Moments of Inertia of Common Geometric Shapes

Shape	Moments of inertia	Figure
Rectangle	$\bar{I}_{x'} = \frac{1}{12}bh^3$ $\bar{I}_{y'} = \frac{1}{12}b^3h$ $I_x = \frac{1}{3}bh^3$ $I_y = \frac{1}{3}b^3h$ $J_C = \frac{1}{12}bh(b^2 + h^2)$	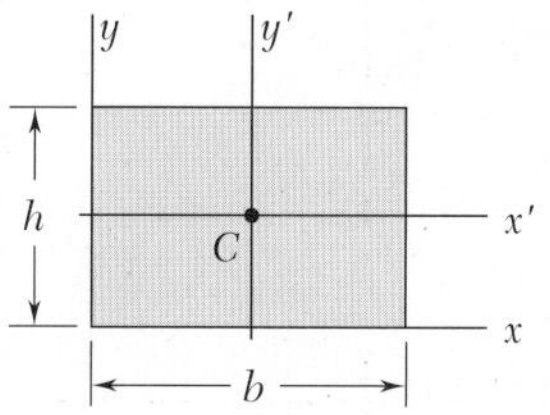
Triangle	$\bar{I}_{x'} = \frac{1}{36}bh^3$ $I_x = \frac{1}{12}bh^3$	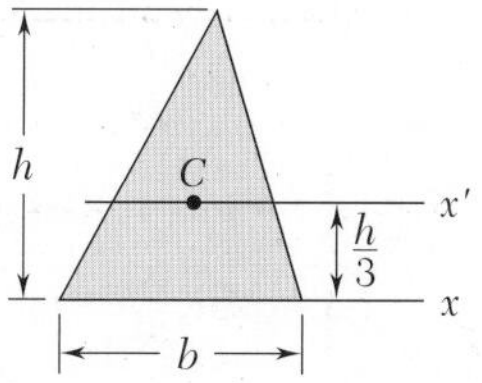
Circle	$\bar{I}_x = \bar{I}_y = \frac{1}{4}\pi r^4$ $J_O = \frac{1}{2}\pi r^4$	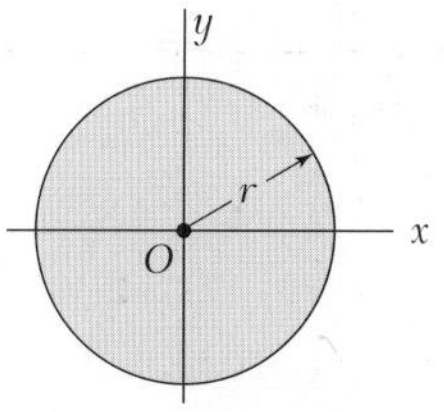
Semicircle	$I_x = I_y = \frac{1}{8}\pi r^4$ $J_O = \frac{1}{4}\pi r^4$	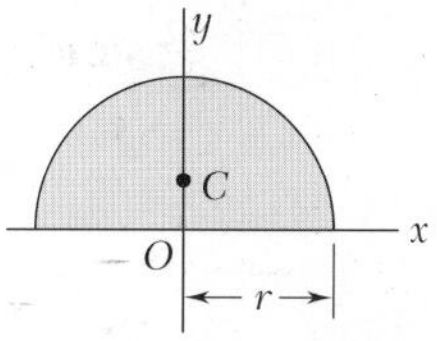
Quarter circle	$I_x = I_y = \frac{1}{16}\pi r^4$ $J_O = \frac{1}{8}\pi r^4$	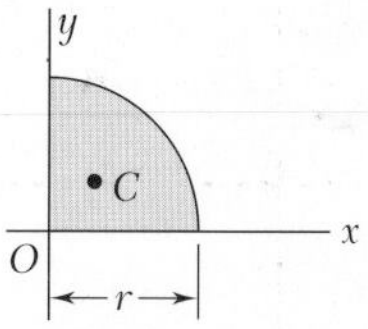
Ellipse	$\bar{I}_x = \frac{1}{4}\pi ab^3$ $\bar{I}_y = \frac{1}{4}\pi a^3b$ $J_O = \frac{1}{4}\pi ab(a^2 + b^2)$	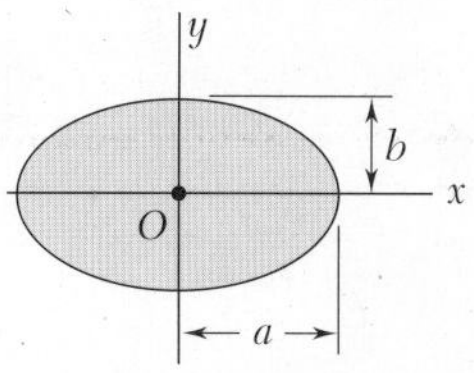

Mass Moments of Inertia of Common Geometric Shapes

Shape	Mass moments of inertia	Figure
Slender rod	$I_y = I_z = \frac{1}{12}mL^2$	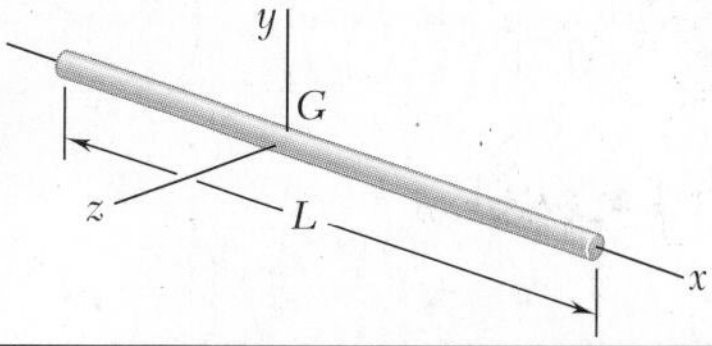
Thin rectangular plate	$I_x = \frac{1}{12}m(b^2 + c^2)$ $I_y = \frac{1}{12}mc^2$ $I_z = \frac{1}{12}mb^2$	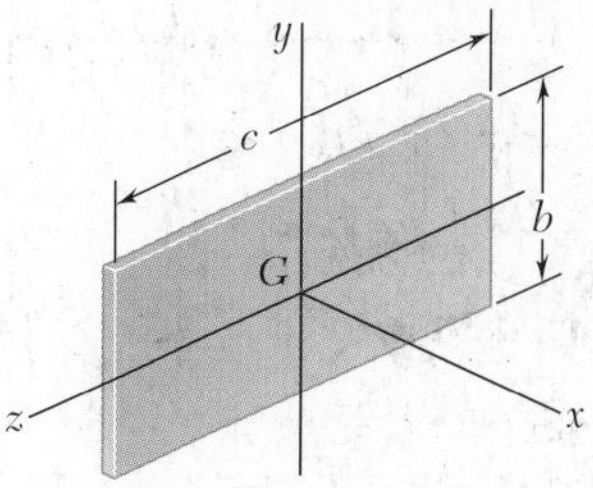
Rectangular prism	$I_x = \frac{1}{12}m(b^2 + c^2)$ $I_y = \frac{1}{12}m(c^2 + a^2)$ $I_z = \frac{1}{12}m(a^2 +$	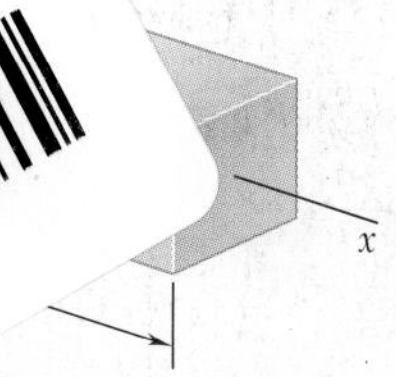
Thin disk	$I_x = \frac{1}{2}mr^2$ $I_y = I_z = \frac{1}{4}mr^2$	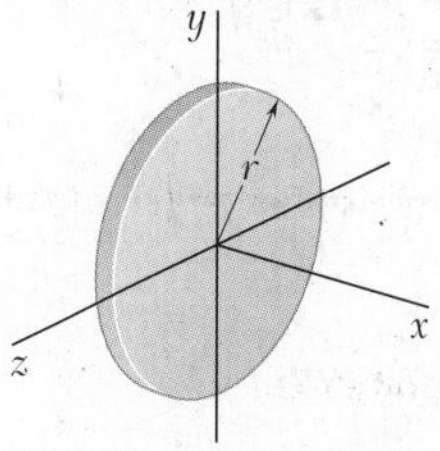
Circular cylinder	$I_x = \frac{1}{2}ma^2$ $I_y = I_z = \frac{1}{12}m(3a^2 + L^2)$	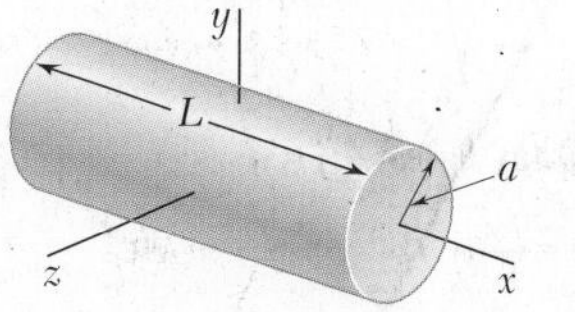
Circular cone	$I_x = \frac{3}{10}ma^2$ $I_y = I_z = \frac{3}{5}m(\frac{1}{4}a^2 + h^2)$	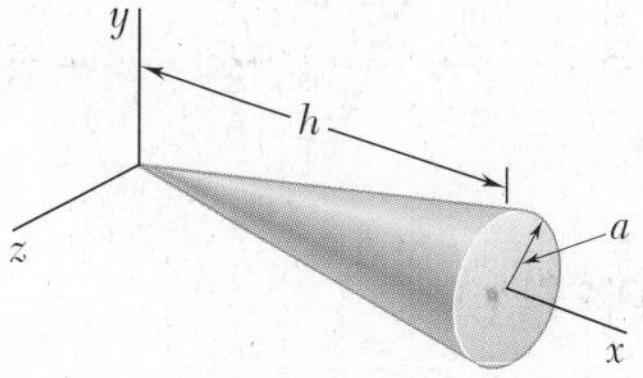
Sphere	$I_x = I_y = I_z = \frac{2}{5}ma^2$	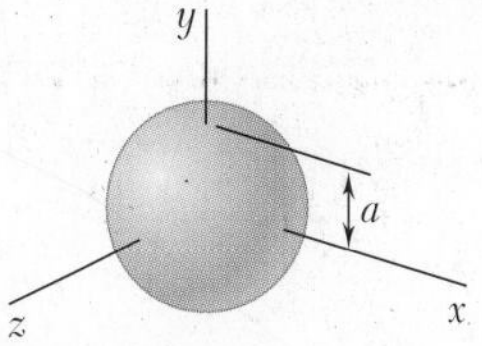

TENTH EDITION IN SI UNITS

VECTOR MECHANICS FOR ENGINEERS

DYNAMICS

Ferdinand P. Beer
Late of Lehigh University

E. Russell Johnston, Jr.
Late of University of Connecticut

Phillip J. Cornwell
Rose-Hulman Institute of Technology

McGraw-Hill Education (Asia)

VECTOR MECHANICS FOR ENGINEERS: DYNAMICS
Tenth Edition in SI Units

Published by McGraw-Hill, a business unit of The McGraw-Hill Companies, Inc., 1221 Avenue of the Americas, New York, NY 10020.

Some ancillaries, including electronic and print components, may not be available to customers outside the United States.

10 9 8 7 6 5 4 3 2 1
CTP SLP
20 16 15 14 13

When ordering this title, use **ISBN 978-1-259-00793-4** or **MHID 1-259-00793-6**

Printed in Singapore

About the Authors

As publishers of the books by Ferd Beer and Russ Johnston we are often asked how did they happen to write their books together with one of them at Lehigh and the other at the University of Connecticut.

The answer to this question is simple. Russ Johnston's first teaching appointment was in the Department of Civil Engineering and Mechanics at Lehigh University. There he met Ferd Beer, who had joined that department two years earlier and was in charge of the courses in mechanics.

Ferd was delighted to discover that the young man who had been hired chiefly to teach graduate structural engineering courses was not only willing but eager to help him reorganize the mechanics courses. Both believed that these courses should be taught from a few basic principles and that the various concepts involved would be best understood and remembered by the students if they were presented to them in a graphic way. Together they wrote lecture notes in statics and dynamics, to which they later added problems they felt would appeal to future engineers, and soon they produced the manuscript of the first edition of *Mechanics for Engineers* that was published in June 1956.

The second edition of *Mechanics for Engineers* and the first edition of *Vector Mechanics for Engineers* found Russ Johnston at Worcester Polytechnic Institute and the next editions at the University of Connecticut. In the meantime, both Ferd and Russ assumed administrative responsibilities in their departments, and both were involved in research, consulting, and supervising graduate students—Ferd in the area of stochastic processes and random vibrations and Russ in the area of elastic stability and structural analysis and design. However, their interest in improving the teaching of the basic mechanics courses had not subsided, and they both taught sections of these courses as they kept revising their texts and began writing the manuscript of the first edition of their *Mechanics of Materials* text.

Their collaboration spanned more than half a century and many successful revisions of all of their textbooks, and Ferd and Russ's contributions to engineering education have earned them a number of honors and awards. They were presented with the Western Electric Fund Award for excellence in the instruction of engineering students by their respective regional sections of the American Society for Engineering Education, and they both received the Distinguished Educator Award from the Mechanics Division of the same society. Starting in 2001, the New Mechanics Educator Award of the Mechanics Division has been named in honor of the Beer and Johnston author team.

Ferdinand P. Beer. Born in France and educated in France and Switzerland, Ferd received an M.S. degree from the Sorbonne and an Sc.D. degree in theoretical mechanics from the University of Geneva. He came to the United States after serving in the French army during the early part of World War II and taught for four years at Williams College in the Williams-MIT joint arts and engineering program. Following his service at Williams College, Ferd joined the faculty of Lehigh University where he taught for thirty-seven years. He held several positions, including University Distinguished Professor and Chairman of the Department of Mechanical Engineering and Mechanics, and in 1995 Ferd was awarded an honorary Doctor of Engineering degree by Lehigh University.

E. Russell Johnston, Jr. Born in Philadelphia, Russ holds a B.S. degree in civil engineering from the University of Delaware and an Sc. D. degree in the field of structural engineering from the Massachusetts Institute of Technology. He taught at Lehigh University and Worcester Polytechnic Institute before joining the faculty of the University of Connecticut where he held the position of chairman of the Civil Engineering Department and taught for twenty-six years. In 1991 Russ received the Outstanding Civil Engineer Award from the Connecticut Section of the American Society of Civil Engineers.

Phillip J. Cornwell. Phil holds a B.S. degree in mechanical engineering from Texas Tech University and M.A. and Ph.D. degrees in mechanical and aerospace engineering from Princeton University. He is currently a professor of mechanical engineering and Vice President of Academic Affairs at Rose-Hulman Institute of Technology where he has taught since 1989. Phil received an SAE Ralph R. Teetor Educational Award in 1992, the Dean's Outstanding Teacher Award at Rose-Hulman in 2000, and the Board of Trustees' Outstanding Scholar Award at Rose-Hulman in 2001.

Brief Contents

Contents

12 Kinetics of Particles: Newton's Second Law 694

13 Kinetics of Particles: Energy and Momentum Methods 762

14 Systems of Particles 866

15 Kinematics of Rigid Bodies 926

16 Plane Motion of Rigid Bodies: Forces and Accelerations 1040

17 Plane Motion of Rigid Bodies: Energy and Momentum Methods 1104

18 Kinetics of Rigid Bodies in Three Dimensions 1172

19 Mechanical Vibrations 1240

Preface

OBJECTIVES

The main objective of a first course in mechanics should be to develop in the engineering student the ability to analyze any problem in a simple and logical manner and to apply to its solution a few, well-understood, basic principles. It is hoped that this text, as well as the preceding volume, *Vector Mechanics for Engineers: Statics*, will help the instructor achieve this goal.†

GENERAL APPROACH

Vector algebra was introduced at the beginning of the first volume and is used in the presentation of the basic principles of statics, as well as in the solution of many problems, particularly three-dimensional problems. Similarly, the concept of vector differentiation will be introduced early in this volume, and vector analysis will be used throughout the presentation of dynamics. This approach leads to more concise derivations of the fundamental principles of mechanics. It also makes it possible to analyze many problems in kinematics and kinetics which could not be solved by scalar methods. The emphasis in this text, however, remains on the correct understanding of the principles of mechanics and on their application to the solution of engineering problems, and vector analysis is presented chiefly as a convenient tool.‡

Practical Applications Are Introduced Early. One of the characteristics of the approach used in this book is that mechanics of *particles* is clearly separated from the mechanics of *rigid bodies.* This approach makes it possible to consider simple practical applications at an early stage and to postpone the introduction of the more difficult concepts. For example:

- In *Statics,* the statics of particles is treated first, and the principle of equilibrium of a particle was immediately applied to practical situations involving only concurrent forces. The statics of rigid bodies is considered later, at which time the vector and scalar products of two vectors were introduced and used to define the moment of a force about a point and about an axis.
- In *Dynamics,* the same division is observed. The basic concepts of force, mass, and acceleration, of work and energy, and of impulse and momentum are introduced and first applied to

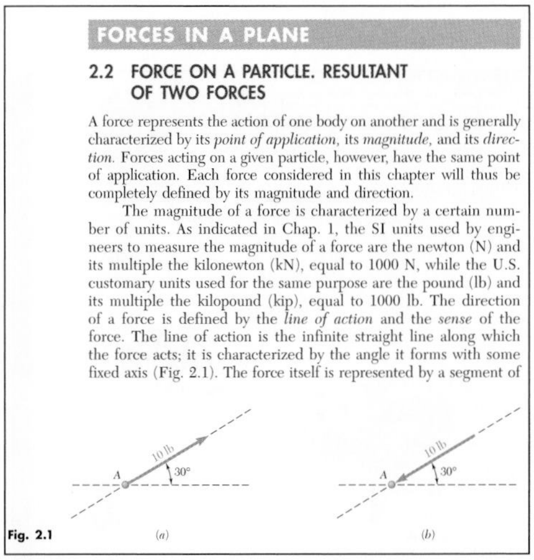
FORCES IN A PLANE

2.2 FORCE ON A PARTICLE. RESULTANT OF TWO FORCES

A force represents the action of one body on another and is generally characterized by its *point of application*, its *magnitude*, and its *direction*. Forces acting on a given particle, however, have the same point of application. Each force considered in this chapter will thus be completely defined by its magnitude and direction.

The magnitude of a force is characterized by a certain number of units. As indicated in Chap. 1, the SI units used by engineers to measure the magnitude of a force are the newton (N) and its multiple the kilonewton (kN), equal to 1000 N, while the U.S. customary units used for the same purpose are the pound (lb) and its multiple the kilopound (kip), equal to 1000 lb. The direction of a force is defined by the *line of action* and the *sense* of the force. The line of action is the infinite straight line along which the force acts; it is characterized by the angle it forms with some fixed axis (Fig. 2.1). The force itself is represented by a segment of

Fig. 2.1

†Both texts also are available in a single volume, *Vector Mechanics for Engineers: Statics and Dynamics,* tenth edition.

‡In a parallel text, *Mechanics for Engineers: Dynamics,* fifth edition, the use of vector algebra is limited to the addition and subtraction of vectors, and vector differentiation is omitted.

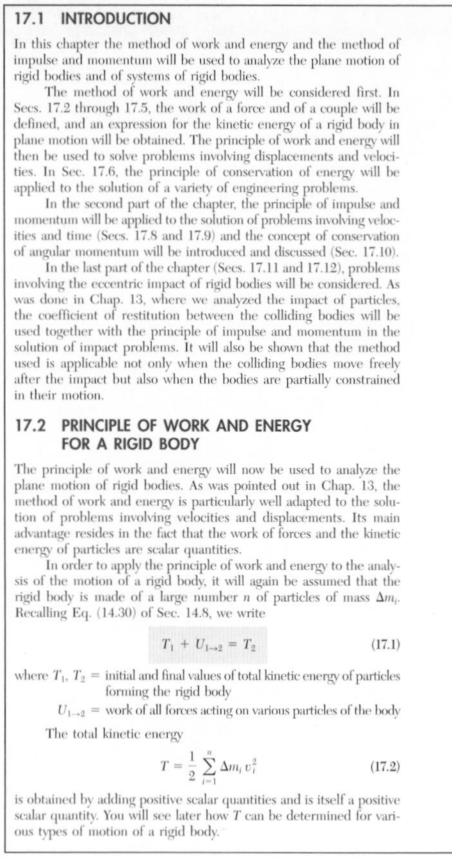

17.1 INTRODUCTION

In this chapter the method of work and energy and the method of impulse and momentum will be used to analyze the plane motion of rigid bodies and of systems of rigid bodies.

The method of work and energy will be considered first. In Secs. 17.2 through 17.5, the work of a force and of a couple will be defined, and an expression for the kinetic energy of a rigid body in plane motion will be obtained. The principle of work and energy will then be used to solve problems involving displacements and velocities. In Sec. 17.6, the principle of conservation of energy will be applied to the solution of a variety of engineering problems.

In the second part of the chapter, the principle of impulse and momentum will be applied to the solution of problems involving velocities and time (Secs. 17.8 and 17.9) and the concept of conservation of angular momentum will be introduced and discussed (Sec. 17.10).

In the last part of the chapter (Secs. 17.11 and 17.12), problems involving the eccentric impact of rigid bodies will be considered. As was done in Chap. 13, where we analyzed the impact of particles, the coefficient of restitution between the colliding bodies will be used together with the principle of impulse and momentum in the solution of impact problems. It will also be shown that the method used is applicable not only when the colliding bodies move freely after the impact but also when the bodies are partially constrained in their motion.

17.2 PRINCIPLE OF WORK AND ENERGY FOR A RIGID BODY

The principle of work and energy will now be used to analyze the plane motion of rigid bodies. As was pointed out in Chap. 13, the method of work and energy is particularly well adapted to the solution of problems involving velocities and displacements. Its main advantage resides in the fact that the work of forces and the kinetic energy of particles are scalar quantities.

In order to apply the principle of work and energy to the analysis of the motion of a rigid body, it will again be assumed that the rigid body is made of a large number n of particles of mass Δm_i. Recalling Eq. (14.30) of Sec. 14.8, we write

$$T_1 + U_{1\rightarrow 2} = T_2 \tag{17.1}$$

where T_1, T_2 = initial and final values of total kinetic energy of particles forming the rigid body

$U_{1\rightarrow 2}$ = work of all forces acting on various particles of the body

The total kinetic energy

$$T = \frac{1}{2}\sum_{i=1}^{n} \Delta m_i\, v_i^2 \tag{17.2}$$

is obtained by adding positive scalar quantities and is itself a positive scalar quantity. You will see later how T can be determined for various types of motion of a rigid body.

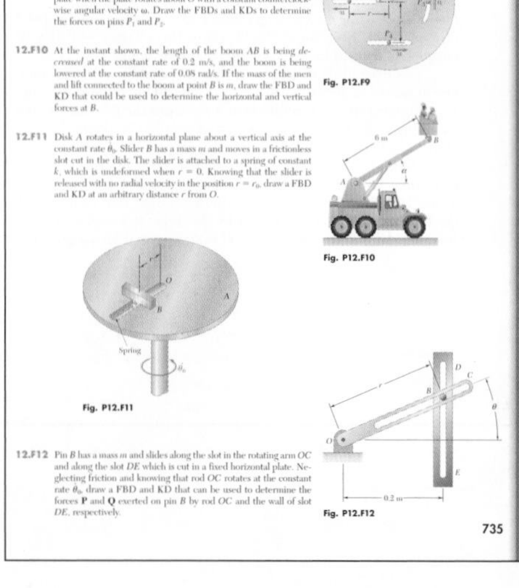

FREE-BODY PRACTICE PROBLEMS

12.F9 Four pins slide in four separate slots cut in a horizontal circular plate as shown. When the plate is at rest, each pin has a velocity directed as shown and of the same constant magnitude u. Each pin has a mass m and maintains the same velocity relative to the plate when the plate rotates about O with a constant counterclockwise angular velocity ω. Draw the FBDs and KDs to determine the forces on pins P_1 and P_2.

12.F10 At the instant shown, the length of the boom AB is being *decreased* at the constant rate of 0.2 m/s, and the boom is being lowered at the constant rate of 0.08 rad/s. If the mass of the men and lift connected to the boom at point B is m, draw the FBD and KD that could be used to determine the horizontal and vertical forces at B.

12.F11 Disk A rotates in a horizontal plane about a vertical axis at the constant rate $\dot{\theta}_0$. Slider B has a mass m and moves in a frictionless slot cut in the disk. The slider is attached to a spring of constant k, which is undeformed when $r = 0$. Knowing that the slider is released with no radial velocity in the position $r = r_0$, draw a FBD and KD at an arbitrary distance r from O.

12.F12 Pin B has a mass m and slides along the slot in the rotating arm OC and along the slot DE which is cut in a fixed horizontal plate. Neglecting friction and knowing that rod OC rotates at the constant rate $\dot{\theta}_0$, draw a FBD and KD that can be used to determine the forces **P** and **Q** exerted on pin B by rod OC and the wall of slot DE, respectively.

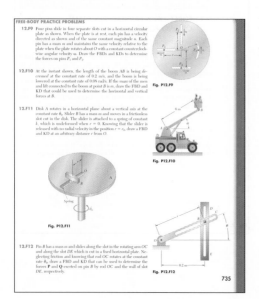

problems involving only particles. Thus, students can familiarize themselves with the three basic methods used in dynamics and learn their respective advantages before facing the difficulties associated with the motion of rigid bodies.

New Concepts Are Introduced in Simple Terms. Since this text is designed for the first course in dynamics, new concepts are presented in simple terms and every step is explained in detail. On the other hand, by discussing the broader aspects of the problems considered, and by stressing methods of general applicability, a definite maturity of approach has been achieved. For example, the concept of potential energy is discussed in the general case of a conservative force. Also, the study of the plane motion of rigid bodies is designed to lead naturally to the study of their general motion in space. This is true in kinematics as well as in kinetics, where the principle of equivalence of external and effective forces is applied directly to the analysis of plane motion, thus facilitating the transition to the study of three-dimensional motion.

Fundamental Principles Are Placed in the Context of Simple Applications. The fact that mechanics is essentially a *deductive* science based on a few fundamental principles is stressed. Derivations have been presented in their logical sequence and with all the rigor warranted at this level. However, the learning process being largely *inductive*, simple applications are considered first. For example:

- The kinematics of particles (Chap. 11) precedes the kinematics of rigid bodies (Chap. 15).
- The fundamental principles of the kinetics of rigid bodies are first applied to the solution of two-dimensional problems (Chaps. 16 and 17), which can be more easily visualized by the student, while three-dimensional problems are postponed until Chap. 18.

The Presentation of the Principles of Kinetics Is Unified. The tenth edition of *Vector Mechanics for Engineers* retains the unified presentation of the principles of kinetics which characterized the previous nine editions. The concepts of linear and angular momentum are introduced in Chap. 12 so that Newton's second law of motion can be presented not only in its conventional form **F** = m**a**, but also as a law relating, respectively, the sum of the forces acting on a particle and the sum of their moments to the rates of change of the linear and angular momentum of the particle. This makes possible an earlier introduction of the principle of conservation of angular momentum and a more meaningful discussion of the motion of a particle under a central force (Sec. 12.9). More importantly, this approach can be readily extended to the study of the motion of a system of particles (Chap. 14) and leads to a more concise and unified treatment of the kinetics of rigid bodies in two and three dimensions (Chaps. 16 through 18).

Free-Body Diagrams Are Used Both to Solve Equilibrium Problems and to Express the Equivalence of Force Systems. Free-body diagrams were introduced early in statics, and their importance was emphasized throughout. They were used not only to solve equilibrium problems but also to express the equivalence of two systems of forces or, more generally, of two systems of vectors. The advantage of this approach becomes apparent in the study of the dynamics of rigid bodies, where it is used to solve three-dimensional as well as two-dimensional problems. By placing the emphasis on "free-body-diagram equations" rather than on the standard algebraic equations of motion, a more intuitive and more complete understanding of the fundamental principles of dynamics can be achieved. This approach, which was first introduced in 1962 in the first edition of *Vector Mechanics for Engineers,* has now gained wide acceptance among mechanics teachers in this country. It is, therefore, used in preference to the method of dynamic equilibrium and to the equations of motion in the solution of all sample problems in this book.

Optional Sections Offer Advanced or Specialty Topics. A large number of optional sections have been included. These sections are indicated by asterisks and thus are easily distinguished from those which form the core of the basic dynamics course. They can be omitted without prejudice to the understanding of the rest of the text.

The topics covered in the optional sections include graphical methods for the solution of rectilinear-motion problems, the trajectory of a particle under a central force, the deflection of fluid streams, problems involving jet and rocket propulsion, the kinematics and kinetics of rigid bodies in three dimensions, damped mechanical vibrations, and electrical analogues. These topics will be found of particular interest when dynamics is taught in the junior year.

The material presented in the text and most of the problems requires no previous mathematical knowledge beyond algebra, trigonometry, elementary calculus, and the elements of vector algebra presented in Chaps. 2 and 3 of the volume on statics.† However, special problems are included, which make use of a more advanced knowledge of calculus, and certain sections, such as Secs. 19.8 and 19.9 on damped vibrations, should be assigned only if students possess the proper mathematical background. In portions of the text using elementary calculus, a greater emphasis is placed on the correct understanding and application of the concepts of differentiation and integration, than on the nimble manipulation of mathematical formulas. In this connection, it should be mentioned that the determination of the centroids of composite areas precedes the calculation of centroids by integration, thus making it possible to establish the concept of moment of area firmly before introducing the use of integration.

†Some useful definitions and properties of vector algebra have been summarized in Appendix A at the end of this volume for the convenience of the reader. Also, Secs. 9.11 through 9.18 of the volume on statics, which deal with the moments of inertia of masses, have been reproduced in Appendix B.

Guided Tour

Chapter Introduction. Each chapter begins with an introductory section setting the purpose and goals of the chapter and describing in simple terms the material to be covered and its application to the solution of engineering problems. New chapter outlines provide students with a preview of chapter topics.

Chapter Lessons. The body of the text is divided into units, each consisting of one or several theory sections, one or several sample problems, and a large number of problems to be assigned. Each unit corresponds to a well-defined topic and generally can be covered in one lesson. In a number of cases, however, the instructor will find it desirable to devote more than one lesson to a given topic. *The Instructor's and Solutions Manual* contains suggestions on the coverage of each lesson.

Sample Problems. The sample problems are set up in much the same form that students will use when solving the assigned problems. They thus serve the double purpose of amplifying the text and demonstrating the type of neat, orderly work that students should cultivate in their own solutions.

Solving Problems on Your Own. A section entitled *Solving Problems on Your Own* is included for each lesson, between the sample problems and the problems to be assigned. The purpose of these sections is to help students organize in their own minds the preceding theory of the text and the solution methods of the sample problems so that they can more successfully solve the homework problems. Also included in these sections are specific suggestions and strategies that will enable the students to more efficiently attack any assigned problems.

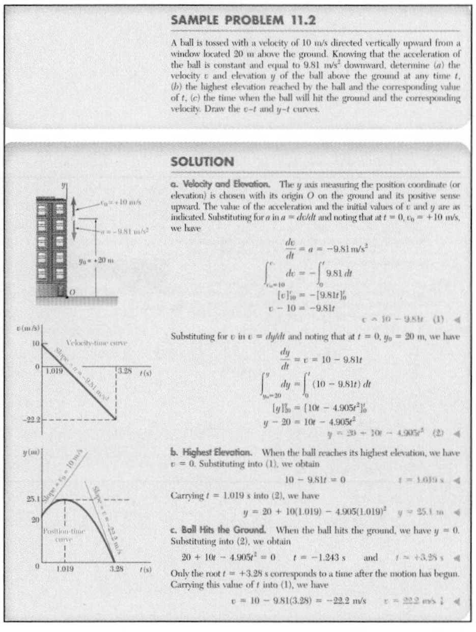

SAMPLE PROBLEM 11.2

A ball is tossed with a velocity of 10 m/s directed vertically upward from a window located 20 m above the ground. Knowing that the acceleration of the ball is constant and equal to 9.81 m/s² downward, determine (*a*) the velocity v and elevation y of the ball above the ground at any time t, (*b*) the highest elevation reached by the ball and the corresponding value of t, (*c*) the time when the ball will hit the ground and the corresponding velocity. Draw the v–t and y–t curves.

SOLUTION

a. Velocity and Elevation. The y axis measuring the position coordinate (or elevation) is chosen with its origin O on the ground and its positive sense upward. The value of the acceleration and the initial values of v and y are as indicated. Substituting for a in $a = dv/dt$ and noting that at $t = 0$, $v_0 = +10$ m/s, we have

$$\frac{dv}{dt} = a = -9.81 \text{ m/s}^2$$
$$\int_{v_0=10}^{v} dv = -\int_0^t 9.81\, dt$$
$$[v]_{10}^{v} = -[9.81t]_0^t$$
$$v - 10 = -9.81t$$
$$v = 10 - 9.81t \quad (1)$$

Substituting for v in $v = dy/dt$ and noting that at $t = 0$, $y_0 = 20$ m, we have

$$\frac{dy}{dt} = v = 10 - 9.81t$$
$$\int_{y_0=20}^{y} dy = \int_0^t (10 - 9.81t)\, dt$$
$$[y]_{20}^{y} = [10t - 4.905t^2]_0^t$$
$$y - 20 = 10t - 4.905t^2$$
$$y = 20 + 10t - 4.905t^2 \quad (2)$$

b. Highest Elevation. When the ball reaches its highest elevation, we have $v = 0$. Substituting into (1), we obtain

$$10 - 9.81t = 0 \qquad t = 1.019 \text{ s}$$

Carrying $t = 1.019$ s into (2), we have

$$y = 20 + 10(1.019) - 4.905(1.019)^2 \qquad y = 25.1 \text{ m}$$

c. Ball Hits the Ground. When the ball hits the ground, we have $y = 0$. Substituting into (2), we obtain

$$20 + 10t - 4.905t^2 = 0 \qquad t = -1.243 \text{ s} \quad \text{and} \quad t = +3.28 \text{ s}$$

Only the root $t = +3.28$ s corresponds to a time after the motion has begun. Carrying this value of t into (1), we have

$$v = 10 - 9.81(3.28) = -22.2 \text{ m/s} \qquad v = 22.2 \text{ m/s} \downarrow$$

Homework Problem Sets. Most of the problems are of a practical nature and should appeal to engineering students. They are primarily designed, however, to illustrate the material presented in the text and to help students understand the principles of mechanics. The problems are grouped according to the portions of material they illustrate and are arranged in order of increasing difficulty. Problems requiring special attention are indicated by asterisks. Answers to 70 percent of the problems are given at the end of the book. Problems for which the answers are given are set in straight type in the text, while problems for which no answer is given are set in italic.

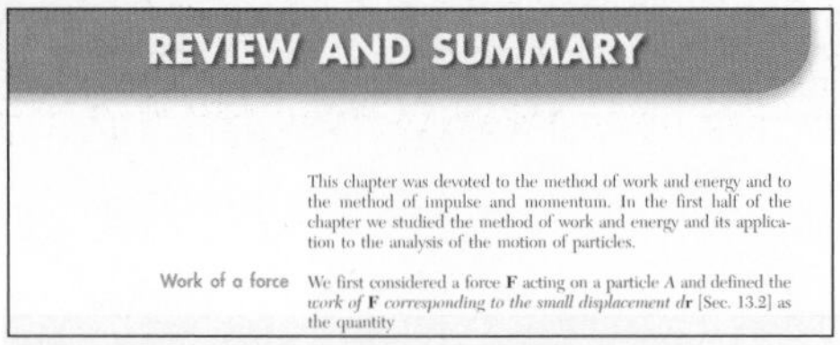

REVIEW AND SUMMARY

This chapter was devoted to the method of work and energy and to the method of impulse and momentum. In the first half of the chapter we studied the method of work and energy and its application to the analysis of the motion of particles.

Work of a force — We first considered a force **F** acting on a particle *A* and defined the *work of* **F** *corresponding to the small displacement d***r** [Sec. 13.2] as the quantity

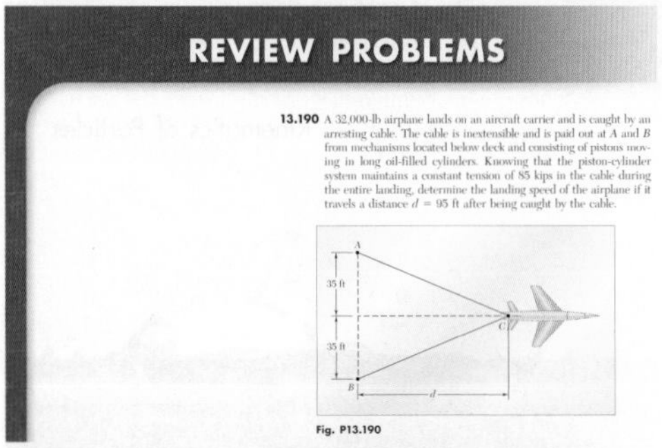

REVIEW PROBLEMS

13.190 A 32,000-lb airplane lands on an aircraft carrier and is caught by an arresting cable. The cable is inextensible and is paid out at *A* and *B* from mechanisms located below deck and consisting of pistons moving in long oil-filled cylinders. Knowing that the piston-cylinder system maintains a constant tension of 85 kips in the cable during the entire landing, determine the landing speed of the airplane if it travels a distance *d* = 95 ft after being caught by the cable.

Fig. P13.190

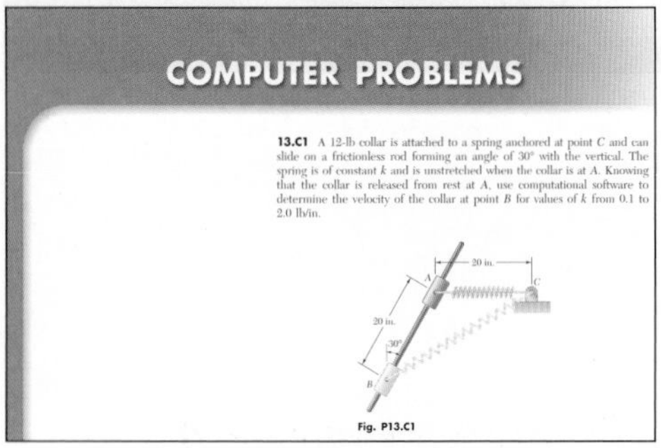

COMPUTER PROBLEMS

13.C1 A 12-lb collar is attached to a spring anchored at point *C* and can slide on a frictionless rod forming an angle of 30° with the vertical. The spring is of constant *k* and is unstretched when the collar is at *A*. Knowing that the collar is released from rest at *A*, use computational software to determine the velocity of the collar at point *B* for values of *k* from 0.1 to 2.0 lb/in.

Fig. P13.C1

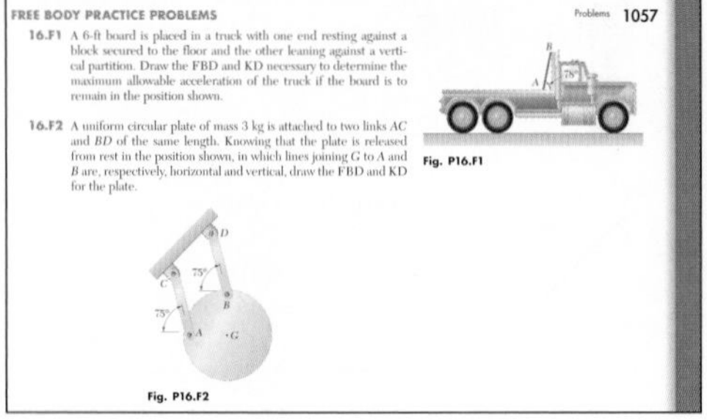

FREE BODY PRACTICE PROBLEMS — Problems 1057

16.F1 A 6-ft board is placed in a truck with one end resting against a block secured to the floor and the other leaning against a vertical partition. Draw the FBD and KD necessary to determine the maximum allowable acceleration of the truck if the board is to remain in the position shown.

16.F2 A uniform circular plate of mass 3 kg is attached to two links *AC* and *BD* of the same length. Knowing that the plate is released from rest in the position shown, in which lines joining *G* to *A* and *B* are, respectively, horizontal and vertical, draw the FBD and KD for the plate.

Fig. P16.F1

Fig. P16.F2

Chapter Review and Summary. Each chapter ends with a review and summary of the material covered in that chapter. Marginal notes are used to help students organize their review work, and cross-references have been included to help them find the portions of material requiring their special attention.

Review Problems. A set of review problems is included at the end of each chapter. These problems provide students further opportunity to apply the most important concepts introduced in the chapter.

Computer Problems. Each chapter includes a set of problems designed to be solved with computational software. Many of these problems provide an introduction to the design process. For example, they may involve the determination of the motion of a particle under initial conditions, the kinematic or kinetic analysis of mechanisms in successive positions, or the numerical integration of various equations of motion. Developing the algorithm required to solve a given mechanics problem will benefit the students in two different ways: (1) It will help them gain a better understanding of the mechanics principles involved; (2) it will provide them with an opportunity to apply their computer skills to the solution of a meaningful engineering problem.

Concept Questions. Educational research has shown that students can often choose appropriate equations and solve algorithmic problems without having a strong conceptual understanding of mechanics principles†. To help assess and develop student conceptual understanding, we have included Concept Questions, which are multiple choice problems that require few, if any, calculations. Each possible incorrect answer typically represents a common misconception (e.g., students often think that a vehicle moving in a curved path at constant speed has zero acceleration). Students are encouraged to solve these problems using the principles and techniques discussed in the text and to use these principles to help them develop their intuition. Mastery and discussion of these Concept Questions will deepen students' conceptual understanding and help them to solve dynamics problems.

Free Body and Impulse-Momentum Practice Problems. Drawing diagrams correctly is a critical step in solving kinetics problems in dynamics. A new type of problem has been added to the text to emphasize the importance of drawing these diagrams. In Chaps. 12 and 16 the Free Body Practice Problems require students to draw a free-body diagram (FBD) showing the applied forces and an equivalent diagram called a "kinetic diagram" (KD) showing $m\mathbf{a}$ or its components and $\overline{\boldsymbol{\alpha}}$. These diagrams provide students with a pictorial representation of Newton's second law and are critical in helping students to correctly solve kinetic problems. In Chaps. 13 and 17 the Impulse-Momentum Practice Problems require students to draw diagrams showing the momenta of the bodies before impact, the impulses exerted on the body during impact, and the final momenta of the bodies. The answers to all of these questions are provided at **www.mhhe.com/beerjohnston.**

†Hestenes, D., Wells, M., and Swakhamer, G (1992). The force concept inventory. *The Physics Teacher*, 30: 141–158.

Streveler, R. A., Litzinger, T. A., Miller, R. L., and Steif P. S. (2008). Learning conceptual knowledge in the engineering sciences: Overview and future research directions, *JEE*, 279–294.

What Resources Support This Textbook?

Instructor's and Solutions Manual. *The Instructor's and Solutions Manual* that accompanies the tenth edition features typeset, one-per-page solutions to the end of chapter problems. This Manual also features a number of tables designed to assist instructors in creating a schedule of assignments for their course. The various topics covered in the text have been listed in Table I and a suggested number of periods to be spent on each topic has been indicated. Table II prepares a brief description of all groups of problems and a classification of the problems in each group according to the units used. Sample lesson schedules are shown in Tables III, IV, and V, together with various alternative lists of assigned homework problems.

For additional resources related to users of this SI edition, please visit **www.mheducation.asia/olc/beerjohnston**.

McGraw-Hill Connect Engineering McGraw-Hill Connect Engineering is a web-based assignment and assessment platform that gives students the means to better connect with their coursework, their instructors, and the important concepts that they will need to know for success now and in the future. With Connect Engineering, instructors can deliver assignments, quizzes, and tests easily online. Students can practice important skills at their own pace and on their own schedule.

Connect Engineering for *Vector Mechanics for Engineers* is available at **www.mhhe.com/beerjohnston** and includes algorithmic problems from the text, Lecture PowerPoints, an image bank, and animations.

Mc Graw Hill connect® |ENGINEERING

Hands-on Mechanics. Hands-on Mechanics is a website designed for instructors who are interested in incorporating three-dimensional, hands-on teaching aids into their lectures. Developed through a partnership between the McGraw-Hill Engineering Team and the Department of Civil and Mechanical Engineering at the United States Military Academy at West Point, this website not only provides detailed instructions for how to build 3-D teaching tools using materials found in any lab or local hardware store, but also provides a community where educators can share ideas, trade best practices, and submit their own original demonstrations for posting on the site. Visit **www.handsonmechanics.com**.

Acknowledgments

A special thanks to Dean Updike of Lehigh University who thoroughly checked the solutions and answers of all problems in this edition and then prepared the solutions for the accompanying *Instructor's and Solutions Manual.*

We are pleased to recognize Dennis Ormond of Fine Line Illustrations for the artful illustrations which contributed much to the effectiveness of the text.

The authors thank the many companies that provided photographs for this edition. We also wish to recognize the determined efforts and patience of our photo researcher Sabina Dowell.

The authors also thank the members of the staff at McGraw-Hill for their support and dedication during the preparation of this new edition.

E. Russell Johnston, Jr.
Phillip J. Cornwell

The authors gratefully acknowledge the many helpful comments and suggestions offered by focus group attendees and users of the previous editions of *Vector Mechanics for Engineers.*

George Adams
Northeastern University

William Altenhof
University of Windsor

Sean B. Anderson
Boston University

Manohar Arora
Colorado School of Mines

Gilbert Baladi
Michigan State University

Francois Barthelat
McGill University

Oscar Barton Jr.
U.S. Naval Academy

M. Asghar Bhatti
University of Iowa

Shaohong Cheng
University of Windsor

Philip Datseris
University of Rhode Island

Timothy A. Doughty
University of Portland

Howard Epstein
University of Conneticut

Asad Esmaeily
Kansas State University,
Civil Engineering Department

David Fleming
Florida Institute of Technology

Jeff Hanson
Texas Tech University

David A. Jenkins
University of Florida

Shaofan Li
University of California, Berkeley

William R. Murray
Cal Poly State University

Eric Musslman
University of Minnesota, Duluth

Masoud Olia
Wentworth Institute of Technology

Renee K. B. Petersen
Washington State University

Amir G Rezaei
California State Polytechnic University, Pomona

Martin Sadd
University of Rhode Island

Stefan Seelecke
North Carolina State University

Yixin Shao
McGill University

Muhammad Sharif
The University of Alabama

Anthony Sinclair
University of Toronto

Lizhi Sun
University of California, lrvine

Jeffrey Thomas
Northwestern University

Jiashi Yang
University of Nebraska

Xiangwa Zeng
Case Western Reserve University

Mcgraw-Hill Tegrity®

Tegrity is a service that makes class time available all the time by automatically capturing every lecture in a searchable format for students to review when they study and complete assignments. With a simple one-click start-and-stop process, you capture all computer screens and corresponding audio. Students replay any part of any class with easy-to-use browser-based viewing on a PC or Mac. Educators know that the more students can see, hear, and experience class resources, the better they learn. With Tegrity, students quickly recall key moments by using Tegrity's unique search feature. This search helps students efficiently find what they need, when they need it across an entire semester of class recordings. Help turn all your students' study time into learning moments immediately supported by your lecture.

To learn more about Tegrity watch a 2-minute Flash demo at http://tegritycampus.mhhe.com.

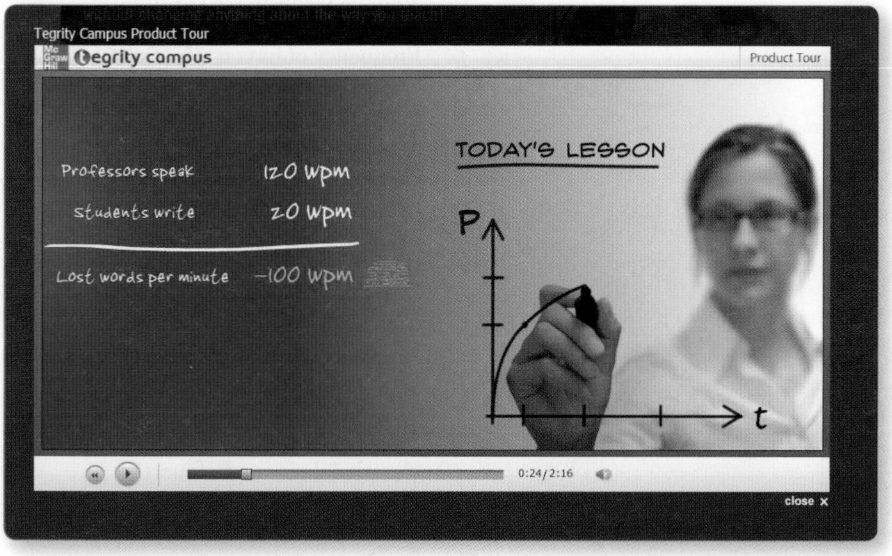

List of Symbols

$\mathbf{a}, a$	Acceleration
a	Constant; radius; distance; semimajor axis of ellipse
$\bar{\mathbf{a}}, \bar{a}$	Acceleration of mass center
$\mathbf{a}_{B/A}$	Acceleration of B relative to frame in translation with A
$\mathbf{a}_{P/\mathcal{F}}$	Acceleration of P relative to rotating frame $\mathcal{F}$
$\mathbf{a}_c$	Coriolis acceleration
$\mathbf{A}, \mathbf{B}, \mathbf{C}, \ldots$	Reactions at supports and connections
$A, B, C, \ldots$	Points
A	Area
b	Width; distance; semiminor axis of ellipse
c	Constant; coefficient of viscous damping
C	Centroid; instantaneous center of rotation; capacitance
d	Distance
$\mathbf{e}_n, \mathbf{e}_t$	Unit vectors along normal and tangent
$\mathbf{e}_r, \mathbf{e}_\theta$	Unit vectors in radial and transverse directions
e	Coefficient of restitution; base of natural logarithms
E	Total mechanical energy; voltage
f	Scalar function
f_f	Frequency of forced vibration
f_n	Natural frequency
$\mathbf{F}$	Force; friction force
g	Acceleration of gravity
G	Center of gravity; mass center; constant of gravitation
h	Angular momentum per unit mass
$\mathbf{H}_O$	Angular momentum about point O
$\dot{\mathbf{H}}_G$	Rate of change of angular momentum $\mathbf{H}_G$ with respect to frame of fixed orientation
$(\dot{\mathbf{H}}_G)_{Gxyz}$	Rate of change of angular momentum $\mathbf{H}_G$ with respect to rotating frame $Gxyz$
$\mathbf{i}, \mathbf{j}, \mathbf{k}$	Unit vectors along coordinate axes
i	Current
$I, I_x, \ldots$	Moments of inertia
$\bar{I}$	Centroidal moment of inertia
$I_{xy}, \ldots$	Products of inertia
J	Polar moment of inertia
k	Spring constant
k_x, k_y, k_O	Radii of gyration
$\bar{k}$	Centroidal radius of gyration
l	Length
$\mathbf{L}$	Linear momentum
L	Length; inductance
m	Mass
m'	Mass per unit length
$\mathbf{M}$	Couple; moment
$\mathbf{M}_O$	Moment about point O
$\mathbf{M}_O^R$	Moment resultant about point O
M	Magnitude of couple or moment; mass of earth
M_{OL}	Moment about axis OL

n	Normal direction
$\mathbf{N}$	Normal component of reaction
O	Origin of coordinates
$\mathbf{P}$	Force; vector
$\dot{\mathbf{P}}$	Rate of change of vector **P** with respect to frame of fixed orientation
q	Mass rate of flow; electric charge
$\mathbf{Q}$	Force; vector
$\dot{\mathbf{Q}}$	Rate of change of vector **Q** with respect to frame of fixed orientation
$(\dot{\mathbf{Q}})_{Oxyz}$	Rate of change of vector **Q** with respect to frame $Oxyz$
$\mathbf{r}$	Position vector
$\mathbf{r}_{B/A}$	Position vector of B relative to A
r	Radius; distance; polar coordinate
$\mathbf{R}$	Resultant force; resultant vector; reaction
R	Radius of earth; resistance
$\mathbf{s}$	Position vector
s	Length of arc
t	Time; thickness; tangential direction
$\mathbf{T}$	Force
T	Tension; kinetic energy
$\mathbf{u}$	Velocity
u	Variable
U	Work
$\mathbf{v}, v$	Velocity
v	Speed
$\overline{\mathbf{v}}, \overline{v}$	Velocity of mass center
$\mathbf{v}_{B/A}$	Velocity of B relative to frame in translation with A
$\mathbf{v}_{P/\mathcal{F}}$	Velocity of P relative to rotating frame $\mathcal{F}$
$\mathbf{V}$	Vector product
V	Volume; potential energy
w	Load per unit length
$\mathbf{W}, W$	Weight; load
x, y, z	Rectangular coordinates; distances
$\dot{x}, \dot{y}, \dot{z}$	Time derivatives of coordinates x, y, z
$\overline{x}, \overline{y}, \overline{z}$	Rectangular coordinates of centroid, center of gravity, or mass center
$\boldsymbol{\alpha}, \alpha$	Angular acceleration
α, β, γ	Angles
γ	Specific weight
δ	Elongation
ε	Eccentricity of conic section or of orbit
$\boldsymbol{\lambda}$	Unit vector along a line
η	Efficiency
θ	Angular coordinate; Eulerian angle; angle; polar coordinate
μ	Coefficient of friction
ρ	Density; radius of curvature
τ	Periodic time
τ_n	Period of free vibration
ϕ	Angle of friction; Eulerian angle; phase angle; angle
φ	Phase difference
ψ	Eulerian angle
$\boldsymbol{\omega}, \omega$	Angular velocity
ω_f	Circular frequency of forced vibration
ω_n	Natural circular frequency
$\boldsymbol{\Omega}$	Angular velocity of frame of reference

The motion of the space shuttle can be described in terms of its *position, velocity,* and *acceleration*. When landing, the pilot of the shuttle needs to consider the wind velocity and the *relative motion* of the shuttle with respect to the wind. The study of motion is known as *kinematics* and is the subject of this chapter.

CHAPTER 11

Kinematics of Particles

11.1 INTRODUCTION TO DYNAMICS

Chapters 1 to 10 were devoted to *statics,* i.e., to the analysis of bodies at rest. We now begin the study of *dynamics,* the part of mechanics that deals with the analysis of bodies in motion.

While the study of statics goes back to the time of the Greek philosophers, the first significant contribution to dynamics was made by Galileo (1564–1642). Galileo's experiments on uniformly accelerated bodies led Newton (1642–1727) to formulate his fundamental laws of motion.

Dynamics includes:

1. *Kinematics,* which is the study of the geometry of motion. Kinematics is used to relate displacement, velocity, acceleration, and time, without reference to the cause of the motion.
2. *Kinetics,* which is the study of the relation existing between the forces acting on a body, the mass of the body, and the motion of the body. Kinetics is used to predict the motion caused by given forces or to determine the forces required to produce a given motion.

Chapters 11 to 14 are devoted to the *dynamics of particles;* in Chap. 11 the *kinematics of particles* will be considered. The use of the word *particles* does not mean that our study will be restricted to small corpuscles; rather, it indicates that in these first chapters the motion of bodies—possibly as large as cars, rockets, or airplanes—will be considered without regard to their size. By saying that the bodies are analyzed as particles, we mean that only their motion as an entire unit will be considered; any rotation about their own mass center will be neglected. There are cases, however, when such a rotation is not negligible; the bodies cannot then be considered as particles. Such motions will be analyzed in later chapters, dealing with the *dynamics of rigid bodies.*

In the first part of Chap. 11, the rectilinear motion of a particle will be analyzed; that is, the position, velocity, and acceleration of a particle will be determined at every instant as it moves along a straight line. First, general methods of analysis will be used to study the motion of a particle; then two important particular cases will be considered, namely, the uniform motion and the uniformly accelerated motion of a particle (Secs. 11.4 and 11.5). In Sec. 11.6 the simultaneous motion of several particles will be considered, and the concept of the relative motion of one particle with respect to another will be introduced. The first part of this chapter concludes with a study of graphical methods of analysis and their application to the solution of various problems involving the rectilinear motion of particles (Secs. 11.7 and 11.8).

In the second part of this chapter, the motion of a particle as it moves along a curved path will be analyzed. Since the position, velocity, and acceleration of a particle will be defined as vector quantities, the concept of the derivative of a vector function will be introduced in Sec. 11.10 and added to our mathematical tools. Applications in which the motion of a particle is defined by the rectangular components of its velocity and acceleration will then be considered; at this point, the motion of a projectile will be analyzed (Sec. 11.11). In Sec. 11.12, the motion of a particle relative to a reference frame in translation will be considered. Finally, the curvilinear motion of a particle will be analyzed in terms of components other than rectangular. The tangential and normal components of a particular velocity and an acceleration will be introduced in Sec. 11.13 and the radial and transverse components of its velocity and acceleration in Sec. 11.14.

RECTILINEAR MOTION OF PARTICLES

11.2 POSITION, VELOCITY, AND ACCELERATION

A particle moving along a straight line is said to be in *rectilinear motion*. At any given instant t, the particle will occupy a certain position on the straight line. To define the position P of the particle, we choose a fixed origin O on the straight line and a positive direction along the line. We measure the distance x from O to P and record it with a plus or minus sign, according to whether P is reached from O by moving along the line in the positive or the negative direction. The distance x, with the appropriate sign, completely defines the position of the particle; it is called the *position coordinate* of the particle considered. For example, the position coordinate corresponding to P in Fig. 11.1*a* is $x = +5$ m; the coordinate corresponding to P' in Fig. 11.1*b* is $x' = -2$ m.

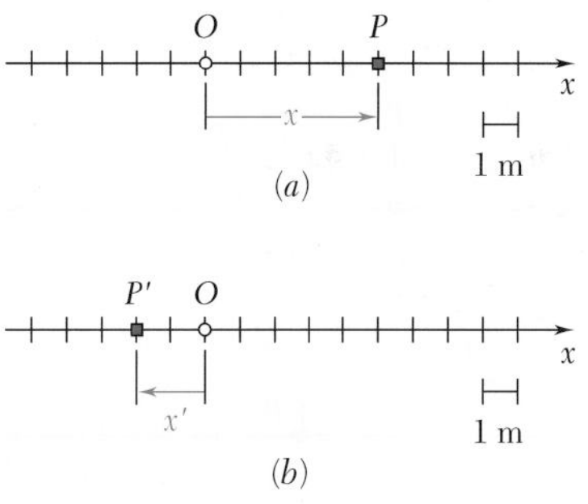

Fig. 11.1

When the position coordinate x of a particle is known for every value of time t, we say that the motion of the particle is known. The "timetable" of the motion can be given in the form of an equation in x and t, such as $x = 6t^2 - t^3$, or in the form of a graph of x versus t as shown in Fig. 11.6. The units most often used to measure the position coordinate x are the meter (m) in the SI system of units† and the foot (ft) in the U.S. customary system of units. Time t is usually measured in seconds (s).

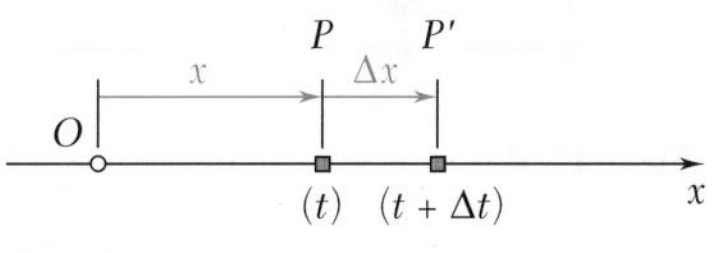

Fig. 11.2

Consider the position P occupied by the particle at time t and the corresponding coordinate x (Fig. 11.2). Consider also the position P' occupied by the particle at a later time $t + \Delta t$; the position coordinate of P' can be obtained by adding to the coordinate x of P the small displacement Δx, which will be positive or negative according to whether P' is to the right or to the left of P. The *average velocity* of the particle over the time interval Δt is defined as the quotient of the displacement Δx and the time interval Δt:

$$\text{Average velocity} = \frac{\Delta x}{\Delta t}$$

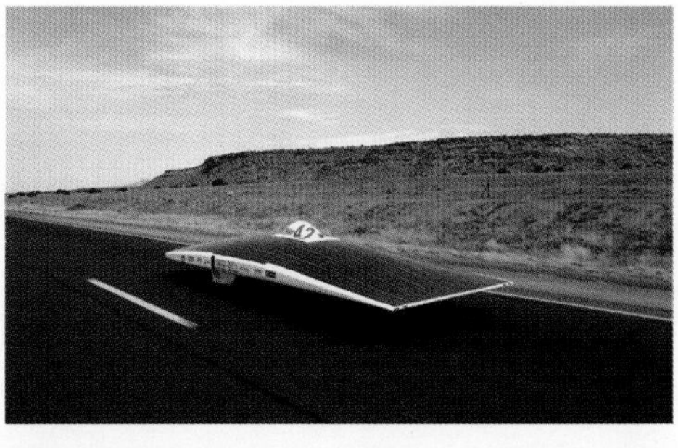

Photo 11.1 The motion of this solar car can be described by its position, velocity, and acceleration.

†Cf. Sec. 1.3.

If SI units are used, Δx is expressed in meters and Δt in seconds; the average velocity will thus be expressed in meters per second (m/s). If U.S. customary units are used, Δx is expressed in feet and Δt in seconds; the average velocity will then be expressed in feet per second (ft/s).

The *instantaneous velocity* v of the particle at the instant t is obtained from the average velocity by choosing shorter and shorter time intervals Δt and displacements Δx:

$$\text{Instantaneous velocity} = v = \lim_{\Delta t \to 0} \frac{\Delta x}{\Delta t}$$

The instantaneous velocity will also be expressed in m/s or ft/s. Observing that the limit of the quotient is equal, by definition, to the derivative of x with respect to t, we write

$$v = \frac{dx}{dt} \tag{11.1}$$

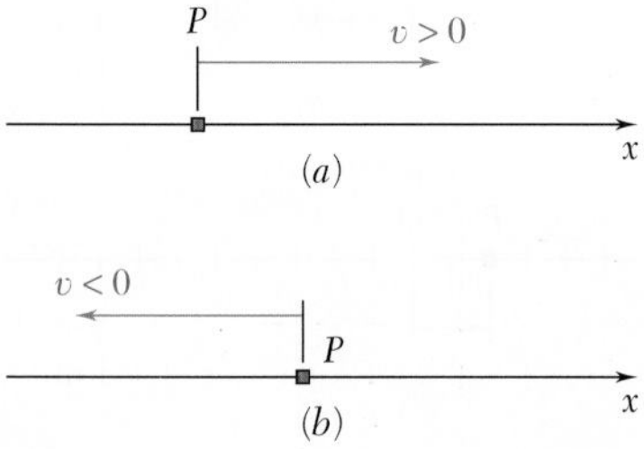

Fig. 11.3

The velocity v is represented by an algebraic number which can be positive or negative.† A positive value of v indicates that x increases, i.e., that the particle moves in the positive direction (Fig. 11.3*a*); a negative value of v indicates that x decreases, i.e., that the particle moves in the negative direction (Fig. 11.3*b*). The magnitude of v is known as the *speed* of the particle.

Consider the velocity v of the particle at time t and also its velocity $v + \Delta v$ at a later time $t + \Delta t$ (Fig. 11.4). The *average acceleration* of the particle over the time interval Δt is defined as the quotient of Δv and Δt:

$$\text{Average acceleration} = \frac{\Delta v}{\Delta t}$$

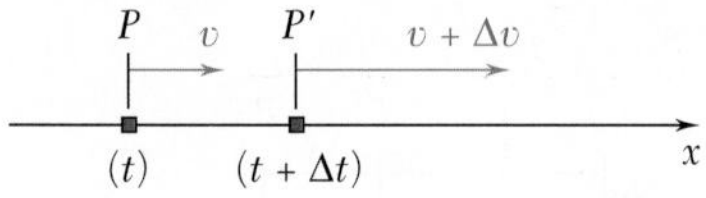

Fig. 11.4

If SI units are used, Δv is expressed in m/s and Δt in seconds; the average acceleration will thus be expressed in m/s^2. If U.S. customary units are used, Δv is expressed in ft/s and Δt in seconds; the average acceleration will then be expressed in ft/s^2.

The *instantaneous acceleration* a of the particle at the instant t is obtained from the average acceleration by choosing smaller and smaller values for Δt and Δv:

$$\text{Instantaneous acceleration} = a = \lim_{\Delta t \to 0} \frac{\Delta v}{\Delta t}$$

The instantaneous acceleration will also be expressed in m/s^2 or ft/s^2. The limit of the quotient, which is by definition the derivative of v

†As you will see in Sec. 11.9, the velocity is actually a vector quantity. However, since we are considering here the rectilinear motion of a particle, where the velocity of the particle has a known and fixed direction, we need only specify the sense and magnitude of the velocity; this can be conveniently done by using a scalar quantity with a plus or minus sign. The same is true of the acceleration of a particle in rectilinear motion.

with respect to t, measures the rate of change of the velocity. We write

$$a = \frac{dv}{dt} \tag{11.2}$$

or, substituting for v from (11.1),

$$a = \frac{d^2x}{dt^2} \tag{11.3}$$

The acceleration a is represented by an algebraic number which can be positive or negative.† A positive value of a indicates that the velocity (i.e., the algebraic number v) increases. This may mean that the particle is moving faster in the positive direction (Fig. 11.5*a*) or that it is moving more slowly in the negative direction (Fig. 11.5*b*); in both cases, Δv is positive. A negative value of a indicates that the velocity decreases; either the particle is moving more slowly in the positive direction (Fig. 11.5*c*) or it is moving faster in the negative direction (Fig. 11.5*d*).

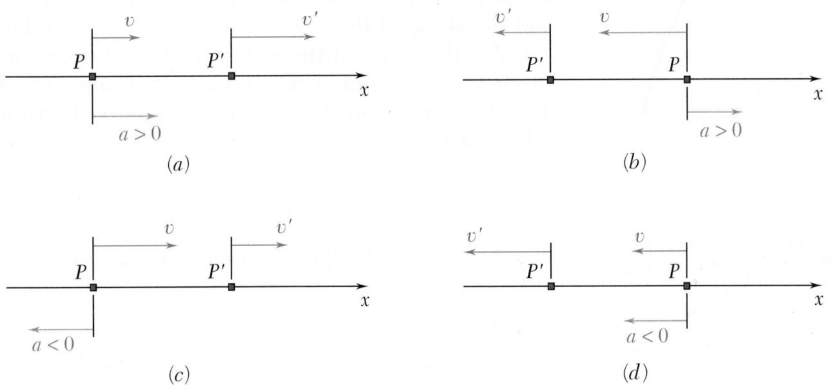

Fig. 11.5

The term *deceleration* is sometimes used to refer to a when the speed of the particle (i.e., the magnitude of v) decreases; the particle is then moving more slowly. For example, the particle of Fig. 11.5 is decelerated in parts b and c; it is truly accelerated (i.e., moves faster) in parts a and d.

Another expression for the acceleration can be obtained by eliminating the differential dt in Eqs. (11.1) and (11.2). Solving (11.1) for dt, we obtain $dt = dx/v$; substituting into (11.2), we write

$$a = v\frac{dv}{dx} \tag{11.4}$$

†See footnote, page 604.

EXAMPLE Consider a particle moving in a straight line, and assume that its position is defined by the equation

$$x = 6t^2 - t^3$$

where t is expressed in seconds and x in meters. The velocity v at any time t is obtained by differentiating x with respect to t:

$$v = \frac{dx}{dt} = 12t - 3t^2$$

The acceleration a is obtained by differentiating again with respect to t:

$$a = \frac{dv}{dt} = 12 - 6t$$

The position coordinate, the velocity, and the acceleration have been plotted against t in Fig. 11.6. The curves obtained are known as *motion curves*. Keep in mind, however, that the particle does not move along any of these curves; the particle moves in a straight line. Since the derivative of a function measures the slope of the corresponding curve, the slope of the x–t curve at any given time is equal to the value of v at that time and the slope of the v–t curve is equal to the value of a. Since $a = 0$ at $t = 2$ s, the slope of the v–t curve must be zero at $t = 2$ s; the velocity reaches a maximum at this instant. Also, since $v = 0$ at $t = 0$ and at $t = 4$ s, the tangent to the x–t curve must be horizontal for both of these values of t.

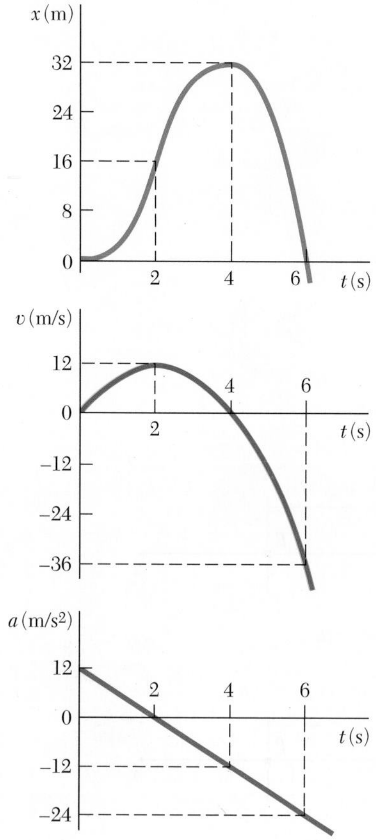

Fig. 11.6

A study of the three motion curves of Fig. 11.6 shows that the motion of the particle from $t = 0$ to $t = \infty$ can be divided into four phases:

1. The particle starts from the origin, $x = 0$, with no velocity but with a positive acceleration. Under this acceleration, the particle gains a positive velocity and moves in the positive direction. From $t = 0$ to $t = 2$ s, x, v, and a are all positive.
2. At $t = 2$ s, the acceleration is zero; the velocity has reached its maximum value. From $t = 2$ s to $t = 4$ s, v is positive, but a is negative; the particle still moves in the positive direction but more and more slowly; the particle is decelerating.
3. At $t = 4$ s, the velocity is zero; the position coordinate x has reached its maximum value. From then on, both v and a are negative; the particle is accelerating and moves in the negative direction with increasing speed.
4. At $t = 6$ s, the particle passes through the origin; its coordinate x is then zero, while the total distance traveled since the beginning of the motion is 64 m. For values of t larger than 6 s, x, v, and a will all be negative. The particle keeps moving in the negative direction, away from O, faster and faster. ■

11.3 DETERMINATION OF THE MOTION OF A PARTICLE

We saw in the preceding section that the motion of a particle is said to be known if the position of the particle is known for every value of the time t. In practice, however, a motion is seldom defined by a relation between x and t. More often, the conditions of the motion will be specified by the type of acceleration that the particle possesses. For example, a freely falling body will have a constant acceleration, directed downward and equal to 9.81 m/s^2, or 32.2 ft/s^2; a mass attached to a spring which has been stretched will have an acceleration proportional to the instantaneous elongation of the spring measured from the equilibrium position, etc. In general, the acceleration of the particle can be expressed as a function of one or more of the variables x, v, and t. In order to determine the position coordinate x in terms of t, it will thus be necessary to perform two successive integrations.

Let us consider three common classes of motion:

1. $a = f(t)$. *The Acceleration Is a Given Function of t.* Solving (11.2) for dv and substituting $f(t)$ for a, we write

$$dv = a\,dt$$
$$dv = f(t)\,dt$$

Integrating both members, we obtain the equation

$$\int dv = \int f(t)\,dt$$

which defines v in terms of t. It should be noted, however, that an arbitrary constant will be introduced as a result of the integration. This is due to the fact that there are many motions which correspond to the given acceleration $a = f(t)$. In order to uniquely define the motion of the particle, it is necessary to specify the *initial conditions* of the motion, i.e., the value v_0 of the velocity and the value x_0 of the position coordinate at $t = 0$. Replacing the indefinite integrals by *definite integrals* with lower limits corresponding to the initial conditions $t = 0$ and $v = v_0$ and upper limits corresponding to $t = t$ and $v = v$, we write

$$\int_{v_0}^{v} dv = \int_0^t f(t)\,dt$$
$$v - v_0 = \int_0^t f(t)\,dt$$

which yields v in terms of t.

Equation (11.1) can now be solved for dx,

$$dx = v\,dt$$

and the expression just obtained substituted for v. Both members are then integrated, the left-hand member with respect to x from $x = x_0$ to $x = x$, and the right-hand member with respect to t from $t = 0$ to $t = t$. The position coordinate x is thus obtained in terms of t; the motion is completely determined.

Two important particular cases will be studied in greater detail in Secs. 11.4 and 11.5: the case when $a = 0$, corresponding to a *uniform motion*, and the case when $a =$ constant, corresponding to a *uniformly accelerated motion*.

2. $a = f(x)$. *The Acceleration Is a Given Function of* x. Rearranging Eq. (11.4) and substituting $f(x)$ for a, we write

$$v\,dv = a\,dx$$
$$v\,dv = f(x)\,dx$$

Since each member contains only one variable, we can integrate the equation. Denoting again by v_0 and x_0, respectively, the initial values of the velocity and of the position coordinate, we obtain

$$\int_{v_0}^{v} v\,dv = \int_{x_0}^{x} f(x)\,dx$$

$$\tfrac{1}{2}v^2 - \tfrac{1}{2}v_0^2 = \int_{x_0}^{x} f(x)\,dx$$

which yields v in terms of x. We now solve (11.1) for dt,

$$dt = \frac{dx}{v}$$

and substitute for v the expression just obtained. Both members can then be integrated to obtain the desired relation between x and t. However, in most cases this last integration cannot be performed analytically and one must resort to a numerical method of integration.

3. $a = f(v)$. *The Acceleration Is a Given Function of* v. We can now substitute $f(v)$ for a in either (11.2) or (11.4) to obtain either of the following relations:

$$f(v) = \frac{dv}{dt} \qquad f(v) = v\frac{dv}{dx}$$

$$dt = \frac{dv}{f(v)} \qquad dx = \frac{v\,dv}{f(v)}$$

Integration of the first equation will yield a relation between v and t; integration of the second equation will yield a relation between v and x. Either of these relations can be used in conjunction with Eq. (11.1) to obtain the relation between x and t which characterizes the motion of the particle.

SAMPLE PROBLEM 11.1

The position of a particle which moves along a straight line is defined by the relation $x = t^3 - 6t^2 - 15t + 40$, where x is expressed in meters and t in seconds. Determine (*a*) the time at which the velocity will be zero, (*b*) the position and distance traveled by the particle at that time, (*c*) the acceleration of the particle at that time, (*d*) the distance traveled by the particle from $t = 4$ s to $t = 6$ s.

SOLUTION

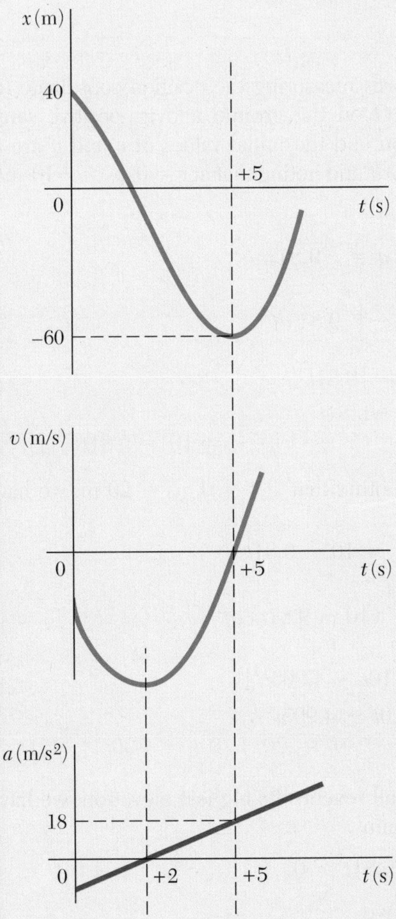

The equations of motion are

$$x = t^3 - 6t^2 - 15t + 40 \quad (1)$$

$$v = \frac{dx}{dt} = 3t^2 - 12t - 15 \quad (2)$$

$$a = \frac{dv}{dt} = 6t - 12 \quad (3)$$

a. Time at Which v = 0. We set $v = 0$ in (2):

$$3t^2 - 12t - 15 = 0 \qquad t = -1 \text{ s} \quad \text{and} \quad t = +5 \text{ s} \;◀$$

Only the root $t = +5$ s corresponds to a time after the motion has begun: for $t < 5$ s, $v < 0$, the particle moves in the negative direction; for $t > 5$ s, $v > 0$, the particle moves in the positive direction.

b. Position and Distance Traveled When v = 0. Carrying $t = +5$ s into (1), we have

$$x_5 = (5)^3 - 6(5)^2 - 15(5) + 40 \qquad x_5 = -60 \text{ m} \;◀$$

The initial position at $t = 0$ was $x_0 = +40$ m. Since $v \neq 0$ during the interval $t = 0$ to $t = 5$ s, we have

$$\text{Distance traveled} = x_5 - x_0 = -60 \text{ m} - 40 \text{ m} = -100 \text{ m}$$

$$\text{Distance traveled} = 100 \text{ m in the negative direction} \;◀$$

c. Acceleration When v = 0. We substitute $t = +5$ s into (3):

$$a_5 = 6(5) - 12 \qquad a_5 = +18 \text{ m/s}^2 \;◀$$

d. Distance Traveled from t = 4 s to t = 6 s. The particle moves in the negative direction from $t = 4$ s to $t = 5$ s and in the positive direction from $t = 5$ s to $t = 6$ s; therefore, the distance traveled during each of these time intervals will be computed separately.

From $t = 4$ s to $t = 5$ s: $x_5 = -60$ m

$$x_4 = (4)^3 - 6(4)^2 - 15(4) + 40 = -52 \text{ m}$$

$$\text{Distance traveled} = x_5 - x_4 = -60 \text{ m} - (-52 \text{ m}) = -8 \text{ m}$$
$$= 8 \text{ m in the negative direction}$$

From $t = 5$ s to $t = 6$ s: $x_5 = -60$ m

$$x_6 = (6)^3 - 6(6)^2 - 15(6) + 40 = -50 \text{ m}$$

$$\text{Distance traveled} = x_6 - x_5 = -50 \text{ m} - (-60 \text{ m}) = +10 \text{ m}$$
$$= 10 \text{ m in the positive direction}$$

Total distance traveled from $t = 4$ s to $t = 6$ s is 8 m + 10 m = 18 m ◀

SAMPLE PROBLEM 11.2

A ball is tossed with a velocity of 10 m/s directed vertically upward from a window located 20 m above the ground. Knowing that the acceleration of the ball is constant and equal to 9.81 m/s² downward, determine (*a*) the velocity v and elevation y of the ball above the ground at any time t, (*b*) the highest elevation reached by the ball and the corresponding value of t, (*c*) the time when the ball will hit the ground and the corresponding velocity. Draw the v–t and y–t curves.

SOLUTION

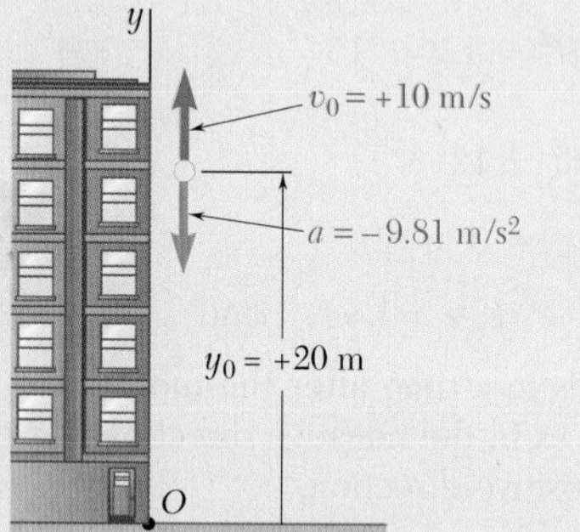

a. Velocity and Elevation. The y axis measuring the position coordinate (or elevation) is chosen with its origin O on the ground and its positive sense upward. The value of the acceleration and the initial values of v and y are as indicated. Substituting for a in $a = dv/dt$ and noting that at $t = 0$, $v_0 = +10$ m/s, we have

$$\frac{dv}{dt} = a = -9.81\,\text{m/s}^2$$

$$\int_{v_0=10}^{v} dv = -\int_0^t 9.81\, dt$$

$$[v]_{10}^{v} = -[9.81t]_0^t$$

$$v - 10 = -9.81t$$

$$v = 10 - 9.81t \quad (1) \blacktriangleleft$$

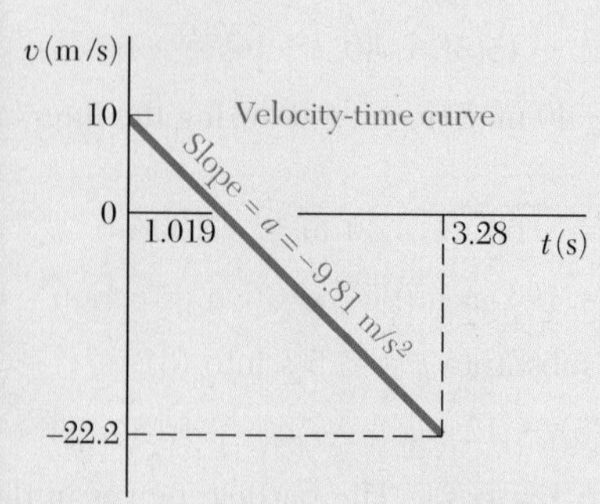

Substituting for v in $v = dy/dt$ and noting that at $t = 0$, $y_0 = 20$ m, we have

$$\frac{dy}{dt} = v = 10 - 9.81t$$

$$\int_{y_0=20}^{y} dy = \int_0^t (10 - 9.81t)\, dt$$

$$[y]_{20}^{y} = [10t - 4.905t^2]_0^t$$

$$y - 20 = 10t - 4.905t^2$$

$$y = 20 + 10t - 4.905t^2 \quad (2) \blacktriangleleft$$

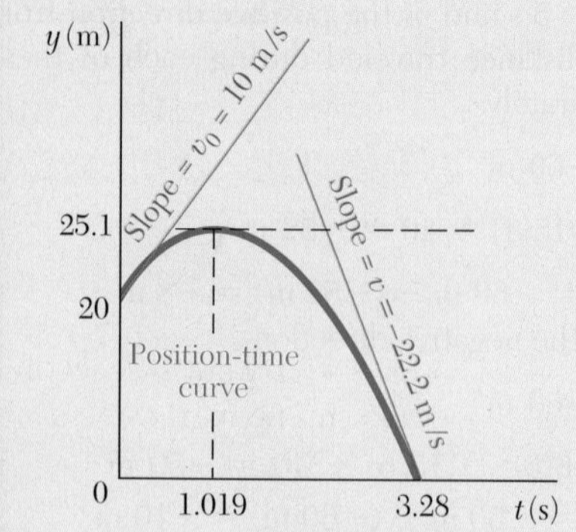

b. Highest Elevation. When the ball reaches its highest elevation, we have $v = 0$. Substituting into (1), we obtain

$$10 - 9.81t = 0 \qquad t = 1.019\text{ s} \blacktriangleleft$$

Carrying $t = 1.019$ s into (2), we have

$$y = 20 + 10(1.019) - 4.905(1.019)^2 \qquad y = 25.1\text{ m} \blacktriangleleft$$

c. Ball Hits the Ground. When the ball hits the ground, we have $y = 0$. Substituting into (2), we obtain

$$20 + 10t - 4.905t^2 = 0 \qquad t = -1.243\text{ s} \quad \text{and} \quad t = +3.28\text{ s} \blacktriangleleft$$

Only the root $t = +3.28$ s corresponds to a time after the motion has begun. Carrying this value of t into (1), we have

$$v = 10 - 9.81(3.28) = -22.2\text{ m/s} \qquad v = 22.2\text{ m/s}\downarrow \blacktriangleleft$$

SAMPLE PROBLEM 11.3

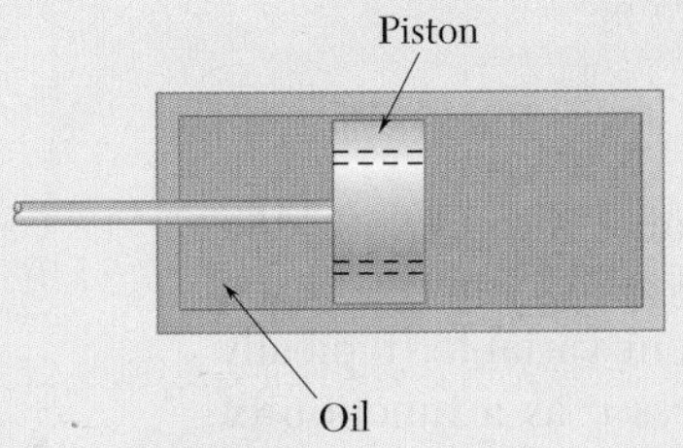

The brake mechanism used to reduce recoil in certain types of guns consists essentially of a piston attached to the barrel and moving in a fixed cylinder filled with oil. As the barrel recoils with an initial velocity v_0, the piston moves and oil is forced through orifices in the piston, causing the piston and the barrel to decelerate at a rate proportional to their velocity; that is, $a = -kv$. Express (*a*) v in terms of t, (*b*) x in terms of t, (*c*) v in terms of x. Draw the corresponding motion curves.

SOLUTION

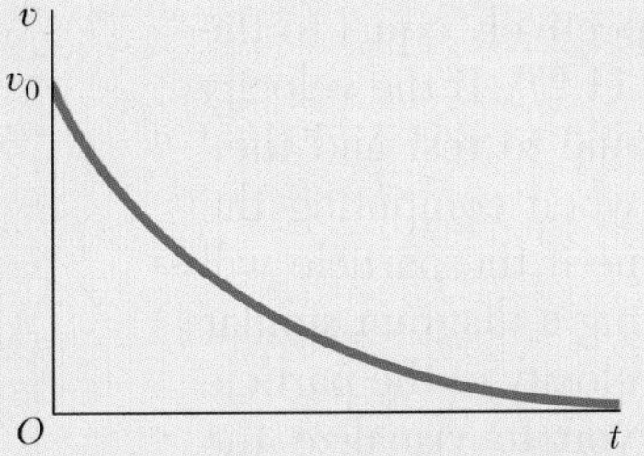

a. *v* in Terms of *t*. Substituting $-kv$ for a in the fundamental formula defining acceleration, $a = dv/dt$, we write

$$-kv = \frac{dv}{dt} \qquad \frac{dv}{v} = -k\,dt \qquad \int_{v_0}^{v} \frac{dv}{v} = -k\int_0^t dt$$

$$\ln\frac{v}{v_0} = -kt \qquad\qquad v = v_0 e^{-kt} \blacktriangleleft$$

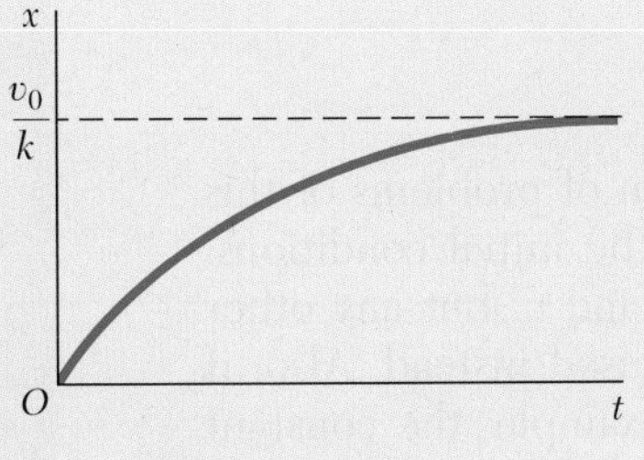

b. *x* in Terms of *t*. Substituting the expression just obtained for v into $v = dx/dt$, we write

$$v_0 e^{-kt} = \frac{dx}{dt}$$

$$\int_0^x dx = v_0 \int_0^t e^{-kt}\,dt$$

$$x = -\frac{v_0}{k}[e^{-kt}]_0^t = -\frac{v_0}{k}(e^{-kt} - 1)$$

$$x = \frac{v_0}{k}(1 - e^{-kt}) \blacktriangleleft$$

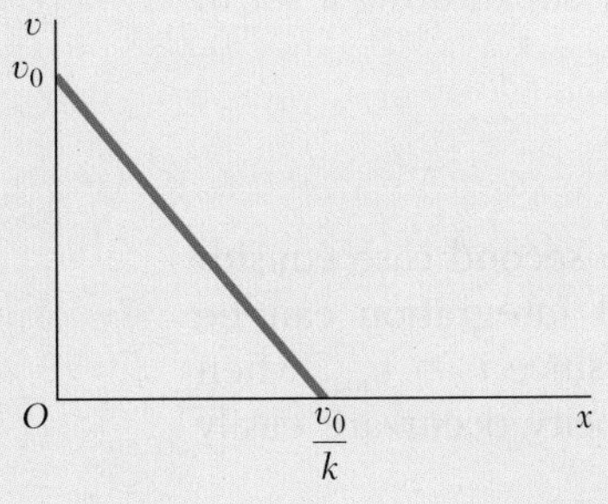

c. *v* in Terms of *x*. Substituting $-kv$ for a in $a = v\,dv/dx$, we write

$$-kv = v\frac{dv}{dx}$$

$$dv = -k\,dx$$

$$\int_{v_0}^{v} dv = -k\int_0^x dx$$

$$v - v_0 = -kx \qquad\qquad v = v_0 - kx \blacktriangleleft$$

Check. Part *c* could have been solved by eliminating t from the answers obtained for parts *a* and *b*. This alternative method can be used as a check. From part *a* we obtain $e^{-kt} = v/v_0$; substituting into the answer of part *b*, we obtain

$$x = \frac{v_0}{k}(1 - e^{-kt}) = \frac{v_0}{k}\left(1 - \frac{v}{v_0}\right) \qquad v = v_0 - kx \qquad \text{(checks)}$$

SOLVING PROBLEMS ON YOUR OWN

In the problems for this lesson, you will be asked to determine the *position*, the *velocity*, or the *acceleration* of a particle in *rectilinear motion*. As you read each problem, it is important that you identify both the independent variable (typically t or x) and what is required (for example, the need to express v as a function of x). You may find it helpful to start each problem by writing down both the given information and a simple statement of what is to be determined.

1. Determining *v*(*t*) and *a*(*t*) for a given *x*(*t*). As explained in Sec. 11.2, the first and the second derivatives of x with respect to t are respectively equal to the velocity and the acceleration of the particle [Eqs. (11.1) and (11.2)]. If the velocity and the acceleration have opposite signs, the particle can come to rest and then move in the opposite direction [Sample Prob. 11.1]. Thus, when computing the total distance traveled by a particle, you should first determine if the particle will come to rest during the specified interval of time. Constructing a diagram similar to that of Sample Prob. 11.1 that shows the position and the velocity of the particle at each critical instant ($v = v_{max}$, $v = 0$, etc.) will help you to visualize the motion.

2. Determining *v*(*t*) and *x*(*t*) for a given *a*(*t*). The solution of problems of this type was discussed in the first part of Sec. 11.3. We used the initial conditions, $t = 0$ and $v = v_0$, for the lower limits of the integrals in t and v, but any other known state (for example, $t = t_1$, $v = v_1$) could have been used instead. Also, if the given function $a(t)$ contains an unknown constant (for example, the constant k if $a = kt$), you will first have to determine that constant by substituting a set of known values of t and a in the equation defining $a(t)$.

3. Determining *v*(*x*) and *x*(*t*) for a given *a*(*x*). This is the second case considered in Sec. 11.3. We again note that the lower limits of integration can be any known state (for example, $x = x_1$, $v = v_1$). In addition, since $v = v_{max}$ when $a = 0$, the positions where the maximum values of the velocity occur are easily determined by writing $a(x) = 0$ and solving for x.

4. Determining *v*(*x*), *v*(*t*), and *x*(*t*) for a given *a*(*v*). This is the last case treated in Sec. 11.3; the appropriate solution techniques for problems of this type are illustrated in Sample Prob. 11.3. All of the general comments for the preceding cases once again apply. Note that Sample Prob. 11.3 provides a summary of how and when to use the equations $v = dx/dt$, $a = dv/dt$, and $a = v\,dv/dx$.

PROBLEMS†

CONCEPT QUESTIONS

11.CQ1 A bus travels the 100 km between A and B at 50 km/h and then another 100 km between B and C at 70 km/h. The average speed of the bus for the entire 200-km trip is:

a. More than 60 km/h.
b. Equal to 60 km/h.
c. Less than 60 km/h.

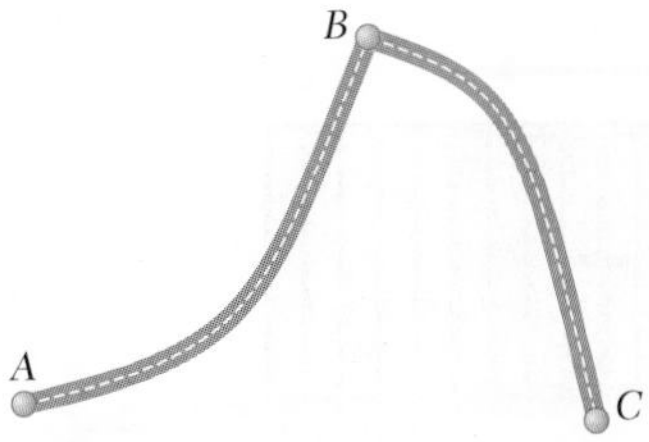

Fig. P11.CQ1

11.CQ2 Two cars A and B race each other down a straight road. The position of each car as a function of time is shown. Which of the following statements are true (more than one answer can be correct)?

a. At time t_2 both cars have traveled the same distance.
b. At time t_1 both cars have the same speed.
c. Both cars have the same speed at some time $t < t_1$.
d. Both cars have the same acceleration at some time $t < t_1$.
e. Both cars have the same acceleration at some time $t_1 < t < t_2$.

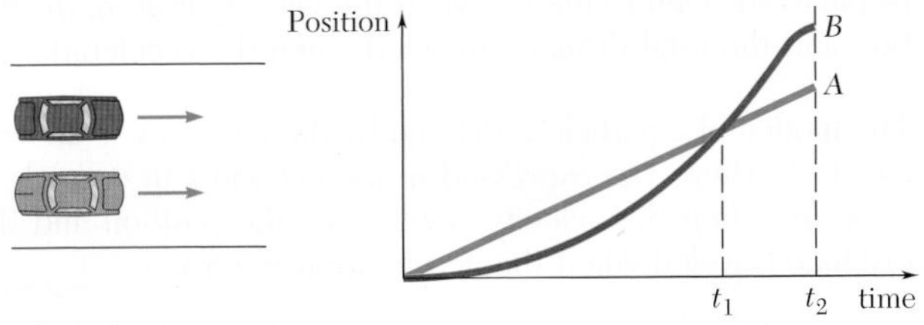

Fig. P11.CQ2

END-OF-SECTION PROBLEMS

11.1 The motion of a particle is defined by the relation $x = t^4 - 10t^2 + 8t + 12$, where x and t are expressed in meters and seconds, respectively. Determine the position, the velocity, and the acceleration of the particle when $t = 1$ s.

11.2 The motion of a particle is defined by the relation $x = 2t^3 - 9t^2 + 12t + 10$, where x and t are expressed in meters and seconds, respectively. Determine the time, the position, and the acceleration of the particle when $v = 0$.

†Answers to all problems set in straight type (such as **11.1**) are given at the end of the book. Answers to problems with a number set in italic type (such as ***11.7***) are not given.

11.3 The vertical motion of mass A is defined by the relation $x = 10 \sin 2t + 15 \cos 2t + 100$, where x and t are expressed in millimeters and seconds, respectively. Determine (*a*) the position, velocity, and acceleration of A when $t = 1$ s, (*b*) the maximum velocity and acceleration of A.

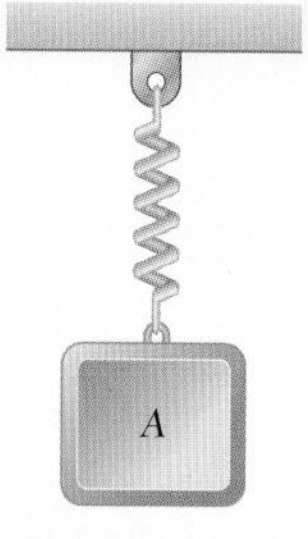

Fig. P11.3

11.4 A loaded railroad car is rolling at a constant velocity when it couples with a spring and dashpot bumper system. After the coupling, the motion of the car is defined by the relation $x = 60e^{-4.8t} \sin 16t$, where x and t are expressed in millimeters and seconds, respectively. Determine the position, the velocity, and the acceleration of the railroad car when (*a*) $t = 0$, (*b*) $t = 0.3$ s.

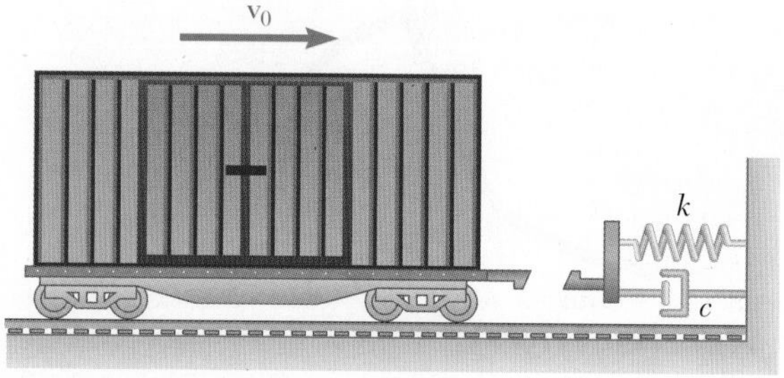

Fig. P11.4

11.5 The motion of a particle is defined by the relation $x = 6t^4 - 2t^3 - 12t^2 + 3t + 3$, where x and t are expressed in meters and seconds, respectively. Determine the time, the position, and the velocity when $a = 0$.

11.6 The motion of a particle is defined by the relation $x = t^3 - 9t^2 + 24t - 8$, where x and t are expressed in meters and seconds, respectively. Determine (*a*) when the velocity is zero, (*b*) the position and the total distance traveled when the acceleration is zero.

11.7 The motion of a particle is defined by the relation $x = 2t^3 - 15t^2 + 24t + 4$, where x is expressed in meters and t in seconds. Determine (*a*) when the velocity is zero, (*b*) the position and the total distance traveled when the acceleration is zero.

11.8 The motion of a particle is defined by the relation $x = t^3 - 6t^2 - 36t - 40$, where x and t are expressed in meters and seconds, respectively. Determine (*a*) when the velocity is zero, (*b*) the velocity, the acceleration, and the total distance traveled when $x = 0$.

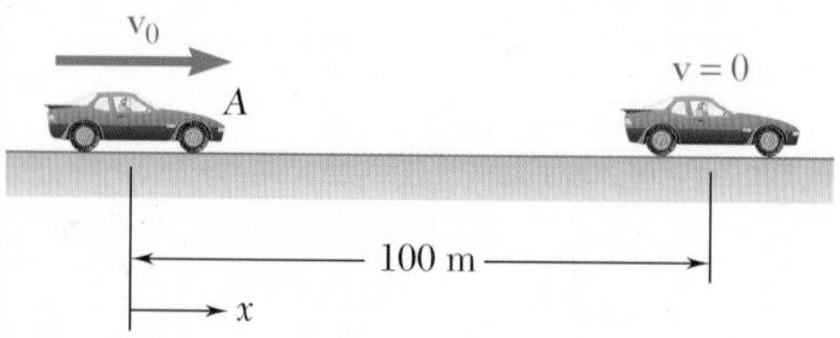

Fig. P11.9

11.9 The brakes of a car are applied, causing it to slow down at a rate of 3 m/s^2. Knowing that the car stops in 100 m, determine (*a*) how fast the car was traveling immediately before the brakes were applied, (*b*) the time required for the car to stop.

11.10 The acceleration of a particle is directly proportional to the time t. At $t = 0$, the velocity of the particle is $v = 400$ mm/s. Knowing that $v = 375$ mm/s and that $x = 500$ mm when $t = 1$ s, determine the velocity, the position, and the total distance traveled when $t = 7$ s.

11.11 The acceleration of a particle is directly proportional to the square of the time t. When $t = 0$, the particle is at $x = 24$ m. Knowing that at $t = 6$ s, $x = 96$ m and $v = 18$ m/s, express x and v in terms of t.

11.12 The acceleration of a particle is defined by the relation $a = kt^2$. (*a*) Knowing that $v = -8$ m/s when $t = 0$ and that $v = +8$ m/s when $t = 2$ s, determine the constant k. (*b*) Write the equations of motion, knowing also that $x = 0$ when $t = 2$ s.

11.13 The acceleration of point A is defined by the relation $a = -1.8 \sin kt$, where a and t are expressed in m/s^2 and seconds, respectively, and $k = 3$ rad/s. Knowing that $x = 0$ and $v = 0.6$ m/s when $t = 0$, determine the velocity and position of point A when $t = 0.5$ s.

11.14 The acceleration of point A is defined by the relation $a = -1.08 \sin kt - 1.44 \cos kt$, where a and t are expressed in m/s^2 and seconds, respectively, and $k = 3$ rad/s. Knowing that $x = 0.16$ m and $v = 0.36$ m/s when $t = 0$, determine the velocity and position of point A when $t = 0.5$ s.

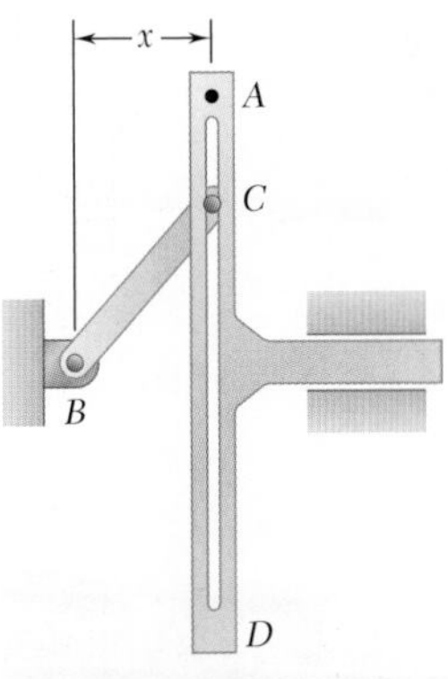

Fig. *P11.13* and *P11.14*

11.15 A piece of electronic equipment that is surrounded by packing material is dropped so that it hits the ground with a speed of 4 m/s. After contact the equipment experiences an acceleration of $a = -kx$, where k is a constant and x is the compression of the packing material. If the packing material experiences a maximum compression of 20 mm, determine the maximum acceleration of the equipment.

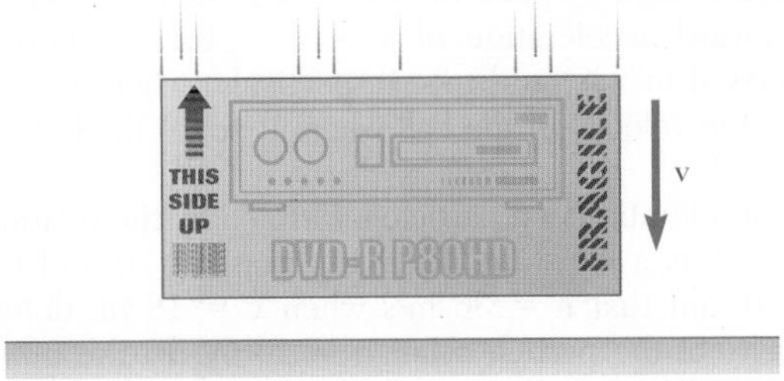

Fig. P11.15

11.16 A projectile enters a resisting medium at $x = 0$ with an initial velocity $\mathbf{v}_0 = 270$ m/s and travels 100 mm before coming to rest. Assuming that the velocity of the projectile is defined by the relation $v = v_0 - kx$, where v is expressed in m/s and x is in meters, determine (*a*) the initial acceleration of the projectile, (*b*) the time required for the projectile to penetrate 97.5 mm into the resisting medium.

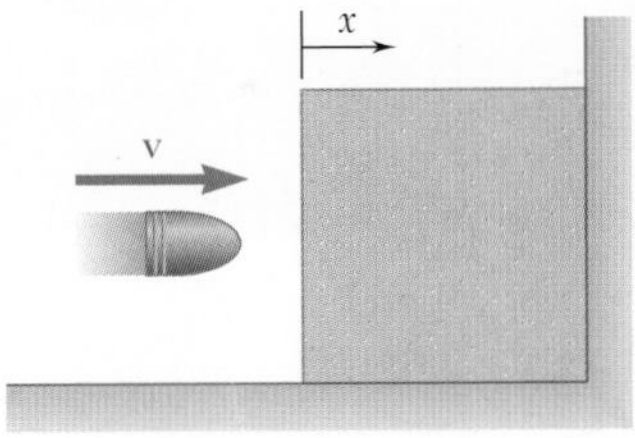

Fig. P11.16

11.17 The acceleration of a particle is defined by the relation $a = -k/x$. It has been experimentally determined that $v = 5$ m/s when $x = 0.2$ m and that $v = 3$ m/s when $x = 0.4$ m. Determine (*a*) the velocity of the particle when $x = 0.5$ m, (*b*) the position of the particle at which its velocity is zero.

11.18 A brass (nonmagnetic) block A and a steel magnet B are in equilibrium in a brass tube under the magnetic repelling force of another steel magnet C located at a distance $x = 0.004$ m from B. The force is inversely proportional to the square of the distance between B and C. If block A is suddenly removed, the acceleration of block B is $a = -9.81 + k/x^2$, where a and x are expressed in m/s^2 and meters, respectively, and $k = 4 \times 10^{-4}$ m^3/s^2. Determine the maximum velocity and acceleration of B.

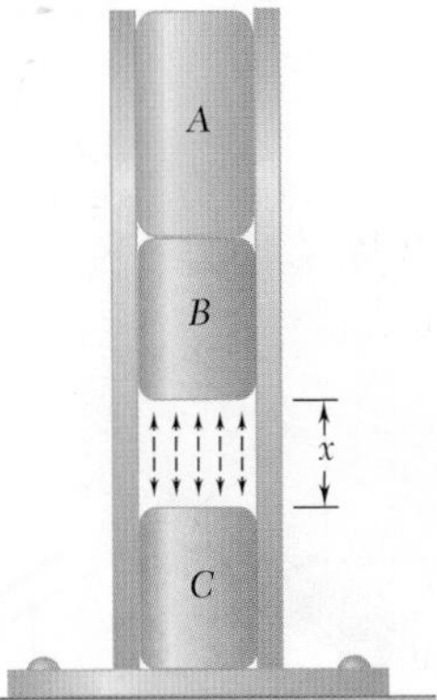

Fig. P11.18

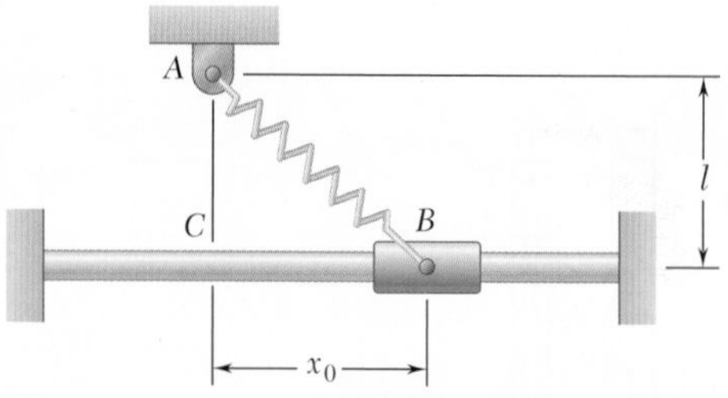

Fig. *P11.20*

11.19 Based on experimental observations, the acceleration of a particle is defined by the relation $a = -(0.1 + \sin x/b)$, where a and x are expressed in m/s^2 and meters, respectively. Knowing that $b = 0.8$ m and that $v = 1$ m/s when $x = 0$, determine (*a*) the velocity of the particle when $x = -1$ m, (*b*) the position where the velocity is maximum, (*c*) the maximum velocity.

11.20 A spring *AB* is attached to a support at *A* and to a collar. The unstretched length of the spring is *l*. Knowing that the collar is released from rest at $x = x_0$ and has an acceleration defined by the relation $a = -100(x - lx/\sqrt{l^2 + x^2})$, determine the velocity of the collar as it passes through point *C*.

11.21 The acceleration of a particle is defined by the relation $a = -0.8v$, where a is expressed in m/s^2 and v in m/s. Knowing that at $t = 0$ the velocity is 1 m/s, determine (*a*) the distance the particle will travel before coming to rest, (*b*) the time required for the particle's velocity to be reduced by 50 percent of its initial value.

11.22 Starting from $x = 0$ with no initial velocity, a particle is given an acceleration $a = 0.8\sqrt{v^2 + 49}$, where a and v are expressed in m/s^2 and m/s, respectively. Determine (*a*) the position of the particle when $v = 24$ m/s, (*b*) the speed of the particle when $x = 40$ m.

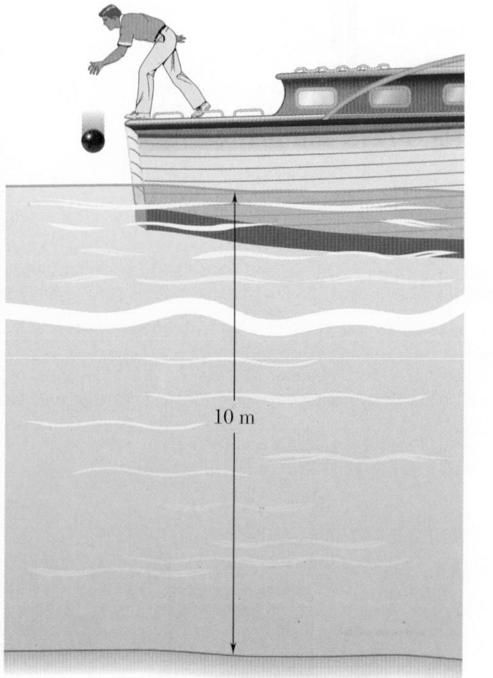

Fig. P11.23

11.23 A bowling ball is dropped from a boat so that it strikes the surface of a lake with a speed of 8 m/s. Assuming the ball experiences a downward acceleration of $a = 3 - 0.1v^2$ (where a and v are expressed in m/s^2 and m/s, respectively) when in the water, determine the velocity of the ball when it strikes the bottom of the lake.

11.24 The acceleration of a particle is defined by the relation $a = -k\sqrt{v}$, where k is a constant. Knowing that $x = 0$ and $v = 81$ m/s at $t = 0$ and that $v = 36$ m/s when $x = 18$ m, determine (*a*) the velocity of the particle when $x = 20$ m, (*b*) the time required for the particle to come to rest.

11.25 A particle is projected to the right from the position $x = 0$ with an initial velocity of 9 m/s. If the acceleration of the particle is defined by the relation $a = -0.6v^{3/2}$, where a and v are expressed in m/s^2 and m/s, respectively, determine (*a*) the distance the particle will have traveled when its velocity is 4 m/s, (*b*) the time when $v = 1$ m/s, (*c*) the time required for the particle to travel 6 m.

11.26 The acceleration of a particle is defined by the relation $a = 0.4(1 - kv)$, where k is a constant. Knowing that at $t = 0$ the particle starts from rest at $x = 4$ m and that when $t = 15$ s, $v = 4$ m/s, determine (*a*) the constant k, (*b*) the position of the particle when $v = 6$ m/s, (*c*) the maximum velocity of the particle.

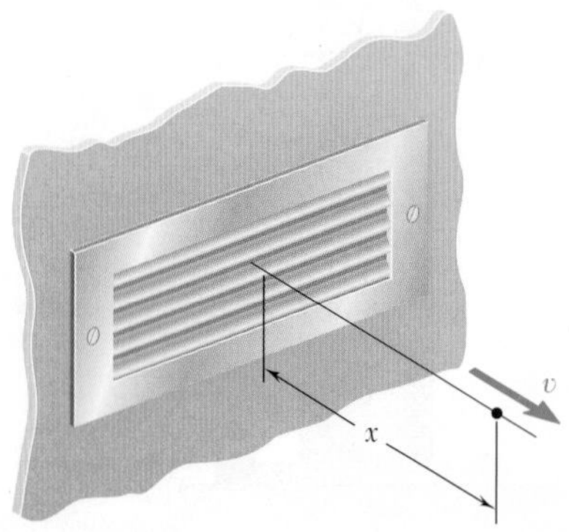

Fig. P11.27

11.27 Experimental data indicate that in a region downstream of a given louvered supply vent the velocity of the emitted air is defined by $v = 0.18v_0/x$, where v and x are expressed in m/s and meters, respectively, and v_0 is the initial discharge velocity of the air. For $v_0 = 3.6$ m/s, determine (*a*) the acceleration of the air at $x = 2$ m, (*b*) the time required for the air to flow from $x = 1$ to $x = 3$ m.

11.28 Based on observations, the speed of a jogger can be approximated by the relation $v = 12(1 - 0.06x)^{0.3}$, where v and x are expressed in km/h and km, respectively. Knowing that $x = 0$ at $t = 0$, determine (*a*) the distance the jogger has run when $t = 1$ h, (*b*) the jogger's acceleration in m/s^2 at $t = 0$, (*c*) the time required for the jogger to run 9 Km.

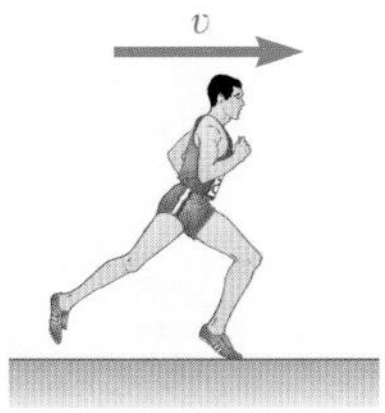

Fig. P11.28

11.29 The acceleration due to gravity at an altitude y above the surface of the earth can be expressed as

$$a = \frac{-9.81}{\left[1 + \left(\dfrac{y}{6.37 \times 10^6}\right)\right]^2}$$

where a and y are expressed in m/s^2 and metre, respectively. Using this expression, compute the height reached by a projectile fired vertically upward from the surface of the earth if its initial velocity is (*a*) 540 m/s, (*b*) 900 m/s, (*c*) 11,180 m/s.

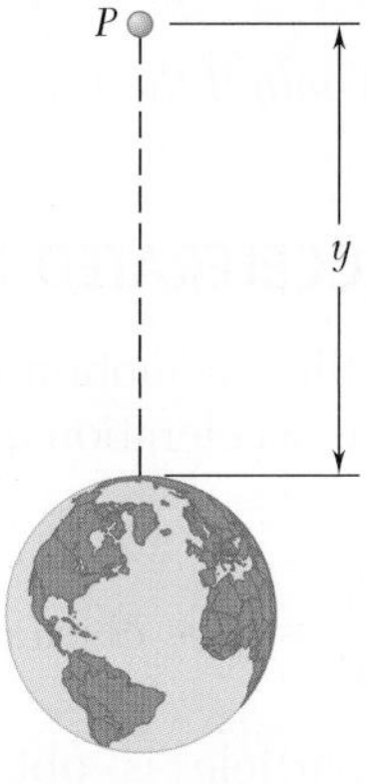

Fig. *P11.29*

11.30 The acceleration due to gravity of a particle falling toward the earth is $a = -gR^2/r^2$, where r is the distance from the *center* of the earth to the particle, R is the radius of the earth, and g is the acceleration due to gravity at the surface of the earth. If $R = 6370$ km, calculate the *escape velocity*, that is, the minimum velocity with which a particle must be projected vertically upward from the surface of the earth if it is not to return to the earth. (*Hint:* $v = 0$ for $r = \infty$.)

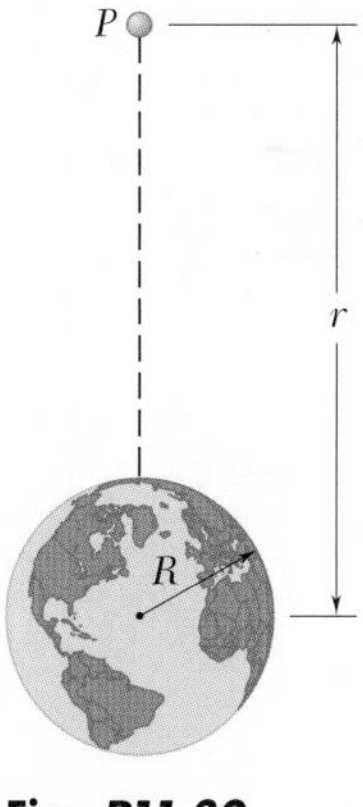

Fig. *P11.30*

11.31 The velocity of a particle is $v = v_0[1 - \sin(\pi t/T)]$. Knowing that the particle starts from the origin with an initial velocity v_0, determine (*a*) its position and its acceleration at $t = 3T$, (*b*) its average velocity during the interval $t = 0$ to $t = T$.

11.32 The velocity of a slider is defined by the relation $v = v'\sin(\omega_n t + \phi)$. Denoting the velocity and the position of the slider at $t = 0$ by v_0 and x_0, respectively, and knowing that the maximum displacement of the slider is $2x_0$, show that (*a*) $v' = (v_0^2 + x_0^2\omega_n^2)/2x_0\omega_n$, (*b*) the maximum value of the velocity occurs when $x = x_0[3 - (v_0/x_0\omega_n)^2]/2$.

11.4 UNIFORM RECTILINEAR MOTION

Uniform rectilinear motion is a type of straight-line motion which is frequently encountered in practical applications. In this motion, the acceleration a of the particle is zero for every value of t. The velocity v is therefore constant, and Eq. (11.1) becomes

$$\frac{dx}{dt} = v = \text{constant}$$

The position coordinate x is obtained by integrating this equation. Denoting by x_0 the initial value of x, we write

$$\int_{x_0}^{x} dv = v \int_0^t dt$$

$$x - x_0 = vt$$

$$x = x_0 + vt \tag{11.5}$$

This equation can be used *only if the velocity of the particle is known to be constant.*

11.5 UNIFORMLY ACCELERATED RECTILINEAR MOTION

Uniformly accelerated rectilinear motion is another common type of motion. In this motion, the acceleration a of the particle is constant, and Eq. (11.2) becomes

$$\frac{dv}{dt} = a = \text{constant}$$

The velocity v of the particle is obtained by integrating this equation:

$$\int_{v_0}^{v} dv = a \int_0^t dt$$

$$v - v_0 = at$$

$$v = v_0 + at \tag{11.6}$$

where v_0 is the initial velocity. Substituting for v in (11.1), we write

$$\frac{dx}{dt} = v_0 + at$$

Denoting by x_0 the initial value of x and integrating, we have

$$\int_{x_0}^{x} dx = \int_0^t (v_0 + at)\, dt$$

$$x - x_0 = v_0 t + \tfrac{1}{2}at^2$$

$$x = x_0 + v_0 t + \tfrac{1}{2}at^2 \tag{11.7}$$

We can also use Eq. (11.4) and write

$$v\frac{dv}{dx} = a = \text{constant}$$

$$v\,dv = a\,dx$$

Integrating both sides, we obtain

$$\int_{v_0}^{v} v\,dv = a\int_{x_0}^{x} dx$$

$$\tfrac{1}{2}(v^2 - v_0^2) = a(x - x_0)$$

$$v^2 = v_0^2 + 2a(x - x_0) \qquad (11.8)$$

The three equations we have derived provide useful relations among position coordinate, velocity, and time in the case of a uniformly accelerated motion, as soon as appropriate values have been substituted for a, v_0, and x_0. The origin O of the x axis should first be defined and a positive direction chosen along the axis; this direction will be used to determine the signs of a, v_0, and x_0. Equation (11.6) relates v and t and should be used when the value of v corresponding to a given value of t is desired, or inversely. Equation (11.7) relates x and t; Eq. (11.8) relates v and x. An important application of uniformly accelerated motion is the motion of a *freely falling body*. The acceleration of a freely falling body (usually denoted by g) is equal to 9.81 m/s^2 or 32.2 ft/s^2.

It is important to keep in mind that the three equations can be used *only when the acceleration of the particle is known to be constant*. If the acceleration of the particle is variable, its motion should be determined from the fundamental equations (11.1) to (11.4) according to the methods outlined in Sec. 11.3.

11.6 MOTION OF SEVERAL PARTICLES

When several particles move independently along the same line, independent equations of motion can be written for each particle. Whenever possible, time should be recorded from the same initial instant for all particles, and displacements should be measured from the same origin and in the same direction. In other words, a single clock and a single measuring tape should be used.

Relative Motion of Two Particles. Consider two particles A and B moving along the same straight line (Fig. 11.7). If the position coordinates x_A and x_B are measured from the same origin, the difference $x_B - x_A$ defines the *relative position coordinate of B with respect to A* and is denoted by $x_{B/A}$. We write

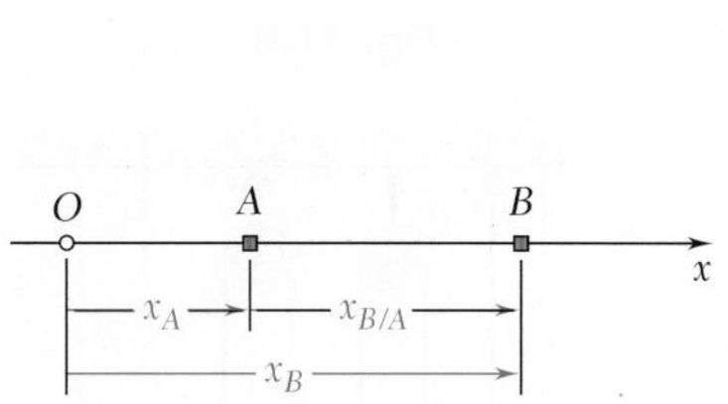

Fig. 11.7

$$x_{B/A} = x_B - x_A \qquad \text{or} \qquad x_B = x_A + x_{B/A} \qquad (11.9)$$

Regardless of the positions of A and B with respect to the origin, a positive sign for $x_{B/A}$ means that B is to the right of A, and a negative sign means that B is to the left of A.

Photo 11.2 Multiple cables and pulleys are used by this shipyard crane.

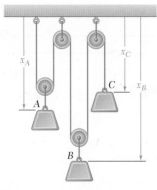

The rate of change of $x_{B/A}$ is known as the *relative velocity of B with respect to A* and is denoted by $v_{B/A}$. Differentiating (11.9), we write

$$v_{B/A} = v_B - v_A \qquad \text{or} \qquad v_B = v_A + v_{B/A} \qquad (11.10)$$

A positive sign for $v_{B/A}$ means that B is *observed from A* to move in the positive direction; a negative sign means that it is observed to move in the negative direction.

The rate of change of $v_{B/A}$ is known as the *relative acceleration of B with respect to A* and is denoted by $a_{B/A}$. Differentiating (11.10), we obtain†

$$a_{B/A} = a_B - a_A \qquad \text{or} \qquad a_B = a_A + a_{B/A} \qquad (11.11)$$

Dependent Motions. Sometimes, the position of a particle will depend upon the position of another particle or of several other particles. The motions are then said to be *dependent*. For example, the position of block B in Fig. 11.8 depends upon the position of block A. Since the rope $ACDEFG$ is of constant length, and since the lengths of the portions of rope CD and EF wrapped around the pulleys remain constant, it follows that the sum of the lengths of the segments AC, DE, and FG is constant. Observing that the length of the segment AC differs from x_A only by a constant and that, similarly, the lengths of the segments DE and FG differ from x_B only by a constant, we write

$$x_A + 2x_B = \text{constant}$$

Since only one of the two coordinates x_A and x_B can be chosen arbitrarily, we say that the system shown in Fig. 11.8 has *one degree of freedom*. From the relation between the position coordinates x_A and x_B, it follows that if x_A is given an increment Δx_A, that is, if block A is lowered by an amount Δx_A, the coordinate x_B will receive an increment $\Delta x_B = -\frac{1}{2}\Delta x_A$. In other words, block B will rise by half the same amount; this can easily be checked directly from Fig. 11.8.

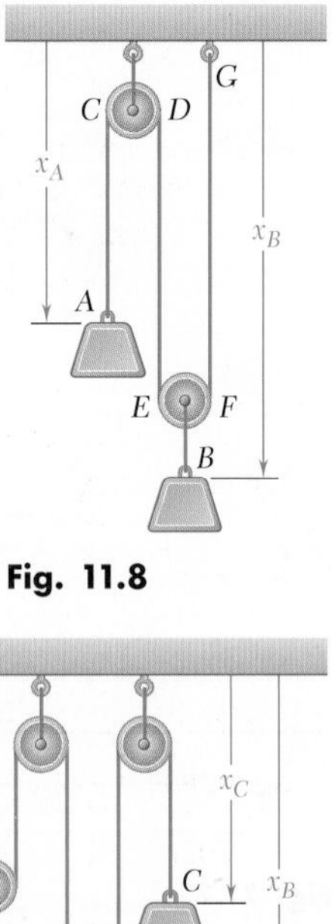

Fig. 11.8

In the case of the three blocks of Fig. 11.9, we can again observe that the length of the rope which passes over the pulleys is constant, and thus the following relation must be satisfied by the position coordinates of the three blocks:

$$2x_A + 2x_B + x_C = \text{constant}$$

Since two of the coordinates can be chosen arbitrarily, we say that the system shown in Fig. 11.9 has *two degrees of freedom*.

When the relation existing between the position coordinates of several particles is *linear*, a similar relation holds between the velocities and between the accelerations of the particles. In the case of the blocks of Fig. 11.9, for instance, we differentiate twice the equation obtained and write

$$2\frac{dx_A}{dt} + 2\frac{dx_B}{dt} + \frac{dx_C}{dt} = 0 \qquad \text{or} \qquad 2v_A + 2v_B + v_C = 0$$

$$2\frac{dv_A}{dt} + 2\frac{dv_B}{dt} + \frac{dv_C}{dt} = 0 \qquad \text{or} \qquad 2a_A + 2a_B + a_C = 0$$

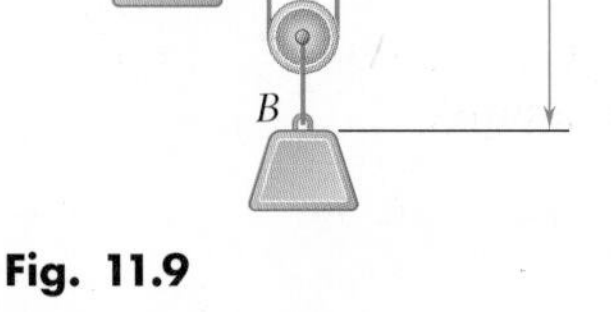

Fig. 11.9

†Note that the product of the subscripts A and B/A used in the right-hand member of Eqs. (11.9), (11.10), and (11.11) is equal to the subscript B used in their left-hand member.

SAMPLE PROBLEM 11.4

A ball is thrown vertically upward from the 12-m level in an elevator shaft with an initial velocity of 18 m/s. At the same instant an open-platform elevator passes the 5-m level, moving upward with a constant velocity of 2 m/s. Determine (*a*) when and where the ball will hit the elevator, (*b*) the relative velocity of the ball with respect to the elevator when the ball hits the elevator.

SOLUTION

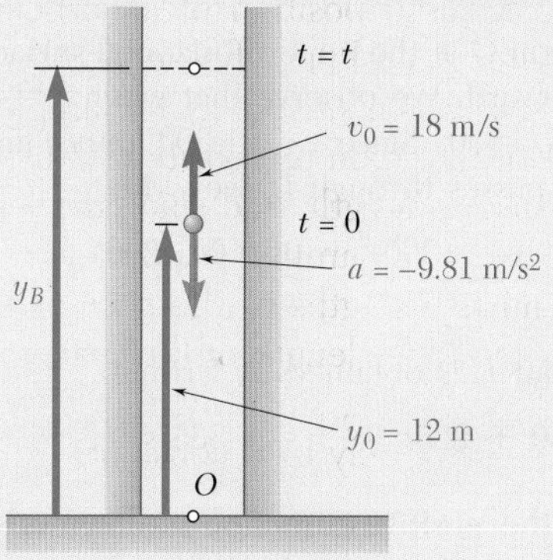

Motion of Ball. Since the ball has a constant acceleration, its motion is *uniformly accelerated*. Placing the origin O of the y axis at ground level and choosing its positive direction upward, we find that the initial position is $y_0 = +12$ m, the initial velocity is $v_0 = +18$ m/s, and the acceleration is $a = -9.81$ m/s². Substituting these values in the equations for uniformly accelerated motion, we write

$$v_B = v_0 + at \qquad v_B = 18 - 9.81t \tag{1}$$
$$y_B = y_0 + v_0t + \tfrac{1}{2}at^2 \qquad y_B = 12 + 18t - 4.905t^2 \tag{2}$$

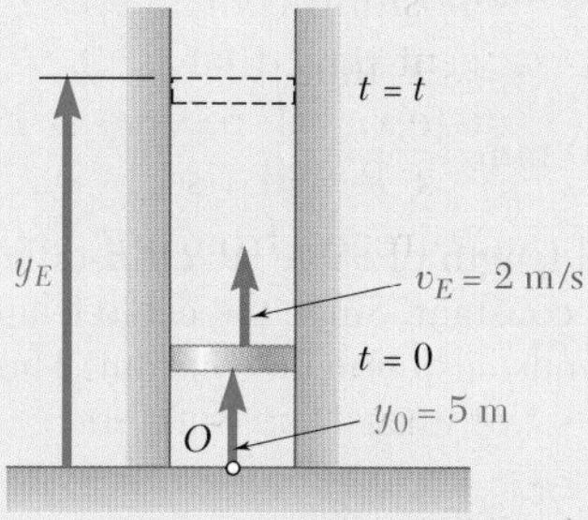

Motion of Elevator. Since the elevator has a constant velocity, its motion is *uniform*. Again placing the origin O at the ground level and choosing the positive direction upward, we note that $y_0 = +5$ m and write

$$v_E = +2 \text{ m/s} \tag{3}$$
$$y_E = y_0 + v_E t \qquad y_E = 5 + 2t \tag{4}$$

Ball Hits Elevator. We first note that the same time t and the same origin O were used in writing the equations of motion of both the ball and the elevator. We see from the figure that when the ball hits the elevator,

$$y_E = y_B \tag{5}$$

Substituting for y_E and y_B from (2) and (4) into (5), we have

$$5 + 2t = 12 + 18t - 4.905t^2$$
$$t = -0.39 \text{ s} \qquad \text{and} \qquad t = 3.65 \text{ s} \quad ◀$$

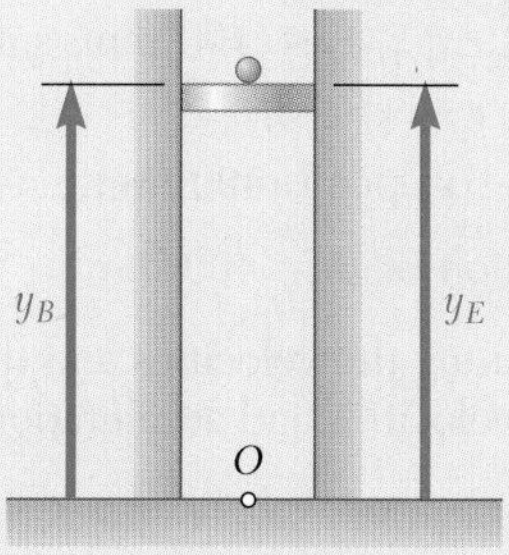

Only the root $t = 3.65$ s corresponds to a time after the motion has begun. Substituting this value into (4), we have

$$y_E = 5 + 2(3.65) = 12.30 \text{ m}$$
$$\text{Elevation from ground} = 12.30 \text{ m} \quad ◀$$

The relative velocity of the ball with respect to the elevator is

$$v_{B/E} = v_B - v_E = (18 - 9.81t) - 2 = 16 - 9.81t$$

When the ball hits the elevator at time $t = 3.65$ s, we have

$$v_{B/E} = 16 - 9.81(3.65) \qquad v_{B/E} = -19.81 \text{ m/s} \quad ◀$$

The negative sign means that the ball is observed from the elevator to be moving in the negative sense (downward).

SAMPLE PROBLEM 11.5

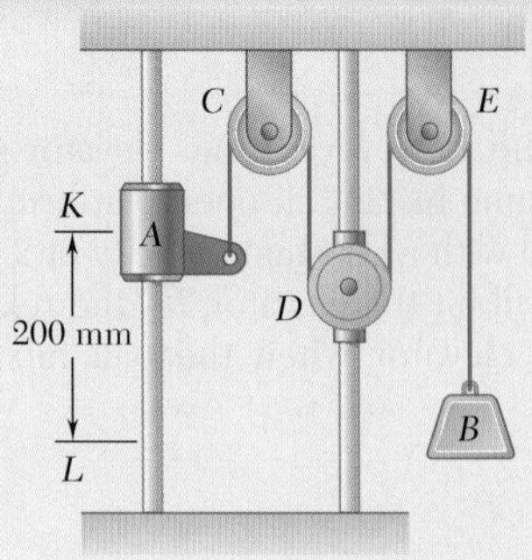

Collar A and block B are connected by a cable passing over three pulleys C, D, and E as shown. Pulleys C and E are fixed, while D is attached to a collar which is pulled downward with a constant velocity of 75 mm/s. At $t = 0$, collar A starts moving downward from position K with a constant acceleration and no initial velocity. Knowing that the velocity of collar A is 300 mm/s as it passes through point L, determine the change in elevation, the velocity, and the acceleration of block B when collar A passes through L.

SOLUTION

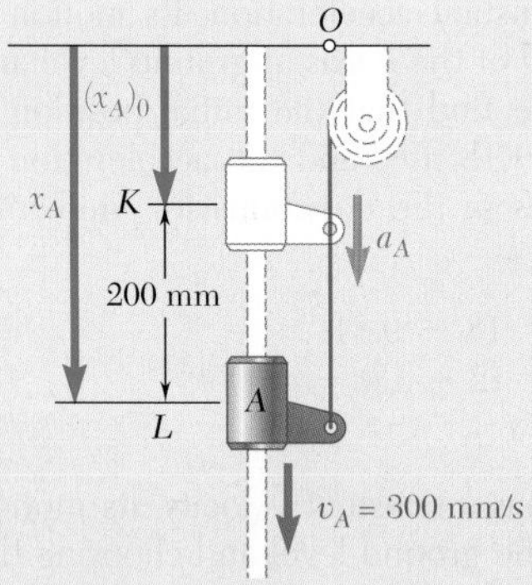

Motion of Collar *A*. We place the origin O at the upper horizontal surface and choose the positive direction downward. We observe that when $t = 0$, collar A is at the position K and $(v_A)_0 = 0$. Since $v_A = 300$ mm/s and $x_A - (x_A)_0 = 200$ mm when the collar passes through L, we write

$$v_A^2 = (v_A)_0^2 + 2a_A[x_A - (x_A)_0] \qquad (300)^2 = 0 + 2a_A(200)$$
$$a_A = 225 \text{ mm/s}^2$$

The time at which collar A reaches point L is obtained by writing

$$v_A = (v_A)_0 + a_A t \qquad 300 = 0 + 225t \qquad t = 1.333 \text{ s}$$

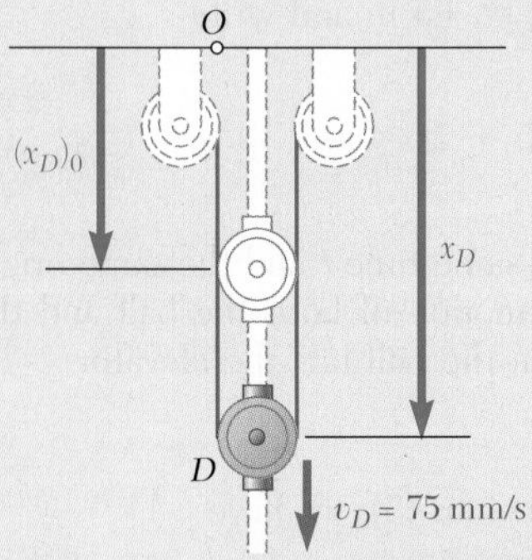

Motion of Pulley *D*. Recalling that the positive direction is downward, we write

$$a_D = 0 \qquad v_D = 75 \text{ mm/s} \qquad x_D = (x_D)_0 + v_D t = (x_D)_0 + 75t$$

When collar A reaches L, at $t = 1.333$ s, we have

$$x_D = (x_D)_0 + 75(1.333) = (x_D)_0 + 100$$

Thus, $\qquad x_D - (x_D)_0 = 100 \text{ mm}$

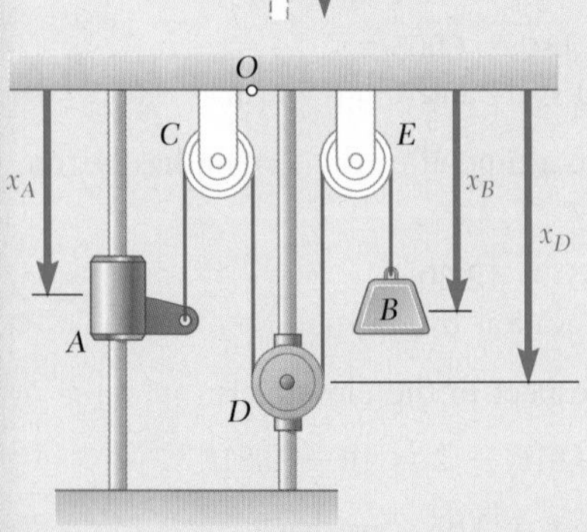

Motion of Block *B*. We note that the total length of cable $ACDEB$ differs from the quantity $(x_A + 2x_D + x_B)$ only by a constant. Since the cable length is constant during the motion, this quantity must also remain constant. Thus, considering the times $t = 0$ and $t = 1.333$ s, we write

$$x_A + 2x_D + x_B = (x_A)_0 + 2(x_D)_0 + (x_B)_0 \qquad (1)$$
$$[x_A - (x_A)_0] + 2[x_D - (x_D)_0] + [x_B - (x_B)_0] = 0 \qquad (2)$$

But we know that $x_A - (x_A)_0 = 200$ mm and $x_D - (x_D)_0 = 100$ mm; substituting these values in (2), we find

$$200 + 2(100) + [x_B - (x_B)_0] = 0 \qquad x_B - (x_B)_0 = -400 \text{ mm}$$

Thus: $\qquad$ Change in elevation of B = 400 mm ↑ ◀

Differentiating (1) twice, we obtain equations relating the velocities and the accelerations of A, B, and D. Substituting for the velocities and accelerations of A and D at $t = 1.333$ s, we have

$$v_A + 2v_D + v_B = 0: \qquad 300 + 2(75) + v_B = 0$$
$$v_B = -450 \text{ mm/s} \qquad v_B = 450 \text{ mm/s} \uparrow \blacktriangleleft$$

$$a_A + 2a_D + a_B = 0: \qquad 225 + 2(0) + a_B = 0$$
$$a_B = -225 \text{ mm/s}^2 \qquad a_B = 225 \text{ mm/s}^2 \uparrow \blacktriangleleft$$

SOLVING PROBLEMS ON YOUR OWN

In this lesson we derived the equations that describe *uniform rectilinear motion* (constant velocity) and *uniformly accelerated rectilinear motion* (constant acceleration). We also introduced the concept of *relative motion*. The equations for relative motion [Eqs. (11.9) to (11.11)] can be applied to the independent or dependent motions of any two particles moving along the same straight line.

A. Independent motion of one or more particles. The solution of problems of this type should be organized as follows:

1. Begin your solution by listing the given information, sketching the system, and selecting the origin and the positive direction of the coordinate axis [Sample Prob. 11.4]. It is always advantageous to have a visual representation of problems of this type.

2. Write the equations that describe the motions of the various particles as well as those that describe how these motions are related [Eq. (5) of Sample Prob. 11.4].

3. Define the initial conditions, i.e., specify the state of the system corresponding to $t = 0$. This is especially important if the motions of the particles begin at different times. In such cases, either of two approaches can be used.

a. Let $t = 0$ be the time when the last particle begins to move. You must then determine the initial position x_0 and the initial velocity v_0 of each of the other particles.

b. Let $t = 0$ be the time when the first particle begins to move. You must then, in each of the equations describing the motion of another particle, replace t with $t - t_0$, where t_0 is the time at which that specific particle begins to move. It is important to recognize that the equations obtained in this way are valid only for $t \geq t_0$.

B. Dependent motion of two or more particles. In problems of this type the particles of the system are connected to each other, typically by ropes or by cables. The method of solution of these problems is similar to that of the preceding group of problems, except that it will now be necessary to describe the *physical connections* between the particles. In the following problems, the connection is provided by one or more cables. For each cable, you will have to write equations similar to the last three equations of Sec. 11.6. We suggest that you use the following procedure:

1. Draw a sketch of the system and select a coordinate system, indicating clearly a positive sense for each of the coordinate axes. For example, in Sample Prob. 11.5 lengths are measured downward from the upper horizontal support. It thus follows that those displacements, velocities, and accelerations which have positive values are directed downward.

2. Write the equation describing the constraint imposed by each cable on the motion of the particles involved. Differentiating this equation twice, you will obtain the corresponding relations among velocities and accelerations.

3. If several directions of motion are involved, you must select a coordinate axis and a positive sense for each of these directions. You should also try to locate the origins of your coordinate axes so that the equations of constraints will be as simple as possible. For example, in Sample Prob. 11.5 it is easier to define the various coordinates by measuring them downward from the upper support than by measuring them upward from the bottom support.

Finally, keep in mind that the method of analysis described in this lesson and the corresponding equations can be used only for particles moving with *uniform* or *uniformly accelerated rectilinear motion*.

PROBLEMS

11.33 A stone is thrown vertically upward from a point on a bridge located 40 m above the water. Knowing that it strikes the water 4 s after release, determine (*a*) the speed with which the stone was thrown upward, (*b*) the speed with which the stone strikes the water.

11.34 A motorist is traveling at 54 km/h when she observes that a traffic light 240 m ahead of her turns red. The traffic light is timed to stay red for 24 s. If the motorist wishes to pass the light without stopping just as it turns green again, determine (*a*) the required uniform deceleration of the car, (*b*) the speed of the car as it passes the light.

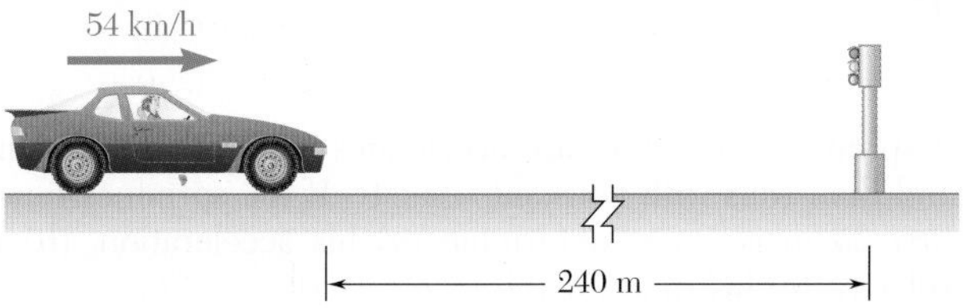

Fig. P11.34

11.35 A motorist enters a freeway at 45 km/h and accelerates uniformly to 99 km/h. From the odometer in the car, the motorist knows that she traveled 0.2 km while accelerating. Determine (*a*) the acceleration of the car, (*b*) the time required to reach 99 km/h.

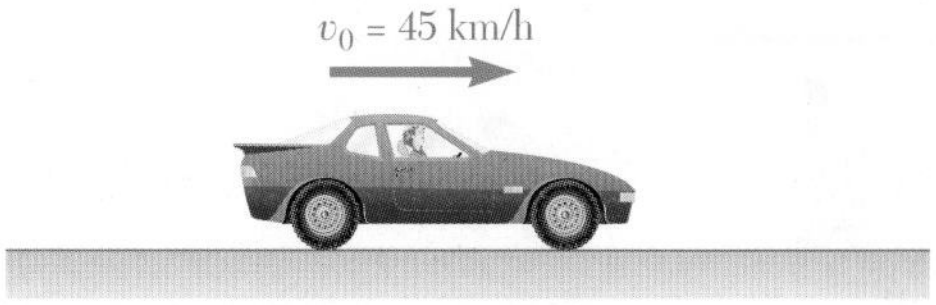

Fig. P11.35

11.36 A group of students launches a model rocket in the vertical direction. Based on tracking data, they determine that the altitude of the rocket was 27 m at the end of the powered portion of the flight and that the rocket landed 16 s later. Knowing that the descent parachute failed to deploy so that the rocket fell freely to the ground after reaching its maximum altitude and assuming that $g = 9.81$ m/s^2, determine (*a*) the speed v_1 of the rocket at the end of powered flight, (*b*) the maximum altitude reached by the rocket.

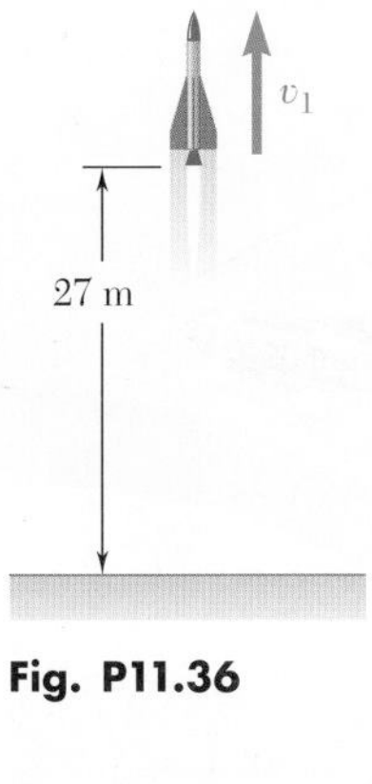

Fig. P11.36

11.37 A small package is released from rest at A and moves along the skate wheel conveyor $ABCD$. The package has a uniform acceleration of 4.8 m/s^2 as it moves down sections AB and CD, and its velocity is constant between B and C. If the velocity of the package at D is 7.2 m/s, determine (a) the distance d between C and D, (b) the time required for the package to reach D.

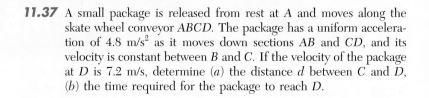

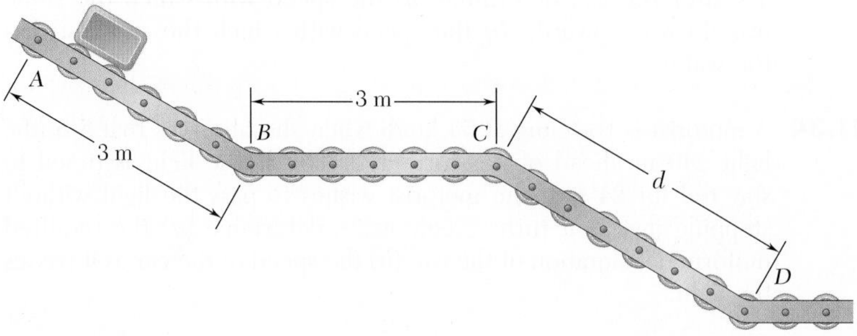

Fig. P11.37

Fig. P11.38

11.38 A sprinter in a 100-m race accelerates uniformly for the first 35 m and then runs with constant velocity. If the sprinter's time for the first 35 m is 5.4 s, determine (a) his acceleration, (b) his final velocity, (c) his time for the race.

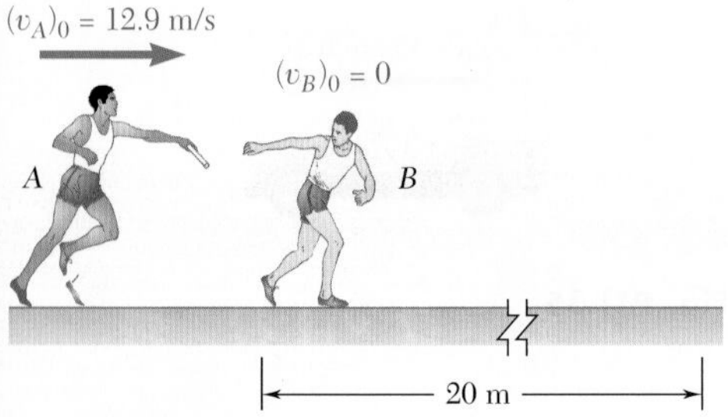

Fig. P11.39

11.39 As relay runner A enters the 20-m-long exchange zone with a speed of 12.9 m/s, he begins to slow down. He hands the baton to runner B 1.82 s later as they leave the exchange zone with the same velocity. Determine (a) the uniform acceleration of each of the runners, (b) when runner B should begin to run.

11.40 In a boat race, boat A is leading boat B by 50 m and both boats are traveling at a constant speed of 180 km/h. At t = 0, the boats accelerate at constant rates. Knowing that when B passes A, t = 8 s and v_A = 225 km/h, determine (a) the acceleration of A, (b) the acceleration of B.

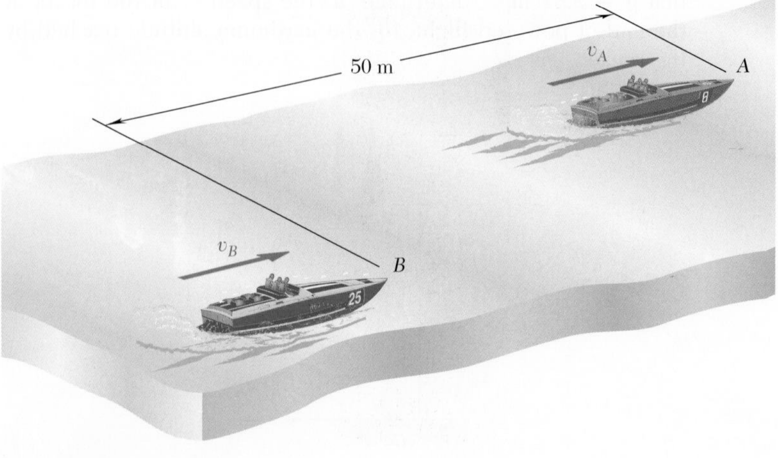

Fig. P11.40

11.41 A police officer in a patrol car parked in a 70 km/h speed zone observes a passing automobile traveling at a slow, constant speed. Believing that the driver of the automobile might be intoxicated, the officer starts his car, accelerates uniformly to 90 km/h in 8 s, and, maintaining a constant velocity of 90 km/h, overtakes the motorist 42 s after the automobile passed him. Knowing that 18 s elapsed before the officer began pursuing the motorist, determine (*a*) the distance the officer traveled before overtaking the motorist, (*b*) the motorist's speed.

11.42 Automobiles *A* and *B* are traveling in adjacent highway lanes and at $t = 0$ have the positions and speeds shown. Knowing that automobile *A* has a constant acceleration of 0.54 m/s^2 and that *B* has a constant deceleration of 0.36 m/s^2, determine (*a*) when and where *A* will overtake *B*, (*b*) the speed of each automobile at that time.

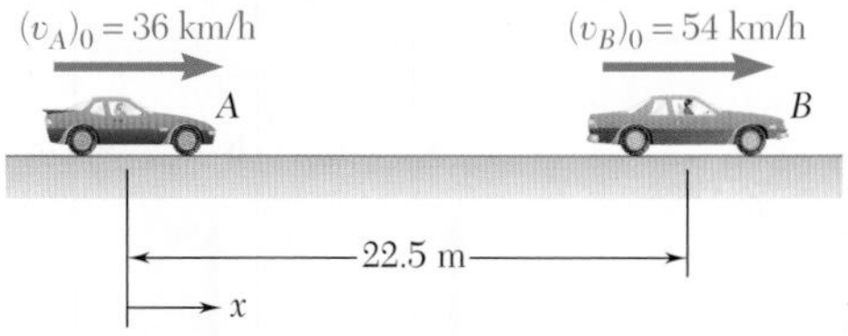

Fig. P11.42

11.43 Two automobiles *A* and *B* are approaching each other in adjacent highway lanes. At $t = 0$, *A* and *B* are 1 km apart, their speeds are $v_A = 108$ km/h and $v_B = 63$ km/h, and they are at points *P* and *Q*, respectively. Knowing that *A* passes point *Q* 40 s after *B* was there and that *B* passes point *P* 42 s after *A* was there, determine (*a*) the uniform accelerations of *A* and *B*, (*b*) when the vehicles pass each other, (*c*) the speed of *B* at that time.

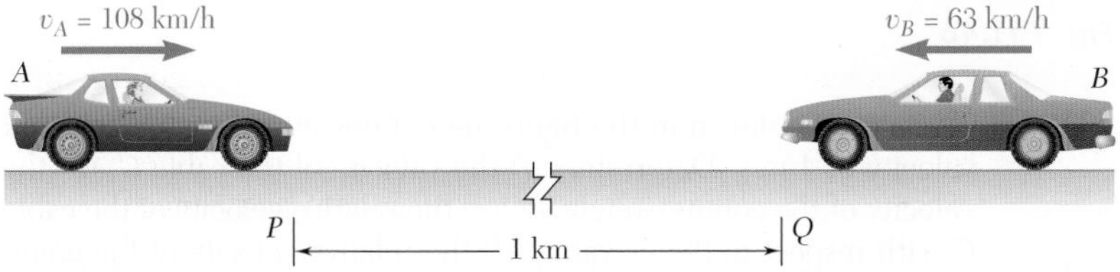

Fig. P11.43

11.44 An elevator is moving upward at a constant speed of 4 m/s. A man standing 10 m above the top of the elevator throws a ball upward with a speed of 3 m/s. Determine (*a*) when the ball will hit the elevator, (*b*) where the ball will hit the elevator with respect to the location of the man.

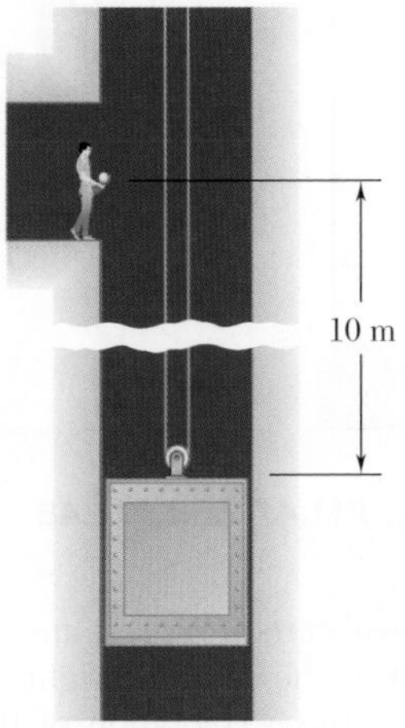

Fig. P11.44

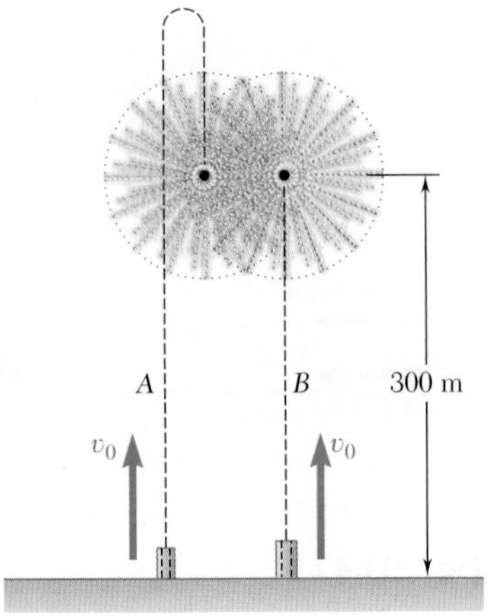

Fig. *P11.45*

11.45 Two rockets are launched at a fireworks display. Rocket A is launched with an initial velocity $v_0 = 100$ m/s and rocket B is launched t_1 s later with the same initial velocity. The two rockets are timed to explode simultaneously at a height of 300 m as A is falling and B is rising. Assuming a constant acceleration $g = 9.81$ m/s^2, determine (*a*) the time t_1, (*b*) the velocity of B relative to A at the time of the explosion.

11.46 Car A is parked along the northbound lane of a highway, and car B is traveling in the southbound lane at a constant speed of 90 km/h. At $t = 0$, A starts and accelerates at a constant rate a_A, while at $t = 5$ s, B begins to slow down with a constant deceleration of magnitude $a_A/6$. Knowing that when the cars pass each other $x = 90$ m and $v_A = v_B$, determine (*a*) the acceleration a_A, (*b*) when the vehicles pass each other, (*c*) the distance d between the vehicles at $t = 0$.

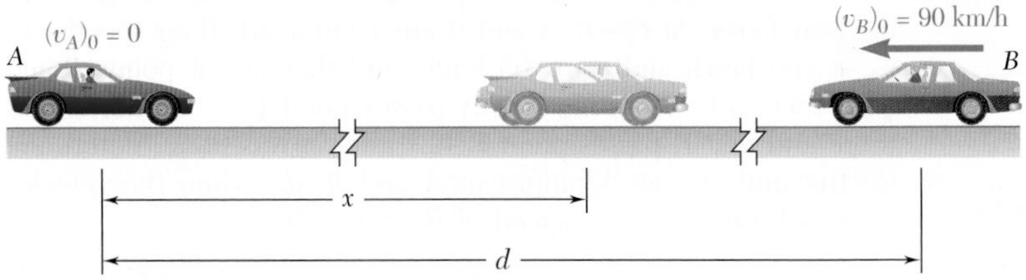

Fig. *P11.46*

11.47 The elevator shown in the figure moves downward with a constant velocity of 4 m/s. Determine (*a*) the velocity of the cable C, (*b*) the velocity of the counterweight W, (*c*) the relative velocity of the cable C with respect to the elevator, (*d*) the relative velocity of the counterweight W with respect to the elevator.

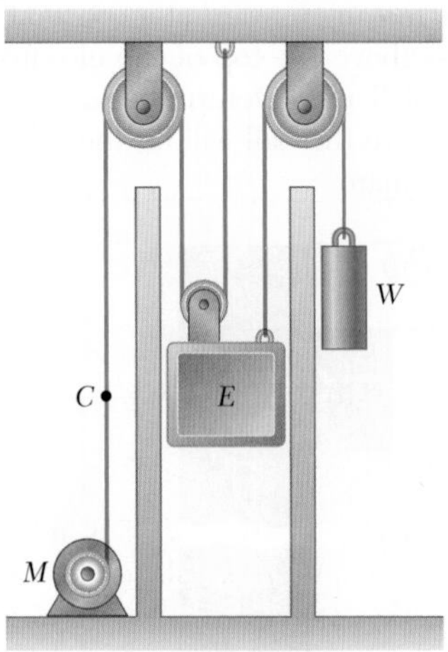

Fig. P11.47 and P11.48

11.48 The elevator shown starts from rest and moves upward with a constant acceleration. If the counterweight W moves through 10 m in 5 s, determine (*a*) the acceleration of the elevator and the cable C, (*b*) the velocity of the elevator after 5 s.

11.49 Slider block A moves to the left with a constant velocity of 6 m/s. Determine (a) the velocity of block B, (b) the velocity of portion D of the cable, (c) the relative velocity of portion C of the cable with respect to portion D.

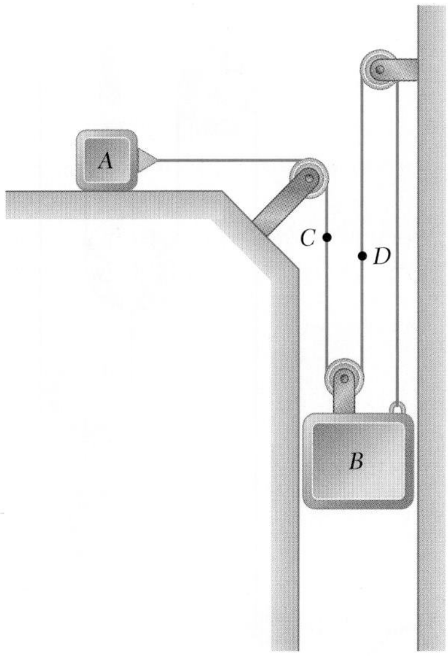

Fig. P11.49 and P11.50

11.50 Block B starts from rest and moves downward with a constant acceleration. Knowing that after slider block A has moved 400 mm its velocity is 4 m/s, determine (a) the accelerations of A and B, (b) the velocity and the change in position of B after 2 s.

11.51 Slider block B moves to the right with a constant velocity of 300 mm/s. Determine (a) the velocity of slider block A, (b) the velocity of portion C of the cable, (c) the velocity of portion D of the cable, (d) the relative velocity of portion C of the cable with respect to slider block A.

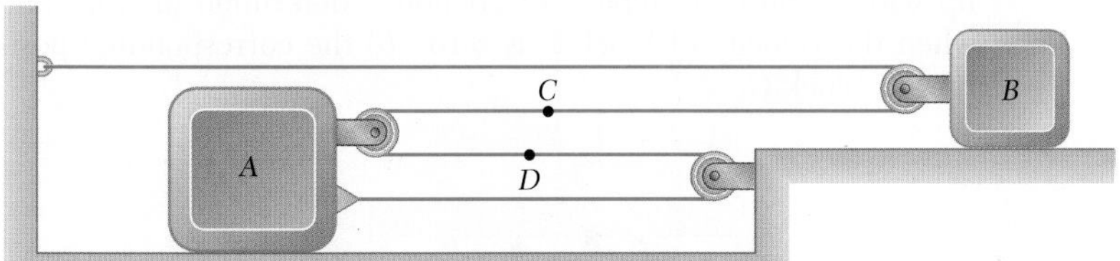

Fig. P11.51 and P11.52

11.52 At the instant shown, slider block B is moving with a constant acceleration, and its speed is 150 mm/s. Knowing that after slider block A has moved 240 mm to the right its velocity is 60 mm/s, determine (a) the accelerations of A and B, (b) the acceleration of portion D of the cable, (c) the velocity and the change in position of slider block B after 4 s.

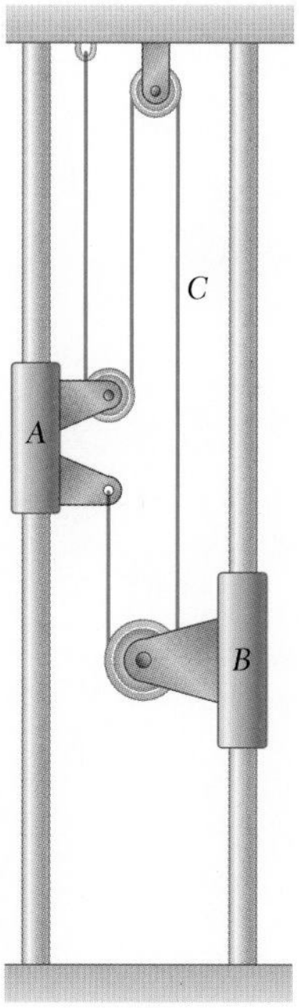

Fig. *P11.53*

11.53 Collar A starts from rest and moves upward with a constant acceleration. Knowing that after 8 s the relative velocity of collar B with respect to collar A is 0.6 m/s, determine (a) the accelerations of A and B, (b) the velocity and the change in position of B after 6 s.

11.54 The motor M reels in the cable at a constant rate of 100 mm/s. Determine (*a*) the velocity of load L, (*b*) the velocity of pulley B with respect to load L.

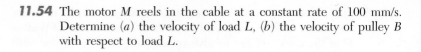

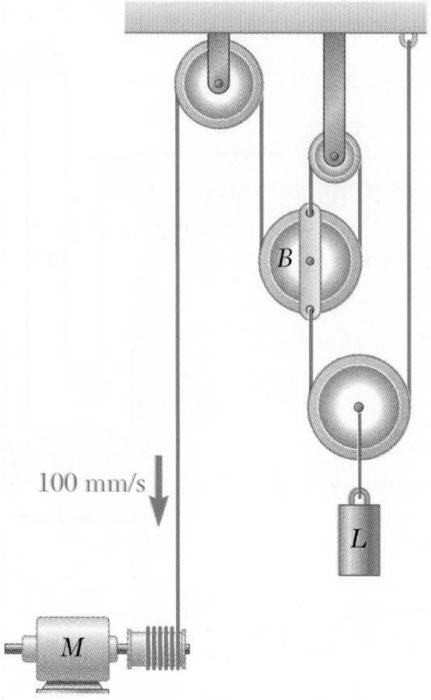

Fig. P11.54

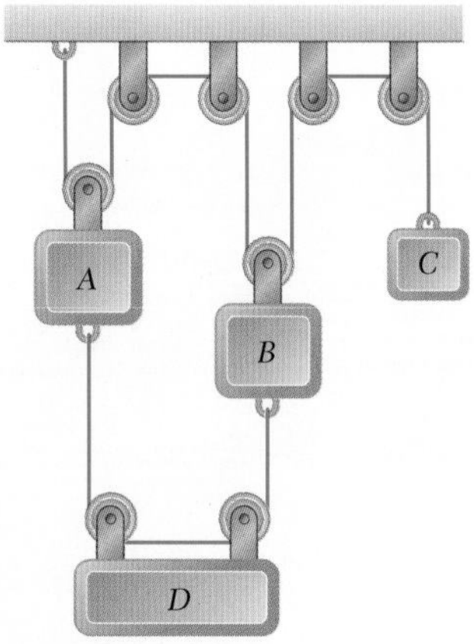

Fig. P11.55

11.55 Block C starts from rest at $t = 0$ and moves downward with a constant acceleration of 100 mm/s^2. Knowing that block B has a constant velocity of 75 mm/s upward, determine (*a*) the time when the velocity of block A is zero, (*b*) the time when the velocity of block A is equal to the velocity of block D, (*c*) the change in position of block A after 5 s.

11.56 Block A starts from rest at $t = 0$ and moves downward with a constant acceleration of 150 mm/s^2. Knowing that block B moves up with a constant velocity of 75 mm/s, determine (*a*) the time when the velocity of block C is zero, (*b*) the corresponding position of block C.

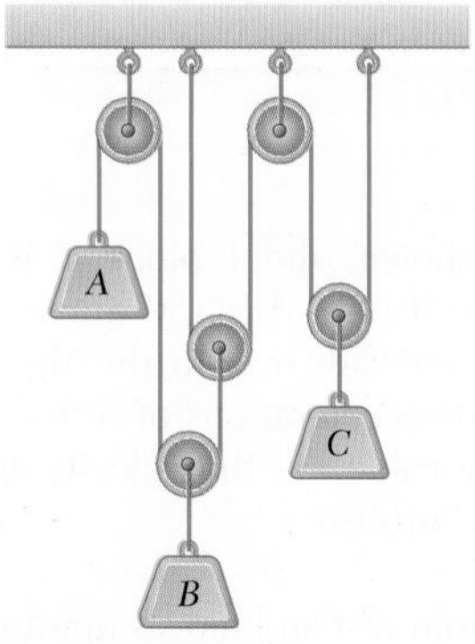

Fig. P11.56

11.57 Block B starts from rest, block A moves with a constant acceleration, and slider block C moves to the right with a constant acceleration of 75 mm/s^2. Knowing that at $t = 2$ s the velocities of B and C are 480 mm/s downward and 280 mm/s to the right, respectively, determine (*a*) the accelerations of A and B, (*b*) the initial velocities of A and C, (*c*) the change in position of slider block C after 3 s.

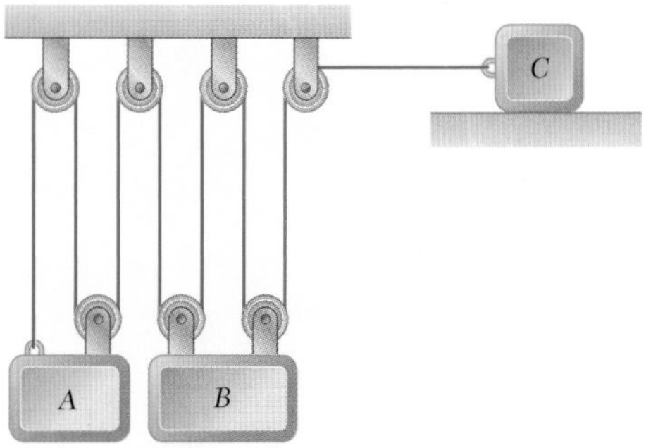

Fig. P11.57 and P11.58

11.58 Block B moves downward with a constant velocity of 20 mm/s. At $t = 0$, block A is moving upward with a constant acceleration, and its velocity is 30 mm/s. Knowing that at $t = 3$ s slider block C has moved 57 mm to the right, determine (*a*) the velocity of slider block C at $t = 0$, (*b*) the accelerations of A and C, (*c*) the change in position of block A after 5 s.

11.59 The system shown starts from rest, and each component moves with a constant acceleration. If the relative acceleration of block C with respect to collar B is 60 mm/s^2 upward and the relative acceleration of block D with respect to block A is 110 mm/s^2 downward, determine (*a*) the velocity of block C after 3 s, (*b*) the change in position of block D after 5 s.

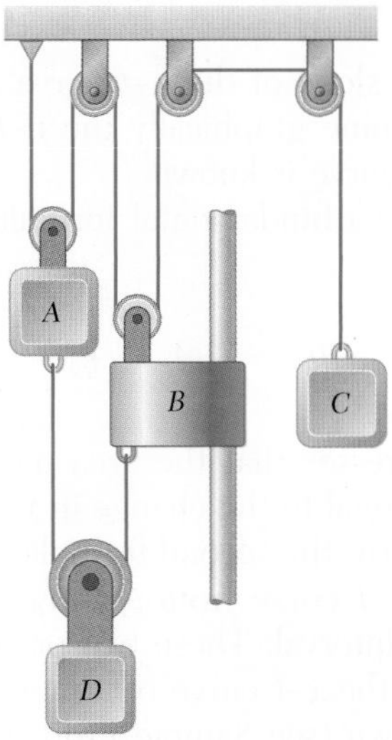

Fig. *P11.59* and *P11.60*

***11.60** The system shown starts from rest, and the length of the upper cord is adjusted so that A, B, and C are initially at the same level. Each component moves with a constant acceleration, and after 2 s the relative change in position of block C with respect to block A is 280 mm upward. Knowing that when the relative velocity of collar B with respect to block A is 80 mm/s downward, the displacements of A and B are 160 mm downward and 320 mm downward, respectively, determine (*a*) the accelerations of A and B if $a_B > 10$ mm/s^2, (*b*) the change in position of block D when the velocity of block C is 600 mm/s upward.

*11.7 GRAPHICAL SOLUTION OF RECTILINEAR-MOTION PROBLEMS

It was observed in Sec. 11.2 that the fundamental formulas

$$v = \frac{dx}{dt} \qquad \text{and} \qquad a = \frac{dv}{dt}$$

have a geometrical significance. The first formula expresses that the velocity at any instant is equal to the slope of the x–t curve at the same instant (Fig. 11.10). The second formula expresses that the acceleration is equal to the slope of the v–t curve. These two properties can be used to determine graphically the v–t and a–t curves of a motion when the x–t curve is known.

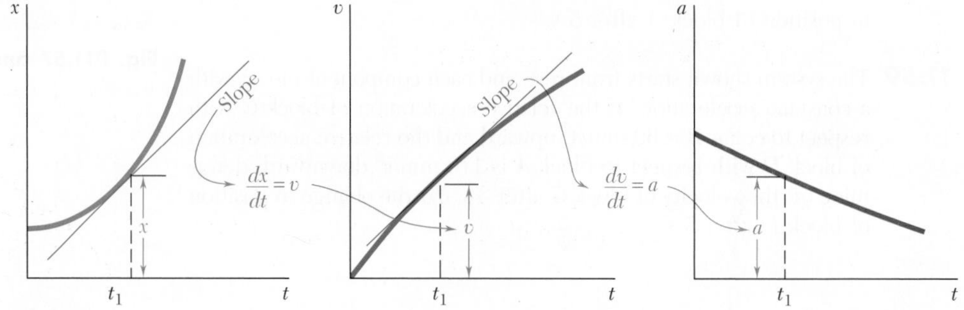

Fig. 11.10

Integrating the two fundamental formulas from a time t_1 to a time t_2, we write

$$x_2 - x_1 = \int_{t_1}^{t_2} v\,dt \qquad \text{and} \qquad v_2 - v_1 = \int_{t_1}^{t_2} a\,dt \tag{11.12}$$

The first formula expresses that the area measured under the v–t curve from t_1 to t_2 is equal to the change in x during that time interval (Fig. 11.11). Similarly, the second formula expresses that the area measured under the a–t curve from t_1 to t_2 is equal to the change in v during that time interval. These two properties can be used to determine graphically the x–t curve of a motion when its v–t curve or its a–t curve is known (see Sample Prob. 11.6).

Graphical solutions are particularly useful when the motion considered is defined from experimental data and when x, v, and a are not analytical functions of t. They can also be used to advantage when the motion consists of distinct parts and when its analysis requires writing a different equation for each of its parts. When using a graphical solution, however, one should be careful to note that (1) the area under the v–t curve measures the *change in* x, not x itself, and similarly, that the area under the a–t curve measures the change in v; (2) an area above the t axis corresponds to an *increase* in x or v, while an area located below the t axis measures a *decrease* in x or v.

It will be useful to remember in drawing motion curves that if the velocity is constant, it will be represented by a horizontal straight line; the position coordinate x will then be a linear function of t and will be represented by an oblique straight line. If the acceleration is

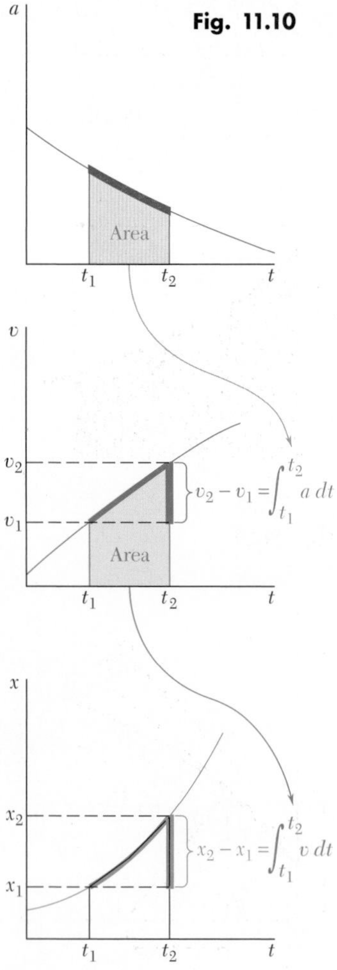

Fig. 11.11

constant and different from zero, it will be represented by a horizontal straight line; v will then be a linear function of t, represented by an oblique straight line, and x will be expressed as a second-degree polynomial in t, represented by a parabola. If the acceleration is a linear function of t, the velocity and the position coordinate will be equal, respectively, to second-degree and third-degree polynomials; a will then be represented by an oblique straight line, v by a parabola, and x by a cubic. In general, if the acceleration is a polynomial of degree n in t, the velocity will be a polynomial of degree $n + 1$ and the position coordinate a polynomial of degree $n + 2$; these polynomials are represented by motion curves of a corresponding degree.

*11.8 OTHER GRAPHICAL METHODS

An alternative graphical solution can be used to determine the position of a particle at a given instant directly from the a–t curve. Denoting the values of x and v at $t = 0$ as x_0 and v_0 and their values at $t = t_1$ as x_1 and v_1, and observing that the area under the v–t curve can be divided into a rectangle of area $v_0 t_1$ and horizontal differential elements of area $(t_1 - t)\,dv$ (Fig. 11.12a), we write

$$x_1 - x_0 = \text{area under } v\text{–}t \text{ curve} = v_0 t_1 + \int_{v_0}^{v_1} (t_1 - t)\,dv$$

Substituting $dv = a\,dt$ in the integral, we obtain

$$x_1 - x_0 = v_0 t_1 + \int_0^{t_1} (t_1 - t)\,a\,dt$$

Referring to Fig. 11.12b, we note that the integral represents the first moment of the area under the a–t curve with respect to the line $t = t_1$ bounding the area on the right. This method of solution is known, therefore, as the *moment-area method*. If the abscissa $\bar{t}$ of the centroid C of the area is known, the position coordinate x_1 can be obtained by writing

$$x_1 = x_0 + v_0 t_1 + (\text{area under } a\text{–}t \text{ curve})(t_1 - \bar{t}) \qquad (11.13)$$

If the area under the a–t curve is a composite area, the last term in (11.13) can be obtained by multiplying each component area by the distance from its centroid to the line $t = t_1$. Areas above the t axis should be considered as positive and areas below the t axis as negative.

Another type of motion curve, the v–x curve, is sometimes used. If such a curve has been plotted (Fig. 11.13), the acceleration a can be obtained at any time by drawing the normal AC to the curve and *measuring the subnormal* BC. Indeed, observing that the angle between AC and AB is equal to the angle θ between the horizontal and the tangent at A (the slope of which is $\tan\theta = dv/dx$), we write

$$BC = AB \tan\theta = v\frac{dv}{dx}$$

and thus, recalling formula (11.4),

$$BC = a$$

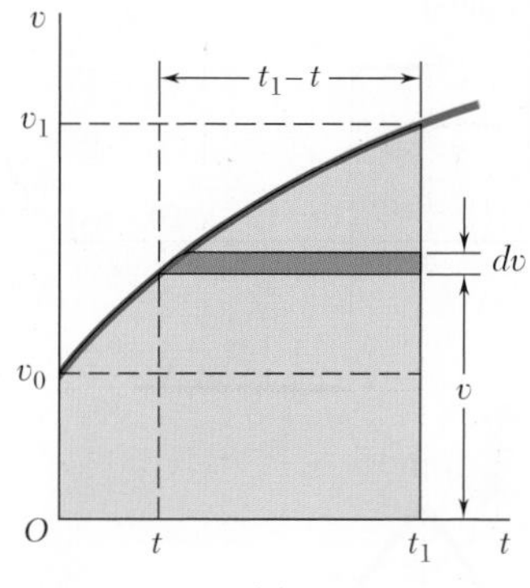

(a)

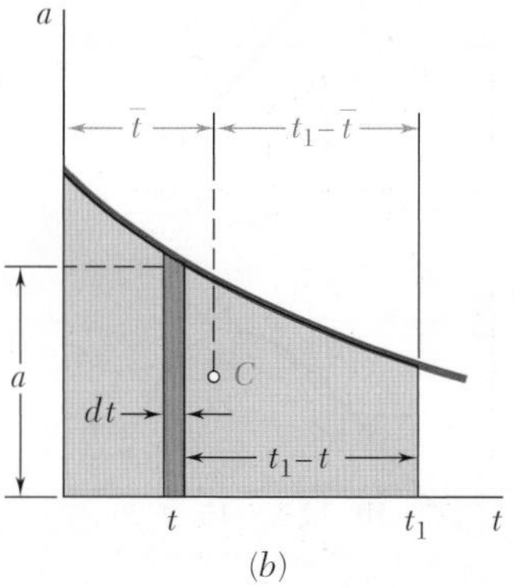

(b)

Fig. 11.12

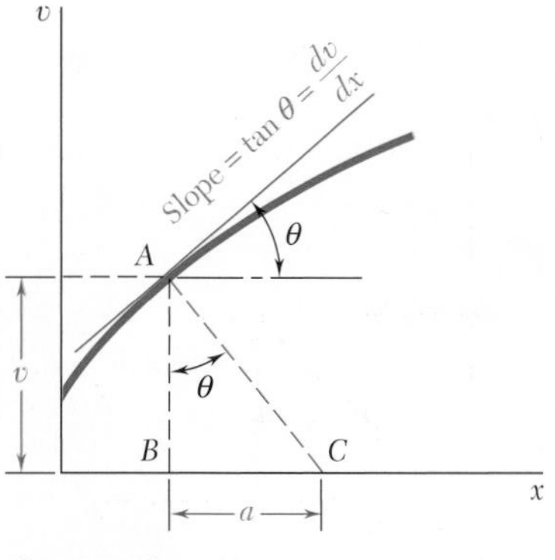

Fig. 11.13

SAMPLE PROBLEM 11.6

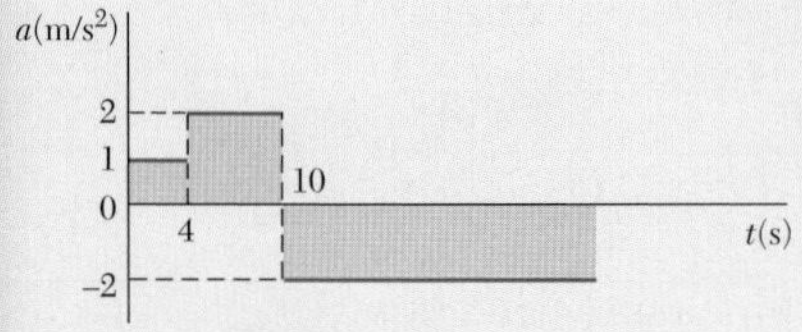

A particle moves in a straight line with the acceleration shown in the figure. Knowing that it starts from the origin with $v_0 = -6$ m/s, (*a*) plot the $v - t$ and $x - t$ curves for $0 < t < 20$ s, (*b*) determine its velocity, its position, and the total distance traveled when $t = 12$ s.

SOLUTION

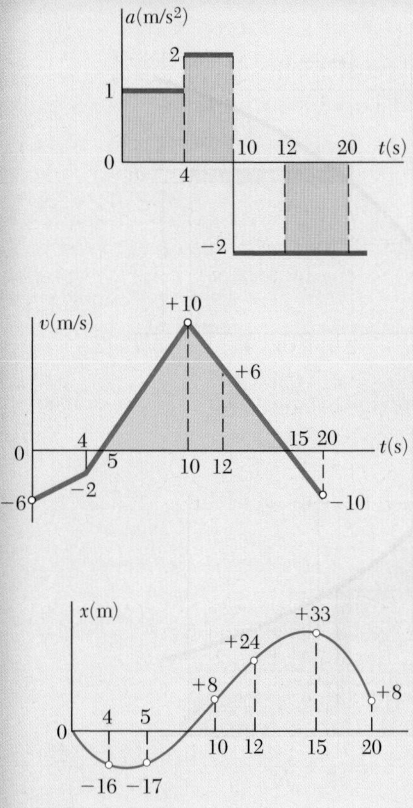

a Acceleration-Time Curve.

Initial conditions: $t = 0,\ v_0 = -6$ m/s, $x_0 = 0$

Change in v = area under a–t curve:

$v_0 = -6$ m/s

$0 < t < 4$s : $v_4 - v_0 = (1\ \text{m/s}^2)(4\text{s}) = +4$ m/s $\qquad v_4 = -2$ m/s

$4\text{s} < t < 10\text{s}$: $v_{10} - v_4 = (2\ \text{m/s}^2)(6\text{s}) = +12$ m/s $\qquad v_{10} = +10$ m/s

$10\text{s} < t < 12\text{s}$: $v_{12} - v_{10} = (-2\ \text{m/s}^2)(2\text{s}) = -4$ m/s $\qquad v_{12} = +6$ m/s

$12\text{s} < t < 20\text{s}$: $v_{20} - v_{12} = (-2\ \text{m/s}^2)(8\text{s}) = -16$ m/s $\qquad v_{20} = -10$ m/s ◀

Change in x = area under $v - t$ curve: $\qquad x_0 = 0$

$0 < t < 4\text{s}$: $x_4 - x_0 = \frac{1}{2}(-6-2)(4) = -16$ m $\qquad x_4 = -16$m

$4\text{s} < t < 5\text{s}$: $x_5 - x_4 = \frac{1}{2}(-2)(1) = -1$ m $\qquad x_5 = -17$m

$5\text{s} < t < 10\text{s}$: $x_{10} - x_5 = \frac{1}{2}(+10)(5) = +25$m $\qquad x_{10} = 8$m

$10\text{s} < t < 12\text{s}$: $x_{12} - x_{10} = \frac{1}{2}(+10+6)(2) = +16$ m $\qquad x_{12} = +24$m

$12\text{s} < t < 15\text{s}$ $\quad x_{15} - x_{12} = \frac{1}{2}(+6)(3) = +9$m $\qquad x_{16} = +33$m

$15\text{s} < t < 20\text{s}$ $\quad x_{20} - x_{15} = \frac{1}{2}(-10)(5) = -25$m $\qquad x_{20} = +8$m

b From above curves, we read

For t=12 s: $\quad v_{12} = +6$ m/s, $x_{12} = +24$ m

Distance traveled $t = 0$ to $t = 12$ s

From $t = 0$ s to $t = 5$ s: Distance traveled = 17 m

From $t = 5$ s to $t = 12$ s: Distance traveled = (17 + 24) = 41 m

Total distance traveled = 58 m ◀

SOLVING PROBLEMS ON YOUR OWN

In this lesson (Secs. 11.7 and 11.8), we reviewed and developed several *graphical techniques for the solution of problems involving rectilinear motion*. These techniques can be used to solve problems directly or to complement analytical methods of solution by providing a visual description, and thus a better understanding, of the motion of a given body. We suggest that you sketch one or more motion curves for several of the problems in this lesson, even if these problems are not part of your homework assignment.

1. Drawing *x–t, v–t*, and *a–t* curves and applying graphical methods. The following properties were indicated in Sec. 11.7 and should be kept in mind as you use a graphical method of solution.

a. The slopes of the *x–t* and *v–t* curves at a time t_1 are respectively equal to the *velocity* and the *acceleration* at time t_1.

b. The areas under the *a–t* and *v–t* curves between the times t_1 and t_2 are respectively equal to the change Δv in the velocity and to the change Δx in the position coordinate during that time interval.

c. If one of the motion curves is known, the fundamental properties we have summarized in paragraphs *a* and *b* will enable you to construct the other two curves. However, when using the properties of paragraph *b*, the velocity and the position coordinate at time t_1 must be known in order to determine the velocity and the position coordinate at time t_2. Thus, in Sample Prob. 11.6, knowing that the initial value of the velocity was zero allowed us to find the velocity at $t = 6$ s: $v_6 = v_0 + \Delta v = 0 + 24 \text{ ft/s} = 24 \text{ ft/s}$.

If you have previously studied the shear and bending-moment diagrams for a beam, you should recognize the analogy that exists between the three motion curves and the three diagrams representing respectively the distributed load, the shear, and the bending moment in the beam. Thus, any techniques that you learned regarding the construction of these diagrams can be applied when drawing the motion curves.

2. Using approximate methods. When the a–t and v–t curves are not represented by analytical functions or when they are based on experimental data, it is often necessary to use approximate methods to calculate the areas under these curves. In those cases, the given area is approximated by a series of rectangles of width Δt. The smaller the value of Δt, the smaller the error introduced by the approximation. The velocity and the position coordinate are obtained by writing

$$v = v_0 + \Sigma a_{\text{ave}}\, \Delta t \qquad x = x_0 + \Sigma v_{\text{ave}}\, \Delta t$$

where a_{ave} and v_{ave} are the heights of an acceleration rectangle and a velocity rectangle, respectively.

(continued)

3. Applying the moment-area method. This graphical technique is used when the a–t curve is given and the change in the position coordinate is to be determined. We found in Sec. 11.8 that the position coordinate x_1 can be expressed as

$$x_1 = x_0 + v_0 t_1 + (\text{area under } a\text{–}t \text{ curve})(t_1 - \bar{t}) \qquad (11.13)$$

Keep in mind that when the area under the a–t curve is a composite area, the same value of t_1 should be used for computing the contribution of each of the component areas.

4. Determining the acceleration from a v–x curve. You saw in Sec. 11.8 that it is possible to determine the acceleration from a v–x curve by direct measurement. It is important to note, however, that this method is applicable only if the same linear scale is used for the v and x axes (for example, 1 in. = 10 ft and 1 in. = 10 ft/s). When this condition is not satisfied, the acceleration can still be determined from the equation

$$a = v\frac{dv}{dx}$$

where the slope dv/dx is obtained as follows: First, draw the tangent to the curve at the point of interest. Next, using appropriate scales, measure along that tangent corresponding increments Δx and Δv. The desired slope is equal to the ratio $\Delta v/\Delta x$.

PROBLEMS

11.61 A subway car leaves station A; it gains speed at the rate of 4 m/s^2 for 6 s and then at the rate of 6 m/s^2 until it has reached the speed of 36 m/s. The car maintains the same speed until it approaches station B; brakes are then applied, giving the car a constant deceleration and bringing it to a stop in 6 s. The total running time from A to B is 40 s. Draw the a–t, v–t, and x–t curves, and determine the distance between stations A and B.

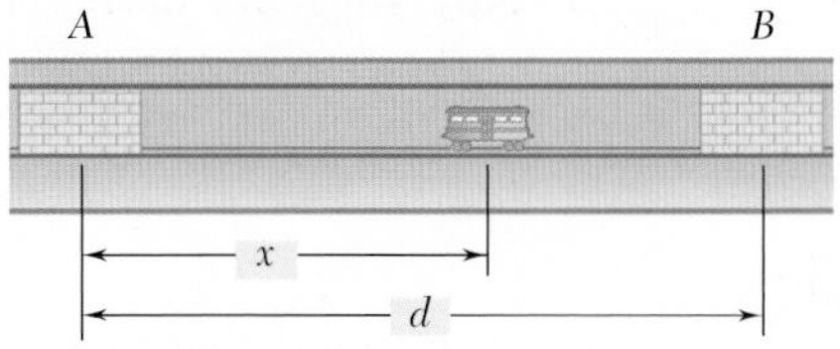

Fig. P11.61

11.62 For the particle and motion of Sample Problem 11.6, plot the v–t and x–t curves for $0 < t < 20$ s and determine (*a*) the maximum value of the velocity of the particle, (*b*) the maximum value of its position coordinate.

11.63 A particle moves in a straight line with the velocity shown in the figure. Knowing that $x = -540$ m at $t = 0$, (*a*) construct the a–t and x–t curves for $0 < t < 50$ s, and determine (*b*) the total distance traveled by the particle when $t = 50$ s, (*c*) the two times at which $x = 0$.

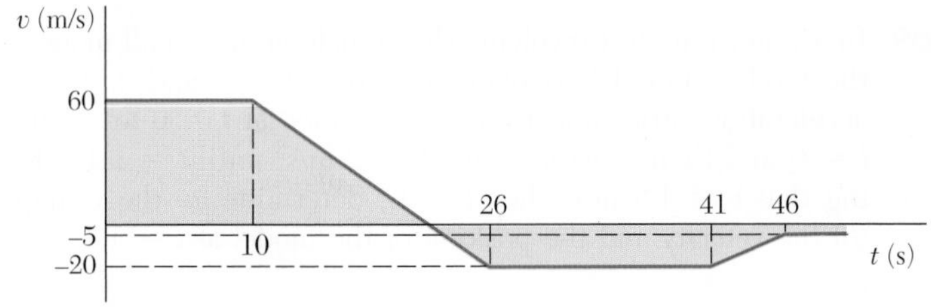

Fig. P11.63 and P11.64

11.64 A particle moves in a straight line with the velocity shown in the figure. Knowing that $x = -540$ m at $t = 0$, (*a*) construct the a–t and x–t curves for $0 < t < 50$ s, and determine (*b*) the maximum value of the position coordinate of the particle, (*c*) the values of t for which the particle is at $x = 100$ m.

11.65 During a finishing operation the bed of an industrial planer moves alternately 750 mm to the right and 750 mm to the left. The velocity of the bed is limited to a maximum value of 150 mm/s to the right and 300 mm/s to the left; the acceleration is successively equal to 150 mm/s^2 to the right, zero, 150 mm/s^2 to the left, zero, etc. Determine the time required for the bed to complete a full cycle, and draw the v–t and x–t curves.

11.66 A parachutist is in free fall at a rate of 200 km/h when he opens his parachute at an altitude of 600 m. Following a rapid and constant deceleration, he then descends at a constant rate of 50 km/h from 586 m to 30 m, where he maneuvers the parachute into the wind to further slow his descent. Knowing that the parachutist lands with a negligible downward velocity, determine (*a*) the time required for the parachutist to land after opening his parachute, (*b*) the initial deceleration.

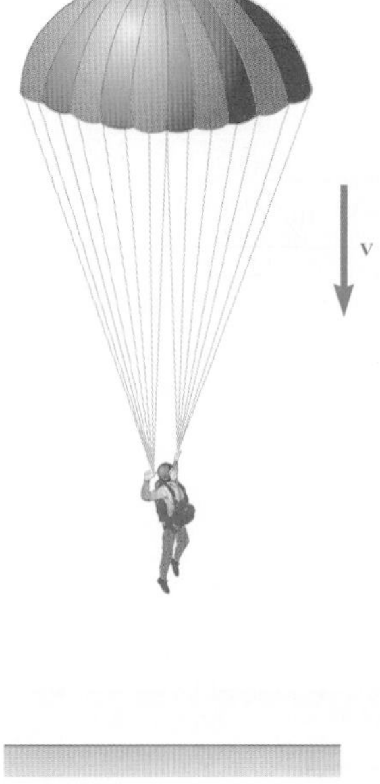

Fig. P11.66

11.67 A commuter train traveling at 60 km/h is 4.5 km from a station. The train then decelerates so that its speed is 30 km/h when it is 0.75 km from the station. Knowing that the train arrives at the station 7.5 min after beginning to decelerate and assuming constant decelerations, determine (*a*) the time required for the train to travel the first 3.75 km, (*b*) the speed of the train as it arrives at the station, (*c*) the final constant deceleration of the train.

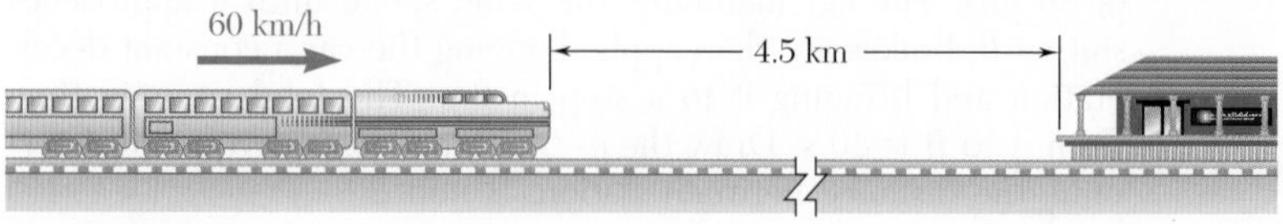

Fig. P11.67

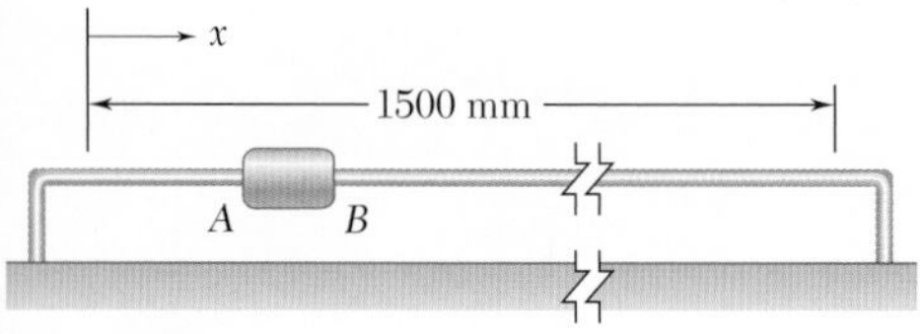

Fig. P11.68

11.68 A temperature sensor is attached to slider *AB* which moves back and forth through 1500 mm. The maximum velocities of the slider are 300 mm/s to the right and 750 mm/s to the left. When the slider is moving to the right, it accelerates and decelerates at a constant rate of 150 mm/s^2; when moving to the left, the slider accelerates and decelerates at a constant rate of 500 mm/s^2. Determine the time required for the slider to complete a full cycle, and construct the v–t and x–t curves of its motion.

11.69 In a water-tank test involving the launching of a small model boat, the model's initial horizontal velocity is 6 m/s and its horizontal acceleration varies linearly from -12 m/s^2 at $t = 0$ to -2 m/s^2 at $t = t_1$ and then remains equal to -2 m/s^2 until $t = 1.4$ s. Knowing that $v = 1.8$ m/s when $t = t_1$, determine (*a*) the value of t_1, (*b*) the velocity and the position of the model at $t = 1.4$ s.

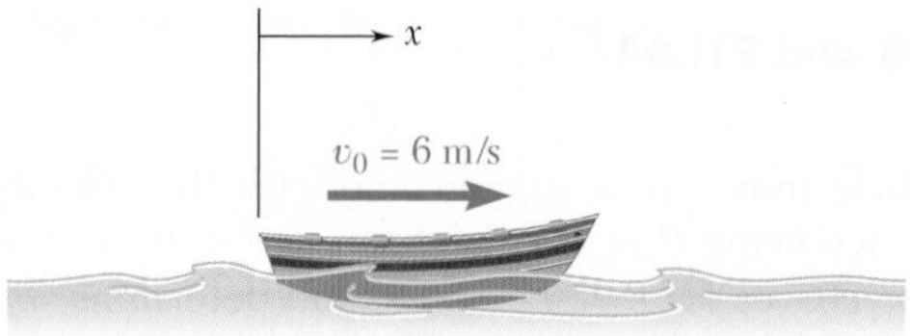

Fig. P11.69

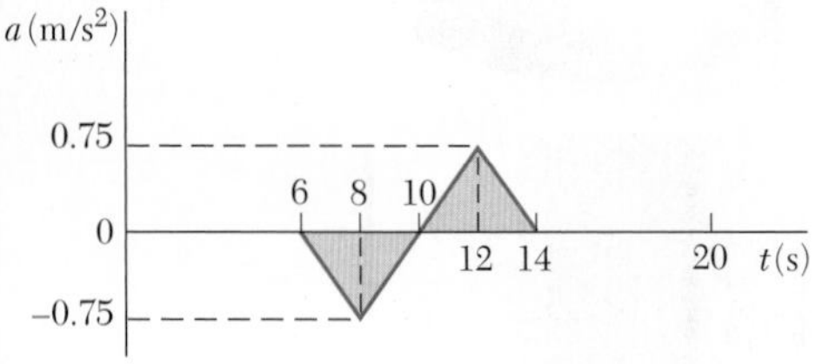

Fig. P11.70

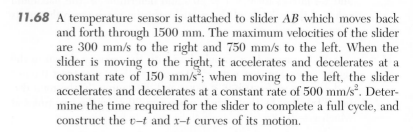

11.70 The acceleration record shown was obtained for a small airplane traveling along a straight course. Knowing that $x = 0$ and $v = 60$ m/s when $t = 0$, determine (*a*) the velocity and position of the plane at $t = 20$ s, (*b*) its average velocity during the interval $6\text{ s} < t < 14\text{ s}$.

11.71 In a 400-m race, runner *A* reaches her maximum velocity v_A in 4 s with constant acceleration and maintains that velocity until she reaches the halfway point with a split time of 25 s. Runner *B* reaches her maximum velocity v_B in 5 s with constant acceleration and maintains that velocity until she reaches the halfway point with a split time of 25.2 s. Both runners then run the second half of the race with the same constant deceleration of 0.1 m/s^2. Determine (*a*) the race times for both runners, (*b*) the position of the winner relative to the loser when the winner reaches the finish line.

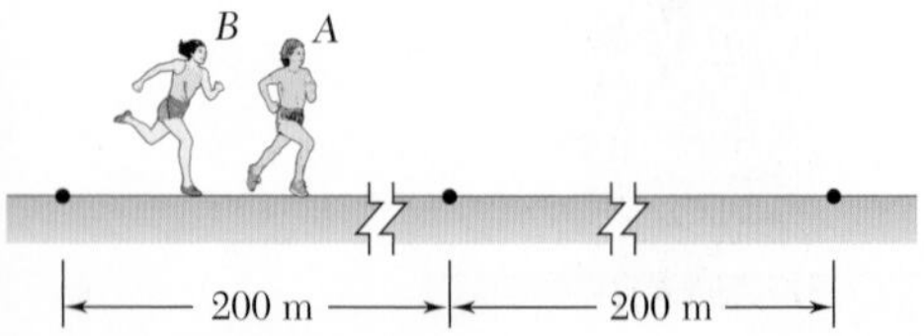

Fig. P11.71

11.72 A car and a truck are both traveling at the constant speed of 50 km/h; the car is 12 m behind the truck. The driver of the car wants to pass the truck, i.e., he wishes to place his car at B, 12 m in front of the truck, and then resume the speed of 50 km/h. The maximum acceleration of the car is 1.5 m/s^2 and the maximum deceleration obtained by applying the brakes is 6 m/s^2. What is the shortest time in which the driver of the car can complete the passing operation if he does not at any time exceed a speed of 75 km/h? Draw the v–t curve.

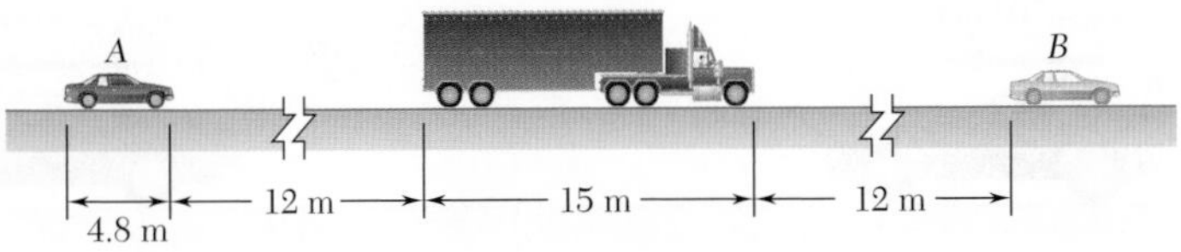

Fig. P11.72

11.73 Solve Prob. 11.72, assuming that the driver of the car does not pay any attention to the speed limit while passing and concentrates on reaching position B and resuming a speed of 50 km/h in the shortest possible time. What is the maximum speed reached? Draw the v–t curve.

11.74 Car A is traveling on a highway at a constant speed $(v_A)_0 = 90$ km/h and is 120 m from the entrance of an access ramp when car B enters the acceleration lane at that point at a speed $(v_B)_0 = 25$ km/h. Car B accelerates uniformly and enters the main traffic lane after traveling 60 m in 5 s. It then continues to accelerate at the same rate until it reaches a speed of 90 km/h, which it then maintains. Determine the final distance between the two cars.

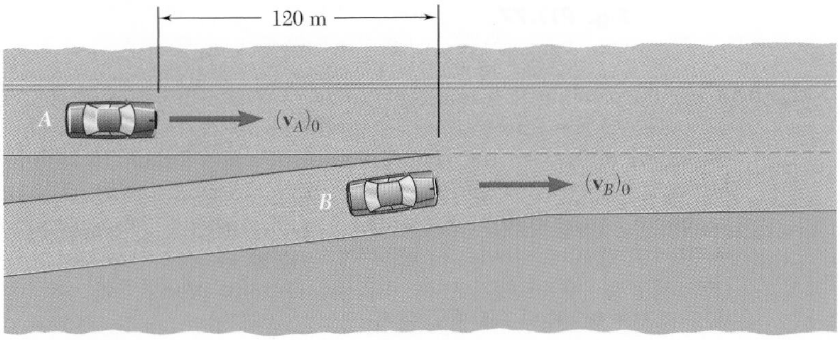

Fig. *P11.74*

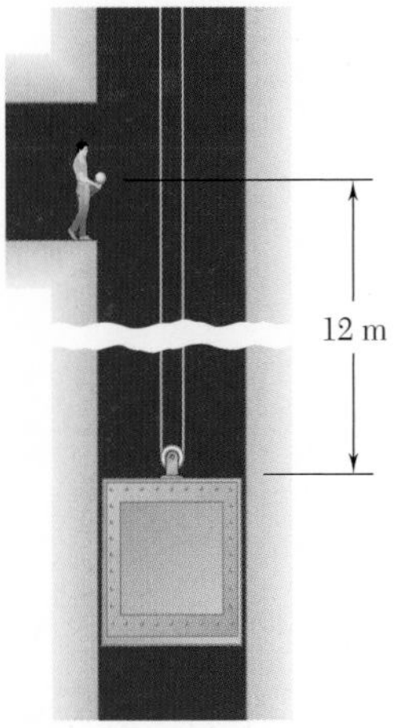

Fig. P11.75

11.75 An elevator starts from rest and moves upward, accelerating at a rate of 1.2 m/s^2 until it reaches a speed of 7.8 m/s, which it then maintains. Two seconds after the elevator begins to move, a man standing 12 m above the initial position of the top of the elevator throws a ball upward with an initial velocity of 20 m/s. Determine when the ball will hit the elevator.

11.76 Car A is traveling at 60 km/h when it enters a 40 km/h speed zone. The driver of car A decelerates at a rate of 5 m/s^2 until reaching a speed of 40 km/h, which she then maintains. When car B, which was initially 20 m behind car A and traveling at a constant speed of 70 km/h, enters the speed zone, its driver decelerates at a rate of 6 m/s^2 until reaching a speed of 35 km/h. Knowing that the driver of car B maintains a speed of 35 km/h, determine (*a*) the closest that car B comes to car A, (*b*) the time at which car A is 25 m in front of car B.

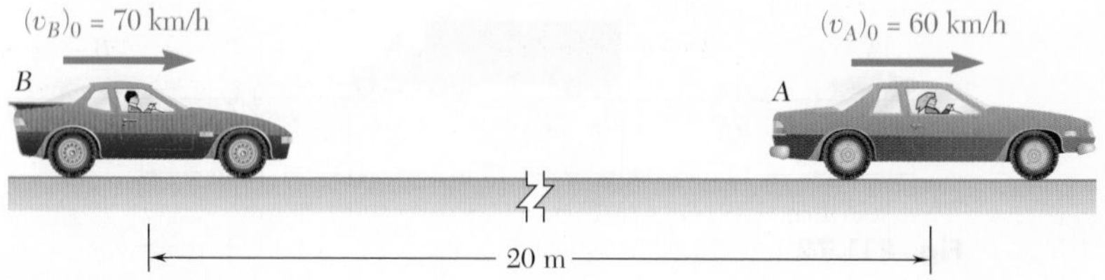

Fig. *P11.76*

11.77 An accelerometer record for the motion of a given part of a mechanism is approximated by an arc of a parabola for 0.2 s and a straight line for the next 0.2 s as shown in the figure. Knowing that $v = 0$ when $t = 0$ and $x = 0.4$ m when $t = 0.4$ s, (*a*) construct the v–t curve for $0 \leq t \leq 0.4$ s, (*b*) determine the position of the part at $t = 0.3$ s and $t = 0.2$ s.

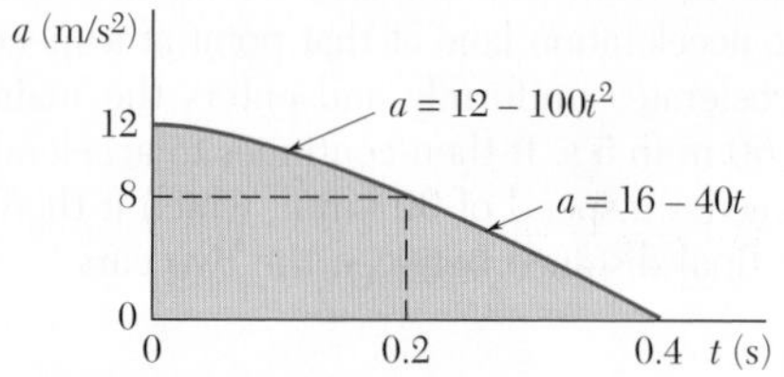

Fig. *P11.77*

11.78 A car is traveling at a constant speed of 54 km/h when its driver sees a child run into the road. The driver applies her brakes until the child returns to the sidewalk and then accelerates to resume her original speed of 54 km/h; the acceleration record of the car is shown in the figure. Assuming $x = 0$ when $t = 0$, determine (*a*) the time t_1 at which the velocity is again 54 km/h, (*b*) the position of the car at that time, (*c*) the average velocity of the car during the interval 1 s $\leq t \leq t_1$.

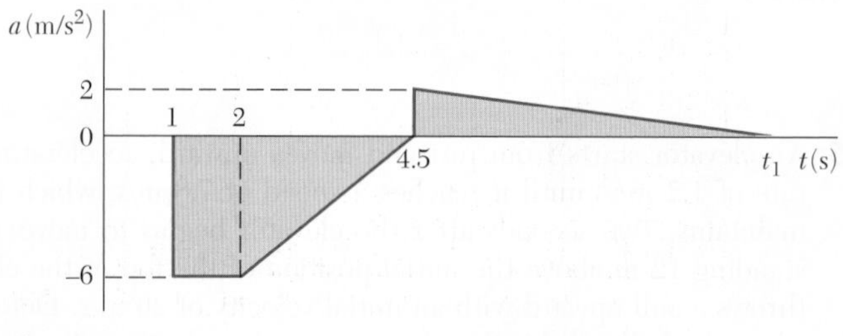

Fig. P11.78

11.79 An airport shuttle train travels between two terminals that are 2.5 km apart. To maintain passenger comfort, the acceleration of the train is limited to ± 1.2 m/s^2, and the jerk, or rate of change of acceleration, is limited to ± 0.24 m/s^2 per second. If the shuttle has a maximum speed of 30 km/h, determine (*a*) the shortest time for the shuttle to travel between the two terminals, (*b*) the corresponding average velocity of the shuttle.

11.80 During a manufacturing process, a conveyor belt starts from rest and travels a total of 400 mm before temporarily coming to rest. Knowing that the jerk, or rate of change of acceleration, is limited to ± 1.5 m/s^2 per second, determine (*a*) the shortest time required for the belt to move 400 mm, (*b*) the maximum and average values of the velocity of the belt during that time.

11.81 Two seconds are required to bring the piston rod of an air cylinder to rest; the acceleration record of the piston rod during the 2 s is as shown. Determine by approximate means (*a*) the initial velocity of the piston rod, (*b*) the distance traveled by the piston rod as it is brought to rest.

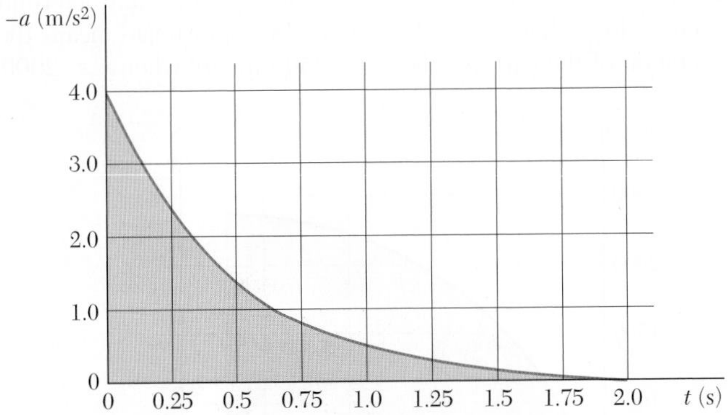

Fig. *P11.81*

11.82 The acceleration record shown was obtained during the speed trials of a sports car. Knowing that the car starts from rest, determine by approximate means (*a*) the velocity of the car at $t = 8$ s, (*b*) the distance the car has traveled at $t = 20$ s.

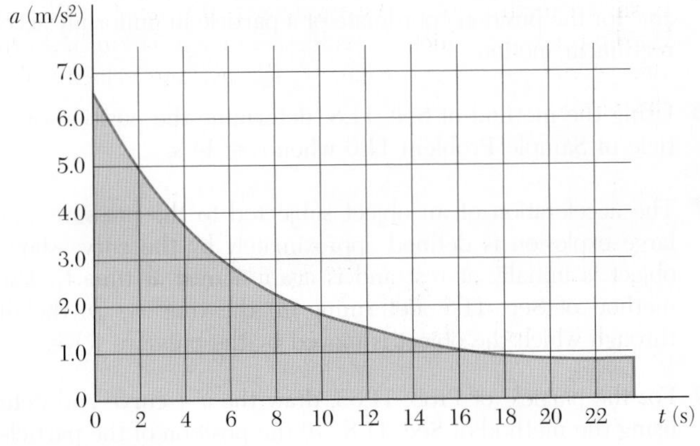

Fig. *P11.82*

11.83 A training airplane has a velocity of 38 m/s when it lands on an aircraft carrier. As the arresting gear of the carrier brings the airplane to rest, the velocity and the acceleration of the airplane are recorded; the results are shown (solid curve) in the figure. Determine by approximate means (*a*) the time required for the airplane to come to rest, (*b*) the distance traveled in that time.

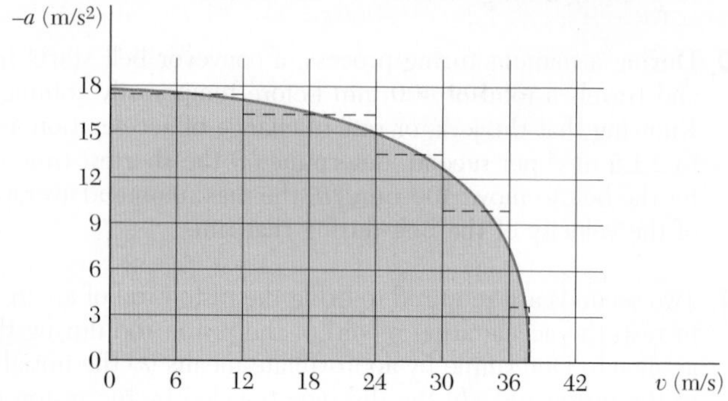

Fig. P11.83

11.84 Shown in the figure is a portion of the experimentally determined v–x curve for a shuttle cart. Determine by approximate means the acceleration of the cart (*a*) when $x = 250$ mm, (*b*) when $v = 2000$ mm/s.

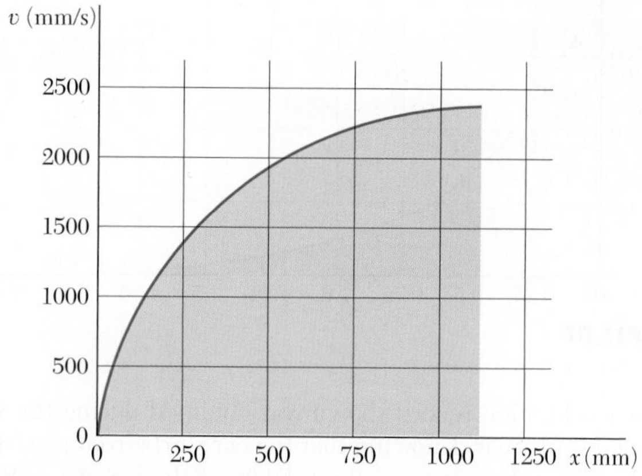

Fig. P11.84

11.85 Using the method of Sec. 11.8, derive the formula $x = x_0 + v_0t + \frac{1}{2}at^2$ for the position coordinate of a particle in uniformly accelerated rectilinear motion.

11.86 Using the method of Sec. 11.8, determine the position of the particle of Sample Problem 11.6 when $t = 14$ s.

11.87 The acceleration of an object subjected to the pressure wave of a large explosion is defined approximately by the curve shown. The object is initially at rest and is again at rest at time t_1. Using the method of Sec. 11.8, determine (*a*) the time t_1, (*b*) the distance through which the object is moved by the pressure wave.

11.88 For the particle of Prob. 11.63, draw the a–t curve and determine, using the method of Sec. 11.8, (*a*) the position of the particle when $t = 52$ s, (*b*) the maximum value of its position coordinate.

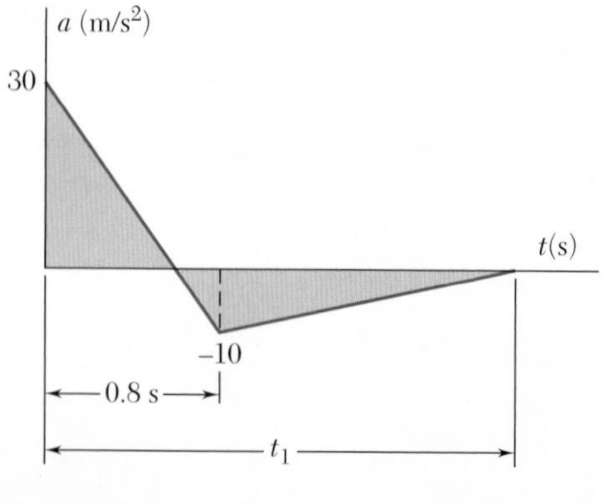

Fig. P11.87

CURVILINEAR MOTION OF PARTICLES

11.9 POSITION VECTOR, VELOCITY, AND ACCELERATION

When a particle moves along a curve other than a straight line, we say that the particle is in *curvilinear motion*. To define the position P occupied by the particle at a given time t, we select a fixed reference system, such as the x, y, z axes shown in Fig. 11.14a, and draw the vector $\mathbf{r}$ joining the origin O and point P. Since the vector $\mathbf{r}$ is characterized by its magnitude r and its direction with respect to the reference axes, it completely defines the position of the particle with respect to those axes; the vector $\mathbf{r}$ is referred to as the *position vector* of the particle at time t.

Consider now the vector $\mathbf{r}'$ defining the position P' occupied by the same particle at a later time $t + \Delta t$. The vector $\Delta\mathbf{r}$ joining P and P' represents the change in the position vector during the time interval Δt since, as we can easily check from Fig. 11.14a, the vector $\mathbf{r}'$ is obtained by adding the vectors $\mathbf{r}$ and $\Delta\mathbf{r}$ according to the triangle rule. We note that $\Delta\mathbf{r}$ represents a change in *direction* as well as a change in *magnitude* of the position vector $\mathbf{r}$. The *average velocity* of the particle over the time interval Δt is defined as the quotient of $\Delta\mathbf{r}$ and Δt. Since $\Delta\mathbf{r}$ is a vector and Δt is a scalar, the quotient $\Delta\mathbf{r}/\Delta t$ is a vector attached at P, of the same direction as $\Delta\mathbf{r}$ and of magnitude equal to the magnitude of $\Delta\mathbf{r}$ divided by Δt (Fig. 11.14b).

The *instantaneous velocity* of the particle at time t is obtained by choosing shorter and shorter time intervals Δt and, correspondingly, shorter and shorter vector increments $\Delta\mathbf{r}$. The instantaneous velocity is thus represented by the vector

$$\mathbf{v} = \lim_{\Delta t \to 0} \frac{\Delta\mathbf{r}}{\Delta t} \tag{11.14}$$

As Δt and $\Delta\mathbf{r}$ become shorter, the points P and P' get closer; the vector $\mathbf{v}$ obtained in the limit must therefore be tangent to the path of the particle (Fig. 11.14c).

Since the position vector $\mathbf{r}$ depends upon the time t, we can refer to it as a *vector function* of the scalar variable t and denote it by $\mathbf{r}(t)$. Extending the concept of derivative of a scalar function introduced in elementary calculus, we will refer to the limit of the quotient $\Delta\mathbf{r}/\Delta t$ as the *derivative* of the vector function $\mathbf{r}(t)$. We write

$$\mathbf{v} = \frac{d\mathbf{r}}{dt} \tag{11.15}$$

The magnitude v of the vector $\mathbf{v}$ is called the *speed* of the particle. It can be obtained by substituting for the vector $\Delta\mathbf{r}$ in formula (11.14) the magnitude of this vector represented by the straight-line segment PP'. But the length of the segment PP' approaches the length Δs of the arc PP' as Δt decreases (Fig. 11.14a), and we can write

$$v = \lim_{\Delta t \to 0} \frac{PP'}{\Delta t} = \lim_{\Delta t \to 0} \frac{\Delta s}{\Delta t} \qquad v = \frac{ds}{dt} \tag{11.16}$$

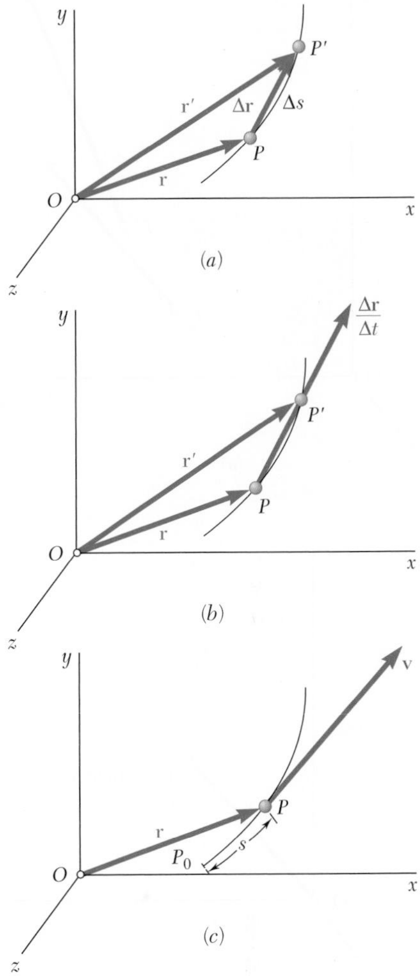

Fig. 11.14

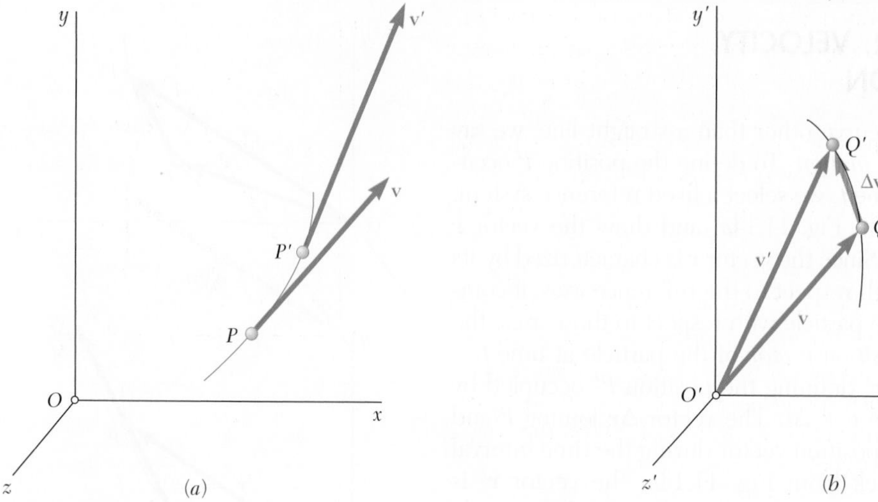

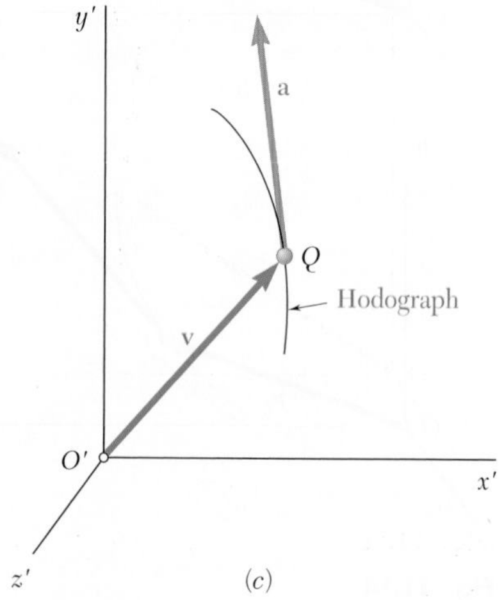

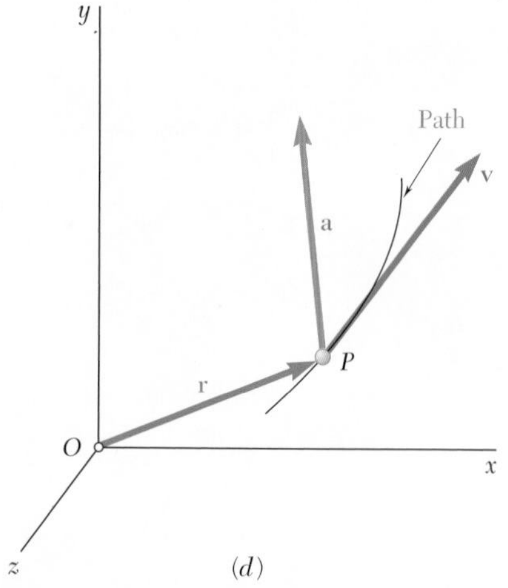

Fig. 11.15

The speed v can thus be obtained by differentiating with respect to t the length s of the arc described by the particle.

Consider the velocity $\mathbf{v}$ of the particle at time t and its velocity $\mathbf{v}'$ at a later time $t + \Delta t$ (Fig. 11.15*a*). Let us draw both vectors $\mathbf{v}$ and $\mathbf{v}'$ from the same origin O' (Fig. 11.15*b*). The vector $\Delta\mathbf{v}$ joining Q and Q' represents the change in the velocity of the particle during the time interval Δt, since the vector $\mathbf{v}'$ can be obtained by adding the vectors $\mathbf{v}$ and $\Delta\mathbf{v}$. We should note that $\Delta\mathbf{v}$ represents a change in the *direction* of the velocity as well as a change in *speed*. The *average acceleration* of the particle over the time interval Δt is defined as the quotient of $\Delta\mathbf{v}$ and Δt. Since $\Delta\mathbf{v}$ is a vector and Δt a scalar, the quotient $\Delta\mathbf{v}/\Delta t$ is a vector of the same direction as $\Delta\mathbf{v}$.

The *instantaneous acceleration* of the particle at time t is obtained by choosing smaller and smaller values for Δt and $\Delta\mathbf{v}$. The instantaneous acceleration is thus represented by the vector

$$\mathbf{a} = \lim_{\Delta t \to 0} \frac{\Delta\mathbf{v}}{\Delta t} \tag{11.17}$$

Noting that the velocity $\mathbf{v}$ is a vector function $\mathbf{v}(t)$ of the time t, we can refer to the limit of the quotient $\Delta\mathbf{v}/\Delta t$ as the derivative of $\mathbf{v}$ with respect to t. We write

$$\mathbf{a} = \frac{d\mathbf{v}}{dt} \tag{11.18}$$

We observe that the acceleration $\mathbf{a}$ is tangent to the curve described by the tip Q of the vector $\mathbf{v}$ when the latter is drawn from a fixed origin O' (Fig. 11.15*c*) and that, in general, the acceleration is *not* tangent to the path of the particle (Fig. 11.15*d*). The curve described by the tip of $\mathbf{v}$ and shown in Fig. 11.15*c* is called the *hodograph* of the motion.

11.10 DERIVATIVES OF VECTOR FUNCTIONS

We saw in the preceding section that the velocity **v** of a particle in curvilinear motion can be represented by the derivative of the vector function **r**(t) characterizing the position of the particle. Similarly, the acceleration **a** of the particle can be represented by the derivative of the vector function **v**(t). In this section, we will give a formal definition of the derivative of a vector function and establish a few rules governing the differentiation of sums and products of vector functions.

Let **P**(u) be a vector function of the scalar variable u. By that we mean that the scalar u completely defines the magnitude and direction of the vector **P**. If the vector **P** is drawn from a fixed origin O and the scalar u is allowed to vary, the tip of **P** will describe a given curve in space. Consider the vectors **P** corresponding, respectively, to the values u and $u + \Delta u$ of the scalar variable (Fig. 11.16*a*). Let $\Delta\mathbf{P}$ be the vector joining the tips of the two given vectors; we write

$$\Delta\mathbf{P} = \mathbf{P}(u + \Delta u) - \mathbf{P}(u)$$

Dividing through by Δu and letting Δu approach zero, *we define the derivative of the vector function* **P**(u):

$$\frac{d\mathbf{P}}{du} = \lim_{\Delta u \to 0} \frac{\Delta\mathbf{P}}{\Delta u} = \lim_{\Delta u \to 0} \frac{\mathbf{P}(u + \Delta u) - \mathbf{P}(u)}{\Delta u} \tag{11.19}$$

As Δu approaches zero, the line of action of $\Delta\mathbf{P}$ becomes tangent to the curve of Fig. 11.16*a*. Thus, the derivative $d\mathbf{P}/du$ of the vector function **P**(u) *is tangent to the curve described by the tip of* **P**(u) (Fig. 11.16*b*).

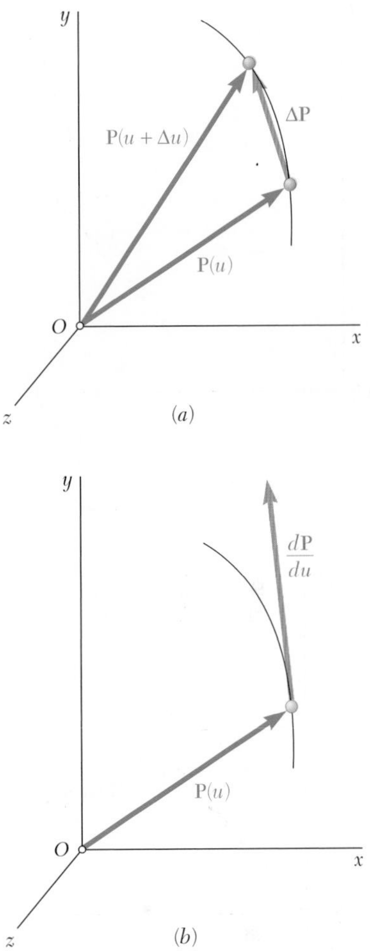

Fig. 11.16

The standard rules for the differentiation of the sums and products of scalar functions can be extended to vector functions. Consider first the *sum of two vector functions* **P**(u) and **Q**(u) of the same scalar variable u. According to the definition given in (11.19), the derivative of the vector **P** + **Q** is

$$\frac{d(\mathbf{P} + \mathbf{Q})}{du} = \lim_{\Delta u \to 0} \frac{\Delta(\mathbf{P} + \mathbf{Q})}{\Delta u} = \lim_{\Delta u \to 0} \left(\frac{\Delta\mathbf{P}}{\Delta u} + \frac{\Delta\mathbf{Q}}{\Delta u} \right)$$

or since the limit of a sum is equal to the sum of the limits of its terms,

$$\frac{d(\mathbf{P} + \mathbf{Q})}{du} = \lim_{\Delta u \to 0} \frac{\Delta\mathbf{P}}{\Delta u} + \lim_{\Delta u \to 0} \frac{\Delta\mathbf{Q}}{\Delta u}$$

$$\frac{d(\mathbf{P} + \mathbf{Q})}{du} = \frac{d\mathbf{P}}{du} + \frac{d\mathbf{Q}}{du} \tag{11.20}$$

The *product of a scalar function* $f(u)$ *and a vector function* **P**(u) of the same scalar variable u will now be considered. The derivative of the vector $f\mathbf{P}$ is

$$\frac{d(f\mathbf{P})}{du} = \lim_{\Delta u \to 0} \frac{(f + \Delta f)(\mathbf{P} + \Delta\mathbf{P}) - f\mathbf{P}}{\Delta u} = \lim_{\Delta u \to 0} \left(\frac{\Delta f}{\Delta u}\mathbf{P} + f\frac{\Delta\mathbf{P}}{\Delta u} \right)$$

or recalling the properties of the limits of sums and products,

$$\frac{d(f\mathbf{P})}{du} = \frac{df}{du}\mathbf{P} + f\frac{d\mathbf{P}}{du} \tag{11.21}$$

The derivatives of the *scalar product* and the *vector product* of two vector functions $\mathbf{P}(u)$ and $\mathbf{Q}(u)$ can be obtained in a similar way. We have

$$\frac{d(\mathbf{P} \cdot \mathbf{Q})}{du} = \frac{d\mathbf{P}}{du} \cdot \mathbf{Q} + \mathbf{P} \cdot \frac{d\mathbf{Q}}{du} \tag{11.22}$$

$$\frac{d(\mathbf{P} \times \mathbf{Q})}{du} = \frac{d\mathbf{P}}{du} \times \mathbf{Q} + \mathbf{P} \times \frac{d\mathbf{Q}}{du} \tag{11.23}$$†

The properties established above can be used to determine the *rectangular components of the derivative of a vector function* $\mathbf{P}(u)$. Resolving $\mathbf{P}$ into components along fixed rectangular axes x, y, z, we write

$$\mathbf{P} = P_x\mathbf{i} + P_y\mathbf{j} + P_z\mathbf{k} \tag{11.24}$$

where P_x, P_y, P_z are the rectangular scalar components of the vector $\mathbf{P}$, and $\mathbf{i}$, $\mathbf{j}$, $\mathbf{k}$ the unit vectors corresponding, respectively, to the x, y, and z axes (Sec. 2.12). By (11.20), the derivative of $\mathbf{P}$ is equal to the sum of the derivatives of the terms in the right-hand member. Since each of these terms is the product of a scalar and a vector function, we should use (11.21). But the unit vectors $\mathbf{i}$, $\mathbf{j}$, $\mathbf{k}$ have a constant magnitude (equal to 1) and fixed directions. Their derivatives are therefore zero, and we write

$$\frac{d\mathbf{P}}{du} = \frac{dP_x}{du}\mathbf{i} + \frac{dP_y}{du}\mathbf{j} + \frac{dP_z}{du}\mathbf{k} \tag{11.25}$$

Noting that the coefficients of the unit vectors are, by definition, the scalar components of the vector $d\mathbf{P}/du$, we conclude that *the rectangular scalar components of the derivative* $d\mathbf{P}/du$ *of the vector function* $\mathbf{P}(u)$ are obtained by differentiating the corresponding scalar components of $\mathbf{P}$.

Rate of Change of a Vector. When the vector $\mathbf{P}$ is a function of the time t, its derivative $d\mathbf{P}/dt$ represents the *rate of change* of $\mathbf{P}$ with respect to the frame $Oxyz$. Resolving $\mathbf{P}$ into rectangular components, we have, by (11.25),

$$\frac{d\mathbf{P}}{dt} = \frac{dP_x}{dt}\mathbf{i} + \frac{dP_y}{dt}\mathbf{j} + \frac{dP_z}{dt}\mathbf{k}$$

or, using dots to indicate differentiation with respect to t,

$$\dot{\mathbf{P}} = \dot{P}_x\mathbf{i} + \dot{P}_y\mathbf{j} + \dot{P}_z\mathbf{k} \tag{11.25'}$$

†Since the vector product is not commutative (Sec. 3.4), the order of the factors in Eq. (11.23) must be maintained.

As you will see in Sec. 15.10, the rate of change of a vector as observed from a *moving frame of reference* is, in general, different from its rate of change as observed from a fixed frame of reference. However, if the moving frame $O'x'y'z'$ is in *translation*, i.e., if its axes remain parallel to the corresponding axes of the fixed frame $Oxyz$ (Fig. 11.17), the same unit vectors **i**, **j**, **k** are used in both frames, and at any given instant the vector **P** has the same components P_x, P_y, P_z in both frames. It follows from (11.25′) that the rate of change $\dot{\mathbf{P}}$ is the same with respect to the frames $Oxyz$ and $O'x'y'z'$. We state, therefore: *The rate of change of a vector is the same with respect to a fixed frame and with respect to a frame in translation.* This property will greatly simplify our work, since we will be concerned mainly with frames in translation.

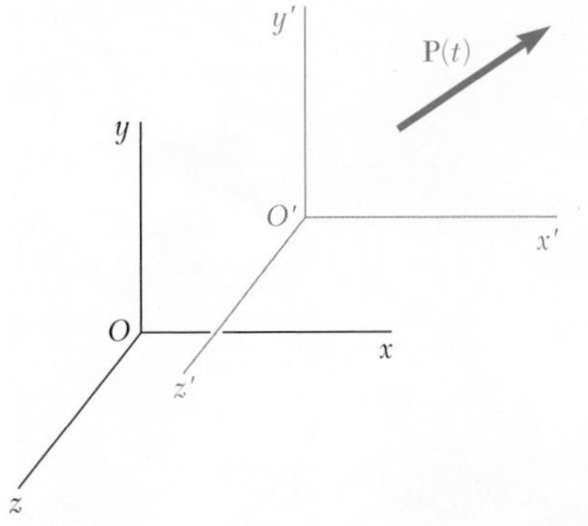

Fig. 11.17

11.11 RECTANGULAR COMPONENTS OF VELOCITY AND ACCELERATION

When the position of a particle P is defined at any instant by its rectangular coordinates x, y, and z, it is convenient to resolve the velocity **v** and the acceleration **a** of the particle into rectangular components (Fig. 11.18).

Resolving the position vector **r** of the particle into rectangular components, we write

$$\mathbf{r} = x\mathbf{i} + y\mathbf{j} + z\mathbf{k} \tag{11.26}$$

where the coordinates x, y, z are functions of t. Differentiating twice, we obtain

$$\mathbf{v} = \frac{d\mathbf{r}}{dt} = \dot{x}\mathbf{i} + \dot{y}\mathbf{j} + \dot{z}\mathbf{k} \tag{11.27}$$

$$\mathbf{a} = \frac{d\mathbf{v}}{dt} = \ddot{x}\mathbf{i} + \ddot{y}\mathbf{j} + \ddot{z}\mathbf{k} \tag{11.28}$$

where $\dot{x}$, $\dot{y}$, $\dot{z}$ and $\ddot{x}$, $\ddot{y}$, $\ddot{z}$ represent, respectively, the first and second derivatives of x, y, and z with respect to t. It follows from (11.27) and (11.28) that the scalar components of the velocity and acceleration are

$$v_x = \dot{x} \qquad v_y = \dot{y} \qquad v_z = \dot{z} \tag{11.29}$$

$$a_x = \ddot{x} \qquad a_y = \ddot{y} \qquad a_z = \ddot{z} \tag{11.30}$$

A positive value for v_x indicates that the vector component $\mathbf{v}_x$ is directed to the right, and a negative value indicates that it is directed to the left. The sense of each of the other vector components can be determined in a similar way from the sign of the corresponding scalar component. If desired, the magnitudes and directions of the velocity and acceleration can be obtained from their scalar components by the methods of Secs. 2.7 and 2.12.

The use of rectangular components to describe the position, the velocity, and the acceleration of a particle is particularly effective when the component a_x of the acceleration depends only upon t, x, and/or v_x, and when, similarly, a_y depends only upon t, y, and/or v_y,

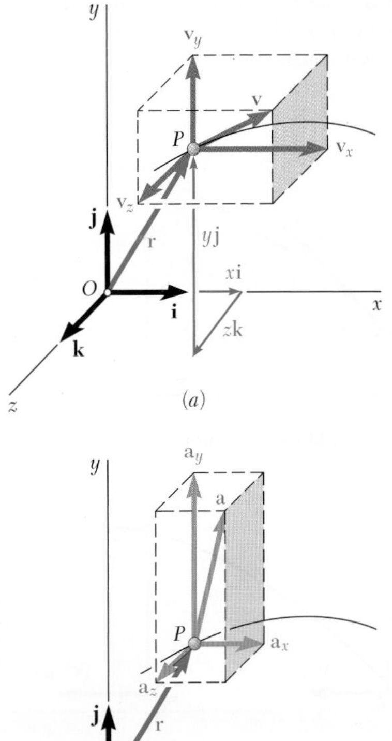

Fig. 11.18

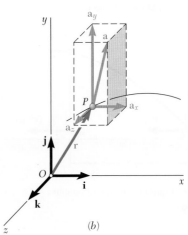

Photo 11.3 The motion of this snowboarder in the air will be a parabola assuming we can neglect air resistance.

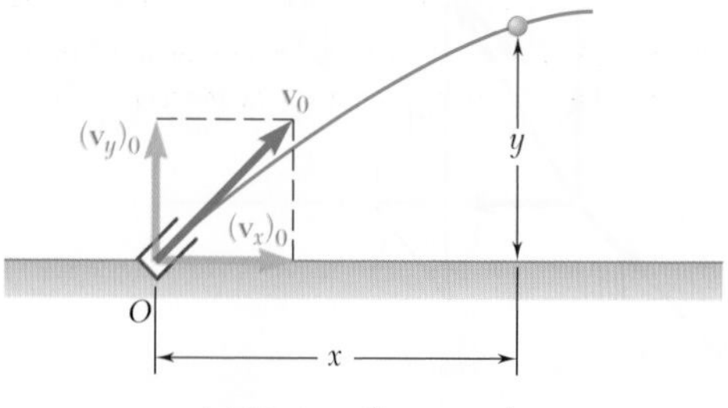

(*a*) Motion of a projectile

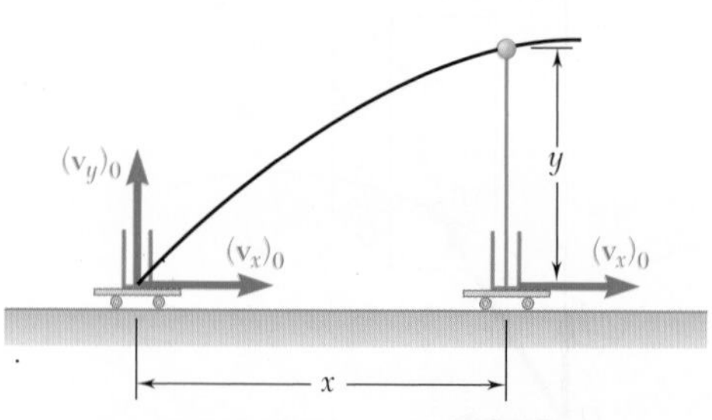

(*b*) Equivalent rectilinear motions

Fig. 11.19

and a_z upon t, z, and/or v_z. Equations (11.30) can then be integrated independently, and so can Eqs. (11.29). In other words, the motion of the particle in the x direction, its motion in the y direction, and its motion in the z direction can be considered separately.

In the case of the *motion of a projectile*, for example, it can be shown (see Sec. 12.5) that the components of the acceleration are

$$a_x = \ddot{x} = 0 \qquad a_y = \ddot{y} = -g \qquad a_z = \ddot{z} = 0$$

if the resistance of the air is neglected. Denoting by x_0, y_0, and z_0 the coordinates of a gun, and by $(v_x)_0$, $(v_y)_0$, and $(v_z)_0$ the components of the initial velocity $\mathbf{v}_0$ of the projectile (a bullet), we integrate twice in t and obtain

$$v_x = \dot{x} = (v_x)_0 \qquad v_y = \dot{y} = (v_y)_0 - gt \qquad v_z = \dot{z} = (v_z)_0$$
$$x = x_0 + (v_x)_0 t \qquad y = y_0 + (v_y)_0 t - \tfrac{1}{2}gt^2 \qquad z = z_0 + (v_z)_0 t$$

If the projectile is fired in the xy plane from the origin O, we have $x_0 = y_0 = z_0 = 0$ and $(v_z)_0 = 0$, and the equations of motion reduce to

$$v_x = (v_x)_0 \qquad v_y = (v_y)_0 - gt \qquad v_z = 0$$
$$x = (v_x)_0 t \qquad y = (v_y)_0 t - \tfrac{1}{2}gt^2 \qquad z = 0$$

These equations show that the projectile remains in the xy plane, that its motion in the horizontal direction is uniform, and that its motion in the vertical direction is uniformly accelerated. The motion of a projectile can thus be replaced by two independent rectilinear motions, which are easily visualized if we assume that the projectile is fired vertically with an initial velocity $(\mathbf{v}_y)_0$ from a platform moving with a constant horizontal velocity $(\mathbf{v}_x)_0$ (Fig. 11.19). The coordinate x of the projectile is equal at any instant to the distance traveled by the platform, and its coordinate y can be computed as if the projectile were moving along a vertical line.

It can be observed that the equations defining the coordinates x and y of a projectile at any instant are the parametric equations of a parabola. Thus, the trajectory of a projectile is *parabolic*. This result, however, ceases to be valid when the resistance of the air or the variation with altitude of the acceleration of gravity is taken into account.

11.12 MOTION RELATIVE TO A FRAME IN TRANSLATION

In the preceding section, a single frame of reference was used to describe the motion of a particle. In most cases this frame was attached to the earth and was considered as fixed. Situations in which it is convenient to use several frames of reference simultaneously will now be analyzed. If one of the frames is attached to the earth, it will be called a *fixed frame of reference*, and the other frames will be referred to as *moving frames of reference*. It should be understood, however, that the selection of a fixed frame of reference is purely arbitrary. Any frame can be designated as "fixed"; all other frames not rigidly attached to this frame will then be described as "moving."

Consider two particles A and B moving in space (Fig. 11.20); the vectors $\mathbf{r}_A$ and $\mathbf{r}_B$ define their positions at any given instant with respect to the fixed frame of reference $Oxyz$. Consider now a system of axes x', y', z' centered at A and parallel to the x, y, z axes. While the origin of these axes moves, their orientation remains the same; the frame of reference $Ax'y'z'$ is in *translation* with respect to $Oxyz$. The vector $\mathbf{r}_{B/A}$ joining A and B defines *the position of B relative to the moving frame $Ax'y'z'$* (or, for short, *the position of B relative to A*).

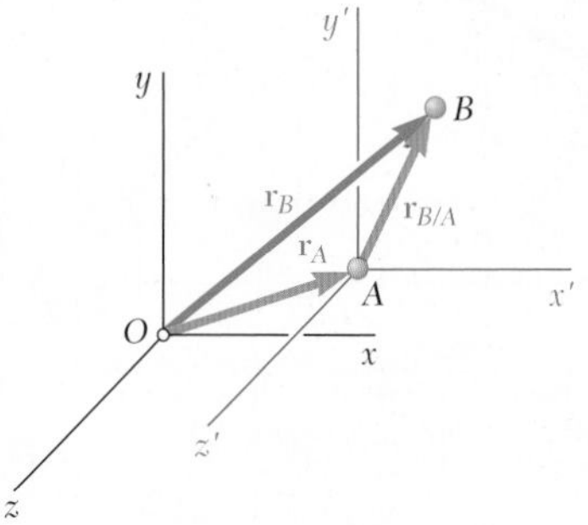

Fig. 11.20

We note from Fig. 11.20 that the position vector $\mathbf{r}_B$ of particle B is the sum of the position vector $\mathbf{r}_A$ of particle A and of the position vector $\mathbf{r}_{B/A}$ of B relative to A; we write

$$\mathbf{r}_B = \mathbf{r}_A + \mathbf{r}_{B/A} \tag{11.31}$$

Differentiating (11.31) with respect to t within the fixed frame of reference, and using dots to indicate time derivatives, we have

$$\dot{\mathbf{r}}_B = \dot{\mathbf{r}}_A + \dot{\mathbf{r}}_{B/A} \tag{11.32}$$

The derivatives $\dot{\mathbf{r}}_A$ and $\dot{\mathbf{r}}_B$ represent, respectively, the velocities $\mathbf{v}_A$ and $\mathbf{v}_B$ of the particles A and B. Since $Ax'y'z'$ is in translation, the derivative $\dot{\mathbf{r}}_{B/A}$ represents the rate of change of $\mathbf{r}_{B/A}$ with respect to the frame $Ax'y'z'$ as well as with respect to the fixed frame (Sec. 11.10). This derivative, therefore, defines *the velocity $\mathbf{v}_{B/A}$ of B relative to the frame $Ax'y'z'$* (or, for short, *the velocity $\mathbf{v}_{B/A}$ of B relative to A*). We write

Photo 11.4 The pilot of a helicopter must take into account the relative motion of the ship when landing.

$$\mathbf{v}_B = \mathbf{v}_A + \mathbf{v}_{B/A} \tag{11.33}$$

Differentiating Eq. (11.33) with respect to t, and using the derivative $\dot{\mathbf{v}}_{B/A}$ to define *the acceleration $\mathbf{a}_{B/A}$ of B relative to the frame $Ax'y'z'$* (or, for short, *the acceleration $\mathbf{a}_{B/A}$ of B relative to A*), we write

$$\mathbf{a}_B = \mathbf{a}_A + \mathbf{a}_{B/A} \tag{11.34}$$

The motion of B with respect to the fixed frame $Oxyz$ is referred to as the *absolute motion of B*. The equations derived in this section show that *the absolute motion of B can be obtained by combining the motion of A and the relative motion of B with respect to the moving frame attached to A*. Equation (11.33), for example, expresses that the absolute velocity $\mathbf{v}_B$ of particle B can be obtained by adding vectorially the velocity of A and the velocity of B relative to the frame $Ax'y'z'$. Equation (11.34) expresses a similar property in terms of the accelerations.† We should keep in mind, however, that *the frame $Ax'y'z'$ is in translation;* that is, while it moves with A, it maintains the same orientation. As you will see later (Sec. 15.14), different relations must be used in the case of a rotating frame of reference.

†Note that the product of the subscripts A and B/A used in the right-hand member of Eqs. (11.31) through (11.34) is equal to the subscript B used in their left-hand member.

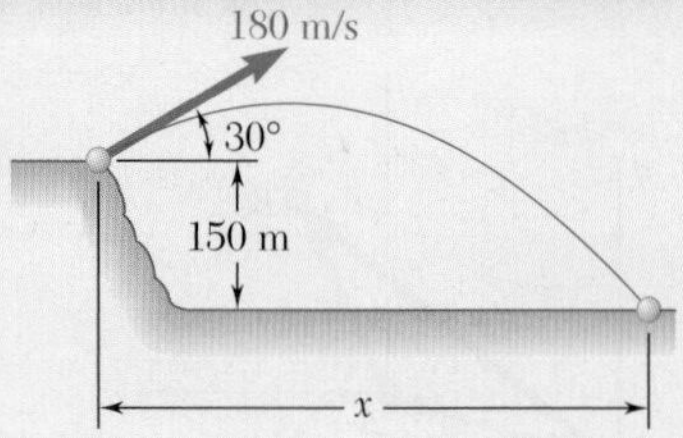

SAMPLE PROBLEM 11.7

A projectile is fired from the edge of a 150-m cliff with an initial velocity of 180 m/s at an angle of 30° with the horizontal. Neglecting air resistance, find (*a*) the horizontal distance from the gun to the point where the projectile strikes the ground, (*b*) the greatest elevation above the ground reached by the projectile.

SOLUTION

The vertical and the horizontal motion will be considered separately.

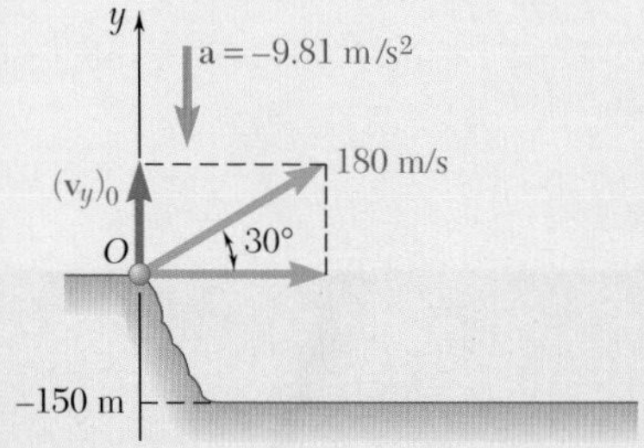

Vertical Motion. ***Uniformly Accelerated Motion.*** Choosing the positive sense of the y axis upward and placing the origin O at the gun, we have

$$(v_y)_0 = (180 \text{ m/s}) \sin 30° = +90 \text{ m/s}$$
$$a = -9.81 \text{ m/s}^2$$

Substituting into the equations of uniformly accelerated motion, we have

$$v_y = (v_y)_0 + at \qquad v_y = 90 - 9.81t \qquad (1)$$
$$y = (v_y)_0 t + \tfrac{1}{2}at^2 \qquad y = 90t - 4.90t^2 \qquad (2)$$
$$v_y^2 = (v_y)_0^2 + 2ay \qquad v_y^2 = 8100 - 19.62y \qquad (3)$$

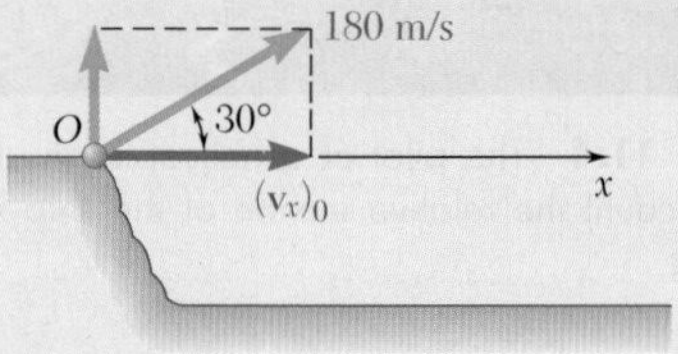

Horizontal Motion. ***Uniform Motion.*** Choosing the positive sense of the x axis to the right, we have

$$(v_x)_0 = (180 \text{ m/s}) \cos 30° = +155.9 \text{ m/s}$$

Substituting into the equation of uniform motion, we obtain

$$x = (v_x)_0 t \qquad x = 155.9t \qquad (4)$$

a. Horizontal Distance. When the projectile strikes the ground, we have

$$y = -150 \text{ m}$$

Carrying this value into Eq. (2) for the vertical motion, we write

$$-150 = 90t - 4.90t^2 \qquad t^2 - 18.37t - 30.6 = 0 \qquad t = 19.91 \text{ s}$$

Carrying $t = 19.91$ s into Eq. (4) for the horizontal motion, we obtain

$$x = 155.9(19.91) \qquad x = 3100 \text{ m} \blacktriangleleft$$

b. Greatest Elevation. When the projectile reaches its greatest elevation, we have $v_y = 0$; carrying this value into Eq. (3) for the vertical motion, we write

$$0 = 8100 - 19.62y \qquad y = 413 \text{ m}$$

Greatest elevation above ground = 150 m + 413 m = 563 m ◀

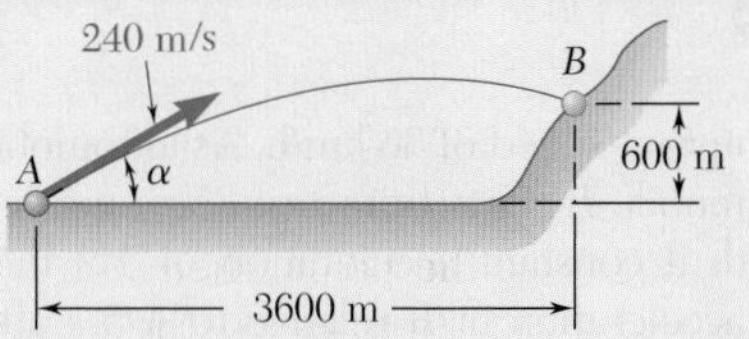

SAMPLE PROBLEM 11.8

A projectile is fired with an initial velocity of 240 m/s at a target B located 600 m above the gun A and at a horizontal distance of 3600 m. Neglecting air resistance, determine the value of the firing angle α.

SOLUTION

The horizontal and the vertical motion will be considered separately.

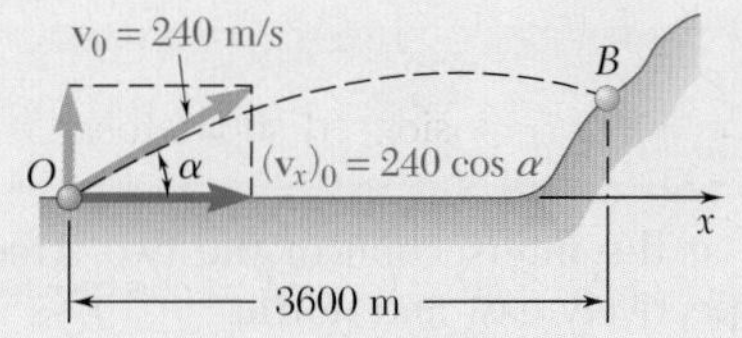

Horizontal Motion. Placing the origin of the coordinate axes at the gun, we have

$$(v_x)_0 = 240 \cos \alpha$$

Substituting into the equation of uniform horizontal motion, we obtain

$$x = (v_x)_0 t \qquad x = (240 \cos \alpha)t$$

The time required for the projectile to move through a horizontal distance of 3600 m is obtained by setting x equal to 3600 m.

$$3600 = (240 \cos \alpha)t$$

$$t = \frac{3600}{240 \cos \alpha} = \frac{15}{\cos \alpha}$$

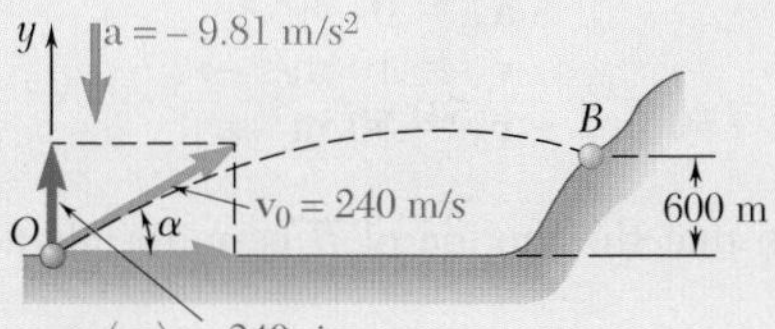

Vertical Motion

$$(v_y)_0 = 240 \sin \alpha \qquad a = -9.81 \text{ m/s}^2$$

Substituting into the equation of uniformly accelerated vertical motion, we obtain

$$y = (v_y)_0 t + \tfrac{1}{2}at^2 \qquad y = (240 \sin \alpha)t - 4.905t^2$$

Projectile Hits Target. When $x = 3600$ m, we must have $y = 600$ m. Substituting for y and setting t equal to the value found above, we write

$$600 = 240 \sin \alpha \frac{15}{\cos \alpha} - 4.905 \left(\frac{15}{\cos \alpha}\right)^2$$

Since $1/\cos^2 \alpha = \sec^2 \alpha = 1 + \tan^2 \alpha$, we have

$$600 = 240(15) \tan \alpha - 4.905(15^2)(1 + \tan^2 \alpha)$$
$$1104 \tan^2 \alpha - 3600 \tan \alpha + 1704 = 0$$

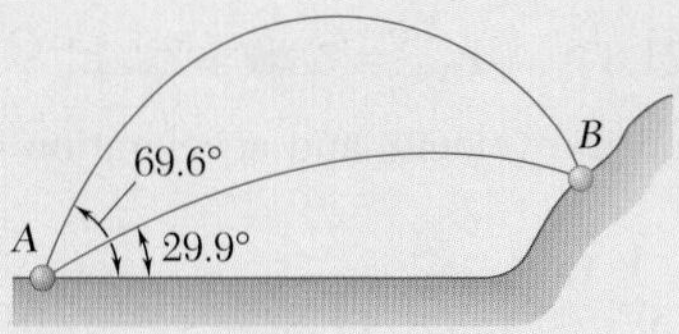

Solving this quadratic equation for $\tan \alpha$, we have

$$\tan \alpha = 0.575 \qquad \text{and} \qquad \tan \alpha = 2.69$$

$$\alpha = 29.9° \qquad \text{and} \qquad \alpha = 69.6° \quad ◀$$

The target will be hit if either of these two firing angles is used (see figure).

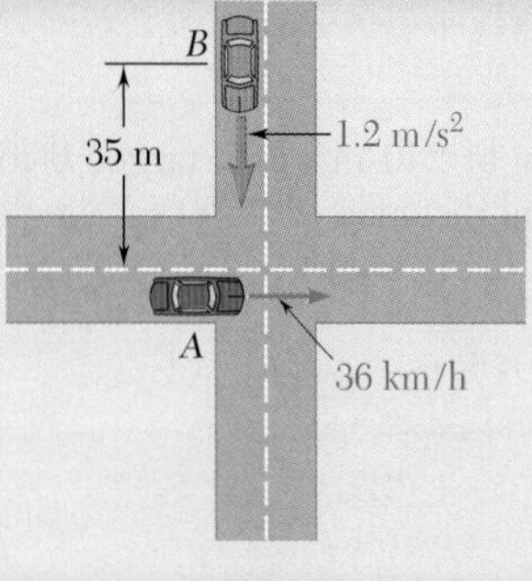

SAMPLE PROBLEM 11.9

Automobile A is traveling east at the constant speed of 36 km/h. As automobile A crosses the intersection shown, automobile B starts from rest 35 m north of the intersection and moves south with a constant acceleration of 1.2 m/s^2. Determine the position, velocity, and acceleration of B relative to A 5 s after A crosses the intersection.

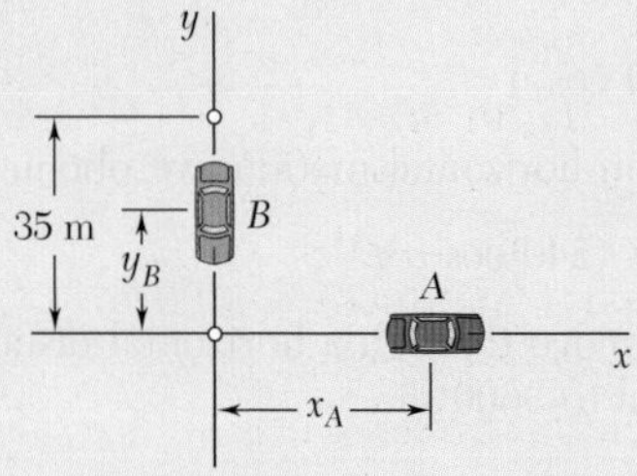

SOLUTION

We choose x and y axes with origin at the intersection of the two streets and with positive senses directed respectively east and north.

Motion of Automobile *A*. First the speed is expressed in m/s:

$$v_A = \left(36\frac{\text{km}}{\text{h}}\right)\left(\frac{1000\text{ m}}{1\text{ km}}\right)\left(\frac{1\text{ h}}{3600\text{ s}}\right) = 10\text{ m/s}$$

Noting that the motion of A is uniform, we write, for any time t,

$$a_A = 0$$
$$v_A = +10\text{ m/s}$$
$$x_A = (x_A)_0 + v_A t = 0 + 10t$$

For $t = 5$ s, we have

$$a_A = 0 \qquad \mathbf{a}_A = 0$$
$$v_A = +10\text{ m/s} \qquad \mathbf{v}_A = 10\text{ m/s} \rightarrow$$
$$x_A = +(10\text{ m/s})(5\text{ s}) = +50\text{ m} \qquad \mathbf{r}_A = 50\text{ m} \rightarrow$$

Motion of Automobile *B*. We note that the motion of B is uniformly accelerated and write

$$a_B = -1.2\text{ m/s}^2$$
$$v_B = (v_B)_0 + at = 0 - 1.2\,t$$
$$y_B = (y_B)_0 + (v_B)_0 t + \tfrac{1}{2}a_B t^2 = 35 + 0 - \tfrac{1}{2}(1.2)t^2$$

For $t = 5$ s, we have

$$a_B = -1.2\text{ m/s}^2 \qquad \mathbf{a}_B = 1.2\text{ m/s}^2 \downarrow$$
$$v_B = -(1.2\text{ m/s}^2)(5\text{ s}) = -6\text{ m/s} \qquad \mathbf{v}_B = 6\text{ m/s} \downarrow$$
$$y_B = 35 - \tfrac{1}{2}(1.2\text{ m/s}^2)(5\text{ s})^2 = +20\text{ m} \qquad \mathbf{r}_B = 20\text{ m} \uparrow$$

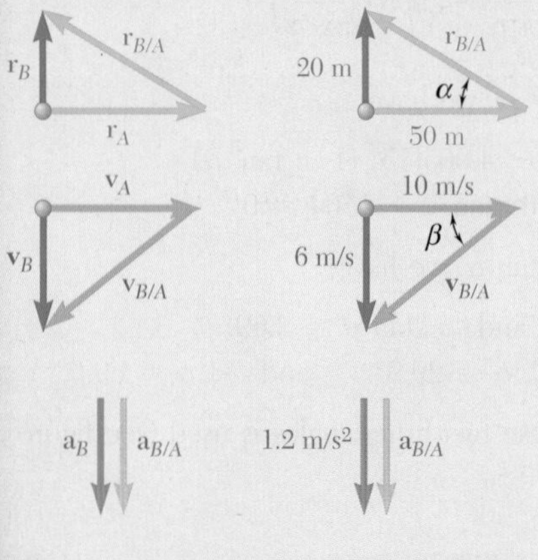

Motion of *B* Relative to *A*. We draw the triangle corresponding to the vector equation $\mathbf{r}_B = \mathbf{r}_A + \mathbf{r}_{B/A}$ and obtain the magnitude and direction of the position vector of B relative to A.

$$r_{B/A} = 53.9\text{ m} \qquad \alpha = 21.8° \qquad \mathbf{r}_{B/A} = 53.9\text{ m} \nwarrow 21.8° \blacktriangleleft$$

Proceeding in a similar fashion, we find the velocity and acceleration of B relative to A.

$$\mathbf{v}_B = \mathbf{v}_A + \mathbf{v}_{B/A}$$
$$v_{B/A} = 11.66\text{ m/s} \qquad \beta = 31.0° \qquad \mathbf{v}_{B/A} = 11.66\text{ m/s} \swarrow 31.0° \blacktriangleleft$$
$$\mathbf{a}_B = \mathbf{a}_A + \mathbf{a}_{B/A} \qquad \mathbf{a}_{B/A} = 1.2\text{ m/s}^2 \downarrow \blacktriangleleft$$

SOLVING PROBLEMS ON YOUR OWN

In the problems for this lesson, you will analyze the *two- and three-dimensional motion* of a particle. While the physical interpretations of the velocity and acceleration are the same as in the first lessons of the chapter, you should remember that these quantities are vectors. In addition, you should understand from your experience with vectors in statics that it will often be advantageous to express position vectors, velocities, and accelerations in terms of their rectangular scalar components [Eqs. (11.27) and (11.28)]. Furthermore, given two vectors **A** and **B**, recall that $\mathbf{A} \cdot \mathbf{B} = 0$ if **A** and **B** are perpendicular to each other, while $\mathbf{A} \times \mathbf{B} = 0$ if **A** and **B** are parallel.

A. Analyzing the motion of a projectile. Many of the following problems deal with the two-dimensional motion of a projectile, where the resistance of the air can be neglected. In Sec. 11.11, we developed the equations which describe this type of motion, and we observed that the horizontal component of the velocity remained constant (uniform motion) while the vertical component of the acceleration was constant (uniformly accelerated motion). We were able to consider separately the horizontal and the vertical motions of the particle. Assuming that the projectile is fired from the origin, we can write the two equations

$$x = (v_x)_0 t \qquad y = (v_y)_0 t - \tfrac{1}{2}gt^2$$

1. If the initial velocity and firing angle are known, the value of y corresponding to any given value of x (or the value of x for any value of y) can be obtained by solving one of the above equations for t and substituting for t into the other, equation [Sample Prob. 11.7].

2. If the initial velocity and the coordinates of a point of the trajectory are known, and you wish to *determine the firing angle* α, begin your solution by expressing the components $(v_x)_0$ and $(v_y)_0$ of the initial velocity as functions of the angle α. These expressions and the known values of x and y are then substituted into the above equations. Finally, solve the first equation for t and substitute that value of t into the second equation to obtain a trigonometric equation in α, which you can solve for that unknown [Sample Prob. 11.8].

(continued)

B. Solving translational two-dimensional relative-motion problems. You saw in Sec. 11.12 that the absolute motion of a particle B can be obtained by combining the motion of a particle A and the *relative motion* of B with respect to a frame attached to A which is in *translation*. The velocity and acceleration of B can then be expressed as shown in Eqs. (11.33) and (11.34), respectively.

1. To visualize the relative motion of B with respect to A, imagine that you are attached to particle A as you observe the motion of particle B. For example, to a passenger in automobile A of Sample Prob. 11.9, automobile B appears to be heading in a southwesterly direction (*south* should be obvious; and *west* is due to the fact that automobile A is moving to the east—automobile B then appears to travel to the west). Note that this conclusion is consistent with the direction of $\mathbf{v}_{B/A}$.

2. To solve a relative-motion problem, first write the vector equations (11.31), (11.33), and (11.34), which relate the motions of particles A and B. You may then use either of the following methods:

a. Construct the corresponding vector triangles and solve them for the desired position vector, velocity, and acceleration [Sample Prob. 11.9].

b. Express all vectors in terms of their rectangular components and solve the two independent sets of scalar equations obtained in that way. If you choose this approach, be sure to select the same positive direction for the displacement, velocity, and acceleration of each particle.

PROBLEMS

CONCEPT QUESTIONS

11.CQ3 Two model rockets are fired simultaneously from a ledge and follow the trajectories shown. Neglecting air resistance, which of the rockets will hit the ground first?

a. A.
b. B.
c. They hit at the same time.
d. The answer depends on h.

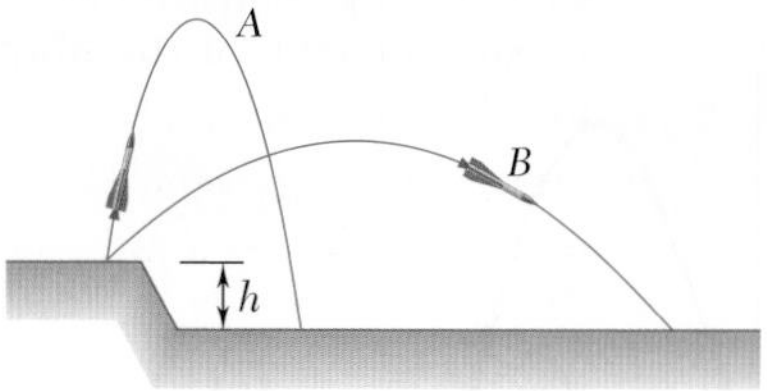

Fig. P11.CQ3

11.CQ4 Ball A is thrown straight up. Which of the following statements about the ball are true at the highest point in its path?

a. The velocity and acceleration are both zero.
b. The velocity is zero, but the acceleration is not zero.
c. The velocity is not zero, but the acceleration is zero.
d. Neither the velocity nor the acceleration is zero.

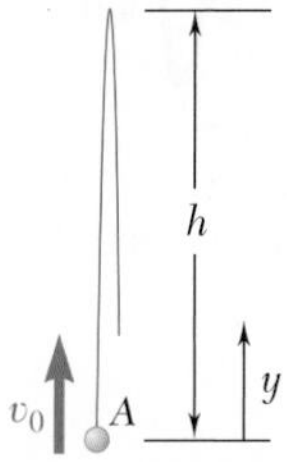

Fig. P11.CQ4

11.CQ5 Ball A is thrown straight up with an initial speed v_0 and reaches a maximum elevation h before falling back down. When A reaches its maximum elevation, a second ball is thrown straight upward with the same initial speed v_0. At what height, y, will the balls cross paths?

a. $y = h$
b. $y > h/2$
c. $y = h/2$
d. $y < h/2$
e. $y = 0$

11.CQ6 Two cars are approaching an intersection at constant speeds as shown. What velocity will car B appear to have to an observer in car A?

a. → **b.** ↘ **c.** ↖ **d.** ↗ **e.** ↙

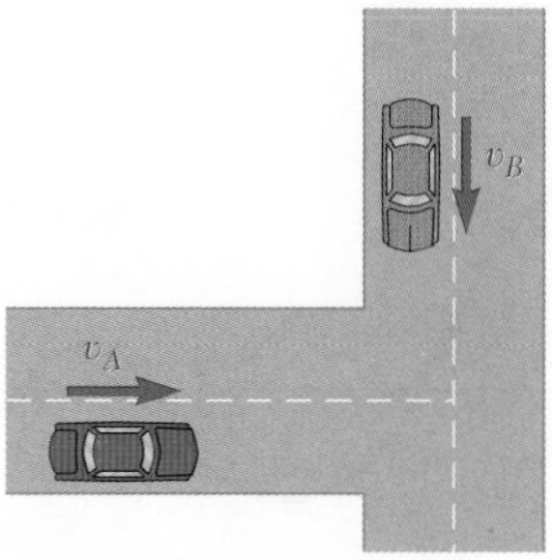

Fig. P11.CQ6

11.CQ7 Blocks A and B are released from rest in the positions shown. Neglecting friction between all surfaces, which figure best indicates the direction α of the acceleration of block B?

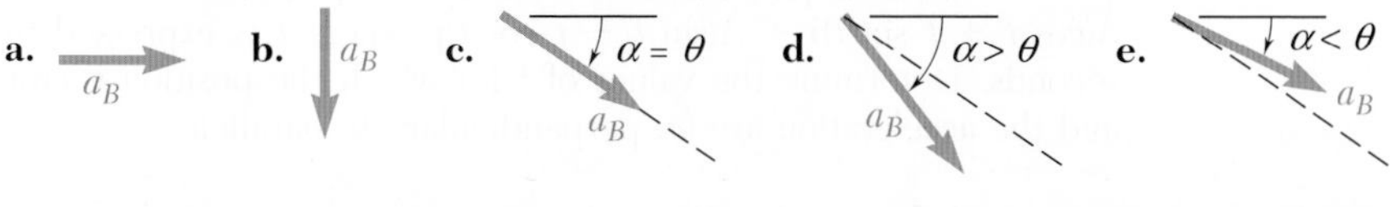

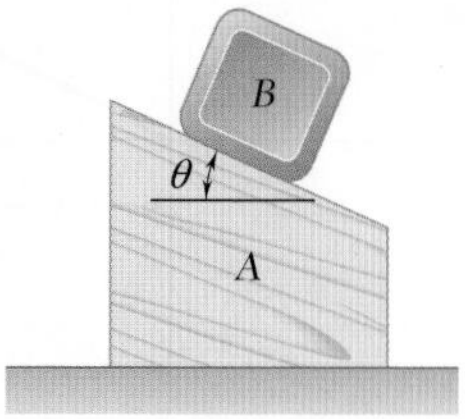

Fig. P11.CQ7

END-OF-SECTION PROBLEMS

11.89 A ball is thrown so that the motion is defined by the equations $x = 5t$ and $y = 2 + 6t - 4.9t^2$, where x and y are expressed in meters and t is expressed in seconds. Determine (*a*) the velocity at $t = 1$ s, (*b*) the horizontal distance the ball travels before hitting the ground.

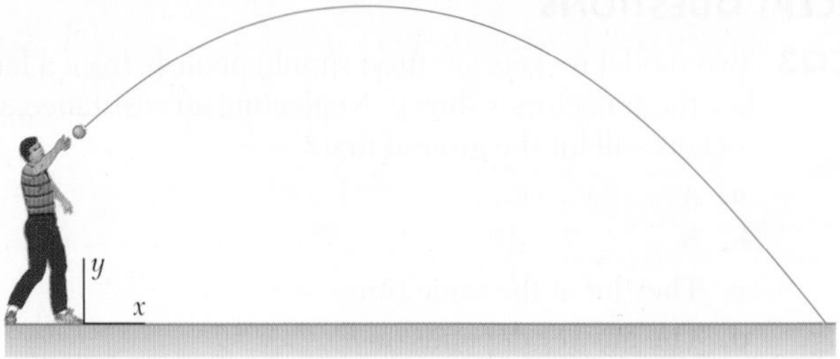

Fig. P11.89

11.90 The motion of a vibrating particle is defined by the position vector $\mathbf{r} = 10(1 - e^{-3t})\mathbf{i} + (4e^{-2t} \sin 15t)\mathbf{j}$, where $\mathbf{r}$ and t are expressed in millimeters and seconds, respectively. Determine the velocity and acceleration when (*a*) $t = 0$, (*b*) $t = 0.5$ s.

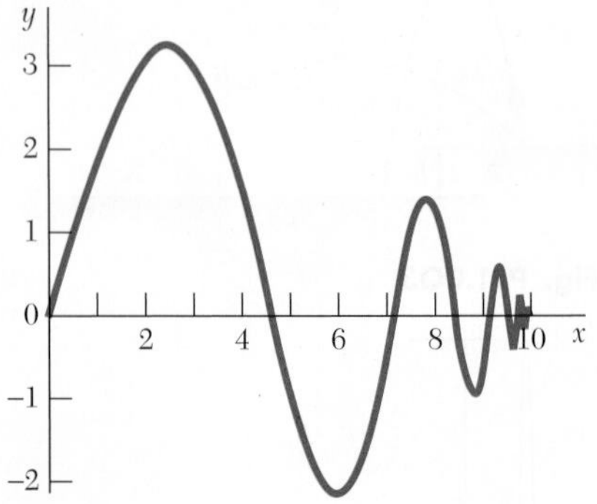

Fig. P11.90

11.91 The motion of a vibrating particle is defined by the position vector $\mathbf{r} = (4 \sin \pi t)\mathbf{i} - (\cos 2\pi t)\mathbf{j}$, where r is expressed in meters and t in seconds. (*a*) Determine the velocity and acceleration when $t = 1$ s. (*b*) Show that the path of the particle is parabolic.

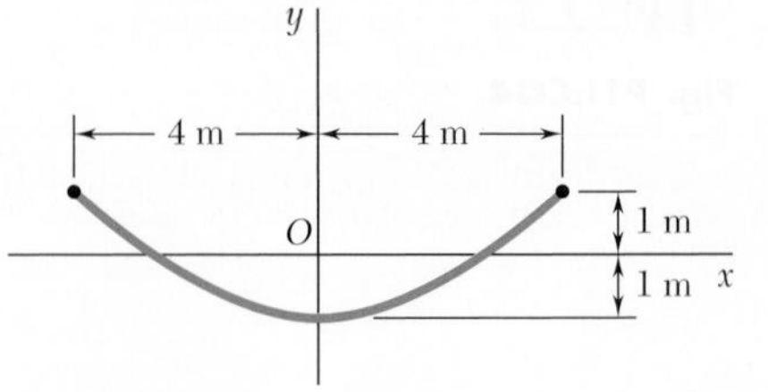

Fig. P11.91

11.92 The motion of a particle is defined by the equations $x = 100t - 50 \sin t$ and $y = 100 - 50 \cos t$, where x and y are expressed in mm and t is expressed in seconds. Sketch the path of the particle, and determine (*a*) the magnitudes of the smallest and largest velocities reached by the particle, (*b*) the corresponding times, positions, and directions of the velocities.

11.93 The damped motion of a vibrating particle is defined by the position vector $\boldsymbol{r} = x_1[1 - 1/(t + 1)]\mathbf{i} + (y_1 e^{-\pi t/2} \cos 2\pi t)\mathbf{j}$, where t is expressed in seconds. For $x_1 = 30$ mm and $y_1 = 20$ mm, determine the position, the velocity, and the acceleration of the particle when (*a*) $t = 0$, (*b*) $t = 1.5$ s.

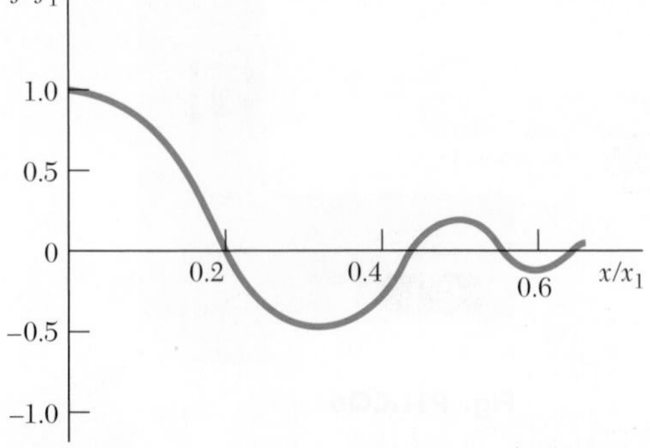

Fig. *P11.93*

11.94 The motion of a particle is defined by the position vector $\boldsymbol{r} = A(\cos t + t \sin t)\mathbf{i} + A(\sin t - t \cos t)\mathbf{j}$, where t is expressed in seconds. Determine the values of t for which the position vector and the acceleration are (*a*) perpendicular, (*b*) parallel.

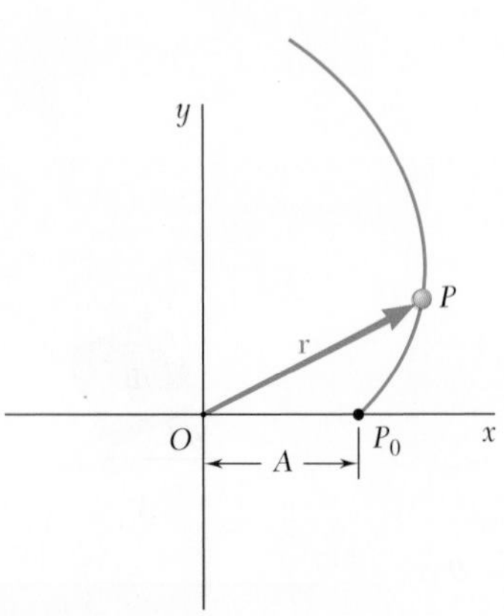

Fig. *P11.94*

11.95 The three-dimensional motion of a particle is defined by the position vector $\boldsymbol{r} = (Rt \cos \omega_n t)\mathbf{i} + ct\mathbf{j} + (Rt \sin \omega_n t)\mathbf{k}$. Determine the magnitudes of the velocity and acceleration of the particle. (The space curve described by the particle is a conic helix.)

***11.96** The three-dimensional motion of a particle is defined by the position vector $\boldsymbol{r} = (At \cos t)\mathbf{i} + (A\sqrt{t^2 + 1})\mathbf{j} + (Bt \sin t)\mathbf{k}$, where r and t are expressed in meters and seconds, respectively. Show that the curve described by the particle lies on the hyperboloid $(y/A)^2 - (x/A)^2 - (z/B)^2 = 1$. For $A = 3$ and $B = 1$, determine (*a*) the magnitudes of the velocity and acceleration when $t = 0$, (*b*) the smallest nonzero value of t for which the position vector and the velocity are perpendicular to each other.

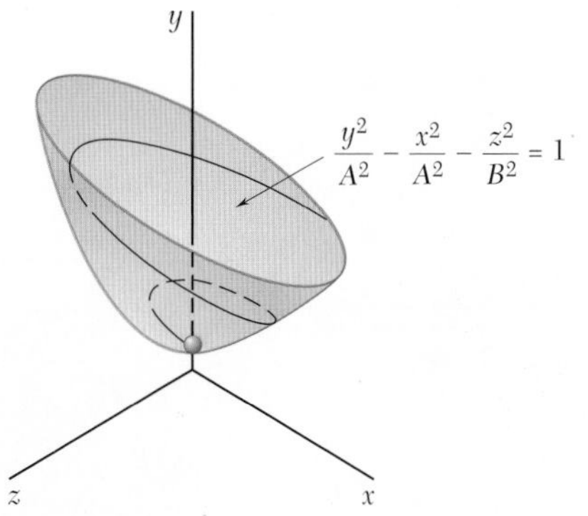

Fig. P11.96

11.97 An airplane used to drop water on brushfires is flying horizontally in a straight line at 315 km/h at an altitude of 80 m. Determine the distance d at which the pilot should release the water so that it will hit the fire at B.

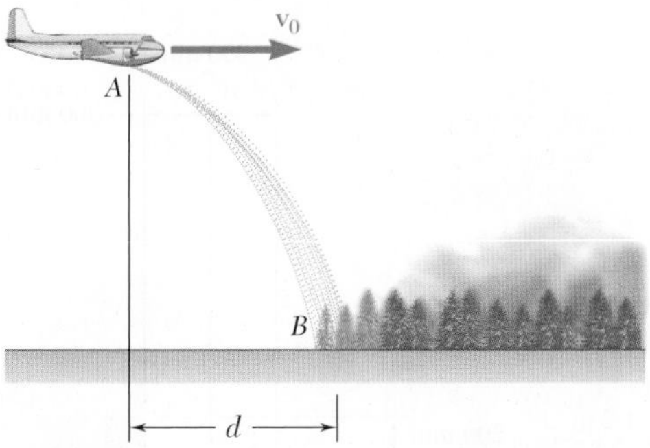

Fig. P11.97

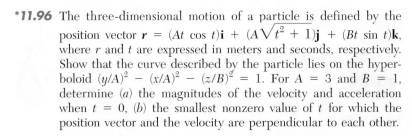

11.98 A helicopter is flying with a constant horizontal velocity of 180 km/h and is directly above point A when a loose part begins to fall. The part lands 6.5 s later at point B on an inclined surface. Determine (*a*) the distance d between points A and B. (*b*) the initial height h.

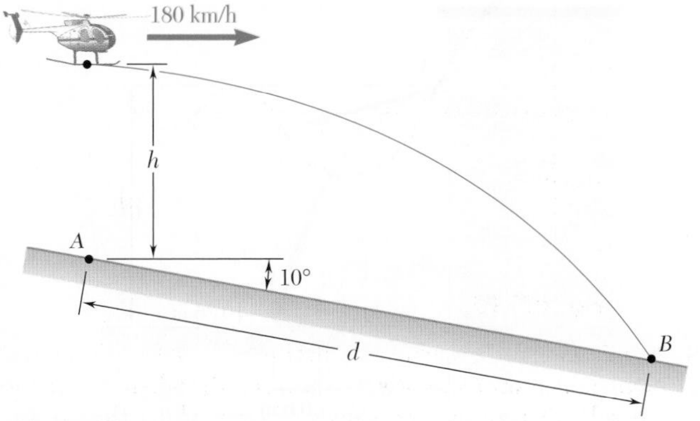

Fig. P11.98

11.99 A baseball pitching machine "throws" baseballs with a horizontal velocity $\mathbf{v}_0$. Knowing that height h varies between 788 mm and 1068 mm, determine (*a*) the range of values of v_0, (*b*) the values of α corresponding to h = 788 mm and h = 1068 mm.

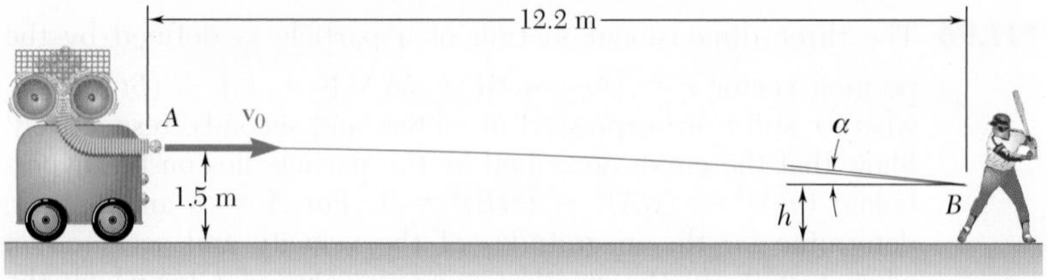

Fig. P11.99

11.100 While delivering newspapers, a girl throws a newspaper with a horizontal velocity $\mathbf{v}_0$. Determine the range of values of v_0 if the newspaper is to land between points B and C.

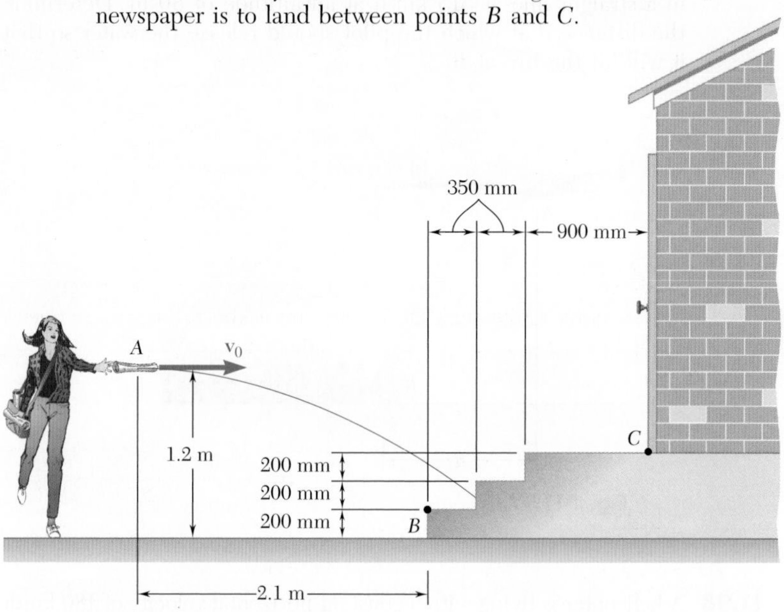

Fig. P11.100

11.101 Water flows from a drain spout with an initial velocity of 0.75 m/s at an angle of 15° with the horizontal. Determine the range of values of the distance d for which the water will enter the trough BC.

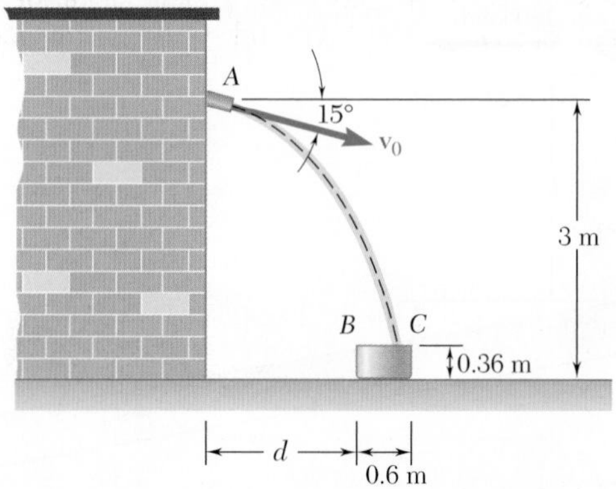

Fig. *P11.101*

11.102 Milk is poured into a glass of height 140 mm and inside diameter 66 mm. If the initial velocity of the milk is 1.2 m/s at an angle of 40° with the horizontal, determine the range of values of the height h for which the milk will enter the glass.

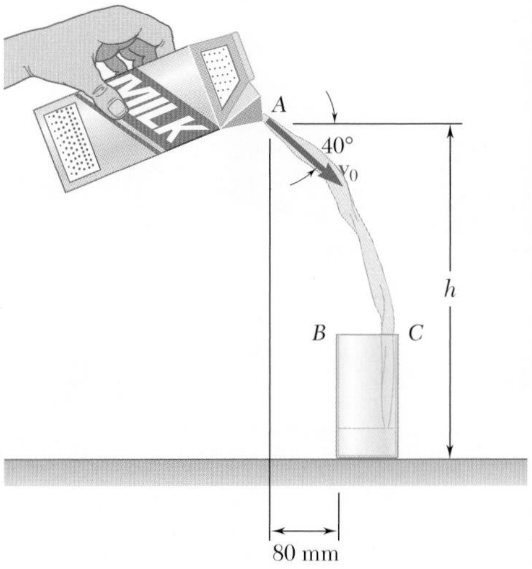

Fig. P11.102

11.103 A volleyball player serves the ball with an initial velocity $\mathbf{v}_0$ of magnitude 13.40 m/s at an angle of 20° with the horizontal. Determine (*a*) if the ball will clear the top of the net, (*b*) how far from the net the ball will land.

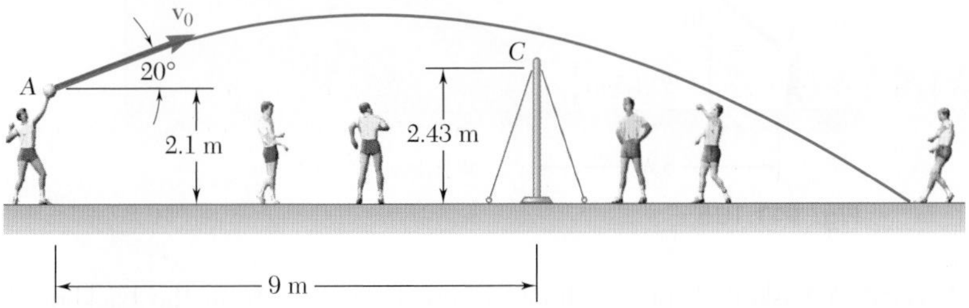

Fig. P11.103

11.104 A golfer hits a golf ball with an initial velocity of 50 m/s at an angle of 25° with the horizontal. Knowing that the fairway slopes downward at an average angle of 5°, determine the distance d between the golfer and point B where the ball first lands.

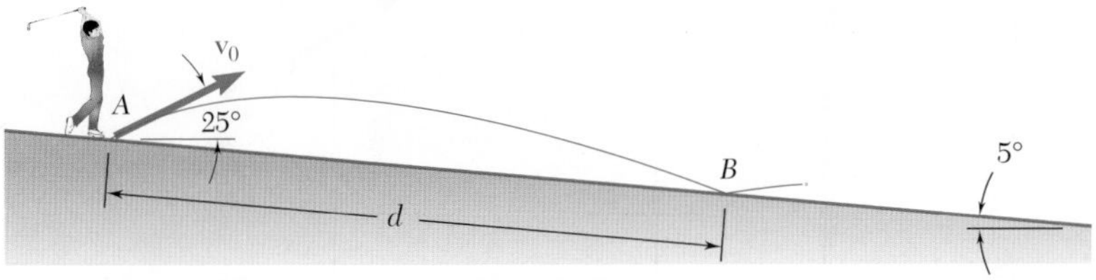

Fig. *P11.104*

11.105 A homeowner uses a snowblower to clear his driveway. Knowing that the snow is discharged at an average angle of 40° with the horizontal, determine the initial velocity v_0 of the snow.

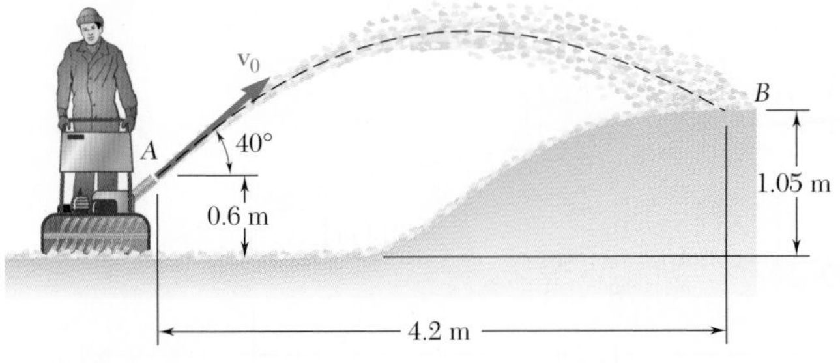

Fig. *P11.105*

11.106 At halftime of a football game souvenir balls are thrown to the spectators with a velocity $\mathbf{v}_0$. Determine the range of values of v_0 if the balls are to land between points B and C.

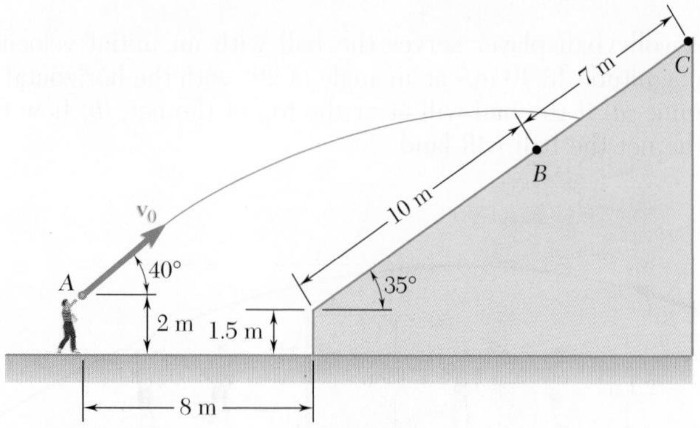

Fig. P11.106

11.107 A basketball player shoots when she is 5 m from the backboard. Knowing that the ball has an initial velocity $\mathbf{v}_0$ at an angle of 30° with the horizontal, determine the value of v_0 when d is equal to (*a*) 225 mm, (*b*) 425 mm.

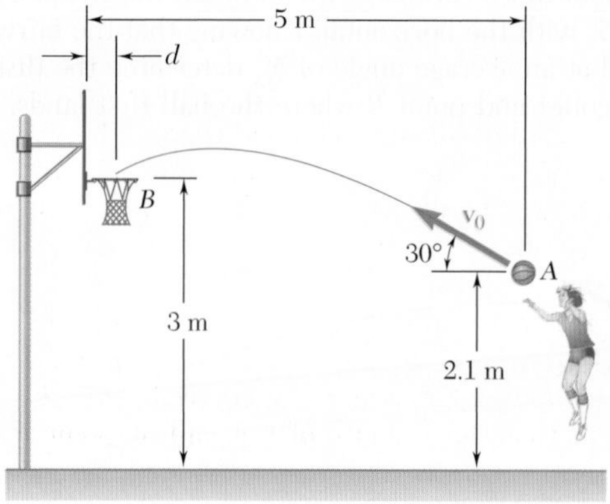

Fig. P11.107

11.108 A tennis player serves the ball at a height $h = 2.5$ m with an initial velocity of $\mathbf{v}_0$ at an angle of 5° with the horizontal. Determine the range of v_0 for which the ball will land in the service area that extends to 6.4 m beyond the net.

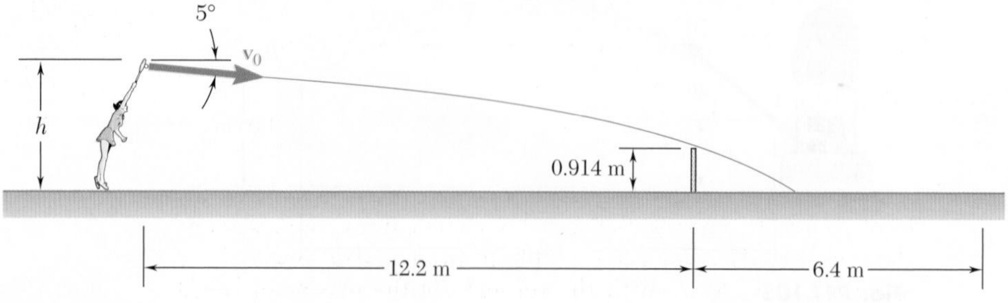

Fig. P11.108

11.109 The nozzle at A discharges cooling water with an initial velocity $\mathbf{v}_0$ at an angle of 6° with the horizontal onto a grinding wheel 350 mm in diameter. Determine the range of values of the initial velocity for which the water will land on the grinding wheel between points B and C.

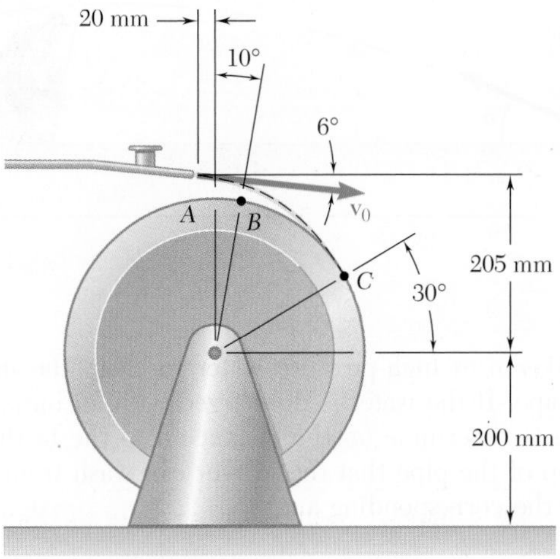

Fig. P11.109

11.110 While holding one of its ends, a worker lobs a coil of rope over the lowest limb of a tree. If he throws the rope with an initial velocity $\mathbf{v}_0$ at an angle of 65° with the horizontal, determine the range of values of v_0 for which the rope will go over only the lowest limb.

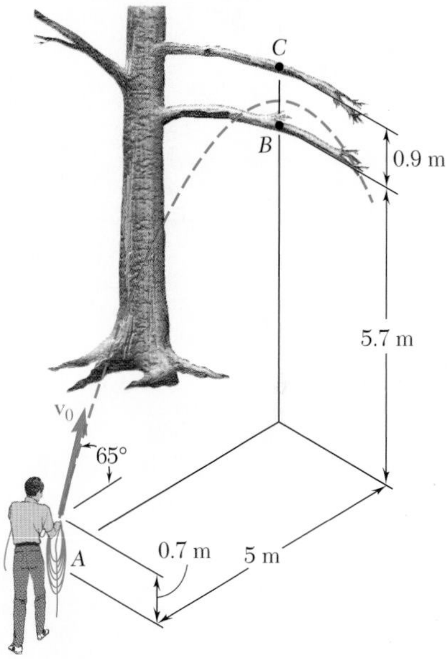

Fig. P11.110

11.111 The pitcher in a softball game throws a ball with an initial velocity $\mathbf{v}_0$ of 72 km/h at an angle α with the horizontal. If the height of the ball at point B is 0.68 m, determine (*a*) the angle α, (*b*) the angle θ that the velocity of the ball at point B forms with the horizontal.

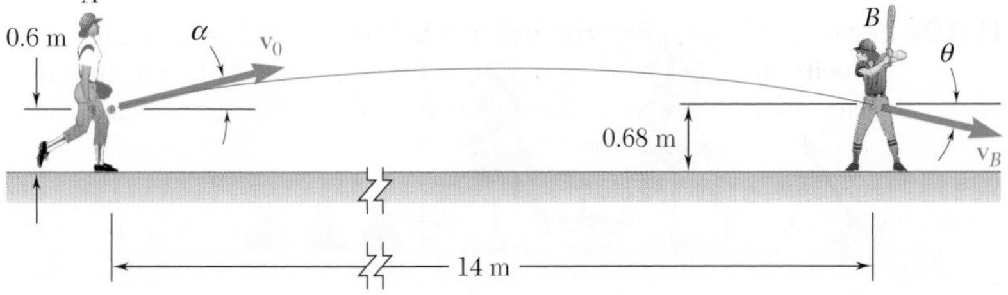

Fig. P11.111

11.112 A model rocket is launched from point A with an initial velocity $\mathbf{v}_0$ of 75 m/s. If the rocket's descent parachute does not deploy and the rocket lands a distance d = 100 m from A, determine (*a*) the angle α that $\mathbf{v}_0$ forms with the vertical, (*b*) the maximum height above point A reached by the rocket, (*c*) the duration of the flight.

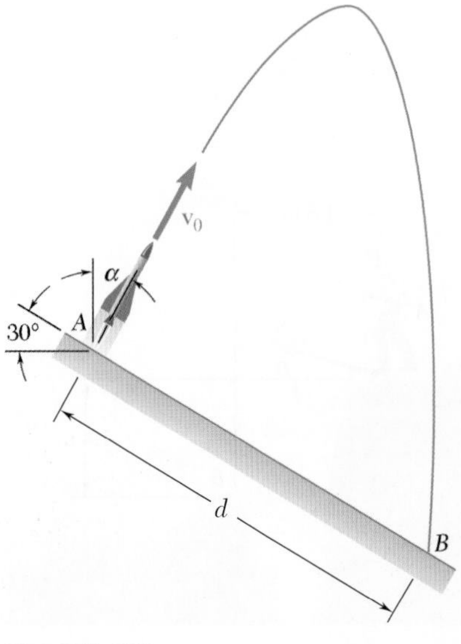

Fig. P11.112

11.113 The initial velocity $\mathbf{v}_0$ of a hockey puck is 160 km/h. Determine (*a*) the largest value (less than 45°) of the angle α for which the puck will enter the net, (*b*) the corresponding time required for the puck to reach the net.

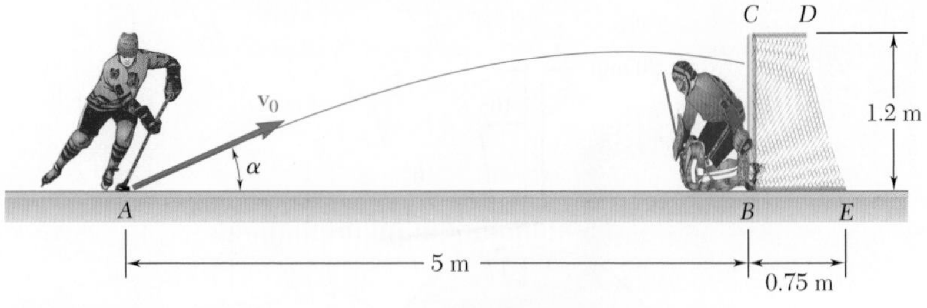

Fig. P11.113

11.114 A worker uses high-pressure water to clean the inside of a long drainpipe. If the water is discharged with an initial velocity $\mathbf{v}_0$ of 11.5 m/s, determine (*a*) the distance *d* to the farthest point *B* on the top of the pipe that the worker can wash from his position at *A*, (*b*) the corresponding angle α.

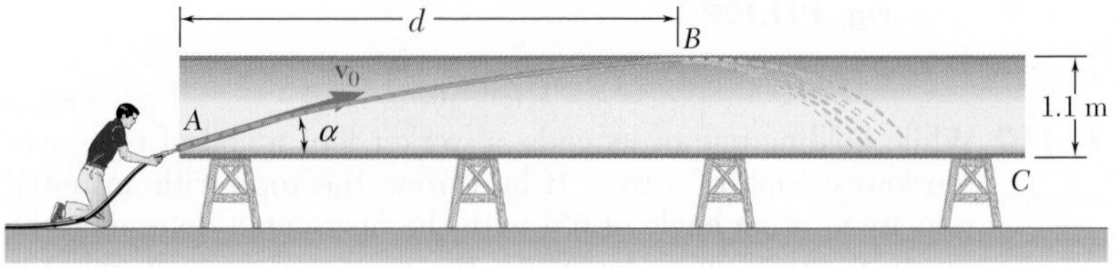

Fig. P11.114

11.115 An oscillating garden sprinkler which discharges water with an initial velocity $\mathbf{v}_0$ of 8 m/s is used to water a vegetable garden. Determine the distance *d* to the farthest point *B* that will be watered and the corresponding angle α when (*a*) the vegetables are just beginning to grow, (*b*) the height *h* of the corn is 1.8 m.

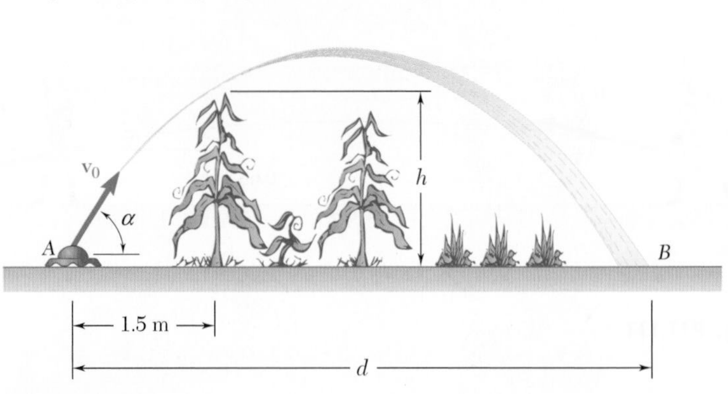

Fig. *P11.115*

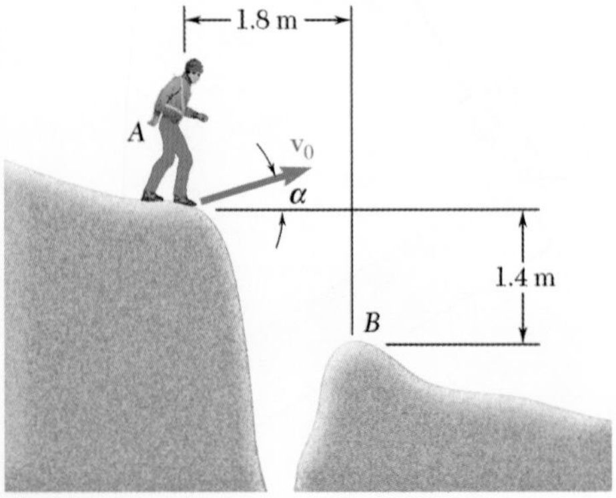

Fig. *P11.116*

****11.116*** A mountain climber plans to jump from *A* to *B* over a crevasse. Determine the smallest value of the climber's initial velocity $\mathbf{v}_0$ and the corresponding value of angle α so that he lands at *B*.

11.117 The velocities of skiers A and B are as shown. Determine the velocity of A with respect to B.

Fig. P11.117

11.118 The three blocks shown move with constant velocities. Find the velocity of each block, knowing that the relative velocity of A with respect to C is 300 mm/s upward and that the relative velocity of B with respect to A is 200 mm/s downward.

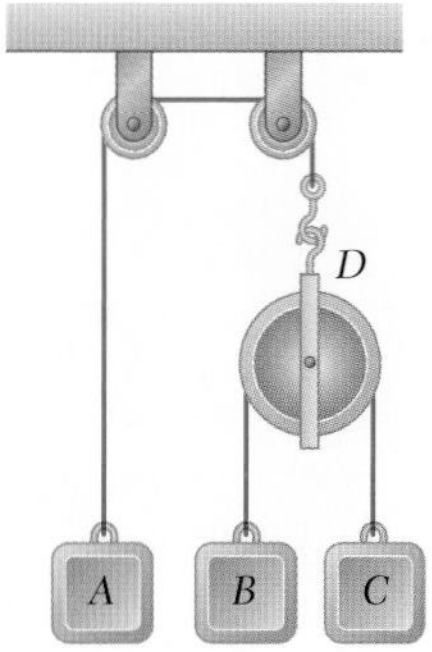

Fig. P11.118

11.119 Three seconds after automobile B passes through the intersection shown, automobile A passes through the same intersection. Knowing that the speed of each automobile is constant, determine (*a*) the relative velocity of B with respect to A, (*b*) the change in position of B with respect to A during a 4-s interval, (*c*) the distance between the two automobiles 2 s after A has passed through the intersection.

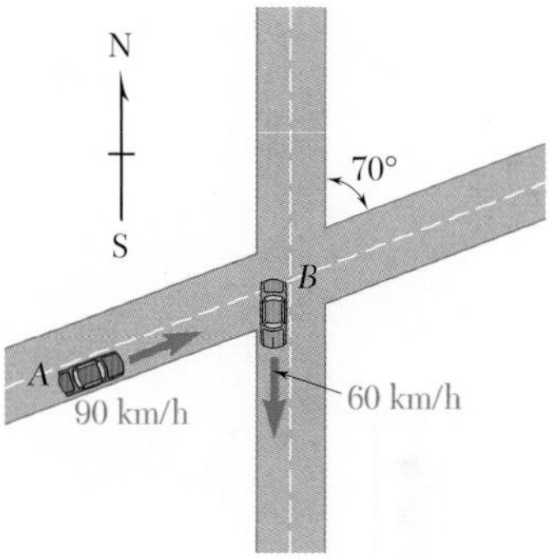

Fig. P11.119

11.120 Shore-based radar indicates that a ferry leaves its slip with a velocity $\mathbf{v} = 18$ km/h $\measuredangle 70°$, while instruments aboard the ferry indicate a speed of 18.4 km/h and a heading of 30° west of south relative to the river. Determine the velocity of the river.

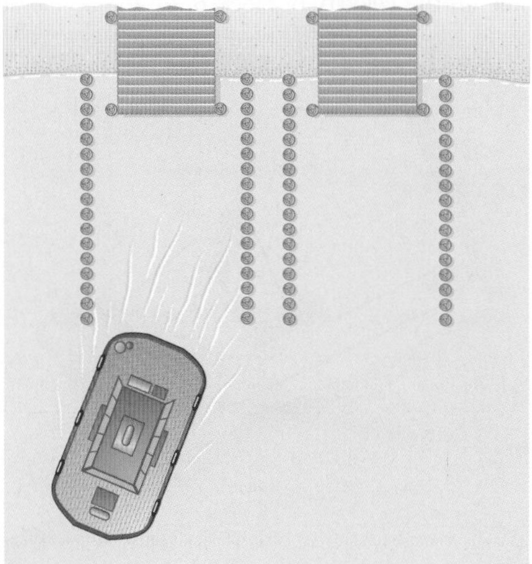

Fig. P11.120

11.121 Airplanes A and B are flying at the same altitude and are tracking the eye of hurricane C. The relative velocity of C with respect to A is $\mathbf{v}_{C/A} = 350$ km/h $\angle 75°$, and the relative velocity of C with respect to B is $\mathbf{v}_{C/B} = 400$ km/h $\angle 40°$. Determine (*a*) the relative velocity of B with respect to A, (*b*) the velocity of A if ground-based radar indicates that the hurricane is moving at a speed of 30 km/h due north, (*c*) the change in position of C with respect to B during a 15-min interval.

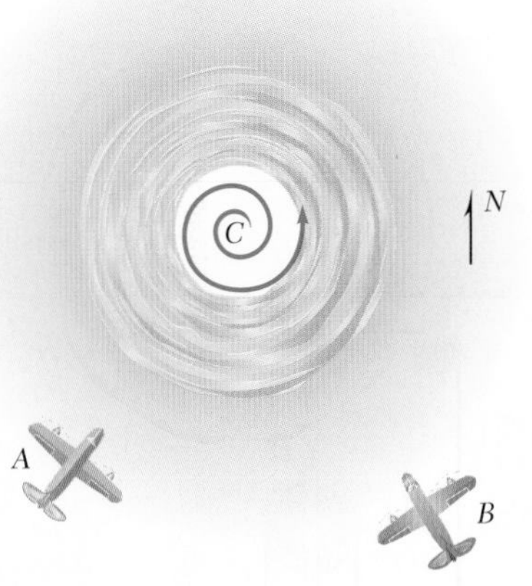

Fig. *P11.121*

11.122 Pin P moves at a constant speed of 150 mm/s in a counterclockwise sense along a circular slot which has been milled in the slider block A shown. Knowing that the block moves downward at a constant speed of 100 mm/s, determine the velocity of pin P when (*a*) $\theta = 30°$, (*b*) $\theta = 120°$.

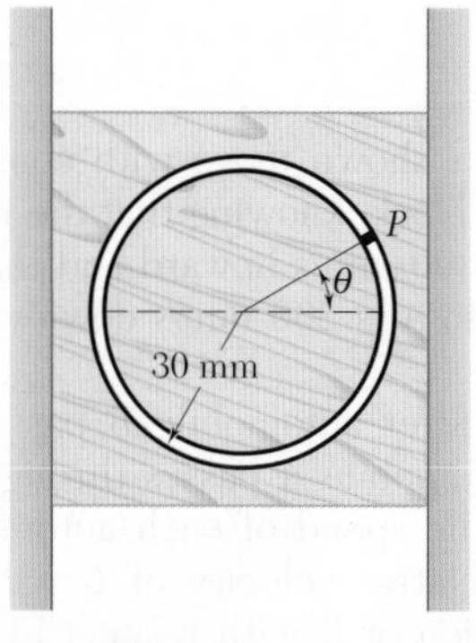

Fig. *P11.122*

11.123 Knowing that at the instant shown assembly A has a velocity of 225 mm/s^2 and an acceleration of 375 mm/s^2 both directed downwards, determine (*a*) the velocity of block B, (*b*) the acceleration of block B.

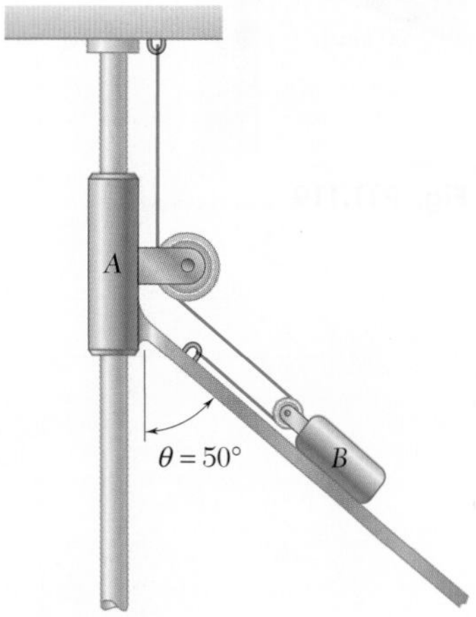

Fig. P11.123

11.124 Knowing that at the instant shown block A has a velocity of 200 mm/s and an acceleration of 150 mm/s^2 both directed down the incline, determine (*a*) the velocity of block B, (*b*) the acceleration of block B.

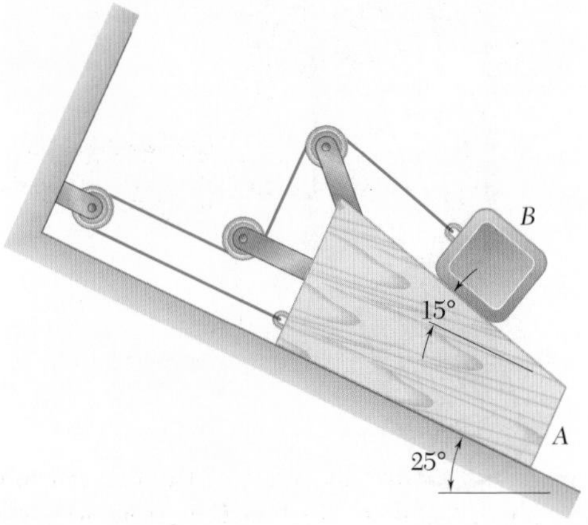

Fig. P11.124

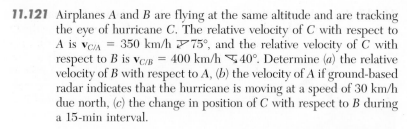

11.125 A boat is moving to the right with a constant deceleration of 0.3 m/s^2 when a boy standing on the deck D throws a ball with an initial velocity relative to the deck which is vertical. The ball rises to a maximum height of 8 m above the release point and the boy must step forward a distance d to catch it at the same height as the release point. Determine (*a*) the distance d, (*b*) the relative velocity of the ball with respect to the deck when the ball is caught.

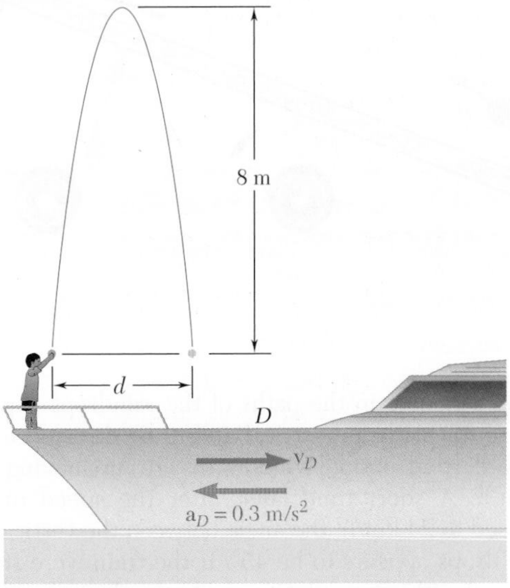

Fig. P11.125

11.126 The assembly of rod A and wedge B starts from rest and moves to the right with a constant acceleration of 2 mm/s^2. Determine (*a*) the acceleration of wedge C, (*b*) the velocity of wedge C when $t = 10$ s.

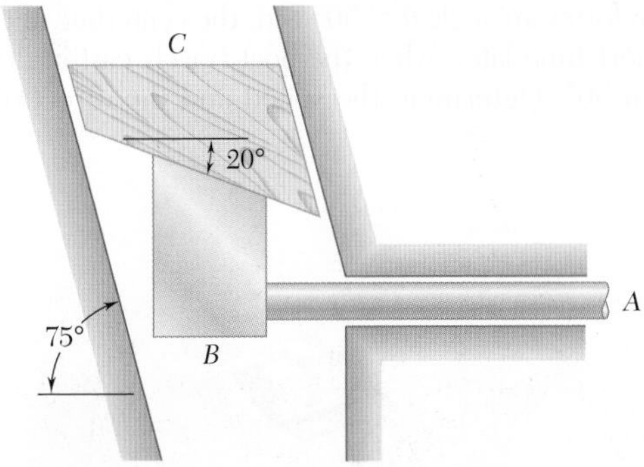

Fig. P11.126

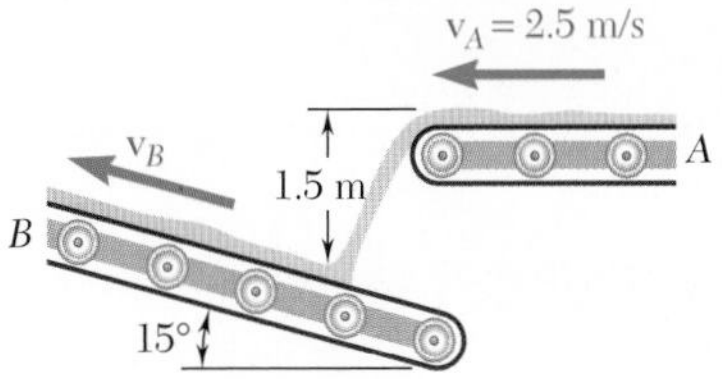

Fig. P11.127

11.127 Determine the required velocity of the belt B if the relative velocity with which the sand hits belt B is to be (*a*) vertical, (*b*) as small as possible.

11.128 Conveyor belt A, which forms a 20° angle with the horizontal, moves at a constant speed of 1.2 m/s and is used to load an airplane. Knowing that a worker tosses duffel bag B with an initial velocity of 0.75 m/s at an angle of 30° with the horizontal, determine the velocity of the bag relative to the belt as it lands on the belt.

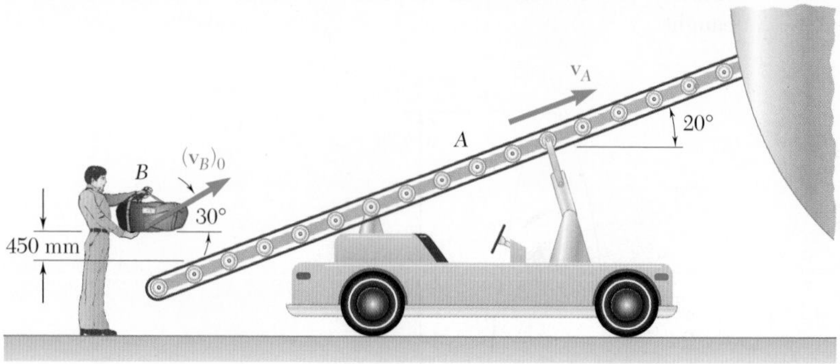

Fig. P11.128

11.129 During a rainstorm the paths of the raindrops appear to form an angle of 30° with the vertical and to be directed to the left when observed from a side window of a train moving at a speed of 15 km/h. A short time later, after the speed of the train has increased to 24 km/h, the angle between the vertical and the paths of the drops appears to be 45°. If the train were stopped, at what angle and with what velocity would the drops be observed to fall?

11.130 As observed from a ship moving due east at 9 km/h, the wind appears to blow from the south. After the ship has changed course and speed, and as it is moving north at 6 km/h, the wind appears to blow from the southwest. Assuming that the wind velocity is constant during the period of observation, determine the magnitude and direction of the true wind velocity.

11.131 When a small boat travels north at 5 km/h, a flag mounted on its stern forms an angle $\theta = 50°$ with the centerline of the boat as shown. A short time later, when the boat travels east at 20 km/h, angle θ is again 50°. Determine the speed and the direction of the wind.

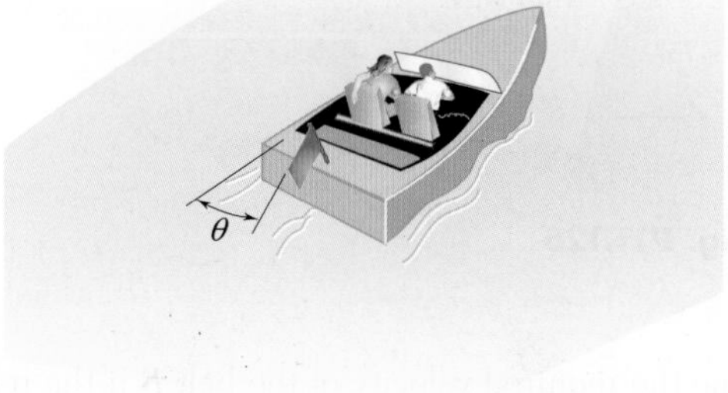

Fig. P11.131

11.132 As part of a department store display, a model train D runs on a slight incline between the store's up and down escalators. When the train and shoppers pass point A, the train appears to a shopper on the up escalator B to move downward at an angle of 22° with the horizontal, and to a shopper on the down escalator C to move upward at an angle of 23° with the horizontal and to travel to the left. Knowing that the speed of the escalators is 1 m/s, determine the speed and the direction of the train.

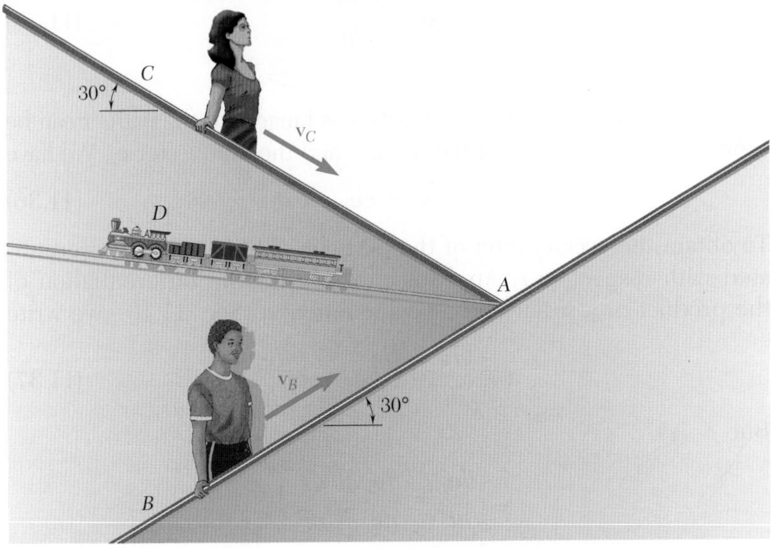

Fig. P11.132

11.13 TANGENTIAL AND NORMAL COMPONENTS

We saw in Sec. 11.9 that the velocity of a particle is a vector tangent to the path of the particle but that, in general, the acceleration is not tangent to the path. It is sometimes convenient to resolve the acceleration into components directed, respectively, along the tangent and the normal to the path of the particle.

Plane Motion of a Particle. First, let us consider a particle which moves along a curve contained in the plane of the figure. Let P be the position of the particle at a given instant. We attach at P a unit vector $\mathbf{e}_t$ tangent to the path of the particle and pointing in the direction of motion (Fig. 11.21*a*). Let $\mathbf{e}'_t$ be the unit vector corresponding to the position P' of the particle at a later instant. Drawing both vectors from the same origin O', we define the vector $\Delta\mathbf{e}_t = \mathbf{e}'_t - \mathbf{e}_t$ (Fig. 11.21*b*). Since $\mathbf{e}_t$ and $\mathbf{e}'_t$ are of unit length, their tips lie on a circle of radius 1. Denoting by $\Delta\theta$ the angle formed by $\mathbf{e}_t$ and $\mathbf{e}'_t$, we find that the magnitude of $\Delta\mathbf{e}_t$ is $2 \sin (\Delta\theta/2)$. Considering now the vector $\Delta\mathbf{e}_t/\Delta\theta$, we note that as $\Delta\theta$ approaches zero, this vector becomes tangent to the unit circle of Fig. 11.21*b*, i.e., perpendicular to $\mathbf{e}_t$, and that its magnitude approaches

$$\lim_{\Delta\theta \to 0} \frac{2 \sin(\Delta\theta/2)}{\Delta\theta} = \lim_{\Delta\theta \to 0} \frac{\sin(\Delta\theta/2)}{\Delta\theta/2} = 1$$

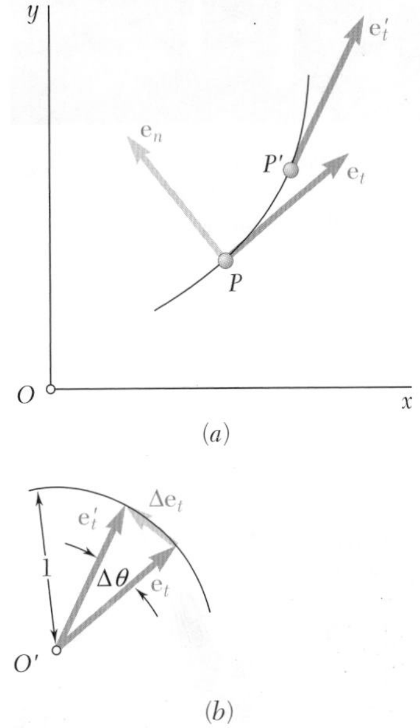

Fig. 11.21

Thus, the vector obtained in the limit is a unit vector along the normal to the path of the particle, in the direction toward which $\mathbf{e}_t$ turns. Denoting this vector by $\mathbf{e}_n$, we write

$$\mathbf{e}_n = \lim_{\Delta\theta \to 0} \frac{\Delta \mathbf{e}_t}{\Delta\theta}$$

$$\mathbf{e}_n = \frac{d\mathbf{e}_t}{d\theta} \tag{11.35}$$

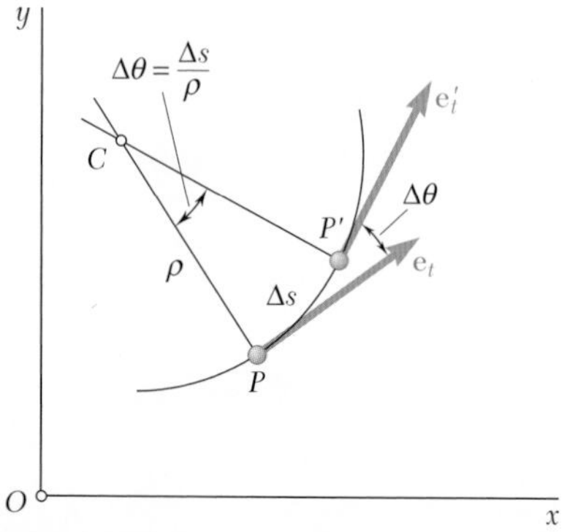

Fig. 11.22

Photo 11.5 The passengers in a train traveling around a curve will experience a normal acceleration toward the center of curvature of the path.

Since the velocity $\mathbf{v}$ of the particle is tangent to the path, it can be expressed as the product of the scalar v and the unit vector $\mathbf{e}_t$. We have

$$\mathbf{v} = v\mathbf{e}_t \tag{11.36}$$

To obtain the acceleration of the particle, (11.36) will be differentiated with respect to t. Applying the rule for the differentiation of the product of a scalar and a vector function (Sec. 11.10), we write

$$\mathbf{a} = \frac{d\mathbf{v}}{dt} = \frac{dv}{dt}\mathbf{e}_t + v\frac{d\mathbf{e}_t}{dt} \tag{11.37}$$

But

$$\frac{d\mathbf{e}_t}{dt} = \frac{d\mathbf{e}_t}{d\theta}\frac{d\theta}{ds}\frac{ds}{dt}$$

Recalling from (11.16) that $ds/dt = v$, from (11.35) that $d\mathbf{e}_t/d\theta = \mathbf{e}_n$, and from elementary calculus that $d\theta/ds$ is equal to $1/\rho$, where ρ is the radius of curvature of the path at P (Fig. 11.22), we have

$$\frac{d\mathbf{e}_t}{dt} = \frac{v}{\rho}\mathbf{e}_n \tag{11.38}$$

Substituting into (11.37), we obtain

$$\mathbf{a} = \frac{dv}{dt}\mathbf{e}_t + \frac{v^2}{\rho}\mathbf{e}_n \tag{11.39}$$

Thus, the scalar components of the acceleration are

$$a_t = \frac{dv}{dt} \qquad a_n = \frac{v^2}{\rho} \tag{11.40}$$

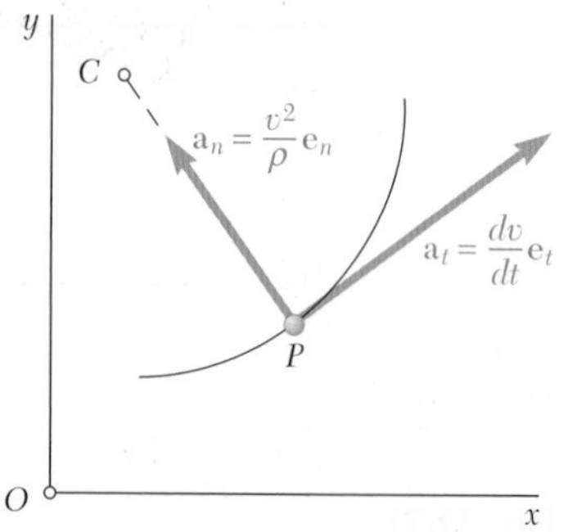

Fig. 11.23

The relations obtained express that the *tangential component* of the acceleration is equal to the *rate of change of the speed of the particle*, while the *normal component* is equal to the *square of the speed divided by the radius of curvature of the path at P*. If the speed of the particle increases, a_t is positive and the vector component $\mathbf{a}_t$ points in the direction of motion. If the speed of the particle decreases, a_t is negative and $\mathbf{a}_t$ points against the direction of motion. The vector component $\mathbf{a}_n$, on the other hand, *is always directed toward the center of curvature C of the path* (Fig. 11.23).

We conclude from the above that the tangential component of the acceleration reflects a change in the speed of the particle, while its normal component reflects a change in the direction of motion of the particle. The acceleration of a particle will be zero only if both its components are zero. Thus, the acceleration of a particle moving with constant speed along a curve will not be zero unless the particle happens to pass through a point of inflection of the curve (where the radius of curvature is infinite) or unless the curve is a straight line.

The fact that the normal component of the acceleration depends upon the radius of curvature of the path followed by the particle is taken into account in the design of structures or mechanisms as widely different as airplane wings, railroad tracks, and cams. In order to avoid sudden changes in the acceleration of the air particles flowing past a wing, wing profiles are designed without any sudden change in curvature. Similar care is taken in designing railroad curves, to avoid sudden changes in the acceleration of the cars (which would be hard on the equipment and unpleasant for the passengers). A straight section of track, for instance, is never directly followed by a circular section. Special transition sections are used to help pass smoothly from the infinite radius of curvature of the straight section to the finite radius of the circular track. Likewise, in the design of high-speed cams, abrupt changes in acceleration are avoided by using transition curves which produce a continuous change in acceleration.

Motion of a Particle in Space. The relations (11.39) and (11.40) still hold in the case of a particle moving along a space curve. However, since there are an infinite number of straight lines which are perpendicular to the tangent at a given point P of a space curve, it is necessary to define more precisely the direction of the unit vector $\mathbf{e}_n$.

Let us consider again the unit vectors $\mathbf{e}_t$ and $\mathbf{e}'_t$ tangent to the path of the particle at two neighboring points P and P' (Fig. 11.24*a*) and the vector $\Delta\mathbf{e}_t$ representing the difference between $\mathbf{e}_t$ and $\mathbf{e}'_t$ (Fig. 11.24*b*). Let us now imagine a plane through P (Fig. 11.24*a*) parallel to the plane defined by the vectors $\mathbf{e}_t$, $\mathbf{e}'_t$, and $\Delta\mathbf{e}_t$ (Fig. 11.24*b*). This plane contains the tangent to the curve at P and is parallel to the tangent at P'. If we let P' approach P, we obtain in the limit the plane which fits the curve most closely in the neighborhood of P. This plane is called the *osculating plane* at P.† It follows from this

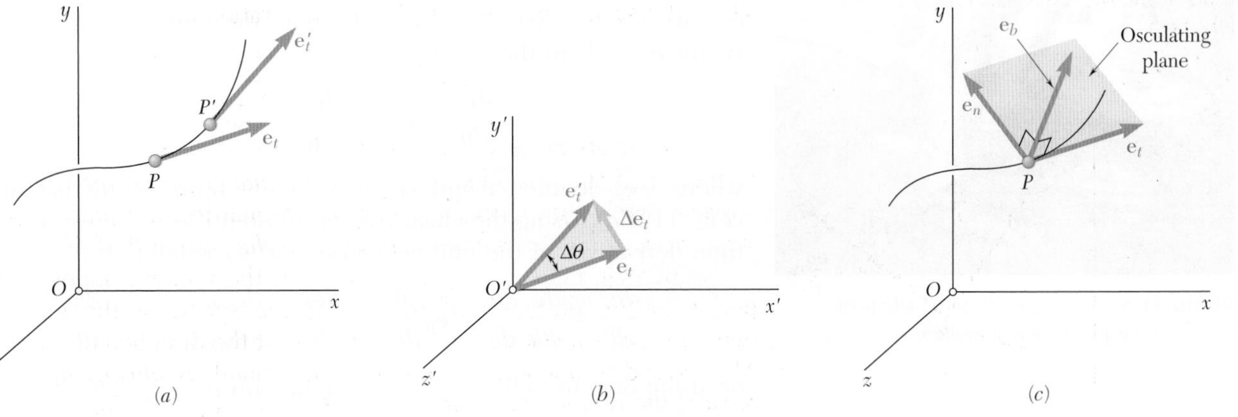

Fig. 11.24

†From the Latin *osculari*, to kiss.

definition that the osculating plane contains the unit vector $\mathbf{e}_n$, since this vector represents the limit of the vector $\Delta\mathbf{e}_t/\Delta\theta$. The normal defined by $\mathbf{e}_n$ is thus contained in the osculating plane; it is called the *principal normal* at P. The unit vector $\mathbf{e}_b = \mathbf{e}_t \times \mathbf{e}_n$ which completes the right-handed triad $\mathbf{e}_t$, $\mathbf{e}_n$, $\mathbf{e}_b$ (Fig. 11.24*c*) defines the *binormal* at P. The binormal is thus perpendicular to the osculating plane. We conclude that the acceleration of the particle at P can be resolved into two components, one along the tangent, the other along the principal normal at P, as indicated in Eq. (11.39). Note that the acceleration has no component along the binormal.

11.14 RADIAL AND TRANSVERSE COMPONENTS

In certain problems of plane motion, the position of the particle P is defined by its polar coordinates r and θ (Fig. 11.25*a*). It is then convenient to resolve the velocity and acceleration of the particle into components parallel and perpendicular, respectively, to the line OP. These components are called *radial* and *transverse components*.

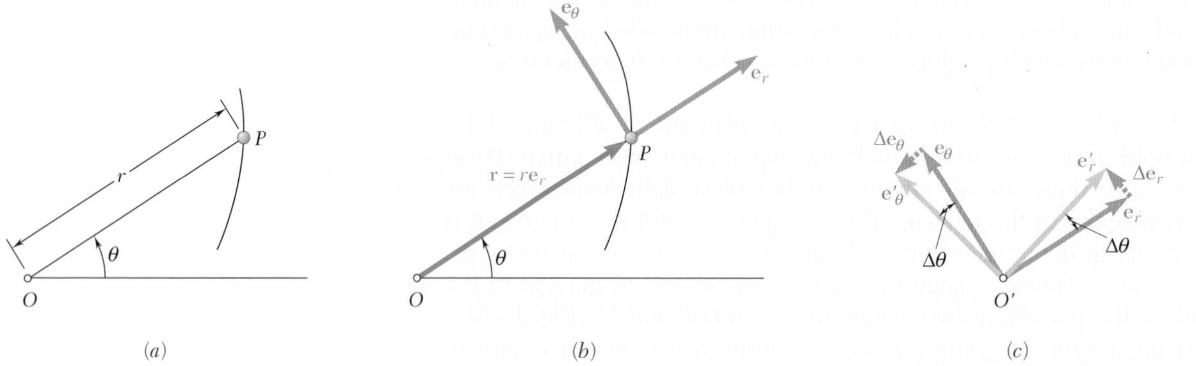

Fig. 11.25

We attach at P two unit vectors, $\mathbf{e}_r$ and $\mathbf{e}_\theta$ (Fig. 11.25*b*). The vector $\mathbf{e}_r$ is directed along OP and the vector $\mathbf{e}_\theta$ is obtained by rotating $\mathbf{e}_r$ through 90° counterclockwise. The unit vector $\mathbf{e}_r$ defines the *radial* direction, i.e., the direction in which P would move if r were increased and θ were kept constant; the unit vector $\mathbf{e}_\theta$ defines the *transverse* direction, i.e., the direction in which P would move if θ were increased and r were kept constant. A derivation similar to the one we used in Sec. 11.13 to determine the derivative of the unit vector $\mathbf{e}_t$ leads to the relations

$$\frac{d\mathbf{e}_r}{d\theta} = \mathbf{e}_\theta \qquad \frac{d\mathbf{e}_\theta}{d\theta} = -\mathbf{e}_r \tag{11.41}$$

where $-\mathbf{e}_r$ denotes a unit vector of sense opposite to that of $\mathbf{e}_r$ (Fig. 11.25*c*). Using the chain rule of differentiation, we express the time derivatives of the unit vectors $\mathbf{e}_r$ and $\mathbf{e}_\theta$ as follows:

$$\frac{d\mathbf{e}_r}{dt} = \frac{d\mathbf{e}_r}{d\theta}\frac{d\theta}{dt} = \mathbf{e}_\theta\frac{d\theta}{dt} \qquad \frac{d\mathbf{e}_\theta}{dt} = \frac{d\mathbf{e}_\theta}{d\theta}\frac{d\theta}{dt} = -\mathbf{e}_r\frac{d\theta}{dt}$$

or, using dots to indicate differentiation with respect to t,

$$\dot{\mathbf{e}}_r = \dot{\theta}\mathbf{e}_\theta \qquad \dot{\mathbf{e}}_\theta = -\dot{\theta}\mathbf{e}_r \tag{11.42}$$

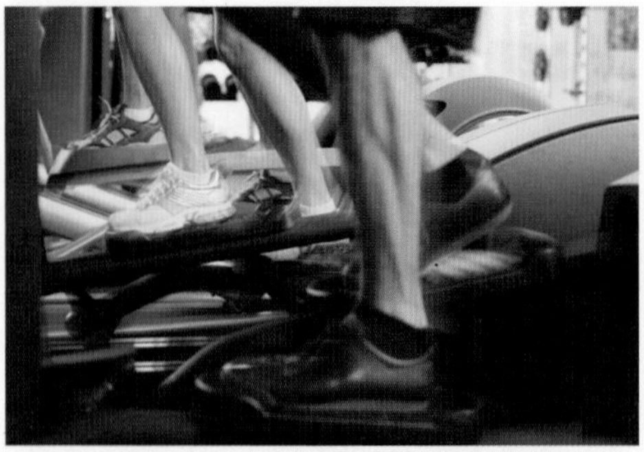

Photo 11.6 The footpads on an elliptical trainer undergo curvilinear motion.

To obtain the velocity **v** of the particle P, we express the position vector **r** of P as the product of the scalar r and the unit vector $\mathbf{e}_r$ and differentiate with respect to t:

$$\mathbf{v} = \frac{d}{dt}(r\mathbf{e}_r) = \dot{r}\mathbf{e}_r + r\dot{\mathbf{e}}_r$$

or, recalling the first of the relations (11.42),

$$\mathbf{v} = \dot{r}\mathbf{e}_r + r\dot{\theta}\mathbf{e}_\theta \tag{11.43}$$

Differentiating again with respect to t to obtain the acceleration, we write

$$\mathbf{a} = \frac{d\mathbf{v}}{dt} = \ddot{r}\mathbf{e}_r + \dot{r}\dot{\mathbf{e}}_r + \dot{r}\dot{\theta}\mathbf{e}_\theta + r\ddot{\theta}\mathbf{e}_\theta + r\dot{\theta}\dot{\mathbf{e}}_\theta$$

or, substituting for $\dot{\mathbf{e}}_r$ and $\dot{\mathbf{e}}_\theta$ from (11.42) and factoring $\mathbf{e}_r$ and $\mathbf{e}_\theta$,

$$\mathbf{a} = (\ddot{r} - r\dot{\theta}^2)\mathbf{e}_r + (r\ddot{\theta} + 2\dot{r}\dot{\theta})\mathbf{e}_\theta \tag{11.44}$$

The scalar components of the velocity and the acceleration in the radial and transverse directions are, therefore,

$$v_r = \dot{r} \qquad v_\theta = r\dot{\theta} \tag{11.45}$$

$$a_r = \ddot{r} - r\dot{\theta}^2 \qquad a_\theta = r\ddot{\theta} + 2\dot{r}\dot{\theta} \tag{11.46}$$

It is important to note that a_r is *not* equal to the time derivative of v_r and that a_θ is *not* equal to the time derivative of v_θ.

In the case of a particle moving along a circle of center O, we have r = constant and $\dot{r} = \ddot{r} = 0$, and the formulas (11.43) and (11.44) reduce, respectively, to

$$\mathbf{v} = r\dot{\theta}\mathbf{e}_\theta \qquad \mathbf{a} = -r\dot{\theta}^2\mathbf{e}_r + r\ddot{\theta}\mathbf{e}_\theta \tag{11.47}$$

Extension to the Motion of a Particle in Space: Cylindrical Coordinates. The position of a particle P in space is sometimes defined by its cylindrical coordinates R, θ, and z (Fig. 11.26*a*). It is then convenient to use the unit vectors $\mathbf{e}_R$, $\mathbf{e}_\theta$, and **k** shown in Fig. 11.26*b*. Resolving the position vector **r** of the particle P into components along the unit vectors, we write

$$\mathbf{r} = R\mathbf{e}_R + z\mathbf{k} \tag{11.48}$$

Observing that $\mathbf{e}_R$ and $\mathbf{e}_\theta$ define, respectively, the radial and transverse directions in the horizontal xy plane, and that the vector **k**, which defines the *axial* direction, is constant in direction as well as in magnitude, we easily verify that

$$\mathbf{v} = \frac{d\mathbf{r}}{dt} = \dot{R}\mathbf{e}_R + R\dot{\theta}\mathbf{e}_\theta + \dot{z}\mathbf{k} \tag{11.49}$$

$$\mathbf{a} = \frac{d\mathbf{v}}{dt} = (\ddot{R} - R\dot{\theta}^2)\mathbf{e}_R + (R\ddot{\theta} + 2\dot{R}\dot{\theta})\mathbf{e}_\theta + \ddot{z}\mathbf{k} \tag{11.50}$$

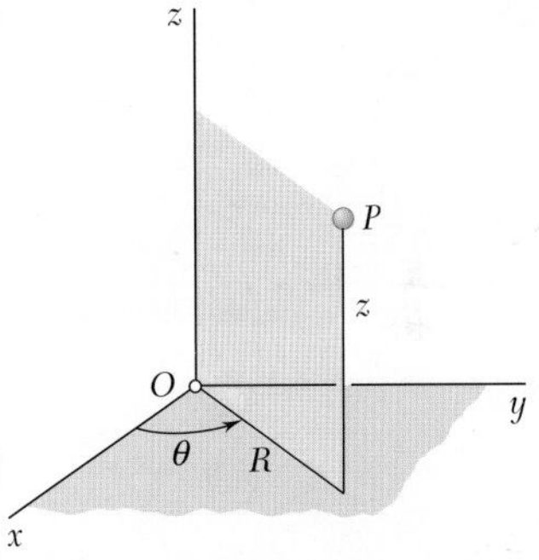

(*a*)

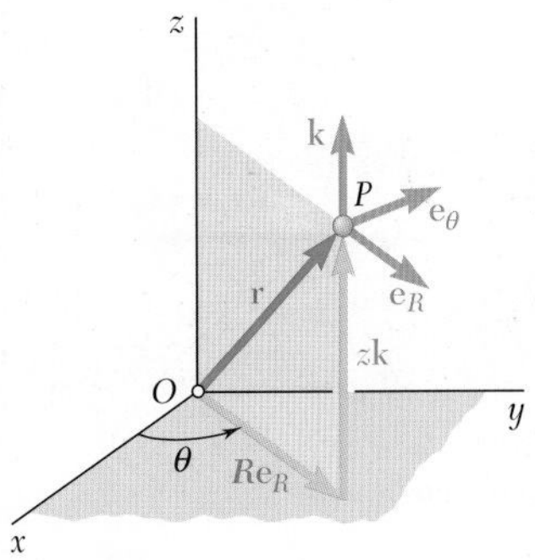

(*b*)

Fig. 11.26

SAMPLE PROBLEM 11.10

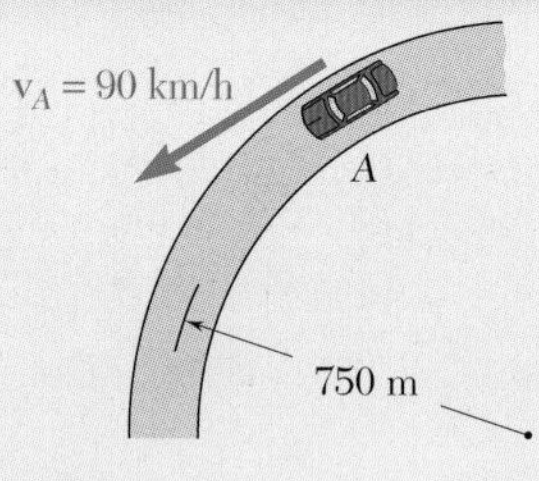

A motorist is traveling on a curved section of highway of radius 750 m at the speed of 90 km/h. The motorist suddenly applies the brakes, causing the automobile to slow down at a constant rate. Knowing that after 8 s the speed has been reduced to 72 km/h, determine the acceleration of the automobile immediately after the brakes have been applied.

SOLUTION

Tangential Component of Acceleration. First the speeds are expressed in m/s.

$$90\,\text{km/h} = \left(90\frac{\text{km}}{\text{h}}\right)\left(\frac{1000\text{ m}}{1\text{ km}}\right)\left(\frac{1\text{ h}}{3600\text{ s}}\right) = 25\text{ m/s}$$

$$72\text{ km/h} = 20\text{ m/s}$$

Since the automobile slows down at a constant rate, we have

$$a_t = \text{average } a_t = \frac{\Delta v}{\Delta t} = \frac{20\text{ m/s} - 25\text{ m/s}}{8\text{ s}} = -0.625\text{ m/s}^2$$

Normal Component of Acceleration. Immediately after the brakes have been applied, the speed is still 25 m/s, and we have

$$a_n = \frac{v^2}{\rho} = \frac{(25\text{ m/s})^2}{750\text{ m}} = 0.833\text{ m/s}^2$$

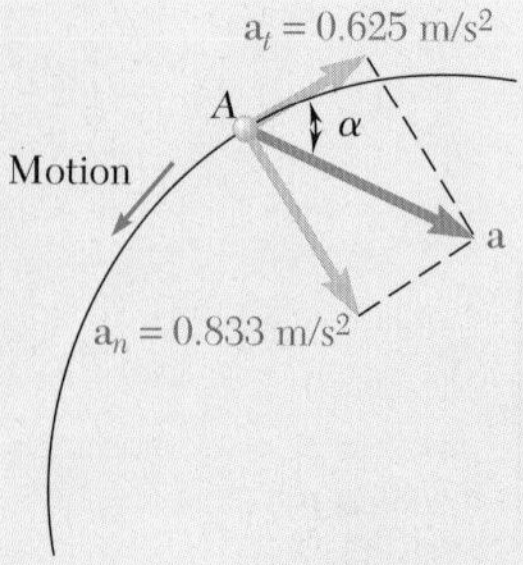

Magnitude and Direction of Acceleration. The magnitude and direction of the resultant **a** of the components $\mathbf{a}_n$ and $\mathbf{a}_t$ are

$$\tan\alpha = \frac{a_n}{a_t} = \frac{0.833\text{ m/s}^2}{0.625\text{ m/s}^2} \qquad \alpha = 53.1° ◀$$

$$a = \frac{a_n}{\sin\alpha} = \frac{0.833\text{ m/s}^2}{\sin 53.1°} \qquad a = 1.041\text{ m/s}^2 ◀$$

SAMPLE PROBLEM 11.11

Determine the minimum radius of curvature of the trajectory described by the projectile considered in Sample Prob. 11.7.

SOLUTION

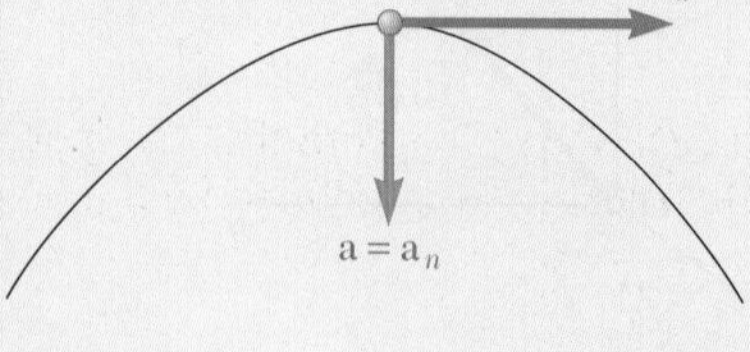

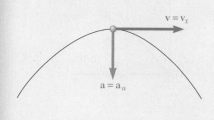

Since $a_n = v^2/\rho$, we have $\rho = v^2/a_n$. The radius will be small when v is small or when a_n is large. The speed v is minimum at the top of the trajectory since $v_y = 0$ at that point; a_n is maximum at that same point, since the direction of the vertical coincides with the direction of the normal. Therefore, the minimum radius of curvature occurs at the top of the trajectory. At this point, we have

$$v = v_x = 155.9\text{ m/s} \qquad a_n = a = 9.81\text{ m/s}^2$$

$$\rho = \frac{v^2}{a_n} = \frac{(155.9\text{ m/s})^2}{9.81\text{ m/s}^2} \qquad \rho = 2480\text{ m} ◀$$

SAMPLE PROBLEM 11.12

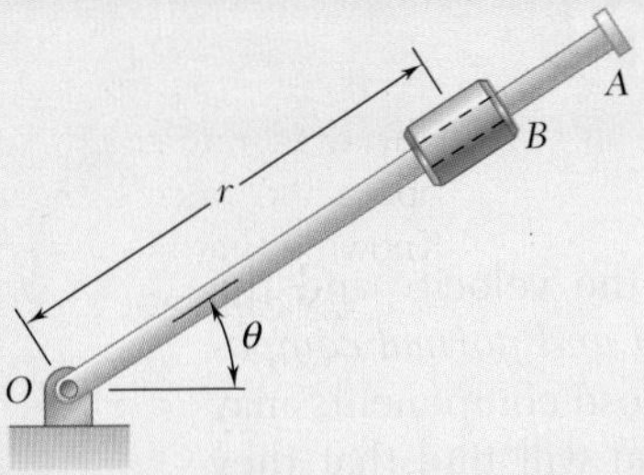

The rotation of the 0.9-m arm OA about O is defined by the relation $\theta = 0.15t^2$, where θ is expressed in radians and t in seconds. Collar B slides along the arm in such a way that its distance from O is $r = 0.9 - 0.12t^2$, where r is expressed in meters and t in seconds. After the arm OA has rotated through 30°, determine (*a*) the total velocity of the collar, (*b*) the total acceleration of the collar, (*c*) the relative acceleration of the collar with respect to the arm.

SOLUTION

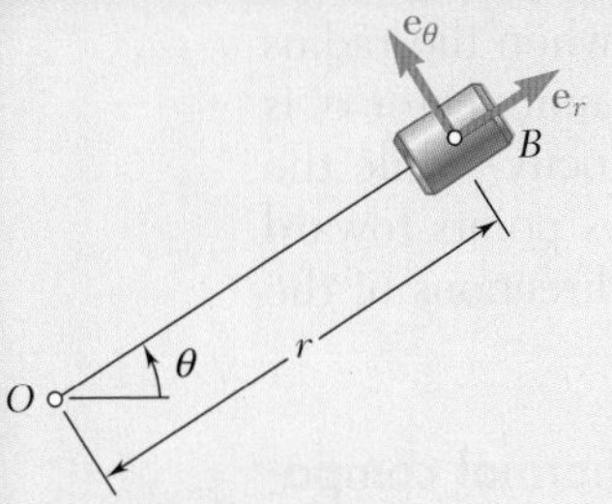

$\mathbf{v} = v_r\mathbf{e}_r + v_\theta\mathbf{e}_\theta$

$\mathbf{a} = a_r\mathbf{e}_r + a_\theta\mathbf{e}_\theta$

Time *t* at which $\theta = 30°$. Substituting $\theta = 30° = 0.524$ rad into the expression for θ, we obtain

$$\theta = 0.15t^2 \qquad 0.524 = 0.15t^2 \qquad t = 1.869 \text{ s}$$

Equations of Motion. Substituting $t = 1.869$ s in the expressions for r, θ, and their first and second derivatives, we have

$$r = 0.9 - 0.12t^2 = 0.481 \text{ m} \qquad \theta = 0.15t^2 = 0.524 \text{ rad}$$
$$\dot{r} = -0.24t = -0.449 \text{ m/s} \qquad \dot{\theta} = 0.30t = 0.561 \text{ rad/s}$$
$$\ddot{r} = -0.24 = -0.240 \text{ m/s}^2 \qquad \ddot{\theta} = 0.30 = 0.300 \text{ rad/s}^2$$

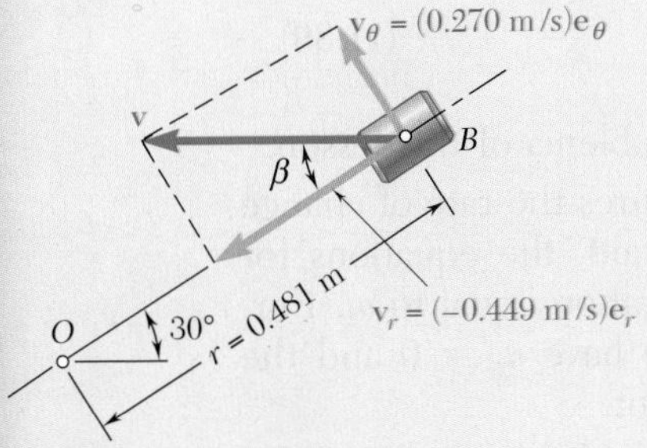

a. Velocity of *B*. Using Eqs. (11.45), we obtain the values of v_r and v_θ when $t = 1.869$ s.

$$v_r = \dot{r} = -0.449 \text{ m/s}$$
$$v_\theta = r\dot{\theta} = 0.481(0.561) = 0.270 \text{ m/s}$$

Solving the right triangle shown, we obtain the magnitude and direction of the velocity,

$$v = 0.524 \text{ m/s} \qquad \beta = 31.0° \quad ◀$$

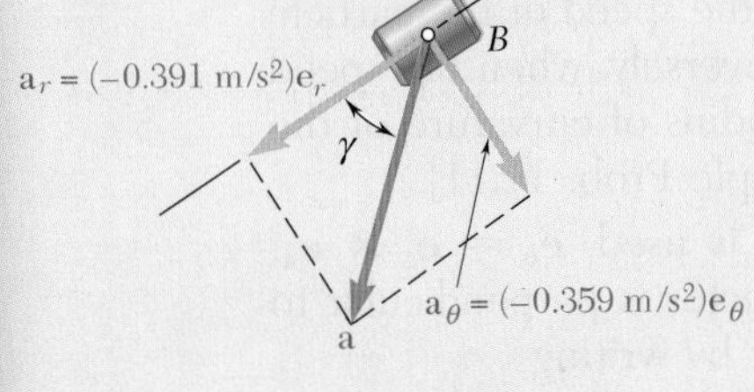

b. Acceleration of *B*. Using Eqs. (11.46), we obtain

$$a_r = \ddot{r} - r\dot{\theta}^2$$
$$= -0.240 - 0.481(0.561)^2 = -0.391 \text{ m/s}^2$$
$$a_\theta = r\ddot{\theta} + 2\dot{r}\dot{\theta}$$
$$= 0.481(0.300) + 2(-0.449)(0.561) = -0.359 \text{ m/s}^2$$
$$a = 0.531 \text{ m/s}^2 \qquad \gamma = 42.6° \quad ◀$$

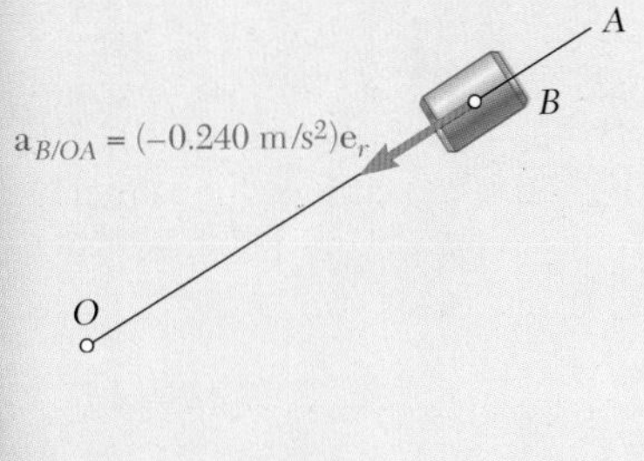

c. Acceleration of *B* with Respect to Arm *OA*. We note that the motion of the collar with respect to the arm is rectilinear and defined by the coordinate r. We write

$$a_{B/OA} = \ddot{r} = -0.240 \text{ m/s}^2$$
$$a_{B/OA} = 0.240 \text{ m/s}^2 \text{ toward } O. \quad ◀$$

SOLVING PROBLEMS ON YOUR OWN

You will be asked in the following problems to express the velocity and the acceleration of particles in terms of either their *tangential and normal components* or their *radial and transverse components*. Although those components may not be as familiar to you as the rectangular components, you will find that they can simplify the solution of many problems and that certain types of motion are more easily described when they are used.

1. Using tangential and normal components. These components are most often used when the particle of interest travels along a circular path or when the radius of curvature of the path is to be determined. Remember that the unit vector $\mathbf{e}_t$ is tangent to the path of the particle (and thus aligned with the velocity) while the unit vector $\mathbf{e}_n$ is directed along the normal to the path and always points toward its center of curvature. It follows that, as the particle moves, the directions of the two unit vectors are constantly changing.

2. Expressing the acceleration in terms of its tangential and normal components. We derived in Sec. 11.13 the following equation, applicable to both the two-dimensional and the three-dimensional motion of a particle:

$$\mathbf{a} = \frac{dv}{dt}\mathbf{e}_t + \frac{v^2}{\rho}\mathbf{e}_n \tag{11.39}$$

The following observations may help you in solving the problems of this lesson.

a. The tangential component of the acceleration measures the rate of change of the speed: $a_t = dv/dt$. It follows that when a_t is constant, the equations for uniformly accelerated motion can be used with the acceleration equal to a_t. Furthermore, when a particle moves at a constant speed, we have $a_t = 0$ and the acceleration of the particle reduces to its normal component.

b. The normal component of the acceleration is always directed toward the center of curvature of the path of the particle, and its magnitude is $a_n = v^2/\rho$. Thus, the normal component can be easily determined if the speed of the particle and the radius of curvature ρ of the path are known. Conversely, when the speed and normal acceleration of the particle are known, the radius of curvature of the path can be obtained by solving this equation for ρ [Sample Prob. 11.11].

c. In three-dimensional motion, a third unit vector is used, $\mathbf{e}_b = \mathbf{e}_t \times \mathbf{e}_n$, which defines the direction of the *binormal*. Since this vector is perpendicular to both the velocity and the acceleration, it can be obtained by writing

$$\mathbf{e}_b = \frac{\mathbf{v} \times \mathbf{a}}{|\mathbf{v} \times \mathbf{a}|}$$

3. Using radial and transverse components. These components are used to analyze the plane motion of a particle P, when the position of P is defined by its polar coordinates r and θ. As shown in Fig. 11.25, the unit vector $\mathbf{e}_r$, which defines the *radial* direction, is attached to P and points away from the fixed point O, while the unit vector $\mathbf{e}_\theta$, which defines the *transverse* direction, is obtained by rotating $\mathbf{e}_r$ *counterclockwise* through 90°. The velocity and the acceleration of a particle were expressed in terms of their radial and transverse components in Eqs. (11.43) and (11.44), respectively. You will note that the expressions obtained contain the first and second derivatives with respect to t of both coordinates r and θ.

In the problems of this lesson, you will encounter the following types of problems involving radial and transverse components:

a. Both r and θ are known functions of t. In this case, you will compute the first and second derivatives of r and θ and substitute the expressions obtained into Eqs. (11.43) and (11.44).

b. A certain relationship exists between r and θ. First, you should determine this relationship from the geometry of the given system and use it to express r as a function of θ. Once the function $r = f(\theta)$ is known, you can apply the chain rule to determine $\dot{r}$ in terms of θ and $\dot{\theta}$, and $\ddot{r}$ in terms of θ, $\dot{\theta}$, $\ddot{\theta}$:

$$\dot{r} = f'(\theta)\dot{\theta}$$

$$\ddot{r} = f''(\theta)\dot{\theta}^2 + f'(\theta)\ddot{\theta}$$

The expressions obtained can then be substituted into Eqs. (11.43) and (11.44).

c. The three-dimensional motion of a particle, as indicated at the end of Sec. 11.14, can often be effectively described in terms of the *cylindrical coordinates* R, θ, and z (Fig. 11.26). The unit vectors then should consist of $\mathbf{e}_R$, $\mathbf{e}_\theta$, and $\mathbf{k}$. The corresponding components of the velocity and the acceleration are given in Eqs. (11.49) and (11.50). Please note that the radial distance R is always measured in a plane parallel to the xy plane, and be careful not to confuse the position vector $\mathbf{r}$ with its radial component $R\mathbf{e}_R$.

PROBLEMS

CONCEPT QUESTIONS

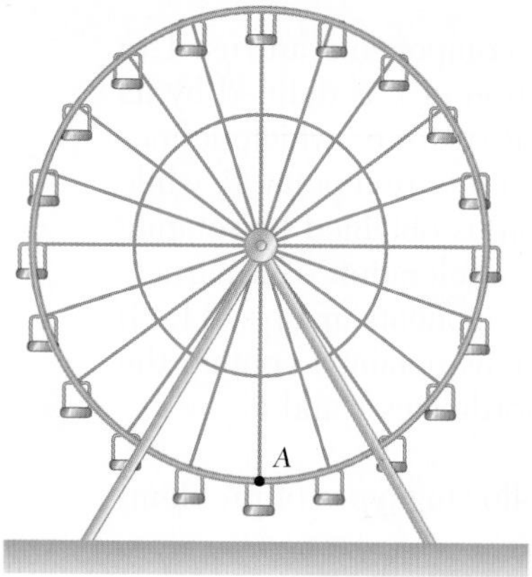

Fig. P11.CQ8

11.CQ8 The Ferris wheel is rotating with a constant angular velocity ω. What is the direction of the acceleration of point A?

a. $\rightarrow$ **b.** $\uparrow$ **c.** $\downarrow$ **d.** $\leftarrow$ **e.** The acceleration is zero.

11.CQ9 A race car travels around the track shown at a constant speed. At which point will the race car have the largest acceleration?

a. *A.* **b.** *B.* **c.** *C.* **d.** *D.* **e.** The acceleration will be zero at all the points.

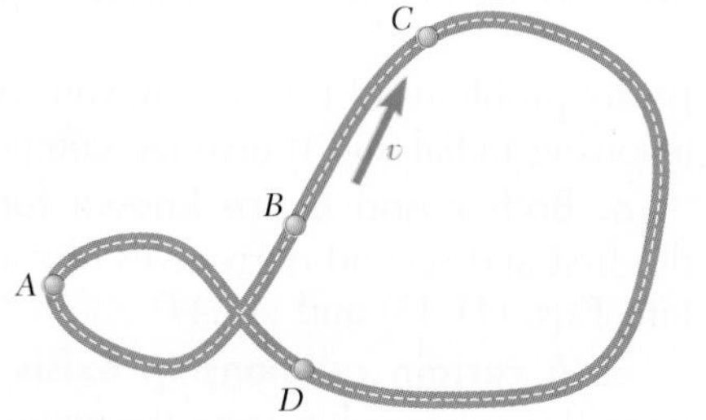

Fig. P11.CQ9

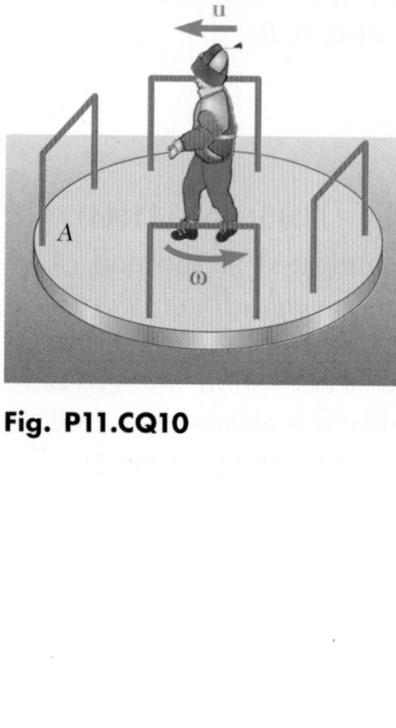

Fig. P11.CQ10

11.CQ10 A child walks across merry-go-round A with a constant speed u relative to A. The merry-go-round undergoes fixed-axis rotation about its center with a constant angular velocity ω counterclockwise. When the child is at the center of A, as shown, what is the direction of his acceleration when viewed from above?

a. $\rightarrow$ **b.** $\leftarrow$ **c.** $\uparrow$ **d.** $\downarrow$ **e.** The acceleration is zero.

END-OF-SECTION PROBLEMS

11.133 Determine the smallest radius that should be used for a highway if the normal component of the acceleration of a car traveling at 72 km/h is not to exceed 0.8 m/s^2.

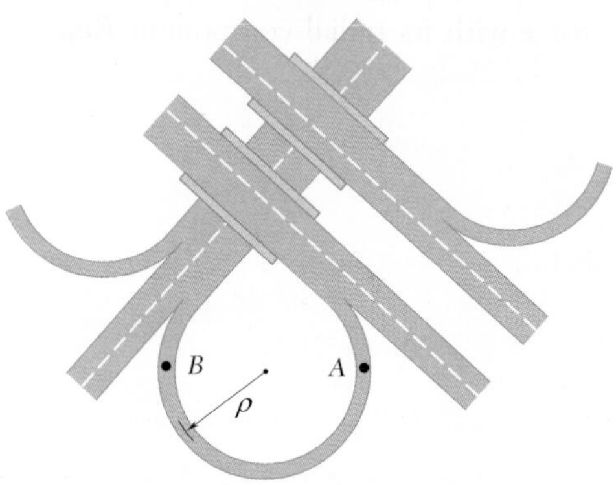

Fig. P11.133

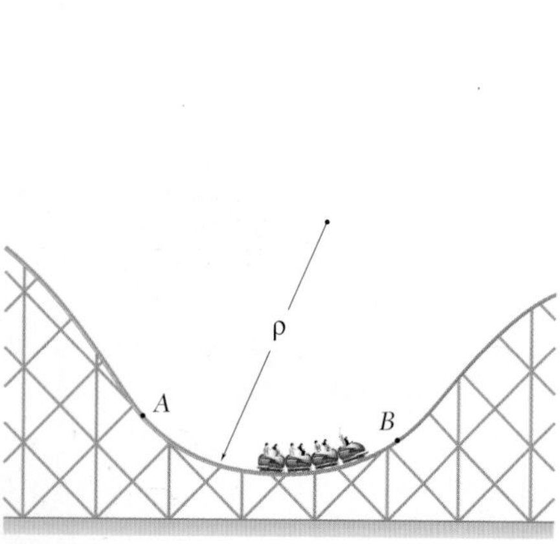

Fig. P11.134

11.134 Determine the maximum speed that the cars of the roller-coaster can reach along the circular portion AB of the track if $\rho = 25$ m and the normal component of their acceleration cannot exceed 3g.

11.135 A bull-roarer is a piece of wood that produces a roaring sound when attached to the end of a string and whirled around in a circle. Determine the magnitude of the normal acceleration of a bull-roarer when it is spun in a circle of radius 0.9 m at a speed of 20 m/s.

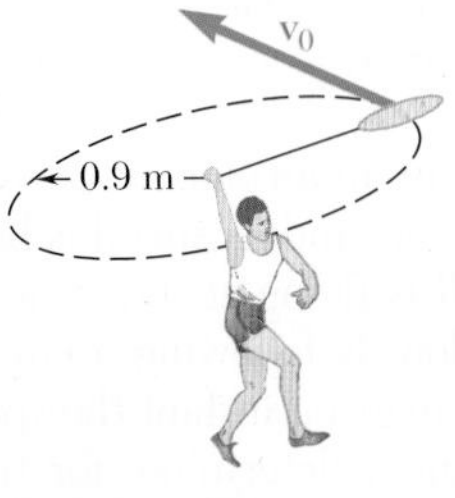

Fig. P11.135

11.136 To test its performance, an automobile is driven around a circular test track of diameter d. Determine (*a*) the value of d if when the speed of the automobile is 72 km/h, the normal component of the acceleration is 3.2 m/s^2, (*b*) the speed of the automobile if d = 180 m and the normal component of the acceleration is measured to be 0.6g.

11.137 An outdoor track is 125 m in diameter. A runner increases her speed at a constant rate from 4 m/s to 7 m/s over a distance of 30 m. Determine the total acceleration of the runner 2 s after she begins to increase her speed.

Fig. P11.137

11.138 A robot arm moves so that P travels in a circle about point B, which is not moving. Knowing that P starts from rest, and its speed increases at a constant rate of 10 mm/s^2, determine (*a*) the magnitude of the acceleration when t = 4 s, (*b*) the time for the magnitude of the acceleration to be 80 mm/s^2.

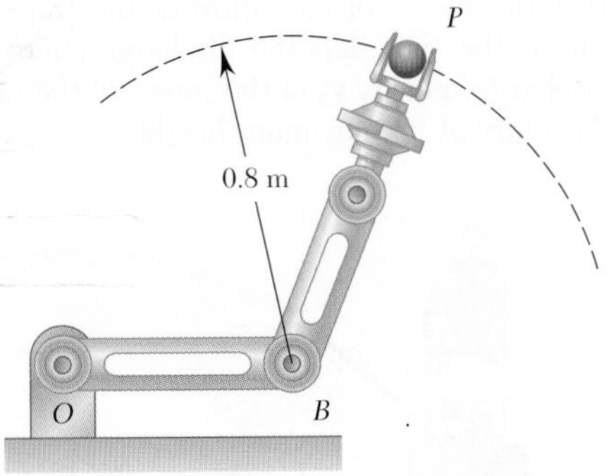

Fig. P11.138

11.139 A monorail train starts from rest on a curve of radius 400 m and accelerates at the constant rate a_t. If the maximum total acceleration of the train must not exceed 1.5 m/s^2, determine (*a*) the shortest distance in which the train can reach a speed of 72 km/h, (*b*) the corresponding constant rate of acceleration a_t.

11.140 A motorist starts from rest at point A on a circular entrance ramp when t = 0, increases the speed of her automobile at a constant rate and enters the highway at point B. Knowing that her speed continues to increase at the same rate until it reaches 100 km/h at point C, determine (*a*) the speed at point B, (*b*) the magnitude of the total acceleration when t = 20 s.

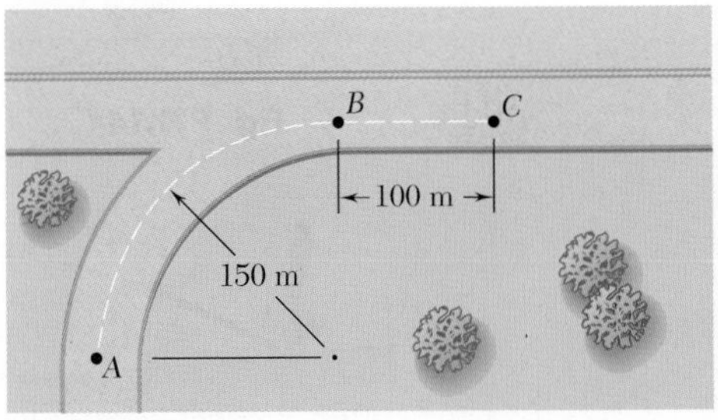

Fig. P11.140

11.141 Race car A is traveling on a straight portion of the track while race car B is traveling on a circular portion of the track. At the instant shown, the speed of A is increasing at the rate of 10 m/s^2, and the speed of B is decreasing at the rate of 6 m/s^2. For the position shown, determine (*a*) the velocity of B relative to A, (*b*) the acceleration of B relative to A.

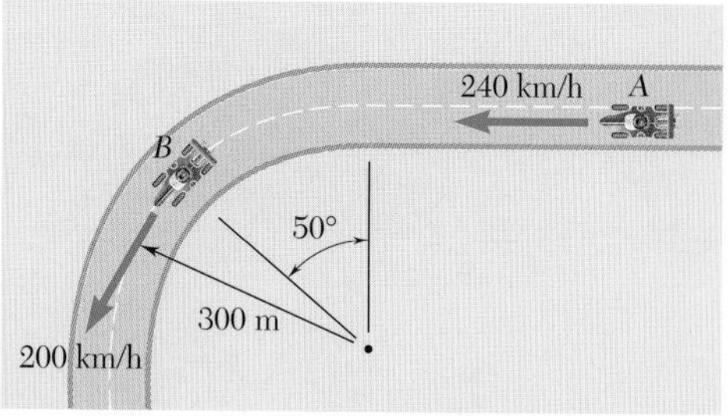

Fig. P11.141

11.142 At a given instant in an airplane race, airplane A is flying horizontally in a straight line, and its speed is being increased at the rate of 8 m/s^2. Airplane B is flying at the same altitude as airplane A and, as it rounds a pylon, is following a circular path of 300-m radius. Knowing that at the given instant the speed of B is being decreased at the rate of 3 m/s^2, determine, for the positions shown, (*a*) the velocity of B relative to A, (*b*) the acceleration of B relative to A.

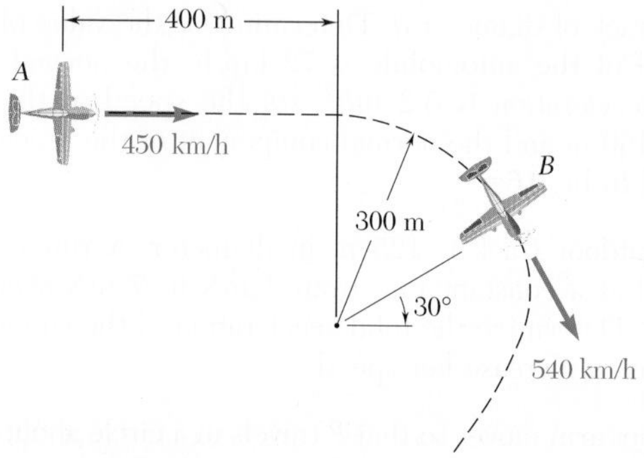

Fig. P11.142

11.143 From a photograph of a homeowner using a snowblower, it is determined that the radius of curvature of the trajectory of the snow was 8.5 m as the snow left the discharge chute at A. Determine (*a*) the discharge velocity $\mathbf{v}_A$ of the snow, (*b*) the radius of curvature of the trajectory at its maximum height.

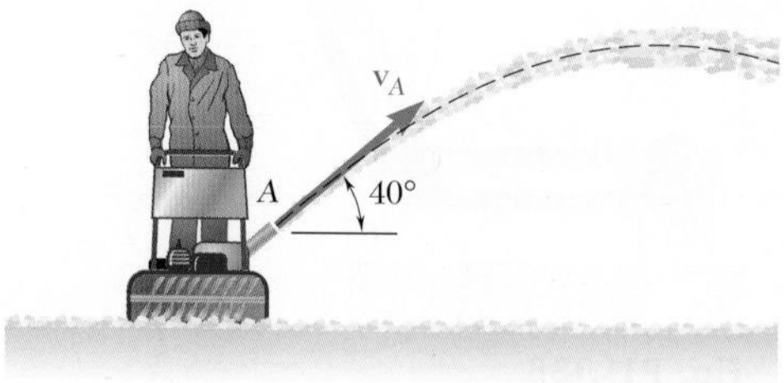

Fig. P11.143

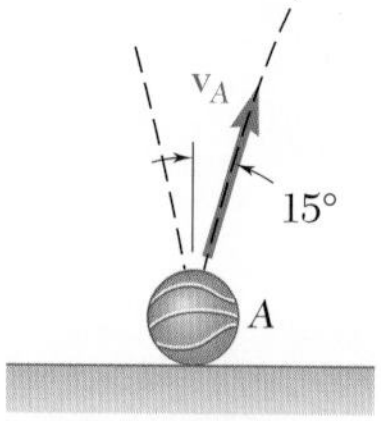

Fig. P11.144

11.144 A basketball is bounced on the ground at point A and rebounds with a velocity $\mathbf{v}_A$ of magnitude 2 m/s as shown. Determine the radius of curvature of the trajectory described by the ball (*a*) at point A, (*b*) at the highest point of the trajectory.

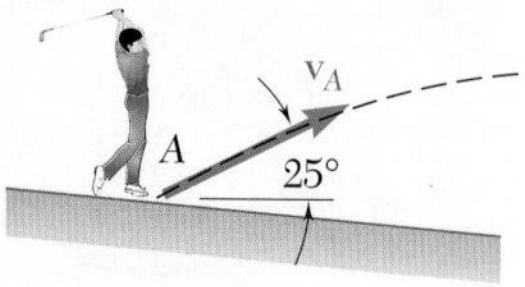

Fig. P11.145

11.145 A golfer hits a golf ball from point A with an initial velocity of 50 m/s at an angle of 25° with the horizontal. Determine the radius of curvature of the trajectory described by the ball (*a*) at point A, (*b*) at the highest point of the trajectory.

11.146 Three children are throwing snowballs at each other. Child *A* throws a snowball with a horizontal velocity $\mathbf{v}_0$. If the snowball just passes over the head of child *B* and hits child *C*, determine the radius of curvature of the trajectory described by the snowball (*a*) at point *B*, (*b*) at point *C*.

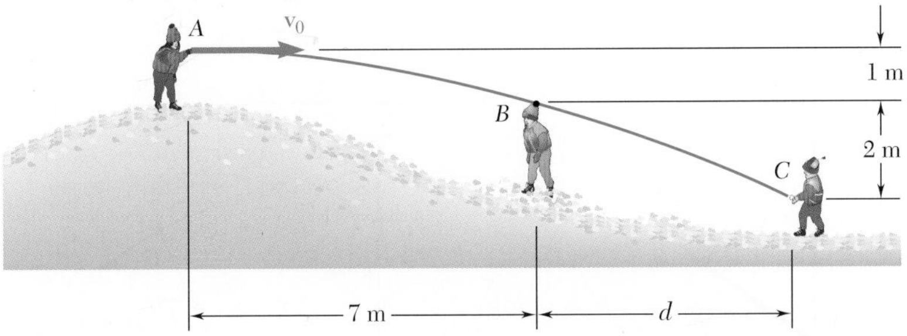

Fig. P11.146

11.147 Coal is discharged from the tailgate *A* of a dump truck with an initial velocity $\mathbf{v}_A = 2$ m/s ⦪ 50°. Determine the radius of curvature of the trajectory described by the coal (*a*) at point *A*, (*b*) at the point of the trajectory 1 m below point *A*.

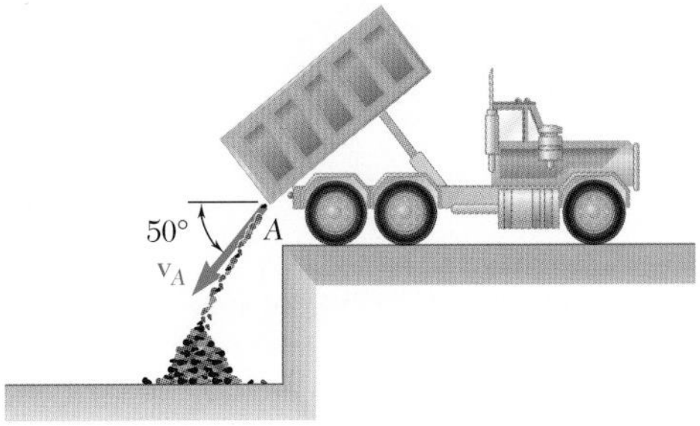

Fig. P11.147

11.148 From measurements of a photograph, it has been found that as the stream of water shown left the nozzle at *A*, it had a radius of curvature of 25 m. Determine (*a*) the initial velocity $\mathbf{v}_A$ of the stream, (*b*) the radius of curvature of the stream as it reaches its maximum height at *B*.

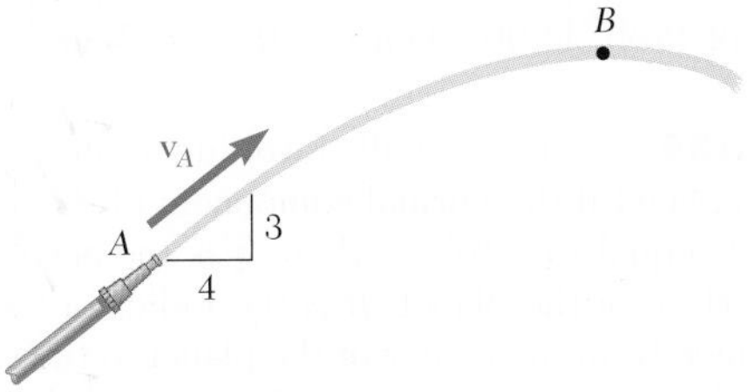

Fig. *P11.148*

11.149 A child throws a ball from point *A* with an initial velocity $\mathbf{v}_A$ of 20 m/s at an angle of 25° with the horizontal. Determine the velocity of the ball at the points of the trajectory described by the ball where the radius of curvature is equal to three-quarters of its value at *A*.

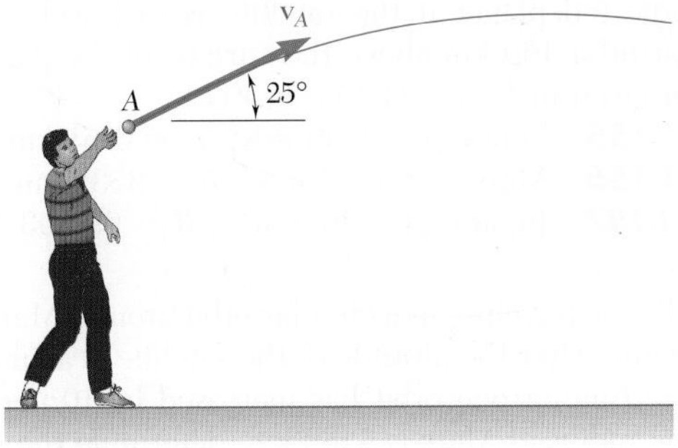

Fig. P11.149

11.150 A projectile is fired from point A with an initial velocity $\mathbf{v}_0$. (*a*) Show that the radius of curvature of the trajectory of the projectile reaches its minimum value at the highest point B of the trajectory. (*b*) Denoting by θ the angle formed by the trajectory and the horizontal at a given point C, show that the radius of curvature of the trajectory at C is $\rho = \rho_{min}/\cos^3\theta$.

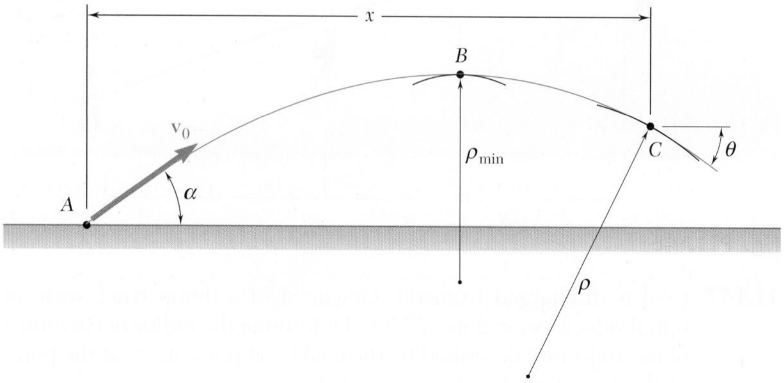

Fig. P11.150

***11.151** Determine the radius of curvature of the path described by the particle of Prob. 11.95 when $t = 0$.

***11.152** Determine the radius of curvature of the path described by the particle of Prob. 11.96 when $t = 0$, $A = 3$, and $B = 1$.

11.153 and 11.154 A satellite will travel indefinitely in a circular orbit around a planet if the normal component of the acceleration of the satellite is equal to $g(R/r)^2$, where g is the acceleration of gravity at the surface of the planet, R is the radius of the planet, and r is the distance from the center of the planet to the satellite. Knowing that the diameter of the sun is 1.39 Gm and that the acceleration of gravity at its surface is 274 m/s^2, determine the radius of the orbit of the indicated planet around the sun assuming that the orbit is circular.

11.153 Earth: $(v_{mean})_{orbit} = 107$ Mm/h.
11.154 Saturn: $(v_{mean})_{orbit} = 34.7$ Mm/h.

11.155 through *11.157* Determine the speed of a satellite relative to the indicated planet if the satellite is to travel indefinitely in a circular orbit 160 km above the surface of the planet. (See information given in Probs. 11.153–11.154.)

11.155 Venus: $g = 8.53$ m/s^2, $R = 6161$ km.
11.156 Mars: $g = 3.83$ m/s^2, $R = 3332$ km.
11.157 Jupiter: $g = 26.0$ m/s^2, $R = 69\,893$ km.

11.158 A satellite is traveling in a circular orbit around Mars at an altitude of 300 km. After the altitude of the satellite is adjusted, it is found that the time of one orbit has increased by 10 percent. Knowing that the radius of Mars is 3382 km, determine the new altitude of the satellite. (See information given in Probs. 11.153–11.154).

11.159 Knowing that the radius of the earth is 6370 km, determine the time of one orbit of the Hubble Space Telescope knowing that the telescope travels in a circular orbit 590 km above the surface of the earth. (See information given in Probs. 11.153–11.154.)

11.160 Satellites A and B are traveling in the same plane in circular orbits around the earth at altitudes of 180 km and 300 km, respectively. If at $t = 0$ the satellites are aligned as shown and knowing that the radius of the earth is $R = 6370$ km, determine when the satellites will next be radially aligned. (See information given in Probs. 11.153–11.155.)

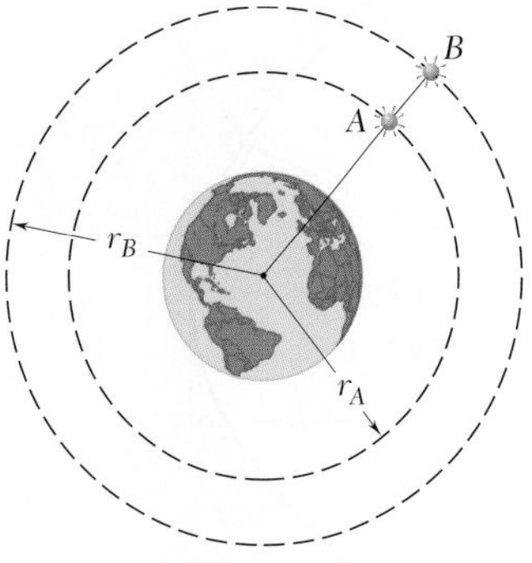

Fig. *P11.160*

11.161 The oscillation of rod OA about O is defined by the relation $\theta = (2/\pi)(\sin \pi t)$, where θ and t are expressed in radians and seconds, respectively. Collar B slides along the rod so that its distance from O is $r = 625/(t + 4)$ where r and t are expressed in mm and seconds, respectively. When $t = 1$ s, determine (*a*) the velocity of the collar, (*b*) the total acceleration of the collar, (*c*) the acceleration of the collar relative to the rod.

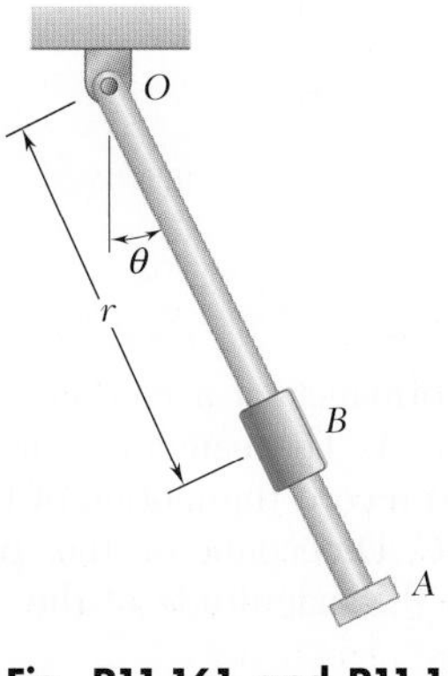

Fig. P11.161 and P11.162

11.162 The rotation of rod OA about O is defined by the relation $\theta = \pi(4t^2 - 8t)$, where θ and t are expressed in radians and seconds, respectively. Collar B slides along the rod so that its distance from O is $r = 250 + 150 \sin \pi t$, where r and t are expressed in mm and seconds, respectively. When $t = 1$ s, determine (*a*) the velocity of the collar, (*b*) the total acceleration of the collar, (*c*) the acceleration of the collar relative to the rod.

11.163 The path of particle P is the ellipse defined by the relations $r = 2/(2 - \cos \pi t)$ and $\theta = \pi t$, where r is expressed in meters, t is in seconds, and θ is in radians. Determine the velocity and the acceleration of the particle when (*a*) $t = 0$, (*b*) $t = 0.5$ s.

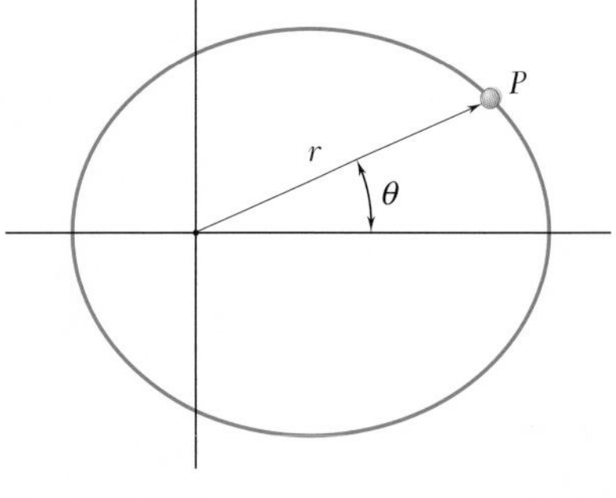

Fig. P11.163

11.164 The two-dimensional motion of a particle is defined by the relations $r = 2a \cos \theta$ and $\theta = bt^2/2$, where a and b are constants. Determine (*a*) the magnitudes of the velocity and acceleration at any instant, (*b*) the radius of curvature of the path. What conclusion can you draw regarding the path of the particle?

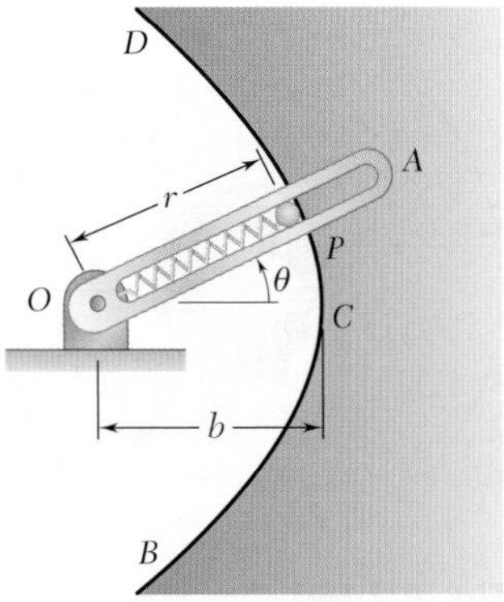

Fig. P11.165

11.165 As rod OA rotates, pin P moves along the parabola BCD. Knowing that the equation of this parabola is $r = 2b/(1 + \cos\theta)$ and that $\theta = kt$, determine the velocity and acceleration of P when (*a*) $\theta = 0$, (*b*) $\theta = 90°$.

11.166 The pin at B is free to slide along the circular slot DE and along the rotating rod OC. Assuming that the rod OC rotates at a constant rate $\dot{\theta}$, (*a*) show that the acceleration of pin B is of constant magnitude, (*b*) determine the direction of the acceleration of pin B.

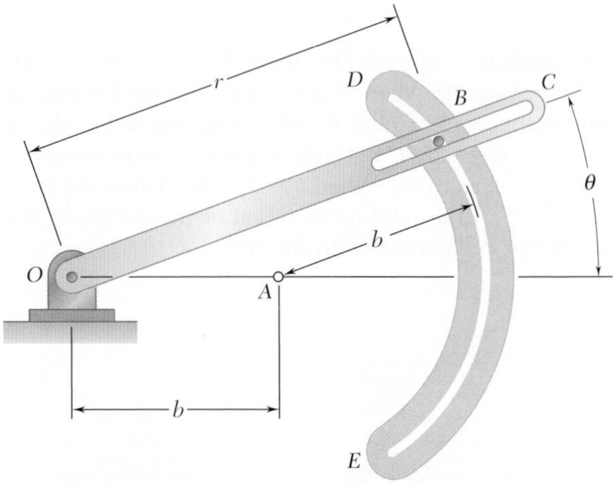

Fig. P11.166

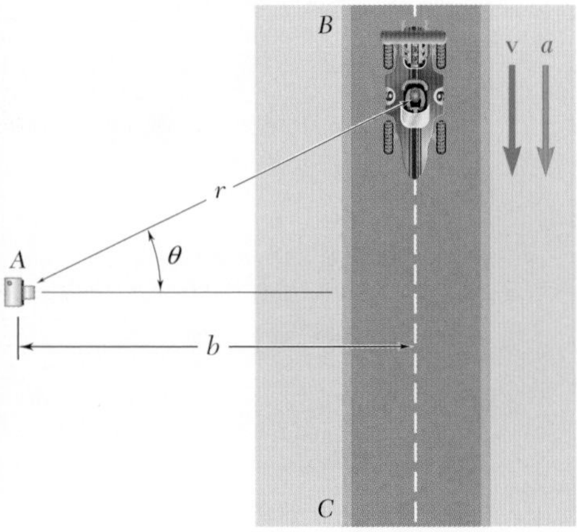

Fig. *P11.167*

11.167 To study the performance of a race car, a high-speed camera is positioned at point A. The camera is mounted on a mechanism which permits it to record the motion of the car as the car travels on straightaway BC. Determine (*a*) the speed of the car in terms of b, θ, and $\dot{\theta}$, (*b*) the magnitude of the acceleration in terms of b, θ, $\dot{\theta}$, and $\ddot{\theta}$.

11.168 After taking off, a helicopter climbs in a straight line at a constant angle β. Its flight is tracked by radar from point A. Determine the speed of the helicopter in terms of d, β, θ, and $\dot{\theta}$.

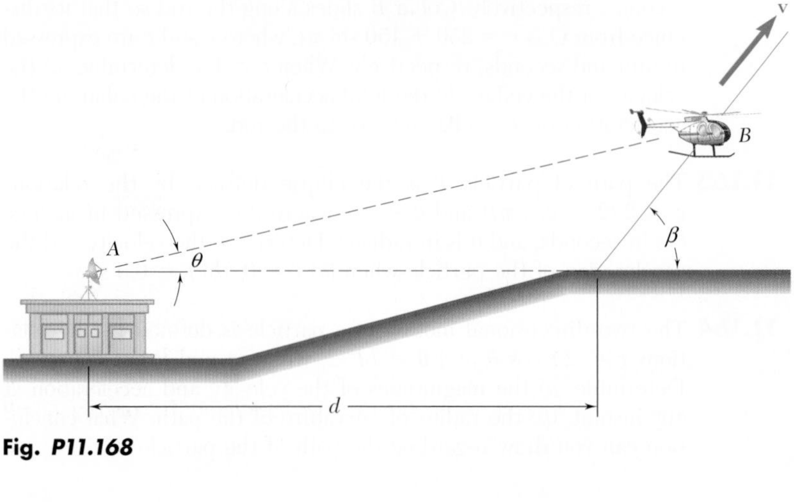

Fig. *P11.168*

11.169 At the bottom of a loop in the vertical plane an airplane has a horizontal velocity of 150 m/s and is speeding up at a rate of 25 m/s^2. The radius of curvature of the loop is 2000 m. The plane is being tracked by radar at O. What are the recorded values of $\dot{r}$, $\ddot{r}$, $\dot{\theta}$ and $\ddot{\theta}$ for this instant?

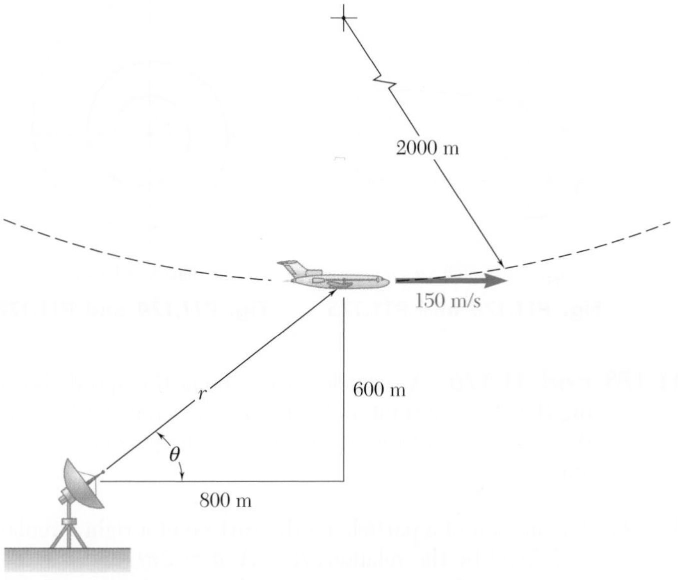

Fig. P11.169

11.170 Pin C is attached to rod BC and slides freely in the slot of rod OA which rotates at the constant rate ω. At the instant when $\beta = 60°$, determine (*a*) $\dot{r}$ and $\dot{\theta}$, (*b*) $\ddot{r}$ and $\ddot{\theta}$. Express your answers in terms of d and ω.

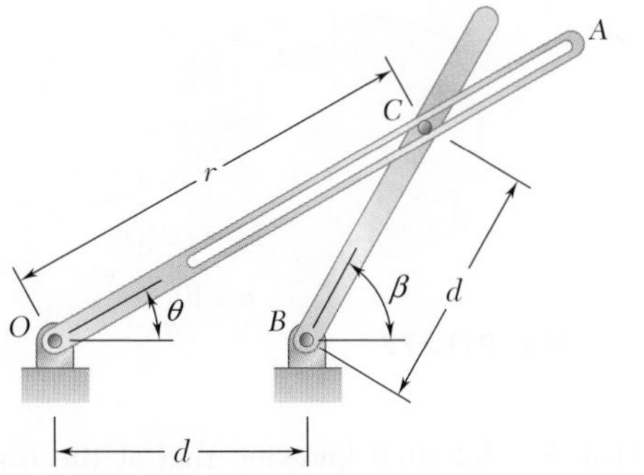

Fig. P11.170

11.171 For the race car of Prob. 11.167, it was found that it took 0.5 s for the car to travel from the position $\theta = 60°$ to the position $\theta = 35°$. Knowing that $b = 25$ m, determine the average speed of the car during the 0.5-s interval.

11.172 For the helicopter of Prob. 11.169, it was found that when the helicopter was at B, the distance and the angle of elevation of the helicopter were $r = 1000$ m and $\theta = 20°$, respectively. Four seconds later, the radar station sighted the helicopter at $r = 1100$ m and $\theta = 23.1°$. Determine the average speed and the angle of climb β of the helicopter during the 4-s interval.

11.173 and 11.174 A particle moves along the spiral shown; determine the magnitude of the velocity of the particle in terms of b, θ, and $\dot{\theta}$.

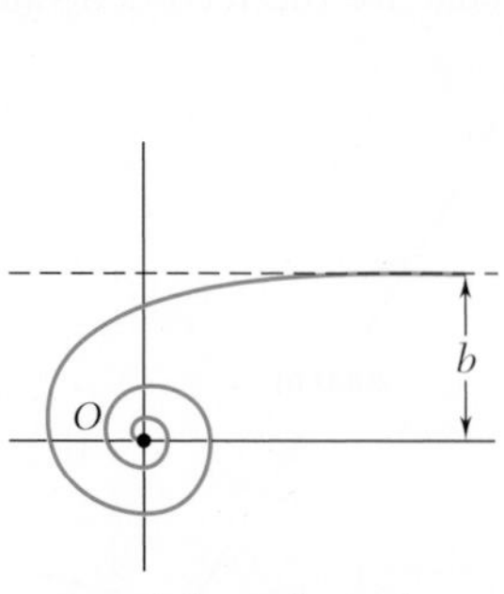

Fig. P11.173 and P11.175

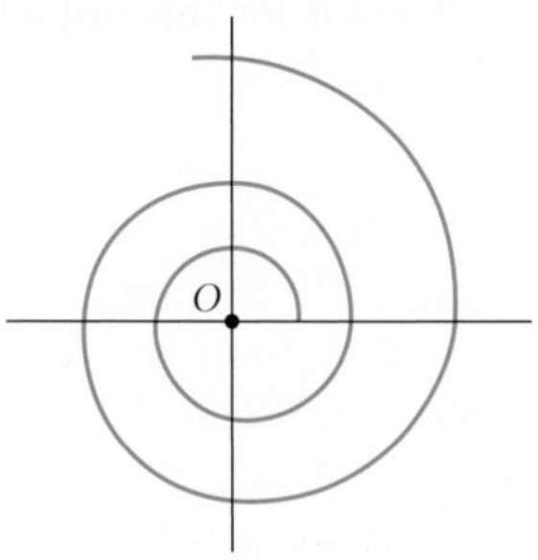

Fig. P11.174 and P11.176

11.175 and 11.176 A particle moves along the spiral shown. Knowing that $\dot{\theta}$ is constant and denoting this constant by ω, determine the magnitude of the acceleration of the particle in terms of b, θ, and ω.

11.177 The motion of a particle on the surface of a right circular cylinder is defined by the relations $R = A$, $\theta = 2\pi t$, and $z = B \sin 2\pi nt$, where A and B are constants and n is an integer. Determine the magnitudes of the velocity and acceleration of the particle at any time t.

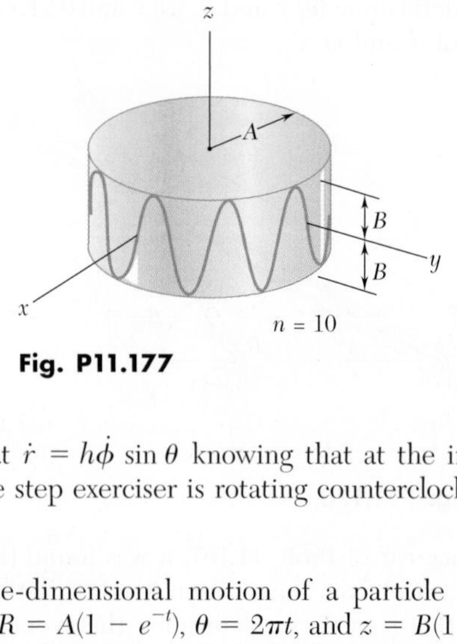

Fig. P11.177

Fig. P11.178

11.178 Show that $\dot{r} = h\dot{\phi} \sin \theta$ knowing that at the instant shown, step AB of the step exerciser is rotating counterclockwise at a constant rate $\dot{\phi}$.

11.179 The three-dimensional motion of a particle is defined by the relations $R = A(1 - e^{-t})$, $\theta = 2\pi t$, and $z = B(1 - e^{-t})$. Determine the magnitudes of the velocity and acceleration when (*a*) $t = 0$, (*b*) $t = \infty$.

***11.180** For the conic helix of Prob. 11.95, determine the angle that the osculating plane forms with the y axis.

***11.181** Determine the direction of the binormal of the path described by the particle of Prob. 11.96 when (*a*) $t = 0$, (*b*) $t = \pi/2$ s.

REVIEW AND SUMMARY

Position coordinate of a particle in rectilinear motion

In the first half of the chapter, we analyzed the *rectilinear motion of a particle,* i.e., the motion of a particle along a straight line. To define the position P of the particle on that line, we chose a fixed origin O and a positive direction (Fig. 11.27). The distance x from O to P, with the appropriate sign, completely defines the position of the particle on the line and is called the *position coordinate* of the particle [Sec. 11.2].

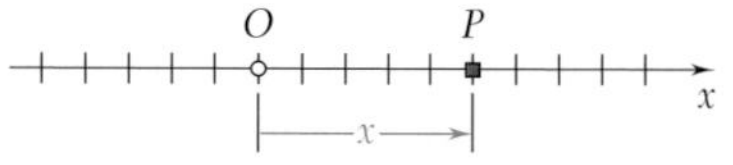

Fig. 11.27

Velocity and acceleration in rectilinear motion

The *velocity* v of the particle was shown to be equal to the time derivative of the position coordinate x,

$$v = \frac{dx}{dt} \tag{11.1}$$

and the *acceleration* a was obtained by differentiating v with respect to t,

$$a = \frac{dv}{dt} \tag{11.2}$$

or

$$a = \frac{d^2x}{dt^2} \tag{11.3}$$

We also noted that a could be expressed as

$$a = v\frac{dv}{dx} \tag{11.4}$$

We observed that the velocity v and the acceleration a were represented by algebraic numbers which can be positive or negative. A positive value for v indicates that the particle moves in the positive direction, and a negative value that it moves in the negative direction. A positive value for a, however, may mean that the particle is truly accelerated (i.e., moves faster) in the positive direction, or that it is decelerated (i.e., moves more slowly) in the negative direction. A negative value for a is subject to a similar interpretation [Sample Prob. 11.1].

Determination of the velocity and acceleration by integration

In most problems, the conditions of motion of a particle are defined by the type of acceleration that the particle possesses and by the initial conditions [Sec. 11.3]. The velocity and position of the particle can then be obtained by integrating two of the equations (11.1) to (11.4). Which of these equations should be selected depends upon the type of acceleration involved [Sample Probs. 11.2 and 11.3].

Uniform rectilinear motion

Two types of motion are frequently encountered: the *uniform rectilinear motion* [Sec. 11.4], in which the velocity v of the particle is constant and

$$x = x_0 + vt \tag{11.5}$$

Uniformly accelerated rectilinear motion

and the *uniformly accelerated rectilinear motion* [Sec. 11.5], in which the acceleration a of the particle is constant and we have

$$v = v_0 + at \tag{11.6}$$
$$x = x_0 + v_0 t + \tfrac{1}{2}at^2 \tag{11.7}$$
$$v^2 = v_0^2 + 2a(x - x_0) \tag{11.8}$$

Relative motion of two particles

When two particles A and B move along the same straight line, we may wish to consider the *relative motion* of B with respect to A

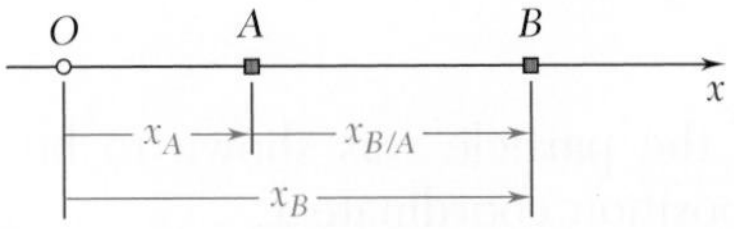

Fig. 11.28

[Sec. 11.6]. Denoting by $x_{B/A}$ the *relative position coordinate* of B with respect to A (Fig. 11.28), we had

$$x_B = x_A + x_{B/A} \tag{11.9}$$

Differentiating Eq. (11.9) twice with respect to t, we obtained successively

$$v_B = v_A + v_{B/A} \tag{11.10}$$
$$a_B = a_A + a_{B/A} \tag{11.11}$$

where $v_{B/A}$ and $a_{B/A}$ represent, respectively, the *relative velocity* and the *relative acceleration* of B with respect to A.

Blocks connected by inextensible cords

When several blocks are *connected by inextensible cords*, it is possible to write a *linear relation* between their position coordinates. Similar relations can then be written between their velocities and between their accelerations and can be used to analyze their motion [Sample Prob. 11.5].

Graphical solutions

It is sometimes convenient to use a *graphical solution* for problems involving the rectilinear motion of a particle [Secs. 11.7 and 11.8]. The graphical solution most commonly used involves the x–t, v–t, and a–t curves [Sec. 11.7; Sample Prob. 11.6]. It was shown that, at any given time t,

$$v = \text{slope of } x\text{–}t \text{ curve}$$
$$a = \text{slope of } v\text{–}t \text{ curve}$$

while, over any given time interval from t_1 to t_2,

$$v_2 - v_1 = \text{area under } a\text{–}t \text{ curve}$$
$$x_2 - x_1 = \text{area under } v\text{–}t \text{ curve}$$

Position vector and velocity in curvilinear motion

In the second half of the chapter, we analyzed the *curvilinear motion of a particle*, i.e., the motion of a particle along a curved path. The position P of the particle at a given time [Sec. 11.9] was defined by the *position vector* **r** joining the *O* of the coordinates and point *P* (Fig. 11.29). The *velocity* **v** of the particle was defined by the relation

$$\mathbf{v} = \frac{d\mathbf{r}}{dt} \tag{11.15}$$

and was found to be a *vector tangent to the path of the particle* and of magnitude v (called the *speed* of the particle) equal to the time derivative of the length *s* of the arc described by the particle:

$$v = \frac{ds}{dt} \tag{11.16}$$

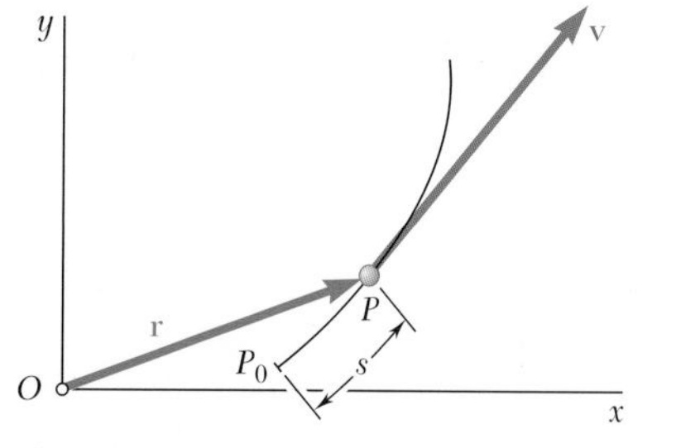

Fig. 11.29

Acceleration in curvilinear motion

The *acceleration* **a** of the particle was defined by the relation

$$\mathbf{a} = \frac{d\mathbf{v}}{dt} \tag{11.18}$$

and we noted that, in general, *the acceleration is not tangent to the path of the particle.*

Derivative of a vector function

Before proceeding to the consideration of the components of velocity and acceleration, we reviewed the formal definition of the derivative of a vector function and established a few rules governing the differentiation of sums and products of vector functions. We then showed that the rate of change of a vector is the same with respect to a fixed frame and with respect to a frame in translation [Sec. 11.10].

Rectangular components of velocity and acceleration

Denoting by x, y, and z the rectangular coordinates of a particle *P*, we found that the rectangular components of the velocity and acceleration of *P* equal, respectively, the first and second derivatives with respect to t of the corresponding coordinates:

$$v_x = \dot{x} \qquad v_y = \dot{y} \qquad v_z = \dot{z} \tag{11.29}$$
$$a_x = \ddot{x} \qquad a_y = \ddot{y} \qquad a_z = \ddot{z} \tag{11.30}$$

Component motions

When the component a_x of the acceleration depends only upon t, x, and/or v_x, and when similarly a_y depends only upon t, y, and/or v_y, and a_z upon t, z, and/or v_z, Eq. (11.30) can be integrated independently. The analysis of the given curvilinear motion can thus be reduced to the analysis of three independent rectilinear component motions [Sec. 11.11]. This approach is particularly effective in the study of the motion of projectiles [Sample Probs. 11.7 and 11.8].

Relative motion of two particles

For two particles *A* and *B* moving in space (Fig. 11.30), we considered the relative motion of *B* with respect to *A*, or more precisely, with respect to a moving frame attached to *A* and in translation with *A* [Sec. 11.12]. Denoting by $\mathbf{r}_{B/A}$ the *relative position vector* of *B* with respect to *A* (Fig. 11.30), we had

$$\mathbf{r}_B = \mathbf{r}_A + \mathbf{r}_{B/A} \tag{11.31}$$

Denoting by $\mathbf{v}_{B/A}$ and $\mathbf{a}_{B/A}$, respectively, the *relative velocity* and the *relative acceleration* of *B* with respect to *A*, we also showed that

$$\mathbf{v}_B = \mathbf{v}_A + \mathbf{v}_{B/A} \tag{11.33}$$

and

$$\mathbf{a}_B = \mathbf{a}_A + \mathbf{a}_{B/A} \tag{11.34}$$

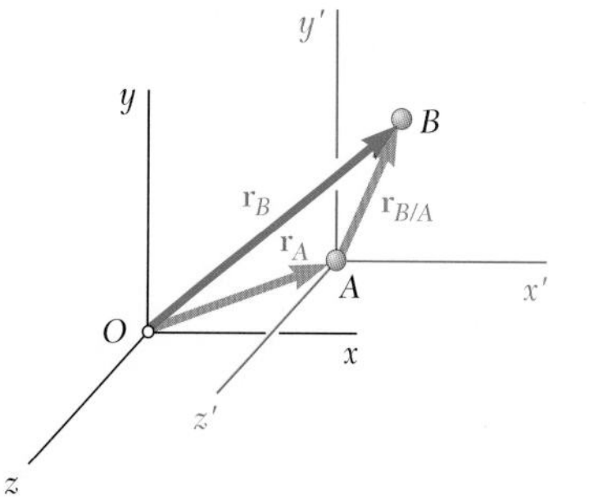

Fig. 11.30

It is sometimes convenient to resolve the velocity and acceleration of a particle P into components other than the rectangular x, y, and z components. For a particle P moving along a path contained in a plane, we attached to P unit vectors $\mathbf{e}_t$ tangent to the path and $\mathbf{e}_n$ normal to the path and directed toward the center of curvature of the path [Sec. 11.13]. We then expressed the velocity and acceleration of the particle in terms of tangential and normal components. We wrote

Tangential and normal components

$$\mathbf{v} = v\mathbf{e}_t \tag{11.36}$$

and

$$\mathbf{a} = \frac{dv}{dt}\mathbf{e}_t + \frac{v^2}{\rho}\mathbf{e}_n \tag{11.39}$$

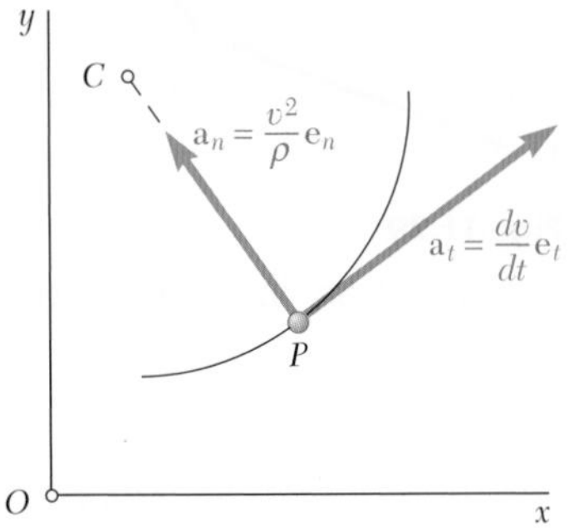

Fig. 11.31

where v is the speed of the particle and ρ the radius of curvature of its path [Sample Probs. 11.10 and 11.11]. We observed that while the velocity $\mathbf{v}$ is directed along the tangent to the path, the acceleration $\mathbf{a}$ consists of a component $\mathbf{a}_t$ directed along the tangent to the path and a component $\mathbf{a}_n$ directed toward the center of curvature of the path (Fig. 11.31).

Motion along a space curve

For a particle P moving along a space curve, we defined the plane which most closely fits the curve in the neighborhood of P as the *osculating plane*. This plane contains the unit vectors $\mathbf{e}_t$ and $\mathbf{e}_n$ which define, respectively, the tangent and principal normal to the curve. The unit vector $\mathbf{e}_b$ which is perpendicular to the osculating plane defines the *binormal*.

Radial and transverse components

When the position of a particle P moving in a plane is defined by its polar coordinates r and θ, it is convenient to use radial and transverse components directed, respectively, along the position vector $\mathbf{r}$ of the particle and in the direction obtained by rotating $\mathbf{r}$ through 90° counterclockwise [Sec. 11.14]. We attached to P unit vectors $\mathbf{e}_r$ and $\mathbf{e}_\theta$ directed, respectively, in the radial and transverse directions (Fig. 11.32). We then expressed the velocity and acceleration of the particle in terms of radial and transverse components

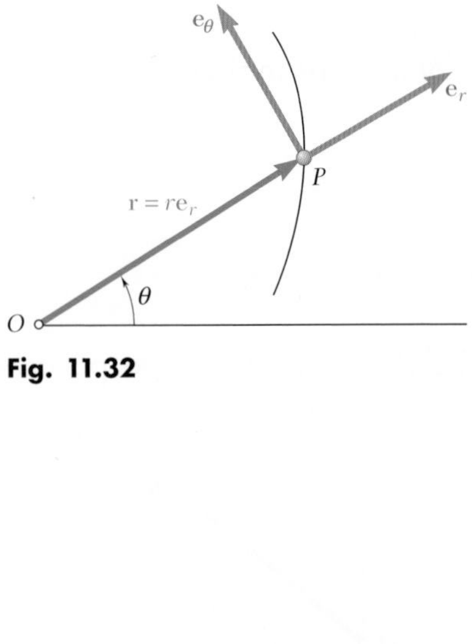

Fig. 11.32

$$\mathbf{v} = \dot{r}\mathbf{e}_r + r\dot{\theta}\mathbf{e}_\theta \tag{11.43}$$
$$\mathbf{a} = (\ddot{r} - r\dot{\theta}^2)\mathbf{e}_r + (r\ddot{\theta} + 2\dot{r}\dot{\theta})\mathbf{e}_\theta \tag{11.44}$$

where dots are used to indicate differentiation with respect to time. The scalar components of the velocity and acceleration in the radial and transverse directions are therefore

$$v_r = \dot{r} \qquad v_\theta = r\dot{\theta} \tag{11.45}$$
$$a_r = \ddot{r} - r\dot{\theta}^2 \qquad a_\theta = r\ddot{\theta} + 2\dot{r}\dot{\theta} \tag{11.46}$$

It is important to note that a_r is *not* equal to the time derivative of v_r, and that a_θ is *not* equal to the time derivative of v_θ [Sample Prob. 11.12].

The chapter ended with a discussion of the use of cylindrical coordinates to define the position and motion of a particle in space.

REVIEW PROBLEMS

11.182 The motion of a particle is defined by the relation $x = 2t^3 - 15t^2 + 24t + 4$, where x and t are expressed in meters and seconds, respectively. Determine (*a*) when the velocity is zero, (*b*) the position and the total distance traveled when the acceleration is zero.

11.183 A particle starting from rest at $x = 1$ m is accelerated so that its velocity doubles in magnitude between $x = 2$ m and $x = 8$ m. Knowing that the acceleration of the particle is defined by the relation $a = k[x - (A/x)]$, determine the values of the constants A and k if the particle has a velocity of 29 m/s when $x = 16$ m.

11.184 A particle moves in a straight line with the acceleration shown in the figure. Knowing that the particle starts from the origin with $v_0 = -2$ m/s, (*a*) construct the v–t and x–t curves for $0 < t < 18$ s, (*b*) determine the position and the velocity of the particle and the total distance traveled when $t = 18$ s.

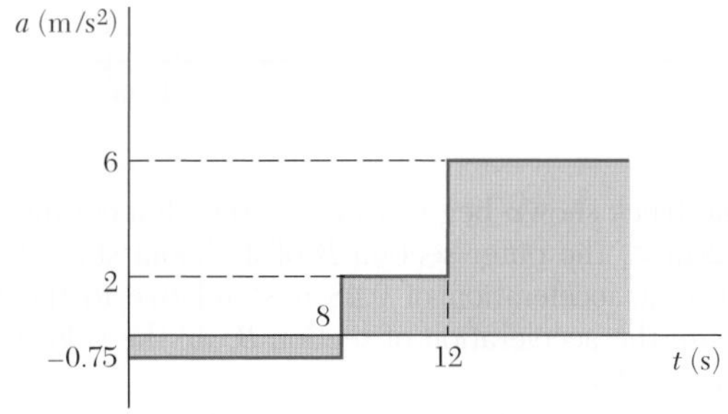

Fig. *P11.184*

11.185 The velocities of commuter trains A and B are as shown. Knowing that the speed of each train is constant and that B reaches the crossing 10 min after A passed through the same crossing, determine (*a*) the relative velocity of B with respect to A, (*b*) the distance between the fronts of the engines 3 min after A passed through the crossing.

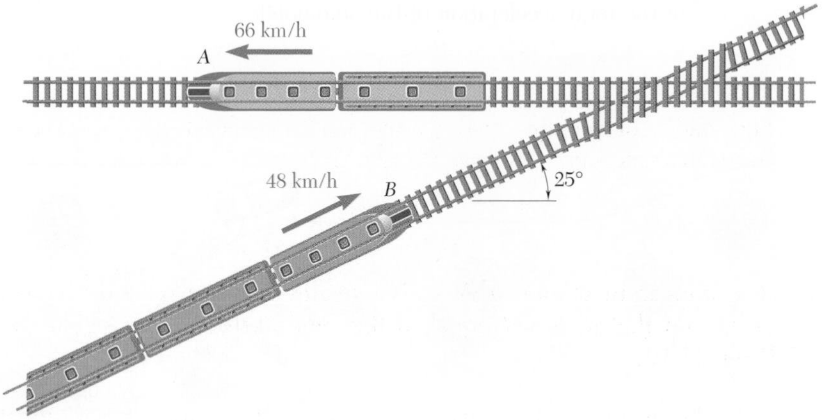

Fig. P11.185

11.186 Slider block B starts from rest and moves to the right with a constant acceleration of 300 mm/s^2. Determine (*a*) the relative acceleration of portion C of the cable with respect to slider block A, (*b*) the velocity of portion C of the cable after 2 s.

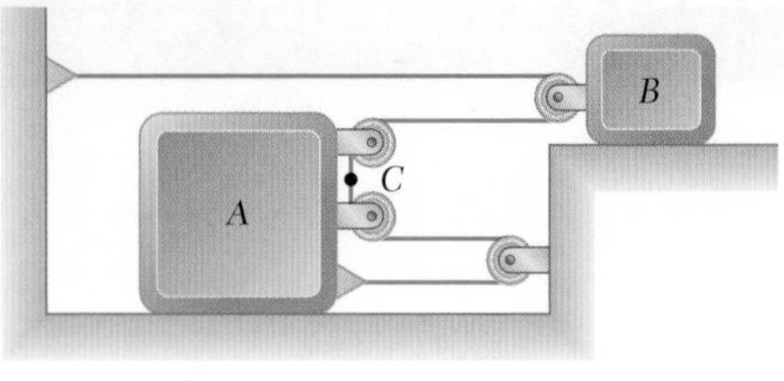

Fig. *P11.186*

11.187 Collar A starts from rest at $t = 0$ and moves downward with a constant acceleration of 175 mm/s^2. Collar B moves upward with a constant acceleration, and its initial velocity is 200 mm/s. Knowing that collar B moves through 500 mm between $t = 0$ and $t = 2$ s, determine (*a*) the accelerations of collar B and block C, (*b*) the time at which the velocity of block C is zero, (*c*) the distance through which block C will have moved at that time.

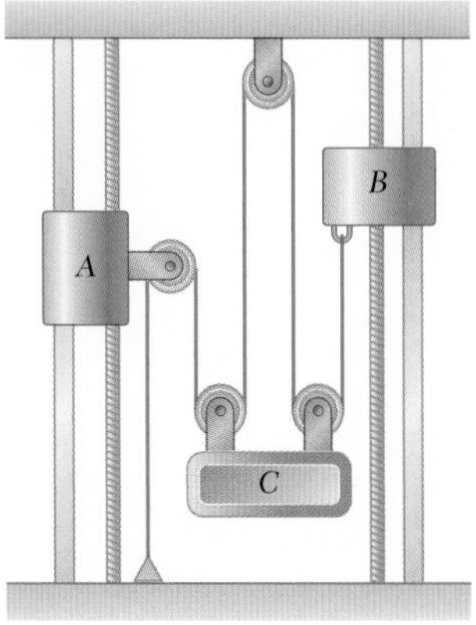

Fig. P11.187

11.188 A golfer hits a ball with an initial velocity of magnitude v_0 at an angle α with the horizontal. Knowing that the ball must clear the tops of two trees and land as close as possible to the flag, determine v_0 and the distance d when the golfer uses (*a*) a six-iron with $\alpha = 31°$, (*b*) a five-iron with $\alpha = 27°$.

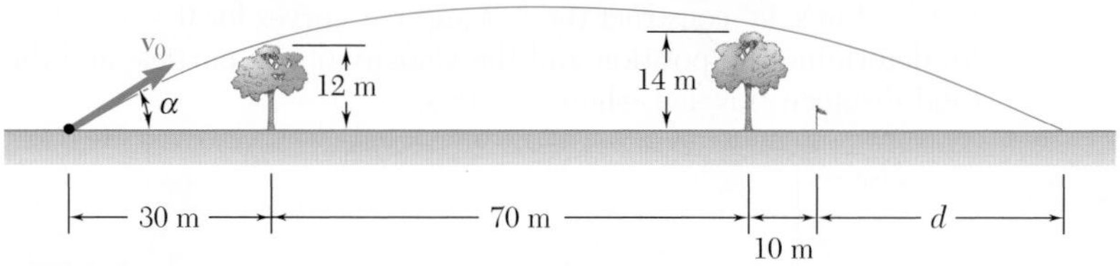

Fig. P11.188

11.189 As the truck shown begins to back up with a constant acceleration of 1.2 m/s^2, the outer section B of its boom starts to retract with a constant acceleration of 0.48 m/s^2 relative to the truck. Determine (*a*) the acceleration of section B, (*b*) the velocity of section B when $t = 2$ s.

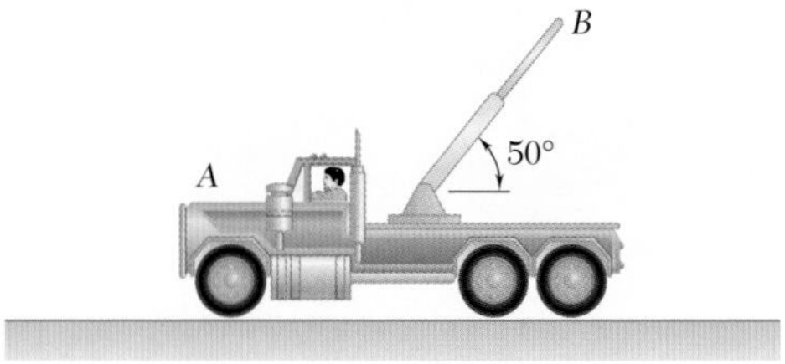

Fig. P11.189

11.190 A motorist traveling along a straight portion of a highway is decreasing the speed of his automobile at a constant rate before exiting from the highway onto a circular exit ramp with a radius of 170 m. He continues to decelerate at the same constant rate so that 10 s after entering the ramp, his speed has decreased to 30 km/h, a speed which he then maintains. Knowing that at this constant speed the total acceleration of the automobile is equal to one-quarter of its value prior to entering the ramp, determine the maximum value of the total acceleration of the automobile.

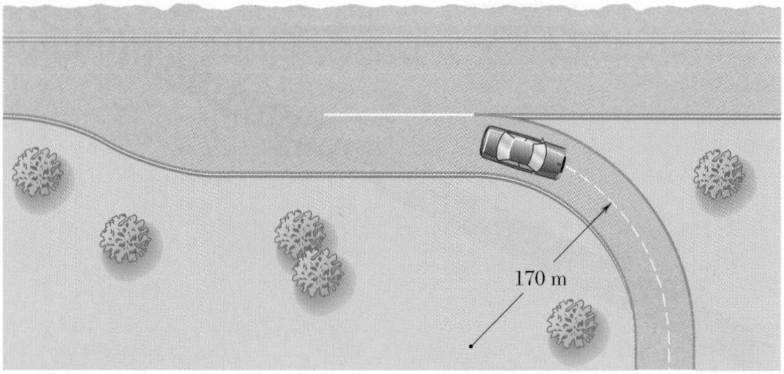

Fig. P11.190

11.191 Sand is discharged at A from a conveyor belt and falls onto the top of a stockpile at B. Knowing that the conveyor belt forms an angle $\alpha = 25°$ with the horizontal, determine (*a*) the speed v_0 of the belt, (*b*) the radius of curvature of the trajectory described by the sand at point B.

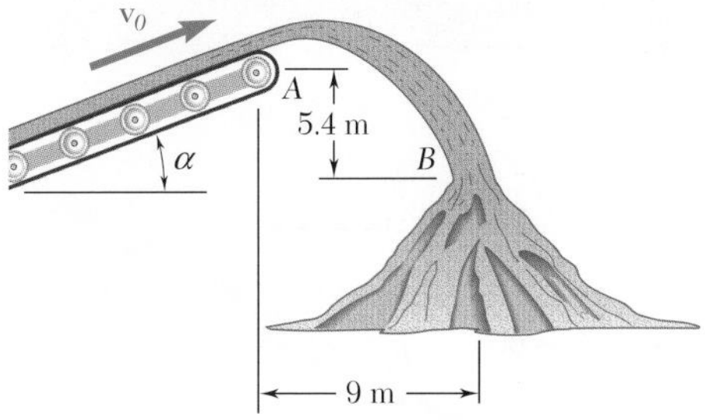

Fig. P11.191

11.192 The end point B of a boom is originally 5 m from fixed point A when the driver starts to retract the boom with a constant radial acceleration of $\ddot{r} = -1.0\ \text{m/s}^2$ and lower it with a constant angular acceleration $\ddot{\theta} = -0.5\ \text{rad/s}^2$. At $t = 2$ s, determine (*a*) the velocity of point B, (*b*) the acceleration of point B, (*c*) the radius of curvature of the path.

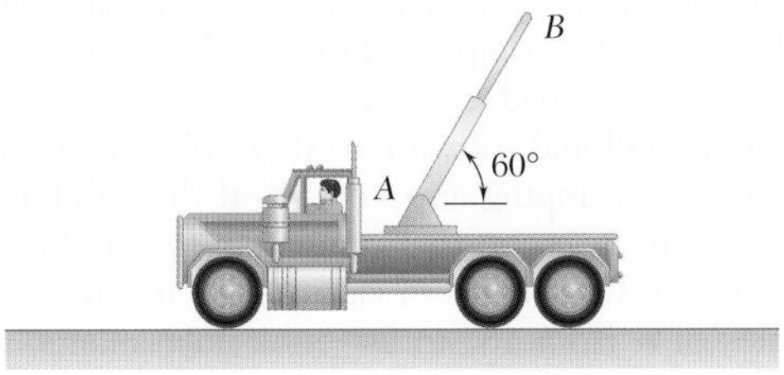

Fig. *P11.192*

11.193 A telemetry system is used to quantify kinematic values of a ski jumper immediately before she leaves the ramp. According to the system $r = 150$ m/s, $\dot{r} = -31.5$ m/s $\ddot{r} = -3\ \text{m/s}^2$, $\theta = 25°$, $\dot{\theta} = 0.07$ rad/s, $\ddot{\theta} = 0.06\ \text{rad/s}^2$. Determine (*a*) the velocity of the skier immediately before she leaves the jump, (*b*) the acceleration of the skier at this instant, (*c*) the distance of the jump d neglecting lift and air resistance.

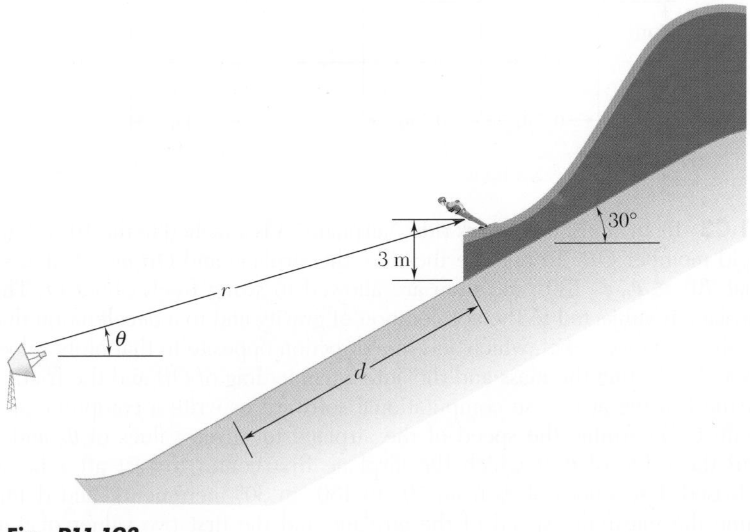

Fig. *P11.193*

COMPUTER PROBLEMS

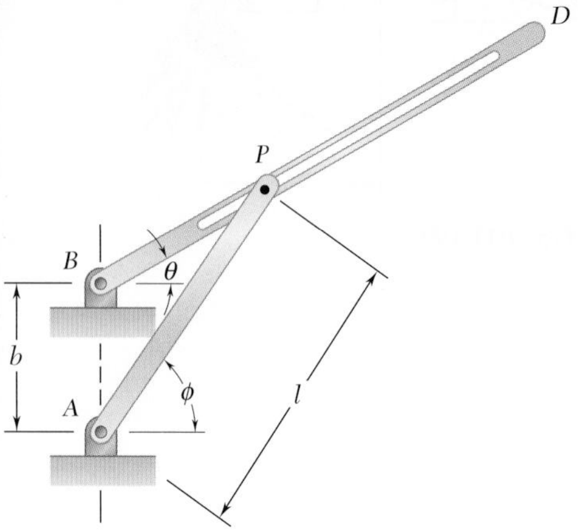

Fig. P11.C1

11.C1 The mechanism shown is known as a Whitworth quick-return mechanism. The input rod AP rotates at a constant rate ϕ, and the pin P is free to slide in the slot of the output rod BD. Plot θ versus ϕ and θ versus ϕ for one revolution of rod AP. Assume $\dot{\phi} = 1$ red/s, $l = 100$ mm, and (*a*) $b = 62.5$ mm, (*b*) $b = 75$ mm, (*c*) $b = 87.5$ mm.

11.C2 A ball is dropped with a velocity $\mathbf{v}_0$ at an angle α with the vertical onto the top step of a flight of stairs consisting of 8 steps. The ball rebounds and bounces down the steps as shown. Each time the ball bounces, at points $A, B, C, \ldots$, the horizontal component of its velocity remains constant and the magnitude of the vertical component of its velocity is reduced by k percent. Use computational software to determine (*a*) if the ball bounces down the steps without skipping any step, (*b*) if the ball bounces down the steps without bouncing twice on the same step, (*c*) the first step on which the ball bounces twice. Use values of v_0 from 1.8 m/s to 3.0 m/s in 0.6-m/s increments, values of α from 18° to 26° in 4° increments, and values of k equal to 40 and 50.

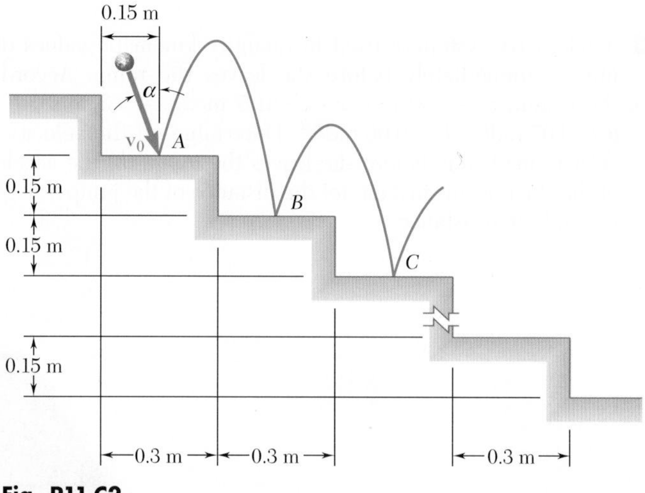

Fig. P11.C2

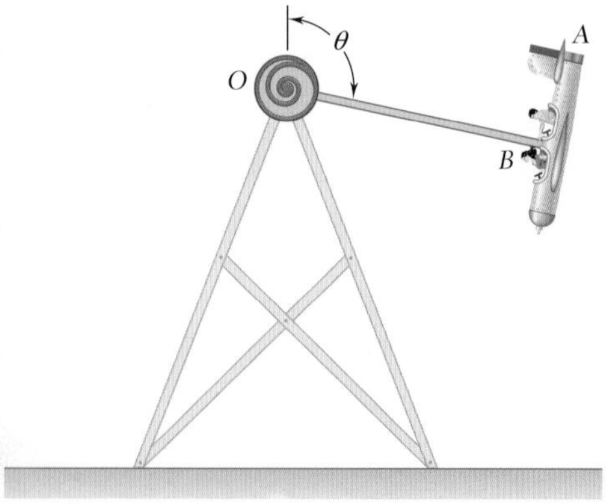

Fig. P11.C3

11.C3 In an amusement park ride, "airplane" A is attached to the 10-m-long rigid member OB. To operate the ride, the airplane and OB are rotated so that $70° \leq \theta_0 \leq 130°$ and then are allowed to swing freely about O. The airplane is subjected to the acceleration of gravity and to a deceleration due to air resistance, $-kv^2$, which acts in a direction opposite to that of its velocity $\mathbf{v}$. Neglecting the mass and the aerodynamic drag of OB and the friction in the bearing at O, use computational software or write a computer program to determine the speed of the airplane for given values of θ_0 and θ and the value of θ at which the airplane first comes to rest after being released. Use values of θ_0 from 70° to 130° in 30° increments, and determine the maximum speed of the airplane and the first two values of θ at which $v = 0$. For each value of θ_0, let (*a*) $k = 0$, (*b*) $k = 2 \times 10^{-4}$ m^{-1}, (*c*) $k = 4 \times 10^{-2}$ m^{-1}. (*Hint:* Express the tangential acceleration of the airplane in terms of g, k, and θ. Recall that $v_\theta = r\dot{\theta}$.)

11.C4 A motorist traveling on a highway at a speed of 90 km/h exits onto an ice-covered exit ramp. Wishing to stop, he applies his brakes until his automobile comes to rest. Knowing that the magnitude of the total acceleration of the automobile cannot exceed 3 m/s^2, use computational software to determine the minimum time required for the automobile to come to rest and the distance it travels on the exit ramp during that time if the exit ramp (*a*) is straight, (*b*) has a constant radius of curvature of 240 m. Solve each part assuming that the driver applies his brakes so that dv/dt, during each time interval, (1) remains constant, (2) varies linearly.

11.C5 An oscillating garden sprinkler discharges water with an initial velocity $\mathbf{v}_0$ of 10 m/s. (*a*) Knowing that the sides but not the top of arbor *BCDE* are open, use computational software to calculate the distance *d* to the point *F* that will be watered for values of α from 20° to 80°. (*b*) Determine the maximum value of *d* and the corresponding value of α.

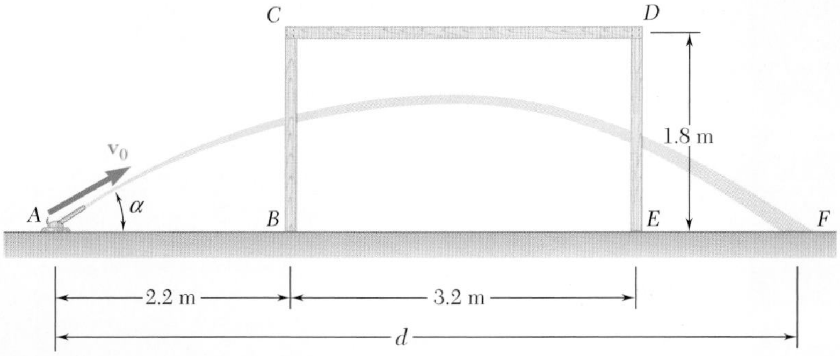

Fig. P11.C5

The forces experienced by the passengers on a roller coaster will depend on whether the roller-coaster car is traveling up a hill or down a hill, in a straight line, or along a horizontal or vertical curved path. The relation existing among force, mass, and acceleration will be studied in this chapter.

C H A P T E R 12

Kinetics of Particles: Newton's Second Law

Chapter 12 Kinetics of Particles: Newton's Second Law

12.1 INTRODUCTION

Newton's first and third laws of motion were used extensively in statics to study bodies at rest and the forces acting upon them. These two laws are also used in dynamics; in fact, they are sufficient for the study of the motion of bodies which have no acceleration. However, when bodies are accelerated, i.e., when the magnitude or the direction of their velocity changes, it is necessary to use Newton's second law of motion to relate the motion of the body with the forces acting on it.

In this chapter we will discuss Newton's second law and apply it to the analysis of the motion of particles. As we state in Sec. 12.2, if the resultant of the forces acting on a particle is not zero, the particle will have an acceleration proportional to the magnitude of the resultant and in the direction of this resultant force. Moreover, the ratio of the magnitudes of the resultant force and of the acceleration can be used to define the *mass* of the particle.

In Sec. 12.3, the *linear momentum* of a particle is defined as the product $\mathbf{L} = m\mathbf{v}$ of the mass m and velocity $\mathbf{v}$ of the particle, and it is demonstrated that Newton's second law can be expressed in an alternative form relating the rate of change of the linear momentum with the resultant of the forces acting on that particle.

Section 12.4 stresses the need for consistent units in the solution of dynamics problems and provides a review of the International System of Units (SI units) and the system of U.S. customary units.

In Secs. 12.5 and 12.6 and in the Sample Problems which follow, Newton's second law is applied to the solution of engineering problems, using either rectangular components or tangential and normal components of the forces and accelerations involved. We recall that an actual body—including bodies as large as a car, rocket, or airplane—can be considered as a particle for the purpose of analyzing its motion as long as the effect of a rotation of the body about its mass center can be ignored.

The second part of the chapter is devoted to the solution of problems in terms of radial and transverse components, with particular emphasis on the motion of a particle under a central force. In Sec. 12.7, the *angular momentum* $\mathbf{H}_O$ of a particle about a point O is defined as the moment about O of the linear momentum of the particle: $\mathbf{H}_O = \mathbf{r} \times m\mathbf{v}$. It then follows from Newton's second law that the rate of change of the angular momentum $\mathbf{H}_O$ of a particle is equal to the sum of the moments about O of the forces acting on that particle.

Section 12.9 deals with the motion of a particle under a *central force*, i.e., under a force directed toward or away from a fixed point O. Since such a force has zero moment about O, it follows that the angular momentum of the particle about O is conserved. This property greatly simplifies the analysis of the motion of a particle under a central force; in Sec. 12.10 it is applied to the solution of problems involving the orbital motion of bodies under gravitational attraction.

Sections 12.11 through 12.13 are optional. They present a more extensive discussion of orbital motion and contain a number of problems related to space mechanics.

12.2 NEWTON'S SECOND LAW OF MOTION

Newton's second law can be stated as follows:

If the resultant force acting on a particle is not zero, the particle will have an acceleration proportional to the magnitude of the resultant and in the direction of this resultant force.

Newton's second law of motion is best understood by imagining the following experiment: A particle is subjected to a force $\mathbf{F}_1$ of constant direction and constant magnitude F_1. Under the action of that force, the particle is observed to move in a straight line and *in the direction of the force* (Fig. 12.1*a*). By determining the position of the particle at various instants, we find that its acceleration has a constant magnitude a_1. If the experiment is repeated with forces $\mathbf{F}_2$, $\mathbf{F}_3$, . . . , of different magnitude or direction (Fig. 12.1*b* and *c*), we find each time that the particle moves in the direction of the force acting on it and that the magnitudes $a_1, a_2, a_3, \ldots$, of the accelerations are proportional to the magnitudes $F_1, F_2, F_3, \ldots$, of the corresponding forces:

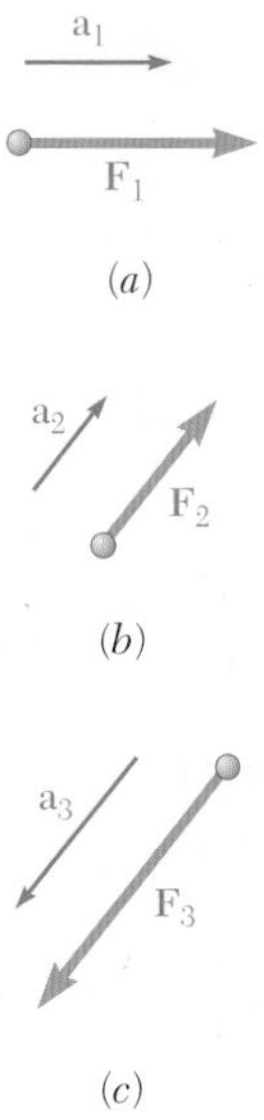

Fig. 12.1

$$\frac{F_1}{a_1} = \frac{F_2}{a_2} = \frac{F_3}{a_3} = \cdots = \text{constant}$$

The constant value obtained for the ratio of the magnitudes of the forces and accelerations is a characteristic of the particle under consideration; it is called the *mass* of the particle and is denoted by m. When a particle of mass m is acted upon by a force $\mathbf{F}$, the force $\mathbf{F}$ and the acceleration $\mathbf{a}$ of the particle must therefore satisfy the relation

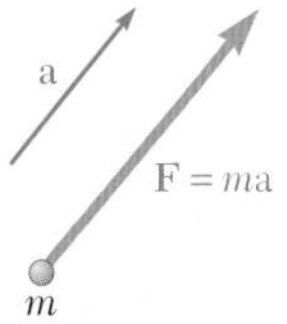

Fig. 12.2

$$\mathbf{F} = m\mathbf{a} \tag{12.1}$$

This relation provides a complete formulation of Newton's second law; it expresses not only that the magnitudes of $\mathbf{F}$ and $\mathbf{a}$ are proportional but also (since m is a positive scalar) that the vectors $\mathbf{F}$ and $\mathbf{a}$ have the same direction (Fig. 12.2). We should note that Eq. (12.1) still holds when $\mathbf{F}$ is not constant but varies with time in magnitude or direction. The magnitudes of $\mathbf{F}$ and $\mathbf{a}$ remain proportional, and the two vectors have the same direction at any given instant. However, they will not, in general, be tangent to the path of the particle.

When a particle is subjected simultaneously to several forces, Eq. (12.1) should be replaced by

$$\Sigma\mathbf{F} = m\mathbf{a} \tag{12.2}$$

where $\Sigma\mathbf{F}$ represents the sum, or resultant, of all the forces acting on the particle.

It should be noted that the system of axes with respect to which the acceleration $\mathbf{a}$ is determined is not arbitrary. These axes must have a constant orientation with respect to the stars, and their origin must either be attached to the sun† or move with a constant velocity

Photo 12.1 When the racecar accelerates forward the rear tires have a friction force acting on them in the direction the car is moving.

†More accurately, to the mass center of the solar system.

with respect to the sun. Such a system of axes is called a *newtonian frame of reference*.† A system of axes attached to the earth does *not* constitute a newtonian frame of reference, since the earth rotates with respect to the stars and is accelerated with respect to the sun. However, in most engineering applications, the acceleration **a** can be determined with respect to axes attached to the earth and Eqs. (12.1) and (12.2) used without any appreciable error. On the other hand, these equations do not hold if **a** represents a relative acceleration measured with respect to moving axes, such as axes attached to an accelerated car or to a rotating piece of machinery.

We observe that if the resultant $\Sigma\mathbf{F}$ of the forces acting on the particle is zero, it follows from Eq. (12.2) that the acceleration **a** of the particle is also zero. If the particle is initially at rest ($\mathbf{v}_0 = 0$) with respect to the newtonian frame of reference used, it will thus remain at rest ($\mathbf{v} = 0$). If originally moving with a velocity $\mathbf{v}_0$, the particle will maintain a constant velocity $\mathbf{v} = \mathbf{v}_0$; that is, it will move with the constant speed v_0 in a straight line. This, we recall, is the statement of Newton's first law (Sec. 2.10). Thus, Newton's first law is a particular case of Newton's second law and can be omitted from the fundamental principles of mechanics.

12.3 LINEAR MOMENTUM OF A PARTICLE. RATE OF CHANGE OF LINEAR MOMENTUM

Replacing the acceleration **a** by the derivative $d\mathbf{v}/dt$ in Eq. (12.2), we write

$$\Sigma\mathbf{F} = m\frac{d\mathbf{v}}{dt}$$

or, since the mass m of the particle is constant,

$$\Sigma\mathbf{F} = \frac{d}{dt}(m\mathbf{v}) \tag{12.3}$$

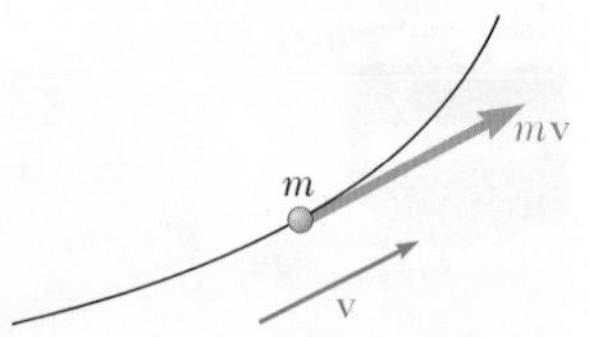

Fig. 12.3

The vector $m\mathbf{v}$ is called the *linear momentum*, or simply the *momentum*, of the particle. It has the same direction as the velocity of the particle, and its magnitude is equal to the product of the mass m and the speed v of the particle (Fig. 12.3). Equation (12.3) expresses that *the resultant of the forces acting on the particle is equal to the rate of change of the linear momentum of the particle.* It is in this form that the second law of motion was originally stated by Newton. Denoting by **L** the linear momentum of the particle,

$$\mathbf{L} = m\mathbf{v} \tag{12.4}$$

and by $\dot{\mathbf{L}}$ its derivative with respect to t, we can write Eq. (12.3) in the alternative form

$$\Sigma\mathbf{F} = \dot{\mathbf{L}} \tag{12.5}$$

†Since stars are not actually fixed, a more rigorous definition of a newtonian frame of reference (also called an *inertial system*) is one *with respect to which Eq. (12.2) holds.*

It should be noted that the mass m of the particle is assumed to be constant in Eqs. (12.3) to (12.5). Equation (12.3) or (12.5) should therefore not be used to solve problems involving the motion of bodies, such as rockets, which gain or lose mass. Problems of that type will be considered in Sec. 14.12.†

It follows from Eq. (12.3) that the rate of change of the linear momentum $m\mathbf{v}$ is zero when $\Sigma\mathbf{F} = 0$. Thus, *if the resultant force acting on a particle is zero, the linear momentum of the particle remains constant, in both magnitude and direction.* This is the principle of *conservation of linear momentum* for a particle, which can be recognized as an alternative statement of Newton's first law (Sec. 2.10).

12.4 SYSTEMS OF UNITS

In using the fundamental equation $\mathbf{F} = m\mathbf{a}$, the units of force, mass, length, and time cannot be chosen arbitrarily. If they are, the magnitude of the force $\mathbf{F}$ required to give an acceleration $\mathbf{a}$ to the mass m will *not* be numerically equal to the product ma; it will be only proportional to this product. Thus, we can choose three of the four units arbitrarily but must choose the fourth unit so that the equation $\mathbf{F} = m\mathbf{a}$ is satisfied. The units are then said to form a system of consistent kinetic units.

Two systems of consistent kinetic units are currently used by engineers, the International System of Units (SI units‡) and the system of U.S. customary units. Both systems were discussed in detail in Sec. 1.3 and are described only briefly in this section.

International System of Units (SI Units). In this system, the base units are the units of length, mass, and time, and are called, respectively, the *meter* (m), the *kilogram* (kg), and the *second* (s). All three are arbitrarily defined (Sec. 1.3). The unit of force is a derived unit. It is called the *newton* (N) and is defined as the force which gives an acceleration of 1 m/s² to a mass of 1 kg (Fig. 12.4). From Eq. (12.1) we write

Fig. 12.4

$$1 \text{ N} = (1 \text{ kg})(1 \text{ m/s}^2) = 1 \text{ kg} \cdot \text{m/s}^2$$

The SI units are said to form an *absolute* system of units. This means that the three base units chosen are independent of the location where measurements are made. The meter, the kilogram, and the second may be used anywhere on the earth; they may even be used on another planet. They will always have the same significance.

The *weight* $\mathbf{W}$ of a body, or *force of gravity* exerted on that body, should, like any other force, be expressed in newtons. Since a body subjected to its own weight acquires an acceleration equal to the acceleration of gravity g, it follows from Newton's second law that the magnitude W of the weight of a body of mass m is

$$W = mg \tag{12.6}$$

†On the other hand, Eqs. (12.3) and (12.5) do hold in *relativistic mechanics*, where the mass m of the particle is assumed to vary with the speed of the particle.

‡SI stands for *Système International d'Unités* (French).

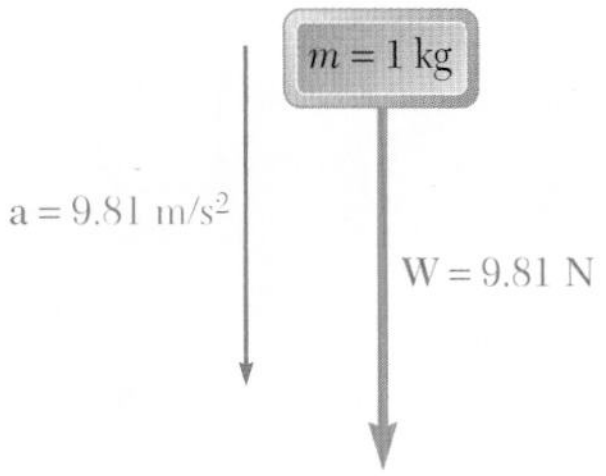

Fig. 12.5

Recalling that $g = 9.81$ m/s^2, we find that the weight of a body of mass 1 kg (Fig. 12.5) is

$$W = (1 \text{ kg})(9.81 \text{ m/s}^2) = 9.81 \text{ N}$$

Multiples and submultiples of the units of length, mass, and force are frequently used in engineering practice. They are, respectively, the *kilometer* (km) and the *millimeter* (mm); the *megagram*† (Mg) and the *gram* (g); and the *kilonewton* (kN). By definition,

$$1 \text{ km} = 1000 \text{ m} \qquad 1 \text{ mm} = 0.001 \text{ m}$$
$$1 \text{ Mg} = 1000 \text{ kg} \qquad 1 \text{ g} = 0.001 \text{ kg}$$
$$1 \text{ kN} = 1000 \text{ N}$$

The conversion of these units to meters, kilograms, and newtons, respectively, can be effected simply by moving the decimal point three places to the right or to the left.

Units other than the units of mass, length, and time can all be expressed in terms of these three base units. For example, the unit of linear momentum can be obtained by recalling the definition of linear momentum and writing

$$mv = (\text{kg})(\text{m/s}) = \text{kg} \cdot \text{m/s}$$

U.S. Customary Units. Most practicing American engineers still commonly use a system in which the base units are the units of length, force, and time. These units are, respectively, the *foot* (ft), the *pound* (lb), and the *second* (s). The second is the same as the corresponding SI unit. The foot is defined as 0.3048 m. The pound is defined as the *weight* of a platinum standard, called the *standard pound*, which is kept at the National Institute of Standards and Technology outside Washington and the mass of which is 0.453 592 43 kg. Since the weight of a body depends upon the gravitational attraction of the earth, which varies with location, it is specified that the standard pound should be placed at sea level and at a latitude of 45° to properly define a force of 1 lb. Clearly, the U.S. customary units do not form an absolute system of units. Because of their dependence upon the gravitational attraction of the earth, they are said to form a *gravitational* system of units.

While the standard pound also serves as the unit of mass in commercial transactions in the United States, it cannot be so used in engineering computations since such a unit would not be consistent with the base units defined in the preceding paragraph. Indeed, when acted upon by a force of 1 lb, that is, when subjected to its own weight, the standard pound receives the acceleration of gravity, $g = 32.2$ ft/s^2 (Fig. 12.6), and not the unit acceleration required by Eq. (12.1). The unit of mass consistent with the foot, the pound, and the second is the mass which receives an acceleration of 1 ft/s^2 when a force of 1 lb is applied to it (Fig. 12.7). This unit, sometimes called a *slug*, can be derived from the equation $F = ma$ after substituting 1 lb and 1 ft/s^2 for F and a, respectively. We write

$$F = ma \qquad 1 \text{ lb} = (1 \text{ slug})(1 \text{ ft/s}^2)$$

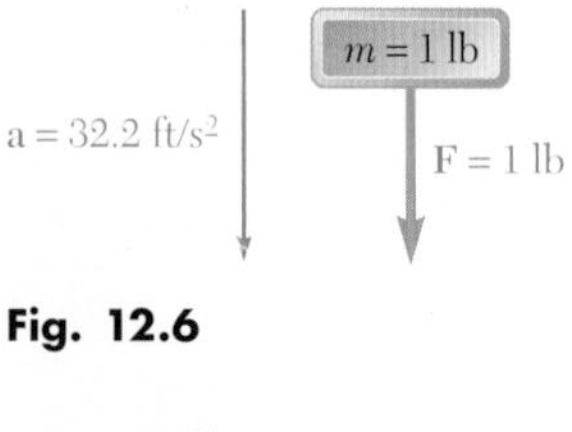

Fig. 12.6

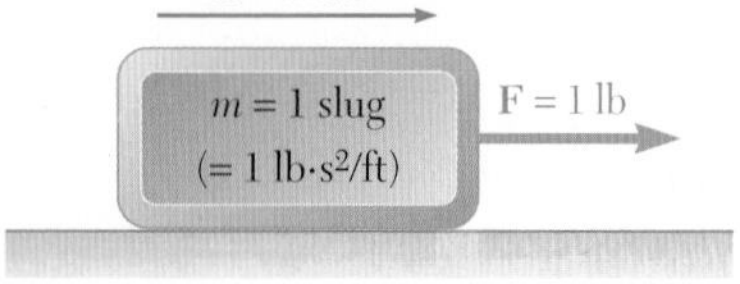

Fig. 12.7

†Also known as a *metric ton*.

and obtain

$$1 \text{ slug} = \frac{1 \text{ lb}}{1 \text{ ft/s}^2} = 1 \text{ lb} \cdot \text{s}^2\text{/ft}$$

Comparing Figs. 12.6 and 12.7, we conclude that the slug is a mass 32.2 times larger than the mass of the standard pound.

The fact that bodies are characterized in the U.S. customary system of units by their weight in pounds rather than by their mass in slugs was a convenience in the study of statics, where we were dealing for the most part with weights and other forces and only seldom with masses. However, in the study of kinetics, which involves forces, masses, and accelerations, it will be repeatedly necessary to express in slugs the mass m of a body, the weight W of which has been given in pounds. Recalling Eq. (12.6), we will write

$$m = \frac{W}{g} \tag{12.7}$$

where g is the acceleration of gravity ($g = 32.2 \text{ ft/s}^2$).

Units other than the units of force, length, and time can all be expressed in terms of these three base units. For example, the unit of linear momentum can be obtained by using the definition of linear momentum to write

$$mv = (\text{lb} \cdot \text{s}^2\text{/ft})(\text{ft/s}) = \text{lb} \cdot \text{s}$$

Conversion from One System of Units to Another. The conversion from U.S. customary units to SI units, and vice versa, was discussed in Sec. 1.4. You will recall that the conversion factors obtained for the units of length, force, and mass are, respectively,

Length:	$1 \text{ ft} = 0.3048 \text{ m}$
Force:	$1 \text{ lb} = 4.448 \text{ N}$
Mass:	$1 \text{ slug} = 1 \text{ lb} \cdot \text{s}^2\text{/ft} = 14.59 \text{ kg}$

Although it cannot be used as a consistent unit of mass, the mass of the standard pound is, by definition,

$$1 \text{ pound-mass} = 0.4536 \text{ kg}$$

This constant can be used to determine the *mass* in SI units (kilograms) of a body which has been characterized by its *weight* in U.S. customary units (pounds).

12.5 EQUATIONS OF MOTION

Consider a particle of mass m acted upon by several forces. We recall from Sec. 12.2 that Newton's second law can be expressed by the equation

$$\Sigma\mathbf{F} = m\mathbf{a} \tag{12.2}$$

which relates the forces acting on the particle and the vector $m\mathbf{a}$ (Fig. 12.8). In order to solve problems involving the motion of a particle, however, it will be found more convenient to replace Eq. (12.2) by equivalent equations involving scalar quantities.

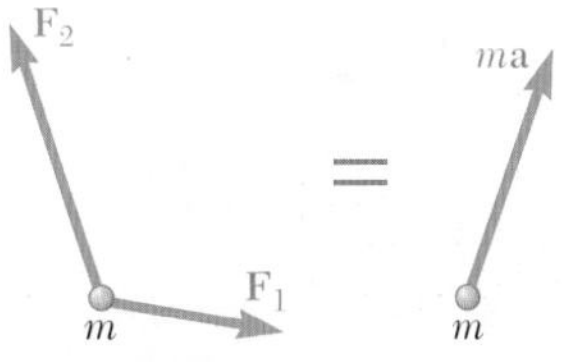

Fig. 12.8

Photo 12.2 The pilot of a fighter aircraft will experience very large normal forces when taking a sharp turn.

Rectangular Components. Resolving each force **F** and the acceleration **a** into rectangular components, we write

$$\Sigma(F_x\mathbf{i} + F_y\mathbf{j} + F_z\mathbf{k}) = m(a_x\mathbf{i} + a_y\mathbf{j} + a_z\mathbf{k})$$

from which it follows that

$$\Sigma F_x = ma_x \qquad \Sigma F_y = ma_y \qquad \Sigma F_z = ma_z \tag{12.8}$$

Recalling from Sec. 11.11 that the components of the acceleration are equal to the second derivatives of the coordinates of the particle, we have

$$\Sigma F_x = m\ddot{x} \qquad \Sigma F_y = m\ddot{y} \qquad \Sigma F_z = m\ddot{z} \tag{12.8$'$}$$

Consider, as an example, the motion of a projectile. If the resistance of the air is neglected, the only force acting on the projectile after it has been fired is its weight $\mathbf{W} = -W\mathbf{j}$. The equations defining the motion of the projectile are therefore

$$m\ddot{x} = 0 \qquad m\ddot{y} = -W \qquad m\ddot{z} = 0$$

and the components of the acceleration of the projectile are

$$\ddot{x} = 0 \qquad \ddot{y} = -\frac{W}{m} = -g \qquad \ddot{z} = 0$$

where g is 9.81 m/s^2 or 32.2 ft/s^2. The equations obtained can be integrated independently, as shown in Sec. 11.11, to obtain the velocity and displacement of the projectile at any instant.

When a problem involves two or more bodies, equations of motion should be written for each of the bodies (see Sample Probs. 12.3 and 12.4). You will recall from Sec. 12.2 that all accelerations should be measured with respect to a newtonian frame of reference. In most engineering applications, accelerations can be determined with respect to axes attached to the earth, but relative accelerations measured with respect to moving axes, such as axes attached to an accelerated body, cannot be substituted for **a** in the equations of motion.

Tangential and Normal Components. Resolving the forces and the acceleration of the particle into components along the tangent to the path (in the direction of motion) and the normal (toward the inside of the path) (Fig. 12.9), and substituting into Eq. (12.2), we obtain the two scalar equations

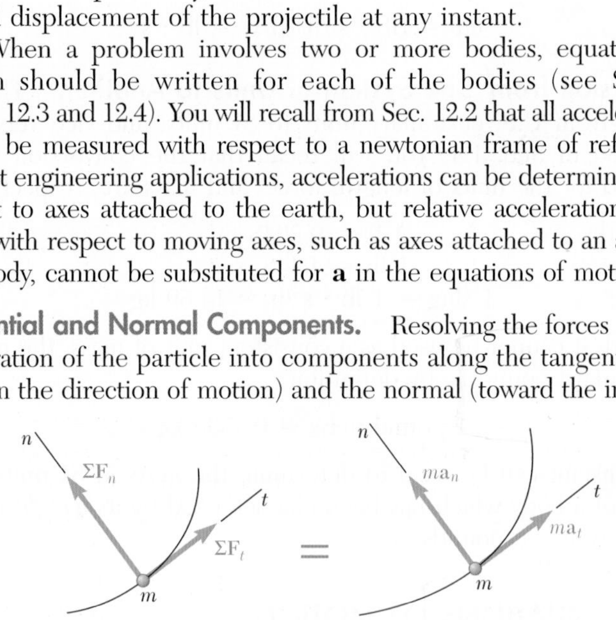

Fig. 12.9

$$\Sigma F_t = ma_t \qquad \Sigma F_n = ma_n \tag{12.9}$$

Substituting for a_t and a_n from Eqs. (11.40), we have

$$\Sigma F_t = m\frac{dv}{dt} \qquad \Sigma F_n = m\frac{v^2}{\rho} \tag{12.9$'$}$$

The equations obtained may be solved for two unknowns.

12.6 DYNAMIC EQUILIBRIUM

Returning to Eq. (12.2) and transposing the right-hand member, we write Newton's second law in the alternative form

$$\Sigma \mathbf{F} - m\mathbf{a} = 0 \tag{12.10}$$

which expresses that if we add the vector $-m\mathbf{a}$ to the forces acting on the particle, *we obtain a system of vectors equivalent to zero* (Fig. 12.10). The vector $-m\mathbf{a}$, of magnitude ma and of *direction opposite* to that of the acceleration, is called an *inertia vector*. The particle may thus be considered to be in equilibrium under the given forces and the inertia vector. The particle is said to be in *dynamic equilibrium*, and the problem under consideration can be solved by the methods developed earlier in statics.

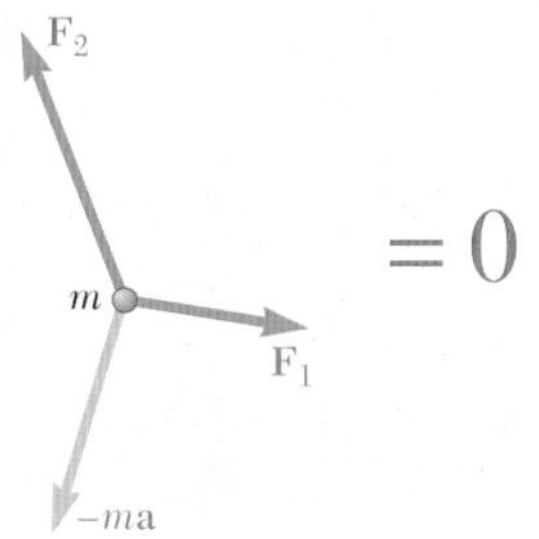

Fig. 12.10

In the case of coplanar forces, all the vectors shown in Fig. 12.10, *including the inertia vector*, can be drawn tip-to-tail to form a closed-vector polygon. Or the sums of the components of all the vectors in Fig. 12.10, again including the inertia vector, can be equated to zero. Using rectangular components, we therefore write

$$\Sigma F_x = 0 \qquad \Sigma F_y = 0 \qquad \textbf{\textit{including inertia vector}} \tag{12.11}$$

When tangential and normal components are used, it is more convenient to represent the inertia vector by its two components $-m\mathbf{a}_t$ and $-m\mathbf{a}_n$ in the sketch itself (Fig. 12.11). The tangential component of the inertia vector provides a measure of the resistance the particle offers to a change in speed, while its normal component (also called *centrifugal force*) represents the tendency of the particle to leave its curved path. We should note that either of these two components may be zero under special conditions: (1) If the particle starts from rest, its initial velocity is zero and the normal component of the inertia vector is zero at $t = 0$; (2) if the particle moves at constant speed along its path, the tangential component of the inertia vector is zero and only its normal component needs to be considered.

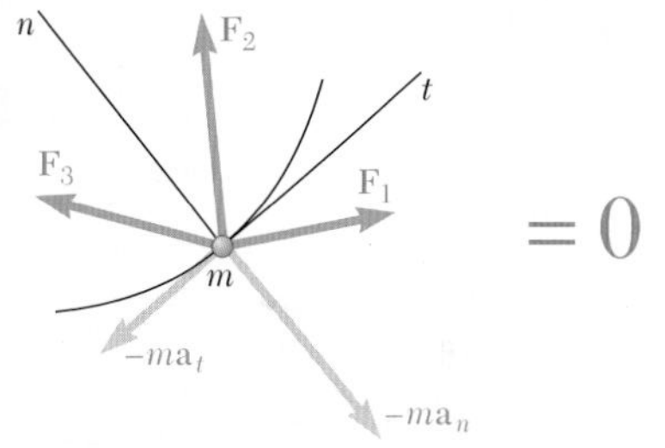

Fig. 12.11

Because they measure the resistance that particles offer when we try to set them in motion or when we try to change the conditions of their motion, inertia vectors are often called *inertia forces*. The inertia forces, however, are not forces like the forces found in statics, which are either contact forces or gravitational forces (weights). Many people, therefore, object to the use of the word "force" when referring to the vector $-m\mathbf{a}$ or even avoid altogether the concept of dynamic equilibrium. Others point out that inertia forces and actual forces, such as gravitational forces, affect our senses in the same way and cannot be distinguished by physical measurements. A man riding in an elevator which is accelerated upward will have the feeling that his weight has suddenly increased; and no measurement made within the elevator could establish whether the elevator is truly accelerated or whether the force of attraction exerted by the earth has suddenly increased.

Sample problems have been solved in this text by the direct application of Newton's second law, as illustrated in Figs. 12.8 and 12.9, rather than by the method of dynamic equilibrium.

Photo 12.3 The angle each rider is with respect to the horizontal will depend on the weight of the rider and the speed of rotation.

SAMPLE PROBLEM 12.1

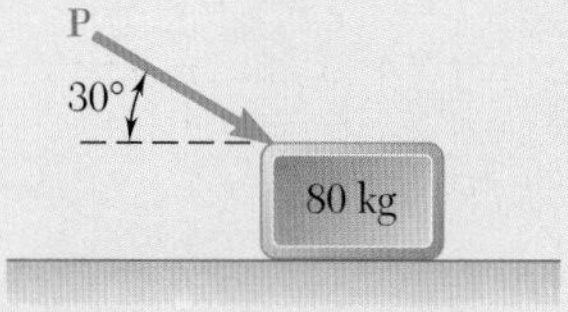

An 80-kg block rests on a horizontal plane. Find the magnitude of the force **P** required to give the block an acceleration of 2.5 m/s^2 to the right. The coefficient of kinetic friction between the block and the plane is $\mu_k = 0.25$.

SOLUTION

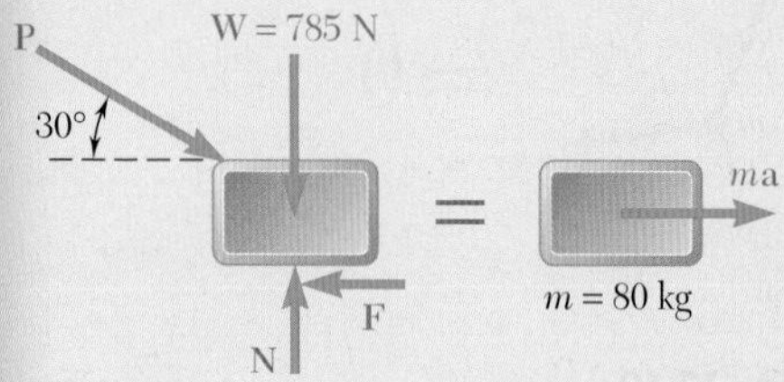

The weight of the block is

$$W = mg = (80 \text{ kg})(9.81 \text{ m/s}^2) = 785 \text{ N}$$

We note that $F = \mu_k N = 0.25N$ and that $a = 2.5$ m/s^2. Expressing that the forces acting on the block are equivalent to the vector $m\mathbf{a}$, we write

$$\overset{+}{\rightarrow}\Sigma F_x = ma: \quad P \cos 30° - 0.25N = (80 \text{ kg})(2.5 \text{ m/s}^2)$$

$$P \cos 30° - 0.25N = 200 \text{ N} \qquad (1)$$

$$+\uparrow \Sigma F_y = 0: \quad N - P \sin 30° - 785 \text{ N} = 0 \qquad (2)$$

Solving (2) for N and substituting the result into (1), we obtain

$$N = P \sin 30° + 785 \text{ N}$$

$$P \cos 30° - 0.25(P \sin 30° + 785 \text{ N}) = 200 \text{ N} \qquad P = 535 \text{ N} \blacktriangleleft$$

SAMPLE PROBLEM 12.2

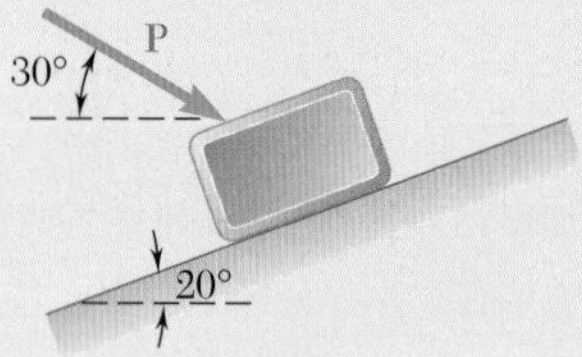

A 20-kg package is at rest on an incline when a force **P** is applied to it. Determine the magnitude of **P** if 10 s is required for the package to travel 5 m up the incline. The static and kinetic coefficients of friction between the package and the incline are 0.4 and 0.3 respectively.

SOLUTION

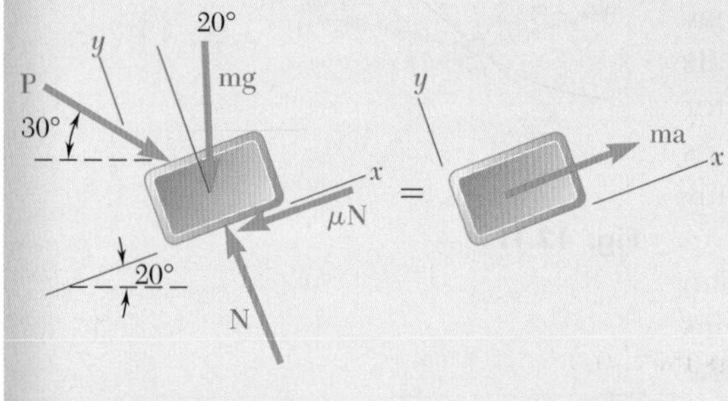

Kinematics: Uniformly accelerated motion. $(x_0 = 0,\ v_0 = 0)$

$$x = x_0 + v_0 t + \frac{1}{2}at^2,$$

or

$$a = \frac{2x}{t^2} = \frac{(2)(5)}{(10)^2} = 0.100 \text{ m/s}^2$$

$$+\nwarrow \Sigma F_y = 0: \quad N - P \sin 50° - mg \cos 20° = 0$$

$$N = P \sin 50° + mg \cos 20°$$

$$+\nearrow \Sigma F_x = ma: \quad P \cos 50° - mg \sin 20° - \mu N = ma$$

or

$$P \cos 50° - mg \sin 20° - \mu(P \sin 50° + mg \cos 20°) = ma$$

$$P = \frac{ma + mg(\sin 20° + \mu \cos 20°)}{\cos 50° - \mu \sin 50°}$$

For motion impending, set $a = 0$ and $\mu = \mu_s = 0.4$

$$P = \frac{(20)(0) + (20)(9.81)(\sin 20° + 0.4 \cos 20°)}{\cos 50° - 0.4 \sin 50°}$$

$$= 419 \text{ N} \blacktriangleleft$$

For motion with $a = 0.100$ m/s^2, use $\mu = \mu_k = 0.3$.

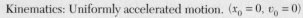

$$P = \frac{(20)(0.100) + (20)(9.81)(\sin 20° + 0.3 \cos 20°)}{\cos 50° - 0.3 \sin 50°} \qquad P = 301 \text{ N} \blacktriangleleft$$

SAMPLE PROBLEM 12.3

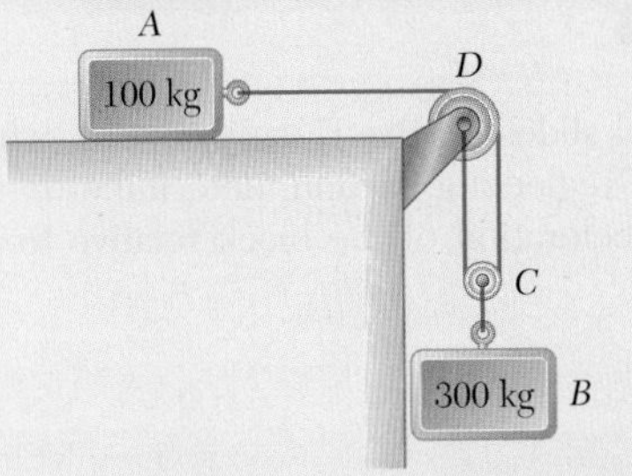

The two blocks shown start from rest. The horizontal plane and the pulley are frictionless, and the pulley is assumed to be of negligible mass. Determine the acceleration of each block and the tension in each cord.

SOLUTION

Kinematics. We note that if block A moves through x_A to the right, block B moves down through

$$x_B = \tfrac{1}{2}x_A$$

Differentiating twice with respect to t, we have

$$a_B = \tfrac{1}{2}a_A \qquad (1)$$

Kinetics. We apply Newton's second law successively to block A, block B, and pulley C.

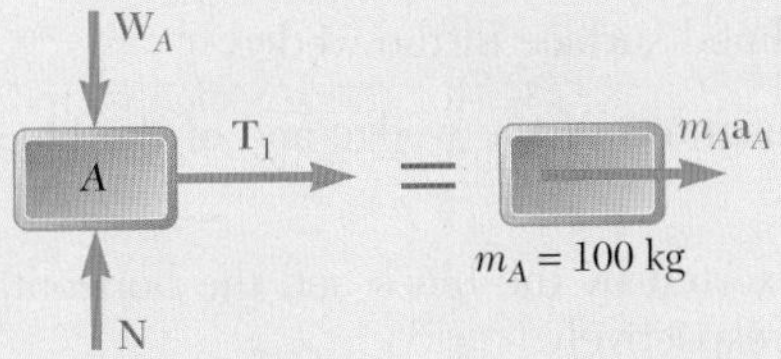

Block A. Denoting by T_1 the tension in cord ACD, we write

$$\xrightarrow{+}\Sigma F_x = m_A a_A: \qquad T_1 = 100a_A \qquad (2)$$

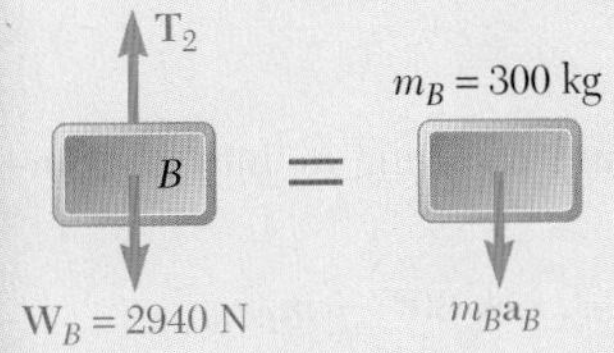

Block B. Observing that the weight of block B is

$$W_B = m_B g = (300 \text{ kg})(9.81 \text{ m/s}^2) = 2940 \text{ N}$$

and denoting by T_2 the tension in cord BC, we write

$$+\downarrow\Sigma F_y = m_B a_B: \qquad 2940 - T_2 = 300a_B$$

or, substituting for a_B from (1),

$$2940 - T_2 = 300(\tfrac{1}{2}a_A)$$
$$T_2 = 2940 - 150a_A \qquad (3)$$

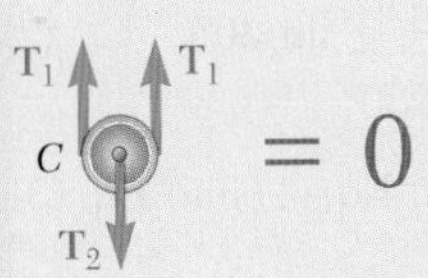

Pulley C. Since m_C is assumed to be zero, we have

$$+\downarrow\Sigma F_y = m_C a_C = 0: \qquad T_2 - 2T_1 = 0 \qquad (4)$$

Substituting for T_1 and T_2 from (2) and (3), respectively, into (4) we write

$$2940 - 150a_A - 2(100a_A) = 0$$
$$2940 - 350a_A = 0 \qquad a_A = 8.40 \text{ m/s}^2 \blacktriangleleft$$

Substituting the value obtained for a_A into (1) and (2), we have

$$a_B = \tfrac{1}{2}a_A = \tfrac{1}{2}(8.40 \text{ m/s}^2) \qquad a_B = 4.20 \text{ m/s}^2 \blacktriangleleft$$
$$T_1 = 100a_A = (100 \text{ kg})(8.40 \text{ m/s}^2) \qquad T_1 = 840 \text{ N} \blacktriangleleft$$

Recalling (4), we write

$$T_2 = 2T_1 \qquad T_2 = 2(840 \text{ N}) \qquad T_2 = 1680 \text{ N} \blacktriangleleft$$

We note that the value obtained for T_2 is *not* equal to the weight of block B.

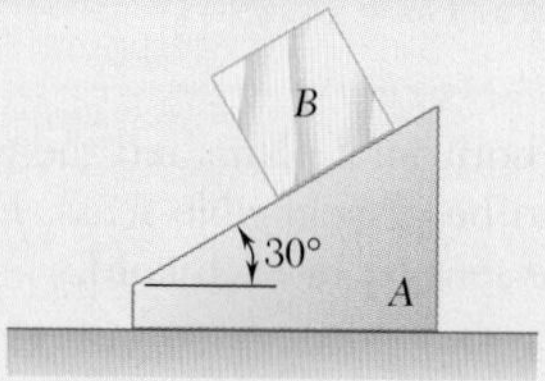

SAMPLE PROBLEM 12.4

The 6-kg block B starts from rest and slides on the 15-kg wedge A, which is supported by a horizontal surface. Neglecting friction, determine (a) the acceleration of the wedge, (b) the acceleration of the block relative to the wedge.

SOLUTION

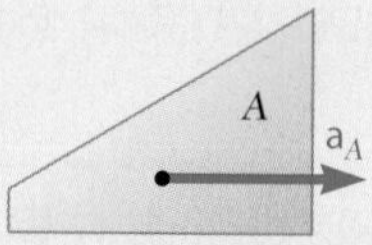

Kinematics. We first examine the acceleration of the wedge and the acceleration of the block.

Wedge A. Since the wedge is constrained to move on the horizontal surface, its acceleration $\mathbf{a}_A$ is horizontal. We will assume that it is directed to the right.

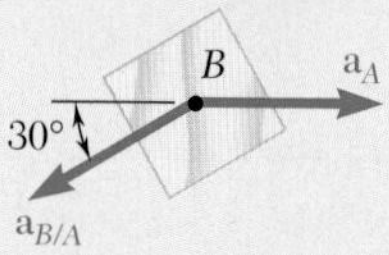

Block B. The acceleration $\mathbf{a}_B$ of block B can be expressed as the sum of the acceleration of A and the acceleration of B relative to A. We have

$$\mathbf{a}_B = \mathbf{a}_A + \mathbf{a}_{B/A}$$

where $\mathbf{a}_{B/A}$ is directed along the inclined surface of the wedge.

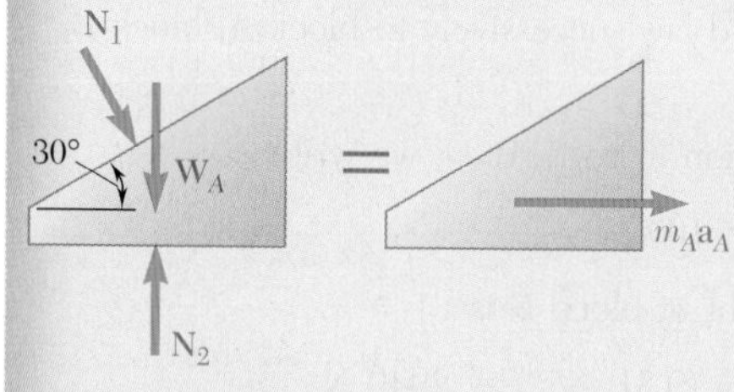

Kinetics. We draw the free-body diagrams of the wedge and of the block and apply Newton's second law.

Wedge A. We denote the forces exerted by the block and the horizontal surface on wedge A by $\mathbf{N}_1$ and $\mathbf{N}_2$, respectively.

$$\overset{+}{\rightarrow}\Sigma F_x = m_A a_A: \qquad N_1 \sin 30° = m_A a_A$$

$$0.5N_1 = m_A a_A \qquad (1)$$

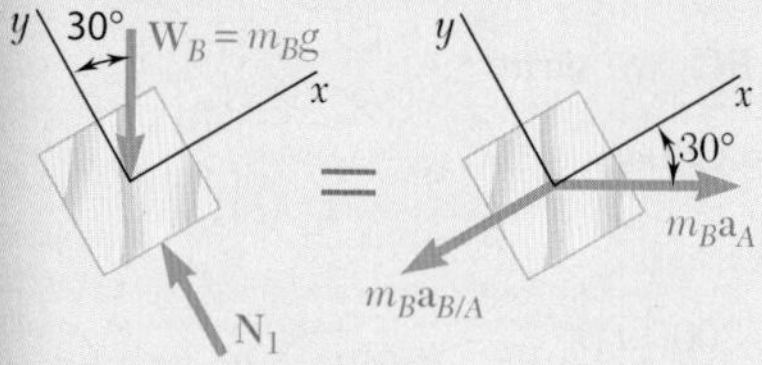

Block B. Using the coordinate axes shown and resolving $\mathbf{a}_B$ into its components $\mathbf{a}_A$ and $\mathbf{a}_{B/A}$, we write

$$+\nearrow\Sigma F_x = m_B a_x: \qquad -m_B g \sin 30° = m_B a_A \cos 30° - m_B a_{B/A}$$

$$-m_B g \sin 30° = m_B(a_A \cos 30° - a_{B/A})$$

$$a_{B/A} = a_A \cos 30° + g \sin 30° \qquad (2)$$

$$+\nwarrow\Sigma F_y = m_B a_y: \qquad N_1 - m_B g \cos 30° = -m_B a_A \sin 30°$$

a. Acceleration of Wedge A. Substituting for N_1 from Eq. (1) into Eq. (3), and noting that $W_A = m_A g$,

$$2m_A a_A - m_B g \cos 30° = -\, m_B a_A \sin 30°$$

Solving for a_A and substituting the numerical data, we write

$$a_A = \frac{m_B g \cos 30°}{2m_A + m_B \sin 30°} = \frac{(6\text{ kg})\ (9.81\text{ m/s}^2)\cos 30°}{2(15\text{ kg}) + (6\text{ kg})\sin 30°}$$

$$a_A = +1.545\text{ m/s}^2 \qquad \mathbf{a}_A = 1.545\text{ m/s}^2 \rightarrow \blacktriangleleft$$

b. Acceleration of Block *B* Relative to *A*. Substituting the value obtained for a_A into Eq. (2), we have

$$a_{B/A} = (1.545\text{ m/s}^2) \cos 30° + (9.81\text{ m/s}^2) \sin 30°$$

$$a_{B/A} = +6.24\text{ m/s}^2 \qquad \mathbf{a}_{B/A} = 6.24\text{ m/s}^2 \swarrow 30° \blacktriangleleft$$

SAMPLE PROBLEM 12.5

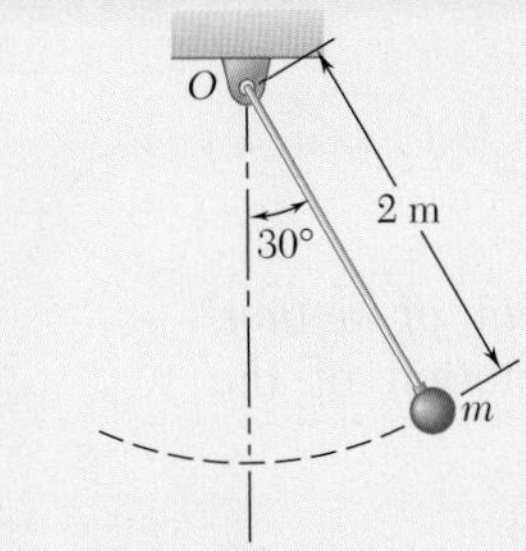

The bob of a 2-m pendulum describes an arc of circle in a vertical plane. If the tension in the cord is 2.5 times the weight of the bob for the position shown, find the velocity and the acceleration of the bob in that position.

SOLUTION

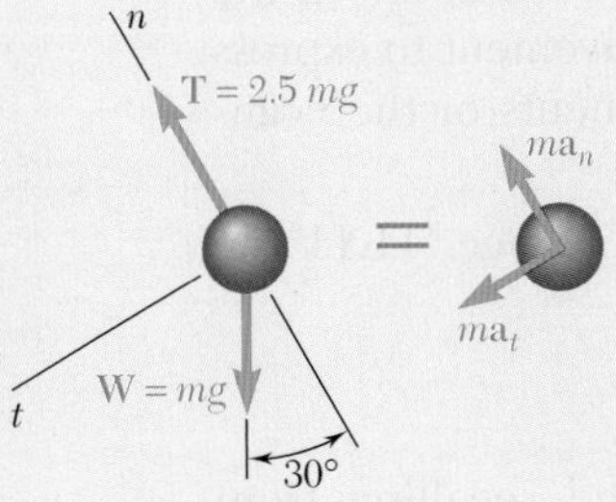

The weight of the bob is $W = mg$; the tension in the cord is thus 2.5 mg. Recalling that $\mathbf{a}_n$ is directed toward O and assuming $\mathbf{a}_t$ as shown, we apply Newton's second law and obtain

$+\swarrow \Sigma F_t = ma_t$: $\quad mg \sin 30° = ma_t$

$$a_t = g \sin 30° = +4.90 \text{ m/s}^2 \qquad \mathbf{a}_t = 4.90 \text{ m/s}^2 \swarrow \blacktriangleleft$$

$+\nwarrow \Sigma F_n = ma_n$: $\quad 2.5\ mg - mg \cos 30° = ma_n$

$$a_n = 1.634\ g = +16.03 \text{ m/s}^2 \qquad \mathbf{a}_n = 16.03 \text{ m/s}^2 \nwarrow \blacktriangleleft$$

Since $a_n = v^2/\rho$, we have $v^2 = \rho a_n = (2 \text{ m})(16.03 \text{ m/s}^2)$

$$v = \pm 5.66 \text{ m/s} \qquad \mathbf{v} = 5.66 \text{ m/s} \swarrow\!\!\nearrow \text{(up or down)} \blacktriangleleft$$

SAMPLE PROBLEM 12.6

Determine the rated speed of a highway curve of radius $\rho = 120$ m banked through an angle $\theta = 18°$. The *rated speed* of a banked highway curve is the speed at which a car should travel if no lateral friction force is to be exerted on its wheels.

SOLUTION

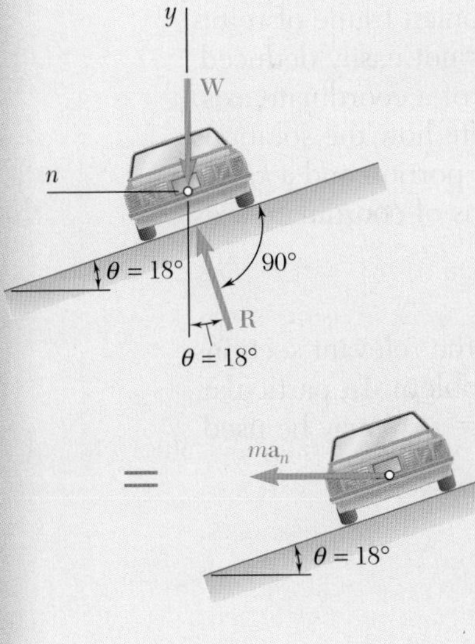

The car travels in a *horizontal* circular path of radius ρ. The normal component $\mathbf{a}_n$ of the acceleration is directed toward the center of the path; its magnitude is $a_n = v^2/\rho$, where v is the speed of the car in m/s. The mass m of the car is W/g, where W is the weight of the car. Since no lateral friction force is to be exerted on the car, the reaction $\mathbf{R}$ of the road is shown perpendicular to the roadway. Applying Newton's second law, we write

$+\uparrow \Sigma F_y = 0$: $\qquad R \cos \theta - W = 0 \qquad R = \dfrac{W}{\cos\theta} \qquad (1)$

$\overset{+}{\leftarrow} \Sigma F_n = ma_n$: $\qquad R \sin \theta = \dfrac{W}{g} a_n \qquad (2)$

Substituting for R from (1) into (2), and recalling that $a_n = v^2/\rho$,

$$\frac{W}{\cos \theta} \sin \theta = \frac{W}{g}\frac{v^2}{\rho} \qquad v^2 = g\rho \tan \theta$$

Substituting $\rho = 120$ m and $\theta = 18°$ into this equation, we obtain

$$v^2 = (9.81 \text{ m/s}^2)(120 \text{ m})(\tan 18°)$$
$$v = 19.56 \text{ m/s} \qquad v = 70.4 \text{ km/h} \blacktriangleleft$$

SOLVING PROBLEMS ON YOUR OWN

In the problems for this lesson, you will apply *Newton's second law of motion,* $\Sigma\mathbf{F} = m\mathbf{a}$, to relate the forces acting on a particle to the motion of the particle.

1. Writing the equations of motion. When applying Newton's second law to the types of motion discussed in this lesson, you will find it most convenient to express the vectors **F** and **a** in terms of either their rectangular components or their tangential and normal components.

a. When using rectangular components, and recalling from Sec. 11.11 the expressions found for a_x, a_y, and a_z, you will write

$$\Sigma F_x = m\ddot{x} \qquad \Sigma F_y = m\ddot{y} \qquad \Sigma F_z = m\ddot{z}$$

b. When using tangential and normal components, and recalling from Sec. 11.13 the expressions found for a_t and a_n, you will write

$$\Sigma F_t = m\frac{dv}{dt} \qquad \Sigma F_n = m\frac{v^2}{\rho}$$

2. Drawing a free-body diagram showing the applied forces *and an equivalent diagram* showing the vector $m\mathbf{a}$ or its components will provide you with a pictorial representation of Newton's second law [Sample Probs. 12.1 through 12.6]. These diagrams will be of great help to you when writing the equations of motion. Note that when a problem involves two or more bodies, it is usually best to consider each body separately.

3. Applying Newton's second law. As we observed in Sec. 12.2, the acceleration used in the equation $\Sigma\mathbf{F} = m\mathbf{a}$ should always be the *absolute acceleration* of the particle (that is, it should be measured with respect to a newtonian frame of reference). Also, *if the sense of the acceleration* **a** *is unknown* or is not easily deduced, assume an arbitrary sense for **a** (usually the positive direction of a coordinate axis) and then let the solution provide the correct sense. Finally, note how the solutions of Sample Probs. 12.3 and 12.4 were divided into a *kinematics* portion and a *kinetics* portion, and how in Sample Prob. 12.4 we used two systems of coordinate axes to simplify the equations of motion.

4. When a problem involves dry friction, be sure to review the relevant sections of *Statics* [Secs. 8.1 to 8.3] before attempting to solve that problem. In particular, you should know when each of the equations $F = \mu_s N$ and $F = \mu_k N$ may be used.

You should also recognize that if the motion of a system is not specified, it is necessary first to assume a possible motion and then to check the validity of that assumption.

5. Solving problems involving relative motion. When a body B moves with respect to a body A, as in Sample Prob. 12.4, it is often convenient to express the acceleration of B as

$$\mathbf{a}_B = \mathbf{a}_A + \mathbf{a}_{B/A}$$

where $\mathbf{a}_{B/A}$ is the acceleration of B relative to A, that is, the acceleration of B as observed from a frame of reference attached to A and in translation. If B is observed to move in a straight line, $\mathbf{a}_{B/A}$ will be directed along that line. On the other hand, if B is observed to move along a circular path, the relative acceleration $\mathbf{a}_{B/A}$ should be resolved into components tangential and normal to that path.

6. Finally, always consider the implications of any assumption you make. Thus, in a problem involving two cords, if you assume that the tension in one of the cords is equal to its maximum allowable value, check whether any requirements set for the other cord will then be satisfied. For instance, will the tension T in that cord satisfy the relation $0 \leq T \leq T_{max}$? That is, will the cord remain taut and will its tension be less than its maximum allowable value?

PROBLEMS

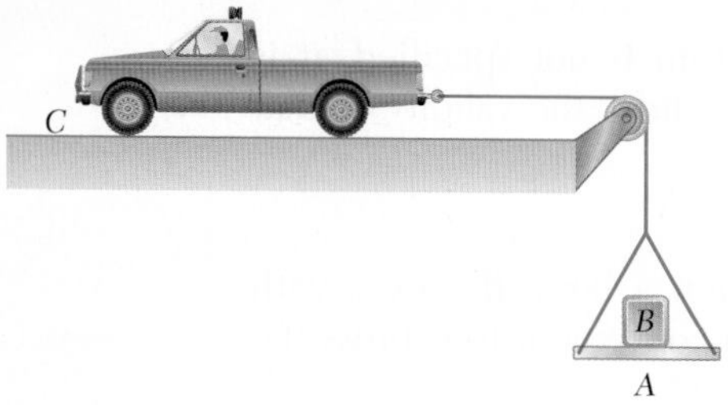

Fig. P12.CQ1

CONCEPT QUESTIONS

12.CQ1 A 1000-N boulder B is resting on a 200-N platform A when truck C accelerates to the left with a constant acceleration. Which of the following statements are true (more than one may be true)?

a. The tension in the cord connected to the truck is 200 N.
b. The tension in the cord connected to the truck is 1000 N.
c. The tension in the cord connected to the truck is greater than 1200 N.
d. The normal force between A and B is 1000 N.
e. The normal force between A and B is 1200 N.
f. None of the above are true.

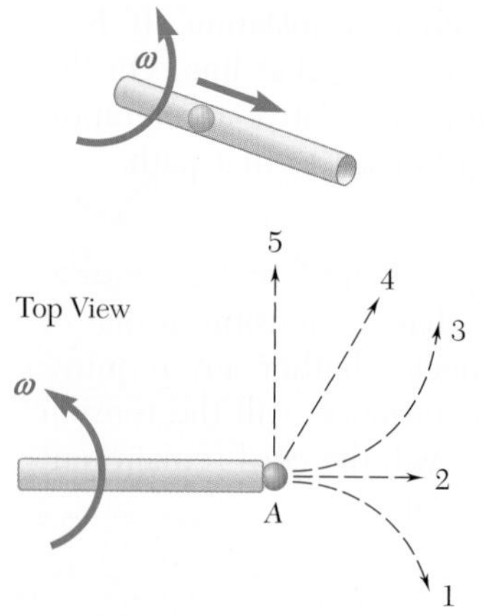

Fig. P12.CQ2

12.CQ2 Marble A is placed in a hollow tube, and the tube is swung in a horizontal plane causing the marble to be thrown out. As viewed from the top, which of the following choices best describes the path of the marble after leaving the tube?

a. 1 **b.** 2 **c.** 3 **d.** 4 **e.** 5

12.CQ3 The two systems shown start from rest. On the left, two 200-N weights are connected by an inextensible cord, and on the right, a constant 200-N force pulls on the cord. Neglecting all frictional forces, which of the following statements is true?

a. Blocks A and C will have the same acceleration.
b. Block C will have a larger acceleration than block A.
c. Block A will have a larger acceleration than block C.
d. Block A will not move.
e. None of the above are true.

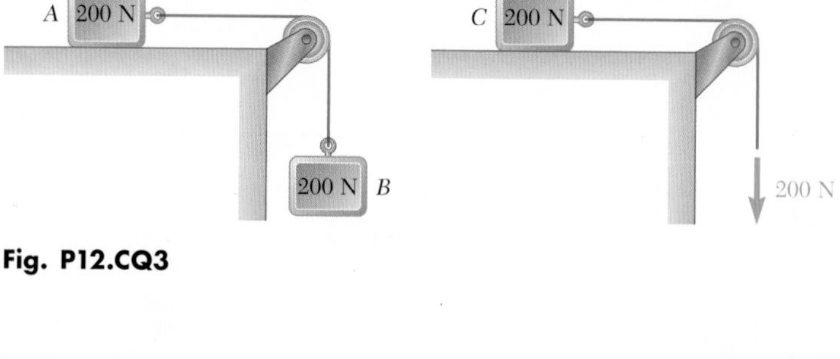

Fig. P12.CQ3

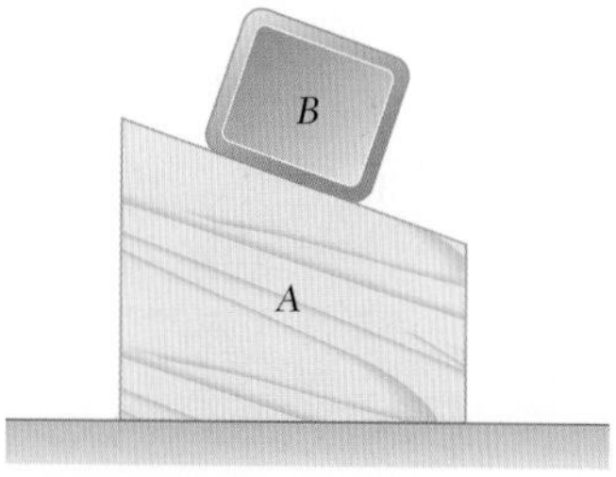

Fig. P12.CQ4

12.CQ4 Blocks A and B are released from rest in the position shown. Neglecting friction, the normal force between block A and the ground is:

a. Less than the weight of A plus the weight of B.
b. Equal to the weight of A plus the weight of B.
c. Greater than the weight of A plus the weight of B.

12.CQ5 People sit on a Ferris wheel at points A, B, C, and D. The Ferris wheel travels at a constant angular velocity. At the instant shown, which person experiences the largest force from his or her chair (back and seat)? Assume you can neglect the size of the chairs—that is, the people are located the same distance from the axis of rotation.

a. A

b. B

c. C

d. D

e. The force is the same for all the passengers.

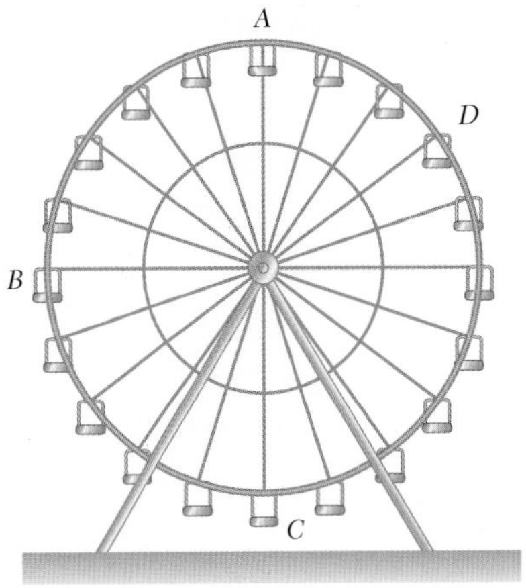

Fig. P12.CQ5

FREE BODY PRACTICE PROBLEMS

12.F1 Crate A is gently placed with zero initial velocity onto a moving conveyor belt. The coefficient of kinetic friction between the crate and the belt is μ_k. Draw the FBD and KD for A immediately after it contacts the belt.

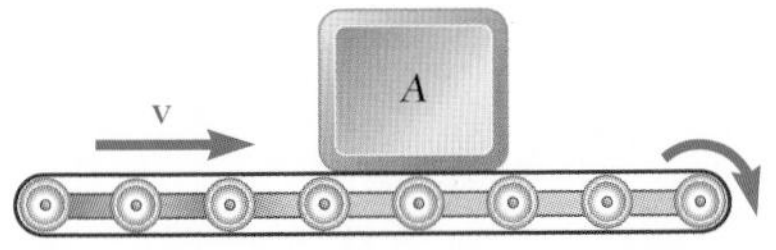

Fig. P12.F1

12.F2 Two blocks weighing W_A and W_B are at rest on a conveyor that is initially at rest. The belt is suddenly started in an upward direction so that slipping occurs between the belt and the boxes. Assuming the coefficient of friction between the boxes and the belt is μ_k, draw the FBDs and KDs for blocks A and B. How would you determine if A and B remain in contact?

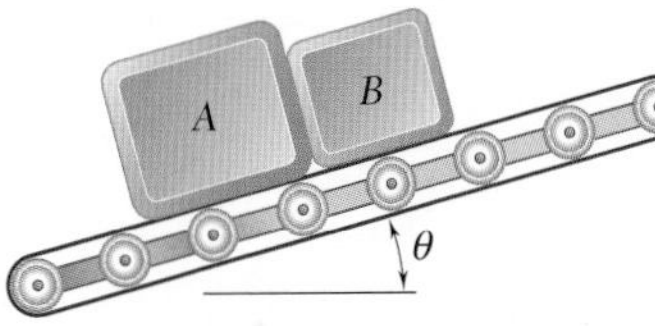

Fig. P12.F2

12.F3 Objects A, B, and C have masses m_A, m_B, and m_C, respectively. The coefficient of kinetic friction between A and B is μ_k, and the friction between A and the ground is negligible and the pulleys are massless and frictionless. Assuming B slides on A, draw the FBD and KD for each of the three masses A, B, and C.

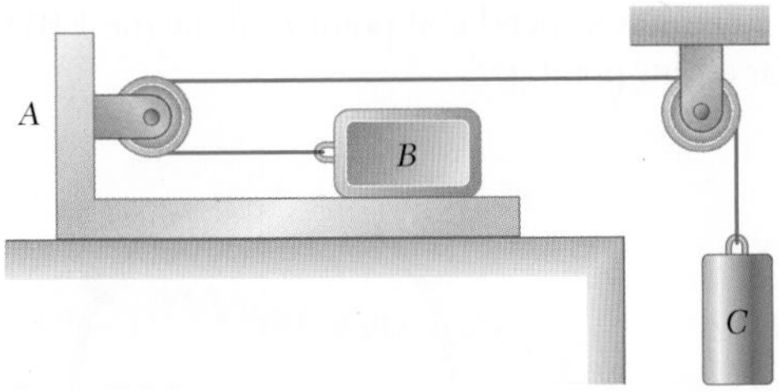

Fig. P12.F3

12.F4 Blocks A and B have masses m_A and m_B, respectively. Neglecting friction between all surfaces, draw the FBD and KD for each mass.

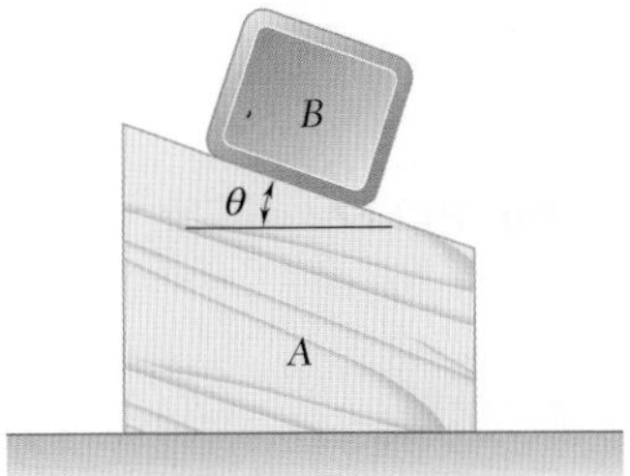

Fig. P12.F4

12.F5 Blocks A and B have masses m_A and m_B, respectively. Neglecting friction between all surfaces, draw the FBD and KD for the two systems shown.

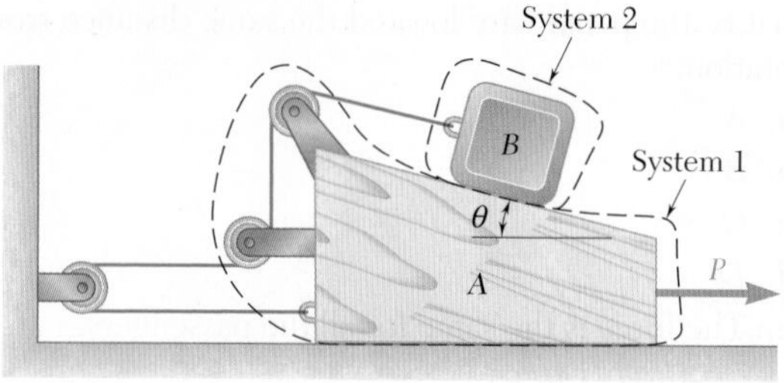

Fig. P12.F5

12.F6 A pilot of mass m flies a jet in a half-vertical loop of radius R so that the speed of the jet, v, remains constant. Draw a FBD and KD of the pilot at points A, B, and C.

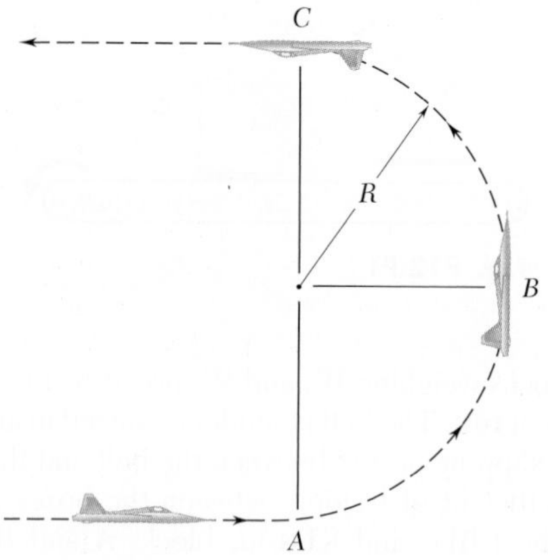

Fig. P12.F6

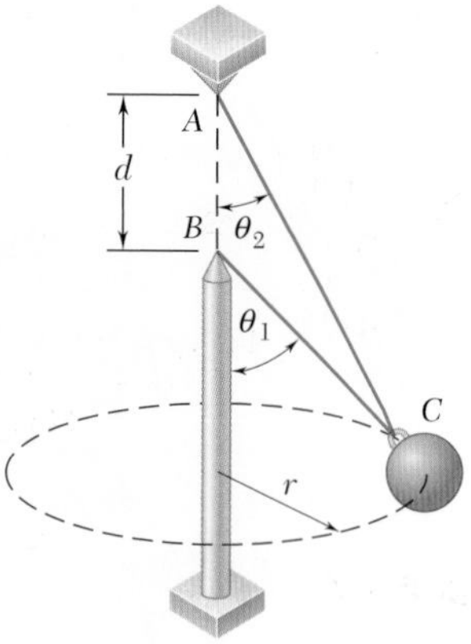

Fig. P12.F7

12.F7 Wires AC and BC are attached to a sphere which revolves at a constant speed v in the horizontal circle of radius r as shown. Draw a FBD and KD of C.

12.F8 A collar of mass m is attached to a spring and slides without friction along a circular rod in a vertical plane. The spring has an undeformed length of 125 mm. and a constant k. Knowing that the collar has a speed v at point B, draw the FBD and KD of the collar at this point.

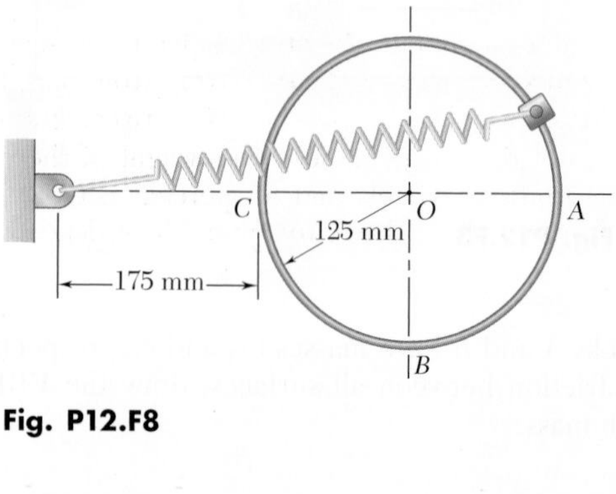

Fig. P12.F8

12.1 The acceleration due to gravity on the moon is 1.62 m/s^2. Determine (*a*) the weight in newtons, (*b*) the mass in kilograms, on the moon, of a gold bar, the mass of which has been officially designated as 2 kg.

12.2 The value of g at any latitude ϕ may be obtained from the formula

$$g = 9.78(1+0.0053 \sin^2\phi)\text{m/s}^2$$

which takes into account the effect of the rotation of the earth, as well as the fact that the earth is not truly spherical. Determine to four significant figures (*a*) the weight in newtons, (*b*) the mass in kg at the latitudes of 0°, 45°, 60°, of a silver bar, the mass of which has been officially designated as 5 kg.

12.3 A 400-kg satellite has been placed in a circular orbit 1500 km above the surface of the earth. The acceleration of gravity at this elevation is 6.43 m/s^2. Determine the linear momentum of the satellite, knowing that its orbital speed is 25.6×10^3 km/h.

12.4 A spring scale A and a lever scale B having equal lever arms are fastened to the roof of an elevator, and identical packages are attached to the scales as shown. Knowing that when the elevator moves downward with an acceleration of 1 m/s^2 the spring scale indicates a load of 60 N, determine (*a*) the weight of the packages, (*b*) the load indicated by the spring scale and the mass needed to balance the lever scale when the elevator moves upward with an acceleration of 1 m/s^2.

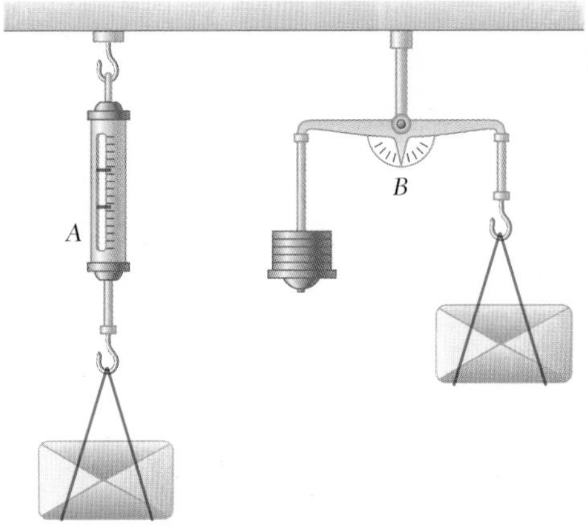

Fig. *P12.4*

12.5 In anticipation of a long 7° upgrade, a bus driver accelerates at a constant rate of 0.9 m/s^2 while still on a level section of the highway. Knowing that the speed of the bus is 100 km/h as it begins to climb the grade and that the driver does not change the setting of his throttle or shift gears, determine the distance traveled by the bus up the grade when its speed has decreased to 80 km/h.

12.6 A hockey player hits a puck so that it comes to rest in 9 s after sliding 30 m on the ice. Determine (*a*) the initial velocity of the puck, (*b*) the coefficient of friction between the puck and the ice.

12.7 The acceleration of a package sliding at point A is 3 m/s^2. Assuming that the coefficient of kinetic friction is the same for each section, determine the acceleration of the package at point B.

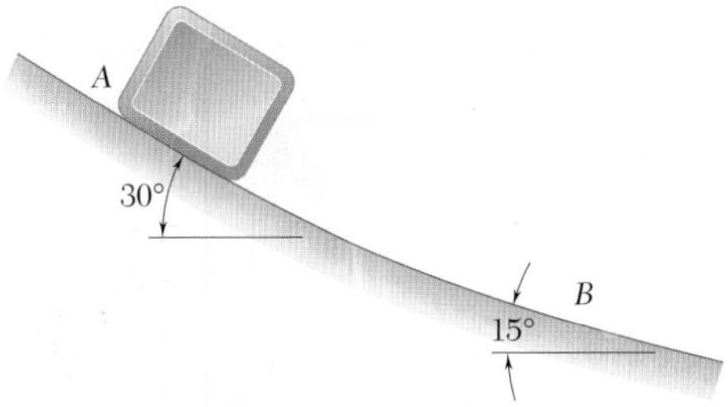

Fig. P12.7

12.8 Determine the maximum theoretical speed that may be achieved over a distance of 60 m by a car starting from rest, knowing that the coefficient of static friction is 0.80 between the tires and the pavement and that 60 percent of the weight of the car is distributed over its front wheels and 40 percent over its rear wheels. Assume (*a*) four-wheel drive, (*b*) front-wheel drive, (*c*) rear-wheel drive.

12.9 If an automobile's braking distance from 90 km/h is 45 m on level pavement, determine the automobile's braking distance from 90 km/h when it is (*a*) going up a 5° incline, (*b*) going down a 3-percent incline. Assume the braking force is independent of grade.

12.10 A mother and her child are skiing together, and the mother is holding the end of a rope tied to the child's waist. They are moving at a speed of 7.2 km/h on a gently sloping portion of the ski slope when the mother observes that they are approaching a steep descent. She pulls on the rope with an average force of 7 N. Knowing the coefficient of friction between the child and the ground is 0.1 and the angle of the rope does not change, determine (*a*) the time required for the child's speed to be cut in half, (*b*) the distance traveled in this time.

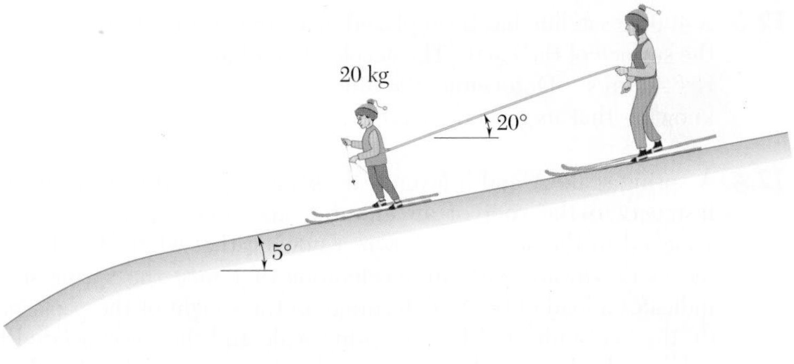

Fig. P12.10

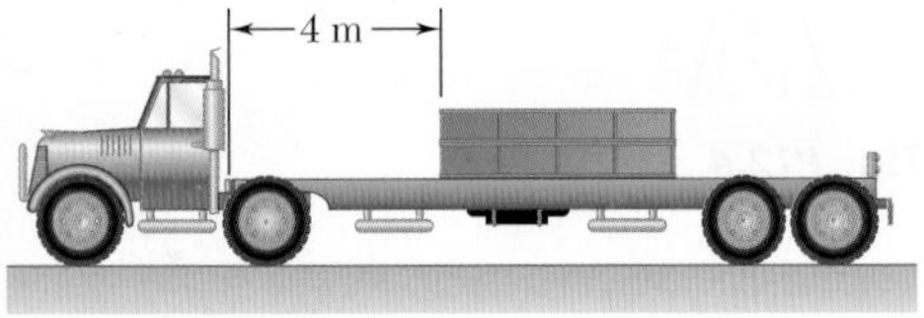

Fig. P12.11

12.11 The coefficients of friction between the load and the flatbed trailer shown are $\mu_s = 0.40$ and $\mu_k = 0.30$. Knowing that the speed of the rig is 72 km/h, determine the shortest distance in which the rig can be brought to a stop if the load is not to shift.

12.12 A light train made up of two cars is traveling at 90 km/h when the brakes are applied to both cars. Knowing that car *A* has a mass of 25 Mg and car *B* a mass of 20 Mg, and that the braking force is 30 kN on each car, determine (*a*) the distance traveled by the train before it comes to a stop, (*b*) the force in the coupling between the cars while the train is slowing down.

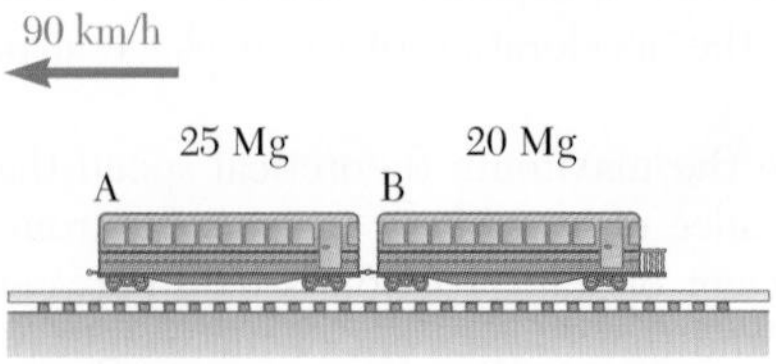

Fig. P12.12

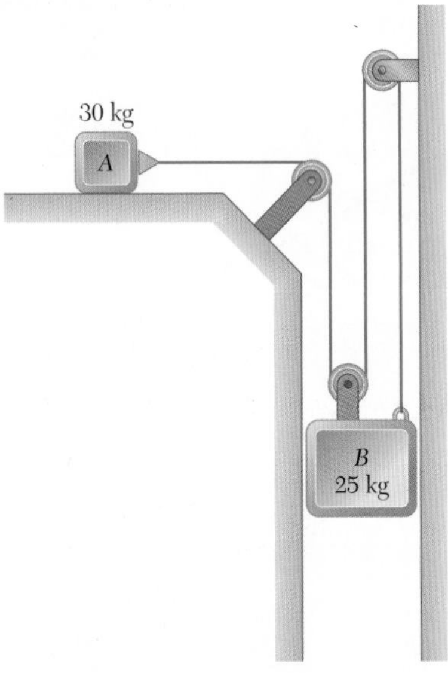

Fig. P12.13 and P12.14

12.13 The two blocks shown are originally at rest. Neglecting the masses of the pulleys and the effect of friction in the pulleys and between block *A* and the horizontal surface, determine (*a*) the acceleration of each block, (*b*) the tension in the cable.

12.14 The two blocks shown are originally at rest. Neglecting the masses of the pulleys and the effect of friction in the pulleys and assuming that the coefficients of friction between block A and the horizontal surface are $\mu_s = 0.25$ and $\mu_k = 0.20$, determine (*a*) the acceleration of each block, (*b*) the tension in the cable.

12.15 Each of the systems shown is initially at rest. Neglecting axle friction and the masses of the pulleys, determine for each system (*a*) the acceleration of block *A*, (*b*) the velocity of block *A* after it has moved through 3 m, (*c*) the time required for block *A* to reach a velocity of 6 m/s.

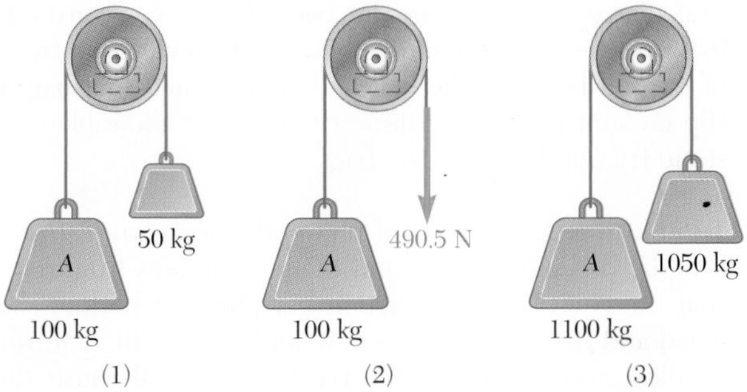

Fig. P12.15

12.16 Boxes *A* and *B* are at rest on a conveyor belt that is initially at rest. The belt is suddenly started in an upward direction so that slipping occurs between the belt and the boxes. Knowing that the coefficients of kinetic friction between the belt and the boxes are $(\mu_k)_A = 0.30$ and $(\mu_k)_B = 0.32$, determine the initial acceleration of each box.

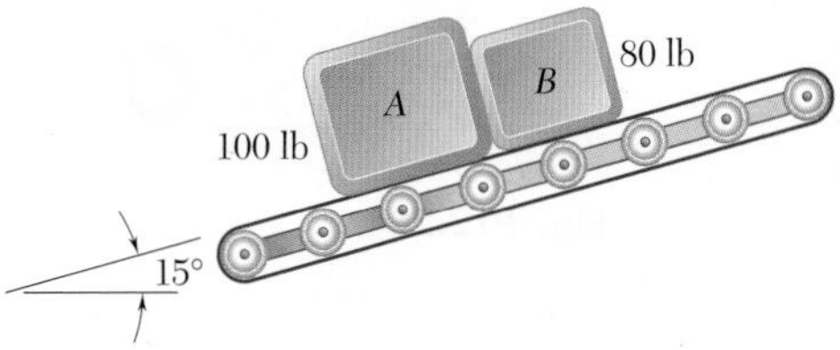

Fig. P12.16

12.17 A 2500-kg truck is being used to lift a 500-kg boulder *B* that is on a 100-kg pallet *A*. Knowing the acceleration of the truck is 0.3 m/s², determine (*a*) the horizontal force between the tires and the ground, (*b*) the force between the boulder and the pallet.

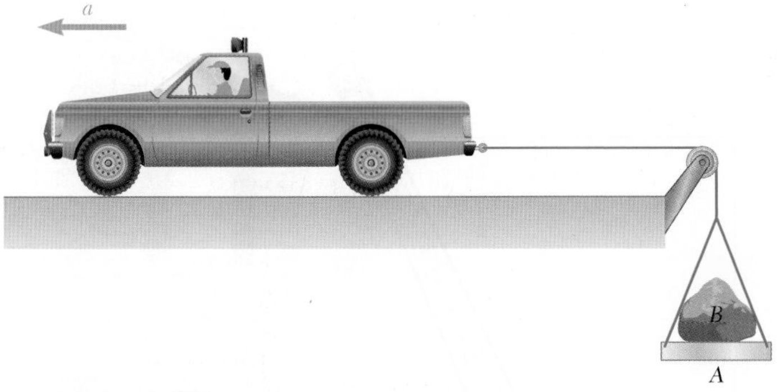

Fig. P12.17

12.18 Block *A* has a mass of 40 kg, and block *B* has a mass of 8 kg. The coefficients of friction between all surfaces of contact are $\mu_s = 0.20$ and $\mu_k = 0.15$. If $P = 0$, determine (*a*) the acceleration of block *B*, (*b*) the tension in the cord.

12.19 Block *A* has a mass of 40 kg, and block *B* has a mass of 8 kg. The coefficients of friction between all surfaces of contact are $\mu_s = 0.20$ and $\mu_k = 0.15$. If $P = 40\text{ N} \rightarrow$, determine (*a*) the acceleration of block *B*, (*b*) the tension in the cord.

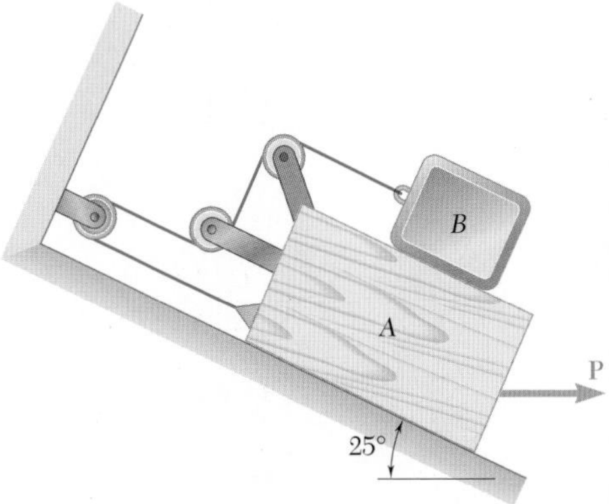

Fig. P12.18 and P12.19

12.20 A package is at rest on a conveyor belt which is initially at rest. The belt is started and moves to the right for 1.3 s with a constant acceleration of 2 m/s^2. The belt then moves with a constant deceleration $\mathbf{a}_2$ and comes to a stop after a total displacement of 2.2 m. Knowing that the coefficients of friction between the package and the belt are $\mu_s = 0.35$ and $\mu_k = 0.25$, determine (*a*) the deceleration $\mathbf{a}_2$ of the belt, (*b*) the displacement of the package relative to the belt as the belt comes to a stop.

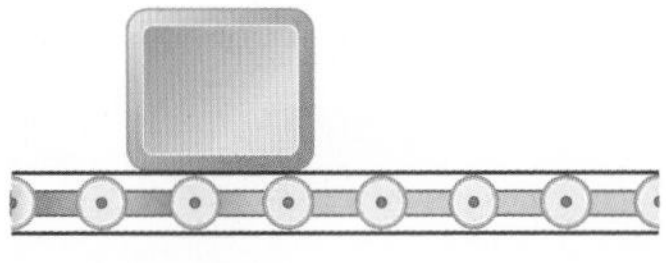

Fig. P12.20

12.21 A baggage conveyor is used to unload luggage from an airplane. The 10-kg duffel bag *A* is sitting on top of the 20-kg suitcase *B*. The conveyor is moving the bags down at a constant speed of 0.5 m/s when the belt suddenly stops. Knowing that the coefficient of friction between the belt and *B* is 0.3 and that bag *A* does not slip on suitcase *B*, determine the smallest allowable coefficient of static friction between the bags.

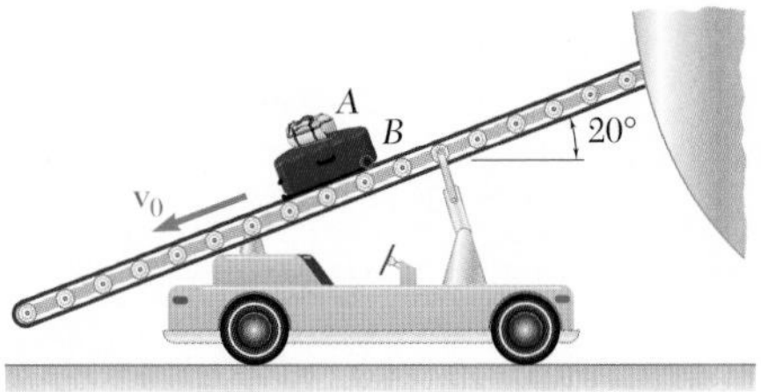

Fig. *P12.21*

12.22 To unload a bound stack of plywood from a truck, the driver first tilts the bed of the truck and then accelerates from rest. Knowing that the coefficients of friction between the bottom sheet of plywood and the bed are $\mu_s = 0.40$ and $\mu_k = 0.30$, determine (*a*) the smallest acceleration of the truck which will cause the stack of plywood to slide, (*b*) the acceleration of the truck which causes corner *A* of the stack to reach the end of the bed in 0.9 s.

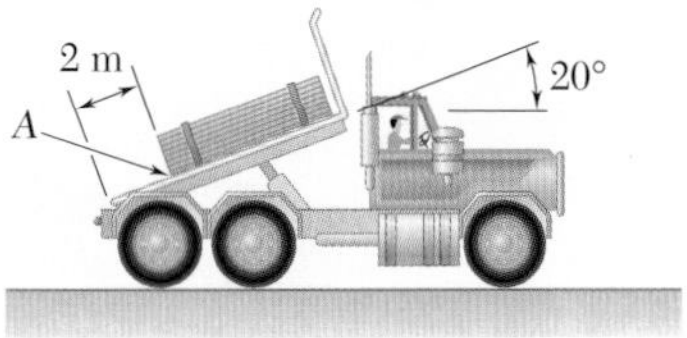

Fig. *P12.22*

12.23 To transport a series of bundles of shingles *A* to a roof, a contractor uses a motor-driven lift consisting of a horizontal platform *BC* which rides on rails attached to the sides of a ladder. The lift starts from rest and initially moves with a constant acceleration $\mathbf{a}_1$ as shown. The lift then decelerates at a constant rate $\mathbf{a}_2$ and comes to rest at *D*, near the top of the ladder. Knowing that the coefficient of static friction between a bundle of shingles and the horizontal platform is 0.30, determine the largest allowable acceleration $\mathbf{a}_1$ and the largest allowable deceleration $\mathbf{a}_2$ if the bundle is not to slide on the platform.

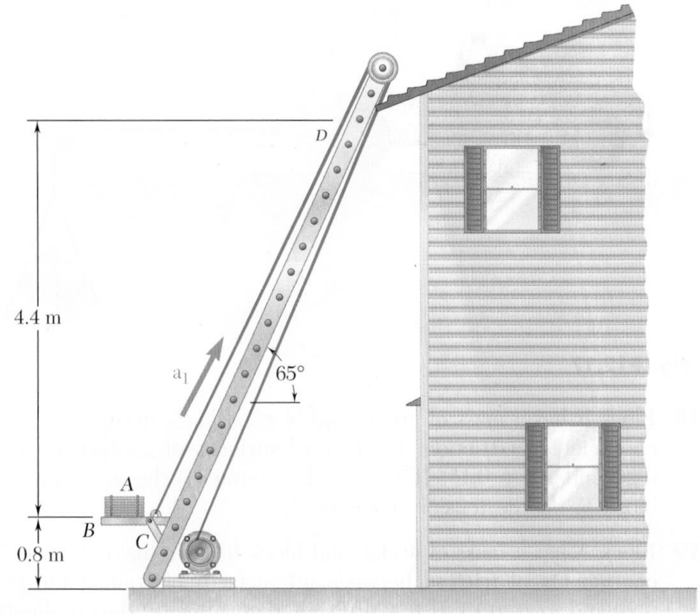

Fig. P12.23

12.24 An airplane has a mass of 25 Mg and its engines develop a total thrust of 40 kN during take-off. If the drag **D** exerted on the plane has a magnitude $D = 2.25\ v^2$, where v is expressed in meters per second and D in newtons, and if the plane becomes airborne at a speed of 240 km/h, determine the length of runway required for the plane to take off.

12.25 The propellers of a ship of weight W can produce a propulsive force $\mathbf{F}_0$; they produce a force of the same magnitude but of opposite direction when the engines are reversed. Knowing that the ship was proceeding forward at its maximum speed v_0 when the engines were put into reverse, determine the distance the ship travels before coming to a stop. Assume that the frictional resistance of the water varies directly with the square of the velocity.

12.26 A constant force **P** is applied to a piston and rod of total mass m to make them move in a cylinder filled with oil. As the piston moves, the oil is forced through orifices in the piston and exerts on the piston a force of magnitude kv in a direction opposite to the motion of the piston. Knowing that the piston starts from rest at $t = 0$ and $x = 0$, show that the equation relating x, v, and t, where x is the distance traveled by the piston and v is the speed of the piston, is linear in each of these variables.

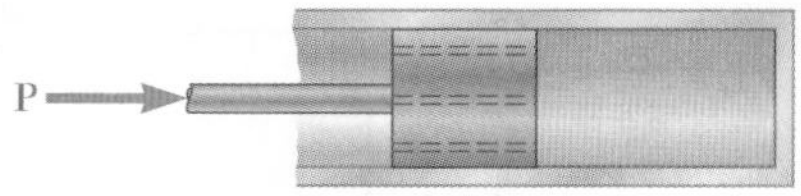

Fig. P12.26

12.27 A spring AB of constant k is attached to a support at A and to a collar of mass m. The unstretched length of the spring is l. Knowing that the collar is released from rest at $x = x_0$ and neglecting friction between the collar and the horizontal rod, determine the magnitude of the velocity of the collar as it passes through point C.

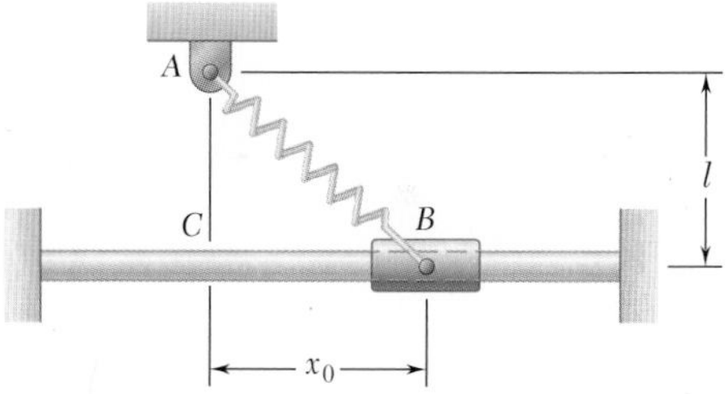

Fig. P12.27

12.28 Block A has a mass of 10 kg, and blocks B and C have masses of 5 kg each. Knowing that the blocks are initially at rest and that B moves through 3 m in 2 s, determine (*a*) the magnitude of the force **P**, (*b*) the tension in the cord AD. Neglect the masses of the pulleys and axle friction.

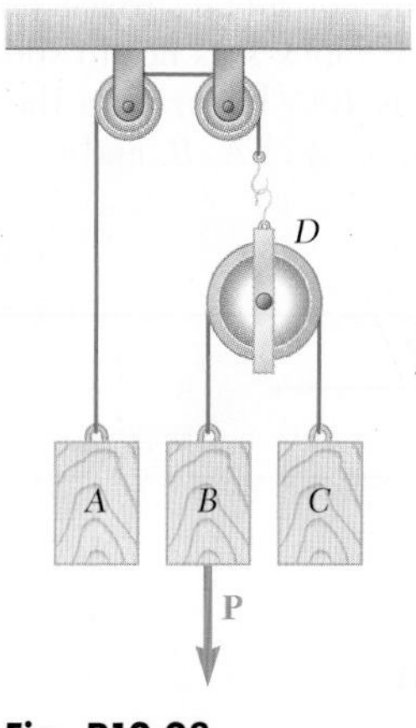

Fig. P12.28

12.29 A 20-kg sliding panel is supported by rollers at B and C. A 12.5-kg counterweight A is attached to a cable as shown and, in cases a and c, is initially in contact with a vertical edge of the panel. Neglecting friction, determine in each case shown the acceleration of the panel and the tension in the cord immediately after the system is released from rest.

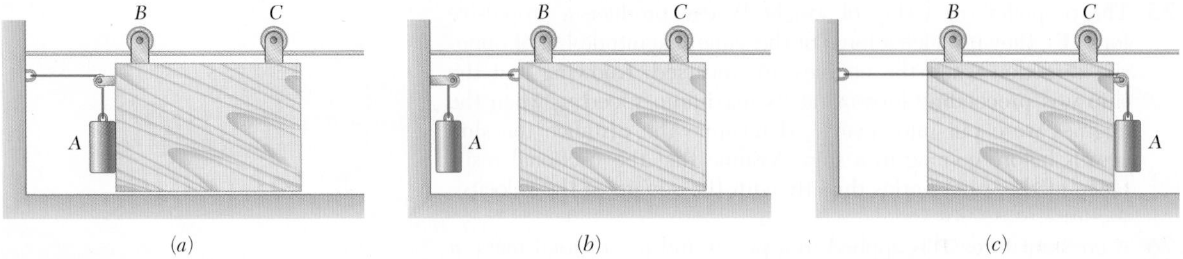

Fig. P12.29

12.30 The coefficients of friction between blocks A and C and the horizontal surfaces are $\mu_s = 0.24$ and $\mu_k = 0.20$. Knowing that $m_A = 5$ kg, $m_B = 10$ kg, and $m_C = 10$ kg, determine (a) the tension in the cord, (b) the acceleration of each block.

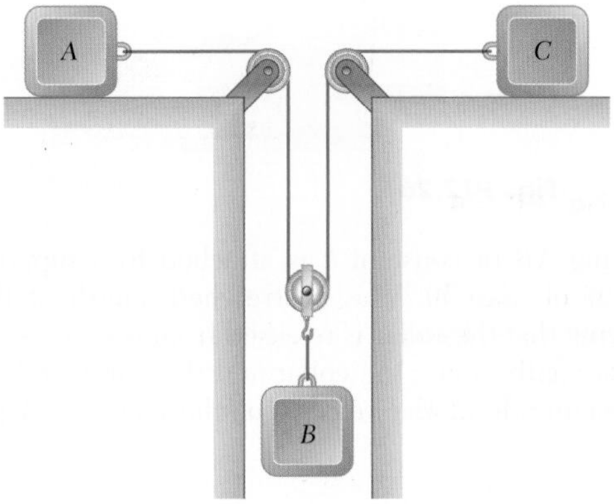

Fig. P12.30

12.31 A 6-kg block B rests as shown on a 10-kg bracket A. The coefficients of friction are $\mu_s = 0.30$ and $\mu_k = 0.25$ between block B and bracket A, and there is no friction in the pulley or between the bracket and the horizontal surface. (a) Determine the maximum mass of block C if block B is not to slide on bracket A. (b) If the mass of block C is 10% larger than the answer found in a determine the accelerations of A, B, and C.

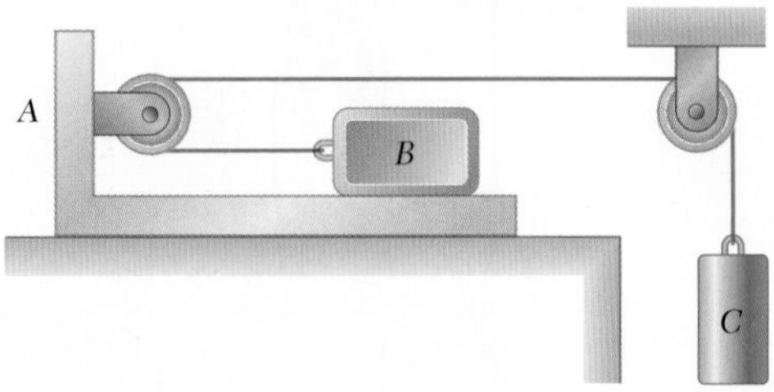

Fig. P12.31

12.32 The masses of blocks A, B, C, and D are 9 kg, 9 kg, 6 kg, and 7 kg, respectively. Knowing that a downward force of magnitude 120 N is applied to block D, determine (*a*) the acceleration of each block, (*b*) the tension in cord ABC. Neglect the weights of the pulleys and the effect of friction.

12.33 The masses of blocks A, B, C, and D are 9 kg, 9 kg, 6 kg, and 7 kg, respectively. Knowing that a downward force of magnitude 50 N is applied to block B and that the system starts from rest, determine at $t = 3$ s the velocity (*a*) of D relative to A, (*b*) of C relative to D. Neglect the weights of the pulleys and the effect of friction.

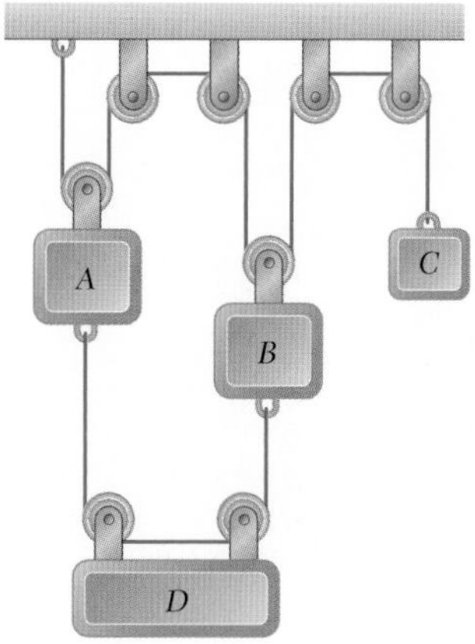

Fig. *P12.32* and *P12.33*

12.34 The 15-kg block B is supported by the 25-kg block A and is attached to a cord to which a 225-N horizontal force is applied as shown. Neglecting friction, determine (*a*) the acceleration of block A, (*b*) the acceleration of block B relative to A.

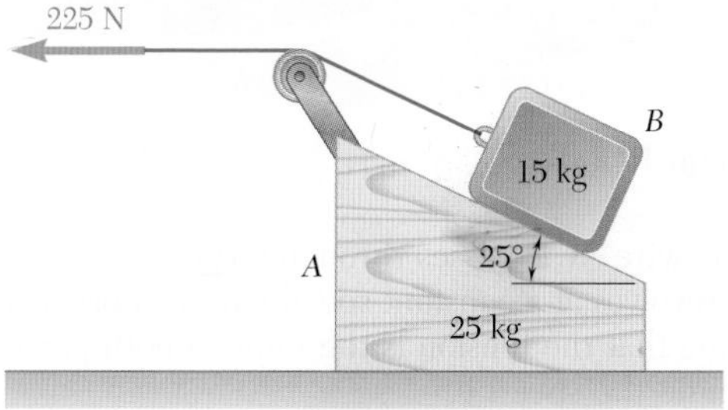

Fig. P12.34

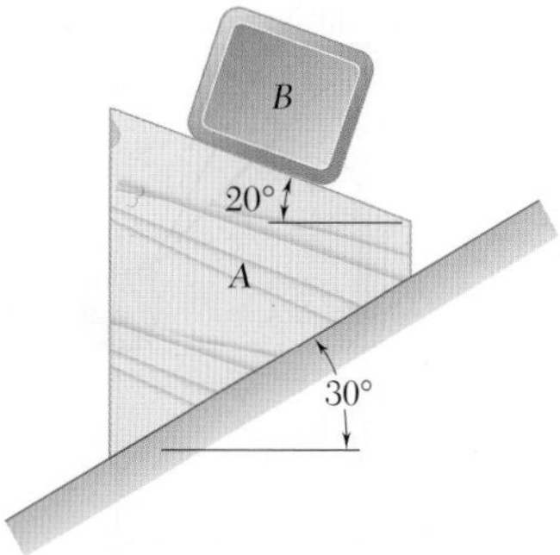

Fig. P12.35

12.35 Block B of mass 10 kg rests as shown on the upper surface of a 22-kg wedge A. Knowing that the system is released from rest and neglecting friction, determine (*a*) the acceleration of B, (*b*) the velocity of B relative to A at $t = 0.5$ s.

12.36 A 450-g tetherball A is moving along a horizontal circular path at a constant speed of 4 m/s. Determine (*a*) the angle θ that the cord forms with pole BC, (*b*) the tension in the cord.

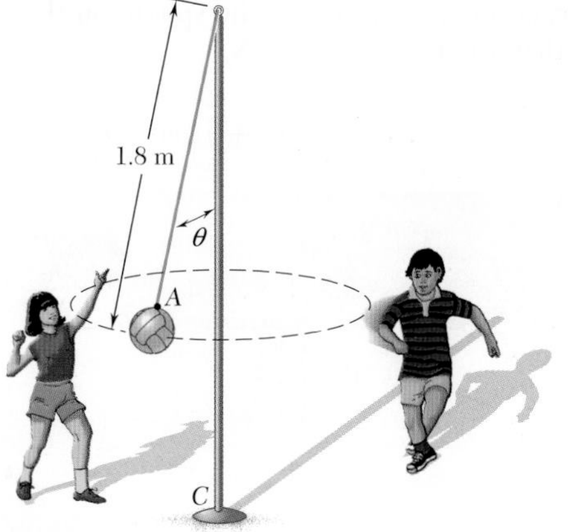

Fig. P12.36

12.37 During a hammer thrower's practice swings, the 7.1-kg head A of the hammer revolves at a constant speed v in a horizontal circle as shown. If $\rho = 0.93$ m and $\theta = 60°$, determine (*a*) the tension in wire BC, (*b*) the speed of the hammer's head.

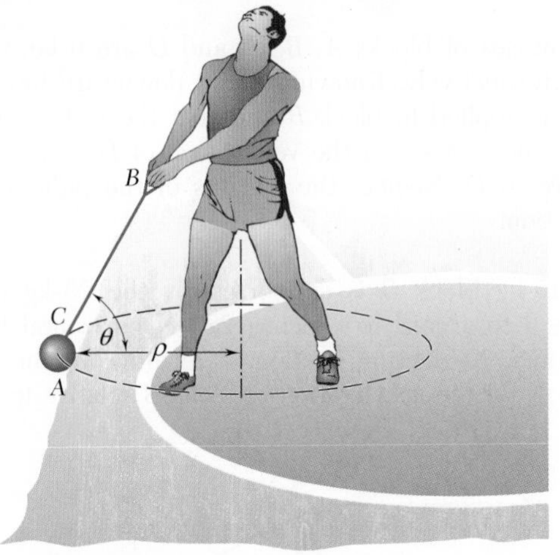

Fig. P12.37

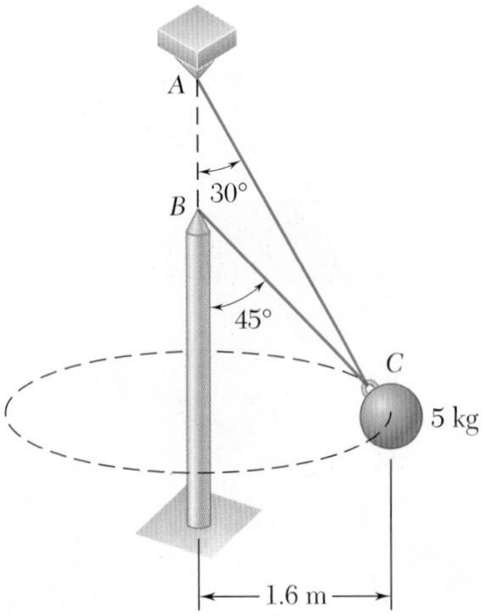

Fig. P12.38, P12.39, and P12.40

12.38 A single wire ACB passes through a ring at C attached to a sphere which revolves at a constant speed v in the horizontal circle shown. Knowing that the tension is the same in both portions of the wire, determine the speed v.

12.39 Two wires AC and BC are tied at C to a sphere which revolves at a constant speed v in the horizontal circle shown. Determine the range of values of v for which both wires remain taut.

***12.40** Two wires AC and BC are tied at C to a sphere which revolves at a constant speed v in the horizontal circle shown. Determine the range of the allowable values of v if both wires are to remain taut and if the tension in either of the wires is not to exceed 60 N.

12.41 A 100-g sphere D is at rest relative to drum ABC which rotates at a constant rate. Neglecting friction, determine the range of the allowable values of the velocity v of the sphere if neither of the normal forces exerted by the sphere on the inclined surfaces of the drum is to exceed 1.1 N.

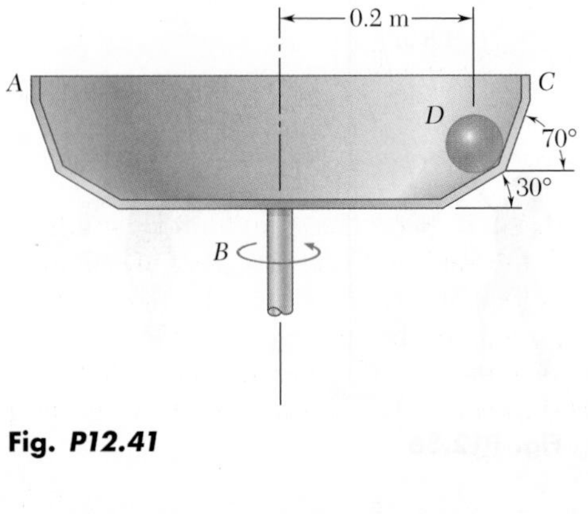

Fig. *P12.41*

***12.42** As part of an outdoor display, a 6-kg model C of the earth is attached to wires AC and BC and revolves at a constant speed v in the horizontal circle shown. Determine the range of the allowable values of v if both wires are to remain taut and if the tension in either of the wires is not to exceed 125-N.

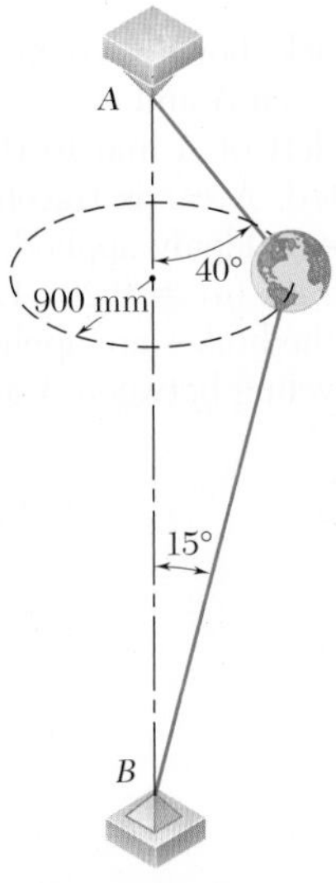

Fig. P12.42

***12.43** The 600-g flyballs of a centrifugal governor revolve at a constant speed v in the horizontal circle of 150 mm radius shown. Neglecting the weights of links AB, BC, AD, and DE and requiring that the links support only tensile forces, determine the range of the allowable values of v so that the magnitudes of the forces in the links do not exceed 80 N.

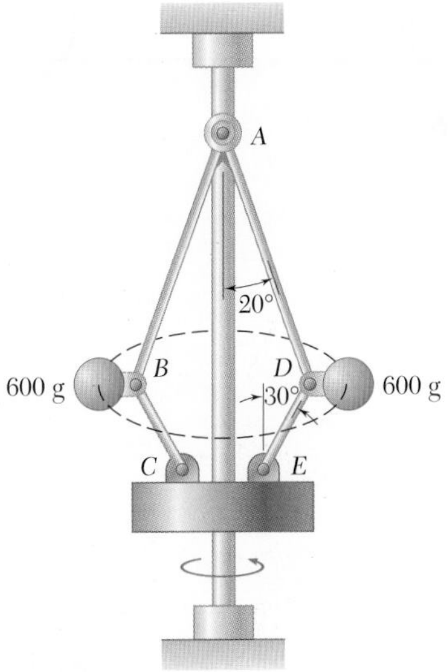

Fig. P12.43

12.44 A 60-kg wrecking ball B is attached to a 15-m-long steel cable AB and swings in the vertical arc shown. Determine the tension in the cable (*a*) at the top C of the swing, (*b*) at the bottom D of the swing, where the speed of B is 4.2 m/s.

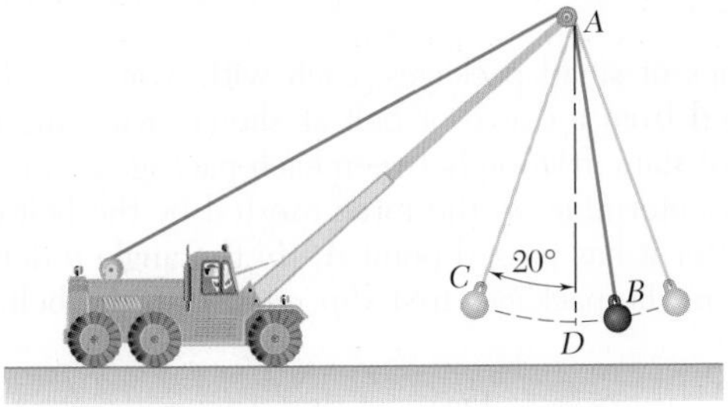

Fig. P12.44

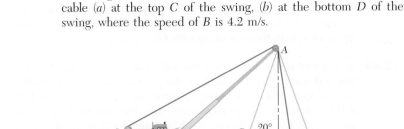

12.45 During a high-speed chase, a 1000-kg sports car traveling at a speed of 160 km/h just loses contact with the road as it reaches the crest A of a hill. (*a*) Determine the radius of curvature ρ of the vertical profile of the road at A. (*b*) Using the value of ρ found in part *a*, determine the force exerted on a 80-kg driver by the seat of his 1200-kg car as the car, traveling at a constant speed of 75 km/h, passes through A.

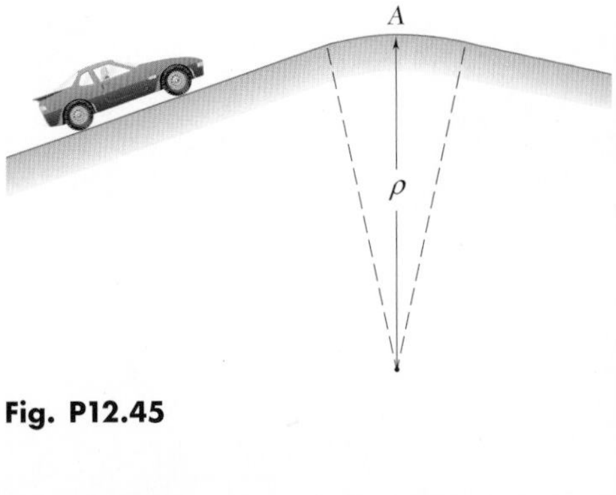

Fig. P12.45

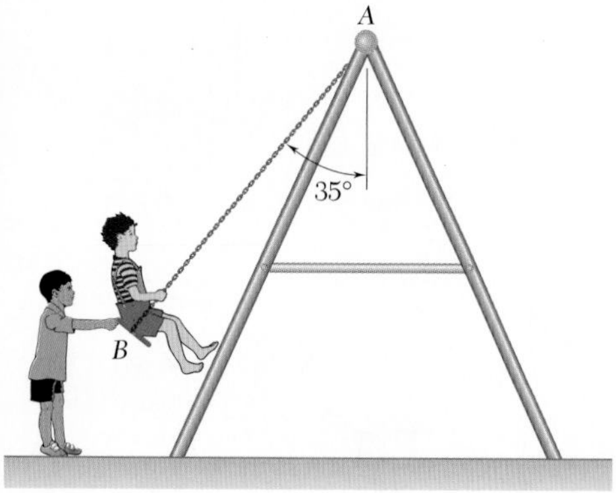

Fig. P12.46

12.46 A child having a mass of 22 kg sits on a swing and is held in the position shown by a second child. Neglecting the mass of the swing, determine the tension in rope AB (*a*) while the second child holds the swing with his arms outstretched horizontally, (*b*) immediately after the swing is released.

12.47 The roller-coaster track shown is contained in a vertical plane. The portion of track between A and B is straight and horizontal, while the portions to the left of A and to the right of B have radii of curvature as indicated. A car is traveling at a speed of 72 km/h when the brakes are suddenly applied, causing the wheels of the car to slide on the track ($\mu_k = 0.20$). Determine the initial deceleration of the car if the brakes are applied as the car (*a*) has almost reached A, (*b*) is traveling between A and B, (*c*) has just passed B.

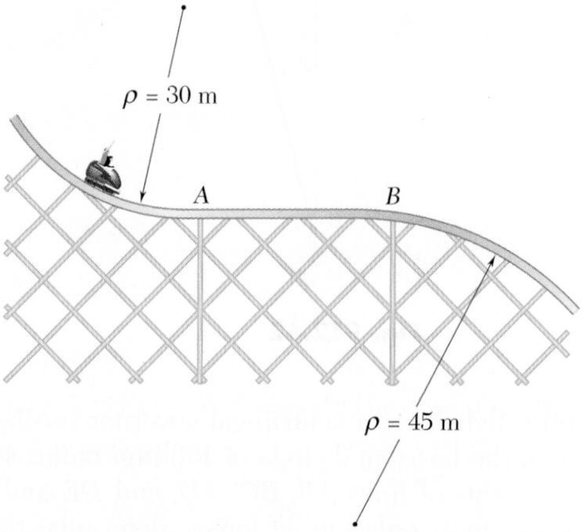

Fig. P12.47

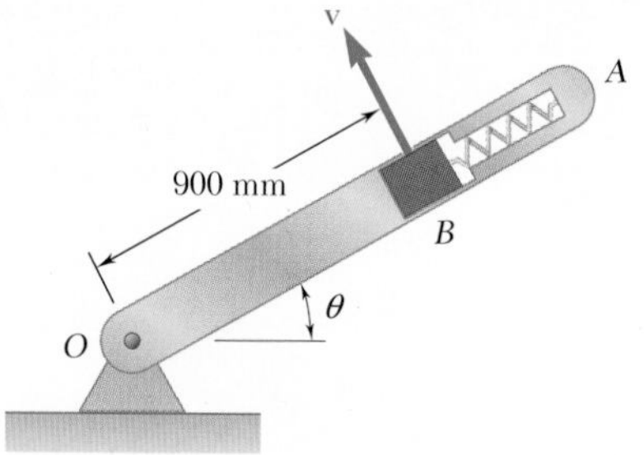

Fig. P12.48

12.48 A 250-g block fits inside a small cavity cut in arm OA, which rotates in the vertical plane at a constant rate such that $v = 3$ m/s. Knowing that the spring exerts on block B a force of magnitude $P = 1.5$ N and neglecting the effect of friction, determine the range of values of θ for which block B is in contact with the face of the cavity closest to the axis of rotation O.

12.49 A series of small packages, each with a mass of 0.5 kg, are discharged from a conveyor belt as shown. Knowing that the coefficient of static friction between each package and the conveyor belt is 0.4, determine (*a*) the force exerted by the belt on the package just after it has passed point A, (*b*) the angle θ defining the point B where the packages first *slip* relative to the belt.

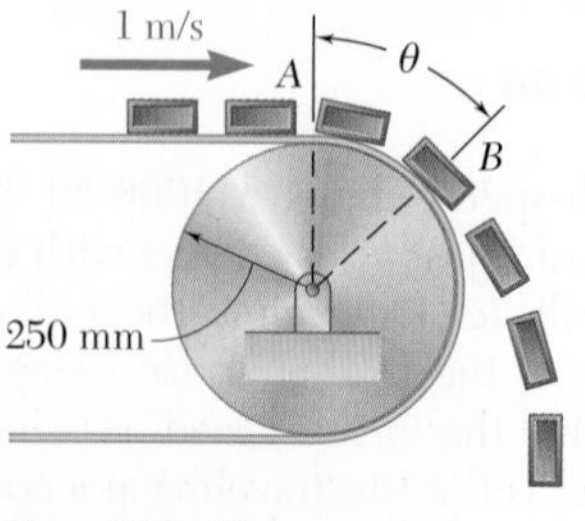

Fig. P12.49

12.50 A 54-kg pilot flies a jet trainer in a half-vertical loop of 1200-m radius so that the speed of the trainer decreases at a constant rate. Knowing that the pilot's apparent weights at points A and C are 1680 N and 350 N, respectively, determine the force exerted on her by the seat of the trainer when the trainer is at point B.

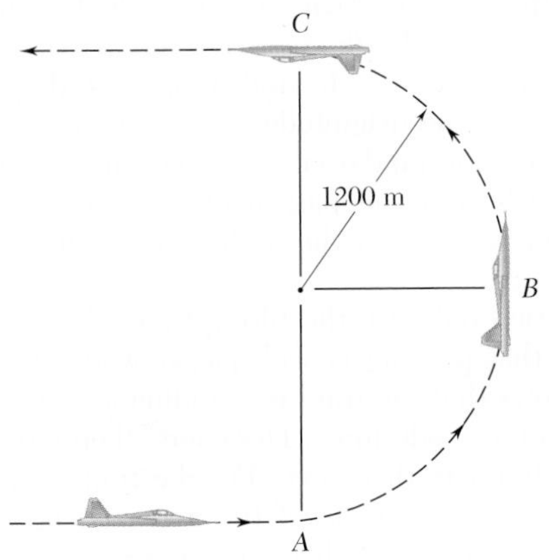

Fig. P12.50

12.51 A carnival ride is designed to allow the general public to experience high-acceleration motion. The ride rotates about point O in a horizontal circle such that the rider has a speed v_0. The rider reclines on a platform A which rides on rollers such that friction is negligible. A mechanical stop prevents the platform from rolling down the incline. Determine (*a*) the speed v_0 at which the platform A begins to roll upward, (*b*) the normal force experienced by an 80-kg rider at this speed.

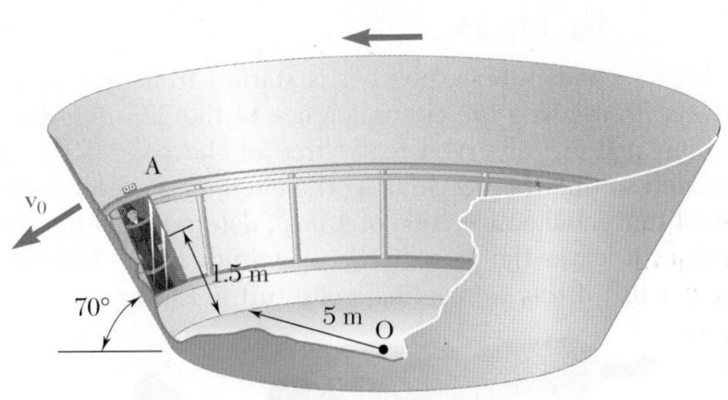

Fig. P12.51

12.52 A curve in a speed track has a radius of 300-m and a rated speed of 190 km/h. (See Sample Prob. 12.5 for the definition of rated speed.) Knowing that a racing car starts skidding on the curve when traveling at a speed of 290 km/h, determine (*a*) the banking angle θ, (*b*) the coefficient of static friction between the tires and the track under the prevailing conditions, (*c*) the minimum speed at which the same car could negotiate the curve.

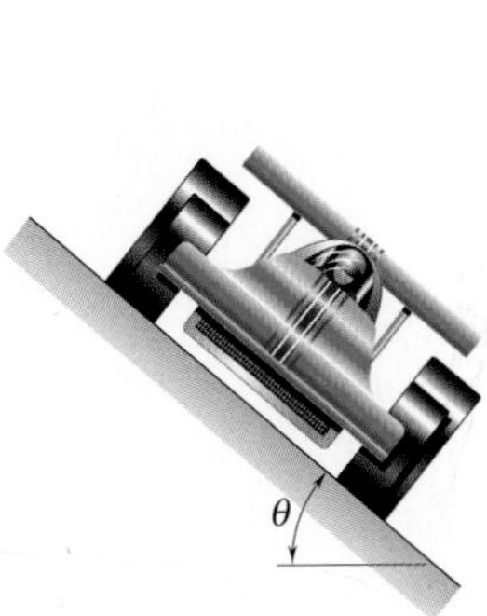

Fig. P12.52

Fig. P12.53 and *P12.54*

12.53 Tilting trains, such as the *American Flyer* which will run from Washington to New York and Boston, are designed to travel safely at high speeds on curved sections of track which were built for slower, conventional trains. As it enters a curve, each car is tilted by hydraulic actuators mounted on its trucks. The tilting feature of the cars also increases passenger comfort by eliminating or greatly reducing the side force $\mathbf{F}_s$ (parallel to the floor of the car) to which passengers feel subjected. For a train traveling at 160 km/h on a curved section of track banked through an angle $\theta = 6°$ and with a rated speed of 100 km/h, determine (*a*) the magnitude of the side force felt by a passenger of weight W in a standard car with no tilt ($\phi = 0$), (*b*) the required angle of tilt ϕ if the passenger is to feel no side force. (See Sample Prob. 12.5 for the definition of rated speed.)

12.54 Tests carried out with the tilting trains described in Prob. 12.53 revealed that passengers feel queasy when they see through the car windows that the train is rounding a curve at high speed, yet do not feel any side force. Designers, therefore, prefer to reduce, but not eliminate that force. For the train of Prob. 12.53, determine the required angle of tilt ϕ if passengers are to feel side forces equal to 10% of their weights.

12.55 A 3-kg block is at rest relative to a parabolic dish which rotates at a constant rate about a vertical axis. Knowing that the coefficient of static friction is 0.5 and that $r = 2$ m, determine the maximum allowable velocity v of the block.

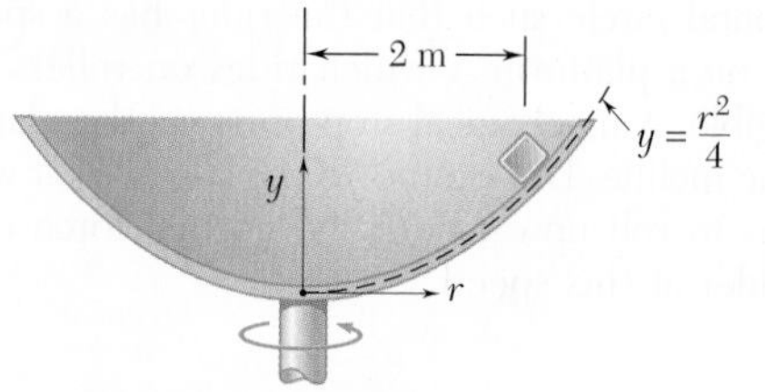

Fig. P12.55

12.56 Three seconds after a polisher is started from rest, small tufts of fleece from along the circumference of the 225-mm-diameter polishing pad are observed to fly free of the pad. If the polisher is started so that the fleece along the circumference undergoes a constant tangential acceleration of 4 m/s^2, determine (*a*) the speed v of a tuft as it leaves the pad, (*b*) the magnitude of the force required to free a tuft if the average mass of a tuft is 1.6 mg.

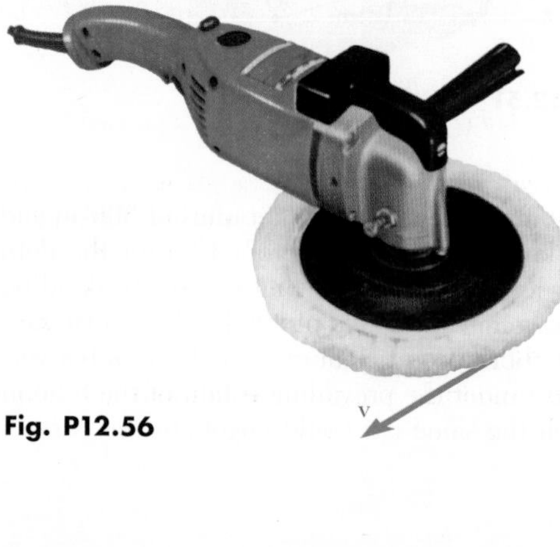

Fig. P12.56

12.57 A turntable A is built into a stage for use in a theatrical production. It is observed during a rehearsal that a trunk B starts to slide on the turntable 10 s after the turntable begins to rotate. Knowing that the trunk undergoes a constant tangential acceleration of 0.24 m/s^2, determine the coefficient of static friction between the trunk and the turntable.

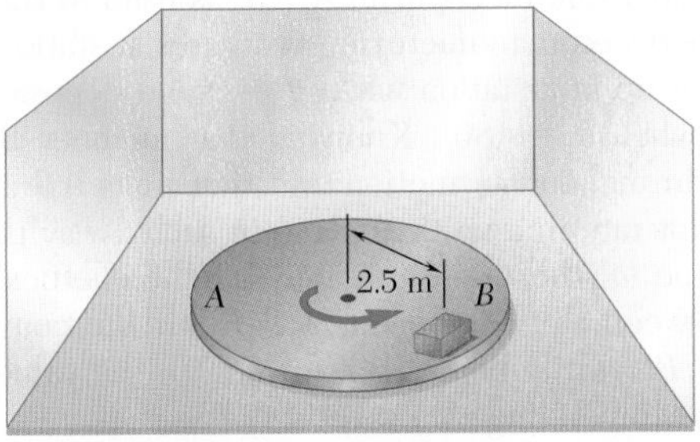

Fig. P12.57

12.58 A small, 300-g collar D can slide on portion AB of a rod which is bent as shown. Knowing that $\alpha = 40°$ and that the rod rotates about the vertical AC at a constant rate of 5 rad/s, determine the value of r for which the collar will not slide on the rod if the effect of friction between the rod and the collar is neglected.

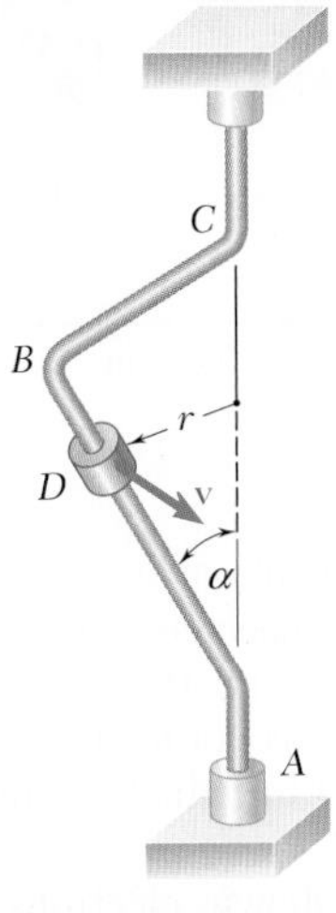

Fig. P12.58 and *P12.59*

12.59 A small, 200-g collar D can slide on portion AB of a rod which is bent as shown. Knowing that the rod rotates about the vertical AC at a constant rate and that $\alpha = 30°$ and $r = 600$ mm, determine the range of values of the speed v for which the collar will not slide on the rod if the coefficient of static friction between the rod and the collar is 0.30.

12.60 A semicircular slot of 250-mm radius is cut in a flat plate which rotates about the vertical AD at a constant rate of 14 rad/s. A small, 400-g block E is designed to slide in the slot as the plate rotates. Knowing that the coefficients of friction are $\mu_s = 0.35$ and $\mu_k = 0.25$, determine whether the block will slide in the slot if it is released in the position corresponding to (*a*) $\theta = 80°$, (*b*) $\theta = 40°$. Also determine the magnitude and the direction of the friction force exerted on the block immediately after it is released.

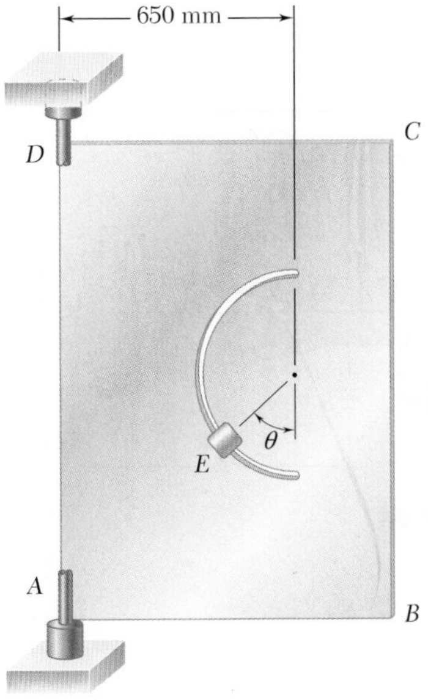

Fig. P12.60

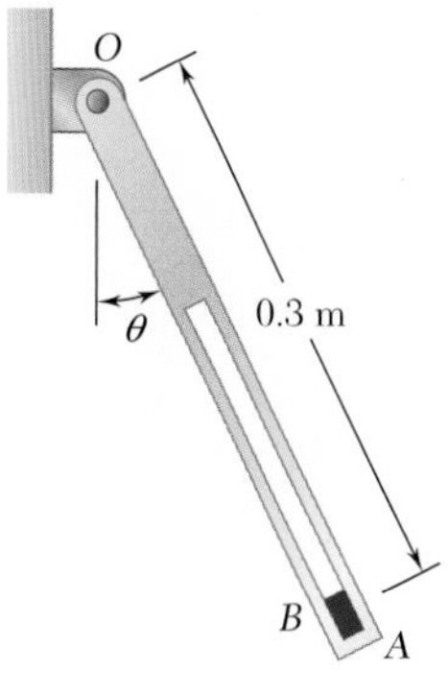

Fig. P12.61

12.61 A small block B fits inside a slot cut in arm OA which rotates in a vertical plane at a constant rate. The block remains in contact with the end of the slot closest to A and its speed is 1.4 m/s for $0 \leq \theta \leq 150°$. Knowing that the block begins to slide when $\theta = 150°$, determine the coefficient of static friction between the block and the slot.

12.62 The parallel-link mechanism $ABCD$ is used to transport a component I between manufacturing processes at stations E, F, and G by picking it up at a station when $\theta = 0$ and depositing it at the next station when $\theta = 180°$. Knowing that member BC remains horizontal throughout its motion and that links AB and CD rotate at a constant rate in a vertical plane in such a way that $v_B = 0.7$ m/s, determine (*a*) the minimum value of the coefficient of static friction between the component and BC if the component is not to slide on BC while being transferred, (*b*) the values of θ for which sliding is impending.

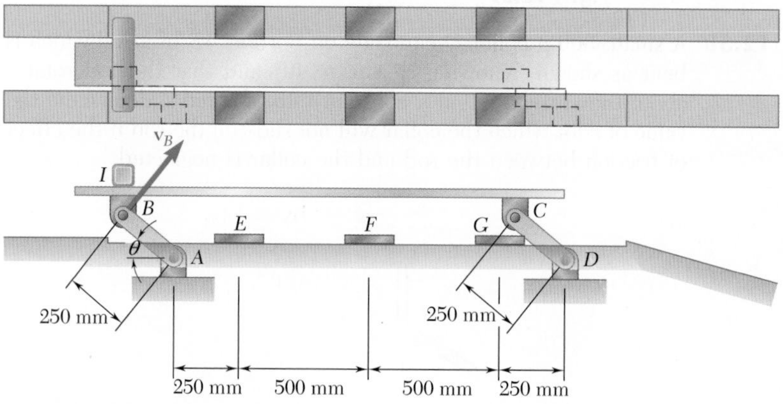

Fig. P12.62

12.63 Knowing that the coefficients of friction between the component I and member BC of the mechanism of Prob. 12.62 are $\mu_s = 0.35$ and $\mu_k = 0.25$, determine (*a*) the maximum allowable constant speed v_B if the component is not to slide on BC while being transferred, (*b*) the values of θ for which sliding is impending.

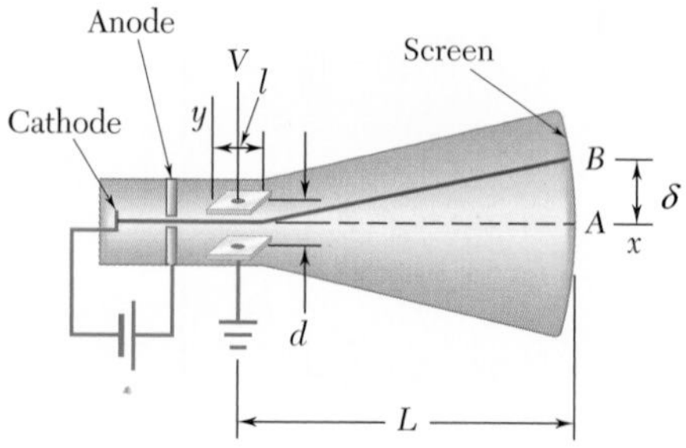

Fig. P12.64

12.64 In the cathode-ray tube shown, electrons emitted by the cathode and attracted by the anode pass through a small hole in the anode and then travel in a straight line with a speed v_0 until they strike the screen at A. However, if a difference of potential V is established between the two parallel plates, the electrons will be subjected to a force **F** perpendicular to the plates while they travel between the plates and will strike the screen at point B, which is at a distance δ from A. The magnitude of the force **F** is $F = eV/d$, where $-e$ is the charge of an electron and d is the distance between the plates. Derive an expression for the deflection d in terms of V, v_0, the charge $-e$ and the mass m of an electron, and the dimensions d, l, and L.

12.65 In Prob. 12.64, determine the smallest allowable value of the ratio d/l in terms of e, m, v_0, and V if at $x = l$ the minimum permissible distance between the path of the electrons and the positive plate is $0.05d$.

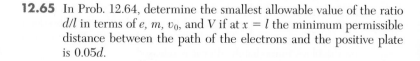

12.7 ANGULAR MOMENTUM OF A PARTICLE. RATE OF CHANGE OF ANGULAR MOMENTUM

Consider a particle P of mass m moving with respect to a newtonian frame of reference $Oxyz$. As we saw in Sec. 12.3, the linear momentum of the particle at a given instant is defined as the vector $m\mathbf{v}$ obtained by multiplying the velocity $\mathbf{v}$ of the particle by its mass m. The moment about O of the vector $m\mathbf{v}$ is called the *moment of momentum,* or the *angular momentum,* of the particle about O at that instant and is denoted by $\mathbf{H}_O$. Recalling the definition of the moment of a vector (Sec. 3.6) and denoting by $\mathbf{r}$ the position vector of P, we write

$$\mathbf{H}_O = \mathbf{r} \times m\mathbf{v} \qquad (12.12)$$

and note that $\mathbf{H}_O$ is a vector perpendicular to the plane containing $\mathbf{r}$ and $m\mathbf{v}$ and of magnitude

$$H_O = rmv \sin \phi \qquad (12.13)$$

where ϕ is the angle between $\mathbf{r}$ and $m\mathbf{v}$ (Fig. 12.12). The sense of $\mathbf{H}_O$ can be determined from the sense of $m\mathbf{v}$ by applying the right-hand rule. The unit of angular momentum is obtained by multiplying the units of length and of linear momentum (Sec. 12.4). With SI units, we have

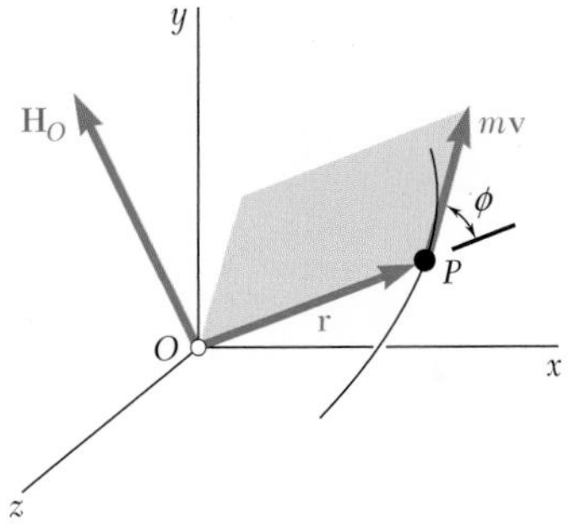

Fig. 12.12

$$(\text{m})(\text{kg} \cdot \text{m/s}) = \text{kg} \cdot \text{m}^2/\text{s}$$

With U.S. customary units, we write

$$(\text{ft})(\text{lb} \cdot \text{s}) = \text{ft} \cdot \text{lb} \cdot \text{s}$$

Resolving the vectors $\mathbf{r}$ and $m\mathbf{v}$ into components and applying formula (3.10), we write

$$\mathbf{H}_O = \begin{vmatrix} \mathbf{i} & \mathbf{j} & \mathbf{k} \\ x & y & z \\ mv_x & mv_y & mv_z \end{vmatrix} \qquad (12.14)$$

The components of $\mathbf{H}_O$, which also represent the moments of the linear momentum $m\mathbf{v}$ about the coordinate axes, can be obtained by expanding the determinant in (12.14). We have

$$\begin{aligned} H_x &= m(yv_z - zv_y) \\ H_y &= m(zv_x - xv_z) \\ H_z &= m(xv_y - yv_x) \end{aligned} \qquad (12.15)$$

In the case of a particle moving in the xy plane, we have $z = v_z = 0$ and the components H_x and H_y reduce to zero. The angular momentum is thus perpendicular to the xy plane; it is then completely defined by the scalar

$$H_O = H_z = m(xv_y - yv_x) \qquad (12.16)$$

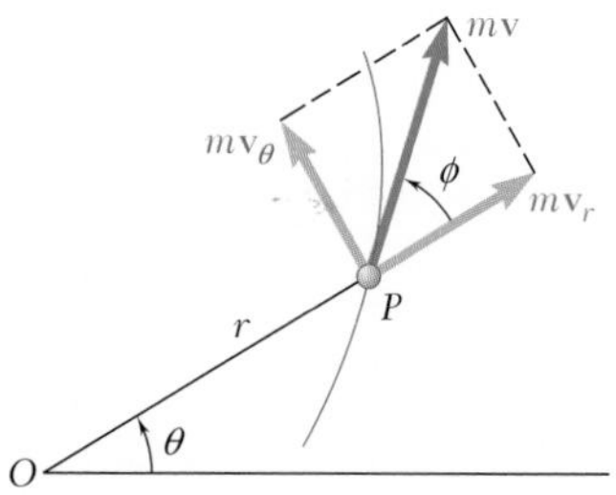

Fig. 12.13

which will be positive or negative according to the sense in which the particle is observed to move from O. If polar coordinates are used, we resolve the linear momentum of the particle into radial and transverse components (Fig. 12.13) and write

$$H_O = rmv \sin \phi = rmv_\theta \tag{12.17}$$

or, recalling from (11.45) that $v_\theta = r\dot{\theta}$,

$$H_O = mr^2\dot{\theta} \tag{12.18}$$

Let us now compute the derivative with respect to t of the angular momentum $\mathbf{H}_O$ of a particle P moving in space. Differentiating both members of Eq. (12.12), and recalling the rule for the differentiation of a vector product (Sec. 11.10), we write

$$\dot{\mathbf{H}}_O = \dot{\mathbf{r}} \times m\mathbf{v} + \mathbf{r} \times m\dot{\mathbf{v}} = \mathbf{v} \times m\mathbf{v} + \mathbf{r} \times m\mathbf{a}$$

Since the vectors $\mathbf{v}$ and $m\mathbf{v}$ are collinear, the first term of the expression obtained is zero; and, by Newton's second law, $m\mathbf{a}$ is equal to the sum $\Sigma\mathbf{F}$ of the forces acting on P. Noting that $\mathbf{r} \times \Sigma\mathbf{F}$ represents the sum $\Sigma\mathbf{M}_O$ of the moments about O of these forces, we write

$$\Sigma\mathbf{M}_O = \dot{\mathbf{H}}_O \tag{12.19}$$

Equation (12.19), which results directly from Newton's second law, states that *the sum of the moments about O of the forces acting on the particle is equal to the rate of change of the moment of momentum, or angular momentum, of the particle about O.*

12.8 EQUATIONS OF MOTION IN TERMS OF RADIAL AND TRANSVERSE COMPONENTS

Consider a particle P, of polar coordinates r and θ, which moves in a plane under the action of several forces. Resolving the forces and the acceleration of the particle into radial and transverse components (Fig. 12.14) and substituting into Eq. (12.2), we obtain the two scalar equations

$$\Sigma F_r = ma_r \qquad \Sigma F_\theta = ma_\theta \tag{12.20}$$

Substituting for a_r and a_θ from Eqs. (11.46), we have

$$\Sigma F_r = m(\ddot{r} - r\dot{\theta}^2) \tag{12.21}$$

$$\Sigma F_\theta = m(r\ddot{\theta} + 2\dot{r}\dot{\theta}) \tag{12.22}$$

The equations obtained can be solved for two unknowns.

Photo 12.4 The forces on the specimens used in a high speed centrifuge can be described in terms of radial and transverse components.

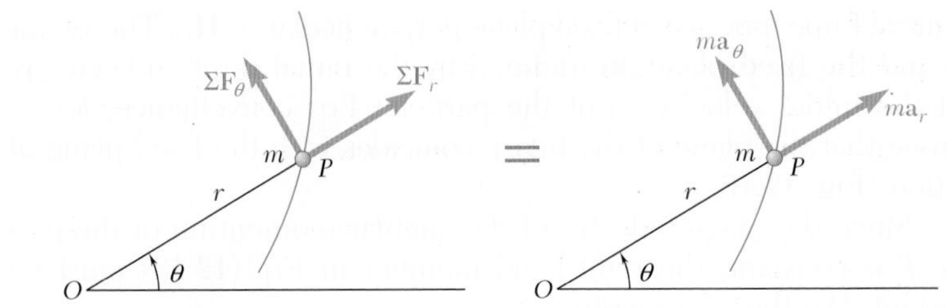

Fig. 12.14

Equation (12.22) could have been derived from Eq. (12.19). Recalling (12.18) and noting that $\Sigma M_O = r\Sigma F_\theta$, Eq. (12.19) yields

$$\begin{aligned} r\Sigma F_\theta &= \frac{d}{dt}(mr^2\dot{\theta}) \\ &= m(r^2\ddot{\theta} + 2r\dot{r}\dot{\theta}) \end{aligned}$$

and, after dividing both members by r,

$$\Sigma F_\theta = m(r\ddot{\theta} + 2\dot{r}\dot{\theta}) \qquad (12.22)$$

12.9 MOTION UNDER A CENTRAL FORCE. CONSERVATION OF ANGULAR MOMENTUM

When the only force acting on a particle P is a force **F** directed toward or away from a fixed point O, the particle is said to be moving *under a central force,* and the point O is referred to as the *center of force* (Fig. 12.15). Since the line of action of **F** passes through O, we must have $\Sigma \mathbf{M}_O = 0$ at any given instant. Substituting into Eq. (12.19), we therefore obtain

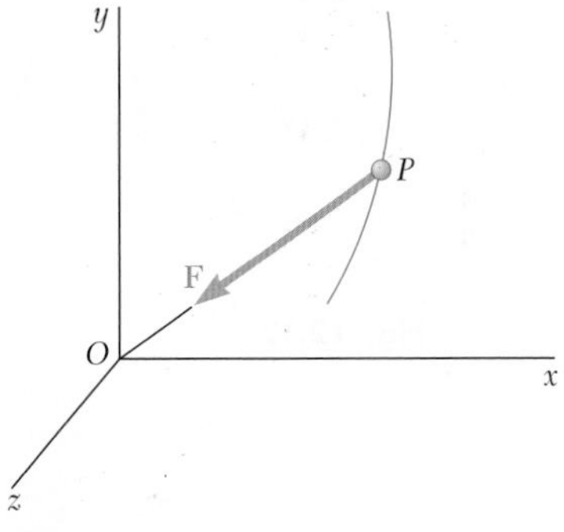

Fig. 12.15

$$\dot{\mathbf{H}}_O = 0$$

for all values of t and, integrating in t,

$$\mathbf{H}_O = \text{constant} \qquad (12.23)$$

We thus conclude that *the angular momentum of a particle moving under a central force is constant, in both magnitude and direction.*

Recalling the definition of the angular momentum of a particle (Sec. 12.7), we write

$$\mathbf{r} \times m\mathbf{v} = \mathbf{H}_O = \text{constant} \qquad (12.24)$$

from which it follows that the position vector **r** of the particle P must be perpendicular to the constant vector $\mathbf{H}_O$. Thus, a particle under

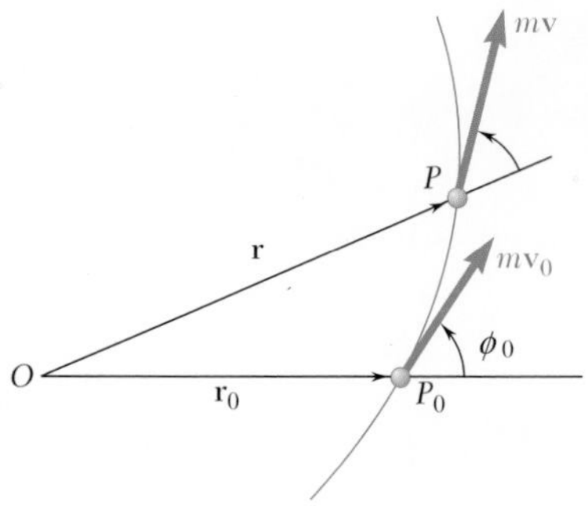

Fig. 12.16

a central force moves in a fixed plane perpendicular to $\mathbf{H}_O$. The vector $\mathbf{H}_O$ and the fixed plane are defined by the initial position vector $\mathbf{r}_0$ and the initial velocity $\mathbf{v}_0$ of the particle. For convenience, let us assume that the plane of the figure coincides with the fixed plane of motion (Fig. 12.16).

Since the magnitude H_O of the angular momentum of the particle P is constant, the right-hand member in Eq. (12.13) must be constant. We therefore write

$$rmv \sin \phi = r_0 m v_0 \sin \phi_0 \tag{12.25}$$

This relation applies to the motion of any particle under a central force. Since the gravitational force exerted by the sun on a planet is a central force directed toward the center of the sun, Eq. (12.25) is fundamental to the study of planetary motion. For a similar reason, it is also fundamental to the study of the motion of space vehicles in orbit about the earth.

Alternatively, recalling Eq. (12.18), we can express the fact that the magnitude H_O of the angular momentum of the particle P is constant by writing

$$mr^2\dot{\theta} = H_O = \text{constant} \tag{12.26}$$

or, dividing by m and denoting by h the angular momentum per unit mass H_O/m,

$$r^2\dot{\theta} = h \tag{12.27}$$

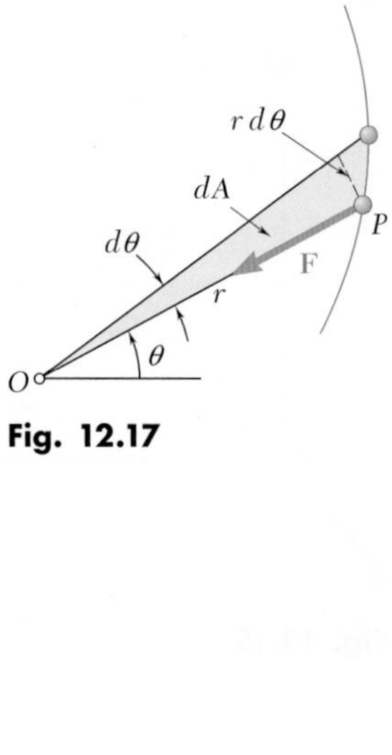

Fig. 12.17

Equation (12.27) can be given an interesting geometric interpretation. Observing from Fig. 12.17 that the radius vector OP sweeps an infinitesimal area $dA = \frac{1}{2}r^2\, d\theta$ as it rotates through an angle $d\theta$, and defining the *areal velocity* of the particle as the quotient dA/dt, we note that the left-hand member of Eq. (12.27) represents twice the areal velocity of the particle. We thus conclude that *when a particle moves under a central force, its areal velocity is constant.*

12.10 NEWTON'S LAW OF GRAVITATION

As you saw in the preceding section, the gravitational force exerted by the sun on a planet or by the earth on an orbiting satellite is an important example of a central force. In this section, you will learn how to determine the magnitude of a gravitational force.

In his *law of universal gravitation,* Newton states that two particles of masses M and m at a distance r from each other attract each other with equal and opposite forces $\mathbf{F}$ and $-\mathbf{F}$ directed along the line joining the particles (Fig. 12.18). The common magnitude F of the two forces is

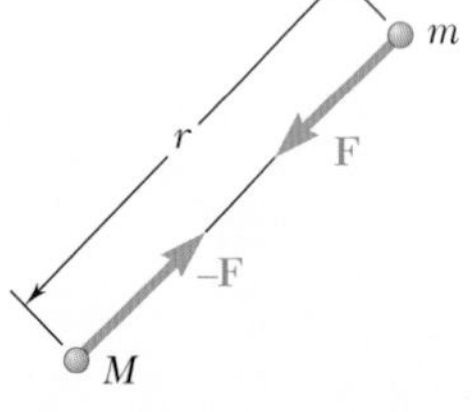

Fig. 12.18

$$F = G\frac{Mm}{r^2} \tag{12.28}$$

where G is a universal constant, called the *constant of gravitation*. Experiments show that the value of G is $(66.73 \pm 0.03) \times 10^{-12}$ m^3/kg · s^2 in SI units or approximately 34.4×10^{-9} ft^4/lb · s^4 in U.S. customary units. Gravitational forces exist between any pair of bodies, but their effect is appreciable only when one of the bodies has a very large mass. The effect of gravitational forces is apparent in the cases of the motion of a planet about the sun, of satellites orbiting about the earth, or of bodies falling on the surface of the earth.

Since the force exerted by the earth on a body of mass m located on or near its surface is defined as the weight **W** of the body, we can substitute the magnitude $W = mg$ of the weight for F, and the radius R of the earth for r, in Eq. (12.28). We obtain

$$W = mg = \frac{GM}{R^2}m \qquad \text{or} \qquad g = \frac{GM}{R^2} \tag{12.29}$$

where M is the mass of the earth. Since the earth is not truly spherical, the distance R from the center of the earth depends upon the point selected on its surface, and the values of W and g will thus vary with the altitude and latitude of the point considered. Another reason for the variation of W and g with latitude is that a system of axes attached to the earth does not constitute a newtonian frame of reference (see Sec. 12.2). A more accurate definition of the weight of a body should therefore include a component representing the centrifugal force due to the rotation of the earth. Values of g at sea level vary from 9.781 m/s^2, or 32.09 ft/s^2, at the equator to 9.833 m/s^2, or 32.26 ft/s^2, at the poles.†

The force exerted by the earth on a body of mass m located in space at a distance r from its center can be found from Eq. (12.28). The computations will be somewhat simplified if we note that according to Eq. (12.29), the product of the constant of gravitation G and the mass M of the earth can be expressed as

$$GM = gR^2 \tag{12.30}$$

where g and the radius R of the earth will be given their average values $g = 9.81$ m/s^2 and $R = 6.37 \times 10^6$ m in SI units‡ and $g =$ 32.2 ft/s^2 and $R =$ (3960 mi)(5280 ft/mi) in U.S. customary units.

The discovery of the law of universal gravitation has often been attributed to the belief that, after observing an apple falling from a tree, Newton had reflected that the earth must attract an apple and the moon in much the same way. While it is doubtful that this incident actually took place, it may be said that Newton would not have formulated his law if he had not first perceived that the acceleration of a falling body must have the same cause as the acceleration which keeps the moon in its orbit. This basic concept of the continuity of gravitational attraction is more easily understood today, when the gap between the apple and the moon is being filled with artificial earth satellites.

†A formula expressing g in terms of the latitude ϕ was given in Prob. 12.2.

‡The value of R is easily found if one recalls that the circumference of the earth is $2\pi R = 40 \times 10^6$ m.

SAMPLE PROBLEM 12.7

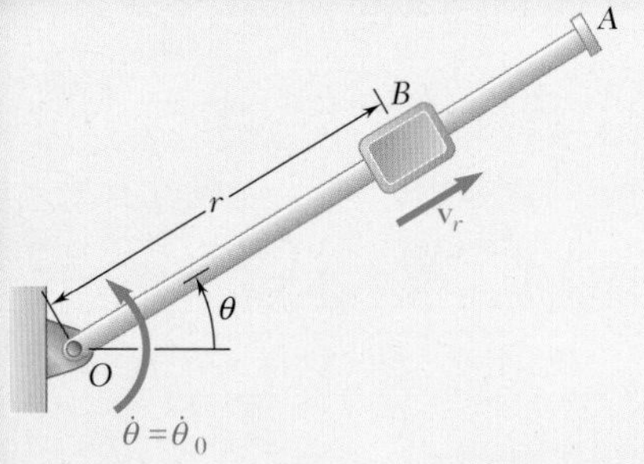

A block B of mass m can slide freely on a frictionless arm OA which rotates in a horizontal plane at a constant rate $\dot{\theta}_0$. Knowing that B is released at a distance r_0 from O, express as a function of r, (*a*) the component v_r of the velocity of B along OA, (*b*) the magnitude of the horizontal force $\mathbf{F}$ exerted on B by the arm OA.

SOLUTION

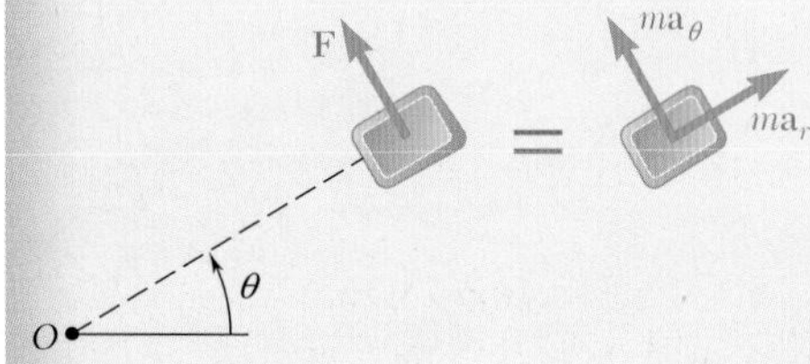

Since all other forces are perpendicular to the plane of the figure, the only force shown acting on B is the force $\mathbf{F}$ perpendicular to OA.

Equations of Motion. Using radial and transverse components,

$$+\nearrow \Sigma F_r = ma_r: \qquad 0 = m(\ddot{r} - r\dot{\theta}^2) \qquad (1)$$

$$+\nwarrow \Sigma F_\theta = ma_\theta: \qquad F = m(r\ddot{\theta} + 2\dot{r}\dot{\theta}) \qquad (2)$$

a. Component v_r of Velocity. Since $v_r = \dot{r}$, we have

$$\ddot{r} = \dot{v}_r = \frac{dv_r}{dt} = \frac{dv_r}{dr}\frac{dr}{dt} = v_r \frac{dv_r}{dr}$$

Substituting for $\ddot{r}$ in (1), recalling that $\dot{\theta} = \dot{\theta}_0$, and separating the variables,

$$v_r \, dv_r = \dot{\theta}_0^2 r \, dr$$

Multiplying by 2, and integrating from 0 to v_r and from r_0 to r,

$$v_r^2 = \dot{\theta}_0^2(r^2 - r_0^2) \qquad v_r = \dot{\theta}_0(r^2 - r_0^2)^{1/2} \quad ◀$$

b. Horizontal Force F. Setting $\dot{\theta} = \dot{\theta}_0$, $\ddot{\theta} = 0$, $\dot{r} = v_r$ in Eq. (2), and substituting for v_r the expression obtained in part *a*,

$$F = 2m\dot{\theta}_0(r^2 - r_0^2)^{1/2}\dot{\theta}_0 \qquad F = 2m\dot{\theta}_0^2(r^2 - r_0^2)^{1/2} \quad ◀$$

SAMPLE PROBLEM 12.8

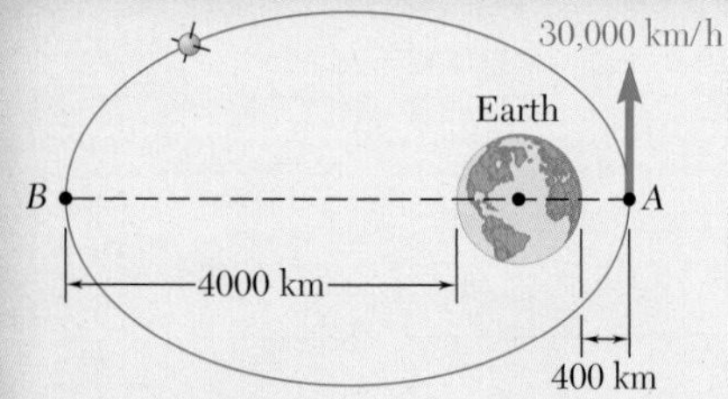

A satellite is launched in a direction parallel to the surface of the earth with a velocity of 30,000 km/h from an altitude of 400 km. Determine the velocity of the satellite as it reaches its maximum altitude of 4000 km. It is recalled that the radius of the earth is 6370 km.

SOLUTION

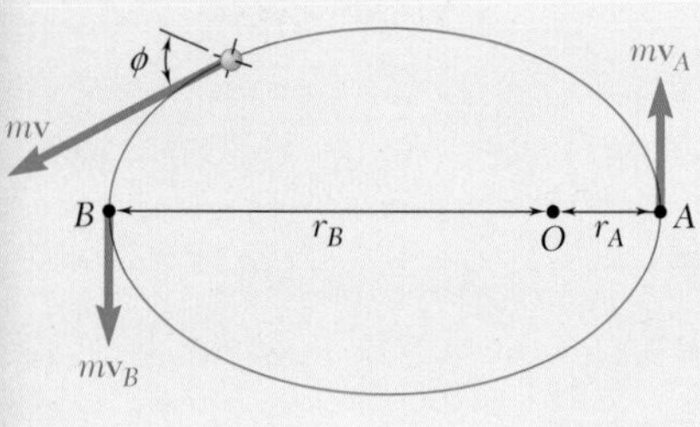

Since the satellite is moving under a central force directed toward the center O of the earth, its angular momentum $\mathbf{H}_O$ is constant. From Eq. (12.13) we have

$$rmv \sin \phi = H_O = \text{constant}$$

which shows that v is minimum at B, where both r and $\sin \phi$ are maximum. Expressing conservation of angular momentum between A and B,

$$r_A m v_A = r_B m v_B$$

$$v_B = v_A \frac{r_A}{r_B} = (30{,}000 \text{ km/h}) \left(\frac{6370 \text{ km} + 400 \text{ km}}{6370 \text{ km} + 4000 \text{ km}} \right)$$

$$v_B = 19{,}590 \text{ km/h} \quad ◀$$

Note: It may be observed that r is the distance from the centre of the earth and is expressed as $r = R_{earth}$ + altitude.

SOLVING PROBLEMS ON YOUR OWN

In this lesson we continued our study of Newton's second law by expressing the force and the acceleration in terms of their *radial and transverse components*, where the corresponding equations of motion are

$$\Sigma F_r = ma_r: \qquad \Sigma F_r = m(\ddot{r} - r\dot{\theta}^2)$$
$$\Sigma F_\theta = ma_\theta: \qquad \Sigma F_\theta = m(r\ddot{\theta} + 2\dot{r}\dot{\theta})$$

We introduced the *moment of the momentum*, or the *angular momentum*, $\mathbf{H}_O$ of a particle about O:

$$\mathbf{H}_O = \mathbf{r} \times m\mathbf{v} \tag{12.12}$$

and found that $\mathbf{H}_O$ is constant when the particle moves under a *central force* with its center located at O.

1. Using radial and transverse components. Radial and transverse components were introduced in the last lesson of Chap. 11 [Sec. 11.14]; you should review that material before attempting to solve the following problems. Also, our comments in the preceding lesson regarding the application of Newton's second law (drawing a free-body diagram and a $m\mathbf{a}$ diagram, etc.) still apply [Sample Prob. 12.7]. Finally, note that the solution of that sample problem depends on the application of techniques developed in Chap. 11—you will need to use similar techniques to solve some of the problems of this lesson.

2. Solving problems involving the motion of a particle under a central force. In problems of this type, the angular momentum $\mathbf{H}_O$ of the particle about the center of force O is conserved. You will find it convenient to introduce the constant $h = H_O/m$ representing the angular momentum per unit mass. Conservation of the angular momentum of the particle P about O can then be expressed by either of the following equations

$$rv \sin \phi = h \qquad \text{or} \qquad r^2\dot{\theta} = h$$

where r and θ are the polar coordinates of P, and ϕ is the angle that the velocity $\mathbf{v}$ of the particle forms with the line OP (Fig. 12.16). The constant h can be determined from the initial conditions and either of the above equations can be solved for one unknown.

(continued)

3. In space-mechanics problems involving the orbital motion of a planet about the sun, or a satellite about the earth, the moon, or some other planet, the central force $\mathbf{F}$ is the force of gravitational attraction; it is directed *toward* the center of force O and has the magnitude

$$F = G\frac{Mm}{r^2} \qquad (12.28)$$

Note that in the particular case of the gravitational force exerted by the earth, the product GM can be replaced by gR^2, where R is the radius of the earth [Eq. 12.30].

The following two cases of orbital motion are frequently encountered:

a. For a satellite in a circular orbit, the force $\mathbf{F}$ is normal to the orbit and you can write $F = ma_n$; substituting for F from Eq. (12.28) and observing that $a_n = v^2/\rho = v^2/r$, you will obtain

$$G\frac{Mm}{r^2} = m\frac{v^2}{r} \qquad \text{or} \qquad v^2 = \frac{GM}{r}$$

b. For a satellite in an elliptic orbit, the radius vector $\mathbf{r}$ and the velocity $\mathbf{v}$ of the satellite are perpendicular to each other at the points A and B which are, respectively, farthest and closest to the center of force O [Sample Prob. 12.8]. Thus, conservation of angular momentum of the satellite between these two points can be expressed as

$$r_A m v_A = r_B m v_B$$

FREE-BODY PRACTICE PROBLEMS

12.F9 Four pins slide in four separate slots cut in a horizontal circular plate as shown. When the plate is at rest, each pin has a velocity directed as shown and of the same constant magnitude u. Each pin has a mass m and maintains the same velocity relative to the plate when the plate rotates about O with a constant counterclockwise angular velocity ω. Draw the FBDs and KDs to determine the forces on pins P_1 and P_2.

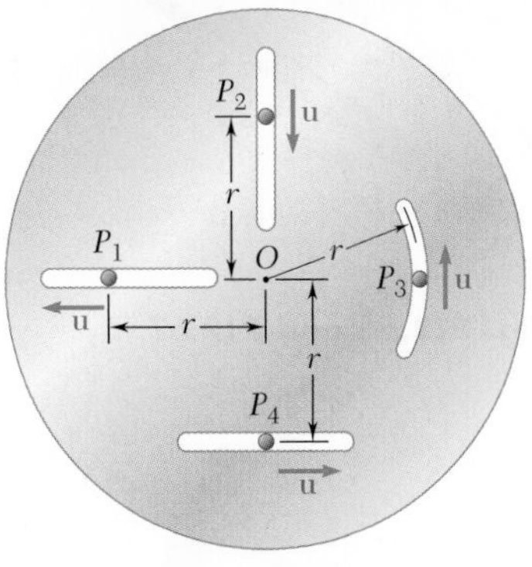

Fig. P12.F9

12.F10 At the instant shown, the length of the boom AB is being *decreased* at the constant rate of 0.2 m/s, and the boom is being lowered at the constant rate of 0.08 rad/s. If the mass of the men and lift connected to the boom at point B is m, draw the FBD and KD that could be used to determine the horizontal and vertical forces at B.

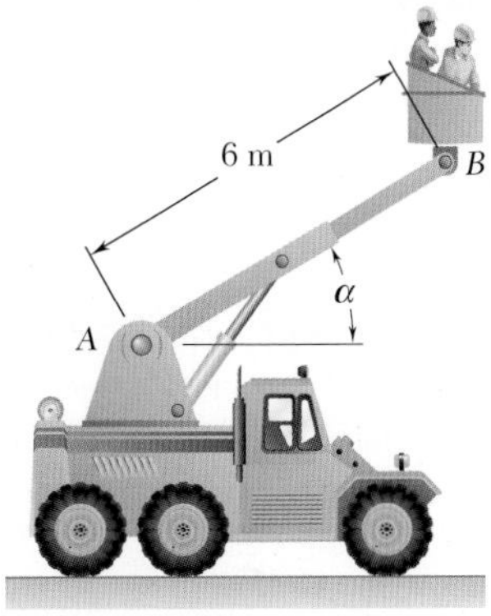

Fig. P12.F10

12.F11 Disk A rotates in a horizontal plane about a vertical axis at the constant rate $\dot{\theta}_0$. Slider B has a mass m and moves in a frictionless slot cut in the disk. The slider is attached to a spring of constant k, which is undeformed when $r = 0$. Knowing that the slider is released with no radial velocity in the position $r = r_0$, draw a FBD and KD at an arbitrary distance r from O.

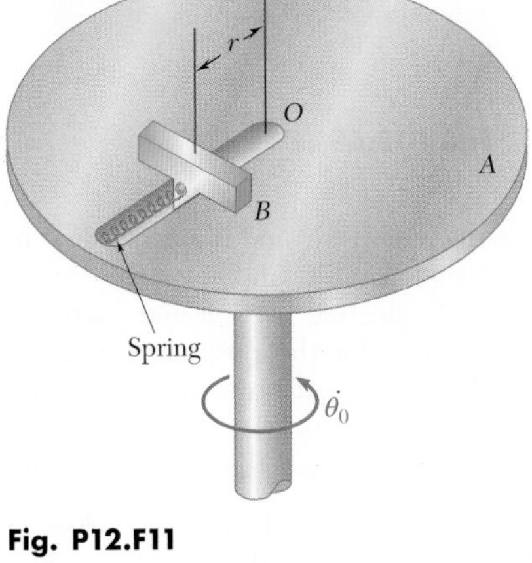

Fig. P12.F11

12.F12 Pin B has a mass m and slides along the slot in the rotating arm OC and along the slot DE which is cut in a fixed horizontal plate. Neglecting friction and knowing that rod OC rotates at the constant rate $\dot{\theta}_0$, draw a FBD and KD that can be used to determine the forces **P** and **Q** exerted on pin B by rod OC and the wall of slot DE, respectively.

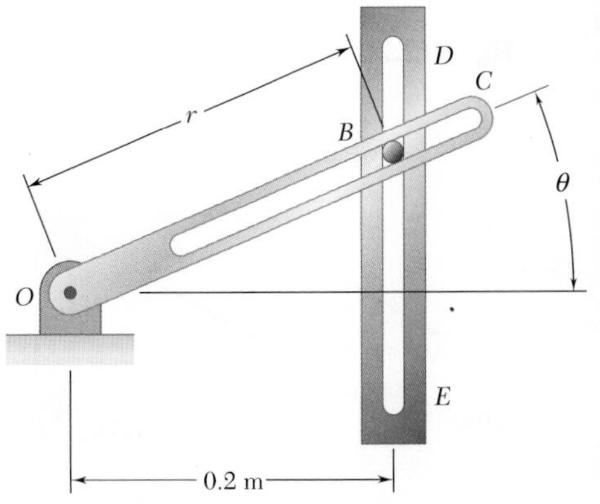

Fig. P12.F12

END-OF-SECTION PROBLEMS

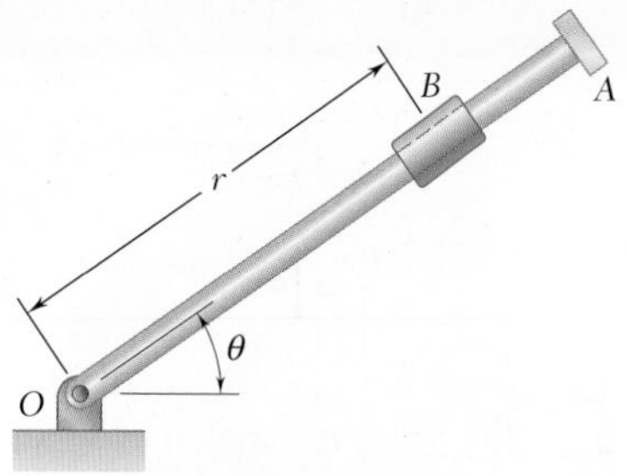

Fig. P12.66 and P12.67

12.66 Rod OA rotates about O in a horizontal plane. The motion of the 300-g collar B is defined by the relations $r = 300 + 100 \cos (0.5\, \pi t)$ and $\theta = \pi(t^2 - 3t)$, where r is expressed in millimeters, t in seconds, and θ in radians. Determine the radial and transverse components of the force exerted on the collar when (a) $t = 0$, (b) $t = 0.5$ s.

12.67 Rod OA oscillates about O in a horizontal plane. The motion of the 2.5 kg collar B is defined by the relations $r = 250/(t + 4)$ and $\theta = (2/\pi) \sin \pi t$, where r is expressed in millimeters, t in seconds, and θ in radians. Determine the radial and transverse components of the force exerted on the collar when (a) $t = 1$ s, (b) $t = 6$ s.

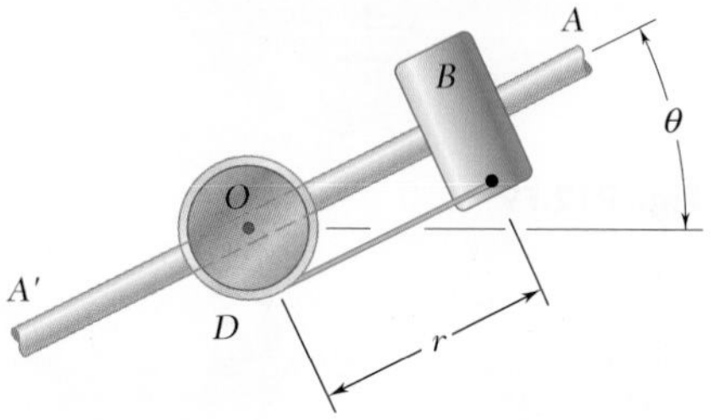

Fig. P12.68

12.68 The 3-kg collar B slides on the frictionless arm AA'. The arm is attached to drum D and rotates about O in a horizontal plane at the rate $\dot{\theta} = 0.75t$, where $\dot{\theta}$ and t are expressed in rad/s and seconds, respectively. As the arm-drum assembly rotates, a mechanism within the drum releases cord so that the collar moves outward from O with a constant speed of 0.5 m/s. Knowing that at $t = 0$, $r = 0$, determine the time at which the tension in the cord is equal to the magnitude of the horizontal force exerted on B by arm AA'.

12.69 The horizontal rod OA rotates about a vertical shaft according to the relation $\dot{\theta} = 10t$, where $\dot{\theta}$ and t are expressed in rad/s and seconds, respectively. A 250-g collar B is held by a cord with a breaking strength of 18 N. Neglecting friction, determine, immediately after the cord breaks, (a) the relative acceleration of the collar with respect to the rod, (b) the magnitude of the horizontal force exerted on the collar by the rod.

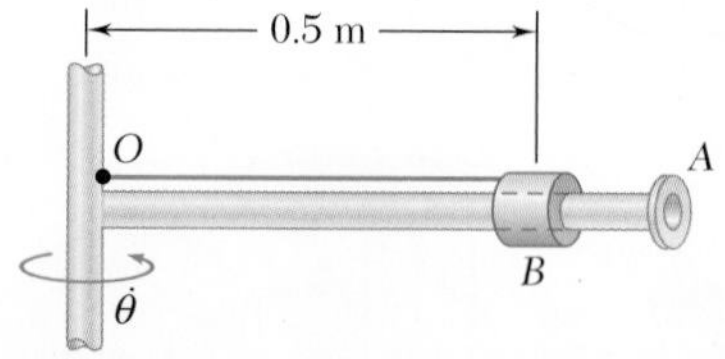

Fig. P12.69

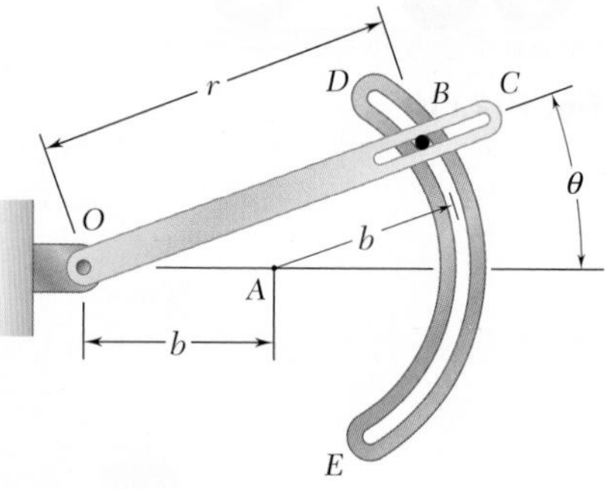

Fig. P12.70

12.70 Pin B has a mass of 100 g and is free to slide in a horizontal plane along the rotating arm OC and along the circular slot DE of radius $b = 500$ mm. Neglecting friction and assuming that $\dot{\theta} = 15$ rad/s and $\ddot{\theta} = 250$ rad/s^2 for the position $\theta = 20°$, determine for that position (a) the radial and transverse components of the resultant force exerted on pin B, (b) the forces $\mathbf{P}$ and $\mathbf{Q}$ exerted on pin B, respectively, by rod OC and the wall of slot DE.

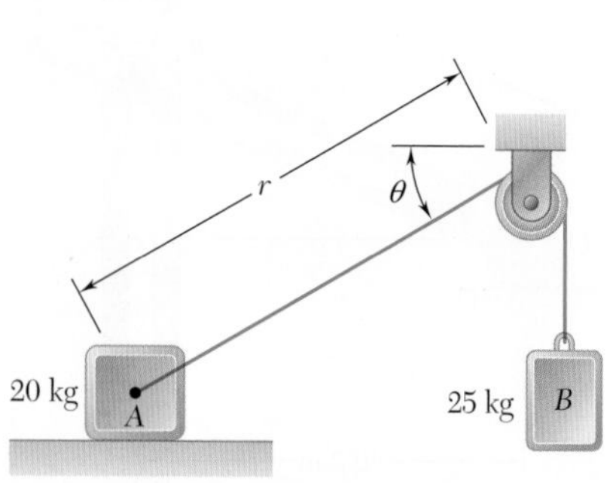

Fig. P12.71 and P12.72

12.71 The two blocks are released from rest when $r = 0.8$ m and $\theta = 30°$. Neglecting the mass of the pulley and the effect of friction in the pulley and between block A and the horizontal surface, determine (a) the initial tension in the cable, (b) the initial acceleration of block A, (c) the initial acceleration of block B.

12.72 The velocity of block A is 2 m/s to the right at the instant when $r = 0.8$ m and $\theta = 30°$. Neglecting the mass of the pulley and the effect of friction in the pulley and between block A and the horizontal surface, determine, at this instant, (a) the tension in the cable, (b) the acceleration of block A, (c) the acceleration of block B.

***12.73** Slider C has a mass of 250-g and may move in a slot cut in arm AB, which rotates at the constant rate $\dot{\theta}_0 = 10$ rad/s in a horizontal plane. The slider is attached to a spring of constant $k = 50$ N/m, which is unstretched when $r = 0$. Knowing that the slider is released from rest with no radial velocity in the position $r = 500$ mm and neglecting friction, determine for the position $r = 300$ mm (*a*) the radial and transverse components of the velocity of the slider, (*b*) the radial and transverse components of its acceleration, (*c*) the horizontal force exerted on the slider by arm AB.

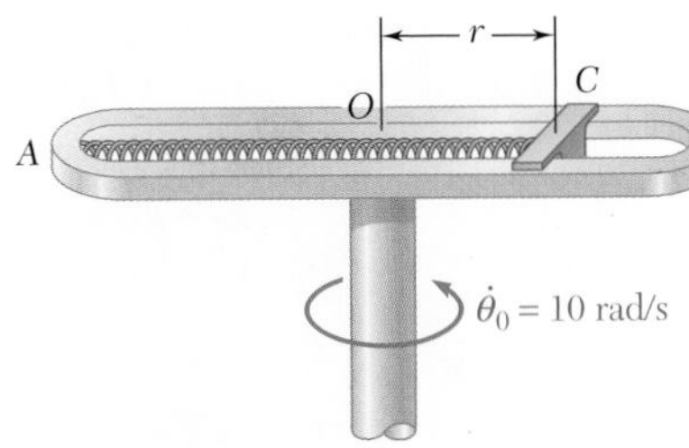

Fig. *P12.73*

12.74 A particle of mass m is projected from point A with an initial velocity $\mathbf{v}_0$ perpendicular to line OA and moves under a central force $\mathbf{F}$ directed away from the center of force O. Knowing that the particle follows a path defined by the equation $r = r_0/\sqrt{\cos 2\theta}$ and using Eq. (12.27), express the radial and transverse components of the velocity $\mathbf{v}$ of the particle as functions of θ.

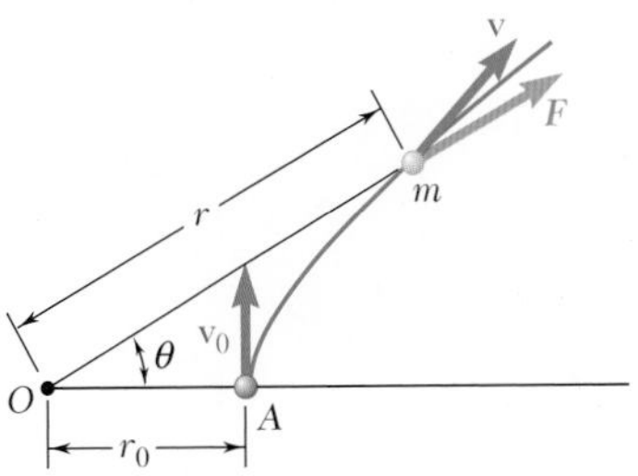

Fig. P12.74

12.75 For the particle of Prob. 12.74, show (*a*) that the velocity of the particle and the central force $\mathbf{F}$ are proportional to the distance r from the particle to the center of force O, (*b*) that the radius of curvature of the path is proportional to r^3.

12.76 A particle of mass m is projected from point A with an initial velocity $\mathbf{v}_0$ perpendicular to line OA and moves under a central force $\mathbf{F}$ along a semicircular path of diameter OA. Observing that $r = r_0 \cos\theta$ and using Eq. (12.27), show that the speed of the particle is $v = v_0/\cos^2\theta$.

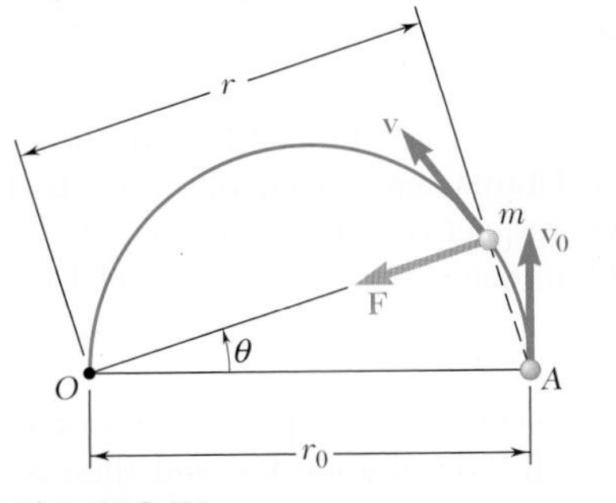

Fig. P12.76

12.77 For the particle of Prob. 12.76, determine the tangential component F_t of the central force $\mathbf{F}$ along the tangent to the path of the particle for (*a*) $\theta = 0$, (*b*) $\theta = 45°$.

12.78 Determine the mass of the earth knowing that the mean radius of the moon's orbit about the earth is 384,000 km and that the moon requires 27.32 days to complete one full revolution about the earth.

12.79 Show that the radius r of the moon's orbit can be determined from the radius R of the earth, the acceleration of gravity g at the surface of the earth, and the time τ required for the moon to complete one full revolution about the earth. Compute r knowing that $\tau = 27.3$ days.

12.80 Communication satellites are placed in a geosynchronous orbit, i.e., in a circular orbit such that they complete one full revolution about the earth in one sidereal day (23.934 h), and thus appear stationary with respect to the ground. Determine (*a*) the altitude of these satellites above the surface of the earth, (*b*) the velocity with which they describe their orbit.

12.81 Show that the radius r of the orbit of a moon of a given planet can be determined from the radius R of the planet, the acceleration of gravity at the surface of the planet, and the time τ required by the moon to complete one full revolution about the planet. Determine the acceleration of gravity at the surface of the planet Jupiter knowing that $R = 71\ 492$ km and that $\tau = 3.551$ days and $r = 670.9 \times 10^3$ km for its moon Europa.

12.82 The orbit of the planet Venus is nearly circular with an orbital velocity of 126.5×10^3 km/h. Knowing that the mean distance from the center of the sun to the center of Venus is 108×10^6 km and that the radius of the sun is 695.5×10^3 km, determine (*a*) the mass of the sun, (*b*) the acceleration of gravity at the surface of the sun.

12.83 A satellite is placed into a circular orbit about the planet Saturn at an altitude of 3400 km. The satellite describes its orbit with a velocity of 87,500 km/h. Knowing that the radius of the orbit about Saturn and the periodic time of Atlas, one of Saturn's moons, are 136.9×10^3 km and 0.6017 days, respectively, determine (*a*) the radius of Saturn, (*b*) the mass of Saturn. (The *periodic time* of a satellite is the time it requires to complete one full revolution about the planet.)

12.84 The periodic times (see Prob. 12.83) of the planet Uranus's moons Juliet and Titania have been observed to be 0.4931 days and 8.706 days, respectively. Knowing that the radius of Juliet's orbit is 64 360 km, determine (*a*) the mass of Uranus, (*b*) the radius of Titania's orbit.

12.85 A 500-kg spacecraft first is placed into a circular orbit about the earth at an altitude of 4500 km and then is transferred to a circular orbit about the moon. Knowing that the mass of the moon is 0.01230 times the mass of the earth and that the radius of the moon is 1737 km, determine (*a*) the gravitational force exerted on the spacecraft as it was orbiting the earth, (*b*) the required radius of the orbit of the spacecraft about the moon if the periodic times (see Prob. 12.83) of the two orbits are to be equal, (*c*) the acceleration of gravity at the surface of the moon.

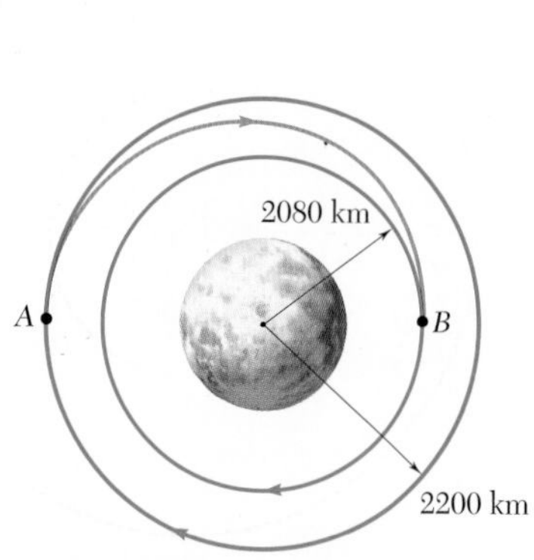

Fig. P12.86

12.86 A space vehicle is in a circular orbit of 2200-km radius around the moon. To transfer it to a smaller circular orbit of 2080-km radius, the vehicle is first placed on an elliptic path AB by reducing its speed by 26.3 m/s as it passes through A. Knowing that the mass of the moon is 73.49×10^{21} kg, determine (*a*) the speed of the vehicle as it approaches B on the elliptic path, (*b*) the amount by which its speed should be reduced as it approaches B to insert it into the smaller circular orbit.

12.87 Plans for an unmanned landing mission on the planet Mars called for the earth-return vehicle to first describe a circular orbit at an altitude $d_A = 2200$ km above the surface of the planet with a velocity of 2771 m/s. As it passed through point A, the vehicle was to be inserted into an elliptic transfer orbit by firing its engine and increasing its speed by $\Delta v_A = 1046$ m/s. As it passed through point B, at an altitude $d_B = 100\ 000$ km, the vehicle was to be inserted into a second transfer orbit located in a slightly different plane, by changing the direction of its velocity and reducing its speed by $\Delta v_B = -22.0$ m/s. Finally, as the vehicle passed through point C, at an altitude $d_C = 1000$ km, its speed was to be increased by $\Delta v_C = 660$ m/s to insert it into its return trajectory. Knowing that the radius of the planet Mars is $R = 3400$ km, determine the velocity of the vehicle after completion of the last maneuver.

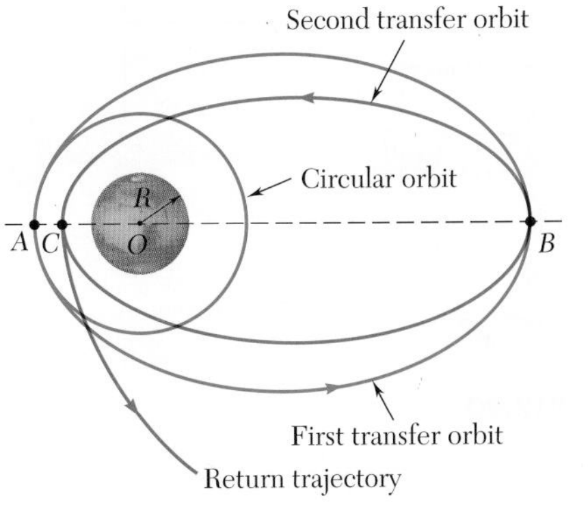

Fig. P12.87

12.88 To place a communications satellite into a geosynchronous orbit (see Prob. 12.80) at an altitude of 35,800 km above the surface of the earth, the satellite first is released from a space shuttle, which is in a circular orbit at an altitude of 300 km, and then is propelled by an upper-stage booster to its final altitude. As the satellite passes through A, the booster's motor is fired to insert the satellite into an elliptic transfer orbit. The booster is again fired at B to insert the satellite into a geosynchronous orbit. Knowing that the second firing increases the speed of the satellite by 1440 m/s, determine (*a*) the speed of the satellite as it approaches B on the elliptic transfer orbit, (*b*) the increase in speed resulting from the first firing at A.

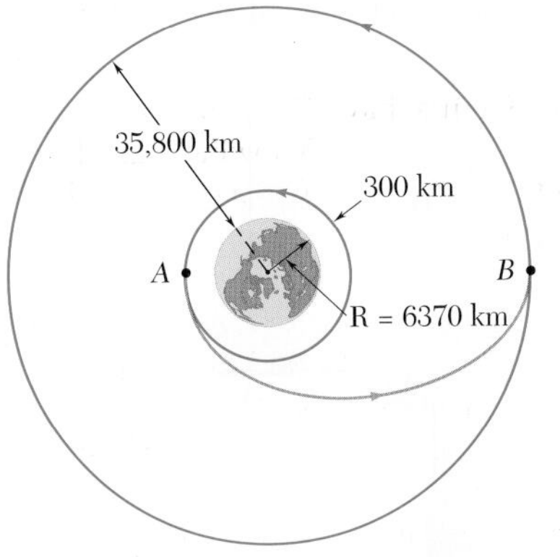

Fig. P12.88

12.89 A space shuttle S and a satellite A are in the circular orbits shown. In order for the shuttle to recover the satellite, the shuttle is first placed in an elliptic path BC by increasing its speed by $\Delta v_B = 84$ m/s as it passes through B. As the shuttle approaches C, its speed is increased by $\Delta v_C = 78$ m/s to insert it into a second elliptic transfer orbit CD. Knowing that the distance from O to C is 6860 km, determine the amount by which the speed of the shuttle should be increased as it approaches D to insert it into the circular orbit of the satellite.

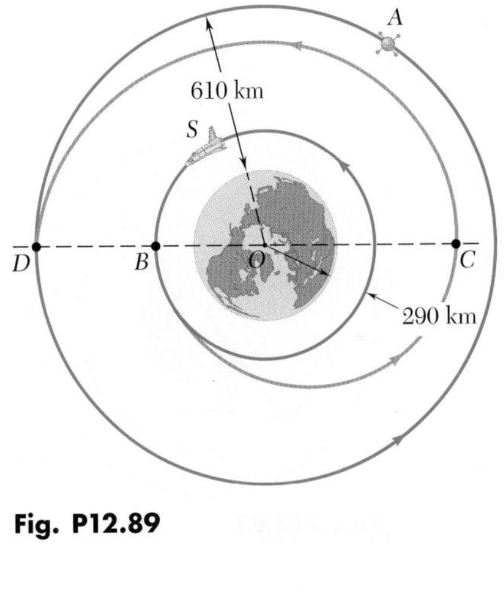

Fig. P12.89

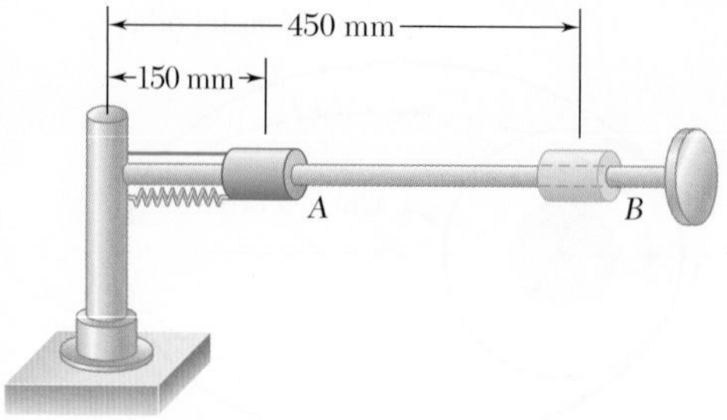

Fig. P12.90

12.90 A 1-kg collar can slide on a horizontal rod which is free to rotate about a vertical shaft. The collar is initially held at A by a cord attached to the shaft. A spring of constant 30 N/m is attached to the collar and to the shaft and is undeformed when the collar is at A. As the rod rotates at the rate $\dot{\theta} = 16$ rad/s, the cord is cut and the collar moves out along the rod. Neglecting friction and the mass of the rod, determine (*a*) the radial and transverse components of the acceleration of the collar at A, (*b*) the acceleration of the collar relative to the rod at A, (*c*) the transverse component of the velocity of the collar at B.

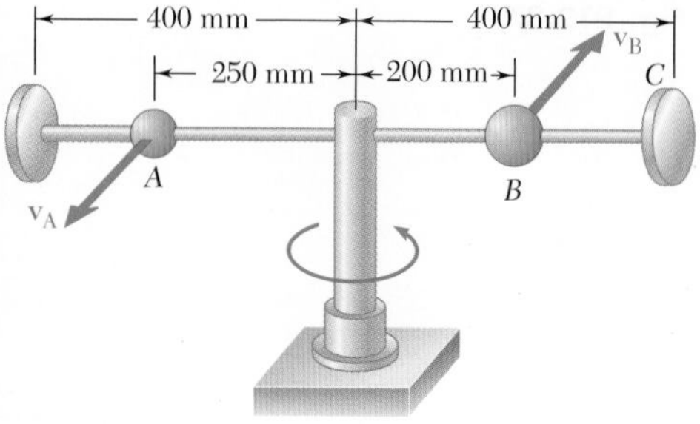

Fig. P12.91

12.91 A 200-g ball A and a 400-g ball B are mounted on a horizontal rod which rotates freely about a vertical shaft. The balls are held in the positions shown by pins. The pin holding B is suddenly removed and the ball moves to position C as the rod rotates. Neglecting friction and the mass of the rod and knowing that the initial speed of A is $v_A = 2.5$ m/s, determine (*a*) the radial and transverse components of the acceleration of ball B immediately after the pin is removed, (*b*) the acceleration of ball B relative to the rod at that instant, (*c*) the speed of ball A after ball B has reached the stop at C.

12.92 Two 1.25-kg collars A and B can slide without friction on a frame, consisting of the horizontal rod OE and the vertical rod CD, which is free to rotate about CD. The two collars are connected by a cord running over a pulley that is attached to the frame at O and a stop prevents collar B from moving. The frame is rotating at the rate $\dot{\theta} =$ 12 rad/s and $r = 200$ mm when the stop is removed allowing collar A to move out along rod OE. Neglecting friction and the mass of the frame, determine, for the position $r = 400$ mm, (*a*) the transverse component of the velocity of collar A, (*b*) the tension in the cord and the acceleration of collar A relative to the rod OE.

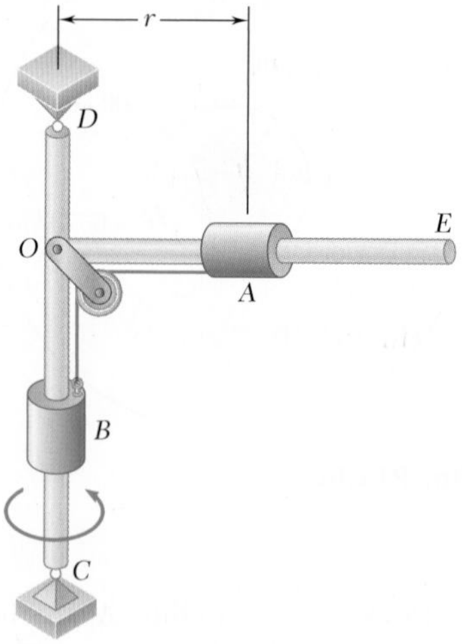

Fig. P12.92

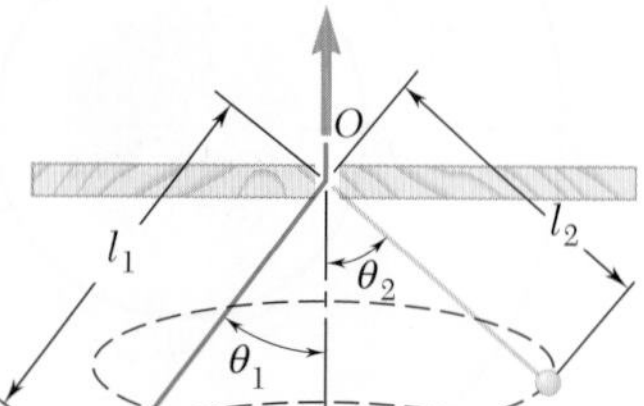

Fig. P12.93

12.93 A small ball swings in a horizontal circle at the end of a cord of length l_1, which forms an angle θ_1 with the vertical. The cord is then slowly drawn through the support at O until the length of the free end is l_2. (*a*) Derive a relation among l_1, l_2, θ_1, and θ_2. (*b*) If the ball is set in motion so that initially $l_1 = 0.8$ m and $\theta_1 = 35°$, determine the angle θ_2 when $l_2 = 0.6$ m.

*12.11 TRAJECTORY OF A PARTICLE UNDER A CENTRAL FORCE

Consider a particle P moving under a central force **F**. We propose to obtain the differential equation which defines its trajectory.

Assuming that the force **F** is directed toward the center of force O, we note that ΣF_r and ΣF_θ reduce, respectively, to $-F$ and zero in Eqs. (12.21) and (12.22). We therefore write

$$m(\ddot{r} - r\dot{\theta}^2) = -F \qquad (12.31)$$

$$m(r\ddot{\theta} + 2\dot{r}\dot{\theta}) = 0 \qquad (12.32)$$

These equations define the motion of P. We will, however, replace Eq. (12.32) by Eq. (12.27), which is equivalent to Eq. (12.32), as can easily be checked by differentiating it with respect to t, but which is more convenient to use. We write

$$r^2\dot{\theta} = h \qquad \text{or} \qquad r^2\frac{d\theta}{dt} = h \qquad (12.33)$$

Equation (12.33) can be used to eliminate the independent variable t from Eq. (12.31). Solving Eq. (12.33) for $\dot{\theta}$ or $d\theta/dt$, we have

$$\dot{\theta} = \frac{d\theta}{dt} = \frac{h}{r^2} \qquad (12.34)$$

from which it follows that

$$\dot{r} = \frac{dr}{dt} = \frac{dr}{d\theta}\frac{d\theta}{dt} = \frac{h}{r^2}\frac{dr}{d\theta} = -h\frac{d}{d\theta}\left(\frac{1}{r}\right) \qquad (12.35)$$

$$\ddot{r} = \frac{d\dot{r}}{dt} = \frac{d\dot{r}}{d\theta}\frac{d\theta}{dt} = \frac{h}{r^2}\frac{d\dot{r}}{d\theta}$$

or, substituting for $\dot{r}$ from (12.35),

$$\ddot{r} = \frac{h}{r^2}\frac{d}{d\theta}\left[-h\frac{d}{d\theta}\left(\frac{1}{r}\right)\right]$$

$$\ddot{r} = -\frac{h^2}{r^2}\frac{d^2}{d\theta^2}\left(\frac{1}{r}\right) \qquad (12.36)$$

Substituting for θ and $\ddot{r}$ from (12.34) and (12.36), respectively, in Eq. (12.31) and introducing the function $u = 1/r$, we obtain after reductions

$$\frac{d^2u}{d\theta^2} + u = \frac{F}{mh^2u^2} \qquad (12.37)$$

In deriving Eq. (12.37), the force **F** was assumed directed toward O. The magnitude F should therefore be positive if **F** is actually directed toward O (attractive force) and negative if **F** is directed away from O (repulsive force). If F is a known function of r and thus of u, Eq. (12.37) is a differential equation in u and θ. This differential equation defines the trajectory followed by the particle under the central force **F**. The equation of the trajectory can be obtained by solving the differential equation (12.37) for u as a function of θ and determining the constants of integration from the initial conditions.

Photo 12.5 The Hubble telescope was carried into orbit by the space shuttle in 1990 (first geosynchronous from NASA).

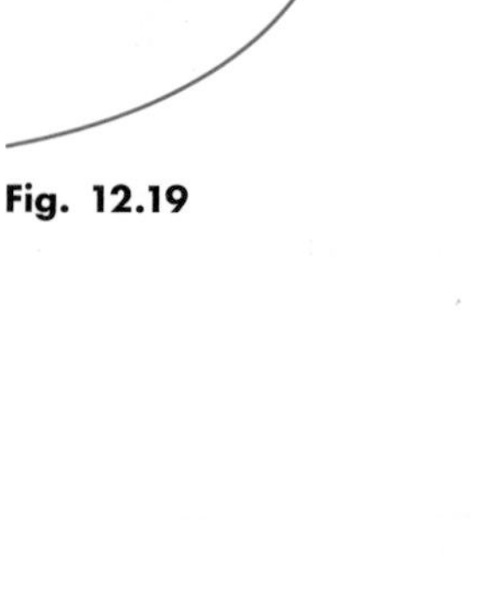

Fig. 12.19

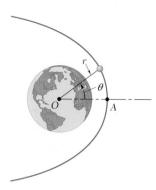

*12.12 APPLICATION TO SPACE MECHANICS

After the last stages of their launching rockets have burned out, earth satellites and other space vehicles are subjected to only the gravitational pull of the earth. Their motion can therefore be determined from Eqs. (12.33) and (12.37), which govern the motion of a particle under a central force, after F has been replaced by the expression obtained for the force of gravitational attraction.† Setting in Eq. (12.37)

$$F = \frac{GMm}{r^2} = GMmu^2$$

where M = mass of earth
m = mass of space vehicle
r = distance from center of earth to vehicle
$u = 1/r$

we obtain the differential equation

$$\frac{d^2u}{d\theta^2} + u = \frac{GM}{h^2} \tag{12.38}$$

where the right-hand member is observed to be a constant.

The solution of the differential equation (12.38) is obtained by adding the particular solution $u = GM/h^2$ to the general solution $u = C \cos(\theta - \theta_0)$ of the corresponding homogeneous equation (i.e., the equation obtained by setting the right-hand member equal to zero). Choosing the polar axis so that $\theta_0 = 0$, we write

$$\frac{1}{r} = u = \frac{GM}{h^2} + C \cos\theta \tag{12.39}$$

Equation (12.39) is the equation of a *conic section* (ellipse, parabola, or hyperbola) in the polar coordinates r and θ. The origin O of the coordinates, which is located at the center of the earth, is a *focus* of this conic section, and the polar axis is one of its axes of symmetry (Fig. 12.19).

The ratio of the constants C and GM/h^2 defines the *eccentricity* ε of the conic section; letting

$$\varepsilon = \frac{C}{GM/h^2} = \frac{Ch^2}{GM} \tag{12.40}$$

we can write Eq. (12.39) in the form

$$\frac{1}{r} = \frac{GM}{h^2}(1 + \varepsilon \cos\theta) \tag{12.39'}$$

This equation represents three possible trajectories.

1. $\varepsilon > 1$, or $C > GM/h^2$: There are two values θ_1 and $-\theta_1$ of the polar angle, defined by $\cos\theta_1 = -GM/Ch^2$, for which the

†It is assumed that the space vehicles considered here are attracted by the earth only and that their mass is negligible compared with the mass of the earth. If a vehicle moves very far from the earth, its path may be affected by the attraction of the sun, the moon, or another planet.

right-hand member of Eq. (12.39) becomes zero. For both these values, the radius vector r becomes infinite; the conic section is a *hyperbola* (Fig. 12.20).

2. $\varepsilon = 1$, or $C = GM/h^2$: The radius vector becomes infinite for $\theta = 180°$; the conic section is a *parabola*.

3. $\varepsilon < 1$, or $C < GM/h^2$: The radius vector remains finite for every value of θ; the conic section is an *ellipse*. In the particular case when $\varepsilon = C = 0$, the length of the radius vector is constant; the conic section is a circle.

Let us now see how the constants C and GM/h^2, which characterize the trajectory of a space vehicle, can be determined from the vehicle's position and velocity at the beginning of its free flight. We will assume that, as is generally the case, the powered phase of its flight has been programmed in such a way that as the last stage of the launching rocket burns out, the vehicle has a velocity parallel to the surface of the earth (Fig. 12.21). In other words, we will assume that the space vehicle begins its free flight at the vertex A of its trajectory.†

Denoting the radius vector and speed of the vehicle at the beginning of its free flight by r_0 and v_0, respectively, we observe that the velocity reduces to its transverse component and, thus, that $v_0 = r_0\dot{\theta}_0$. Recalling Eq. (12.27), we express the angular momentum per unit mass h as

$$h = r_0^2\dot{\theta}_0 = r_0 v_0 \qquad (12.41)$$

The value obtained for h can be used to determine the constant GM/h^2. We also note that the computation of this constant will be simplified if we use the relation obtained in Sec. 12.10.

$$GM = gR^2 \qquad (12.30)$$

where R is the radius of the earth ($R = 6.37 \times 10^6$ m or 3960 mi) and g is the acceleration of gravity at the surface of the earth.

The constant C is obtained by setting $\theta = 0$, $r = r_0$ in (12.39):

$$C = \frac{1}{r_0} - \frac{GM}{h^2} \qquad (12.42)$$

Substituting for h from (12.41), we can then easily express C in terms of r_0 and v_0.

Let us now determine the initial conditions corresponding to each of the three fundamental trajectories indicated above. Considering first the parabolic trajectory, we set C equal to GM/h^2 in Eq. (12.42) and eliminate h between Eqs. (12.41) and (12.42). Solving for v_0, we obtain

$$v_0 = \sqrt{\frac{2GM}{r_0}}$$

We can easily check that a larger value of the initial velocity corresponds to a hyperbolic trajectory and a smaller value corresponds to an elliptic orbit. Since the value of v_0 obtained for the parabolic trajectory

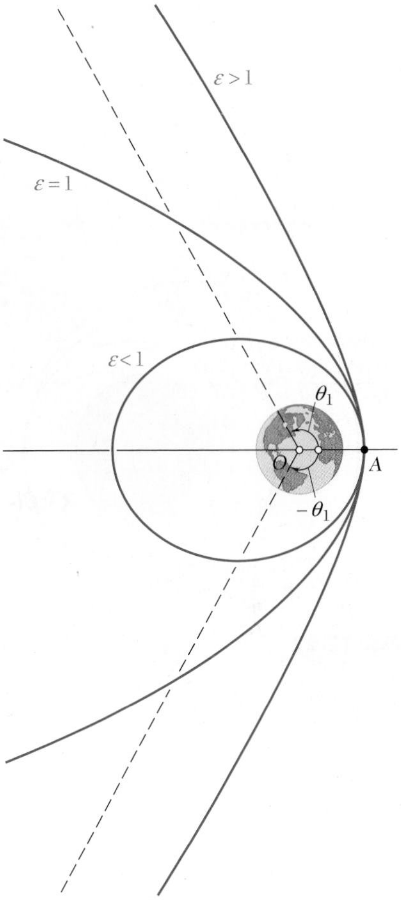

Fig. 12.20

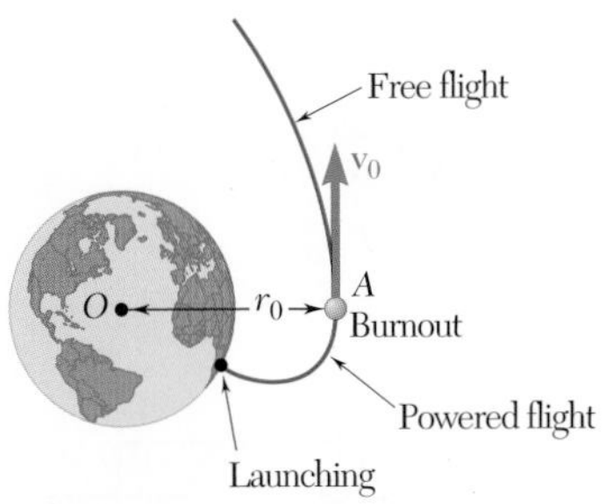

Fig. 12.21

†Problems involving oblique launchings will be considered in Sec. 13.9.

is the smallest value for which the space vehicle does not return to its starting point, it is called the *escape velocity*. We write therefore

$$v_{esc} = \sqrt{\frac{2GM}{r_0}} \qquad \text{or} \qquad v_{esc} = \sqrt{\frac{2gR^2}{r_0}} \tag{12.43}$$

if we make use of Eq. (12.30). We note that the trajectory will be (1) hyperbolic if $v_0 > v_{esc}$, (2) parabolic if $v_0 = v_{esc}$, and (3) elliptic if $v_0 < v_{esc}$.

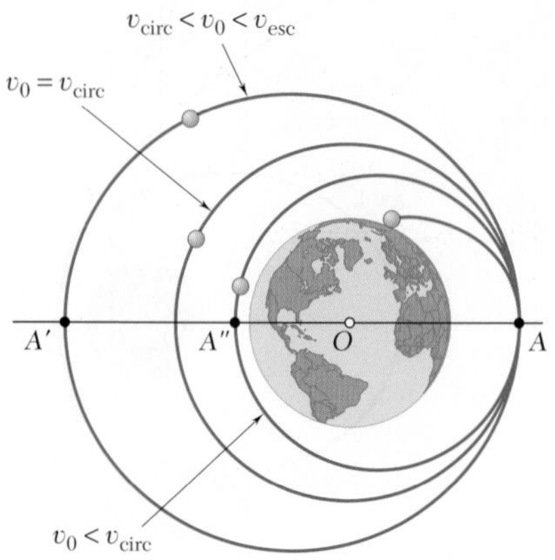

Fig. 12.22

Among the various possible elliptic orbits, the one obtained when $C = 0$, the *circular orbit*, is of special interest. The value of the initial velocity corresponding to a circular orbit is easily found to be

$$v_{circ} = \sqrt{\frac{GM}{r_0}} \qquad \text{or} \qquad v_{circ} = \sqrt{\frac{gR^2}{r_0}} \tag{12.44}$$

if Eq. (12.30) is taken into account. We note from Fig. 12.22 that for values of v_0 larger than v_{circ} but smaller than v_{esc}, point A where free flight begins is the point of the orbit closest to the earth; this point is called the *perigee*, while point A', which is farthest away from the earth, is known as the *apogee*. For values of v_0 smaller than v_{circ}, point A is the apogee, while point A'', on the other side of the orbit, is the perigee. For values of v_0 much smaller than v_{circ}, the trajectory of the space vehicle intersects the surface of the earth; in such a case, the vehicle does not go into orbit.

Ballistic missiles, which were designed to hit the surface of the earth, also travel along elliptic trajectories. In fact, we should now realize that any object projected in vacuum with an initial velocity v_0 smaller than v_{esc} will move along an elliptic path. It is only when the distances involved are small that the gravitational field of the earth can be assumed uniform and that the elliptic path can be approximated by a parabolic path, as was done earlier (Sec. 11.11) in the case of conventional projectiles.

Periodic Time. An important characteristic of the motion of an earth satellite is the time required by the satellite to describe its orbit. This time, known as the *periodic time* of the satellite, is denoted by τ. We first observe, in view of the definition of areal velocity (Sec. 12.9), that τ can be obtained by dividing the area inside the orbit by the areal velocity. Noting that the area of an ellipse is equal to πab, where a and b denote the semimajor and semiminor axes, respectively, and that the areal velocity is equal to $h/2$, we write

$$\tau = \frac{2\pi ab}{h} \tag{12.45}$$

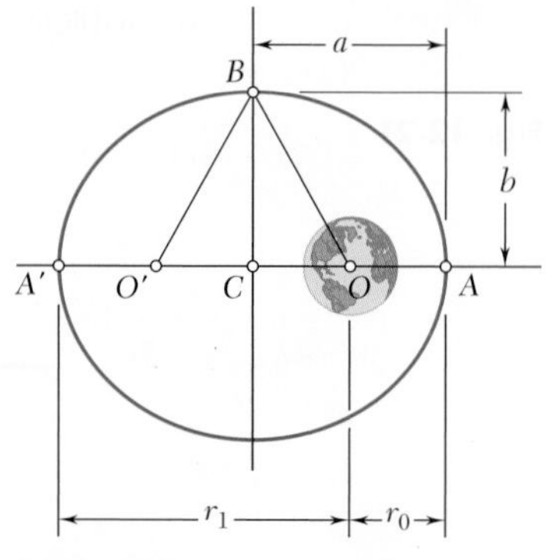

Fig. 12.23

While h can be readily determined from r_0 and v_0 in the case of a satellite launched in a direction parallel to the surface of the earth, the semiaxes a and b are not directly related to the initial conditions. Since, on the other hand, the values r_0 and r_1 of r corresponding to the perigee and apogee of the orbit can easily be determined from Eq. (12.39), we will express the semiaxes a and b in terms of r_0 and r_1.

Consider the elliptic orbit shown in Fig. 12.23. The earth's center is located at O and coincides with one of the two foci of the ellipse, while the points A and A' represent, respectively, the perigee and apogee of the orbit. We easily check that

$$r_0 + r_1 = 2a$$

and thus

$$a = \tfrac{1}{2}(r_0 + r_1) \qquad (12.46)$$

Recalling that the sum of the distances from each of the foci to any point of the ellipse is constant, we write

$$O'B + BO = O'A + OA = 2a \qquad \text{or} \qquad BO = a$$

On the other hand, we have $CO = a - r_0$. We can therefore write

$$b^2 = (BC)^2 = (BO)^2 - (CO)^2 = a^2 - (a - r_0)^2$$
$$b^2 = r_0(2a - r_0) = r_0 r_1$$

and thus

$$b = \sqrt{r_0 r_1} \qquad (12.47)$$

Formulas (12.46) and (12.47) indicate that the semimajor and semiminor axes of the orbit are equal, respectively, to the arithmetic and geometric means of the maximum and minimum values of the radius vector. Once r_0 and r_1 have been determined, the lengths of the semiaxes can be easily computed and substituted for a and b in formula (12.45).

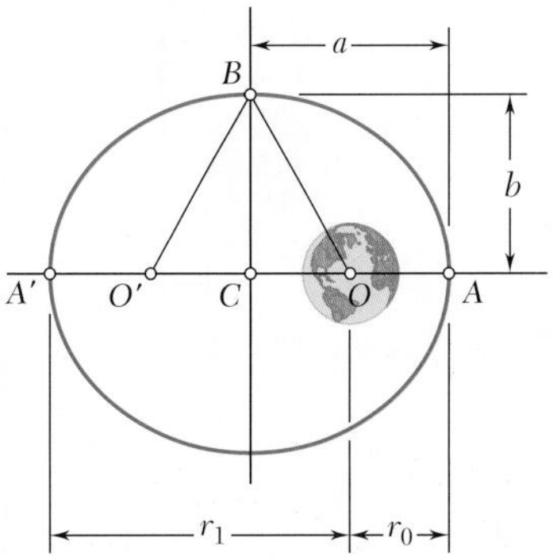

Fig. 12.23 *(repeated)*

*12.13 KEPLER'S LAWS OF PLANETARY MOTION

The equations governing the motion of an earth satellite can be used to describe the motion of the moon around the earth. In that case, however, the mass of the moon is not negligible compared with the mass of the earth, and the results obtained are not entirely accurate.

The theory developed in the preceding sections can also be applied to the study of the motion of the planets around the sun. Although another error is introduced by neglecting the forces exerted by the planets on one another, the approximation obtained is excellent. Indeed, even before Newton had formulated his fundamental theory, the properties expressed by Eq. (12.39), where M now represents the mass of the sun, and by Eq. (12.33) had been discovered by the German astronomer Johann Kepler (1571–1630) from astronomical observations of the motion of the planets.

Kepler's three *laws of planetary motion* can be stated as follows:

1. Each planet describes an ellipse, with the sun located at one of its foci.
2. The radius vector drawn from the sun to a planet sweeps equal areas in equal times.
3. The squares of the periodic times of the planets are proportional to the cubes of the semimajor axes of their orbits.

The first law states a particular case of the result established in Sec. 12.12, and the second law expresses that the areal velocity of each planet is constant (see Sec. 12.9). Kepler's third law can also be derived from the results obtained in Sec. 12.12.†

†See Prob. 12.120.

SAMPLE PROBLEM 12.9

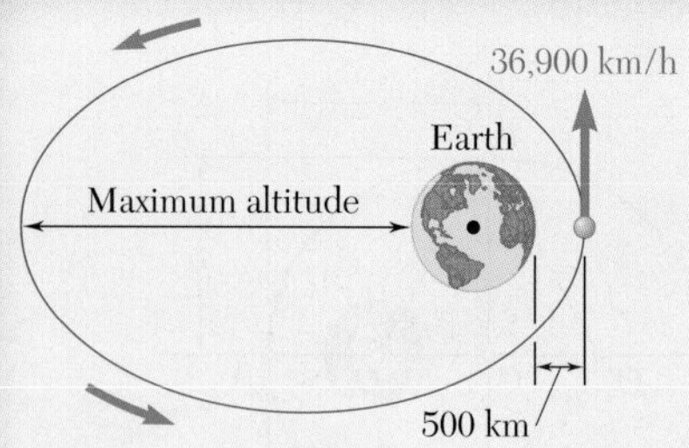

A satellite is launched in a direction parallel to the surface of the earth with a velocity of 36 900 km/h from an altitude of 500 km. Determine *(a)* the maximum altitude reached by the satellite, *(b)* the periodic time of the satellite.

SOLUTION

a. Maximum Altitude. After the satellite is launched, it is subjected only to the gravitational attraction of the earth; its motion is thus governed by Eq. (12.39),

$$\frac{1}{r} = \frac{GM}{h^2} + C \cos \theta \qquad (1)$$

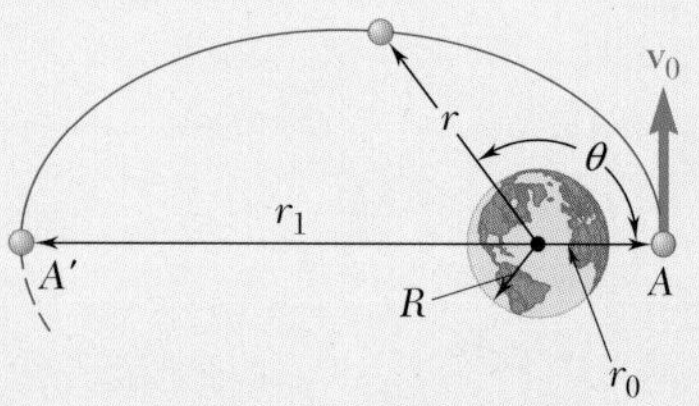

Since the radial component of the velocity is zero at the point of launching A, we have $h = r_0 v_0$. Recalling that for the earth $R = 6370$ km, we compute

$$r_0 = 6370 \text{ km} + 500 \text{ km} = 6870 \text{ km} = 6.87 \times 10^6 \text{ m}$$

$$v_0 = 36\,900 \text{ km/h} = \frac{36.9 \times 10^6 \text{ m}}{3.6 \times 10^3 \text{ s}} = 10.25 \times 10^3 \text{ m/s}$$

$$h = r_0 v_0 = (6.87 \times 10^6 \text{ m})(10.25 \times 10^3 \text{ m/s}) = 70.4 \times 10^9 \text{ m}^2/\text{s}$$

$$h^2 = 4.96 \times 10^{21} \text{ m}^4/\text{s}^2$$

Since $GM = gR^2$, where R is the radius of the earth, we have

$$GM = gR^2 = (9.81 \text{ m/s}^2)(6.37 \times 10^6 \text{ m})^2 = 398 \times 10^{12} \text{ m}^3/\text{s}^2$$

$$\frac{GM}{h^2} = \frac{398 \times 10^{12} \text{ m}^3/\text{s}^2}{4.96 \times 10^{21} \text{ m}^4/\text{s}^2} = 80.3 \times 10^{-9} \text{ m}^{-1}$$

Substituting this value into (1), we obtain

$$\frac{1}{r} = 80.3 \times 10^{-9} \text{ m}^{-1} + C \cos \theta \qquad (2)$$

Noting that at point A we have $\theta = 0$ and $r = r_0 = 6.87 \times 10^6$ m, we compute the constant C:

$$\frac{1}{6.87 \times 10^6 \text{ m}} = 80.3 \times 10^{-9} \text{ m}^{-1} + C \cos 0° \qquad C = 65.3 \times 10^{-9} \text{ m}^{-1}$$

At A', the point on the orbit farthest from the earth, we have $\theta = 180°$. Using (2), we compute the corresponding distance r_1:

$$\frac{1}{r_1} = 80.3 \times 10^{-9} \text{ m}^{-1} + (65.3 \times 10^{-9} \text{ m}^{-1}) \cos 180°$$

$$r_1 = 66.7 \times 10^6 \text{ m} = 66\,700 \text{ km}$$

Maximum altitude = 66 700 km − 6370 km = 60 300 km ◀

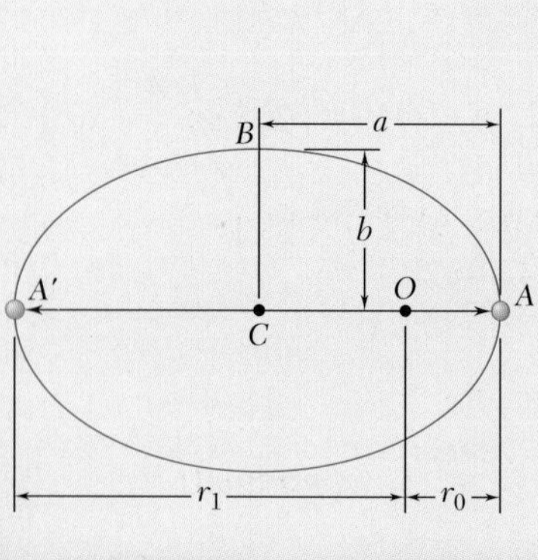

b. Periodic Time. Since A and A' are the perigee and apogee, respectively, of the elliptic orbit, we use Eqs. (12.46) and (12.47) and compute the semimajor and semiminor axes of the orbit:

$$a = \tfrac{1}{2}(r_0 + r_1) = \tfrac{1}{2}(6.87 + 66.7)(10^6) \text{ m} = 36.8 \times 10^6 \text{ m}$$

$$b = \sqrt{r_0 r_1} = \sqrt{(6.87)(66.7)} \times 10^6 \text{ m} = 21.4 \times 10^6 \text{ m}$$

$$\tau = \frac{2\pi ab}{h} = \frac{2\pi(36.8 \times 10^6 \text{m})(21.4 \times 10^6 \text{m})}{70.4 \times 10^9 \text{ m}^2/\text{s}}$$

$$\tau = 70.3 \times 10^3 \text{ s} = 1171 \text{ min} = 19 \text{ h } 31 \text{ min}$$ ◀

SOLVING PROBLEMS ON YOUR OWN

In this lesson, we continued our study of the motion of a particle under a central force and applied the results to problems in space mechanics. We found that the trajectory of a particle under a central force is defined by the differential equation

$$\frac{d^2u}{d\theta^2} + u = \frac{F}{mh^2u^2} \tag{12.37}$$

where u is the reciprocal of the distance r of the particle to the center of force ($u = 1/r$), F is the magnitude of the central force $\mathbf{F}$, and h is a constant equal to the angular momentum per unit mass of the particle. In space-mechanics problems, $\mathbf{F}$ is the force of gravitational attraction exerted on the satellite or spacecraft by the sun, earth, or other planet about which it travels. Substituting $F = GMm/r^2 = GMmu^2$ into Eq. (12.37), we obtain for that case

$$\frac{d^2u}{d\theta^2} + u = \frac{GM}{h^2} \tag{12.38}$$

where the right-hand member is a constant.

1. Analyzing the motion of satellites and spacecraft. The solution of the differential equation (12.38) defines the trajectory of a satellite or spacecraft. It was obtained in Sec. 12.12 and was given in the alternative forms

$$\frac{1}{r} = \frac{GM}{h^2} + C\cos\theta \qquad \text{or} \qquad \frac{1}{r} = \frac{GM}{h^2}(1 + \varepsilon\cos\theta) \qquad (12.39, 12.39')$$

Remember when applying these equations that $\theta = 0$ always corresponds to the perigee (the point of closest approach) of the trajectory (Fig. 12.19) and that h is a constant for a given trajectory. Depending on the value of the eccentricity ε, the trajectory will be a hyperbola, a parabola, or an ellipse.

a. $\varepsilon > 1$: The trajectory is a hyperbola, so that for this case the spacecraft never returns to its starting point.

b. $\varepsilon = 1$: The trajectory is a parabola. This is the limiting case between open (hyperbolic) and closed (elliptic) trajectories. We had observed for this case that the velocity v_0 at the perigee is equal to the escape velocity v_{esc},

$$v_0 = v_{esc} = \sqrt{\frac{2GM}{r_0}} \tag{12.43}$$

Note that the escape velocity is the smallest velocity for which the spacecraft does not return to its starting point.

c. $\varepsilon < 1$: The trajectory is an elliptic orbit. For problems involving elliptic orbits, you may find that the relation derived in Prob. 12.102,

$$\frac{1}{r_0} + \frac{1}{r_1} = \frac{2GM}{h^2}$$

(continued)

will be useful in the solution of subsequent problems. When you apply this equation, remember that r_0 and r_1 are the distances from the center of force to the perigee ($\theta = 0$) and apogee ($\theta = 180°$), respectively; that $h = r_0 v_0 = r_1 v_1$; and that, for a satellite orbiting the earth, $GM_{\text{earth}} = gR^2$, where R is the radius of the earth. Also recall that the trajectory is a circle when $\varepsilon = 0$.

2. Determining the point of impact of a descending spacecraft. For problems of this type, you may assume that the trajectory is elliptic and that the initial point of the descent trajectory is the apogee of the path (Fig. 12.22). Note that at the point of impact, the distance r in Eqs. (12.39) and (12.39′) is equal to the radius R of the body on which the spacecraft lands or crashes. In addition, we have $h = Rv_I \sin \phi_I$, where v_I is the speed of the spacecraft at impact and ϕ_I is the angle that its path forms with the vertical at the point of impact.

3. Calculating the time to travel between two points on a trajectory. For central force motion, the time t required for a particle to travel along a portion of its trajectory can be determined by recalling from Sec. 12.9 that the rate at which area is swept per unit time by the position vector **r** is equal to one-half of the angular momentum per unit mass h of the particle: $dA/dt = h/2$. It follows, since h is a constant for a given trajectory, that

$$t = \frac{2A}{h}$$

where A is the total area swept in the time t.

a. In the case of an elliptic trajectory, the time required to complete one orbit is called the *periodic time* and is expressed as

$$\tau = \frac{2(\pi ab)}{h} \tag{12.45}$$

where a and b are the semimajor and semiminor axes, respectively, of the ellipse and are related to the distances r_0 and r_1 by

$$a = \tfrac{1}{2}(r_0 + r_1) \qquad \text{and} \qquad b = \sqrt{r_0 r_1} \qquad (12.46, 12.47)$$

b. Kepler's third law provides a convenient relation between the periodic times of two satellites describing elliptic orbits about the same body [Sec. 12.13]. Denoting the semimajor axes of the two orbits by a_1 and a_2, respectively, and the corresponding periodic times by τ_1 and τ_2, we have

$$\frac{\tau_1^2}{\tau_2^2} = \frac{a_1^3}{a_2^3}$$

c. In the case of a parabolic trajectory, you may be able to use the expression given on the inside of the front cover of the book for a parabolic or a semiparabolic area to calculate the time required to travel between two points of the trajectory.

PROBLEMS

CONCEPTS QUESTIONS

12.CQ6 A uniform crate C with mass m_C is being transported to the left by a forklift with a constant speed v_1. What is the magnitude of the angular momentum of the crate about point D, that is, the upper left corner of the crate?

a. 0
b. mv_1a
c. mv_1b
d. $mv_1\sqrt{a^2 + b^2}$

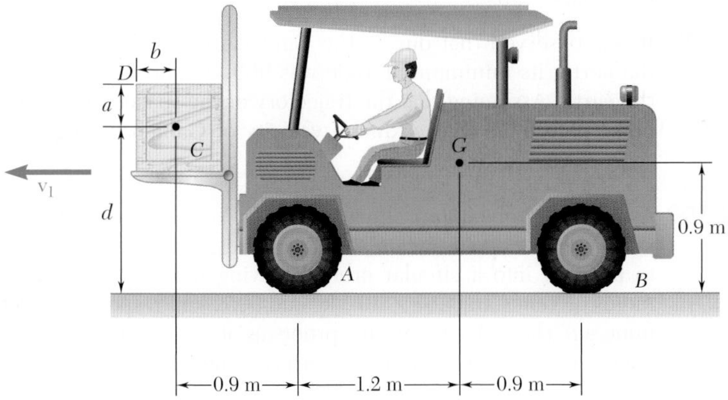

Fig. P12.CQ6 and P12.CQ7

12.CQ7 A uniform crate C with mass m_C is being transported to the left by a forklift with a constant speed v_1. What is the magnitude of the angular momentum of the crate about point A, that is, the point of contact between the front tire of the forklift and the ground?

a. 0
b. mv_1d
c. $3mv_1$
d. $mv_1\sqrt{3^2 + d^2}$

END-OF-SECTION PROBLEMS

12.94 A particle of mass m is projected from point A with an initial velocity $\mathbf{v}_0$ perpendicular to OA and moves under a central force $\mathbf{F}$ along an elliptic path defined by the equation $r = r_0/(2 - \cos\theta)$. Using Eq. (12.37), show that $\mathbf{F}$ is inversely proportional to the square of the distance r from the particle to the center of force O.

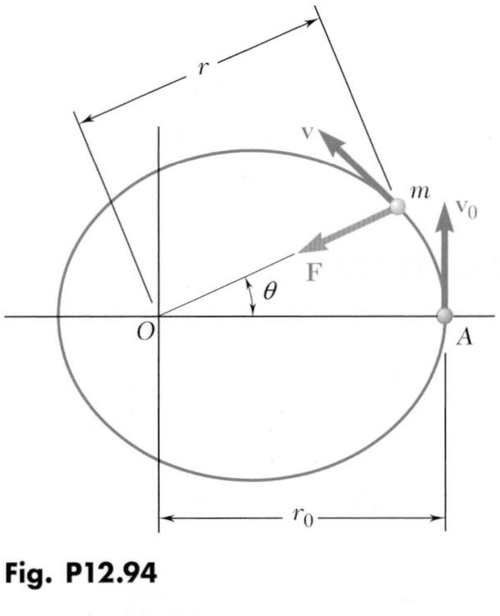

Fig. P12.94

12.95 A particle of mass m describes the logarithmic spiral $r = r_0 e^{b\theta}$ under a central force **F** directed toward the center of force O. Using Eq. (12.37), show that **F** is inversely proportional to the cube of the distance r from the particle to O.

12.96 For the particle of Prob. 12.74, and using Eq. (12.37), show that the central force **F** is proportional to the distance r from the particle to the center of force O.

12.97 A particle of mass m describes the path defined by the equation $r = r_0 \sin \theta$ under a central force **F** directed toward the center of force O. Using Eq. (12.37), show that **F** is inversely proportional to the fifth power of the distance r from the particle to O.

12.98 It was observed that during its second flyby of the earth, the Galileo spacecraft had a velocity of 14.1 km/s as it reached its minimum altitude of 303 km above the surface of the earth. Determine the eccentricity of the trajectory of the spacecraft during this portion of its flight.

12.99 It was observed that during the Galileo spacecraft's first flyby of the earth, its minimum altitude was 960 km above the surface of the earth. Assuming that the trajectory of the spacecraft was parabolic, determine the maximum velocity of Galileo during its first flyby of the earth.

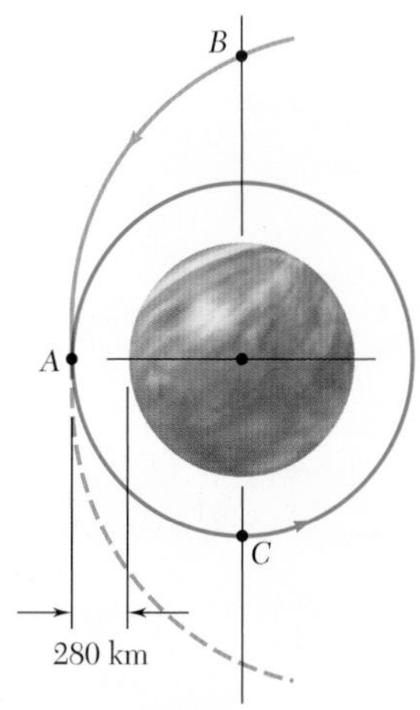

Fig. P12.100

12.100 As a space probe approaching the planet Venus on a parabolic trajectory reaches point A closest to the planet, its velocity is decreased to insert it into a circular orbit. Knowing that the mass and the radius of Venus are 4.87×10^{24} kg and 6052 km, respectively, determine (a) the velocity of the probe as it approaches A, (b) the decrease in velocity required to insert it into the circular orbit.

12.101 It was observed that as the Voyager I spacecraft reached the point of its trajectory closest to the planet Saturn, it was at a distance of 185×10^3 km from the center of the planet and had a velocity of 21.0 km/s. Knowing that Tethys, one of Saturn's moons, describes a circular orbit of radius 295×10^3 km at a speed of 11.35 km/s, determine the eccentricity of the trajectory of Voyager I on its approach to Saturn.

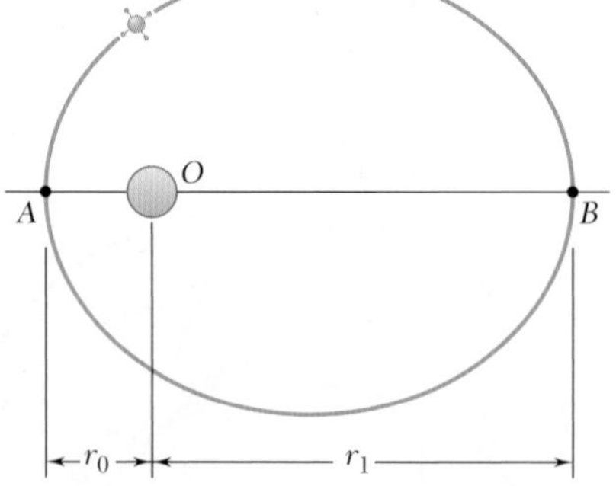

Fig. P12.102

12.102 A satellite describes an elliptic orbit about a planet of mass M. Denoting by r_0 and r_1, respectively, the minimum and maximum values of the distance r from the satellite to the center of the planet, derive the relation

$$\frac{1}{r_0} + \frac{1}{r_1} = \frac{2GM}{h^2}$$

where h is the angular momentum per unit mass of the satellite.

12.103 A space probe is describing a circular orbit about a planet of radius R. The altitude of the probe above the surface of the planet is αR and its speed is v_0. To place the probe in an elliptic orbit which will bring it closer to the planet, its speed is reduced from v_0 to βv_0, where $\beta < 1$, by firing its engine for a short interval of time. Determine the smallest permissible value of β if the probe is not to crash on the surface of the planet.

12.104 At main engine cutoff of its thirteenth flight, the space shuttle Discovery was in an elliptic orbit of minimum altitude 60 km and maximum altitude 500 km above the surface of the earth. Knowing that at point A the shuttle had a velocity $\mathbf{v}_0$ parallel to the surface of the earth and that the shuttle was transferred to a circular orbit as it passed through point B, determine (*a*) the speed v_0 of the shuttle at A, (*b*) the increase in speed required at B to insert the shuttle into the circular orbit.

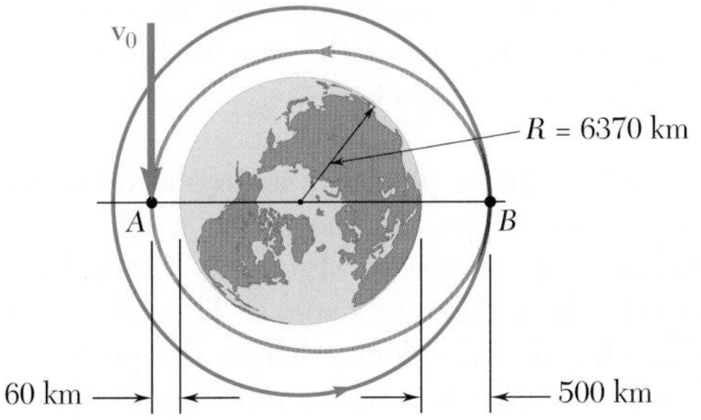

Fig. P12.104

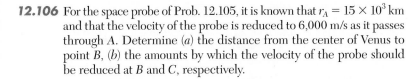

12.105 A space probe is to be placed in a circular orbit of 9000-km radius about the planet Venus in a specified plane. As the probe reaches A, the point of its original trajectory closest to Venus, it is inserted in a first elliptic transfer orbit by reducing its speed of Δv_A. This orbit brings it to point B with a much reduced velocity. There the probe is inserted in a second transfer orbit located in the specified plane by changing the direction of its velocity and further reducing its speed by Δv_B. Finally, as the probe reaches point C, it is inserted in the desired circular orbit by reducing its speed by Δv_C. Knowing that the mass of Venus is 0.82 times the mass of the earth, that $r_A = 15 \times 10^3$ km and $r_B = 300 \times 10^3$ km, and that the probe approaches A on a parabolic trajectory, determine by how much the velocity of the probe should be reduced (*a*) at A, (*b*) at B, (*c*) at C.

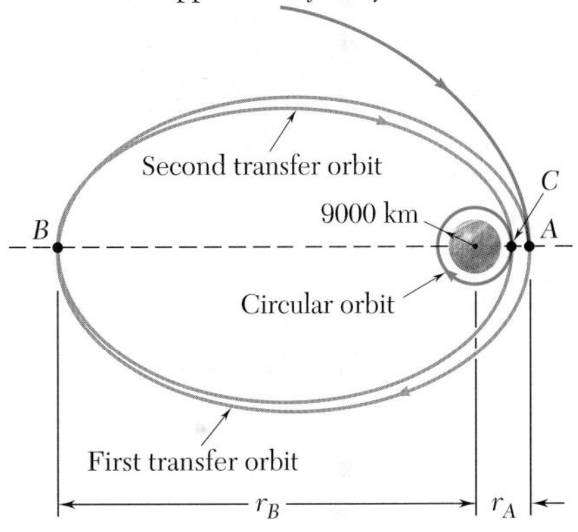

Fig. *P12.105*

12.106 For the space probe of Prob. 12.105, it is known that $r_A = 15 \times 10^3$ km and that the velocity of the probe is reduced to 6,000 m/s as it passes through A. Determine (*a*) the distance from the center of Venus to point B, (*b*) the amounts by which the velocity of the probe should be reduced at B and C, respectively.

12.107 As it describes an elliptic orbit about the sun, a spacecraft reaches a maximum distance of 320×10^6 km from the center of the sun at point A (called the aphelion) and a minimum distance of 150×10^6 km at point B (called the perihelion). To place the spacecraft in a smaller elliptic orbit with aphelion at A' and perihelion at B', where A' and B' are located 260×10^6 km and 135×10^6 km, respectively, from the center of the sun, the speed of the spacecraft is first reduced as it passes through A and then is further reduced as it passes through B'. Knowing that the mass of the sun is 332.8×10^3 times the mass of the earth, determine (*a*) the speed of the spacecraft at A, (*b*) the amounts by which the speed of the spacecraft should be reduced at A and B' to insert it into the desired elliptic orbit.

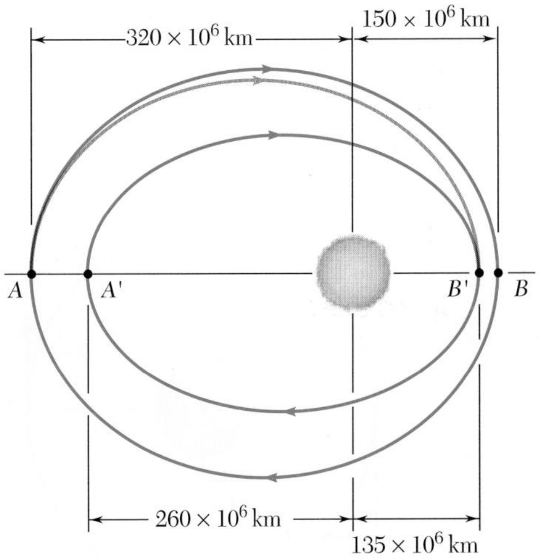

Fig. P12.107

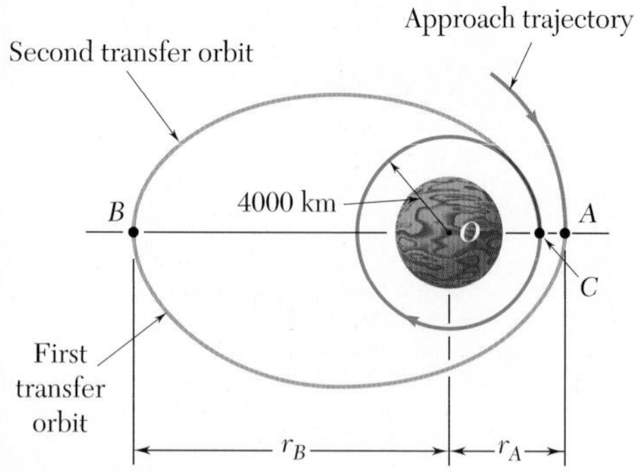

Fig. P12.110

12.108 Halley's comet travels in an elongated elliptic orbit for which the minimum distance from the sun is approximately $\frac{1}{2} r_E$, where $r_E = 150 \times 10^6$ km is the mean distance from the sun to the earth. Knowing that the periodic time of Halley's comet is about 76 years, determine the maximum distance from the sun reached by the comet.

12.109 Based on observations made during the 1996 sighting of comet Hyakutake, it was concluded that the trajectory of the comet is a highly elongated ellipse for which the eccentricity is approximately $\varepsilon = 0.999887$. Knowing that for the 1996 sighting the minimum distance between the comet and the sun was $0.230R_E$, where R_E is the mean distance from the sun to the earth, determine the periodic time of the comet.

12.110 A space probe is to be placed in a circular orbit of radius 4000 km about the planet Mars. As the probe reaches *A*, the point of its original trajectory closest to Mars, it is inserted into a first elliptic transfer orbit by reducing its speed. This orbit brings it to point *B* with a much-reduced velocity. There the probe is inserted into a second transfer orbit by further reducing its speed. Knowing that the mass of Mars is 0.1074 times the mass of the earth, that $r_A = 9000$ km and $r_B = 180\ 000$ km, and that the probe approaches *A* on a parabolic trajectory, determine the time needed for the space probe to travel from *A* to *B* on its first transfer orbit.

12.111 A space shuttle is in an elliptic orbit of eccentricity 0.0356 and a minimum altitude of 300 km above the surface of the earth. Knowing that the radius of the earth is 6370 km, determine the periodic time for the orbit.

12.112 The Clementine spacecraft described an elliptic orbit of minimum altitude $h_A = 400$ km and maximum altitude $h_B = 2940$ km above the surface of the moon. Knowing that the radius of the moon is 1737 km and that the mass of the moon is 0.01230 times the mass of the earth, determine the periodic time of the spacecraft.

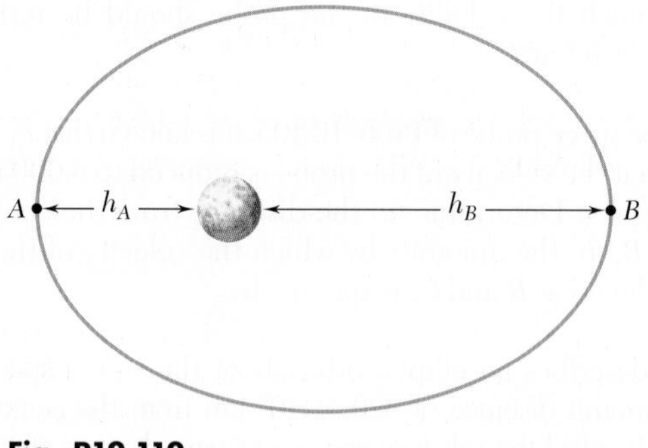

Fig. P12.112

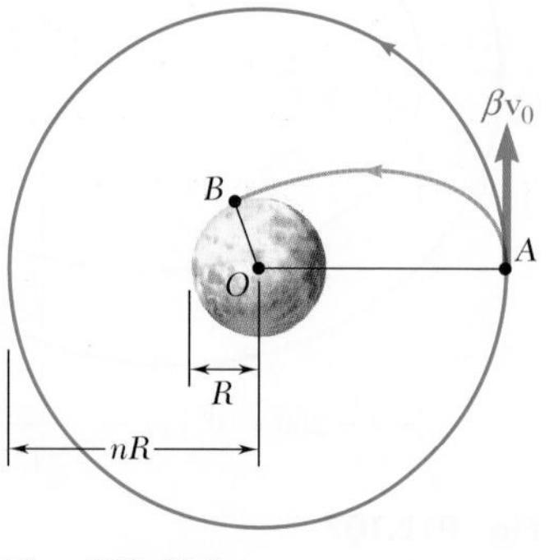

Fig. P12.114

12.113 Determine the time needed for the space probe of Prob. 12.100 to travel from *B* to *C*.

12.114 A space probe is describing a circular orbit of radius *nR* with a velocity v_0 about a planet of radius *R* and center *O*. As the probe passes through point *A*, its velocity is reduced from v_0 to βv_0, where $\beta < 1$, to place the probe on a crash trajectory. Express in terms of *n* and β the angle *AOB*, where *B* denotes the point of impact of the probe on the planet.

12.115 A long-range ballistic trajectory between points A and B on the earth's surface consists of a portion of an ellipse with the apogee at point C. Knowing that point C is 1500 km above the surface of the earth and the range $R\phi$ of the trajectory is 6000 km, determine (*a*) the velocity of the projectile at C, (*b*) the eccentricity ε of the trajectory.

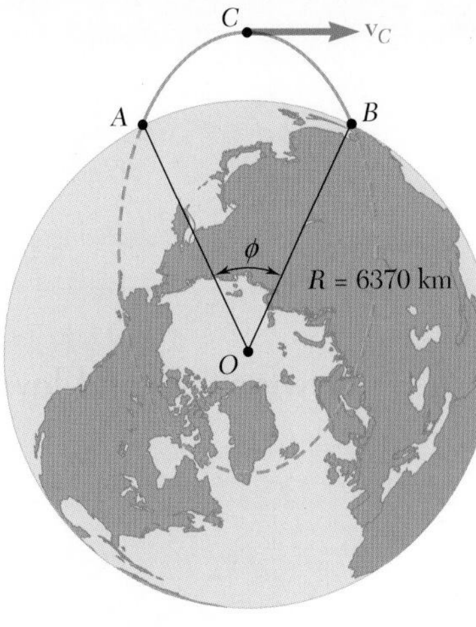

Fig. P12.115

12.116 A space shuttle is describing a circular orbit at an altitude of 563 km above the surface of the earth. As it passes through point A, it fires its engine for a short interval of time to reduce its speed by 152 m/s and begin its descent toward the earth. Determine the angle AOB so that the altitude of the shuttle at point B is 121 km. (*Hint:* Point A is the apogee of the elliptic descent trajectory.)

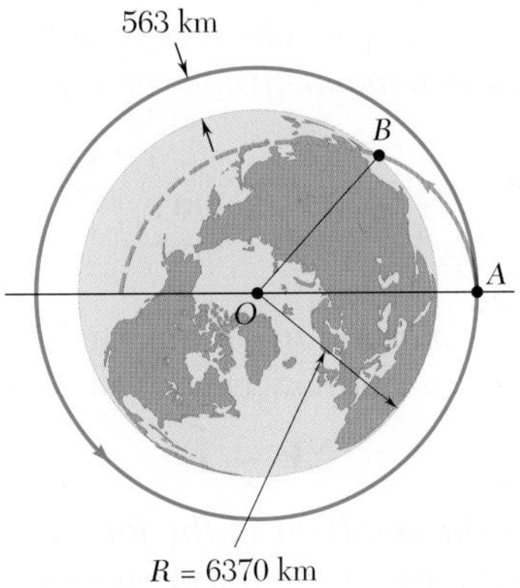

Fig. *P12.116*

12.117 As a spacecraft approaches the planet Jupiter, it releases a probe which is to enter the planet's atmosphere at point B at an altitude of 450 km above the surface of the planet. The trajectory of the probe is a hyperbola of eccentricity $\varepsilon = 1.031$. Knowing that the radius and the mass of Jupiter are 71.429×10^3 km and 1.9×10^{27} kg, respectively, and that the velocity $\mathbf{v}_B$ of the probe at B forms an angle of 82.9° with the direction of OA, determine (*a*) the angle AOB, (*b*) the speed v_B of the probe at B.

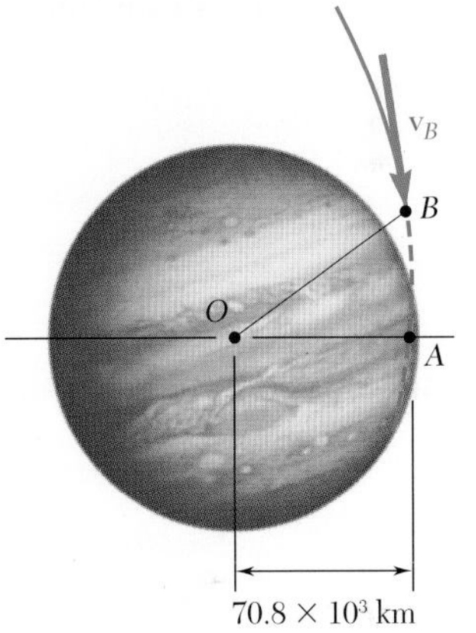

Fig. *P12.117*

12.118 A satellite describes an elliptic orbit about a planet. Denoting by r_0 and r_1 the distances corresponding, respectively, to the perigee and apogee of the orbit, show that the curvature of the orbit at each of these two points can be expressed as

$$\frac{1}{\rho} = \frac{1}{2}\left(\frac{1}{r_0} + \frac{1}{r_1}\right)$$

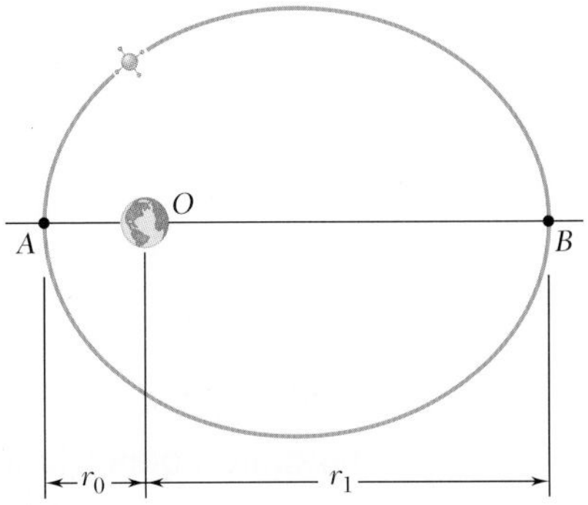

Fig. P12.118 and *P12.119*

12.119 (*a*) Express the eccentricity ε of the elliptic orbit described by a satellite about a planet in terms of the distances r_0 and r_1 corresponding, respectively, to the perigee and apogee of the orbit. (*b*) Use the result obtained in part *a* and the data given in Prob. 12.109, where $R_E = 149.6 \times 10^6$ km, to determine the approximate maximum distance from the sun reached by comet Hyakutake.

12.120 Derive Kepler's third law of planetary motion from Eqs. (12.39) and (12.45).

12.121 Show that the angular momentum per unit mass h of a satellite describing an elliptic orbit of semimajor axis a and eccentricity ε about a planet of mass M can be expressed as

$$h = \sqrt{GMa(1 - \varepsilon^2)}$$

REVIEW AND SUMMARY

This chapter was devoted to Newton's second law and its application to the analysis of the motion of particles.

Newton's second law

Denoting by m the mass of a particle, by $\Sigma\mathbf{F}$ the sum, or resultant, of the forces acting on the particle, and by $\mathbf{a}$ the acceleration of the particle relative to a *newtonian frame of reference* [Sec. 12.2], we wrote

$$\Sigma\mathbf{F} = m\mathbf{a} \tag{12.2}$$

Linear momentum

Introducing the *linear momentum* of a particle, $\mathbf{L} = m\mathbf{v}$ [Sec. 12.3], we saw that Newton's second law can also be written in the form

$$\Sigma\mathbf{F} = \dot{\mathbf{L}} \tag{12.5}$$

which expresses that *the resultant of the forces acting on a particle is equal to the rate of change of the linear momentum of the particle.*

Consistent systems of units

Equation (12.2) holds only if a consistent system of units is used. With SI units, the forces should be expressed in newtons, the masses in kilograms, and the accelerations in m/s^2; with U.S. customary units, the forces should be expressed in pounds, the masses in $lb \cdot s^2/ft$ (also referred to as *slugs*), and the accelerations in ft/s^2 [Sec. 12.4].

Equations of motion for a particle

To solve a problem involving the motion of a particle, Eq. (12.2) should be replaced by equations containing scalar quantities [Sec. 12.5]. Using *rectangular components* of $\mathbf{F}$ and $\mathbf{a}$, we wrote

$$\Sigma F_x = ma_x \qquad \Sigma F_y = ma_y \qquad \Sigma F_z = ma_z \tag{12.8}$$

Using *tangential and normal components*, we had

$$\Sigma F_t = m\frac{dv}{dt} \qquad \Sigma F_n = m\frac{v^2}{\rho} \tag{12.9'}$$

Dynamic equilibrium

We also noted [Sec. 12.6] that the equations of motion of a particle can be replaced by equations similar to the equilibrium equations used in statics if a vector $-m\mathbf{a}$ of magnitude ma but of sense opposite to that of the acceleration is added to the forces applied to the particle; the particle is then said to be in *dynamic equilibrium*. For the sake of uniformity, however, all the Sample Problems were solved by using the equations of motion, first with rectangular components [Sample Probs. 12.1 through 12.4], then with tangential and normal components [Sample Probs. 12.5 and 12.6].

Angular momentum

In the second part of the chapter, we defined the *angular momentum* $\mathbf{H}_O$ of a particle about a point O as the moment about O of the linear momentum $m\mathbf{v}$ of that particle [Sec. 12.7]. We wrote

$$\mathbf{H}_O = \mathbf{r} \times m\mathbf{v} \tag{12.12}$$

and noted that $\mathbf{H}_O$ is a vector perpendicular to the plane containing $\mathbf{r}$ and $m\mathbf{v}$ (Fig. 12.24) and of magnitude

$$H_O = rmv \sin \phi \tag{12.13}$$

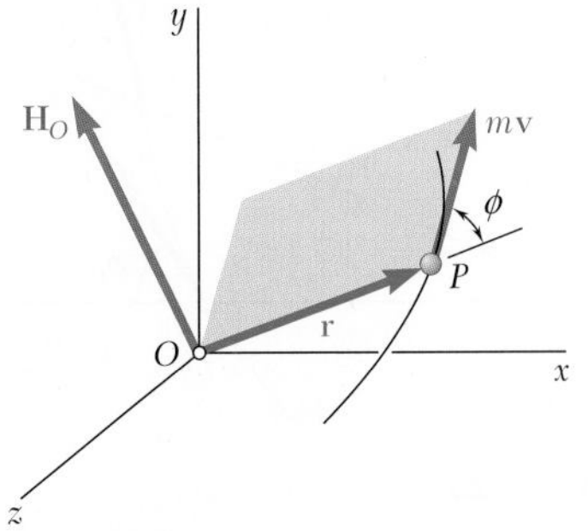

Fig. 12.24

Resolving the vectors $\mathbf{r}$ and $m\mathbf{v}$ into rectangular components, we expressed the angular momentum $\mathbf{H}_O$ in the determinant form

$$\mathbf{H}_O = \begin{vmatrix} \mathbf{i} & \mathbf{j} & \mathbf{k} \\ x & y & z \\ mv_x & mv_y & mv_z \end{vmatrix} \tag{12.14}$$

In the case of a particle moving in the xy plane, we have $z = v_z = 0$. The angular momentum is perpendicular to the xy plane and is completely defined by its magnitude. We wrote

$$H_O = H_z = m(xv_y - yv_x) \tag{12.16}$$

Rate of change of angular momentum

Computing the rate of change $\dot{\mathbf{H}}_O$ of the angular momentum $\mathbf{H}_O$, and applying Newton's second law, we wrote the equation

$$\Sigma\mathbf{M}_O = \dot{\mathbf{H}}_O \tag{12.19}$$

which states that *the sum of the moments about O of the forces acting on a particle is equal to the rate of change of the angular momentum of the particle about O.*

Radial and transverse components

In many problems involving the plane motion of a particle, it is found convenient to use *radial and transverse components* [Sec. 12.8, Sample Prob. 12.7] and to write the equations

$$\Sigma F_r = m(\ddot{r} - r\dot{\theta}^2) \tag{12.21}$$
$$\Sigma F_\theta = m(r\ddot{\theta} + 2\dot{r}\dot{\theta}) \tag{12.22}$$

Motion under a central force

When the only force acting on a particle P is a force $\mathbf{F}$ directed toward or away from a fixed point O, the particle is said to be moving *under a central force* [Sec. 12.9]. Since $\Sigma\mathbf{M}_O = 0$ at any given instant, it follows from Eq. (12.19) that $\dot{\mathbf{H}}_O = 0$ for all values of t and, thus, that

$$\mathbf{H}_O = \text{constant} \tag{12.23}$$

We concluded that *the angular momentum of a particle moving under a central force is constant, both in magnitude and direction*, and that the particle moves in a plane perpendicular to the vector $\mathbf{H}_O$.

Recalling Eq. (12.13), we wrote the relation

$$rmv \sin \phi = r_0 m v_0 \sin \phi_0 \tag{12.25}$$

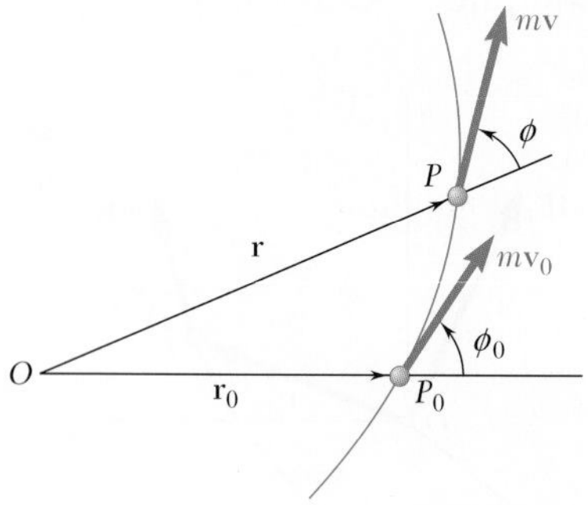

Fig. 12.25

for the motion of any particle under a central force (Fig. 12.25). Using polar coordinates and recalling Eq. (12.18), we also had

$$r^2\dot{\theta} = h \tag{12.27}$$

where h is a constant representing the angular momentum per unit mass, H_O/m, of the particle. We observed (Fig. 12.26) that the infinitesimal area dA swept by the radius vector OP as it rotates through $d\theta$ is equal to $\frac{1}{2}r^2 d\theta$ and, thus, that the left-hand member of Eq. (12.27) represents twice the *areal velocity* dA/dt of the particle. Therefore, *the areal velocity of a particle moving under a central force is constant.*

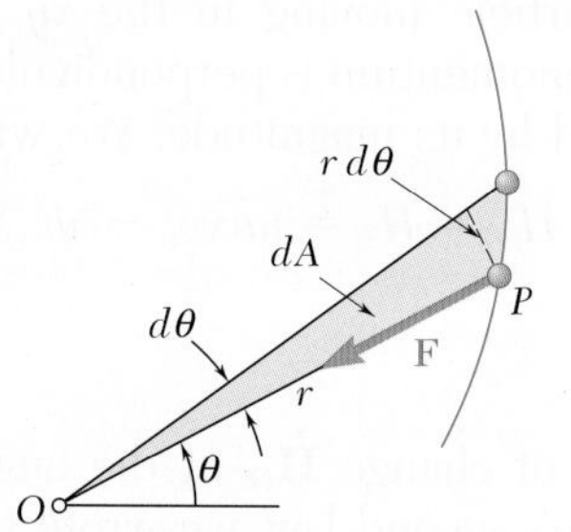

Fig. 12.26

Newton's law of universal gravitation

An important application of the motion under a central force is provided by the orbital motion of bodies under gravitational attraction [Sec. 12.10]. According to *Newton's law of universal gravitation,* two particles at a distance r from each other and of masses M and m, respectively, attract each other with equal and opposite forces **F** and −**F** directed along the line joining the particles (Fig. 12.27). The common magnitude F of the two forces is

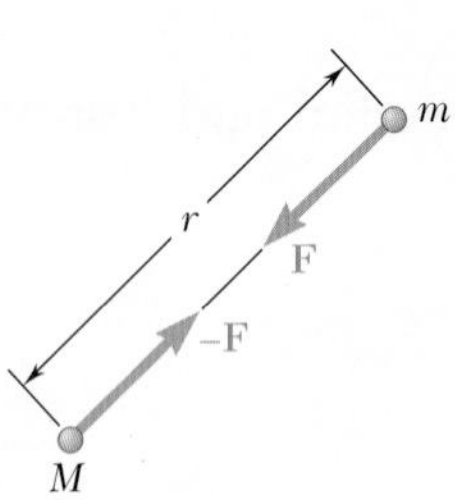

Fig. 12.27

$$F = G\frac{Mm}{r^2} \tag{12.28}$$

where G is the *constant of gravitation*. In the case of a body of mass m subjected to the gravitational attraction of the earth, the product GM, where M is the mass of the earth, can be expressed as

$$GM = gR^2 \tag{12.30}$$

where $g = 9.81 \text{ m/s}^2 = 32.2 \text{ ft/s}^2$ and R is the radius of the earth.

Orbital motion

It was shown in Sec. 12.11 that a particle moving under a central force describes a trajectory defined by the differential equation

$$\frac{d^2u}{d\theta^2} + u = \frac{F}{mh^2u^2} \tag{12.37}$$

where $F > 0$ corresponds to an attractive force and $u = 1/r$. In the case of a particle moving under a force of gravitational attraction [Sec. 12.12], we substituted for F the expression given in Eq. (12.28). Measuring θ from the axis OA joining the focus O to the point A of the trajectory closest to O (Fig. 12.28), we found that the solution to Eq. (12.37) was

$$\frac{1}{r} = u = \frac{GM}{h^2} + C\cos\theta \qquad (12.39)$$

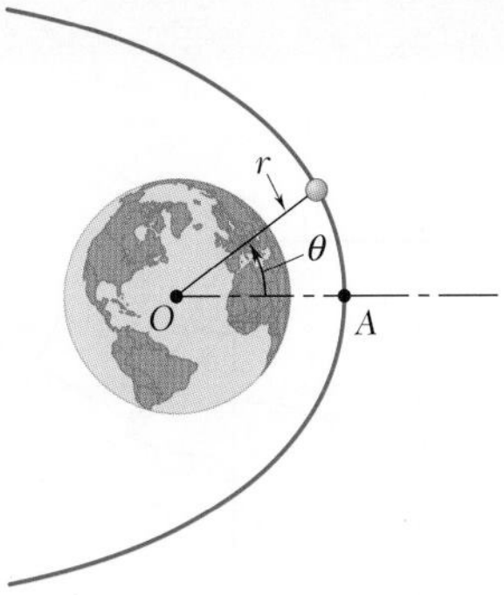

Fig. 12.28

This is the equation of a conic of eccentricity $\varepsilon = Ch^2/GM$. The conic is an *ellipse* if $\varepsilon < 1$, a *parabola* if $\varepsilon = 1$, and a hyperbola if $\varepsilon > 1$. The constants C and h can be determined from the initial conditions; if the particle is projected from point A ($\theta = 0$, $r = r_0$) with an initial velocity $\mathbf{v}_0$ perpendicular to OA, we have $h = r_0 v_0$ [Sample Prob. 12.9].

Escape velocity

It was also shown that the values of the initial velocity corresponding, respectively, to a parabolic and a circular trajectory were

$$v_{esc} = \sqrt{\frac{2GM}{r_0}} \qquad (12.43)$$

$$v_{circ} = \sqrt{\frac{GM}{r_0}} \qquad (12.44)$$

and that the first of these values, called the *escape velocity,* is the smallest value of v_0 for which the particle will not return to its starting point.

Periodic time

The *periodic time* τ of a planet or satellite was defined as the time required by that body to describe its orbit. It was shown that

$$\tau = \frac{2\pi ab}{h} \qquad (12.45)$$

where $h = r_0 v_0$ and where a and b represent the semimajor and semiminor axes of the orbit. It was further shown that these semiaxes are respectively equal to the arithmetic and geometric means of the maximum and minimum values of the radius vector r.

Kepler's laws

The last section of the chapter [Sec. 12.13] presented *Kepler's laws of planetary motion* and showed that these empirical laws, obtained from early astronomical observations, confirm Newton's laws of motion as well as his law of gravitation.

REVIEW PROBLEMS

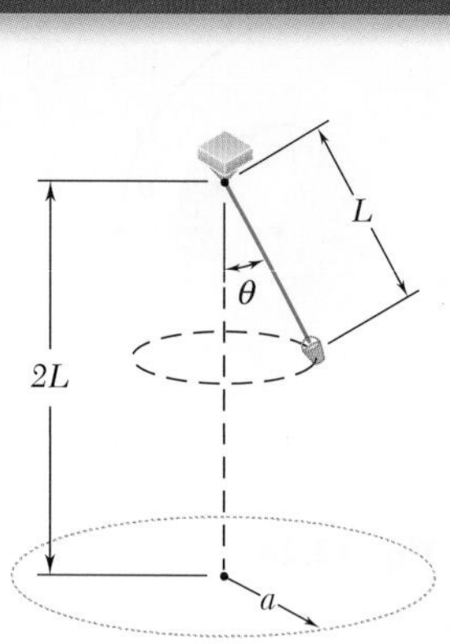

Fig. *P12.123*

12.122 Knowing that the system shown starts from rest, find the velocity at $t = 1.2$ s of (*a*) collar *A*, (*b*) collar *B*. Neglect the masses of the pulleys and the effect of friction.

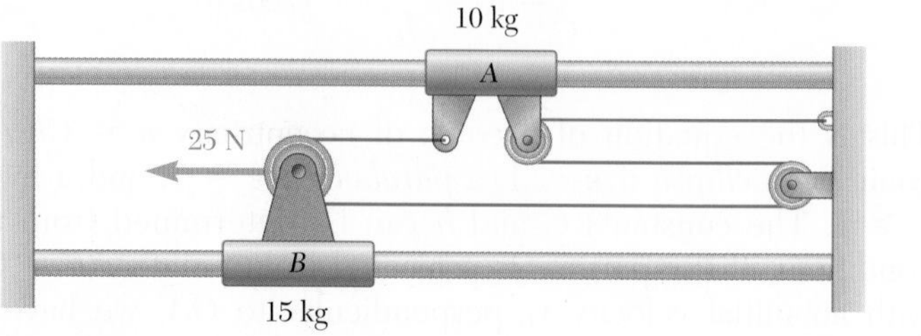

Fig. *P12.122*

12.123 A bucket is attached to a rope of length $L = 1.2$ m and is made to revolve in a horizontal circle. Drops of water leaking from the bucket fall and strike the floor along the perimeter of a circle of radius *a*. Determine the radius *a* when $\theta = 30°$.

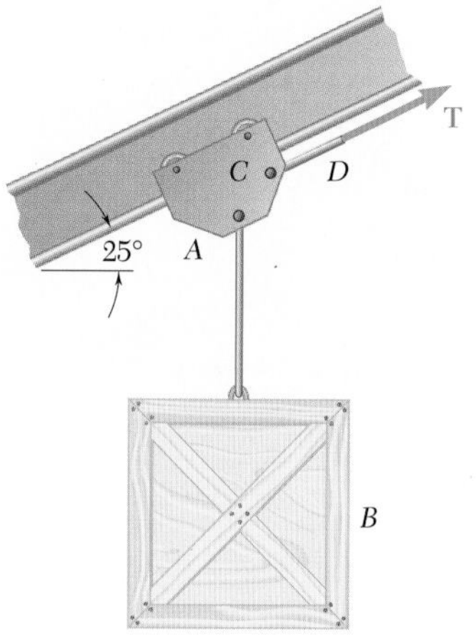

Fig. P12.125

12.124 A 6-kg block *B* rests as shown on the upper surface of a 15-kg wedge *A*. Neglecting friction, determine immediately after the system is released from rest (*a*) the acceleration of *A*, (*b*) the acceleration of *B* relative to *A*.

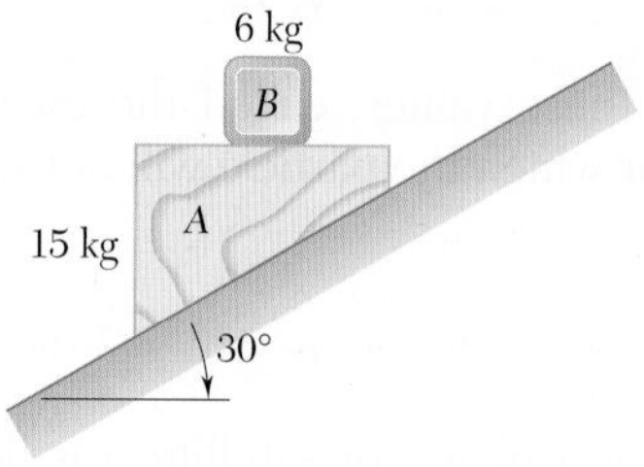

Fig. P12.124

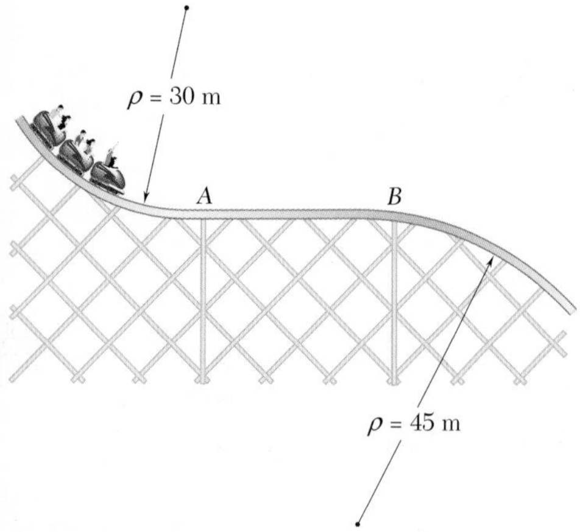

Fig. P12.126

12.125 A 250-kg crate *B* is suspended from a cable attached to a 20-kg trolley *A* which rides on an inclined I-beam as shown. Knowing that at the instant shown the trolley has an acceleration of 0.4 m/s^2 up and to the right, determine (*a*) the acceleration of *B* relative to *A*, (*b*) the tension in cable *CD*.

12.126 The roller-coaster track shown is contained in a vertical plane. The portion of track between *A* and *B* is straight and horizontal, while the portions to the left of *A* and to the right of *B* have radii of curvature as indicated. A car is traveling at a speed of 72 km/h when the brakes are suddenly applied, causing the wheels of the car to slide on the track ($\mu_k = 0.25$). Determine the initial deceleration of the car if the brakes are applied as the car (*a*) has almost reached *A*, (*b*) is traveling between *A* and *B*, (*c*) has just passed *B*.

12.127 The 100-g pin B slides along the slot in the rotating arm OC and along the slot DE which is cut in a fixed horizontal plate. Neglecting friction and knowing that rod OC rotates at the constant rate $\dot{\theta}_0 = 12$ rad/s, determine for any given value of θ (*a*) the radial and transverse components of the resultant force $\mathbf{F}$ exerted on pin B, (*b*) the forces $\mathbf{P}$ and $\mathbf{Q}$ exerted on pin B by rod OC and the wall of slot DE, respectively.

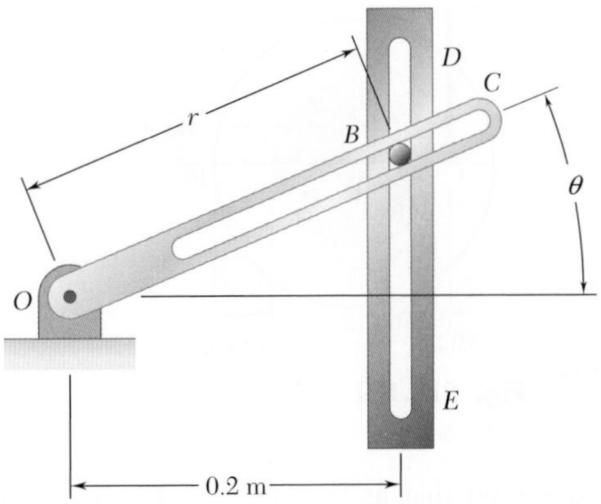

Fig. P12.127

12.128 A small 200-g collar C can slide on a semicircular rod which is made to rotate about the vertical AB at the constant rate of 6 rad/s. Determine the minimum required value of the coefficient of static friction between the collar and the rod if the collar is not to slide when (*a*) $\theta = 90°$, (*b*) $\theta = 75°$, (*c*) $\theta = 45°$. Indicate in each case the direction of the impending motion.

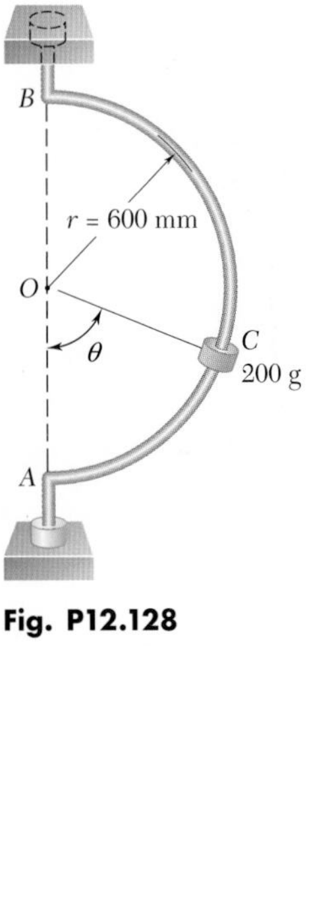

Fig. P12.128

12.129 Telemetry technology is used to quantify kinematic values of a 200-kg roller-coaster cart as it passes overhead. According to the system, $r = 25$ m, $\dot{r} = -10$ m/s, $\ddot{r} = -2$ m/s^2, $\theta = 90°$, $\dot{\theta} = -0.4$ rad/s, $\ddot{\theta} = -0.32$ rad/s^2. At this instant, determine (*a*) the normal force between the cart and the track, (*b*) the radius of curvature of the track.

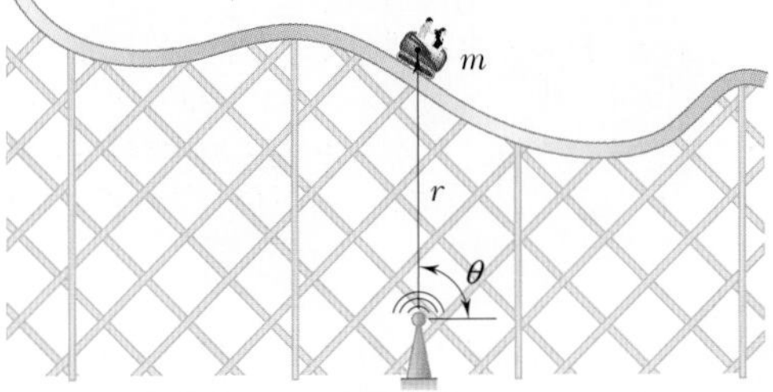

Fig. P12.129

12.130 The radius of the orbit of a moon of a given planet is equal to twice the radius of that planet. Denoting by ρ the mean density of the planet, show that the time required by the moon to complete one full revolution about the planet is $(24\pi/G\rho)^{1/2}$, where G is the constant of gravitation.

12.131 At engine burnout on a mission, a shuttle had reached point A at an altitude of 80 km above the surface of the earth and had a horizontal velocity $\mathbf{v}_0$. Knowing that its first orbit was elliptic and that the shuttle was transferred to a circular orbit as it passed through point B at an altitude of 270 km, determine (*a*) the time needed for the shuttle to travel from A to B on its original elliptic orbit, (*b*) the periodic time of the shuttle on its final circular orbit.

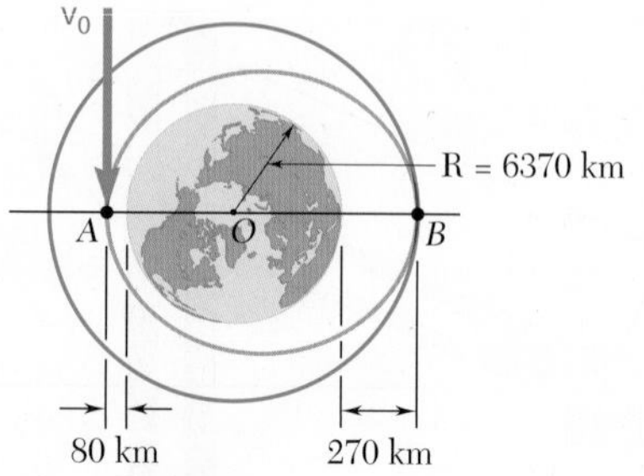

Fig. *P12.131*

12.132 It was observed that as the Galileo spacecraft reached the point on its trajectory closest to Io, a moon of the planet Jupiter, it was at a distance of 2800 km from the center of Io and had a velocity of 15×10^3 m/s. Knowing that the mass of Io is 0.01496 times the mass of the earth, determine the eccentricity of the trajectory of the spacecraft as it approached Io.

***12.133** Disk A rotates in a horizontal plane about a of vertical axis at the constant rate $\dot{\theta}_0 = 10$ rad/s. Slider B has mass 1 kg and moves in a frictionless slot cut in the disk. The slider is attached to a spring of constant k, which is undeformed when $r = 0$. Knowing that the slider is released with no radial velocity in the position $r = 500$ mm, determine the position of the slider and the horizontal force exerted on it by the disk at $t = 0.1$ s for (*a*) $k = 100$ N/m, (*b*) $k = 200$ N/m.

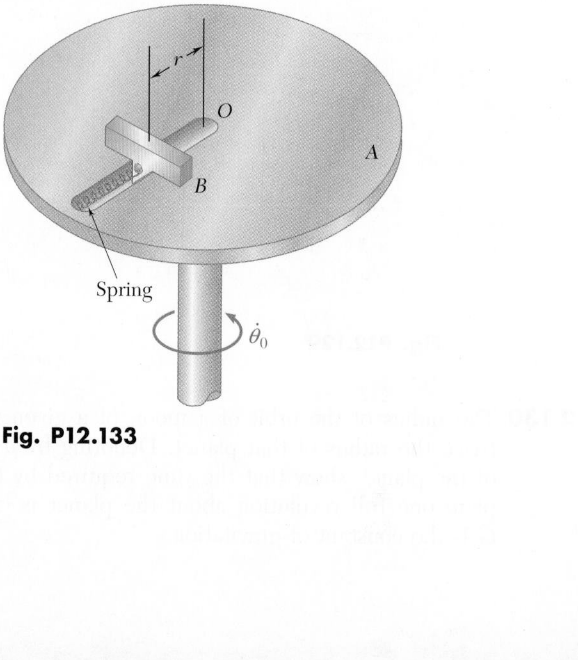

Fig. P12.133

COMPUTER PROBLEMS

12.C1 Block B of mass 10 kg is initially at rest as shown on the upper surface of a 20-kg wedge A which is supported by a horizontal surface. A 2-kg block C is connected to block B by a cord which passes over a pulley of negligible mass. Using computational software and denoting by μ the coefficient of friction at all surfaces, use this program to determine the accelerations for values of $\mu \geq 0$. Use 0.01 increments for μ until the wedge does not move and then use 0.1 increments until no motion occurs.

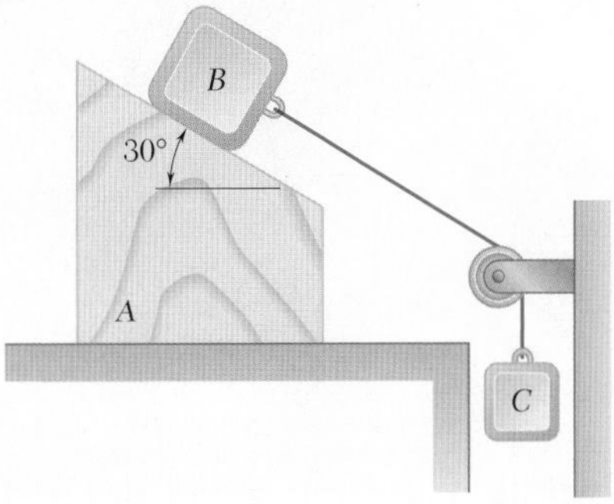

Fig. P12.C1

12.C2 A small, 500-g block is at rest at the top of a cylindrical surface. The block is given an initial velocity $\mathbf{v}_0$ to the right of magnitude 3 m/s, which causes it to slide on the cylindrical surface. Using computational software calculate and plot the values of θ at which the block leaves the surface for values of μ_k, the coefficient of kinetic friction between the block and the surface, from 0 to 0.4.

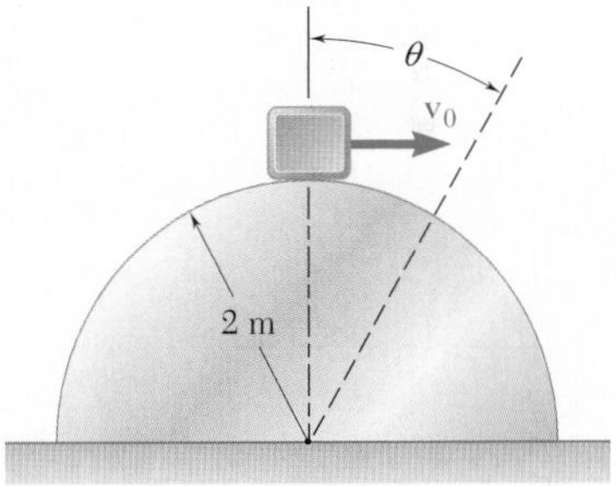

Fig. P12.C2

12.C3 A block of mass m is attached to a spring of constant k. The block is released from rest when the spring is in a horizontal and undeformed position. Use computational software to determine, for various selected values of k/m and r_0, (*a*) the length of the spring and the magnitude and direction of the velocity of the block as the block passes directly under the point of suspension of the spring, (*b*) the value of k/m when $r_0 = 1$ m for which that velocity is horizontal.

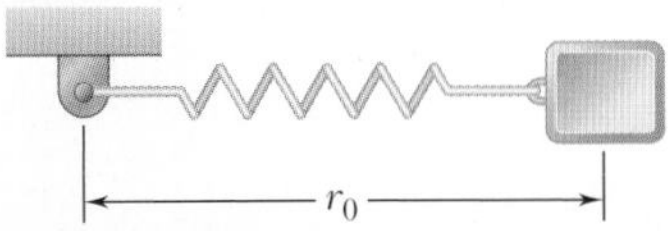

Fig. P12.C3

12.C4 Use computational software to determine the ranges of values of θ for which the block E of Prob. 12.60 will not slide in the semicircular slot of the flat plate. Assuming a coefficient of static friction of 0.35, determine the ranges of values when the constant rate of rotation of the plate is (*a*) 14 rad/s, (*b*) 2 rad/s.

A golf ball will deform upon impact as shown by this high-speed photo. The maximum deformation will occur when the club head velocity and the ball velocity are the same. In this chapter impacts will be analyzed using the coefficient of restitution and conservation of linear momentum. The kinetics of particles using energy and momentum methods is the subject of this chapter.

CHAPTER 13

Kinetics of Particles: Energy and Momentum Methods

13.1 INTRODUCTION

In the preceding chapter, most problems dealing with the motion of particles were solved through the use of the fundamental equation of motion $\mathbf{F} = m\mathbf{a}$. Given a particle acted upon by a force $\mathbf{F}$, we could solve this equation for the acceleration $\mathbf{a}$; then, by applying the principles of kinematics, we could determine from $\mathbf{a}$ the velocity and position of the particle at any time.

Using the equation $\mathbf{F} = m\mathbf{a}$ together with the principles of kinematics allows us to obtain two additional methods of analysis, the *method of work and energy* and the *method of impulse and momentum*. The advantage of these methods lies in the fact that they make the determination of the acceleration unnecessary. Indeed, the method of work and energy directly relates force, mass, velocity, and displacement, while the method of impulse and momentum relates force, mass, velocity, and time.

The method of work and energy will be considered first. In Secs. 13.2 through 13.4, the *work of a force* and the *kinetic energy of a particle* are discussed and the principle of work and energy is applied to the solution of engineering problems. The concepts of *power* and *efficiency* of a machine are introduced in Sec. 13.5.

Sections 13.6 through 13.8 are devoted to the concept of *potential energy* of a conservative force and to the application of the principle of conservation of energy to various problems of practical interest. In Sec. 13.9, the principles of conservation of energy and of conservation of angular momentum are used jointly to solve problems of space mechanics.

The second part of the chapter is devoted to the *principle of impulse and momentum* and to its application to the study of the motion of a particle. As you will see in Sec. 13.11, this principle is particularly effective in the study of the *impulsive motion* of a particle, where very large forces are applied for a very short time interval.

In Secs. 13.12 through 13.14, the *central impact* of two bodies will be considered. It will be shown that a certain relation exists between the relative velocities of the two colliding bodies before and after impact. This relation, together with the fact that the total momentum of the two bodies is conserved, can be used to solve a number of problems of practical interest.

Finally, in Sec. 13.15, you will learn to select from the three fundamental methods presented in Chaps. 12 and 13 the method best suited for the solution of a given problem. You will also see how the principle of conservation of energy and the method of impulse and momentum can be combined to solve problems involving only conservative forces, except for a short impact phase during which impulsive forces must also be taken into consideration.

13.2 WORK OF A FORCE

We will first define the terms *displacement* and *work* as they are used in mechanics.† Consider a particle which moves from a point

†The definition of work was given in Sec. 10.2, and the basic properties of the work of a force were outlined in Secs. 10.2 and 10.6. For convenience, we repeat here the portions of this material which relate to the kinetics of particles.

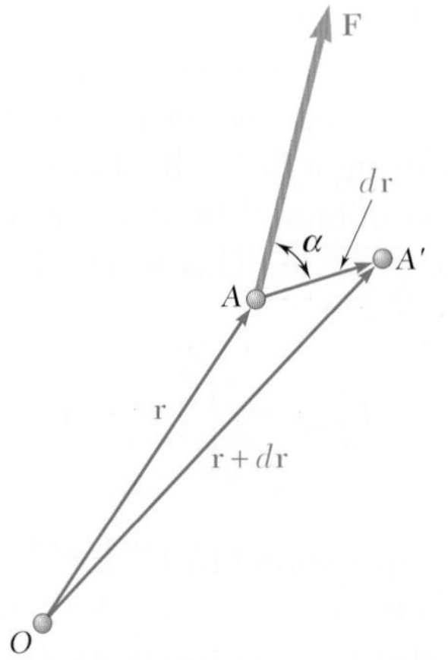

Fig. 13.1

A to a neighboring point A' (Fig. 13.1). If **r** denotes the position vector corresponding to point A, the small vector joining A and A' can be denoted by the differential $d\mathbf{r}$; the vector $d\mathbf{r}$ is called the *displacement* of the particle. Now, let us assume that a force **F** is acting on the particle. The *work of the force* **F** *corresponding to the displacement* $d\mathbf{r}$ is defined as the quantity

$$dU = \mathbf{F} \cdot d\mathbf{r} \tag{13.1}$$

obtained by forming the scalar product of the force **F** and the displacement $d\mathbf{r}$. Denoting by F and ds, respectively, the magnitudes of the force and of the displacement, and by α the angle formed by **F** and $d\mathbf{r}$, and recalling the definition of the scalar product of two vectors (Sec. 3.9), we write

$$dU = F\,ds\cos\alpha \tag{13.1$'$}$$

Using formula (3.30), we can also express the work dU in terms of the rectangular components of the force and of the displacement:

$$dU = F_x\,dx + F_y\,dy + F_z\,dz \tag{13.1$''$}$$

Being a *scalar quantity*, work has a magnitude and a sign but no direction. We also note that work should be expressed in units obtained by multiplying units of length by units of force. Thus, if U.S. customary units are used, work should be expressed in ft · lb or in · lb. If SI units are used, work should be expressed in N · m. The unit of work N · m is called a *joule* (J).† Recalling the conversion factors indicated in Sec. 12.4, we write

$$1\text{ ft}\cdot\text{lb} = (1\text{ ft})(1\text{ lb}) = (0.3048\text{ m})(4.448\text{ N}) = 1.356\text{ J}$$

It follows from (13.1′) that the work dU is positive if the angle α is acute and negative if α is obtuse. Three particular cases are of special

†The joule (J) is the SI unit of *energy*, whether in mechanical form (work, potential energy, kinetic energy) or in chemical, electrical, or thermal form. We should note that even though N · m = J, the moment of a force must be expressed in N · m and not in joules, since the moment of a force is not a form of energy.

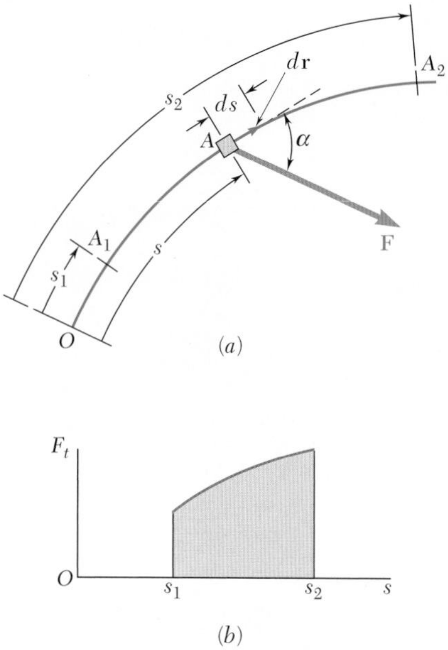

Fig. 13.2

interest. If the force **F** has the same direction as $d\mathbf{r}$, the work dU reduces to $F\,ds$. If **F** has a direction opposite to that of $d\mathbf{r}$, the work is $dU = -F\,ds$. Finally, if **F** is perpendicular to $d\mathbf{r}$, the work dU is zero.

The work of **F** during a *finite* displacement of the particle from A_1 to A_2 (Fig. 13.2*a*) is obtained by integrating Eq. (13.1) along the path described by the particle. This work, denoted by $U_{1\rightarrow2}$, is

$$U_{1\rightarrow2} = \int_{A_1}^{A_2} \mathbf{F} \cdot d\mathbf{r} \tag{13.2}$$

Using the alternative expression (13.1′) for the elementary work dU, and observing that $F \cos \alpha$ represents the tangential component F_t of the force, we can also express the work $U_{1\rightarrow2}$ as

$$U_{1\rightarrow2} = \int_{s_1}^{s_2} (F \cos \alpha)\, ds = \int_{s_1}^{s_2} F_t\, ds \tag{13.2′}$$

where the variable of integration s measures the distance traveled by the particle along the path. The work $U_{1\rightarrow2}$ is represented by the area under the curve obtained by plotting $F_t = F \cos \alpha$ against s (Fig. 13.2*b*).

When the force **F** is defined by its rectangular components, the expression (13.1″) can be used for the elementary work. We then write

$$U_{1\rightarrow2} = \int_{A_1}^{A_2} (F_x\, dx + F_y\, dy + F_z\, dz) \tag{13.2″}$$

where the integration is to be performed along the path described by the particle.

Work of a Constant Force in Rectilinear Motion. When a particle moving in a straight line is acted upon by a force **F** of constant magnitude and of constant direction (Fig. 13.3), formula (13.2′) yields

$$U_{1\rightarrow2} = (F \cos \alpha)\, \Delta x \tag{13.3}$$

where α = angle the force forms with direction of motion
Δx = displacement from A_1 to A_2

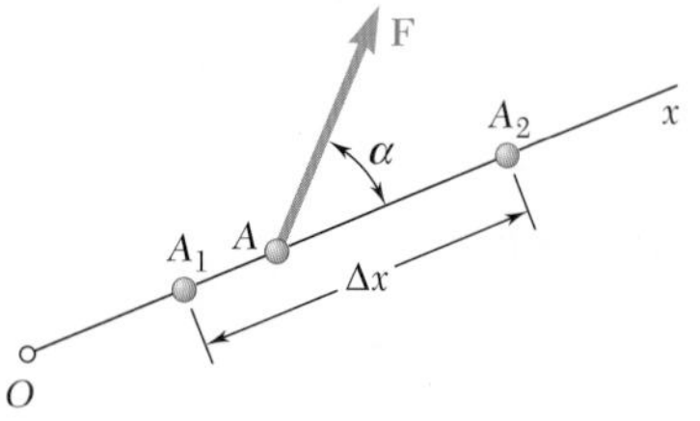

Fig. 13.3

Work of the Force of Gravity. The work of the weight **W** of a body, i.e., of the force of gravity exerted on that body, is obtained by substituting the components of **W** into (13.1″) and (13.2″). With the y axis chosen upward (Fig. 13.4), we have $F_x = 0$, $F_y = -W$, and $F_z = 0$, and we write

$$dU = -W\, dy$$

$$U_{1\rightarrow2} = -\int_{y_1}^{y_2} W\, dy = Wy_1 - Wy_2 \tag{13.4}$$

or

$$U_{1\rightarrow2} = -W(y_2 - y_1) = -W\, \Delta y \tag{13.4′}$$

where Δy is the vertical displacement from A_1 to A_2. The work of the weight **W** is thus equal to *the product of W and the vertical*

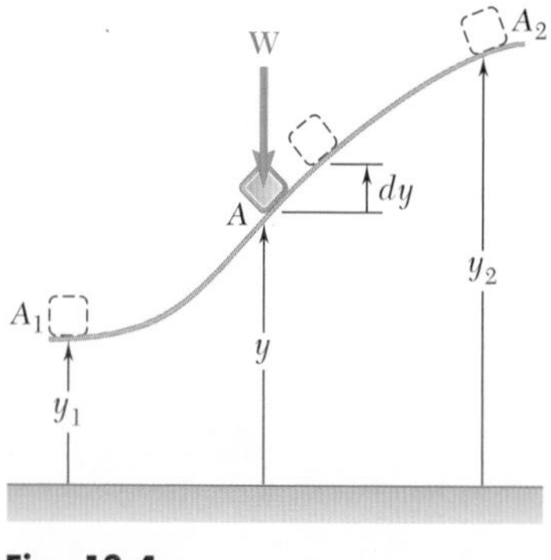

Fig. 13.4

displacement of the center of gravity of the body. The work is *positive* when $\Delta y < 0$, that is, *when the body moves down*.

Work of the Force Exerted by a Spring. Consider a body A attached to a fixed point B by a spring; it is assumed that the spring is undeformed when the body is at A_0 (Fig. 13.5*a*). Experimental evidence shows that the magnitude of the force $\mathbf{F}$ exerted by the spring on body A is proportional to the deflection x of the spring measured from the position A_0. We have

$$F = kx \tag{13.5}$$

where k is the *spring constant*, expressed in N/m or kN/m if SI units are used and in lb/ft or lb/in. if U.S. customary units are used.†

The work of the force $\mathbf{F}$ exerted by the spring during a finite displacement of the body from $A_1(x = x_1)$ to $A_2(x = x_2)$ is obtained by writing

$$dU = -F\,dx = -kx\,dx$$

$$U_{1\to2} = -\int_{x_1}^{x_2} kx\,dx = \tfrac{1}{2}kx_1^2 - \tfrac{1}{2}kx_2^2 \tag{13.6}$$

Care should be taken to express k and x in consistent units. For example, if U.S. customary units are used, k should be expressed in lb/ft and x in feet, or k in lb/in. and x in inches; in the first case, the work is obtained in ft · lb, in the second case, in in · lb. We note that the work of the force $\mathbf{F}$ exerted by the spring on the body is *positive* when $x_2 < x_1$, that is, *when the spring is returning to its undeformed position*.

Since Eq. (13.5) is the equation of a straight line of slope k passing through the origin, the work $U_{1\to2}$ of $\mathbf{F}$ during the displacement from A_1 to A_2 can be obtained by evaluating the area of the trapezoid shown in Fig. 13.5*b*. This is done by computing F_1 and F_2 and multiplying the base Δx of the trapezoid by its mean height $\frac{1}{2}(F_1 + F_2)$. Since the work of the force $\mathbf{F}$ exerted by the spring is positive for a negative value of Δx, we write

$$U_{1\to2} = -\tfrac{1}{2}(F_1 + F_2)\,\Delta x \tag{13.6$'$}$$

Formula (13.6′) is usually more convenient to use than (13.6) and affords fewer chances of confusing the units involved.

Work of a Gravitational Force. We saw in Sec. 12.10 that two particles of mass M and m at a distance r from each other attract each other with equal and opposite forces $\mathbf{F}$ and $-\mathbf{F}$, directed along the line joining the particles and of magnitude

$$F = G\frac{Mm}{r^2}$$

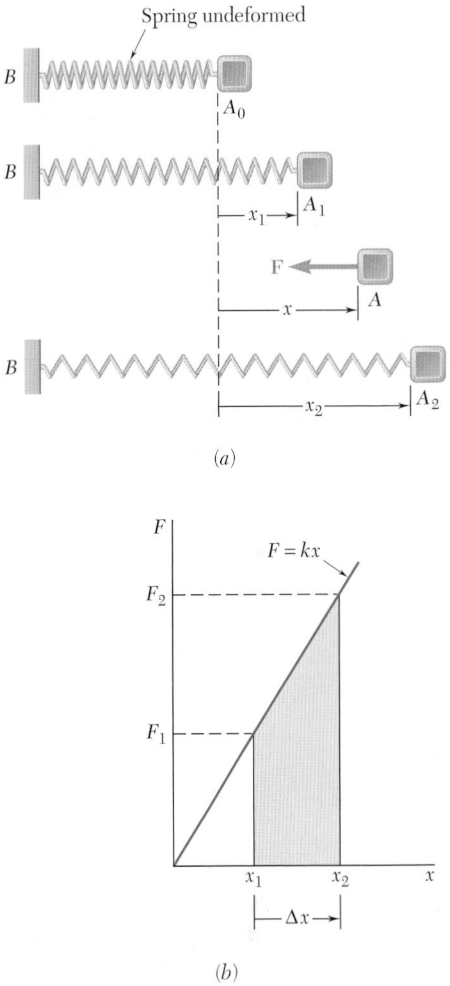

Fig. 13.5

†The relation $F = kx$ is correct under static conditions only. Under dynamic conditions, formula (13.5) should be modified to take the inertia of the spring into account. However, the error introduced by using the relation $F = kx$ in the solution of kinetics problems is small if the mass of the spring is small compared with the other masses in motion.

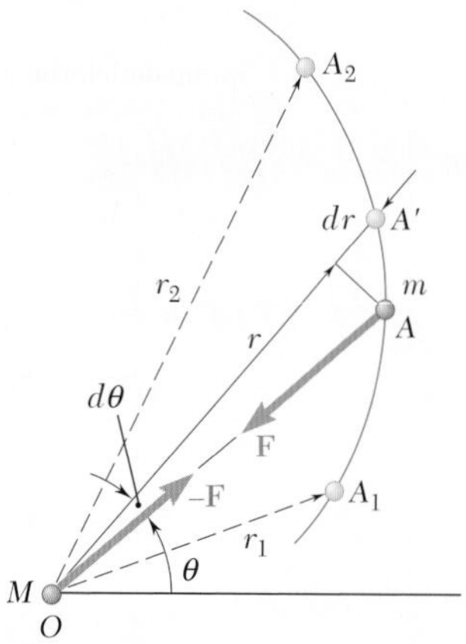

Fig. 13.6

Let us assume that the particle M occupies a fixed position O while the particle m moves along the path shown in Fig. 13.6. The work of the force $\mathbf{F}$ exerted on the particle m during an infinitesimal displacement of the particle from A to A' can be obtained by multiplying the magnitude F of the force by the radial component dr of the displacement. Since $\mathbf{F}$ is directed toward O, the work is negative and we write

$$dU = -F\,dr = -G\frac{Mm}{r^2}dr$$

The work of the gravitational force $\mathbf{F}$ during a finite displacement from $A_1(r = r_1)$ to $A_2(r = r_2)$ is therefore

$$U_{1\rightarrow 2} = -\int_{r_1}^{r_2} \frac{GMm}{r^2}dr = \frac{GMm}{r_2} - \frac{GMm}{r_1} \tag{13.7}$$

where M is the mass of the earth. This formula can be used to determine the work of the force exerted by the earth on a body of mass m at a distance r from the center of the earth, when r is larger than the radius R of the earth. Recalling the first of the relations (12.29), we can replace the product GMm in Eq. (13.7) by WR^2, where R is the radius of the earth ($R = 6.37 \times 10^6$ m or 3960 mi) and W is the weight of the body at the surface of the earth.

A number of forces frequently encountered in problems of kinetics *do no work*. They are forces applied to fixed points ($ds = 0$) or acting in a direction perpendicular to the displacement ($\cos\alpha = 0$). Among the forces which do no work are the following: the reaction at a frictionless pin when the body supported rotates about the pin, the reaction at a frictionless surface when the body in contact moves along the surface, the reaction at a roller moving along its track, and the weight of a body when its center of gravity moves horizontally.

13.3 KINETIC ENERGY OF A PARTICLE. PRINCIPLE OF WORK AND ENERGY

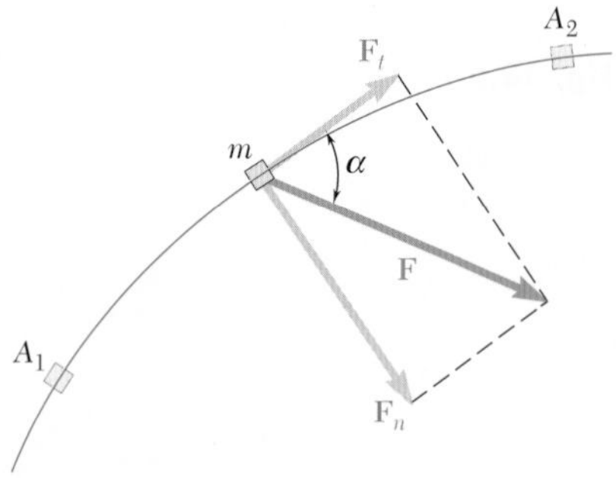

Fig. 13.7

Consider a particle of mass m acted upon by a force $\mathbf{F}$ and moving along a path which is either rectilinear or curved (Fig. 13.7). Expressing Newton's second law in terms of the tangential components of the force and of the acceleration (see Sec. 12.5), we write

$$F_t = ma_t \qquad \text{or} \qquad F_t = m\frac{dv}{dt}$$

where v is the speed of the particle. Recalling from Sec. 11.9 that $v = ds/dt$, we obtain

$$F_t = m\frac{dv}{ds}\frac{ds}{dt} = mv\frac{dv}{ds}$$
$$F_t\,ds = mv\,dv$$

Integrating from A_1, where $s = s_1$ and $v = v_1$, to A_2, where $s = s_2$ and $v = v_2$, we write

$$\int_{s_1}^{s_2} F_t\,ds = m\int_{v_1}^{v_2} v\,dv = \tfrac{1}{2}mv_2^2 - \tfrac{1}{2}mv_1^2 \tag{13.8}$$

The left-hand member of Eq. (13.8) represents the work $U_{1\rightarrow 2}$ of the force $\mathbf{F}$ exerted on the particle during the displacement from A_1 to

A_2; as indicated in Sec. 13.2, the work $U_{1\to2}$ is a scalar quantity. The expression $\frac{1}{2}mv^2$ is also a scalar quantity; it is defined as the kinetic energy of the particle and is denoted by T. We write

$$T = \tfrac{1}{2}mv^2 \tag{13.9}$$

Substituting into (13.8), we have

$$U_{1\to2} = T_2 - T_1 \tag{13.10}$$

which expresses that, when a particle moves from A_1 to A_2 under the action of a force **F**, *the work of the force* **F** *is equal to the change in kinetic energy of the particle*. This is known as the *principle of work and energy*. Rearranging the terms in (13.10), we write

$$T_1 + U_{1\to2} = T_2 \tag{13.11}$$

Thus, *the kinetic energy of the particle at* A_2 *can be obtained by adding to its kinetic energy at* A_1 *the work done during the displacement from* A_1 *to* A_2 *by the force* **F** *exerted on the particle*. Like Newton's second law from which it is derived, the principle of work and energy applies only with respect to a newtonian frame of reference (Sec. 12.2). The speed v used to determine the kinetic energy T should therefore be measured with respect to a newtonian frame of reference.

Since both work and kinetic energy are scalar quantities, their sum can be computed as an ordinary algebraic sum, the work $U_{1\to2}$ being considered as positive or negative according to the direction of **F**. When several forces act on the particle, the expression $U_{1\to2}$ represents the total work of the forces acting on the particle; it is obtained by adding algebraically the work of the various forces.

As noted above, the kinetic energy of a particle is a scalar quantity. It further appears from the definition $T = \frac{1}{2}mv^2$ that regardless of the direction of motion of the particle the kinetic energy is always positive. Considering the particular case when $v_1 = 0$ and $v_2 = v$, and substituting $T_1 = 0$ and $T_2 = T$ into (13.10), we observe that the work done by the forces acting on the particle is equal to T. Thus, the kinetic energy of a particle moving with a speed v represents the work which must be done to bring the particle from rest to the speed v. Substituting $T_1 = T$ and $T_2 = 0$ into (13.10), we also note that when a particle moving with a speed v is brought to rest, the work done by the forces acting on the particle is $-T$. Assuming that no energy is dissipated into heat, we conclude that the work done by the forces exerted *by the particle* on the bodies which cause it to come to rest is equal to T. Thus, the kinetic energy of a particle also represents *the capacity to do work associated with the speed of the particle*.

The kinetic energy is measured in the same units as work, i.e., in joules if SI units are used and in ft · lb if U.S. customary units are used. We check that, in SI units,

$$T = \tfrac{1}{2}mv^2 = \text{kg}(\text{m/s})^2 = (\text{kg}\cdot\text{m/s}^2)\text{m} = \text{N}\cdot\text{m} = \text{J}$$

while, in customary units,

$$T = \tfrac{1}{2}mv^2 = (\text{lb}\cdot\text{s}^2/\text{ft})(\text{ft/s})^2 = \text{ft}\cdot\text{lb}$$

13.4 APPLICATIONS OF THE PRINCIPLE OF WORK AND ENERGY

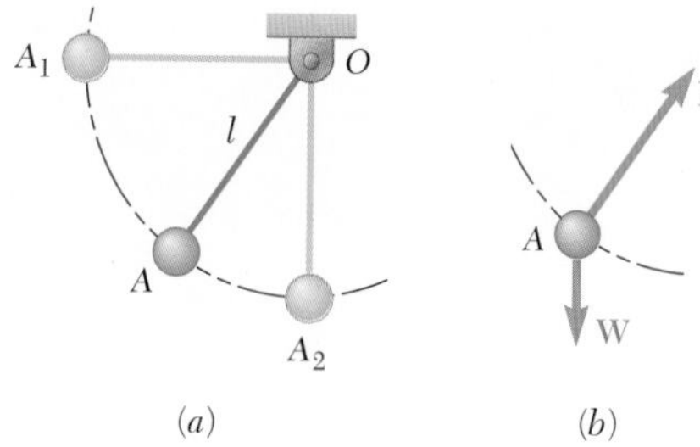

Fig. 13.8

The application of the principle of work and energy greatly simplifies the solution of many problems involving forces, displacements, and velocities. Consider, for example, the pendulum OA consisting of a bob A of weight W attached to a cord of length l (Fig. 13.8*a*). The pendulum is released with no initial velocity from a horizontal position OA_1 and allowed to swing in a vertical plane. We wish to determine the speed of the bob as it passes through A_2, directly under O.

We first determine the work done during the displacement from A_1 to A_2 by the forces acting on the bob. We draw a free-body diagram of the bob, showing all the *actual* forces acting on it, i.e., the weight $\mathbf{W}$ and the force $\mathbf{P}$ exerted by the cord (Fig. 13.8*b*). (An inertia vector is not an actual force and *should not* be included in the free-body diagram.) We note that the force $\mathbf{P}$ does no work, since it is normal to the path; the only force which does work is thus the weight $\mathbf{W}$. The work of $\mathbf{W}$ is obtained by multiplying its magnitude W by the vertical displacement l (Sec. 13.2); since the displacement is downward, the work is positive. We therefore write $U_{1\rightarrow 2} = Wl$.

Now considering the kinetic energy of the bob, we find $T_1 = 0$ at A_1 and $T_2 = \frac{1}{2}(W/g)v_2^2$ at A_2. We can now apply the principle of work and energy; recalling formula (13.11), we write

$$T_1 + U_{1\rightarrow 2} = T_2 \qquad 0 + Wl = \frac{1}{2}\frac{W}{g}v_2^2$$

Solving for v_2, we find $v_2 = \sqrt{2gl}$. We note that the speed obtained is that of a body falling freely from a height l.

The example we have considered illustrates the following advantages of the method of work and energy:

1. In order to find the speed at A_2, there is no need to determine the acceleration in an intermediate position A and to integrate the expression obtained from A_1 to A_2.
2. All quantities involved are scalars and can be added directly, without using x and y components.
3. Forces which do no work are eliminated from the solution of the problem.

What is an advantage in one problem, however, may be a disadvantage in another. It is evident, for instance, that the method of work and energy cannot be used to directly determine an acceleration. It is also evident that in determining a force which is normal to the path of the particle, a force which does no work, the method of work and energy must be supplemented by the direct application of Newton's second law. Suppose, for example, that we wish to determine the tension in the cord of the pendulum of Fig. 13.8*a* as the bob passes through A_2. We draw a free-body diagram of the bob in that position (Fig. 13.9) and express Newton's second law in terms of tangential and normal components. The equations $\Sigma F_t = ma_t$ and $\Sigma F_n = ma_n$ yield, respectively, $a_t = 0$ and

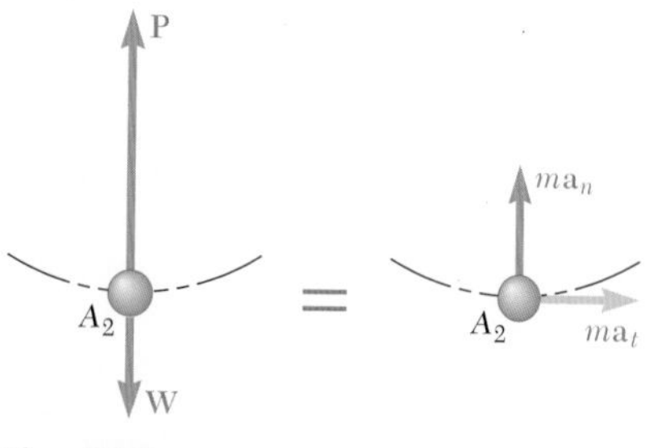

Fig. 13.9

$$P - W = ma_n = \frac{W}{g}\frac{v_2^2}{l}$$

But the speed at A_2 was determined earlier by the method of work and energy. Substituting $v_2^2 = 2gl$ and solving for P, we write

$$P = W + \frac{W}{g}\frac{2gl}{l} = 3W$$

When a problem involves two particles or more, the principle of work and energy can be applied to each particle separately. Adding the kinetic energies of the various particles, and considering the work of all the forces acting on them, we can also write a single equation of work and energy for all the particles involved. We have

$$T_1 + U_{1\rightarrow 2} = T_2 \tag{13.11}$$

where T represents the arithmetic sum of the kinetic energies of the particles involved (all terms are positive) and $U_{1\rightarrow 2}$ is the work of all the forces acting on the particles, *including the forces of action and reaction exerted by the particles on each other*. In problems involving bodies connected by *inextensible cords or links*, however, the work of the forces exerted by a given cord or link on the two bodies it connects cancels out, since the points of application of these forces move through equal distances (see Sample Prob. 13.2).†

Since friction forces have a direction opposite of that of the displacement of the body on which they act, *the work of friction forces is always negative*. This work represents energy dissipated into heat and always results in a decrease in the kinetic energy of the body involved (see Sample Prob. 13.3).

13.5 POWER AND EFFICIENCY

Power is defined as the time rate at which work is done. In the selection of a motor or engine, power is a much more important criterion than is the actual amount of work to be performed. Either a small motor or a large power plant can be used to do a given amount of work; but the small motor may require a month to do the work done by the power plant in a matter of minutes. If ΔU is the work done during the time interval Δt, then the average power during that time interval is

$$\text{Average power} = \frac{\Delta U}{\Delta t}$$

Letting Δt approach zero, we obtain at the limit

$$\text{Power} = \frac{dU}{dt} \tag{13.12}$$

†The application of the method of work and energy to a system of particles is discussed in detail in Chap. 14.

Substituting the scalar product $\mathbf{F} \cdot d\mathbf{r}$ for dU, we can also write

$$\text{Power} = \frac{dU}{dt} = \frac{\mathbf{F} \cdot d\mathbf{r}}{dt}$$

and, recalling that $d\mathbf{r}/dt$ represents the velocity $\mathbf{v}$ of the point of application of $\mathbf{F}$,

$$\text{Power} = \mathbf{F} \cdot \mathbf{v} \tag{13.13}$$

Since power was defined as the time rate at which work is done, it should be expressed in units obtained by dividing units of work by the unit of time. Thus, if SI units are used, power should be expressed in J/s; this unit is called a *watt* (W). We have

$$1 \text{ W} = 1 \text{ J/s} = 1 \text{ N} \cdot \text{m/s}$$

If U.S. customary units are used, power should be expressed in ft · lb/s or in *horsepower* (hp), with the latter defined as

$$1 \text{ hp} = 550 \text{ ft} \cdot \text{lb/s}$$

Recalling from Sec. 13.2 that 1 ft · lb = 1.356 J, we verify that

$$1 \text{ ft} \cdot \text{lb/s} = 1.356 \text{ J/s} = 1.356 \text{ W}$$
$$1 \text{ hp} = 550(1.356 \text{ W}) = 746 \text{ W} = 0.746 \text{ kW}$$

The *mechanical efficiency* of a machine was defined in Sec. 10.5 as the ratio of the output work to the input work:

$$\eta = \frac{\text{output work}}{\text{input work}} \tag{13.14}$$

This definition is based on the assumption that work is done at a constant rate. The ratio of the output to the input work is therefore equal to the ratio of the rates at which output and input work are done, and we have

$$\eta = \frac{\text{power output}}{\text{power input}} \tag{13.15}$$

Because of energy losses due to friction, the output work is always smaller than the input work, and consequently the power output is always smaller than the power input. The mechanical efficiency of a machine is therefore always less than 1.

When a machine is used to transform mechanical energy into electric energy, or thermal energy into mechanical energy, its *overall efficiency* can be obtained from formula (13.15). The overall efficiency of a machine is always less than 1; it provides a measure of all the various energy losses involved (losses of electric or thermal energy as well as frictional losses). Note that it is necessary to express the power output and the power input in the same units before using formula (13.15).

SAMPLE PROBLEM 13.1

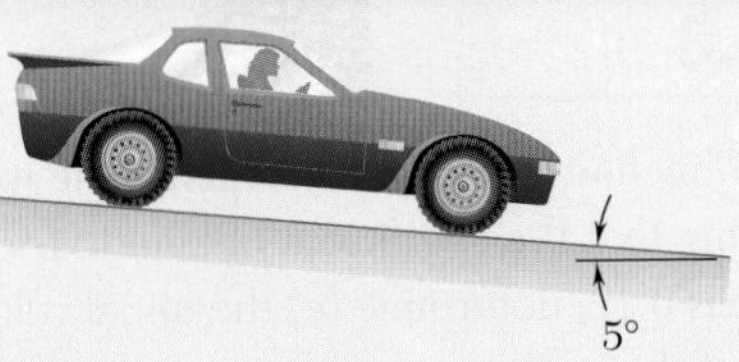

An automobile weighing 1000 kg is driven down a 5° incline at a speed of 72 km/h when the brakes are applied, causing a constant total braking force (applied by the road on the tires) of 5000-N. Determine the distance traveled by the automobile as it comes to a stop.

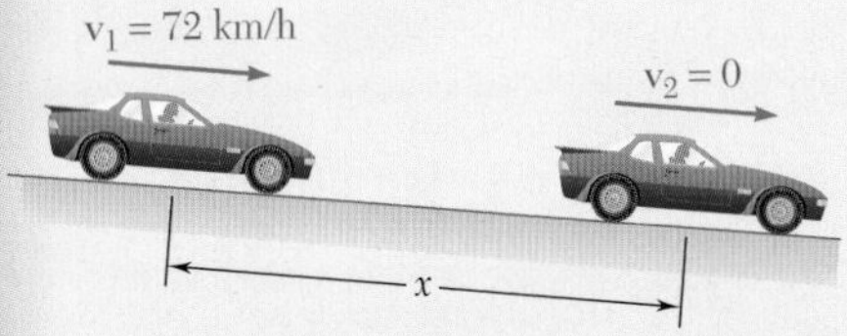

SOLUTION

Kinetic Energy

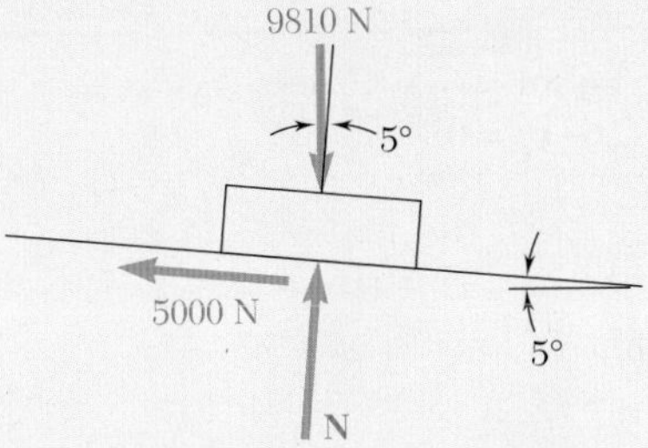

Position 1: $v_1 = \left(72 \dfrac{\text{km}}{\text{h}}\right)\left(\dfrac{1000 \text{ m}}{1 \text{ km}}\right)\left(\dfrac{1 \text{ h}}{3600 \text{ s}}\right) = 20 \text{ m/s}$

$$T_1 = \tfrac{1}{2}mv_1^2 = \tfrac{1}{2}(1000 \text{ kg})(20 \text{ m/s})^2 = 20{,}000 \text{ J}$$

Position 2: $v_2 = 0 \qquad T_2 = 0$

Work $\quad U_{1\to 2} = -5000x + (1000 \text{ kg})(9.81 \text{ m/s}^2)(\sin 5°)x = -4145x$

Principle of Work and Energy

$$T_1 + U_{1\to 2} = T_2$$
$$200{,}000 - 4145x = 0 \qquad x = 48.25 \text{ m} \blacktriangleleft$$

SAMPLE PROBLEM 13.2

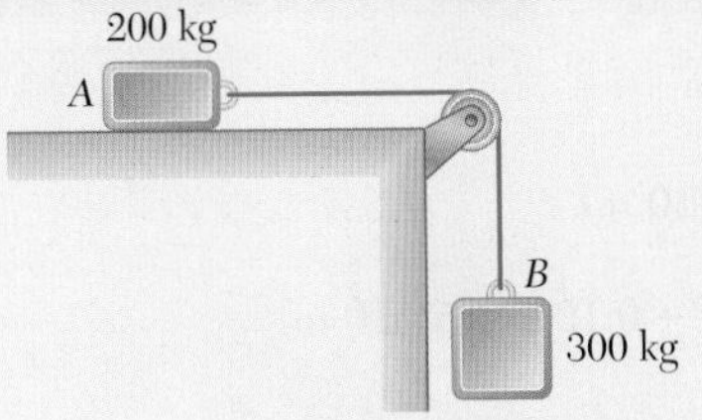

Two blocks are joined by an inextensible cable as shown. If the system is released from rest, determine the velocity of block A after it has moved 2 m. Assume that the coefficient of kinetic friction between block A and the plane is $\mu_k = 0.25$ and that the pulley is weightless and frictionless.

SOLUTION

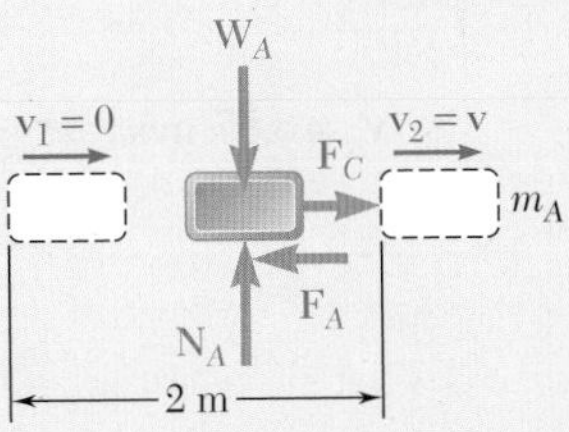

Work and Energy for Block *A*. We denote the friction force by $\mathbf{F}_A$ and the force exerted by the cable by $\mathbf{F}_C$, and write

$$m_A = 200 \text{ kg} \qquad W_A = (200 \text{ kg})(9.81 \text{ m/s}^2) = 1962 \text{ N}$$
$$F_A = \mu_k N_A = \mu_k W_A = 0.25(1962 \text{ N}) = 490 \text{ N}$$
$$T_1 + U_{1\to 2} = T_2: \quad 0 + F_C(2 \text{ m}) - F_A(2 \text{ m}) = \tfrac{1}{2}m_A v^2$$
$$F_C(2 \text{ m}) - (490 \text{ N})(2 \text{ m}) = \tfrac{1}{2}(200 \text{ kg})v^2 \qquad (1)$$

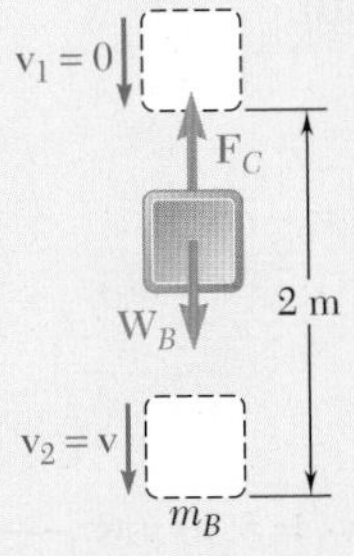

Work and Energy for Block *B*. We write

$$m_B = 300 \text{ kg} \qquad W_B = (300 \text{ kg})(9.81 \text{ m/s}^2) = 2940 \text{ N}$$
$$T_1 + U_{1\to 2} = T_2: \quad 0 + W_B(2 \text{ m}) - F_C(2 \text{ m}) = \tfrac{1}{2}m_B v^2$$
$$(2940 \text{ N})(2 \text{ m}) - F_C(2 \text{ m}) = \tfrac{1}{2}(300 \text{ kg})v^2 \qquad (2)$$

Adding the left-hand and right-hand members of (1) and (2), we observe that the work of the forces exerted by the cable on A and B cancels out:

$$(2940 \text{ N})(2 \text{ m}) - (490 \text{ N})(2 \text{ m}) = \tfrac{1}{2}(200 \text{ kg} + 300 \text{ kg})v^2$$
$$4900 \text{ J} = \tfrac{1}{2}(500 \text{ kg})v^2 \qquad v = 4.43 \text{ m/s} \blacktriangleleft$$

SAMPLE PROBLEM 13.3

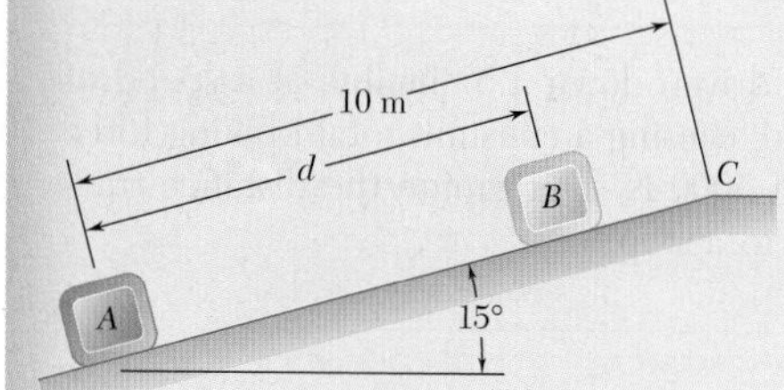

A package is projected 10 m up a 15° incline so that it just reaches the top of the incline with zero velocity. Knowing that the coefficient of kinetic friction between the package and the incline is 0.12, determine (*a*) the initial velocity of the package at *A*, (*b*) the velocity of the package as it returns to its original position.

SOLUTION

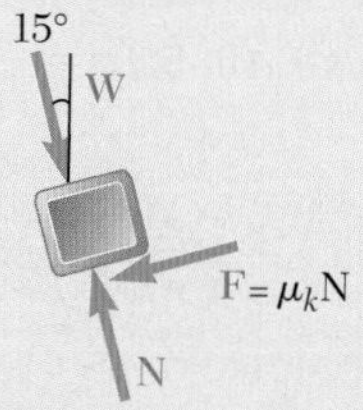

(*a*) Motion up the plane, from *A* to *C*, $-v_C = 0$.

$$T_A = \frac{1}{2} m v_A^2, \quad T_C = 0$$

$$U_{A-C} = (-W \sin 15° - F)(10 \text{ m})$$

$$\Sigma F = 0\text{:} \quad N - W \cos 15° = 0$$

$$N = W \cos 15°$$

$$F = \mu_k N = 0.12\, W \cos 15°$$

$$U_{A-C} = -W(\sin 15° + 0.12 \cos 15°)(10 \text{ m})$$

$$T_A + U_{A-C} = T_C \qquad \frac{1}{2}\frac{W}{g} v_A^2 - W(\sin 15° + 0.12 \cos 15°)(10 \text{ m})$$

$$v_A^2 = (2)(9.81)(\sin 15° + 0.12 \cos 15°)(10 \text{ m})$$

$$v_A^2 = 73.5$$

$\mathbf{v}_A = 8.57$ m/s ∠ 15° ◀

(*b*) Motion down the plane from *C* to *A*.

$$T_C = 0 \quad T_A = \frac{1}{2} m v_A^2 \quad U_{C-A} = (W \sin 15° - F)10$$

(*F* reverses direction.)

$$T_C + U_{C-A} = T_A \qquad 0 + W(\sin 15° - 0.12 \cos 15°)(10 \text{ m}) = \frac{1}{2} m v_A^2$$

$$v_A^2 = (2)(9.81)(\sin 15° - 0.12 \cos 15°)(10 \text{ m})$$

$$v_A^2 = 28.039$$

$\mathbf{v}_A = 5.30$ m/s ∠ 15° ◀

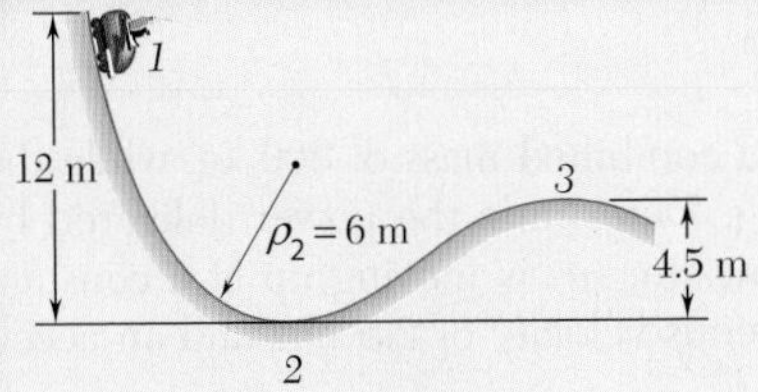

SAMPLE PROBLEM 13.4

A 1000-kg car starts from rest at point *1* and moves without friction down the track shown. (*a*) Determine the force exerted by the track on the car at point 2, where the radius of curvature of the track is 6 m (*b*) Determine the minimum safe value of the radius of curvature at point 3.

SOLUTION

a. Force Exerted by the Track at Point 2. The principle of work and energy is used to determine the velocity of the car as it passes through point 2.

Kinetic Energy. $T_1 = 0 \qquad T_2 = \frac{1}{2}mv_2^2$

Work. The only force which does work is the weight **W**. Since the vertical displacement from point *1* to point 2 is 12 m downward, the work of the weight is

$$U_{1\rightarrow 2} = +W(12 \text{ m}) = \text{mg}(12 \text{ m})$$

Principle of Work and Energy

$$T_1 + U_{1\rightarrow 2} = T_2 \qquad 0 + mg(12 \text{ m}) = \frac{1}{2}\text{m}v_2^2$$

$$v_2^2 = 24g = (24\text{m})\left(9.81\frac{\text{m}}{\text{s}^2}\right) \qquad v_2 = 15.34 \text{ m/s}$$

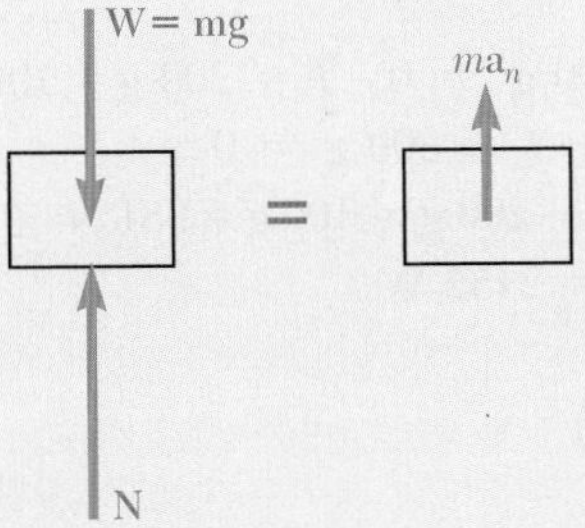

Newton's Second Law at Point 2. The acceleration $\mathbf{a}_n$ of the car at point 2 has a magnitude $a_n = v_2^2/\rho$ and is directed upward. Since the external forces acting on the car are **W** and **N**, we write

$$+\uparrow\Sigma F_n = ma_n: \qquad -W + N = ma_n$$

$$= \frac{mv_2^2}{\rho}$$

$$= m\frac{24\text{g}}{6} = 4\text{mg} = 4W$$

$$N = 5W = 5(1000 \text{ kg})(9.81 \text{ m/s}^2)\text{ N} = 49.05 \text{ kN}\uparrow \blacktriangleleft$$

b. Minimum Value of ρ at Point 3. ***Principle of Work and Energy.*** Applying the principle of work and energy between point *1* and point 3, we obtain

$$T_1 + U_{1\rightarrow 3} = T_3 \qquad 0 + mg(7.5 \text{ m}) = \frac{1}{2}mv_3^2$$

$$v_3^2 = 15\text{g} = (15\text{m})(9.81 \text{m/s}^2) \qquad v_3 = 12.13 \text{ m/s}$$

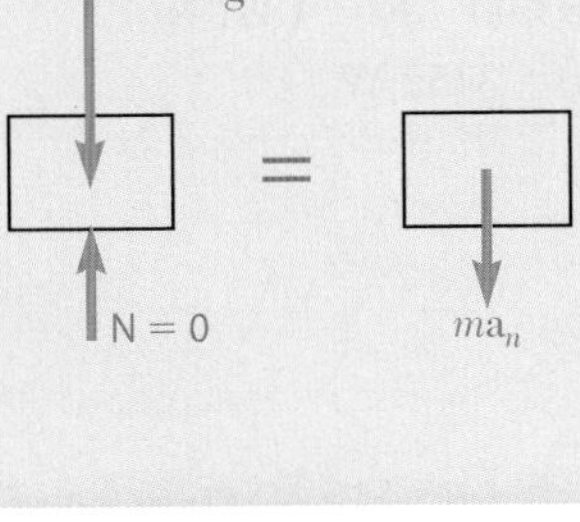

Newton's Second Law at Point 3. The minimum safe value of ρ occurs when $\mathbf{N} = 0$. In this case, the acceleration $\mathbf{a}_n$, of magnitude $a_n = v_3^2/\rho$, is directed downward, and we write

$$+\downarrow\Sigma F_n = ma_n: \qquad \text{mg} = m\frac{v_3^2}{\rho}$$

$$\rho = \frac{v_3^2}{\text{g}} = \frac{15\text{g}}{\text{g}} \qquad \rho = 15 \text{ m} \blacktriangleleft$$

SAMPLE PROBLEM 13.5

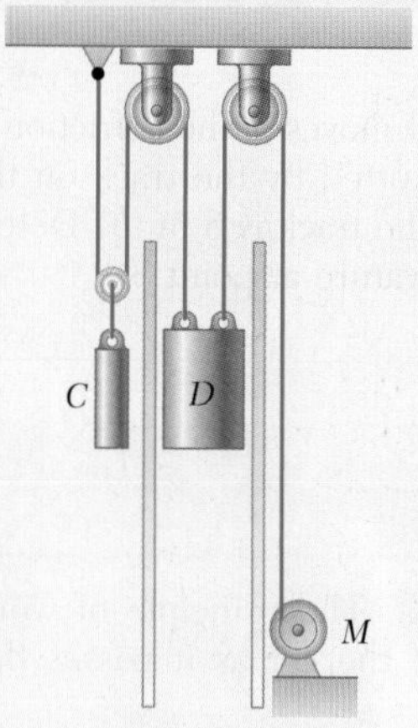

The dumbwaiter D and its load have a combined mass of 300 kg, while the counterweight C has a mass of 400 kg. Determine the power delivered by the electric motor M when the dumbwaiter (*a*) is moving up at a constant speed of 2.5 m/s, (*b*) has an instantaneous velocity of 2.5 m/s and an acceleration of 1 m/s², both directed upward.

SOLUTION

Since the force **F** exerted by the motor cable has the same direction as the velocity $\mathbf{v}_D$ of the dumbwaiter, the power is equal to Fv_D, where $v_D = 2.5$ m/s. To obtain the power, we must first determine **F** in each of the two given situations.

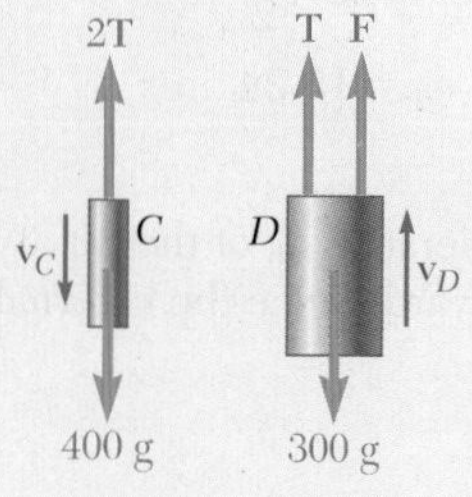

a. Uniform Motion. We have $\mathbf{a}_C = \mathbf{a}_D = 0$; both bodies are in equilibrium.

Free Body C: $+\uparrow\Sigma F_y = 0$: $2T - 400\text{ g} = 0 \quad T = 200\text{ g} = 1962\text{ N}$

Free Body D: $+\uparrow\Sigma F_y = 0$: $F + T - 300\text{ g} = 0$

$$F = 300\text{ g} - T = 300\text{ g} - 200\text{ g} = 100\text{ g} = 981\text{ N}$$

$$Fv_D = (981\text{ N})(2.5\text{ m/s}) = 2452\text{ W}$$

Power = 2450 W ◀

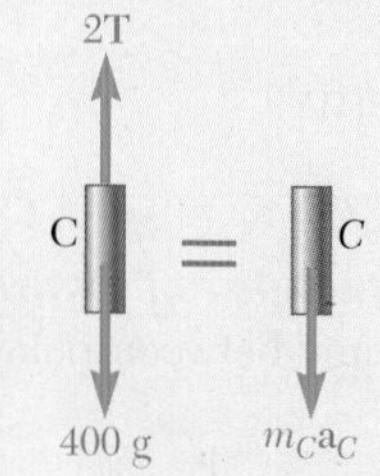

b. Accelerated Motion. We have

$$\mathbf{a}_D = 1\text{ m/s}^2\uparrow \qquad \mathbf{a}_C = -\tfrac{1}{2}\mathbf{a}_D = 0.5\text{ m/s}^2\downarrow$$

The equations of motion are

Free Body C: $+\downarrow\Sigma F_y = m_C a_C$: $400\text{ g} - 2T = 400\,(0.5)$

$$T = \frac{(400)(9.81) - 400\,(0.5)}{2} = 1862\text{ N}$$

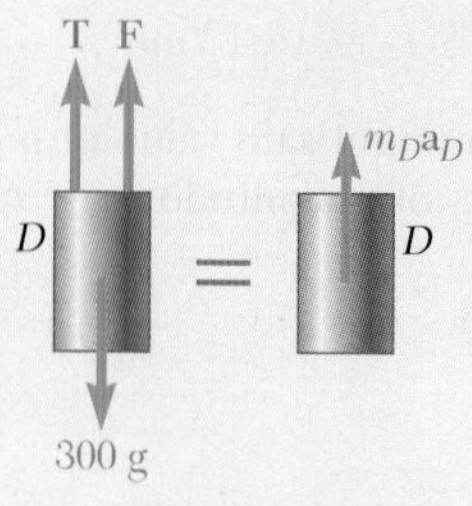

Free Body D: $+\uparrow\Sigma F_y = m_D a_D$: $F + T - 300\text{ g} = 300\,(1)$

$$F + 1862 - 300\,(9.81) = 300 \qquad F = 1381\text{ N}$$

$$Fv_D = (1381\text{ N})(2.5\text{ m/s}) = 3452\text{ W}$$

Power = 3450 W ◀

SOLVING PROBLEMS ON YOUR OWN

In the preceding chapter, you solved problems dealing with the motion of a particle by using the fundamental equation $\mathbf{F} = m\mathbf{a}$ to determine the acceleration $\mathbf{a}$. By applying the principles of kinematics you were then able to determine from $\mathbf{a}$ the velocity and displacement of the particle at any time. In this lesson we combined $\mathbf{F} = m\mathbf{a}$ and the principles of kinematics to obtain an additional method of analysis called the *method of work and energy*. This eliminates the need to calculate the acceleration and will enable you to relate the velocities of the particle at two points along its path of motion. To solve a problem by the method of work and energy you will follow these steps:

1. Computing the work of each of the forces. The work $U_{1\rightarrow 2}$ of a given force $\mathbf{F}$ during the finite displacement of the particle from A_1 to A_2 is defined as

$$U_{1\rightarrow 2} = \int \mathbf{F} \cdot d\mathbf{r} \qquad \text{or} \qquad U_{1\rightarrow 2} = \int (F \cos \alpha)\, ds \qquad (13.2,\ 13.2')$$

where α is the angle between $\mathbf{F}$ and the displacement $d\mathbf{r}$. The work $U_{1\rightarrow 2}$ is a scalar quantity and is expressed in ft · lb or in · lb in the U.S. customary system of units and in N · m or joules (J) in the SI system of units. Note that the work done is zero for a force perpendicular to the displacement ($\alpha = 90°$). Negative work is done for $90° < \alpha < 180°$ and in particular for a friction force, which is always opposite in direction to the displacement ($\alpha = 180°$).

The work $U_{1\rightarrow 2}$ can be easily evaluated in the following cases that you will encounter:

a. Work of a constant force in rectilinear motion

$$U_{1\rightarrow 2} = (F \cos \alpha)\, \Delta x \qquad (13.3)$$

where α = angle the force forms with the direction of motion
Δx = displacement from A_1 to A_2 (Fig. 13.3)

b. Work of the force of gravity

$$U_{1\rightarrow 2} = -W\, \Delta y \qquad (13.4')$$

where Δy is the vertical displacement of the center of gravity of the body of weight W. Note that the work is positive when Δy is negative, that is, when the body moves down (Fig. 13.4).

c. Work of the force exerted by a spring

$$U_{1\rightarrow 2} = \tfrac{1}{2}kx_1^2 - \tfrac{1}{2}kx_2^2 \qquad (13.6)$$

where k is the spring constant and x_1 and x_2 are the elongations of the spring corresponding to the positions A_1 and A_2 (Fig. 13.5).

(continued)

d. Work of a gravitational force

$$U_{1\rightarrow 2} = \frac{GMm}{r_2} - \frac{GMm}{r_1} \qquad (13.7)$$

for a displacement of the body from $A_1(r = r_1)$ to $A_2(r = r_2)$ (Fig. 13.6).

2. Calculate the kinetic energy at A_1 and A_2. The kinetic energy T is

$$T = \tfrac{1}{2}mv^2 \qquad (13.9)$$

where m is the mass of the particle and v is the magnitude of its velocity. The units of kinetic energy are the same as the units of work, that is, ft · lb or in · lb if U.S. customary units are used and N · m or joules (J) if SI units are used.

3. Substitute the values for the work done $U_{1\rightarrow 2}$ and the kinetic energies T_1 and T_2 into the equation

$$T_1 + U_{1\rightarrow 2} = T_2 \qquad (13.11)$$

You will now have *one equation* which you can solve for *one unknown*. Note that this equation does not yield the time of travel or the acceleration directly. However, if you know the radius of curvature ρ of the path of the particle at a point where you have obtained the velocity v, you can express the normal component of the acceleration as $a_n = v^2/\rho$ and obtain the normal component of the force exerted on the particle by writing $F_n = mv^2/\rho$.

4. Power was introduced in this lesson as the time rate at which work is done, $P = dU/dt$. Power is measured in ft · lb/s or *horsepower* (hp) in U.S. customary units and in J/s or *watts* (W) in the SI system of units. To calculate the power, you can use the equivalent formula,

$$P = \mathbf{F} \cdot \mathbf{v} \qquad (13.13)$$

where **F** and **v** denote the force and the velocity, respectively, at a given time [Sample Prob. 13.5]. In some problems [see, e.g., Prob. 13.47], you will be asked for the *average power*, which can be obtained by dividing the total work by the time interval during which the work is done.

PROBLEMS

CONCEPT QUESTION

13.CQ1 Block A is traveling with a speed v_0 on a smooth surface when the surface suddenly becomes rough with a coefficient of friction of μ causing the block to stop after a distance d. If block A were traveling twice as fast, that is, at a speed $2v_0$, how far will it travel on the rough surface before stopping?

a. $d/2$
b. d
c. $\sqrt{2}d$
d. $2d$
e. $4d$

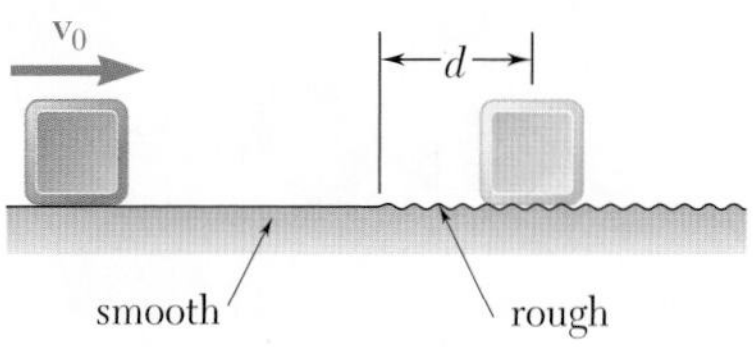

Fig. P13.CQ1

END-OF-SECTION PROBLEMS

13.1 A 400-kg satellite was placed in a circular orbit 1500 km above the surface of the earth. At this elevation the acceleration of gravity is 6.43 m/s^2. Determine the kinetic energy of the satellite, knowing that its orbital speed is 25.6×10^3 km/h.

13.2 A 0.5-kg stone is dropped down the "bottomless pit" at Carlsbad Caverns and strikes the ground with a speed of 30 m/s. Neglecting air resistance, (*a*) determine the kinetic energy of the stone as it strikes the ground and the height h from which it was dropped. (*b*) Solve part *a* assuming that the same stone is dropped down a hole on the moon. (Acceleration of gravity on the moon = 1.63 m/s^2.)

Fig. P13.2

13.3 A baseball player hits a 160-g baseball with an initial velocity of 40 m/s at an angle of 40° with the horizontal as shown. Determine (*a*) the kinetic energy of the ball immediately after it is hit, (*b*) the kinetic energy of the ball when it reaches its maximum height, (*c*) the maximum height above the ground reached by the ball.

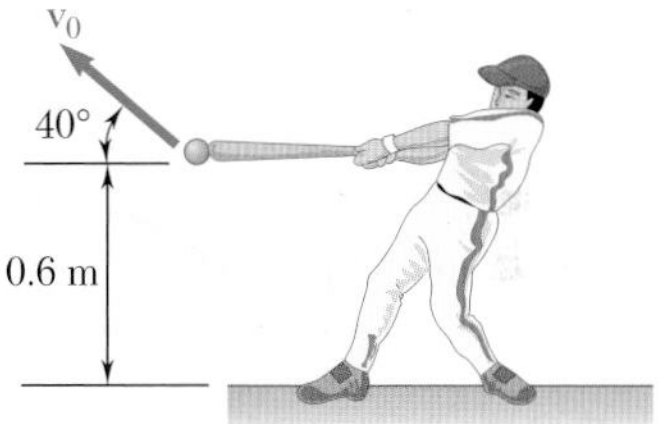

Fig. P13.3

13.4 A 500-kg communications satellite is in a circular geosynchronous orbit and completes one revolution about the earth in 23 h and 56 min at an altitude of 35 800 km above the surface of the earth. Knowing that the radius of the earth is 6370 km, determine the kinetic energy of the satellite.

13.5 In an ore-mixing operation, a bucket full of ore is suspended from a traveling crane which moves along a stationary bridge. The bucket is to swing no more than 4 m horizontally when the crane is brought to a sudden stop. Determine the maximum allowable speed v of the crane.

13.6 In an ore-mixing operation, a bucket full of ore is suspended from a traveling crane which moves along a stationary bridge. The crane is traveling at a speed of 3 m/s when it is brought to a sudden stop. Determine the maximum horizontal distance through which the bucket will swing.

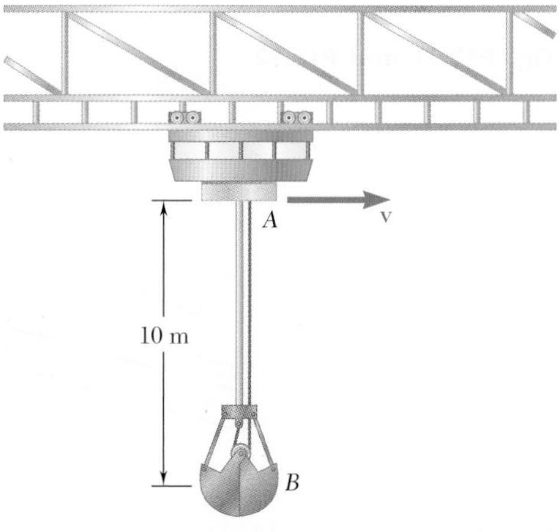

Fig. P13.5 and P13.6

13.7 Determine the maximum theoretical speed that may be achieved over a distance of 110 m by a car starting from rest assuming there is no slipping. The coefficient of static friction between the tires and pavement is 0.75, and 60 percent of the weight of the car is distributed over its front wheels and 40 percent over its rear wheels. Assume (*a*) front-wheel drive, (*b*) rear-wheel drive.

Fig. P13.8

13.8 Skid marks on a drag racetrack indicate that the rear (drive) wheels of a car slip for the first 20 m of the 400-m track. (*a*) Knowing that the coefficient of kinetic friction is 0.60, determine the speed of the car at the end of the first 20-m portion of the track if it starts from rest and the front wheels are just off the ground. (*b*) What is the maximum theoretical speed of the car at the finish line if, after skidding for 20 m, it is driven without the wheels slipping for the remainder of the race? Assume that while the car is rolling without slipping, 60 percent of the weight of the car is on the rear wheels and the coefficient of static friction is 0.75. Ignore air resistance and rolling resistance.

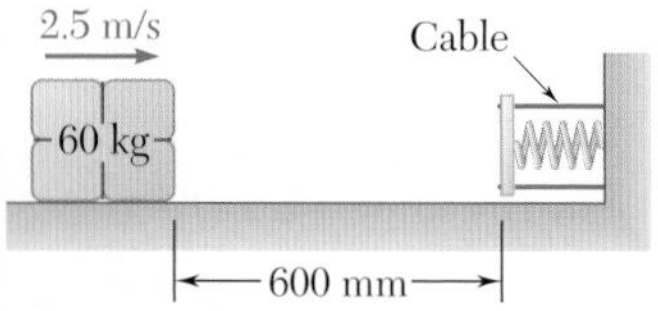

Fig. P13.9

13.9 A spring is used to stop a 60-kg package which is sliding on a horizontal surface. The spring has a constant $k = 20$ kN/m and is held by cables so that it is initially compressed 120 mm. Knowing that the package has a velocity of 2.5 m/s in the position shown and that the maximum additional deflection of the spring is 40 mm, determine (*a*) the coefficient of kinetic friction between the package and the surface, (*b*) the velocity of the package as it passes again through the position shown.

13.10 A 1.4-kg model rocket is launched vertically from rest with a constant thrust of 25 N until the rocket reaches an altitude of 15 m and the thrust ends. Neglecting air resistance, determine (*a*) the speed of the rocket when the thrust ends, (*b*) the maximum height reached by the rocket, (*c*) the speed of the rocket when it returns to the ground.

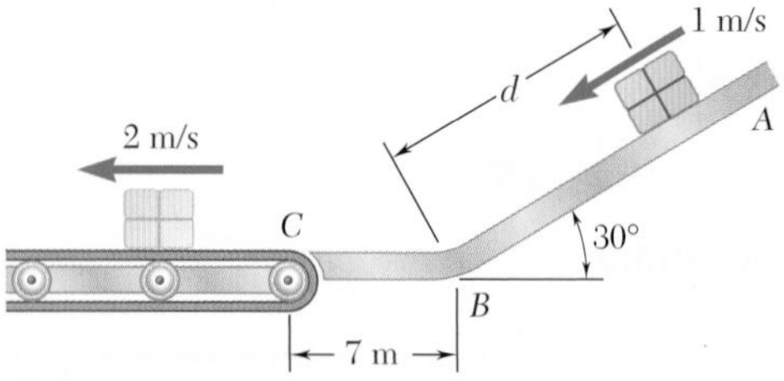

Fig. P13.11 and P13.12

13.11 Packages are thrown down an incline at A with a velocity of 1 m/s. The packages slide along the surface ABC to a conveyor belt which moves with a velocity of 2 m/s. Knowing that $\mu_k = 0.25$ between the packages and the surface ABC, determine the distance d if the packages are to arrive at C with a velocity of 2 m/s.

13.12 Packages are thrown down an incline at A with a velocity of 1 m/s. The packages slide along the surface ABC to a conveyor belt which moves with a velocity of 2 m/s. Knowing that $d = 7.5$ m and $\mu_k = 0.25$ between the packages and all surfaces, determine (*a*) the speed of the package at C, (*b*) the distance a package will slide on the conveyor belt before it comes to rest relative to the belt.

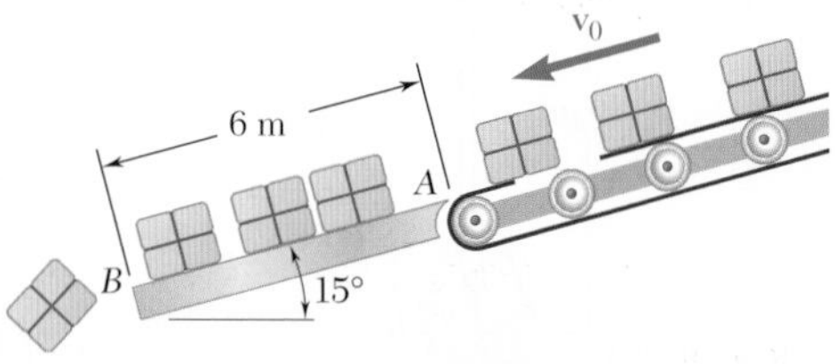

Fig. *P13.13* and *P13.14*

13.13 Boxes are transported by a conveyor belt with a velocity $\mathbf{v}_0$ to a fixed incline at A where they slide and eventually fall off at B. Knowing that $\mu_k = 0.40$, determine the velocity of the conveyor belt if the boxes leave the incline at B with a velocity of 2.5 m/s.

13.14 Boxes are transported by a conveyor belt with a velocity $\mathbf{v}_0$ to a fixed incline at A where they slide and eventually fall off at B. Knowing that $\mu_k = 0.40$, determine the velocity of the conveyor belt if the boxes are to have zero velocity at B.

13.15 A 1200-kg trailer is hitched to a 1400-kg car. The car and trailer are traveling at 72 km/h when the driver applies the brakes on both the car and the trailer. Knowing that the braking forces exerted on the car and the trailer are 5000 N and 4000 N, respectively, determine (*a*) the distance traveled by the car and trailer before they come to a stop, (*b*) the horizontal component of the force exerted by the trailer hitch on the car.

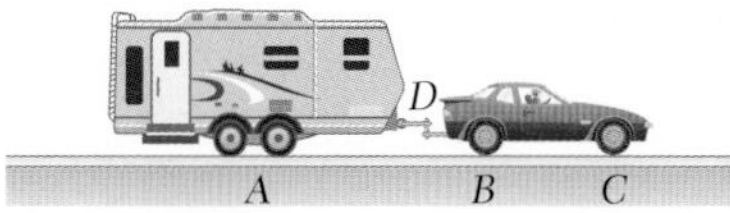

Fig. P13.15

13.16 A trailer truck enters a 2 percent uphill grade traveling at 72 km/h and reaches a speed of 108 km/h in 300 m. The cab has a mass of 1800 kg and the trailer 5400 kg. Determine (*a*) the average force at the wheels of the cab, (*b*) the average force in the coupling between the cab and the trailer.

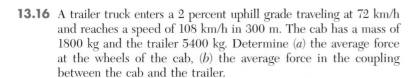

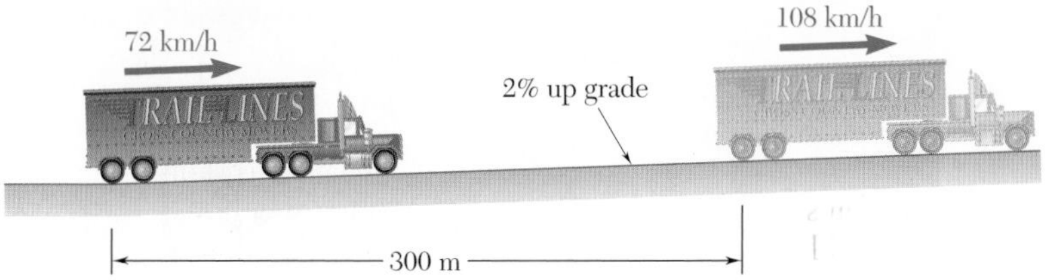

Fig. P13.16

13.17 The subway train shown is traveling at a speed of 50 km/h when the brakes are fully applied on the wheels of cars *B* and *C*, causing them to slide on the track, but are not applied on the wheels of car *A*. Knowing that the coefficient of kinetic friction is 0.35 between the wheels and the track, determine (*a*) the distance required to bring the train to a stop, (*b*) the force in each coupling.

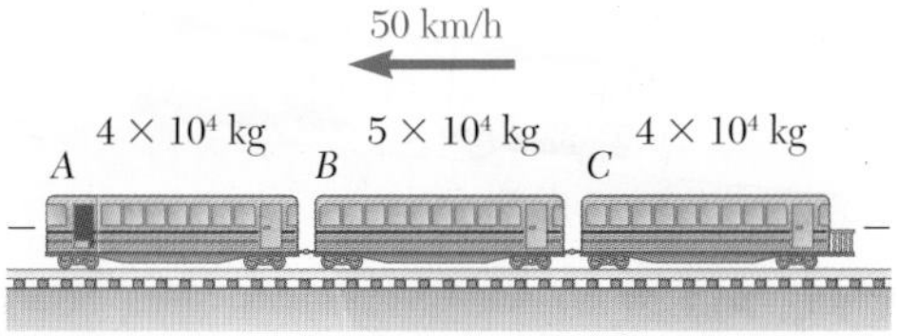

Fig. P13.17

13.18 Solve Prob. 13.17 assuming that the brakes are applied only on the wheels of car *A*.

13.19 Blocks *A* and *B* have masses of 11 kg and 5 kg, respectively, and they are both at a height $h = 2$ m above the ground when the system is released from rest. Just before hitting the ground block *A* is moving at a speed of 3 m/s. Determine (*a*) the amount of energy dissipated in friction by the pulley, (*b*) the tension in each portion of the cord during the motion.

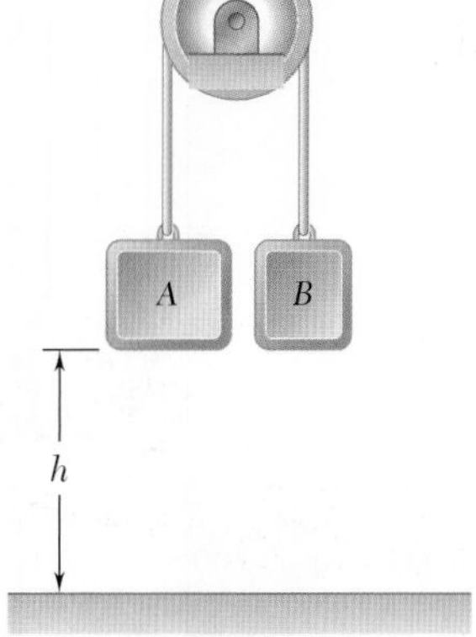

Fig. P13.19

13.20 The system shown is at rest when a constant 150-N force is applied to collar B. (a) If the force acts through the entire motion, determine the speed of collar B as it strikes the support at C. (b) After what distance d should the 150-N force be removed if the collar is to reach support C with zero velocity?

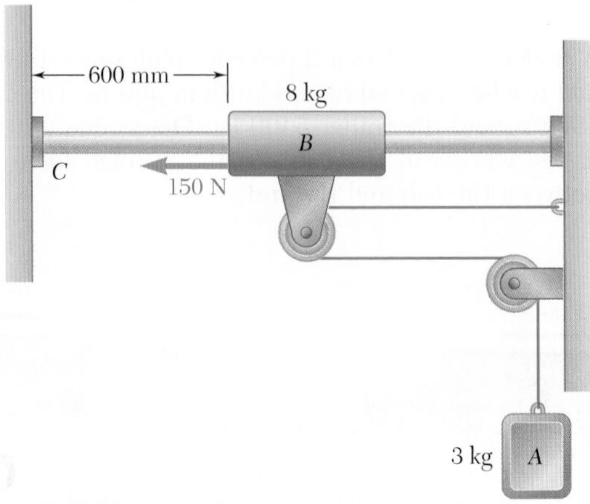

Fig. P13.20

13.21 Car B is towing car A at a constant speed of 10 m/s on an uphill grade when the brakes of car A are fully applied causing all four wheels to skid. The driver of car B does not change the throttle setting or change gears. The masses of the cars A and B are 1400 kg and 1200 kg, respectively, and the coefficient of kinetic friction is 0.8. Neglecting air resistance and rolling resistance, determine (a) the distance traveled by the cars before they come to a stop, (b) the tension in the cable.

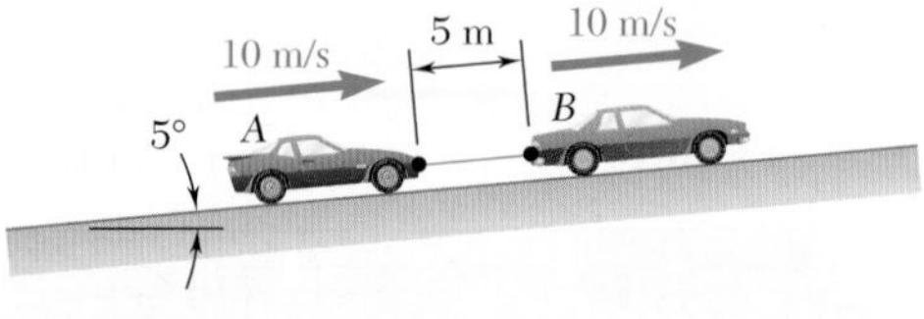

Fig. *P13.21*

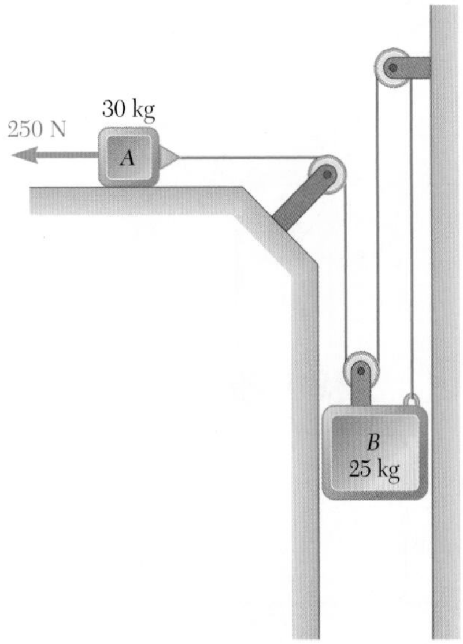

Fig. *P13.22* and P13.23

13.22 The system shown is at rest when a constant 250-N force is applied to block A. Neglecting the masses of the pulleys and the effect of friction in the pulleys and between block A and the horizontal surface, determine (a) the velocity of block B after block A has moved 2 m, (b) the tension in the cable.

13.23 The system shown is at rest when a constant 250-N force is applied to block A. Neglecting the masses of the pulleys and the effect of friction in the pulleys and assuming that the coefficients of friction between block A and the horizontal surface are $\mu_s = 0.25$ and $\mu_k = 0.20$, determine (a) the velocity of block B after block A has moved 2 m, (b) the tension in the cable.

13.24 Two blocks A and B, of mass 4 kg and 5 kg, respectively, are connected by a cord which passes over pulleys as shown. A 3-kg collar C is placed on block A and the system is released from rest. After the blocks have moved 0.9 m, collar C is removed and blocks A and B continue to move. Determine the speed of block A just before it strikes the ground.

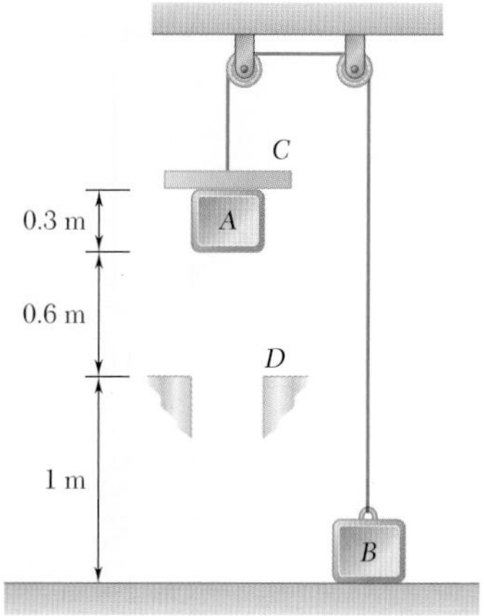

Fig. P13.24

13.25 Four packages, each of mass 3 kg, are held in place by friction on a conveyor which is disengaged from its drive motor. When the system is released from rest, package *1* leaves the belt at A just as package *4* comes onto the inclined portion of the belt at B. Determine (*a*) the speed of package *2* as it leaves the belt at A, (*b*) the speed of package *3* as it leaves the belt at A. Neglect the mass of the belt and rollers.

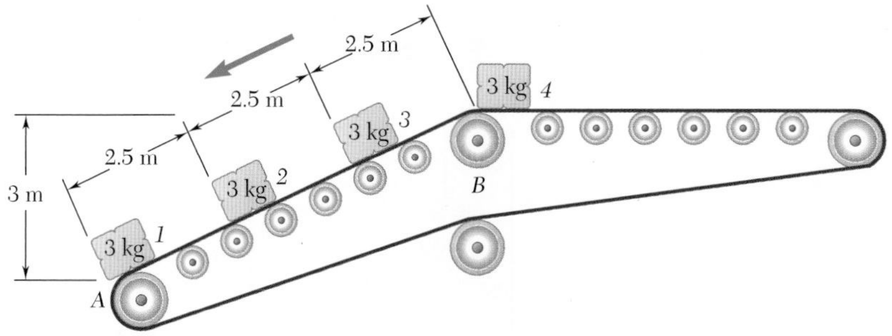

Fig. P13.25

13.26 A 3-kg block rests on top of a 2-kg block supported by but not attached to a spring of constant 40 N/m. The upper block is suddenly removed. Determine (*a*) the maximum speed reached by the 2-kg block, (*b*) the maximum height reached by the 2-kg block.

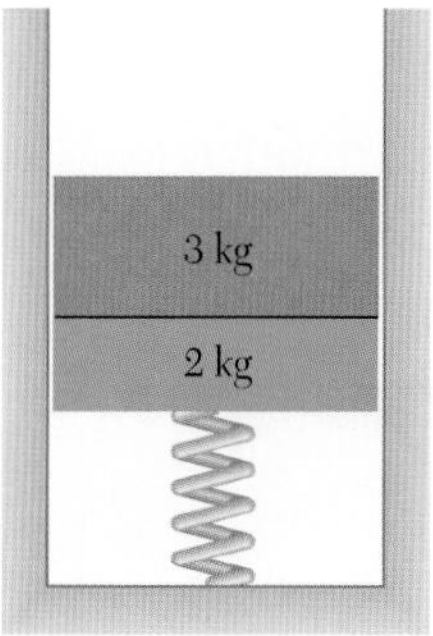

Fig. P13.26

13.27 Solve Prob. 13.26, assuming that the 2-kg block is attached to the spring.

13.28 A 4-kg collar C slides on a horizontal rod between springs A and B. If the collar is pushed to the right until spring B is compressed 50 mm and released, determine the distance through which the collar will travel assuming (*a*) no friction between the collar and the rod, (*b*) a coefficient of friction $\mu_k = 0.35$.

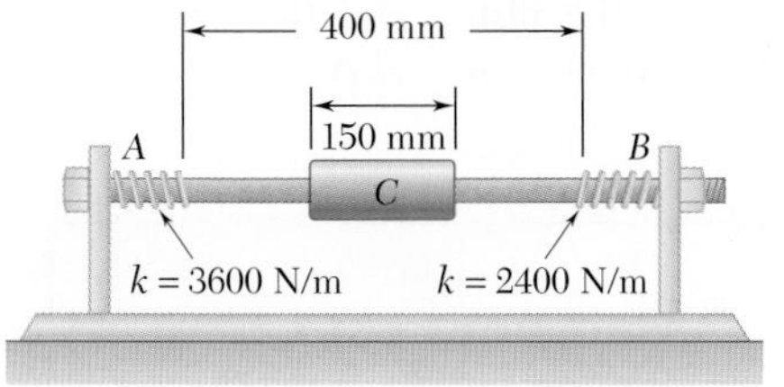

Fig. P13.28

13.29 A 3-kg block is attached to a cable and to a spring as shown. The constant of the spring is $k = 1600$ N/m and the tension in the cable is 15-N. If the cable is cut, determine (*a*) the maximum displacement of the block, (*b*) the maximum speed of the block.

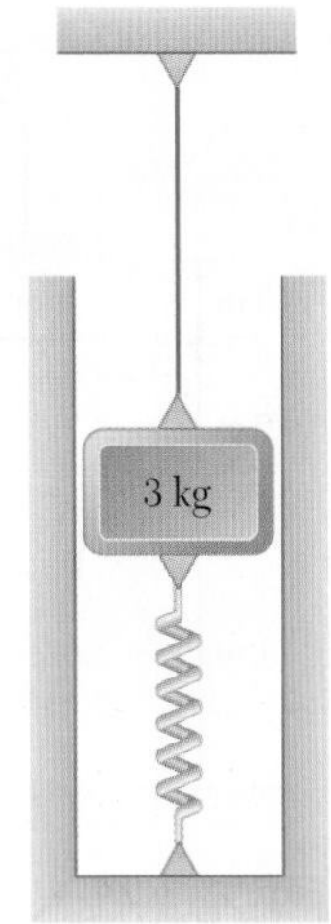

Fig. P13.29

13.30 A 10-kg block is attached to spring *A* and connected to spring *B* by a cord and pulley. The block is held in the position shown with both springs unstretched when the support is removed and the block is released with no initial velocity. Knowing that the constant of each spring is 2 kN/m, determine (*a*) the velocity of the block after it has moved down 50 mm, (*b*) the maximum velocity achieved by the block.

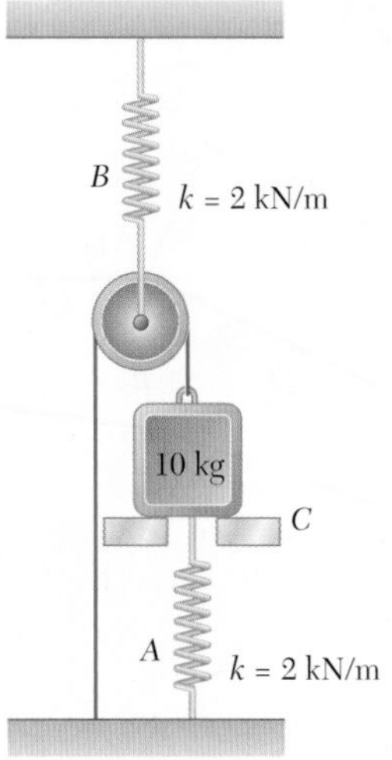

Fig. *P13.30*

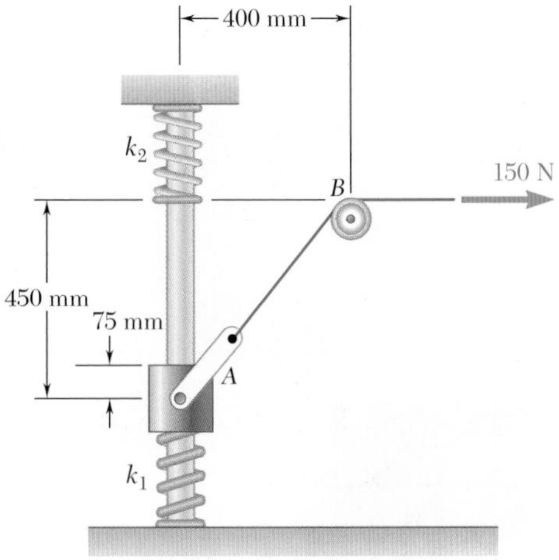

Fig. *P13.31*

13.31 A 5-kg collar *A* is at rest on top of, but not attached to, a spring with stiffness $k_1 = 400$ N/m when a constant 150-N force is applied to the cable. Knowing *A* has a speed of 1 m/s when the upper spring is compressed 75 mm, determine the spring stiffness k_2. Ignore friction and the mass of the pulley.

13.32 A piston of mass *m* and cross-sectional area *A* is in equilibrium under the pressure *p* at the center of a cylinder closed at both ends. Assuming that the piston is moved to the left a distance *a*/2 and released, and knowing that the pressure on each side of the piston varies inversely with the volume, determine the velocity of the piston as it again reaches the center of the cylinder. Neglect friction between the piston and the cylinder and express your answer in terms of *m*, *a*, *p*, and *A*.

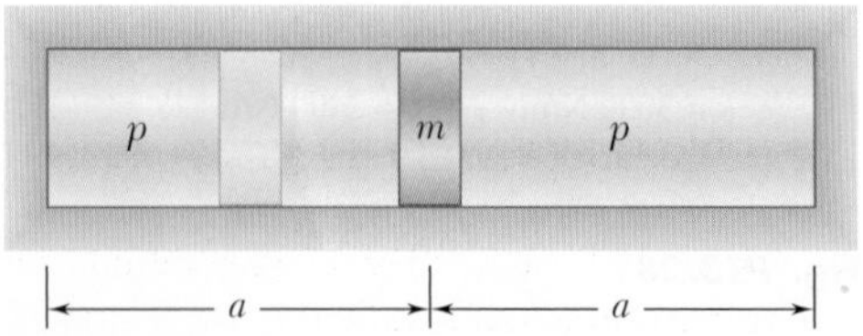

Fig. P13.32

13.33 An uncontrolled automobile traveling at 100 km/h strikes squarely a highway crash cushion of the type shown in which the automobile is brought to rest by successively crushing steel barrels. The magnitude F of the force required to crush the barrels is shown as a function of the distance x the automobile has moved into the cushion. Knowing that the mass of the automobile is 1000 kg and neglecting the effect of friction, determine (*a*) the distance the automobile will move into the cushion before it comes to rest, (*b*) the maximum deceleration of the automobile.

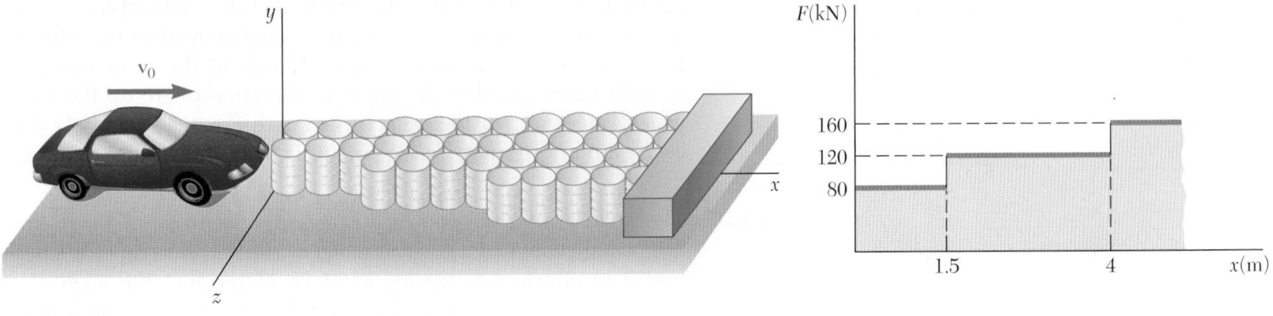

Fig. P13.33

13.34 A 300-g brass (nonmagnetic) block A and a 200-g steel magnet B are in equilibrium in a brass tube under the magnetic repelling force of another steel magnet C located at a distance $x = 4$ mm from B. The force is inversely proportional to the square of the distance between B and C. If block A is suddenly removed, determine (*a*) the maximum velocity of B, (*b*) the maximum acceleration of B. Assume that air resistance and friction are negligible.

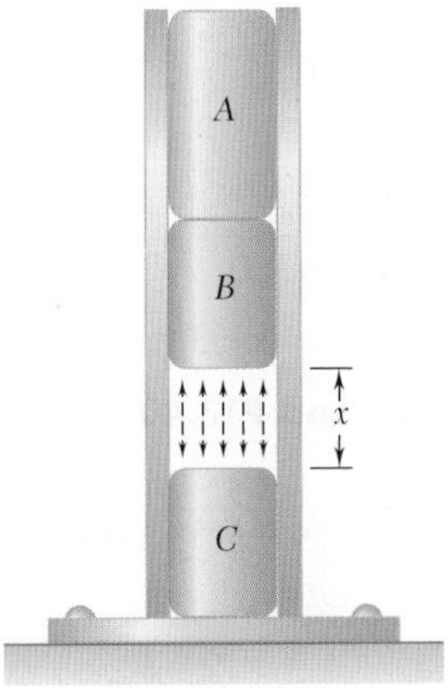

Fig. P13.34

13.35 Nonlinear springs are classified as hard or soft, depending upon the curvature of their force-deflection curve (see figure). If a delicate instrument having a mass of 5 kg is placed on a spring of length l so that its base is just touching the undeformed spring and then inadvertently released from that position, determine the maximum deflection x_m of the spring and the maximum force F_m exerted by the spring, assuming (*a*) a linear spring of constant $k = 3$ kN/m, (*b*) a hard, nonlinear spring, for which $F = (3 \text{ kN/m})(x + 160x^3)$.

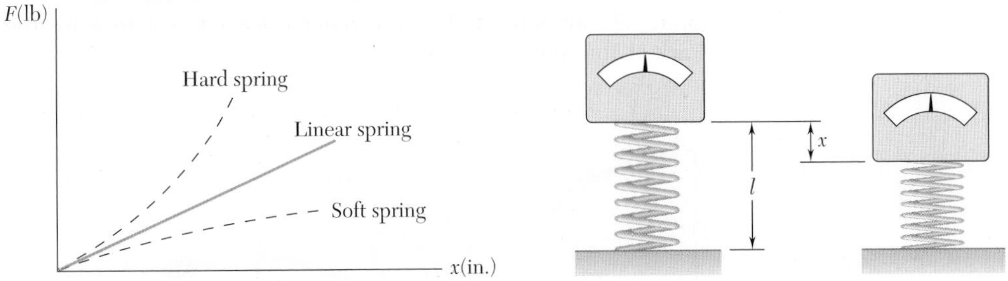

Fig. P13.35

13.36 A rocket is fired vertically from the surface of the moon with a speed v_0. Derive a formula for the ratio h_n/h_u of heights reached with a speed v, if Newton's law of gravitation is used to calculate h_n and a uniform gravitational field is used to calculate h_u. Express your answer in terms of the acceleration of gravity g_m on the surface of the moon, the radius R_m of the moon, and the speeds v and v_0.

13.37 Express the acceleration of gravity g_h at an altitude h above the surface of the earth in terms of the acceleration of gravity g_0 at the surface of the earth, the altitude h, and the radius R of the earth. Determine the percent error if the weight that an object has on the surface of earth is used as its weight at an altitude of (*a*) 1 km, (*b*) 1000 km.

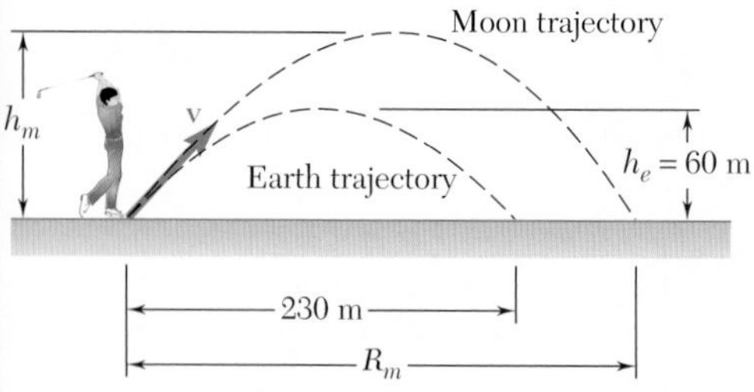

Fig. P13.38

13.38 A golf ball struck on earth rises to a maximum height of 60 m and hits the ground 230 m away. How high will the same golf ball travel on the moon if the magnitude and direction of its velocity are the same as they were on earth immediately after the ball was hit? Assume that the ball is hit and lands at the same elevation in both cases and that the effect of the atmosphere on the earth is neglected, so that the trajectory in both cases is a parabola. The acceleration of gravity on the moon is 0.165 times that on earth.

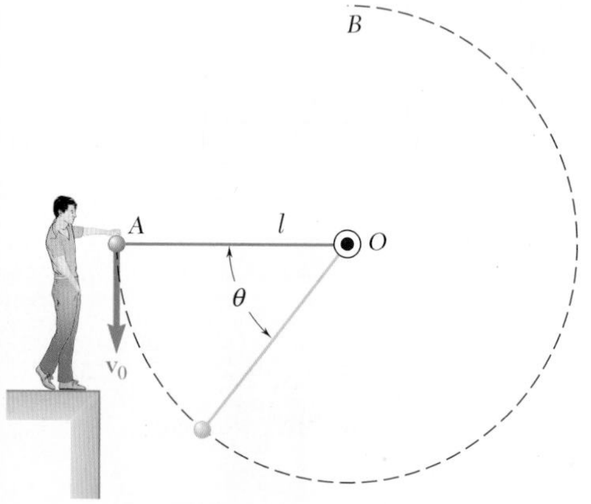

Fig. P13.39 and P13.40

13.39 The sphere at A is given a downward velocity $\mathbf{v}_0$ of magnitude 5 m/s and swings in a vertical plane at the end of a rope of length l = 2 m attached to a support at O. Determine the angle θ at which the rope will break, knowing that it can withstand a maximum tension equal to twice the weight of the sphere.

13.40 The sphere at A is given a downward velocity $\mathbf{v}_0$ and swings in a vertical circle of radius l and center O. Determine the smallest velocity $\mathbf{v}_0$ for which the sphere will reach point B as it swings about point O (*a*) if AO is a rope, (*b*) if AO is a slender rod of negligible mass.

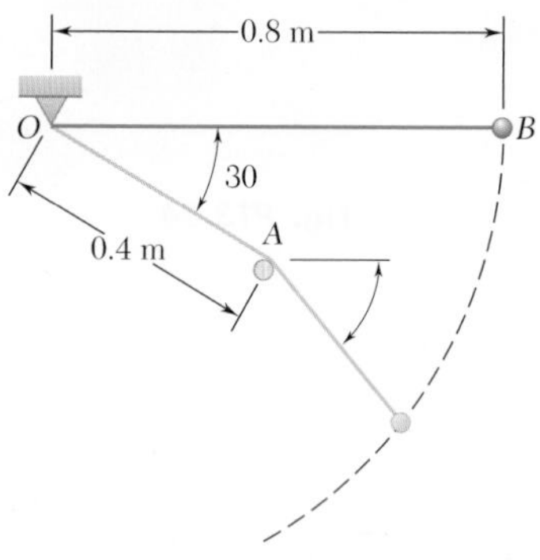

Fig. P13.41

13.41 A small sphere B of mass m is released from rest in the position shown and swings freely in a vertical plane, first about O and then about the peg A after the cord comes in contact with the peg. Determine the tension in the cord (*a*) just before the sphere comes in contact with the peg, (*b*) just after it comes in contact with the peg.

13.42 A roller coaster starts from rest at A, rolls down the track to B, describes a circular loop of 12-m diameter, and moves up and down past point E. Knowing that h = 18 m and assuming no energy loss due to friction, determine (*a*) the force exerted by his seat on a 80-kg rider at B and D, (*b*) the minimum value of the radius of curvature at E if the roller coaster is not to leave the track at that point.

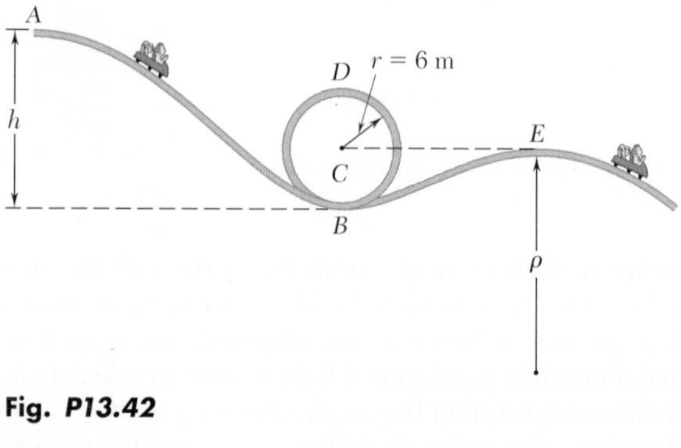

Fig. P13.42

13.43 In Prob. 13.42, determine the range of values of h for which the roller coaster will not leave the track at D or E, knowing that the radius of curvature at E is $\rho = 22.5$ m. Assume no energy loss due to friction.

13.44 A small block slides at a speed v on a horizontal surface. Knowing that $h = 0.9$ m, determine the required speed of the block if it is to leave the cylindrical surface BCD when $\theta = 30°$.

13.45 A small block slides at a speed $v = 2.5$ m/s on a horizontal surface at a height $h = 1$ m above the ground. Determine (*a*) the angle θ at which it will leave the cylindrical surface BCD, (*b*) the distance x at which it will hit the ground. Neglect friction and air resistance.

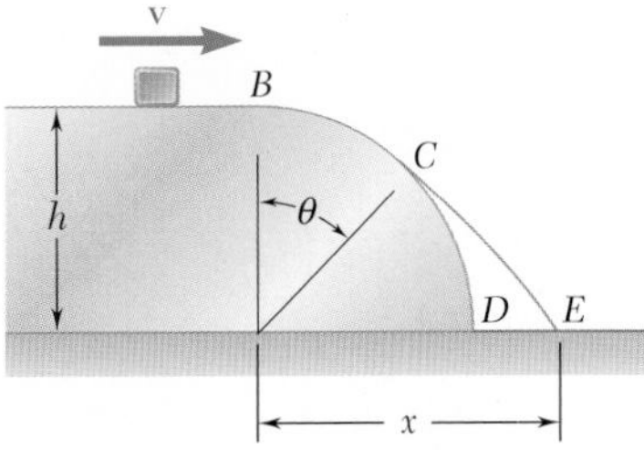

Fig. P13.44 and P13.45

13.46 A chair-lift is designed to transport 1000 skiers per hour from the base A to the summit B. The average mass of a skier is 70 kg and the average speed of the lift is 75 m/min. Determine (*a*) the average power required, (*b*) the required capacity of the motor if the mechanical efficiency is 85 percent and if a 300-percent overload is to be allowed.

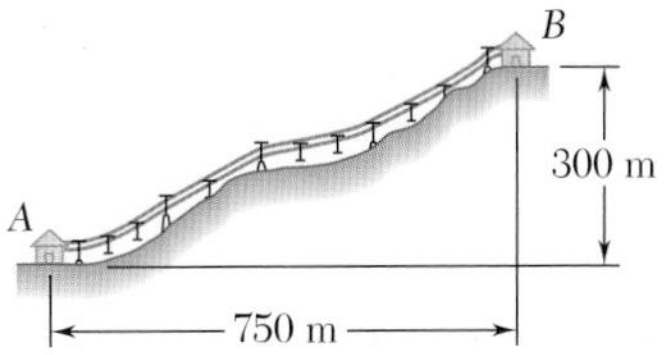

Fig. P13.46

13.47 It takes 15 s to raise a 1200-kg car and the supporting 300-kg hydraulic car-lift platform to a height of 2.8 m. Determine (*a*) the average output power delivered by the hydraulic pump to lift the system, (*b*) the average electric power required, knowing that the overall conversion efficiency from electric to mechanical power for the system is 82 percent.

Fig. P13.47

13.48 The velocity of the lift of Prob. 13.47 increases uniformly from zero to its maximum value at mid-height in 7.5 s and then decreases uniformly to zero in 7.5 s. Knowing that the peak power output of the hydraulic pump is 6 kW when the velocity is maximum, determine the maximum lift force provided by the pump.

13.49 (*a*) A 50-kg woman rides a 7.5-kg bicycle up a 3-percent slope at a constant speed of 1.5 m/s. How much power must be developed by the woman? (*b*) A 75-kg man on a 9-kg bicycle starts down the same slope and maintains a constant speed of 6 m/s by braking. How much power is dissipated by the brakes? Ignore air resistance and rolling resistance.

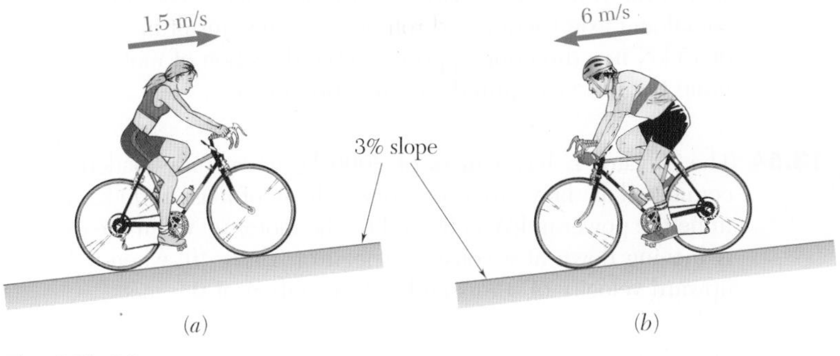

Fig. P13.46

13.50 A power specification formula is to be derived for electric motors which drive conveyor belts moving solid material at different rates to different heights and distances. Denoting the efficiency of a motor by η and neglecting the power needed to drive the belt itself, derive a formula in the SI system of units for the power P in kW, in terms of the mass flow rate m in kg/h, the height b and horizontal distance l in meters.

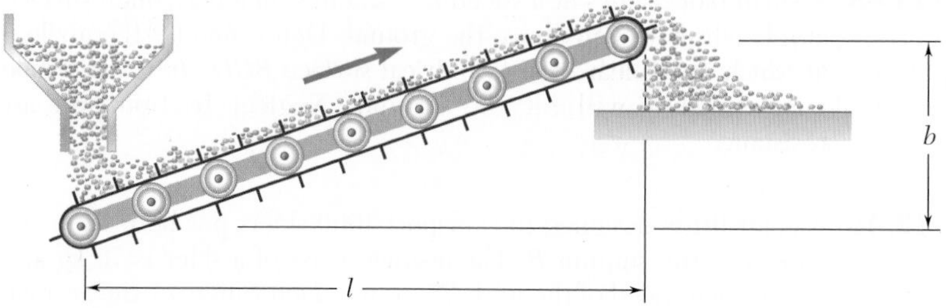

Fig. P13.50

Fig. P13.51

13.51 In an automobile drag race, the rear (drive) wheels of a 1000-kg car skid for the first 20 m and roll with sliding impending during the remaining 380 m. The front wheels of the car are just off the ground for the first 20 m, and for the remainder of the race 80 percent of the weight is on the rear wheels. Knowing that the coefficients of friction are $\mu_s = 0.90$ and $\mu_k = 0.68$, determine the power developed by the car at the drive wheels (*a*) at the end of the 20-m portion of the race, (*b*) at the end of the race. Give your answer in kW and in hp. Ignore the effect of air resistance and rolling friction.

13.52 The frictional resistance of a ship is known to vary directly as the 1.75 power of the speed v of the ship. A single tugboat at full power can tow the ship at a constant speed of 4.5 km/h by exerting a constant force of 300 kN. Determine (*a*) the power developed by the tugboat, (*b*) the maximum speed at which two tugboats, capable of delivering the same power, can tow the ship.

13.53 A train of total mass equal to 500 Mg starts from rest and accelerates uniformely to a speed of 90 km/h in 50 s. After reaching this speed, the train travels with a constant velocity. The track is horizontal and axle friction and rolling resistance result in a total force of 15 kN in a direction opposite to the direction of motion. Determine the power required as a function of time.

13.54 The elevator E has a mass of 3000 kg when fully loaded and is connected as shown to a counterweight W of mass 1000 kg. Determine the power in kW delivered by the motor (*a*) when the elevator is moving down at a constant speed of 3 m/s, (*b*) when it has an upward velocity of 3 m/s and a deceleration of 0.5 m/s^2.

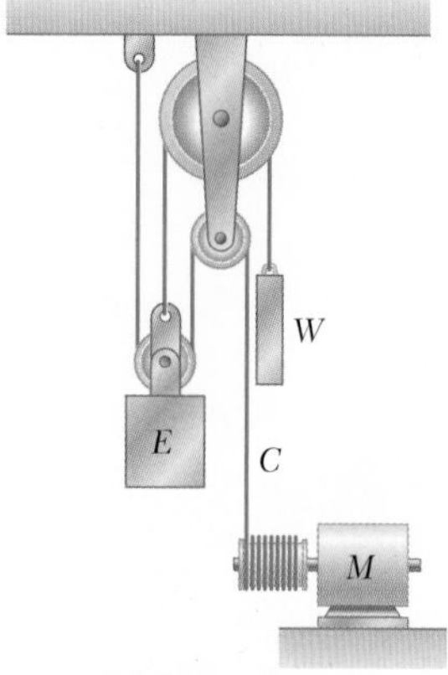

Fig. P13.54

13.6 POTENTIAL ENERGY†

Let us consider again a body of weight **W** which moves along a curved path from a point A_1 of elevation y_1 to a point A_2 of elevation y_2 (Fig. 13.4). We recall from Sec. 13.2 that the work of the force of gravity **W** during this displacement is

$$U_{1\rightarrow 2} = Wy_1 - Wy_2 \qquad (13.4)$$

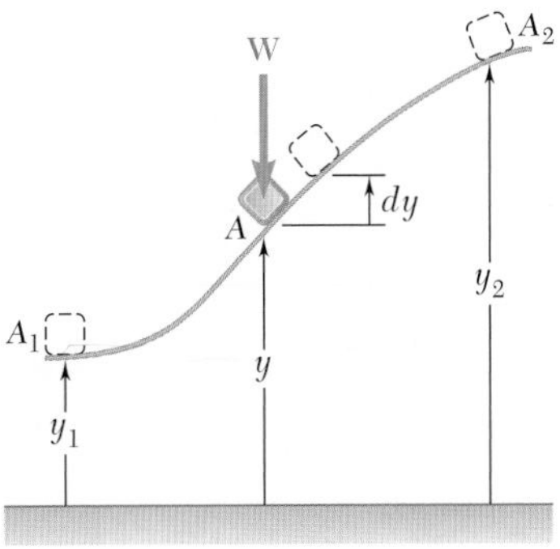

Fig. 13.4 *(repeated)*

The work of **W** may thus be obtained by subtracting the value of the function Wy corresponding to the second position of the body from its value corresponding to the first position. The work of **W** is independent of the actual path followed; it depends only upon the initial and final values of the function Wy. This function is called the *potential energy* of the body with respect to the *force of gravity* **W** and is denoted by V_g. We write

$$U_{1\rightarrow 2} = (V_g)_1 - (V_g)_2 \qquad \text{with } V_g = Wy \qquad (13.16)$$

We note that if $(V_g)_2 > (V_g)_1$, that is, *if the potential energy increases* during the displacement (as in the case considered here), *the work* $U_{1\rightarrow 2}$ *is negative*. If, on the other hand, the work of **W** is positive, the potential energy decreases. Therefore, the potential energy V_g of the body provides a measure of the work which can be done by its weight **W**. Since only the *change* in potential energy, and not the actual value of V_g, is involved in formula (13.16), an arbitrary constant can be added to the expression obtained for V_g. In other words, the level, or datum, from which the elevation y is measured can be chosen arbitrarily. Note that potential energy is expressed in the same units as work, i.e., in joules if SI units are used and in ft · lb or in · lb if U.S. customary units are used.

It should be noted that the expression just obtained for the potential energy of a body with respect to gravity is valid only as long as the weight **W** of the body can be assumed to remain constant, i.e., as long as the displacements of the body are small compared with the radius of the earth. In the case of a space vehicle, however, we should take into consideration the variation of the force of gravity with the distance r from the center of the earth. Using the expression obtained in Sec. 13.2 for the work of a gravitational force, we write (Fig. 13.6)

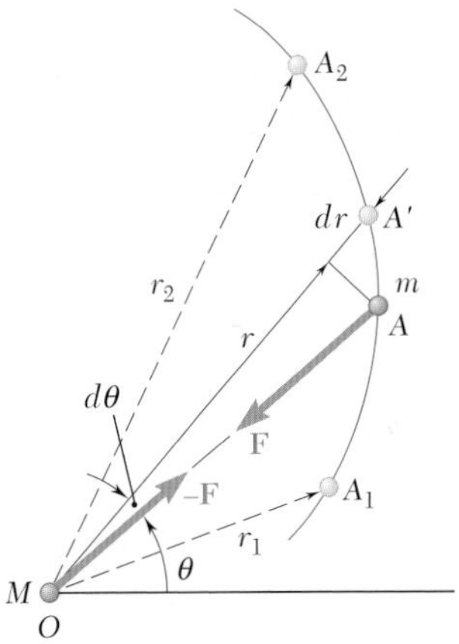

Fig. 13.6 *(repeated)*

$$U_{1\rightarrow 2} = \frac{GMm}{r_2} - \frac{GMm}{r_1} \qquad (13.7)$$

The work of the force of gravity can therefore be obtained by subtracting the value of the function $-GMm/r$ corresponding to the second position of the body from its value corresponding to the first position. Thus, the expression which should be used for the potential energy V_g when the variation in the force of gravity cannot be neglected is

$$V_g = -\frac{GMm}{r} \qquad (13.17)$$

†Some of the material in this section has already been considered in Sec. 10.7.

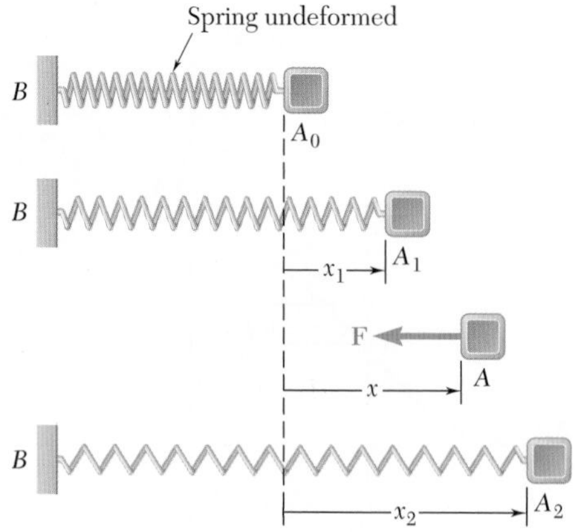

Fig. 13.5 *(repeated)*

Taking the first of the relations (12.29) into account, we write V_g in the alternative form

$$V_g = -\frac{WR^2}{r} \tag{13.17'}$$

where R is the radius of the earth and W is the value of the weight of the body at the surface of the earth. When either of the relations (13.17) or (13.17′) is used to express V_g, the distance r should, of course, be measured from the center of the earth.† Note that V_g is always negative and that it approaches zero for very large values of r.

Consider now a body attached to a spring and moving from a position A_1, corresponding to a deflection x_1 of the spring, to a position A_2, corresponding to a deflection x_2 of the spring (Fig. 13.5). We recall from Sec. 13.2 that the work of the force **F** exerted by the spring on the body is

$$U_{1\rightarrow 2} = \tfrac{1}{2}kx_1^2 - \tfrac{1}{2}kx_2^2 \tag{13.6}$$

The work of the elastic force is thus obtained by subtracting the value of the function $\frac{1}{2}kx^2$ corresponding to the second position of the body from its value corresponding to the first position. This function is denoted by V_e and is called the *potential energy* of the body with respect to the *elastic force* **F**. We write

$$U_{1\rightarrow 2} = (V_e)_1 - (V_e)_2 \qquad \text{with } V_e = \tfrac{1}{2}kx^2 \tag{13.18}$$

and observe that during the displacement considered, the work of the force **F** exerted by the spring on the body is negative and the potential energy V_e increases. You should note that the expression obtained for V_e is valid only if the deflection of the spring is measured from its undeformed position. On the other hand, formula (13.18) can be used even when the spring is rotated about its fixed end (Fig. 13.10*a*). The work of the elastic force depends only upon the initial and final deflections of the spring (Fig. 13.10*b*).

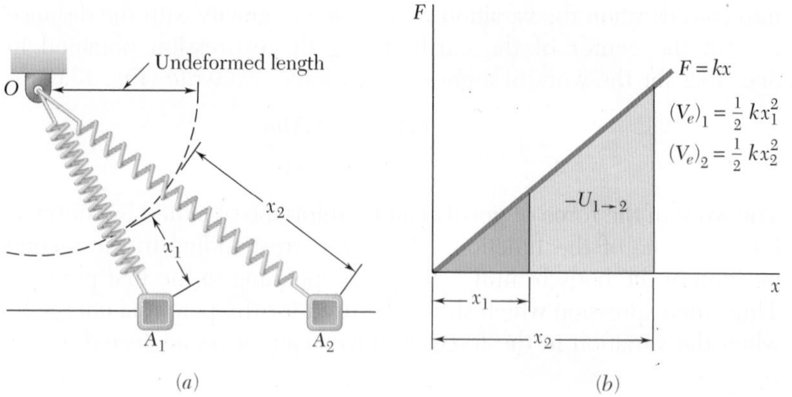

Fig. 13.10

†The expressions given for V_g in (13.17) and (13.17′) are valid only when $r \geq R$, that is, when the body considered is above the surface of the earth.

The concept of potential energy can be used when forces other than gravity forces and elastic forces are involved. Indeed, it remains valid as long as the work of the force considered is independent of the path followed by its point of application as this point moves from a given position A_1 to a given position A_2. Such forces are said to be *conservative forces;* the general properties of conservative forces are studied in the following section.

*13.7 CONSERVATIVE FORCES

As indicated in the preceding section, a force **F** acting on a particle A is said to be conservative *if its work $U_{1\rightarrow2}$ is independent of the path followed by the particle A as it moves from A_1 to A_2* (Fig. 13.11*a*). We can then write

$$U_{1\rightarrow2} = V(x_1, y_1, z_1) - V(x_2, y_2, z_2) \tag{13.19}$$

or, for short,

$$U_{1\rightarrow2} = V_1 - V_2 \tag{13.19'}$$

The function $V(x, y, z)$ is called the potential energy, or *potential function,* of **F**.

We note that if A_2 is chosen to coincide with A_1, that is, if the particle describes a closed path (Fig. 13.11*b*), we have $V_1 = V_2$ and the work is zero. Thus for any conservative force **F** we can write

$$\oint \mathbf{F} \cdot d\mathbf{r} = 0 \tag{13.20}$$

where the circle on the integral sign indicates that the path is closed.

Let us now apply (13.19) between two neighboring points $A(x, y, z)$ and $A'(x + dx, y + dy, z + dz)$. The elementary work dU corresponding to the displacement $d\mathbf{r}$ from A to A' is

$$dU = V(x, y, z) - V(x + dx, y + dy, z + dz)$$

or

$$dU = -dV(x, y, z) \tag{13.21}$$

Thus, the elementary work of a conservative force is an *exact differential.*

Substituting for dU in (13.21) the expression obtained in (13.1″) and recalling the definition of the differential of a function of several variables, we write

$$F_x\, dx + F_y\, dy + F_z\, dz = -\left(\frac{\partial V}{\partial x}dx + \frac{\partial V}{\partial y}dy + \frac{\partial V}{\partial z}dz\right).$$

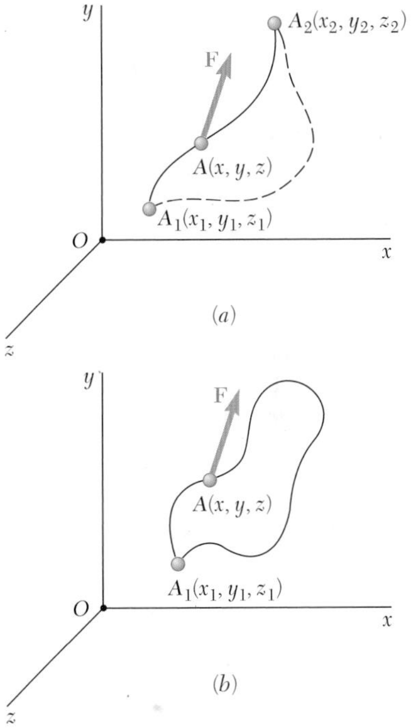

Fig. 13.11

from which it follows that

$$F_x = -\frac{\partial V}{\partial x} \qquad F_y = -\frac{\partial V}{\partial y} \qquad F_z = -\frac{\partial V}{\partial z} \tag{13.22}$$

It is clear that the components of **F** must be functions of the coordinates x, y, and z. Thus, a *necessary* condition for a conservative force is that it depend only upon the position of its point of application. The relations (13.22) can be expressed more concisely if we write

$$\mathbf{F} = F_x\mathbf{i} + F_y\mathbf{j} + F_z\mathbf{k} = -\left(\frac{\partial V}{\partial x}\mathbf{i} + \frac{\partial V}{\partial y}\mathbf{j} + \frac{\partial V}{\partial z}\mathbf{k}\right)$$

The vector in parentheses is known as the *gradient of the scalar function* V and is denoted by **grad** V. We thus write for any conservative force

$$\mathbf{F} = -\mathbf{grad}\ V \tag{13.23}$$

The relations (13.19) to (13.23) were shown to be satisfied by any conservative force. It can also be shown that if a force **F** satisfies one of these relations, **F** must be a conservative force.

13.8 CONSERVATION OF ENERGY

We saw in the preceding two sections that the work of a conservative force, such as the weight of a particle or the force exerted by a spring, can be expressed as a change in potential energy. When a particle moves under the action of conservative forces, the principle of work and energy stated in Sec. 13.3 can be expressed in a modified form. Substituting for $U_{1\rightarrow 2}$ from (13.19′) into (13.10), we write

$$V_1 - V_2 = T_2 - T_1$$

$$T_1 + V_1 = T_2 + V_2 \tag{13.24}$$

Formula (13.24) indicates that when a particle moves under the action of conservative forces, *the sum of the kinetic energy and of the potential energy of the particle remains constant*. The sum $T + V$ is called the *total mechanical energy* of the particle and is denoted by E.

Consider, for example, the pendulum analyzed in Sec. 13.4, which is released with no velocity from A_1 and allowed to swing in a vertical plane (Fig. 13.12). Measuring the potential energy from the level of A_2, we have, at A_1,

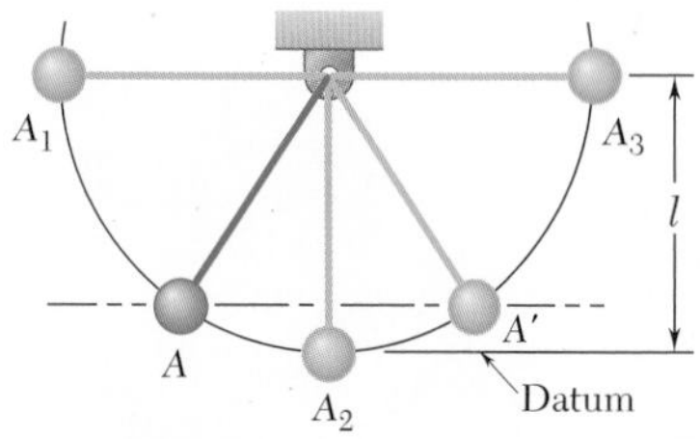

Fig. 13.12

$$T_1 = 0 \qquad V_1 = Wl \qquad T_1 + V_1 = Wl$$

Recalling that at A_2 the speed of the pendulum is $v_2 = \sqrt{2gl}$, we have

$$T_2 = \tfrac{1}{2}mv_2^2 = \frac{1}{2}\frac{W}{g}(2gl) = Wl \qquad V_2 = 0$$
$$T_2 + V_2 = Wl$$

We thus check that the total mechanical energy $E = T + V$ of the pendulum is the same at A_1 and A_2. Whereas the energy is entirely potential at A_1, it becomes entirely kinetic at A_2, and as the pendulum keeps swinging to the right, the kinetic energy is transformed back into potential energy. At A_3, $T_3 = 0$ and $V_3 = Wl$.

Since the total mechanical energy of the pendulum remains constant and since its potential energy depends only upon its elevation, the kinetic energy of the pendulum will have the same value at any two points located on the same level. Thus, the speed of the pendulum is the same at A and at A' (Fig. 13.12). This result can be extended to the case of a particle moving along any given path, regardless of the shape of the path, as long as the only forces acting on the particle are its weight and the normal reaction of the path. The particle of Fig. 13.13, for example, which slides in a vertical plane along a frictionless track, will have the same speed at A, A', and A''.

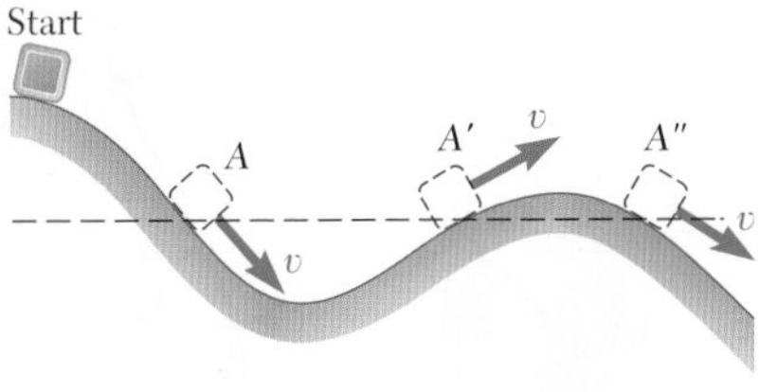

Fig. 13.13

While the weight of a particle and the force exerted by a spring are conservative forces, *friction forces are nonconservative forces*. In other words, *the work of a friction force cannot be expressed as a change in potential energy*. The work of a friction force depends upon the path followed by its point of application; and while the work $U_{1\rightarrow 2}$ defined by (13.19) is positive or negative according to the sense of motion, *the work of a friction force*, as we noted in Sec. 13.4, *is always negative*. It follows that when a mechanical system involves friction, its total mechanical energy does not remain constant but decreases. The energy of the system, however, is not lost; it is transformed into heat, and the sum of the *mechanical energy* and of the *thermal energy* of the system remains constant.

Other forms of energy can also be involved in a system. For instance, a generator converts mechanical energy into *electric energy;* a gasoline engine converts *chemical energy* into mechanical energy; a nuclear reactor converts *mass* into thermal energy. If all forms of energy are considered, the energy of any system can be considered as constant and the principle of conservation of energy remains valid under all conditions.

13.9 MOTION UNDER A CONSERVATIVE CENTRAL FORCE. APPLICATION TO SPACE MECHANICS

We saw in Sec. 12.9 that when a particle P moves under a central force $\mathbf{F}$, the angular momentum $\mathbf{H}_O$ of the particle about the center of force O is constant. If the force $\mathbf{F}$ is also conservative, there exists a potential energy V associated with $\mathbf{F}$, and the total energy $E = T + V$ of the particle is constant (Sec. 13.8). Thus, when a particle moves under a conservative central force, both the principle of conservation of angular momentum and the principle of conservation of energy can be used to study its motion.

Consider, for example, a space vehicle of mass m moving under the earth's gravitational force. Let us assume that it begins its free flight at point P_0 at a distance r_0 from the center of the earth, with a velocity $\mathbf{v}_0$ forming an angle ϕ_0 with the radius vector OP_0

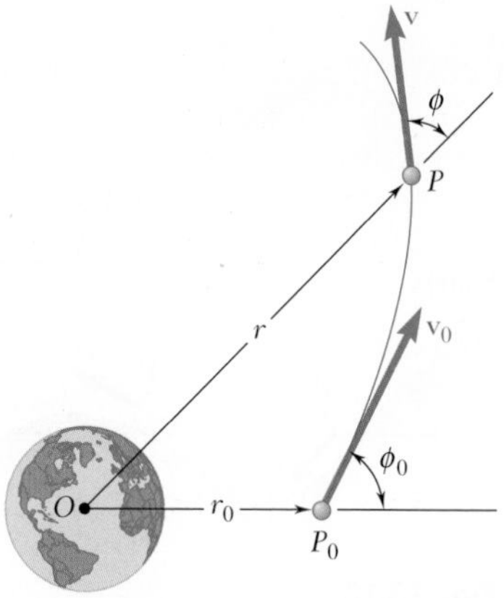

Fig. 13.14

(Fig. 13.14). Let P be a point of the trajectory described by the vehicle; we denote by r the distance from O to P, by $\mathbf{v}$ the velocity of the vehicle at P, and by ϕ the angle formed by $\mathbf{v}$ and the radius vector OP. Applying the principle of conservation of angular momentum about O between P_0 and P (Sec. 12.9), we write

$$r_0 m v_0 \sin \phi_0 = r m v \sin \phi \tag{13.25}$$

Recalling the expression (13.17) obtained for the potential energy due to a gravitational force, we apply the principle of conservation of energy between P_0 and P and write

$$T_0 + V_0 = T + V$$

$$\tfrac{1}{2}mv_0^2 - \frac{GMm}{r_0} = \tfrac{1}{2}mv^2 - \frac{GMm}{r} \tag{13.26}$$

where M is the mass of the earth.

Equation (13.26) can be solved for the magnitude v of the velocity of the vehicle at P when the distance r from O to P is known; Eq. (13.25) can then be used to determine the angle ϕ that the velocity forms with the radius vector OP.

Equations (13.25) and (13.26) can also be used to determine the maximum and minimum values of r in the case of a satellite launched from P_0 in a direction forming an angle ϕ_0 with the vertical OP_0 (Fig. 13.15). The desired values of r are obtained by making $\phi = 90°$ in (13.25) and eliminating v between Eqs. (13.25) and (13.26).

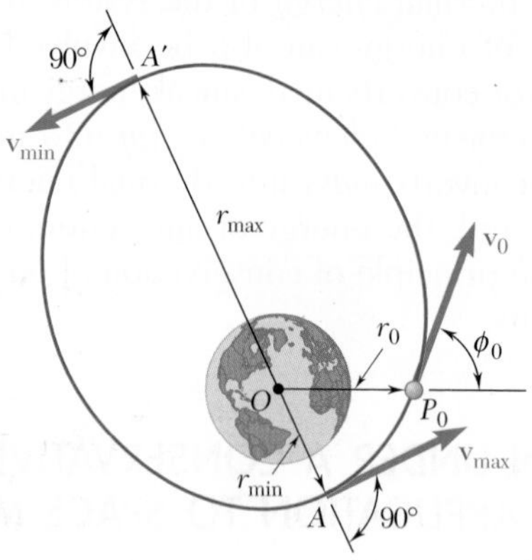

Fig. 13.15

It should be noted that the application of the principles of conservation of energy and of conservation of angular momentum leads to a more fundamental formulation of the problems of space mechanics than does the method indicated in Sec. 12.12. In all cases involving oblique launchings, it will also result in much simpler computations. And while the method of Sec. 12.12 must be used when the actual trajectory or the periodic time of a space vehicle is to be determined, the calculations will be simplified if the conservation principles are first used to compute the maximum and minimum values of the radius vector r.

SAMPLE PROBLEM 13.6

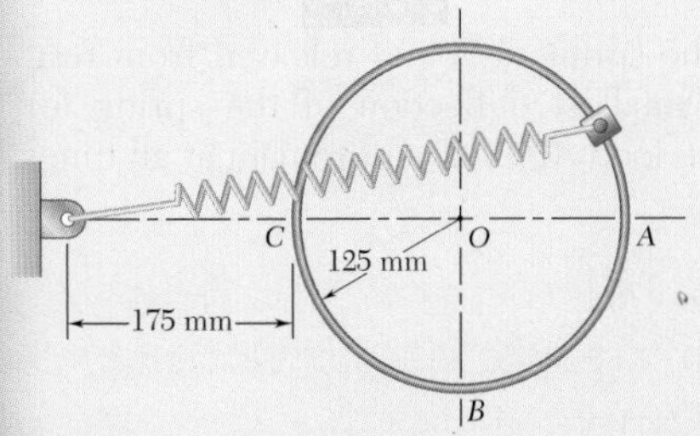

A 1.5-kg collar is attached to a spring and slides without friction along a circular rod in a horizontal plane. The spring has an undeformed length of 150 mm and a constant k = 400 N/m. Knowing that the collar is in equilibrium at A and is given a slight push to get it moving, determine the velocity of the collar (a) as it passes through B, (b) as it passes through C.

SOLUTION

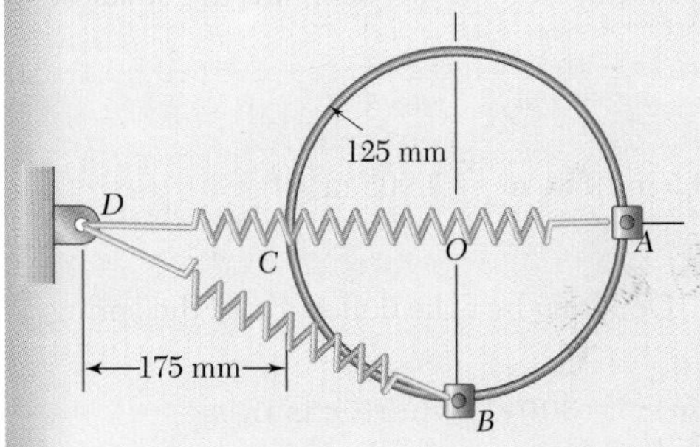

Position 1. ***Kinetic Energy:***

$$v_A = 0 \; T_A = 0$$

Potential Energy:

$$\Delta L_{AD} = \Delta L_{AD} - L_O$$

$$\Delta L_{AD} = (175 \text{ mm} + 250 \text{ mm}) - (150 \text{ mm})$$

$$= 275 \text{ mm} = 0.275 \text{ m}$$

$$V_A = \frac{1}{2}\, k(\Delta L_{AD})^2$$

$$V_A = \frac{1}{2}\,(400 \text{ N/m})\,(0.275 \text{ m})^2$$

$$V_A = 15.125 \text{ J}$$

Position B. ***Kinetic Energy:***

$$T_B = \tfrac{1}{2}mv_B^2 = \left(\frac{1.5}{2}\text{ kg}\right)(v_B^2) = (0.75)(v_B^2)$$

Potential Energy:

$$L_{BD} = (300^2 \text{ mm} + 125^2 \text{ mm})^{1/2}$$
$$= 325 \text{ mm}$$
$$\Delta_{BD} = L_{BD} - L_O$$
$$= (325 \text{ mm} - 150 \text{ m})$$
$$= 175 \text{ mm} = 0.175 \text{ m}$$

$$V_B = \frac{1}{2}k(\Delta L_{BD})^2$$

$$= \frac{1}{2}(400 \text{ N/m})\,(0.175 \text{ m})^2$$

$$= 6.125 \text{ J}$$

Consentation of Energy:

$$T_A + V_A = T_B + V_B \qquad 0 + 15.125 = 0.75v_B^2 + 6.125$$

$$v_B^2 = \left(\frac{(15.125 - 6.125)}{(0.75)}\right) = 12.00\text{m}^2/\text{s}^2$$

$$v_B = 3.46 \text{ m/s} \;\blacktriangleleft$$

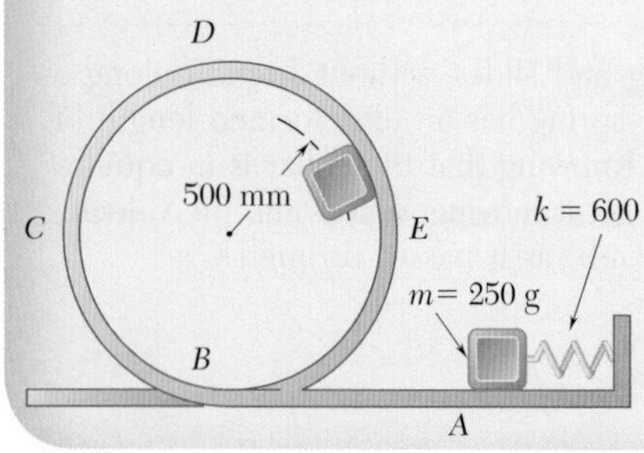

SAMPLE PROBLEM 13.7

The 250 g pellet is pushed against the spring at A and released from rest. Neglecting friction, determine the smallest deflection of the spring for which the pellet will travel around the loop $ABCDE$ and remain at all times in contact with the loop.

SOLUTION

Required Speed at Point D. As the pellet passes through the highest point D, its potential energy will respect to gravity is maximum and thus, its kinetic energy and speed are minimum. Since the pellet must remain in contact with the loop, the force **N** exerted on the pellet by the loop must be equal to or greater than zero. Setting **N** = 0, we compute the smallest possible speed v_D.

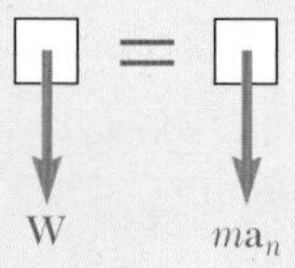

$$+\downarrow \Sigma F_n = ma_n: \qquad W = ma_n \qquad mg = ma_n \qquad a_n = g$$

$$a_n = \frac{v_D^2}{r}: \quad v_D^2 = ra_n = rg = (0.5\text{ m})(9.81\ \frac{\text{m}}{\text{s}^2}) = 4.905\text{ m}^2/\text{s}^2$$

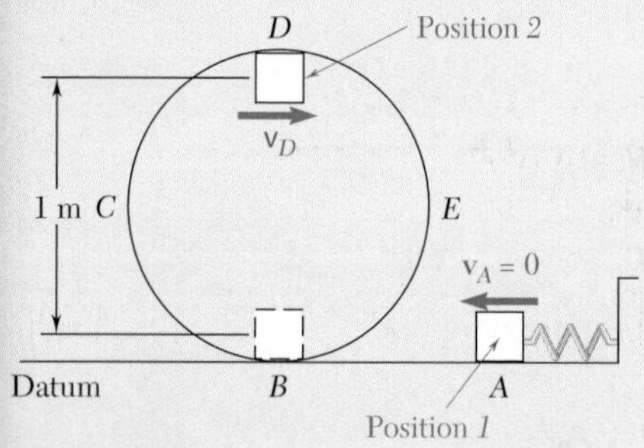

Position 1. ***Potential Energy.*** Denoting by x the deflection of the spring and noting that $k = 600$ N/m.

$$V_e = \tfrac{1}{2}kx^2 = \tfrac{1}{2}(600\text{ N/m})x^2 = 300x^2, \text{ where } x \text{ is in } m$$

Choosing the datum at A, we have $V_g = 0$; therefore

$$V_1 = V_e + V_g = 300x^2$$

Kinetic Energy. Since the pellet is released from rest, $v_A = 0$ and we have $T_1 = 0$.

Position 2. ***Potential Energy.*** The spring is now undeformed; thus $V_e = 0$. Since the pellet is 1 m above the datum, we have

$$V_g = mgy = (0.25\text{ kg})\ (9.81\text{ m/s}^2)\ (1\text{ m}) = 2.45\text{ J}$$
$$V_2 = V_e + V_g = 2.45\text{ J}$$

Kinetic Energy. Using the value of v_D^2 obtained above, we write

$$T_2 = \tfrac{1}{2}mv_D^2 = \frac{1}{2} \times (0.25\text{ kg})\left(4.905\ \frac{\text{m}^2}{\text{s}^2}\right) = 0.613\text{ J}$$

Conservation of Energy. Applying the principle of conservation of energy between positions 1 and 2, we write

$$T_1 + V_1 = T_2 + V_2$$
$$0 + 300x^2 = 2.45\text{ J} + 0.613\text{ J}$$
$$x = 0.101\text{ m} \qquad\qquad x = 101\text{ mm} \blacktriangleleft$$

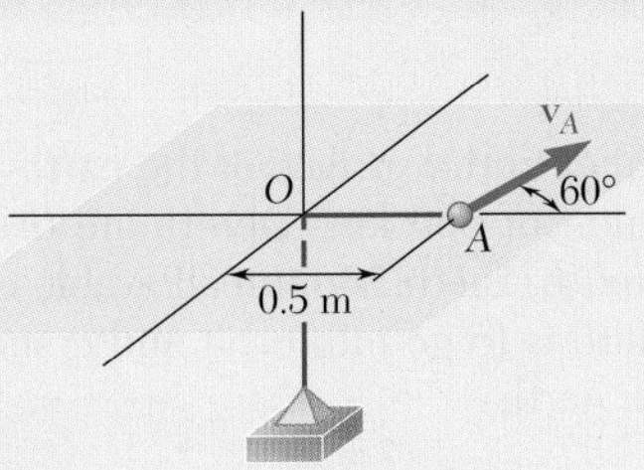

SAMPLE PROBLEM 13.8

A sphere of mass $m = 0.6$ kg is attached to an elastic cord of constant $k = 100$ N/m, which is undeformed when the sphere is located at the origin O. Knowing that the sphere may slide without friction on the horizontal surface and that in the position shown its velocity $\mathbf{v}_A$ has a magnitude of 20 m/s, determine (*a*) the maximum and minimum distances from the sphere to the origin O, (*b*) the corresponding values of its speed.

SOLUTION

The force exerted by the cord on the sphere passes through the fixed point O, and its work can be expressed as a change in potential energy. It is therefore a conservative central force, and both the total energy of the sphere and its angular momentum about O are conserved.

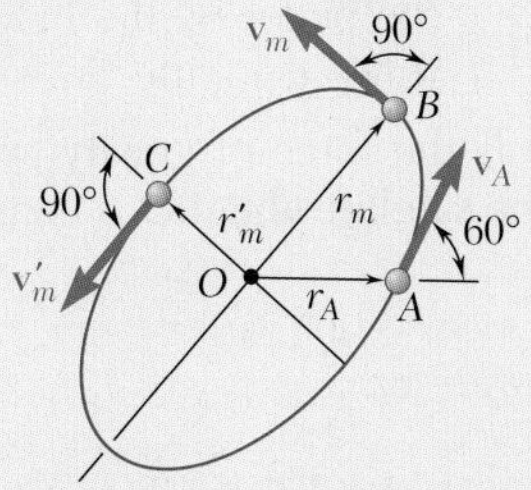

Conservation of Angular Momentum About O. At point B, where the distance from O is maximum, the velocity of the sphere is perpendicular to OB and the angular momentum is $r_m m v_m$. A similar property holds at point C, where the distance from O is minimum. Expressing conservation of angular momentum between A and B, we write

$$r_A m v_A \sin 60° = r_m m v_m$$

$$(0.5 \text{ m})(0.6 \text{ kg})(20 \text{ m/s}) \sin 60° = r_m(0.6 \text{ kg})v_m$$

$$v_m = \frac{8.66}{r_m} \qquad (1)$$

Conservation of Energy

At point A: $\quad T_A = \frac{1}{2}mv_A^2 = \frac{1}{2}(0.6 \text{ kg})(20 \text{ m/s})^2 = 120 \text{ J}$

$\quad V_A = \frac{1}{2}kr_A^2 = \frac{1}{2}(100 \text{ N/m})(0.5 \text{ m})^2 = 12.5 \text{ J}$

At point B: $\quad T_B = \frac{1}{2}mv_m^2 = \frac{1}{2}(0.6 \text{kg})v_m^2 = 0.3v_m^2$

$\quad V_B = \frac{1}{2}kr_m^2 = \frac{1}{2}(100 \text{ N/m})r_m^2 = 50r_m^2$

Applying the principle of conservation of energy between points A and B, we write

$$T_A + V_A = T_B + V_B$$

$$120 + 12.5 = 0.3v_m^2 + 50r_m^2 \qquad (2)$$

a. Maximum and Minimum Values of Distance. Substituting for v_m from Eq. (1) into Eq. (2) and solving for r_m^2, we obtain

$$r_m^2 = 2.468 \text{ or } 0.1824 \qquad r_m = 1.571 \text{ m}, \; r'_m = 0.427 \text{ m} \blacktriangleleft$$

b. Corresponding Values of Speed. Substituting the values obtained for r_m and r'_m into Eq. (1), we have

$$v_m = \frac{8.66}{1.571} \qquad v_m = 5.51 \text{ m/s} \blacktriangleleft$$

$$v'_m = \frac{8.66}{0.427} \qquad v'_m = 20.3 \text{ m/s} \blacktriangleleft$$

Note. It can be shown that the path of the sphere is an ellipse of *center O*.

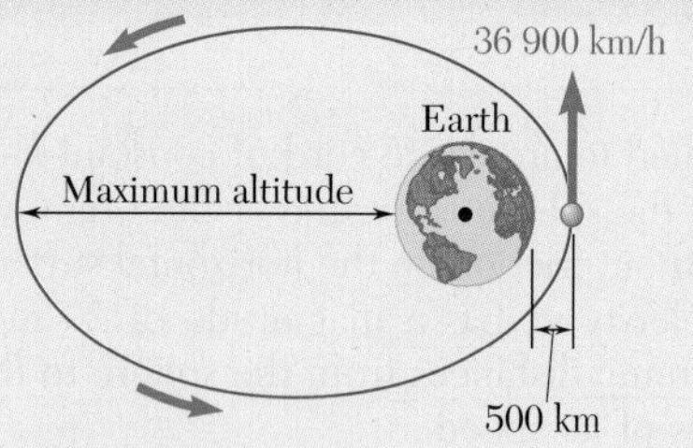

SAMPLE PROBLEM 13.9

A satellite is launched in a direction parallel to the surface of the earth with a velocity of 36 900 km/h from an altitude of 500 km. Determine (*a*) the maximum altitude reached by the satellite, (*b*) the maximum allowable error in the direction of launching if the satellite is to go into orbit and come no closer than 200 km to the surface of the earth.

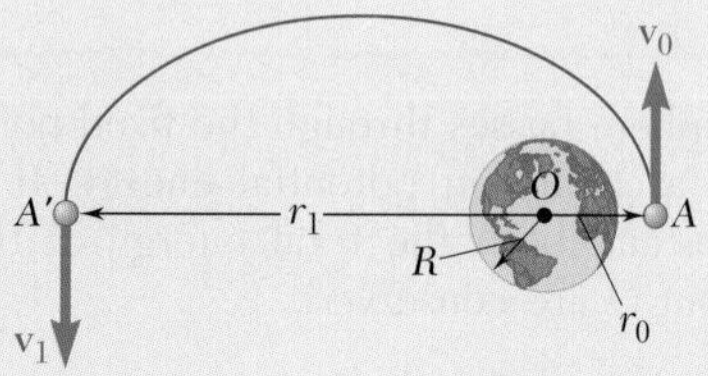

SOLUTION

a. Maximum Altitude. We denote by A' the point of the orbit farthest from the earth and by r_1 the corresponding distance from the center of the earth. Since the satellite is in free flight between A and A', we apply the principle of conservation of energy:

$$T_A + V_A = T_{A'} + V_{A'}$$

$$\tfrac{1}{2}mv_0^2 - \frac{GMm}{r_0} = \tfrac{1}{2}mv_1^2 - \frac{GMm}{r_1} \qquad (1)$$

Since the only force acting on the satellite is the force of gravity, which is a central force, the angular momentum of the satellite about O is conserved. Considering points A and A', we write

$$r_0 m v_0 = r_1 m v_1 \qquad v_1 = v_0 \frac{r_0}{r_1} \qquad (2)$$

Substituting this expression for v_1 into Eq. (1), dividing each term by the mass m, and rearranging the terms, we obtain

$$\tfrac{1}{2}v_0^2\left(1 - \frac{r_0^2}{r_1^2}\right) = \frac{GM}{r_0}\left(1 - \frac{r_0}{r_1}\right) \qquad 1 + \frac{r_0}{r_1} = \frac{2GM}{r_0 v_0^2} \qquad (3)$$

Recalling that the radius of the earth is $R = 6370$ km, we compute

$$r_0 = 6370 \text{ km} + 500 \text{ km} = 6870 \text{ km} = 6.87 \times 10^6 \text{ m}$$
$$v_0 = 36\,900 \text{ km/h} = (36.9 \times 10^6 \text{ m})/(3.6 \times 10^3 \text{ s}) = 10.25 \times 10^3 \text{ m/s}$$
$$GM = gR^2 = (9.81 \text{ m/s}^2)(6.37 \times 10^6 \text{ m})^2 = 398 \times 10^{12} \text{ m}^3/\text{s}^2$$

Substituting these values into (3), we obtain $r_1 = 66.8 \times 10^6$ m.

Maximum altitude $= 66.8 \times 10^6 \text{ m} - 6.37 \times 10^6 \text{ m} = 60.4 \times 10^6 \text{ m} =$ 60 400 km ◀

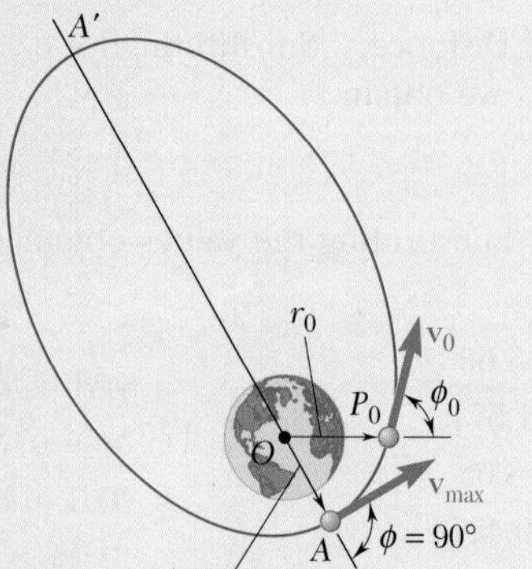

b. Allowable Error in Direction of Launching. The satellite is launched from P_0 in a direction forming an angle ϕ_0 with the vertical OP_0. The value of ϕ_0 corresponding to $r_{min} = 6370 \text{ km} + 200 \text{ km} = 6570$ km is obtained by applying the principles of conservation of energy and of conservation of angular momentum between P_0 and A:

$$\tfrac{1}{2}mv_0^2 - \frac{GMm}{r_0} = \tfrac{1}{2}mv_{max}^2 - \frac{GMm}{r_{min}} \qquad (4)$$

$$r_0 m v_0 \sin\phi_0 = r_{min} m v_{max} \qquad (5)$$

Solving (5) for v_{max} and then substituting for v_{max} into (4), we can solve (4) for $\sin\phi_0$. Using the values of v_0 and GM computed in part *a* and noting that $r_0/r_{min} = 6870/6570 = 1.0457$, we find

$\sin\phi_0 = 0.9801 \qquad \phi_0 = 90° \pm 11.5° \qquad$ Allowable error $= \pm 11.5°$ ◀

SOLVING PROBLEMS ON YOUR OWN

In this lesson you learned that when the work done by a force **F** acting on a particle A *is independent of the path followed by the particle* as it moves from a given position A_1 to a given position A_2 (Fig. 13.11a), then a function V, called *potential energy,* can be defined for the force **F**. Such forces are said to be *conservative forces,* and you can write

$$U_{1\rightarrow 2} = V(x_1, y_1, z_1) - V(x_2, y_2, z_2) \qquad (13.19)$$

or, for short,

$$U_{1\rightarrow 2} = V_1 - V_2 \qquad (13.19')$$

Note that the work is negative when the change in the potential energy is positive, i.e., when $V_2 > V_1$.

Substituting the above expression into the equation for work and energy, you can write

$$T_1 + V_1 = T_2 + V_2 \qquad (13.24)$$

which shows that when a particle moves under the action of a conservative force *the sum of the kinetic and potential energies of the particle remains constant.*

Your solution of problems using the above formula will consist of the following steps.

1. Determine whether all the forces involved are conservative. If some of the forces are not conservative, for example if friction is involved, you must use the method of work and energy from the previous lesson, since the work done by such forces depends upon the path followed by the particle and a potential function does not exist. If there is no friction and if all the forces are conservative, you can proceed as follows.

2. Determine the kinetic energy $T = \frac{1}{2}mv^2$ at each end of the path.

3. Compute the potential energy for all the forces involved at each end of the path. You will recall that the following expressions for the potential energy were derived in this lesson.

a. The potential energy of a weight W close to the surface of the earth and at a height y above a given datum,

$$V_g = Wy \tag{13.16}$$

b. The potential energy of a mass m located at a distance r from the center of the earth, large enough so that the variation of the force of gravity must be taken into account,

$$V_g = -\frac{GMm}{r} \tag{13.17}$$

where the distance r is measured from the center of the earth and V_g is equal to zero at $r = \infty$.

c. The potential energy of a body with respect to an elastic force $F = kx$,

$$V_e = \tfrac{1}{2}kx^2 \tag{13.18}$$

where the distance x is the deflection of the elastic spring measured from its *undeformed* position and k is the spring constant. Note that V_e *depends only upon the deflection* x and not upon the path of the body attached to the spring. Also, V_e is always positive, whether the spring is compressed or elongated.

4. Substitute your expressions for the kinetic and potential energies into Eq. (13.24). You will be able to solve this equation for one unknown, for example, for a velocity [Sample Prob. 13.6]. If more than one unknown is involved, you will have to search for another condition or equation, such as the minimum speed [Sample Prob. 13.7] or the minimum potential energy of the particle. For problems involving a central force, a second equation can be obtained by using conservation of angular momentum [Sample Prob. 13.8]. This is especially useful in applications to space mechanics [Sec. 13.9].

PROBLEMS

CONCEPT QUESTIONS

13.CQ2 Two small balls A and B with masses $2m$ and m, respectively, are released from rest at a height h above the ground. Neglecting air resistance, which of the following statements is true when the two balls hit the ground?

a. The kinetic energy of A is the same as the kinetic energy of B.
b. The kinetic energy of A is half the kinetic energy of B.
c. The kinetic energy of A is twice the kinetic energy of B.
d. The kinetic energy of A is four times the kinetic energy of B.

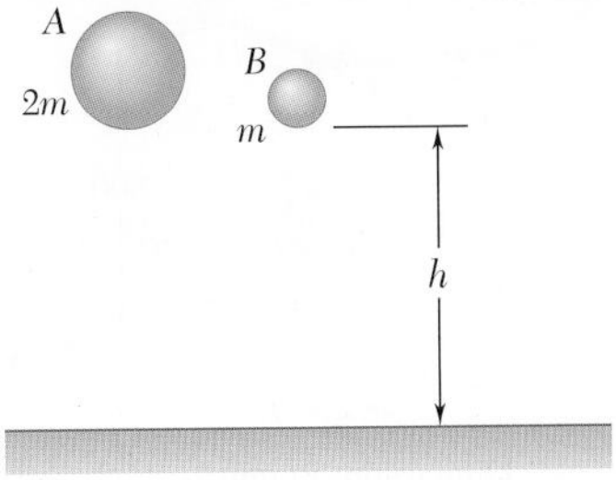

Fig. P13.CQ2

13.CQ3 A small block A is released from rest and slides down the frictionless ramp to the loop. The maximum height h of the loop is the same as the initial height of the block. Will A make it completely around the loop without losing contact with the track?

a. Yes
b. No
c. Need more information

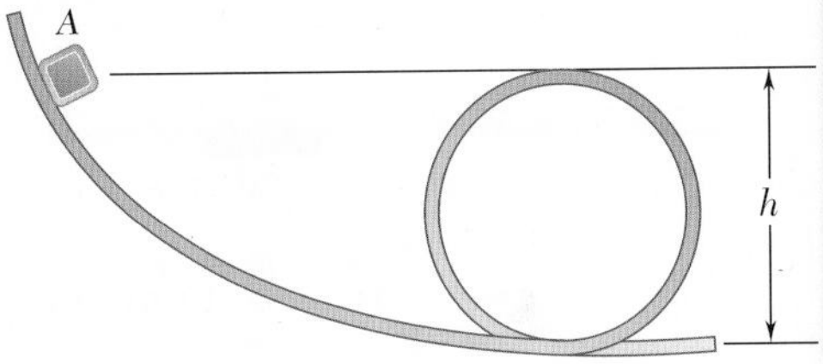

Fig. P13.CQ3

END-OF-SECTION PROBLEMS

13.55 A force **P** is slowly applied to a plate that is attached to two springs and causes a deflection x_0. In each of the two cases shown, derive an expression for the constant k_e, in terms of k_1 and k_2, of the single spring equivalent to the given system, that is, of the single spring which will undergo the same deflection x_0 when subjected to the same force **P**.

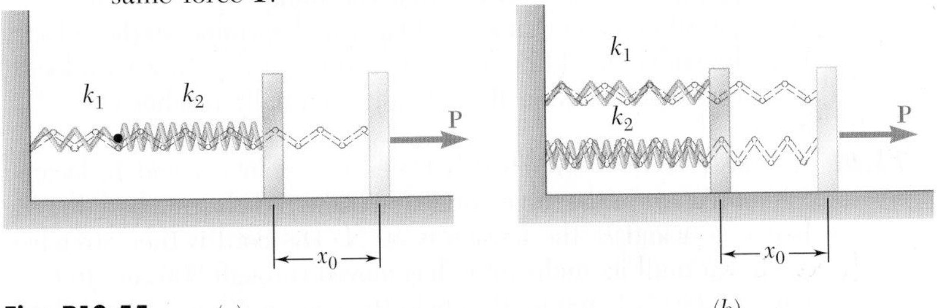

Fig. P13.55 (*a*) (*b*)

13.56 A loaded railroad car of mass m is rolling at a constant velocity $\mathbf{v}_0$ when it couples with a massless bumper system. Determine the maximum deflection of the bumper assuming the two springs are (*a*) in series (as shown), (*b*) in parallel.

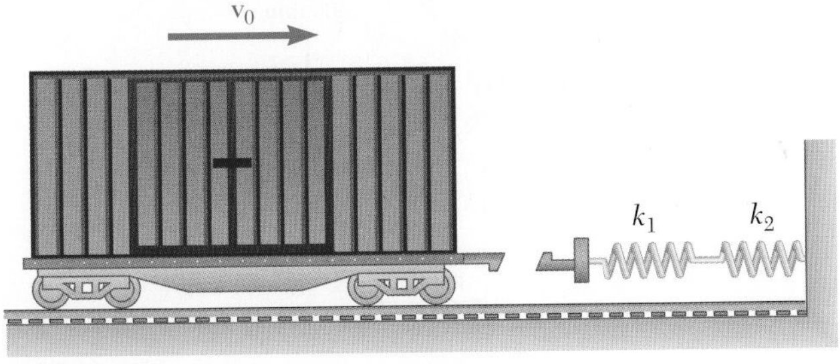

Fig. P13.56

13.57 A 600-g collar C may slide along a horizontal, semicircular rod ABD. The spring CE has an undeformed length of 250 mm and a spring constant of 135 N/m. Knowing that the collar is released from rest at A and neglecting friction, determine the speed of the collar (*a*) at B, (*b*) at D.

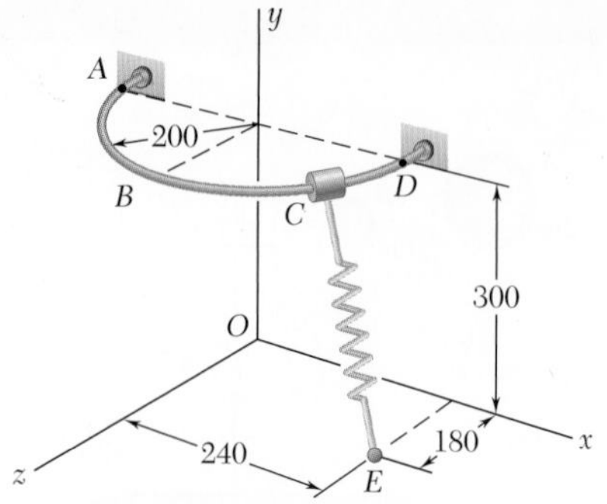

Dimensions in mm

Fig. P13.57

13.58 A 10-kg collar slides without friction along a vertical rod as shown. The spring attached to the collar has an undeformed length of 100 mm and a constant of 600 N/m. If the collar is released from rest in position 1, determine its velocity after it has moved 150 mm to position 2.

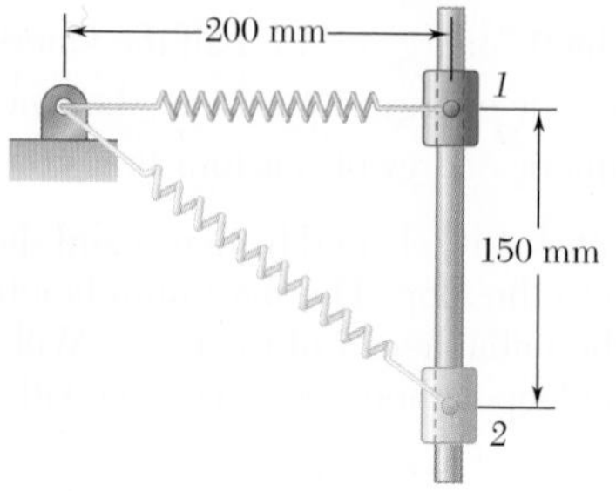

Fig. *P13.58*

13.59 A 1.2-kg collar C may slide without friction along a horizontal rod. It is attached to three springs, each of constant $k = 400$ N/m and 150-mm undeformed length. Knowing that the collar is released from rest in the position shown, determine the maximum speed it will reach in the ensuing motion.

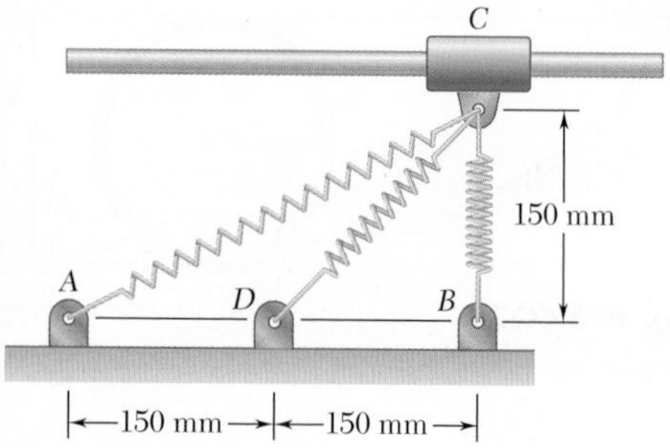

Fig. P13.59

13.60 A 500-g collar can slide without friction on the curved rod BC in a *horizontal* plane. Knowing that the undeformed length of the spring is 80 mm and that $k = 400$ kN/m, determine (*a*) the velocity that the collar should be given at A to reach B with zero velocity, (*b*) the velocity of the collar when it eventually reaches C.

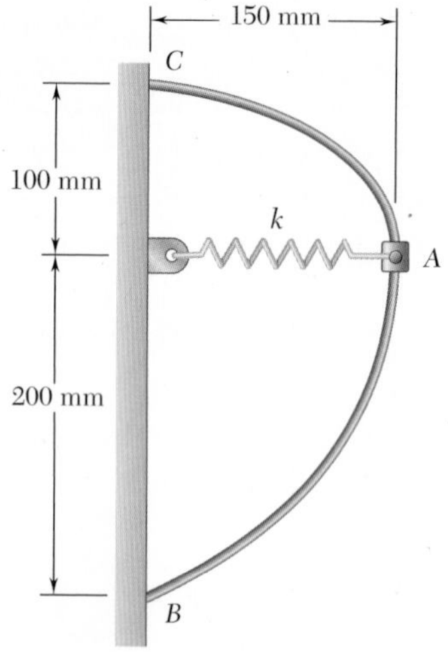

Fig. *P13.60*

13.61 An elastic cord is stretched between two points A and B, located 800 mm apart in the same horizontal plane. When stretched directly between A and B, the tension is 40 N. The cord is then stretched as shown until its midpoint C has moved through 300 mm to C'; a force of 240 N is required to hold the cord at C'. A 0.1-kg pellet is placed at C', and the cord is released. Determine the speed of the pellet as it passes through C.

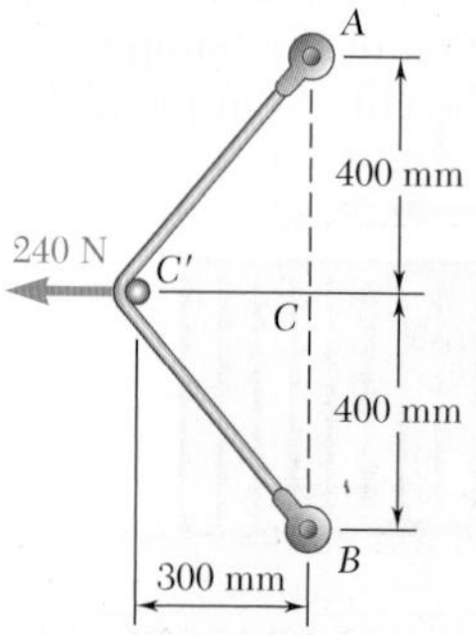

Fig. *P13.61*

13.62 An elastic cable is to be designed for bungee jumping from a tower 40 m high. The specifications call for the cable to be 25 m long when unstretched, and to stretch to a total length of 30 m when a 300-kg mass is attached to it and dropped from the tower. Determine (*a*) the required spring constant k of the cable, (*b*) how close to the ground a 90-kg man will come if he uses this cable to jump from the tower.

Fig. P13.62

13.63 It is shown in mechanics of materials that when an elastic beam AB supports a block of weight W at a given point B, the deflection y_{st} (called the static deflection) is proportional to W. Show that if the same block is dropped from a height h onto the end B of a cantilever beam AB and does not bounce off, the maximum deflection y_m in the ensuing motion can be expressed as $y_m = y_{st}(1 + \sqrt{1 + 2h/y_{st}})$ Note that this formula is approximate, since it is based on the assumption that there is no energy dissipated in the impact and that the weight of the beam is small compared to the weight of the block.

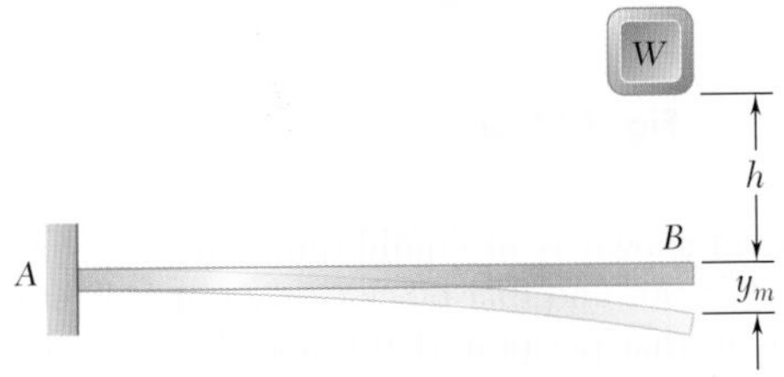

Fig. P13.63

13.64 A 2-kg collar is attached to a spring and slides without friction in a vertical plane along the curved rod ABC. The spring is undeformed when the collar is at C and its constant is 600 N/m. If the collar is released at A with no initial velocity, determine its velocity (*a*) as it passes through B, (*b*) as it reaches C.

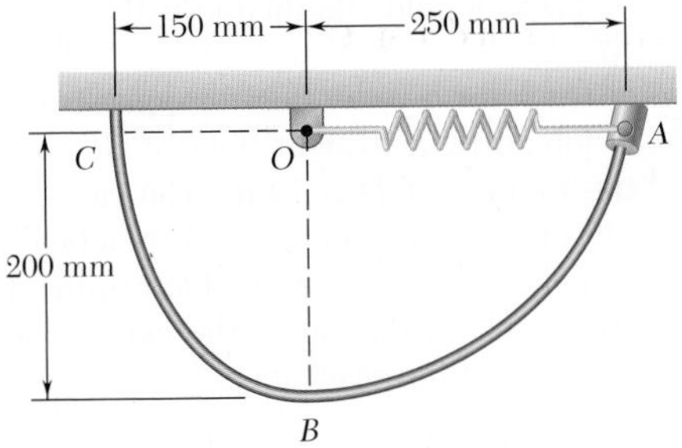

Fig. P13.64

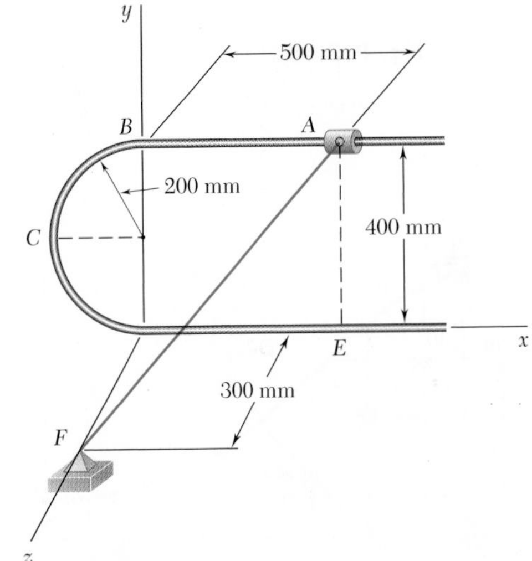

Fig. P13.65

13.65 A 1-kg collar can slide along the rod shown. It is attached to an elastic cord anchored at F, which has an undeformed length of 250 mm and spring constant of 75 N/m. Knowing that the collar is released from rest at A and neglecting friction, determine the speed of the collar (*a*) at B, (*b*) at E.

13.66 A thin circular rod is supported in a *vertical plane* by a bracket at A. Attached to the bracket and loosely wound around the rod is a spring of constant $k = 50$ N/m and undeformed length equal to the arc of circle AB. A 250-g collar C, not attached to the spring, can slide without friction along the rod. Knowing that the collar is released from rest at an angle θ with the vertical, determine (*a*) the smallest value of θ for which the collar will pass through D and reach point A, (*b*) the velocity of the collar as it reaches point A.

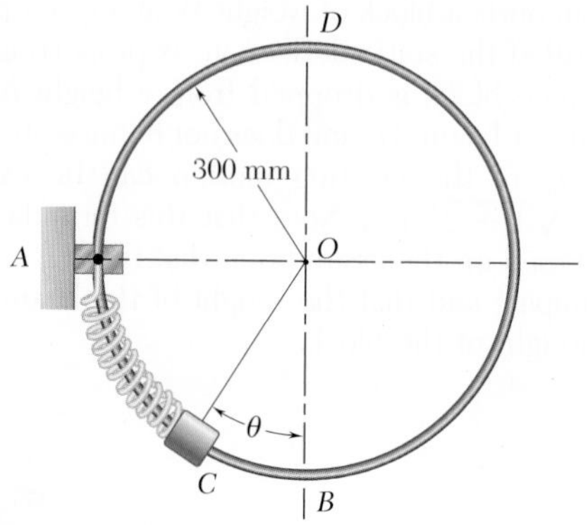

Fig. P13.66

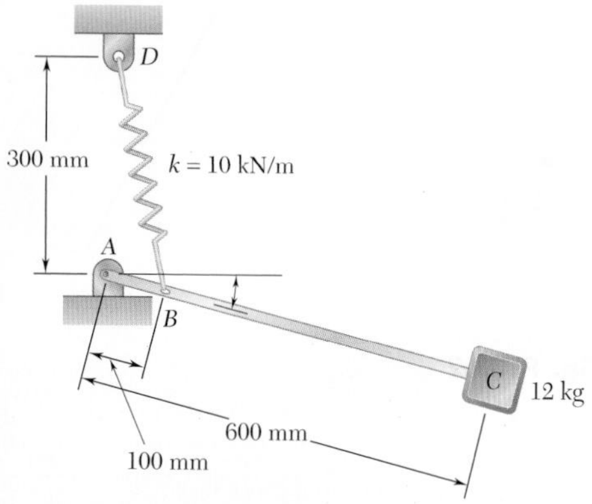

Fig. P13.67

13.67 The system shown is in equilibrium when $\phi = 0$. Knowing that initially $\phi = 90°$ and that block C is given a slight nudge when the system is in that position, determine the speed of the block as it passes through the equilibrium position $\phi = 0$. Neglect the weight of the rod.

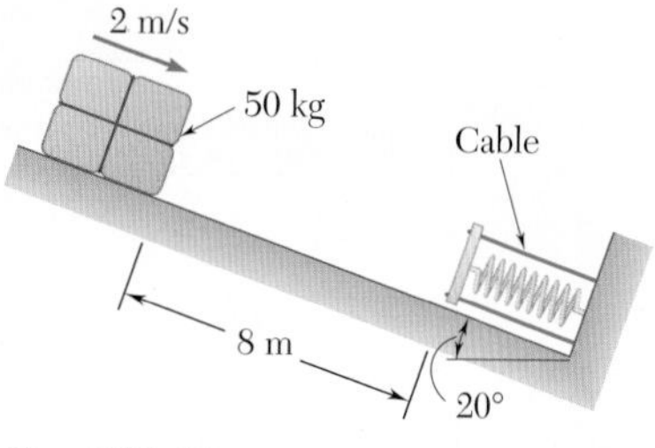

Fig. P13.68

13.68 A spring is used to stop a 50-kg package which is moving down a 20° incline. The spring has a constant $k = 30$ kN/m and is held by cables so that it is initially compressed 50 mm. Knowing that the velocity of the package is 2 m/s when it is 8 m from the spring and neglecting friction, determine the maximum additional deformation of the spring in bringing the package to rest.

13.69 Solve Prob. 13.68 assuming the kinetic coefficient of friction between the package and the incline is 0.2.

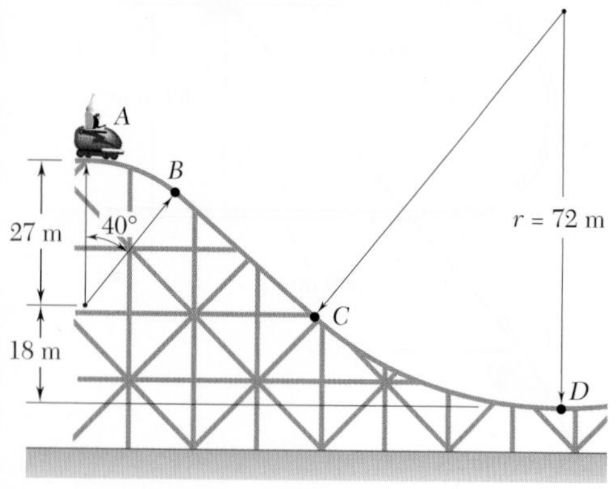

Fig. P13.70 and P13.71

13.70 A section of track for a roller coaster consists of two circular arcs AB and CD joined by a straight portion BC. The radius of AB is 27 m and the radius of CD is 72 m. The car and its occupants, of total mass 250 kg, reach point A with practically no velocity and then drop freely along the track. Determine the normal force exerted by the track on the car as the car reaches point B. Ignore air resistance and rolling resistance.

13.71 A section of track for a roller coaster consists of two circular arcs AB and CD joined by a straight portion BC. The radius of AB is 27 m and the radius of CD is 72 m. The car and its occupants, of total mass 250 kg, reach point A with practically no velocity and then drop freely along the track. Determine the maximum and minimum values of the normal force exerted by the track on the car as the car travels from A to D. Ignore air resistance and rolling resistance.

13.72 A 500-g collar is attached to a spring and slides without friction along a circular rod in a *vertical* plane. The spring has an undeformed length of 125 mm and a constant $k = 150$ N/m. Knowing that the collar is released from being held at A determine the speed of the collar and the normal force between the collar and the rod as the collar passes through B.

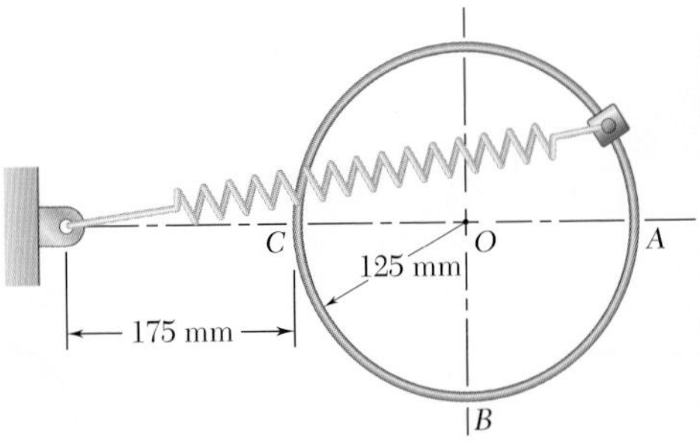

Fig. P13.72

13.73 A 600-g collar can slide without friction along the semicircular rod BCD. The spring is of constant 360 N/m and its undeformed length is 200 mm. Knowing that the collar is released from rest at B, determine (*a*) the speed of the collar as it passes through C, (*b*) the force exerted by the rod on the collar at C

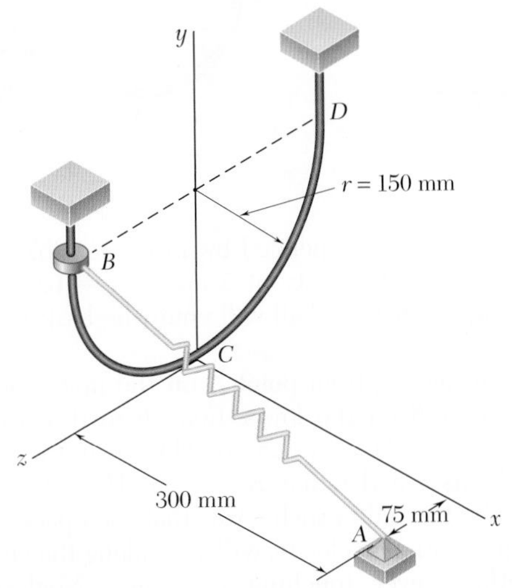

Fig. P13.73

13.74 A 200-g package is projected upward with a velocity $\mathbf{v}_0$ by a spring at A; it moves around a frictionless loop and is deposited at C. For each of the two loops shown, determine (*a*) the smallest velocity $\mathbf{v}_0$ for which the package will reach C, (*b*) the corresponding force exerted by the package on the loop just before the package leaves the loop at C.

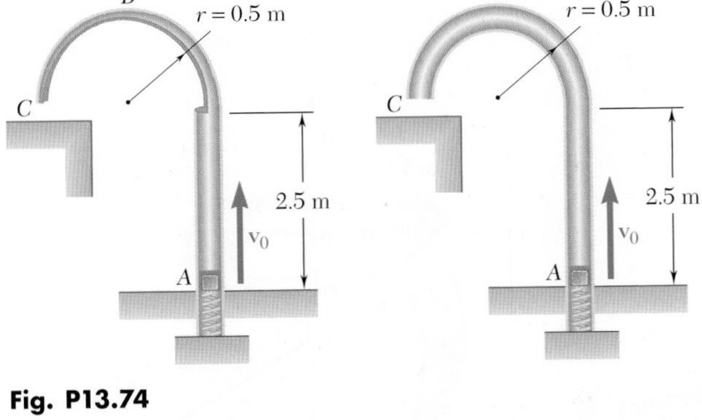

Fig. P13.74

13.75 If the package of Prob. 13.74 is not to hit the horizontal surface at C with a speed greater than 3.5 m/s, (*a*) show that this requirement can be satisfied only by the second loop, (*b*) determine the largest allowable initial velocity $\mathbf{v}_0$ when the second loop is used.

13.76 A small package of weight W is projected into a vertical return loop at A with a velocity $\mathbf{v}_0$. The package travels without friction along a circle of radius r and is deposited on a horizontal surface at C. For each of the two loops shown, determine (*a*) the smallest velocity $\mathbf{v}_0$ for which the package will reach the horizontal surface at C, (*b*) the corresponding force exerted by the loop on the package as it passes point B.

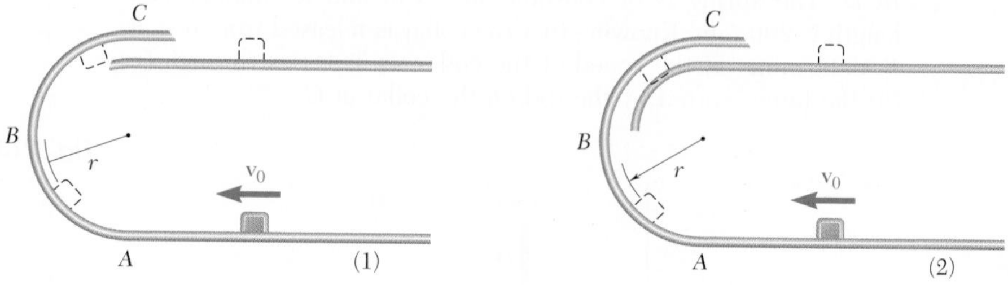

Fig. P13.76

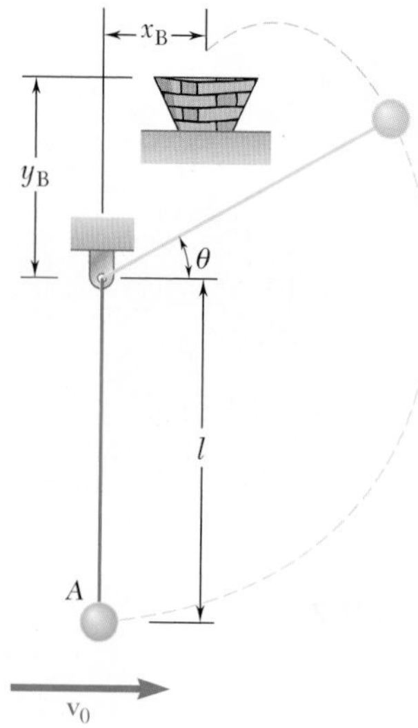

Fig. P13.77

13.77 The 1-kg ball at A is suspended by an inextensible cord and given an initial horizontal velocity of 5 m/s. If $l = 0.6$ m and $x_B = 0$, determine y_B so that the ball will enter the basket.

***13.78** Packages are moved from point A on the upper floor of a warehouse to point B on the lower floor, 6 m directly below A, by means of a chute, the centerline of which is in the shape of a helix of vertical axis y and radius $R = 4$ m. The cross section of the chute is to be banked in such a way that each package, after being released at A with no velocity, will slide along the centerline of the chute without ever touching its edges. Neglecting friction, (*a*) express as a function of the elevation y of a given point P of the centerline the angle ϕ formed by the normal to the surface of the chute at P and the principal normal of the centerline at that point, (*b*) determine the magnitude and direction of the force exerted by the chute on a 10-kg package as it reaches point B. *Hint:* The principal normal to the helix at any point P is horizontal and directed toward the y axis, and the radius of curvature of the helix is $\rho = R[1 + (h/2\pi R)^2]$.

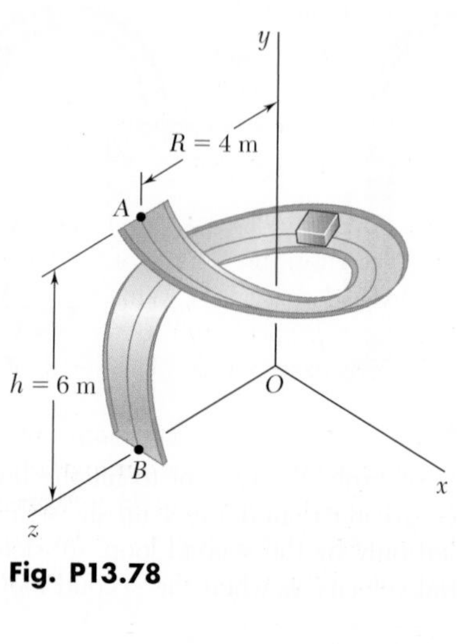

Fig. P13.78

***13.79** Prove that a force $F(x, y, z)$ is conservative if, and only if, the following relations are satisfied:

$$\frac{\partial F_x}{\partial y} = \frac{\partial F_y}{\partial x} \qquad \frac{\partial F_y}{\partial z} = \frac{\partial F_z}{\partial y} \qquad \frac{\partial F_z}{\partial x} = \frac{\partial F_x}{\partial z}$$

13.80 The force $\mathbf{F} = (yz\mathbf{i} + zx\mathbf{j} + xy\mathbf{k})/xyz$ acts on the particle $P(x, y, z)$ which moves in space. (*a*) Using the relation derived in Prob. 13.79, show that this force is a conservative force. (*b*) Determine the potential function associated with **F**.

***13.81** A force F acts on a particle $P(x, y)$ which moves in the xy plane. Determine whether **F** is a conservative force and compute the work of **F** when P describes in a clockwise sense the path A, B, C, A including the quarter circle $x^2 + y^2 = a^2$, if (*a*) $\mathbf{F} = ky\mathbf{i}$, (*b*) $\mathbf{F} = k(y\mathbf{i} + x\mathbf{j})$.

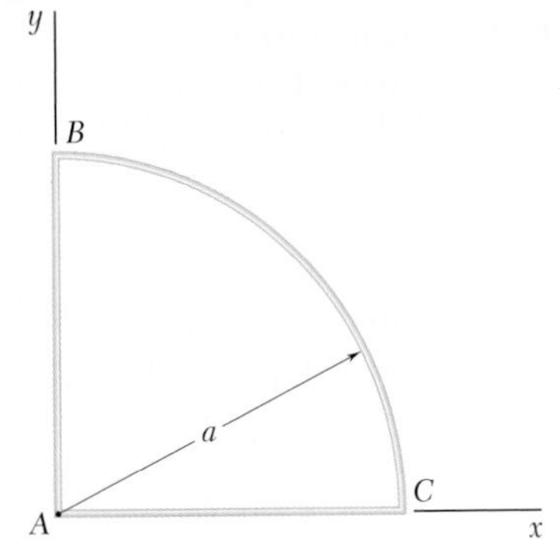

Fig. P13.81

***13.82** The potential function associated with a force **P** in space is known to be $V(x, y, z) = -(x^2 + y^2 + z^2)^{1/2}$. (*a*) Determine the x, y, and z components of **P**. (*b*) Calculate the work done by **P** from O to D by integrating along the path $OABD$, and show that it is equal to the negative of the change in potential from O to D.

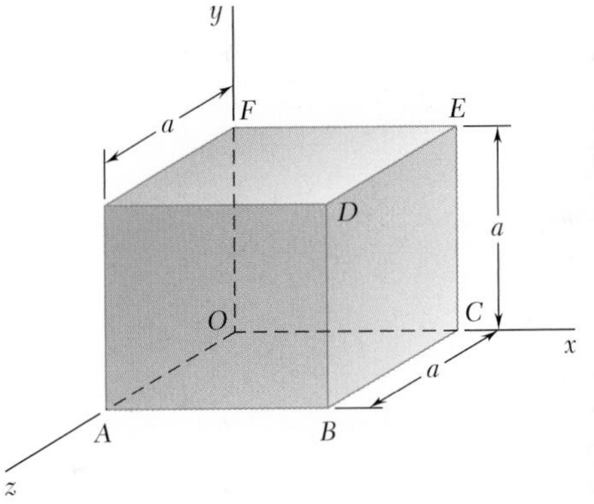

Fig. P13.82

***13.83** (*a*) Calculate the work done from D to O by the force **P** of Prob. 13.82 by integrating along the diagonal of the cube. (*b*) Using the result obtained and the answer to part *b* of Prob. 13.82, verify that the work done by a conservative force around the closed path $OABDO$ is zero.

***13.84** The force $\mathbf{F} = (x\mathbf{i} + y\mathbf{j} + z\mathbf{k})/(x^2 + y^2 + z^2)^{3/2}$ acts on the particle $P(x, y, z)$ which moves in space. (*a*) Using the relations derived in Prob. 13.79, prove that **F** is a conservative force. (*b*) Determine the potential function $V(x, y, z)$ associated with **F**.

13.85 (*a*) Determine the kinetic energy per unit mass which a missile must have after being fired from the surface of the earth if it is to reach an infinite distance from the earth. (*b*) What is the initial velocity of the missile (called the *escape velocity*)? Give your answers in SI units and show that the answer to part *b* is independent of the firing angle.

13.86 A satellite describes an elliptic orbit of minimum altitude 606 km above the surface of the earth. The semimajor and semiminor axes are 17 440 km and 13 950 km, respectively. Knowing that the speed of the satellite at point C is 4.78 km/s, determine (*a*) the speed at point A, the perigee, (*b*) the speed at point B, the apogee.

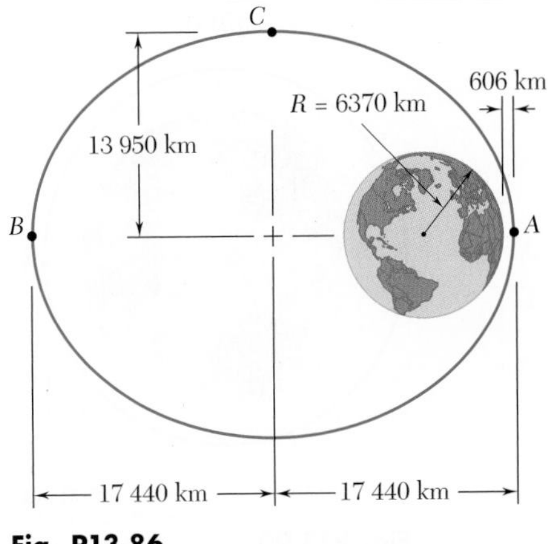

Fig. P13.86

13.87 While describing a circular orbit 300 km above the earth a space vehicle launches a 3600-kg communications satellite. Determine (*a*) the additional energy required to place the satellite in a geosynchronous orbit at an altitude of 35 770 km above the surface of the earth, (*b*) the energy required to place the satellite in the same orbit by launching it from the surface of the earth, excluding the energy needed to overcome air resistance. (A *geosynchronous orbit* is a circular orbit in which the satellite appears stationary with respect to the ground.)

13.88 A lunar excursion module (LEM) was used in the Apollo moon-landing missions to save fuel by making it unnecessary to launch the entire Apollo spacecraft from the moon's surface on its return trip to earth. Check the effectiveness of this approach by computing the energy per kg required for a spacecraft (as weighed on the earth) to escape the moon's gravitational field if the spacecraft starts from (*a*) the moon's surface, (*b*) a circular orbit 80 km above the moon's surface. Neglect the effect of the earth's gravitational field. (The radius of the moon is 1740 km and its mass is 0.0123 times the mass of the earth.)

13.89 Knowing that the velocity of an experimental space probe fired from the earth has a magnitude v_A = 32.5 Mm/h at point A, determine the speed of the probe as it passes through point B.

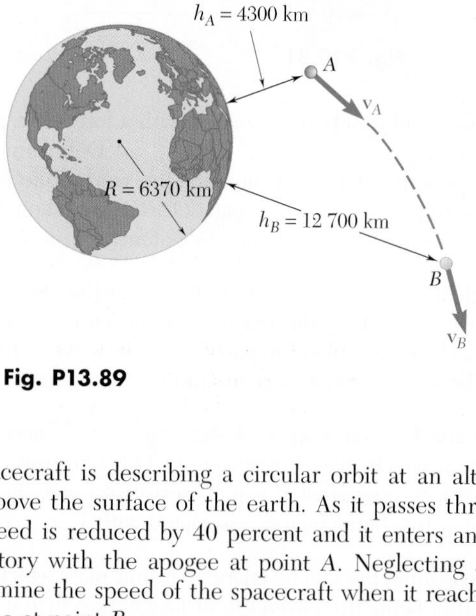

Fig. P13.89

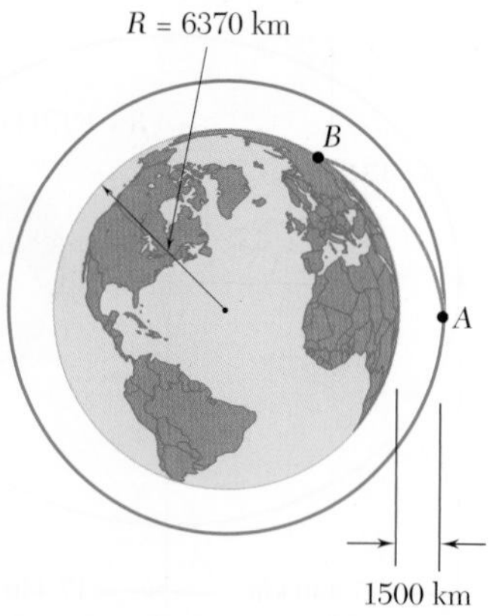

Fig. P13.90

13.90 A spacecraft is describing a circular orbit at an altitude of 1500 km above the surface of the earth. As it passes through point A, its speed is reduced by 40 percent and it enters an elliptic crash trajectory with the apogee at point A. Neglecting air resistance, determine the speed of the spacecraft when it reaches the earth's surface at point B.

13.91 Observations show that a celestial body traveling at 2×10^6 km/h appears to be describing about point B a circle of radius equal to 60 light years. Point B is suspected of being a very dense concentration of mass called a black hole. Determine the ratio M_B/M_S of the mass at B to the mass of the sun. (The mass of the sun is 330,000 times the mass of the earth, and a light year is the distance traveled by light in one year at 3×10^5 km/s.)

13.92 (*a*) Show that, by setting $r = R + y$ in the right-hand member of Eq. (13.17′) and expanding that member in a power series in y/R, the expression in Eq. (13.16) for the potential energy V_g due to gravity is a first-order approximation for the expression given in Eq. (13.17′). (*b*) Using the same expansion, derive a second-order approximation for V_g.

13.93 Collar *A* has a mass of 3 kg and is attached to a spring of constant 1200 N/m and of undeformed length equal to 0.5 m. The system is set in motion with $r = 0.3$ m, $v_\theta = 2$ m/s, and $v_r = 0$. Neglecting the mass of the rod and the effect of friction, determine the radial and transverse components of the velocity of the collar when $r = 0.6$ m.

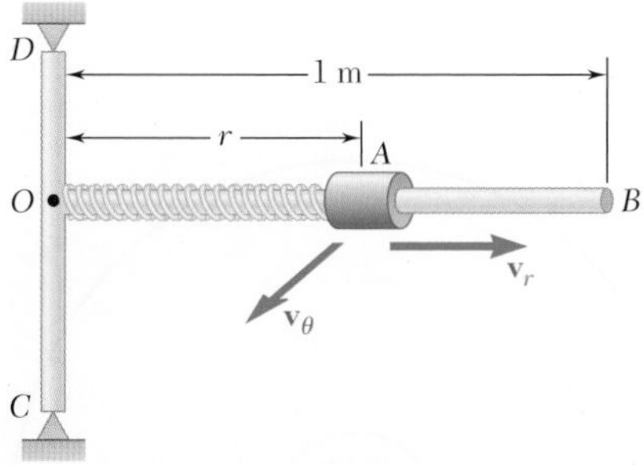

Fig. P13.93 and P13.94

13.94 Collar *A* has a mass of 3 kg and is attached to a spring of constant 1200 N/m and of undeformed length equal to 0.5 m. The system is set in motion with $r = 0.3$ m, $v_\theta = 2$ m/s, and $v_r = 0$. Neglecting the mass of the rod and the effect of friction, determine (*a*) the maximum distance between the origin and the collar, (*b*) the corresponding speed. (*Hint:* Solve the equation obtained for *r* by trial and error.)

13.95 A 1.8-kg collar *A* and a 0.7-kg collar *B* can slide without friction on a frame, consisting of the horizontal rod *OE* and the vertical rod *CD*, which is free to rotate about *CD*. The two collars are connected by a cord running over a pulley that is attached to the frame at *O*. At the instant shown, the velocity $\mathbf{v}_A$ of collar *A* has a magnitude of 2.1 m/s and a stop prevents collar *B* from moving. If the stop is suddenly removed, determine (*a*) the velocity of collar *A* when it is 0.2 m from *O*, (*b*) the velocity of collar *A* when collar *B* comes to rest. (Assume that collar *B* does not hit *O*, that collar *A* does not come off rod *OE*, and that the mass of the frame is negligible.)

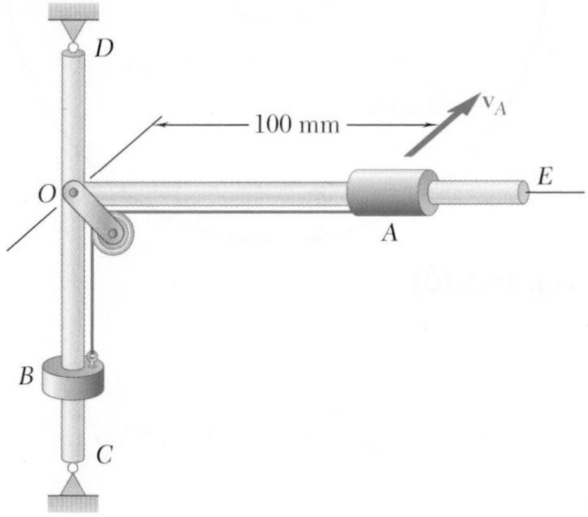

Fig. *P13.95*

13.96 A 200-g ball may slide on a horizontal frictionless surface and is attached to a fixed point *O* by means of an elastic cord of constant $k = 150$ N/m and undeformed length equal to 600 mm. The ball is placed at point *A*, 900 mm from *O*, and is given an initial velocity $\mathbf{v}_A$ in a direction perpendicular to *OA*. Knowing that the ball passes a distance $d = 100$ m from *O*, determine (*a*) the initial speed v_A of the ball, (*b*) its speed *v* after the cord has become slack.

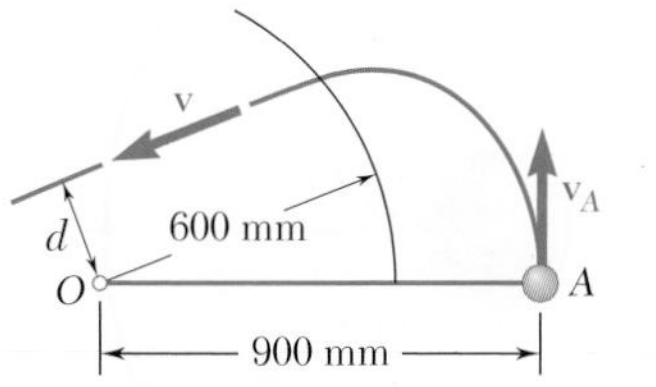

Fig. P13.96

13.97 For the ball of Prob. 13.96, determine (*a*) the smallest magnitude of the initial velocity $\mathbf{v}_A$ for which the elastic cord remains taut at all times, (*b*) the corresponding maximum speed reached by the ball.

13.98 Using the principles of conservation of energy and conservation of angular momentum, solve part *a* of Sample Prob. 12.9.

13.99 Solve Sample Prob. 13.8, assuming that the elastic cord is replaced by a central force **F** of magnitude $(80/r^2)$ N directed toward *O*.

13.100 A spacecraft is describing an elliptic orbit of minimum altitude $h_A = 2400$ km and maximum altitude $h_B = 9600$ km above the surface of the earth. Determine the speed of the spacecraft at A.

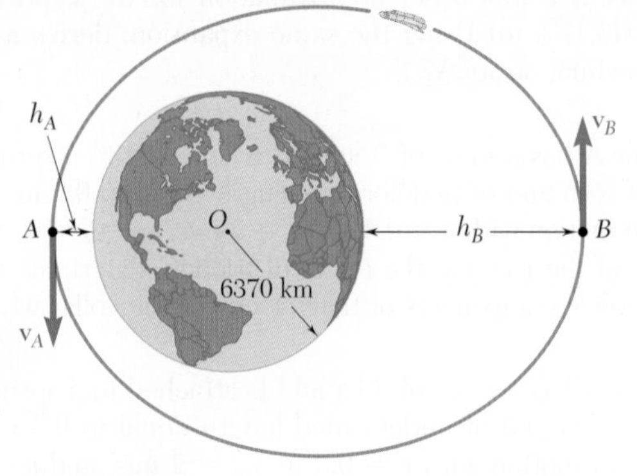

Fig. P13.100

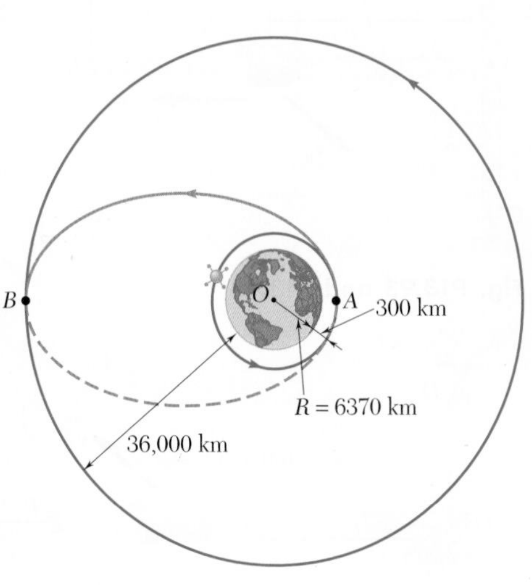

Fig. P13.101

13.101 While describing a circular orbit, 300 km above the surface of the earth, a space shuttle ejects at point A an inertial upper stage (IUS) carrying a communications satellite to be placed in a geosynchronous orbit (see Prob. 13.87) at an altitude of 36,000 km above the surface of the earth. Determine (*a*) the velocity of the IUS relative to the shuttle after its engine has been fired at A, (*b*) the increase in velocity required at B to place the satellite in its final orbit.

13.102 A spacecraft approaching the planet Saturn reaches point A with a velocity $\mathbf{v}_A$ of magnitude 20×10^3 m/s. It is to be placed in an elliptic orbit about Saturn so that it will be able to periodically examine Tethys, one of Saturn's moons. Tethys is in a circular orbit of radius 300×10^3 km about the center of Saturn, traveling at a speed of 11.1×10^3 m/s. Determine (*a*) the decrease in speed required by the spacecraft at A to achieve the desired orbit, (*b*) the speed of the spacecraft when it reaches the orbit of Tethys at B.

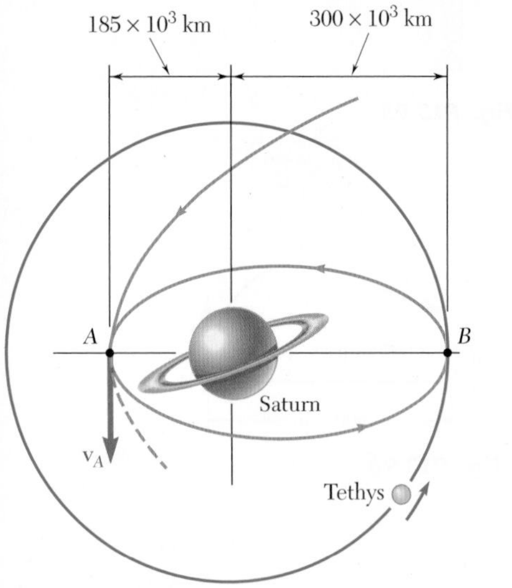

Fig. P13.102

13.103 A spacecraft traveling along a parabolic path toward the planet Jupiter is expected to reach point A with a velocity $\mathbf{v}_A$ of magnitude 26.9 km/s. Its engines will then be fired to slow it down, placing it into an elliptic orbit which will bring it to within 100×10^3 km of Jupiter. Determine the decrease in speed Δv at point A which will place the spacecraft into the required orbit. The mass of Jupiter is 319 times the mass of the earth.

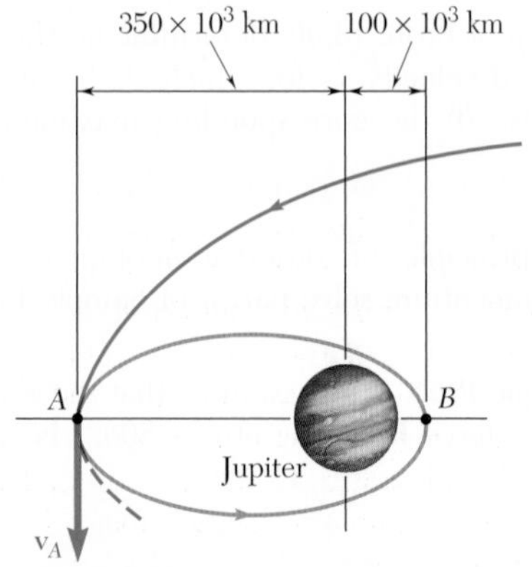

Fig. P13.103

13.104 As a first approximation to the analysis of a space flight from the earth to Mars, it is assumed that the orbits of the earth and Mars are circular and coplanar. The mean distances from the sun to the earth and to Mars are 149.6×10^6 km and 227.8×10^6 km, respectively. To place the spacecraft into an elliptical transfer orbit at point A, its speed is increased over a short interval of time to v_A which is faster than the earth's orbital speed. When the spacecraft reaches point B on the elliptical transfer orbit, its speed v_B is increased to the orbital speed of Mars. Knowing that the mass of the sun is 332.8×10^3 times the mass of the earth, determine the increase in velocity required (*a*) at A, (*b*) at B.

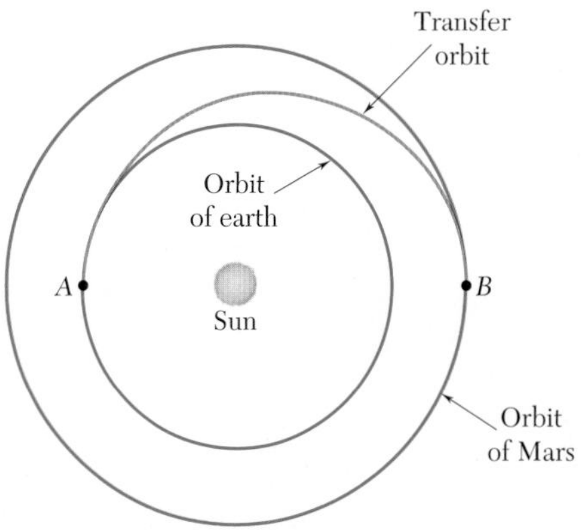

Fig. P13.104

13.105 The optimal way of transferring a space vehicle from an inner circular orbit to an outer coplanar circular orbit is to fire its engines as it passes through A to increase its speed and place it in an elliptic transfer orbit. Another increase in speed as it passes through B will place it in the desired circular orbit. For a vehicle in a circular orbit about the earth at an altitude $h_1 = 300$ km, which is to be transferred to a circular orbit at an altitude $h_2 = 800$ km, determine (*a*) the required increases in speed at A and at B, (*b*) the total energy per unit mass required to execute the transfer.

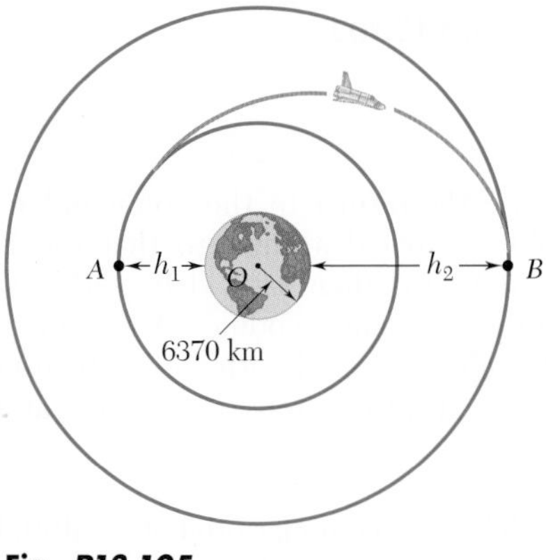

Fig. P13.105

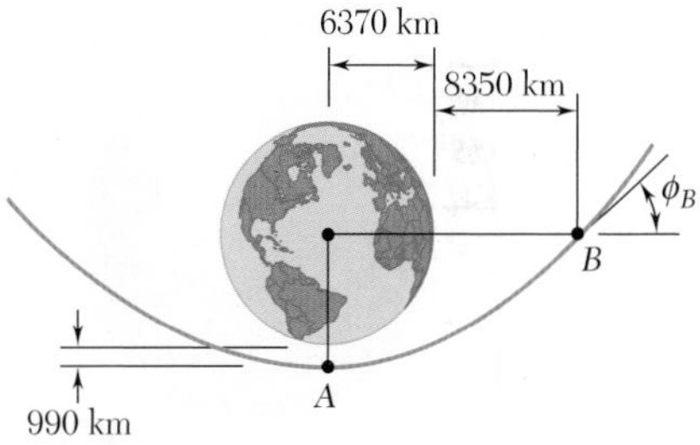

Fig. P13.106

13.106 During a flyby of the earth, the velocity of a spacecraft is 10.4 km/s as it reaches its minimum altitude of 990 km above the surface at point A. At point B the spacecraft is observed to have an altitude of 8350 km. Determine (*a*) the magnitude of the velocity at point B, (*b*) the angle ϕ_B.

13.107 A space platform is in a circular orbit about the earth at an altitude of 300 km. As the platform passes through A, a rocket carrying a communications satellite is launched from the platform with a relative velocity of magnitude 3.44 km/s in a direction tangent to the orbit of the platform. This was intended to place the rocket in an elliptic transfer orbit bringing it to point B, where the rocket would again be fired to place the satellite in a geosynchronous orbit of radius 42 140 km. After launching, it was discovered that the relative velocity imparted to the rocket was too large. Determine the angle γ at which the rocket will cross the intended orbit at point C.

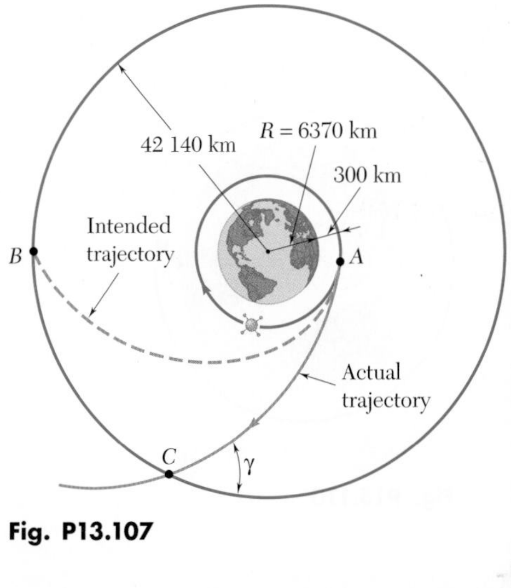

Fig. P13.107

13.108 A satellite is projected into space with a velocity $\mathbf{v}_0$ at a distance r_0 from the center of the earth by the last stage of its launching rocket. The velocity $\mathbf{v}_0$ was designed to send the satellite into a circular orbit of radius r_0. However, owing to a malfunction of control, the satellite is not projected horizontally but at an angle α with the horizontal and, as a result, is propelled into an elliptic orbit. Determine the maximum and minimum values of the distance from the center of the earth to the satellite.

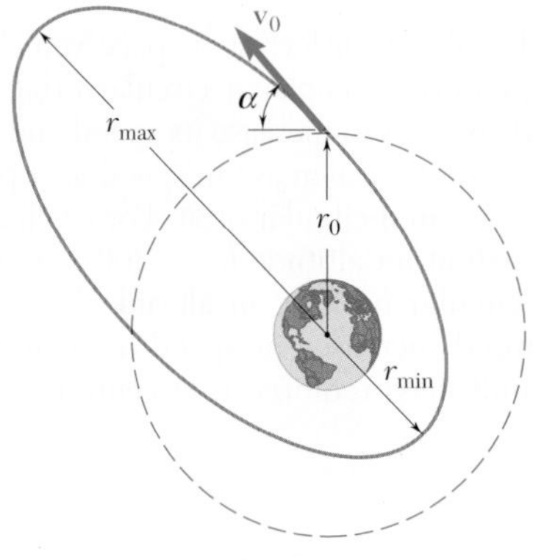

Fig. P13.108

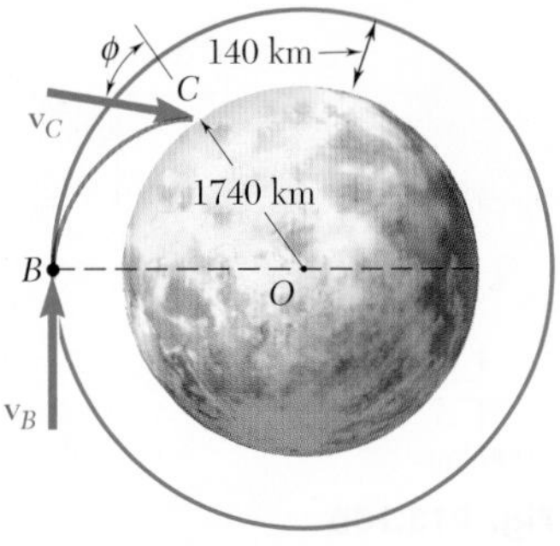

Fig. P13.109

13.109 Upon the LEM's return to the command module, the Apollo spacecraft was turned around so that the LEM faced to the rear. The LEM was then cast adrift with a velocity of 200 m/s relative to the command module. Determine the magnitude and direction (angle ϕ formed with the vertical OC) of the velocity $\mathbf{v}_C$ of the LEM just before it crashed at C on the moon's surface.

13.110 A space vehicle is in a circular orbit at an altitude of 360 km above the earth. To return to earth, it decreases its speed as it passes through A by firing its engine for a short interval of time in a direction opposite to the direction of its motion. Knowing that the velocity of the space vehicle should form an angle $\phi_B = 60°$ with the vertical as it reaches point B at an altitude of 60 km, determine (*a*) the required speed of the vehicle as it leaves its circular orbit at A, (*b*) its speed at point B.

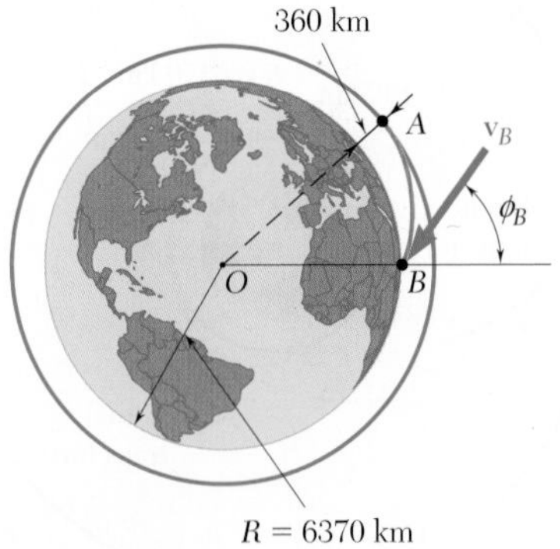

Fig. P13.110

***13.111** In Prob. 13.110, the speed of the space vehicle was decreased as it passed through A by firing its engine in a direction opposite to the direction of motion. An alternative strategy for taking the space vehicle out of its circular orbit would be to turn it around so that its engine would point away from the earth and then give it an incremental velocity $\Delta\mathbf{v}_A$ toward the center O of the earth. This would likely require a smaller expenditure of energy when firing the engine at A, but might result in too fast a descent at B. Assuming this strategy is used with only 50 percent of the energy expenditure used in Prob. 13.110, determine the resulting values of ϕ_B and v_B.

13.112 Show that the values v_A and v_P of the speed of an earth satellite at the apogee A and the perigee P of an elliptic orbit are defined by the relations

$$v_A^2 = \frac{2GM}{r_A + r_P}\frac{r_P}{r_A} \qquad v_P^2 = \frac{2GM}{r_A + r_P}\frac{r_A}{r_P}$$

where M is the mass of the earth, and r_A and r_P represent, respectively, the maximum and minimum distances of the orbit to the center of the earth.

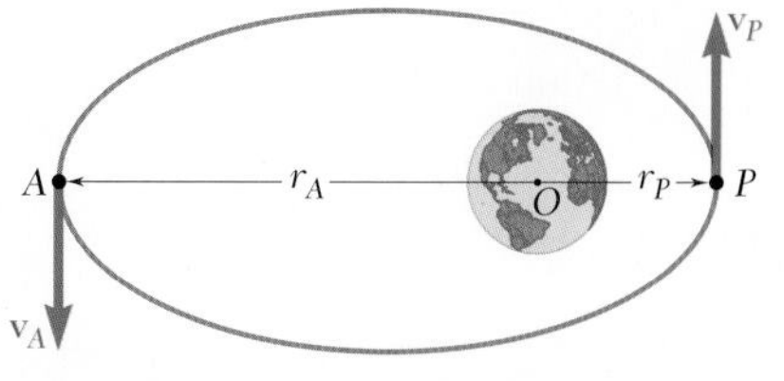

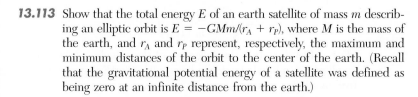

Fig. *P13.112* and *P13.113*

13.113 Show that the total energy E of an earth satellite of mass m describing an elliptic orbit is $E = -GMm/(r_A + r_P)$, where M is the mass of the earth, and r_A and r_P represent, respectively, the maximum and minimum distances of the orbit to the center of the earth. (Recall that the gravitational potential energy of a satellite was defined as being zero at an infinite distance from the earth.)

***13.114** A space probe describes a circular orbit of radius nR with a velocity $\mathbf{v}_0$ about a planet of radius R and center O. Show that (*a*) in order for the probe to leave its orbit and hit the planet at an angle θ with the vertical, its velocity must be reduced to $\alpha\mathbf{v}_0$, where

$$\alpha = \sin\theta\sqrt{\frac{2(n-1)}{n^2 - \sin^2\theta}}$$

(*b*) the probe will not hit the planet if α is larger than $\sqrt{2/(1+n)}$.

13.115 A missile is fired from the ground with an initial velocity $\mathbf{v}_0$ forming an angle ϕ_0 with the vertical. If the missile is to reach a maximum altitude equal to αR, where R is the radius of the earth, (*a*) show that the required angle ϕ_0 is defined by the relation

$$\sin\phi_0 = (1+\alpha)\sqrt{1 - \frac{\alpha}{1+\alpha}\left(\frac{v_{esc}}{v_0}\right)^2}$$

where v_{esc} is the escape velocity, (*b*) determine the range of allowable values of v_0.

13.116 A spacecraft of mass m describes a circular orbit of radius r_1 around the earth. (*a*) Show that the additional energy ΔE which must be imparted to the spacecraft to transfer it to a circular orbit of larger radius r_2 is

$$\Delta E = \frac{GMm(r_2 - r_1)}{2r_1 r_2}$$

where M is the mass of the earth. (*b*) Further show that if the transfer from one circular orbit to the other is executed by placing the spacecraft on a transitional semielliptic path AB, the amounts of energy ΔE_A and ΔE_B which must be imparted at A and B are, respectively, proportional to r_2 and r_1:

$$\Delta E_A = \frac{r_2}{r_1 + r_2}\Delta E \qquad \Delta E_B = \frac{r_1}{r_1 + r_2}\Delta E$$

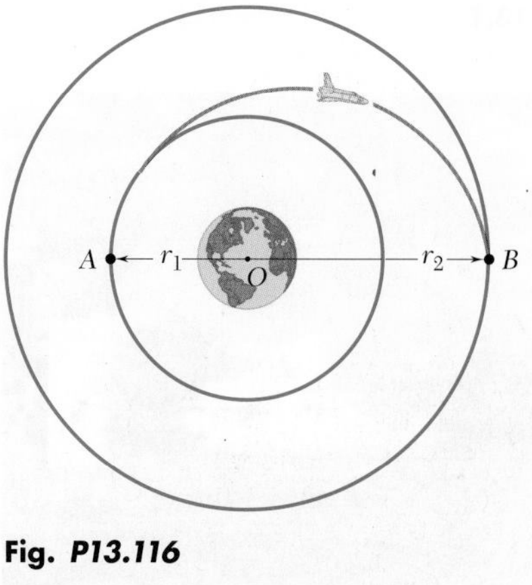

Fig. *P13.116*

***13.117** Using the answers obtained in Prob. 13.108, show that the intended circular orbit and the resulting elliptic orbit intersect at the ends of the minor axis of the elliptic orbit.

***13.118** (*a*) Express in terms of r_{min} and v_{max} the angular momentum per unit mass, h, and the total energy per unit mass, E/m, of a space vehicle moving under the gravitational attraction of a planet of mass M (Fig. 13.15). (*b*) Eliminating v_{max} between the equations obtained, derive the formula

$$\frac{1}{r_{min}} = \frac{GM}{h^2}\left[1 + \sqrt{1 + \frac{2E}{m}\left(\frac{h}{GM}\right)^2}\,\right]$$

(*c*) Show that the eccentricity ε of the trajectory of the vehicle can be expressed as

$$\varepsilon = \sqrt{1 + \frac{2E}{m}\left(\frac{h}{GM}\right)^2}$$

(*d*) Further show that the trajectory of the vehicle is a hyperbola, an ellipse, or a parabola, depending on whether E is positive, negative, or zero.

13.10 PRINCIPLE OF IMPULSE AND MOMENTUM

A third basic method for the solution of problems dealing with the motion of particles will be considered now. This method is based on the principle of impulse and momentum and can be used to solve problems involving force, mass, velocity, and time. It is of particular interest in the solution of problems involving impulsive motion and problems involving impact (Secs. 13.11 and 13.12).

Photo 13.1

Consider a particle of mass m acted upon by a force $\mathbf{F}$. As we saw in Sec. 12.3, Newton's second law can be expressed in the form

$$\mathbf{F} = \frac{d}{dt}(m\mathbf{v}) \tag{13.27}$$

where $m\mathbf{v}$ is the linear momentum of the particle. Multiplying both sides of Eq. (13.27) by dt and integrating from a time t_1 to a time t_2, we write

$$\mathbf{F}\,dt = d(m\mathbf{v})$$

$$\int_{t_1}^{t_2} \mathbf{F}\,dt = m\mathbf{v}_2 - m\mathbf{v}_1$$

or, transposing the last term,

$$m\mathbf{v}_1 + \int_{t_1}^{t_2} \mathbf{F}\,dt = m\mathbf{v}_2 \tag{13.28}$$

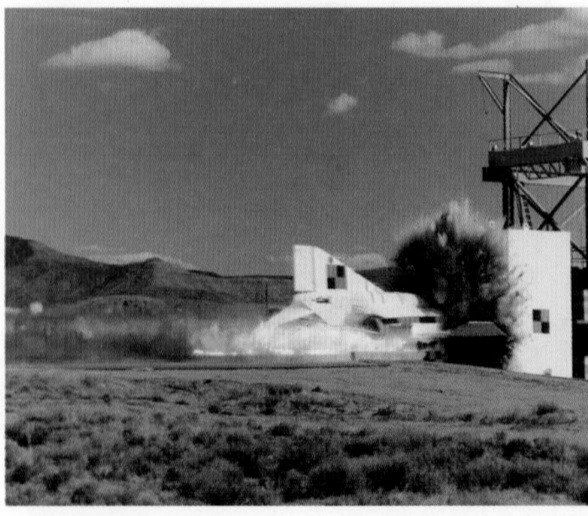

Photo 13.2 This impact test between an F-4 Phantom and a rigid reinforced target was to determine the impact force as a function of time.

The integral in Eq. (13.28) is a vector known as the *linear impulse*, or simply the *impulse*, of the force $\mathbf{F}$ during the interval of time considered. Resolving $\mathbf{F}$ into rectangular components, we write

$$\mathbf{Imp}_{1\to 2} = \int_{t_1}^{t_2} \mathbf{F}\,dt$$

$$= \mathbf{i}\int_{t_1}^{t_2} F_x\,dt + \mathbf{j}\int_{t_1}^{t_2} F_y\,dt + \mathbf{k}\int_{t_1}^{t_2} F_z\,dt \tag{13.29}$$

and note that the components of the impulse of the force **F** are, respectively, equal to the areas under the curves obtained by plotting the components F_x, F_y, and F_z against t (Fig. 13.16). In the case of a force **F** of constant magnitude and direction, the impulse is represented by the vector $\mathbf{F}(t_2 - t_1)$, which has the same direction as **F**.

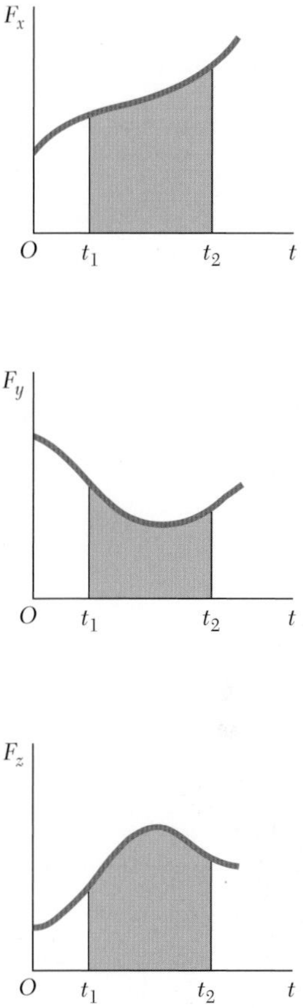

Fig. 13.16

If SI units are used, the magnitude of the impulse of a force is expressed in N · s. But, recalling the definition of the newton, we have

$$\mathrm{N} \cdot \mathrm{s} = (\mathrm{kg} \cdot \mathrm{m/s^2}) \cdot \mathrm{s} = \mathrm{kg} \cdot \mathrm{m/s}$$

which is the unit obtained in Sec. 12.4 for the linear momentum of a particle. We thus check that Eq. (13.28) is dimensionally correct. If U.S. customary units are used, the impulse of a force is expressed in lb · s, which is also the unit obtained in Sec. 12.4 for the linear momentum of a particle.

Equation (13.28) expresses that when a particle is acted upon by a force **F** during a given time interval, *the final momentum* $m\mathbf{v}_2$ *of the particle can be obtained by adding vectorially its initial momentum* $m\mathbf{v}_1$ *and the impulse of the force* **F** *during the time interval considered*

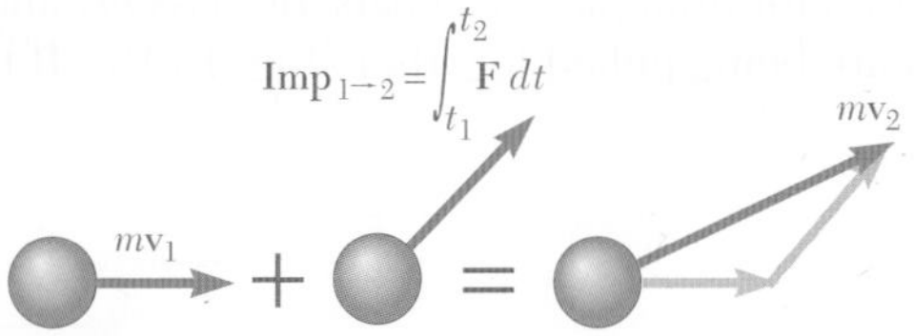

Fig. 13.17

(Fig. 13.17). We write

$$m\mathbf{v}_1 + \mathbf{Imp}_{1\rightarrow 2} = m\mathbf{v}_2 \tag{13.30}$$

We note that while kinetic energy and work are scalar quantities, momentum and impulse are vector quantities. To obtain an analytic solution, it is thus necessary to replace Eq. (13.30) by the corresponding component equations

$$\begin{aligned} (mv_x)_1 + \int_{t_1}^{t_2} F_x\,dt &= (mv_x)_2 \\ (mv_y)_1 + \int_{t_1}^{t_2} F_y\,dt &= (mv_y)_2 \\ (mv_z)_1 + \int_{t_1}^{t_2} F_z\,dt &= (mv_z)_2 \end{aligned} \tag{13.31}$$

When several forces act on a particle, the impulse of each of the forces must be considered. We have

$$m\mathbf{v}_1 + \Sigma\,\mathbf{Imp}_{1\rightarrow 2} = m\mathbf{v}_2 \tag{13.32}$$

Again, the equation obtained represents a relation between vector quantities; in the actual solution of a problem, it should be replaced by the corresponding component equations.

When a problem involves two particles or more, each particle can be considered separately and Eq. (13.32) can be written for each particle. We can also add vectorially the momenta of all the particles and the impulses of all the forces involved. We write then

$$\Sigma m\mathbf{v}_1 + \Sigma\ \mathbf{Imp}_{1\rightarrow 2} = \Sigma m\mathbf{v}_2 \tag{13.33}$$

Since the forces of action and reaction exerted by the particles on each other form pairs of equal and opposite forces, and since the time interval from t_1 to t_2 is common to all the forces involved, the impulses of the forces of action and reaction cancel out and only the impulses of the external forces need be considered.†

If no external force is exerted on the particles or, more generally, if the sum of the external forces is zero, the second term in Eq. (13.33) vanishes and Eq. (13.33) reduces to

$$\Sigma m\mathbf{v}_1 = \Sigma m\mathbf{v}_2 \tag{13.34}$$

which expresses that *the total momentum of the particles is conserved*. Consider, for example, two boats, of mass m_A and m_B, initially at rest, which are being pulled together (Fig. 13.18). If the resistance

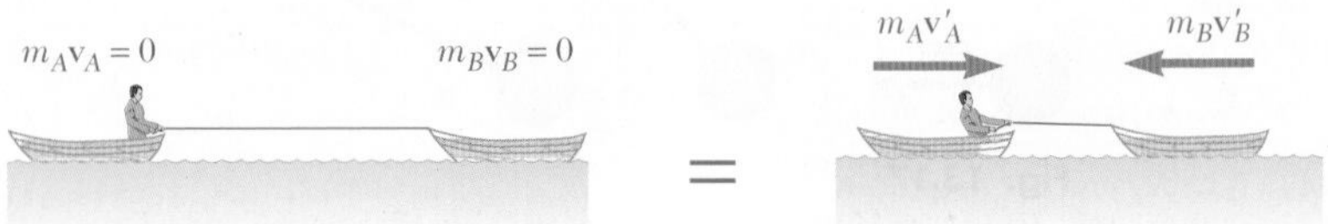

Fig. 13.18

of the water is neglected, the only external forces acting on the boats are their weights and the buoyant forces exerted on them. Since these forces are balanced, we write

$$\Sigma m\mathbf{v}_1 = \Sigma m\mathbf{v}_2$$
$$0 = m_A\mathbf{v}'_A + m_B\mathbf{v}'_B$$

where $\mathbf{v}'_A$ and $\mathbf{v}'_B$ represent the velocities of the boats after a finite interval of time. The equation obtained indicates that the boats move in opposite directions (toward each other) with velocities inversely proportional to their masses.‡

†We should note the difference between this statement and the corresponding statement made in Sec. 13.4 regarding the work of the forces of action and reaction between several particles. While the sum of the impulses of these forces is always zero, the sum of their work is zero only under special circumstances, e.g., when the various bodies involved are connected by inextensible cords or links and are thus constrained to move through equal distances.

‡Blue equals signs are used in Fig. 13.18 and throughout the remainder of this chapter to express that two systems of vectors are *equipollent*, i.e., that they have the same resultant and moment resultant (cf. Sec. 3.19). Red equals signs will continue to be used to indicate that two systems of vectors are *equivalent*, i.e., that they have the same effect. This and the concept of conservation of momentum for a system of particles will be discussed in greater detail in Chap. 14.

13.11 IMPULSIVE MOTION

A force acting on a particle during a very short time interval that is large enough to produce a definite change in momentum is called an *impulsive force* and the resulting motion is called an *impulsive motion*. For example, when a baseball is struck, the contact between bat and ball takes place during a very short time interval Δt. But the average value of the force $\mathbf{F}$ exerted by the bat on the ball is very large, and the resulting impulse $\mathbf{F}\ \Delta t$ is large enough to change the sense of motion of the ball (Fig. 13.19).

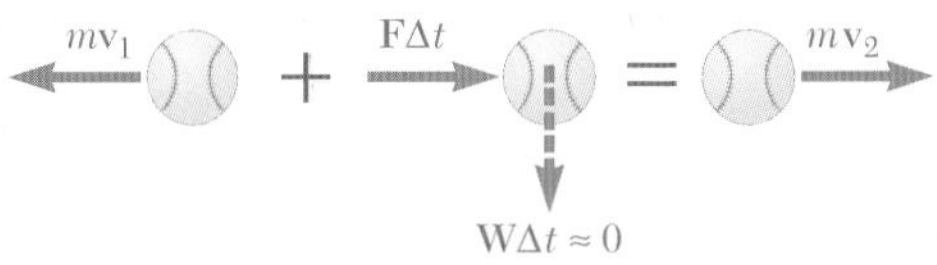

Fig. 13.19

When impulsive forces act on a particle, Eq. (13.32) becomes

$$m\mathbf{v}_1 + \Sigma\mathbf{F}\ \Delta t = m\mathbf{v}_2 \tag{13.35}$$

Any force which is not an impulsive force may be neglected, since the corresponding impulse $\mathbf{F}\ \Delta t$ is very small. *Nonimpulsive forces* include the weight of the body, the force exerted by a spring, or any other force which is *known* to be small compared with an impulsive force. Unknown reactions may or may not be impulsive; their impulses should therefore be included in Eq. (13.35) as long as they have not been proved negligible. The impulse of the weight of the baseball considered above, for example, may be neglected. If the motion of the bat is analyzed, the impulse of the weight of the bat can also be neglected. The impulses of the reactions of the player's hands on the bat, however, should be included; these impulses will not be negligible if the ball is incorrectly hit.

We note that the method of impulse and momentum is particularly effective in the analysis of the impulsive motion of a particle, since it involves only the initial and final velocities of the particle and the impulses of the forces exerted on the particle. The direct application of Newton's second law, on the other hand, would require the determination of the forces as functions of the time and the integration of the equations of motion over the time interval Δt.

In the case of the impulsive motion of several particles, Eq. (13.33) can be used. It reduces to

$$\Sigma m\mathbf{v}_1 + \Sigma\mathbf{F}\ \Delta t = \Sigma m\mathbf{v}_2 \tag{13.36}$$

where the second term involves only impulsive, external forces. If all the external forces acting on the various particles are nonimpulsive, the second term in Eq. (13.36) vanishes and this equation reduces to Eq. (13.34). We write

$$\Sigma m\mathbf{v}_1 = \Sigma m\mathbf{v}_2 \tag{13.34}$$

which expresses that the total momentum of the particles is conserved. This situation occurs, for example, when two particles which are moving freely collide with one another. We should note, however, that while the total momentum of the particles is conserved, their total energy is generally *not* conserved. Problems involving the collision or *impact* of two particles will be discussed in detail in Secs. 13.12 through 13.14.

SAMPLE PROBLEM 13.10

An automobile weighing 1800 kg is driven down a 5° incline at a speed of 100 km/h when the brakes are applied, causing a constant total braking force (applied by the road on the tires) of 7000-N. Determine the time required for the automobile to come to a stop.

SOLUTION

We apply the principle of impulse and momentum. Since each force is constant in magnitude and direction, each corresponding impulse is equal to the product of the force and of the time interval t.

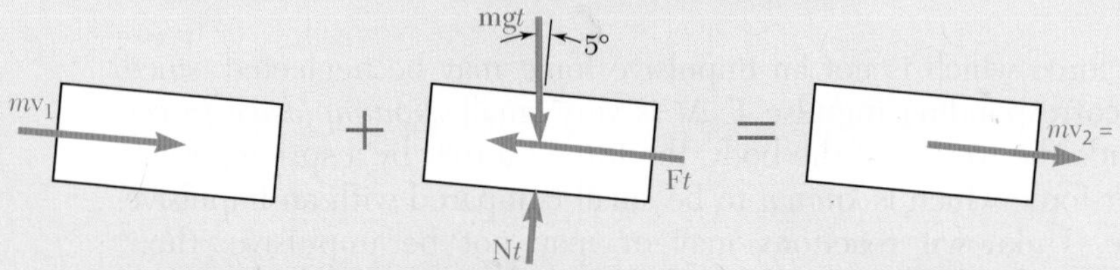

$$m\mathbf{v}_1 + \Sigma\,\mathbf{Imp}_{1\rightarrow 2} = m\mathbf{v}_2$$

$+\searrow$ components: $mv_1 + (\text{mg} \sin 5°)t - Ft = 0$

$$100 \text{ km/h} = 100 \times \frac{1000 \text{ m}}{1 \text{ km}} \times \frac{1 \text{ h}}{3600 \text{ s}} = 27.78 \text{ m/s}$$

$$(1800 \text{ kg})(27.78 \text{ m/s}) + (1800 \text{ kg})(9.81 \text{ m/s}^2) \sin 5° t - (7000 \text{ N})t = 0$$

$$t = 9.16 \text{ s} \blacktriangleleft$$

SAMPLE PROBLEM 13.11

A 120-g baseball is pitched with a velocity of 24 m/s toward a batter. After the ball is hit by the bat B, it has a velocity of 36 m/s in the direction shown. If the bat and ball are in contact 0.015 s, determine the average impulsive force exerted on the ball during the impact.

SOLUTION

We apply the principle of impulse and momentum to the ball. Since the weight of the ball is a nonimpulsive force, it can be neglected.

$$m\mathbf{v}_1 + \Sigma\,\mathbf{Imp}_{1\rightarrow 2} = m\mathbf{v}_2$$

$\xrightarrow{+} x$ components: $-mv_1 + F_x \Delta t = mv_2 \cos 40°$

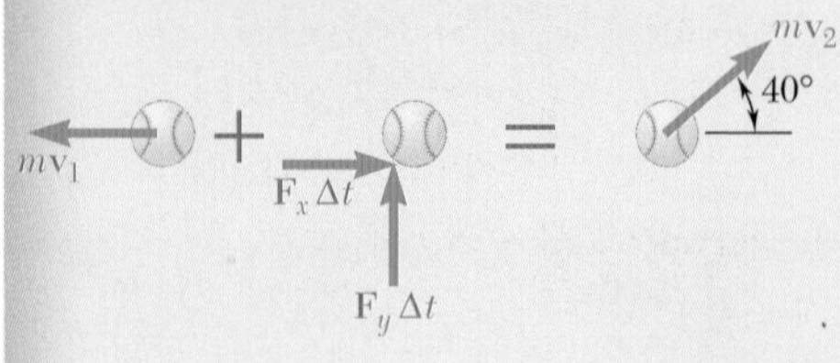

$$-(0.12 \text{ kg})(24 \text{ m/s}) + F_x(0.015 \text{ s}) = (0.12 \text{ kg})(36 \text{ m/s}) \cos 40°$$

$$F_x = +412.6 \text{ N}$$

$+\uparrow y$ components: $0 + F_y \Delta t = mv_2 \sin 40°$

$$F_y(0.015 \text{ s}) = (0.12 \text{ kg})(36 \text{ m/s}) \sin 40°$$

$$F_y = +185.1 \text{N}$$

From its components F_x and F_y we determine the magnitude and direction of the force $\mathbf{F}$:

$$\mathbf{F} = 452 \text{ N} \measuredangle 24.2° \blacktriangleleft$$

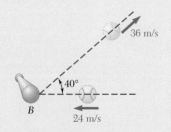

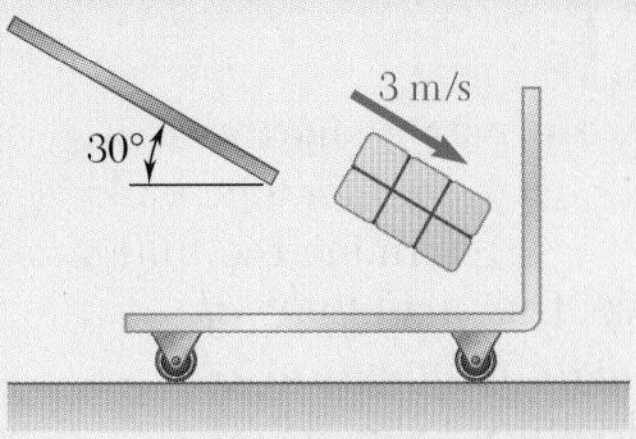

SAMPLE PROBLEM 13.12

A 10-kg package drops from a chute into a 25-kg cart with a velocity of 3 m/s. Knowing that the cart is initially at rest and can roll freely, determine (*a*) the final velocity of the cart, (*b*) the impulse exerted by the cart on the package, (*c*) the fraction of the initial energy lost in the impact.

SOLUTION

We first apply the principle of impulse and momentum to the package-cart system to determine the velocity $\mathbf{v}_2$ of the cart and package. We then apply the same principle to the package alone to determine the impulse $\mathbf{F}\,\Delta t$ exerted on it.

a. Impulse-Momentum Principle: Package and Cart

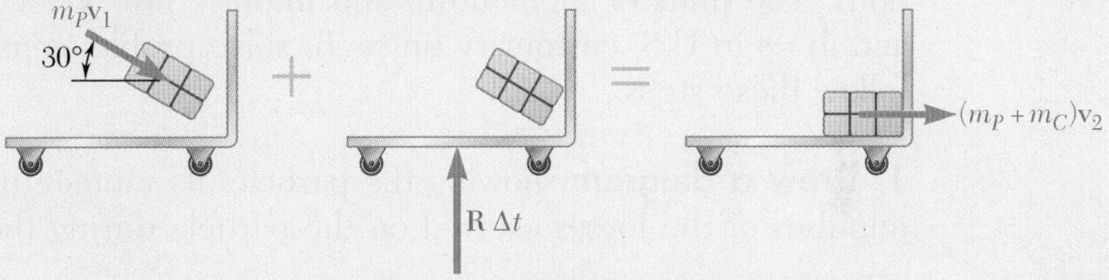

$$m_P\mathbf{v}_1 + \Sigma\,\mathbf{Imp}_{1\rightarrow 2} = (m_P + m_C)\mathbf{v}_2$$

$\overset{+}{\rightarrow} x$ components:

$$m_P v_1 \cos 30° + 0 = (m_P + m_C)v_2$$

$$(10\text{ kg})(3\text{ m/s})\cos 30° = (10\text{ kg} + 25\text{ kg})v_2$$

$$\mathbf{v}_2 = 0.742\text{ m/s}\rightarrow \quad ◀$$

We note that the equation used expresses conservation of momentum in the x direction.

b. Impulse-Momentum Principle: Package

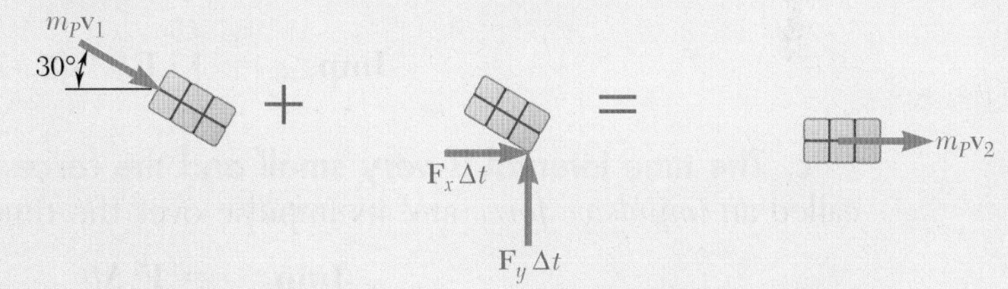

$$m_P\mathbf{v}_1 + \Sigma\,\mathbf{Imp}_{1\rightarrow 2} = m_P\mathbf{v}_2$$

$\overset{+}{\rightarrow} x$ components:

$$(10\text{ kg})(3\text{ m/s})\cos 30° + F_x\,\Delta t = (10\text{ kg})(0.742\text{ m/s})$$

$$F_x\,\Delta t = -18.56\text{ N}\cdot\text{s}$$

$+\uparrow y$ components:

$$-m_P v_1 \sin 30° + F_y\,\Delta t = 0$$

$$-(10\text{ kg})(3\text{ m/s})\sin 30° + F_y\,\Delta t = 0$$

$$F_y\,\Delta t = +15\text{ N}\cdot\text{s}$$

The impulse exerted on the package is $\quad \mathbf{F}\,\Delta t = 23.9\text{ N}\cdot\text{s} \measuredangle 38.9° \quad ◀$

c. Fraction of Energy Lost. The initial and final energies are

$$T_1 = \tfrac{1}{2}m_P v_1^2 = \tfrac{1}{2}(10\text{ kg})(3\text{ m/s})^2 = 45\text{ J}$$

$$T_2 = \tfrac{1}{2}(m_P + m_C)v_2^2 = \tfrac{1}{2}(10\text{ kg} + 25\text{ kg})(0.742\text{ m/s})^2 = 9.63\text{ J}$$

The fraction of energy lost is $\quad \dfrac{T_1 - T_2}{T_1} = \dfrac{45\text{J} - 9.63\text{J}}{45\text{J}} = 0.786 \quad ◀$

SOLVING PROBLEMS ON YOUR OWN

In this lesson we integrated Newton's second law to derive the *principle of impulse and momentum* for a particle. Recalling that the *linear momentum* of a particle was defined as the product of its mass m and its velocity $\mathbf{v}$ [Sec. 12.3], we wrote

$$m\mathbf{v}_1 + \Sigma\ \mathbf{Imp}_{1\to 2} = m\mathbf{v}_2 \tag{13.32}$$

This equation expresses that the linear momentum $m\mathbf{v}_2$ of a particle at time t_2 can be obtained by adding to its linear momentum $m\mathbf{v}_1$ at time t_1 the *impulses* of the forces exerted on the particle during the time interval t_1 to t_2. For computing purposes, the momenta and impulses may be expressed in terms of their rectangular components, and Eq. (13.32) can be replaced by the equivalent scalar equations. The units of momentum and impulse are N · s in the SI system of units and lb · s in U.S. customary units. To solve problems using this equation you can follow these steps:

1. Draw a diagram showing the particle, its momentum at t_1 and at t_2, and the impulses of the forces exerted on the particle during the time interval t_1 to t_2.

2. Calculate the impulse of each force, expressing it in terms of its rectangular components if more than one direction is involved. You may encounter the following cases:

a. The time interval is finite and the force is constant.

$$\mathbf{Imp}_{1\to 2} = \mathbf{F}(t_2 - t_1)$$

b. The time interval is finite and the force is a function of *t*.

$$\mathbf{Imp}_{1\to 2} = \int_{t_1}^{t_2} \mathbf{F}(t)\, dt$$

c. The time interval is very small and the force is very large. The force is called an *impulsive force* and its impulse over the time interval $t_2 - t_1 = \Delta t$ is

$$\mathbf{Imp}_{1\to 2} = \mathbf{F}\ \Delta t$$

Note that this impulse is *zero for a nonimpulsive force* such as the *weight* of a body, the force exerted by a *spring*, or any other force which is known to be small by comparison with the impulsive forces. Unknown reactions, however, *cannot be assumed* to be nonimpulsive and their impulses should be taken into account.

3. Substitute the values obtained for the impulses into Eq. (13.32) or into the equivalent scalar equations. You will find that the forces and velocities in the problems of this lesson are contained in a plane. You will, therefore, write two scalar equations and solve these equations for *two unknowns*. These unknowns may be a *time* [Sample Prob. 13.10], a *velocity* and an *impulse* [Sample Prob. 13.12], or an *average impulsive force* [Sample Prob. 13.11].

4. When several particles are involved, a separate diagram should be drawn for each particle, showing the initial and final momentum of the particle, as well as the impulses of the forces exerted on the particle.

a. It is usually convenient, however, to first consider a diagram including all the particles. This diagram leads to the equation

$$\Sigma m\mathbf{v}_1 + \Sigma\ \mathbf{Imp}_{1\rightarrow 2} = \Sigma m\mathbf{v}_2 \qquad (13.33)$$

where the impulses of *only the forces external to the system* need be considered. Therefore, the two equivalent scalar equations will not contain any of the impulses of the unknown internal forces.

b. If the sum of the impulses of the external forces is zero, Eq. (13.33) reduces to

$$\Sigma m\mathbf{v}_1 = \Sigma m\mathbf{v}_2 \qquad (13.34)$$

which expresses that *the total momentum of the particles is conserved*. This occurs either if the resultant of the external forces is zero or, when the time interval Δt is very short (impulsive motion), if all the external forces are nonimpulsive. Keep in mind, however, that the total momentum may be conserved *in one direction*, but not in another [Sample Prob. 13.12].

PROBLEMS

CONCEPT QUESTIONS

13.CQ4 A large insect impacts the front windshield of a sports car traveling down a road. Which of the following statements is true during the collision?

a. The car exerts a greater force on the insect than the insect exerts on the car.

b. The insect exerts a greater force on the car than the car exerts on the insect.

c. The car exerts a force on the insect, but the insect does not exert a force on the car.

d. The car exerts the same force on the insect as the insect exerts on the car.

e. Neither exerts a force on the other; the insect gets smashed simply because it gets in the way of the car.

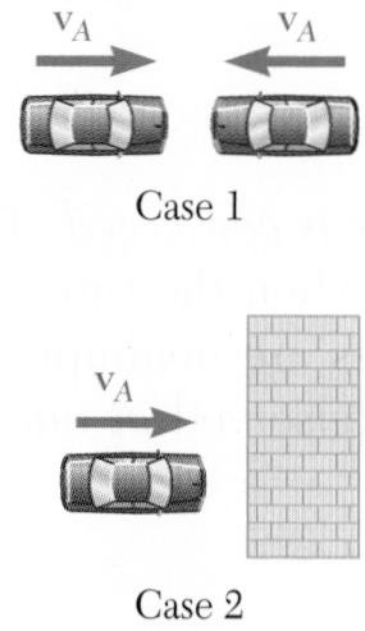

Fig. P13.CQ5

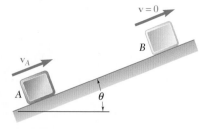

13.CQ5 The expected damages associated with two types of perfectly plastic collisions are to be compared. In the first case, two identical cars traveling at the same speed impact each other head-on. In the second case, the car impacts a massive concrete wall. In which case would you expect the car to be more damaged?

a. Case 1

b. Case 2

c. The same damage in each case

IMPULSE-MOMENTUM PRACTICE PROBLEMS

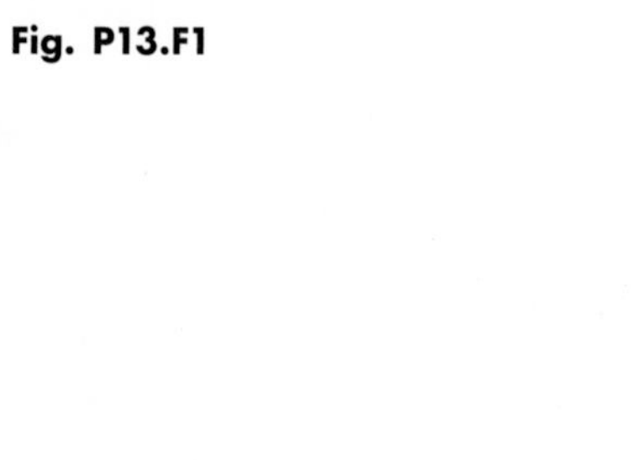

Fig. P13.F1

13.F1 The initial velocity of the block in position A is 10 m/s. The coefficient of kinetic friction between the block and the plane is $\mu_k = 0.30$. Draw the impulse-momentum diagram that can be used to determine the time it takes for the block to reach B with zero velocity, if $\theta = 20°$.

13.F2 A 2-kg collar which can slide on a frictionless vertical rod is acted upon by a force **P** which varies in magnitude as shown. Knowing that the collar is initially at rest, draw the impulse-momentum diagram that can be used to determine its velocity at $t = 3$ s.

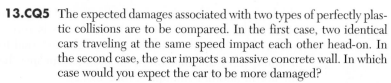

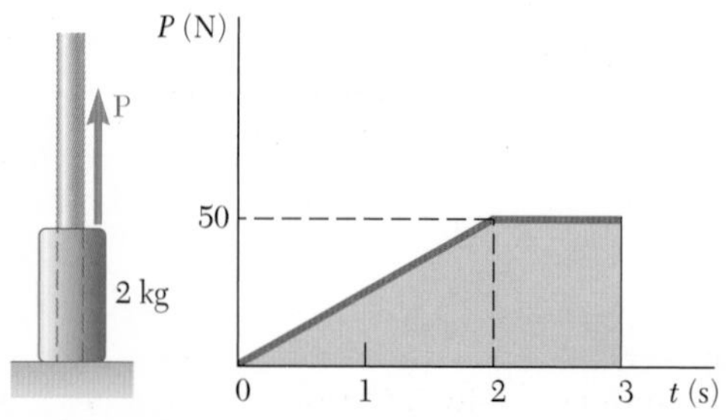

Fig. P13.F2

13.F3 The 15-kg suitcase A has been propped up against one end of a 40-kg luggage carrier B and is prevented from sliding down by other luggage. When the luggage is unloaded and the last heavy trunk is removed from the carrier, the suitcase is free to slide down, causing the 40-kg carrier to move to the left with a velocity v_B of magnitude 0.8 m/s. Neglecting friction, draw the impulse-momentum diagrams that can be used to determine (*a*) the velocity of A as it rolls on the carrier, (*b*) the velocity of the carrier after the suitcase hits the right side of the carrier without bouncing back.

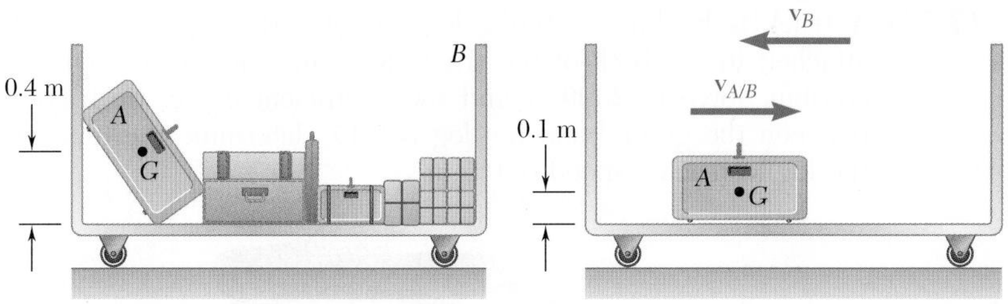

Fig. P13.F3

13.F4 Car A was traveling west at a speed of 15 m/s and car B was traveling north at an unknown speed when they slammed into each other at an intersection. Upon investigation it was found that after the crash the two cars got stuck and skidded off at an angle of 50° north of east. Knowing the masses of A and B are m_A and m_B, respectively, draw the impulse-momentum diagram that can be used to determine the velocity of B before impact.

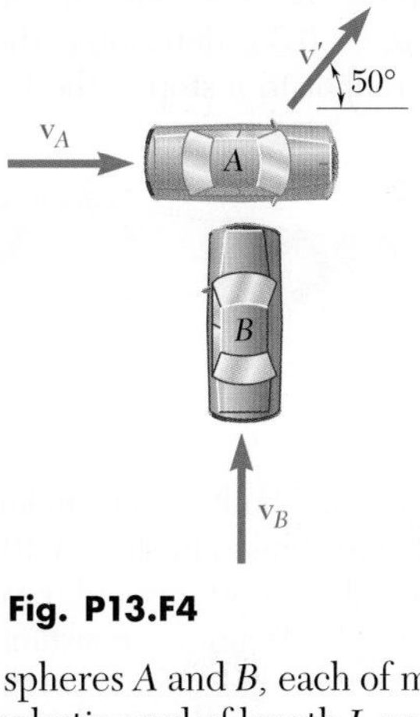

Fig. P13.F4

13.F5 Two identical spheres A and B, each of mass m, are attached to an inextensible inelastic cord of length L and are resting at a distance a from each other on a frictionless horizontal surface. Sphere B is given a velocity $\mathbf{v}_0$ in a direction perpendicular to line AB and moves it without friction until it reaches B' where the cord becomes taut. Draw the impulse-momentum diagram that can be used to determine the magnitude of the velocity of each sphere immediately after the cord has become taut.

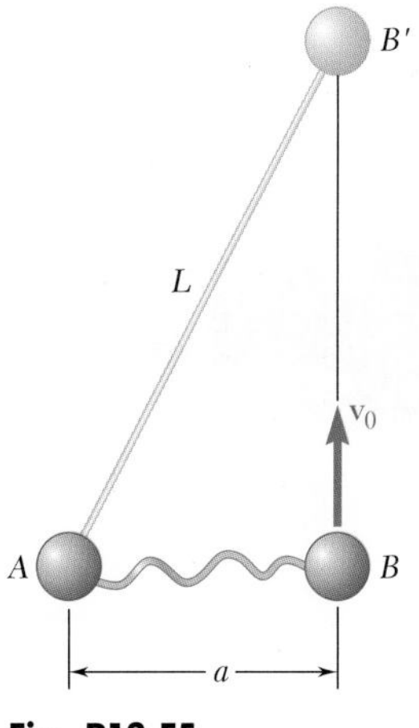

Fig. P13.F5

END-OF-SECTION PROBLEMS

13.119 A 35 000-Mg ocean liner has an initial velocity of 4 km/h. Neglecting the frictional resistance of the water, determine the time required to bring the liner to rest by using a single tugboat which exerts a constant force of 150 kN.

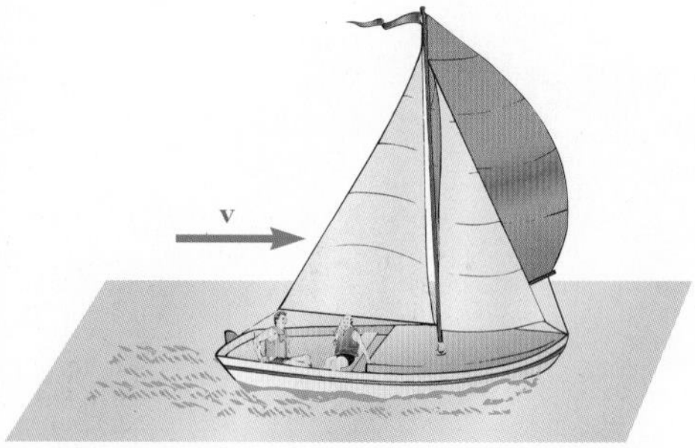

Fig. P13.121

13.120 A 1200-kg automobile is moving at a speed of 90 km/h when the brakes are fully applied, causing all four wheels to skid. Determine the time required to stop the automobile (*a*) on dry pavement ($\mu_k = 0.75$), (*b*) on an icy road ($\mu_k = 0.10$).

13.121 A sailboat of mass 500 kg with its occupants is running down wind at 12 km/h when its spinnaker is raised to increase its speed. Determine the net force provided by the spinnaker over the 10-s interval that it takes for the boat to reach a speed of 18 km/h.

13.122 A truck is hauling a 300-kg log out of a ditch using a winch attached to the back of the truck. Knowing the winch applies a constant force of 2500 N and the coefficient of kinetic friction between the ground and the log is 0.45, determine the time for the log to reach a speed of 0.5 m/s.

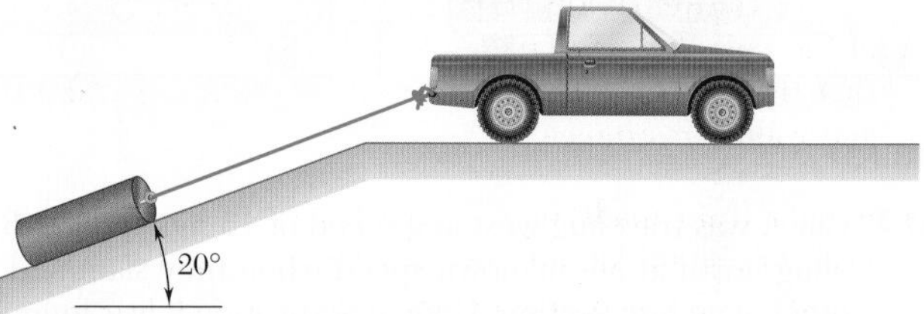

Fig. *P13.122*

13.123 A truck is traveling down a road with a 3-percent grade at a speed of 72 km/h when the brakes are applied. Knowing the coefficients of friction between the load and the flatbed trailer shown are $\mu_s = 0.40$ and $\mu_k = 0.35$, determine the shortest time in which the rig can be brought to a stop if the load is not to shift.

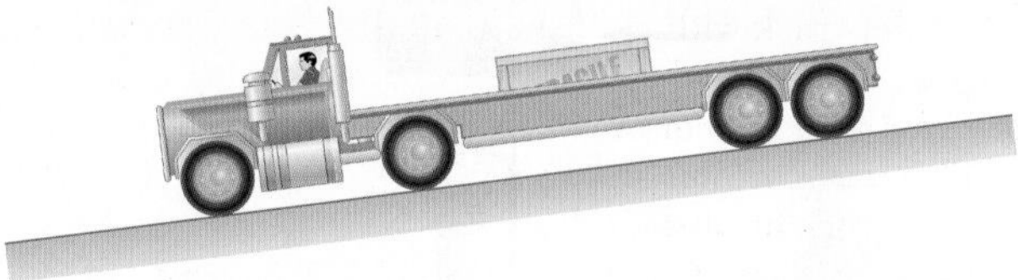

Fig. P13.123

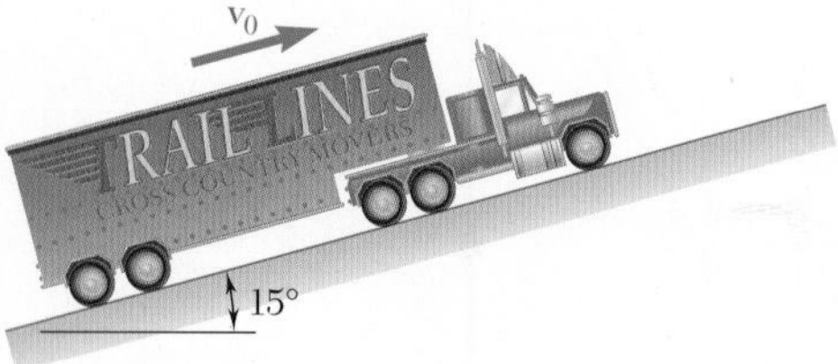

Fig. P13.124

13.124 Steep safety ramps are built beside mountain highways to enable vehicles with defective brakes to stop. A 10,000 kg truck enters a 15° ramp at a high speed $v_0 = 30$ m/s and travels for 6 s before its speed is reduced to 10 m/s. Assuming constant deceleration, determine (*a*) the magnitude of the braking force, (*b*) the additional time required for the truck to stop. Neglect air resistance and rolling resistance.

13.125 Baggage on the floor of the baggage car of a high-speed train is not prevented from moving other than by friction. Determine the smallest allowable value of the coefficient of static friction between a trunk and the floor of the car if the trunk is not to slide when the train decreases its speed at a constant rate from 200 km/h to 90 km/h in a time interval of 12 s.

13.126 A 2-kg particle is acted upon by the force, expressed in newtons, $\mathbf{F} = (8 - 6t)\mathbf{i} + (4 - t^2)\mathbf{j} + (4 + t)\mathbf{k}$. Knowing that the velocity of the particle is $\mathbf{v} = (150 \text{ m/s})\mathbf{i} + (100 \text{ m/s})\mathbf{j} - (250 \text{ m/s})\mathbf{k}$ at $t = 0$, determine (*a*) the time at which the velocity of the particle is parallel to the *yz* plane, (*b*) the corresponding velocity of the particle.

13.127 A truck is traveling down a road with a 4-percent grade at a speed of 80 km/h when its brakes are applied to slow it down to 30 km/h. An antiskid braking system limits the braking force to a value at which the wheels of the truck are just about to slide. Knowing that the coefficient of static friction between the road and the wheels is 0.60, determine the shortest time needed for the truck to slow down.

13.128 Skid marks on a drag race track indicate that the rear (drive) wheels of a car slip for the first 20 m of the 400-m track. (*a*) Knowing that the coefficient of kinetic friction is 0.60, determine the shortest possible time for the car to travel the initial 20-m portion of the track if it starts from rest with its front wheels just off the ground. (*b*) Determine the minimum time for the car to run the whole race if, after skidding for 20 m, the wheels roll without sliding for the remainder of the race. Assume for the rolling portion of the race that 65 percent of the weight is on the rear wheels and that the coefficient of static friction is 0.85. Ignore air resistance and rolling resistance.

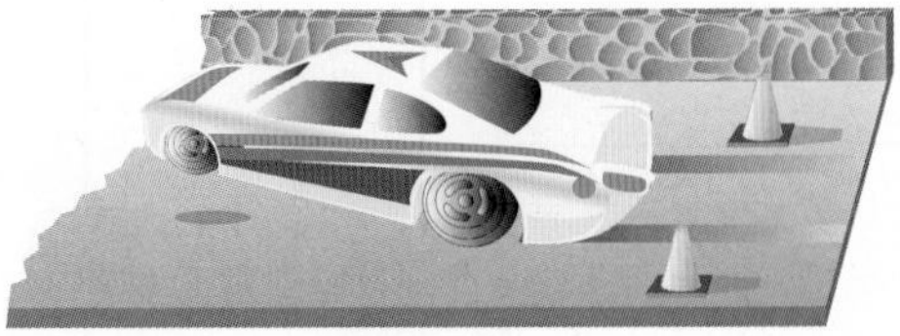

Fig. P13.128

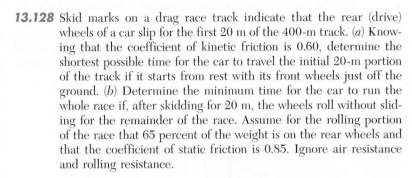

13.129 A light train made of two cars travels at 72 km/h. Car A has a mass of 18,000 kg, and car B has a mass of 13,000 kg. When the brakes are applied, a constant braking force of 21.5 kN is applied to each car. Determine (*a*) the time required for the train to stop after the brakes are applied, (*b*) the force in the coupling between the cars while the train is slowing down.

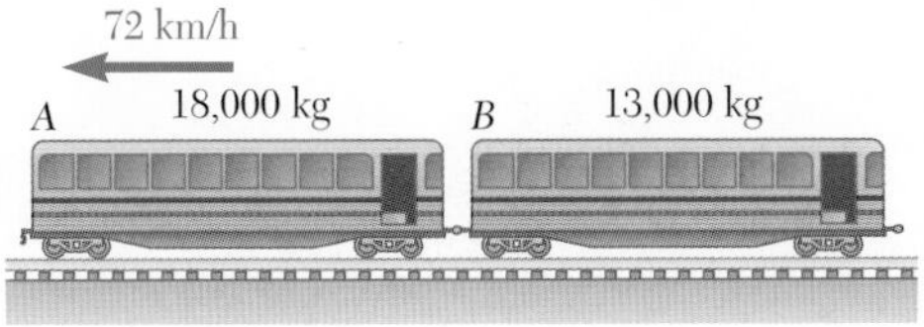

Fig. P13.129

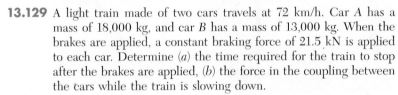

13.130 Solve Prob. 13.129, assuming that a constant braking force of 21.5 kN is applied to car B but that the brakes on car A are not applied.

13.131 A trailer truck with a 2000-kg cab and an 8000-kg trailer is traveling on a level road at 90 km/h. The brakes on the trailer fail and the antiskid system of the cab provides the largest possible force which will not cause the wheels of the cab to slide. Knowing that the coefficient of static friction is 0.65, determine (*a*) the shortest time for the rig to come to a stop, (*b*) the force in the coupling during that time.

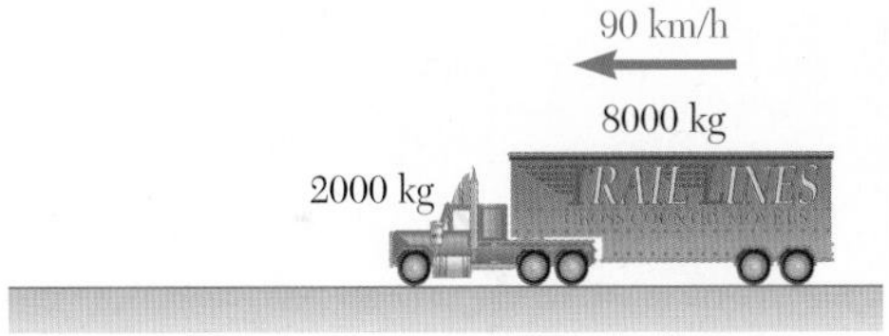

Fig. P13.131

13.132 The system shown is at rest when a constant 150-N force is applied to collar B. Neglecting the effect of friction, determine (*a*) the time at which the velocity of collar B will be 2.5 m/s to the left, (*b*) the corresponding tension in the cable.

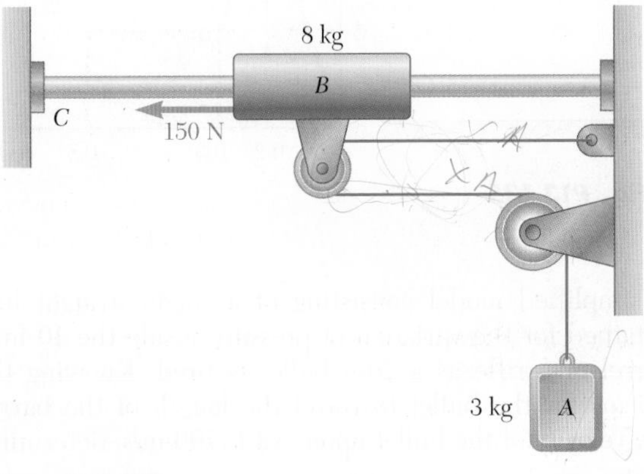

Fig. P13.132

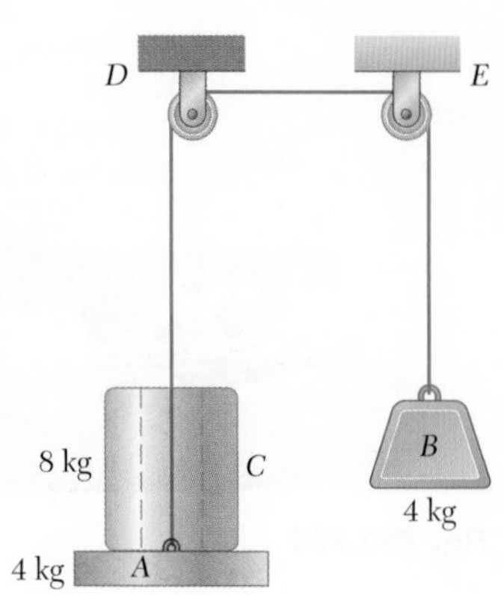

Fig. P13.133

13.133 An 8-kg cylinder C rests on a 4-kg platform A supported by a cord which passes over the pulleys D and E and is attached to a 4-kg block B. Knowing that the system is released from rest, determine (*a*) the velocity of block B after 0.8 s, (*b*) the force exerted by the cylinder on the platform.

13.134 An estimate of the expected load on over-the-shoulder seat belts is to be made before designing prototype belts that will be evaluated in automobile crash tests. Assuming that an automobile traveling at 72 km/h is brought to a stop in 110 ms, determine (*a*) the average impulsive force exerted by a 100-kg man on the belt, (*b*) the maximum force F_m exerted on the belt if the force-time diagram has the shape shown.

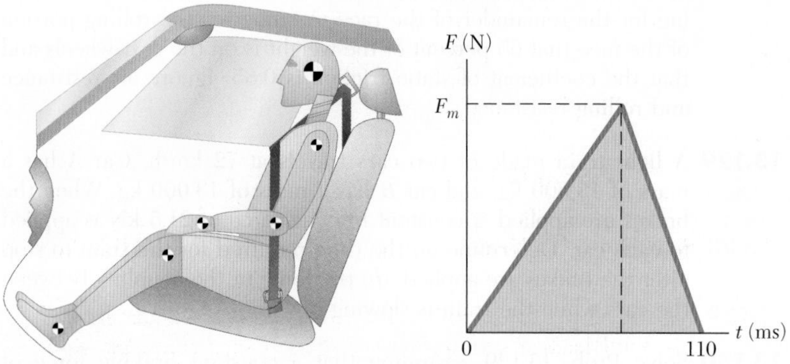

Fig. P13.142

13.135 A 60-g model rocket is fired vertically. The engine applies a thrust **P** which varies in magnitude as shown. Neglecting air resistance and the change in mass of the rocket, determine (*a*) the maximum speed of the rocket as it goes up, (*b*) the time for the rocket to reach its maximum elevation.

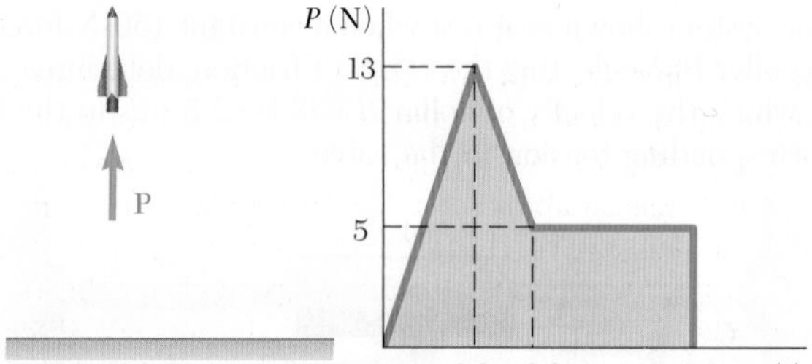

Fig. P13.135

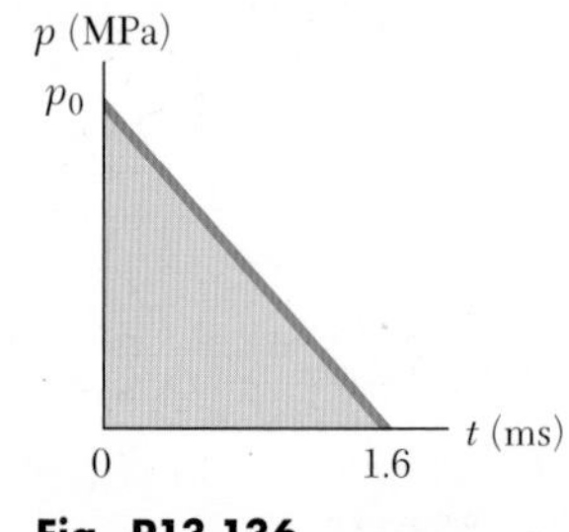

Fig. P13.136

13.136 A simplified model consisting of a single straight line is to be obtained for the variation of pressure inside the 10-mm-diameter barrel of a rifle as a 20-g bullet is fired. Knowing that it takes 1.6 ms for the bullet to travel the length of the barrel and that the velocity of the bullet upon exit is 700 m/s, determine the value of p_0.

13.137 A 60-kg block initially at rest is acted upon by a force **P** which varies as shown. Knowing that the coefficients of friction between the block and the horizontal surface are $\mu_s = 0.50$ and $\mu_k = 0.40$, determine (*a*) the time at which the block will start moving, (*b*) the maximum speed reached by the block, (*c*) the time at which the block will stop moving.

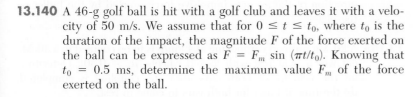

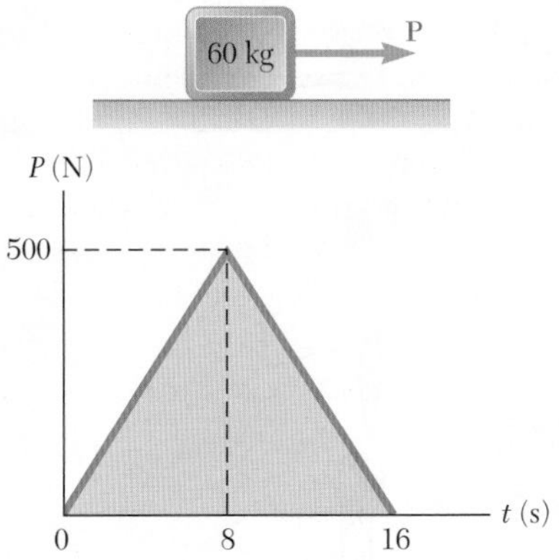

Fig. P13.137

13.138 Solve Prob. 13.137, assuming that the mass of the block is 90 kg.

13.139 A baseball player catching a ball can soften the impact by pulling his hand back. Assuming that a 150-g ball reaches his glove at 140 km/h and that the player pulls his hand back during the impact at an average speed of 9 m/s over a distance of 150 mm, bringing the ball to a stop, determine the average impulsive force exerted on the player's hand.

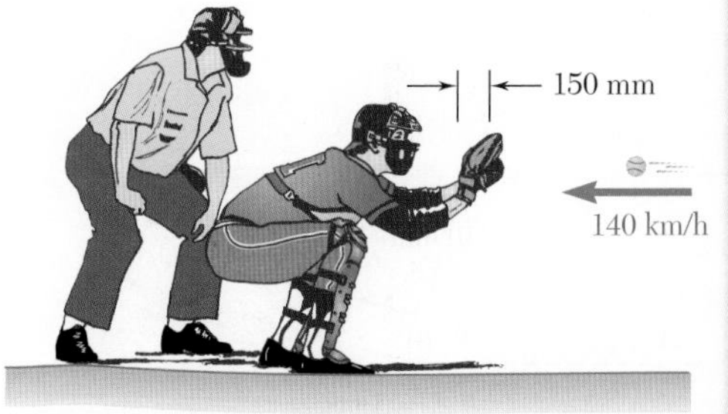

Fig. P13.139

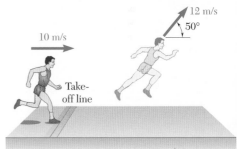

13.140 A 46-g golf ball is hit with a golf club and leaves it with a velocity of 50 m/s. We assume that for $0 \le t \le t_0$, where t_0 is the duration of the impact, the magnitude F of the force exerted on the ball can be expressed as $F = F_m \sin(\pi t/t_0)$. Knowing that $t_0 = 0.5$ ms, determine the maximum value F_m of the force exerted on the ball.

13.141 The triple jump is a track-and-field event in which an athlete gets a running start and tries to leap as far as he can with a hop, step, and jump. Shown in the figure is the initial hop of the athlete. Assuming that he approaches the takeoff line from the left with a horizontal velocity of 10 m/s, remains in contact with the ground for 0.18 s, and takes off at a 50° angle with a velocity of 12 m/s, determine the vertical component of the average impulsive force exerted by the ground on his foot. Give your answer in terms of the weight W of the athlete.

Fig. P13.141

13.142 The last segment of the triple jump track-and-field event is the jump, in which the athlete makes a final leap, landing in a sand-filled pit. Assuming that the velocity of a 80-kg athlete just before landing is 9 m/s at an angle of 35° with the horizontal and that the athlete comes to a complete stop in 0.22 s after landing, determine the horizontal component of the average impulsive force exerted on his feet during landing.

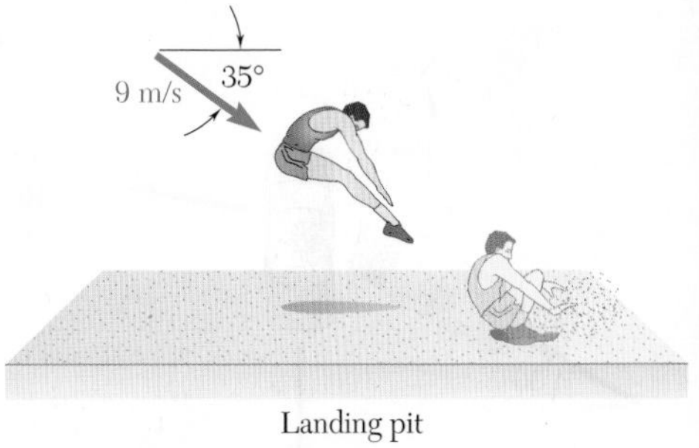

Fig. P13.142

13.143 The design for a new cementless hip implant is to be studied using an instrumented implant and a fixed simulated femur. Assuming the punch applies an average force of 2 kN over a time of 2 ms to the 200-g implant, determine (*a*) the velocity of the implant immediately after impact, (*b*) the average resistance of the implant to penetration if the implant moves 1 mm before coming to rest.

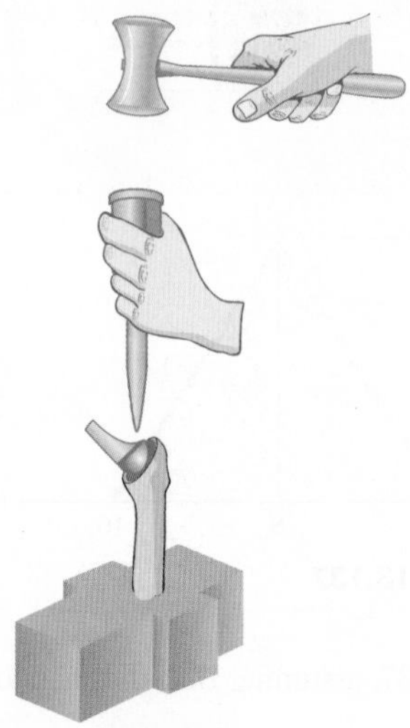

Fig. *P13.143*

13.144 A 25-g steel-jacketed bullet is fired horizontally with a velocity of 600 m/s and ricochets off a steel plate along the path *CD* with a velocity of 400 m/s. Knowing that the bullet leaves a 10-mm scratch on the plate and assuming that its average speed is 500 m/s while it is in contact with the plate, determine the magnitude and direction of the average impulsive force exerted by the bullet on the plate.

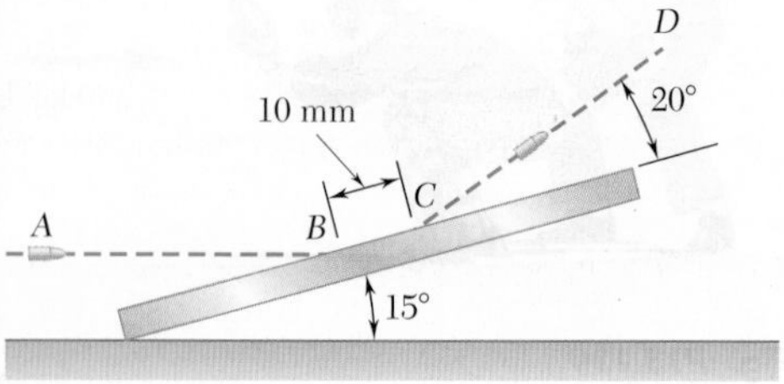

Fig. *P13.144*

13.145 A 20-Mg railroad car moving at 4 km/h is to be coupled to a 40-Mg car which is at rest with locked wheels ($\mu_k = 0.30$). Determine (*a*) the velocity of both cars after the coupling is completed, (*b*) the time it takes for both cars to come to rest.

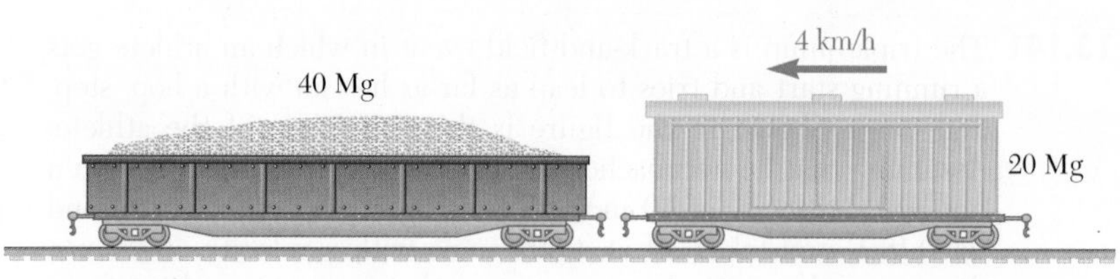

Fig. P13.145

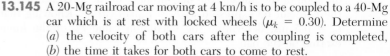

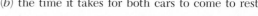

13.146 At an intersection car *B* was traveling south and car *A* was traveling 30° north of east when they slammed into each other. Upon investigation it was found that after the crash the two cars got stuck and skidded off at an angle of 10° north of east. Each driver claimed that he was going at the speed limit of 50 km/h and that he tried to slow down but couldn't avoid the crash because the other driver was going a lot faster. Knowing that the masses of cars *A* and *B* were 1500 kg and 1200 kg, respectively, determine (*a*) which car was going faster, (*b*) the speed of the faster of the two cars if the slower car was traveling at the speed limit.

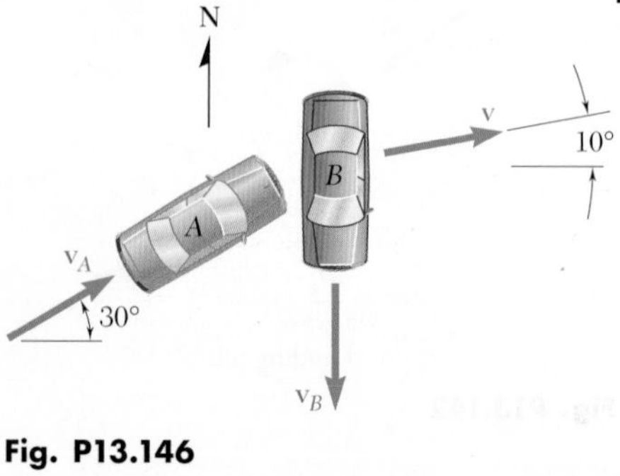

Fig. P13.146

13.147 The 650-kg hammer of a drop-hammer pile driver falls from a height of 1.2 m onto the top of a 140-kg pile, driving it 110 mm into the ground. Assuming perfectly plastic impact ($e = 0$), determine the average resistance of the ground to penetration.

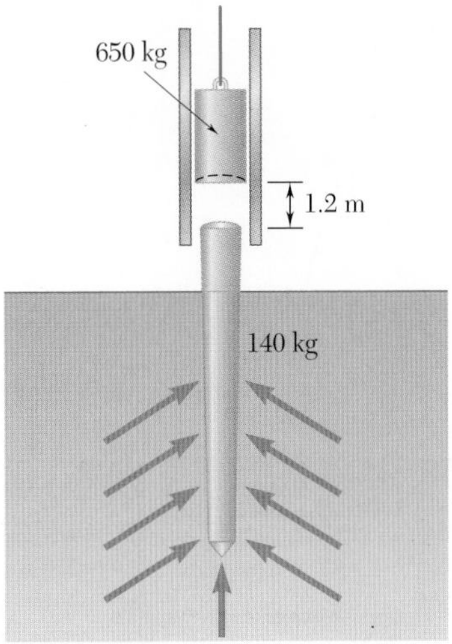

Fig. P13.147

13.148 A small rivet connecting two pieces of sheet metal is being clinched by hammering. Determine the impulse exerted on the rivet and the energy absorbed by the rivet under each blow, knowing that the head of the hammer has a mass of 750 g and that it strikes the rivet with a velocity of 6 m/s. Assume that the hammer does not rebound and that the anvil is supported by springs and (*a*) has an infinite mass (rigid support), (*b*) has a mass of 4.5 kg.

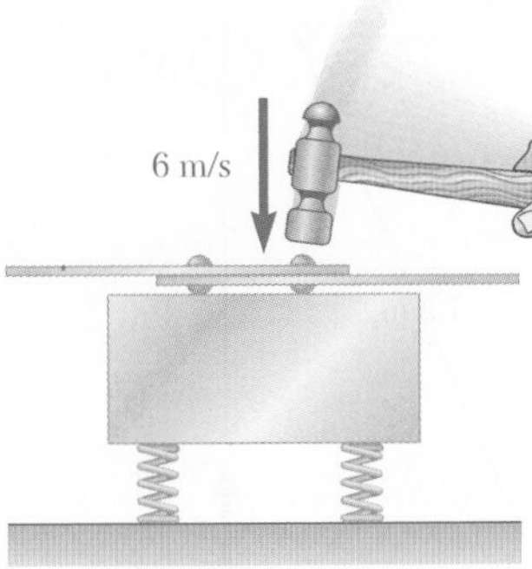

Fig. P13.148

13.149 Bullet *B* has a mass 15 g and blocks *A* and *C* both have a mass 2 kg. The coefficient of friction between the blocks and the plane is $\mu_k = 0.25$. Initially the bullet is moving at v_0 and blocks *A* and *C* are at rest (Fig. 1). After the bullet passes through *A* it becomes embedded in block *C* and all three objects come to stop in the positions shown (Fig. 2). Determine the initial speed of the bullet v_0.

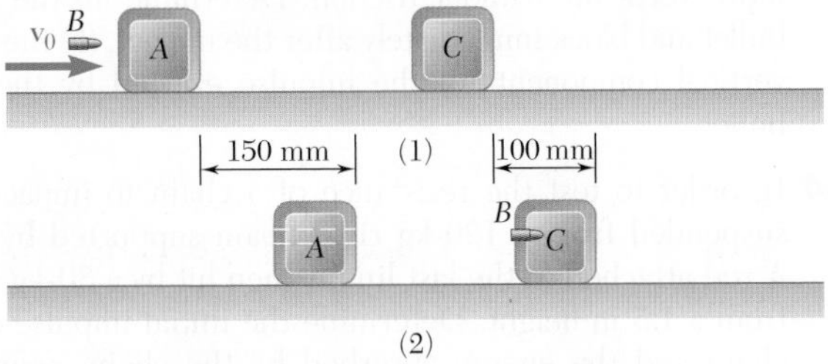

Fig. P13.149

13.150 A 90-kg man and a 60-kg woman stand at opposite ends of a 150-kg boat, ready to dive, each with a 5-m/s velocity relative to the boat. Determine the velocity of the boat after they have both dived, if (*a*) the woman dives first, (*b*) the man dives first.

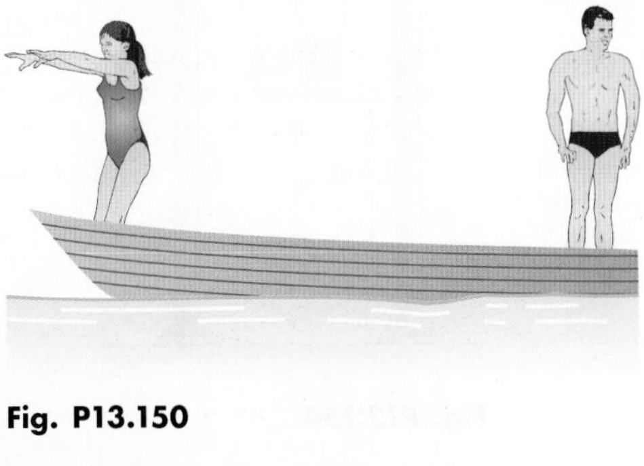

Fig. P13.150

13.151 A 75-g ball is projected from a height of 1.6 m with a horizontal velocity of 2 m/s and bounces from a 400-g smooth plate supported by springs. Knowing that the height of the rebound is 0.6 m, determine (*a*) the velocity of the plate immediately after the impact, (*b*) the energy lost due to the impact.

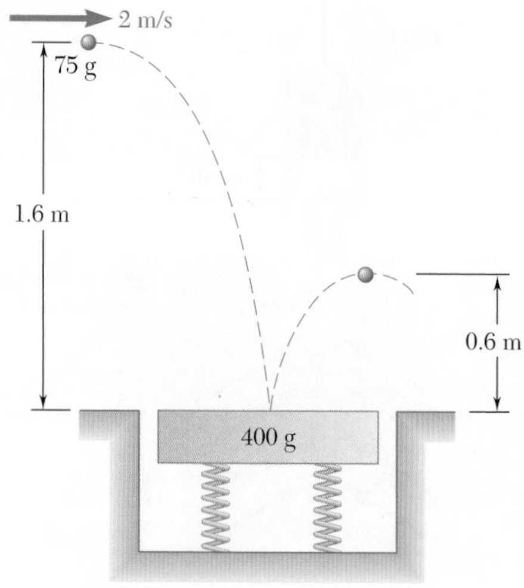

Fig. P13.151

13.152 A 2-kg sphere A is connected to a fixed point O by an inextensible cord of length 1.2 m. The sphere is resting on a frictionless horizontal surface at a distance of 0.5 m from O when it is given a velocity $\mathbf{v}_0$ in a direction perpendicular to line OA. It moves freely until it reaches position A', when the cord becomes taut. Determine the maximum allowable velocity $\mathbf{v}_0$ if the impulse of the force exerted on the cord is not to exceed 3 N · S.

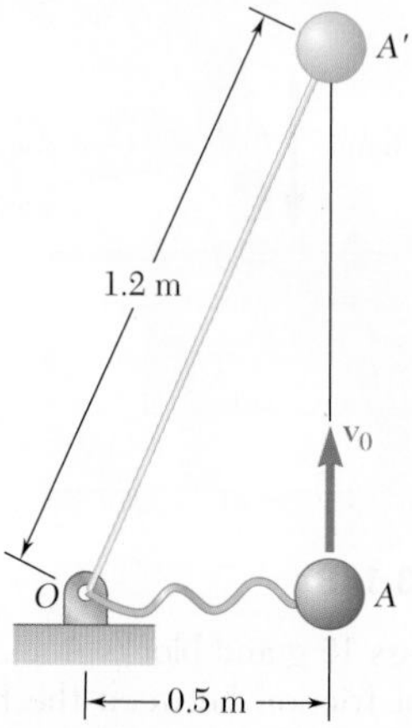

Fig. P13.152

13.153 A 25-g bullet is traveling with a velocity of 425 m/s when it impacts and becomes embedded in a 2.5-kg wooden block. The block can move vertically without friction. Determine (*a*) the velocity of the bullet and block immediately after the impact, (*b*) the horizontal and vertical components of the impulse exerted by the block on the bullet.

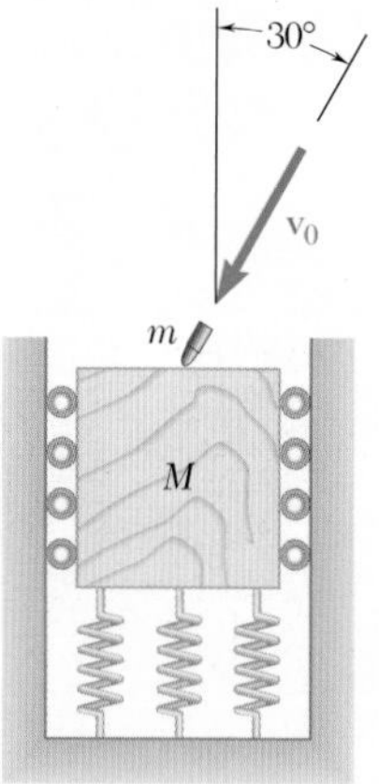

Fig. P13.153

13.154 In order to test the resistance of a chain to impact, the chain is suspended from a 120-kg rigid beam supported by two columns. A rod attached to the last link is then hit by a 30-kg block dropped from a 1.5 m height. Determine the initial impulse exerted on the chain and the energy absorbed by the chain, assuming that the block does not rebound from the rod and that the columns supporting the beam are (*a*) perfectly rigid, (*b*) equivalent to two perfectly elastic springs.

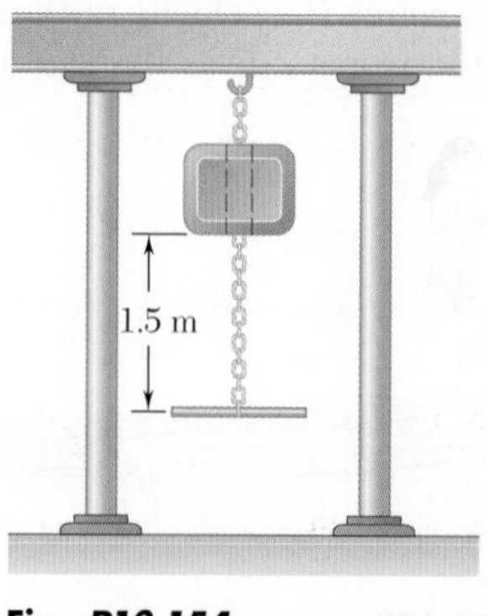

Fig. P13.154

13.12 IMPACT

A collision between two bodies which occurs in a very small interval of time and during which the two bodies exert relatively large forces on each other is called an *impact*. The common normal to the surfaces in contact during the impact is called the *line of impact*. If the mass centers on the two colliding bodies are located on this line, the impact is a *central impact*. Otherwise, the impact is said to be *eccentric*. Our present study will be limited to the central impact of two particles. The analysis of the eccentric impact of two rigid bodies will be considered later, in Sec. 17.12.

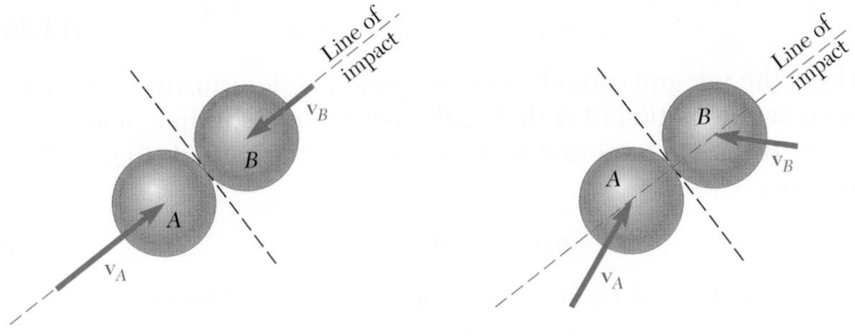

(*a*) Direct central impact (*b*) Oblique central impact

Fig. 13.20

If the velocities of the two particles are directed along the line of impact, the impact is said to be a *direct impact* (Fig. 13.20*a*). If either or both particles move along a line other than the line of impact, the impact is said to be an *oblique impact* (Fig. 13.20*b*).

13.13 DIRECT CENTRAL IMPACT

Consider two particles *A* and *B*, of mass m_A and m_B, which are moving in the same straight line and to the right with known velocities $\mathbf{v}_A$ and $\mathbf{v}_B$ (Fig. 13.21*a*). If $\mathbf{v}_A$ is larger than $\mathbf{v}_B$, particle *A* will eventually strike particle *B*. Under the impact, the two particles will *deform* and, at the end of the period of deformation, they will have the same velocity $\mathbf{u}$ (Fig. 13.21*b*). A period of *restitution* will then take place, at the end of which, depending upon the magnitude of the impact forces and upon the materials involved, the two particles either will have regained their original shape or will stay permanently deformed. Our purpose here is to determine the velocities $\mathbf{v}'_A$ and $\mathbf{v}'_B$ of the particles at the end of the period of restitution (Fig. 13.21*c*).

Considering first the two particles as a single system, we note that there is no impulsive, external force. Thus, the total momentum of the two particles is conserved, and we write

$$m_A\mathbf{v}_A + m_B\mathbf{v}_B = m_A\mathbf{v}'_A + m_B\mathbf{v}'_B$$

Since all the velocities considered are directed along the same axis, we can replace the equation obtained by the following relation involving only scalar components:

$$m_Av_A + m_Bv_B = m_Av'_A + m_Bv'_B \qquad (13.37)$$

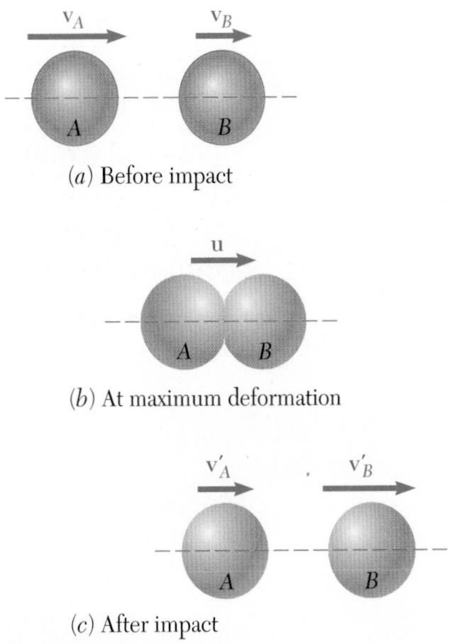

(*a*) Before impact
(*b*) At maximum deformation
(*c*) After impact

Fig. 13.21

A positive value for any of the scalar quantities v_A, v_B, v'_A, or v'_B means that the corresponding vector is directed to the right; a negative value indicates that the corresponding vector is directed to the left.

To obtain the velocities $\mathbf{v}'_A$ and $\mathbf{v}'_B$, it is necessary to establish a second relation between the scalars v'_A and v'_B. For this purpose, let us now consider the motion of particle A during the period of deformation and apply the principle of impulse and momentum. Since the only impulsive force acting on A during this period is the force $\mathbf{P}$ exerted by B (Fig. 13.22a), we write, using again scalar components,

$$m_A v_A - \int P\,dt = m_A u \tag{13.38}$$

where the integral extends over the period of deformation. Considering now the motion of A during the period of restitution, and denoting by $\mathbf{R}$ the force exerted by B on A during this period (Fig. 13.22b), we write

$$m_A u - \int R\,dt = m_A v'_A \tag{13.39}$$

where the integral extends over the period of restitution.

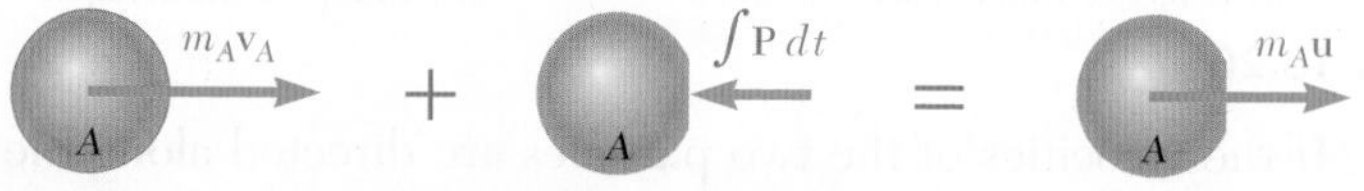

(a) Period of deformation

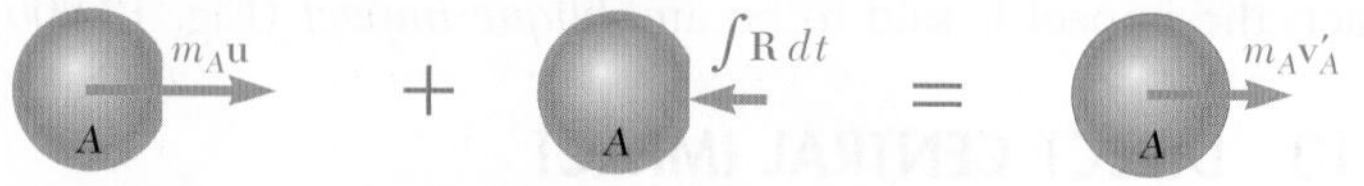

(b) Period of restitution

Fig. 13.22

In general, the force $\mathbf{R}$ exerted on A during the period of restitution differs from the force $\mathbf{P}$ exerted during the period of deformation, and the magnitude $\int R\,dt$ of its impulse is smaller than the magnitude $\int P\,dt$ of the impulse of $\mathbf{P}$. The ratio of the magnitudes of the impulses corresponding, respectively, to the period of restitution and to the period of deformation is called the *coefficient of restitution* and is denoted by e. We write

$$e = \frac{\int R\,dt}{\int P\,dt} \tag{13.40}$$

The value of the coefficient e is always between 0 and 1. It depends to a large extent on the two materials involved, but it also varies considerably with the impact velocity and the shape and size of the two colliding bodies.

Solving Eqs. (13.38) and (13.39) for the two impulses and substituting into (13.40), we write

$$e = \frac{u - v'_A}{v_A - u} \tag{13.41}$$

A similar analysis of particle B leads to the relation

$$e = \frac{v'_B - u}{u - v_B} \tag{13.42}$$

Since the quotients in (13.41) and (13.42) are equal, they are also equal to the quotient obtained by adding, respectively, their numerators and their denominators. We have, therefore,

$$e = \frac{(u - v'_A) + (v'_B - u)}{(v_A - u) + (u - v_B)} = \frac{v'_B - v'_A}{v_A - v_B}$$

and

$$v'_B - v'_A = e(v_A - v_B) \tag{13.43}$$

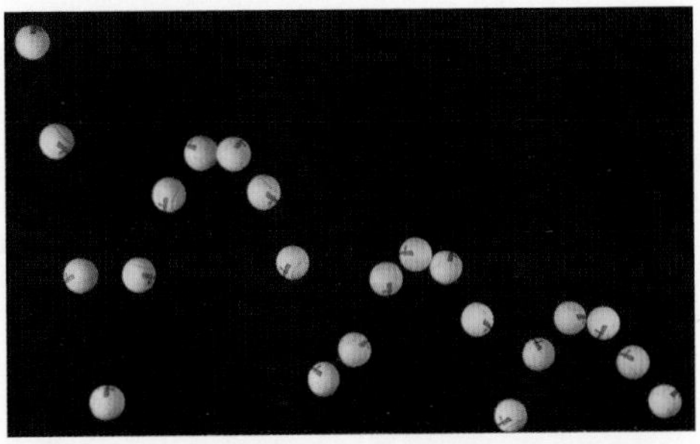

Photo 13.3 The height the tennis ball bounces decreases after each impact because it has a coefficient of restitution less than one and energy is lost with each bounce.

Since $v'_B - v'_A$ represents the relative velocity of the two particles after impact and $v_A - v_B$ represents their relative velocity before impact, formula (13.43) expresses that *the relative velocity of the two particles after impact can be obtained by multiplying their relative velocity before impact by the coefficient of restitution*. This property is used to determine experimentally the value of the coefficient of restitution of two given materials.

The velocities of the two particles after impact can now be obtained by solving Eqs. (13.37) and (13.43) simultaneously for v'_A and v'_B. It is recalled that the derivation of Eqs. (13.37) and (13.43) was based on the assumption that particle B is located to the right of A, and that both particles are initially moving to the right. If particle B is initially moving to the left, the scalar v_B should be considered negative. The same sign convention holds for the velocities after impact: A positive sign for v'_A will indicate that particle A moves to the right after impact, and a negative sign will indicate that it moves to the left.

Two particular cases of impact are of special interest:

1. $e = 0$, *Perfectly Plastic Impact*. When $e = 0$, Eq. (13.43) yields $v'_B = v'_A$. There is no period of restitution, and both particles stay together after impact. Substituting $v'_B = v'_A = v'$ into Eq. (13.37), which expresses that the total momentum of the particles is conserved, we write

$$m_A v_A + m_B v_B = (m_A + m_B)v' \tag{13.44}$$

This equation can be solved for the common velocity v' of the two particles after impact.

2. $e = 1$, *Perfectly Elastic Impact*. When $e = 1$, Eq. (13.43) reduces to

$$v'_B - v'_A = v_A - v_B \tag{13.45}$$

which expresses that the relative velocities before and after impact are equal. The impulses received by each particle during the period of deformation and during the period of restitution are equal. The particles move away from each other after impact with the same velocity with which they approached each other before impact. The velocities v'_A and v'_B can be obtained by solving Eqs. (13.37) and (13.45) simultaneously.

It is worth noting that *in the case of a perfectly elastic impact, the total energy of the two particles,* as well as their total momentum, *is conserved.* Equations (13.37) and (13.45) can be written as follows:

$$m_A(v_A - v'_A) = m_B(v'_B - v_B) \qquad (13.37')$$

$$v_A + v'_A = v_B + v'_B \qquad (13.45')$$

Multiplying (13.37′) and (13.45′) member by member, we have

$$m_A(v_A - v'_A)(v_A + v'_A) = m_B(v'_B - v_B)(v'_B + v_B)$$

$$m_A v_A^2 - m_A(v'_A)^2 = m_B(v'_B)^2 - m_B v_B^2$$

Rearranging the terms in the equation obtained and multiplying by $\frac{1}{2}$, we write

$$\tfrac{1}{2}m_A v_A^2 + \tfrac{1}{2}m_B v_B^2 = \tfrac{1}{2}m_A(v'_A)^2 + \tfrac{1}{2}m_B(v'_B)^2 \qquad (13.46)$$

which expresses that the kinetic energy of the particles is conserved. It should be noted, however, that *in the general case of impact,* i.e., when e is not equal to 1, *the total energy of the particles is not conserved.* This can be shown in any given case by comparing the kinetic energies before and after impact. The lost kinetic energy is in part transformed into heat and in part spent in generating elastic waves within the two colliding bodies.

13.14 OBLIQUE CENTRAL IMPACT

Let us now consider the case when the velocities of the two colliding particles are *not* directed along the line of impact (Fig. 13.23). As indicated in Sec. 13.12, the impact is said to be *oblique.* Since the

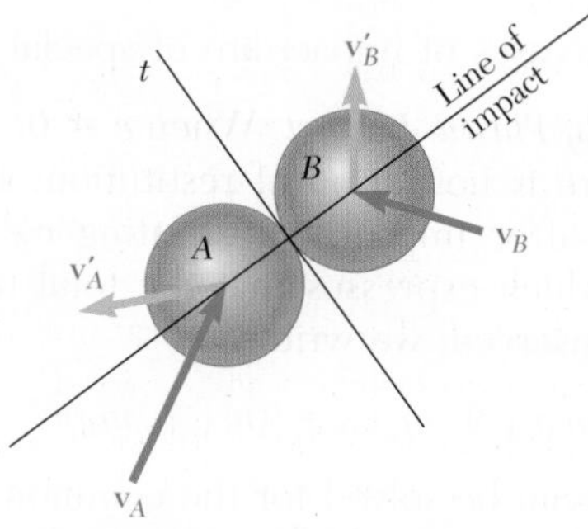

Fig. 13.23

Photo 13.4 When pool balls strike each other there is a transfer of momentum.

velocities $\mathbf{v}'_A$ and $\mathbf{v}'_B$ of the particles after impact are unknown in direction as well as in magnitude, their determination will require the use of four independent equations.

We choose as coordinate axes the n axis along the line of impact, i.e., along the common normal to the surfaces in contact, and the t axis along their common tangent. Assuming that the particles are perfectly *smooth and frictionless,* we observe that the only impulses

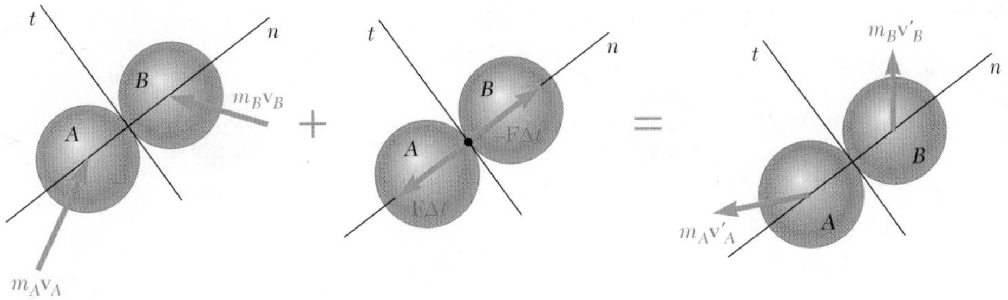

Fig. 13.24

exerted on the particles during the impact are due to internal forces directed along the line of impact, i.e., along the n axis (Fig. 13.24). It follows that

1. The component along the t axis of the momentum of each particle, considered separately, is conserved; hence the t component of the velocity of each particle remains unchanged. We write

$$(v_A)_t = (v'_A)_t \qquad (v_B)_t = (v'_B)_t \tag{13.47}$$

2. The component along the n axis of the total momentum of the two particles is conserved. We write

$$m_A(v_A)_n + m_B(v_B)_n = m_A(v'_A)_n + m_B(v'_B)_n \tag{13.48}$$

3. The component along the n axis of the relative velocity of the two particles after impact is obtained by multiplying the n component of their relative velocity before impact by the coefficient of restitution. Indeed, a derivation similar to that given in Sec. 13.13 for direct central impact yields

$$(v'_B)_n - (v'_A)_n = e[(v_A)_n - (v_B)_n] \tag{13.49}$$

We have thus obtained four independent equations which can be solved for the components of the velocities of A and B after impact. This method of solution is illustrated in Sample Prob. 13.15.

Our analysis of the oblique central impact of two particles has been based so far on the assumption that both particles moved freely before and after the impact. Let us now examine the case when one or both of the colliding particles is constrained in its motion. Consider, for instance, the collision between block A, which is constrained to move on a horizontal surface, and ball B, which is free to move in the plane of the figure (Fig. 13.25). Assuming no friction between the block and the ball, or between the block and the horizontal surface, we note that the impulses exerted on the system consist of the impulses of the internal forces $\mathbf{F}$ and $-\mathbf{F}$ directed along the line of impact, i.e., along the n axis, and of the impulse of the external force $\mathbf{F}_{ext}$ exerted by the horizontal surface on block A and directed along the vertical (Fig. 13.26).

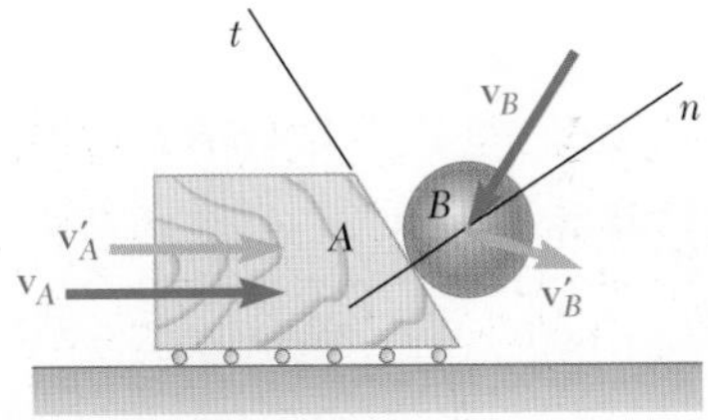

Fig. 13.25

The velocities of block A and ball B immediately after the impact are represented by three unknowns: the magnitude of the velocity $\mathbf{v}'_A$ of block A, which is known to be horizontal, and the magnitude and

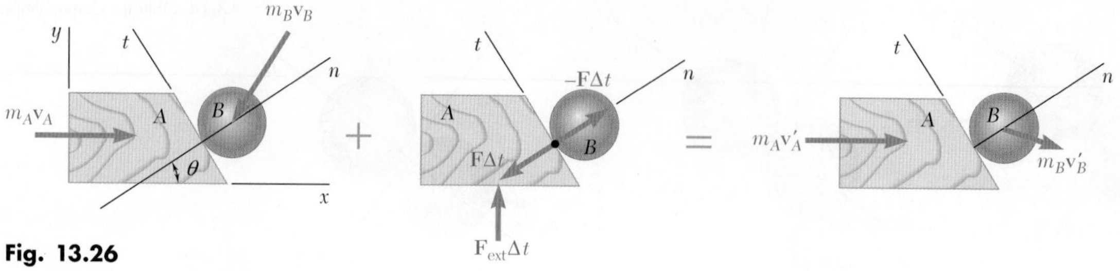

Fig. 13.26

direction of the velocity $\mathbf{v}'_B$ of ball B. We must therefore write three equations by expressing that

1. The component along the t axis of the momentum of ball B is conserved; hence the t component of the velocity of ball B remains unchanged. We write

$$(v_B)_t = (v'_B)_t \tag{13.50}$$

2. The component along the horizontal x axis of the total momentum of block A and ball B is conserved. We write

$$m_A v_A + m_B(v_B)_x = m_A v'_A + m_B(v'_B)_x \tag{13.51}$$

3. The component along the n axis of the relative velocity of block A and ball B after impact is obtained by multiplying the n component of their relative velocity before impact by the coefficient of restitution. We write again

$$(v'_B)_n - (v'_A)_n = e[(v_A)_n - (v_B)_n] \tag{13.49}$$

We should note, however, that in the case considered here, the validity of Eq. (13.49) cannot be established through a mere extension of the derivation given in Sec. 13.13 for the direct central impact of two particles moving in a straight line. Indeed, these particles were not subjected to any external impulse, while block A in the present analysis is subjected to the impulse exerted by the horizontal surface. To prove that Eq. (13.49) is still valid, we will first apply the principle of impulse and momentum to block A over the period of deformation (Fig. 13.27). Considering only the horizontal components, we write

$$m_A v_A - (\textstyle\int P\,dt) \cos \theta = m_A u \tag{13.52}$$

where the integral extends over the period of deformation and where $\mathbf{u}$ represents the velocity of block A at the end of that period. Considering now the period of restitution, we write in a similar way

$$m_A u - (\textstyle\int R\,dt) \cos \theta = m_A v'_A \tag{13.53}$$

where the integral extends over the period of restitution.

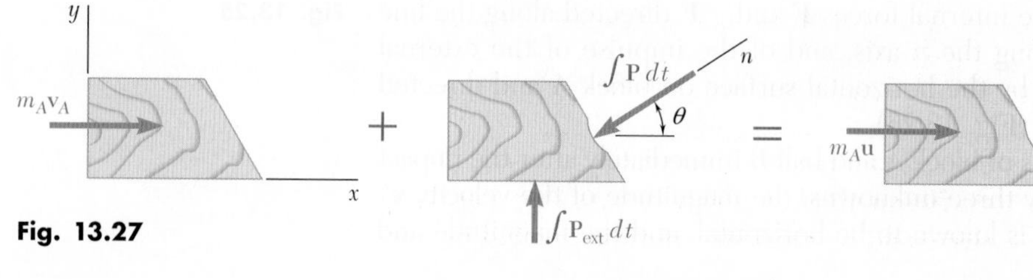

Fig. 13.27

Recalling from Sec. 13.13 the definition of the coefficient of restitution, we write

$$e = \frac{\int R\,dt}{\int P\,dt} \tag{13.40}$$

Solving Eqs. (13.52) and (13.53) for the integrals $\int P\,dt$ and $\int R\,dt$, and substituting into Eq. (13.40), we have, after reductions,

$$e = \frac{u - v'_A}{v_A - u}$$

or, multiplying all velocities by cos θ to obtain their projections on the line of impact.

$$e = \frac{u_n - (v'_A)_n}{(v_A)_n - u_n} \tag{13.54}$$

We note that Eq. (13.54) is identical to Eq. (13.41) of Sec. 13.13, except for the subscripts n which are used here to indicate that we are considering velocity components along the line of impact. Since the motion of ball B is unconstrained, the proof of Eq. (13.49) can be completed in the same manner as the derivation of Eq. (13.43) of Sec. 13.13. Thus, we conclude that the relation (13.49) between the components along the line of impact of the relative velocities of two colliding particles remains valid when one of the particles is constrained in its motion. The validity of this relation is easily extended to the case when both particles are constrained in their motion.

13.15 PROBLEMS INVOLVING ENERGY AND MOMENTUM

You now have at your disposal three different methods for the solution of kinetics problems: the direct application of Newton's second law, $\Sigma\mathbf{F} = m\mathbf{a}$; the method of work and energy; and the method of impulse and momentum. To derive maximum benefit from these three methods, you should be able to choose the method best suited for the solution of a given problem. You should also be prepared to use different methods for solving the various parts of a problem when such a procedure seems advisable.

You have already seen that the method of work and energy is in many cases more expeditious than the direct application of Newton's second law. As indicated in Sec. 13.4, however, the method of work and energy has limitations, and it must sometimes be supplemented by the use of $\Sigma\mathbf{F} = m\mathbf{a}$. This is the case, for example, when you wish to determine an acceleration or a normal force.

For the solution of problems involving no impulsive forces, it will usually be found that the equation $\Sigma\mathbf{F} = m\mathbf{a}$ yields a solution just as fast as the method of impulse and momentum and that the method of work and energy, if it applies, is more rapid and more convenient. However, in problems of impact, the method of impulse and momentum is the only practicable method. A solution based on the direct application of $\Sigma\mathbf{F} = m\mathbf{a}$ would be unwieldy, and the method of work and energy cannot be used since impact (unless perfectly elastic) involves a loss of mechanical energy.

Many problems involve only conservative forces, except for a short impact phase during which impulsive forces act. The solution of such problems can be divided into several parts. The part corresponding to the impact phase calls for the use of the method of impulse and momentum and of the relation between relative velocities, and the other parts can usually be solved by the method of work and energy. If the problem involves the determination of a normal force, however, the use of $\Sigma\mathbf{F} = m\mathbf{a}$ is necessary.

Consider, for example, a pendulum A, of mass m_A and length l, which is released with no velocity from a position A_1 (Fig. 13.28*a*). The pendulum swings freely in a vertical plane and hits a second pendulum B, of mass m_B and same length l, which is initially at rest. After the impact (with coefficient of restitution e), pendulum B swings through an angle θ that we wish to determine.

The solution of the problem can be divided into three parts:

1. *Pendulum A Swings from A_1 to A_2.* The principle of conservation of energy can be used to determine the velocity $(\mathbf{v}_A)_2$ of the pendulum at A_2 (Fig. 13.28*b*).
2. *Pendulum A Hits Pendulum B.* Using the fact that the total momentum of the two pendulums is conserved and the relation between their relative velocities, we determine the velocities $(\mathbf{v}_A)_3$ and $(\mathbf{v}_B)_3$ of the two pendulums after impact (Fig. 13.28*c*).
3. *Pendulum B Swings from B_3 to B_4.* Applying the principle of conservation of energy to pendulum B, we determine the maximum elevation y_4 reached by that pendulum (Fig. 13.28*d*). The angle θ can then be determined by trigonometry.

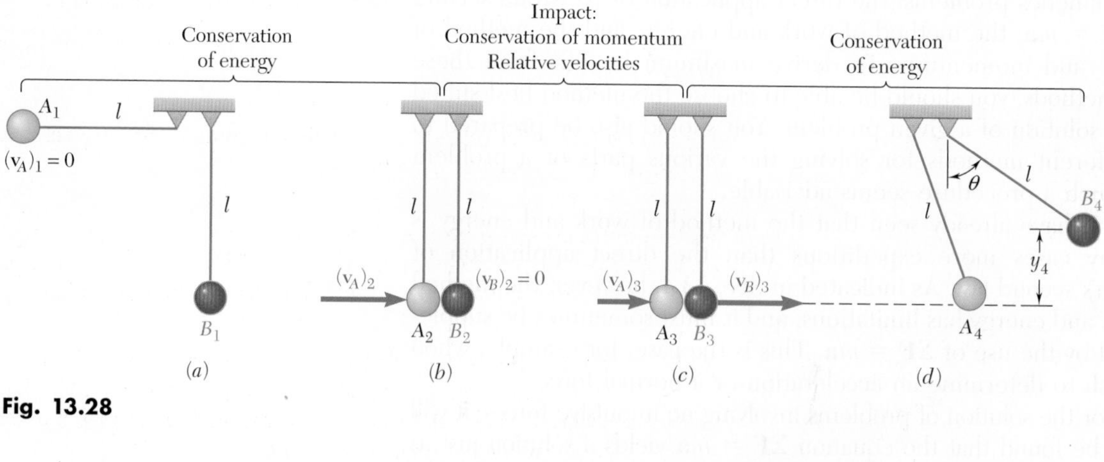

Fig. 13.28

We note that if the tensions in the cords holding the pendulums are to be determined, the method of solution just described should be supplemented by the use of $\Sigma\mathbf{F} = m\mathbf{a}$.

SAMPLE PROBLEM 13.13

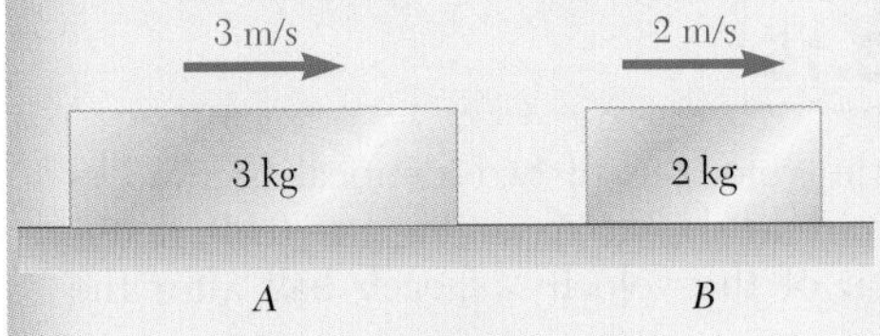

Two steel blocks are sliding on a frictionless horizontal surface with the velocities shown. Knowing that after impact the velocity of *B* is observed to be 3.1 m/s to the right, determine the coefficient of restitution between the two blocks.

SOLUTION

We express that the total momentum of the two blocks is conserved.

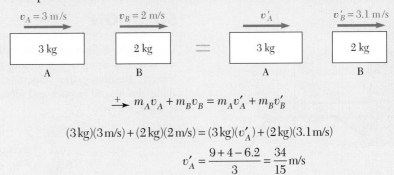

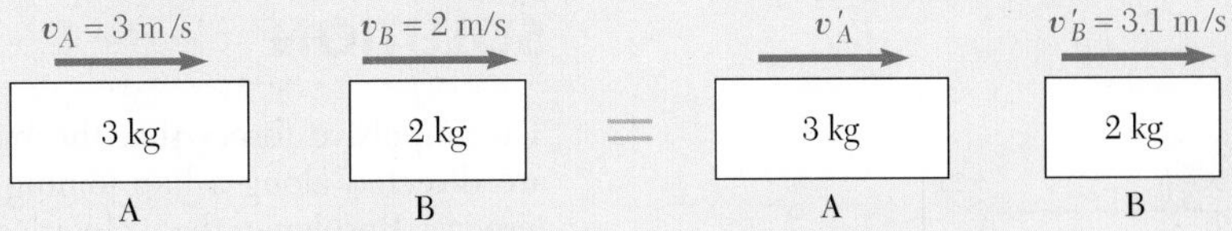

$$\xrightarrow{+} \quad m_A v_A + m_B v_B = m_A v'_A + m_B v'_B$$

$$(3\,\text{kg})(3\,\text{m/s}) + (2\,\text{kg})(2\,\text{m/s}) = (3\,\text{kg})(v'_A) + (2\,\text{kg})(3.1\,\text{m/s})$$

$$v'_A = \frac{9+4-6.2}{3} = \frac{34}{15}\,\text{m/s}$$

The coefficient of restitution is obtained by writing.

$$e = \frac{v'_B - v'_A}{v_A - v_B} = \frac{3.1 - \frac{34}{15}}{3-2} = \frac{5}{6} \qquad e = 0.833 \blacktriangleleft$$

SAMPLE PROBLEM 13.14

A ball is thrown against a frictionless, vertical wall. Immediately before the ball strikes the wall, its velocity has a magnitude v and forms an angle of 30° with the horizontal. Knowing that $e = 0.90$, determine the magnitude and direction of the velocity of the ball as it rebounds from the wall.

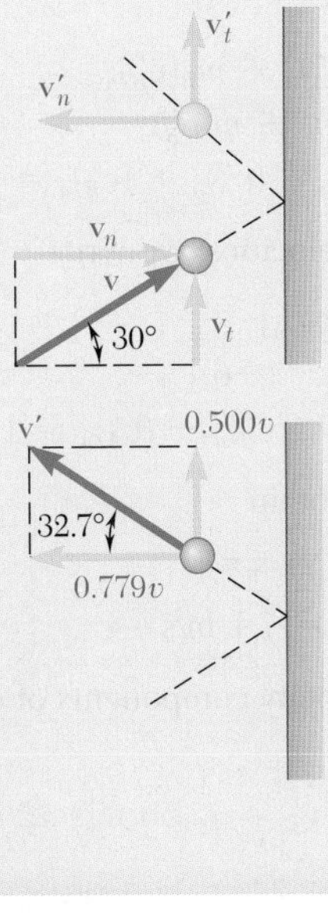

SOLUTION

We resolve the initial velocity of the ball into components respectively perpendicular and parallel to the wall:

$$v_n = v \cos 30° = 0.866v \qquad v_t = v \sin 30° = 0.500v$$

Motion Parallel to the Wall. Since the wall is frictionless, the impulse it exerts on the ball is perpendicular to the wall. Thus, the component parallel to the wall of the momentum of the ball is conserved and we have

$$\mathbf{v}'_t = \mathbf{v}_t = 0.500v \uparrow$$

Motion Perpendicular to the Wall. Since the mass of the wall (and earth) is essentially infinite, expressing that the total momentum of the ball and wall is conserved would yield no useful information. Using the relation (13.49) between relative velocities, we write

$$0 - v'_n = e(v_n - 0)$$

$$v'_n = -0.90(0.866v) = -0.779v \qquad \mathbf{v}'_n = 0.779v \leftarrow$$

Resultant Motion. Adding vectorially the components $\mathbf{v}'_n$ and $\mathbf{v}'_t$,

$$\mathbf{v}' = 0.926v \measuredangle 32.7° \blacktriangleleft$$

SAMPLE PROBLEM 13.15

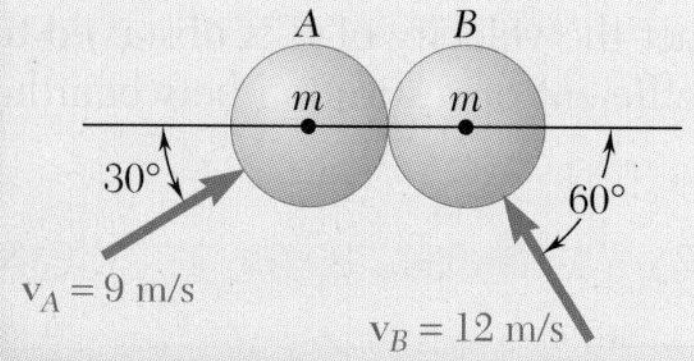

The magnitude and direction of the velocities of two identical frictionless balls before they strike each other are as shown. Assuming $e = 0.90$, determine the magnitude and direction of the velocity of each ball after the impact.

SOLUTION

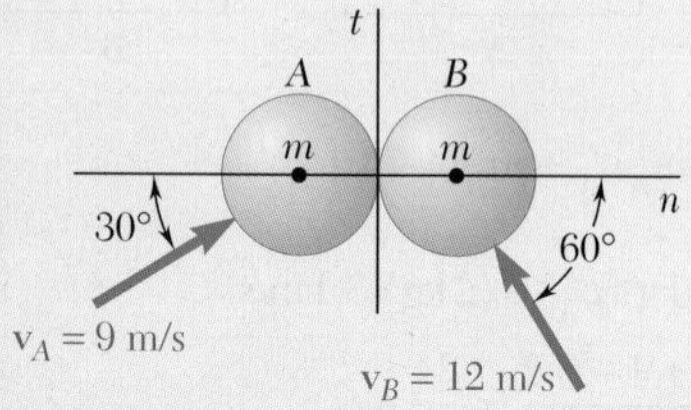

The impulsive forces that the balls exert on each other during the impact are directed along a line joining the centers of the balls called the *line of impact*. Resolving the velocities into components directed, respectively, along the line of impact and along the common tangent to the surfaces in contact, we write

$$(v_A)_n = v_A \cos 30° = +\ 7.79 \text{ m/s}$$
$$(v_A)_t = v_A \sin 30° = +\ 4.5 \text{ m/s}$$
$$(v_B)_n = -v_B \cos 60° = -\ 6 \text{ m/s}$$
$$(v_B)_t = v_B \sin 60° = +10.39 \text{ m/s}$$

Principle of Impulse and Momentum. In the adjoining sketches we show in turn the initial momenta, the impulses, and the final momenta.

Motion along the Common Tangent. Considering only the t components, we apply the principle of impulse and momentum to each ball *separately*. Since the impulsive forces are directed along the line of impact, the t component of the momentum, and hence the t component of the velocity of each ball, is unchanged. We have

$$(\mathbf{v}'_A)_t = 4.5 \text{ m/s} \uparrow \qquad (\mathbf{v}'_B)_t = 10.39 \text{ m/s} \uparrow$$

Motion along the Line of Impact. In the n direction, we consider the two balls as a single system and note that by Newton's third law, the internal impulses are, respectively, $\mathbf{F}\,\Delta t$ and $-\mathbf{F}\,\Delta t$ and cancel. We thus write that the total momentum of the balls is conserved:

$$m_A(v_A)_n + m_B(v_B)_n = m_A(v'_A)_n + m_B(v'_B)_n$$
$$m(7.79) + m(-6) = m(v'_A)_n + m(v'_B)_n$$
$$(v'_A)_n + (v'_B)_n = 1.79 \qquad (1)$$

Using the relation (13.49) between relative velocities, we write

$$(v'_B)_n - (v'_A)_n = e[(v_A)_n - (v_B)_n]$$
$$(v'_B)_n - (v'_A)_n = (0.90)[7.79 - (-6)]$$
$$(v'_B)_n - (v'_A)_n = 12.41 \qquad (2)$$

Solving Eqs. (1) and (2) simultaneously, we obtain

$$(v'_A)_n = -\ 5.31 \qquad (v'_B)_n = +7.1$$
$$(\mathbf{v}'_A)_n = 5.31 \text{ m/s} \leftarrow \qquad (\mathbf{v}'_B)_n = 7.1 \text{ m/s} \rightarrow$$

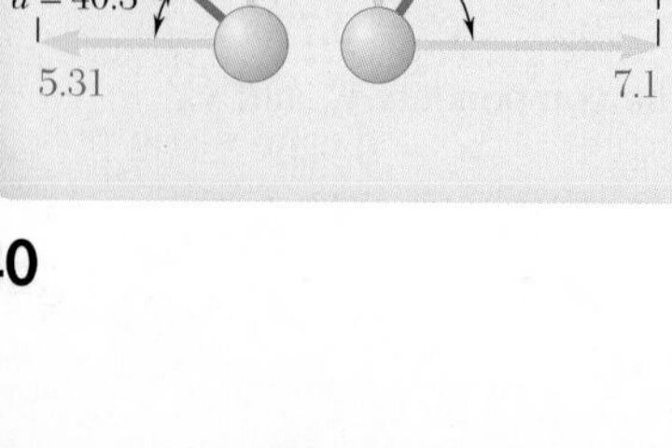

Resultant Motion. Adding vectorially the velocity components of each ball, we obtain

$$\mathbf{v}'_A = 6.96 \text{ m/s} \nwarrow 40.3° \qquad \mathbf{v}'_B = 12.58 \text{ m/s} \nearrow 55.6° \blacktriangleleft$$

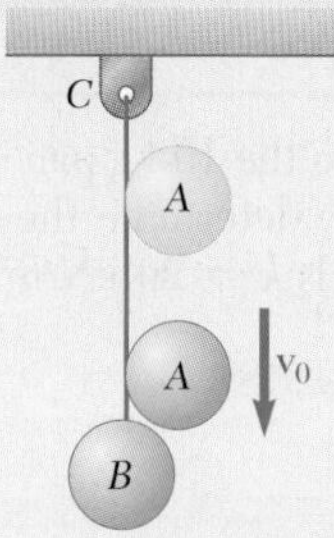

SAMPLE PROBLEM 13.16

Ball B is hanging from an inextensible cord BC. An identical ball A is released from rest when it is just touching the cord and acquires a velocity $\mathbf{v}_0$ before striking ball B. Assuming perfectly elastic impact ($e = 1$) and no friction, determine the velocity of each ball immediately after impact.

SOLUTION

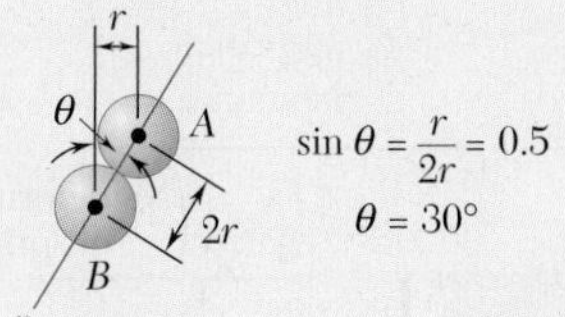

Since ball B is constrained to move in a circle of center C, its velocity $\mathbf{v}_B$ after impact must be horizontal. Thus the problem involves three unknowns: the magnitude v'_B of the velocity of B, and the magnitude and direction of the velocity $\mathbf{v}'_A$ of A after impact.

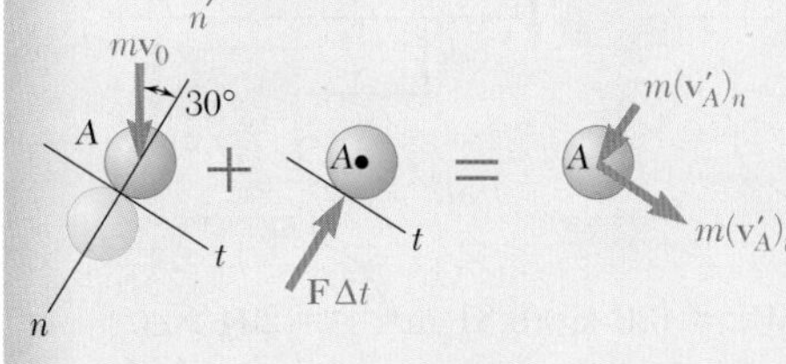

Impulse-Momentum Principle: Ball A

$$m\mathbf{v}_A + \mathbf{F}\,\Delta t = m\mathbf{v}'_A$$

$+\searrow t$ components: $\quad mv_0 \sin 30° + 0 = m(v'_A)_t$

$$(v'_A)_t = 0.5v_0 \qquad (1)$$

We note that the equation used expresses conservation of the momentum of ball A along the common tangent to balls A and B.

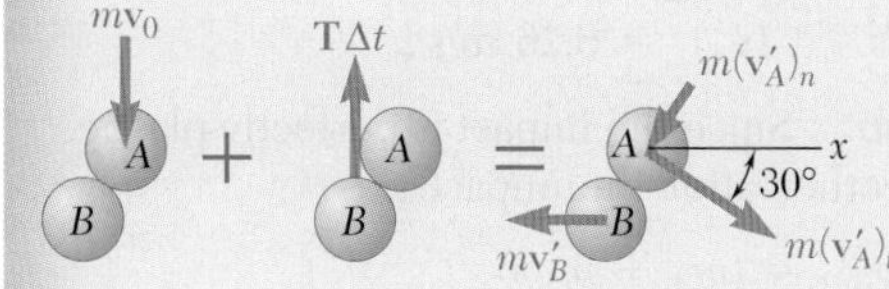

Impulse-Momentum Principle: Balls A and B

$$m\mathbf{v}_A + \mathbf{T}\,\Delta t = m\mathbf{v}'_A + m\mathbf{v}'_B$$

$\overset{+}{\rightarrow} x$ components: $\quad 0 = m(v'_A)_t \cos 30° - m(v'_A)_n \sin 30° - mv'_B$

We note that the equation obtained expresses conservation of the total momentum in the x direction. Substituting for $(v'_A)_t$ from Eq. (1) and rearranging terms, we write

$$0.5(v'_A)_n + v'_B = 0.433v_0 \qquad (2)$$

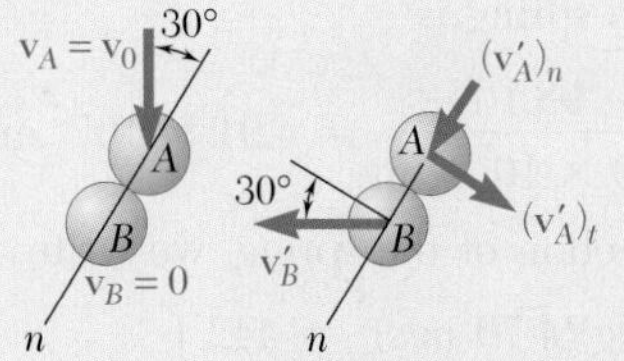

Relative Velocities Along the Line of Impact. Since $e = 1$, Eq. (13.49) yields

$$(v'_B)_n - (v'_A)_n = (v_A)_n - (v_B)_n$$

$$v'_B \sin 30° - (v'_A)_n = v_0 \cos 30° - 0$$

$$0.5v'_B - (v'_A)_n = 0.866v_0 \qquad (3)$$

Solving Eqs. (2) and (3) simultaneously, we obtain

$$(v'_A)_n = -0.520v_0 \qquad v'_B = 0.693v_0$$

$$\mathbf{v}'_B = 0.693v_0 \leftarrow \blacktriangleleft$$

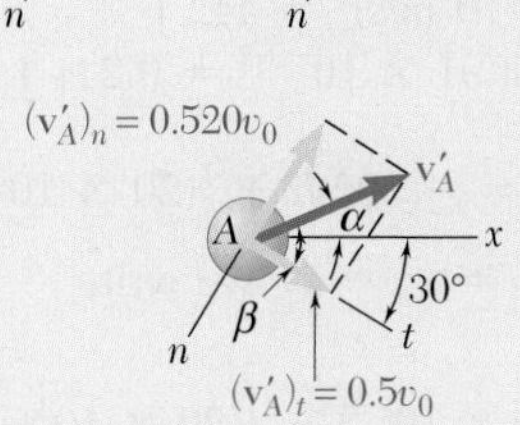

Recalling Eq. (1) we draw the adjoining sketch and obtain by trigonometry

$$v'_A = 0.721v_0 \qquad \beta = 46.1° \qquad \alpha = 46.1° - 30° = 16.1°$$

$$\mathbf{v}'_A = 0.721v_0 \measuredangle 16.1° \blacktriangleleft$$

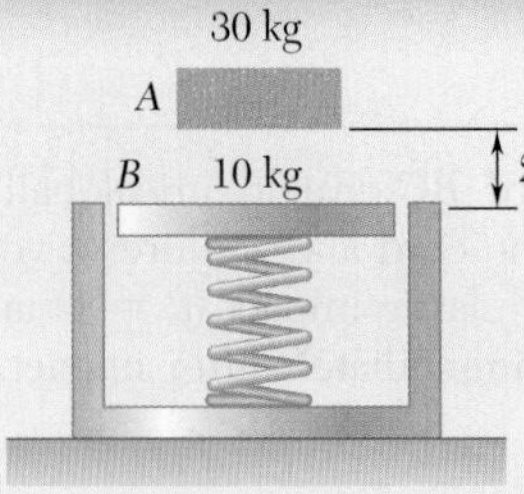

SAMPLE PROBLEM 13.17

A 30-kg block is dropped from a height of 2 m onto the 10-kg pan of a spring scale. Assuming the impact to be perfectly plastic, determine the maximum deflection of the pan. The constant of the spring is $k = 20$ kN/m.

SOLUTION

The impact between the block and the pan *must* be treated separately; therefore we divide the solution into three parts.

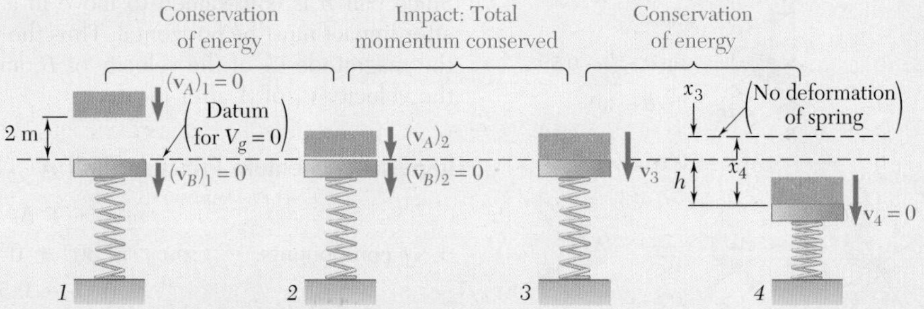

Conservation of Energy. Block: $W_A = (30\text{ kg})(9.81\text{ m/s}^2) = 294\text{ N}$

$$T_1 = \tfrac{1}{2}m_A(v_A)_1^2 = 0 \qquad V_1 = W_A y = (294\text{ N})(2\text{ m}) = 588\text{ J}$$
$$T_2 = \tfrac{1}{2}m_A(v_A)_2^2 = \tfrac{1}{2}(30\text{ kg})(v_A)_2^2 \qquad V_2 = 0$$
$$T_1 + V_1 = T_2 + V_2: \qquad 0 + 588\text{ J} = \tfrac{1}{2}(30\text{ kg})(v_A)_2^2 + 0$$
$$(v_A)_2 = +6.26\text{ m/s} \qquad (\mathbf{v}_A)_2 = 6.26\text{ m/s}\downarrow$$

Impact: Conservation of Momentum. Since the impact is perfectly plastic, $e = 0$; the block and pan move together after the impact.

$$m_A(v_A)_2 + m_B(v_B)_2 = (m_A + m_B)v_3$$
$$(30\text{ kg})(6.26\text{ m/s}) + 0 = (30\text{ kg} + 10\text{ kg})v_3$$
$$v_3 = +4.70\text{ m/s} \qquad \mathbf{v}_3 = 4.70\text{ m/s}\downarrow$$

Conservation of Energy. Initially the spring supports the weight W_B of the pan; thus the initial deflection of the spring is

$$x_3 = \frac{W_B}{k} = \frac{(10\text{kg})(9.81\text{ m/s}^2)}{20 \times 10^3\text{ N/m}} = \frac{98.1\text{ N}}{20 \times 10^3\text{ N/m}} = 4.91 \times 10^{-3}\text{ m}$$

Denoting by x_4 the total maximum deflection of the spring, we write

$$T_3 = \tfrac{1}{2}(m_A + m_B)v_3^2 = \tfrac{1}{2}(30\text{ kg} + 10\text{ kg})(4.70\text{ m/s})^2 = 442\text{ J}$$
$$V_3 = V_g + V_e = 0 + \tfrac{1}{2}kx_3^2 = \tfrac{1}{2}(20 \times 10^3)(4.91 \times 10^{-3})^2 = 0.241\text{ J}$$
$$T_4 = 0$$
$$V_4 = V_g + V_e = (W_A + W_B)(-h) + \tfrac{1}{2}kx_4^2 = -(392)h + \tfrac{1}{2}(20 \times 10^3)x_4^2$$

Noting that the displacement of the pan is $h = x_4 - x_3$, we write

$$T_3 + V_3 = T_4 + V_4:$$
$$442 + 0.241 = 0 - 392(x_4 - 4.91 \times 10^{-3}) + \tfrac{1}{2}(20 \times 10^3)x_4^2$$
$$x_4 = 0.230\text{ m} \qquad h = x_4 - x_3 = 0.230\text{ m} - 4.91 \times 10^{-3}\text{ m}$$
$$h = 0.225\text{ m} \qquad h = 225\text{ mm} \blacktriangleleft$$

SOLVING PROBLEMS ON YOUR OWN

This lesson deals with the *impact of two bodies,* i.e., with a collision occurring in a very small interval of time. You will solve a number of impact problems by expressing that the total momentum of the two bodies is conserved and noting the relationship which exists between the relative velocities of the two bodies before and after impact.

1. As a first step in your solution you should select and draw the following coordinate axes: the t axis, which is tangent to the surfaces of contact of the two colliding bodies, and the n axis, which is normal to the surfaces of contact and defines the *line of impact*. In all the problems of this lesson the line of impact passes through the mass centers of the colliding bodies, and the impact is referred to as a *central impact*.

2. Next you will draw a diagram showing the momenta of the bodies before impact, the impulses exerted on the bodies during impact, and the final momenta of the bodies after impact (Fig. 13.24). You will then observe whether the impact is a *direct central impact* or an *oblique central impact*.

3. Direct central impact. This occurs when the velocities of bodies A and B before impact are *both directed along the line of impact* (Fig. 13.20*a*).

a. Conservation of momentum. Since the impulsive forces are internal to the system, you can write that the *total momentum of A and B is conserved,*

$$m_A v_A + m_B v_B = m_A v'_A + m_B v'_B \tag{13.37}$$

where v_A and v_B denote the velocities of bodies A and B before impact and v'_A and v'_B denote their velocities after impact.

b. Coefficient of restitution. You can also write the following relation between the *relative velocities* of the two bodies before and after impact,

$$v'_B - v'_A = e(v_A - v_B) \tag{13.43}$$

where e is the coefficient of restitution between the two bodies.

Note that Eqs. (13.37) and (13.43) are scalar equations which can be solved for two unknowns. Also, be careful to adopt a consistent sign convention for all velocities.

4. Oblique central impact. This occurs when *one or both* of the initial velocities of the two bodies is *not directed* along the line of impact (Fig. 13.20*b*). To solve problems of this type, you should *first resolve into components* along the t axis and the n axis the momenta and impulses shown in your diagram.

(*continued*)

a. Conservation of momentum. Since the impulsive forces act along the line of impact, i.e., along the n axis, the component along the t axis of the momentum *of each body* is conserved. Therefore, you can write for each body that the t components of its velocity before and after impact are equal,

$$(v_A)_t = (v'_A)_t \qquad (v_B)_t = (v'_B)_t \tag{13.47}$$

Also, the component along the n axis of the *total momentum* of the system is conserved,

$$m_A(v_A)_n + m_B(v_B)_n = m_A(v'_A)_n + m_B(v'_B)_n \tag{13.48}$$

b. Coefficient of restitution. The relation between the relative velocities of the two bodies before and after impact can be written in the n direction only,

$$(v'_B)_n - (v'_A)_n = e[(v_A)_n - (v_B)_n] \tag{13.49}$$

You now have four equations that you can solve for four unknowns. Note that after finding all the velocities, you can determine the impulse exerted by body A on body B by drawing an impulse-momentum diagram for B alone and equating components in the n direction.

c. When the motion of one of the colliding bodies is constrained, you must include the impulses of the external forces in your diagram. You will then observe that some of the above relations do not hold. However, in the example shown in Fig. 13.26 the total momentum of the system is conserved in a direction perpendicular to the external impulse. You should also note that when a body A bounces off a fixed surface B, the only conservation of momentum equation which can be used is the first of Eqs. (13.47) [Sample Prob. 13.14].

5. Remember that energy is lost during most impacts. The only exception is for *perfectly elastic* impacts ($e = 1$), where energy is conserved. Thus, in the general case of impact, where $e < 1$, the energy is not conserved. Therefore, be careful *not to apply* the principle of conservation of energy through an impact situation. Instead, apply this principle separately to the motions preceding and following the impact [Sample Prob. 13.17].

PROBLEMS

CONCEPT QUESTION

13.CQ6 A 5-kg ball A strikes a 1-kg ball B that is initially at rest. Is it possible that after the impact A is not moving and B has a speed of $5v$?

a. Yes

b. No

Explain your answer.

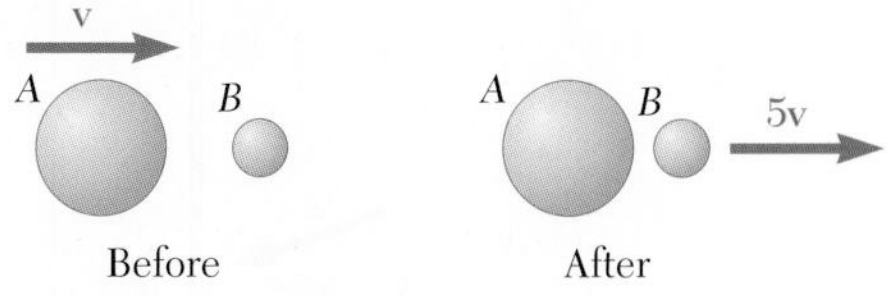

Fig. P13.CQ6

IMPULSE-MOMENTUM PRACTICE PROBLEMS

13.F6 A sphere with a speed v_0 rebounds after striking a frictionless inclined plane as shown. Draw the impulse-momentum diagram that can be used to find the velocity of the sphere after the impact.

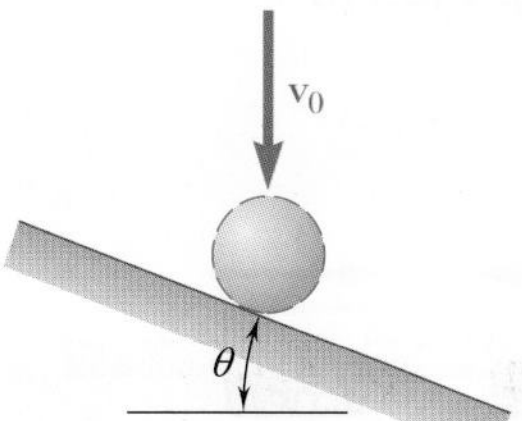

Fig. P13.F6

13.F7 An 80-Mg railroad engine A coasting at 6.5 km/h strikes a 20-Mg flatcar C carrying a 30-Mg load B which can slide along the floor of the car ($\mu_k = 0.25$). The flatcar was at rest with its brakes released. Instead of A and C coupling as expected, it is observed that A rebounds with a speed of 2 km/h after the impact. Draw impulse-momentum diagrams that can be used to determine (a) the coefficient of restitution and the speed of the flatcar immediately after impact, (b) the time it takes the load to slide to a stop relative to the car.

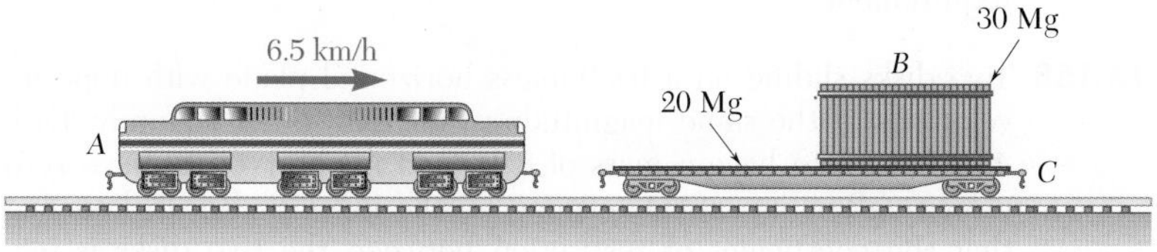

Fig. P13.F7

13.F8 Two frictionless balls strike each other as shown. The coefficient of restitution between the balls is e. Draw the impulse-momentum diagram that could be used to find the velocities of A and B after the impact.

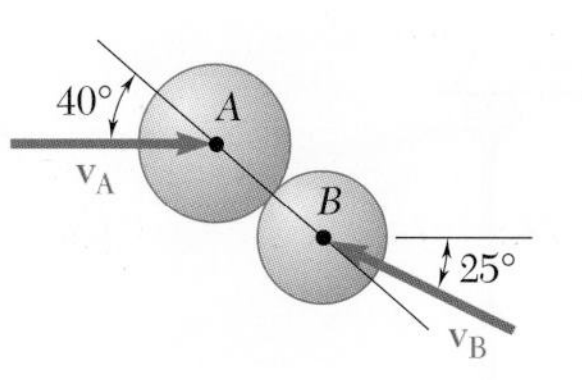

Fig. P13.F8

13.F9 A 10-kg ball A moving horizontally at 12 m/s strikes a 10-kg block B. The coefficient of restitution of the impact is 0.4 and the coefficient of kinetic friction between the block and the inclined surface is 0.5. Draw the impulse-momentum diagram that can be used to determine the speeds of A and B after the impact.

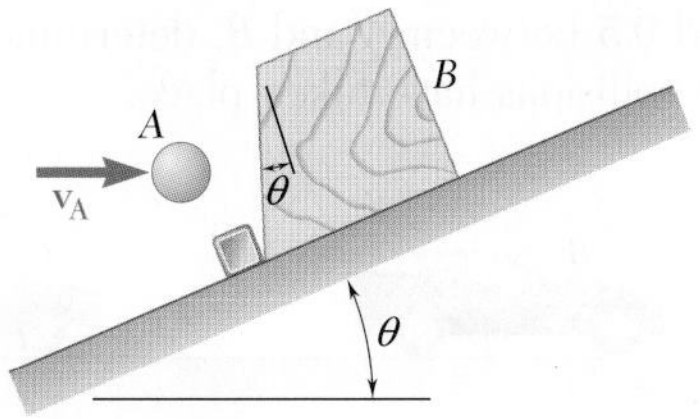

Fig. P13.F9

13.F10 Block A of mass m_A strikes ball B of mass m_B with a speed of v_A as shown. Draw the impulse-momentum diagram that can be used to determine the speeds of A and B after the impact and the impulse during the impact.

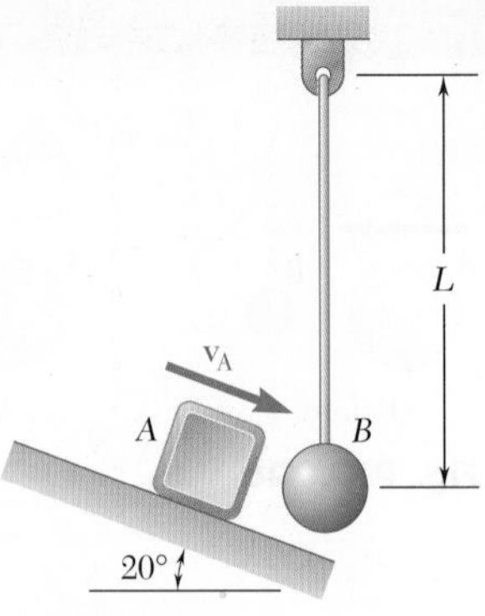

Fig. P13.F10

END-OF-SECTION PROBLEMS

13.155 The coefficient of restitution between the two collars is known to be 0.70. Determine (*a*) their velocities after impact, (*b*) the energy loss during impact.

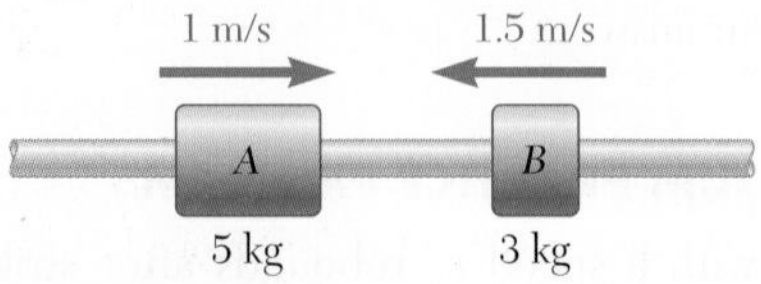

Fig. P13.155

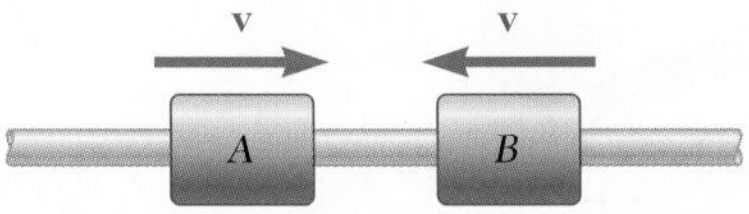

Fig. P13.156

13.156 Collars A and B, of the same mass m, are moving toward each other with identical speeds as shown. Knowing that the coefficient of restitution between the collars is e, determine the energy lost in the impact as a function of m, e, and v.

13.157 One of the requirements for tennis balls to be used in official competition is that, when dropped onto a rigid surface from a height of 2.5 m, the height of the first bounce of the ball must be in the range $1.325 \text{ m} \leq h \leq 1.45 \text{ m}$. Determine the range of the coefficients of restitution of the tennis balls satisfying this requirement.

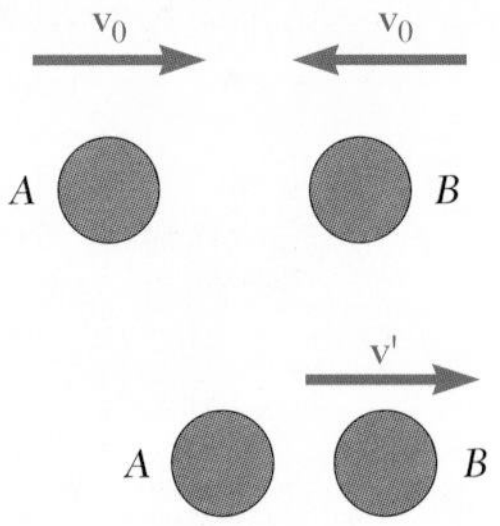

Fig. P13.158

13.158 Two disks sliding on a frictionless horizontal plane with opposite velocities of the same magnitude v_0 hit each other squarely. Disk A is known to have a mass of 3 kg and is observed to have zero velocity after impact. Determine (*a*) the mass of disk B, knowing that the coefficient of restitution between the two disks is 0.5, (*b*) the range of possible values of the mass of disk B if the coefficient of restitution between the two disks is unknown.

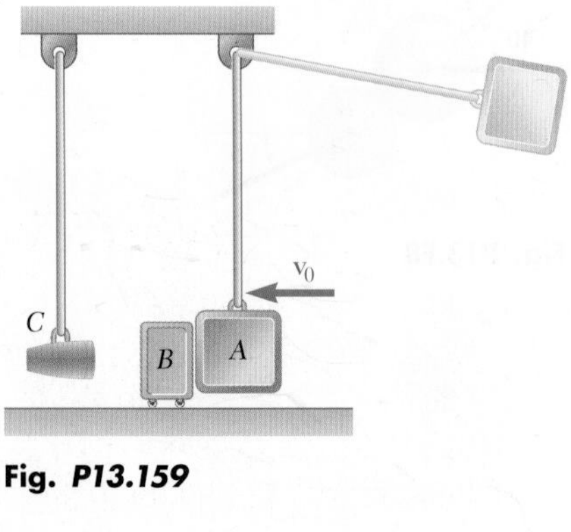

Fig. P13.159

13.159 To apply shock loading to an artillery shell, a 20-kg pendulum A is released from a known height and strikes impactor B at a known velocity $\mathbf{v}_0$. Impactor B then strikes the 1-kg artillery shell C. Knowing the coefficient of restitution between all objects is e, determine the mass of B to maximize the impulse applied to the artillery shell C.

13.160 Two identical cars A and B are at rest on a loading dock with brakes released. Car C, of a slightly different style but of the same weight, has been pushed by dockworkers and hits car B with a velocity of 1.5 m/s. Knowing that the coefficient of restitution is 0.8 between B and C and 0.5 between A and B, determine the velocity of each car after all collisions have taken place.

Fig. P13.160

13.161 Three steel spheres of equal weight are suspended from the ceiling by cords of equal length which are spaced at a distance slightly greater than the diameter of the spheres. After being pulled back and released, sphere A hits sphere B, which then hits sphere C. Denoting by e the coefficient of restitution between the spheres and by $\mathbf{v}_0$ the velocity of A just before it hits B, determine (*a*) the velocities of A and B immediately after the first collision, (*b*) the velocities of B and C immediately after the second collision. (*c*) Assuming now that n spheres are suspended from the ceiling and that the first sphere is pulled back and released as described above, determine the velocity of the last sphere after it is hit for the first time. (*d*) Use the result of part c to obtain the velocity of the last sphere when $n = 5$ and $e = 0.9$.

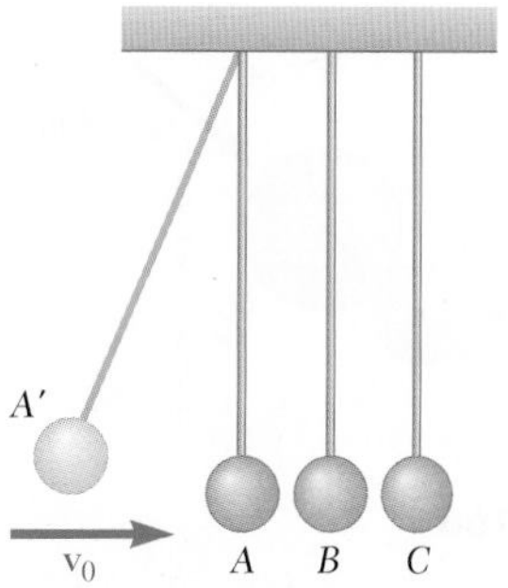

Fig. P13.161

13.162 At an amusement park there are 200-kg bumper cars A, B, and C that have riders with masses of 40 kg, 60 kg, and 35 kg, respectively. Car A is moving to the right with a velocity $\mathbf{v}_A = 2$ m/s and car C has a velocity $\mathbf{v}_B = 1.5$ m/s to the left, but car B is initially at rest. The coefficient of restitution between each car is 0.8. Determine the final velocity of each car, after all impacts, assuming (*a*) cars A and C hit car B at the same time, (*b*) car A hits car B before car C does.

Fig. *P13.162* and P13.163

13.163 At an amusement park there are 200-kg bumper cars A, B, and C that have riders with masses of 40 kg, 60 kg, and 35 kg, respectively. Car A is moving to the right with a velocity $\mathbf{v}_A = 2$ m/s when it hits stationary car B. The coefficient of restitution between each car is 0.8. Determine the velocity of car C so that after car B collides with car C the velocity of car B is zero.

13.164 Two identical billiard balls can move freely on a horizontal table. Ball A has a velocity $\mathbf{v}_0$ as shown and hits ball B, which is at rest, at a point C defined by $\theta = 45°$. Knowing that the coefficient of restitution between the two balls is $e = 0.8$ and assuming no friction, determine the velocity of each ball after impact.

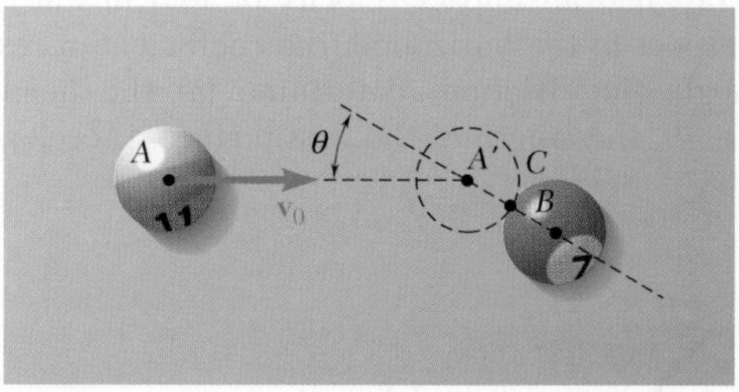

Fig. P13.164

13.165 The coefficient of restitution is 0.9 between the two 60-mm diameter billiard balls A and B. Ball A is moving in the direction shown with a velocity of 1 m/s when it strikes ball B, which is at rest. Knowing that after impact B is moving in the x direction, determine (*a*) the angle θ, (*b*) the velocity of B after impact.

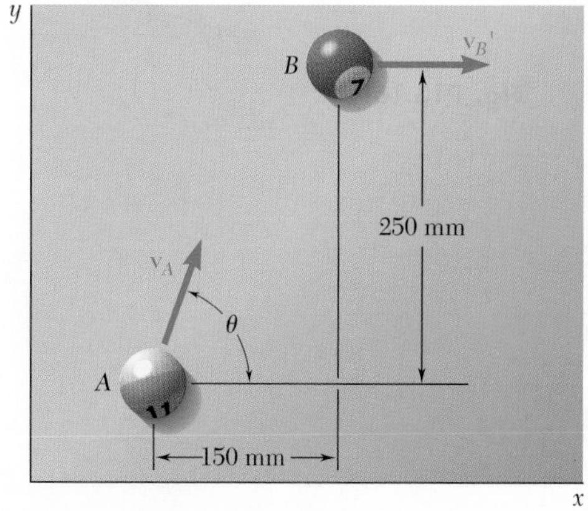

Fig. P13.165

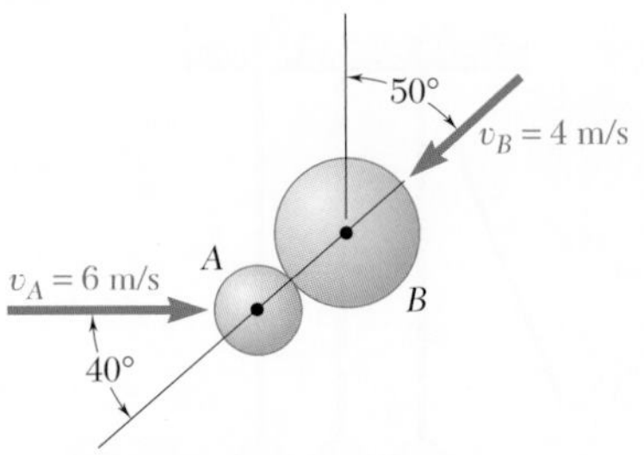

Fig. P13.166

13.166 A 600-g ball A is moving with a velocity of magnitude 6 m/s when it is hit as shown by a 1-kg ball B which has a velocity of magnitude 4 m/s. Knowing that the coefficient of restitution is 0.8 and assuming no friction, determine the velocity of each ball after impact.

13.167 Two identical hockey pucks are moving on a hockey rink at the same speed of 3 m/s and in perpendicular directions when they strike each other as shown. Assuming a coefficient of restitution $e = 0.9$, determine the magnitude and direction of the velocity of each puck after impact.

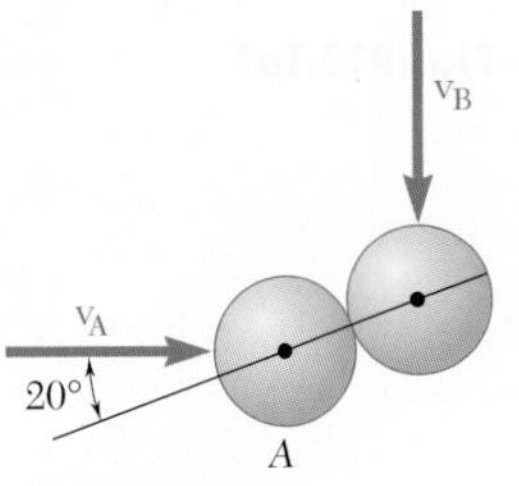

Fig. P13.167

13.168 Two identical pool balls of 57.15 mm diameter may move freely on a pool table. Ball B is at rest and ball A has an initial velocity $\mathbf{v} = v_0\mathbf{i}$. (*a*) Knowing that $b = 50$ mm and $e = 0.7$, determine the velocity of each ball after impact. (*b*) Show that if $e = 1$, the final velocities of the balls form a right angle for all values of b.

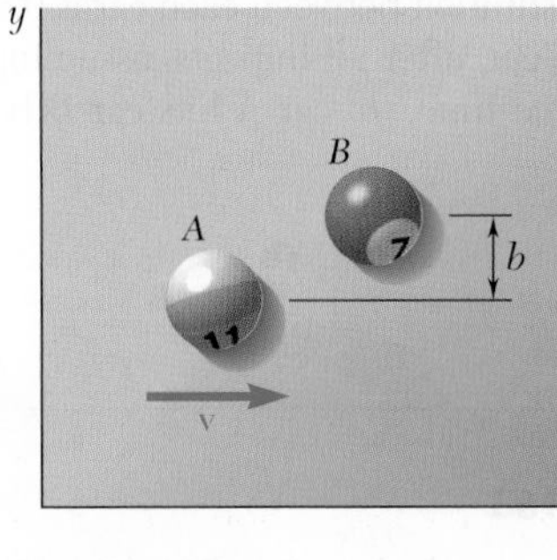

Fig. P13.168

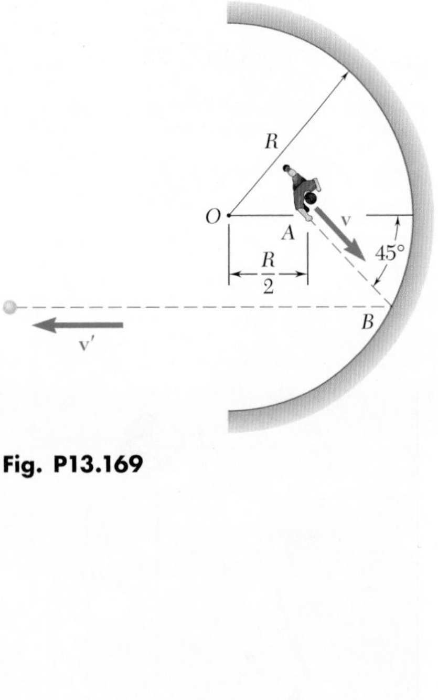

Fig. P13.169

13.169 A boy located at point A halfway between the center O of a semicircular wall and the wall itself throws a ball at the wall in a direction forming an angle of 45° with OA. Knowing that after hitting the wall the ball rebounds in a direction parallel to OA, determine the coefficient of restitution between the ball and the wall.

13.170 The Mars Pathfinder spacecraft used large airbags to cushion its impact with the planet's surface when landing. Assuming the spacecraft had an impact velocity of 18.5 m/s at an angle of 45° with respect to the horizontal, the coefficient of restitution is 0.85 and neglecting friction, determine (*a*) the height of the first bounce, (*b*) the length of the first bounce. (Acceleration of gravity on Mars $= 3.73$ m/s^2.)

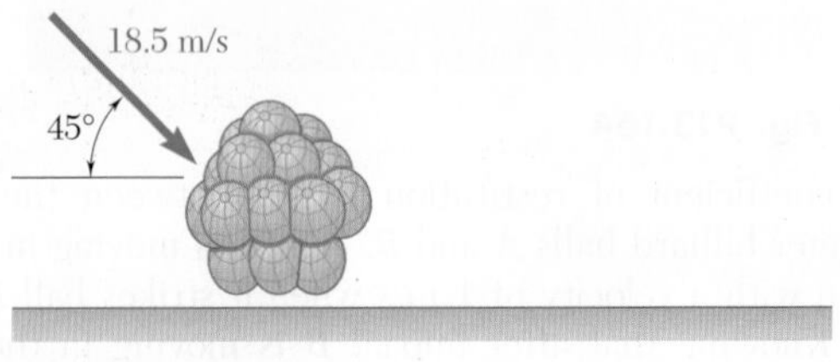

Fig. P13.170

13.171 A girl throws a ball at an inclined wall from a height of 1.2 m, hitting the wall at A with a horizontal velocity $\mathbf{v}_0$ of magnitude 15 m/s. Knowing that the coefficient of restitution between the ball and the wall is 0.9 and neglecting friction, determine the distance d from the foot of the wall to the point B where the ball will hit the ground after bouncing off the wall.

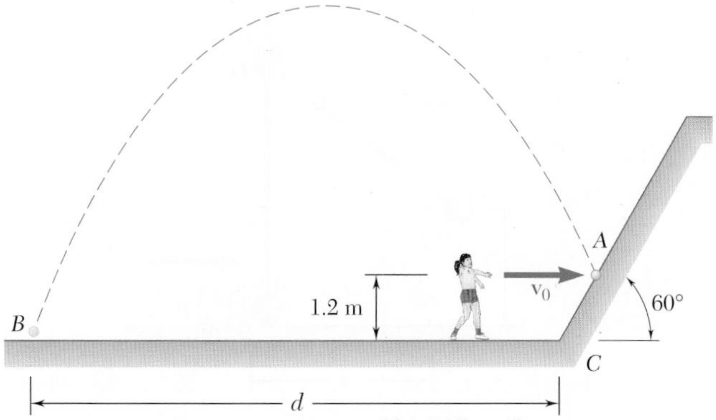

Fig. P13.171

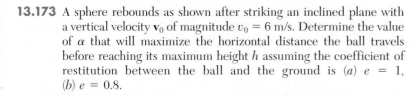

13.172 A sphere rebounds as shown after striking an inclined plane with a vertical velocity $\mathbf{v}_0$ of magnitude $v_0 = 5$ m/s. Knowing that $\alpha = 30°$ and $e = 0.8$ between the sphere and the plane, determine the height h reached by the sphere.

13.173 A sphere rebounds as shown after striking an inclined plane with a vertical velocity $\mathbf{v}_0$ of magnitude $v_0 = 6$ m/s. Determine the value of α that will maximize the horizontal distance the ball travels before reaching its maximum height h assuming the coefficient of restitution between the ball and the ground is (*a*) $e = 1$, (*b*) $e = 0.8$.

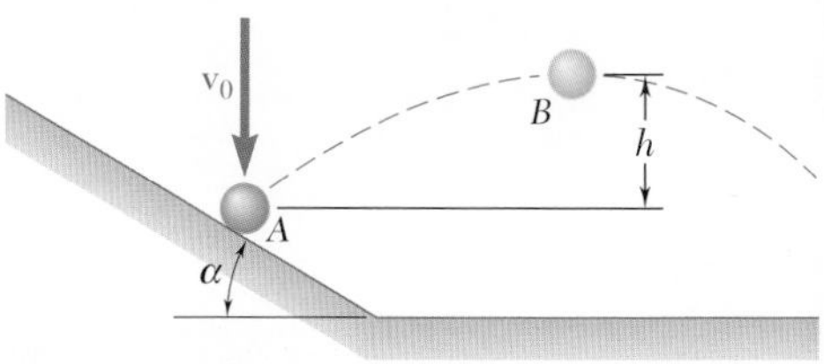

Fig. *P13.172* and P13.173

13.174 Two cars of the same mass run head-on into each other at C. After the collision, the cars skid with their brakes locked and come to a stop in the positions shown in the lower part of the figure. Knowing that the speed of car A just before impact was 8 km/h and that the coefficient of kinetic friction between the pavement and the tires of both cars is 0.30, determine (*a*) the speed of car B just before impact, (*b*) the effective coefficient of restitution between the two cars.

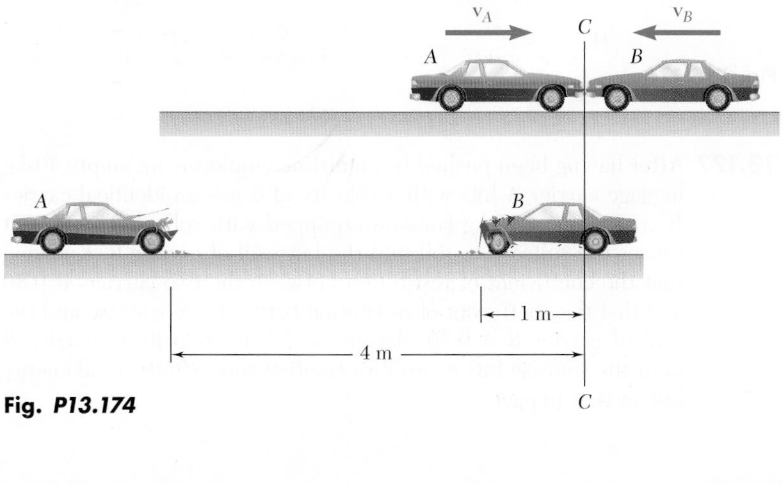

Fig. *P13.174*

13.175 A 1-kg block B is moving with a velocity $\mathbf{v}_0$ of magnitude $v_0 = 2$ m/s as it hits the 0.5-kg sphere A, which is at rest and hanging from a cord attached at O. Knowing that $\mu_k = 0.6$ between the block and the horizontal surface and $e = 0.8$ between the block and the sphere, determine after impact (*a*) the maximum height h reached by the sphere, (*b*) the distance x traveled by the block.

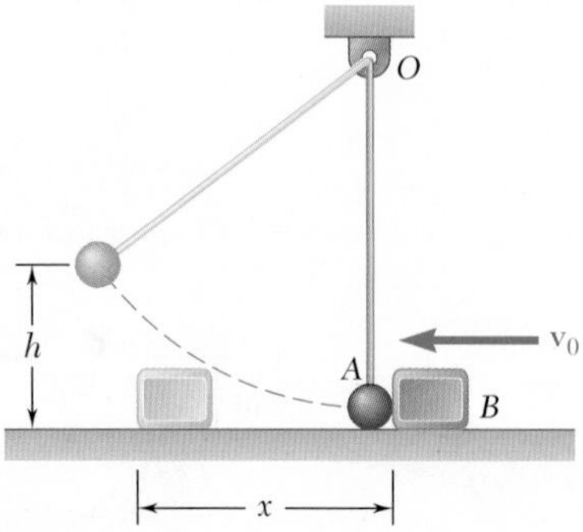

Fig. P13.175

13.176 A 90-g ball thrown with a horizontal velocity $\mathbf{v}_0$ strikes a 720-g plate attached to a vertical wall at a height of 900 mm above the ground. It is observed that after rebounding, the ball hits the ground at a distance of 480 mm from the wall when the plate is rigidly attached to the wall (Fig. 1) and at a distance of 220 mm when a foam-rubber mat is placed between the plate and the wall (Fig. 2). Determine (*a*) the coefficient of restitution e between the ball and the plate, (*b*) the initial velocity $\mathbf{v}_0$ of the ball.

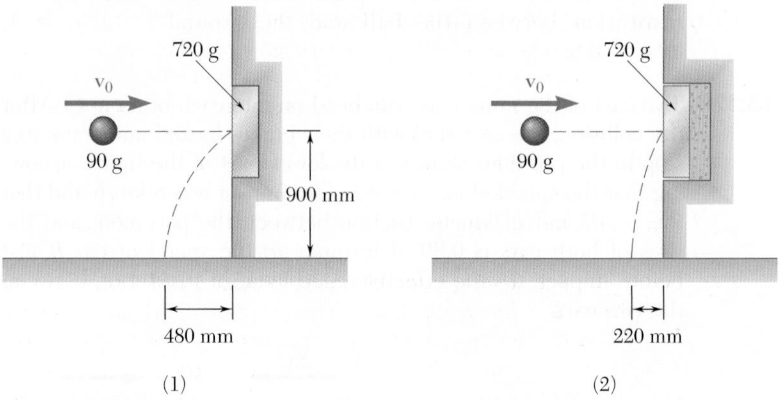

Fig. P13.176

13.177 After having been pushed by an airline employee, an empty 40-kg luggage carrier A hits with a velocity of 5 m/s an identical carrier B containing a 15-kg suitcase equipped with rollers. The impact causes the suitcase to roll into the left wall of carrier B. Knowing that the coefficient of restitution between the two carriers is 0.80 and that the coefficient of restitution between the suitcase and the wall of carrier B is 0.30, determine (*a*) the velocity of carrier B after the suitcase hits its wall for the first time, (*b*) the total energy lost in that impact.

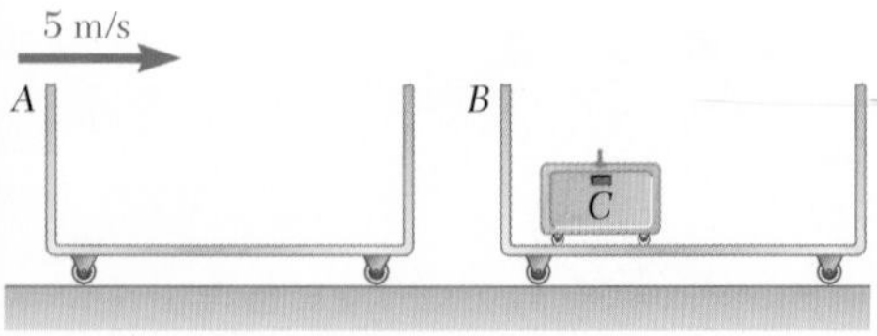

Fig. P13.177

13.178 Blocks A and B each have a mass of 400 g each and block C has a mass of 1.2 kg. The coefficient of friction between the blocks and the plane is $\mu_k = 0.30$. Initially block A is moving at a speed $v_0 = 5$ m/s and blocks B and C are at rest (Fig. 1). After A strikes B and B strikes C, all three blocks come to a stop in the positions shown (Fig. 2). Determine *(a)* the coefficients of restitution between A and B and between B and C, *(b)* the displacement x of block C.

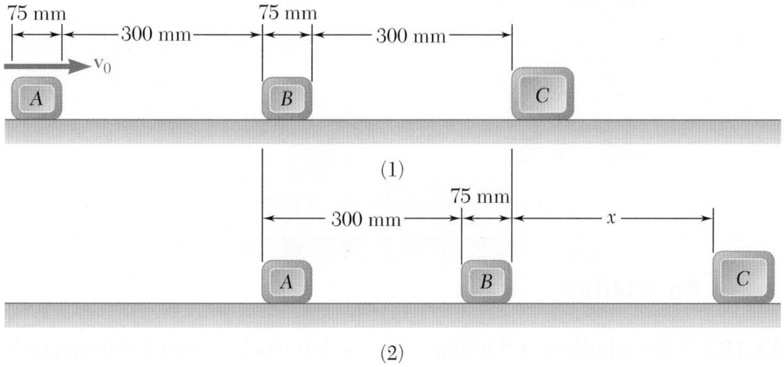

Fig. *P13.178*

13.179 A 0.5-kg sphere A is dropped from a height of 0.6 m onto a 1.0-kg plate B, which is supported by a nested set of springs and is initially at rest. Knowing that the coefficient of restitution between the sphere and the plate is $e = 0.8$, determine *(a)* the height h reached by the sphere after rebound, *(b)* the constant k of the single spring equivalent to the given set if the maximum deflection of the plate is observed to be equal to $3h$.

13.180 A 0.5-kg sphere A is dropped from a height of 0.6 m onto 1.0-kg plate B, which is supported by a nested set of springs and is initially at rest. Knowing that the set of springs is equivalent to a single spring of constant $k = 900$ N/m, determine *(a)* the value of the coefficient of restitution between the sphere and the plate for which the height h reached by the sphere after rebound is maximum, *(b)* the corresponding value of h, (c) the corresponding value of the maximum deflection of the plate.

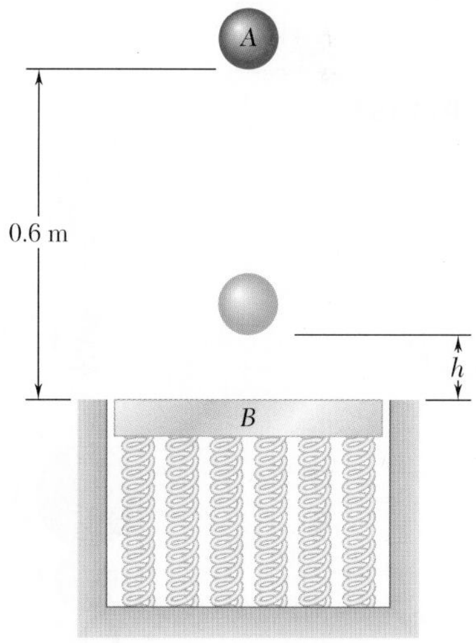

Fig. P13.179 and P13.180

13.181 The three blocks shown are identical. Blocks B and C are at rest when block B is hit by block A, which is moving with a velocity $\mathbf{v}_A$ of 1 m/s. After the impact, which is assumed to be perfectly plastic ($e = 0$), the velocity of blocks A and B decreases due to friction, while block C picks up speed, until all three blocks are moving with the same velocity **v**. Knowing that the coefficient of kinetic friction between all surfaces is $\mu_k = 0.20$, determine *(a)* the time required for the three blocks to reach the same velocity, *(b)* the total distance traveled by each block during that time.

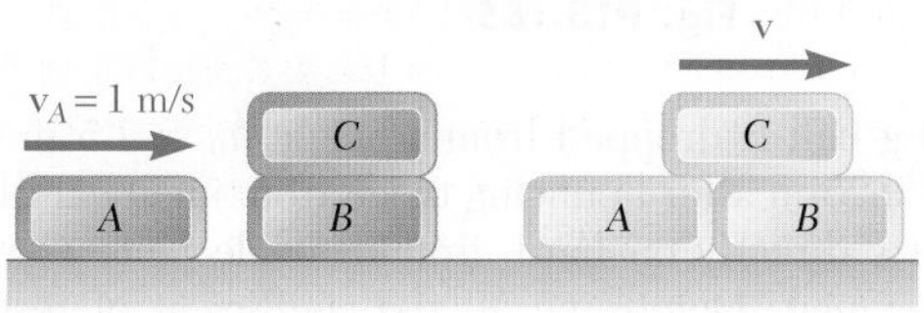

Fig. *P13.181*

13.182 Block A is released from rest and slides down the frictionless surface of B until it hits a bumper on the right end of B. Block A has a mass of 10 kg and object B has a mass of 30 kg and B can roll freely on the ground. Determine the velocities of A and B immediately after impact when (*a*) $e = 0$, (*b*) $e = 0.7$.

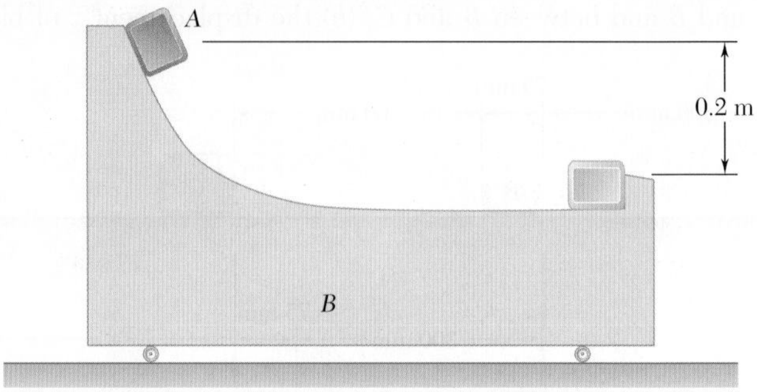

Fig. P13.182

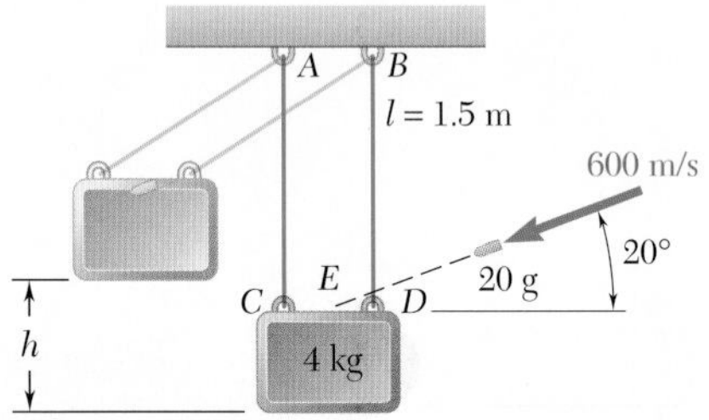

Fig. P13.183

13.183 A 20-g bullet fired into a 4-kg wooden block suspended from cords AC and BD penetrates the block at point E, halfway between C and D, without hitting cord BD. Determine (*a*) the maximum height h to which the block and the embedded bullet will swing after impact, (*b*) the total impulse exerted on the block by the two cords during the impact.

13.184 A 1-kg ball A is suspended from a spring of constant 2000 N/m. and is initially at rest when it is struck by 0.5-kg ball B as shown. Neglecting friction and knowing the coefficient of restitution between the balls is 0.6, determine (*a*) the velocities of A and B after the impact, (*b*) the maximum height reached by A.

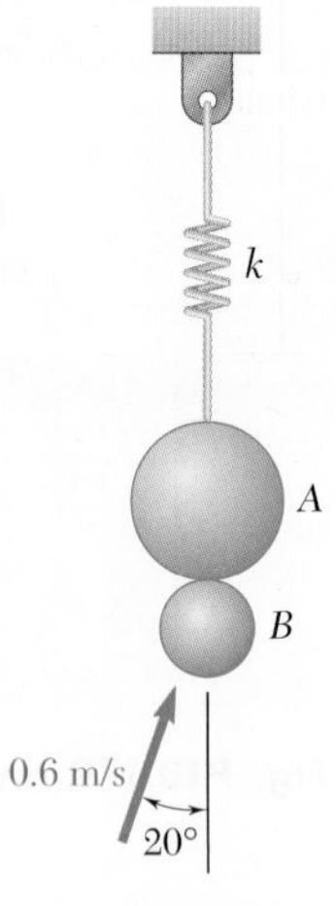

Fig. P13.184

13.185 Ball B is hanging from an inextensible cord. An identical ball A is released from rest when it is just touching the cord and drops through the vertical distance $h_A = 200$ mm before striking ball B. Assuming $e = 0.9$ and no friction, determine the resulting maximum vertical displacement h_B of the ball B.

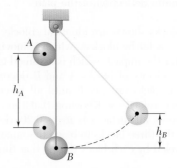

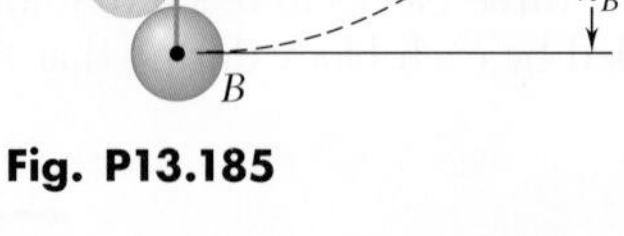

Fig. P13.185

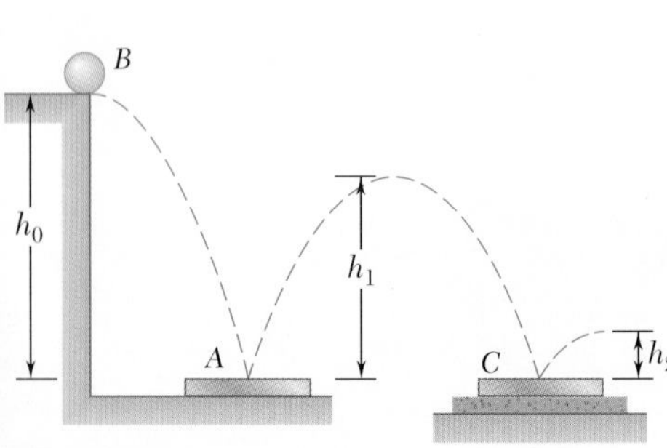

Fig. P13.186

13.186 A 70-g ball B dropped from a height $h_0 = 1.5$ m reaches a height $h_2 = 0.25$ m after bouncing twice from identical 210-g plates. Plate A rests directly on hard ground, while plate C rests on a foam-rubber mat. Determine (*a*) the coefficient of restitution between the ball and the plates, (*b*) the height h_1 of the ball's first bounce.

13.187 A 700-g sphere A moving with a velocity $\mathbf{v}_0$ parallel to the ground strikes the inclined face of a 2.1-kg wedge B which can roll freely on the ground and is initially at rest. After impact the sphere is observed from the ground to be moving straight up. Knowing that the coefficient of restitution between the sphere and the wedge is $e = 0.6$, determine (*a*) the angle θ that the inclined face of the wedge makes with the horizontal, (*b*) the energy lost due to the impact.

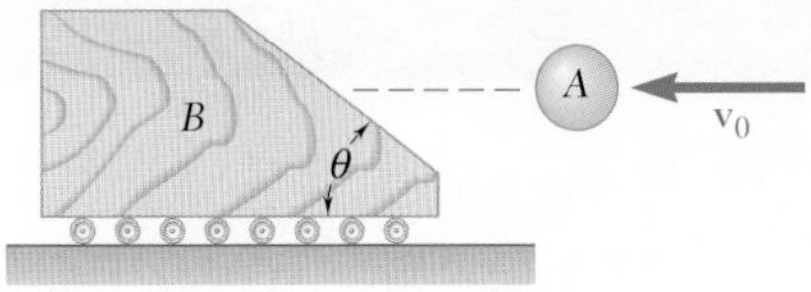

Fig. *P13.187*

13.188 When the rope is at an angle of $\alpha = 30°$ the 0.5-kg sphere A has a speed $v_0 = 1.2$ m/s. The coefficient of restitution between A and the 0.9-kg wedge B is 0.7 and the length of rope $l = 0.8$ m. The spring constant has a value of 500 N/m and $\theta = 20°$. Determine the velocity of A and B immediately after the impact.

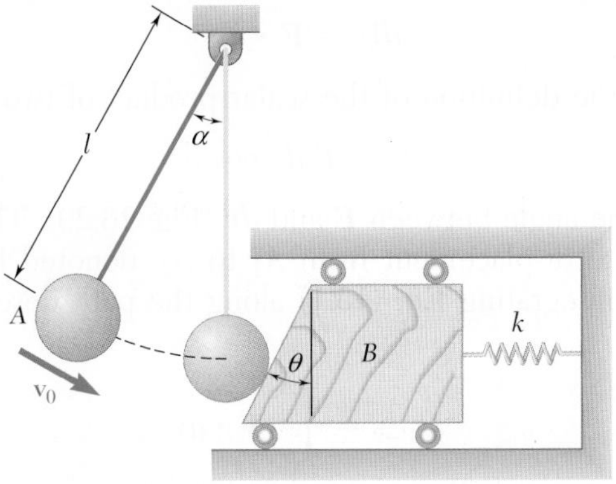

Fig. P13.188 and *P13.189*

13.189 When the rope is at an angle of $\alpha = 30°$ the 1-kg sphere A has a speed $v_0 = 0.6$ m/s. The coefficient of restitution between A and the 2-kg wedge B is 0.8 and the length of rope $l = 0.9$ m. The spring constant has a value of 1500 N/m and $\theta = 20°$. Determine (*a*) the velocities of A and B immediately after the impact, (*b*) the maximum deflection of the spring assuming A does not strike B again before this point.

REVIEW AND SUMMARY

This chapter was devoted to the method of work and energy and to the method of impulse and momentum. In the first half of the chapter we studied the method of work and energy and its application to the analysis of the motion of particles.

Work of a force

We first considered a force $\mathbf{F}$ acting on a particle A and defined the *work of* $\mathbf{F}$ *corresponding to the small displacement* $d\mathbf{r}$ [Sec. 13.2] as the quantity

$$dU = \mathbf{F} \cdot d\mathbf{r} \tag{13.1}$$

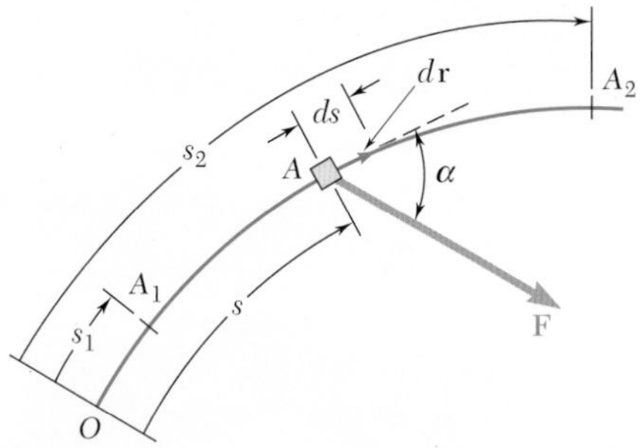

Fig. 13.29

or, recalling the definition of the scalar product of two vectors,

$$dU = F\, ds \cos \alpha \tag{13.1'}$$

where α is the angle between $\mathbf{F}$ and $d\mathbf{r}$ (Fig. 13.29). The work of $\mathbf{F}$ during a finite displacement from A_1 to A_2, denoted by $U_{1\rightarrow 2}$, was obtained by integrating Eq. (13.1) along the path described by the particle:

$$U_{1\rightarrow 2} = \int_{A_1}^{A_2} \mathbf{F} \cdot d\mathbf{r} \tag{13.2}$$

For a force defined by its rectangular components, we wrote

$$U_{1\rightarrow 2} = \int_{A_1}^{A_2} (F_x\, dx + F_y\, dy + F_z\, dz) \tag{13.2''}$$

Work of a weight

The work of the weight $\mathbf{W}$ of a body as its center of gravity moves from the elevation y_1 to y_2 (Fig. 13.30) was obtained by substituting $F_x = F_z = 0$ and $F_y = -W$ into Eq. (13.2″) and integrating. We found

$$U_{1\rightarrow 2} = -\int_{y_1}^{y_2} W\, dy = Wy_1 - Wy_2 \tag{13.4}$$

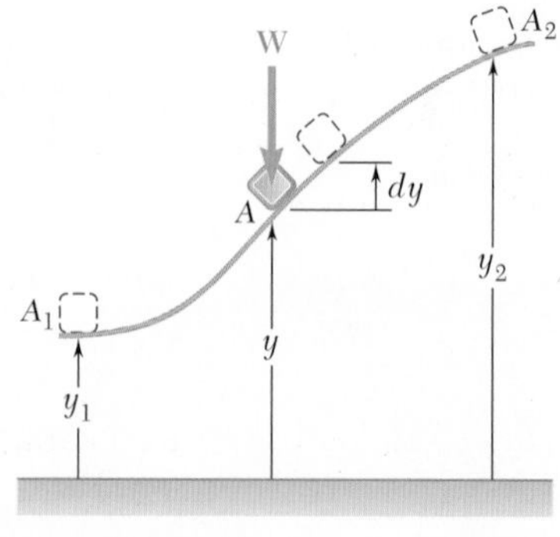

Fig. 13.30

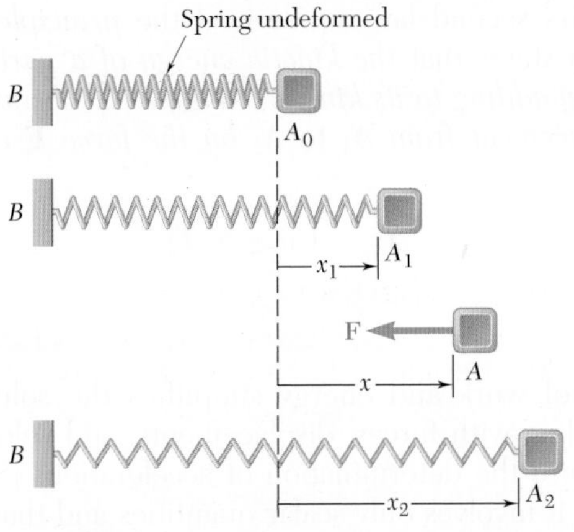

Fig. 13.31

Work of the force exerted by a spring

The work of a force **F** exerted by a spring on a body A during a finite displacement of the body (Fig. 13.31) from $A_1(x = x_1)$ to $A_2(x = x_2)$ was obtained by writing

$$dU = -F\,dx = -kx\,dx$$

$$U_{1\rightarrow 2} = -\int_{x_1}^{x_2} kx\,dx = \tfrac{1}{2}kx_1^2 - \tfrac{1}{2}kx_2^2 \qquad (13.6)$$

The work of **F** is therefore positive *when the spring is returning to its undeformed position.*

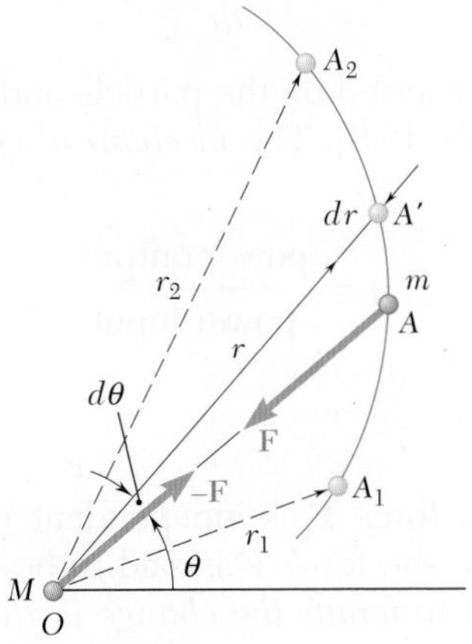

Fig. 13.32

Work of the gravitational force

The *work of the gravitational force* **F** exerted by a particle of mass M located at O on a particle of mass m as the latter moves from A_1 to A_2 (Fig. 13.32) was obtained by recalling from Sec. 12.10 the expression for the magnitude of **F** and writing

$$U_{1\rightarrow 2} = -\int_{r_1}^{r_2} \frac{GMm}{r^2}\,dr = \frac{GMm}{r_2} - \frac{GMm}{r_1} \qquad (13.7)$$

Kinetic energy of a particle

The *kinetic energy of a particle* of mass m moving with a velocity **v** [Sec. 13.3] was defined as the scalar quantity

$$T = \tfrac{1}{2}mv^2 \qquad (13.9)$$

Principle of work and energy

From Newton's second law we derived the *principle of work and energy*, which states that *the kinetic energy of a particle at A_2 can be obtained by adding to its kinetic energy at A_1 the work done during the displacement from A_1 to A_2 by the force* **F** *exerted on the particle:*

$$T_1 + U_{1\rightarrow 2} = T_2 \tag{13.11}$$

Method of work and energy

The method of work and energy simplifies the solution of many problems dealing with forces, displacements, and velocities, since it does not require the determination of accelerations [Sec. 13.4]. We also note that it involves only scalar quantities and that forces which do no work need not be considered [Sample Probs. 13.1 and 13.3]. However, this method should be supplemented by the direct application of Newton's second law to determine a force normal to the path of the particle [Sample Prob. 13.4].

Power and mechanical efficiency

The power developed by a machine and its mechanical efficiency were discussed in Sec. 13.5. Power was defined as the time rate at which work is done:

$$\text{Power} = \frac{dU}{dt} = \mathbf{F} \cdot \mathbf{v} \qquad (13.12,\ 13.13)$$

where **F** is the force exerted on the particle and **v** the velocity of the particle [Sample Prob. 13.5]. The *mechanical efficiency*, denoted by η, was expressed as

$$\eta = \frac{\text{power output}}{\text{power input}} \tag{13.15}$$

Conservative force. Potential energy

When the work of a force **F** is independent of the path followed [Secs. 13.6 and 13.7], the force **F** is said to be a *conservative force*, and its work is equal to *minus the change in the potential energy* V associated with **F**:

$$U_{1\rightarrow 2} = V_1 - V_2 \tag{13.19$'$}$$

The following expressions were obtained for the potential energy associated with each of the forces considered earlier:

Force of gravity (weight): $\quad V_g = Wy \qquad (13.16)$

Gravitational force: $\quad V_g = -\dfrac{GMm}{r} \qquad (13.17)$

Elastic force exerted by a spring: $\quad V_e = \frac{1}{2}kx^2 \qquad (13.18)$

Substituting for $U_{1\to2}$ from Eq. (13.19′) into Eq. (13.11) and rearranging the terms [Sec. 13.8], we obtained

$$T_1 + V_1 = T_2 + V_2 \qquad (13.24)$$

Principle of conservation of energy

This is the *principle of conservation of energy,* which states that when a particle moves under the action of conservative forces, *the sum of its kinetic and potential energies remains constant.* The application of this principle facilitates the solution of problems involving only conservative forces [Sample Probs. 13.6 and 13.7].

Motion under a gravitational force

Recalling from Sec. 12.9 that, when a particle moves under a central force **F**, its angular momentum about the center of force O remains constant, we observed [Sec. 13.9] that, if the central force **F** is also conservative, the principles of conservation of angular momentum and of conservation of energy can be used jointly to analyze the motion of the particle [Sample Prob. 13.8]. Since the gravitational force exerted by the earth on a space vehicle is both central and conservative, this approach was used to study the motion of such vehicles [Sample Prob. 13.9] and was found particularly effective in the case of an *oblique launching.* Considering the initial position P_0 and an arbitrary position P of the vehicle (Fig. 13.33), we wrote

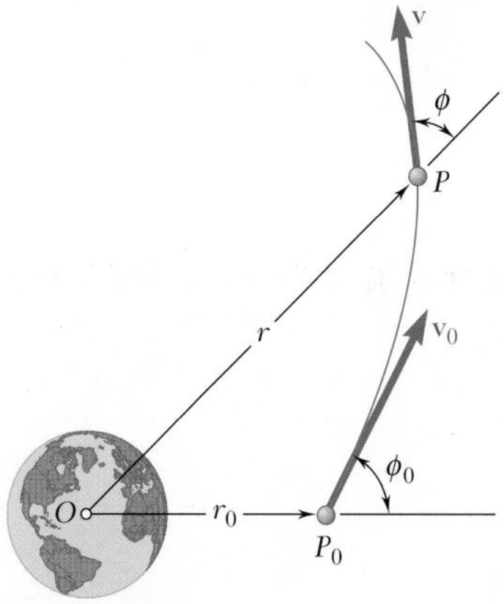

Fig. 13.33

$$(H_O)_0 = H_O: \qquad r_0 m v_0 \sin \phi_0 = r m v \sin \phi \qquad (13.25)$$

$$T_0 + V_0 = T + V: \qquad \tfrac{1}{2} m v_0^2 - \frac{GMm}{r_0} = \tfrac{1}{2} m v^2 - \frac{GMm}{r} \qquad (13.26)$$

where m was the mass of the vehicle and M the mass of the earth.

Principle of impulse and momentum for a particle

The second half of the chapter was devoted to the method of impulse and momentum and to its application to the solution of various types of problems involving the motion of particles.

The *linear momentum of a particle* was defined [Sec. 13.10] as the product $m\mathbf{v}$ of the mass m of the particle and its velocity **v**. From Newton's second law, $\mathbf{F} = m\mathbf{a}$, we derived the relation

$$m\mathbf{v}_1 + \int_{t_1}^{t_2} \mathbf{F}\, dt = m\mathbf{v}_2 \qquad (13.28)$$

where $m\mathbf{v}_1$ and $m\mathbf{v}_2$ represent the momentum of the particle at a time t_1 and a time t_2, respectively, and where the integral defines the *linear impulse of the force* **F** during the corresponding time interval. We wrote therefore

$$m\mathbf{v}_1 + \mathbf{Imp}_{1\to2} = m\mathbf{v}_2 \qquad (13.30)$$

which expresses the principle of impulse and momentum for a particle.

When the particle considered is subjected to several forces, the sum of the impulses of these forces should be used; we had

$$m\mathbf{v}_1 + \Sigma\ \mathbf{Imp}_{1\rightarrow 2} = m\mathbf{v}_2 \tag{13.32}$$

Since Eqs. (13.30) and (13.32) involve *vector quantities,* it is necessary to consider their x and y components separately when applying them to the solution of a given problem [Sample Probs. 13.10 and 13.11].

Impulsive motion

The method of impulse and momentum is particularly effective in the study of the *impulsive motion* of a particle, when very large forces, called *impulsive forces,* are applied for a very short interval of time Δt, since this method involves the impulses $\mathbf{F}\ \Delta t$ of the forces, rather than the forces themselves [Sec. 13.11]. Neglecting the impulse of any nonimpulsive force, we wrote

$$m\mathbf{v}_1 + \Sigma\mathbf{F}\ \Delta t = m\mathbf{v}_2 \tag{13.35}$$

In the case of the impulsive motion of several particles, we had

$$\Sigma m\mathbf{v}_1 + \Sigma\mathbf{F}\ \Delta t = \Sigma m\mathbf{v}_2 \tag{13.36}$$

where the second term involves only impulsive, external forces [Sample Prob. 13.12].

In the particular case *when the sum of the impulses of the external forces is zero,* Eq. (13.36) reduces to $\Sigma m\mathbf{v}_1 = \Sigma m\mathbf{v}_2$; that is, *the total momentum of the particles is conserved.*

Direct central impact

In Secs. 13.12 through 13.14, we considered the *central impact* of two colliding bodies. In the case of a *direct central impact* [Sec. 13.13], the two colliding bodies A and B were moving along the *line of impact* with velocities $\mathbf{v}_A$ and $\mathbf{v}_B$, respectively (Fig. 13.34). Two equations could be used to determine their velocities $\mathbf{v}'_A$ and $\mathbf{v}'_B$ after the impact.

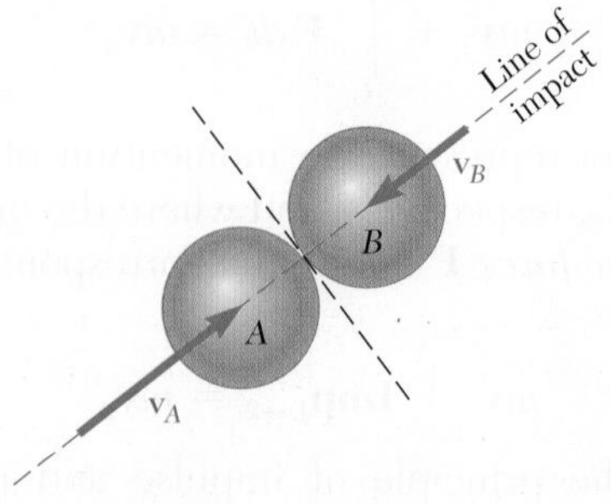

Fig. 13.34

The first expressed conservation of the total momentum of the two bodies,

$$m_A v_A + m_B v_B = m_A v'_A + m_B v'_B \qquad (13.37)$$

where a positive sign indicates that the corresponding velocity is directed to the right, while the second related the *relative velocities* of the two bodies before and after the impact,

$$v'_B - v'_A = e(v_A - v_B) \qquad (13.43)$$

The constant e is known as the *coefficient of restitution;* its value lies between 0 and 1 and depends in a large measure on the materials involved. When $e = 0$, the impact is said to be *perfectly plastic;* when $e = 1$, it is said to be *perfectly elastic* [Sample Prob. 13.13].

Oblique central impact

In the case of an *oblique central impact* [Sec. 13.14], the velocities of the two colliding bodies before and after the impact were resolved into n components along the line of impact and t components along the common tangent to the surfaces in contact (Fig. 13.35). We observed that the t component of the velocity of each body remained

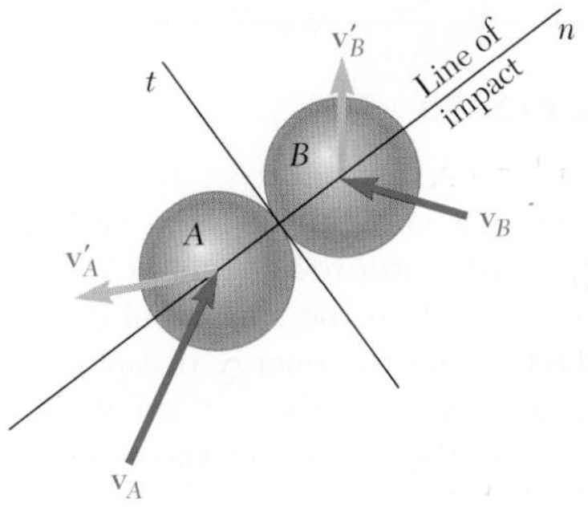

Fig. 13.35

unchanged, while the n components satisfied equations similar to Eqs. (13.37) and (13.43) [Sample Probs. 13.14 and 13.15]. It was shown that although this method was developed for bodies moving freely before and after the impact, it could be extended to the case when one or both of the colliding bodies is constrained in its motion [Sample Prob. 13.16].

Using the three fundamental methods of kinetic analysis

In Sec. 13.15, we discussed the relative advantages of the three fundamental methods presented in this chapter and the preceding one, namely, Newton's second law, work and energy, and impulse and momentum. We noted that the method of work and energy and the method of impulse and momentum can be combined to solve problems involving a short impact phase during which impulsive forces must be taken into consideration [Sample Prob. 13.17].

REVIEW PROBLEMS

13.190 A section of track for a roller coaster consists of two circular arcs AB and CD joined by a straight portion BC. The radius of AB is 30 m and the radius of CD is 80 m. The car and its occupants, of total mass 300 kg reach point A with practically no velocity and then drop freely along the track. Determine the normal force exerted by the track on the car as the car reaches point B. Ignore air resistance and rolling resistance.

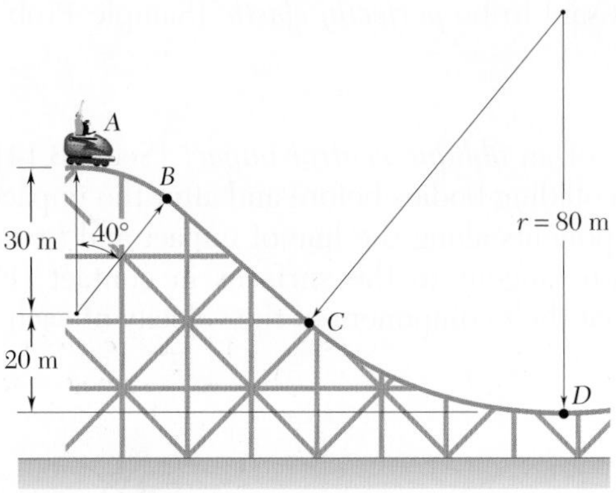

Fig. P13.190

13.191 A 60-g pellet shot vertically from a spring-loaded pistol on the surface of the earth rises to a height of 90 m. The same pellet shot from the same pistol on the surface of the moon rises to a height of 570 m. Determine the energy dissipated by aerodynamic drag when the pellet is shot on the surface of the earth. (The acceleration of gravity on the surface of the moon is 0.165 times that on the surface of the earth.)

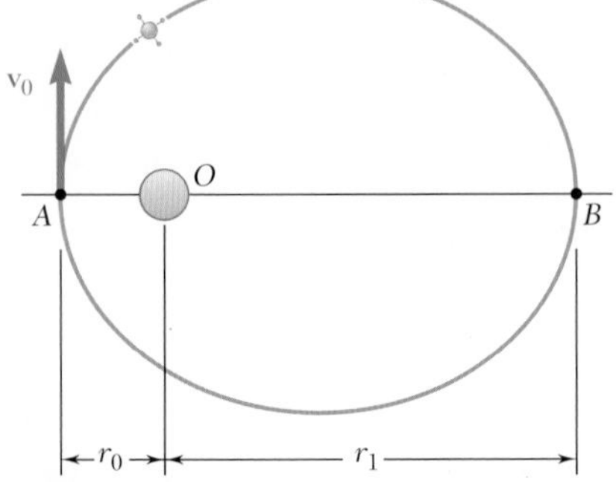

Fig. P13.192

13.192 A satellite describes an elliptic orbit about a planet of mass M. The minimum and maximum values of the distance r from the satellite to the center of the planet are, respectively, r_0 and r_1. Use the principles of conservation of energy and conservation of angular momentum to derive the relation

$$\frac{1}{r_0} + \frac{1}{r_1} = \frac{2GM}{h^2}$$

where h is the angular momentum per unit mass of the satellite and G is the constant of gravitation.

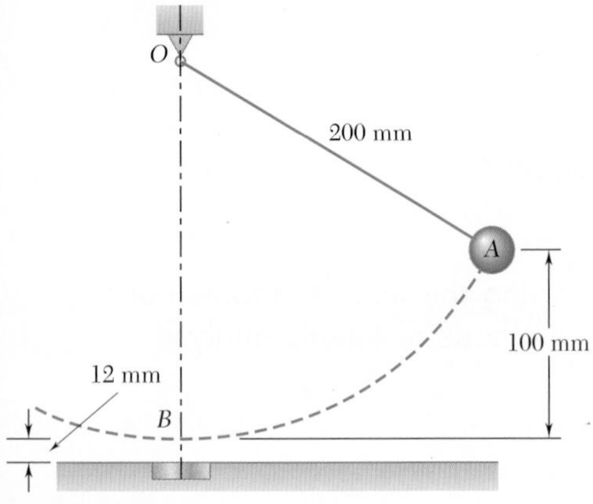

Fig. *P13.193*

13.193 A 60-g steel sphere attached to a 200-mm cord can swing about point O in a vertical plane. It is subjected to its own weight and to a force **F** exerted by a small magnet embedded in the ground. The magnitude of that force expressed in newtons is $F = 3000/r^2$, where r is the distance from the magnet to the sphere expressed in millimeters. Knowing that the sphere is released from rest at A, determine its speed as it passes through point B.

13.194 A shuttle is to rendezvous with a space station which is in a circular orbit at an altitude of 400 km above the surface of the earth. The shuttle has reached an altitude of 60 km when its engine is turned off at point B. Knowing that at that time the velocity $\mathbf{v}_0$ of the shuttle forms an angle $\phi_0 = 55°$ with the vertical, determine the required magnitude of $\mathbf{v}_0$ if the trajectory of the shuttle is to be tangent at A to the orbit of the space station.

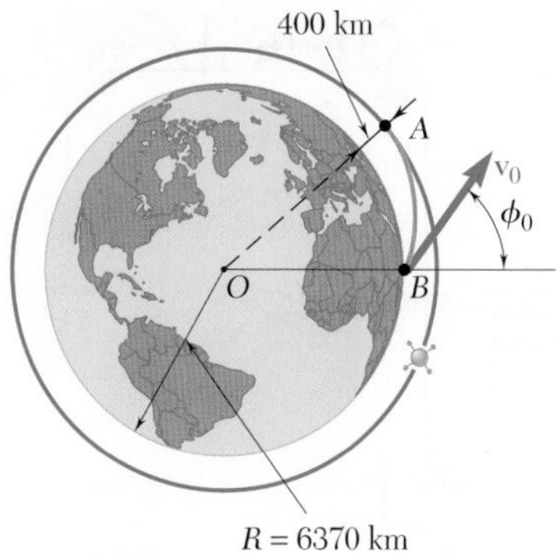

Fig. P13.194

13.195 A 300-g block is released from rest after a spring of constant $k = 600$ N/m has been compressed 160 mm. Determine the force exerted by the loop $ABCD$ on the block as the block passes through (*a*) point A, (*b*) point B, (*c*) point C. Assume no friction.

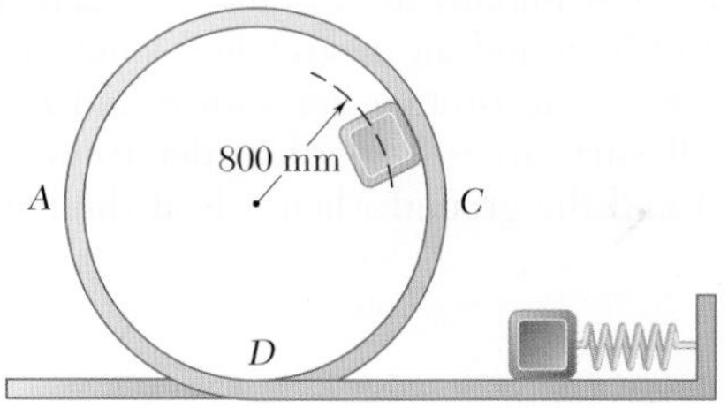

Fig. P13.195

13.196 A small sphere B of mass m is attached to an inextensible cord of length $2a$, which passes around the fixed peg A and is attached to a fixed support at O. The sphere is held close to the support at O and released with no initial velocity. It drops freely to point C, where the cord becomes taut, and swings in a vertical plane, first about A and then about O. Determine the vertical distance from line OD to the highest point C'' that the sphere will reach.

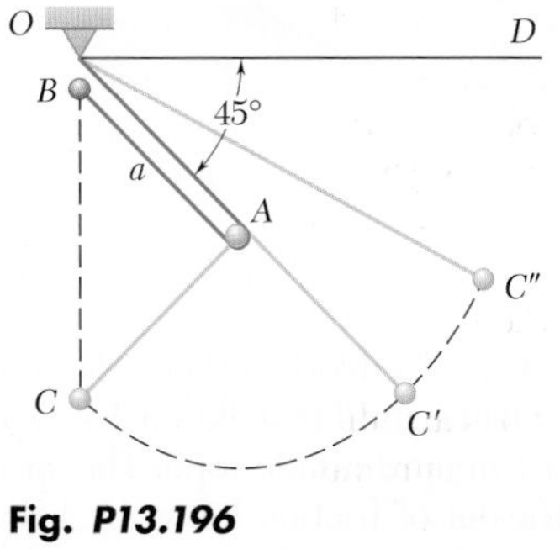

Fig. P13.196

13.197 A 300-g collar A is released from rest, slides down a frictionless rod, and strikes a 900-g collar B which is at rest and supported by a spring of constant 500 N/m. Knowing that the coefficient of restitution between the two collars is 0.9, determine (*a*) the maximum distance collar A moves up the rod after impact, (*b*) the maximum distance collar B moves down the rod after impact.

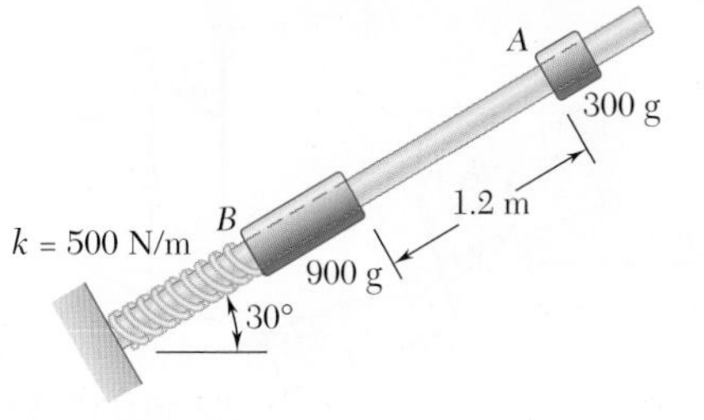

Fig. P13.197

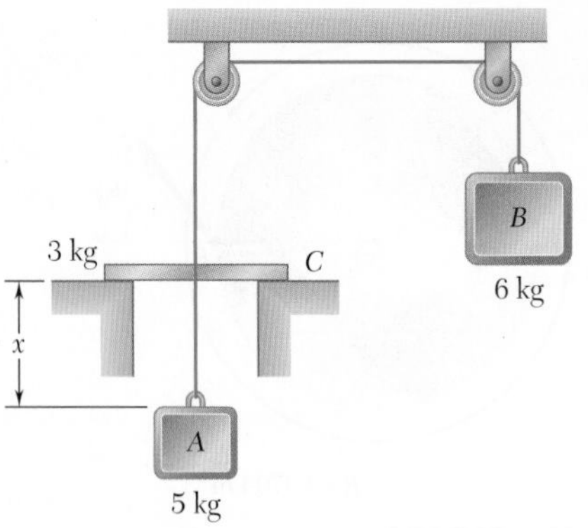

Fig. P13.198

13.198 Blocks A and B are connected by a cord which passes over pulleys and through a collar C. The system is released from rest when $x = 1.7$ m. As block A rises, it strikes collar C with perfectly plastic impact ($e = 0$). After impact, the two blocks and the collar keep moving until they come to a stop and reverse their motion. As A and C move down, C hits the ledge and blocks A and B keep moving until they come to another stop. Determine (*a*) the velocity of the blocks and collar immediately after A hits C, (*b*) the distance the blocks and collar move after the impact before coming to a stop, (*c*) the value of x at the end of one complete cycle.

13.199 A 2-kg ball B is traveling horizontally at 10 m/s when it strikes 2-kg ball A. Ball A is initially at rest and is attached to a spring with constant 100 N/m and an unstretched length of 1.2 m. Knowing the coefficient of restitution between A and B is 0.8 and friction between all surfaces is negligible, determine the normal force between A and the ground when it is at the bottom of the hill.

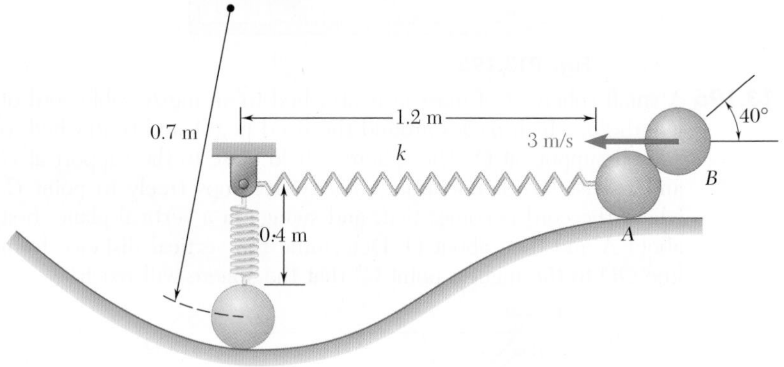

Fig. P13.199

13.200 A 2-kg block A is pushed up against a spring compressing it a distance $x = 0.1$ m. The block is then released from rest and slides down the 20° incline until it strikes a 1-kg sphere B which is suspended from a 1-m inextensible rope. The spring constant $k = 800$ N/m, the coefficient of friction between A and the ground is 0.2, the distance A slides from the unstretched length of the spring $d = 1.5$ m, and the coefficient of restitution between A and B is 0.8. When $\alpha = 40°$, determine (*a*) the speed of B, (*b*) the tension in the rope.

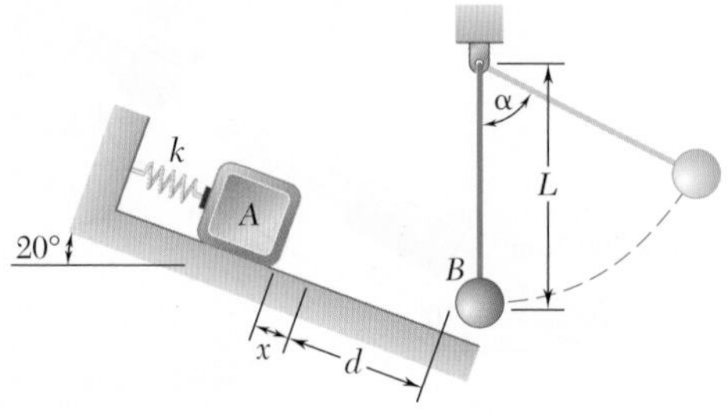

Fig. P13.200

*13.201 The 1-kg ball at A is suspended by an inextensible cord and given an initial horizontal velocity of $\mathbf{v}_0$. If $l = 600$ mm, $x_B = 90$ mm, and $y_B = 120$ mm, determine the initial velocity $\mathbf{v}_0$ so that the ball will enter the basket. (*Hint:* Use a computer to solve the resulting set of equations.)

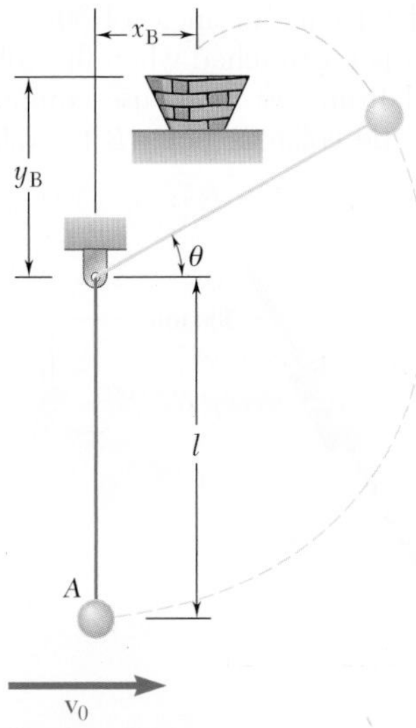

Fig. *P13.201*

COMPUTER PROBLEMS

13.C1 A 6-kg collar is attached to a spring anchored at point C and can slide on a frictionless rod forming an angle of 30° with the vertical. The spring is of constant k and is unstretched when the collar is at A. Knowing that the collar is released from rest at A, use computational software to determine the velocity of the collar at point B for values of k from 20 to 400 N/m.

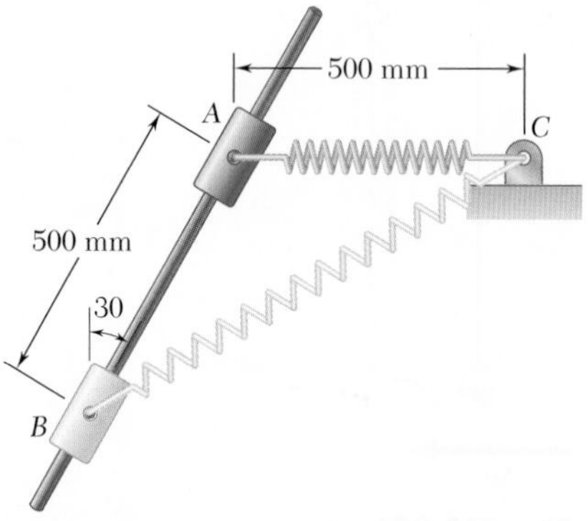

Fig. P13.C1

13.C2 Skid marks on a drag race track indicate that the rear (drive) wheels of a 1000-kg car slip with the front wheels just off the ground for the first 20 m of the 500-m track. The car is driven with slipping impending, with 60 percent of its weight on the rear wheels, for the remaining 480 m of the race. Knowing that the coefficients of kinetic and static friction are 0.60 and 0.85, respectively, and that the force due to the aerodynamic drag is $F_d = 0.544v^2$, where the speed v is expressed in m/s and the force F_d in N, use computational software to determine the time elapsed and the speed of the car at various points along the track, (a) taking the force F_d into account, (b) ignoring the force F_d. If you write a computer program use increments of distance $\Delta x = 0.025$ m in your calculations, and tabulate your results every 2 m for the first 20 m and every 20 m for the remaining 480 m. [*Hint:* The time Δt_i required for the car to move through the increment of distance Δx_i can be obtained by dividing Δx_i by the average velocity $\frac{1}{2}(v_i + v_{i+1})$ of the car over Δx_i if the acceleration of the car is assumed to remain constant over Δx.]

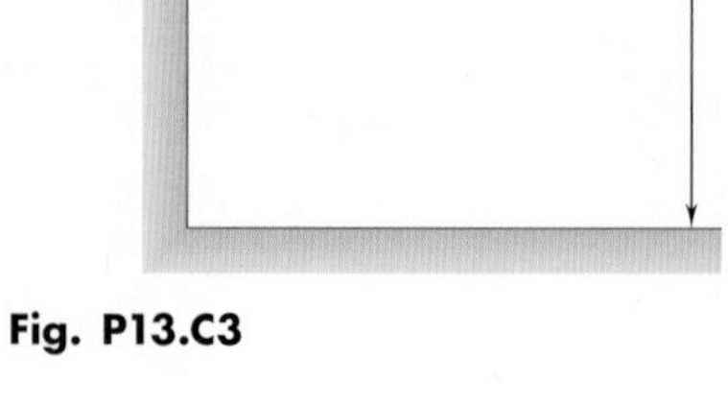

Fig. P13.C3

13.C3 A 5-kg bag is gently pushed off the top of a wall and swings in a vertical plane at the end of a 2.4-m rope which can withstand a maximum tension F_m. For F_m from 40 to 140 N use computational software to determine (a) the difference in elevation h between point A and point B where the rope will break, (b) the distance d from the vertical wall to the point where the bag will strike the floor.

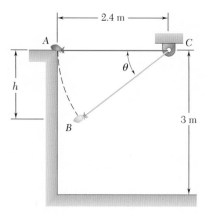

13.C4 Use computational software to determine (a) the time required for the system of Prob. 13.198 to complete 10 successive cycles of the motion described in that problem, starting with $x = 1.7$ m, (b) the value of x at the end of the tenth cycle.

13.C5 A 700-g ball B is hanging from an inextensible cord attached to a support at C. A 350-g ball A strikes B with a velocity $\mathbf{v}_0$ at an angle θ_0 with the vertical. Assuming no friction and denoting by e the coefficient of restitution, use computational software to determine the magnitudes v'_A and v'_B of the velocities of the balls immediately after impact and the percentage of energy lost in the collision for $v_0 = 6$ m/s and values of θ_0 from 20° to 150°, assuming (*a*) $e = 1$, (*b*) $e = 0.75$, (*c*) $e = 0$.

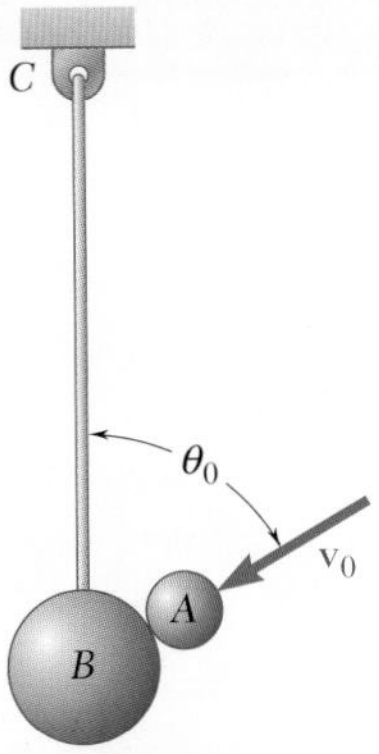

Fig. P13.C5

13.C6 In Prob. 13.110, a space vehicle was in a circular orbit at an altitude of 360 km above the surface of the earth. To return to earth it decreased its speed as it passed through A by firing its engine for a short interval of time in a direction opposite to the direction of its motion. Its resulting velocity as it reached point B at an altitude of 60 km formed an angle $\phi_B = 60°$ with the vertical. An alternative strategy for taking the space vehicle out of its circular orbit would be to turn it around so that its engine pointed away from the earth and then give it an incremental velocity $\Delta\mathbf{v}_A$ toward the center O of the earth. This would likely require a smaller expenditure of energy when firing the engine at A, but might result in too fast a descent at B. Assuming that this strategy is used, use computational software to determine the values of ϕ_B and v_B for an energy expenditure ranging from 5 to 100 percent of that needed in Prob. 13.110.

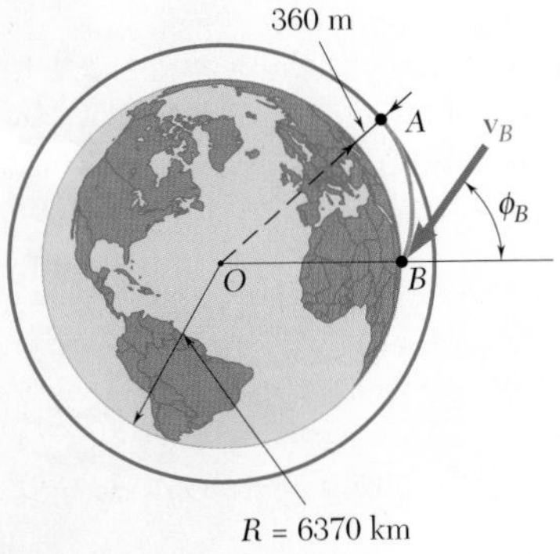

Fig. P13.C6

The thrust for this XR-5M15 prototype engine is produced by gas particles being ejected at a high velocity. The determination of the forces on the test stand is based on the analysis of the motion of a *variable system of particles,* i.e., the motion of a large number of air particles considered together rather than separately.

CHAPTER 14

Systems of Particles

14.1 INTRODUCTION

In this chapter you will study the motion of *systems of particles,* i.e., the motion of a large number of particles considered together. The first part of the chapter is devoted to systems consisting of well-defined particles; the second part considers the motion of variable systems, i.e., systems which are continually gaining or losing particles, or doing both at the same time.

In Sec. 14.2, Newton's second law will first be applied to each particle of the system. Defining the *effective force* of a particle as the product $m_i\mathbf{a}_i$ of its mass m_i and its acceleration $\mathbf{a}_i$, we will show that the *external forces* acting on the various particles form a system equipollent to the system of the effective forces, i.e., both systems have the same resultant and the same moment resultant about any given point. In Sec. 14.3, it will be further shown that the resultant and moment resultant of the external forces are equal, respectively, to the rate of change of the total linear momentum and of the total angular momentum of the particles of the system.

In Sec. 14.4, the *mass center* of a system of particles is defined and the motion of that point is described, while in Sec. 14.5 the motion of the particles about their mass center is analyzed. The conditions under which the linear momentum and the angular momentum of a system of particles are conserved are discussed in Sec. 14.6, and the results obtained in that section are applied to the solution of various problems.

Sections 14.7 and 14.8 deal with the application of the work-energy principle to a system of particles, and Sec. 14.9 with the application of the impulse-momentum principle. These sections also contain a number of problems of practical interest.

It should be noted that while the derivations given in the first part of this chapter are carried out for a system of independent particles, they remain valid when the particles of the system are rigidly connected, i.e., when they form a rigid body. In fact, the results obtained here will form the foundation of our discussion of the kinetics of rigid bodies in Chaps. 16 through 18.

The second part of this chapter is devoted to the study of variable systems of particles. In Sec. 14.11 you will consider steady streams of particles, such as a stream of water diverted by a fixed vane, or the flow of air through a jet engine, and learn to determine the force exerted by the stream on the vane and the thrust developed by the engine. Finally, in Sec. 14.12, you will learn how to analyze systems which gain mass by continually absorbing particles or lose mass by continually expelling particles. Among the various practical applications of this analysis will be the determination of the thrust developed by a rocket engine.

14.2 APPLICATION OF NEWTON'S LAWS TO THE MOTION OF A SYSTEM OF PARTICLES. EFFECTIVE FORCES

In order to derive the equations of motion for a system of n particles, let us begin by writing Newton's second law for each individual particle of the system. Consider the particle P_i, where $1 \le i \le n$. Let

m_i be the mass of P_i and $\mathbf{a}_i$ its acceleration with respect to the newtonian frame of reference $Oxyz$. The force exerted on P_i by another particle P_j of the system (Fig. 14.1), called an *internal force,* will be denoted by $\mathbf{f}_{ij}$. The resultant of the internal forces exerted on P_i by all the other particles of the system is thus $\sum_{j=1}^{n} \mathbf{f}_{ij}$ (where $\mathbf{f}_{ii}$ has no meaning and is assumed to be equal to zero). Denoting, on the other hand, by $\mathbf{F}_i$ the resultant of all the *external forces* acting on P_i, we write Newton's second law for the particle P_i as follows:

$$\mathbf{F}_i + \sum_{j=1}^{n} \mathbf{f}_{ij} = m_i\mathbf{a}_i \tag{14.1}$$

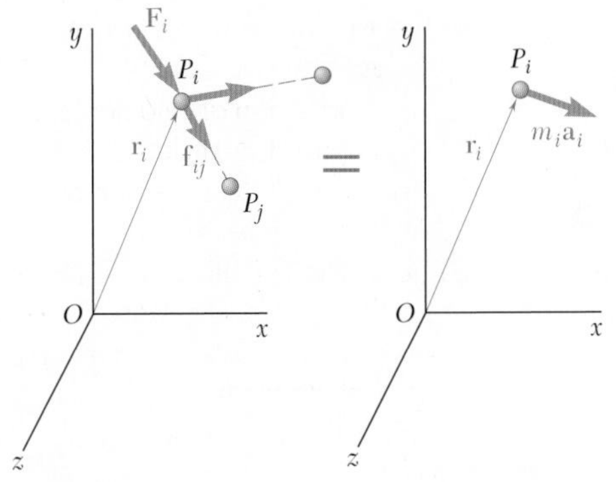

Fig. 14.1

Denoting by $\mathbf{r}_i$ the position vector of P_i and taking the moments about O of the various terms in Eq. (14.1), we also write

$$\mathbf{r}_i \times \mathbf{F}_i + \sum_{j=1}^{n} (\mathbf{r}_i \times \mathbf{f}_{ij}) = \mathbf{r}_i \times m_i\mathbf{a}_i \tag{14.2}$$

Repeating this procedure for each particle P_i of the system, we obtain n equations of the type (14.1) and n equations of the type (14.2), where i takes successively the values 1, 2, . . . , n. The vectors $m_i\mathbf{a}_i$ are referred to as the *effective forces* of the particles.† Thus the equations obtained express the fact that the external forces $\mathbf{F}_i$ and the internal forces $\mathbf{f}_{ij}$ acting on the various particles form a system equivalent to the system of the effective forces $m_i\mathbf{a}_i$ (i.e., one system may be replaced by the other) (Fig. 14.2).

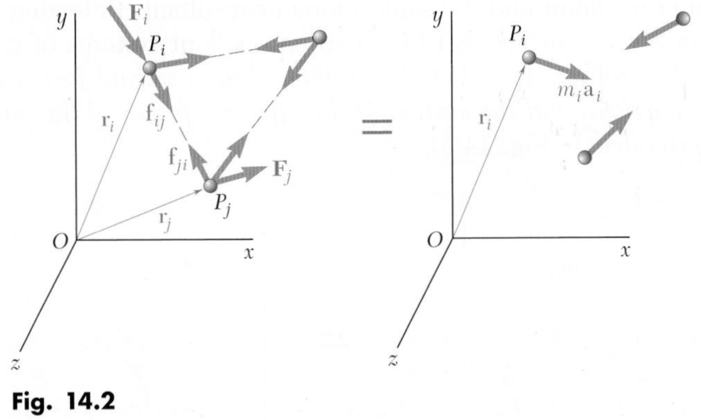

Fig. 14.2

Before proceeding further with our derivation, let us examine the internal forces $\mathbf{f}_{ij}$. We note that these forces occur in pairs $\mathbf{f}_{ij}$, $\mathbf{f}_{ji}$, where $\mathbf{f}_{ij}$ represents the force exerted by the particle P_j on the particle P_i and $\mathbf{f}_{ji}$ represents the force exerted by P_i on P_j (Fig. 14.2). Now, according to Newton's third law (Sec. 6.1), as extended by Newton's law of gravitation to particles acting at a distance (Sec. 12.10), the forces $\mathbf{f}_{ij}$ and $\mathbf{f}_{ji}$ are equal and opposite and have the same line of action. Their sum is therefore $\mathbf{f}_{ij} + \mathbf{f}_{ji} = 0$, and the sum of their moments about O is

$$\mathbf{r}_i \times \mathbf{f}_{ij} + \mathbf{r}_j \times \mathbf{f}_{ji} = \mathbf{r}_i \times (\mathbf{f}_{ij} + \mathbf{f}_{ji}) + (\mathbf{r}_j - \mathbf{r}_i) \times \mathbf{f}_{ji} = 0$$

†Since these vectors represent the resultants of the forces acting on the various particles of the system, they can truly be considered as forces.

since the vectors $\mathbf{r}_j - \mathbf{r}_i$ and $\mathbf{f}_{ji}$ in the last term are collinear. Adding all the internal forces of the system and summing their moments about O, we obtain the equations

$$\sum_{i=1}^{n}\sum_{j=1}^{n} \mathbf{f}_{ij} = 0 \qquad \sum_{i=1}^{n}\sum_{j=1}^{n} (\mathbf{r}_i \times \mathbf{f}_{ij}) = 0 \tag{14.3}$$

which express the fact that the resultant and the moment resultant of the internal forces of the system are zero.

Returning now to the n equations (14.1), where $i = 1, 2, \ldots, n$, we sum their left-hand members and sum their right-hand members. Taking into account the first of Eqs. (14.3), we obtain

$$\sum_{i=1}^{n} \mathbf{F}_i = \sum_{i=1}^{n} m_i\mathbf{a}_i \tag{14.4}$$

Proceeding similarly with Eq. (14.2) and taking into account the second of Eqs. (14.3), we have

$$\sum_{i=1}^{n} (\mathbf{r}_i \times \mathbf{F}_i) = \sum_{i=1}^{n} (\mathbf{r}_i \times m_i\mathbf{a}_i) \tag{14.5}$$

Equations (14.4) and (14.5) express the fact that the system of the external forces $\mathbf{F}_i$ and the system of the effective forces $m_i\mathbf{a}_i$ have the same resultant and the same moment resultant. Referring to the definition given in Sec. 3.19 for two equipollent systems of vectors, we can therefore state that *the system of the external forces acting on the particles and the system of the effective forces of the particles are equipollent*† (Fig. 14.3).

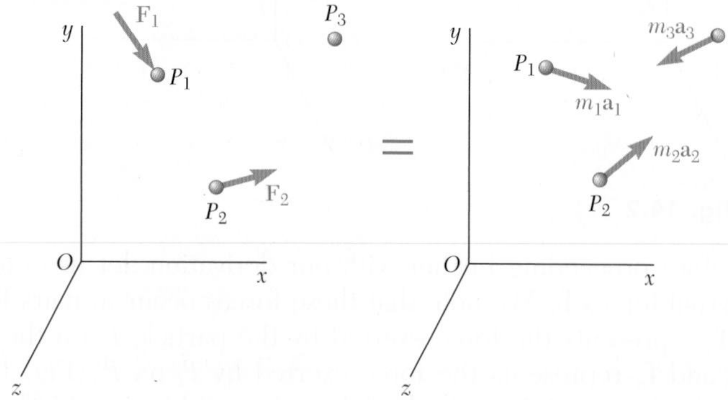

Fig. 14.3

†The result just obtained is often referred to as *d'Alembert's principle*, after the French mathematician Jean le Rond d'Alembert (1717–1783). However, d'Alembert's original statement refers to the motion of a system of connected bodies, with $\mathbf{f}_{ij}$ representing constraint forces which if applied by themselves will not cause the system to move. Since, as it will now be shown, this is in general not the case for the internal forces acting on a system of free particles, the consideration of d'Alembert's principle will be postponed until the motion of rigid bodies is considered (Chap. 16).

Equations (14.3) express the fact that the system of the internal forces $\mathbf{f}_{ij}$ is equipollent to zero. Note, however, that it does *not* follow that the internal forces have no effect on the particles under consideration. Indeed, the gravitational forces that the sun and the planets exert on one another are internal to the solar system and equipollent to zero. Yet these forces are alone responsible for the motion of the planets about the sun.

Similarly, it does not follow from Eqs. (14.4) and (14.5) that two systems of external forces which have the same resultant and the same moment resultant will have the same effect on a given system of particles. Clearly, the systems shown in Figs. 14.4*a* and 14.4*b* have

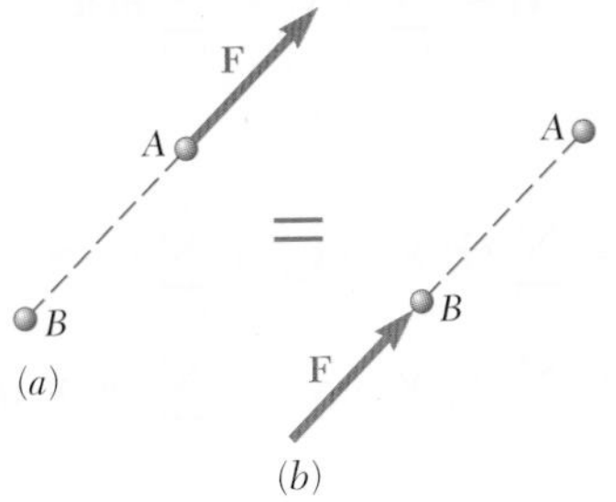

Fig. 14.4

the same resultant and the same moment resultant; yet the first system accelerates particle *A* and leaves particle *B* unaffected, while the second accelerates *B* and does not affect *A*. It is important to recall that when we stated in Sec. 3.19 that two equipollent systems of forces acting on a rigid body are also equivalent, we specifically noted that this property could *not* be extended to a system of forces acting on a set of independent particles such as those considered in this chapter.

In order to avoid any confusion, blue equals signs are used to connect equipollent systems of vectors, such as those shown in Figs. 14.3 and 14.4. These signs indicate that the two systems of vectors have the same resultant and the same moment resultant. Red equals signs will continue to be used to indicate that two systems of vectors are equivalent, i.e., that one system can actually be replaced by the other (Fig. 14.2).

14.3 LINEAR AND ANGULAR MOMENTUM OF A SYSTEM OF PARTICLES

Equations (14.4) and (14.5), obtained in the preceding section for the motion of a system of particles, can be expressed in a more condensed form if we introduce the linear and the angular momentum of the system of particles. Defining the linear momentum **L** of the system of particles as the sum of the linear momenta of the various particles of the system (Sec. 12.3), we write

$$\mathbf{L} = \sum_{i=1}^{n} m_i \mathbf{v}_i \tag{14.6}$$

Defining the angular momentum $\mathbf{H}_O$ about O of the system of particles in a similar way (Sec. 12.7), we have

$$\mathbf{H}_O = \sum_{i=1}^{n} (\mathbf{r}_i \times m_i\mathbf{v}_i) \tag{14.7}$$

Differentiating both members of Eqs. (14.6) and (14.7) with respect to t, we write

$$\dot{\mathbf{L}} = \sum_{i=1}^{n} m_i\dot{\mathbf{v}}_i = \sum_{i=1}^{n} m_i\mathbf{a}_i \tag{14.8}$$

and

$$\begin{aligned}\dot{\mathbf{H}}_O &= \sum_{i=1}^{n} (\dot{\mathbf{r}}_i \times m_i\mathbf{v}_i) + \sum_{i=1}^{n} (\mathbf{r}_i \times m_i\dot{\mathbf{v}}_i) \\ &= \sum_{i=1}^{n} (\mathbf{v}_i \times m_i\mathbf{v}_i) + \sum_{i=1}^{n} (\mathbf{r}_i \times m_i\mathbf{a}_i)\end{aligned}$$

which reduces to

$$\dot{\mathbf{H}}_O = \sum_{i=1}^{n} (\mathbf{r}_i \times m_i\mathbf{a}_i) \tag{14.9}$$

since the vectors $\mathbf{v}_i$ and $m_i\mathbf{v}_i$ are collinear.

We observe that the right-hand members of Eqs. (14.8) and (14.9) are respectively identical with the right-hand members of Eqs. (14.4) and (14.5). It follows that the left-hand members of these equations are respectively equal. Recalling that the left-hand member of Eq. (14.5) represents the sum of the moments $\mathbf{M}_O$ about O of the external forces acting on the particles of the system, and omitting the subscript i from the sums, we write

$$\Sigma\mathbf{F} = \dot{\mathbf{L}} \tag{14.10}$$

$$\Sigma\mathbf{M}_O = \dot{\mathbf{H}}_O \tag{14.11}$$

These equations express that *the resultant and the moment resultant about the fixed point O of the external forces are respectively equal to the rates of change of the linear momentum and of the angular momentum about O of the system of particles.*

14.4 MOTION OF THE MASS CENTER OF A SYSTEM OF PARTICLES

Equation (14.10) may be written in an alternative form if the *mass center* of the system of particles is considered. The mass center of the system is the point G defined by the position vector $\bar{\mathbf{r}}$, which satisfies the relation

$$m\bar{\mathbf{r}} = \sum_{i=1}^{n} m_i\mathbf{r}_i \tag{14.12}$$

where m represents the total mass $\sum_{i=1}^{n} m_i$ of the particles. Resolving the position vectors $\bar{\mathbf{r}}$ and $\mathbf{r}_i$ into rectangular components, we obtain the following three scalar equations, which can be used to determine the coordinates $\bar{x}, \bar{y}, \bar{z}$ of the mass center:

$$m\bar{x} = \sum_{i=1}^{n} m_i x_i \qquad m\bar{y} = \sum_{i=1}^{n} m_i y_i \qquad m\bar{z} = \sum_{i=1}^{n} m_i z_i \tag{14.12$'$}$$

Since $m_i g$ represents the weight of the particle P_i, and mg the total weight of the particles, G is also the center of gravity of the system of particles. However, in order to avoid any confusion, G will be referred to as the *mass center* of the system of particles when properties associated with the *mass* of the particles are being discussed, and as the *center of gravity* of the system when properties associated with the *weight* of the particles are being considered. Particles located outside the gravitational field of the earth, for example, have a mass but no weight. We can then properly refer to their mass center, but obviously not to their center of gravity.†

Differentiating both members of Eq. (14.12) with respect to t, we write

$$m\dot{\bar{\mathbf{r}}} = \sum_{i=1}^{n} m_i\dot{\mathbf{r}}_i$$

or

$$m\bar{\mathbf{v}} = \sum_{i=1}^{n} m_i\mathbf{v}_i \tag{14.13}$$

where $\bar{\mathbf{v}}$ represents the velocity of the mass center G of the system of particles. But the right-hand member of Eq. (14.13) is, by definition, the linear momentum $\mathbf{L}$ of the system (Sec. 14.3). We therefore have

$$\mathbf{L} = m\bar{\mathbf{v}} \tag{14.14}$$

and, differentiating both members with respect to t,

$$\dot{\mathbf{L}} = m\bar{\mathbf{a}} \tag{14.15}$$

†It may also be pointed out that the mass center and the center of gravity of a system of particles do not exactly coincide, since the weights of the particles are directed toward the center of the earth and thus do not truly form a system of parallel forces.

where $\bar{\mathbf{a}}$ represents the acceleration of the mass center G. Substituting for $\dot{\mathbf{L}}$ from (14.15) into (14.10), we write the equation

$$\Sigma \mathbf{F} = m\bar{\mathbf{a}} \tag{14.16}$$

which defines the motion of the mass center G of the system of particles.

We note that Eq. (14.16) is identical with the equation we would obtain for a particle of mass m equal to the total mass of the particles of the system, acted upon by all the external forces. We therefore state that *the mass center of a system of particles moves as if the entire mass of the system and all the external forces were concentrated at that point.*

This principle is best illustrated by the motion of an exploding shell. We know that if air resistance is neglected, it can be assumed that a shell will travel along a parabolic path. After the shell has exploded, the mass center G of the fragments of shell will continue to travel along the same path. Indeed, point G must move as if the mass and the weight of all fragments were concentrated at G; it must, therefore, move as if the shell had not exploded.

It should be noted that the preceding derivation does not involve the moments of the external forces. Therefore, *it would be wrong to assume* that the external forces are equipollent to a vector $m\bar{\mathbf{a}}$ attached at the mass center G. This is not in general the case since, as you will see in the next section, the sum of the moments about G of the external forces is not in general equal to zero.

14.5 ANGULAR MOMENTUM OF A SYSTEM OF PARTICLES ABOUT ITS MASS CENTER

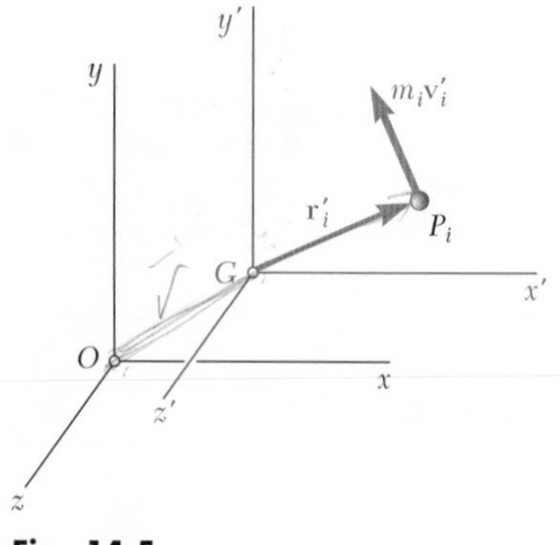

Fig. 14.5

In some applications (for example, in the analysis of the motion of a rigid body) it is convenient to consider the motion of the particles of the system with respect to a centroidal frame of reference $Gx'y'z'$ which translates with respect to the newtonian frame of reference $Oxyz$ (Fig. 14.5). Although a centroidal frame is not, in general, a newtonian frame of reference, it will be seen that the fundamental relation (14.11) holds when the frame $Oxyz$ is replaced by $Gx'y'z'$.

Denoting, respectively, by $\mathbf{r}'_i$ and $\mathbf{v}'_i$ the position vector and the velocity of the particle P_i relative to the moving frame of reference $Gx'y'z'$, we define the *angular momentum* $\mathbf{H}'_G$ of the system of particles *about the mass center* G as follows:

$$\mathbf{H}'_G = \sum_{i=1}^{n} (\mathbf{r}'_i \times m_i\mathbf{v}'_i) \tag{14.17}$$

We now differentiate both members of Eq. (14.17) with respect to t. This operation is similar to that performed in Sec. 14.3 on Eq. (14.7), and so we write immediately

$$\dot{\mathbf{H}}'_G = \sum_{i=1}^{n} (\mathbf{r}'_i \times m_i\mathbf{a}'_i) \tag{14.18}$$

where $\mathbf{a}_i'$ denotes the acceleration of P_i relative to the moving frame of reference. Referring to Sec. 11.12, we write

$$\mathbf{a}_i = \overline{\mathbf{a}} + \mathbf{a}_i'$$

where $\mathbf{a}_i$ and $\overline{\mathbf{a}}$ denote, respectively, the accelerations of P_i and G relative to the frame $Oxyz$. Solving for $\mathbf{a}_i'$ and substituting into (14.18), we have

$$\dot{\mathbf{H}}_G' = \sum_{i=1}^{n} (\mathbf{r}_i' \times m_i\mathbf{a}_i) - \left(\sum_{i=1}^{n} m_i\mathbf{r}_i'\right) \times \overline{\mathbf{a}} \qquad (14.19)$$

But, by (14.12), the second sum in Eq. (14.19) is equal to $m\overline{\mathbf{r}}'$ and thus to zero, since the position vector $\overline{\mathbf{r}}'$ of G relative to the frame $Gx'y'z'$ is clearly zero. On the other hand, since $\mathbf{a}_i$ represents the acceleration of P_i relative to a newtonian frame, we can use Eq. (14.1) and replace $m_i\mathbf{a}_i$ by the sum of the internal forces $\mathbf{f}_{ij}$ and of the resultant $\mathbf{F}_i$ of the external forces acting on P_i. But a reasoning similar to that used in Sec. 14.2 shows that the moment resultant about G of the internal forces $\mathbf{f}_{ij}$ of the entire system is zero. The first sum in Eq. (14.19) therefore reduces to the moment resultant about G of the external forces acting on the particles of the system, and we write

$$\Sigma\mathbf{M}_G = \dot{\mathbf{H}}_G' \qquad (14.20)$$

which expresses that *the moment resultant about G of the external forces is equal to the rate of change of the angular momentum about G of the system of particles.*

It should be noted that in Eq. (14.17) we defined the angular momentum $\mathbf{H}_G'$ as the sum of the moments about G of the momenta of the particles $m_i\mathbf{v}_i'$ *in their motion relative to the centroidal frame of reference* $Gx'y'z'$. We may sometimes want to compute the sum $\mathbf{H}_G$ of the moments about G of the momenta of the particles $m_i\mathbf{v}_i$ *in their absolute motion,* i.e., in their motion as observed from the newtonian frame of reference $Oxyz$ (Fig. 14.6):

$$\mathbf{H}_G = \sum_{i=1}^{n} (\mathbf{r}_i' \times m_i\mathbf{v}_i) \qquad (14.21)$$

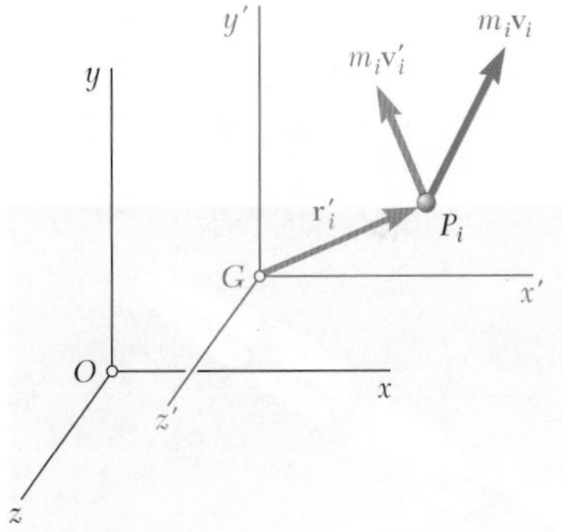

Fig. 14.6

Remarkably, the angular momenta $\mathbf{H}_G'$ and $\mathbf{H}_G$ are identically equal. This can be verified by referring to Sec. 11.12 and writing

$$\mathbf{v}_i = \overline{\mathbf{v}} + \mathbf{v}_i' \qquad (14.22)$$

Substituting for $\mathbf{v}_i$ from (14.22) into Eq. (14.21), we have

$$\mathbf{H}_G = \left(\sum_{i=1}^{n} m_i\mathbf{r}_i'\right) \times \overline{\mathbf{v}} + \sum_{i=1}^{n} (\mathbf{r}_i' \times m_i\mathbf{v}_i')$$

But, as observed earlier, the first sum is equal to zero. Thus $\mathbf{H}_G$ reduces to the second sum, which, by definition, is equal to $\mathbf{H}_G'$.†

†Note that this property is peculiar to the centroidal frame $Gx'y'z'$ and does not, in general, hold for other frames of reference (see Prob. 14.29).

Taking advantage of the property we have just established, we simplify our notation by dropping the prime (′) from Eq. (14.20) and writing

$$\Sigma \mathbf{M}_G = \dot{\mathbf{H}}_G \tag{14.23}$$

where it is understood that the angular momentum $\mathbf{H}_G$ can be computed by forming the moments about G of the momenta of the particles in their motion with respect to either the newtonian frame $Oxyz$ or the centroidal frame $Gx'y'z'$:

$$\mathbf{H}_G = \sum_{i=1}^{n} (\mathbf{r}'_i \times m_i\mathbf{v}_i) = \sum_{i=1}^{n} (\mathbf{r}'_i \times m_i\mathbf{v}'_i) \tag{14.24}$$

14.6 CONSERVATION OF MOMENTUM FOR A SYSTEM OF PARTICLES

If no external force acts on the particles of a system, the left-hand members of Eqs. (14.10) and (14.11) are equal to zero and these equations reduce to $\dot{\mathbf{L}} = 0$ and $\dot{\mathbf{H}}_O = 0$. We conclude that

$$\mathbf{L} = \text{constant} \qquad \mathbf{H}_O = \text{constant} \tag{14.25}$$

The equations obtained express that the linear momentum of the system of particles and its angular momentum about the fixed point O are conserved.

Photo 14.1 If no external forces are acting on the two stages of this rocket, the linear and angular momentum of the system will be conserved.

In some applications, such as problems involving central forces, the moment about a fixed point O of each of the external forces can be zero without any of the forces being zero. In such cases, the second of Eqs. (14.25) still holds; the angular momentum of the system of particles about O is conserved.

The concept of conservation of momentum can also be applied to the analysis of the motion of the mass center G of a system of particles and to the analysis of the motion of the system about G. For example, if the sum of the external forces is zero, the first of Eqs. (14.25) applies. Recalling Eq. (14.14), we write

$$\overline{\mathbf{v}} = \text{constant} \tag{14.26}$$

which expresses that the mass center G of the system moves in a straight line and at a constant speed. On the other hand, if the sum of the moments about G of the external forces is zero, it follows from Eq. (14.23) that the angular momentum of the system about its mass center is conserved:

$$\mathbf{H}_G = \text{constant} \tag{14.27}$$

SAMPLE PROBLEM 14.1

A 200-kg space vehicle is observed at $t = 0$ to pass through the origin of a newtonian reference frame $Oxyz$ with velocity $\mathbf{v}_0 = (150 \text{ m/s})\mathbf{i}$ relative to the frame. Following the detonation of explosive charges, the vehicle separates into three parts A, B, and C, of mass 100 kg, 60 kg, and 40 kg, respectively. Knowing that at $t = 2.5$ s the positions of parts A and B are observed to be $A(555, -180, 240)$ and $B(255, 0, -120)$, where the coordinates are expressed in meters, determine the position of part C at that time.

SOLUTION

Since there is no external force, the mass center G of the system moves with the constant velocity $\mathbf{v}_0 = (150 \text{ m/s})\mathbf{i}$. At $t = 2.5$ s, its position is

$$\bar{\mathbf{r}} = \mathbf{v}_0 t = (150 \text{ m/s})\mathbf{i}(2.5 \text{ s}) = (375 \text{ m})\mathbf{i}$$

Recalling Eq. (14.12), we write

$$m\bar{\mathbf{r}} = m_A\mathbf{r}_A + m_B\mathbf{r}_B + m_C\mathbf{r}_C$$

$$(200 \text{ kg})(375 \text{ m})\mathbf{i} = (100 \text{ kg})[(555 \text{ m})\mathbf{i} - (180 \text{ m})\mathbf{j} + (240 \text{ m})\mathbf{k}] + (60 \text{ kg})[(255 \text{ m})\mathbf{i} - (120 \text{ m})\mathbf{k}] + (40 \text{ kg})\mathbf{r}_C$$

$$\mathbf{r}_C = (105 \text{ m})\mathbf{i} + (450 \text{ m})\mathbf{j} - (420 \text{ m})\mathbf{k} \quad ◀$$

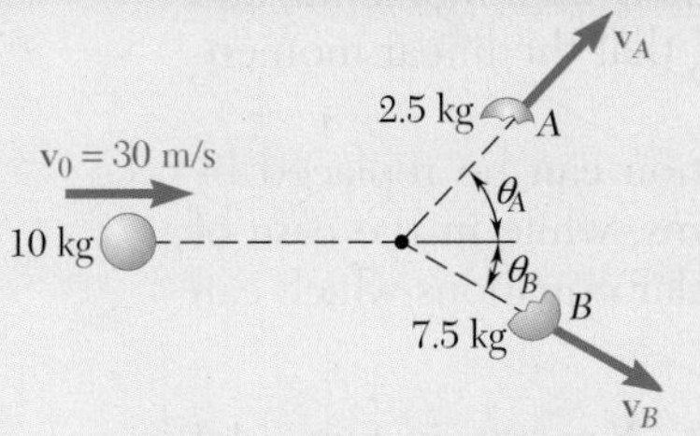

SAMPLE PROBLEM 14.2

A 10-kg projectile is moving with a velocity of 30 m/s when it explodes into two fragments A and B, weighing 2.5 kg and 7.5 kg, respectively. Knowing that immediately after the explosion, fragments A and B travel in directions defined respectively by $\theta_A = 45°$ and $\theta_B = 30°$, determine the velocity of each fragment.

SOLUTION

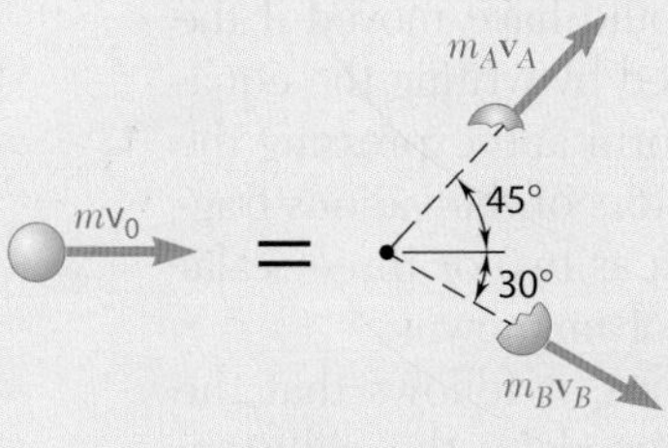

Since there is no external force, the linear momentum of the system is conserved, and we write

$$m_A\mathbf{v}_A + m_B\mathbf{v}_B = m\mathbf{v}_0$$

$$2.5\,\mathbf{v}_A + 7.5\,\mathbf{v}_B = 10\,\mathbf{v}_0$$

$\overset{+}{\rightarrow} x$ components: $\quad 2.5v_A \cos 45° + 7.5v_B \cos 30° = 10(30)$

$+\uparrow y$ components: $\quad 2.5v_A \sin 45° - 7.5v_B \sin 30° = 0$

Solving simultaneously the two equations for v_A and v_B, we have

$$v_A = 62.2 \text{ m/s} \qquad v_B = 29.3 \text{ m/s}$$

$$\mathbf{v}_A = 62.2 \text{ m/s} \measuredangle 45° \qquad \mathbf{v}_B = 29.3 \text{ m/s} \measuredangle 30° \quad ◀$$

SOLVING PROBLEMS ON YOUR OWN

This chapter deals with the motion of *systems of particles,* that is, with the motion of a large number of particles considered together, rather than separately. In this first lesson you learned to compute the *linear momentum* and the *angular momentum* of a system of particles. We defined the linear momentum **L** of a system of particles as the sum of the linear momenta of the particles and we defined the angular momentum $\mathbf{H}_O$ of the system as the sum of the angular momenta of the particles about O:

$$\mathbf{L} = \sum_{i=1}^{n} m_i\mathbf{v}_i \qquad \mathbf{H}_O = \sum_{i=1}^{n} (\mathbf{r}_i \times m_i\mathbf{v}_i) \qquad (14.6,\ 14.7)$$

In this lesson, you will solve a number of problems of practical interest, either by observing that the linear momentum of a system of particles is conserved or by considering the motion of the mass center of a system of particles.

1. Conservation of the linear momentum of a system of particles. This occurs *when the resultant of the external forces acting on the particles of the system is zero*. You may encounter such a situation in the following types of problems.

a. Problems involving the rectilinear motion of objects such as colliding automobiles and railroad cars. After you have checked that the resultant of the external forces is zero, equate the algebraic sums of the initial momenta and final momenta to obtain an equation which can be solved for one unknown.

b. Problems involving the two-dimensional or three-dimensional motion of objects such as exploding shells, or colliding aircraft, automobiles, or billiard balls. After you have checked that the resultant of the external forces is zero, add vectorially the initial momenta of the objects, add vectorially their final momenta, and equate the two sums to obtain a vector equation expressing that the linear momentum of the system is conserved.

In the case of a two-dimensional motion, this equation can be replaced by two scalar equations which can be solved for two unknowns, while in the case of a three-dimensional motion it can be replaced by three scalar equations which can be solved for three unknowns.

2. Motion of the mass center of a system of particles. You saw in Sec. 14.4 that *the mass center of a system of particles moves as if the entire mass of the system and all of the external forces were concentrated at that point.*

a. In the case of a body exploding while in motion, it follows that the mass center of the resulting fragments moves as the body itself would have moved if the explosion had not occurred. Problems of this type can be solved by writing the equation of motion of the mass center of the system in vectorial form and expressing the position vector of the mass center in terms of the position vectors of the various fragments [Eq. (14.12)]. You can then rewrite the vector equation as two or three scalar equations and solve the equations for an equivalent number of unknowns.

b. In the case of the collision of several moving bodies, it follows that the motion of the mass center of the various bodies is unaffected by the collision. Problems of this type can be solved by writing the equation of motion of the mass center of the system in vectorial form and expressing its position vector before and after the collision in terms of the position vectors of the relevant bodies [Eq. (14.12)]. You can then rewrite the vector equation as two or three scalar equations and solve these equations for an equivalent number of unknowns.

PROBLEMS

14.1 A 30-g bullet is fired with a horizontal velocity of 450 m/s and becomes embedded in block *B* which has a mass of 3 kg. After the impact, block *B* slides on 30-kg carrier *C* until it impacts the end of the carrier. Knowing the impact between *B* and *C* is perfectly plastic and the coefficient of kinetic friction between *B* and *C* is 0.2, determine (*a*) the velocity of the bullet and *B* after the first impact, (*b*) the final velocity of the carrier.

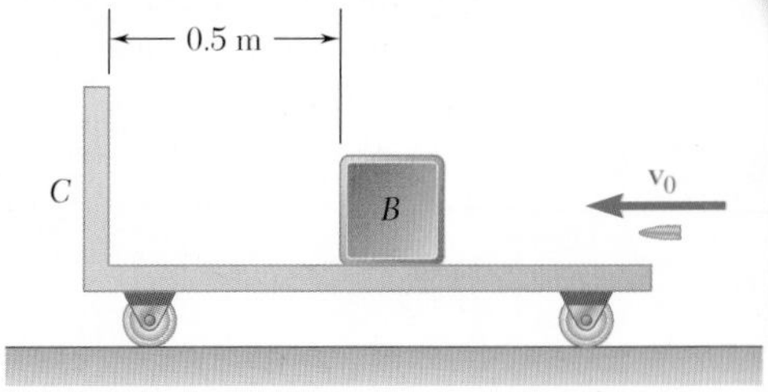

Fig. P14.1 and P14.2

14.2 A 30-g bullet is fired with a horizontal velocity of 450 m/s through 3-kg block *B* and becomes embedded in carrier *C* which has a mass of 30 kg. After the impact, block *B* slides 0.3 m on *C* before coming to rest relative to the carrier. Knowing the coefficient of kinetic friction between *B* and *C* is 0.2, determine (*a*) the velocity of the bullet immediately after passing through *B*, (*b*) the final velocity of the carrier.

14.3 An airline employee tosses two suitcases, of mass 15 kg and 20 kg, respectively, onto a 25-kg baggage carrier in rapid succession. Knowing that the carrier is initially at rest and that the employee imparts a 3-m/s horizontal velocity to the 15-kg suitcase and a 2-m/s horizontal velocity to the 20-kg suitcase, determine the final velocity of the baggage carrier if the first suitcase tossed onto the carrier is (*a*) the 15-kg suitcase, (*b*) the 20-kg suitcase.

Fig. P14.3

14.4 A bullet is fired with a horizontal velocity of 450 m/s through a 3-kg block *A* and becomes embedded in a 2.5-kg block *B*. Knowing that blocks *A* and *B* start moving with velocities of 1.5 m/s and 2.7 m/s, respectively, determine (*a*) the mass of the bullet, (*b*) its velocity as it travels from block *A* to block *B*.

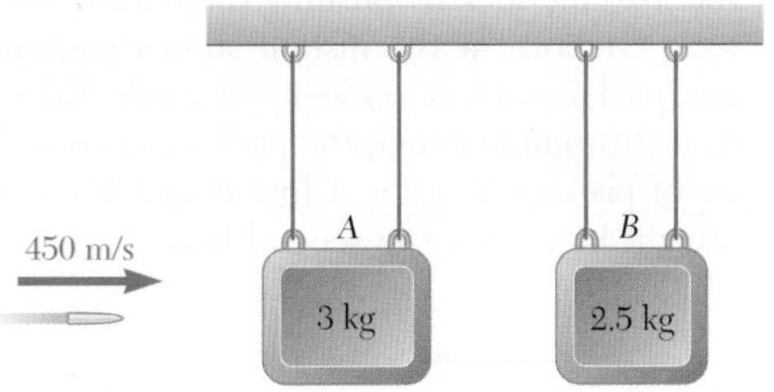

Fig. *P14.4*

14.5 A 90-kg man and a 60-kg woman stand at opposite ends of a 150-kg boat, ready to dive, each with a 5 m/s velocity relative to the boat. Determine the velocity of the boat after they have both dived, if (*a*) the woman dives first, (*b*) the man dives first.

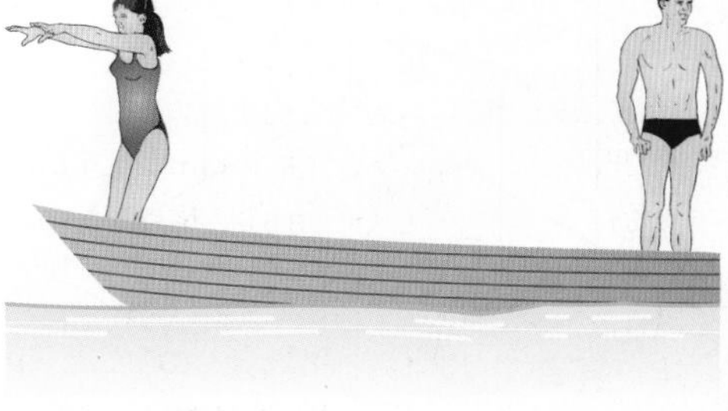

Fig. P14.5

14.6 A 90-kg man and a 60-kg woman stand side by side at the same end of a 150-kg boat, ready to dive, each with a 5 m/s velocity relative to the boat. Determine the velocity of the boat after they have both dived, if (*a*) the woman dives first, (*b*) the man dives first.

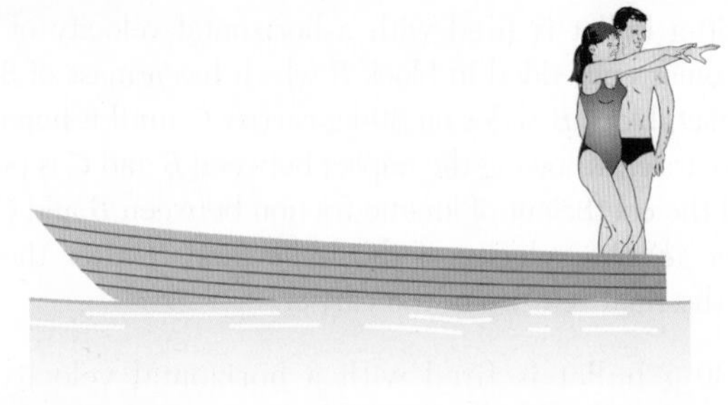

Fig. P14.6

14.7 A 40-Mg boxcar *A* is moving in a railroad switchyard with a velocity of 9 km/h toward cars *B* and *C*, which are both at rest with their brakes off at a short distance from each other. Car *B* is a 25-Mg flatcar supporting a 30-Mg container, and car *C* is a 35-Mg boxcar. As the cars hit each other they get automatically and tightly coupled. Determine the velocity of car *A* immediately after each of the two couplings, assuming that the container (*a*) does not slide on the flatcar, (*b*) slides after the first coupling but hits a stop before the second coupling occurs, (*c*) slides and hits the stop only after the second coupling has occurred.

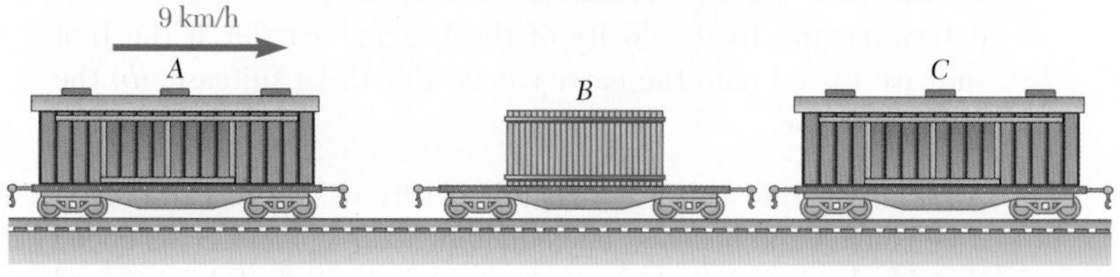

Fig. P14.7

14.8 Packages in an automobile parts supply house are transported to the loading dock by pushing them along on a roller track with very little friction. At the instant shown packages *B* and *C* are at rest and package *A* has a velocity of 2 m/s. Knowing that the coefficient of restitution between the packages is 0.3, determine (*a*) the velocity of package *C* after *A* hits *B* and *B* hits *C*, (*b*) the velocity of *A* after it hits *B* for the second time.

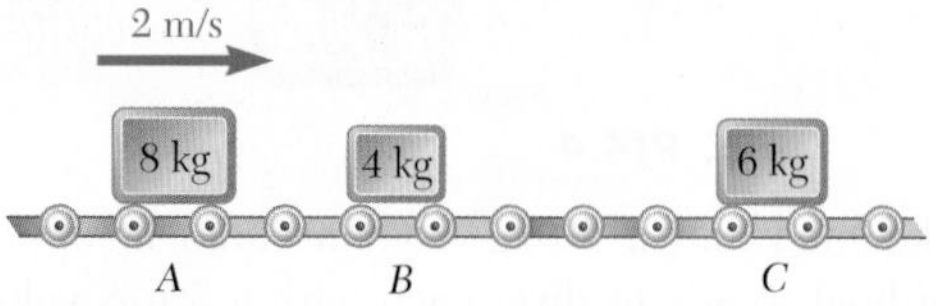

Fig. P14.8

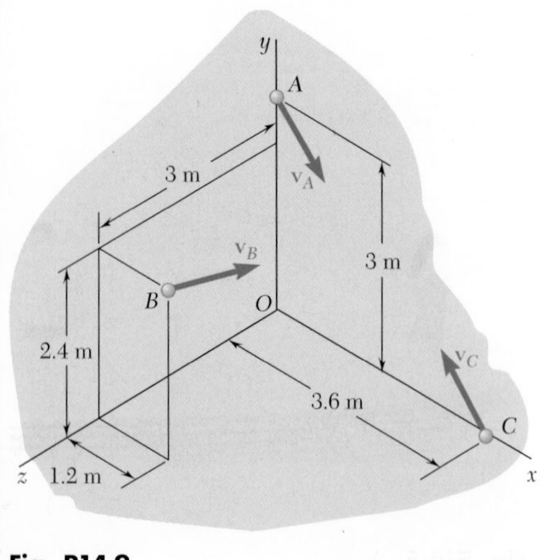

Fig. P14.9

14.9 A system consists of three particles *A*, *B*, and *C*. We know that $m_A = 3$ kg, $m_B = 2$ kg, and $m_C = 4$ kg and that the velocities of the particles expressed in m/s are, respectively, $\mathbf{v}_A = 4\mathbf{i} + 2\mathbf{j} + 2\mathbf{k}$, $\mathbf{v}_B = 4\mathbf{i} + 3\mathbf{j}$, and $\mathbf{v}_C = -2\mathbf{i} + 4\mathbf{j} + 2\mathbf{k}$. Determine the angular momentum $\mathbf{H}_O$ of the system about *O*.

14.10 For the system of particles of Prob. 14.9, determine (*a*) the position vector $\bar{\mathbf{r}}$ of the mass center G of the system, (*b*) the linear momentum $m\bar{\mathbf{v}}$ of the system, (*c*) the angular momentum $\mathbf{H}_G$ of the system about G. Also verify that the answers to this problem and to Prob. 14.9 satisfy the equation given in Prob. 14.27.

14.11 A system consists of three particles A, B, and C. We know that $m_A = 2.5$ kg, $m_B = 2$ kg, and $m_C = 1.5$ kg and that the velocities of the particles expressed in m/s are, respectively, $\mathbf{v}_A = \mathbf{i} + 1.5\mathbf{j} - \mathbf{k}$, $\mathbf{v}_B = v_x\mathbf{i} + v_y\mathbf{j} + v_z\mathbf{k}$, and $\mathbf{v}_C = -1.5\mathbf{i} - \mathbf{j} + 0.5\mathbf{k}$. Determine (*a*) the components v_x and v_y of the velocity of particle B for which the angular momentum $\mathbf{H}_O$ of the system about O is parallel to the x axis, (*b*) the value of $\mathbf{H}_O$.

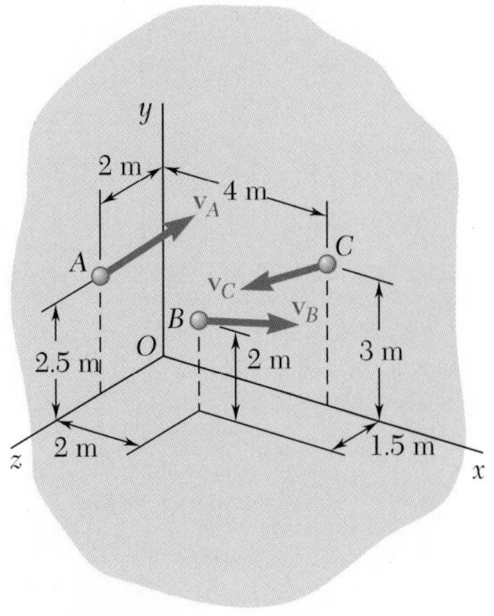

Fig. P14.11

14.12 For the system of particles of Prob. 14.11, determine (*a*) the components v_x and v_z of the velocity of particle B for which the angular momentum $\mathbf{H}_O$ of the system about O is parallel to the z axis, (*b*) the value of $\mathbf{H}_O$.

14.13 A system consists of three particles A, B, and C. We know that $m_A = 3$ kg, $m_B = 4$ kg, and $m_c = 5$ kg and that the velocities of the particles expressed in m/s are, respectively, $\mathbf{v}_A = -4\mathbf{i} + 4\mathbf{j} + 6\mathbf{k}$, $\mathbf{v}_B = -6\mathbf{i} + 8\mathbf{j} + 4\mathbf{k}$, and $\mathbf{v}_C = 2\mathbf{i} - 6\mathbf{j} - 4\mathbf{k}$. Determine the angular momentum $\mathbf{H}_O$ of the system about O.

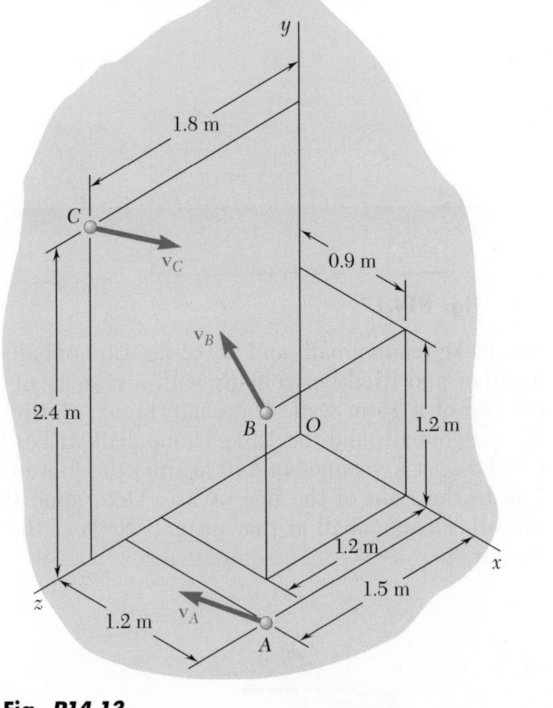

Fig. *P14.13*

14.14 For the system of particles of Prob. 14.13, determine (*a*) the position vector $\bar{\mathbf{r}}$ of the mass center G of the system, (*b*) the linear momentum $m\bar{\mathbf{v}}$ of the system, (*c*) the angular momentum $\mathbf{H}_G$ of the system about G. Also verify that the answers to this problem and to Prob. 14.13 satisfy the equation given in Prob. 14.27.

14.15 A 13-kg projectile is passing through the origin O with a velocity $\mathbf{v}_0 = (35\ \text{m/s})\mathbf{i}$ when it explodes into two fragments A and B, of mass 5 kg and 8 kg, respectively. Knowing that 3 s later the position of fragment A is (90 m, 7 m, −14 m), determine the position of fragment B at the same instant. Assume $a_y = -g = -9.81\ \text{m/s}^2$ and neglect air resistance.

14.16 A 300-kg space vehicle traveling with a velocity $\mathbf{v}_0 = (360\ \text{m/s})\mathbf{i}$ passes through the origin O at $t = 0$. Explosive charges then separate the vehicle into three parts A, B, and C, with mass, respectively, 150 kg, 100 kg, and 50 kg. Knowing that at $t = 4$ s, the positions of parts A and B are observed to be A (1170 m, −290 m, –585 m) and B (1975 m, 365 m, 800 m), determine the corresponding position of part C. Neglect the effect of gravity.

14.17 A 2-kg model rocket is launched vertically and reaches an altitude of 70 m with a speed of 30 m/s at the end of powered flight, time $t = 0$. As the rocket approaches its maximum altitude it explodes into two parts of masses $m_A = 0.7$ kg and $m_B = 1.3$ kg. Part A is observed to strike the ground 80 m west of the launch point at $t = 6$ s. Determine the position of part B at that time.

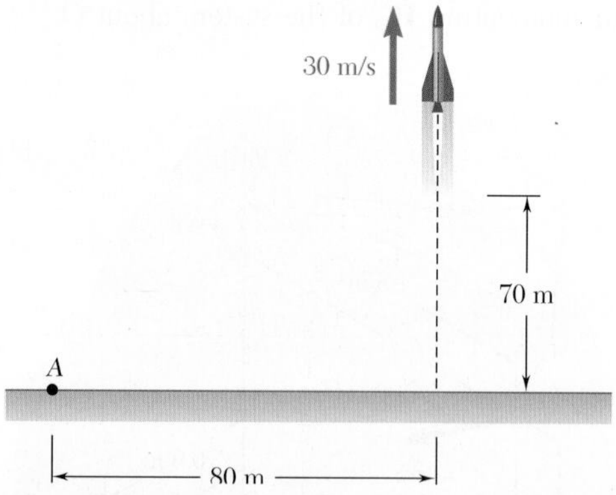

Fig. *P14.17*

14.18 An 18-kg cannonball and a 12-kg cannonball are chained together and fired horizontally with a velocity of 165 m/s from the top of a 15-m wall. The chain breaks during the flight of the cannonballs and the 12-kg cannonball strikes the ground at $t = 1.5$ s, at a distance of 240 m from the foot of the wall, and 7 m to the right of the line of fire. Determine the position of the other cannonball at that instant. Neglect the resistance of the air.

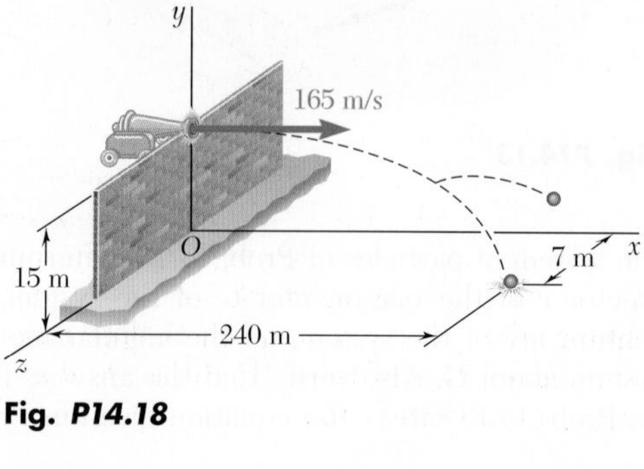

Fig. *P14.18*

14.19 and 14.20 Car A was traveling east at high speed when it collided at point O with car B, which was traveling north at 72 km/h. Car C, which was traveling west at 90 km/h, was 10 m east and 3 m north of point O at the time of the collision. Because the pavement was wet, the driver of car C could not prevent his car from sliding into the other two cars, and the three cars, stuck together, kept sliding until they hit the utility pole P. Knowing that the masses of cars A, B, and C are, respectively, 1500 kg, 1300 kg, and 1200 kg, and neglecting the forces exerted on the cars by the wet pavement solve the problems indicated.

14.19 Knowing that the speed of car A was 129.6 km/h and that the time elapsed from the first collision to the stop at P was 2.4 s, determine the coordinates of the utility pole P.

14.20 Knowing that the coordinates of the utility pole are $x_p = 18$ m and $y_p = 13.9$ m, determine (*a*) the time elapsed from the first collision to the stop at P, (*b*) the speed of car A.

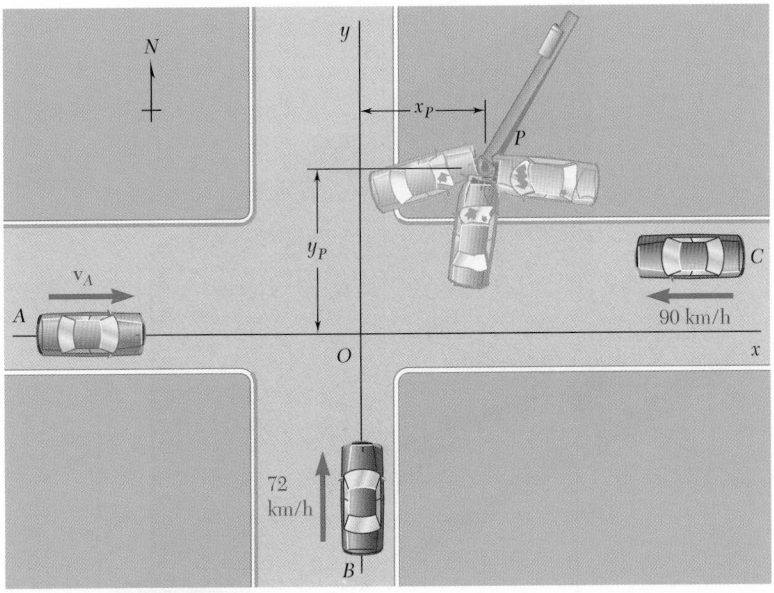

Fig. P14.19 and P14.20

14.21 An expert archer demonstrates his ability by hitting tennis balls thrown by an assistant. A 58-g tennis ball has a velocity of (10 m/s)**i** − (2 m/s)**j** and is 10 m above the ground when it is hit by a 40-g arrow traveling with a velocity of (50 m/s)**j** + (70 m/s)**k** where **j** is directed upwards. Determine the position P where the ball and arrow will hit the ground, relative to point O located directly under the point of impact.

14.22 Two spheres, each of mass m, can slide freely on a frictionless, horizontal surface. Sphere A is moving at a speed $v_0 = 5$ m/s when it strikes sphere B which is at rest and the impact causes sphere B to break into two pieces, each of mass $m/2$. Knowing that 0.7 s after the collision one piece reaches point C and 0.9 s after the collision the other piece reaches point D, determine (*a*) the velocity of sphere A after the collision, (*b*) the angle θ and the speeds of the two pieces after the collision.

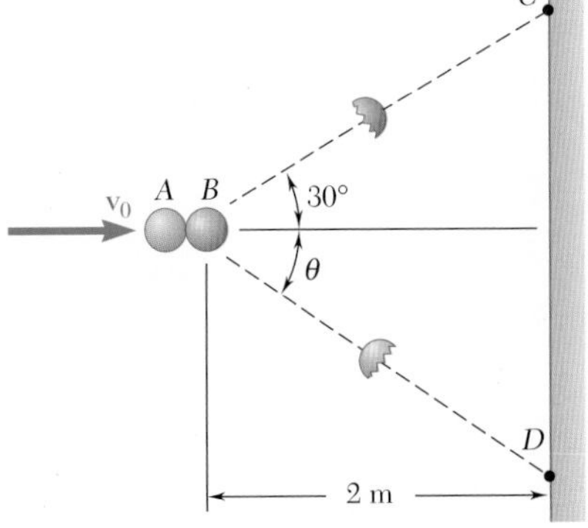

Fig. P14.22

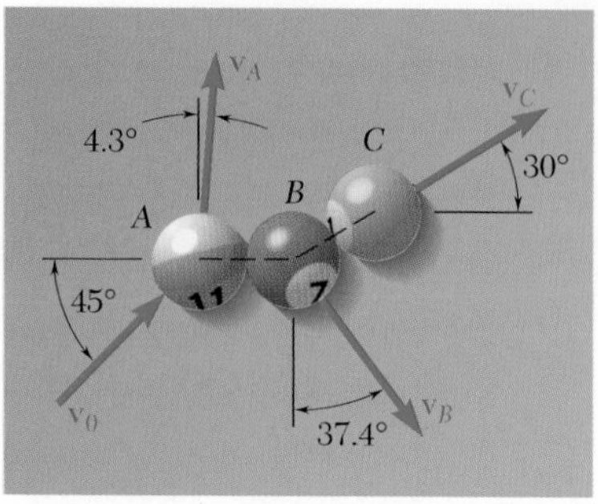

Fig. P14.23

14.23 In a game of pool, ball *A* is moving with a velocity $\mathbf{v}_0$ when it strikes balls *B* and *C* which are at rest and aligned as shown. Knowing that after the collision the three balls move in the directions indicated and that $v_0 = 4$ m/s and $v_C = 2$ m/s, determine the magnitude of the velocity of (*a*) ball *A*, (*b*) ball *B*.

14.24 A 6-kg shell moving with a velocity $\mathbf{v}_0 = (12 \text{ m/s})\mathbf{i} - (9 \text{ m/s})\mathbf{j} - (360 \text{ m/s})\mathbf{k}$ explodes at point *D* into three fragments *A*, *B*, and *C* of mass, respectively, 3 kg, 2 kg, and 1 kg. Knowing that the fragments hit the vertical wall at the points indicated, determine the speed of each fragment immediately after the explosion. Assume that elevation changes due to gravity may be neglected.

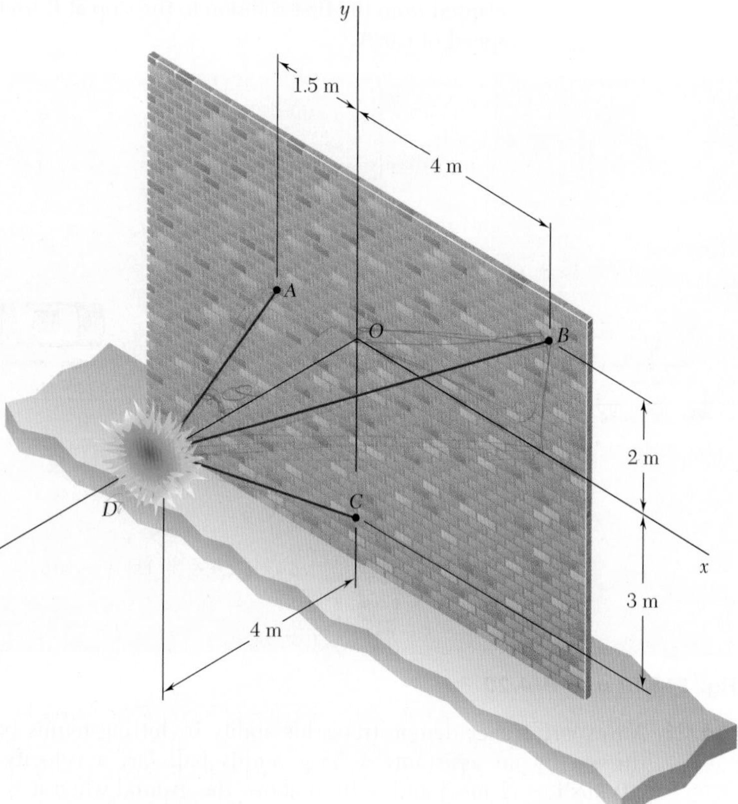

Fig. P14.24 and P14.25

14.25 A 6-kg shell moving with a velocity $\mathbf{v}_0 = (12 \text{ m/s})\mathbf{i} - (9 \text{ m/s})\mathbf{j} - (360 \text{ m/s})\mathbf{k}$ explodes at point *D* into three fragments *A*, *B*, and *C* of mass, respectively, 2 kg, 1 kg, and 3 kg. Knowing that the fragments hit the vertical wall at the points indicated, determine the speed of each fragment immediately after the explosion. Assume that elevation changes due to gravity may be neglected.

14.26 In a scattering experiment, an alpha particle A is projected with the velocity $\mathbf{u}_0 = -(600 \text{ m/s})\mathbf{i} + (750 \text{ m/s})\mathbf{j} - (800 \text{ m/s})\mathbf{k}$ into a stream of oxygen nuclei moving with a common velocity $\mathbf{v}_0 = (600 \text{ m/s})\mathbf{j}$. After colliding successively with the nuclei B and C, particle A is observed to move along the path defined by the points A_1 (280, 240, 120) and A_2 (360, 320, 160), while nuclei B and C are observed to move along paths defined, respectively, by B_1 (147, 220, 130) and B_2 (114, 290, 120), and by C_1 (240, 232, 90) and C_2 (240, 280, 75). All paths are along straight lines and all coordinates are expressed in millimeters. Knowing that the mass of an oxygen nucleus is four times that of an alpha particle, determine the speed of each of the three particles after the collisions.

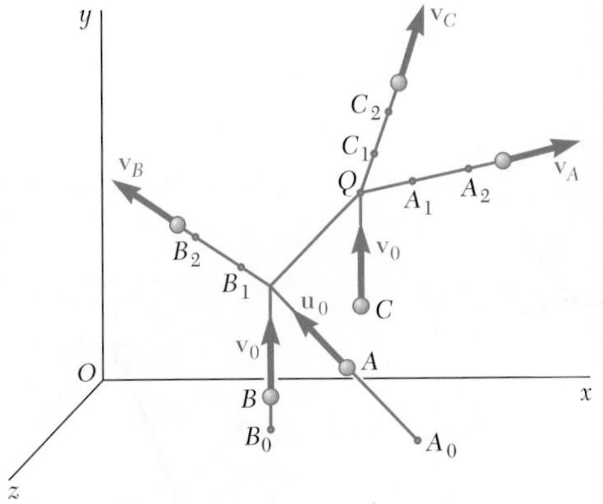

Fig. P14.26

14.27 Derive the relation

$$\mathbf{H}_O = \bar{\mathbf{r}} \times m\bar{\mathbf{v}} + H_G$$

between the angular momenta $\mathbf{H}_O$ and $\mathbf{H}_G$ defined in Eqs. (14.7) and (14.24), respectively. The vectors $\bar{\mathbf{r}}$ and $\bar{\mathbf{v}}$ define, respectively, the position and velocity of the mass center G of the system of particles relative to the newtonian frame of reference $Oxyz$, and m represents the total mass of the system.

14.28 Show that Eq. (14.23) may be derived directly from Eq. (14.11) by substituting for $\mathbf{H}_O$ the expression given in Prob. 14.27.

14.29 Consider the frame of reference $Ax'y'z'$ in translation with respect to the newtonian frame of reference $Oxyz$. We define the angular momentum $\mathbf{H}'_A$ of a system of n particles about A as the sum

$$\mathbf{H}'_A = \sum_{i=1}^{n} \mathbf{r}'_i \times m_i\mathbf{v}'_i \qquad (1)$$

of the moments about A of the momenta $m_i\mathbf{v}'_i$ of the particles in their motion relative to the frame $Ax'y'z'$. Denoting by $\mathbf{H}_A$ the sum

$$\mathbf{H}_A = \sum_{i=1}^{n} \mathbf{r}'_i \times m_i\mathbf{v}_i$$

of the moments about A of the momenta $m_i\mathbf{v}_i$ of the particles in their motion relative to the newtonian frame $Oxyz$, show that $\mathbf{H}_A = \mathbf{H}'_A$ at a given instant if, and only if, one of the following conditions is satisfied at that instant: (*a*) A has zero velocity with respect to the frame $Oxyz$, (*b*) A coincides with the mass center G of the system, (*c*) the velocity $\mathbf{v}_A$ relative to $Oxyz$ is directed along the line AG.

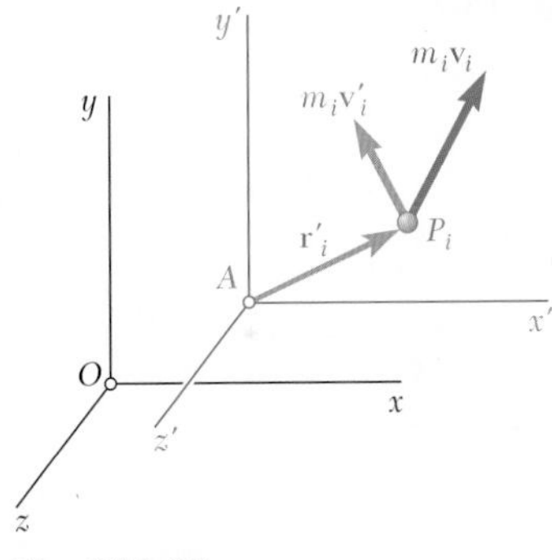

Fig. P14.29

14.30 Show that the relation $\Sigma\mathbf{M}_A = \dot{\mathbf{H}}'_A$, where $\mathbf{H}'_A$ is defined by Eq. (1) of Prob. 14.29 and where $\Sigma\mathbf{M}_A$ represents the sum of the moments about A of the external forces acting on the system of particles, is valid if, and only if, one of the following conditions is satisfied: (*a*) the frame $Ax'y'z'$ is itself a newtonian frame of reference, (*b*) A coincides with the mass center G, (*c*) the acceleration $\mathbf{a}_A$ of A relative to $Oxyz$ is directed along the line AG.

14.7 KINETIC ENERGY OF A SYSTEM OF PARTICLES

The kinetic energy T of a system of particles is defined as the sum of the kinetic energies of the various particles of the system. Referring to Sec. 13.3, we therefore write

$$T = \frac{1}{2}\sum_{i=1}^{n} m_i v_i^2 \tag{14.28}$$

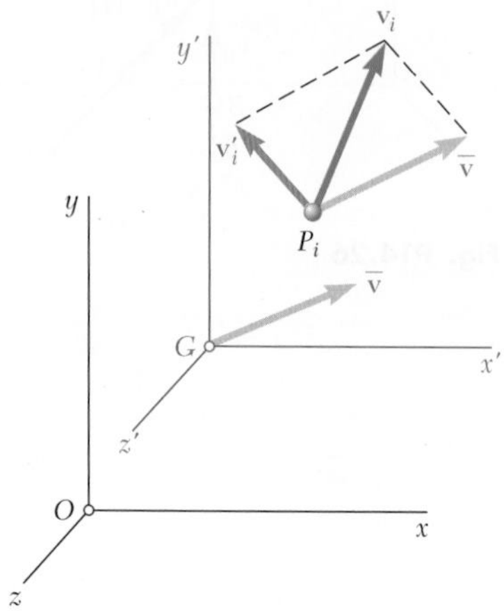

Fig. 14.7

Using a Centroidal Frame of Reference. It is often convenient when computing the kinetic energy of a system comprising a large number of particles (as in the case of a rigid body) to consider separately the motion of the mass center G of the system and the motion of the system relative to a moving frame attached to G.

Let P_i be a particle of the system, $\mathbf{v}_i$ its velocity relative to the newtonian frame of reference $Oxyz$, and $\mathbf{v}'_i$ its velocity relative to the moving frame $Gx'y'z'$ which is in translation with respect to $Oxyz$ (Fig. 14.7). We recall from the preceding section that

$$\mathbf{v}_i = \overline{\mathbf{v}} + \mathbf{v}'_i \tag{14.22}$$

where $\overline{\mathbf{v}}$ denotes the velocity of the mass center G relative to the newtonian frame $Oxyz$. Observing that v_i^2 is equal to the scalar product $\mathbf{v}_i \cdot \mathbf{v}_i$, we express the kinetic energy T of the system relative to the newtonian frame $Oxyz$ as follows:

$$T = \frac{1}{2}\sum_{i=1}^{n} m_i v_i^2 = \frac{1}{2}\sum_{i=1}^{n} (m_i \mathbf{v}_i \cdot \mathbf{v}_i)$$

or, substituting for $\mathbf{v}_i$ from (14.22),

$$\begin{aligned} T &= \frac{1}{2}\sum_{i=1}^{n} [m_i(\overline{\mathbf{v}} + \mathbf{v}'_i) \cdot (\overline{\mathbf{v}} + \mathbf{v}'_i)] \\ &= \frac{1}{2}\left(\sum_{i=1}^{n} m_i\right)\overline{v}^2 + \overline{\mathbf{v}} \cdot \sum_{i=1}^{n} m_i \mathbf{v}'_i + \frac{1}{2}\sum_{i=1}^{n} m_i v_i'^2 \end{aligned}$$

The first sum represents the total mass m of the system. Recalling Eq. (14.13), we note that the second sum is equal to $m\overline{\mathbf{v}}'$ and thus to zero, since $\overline{\mathbf{v}}'$, which represents the velocity of G relative to the frame $Gx'y'z'$, is clearly zero. We therefore write

$$T = \tfrac{1}{2} m\overline{v}^2 + \frac{1}{2}\sum_{i=1}^{n} m_i v_i'^2 \tag{14.29}$$

This equation shows that the kinetic energy T of a system of particles can be obtained *by adding the kinetic energy of the mass center G* (assuming the entire mass concentrated at G) *and the kinetic energy of the system in its motion relative to the frame $Gx'y'z'$*.

14.8 WORK-ENERGY PRINCIPLE. CONSERVATION OF ENERGY FOR A SYSTEM OF PARTICLES

The principle of work and energy can be applied to each particle P_i of a system of particles. We write

$$T_1 + U_{1\rightarrow 2} = T_2 \tag{14.30}$$

for each particle P_i, where $U_{1\rightarrow 2}$ represents the work done by the internal forces $\mathbf{f}_{ij}$ and the resultant external force $\mathbf{F}_i$ acting on P_i. Adding the kinetic energies of the various particles of the system and considering the work of all the forces involved, we can apply Eq. (14.30) to the entire system. The quantities T_1 and T_2 now represent the kinetic energy of the entire system and can be computed from either Eq. (14.28) or Eq. (14.29). The quantity $U_{1\rightarrow 2}$ represents the work of all the forces acting on the particles of the system. Note that while the internal forces $\mathbf{f}_{ij}$ and $\mathbf{f}_{ji}$ are equal and opposite, the work of these forces will not, in general, cancel out, since the particles P_i and P_j on which they act will, in general, undergo different displacements. Therefore, in computing $U_{1\rightarrow 2}$, *we must consider the work of the internal forces* $\mathbf{f}_{ij}$ *as well as the work of the external forces* $\mathbf{F}_i$.

If all the forces acting on the particles of the system are conservative, Eq. (14.30) can be replaced by

$$T_1 + V_1 = T_2 + V_2 \tag{14.31}$$

where V represents the potential energy associated with the internal and external forces acting on the particles of the system. Equation (14.31) expresses the principle of *conservation of energy* for the system of particles.

14.9 PRINCIPLE OF IMPULSE AND MOMENTUM FOR A SYSTEM OF PARTICLES

Integrating Eqs. (14.10) and (14.11) in t from t_1 to t_2, we write

$$\Sigma \int_{t_1}^{t_2} \mathbf{F}\, dt = \mathbf{L}_2 - \mathbf{L}_1 \tag{14.32}$$

$$\Sigma \int_{t_1}^{t_2} \mathbf{M}_O\, dt = (\mathbf{H}_O)_2 - (\mathbf{H}_O)_1 \tag{14.33}$$

Recalling the definition of the linear impulse of a force given in Sec. 13.10, we observe that the integrals in Eq. (14.32) represent the linear impulses of the external forces acting on the particles of the system. We shall refer in a similar way to the integrals in Eq. (14.33) as the *angular impulses* about O of the external forces. Thus, Eq. (14.32) expresses that the sum of the linear impulses of the external forces acting on the system is equal to the change in linear momentum of the system. Similarly, Eq. (14.33) expresses that the sum of the angular impulses about O of the external forces is equal to the change in angular momentum about O of the system.

Photo 14.2 When a golf ball is hit out of a sand trap, some of the momentum of the club is transferred to the golf ball and any sand that is hit.

In order to make clear the physical significance of Eqs. (14.32) and (14.33), we will rearrange the terms in these equations and write

$$\mathbf{L}_1 + \sum \int_{t_1}^{t_2} \mathbf{F}\, dt = \mathbf{L}_2 \tag{14.34}$$

$$(\mathbf{H}_O)_1 + \sum \int_{t_1}^{t_2} \mathbf{M}_O\, dt = (\mathbf{H}_O)_2 \tag{14.35}$$

In parts *a* and *c* of Fig. 14.8 we have sketched the momenta of the particles of the system at times t_1 and t_2, respectively. In part *b* we have shown a vector equal to the sum of the linear impulses of the external forces and a couple of moment equal to the sum of the angular impulses about *O* of the external forces. For simplicity, the particles have been

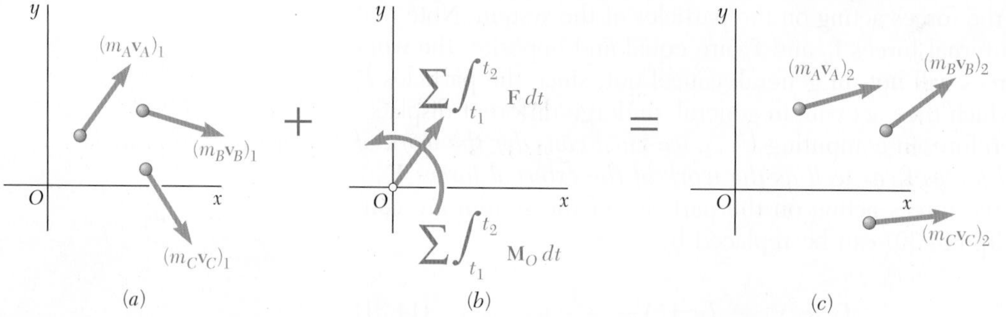

Fig. 14.8

assumed to move in the plane of the figure, but the present discussion remains valid in the case of particles moving in space. Recalling from Eq. (14.6) that **L**, by definition, is the resultant of the momenta $m_i\mathbf{v}_i$, we note that Eq. (14.34) expresses that the resultant of the vectors shown in parts *a* and *b* of Fig. 14.8 is equal to the resultant of the vectors shown in part *c* of the same figure. Recalling from Eq. (14.7) that $\mathbf{H}_O$ is the moment resultant of the momenta $m_i\mathbf{v}_i$, we note that Eq. (14.35) similarly expresses that the moment resultant of the vectors in parts *a* and *b* of Fig. 14.8 is equal to the moment resultant of the vectors in part *c*. Together, Eqs. (14.34) and (14.35) thus express that *the momenta of the particles at time* t_1 *and the impulses of the external forces from* t_1 *to* t_2 *form a system of vectors equipollent to the system of the momenta of the particles at time* t_2. This has been indicated in Fig. 14.8 by the use of blue plus and equals signs.

If no external force acts on the particles of the system, the integrals in Eqs. (14.34) and (14.35) are zero, and these equations yield

$$\mathbf{L}_1 = \mathbf{L}_2 \tag{14.36}$$

$$(\mathbf{H}_O)_1 = (\mathbf{H}_O)_2 \tag{14.37}$$

We thus check the result obtained in Sec. 14.6: If no external force acts on the particles of a system, the linear momentum and the angular momentum about *O* of the system of particles are conserved. The system of the initial momenta is equipollent to the system of the final momenta, and it follows that the angular momentum of the system of particles about *any* fixed point is conserved.

SAMPLE PROBLEM 14.3

For the 200-kg space vehicle of Sample Prob. 14.1, it is known that at t = 2.5 s, the velocity of part A is $\mathbf{v}_A = (270 \text{ m/s})\mathbf{i} - (120 \text{ m/s})\mathbf{j} + (160 \text{ m/s})\mathbf{k}$ and the velocity of part B is parallel to the xz plane. Determine the velocity of part C.

SOLUTION

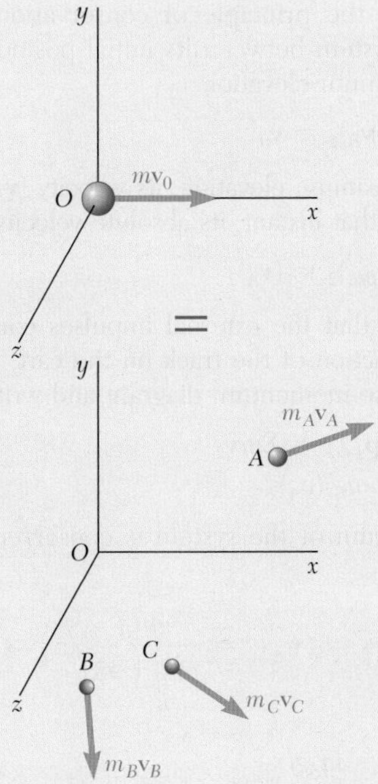

Since there is no external force, the initial momentum $m\mathbf{v}_0$ is equipollent to the system of the final momenta. Equating first the sums of the vectors in both parts of the adjoining sketch, and then the sums of their moments about O, we write

$$\mathbf{L}_1 = \mathbf{L}_2: \qquad m\mathbf{v}_0 = m_A\mathbf{v}_A + m_B\mathbf{v}_B + m_C\mathbf{v}_C \tag{1}$$

$$(\mathbf{H}_O)_1 = (\mathbf{H}_O)_2: \qquad 0 = \mathbf{r}_A \times m_A\mathbf{v}_A + \mathbf{r}_B \times m_B\mathbf{v}_B + \mathbf{r}_C \times m_C\mathbf{v}_C \tag{2}$$

Recalling from Sample Prob. 14.1 that $\mathbf{v}_0 = (150 \text{ m/s})\mathbf{i}$,

$$m_A = 100 \text{ kg} \qquad m_B = 60 \text{ kg} \qquad m_C = 40 \text{ kg}$$

$$\mathbf{r}_A = (555 \text{ m})\mathbf{i} - (180 \text{ m})\mathbf{j} + (240 \text{ m})\mathbf{k}$$

$$\mathbf{r}_B = (255 \text{ m})\mathbf{i} - (120 \text{ m})\mathbf{k}$$

$$\mathbf{r}_C = (105 \text{ m})\mathbf{i} + (450 \text{ m})\mathbf{j} - (420 \text{ m})\mathbf{k}$$

and using the information given in the statement of this problem, we rewrite Eqs. (1) and (2) as follows:

$$200(150\mathbf{i}) = 100(270\mathbf{i} - 120\mathbf{j} + 160\mathbf{k}) + 60[(v_B)_x\mathbf{i} + (v_B)_z\mathbf{k}] + 40[(v_C)_x\mathbf{i} + (v_C)_y\mathbf{j} + (v_C)_z\mathbf{k}] \tag{1'}$$

$$0 = 100\begin{vmatrix} \mathbf{i} & \mathbf{j} & \mathbf{k} \\ 555 & -180 & 240 \\ 270 & -120 & 160 \end{vmatrix} + 60\begin{vmatrix} \mathbf{i} & \mathbf{j} & \mathbf{k} \\ 255 & 0 & -120 \\ (v_B)_x & 0 & (v_B)_z \end{vmatrix} + 40\begin{vmatrix} \mathbf{i} & \mathbf{j} & \mathbf{k} \\ 105 & 450 & -420 \\ (v_C)_x & (v_C)_y & (v_C)_z \end{vmatrix} \tag{2'}$$

Equating to zero the coefficient of $\mathbf{j}$ in (1′) and the coefficients of $\mathbf{i}$ and $\mathbf{k}$ in (2′), we write, after reductions, the three scalar equations

$$(v_C)_y - 300 = 0$$

$$450(v_C)_z + 420(v_C)_y = 0$$

$$105(v_C)_y - 450(v_C)_x - 45\,000 = 0$$

which yield, respectively,

$$(v_C)_y = 300 \qquad (v_C)_z = -280 \qquad (v_C)_x = -30$$

The velocity of part C is thus

$$\mathbf{v}_C = -(30 \text{ m/s})\mathbf{i} + (300 \text{ m/s})\mathbf{j} - (280 \text{ m/s})\mathbf{k} \quad ◀$$

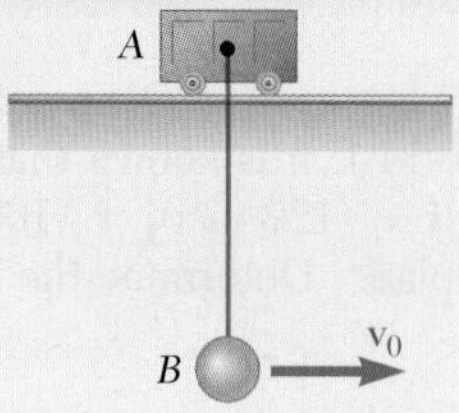

SAMPLE PROBLEM 14.4

Ball B, of mass m_B, is suspended from a cord of length l attached to cart A, of mass m_A, which can roll freely on a frictionless horizontal track. If the ball is given an initial horizontal velocity $\mathbf{v}_0$ while the cart is at rest, determine (*a*) the velocity of B as it reaches its maximum elevation, (*b*) the maximum vertical distance h through which B will rise. (It is assumed that $v_0^2 < 2gl$.)

SOLUTION

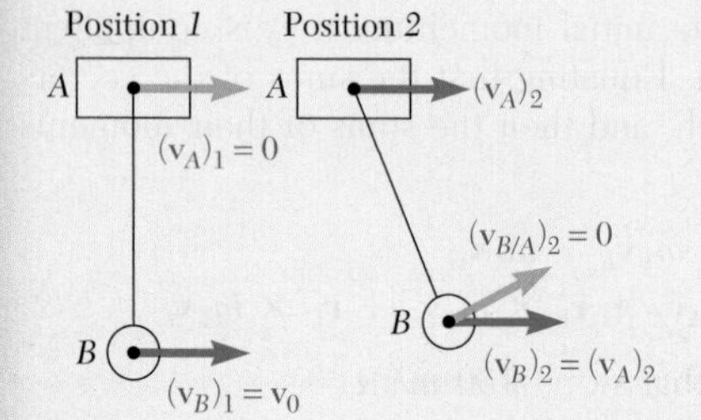

The impulse-momentum principle and the principle of conservation of energy will be applied to the cart-ball system between its initial position 1 and position 2, when B reaches its maximum elevation.

Velocities ***Position 1:*** $(\mathbf{v}_A)_1 = 0 \qquad (\mathbf{v}_B)_1 = \mathbf{v}_0$ (1)

Position 2: When ball B reaches its maximum elevation, its velocity $(\mathbf{v}_{B/A})_2$ relative to its support A is zero. Thus, at that instant, its absolute velocity is

$$(\mathbf{v}_B)_2 = (\mathbf{v}_A)_2 + (\mathbf{v}_{B/A})_2 = (\mathbf{v}_A)_2 \tag{2}$$

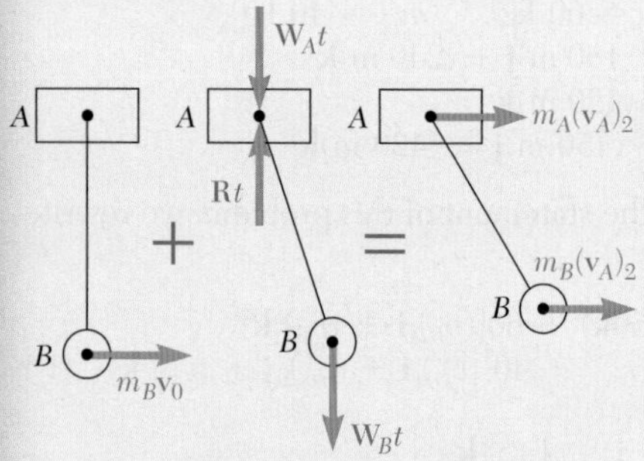

Impulse-Momentum Principle. Noting that the external impulses consist of $\mathbf{W}_At$, $\mathbf{W}_Bt$, and $\mathbf{R}t$, where $\mathbf{R}$ is the reaction of the track on the cart, and recalling (1) and (2), we draw the impulse-momentum diagram and write

$$\Sigma m\mathbf{v}_1 + \Sigma \textbf{ Ext Imp}_{1\rightarrow 2} = \Sigma m\mathbf{v}_2$$

$\xrightarrow{+} x$ components: $\qquad m_Bv_0 = (m_A + m_B)(v_A)_2$

which expresses that the linear momentum of the system is conserved in the horizontal direction. Solving for $(v_A)_2$:

$$(v_A)_2 = \frac{m_B}{m_A + m_B}v_0 \qquad (\mathbf{v}_B)_2 = (\mathbf{v}_A)_2 = \frac{m_B}{m_A + m_B}v_0 \rightarrow \blacktriangleleft$$

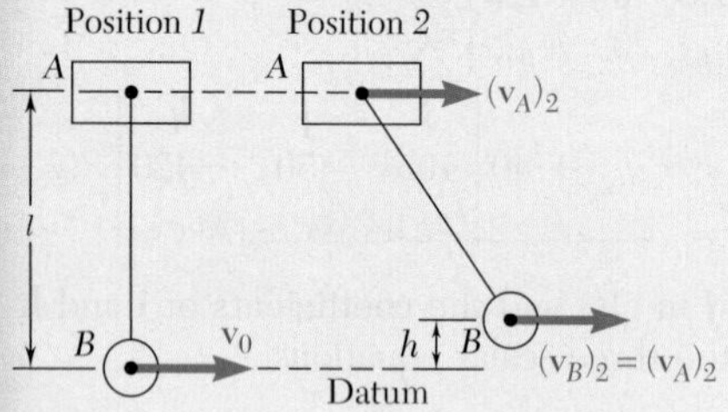

Conservation of Energy

Position 1.	***Potential Energy:***	$V_1 = m_Agl$
	Kinetic Energy:	$T_1 = \frac{1}{2}m_Bv_0^2$
Position 2.	***Potential Energy:***	$V_2 = m_Agl + m_Bgh$
	Kinetic Energy:	$T_2 = \frac{1}{2}(m_A + m_B)(v_A)_2^2$

$$T_1 + V_1 = T_2 + V_2: \quad \tfrac{1}{2}m_Bv_0^2 + m_Agl = \tfrac{1}{2}(m_A + m_B)(v_A)_2^2 + m_Agl + m_Bgh$$

Solving for h, we have

$$h = \frac{v_0^2}{2g} - \frac{m_A + m_B}{m_B}\frac{(v_A)_2^2}{2g}$$

or, substituting for $(v_A)_2$ the expression found above,

$$h = \frac{v_0^2}{2g} - \frac{m_B}{m_A + m_B}\frac{v_0^2}{2g} \qquad h = \frac{m_A}{m_A + m_B}\frac{v_0^2}{2g} \blacktriangleleft$$

Remarks. (1) Recalling that $v_0^2 < 2gl$, it follows from the last equation that $h < l$; we thus check that B stays below A as assumed in our solution.

(2) For $m_A \gg m_B$, the answers obtained reduce to $(\mathbf{v}_B)_2 = (\mathbf{v}_A)_2 = 0$ and $h = v_0^2/2g$; B oscillates as a simple pendulum with A fixed. For $m_A \ll m_B$, they reduce to $(\mathbf{v}_B)_2 = (\mathbf{v}_A)_2 = \mathbf{v}_0$ and $h = 0$; A and B move with the same constant velocity $\mathbf{v}_0$.

SAMPLE PROBLEM 14.5

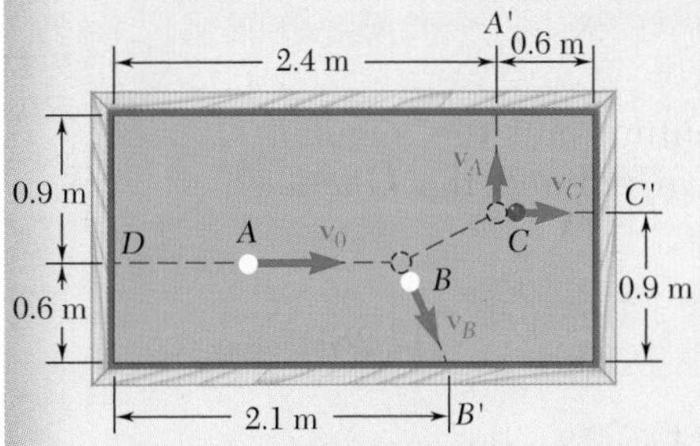

In a game of billiards, ball A is given an initial velocity $\mathbf{v}_0$ of magnitude $v_0 = 3$ m/s along line DA parallel to the axis of the table. It hits ball B and then ball C, which are both at rest. Knowing that A and C hit the sides of the table squarely at points A' and C', respectively, that B hits the side obliquely at B', and assuming frictionless surfaces and perfectly elastic impacts, determine the velocities $\mathbf{v}_A$, $\mathbf{v}_B$, and $\mathbf{v}_C$ with which the balls hit the sides of the table. (*Remark:* In this sample problem and in several of the problems which follow, the billiard balls are assumed to be particles moving freely in a horizontal plane, rather than the rolling and sliding spheres they actually are.)

SOLUTION

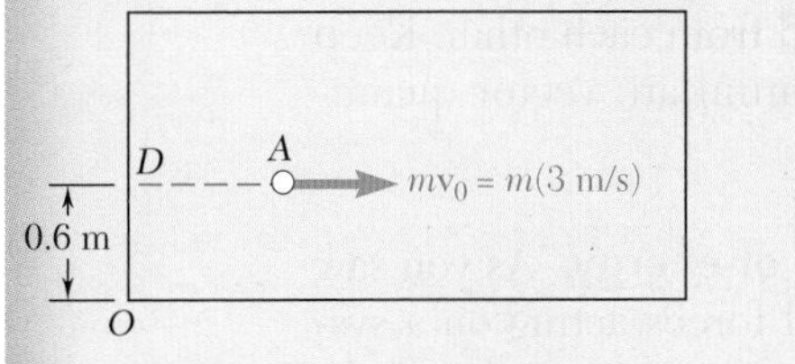

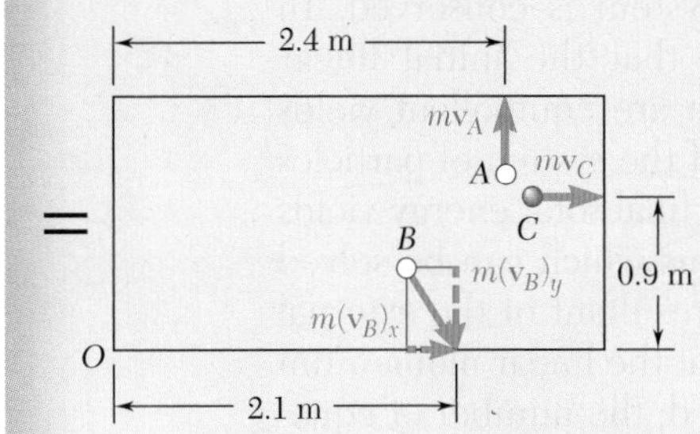

Conservation of Momentum. Since there is no external force, the initial momentum $m\mathbf{v}_0$ is equipollent to the system of momenta after the two collisions (and before any of the balls hits the side of the table). Referring to the adjoining sketch, we write

$$\xrightarrow{+} x \text{ components:} \quad m(3 \text{ m/s}) = m(v_B)_x + mv_C \tag{1}$$

$$+\uparrow y \text{ components:} \quad 0 = mv_A - m(v_B)_y \tag{2}$$

$$+\circlearrowleft \text{ moments about } O\text{:} \quad -(0.6 \text{ m})m(3 \text{ m/s}) = (2.4 \text{ m})mv_A - (2.1 \text{ m})m(v_B)_y - (0.9 \text{ m})mv_C \tag{3}$$

Solving the three equations for v_A, $(v_B)_x$, and $(v_B)_y$ in terms of v_C,

$$v_A = (v_B)_y = 3v_C - 6 \qquad (v_B)_x = 3 - v_C \tag{4}$$

Conservation of Energy. Since the surfaces are frictionless and the impacts are perfectly elastic, the initial kinetic energy $\frac{1}{2}mv_0^2$ is equal to the final kinetic energy of the system:

$$\tfrac{1}{2}mv_0^2 = \tfrac{1}{2}m_Av_A^2 + \tfrac{1}{2}m_Bv_B^2 + \tfrac{1}{2}m_Cv_C^2$$

$$v_A^2 + (v_B)_x^2 + (v_B)_y^2 + v_C^2 = (3 \text{ m/s})^2 \tag{5}$$

Substituting for v_A, $(v_B)_x$, and $(v_B)_y$ from (4) into (5), we have

$$2(3v_C - 6)^2 + (3 - v_C)^2 + v_C^2 = 9$$

$$20v_C^2 - 78v_C + 72 = 0$$

Solving for v_C, we find $v_C = 1.5$ m/s and $v_C = 2.4$ m/s. Since only the second root yields a positive value for v_A after substitution into Eqs. (4), we conclude that $v_C = 2.4$ m/s and

$$v_A = (v_B)_y = 3(2.4) - 6 = 1.2 \text{ m/s} \qquad (v_B)_x = 3 - 2.4 = 0.6 \text{ m/s}$$

$$\mathbf{v}_A = 1.2 \text{ m/s}\uparrow \qquad \mathbf{v}_B = 1.342 \text{ m/s} \measuredangle 63.4° \qquad \mathbf{v}_C = 2.4 \text{ m/s} \rightarrow$$ ◀

SOLVING PROBLEMS ON YOUR OWN

In the preceding lesson we defined the linear momentum and the angular momentum of a system of particles. In this lesson we defined the *kinetic energy* T of a system of particles:

$$T = \frac{1}{2} \sum_{i=1}^{n} m_i v_i^2 \qquad (14.28)$$

The solutions of the problems in the preceding lesson were based on the conservation of the linear momentum of a system of particles or on the observation of the motion of the mass center of a system of particles. In this lesson you will solve problems involving the following:

1. Computation of the kinetic energy lost in collisions. The kinetic energy T_1 of the system of particles before the collisions and its kinetic energy T_2 after the collisions are computed from Eq. (14.28) and are subtracted from each other. Keep in mind that, while linear momentum and angular momentum are vector quantities, kinetic energy is a *scalar* quantity.

2. Conservation of linear momentum and conservation of energy. As you saw in the preceding lesson, when the resultant of the external forces acting on a system of particles is zero, the linear momentum of the system is conserved. In problems involving two-dimensional motion, expressing that the initial linear momentum and the final linear momentum of the system are equipollent yields two algebraic equations. Equating the initial total energy of the system of particles (including potential energy as well as kinetic energy) to its final total energy yields an additional equation. Thus, you can write three equations which can be solved for three unknowns [Sample Prob. 14.5]. Note that if the resultant of the external forces is not zero but has a fixed direction, the component of the linear momentum in a direction perpendicular to the resultant is still conserved; the number of equations which can be used is then reduced to two [Sample Prob. 14.4].

3. Conservation of linear and angular momentum. When no external forces act on a system of particles, both the linear momentum of the system and its angular momentum about some arbitrary point are conserved. In the case of three-dimensional motion, this will enable you to write as many as six equations, although you may need to solve only some of them to obtain the desired answers [Sample Prob. 14.3]. In the case of two-dimensional motion, you will be able to write three equations which can be solved for three unknowns.

4. Conservation of linear and angular momentum and conservation of energy. In the case of the two-dimensional motion of a system of particles which are not subjected to any external forces, you will obtain two algebraic equations by expressing that the linear momentum of the system is conserved, one equation by writing that the angular momentum of the system about some arbitrary point is conserved, and a fourth equation by expressing that the total energy of the system is conserved. These equations can be solved for four unknowns.

PROBLEMS

14.31 Determine the energy lost due to friction and the impacts for Prob. 14.1.

14.32 In Prob. 14.4, determine the energy lost as the bullet (*a*) passes through block *A*, (*b*) becomes embedded in block *B*.

14.33 In Prob. 14.6, determine the work done by the woman and by the man as each dives from the boat, assuming that the woman dives first.

14.34 Determine the energy lost as a result of the series of collisions described in Prob. 14.8.

14.35 Two automobiles *A* and *B*, of mass m_A and m_B, respectively, are traveling in opposite directions when they collide head on. The impact is assumed perfectly plastic, and it is further assumed that the energy absorbed by each automobile is equal to its loss of kinetic energy with respect to a moving frame of reference attached to the mass center of the two-vehicle system. Denoting by E_A and E_B, respectively, the energy absorbed by automobile *A* and by automobile *B*, (*a*) show that $E_A/E_B = m_B/m_A$, that is, the amount of energy absorbed by each vehicle is inversely proportional to its mass, (*b*) compute E_A and E_B, knowing that $m_A = 1600$ kg and $m_B = 900$ kg and that the speeds of *A* and *B* are, respectively, 90 km/h and 60 km/h.

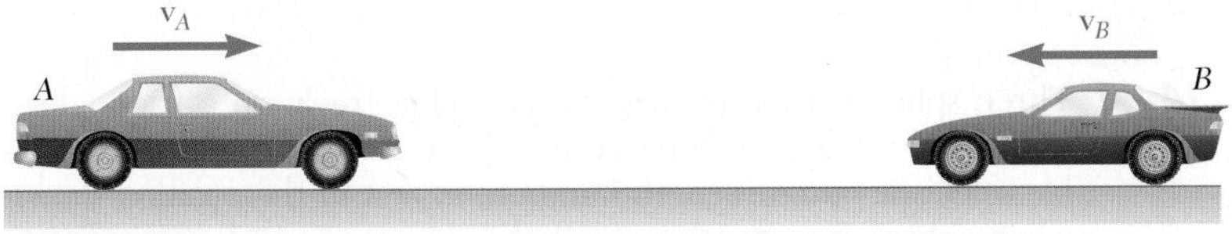

Fig. P14.35

14.36 It is assumed that each of the two automobiles involved in the collision described in Prob. 14.35 had been designed to safely withstand a test in which it crashed into a solid, immovable wall at the speed v_0. The severity of the collision of Prob. 14.35 may then be measured for each vehicle by the ratio of the energy it absorbed in the collision to the energy it absorbed in the test. On that basis, show that the collision described in Prob. 14.35 is $(m_A/m_B)^2$ times more severe for automobile *B* than for automobile *A*.

14.37 Solve Sample Prob. 14.4, assuming that cart *A* is given an initial horizontal velocity $\mathbf{v}_0$ while ball *B* is at rest.

14.38 Two hemispheres are held together by a cord which maintains a spring under compression (the spring is not attached to the hemispheres). The potential energy of the compressed spring is 120 J and the assembly has an initial velocity $\mathbf{v}_0$ of magnitude $v_0 = 8$ m/s. Knowing that the cord is severed when $\theta = 30°$, causing the hemispheres to fly apart, determine the resulting velocity of each hemisphere.

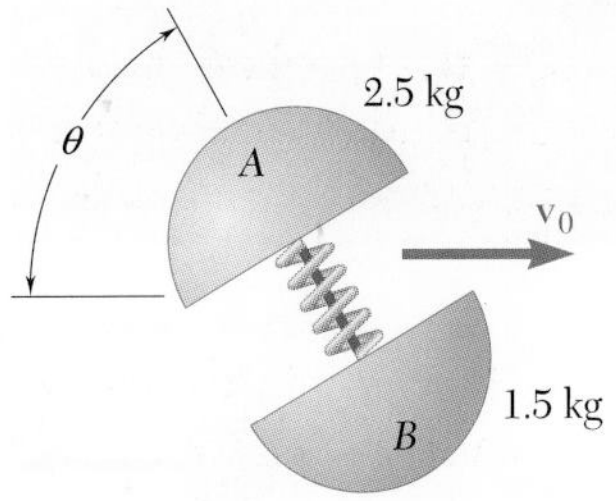

Fig. P14.38

14.39 A 9-kg block *B* starts from rest and slides down the inclined surface of a 15-kg wedge *A* which is supported by a horizontal surface. Neglecting friction, determine (*a*) the velocity of *B* relative to *A* after it has slid 0.6 m down the surface of the wedge, (*b*) the corresponding velocity of the wedge.

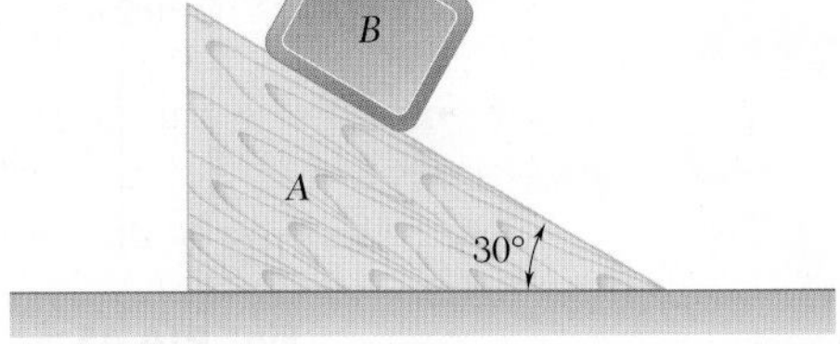

Fig. *P14.39*

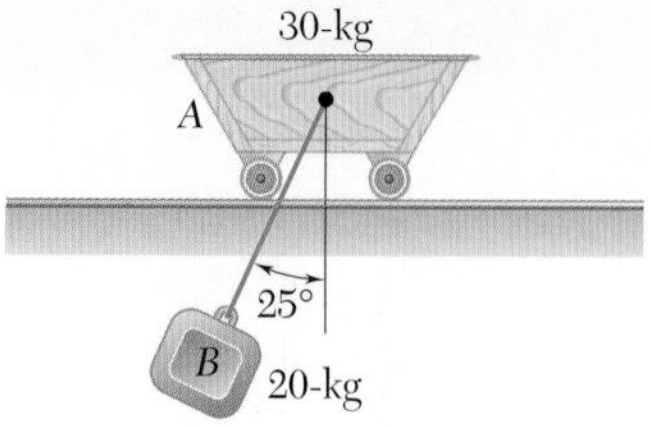

Fig. *P14.40*

14.40 A 20-kg block B is suspended from a 2-m cord attached to a 30-kg cart A, which may roll freely on a frictionless, horizontal track. If the system is released from rest in the position shown, determine the velocities of A and B as B passes directly under A.

14.41 and 14.42 In a game of pool, ball A is moving with a velocity $\mathbf{v}_0$ of magnitude $v_0 = 5$ m/s when it strikes balls B and C, which are at rest and aligned as shown. Knowing that after the collision the three balls move in the directions indicated and assuming frictionless surfaces and perfectly elastic impact (i.e., conservation of energy), determine the magnitudes of the velocities $\mathbf{v}_A$, $\mathbf{v}_B$, and $\mathbf{v}_C$.

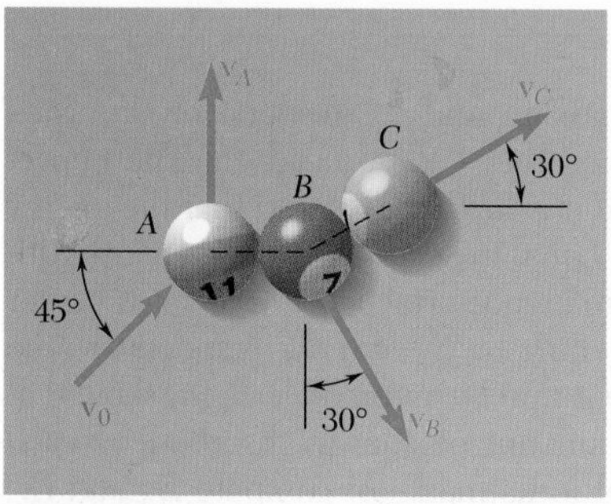

Fig. P14.41

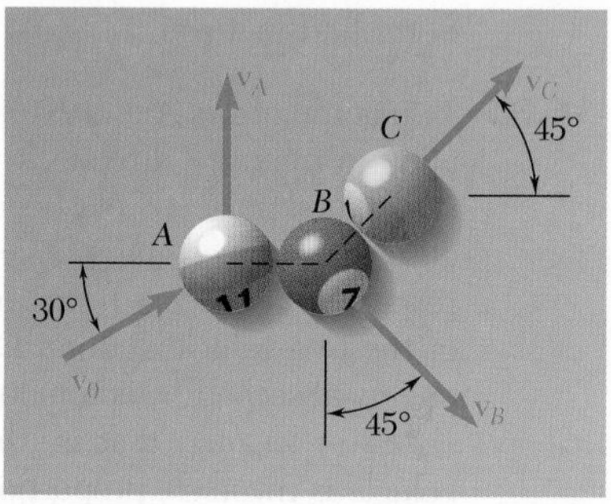

Fig. P14.42

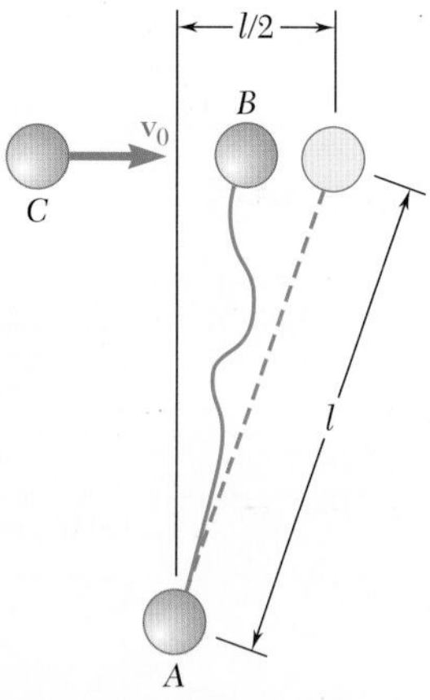

Fig. *P14.43*

14.43 Three spheres, each of mass m, can slide freely on a frictionless, horizontal surface. Spheres A and B are attached to an inextensible, inelastic cord of length l and are at rest in the position shown when sphere B is struck squarely by sphere C which is moving to the right with a velocity $\mathbf{v}_0$. Knowing that the cord is slack when sphere B is struck by sphere C and assuming perfectly elastic impact between B and C, determine (*a*) the velocity of each sphere immediately after the cord becomes taut, (*b*) the fraction of the initial kinetic energy of the system which is dissipated when the cord becomes taut.

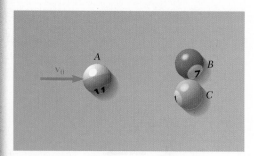

Fig. *P14.44*

14.44 In a game of pool, ball A is moving with the velocity $\mathbf{v}_0 = v_0\mathbf{i}$ when it strikes balls B and C, which are at rest side by side. Assuming frictionless surfaces and perfectly elastic impact (i.e., conservation of energy), determine the final velocity of each ball, assuming that the path of A is (*a*) perfectly centered and that A strikes B and C simultaneously, (*b*) not perfectly centered and that A strikes B slightly before it strikes C.

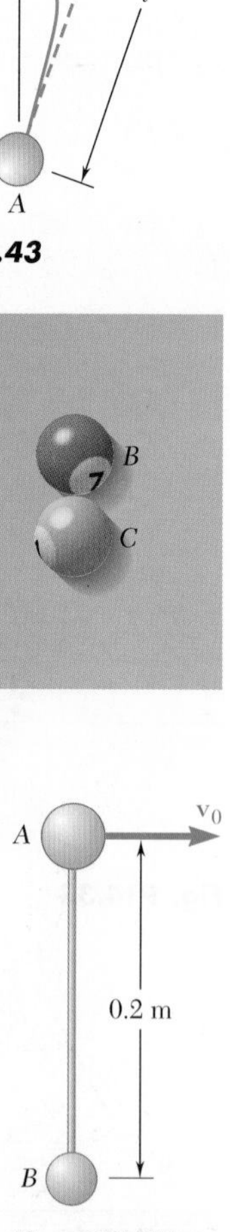

Fig. P14.45

14.45 Two small spheres A and B, of mass 2.5 kg and 1 kg, respectively, are connected by a rigid rod of negligible mass. The two spheres are resting on a horizontal, frictionless surface when A is suddenly given the velocity $\mathbf{v}_0 = (3.5 \text{ m/s})\mathbf{i}$. Determine (*a*) the linear momentum of the system and its angular momentum about its mass center G, (*b*) the velocities of A and B after the rod AB has rotated through 180°.

14.46 A 360-kg space vehicle traveling with a velocity $\mathbf{v}_0 = (450 \text{ m/s})\mathbf{k}$ passes through the origin O. Explosive charges then separate the vehicle into three parts A, B, and C, with masses of 60 kg, 120 kg, and 180 kg, respectively. Knowing that shortly thereafter the positions of the three parts are, respectively, $A(72, 72, 648)$, $B(180, 396, 972)$, and $C(-144, -288, 576)$, where the coordinates are expressed in meters, that the velocity of B is $\mathbf{v}_B = (150\text{m/s})\mathbf{i} + (330 \text{ m/s})\mathbf{j} + (660 \text{ m/s})\mathbf{k}$, and that the x component of the velocity of C is -120 m/s, determine the velocity of part A.

14.47 Four small disks A, B, C, and D can slide freely on a frictionless horizontal surface. Disks B, C, and D are connected by light rods and are at rest in the position shown when disk B is struck squarely by disk A which is moving to the right with a velocity $\mathbf{v}_0 = (12 \text{ m/s})\mathbf{i}$. The masses of the disks are $m_A = m_B = m_C = 7.5$ kg, and $m_D = 15$ kg. Knowing that the velocities of the disks immediately after the impact are $\mathbf{v}_A = \mathbf{v}_B = (2.5 \text{ m/s})\mathbf{i}$, $\mathbf{v}_C = v_C\mathbf{i}$, and $\mathbf{v}_D = v_D\mathbf{i}$, determine (*a*) the speeds v_C and v_D, (*b*) the fraction of the initial kinetic energy of the system which is dissipated during the collision.

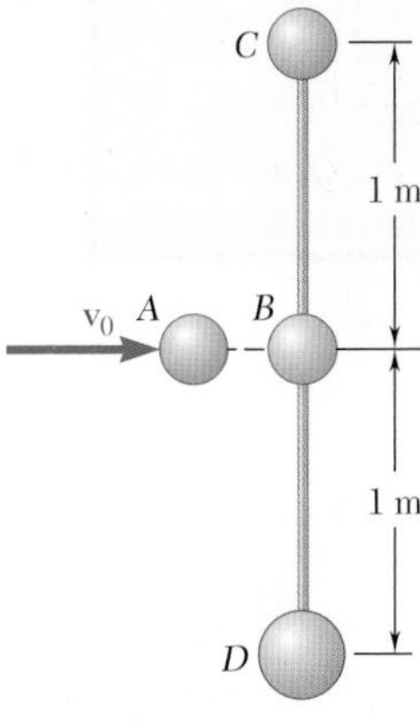

Fig. P14.47

14.48 In the scattering experiment of Prob. 14.26, it is known that the alpha particle is projected from $A_0(300, 0, 300)$ and that it collides with the oxygen nucleus C at $Q(240, 200, 100)$, where all coordinates are expressed in millimeters. Determine the coordinates of point B_0 where the original path of nucleus B intersects the zx plane. (*Hint.* Express that the angular momentum of the three particles about Q is conserved.)

14.49 Three identical small spheres, each of mass 1 kg, can slide freely on a horizontal frictionless surface. Spheres B and C are connected by a light rod and are at rest in the position shown when sphere B is struck squarely by sphere A which is moving to the right with a velocity $\mathbf{v}_0 = (2.4 \text{ m/s})\mathbf{i}$. Knowing that $\theta = 45°$ and that the velocities of spheres A and B immediately after the impact are $\mathbf{v}_A = 0$ and $\mathbf{v}_B = (1.8 \text{ m/s})\mathbf{i} + (\mathbf{v}_B)_y\mathbf{j}$, determine $(v_B)_y$ and the velocity of C immediately after impact.

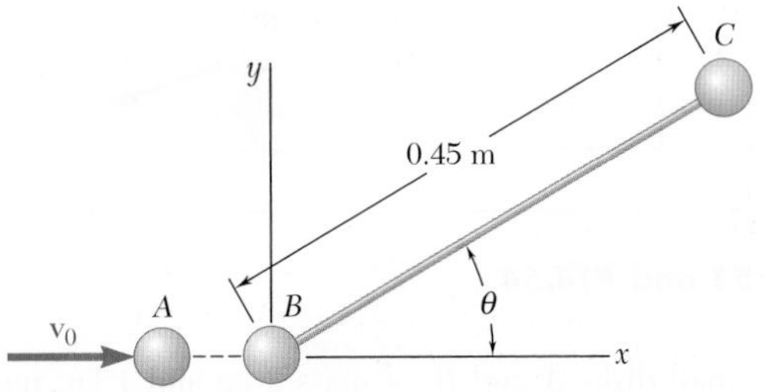

Fig. P14.49

14.50 Three small spheres A, B, and C, each of mass m, are connected to a small ring D of negligible mass by means of three inextensible, inelastic cords of length l. The spheres can slide freely on a frictionless horizontal surface and are rotating initially at a speed v_0 about ring D which is at rest. Suddenly the cord CD breaks. After the other two cords have again become taut, determine (*a*) the speed of ring D, (*b*) the relative speed at which spheres A and B rotate about D, (*c*) the fraction of the original energy of spheres A and B which is dissipated when cords AD and BD again became taut.

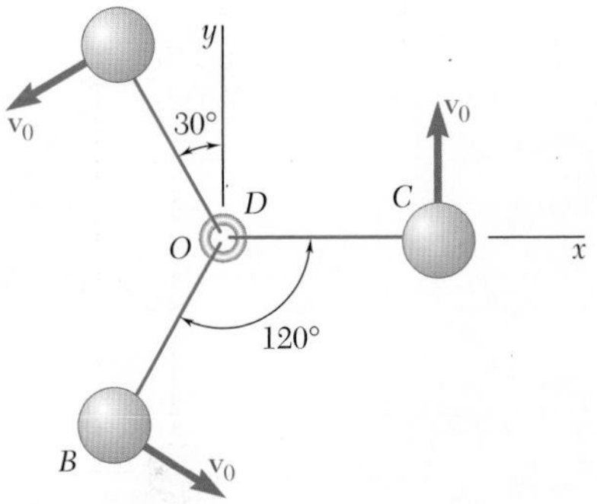

Fig. P14.50

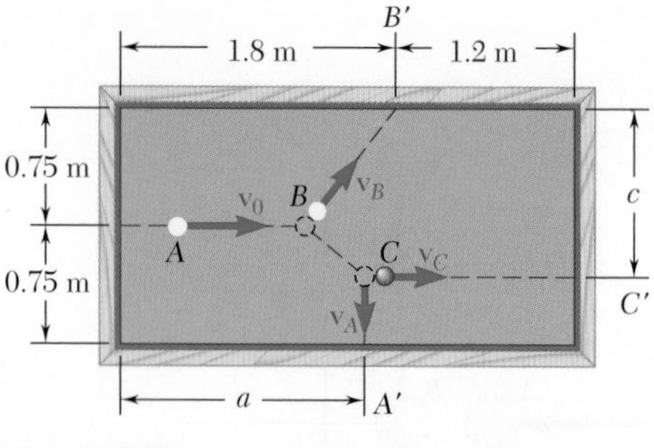

Fig. P14.51

14.51 In a game of billiards, ball A is given an initial velocity $\mathbf{v}_0$ along the longitudinal axis of the table. It hits ball B and then ball C, which are both at rest. Balls A and C are observed to hit the sides of the table squarely at A' and C', respectively, and ball B is observed to hit the side obliquely at B'. Knowing that $v_0 = 4$ m/s, $v_A = 1.92$ m/s, and $a = 1.65$ m, determine (*a*) the velocities $\mathbf{v}_B$ and $\mathbf{v}_C$ of balls B and C, (*b*) the point C' where ball C hits the side of the table. Assume frictionless surfaces and perfectly elastic impacts (i.e., conservation of energy).

14.52 For the game of billiards of Prob. 14.51, it is now assumed that $v_0 = 5$ m/s, $v_C = 3.2$ m/s, and $c = 1.22$ m. Determine (*a*) the velocities $\mathbf{v}_A$ and $\mathbf{v}_B$ of balls A and B, (*b*) the point A' where ball A hits the side of the table.

14.53 Two small disks A and B, of mass 3 kg and 1.5 kg, respectively, may slide on a horizontal, frictionless surface. They are connected by a cord, 600 mm long, and spin counterclockwise about their mass center G at the rate of 10 rad/s. At $t = 0$, the coordinates of G are $\bar{x}_0 = 0$, $\bar{y}_0 = 2$ m, and its velocity $\bar{\mathbf{v}}_0 = (1.2\text{ m/s})\mathbf{i} + (0.96\text{ m/s})\mathbf{j}$. Shortly thereafter the cord breaks; disk A is then observed to move along a path parallel to the y axis and disk B along a path which intersects the x axis at a distance $b = 7.5$ m from O. Determine (*a*) the velocities of A and B after the cord breaks, (*b*) the distance a from the y axis to the path of A.

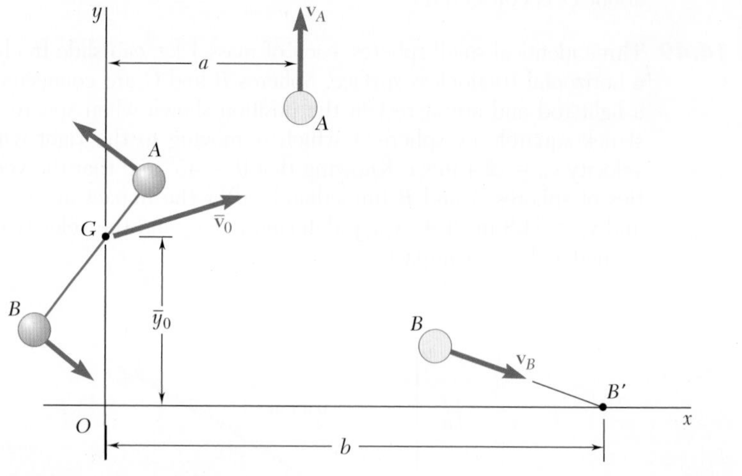

Fig. *P14.53 and P14.54*

14.54 Two small disks A and B, of mass 2 kg and 1 kg, respectively, may slide on a horizontal and frictionless surface. They are connected by a cord of negligible mass and spin about their mass center G. At $t = 0$, G is moving with the velocity $\bar{\mathbf{v}}_0$ and its coordinates are $\bar{x}_0 = 0$, $\bar{y}_0 = 1.89$ m. Shortly thereafter, the cord breaks and disk A is observed to move with a velocity $\mathbf{v}_A = (5\text{ m/s})\mathbf{j}$ in a straight line and at a distance $a = 2.56$ m from the y axis, while B moves with a velocity $\mathbf{v}_B = (7.2\text{ m/s})\mathbf{i} - (4.6\text{ m/s})\mathbf{j}$ along a path intersecting the x axis at a distance $b = 7.48$ m from the origin O. Determine (*a*) the initial velocity $\bar{\mathbf{v}}_0$ of the mass center G of the two disks, (*b*) the length of the cord initially connecting the two disks, (*c*) the rate in rad/s at which the disks were spinning about G.

14.55 Three small identical spheres A, B, and C, which can slide on a horizontal, frictionless surface, are attached to three 200-mm-long strings, which are tied to a ring G. Initially the spheres rotate clockwise about the ring with a relative velocity of 0.8 m/s and the ring moves along the x axis with a velocity $\mathbf{v}_0 = (0.4 \text{ m/s})\mathbf{i}$. Suddenly the ring breaks and the three spheres move freely in the xy plane with A and B following paths parallel to the y axis at a distance $a = 346$ mm from each other and C following a path parallel to the x axis. Determine (*a*) the velocity of each sphere, (*b*) the distance d.

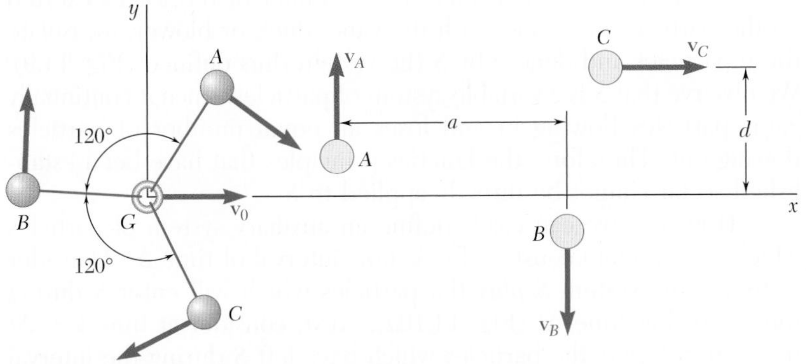

Fig. P14.55 and P14.56

14.56 Three small identical spheres A, B, and C, which can slide on a horizontal, frictionless surface, are attached to three strings of length l which are tied to a ring G. Initially the spheres rotate clockwise about the ring which moves along the x axis with a velocity $\mathbf{v}_0$. Suddenly the ring breaks and the three spheres move freely in the xy plane. Knowing that $\mathbf{v}_A = (1.039 \text{ m/s})\mathbf{j}$, $\mathbf{v}_C = (1.8 \text{ m/s})\mathbf{i}$, $a = 416$ mm, and $d = 240$ mm, determine (*a*) the initial velocity of the ring, (*b*) the length l of the strings, (*c*) the rate in rad/s at which the spheres were rotating about G.

*14.10 VARIABLE SYSTEMS OF PARTICLES

All the systems of particles considered so far consisted of well-defined particles. These systems did not gain or lose any particles during their motion. In a large number of engineering applications, however, it is necessary to consider *variable systems of particles*, i.e., systems which are continually gaining or losing particles, or doing both at the same time. Consider, for example, a hydraulic turbine. Its analysis involves the determination of the forces exerted by a stream of water on rotating blades, and we note that the particles of water in contact with the blades form an everchanging system which continually acquires and loses particles. Rockets furnish another example of variable systems, since their propulsion depends upon the continual ejection of fuel particles.

We recall that all the kinetics principles established so far were derived for constant systems of particles, which neither gain nor lose particles. We must therefore find a way to reduce the analysis of a variable system of particles to that of an auxiliary constant system. The procedure to follow is indicated in Secs. 14.11 and 14.12 for two broad categories of applications: a steady stream of particles and a system that is gaining or losing mass.

*14.11 STEADY STREAM OF PARTICLES

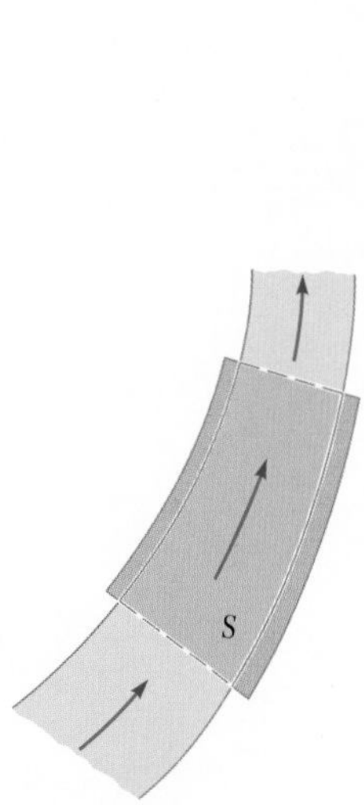

Fig. 14.9

Consider a steady stream of particles, such as a stream of water diverted by a fixed vane or a flow of air through a duct or through a blower. In order to determine the resultant of the forces exerted on the particles in contact with the vane, duct, or blower, we isolate these particles and denote by S the system thus defined (Fig. 14.9). We observe that S is a variable system of particles, since it continually gains particles flowing in and loses an equal number of particles flowing out. Therefore, the kinetics principles that have been established so far cannot be directly applied to S.

However, we can easily define an auxiliary system of particles which does remain constant for a short interval of time Δt. Consider at time t the system S *plus* the particles which will enter S during the interval at time Δt (Fig. 14.10*a*). Next, consider at time $t + \Delta t$ the system S *plus* the particles which have left S during the interval Δt (Fig. 14.10*c*). Clearly, *the same particles are involved in both cases,* and we can apply to those particles the principle of impulse and momentum. Since the total mass m of the system S remains constant, the particles entering the system and those leaving the system in the time Δt must have the same mass Δm. Denoting by $\mathbf{v}_A$ and $\mathbf{v}_B$, respectively, the velocities of the particles entering S at A and leaving S at B, we represent the momentum of the particles entering S by $(\Delta m)\mathbf{v}_A$ (Fig. 14.10*a*) and the momentum of the particles leaving S by $(\Delta m)\mathbf{v}_B$ (Fig. 14.10*c*). We also represent by appropriate vectors the momenta $m_i\mathbf{v}_i$ of the particles forming S and the impulses of the forces exerted on S and indicate by blue plus and equals signs that the system of the momenta and impulses in parts a and b of Fig. 14.10 is equipollent to the system of the momenta in part c of the same figure.

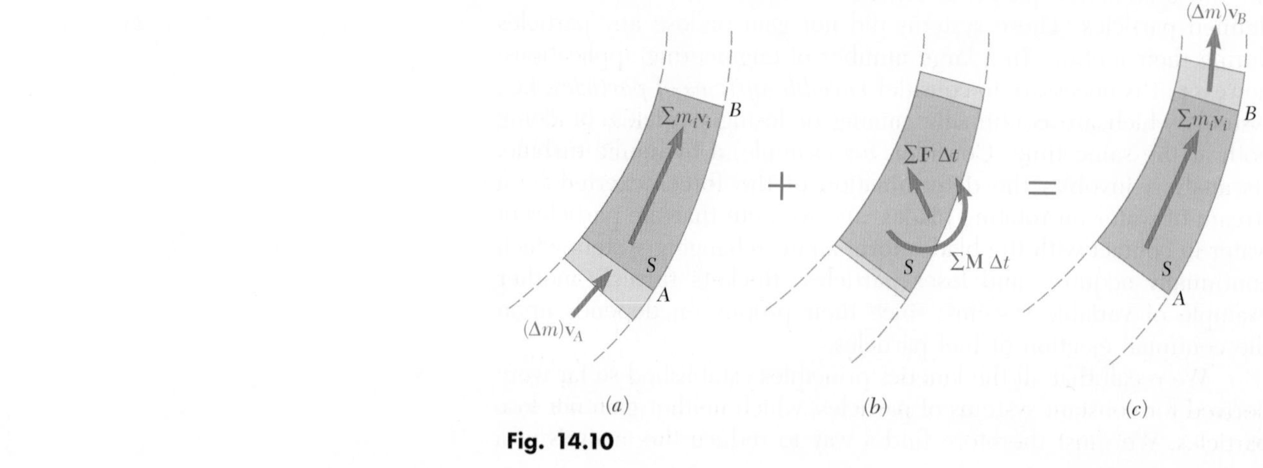

Fig. 14.10

The resultant $\Sigma m_i\mathbf{v}_i$ of the momenta of the particles of S is found on both sides of the equals sign and can thus be omitted. We conclude that *the system formed by the momentum* $(\Delta m)\mathbf{v}_A$ *of the particles entering S in the time* Δt *and the impulses of the forces exerted on S during that time is equipollent to the momentum* $(\Delta m)\mathbf{v}_B$ *of the particles leaving S in the same time* Δt. We can therefore write

$$(\Delta m)\mathbf{v}_A + \Sigma\mathbf{F}\,\Delta t = (\Delta m)\mathbf{v}_B \tag{14.38}$$

A similar equation can be obtained by taking the moments of the vectors involved (see Sample Prob. 14.5). Dividing all terms of Eq. (14.38) by Δt and letting Δt approach zero, we obtain at the limit

$$\Sigma\mathbf{F} = \frac{dm}{dt}(\mathbf{v}_B - \mathbf{v}_A) \tag{14.39}$$

where $\mathbf{v}_B - \mathbf{v}_A$ represents the difference between the *vector* $\mathbf{v}_B$ and the *vector* $\mathbf{v}_A$.

If SI units are used, dm/dt is expressed in kg/s and the velocities in m/s; we check that both members of Eq. (14.39) are expressed in the same units (newtons). If U.S. customary units are used, dm/dt must be expressed in slugs/s and the velocities in ft/s; we check again that both members of the equation are expressed in the same units (pounds).†

The principle we have established can be used to analyze a large number of engineering applications. Some of the more common of these applications will be considered next.

Fluid Stream Diverted by a Vane. If the vane is fixed, the method of analysis given above can be applied directly to find the force $\mathbf{F}$ exerted by the vane on the stream. We note that $\mathbf{F}$ is the only force which needs to be considered since the pressure in the stream is constant (atmospheric pressure). The force exerted by the stream on the vane will be equal and opposite to $\mathbf{F}$. If the vane moves with a constant velocity, the stream is not steady. However, it will appear steady to an observer moving with the vane. We should therefore choose a system of axes moving with the vane. Since this system of axes is not accelerated, Eq. (14.38) can still be used, but $\mathbf{v}_A$ and $\mathbf{v}_B$ must be replaced by the *relative velocities* of the stream with respect to the vane (see Sample Prob. 14.7).

Fluid Flowing Through a Pipe. The force exerted by the fluid on a pipe transition such as a bend or a contraction can be determined by considering the system of particles S in contact with the transition. Since, in general, the pressure in the flow will vary, the forces exerted on S by the adjoining portions of the fluid should also be considered.

†It is often convenient to express the mass rate of flow dm/dt as the product ρQ, where ρ is the density of the stream (mass per unit volume) and Q its volume rate of flow (volume per unit time). If SI units are used, ρ is expressed in kg/m^3 (for instance, $\rho = 1000$ kg/m^3 for water) and Q in m^3/s. However, if U.S. customary units are used, ρ will generally have to be computed from the corresponding specific weight γ (weight per unit volume), $\rho = \gamma/g$. Since γ is expressed in lb/ft^3 (for instance, $\gamma = 62.4$ lb/ft^3 for water), ρ is obtained in slugs/ft^3. The volume rate of flow Q is expressed in ft^3/s.

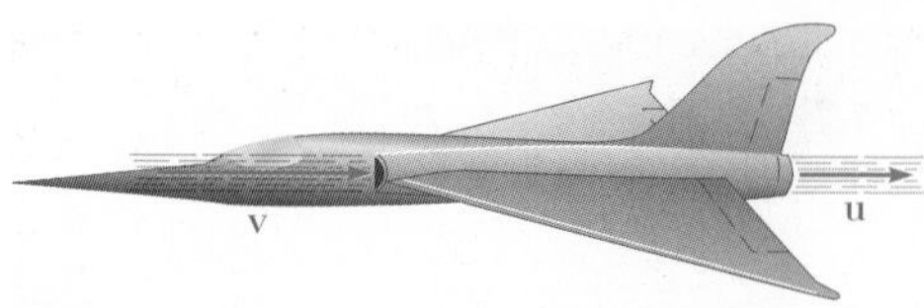

Fig. 14.11

Jet Engine. In a jet engine, air enters with no velocity through the front of the engine and leaves through the rear with a high velocity. The energy required to accelerate the air particles is obtained by burning fuel. The mass of the burned fuel in the exhaust gases will usually be small enough compared with the mass of the air flowing through the engine that it can be neglected. Thus, the analysis of a jet engine reduces to that of an airstream. This stream can be considered as a steady stream if all velocities are measured with respect to the airplane. It will be assumed, therefore, that the airstream enters the engine with a velocity $\mathbf{v}$ of magnitude equal to the speed of the airplane and leaves with a velocity $\mathbf{u}$ equal to the relative velocity of the exhaust gases (Fig. 14.11). Since the intake and exhaust pressures are nearly atmospheric, the only external force which needs to be considered is the force exerted by the engine on the airstream. This force is equal and opposite to the thrust.†

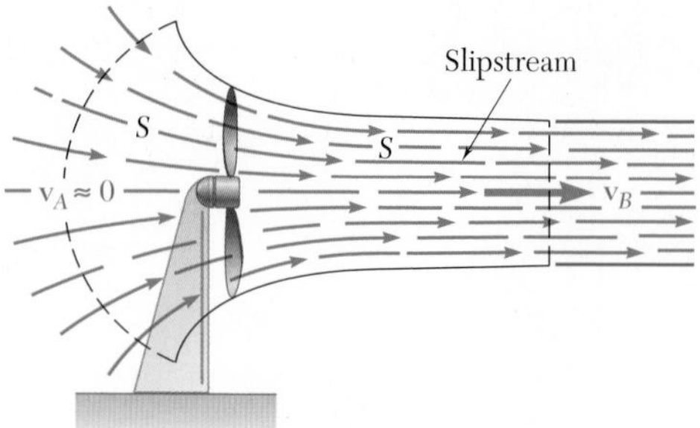

Fig. 14.12

Fan. We consider the system of particles S shown in Fig. 14.12. The velocity $\mathbf{v}_A$ of the particles entering the system is assumed equal to zero, and the velocity $\mathbf{v}_B$ of the particles leaving the system is the velocity of the *slipstream*. The rate of flow can be obtained by multiplying v_B by the cross-sectional area of the slipstream. Since the pressure all around S is atmospheric, the only external force acting on S is the thrust of the fan.

Helicopter. The determination of the thrust created by the rotating blades of a hovering helicopter is similar to the determination of the thrust of a fan. The velocity $\mathbf{v}_A$ of the air particles as they approach the blades is assumed to be zero, and the rate of flow is obtained by multiplying the magnitude of the velocity $\mathbf{v}_B$ of the slipstream by its cross-sectional area.

*14.12 SYSTEMS GAINING OR LOSING MASS

Let us now analyze a different type of variable system of particles, namely, a system which gains mass by continually absorbing particles or loses mass by continually expelling particles. Consider the system S shown in Fig. 14.13. Its mass, equal to m at the instant t, increases by Δm in the interval of time Δt. In order to apply the principle of impulse and momentum to the analysis of this system, we must consider at time t the system S *plus* the particles of mass Δm which S absorbs during the time interval Δt. The velocity of S at time t is denoted by $\mathbf{v}$, the velocity of S at time $t + \Delta t$ is denoted by $\mathbf{v} + \Delta\mathbf{v}$, and the absolute velocity of the particles absorbed is denoted by $\mathbf{v}_a$. Applying the principle of impulse and momentum, we write

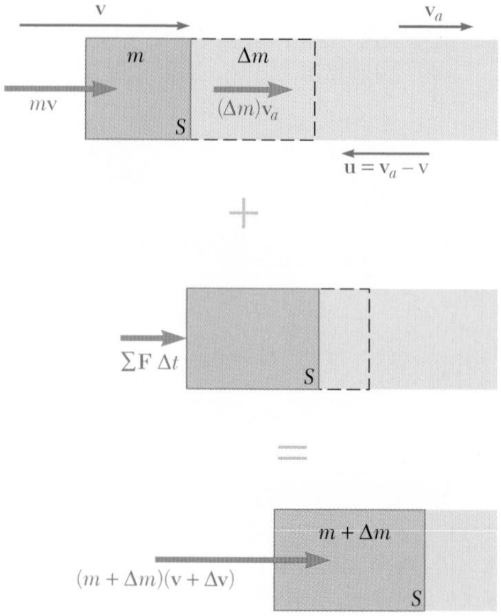

Fig. 14.13

$$m\mathbf{v} + (\Delta m)\mathbf{v}_a + \Sigma\mathbf{F}\,\Delta t = (m + \Delta m)(\mathbf{v} + \Delta\mathbf{v}) \qquad (14.40)$$

†Note that if the airplane is accelerated, it cannot be used as a newtonian frame of reference. The same result will be obtained for the thrust, however, by using a reference frame at rest with respect to the atmosphere, since the air particles will then be observed to enter the engine with no velocity and to leave it with a velocity of magnitude $u - v$.

Solving for the sum $\Sigma\mathbf{F}\,\Delta t$ of the impulses of the external forces acting on S (excluding the forces exerted by the particles being absorbed), we have

$$\Sigma\mathbf{F}\,\Delta t = m\Delta\mathbf{v} + \Delta m(\mathbf{v} - \mathbf{v}_a) + (\Delta m)(\Delta\mathbf{v}) \qquad (14.41)$$

Introducing the *relative velocity* $\mathbf{u}$ with respect to S of the particles which are absorbed, we write $\mathbf{u} = \mathbf{v}_a - \mathbf{v}$ and note, since $v_a < v$, that the relative velocity $\mathbf{u}$ is directed to the left, as shown in Fig. 14.13. Neglecting the last term in Eq. (14.41), which is of the second order, we write

$$\Sigma\mathbf{F}\,\Delta t = m\,\Delta\mathbf{v} - (\Delta m)\mathbf{u}$$

Dividing through by Δt and letting Δt approach zero, we have at the limit†

$$\Sigma\mathbf{F} = m\frac{d\mathbf{v}}{dt} - \frac{dm}{dt}\mathbf{u} \qquad (14.42)$$

Rearranging the terms and recalling that $d\mathbf{v}/dt = \mathbf{a}$, where $\mathbf{a}$ is the acceleration of the system S, we write

$$\Sigma\mathbf{F} + \frac{dm}{dt}\mathbf{u} = m\mathbf{a} \qquad (14.43)$$

which shows that the action on S of the particles being absorbed is equivalent to a thrust

$$\mathbf{P} = \frac{dm}{dt}\mathbf{u} \qquad (14.44)$$

which tends to slow down the motion of S, since the relative velocity $\mathbf{u}$ of the particles is directed to the left. If SI units are used, dm/dt is expressed in kg/s, the relative velocity u in m/s, and the corresponding thrust in newtons. If U.S. customary units are used, dm/dt must be expressed in slugs/s, u in ft/s, and the corresponding thrust in pounds.‡

The equations obtained can also be used to determine the motion of a system S losing mass. In this case, the rate of change of mass is negative, and the action on S of the particles being expelled is equivalent to a thrust in the direction of $-\mathbf{u}$, that is, in the direction opposite to that in which the particles are being expelled. A *rocket* represents a typical case of a system continually losing mass (see Sample Prob. 14.8).

Photo 14.3 As the shuttle's booster rockets are fired, the gas particles they eject provide the thrust required for liftoff.

†When the absolute velocity $\mathbf{v}_a$ of the particles absorbed is zero, $\mathbf{u} = -\mathbf{v}$, and formula (14.42) becomes

$$\Sigma\mathbf{F} = \frac{d}{dt}(m\mathbf{v})$$

Comparing the formula obtained to Eq. (12.3) of Sec. 12.3, we observe that Newton's second law can be applied to a system gaining mass, *provided that the particles absorbed are initially at rest.* It may also be applied to a system losing mass, *provided that the velocity of the particles expelled is zero* with respect to the frame of reference selected.

‡See footnote on page 899.

SAMPLE PROBLEM 14.6

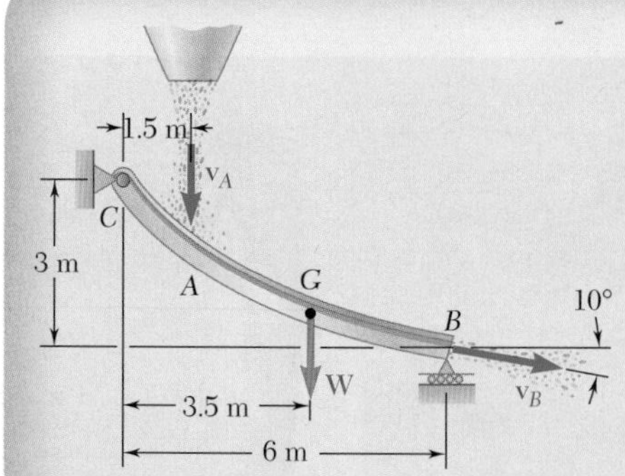

Grain falls from a hopper onto a chute CB at the rate of 120 kg/s. It hits the chute at A with a velocity of 10 m/s and leaves at B with a velocity of 7.5 m/s, forming an angle of 10° with the horizontal. Knowing that the combined weight of the chute and of the grain it supports is a force **W** of magnitude 3000 N applied at G, determine the reaction at the roller support B and the components of the reaction at the hinge C.

SOLUTION

We apply the principle of impulse and momentum for the time interval Δt to the system consisting of the chute, the grain it supports, and the amount of grain which hits the chute in the interval Δt. Since the chute does not move, it has no momentum. We also note that the sum $\Sigma m_i \mathbf{v}_i$ of the momenta of the particles supported by the chute is the same at t and $t + \Delta t$ and can thus be omitted.

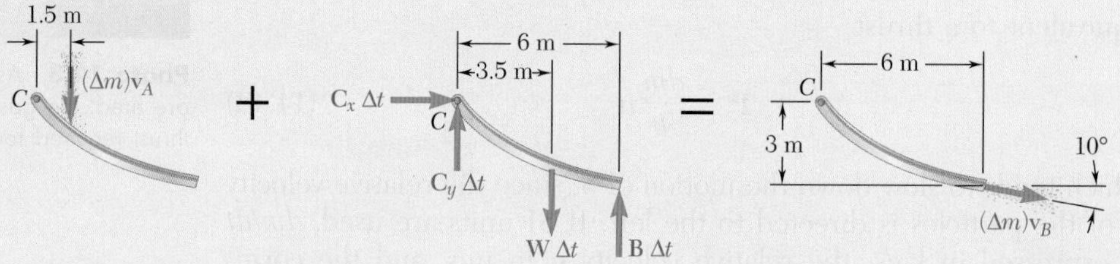

Since the system formed by the momentum $(\Delta m)\mathbf{v}_A$ and the impulses is equipollent to the momentum $(\Delta m)\mathbf{v}_B$, we write

$\xrightarrow{+} x$ components: $\quad C_x\,\Delta t = (\Delta m)v_B \cos 10° \qquad (1)$

$+\uparrow y$ components: $\quad -(\Delta m)v_A + C_y\,\Delta t - W\,\Delta t + B\,\Delta t$
$\qquad = -(\Delta m)v_B \sin 10° \qquad (2)$

$+\circlearrowleft$ moments about C: $\quad -1.5(\Delta m)v_A - 3.5(W\,\Delta t) + 6(B\,\Delta t)$
$\qquad = 3(\Delta m)v_B \cos 10° - 6(\Delta m)v_B \sin 10° \qquad (3)$

Using the given data, $W = 3000$ N, $v_A = 10$ m/s, $v_B = 7.5$ m/s, and $\Delta m/\Delta t = 120$ kg/s, and solving Eq. (3) for B and Eq. (1) for C_x,

$$6B = 3.5(3000) + 1.5(120)(10) + 3(120)(7.5)(\cos 10° - 2 \sin 10°)$$
$$B = 2340 \text{ N} \qquad \mathbf{B} = 2340 \text{ N}\uparrow \blacktriangleleft$$

$$C_x = (120)(7.5)\cos 10° = 886 \text{ N} \qquad \mathbf{C}_x = 886 \text{ N} \rightarrow \blacktriangleleft$$

Substituting for B and solving Eq. (2) for C_y,

$$C_y = 3000 - 2340 + (120)(10 - 7.5 \sin 10°) = 1704 \text{ N}$$
$$\mathbf{C}_y = 1704 \text{ N}\uparrow \blacktriangleleft$$

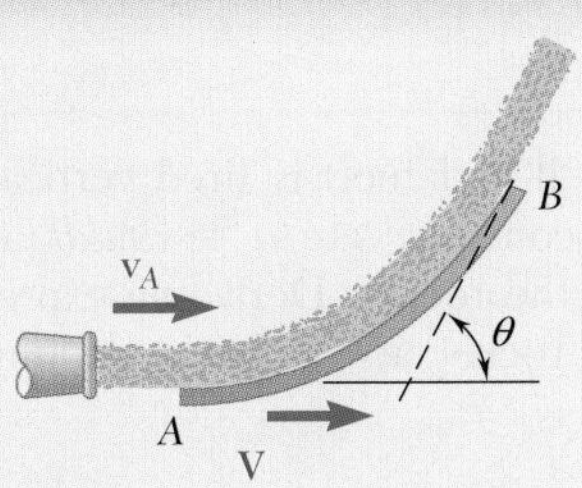

SAMPLE PROBLEM 14.7

A nozzle discharges a stream of water of cross-sectional area A with a velocity $\mathbf{v}_A$. The stream is deflected by a *single* blade which moves to the right with a constant velocity $\mathbf{V}$. Assuming that the water moves along the blade at constant speed, determine (*a*) the components of the force $\mathbf{F}$ exerted by the blade on the stream, (*b*) the velocity $\mathbf{V}$ for which maximum power is developed.

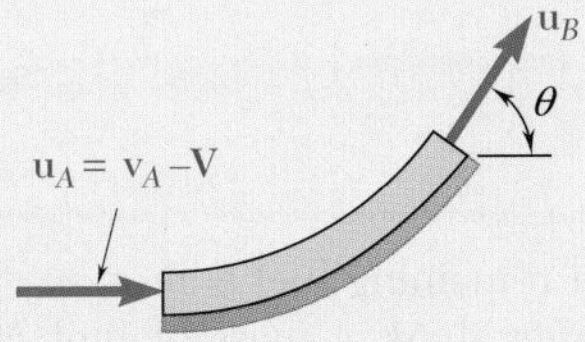

SOLUTION

a. Components of Force Exerted on Stream. We choose a coordinate system which moves with the blade at a constant velocity $\mathbf{V}$. The particles of water strike the blade with a relative velocity $\mathbf{u}_A = \mathbf{v}_A - \mathbf{V}$ and leave the blade with a relative velocity $\mathbf{u}_B$. Since the particles move along the blade at a constant speed, the relative velocities $\mathbf{u}_A$ and $\mathbf{u}_B$ have the same magnitude u. Denoting the density of water by ρ, the mass of the particles striking the blade during the time interval Δt is $\Delta m = A\rho(v_A - V)\,\Delta t$; an equal mass of particles leaves the blade during Δt. We apply the principle of impulse and momentum to the system formed by the particles in contact with the blade and the particles striking the blade in the time Δt.

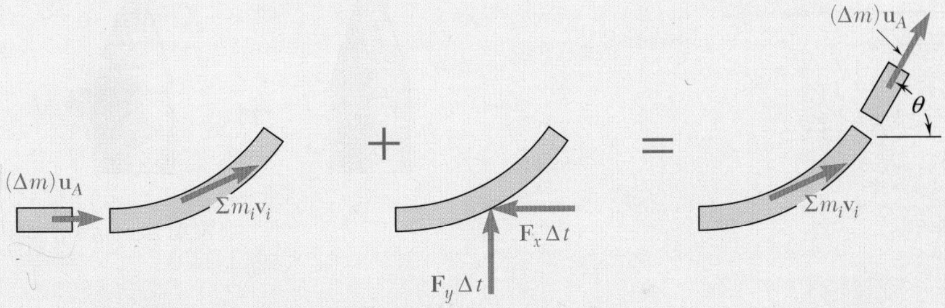

Recalling that $\mathbf{u}_A$ and $\mathbf{u}_B$ have the same magnitude u, and omitting the momentum $\Sigma m_i\mathbf{v}_i$ which appears on both sides, we write

$\xrightarrow{+} x$ components: $\quad (\Delta m)u - F_x\,\Delta t = (\Delta m)u\cos\theta$

$+\uparrow y$ components: $\quad +F_y\,\Delta t = (\Delta m)u\sin\theta$

Substituting $\Delta m = A\rho(v_A - V)\,\Delta t$ and $u = v_A - V$, we obtain

$$\mathbf{F}_x = A\rho(v_A - V)^2(1 - \cos\theta) \leftarrow \qquad \mathbf{F}_y = A\rho(v_A - V)^2\sin\theta\uparrow \quad ◀$$

b. Velocity of Blade for Maximum Power. The power is obtained by multiplying the velocity V of the blade by the component F_x of the force exerted by the stream on the blade.

$$\text{Power} = F_xV = A\rho(v_A - V)^2(1 - \cos\theta)V$$

Differentiating the power with respect to V and setting the derivative equal to zero, we obtain

$$\frac{d(\text{power})}{dV} = A\rho(v_A^2 - 4v_AV + 3V^2)(1 - \cos\theta) = 0$$

$$V = v_A \qquad V = \tfrac{1}{3}v_A \qquad \text{For maximum power } \mathbf{V} = \tfrac{1}{3}v_A \rightarrow \quad ◀$$

Note. These results are valid only when a *single* blade deflects the stream. Different results are obtained when a series of blades deflects the stream, as in a Pelton-wheel turbine. (See Prob. 14.81.)

SAMPLE PROBLEM 14.8

A rocket of initial mass m_0 (including shell and fuel) is fired vertically at time $t = 0$. The fuel is consumed at a constant rate $q = dm/dt$ and is expelled at a constant speed u relative to the rocket. Derive an expression for the magnitude of the velocity of the rocket at time t, neglecting the resistance of the air.

SOLUTION

At time t, the mass of the rocket shell and remaining fuel is $m = m_0 - qt$, and the velocity is $\mathbf{v}$. During the time interval Δt, a mass of fuel $\Delta m = q\ \Delta t$ is expelled with a speed u relative to the rocket. Denoting by $\mathbf{v}_e$ the absolute velocity of the expelled fuel, we apply the principle of impulse and momentum between time t and time $t + \Delta t$.

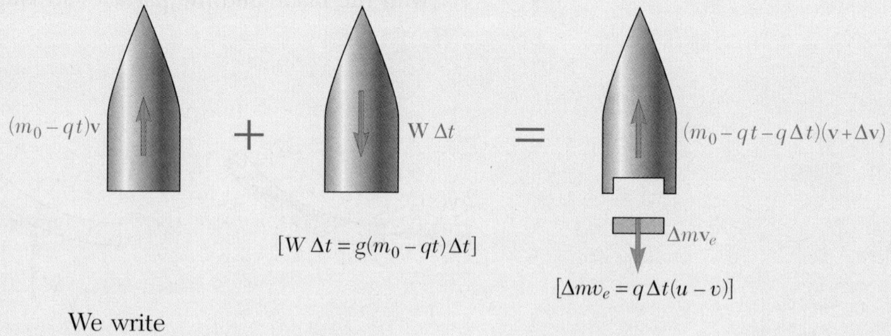

We write

$$(m_0 - qt)v - g(m_0 - qt)\ \Delta t = (m_0 - qt - q\ \Delta t)(v + \Delta v) - q\ \Delta t(u - v)$$

Dividing through by Δt and letting Δt approach zero, we obtain

$$-g(m_0 - qt) = (m_0 - qt)\frac{dv}{dt} - qu$$

Separating variables and integrating from $t = 0$, $v = 0$ to $t = t$, $v = v$,

$$dv = \left(\frac{qu}{m_0 - qt} - g\right)dt \qquad \int_0^v dv = \int_0^t \left(\frac{qu}{m_0 - qt} - g\right)dt$$

$$v = [-u \ln(m_0 - qt) - gt]_0^t \qquad v = u \ln \frac{m_0}{m_0 - qt} - gt \quad ◀$$

Remark. The mass remaining at time t_f, after all the fuel has been expended, is equal to the mass of the rocket shell $m_s = m_0 - qt_f$, and the maximum velocity attained by the rocket is $v_m = u \ln(m_0/m_s) - gt_f$. Assuming that the fuel is expelled in a relatively short period of time, the term gt_f is small and we have $v_m \approx u \ln(m_0/m_s)$. In order to escape the gravitational field of the earth, a rocket must reach a velocity of 11.18 km/s. Assuming $u = 2200$ m/s and $v_m = 11.18$ km/s, we obtain $m_0/m_s = 161$. Thus, to project each kilogram of the rocket shell into space, it is necessary to consume more than 161 kg of fuel if a propellant yielding $u = 2200$ m/s is used.

SOLVING PROBLEMS ON YOUR OWN

This lesson is devoted to the study of the motion of *variable systems of particles*, i.e., systems which are continually *gaining or losing particles* or doing both at the same time. The problems you will be asked to solve will involve (1) *steady streams of particles* and (2) *systems gaining or losing mass*.

1. To solve problems involving a steady stream of particles, you will consider a portion S of the stream and express that the system formed by the momentum of the particles entering S at A in the time Δt and the impulses of the forces exerted on S during that time is equipollent to the momentum of the particles leaving S at B in the same time Δt (Fig. 14.10). Considering only the resultants of the vector systems involved, you can write the vector equation

$$(\Delta m)\mathbf{v}_A + \Sigma\mathbf{F}\,\Delta t = (\Delta m)\mathbf{v}_B \qquad (14.38)$$

You may want to consider as well the moments about a given point of the vector systems involved to obtain an additional equation [Sample Prob. 14.6], but many problems can be solved using Eq. (14.38) or the equation obtained by dividing all terms by Δt and letting Δt approach zero,

$$\Sigma\mathbf{F} = \frac{dm}{dt}(\mathbf{v}_B - \mathbf{v}_A) \qquad (14.39)$$

where $\mathbf{v}_B - \mathbf{v}_A$ represents a *vector subtraction* and where the mass rate of flow dm/dt can be expressed as the product ρQ of the density ρ of the stream (mass per unit volume) and the volume rate of flow Q (volume per unit time). If U.S. customary units are used, ρ is expressed as the ratio γ/g, where γ is the specific weight of the stream and g is the acceleration of gravity.

Typical problems involving a steady stream of particles have been described in Sec. 14.11. You may be asked to determine the following:

a. Thrust caused by a diverted flow. Equation (14.39) is applicable, but you will get a better understanding of the problem if you use a solution based on Eq. (14.38).

b. Reactions at supports of vanes or conveyor belts. First draw a diagram showing on one side of the equals sign the momentum $(\Delta m)\mathbf{v}_A$ of the particles impacting the vane or belt in the time Δt, as well as the impulses of the loads and reactions at the supports during that time, and showing on the other side the momentum $(\Delta m)\mathbf{v}_B$ of the particles leaving the vane or belt in the time Δt [Sample Prob. 14.6]. Equating the x components, y components, and moments of the quantities on both sides of the equals sign will yield three scalar equations which can be solved for three unknowns.

c. Thrust developed by a jet engine, a propeller, or a fan. In most cases, a single unknown is involved, and that unknown can be obtained by solving the scalar equation derived from Eq. (14.38) or Eq. (14.39).

(continued)

2. To solve problems involving systems gaining mass, you will consider the system S, which has a mass m and is moving with a velocity $\mathbf{v}$ at time t, and the particles of mass Δm with velocity $\mathbf{v}_a$ that S will absorb in the time interval Δt (Fig. 14.13). You will then express that the total momentum of S and of the particles that will be absorbed, *plus* the impulse of the external forces exerted on S, are equipollent to the momentum of S at time $t + \Delta t$. Noting that the mass of S and its velocity at that time are, respectively, $m + \Delta m$ and $\mathbf{v} + \Delta\mathbf{v}$, you will write the vector equation

$$m\mathbf{v} + (\Delta m)\mathbf{v}_a + \Sigma\mathbf{F}\,\Delta t = (m + \Delta m)(\mathbf{v} + \Delta\mathbf{v}) \qquad (14.40)$$

As was shown in Sec. 14.12, if you introduce the relative velocity $\mathbf{u} = \mathbf{v}_a - \mathbf{v}$ of the particles being absorbed, you obtain the following expression for the resultant of the external forces applied to S:

$$\Sigma\mathbf{F} = m\frac{d\mathbf{v}}{dt} - \frac{dm}{dt}\mathbf{u} \qquad (14.42)$$

Furthermore, it was shown that the action on S of the particles being absorbed is equivalent to a thrust

$$\mathbf{P} = \frac{dm}{dt}\mathbf{u} \qquad (14.44)$$

exerted in the direction of the relative velocity of the particles being absorbed.

Examples of systems gaining mass are conveyor belts and moving railroad cars being loaded with gravel or sand, and chains being pulled out of a pile.

3. To solve problems involving systems losing mass, such as rockets and rocket engines, you can use Eqs. (14.40) through (14.44), provided that you give negative values to the increment of mass Δm and to the rate of change of mass dm/dt. It follows that the thrust defined by Eq. (14.44) will be exerted in a direction opposite to the direction of the relative velocity of the particles being ejected.

PROBLEMS

14.57 A stream of water of cross-section area A_1 and velocity $\mathbf{v}_1$ strikes a circular plate which is held motionless by a force **P**. A hole in the circular plate of area A_2 results in a discharge jet having a velocity $\mathbf{v}_1$. Determine the magnitude of **P**.

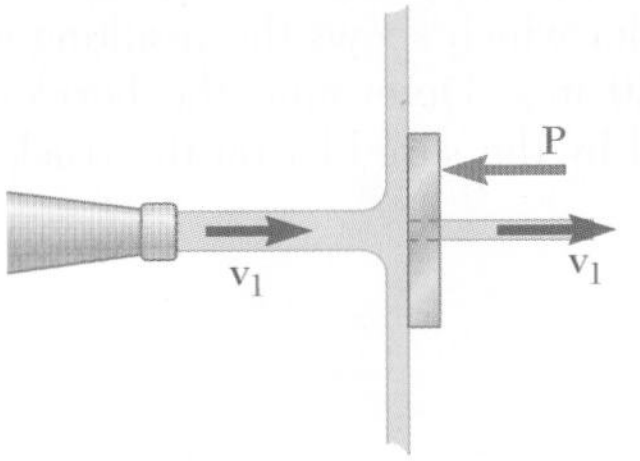

Fig. P14.57

14.58 A jet ski is placed in a channel and is tethered so that it is stationary. Water enters the jet ski with velocity $\mathbf{v}_1$ and exits with velocity $\mathbf{v}_2$. Knowing the inlet area is A_1 and the exit area is A_2, determine the tension in the tether.

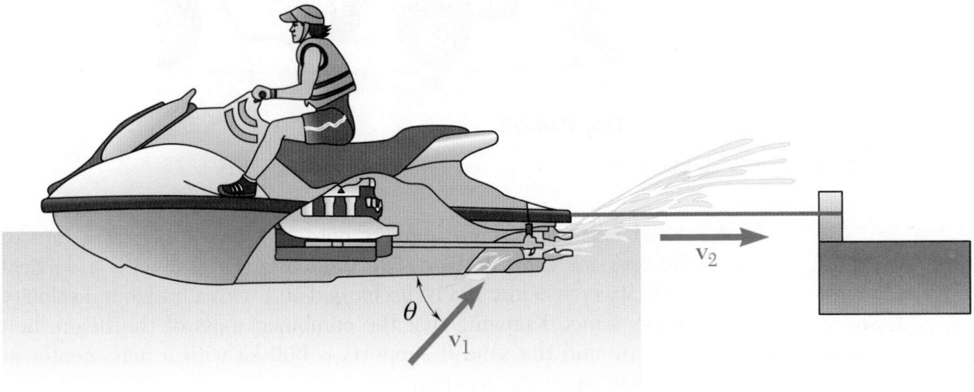

Fig. P14.58

14.59 A stream of water of cross-section area A and velocity $\mathbf{v}_1$ strikes a plate which is held motionless by a force **P**. Determine the magnitude of **P**, knowing that $A = 500 \text{ mm}^2$, $v_1 = 25$ m/s and $V = 0$.

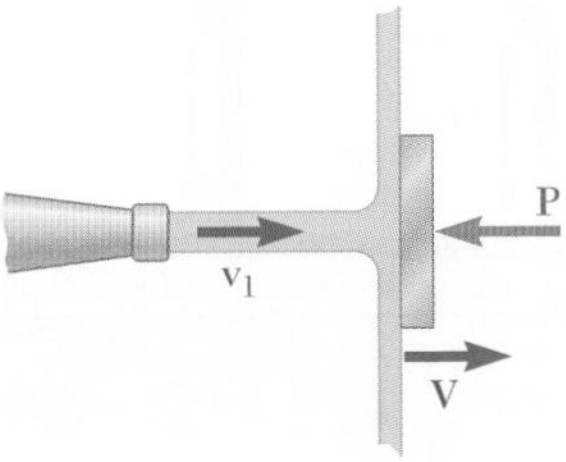

Fig. P14.59 and P14.60

14.60 A stream of water of cross-section area A and velocity $\mathbf{v}_1$ strikes a plate which moves to the right with a velocity **V**. Determine the magnitude of **V**, knowing that $A = 600 \text{ mm}^2$, $v_1 = 25$ m/s and $P = 400$ N.

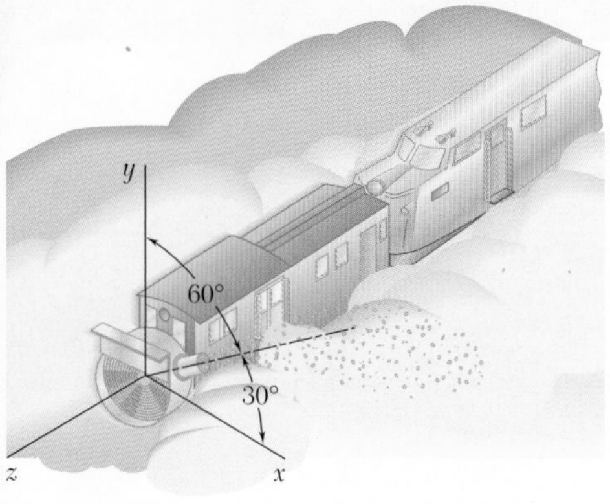

Fig. P14.61

14.61 A rotary power plow is used to remove snow from a level section of railroad track. The plow car is placed ahead of an engine which propels it at a constant speed of 20 km/h. The plow car clears 160 Mg of snow per minute, projecting it in the direction shown with a velocity of 12 m/s relative to the plow car. Neglecting friction, determine (*a*) the force exerted by the engine on the plow car, (*b*) the lateral force exerted by the track on the plow.

14.62 Tree limbs and branches are being fed at *A* at the rate of 5 kg/s into a shredder which spews the resulting wood chips at *C* with a velocity of 20 m/s. Determine the horizontal component of the force exerted by the shredder on the truck hitch at *D*.

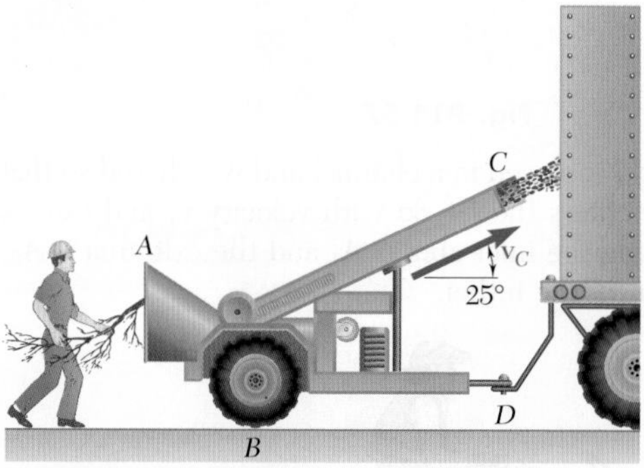

Fig. P14.62

14.63 Sand falls from three hoppers onto a conveyor belt at a rate of 40 kg/s for each hopper. The sand hits the belt with a vertical velocity $v_1 = 3$ m/s and is discharged at *A* with a horizontal velocity $v_2 = 4$ m/s. Knowing that the combined mass of the beam, belt system, and the sand it supports is 600 kg with a mass center at *G*, determine the reaction at *E*.

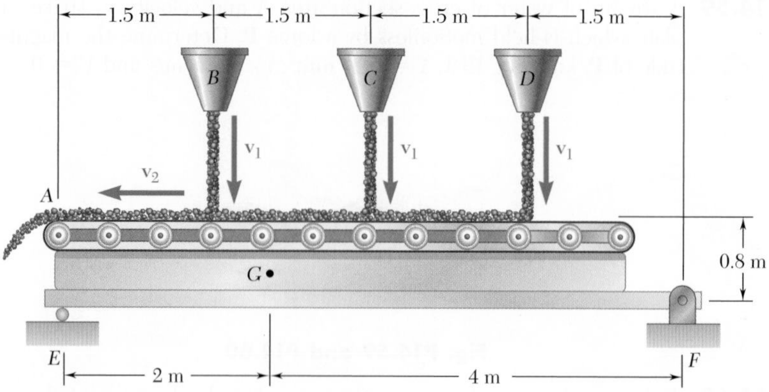

Fig. P14.63

14.64 The stream of water shown flows at a rate of 550 L/min and moves with a velocity of magnitude 18 m/s at both *A* and *B*. The vane is supported by a pin and bracket at *C* and by a load cell at *D* which can exert only a horizontal force. Neglecting the weight of the vane, determine the components of the reactions at *C* and *D*.

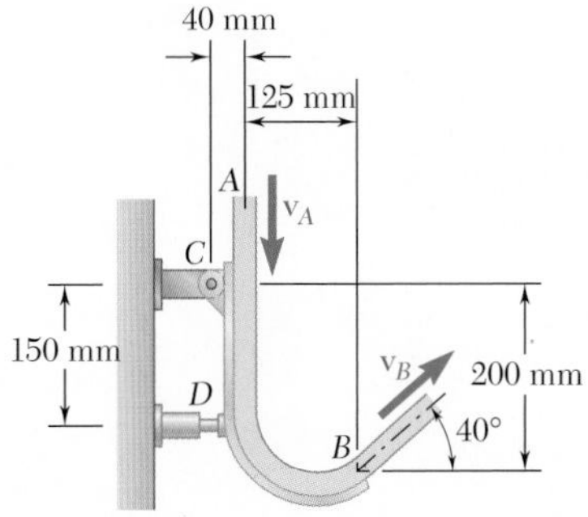

Fig. P14.64

14.65 The nozzle discharges water at the rate of 1.3 m^3/min. Knowing the velocity of the water at both *A* and *B* has a magnitude of 20 m/s and neglecting the weight of the vane, determine the components of the reactions at *C* and *D*.

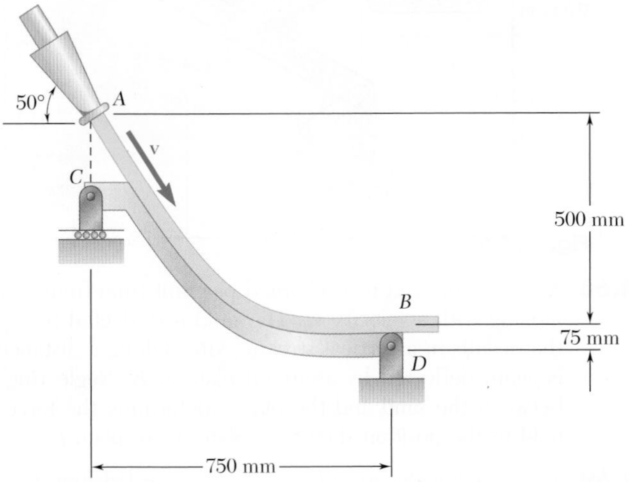

Fig. P14.65

14.66 A high-speed jet of air issues from nozzle *A* with a velocity of $\mathbf{v}_A$ and mass flow rate of 0.36 kg/s. The air impinges on a vane causing it to rotate to the position shown. The vane has a mass of 6 kg. Knowing that the magnitude of the air velocity is equal at *A* and *B*, determine (*a*) the magnitude of the velocity at *A*, (*b*) the components of the reactions at *O*.

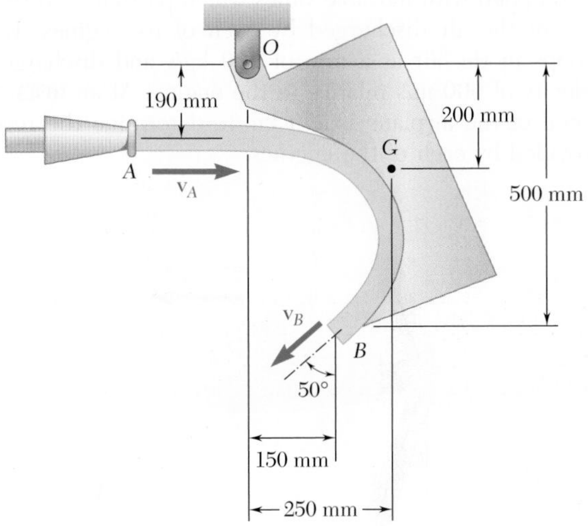

Fig. P14.66

14.67 Coal is being discharged from a first conveyor belt at the rate of 120 kg/s. It is received at A by a second belt which discharges it again at B. Knowing that $v_1 = 3$ m/s and $v_2 = 4.25$ m/s and that the second belt assembly and the coal it supports have a total mass of 472 kg, determine the components of the reactions at C and D.

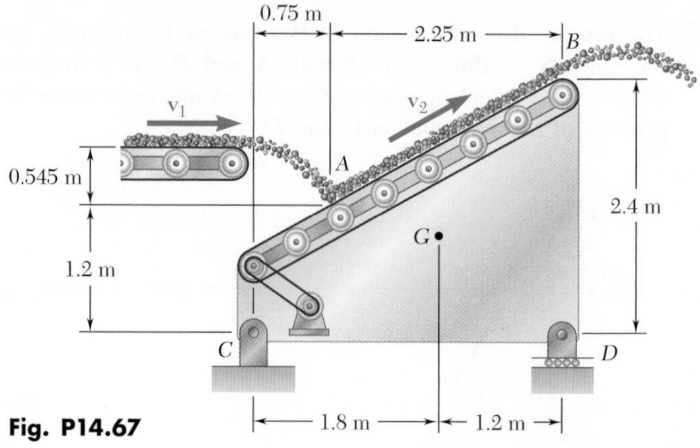

Fig. P14.67

14.68 A mass q of sand is discharged per unit time from a conveyor belt moving with a velocity $\mathbf{v}_0$. The sand is deflected by a plate at A so that it falls in a vertical stream. After falling a distance h the sand is again deflected by a curved plate at B. Neglecting the friction between the sand and the plates, determine the force required to hold in the position shown (*a*) plate A, (*b*) plate B.

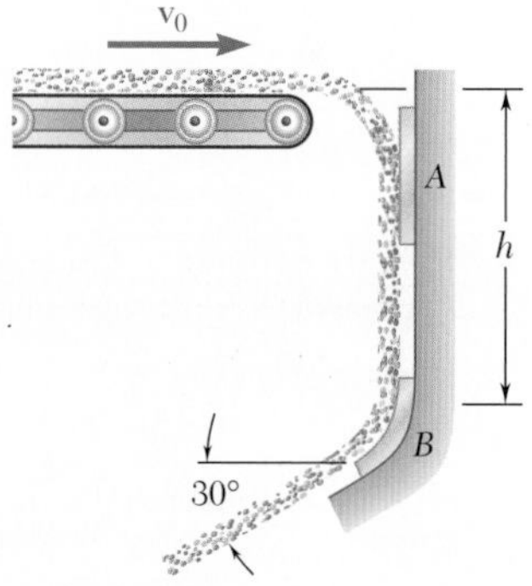

Fig. P14.68

14.69 The total drag due to air friction on a jet airplane traveling at 900 km/h is 35 kN. Knowing that the exhaust velocity is 600 m/s relative to the airplane, determine the mass of air which must pass through the engine per second to maintain the speed of 900 km/h in level flight.

14.70 While cruising in level flight at a speed of 900 km/h, a jet plane scoops in air at the rate of 90 kg/s and discharges it with a velocity of 660 m/s relative to the airplane. Determine the total drag due to air friction on the airplane.

14.71 In order to shorten the distance required for landing, a jet airplane is equipped with movable vanes which partially reverse the direction of the air discharged by each of its engines. Each engine scoops in the air at a rate of 120 kg/s and discharges it with a velocity of 600 m/s relative to the engine. At an instant when the speed of the airplane is 270 km/h, determine the reverse thrust provided by each of the engines.

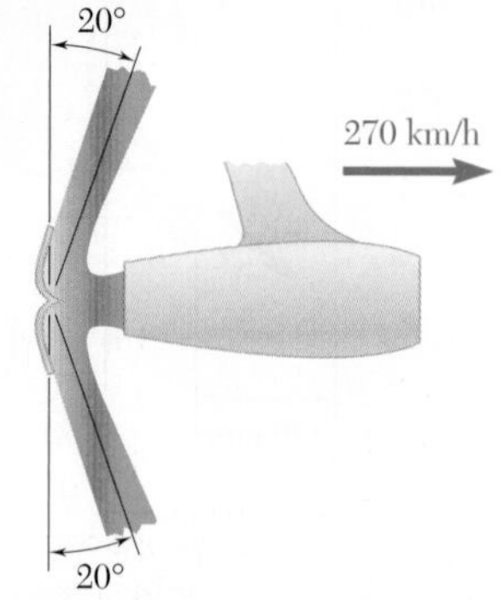

Fig. P14.71

14.72 The helicopter shown can produce a maximum downward air speed of 25 m/s in a 10 m-diameter slipstream. Knowing that the weight of the helicopter and its crew is 18 kN and assuming $\rho = 1.21\ \text{kg/m}^3$ for air, determine the maximum load that the helicopter can lift while hovering in midair.

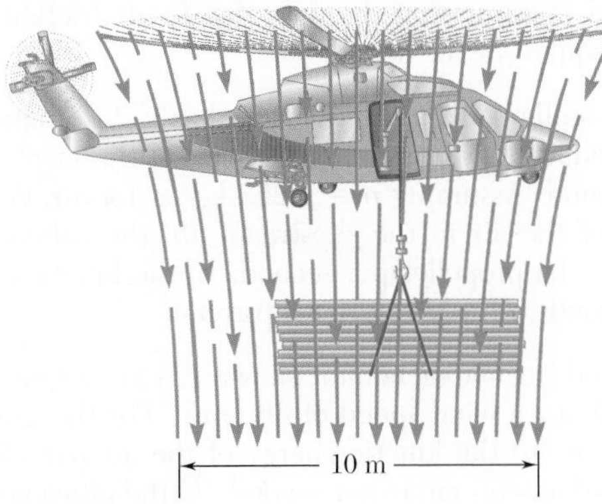

Fig. P14.72

14.73 A floor fan designed to deliver air at a maximum velocity of 6 m/s in a 400-mm-diameter slipstream is supported by a 200-mm-diameter circular base plate. Knowing that the total weight of the assembly is 60 N and that its center of gravity is located directly above the center of the base plate, determine the maximum height h at which the fan may be operated if it is not to tip over. Assume $\rho = 1.21\ \text{kg/m}^3$ for air and neglect the approach velocity of the air.

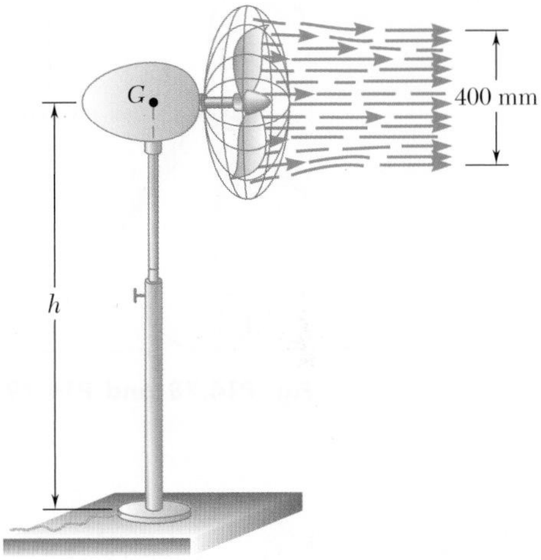

Fig. P14.73

14.74 The jet engine shown scoops in air at A at a rate of 100 kg/s and discharges it at B with a velocity of 600 m/s relative to the airplane. Determine the magnitude and line of action of the propulsive thrust developed by the engine when the speed of the airplane is (*a*) 500 km/h, (*b*) 1000 km/h.

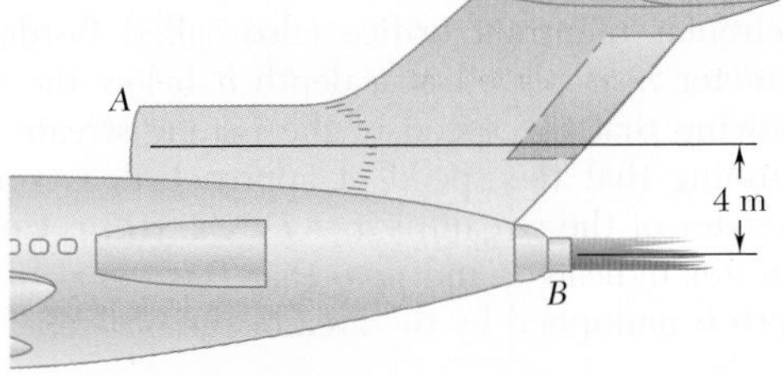

Fig. P14.74

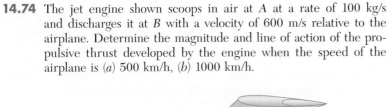

14.75 A jet airliner is cruising at a speed of 900 km/h with each of its three engines discharging air with a velocity of 800 m/s relative to the plane. Determine the speed of the airliner after it has lost the use of (*a*) one of its engines, (*b*) two of its engines. Assume that the drag due to air friction is proportional to the square of the speed and that the remaining engines keep operating at the same rate.

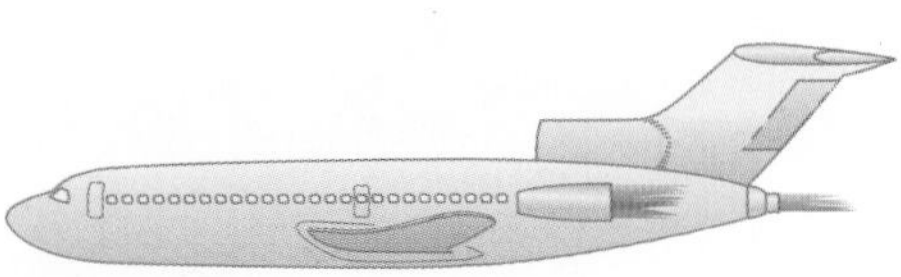

Fig. P14.75

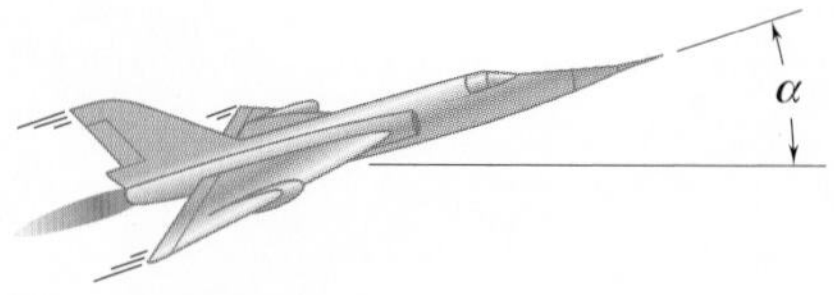

Fig. P14.76

14.76 A 16-Mg jet airplane maintains a constant speed of 774 km/h while climbing at an angle $\alpha = 18°$. The airplane scoops in air at a rate of 300 kg/s and discharges it with a velocity of 665 m/s relative to the airplane. If the pilot changes to a horizontal flight while maintaining the same engine setting, determine (*a*) the initial acceleration of the plane, (*b*) the maximum horizontal speed that will be attained. Assume that the drag due to air friction is proportional to the square of the speed.

14.77 The propeller of a small airplane has a 2-m-diameter slipstream and produces a thrust of 3600 N when the airplane is at rest on the ground. Assuming $\rho = 1.225$ kg/m^3 for air, determine (*a*) the speed of the air in the slipstream, (*b*) the volume of air passing through the propeller per second, (*c*) the kinetic energy imparted per second to the air in the slipstream.

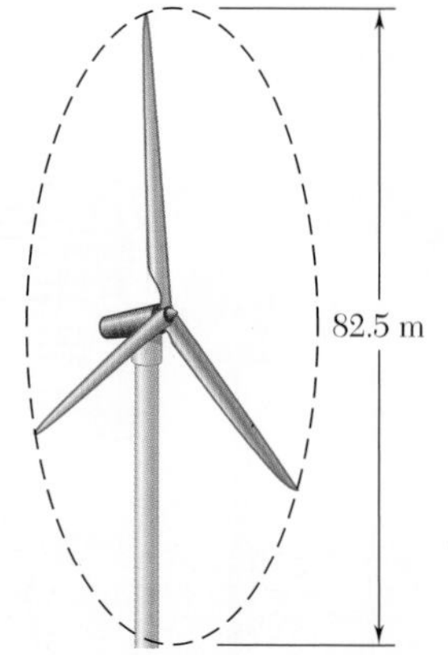

Fig. P14.78 and P14.79

14.78 The wind turbine–generator shown has an output-power rating of 1.5 MW for a wind speed of 36 km/h. For the given wind speed, determine (*a*) the kinetic energy of the air particles entering the 82.5-m-diameter circle per second, (*b*) the efficiency of this energy conversion system. Assume $\rho = 1.21$ kg/m^3 for air.

14.79 A wind turbine-generator system having a diameter of 82.5 m produces 1.5 MW at a wind speed of 12 m/s. Determine the diameter of blade necessary to produce 10 MW of power assuming the efficiency is the same for both designs and $\rho = 1.21$ kg/m^3 for air.

14.80 While cruising in level flight at a speed of 900 km/h, a jet airplane scoops in air at a rate of 120 kg/s and discharges it with a velocity of 650 m/s relative to the airplane. Determine (*a*) the power actually used to propel the airplane, (*b*) the total power developed by the engine, (*c*) the mechanical efficiency of the airplane.

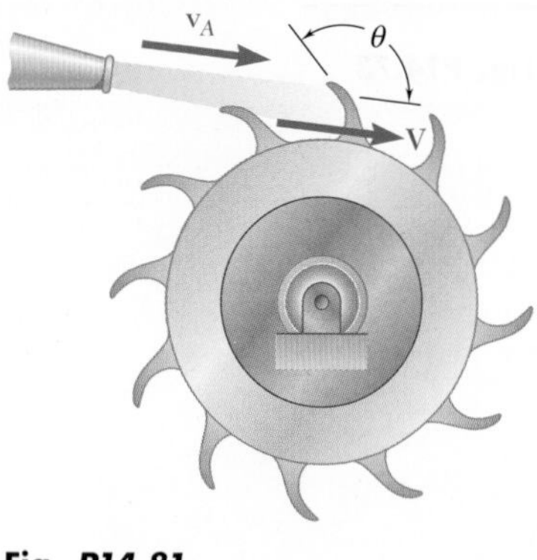

Fig. *P14.81*

14.81 In a Pelton-wheel turbine, a stream of water is deflected by a series of blades so that the rate at which water is deflected by the blades is equal to the rate at which water issues from the nozzle ($\Delta m/\Delta t = A\rho v_A$). Using the same notation as in Sample Prob. 14.7, (*a*) determine the velocity **V** of the blades for which maximum power is developed, (*b*) derive an expression for the maximum power, (*c*) derive an expression for the mechanical efficiency.

14.82 A circular reentrant orifice (also called Borda's mouthpiece) of diameter D is placed at a depth h below the surface of a tank. Knowing that the speed of the issuing stream is $v = \sqrt{2gh}$ and assuming that the speed of approach v_1 is zero, show that the diameter of the stream is $d = D/\sqrt{2}$. (*Hint:* Consider the section of water indicated, and note that P is equal to the pressure at a depth h multiplied by the area of the orifice.)

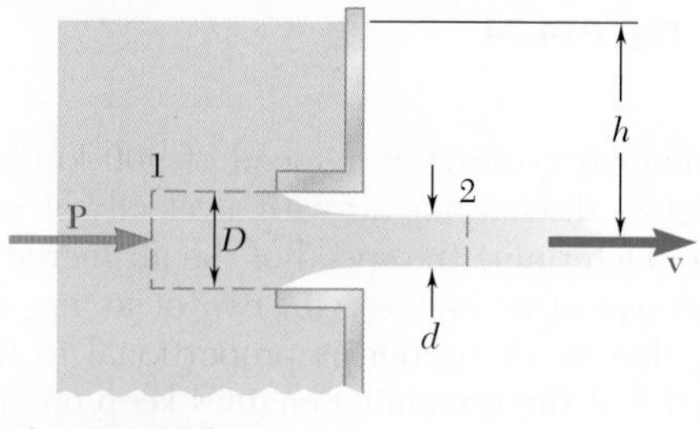

Fig. *P14.82*

14.83 Gravel falls with practically zero velocity onto a conveyor belt at the constant rate $q = dm/dt$. (*a*) Determine the magnitude of the force **P** required to maintain a constant belt speed v. (*b*) Show that the kinetic energy acquired by the gravel in a given time interval is equal to half the work done in that interval by the force **P**. Explain what happens to the other half of the work done by **P**.

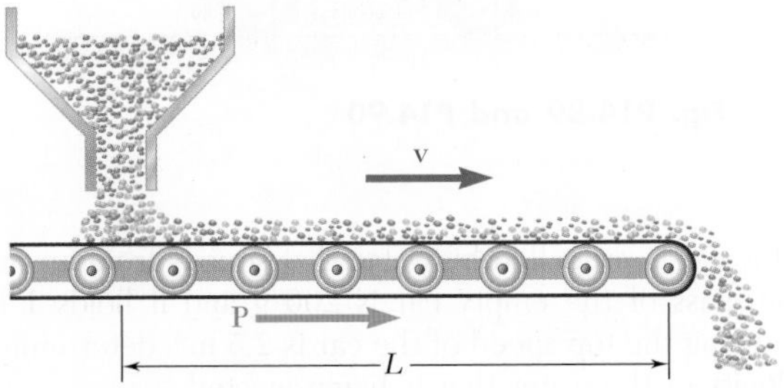

Fig. P14.83

***14.84** The depth of water flowing in a rectangular channel of width b at a speed v_1 and a depth d_1 increases to a depth d_2 at a *hydraulic jump*. Express the rate of flow Q in terms of b, d_1, and d_2.

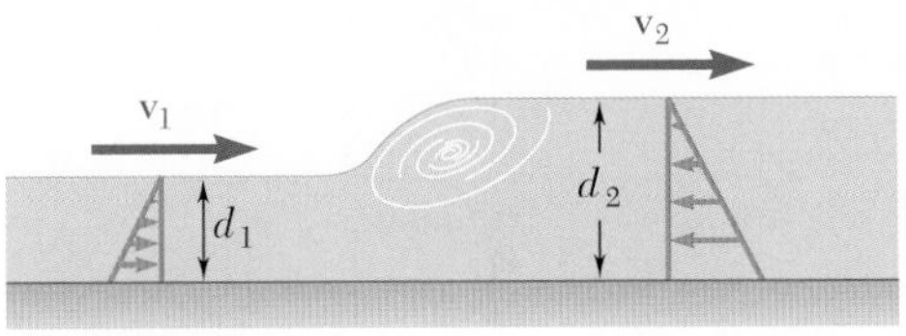

Fig. P14.84

***14.85** Determine the rate of flow in the channel of Prob. 14.84, knowing that $b = 3.6$ m, $d_1 = 1.2$ m, and $d_2 = 1.5$ m.

14.86 A chain of length l and mass m lies in a pile on the floor. If its end A is raised vertically at a constant speed v, express in terms of the length y of chain which is off the floor at any given instant (*a*) the magnitude of the force **P** applied to A, (*b*) the reaction of the floor.

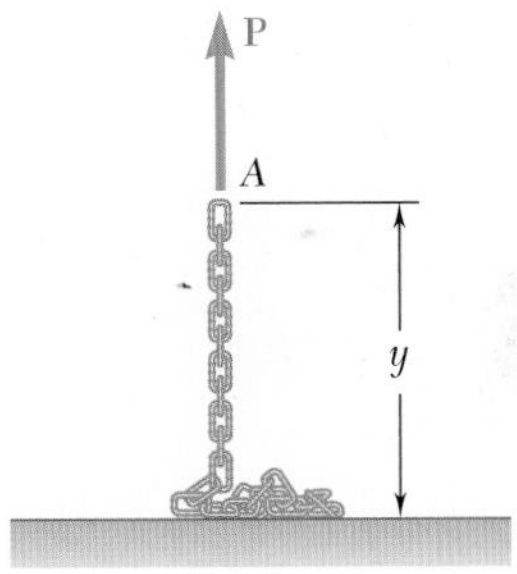

Fig. P14.86

14.87 Solve Prob. 14.86, assuming that the chain is being *lowered* to the floor at a constant speed v.

14.88 The ends of a chain lie in piles at A and C. When released from rest at time $t = 0$, the chain moves over the pulley at B, which has a negligible mass. Denoting by L the length of chain connecting the two piles and neglecting friction, determine the speed v of the chain at time t.

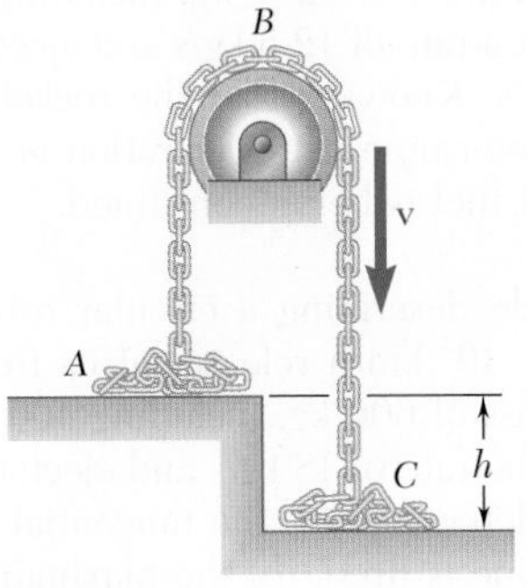

Fig. P14.88

14.89 A toy car is propelled by water that squirts from an internal tank at a constant 2 m/s relative to the car. The mass of the empty car is 200 g and it holds 1 kg of water. Neglecting other tangential forces determine the top speed of the car.

Fig. P14.89 and P14.90

14.90 A toy car is propelled by water that squirts from an internal tank. The mass of the empty car is 200 g and it holds 1 kg of water. Knowing the top speed of the car is 2.5 m/s determine the relative velocity of the water that is being ejected.

14.91 The main propulsion system of a space shuttle consists of three identical rocket engines which provide a total thrust of 6 MN. Determine the rate at which the hydrogen-oxygen propellant is burned by each of the three engines, knowing that it is ejected with a relative velocity of 3750 m/s.

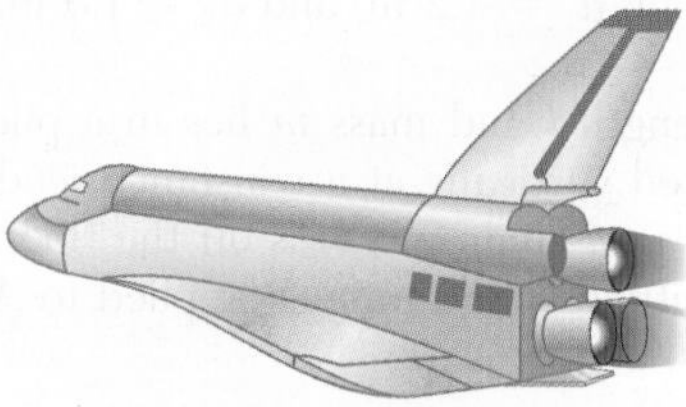

Fig. P14.91 and *P14.92*

14.92 The main propulsion system of a space shuttle consists of three identical rocket engines, each of which burns the hydrogen-oxygen propellant at the rate of 340 kg/s and ejects it with a relative velocity of 3750 m/s. Determine the total thrust provided by the three engines.

14.93 A rocket has a mass of 1200 kg, including 1000 kg of fuel, which is consumed at a rate of 12.5 kg/s and ejected with a relative velocity of 4000 m/s. Knowing that the rocket is fired vertically from the ground, determine its acceleration (*a*) as it is fired, (*b*) as the last particle of fuel is being consumed.

14.94 A space vehicle describing a circular orbit about the earth at a speed of 24×10^3 km/h releases at its front end a capsule which has a gross mass of 600 kg, including 400 kg of fuel. If the fuel is consumed at the rate of 18 kg/s and ejected with a relative velocity of 3000 m/s, determine (*a*) the tangential acceleration of the capsule as its engine is fired, (*b*) the maximum speed attained by the capsule.

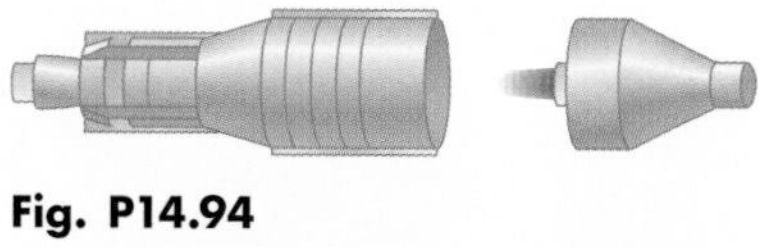

Fig. P14.94

14.95 A 540-kg spacecraft is mounted on top of a rocket with a mass of 19 Mg, including 17.8 Mg of fuel. Knowing that the fuel is consumed at a rate of 225 kg/s and ejected with a relative velocity of 3600 m/s, determine the maximum speed imparted to the spacecraft if the rocket is fired vertically from the ground.

Fig. P14.95

Fig. P14.96

14.96 The rocket used to launch the 540-kg spacecraft of Prob. 14.95 is redesigned to include two stages A and B, each of mass 9.5 Mg, including 8.9 Mg of fuel. The fuel is again consumed at a rate of 225 kg/s and ejected with a relative velocity of 3600 m/s. Knowing that when stage A expels its last particle of fuel, its casing is released and jettisoned, determine (*a*) the speed of the rocket at that instant, (*b*) the maximum speed imparted to the spacecraft.

14.97 A communication satellite weighing 50 kN, including fuel, has been ejected from a space shuttle describing a low circular orbit around the earth. After the satellite has slowly drifted to a safe distance from the shuttle, its engine is fired to increase its velocity by 2500 m/s as a first step to its transfer to a geosynchronous orbit. Knowing that the fuel is ejected with a relative velocity of 4000 m/s, determine the weight of fuel consumed in this maneuver.

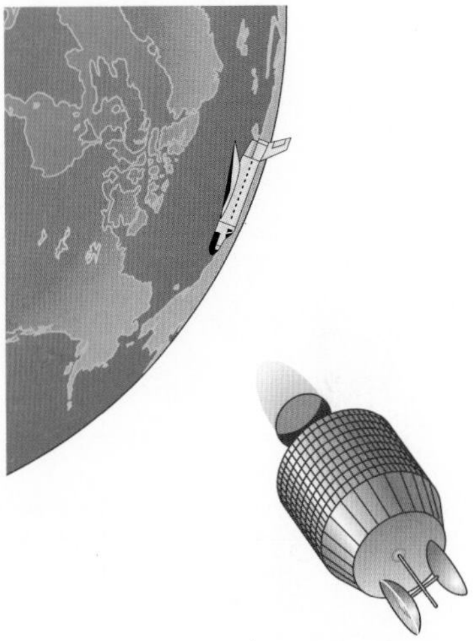

Fig. *P14.97*

14.98 Determine the increase in velocity of the communication satellite of Prob. 14.97 after 12.5 kN of fuel has been consumed.

14.99 Determine the distance separating the communication satellite of Prob. 14.97 from the space shuttle 60 s after its engine has been fired, knowing that the fuel is consumed at a rate of 20 kg/s.

14.100 For the rocket of Prob. 14.93, determine (*a*) the altitude at which all of the fuel has been consumed, (*b*) the velocity of the rocket at this time.

14.101 Determine the altitude reached by the spacecraft of Prob. 14.95 when all the fuel of its launching rocket has been consumed.

14.102 For the spacecraft and the two-stage launching rocket of Prob. 14.96, determine the altitude at which (*a*) stage A of the rocket is released, (*b*) the fuel of both stages has been consumed.

14.103 In a jet airplane, the kinetic energy imparted to the exhaust gases is wasted as far as propelling the airplane is concerned. The useful power is equal to the product of the force available to propel the airplane and the speed of the airplane. If v is the speed of the airplane and u is the relative speed of the expelled gases, show that the mechanical efficiency of the airplane is $\eta = 2v/(u + v)$. Explain why $\eta = 1$ when $u = v$.

14.104 In a rocket, the kinetic energy imparted to the consumed and ejected fuel is wasted as far as propelling the rocket is concerned. The useful power is equal to the product of the force available to propel the rocket and the speed of the rocket. If v is the speed of the rocket and u is the relative speed of the expelled fuel, show that the mechanical efficiency of the rocket is $\eta = 2uv/(u^2 + v^2)$. Explain why $\eta = 1$ when $u = v$.

REVIEW AND SUMMARY

In this chapter we analyzed the motion of *systems of particles*, i.e., the motion of a large number of particles considered together. In the first part of the chapter we considered systems consisting of well-defined particles, while in the second part we analyzed systems which are continually gaining or losing particles, or doing both at the same time.

Effective forces

We first defined the *effective force* of a particle P_i of a given system as the product $m_i\mathbf{a}_i$ of its mass m_i and its acceleration $\mathbf{a}_i$ with respect to a newtonian frame of reference centered at O [Sec. 14.2]. We then showed that *the system of the external forces acting on the particles and the system of the effective forces of the particles are equipollent;* i.e., both systems have the *same resultant* and the *same moment resultant* about O:

$$\sum_{i=1}^{n} \mathbf{F}_i = \sum_{i=1}^{n} m_i\mathbf{a}_i \tag{14.4}$$

$$\sum_{i=1}^{n} (\mathbf{r}_i \times \mathbf{F}_i) = \sum_{i=1}^{n} (\mathbf{r}_i \times m_i\mathbf{a}_i) \tag{14.5}$$

Linear and angular momentum of a system of particles

Defining the *linear momentum* $\mathbf{L}$ and the *angular momentum* $\mathbf{H}_O$ *about point* O of the system of particles [Sec. 14.3] as

$$\mathbf{L} = \sum_{i=1}^{n} m_i\mathbf{v}_i \qquad \mathbf{H}_O = \sum_{i=1}^{n} (\mathbf{r}_i \times m_i\mathbf{v}_i) \tag{14.6, 14.7}$$

we showed that Eqs. (14.4) and (14.5) can be replaced by the equations

$$\Sigma\mathbf{F} = \dot{\mathbf{L}} \qquad \Sigma\mathbf{M}_O = \dot{\mathbf{H}}_O \tag{14.10, 14.11}$$

which express that *the resultant and the moment resultant about O of the external forces are, respectively, equal to the rates of change of the linear momentum and of the angular momentum about O of the system of particles.*

Motion of the mass center of a system of particles

In Sec. 14.4, we defined the mass center of a system of particles as the point G whose position vector $\bar{\mathbf{r}}$ satisfies the equation

$$m\bar{\mathbf{r}} = \sum_{i=1}^{n} m_i\mathbf{r}_i \tag{14.12}$$

where m represents the total mass $\sum_{i=1}^{n} m_i$ of the particles. Differentiating both members of Eq. (14.12) twice with respect to t, we obtained the relations

$$\mathbf{L} = m\bar{\mathbf{v}} \qquad \dot{\mathbf{L}} = m\bar{\mathbf{a}} \qquad (14.14, 14.15)$$

where $\bar{\mathbf{v}}$ and $\bar{\mathbf{a}}$ represent, respectively, the velocity and the acceleration of the mass center G. Substituting for $\dot{\mathbf{L}}$ from (14.15) into (14.10), we obtained the equation

$$\Sigma\mathbf{F} = m\bar{\mathbf{a}} \qquad (14.16)$$

from which we concluded that *the mass center of a system of particles moves as if the entire mass of the system and all the external forces were concentrated at that point* [Sample Prob. 14.1].

Angular momentum of a system of particles about its mass center

In Sec. 14.5 we considered the motion of the particles of a system with respect to a centroidal frame $Gx'y'z'$ attached to the mass center G of the system and in translation with respect to the newtonian frame $Oxyz$ (Fig. 14.14). We defined the *angular momentum* of the system *about its mass center* G as the sum of the moments about G of the momenta $m_i\mathbf{v}'_i$ of the particles in their motion relative to the frame $Gx'y'z'$. We also noted that the same result can be obtained by considering the moments about G of the momenta $m_i\mathbf{v}_i$ of the particles in their absolute motion. We therefore wrote

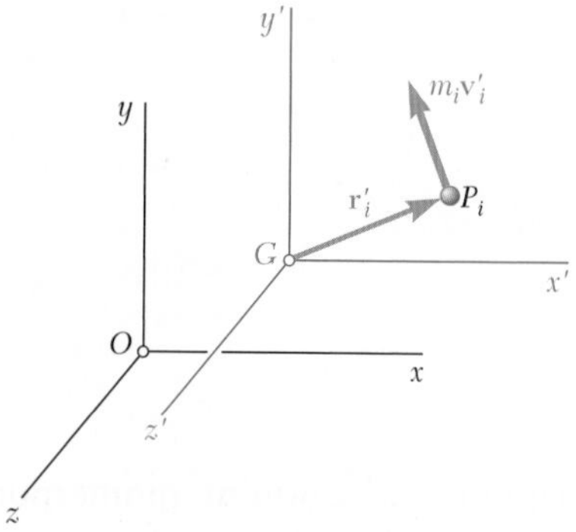

Fig. 14.14

$$\mathbf{H}_G = \sum_{i=1}^{n} (\mathbf{r}'_i \times m_i\mathbf{v}_i) = \sum_{i=1}^{n} (\mathbf{r}'_i \times m_i\mathbf{v}'_i) \qquad (14.24)$$

and derived the relation

$$\Sigma\mathbf{M}_G = \dot{\mathbf{H}}_G \qquad (14.23)$$

which expresses that *the moment resultant about G of the external forces is equal to the rate of change of the angular momentum about G of the system of particles*. As will be seen later, this relation is fundamental to the study of the motion of rigid bodies.

Conservation of momentum

When no external force acts on a system of particles [Sec. 14.6], it follows from Eqs. (14.10) and (14.11) that the linear momentum $\mathbf{L}$ and the angular momentum $\mathbf{H}_O$ of the system are conserved [Sample Probs. 14.2 and 14.3]. In problems involving central forces, the angular momentum of the system about the center of force O will also be conserved.

Kinetic energy of a system of particles

The kinetic energy T of a system of particles was defined as the sum of the kinetic energies of the particles [Sec. 14.7]:

$$T = \frac{1}{2}\sum_{i=1}^{n} m_i v_i^2 \qquad (14.28)$$

Using the centroidal frame of reference $Gx'y'z'$ of Fig. 14.14, we noted that the kinetic energy of the system can also be obtained by adding the kinetic energy $\frac{1}{2}m\bar{v}^2$ associated with the motion of the mass center G and the kinetic energy of the system in its motion relative to the frame $Gx'y'z'$:

$$T = \tfrac{1}{2}m\bar{v}^2 + \frac{1}{2}\sum_{i=1}^{n} m_i v_i'^2 \qquad (14.29)$$

Principle of work and energy

The *principle of work and energy* can be applied to a system of particles as well as to individual particles [Sec. 14.8]. We wrote

$$T_1 + U_{1\rightarrow 2} = T_2 \qquad (14.30)$$

and noted that $U_{1\rightarrow 2}$ represents the work of *all* the forces acting on the particles of the system, internal as well as external.

Conservation of energy

If all the forces acting on the particles of the system are *conservative*, we can determine the potential energy V of the system and write

$$T_1 + V_1 = T_2 + V_2 \qquad (14.31)$$

which expresses the *principle of conservation of energy* for a system of particles.

Principle of impulse and momentum

We saw in Sec. 14.9 that the *principle of impulse and momentum* for a system of particles can be expressed graphically as shown in Fig. 14.15. It states that the momenta of the particles at time t_1 and the impulses of the external forces from t_1 to t_2 form a system of vectors equipollent to the system of the momenta of the particles at time t_2.

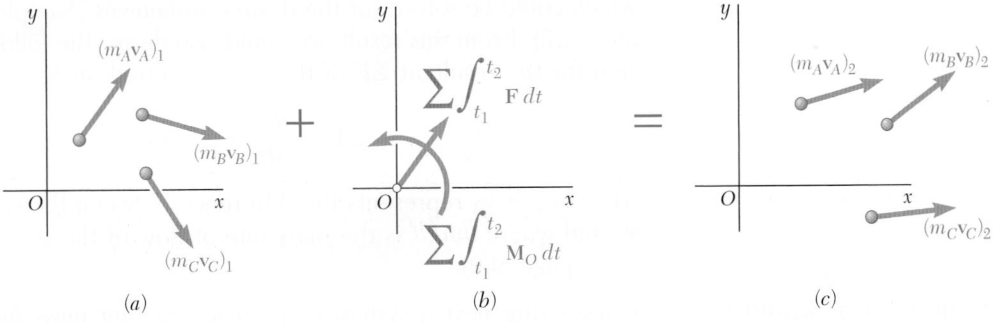

Fig. 14.15

If no external force acts on the particles of the system, the systems of momenta shown in parts *a* and *c* of Fig. 14.15 are equipollent and we have

$$\mathbf{L}_1 = \mathbf{L}_2 \qquad (\mathbf{H}_O)_1 = (\mathbf{H}_O)_2 \qquad (14.36, 14.37)$$

Use of conservation principles in the solution of problems involving systems of particles

Many problems involving the motion of systems of particles can be solved by applying simultaneously the principle of impulse and momentum and the principle of conservation of energy [Sample Prob. 14.4] or by expressing that the linear momentum, angular momentum, and energy of the system are conserved [Sample Prob. 14.5].

Variable systems of particles
Steady stream of particles

In the second part of the chapter, we considered *variable systems of particles*. First we considered a *steady stream of particles,* such as a stream of water diverted by a fixed vane or the flow of air through a jet engine [Sec. 14.11]. Applying the principle of impulse and momentum to a system S of particles during a time interval Δt, and including the particles which enter the system at A during that time interval and those (of the same mass Δm) which leave the system at B, we concluded that *the system formed by the momentum* $(\Delta m)\mathbf{v}_A$ *of the particles entering S in the time* Δt *and the impulses of the forces exerted on S during that time is equipollent to the momentum* $(\Delta m)\mathbf{v}_B$ *of the particles leaving S in the same time* Δt (Fig. 14.16). Equating

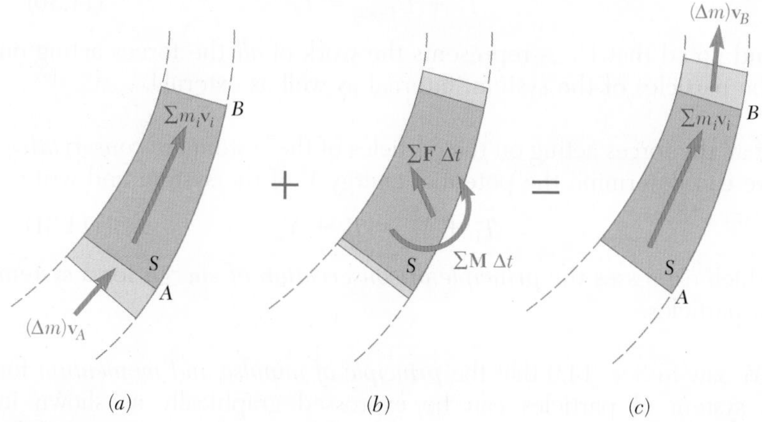

Fig. 14.16

the x components, y components, and moments about a fixed point of the vectors involved, we could obtain as many as three equations, which could be solved for the desired unknowns [Sample Probs. 14.6 and 14.7]. From this result, we could also derive the following expression for the resultant $\Sigma\mathbf{F}$ of the forces exerted on S,

$$\Sigma\mathbf{F} = \frac{dm}{dt}(\mathbf{v}_B - \mathbf{v}_A) \qquad (14.39)$$

where $\mathbf{v}_B - \mathbf{v}_A$ represents the difference between the *vectors* $\mathbf{v}_B$ and $\mathbf{v}_A$ and where dm/dt is the mass rate of flow of the stream (see footnote, page 899).

Systems gaining or losing mass

Considering next a system of particles gaining mass by continually absorbing particles or losing mass by continually expelling particles [Sec. 14.12], as in the case of a rocket, we applied the principle of impulse and momentum to the system during a time interval Δt, being careful to include the particles gained or lost during that time interval [Sample Prob. 14.8]. We also noted that the action on a system S of the particles being *absorbed* by S was equivalent to a thrust

$$\mathbf{P} = \frac{dm}{dt}\mathbf{u} \qquad (14.44)$$

where dm/dt is the rate at which mass is being absorbed, and $\mathbf{u}$ is the velocity of the particles *relative to S*. In the case of particles being *expelled* by S, the rate dm/dt is negative and the thrust $\mathbf{P}$ is exerted in a direction opposite to that in which the particles are being expelled.

REVIEW PROBLEMS

14.105 Three identical freight cars have the velocities indicated. Assuming that car B is first hit by car A, determine the velocity of each car after all the collisions have taken place if (*a*) all three cars get automatically coupled, (*b*) cars A and B get automatically coupled while cars B and C bounce off each other with a coefficient of restitution $e = 0.8$.

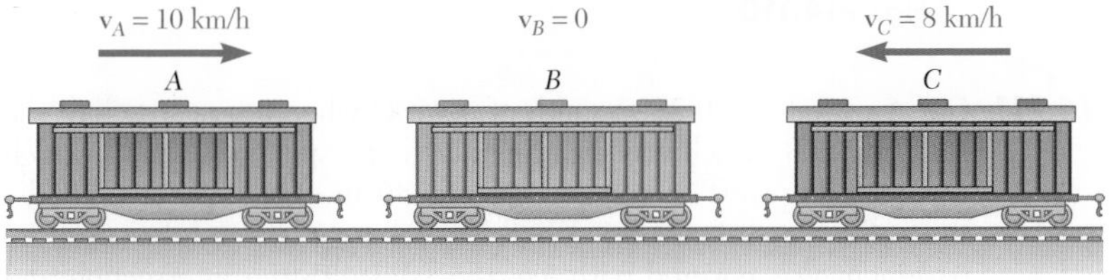

Fig. P14.105

14.106 A 30-g bullet is fired with a velocity of 480 m/s into block A, which has a mass of 5 kg. The coefficient of kinetic friction between block A and cart BC is 0.50. Knowing that the cart has a mass of 4 kg and can roll freely, determine (*a*) the final velocity of the cart and block, (*b*) the final position of the block on the cart.

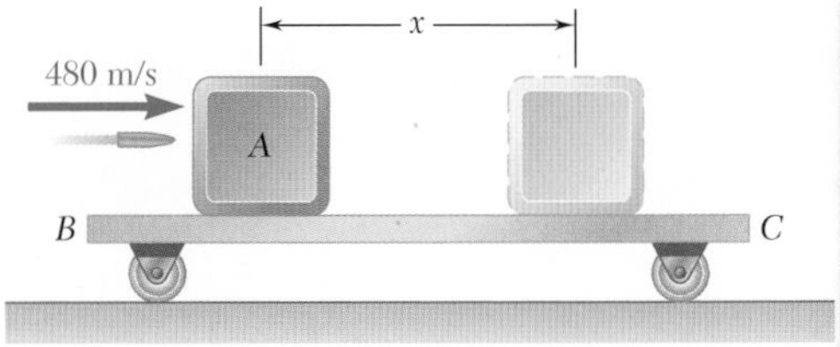

Fig. P14.106

14.107 An 80-Mg railroad engine A coasting at 6.5 km/h strikes a 20-Mg flatcar C carrying a 30-Mg load B which can slide along the floor of the car ($\mu_k = 0.25$). Knowing that the car was at rest with its brakes released and that it automatically coupled with the engine upon impact, determine the velocity of the car (*a*) immediately after impact, (*b*) after the load has slid to a stop relative to the car.

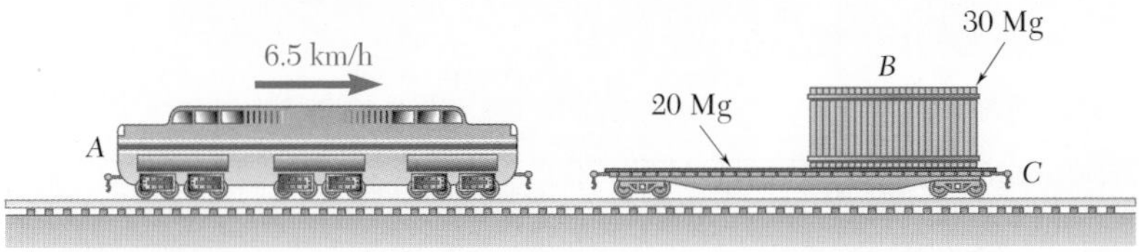

Fig. P14.107

14.108 A 4500-kg helicopter A was traveling due east in level flight at a speed of 120 km/h and at an altitude of 750 m when it was hit by a 6000-kg helicopter B. As a result of the collision, both helicopters lost their lift, and their entangled wreckage fell to the ground in 12 s at a point located 450 m east and 115.2 m south of the point of impact. Neglecting air resistance, determine the velocity components of helicopter B just before the collision.

14.109 Mass C, which has a mass of 4 kg, is suspended from a cord attached to cart A, which has a mass of 5 kg and can roll freely on a frictionless horizontal track. A 60-g bullet is fired with a speed $v_0 = 500$ m/s and gets lodged in block C. Determine (*a*) the velocity of C as it reaches its maximum elevation, (*b*) the maximum vertical distance h through which C will rise.

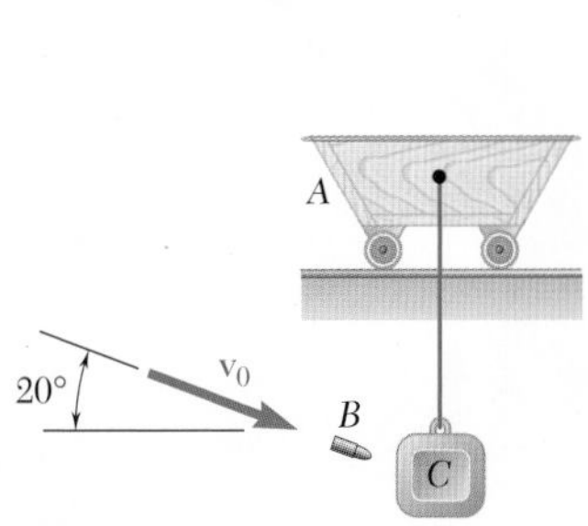

Fig. P14.109

14.110 A 7.5-kg block B is at rest and a spring of constant $k = 15000$ N/m is held compressed 75 mm by a cord. After 2.5-kg block A is placed against the end of the spring the cord is cut causing A and B to move. Neglecting friction, determine the velocities of blocks A and B immediately after A leaves B.

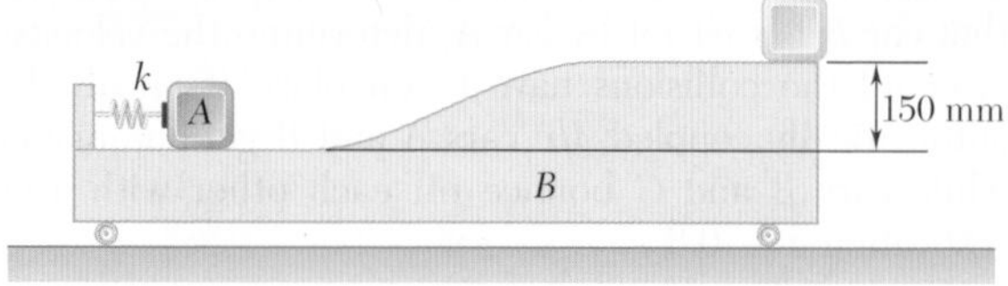

Fig. P14.110

14.111 Car A was at rest 9.28 m south of point O when it was struck in the rear by car B, which was traveling north at a speed v_B. Car C, which was traveling west at a speed v_C, was 40 m east of point O at the time of the collision. Cars A and B stuck together and, because the pavement was covered with ice, they slid into the intersection and were struck by car C which had not changed its speed. Measurements based on a photograph taken from a traffic helicopter shortly after the second collision indicated that the positions of the cars, expressed in meters, were $\mathbf{r}_A = -10.1\mathbf{i} + 16.9\mathbf{j}$, $\mathbf{r}_B = -10.1\mathbf{i} + 20.4\mathbf{j}$, and $\mathbf{r}_C = -19.8\mathbf{i} - 15.2\mathbf{j}$. Knowing that the masses of cars A, B, and C are, respectively, 1400 kg, 1800 kg, and 1600 kg, and that the time elapsed between the first collision and the time the photograph was taken was 3.4 s, determine the initial speeds of cars B and C.

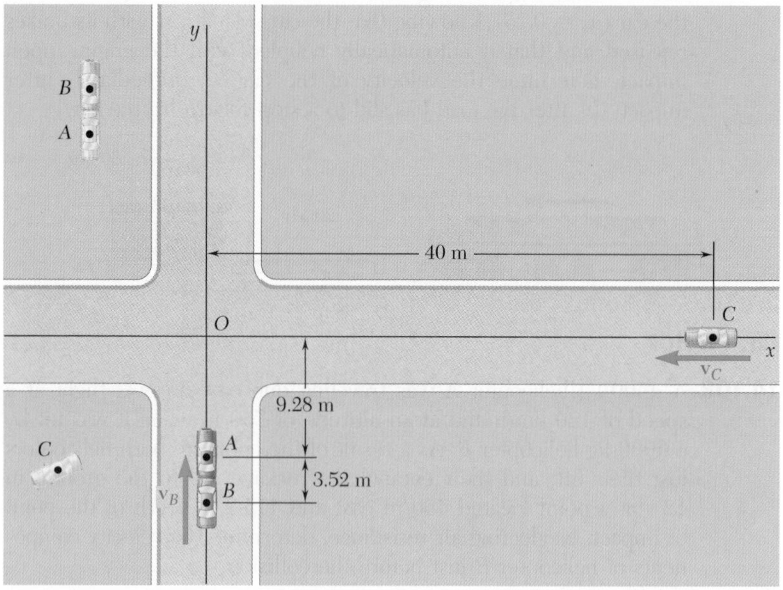

Fig. P14.111

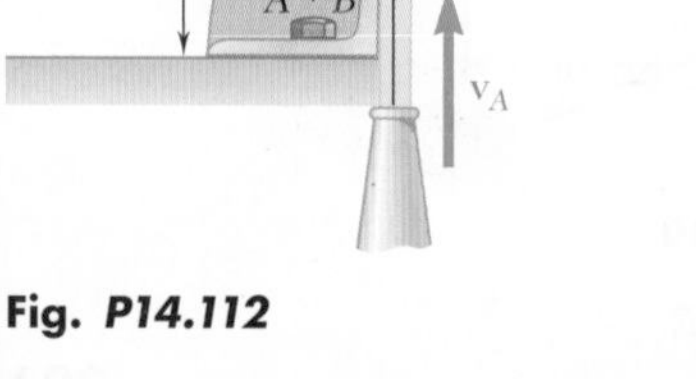

Fig. P14.112

14.112 The nozzle shown discharges water at the rate of 800 L/min. Knowing that at both B and C the stream of water moves with a velocity of magnitude 30 m/s, and neglecting the weight of the vane, determine the force-couple system which must be applied at A to hold the vane in place.

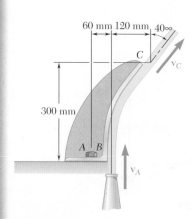

14.113 A possible method for reducing the speed of a training plane as it lands on an aircraft carrier consists in having the tail of the plane hook into the end of a heavy chain of length l which lies in a pile below deck. Denoting by m the mass of the plane and by v_0 its speed at touchdown, and assuming no other retarding force, determine (*a*) the required mass of the chain if the speed of the plane is to be reduced to βv_0, where $\beta < 1$, (*b*) the maximum value of the force exerted by the chain on the plane.

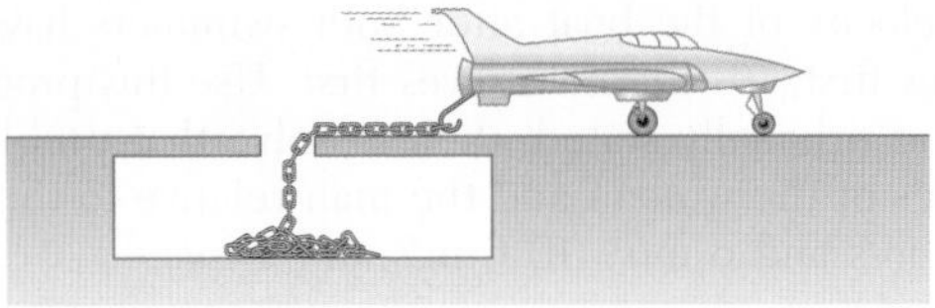

Fig. P14.113

14.114 A railroad car of length L and mass m_0 when empty is moving freely on a horizontal track while being loaded with sand from a stationary chute at a rate $dm/dt = q$. Knowing that the car was approaching the chute at a speed v_0, determine (*a*) the mass of the car and its load after the car has cleared the chute, (*b*) the speed of the car at that time.

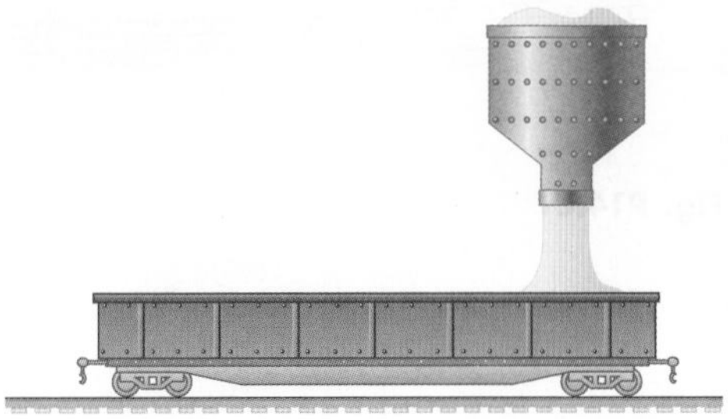

Fig. P14.114

14.115 A garden sprinkler has four rotating arms, each of which consists of two horizontal straight sections of pipe forming an angle of 120° with each other. Each arm discharges water at a rate of 20 L/min with a velocity of 18 m/s relative to the arm. Knowing that the friction between the moving and stationary parts of the sprinkler is equivalent to a couple of magnitude $M = 0.375$ N · m, determine the constant rate at which the sprinkler rotates.

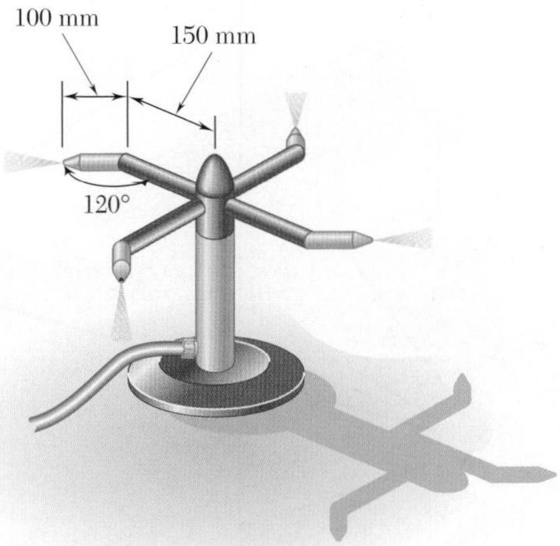

Fig. P14.115

14.116 A chain of length l and mass m falls through a small hole in a plate. Initially, when y is very small, the chain is at rest. In each case shown, determine (*a*) the acceleration of the first link A as a function of y, (*b*) the velocity of the chain as the last link passes through the hole. In case *1* assume that the individual links are at rest until they fall through the hole; in case 2 assume that at any instant all links have the same speed. Ignore the effect of friction.

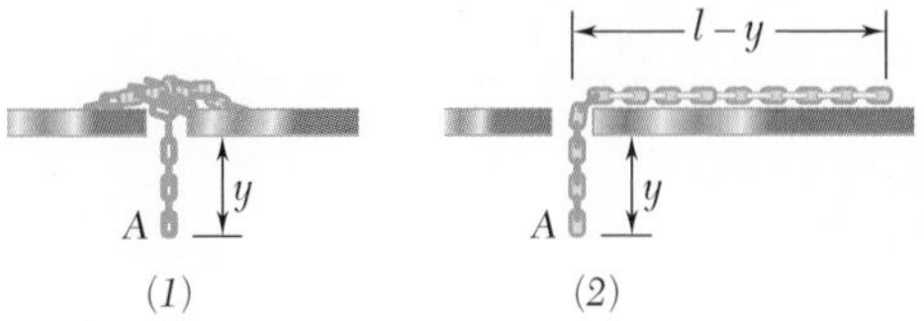

Fig. P14.116

COMPUTER PROBLEMS

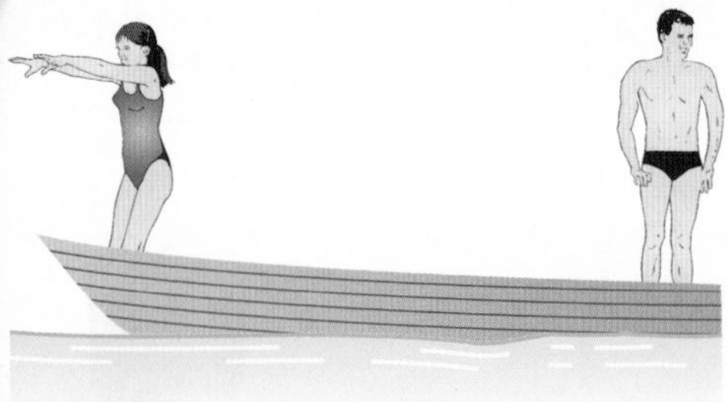

Fig. P14.C1

14.C1 A man and a woman, of weights W_m and W_w, stand at opposite ends of a stationary boat of weight W_b, ready to dive with velocities v_m and v_w, respectively, relative to the boat. Use computational software to determine the velocity of the boat after both swimmers have dived if (*a*) the woman dives first, (*b*) the man dives first. Use this program first to solve Prob. 14.5 as originally stated, then to solve that problem assuming that the velocities of the woman and the man relative to the boat are, respectively, (i) 4 m/s and 6 m/s, (ii) 6 m/s and 4 m/s.

14.C2 A system of particles consists of n particles A_i of mass m_i and coordinates x_i, y_i, and z_i, having velocities of components $(v_x)_i$, $(v_y)_i$, and $(v_z)_i$. Derive expressions for the components of the angular momentum of the system about the origin O of the coordinates. Use computational software to solve Probs. 14.11 and 14.13.

14.C3 A shell moving with a velocity of known components v_x, v_y, and v_z explodes into three fragments of weights W_1, W_2, and W_3 at point A_0 at a distance d from a vertical wall. Use computational software to determine the speed of each fragment immediately after the explosion, knowing the coordinates x_i and y_i of the points A_i $(i = 1, 2, 3)$ where the fragments hit the wall. Use this program to solve (*a*) Prob. 14.24, (*b*) Prob. 14.25.

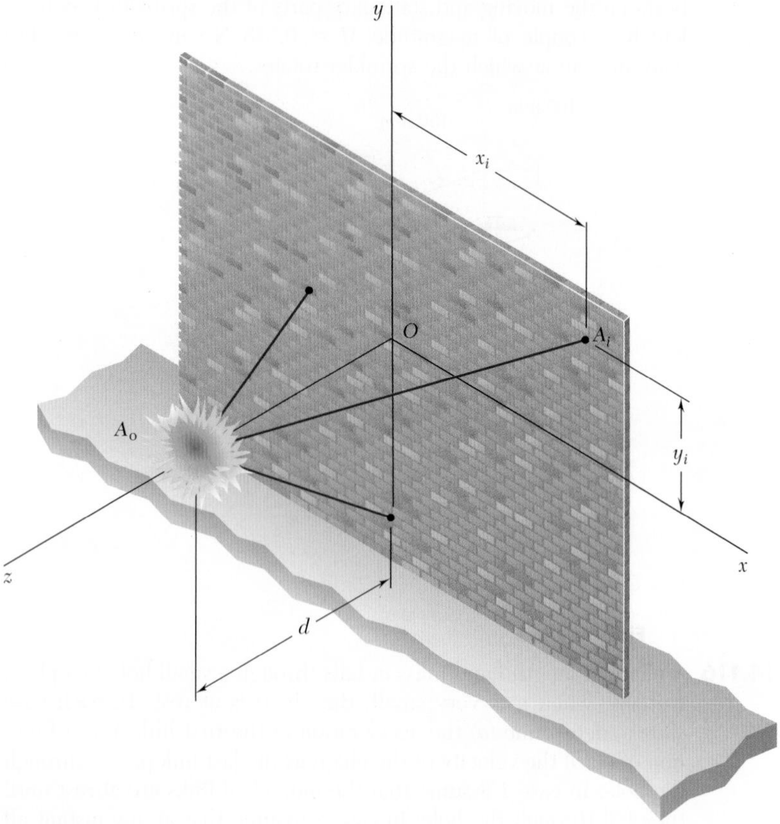

Fig. P14.C3

14.C4 As a 6000-kg training plane lands on an aircraft carrier at a speed of 180 km/h, its tail hooks into the end of an 80-m long chain which lies in a pile below deck. Knowing that the chain has a mass per unit length of 50 kg/m and assuming no other retarding force, use computational software to determine the distance traveled by the plane while the chain is being pulled out and the corresponding values of the time and of the velocity and deceleration of the plane.

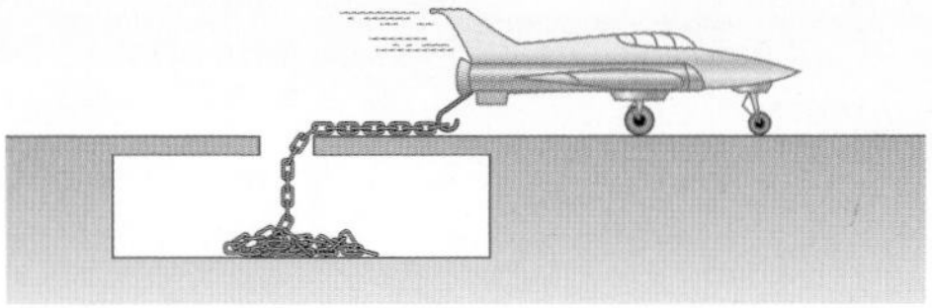

Fig. P14.C4

14.C5 A 16-Mg jet airplane maintains a constant speed of 774 km/h while climbing at an angle $\alpha = 18°$. The airplane scoops in air at a rate of 300 kg/s and discharges it with a velocity of 665 m/s relative to the airplane. Knowing that the pilot changes the angle of climb α while maintaining the same engine setting, use computational software to calculate and plot values of α from 0 to 20° (*a*) the initial acceleration of the plane, (*b*) the maximum speed that will be attained. Assume that the drag due to air friction is proportional to the square of the speed.

14.C6 A rocket has a mass of 1200 kg, including 1000 kg of fuel, which is consumed at the rate of 12.5 kg/s and ejected with a relative velocity of 3600 m/s. Knowing that the rocket is fired vertically from the ground, assuming a constant value for the acceleration of gravity, and using 4-s time intervals, use computational software to determine and plot from the time of ignition to the time when the last particle of fuel is being consumed (*a*) the acceleration a of the rocket in m/s, (*b*) its velocity v in m/s, (*c*) its elevation h above the ground in km. (*Hint:* Use for v the expression derived in Sample Prob. 14.8, and integrate this expression analytically to obtain h.)

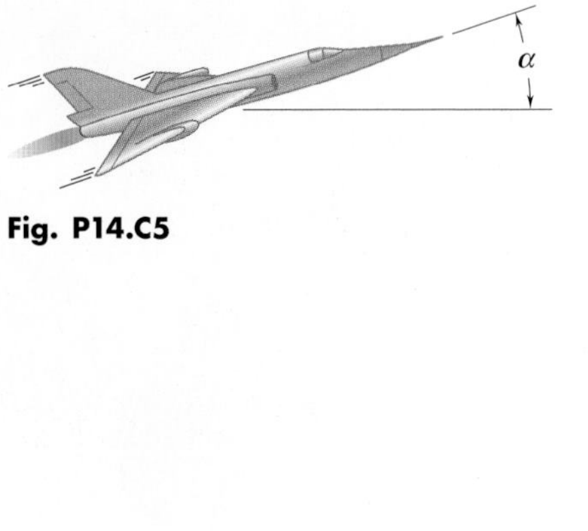

Fig. P14.C5

This huge crank belongs to a Wartsila-Sulzer RTA96-C turbocharged two-stroke diesel engine. In this chapter you will learn to perform the *kinematic* analysis of rigid bodies that undergo *translation, fixed axis rotation,* and *general plane motion.*

CHAPTER 15

Kinematics of Rigid Bodies

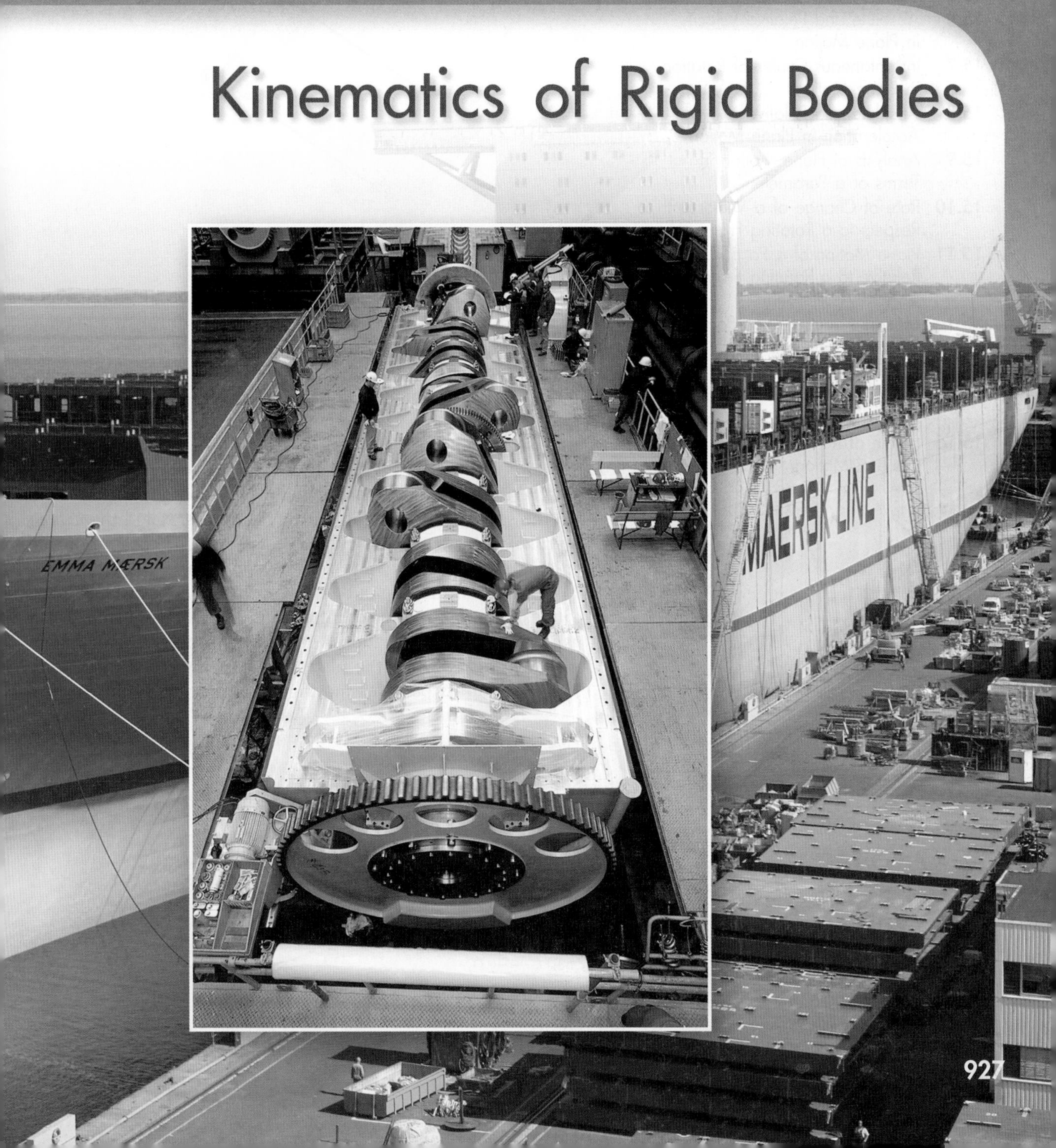

Chapter 15 Kinematics of Rigid Bodies

15.1 INTRODUCTION

In this chapter, the kinematics of *rigid bodies* will be considered. You will investigate the relations existing between the time, the positions, the velocities, and the accelerations of the various particles forming a rigid body. As you will see, the various types of rigid-body motion can be conveniently grouped as follows:

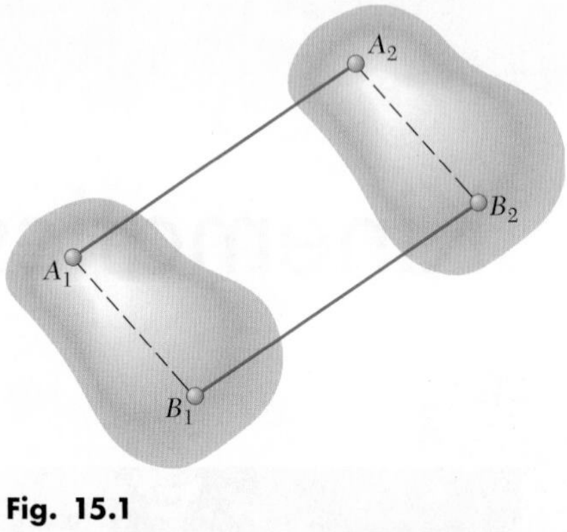

Fig. 15.1

1. *Translation*. A motion is said to be a translation if any straight line inside the body keeps the same direction during the motion. It can also be observed that in a translation all the particles forming the body move along parallel paths. If these paths are straight lines, the motion is said to be a *rectilinear translation* (Fig. 15.1); if the paths are curved lines, the motion is a *curvilinear translation* (Fig. 15.2).
2. *Rotation About a Fixed Axis*. In this motion, the particles forming the rigid body move in parallel planes along circles centered on the same fixed axis (Fig. 15.3). If this axis, called the *axis of rotation*, intersects the rigid body, the particles located on the axis have zero velocity and zero acceleration.

   Rotation should not be confused with certain types of curvilinear translation. For example, the plate shown in Fig. 15.4*a* is in curvilinear translation, with all its particles moving along *parallel* circles, while the plate shown in Fig. 15.4*b* is in rotation, with all its particles moving along *concentric* circles.

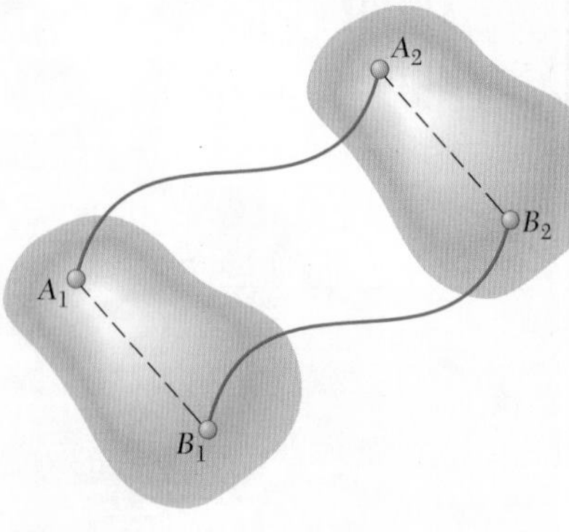

Fig. 15.2

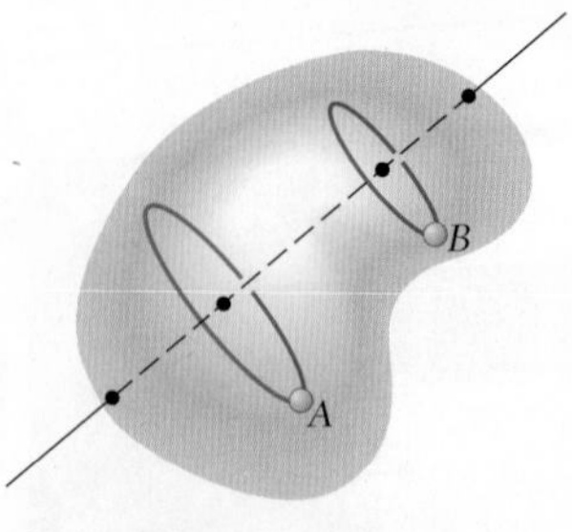

Fig. 15.3

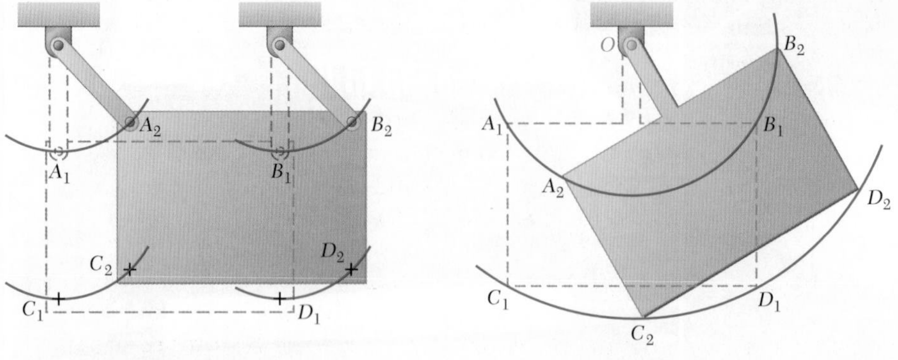

(*a*) Curvilinear translation (*b*) Rotation

Fig. 15.4

In the first case, any given straight line drawn on the plate will maintain the same direction, whereas in the second case, point O remains fixed.

Because each particle moves in a given plane, the rotation of a body about a fixed axis is said to be a *plane motion*.

3. *General Plane Motion*. There are many other types of plane motion, i.e., motions in which all the particles of the body move in parallel planes. Any plane motion that is neither a rotation nor a translation is referred to as a general plane motion. Two examples of general plane motion are given in Fig. 15.5.

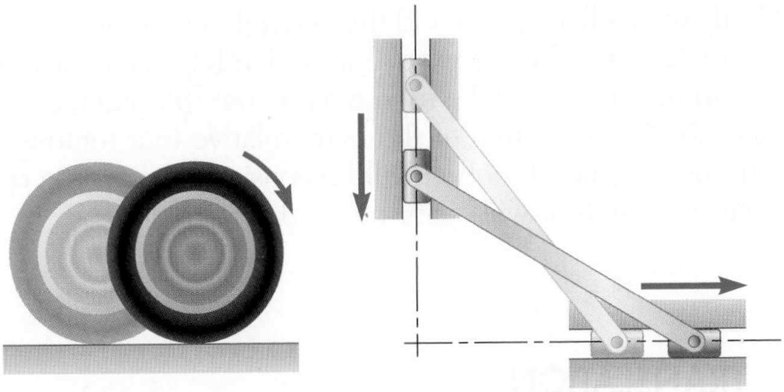

(*a*) Rolling wheel (*b*) Sliding rod

Fig. 15.5

4. *Motion About a Fixed Point*. The three-dimensional motion of a rigid body attached at a fixed point O, e.g., the motion of a top on a rough floor (Fig. 15.6), is known as motion about a fixed point.
5. *General Motion*. Any motion of a rigid body that does not fall in any of the categories above is referred to as a general motion.

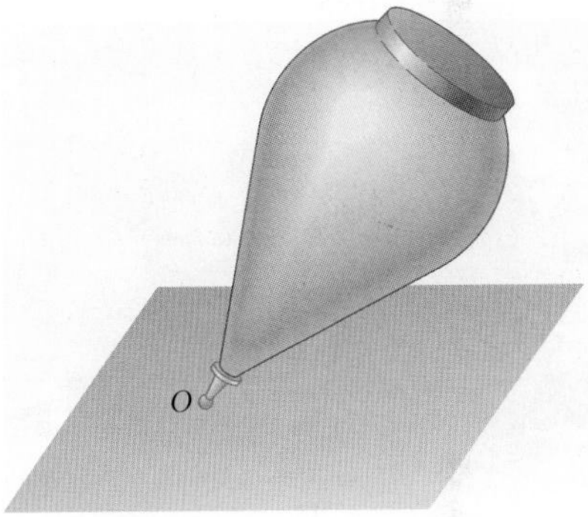

Fig. 15.6

After a brief discussion in Sec. 15.2 of the motion of translation, the rotation of a rigid body about a fixed axis is considered in Sec. 15.3. The *angular velocity* and the *angular acceleration* of a rigid body about a fixed axis will be defined, and you will learn to express the velocity and the acceleration of a given point of the body in terms of its position vector and the angular velocity and angular acceleration of the body.

The following sections are devoted to the study of the general plane motion of a rigid body and to its application to the analysis of mechanisms such as gears, connecting rods, and pin-connected linkages. Resolving the plane motion of a slab into a translation and a rotation (Secs. 15.5 and 15.6), we will then express the velocity of a point B of the slab as the sum of the velocity of a reference point A and of the velocity of B relative to a frame of reference translating with A (i.e., moving with A but not rotating). The same approach is used later in Sec. 15.8 to express the acceleration of B in terms of the acceleration of A and of the acceleration of B relative to a frame translating with A.

An alternative method for the analysis of velocities in plane motion, based on the concept of *instantaneous center of rotation,* is given in Sec. 15.7; and still another method of analysis, based on the use of parametric expressions for the coordinates of a given point, is presented in Sec. 15.9.

The motion of a particle relative to a rotating frame of reference and the concept of *Coriolis acceleration* are discussed in Secs. 15.10 and 15.11, and the results obtained are applied to the analysis of the plane motion of mechanisms containing parts which slide on each other.

The remaining part of the chapter is devoted to the analysis of the three-dimensional motion of a rigid body, namely, the motion of a rigid body with a fixed point and the general motion of a rigid body. In Secs. 15.12 and 15.13, a fixed frame of reference or a frame of reference in translation will be used to carry out this analysis; in Secs. 15.14 and 15.15, the motion of the body relative to a rotating frame or to a frame in general motion will be considered, and the concept of Coriolis acceleration will again be used.

Photo 15.1 This replica of a battering ram at *Château des Baux, France* undergoes curvilinear translation.

15.2 TRANSLATION

Consider a rigid body in translation (either rectilinear or curvilinear translation), and let A and B be any two of its particles (Fig. 15.7*a*). Denoting, respectively, by $\mathbf{r}_A$ and $\mathbf{r}_B$ the position vectors of A and B with respect to a fixed frame of reference and by $\mathbf{r}_{B/A}$ the vector joining A and B, we write

$$\mathbf{r}_B = \mathbf{r}_A + \mathbf{r}_{B/A} \tag{15.1}$$

Let us differentiate this relation with respect to t. We note that from the very definition of a translation, the vector $\mathbf{r}_{B/A}$ must maintain a constant direction; its magnitude must also be constant, since A and B

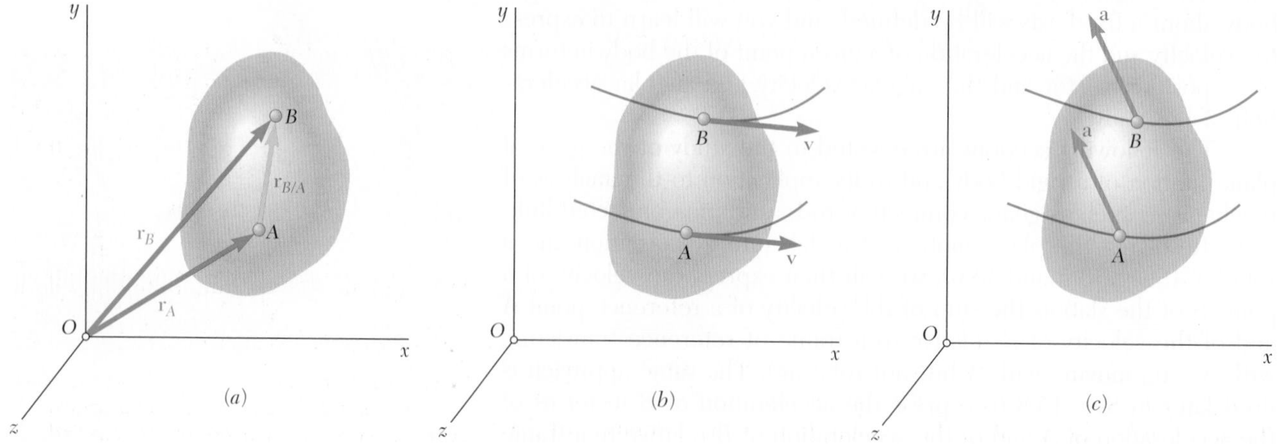

Fig. 15.7

belong to the same rigid body. Thus, the derivative of $\mathbf{r}_{B/A}$ is zero and we have

$$\mathbf{v}_B = \mathbf{v}_A \tag{15.2}$$

Differentiating once more, we write

$$\mathbf{a}_B = \mathbf{a}_A \tag{15.3}$$

Thus, *when a rigid body is in translation, all the points of the body have the same velocity and the same acceleration at any given instant* (Fig. 15.7*b* and *c*). In the case of curvilinear translation, the velocity and acceleration change in direction as well as in magnitude at every instant. In the case of rectilinear translation, all particles of the body move along parallel straight lines, and their velocity and acceleration keep the same direction during the entire motion.

15.3 ROTATION ABOUT A FIXED AXIS

Consider a rigid body which rotates about a fixed axis AA'. Let P be a point of the body and $\mathbf{r}$ its position vector with respect to a fixed frame of reference. For convenience, let us assume that the frame is centered at point O on AA' and that the z axis coincides with AA' (Fig. 15.8). Let B be the projection of P on AA'; since P must remain at a constant distance from B, it will describe a circle of center B and of radius $r \sin \phi$, where ϕ denotes the angle formed by $\mathbf{r}$ and AA'.

The position of P and of the entire body is completely defined by the angle θ the line BP forms with the zx plane. The angle θ is known as the *angular coordinate* of the body and is defined as positive when viewed as counterclockwise from A'. The angular coordinate will be expressed in radians (rad) or, occasionally, in degrees (°) or revolutions (rev). We recall that

$$1 \text{ rev} = 2\pi \text{ rad} = 360°$$

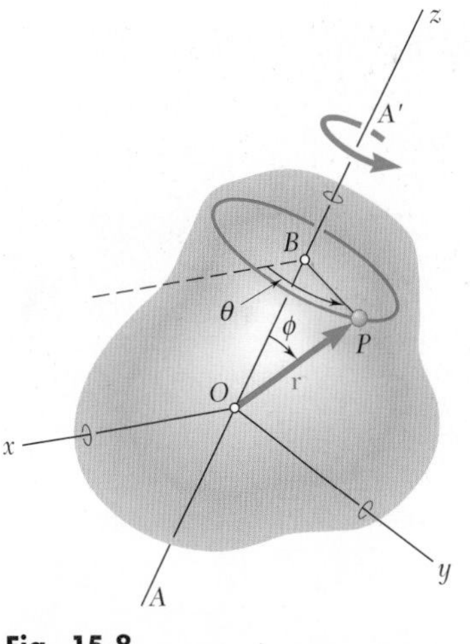

Fig. 15.8

We recall from Sec. 11.9 that the velocity $\mathbf{v} = d\mathbf{r}/dt$ of a particle P is a vector tangent to the path of P and of magnitude $v = ds/dt$. Observing that the length Δs of the arc described by P when the body rotates through $\Delta\theta$ is

$$\Delta s = (BP)\,\Delta\theta = (r \sin \phi)\,\Delta\theta$$

and dividing both members by Δt, we obtain at the limit, as Δt approaches zero,

$$v = \frac{ds}{dt} = r\dot{\theta} \sin \phi \tag{15.4}$$

where $\dot{\theta}$ denotes the time derivative of θ. (Note that the angle θ depends on the position of P within the body, but the rate of change $\dot{\theta}$ is itself independent of P.) We conclude that the velocity $\mathbf{v}$ of P is a vector perpendicular to the plane containing AA' and $\mathbf{r}$, and of

Photo 15.2 For the central gear rotating about a fixed axis, the angular velocity and angular acceleration of that gear are vectors directed along the vertical axis of rotation.

magnitude v defined by (15.4). But this is precisely the result we would obtain if we drew along AA' a vector $\boldsymbol{\omega} = \dot{\theta}\mathbf{k}$ and formed the vector product $\boldsymbol{\omega} \times \mathbf{r}$ (Fig. 15.9). We thus write

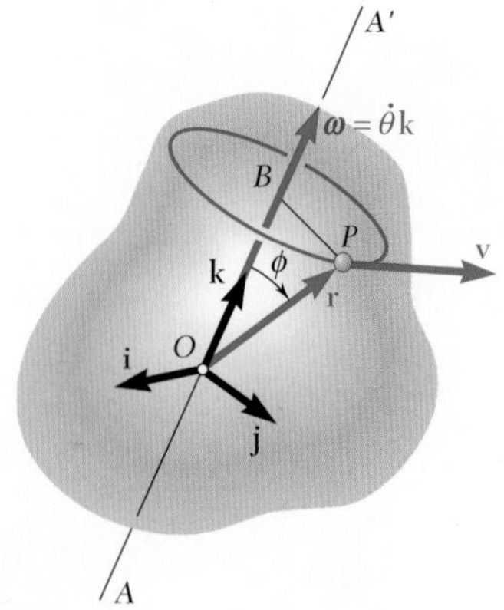

Fig. 15.9

$$\mathbf{v} = \frac{d\mathbf{r}}{dt} = \boldsymbol{\omega} \times \mathbf{r} \qquad (15.5)$$

The vector

$$\boldsymbol{\omega} = \omega\mathbf{k} = \dot{\theta}\mathbf{k} \qquad (15.6)$$

which is directed along the axis of rotation, is called the *angular velocity* of the body and is equal in magnitude to the rate of change $\dot{\theta}$ of the angular coordinate; its sense may be obtained by the right-hand rule (Sec. 3.6) from the sense of rotation of the body.†

The acceleration $\mathbf{a}$ of the particle P will now be determined. Differentiating (15.5) and recalling the rule for the differentiation of a vector product (Sec. 11.10), we write

$$\begin{aligned} \mathbf{a} &= \frac{d\mathbf{v}}{dt} = \frac{d}{dt}(\boldsymbol{\omega} \times \mathbf{r}) \\ &= \frac{d\boldsymbol{\omega}}{dt} \times \mathbf{r} + \boldsymbol{\omega} \times \frac{d\mathbf{r}}{dt} \\ &= \frac{d\boldsymbol{\omega}}{dt} \times \mathbf{r} + \boldsymbol{\omega} \times \mathbf{v} \end{aligned} \qquad (15.7)$$

The vector $d\boldsymbol{\omega}/dt$ is denoted by $\boldsymbol{\alpha}$ and is called the *angular acceleration* of the body. Substituting also for $\mathbf{v}$ from (15.5), we have

$$\mathbf{a} = \boldsymbol{\alpha} \times \mathbf{r} + \boldsymbol{\omega} \times (\boldsymbol{\omega} \times \mathbf{r}) \qquad (15.8)$$

Differentiating (15.6) and recalling that $\mathbf{k}$ is constant in magnitude and direction, we have

$$\boldsymbol{\alpha} = \alpha\mathbf{k} = \dot{\omega}\mathbf{k} = \ddot{\theta}\mathbf{k} \qquad (15.9)$$

Thus, the angular acceleration of a body rotating about a fixed axis is a vector directed along the axis of rotation, and is equal in magnitude to the rate of change $\dot{\omega}$ of the angular velocity. Returning to (15.8), we note that the acceleration of P is the sum of two vectors. The first vector is equal to the vector product $\boldsymbol{\alpha} \times \mathbf{r}$; it is tangent to the circle described by P and therefore represents the tangential component of the acceleration. The second vector is equal to the *vector triple product* $\boldsymbol{\omega} \times (\boldsymbol{\omega} \times \mathbf{r})$ obtained by forming the vector product of $\boldsymbol{\omega}$ and $\boldsymbol{\omega} \times \mathbf{r}$; since $\boldsymbol{\omega} \times \mathbf{r}$ is tangent to the circle described by P, the vector triple product is directed toward the center B of the circle and therefore represents the normal component of the acceleration.

†It will be shown in Sec. 15.12 in the more general case of a rigid body rotating simultaneously about axes having different directions that angular velocities obey the parallelogram law of addition and thus are actually vector quantities.

Rotation of a Representative Slab. The rotation of a rigid body about a fixed axis can be defined by the motion of a representative slab in a reference plane perpendicular to the axis of rotation. Let us choose the xy plane as the reference plane and assume that it coincides with the plane of the figure, with the z axis pointing out of the paper (Fig. 15.10). Recalling from (15.6) that $\boldsymbol{\omega} = \omega\mathbf{k}$, we

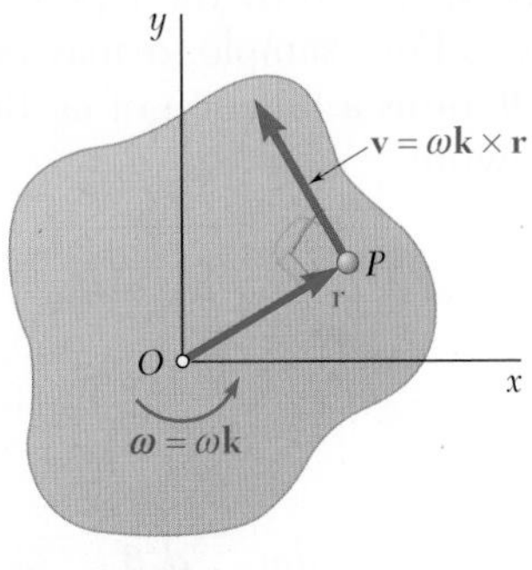

Fig. 15.10

note that a positive value of the scalar ω corresponds to a counterclockwise rotation of the representative slab, and a negative value to a clockwise rotation. Substituting $\omega\mathbf{k}$ for $\boldsymbol{\omega}$ into Eq. (15.5), we express the velocity of any given point P of the slab as

$$\mathbf{v} = \omega\mathbf{k} \times \mathbf{r} \tag{15.10}$$

Since the vectors $\mathbf{k}$ and $\mathbf{r}$ are mutually perpendicular, the magnitude of the velocity $\mathbf{v}$ is

$$v = r\omega \tag{15.10$'$}$$

and its direction can be obtained by rotating $\mathbf{r}$ through 90° in the sense of rotation of the slab.

Substituting $\boldsymbol{\omega} = \omega\mathbf{k}$ and $\boldsymbol{\alpha} = \alpha\mathbf{k}$ into Eq. (15.8), and observing that cross-multiplying $\mathbf{r}$ twice by $\mathbf{k}$ results in a 180° rotation of the vector $\mathbf{r}$, we express the acceleration of point P as

$$\mathbf{a} = \alpha\mathbf{k} \times \mathbf{r} - \omega^2\mathbf{r} \tag{15.11}$$

Resolving $\mathbf{a}$ into tangential and normal components (Fig. 15.11), we write

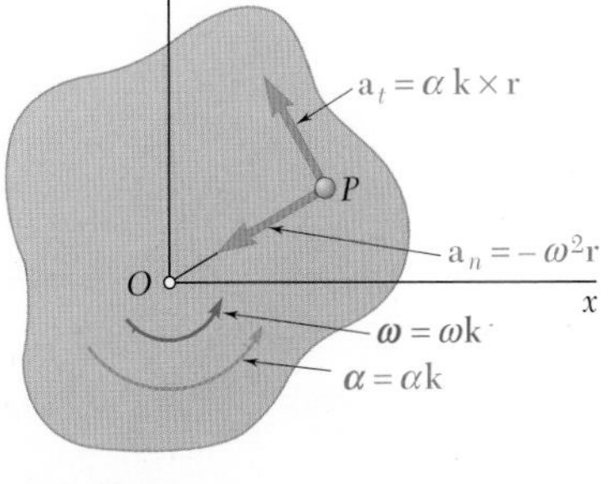

Fig. 15.11

$$\begin{aligned} \mathbf{a}_t &= \alpha\mathbf{k} \times \mathbf{r} \qquad & a_t &= r\alpha \\ \mathbf{a}_n &= -\omega^2\mathbf{r} \qquad & a_n &= r\omega^2 \end{aligned} \tag{15.11$'$}$$

The tangential component $\mathbf{a}_t$ points in the counterclockwise direction if the scalar α is positive, and in the clockwise direction if α is negative. The normal component $\mathbf{a}_n$ always points in the direction opposite to that of $\mathbf{r}$, that is, toward O.

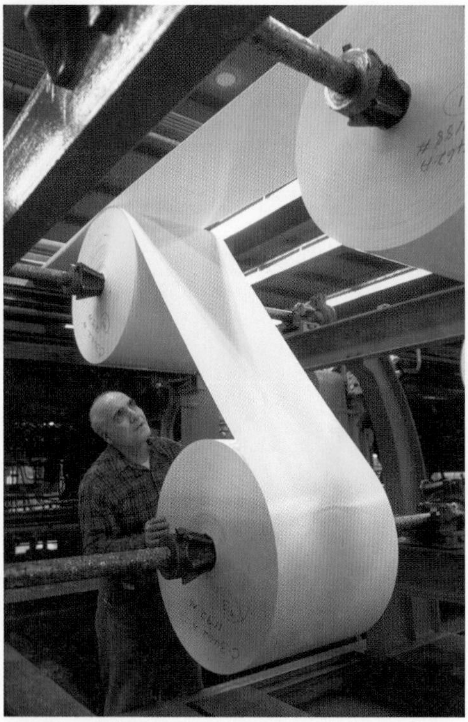

Photo 15.3 If the lower roll has a constant angular velocity, the speed of the paper being wound onto it increases as the radius of the roll increases.

15.4 EQUATIONS DEFINING THE ROTATION OF A RIGID BODY ABOUT A FIXED AXIS

The motion of a rigid body rotating about a fixed axis AA' is said to be *known* when its angular coordinate θ can be expressed as a known function of t. In practice, however, the rotation of a rigid body is seldom defined by a relation between θ and t. More often, the conditions of motion will be specified by the type of angular acceleration that the body possesses. For example, α may be given as a function of t, as a function of θ, or as a function of ω. Recalling the relations (15.6) and (15.9), we write

$$\omega = \frac{d\theta}{dt} \tag{15.12}$$

$$\alpha = \frac{d\omega}{dt} = \frac{d^2\theta}{dt^2} \tag{15.13}$$

or, solving (15.12) for dt and substituting into (15.13),

$$\alpha = \omega\frac{d\omega}{d\theta} \tag{15.14}$$

Since these equations are similar to those obtained in Chap. 11 for the rectilinear motion of a particle, their integration can be performed by following the procedure outlined in Sec. 11.3.

Two particular cases of rotation are frequently encountered:

1. *Uniform Rotation.* This case is characterized by the fact that the angular acceleration is zero. The angular velocity is thus constant, and the angular coordinate is given by the formula

$$\theta = \theta_0 + \omega t \tag{15.15}$$

2. *Uniformly Accelerated Rotation.* In this case, the angular acceleration is constant. The following formulas relating angular velocity, angular coordinate, and time can then be derived in a manner similar to that described in Sec. 11.5. The similarity between the formulas derived here and those obtained for the rectilinear uniformly accelerated motion of a particle is apparent.

$$\begin{aligned} \omega &= \omega_0 + \alpha t \\ \theta &= \theta_0 + \omega_0 t + \tfrac{1}{2}\alpha t^2 \\ \omega^2 &= \omega_0^2 + 2\alpha(\theta - \theta_0) \end{aligned} \tag{15.16}$$

It should be emphasized that formula (15.15) can be used only when $\alpha = 0$, and formulas (15.16) can be used only when $\alpha =$ constant. In any other case, the general formulas (15.12) to (15.14) should be used.

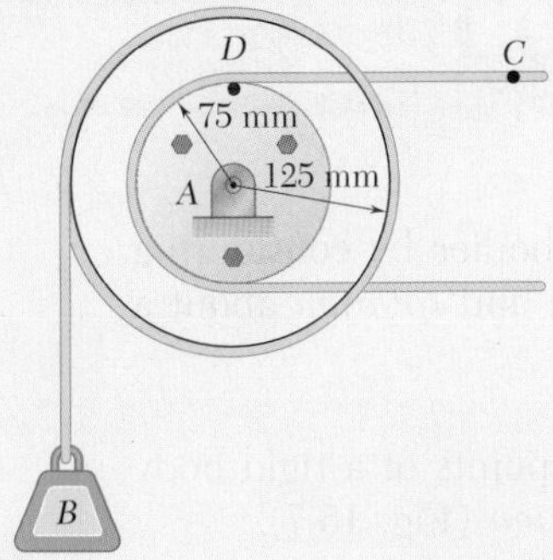

SAMPLE PROBLEM 15.1

Load B is connected to a double pulley by one of the two inextensible cables shown. The motion of the pulley is controlled by cable C, which has a constant acceleration of 225 mm/s² and an initial velocity of 300 mm/s, both directed to the right. Determine (*a*) the number of revolutions executed by the pulley in 2 s, (*b*) the velocity and change in position of the load B after 2 s, and (*c*) the acceleration of point D on the rim of the inner pulley at $t = 0$.

SOLUTION

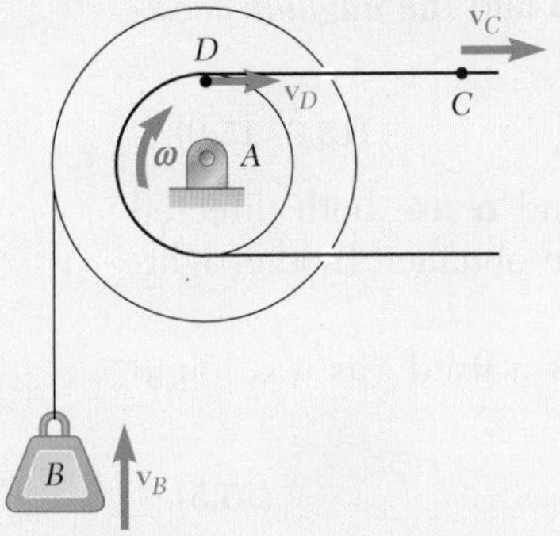

a. Motion of Pulley. Since the cable is inextensible, the velocity of point D is equal to the velocity of point C and the tangential component of the acceleration of D is equal to the acceleration of C.

$$(\mathbf{v}_D)_0 = (\mathbf{v}_C)_0 = 300\ \text{mm/s} \rightarrow \qquad (\mathbf{a}_D)_t = \mathbf{a}_C = 225\ \text{mm/s}^2 \rightarrow$$

Noting that the distance from D to the center of the pulley is 75 mm, we write

$$(v_D)_0 = r\omega_0 \qquad 300\ \text{mm/s} = (75\ \text{mm})\omega_0 \qquad \boldsymbol{\omega}_0 = 4\ \text{rad/s} \circlearrowright$$
$$(a_D)_t = r\alpha \qquad 225\ \text{mm/s}^2 = (75\ \text{mm})\alpha \qquad \boldsymbol{\alpha} = 3\ \text{rad/s}^2 \circlearrowright$$

Using the equations of uniformly accelerated motion, we obtain, for $t = 2$ s,

$$\omega = \omega_0 + \alpha t = 4\ \text{rad/s} + (3\ \text{rad/s}^2)(2\ \text{s}) = 10\ \text{rad/s}$$
$$\boldsymbol{\omega} = 10\ \text{rad/s} \circlearrowright$$
$$\theta = \omega_0 t + \tfrac{1}{2}\alpha t^2 = (4\ \text{rad/s})(2\ \text{s}) + \tfrac{1}{2}(3\ \text{rad/s}^2)(2\ \text{s})^2 = 14\ \text{rad}$$
$$\theta = 14\ \text{rad} \circlearrowright$$

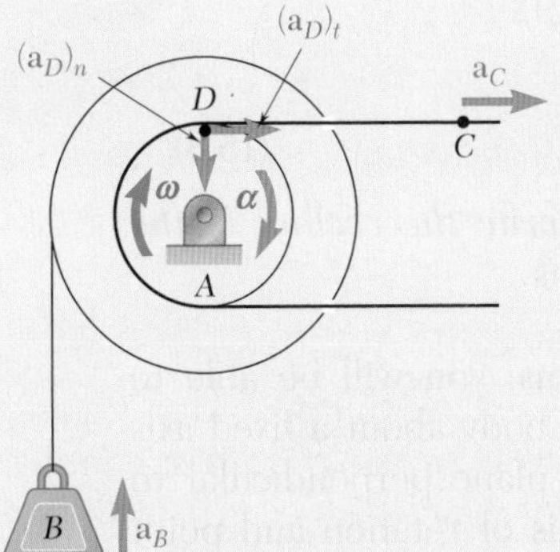

$$\text{Number of revolutions} = (14\ \text{rad})\left(\frac{1\text{rev}}{2\pi\ \text{rad}}\right) = 2.23\ \text{rev} \blacktriangleleft$$

b. Motion of Load B. Using the following relations between linear and angular motion, with $r = 125$ mm, we write

$$v_B = r\omega = (125\ \text{mm})(10\ \text{rad/s}) = 1250\ \text{mm/s} \qquad \mathbf{v}_B = 1.25\ \text{m/s} \uparrow \blacktriangleleft$$
$$\Delta y_B = r\theta = (125\ \text{mm})(14\ \text{rad}) = 1750\ \text{mm} \quad \Delta y_B = 1.75\ \text{m upward} \blacktriangleleft$$

c. Acceleration of Point D at $t = 0$. The tangential component of the acceleration is

$$(\mathbf{a}_D)_t = \mathbf{a}_C = 225\ \text{mm/s}^2 \rightarrow$$

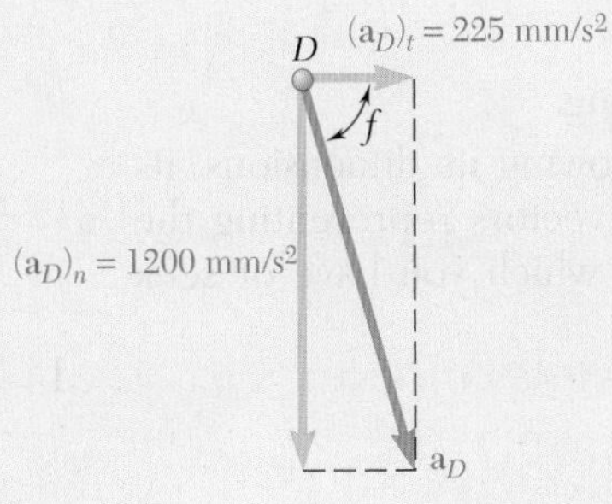

Since, at $t = 0$, $\omega_0 = 4$ rad/s, the normal component of the acceleration is

$$(a_D)_n = r_D\omega_0^2 = (75\ \text{mm})(4\ \text{rad/s})^2 = 1200\ \text{mm/s}^2 \qquad (\mathbf{a}_D)_n = 1200\ \text{mm/s}^2 \downarrow$$

The magnitude and direction of the total acceleration can be obtained by writing

$$\tan\phi = (1200\ \text{mm/s}^2)/(225\ \text{mm/s}^2) \qquad \phi = 79.4^\circ$$
$$a_D \sin 79.4^\circ = 1200\ \text{mm/s}^2 \qquad a_D = 1220\ \text{mm/s}^2$$
$$\mathbf{a}_D = 1.22\ \text{m/s}^2 \measuredangle 79.4^\circ \blacktriangleleft$$

SOLVING PROBLEMS ON YOUR OWN

In this lesson we began the study of the motion of rigid bodies by considering two particular types of motion of rigid bodies: *translation* and *rotation* about a *fixed axis*.

1. Rigid body in translation. At any given instant, all the points of a rigid body in translation have the *same velocity* and the *same acceleration* (Fig. 15.7).

2. Rigid body rotating about a fixed axis. The position of a rigid body rotating about a fixed axis was defined at any given instant by the *angular coordinate* θ, which is usually measured in *radians*. Selecting the unit vector $\mathbf{k}$ along the fixed axis and in such a way that the rotation of the body appears counterclockwise as seen from the tip of $\mathbf{k}$, we defined the *angular velocity* $\boldsymbol{\omega}$ and the *angular acceleration* $\boldsymbol{\alpha}$ of the body:

$$\boldsymbol{\omega} = \dot{\theta}\mathbf{k} \qquad \boldsymbol{\alpha} = \ddot{\theta}\mathbf{k} \tag{15.6, 15.9}$$

In solving problems, keep in mind that the vectors $\boldsymbol{\omega}$ and $\boldsymbol{\alpha}$ are both directed along the fixed axis of rotation and that their sense can be obtained by the right-hand rule.

a. The velocity of a point *P* of a body rotating about a fixed axis was found to be

$$\mathbf{v} = \boldsymbol{\omega} \times \mathbf{r} \tag{15.5}$$

where $\boldsymbol{\omega}$ is the angular velocity of the body and $\mathbf{r}$ is the position vector drawn from any point on the axis of rotation to point P (Fig. 15.9).

b. The acceleration of point *P* was found to be

$$\mathbf{a} = \boldsymbol{\alpha} \times \mathbf{r} + \boldsymbol{\omega} \times (\boldsymbol{\omega} \times \mathbf{r}) \tag{15.8}$$

Since vector products are not commutative, *be sure to write the vectors in the order shown* when using either of the above two equations.

3. Rotation of a representative slab. In many problems, you will be able to reduce the analysis of the rotation of a three-dimensional body about a fixed axis to the study of the rotation of a representative slab in a plane perpendicular to the fixed axis. The z axis should be directed along the axis of rotation and point out of the paper. Thus, the representative slab will be rotating in the xy plane about the origin O of the coordinate system (Fig. 15.10).

To solve problems of this type you should do the following:

a. Draw a diagram of the representative slab, showing its dimensions, its angular velocity and angular acceleration, as well as the vectors representing the velocities and accelerations of the points of the slab for which you have or seek information.

b. Relate the rotation of the slab and the motion of points of the slab by writing the equations

$$v = r\omega \qquad (15.10')$$

$$a_t = r\alpha \qquad a_n = r\omega^2 \qquad (15.11')$$

Remember that the velocity $\mathbf{v}$ and the component $\mathbf{a}_t$ of the acceleration of a point P of the slab are tangent to the circular path described by P. The directions of $\mathbf{v}$ and $\mathbf{a}_t$ are found by rotating the position vector $\mathbf{r}$ through 90° in the sense indicated by $\boldsymbol{\omega}$ and $\boldsymbol{\alpha}$, respectively. The normal component $\mathbf{a}_n$ of the acceleration of P is always directed toward the axis of rotation.

4. Equations defining the rotation of a rigid body. You must have been pleased to note the similarity existing between the equations defining the rotation of a rigid body about a fixed axis [Eqs. (15.12) through (15.16)] and those in Chap. 11 defining the rectilinear motion of a particle [Eqs. (11.1) through (11.8)]. All you have to do to obtain the new set of equations is to substitute θ, ω, and α for x, v, and a in the equations of Chap. 11.

PROBLEMS

CONCEPT QUESTIONS

15.CQ1 A rectangular plate swings from arms of equal length as shown. What is the magnitude of the angular velocity of the plate?

a. 0 rad/s
b. 1 rad/s
c. 2 rad/s
d. 3 rad/s
e. Need to know the location of the center of gravity.

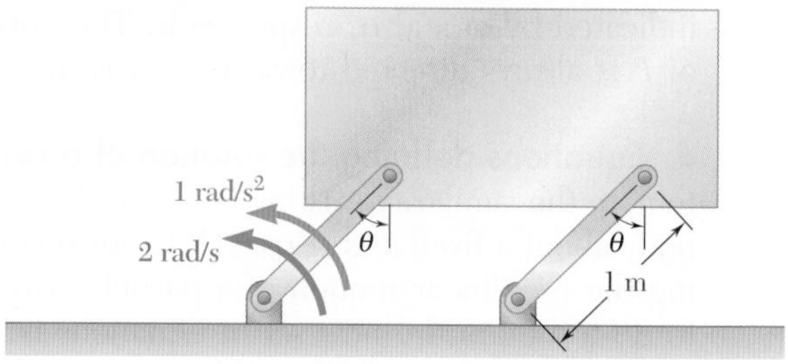

Fig. P15.CQ1

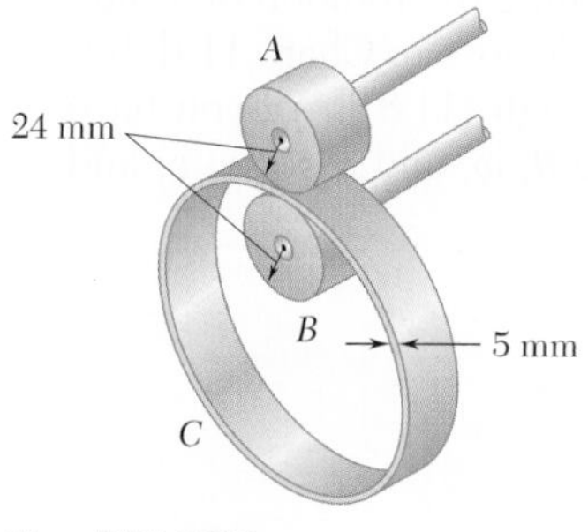

Fig. P15.CQ2

15.CQ2 Knowing that wheel A rotates with a constant angular velocity and that no slipping occurs between ring C and wheel A and wheel B, which of the following statements concerning the angular speeds of the three objects is true?

a. $\omega_a = \omega_b$
b. $\omega_a > \omega_b$
c. $\omega_a < \omega_b$
d. $\omega_a = \omega_c$
e. The contact points between A and C have the same acceleration.

END-OF-SECTION PROBLEMS

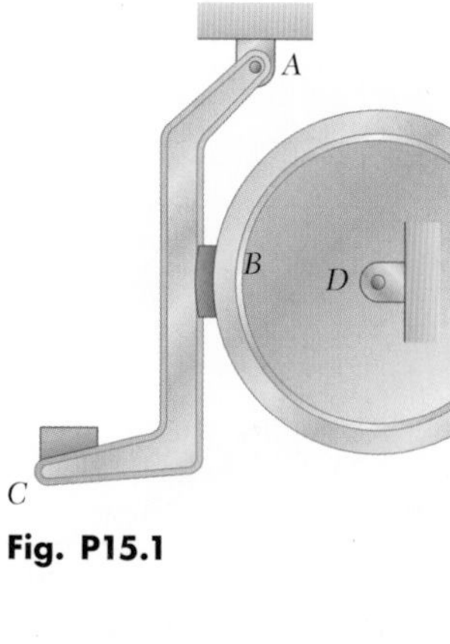

Fig. P15.1

15.1 The brake drum is attached to a larger flywheel that is not shown. The motion of the brake drum is defined by the relation $\theta = 36t - 1.6t^2$, where θ is expressed in radians and t in seconds. Determine (*a*) the angular velocity at $t = 2$ s, (*b*) the number of revolutions executed by the brake drum before coming to rest.

15.2 The motion of an oscillating crank is defined by the relation $\theta = \theta_0 \sin(\pi t/T) - (0.5\theta_0)\sin(2\pi t/T)$, where θ is expressed in radians and t in seconds. Knowing that $\theta_0 = 6$ rad and $T = 4$ s, determine the angular coordinate, the angular velocity, and the angular acceleration of the crank when (*a*) $t = 0$, (*b*) $t = 2$ s.

15.3 The motion of a disk rotating in an oil bath is defined by the relation $\theta = \theta_0(1 - e^{-t/4})$, where θ is expressed in radians and t in seconds. Knowing that $\theta_0 = 0.40$ rad, determine the angular coordinate, velocity, and acceleration of the disk when (*a*) $t = 0$, (*b*) $t = 3$ s, (*c*) $t = \infty$.

15.4 The rotor of a gas turbine is rotating at a speed of 6900 rpm when the turbine is shut down. It is observed that 4 min is required for the rotor to coast to rest. Assuming uniformly accelerated motion, determine (*a*) the angular acceleration, (*b*) the number of revolutions that the rotor executes before coming to rest.

15.5 A small grinding wheel is attached to the shaft of an electric motor which has a rated speed of 3600 rpm. When the power is turned on, the unit reaches its rated speed in 5 s, and when the power is turned off, the unit coasts to rest in 70 s. Assuming uniformly accelerated motion, determine the number of revolutions that the motor executes (*a*) in reaching its rated speed, (*b*) in coasting to rest.

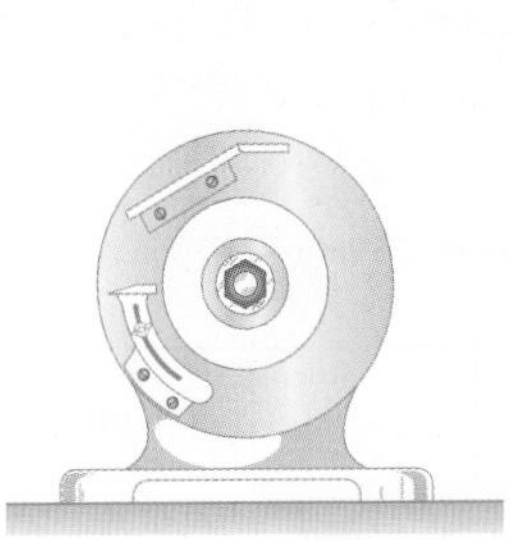

Fig. P15.5

15.6 A connecting rod is supported by a knife-edge at point *A*. For small oscillations the angular acceleration of the connecting rod is governed by the relation $\alpha = -6\theta$ where α is expressed in rad/s^2 and θ in radians. Knowing that the connecting rod is released from rest when $\theta = 20°$, determine (*a*) the maximum angular velocity, (*b*) the angular position when $t = 2$ s.

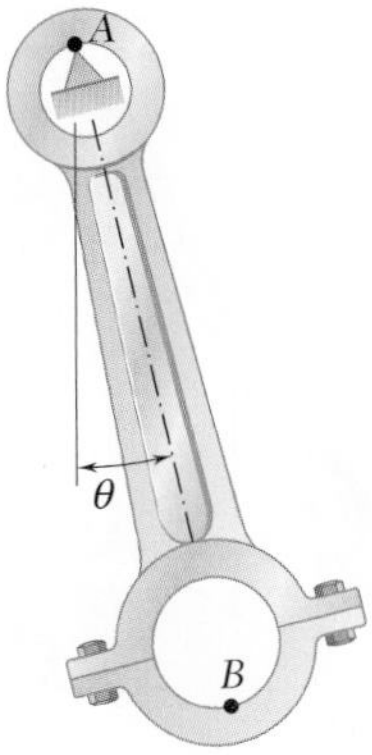

Fig. P15.6

15.7 When studying whiplash resulting from rear-end collisions, the rotation of the head is of primary interest. An impact test was performed, and it was found that the angular acceleration of the head is defined by the relation $\alpha = 700 \cos \theta + 70 \sin \theta$, where α is expressed in rad/s^2 and θ in radians. Knowing that the head is initially at rest, determine the angular velocity of the head when $\theta = 30°$.

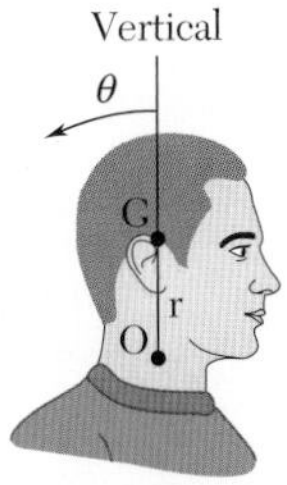

Fig. *P15.7*

15.8 The angular acceleration of an oscillating disk is defined by the relation $\alpha = -k\theta$. Determine (*a*) the value of *k* for which $\omega = 8$ rad/s when $\theta = 0$ and $\theta = 4$ rad when $\omega = 0$, (*b*) the angular velocity of the disk when $\theta = 3$ rad.

15.9 The angular acceleration of a shaft is defined by the relation $\alpha = -0.25\omega$, where α is expressed in rad/s^2 and ω in rad/s. Knowing that at $t = 0$ the angular velocity of the shaft is 20 rad/s, determine (*a*) the number of revolutions the shaft will execute before coming to rest, (*b*) the time required for the shaft to come to rest, (*c*) the time required for the angular velocity of the shaft to be reduced to 1 percent of its initial value.

15.10 The bent rod *ABCDE* rotates about a line joining points *A* and *E* with a constant angular velocity of 9 rad/s. Knowing that the rotation is clockwise as viewed from *E*, determine the velocity and acceleration of corner *C*.

15.11 In Prob. 15.10, determine the velocity and acceleration of corner *B*, assuming that the angular velocity is 9 rad/s and increases at the rate of 45 rad/s^2.

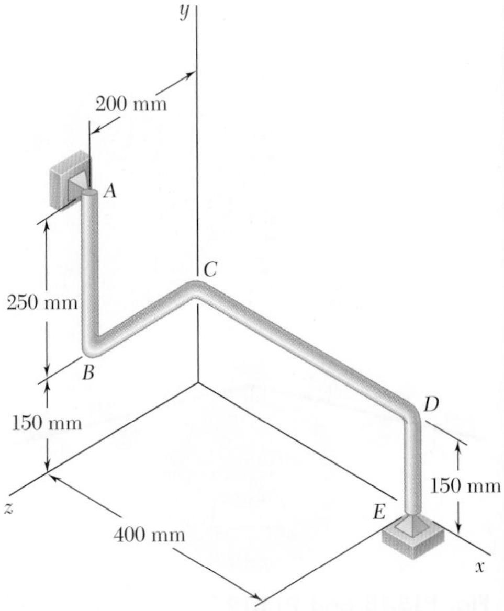

Fig. P15.10

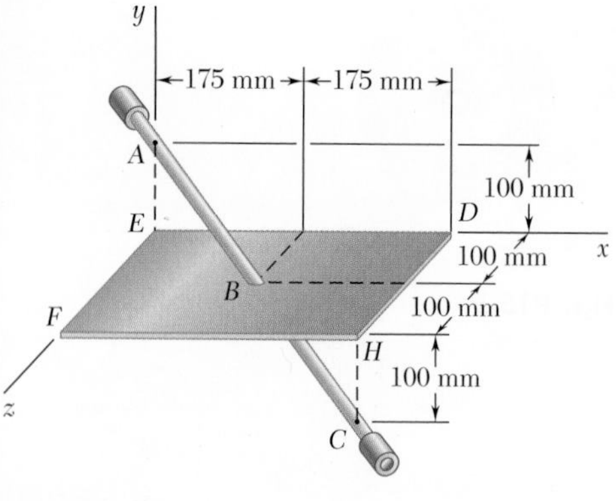

Fig. P15.12

15.12 The assembly shown consists of the straight rod *ABC* which passes through and is welded to the rectangular plate *DEFH*. The assembly rotates about the axis *AC* with a constant angular velocity of 9 rad/s. Knowing that the motion when viewed from *C* is counterclockwise, determine the velocity and acceleration of corner *F*.

15.13 In Prob. 15.12, determine the acceleration of corner *H*, assuming that the angular velocity is 9 rad/s and decreases at a rate of 18 rad/s^2.

15.14 A circular plate of 120-mm radius is supported by two bearings *A* and *B* as shown. The plate rotates about the rod joining *A* and *B* with a constant angular velocity of 26 rad/s. Knowing that, at the instant considered, the velocity of point *C* is directed to the right, determine the velocity and acceleration of point *E*.

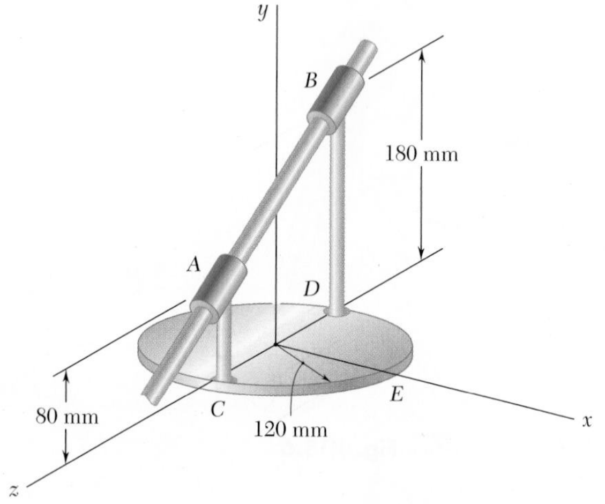

Fig. P15.14

15.15 In Prob. 15.14, determine the velocity and acceleration of point *E*, assuming that the angular velocity is 26 rad/s and increases at the rate of 65 rad/s^2.

15.16 The earth makes one complete revolution around the sun in 365.24 days. Assuming that the orbit of the earth is circular and has a radius of 15×10^6 km, determine the velocity and acceleration of the earth.

15.17 The earth makes one complete revolution on its axis in 23 h 56 min. Knowing that the mean radius of the earth is 6370 km, determine the linear velocity and acceleration of a point on the surface of the earth (*a*) at the equator, (*b*) at Philadelphia, latitude 40° north, (*c*) at the North Pole.

15.18 A series of small machine components being moved by a conveyor belt pass over a 120-mm-radius idler pulley. At the instant shown, the velocity of point *A* is 300 mm/s to the left and its acceleration is 180 mm/s^2 to the right. Determine (*a*) the angular velocity and angular acceleration of the idler pulley, (*b*) the total acceleration of the machine component at *B*.

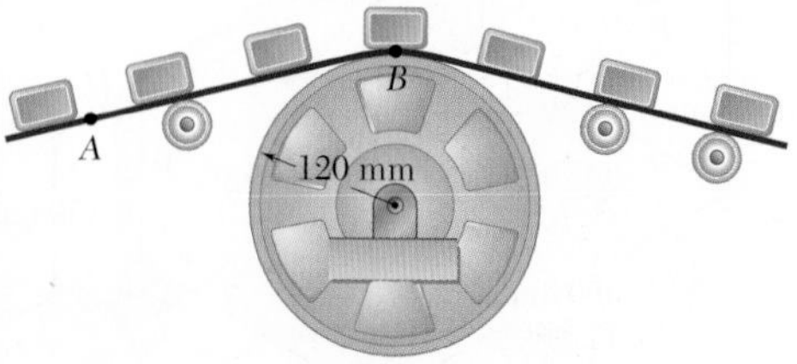

Fig. P15.18 and P15.19

15.19 A series of small machine components being moved by a conveyor belt pass over a 120-mm-radius idler pulley. At the instant shown, the angular velocity of the idler pulley is 4 rad/s clockwise. Determine the angular acceleration of the pulley for which the magnitude of the total acceleration of the machine component at *B* is 2400 mm/s^2.

15.20 The belt sander shown is initially at rest. If the driving drum B has a constant angular acceleration of 120 rad/s^2 counterclockwise, determine the magnitude of the acceleration of the belt at point C when (*a*) $t = 0.5$ s, (*b*) $t = 2$ s.

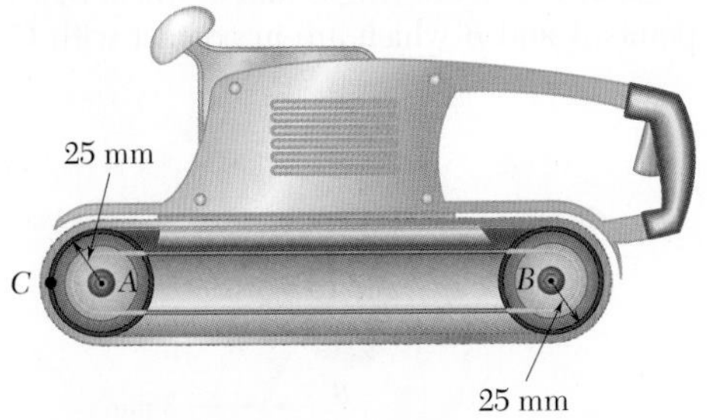

Fig. P15.20 and P15.21

15.21 The rated speed of drum B of the belt sander shown is 2400 rpm. When the power is turned off, it is observed that the sander coasts from its rated speed to rest in 10 s. Assuming uniformly decelerated motion, determine the velocity and acceleration of point C of the belt, (*a*) immediately before the power is turned off, (*b*) 9 s later.

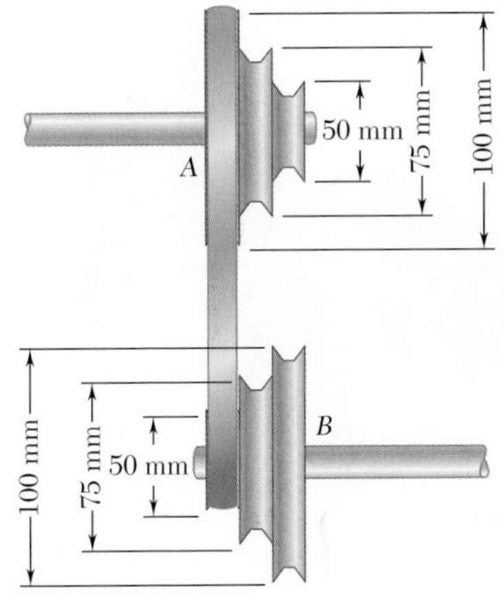

Fig. P15.22

15.22 The two pulleys shown may be operated with the V belt in any of three positions. If the angular acceleration of shaft A is 6 rad/s^2 and if the system is initially at rest, determine the time required for shaft B to reach a speed of 400 rpm with the belt in each of the three positions.

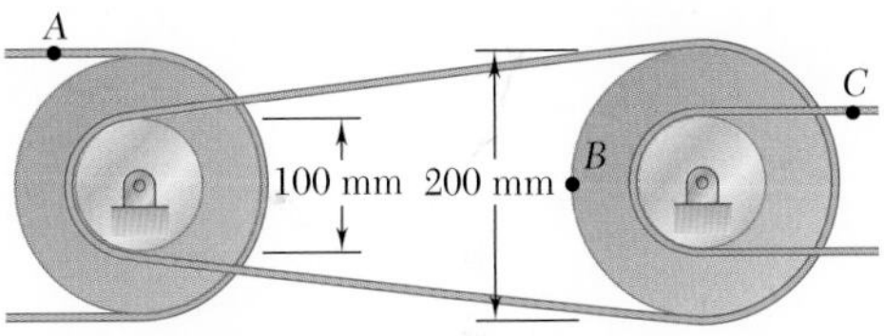

Fig. P15.23

15.23 Three belts move over two pulleys without slipping in the speed reduction system shown. At the instant shown, the velocity of point A on the input belt is 0.6 m/s to the right, decreasing at the rate of 1.8 m/s^2. Determine, at this instant, (*a*) the velocity and acceleration of point C on the output belt, (*b*) the acceleration of point B on the output pulley.

15.24 A gear reduction system consists of three gears A, B, and C. Knowing that gear A rotates clockwise with a constant angular velocity $\omega_A = 600$ rpm, determine (*a*) the angular velocities of gears B and C, (*b*) the accelerations of the points on gears B and C which are in contact.

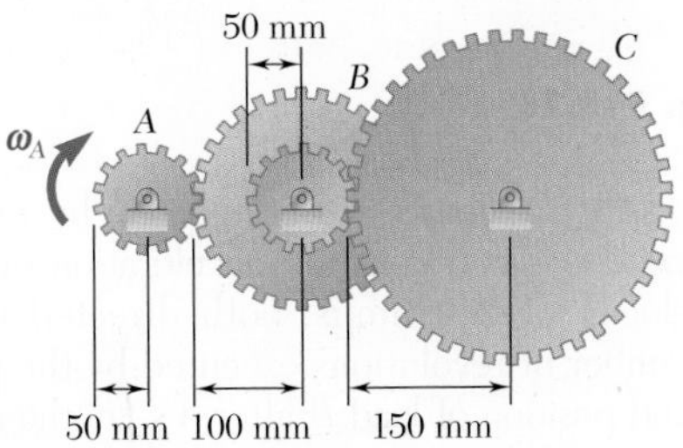

Fig. P15.24

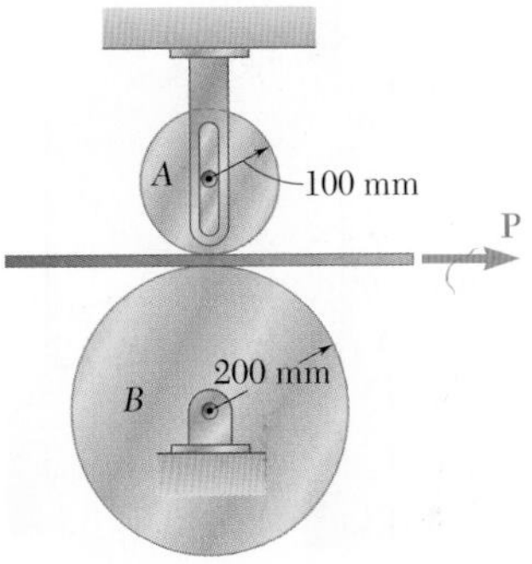

Fig. P15.25

15.25 A belt is pulled to the right between cylinders A and B. Knowing that the speed of the belt is a constant 1.5 m/s and no slippage occurs, determine (*a*) the angular velocities of A and B, (*b*) the accelerations of the points which are in contact with the belt.

15.26 Ring C has an inside radius of 55 mm and an outside radius of 60 mm and is positioned between two wheels A and B, each of 24-mm outside radius. Knowing that wheel A rotates with a constant angular velocity of 300 rpm and that no slipping occurs, determine (*a*) the angular velocity of ring C and of wheel B, (*b*) the acceleration of the points A and B which are in contact with C.

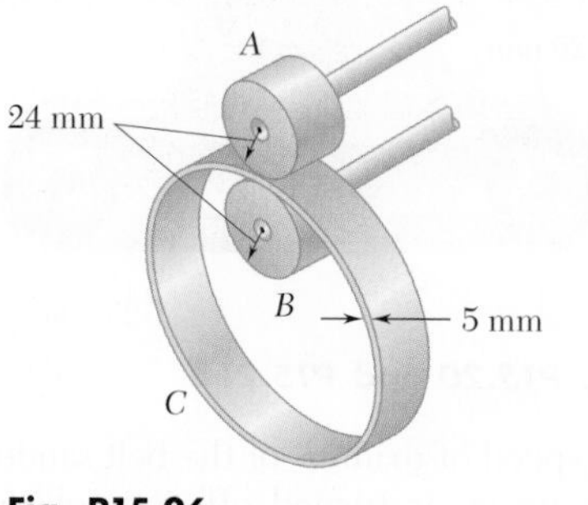

Fig. P15.26

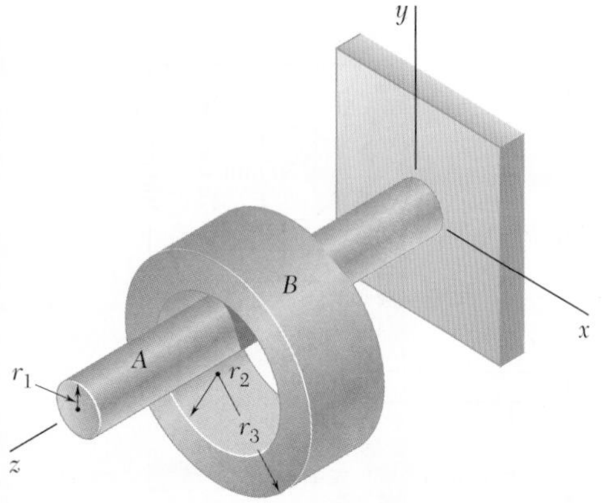

Fig. P15.27

15.27 Ring B has an inside radius r_2 and hangs from the horizontal shaft A as shown. Shaft A rotates with a constant angular velocity of 25 rad/s and no slipping occurs. Knowing that $r_1 = 12$ mm, $r_2 = 30$ mm, and $r_3 = 40$ mm, determine (*a*) the angular velocity of ring B, (*b*) the accelerations of the points of shaft A and ring B which are in contact, (*c*) the magnitude of the acceleration of a point on the outside surface of ring B.

15.28 A plastic film moves over two drums. During a 4-s interval the speed of the tape is increased uniformly from $v_0 = 0.6$ m/s to $v_1 = 1.2$ m/s. Knowing that the tape does not slip on the drums, determine (*a*) the angular acceleration of drum B, (*b*) the number of revolutions executed by drum B during the 4-s interval.

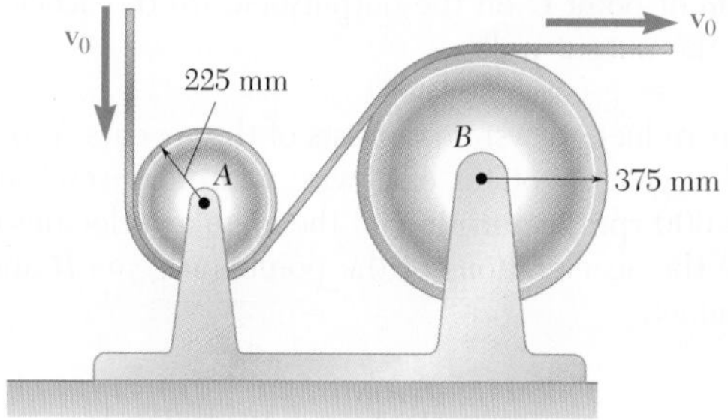

Fig. P15.28

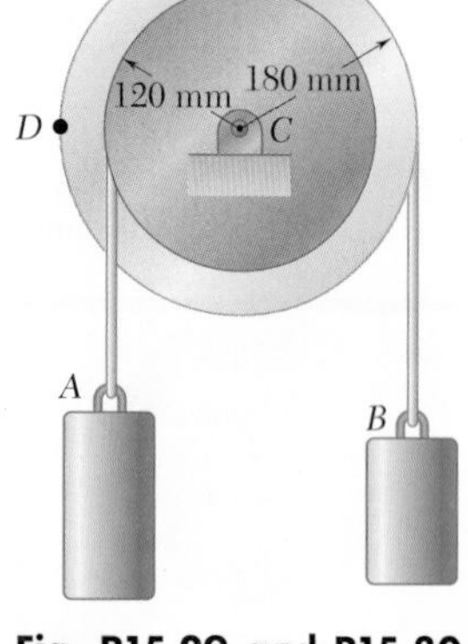

Fig. P15.29 and P15.30

15.29 A pulley and two loads are connected by inextensible cords as shown. Load A has a constant acceleration of 300 mm/s^2 and an initial velocity of 240 mm/s, both directed upward. Determine (*a*) the number of revolutions executed by the pulley in 3 s, (*b*) the velocity and position of load B after 3 s, (*c*) the acceleration of point D on the rim of the pulley at $t = 0$.

15.30 A pulley and two loads are connected by inextensible cords as shown. The pulley starts from rest at $t = 0$ and is accelerated at the uniform rate of 2.4 rad/s^2 clockwise. At $t = 4$ s, determine the velocity and position (*a*) of load A, (*b*) of load B.

15.31 A load is to be raised 6 m by the hoisting system shown. Assuming gear *A* is initially at rest, accelerates uniformly to a speed of 120 rpm in 5 s, and then maintains a constant speed of 120 rpm, determine (*a*) the number of revolutions executed by gear *A* in raising the load, (*b*) the time required to raise the load.

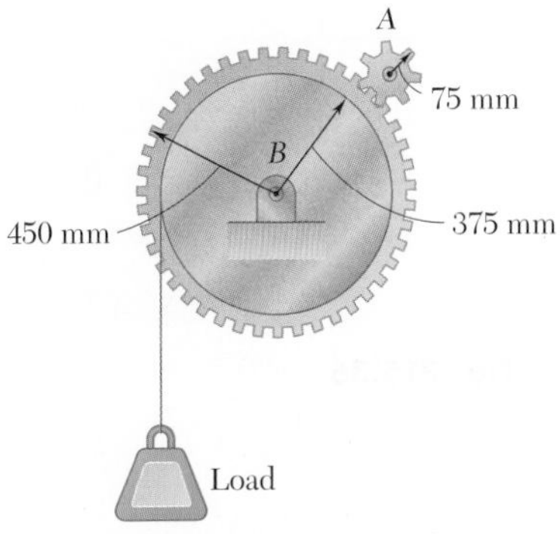

Fig. P15.31

15.32 Disk *B* is at rest when it is brought into contact with disk *A* which is rotating freely at 450 rpm clockwise. After 6 s of slippage, during which each disk has a constant angular acceleration, disk *A* reaches a final angular velocity of 140 rpm clockwise. Determine the angular acceleration of each disk during the period of slippage.

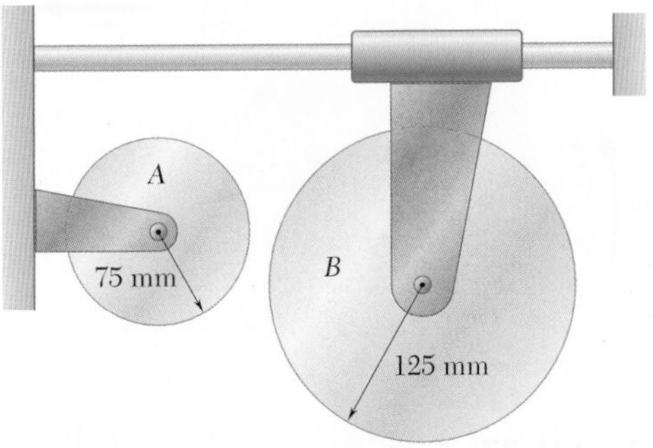

Fig. P15.32 and P15.33

15.33 and *15.34* A simple friction drive consists of two disks *A* and *B*. Initially, disk *A* has a clockwise angular velocity of 500 rpm and disk *B* is at rest. It is known that disk *A* will coast to rest in 60 s. However, rather than waiting until both disks are at rest to bring them together, disk *B* is given a constant angular acceleration of 2.5 rad/s^2 counterclockwise. Determine (*a*) at what time the disks can be brought together if they are not to slip, (*b*) the angular velocity of each disk as contact is made.

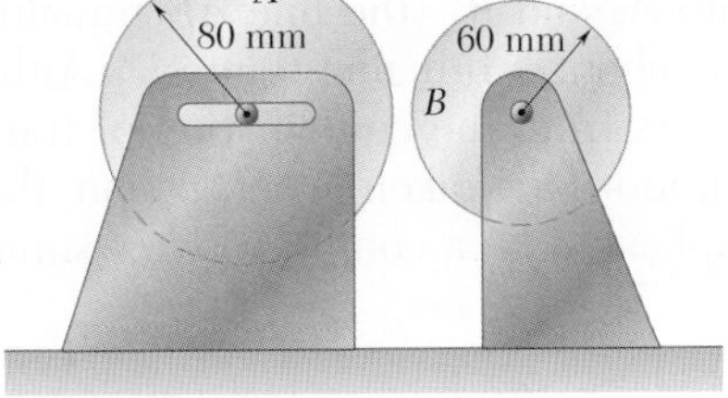

Fig. *P15.34* and *P15.35*

15.35 Two friction disks *A* and *B* are both rotating freely at 240 rpm counterclockwise when they are brought into contact. After 8 s of slippage, during which each disk has a constant angular acceleration, disk *A* reaches a final angular velocity of 60 rpm counterclockwise. Determine (*a*) the angular acceleration of each disk during the period of slippage, (*b*) the time at which the angular velocity of disk *B* is equal to zero.

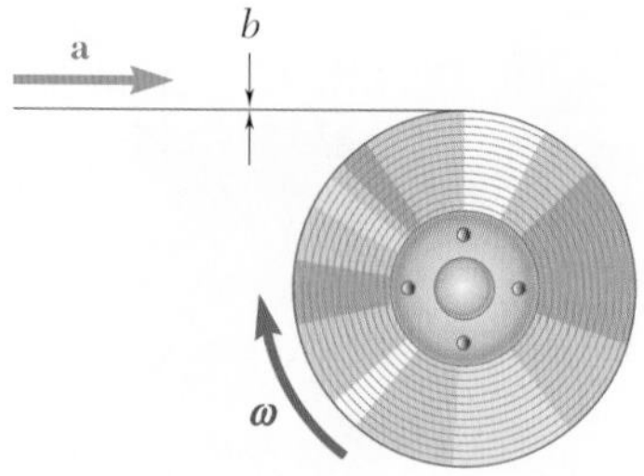

Fig. P15.36

*15.36 Steel tape is being wound onto a spool which rotates with a constant angular velocity $\boldsymbol{\omega}_0$. Denoting by r the radius of the spool and tape at any given time and by b the thickness of the tape, derive an expression for the acceleration of the tape as it approaches the spool.

*15.37 In a continuous printing process, paper is drawn into the presses at a constant speed v. Denoting by r the radius of the paper roll at any given time and by b the thickness of the paper, derive an expression for the angular acceleration of the paper roll.

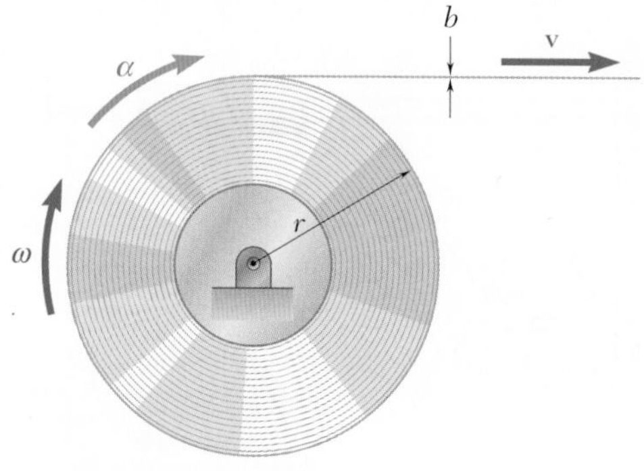

Fig. P15.37

15.5 GENERAL PLANE MOTION

As indicated in Sec. 15.1, we understand by general plane motion a plane motion which is neither a translation nor a rotation. As you will presently see, however, *a general plane motion can always be considered as the sum of a translation and a rotation.*

Consider, for example, a wheel rolling on a straight track (Fig. 15.12). Over a certain interval of time, two given points A and B will have moved, respectively, from A_1 to A_2 and from B_1 to B_2. The same result could be obtained through a translation which would bring A and B into A_2 and B_1' (the line AB remaining vertical), followed by a rotation about A bringing B into B_2. Although the original rolling motion differs from the combination of translation and rotation when these motions are taken in succession, the original motion can be exactly duplicated by a combination of simultaneous translation and rotation.

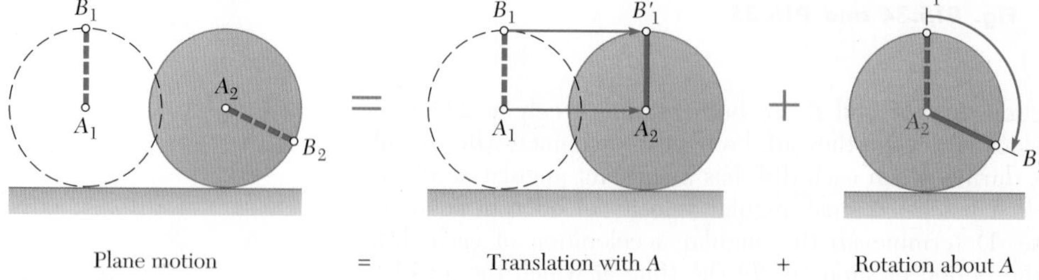

Fig. 15.12

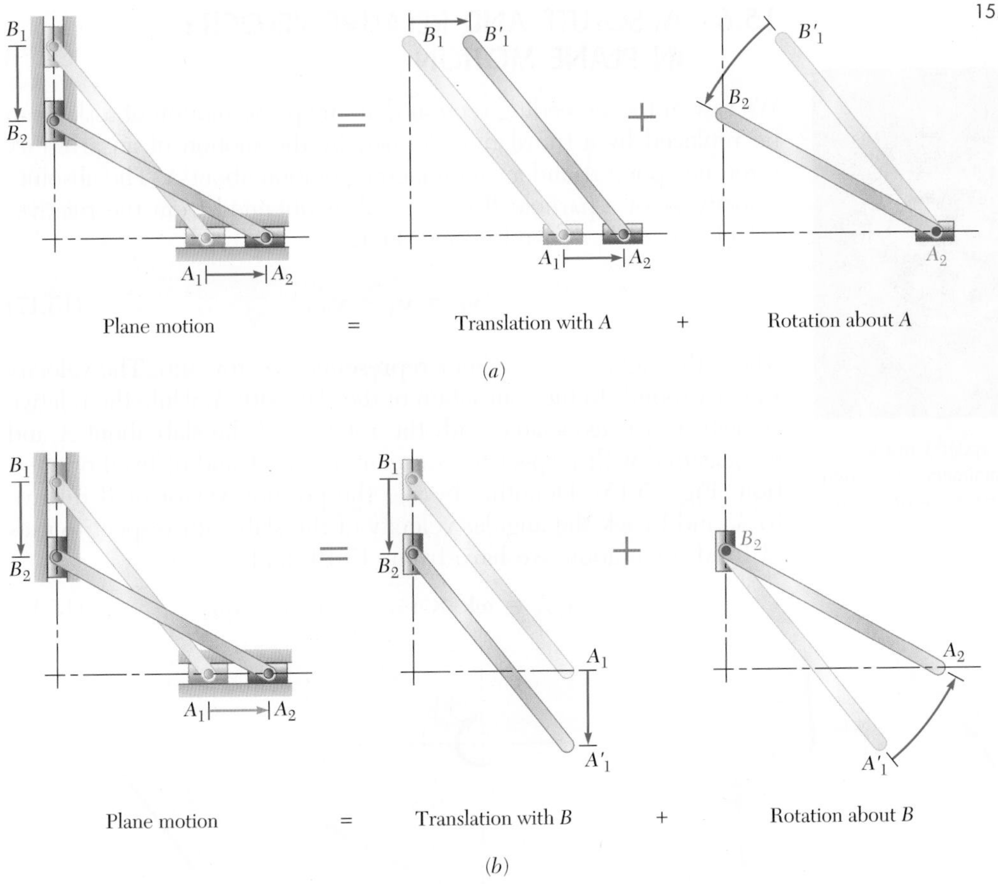

Fig. 15.13

Another example of plane motion is given in Fig. 15.13, which represents a rod whose extremities slide along a horizontal and a vertical track, respectively. This motion can be replaced by a translation in a horizontal direction and a rotation about A (Fig. 15.13*a*) or by a translation in a vertical direction and a rotation about B (Fig. 15.13*b*).

In the general case of plane motion, we will consider a small displacement which brings two particles A and B of a representative slab, respectively, from A_1 and B_1 into A_2 and B_2 (Fig. 15.14). This displacement can be divided into two parts: in one, the particles move into A_2 and B'_1 while the line AB maintains the same direction; in the other, B moves into B_2 while A remains fixed. The first part of the motion is clearly a translation and the second part a rotation about A.

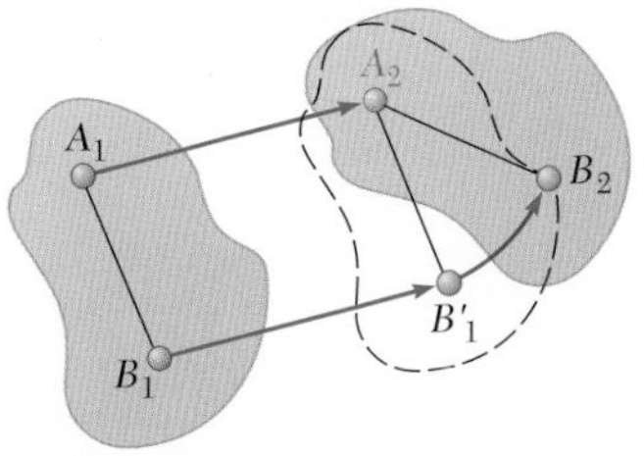

Fig. 15.14

Recalling from Sec. 11.12 the definition of the relative motion of a particle with respect to a moving frame of reference—as opposed to its absolute motion with respect to a fixed frame of reference—we can restate as follows the result obtained above: Given two particles A and B of a rigid slab in plane motion, the relative motion of B with respect to a frame attached to A and of fixed orientation is a rotation. To an observer moving with A but not rotating, particle B will appear to describe an arc of circle centered at A.

Photo 15.4 Planetary gear systems are used to high reduction ratios with minimum space and weight. The small gears undergo general plane motion.

15.6 ABSOLUTE AND RELATIVE VELOCITY IN PLANE MOTION

We saw in the preceding section that any plane motion of a slab can be replaced by a translation defined by the motion of an arbitrary reference point A and a simultaneous rotation about A. The absolute velocity $\mathbf{v}_B$ of a particle B of the slab is obtained from the relative-velocity formula derived in Sec. 11.12,

$$\mathbf{v}_B = \mathbf{v}_A + \mathbf{v}_{B/A} \tag{15.17}$$

where the right-hand member represents a vector sum. The velocity $\mathbf{v}_A$ corresponds to the translation of the slab with A, while the relative velocity $\mathbf{v}_{B/A}$ is associated with the rotation of the slab about A and is measured with respect to axes centered at A and of fixed orientation (Fig. 15.15). Denoting by $\mathbf{r}_{B/A}$ the position vector of B relative to A, and by $\omega\mathbf{k}$ the angular velocity of the slab with respect to axes of fixed orientation, we have from (15.10) and (15.10′)

$$\mathbf{v}_{B/A} = \omega\mathbf{k} \times \mathbf{r}_{B/A} \qquad v_{B/A} = r\omega \tag{15.18}$$

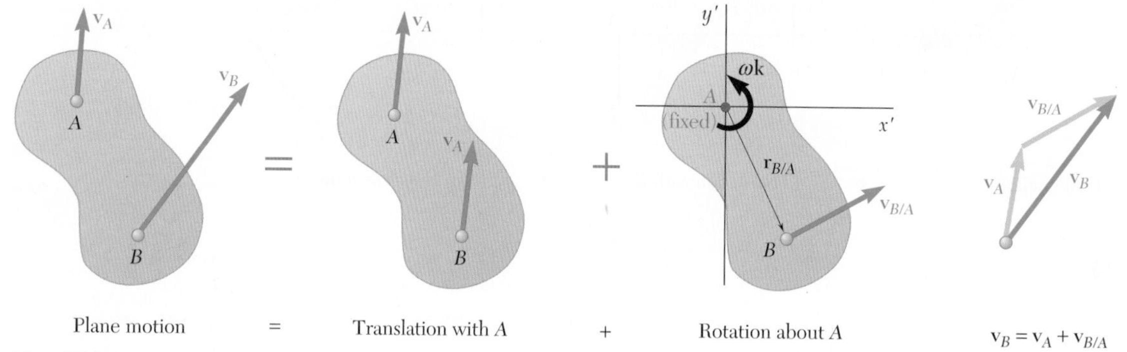

Fig. 15.15

where r is the distance from A to B. Substituting for $\mathbf{v}_{B/A}$ from (15.18) into (15.17), we can also write

$$\mathbf{v}_B = \mathbf{v}_A + \omega\mathbf{k} \times \mathbf{r}_{B/A} \tag{15.17′}$$

As an example, let us again consider the rod AB of Fig. 15.13. Assuming that the velocity $\mathbf{v}_A$ of end A is known, we propose to find the velocity $\mathbf{v}_B$ of end B and the angular velocity $\boldsymbol{\omega}$ of the rod, in terms of the velocity $\mathbf{v}_A$, the length l, and the angle θ. Choosing A as a reference point, we express that the given motion is equivalent to a translation with A and a simultaneous rotation about A (Fig. 15.16). The absolute velocity of B must therefore be equal to the vector sum

$$\mathbf{v}_B = \mathbf{v}_A + \mathbf{v}_{B/A} \tag{15.17}$$

We note that while the direction of $\mathbf{v}_{B/A}$ is known, its magnitude $l\omega$ is unknown. However, this is compensated for by the fact that the direction of $\mathbf{v}_B$ is known. We can therefore complete the diagram of Fig. 15.16. Solving for the magnitudes v_B and ω, we write

$$v_B = v_A \tan\theta \qquad \omega = \frac{v_{B/A}}{l} = \frac{v_A}{l\cos\theta} \tag{15.19}$$

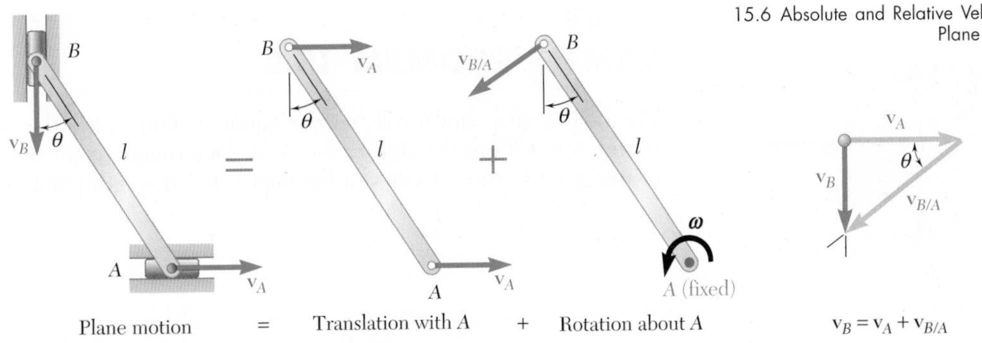

Fig. 15.16

The same result can be obtained by using B as a point of reference. Resolving the given motion into a translation with B and a simultaneous rotation about B (Fig. 15.17), we write the equation

$$\mathbf{v}_A = \mathbf{v}_B + \mathbf{v}_{A/B} \tag{15.20}$$

which is represented graphically in Fig. 15.17. We note that $\mathbf{v}_{A/B}$ and $\mathbf{v}_{B/A}$ have the same magnitude $l\omega$ but opposite sense. The sense of the relative velocity depends, therefore, upon the point of reference which has been selected and should be carefully ascertained from the appropriate diagram (Fig. 15.16 or 15.17).

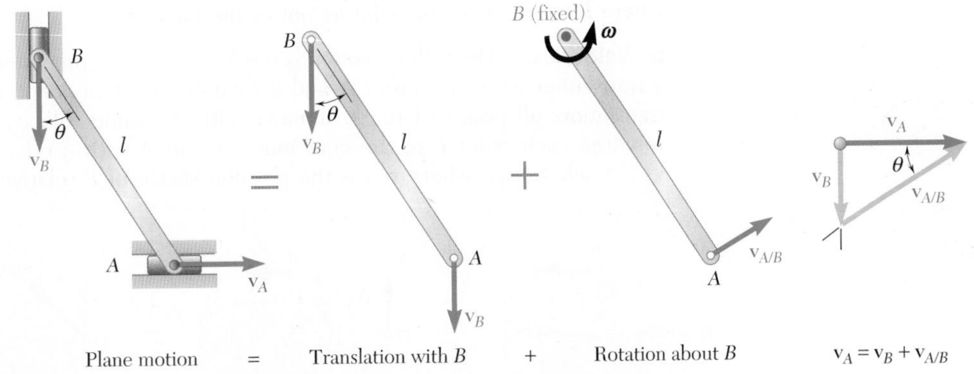

Fig. 15.17

Finally, we observe that the angular velocity $\boldsymbol{\omega}$ of the rod in its rotation about B is the same as in its rotation about A. It is measured in both cases by the rate of change of the angle θ. This result is quite general; we should therefore bear in mind that *the angular velocity* $\boldsymbol{\omega}$ *of a rigid body in plane motion is independent of the reference point.*

Most mechanisms consist not of one but of *several* moving parts. When the various parts of a mechanism are pin-connected, the analysis of the mechanism can be carried out by considering each part as a rigid body, keeping in mind that the points where two parts are connected must have the same absolute velocity. A similar analysis can be used when gears are involved, since the teeth in contact must also have the same absolute velocity. However, when a mechanism contains parts which slide on each other, the relative velocity of the parts in contact must be taken into account (see Secs. 15.10 and 15.11).

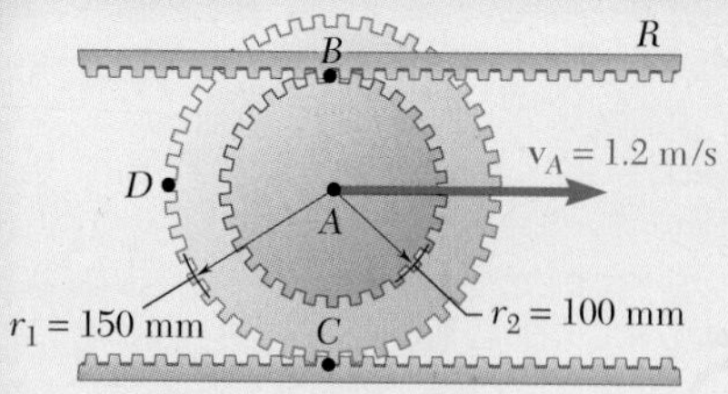

SAMPLE PROBLEM 15.2

The double gear shown rolls on the stationary lower rack; the velocity of its center A is 1.2 m/s directed to the right. Determine (*a*) the angular velocity of the gear, (*b*) the velocities of the upper rack R and of point D of the gear.

SOLUTION

a. Angular Velocity of the Gear. Since the gear rolls on the lower rack, its center A moves through a distance equal to the outer circumference $2\pi r_1$ for each full revolution of the gear. Noting that 1 rev = 2π rad, and that when A moves to the right ($x_A > 0$) the gear rotates clockwise ($\theta < 0$), we write

$$\frac{x_A}{2\pi r_1} = -\frac{\theta}{2\pi} \qquad x_A = -r_1\theta$$

Differentiating with respect to the time t and substituting the known values $v_A = 1.2$ m/s and $r_1 = 150$ mm $= 0.150$ m, we obtain

$$v_A = -r_1\omega \qquad 1.2 \text{ m/s} = -(0.150 \text{ m})\omega \qquad \omega = -8 \text{ rad/s}$$

$$\boldsymbol{\omega} = \omega\mathbf{k} = -(8 \text{ rad/s})\mathbf{k} \quad ◀$$

where $\mathbf{k}$ is a unit vector pointing out of the paper.

b. Velocities. The rolling motion is resolved into two component motions: a translation with the center A and a rotation about the center A. In the translation, all points of the gear move with the same velocity $\mathbf{v}_A$. In the rotation, each point P of the gear moves about A with a relative velocity $\mathbf{v}_{P/A} = \omega\mathbf{k} \times \mathbf{r}_{P/A}$, where $\mathbf{r}_{P/A}$ is the position vector of P relative to A.

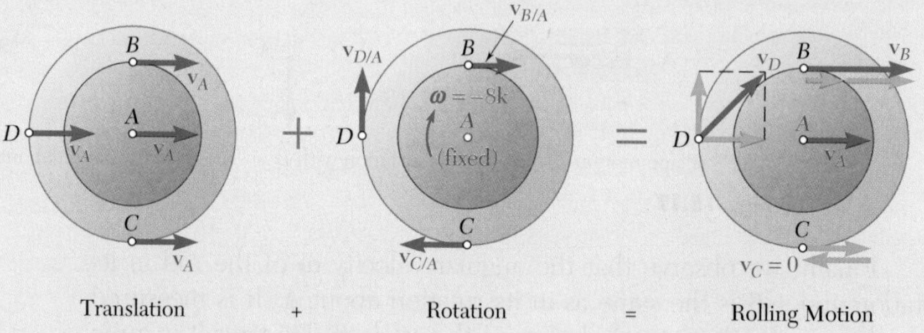

Velocity of Upper Rack. The velocity of the upper rack is equal to the velocity of point B; we write

$$\mathbf{v}_R = \mathbf{v}_B = \mathbf{v}_A + \mathbf{v}_{B/A} = \mathbf{v}_A + \omega\mathbf{k} \times \mathbf{r}_{B/A}$$
$$= (1.2 \text{ m/s})\mathbf{i} - (8 \text{ rad/s})\mathbf{k} \times (0.100 \text{ m})\mathbf{j}$$
$$= (1.2 \text{ m/s})\mathbf{i} + (0.8 \text{ m/s})\mathbf{i} = (2 \text{ m/s})\mathbf{i}$$

$$\mathbf{v}_R = 2 \text{ m/s} \rightarrow \quad ◀$$

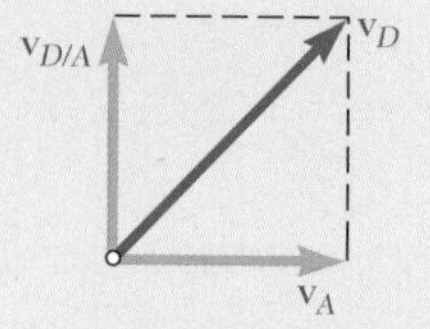

Velocity of Point D

$$\mathbf{v}_D = \mathbf{v}_A + \mathbf{v}_{D/A} = \mathbf{v}_A + \omega\mathbf{k} \times \mathbf{r}_{D/A}$$
$$= (1.2 \text{ m/s})\mathbf{i} - (8 \text{ rad/s})\mathbf{k} \times (-0.150 \text{ m})\mathbf{i}$$
$$= (1.2 \text{ m/s})\mathbf{i} + (1.2 \text{ m/s})\mathbf{j}$$

$$\mathbf{v}_D = 1.697 \text{ m/s} \measuredangle 45° \quad ◀$$

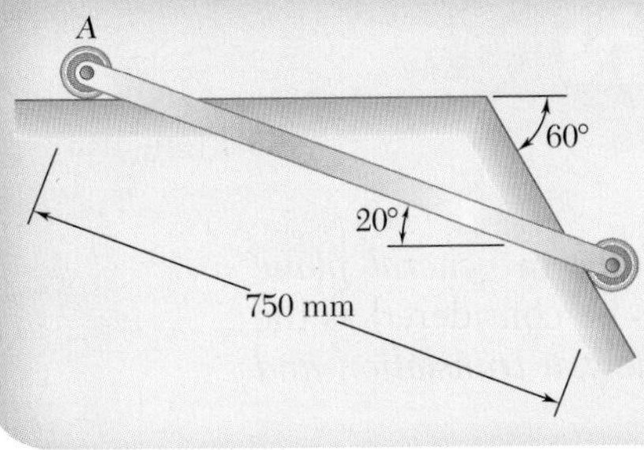

SAMPLE PROBLEM 15.3

Small wheels have been attached to the ends of rod AB and roll freely along the surfaces shown. Knowing that wheel A moves to the left with a constant velocity of 1.5 m/s, determine (a) the angular velocity of the rod, (b) the velocity of end B of the rod.

SOLUTION

Motion of part B: The velocity of point A is horizontal (to the left) where as velocity of part B is upward along the incline (at an angle of 60° from the horizontal). Resolving the reaction of AB into a translation with A and a rotation about A, we obtain

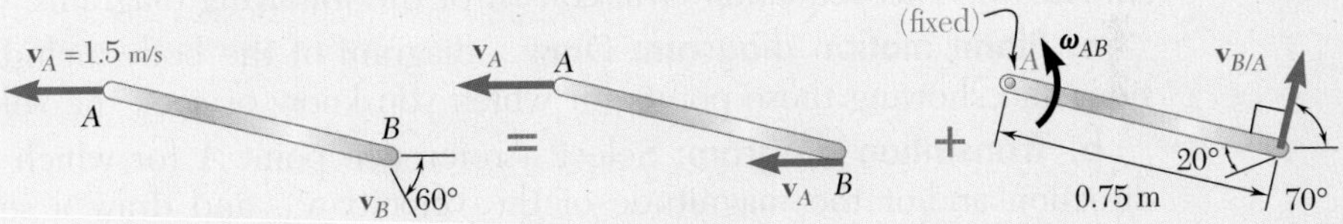

Expressing the relation between the velocities $\mathbf{v}_B$, $\mathbf{v}_A$ and $\mathbf{v}_{B/A}$

$$\mathbf{v}_B = \mathbf{v}_A + \mathbf{v}_{B/A}$$

$$[v_B \searrow 60°] = [1.5 \text{ m/s} \leftarrow] + [v_{B/A} \angle 70°]$$

We draw a vector diagram corresponding to this equation.

Law of sines.

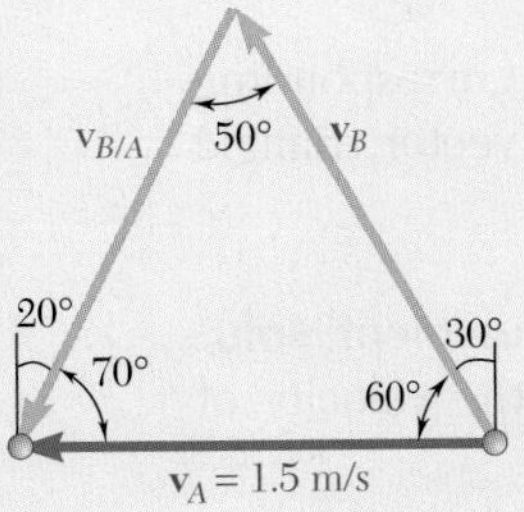

$$\frac{v_B}{\sin 70°} = \frac{v_{B/A}}{\sin 60°} = \frac{1.5 \text{ m/s}}{\sin 50°}$$

$$\mathbf{V}_B = 1.840 \text{ m/s} \searrow 60° \quad \blacktriangleleft$$

$$\mathbf{V}_{B/A} = 1.696 \text{ m/s} \angle 70°$$

(b)

$$v_{B/A} = (AB)\omega_{AB}$$

$$1.696 \text{ m/s} = (0.75 \text{ m})\omega_{AB}$$

(a) $\omega_{AB} = 2.261$ rad/s $\qquad \boldsymbol{\omega}_{AB} = 2.26 \text{ rad/s} \circlearrowleft \quad \blacktriangleleft$

SOLVING PROBLEMS ON YOUR OWN

In this lesson you learned to analyze the velocity of bodies in *general plane motion*. You found that a general plane motion can always be considered as the sum of the two motions you studied in the last lesson, namely, *a translation and a rotation*.

To solve a problem involving the velocity of a body in plane motion you should take the following steps.

1. Whenever possible determine the velocity of the points of the body where the body is connected to another body whose motion is known. That other body may be an arm or crank rotating with a given angular velocity [Sample Prob. 15.3].

2. Next start drawing a "diagram equation" to use in your solution (Figs. 15.15 and 15.16). This "equation" will consist of the following diagrams.

a. Plane motion diagram: Draw a diagram of the body including all dimensions and showing those points for which you know or seek the velocity.

b. Translation diagram: Select a reference point A for which you know the direction and/or the magnitude of the velocity $\mathbf{v}_A$, and draw a second diagram showing the body in translation with all of its points having the same velocity $\mathbf{v}_A$.

c. Rotation diagram: Consider point A as a fixed point and draw a diagram showing the body in rotation about A. Show the angular velocity $\boldsymbol{\omega} = \omega\mathbf{k}$ of the body and the relative velocities with respect to A of the other points, such as the velocity $\mathbf{v}_{B/A}$ of B relative to A.

3. Write the relative-velocity formula

$$\mathbf{v}_B = \mathbf{v}_A + \mathbf{v}_{B/A}$$

While you can solve this vector equation analytically by writing the corresponding scalar equations, you will usually find it easier to solve it by using a vector triangle (Fig. 15.16).

4. A different reference point can be used to obtain an equivalent solution. For example, if point B is selected as the reference point, the velocity of point A is expressed as

$$\mathbf{v}_A = \mathbf{v}_B + \mathbf{v}_{A/B}$$

Note that the relative velocities $\mathbf{v}_{B/A}$ and $\mathbf{v}_{A/B}$ have the same magnitude but opposite sense. Relative velocities, therefore, depend upon the reference point that has been selected. The angular velocity, however, is independent of the choice of reference point.

PROBLEMS

CONCEPT QUESTIONS

15.CQ3 The ball rolls without slipping on the fixed surface as shown. What is the direction of the velocity of point A?

a. $\rightarrow$ **b.** $\nearrow$ **c.** $\uparrow$ **d.** $\downarrow$ **e.** $\searrow$

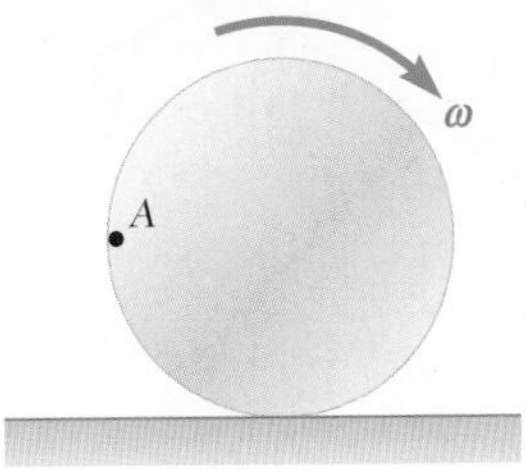

Fig. P15.CQ3

15.CQ4 Three uniform rods—ABC, DCE, and FGH—are connected as shown. Which of the following statements concerning the angular speed of the three objects is true?

a. $\omega_{ABC} = \omega_{DCE} = \omega_{FGH}$

b. $\omega_{DCE} > \omega_{ABC} > \omega_{FGH}$

c. $\omega_{DCE} < \omega_{ABC} < \omega_{FGH}$

d. $\omega_{ABC} > \omega_{DCE} > \omega_{FGH}$

e. $\omega_{FGH} = \omega_{DCE} < \omega_{ABC}$

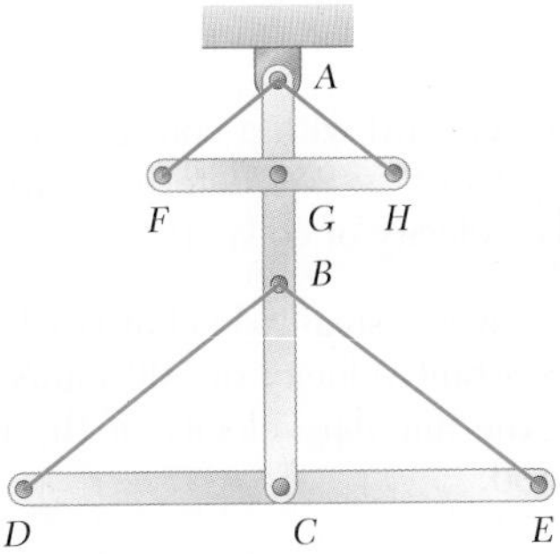

Fig. P15.CQ4

END-OF-SECTION PROBLEMS

15.38 An automobile travels to the right at a constant speed of 80 km/h. If the diameter of a wheel is 500 mm, determine the velocities of points B, C, D, and E on the rim of the wheel.

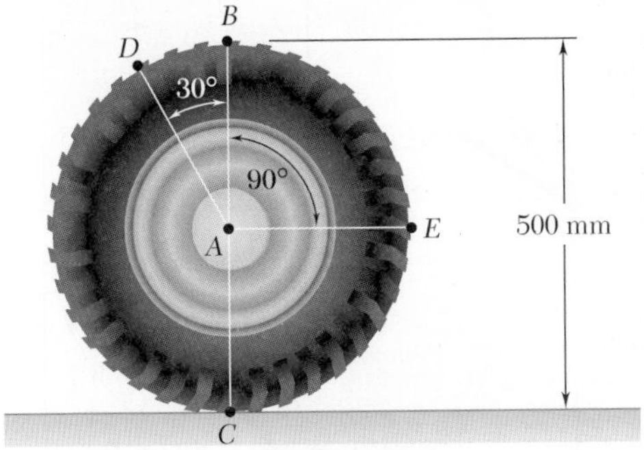

Fig. P15.38

15.39 The motion of rod AB is guided by pins attached at A and B which slide in the slots shown. At the instant shown, $\theta = 40°$ and the pin at B moves upward to the left with a constant velocity of 150 mm/s. Determine (*a*) the angular velocity of the rod, (*b*) the velocity of the pin at end A.

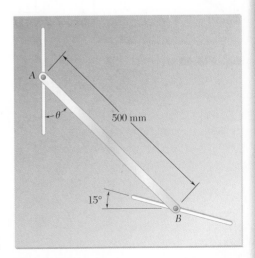

Fig. P15.39

15.40 Collar B moves upward with a constant velocity of 1.5 m/s. At the instant when $\theta = 50°$, determine (*a*) the angular velocity of rod AB, (*b*) the velocity of end A of the rod.

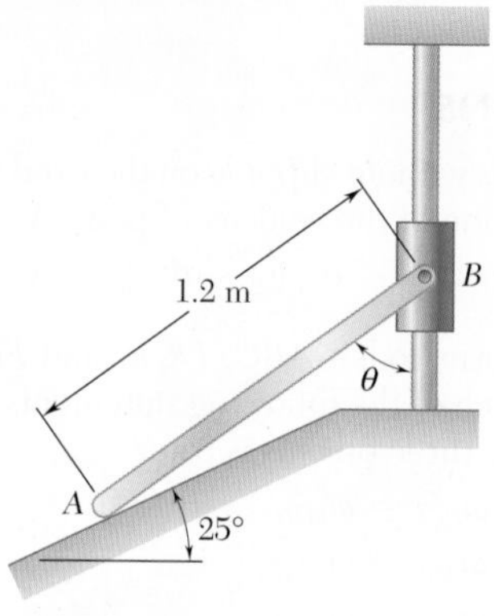

Fig. P15.40

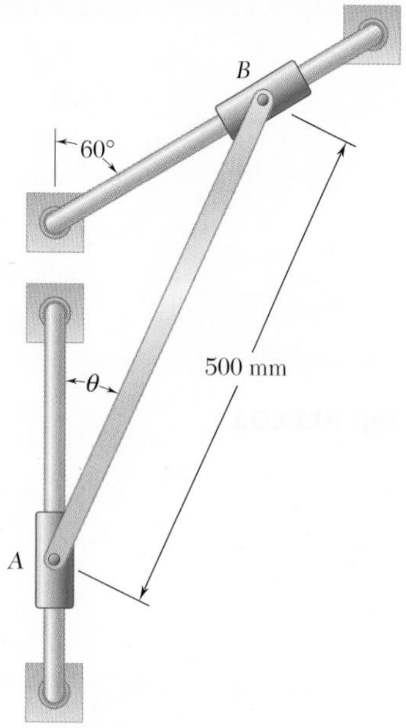

Fig. P15.41 and P15.42

15.41 Collar B moves downward to the left with a constant velocity of 1.6 m/s. At the instant shown when $\theta = 40°$, determine (*a*) the angular velocity of rod AB, (*b*) the velocity of collar A.

15.42 Collar A moves upward with a constant velocity of 1.2 m/s. At the instant shown when $\theta = 25°$, determine (*a*) the angular velocity of rod AB, (*b*) the velocity of collar B.

15.43 Rod AB moves over a small wheel at C while end A moves to the right with a constant velocity of 500 mm/s. At the instant shown, determine (*a*) the angular velocity of the rod, (*b*) the velocity of end B of the rod.

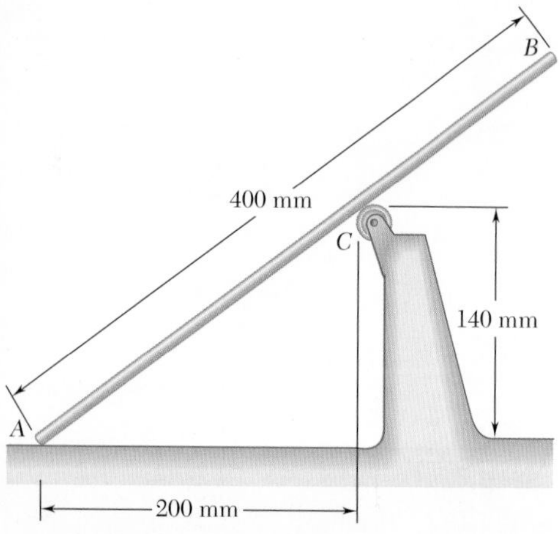

Fig. P15.43

15.44 The plate shown moves in the xy plane. Knowing that $(v_A)_x = 120$ mm/s, $(v_B)_y = 300$ mm/s, and $(v_C)_y = -60$ mm/s, determine (*a*) the angular velocity of the plate, (*b*) the velocity of point A.

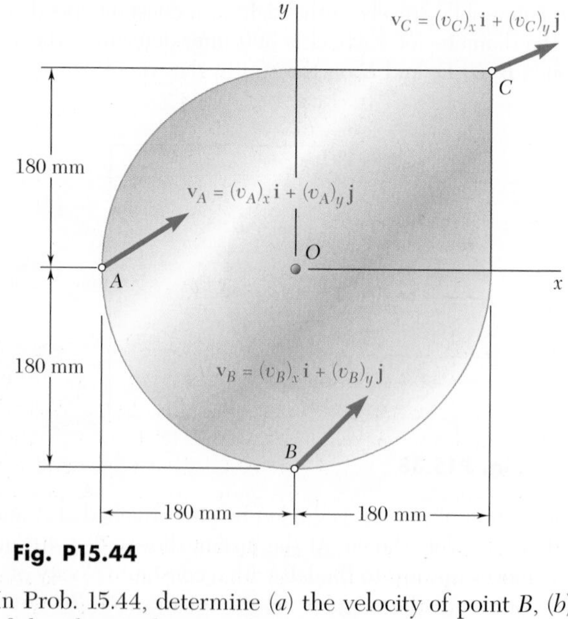

Fig. P15.44

15.45 In Prob. 15.44, determine (*a*) the velocity of point B, (*b*) the point of the plate with zero velocity.

15.46 The plate shown moves in the xy plane. Knowing that $(v_A)_x = 250$ mm/s, $(v_B)_y = -450$ mm/s, and $(v_C)_x = -500$ mm/s, determine (*a*) the angular velocity of the plate, (*b*) the velocity of point A.

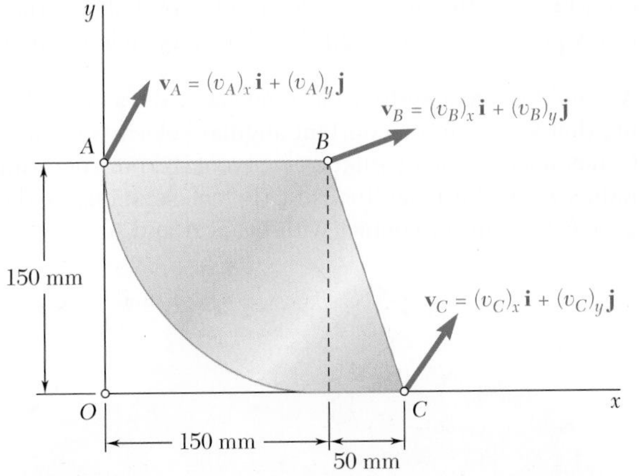

Fig. P15.46

15.47 The plate shown moves in the xy plane. Knowing that $(v_A)_x = 300$ mm/s, $(v_B)_x = -100$ mm/s, and $(v_C)_y = -600$ mm/s, determine (*a*) the angular velocity of the plate, (*b*) the velocity of point B.

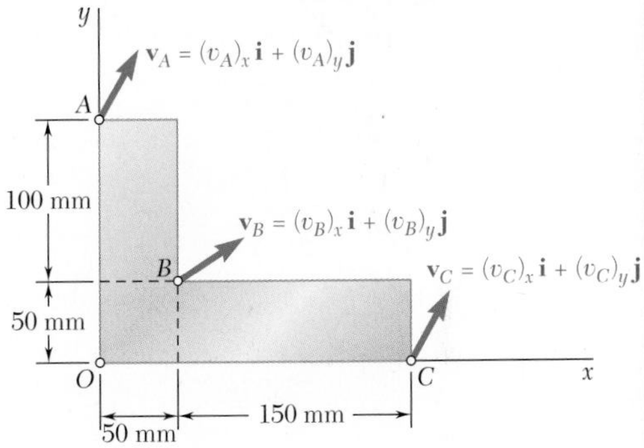

Fig. P15.47

15.48 In the planetary gear system shown, the radius of gears A, B, C, and D is a and the radius of the outer gear E is $3a$. Knowing that the angular velocity of gear A is ω_A clockwise and that the outer gear E is stationary, determine (*a*) the angular velocity of each planetary gear, (*b*) the angular velocity of the spider connecting the planetary gears.

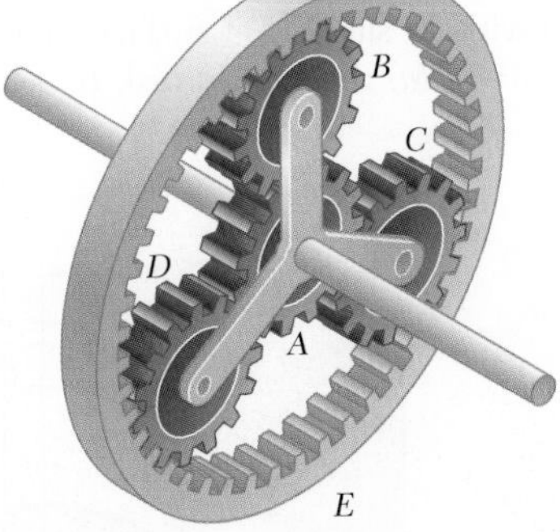

Fig. P15.48 and P15.49

15.49 In the planetary gear system shown, the radius of gears A, B, C, and D is 30 mm and the radius of the outer gear E is 90 mm. Knowing that gear E has an angular velocity of 180 rpm clockwise and that the central gear A has an angular velocity of 240 rpm clockwise, determine (*a*) the angular velocity of each planetary gear, (*b*) the angular velocity of the spider connecting the planetary gears.

15.50 Arm AB rotates with an angular velocity of 20 rad/s counterclockwise. Knowing that the outer gear C is stationary, determine (*a*) the angular velocity of gear B, (*b*) the velocity of the gear tooth located at point D.

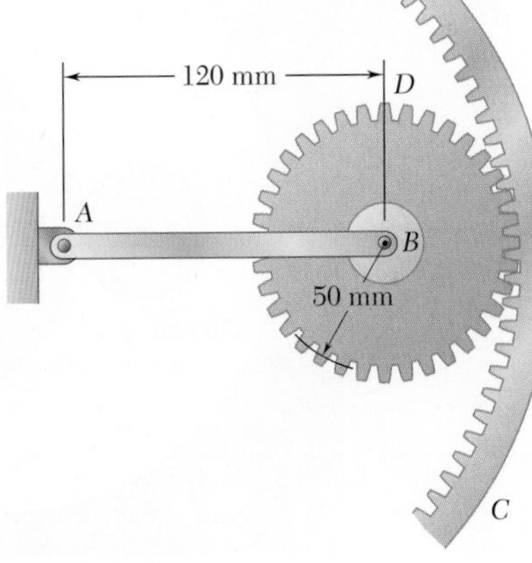

Fig. P15.50

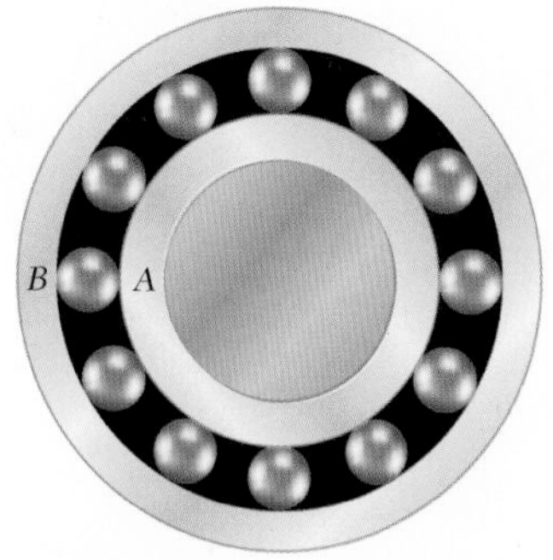

Fig. P15.51

15.51 In the simplified sketch of a ball bearing shown, the diameter of the inner race A is 60 mm and the diameter of each ball is 12 mm. The outer race B is stationary while the inner race has an angular velocity of 3600 rpm. Determine (*a*) the speed of the center of each ball, (*b*) the angular velocity of each ball, (*c*) the number of times per minute each ball describes a complete circle.

15.52 A simplified gear system for a mechanical watch is shown. Knowing that gear A has a constant angular velocity of 1 rev/h and gear C has a constant angular velocity of 1 rpm, determine (*a*) the radius r, (*b*) the magnitudes of the accelerations of the points on gear B that are in contact with gears A and C.

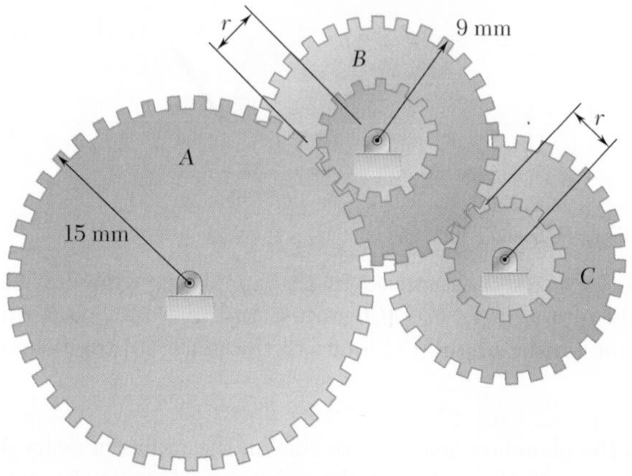

Fig. *P15.52*

15.53 and *15.54* Arm ACB rotates about point C with an angular velocity of 40 rad/s counterclockwise. Two friction disks A and B are pinned at their centers to arm ACB as shown. Knowing that the disks roll without slipping at surfaces of contact, determine the angular velocity of (*a*) disk A, (*b*) disk B.

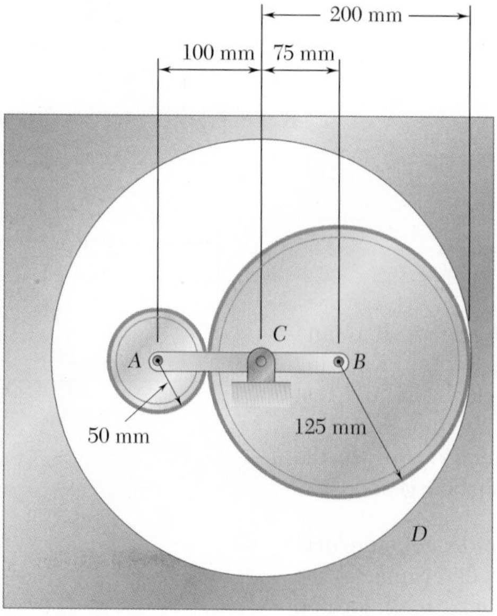

Fig. P15.53

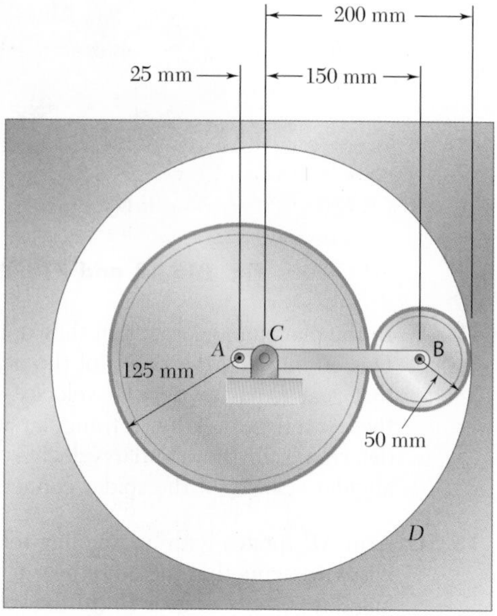

Fig. *P15.54*

15.55 Knowing that at the instant shown the velocity of collar A is 900 mm/s to the left, determine (*a*) the angular velocity of rod ADB, (*b*) the velocity of point B.

15.56 Knowing that at the instant shown the angular velocity of rod DE is 2.4 rad/s clockwise, determine (*a*) the velocity of collar A, (*b*) the velocity of point B.

15.57 A straight rack rests on a gear of radius r and is attached to a block B as shown. Denoting by ω_D the clockwise angular velocity of gear D and by θ the angle formed by the rack and the horizontal, derive expressions for the velocity of block B and the angular velocity of the rack in terms of r, θ, and ω_D.

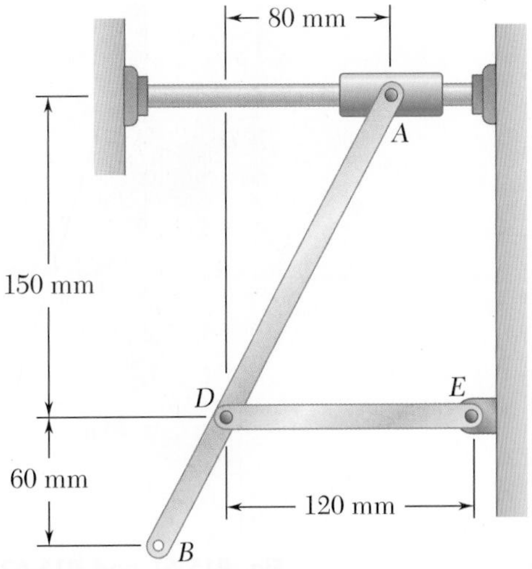

Fig. P15.55 and P15.56

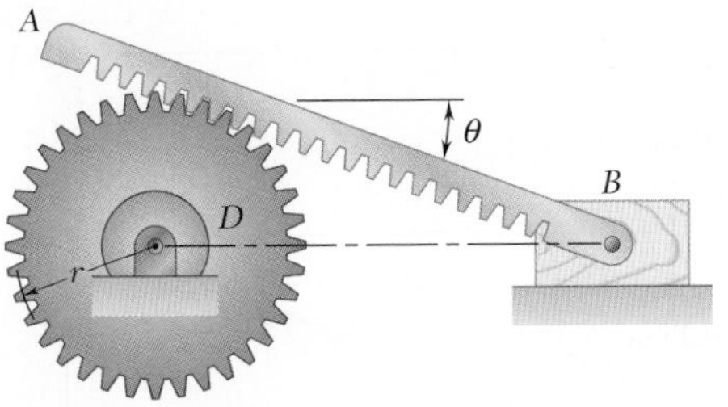

Fig. P15.57 and P15.58

15.58 A straight rack rests on a gear of radius $r = 60$ mm and is attached to a block B as shown. Knowing that at the instant shown the velocity of block B is 200 mm/s to the right and $\theta = 25°$, determine (*a*) the angular velocity of gear D, (*b*) the angular velocity of the rack.

15.59 Knowing that at the instant shown the angular velocity of crank AB is 2.7 rad/s clockwise, determine (*a*) the angular velocity of link BD, (*b*) the velocity of collar D, (*c*) the velocity of the midpoint of link BD.

15.60 In the eccentric shown, a disk of 40-mm-radius revolves about shaft O that is located 10 mm from the center A of the disk. The distance between the center A of the disk and the pin at B is 160 mm. Knowing that the angular velocity of the disk is 900 rpm clockwise, determine the velocity of the block when $\theta = 30°$.

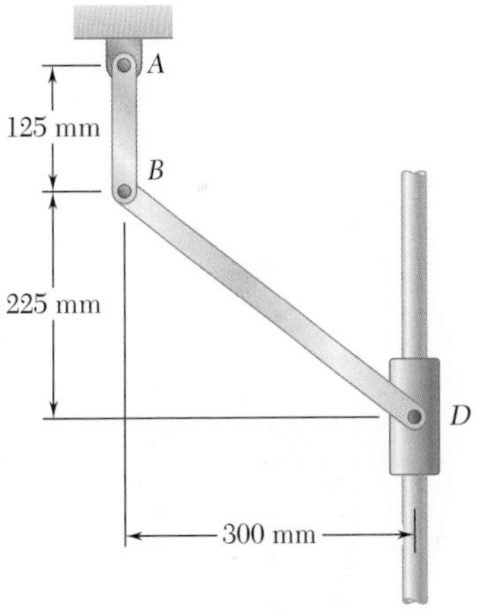

Fig. P15.59

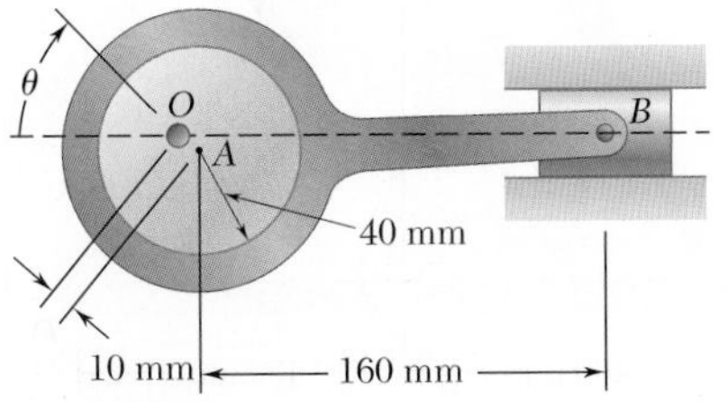

Fig. *P15.60*

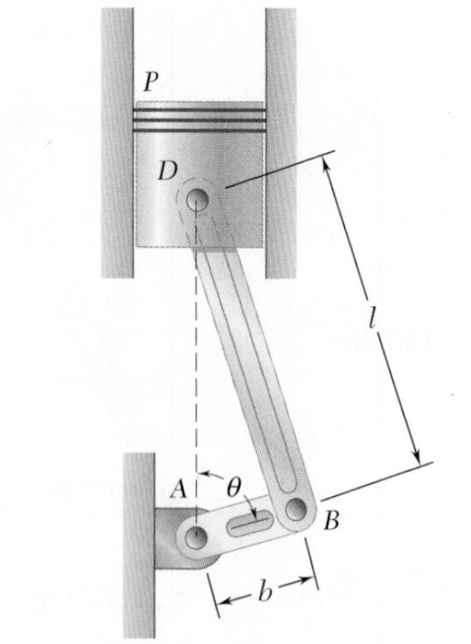

Fig. P15.61 and P15.62

15.61 In the engine system shown, $l = 160$ mm and $b = 60$ mm. Knowing that the crank AB rotates with a constant angular velocity of 1000 rpm clockwise, determine the velocity of the piston P and the angular velocity of the connecting rod when (*a*) $\theta = 0$, (*b*) $\theta = 90°$.

15.62 In the engine system shown, $l = 160$ mm and $b = 60$ mm. Knowing that crank AB rotates with a constant angular velocity of 1000 rpm clockwise, determine the velocity of the piston P and the angular velocity of the connecting rod when $\theta = 60°$.

15.63 Knowing that at the instant shown the angular velocity of rod AB is 15 rad/s clockwise, determine (*a*) the angular velocity of rod BD, (*b*) the velocity of the midpoint of rod BD.

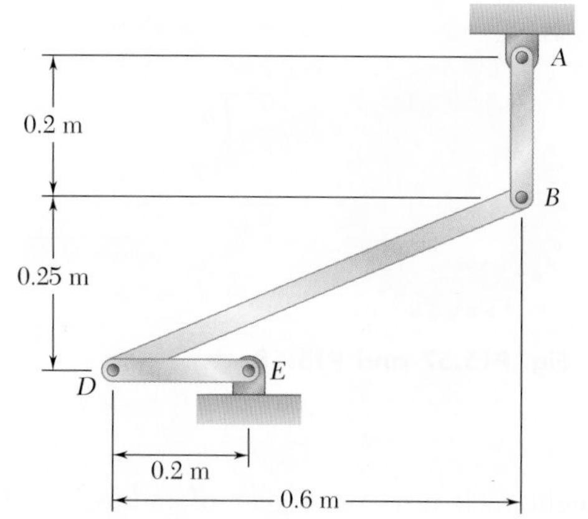

Fig. P15.63

15.64 and 15.65 In the position shown, bar AB has an angular velocity of 4 rad/s clockwise. Determine the angular velocity of bars BD and DE.

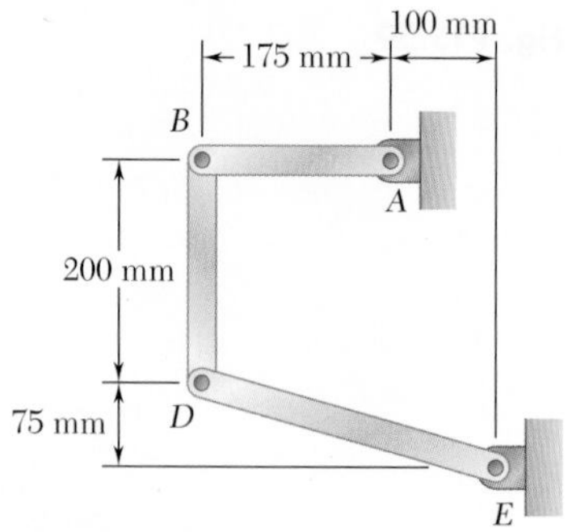

Fig. P15.64

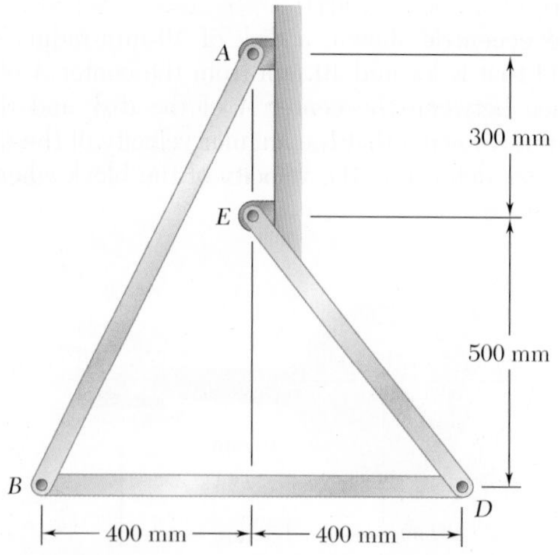

Fig. P15.65

15.66 Roberts linkage is named after Richard Roberts (1789–1864) and can be used to draw a close approximation to a straight line by locating a pen at point *F*. The distance *AB* is the same as *BF*, *DF*, and *DE*. Knowing that the angular velocity of bar *AB* is 5 rad/s clockwise in the position shown, determine (*a*) the angular velocity of bar *DE*, (*b*) the velocity of point *F*.

15.67 Roberts linkage is named after Richard Roberts (1789–1864) and can be used to draw a close approximation to a straight line by locating a pen at point *F*. The distance *AB* is the same as *BF*, *DF*, and *DE*. Knowing that the angular velocity of plate *BDF* is 2 rad/s counterclockwise when $\theta = 90°$, determine (*a*) the angular velocities of bars *AB* and *DE*, (*b*) the velocity of point *F*. When $\theta = 90°$, point *F* may be assumed to coincide with point *E*, with negligible error in the velocity analysis.

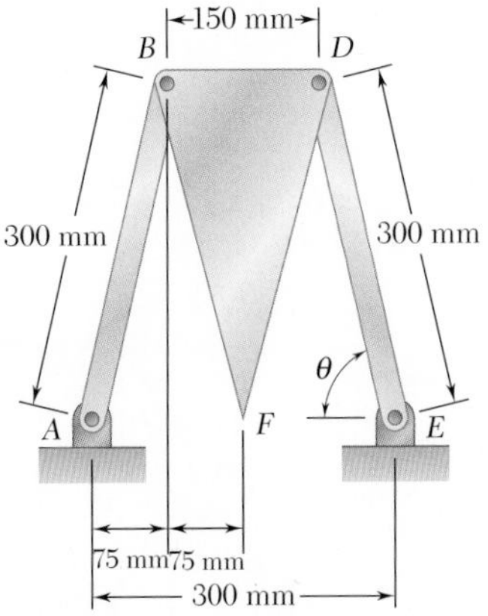

Fig. P15.66 and P15.67

15.68 In the position shown, bar *DE* has a constant angular velocity of 10 rad/s clockwise. Knowing that $h = 500$ mm, determine (*a*) the angular velocity of bar *FBD*, (*b*) the velocity of point *F*.

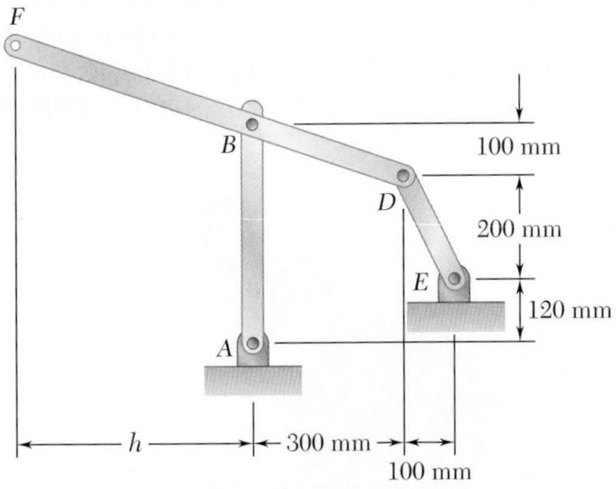

Fig. P15.68 and P15.69

15.69 In the position shown, bar *DE* has a constant angular velocity of 10 rad/s clockwise. Determine (*a*) the distance *h* for which the velocity of point *F* is vertical, (*b*) the corresponding velocity of point *F*.

15.70 Both 150-mm.-radius wheels roll without slipping on the horizontal surface. Knowing that the distance *AD* is 125 mm., the distance *BE* is 100 mm., and *D* has a velocity of 150 mm/s to the right, determine the velocity of point *E*.

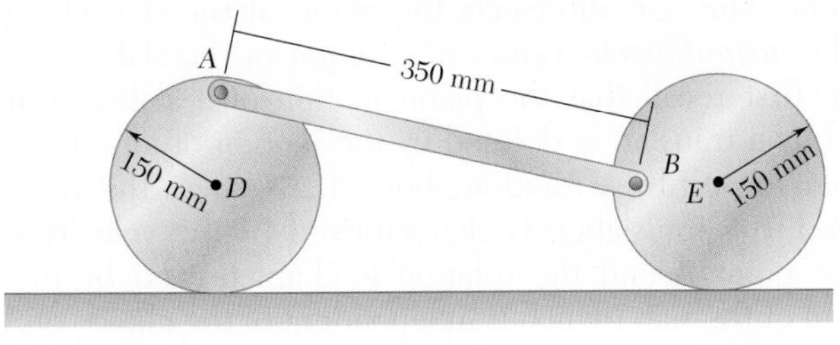

Fig. P15.70

15.71 The 80-mm-radius wheel shown rolls to the left with a velocity of 900 mm/s. Knowing that the distance AD is 50 mm, determine the velocity of the collar and the angular velocity of rod AB when (*a*) $\beta = 0$, (*b*) $\beta = 90°$.

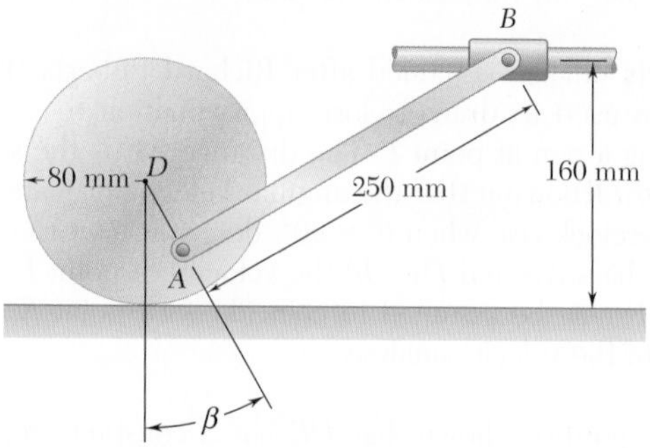

Fig. P15.71

***15.72** For the gearing shown, derive an expression for the angular velocity ω_C of gear C and show that ω_C is independent of the radius of gear B. Assume that point A is fixed and denote the angular velocities of rod ABC and gear A by ω_{ABC} and ω_A, respectively.

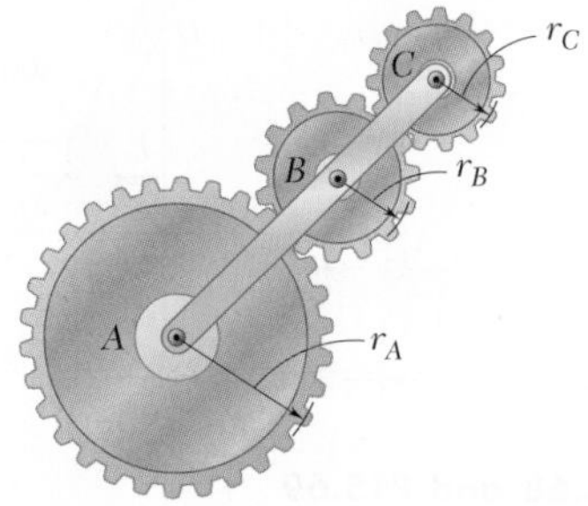

Fig. P15.72

15.7 INSTANTANEOUS CENTER OF ROTATION IN PLANE MOTION

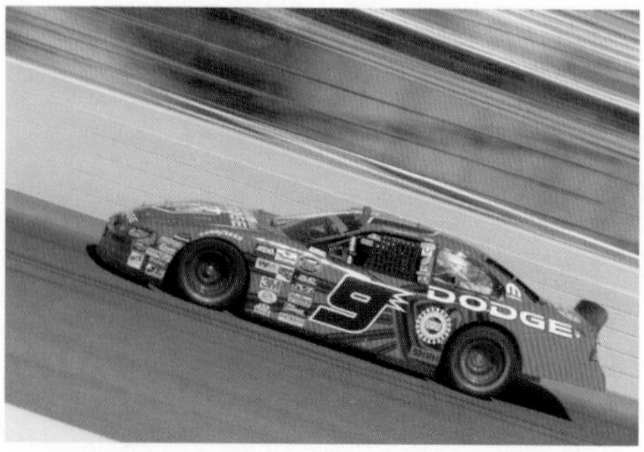

Photo 15.5 If the tires of this car are rolling without sliding, the instantaneous center of rotation of a tire is the point of contact between the road and the tire.

Consider the general plane motion of a slab. We propose to show that at any given instant the velocities of the various particles of the slab are the same as if the slab were rotating about a certain axis perpendicular to the plane of the slab, called the *instantaneous axis of rotation*. This axis intersects the plane of the slab at a point C, called the *instantaneous center of rotation* of the slab.

We first recall that the plane motion of a slab can always be replaced by a translation defined by the motion of an arbitrary reference point A and by a rotation about A. As far as the velocities are concerned, the translation is characterized by the velocity $\mathbf{v}_A$ of the reference point A and the rotation is characterized by the angular velocity $\boldsymbol{\omega}$ of the slab (which is independent of the choice of A). Thus, the velocity $\mathbf{v}_A$ of point A and the angular velocity $\boldsymbol{\omega}$ of the slab define

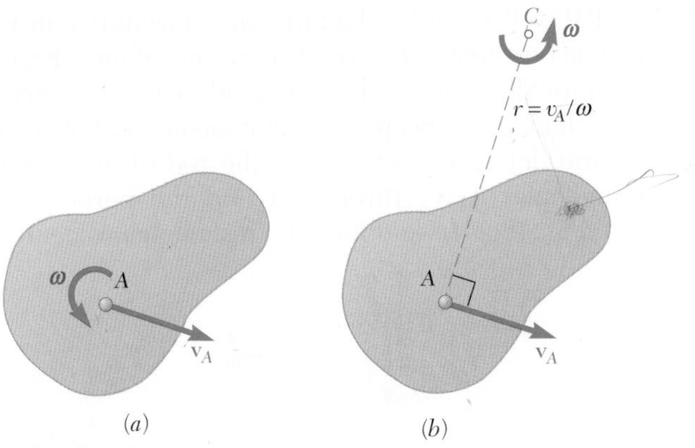

Fig. 15.18

completely the velocities of all the other particles of the slab (Fig. 15.18*a*). Now let us assume that $\mathbf{v}_A$ and $\boldsymbol{\omega}$ are known and that they are both different from zero. (If $\mathbf{v}_A = 0$, point A is itself the instantaneous center of rotation, and if $\boldsymbol{\omega} = 0$, all the particles have the same velocity $\mathbf{v}_A$.) These velocities could be obtained by letting the slab rotate with the angular velocity $\boldsymbol{\omega}$ about a point C located on the perpendicular to $\mathbf{v}_A$ at a distance $r = v_A/\omega$ from A as shown in Fig. 15.18*b*. We check that the velocity of A would be perpendicular to AC and that its magnitude would be $r\omega = (v_A/\omega)\omega = v_A$. Thus the velocities of all the other particles of the slab would be the same as originally defined. Therefore, *as far as the velocities are concerned, the slab seems to rotate about the instantaneous center* C at the instant considered.

The position of the instantaneous center can be defined in two other ways. If the directions of the velocities of two particles A and B of the slab are known and if they are different, the instantaneous center C is obtained by drawing the perpendicular to $\mathbf{v}_A$ through A and the perpendicular to $\mathbf{v}_B$ through B and determining the point in which these two lines intersect (Fig. 15.19*a*). If the velocities $\mathbf{v}_A$ and $\mathbf{v}_B$ of two particles A and B are perpendicular to the line AB and if their magnitudes are known, the instantaneous center can be found by intersecting the line AB with the line joining the extremities of the vectors $\mathbf{v}_A$ and $\mathbf{v}_B$ (Fig. 15.19*b*). Note that if $\mathbf{v}_A$ and $\mathbf{v}_B$ were parallel

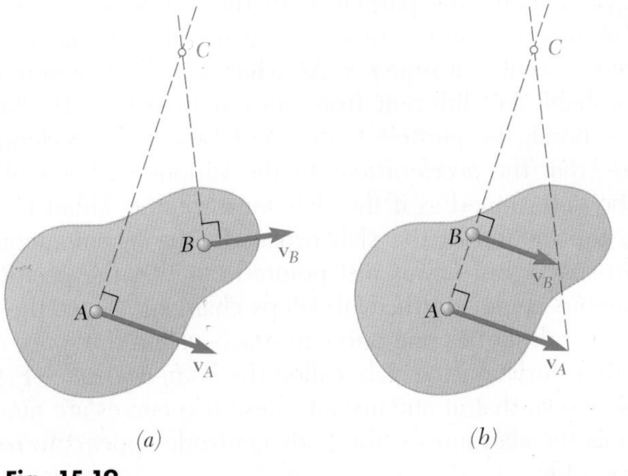

Fig. 15.19

in Fig. 15.19*a* or if $\mathbf{v}_A$ and $\mathbf{v}_B$ had the same magnitude in Fig. 15.19*b*, the instantaneous center C would be at an infinite distance and $\boldsymbol{\omega}$ would be zero; all points of the slab would have the same velocity.

To see how the concept of instantaneous center of rotation can be put to use, let us consider again the rod of Sec. 15.6. Drawing the perpendicular to $\mathbf{v}_A$ through A and the perpendicular to $\mathbf{v}_B$ through B (Fig. 15.20), we obtain the instantaneous center C. At the

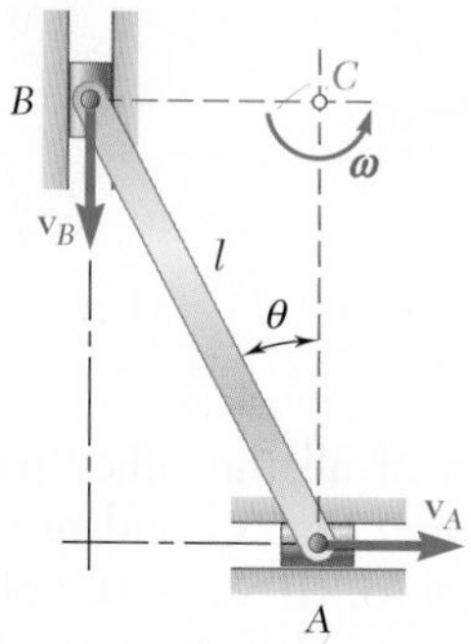

Fig. 15.20

instant considered, the velocities of all the particles of the rod are thus the same as if the rod rotated about C. Now, if the magnitude v_A of the velocity of A is known, the magnitude ω of the angular velocity of the rod can be obtained by writing

$$\omega = \frac{v_A}{AC} = \frac{v_A}{l \cos \theta}$$

The magnitude of the velocity of B can then be obtained by writing

$$v_B = (BC)\omega = l \sin \theta \frac{v_A}{l \cos \theta} = v_A \tan \theta$$

Note that only *absolute* velocities are involved in the computation.

The instantaneous center of a slab in plane motion can be located either on the slab or outside the slab. If it is located on the slab, the particle C coinciding with the instantaneous center at a given instant t must have zero velocity at that instant. However, it should be noted that the instantaneous center of rotation is valid only at a given instant. Thus, the particle C of the slab which coincides with the instantaneous center at time t will generally not coincide with the instantaneous center at time $t + \Delta t$; while its velocity is zero at time t, it will probably be different from zero at time $t + \Delta t$. This means that, in general, the particle C *does not have zero acceleration* and, therefore, that the *accelerations* of the various particles of the slab *cannot* be determined as if the slab were rotating about C.

As the motion of the slab proceeds, the instantaneous center moves in space. But it was just pointed out that the position of the instantaneous center on the slab keeps changing. Thus, the instantaneous center describes one curve in space, called the *space centrode*, and another curve on the slab, called the *body centrode* (Fig. 15.21). It can be shown that at any instant, these two curves are tangent at C and that as the slab moves, the body centrode appears to *roll* on the space centrode.

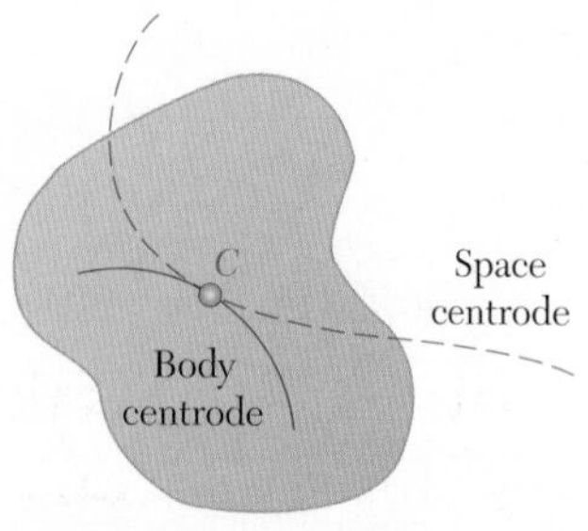

Fig. 15.21

SAMPLE PROBLEM 15.4

Solve Sample Prob. 15.2, using the method of the instantaneous center of rotation.

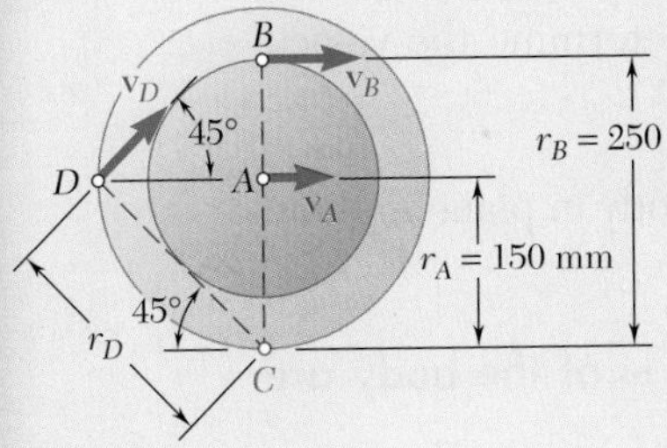

SOLUTION

a. Angular Velocity of the Gear. Since the gear rolls on the stationary lower rack, the point of contact C of the gear with the rack has no velocity; point C is therefore the instantaneous center of rotation. We write

$$v_A = r_A\omega \qquad 1.2 \text{ m/s} = (0.150 \text{ m})\omega$$

$$\boldsymbol{\omega} = 8 \text{ rad/s} \circlearrowright \blacktriangleleft$$

b. Velocities. As far as velocities are concerned, all points of the gear seem to rotate about the instantaneous center.

Velocity of Upper Rack. Recalling that $v_R = v_B$, we write

$$v_R = v_B = r_B\omega \qquad v_R = (0.250 \text{ m})(8 \text{ rad/s}) = 2 \text{ m/s}$$

$$\mathbf{v}_R = 2 \text{ m/s} \rightarrow \blacktriangleleft$$

Velocity of Point D. Since $r_D = (0.150 \text{ m})\sqrt{2} = 0.2121$ m, we write

$$v_D = r_D\omega \qquad v_D = (0.2121 \text{ m})(8 \text{ rad/s}) = 1.697 \text{ m/s}$$

$$\mathbf{v}_D = 1.697 \text{ m/s} \angle 45° \blacktriangleleft$$

SAMPLE PROBLEM 15.5

Solve Sample Prob. 15.3 using the method of the instantaneous center of rotation.

SOLUTION

Given the geometry of rod AB and the velocity of A = 1500 mm/s to the left ←. For the rigid body AB, the direction of velocity of A and velocity of B are known. So we draw perpendiculars to these directions which intersect C, the instantaneous centre. Since the magnitude of $\mathbf{V}_A$ is also known, we can find ω as the rigid body AB at this instant is moving in a circular motion about C.

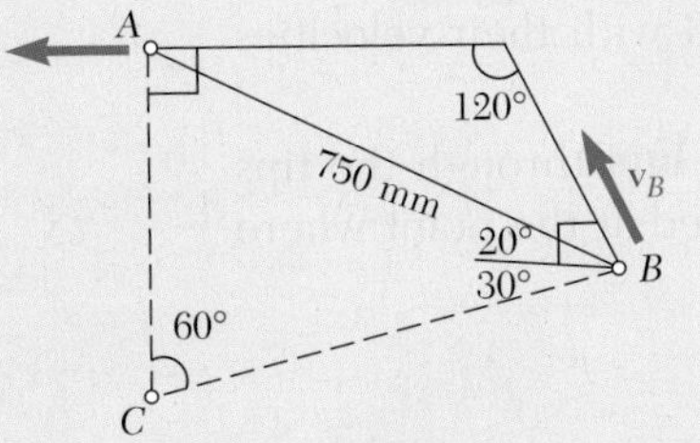

From the given geometry we observe that $\angle ACB = 60°$, and $\angle ABC = 20° + 30° = 50°$. Thus $\angle BAC = 70°$ using law of sines for $\Delta\, ABC$

$$\frac{AC}{\sin 50°} = \frac{750}{\sin 60°} = \frac{BC}{\sin 70°}$$

$$AC = 663.4 \text{ mm}$$

$$BC = 813.8 \text{ mm}$$

$$\mathbf{V}_A = AC(\omega_{AB}) \Rightarrow \omega_{AB} = \frac{\mathbf{V}_A}{AC} = \frac{1500}{663.4 \text{ mm}} = 2.26 \text{ rad/s} \circlearrowleft$$

$$V_B = BC(\omega_{AB}) = (813.8 \text{ mm})(2.26 \text{ rad/s}) = 1840 \frac{\text{mm}}{\text{s}}$$

$$= 1.84 \text{ m/s } 60° \blacktriangleleft$$

SOLVING PROBLEMS ON YOUR OWN

In this lesson we introduced the *instantaneous center of rotation* in plane motion. This provides us with an alternative way for solving problems involving the *velocities* of the various points of a body in plane motion.

As its name suggests, the *instantaneous center of rotation* is the point about which you can assume a body is rotating at a given instant, as you determine the velocities of the points of the body at that instant.

A. To determine the instantaneous center of rotation of a body in plane motion, you should use one of the following procedures.

1. If the velocity $\mathbf{v}_A$ of a point *A* and the angular velocity ω of the body are both known (Fig. 15.18):

a. Draw a sketch of the body, showing point *A*, its velocity $\mathbf{v}_A$, and the angular velocity $\boldsymbol{\omega}$ of the body.

b. From *A* draw a line perpendicular to $\mathbf{v}_A$ on the side of $\mathbf{v}_A$ from which this velocity is viewed as having *the same sense as* $\boldsymbol{\omega}$.

c. Locate the instantaneous center *C* on this line, at a distance $r = v_A/\omega$ from point *A*.

2. If the directions of the velocities of two points *A* and *B* are known and are different (Fig. 15.19*a*):

a. Draw a sketch of the body, showing points *A* and *B* and their velocities $\mathbf{v}_A$ and $\mathbf{v}_B$.

b. From *A* and *B* draw lines perpendicular to $\mathbf{v}_A$ and $\mathbf{v}_B$, respectively. The instantaneous center *C* is located at the point where the two lines intersect.

c. If the velocity of one of the two points is known, you can determine the angular velocity of the body. For example, if you know $\boldsymbol{v}_A$, you can write $\omega = v_A/AC$, where *AC* is the distance from point *A* to the instantaneous center *C*.

3. If the velocities of two points *A* and *B* are known and are both perpendicular to the line *AB* (Fig. 15.19*b*):

a. Draw a sketch of the body, showing points *A* and *B* with their velocities $\mathbf{v}_A$ and $\mathbf{v}_B$ *drawn to scale.*

b. Draw a line through points *A* and *B*, and another line through the tips of the vectors $\mathbf{v}_A$ and $\mathbf{v}_B$. The instantaneous center *C* is located at the point where the two lines intersect.

c. The angular velocity of the body is obtained by either dividing $\mathbf{v}_A$ by AC or $\mathbf{v}_B$ by BC.

d. If the velocities $\mathbf{v}_A$ and $\mathbf{v}_B$ have the same magnitude, the two lines drawn in part *b* do not intersect; the instantaneous center C is at an infinite distance. The angular velocity $\boldsymbol{\omega}$ is zero and *the body is in translation.*

B. Once you have determined the instantaneous center and the angular velocity of a body, you can determine the velocity $\mathbf{v}_P$ of any point P of the body in the following way.

1. Draw a sketch of the body, showing point P, the instantaneous center of rotation C, and the angular velocity $\boldsymbol{\omega}$.

2. Draw a line from P to the instantaneous center C and measure or calculate the distance from P to C.

3. The velocity $\mathbf{v}_P$ is a vector perpendicular to the line PC, of the same sense as $\boldsymbol{\omega}$, and of magnitude $v_P = (PC)\omega$.

Finally, keep in mind that the instantaneous center of rotation can be used *only* to determine velocities. *It cannot be used to determine accelerations.*

PROBLEMS

CONCEPT QUESTIONS

15.CQ5 The disk rolls without sliding on the fixed horizontal surface. At the instant shown, the instantaneous center of zero velocity for rod AB would be located in which region?

a. Region 1
b. Region 2
c. Region 3
d. Region 4
e. Region 5
f. Region 6

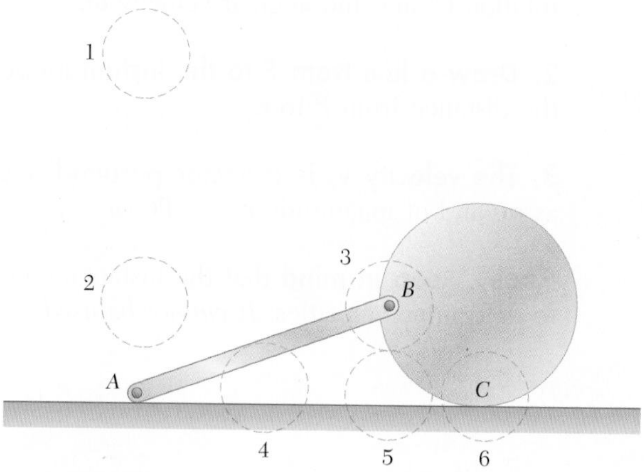

Fig. P15.CQ5

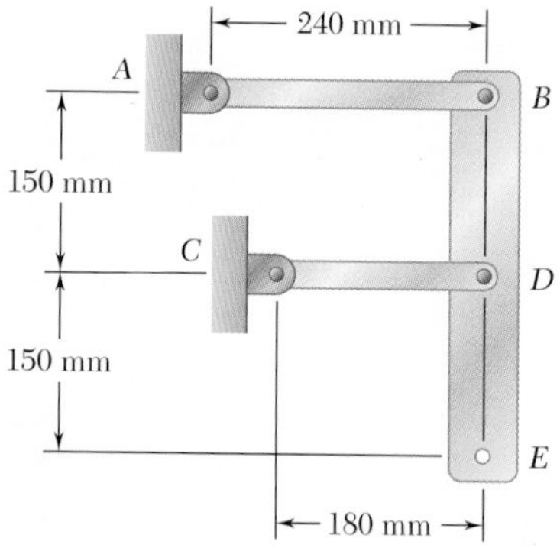

Fig. P15.CQ6

15.CQ6 Bar BDE is pinned to two links, AB and CD. At the instant shown, the angular velocities of link AB, link CD, and bar BDE are ω_{AB}, ω_{CD}, and ω_{BDE}, respectively. Which of the following statements concerning the angular speeds of the three objects is true at this instant?

a. $\omega_{AB} = \omega_{CD} = \omega_{BDE}$
b. $\omega_{BDE} > \omega_{AB} > \omega_{CD}$
c. $\omega_{AB} = \omega_{CD} > \omega_{BDE}$
d. $\omega_{AB} > \omega_{CD} > \omega_{BDE}$
e. $\omega_{CD} > \omega_{AB} > \omega_{BDE}$

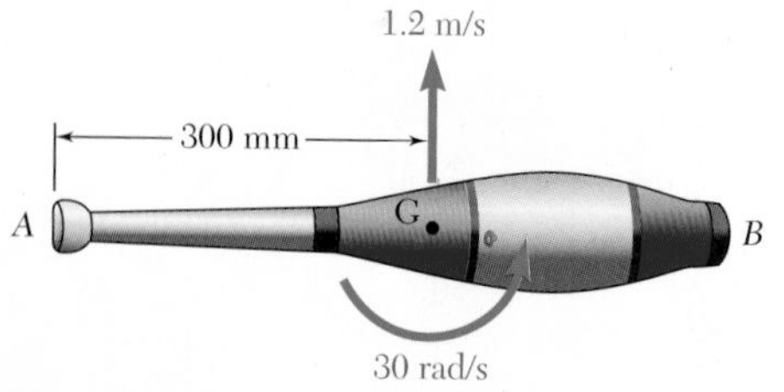

Fig. *P15.73*

END-OF-SECTION PROBLEMS

15.73 A juggling club is thrown vertically into the air. The center of gravity G of the 500-mm. club is located 300-mm. from the knob. Knowing that at the instant shown, G has a velocity of 1.2 m/s upwards and the club has an angular velocity of 30 rad/s counterclockwise, determine (a) the speeds of points A and B, (b) the location of the instantaneous center of rotation.

15.74 A 2.5 m beam AE is being lowered by means of two overhead cranes. At the instant shown it is known that the velocity of point D is 600 mm/s downward and the velocity of point E is 900 mm/s downward. Determine (*a*) the instantaneous center of rotation of the beam, (*b*) the velocity of point A.

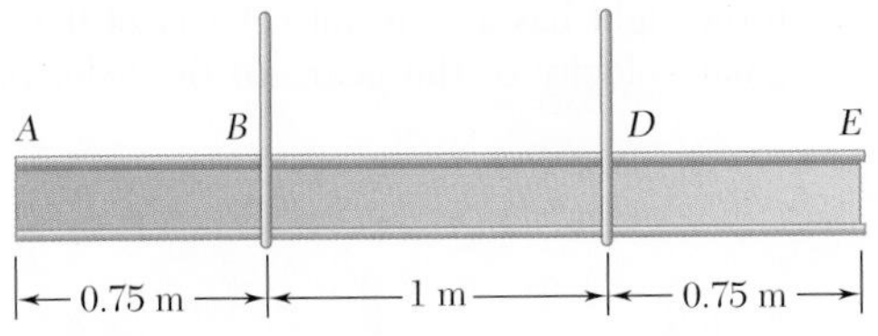

Fig. P15.74

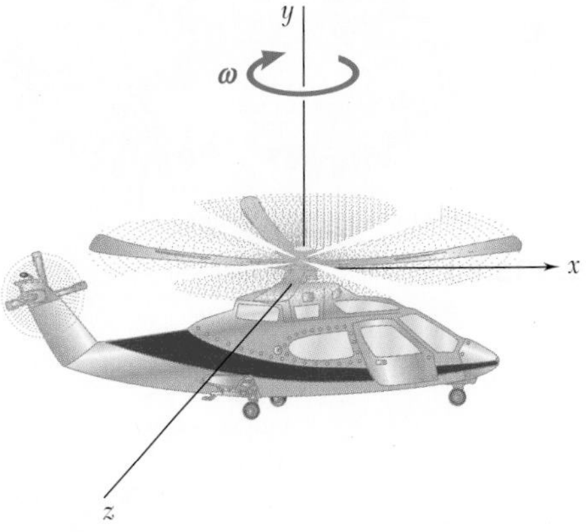

Fig. P15.75

15.75 A helicopter moves horizontally in the x direction at a speed of 200 km/h. Knowing that the main blades rotate clockwise with an angular velocity of 180 rpm, determine the instantaneous axis of rotation of the main blades.

15.76 and 15.77 A 60-mm-radius drum is rigidly attached to a 100-mm-radius drum as shown. One of the drums rolls without sliding on the surface shown, and a cord is wound around the other drum. Knowing that end E of the cord is pulled to the left with a velocity of 120 mm/s, determine (*a*) the angular velocity of the drums, (*b*) the velocity of the center of the drums, (*c*) the length of cord wound or unwound per second.

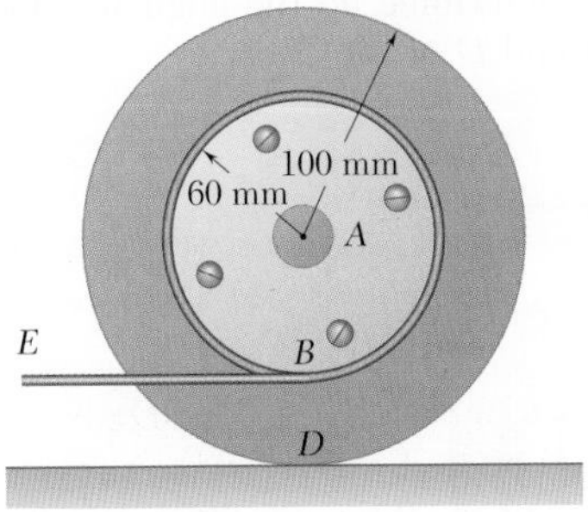

Fig. P15.76

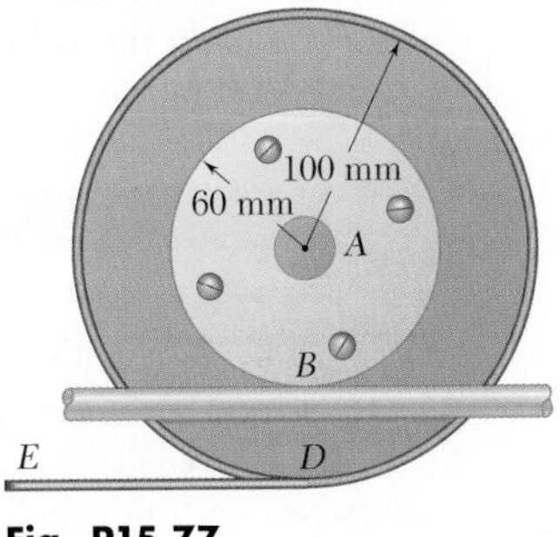

Fig. P15.77

15.78 The spool of tape shown and its frame assembly are pulled upward at a speed v_A = 750 mm/s. Knowing that the 80-mm-radius spool has an angular velocity of 15 rad/s clockwise and that at the instant shown the total thickness of the tape on the spool is 20 mm, determine (*a*) the instantaneous center of rotation of the spool, (*b*) the velocities of points B and D.

15.79 The spool of tape shown and its frame assembly are pulled upward at a speed v_A = 100 mm/s. Knowing that end B of the tape is pulled downward with a velocity of 300 mm/s and that at the instant shown the total thickness of the tape on the spool is 20 mm, determine (*a*) the instantaneous center of rotation of the spool, (*b*) the velocity of point D of the spool.

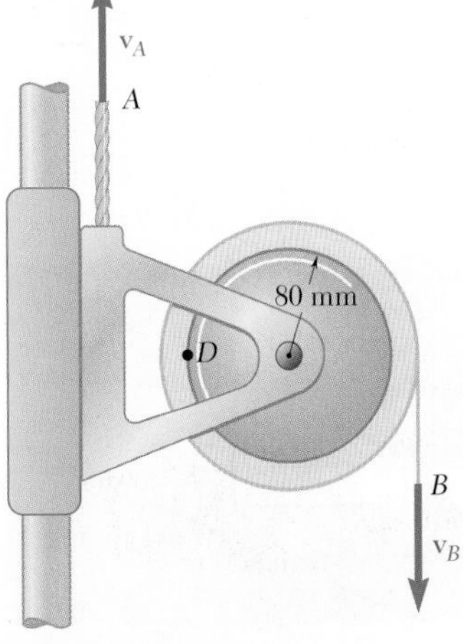

Fig. P15.78 and P15.79

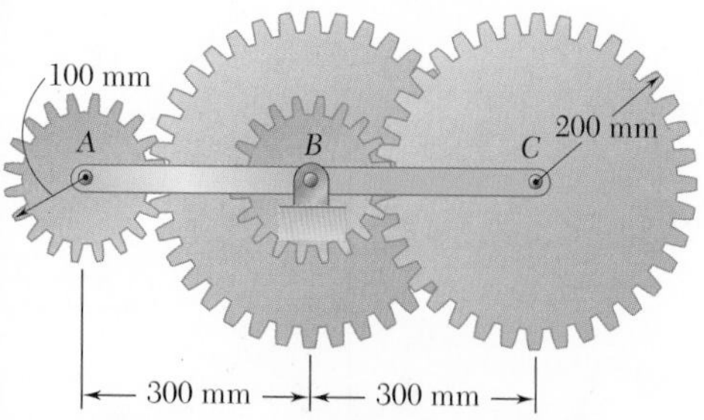

Fig. P15.80

15.80 The arm *ABC* rotates with an angular velocity of 4 rad/s counterclockwise. Knowing that the angular velocity of the intermediate gear *B* is 8 rad/s counterclockwise, determine (*a*) the instantaneous centers of rotation of gears *A* and *C*, (*b*) the angular velocities of gears *A* and *C*.

15.81 The double gear rolls on the stationary left rack *R*. Knowing that the rack on the right has a constant velocity of 0.6 m/s, determine (*a*) the angular velocity of the gear, (*b*) the velocities of points *A* and *D*.

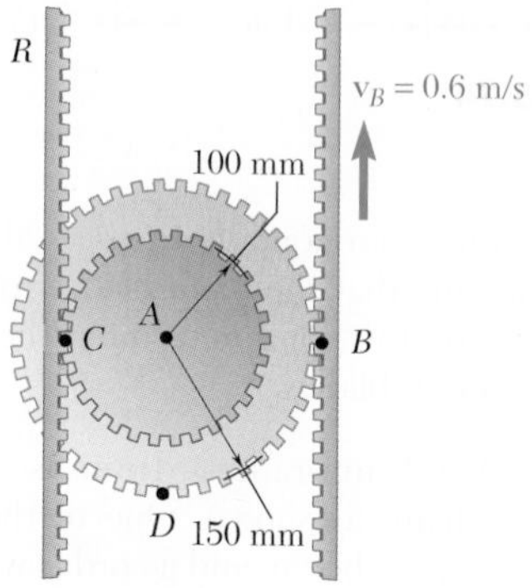

Fig. P15.81

15.82 An overhead door is guided by wheels at *A* and *B* that roll in horizontal and vertical tracks. Knowing that when $\theta = 40°$ the velocity of wheel *B* is 0.5 m/s upward, determine (*a*) the angular velocity of the door, (*b*) the velocity of end *D* of the door.

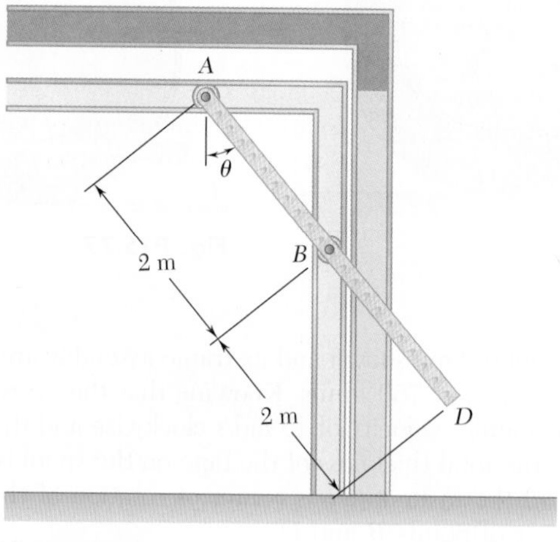

Fig. P15.82

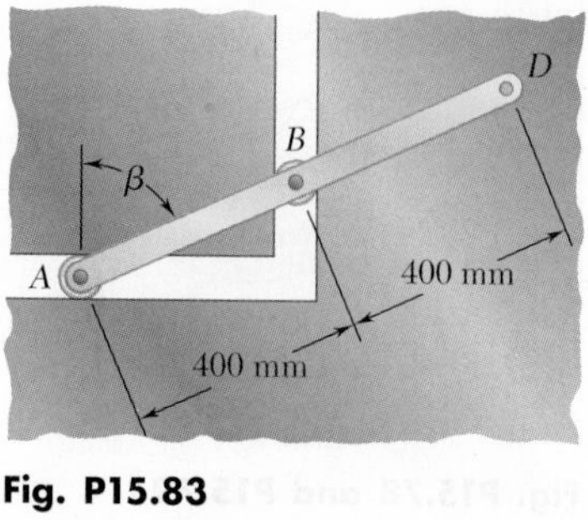

Fig. P15.83

15.83 Rod *ABD* is guided by wheels at *A* and *B* that roll in horizontal and vertical tracks. Knowing that at the instant $\beta = 60°$ and the velocity of wheel *B* is 1 m/s downward, determine (*a*) the angular velocity of the rod, (*b*) the velocity of point *D*.

15.84 Rod BDE is partially guided by a roller at D which moves in a vertical track. Knowing that at the instant shown the angular velocity of crank AB is 5 rad/s clockwise and that $\beta = 25°$, determine (*a*) the angular velocity of the rod, (*b*) the velocity of point E.

15.85 Rod BDE is partially guided by a roller at D which moves in a vertical track. Knowing that at the instant shown $\beta = 30°$, point E has a velocity of 2 m/s down and to the right, determine the angular velocities of rod BDE and crank AB.

15.86 Knowing that at the instant shown, the velocity of collar D is 1.6 m/s upward, determine (*a*) the angular velocity of rod AD, (*b*) the velocity of point B, (*c*) the velocity of point A.

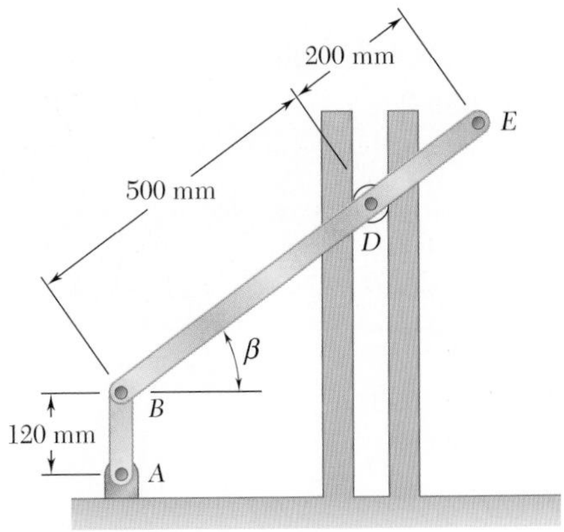

Fig. P15.84 and P15.85

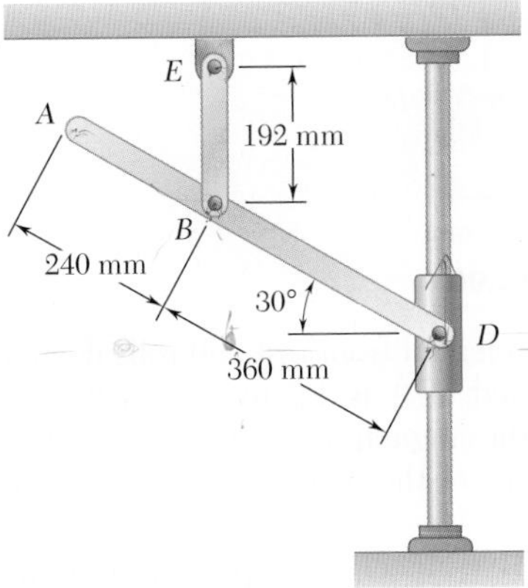

Fig. P15.86 and P15.87

15.87 Knowing that at the instant shown, the angular velocity of rod BE is 4 rad/s counterclockwise, determine (*a*) the angular velocity of rod AD, (*b*) the velocity of collar D, (*c*) the velocity of point A.

15.88 Rod AB can slide freely along the floor and the inclined plane. Denoting by $\mathbf{v}_A$ the velocity of point A, derive an expression for (*a*) the angular velocity of the rod, (*b*) the velocity of end B.

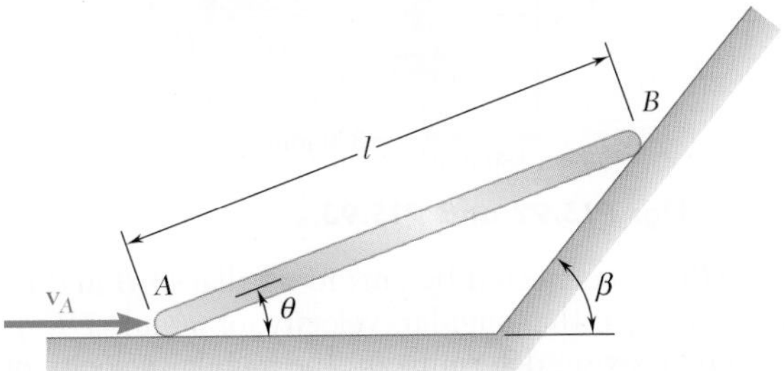

Fig. P15.88 and P15.89

15.89 Rod AB can slide freely along the floor and the inclined plane. Knowing that $\theta = 20°$, $\beta = 50°$, $l = 0.6$ m, and $v_A = 3$ m/s, determine (*a*) the angular velocity of the rod, (*b*) the velocity of end B.

15.90 Two slots have been cut in plate *FG* and the plate has been placed so that the slots fit two fixed pins *A* and *B*. Knowing that at the instant shown the angular velocity of crank *DE* is 6 rad/s clockwise, determine (*a*) the velocity of point *F*, (*b*) the velocity of point *G*.

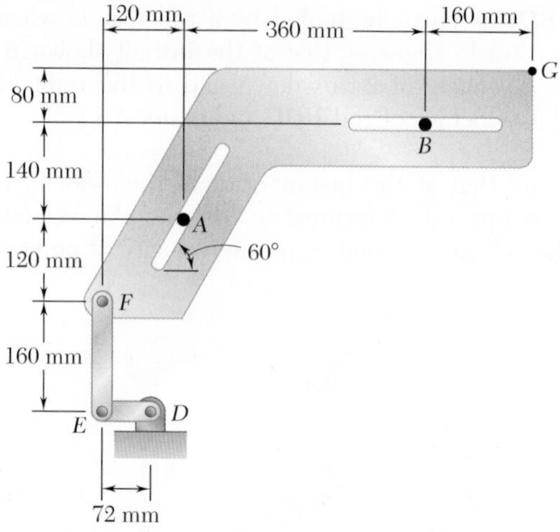

Fig. P15.90

15.91 The disk is released from rest and rolls down the incline. Knowing that the speed of *A* is 1.2 m/s when $\theta = 0°$, determine at that instant (*a*) the angular velocity of the rod, (*b*) the velocity of *B*. (Only portions of the two tracks are shown.)

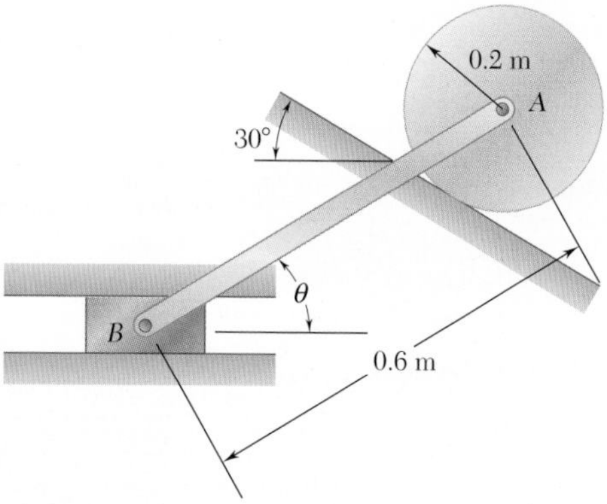

Fig. *P15.91*

15.92 Arm *ABD* is connected by pins to a collar at *B* and to crank *DE*. Knowing that the velocity of collar *B* is 400 mm/s upward, determine (*a*) the angular velocity of arm *ABD*, (*b*) the velocity of point *A*.

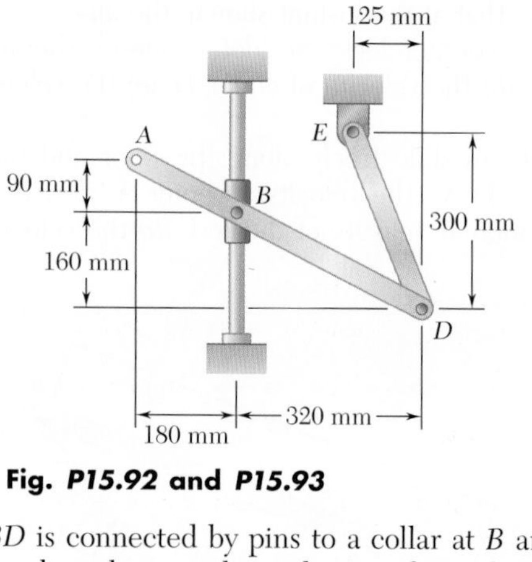

Fig. *P15.92* and *P15.93*

15.93 Arm *ABD* is connected by pins to a collar at *B* and to crank *DE*. Knowing that the angular velocity of crank *DE* is 1.2 rad/s counterclockwise, determine (*a*) the angular velocity of arm *ABD*, (*b*) the velocity of point *A*.

15.94 Two links *AB* and *BD*, each 625 mm. long, are connected at *B* and guided by hydraulic cylinders attached at *A* and *D*. Knowing that *D* is stationary and that the velocity of *A* is 750 mm/s to the right, determine at the instant shown (*a*) the angular velocity of each link, (*b*) the velocity of *B*.

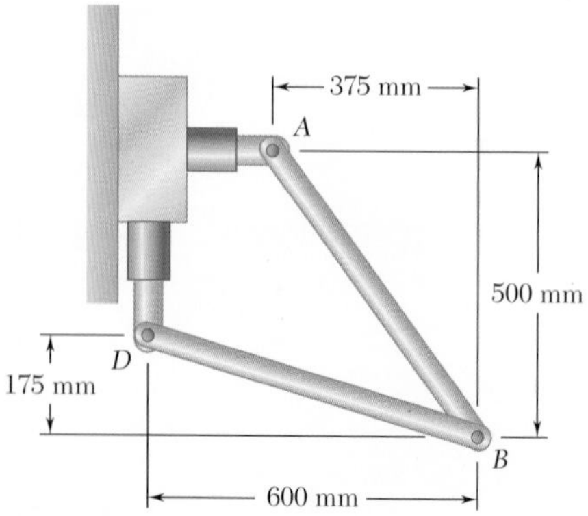

Fig. P15.94

15.95 Two 500-mm rods are pin-connected at D as shown. Knowing that B moves to the left with a constant velocity of 360 mm/s, determine at the instant shown (*a*) the angular velocity of each rod, (*b*) the velocity of E.

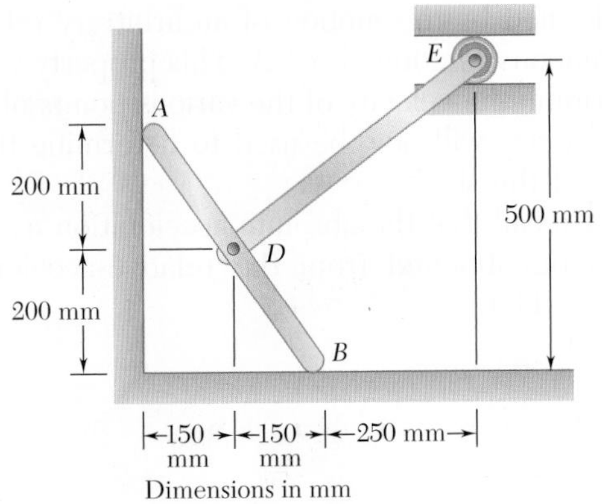

Fig. P15.95

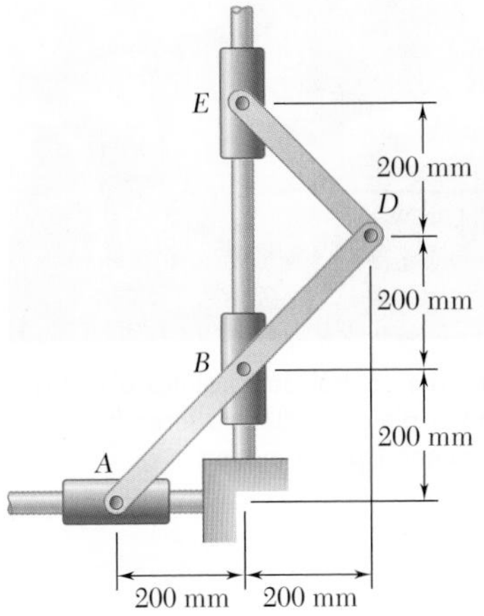

Fig. P15.96

15.96 Two rods ABD and DE are connected to three collars as shown. Knowing that the angular velocity of ABD is 5 rad/s clockwise, determine at the instant shown (*a*) the angular velocity of DE, (*b*) the velocity of collar E.

15.97 Two collars C and D move along the vertical rod shown. Knowing that the velocity of collar C is 660 mm/s downward, determine (*a*) the velocity of collar D, (*b*) the angular velocity of member AB.

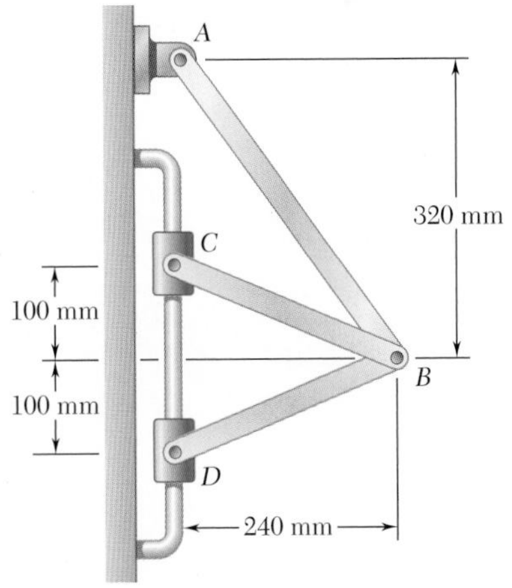

Fig. P15.97

15.98 Two rods AB and DE are connected as shown. Knowing that point D moves to the left with a velocity of 1 m/s, determine (*a*) the angular velocity of each rod, (*b*) the velocity of point A.

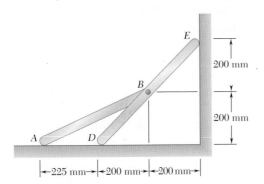

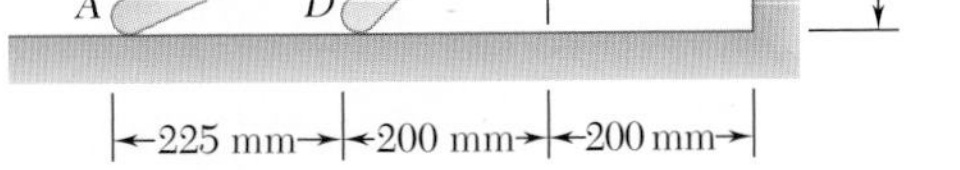

Fig. P15.98

15.99 Describe the space centrode and the body centrode of rod ABD of Prob. 15.83. (*Hint:* The body centrode need not lie on a physical portion of the rod.)

15.100 Describe the space centrode and the body centrode of the gear of Sample Prob. 15.2 as the gear rolls on the stationary horizontal rack.

15.101 Using the method of Sec. 15.7, solve Prob. 15.60.

15.102 Using the method of Sec. 15.7, solve Prob. 15.64.

15.103 Using the method of Sec. 15.7, solve Prob. 15.65.

15.104 Using the method of Sec. 15.7, solve Prob. 15.38.

Photo 15.6 The central gear rotates about a fixed axis and is pin-connected to three bars which are in general plane motion.

15.8 ABSOLUTE AND RELATIVE ACCELERATION IN PLANE MOTION

We saw in Sec. 15.5 that any plane motion can be replaced by a translation defined by the motion of an arbitrary reference point A and a simultaneous rotation about A. This property was used in Sec. 15.6 to determine the velocity of the various points of a moving slab. The same property will now be used to determine the acceleration of the points of the slab.

We first recall that the absolute acceleration $\mathbf{a}_B$ of a particle of the slab can be obtained from the relative-acceleration formula derived in Sec. 11.12,

$$\mathbf{a}_B = \mathbf{a}_A + \mathbf{a}_{B/A} \tag{15.21}$$

where the right-hand member represents a vector sum. The acceleration $\mathbf{a}_A$ corresponds to the translation of the slab with A, while the relative acceleration $\mathbf{a}_{B/A}$ is associated with the rotation of the slab about A and is measured with respect to axes centered at A and of fixed orientation. We recall from Sec. 15.3 that the relative acceleration $\mathbf{a}_{B/A}$ can be resolved into two components, a *tangential component* $(\mathbf{a}_{B/A})_t$ perpendicular to the line AB, and a *normal component* $(\mathbf{a}_{B/A})_n$ directed toward A (Fig. 15.22). Denoting by $\mathbf{r}_{B/A}$ the position vector of B relative to A and, respectively, by $\omega\mathbf{k}$ and $\alpha\mathbf{k}$ the angular velocity and angular acceleration of the slab with respect to axes of fixed orientation, we have

$$\begin{aligned} (\mathbf{a}_{B/A})_t &= \alpha\mathbf{k} \times \mathbf{r}_{B/A} \qquad & (a_{B/A})_t &= r\alpha \\ (\mathbf{a}_{B/A})_n &= -\omega^2\mathbf{r}_{B/A} \qquad & (a_{B/A})_n &= r\omega^2 \end{aligned} \tag{15.22}$$

where r is the distance from A to B. Substituting into (15.21) the expressions obtained for the tangential and normal components of $\mathbf{a}_{B/A}$, we can also write

$$\mathbf{a}_B = \mathbf{a}_A + \alpha\mathbf{k} \times \mathbf{r}_{B/A} - \omega^2\mathbf{r}_{B/A} \tag{15.21'}$$

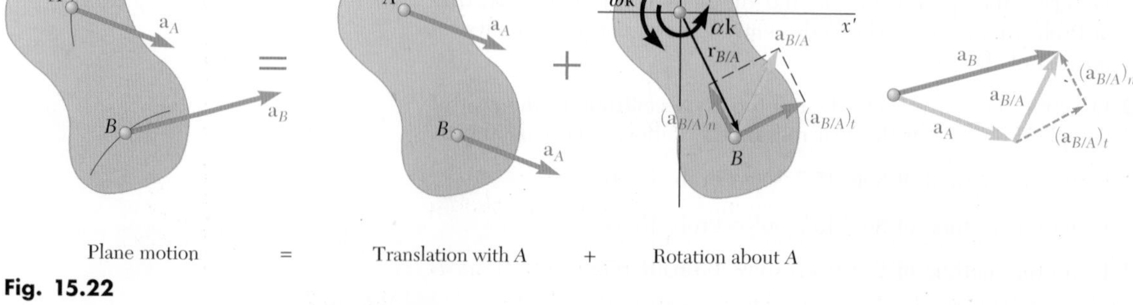

Fig. 15.22

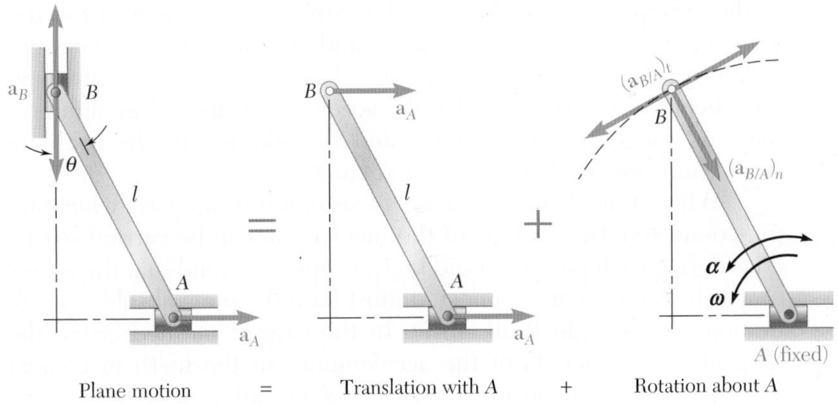

Fig. 15.23

As an example, let us again consider the rod AB whose extremities slide, respectively, along a horizontal and a vertical track (Fig. 15.23). Assuming that the velocity $\mathbf{v}_A$ and the acceleration $\mathbf{a}_A$ of A are known, we propose to determine the acceleration $\mathbf{a}_B$ of B and the angular acceleration $\boldsymbol{\alpha}$ of the rod. Choosing A as a reference point, we express that the given motion is equivalent to a translation with A and a rotation about A. The absolute acceleration of B must be equal to the sum

$$\begin{aligned} \mathbf{a}_B &= \mathbf{a}_A + \mathbf{a}_{B/A} \\ &= \mathbf{a}_A + (\mathbf{a}_{B/A})_n + (\mathbf{a}_{B/A})_t \end{aligned} \tag{15.23}$$

where $(\mathbf{a}_{B/A})_n$ has the magnitude $l\omega^2$ and is *directed toward A,* while $(\mathbf{a}_{B/A})_t$ has the magnitude $l\alpha$ and is perpendicular to AB. Students should note that there is no way to tell whether the tangential component $(\mathbf{a}_{B/A})_t$ is directed to the left or to the right, and therefore both possible directions for this component are indicated in Fig. 15.23. Similarly, both possible senses for $\mathbf{a}_B$ are indicated, since it is not known whether point B is accelerated upward or downward.

Equation (15.23) has been expressed geometrically in Fig. 15.24. Four different vector polygons can be obtained, depending upon the sense of $\mathbf{a}_A$ and the relative magnitude of a_A and $(a_{B/A})_n$. If we are to determine a_B and α from one of these diagrams, we must know not only a_A and θ but also ω. The angular velocity of the rod should therefore be separately determined by one of the methods indicated in Secs. 15.6 and 15.7. The values of a_B and α can then be obtained by considering successively the x and y components of the vectors shown in Fig. 15.24. In the case of polygon a, for example, we write

$$\overset{+}{\rightarrow} x \text{ components:} \qquad 0 = a_A + l\omega^2 \sin\theta - l\alpha\cos\theta$$
$$+\!\uparrow y \text{ components:} \qquad -a_B = -l\omega^2\cos\theta - l\alpha\sin\theta$$

and solve for a_B and α. The two unknowns can also be obtained by direct measurement on the vector polygon. In that case, care should be taken to draw first the known vectors $\mathbf{a}_A$ and $(\mathbf{a}_{B/A})_n$.

It is quite evident that the determination of accelerations is considerably more involved than the determination of velocities. Yet

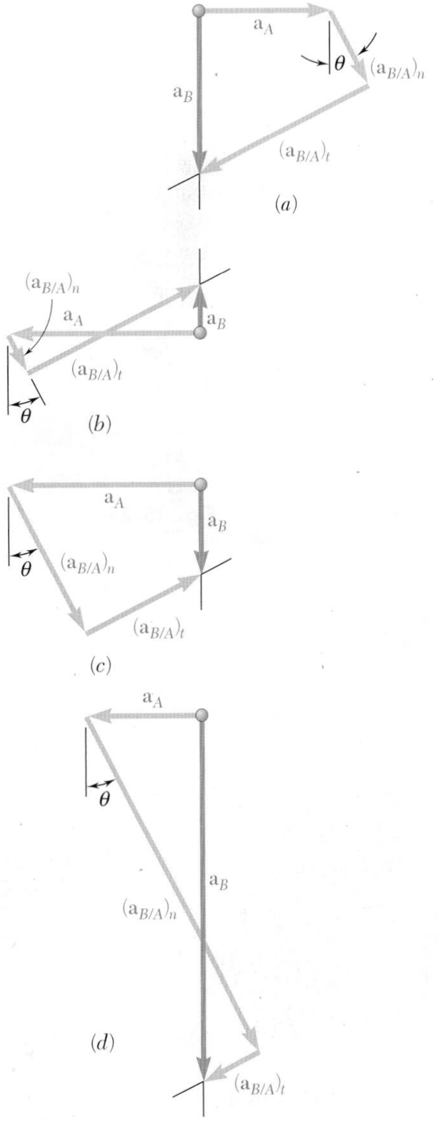

Fig. 15.24

in the example considered here, the extremities A and B of the rod were moving along straight tracks, and the diagrams drawn were relatively simple. If A and B had moved along curved tracks, it would have been necessary to resolve the accelerations $\mathbf{a}_A$ and $\mathbf{a}_B$ into normal and tangential components and the solution of the problem would have involved six different vectors.

When a mechanism consists of several moving parts which are pin-connected, the analysis of the mechanism can be carried out by considering each part as a rigid body, keeping in mind that the points at which two parts are connected must have the same absolute acceleration (see Sample Prob. 15.7). In the case of meshed gears, the tangential components of the accelerations of the teeth in contact are equal, but their normal components are different.

*15.9 ANALYSIS OF PLANE MOTION IN TERMS OF A PARAMETER

In the case of certain mechanisms, it is possible to express the coordinates x and y of all the significant points of the mechanism by means of simple analytic expressions containing a single parameter. It is sometimes advantageous in such a case to determine the absolute velocity and the absolute acceleration of the various points of the mechanism directly, since the components of the velocity and of the acceleration of a given point can be obtained by differentiating the coordinates x and y of that point.

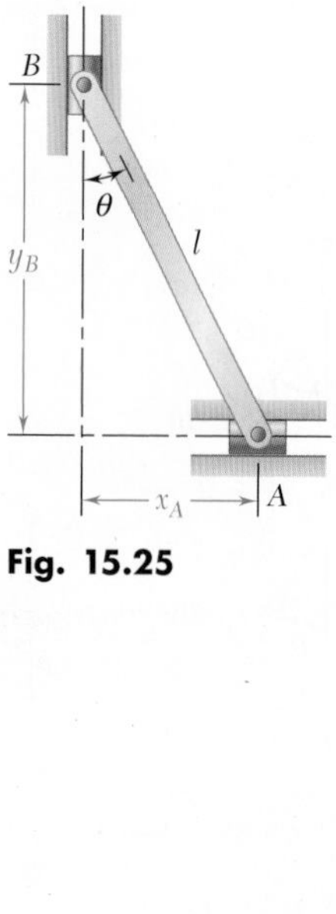

Fig. 15.25

Let us consider again the rod AB whose extremities slide, respectively, in a horizontal and a vertical track (Fig. 15.25). The coordinates x_A and y_B of the extremities of the rod can be expressed in terms of the angle θ the rod forms with the vertical:

$$x_A = l \sin \theta \qquad y_B = l \cos \theta \tag{15.24}$$

Differentiating Eqs. (15.24) twice with respect to t, we write

$$v_A = \dot{x}_A = l\dot{\theta} \cos \theta$$
$$a_A = \ddot{x}_A = -l\dot{\theta}^2 \sin \theta + l\ddot{\theta} \cos \theta$$

$$v_B = \dot{y}_B = -l\dot{\theta} \sin \theta$$
$$a_B = \ddot{y}_B = -l\dot{\theta}^2 \cos \theta - l\ddot{\theta} \sin \theta$$

Recalling that $\dot{\theta} = \omega$ and $\ddot{\theta} = \alpha$, we obtain

$$v_A = l\omega \cos \theta \qquad v_B = -l\omega \sin \theta \tag{15.25}$$

$$a_A = -l\omega^2 \sin \theta + l\alpha \cos \theta \qquad a_B = -l\omega^2 \cos \theta - l\alpha \sin \theta \tag{15.26}$$

We note that a positive sign for v_A or a_A indicates that the velocity $\mathbf{v}_A$ or the acceleration $\mathbf{a}_A$ is directed to the right; a positive sign for v_B or a_B indicates that $\mathbf{v}_B$ or $\mathbf{a}_B$ is directed upward. Equations (15.25) can be used, for example, to determine v_B and ω when v_A and θ are known. Substituting for ω in (15.26), we can then determine a_B and α if a_A is known.

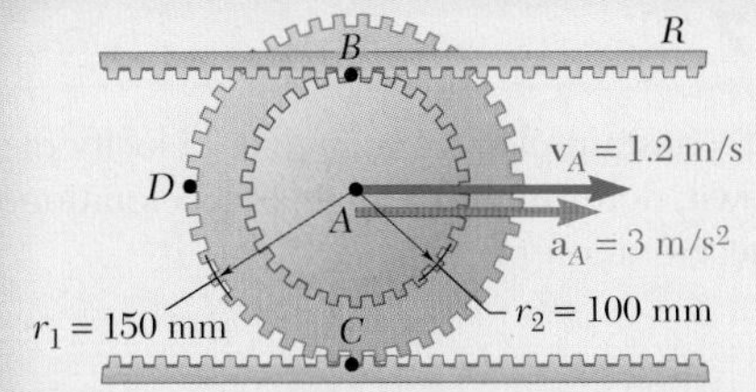

SAMPLE PROBLEM 15.6

The center of the double gear of Sample Prob. 15.2 has a velocity of 1.2 m/s to the right and an acceleration of 3 m/s^2 to the right. Recalling that the lower rack is stationary, determine (*a*) the angular acceleration of the gear, (*b*) the acceleration of points *B*, *C*, and *D* of the gear.

SOLUTION

a. Angular Acceleration of the Gear. In Sample Prob. 15.2, we found that $x_A = -r_1\theta$ and $v_A = -r_1\omega$. Differentiating the latter with respect to time, we obtain $a_A = -r_1\alpha$.

$$v_A = -r_1\omega \qquad 1.2 \text{ m/s} = -(0.150 \text{ m})\omega \qquad \omega = -8 \text{ rad/s}$$
$$a_A = -r_1\alpha \qquad 3 \text{ m/s}^2 = -(0.150 \text{ m})\alpha \qquad \alpha = -20 \text{ rad/s}^2$$
$$\boldsymbol{\alpha} = \alpha\mathbf{k} = -(20 \text{ rad/s}^2)\mathbf{k} \quad ◀$$

b. Accelerations. The rolling motion of the gear is resolved into a translation with *A* and a rotation about *A*.

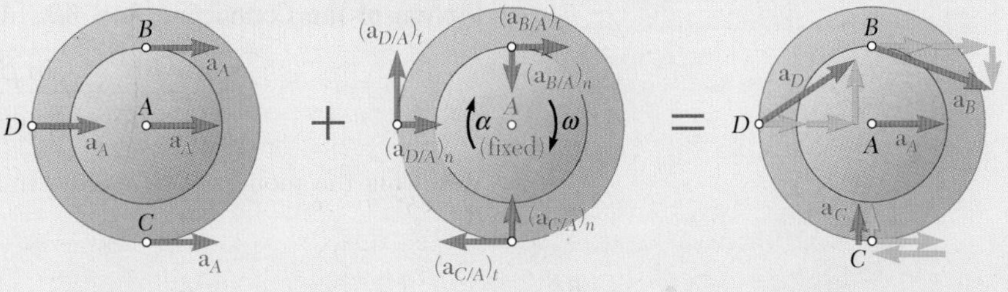

Translation + Rotation = Rolling motion

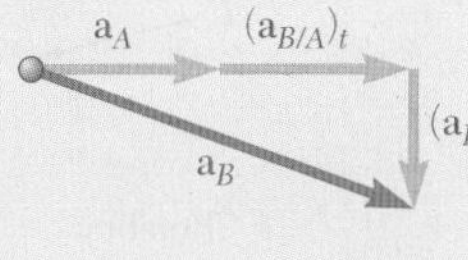

Acceleration of Point *B*. Adding vectorially the accelerations corresponding to the translation and to the rotation, we obtain

$$\begin{aligned}\mathbf{a}_B &= \mathbf{a}_A + \mathbf{a}_{B/A} = \mathbf{a}_A + (\mathbf{a}_{B/A})_t + (\mathbf{a}_{B/A})_n \\ &= \mathbf{a}_A + \alpha\mathbf{k} \times \mathbf{r}_{B/A} - \omega^2\mathbf{r}_{B/A} \\ &= (3 \text{ m/s}^2)\mathbf{i} - (20 \text{ rad/s}^2)\mathbf{k} \times (0.100 \text{ m})\mathbf{j} - (8 \text{ rad/s})^2(0.100 \text{ m})\mathbf{j} \\ &= (3 \text{ m/s}^2)\mathbf{i} + (2 \text{ m/s}^2)\mathbf{i} - (6.40 \text{ m/s}^2)\mathbf{j}\end{aligned}$$
$$\mathbf{a}_B = 8.12 \text{ m/s}^2 \measuredangle 52.0° \quad ◀$$

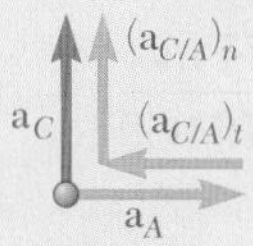

Acceleration of Point *C*

$$\begin{aligned}\mathbf{a}_C &= \mathbf{a}_A + \mathbf{a}_{C/A} = \mathbf{a}_A + \alpha\mathbf{k} \times \mathbf{r}_{C/A} - \omega^2\mathbf{r}_{C/A} \\ &= (3 \text{ m/s}^2)\mathbf{i} - (20 \text{ rad/s}^2)\mathbf{k} \times (-0.150 \text{ m})\mathbf{j} - (8 \text{ rad/s})^2(-0.150 \text{ m})\mathbf{j} \\ &= (3 \text{ m/s}^2)\mathbf{i} - (3 \text{ m/s}^2)\mathbf{i} + (9.60 \text{ m/s}^2)\mathbf{j}\end{aligned}$$
$$\mathbf{a}_C = 9.60 \text{ m/s}^2 \uparrow \quad ◀$$

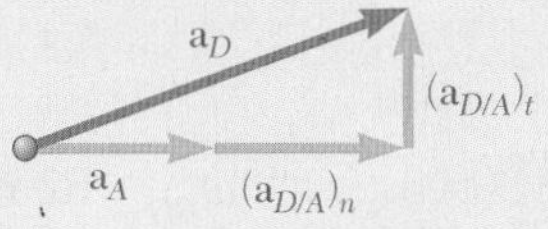

Acceleration of Point *D*

$$\begin{aligned}\mathbf{a}_D &= \mathbf{a}_A + \mathbf{a}_{D/A} = \mathbf{a}_A + \alpha\mathbf{k} \times \mathbf{r}_{D/A} - \omega^2\mathbf{r}_{D/A} \\ &= (3 \text{ m/s}^2)\mathbf{i} - (20 \text{ rad/s}^2)\mathbf{k} \times (-0.150 \text{ m})\mathbf{i} - (8 \text{ rad/s})^2(-0.150 \text{ m})\mathbf{i} \\ &= (3 \text{ m/s}^2)\mathbf{i} + (3 \text{ m/s}^2)\mathbf{j} + (9.60 \text{ m/s}^2)\mathbf{i}\end{aligned}$$
$$\mathbf{a}_D = 12.95 \text{ m/s}^2 \measuredangle 13.4° \quad ◀$$

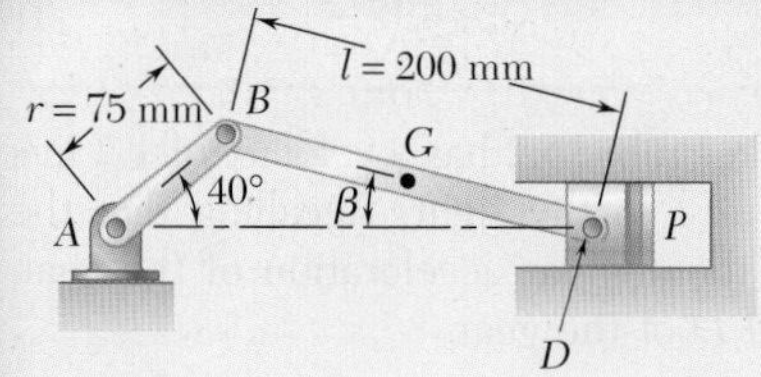

SAMPLE PROBLEM 15.7

Crank AB of the engine system has a constant clockwise angular velocity of 2000 rpm. For the crank position shown, determine the angular acceleration of the connecting rod BD and the acceleration of point D.

SOLUTION

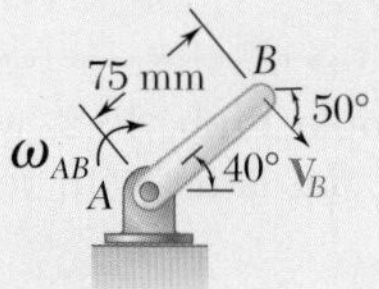

We first determine ω_{BD}

Motion of Crank *AB*. The crank AB rotates about point A. Expressing ω_{AB} in rad/s and writing $v_B = r\omega_{AB}$.

$$\omega_{AB} = \left(2000\frac{\text{rev}}{\text{min}}\right)\left(\frac{1\text{ min}}{60\text{ s}}\right)\left(\frac{2\pi\text{ rad}}{1\text{ rev}}\right) = 209.4\text{ rad/s}$$

$$v_B = (AB)\omega_{AB} = (0.075\text{ m})(209.4\text{ rad/s}) = 15.71\text{ m/s}$$

$$\mathbf{v}_B = 15.71\text{ m/s} \searrow 50°$$

Motion of the Connecting Rod *BD*. Using the law of sines,

$$\frac{\sin 40°}{200\text{ mm}} = \frac{\sin\beta}{75\text{ mm}} \qquad \beta = 13.95°$$

Resolving the motion of BD into a translation with B and a rotation about B,

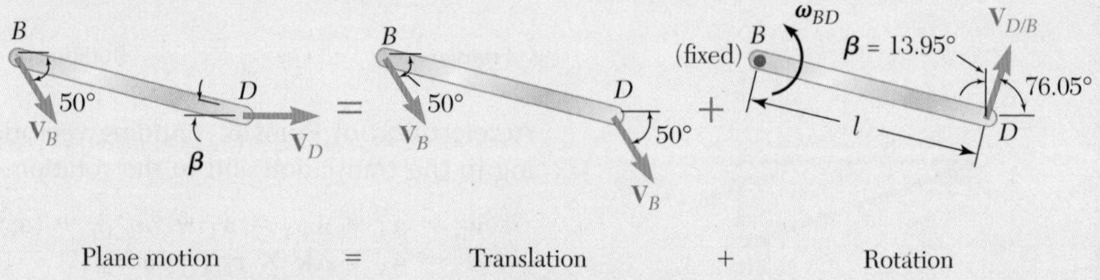

$$\mathbf{v}_D = \mathbf{v}_B + \mathbf{v}_{D/B}$$

The vector diagram corresponding to this equation is drawn.

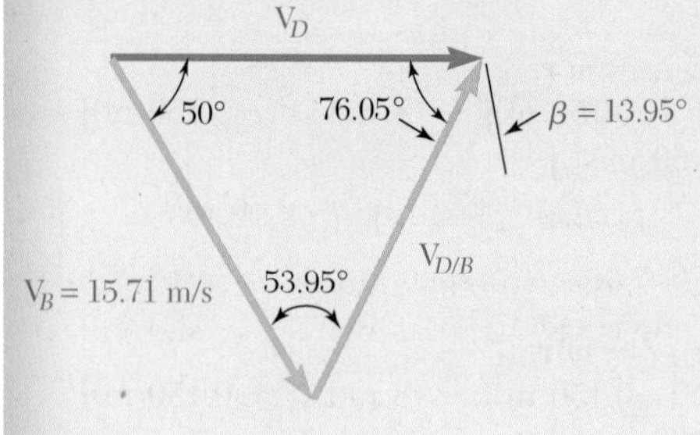

$$\frac{v_D}{\sin 53.95°} = \frac{v_{D/B}}{\sin 50°} = \frac{15.71\text{ m/s}}{\sin 76.05°}$$

$$v_{D/B} = 12.40\text{ m/s} \qquad \mathbf{v}_{D/B} = 12.40\text{ m/s} \angle 76.05$$

Since $v_{D/B} = l\omega_{BD}$, we have

$$12.40\text{ m/s} = (0.2\text{ m})\omega_{BD}$$

$$\omega_{BD} = 62\text{ rad/s} \circlearrowleft$$

We now determine the angular acceleration of BD and acceleration of point D.

Motion of Crank AB. Since the crank rotates about A with constant we have $\alpha_{AB} = 0$. The acceleration of B is therefore directed toward A and has a magnitude

$$a_B = r\omega_{AB}^2 = (\frac{75}{1000}\text{ m})(209.4 \text{ rad/s})^2 = 3289 \text{ m/s}^2$$

$$\mathbf{a}_B = 32{,}89 \text{ m/s}^2 \measuredangle 40°$$

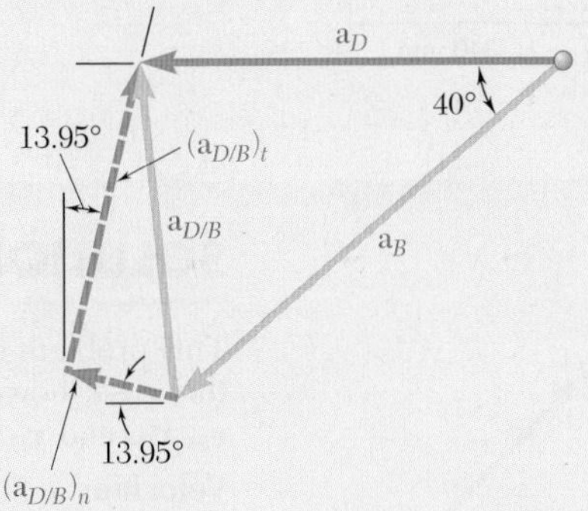

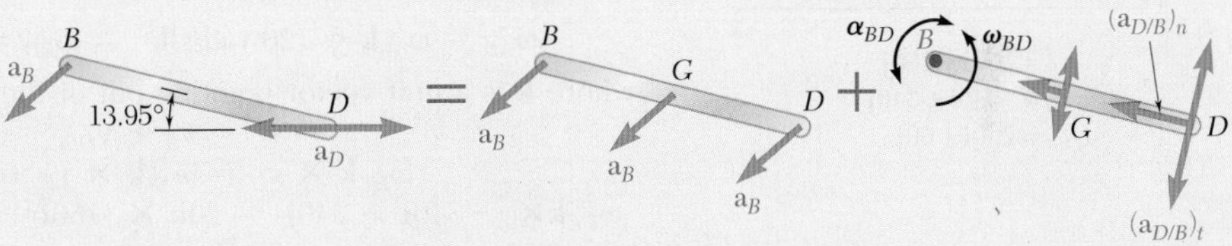

Plane motion = Translation + Rotation

The motion of BD is resolved into a translation with B and a rotation about B. The relative acceleration $\mathbf{a}_{D/B}$ is resolved into normal and tangential components:

$$(a_{D/B})_n = (BD)\omega_{BD}^2 = (\frac{200}{1000}\text{m})(62.0 \text{ rad/s})^2 = 768.8 \text{m/s}^2$$

$$(\mathbf{a}_{D/B})_n = 768.8 \text{m/s}^2 \measuredangle 13.95°$$

$$(a_{D/B})_t = (BD)\alpha_{BD} = (\frac{200}{1000}\text{m}) \quad \alpha_{BD} = 0.2\,\alpha_{BD}$$

$$(\mathbf{a}_{D/B})_t = 0.2\,\alpha_{BD} \measuredangle 76.05°$$

While $(\mathbf{a}_{D/B})_t$ must be perpendicular to BD, its sense is not known.

Noting that the acceleration $\mathbf{a}_D$ must be horizontal, we write

$$\mathbf{a}_D = \mathbf{a}_B + \mathbf{a}_{D/B} = \mathbf{a}_B + (\mathbf{a}_{D/B})_n + (\mathbf{a}_{D/B})_t$$

$$[a_D \leftrightarrow] = [3289 \measuredangle 40°] + [768.8 \measuredangle 13.95°] + [0.2\alpha_{BD} \measuredangle 76.05°]$$

Equating x and y components, we obtain the following scalar equations:

$\overset{+}{\rightarrow} x$ components:

$$-a_D = -3289 \cos 40° - 768.8 \cos 13.95° + 0.2\alpha_{BD} \sin 13.95°$$

$+\uparrow y$ components:

$$0 = -3289 \sin 40° + 768.8 \sin 13.95° + 0.2\alpha_{BD} \cos 13.95°$$

Solving the equations simultaneously, we obtain $\alpha_{BD} = +9940 \text{ rad/s}^2$ and $a_D = +9290 \text{ ft/s}^2$. The positive signs indicate that the senses shown on the vector polygon are correct; we write

$$\alpha_{BD} = 9940 \text{ rad/s}^2 \circlearrowleft \blacktriangleleft$$

$$\mathbf{a}_D = 2790 \text{ m/s}^2 \leftarrow \blacktriangleleft$$

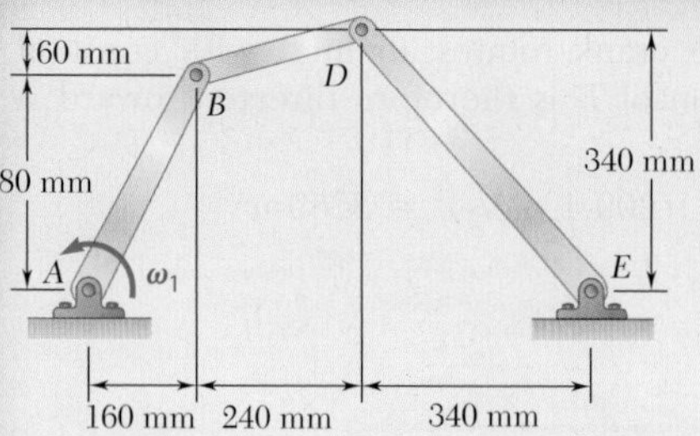

SAMPLE PROBLEM 15.8

The linkage *ABDE* moves in the vertical plane. Knowing that in the position shown crank *AB* has a constant angular velocity $\boldsymbol{\omega}_1$ of 20 rad/s counterclockwise, determine the angular velocities and angular accelerations of the connecting rod *BD* and of the crank *DE*.

SOLUTION

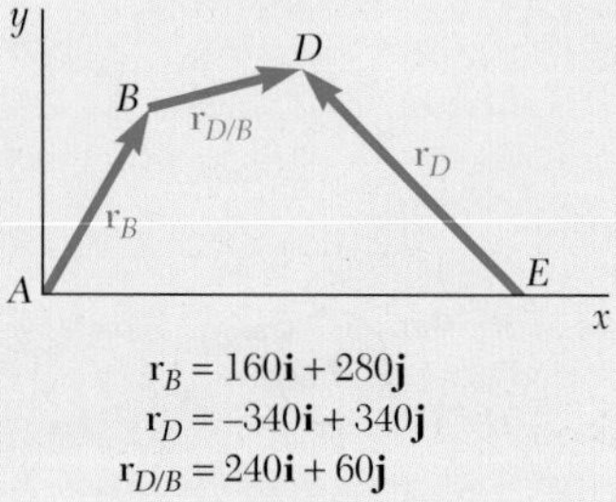

This problem could be solved by the method used in Sample Prob. 15.7. In this case, however, the vector approach will be used. The position vectors $\mathbf{r}_B$, $\mathbf{r}_D$, and $\mathbf{r}_{D/B}$ are chosen as shown in the sketch.

Velocities. Since the motion of each element of the linkage is contained in the plane of the figure, we have

$$\boldsymbol{\omega}_{AB} = \omega_{AB}\mathbf{k} = (20 \text{ rad/s})\mathbf{k} \qquad \boldsymbol{\omega}_{BD} = \omega_{BD}\mathbf{k} \qquad \boldsymbol{\omega}_{DE} = \omega_{DE}\mathbf{k}$$

where $\mathbf{k}$ is a unit vector pointing out of the paper. We now write

$$\mathbf{v}_D = \mathbf{v}_B + \mathbf{v}_{D/B}$$
$$\omega_{DE}\mathbf{k} \times \mathbf{r}_D = \omega_{AB}\mathbf{k} \times \mathbf{r}_B + \omega_{BD}\mathbf{k} \times \mathbf{r}_{D/B}$$
$$\omega_{DE}\mathbf{k}\times (-340\mathbf{i} + 340\mathbf{j}) = 20\mathbf{k} \times (160\mathbf{i} + 280\mathbf{j}) + \omega_{BD}\mathbf{k}\times (240\mathbf{i} + 60\mathbf{j})$$

Dividing each term by 20 we get

$$-17\omega_{DE}\mathbf{j} - 17\omega_{DE}\mathbf{i} = 160\mathbf{j} - 280\mathbf{i} + 12\omega_{BD}\mathbf{j} - 3\omega_{BD}\mathbf{i}$$

Equating the coefficients of the unit vectors $\mathbf{i}$ and $\mathbf{j}$, we obtain the following two scalar equations:

$$-17\omega_{DE} = -280 - 3\omega_{BD}$$
$$-17\omega_{DE} = +160 + 12\omega_{BD}$$
$$\boldsymbol{\omega}_{BD} = -(29.33 \text{ rad/s})\mathbf{k} \qquad \boldsymbol{\omega}_{DE} = (11.29 \text{ rad/s})\mathbf{k} \quad ◀$$

Expressing $\mathbf{r}$ in m we have

$$\mathbf{r}_B = 0.16\mathbf{i} + 0.28\mathbf{j}$$
$$\mathbf{r}_D = -0.34\mathbf{i} + 0.34\mathbf{j}$$
$$\mathbf{r}_{D/B} = 0.24\mathbf{i} + 0.06\mathbf{j}$$

Accelerations. Noting that at the instant considered crank *AB* has a constant angular velocity, we write

$$\boldsymbol{\alpha}_{AB} = 0 \qquad \boldsymbol{\alpha}_{BD} = \alpha_{BD}\mathbf{k} \qquad \boldsymbol{\alpha}_{DE} = \alpha_{DE}\mathbf{k}$$
$$\mathbf{a}_D = \mathbf{a}_B + \mathbf{a}_{D/B} \tag{1}$$

Each term of Eq. (1) is evaluated separately:

$$\mathbf{a}_D = \alpha_{DE}\mathbf{k} \times \mathbf{r}_D - \omega_{DE}^2\mathbf{r}_D$$
$$= \alpha_{DE}\mathbf{k}\times (-0.34\mathbf{i} + 0.34\mathbf{j}) - (11.29)^2(-0.34\mathbf{i} + 0.34\mathbf{j})$$
$$= -0.34\alpha_{DE}\mathbf{j} - 0.34\alpha_{DE}\mathbf{i} + 43.33\mathbf{i} - 43.33\mathbf{j}$$
$$\mathbf{a}_B = \alpha_{AB}\mathbf{k} \times \mathbf{r}_B - \omega_{AB}^2\mathbf{r}_B = 0 - (20)^2(16\mathbf{i} + 0.28\mathbf{j})$$
$$= -64\mathbf{i} - 112\mathbf{j}$$
$$\mathbf{a}_{D/B} = \alpha_{BD}\mathbf{k} \times \mathbf{r}_{D/B} - \omega_{BD}^2\mathbf{r}_{D/B}$$
$$= \alpha_{BD}\mathbf{k} \times (0.24\mathbf{i} + 0.06\mathbf{j}) - (29.33)^2(0.24\mathbf{i} + 0.06\mathbf{j})$$
$$= 0.24\alpha_{BD}\mathbf{j} - 0.06\alpha_{BD}\mathbf{i} - 206.4\mathbf{i} - 51.61\mathbf{j}$$

Substituting into Eq. (1) and equating the coefficients of $\mathbf{i}$ and $\mathbf{j}$, we obtain

$$-0.34\alpha_{DE} + 0.06\alpha_{BD} = -313.7$$
$$-0.34\alpha_{DE} - 0.24\alpha_{BD} = -120.28$$
$$\boldsymbol{\alpha}_{BD} = -(645 \text{ rad/s}^2)\mathbf{k} \qquad \boldsymbol{\alpha}_{DE} = (809 \text{ rad/s}^2)\mathbf{k} \quad ◀$$

SOLVING PROBLEMS ON YOUR OWN

This lesson was devoted to the determination of the *accelerations* of the points of a *rigid body in plane motion.* As you did previously for velocities, you will again consider the plane motion of a rigid body as the sum of two motions, namely, *a translation and a rotation.*

To solve a problem involving accelerations in plane motion you should use the following steps:

1. Determine the angular velocity of the body. To find $\boldsymbol{\omega}$ you can either

a. Consider the motion of the body as the sum of a translation and a rotation as you did in Sec. 15.6, or

b. Use the instantaneous center of rotation of the body as you did in Sec. 15.7. However, *keep in mind that you cannot use the instantaneous center to determine accelerations.*

2. Start drawing a "diagram equation" to use in your solution. This "equation" will involve the following diagrams (Fig. 15.22).

a. Plane motion diagram. Draw a sketch of the body, including all dimensions, as well as the angular velocity $\boldsymbol{\omega}$. Show the angular acceleration $\boldsymbol{\alpha}$ with its magnitude and sense if you know them. Also show those points for which you know or seek the accelerations, indicating all that you know about these accelerations.

b. Translation diagram. Select a reference point A for which you know the direction, the magnitude, or a component of the acceleration $\mathbf{a}_A$. Draw a second diagram showing the body in translation with each point having the same acceleration as point A.

c. Rotation diagram. Considering point A as a fixed reference point, draw a third diagram showing the body in rotation about A. Indicate the normal and tangential components of the relative accelerations of other points, such as the components $(\mathbf{a}_{B/A})_n$ and $(\mathbf{a}_{B/A})_t$ of the acceleration of point B with respect to point A.

3. Write the relative-acceleration formula

$$\mathbf{a}_B = \mathbf{a}_A + \mathbf{a}_{B/A} \qquad \text{or} \qquad \mathbf{a}_B = \mathbf{a}_A + (\mathbf{a}_{B/A})_n + (\mathbf{a}_{B/A})_t$$

The sample problems illustrate three different ways to use this vector equation:

a. If α is given or can easily be determined, you can use this equation to determine the accelerations of various points of the body [Sample Prob. 15.6].

b. If α cannot easily be determined, select for point B a point for which you know the direction, the magnitude, or a component of the acceleration $\mathbf{a}_B$ and draw a vector diagram of the equation. Starting at the same point, draw all known acceleration components in tip-to-tail fashion for each member of the equation. Complete the diagram by drawing the two remaining vectors in appropriate directions and in such a way that the two sums of vectors end at a common point.

The magnitudes of the two remaining vectors can be found either graphically or analytically. Usually an analytic solution will require the solution of two simultaneous equations [Sample Prob. 15.7]. However, by first considering the components of the various vectors in a direction perpendicular to one of the unknown vectors, you may be able to obtain an equation in a single unknown.

One of the two vectors obtained by the method just described will be $(\mathbf{a}_{B/A})_t$, from which you can compute α. Once α has been found, the vector equation can be used to determine the acceleration of any other point of the body.

c. A full vector approach can also be used to solve the vector equation. This is illustrated in Sample Prob. 15.8.

4. The analysis of plane motion in terms of a parameter completed this lesson. This method should be used *only if it is possible* to express the coordinates x and y of all significant points of the body in terms of a single parameter (Sec. 15.9). By differentiating twice with respect to t the coordinates x and y of a given point, you can determine the rectangular components of the absolute velocity and absolute acceleration of that point.

PROBLEMS

CONCEPT QUESTION

15.CQ7 A rear-wheel-drive car starts from rest and accelerates to the left so that the tires do not slip on the road. What is the direction of the acceleration of the point on the tire in contact with the road, that is, point A?

a. ← **b.** ↖ **c.** ↑ **d.** ↓ **e.** ↙

Fig. P15.CQ7

END-OF-SECTION PROBLEMS

15.105 A 3.5-m steel beam is lowered by means of two cables unwinding at the same speed from overhead cranes. As the beam approaches the ground, the crane operators apply brakes to slow down the unwinding motion. At the instant considered, the deceleration of the cable attached at A is 4 m/s^2, while that of the cable at B is 1.5 m/s^2. Determine (*a*) the angular acceleration of the beam, (*b*) the acceleration of point C.

15.106 The acceleration of point C is 0.3 m/s^2 downward and the angular acceleration of the beam is 0.8 rad/s^2 clockwise. Knowing that the angular velocity of the beam is zero at the instant considered, determine the acceleration of each cable.

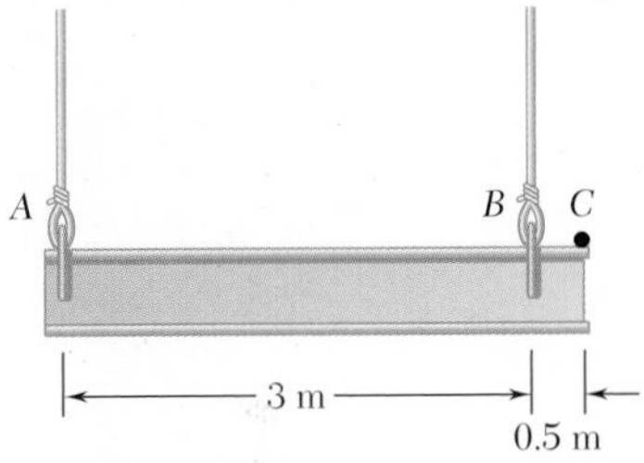

Fig. P15.105 and P15.106

15.107 A 900-mm rod rests on a horizontal table. A force **P** applied as shown produces the following accelerations: $\mathbf{a}_A = 3.6$ m/s^2 to the right, $\alpha = 6$ rad/s^2 counterclockwise as viewed from above. Determine the acceleration (*a*) of point G, (*b*) of point B.

15.108 In Prob. 15.107, determine the point of the rod that (*a*) has no acceleration, (*b*) has an acceleration of 2.4 m/s^2 to the right.

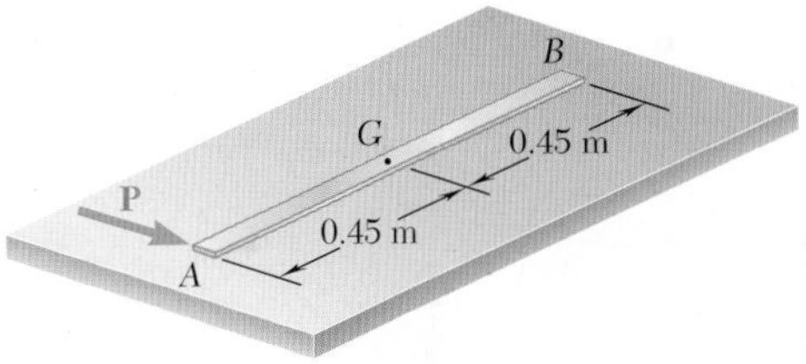

Fig. P15.107 and P15.108

15.109 Knowing that at the instant shown crank BC has a constant angular velocity of 45 rpm clockwise, determine the acceleration (*a*) of point A, (*b*) of point D.

15.110 Bar BDE is attached to two links AB and CD. Knowing that at the instant shown link AB has zero angular acceleration and an angular velocity of 3 rad/s clockwise, determine the acceleration (*a*) of point D, (*b*) of point E.

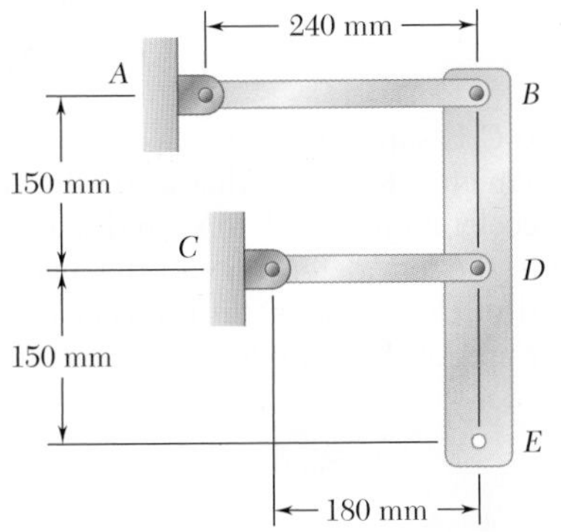

Fig. P15.110

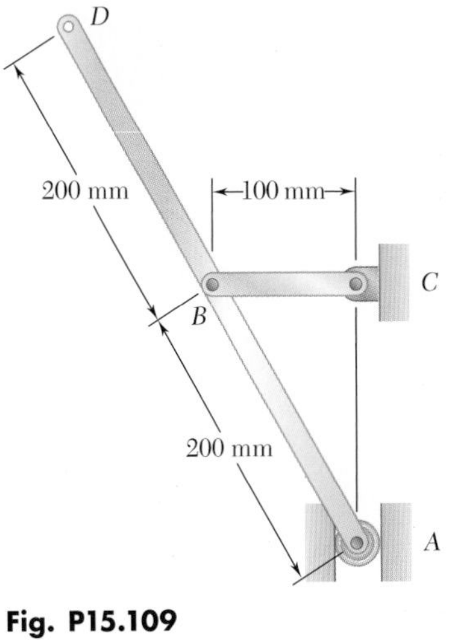

Fig. P15.109

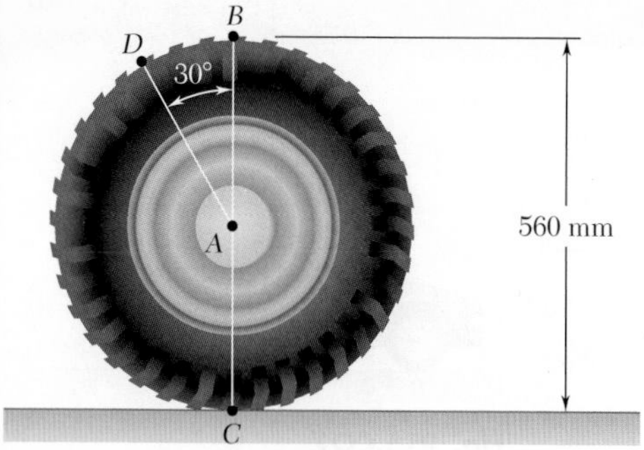

Fig. P15.111

15.111 An automobile travels to the left at a constant speed of 72 km/h. Knowing that the diameter of the wheel is 560 mm, determine the acceleration (*a*) of point *B*, (*b*) of point *C*, (*c*) of point *D*.

15.112 The 500-mm-radius flywheel is rigidly attached to a 40-mm-radius shaft that can roll along parallel rails. Knowing that at the instant shown the center of the shaft has a velocity of 30 mm/s and an acceleration of 10 mm/s^2, both directed down to the left, determine the acceleration (*a*) of point *A*, (*b*) of point *B*.

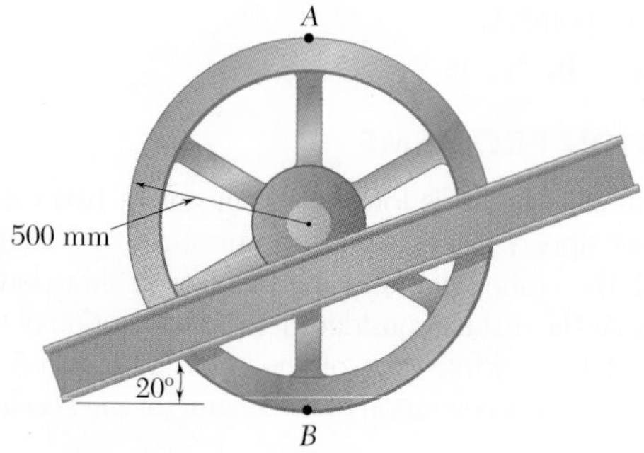

Fig. P15.112

15.113 and 15.114 A 75-mm-radius drum is rigidly attached to a 125-mm-radius drum as shown. One of the drums rolls without sliding on the surface shown, and a cord is wound around the other drum. Knowing that at the instant shown end *D* of the cord has a velocity of 200 mm/s and an acceleration of 750 mm/s^2, both directed to the left, determine the accelerations of points *A*, *B*, and *C* of the drums.

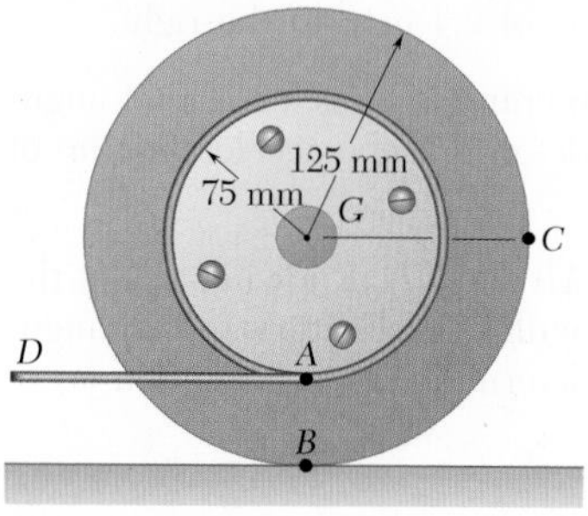

Fig. P15.113

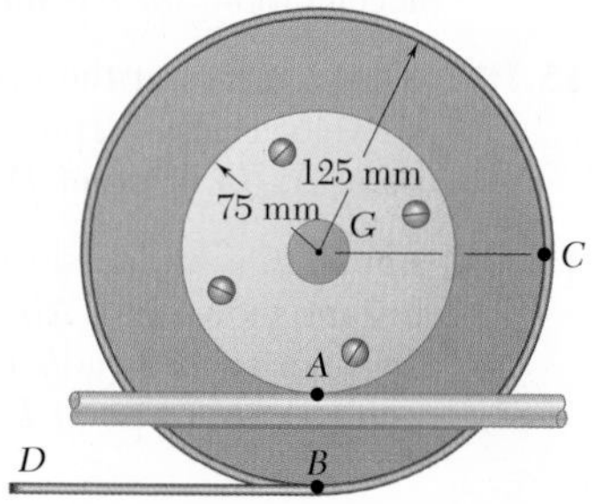

Fig. P15.114

15.115 A carriage *C* is supported by a caster *A* and a cylinder *B*, each of 50-mm diameter. Knowing that at the instant shown the carriage has an acceleration of 2.4 m/s^2 and a velocity of 1.5 m/s, both directed to the left, determine (*a*) the angular accelerations of the caster and of the cylinder, (*b*) the accelerations of the centers of the caster and of the cylinder.

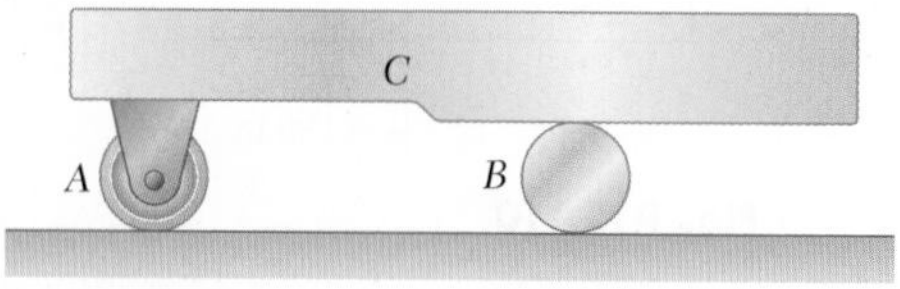

Fig. P15.115

15.116 A wheel rolls without slipping on a fixed cylinder. Knowing that at the instant shown the angular velocity of the wheel is 10 rad/s clockwise and its angular acceleration is 30 rad/s^2 counterclockwise, determine the acceleration of (*a*) point *A*, (*b*) point *B*, (*c*) point *C*.

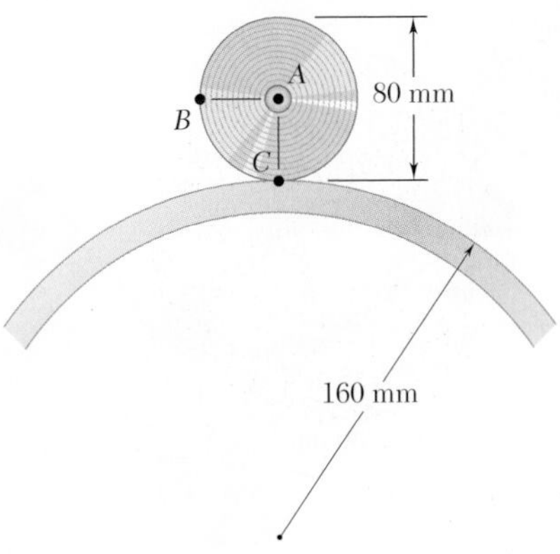

Fig. *P15.116*

15.117 The 100-mm-radius drum rolls without slipping on a portion of a belt which moves downward to the left with a constant velocity of 120 mm/s. Knowing that at a given instant the velocity and acceleration of the center *A* of the drum are as shown, determine the acceleration of point *D*.

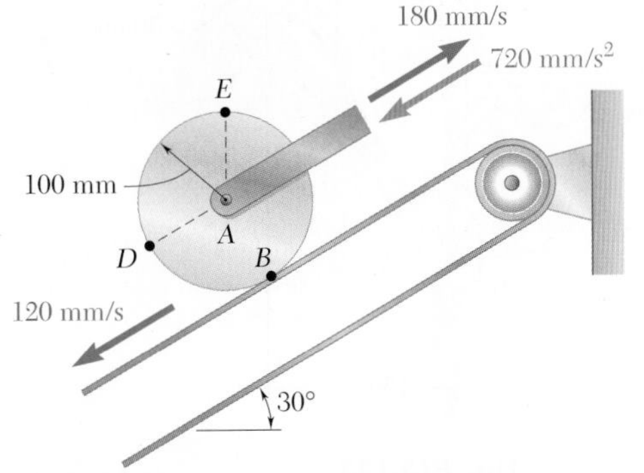

Fig. *P15.117*

15.118 In the planetary gear system shown the radius of gears *A*, *B*, *C*, and *D* is 100 mm and the radius of the outer gear *E* is 300 mm. Knowing that gear *A* has a constant angular velocity of 150 rpm clockwise and that the outer gear *E* is stationary, determine the magnitude of the acceleration of the tooth of gear *D* that is in contact with (*a*) gear *A*, (*b*) gear *E*.

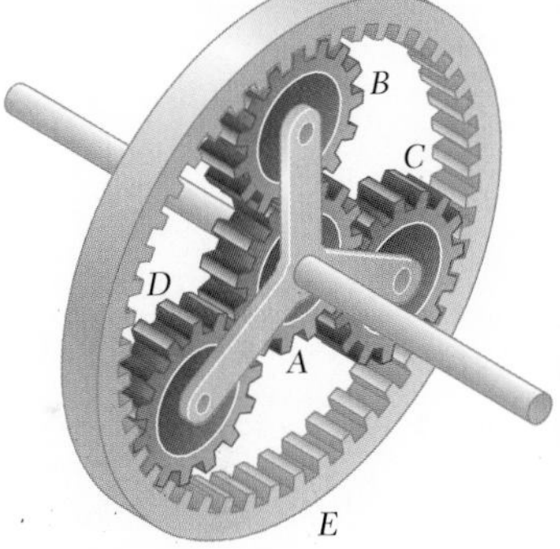

Fig. P15.118

15.119 The 200-mm-radius disk rolls without sliding on the surface shown. Knowing that the distance *BG* is 160 mm and that at the instant shown the disk has an angular velocity of 8 rad/s counterclockwise and an angular acceleration of 2 rad/s^2 clockwise, determine the acceleration of *A*.

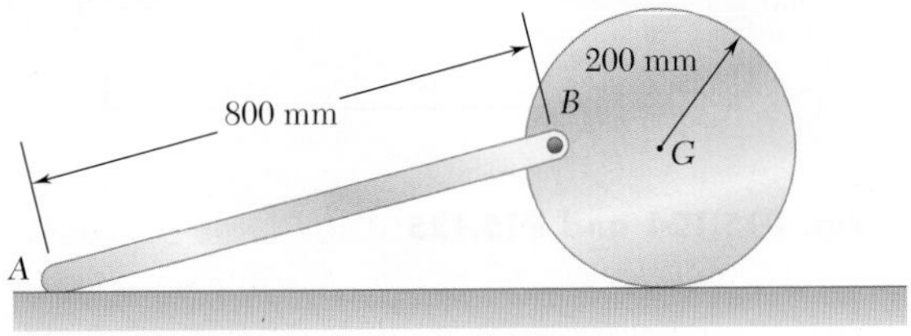

Fig. *P15.119*

15.120 Knowing that crank *AB* rotates about point *A* with a constant angular velocity of 900 rpm clockwise, determine the acceleration of the piston *P* when $\theta = 60°$.

15.121 Knowing that crank *AB* rotates about point *A* with a constant angular velocity of 900 rpm clockwise, determine the acceleration of the piston *P* when $\theta = 120°$.

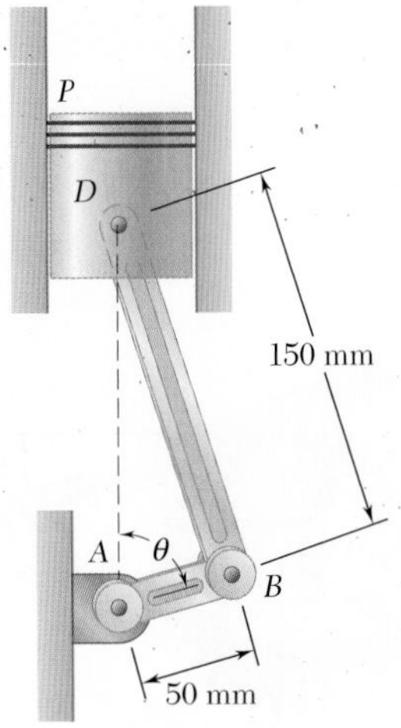

Fig. P15.120 and P15.121

15.122 In the two-cylinder air compressor shown the connecting rods BD and BE are each 190 mm long and crank AB rotates about the fixed point A with a constant angular velocity of 1500 rpm clockwise. Determine the acceleration of each piston when $\theta = 0$.

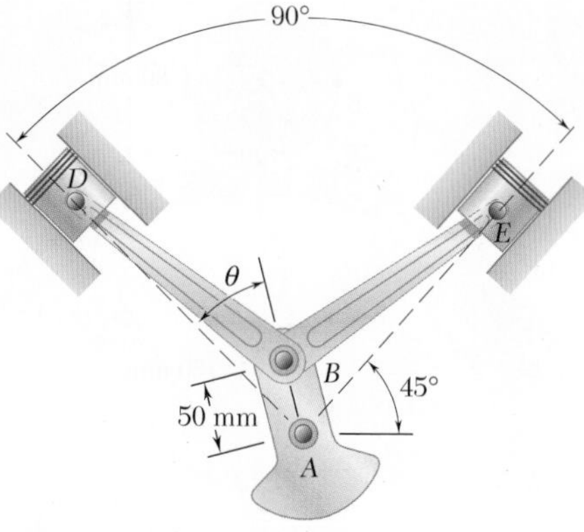

Fig. P15.122

15.123 The disk shown has a constant angular velocity of 500 rpm counterclockwise. Knowing that rod BD is 250 mm long, determine the acceleration of collar D when (*a*) $\theta = 90°$, (*b*) $\theta = 180°$.

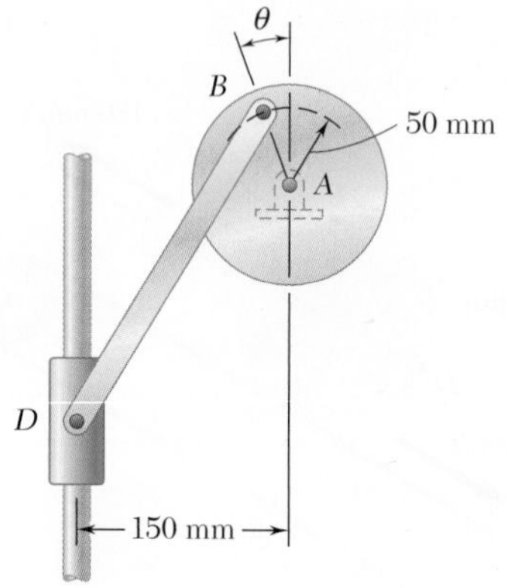

Fig. *P15.123*

15.124 Arm AB has a constant angular velocity of 16 rad/s counterclockwise. At the instant when $\theta = 90°$, determine the acceleration (*a*) of collar D, (*b*) of the midpoint G of bar BD.

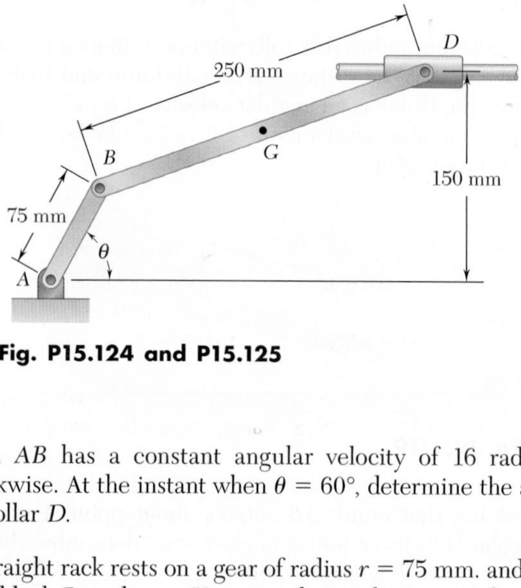

Fig. P15.124 and P15.125

15.125 Arm AB has a constant angular velocity of 16 rad/s counterclockwise. At the instant when $\theta = 60°$, determine the acceleration of collar D.

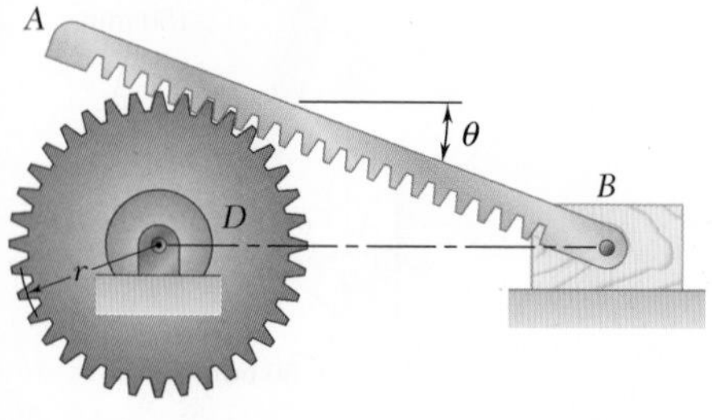

Fig. *P15.126*

15.126 A straight rack rests on a gear of radius $r = 75$ mm. and is attached to a block B as shown. Knowing that at the instant shown $\theta = 20°$, the angular velocity of gear D is 3 rad/s clockwise, and it is speeding up at a rate of 2 rad/s^2, determine (*a*) the angular acceleration of AB, (*b*) the acceleration of block B.

15.127 Knowing that at the instant shown rod *AB* has a constant angular velocity of 6 rad/s clockwise, determine the acceleration of point *D*.

15.128 Knowing that at the instant shown rod *AB* has a constant angular velocity of 6 rad/s clockwise, determine (*a*) the angular acceleration of member *BDE*, (*b*) the acceleration of point *E*.

15.129 Knowing that at the instant shown rod *AB* has zero angular acceleration and an angular velocity ω_0 clockwise, determine (*a*) the angular acceleration of arm *DE*, (*b*) the acceleration of point *D*.

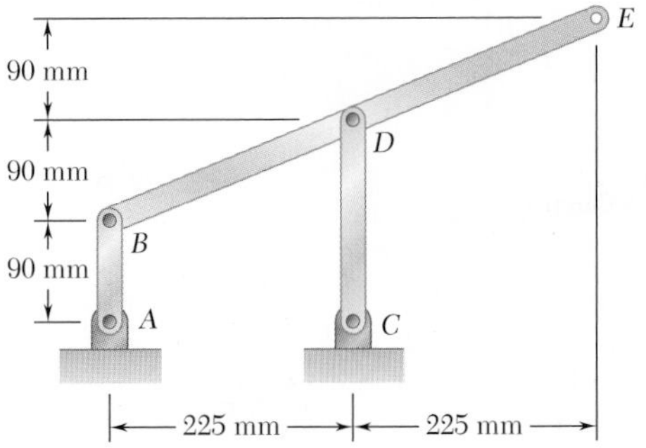

Fig. P15.127 and P15.128

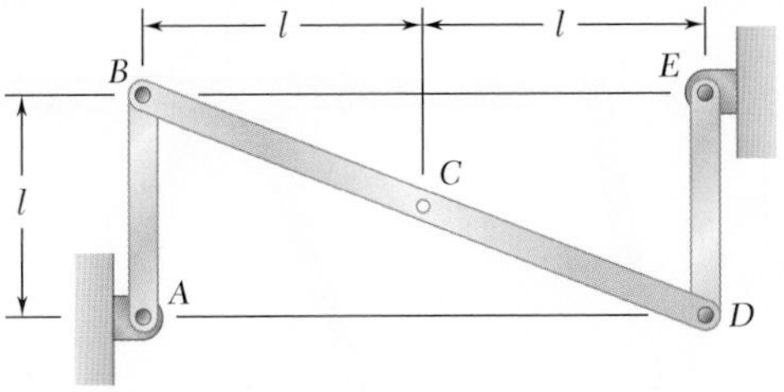

Fig. P15.129 and P15.130

15.130 At the instant shown rod *AB* has zero angular acceleration and an angular velocity of 8 rad/s clockwise. Knowing that $l = 0.3$ m, determine the acceleration of the midpoint *C* of member *BD*.

15.131 and 15.132 Knowing that at the instant shown bar *AB* has a constant angular velocity of 4 rad/s clockwise, determine the angular acceleration (*a*) of bar *BD*, (*b*) of bar *DE*.

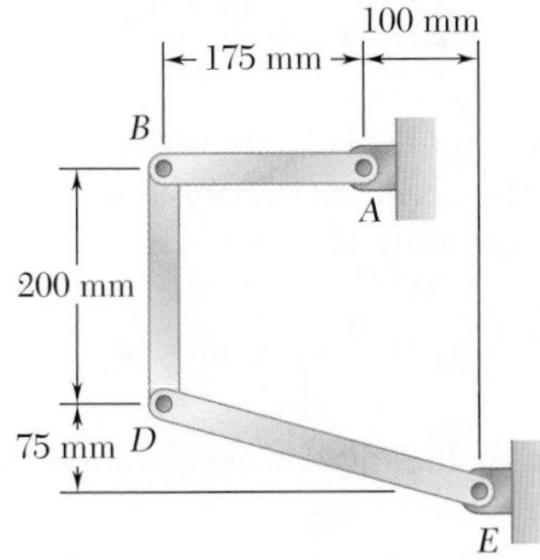

Fig. P15.132 and P15.134

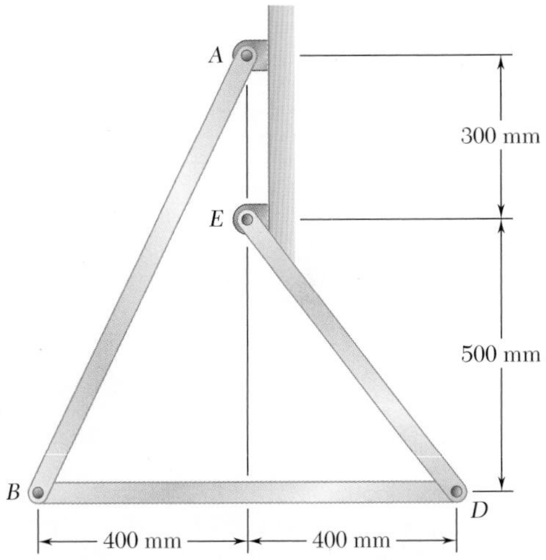

Fig. P15.131 and P15.133

15.133 and 15.134 Knowing that at the instant shown bar *AB* has an angular velocity of 4 rad/s and an angular acceleration of 2 rad/s^2, both clockwise, determine the angular acceleration (*a*) of bar *BD*, (*b*) of bar *DE* by using the vector approach as is done in Sample Prob. 15.8.

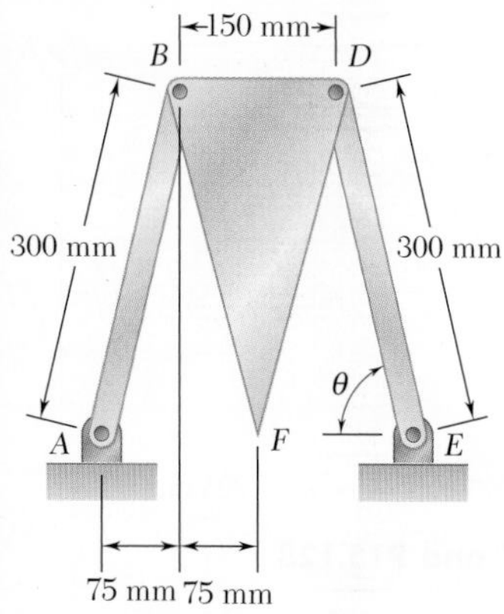

Fig. P15.135

15.135 Roberts linkage is named after Richard Roberts (1789–1864) and can be used to draw a close approximation to a straight line by locating a pen at point F. The distance AB is the same as BF, DF, and DE. Knowing that at the instant shown, bar AB has a constant angular velocity of 4 rad/s clockwise, determine (*a*) the angular acceleration of bar DE, (*b*) the acceleration of point F.

15.136 For the oil pump rig shown, link AB causes the beam BCE to oscillate as the crank OA revolves. Knowing that OA has a radius of 0.6 m and a constant clockwise angular velocity of 20 rpm, determine the velocity and acceleration of point D at the instant shown.

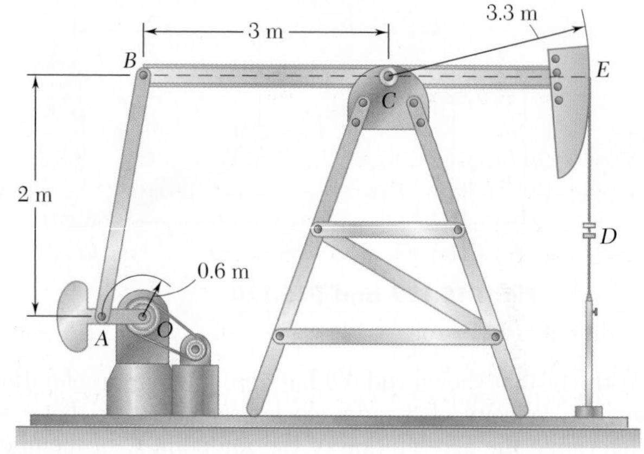

Fig. P15.136

15.137 Denoting by $\mathbf{r}_A$ the position vector of a point A of a rigid slab that is in plane motion, show that (*a*) the position vector $\mathbf{r}_C$ of the instantaneous center of rotation is

$$\mathbf{r}_C = \mathbf{r}_A + \frac{\boldsymbol{\omega} \times \mathbf{v}_A}{\omega^2}$$

where $\boldsymbol{\omega}$ is the angular velocity of the slab and $\mathbf{v}_A$ is the velocity of point A, (*b*) the acceleration of the instantaneous center of rotation is zero if, and only if,

$$\mathbf{a}_A = \frac{\alpha}{\omega}\mathbf{v}_A + \boldsymbol{\omega} \times \mathbf{v}_A$$

where $\boldsymbol{\alpha} = \alpha\mathbf{k}$ is the angular acceleration of the slab.

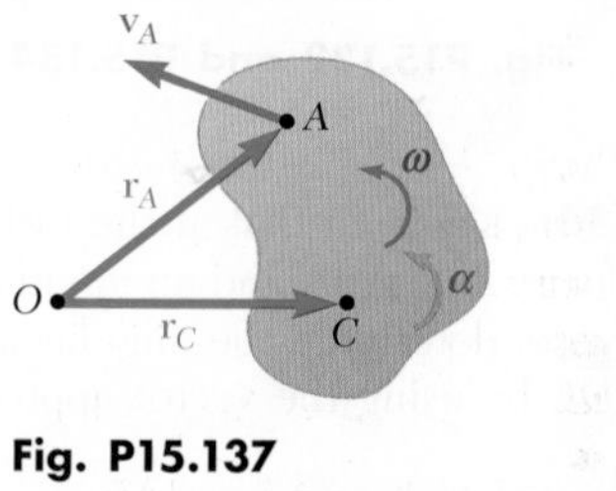

Fig. P15.137

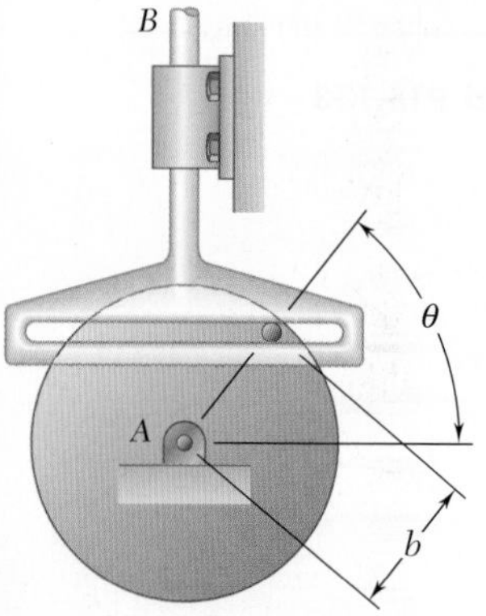

Fig. P15.138

***15.138** The drive disk of the Scotch crosshead mechanism shown has an angular velocity $\boldsymbol{\omega}$ and an angular acceleration $\boldsymbol{\alpha}$, both directed counterclockwise. Using the method of Sec. 15.9, derive expressions for the velocity and acceleration of point B.

***15.139** The wheels attached to the ends of rod AB roll along the surfaces shown. Using the method of Sec. 15.9, derive an expression for the angular velocity of the rod in terms of v_B, θ, l, and β.

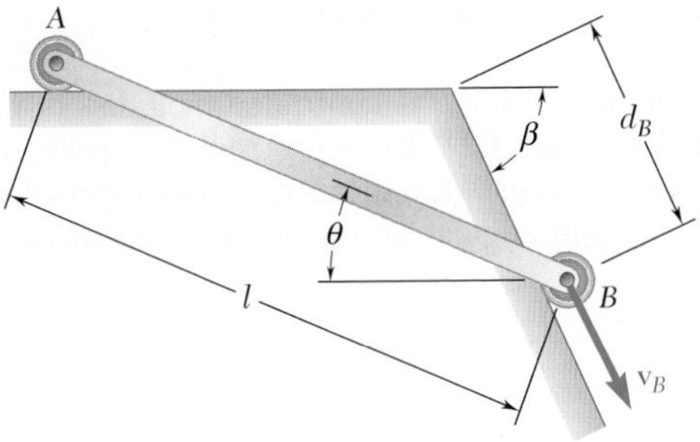

Fig. P15.139 and P15.140

***15.140** The wheels attached to the ends of rod AB roll along the surfaces shown. Using the method of Sec. 15.9 and knowing that the acceleration of wheel B is zero, derive an expression for the angular acceleration of the rod in terms of v_B, θ, l, and β.

***15.141** A disk of radius r rolls to the right with a constant velocity $\mathbf{v}$. Denoting by P the point of the rim in contact with the ground at $t = 0$, derive expressions for the horizontal and vertical components of the velocity of P at any time t.

***15.142** Rod AB moves over a small wheel at C while end A moves to the right with a constant velocity $\mathbf{v}_A$. Using the method of Sec. 15.9, derive expressions for the angular velocity and angular acceleration of the rod.

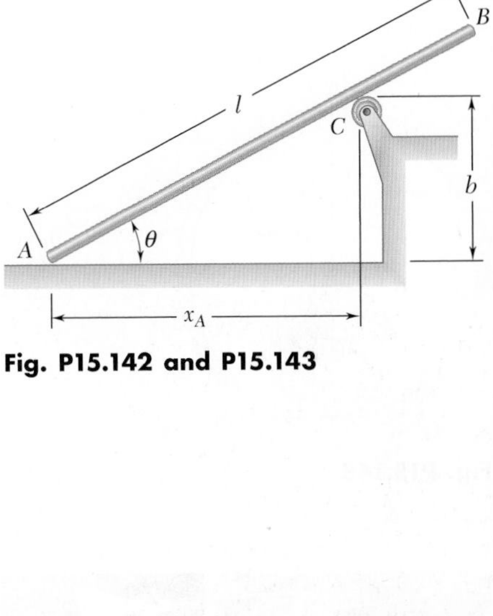

Fig. P15.142 and P15.143

***15.143** Rod AB moves over a small wheel at C while end A moves to the right with a constant velocity $\mathbf{v}_A$. Using the method of Sec. 15.9, derive expressions for the horizontal and vertical components of the velocity of point B.

15.144 Crank AB rotates with a constant clockwise angular velocity $\boldsymbol{\omega}$. Using the method of Sec. 15.9, derive expressions for the angular velocity of rod BD and the velocity of the point on the rod coinciding with point E in terms of θ, ω, b, and l.

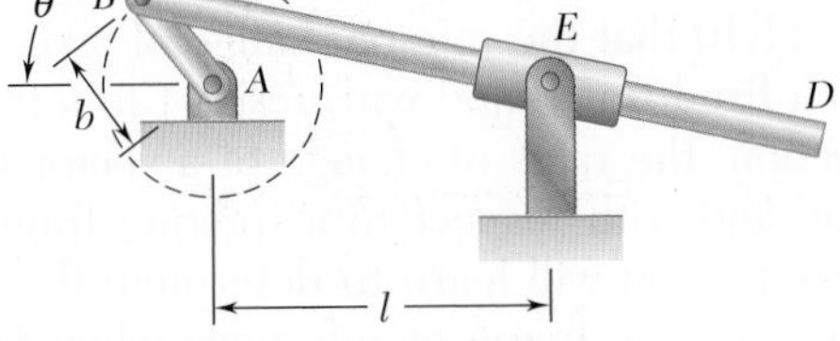

Fig. P15.144 and P15.145

15.145 Crank AB rotates with a constant clockwise angular velocity $\boldsymbol{\omega}$. Using the method of Sec. 15.9, derive an expression for the angular acceleration of rod BD in terms of θ, ω, b, and l.

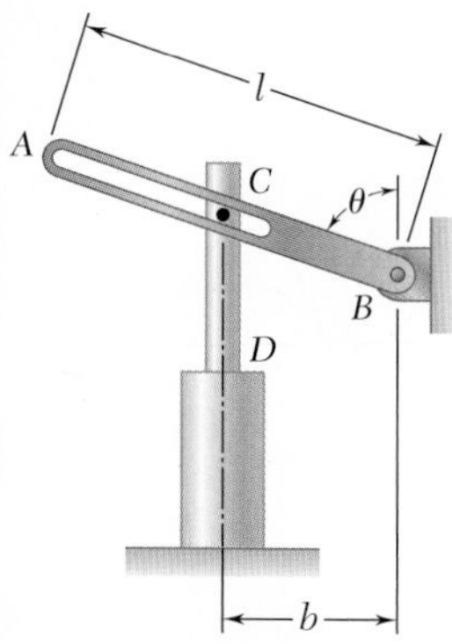

Fig. P15.146

15.146 Pin C is attached to rod CD and slides in a slot cut in arm AB. Knowing that rod CD moves vertically upward with a constant velocity $\mathbf{v}_0$, derive an expression for (*a*) the angular velocity of arm AB, (*b*) the components of the velocity of point A, (*c*) an expression for the angular acceleration of arm AB.

***15.147** The position of rod AB is controlled by a disk of radius r which is attached to yoke CD. Knowing that the yoke moves vertically upward with a constant velocity $\mathbf{v}_0$, derive expressions for the angular velocity and angular acceleration of rod AB.

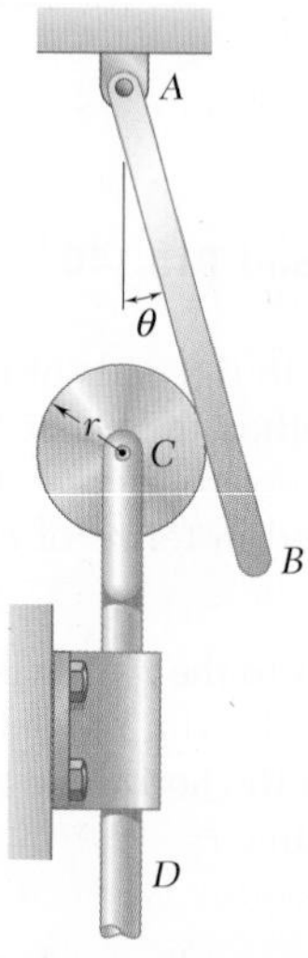

Fig. P15.147

***15.148** A wheel of radius r rolls without slipping along the inside of a fixed cylinder of radius R with a constant angular velocity $\boldsymbol{\omega}$. Denoting by P the point of the wheel in contact with the cylinder at $t = 0$, derive expressions for the horizontal and vertical components of the velocity of P at any time t. (The curve described by point P is a *hypocycloid*.)

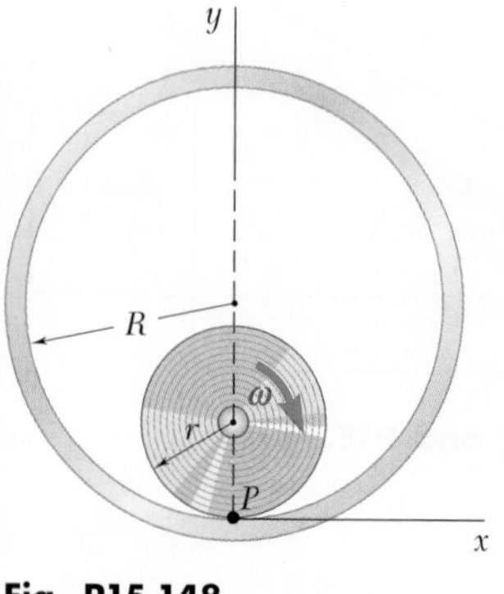

Fig. P15.148

***15.149** In Prob. 15.148, show that the path of P is a vertical straight line when $r = R/2$. Derive expressions for the corresponding velocity and acceleration of P at any time t.

Photo 15.7 A Geneva mechanism is used to convert rotary motion into intermittent motion.

15.10 RATE OF CHANGE OF A VECTOR WITH RESPECT TO A ROTATING FRAME

We saw in Sec. 11.10 that the rate of change of a vector is the same with respect to a fixed frame and with respect to a frame in translation. In this section, the rates of change of a vector $\mathbf{Q}$ with respect to a fixed frame and with respect to a rotating frame of reference will be considered.† You will learn to determine the rate of change of $\mathbf{Q}$ with respect to one frame of reference when $\mathbf{Q}$ is defined by its components in another frame.

†It is recalled that the selection of a fixed frame of reference is arbitrary. Any frame may be designated as "fixed"; all others will then be considered as moving.

Consider two frames of reference centered at O, a fixed frame $OXYZ$ and a frame $Oxyz$ which rotates about the fixed axis OA; let $\boldsymbol{\Omega}$ denote the angular velocity of the frame $Oxyz$ at a given instant (Fig. 15.26). Consider now a vector function $\mathbf{Q}(t)$ represented by the vector $\mathbf{Q}$ attached at O; as the time t varies, both the direction and the magnitude of $\mathbf{Q}$ change. Since the variation of $\mathbf{Q}$ is viewed differently by an observer using $OXYZ$ as a frame of reference and by an observer using $Oxyz$, we should expect the rate of change of $\mathbf{Q}$ to depend upon the frame of reference which has been selected. Therefore, the rate of change of $\mathbf{Q}$ with respect to the fixed frame $OXYZ$ will be denoted by $(\dot{\mathbf{Q}})_{OXYZ}$, and the rate of change of $\mathbf{Q}$ with respect to the rotating frame $Oxyz$ will be denoted by $(\dot{\mathbf{Q}})_{Oxyz}$. We propose to determine the relation existing between these two rates of change.

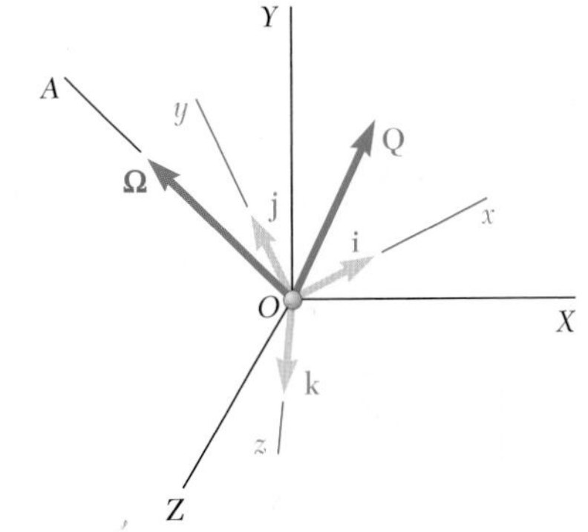

Fig. 15.26

Let us first resolve the vector $\mathbf{Q}$ into components along the x, y, and z axes of the rotating frame. Denoting by $\mathbf{i}$, $\mathbf{j}$, and $\mathbf{k}$ the corresponding unit vectors, we write

$$\mathbf{Q} = Q_x\mathbf{i} + Q_y\mathbf{j} + Q_z\mathbf{k} \tag{15.27}$$

Differentiating (15.27) with respect to t and considering the unit vectors $\mathbf{i}$, $\mathbf{j}$, $\mathbf{k}$ as fixed, we obtain the rate of change of $\mathbf{Q}$ *with respect to the rotating frame Oxyz:*

$$(\dot{\mathbf{Q}})_{Oxyz} = \dot{Q}_x\mathbf{i} + \dot{Q}_y\mathbf{j} + \dot{Q}_z\mathbf{k} \tag{15.28}$$

To obtain the rate of change of $\mathbf{Q}$ *with respect to the fixed frame OXYZ*, we must consider the unit vectors $\mathbf{i}$, $\mathbf{j}$, $\mathbf{k}$ as variable when differentiating (15.27). We therefore write

$$(\dot{\mathbf{Q}})_{OXYZ} = \dot{Q}_x\mathbf{i} + \dot{Q}_y\mathbf{j} + \dot{Q}_z\mathbf{k} + Q_x\frac{d\mathbf{i}}{dt} + Q_y\frac{d\mathbf{j}}{dt} + Q_z\frac{d\mathbf{k}}{dt} \tag{15.29}$$

Recalling (15.28), we observe that the sum of the first three terms in the right-hand member of (15.29) represents the rate of change $(\dot{\mathbf{Q}})_{Oxyz}$. We note, on the other hand, that the rate of change $(\dot{\mathbf{Q}})_{OXYZ}$ would reduce to the last three terms in (15.29) if the vector $\mathbf{Q}$ were fixed within the frame $Oxyz$, since $(\dot{\mathbf{Q}})_{Oxyz}$ would then be zero. But in that case, $(\dot{\mathbf{Q}})_{OXYZ}$ would represent the velocity of a particle located at the tip of $\mathbf{Q}$ and belonging to a body rigidly attached to the frame $Oxyz$. Thus, the last three terms in (15.29) represent the velocity of that particle; since the frame $Oxyz$ has an angular velocity $\boldsymbol{\Omega}$ with respect to $OXYZ$ at the instant considered, we write, by (15.5),

$$Q_x\frac{d\mathbf{i}}{dt} + Q_y\frac{d\mathbf{j}}{dt} + Q_z\frac{d\mathbf{k}}{dt} = \boldsymbol{\Omega} \times \mathbf{Q} \tag{15.30}$$

Substituting from (15.28) and (15.30) into (15.29), we obtain the fundamental relation

$$(\dot{\mathbf{Q}})_{OXYZ} = (\dot{\mathbf{Q}})_{Oxyz} + \boldsymbol{\Omega} \times \mathbf{Q} \tag{15.31}$$

We conclude that the rate of change of the vector $\mathbf{Q}$ with respect to the fixed frame $OXYZ$ is made of two parts: The first part represents the rate of change of $\mathbf{Q}$ with respect to the rotating frame $Oxyz$; the second part, $\boldsymbol{\Omega} \times \mathbf{Q}$, is induced by the rotation of the frame $Oxyz$.

The use of relation (15.31) simplifies the determination of the rate of change of a vector **Q** with respect to a fixed frame of reference *OXYZ* when the vector **Q** is defined by its components along the axes of a rotating frame *Oxyz*, since this relation does not require the separate computation of the derivatives of the unit vectors defining the orientation of the rotating frame.

15.11 PLANE MOTION OF A PARTICLE RELATIVE TO A ROTATING FRAME. CORIOLIS ACCELERATION

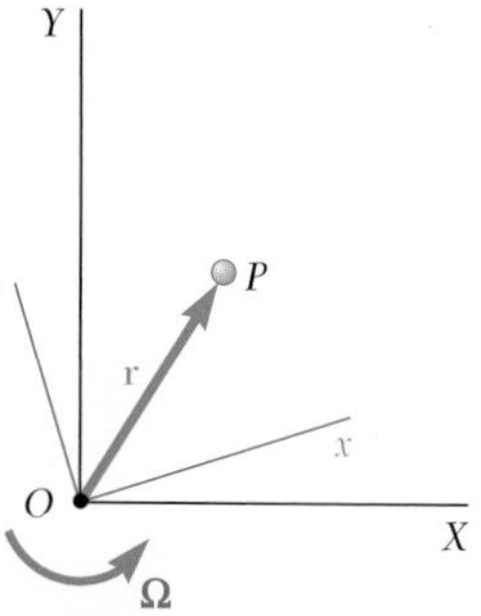

Fig. 15.27

Consider two frames of reference, both centered at *O* and both in the plane of the figure, a fixed frame *OXY* and a rotating frame *Oxy* (Fig. 15.27). Let *P* be a particle moving in the plane of the figure. The position vector **r** of *P* is the same in both frames, but its rate of change depends upon the frame of reference which has been selected.

The absolute velocity $\mathbf{v}_P$ of the particle is defined as the velocity observed from the fixed frame *OXY* and is equal to the rate of change $(\dot{\mathbf{r}})_{OXY}$ of **r** with respect to that frame. We can, however, express $\mathbf{v}_P$ in terms of the rate of change $(\dot{\mathbf{r}})_{Oxy}$ observed from the rotating frame if we make use of Eq. (15.31). Denoting by $\boldsymbol{\Omega}$ the angular velocity of the frame *Oxy* with respect to *OXY* at the instant considered, we write

$$\mathbf{v}_P = (\dot{\mathbf{r}})_{OXY} = \boldsymbol{\Omega} \times \mathbf{r} + (\dot{\mathbf{r}})_{Oxy} \tag{15.32}$$

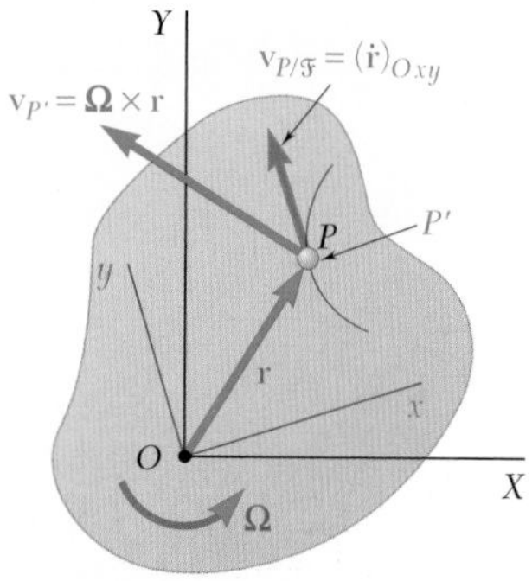

Fig. 15.28

But $(\dot{\mathbf{r}})_{Oxy}$ defines the velocity of the particle *P* relative to the rotating frame *Oxy*. Denoting the rotating frame by $\mathcal{F}$ for short, we represent the velocity $(\dot{\mathbf{r}})_{Oxy}$ of *P* relative to the rotating frame by $\mathbf{v}_{P/\mathcal{F}}$. Let us imagine that a rigid slab has been attached to the rotating frame. Then $v_{P/\mathcal{F}}$ represents the velocity of *P* along the path that it describes on that slab (Fig. 15.28), and the term $\boldsymbol{\Omega} \times \mathbf{r}$ in (15.32) represents the velocity $\mathbf{v}_{P'}$ of the point *P'* of the slab—or rotating frame—which coincides with *P* at the instant considered. Thus, we have

$$\mathbf{v}_P = \mathbf{v}_{P'} + \mathbf{v}_{P/\mathcal{F}} \tag{15.33}$$

where $\mathbf{v}_P$ = absolute velocity of particle *P*
$\mathbf{v}_{P'}$ = velocity of point *P'* of moving frame $\mathcal{F}$ coinciding with *P*
$\mathbf{v}_{P/\mathcal{F}}$ = velocity of *P* relative to moving frame $\mathcal{F}$

The absolute acceleration $\mathbf{a}_P$ of the particle is defined as the rate of change of $\mathbf{v}_P$ with respect to the fixed frame *OXY*. Computing the rates of change with respect to *OXY* of the terms in (15.32), we write

$$\mathbf{a}_P = \dot{\mathbf{v}}_P = \dot{\boldsymbol{\Omega}} \times \mathbf{r} + \boldsymbol{\Omega} \times \dot{\mathbf{r}} + \frac{d}{dt}[(\dot{\mathbf{r}})_{Oxy}] \tag{15.34}$$

where all derivatives are defined with respect to *OXY*, except where indicated otherwise. Referring to Eq. (15.31), we note that the last term in (15.34) can be expressed as

$$\frac{d}{dt}[(\dot{\mathbf{r}})_{Oxy}] = (\ddot{\mathbf{r}})_{Oxy} + \boldsymbol{\Omega} \times (\dot{\mathbf{r}})_{Oxy}$$

On the other hand, $\dot{\mathbf{r}}$ represents the velocity $\mathbf{v}_P$ and can be replaced by the right-hand member of Eq. (15.32). After completing these two substitutions into (15.34), we write

$$\mathbf{a}_P = \dot{\boldsymbol{\Omega}} \times \mathbf{r} + \boldsymbol{\Omega} \times (\boldsymbol{\Omega} \times \mathbf{r}) + 2\boldsymbol{\Omega} \times (\dot{\mathbf{r}})_{Oxy} + (\ddot{\mathbf{r}})_{Oxy} \tag{15.35}$$

Referring to the expression (15.8) obtained in Sec. 15.3 for the acceleration of a particle in a rigid body rotating about a fixed axis, we note that the sum of the first two terms represents the acceleration $\mathbf{a}_{P'}$ of the point P' of the rotating frame which coincides with P at the instant considered. On the other hand, the last term defines the acceleration $\mathbf{a}_{P/\mathcal{F}}$ of P relative to the rotating frame. If it were not for the third term, which has not been accounted for, a relation similar to (15.33) could be written for the accelerations, and $\mathbf{a}_P$ could be expressed as the sum of $\mathbf{a}_{P'}$ and $\mathbf{a}_{P/\mathcal{F}}$. However, it is clear that *such a relation would be incorrect* and that we must include the additional term. This term, which will be denoted by $\mathbf{a}_c$, is called the *complementary acceleration*, or *Coriolis acceleration*, after the French mathematician de Coriolis (1792–1843). We write

$$\mathbf{a}_P = \mathbf{a}_{P'} + \mathbf{a}_{P/\mathcal{F}} + \mathbf{a}_c \tag{15.36}$$

where $\mathbf{a}_P$ = absolute acceleration of particle P
$\mathbf{a}_{P'}$ = acceleration of point P' of moving frame $\mathcal{F}$ coinciding with P
$\mathbf{a}_{P/\mathcal{F}}$ = acceleration of P relative to moving frame $\mathcal{F}$
$\mathbf{a}_c = 2\boldsymbol{\Omega} \times (\dot{\mathbf{r}})_{Oxy} = 2\boldsymbol{\Omega} \times \mathbf{v}_{P/\mathcal{F}}$
= complementary, or Coriolis, acceleration†

We note that since point P' moves in a circle about the origin O, its acceleration $\mathbf{a}_{P'}$ has, in general, two components: a component $(\mathbf{a}_{P'})_t$ tangent to the circle, and a component $(\mathbf{a}_{P'})_n$ directed toward O. Similarly, the acceleration $\mathbf{a}_{P/\mathcal{F}}$ generally has two components: a component $(\mathbf{a}_{P/\mathcal{F}})_t$ tangent to the path that P describes on the rotating slab, and a component $(\mathbf{a}_{P/\mathcal{F}})_n$ directed toward the center of curvature of that path. We further note that since the vector $\boldsymbol{\Omega}$ is perpendicular to the plane of motion, and thus to $\mathbf{v}_{P/\mathcal{F}}$, the magnitude of the Coriolis acceleration $\mathbf{a}_c = 2\boldsymbol{\Omega} \times \mathbf{v}_{P/\mathcal{F}}$ is equal to $2\Omega v_{P/\mathcal{F}}$, and its direction can be obtained by rotating the vector $\mathbf{v}_{P/\mathcal{F}}$ through 90° in the sense of rotation of the moving frame (Fig. 15.29). The Coriolis acceleration reduces to zero when either $\boldsymbol{\Omega}$ or $\mathbf{v}_{P/\mathcal{F}}$ is zero.

The following example will help in understanding the physical meaning of the Coriolis acceleration. Consider a collar P which is

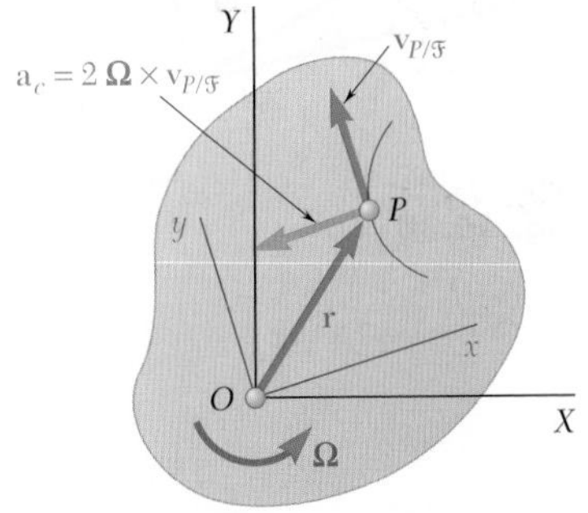

Fig. 15.29

†It is important to note the difference between Eq. (15.36) and Eq. (15.21) of Sec. 15.8. When we wrote

$$\mathbf{a}_B = \mathbf{a}_A + \mathbf{a}_{B/A} \tag{15.21}$$

in Sec. 15.8, we were expressing the absolute acceleration of point B as the sum of its acceleration $\mathbf{a}_{B/A}$ relative to a *frame in translation* and of the acceleration $\mathbf{a}_A$ of a point of that frame. We are now trying to relate the absolute acceleration of point P to its acceleration $\mathbf{a}_{P/\mathcal{F}}$ relative to a *rotating frame* $\mathcal{F}$ and to the acceleration $\mathbf{a}_{P'}$ of the point P' of that frame which coincides with P; Eq. (15.36) shows that because the frame is rotating, it is necessary to include an additional term representing the Coriolis acceleration $\mathbf{a}_c$.

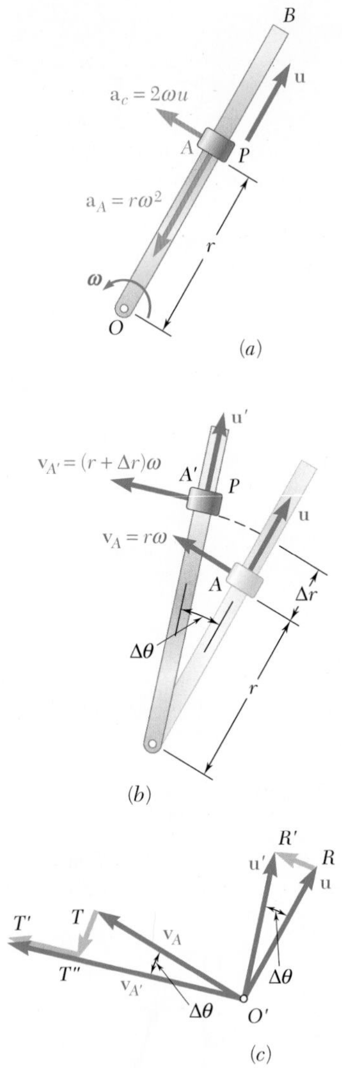

Fig. 15.30

made to slide at a constant relative speed u along a rod OB rotating at a constant angular velocity $\boldsymbol{\omega}$ about O (Fig. 15.30*a*). According to formula (15.36), the absolute acceleration of P can be obtained by adding vectorially the acceleration $\mathbf{a}_A$ of the point A of the rod coinciding with P, the relative acceleration $\mathbf{a}_{P/OB}$ of P with respect to the rod, and the Coriolis acceleration $\mathbf{a}_c$. Since the angular velocity $\boldsymbol{\omega}$ of the rod is constant, $\mathbf{a}_A$ reduces to its normal component $(\mathbf{a}_A)_n$ of magnitude $r\omega^2$; and since u is constant, the relative acceleration $\mathbf{a}_{P/OB}$ is zero. According to the definition given above, the Coriolis acceleration is a vector perpendicular to OB, of magnitude $2\omega u$, and directed as shown in the figure. The acceleration of the collar P consists, therefore, of the two vectors shown in Fig. 15.30*a*. Note that the result obtained can be checked by applying the relation (11.44).

To understand better the significance of the Coriolis acceleration, let us consider the absolute velocity of P at time t and at time $t + \Delta t$ (Fig. 15.30*b*). The velocity at time t can be resolved into its components $\mathbf{u}$ and $\mathbf{v}_A$; the velocity at time $t + \Delta t$ can be resolved into its components $\mathbf{u}'$ and $\mathbf{v}_{A'}$. Drawing these components from the same origin (Fig. 15.30*c*), we note that the change in velocity during the time Δt can be represented by the sum of three vectors, $\overrightarrow{RR'}$, $\overrightarrow{TT''}$, and $\overrightarrow{T''T'}$. The vector $\overrightarrow{TT''}$ measures the change in direction of the velocity $\mathbf{v}_A$, and the quotient $\overrightarrow{TT''}/\Delta t$ represents the acceleration $\mathbf{a}_A$ when Δt approaches zero. We check that the direction of $\overrightarrow{TT''}$ is that of $\mathbf{a}_A$ when Δt approaches zero and that

$$\lim_{\Delta t \to 0} \frac{TT''}{\Delta t} = \lim_{\Delta t \to 0} v_A \frac{\Delta\theta}{\Delta t} = r\omega\omega = r\omega^2 = a_A$$

The vector $\overrightarrow{RR'}$ measures the change in direction of $\mathbf{u}$ due to the rotation of the rod; the vector $\overrightarrow{T''T'}$ measures the change in magnitude of $\mathbf{v}_A$ due to the motion of P on the rod. The vectors $\overrightarrow{RR'}$ and $\overrightarrow{T''T'}$ result from the *combined effect* of the relative motion of P and of the rotation of the rod; they would vanish if *either* of these two motions stopped. It is easily verified that the sum of these two vectors defines the Coriolis acceleration. Their direction is that of $\mathbf{a}_c$ when Δt approaches zero, and since $RR' = u\,\Delta\theta$ and $T''T' = v_{A'} - v_A = (r + \Delta r)\omega - r\omega = \omega\,\Delta r$, we check that a_c is equal to

$$\lim_{\Delta t \to 0}\left(\frac{RR'}{\Delta t} + \frac{T''T'}{\Delta t}\right) = \lim_{\Delta t \to 0}\left(u\frac{\Delta\theta}{\Delta t} + \omega\frac{\Delta r}{\Delta t}\right) = u\omega + \omega u = 2\omega u$$

Formulas (15.33) and (15.36) can be used to analyze the motion of mechanisms which contain parts sliding on each other. They make it possible, for example, to relate the absolute and relative motions of sliding pins and collars (see Sample Probs. 15.9 and 15.10). The concept of Coriolis acceleration is also very useful in the study of long-range projectiles and of other bodies whose motions are appreciably affected by the rotation of the earth. As was pointed out in Sec. 12.2, a system of axes attached to the earth does not truly constitute a newtonian frame of reference; such a system of axes should actually be considered as rotating. The formulas derived in this section will therefore facilitate the study of the motion of bodies with respect to axes attached to the earth.

SAMPLE PROBLEM 15.9

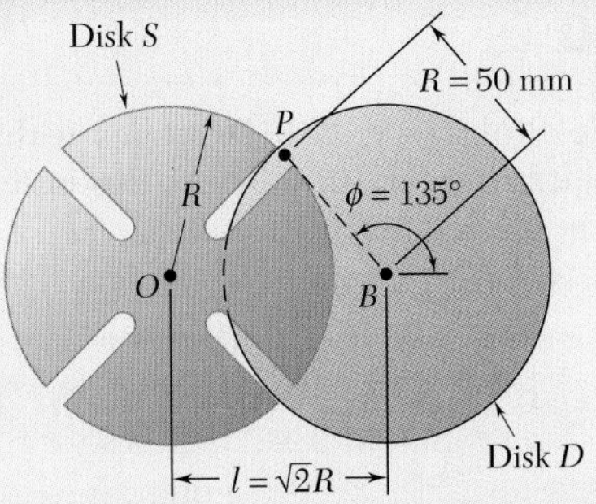

The Geneva mechanism shown is used in many counting instruments and in other applications where an intermittent rotary motion is required. Disk D rotates with a constant counterclockwise angular velocity $\boldsymbol{\omega}_D$ of 10 rad/s. A pin P is attached to disk D and slides along one of several slots cut in disk S. It is desirable that the angular velocity of disk S be zero as the pin enters and leaves each slot; in the case of four slots, this will occur if the distance between the centers of the disks is $l = \sqrt{2}\,R$.

At the instant when $\phi = 150°$, determine (*a*) the angular velocity of disk S, (*b*) the velocity of pin P relative to disk S.

SOLUTION

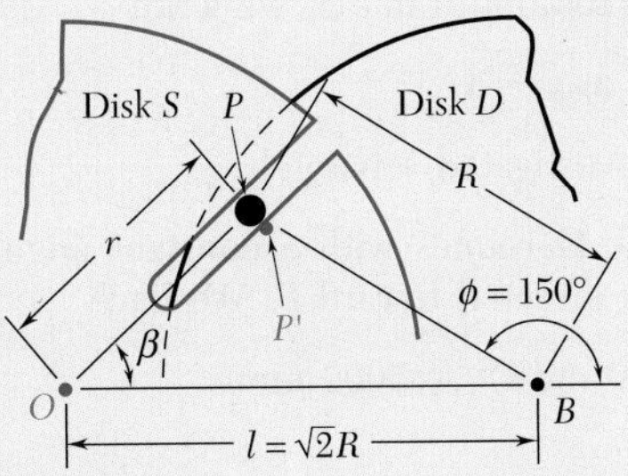

We solve triangle OPB, which corresponds to the position $\phi = 150°$. Using the law of cosines, we write

$$r^2 = R^2 + l^2 - 2Rl\cos 30° = 0.551R^2 \qquad r = 0.742R = 37.1 \text{ mm}$$

From the law of sines,

$$\frac{\sin\beta}{R} = \frac{\sin 30°}{r} \qquad \sin\beta = \frac{\sin 30°}{0.742} \qquad \beta = 42.4°$$

Since pin P is attached to disk D, and since disk D rotates about point B, the magnitude of the absolute velocity of P is

$$v_P = R\omega_D = (50 \text{ mm})(10 \text{ rad/s}) = 500 \text{ mm/s}$$
$$\mathbf{v}_P = 500 \text{ mm/s} \nearrow 60°$$

We consider now the motion of pin P along the slot in disk S. Denoting by P' the point of disk S which coincides with P at the instant considered and selecting a rotating frame $\mathcal{S}$ attached to disk S, we write

$$\mathbf{v}_P = \mathbf{v}_{P'} + \mathbf{v}_{P/\mathcal{S}}$$

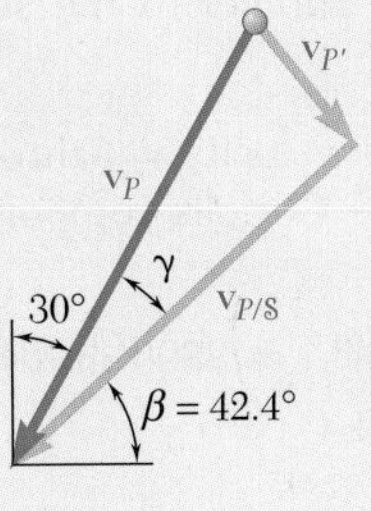

Noting that $\mathbf{v}_{P'}$ is perpendicular to the radius OP and that $\mathbf{v}_{P/\mathcal{S}}$ is directed along the slot, we draw the velocity triangle corresponding to the equation above. From the triangle, we compute

$$\gamma = 90° - 42.4° - 30° = 17.6°$$
$$v_{P'} = v_P \sin\gamma = (500 \text{ mm/s}) \sin 17.6°$$
$$\mathbf{v}_{P'} = 151.2 \text{ mm/s} \searrow 42.4°$$
$$v_{P/\mathcal{S}} = v_P \cos\gamma = (500 \text{ mm/s}) \cos 17.6°$$
$$\mathbf{v}_{P/S} = \mathbf{v}_{P/\mathcal{S}} = 477 \text{ mm/s} \nearrow 42.4° \quad ◀$$

Since $\mathbf{v}_{P'}$ is perpendicular to the radius OP, we write

$$v_{P'} = r\omega_{\mathcal{S}} \qquad 151.2 \text{ mm/s} = (37.1 \text{ mm})\omega_{\mathcal{S}}$$
$$\boldsymbol{\omega}_S = \boldsymbol{\omega}_{\mathcal{S}} = 4.08 \text{ rad/s} \circlearrowright \quad ◀$$

SAMPLE PROBLEM 15.10

In the Geneva mechanism of Sample Prob. 15.9, disk D rotates with a constant counterclockwise angular velocity $\boldsymbol{\omega}_D$ of 10 rad/s. At the instant when $\phi = 150°$, determine the angular acceleration of disk S.

SOLUTION

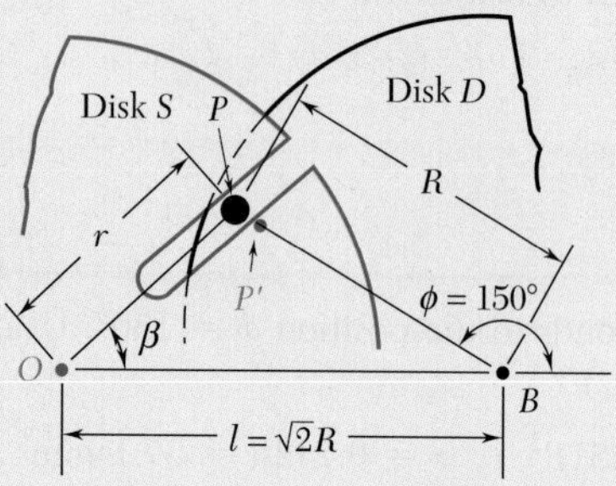

Referring to Sample Prob. 15.9, we obtain the angular velocity of the frame $\mathcal{S}$ attached to disk S and the velocity of the pin relative to $\mathcal{S}$:

$$\omega_{\mathcal{S}} = 4.08 \text{ rad/s} \circlearrowright$$

$$\beta = 42.4° \qquad \mathbf{v}_{P/\mathcal{S}} = 477 \text{ mm/s} \nearrow 42.4°$$

Since pin P moves with respect to the rotating frame $\mathcal{S}$, we write

$$\mathbf{a}_P = \mathbf{a}_{P'} + \mathbf{a}_{P/\mathcal{S}} + \mathbf{a}_c \qquad (1)$$

Each term of this vector equation is investigated separately.

Absolute Acceleration $\mathbf{a}_P$. Since disk D rotates with a constant angular velocity, the absolute acceleration $\mathbf{a}_P$ is directed toward B. We have

$$a_P = R\omega_D^2 = (500 \text{ mm})(10 \text{ rad/s})^2 = 5000 \text{ mm/s}^2$$

$$\mathbf{a}_P = 5000 \text{ mm/s}^2 \searrow 30°$$

Acceleration $\mathbf{a}_{P'}$ of the Coinciding Point P'. The acceleration $\mathbf{a}_{P'}$ of the point P' of the frame $\mathcal{S}$ which coincides with P at the instant considered is resolved into normal and tangential components. (We recall from Sample Prob. 15.9 that $r = 37.1$ mm.)

$$(a_{P'})_n = r\omega_{\mathcal{S}}^2 = (37.1 \text{ mm})(4.08 \text{ rad/s})^2 = 618 \text{ mm/s}^2$$

$$(\mathbf{a}_{P'})_n = 618 \text{ mm/s}^2 \nearrow 42.4°$$

$$(a_{P'})_t = r\alpha_{\mathcal{S}} = 37.1\alpha_{\mathcal{S}} \qquad (\mathbf{a}_{P'})_t = 37.1\alpha_{\mathcal{S}} \searrow 42.4°$$

Relative Acceleration $\mathbf{a}_{P/\mathcal{S}}$. Since the pin P moves in a straight slot cut in disk S, the relative acceleration $\mathbf{a}_{P/\mathcal{S}}$ must be parallel to the slot; i.e., its direction must be $\swarrow$ 42.4°.

Coriolis Acceleration $\mathbf{a}_c$. Rotating the relative velocity $\mathbf{v}_{P/\mathcal{S}}$ through 90° in the sense of $\boldsymbol{\omega}_{\mathcal{S}}$, we obtain the direction of the Coriolis component of the acceleration: $\nwarrow$ 42.4°. We write

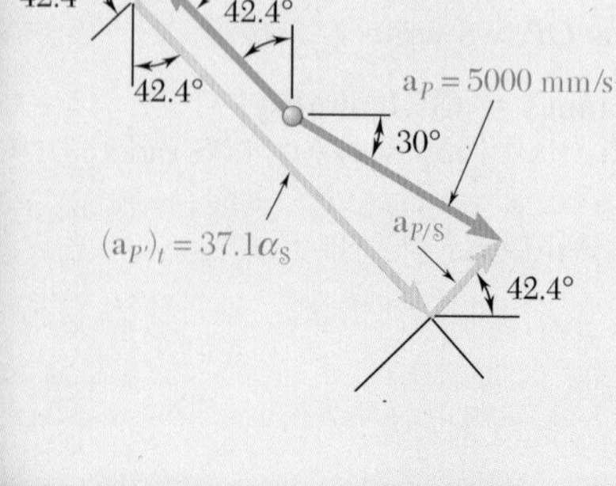

$$a_c = 2\omega_{\mathcal{S}} v_{P/\mathcal{S}} = 2(4.08 \text{ rad/s})(477 \text{ mm/s}) = 3890 \text{ mm/s}^2$$

$$\mathbf{a}_c = 3890 \text{ mm/s}^2 \nwarrow 42.4°$$

We rewrite Eq. (1) and substitute the accelerations found above:

$$\mathbf{a}_P = (\mathbf{a}_{P'})_n + (\mathbf{a}_{P'})_t + \mathbf{a}_{P/\mathcal{S}} + \mathbf{a}_c$$

$$[5000 \searrow 30°] = [618 \nearrow 42.4°] + [37.1\alpha_{\mathcal{S}} \searrow 42.4°] + [a_{P/\mathcal{S}} \swarrow 42.4°] + [3890 \nwarrow 42.4°]$$

Equating components in a direction perpendicular to the slot,

$$5000 \cos 17.6° = 37.1\alpha_{\mathcal{S}} - 3890$$

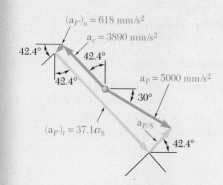

$$\boldsymbol{\alpha}_S = \boldsymbol{\alpha}_{\mathcal{S}} = 233 \text{ rad/s}^2 \circlearrowright \quad ◀$$

SOLVING PROBLEMS ON YOUR OWN

In this lesson you studied the rate of change of a vector with respect to a rotating frame and then applied your knowledge to the analysis of the plane motion of a particle relative to a rotating frame.

1. Rate of change of a vector with respect to a fixed frame and with respect to a rotating frame. Denoting by $(\dot{\mathbf{Q}})_{OXYZ}$ the rate of change of a vector $\mathbf{Q}$ with respect to a fixed frame $OXYZ$ and by $(\dot{\mathbf{Q}})_{Oxyz}$ its rate of change with respect to a rotating frame $Oxyz$, we obtained the fundamental relation

$$(\dot{\mathbf{Q}})_{OXYZ} = (\dot{\mathbf{Q}})_{Oxyz} + \boldsymbol{\Omega} \times \mathbf{Q} \tag{15.31}$$

where $\boldsymbol{\Omega}$ is the angular velocity of the rotating frame.

This fundamental relation will now be applied to the solution of two-dimensional problems.

2. Plane motion of a particle relative to a rotating frame. Using the above fundamental relation and designating by $\mathcal{F}$ the rotating frame, we obtained the following expressions for the velocity and the acceleration of a particle P:

$$\mathbf{v}_P = \mathbf{v}_{P'} + \mathbf{v}_{P/\mathcal{F}} \tag{15.33}$$

$$\mathbf{a}_P = \mathbf{a}_{P'} + \mathbf{a}_{P/\mathcal{F}} + \mathbf{a}_c \tag{15.36}$$

In these equations:

a. The subscript *P* refers to the absolute motion of the particle P, that is, to its motion with respect to a fixed frame of reference OXY.

b. The subscript *P′* refers to the motion of the point P' of the rotating frame $\mathcal{F}$ which coincides with P at the instant considered.

c. The subscript *P/𝓕* refers to the motion of the particle P relative to the rotating frame $\mathcal{F}$.

d. The term $\mathbf{a}_c$ represents the Coriolis acceleration of point *P*. Its magnitude is $2\Omega v_{P/\mathcal{F}}$, and its direction is found by rotating $\mathbf{v}_{P/\mathcal{F}}$ through 90° in the sense of rotation of the frame $\mathcal{F}$.

You should keep in mind that the Coriolis acceleration should be taken into account whenever a part of the mechanism you are analyzing is moving with respect to another part that is rotating. The problems you will encounter in this lesson involve collars that slide on rotating rods, booms that extend from cranes rotating in a vertical plane, etc.

When solving a problem involving a rotating frame, you will find it convenient to draw vector diagrams representing Eqs. (15.33) and (15.36), respectively, and use these diagrams to obtain either an analytical or a graphical solution.

PROBLEMS

CONCEPT QUESTION

15.CQ8 A person walks radially inward on a platform that is rotating counterclockwise about its center. Knowing that the platform has a constant angular velocity $\boldsymbol{\omega}$ and the person walks with a constant speed $\mathbf{u}$ relative to the platform, what is the direction of the acceleration of the person at the instant shown?

a. Negative x
b. Negative y
c. Negative x and positive y
d. Positive x and positive y
e. Negative x and negative y

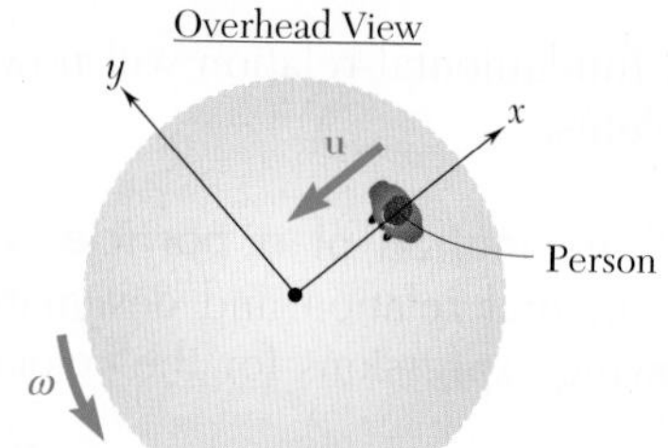

Fig. P15.CQ8

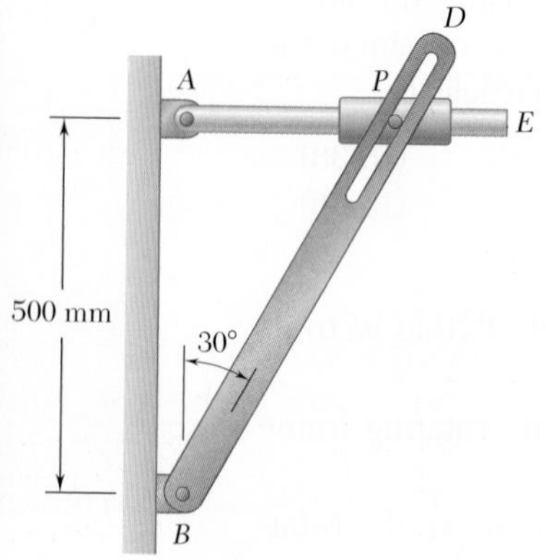

Fig. P15.150 and P15.151

END-OF-SECTION PROBLEMS

15.150 and 15.151 Pin P is attached to the collar shown; the motion of the pin is guided by a slot cut in rod BD and by the collar that slides on rod AE. Knowing that at the instant considered the rods rotate clockwise with constant angular velocities, determine for the given data the velocity of pin P.

15.150 $\omega_{AE} = 8$ rad/s, $\omega_{BD} = 3$ rad/s
15.151 $\omega_{AE} = 7$ rad/s, $\omega_{BD} = 4.8$ rad/s

15.152 and 15.153 Two rotating rods are connected by slider block P. The rod attached at A rotates with a constant clockwise angular velocity ω_A. For the given data, determine for the position shown (*a*) the angular velocity of the rod attached at B, (*b*) the relative velocity of slider block P with respect to the rod on which it slides.

15.152 $b = 200$ mm, $\omega_A = 6$ rad/s.
15.153 $b = 300$ mm, $\omega_A = 10$ rad/s

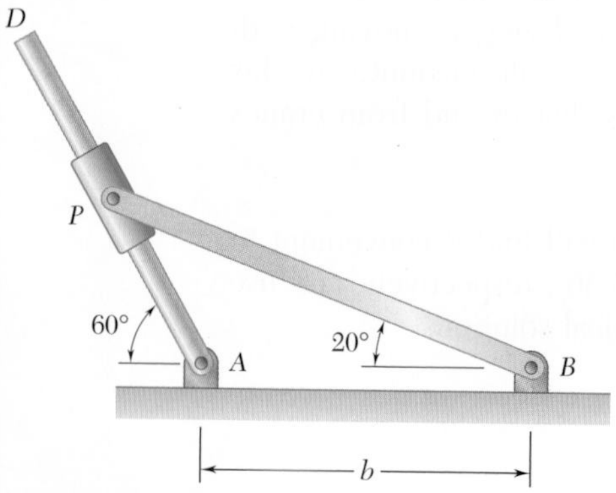

Fig. P15.153

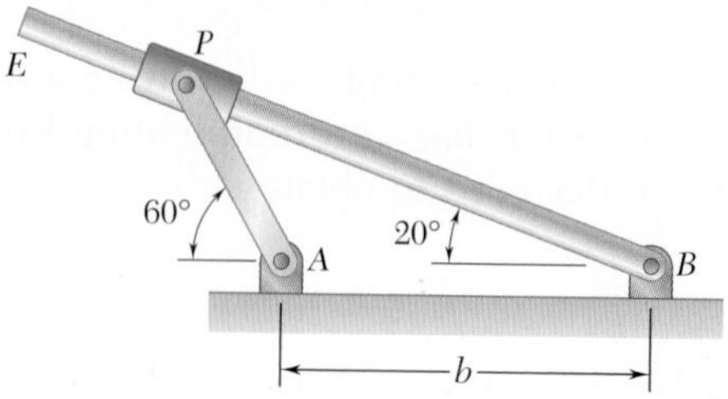

Fig. P15.152

15.154 Pin P is attached to the wheel shown and slides in a slot cut in bar BD. The wheel rolls to the right without slipping with a constant angular velocity of 20 rad/s. Knowing that $x = 480$ mm when $\theta = 0$, determine the angular velocity of the bar and the relative velocity of pin P with respect to the rod when (a) $\theta = 0$, (b) $\theta = 90°$.

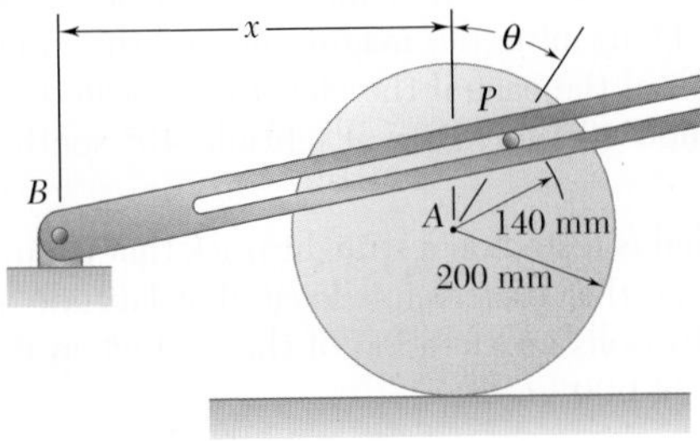

Fig. P15.154

15.155 Bar AB rotates clockwise with a constant angular velocity of 8 rad/s and rod EF rotates clockwise with a constant angular velocity of 6 rad/s. Determine at the instant shown (a) the angular velocity of bar BD, (b) the relative velocity of collar D with respect to rod EF.

15.156 Bar AB rotates clockwise with a constant angular velocity of 4 rad/s. Knowing that the magnitude of the velocity of collar D is 6 m/s and that the angular velocity of bar BD is counterclockwise at the instant shown, determine (a) the angular velocity of bar EF, (b) the relative velocity of collar D with respect to rod EF.

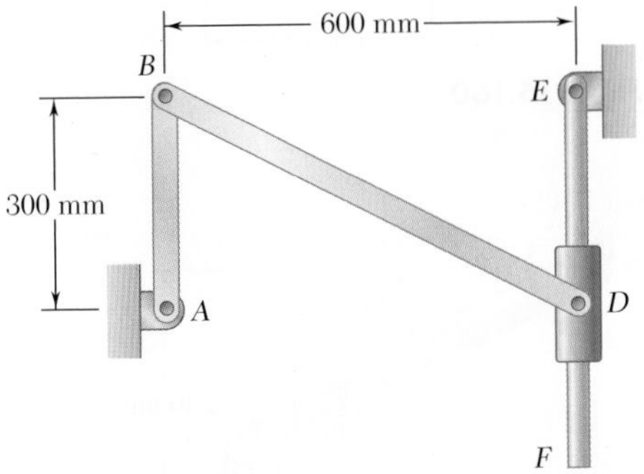

Fig. P15.155 and *P15.156*

15.157 The motion of pin P is guided by slots cut in rods AD and BE. Knowing that bar AD has a constant angular velocity of 4 rad/s clockwise and bar BE has an angular velocity of 5 rad/s counterclockwise and is slowing down at a rate of 2 rad/s^2, determine the velocity of P for the position shown.

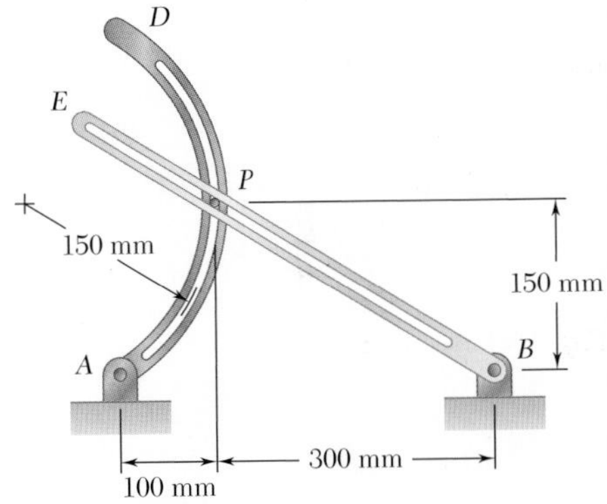

Fig. P15.157

15.158 Four pins slide in four separate slots cut in a circular plate as shown. When the plate is at rest, each pin has a velocity directed as shown and of the same constant magnitude u. If each pin maintains the same velocity relative to the plate when the plate rotates about O with a constant counterclockwise angular velocity $\boldsymbol{\omega}$, determine the acceleration of each pin.

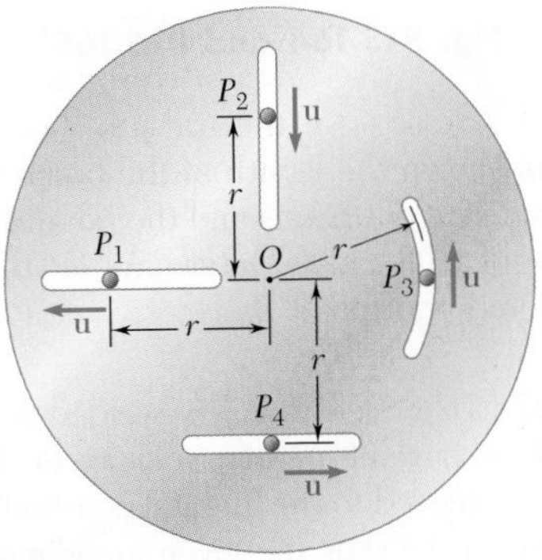

Fig. P15.158

15.159 Solve Prob. 15.158, assuming that the plate rotates about O with a constant clockwise angular velocity $\boldsymbol{\omega}$.

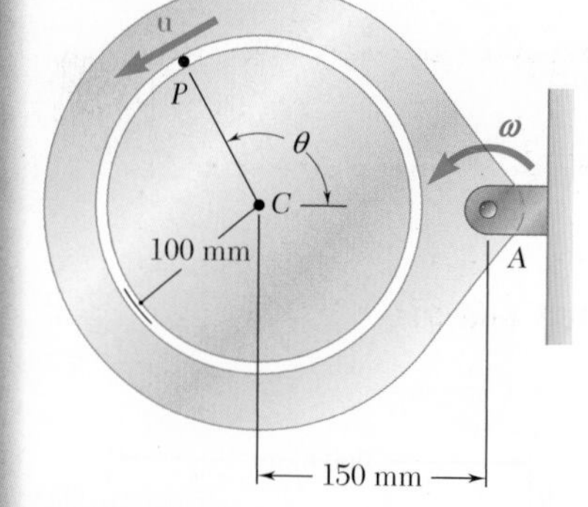

Fig. P15.160

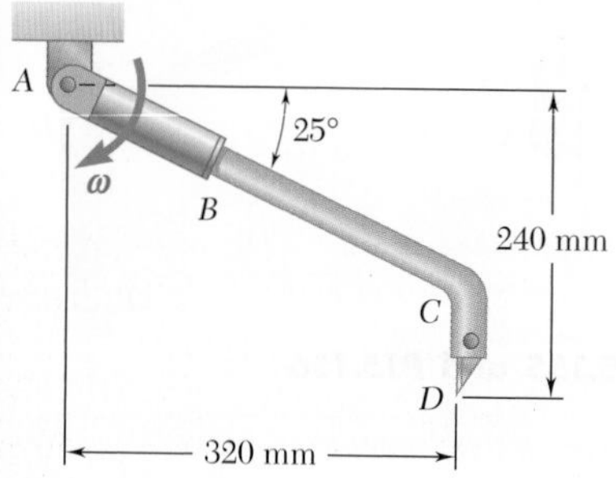

Fig. P15.163

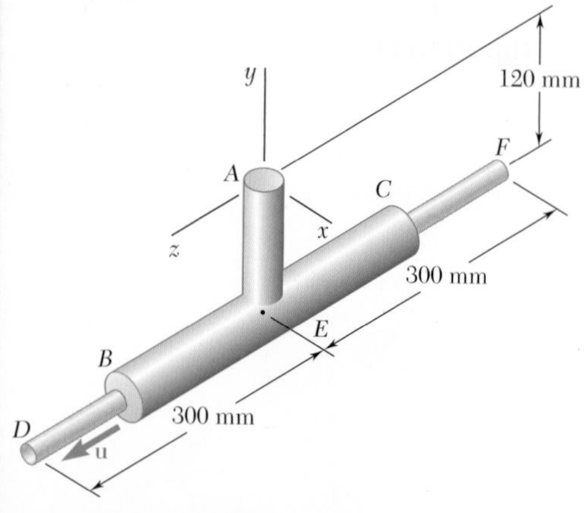

Fig. P15.166 and P15.167

15.160 Pin P slides in the circular slot cut in the plate shown at a constant relative speed $u = 500$ mm/s. Assuming that at the instant shown the angular velocity of the plate is 6 rad/s and is increasing at the rate of 20 rad/s^2, determine the acceleration of pin P when $\theta = 90°$.

15.161 The cage of a mine elevator moves downward at a constant speed of 12 m/s. Determine the magnitude and direction of the Coriolis acceleration of the cage if the elevator is located (*a*) at the equator, (*b*) at latitude 40° north, (*c*) at latitude 40° south.

15.162 A rocket sled is tested on a straight track that is built along a meridian. Knowing that the track is located at latitude 40° north, determine the Coriolis acceleration of the sled when it is moving north at a speed of 900 km/h.

15.163 The motion of blade D is controlled by the robot arm ABC. At the instant shown the arm is rotating clockwise at the constant rate $\omega = 1.8$ rad/s and the length of portion BC of the arm is being decreased at the constant rate of 250 mm/s. Determine (*a*) the velocity of D, (*b*) the acceleration of D.

15.164 At the instant shown the length of the boom AB is being *decreased* at the constant rate of 0.2 m/s and the boom is being lowered at the constant rate of 0.08 rad/s. Determine (*a*) the velocity of point B, (*b*) the acceleration of point B.

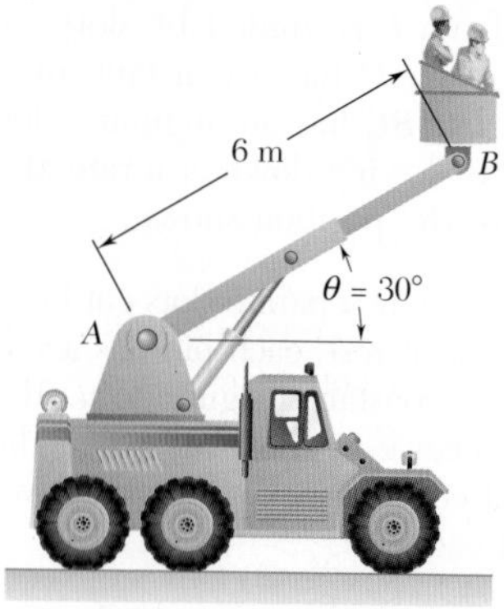

Fig. P15.164 and P15.165

15.165 At the instant shown the length of the boom AB is being *increased* at the constant rate of 0.2 m/s and the boom is being lowered at the constant rate of 0.08 rad/s. Determine (*a*) the velocity of point B, (*b*) the acceleration of point B.

15.166 and 15.167 The sleeve BC is welded to an arm that rotates about A with a constant angular velocity $\boldsymbol{\omega}$. In the position shown rod DF is being moved to the left at a constant speed $u = 400$ mm/s relative to the sleeve. For the given angular velocity $\boldsymbol{\omega}$, determine the acceleration (*a*) of point D, (*b*) of the point of rod DF that coincides with point E.

15.166 $\boldsymbol{\omega} = (3 \text{ rad/s})\,\mathbf{i}$
15.167 $\boldsymbol{\omega} = (3 \text{ rad/s})\,\mathbf{j}$

15.168 and 15.169 A chain is looped around two gears of radius 40 mm that can rotate freely with respect to the 320-mm arm AB. The chain moves about arm AB in a clockwise direction at the constant rate of 80 mm/s relative to the arm. Knowing that in the position shown arm AB rotates clockwise about A at the constant rate $\omega = 0.75$ rad/s, determine the acceleration of each of the chain links indicated.

15.168 Links 1 and 2
15.169 Links 3 and 4

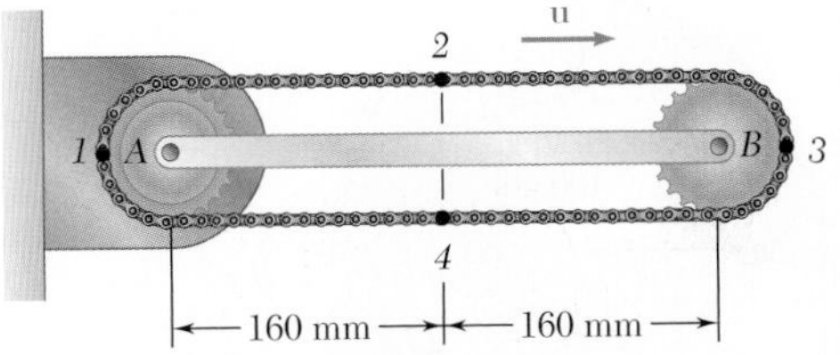

Fig. P15.168 and P15.169

15.170 A basketball player shoots a free throw in such a way that his shoulder can be considered a pin joint at the moment of release as shown. Knowing that at the instant shown the upper arm SE has a constant angular velocity of 2 rad/s counterclockwise and the forearm EW has a constant clockwise angular velocity of 4 rad/s with respect to SE, determine the velocity and acceleration of the wrist W.

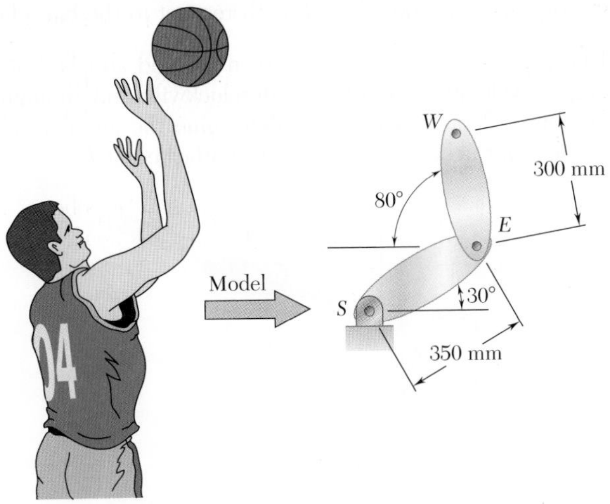

Fig. P15.170

15.171 The human leg can be crudely approximated as two rigid bars (the femur and the tibia) connected with a pin joint. At the instant shown, the velocity of the ankle A is zero, the tibia AK has an angular velocity of 1.5 rad/s counterclockwise and an angular acceleration of 1 rad/s^2 counterclockwise. Determine the relative angular velocity and relative angular acceleration of the femur KH with respect to AK so that the velocity and acceleration of H are both straight up at this instant.

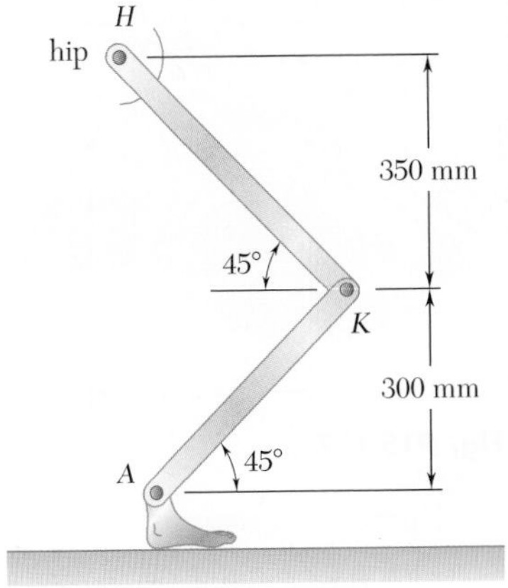

Fig. P15.171

15.172 The collar P slides outward at a constant relative speed u along rod AB, which rotates counterclockwise with a constant angular velocity of 20 rpm. Knowing that $r = 250$ mm when $\theta = 0$ and that the collar reaches B when $\theta = 90°$, determine the magnitude of the acceleration of the collar P just as it reaches B.

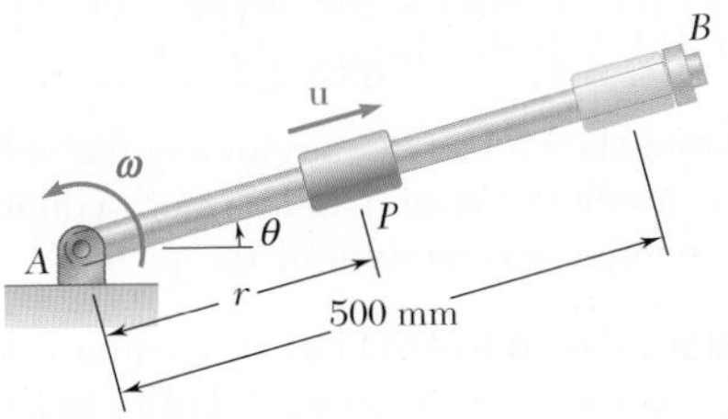

Fig. *P15.172*

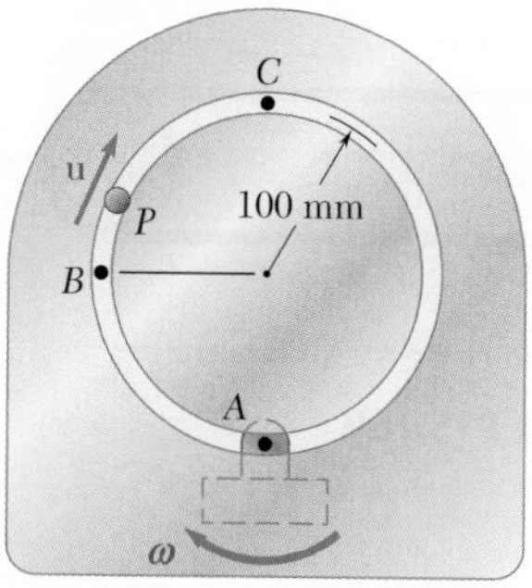

Fig. P15.173 and P15.174

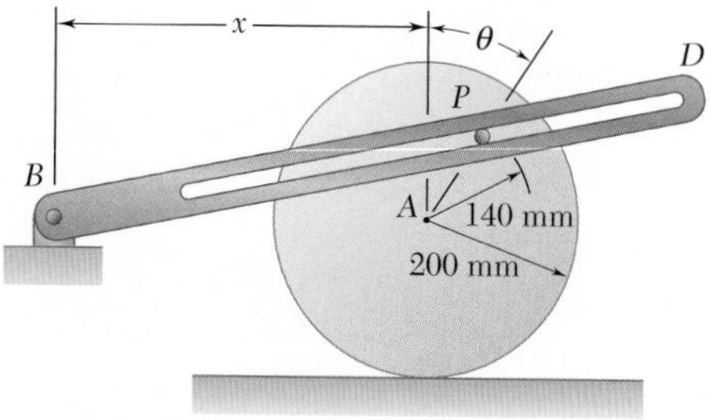

Fig. P15.175

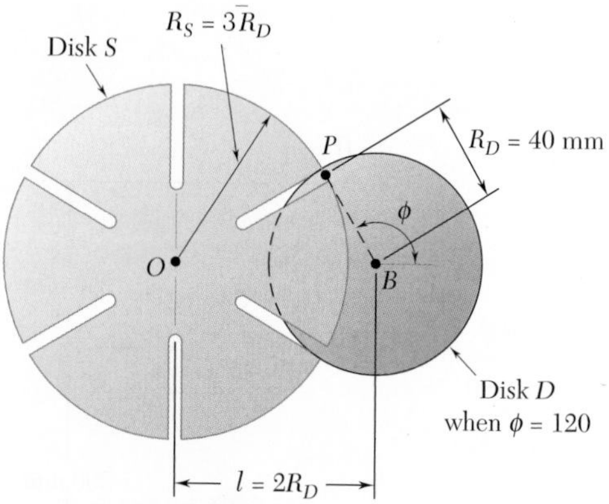

Fig. P15.177

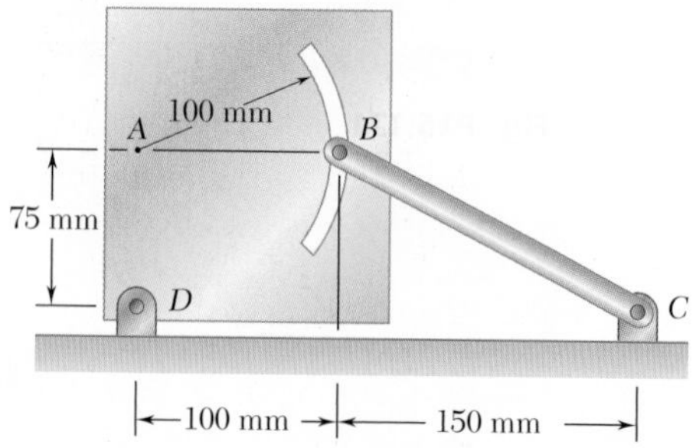

Fig. P15.179 and P15.180

15.173 Pin P slides in a circular slot cut in the plate shown at a constant relative speed u = 90 mm/s. Knowing that at the instant shown the plate rotates clockwise about A at the constant rate ω = 3 rad/s, determine the acceleration of the pin if it is located at (*a*) point A, (*b*) point B, (*c*) point C.

15.174 Pin P slides in a circular slot cut in the plate shown at a constant relative speed u = 90 mm/s. Knowing that at the instant shown the angular velocity $\boldsymbol{\omega}$ of the plate is 3 rad/s clockwise and is decreasing at the rate of 5 rad/s^2, determine the acceleration of the pin if it is located at (*a*) point A, (*b*) point B, (*c*) point C.

15.175 Pin P is attached to the wheel shown and slides in a slot cut in bar BD. The wheel rolls to the right without slipping with a constant angular velocity of 20 rad/s. Knowing that x = 480 mm when $\theta = 0$, determine (*a*) the angular acceleration of the bar, (*b*) the relative acceleration of pin P with respect to the bar when $\theta = 0$.

15.176 Knowing that at the instant shown the rod attached at A has an angular velocity of 5 rad/s counterclockwise and an angular acceleration of 2 rad/s^2 clockwise, determine the angular velocity and the angular acceleration of the rod attached at B.

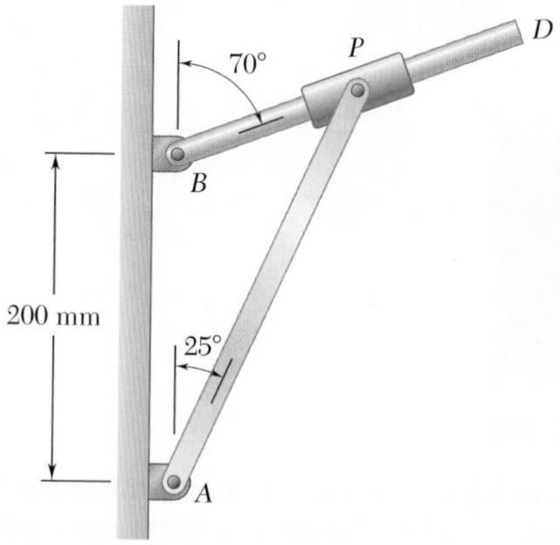

Fig. P15.176

15.177 The Geneva mechanism shown is used to provide an intermittent rotary motion of disk S. Disk D rotates with a constant counterclockwise angular velocity $\boldsymbol{\omega}_D$ of 8 rad/s. A pin P is attached to disk D and can slide in one of the six equally spaced slots cut in disk S. It is desirable that the angular velocity of disk S be zero as the pin enters and leaves each of the six slots; this will occur if the distance between the centers of the disks and the radii of the disks are related as shown. Determine the angular velocity and angular acceleration of disk S at the instant when $\phi = 150°$.

15.178 In Prob. 15.177, determine the angular velocity and angular acceleration of disk S at the instant when $\phi = 135°$.

15.179 At the instant shown bar BC has an angular velocity of 3 rad/s and an angular acceleration of 2 rad/s^2, both counterclockwise, determine the angular acceleration of the plate.

15.180 At the instant shown bar BC has an angular velocity of 3 rad/s and an angular acceleration of 2 rad/s^2, both clockwise, determine the angular acceleration of the plate.

***15.181** Rod AB passes through a collar which is welded to link DE. Knowing that at the instant shown block A moves to the right at a constant speed of 2 m/s, determine (*a*) the angular velocity of rod AB, (*b*) the velocity relative to the collar of the point of the rod in contact with the collar, (*c*) the acceleration of the point of the rod in contact with the collar. (*Hint:* Rod AB and link DE have the same $\boldsymbol{\omega}$ and the same $\boldsymbol{\alpha}$.)

***15.182** Solve Prob. 15.182 assuming block A moves to the left at a constant speed of 2 m/s.

***15.183** In Prob. 15.157, determine the acceleration of pin P.

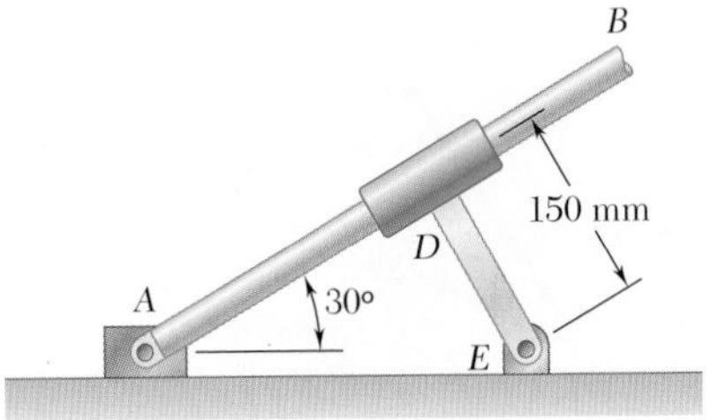

Fig. P15.181

*15.12 MOTION ABOUT A FIXED POINT

In Sec. 15.3 the motion of a rigid body constrained to rotate about a fixed axis was considered. The more general case of the motion of a rigid body which has a fixed point O will now be examined.

First, it will be proved that *the most general displacement of a rigid body with a fixed point O is equivalent to a rotation of the body about an axis through O.*† Instead of considering the rigid body itself, we can detach a sphere of center O from the body and analyze the motion of that sphere. Clearly, the motion of the sphere completely characterizes the motion of the given body. Since three points define the position of a solid in space, the center O and two points A and B on the surface of the sphere will define the position of the sphere and thus the position of the body. Let A_1 and B_1 characterize the position of the sphere at one instant, and let A_2 and B_2 characterize its position at a later instant (Fig. 15.31*a*). Since the sphere is rigid, the lengths of the arcs of great circle A_1B_1 and A_2B_2 must be equal, but except for this requirement, the positions of A_1, A_2, B_1, and B_2 are arbitrary. We propose to prove that the points A and B can be brought, respectively, from A_1 and B_1 into A_2 and B_2 by a single rotation of the sphere about an axis.

For convenience, and without loss of generality, we select point B so that its initial position coincides with the final position of A; thus, $B_1 = A_2$ (Fig. 15.31*b*). We draw the arcs of great circle A_1A_2, A_2B_2 and the arcs bisecting, respectively, A_1A_2 and A_2B_2. Let C be the point of intersection of these last two arcs; we complete the construction by drawing A_1C, A_2C, and B_2C. As pointed out above, because of the rigidity of the sphere, $A_1B_1 = A_2B_2$. Since C is by construction equidistant from A_1, A_2, and B_2, we also have $A_1C = A_2C = B_2C$. As a result, the spherical triangles A_1CA_2 and B_1CB_2 are congruent and the angles A_1CA_2 and B_1CB_2 are equal. Denoting by θ the common value of these angles, we conclude that the sphere can be brought from its initial position into its final position by a single rotation through θ about the axis OC.

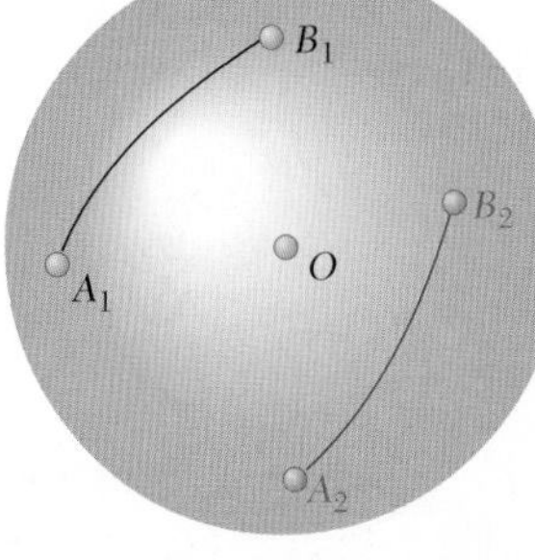

(*a*)

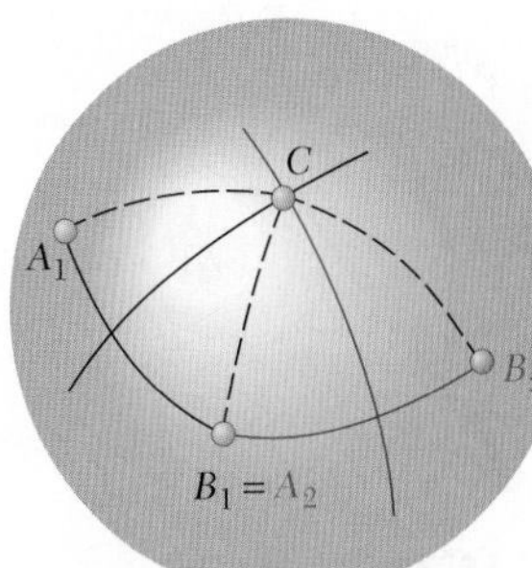

(*b*)

Fig. 15.31

†This is known as *Euler's theorem.*

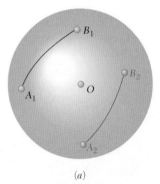

It follows that the motion during a time interval Δt of a rigid body with a fixed point O can be considered as a rotation through $\Delta\theta$ about a certain axis. Drawing along that axis a vector of magnitude $\Delta\theta/\Delta t$ and letting Δt approach zero, we obtain at the limit the *instantaneous axis of rotation* and the angular velocity $\boldsymbol{\omega}$ of the body at the instant considered (Fig. 15.32). The velocity of a particle P of the body can then be obtained, as in Sec. 15.3, by forming the vector product of $\boldsymbol{\omega}$ and of the position vector $\mathbf{r}$ of the particle:

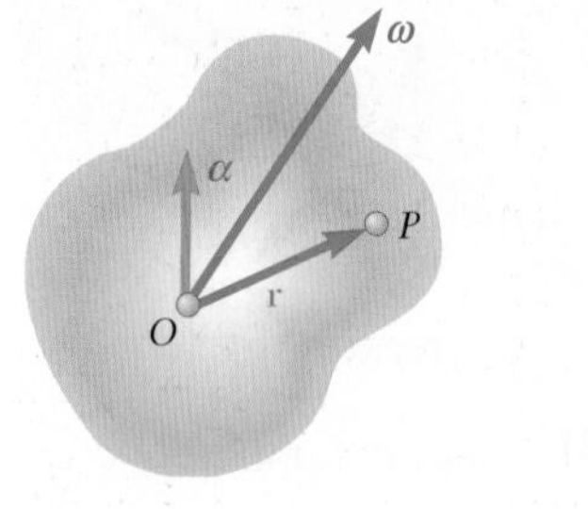

Fig. 15.32

$$\mathbf{v} = \frac{d\mathbf{r}}{dt} = \boldsymbol{\omega} \times \mathbf{r} \tag{15.37}$$

The acceleration of the particle is obtained by differentiating (15.37) with respect to t. As in Sec. 15.3 we have

$$\mathbf{a} = \boldsymbol{\alpha} \times \mathbf{r} + \boldsymbol{\omega} \times (\boldsymbol{\omega} \times \mathbf{r}) \tag{15.38}$$

where the angular acceleration $\boldsymbol{\alpha}$ is defined as the derivative

$$\boldsymbol{\alpha} = \frac{d\boldsymbol{\omega}}{dt} \tag{15.39}$$

of the angular velocity $\boldsymbol{\omega}$.

In the case of the motion of a rigid body with a fixed point, the direction of $\boldsymbol{\omega}$ and of the instantaneous axis of rotation changes from one instant to the next. The angular acceleration $\boldsymbol{\alpha}$ therefore reflects the change in direction of $\boldsymbol{\omega}$ as well as its change in magnitude and, in general, *is not directed along the instantaneous axis of rotation.* While the particles of the body located on the instantaneous axis of rotation have zero velocity at the instant considered, they do not have zero acceleration. Also, the accelerations of the various particles of the body *cannot* be determined as if the body were rotating permanently about the instantaneous axis.

Recalling the definition of the velocity of a particle with position vector $\mathbf{r}$, we note that the angular acceleration $\boldsymbol{\alpha}$, as expressed in (15.39), represents the velocity of the tip of the vector $\boldsymbol{\omega}$. This property may be useful in the determination of the angular acceleration of a rigid body. For example, it follows that the vector $\boldsymbol{\alpha}$ is tangent to the curve described in space by the tip of the vector $\boldsymbol{\omega}$.

We should note that the vector $\boldsymbol{\omega}$ moves within the body, as well as in space. It thus generates two cones called, respectively, the *body cone* and the *space cone* (Fig. 15.33).† It can be shown that at any given instant, the two cones are tangent along the instantaneous axis of rotation and that as the body moves, the body cone appears to *roll* on the space cone.

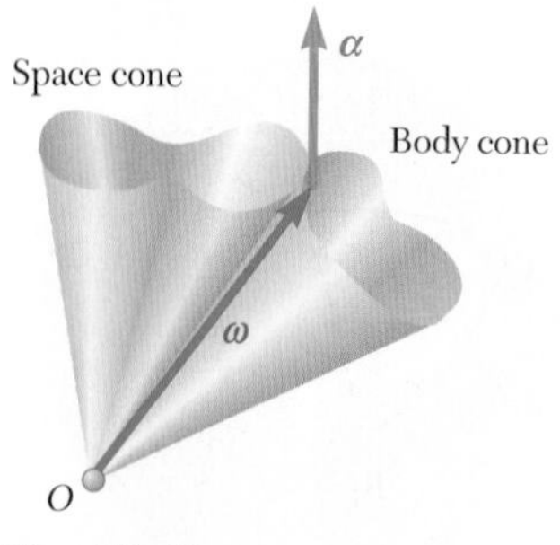

Fig. 15.33

†It is recalled that a *cone* is, by definition, a surface generated by a straight line passing through a fixed point. In general, the cones considered here *will not be circular cones.*

Before concluding our analysis of the motion of a rigid body with a fixed point, we should prove that angular velocities are actually vectors. As indicated in Sec. 2.3, some quantities, such as the *finite rotations* of a rigid body, have magnitude and direction but do not obey the parallelogram law of addition; these quantities cannot be considered as vectors. In contrast, angular velocities (and also *infinitesimal rotations*), as will be demonstrated presently, *do obey* the parallelogram law and thus are truly vector quantities.

Photo 15.8 When the ladder rotates about its fixed base, its angular velocity can be obtained by adding the angular velocities which correspond to simultaneous rotations about two different axes.

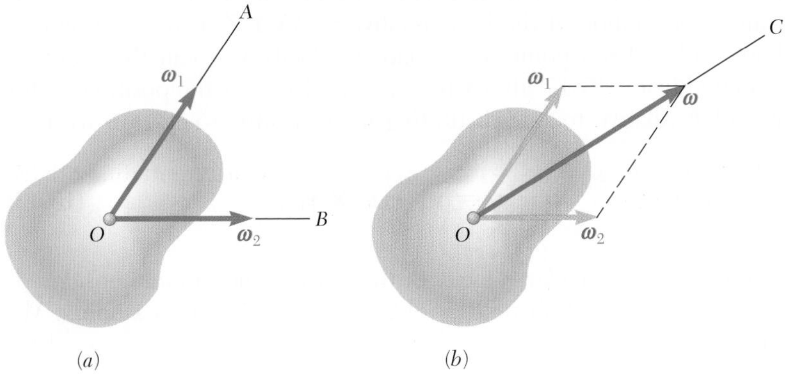

Fig. 15.34

Consider a rigid body with a fixed point O which at a given instant rotates simultaneously about the axes OA and OB with angular velocities $\boldsymbol{\omega}_1$ and $\boldsymbol{\omega}_2$ (Fig. 15.34*a*). We know that this motion must be equivalent at the instant considered to a single rotation of angular velocity $\boldsymbol{\omega}$. We propose to show that

$$\boldsymbol{\omega} = \boldsymbol{\omega}_1 + \boldsymbol{\omega}_2 \tag{15.40}$$

i.e., that the resulting angular velocity can be obtained by adding $\boldsymbol{\omega}_1$ and $\boldsymbol{\omega}_2$ by the parallelogram law (Fig. 15.34*b*).

Consider a particle P of the body, defined by the position vector $\mathbf{r}$. Denoting, respectively, by $\mathbf{v}_1$, $\mathbf{v}_2$, and $\mathbf{v}$ the velocity of P when the body rotates about OA only, about OB only, and about both axes simultaneously, we write

$$\mathbf{v} = \boldsymbol{\omega} \times \mathbf{r} \qquad \mathbf{v}_1 = \boldsymbol{\omega}_1 \times \mathbf{r} \qquad \mathbf{v}_2 = \boldsymbol{\omega}_2 \times \mathbf{r} \tag{15.41}$$

But the vectorial character of *linear* velocities is well established (since they represent the derivatives of position vectors). We therefore have

$$\mathbf{v} = \mathbf{v}_1 + \mathbf{v}_2$$

where the plus sign indicates vector addition. Substituting from (15.41), we write

$$\boldsymbol{\omega} \times \mathbf{r} = \boldsymbol{\omega}_1 \times \mathbf{r} + \boldsymbol{\omega}_2 \times \mathbf{r}$$
$$\boldsymbol{\omega} \times \mathbf{r} = (\boldsymbol{\omega}_1 + \boldsymbol{\omega}_2) \times \mathbf{r}$$

where the plus sign still indicates vector addition. Since the relation obtained holds for an arbitrary $\mathbf{r}$, we conclude that (15.40) must be true.

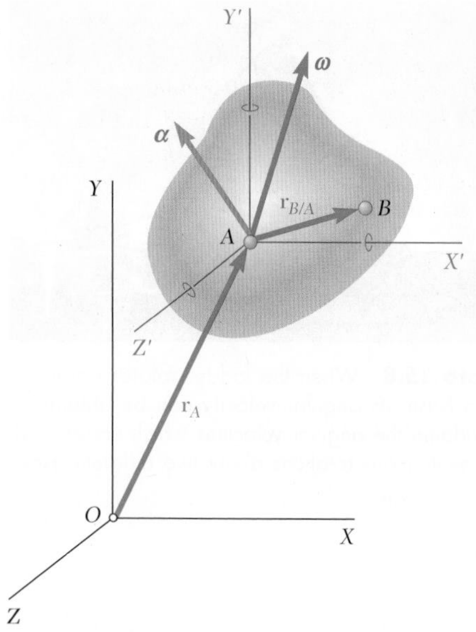

Fig. 15.35

*15.13 GENERAL MOTION

The most general motion of a rigid body in space will now be considered. Let A and B be two particles of the body. We recall from Sec. 11.12 that the velocity of B with respect to the fixed frame of reference $OXYZ$ can be expressed as

$$\mathbf{v}_B = \mathbf{v}_A + \mathbf{v}_{B/A} \tag{15.42}$$

where $\mathbf{v}_{B/A}$ is the velocity of B relative to a frame $AX'Y'Z'$ attached to A and of fixed orientation (Fig. 15.35). Since A is fixed in this frame, the motion of the body relative to $AX'Y'Z'$ is the motion of a body with a fixed point. The relative velocity $\mathbf{v}_{B/A}$ can therefore be obtained from (15.37) after $\mathbf{r}$ has been replaced by the position vector $\mathbf{r}_{B/A}$ of B relative to A. Substituting for $\mathbf{v}_{B/A}$ into (15.42), we write

$$\mathbf{v}_B = \mathbf{v}_A + \boldsymbol{\omega} \times \mathbf{r}_{B/A} \tag{15.43}$$

where $\boldsymbol{\omega}$ is the angular velocity of the body at the instant considered.

The acceleration of B is obtained by a similar reasoning. We first write

$$\mathbf{a}_B = \mathbf{a}_A + \mathbf{a}_{B/A}$$

and, recalling Eq. (15.38),

$$\mathbf{a}_B = \mathbf{a}_A + \boldsymbol{\alpha} \times \mathbf{r}_{B/A} + \boldsymbol{\omega} \times (\boldsymbol{\omega} \times \mathbf{r}_{B/A}) \tag{15.44}$$

where $\boldsymbol{\alpha}$ is the angular acceleration of the body at the instant considered.

Equations (15.43) and (15.44) show that *the most general motion of a rigid body is equivalent, at any given instant, to the sum of a translation,* in which all the particles of the body have the same velocity and acceleration as a reference particle A, *and of a motion in which particle A is assumed to be fixed.*†

It is easily shown, by solving (15.43) and (15.44) for $\mathbf{v}_A$ and $\mathbf{a}_A$, that the motion of the body with respect to a frame attached to B would be characterized by the same vectors $\boldsymbol{\omega}$ and $\boldsymbol{\alpha}$ as its motion relative to $AX'Y'Z'$. The angular velocity and angular acceleration of a rigid body at a given instant are thus independent of the choice of reference point. On the other hand, one should keep in mind that whether the moving frame is attached to A or to B, it should maintain a fixed orientation; that is, it should remain parallel to the fixed reference frame $OXYZ$ throughout the motion of the rigid body. In many problems it will be more convenient to use a moving frame which is allowed to rotate as well as to translate. The use of such moving frames will be discussed in Secs. 15.14 and 15.15.

†It is recalled from Sec. 15.12 that, in general, the vectors $\boldsymbol{\omega}$ and $\boldsymbol{\alpha}$ are not collinear, and that the accelerations of the particles of the body in their motion relative to the frame $AX'Y'Z'$ cannot be determined as if the body were rotating permanently about the instantaneous axis through A.

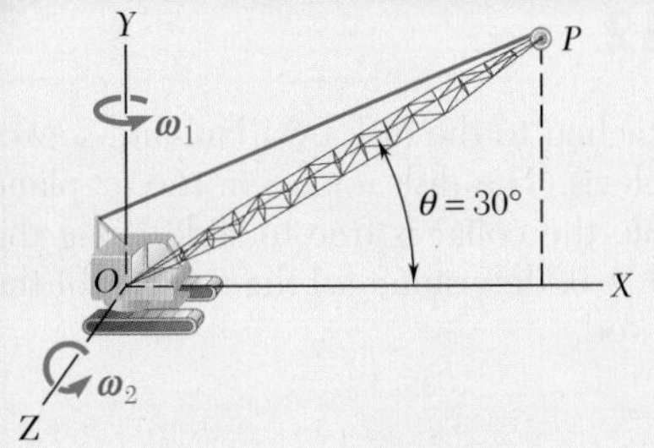

SAMPLE PROBLEM 15.11

The crane shown rotates with a constant angular velocity $\boldsymbol{\omega}_1$ of 0.30 rad/s. Simultaneously, the boom is being raised with a constant angular velocity $\boldsymbol{\omega}_2$ of 0.50 rad/s relative to the cab. Knowing that the length of the boom OP is $l = 12$ m, determine (*a*) the angular velocity $\boldsymbol{\omega}$ of the boom, (*b*) the angular acceleration $\boldsymbol{\alpha}$ of the boom, (*c*) the velocity **v** of the tip of the boom, (*d*) the acceleration **a** of the tip of the boom.

SOLUTION

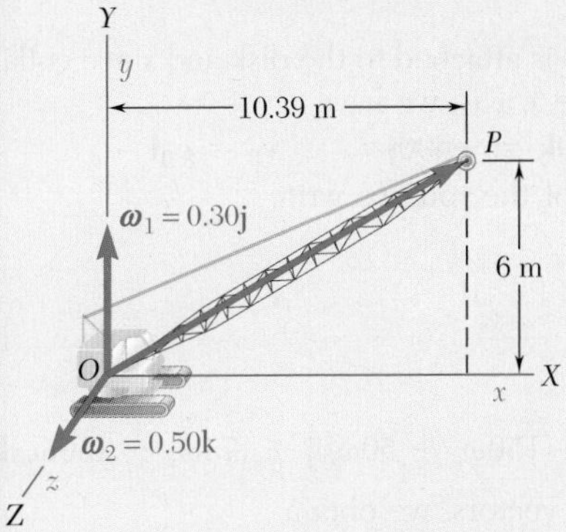

a. Angular Velocity of Boom. Adding the angular velocity $\boldsymbol{\omega}_1$ of the cab and the angular velocity $\boldsymbol{\omega}_2$ of the boom relative to the cab, we obtain the angular velocity $\boldsymbol{\omega}$ of the boom at the instant considered:

$$\boldsymbol{\omega} = \boldsymbol{\omega}_1 + \boldsymbol{\omega}_2 \qquad \boldsymbol{\omega} = (0.30 \text{ rad/s})\mathbf{j} + (0.50 \text{ rad/s})\mathbf{k} \quad ◀$$

b. Angular Acceleration of Boom. The angular acceleration $\boldsymbol{\alpha}$ of the boom is obtained by differentiating $\boldsymbol{\omega}$. Since the vector $\boldsymbol{\omega}_1$ is constant in magnitude and direction, we have

$$\boldsymbol{\alpha} = \dot{\boldsymbol{\omega}} = \dot{\boldsymbol{\omega}}_1 + \dot{\boldsymbol{\omega}}_2 = 0 + \dot{\boldsymbol{\omega}}_2$$

where the rate of change $\dot{\boldsymbol{\omega}}_2$ is to be computed with respect to the fixed frame $OXYZ$. However, it is more convenient to use a frame $Oxyz$ attached to the cab and rotating with it, since the vector $\boldsymbol{\omega}_2$ also rotates with the cab and therefore has zero rate of change with respect to that frame. Using Eq. (15.31) with $\mathbf{Q} = \boldsymbol{\omega}_2$ and $\boldsymbol{\Omega} = \boldsymbol{\omega}_1$, we write

$$(\dot{\mathbf{Q}})_{OXYZ} = (\dot{\mathbf{Q}})_{Oxyz} + \boldsymbol{\Omega} \times \mathbf{Q}$$
$$(\dot{\boldsymbol{\omega}}_2)_{OXYZ} = (\dot{\boldsymbol{\omega}}_2)_{Oxyz} + \boldsymbol{\omega}_1 \times \boldsymbol{\omega}_2$$
$$\boldsymbol{\alpha} = (\dot{\boldsymbol{\omega}}_2)_{OXYZ} = 0 + (0.30 \text{ rad/s})\mathbf{j} \times (0.50 \text{ rad/s})\mathbf{k}$$

$$\boldsymbol{\alpha} = (0.15 \text{ rad/s}^2)\mathbf{i} \quad ◀$$

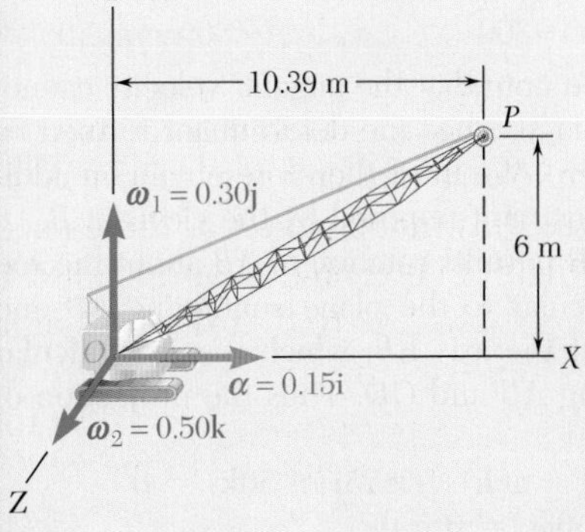

c. Velocity of Tip of Boom. Noting that the position vector of point P is $\mathbf{r} = (10.39 \text{ m})\mathbf{i} + (6 \text{ m})\mathbf{j}$ and using the expression found for $\boldsymbol{\omega}$ in part *a*, we write

$$\mathbf{v} = \boldsymbol{\omega} \times \mathbf{r} = \begin{vmatrix} \mathbf{i} & \mathbf{j} & \mathbf{k} \\ 0 & 0.30 \text{ rad/s} & 0.50 \text{ rad/s} \\ 10.39 \text{ m} & 6 \text{ m} & 0 \end{vmatrix}$$

$$\mathbf{v} = -(3 \text{ m/s})\mathbf{i} + (5.20 \text{ m/s})\mathbf{j} - (3.12 \text{ m/s})\mathbf{k} \quad ◀$$

d. Acceleration of Tip of Boom. Recalling that $\mathbf{v} = \boldsymbol{\omega} \times \mathbf{r}$, we write

$$\mathbf{a} = \boldsymbol{\alpha} \times \mathbf{r} + \boldsymbol{\omega} \times (\boldsymbol{\omega} \times \mathbf{r}) = \boldsymbol{\alpha} \times \mathbf{r} + \boldsymbol{\omega} \times \mathbf{v}$$

$$\mathbf{a} = \begin{vmatrix} \mathbf{i} & \mathbf{j} & \mathbf{k} \\ 0.15 & 0 & 0 \\ 10.39 & 6 & 0 \end{vmatrix} + \begin{vmatrix} \mathbf{i} & \mathbf{j} & \mathbf{k} \\ 0 & 0.30 & 0.50 \\ -3 & 5.20 & -3.12 \end{vmatrix}$$
$$= 0.90\mathbf{k} - 0.94\mathbf{i} - 2.60\mathbf{i} - 1.50\mathbf{j} + 0.90\mathbf{k}$$

$$\mathbf{a} = -(3.54 \text{ m/s}^2)\mathbf{i} - (1.50 \text{ m/s}^2)\mathbf{j} + (1.80 \text{ m/s}^2)\mathbf{k} \quad ◀$$

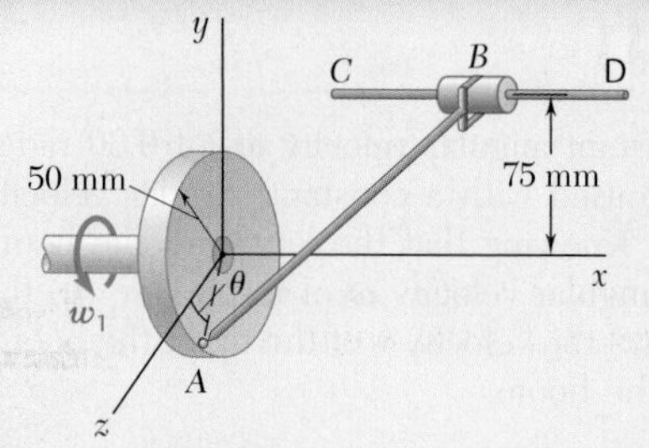

SAMPLE PROBLEM 15.12

The rod AB, of length 175 mm, is attached to the disk by a ball-and-socket connection and to the collar B by a clevis. The disk rotates in the yz plane at a constant rate $\omega_1 = 12$ rad/s, while the collar is free to slide along the horizontal rod CD. For the position $\theta = 0$, determine (*a*) the velocity of the collar, (*b*) the angular velocity of the rod.

SOLUTION

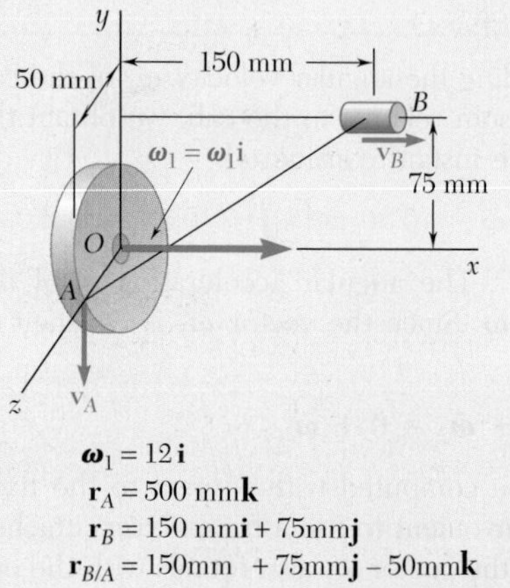

As the length AB = 175 mm, it can be worked out that the coordinates of B are (150 mm, 75 mm, 0)

a. Velocity of Collar. Since point A is attached to the disk and since collar B moves in a direction parallel to the x axis, we have

$$\mathbf{v}_A = \boldsymbol{\omega}_1 \times \mathbf{r}_A = 12\mathbf{i} \times 50\mathbf{k} = -600\mathbf{j} \qquad \mathbf{v}_B = v_B\mathbf{i}$$

Denoting by $\boldsymbol{\omega}$ the angular velocity of the rod, we write

$$v_B\mathbf{i} = -600\mathbf{j} + \begin{vmatrix} \mathbf{i} & \mathbf{j} & \mathbf{k} \\ \omega_x & \omega_y & \omega_z \\ 150 & 75 & -50 \end{vmatrix}$$

$$v_B\mathbf{i} = -600\mathbf{j} + (-50\omega_y - 75\omega_z)\mathbf{i} + (150\omega_z + 50\omega_x)\mathbf{j} + (75\omega_x - 150\omega_y)\mathbf{k}$$

Equating the coefficients of the unit vectors, we obtain

$$v_B = -50\omega_y - 75\omega_z \qquad (1)$$
$$600 = 50\omega_x + 150\omega_z \qquad (2)$$
$$0 = 75\omega_x - 150\omega_y \qquad (3)$$

Clearly the three equations obtained cannot be solved for the four unknown v_B, ω_x, ω_y, and ω_z. An additional equation reflecting the type of correction will be obtained in part b. However, since the velocity of B does not depend on the type of correction, we should be able to obtain v_B by eliminating the other olumes from Eqs. (1), (2) and (3).

Multiplying Eqs. (1), (2), (3), respectively, by 6, 3, −2 and adding, we write

$$6v_B + 1800 = 0 \qquad v_B = -300 \qquad \mathbf{v}_B = -(300 \text{ mm/s})\mathbf{i}$$

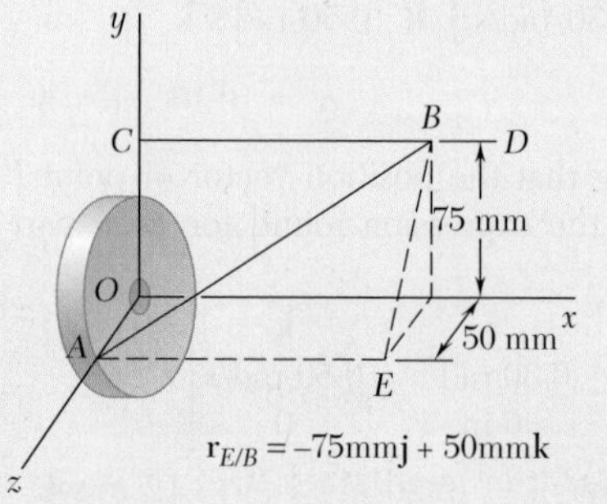

b. Angular Velocity of Rod *AB*. We note that the angular velocity cannot be determined from Eqs. (1), (2), and (3), since the determinant formed by the coefficients of ω_x, ω_y, and ω_z is zero. We must therefore obtain an additional equation by considering the constraint imposed by the clevis at B.

The collar-clevis connection at B permits rotation of AB about the rod CD and also about an axis perpendicular to the plane containing AB and CD. It prevents rotation of AB about the axis EB, which is perpendicular to CD and lies in the plane containing AB and CD. Thus the projection of $\boldsymbol{\omega}$ on $\mathbf{r}_{E/B}$ must be zero and we write†

$$\boldsymbol{\omega} \cdot \mathbf{r}_{E/B} = 0 \qquad (\omega_x\mathbf{i} + \omega_y\mathbf{j} + \omega_z\mathbf{k}) \cdot (-75\mathbf{j} + 50\mathbf{k}) = 0$$
$$-75\omega_y + 50\omega_z = 0 \qquad (4)$$

Solving Eqs. (1) through (4) simultaneously, we obtain

$$v_B = -300 \qquad \omega_x = 3.69 \qquad \omega_y = 1.846 \qquad \omega_z = 2.77$$

$$\boldsymbol{\omega} = (3.69 \text{ rad/s})\mathbf{i} + (1.846 \text{ rad/s})\mathbf{j} + (2.77 \text{ rad/s})\mathbf{k}$$

†We could also note that the direction of EB is that of the vector triple product $\mathbf{r}_{B/C} \times (\mathbf{r}_{B/C} \times \mathbf{r}_{B/A})$ and write $\boldsymbol{\omega} \cdot [\mathbf{r}_{B/C} \times (\mathbf{r}_{B/C} \times \mathbf{r}_{B/A})] = 0$. This formulation would be particularly useful if the rod CD were skew.

SOLVING PROBLEMS ON YOUR OWN

In this lesson you started the study of the *kinematics of rigid bodies in three dimensions*. You first studied the *motion of a rigid body about a fixed point* and then the *general motion of a rigid body*.

A. Motion of a rigid body about a fixed point. To analyze the motion of a point B of a body rotating about a fixed point O you may have to take some or all of the following steps.

1. Determine the position vector r connecting the fixed point O to point B.

2. Determine the angular velocity ω of the body with respect to a fixed frame of reference. The angular velocity $\boldsymbol{\omega}$ will often be obtained by adding two component angular velocities $\boldsymbol{\omega}_1$ and $\boldsymbol{\omega}_2$ [Sample Prob. 15.11].

3. Compute the velocity of B by using the equation

$$\mathbf{v} = \boldsymbol{\omega} \times \mathbf{r} \tag{15.37}$$

Your computation will usually be facilitated if you express the vector product as a determinant.

4. Determine the angular acceleration α of the body. The angular acceleration $\boldsymbol{\alpha}$ represents the rate of change $(\dot{\boldsymbol{\omega}})_{OXYZ}$ of the vector $\boldsymbol{\omega}$ *with respect to a fixed frame of reference OXYZ* and reflects both *a change in magnitude and a change in direction* of the angular velocity. However, when computing $\boldsymbol{\alpha}$ you may find it convenient to first compute the rate of change $(\dot{\boldsymbol{\omega}})_{Oxyz}$ of $\boldsymbol{\omega}$ with respect to a rotating frame of reference $Oxyz$ of your choice and use Eq. (15.31) of the preceding lesson to obtain $\boldsymbol{\alpha}$. You will write

$$\boldsymbol{\alpha} = (\dot{\boldsymbol{\omega}})_{OXYZ} = (\dot{\boldsymbol{\omega}})_{Oxyz} + \boldsymbol{\Omega} \times \boldsymbol{\omega}$$

where $\boldsymbol{\Omega}$ is the angular velocity of the rotating frame $Oxyz$ [Sample Prob. 15.11].

5. Compute the acceleration of B by using the equation

$$\mathbf{a} = \boldsymbol{\alpha} \times \mathbf{r} + \boldsymbol{\omega} \times (\boldsymbol{\omega} \times \mathbf{r}) \tag{15.38}$$

Note that the vector product $(\boldsymbol{\omega} \times \mathbf{r})$ represents the velocity of point B and was computed in step 3. Also, the computation of the first vector product in (15.38) will be facilitated if you express this product in determinant form. Remember that, as was the case with the plane motion of a rigid body, the instantaneous axis of rotation *cannot* be used to determine accelerations.

B. General motion of a rigid body. The general motion of a rigid body may be considered as *the sum of a translation and a rotation.* Keep the following in mind:

a. In the translation part of the motion, all the points of the body have the *same velocity* $\mathbf{v}_A$ *and the same acceleration* $\mathbf{a}_A$ as the point A of the body that has been selected as the reference point.

b. In the rotation part of the motion, the same reference point A is assumed to be a *fixed point.*

1. To determine the velocity of a point *B* of the rigid body when you know the velocity $\mathbf{v}_A$ of the reference point A and the angular velocity $\boldsymbol{\omega}$ of the body, you simply add $\mathbf{v}_A$ to the velocity $\mathbf{v}_{B/A} = \boldsymbol{\omega} \times \mathbf{r}_{B/A}$ of B in its rotation about A:

$$\mathbf{v}_B = \mathbf{v}_A + \boldsymbol{\omega} \times \mathbf{r}_{B/A} \qquad (15.43)$$

As indicated earlier, the computation of the vector product will usually be facilitated if you express this product in determinant form.

Equation (15.43) can also be used to determine the magnitude of $\mathbf{v}_B$ when its direction is known, even if $\boldsymbol{\omega}$ is not known. While the corresponding three scalar equations are linearly dependent and the components of $\boldsymbol{\omega}$ are indeterminate, these components can be eliminated and $\mathbf{v}_A$ can be found by using an appropriate linear combination of the three equations [Sample Prob. 15.12, part *a*]. Alternatively, you can assign an arbitrary value to one of the components of $\boldsymbol{\omega}$ and solve the equations for $\mathbf{v}_A$. However, an additional equation must be sought in order to determine the true values of the components of $\boldsymbol{\omega}$ [Sample Prob. 15.12, part *b*].

2. To determine the acceleration of a point *B* of the rigid body when you know the acceleration $\mathbf{a}_A$ of the reference point A and the angular acceleration $\boldsymbol{\alpha}$ of the body, you simply add $\mathbf{a}_A$ to the acceleration of B in its rotation about A, as expressed by Eq. (15.38):

$$\mathbf{a}_B = \mathbf{a}_A + \boldsymbol{\alpha} \times \mathbf{r}_{B/A} + \boldsymbol{\omega} \times (\boldsymbol{\omega} \times \mathbf{r}_{B/A}) \qquad (15.44)$$

Note that the vector product $(\boldsymbol{\omega} \times \mathbf{r}_{B/A})$ represents the velocity $\mathbf{v}_{B/A}$ of B relative to A and may already have been computed as part of your calculation of $\mathbf{v}_B$. We also remind you that the computation of the other two vector products will be facilitated if you express these products in determinant form.

The three scalar equations associated with Eq. (15.44) can also be used to determine the magnitude of $\mathbf{a}_B$ when its direction is known, even if $\boldsymbol{\omega}$ and $\boldsymbol{\alpha}$ are not known. While the components of $\boldsymbol{\omega}$ and $\boldsymbol{\alpha}$ are indeterminate, you can assign arbitrary values to one of the components of $\boldsymbol{\omega}$ and to one of the components of $\boldsymbol{\alpha}$ and solve the equations for $\mathbf{a}_B$.

PROBLEMS

END-OF-SECTION PROBLEMS

15.184 At the instant considered the radar antenna shown rotates about the origin of coordinates with an angular velocity $\boldsymbol{\omega} = \omega_x\mathbf{i} + \omega_y\mathbf{j} + \omega_z\mathbf{k}$. Knowing that $(v_A)_y = 300$ mm/s, $(v_B)_y = 180$ mm/s, and $(v_B)_z = 360$ mm/s, determine (*a*) the angular velocity of the antenna, (*b*) the velocity of point *A*.

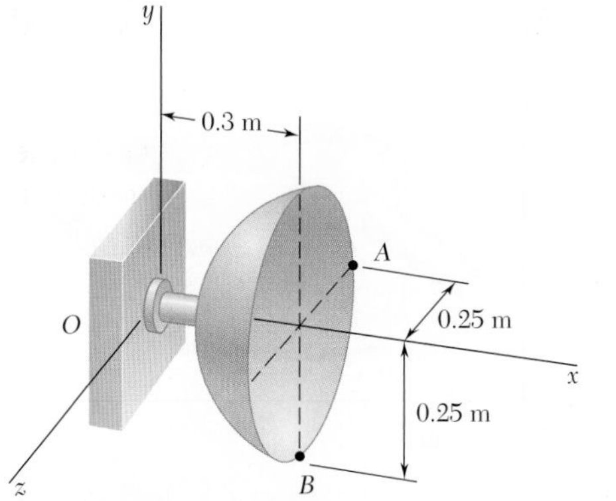

Fig. P15.184 and P15.185

15.185 At the instant considered the radar antenna shown rotates about the origin of coordinates with an angular velocity $\boldsymbol{\omega} = \omega_x\mathbf{i} + \omega_y\mathbf{j} + \omega_z\mathbf{k}$. Knowing that $(v_A)_x = 100$ mm/s, $(v_A)_y = -90$ mm/s, and $(v_B)_z = 120$ mm/s, determine (*a*) the angular velocity of the antenna, (*b*) the velocity of point *A*.

15.186 Plate *ABD* and rod *OB* are rigidly connected and rotate about the ball-and-socket joint *O* with an angular velocity $\boldsymbol{\omega} = \omega_x\mathbf{i} + \omega_y\mathbf{j} + \omega_z\mathbf{k}$. Knowing that $\mathbf{v}_A = (80 \text{ mm/s})\mathbf{i} + (360 \text{ mm/s})\mathbf{j} + (v_A)_z\mathbf{k}$ and $\omega_x = 1.5$ rad/s, determine (*a*) the angular velocity of the assembly, (*b*) the velocity of point *D*.

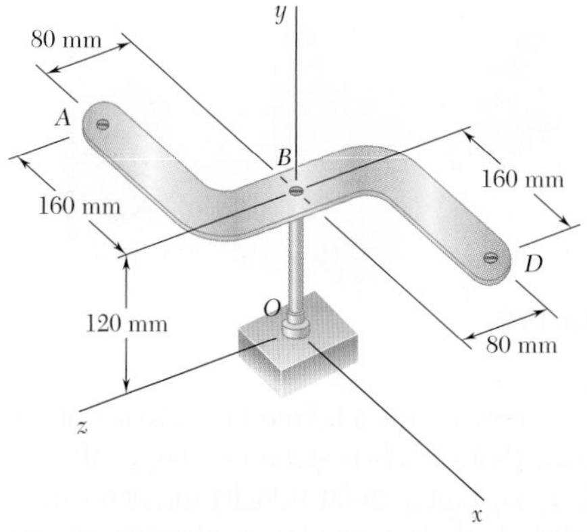

Fig. P15.186

15.187 The blade assembly of an oscillating fan rotates with a constant angular velocity $\boldsymbol{\omega}_1 = -(360 \text{ rpm})\mathbf{i}$ with respect to the motor housing. Determine the angular acceleration of the blade assembly, knowing that at the instant shown the angular velocity and angular acceleration of the motor housing are, respectively, $\boldsymbol{\omega}_2 = -(2.5 \text{ rpm})\mathbf{j}$ and $\boldsymbol{\alpha}_2 = 0$.

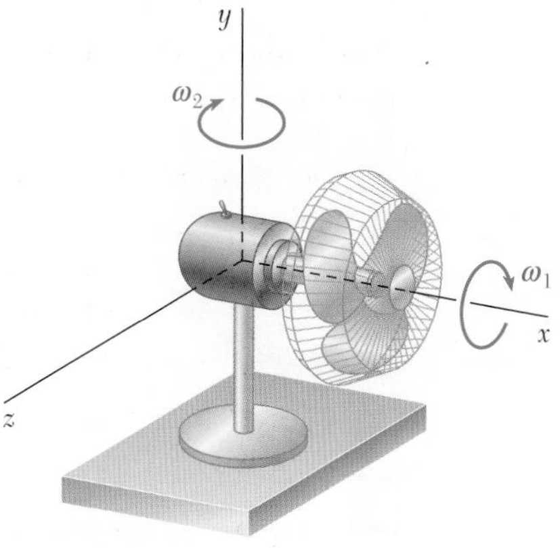

Fig. P15.187

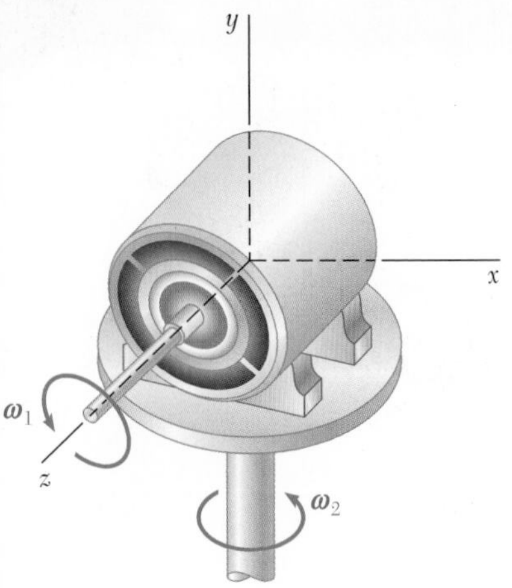

Fig. P15.188

15.188 The rotor of an electric motor rotates at the constant rate $\omega_1 =$ 1800 rpm. Determine the angular acceleration of the rotor as the motor is rotated about the y axis with a constant angular velocity $\boldsymbol{\omega}_2$ of 6 rpm counterclockwise when viewed from the positive y axis.

15.189 The disk of a portable sander rotates at the constant rate $\omega_1 =$ 4400 rpm as shown. Determine the angular acceleration of the disk as a worker rotates the sander about the z axis with an angular velocity of 0.5 rad/s and an angular acceleration of 2.5 rad/s^2, both clockwise when viewed from the positive z axis.

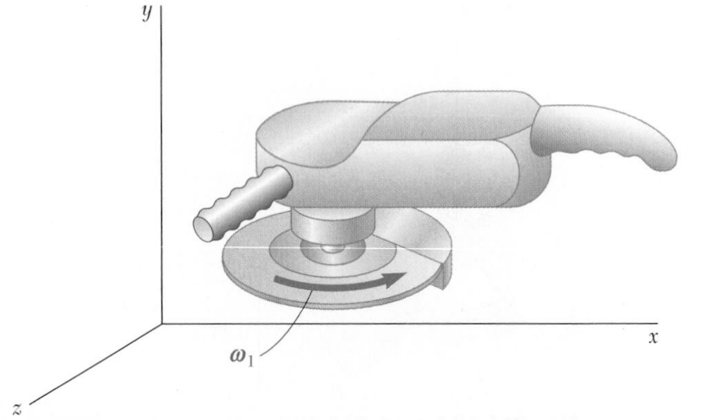

Fig. P15.189

15.190 Knowing that the turbine rotor shown rotates at a constant rate $\omega_1 = 9000$ rpm, determine the angular acceleration of the rotor if the turbine housing has a constant angular velocity of 2.4 rad/s clockwise as viewed from (*a*) the positive y axis, (*b*) the positive z axis.

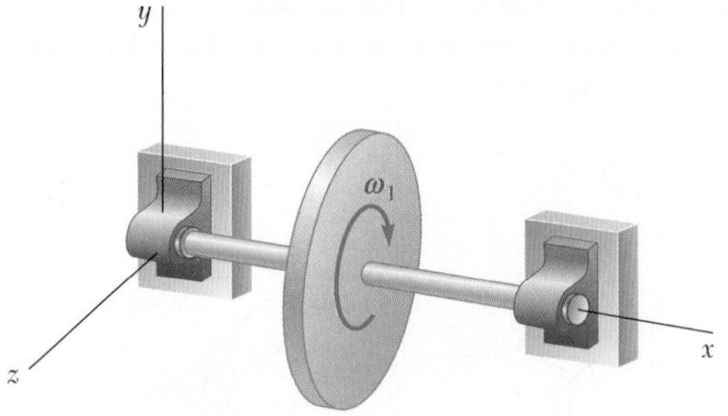

Fig. P15.190

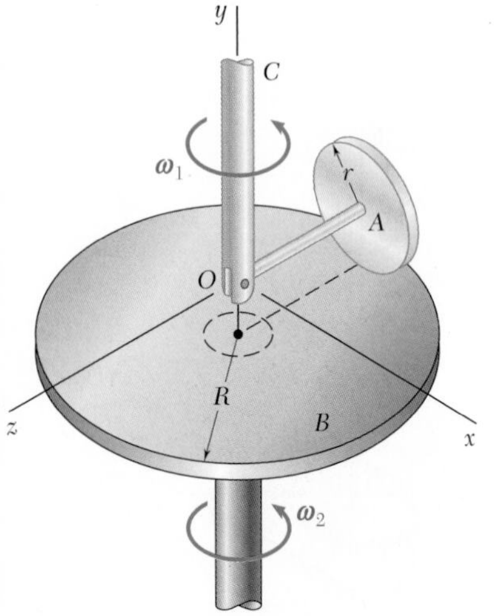

Fig. *P15.191* and *P15.192*

15.191 In the system shown, disk *A* is free to rotate about the horizontal rod *OA*. Assuming that disk *B* is stationary ($\omega_2 = 0$), and that shaft *OC* rotates with a constant angular velocity $\boldsymbol{\omega}_1$, determine (*a*) the angular velocity of disk *A*, (*b*) the angular acceleration of disk *A*.

15.192 In the system shown, disk *A* is free to rotate about the horizontal rod *OA*. Assuming that shaft *OC* and disk *B* rotate with constant angular velocities $\boldsymbol{\omega}_1$ and $\boldsymbol{\omega}_2$, respectively, both counterclockwise, determine (*a*) the angular velocity of disk *A*, (*b*) the angular acceleration of disk *A*.

15.193 The L-shaped arm BCD rotates about the z axis with a constant angular velocity $\omega_1 = 5$ rad/s. Knowing that the 150-mm-radius disk rotates about BC with a constant angular velocity $\omega_2 = 4$ rad/s, determine (*a*) the velocity of point A, (*b*) the acceleration of point A.

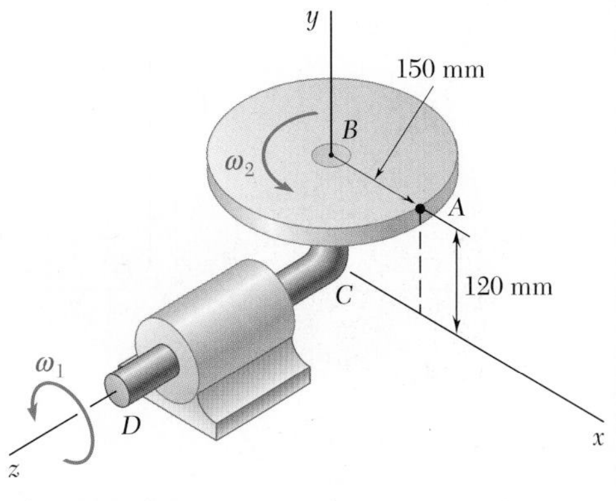

Fig. P15.193

15.194 A gun barrel of length $OP = 4$ m is mounted on a turret as shown. To keep the gun aimed at a moving target the azimuth angle β is being increased at the rate $d\beta/dt = 30°/s$ and the elevation angle γ is being increased at the rate $d\gamma/dt = 10°/s$. For the position $\beta = 90°$ and $\gamma = 30°$, determine (*a*) the angular velocity of the barrel, (*b*) the angular acceleration of the barrel, (*c*) the velocity and acceleration of point P.

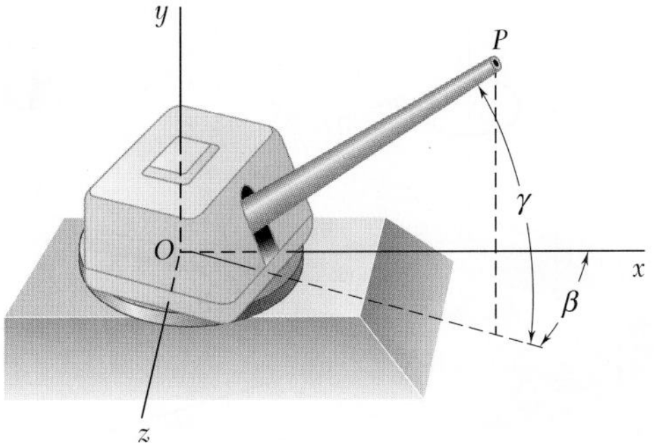

Fig. *P15.194*

15.195 A 100-mm-radius disk spins at the constant rate $\omega_2 = 4$ rad/s about an axis held by a housing attached to a horizontal rod that rotates at the constant rate $\omega_1 = 5$ rad/s. For the position shown, determine (*a*) the angular acceleration of the disk, (*b*) the acceleration of point P on the rim of the disk if $\theta = 0$, (*c*) the acceleration of point P on the rim of the disk if $\theta = 90°$.

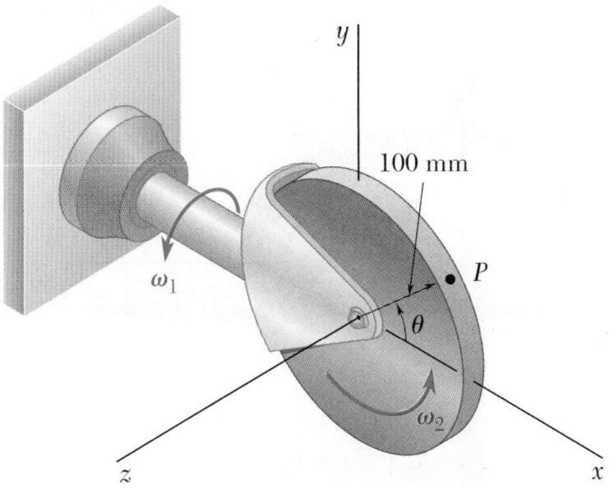

Fig. P15.195 and P15.196

15.196 A 100-mm-radius disk spins at the constant rate $\omega_2 = 4$ rad/s about an axis held by a housing attached to a horizontal rod that rotates at the constant rate $\omega_1 = 5$ rad/s. Knowing that $\theta = 30°$, determine the acceleration of point P on the rim of the disk.

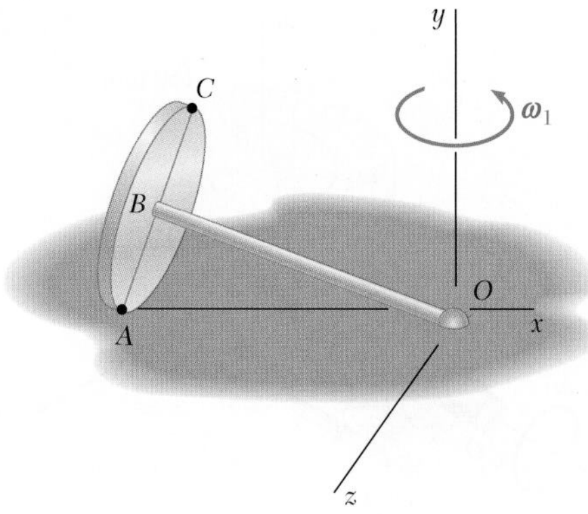

Fig. P15.197

15.197 A 30-mm-radius wheel is mounted on an axle OB of length 100 mm. The wheel rolls without sliding on the horizontal floor, and the axle is perpendicular to the plane of the wheel. Knowing that the system rotates about the y axis at a constant rate $\omega_1 = 2.4$ rad/s, determine (*a*) the angular velocity of the wheel, (*b*) the angular acceleration of the wheel, (*c*) the acceleration of point C located at the highest point on the rim of the wheel.

15.198 At the instant shown, the robotic arm ABC is being rotated simultaneously at the constant rate $\omega_1 = 0.15$ rad/s about the y axis, and at the constant rate $\omega_2 = 0.25$ rad/s about the z axis. Knowing that the length of arm ABC is 1 m, determine (*a*) the angular acceleration of the arm, (*b*) the velocity of point C, (*c*) the acceleration of point C.

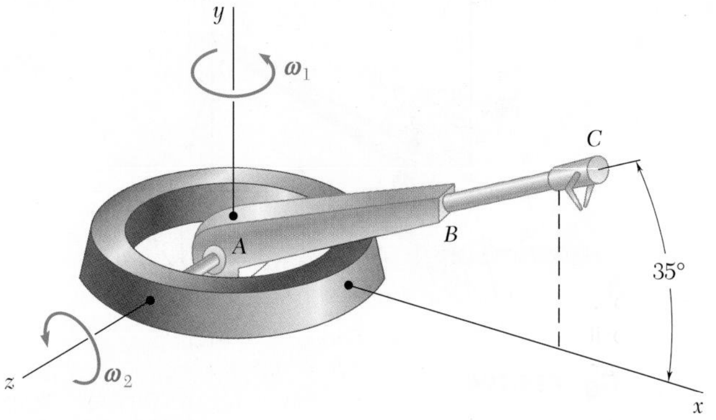

Fig. P15.198

15.199 In the planetary gear system shown, gears A and B are rigidly connected to each other and rotate as a unit about the inclined shaft. Gears C and D rotate with constant angular velocities of 30 rad/s and 20 rad/s, respectively (both counterclockwise when viewed from the right). Choosing the x axis to the right, the y axis upward, and the z axis pointing out of the plane of the figure, determine (*a*) the common angular velocity of gears A and B, (*b*) the angular velocity of shaft FH, which is rigidly attached to the inclined shaft.

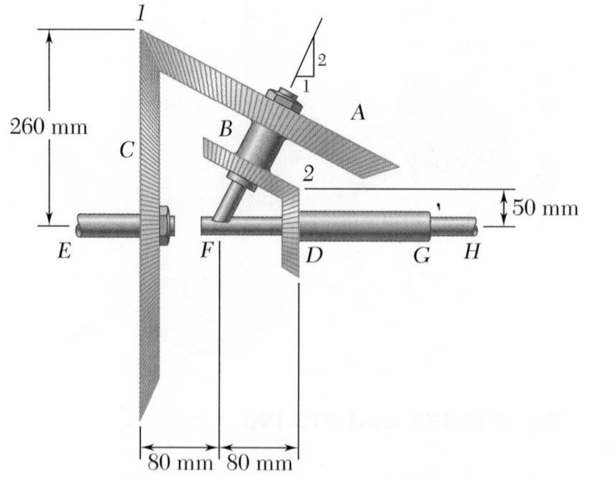

Fig. P15.199

15.200 In Prob. 15.199, determine (*a*) the common angular acceleration of gears A and B, (*b*) the acceleration of the tooth of gear A which is in contact with gear C at point I.

15.201 Several rods are brazed together to form the robotic guide arm shown which is attached to a ball-and-socket joint at O. Rod OA slides in a straight inclined slot while rod OB slides in a slot parallel to the z-axis. Knowing that at the instant shown $\mathbf{v}_B = (180\text{ mm/s})\,\mathbf{k}$, determine (*a*) the angular velocity of the guide arm, (*b*) the velocity of point A, (*c*) the velocity of point C.

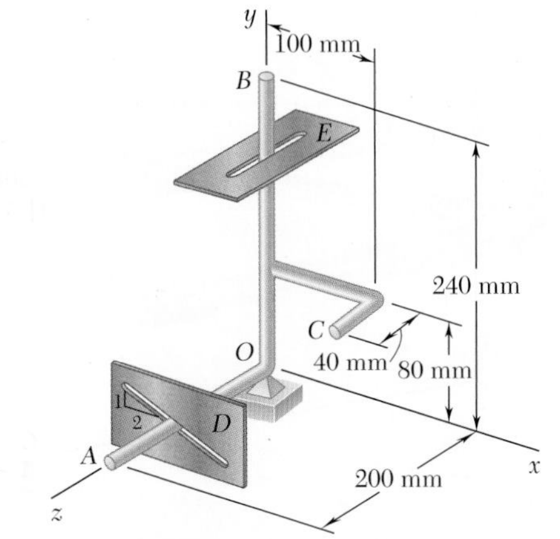

Fig. *P15.201*

15.202 In Prob. 15.199 the speed of point B is known to be constant. For the position shown, determine (*a*) the angular acceleration of the guide arm, (*b*) the acceleration of point C.

15.203 Rod AB of length 500 mm is connected by ball-and-socket joints to collars A and B, which slide along the two rods shown. Knowing that collar B moves toward point E at a constant speed of 400 mm/s, determine the velocity of collar A as collar B passes through point D.

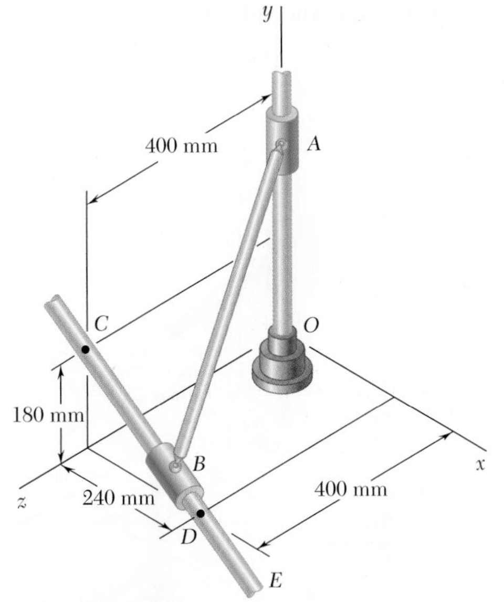

Fig. P15.203

15.204 Rod AB of length 275 mm is connected by ball-and-socket joints to collars A and B, which slide along the two rods shown. Knowing that collar B moves toward the origin O at a constant speed of 180 mm/s, determine the velocity of collar A when $c = 175$ mm.

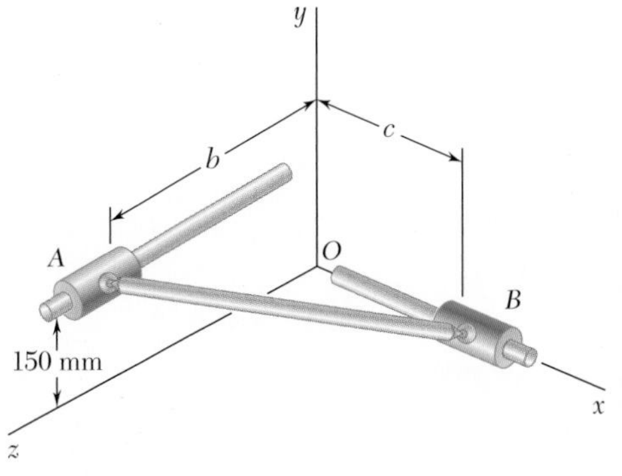

Fig. P15.204

15.205 Rod BC and BD are each 840 mm long and are connected by ball-and-socket joints to collars which may slide on the fixed rods shown. Knowing that collar B moves toward A at a constant speed of 390 mm/s, determine the velocity of collar C for the position shown.

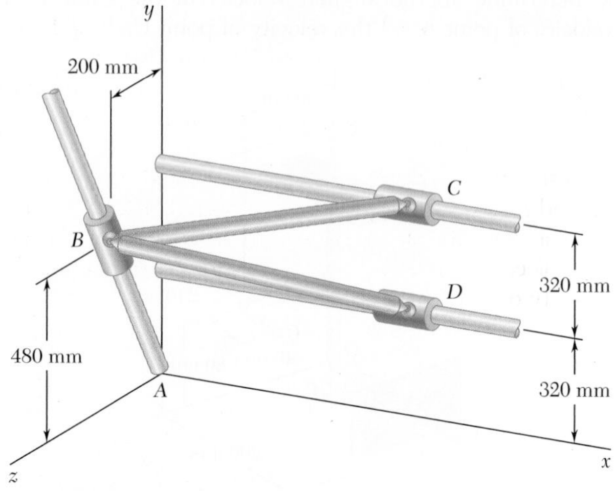

Fig. P15.205

15.206 Rod AB is connected by ball-and-socket joints to collar A and to the 400-mm-diameter disk C. Knowing that disk C rotates counterclockwise at the constant rate $\omega_0 = 3$ rad/s in the zx plane, determine the velocity of collar A for the position shown.

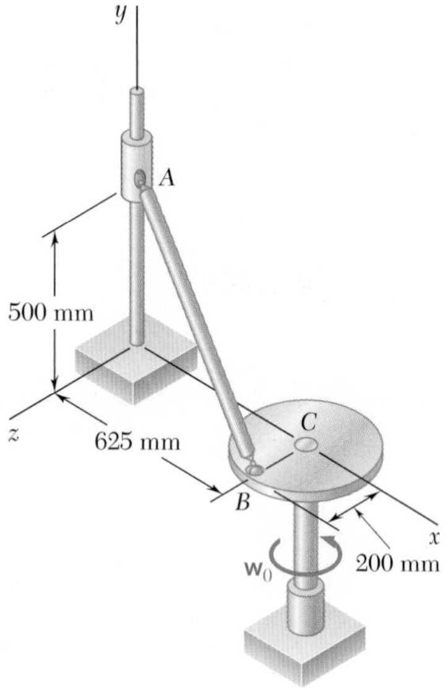

Fig. P15.206

15.207 Rod AB of length 580 mm is connected by ball-and-socket joints to the rotating crank BC and to the collar A. Crank BC is of length 160 mm and rotates in the horizontal xy plane at the constant rate $\omega_0 = 10$ rad/s. At the instant shown, when crank BC is parallel to the z axis, determine the velocity of collar A.

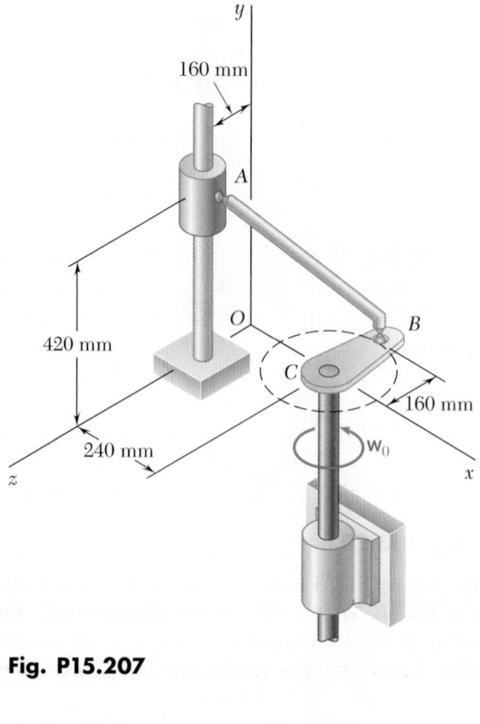

Fig. P15.207

15.208 Rod AB of length 300 mm is connected by ball-and-socket joints to collars A and B, which slide along the two rods shown. Knowing that collar B moves toward point D at a constant speed of 50 mm/s, determine the velocity of collar A when $c = 80$ mm.

15.209 Rod AB of length 300 mm is connected by ball-and-socket joints to collars A and B, which slide along the two rods shown. Knowing that collar B moves toward point D at a constant speed of 50 mm/s, determine the velocity of collar A when $c = 120$ mm.

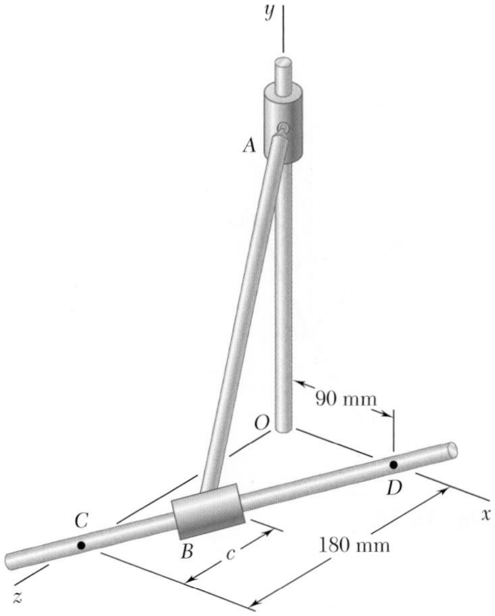

Fig. *P15.208* and *P15.209*

15.210 Two shafts AC and EG, which lie in the vertical yz plane, are connected by a universal joint at D. Shaft AC rotates with a constant angular velocity $\boldsymbol{\omega}_1$ as shown. At a time when the arm of the crosspiece attached to shaft AC is vertical, determine the angular velocity of shaft EG.

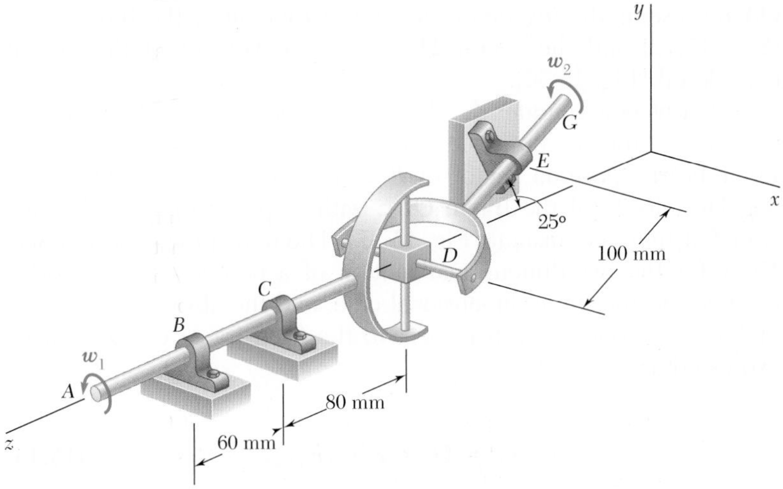

Fig. P15.210

15.211 Solve Prob. 15.210, assuming that the arm of the crosspiece attached to the shaft AC is horizontal.

15.212 In Prob. 15.206, the ball-and-socket joint between the rod and collar A is replaced by the clevis shown. Determine (*a*) the angular velocity of the rod, (*b*) the velocity of collar A.

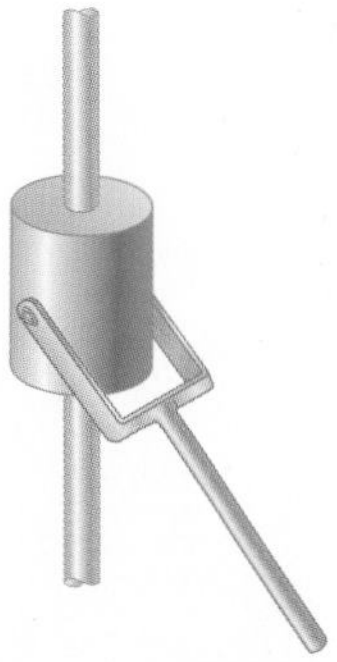

Fig. P15.212

15.213 In Prob. 15.205, the ball-and-socket joint between the rod and collar C is replaced by the clevis connection shown. Determine (*a*) the angular velocity of the rod, (*b*) the velocity of collar C.

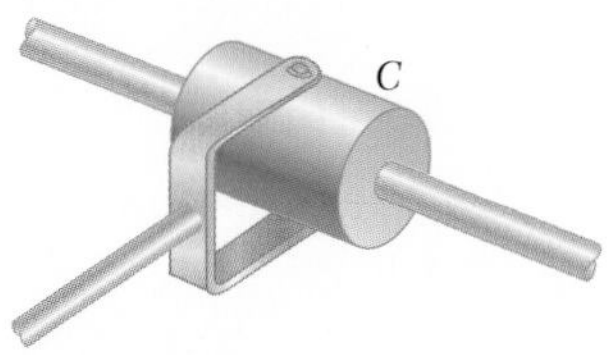

Fig. P15.213

15.214 Mechanism of Prob. 15.204, determine the acceleration of collar A.

***15.215** In Prob. 15.205, determine the acceleration of collar C.

15.216 Mechanism of Prob. 15.206, determine the acceleration of collar A.

15.217 Mechanism of Prob. 15.207, determine the acceleration of collar A.

15.218 In Prob. 15.208, determine the acceleration of collar A.

15.219 In Prob. 15.209, determine the acceleration of collar A.

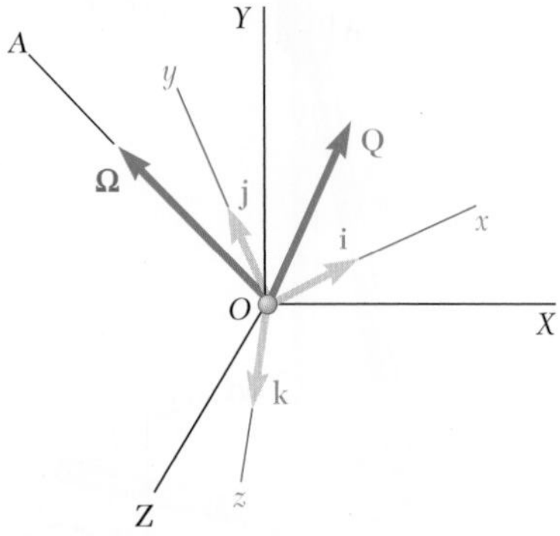

Fig. 15.36

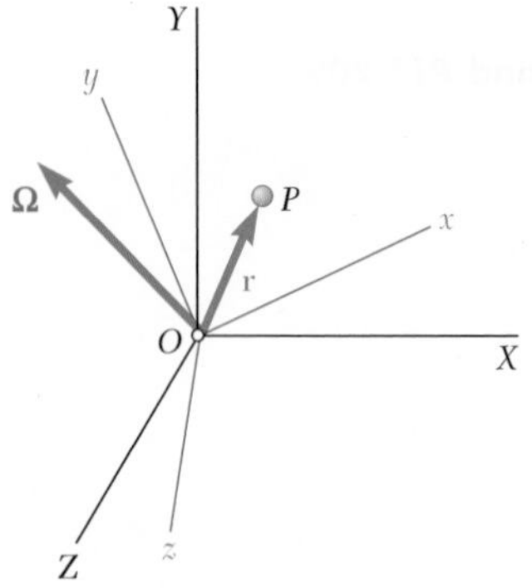

Fig. 15.37

*15.14 THREE-DIMENSIONAL MOTION OF A PARTICLE RELATIVE TO A ROTATING FRAME. CORIOLIS ACCELERATION

We saw in Sec. 15.10 that given a vector function $\mathbf{Q}(t)$ and two frames of reference centered at O—a fixed frame $OXYZ$ and a rotating frame $Oxyz$—the rates of change of $\mathbf{Q}$ with respect to the two frames satisfy the relation

$$(\dot{\mathbf{Q}})_{OXYZ} = (\dot{\mathbf{Q}})_{Oxyz} + \boldsymbol{\Omega} \times \mathbf{Q} \tag{15.31}$$

We had assumed at the time that the frame $Oxyz$ was constrained to rotate about a fixed axis OA. However, the derivation given in Sec. 15.10 remains valid when the frame $Oxyz$ is constrained only to have a fixed point O. Under this more general assumption, the axis OA represents the *instantaneous* axis of rotation of the frame $Oxyz$ (Sec. 15.12) and the vector $\boldsymbol{\Omega}$, its angular velocity at the instant considered (Fig. 15.36).

Let us now consider the three-dimensional motion of a particle P relative to a rotating frame $Oxyz$ constrained to have a fixed origin O. Let $\mathbf{r}$ be the position vector of P at a given instant and $\boldsymbol{\Omega}$ be the angular velocity of the frame $Oxyz$ with respect to the fixed frame $OXYZ$ at the same instant (Fig. 15.37). The derivations given in Sec. 15.11 for the two-dimensional motion of a particle can be readily extended to the three-dimensional case, and the absolute velocity $\mathbf{v}_P$ of P (i.e., its velocity with respect to the fixed frame $OXYZ$) can be expressed as

$$\mathbf{v}_P = \boldsymbol{\Omega} \times \mathbf{r} + (\dot{\mathbf{r}})_{Oxyz} \tag{15.45}$$

Denoting by $\mathcal{F}$ the rotating frame $Oxyz$, we write this relation in the alternative form

$$\mathbf{v}_P = \mathbf{v}_{P'} + \mathbf{v}_{P/\mathcal{F}} \tag{15.46}$$

where $\mathbf{v}_P$ = absolute velocity of particle P
$\mathbf{v}_{P'}$ = velocity of point P' of moving frame $\mathcal{F}$ coinciding with P
$\mathbf{v}_{P/\mathcal{F}}$ = velocity of P relative to moving frame $\mathcal{F}$

The absolute acceleration $\mathbf{a}_P$ of P can be expressed as

$$\mathbf{a}_P = \dot{\boldsymbol{\Omega}} \times \mathbf{r} + \boldsymbol{\Omega} \times (\boldsymbol{\Omega} \times \mathbf{r}) + 2\boldsymbol{\Omega} \times (\dot{\mathbf{r}})_{Oxyz} + (\ddot{\mathbf{r}})_{Oxyz} \tag{15.47}$$

An alternative form is

$$\mathbf{a}_P = \mathbf{a}_{P'} + \mathbf{a}_{P/\mathcal{F}} + \mathbf{a}_c \tag{15.48}$$

where $\mathbf{a}_P$ = absolute acceleration of particle P
$\mathbf{a}_{P'}$ = acceleration of point P' of moving frame $\mathcal{F}$ coinciding with P
$\mathbf{a}_{P/\mathcal{F}}$ = acceleration of P relative to moving frame $\mathcal{F}$
$\mathbf{a}_c = 2\boldsymbol{\Omega} \times (\dot{\mathbf{r}})_{Oxyz} = 2\boldsymbol{\Omega} \times \mathbf{v}_{P/\mathcal{F}}$
= complementary, or Coriolis, acceleration†

We note that the Coriolis acceleration is perpendicular to the vectors $\boldsymbol{\Omega}$ and $\mathbf{v}_{P/\mathcal{F}}$. However, since these vectors are usually not perpendicular to each other, the magnitude of $\mathbf{a}_c$ is in general *not* equal to $2\Omega v_{P/\mathcal{F}}$, as was the case for the plane motion of a particle. We further note that the Coriolis acceleration reduces to zero when the vectors $\boldsymbol{\Omega}$ and $\mathbf{v}_{P/\mathcal{F}}$ are parallel, or when either of them is zero.

Rotating frames of reference are particularly useful in the study of the three-dimensional motion of rigid bodies. If a rigid body has a fixed point O, as was the case for the crane of Sample Prob. 15.11, we can use a frame $Oxyz$ which is neither fixed nor rigidly attached to the rigid body. Denoting by $\boldsymbol{\Omega}$ the angular velocity of the frame $Oxyz$, we then resolve the angular velocity $\boldsymbol{\omega}$ of the body into the components $\boldsymbol{\Omega}$ and $\boldsymbol{\omega}_{B/\mathcal{F}}$, where the second component represents the angular velocity of the body relative to the frame $Oxyz$ (see Sample Prob. 15.14). An appropriate choice of the rotating frame often leads to a simpler analysis of the motion of the rigid body than would be possible with axes of fixed orientation. This is especially true in the case of the general three-dimensional motion of a rigid body, i.e., when the rigid body under consideration has no fixed point (see Sample Prob. 15.15).

*15.15 FRAME OF REFERENCE IN GENERAL MOTION

Consider a fixed frame of reference $OXYZ$ and a frame $Axyz$ which moves in a known, but arbitrary, fashion with respect to $OXYZ$ (Fig. 15.38). Let P be a particle moving in space. The position of P is defined at any instant by the vector $\mathbf{r}_P$ in the fixed frame, and by the vector $\mathbf{r}_{P/A}$ in the moving frame. Denoting by $\mathbf{r}_A$ the position vector of A in the fixed frame, we have

$$\mathbf{r}_P = \mathbf{r}_A + \mathbf{r}_{P/A} \qquad (15.49)$$

The absolute velocity $\mathbf{v}_P$ of the particle is obtained by writing

$$\mathbf{v}_P = \dot{\mathbf{r}}_P = \dot{\mathbf{r}}_A + \dot{\mathbf{r}}_{P/A} \qquad (15.50)$$

where the derivatives are defined with respect to the fixed frame $OXYZ$. The first term in the right-hand member of (15.50) thus represents the velocity $\mathbf{v}_A$ of the origin A of the moving axes. On the other hand, since the rate of change of a vector is the same with respect to a fixed frame and with respect to a frame in translation (Sec. 11.10), the second term can be regarded as the velocity $\mathbf{v}_{P/A}$ of

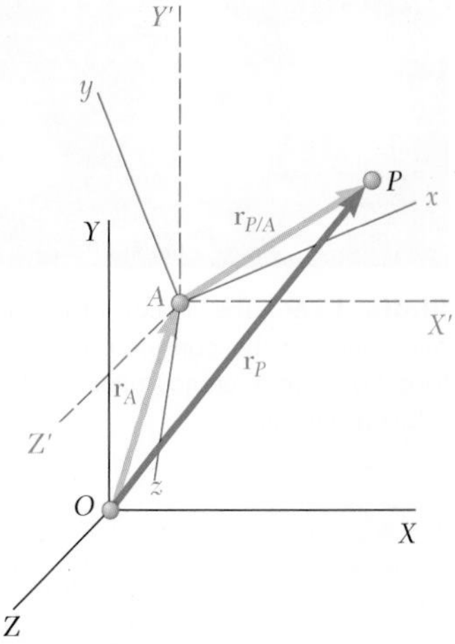

Fig. 15.38

†It is important to note the difference between Eq. (15.48) and Eq. (15.21) of Sec. 15.8. See the footnote on page 988.

P relative to the frame $AX'Y'Z'$ of the same orientation as $OXYZ$ and the same origin as $Axyz$. We therefore have

$$\mathbf{v}_P = \mathbf{v}_A + \mathbf{v}_{P/A} \tag{15.51}$$

But the velocity $\mathbf{v}_{P/A}$ of P relative to $AX'Y'Z'$ can be obtained from (15.45) by substituting $\mathbf{r}_{P/A}$ for $\mathbf{r}$ in that equation. We write

$$\mathbf{v}_P = \mathbf{v}_A + \mathbf{\Omega} \times \mathbf{r}_{P/A} + (\dot{\mathbf{r}}_{P/A})_{Axyz} \tag{15.52}$$

where $\mathbf{\Omega}$ is the angular velocity of the frame $Axyz$ at the instant considered.

The absolute acceleration $\mathbf{a}_P$ of the particle is obtained by differentiating (15.51) and writing

$$\mathbf{a}_P = \dot{\mathbf{v}}_P = \dot{\mathbf{v}}_A + \dot{\mathbf{v}}_{P/A} \tag{15.53}$$

where the derivatives are defined with respect to either of the frames $OXYZ$ or $AX'Y'Z'$. Thus, the first term in the right-hand member of (15.53) represents the acceleration $\mathbf{a}_A$ of the origin A of the moving axes and the second term represents the acceleration $\mathbf{a}_{P/A}$ of P relative to the frame $AX'Y'Z'$. This acceleration can be obtained from (15.47) by substituting $\mathbf{r}_{P/A}$ for $\mathbf{r}$. We therefore write

$$\begin{aligned}\mathbf{a}_P = \mathbf{a}_A + \dot{\mathbf{\Omega}} \times \mathbf{r}_{P/A} + \mathbf{\Omega} \times (\mathbf{\Omega} \times \mathbf{r}_{P/A})\\ + 2\mathbf{\Omega} \times (\dot{\mathbf{r}}_{P/A})_{Axyz} + (\ddot{\mathbf{r}}_{P/A})_{Axyz}\end{aligned} \tag{15.54}$$

Photo 15.9 The motion of air particles in a hurricane can be considered as motion relative to a frame of reference attached to the Earth and rotating with it.

Formulas (15.52) and (15.54) make it possible to determine the velocity and acceleration of a given particle with respect to a fixed frame of reference, when the motion of the particle is known with respect to a moving frame. These formulas become more significant, and considerably easier to remember, if we note that the sum of the first two terms in (15.52) represents the velocity of the point P' of the moving frame which coincides with P at the instant considered, and that the sum of the first three terms in (15.54) represents the acceleration of the same point. Thus, the relations (15.46) and (15.48) of the preceding section are still valid in the case of a reference frame in general motion, and we can write

$$\mathbf{v}_P = \mathbf{v}_{P'} + \mathbf{v}_{P/\mathcal{F}} \tag{15.46}$$

$$\mathbf{a}_P = \mathbf{a}_{P'} + \mathbf{a}_{P/\mathcal{F}} + \mathbf{a}_c \tag{15.48}$$

where the various vectors involved have been defined in Sec. 15.14.

It should be noted that if the moving reference frame $\mathcal{F}$ (or $Axyz$) is in translation, the velocity and acceleration of the point P' of the frame which coincides with P become, respectively, equal to the velocity and acceleration of the origin A of the frame. On the other hand, since the frame maintains a fixed orientation, $\mathbf{a}_c$ is zero, and the relations (15.46) and (15.48) reduce, respectively, to the relations (11.33) and (11.34) derived in Sec. 11.12.

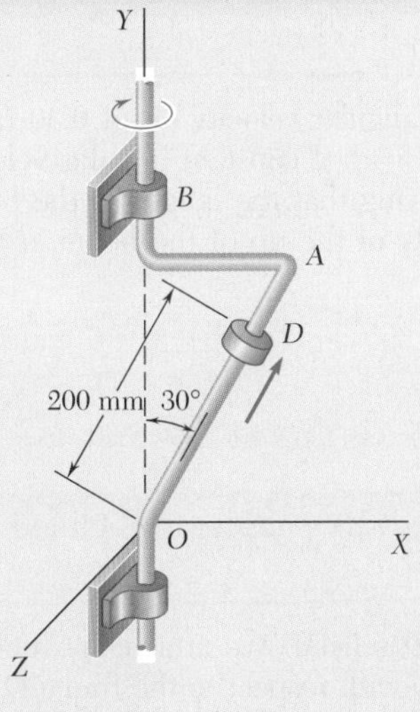

SAMPLE PROBLEM 15.13

The bent rod OAB rotates about the vertical OB. At the instant considered, its angular velocity and angular acceleration are, respectively, 20 rad/s and 200 rad/s^2, both clockwise when viewed from the positive Y axis. The collar D moves along the rod, and at the instant considered, OD = 200 mm. The velocity and acceleration of the collar relative to the rod are, respectively, 1.25 m/s and 15 m/s^2, both upward. Determine (*a*) the velocity of the collar, (*b*) the acceleration of the collar.

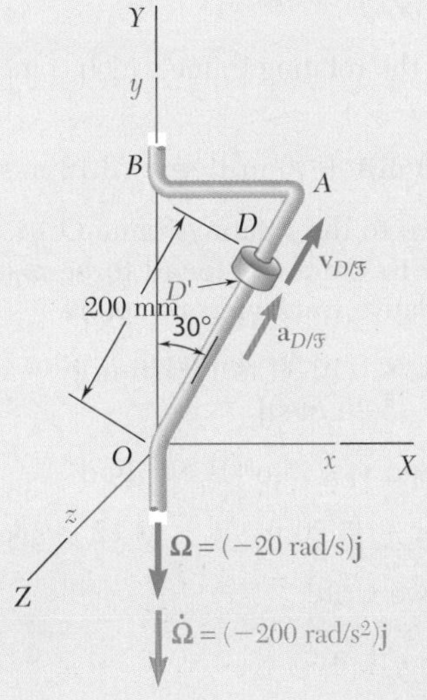

SOLUTION

Frames of Reference. The frame $OXYZ$ is fixed. We attach the rotating frame $Oxyz$ to the bent rod. Its angular velocity and angular acceleration relative to $OXYZ$ are therefore $\boldsymbol{\Omega} = (-20 \text{ rad/s})\mathbf{j}$ and $\dot{\boldsymbol{\Omega}} = (-200 \text{ rad/s}^2)\mathbf{j}$, respectively. The position vector of D is

$$\mathbf{r} = (200 \text{ mm})(\sin 30°\mathbf{i} + \cos 30°\mathbf{j}) = (100 \text{ mm})\mathbf{i} + (173.25 \text{ mm})\mathbf{j}$$

a. Velocity $\mathbf{v}_D$. Denoting by D' the point of the rod which coincides with D and by $\mathcal{F}$ the rotating frame $Oxyz$, we write from Eq. (15.46)

$$\mathbf{v}_D = \mathbf{v}_{D'} + \mathbf{v}_{D/\mathcal{F}} \qquad (1)$$

where

$$\mathbf{v}_{D'} = \boldsymbol{\Omega} \times \mathbf{r} = (-20 \text{ rad/s})\mathbf{j} \times [(100 \text{ mm})\mathbf{i} + (173.25 \text{ mm})\mathbf{j}] = (2000 \text{ mm/s})\mathbf{k}$$
$$\mathbf{v}_{D/\mathcal{F}} = (1250 \text{ mm/s})(\sin 30°\mathbf{i} + \cos 30°\mathbf{j}) = (625 \text{ mm/s})\mathbf{i} + (1083 \text{ mm/s})\mathbf{j}$$

Substituting the values obtained for $\mathbf{v}_{D'}$ and $\mathbf{v}_{D/\mathcal{F}}$ into (1), we find

$$\mathbf{v}_D = (625 \text{ mm/s})\mathbf{i} + (1083 \text{ mm/s})\mathbf{j} + (2000 \text{ mm/s})\mathbf{k} \quad ◀$$
$$= (0.625 \text{ m/s})\mathbf{i} + (1.083 \text{ m/s})\mathbf{j} + (2 \text{ m/s})\mathbf{k}$$

b. Acceleration $\mathbf{a}_D$. From Eq. (15.48) we write

$$\mathbf{a}_D = \mathbf{a}_{D'} + \mathbf{a}_{D/\mathcal{F}} + \mathbf{a}_c \qquad (2)$$

where

$$\mathbf{a}_{D'} = \dot{\boldsymbol{\Omega}} \times \mathbf{r} + \boldsymbol{\Omega} \times (\boldsymbol{\Omega} \times \mathbf{r})$$
$$= (-200 \text{ rad/s}^2)\mathbf{j} \times [(100 \text{ mm})\mathbf{i} + (173.25 \text{ mm})\mathbf{j}] - (20 \text{ red/s})\mathbf{j} \times (2000 \text{ mm/s})\mathbf{k}$$
$$= +(20{,}000 \text{ mm/s}^2)\mathbf{k} - (40{,}000 \text{ mm/s}^2)\mathbf{i}$$
$$\mathbf{a}_{D/\mathcal{F}} = (15{,}000 \text{ mm/s}^2)(\sin 30°\mathbf{i} + \cos 30°\mathbf{j}) = (7500 \text{ mm/s}^2)\mathbf{i} + (12{,}990 \text{ mm/s}^2)\mathbf{j}$$
$$\mathbf{a}_c = 2\boldsymbol{\Omega} \times \mathbf{v}_{D/\mathcal{F}}$$
$$= 2(-20 \text{ rad/s})\mathbf{j} \times [(625 \text{ mm/s})\mathbf{i} + (1083 \text{ mm/s})\mathbf{j}] = (25{,}000 \text{ mm/s}^2)\mathbf{k}$$

Substituting the values obtained for $\mathbf{a}_{D'}$, $\mathbf{a}_{D/\mathcal{F}}$, and $\mathbf{a}_c$ into (2),

$$\mathbf{a}_D = -(32{,}5000 \text{ mm/s}^2)\mathbf{i} + (12{,}990 \text{ mm/s}^2)\mathbf{j} + (45{,}000 \text{ mm/s}^2)\mathbf{k} \quad ◀$$
$$= -(32.5 \text{ m/s}^2)\mathbf{i} + (12.99 \text{ m/s}^2)\mathbf{j} + (45 \text{ m/s}^2)\mathbf{k}$$

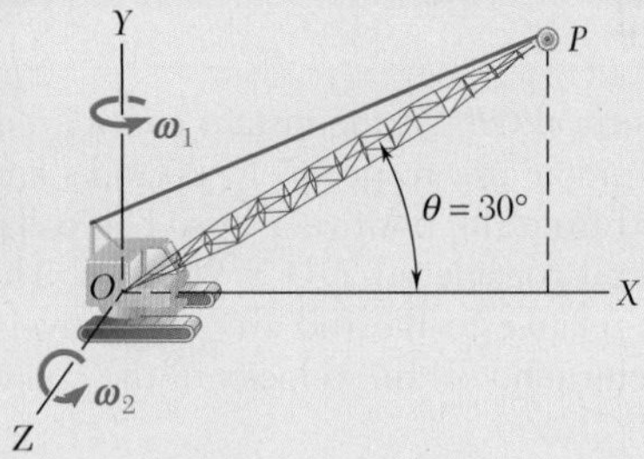

SAMPLE PROBLEM 15.14

The crane shown rotates with a constant angular velocity $\boldsymbol{\omega}_1$ of 0.30 rad/s. Simultaneously, the boom is being raised with a constant angular velocity $\boldsymbol{\omega}_2$ of 0.50 rad/s relative to the cab. Knowing that the length of the boom OP is $l = 12$ m, determine (*a*) the velocity of the tip of the boom, (*b*) the acceleration of the tip of the boom.

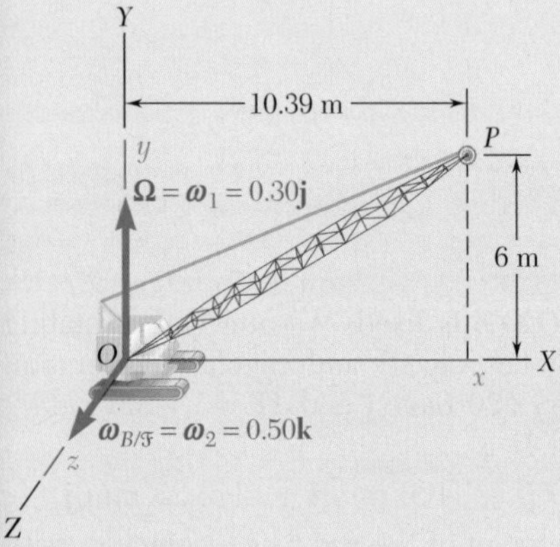

SOLUTION

Frames of Reference. The frame $OXYZ$ is fixed. We attach the rotating frame $Oxyz$ to the cab. Its angular velocity with respect to the frame $OXYZ$ is therefore $\boldsymbol{\Omega} = \boldsymbol{\omega}_1 = (0.30 \text{ rad/s})\mathbf{j}$. The angular velocity of the boom relative to the cab and the rotating frame $Oxyz$ (or $\mathcal{F}$, for short) is $\boldsymbol{\omega}_{B/\mathcal{F}} = \boldsymbol{\omega}_2 = (0.50 \text{ rad/s})\mathbf{k}$.

a. Velocity $\mathbf{v}_P$. From Eq. (15.46) we write

$$\mathbf{v}_P = \mathbf{v}_{P'} + \mathbf{v}_{P/\mathcal{F}} \tag{1}$$

where $\mathbf{v}_{P'}$ is the velocity of the point P' of the rotating frame which coincides with P:

$$\mathbf{v}_{P'} = \boldsymbol{\Omega} \times \mathbf{r} = (0.30 \text{ rad/s})\mathbf{j} \times [(10.39 \text{ m})\mathbf{i} + (6 \text{ m})\mathbf{j}] = -(3.12 \text{ m/s})\mathbf{k}$$

and where $\mathbf{v}_{P/\mathcal{F}}$ is the velocity of P relative to the rotating frame $Oxyz$. But the angular velocity of the boom relative to $Oxyz$ was found to be $\boldsymbol{\omega}_{B/\mathcal{F}} = (0.50 \text{ rad/s})\mathbf{k}$. The velocity of its tip P relative to $Oxyz$ is therefore

$$\begin{aligned} \mathbf{v}_{P/\mathcal{F}} = \boldsymbol{\omega}_{B/\mathcal{F}} \times \mathbf{r} &= (0.50 \text{ rad/s})\mathbf{k} \times [(10.39 \text{ m})\mathbf{i} + (6 \text{ m})\mathbf{j}] \\ &= -(3 \text{ m/s})\mathbf{i} + (5.20 \text{ m/s})\mathbf{j} \end{aligned}$$

Substituting the values obtained for $\mathbf{v}_{P'}$ and $\mathbf{v}_{P/\mathcal{F}}$ into (1), we find

$$\mathbf{v}_P = -(3 \text{ m/s})\mathbf{i} + (5.20 \text{ m/s})\mathbf{j} - (3.12 \text{ m/s})\mathbf{k} \quad ◀$$

b. Acceleration $\mathbf{a}_P$. From Eq. (15.48) we write

$$\mathbf{a}_P = \mathbf{a}_{P'} + \mathbf{a}_{P/\mathcal{F}} + \mathbf{a}_c \tag{2}$$

Since $\boldsymbol{\Omega}$ and $\boldsymbol{\omega}_{B/\mathcal{F}}$ are both constant, we have

$$\begin{aligned} \mathbf{a}_{P'} &= \boldsymbol{\Omega} \times (\boldsymbol{\Omega} \times \mathbf{r}) = (0.30 \text{ rad/s})\mathbf{j} \times (-3.12 \text{ m/s})\mathbf{k} = -(0.94 \text{ m/s}^2)\mathbf{i} \\ \mathbf{a}_{P/\mathcal{F}} &= \boldsymbol{\omega}_{B/\mathcal{F}} \times (\boldsymbol{\omega}_{B/\mathcal{F}} \times \mathbf{r}) \\ &= (0.50 \text{ rad/s})\mathbf{k} \times [-(3 \text{ m/s})\mathbf{i} + (5.20 \text{ m/s})\mathbf{j}] \\ &= -(1.50 \text{ m/s}^2)\mathbf{j} - (2.60 \text{ m/s}^2)\mathbf{i} \\ \mathbf{a}_c &= 2\boldsymbol{\Omega} \times \mathbf{v}_{P/\mathcal{F}} \\ &= 2(0.30 \text{ rad/s})\mathbf{j} \times [-(3 \text{ m/s})\mathbf{i} + (5.20 \text{ m/s})\mathbf{j}] = (1.80 \text{ m/s}^2)\mathbf{k} \end{aligned}$$

Substituting for $\mathbf{a}_{P'}$, $\mathbf{a}_{P/\mathcal{F}}$, and $\mathbf{a}_c$ into (2), we find

$$\mathbf{a}_P = -(3.54 \text{ m/s}^2)\mathbf{i} - (1.50 \text{ m/s}^2)\mathbf{j} + (1.80 \text{ m/s}^2)\mathbf{k} \quad ◀$$

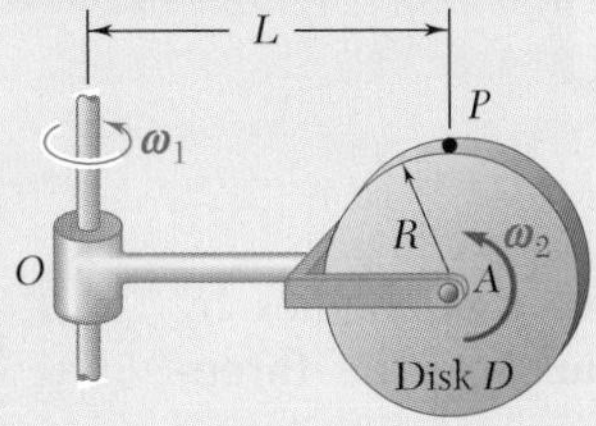

SAMPLE PROBLEM 15.15

Disk D, of radius R, is pinned to end A of the arm OA of length L located in the plane of the disk. The arm rotates about a vertical axis through O at the constant rate ω_1, and the disk rotates about A at the constant rate ω_2. Determine (*a*) the velocity of point P located directly above A, (*b*) the acceleration of P, (*c*) the angular velocity and angular acceleration of the disk.

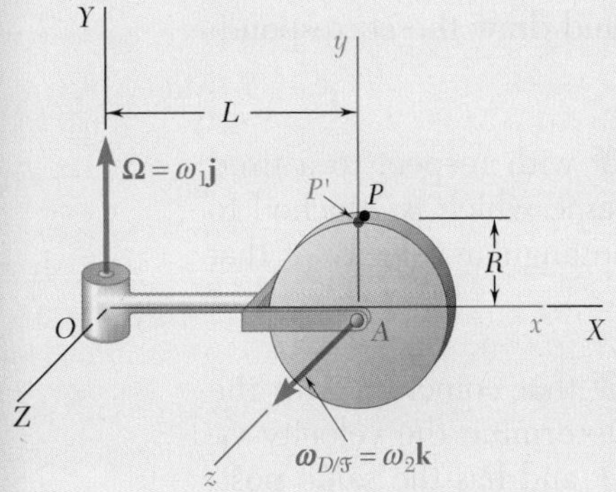

SOLUTION

Frames of Reference. The frame $OXYZ$ is fixed. We attach the moving frame $Axyz$ to the arm OA. Its angular velocity with respect to the frame $OXYZ$ is therefore $\boldsymbol{\Omega} = \omega_1\mathbf{j}$. The angular velocity of disk D relative to the moving frame $Axyz$ (or $\mathcal{F}$, for short) is $\boldsymbol{\omega}_{D/\mathcal{F}} = \omega_2\mathbf{k}$. The position vector of P relative to O is $\mathbf{r} = L\mathbf{i} + R\mathbf{j}$, and its position vector relative to A is $\mathbf{r}_{P/A} = R\mathbf{j}$.

a. Velocity $\mathbf{v}_P$. Denoting by P' the point of the moving frame which coincides with P, we write from Eq. (15.46)

$$\mathbf{v}_P = \mathbf{v}_{P'} + \mathbf{v}_{P/\mathcal{F}} \qquad (1)$$

where $\mathbf{v}_{P'} = \boldsymbol{\Omega} \times \mathbf{r} = \omega_1\mathbf{j} \times (L\mathbf{i} + R\mathbf{j}) = -\omega_1 L\mathbf{k}$

$$\mathbf{v}_{P/\mathcal{F}} = \boldsymbol{\omega}_{D/\mathcal{F}} \times \mathbf{r}_{P/A} = \omega_2\mathbf{k} \times R\mathbf{j} = -\omega_2 R\mathbf{i}$$

Substituting the values obtained for $\mathbf{v}_{P'}$ and $\mathbf{v}_{P/\mathcal{F}}$ into (1), we find

$$\mathbf{v}_P = -\omega_2 R\mathbf{i} - \omega_1 L\mathbf{k} \quad ◀$$

b. Acceleration $\mathbf{a}_P$. From Eq. (15.48) we write

$$\mathbf{a}_P = \mathbf{a}_{P'} + \mathbf{a}_{P/\mathcal{F}} + \mathbf{a}_c \qquad (2)$$

Since $\boldsymbol{\Omega}$ and $\boldsymbol{\omega}_{D/\mathcal{F}}$ are both constant, we have

$$\mathbf{a}_{P'} = \boldsymbol{\Omega} \times (\boldsymbol{\Omega} \times \mathbf{r}) = \omega_1\mathbf{j} \times (-\omega_1 L\mathbf{k}) = -\omega_1^2 L\mathbf{i}$$
$$\mathbf{a}_{P/\mathcal{F}} = \boldsymbol{\omega}_{D/\mathcal{F}} \times (\boldsymbol{\omega}_{D/\mathcal{F}} \times \mathbf{r}_{P/A}) = \omega_2\mathbf{k} \times (-\omega_2 R\mathbf{i}) = -\omega_2^2 R\mathbf{j}$$
$$\mathbf{a}_c = 2\boldsymbol{\Omega} \times \mathbf{v}_{P/\mathcal{F}} = 2\omega_1\mathbf{j} \times (-\omega_2 R\mathbf{i}) = 2\omega_1\omega_2 R\mathbf{k}$$

Substituting the values obtained into (2), we find

$$\mathbf{a}_P = -\omega_1^2 L\mathbf{i} - \omega_2^2 R\mathbf{j} + 2\omega_1\omega_2 R\mathbf{k} \quad ◀$$

c. Angular Velocity and Angular Acceleration of Disk.

$$\boldsymbol{\omega} = \boldsymbol{\Omega} + \boldsymbol{\omega}_{D/\mathcal{F}} \qquad \boldsymbol{\omega} = \omega_1\mathbf{j} + \omega_2\mathbf{k} \quad ◀$$

Using Eq. (15.31) with $\mathbf{Q} = \boldsymbol{\omega}$, we write

$$\boldsymbol{\alpha} = (\dot{\boldsymbol{\omega}})_{OXYZ} = (\dot{\boldsymbol{\omega}})_{Axyz} + \boldsymbol{\Omega} \times \boldsymbol{\omega}$$
$$= 0 + \omega_1\mathbf{j} \times (\omega_1\mathbf{j} + \omega_2\mathbf{k})$$

$$\boldsymbol{\alpha} = \omega_1\omega_2\mathbf{i} \quad ◀$$

SOLVING PROBLEMS ON YOUR OWN

In this lesson you concluded your study of the kinematics of rigid bodies by learning how to use an auxiliary frame of reference $\mathcal{F}$ to analyze the three-dimensional motion of a rigid body. This auxiliary frame may be a *rotating frame* with a fixed origin O, or it may be a *frame in general motion*.

A. Using a rotating frame of reference. As you approach a problem involving the use of a rotating frame $\mathcal{F}$ you should take the following steps.

1. Select the rotating frame $\mathcal{F}$ that you wish to use and draw the corresponding coordinate axes x, y, and z from the fixed point O.

2. Determine the angular velocity Ω of the frame $\mathcal{F}$ with respect to a fixed frame $OXYZ$. In most cases, you will have selected a frame which is attached to some rotating element of the system; $\boldsymbol{\Omega}$ will then be the angular velocity of that element.

3. Designate as P' the point of the rotating frame $\mathcal{F}$ that coincides with the point P of interest at the instant you are considering. Determine the velocity $\mathbf{v}_{P'}$ and the acceleration $\mathbf{a}_{P'}$ of point P'. Since P' is part of $\mathcal{F}$ and has the same position vector $\mathbf{r}$ as P, you will find that

$$\mathbf{v}_{P'} = \boldsymbol{\Omega} \times \mathbf{r} \qquad \text{and} \qquad \mathbf{a}_{P'} = \boldsymbol{\alpha} \times \mathbf{r} + \boldsymbol{\Omega} \times (\boldsymbol{\Omega} \times \mathbf{r})$$

where $\boldsymbol{\alpha}$ is the angular acceleration of $\mathcal{F}$. However, in many of the problems that you will encounter, the angular velocity of $\mathcal{F}$ is constant in both magnitude and direction, and $\boldsymbol{\alpha} = 0$.

4. Determine the velocity and acceleration of point P with respect to the frame $\mathcal{F}$. As you are trying to determine $\mathbf{v}_{P/\mathcal{F}}$ and $\mathbf{a}_{P/\mathcal{F}}$ you will find it useful to visualize the motion of P on frame $\mathcal{F}$ when the frame is not rotating. If P is a point of a rigid body $\mathcal{B}$ which has an angular velocity $\boldsymbol{\omega}_{\mathcal{B}}$ and an angular acceleration $\boldsymbol{\alpha}_{\mathcal{B}}$ relative to $\mathcal{F}$ [Sample Prob. 15.14], you will find that

$$\mathbf{v}_{P/\mathcal{F}} = \boldsymbol{\omega}_{\mathcal{B}} \times \mathbf{r} \qquad \text{and} \qquad \mathbf{a}_{P/\mathcal{F}} = \boldsymbol{\alpha}_{\mathcal{B}} \times \mathbf{r} + \boldsymbol{\omega}_{\mathcal{B}} \times (\boldsymbol{\omega}_{\mathcal{B}} \times \mathbf{r})$$

In many of the problems that you will encounter, the angular velocity of body $\mathcal{B}$ relative to frame $\mathcal{F}$ is constant in both magnitude and direction, and $\boldsymbol{\alpha}_{\mathcal{B}} = 0$.

5. Determine the Coriolis acceleration. Considering the angular velocity $\boldsymbol{\Omega}$ of frame $\mathcal{F}$ and the velocity $\mathbf{v}_{P/\mathcal{F}}$ of point P relative to that frame, which was computed in the previous step, you write

$$\mathbf{a}_c = 2\boldsymbol{\Omega} \times \mathbf{v}_{P/\mathcal{F}}$$

(continued)

6. The velocity and the acceleration of *P* with respect to the fixed frame *OXYZ* can now be obtained by adding the expressions you have determined:

$$\mathbf{v}_P = \mathbf{v}_{P'} + \mathbf{v}_{P/\mathcal{F}} \qquad (15.46)$$
$$\mathbf{a}_P = \mathbf{a}_{P'} + \mathbf{a}_{P/\mathcal{F}} + \mathbf{a}_c \qquad (15.48)$$

B. Using a frame of reference in general motion. The steps that you will take differ only slightly from those listed under A. They consist of the following:

1. Select the frame $\mathcal{F}$ that you wish to use and a reference point *A* in that frame, from which you will draw the coordinate axes, x, y, and z defining that frame. You will consider the motion of the frame as the sum of a *translation with A and a rotation about A.*

2. Determine the velocity $\mathbf{v}_A$ of point *A* and the angular velocity $\boldsymbol{\Omega}$ of the frame. In most cases, you will have selected a frame which is attached to some element of the system; $\boldsymbol{\Omega}$ will then be the angular velocity of that element.

3. Designate as *P*′ the point of frame $\mathcal{F}$ that coincides with the point P of interest at the instant you are considering, and determine the velocity $\mathbf{v}_{P'}$ and the acceleration $\mathbf{a}_{P'}$ of that point. In some cases, this can be done by visualizing the motion of P if that point were prevented from moving with respect to $\mathcal{F}$ [Sample Prob. 15.15]. A more general approach is to recall that the motion of P' is the sum of a translation with the reference point A and a rotation about A. The velocity $\mathbf{v}_{P'}$ and the acceleration $\mathbf{a}_{P'}$ of P', therefore, can be obtained by adding $\mathbf{v}_A$ and $\mathbf{a}_A$, respectively, to the expressions found in paragraph A3 and replacing the position vector $\mathbf{r}$ by the vector $\mathbf{r}_{P/A}$ drawn from A to P:

$$\mathbf{v}_{P'} = \mathbf{v}_A + \boldsymbol{\Omega} \times \mathbf{r}_{P/A} \qquad \mathbf{a}_{P'} = \mathbf{a}_A + \boldsymbol{\alpha} \times \mathbf{r}_{P/A} + \boldsymbol{\Omega} \times (\boldsymbol{\Omega} \times \mathbf{r}_{P/A})$$

Steps 4, 5, and 6 are the same as in Part *A*, except that the vector $\mathbf{r}$ should again be replaced by $\mathbf{r}_{P/A}$. Thus, Eqs. (15.46) and (15.48) can still be used to obtain the velocity and the acceleration of P with respect to the fixed frame of reference *OXYZ.*

PROBLEMS

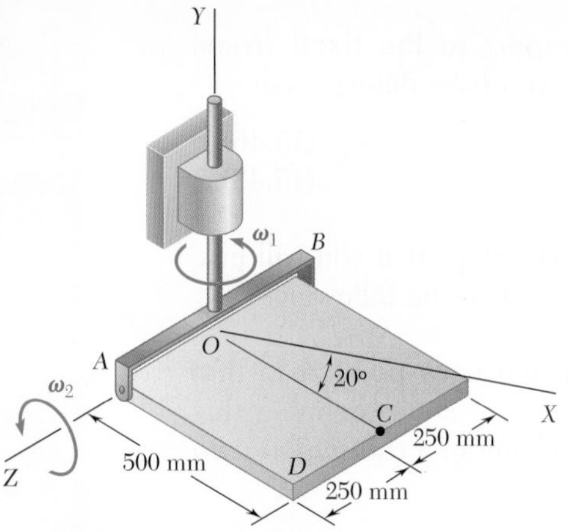

Fig. P15.220 and P15.221

END-OF-SECTION PROBLEMS

15.220 A square plate of side 500 mm. is hinged at A and B to a clevis. The plate rotates at the constant rate $\omega_2 = 4$ rad/s with respect to the clevis, which itself rotates at the constant rate $\omega_1 = 3$ rad/s about the Y axis. For the position shown, determine (*a*) the velocity of point C, (*b*) the acceleration of point C.

15.221 A square plate of side 500 mm is hinged at A and B to a clevis. The plate rotates at the constant rate $\omega_2 = 4$ rad/s with respect to the clevis, which itself rotates at the constant rate $\omega_1 = 3$ rad/s about the Y axis. For the position shown, determine (*a*) the velocity of corner D, (*b*) the acceleration of corner D.

15.222 and 15.223 The rectangular plate shown rotates at the constant rate $\omega_2 = 12$ rad/s with respect to arm AE, which itself rotates at the constant rate $\omega_1 = 9$ rad/s about the Z axis. For the position shown, determine the velocity and acceleration of the point of the plate indicated.

15.222 Corner B
15.223 Corner C

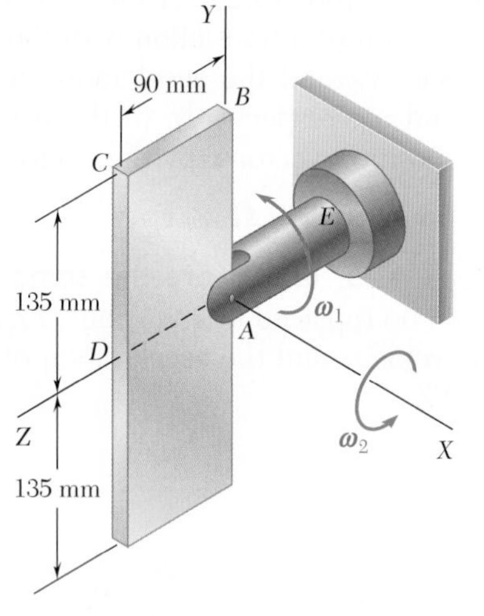

Fig. P15.222 and P15.223

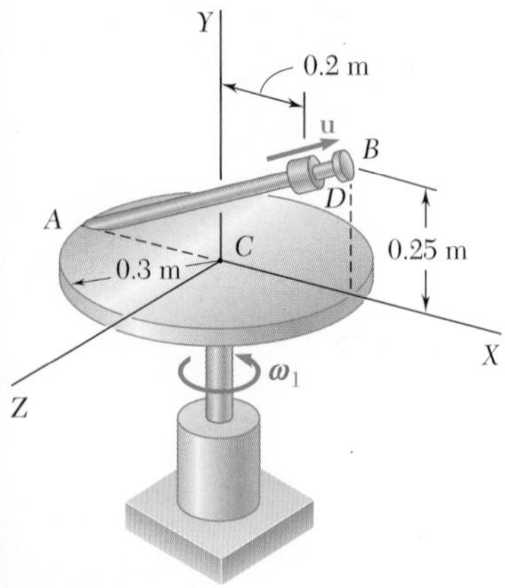

Fig. P15.224

15.224 Rod AB is welded to the 0.3-m-radius plate which rotates at the constant rate $\omega_1 = 6$ rad/s. Knowing that collar D moves toward end B of the rod at a constant speed $u = 1.3$ m/s, determine, for the position shown, (*a*) the velocity of D, (*b*) the acceleration of D.

15.225 The bent rod shown rotates at the constant rate $\omega_1 = 3$ rad/s. Knowing that collar C moves toward point D at a constant relative speed $u = 1$ m/s, determine, for the position shown, the velocity and acceleration of C if (*a*) $x = 100$ mm (*b*) $x = 400$ mm.

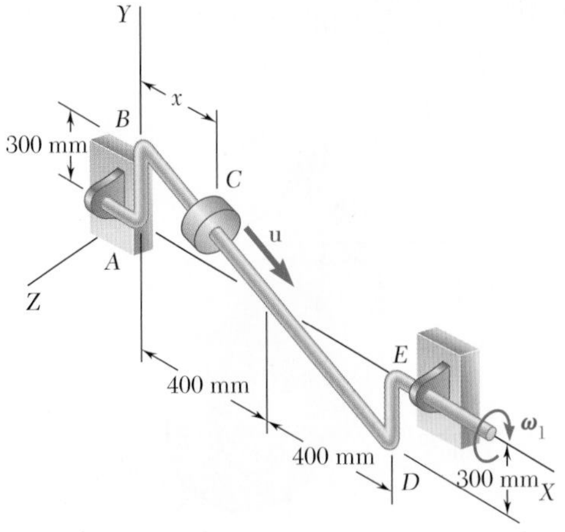

Fig. P15.225

15.226 The bent pipe shown rotates at the constant rate $\omega_1 = 10$ rad/s. Knowing that a ball bearing D moves in portion BC of the pipe toward end C at a constant relative speed $u = 0.6$ m/s, determine at the instant shown (*a*) the velocity of D, (*b*) the acceleration of D.

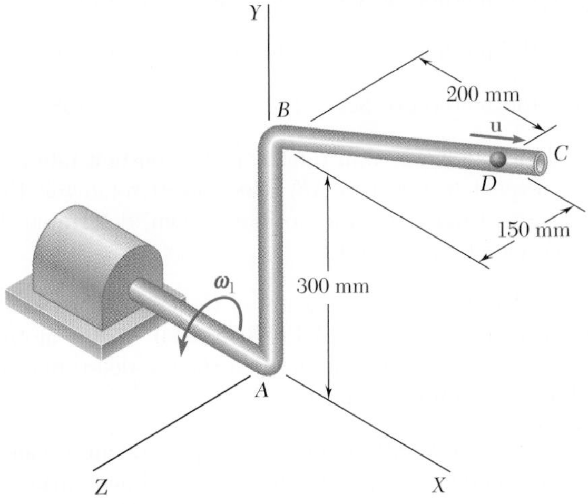

Fig. P15.226

15.227 The circular plate shown rotates about its vertical diameter at the constant rate $\omega_1 = 10$ rad/s. Knowing that in the position shown the disk lies in the XY plane and point D of strap CD moves upward at a constant relative speed $u = 1.5$ m/s, determine (*a*) the velocity of D, (b) the acceleration of D.

15.228 Manufactured items are spray-painted as they pass through the automated work station shown. Knowing that the bent pipe ACE rotates at the constant rate $\omega_1 = 0.4$ rad/s and that at point D the paint moves through the pipe at a constant relative speed $u = 150$ mm/s, determine, for the position shown, (*a*) the velocity of the paint at D, (*b*) the acceleration of the paint at D.

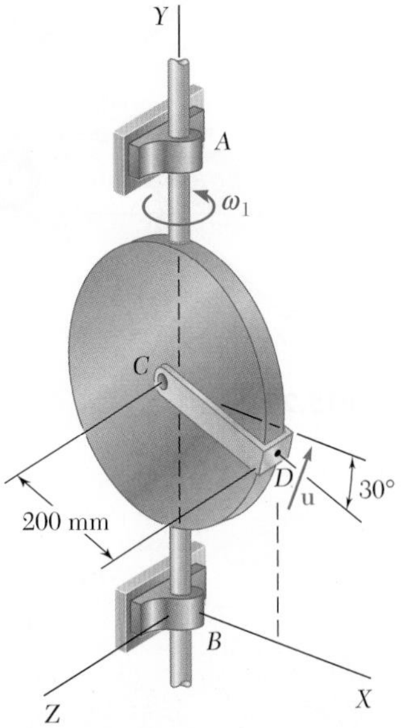

Fig. P15.227

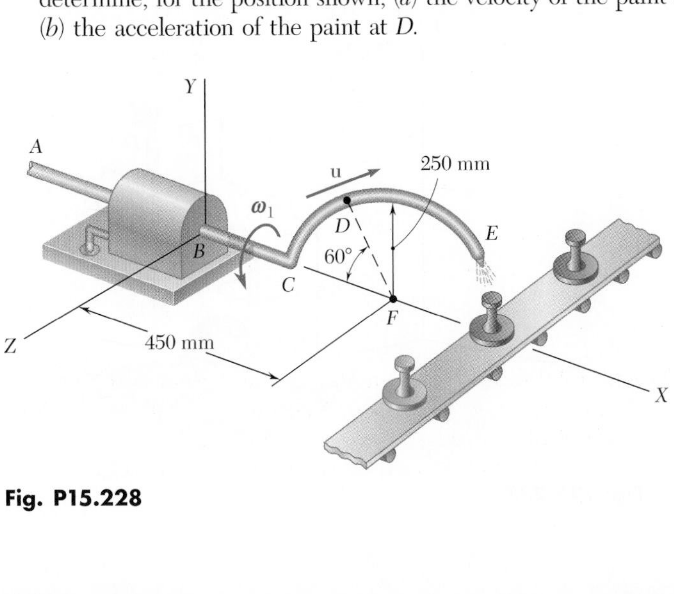

Fig. P15.228

15.229 Solve Prob. 15.227, assuming that at the instant shown the angular velocity ω_1 of the plate is 10 rad/s and is decreasing at the rate of 25 rad/s^2, while the relative speed u of point D of strap CD is 1.5 m/s and is decreasing at the rate of 3 m/s^2.

15.230 Solve Prob. 15.225, assuming that at the instant shown the angular velocity $\boldsymbol{\omega}_1$ of the rod is 3 rad/s and is increasing at the rate of 12 rad/s^2, while the relative speed u of the collar is 1m/s and is decreasing at the rate of 2 m/s^2.

15.231 Using the method of Sec. 15.14, solve Prob. 15.192.

15.232 Using the method of Sec. 15.14, solve Prob. 15.196.

15.233 Using the method of Sec. 15.14, solve Prob. 15.198.

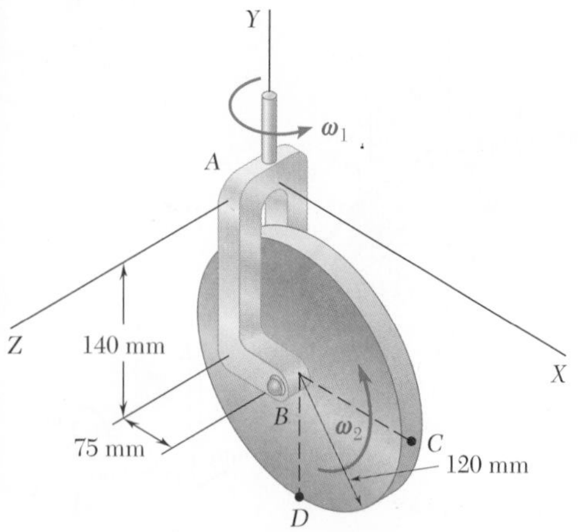

Fig. P15.234 and P15.235

15.234 A disk of radius 120 mm rotates at the constant rate $\omega_2 = 5$ rad/s with respect to the arm AB, which itself rotates at the constant rate $\omega_1 = 3$ rad/s. For the position shown, determine the velocity and acceleration of point C.

15.235 A disk of radius 120 mm rotates at the constant rate $\omega_2 = 5$ rad/s with respect to the arm AB, which itself rotates at the constant rate $\omega_1 = 3$ rad/s. For the position shown, determine the velocity and acceleration of point D.

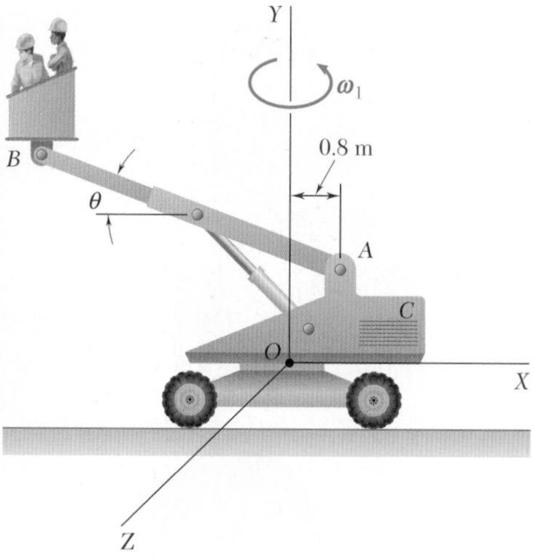

Fig. P15.236

15.236 The arm AB of length 5 m is used to provide an elevated platform for construction workers. In the position shown, arm AB is being raised at the constant rate $d\theta/dt = 0.25$ rad/s; simultaneously, the unit is being rotated about the Y axis at the constant rate $\omega_1 = 0.15$ rad/s. Knowing that $\theta = 20°$, determine the velocity and acceleration of point B.

15.237 The remote manipulator system (RMS) shown is used to deploy payloads from the cargo bay of space shuttles. At the instant shown, the whole RMS is rotating at the constant rate $\omega_1 = 0.03$ rad/s about the axis AB. At the same time, portion BCD rotates as a rigid body at the constant rate $\omega_2 = d\beta/dt = 0.04$ rad/s about an axis through B parallel to the X axis. Knowing that $\beta = 30°$, determine (*a*) the angular acceleration of BCD, (*b*) the velocity of D, (*c*) the acceleration of D.

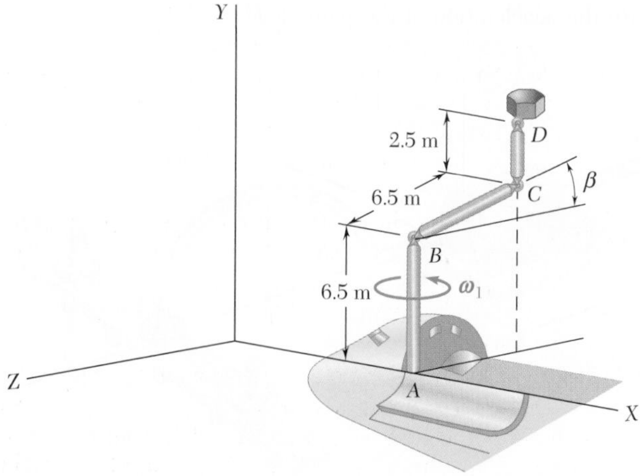

Fig. P15.237

15.238 The body AB and rod BC of the robotic component shown rotate at the constant rate $\omega_1 = 0.60$ rad/s about the Y axis. Simultaneously a wire-and-pulley control causes arm CD to rotate about C at the constant rate $\omega = d\beta/dt = 0.45$ rad/s. Knowing $\beta = 120°$, determine (*a*) the angular acceleration of arm CD, (*b*) the velocity of D, (*c*) the acceleration of D.

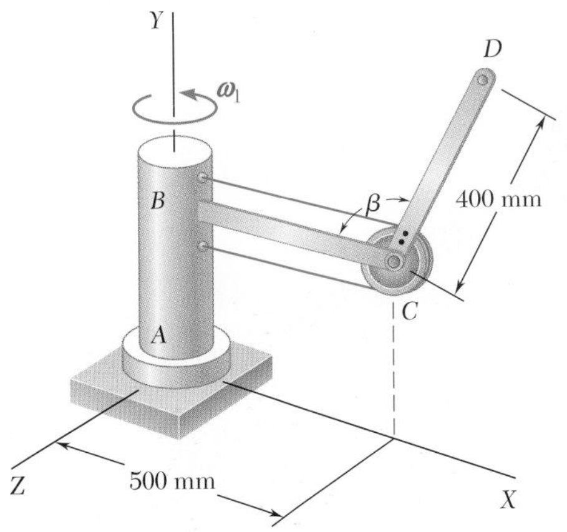

Fig. P15.238

15.239 The crane shown rotates at the constant rate $\omega_1 = 0.25$ rad/s; simultaneously, the telescoping boom is being lowered at the constant rate $\omega_2 = 0.40$ rad/s. Knowing that at the instant shown the length of the boom is 6 m and is increasing at the constant rate $u = 0.5$ m/s, determine the velocity and acceleration of point B.

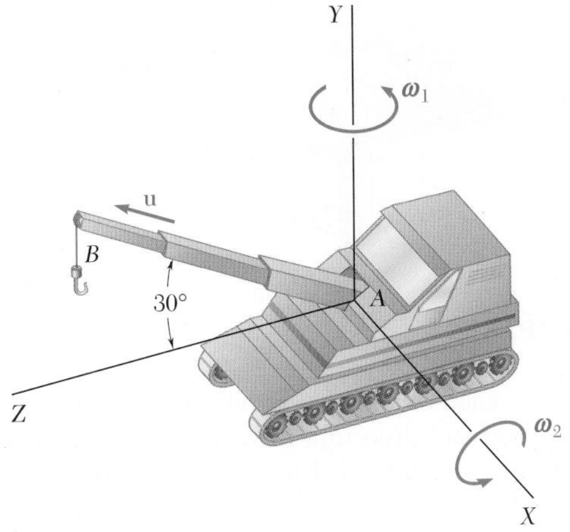

Fig. P15.239

15.240 The vertical plate shown is welded to arm EFG, and the entire unit rotates at the constant rate $\omega_1 = 2$ rad/s about the Y axis. At the same time, a continuous link belt moves around the perimeter of the plate at a constant speed $u = 100$ mm/s. For the position shown, determine the acceleration of the link of the belt located (*a*) at point A, (*b*) at point B.

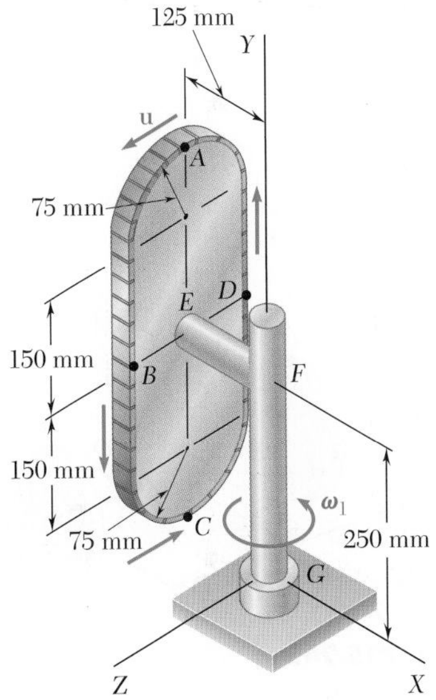

Fig. P15.240 and P15.241

15.241 The vertical plate shown is welded to arm EFG, and the entire unit rotates at the constant rate $\omega_1 = 2$ rad/s about the Y axis. At the same time, a continuous link belt moves around the perimeter of the plate at a constant speed $u = 100$ mm/s. For the position shown, determine the acceleration of the link of the belt located (*a*) at point C, (*b*) at point D.

15.242 A disk of 180-mm radius rotates at the constant rate $\omega_2 = 12$ rad/s with respect to arm CD, which itself rotates at the constant rate $\omega_1 = 8$ rad/s about the Y axis. Determine at the instant shown the velocity and acceleration of point A on the rim of the disk.

15.243 A disk of 180-mm radius rotates at the constant rate $\omega_2 = 12$ rad/s with respect to arm CD, which itself rotates at the constant rate $\omega_1 = 8$ rad/s about the Y axis. Determine at the instant shown the velocity and acceleration of point B on the rim of the disk.

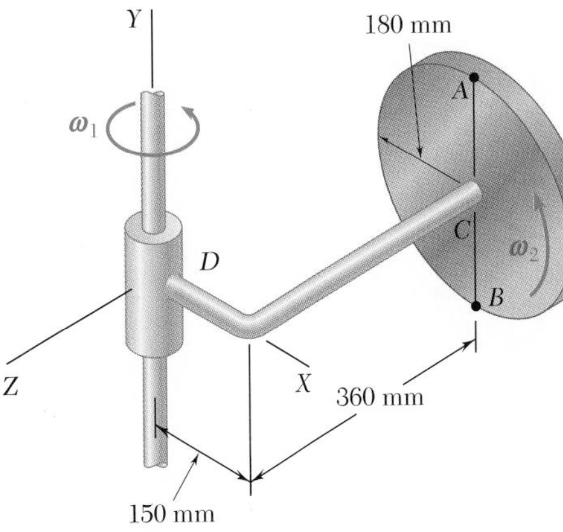

Fig. P15.242 and P15.243

15.244 A square plate of side $2r$ is welded to a vertical shaft which rotates with a constant angular velocity $\boldsymbol{\omega}_1$. At the same time, rod AB of length r rotates about the center of the plate with a constant angular velocity $\boldsymbol{\omega}_2$ with respect to the plate. For the position of the plate shown, determine the acceleration of end B of the rod if (*a*) $\theta = 0$, (*b*) $\theta = 90°$, (*c*) $\theta = 180°$.

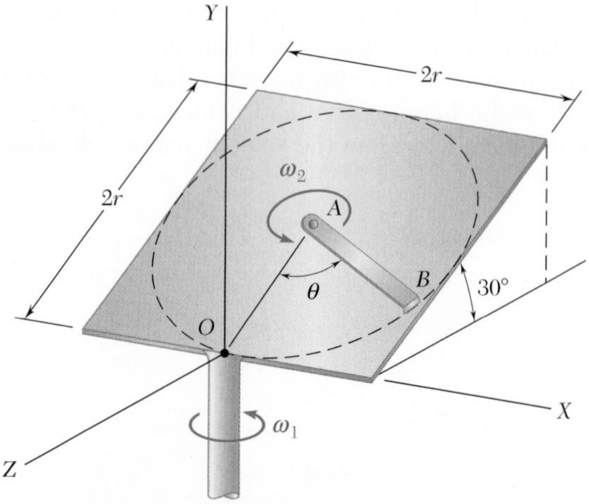

Fig. P15.244

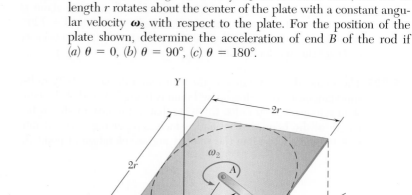

15.245 Two disks, each of 130-mm radius, are welded to the 500-mm rod CD. The rod-and-disks unit rotates at the constant rate $\omega_2 = 3$ rad/s with respect to arm AB. Knowing that at the instant shown $\omega_1 = 4$ rad/s, determine the velocity and acceleration of (*a*) point E, (*b*) point F.

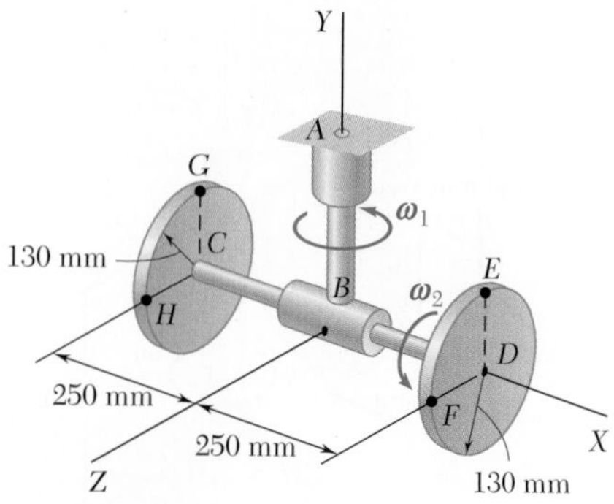

Fig. P15.245

15.246 In Prob. 15.245, determine the velocity and acceleration of (*a*) point G, (*b*) point H.

15.247 The position of the stylus tip A is controlled by the robot shown. In the position shown, the stylus moves at a constant speed $u = 180$ mm/s relative to the solenoid BC. At the same time, arm CD rotates at the constant rate $\omega_2 = 1.6$ rad/s with respect to component DEG. Knowing that the entire robot rotates about the X axis at the constant rate $\omega_1 = 1.2$ rad/s, determine (*a*) the velocity of A, (*b*) the acceleration of A.

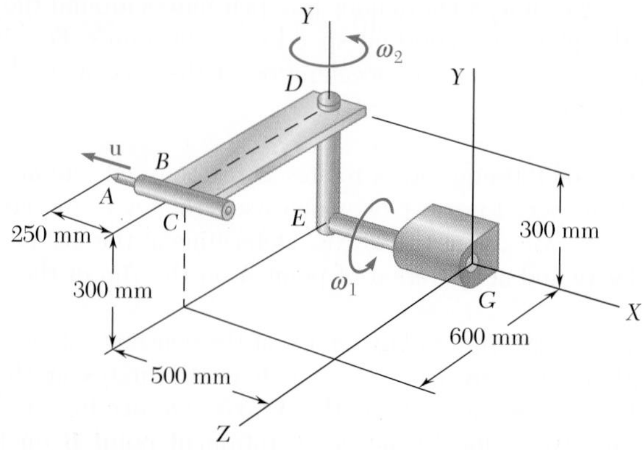

Fig. P15.247

REVIEW AND SUMMARY

This chapter was devoted to the study of the kinematics of rigid bodies.

Rigid body in translation

We first considered the *translation* of a rigid body [Sec. 15.2] and observed that in such a motion, *all points of the body have the same velocity and the same acceleration at any given instant.*

Rigid body in rotation about a fixed axis

We next considered the *rotation* of a rigid body about a fixed axis [Sec. 15.3]. The position of the body is defined by the angle θ that the line BP, drawn from the axis of rotation to a point P of the body, forms with a fixed plane (Fig. 15.39). We found that the magnitude of the velocity of P is

$$v = \frac{ds}{dt} = r\dot{\theta} \sin \phi \qquad (15.4)$$

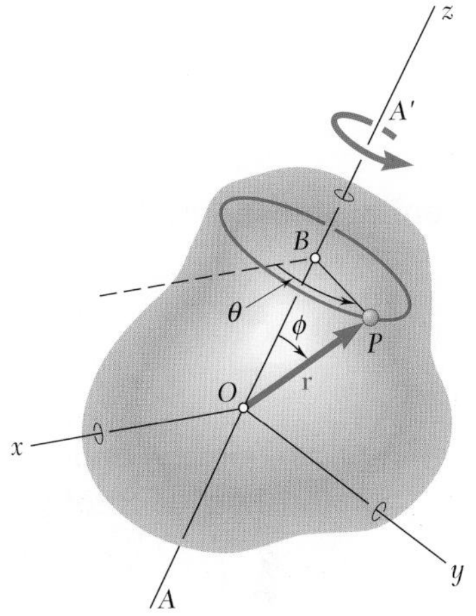

Fig. 15.39

where $\dot{\theta}$ is the time derivative of θ. We then expressed the velocity of P as

$$\mathbf{v} = \frac{d\mathbf{r}}{dt} = \boldsymbol{\omega} \times \mathbf{r} \qquad (15.5)$$

where the vector

$$\boldsymbol{\omega} = \omega\mathbf{k} = \dot{\theta}\mathbf{k} \qquad (15.6)$$

is directed along the fixed axis of rotation and represents the *angular velocity* of the body.

Denoting by $\boldsymbol{\alpha}$ the derivative $d\boldsymbol{\omega}/dt$ of the angular velocity, we expressed the acceleration of P as

$$\mathbf{a} = \boldsymbol{\alpha} \times \mathbf{r} + \boldsymbol{\omega} \times (\boldsymbol{\omega} \times \mathbf{r}) \qquad (15.8)$$

Differentiating (15.6), and recalling that $\mathbf{k}$ is constant in magnitude and direction, we found that

$$\boldsymbol{\alpha} = \alpha\mathbf{k} = \dot{\omega}\mathbf{k} = \ddot{\theta}\mathbf{k} \qquad (15.9)$$

The vector $\boldsymbol{\alpha}$ represents the *angular acceleration* of the body and is directed along the fixed axis of rotation.

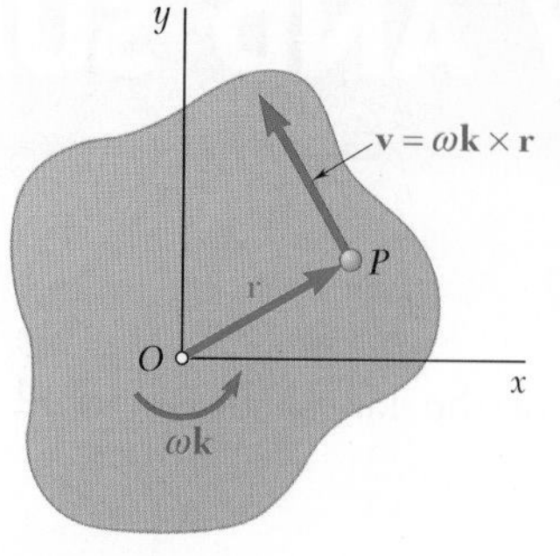

Fig. 15.40

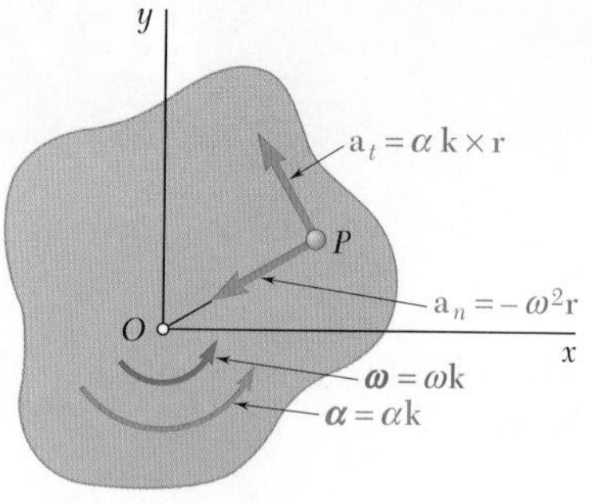

Fig. 15.41

Rotation of a representative slab

Next we considered the motion of a representative slab located in a plane perpendicular to the axis of rotation of the body (Fig. 15.40). Since the angular velocity is perpendicular to the slab, the velocity of a point P of the slab was expressed as

$$\mathbf{v} = \omega\mathbf{k} \times \mathbf{r} \tag{15.10}$$

Tangential and normal components

where $\mathbf{v}$ is contained in the plane of the slab. Substituting $\boldsymbol{\omega} = \omega\mathbf{k}$ and $\boldsymbol{\alpha} = \alpha\mathbf{k}$ into (15.8), we found that the acceleration of P could be resolved into tangential and normal components (Fig. 15.41) respectively equal to

$$\mathbf{a}_t = \alpha\mathbf{k} \times \mathbf{r} \qquad a_t = r\alpha$$
$$\mathbf{a}_n = -\omega^2\mathbf{r} \qquad a_n = r\omega^2 \tag{15.11'}$$

Angular velocity and angular acceleration of rotating slab

Recalling Eqs. (15.6) and (15.9), we obtained the following expressions for the *angular velocity* and the *angular acceleration* of the slab [Sec. 15.4]:

$$\omega = \frac{d\theta}{dt} \tag{15.12}$$

$$\alpha = \frac{d\omega}{dt} = \frac{d^2\theta}{dt^2} \tag{15.13}$$

or

$$\alpha = \omega\frac{d\omega}{d\theta} \tag{15.14}$$

We noted that these expressions are similar to those obtained in Chap. 11 for the rectilinear motion of a particle.

Two particular cases of rotation are frequently encountered: *uniform rotation* and *uniformly accelerated rotation.* Problems involving either of these motions can be solved by using equations similar to those used in Secs. 11.4 and 11.5 for the uniform rectilinear motion and the uniformly accelerated rectilinear motion of a particle, but where x, v, and a are replaced by θ, ω, and α, respectively [Sample Prob. 15.1].

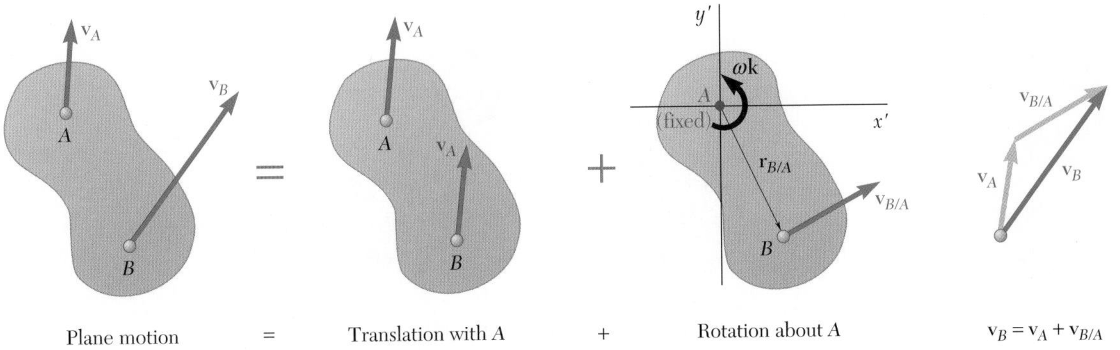

Fig. 15.42

Velocities in plane motion

The *most general plane motion* of a rigid slab can be considered as the *sum of a translation and a rotation* [Sec. 15.5]. For example, the slab shown in Fig. 15.42 can be assumed to translate with point A, while simultaneously rotating about A. It follows [Sec. 15.6] that the velocity of any point B of the slab can be expressed as

$$\mathbf{v}_B = \mathbf{v}_A + \mathbf{v}_{B/A} \tag{15.17}$$

where $\mathbf{v}_A$ is the velocity of A and $\mathbf{v}_{B/A}$ the relative velocity of B with respect to A or, more precisely, with respect to axes $x'y'$ translating with A. Denoting by $\mathbf{r}_{B/A}$ the position vector of B relative to A, we found that

$$\mathbf{v}_{B/A} = \omega\mathbf{k} \times \mathbf{r}_{B/A} \qquad v_{B/A} = r\omega \tag{15.18}$$

The fundamental equation (15.17) relating the absolute velocities of points A and B and the relative velocity of B with respect to A was expressed in the form of a vector diagram and used to solve problems involving the motion of various types of mechanisms [Sample Probs. 15.2 and 15.3].

Instantaneous center of rotation

Another approach to the solution of problems involving the velocities of the points of a rigid slab in plane motion was presented in Sec. 15.7 and used in Sample Probs. 15.4 and 15.5. It is based on the determination of the *instantaneous center of rotation* C of the slab (Fig. 15.43).

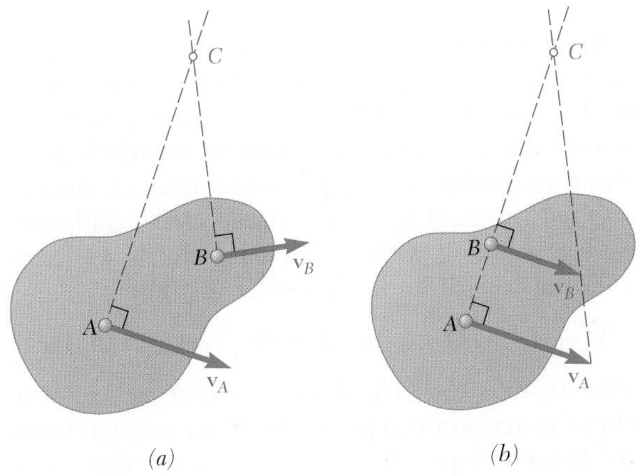

Fig. 15.43

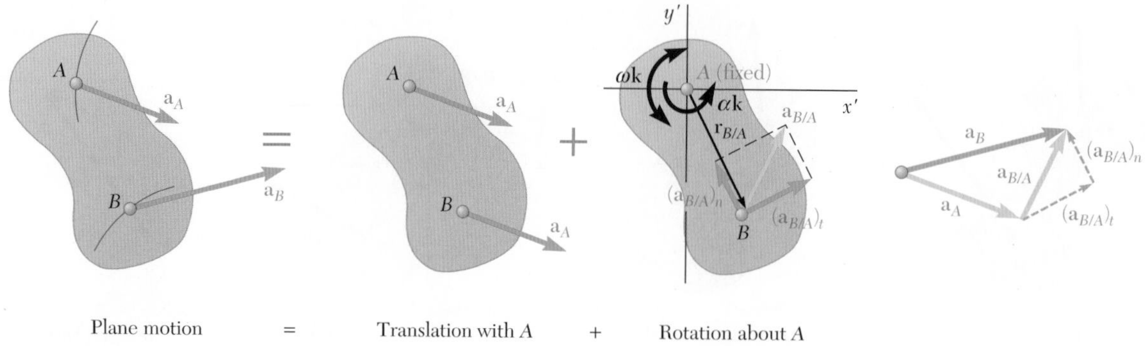

Fig. 15.44

Accelerations in plane motion

The fact that any plane motion of a rigid slab can be considered as the sum of a translation of the slab with a reference point A and a rotation about A was used in Sec. 15.8 to relate the absolute accelerations of any two points A and B of the slab and the relative acceleration of B with respect to A. We had

$$\mathbf{a}_B = \mathbf{a}_A + \mathbf{a}_{B/A} \tag{15.21}$$

where $\mathbf{a}_{B/A}$ consisted of a *normal component* $(\mathbf{a}_{B/A})_n$ of magnitude $r\omega^2$ directed toward A, and a *tangential component* $(\mathbf{a}_{B/A})_t$ of magnitude $r\alpha$ perpendicular to the line AB (Fig. 15.44). The fundamental relation (15.21) was expressed in terms of vector diagrams or vector equations and used to determine the accelerations of given points of various mechanisms [Sample Probs. 15.6 through 15.8]. It should be noted that the instantaneous center of rotation C considered in Sec. 15.7 cannot be used for the determination of accelerations, since point C, in general, does *not* have zero acceleration.

Coordinates expressed in terms of a parameter

In the case of certain mechanisms, it is possible to express the coordinates x and y of all significant points of the mechanism by means of simple analytic expressions containing a *single parameter.* The components of the absolute velocity and acceleration of a given point are then obtained by differentiating twice with respect to the time t the coordinates x and y of that point [Sec. 15.9].

Rate of change of a vector with respect to a rotating frame

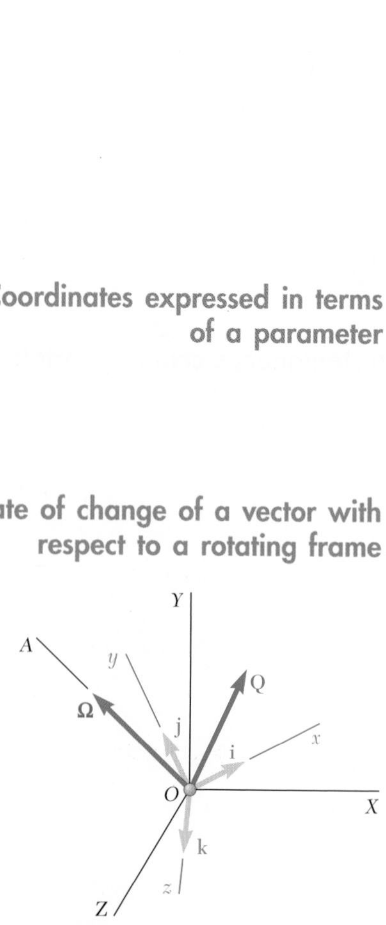

Fig. 15.45

While the rate of change of a vector is the same with respect to a fixed frame of reference and with respect to a frame in translation, the rate of change of a vector with respect to a rotating frame is different. Therefore, in order to study the motion of a particle relative to a rotating frame we first had to compare the rates of change of a general vector $\mathbf{Q}$ with respect to a fixed frame $OXYZ$ and with respect to a frame $Oxyz$ rotating with an angular velocity $\boldsymbol{\Omega}$ [Sec. 15.10] (Fig. 15.45). We obtained the fundamental relation

$$(\dot{\mathbf{Q}})_{OXYZ} = (\dot{\mathbf{Q}})_{Oxyz} + \boldsymbol{\Omega} \times \mathbf{Q} \tag{15.31}$$

and we concluded that the rate of change of the vector $\mathbf{Q}$ with respect to the fixed frame $OXYZ$ is made of two parts: The first part represents the rate of change of $\mathbf{Q}$ with respect to the rotating frame $Oxyz$; the second part, $\boldsymbol{\Omega} \times \mathbf{Q}$, is induced by the rotation of the frame $Oxyz$.

Plane motion of a particle relative to a rotating frame

The next part of the chapter [Sec. 15.11] was devoted to the two-dimensional kinematic analysis of a particle P moving with respect to a frame $\mathcal{F}$ rotating with an angular velocity $\mathbf{\Omega}$ about a fixed axis (Fig. 15.46). We found that the absolute velocity of P could be expressed as

$$\mathbf{v}_P = \mathbf{v}_{P'} + \mathbf{v}_{P/\mathcal{F}} \tag{15.33}$$

where $\mathbf{v}_P$ = absolute velocity of particle P
$\mathbf{v}_{P'}$ = velocity of point P' of moving frame $\mathcal{F}$ coinciding with P
$\mathbf{v}_{P/\mathcal{F}}$ = velocity of P relative to moving frame $\mathcal{F}$

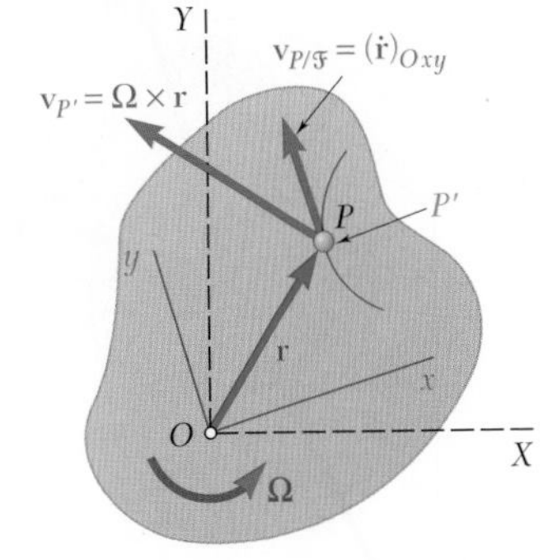

Fig. 15.46

We noted that the same expression for $\mathbf{v}_P$ is obtained if the frame is in translation rather than in rotation. However, when the frame is in rotation, the expression for the acceleration of P is found to contain an additional term $\mathbf{a}_c$ called the *complementary acceleration* or *Coriolis acceleration*. We wrote

$$\mathbf{a}_P = \mathbf{a}_{P'} + \mathbf{a}_{P/\mathcal{F}} + \mathbf{a}_c \tag{15.36}$$

where $\mathbf{a}_P$ = absolute acceleration of particle P
$\mathbf{a}_{P'}$ = acceleration of point P' of moving frame $\mathcal{F}$ coinciding with P
$\mathbf{a}_{P/\mathcal{F}}$ = acceleration of P relative to moving frame $\mathcal{F}$
$\mathbf{a}_c = 2\mathbf{\Omega} \times (\dot{\mathbf{r}})_{Oxy} = 2\mathbf{\Omega} \times \mathbf{v}_{P/\mathcal{F}}$
= complementary, or Coriolis, acceleration

Since $\mathbf{\Omega}$ and $\mathbf{v}_{P/\mathcal{F}}$ are perpendicular to each other in the case of plane motion, the Coriolis acceleration was found to have a magnitude $a_c = 2\Omega v_{P/\mathcal{F}}$ and to point in the direction obtained by rotating the vector $\mathbf{v}_{P/\mathcal{F}}$ through 90° in the sense of rotation of the moving frame. Formulas (15.33) and (15.36) can be used to analyze the motion of mechanisms which contain parts sliding on each other [Sample Probs. 15.9 and 15.10].

Motion of a rigid body with a fixed point

The last part of the chapter was devoted to the study of the kinematics of rigid bodies in three dimensions. We first considered the motion of a rigid body with a fixed point [Sec. 15.12]. After proving that the most general displacement of a rigid body with a fixed point O is equivalent to a rotation of the body about an axis through O, we were able to define the angular velocity $\boldsymbol{\omega}$ and the *instantaneous axis of rotation* of the body at a given instant. The velocity of a point P of the body (Fig. 15.47) could again be expressed as

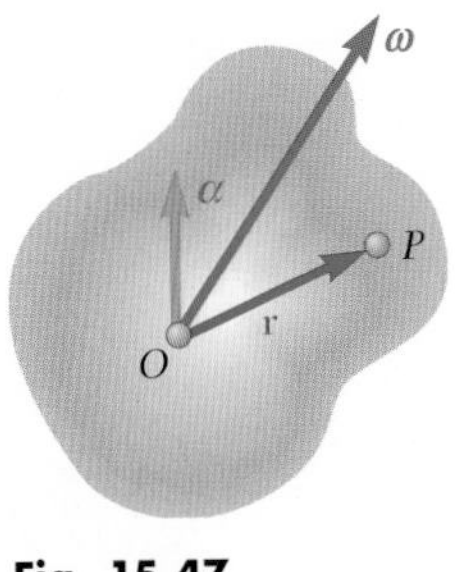

Fig. 15.47

$$\mathbf{v} = \frac{d\mathbf{r}}{dt} = \boldsymbol{\omega} \times \mathbf{r} \tag{15.37}$$

Differentiating this expression, we also wrote

$$\mathbf{a} = \boldsymbol{\alpha} \times \mathbf{r} + \boldsymbol{\omega} \times (\boldsymbol{\omega} \times \mathbf{r}) \tag{15.38}$$

However, since the direction of $\boldsymbol{\omega}$ changes from one instant to the next, the angular acceleration $\boldsymbol{\alpha}$ is, in general, not directed along the instantaneous axis of rotation [Sample Prob. 15.11].

General motion in space

It was shown in Sec. 15.13 that *the most general motion of a rigid body in space is equivalent, at any given instant, to the sum of a translation and a rotation.* Considering two particles A and B of the body, we found that

$$\mathbf{v}_B = \mathbf{v}_A + \mathbf{v}_{B/A} \tag{15.42}$$

where $\mathbf{v}_{B/A}$ is the velocity of B relative to a frame $AX'Y'Z'$ attached to A and of fixed orientation (Fig. 15.48). Denoting by $\mathbf{r}_{B/A}$ the position vector of B relative to A, we wrote

$$\mathbf{v}_B = \mathbf{v}_A + \boldsymbol{\omega} \times \mathbf{r}_{B/A} \tag{15.43}$$

where $\boldsymbol{\omega}$ is the angular velocity of the body at the instant considered [Sample Prob. 15.12]. The acceleration of B was obtained by a similar reasoning. We first wrote

$$\mathbf{a}_B = \mathbf{a}_A + \mathbf{a}_{B/A}$$

and, recalling Eq. (15.38),

$$\mathbf{a}_B = \mathbf{a}_A + \boldsymbol{\alpha} \times \mathbf{r}_{B/A} + \boldsymbol{\omega} \times (\boldsymbol{\omega} \times \mathbf{r}_{B/A}) \tag{15.44}$$

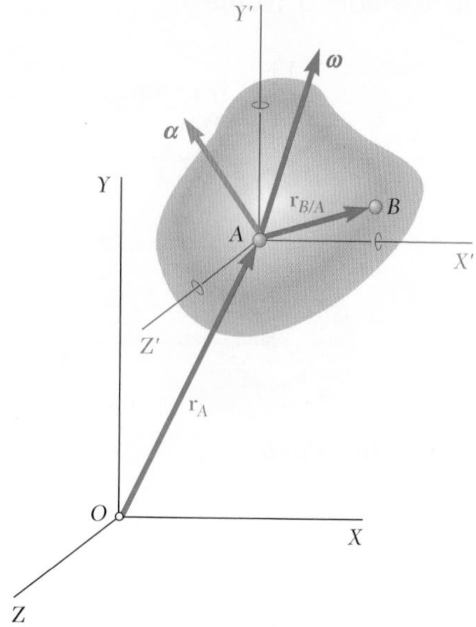

Fig. 15.48

Three-dimensional motion of a particle relative to a rotating frame

In the final two sections of the chapter we considered the three-dimensional motion of a particle P relative to a frame $Oxyz$ rotating with an angular velocity $\boldsymbol{\Omega}$ with respect to a fixed frame $OXYZ$ (Fig. 15.49). In Sec. 15.14 we expressed the absolute velocity $\mathbf{v}_P$ of P as

$$\mathbf{v}_P = \mathbf{v}_{P'} + \mathbf{v}_{P/\mathcal{F}} \tag{15.46}$$

where $\mathbf{v}_P$ = absolute velocity of particle P
$\mathbf{v}_{P'}$ = velocity of point P' of moving frame $\mathcal{F}$ coinciding with P
$\mathbf{v}_{P/\mathcal{F}}$ = velocity of P relative to moving frame $\mathcal{F}$

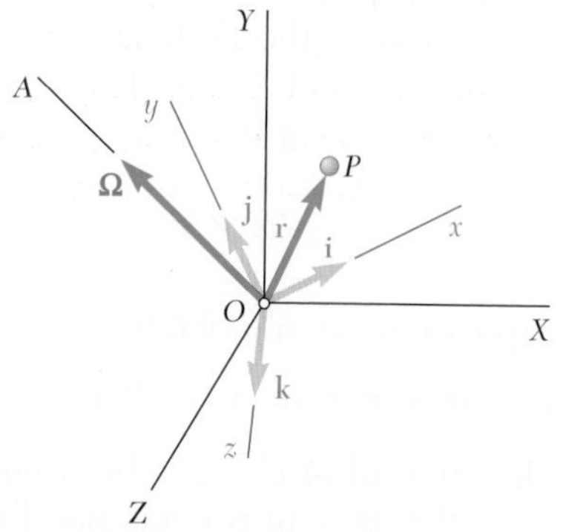

Fig. 15.49

The absolute acceleration $\mathbf{a}_P$ of P was then expressed as

$$\mathbf{a}_P = \mathbf{a}_{P'} + \mathbf{a}_{P/\mathcal{F}} + \mathbf{a}_c \qquad (15.48)$$

where $\mathbf{a}_P$ = absolute acceleration of particle P
$\mathbf{a}_{P'}$ = acceleration of point P' of moving frame $\mathcal{F}$ coinciding with P
$\mathbf{a}_{P/\mathcal{F}}$ = acceleration of P relative to moving frame $\mathcal{F}$
$\mathbf{a}_c = 2\boldsymbol{\Omega} \times (\dot{\mathbf{r}})_{Oxyz} = 2\boldsymbol{\Omega} \times \mathbf{v}_{P/\mathcal{F}}$
= complementary, or Coriolis, acceleration

It was noted that the magnitude a_c of the Coriolis acceleration is not equal to $2\Omega v_{P/\mathcal{F}}$ [Sample Prob. 15.13] except in the special case when $\boldsymbol{\Omega}$ and $\mathbf{v}_{P/\mathcal{F}}$ are perpendicular to each other.

Frame of reference in general motion

We also observed [Sec. 15.15] that Eqs. (15.46) and (15.48) remain valid when the frame $Axyz$ moves in a known, but arbitrary, fashion with respect to the fixed frame $OXYZ$ (Fig. 15.50), provided that the motion of A is included in the terms $\mathbf{v}_{P'}$ and $\mathbf{a}_{P'}$ representing the absolute velocity and acceleration of the coinciding point P'.

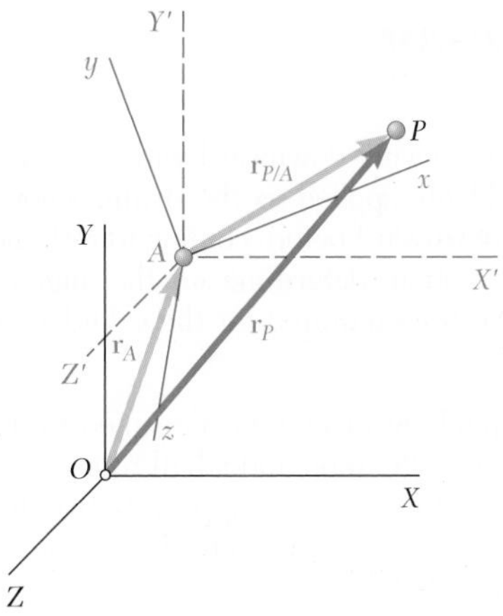

Fig. 15.50

Rotating frames of reference are particularly useful in the study of the three-dimensional motion of rigid bodies. Indeed, there are many cases where an appropriate choice of the rotating frame will lead to a simpler analysis of the motion of the rigid body than would be possible with axes of fixed orientation [Sample Probs. 15.14 and 15.15].

REVIEW PROBLEMS

15.248 The angular acceleration of the 600-mm-radius circular plate shown is defined by the relation $\alpha = \alpha_0 e^{-t}$. Knowing that the plate is at rest when $t = 0$ and that $\alpha_0 = 10 \text{ rad/s}^2$, determine the magnitude of the total acceleration of point B when (*a*) $t = 0$, (*b*) $t = 0.5$ s, (*c*) $t = \infty$.

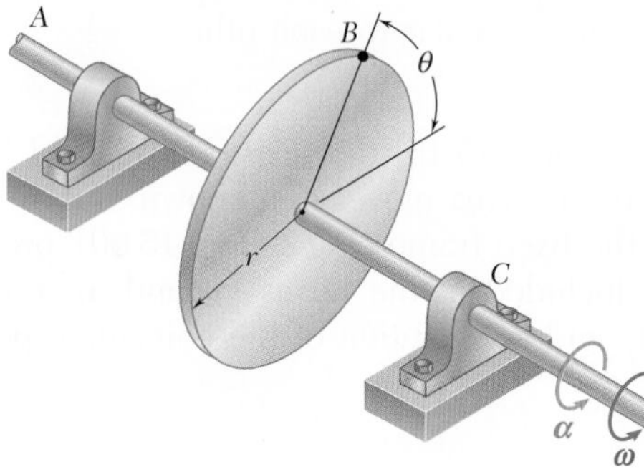

Fig. P15.248

15.249 Cylinder A is moving downward with a velocity of 3 m/s when the brake is suddenly applied to the drum. Knowing that the cylinder moves 6 m downward before coming to rest and assuming uniformly accelerated motion, determine (*a*) the angular acceleration of the drum, (*b*) the time required for the cylinder to come to rest.

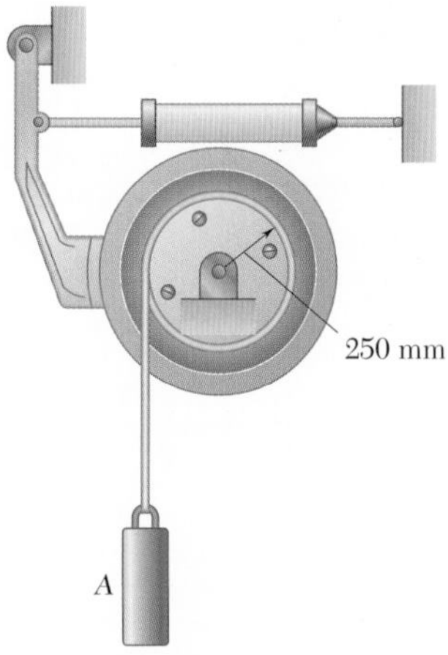

Fig. P15.249

15.250 A baseball pitching machine is designed to deliver a baseball with a ball speed of 108 kmph and a ball rotation of 300 rpm clockwise. Knowing that there is no slipping between the wheels and the baseball during the ball launch, determine the angular velocities of wheels A and B.

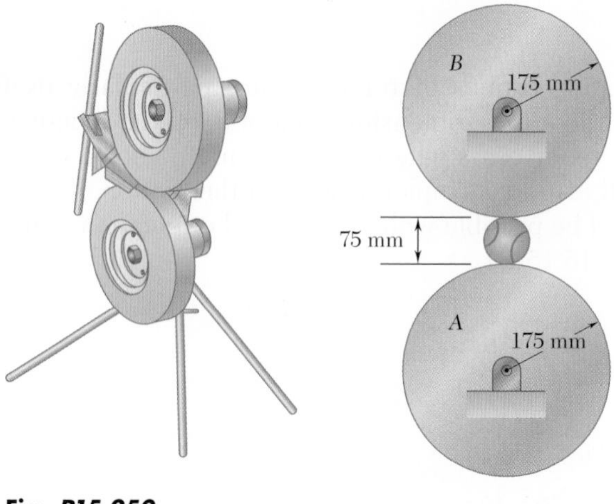

Fig. *P15.250*

15.251 Knowing that inner gear A is stationary and outer gear C starts from rest and has a constant angular acceleration of 4 rad/s^2 clockwise, determine at $t = 5$ s (*a*) the angular velocity of arm AB, (*b*) the angular velocity of gear B, (*c*) the acceleration of the point on gear B that is in contact with gear A.

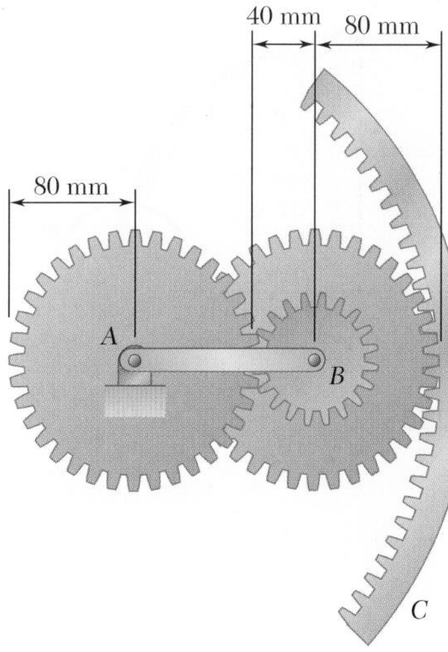

Fig. P15.251

15.252 Knowing that at the instant shown bar AB has an angular velocity of 10 rad/s clockwise and it is slowing down at a rate of 2 rad/s^2, determine the angular accelerations of bar BD and bar DE.

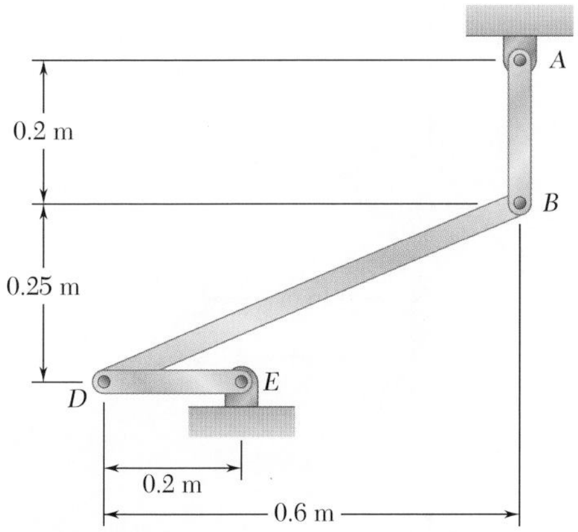

Fig. P15.252

15.253 Knowing that at the instant shown rod AB has zero angular acceleration and an angular velocity of 15 rad/s counterclockwise, determine (*a*) the angular acceleration of arm DE, (*b*) the acceleration of point D.

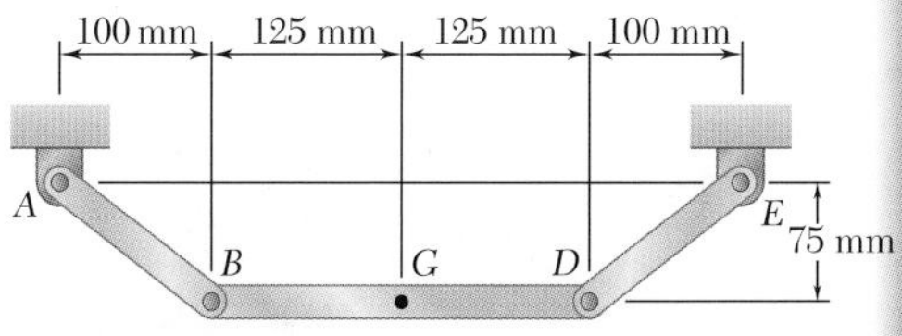

Fig. P15.253

15.254 Rod AB is attached to a collar at A and is fitted with a wheel at B that has a radius $r = 15$ mm. Knowing that when $\theta = 60°$ the collar has a velocity of 250 mm/s upward and it is slowing down at a rate of 150 mm/s^2, determine (*a*) the angular acceleration of rod AB, (*b*) the angular acceleration of the wheel.

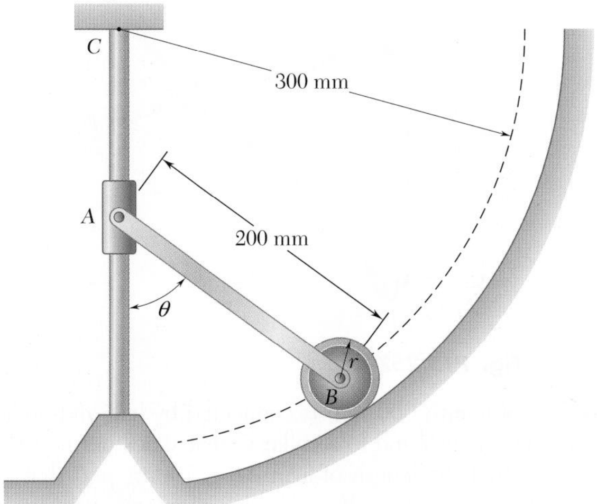

Fig. P15.254

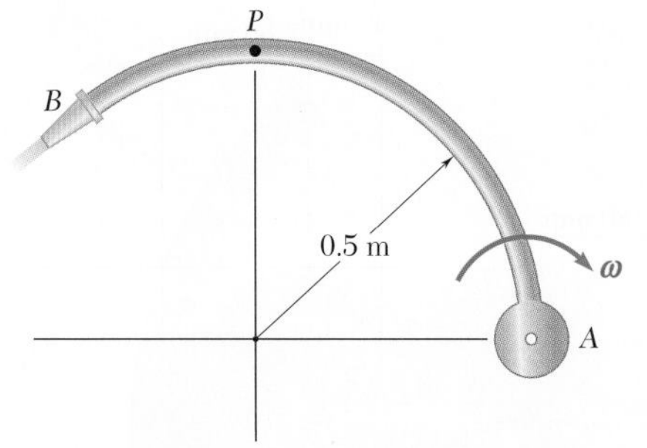

Fig. P15.255

15.255 Water flows through a curved pipe *AB* that rotates with a constant clockwise angular velocity of 90 rpm. If the velocity of the water relative to the pipe is 8 m/s, determine the total acceleration of a particle of water at point *P*.

15.256 A disk of 0.15-m radius rotates at the constant rate ω_2 with respect to plate *BC*, which itself rotates at the constant rate ω_1 about the *y* axis. Knowing that $\omega_1 = \omega_2 = 3$ rad/s, determine, for the position shown, the velocity and acceleration (*a*) of point *D*, (*b*) of point *F*.

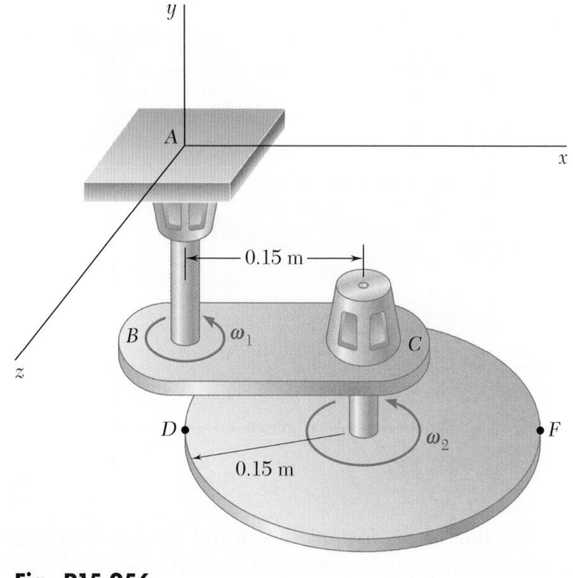

Fig. P15.256

15.257 Two rods *AE* and *BD* pass through holes drilled into a hexagonal block. (The holes are drilled in different planes so that the rods will not touch each other.) Knowing that rod *AE* has an angular velocity of 20 rad/s clockwise and an angular acceleration of 4 rad/s² counterclockwise when $\theta = 90°$, determine (*a*) the relative velocity of the block with respect to each rod, (*b*) the relative acceleration of the block with respect to each rod.

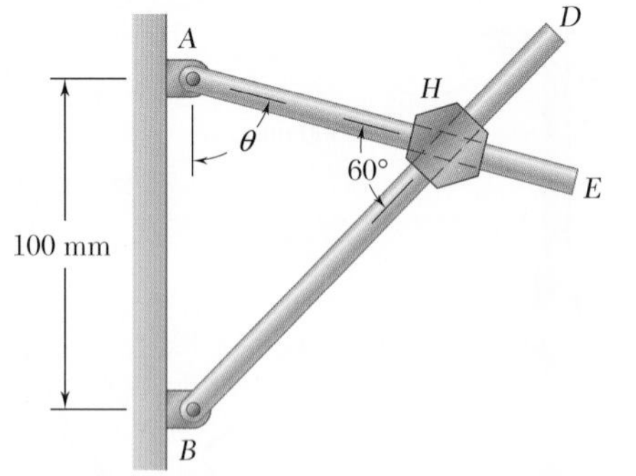

Fig. P15.257

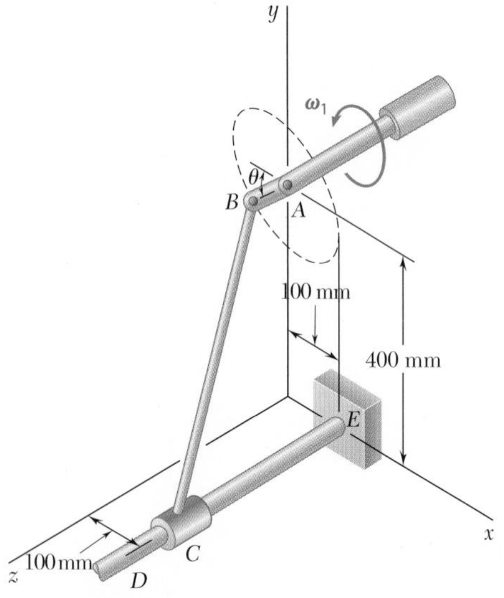

Fig. P15.258

15.258 Rod *BC* of length 600 mm is connected by ball-and-socket joints to a rotating arm *AB* and to a collar *C* that slides on the fixed rod *DE*. Knowing that the length of arm *AB* is 100 in. and that it rotates at the constant rate $\omega_1 = 10$ rad/s, determine the velocity of collar *C* when $\theta = 0$.

15.259 In the position shown the thin rod moves at a constant speed u = 100 mm/s out of the tube BC. At the same time tube BC rotates at the constant rate ω_2 = 2 rad/s with respect to arm CD. Knowing that the entire assembly rotates about the X axis at the constant rate ω_1 = 1 rad/s, determine the velocity and acceleration of end A of the rod.

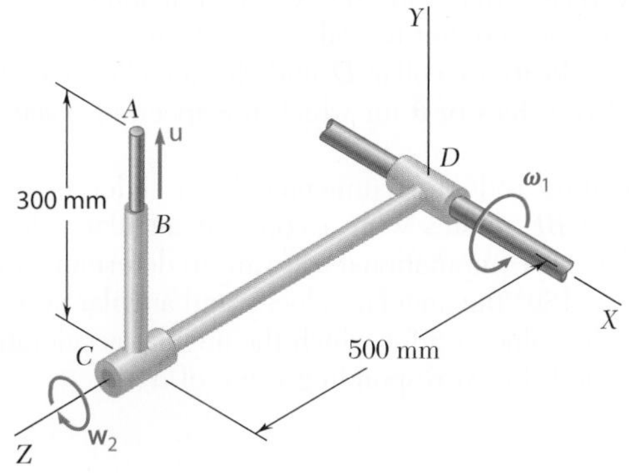

Fig. P15.259

COMPUTER PROBLEMS

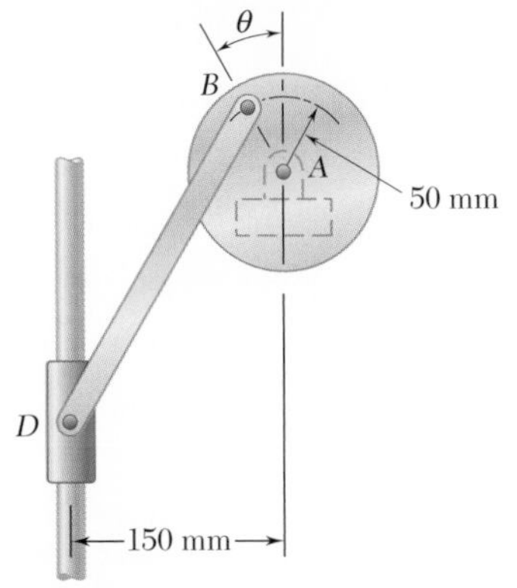

Fig. P15.C1

15.C1 The disk shown has a constant angular velocity of 500 rpm counterclockwise. Knowing that rod BD is 250 mm long, use computational software to determine and plot for values of θ from 0 to 360° and using 30° increments, the velocity of collar D and the angular velocity of rod BD. Determine the two values of θ for which the speed of collar D is zero.

15.C2 Two rotating rods are connected by a slider block P as shown. Knowing that rod BP rotates with a constant angular velocity of 6 rad/s counterclockwise, use computational software to determine and plot for values of θ from 0 to 180° the angular velocity and angular acceleration of rod AE. Determine the value of θ for which the angular acceleration α_{AE} of rod AE is maximum and the corresponding value of α_{AE}.

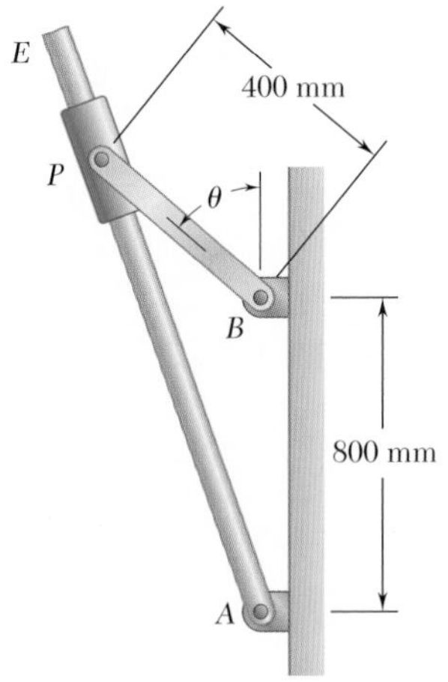

Fig. P15.C2

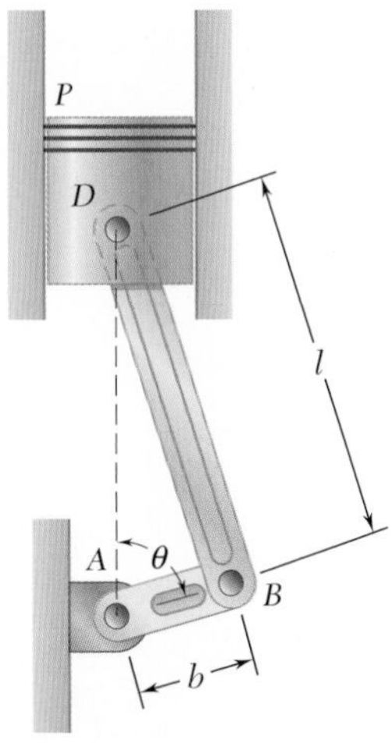

Fig. P15.C3

15.C3 In the engine system shown, $l = 160$ mm and $b = 60$ mm. Knowing that crank AB rotates with a constant angular velocity of 1000 rpm clockwise, use computational software to determine and plot for values of θ from 0 to 180° and using 10° increments, (*a*) the angular velocity and angular acceleration of rod BD, (*b*) the velocity and acceleration of the piston P.

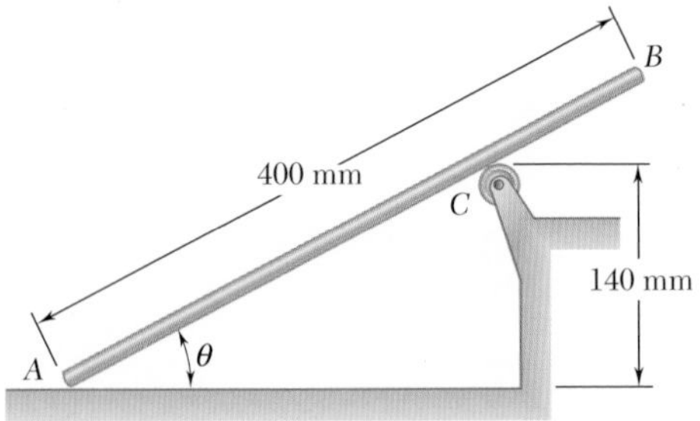

Fig. P15.C4

15.C4 Rod AB moves over a small wheel at C while end A moves to the right with a constant velocity of 180 mm/s. Use computational software to determine and plot for values of θ from 20° to 90° and using 5° increments, the velocity of point B and the angular acceleration of the rod. Determine the value of θ for which the angular acceleration α of the rod is maximum and the corresponding value of α.

15.C5 Rod BC of length 600 mm is connected by ball-and-socket joints to the rotating arm AB and to collar C that slides on the fixed rod DE. Arm AB of length 100 mm rotates in the XY plane with a constant angular velocity of 10 rad/s. Use computational software to determine and plot for values of θ from 0 to 360° the velocity of collar C. Determine the two values of θ for which the velocity of collar C is zero.

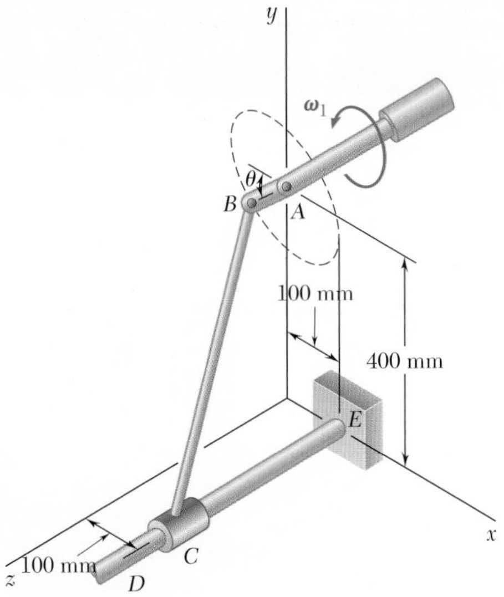

Fig. P15.C5

15.C6 Rod AB of length 625 mm is connected by ball-and-socket joints to collars A and B, which slide along the two rods shown. Collar B moves toward support E at a constant speed of 500 mm/s. Denoting by d the distance from point C to collar B, use computational software to determine and plot the velocity of collar A for values of d from 0 to 375 mm.

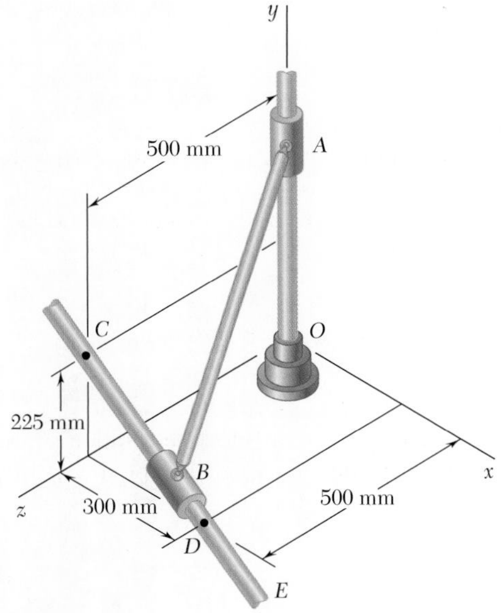

Fig. P15.C6

Three-bladed wind turbines, similar to the ones shown in this picture of a wind farm, are currently the most common design. In this chapter you will learn to analyze the motion of a rigid body by considering the motion of its mass center, the motion relative to its mass center, and the external forces acting on it.

CHAPTER 16

Plane Motion of Rigid Bodies: Forces and Accelerations

16.1 INTRODUCTION

In this chapter and in Chaps. 17 and 18, you will study the *kinetics of rigid bodies,* i.e., the relations existing between the forces acting on a rigid body, the shape and mass of the body, and the motion produced. In Chaps. 12 and 13, you studied similar relations, assuming then that the body could be considered as a particle, i.e., that its mass could be concentrated in one point and that all forces acted at that point. The shape of the body, as well as the exact location of the points of application of the forces, will now be taken into account. You will also be concerned not only with the motion of the body as a whole but also with the motion of the body about its mass center.

Our approach will be to consider rigid bodies as made of large numbers of particles and to use the results obtained in Chap. 14 for the motion of systems of particles. Specifically, two equations from Chap. 14 will be used: Eq. (14.16), $\Sigma\mathbf{F} = m\bar{\mathbf{a}}$, which relates the resultant of the external forces and the acceleration of the mass center G of the system of particles, and Eq. (14.23), $\Sigma\mathbf{M}_G = \dot{\mathbf{H}}_G$, which relates the moment resultant of the external forces and the angular momentum of the system of particles about G.

Except for Sec. 16.2, which applies to the most general case of the motion of a rigid body, the results derived in this chapter will be limited in two ways: (1) They will be restricted to the *plane motion* of rigid bodies, i.e., to a motion in which each particle of the body remains at a constant distance from a fixed reference plane. (2) The rigid bodies considered will consist only of plane slabs and of bodies which are symmetrical with respect to the reference plane.† The study of the plane motion of nonsymmetrical three-dimensional bodies and, more generally, the motion of rigid bodies in three-dimensional space will be postponed until Chap. 18.

In Sec. 16.3, we define the angular momentum of a rigid body in plane motion and show that the rate of change of the angular momentum $\dot{\mathbf{H}}_G$ about the mass center is equal to the product $\bar{I}\boldsymbol{\alpha}$ of the centroidal mass moment of inertia $\bar{I}$ and the angular acceleration $\boldsymbol{\alpha}$ of the body. D'Alembert's principle, introduced in Sec. 16.4, is used to prove that the external forces acting on a rigid body are equivalent to a vector $m\bar{\mathbf{a}}$ attached at the mass center and a couple of moment $\bar{I}\boldsymbol{\alpha}$.

In Sec. 16.5, we derive the principle of transmissibility using only the parallelogram law and Newton's laws of motion, allowing us to remove this principle from the list of axioms (Sec. 1.2) required for the study of the statics and dynamics of rigid bodies.

Free-body-diagram equations are introduced in Sec. 16.6 and will be used in the solution of all problems involving the plane motion of rigid bodies.

After considering the plane motion of connected rigid bodies in Sec. 16.7, you will be prepared to solve a variety of problems involving the translation, centroidal rotation, and unconstrained motion of rigid bodies. In Sec. 16.8 and in the remaining part of the chapter, the solution of problems involving noncentroidal rotation, rolling motion, and other partially constrained plane motions of rigid bodies will be considered.

†Or, more generally, bodies which have a principal centroidal axis of inertia perpendicular to the reference plane.

16.2 EQUATIONS OF MOTION FOR A RIGID BODY

Consider a rigid body acted upon by several external forces $\mathbf{F}_1$, $\mathbf{F}_2$, $\mathbf{F}_3$, . . . (Fig. 16.1). We can assume that the body is made of a large number n of particles of mass Δm_i ($i = 1, 2, \ldots, n$) and apply the results obtained in Chap. 14 for a system of particles (Fig. 16.2). Considering first the motion of the mass center G of the body with respect to the newtonian frame of reference $Oxyz$, we recall Eq. (14.16) and write

$$\Sigma \mathbf{F} = m\bar{\mathbf{a}} \tag{16.1}$$

where m is the mass of the body and $\bar{\mathbf{a}}$ is the acceleration of the mass center G. Turning now to the motion of the body relative to the centroidal frame of reference $Gx'y'z'$, we recall Eq. (14.23) and write

$$\Sigma \mathbf{M}_G = \dot{\mathbf{H}}_G \tag{16.2}$$

where $\dot{\mathbf{H}}_G$ represents the rate of change of $\mathbf{H}_G$, the angular momentum about G of the system of particles forming the rigid body. In the following, $\mathbf{H}_G$ will simply be referred to as the *angular momentum of the rigid body about its mass center G*. Together Eqs. (16.1) and (16.2) express that *the system of the external forces is equipollent to the system consisting of the vector $m\bar{\mathbf{a}}$ attached at G and the couple of moment $\dot{\mathbf{H}}_G$* (Fig. 16.3).†

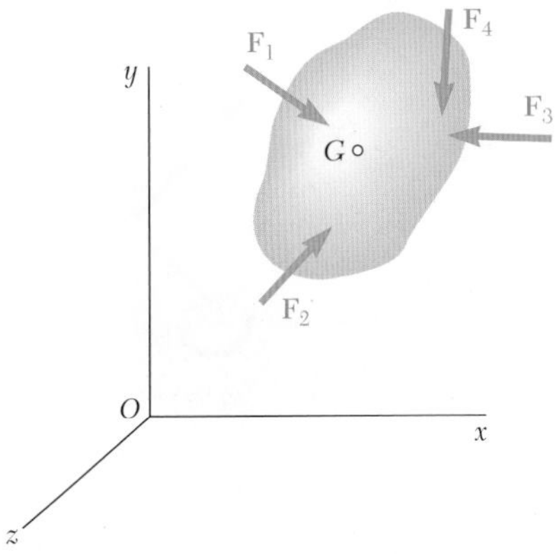

Fig. 16.1

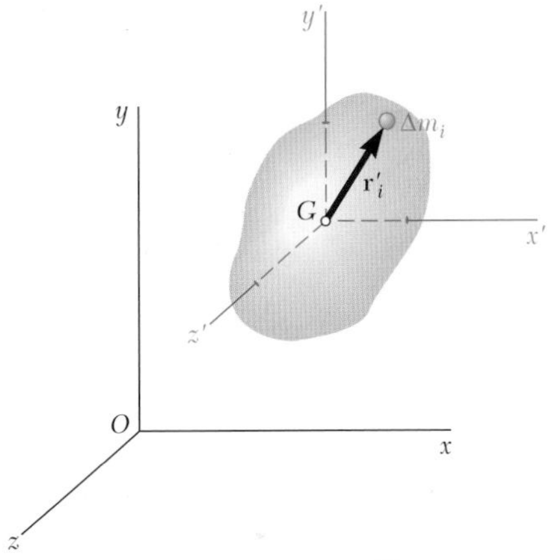

Fig. 16.2

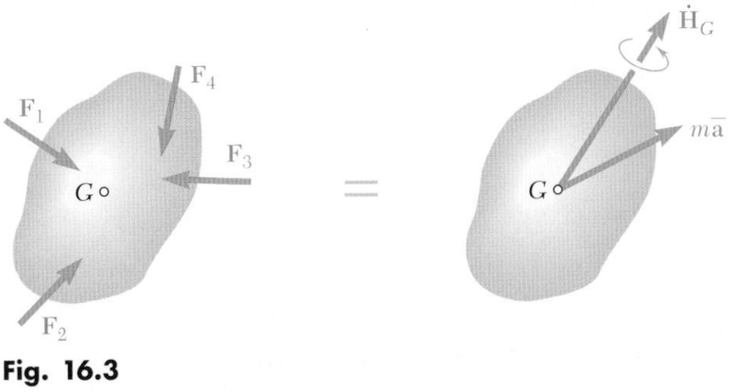

Fig. 16.3

Equations (16.1) and (16.2) apply in the most general case of the motion of a rigid body. In the rest of this chapter, however, our analysis will be limited to the *plane motion* of rigid bodies, i.e., to a motion in which each particle remains at a constant distance from a fixed reference plane, and it will be assumed that the rigid bodies considered consist only of plane slabs and of bodies which are symmetrical with respect to the reference plane. Further study of the plane motion of nonsymmetrical three-dimensional bodies and of the motion of rigid bodies in three-dimensional space will be postponed until Chap. 18.

Photo 16.1 The system of external forces acting on the man and wakeboard includes the weights, the tension in the tow rope, and the forces exerted by the water and the air.

†Since the systems involved act on a rigid body, we could conclude at this point, by referring to Sec. 3.19, that the two systems are *equivalent* as well as equipollent and use red rather than blue equals signs in Fig. 16.3. However, by postponing this conclusion, we will be able to arrive at it independently (Secs. 16.4 and 18.5), thereby eliminating the necessity of including the principle of transmissibility among the axioms of mechanics (Sec. 16.5).

16.3 ANGULAR MOMENTUM OF A RIGID BODY IN PLANE MOTION

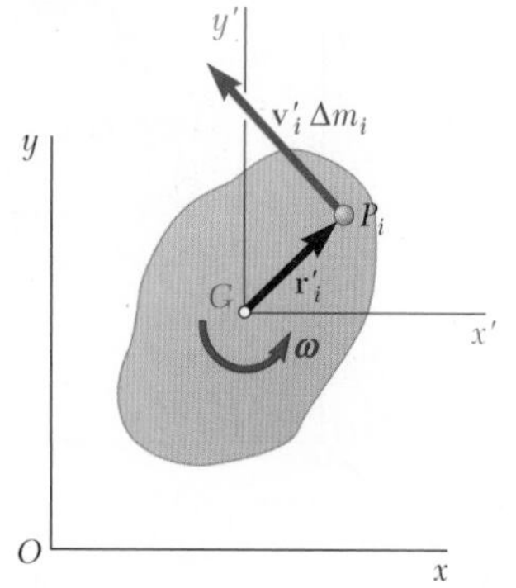

Fig. 16.4

Consider a rigid slab in plane motion. Assuming that the slab is made of a large number n of particles P_i of mass Δm_i and recalling Eq. (14.24) of Sec. 14.5, we note that the angular momentum $\mathbf{H}_G$ of the slab about its mass center G can be computed by taking the moments about G of the momenta of the particles of the slab in their motion with respect to either of the frames Oxy or $Gx'y'$ (Fig. 16.4). Choosing the latter course, we write

$$\mathbf{H}_G = \sum_{i=1}^{n} (\mathbf{r}'_i \times \mathbf{v}'_i\,\Delta m_i) \tag{16.3}$$

where $\mathbf{r}'_i$ and $\mathbf{v}'_i\,\Delta m_i$ denote, respectively, the position vector and the linear momentum of the particle P_i relative to the centroidal frame of reference $Gx'y'$. But since the particle belongs to the slab, we have $\mathbf{v}'_i = \boldsymbol{\omega} \times \mathbf{r}'_i$, where $\boldsymbol{\omega}$ is the angular velocity of the slab at the instant considered. We write

$$\mathbf{H}_G = \sum_{i=1}^{n} [\mathbf{r}'_i \times (\boldsymbol{\omega} \times \mathbf{r}'_i)\,\Delta m_i]$$

Referring to Fig. 16.4, we easily verify that the expression obtained represents a vector of the same direction as $\boldsymbol{\omega}$ (i.e., perpendicular to the slab) and of magnitude equal to $\omega\Sigma r_i'^2\,\Delta m_i$. Recalling that the sum $\Sigma r_i'^2\,\Delta m_i$ represents the moment of inertia $\bar{I}$ of the slab about a centroidal axis perpendicular to the slab, we conclude that the angular momentum $\mathbf{H}_G$ of the slab about its mass center is

$$\mathbf{H}_G = \bar{I}\boldsymbol{\omega} \tag{16.4}$$

Differentiating both members of Eq. (16.4) we obtain

$$\dot{\mathbf{H}}_G = \bar{I}\dot{\boldsymbol{\omega}} = \bar{I}\boldsymbol{\alpha} \tag{16.5}$$

Thus the rate of change of the angular momentum of the slab is represented by a vector of the same direction as $\boldsymbol{\alpha}$ (i.e., perpendicular to the slab) and of magnitude $\bar{I}\alpha$.

It should be kept in mind that the results obtained in this section have been derived for a rigid slab in plane motion. As you will see in Chap. 18, they remain valid in the case of the plane motion of rigid bodies which are symmetrical with respect to the reference plane.† However, they do not apply in the case of nonsymmetrical bodies or in the case of three-dimensional motion.

Photo 16.2 The hard disk and pick-up arms of a hard disk computer undergo centroidal rotation.

†Or, more generally, bodies which have a principal centroidal axis of inertia perpendicular to the reference plane.

16.4 PLANE MOTION OF A RIGID BODY. D'ALEMBERT'S PRINCIPLE

Consider a rigid slab of mass m moving under the action of several external forces $\mathbf{F}_1$, $\mathbf{F}_2$, $\mathbf{F}_3$, . . . , contained in the plane of the slab (Fig. 16.5). Substituting for $\dot{\mathbf{H}}_G$ from Eq. (16.5) into Eq. (16.2) and writing the fundamental equations of motion (16.1) and (16.2) in scalar form, we have

$$\Sigma F_x = m\bar{a}_x \qquad \Sigma F_y = m\bar{a}_y \qquad \Sigma M_G = \bar{I}\alpha \tag{16.6}$$

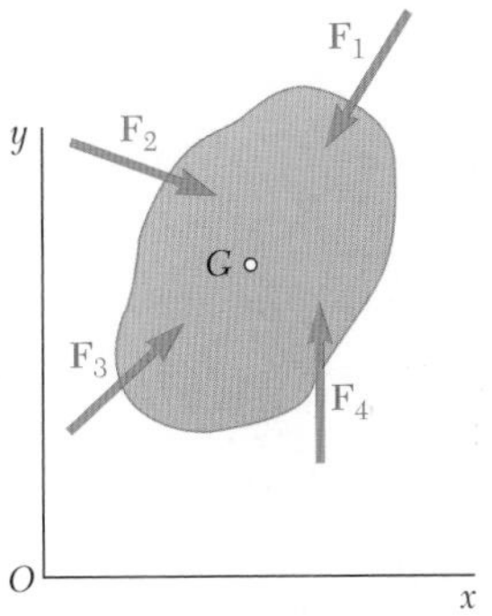

Fig. 16.5

Equations (16.6) show that the acceleration of the mass center G of the slab and its angular acceleration $\boldsymbol{\alpha}$ are easily obtained once the resultant of the external forces acting on the slab and their moment resultant about G have been determined. Given appropriate initial conditions, the coordinates $\bar{x}$ and $\bar{y}$ of the mass center and the angular coordinate θ of the slab can then be obtained by integration at any instant t. Thus *the motion of the slab is completely defined by the resultant and moment resultant about G of the external forces acting on it.*

This property, which will be extended in Chap. 18 to the case of the three-dimensional motion of a rigid body, is characteristic of the motion of a rigid body. Indeed, as we saw in Chap. 14, the motion of a system of particles which are not rigidly connected will in general depend upon the specific external forces acting on the various particles, as well as upon the internal forces.

Since the motion of a rigid body depends only upon the resultant and moment resultant of the external forces acting on it, it follows that *two systems of forces which are equipollent,* i.e., which have the same resultant and the same moment resultant, *are also equivalent;* that is, they have exactly the same effect on a given rigid body.†

Consider in particular the system of the external forces acting on a rigid body (Fig. 16.6*a*) and the system of the effective forces associated with the particles forming the rigid body (Fig. 16.6*b*). It was shown in Sec. 14.2 that the two systems thus defined are equipollent. But since the particles considered now form a rigid body, it follows from the discussion above that the two systems are also equivalent. We can thus state that *the external forces acting on a rigid body are equivalent to the effective forces of the various particles forming the body.* This statement is referred to as *d'Alembert's principle* after the French mathematician Jean le Rond d'Alembert (1717–1783), even though d'Alembert's original statement was written in a somewhat different form.

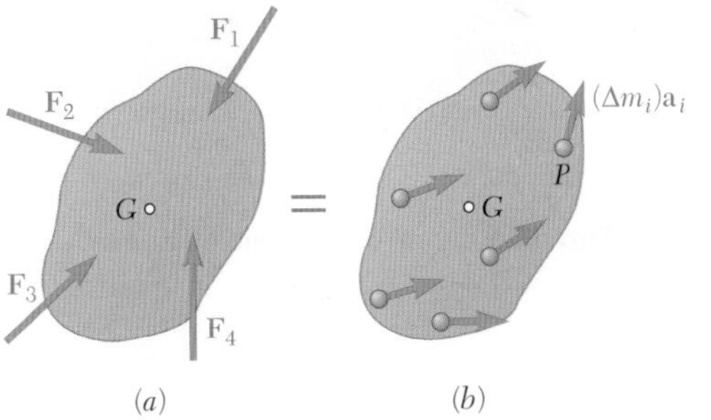

Fig. 16.6

The fact that the system of external forces is *equivalent* to the system of the effective forces has been emphasized by the use of a red equals sign in Fig. 16.6 and also in Fig. 16.7, where using results obtained earlier in this section, we have replaced the effective forces by a vector $m\bar{\mathbf{a}}$ attached at the mass center G of the slab and a couple of moment $\bar{I}\boldsymbol{\alpha}$.

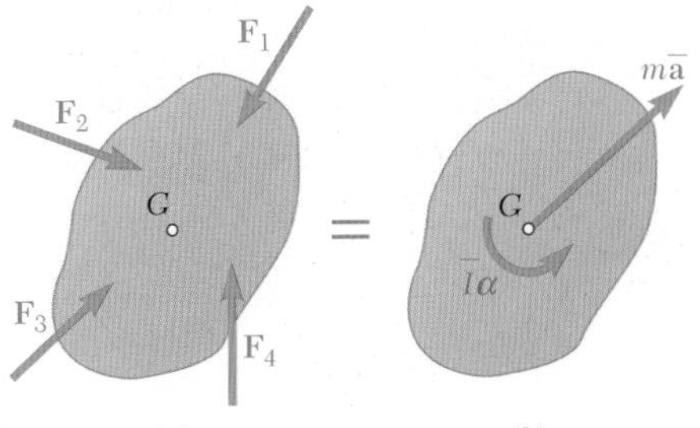

Fig. 16.7

†This result has already been derived in Sec. 3.19 from the principle of transmissibility (Sec. 3.3). The present derivation is independent of that principle, however, and will make possible its elimination from the axioms of mechanics (Sec. 16.5).

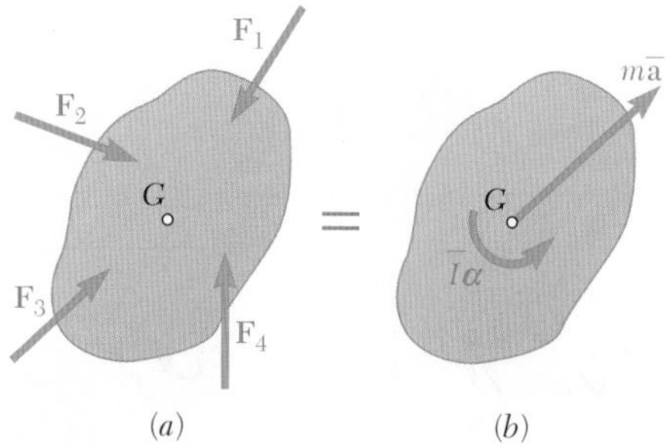

Fig. 16.7 (repeated)

Translation. In the case of a body in translation, the angular acceleration of the body is identically equal to zero and its effective forces reduce to the vector $m\bar{\mathbf{a}}$ attached at G (Fig. 16.8). Thus, the resultant of the external forces acting on a rigid body in translation passes through the mass center of the body and is equal to $m\bar{\mathbf{a}}$.

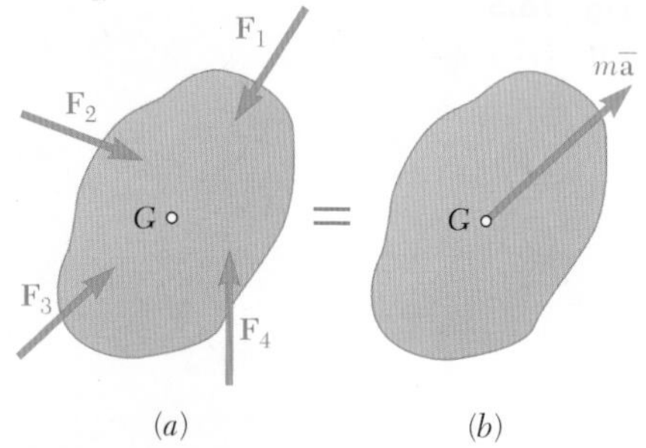

Fig. 16.8 Translation.

Centroidal Rotation. When a slab, or, more generally, a body symmetrical with respect to the reference plane, rotates about a fixed axis perpendicular to the reference plane and passing through its mass center G, we say that the body is in *centroidal rotation*. Since the acceleration $\bar{\mathbf{a}}$ is identically equal to zero, the effective forces of the body reduce to the couple $\bar{I}\boldsymbol{\alpha}$ (Fig. 16.9). Thus, the external forces acting on a body in centroidal rotation are equivalent to a couple of moment $\bar{I}\boldsymbol{\alpha}$.

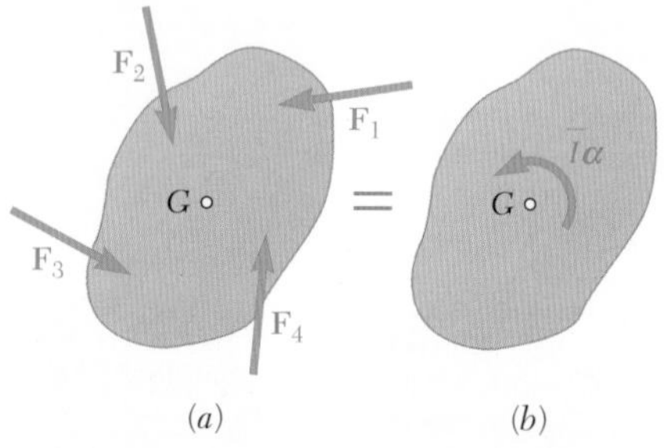

Fig. 16.9 Centroidal rotation.

General Plane Motion. Comparing Fig. 16.7 with Figs. 16.8 and 16.9, we observe that from the point of view of *kinetics*, the most general plane motion of a rigid body symmetrical with respect to the reference plane can be replaced by the sum of a translation and a centroidal rotation. We should note that this statement is more restrictive than the similar statement made earlier from the point of view of *kinematics* (Sec. 15.5), since we now require that the mass center of the body be selected as the reference point.

Referring to Eqs. (16.6), we observe that the first two equations are identical with the equations of motion of a particle of mass m acted upon by the given forces $\mathbf{F}_1$, $\mathbf{F}_2$, $\mathbf{F}_3$, . . . We thus check that *the mass center G of a rigid body in plane motion moves as if the entire mass of the body were concentrated at that point, and as if all the external forces acted on it*. We recall that this result has already been obtained in Sec. 14.4 in the general case of a system of particles, the particles being not necessarily rigidly connected. We also note, as we did in Sec. 14.4, that the system of the external forces does not, in general, reduce to a single vector $m\bar{\mathbf{a}}$ attached at G. Therefore, in the general case of the plane motion of a rigid body, *the resultant of the external forces acting on the body does not pass through the mass center of the body*.

Finally, it should be observed that the last of Eqs. (16.6) would still be valid if the rigid body, while subjected to the same applied forces, were constrained to rotate about a fixed axis through G. Thus, *a rigid body in plane motion rotates about its mass center as if this point were fixed*.

*16.5 A REMARK ON THE AXIOMS OF THE MECHANICS OF RIGID BODIES

The fact that two equipollent systems of external forces acting on a rigid body are also equivalent, i.e., have the same effect on that rigid body, has already been established in Sec. 3.19. But there it was derived from the *principle of transmissibility*, one of the axioms used in our study of the statics of rigid bodies. It should be noted that this axiom has not been used in the present chapter because Newton's second and third laws of motion make its use unnecessary in the study of the dynamics of rigid bodies.

In fact, the principle of transmissibility may now be *derived* from the other axioms used in the study of mechanics. This principle stated, without proof (Sec. 3.3), that the conditions of equilibrium or motion of a rigid body remain unchanged if a force **F** acting at a given point of the rigid body is replaced by a force **F′** of the same magnitude and same direction, but acting at a different point, provided that the two forces have the same line of action. But since **F** and **F′** have the same moment about any given point, it is clear that they form two equipollent systems of external forces. Thus, we may now *prove,* as a result of what we established in the preceding section, that **F** and **F′** have the same effect on the rigid body (Fig. 3.3).

The principle of transmissibility can therefore be removed from the list of axioms required for the study of the mechanics of rigid bodies. These axioms are reduced to the parallelogram law of addition of vectors and to Newton's laws of motion.

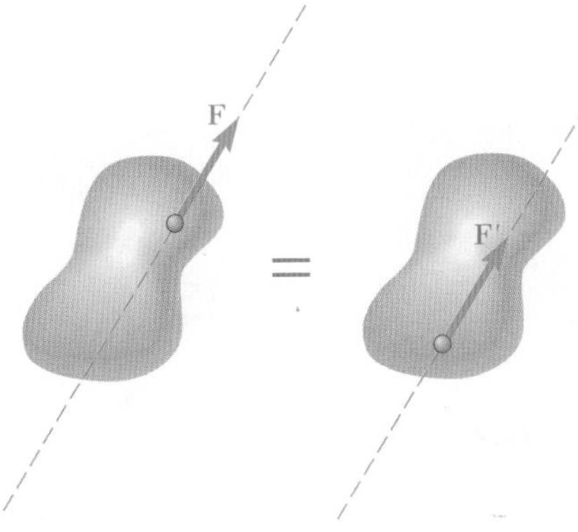

Fig. 3.3 (repeated)

16.6 SOLUTION OF PROBLEMS INVOLVING THE MOTION OF A RIGID BODY

We saw in Sec. 16.4 that when a rigid body is in plane motion, there exists a fundamental relation between the forces $\mathbf{F}_1$, $\mathbf{F}_2$, $\mathbf{F}_3$, . . . , acting on the body, the acceleration $\overline{\mathbf{a}}$ of its mass center, and the angular acceleration $\boldsymbol{\alpha}$ of the body. This relation, which is represented in Fig. 16.7 in the form of a *free-body-diagram equation,* can be used to determine the acceleration $\overline{\mathbf{a}}$ and the angular acceleration $\boldsymbol{\alpha}$ produced by a given system of forces acting on a rigid body or, conversely, to determine the forces which produce a given motion of the rigid body.

The three algebraic equations (16.6) can be used to solve problems of plane motion.† However, our experience in statics suggests that the solution of many problems involving rigid bodies could be simplified by an appropriate choice of the point about which the moments of the forces are computed. It is therefore preferable to remember the relation existing between the forces and the accelerations in the pictorial form shown in Fig. 16.7 and to derive from this fundamental relation the component or moment equations which fit best the solution of the problem under consideration.

The fundamental relation shown in Fig. 16.7 can be presented in an alternative form if we add to the external forces an inertia vector $-m\overline{\mathbf{a}}$ of sense opposite to that of $\overline{\mathbf{a}}$, attached at G, and an inertia couple $-\overline{I}\boldsymbol{\alpha}$ of moment equal in magnitude to $\overline{I}\alpha$ and of sense opposite to that of $\boldsymbol{\alpha}$ (Fig. 16.10). The system obtained is equivalent to zero, and the rigid body is said to be in *dynamic equilibrium.*

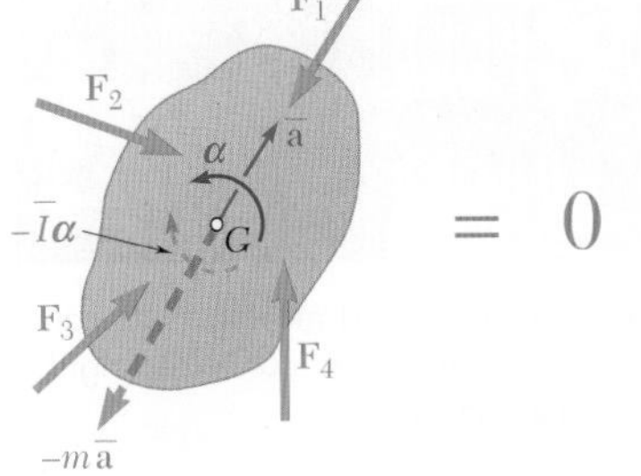

Fig. 16.10

Whether the principle of equivalence of external and effective forces is directly applied, as in Fig. 16.7, or whether the concept of dynamic equilibrium is introduced, as in Fig. 16.10, the use of free-body-diagram equations showing vectorially the relationship existing between the forces applied on the rigid body and the resulting linear and angular accelerations presents considerable advantages over the blind application of formulas (16.6). These advantages can be summarized as follows:

1. The use of a pictorial representation provides a much clearer understanding of the effect of the forces on the motion of the body.

†We recall that the last of Eqs. (16.6) is valid only in the case of the plane motion of a rigid body symmetrical with respect to the reference plane. In all other cases, the methods of Chap. 18 should be used.

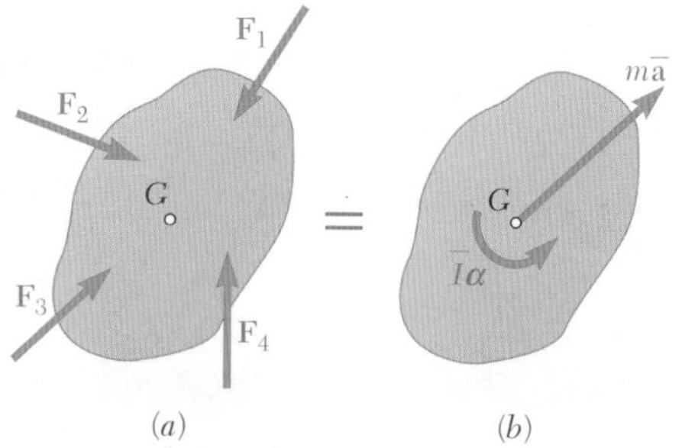

Fig. 16.7 (repeated)

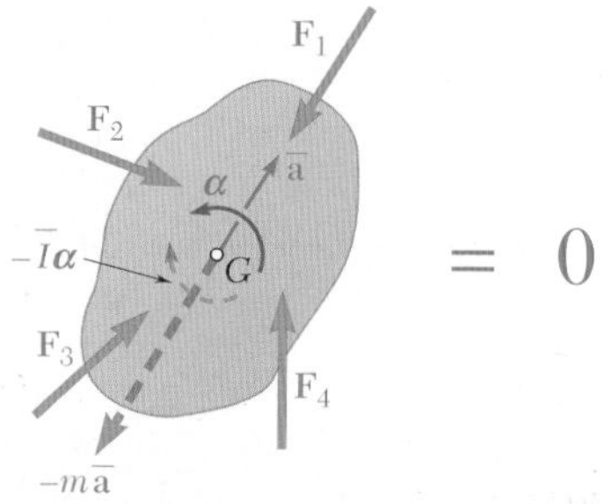

Fig. 16.10 (repeated)

Photo 16.3 The forklift and moving load can be analyzed as a system of two connected rigid bodies in plane motion.

2. This approach makes it possible to divide the solution of a dynamics problem into two parts: In the first part, the analysis of the kinematic and kinetic characteristics of the problem leads to the free-body diagrams of Fig. 16.7 or 16.10; in the second part, the diagram obtained is used to analyze the various forces and vectors involved by the methods of Chap. 3.
3. A unified approach is provided for the analysis of the plane motion of a rigid body, regardless of the particular type of motion involved. While the kinematics of the various motions considered may vary from one case to the other, the approach to the kinetics of the motion is consistently the same. In every case a diagram will be drawn showing the external forces, the vector $m\bar{\mathbf{a}}$ associated with the motion of G, and the couple $\bar{I}\boldsymbol{\alpha}$ associated with the rotation of the body about G.
4. The resolution of the plane motion of a rigid body into a translation and a centroidal rotation, which is used here, is a basic concept which can be applied effectively throughout the study of mechanics. It will be used again in Chap. 17 with the method of work and energy and the method of impulse and momentum.
5. As you will see in Chap. 18, this approach can be extended to the study of the general three-dimensional motion of a rigid body. The motion of the body will again be resolved into a translation and a rotation about the mass center, and free-body-diagram equations will be used to indicate the relationship existing between the external forces and the rates of change of the linear and angular momentum of the body.

16.7 SYSTEMS OF RIGID BODIES

The method described in the preceding section can also be used in problems involving the plane motion of several connected rigid bodies. For each part of the system, a diagram similar to Fig. 16.7 or Fig. 16.10 can be drawn. The equations of motion obtained from these diagrams are solved simultaneously.

In some cases, as in Sample Prob. 16.3, a single diagram can be drawn for the entire system. This diagram should include all the external forces, as well as the vectors $m\bar{\mathbf{a}}$ and the couples $\bar{I}\boldsymbol{\alpha}$ associated with the various parts of the system. However, internal forces such as the forces exerted by connecting cables, can be omitted since they occur in pairs of equal and opposite forces and are thus equipollent to zero. The equations obtained by expressing that the system of the external forces is equipollent to the system of the effective forces can be solved for the remaining unknowns.†

It is not possible to use this second approach in problems involving more than three unknowns, since only three equations of motion are available when a single diagram is used. We need not elaborate upon this point, since the discussion involved would be completely similar to that given in Sec. 6.11 in the case of the equilibrium of a system of rigid bodies.

†Note that we cannot speak of *equivalent* systems since we are not dealing with a single rigid body.

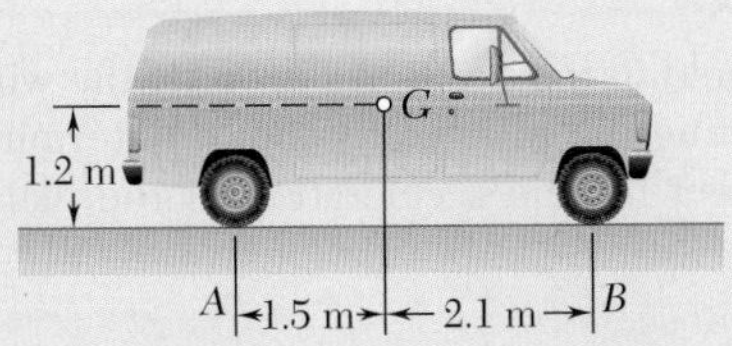

SAMPLE PROBLEM 16.1

When the forward speed of the truck shown was 10 m/s, the brakes were suddenly applied, causing all four wheels to stop rotating. It was observed that the truck skidded to rest in 7 m. Determine the magnitude of the normal reaction and of the friction force at each wheel as the truck skidded to rest. The weight of the truck is W N.

SOLUTION

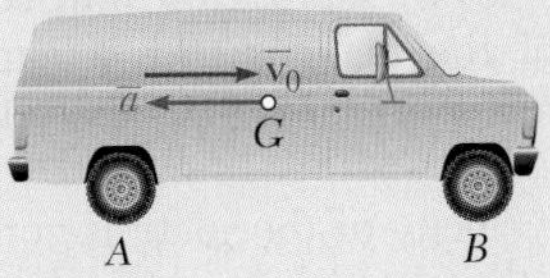

Kinematics of Motion. Choosing the positive sense to the right and using the equations of uniformly accelerated motion, we write

$$\bar{v}_0 = +10 \text{ m/s} \qquad \bar{v}^2 = \bar{v}_0^2 + 2\,\bar{a}\,\bar{x} \qquad 0 = (10)^2 + 2\bar{a}(7)$$

$$\bar{a} = -7.14 \text{ m/s}^2 \qquad \bar{\mathbf{a}} = 7.14 \text{ m/s}^2 \leftarrow$$

The negative sign indicates that the direction of acceleration is opposite to the shown deraction.

Equations of Motion. The external forces consist of the weight **W** of the truck and of the normal reactions and friction forces at the wheels. (The vectors $\mathbf{N}_A$ and $\mathbf{F}_A$ represent the sum of the reactions at the rear wheels, while $\mathbf{N}_B$ and $\mathbf{F}_B$ represent the sum of the reactions at the front wheels.) Since the truck is in translation, the effective forces reduce to the vector $m\bar{\mathbf{a}}$ attached at G. Three equations of motion are obtained by expressing that the system of the external forces is equivalent to the system of the effective forces.

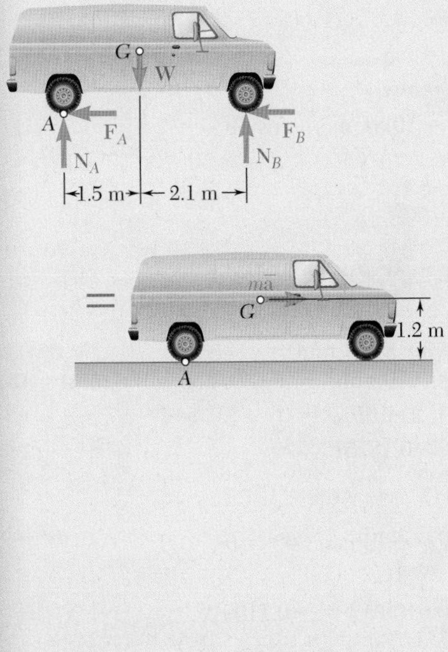

$$+\uparrow \Sigma F_y = \Sigma (F_y)_{\text{eff}}: \qquad N_A + N_B - W = 0$$

Since $F_A = \mu_k N_A$ and $F_B = \mu_k N_B$, where μ_k is the coefficient of kinetic friction, we find that

$$F_A + F_B = \mu_k (N_A + N_B) = \mu_k W$$

$$\xrightarrow{+} \Sigma F_x = \Sigma (F_x)_{\text{eff}}: \qquad -(F_A + F_B) = -m\bar{a}$$

$$-\mu_k \,\text{mg} = -\text{m}(-7.14)$$

$$\mu_k = \frac{7.14}{9.81} = 0.728$$

$$+\circlearrowleft \Sigma M_A = \Sigma (M_A)_{\text{eff}}: \qquad -W(1.5 \text{ m}) + N_B(3.6 \text{ m}) = -m\bar{a}(1.2 \text{ m})$$

$$-W(1.5 \text{ m}) + N_B(3.6 \text{ m}) = \frac{-W}{9.81 \text{m/s}^2}(-7.14 \text{ m/s}^2)(1.2 \text{ m})$$

$$N_B = 0.659W$$

$$F_B = \mu_k N_B = (0.728)(0.659W) \qquad F_B = 0.48\,W$$

$$+\uparrow \Sigma F_y = \Sigma (F_y)_{\text{eff}}: \qquad N_A + N_B - W = 0$$

$$N_A + 0.659W - W = 0$$

$$N_A = 0.341W$$

$$F_A = \mu_k N_A = (0.728)(0.341\,W) \qquad F_A = 0.248\,W$$

Reactions at Each Wheel. Recalling that the values computed above represent the sum of the reactions at the two front wheels or the two rear wheels, we obtain the magnitude of the reactions at each wheel by writing

$$N_{\text{front}} = \tfrac{1}{2}N_B = 0.3295W \qquad N_{\text{rear}} = \tfrac{1}{2}N_A = 0.1705W \quad ◀$$

$$F_{\text{front}} = \tfrac{1}{2}F_B = 0.24W \qquad F_{\text{rear}} = \tfrac{1}{2}F_A = 0.124W \quad ◀$$

SAMPLE PROBLEM 16.2

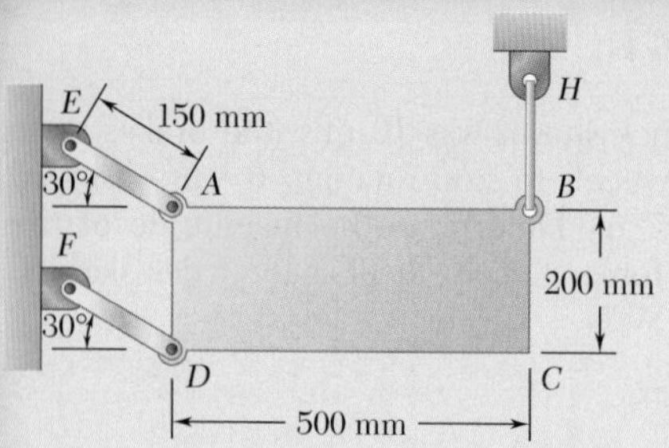

The thin plate *ABCD* of mass 8 kg is held in the position shown by the wire *BH* and two links *AE* and *DF*. Neglecting the mass of the links, determine immediately after wire *BH* has been cut (*a*) the acceleration of the plate, (*b*) the force in each link.

SOLUTION

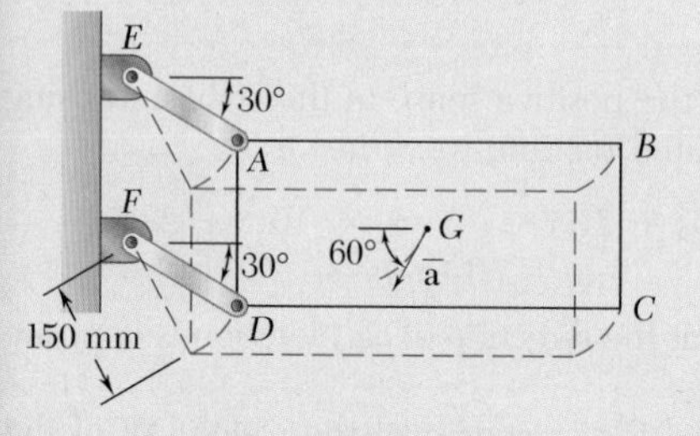

Kinematics of Motion. After wire *BH* has been cut, we observe that corners *A* and *D* move along parallel circles of radius 150 mm centered, respectively, at *E* and *F*. The motion of the plate is thus a curvilinear translation; the particles forming the plate move along parallel circles of radius 150 mm.

At the instant wire *BH* is cut, the velocity of the plate is zero. Thus the acceleration $\bar{\mathbf{a}}$ of the mass center *G* of the plate is tangent to the circular path which will be described by *G*.

Equations of Motion. The external forces consist of the weight **W** and the forces $\mathbf{F}_{AE}$ and $\mathbf{F}_{DF}$ exerted by the links. Since the plate is in translation, the effective forces reduce to the vector $m\bar{\mathbf{a}}$ attached at *G* and directed along the *t* axis. A free-body-diagram equation is drawn to show that the system of the external forces is equivalent to the system of the effective forces.

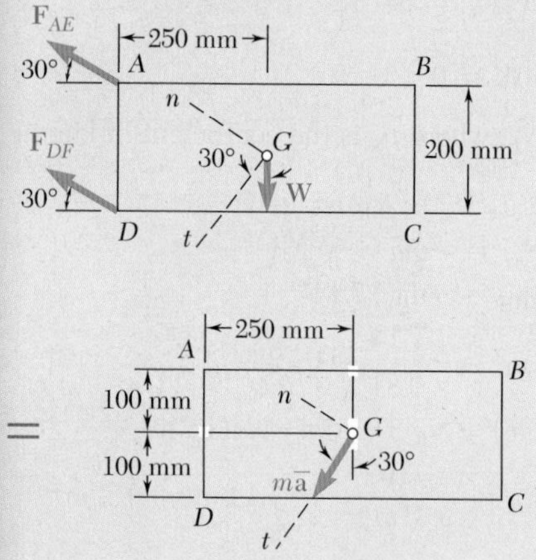

a. Acceleration of the Plate.

$+\swarrow \Sigma F_t = \Sigma (F_t)_{\text{eff}}$:

$$W \cos 30° = m\bar{a}$$

$$mg \cos 30° = m\bar{a}$$

$$\bar{a} = g \cos 30° = (9.81 \text{ m/s}^2) \cos 30° \qquad (1)$$

$$\bar{\mathbf{a}} = 8.50 \text{ m/s}^2 \nearrow 60° \quad \blacktriangleleft$$

b. Forces in Links *AE* and *DF*.

$+\nwarrow \Sigma F_n = \Sigma (F_n)_{\text{eff}}$: $\quad F_{AE} + F_{DF} - W \sin 30° = 0 \qquad (2)$

$+\circlearrowright \Sigma M_G = \Sigma (M_G)_{\text{eff}}$:

$$(F_{AE} \sin 30°)(250 \text{ mm}) - (F_{AE} \cos 30°)(100 \text{ mm})$$
$$+ (F_{DF} \sin 30°)(250 \text{ mm}) + (F_{DF} \cos 30°)(100 \text{ mm}) = 0$$

$$38.4F_{AE} + 211.6F_{DF} = 0$$

$$F_{DF} = -0.1815F_{AE} \qquad (3)$$

Substituting for F_{DF} from (3) into (2), we write

$$F_{AE} - 0.1815F_{AE} - W \sin 30° = 0$$

$$F_{AE} = 0.6109W$$

$$F_{DF} = -0.1815(0.6109W) = -0.1109W$$

Noting that $W = mg = (8 \text{ kg})(9.81 \text{ m/s}^2) = 78.48$ N, we have

$$F_{AE} = 0.6109(78.48 \text{ N}) \qquad F_{AE} = 47.9 \text{ N } T \quad \blacktriangleleft$$

$$F_{DF} = -0.1109(78.48 \text{ N}) \qquad F_{DF} = 8.70 \text{ N } C \quad \blacktriangleleft$$

SAMPLE PROBLEM 16.3

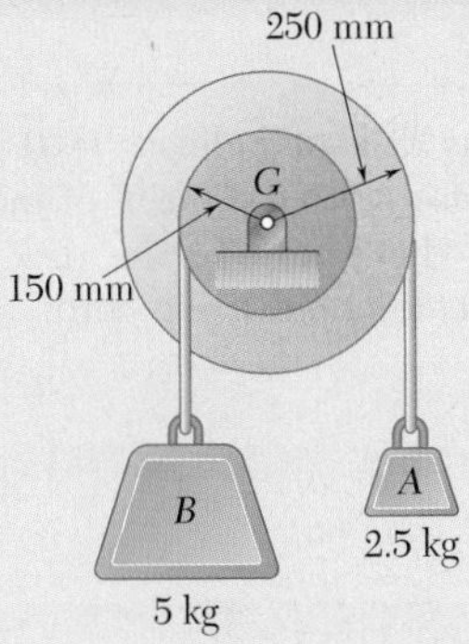

A pulley of mass 6 kg and having a radius of gyration of 200 mm is connected to two blocks as shown. Assuming no axle friction, determine the angular acceleration of the pulley and the acceleration of each block.

SOLUTION

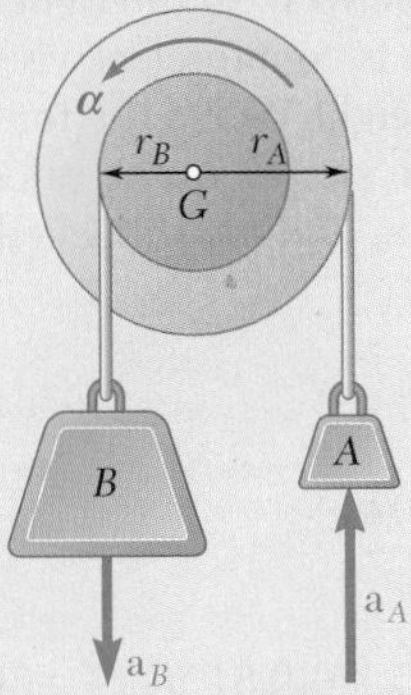

Although an arbitrary sense of motion can be assumed (since no friction forces are involved) and later checked by the sign of the answer, we may prefer to determine the actual sense of rotation of the pulley first. The weight of block *B* required to maintain the equilibrium of the pulley when it is acted upon by the 2.5-kg block *A* is first determined. We write

$$+\circlearrowleft \Sigma M_G = 0: \quad m_B g(150 \text{ mm}) - (2.5 \text{ kg})g\ (250 \text{ mm}) = 0 \quad m_B = 4.167 \text{ kg}$$

Since block *B* actually weighs 5 kg, the pulley will rotate counterclockwise.

Kinematics of Motion. Assuming $\boldsymbol{\alpha}$ counterclockwise and noting that $a_A = r_A\alpha$ and $a_B = r_B\alpha$, we obtain

$$\mathbf{a}_A = (0.25 \text{ m})\alpha \uparrow \qquad \mathbf{a}_B = (0.15 \text{ m})\alpha \downarrow$$

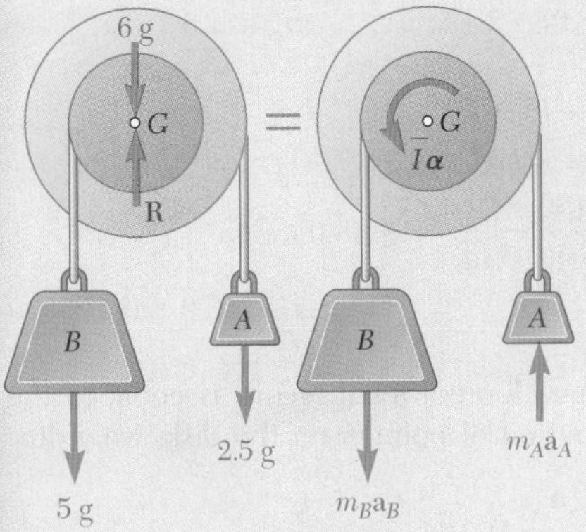

Equations of Motion. A single system consisting of the pulley and the two blocks is considered. Forces external to this system consist of the weights of the pulley and the two blocks and of the reaction at *G*. (The forces exerted by the cables on the pulley and on the blocks are internal to the system considered and cancel out.) Since the motion of the pulley is a centroidal rotation and the motion of each block is a translation, the effective forces reduce to the couple $\bar{I}\boldsymbol{\alpha}$ and the two vectors $m\mathbf{a}_A$ and $m\mathbf{a}_B$. The centroidal moment of inertia of the pulley is

$$\bar{I} = m\bar{k}^2 = (6 \text{ kg})(0.2 \text{ m})^2 = 0.24 \text{ kg}\cdot\text{m}^2$$

Since the system of the external forces is equipollent to the system of the effective forces, we write

$$+\circlearrowleft \Sigma M_G = \Sigma (M_G)_{\text{eff}}:$$

$$(5 \text{ kg})(9.81 \text{ m/s}^2)(0.15 \text{ m}) - (2.5 \text{ kg})(9.81 \text{ m/s}^2)(0.25\text{m}) = +\bar{I}\alpha + m_B a_B(0.15 \text{ m}) + m_A a_A(0.25 \text{ m})$$

$$7.3575 - 6.1312 = 0.24\,\alpha + 5(0.15\,\alpha)(0.15) + 2.5(0.25\,\alpha)(0.25)$$

$$\alpha = +2.41 \text{ rad/s}^2 \qquad \boldsymbol{\alpha} = 2.41 \text{ rad/s}^2 \circlearrowleft \blacktriangleleft$$

$$a_A = r_A\alpha = (0.25 \text{ m})(2.41 \text{rad/s}^2) \qquad \mathbf{a}_A = 0.603 \text{ m/s}^2 \uparrow \blacktriangleleft$$

$$a_B = r_B\alpha = (0.15 \text{ m})(2.41 \text{rad/s}^2) \qquad \mathbf{a}_B = 0.362 \text{ m/s}^2 \downarrow \blacktriangleleft$$

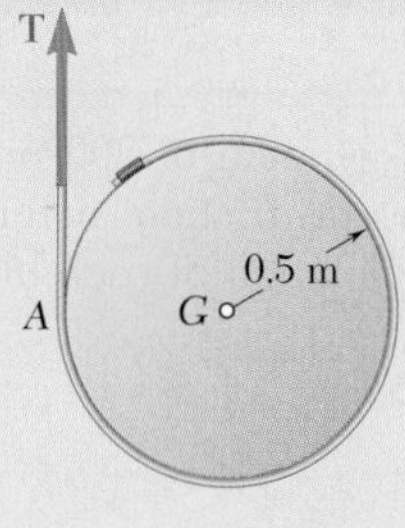

SAMPLE PROBLEM 16.4

A cord is wrapped around a homogeneous disk of radius $r = 0.5$ m and mass $m = 15$ kg. If the cord is pulled upward with a force $\mathbf{T}$ of magnitude 180 N, determine (*a*) the acceleration of the center of the disk, (*b*) the angular acceleration of the disk, (*c*) the acceleration of the cord.

SOLUTION

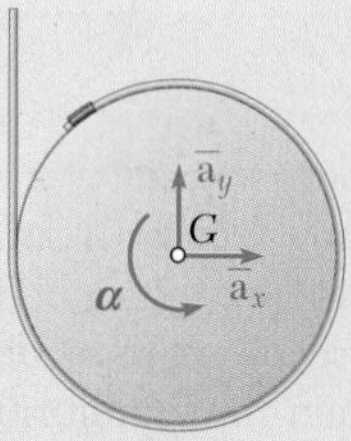

Equations of Motion. We assume that the components $\bar{\mathbf{a}}_x$ and $\bar{\mathbf{a}}_y$ of the acceleration of the center are directed, respectively, to the right and upward and that the angular acceleration of the disk is counterclockwise. The external forces acting on the disk consist of the weight $\mathbf{W}$ and the force $\mathbf{T}$ exerted by the cord. This system is equivalent to the system of the effective forces, which consists of a vector of components $m\bar{\mathbf{a}}_x$ and $m\bar{\mathbf{a}}_y$ attached at G and a couple $\bar{I}\boldsymbol{\alpha}$. We write

$$\overset{+}{\rightarrow}\Sigma F_x = \Sigma(F_x)_{\text{eff}}: \qquad 0 = m\bar{a}_x \qquad\qquad \bar{\mathbf{a}}_x = 0 \blacktriangleleft$$

$$+\uparrow\Sigma F_y = \Sigma(F_y)_{\text{eff}}: \qquad T - W = m\bar{a}_y$$

$$\bar{a}_y = \frac{T - W}{m}$$

Since $T = 180$ N, $m = 15$ kg, and $W = (15 \text{ kg})(9.81 \text{ m/s}^2) = 147.1$ N, we have

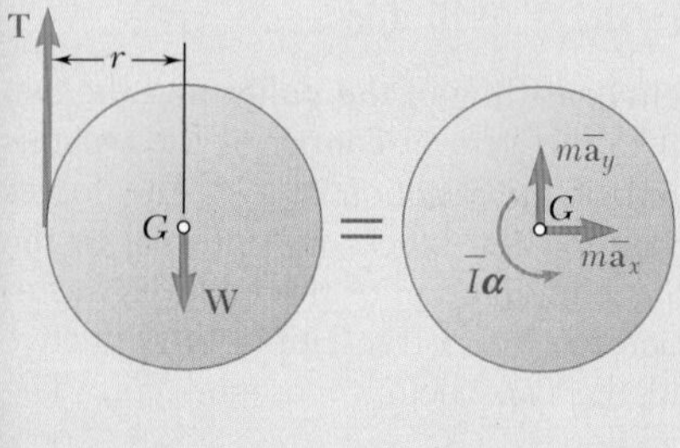

$$\bar{a}_y = \frac{180 \text{ N} - 147.1 \text{ N}}{15 \text{ kg}} = +2.19 \text{ m/s}^2 \qquad \bar{\mathbf{a}}_y = 2.19 \text{ m/s}^2 \uparrow \blacktriangleleft$$

$$+\circlearrowleft\Sigma M_G = \Sigma(M_G)_{\text{eff}}: \qquad -Tr = \bar{I}\alpha$$

$$-Tr = (\tfrac{1}{2}mr^2)\alpha$$

$$\alpha = -\frac{2T}{mr} = -\frac{2(180 \text{ N})}{(15 \text{ kg})(0.5 \text{ m})} = -48.0 \text{ rad/s}^2$$

$$\boldsymbol{\alpha} = 48.0 \text{ rad/s}^2 \circlearrowright \blacktriangleleft$$

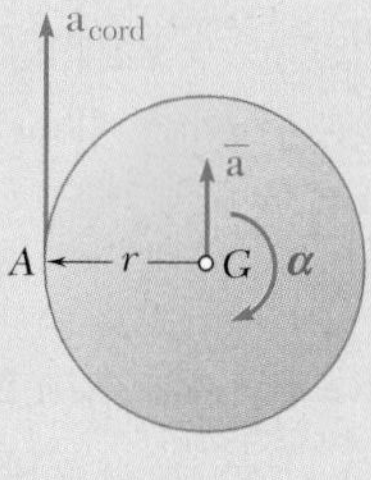

Acceleration of Cord. Since the acceleration of the cord is equal to the tangential component of the acceleration of point A on the disk, we write

$$\mathbf{a}_{\text{cord}} = (\mathbf{a}_A)_t = \bar{\mathbf{a}} + (\mathbf{a}_{A/G})_t$$

$$= [2.19 \text{ m/s}^2 \uparrow] + [(0.5 \text{ m})(48 \text{ rad/s}^2)\uparrow]$$

$$\mathbf{a}_{\text{cord}} = 26.2 \text{ m/s}^2 \uparrow \blacktriangleleft$$

SAMPLE PROBLEM 16.5

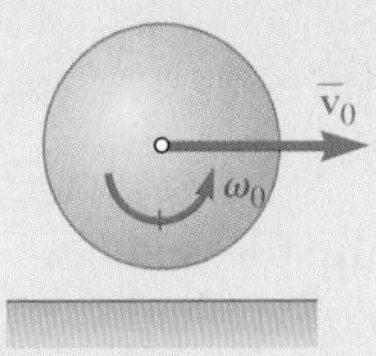

A uniform sphere of mass m and radius r is projected along a rough horizontal surface with a linear velocity $\bar{\mathbf{v}}_0$ and angular velocity $\boldsymbol{\omega}_0$ as shown. Denoting by μ_k the coefficient of kinetic friction between the sphere and the floor, determine (*a*) the required magnitude of $\boldsymbol{\omega}_0$, (*b*) the time t_1 required for the sphere to come to rest, (*c*) the distance the sphere will move before coming to rest.

SOLUTION

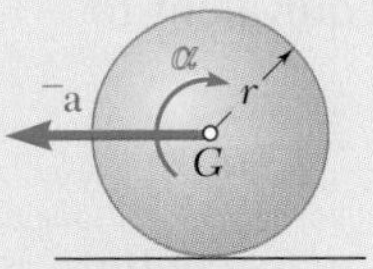

Equations of Motion. The positive sense is chosen to the left for $\bar{\mathbf{a}}$ and clockwise for $\boldsymbol{\alpha}$. The external forces acting on the sphere consist of the weight **W**, the normal reaction **N**, and the friction force **F**. Since the point of the sphere in contact with the surface is sliding to the right, the friction force **F** is directed to the left. While the sphere is sliding, the magnitude of the friction force is $F = \mu_k N$. The effective forces consist of the vector $m\bar{\mathbf{a}}$ attached at G and the couple $\bar{I}\boldsymbol{\alpha}$. Expressing that the system of the external forces is equivalent to the system of the effective forces, we write

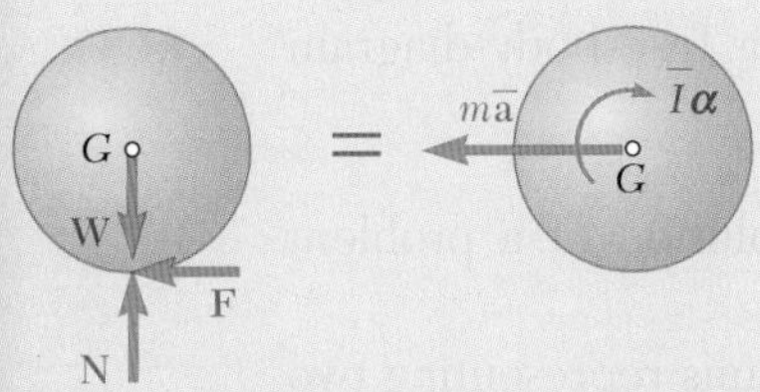

$+\uparrow \Sigma F_y = \Sigma(F_y)_{\text{eff}}$: $\quad N - W = 0$

$$N = W = mg \qquad F = \mu_k N = \mu_k mg$$

$\xrightarrow{+} \Sigma F_x = \Sigma(F_x)_{\text{eff}}$: $\quad F = m\bar{a} \qquad \mu_k mg = m\bar{a} \qquad \bar{a} = \mu_k g$

$+\circlearrowright \Sigma M_G = \Sigma(M_G)_{\text{eff}}$: $\quad Fr = \bar{I}\alpha$

Noting that $\bar{I} = \frac{2}{5}mr^2$ and substituting the value obtained for F, we write

$$(\mu_k mg)r = \tfrac{2}{5}mr^2\alpha \qquad \alpha = \frac{5}{2}\frac{\mu_k g}{r}$$

<u>Kinematics</u>: $\quad \xrightarrow{+} v = v_0 - \bar{a}t$

$$v = v_0 - \mu_k g t$$

For $v = 0$ when $t = t_1$: $\quad 0 = v_0 - \mu_k g t_1; \qquad t_1 = \dfrac{v_0}{\mu_k g} \qquad (1)$

$$\circlearrowleft^{+}\ \omega = \omega_0 - \alpha t$$

$$\omega = \omega_0 - \left(\frac{5}{2}\frac{\mu_k g}{r}\right)t$$

For $\omega = 0$ when $t = t_1$ $\; 0 = \omega_0 - \dfrac{5}{2}\dfrac{\mu_k g}{r}t_1; \qquad t_1 = \dfrac{2r}{5\mu_k g}\omega_0 \qquad (2)$ ◀

Set Eq. (1) = Eq. (2) $\; \dfrac{v_0}{\mu_k g} = \dfrac{2r}{5\mu_k g}\omega_0; \qquad \omega_0 = \dfrac{5}{2}\dfrac{v_0}{r}$ ◀

Distance traveled: $\quad s_1 = v_0 t_1 - \dfrac{1}{2}\bar{a}t_1^2$

$$s_1 = v_0\left(\frac{v_0}{\mu_k g}\right) - \frac{1}{2}(\mu_k g)\left(\frac{v_0}{\mu_k g}\right)^2; \qquad s_1 = \frac{v_0^2}{2\mu_k g}$$ ◀

SOLVING PROBLEMS ON YOUR OWN

This chapter deals with the *plane motion* of rigid bodies, and in this first lesson we considered rigid bodies that are free to move under the action of applied forces.

1. Effective forces. We first recalled that a rigid body consists of a large number of particles. The effective forces of the particles forming the body were found to be equivalent to a vector $m\bar{\mathbf{a}}$ attached at the mass center G of the body and a couple of moment $\bar{I}\boldsymbol{\alpha}$ [Fig. 16.7]. Noting that the applied forces are equivalent to the effective forces, we wrote

$$\Sigma F_x = m\bar{a}_x \qquad \Sigma F_y = m\bar{a}_y \qquad \Sigma M_G = \bar{I}\alpha \qquad (16.5)$$

where $\bar{a}_x$ and $\bar{a}_y$ are the x and y components of the acceleration of the mass center G of the body and α is the angular acceleration of the body. It is important to note that when these equations are used, *the moments of the applied forces must be computed with respect to the mass center of the body*. However, you learned a more efficient method of solution based on the use of a free-body-diagram equation.

2. Free-body-diagram equation. Your first step in the solution of a problem should be to draw a *free-body-diagram equation*.

a. A free-body-diagram equation consists of two diagrams representing two equivalent systems of vectors. *In the first diagram* you should show *the forces exerted on the body,* including the applied forces, the reactions at the supports, and the weight of the body. *In the second diagram* you should show the vector $m\bar{\mathbf{a}}$ and the couple $\bar{I}\boldsymbol{\alpha}$ representing *the effective forces*.

b. Using a free-body-diagram equation allows you to *sum components in any direction and to sum moments about any point.* When writing the three equations of motion needed to solve a given problem, you can therefore select one or more equations involving a single unknown. Solving these equations first and substituting the values obtained for the unknowns into the remaining equation(s) will yield a simpler solution.

3. Plane motion of a rigid body. The problems that you will be asked to solve will fall into one of the following categories.

a. Rigid body in translation. For a body in translation, the angular acceleration is zero. The effective forces reduce to *the vector* $m\bar{\mathbf{a}}$ applied at the mass center [Sample Probs. 16.1 and 16.2].

b. Rigid body in centroidal rotation. For a body in centroidal rotation, the acceleration of the mass center is zero. The effective forces reduce to *the couple* $\bar{I}\boldsymbol{\alpha}$ [Sample Prob. 16.3].

c. Rigid body in general plane motion. You can consider the general plane motion of a rigid body as the sum of a translation and a centroidal rotation. The effective forces are equivalent to the vector $m\bar{\mathbf{a}}$ and the couple $\bar{I}\boldsymbol{\alpha}$ [Sample Probs. 16.4 and 16.5].

4. Plane motion of a system of rigid bodies. You first should draw a free-body-diagram equation that includes all the rigid bodies of the system. A vector $m\bar{\mathbf{a}}$ and a couple $\bar{I}\boldsymbol{\alpha}$ are attached to each body. However, the forces exerted on each other by the various bodies of the system can be omitted, since they occur in pairs of equal and opposite forces.

a. If no more than three unknowns are involved, you can use this free-body-diagram equation and sum components in any direction and sum moments about any point to obtain equations that can be solved for the desired unknowns [Sample Prob. 16.3].

b. If more than three unknowns are involved, you must draw a separate free-body-diagram equation for each of the rigid bodies of the system. Both internal forces and external forces should be included in each of the free-body-diagram equations, and care should be taken to represent with equal and opposite vectors the forces that two bodies exert on each other.

PROBLEMS

CONCEPT QUESTIONS

16.CQ1 Two pendulums, A and B, with the masses and lengths shown are released from rest. Which system has a larger mass moment of inertia about its pivot point?

a. A

b. B

c. They are the same.

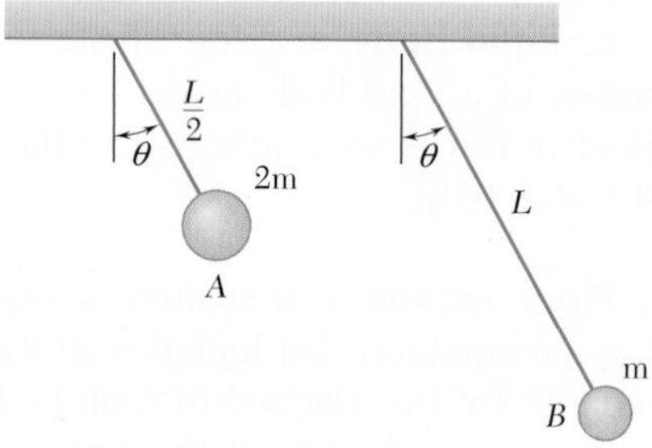

Fig. P16.CQ1 and P16.CQ2

16.CQ2 Two pendulums, A and B, with the masses and lengths shown are released from rest. Which system has a larger angular acceleration immediately after release?

a. A

b. B

c. They are the same.

16.CQ3 Two solid cylinders, A and B, have the same mass m and the radii $2r$ and r, respectively. Each is accelerated from rest with a force applied as shown. In order to impart identical angular accelerations to both cylinders, what is the relationship between F_1 and F_2?

a. $F_1 = 0.5F_2$

b. $F_1 = F_2$

c. $F_1 = 2F_2$

d. $F_1 = 4F_2$

e. $F_1 = 8F_2$

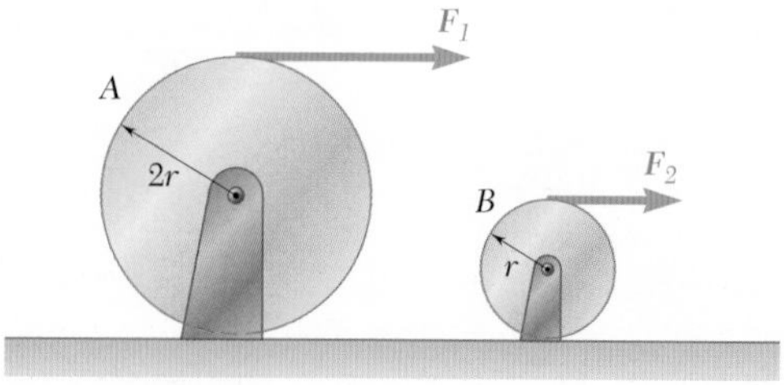

Fig. P16.CQ3

FREE BODY PRACTICE PROBLEMS

16.F1 A 1.8 m board is placed in a truck with one end resting against a block secured to the floor and the other leaning against a vertical partition. Draw the FBD and KD necessary to determine the maximum allowable acceleration of the truck if the board is to remain in the position shown.

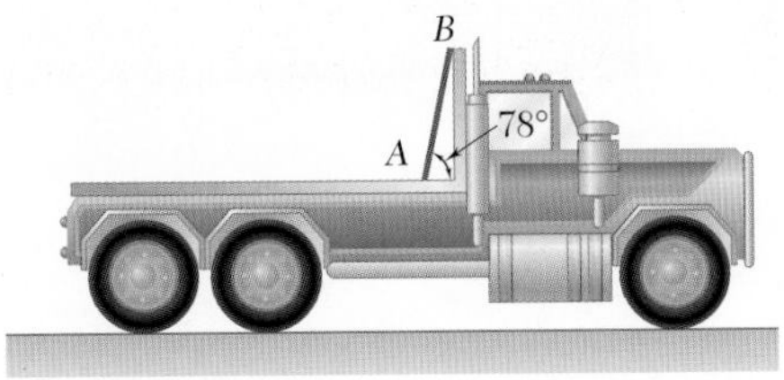

Fig. P16.F1

16.F2 A uniform circular plate of mass 3 kg is attached to two links *AC* and *BD* of the same length. Knowing that the plate is released from rest in the position shown, in which lines joining *G* to *A* and *B* are, respectively, horizontal and vertical, draw the FBD and KD for the plate.

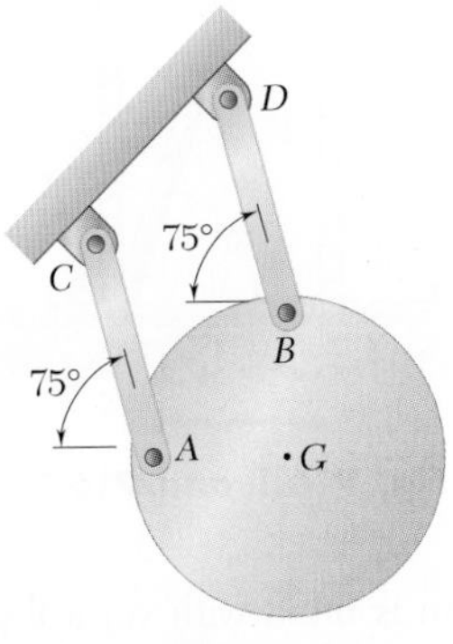

Fig. P16.F2

16.F3 Two uniform disks and two cylinders are assembled as indicated. Disk *A* has a mass of 10 kg and disk *B* has mass of 6 kg. Knowing that the system is released from rest, draw the FBD and KD for the whole system.

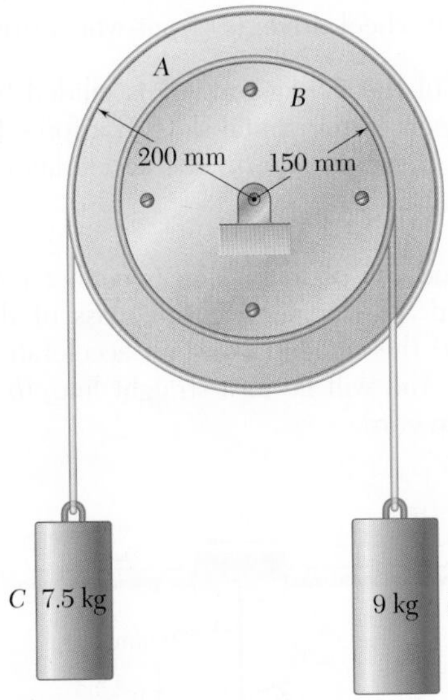

Fig. P16.F3

16.F4 The 200-kg crate shown is lowered by means of two overhead cranes. Knowing the tension in each cable, draw the FBD and KD that can be used to determine the angular acceleration of the crate and the acceleration of the center of gravity.

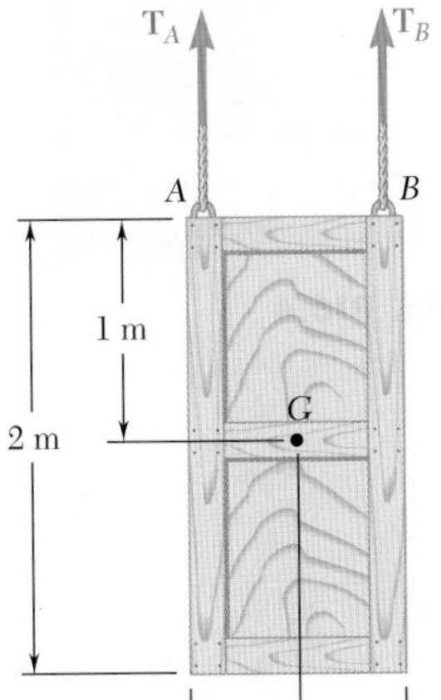

Fig. P16.F4

END-OF-SECTION PROBLEMS

16.1 A conveyor system is fitted with vertical panels, and a 300-mm rod AB of mass 2.5 kg is lodged between two panels as shown. If the rod is to remain in the position shown, determine the maximum allowable acceleration of the system.

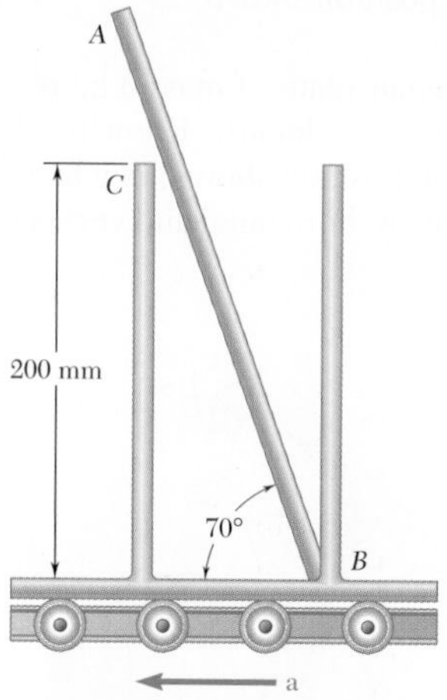

Fig. P16.1 and P16.2

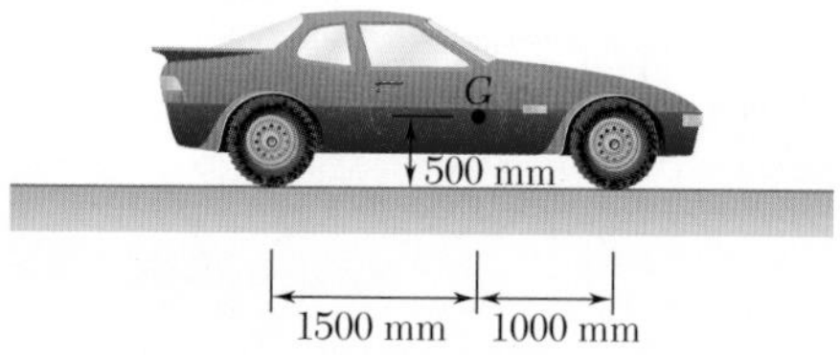

Fig. P16.3

16.2 A conveyor system is fitted with vertical panels, and a 300-mm rod AB of mass 2.5 kg is lodged between two panels as shown. Knowing that the acceleration of the system is 1.5 m/s^2 to the left, determine (*a*) the force exerted on the rod at C, (*b*) the reaction at B.

16.3 Knowing that the coefficient of static friction between the tires and the road is 0.80 for the automobile shown, determine the maximum possible acceleration on a level road, assuming (*a*) four-wheel drive, (*b*) rear-wheel drive, (*c*) front-wheel drive.

16.4 The motion of the 2.5-kg rod AB is guided by two small wheels which roll freely in horizontal slots. If a force **P** of magnitude 8 N is applied at B, determine (*a*) the acceleration of the rod, (*b*) the reactions at A and B.

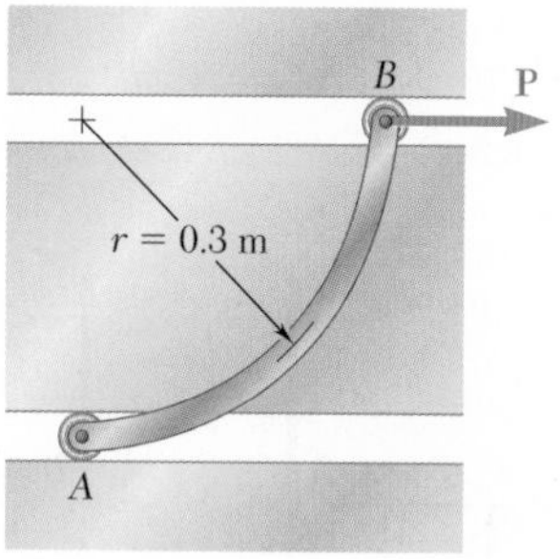

Fig. P16.4

16.5 A uniform rod BC of mass 4 kg is connected to a collar A by a 250-mm cord AB. Neglecting the mass of the collar and cord, determine (*a*) the smallest constant acceleration $\mathbf{a}_A$ for which the cord and the rod will lie in a straight line, (*b*) the corresponding tension in the cord.

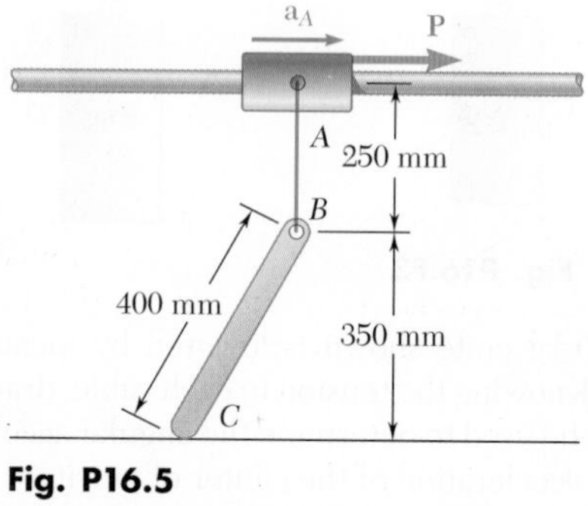

Fig. P16.5

16.6 A 2000-kg truck is being used to lift a 400-kg boulder B that is on a 50-kg pallet A. Knowing the acceleration of the rear-wheel-drive truck is 1 m/s^2, determine (a) the reaction at each of the front wheels, (b) the force between the boulder and the pallet.

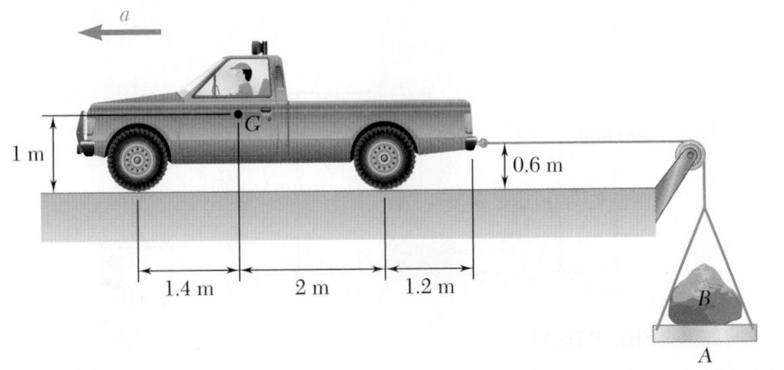

Fig. P16.6

16.7 The support bracket shown is used to transport a cylindrical can from one elevation to another. Knowing that $\mu_s = 0.25$ between the can and the bracket, determine (a) the magnitude of the upward acceleration **a** for which the can will slide on the bracket, (b) the smallest ratio h/d for which the can will tip before it slides.

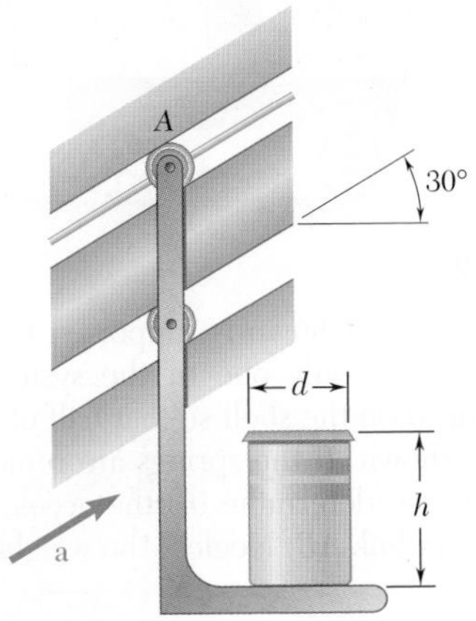

Fig. P16.7

16.8 Solve Prob. 16.7, assuming that the acceleration **a** of the bracket is directed downward.

16.9 A 20-kg cabinet is mounted on casters that allow it to move freely ($\mu = 0$) on the floor. If a 100-N force is applied as shown, determine (a) the acceleration of the cabinet, (b) the range of values of h for which the cabinet will not tip.

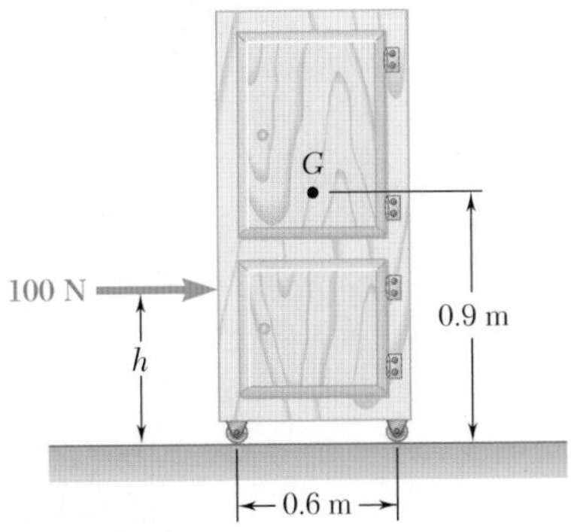

Fig. P16.9

16.10 Solve Prob. 16.9, assuming that the casters are locked and slide on the rough floor ($\mu_k = 0.25$).

16.11 A completely filled barrel and its contents have a combined mass of 90 kg. A cylinder C is connected to the barrel at a height h = 550 mm as shown. Knowing $\mu_s = 0.40$ and $\mu_k = 0.35$, determine the maximum mass of C so the barrel will not tip.

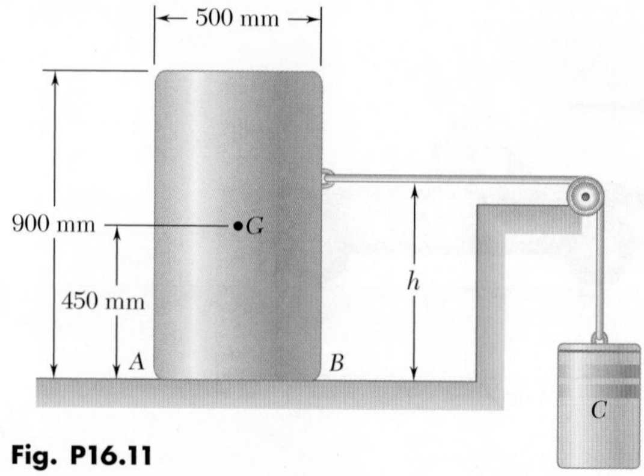

Fig. P16.11

16.12 A 40-kg vase has a 200-mm-diameter base and is being moved using a 100-kg utility cart as shown. The cart moves freely ($\mu = 0$) on the ground. Knowing the coefficient of static friction between the vase and the cart is $\mu_s = 0.4$, determine the maximum force **F** that can be applied if the vase is not to slide or tip.

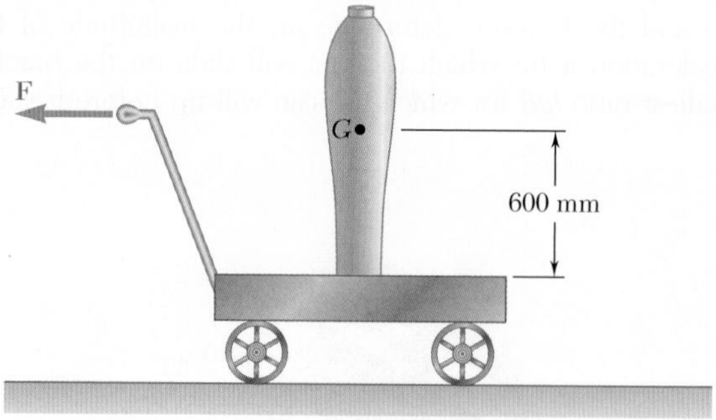

Fig. P16.12

16.13 The retractable shelf shown is supported by two identical linkage-and-spring systems; only one of the systems is shown. A 20-kg machine is placed on the shelf so that half of its weight is supported by the system shown. If the springs are removed and the system is released from rest, determine (*a*) the acceleration of the machine, (*b*) the tension in link AB. Neglect the weight of the shelf and links.

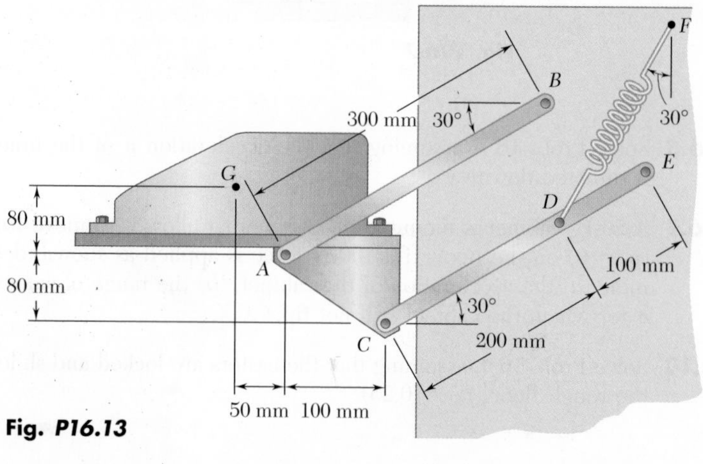

Fig. P16.13

16.14 A uniform rectangular plate has a mass of 5 kg and is held in position by three ropes as shown. Knowing that $\theta = 30°$, determine, immediately after rope CF has been cut, (*a*) the acceleration of the plate, (*b*) the tension in ropes AD and BE.

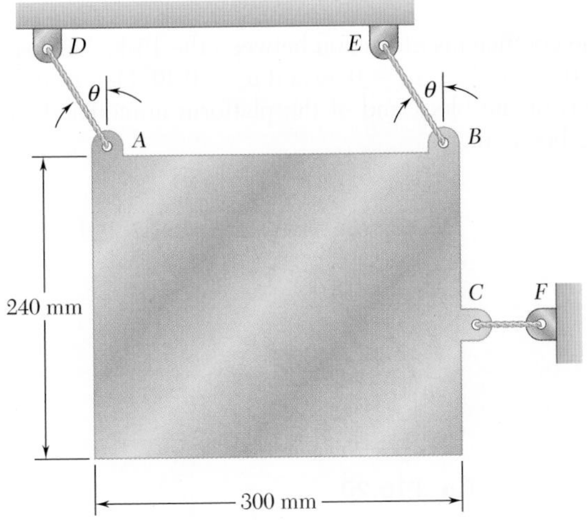

Fig. P16.14 and P16.15

16.15 A uniform rectangular plate has a mass of 5 kg and is held in position by three ropes as shown. Determine the largest value of θ for which both ropes AD and BE remain taut immediately after rope CF has been cut.

16.16 Three bars, each of mass 3 kg, are welded together and pin-connected to two links BE and CF. Neglecting the weight of the links, determine the force in each link immediately after the system is released from rest.

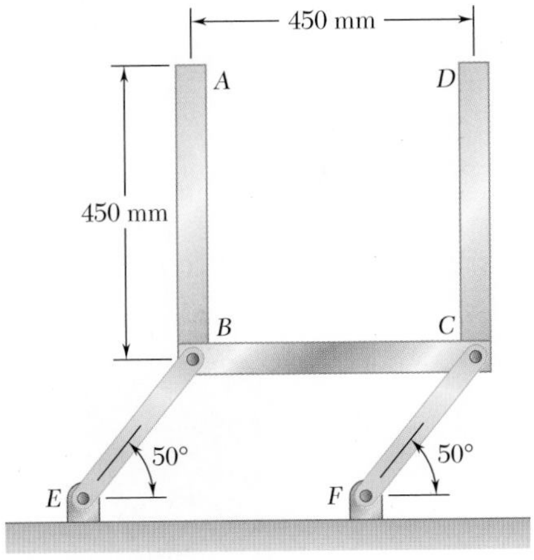

Fig. *P16.16*

16.17 Members ACE and DCB are each 600 mm long and are connected by a pin at C. The mass center of the 10-kg member AB is located at G. Determine (*a*) the acceleration of AB immediately after the system has been released from rest in the position shown, (*b*) the corresponding force exerted by roller A on member AB. Neglect the weight of members ACE and DCB.

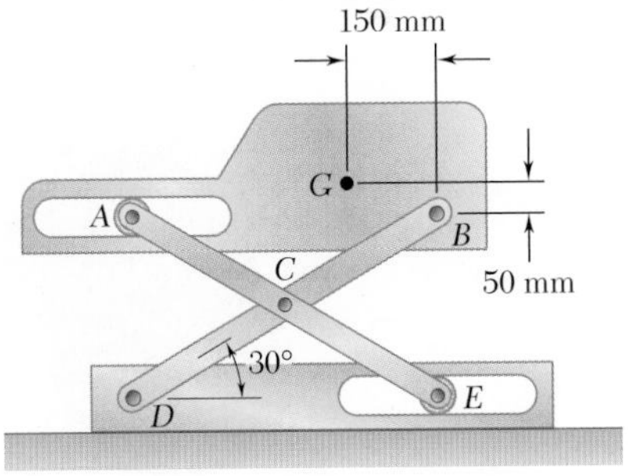

Fig. *P16.17*

16.18 The 7.5-kg rod BC connects a disk centered at A to crank CD. Knowing that the disk is made to rotate at the constant speed of 180 rpm, determine for the position shown the vertical components of the forces exerted on rod BC by pins at B and C.

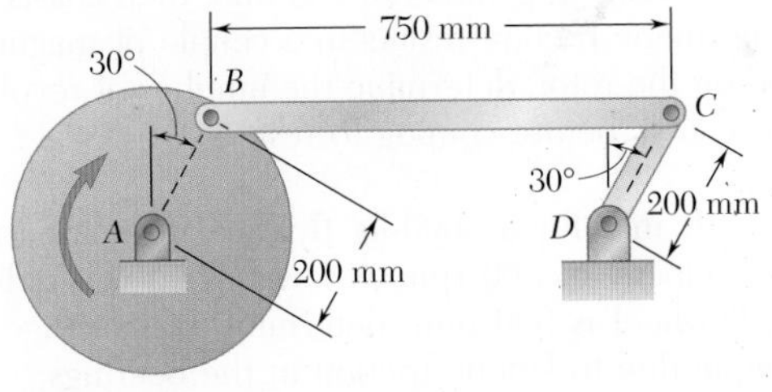

Fig. P16.18

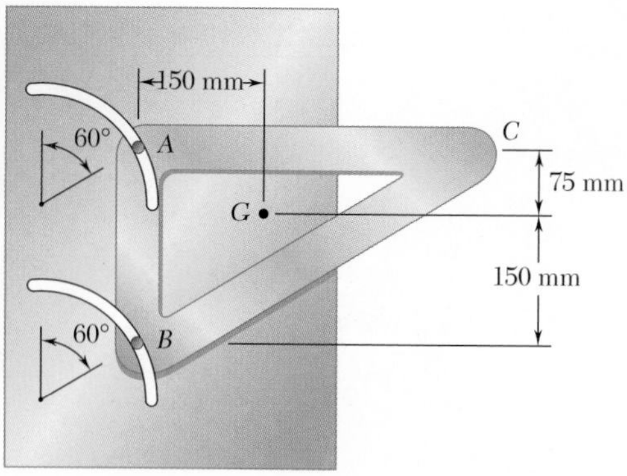

Fig. P16.19

16.19 The triangular weldment ABC is guided by two pins that slide freely in parallel curved slots of radius 150 mm cut in a vertical plate. The weldment has a mass of 8 kg and its mass center is located at point G. Knowing that at the instant shown the velocity of each pin is 750 mm/s downward along the slots, determine (*a*) the acceleration of the weldment, (*b*) the reactions at A and B.

16.20 The coefficients of friction between the 15-kg block and the 2.5-kg platform BD are $\mu_s = 0.50$ and $\mu_k = 0.40$. Determine the accelerations of the block and of the platform immediately after wire AB has been cut.

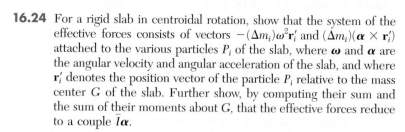

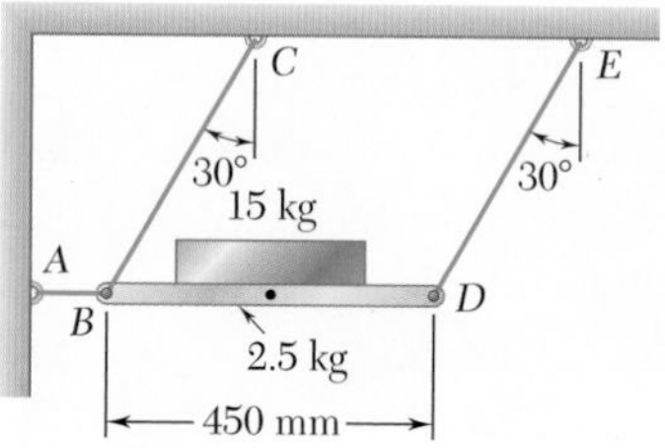

Fig. P16.20

16.21 Draw the shear and bending-moment diagrams for the vertical rod AB of Prob. 16.16.

***16.22** Draw the shear and bending-moment diagrams for the connecting rod BC of Prob. 16.18.

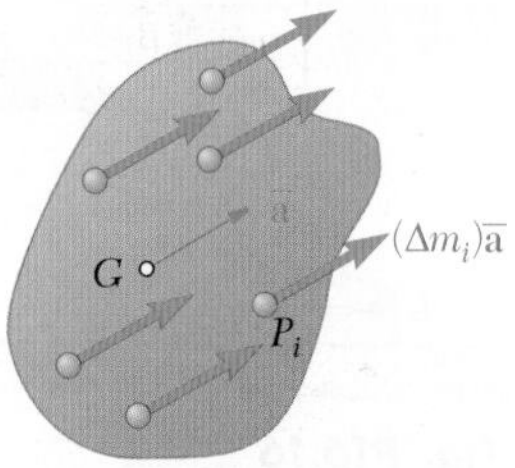

Fig. P16.23

16.23 For a rigid slab in translation, show that the system of the effective forces consists of vectors $(\Delta m_i)\bar{\mathbf{a}}$ attached to the various particles of the slab, where $\bar{\mathbf{a}}$ is the acceleration of the mass center G of the slab. Further show, by computing their sum and the sum of their moments about G, that the effective forces reduce to a single vector $m\bar{\mathbf{a}}$ attached at G.

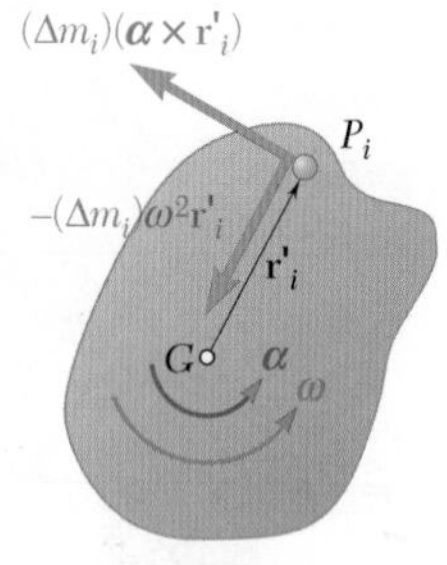

Fig. P16.24

16.24 For a rigid slab in centroidal rotation, show that the system of the effective forces consists of vectors $-(\Delta m_i)\omega^2\mathbf{r}'_i$ and $(\Delta m_i)(\boldsymbol{\alpha} \times \mathbf{r}'_i)$ attached to the various particles P_i of the slab, where $\boldsymbol{\omega}$ and $\boldsymbol{\alpha}$ are the angular velocity and angular acceleration of the slab, and where $\mathbf{r}'_i$ denotes the position vector of the particle P_i relative to the mass center G of the slab. Further show, by computing their sum and the sum of their moments about G, that the effective forces reduce to a couple $\bar{I}\boldsymbol{\alpha}$.

16.25 The rotor of an electric motor has an angular velocity of 3600 rpm when the load and power are cut off. The 50-kg rotor, which has a centroidal radius of gyration of 180 mm, then coasts to rest. Knowing that kinetic friction results in a couple of magnitude 3.5 N · m exerted on the rotor, determine the number of revolutions that the rotor executes before coming to rest.

16.26 It takes 10 min for a 3000-kg flywheel to coast to rest from an angular velocity of 300 rpm. Knowing that the radius of gyration of the flywheel is 900 mm, determine the average magnitude of the couple due to kinetic friction in the bearings.

16.27 The 200-mm-radius brake drum is attached is a larger flywheel that is not shown. The total mass moment of inertia of the drum and the flywheel is 20 kg · m^2 and the coefficient of kinetic friction between the drum and the brake shoe is 0.35. Knowing that the angular velocity of the flywheel is 360 rpm counterclockwise when a force **P** of magnitude 400 N is applied to the pedal C, determine the number of revolutions executed by the flywheel before it comes to rest.

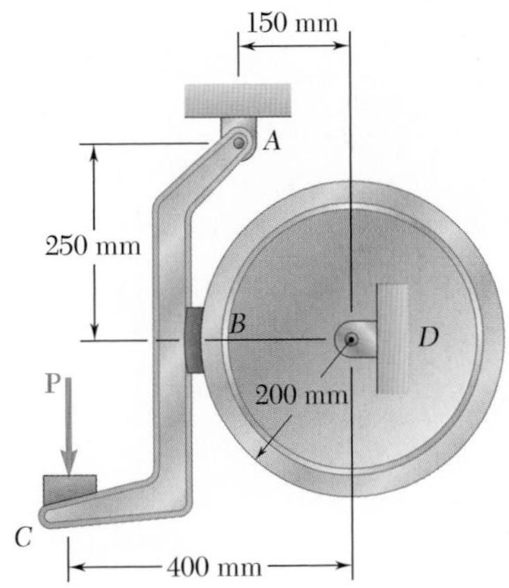

Fig. P16.27

16.28 Solve Prob. 16.27, assuming that the initial angular velocity of the flywheel is 360 rpm clockwise.

16.29 The 100-mm-radius brake drum is attached to a flywheel which is not shown. The drum and flywheel together have a mass of 300 kg and a radius of gyration of 600 mm. The coefficient of kinetic friction between the brake band and the drum is 0.30. Knowing that a force **P** of magnitude 50 N is applied at A when the angular velocity is 180 rpm counterclockwise, determine the time required to stop the flywheel when a = 200 mm and b = 160 mm.

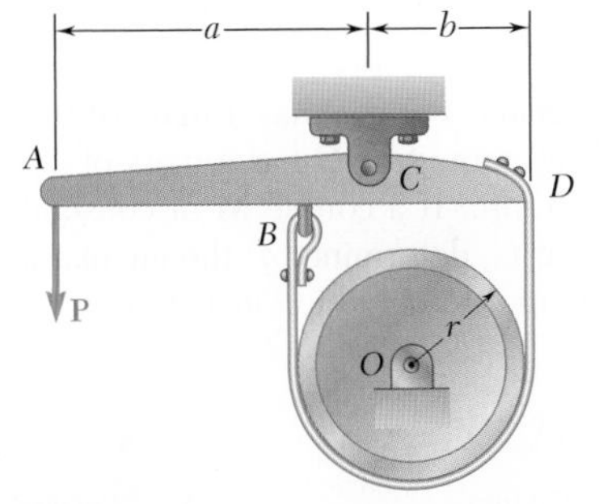

Fig. P16.29

16.30 The 180-mm-radius disk is at rest when it is placed in contact with a belt moving at a constant speed. Neglecting the weight of the link AB and knowing that the coefficient of kinetic friction between the disk and the belt is 0.40, determine the angular acceleration of the disk while slipping occurs.

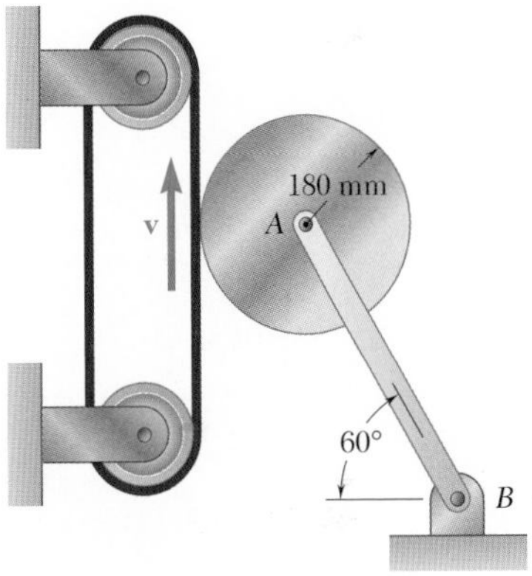

Fig. P16.30

16.31 Solve Prob. 16.30, assuming that the direction of motion of the belt is reversed.

16.32 In order to determine the mass moment of inertia of a flywheel of radius 600 mm, a 12-kg block is attached to a wire that is wrapped around the flywheel. The block is released and is observed to fall 3 m in 4.6 s. To eliminate bearing friction from the computation, a second block of mass 24 kg is used and is observed to fall 3 m in 3.1 s. Assuming that the moment of the couple due to friction remains constant, determine the mass moment of inertia of the flywheel.

16.33 The flywheel shown has a radius of 500 mm, a mass of 120 kg, and a radius of gyration of 375 mm. A 15-kg block A is attached to a wire that is wrapped around the flywheel, and the system is released from rest. Neglecting the effect of friction, determine (*a*) the acceleration of block A, (*b*) the speed of block A after it has moved 1.5 m.

Fig. *P16.32* and P16.33

16.34 Each of the double pulleys shown has a mass moment of inertia of 20 kg·m^2 and is initially at rest. The outside radius is 500 mm, and the inner radius is 250 mm. Determine (*a*) the angular acceleration of each pulley, (*b*) the angular velocity of each pulley after point *A* on the cord has moved 3 m.

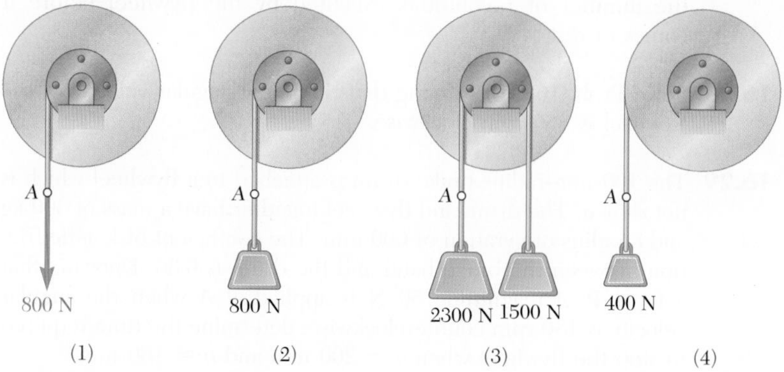

Fig. P16.34

16.35 Each of the gears *A* and *B* has a mass of 9 kg and has a radius of gyration of 200 mm; gear *C* has a mass of 3 kg and has a radius of gyration of 75 mm. If a couple **M** of constant magnitude 5 N-m is applied to gear *C*, determine (*a*) the angular acceleration of gear *A*, (*b*) the tangential force which gear *C* exerts on gear *A*.

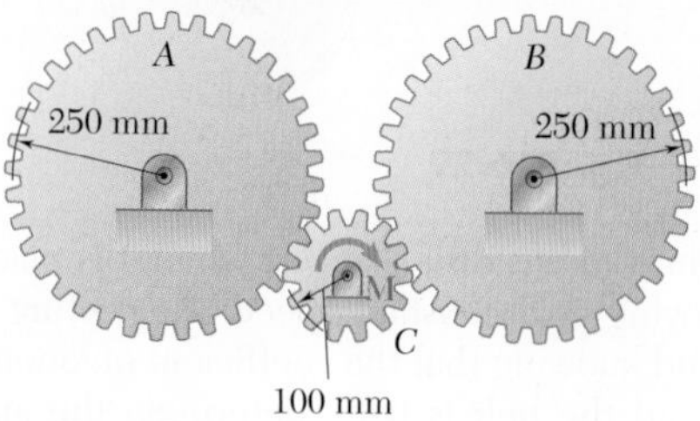

Fig. *P16.35*

16.36 Solve Prob. 16.35, assuming that the couple **M** is applied to gear *A*.

16.37 The 150-mm-radius brake drum is attached to a larger flywheel that is not shown. The total mass moment of inertia of the drum and the flywheel is 75 kg · m^2. A band brake is used to control the motion of the system and the coefficient of kinetic friction between the belt and the drum is 0.25. Knowing that the 100-N force **P** is applied when the initial angular velocity of the system is 240 rpm clockwise, determine the time required for the system to stop. Show that the same result is obtained if the initial angular velocity of the system is 240 rpm counterclockwise.

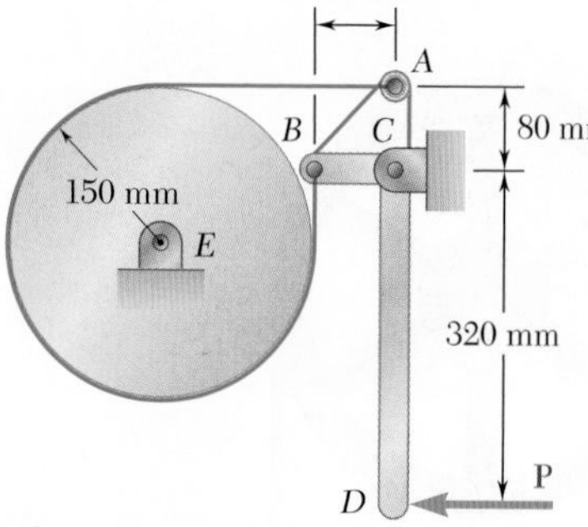

Fig. *P16.37*

16.38 Disks A and B are bolted together, and cylinders D and E are attached to separate cords wrapped on the disks. A single cord passes over disks B and C. Disk A has a mass of 10 kg and disks B and C each have a mass of 6 kg. Knowing that the system is released from rest and that no slipping occurs between the cords and the disks, determine the acceleration (a) of cylinder D, (b) of cylinder E.

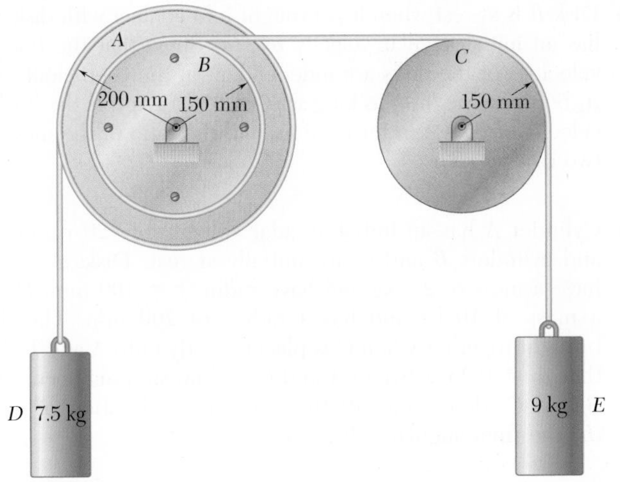

Fig. P16.38

16.39 A belt of negligible mass passes between cylinders A and B and is pulled to the right with a force **P**. Cylinders A and B have a mass of, respectively, 2.5 and 10 kg. The shaft of cylinder A is free to slide in a vertical slot and the coefficients of friction between the belt and each of the cylinders are $\mu_s = 0.50$ and $\mu_k = 0.40$. For $P = 18$ N, determine (a) whether slipping occurs between the belt and either cylinder, (b) the angular acceleration of each cylinder.

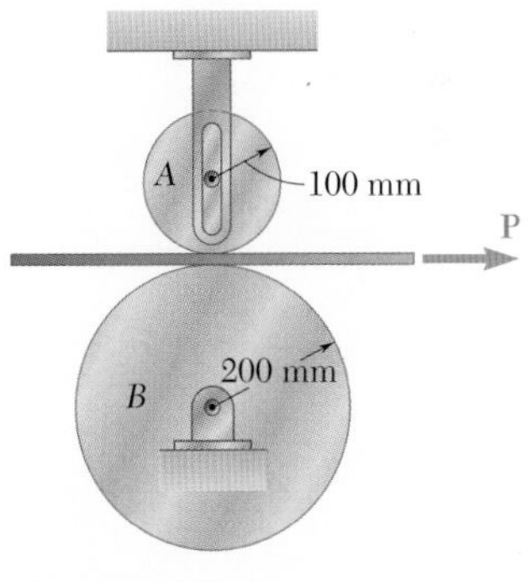

Fig. P16.39

16.40 Solve Prob. 16.39 for $P = 10$ N.

16.41 Disk A has a mass of 6 kg and an initial angular velocity of 360 rpm clockwise; disk B has a mass of 3 kg and is initially at rest. The disks are brought together by applying a horizontal force of magnitude 20 N to the axle of disk A. Knowing that $\mu_k = 0.15$ between the disks and neglecting bearing friction, determine (a) the angular acceleration of each disk, (b) the final angular velocity of each disk.

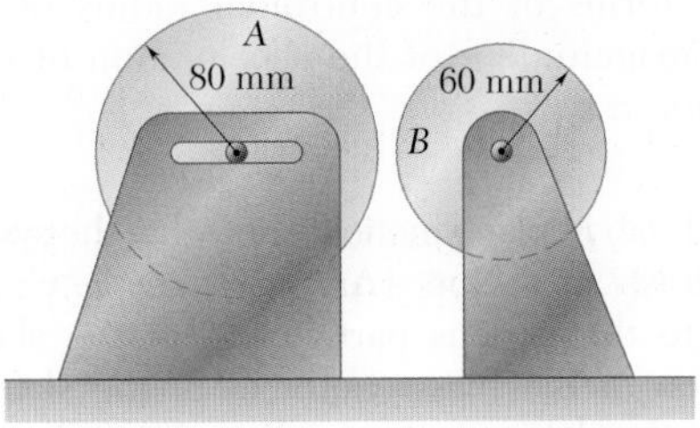

Fig. P16.41

16.42 Solve Prob. 16.41, assuming that initially disk A is at rest and disk B has an angular velocity of 360 rpm clockwise.

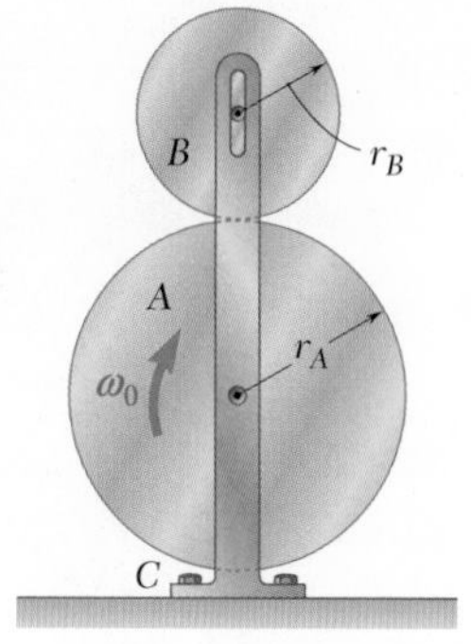

Fig. P16.43 and P16.44

16.43 Disk A has a mass $m_A = 4$ kg, a radius $r_A = 300$ mm, and an initial angular velocity $\boldsymbol{\omega}_0 = 300$ rpm clockwise. Disk B has a mass $m_B = 1.6$ kg, a radius $r_B = 180$ mm, and is at rest when it is brought into contact with disk A. Knowing that $\mu_k = 0.35$ between the disks and neglecting bearing friction, determine (*a*) the angular acceleration of each disk, (*b*) the reaction at the support C.

16.44 Disk B is at rest when it is brought into contact with disk A, which has an initial angular velocity $\boldsymbol{\omega}_0$. (*a*) Show that the final angular velocities of the disks are independent of the coefficient of friction μ_k between the disks as long as $\mu_k \neq 0$. (*b*) Express the final angular velocity of disk A in terms of ω_0 and the ratio of the masses of the two disks m_A/m_B.

16.45 Cylinder A has an initial angular velocity of 720 rpm clockwise, and cylinders B and C are initially at rest. Disks A and B each have a mass of 2.5 kg and have radius $r = 100$ mm. Disk C has a mass of 10 kg and has a radius of 200 mm. The disks are brought together when C is placed gently onto A and B. Knowing that $\mu_k = 0.25$ between A and C and no slipping occurs between B and C, determine (*a*) the angular acceleration of each disk, (*b*) the final angular velocity of each disk.

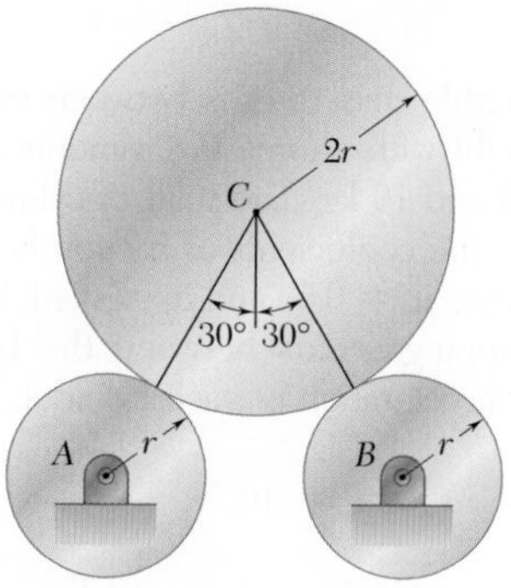

Fig. *P16.45*

16.46 Show that the system of the effective forces for a rigid slab in plane motion reduces to a single vector, and express the distance from the mass center G of the slab to the line of action of this vector in terms of the centroidal radius of gyration $\bar{k}$ of the slab, the magnitude $\bar{a}$ of the acceleration of G, and the angular acceleration α.

16.47 For a rigid slab in plane motion, show that the system of the effective forces consists of vectors $(\Delta m_i)\bar{\mathbf{a}}$, $-(\Delta m_i)\omega^2\mathbf{r}'_i$, and $(\Delta m_i)(\boldsymbol{\alpha} \times \mathbf{r}'_i)$ attached to the various particles P_i of the slab, where $\bar{\mathbf{a}}$ is the acceleration of the mass center G of the slab, $\boldsymbol{\omega}$ is the angular velocity of the slab, $\boldsymbol{\alpha}$ is its angular acceleration, and $\mathbf{r}'_i$ denotes the position vector of the particle P_i, relative to G. Further show, by computing their sum and the sum of their moments about G, that the effective forces reduce to a vector $m\bar{\mathbf{a}}$ attached at G and a couple $\bar{I}\boldsymbol{\alpha}$.

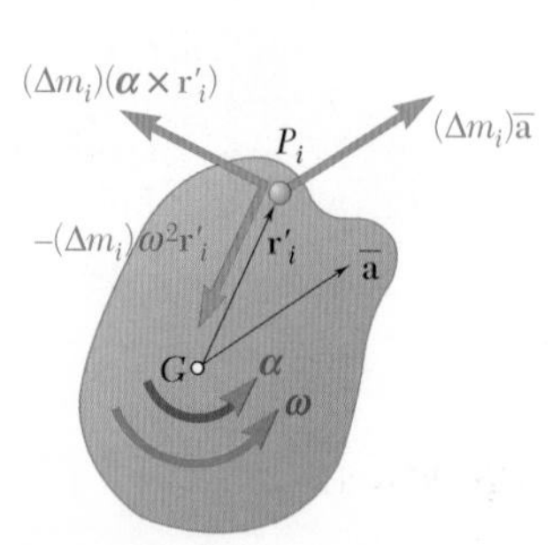

Fig. P16.47

16.48 A uniform slender rod AB rests on a frictionless horizontal surface, and a force **P** of magnitude 1 N is applied at A in a direction perpendicular to the rod. Knowing that the rod weighs 9 N, determine (a) the acceleration of point A, (b) acceleration of point B.

16.49 (a) In Prob. 16.48, determine the point of the rod AB at which the force **P** should be applied if the acceleration of point B is to be zero. (b) Knowing that $P = 1$ N, determine the corresponding acceleration of point A.

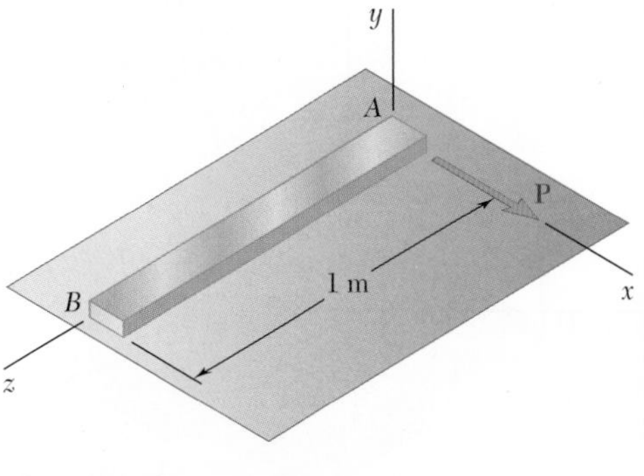

Fig. P16.48

16.50 A force **P** of magnitude 3 N is applied to a tape wrapped around a thin hoop of mass 2.4 kg. Knowing that the body rests on a frictionless horizontal surface, determine the acceleration of (a) point A, (b) point B.

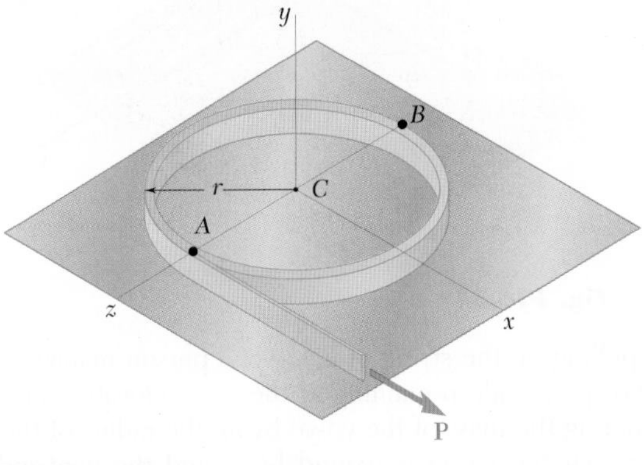

Fig. P16.50

16.51 A force **P** is applied to a tape wrapped around a uniform disk that rests on a frictionless horizontal surface. Show that for each 360° rotation of the disk the center of the disk will move a distance πr.

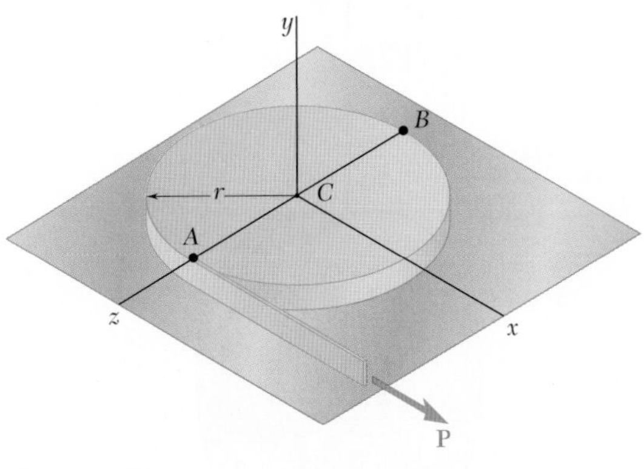

Fig. P16.51

16.52 A 120-kg satellite has a radius of gyration of 600 mm with respect to the y axis and is symmetrical with respect to the zx plane. Its orientation is changed by firing four small rockets A, B, C, and D, each of which produces a 16.20-N thrust **T** directed as shown. Determine the angular acceleration of the satellite and the acceleration of its mass center G (a) when all four rockets are fired, (b) when all rockets except D are fired.

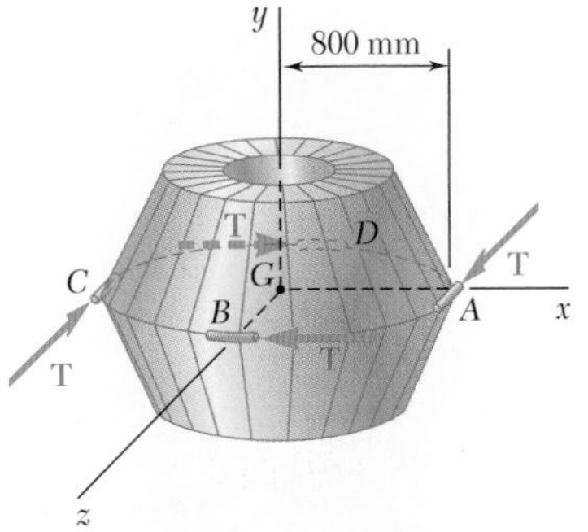

Fig. P16.52

16.53 A rectangular plate of mass 5 kg is suspended from four vertical wires, and a force **P** of magnitude 6 N is applied to corner C as shown. Immediately after **P** is applied, determine the acceleration of (*a*) the midpoint of edge BC, (*b*) corner B.

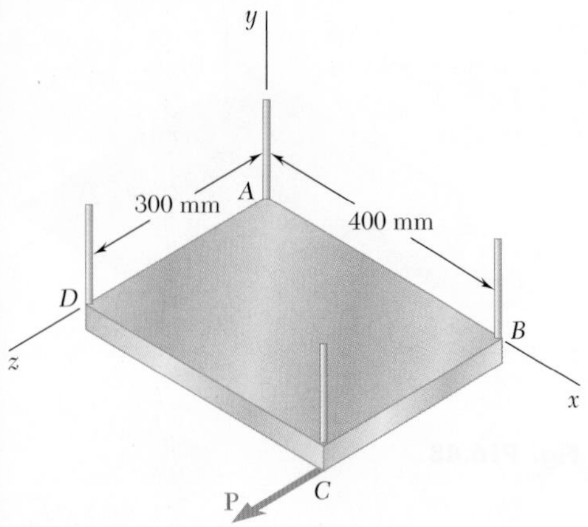

Fig. P16.53

16.54 A uniform slender L-shaped bar ABC is at rest on a horizontal surface when a force **P** of magnitude 4 N is applied at point A. Neglecting friction between the bar and the surface and knowing that the mass of the bar is 2 kg, determine (*a*) the initial angular acceleration of the bar, (*b*) the initial acceleration of point B.

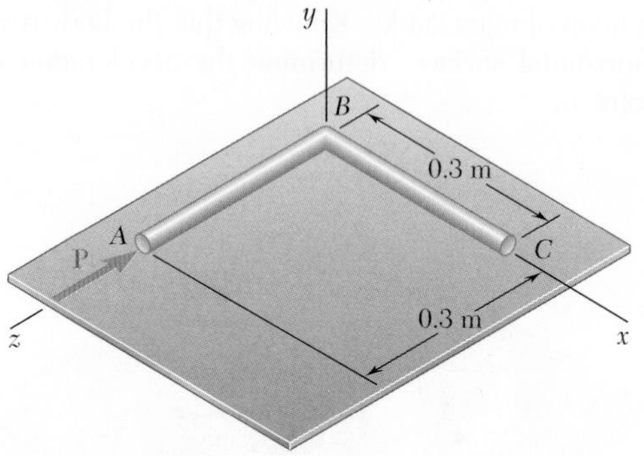

Fig. P16.54

16.55 By pulling on the string of a yo-yo, a person manages to make the yo-yo spin, while remaining at the same elevation above the floor. Denoting the mass of the yo-yo by m, the radius of the inner drum on which the string is wound by r, and the centroidal radius of gyration of the yo-yo by $\bar{k}$, determine the angular acceleration of the yo-yo.

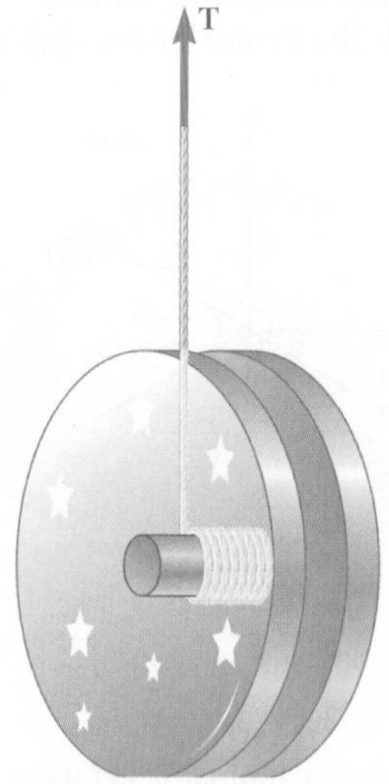

Fig. P16.55 and P16.56

16.56 The 80-g yo-yo shown has a centroidal radius of gyration of 30 mm. The radius of the inner drum on which a string is wound is 6 mm. Knowing that at the instant shown the acceleration of the center of the yo-yo is 1 m/s^2 upward, determine (*a*) the required tension **T** in the string, (*b*) the corresponding angular acceleration of the yo-yo.

16.57 A 3-kg sprocket wheel has a centroidal radius of gyration of 70 mm and is suspended from a chain as shown. Determine the acceleration of points A and B of the chain, knowing that $T_A = 14$ N and $T_B = 18$ N.

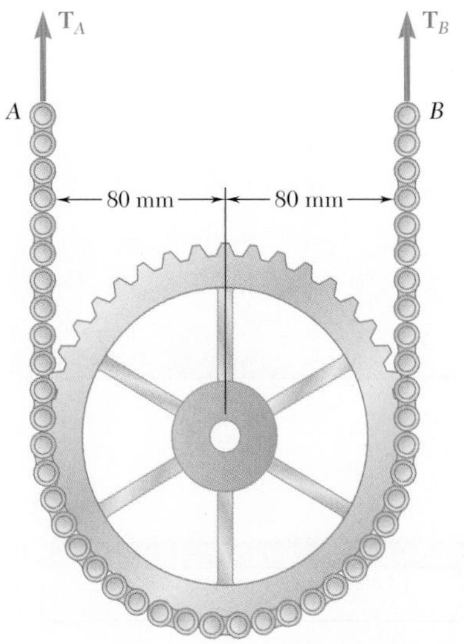

Fig. P16.57

16.58 The steel roll shown has a mass of 1200 kg, a centroidal radius of gyration of 150 mm, and is lifted by two cables looped around its shaft. Knowing that for each cable $T_A = 3100$ N and $T_B = 3300$ N, determine (*a*) the angular acceleration of the roll, (*b*) the acceleration of its mass center.

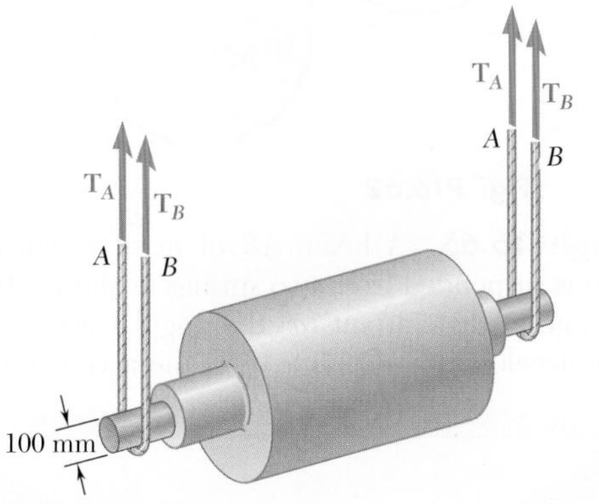

Fig. P16.58 and P16.59

16.59 The steel roll shown has a mass of 1200 kg, has a centroidal radius of gyration of 150 mm, and is lifted by two cables looped around its shaft. Knowing that at the instant shown the acceleration of the roll is 150 mm/s^2 downward and that for each cable $T_A = 3000$ N, determine (*a*) the corresponding tension T_B, (*b*) the angular acceleration of the roll.

16.60 and *16.61* A 5 m beam weighing 2500 N is lowered by means of two cables unwinding from overhead cranes. As the beam approaches the ground, the crane operators apply brakes to slow the unwinding motion. Knowing that the deceleration of cable A is 6 m/s^2 and the deceleration of cable B is 1 m/s^2, determine the tension in each cable.

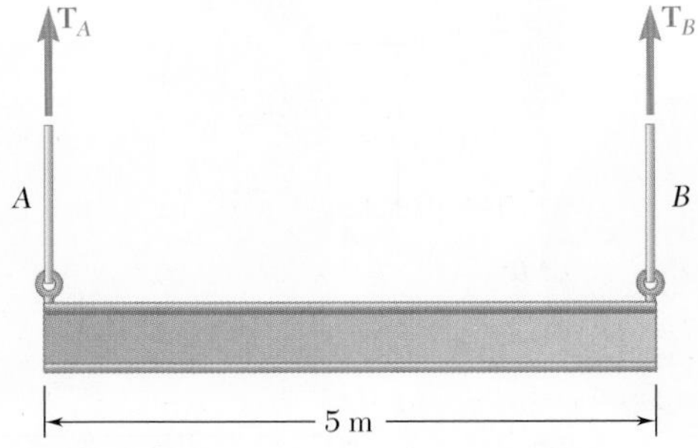

Fig. P16.60

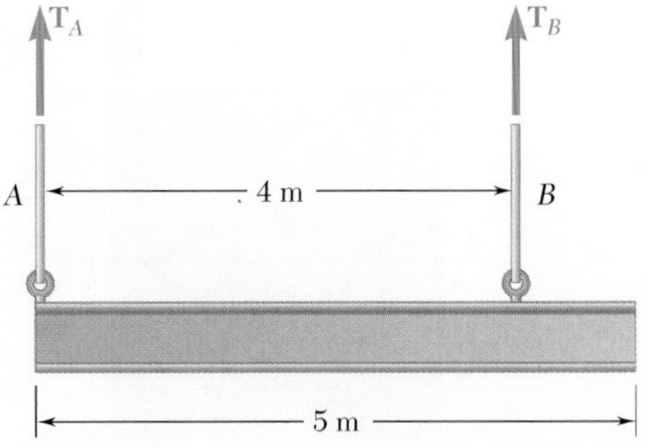

Fig. *P16.61*

16.62 Two uniform cylinders, each of mass 7 kg and radius $r = 125$ mm, are connected by a belt as shown. If the system is released from rest, determine (*a*) the angular acceleration of each cylinder, (*b*) the tension in the portion of belt connecting the two cylinders, (*c*) the velocity of the center of the cylinder *A* after it has moved through 1 m.

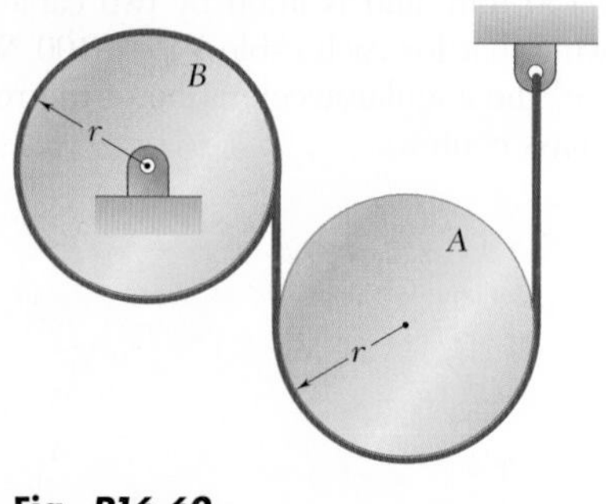

Fig. P16.62

16.63 through 16.65 A beam *AB* of mass *m* and of uniform cross section is suspended from two springs as shown. If spring 2 breaks, determine at that instant (*a*) the angular acceleration of the bar, (*b*) the acceleration of point A, (*c*) the acceleration of point *B*.

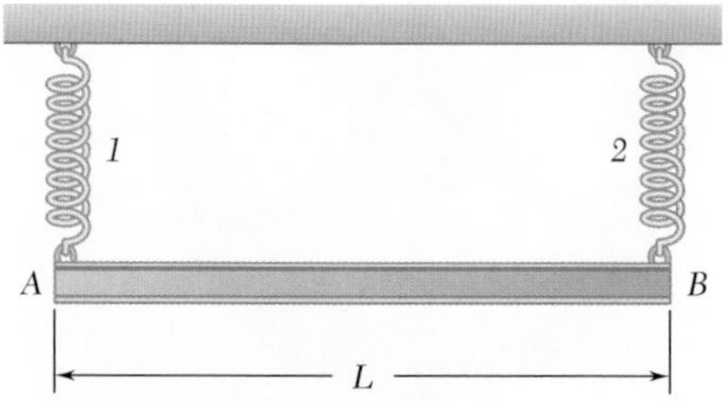

Fig. P16.63

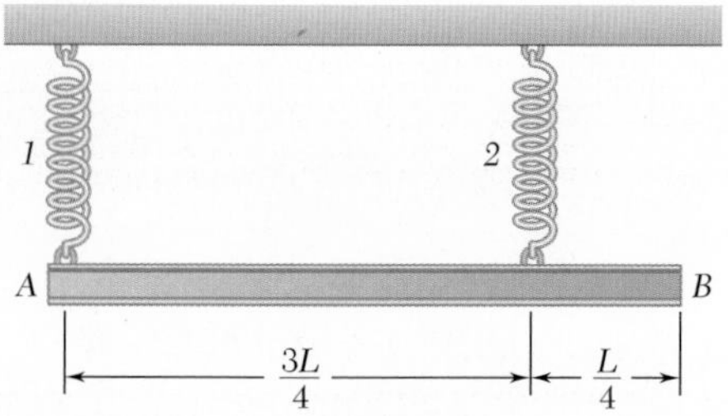

Fig. P16.64

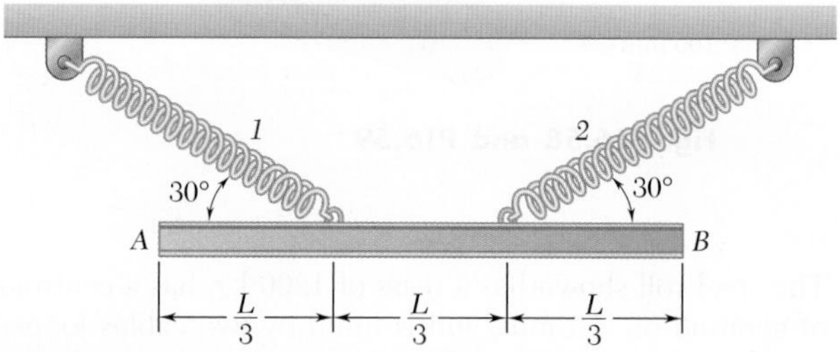

Fig. P16.65

16.66 through 16.68 A thin plate of the shape indicated and of mass *m* is suspended from two springs as shown. If spring 2 breaks, determine the acceleration at that instant (*a*) of point *A*, (*b*) of point *B*.

16.66 A square plate of side *b*
16.67 A circular plate of diameter *b*
16.68 A rectangular plate of height *b* and width *a*

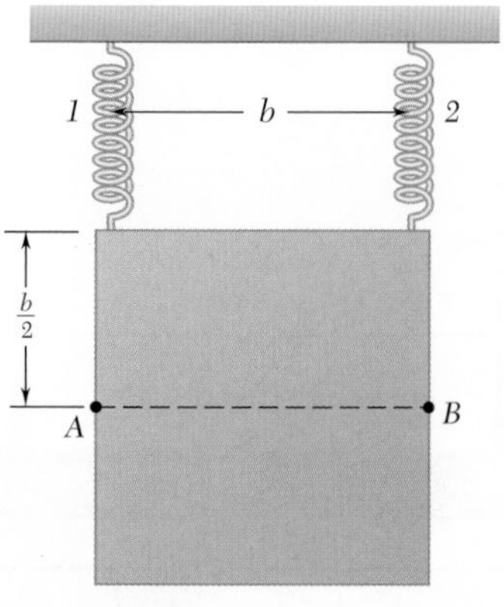

Fig. P16.66

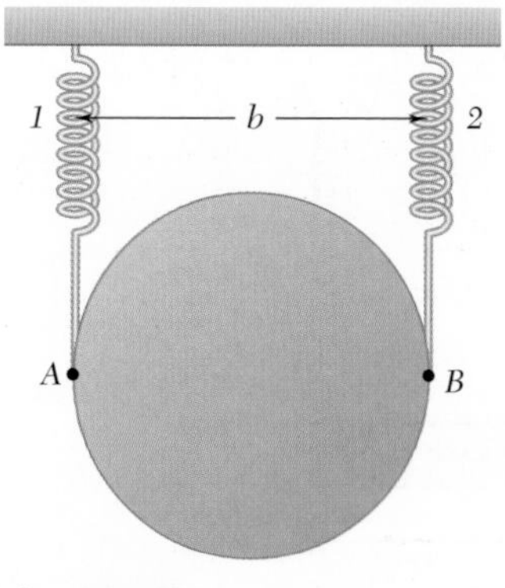

Fig. P16.67

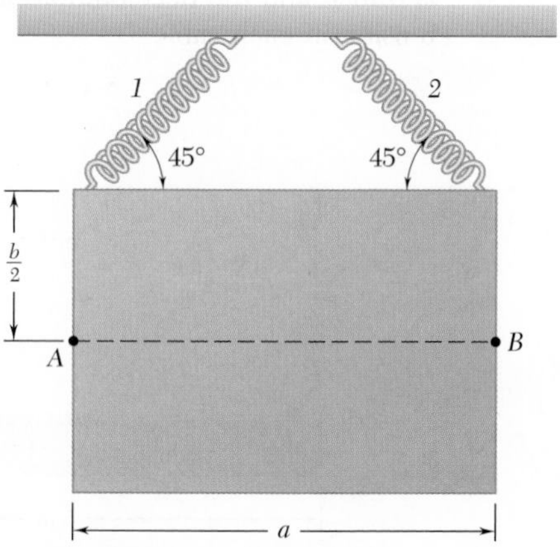

Fig. P16.68

16.69 A uniform sphere of mass m and radius r is projected along a rough horizontal surface with a linear velocity $\bar{\mathbf{v}}_0$ and no angular velocity. Denoting by μ_k the coefficient of kinetic friction between the sphere and the floor, determine (*a*) the required magnitude of $\boldsymbol{\omega}_0$, (*b*) the time t_1 required for the sphere to come to rest, (*c*) the distance the sphere will move before coming to rest.

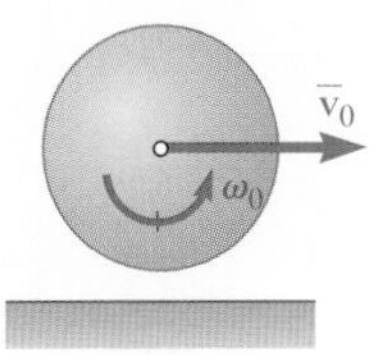

Fig. P16.69

16.70 Solve Sample Problem 16.5, assuming that the sphere is replaced by a uniform thin hoop of radius r and mass m.

16.71 A bowler projects a 200-mm-diameter ball of mass 6 kg along an alley with a forward velocity $\mathbf{v}_0$ of 5 m/s and a backspin $\boldsymbol{\omega}_0$ of 9 rad/s. Knowing that the coefficient of kinetic friction between the ball and the alley is 0.10, determine (*a*) the time t_1 at which the ball will start rolling without sliding, (*b*) the speed of the ball at time t_1, (*c*) the distance the ball will have traveled at time t_1.

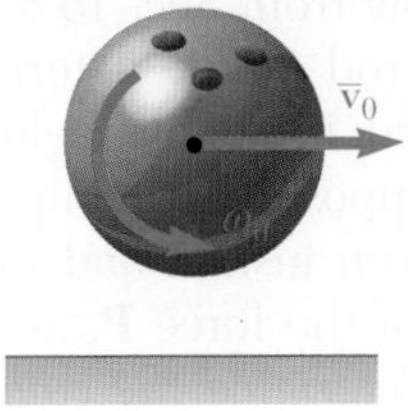

Fig. P16.71

16.72 Solve Prob. 16.71, assuming that the bowler projects the ball with the same forward velocity but with a backspin of 18 rad/s.

16.73 A uniform sphere of radius r and mass m is placed with no initial velocity on a belt that moves to the right with a constant velocity $\mathbf{v}_1$. Denoting by μ_k the coefficient of kinetic friction between the sphere and the belt, determine (*a*) the time t_1 at which the sphere will start rolling without sliding, (*b*) the linear and angular velocities of the sphere at time t_1.

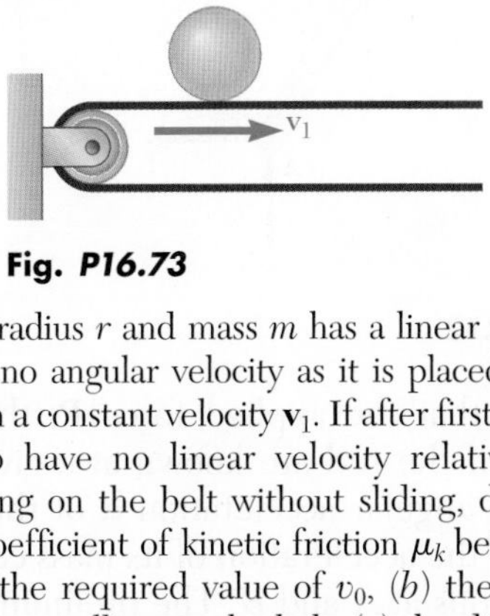

Fig. *P16.73*

16.74 A sphere of radius r and mass m has a linear velocity $\mathbf{v}_0$ directed to the left and no angular velocity as it is placed on a belt moving to the right with a constant velocity $\mathbf{v}_1$. If after first sliding on the belt the sphere is to have no linear velocity relative to the ground as it starts rolling on the belt without sliding, determine in terms of v_1 and the coefficient of kinetic friction μ_k between the sphere and the belt (*a*) the required value of v_0, (*b*) the time t_1 at which the sphere will start rolling on the belt, (*c*) the distance the sphere will have moved relative to the ground at time t_1.

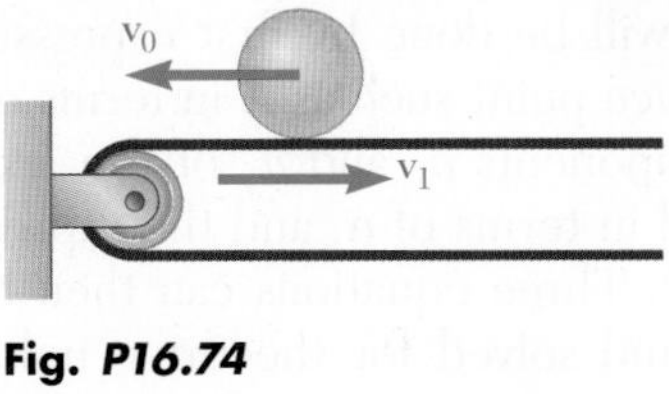

Fig. *P16.74*

16.8 CONSTRAINED PLANE MOTION

Most engineering applications deal with rigid bodies which are moving under given constraints. For example, cranks must rotate about a fixed axis, wheels must roll without sliding, and connecting rods must describe certain prescribed motions. In all such cases, definite relations exist between the components of the acceleration $\mathbf{\bar{a}}$ of the mass center G of the body considered and its angular acceleration $\boldsymbol{\alpha}$; the corresponding motion is said to be a *constrained motion*.

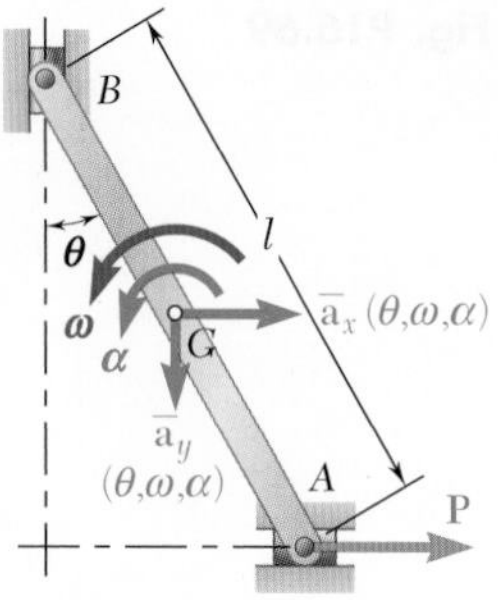

Fig. 16.11

The solution of a problem involving a constrained plane motion calls first for a *kinematic analysis* of the problem. Consider, for example, a slender rod AB of length l and mass m whose extremities are connected to blocks of negligible mass which slide along horizontal and vertical frictionless tracks. The rod is pulled by a force $\mathbf{P}$ applied at A (Fig. 16.11). We know from Sec. 15.8 that the acceleration $\mathbf{\bar{a}}$ of the mass center G of the rod can be determined at any given instant from the position of the rod, its angular velocity, and its angular acceleration at that instant. Suppose, for example, that the values of θ, ω, and α are known at a given instant and that we wish to determine the corresponding value of the force $\mathbf{P}$, as well as the reactions at A and B. We should first *determine the components $\bar{a}_x$ and $\bar{a}_y$ of the acceleration of the mass center G* by the method of Sec. 15.8. We next apply d'Alembert's principle (Fig. 16.12), using the expressions obtained for $\bar{a}_x$ and $\bar{a}_y$. The unknown forces $\mathbf{P}$, $\mathbf{N}_A$, and $\mathbf{N}_B$ can then be determined by writing and solving the appropriate equations.

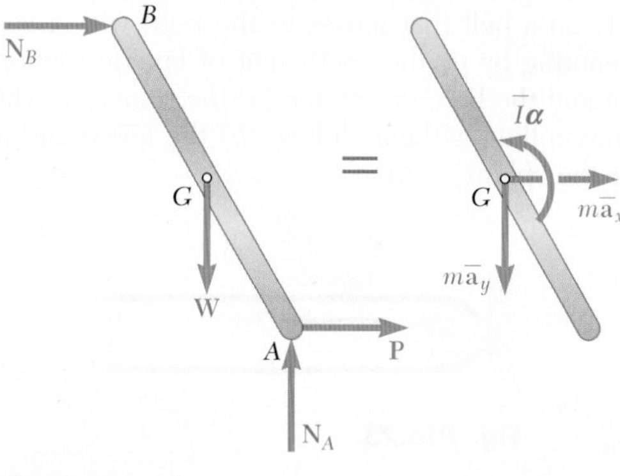

Fig. 16.12

Suppose now that the applied force $\mathbf{P}$, the angle θ, and the angular velocity ω of the rod are known at a given instant and that we wish to find the angular acceleration α of the rod and the components $\bar{a}_x$ and $\bar{a}_y$ of the acceleration of its mass center at that instant, as well as the reactions at A and B. The preliminary kinematic study of the problem will have for its object *to express the components $\bar{a}_x$ and $\bar{a}_y$ of the acceleration of G in terms of the angular acceleration α of the rod*. This will be done by first expressing the acceleration of a suitable reference point such as A in terms of the angular acceleration α. The components $\bar{a}_x$ and $\bar{a}_y$ of the acceleration of G can then be determined in terms of α, and the expressions obtained carried into Fig. 16.12. Three equations can then be derived in terms of α, N_A, and N_B and solved for the three unknowns (see Sample

Prob. 16.10). Note that the method of dynamic equilibrium can also be used to carry out the solution of the two types of problems we have considered (Fig. 16.13).

When a mechanism consists of *several moving parts*, the approach just described can be used with each part of the mechanism. The procedure required to determine the various unknowns is then similar to the procedure followed in the case of the equilibrium of a system of connected rigid bodies (Sec. 6.11).

Earlier, we analyzed two particular cases of constrained plane motion: the translation of a rigid body, in which the angular acceleration of the body is constrained to be zero, and the centroidal rotation, in which the acceleration $\bar{\mathbf{a}}$ of the mass center of the body is constrained to be zero. Two other particular cases of constrained plane motion are of special interest: the *noncentroidal rotation* of a rigid body and the *rolling motion* of a disk or wheel. These two cases can be analyzed by one of the general methods described above. However, in view of the range of their applications, they deserve a few special comments.

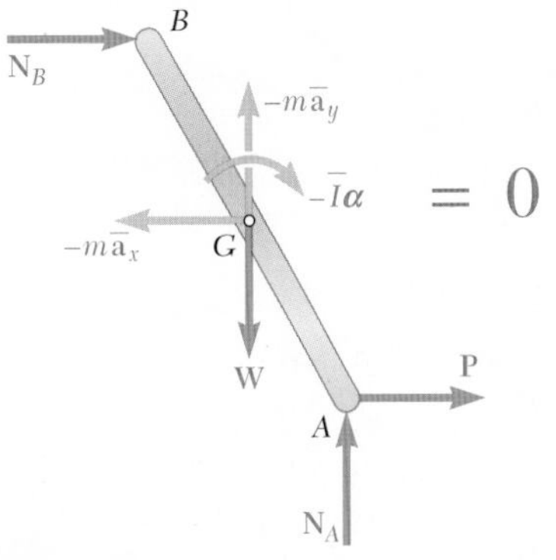

Fig. 16.13

Noncentroidal Rotation. The motion of a rigid body constrained to rotate about a fixed axis which does not pass through its mass center is called *noncentroidal rotation*. The mass center G of the body moves along a circle of radius $\bar{r}$ centered at the point O, where the axis of rotation intersects the plane of reference (Fig. 16.14). Denoting, respectively, by $\boldsymbol{\omega}$ and $\boldsymbol{\alpha}$ the angular velocity and the angular acceleration of the line OG, we obtain the following expressions for the tangential and normal components of the acceleration of G:

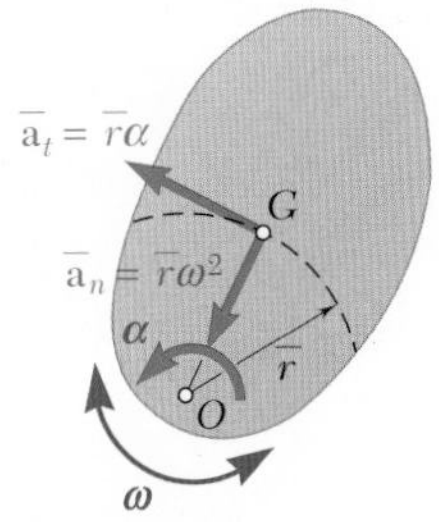

Fig. 16.14

$$\bar{a}_t = \bar{r}\alpha \qquad \bar{a}_n = \bar{r}\omega^2 \tag{16.7}$$

Since line OG belongs to the body, its angular velocity $\boldsymbol{\omega}$ and its angular acceleration $\boldsymbol{\alpha}$ also represent the angular velocity and the angular acceleration of the body in its motion relative to G. Equations (16.7) define, therefore, the kinematic relation existing between the motion of the mass center G and the motion of the body about G. They should be used to eliminate $\bar{a}_t$ and $\bar{a}_n$ from the equations obtained by applying d'Alembert's principle (Fig. 16.15) or the method of dynamic equilibrium (Fig. 16.16).

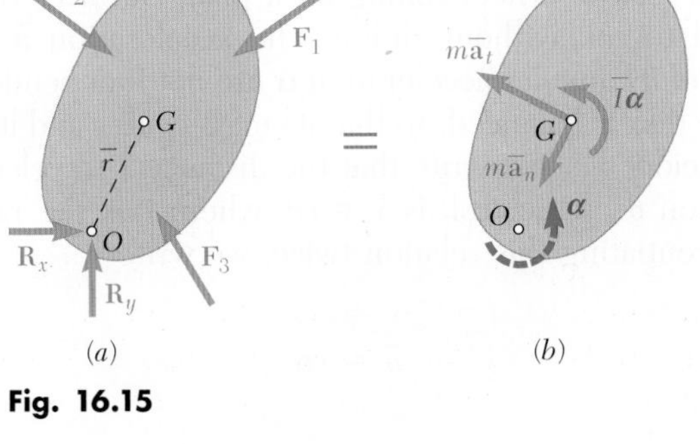

Fig. 16.15

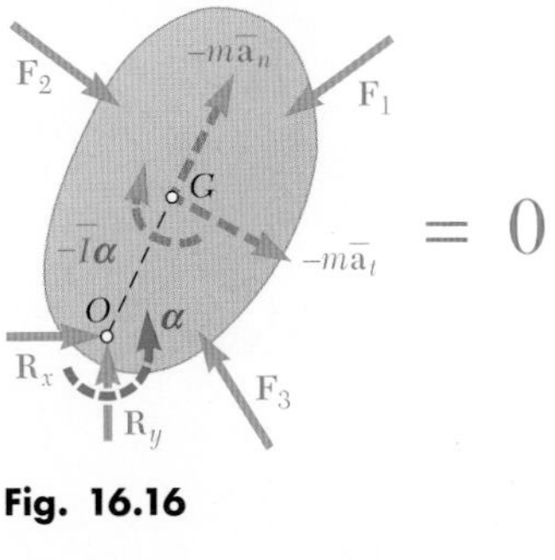

Fig. 16.16

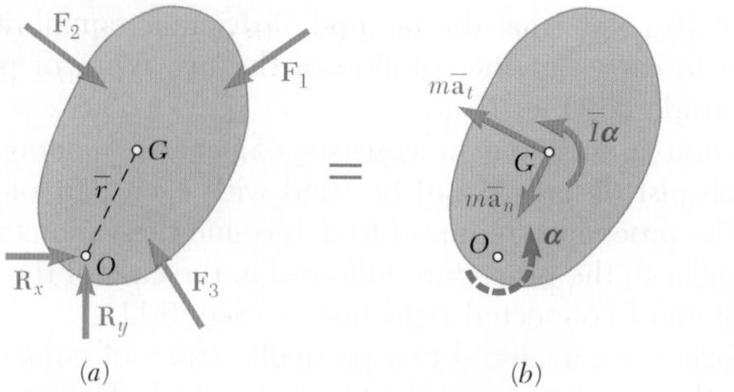

Fig. 16.15 (repeated)

An interesting relation is obtained by equating the moments about the fixed point O of the forces and vectors shown, respectively, in parts a and b of Fig. 16.15. We write

$$+\circlearrowleft \Sigma M_O = \bar{I}\alpha + (m\bar{r}\alpha)\bar{r} = (\bar{I} + m\bar{r}^2)\alpha$$

But according to the parallel-axis theorem, we have $\bar{I} + m\bar{r}^2 = I_O$, where I_O denotes the moment of inertia of the rigid body about the fixed axis. We therefore write

$$\Sigma M_O = I_O\alpha \tag{16.8}$$

Although formula (16.8) expresses an important relation between the sum of the moments of the external forces about the fixed point O and the product $I_O\alpha$, it should be clearly understood that this formula does not mean that the system of the external forces is equivalent to a couple of moment $I_O\alpha$. The system of the effective forces, and thus the system of the external forces, reduces to a couple only when O coincides with G—that is, *only when the rotation is centroidal* (Sec. 16.4). In the more general case of noncentroidal rotation, the system of the external forces does not reduce to a couple.

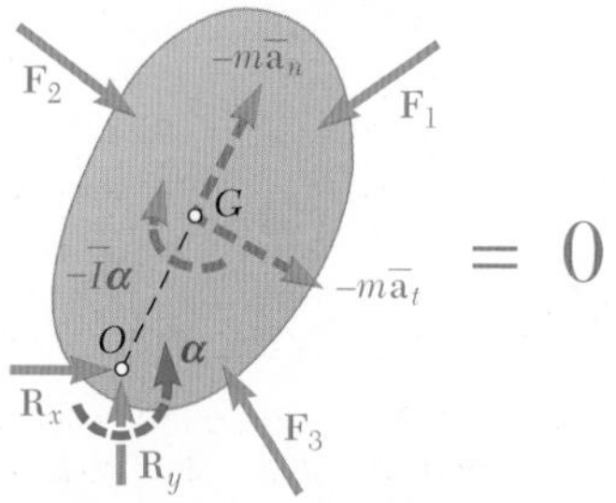

Fig. 16.16 (repeated)

A particular case of noncentroidal rotation is of special interest—the case of *uniform rotation,* in which the angular velocity $\boldsymbol{\omega}$ is constant. Since $\boldsymbol{\alpha}$ is zero, the inertia couple in Fig. 16.16 vanishes and the inertia vector reduces to its normal component. This component (also called *centrifugal force*) represents the tendency of the rigid body to break away from the axis of rotation.

Rolling Motion. Another important case of plane motion is the motion of a disk or wheel rolling on a plane surface. If the disk is constrained to roll without sliding, the acceleration $\bar{\mathbf{a}}$ of its mass center G and its angular acceleration $\boldsymbol{\alpha}$ are not independent. Assuming that the disk is balanced, so that its mass center and its geometric center coincide, we first write that the distance $\bar{x}$ traveled by G during a rotation θ of the disk is $\bar{x} = r\theta$, where r is the radius of the disk. Differentiating this relation twice, we write

$$\bar{a} = r\alpha \tag{16.9}$$

Recalling that the system of the effective forces in plane motion reduces to a vector $m\bar{\mathbf{a}}$ and a couple $\bar{I}\boldsymbol{\alpha}$, we find that in the particular case of the rolling motion of a balanced disk, the effective forces reduce to a vector of magnitude $mr\alpha$ attached at G and to a couple of magnitude $\bar{I}\alpha$. We may thus express that the external forces are equivalent to the vector and couple shown in Fig. 16.17.

When a disk *rolls without sliding,* there is no relative motion between the point of the disk in contact with the ground and the ground itself. Thus, as far as the computation of the friction force **F** is concerned, a rolling disk can be compared with a block at rest on a surface. The magnitude F of the friction force can have any value, as long as this value does not exceed the maximum value $F_m = \mu_s N$, where μ_s is the coefficient of static friction and N is the magnitude of the normal force. In the case of a rolling disk, the magnitude F of the friction force should therefore be determined independently of N by solving the equation obtained from Fig. 16.17.

When *sliding is impending,* the friction force reaches its maximum value $F_m = \mu_s N$ and can be obtained from N.

When the disk *rotates and slides* at the same time, a relative motion exists between the point of the disk which is in contact with the ground and the ground itself, and the force of friction has the magnitude $F_k = \mu_k N$, where μ_k is the coefficient of kinetic friction. In this case, however, the motion of the mass center G of the disk and the rotation of the disk about G are independent, and $\bar{a}$ is not equal to $r\alpha$.

These three different cases can be summarized as follows:

Rolling, no sliding:	$F \leq \mu_s N$	$\bar{a} = r\alpha$
Rolling, sliding impending:	$F = \mu_s N$	$\bar{a} = r\alpha$
Rotating and sliding:	$F = \mu_k N$	$\bar{a}$ *and* α *independent*

When it is not known whether or not a disk slides, it should first be assumed that the disk rolls without sliding. If F is found smaller than or equal to $\mu_s N$, the assumption is proved correct. If F is found larger than $\mu_s N$, the assumption is incorrect and the problem should be started again, assuming rotating and sliding.

When a disk is *unbalanced,* i.e., when its mass center G does not coincide with its geometric center O, the relation (16.9) does not hold between $\bar{a}$ and α. However, a similar relation holds between the magnitude a_O of the acceleration of the geometric center and the angular acceleration α of an unbalanced disk which rolls without sliding. We have

$$a_O = r\alpha \tag{16.10}$$

To determine $\bar{a}$ in terms of the angular acceleration α and the angular velocity ω of the disk, we can use the relative-acceleration formula

$$\begin{aligned}\bar{\mathbf{a}} = \bar{\mathbf{a}}_G &= \mathbf{a}_O + \mathbf{a}_{G/O} \\ &= \mathbf{a}_O + (\mathbf{a}_{G/O})_t + (\mathbf{a}_{G/O})_n\end{aligned} \tag{16.11}$$

where the three component accelerations obtained have the directions indicated in Fig. 16.18 and the magnitudes $a_O = r\alpha$, $(a_{G/O})_t = (OG)\alpha$, and $(a_{G/O})_n = (OG)\omega^2$.

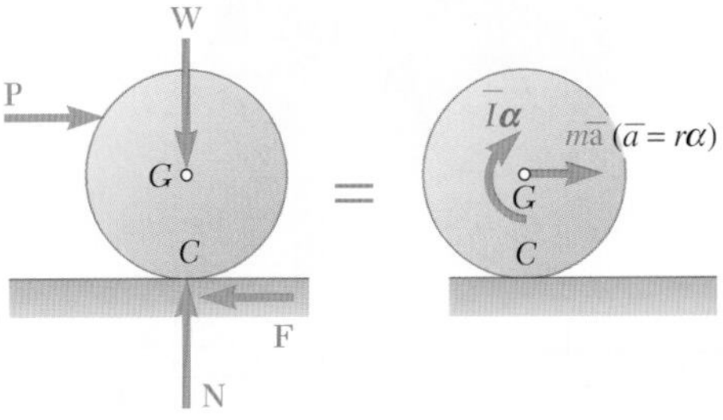

Fig. 16.17

Photo 16.4 As the ball hits the bowling alley, it first spins and slides, then rolls without sliding.

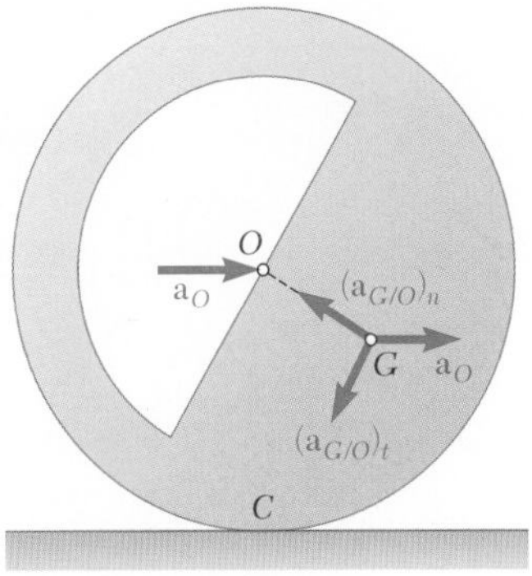

Fig. 16.18

SAMPLE PROBLEM 16.6

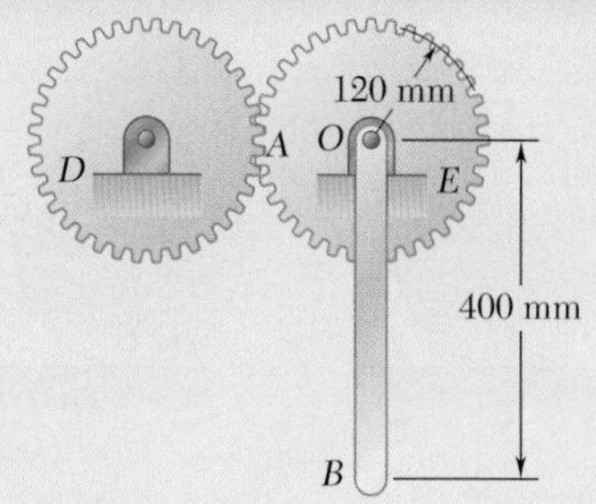

The portion *AOB* of a mechanism consists of a 400-mm steel rod *OB* welded to a gear *E* of radius 120 mm which can rotate about a horizontal shaft *O*. It is actuated by a gear *D* and, at the instant shown, has a clockwise angular velocity of 8 rad/s and a counterclockwise angular acceleration of 40 rad/s². Knowing that rod *OB* has a mass of 3 kg and gear *E* a mass of 4 kg and a radius of gyration of 85 mm, determine (*a*) the tangential force exerted by gear *D* on gear *E*, (*b*) the components of the reaction at shaft *O*.

SOLUTION

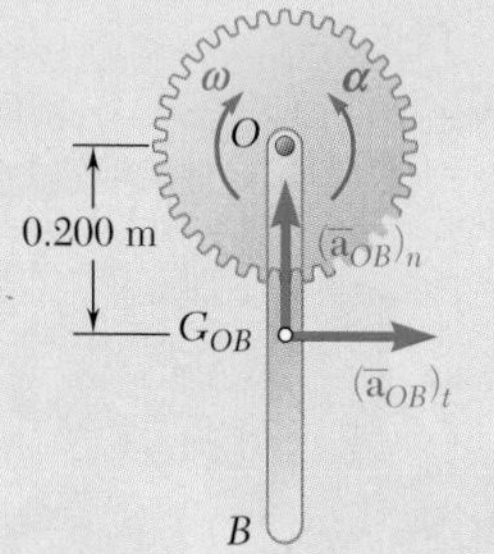

In determining the effective forces of the rigid body *AOB*, gear *E* and rod *OB* will be considered separately. Therefore, the components of the acceleration of the mass center G_{OB} of the rod will be determined first:

$$(\bar{a}_{OB})_t = \bar{r}\alpha = (0.200 \text{ m})(40 \text{ rad/s}^2) = 8 \text{ m/s}^2$$
$$(\bar{a}_{OB})_n = \bar{r}\omega^2 = (0.200 \text{ m})(8 \text{ rad/s})^2 = 12.8 \text{ m/s}^2$$

Equations of Motion. Two sketches of the rigid body *AOB* have been drawn. The first shows the external forces consisting of the weight $\mathbf{W}_E$ of gear *E*, the weight $\mathbf{W}_{OB}$ of the rod *OB*, the force **F** exerted by gear *D*, and the components $\mathbf{R}_x$ and $\mathbf{R}_y$ of the reaction at *O*. The magnitudes of the weights are, respectively,

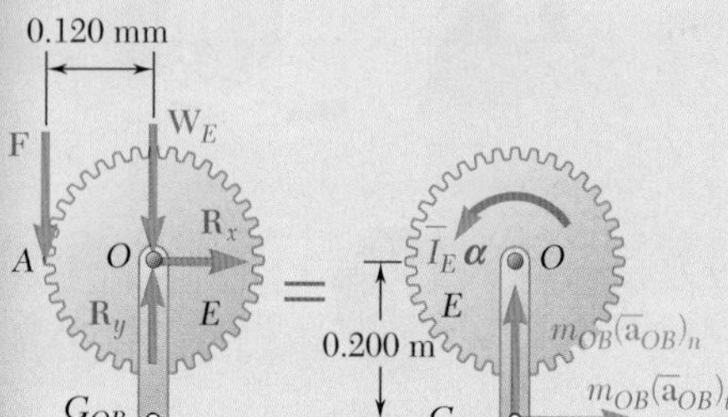

$$W_E = m_E g = (4 \text{ kg})(9.81 \text{ m/s}^2) = 39.2 \text{ N}$$
$$W_{OB} = m_{OB} g = (3 \text{ kg})(9.81 \text{ m/s}^2) = 29.4 \text{ N}$$

The second sketch shows the effective forces, which consist of a couple $\bar{I}_E\boldsymbol{\alpha}$ (since gear *E* is in centroidal rotation) and of a couple and two vector components at the mass center of *OB*. Since the accelerations are known, we compute the magnitudes of these components and couples:

$$\bar{I}_E\alpha = m_E\bar{k}_E^2\alpha = (4 \text{ kg})(0.085 \text{ m})^2(40 \text{ rad/s}^2) = 1.156 \text{ N}\cdot\text{m}$$
$$m_{OB}(\bar{a}_{OB})_t = (3 \text{ kg})(8 \text{ m/s}^2) = 24.0 \text{ N}$$
$$m_{OB}(\bar{a}_{OB})_n = (3 \text{ kg})(12.8 \text{ m/s}^2) = 38.4 \text{ N}$$
$$\bar{I}_{OB}\alpha = (\tfrac{1}{12}m_{OB}L^2)\alpha = \tfrac{1}{12}(3 \text{ kg})(0.400 \text{ m})^2(40 \text{ rad/s}^2) = 1.600 \text{ N}\cdot\text{m}$$

Expressing that the system of the external forces is equivalent to the system of the effective forces, we write the following equations:

$+\circlearrowleft \Sigma M_O = \Sigma(M_O)_{\text{eff}}$:

$$F(0.120 \text{ m}) = \bar{I}_E\alpha + m_{OB}(\bar{a}_{OB})_t(0.200 \text{ m}) + \bar{I}_{OB}\alpha$$
$$F(0.120 \text{ m}) = 1.156 \text{ N}\cdot\text{m} + (24.0 \text{ N})(0.200 \text{ m}) + 1.600 \text{ N}\cdot\text{m}$$

$F = 63.0$ N $\qquad \mathbf{F} = 63.0 \text{ N}\downarrow$ ◀

$\overset{+}{\rightarrow}\Sigma F_x = \Sigma(F_x)_{\text{eff}}$: $\qquad R_x = m_{OB}(\bar{a}_{OB})_t$

$R_x = 24.0$ N $\qquad \mathbf{R}_x = 24.0 \text{ N}\rightarrow$ ◀

$+\uparrow\Sigma F_y = \Sigma(F_y)_{\text{eff}}$: $\qquad R_y - F - W_E - W_{OB} = m_{OB}(\bar{a}_{OB})_n$

$$R_y - 63.0 \text{ N} - 39.2 \text{ N} - 29.4 \text{ N} = 38.4 \text{ N}$$

$R_y = 170.0$ N $\qquad \mathbf{R}_y = 170.0 \text{ N}\uparrow$ ◀

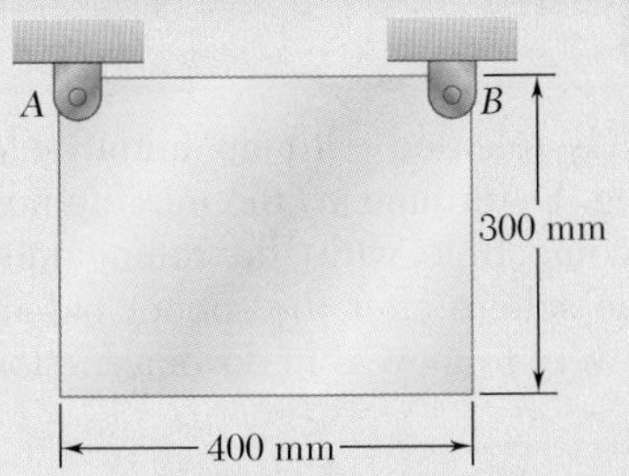

SAMPLE PROBLEM 16.7

A 300 × 400 mm rectangular plate of mass 30 kg is suspended from two pins *A* and *B*. If pin *B* is suddenly removed, determine (*a*) the angular acceleration of the plate, (*b*) the components of the reaction at pin A, immediately after pin *B* has been removed.

SOLUTION

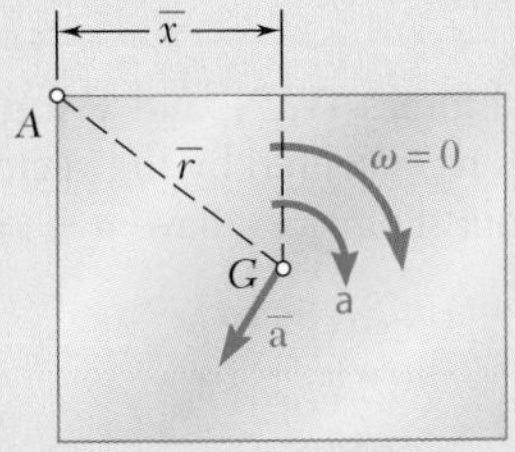

a. Angular Acceleration. We observe that as the plate rotates about point A, its mass center G describes a circle of radius $\bar{r}$ with center at A.

Since the plate is released from rest ($\omega = 0$), the normal component of the acceleration of G is zero. The magnitude of the acceleration $\bar{\mathbf{a}}$ of the mass center G is thus $\bar{a} = \bar{r}\alpha$. We draw the diagram shown to express that the external forces are equivalent to the effective forces:

$$+\circlearrowright \Sigma M_A = \Sigma(M_A)_{\text{eff}}: \qquad W\bar{x} = (m\bar{a})\bar{r} + \bar{I}\alpha$$

Since $\bar{a} = \bar{r}\alpha$, we have

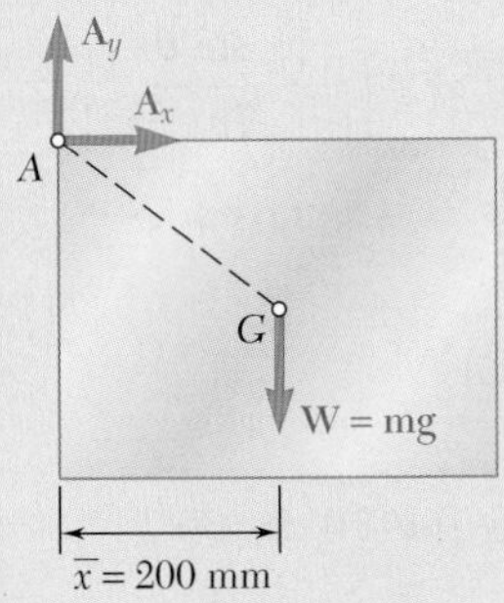

$$W\bar{x} = m(\bar{r}\alpha)\bar{r} + \bar{I}\alpha \qquad \alpha = \frac{mg\bar{x}}{m\bar{r}^2 + \bar{I}} \tag{1}$$

The centroidal moment of inertia of the plate is

$$\bar{I} = \frac{m}{12}(a^2 + b^2) = \frac{(30\text{ kg})\,[(0.4\text{ m})^2 + (0.3\text{ m})^2]}{12}$$
$$= 0.625\text{ kg}\cdot\text{m}^2$$

Substituting this value of $\bar{I}$ together with W = mg = 294.3 N, $\bar{r}$ = 0.25 m, $\bar{x}$ = 0.2 m into Eq. (1), we obtain

$$\alpha = +\,23.54\text{ rad/s}^2 \qquad \boldsymbol{\alpha} = 23.5\text{ rad/s}^2 \circlearrowright \blacktriangleleft$$

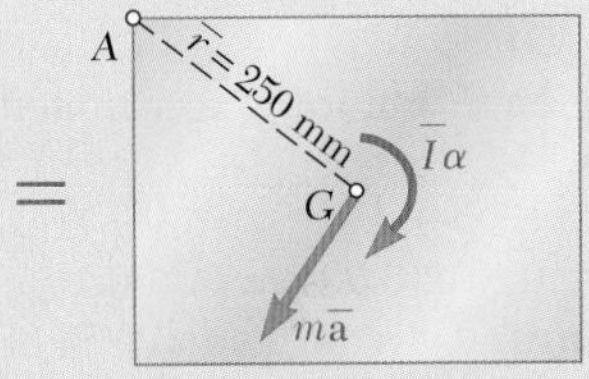

b. Reaction at *A*. Using the computed value of α, we determine the magnitude of the vector $m\bar{\mathbf{a}}$ attached at G.

$$m\bar{a} = m\bar{r}\alpha = (30\text{ kg})(0.25\text{ m})(23.54\text{ rad/s}^2) = 176.6\text{ N}$$

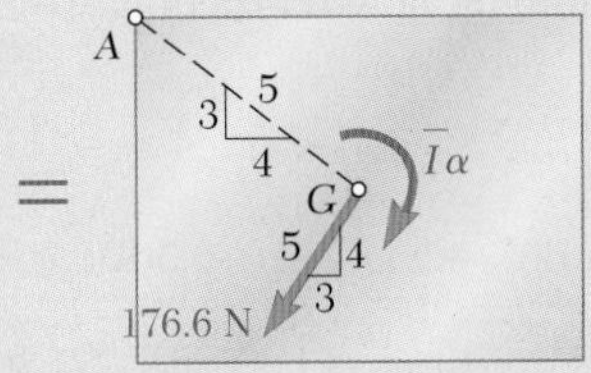

Showing this result on the diagram, we write the equations of motion

$$\xrightarrow{+} \Sigma F_x = \Sigma(F_x)_{\text{eff}}: \qquad A_x = -\tfrac{3}{5}(176.6)$$
$$= -106\text{ N} \qquad \mathbf{A}_x = 106\text{ N} \leftarrow \blacktriangleleft$$

$$+\uparrow \Sigma F_y = \Sigma(F_y)_{\text{eff}}: \qquad A_y - 294.3\text{ N} = -\tfrac{4}{5}(176.6)$$
$$A_y = +153.0\text{ N} \qquad \mathbf{A}_y = 153\text{ N}\uparrow \blacktriangleleft$$

The couple $\bar{I}\boldsymbol{\alpha}$ is not involved in the last two equations; nevertheless, it should be indicated on the diagram.

SAMPLE PROBLEM 16.8

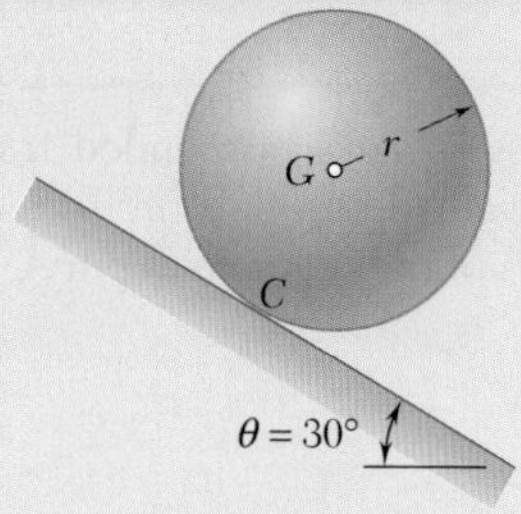

A sphere of radius r and weight W is released with no initial velocity on the incline and rolls without slipping. Determine (*a*) the minimum value of the coefficient of static friction compatible with the rolling motion, (*b*) the velocity of the center G of the sphere after the sphere has rolled 3 m (*c*) the velocity of G if the sphere were to move 3 m down a frictionless 30° incline.

SOLUTION

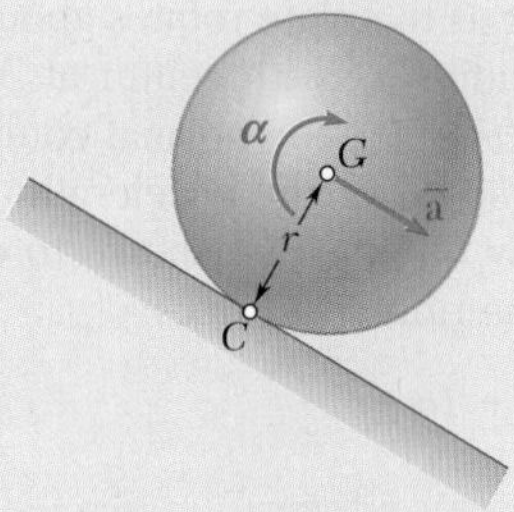

a. Minimum μ_s for Rolling Motion. The external forces **W**, **N**, and **F** form a system equivalent to the system of effective forces represented by the vector $m\bar{\mathbf{a}}$ and the couple $\bar{I}\boldsymbol{\alpha}$. Since the sphere rolls without sliding, we have $\bar{a} = r\alpha$

$$+\circlearrowright \Sigma M_C = \Sigma(M_C)_{\text{eff}}: \qquad (W \sin\theta)r = (m\bar{a})r + \bar{I}\alpha$$
$$(W \sin\theta)r = (mr\alpha)r + \bar{I}\alpha$$

Noting that $m = W/g$ and $\bar{I} = \frac{2}{5}mr^2$, we write

$$(W \sin\theta)r = \left(\frac{W}{g}r\alpha\right)r + \frac{2}{5}\frac{W}{g}r^2\alpha \qquad \alpha = +\frac{5g \sin\theta}{7r}$$

$$\bar{a} = r\alpha = \frac{5g \sin\theta}{7} = \frac{5(9.81 \text{ m/s}^2)\sin 30^\circ}{7} = 3.504 \text{ m/s}^2$$

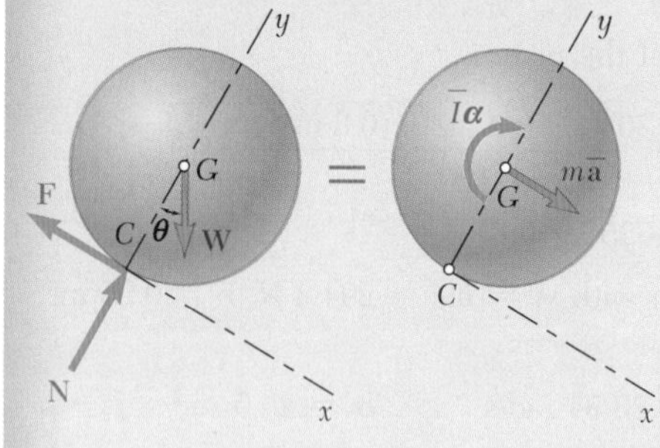

$$+\searrow \Sigma F_x = \Sigma(F_x)_{\text{eff}}: \qquad W \sin\theta - F = m\bar{a}$$
$$W \sin\theta - F = \frac{W}{g}\frac{5g \sin\theta}{7}$$
$$F = +\tfrac{2}{7}W \sin\theta = \tfrac{2}{7}W \sin 30^\circ \qquad \mathbf{F} = 0.143W \measuredangle 30^\circ$$
$$+\nearrow \Sigma F_y = \Sigma(F_y)_{\text{eff}}: \qquad N - W\cos\theta = 0$$
$$N = W\cos\theta = 0.866W \qquad \mathbf{N} = 0.866W \measuredangle 60^\circ$$
$$\mu_s = \frac{F}{N} = \frac{0.143W}{0.866W} \qquad \mu_s = 0.165 \blacktriangleleft$$

b. Velocity of Rolling Sphere. We have uniformly accelerated motion:

$$\bar{v}_0 = 0 \qquad \bar{a} = 3.504 \text{ m/s}^2 \qquad \bar{x} = 3 \text{ m} \qquad \bar{x}_0 = 0$$
$$\bar{v}^2 = \bar{v}_0^2 + 2\bar{a}(\bar{x} - \bar{x}_0) \qquad \bar{v}^2 = 0 + 2(3.504 \text{ m/s}^2)(3\text{m})$$
$$\bar{v} = 4.59 \text{ m/s} \qquad \bar{\mathbf{v}} = 4.59 \text{ m/s} \measuredangle 30^\circ \blacktriangleleft$$

c. Velocity of Sliding Sphere. Assuming now no friction, we have $F = 0$ and obtain

$$+\circlearrowright \Sigma M_G = \Sigma(M_G)_{\text{eff}}: \qquad 0 = \bar{I}\alpha \qquad \alpha = 0$$
$$+\searrow \Sigma F_x = \Sigma(F_x)_{\text{eff}}: \qquad W \sin 30^\circ = m\bar{a} \qquad 0.50W = \frac{W}{g}\bar{a}$$
$$\bar{a} = +4.905 \text{m/s}^2 \qquad \bar{\mathbf{a}} = 4.905 \text{m/s}^2 \measuredangle 30^\circ$$

Substituting $\bar{a} = 4.905$ m/s² into the equations for uniformly accelerated motion, we obtain

$$\bar{v}^2 = \bar{v}_0^2 + 2\bar{a}(\bar{x} - \bar{x}_0) \qquad \bar{v}^2 = 0 + 2(4.905 \text{ m/s}^2)(3 \text{ m})$$
$$\bar{v} = 5.42 \text{ m/s} \qquad \bar{\mathbf{v}} = 5.42 \text{ m/s} \measuredangle 30^\circ \blacktriangleleft$$

SAMPLE PROBLEM 16.9

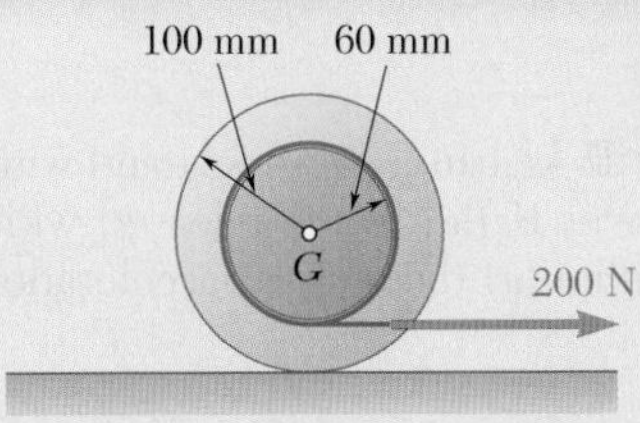

A cord is wrapped around the inner drum of a wheel and pulled horizontally with a force of 200 N. The wheel has a mass of 50 kg and a radius of gyration of 70 mm. Knowing that $\mu_s = 0.20$ and $\mu_k = 0.15$, determine the acceleration of G and the angular acceleration of the wheel.

SOLUTION

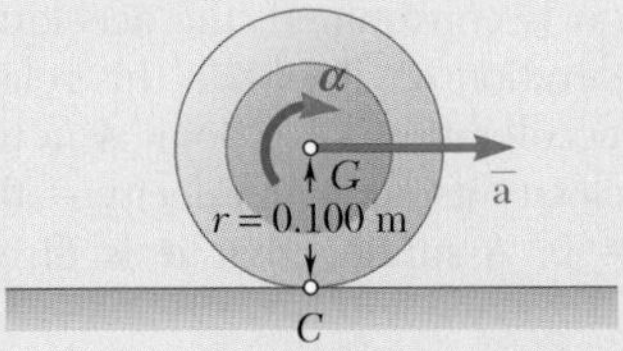

a. Assume Rolling without Sliding. In this case, we have

$$\bar{a} = r\alpha = (0.100 \text{ m})\alpha$$

We can determine whether this assumption is justified by comparing the friction force obtained with the maximum available friction force. The moment of inertia of the wheel is

$$\bar{I} = m\bar{k}^2 = (50 \text{ kg})(0.070 \text{ m})^2 = 0.245 \text{ kg} \cdot \text{m}^2$$

Equations of Motion

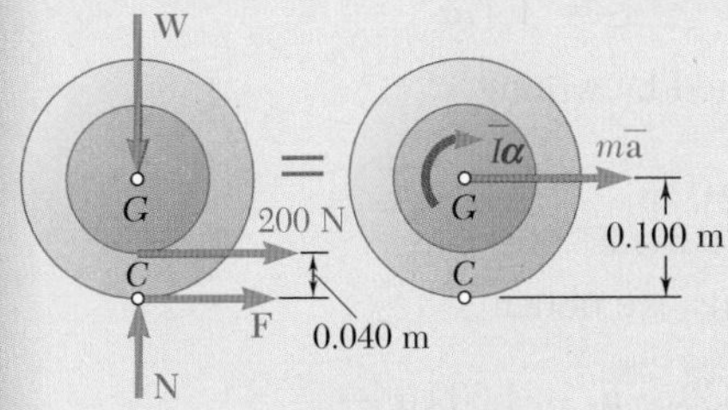

$+\circlearrowright\Sigma M_C = \Sigma(M_C)_{\text{eff}}$: $\quad (200 \text{ N})(0.040 \text{ m}) = m\bar{a}(0.100 \text{ m}) + \bar{I}\alpha$

$$8.00 \text{ N} \cdot \text{m} = (50 \text{ kg})(0.100 \text{ m})\alpha(0.100 \text{ m}) + (0.245 \text{ kg} \cdot \text{m}^2)\alpha$$
$$\alpha = +10.74 \text{ rad/s}^2$$
$$\bar{a} = r\alpha = (0.100 \text{ m})(10.74 \text{ rad/s}^2) = 1.074 \text{ m/s}^2$$

$\overset{+}{\rightarrow}\Sigma F_x = \Sigma(F_x)_{\text{eff}}$: $\quad F + 200 \text{ N} = m\bar{a}$

$$F + 200 \text{ N} = (50 \text{ kg})(1.074 \text{ m/s}^2)$$
$$F = -146.3 \text{ N} \qquad \mathbf{F} = 146.3 \text{ N} \leftarrow$$

$+\uparrow\Sigma F_y = \Sigma(F_y)_{\text{eff}}$:

$$N - W = 0 \qquad N - W = mg = (50 \text{ kg})(9.81 \text{ m/s}^2) = 490.5 \text{ N}$$
$$\mathbf{N} = 490.5 \text{ N} \uparrow$$

Maximum Available Friction Force

$$F_{\text{max}} = \mu_s N = 0.20(490.5 \text{ N}) = 98.1 \text{ N}$$

Since $F > F_{\text{max}}$, the assumed motion is impossible.

b. Rotating and Sliding. Since the wheel must rotate and slide at the same time, we draw a new diagram, where $\bar{\mathbf{a}}$ and $\boldsymbol{\alpha}$ are independent and where

$$F = F_k = \mu_k N = 0.15(490.5 \text{ N}) = 73.6 \text{ N}$$

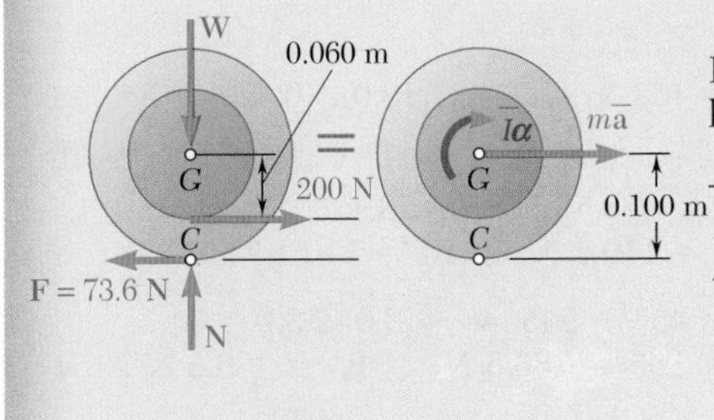

From the computation of part a, it appears that $\mathbf{F}$ should be directed to the left. We write the following equations of motion:

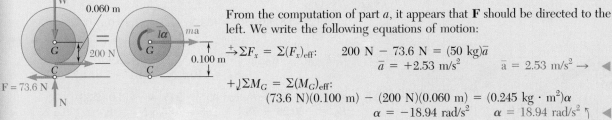

$\overset{+}{\rightarrow}\Sigma F_x = \Sigma(F_x)_{\text{eff}}$: $\quad 200 \text{ N} - 73.6 \text{ N} = (50 \text{ kg})\bar{a}$

$$\bar{a} = +2.53 \text{ m/s}^2 \qquad \bar{\mathbf{a}} = 2.53 \text{ m/s}^2 \rightarrow \blacktriangleleft$$

$+\circlearrowright\Sigma M_G = \Sigma(M_G)_{\text{eff}}$:

$$(73.6 \text{ N})(0.100 \text{ m}) - (200 \text{ N})(0.060 \text{ m}) = (0.245 \text{ kg} \cdot \text{m}^2)\alpha$$
$$\alpha = -18.94 \text{ rad/s}^2 \qquad \boldsymbol{\alpha} = 18.94 \text{ rad/s}^2 \circlearrowleft \blacktriangleleft$$

SAMPLE PROBLEM 16.10

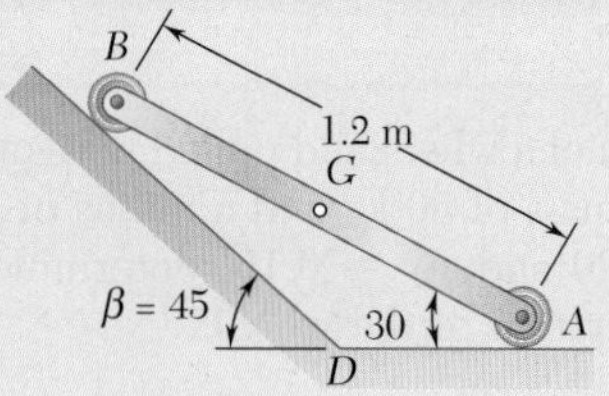

The extremities of a 1.2-m rod weighing 25 kg can move freely and with no friction along two straight tracks as shown. If the rod is released with no velocity from the position shown, determine (*a*) the angular acceleration of the rod, (*b*) the reactions at *A* and *B*.

SOLUTION

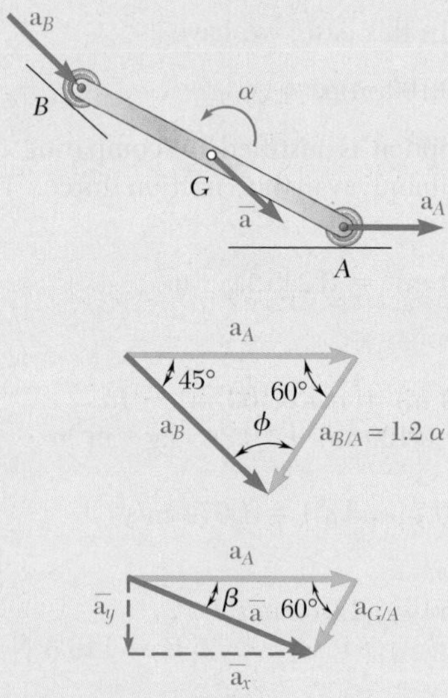

Kinematics of Motion. Since the motion is constrained, the acceleration of *G* must be related to the angular acceleration $\boldsymbol{\alpha}$. To obtain this relation, we first determine the magnitude of the acceleration $\mathbf{a}_A$ of point *A* in terms of α. Since the rod is released with no velocity its angular velecity = 0 and hence the normal component of $\mathbf{a}_{B/A}$ = 0. Assuming that $\boldsymbol{\alpha}$ is directed counterclockwise and noting that $a_{B/A} = 1.2\alpha$, we write

$$\mathbf{a}_B = \mathbf{a}_A + \mathbf{a}_{B/A}$$
$$[a_B \searrow 45°] = [a_A \rightarrow] + [1.2\alpha \nearrow 60°]$$

Noting that $\phi = 75°$ and using the law of sines, we obtain

$$a_A = 1.639\alpha \qquad a_B = 1.47\alpha$$

The acceleration of *G* is now obtained by writing

$$\bar{\mathbf{a}} = \mathbf{a}_G = \mathbf{a}_A + \mathbf{a}_{G/A}$$
$$\bar{\mathbf{a}} = [1.639\,\alpha \rightarrow] + [0.6\,\alpha \nearrow 60°]$$

Resolving $\bar{\mathbf{a}}$ into *x* and *y* components, we obtain

$$\bar{a}_x = 1.639\,\alpha - 0.6\,\alpha\cos 60° = 1.339\,\alpha \qquad \bar{\mathbf{a}}_x = 1.339\,\alpha \rightarrow$$
$$\bar{a}_y = -0.6\,\alpha\sin 60° = -0.520\alpha \qquad \bar{\mathbf{a}}_y = 0.520\,\alpha \downarrow$$

Kinetics of Motion. We draw a free-body-diagram equation expressing that the system of the external forces is equivalent to the system of the effective forces represented by the vector of components $m\bar{\mathbf{a}}_x$ and $m\bar{\mathbf{a}}_y$ attached at *G* and the couple $\bar{I}\boldsymbol{\alpha}$. We compute the following magnitudes:

$$\bar{I} = \tfrac{1}{12}ml^2 = \frac{25}{12}\text{ kg}\,(1.2\text{ m})^2 = 3\text{ kg}\cdot\text{m}^2 \qquad \bar{I}\alpha = 3\,\alpha$$
$$m\bar{a}_x = 25\,(1.339\alpha) = 33.5\,\alpha \qquad m\bar{a}_y = -25(0.520\alpha) = -13.0\,\alpha$$

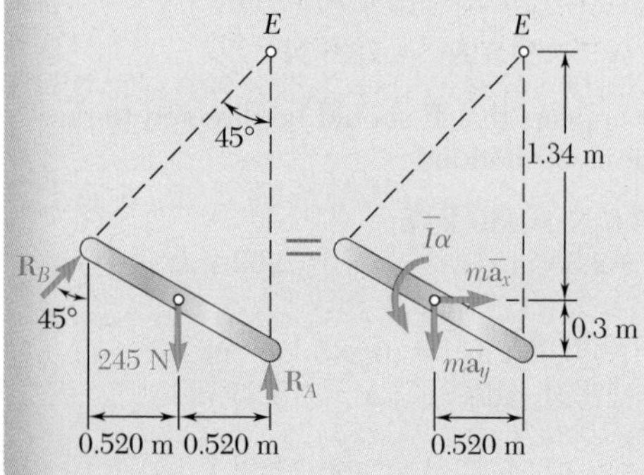

Equations of Motion

$+\circlearrowleft\Sigma M_E = \Sigma(M_E)_{\text{eff}}$:

$$(25)(9.81)(0.520) = (33.5\alpha)(1.34) + (13.0\alpha)(0.520) + 3\alpha$$
$$\alpha = +2.33\text{ rad/s}^2 \qquad \boldsymbol{\alpha} = 2.33\text{ rad/s}^2 \circlearrowleft \blacktriangleleft$$

$\xrightarrow{+}\Sigma F_x = \Sigma(F_x)_{\text{eff}}$: $\quad R_B \sin 45° = (33.5)(2.33) = 78.1$

$$R_B = 110.4\text{ N} \qquad \mathbf{R}_B = 110.4\text{ N} \measuredangle 45° \blacktriangleleft$$

$+\uparrow\Sigma F_y = \Sigma(F_y)_{\text{eff}}$: $\quad R_A + R_B\cos 45° - 245 = -(13.0)(2.33)$

$$R_A = -30.29 - 78.1 + 245 = 136.6\text{ N} \qquad \mathbf{R}_A = 136.6\text{ N}\uparrow \blacktriangleleft$$

SOLVING PROBLEMS ON YOUR OWN

In this lesson we considered the *plane motion of rigid bodies under constraints.* We found that the types of constraints involved in engineering problems vary widely. For example, a rigid body may be constrained to rotate about a fixed axis or to roll on a given surface, or it may be pin-connected to collars or to other bodies.

1. Your solution of a problem involving the constrained motion of a rigid body, will, in general, consist of two steps. First, you will consider the *kinematics of the motion,* and then you will solve the *kinetics portion of the problem.*

2. The kinematic analysis of the motion is done by using the methods you learned in Chap. 15. Due to the constraints, linear and angular accelerations will be related. (They will *not* be independent, as they were in the last lesson.) You should establish *relationships among the accelerations* (angular as well as linear), and your goal should be to express all accelerations in terms of a *single unknown acceleration.* This is the first step taken in the solution of each of the sample problems in this lesson.

a. For a body in noncentroidal rotation, the components of the acceleration of the mass center are $\bar{a}_t = \bar{r}\alpha$ and $\bar{a}_n = \bar{r}\omega^2$, where ω will generally be known [Sample Probs. 16.6 and 16.7].

b. For a rolling disk or wheel, the acceleration of the mass center is $\bar{a} = r\alpha$ [Sample Prob. 16.8].

c. For a body in general plane motion, your best course of action, if neither $\bar{a}$ nor α is known or readily obtainable, is to express $\bar{a}$ in terms of α [Sample Prob. 16.10].

3. The kinetic analysis of the motion is carried out as follows.

a. Start by drawing a free-body-diagram equation. This was done in all the sample problems of this lesson. In each case the left-hand diagram shows the external forces, including the applied forces, the reactions, and the weight of the body. The right-hand diagram shows the vector $m\bar{\mathbf{a}}$ and the couple $\bar{I}\boldsymbol{\alpha}$.

b. Next, reduce the number of unknowns in the free-body-diagram equation by using the relationships among the accelerations that you found in your kinematic analysis. You will then be ready to consider equations that can be written by summing components or moments. Choose first an equation that involves a single unknown. After solving for that unknown, substitute the value obtained into the other equations, which you will then solve for the remaining unknowns.

(continued)

4. When solving problems involving rolling disks or wheels, keep in mind the following.

a. If sliding is impending, the friction force exerted on the rolling body has reached its maximum value, $F_m = \mu_s N$, where N is the normal force exerted on the body and μ_s is the coefficient of *static friction* between the surfaces of contact.

b. If sliding is not impending, the friction force F can have *any value* smaller than F_m and should, therefore, be considered as an independent unknown. After you have determined F, be sure to check that it is smaller than F_m; if it is not, *the body does not roll*, but rotates and slides as described in the next paragraph.

c. If the body rotates and slides at the same time, then the body is *not rolling* and the acceleration $\bar{a}$ of the mass center is *independent* of the angular acceleration α of the body: $\bar{a} \neq r\alpha$. On the other hand, the friction force has a well-defined value, $F = \mu_k N$, where μ_k is the coefficient of *kinetic friction* between the surfaces of contact.

d. For an unbalanced rolling disk or wheel, the relation $\bar{a} = r\alpha$ between the acceleration $\bar{a}$ of the mass center G and the angular acceleration α of the disk or wheel *does not hold anymore*. However, a similar relation holds between the acceleration a_O of the *geometric center* O and the angular acceleration α of the disk or wheel: $a_O = r\alpha$. This relation can be used to express $\bar{a}$ in terms of α and ω (Fig. 16.18).

5. For a system of connected rigid bodies, the goal of your *kinematic analysis* should be to determine all the accelerations from the given data, or to express them all in terms of a single unknown. (For systems with several degrees of freedom, you will need to use as many unknowns as there are degrees of freedom.)

Your *kinetic analysis* will generally be carried out by drawing a free-body-diagram equation for the entire system, as well as for one or several of the rigid bodies involved. In the latter case, both internal and external forces should be included, and care should be taken to represent with equal and opposite vectors the forces that two bodies exert on each other.

PROBLEMS

CONCEPT QUESTIONS

16.CQ4 A cord is attached to a spool when a force **P** is applied to the cord as shown. Assuming the spool rolls without slipping, what direction does the spool move for each case?

Case 1: **a.** left **b.** right **c.** It would not move.

Case 2: **a.** left **b.** right **c.** It would not move.

Case 3: **a.** left **b.** right **c.** It would not move.

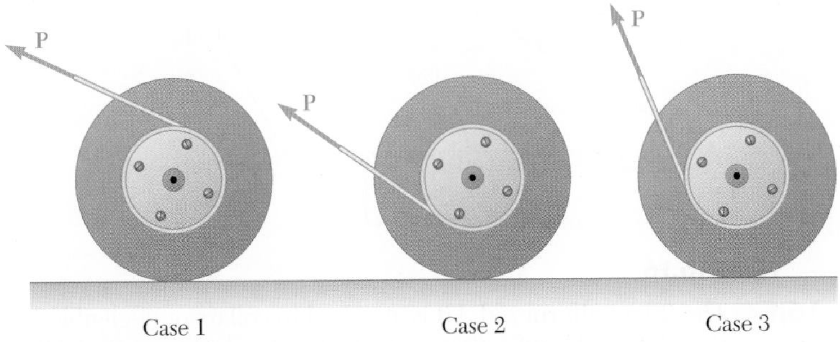

Fig. P16.CQ4 and P16.CQ5

16.CQ5 A cord is attached to a spool when a force **P** is applied to the cord as shown. Assuming the spool rolls without slipping, in what direction does the friction force act for each case?

Case 1: **a.** left **b.** right **c.** The friction force would be zero.

Case 2: **a.** left **b.** right **c.** The friction force would be zero.

Case 3: **a.** left **b.** right **c.** The friction force would be zero.

16.CQ6 A front-wheel-drive car starts from rest and accelerates to the right. Knowing that the tires do not slip on the road, what is the direction of the friction force the road applies to the front tires?

a. left

b. right

c. The friction force is zero.

16.CQ7 A front-wheel-drive car starts from rest and accelerates to the right. Knowing that the tires do not slip on the road, what is the direction of the friction force the road applies to the rear tires?

a. left

b. right

c. The friction force is zero.

FREE BODY PRACTICE PROBLEMS

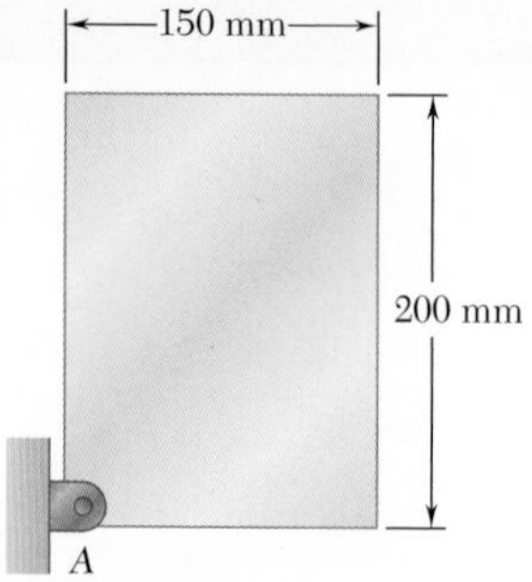

Fig. P16.F5

16.F5 A uniform 150 × 200-mm. rectangular plate of mass m is pinned at A. Knowing the angular velocity of the plate at the instant shown is $\boldsymbol{\omega}$, draw the FBD and KD.

16.F6 Two identical 2-kg slender rods AB and BC are connected by a pin at B and by the cord AC. The assembly rotates in a vertical plane under the combined effect of gravity and a couple **M** applied to rod AB. Knowing that in the position shown the angular velocity of the assembly is $\boldsymbol{\omega}$, draw the FBD and KD that can be used to determine the angular acceleration of the assembly.

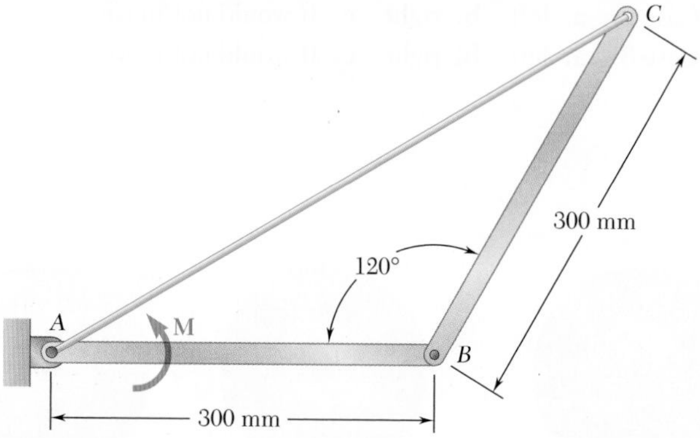

Fig. P16.F6

16.F7 The 2-kg uniform rod AB is attached to collars of negligible mass that slide without friction along the fixed rods shown. Rod AB is at rest in the position $\theta = 25°$ when a horizontal force **P** is applied to collar A causing it to start moving to the left. Draw the FBD and KD for the rod.

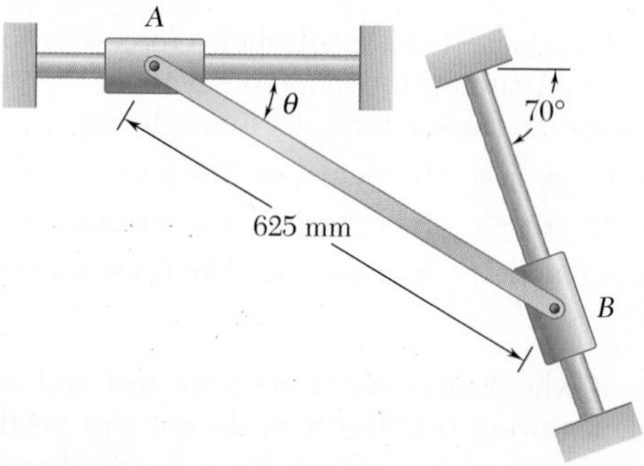

Fig. P16.F7

16.F8 A uniform disk of mass $m = 4$ kg and radius $r = 150$ mm is supported by a belt $ABCD$ that is bolted to the disk at B and C. If the belt suddenly breaks at a point located between A and B, draw the FBD and KD for the disk immediately after the break.

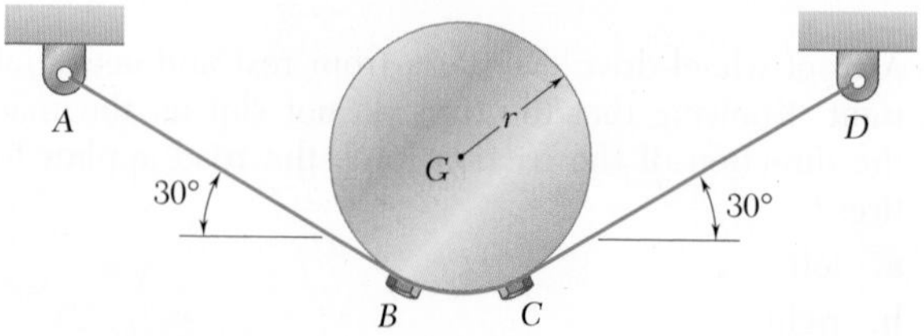

Fig. P16.F8

END-OF-SECTION PROBLEMS

16.75 Show that the couple $\bar{I}\boldsymbol{\alpha}$ of Fig. 16.15 can be eliminated by attaching the vectors $m\bar{\mathbf{a}}_t$ and $m\bar{\mathbf{a}}_n$ at a point P called the *center of percussion*, located on line OG at a distance $GP = \bar{k}^2/\bar{r}$ from the mass center of the body.

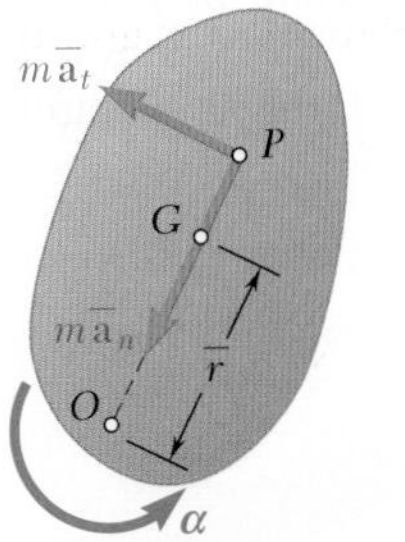

Fig. P16.75

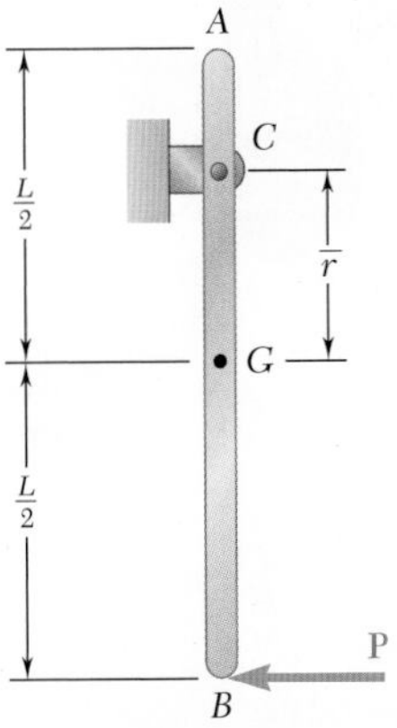

Fig. P16.76

16.76 A uniform slender rod of length $L = 900$ mm and mass $m = 4$ kg is suspended from a hinge at C. A horizontal force $\mathbf{P}$ of magnitude 75 N is applied at end B. Knowing that $\bar{r} = 225$ mm, determine (*a*) the angular acceleration of the rod, (*b*) the components of the reaction at C.

16.77 In Prob. 16.76, determine (*a*) the distance $\bar{r}$ for which the horizontal component of the reaction at C is zero, (*b*) the corresponding angular acceleration of the rod.

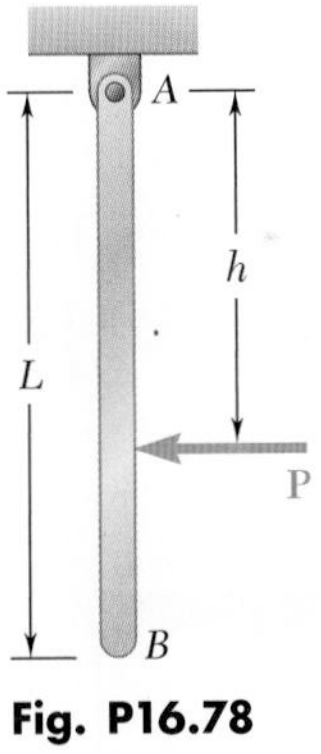

Fig. P16.78

16.78 A uniform slender rod of length $L = 1$ m and mass $m = 2$ kg hangs freely from a hinge at A. If a force $\mathbf{P}$ of magnitude 8 N is applied at B horizontally to the left ($h = L$), determine (*a*) the angular acceleration of the rod, (*b*) the components of the reaction at A.

16.79 In Prob. 16.78, determine (*a*) the distance h for which the horizontal component of the reaction at A is zero, (*b*) the corresponding angular acceleration of the rod.

16.80 The uniform slender rod AB is welded to the hub D, and the system rotates about the vertical axis DE with a constant angular velocity $\boldsymbol{\omega}$. (*a*) Denoting by w the mass per unit length of the rod, express the tension in the rod at a distance z from end A in terms of w, l, z, and $\boldsymbol{\omega}$, (*b*) Determine the tension in the rod for $w = 0.3$ kg/m, $l = 400$ mm, $z = 250$ mm, and $\omega = 150$ rpm.

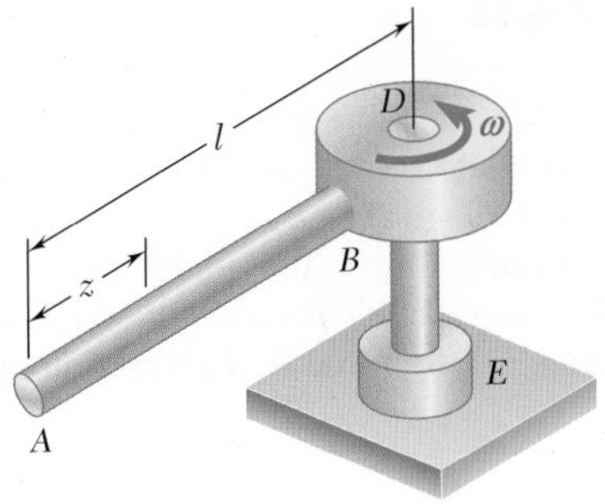

Fig. P16.80

16.81 The shutter shown was formed by removing one quarter of a disk of 20 mm radius and is used to interrupt a beam of light emanating from a lens at C. Knowing that the shutter has a mass of 50 g and rotates at the constant rate of 24 cycles per second, determine the magnitude of the force exerted by the shutter on the shaft at A.

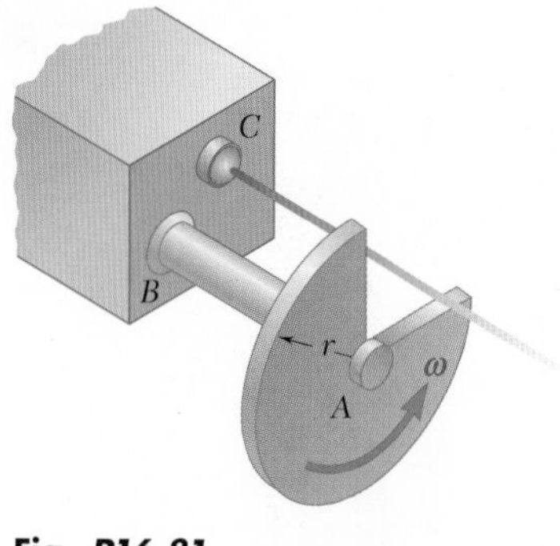

Fig. P16.81

16.82 A slender uniform cone of mass m can swing freely about the horizontal rod AB. If the cone is released from rest in the position shown, determine (a) the acceleration of the tip D, (b) the reaction at C.

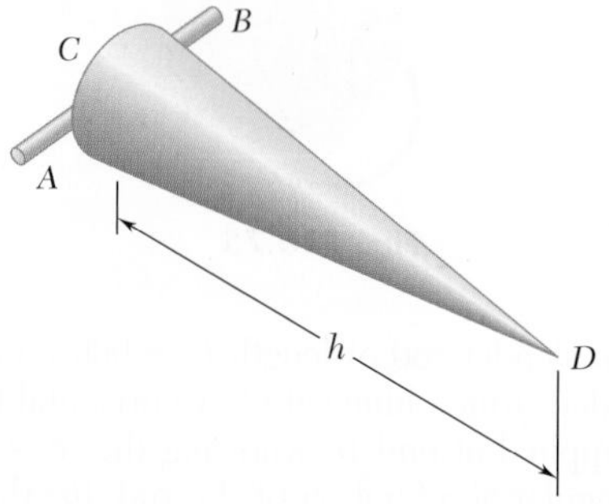

Fig. P16.82

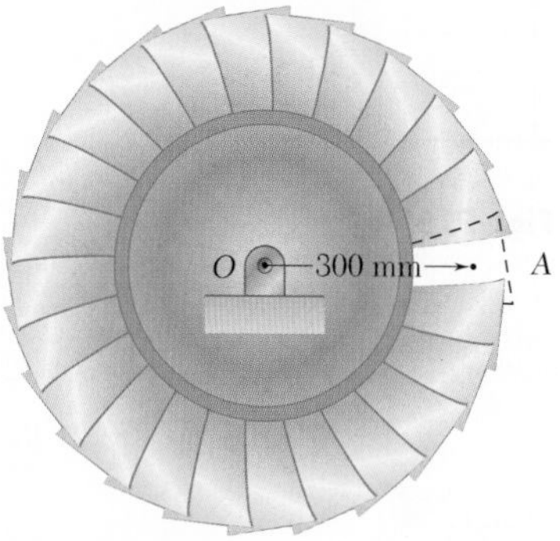

Fig. P16.83

16.83 A turbine disk of mass 26 kg rotates at a constant rate of 9600 rpm. Knowing that the mass center of the disk coincides with the center of rotation O, determine the reaction at O immediately after a single blade at A, of mass 45 g, becomes loose and is thrown off.

16.84 and 16.85 A uniform rod of length L and mass m is supported as shown. If the cable attached at end B suddenly breaks, determine (a) the acceleration of end B, (b) the reaction at the pin support.

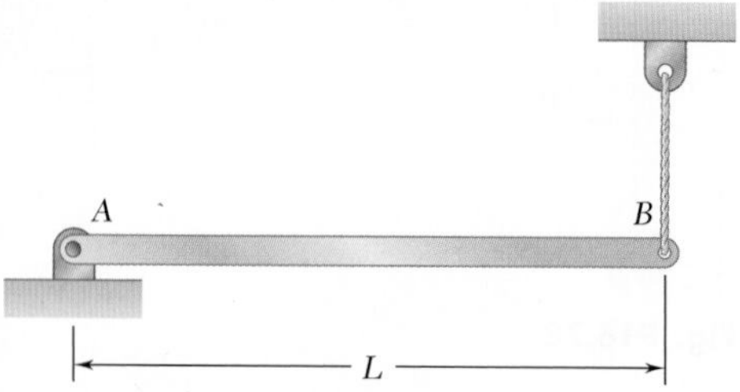

Fig. P16.84

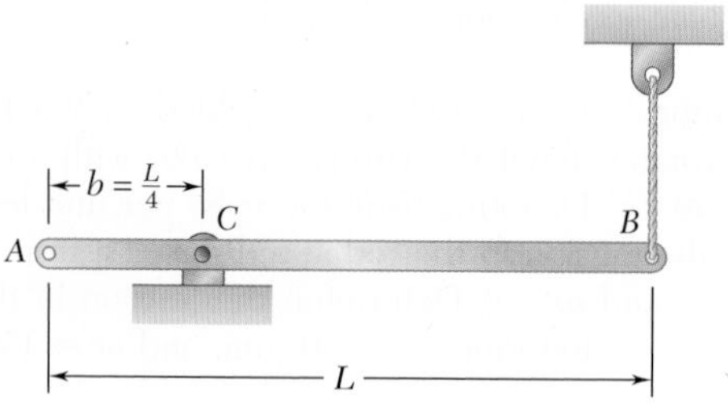

Fig. P16.85

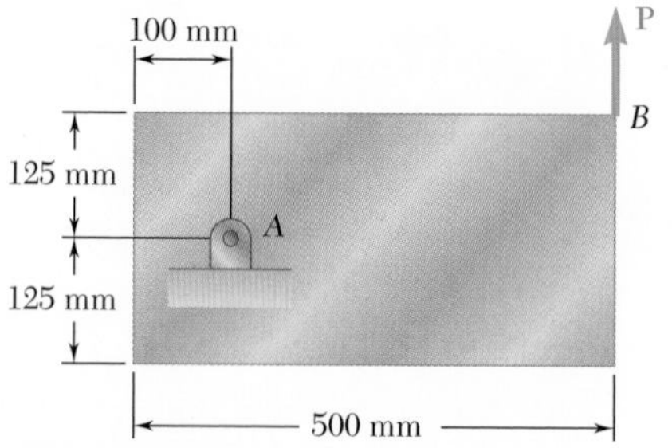

Fig. P16.86

16.86 A 6-kg uniform plate rotates about A in a vertical plane under the combined effect of gravity and of the vertical force **P**. Knowing that at the instant shown the plate has an angular velocity of 20 rad/s and an angular acceleration of 30 rad/s^2 both counterclockwise, determine (a) the force **P**, (b) the components of the reaction at A.

16.87 A 1.5-kg slender rod is welded to a 5-kg uniform disk as shown. The assembly swings freely about *C* in a vertical plane. Knowing that in the position shown the assembly has an angular velocity of 10 rad/s clockwise, determine (*a*) the angular acceleration of the assembly, (*b*) the components of the reaction at *C*.

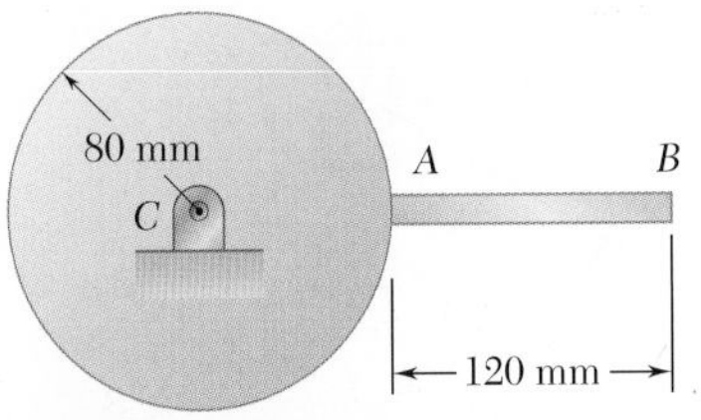

Fig. P16.87

16.88 Two uniform rods, *ABC* of mass 3 kg and *DCE* of mass 4 kg, are connected by a pin at *C* and by two cords *BD* and *BE*. The T-shaped assembly rotates in a vertical plane under the combined effect of gravity and of a couple **M** which is applied to rod *ABC*. Knowing that at the instant shown the tension in cord *BE* is 10 N and the tension in cord *BD* is 2.5 N, determine (*a*) the angular acceleration of the assembly, (*b*) the couple **M**.

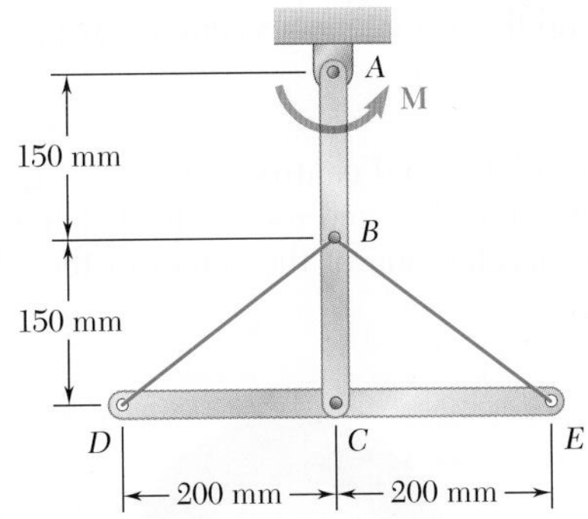

Fig. P16.88

16.89 The object *ABC* consists of two slender rods welded together at point *B*. Rod *AB* has a mass of 1 kg and bar *BC* has a mass of 2 kg. Knowing the magnitude of the angular velocity of *ABC* is 10 rad/s when $\theta = 0$, determine the components of the reaction at point *C* when $\theta = 0$.

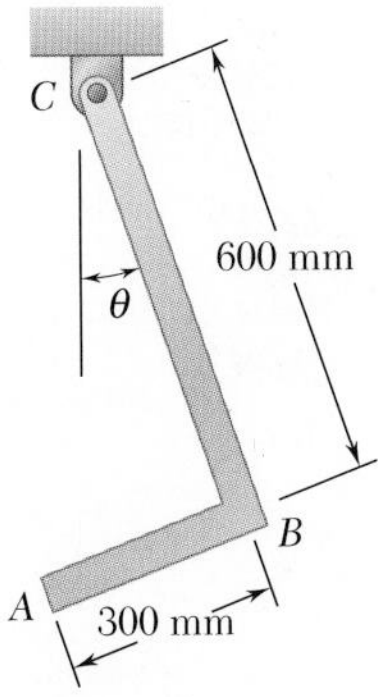

Fig. *P16.89*

16.90 A 3.5-kg slender rod *AB* and a 2-kg slender rod *BC* are connected by a pin at *B* and by the cord *AC*. The assembly can rotate in a vertical plane under the combined effect of gravity and a couple **M** applied to rod *BC*. Knowing that in the position shown the angular velocity of the assembly is zero and the tension in cord *AC* is equal to 25 N, determine (*a*) the angular acceleration of the assembly, (*b*) the magnitude of the couple **M**.

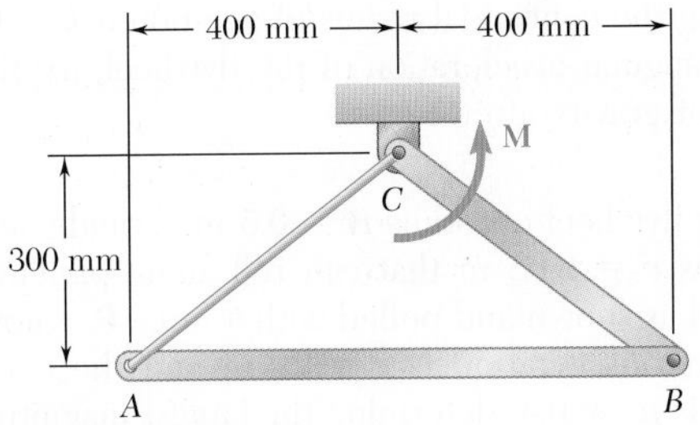

Fig. *P16.90*

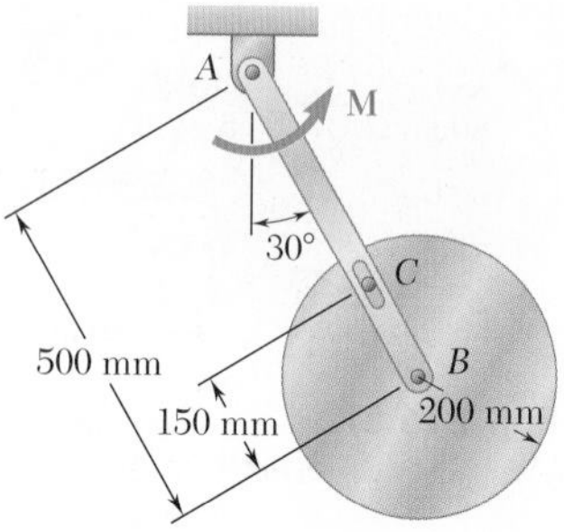

Fig. *P16.91*

16.91 A 9-kg uniform disk is attached to the 5-kg slender rod AB by means of frictionless pins at B and C. The assembly rotates in a vertical plane under the combined effect of gravity and of a couple **M** which is applied to rod AB. Knowing that at the instant shown the assembly has an angular velocity of 6 rad/s and an angular acceleration of 25 rad/s^2, both counterclockwise, determine (a) the couple **M**, (b) the force exerted by pin C on member AB.

16.92 Derive the equation $\Sigma M_C = I_C\alpha$ for the rolling disk of Fig. 16.17, where ΣM_C represents the sum of the moments of the external forces about the instantaneous center C, and I_C is the moment of inertia of the disk about C.

16.93 Show that in the case of an unbalanced disk, the equation derived in Prob. 16.92 is valid only when the mass center G, the geometric center O, and the instantaneous center C happen to lie in a straight line.

16.94 A wheel of radius r and centroidal radius of gyration $\bar{k}$ is released from rest on the incline and rolls without sliding. Derive an expression for the acceleration of the center of the wheel in terms of r, $\bar{k}$, β, and g.

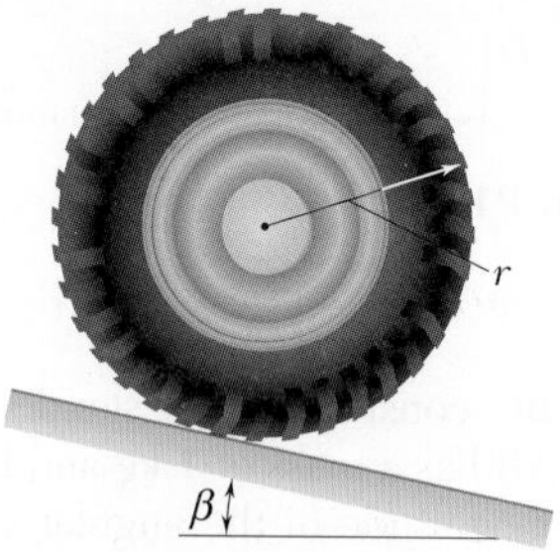

Fig. P16.94

Fig. P16.95

16.95 A homogeneous sphere S, a uniform cylinder C, and a thin pipe P are in contact when they are released from rest on the incline shown. Knowing that all three objects roll without slipping, determine, after 4 s of motion, the clear distance between (a) the pipe and the cylinder, (b) the cylinder and the sphere.

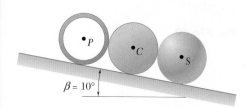

16.96 A 40-kg flywheel of radius $R = 0.5$ m is rigidly attached to a shaft of radius $r = 0.05$ m that can roll along parallel rails. A cord is attached as shown and pulled with a force **P** of magnitude 150 N. Knowing the centroidal radius of gyration is $\bar{k} = 0.4$ m, determine (a) the angular acceleration of the flywheel, (b) the velocity of the center of gravity after 5 s.

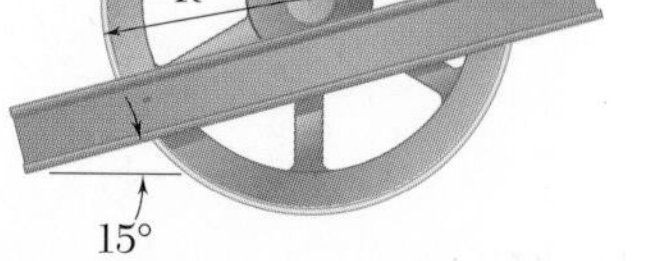

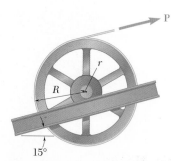

Fig. *P16.96* and *P16.97*

16.97 A 40-kg flywheel of radius $R = 0.5$ m is rigidly attached to a shaft of radius $r = 0.05$ m that can roll along parallel rails. A cord is attached as shown and pulled with a force **P**. Knowing the centroidal radius of gyration is $\bar{k} = 0.4$ m and the coefficient of static friction is $\mu_s = 0.4$, determine the largest magnitude of force **P** for which no slipping will occur.

16.98 through *16.101* A drum of 60-mm radius is attached to a disk of 120-mm radius. The disk and drum have a total mass of 6 kg and a combined radius of gyration of 90 mm. A cord is attached as shown and pulled with a force **P** of magnitude 20 N. Knowing that the disk rolls without sliding, determine (*a*) the angular acceleration of the disk and the acceleration of *G*, (*b*) the minimum value of the coefficient of static friction compatible with this motion.

16.102 through 16.105 A drum of 60-mm radius is attached to a disk of 120-mm radius. The disk and drum have a total mass of 6 kg and a combined radius of gyration of 90 mm. A cord is attached as shown and pulled with a force **P** of magnitude 20 N. Knowing that the disk rolls without sliding, determine (*a*) the angular acceleration of the disk and the acceleration of *G*, (*b*) the minimum value of the coefficient of static friction compatible with this motion.

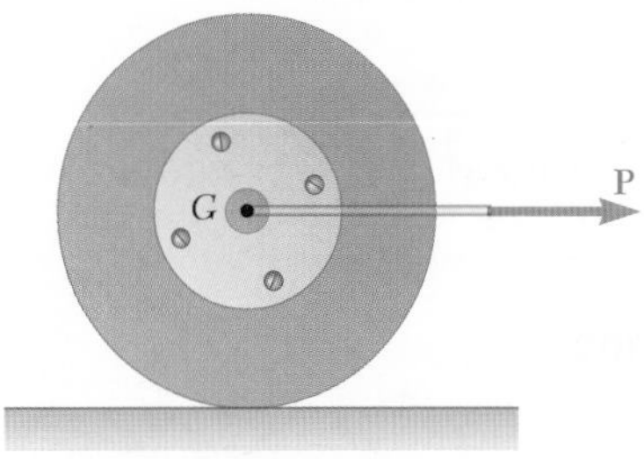

Fig. P16.98 and P16.102

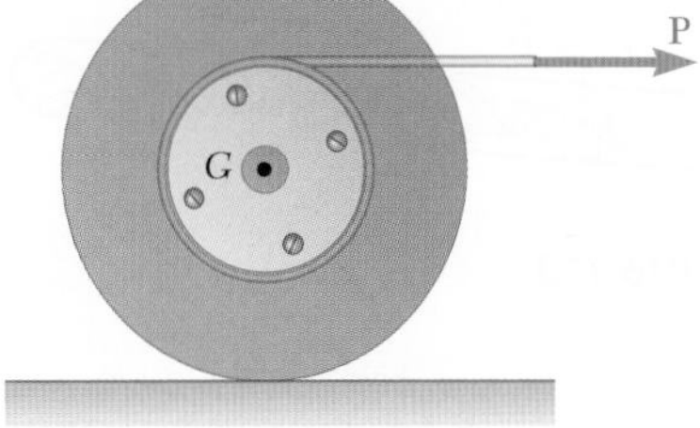

Fig. P16.99 and P16.103

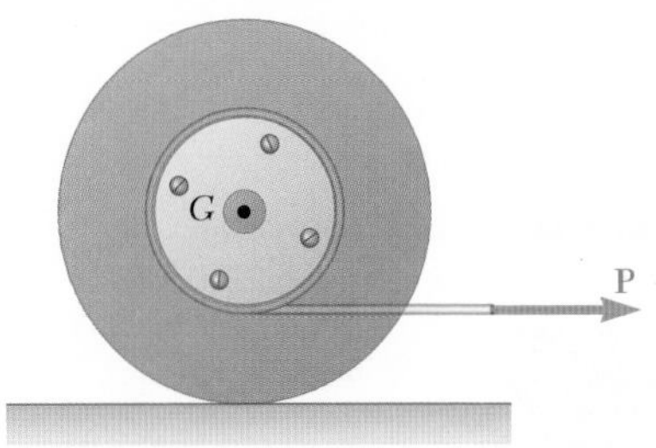

Fig. *P16.100* and P16.104

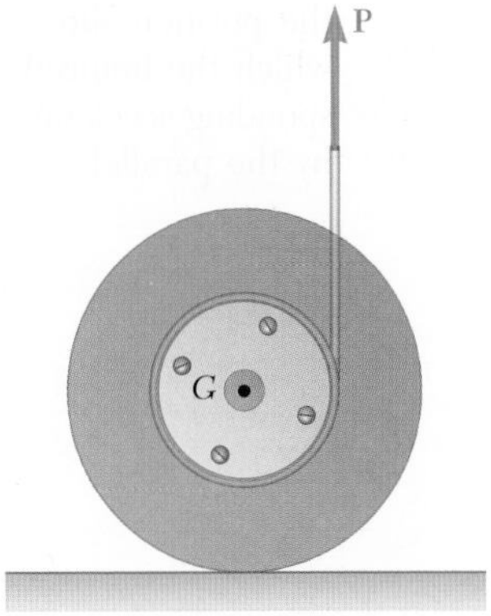

Fig. *P16.101* and P16.105

***16.106* and 16.107** A 300-mm radius cylinder of mass 8 kg rests on a 3-kg carriage. The system is at rest when a force **P** of magnitude 20 N is applied. Knowing that the cylinder rolls without sliding on the carriage and neglecting the mass of the wheels of the carriage, determine (*a*) the acceleration of the carriage, (*b*) the acceleration of point *A*, (*c*) the distance the cylinder has rolled with respect to the carriage after 0.5 s.

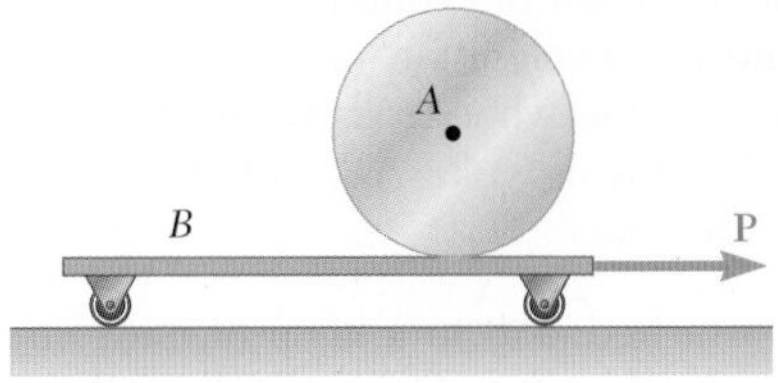

Fig. *P16.106*

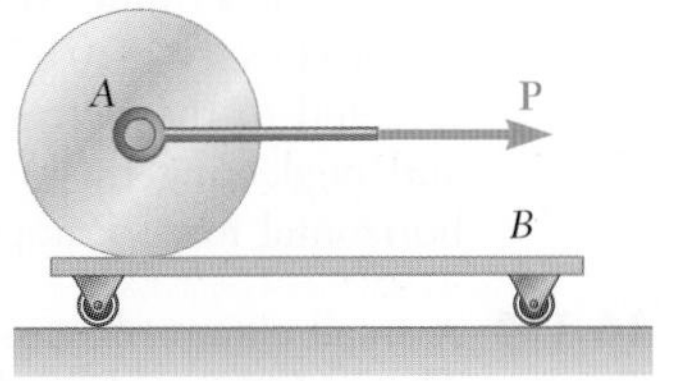

Fig. P16.107

16.108 Gear *C* has a mass of 5 kg and a centroidal radius of gyration of 75 mm. The uniform bar *AB* has a mass of 3 kg and gear *D* is stationary. If the system is released from rest in the position shown, determine (*a*) the angular acceleration of gear *C*, (*b*) the acceleration of point *B*.

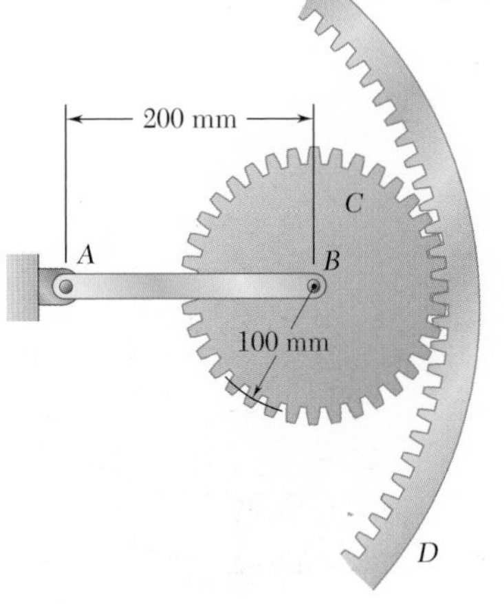

Fig. P16.108

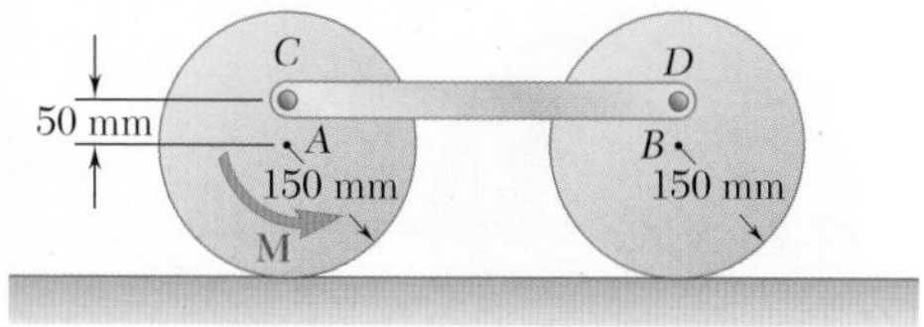

Fig. P16.109

16.109 Two uniform disks A and B, each of mass 2 kg, are connected by a 1.5-kg rod CD as shown. A counterclockwise couple **M** of moment 2.5 N-m is applied to disk A. Knowing that the disks roll without sliding, determine (*a*) the acceleration of the center of each disk, (*b*) the horizontal component of the force exerted on disk B by pin D.

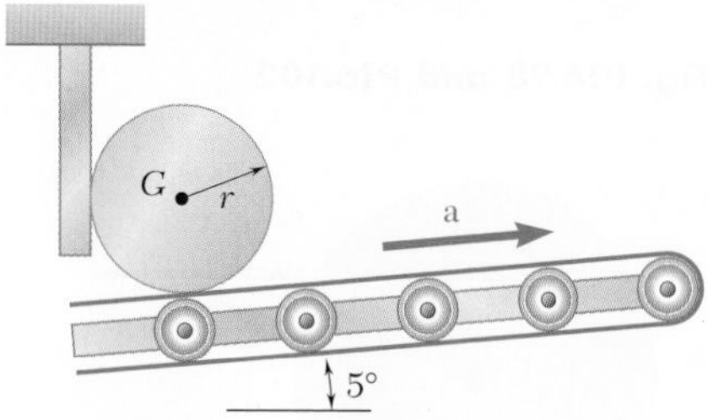

Fig. *P16.110*

16.110 A 5-kg cylinder of radius r = 100 mm is resting on a conveyor belt when the belt is suddenly turned on and it experiences an acceleration of magnitude a = 1.8 m/s^2. The smooth vertical bar holds the cylinder in place when the belt is not moving. Knowing the cylinder rolls without slipping and the friction between the vertical bar and the cylinder is negligible, determine (*a*) the angular acceleration of the cylinder, (*b*) the components of the force the conveyor belt applies to the cylinder.

16.111 A hemisphere of weight W and radius r is released from rest in the position shown. Determine (*a*) the minimum value of μ_s for which the hemisphere starts to roll without sliding, (*b*) the corresponding acceleration of point B [*Hint:* Note that $OG = \frac{3}{8}r$ and that, by the parallel-axis theorem, $\bar{I} = \frac{2}{5}mr^2 - m(OG)^2$.]

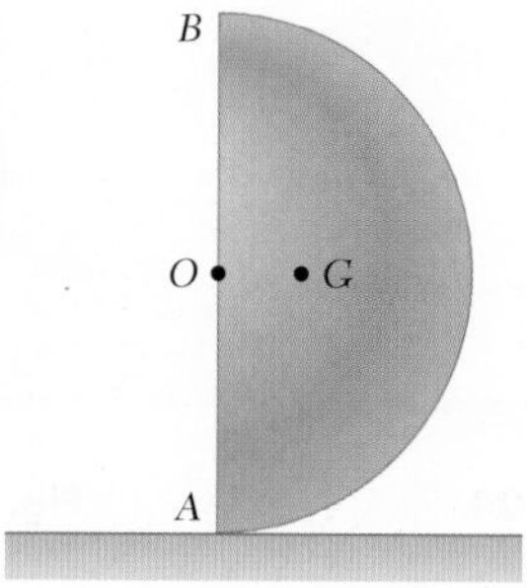

Fig. P16.111

16.112 Solve Prob. 16.111, considering a half cylinder instead of a hemisphere. [*Hint:* Note that $OG = 4r/3\pi$ and that, by the parallel-axis theorem, $\bar{I} = \frac{1}{2}mr^2 - m(OG)^2$.]

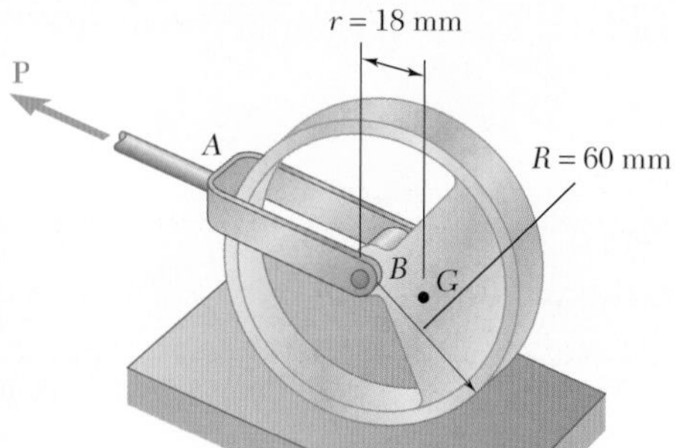

Fig. P16.113

16.113 The center of gravity G of a 1.5-kg unbalanced tracking wheel is located at a distance r = 18 mm from its geometric center B. The radius of the wheel is R = 60 mm and its centroidal radius of gyration is 44 mm. At the instant shown the center B of the wheel has a velocity of 0.35 m/s and an acceleration of 1.2 m/s^2, both directed to the left. Knowing that the wheel rolls without sliding and neglecting the mass of the driving yoke AB, determine the horizontal force **P** applied to the yoke.

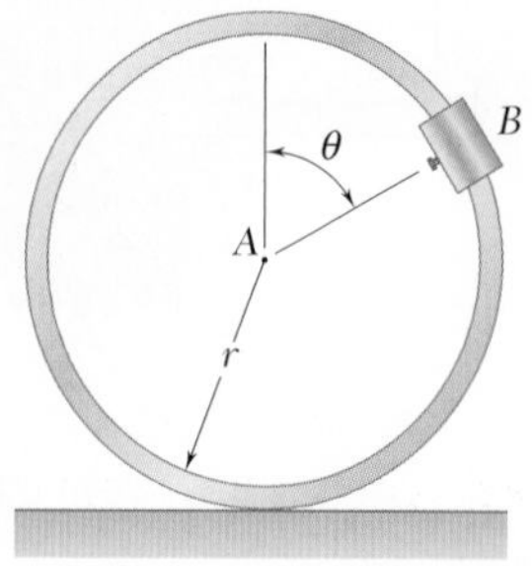

Fig. P16.114 and P16.115

16.114 A small clamp of mass m_B is attached at B to a hoop of mass m_h. The system is released from rest when θ = 90° and rolls without sliding. Knowing that $m_h = 3m_B$, determine (*a*) the angular acceleration of the hoop, (*b*) the horizontal and vertical components of the acceleration of B.

16.115 A small clamp of mass m_B is attached at B to a hoop of mass m_h. Knowing that the system is released from rest and rolls without sliding, derive an expression for the angular acceleration of the hoop in terms of m_B, m_h, r, and θ.

16.116 A 2-kg bar is attached to a 5-kg uniform cylinder by a square pin, P, as shown. Knowing that r =400 mm, h = 200 mm, θ = 20°, L = 500 mm and ω = 2 rad/s at the instant shown, determine the reactions at P at this instant assuming that the cylinder rolls without sliding down the incline.

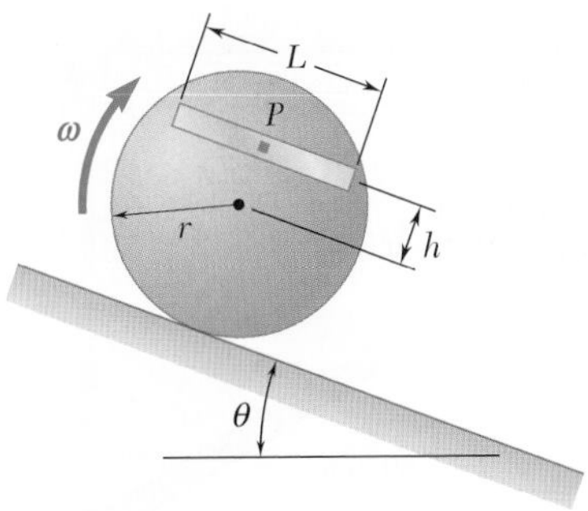

Fig. P16.116

16.117 The ends of the 10-kg uniform rod AB are attached to collars of negligible mass that slide without friction along fixed rods. If the rod is released from rest when θ = 25°, determine immediately after release (*a*) the angular acceleration of the rod, (*b*) the reaction at A, (*b*) the reaction at B.

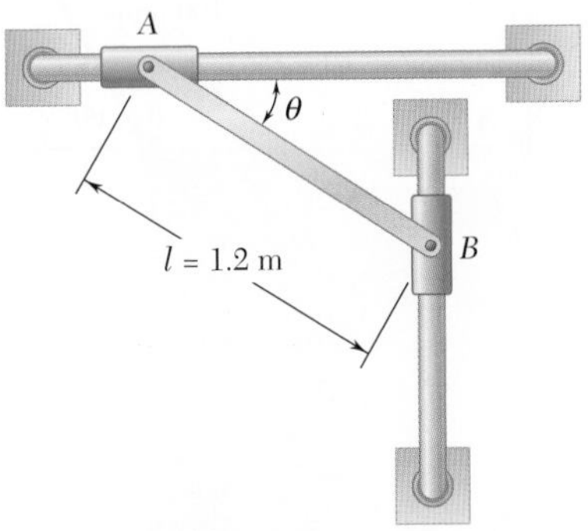

Fig. P16.117 and P16.118

16.118 The ends of the 10-kg uniform rod AB are attached to collars of negligible mass that slide without friction along fixed rods. A vertical force **P** is applied to collar B when θ = 25°, causing the collar to start from rest with an upward acceleration of 12 m/s^2. Determine (*a*) the force **P**, (*b*) the reaction at A.

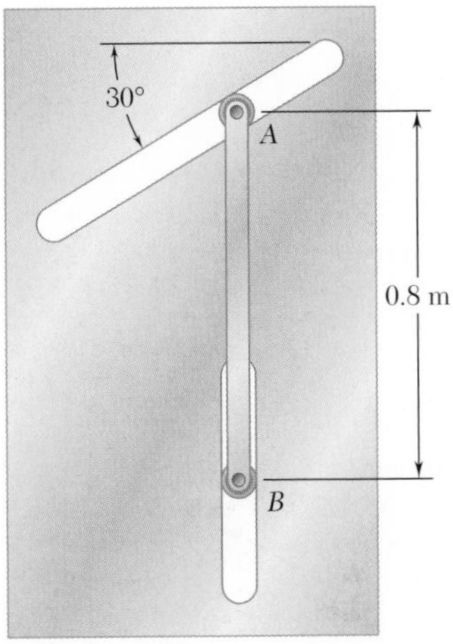

Fig. P16.119

16.119 The motion of the 3-kg uniform rod AB is guided by small wheels of negligible weight that roll along without friction in the slots shown. If the rod is released from rest in the position shown, determine immediately after release (*a*) the angular acceleration of the rod, (*b*) the reaction at B.

16.120 A beam AB of length L and mass m is supported by two cables as shown. If cable BD breaks, determine at that instant the tension in the remaining cable as a function of its initial angular orientation θ.

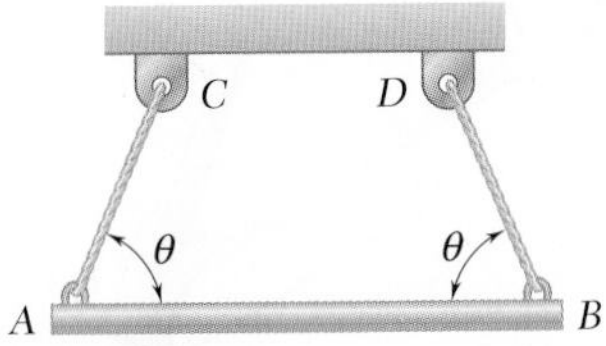

Fig. P16.120

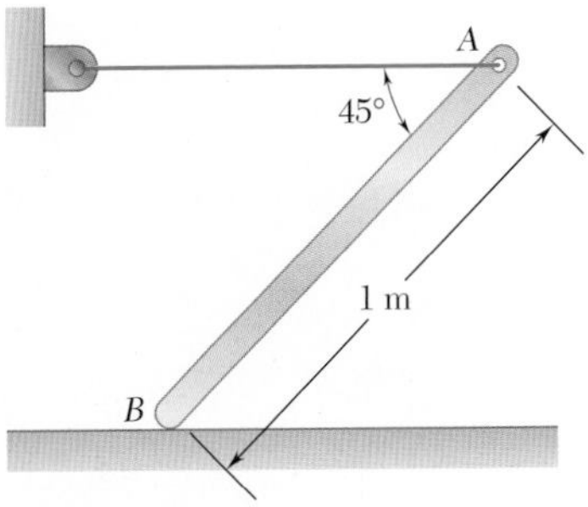

Fig. P16.121

16.121 End A of a uniform 10-kg bar is attached to a horizontal rope and end B contacts a floor with negligible friction. Knowing that the bar is released from rest in the position shown, determine immediately after release (*a*) the angular acceleration of the bar, (*b*) the tension in the rope, (*c*) the reaction at B.

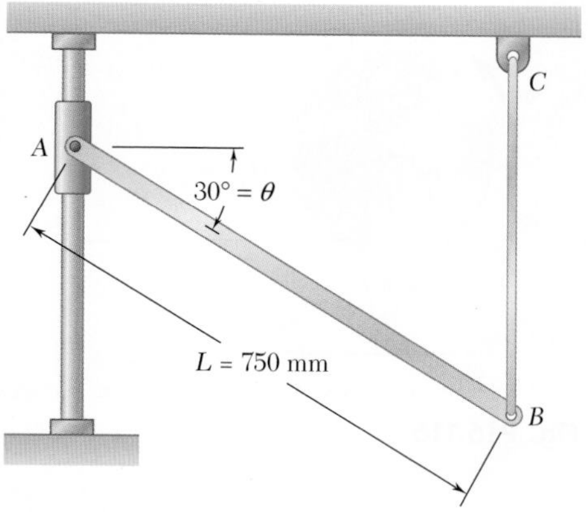

Fig. *P16.122*

16.122 End A of the 8-kg uniform rod AB is attached to a collar that can slide without friction on a vertical rod. End B of the rod is attached to a vertical cable BC. If the rod is released from rest in the position shown, determine immediately after release (*a*) the angular acceleration of the rod, (*b*) the reaction at A.

16.123 A uniform thin plate $ABCD$ has a mass of 8 kg and is held in position by three inextensible cords AE, BF, and CG. If cord AE is cut, determine at that instant (*a*) if the plate is undergoing translation or general plane motion, (*b*) the tension in cords BF and CG.

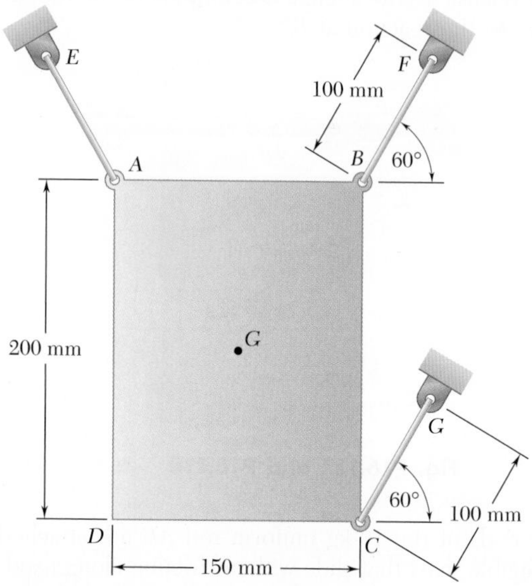

Fig. *P16.123*

16.124 The 4-kg uniform rod ABD is attached to the crank BC and is fitted with a small wheel that can roll without friction along a vertical slot. Knowing that at the instant shown crank BC rotates with an angular velocity of 6 rad/s clockwise and an angular acceleration of 15 rad/s^2 counterclockwise, determine the reaction at A.

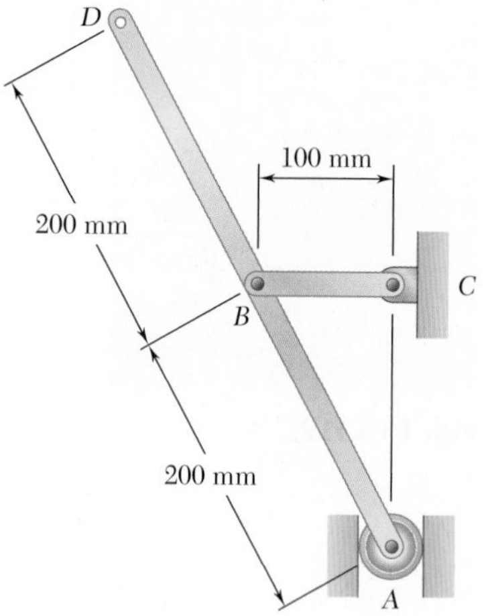

Fig. P16.124

16.125 The 2-kg uniform rod AB is attached to collars of negligible mass which may slide without friction along the fixed rods shown. Rod AB is at rest in the position $\theta = 25°$ when a horizontal force **P** is applied to collar A, causing it to start moving to the left with an acceleration of 4 m/s^2. Determine (*a*) the force **P**, (*b*) the reaction at B.

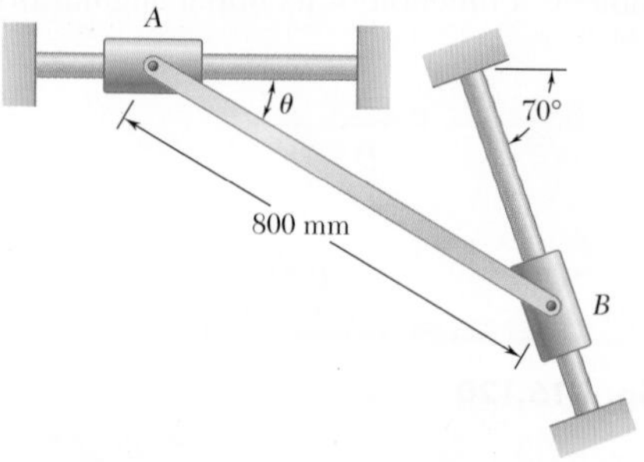

Fig. P16.125 and P16.126

16.126 The 2-kg uniform rod AB is attached to collars of negligible mass which may slide without friction along the fixed rods shown. If rod AB is released from rest in the position $\theta = 25°$, determine immediately after release (*a*) the angular acceleration of the rod, (*b*) the reaction at B.

16.127 The 250-mm uniform rod BD, of mass 5 kg, is connected as shown to disk A and to a collar of negligible mass, that may slide freely along a vertical rod. Knowing that disk A rotates counterclockwise at a constant rate of 500 rpm, determine the reactions at D when $\theta = 0$.

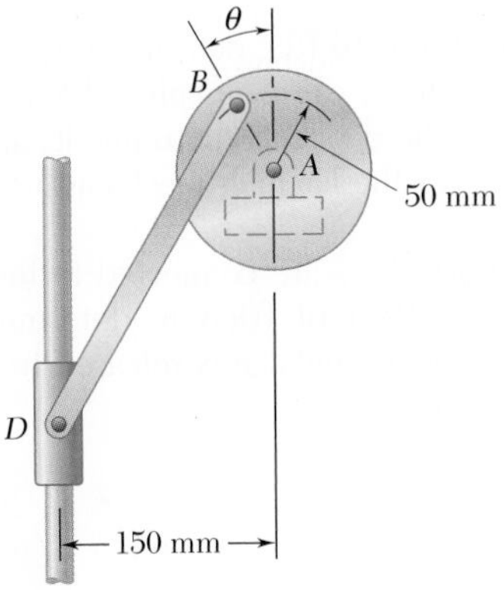

Fig. P16.127

16.128 Solve Prob. 16.127 when $\theta = 90°$.

16.129 The 4-kg uniform slender bar BD is attached to bar AB and a wheel of negligible mass that rolls on a circular surface. Knowing that at the instant shown bar AB has an angular velocity of 6 rad/s and no angular acceleration, determine the reaction at point D.

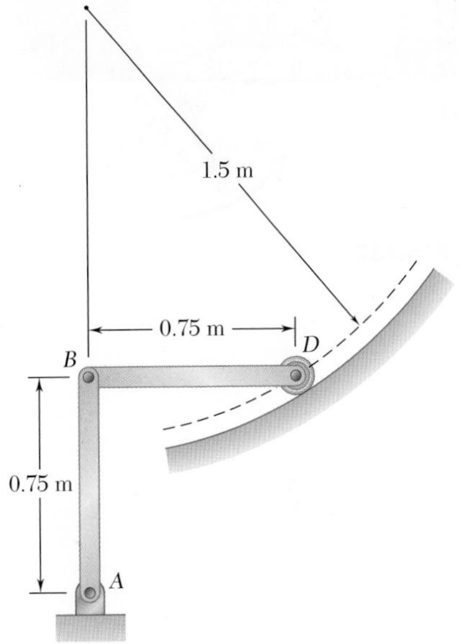

Fig. P16.129

16.130 The motion of the uniform slender rod of length $L = 0.5$ m and mass $m = 3$ kg is guided by pins at A and B that slide freely in frictionless slots, circular and horizontal, cut into a vertical plate as shown. Knowing that at the instant shown the rod has an angular velocity of 3 rad/s counterclockwise and $\theta = 30°$, determine the reactions at points A and B.

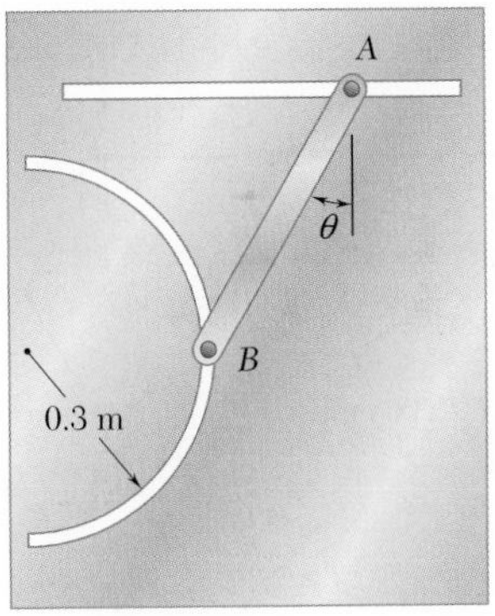

Fig. P16.130

16.131 At the instant shown, the 6-m long, uniform 50-kg pole ABC has an angular velocity of 1 rad/s counterclockwise and point C is sliding to the right. A 500-N horizontal force $\mathbf{P}$ acts at B. Knowing the coefficient of kinetic friction between the pole and the ground is 0.3, determine at this instant (a) the acceleration of the center of gravity, (b) the normal force between the pole and the ground.

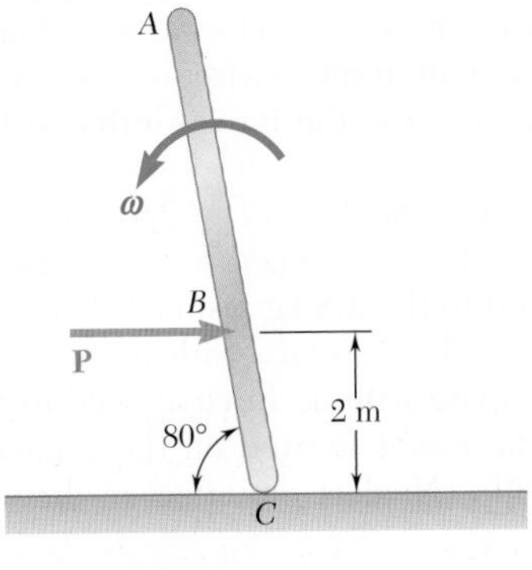

Fig. P16.131

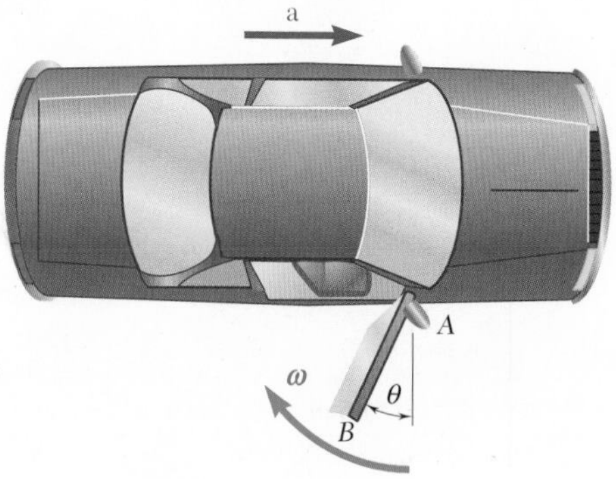

Fig. P16.132

16.132 A driver starts his car with the door on the passenger's side wide open ($\theta = 0$). The 40-kg door has a centroidal radius of gyration $\bar{k} = 300$ mm, and its mass center is located at a distance $r = 500$ mm from its vertical axis of rotation. Knowing that the driver maintains a constant acceleration of 2 m/s^2, determine the angular velocity of the door as it slams shut ($\theta = 90°$).

16.133 For the car of Prob. 16.132, determine the smallest constant acceleration that the driver can maintain if the door is to close and latch, knowing that as the door hits the frame its angular velocity must be at least 2 rad/s for the latching mechanism to operate.

16.134 Two 4-kg uniform bars are connected to form the linkage shown. Neglecting the effect of friction, determine the reaction at D immediately after the linkage is released from rest in the position shown.

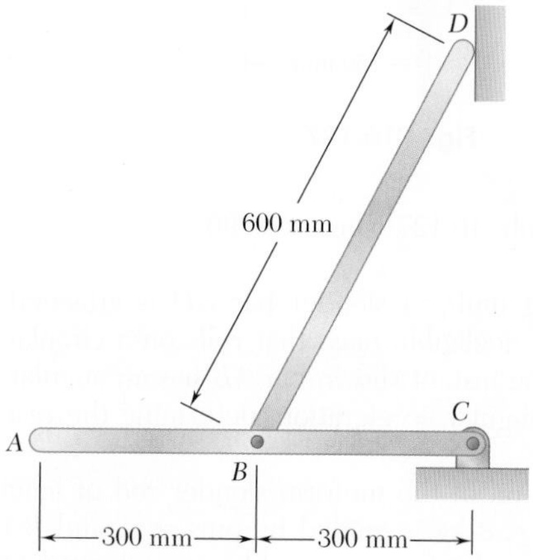

Fig. P16.134

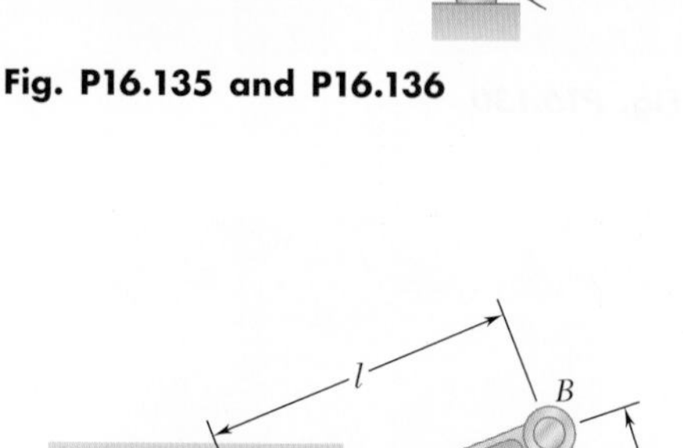

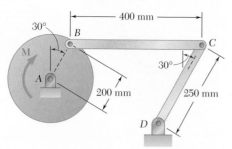

Fig. P16.135 and P16.136

***16.135** The 6-kg rod BC connects a 10-kg disk centered at A to a 5-kg rod CD. The motion of the system is controlled by the couple $\mathbf{M}$ applied to disk A. Knowing that at the instant shown disk A has an angular velocity of 36 rad/s clockwise and no angular acceleration, determine (*a*) the couple $\mathbf{M}$, (*b*) the components of the force exerted at C on rod BC.

***16.136** The 6-kg rod BC connects a 10-kg disk centered at A to a 5-kg rod CD. The motion of the system is controlled by the couple $\mathbf{M}$ applied to disk A. Knowing that at the instant shown disk A has an angular velocity of 36 rad/s clockwise and an angular acceleration of 150 rad/s^2 counterclockwise, determine (*a*) the couple $\mathbf{M}$, (*b*) the components of the force exerted at C on rod BC.

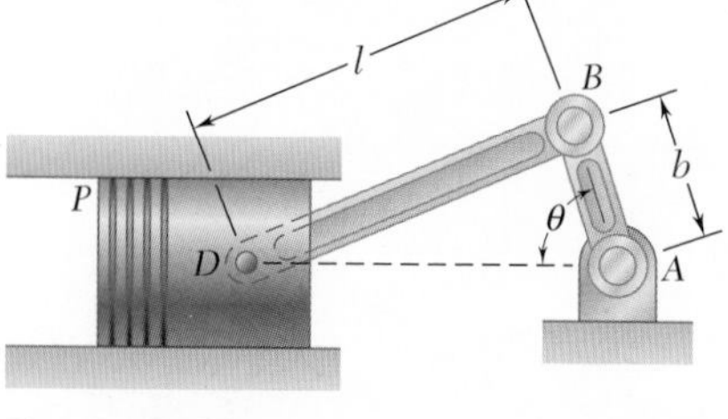

Fig. P16.137

16.137 In the engine system shown $l = 250$ mm and $b = 100$ mm. The connecting rod BD is assumed to be a 1.2-kg uniform slender rod and is attached to the 1.8-kg piston P. During a test of the system, crank AB is made to rotate with a constant angular velocity of 600 rpm clockwise with no force applied to the face of the piston. Determine the forces exerted on the connecting rod at B and D when $\theta = 180°$. (Neglect the effect of the weight of the rod.)

16.138 Solve Prob. 16.137 when $\theta = 90°$.

16.139 The 2-kg rod *AB* and the 3-kg rod *BC* are connected as shown to a disk that is made to rotate in a vertical plane at a constant angular velocity of 6 rad/s clockwise. For the position shown, determine the forces exerted at *A* and *B* on rod *AB*.

16.140 The 2-kg rod *AB* and the 3-kg rod *BC* are connected as shown to a disk that is made to rotate in a vertical plane. Knowing that at the instant shown the disk has an angular acceleration of 18 rad/s^2 clockwise and no angular velocity, determine the components of the forces exerted at *A* and *B* on rod *AB*.

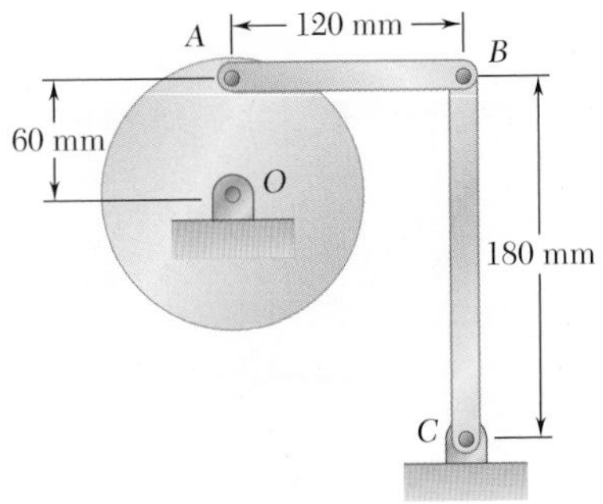

Fig. P16.139 and P16.140

16.141 The linkage *ABCD* is formed by connecting the 3-kg bar *BC* to the 1.5-kg bars *AB* and *CD*. The motion of the linkage is controlled by the couple **M** applied to bar *AB*. Knowing that at the instant shown bar *AB* has an angular velocity of 24 rad/s clockwise and no angular acceleration, determine (*a*) the couple **M**, (*b*) the components of the force exerted at *B* on rod *BC*.

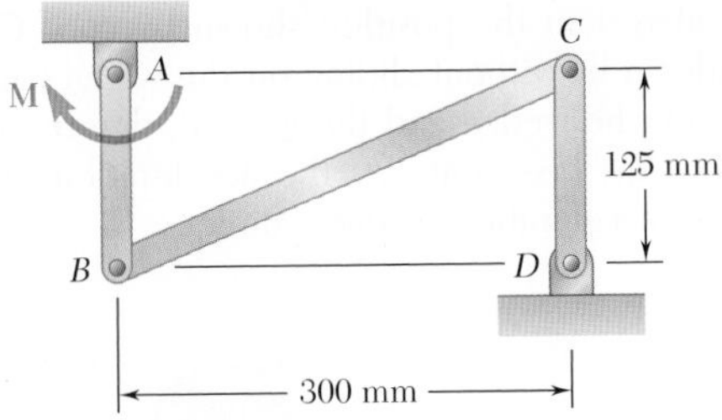

Fig. *P16.141*

16.142 Solve Prob. 16.141, assuming that at the instant shown bar *AB* has an angular velocity of 24 rad/s clockwise and an angular acceleration of 160 rad/s^2 counterclockwise.

***16.143** Draw the shear and bending-moment diagrams for the beam of Prob. 16.84 immediately after the cable at *B* breaks.

***16.144** A uniform slender bar *AB* of mass *m* is suspended as shown from a uniform disk of the same mass *m*. Neglecting the effect of friction, determine the accelerations of points *A* and *B* immediately after a horizontal force **P** has been applied at *B*.

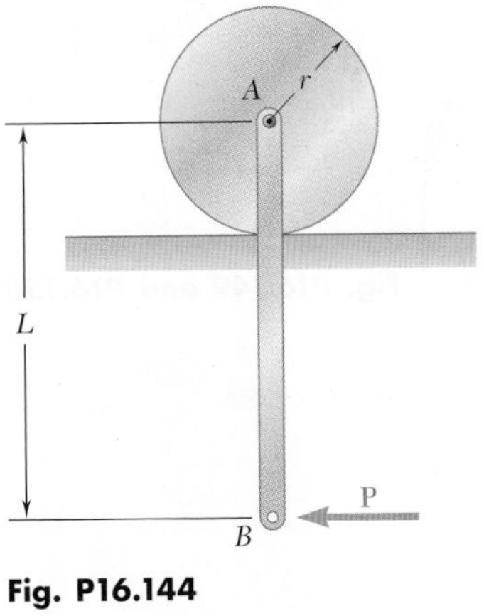

Fig. P16.144

16.145 A uniform rod *AB*, of mass 15 kg and length 1 m, is attached to the 20-kg cart *C*. Neglecting friction, determine immediately after the system has been released from rest, (*a*) the acceleration of the cart, (*b*) the angular acceleration of the rod.

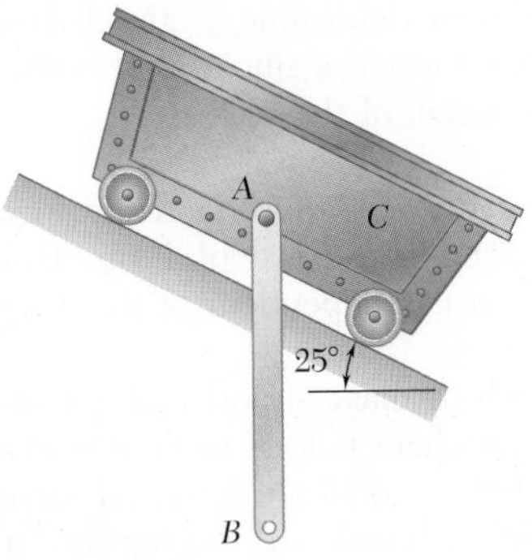

Fig. P16.145

***16.146** The 5-kg slender rod AB is pin-connected to an 8-kg uniform disk as shown. Immediately after the system is released from rest, determine the acceleration of (*a*) point A, (*b*) point B.

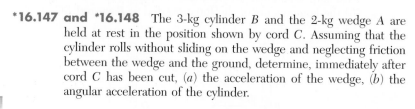

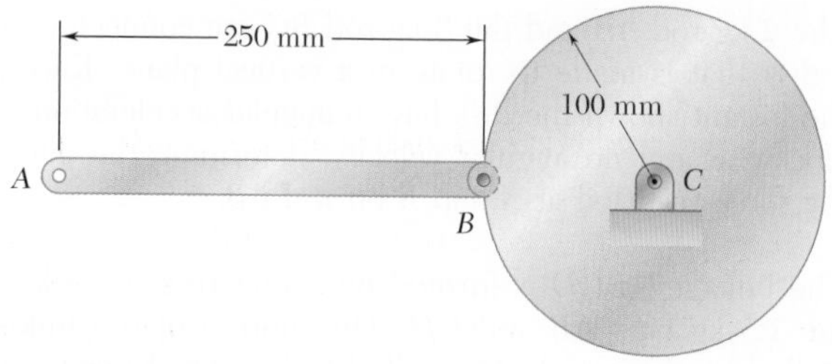

Fig. P16.146

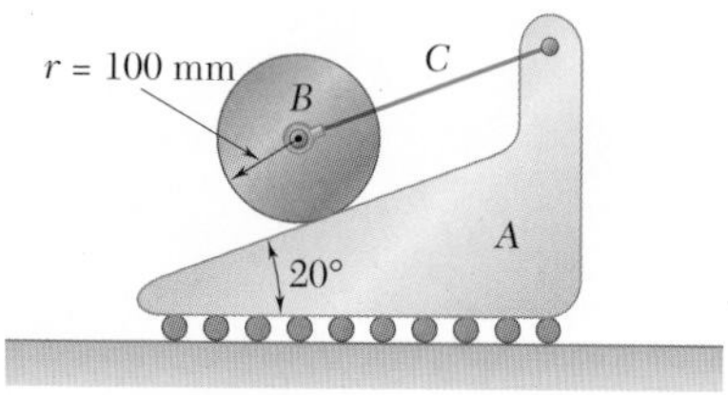

Fig. P16.147

***16.147 and *16.148** The 3-kg cylinder B and the 2-kg wedge A are held at rest in the position shown by cord C. Assuming that the cylinder rolls without sliding on the wedge and neglecting friction between the wedge and the ground, determine, immediately after cord C has been cut, (*a*) the acceleration of the wedge, (*b*) the angular acceleration of the cylinder.

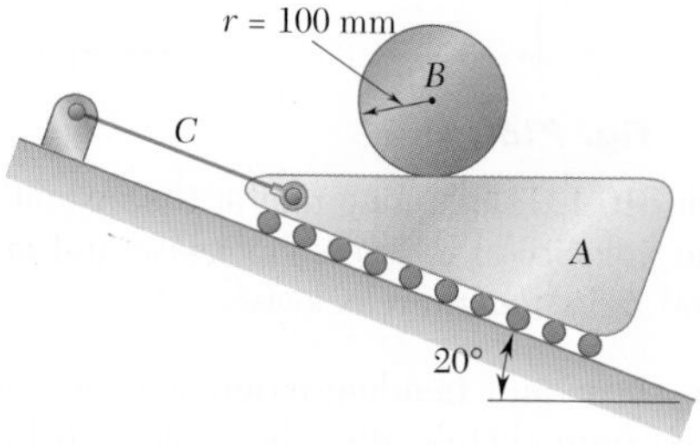

Fig. P16.148

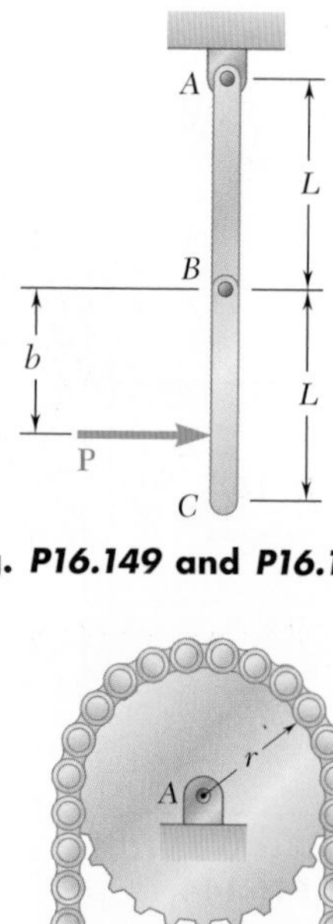

Fig. P16.149 and P16.150

***16.149** Each of the 3-kg bars AB and BC is of length $L = 500$ mm. A horizontal force **P** of magnitude 20 N is applied to bar BC as shown. Knowing that $b = L$ (**P** is applied at C), determine the angular acceleration of each bar.

***16.150** Each of the 3-kg bars AB and BC is of length $L = 500$ mm. A horizontal force **P** of magnitude 20 N is applied to bar BC. For the position shown, determine (*a*) the distance b for which the bars move as if they formed a single rigid body, (*b*) the corresponding angular acceleration of the bars.

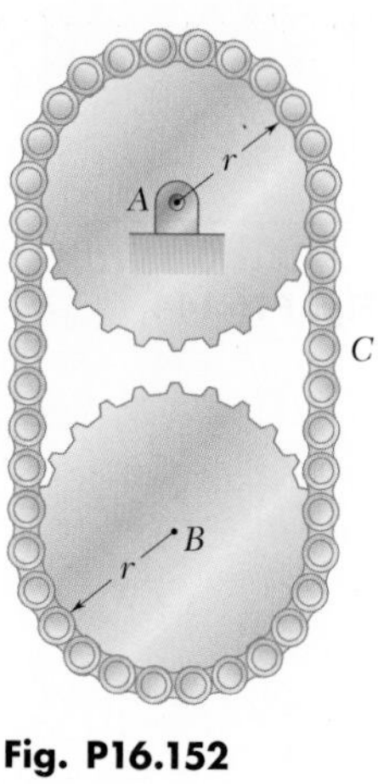

Fig. P16.152

***16.151** (*a*) Determine the magnitude and the location of the maximum bending moment in the rod of Prob. 16.78. (*b*) Show that the answer to part *a* is independent of the weight of the rod.

***16.152** Two disks, each of mass m and radius r are connected as shown by a continuous chain belt of negligible mass. If a pin at point C of the chain belt is suddenly removed, determine (*a*) the angular acceleration of each disk, (*b*) the tension in the left-hand portion of the belt, (*c*) the acceleration of the center of disk B.

REVIEW AND SUMMARY

In this chapter, we studied the *kinetics of rigid bodies,* i.e., the relations existing between the forces acting on a rigid body, the shape and mass of the body, and the motion produced. Except for the first two sections, which apply to the most general case of the motion of a rigid body, our analysis was restricted to the *plane motion of rigid slabs* and rigid bodies symmetrical with respect to the reference plane. The study of the plane motion of nonsymmetrical rigid bodies and of the motion of rigid bodies in three-dimensional space will be considered in Chap. 18.

Fundamental equations of motion for a rigid body

We first recalled [Sec. 16.2] the two fundamental equations derived in Chap. 14 for the motion of a system of particles and observed that they apply in the most general case of the motion of a rigid body. The first equation defines the motion of the mass center G of the body; we have

$$\Sigma \mathbf{F} = m\overline{\mathbf{a}} \tag{16.1}$$

where m is the mass of the body and $\overline{\mathbf{a}}$ the acceleration of G. The second is related to the motion of the body relative to a centroidal frame of reference; we wrote

$$\Sigma \mathbf{M}_G = \dot{\mathbf{H}}_G \tag{16.2}$$

where $\dot{\mathbf{H}}_G$ is the rate of change of the angular momentum $\mathbf{H}_G$ of the body about its mass center G. Together, Eqs. (16.1) and (16.2) express that *the system of the external forces is equipollent to the system consisting of the vector $m\overline{\mathbf{a}}$ attached at G and the couple of moment $\dot{\mathbf{H}}_G$* (Fig. 16.19).

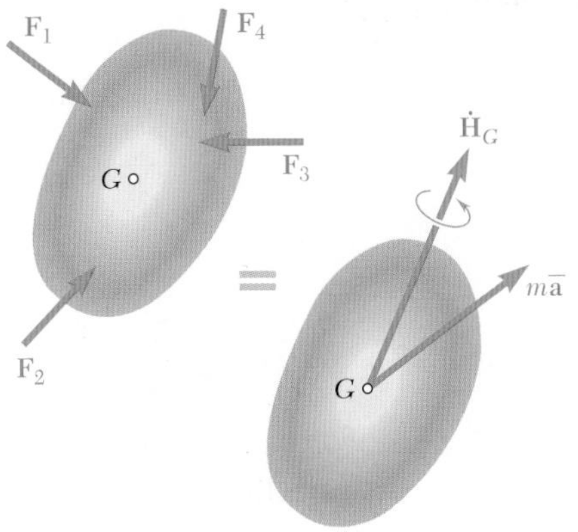

Fig. 16.19

Angular momentum in plane motion

Restricting our analysis at this point and for the rest of the chapter to the plane motion of rigid slabs and rigid bodies symmetrical with respect to the reference plane, we showed [Sec. 16.3] that the angular momentum of the body could be expressed as

$$\mathbf{H}_G = \bar{I}\boldsymbol{\omega} \tag{16.4}$$

where $\bar{I}$ is the moment of inertia of the body about a centroidal axis perpendicular to the reference plane and $\boldsymbol{\omega}$ is the angular velocity of the body. Differentiating both members of Eq. (16.4), we obtained

$$\dot{\mathbf{H}}_G = \bar{I}\dot{\boldsymbol{\omega}} = \bar{I}\boldsymbol{\alpha} \tag{16.5}$$

which shows that in the restricted case considered here, the rate of change of the angular momentum of the rigid body can be represented by a vector of the same direction as $\boldsymbol{\alpha}$ (i.e., perpendicular to the plane of reference) and of magnitude $\bar{I}\alpha$.

Equations for the plane motion of a rigid body

It follows from [Sec. 16.4] that the plane motion of a rigid slab or of a rigid body symmetrical with respect to the reference plane is defined by the three scalar equations

$$\Sigma F_x = m\bar{a}_x \qquad \Sigma F_y = m\bar{a}_y \qquad \Sigma M_G = \bar{I}\alpha \qquad (16.6)$$

D'Alembert's principle

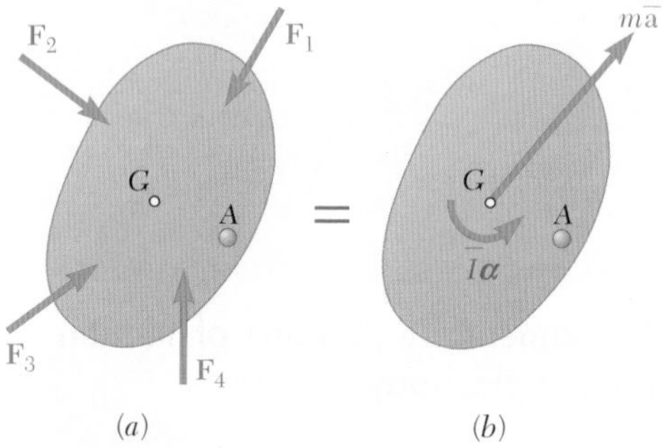

Fig. 16.20

It further follows that *the external forces acting on the rigid body are actually* ***equivalent*** *to the effective forces of the various particles forming the body*. This statement, known as *d'Alembert's principle,* can be expressed in the form of the vector diagram shown in Fig. 16.20, where the effective forces have been represented by a vector $m\bar{\mathbf{a}}$ attached at G and a couple $\bar{I}\boldsymbol{\alpha}$. In the particular case of a slab in *translation,* the effective forces shown in part b of this figure reduce to the single vector $m\bar{\mathbf{a}}$, while in the particular case of a slab in *centroidal rotation,* they reduce to the single couple $\bar{I}\boldsymbol{\alpha}$; in any other case of plane motion, both the vector $m\bar{\mathbf{a}}$ and the couple $\bar{I}\boldsymbol{\alpha}$ should be included.

Free-body-diagram equation

Any problem involving the plane motion of a rigid slab may be solved by drawing a *free-body-diagram equation* similar to that of Fig. 16.20 [Sec. 16.6]. Three equations of motion can then be obtained by equating the x components, y components, and moments about an arbitrary point A, of the forces and vectors involved [Sample Probs. 16.1, 16.2, 16.4, and 16.5]. An alternative solution can be obtained by adding to the external forces an *inertia vector* $-m\bar{\mathbf{a}}$ of sense opposite to that of $\bar{\mathbf{a}}$, attached at G, and an *inertia couple* $-\bar{I}\boldsymbol{\alpha}$ of sense opposite to that of $\boldsymbol{\alpha}$. The system obtained in this way is equivalent to zero, and the slab is said to be in *dynamic equilibrium.*

Connected rigid bodies

The method described above can also be used to solve problems involving the plane motion of several connected rigid bodies [Sec. 16.7]. A free-body-diagram equation is drawn for each part of the system and the equations of motion obtained are solved simultaneously. In some cases, however, a single diagram can be drawn for the entire system, including all the external forces as well as the vectors $m\bar{\mathbf{a}}$ and the couples $\bar{I}\boldsymbol{\alpha}$ associated with the various parts of the system [Sample Prob. 16.3].

Constrained plane motion

In the second part of the chapter, we were concerned with rigid bodies *moving under given constraints* [Sec. 16.8]. While the kinetic analysis of the constrained plane motion of a rigid slab is the same as above, it must be supplemented by a *kinematic analysis* which has for its object to express the components $\bar{a}_x$ and $\bar{a}_y$ of the acceleration of the mass center G of the slab in terms of its angular acceleration α. Problems solved in this way included the *noncentroidal rotation* of rods and plates [Sample Probs. 16.6 and 16.7], the *rolling motion* of spheres and wheels [Sample Probs. 16.8 and 16.9], and the plane motion of *various types of linkages* [Sample Prob. 16.10].

REVIEW PROBLEMS

16.153 A cyclist is riding a bicycle at a speed of 30 kmph on a horizontal road. The distance between the axles is 1050 mm, and the mass center of the cyclist and the bicycle is located 650 mm behind the front axle and 1000 mm above the ground. If the cyclist applies the brakes only on the front wheel, determine the shortest distance in which he can stop without being thrown over the front wheel.

16.154 The forklift truck shown has a mass of 1125 kg and is used to lift a crate of mass $m = 1250$ kg. The truck is moving to the left at a speed of 3 m/s when the brakes are applied on all four wheels. Knowing that the coefficient of static friction between the crate and the fork lift is 0.30, determine the smallest distance in which the truck can be brought to a stop if the crate is not to slide and if the truck is not to tip forward.

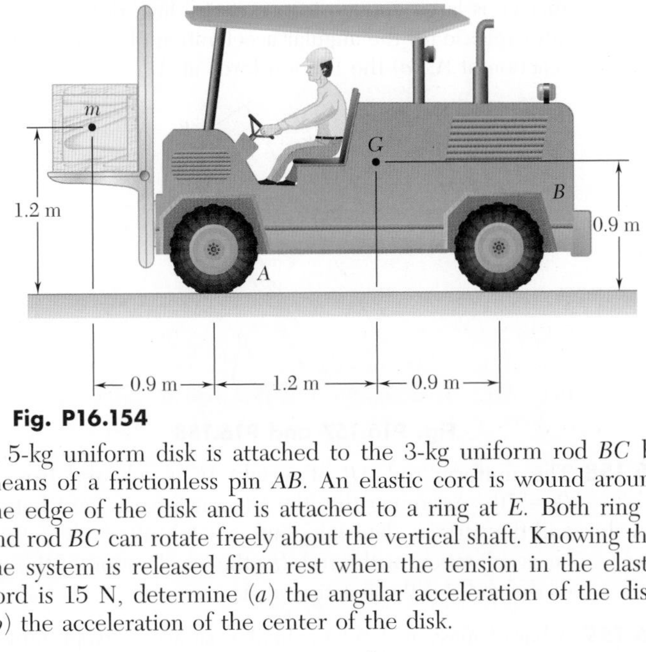

Fig. P16.154

16.155 A 5-kg uniform disk is attached to the 3-kg uniform rod *BC* by means of a frictionless pin *AB*. An elastic cord is wound around the edge of the disk and is attached to a ring at *E*. Both ring *E* and rod *BC* can rotate freely about the vertical shaft. Knowing that the system is released from rest when the tension in the elastic cord is 15 N, determine (*a*) the angular acceleration of the disk, (*b*) the acceleration of the center of the disk.

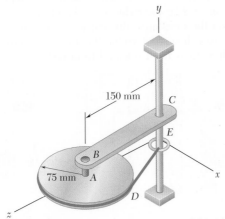

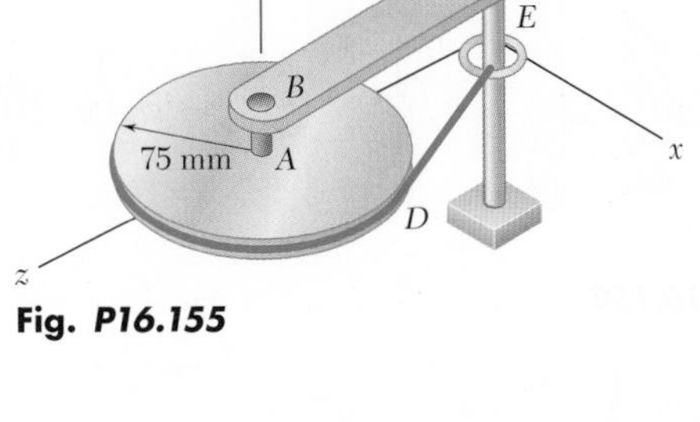

Fig. P16.155

16.156 Identical cylinders of mass m and radius r are pushed by a series of moving arms. Assuming the coefficient of friction between all surfaces to be $\mu < 1$ and denoting by a the magnitude of the acceleration of the arms, derive an expression for (*a*) the maximum allowable value of a if each cylinder is to roll without sliding, (*b*) the minimum allowable value of a if each cylinder is to move to the right without rotating.

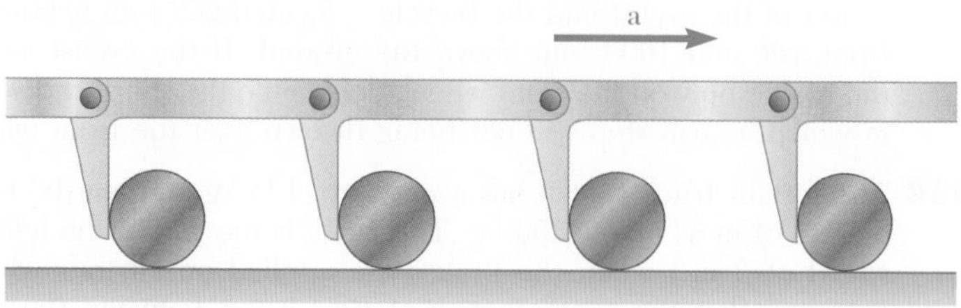

Fig. P16.156

16.157 The uniform rod AB of weight W is released from rest when $\beta = 70°$. Assuming that the friction force between end A and the surface is large enough to prevent sliding, determine immediately after release (*a*) the angular acceleration of the rod, (*b*) the normal reaction at A, (*c*) the friction force at A.

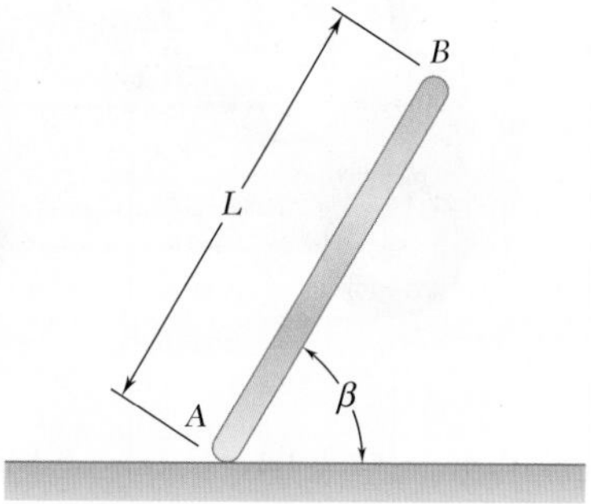

Fig. P16.157 and P16.158

16.158 The uniform rod AB of weight W is released from rest when $\beta = 70°$. Assuming that the friction force is zero between end A and the surface, determine immediately after release (*a*) the angular acceleration of the rod, (*b*) the acceleration of the mass center of the rod, (*c*) the reaction at A.

16.159 A bar of mass $m = 5$ kg is held as shown between four disks, each of mass $m' = 2$ kg and radius $r = 75$ mm. Knowing that the normal forces on the disks are sufficient to prevent any slipping, for each of the cases shown determine the acceleration of the bar immediately after it has been released from rest.

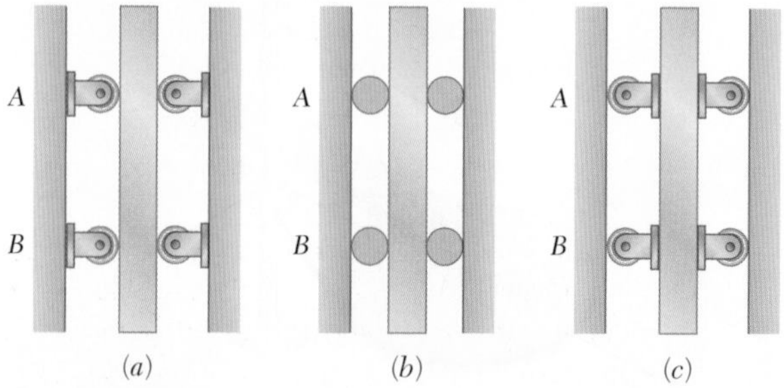

Fig. *P16.159*

16.160 A uniform plate of mass m is suspended in each of the ways shown. For each case determine immediately after the connection B has been released (a) the angular acceleration of the plate, (b) the acceleration of its mass center.

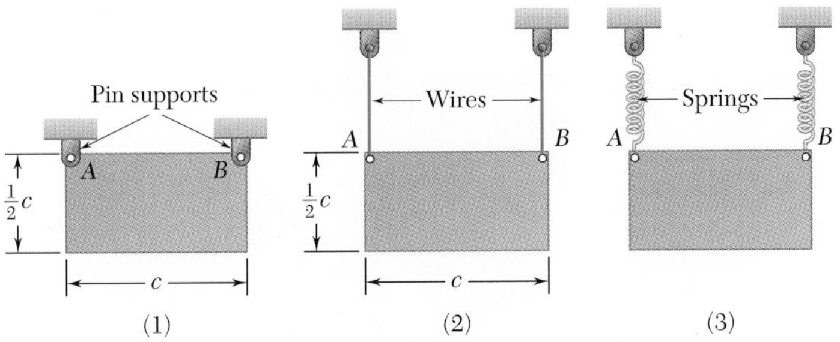

Fig. P16.160

16.161 A cylinder with a circular hole is rolling without slipping on a fixed curved surface as shown. The cylinder would have a mass of 8 kg without the hole, but with the hole it has a mass of 7.5 kg. Knowing that at the instant shown the disk has an angular velocity of 5 rad/s clockwise, determine (a) the angular acceleration of the disk, (b) the components of the reaction force between the cylinder and the ground at this instant.

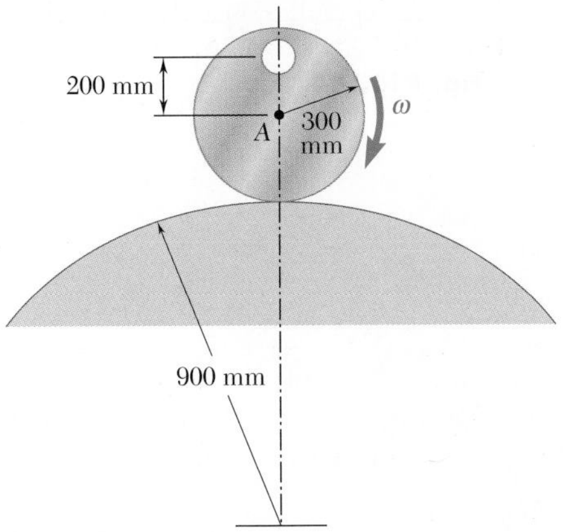

Fig. *P16.161*

16.162 The motion of a square plate of side 150 mm and mass 2.5 kg is guided by pins at corners A and B that slide in slots cut in a vertical wall. Immediately after the plate is released from rest in the position shown, determine (a) the angular acceleration of the plate, (b) the reaction at corner A.

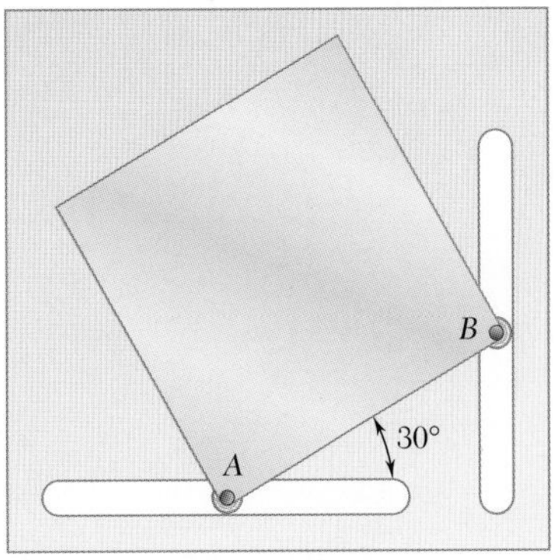

Fig. P16.162

16.163 The motion of a square plate of side 150 mm and mass 2.5 kg is guided by a pin at corner A that slides in a horizontal slot cut in a vertical wall. Immediately after the plate is released from rest in the position shown, determine (*a*) the angular acceleration of the plate, (*b*) the reaction at corner A.

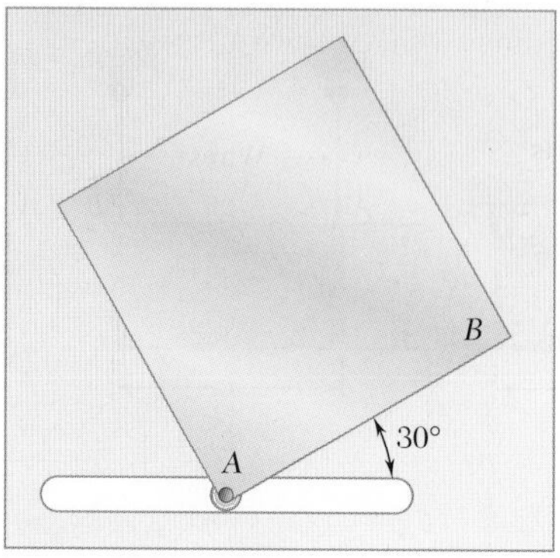

Fig. P16.163

16.164 Two slender rods, each of length l and mass m, are released from rest in the position shown. Knowing that a small knob at end B of rod AB bears on rod CD, determine immediately after release (*a*) the acceleration of end C of rod CD, (*b*) the force exerted on the knob.

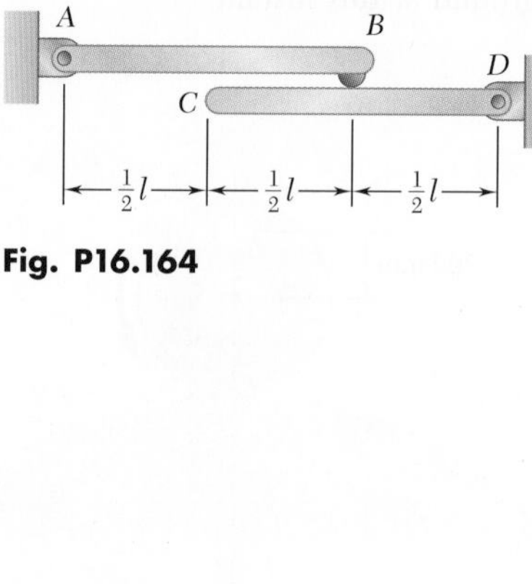

Fig. P16.164

COMPUTER PROBLEMS

16.C1 The 2.5-kg rod AB is released from rest in the position shown. (a) Assuming that the friction force between end A and the surface is large enough to prevent sliding, using software calculate the normal reaction and the friction force at A immediately after release for values of β from 0 to 85°. (b) Knowing that the coefficient of static friction between the rod and the floor is actually equal to 0.50, determine the range of values of β for which the rod will slip immediately after being released from rest.

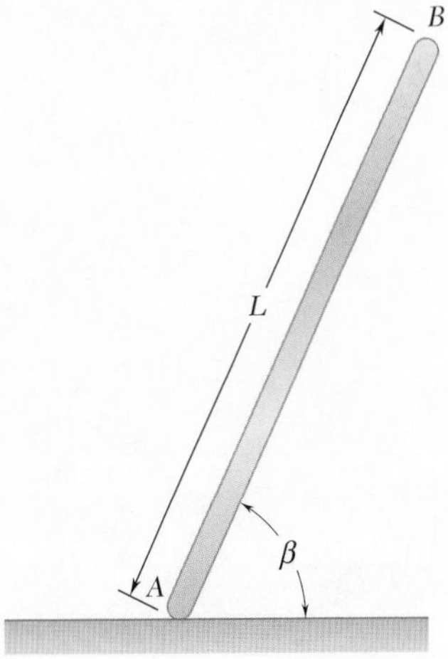

Fig. P16.C1

16.C2 End A of the 5-kg rod AB is moved to the left at a constant speed $v_A = 1.5$ m/s. Using computational software calculate and plot the normal reactions at ends A and B of the rod for values of θ from 0 to 50°. Determine the value of θ at which end B of the rod loses contact with the wall.

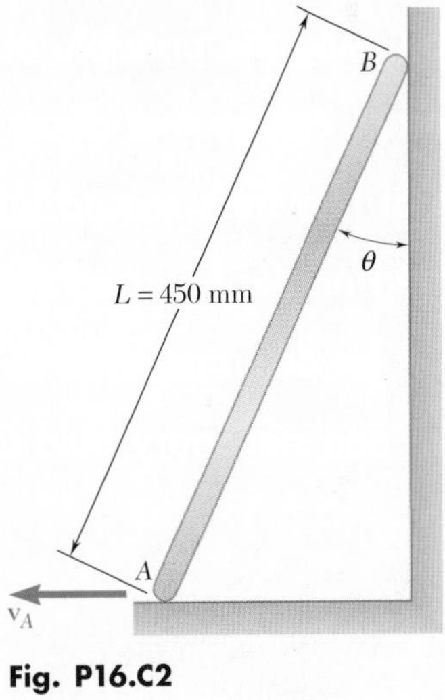

Fig. P16.C2

16.C3 A 15-kg cylinder of diameter $b = 200$ mm and height $h = 150$ mm is placed on a 5-kg platform CD that is held in the position shown by three cables. It is desired to determine the minimum value of μ_s between the cylinder and the platform for which the cylinder does not slip on the platform, immediately after cable AB is cut. Using computational software calculate and plot the minimum allowable value of μ_s for values of θ from 0 to 30°. Knowing that the actual value of μ_s is 0.60, determine the value of θ at which slipping impends.

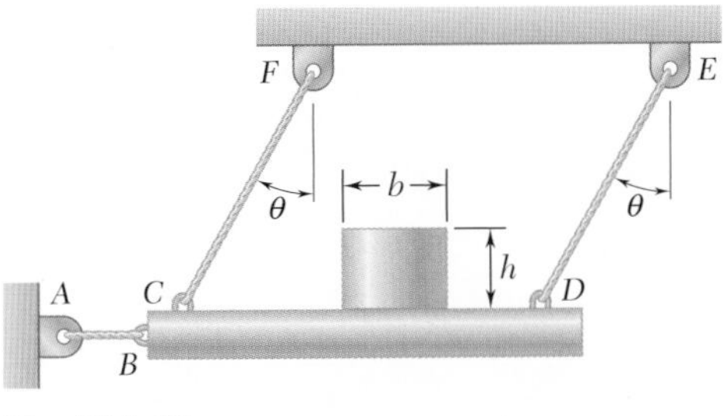

Fig. P16.C3

16.C4 For the engine system of Prob. 15.C3 of Chap. 15, the masses of piston P and the connecting rod BD are 2.5 kg and 3 kg, respectively. Knowing that during a test of the system no force is applied to the face of the piston, use computational software to calculate and plot the horizontal and vertical components of the dynamic reactions exerted on the connecting rod at B and D for values of θ from 0 to 180°.

16.C5 A uniform slender bar AB of mass m is suspended from springs AC and BD as shown. Using computational software calculate and plot the accelerations of ends A and B, immediately after spring AC has broken, for values of θ from 0 to 90°.

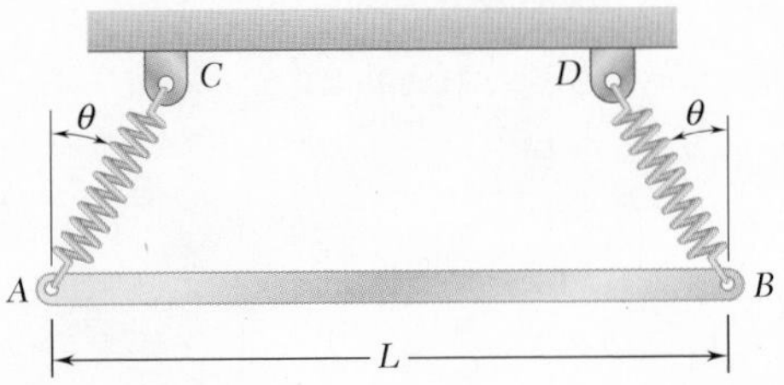

Fig. P16.C5

In this chapter the energy and momentum methods will be added to the tools available for your study of the motion of rigid bodies. For example, by using the principle of conservation of energy and direct application of Newton's second law, the forces exerted on the hands of this gymnast can be determined as he swings from one stationary hold to another.

CHAPTER 17

Plane Motion of Rigid Bodies: Energy and Momentum Methods

Chapter 17 Plane Motion of Rigid Bodies: Energy and Momentum Methods

17.1 INTRODUCTION

In this chapter the method of work and energy and the method of impulse and momentum will be used to analyze the plane motion of rigid bodies and of systems of rigid bodies.

The method of work and energy will be considered first. In Secs. 17.2 through 17.5, the work of a force and of a couple will be defined, and an expression for the kinetic energy of a rigid body in plane motion will be obtained. The principle of work and energy will then be used to solve problems involving displacements and velocities. In Sec. 17.6, the principle of conservation of energy will be applied to the solution of a variety of engineering problems.

In the second part of the chapter, the principle of impulse and momentum will be applied to the solution of problems involving velocities and time (Secs. 17.8 and 17.9) and the concept of conservation of angular momentum will be introduced and discussed (Sec. 17.10).

In the last part of the chapter (Secs. 17.11 and 17.12), problems involving the eccentric impact of rigid bodies will be considered. As was done in Chap. 13, where we analyzed the impact of particles, the coefficient of restitution between the colliding bodies will be used together with the principle of impulse and momentum in the solution of impact problems. It will also be shown that the method used is applicable not only when the colliding bodies move freely after the impact but also when the bodies are partially constrained in their motion.

17.2 PRINCIPLE OF WORK AND ENERGY FOR A RIGID BODY

The principle of work and energy will now be used to analyze the plane motion of rigid bodies. As was pointed out in Chap. 13, the method of work and energy is particularly well adapted to the solution of problems involving velocities and displacements. Its main advantage resides in the fact that the work of forces and the kinetic energy of particles are scalar quantities.

In order to apply the principle of work and energy to the analysis of the motion of a rigid body, it will again be assumed that the rigid body is made of a large number n of particles of mass Δm_i. Recalling Eq. (14.30) of Sec. 14.8, we write

$$T_1 + U_{1\rightarrow 2} = T_2 \tag{17.1}$$

where T_1, T_2 = initial and final values of total kinetic energy of particles forming the rigid body

$U_{1\rightarrow 2}$ = work of all forces acting on various particles of the body

The total kinetic energy

$$T = \frac{1}{2}\sum_{i=1}^{n} \Delta m_i\, v_i^2 \tag{17.2}$$

is obtained by adding positive scalar quantities and is itself a positive scalar quantity. You will see later how T can be determined for various types of motion of a rigid body.

Photo 17.1 The work done by friction reduces the kinetic energy of the wheel.

The expression $U_{1\to2}$ in (17.1) represents the work of all the forces acting on the various particles of the body, whether these forces are internal or external. However, as you will see presently, the total work of the internal forces holding together the particles of a rigid body is zero. Consider two particles A and B of a rigid body and the two equal and opposite forces $\mathbf{F}$ and $-\mathbf{F}$ they exert on each other (Fig. 17.1). While, in general, small displacements $d\mathbf{r}$ and $d\mathbf{r}'$ of the two particles are different, the components of these displacements along AB must be equal; otherwise, the particles would not remain at the same distance from each other and the body would not be rigid. Therefore, the work of $\mathbf{F}$ is equal in magnitude and opposite in sign to the work of $-\mathbf{F}$, and their sum is zero. Thus, the total work of the internal forces acting on the particles of a rigid body is zero, and *the expression $U_{1\to2}$ in Eq. (17.1) reduces to the work of the external forces acting on the body during the displacement considered.*

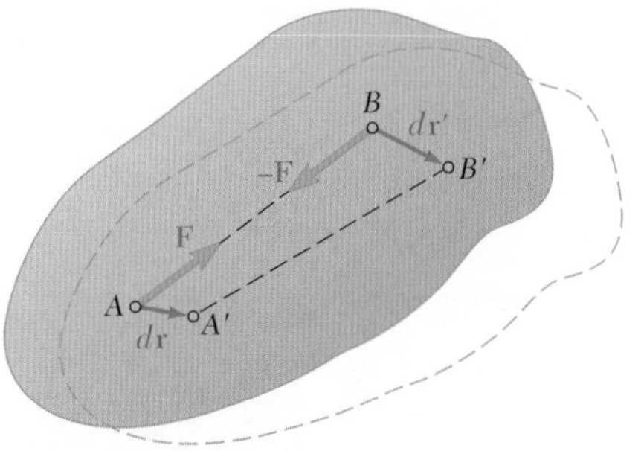

Fig. 17.1

17.3 WORK OF FORCES ACTING ON A RIGID BODY

We saw in Sec. 13.2 that the work of a force $\mathbf{F}$ during a displacement of its point of application from A_1 to A_2 is

$$U_{1\to2} = \int_{A_1}^{A_2} \mathbf{F} \cdot d\mathbf{r} \tag{17.3}$$

or

$$U_{1\to2} = \int_{s_1}^{s_2} (F \cos \alpha)\, ds \tag{17.3'}$$

where F is the magnitude of the force, α is the angle it forms with the direction of motion of its point of application A, and s is the variable of integration which measures the distance traveled by A along its path.

In computing the work of the external forces acting on a rigid body, it is often convenient to determine the work of a couple without considering separately the work of each of the two forces forming the couple. Consider the two forces $\mathbf{F}$ and $-\mathbf{F}$ forming a couple of moment $\mathbf{M}$ and acting on a rigid body (Fig. 17.2). Any small displacement of the rigid body bringing A and B, respectively, into A' and B'' can be divided into two parts: in one part points A and B undergo equal displacements $d\mathbf{r}_1$; in the other part A' remains fixed while B' moves into B'' through a displacement $d\mathbf{r}_2$ of magnitude $ds_2 = r\, d\theta$. In the first part of the motion, the work of $\mathbf{F}$ is equal in magnitude and opposite in sign to the work of $-\mathbf{F}$ and their sum is zero. In the second part of the motion, only force $\mathbf{F}$ works, and its work is $dU = F\, ds_2 = Fr\, d\theta$. But the product Fr is equal to the magnitude M of the moment of the couple. Thus, the work of a couple of moment $\mathbf{M}$ acting on a rigid body is

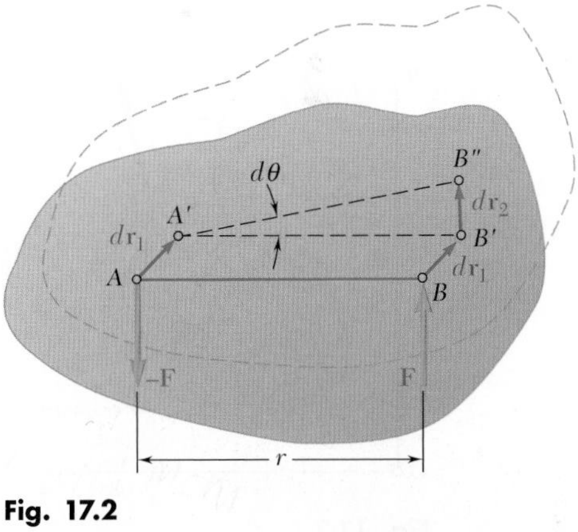

Fig. 17.2

$$dU = M\, d\theta \tag{17.4}$$

where $d\theta$ is the small angle expressed in radians through which the body rotates. We again note that work should be expressed in units obtained by multiplying units of force by units of length. The work of the couple during a finite rotation of the rigid body is obtained by integrating both members of (17.4) from the initial value θ_1 of the angle θ to its final value θ_2. We write

$$U_{1\rightarrow 2} = \int_{\theta_1}^{\theta_2} M\,d\theta \tag{17.5}$$

When the moment **M** *of the couple is constant,* formula (17.5) reduces to

$$U_{1\rightarrow 2} = M(\theta_2 - \theta_1) \tag{17.6}$$

It was pointed out in Sec. 13.2 that a number of forces encountered in problems of kinetics *do no work.* They are forces applied to fixed points or acting in a direction perpendicular to the displacement of their point of application. Among the forces which do no work the following have been listed: the reaction at a frictionless pin when the body supported rotates about the pin, the reaction at a frictionless surface when the body in contact moves along the surface, and the weight of a body when its center of gravity moves horizontally. We can add now that *when a rigid body rolls without sliding on a fixed surface, the friction force* **F** *at the point of contact C does no work.* The velocity $\mathbf{v}_C$ of the point of contact C is zero, and the work of the friction force **F** during a small displacement of the rigid body is

$$dU = F\,ds_C = F(v_C\,dt) = 0$$

17.4 KINETIC ENERGY OF A RIGID BODY IN PLANE MOTION

Consider a rigid body of mass m in plane motion. We recall from Sec. 14.7 that, if the absolute velocity $\mathbf{v}_i$ of each particle P_i of the body is expressed as the sum of the velocity $\overline{\mathbf{v}}$ of the mass center G of the body and of the velocity $\mathbf{v}'_i$ of the particle relative to a frame $Gx'y'$ attached to G and of fixed orientation (Fig. 17.3), the kinetic energy of the system of particles forming the rigid body can be written in the form

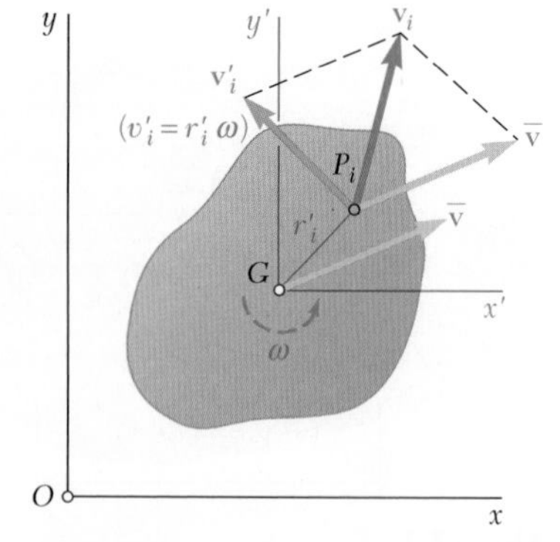

Fig. 17.3

$$T = \tfrac{1}{2}m\overline{v}^2 + \frac{1}{2}\sum_{i=1}^{n} \Delta m_i v_i'^2 \tag{17.7}$$

But the magnitude v'_i of the relative velocity of P_i is equal to the product $r'_i\omega$ of the distance r'_i of P_i from the axis through G perpendicular to the plane of motion and of the magnitude ω of the angular velocity of the body at the instant considered. Substituting into (17.7), we have

$$T = \tfrac{1}{2}m\overline{v}^2 + \frac{1}{2}\left(\sum_{i=1}^{n} r_i'^2\,\Delta m_i\right)w^2 \tag{17.8}$$

or, since the sum represents the moment of inertia $\overline{I}$ of the body about the axis through G,

$$T = \tfrac{1}{2}m\overline{v}^2 + \tfrac{1}{2}\overline{I}\omega^2 \tag{17.9}$$

We note that in the particular case of a body in translation ($\omega = 0$), the expression obtained reduces to $\frac{1}{2}m\bar{v}^2$, while in the case of a centroidal rotation ($\bar{v} = 0$), it reduces to $\frac{1}{2}\bar{I}\omega^2$. We conclude that the kinetic energy of a rigid body in plane motion can be separated into two parts: (1) the kinetic energy $\frac{1}{2}m\bar{v}^2$ associated with the motion of the mass center G of the body, and (2) the kinetic energy $\frac{1}{2}\bar{I}\omega^2$ associated with the rotation of the body about G.

Noncentroidal Rotation. The relation (17.9) is valid for any type of plane motion and can therefore be used to express the kinetic energy of a rigid body rotating with an angular velocity $\boldsymbol{\omega}$ about a fixed axis through O (Fig. 17.4). In that case, however, the kinetic energy of the body can be expressed more directly by noting that the speed v_i of the particle P_i is equal to the product $r_i\omega$ of the distance r_i of P_i from the fixed axis and the magnitude ω of the angular velocity of the body at the instant considered. Substituting into (17.2), we write

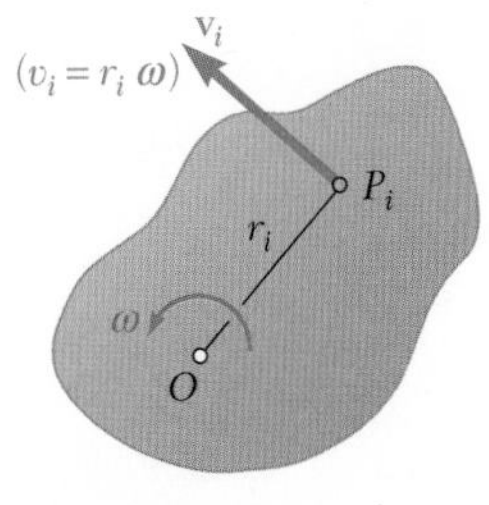

Fig. 17.4

$$T = \frac{1}{2}\sum_{i=1}^{n} \Delta m_i (r_i\omega)^2 = \frac{1}{2}\left(\sum_{i=1}^{n} r_i^2\,\Delta m_i\right)\omega^2$$

or, since the last sum represents the moment of inertia I_O of the body about the fixed axis through O,

$$T = \tfrac{1}{2}I_O\omega^2 \tag{17.10}$$

We note that the results obtained are not limited to the motion of plane slabs or to the motion of bodies which are symmetrical with respect to the reference plane, and can be applied to the study of the plane motion of any rigid body, regardless of its shape. However, since Eq. (17.9) is applicable to any plane motion while Eq. (17.10) is applicable only in cases involving noncentroidal rotation, Eq. (17.9) will be used in the solution of all the sample problems.

17.5 SYSTEMS OF RIGID BODIES

When a problem involves several rigid bodies, each rigid body can be considered separately and the principle of work and energy can be applied to each body. Adding the kinetic energies of all the particles and considering the work of all the forces involved, we can also write the equation of work and energy for the entire system. We have

$$T_1 + U_{1\rightarrow 2} = T_2 \tag{17.11}$$

where T represents the arithmetic sum of the kinetic energies of the rigid bodies forming the system (all terms are positive) and $U_{1\rightarrow 2}$ represents the work of all the forces acting on the various bodies, whether these forces are *internal* or *external* from the point of view of the system as a whole.

The method of work and energy is particularly useful in solving problems involving pin-connected members, blocks and pulleys connected by inextensible cords, and meshed gears. In all these cases, the internal forces occur by pairs of equal and opposite forces, and the points of application of the forces in each pair *move through equal distances* during a small displacement of the system. As a result, the work of the internal forces is zero and $U_{1\rightarrow 2}$ reduces to the work of the *forces external to the system.*

17.6 CONSERVATION OF ENERGY

We saw in Sec. 13.6 that the work of conservative forces, such as the weight of a body or the force exerted by a spring, can be expressed as a change in potential energy. When a rigid body, or a system of rigid bodies, moves under the action of conservative forces, the principle of work and energy stated in Sec. 17.2 can be expressed in a modified form. Substituting for $U_{1\rightarrow 2}$ from (13.19′) into (17.1), we write

$$T_1 + V_1 = T_2 + V_2 \tag{17.12}$$

Formula (17.12) indicates that when a rigid body, or a system of rigid bodies, moves under the action of conservative forces, *the sum of the kinetic energy and of the potential energy of the system remains constant.* It should be noted that in the case of the plane motion of a rigid body, the kinetic energy of the body should include both the *translational* term $\frac{1}{2}m\bar{v}^2$ and the *rotational* term $\frac{1}{2}\bar{I}\omega^2$.

As an example of application of the principle of conservation of energy, let us consider a slender rod AB, of length l and mass m, whose extremities are connected to blocks of negligible mass sliding along horizontal and vertical tracks. We assume that the rod is released with no initial velocity from a horizontal position (Fig. 17.5*a*), and we wish to determine its angular velocity after it has rotated through an angle θ (Fig. 17.5*b*).

Since the initial velocity is zero, we have $T_1 = 0$. Measuring the potential energy from the level of the horizontal track, we write $V_1 = 0$. After the rod has rotated through θ, the center of gravity G of the rod is at a distance $\frac{1}{2}l \sin \theta$ below the reference level and we have

$$V_2 = -\tfrac{1}{2}Wl \sin \theta = -\tfrac{1}{2}mgl \sin \theta$$

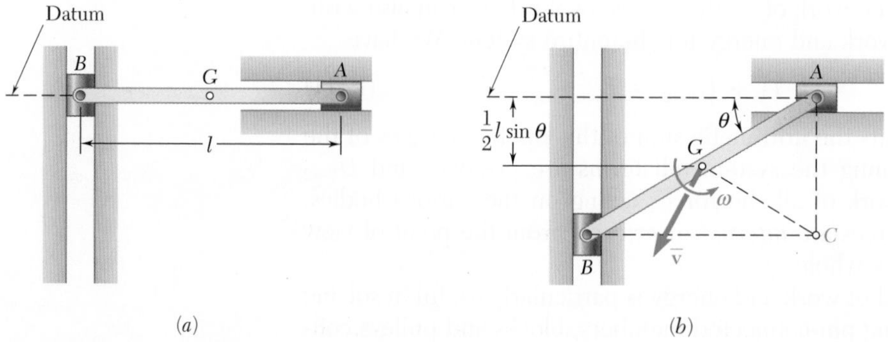

Fig. 17.5

Observing that in this position the instantaneous center of the rod is located at C and that $CG = \frac{1}{2}l$, we write $\bar{v}_2 = \frac{1}{2}l\omega$ and obtain

$$T_2 = \tfrac{1}{2}m\bar{v}_2^2 + \tfrac{1}{2}\bar{I}\omega_2^2 = \tfrac{1}{2}m(\tfrac{1}{2}l\omega)^2 + \tfrac{1}{2}(\tfrac{1}{12}ml^2)\omega^2$$
$$= \frac{1}{2}\frac{ml^2}{3}\omega^2$$

Applying the principle of conservation of energy, we write

$$T_1 + V_1 = T_2 + V_2$$
$$0 = \frac{1}{2}\frac{ml^2}{3}\omega^2 - \tfrac{1}{2}mgl\sin\theta$$
$$\omega = \left(\frac{3g}{l}\sin\theta\right)^{1/2}$$

The advantages of the method of work and energy, as well as its shortcomings, were indicated in Sec. 13.4. Here we should add that the method of work and energy must be supplemented by the application of d'Alembert's principle when reactions at fixed axles, rollers, or sliding blocks are to be determined. For example, in order to compute the reactions at the extremities A and B of the rod of Fig. 17.5b, a diagram should be drawn to express that the system of the external forces applied to the rod is equivalent to the vector $m\bar{\mathbf{a}}$ and the couple $\bar{I}\boldsymbol{\alpha}$. The angular velocity $\boldsymbol{\omega}$ of the rod, however, is determined by the method of work and energy before the equations of motion are solved for the reactions. The complete analysis of the motion of the rod and of the forces exerted on the rod requires, therefore, the combined use of the method of work and energy and of the principle of equivalence of the external and effective forces.

17.7 POWER

Power was defined in Sec. 13.5 as the time rate at which work is done. In the case of a body acted upon by a force $\mathbf{F}$, and moving with a velocity $\mathbf{v}$, the power was expressed as follows:

$$\text{Power} = \frac{dU}{dt} = \mathbf{F}\cdot\mathbf{v} \qquad (13.13)$$

In the case of a rigid body rotating with an angular velocity $\boldsymbol{\omega}$ and acted upon by a couple of moment $\mathbf{M}$ parallel to the axis of rotation, we have, by (17.4),

$$\text{Power} = \frac{dU}{dt} = \frac{M\,d\theta}{dt} = M\omega \qquad (17.13)$$

The various units used to measure power, such as the watt and the horsepower, were defined in Sec. 13.5.

SAMPLE PROBLEM 17.1

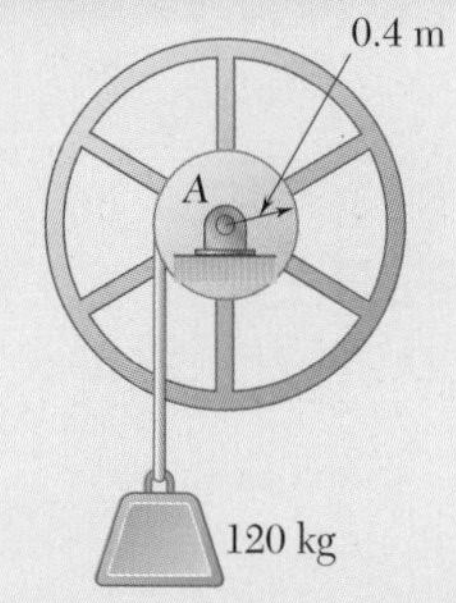

A 120-kg block is suspended from an inextensible cable which is wrapped around a drum of 0.4 m radius rigidly attached to a flywheel. The drum and flywheel have a combined centroidal moment of inertia $\bar{I} = 16\ \text{kg}\cdot\text{m}^2$ At the instant shown, the velocity of the block is 2 m/s directed downward. Knowing that the bearing at A is poorly lubricated and that the bearing friction is equivalent to a couple **M** of magnitude 90 N · m, determine the velocity of the block after it has moved 1.25 m downward.

SOLUTION

We consider the system formed by the flywheel and the block. Since the cable is inextensible, the work done by the internal forces exerted by the cable cancels. The initial and final positions of the system and the external forces acting on the system are as shown.

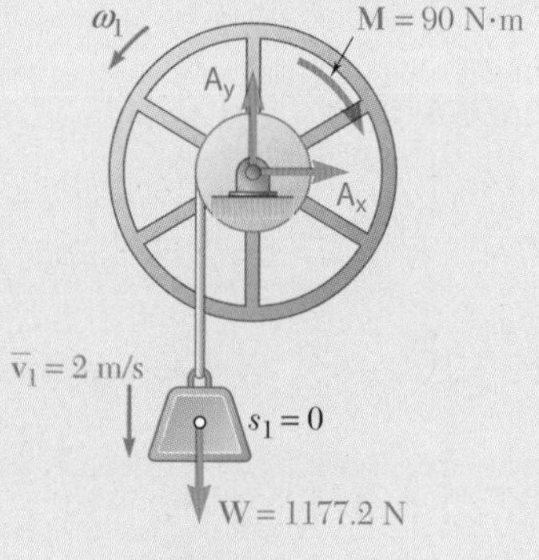

Kinetic Energy. ***Position 1.***

Block: $\bar{v}_1 = 2\ \text{m/s}$

Flywheel: $w_1 = \dfrac{\bar{v}_1}{r} = \dfrac{2\ \text{m/s}}{0.4\ \text{m}} = 5\ \text{rad/s}$

$$T_1 = \tfrac{1}{2}m\bar{v}_1^2 + \tfrac{1}{2}\bar{I}\omega_1^2 \text{ (observing that velocity of centre of mass of drum = 0)}$$

$$= \frac{1}{2}(120\ \text{kg})\ (2\ \text{m/s})^2 + \tfrac{1}{2}(16\ \text{kg}\cdot\text{m}^2)(5\ \text{rad/s})^2$$

$$= 440\ \text{J}$$

Position 2. Noting that $\omega_2 = \bar{v}_2/0.4$ we write

$$T_2 = \tfrac{1}{2}m\bar{v}_2^2 + \tfrac{1}{2}\bar{I}\omega_2^2$$

$$= \frac{1}{2}(120)\ \bar{v}_2^{\,2} + \tfrac{1}{2}(16)\left(\frac{\bar{v}_2}{0.4}\right)^2 = 110\ \bar{v}_2^2$$

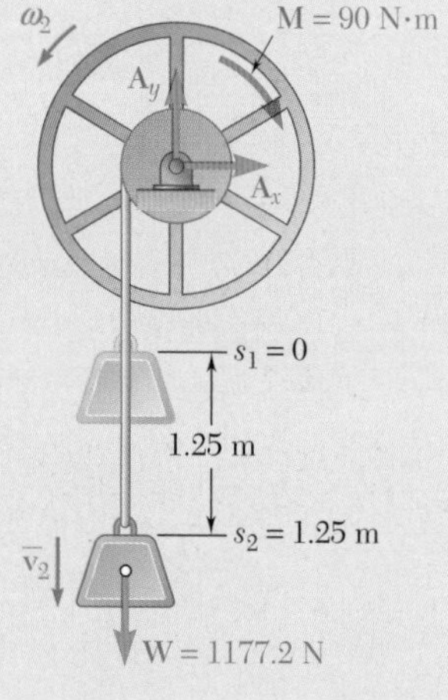

Work. During the motion, only the weight **W** of the block and the friction couple **M** do work. Noting that **W** does positive work and that the friction couple **M** does negative work, we write

$$s_1 = 0 \qquad s_2 = 1.25\ \text{m}$$

$$\theta_1 = 0 \qquad \theta_2 = \frac{s_2}{r} = \frac{1.25\ \text{m}}{0.4\ \text{m}} = 3.125\ \text{rad}$$

$$U_{1\to 2} = W(s_2 - s_1) - M(\theta_2 - \theta_1)$$
$$= (120\ \text{kg})(9.81\ \text{m/s}^2)(1.25\ \text{m}) - (90\ \text{N·m})(3.125\ \text{rad})$$
$$= 1190\ \text{J}$$

Principle of Work and Energy

$$T_1 + U_{1\to 2} = T_2$$
$$(440\ \text{J}) + (1190\ \text{J}) = 110\bar{v}_2^2$$
$$\bar{v}_2 = 3.85\ \text{m/s} \qquad \bar{\mathbf{v}}_2 = 3.85\ \text{m/s}\downarrow$$ ◀

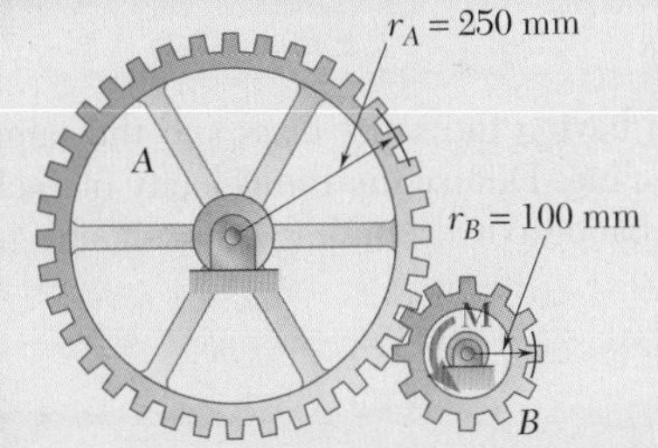

SAMPLE PROBLEM 17.2

Gear A has a mass of 10 kg and a radius of gyration of 200 mm; gear B has a mass of 3 kg and a radius of gyration of 80 mm. The system is at rest when a couple **M** of magnitude 6 N · m is applied to gear B. Neglecting friction, determine (*a*) the number of revolutions executed by gear B before its angular velocity reaches 600 rpm, (*b*) the tangential force which gear B exerts on gear A.

SOLUTION

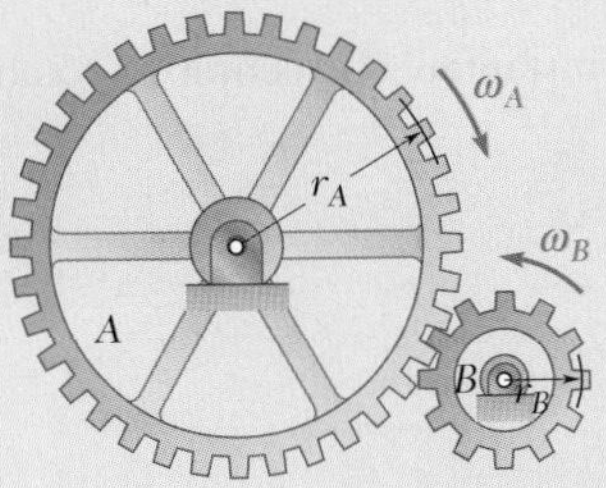

Motion of Entire System. Noting that the peripheral speeds of the gears are equal, we write

$$r_A\omega_A = r_B\omega_B \qquad \omega_A = \omega_B\frac{r_B}{r_A} = \omega_B\frac{100\text{ mm}}{250\text{ mm}} = 0.40\omega_B$$

For $\omega_B = 600$ rpm, we have

$$\omega_B = 62.8\text{ rad/s} \qquad \omega_A = 0.40\omega_B = 25.1\text{ rad/s}$$
$$\bar{I}_A = m_A\bar{k}_A^2 = (10\text{ kg})(0.200\text{ m})^2 = 0.400\text{ kg}\cdot\text{m}^2$$
$$\bar{I}_B = m_B\bar{k}_B^2 = (3\text{ kg})(0.080\text{ m})^2 = 0.0192\text{ kg}\cdot\text{m}^2$$

Kinetic Energy. Since the system is initially at rest, $T_1 = 0$. Adding the kinetic energies of the two gears when $\omega_B = 600$ rpm, we obtain

$$\begin{aligned}T_2 &= \tfrac{1}{2}\bar{I}_A\omega_A^2 + \tfrac{1}{2}\bar{I}_B\omega_B^2\\ &= \tfrac{1}{2}(0.400\text{ kg}\cdot\text{m}^2)(25.1\text{ rad/s})^2 + \tfrac{1}{2}(0.0192\text{ kg}\cdot\text{m}^2)(62.8\text{ rad/s})^2\\ &= 163.9\text{ J}\end{aligned}$$

Work. Denoting by θ_B the angular displacement of gear B, we have

$$U_{1\to2} = M\theta_B = (6\text{N}\cdot\text{m})(\theta_B\text{ rad}) = (6\theta_B)\text{ J}$$

Principle of Work and Energy

$$T_1 + U_{1\to2} = T_2$$
$$0 + (6\theta_B)\text{ J} = 163.9\text{ J}$$
$$\theta_B = 27.32\text{ rad} \qquad \theta_B = 4.35\text{ rev} \blacktriangleleft$$

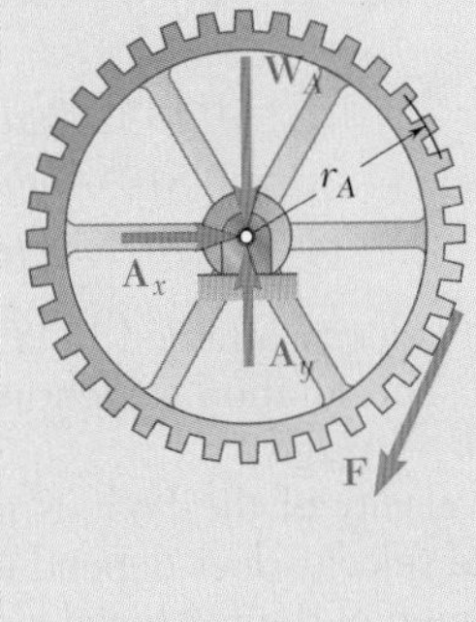

Motion of Gear A. ***Kinetic Energy.*** Initially, gear A is at rest, so $T_1 = 0$. When $\omega_B = 600$ rpm, the kinetic energy of gear A is

$$T_2 = \tfrac{1}{2}\bar{I}_A\omega_A^2 = \tfrac{1}{2}(0.400\text{ kg}\cdot\text{m}^2)(25.1\text{ rad/s})^2 = 126.0\text{ J}$$

Work. The forces acting on gear A are as shown. The tangential force **F** does work equal to the product of its magnitude and of the length $\theta_A r_A$ of the arc described by the point of contact. Since $\theta_A r_A = \theta_B r_B$, we have

$$U_{1\to2} = F(\theta_B r_B) = F(27.3\text{ rad})(0.100\text{ m}) = F(2.73\text{ m})$$

Principle of Work and Energy

$$T_1 + U_{1\to2} = T_2$$
$$0 + F(2.73\text{ m}) = 126.0\text{ J}$$
$$F = +46.2\text{ N} \qquad \mathbf{F} = 46.2\text{ N}\swarrow \blacktriangleleft$$

SAMPLE PROBLEM 17.3

A sphere, a cylinder, and a hoop, each having the same mass and the same radius, are released from rest on an incline. Determine the velocity of each body after it has rolled through a distance corresponding to a change in elevation h.

SOLUTION

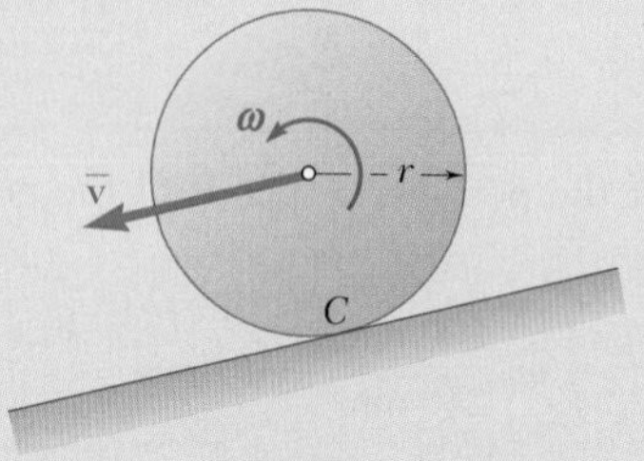

The problem will first be solved in general terms, and then results for each body will be found. We denote the mass by m, the centroidal moment of inertia by $\bar{I}$, the weight by W, and the radius by r.

Kinematics. Since each body rolls, the instantaneous center of rotation is located at C and we write

$$\omega = \frac{\bar{v}}{r}$$

Kinetic Energy

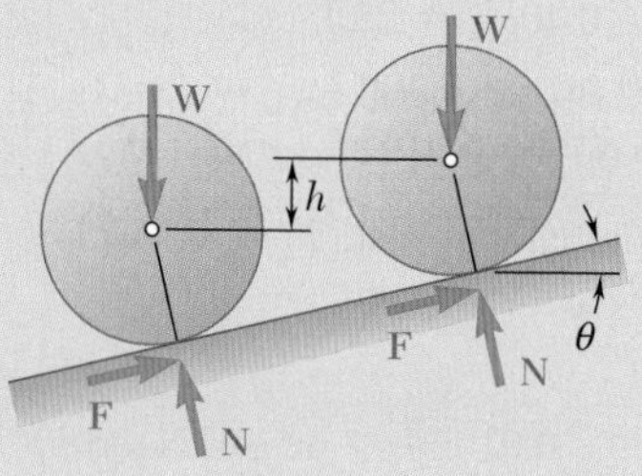

$$T_1 = 0$$
$$T_2 = \tfrac{1}{2}m\bar{v}^2 + \tfrac{1}{2}\bar{I}\omega^2$$
$$= \tfrac{1}{2}m\bar{v}^2 + \tfrac{1}{2}\bar{I}\left(\frac{\bar{v}}{r}\right)^2 = \tfrac{1}{2}\left(m + \frac{\bar{I}}{r^2}\right)\bar{v}^2$$

Work. Since the friction force **F** in rolling motion does no work,

$$U_{1\rightarrow 2} = Wh$$

Principle of Work and Energy

$$T_1 + U_{1\rightarrow 2} = T_2$$
$$0 + Wh = \tfrac{1}{2}\left(m + \frac{\bar{I}}{r^2}\right)\bar{v}^2 \qquad \bar{v}^2 = \frac{2Wh}{m + \bar{I}/r^2}$$

Noting that $W = mg$, we rearrange the result and obtain

$$\bar{v}^2 = \frac{2gh}{1 + \bar{I}/mr^2}$$

Velocities of Sphere, Cylinder, and Hoop. Introducing successively the particular expression for $\bar{I}$, we obtain

Sphere:	$\bar{I} = \frac{2}{5}mr^2$	$\bar{v} = 0.845\sqrt{2gh}$ ◀
Cylinder:	$\bar{I} = \frac{1}{2}mr^2$	$\bar{v} = 0.816\sqrt{2gh}$ ◀
Hoop:	$\bar{I} = mr^2$	$\bar{v} = 0.707\sqrt{2gh}$ ◀

Remark. Let us compare the results with the velocity attained by a frictionless block sliding through the same distance. The solution is identical to the above solution except that $\omega = 0$; we find $\bar{v} = \sqrt{2gh}$.

Comparing the results, we note that the velocity of the body is independent of both its mass and radius. However, the velocity does depend upon the quotient $\bar{I}/mr^2 = \bar{k}^2/r^2$, which measures the ratio of the rotational kinetic energy to the translational kinetic energy. Thus the hoop, which has the largest $\bar{k}$ for a given radius r, attains the smallest velocity, while the sliding block, which does not rotate, attains the largest velocity.

SAMPLE PROBLEM 17.4

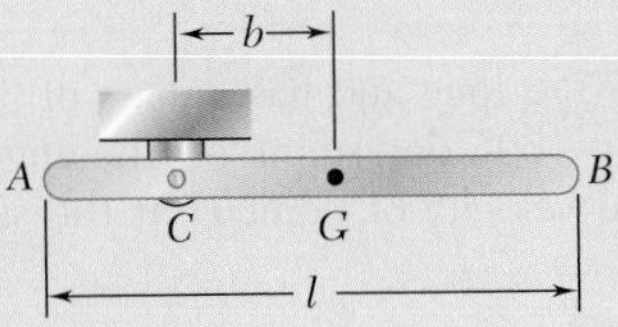

A slender rod of length l is pivoted about a Point C located at a distance b from its center G. It is released from rest in a horizontal position and swings freely. Determine (a) the distance b for which the angular velocity of the rod as it passes through a vertical position is maximum, (b) the corresponding values of its angular velocity and of the reaction at C.

SOLUTION

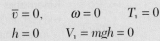

Position 1. Since the rod is released from rest

$$\bar{v} = 0, \qquad \omega = 0 \qquad T_1 = 0$$

Elevation: $h = 0 \qquad V_1 = mgh = 0$

Position 2. Let ω_2 denote the angular velocity of rod at position 2. Since the rod rotates abot c, $\bar{v}_2 = b\omega_2$.

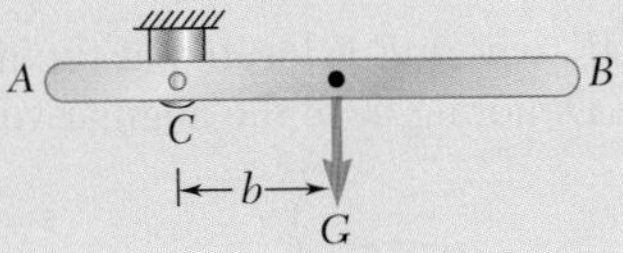

$$\bar{v}_2 = b\omega_2$$

$$\bar{I} = \frac{1}{12}ml^2$$

$$T_2 = \frac{1}{2}m\bar{v}_2^2 + \frac{1}{2}\bar{I}\omega_2^2$$

$$= \frac{1}{2}m\left(b^2 + \frac{1}{12}l^2\right)\omega_2^2$$

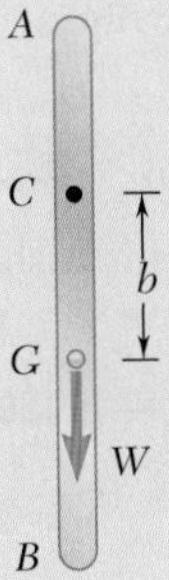

Elevation: $h = -b \qquad V_2 = -mgb$

Principle of conservation of energy.

$$T_1 + V_1 = T_2 + V_2: \quad 0 + 0 = \frac{1}{2}m\left(b^2 + \frac{1}{12}l^2\right)\omega_2^2 - mgb \qquad \omega_2^2 = \frac{2gb}{b^2 + \frac{1}{12}l^2} \qquad (1)$$

(a) Value of b for maximum ω_2.

$$\frac{d}{db}\left(\frac{b}{b^2 + \frac{1}{12}l^2}\right) = \frac{\left(b^2 + \frac{1}{12}l^2\right) - b(2b)}{\left(b^2 + \frac{1}{12}l^2\right)^2} = 0 \quad b^2 = \frac{1}{12}l^2 \quad (2) \qquad b = \frac{l}{\sqrt{12}} \blacktriangleleft$$

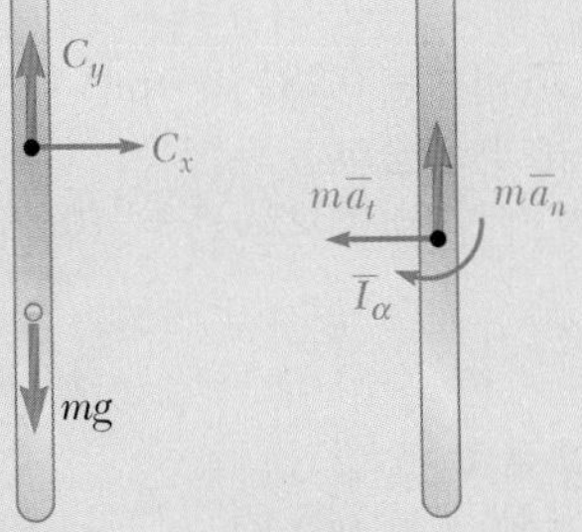

(b) Angular velocity. Substituting the volume of b from (2) in (1)

$$\omega_2^2 = \frac{2g\frac{l}{\sqrt{12}}}{\frac{l^2}{12} + \frac{l^2}{12}} = \sqrt{12}\frac{g}{l} \qquad \omega_2 = 12^{1/4}\sqrt{\frac{g}{l}} \qquad \omega_2 = 1.861\sqrt{\frac{g}{l}} \blacktriangleleft$$

Reaction at C. Since the rod rotates about C, $\bar{a}_n = b\omega_2^2\uparrow$ and $\bar{a}_t = b\alpha\leftarrow$

$$\bar{a}_n = b\omega_2^2 = \frac{l}{\sqrt{12}}\sqrt{12}\frac{g}{l} = g$$

$+\uparrow \Sigma F_y = m\bar{a}_n$: $\quad C_y - mg = mg \quad C_y = 2mg$

$+\circlearrowleft \Sigma M_C = mb\bar{a}_t + \bar{I}\alpha$: $\quad 0 = mb^2\alpha + \bar{I}\alpha = (mb^2 + \bar{I})\alpha$

$$\alpha = 0, \bar{a}_t = 0$$

$\overset{+}{\rightarrow} \Sigma F_x = m\bar{a}_t$: $\quad C_x = -m\bar{a}_t = 0 \qquad$ C = 2mg ↑ ◀

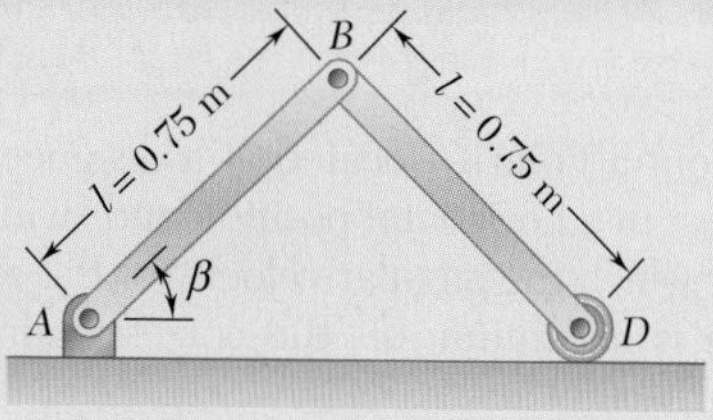

SAMPLE PROBLEM 17.5

Each of the two slender rods shown is 0.75 m long and has a mass of 6 kg. If the system is released from rest with $\beta = 60°$, determine (*a*) the angular velocity of rod *AB* when $\beta = 20°$, (*b*) the velocity of point *D* at the same instant.

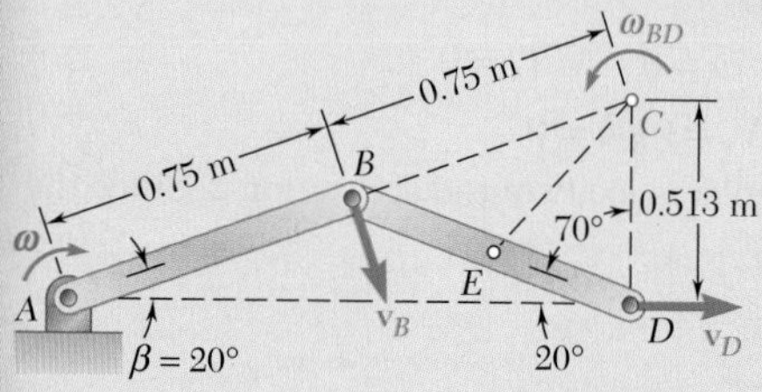

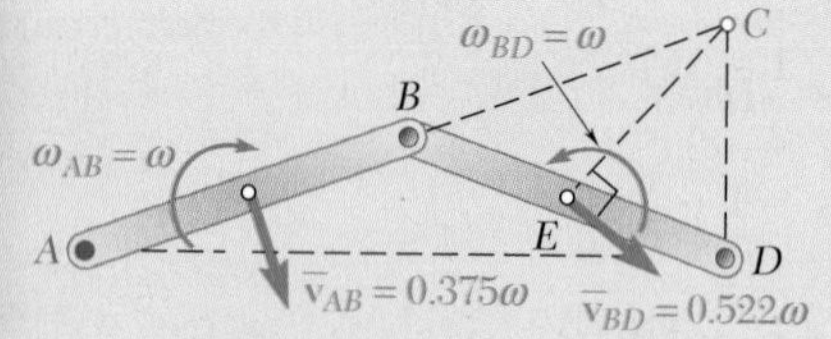

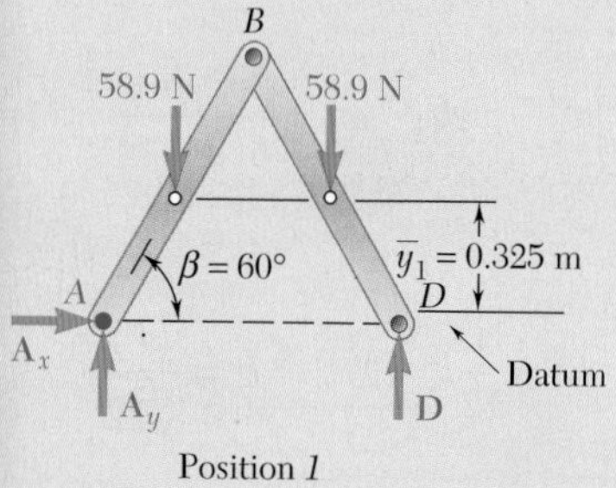

Position *1*

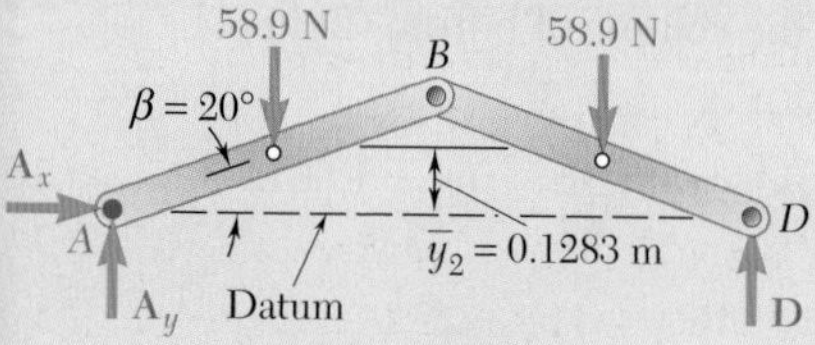

Position 2

SOLUTION

Kinematics of Motion When $\beta = 20°$. Since $\mathbf{v}_B$ is perpendicular to the rod *AB* and $\mathbf{v}_D$ is horizontal, the instantaneous center of rotation of rod *BD* is located at *C*. Considering the geometry of the figure, we obtain

$$BC = 0.75 \text{ m} \qquad CD = 2(0.75 \text{ m}) \sin 20° = 0.513 \text{ m}$$

Applying the law of cosines to triangle *CDE*, where *E* is located at the mass center of rod *BD*, we find $EC = 0.522$ m. Denoting by ω the angular velocity of rod *AB*, we have

$$\bar{v}_{AB} = (0.375 \text{ m})\omega \qquad \bar{\mathbf{v}}_{AB} = 0.375\omega \searrow$$
$$v_B = (0.75 \text{ m})\omega \qquad \mathbf{v}_B = 0.75\omega \searrow$$

Since rod *BD* seems to rotate about point *C*, we write

$$v_B = (BC)\omega_{BD} \qquad (0.75 \text{ m})\omega = (0.75 \text{ m})\omega_{BD} \qquad \boldsymbol{\omega}_{BD} = \omega \circlearrowleft$$
$$\bar{v}_{BD} = (EC)\omega_{BD} = (0.522 \text{ m})\omega \qquad \bar{\mathbf{v}}_{BD} = 0.522\omega \searrow$$

Position *1*. *Potential Energy.* Choosing the datum as shown, and observing that $W = (6 \text{ kg})(9.81 \text{ m/s}^2) = 58.86$ N, we have

$$V_1 = 2W\bar{y}_1 = 2(58.86 \text{ N})(0.325 \text{ m}) = 38.26 \text{ J}$$

Kinetic Energy. Since the system is at rest, $T_1 = 0$.

Position 2. *Potential Energy*

$$V_2 = 2W\bar{y}_2 = 2(58.86 \text{ N})(0.1283 \text{ m}) = 15.10 \text{ J}$$

Kinetic Energy

$$I_{AB} = \bar{I}_{BD} = \tfrac{1}{12}ml^2 = \tfrac{1}{12}(6 \text{ kg})(0.75 \text{ m})^2 = 0.281 \text{ kg} \cdot \text{m}^2$$
$$T_2 = \tfrac{1}{2}m\bar{v}_{AB}^2 + \tfrac{1}{2}\bar{I}_{AB}\omega_{AB}^2 + \tfrac{1}{2}m\bar{v}_{BD}^2 + \tfrac{1}{2}\bar{I}_{BD}\omega_{BD}^2$$
$$= \tfrac{1}{2}(6)(0.375\omega)^2 + \tfrac{1}{2}(0.281)\omega^2 + \tfrac{1}{2}(6)(0.522\omega)^2 + \tfrac{1}{2}(0.281)\omega^2$$
$$= 1.520\omega^2$$

Conservation of Energy

$$T_1 + V_1 = T_2 + V_2$$
$$0 + 38.26 \text{ J} = 1.520\omega^2 + 15.10 \text{ J}$$
$$\omega = 3.90 \text{ rad/s} \qquad \boldsymbol{\omega}_{AB} = 3.90 \text{ rad/s} \circlearrowright \blacktriangleleft$$

Velocity of Point *D*

$$v_D = (CD)\omega = (0.513 \text{ m})(3.90 \text{ rad/s}) = 2.00 \text{ m/s}$$
$$\mathbf{v}_D = 2.00 \text{ m/s} \rightarrow \blacktriangleleft$$

SOLVING PROBLEMS ON YOUR OWN

In this lesson we introduced energy methods to determine the velocity of rigid bodies for various positions during their motion. As you found out previously in Chap. 13, energy methods should be considered for problems involving displacements and velocities.

1. The method of work and energy, when applied to all of the particles forming a rigid body, yields the equation

$$T_1 + U_{1\rightarrow 2} = T_2 \tag{17.1}$$

where T_1 and T_2 are, respectively, the initial and final values of the total kinetic energy of the particles forming the body and $U_{1\rightarrow 2}$ is the *work done by the external forces* exerted on the rigid body.

a. Work of forces and couples. To the expression for the work of a force (Chap. 13), we added the expression for the work of a couple and wrote

$$U_{1\rightarrow 2} = \int_{A_1}^{A_2} \mathbf{F} \cdot d\mathbf{r} \qquad U_{1\rightarrow 2} = \int_{\theta_1}^{\theta_2} M\,d\theta \qquad (17.3, 17.5)$$

When the moment of a couple is constant, the work of the couple is

$$U_{1\rightarrow 2} = M(\theta_2 - \theta_1) \tag{17.6}$$

where θ_1 and θ_2 are expressed in radians [Sample Probs. 17.1 and 17.2].

b. The kinetic energy of a rigid body in plane motion was found by considering the motion of the body as the sum of a translation with its mass center and a rotation about the mass center.

$$T = \tfrac{1}{2} m\bar{v}^2 + \tfrac{1}{2}\bar{I}\omega^2 \tag{17.9}$$

where $\bar{v}$ is the velocity of the mass center and ω is the angular velocity of the body [Sample Probs. 17.3 and 17.4].

2. For a system of rigid bodies we again used the equation

$$T_1 + U_{1\rightarrow 2} = T_2 \tag{17.1}$$

where T is the sum of the kinetic energies of the bodies forming the system and U is the work done by *all the forces acting on the bodies,* internal as well as external. Your computations will be simplified if you keep the following in mind.

a. The forces exerted on each other by pin-connected members or by meshed gears are equal and opposite, and, since they have the same point of application, they undergo equal small displacements. Therefore, *their total work is zero* and can be omitted from your calculations [Sample Prob. 17.2].

(*continued*)

b. The forces exerted by an inextensible cord on the two bodies it connects have the same magnitude and their points of application move through equal distances, but the work of one force is positive and the work of the other is negative. Therefore, *their total work is zero* and can again be omitted from your calculations [Sample Prob. 17.1].

c. The forces exerted by a spring on the two bodies it connects also have the same magnitude, but their points of application will generally move through different distances. Therefore, *their total work is usually not zero* and should be taken into account in your calculations.

3. The principle of conservation of energy can be expressed as

$$T_1 + V_1 = T_2 + V_2 \tag{17.12}$$

where V represents the potential energy of the system. This principle can be used when a body or a system of bodies is acted upon by conservative forces, such as the force exerted by a spring or the force of gravity [Sample Probs. 17.4 and 17.5].

4. The last section of this lesson was devoted to power, which is the time rate at which work is done. For a body acted upon by a couple of moment **M**, the power can be expressed as

$$\text{Power} = M\omega \tag{17.13}$$

where ω is the angular velocity of the body expressed in rad/s. As you did in Chap. 13, you should express power either in watts or in horsepower (1 hp = 550 ft · lb/s).

PROBLEMS

CONCEPT QUESTIONS

17.CQ1 A round object of mass m and radius r is released from rest at the top of a curved surface and rolls without slipping until it leaves the surface with a horizontal velocity as shown. Will a solid sphere, a solid cylinder, or a hoop travel the greatest distance x?

a. Solid sphere

b. Solid cylinder

c. Hoop

d. They will all travel the same distance.

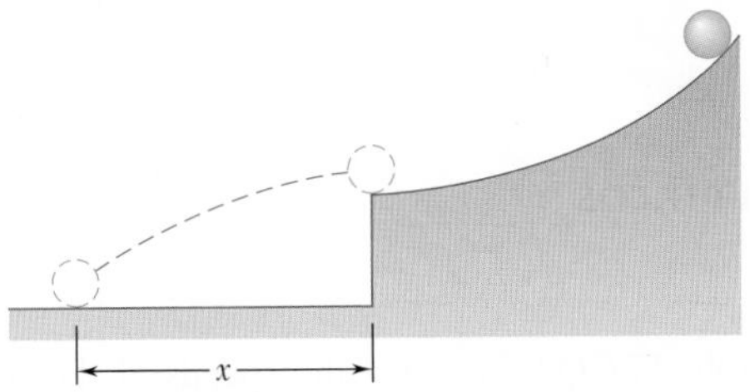

Fig. P17.CQ1

17.CQ2 A solid steel sphere A of radius r and mass m is released from rest and rolls without slipping down an incline as shown. After traveling a distance d, the sphere has a speed v. If a solid steel sphere of radius $2r$ is released from rest on the same incline, what will its speed be after rolling a distance d?

a. $0.25\,v$

b. $0.5\,v$

c. v

d. $2v$

e. $4v$

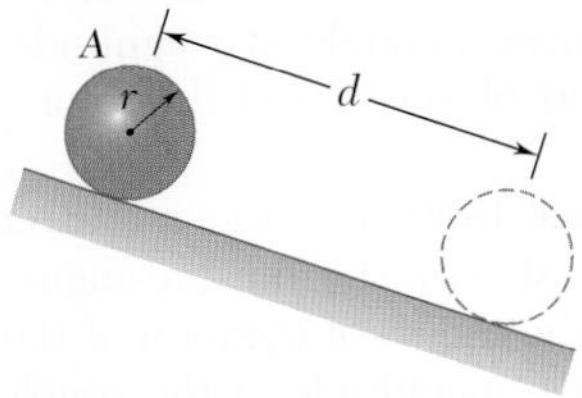

Fig. P17.CQ2

17.CQ3 Slender bar A is rigidly connected to a massless rod BC in Case 1 and two massless cords in Case 2 as shown. The vertical thickness of bar A is negligible compared to L. In both cases A is released from rest at an angle $\theta = \theta_0$. When $\theta = 0°$, which system will have the larger kinetic energy?

a. Case 1

b. Case 2

c. The kinetic energy will be the same.

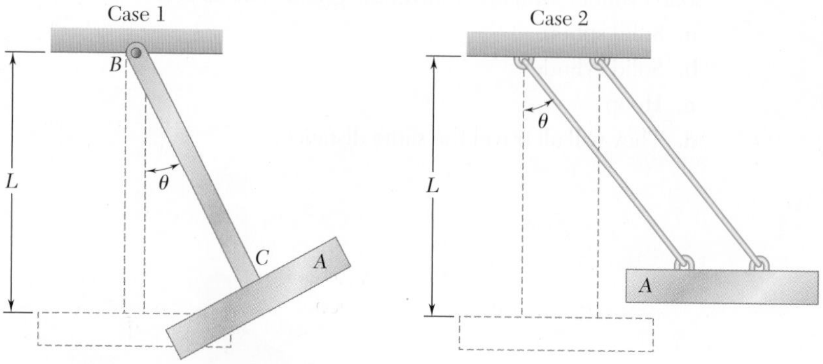

Fig. P17.CQ3 and P17.CQ5

17.CQ4 In Prob. 17.CQ3, how will the speeds of the centers of gravity compare for the two cases when $\theta = 0°$?

a. Case 1 will be larger.

b. Case 2 will be larger.

c. The speeds will be the same.

17.CQ5 Slender bar A is rigidly connected to a massless rod BC in Case 1 and two massless cords in Case 2 as shown. The vertical thickness of bar A is not negligible compared to L. In both cases A is released from rest at an angle $\theta = \theta_0$. When $\theta = 0°$, which system will have the largest kinetic energy?

a. Case 1

b. Case 2

c. The kinetic energy will be the same.

END-OF-SECTION PROBLEMS

17.1 The rotor of an electric motor has an angular velocity of 3600 rpm when the load and power are cut off. The 50-kg rotor then coasts to rest after 5000 revolutions. Knowing that the kinetic friction of the rotor produces a couple of magnitude 4 N · m, determine the centroidal radius of gyration of the rotor.

17.2 It is known that 1500 revolutions are required for the 3000-kg flywheel to coast to rest from an angular velocity of 300 rpm. Knowing that the radius of gyration of the flywheel is 1 m, determine the average magnitude of the couple due to kinetic friction in the bearings.

17.3 Two disks of the same material are attached to a shaft as shown. Disk A has a mass of 15 kg and a radius r = 125 mm. Disk B is three times as thick as disk A. Knowing that a couple **M** of magnitude 20 N · m is to be applied to disk A when the system is at rest, determine the radius nr of disk B if the angular velocity of the system is to be 600 rpm after 4 revolutions.

17.4 Two disks of the same material are attached to a shaft as shown. Disk A is of radius r and has a thickness b, while disk B is of radius nr and thickness $3b$. A couple **M** of constant magnitude is applied when the system is at rest and is removed after the system has executed two revolutions. Determine the value of n which results in the largest final speed for a point on the rim of disk B.

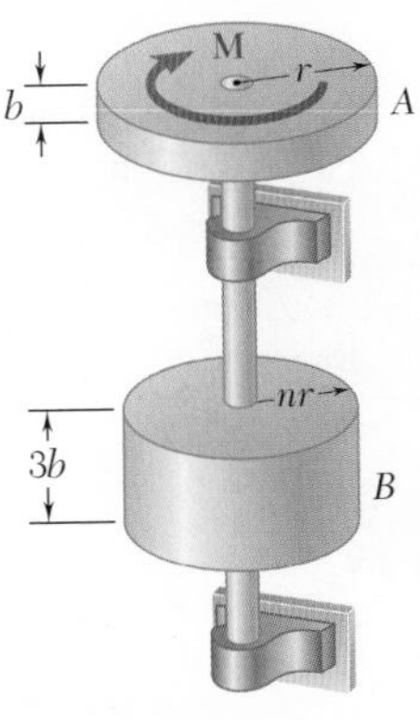

Fig. P17.3 and P17.4

17.5 The flywheel of a small punching machine rotates at 360 rpm. Each punching operation requires 2250 N · m of work and it is desired that the speed of the flywheel after each punching be not less than 95 percent of the original speed. (*a*) Determine the required moment of inertia of the flywheel. (*b*) If a constant 27 N · m couple is applied to the shaft of the flywheel, determine the number of revolutions that must occur between two successive punchings, knowing that the initial velocity is to be 360 rpm at the start of each punching.

17.6 The flywheel of a punching machine has a mass of 300 kg and a radius of gyration of 600 mm. Each punching operation requires 2500 J of work. (*a*) Knowing that the speed of the flywheel is 300 rpm just before a punching, determine the speed immediately after the punching. (*b*) If a constant 25-N · m couple is applied to the shaft of the flywheel, determine the number of revolutions executed before the speed is again 300 rpm.

17.7 Disk A, of mass 5 kg and radius r = 150 mm, is at rest when it is placed in contact with belt BC, which moves to the right with a constant speed v = 12 m/s. Knowing that μ_k = 0.20 between the disk and the belt, determine the number of revolutions executed by the disk before it attains a constant angular velocity.

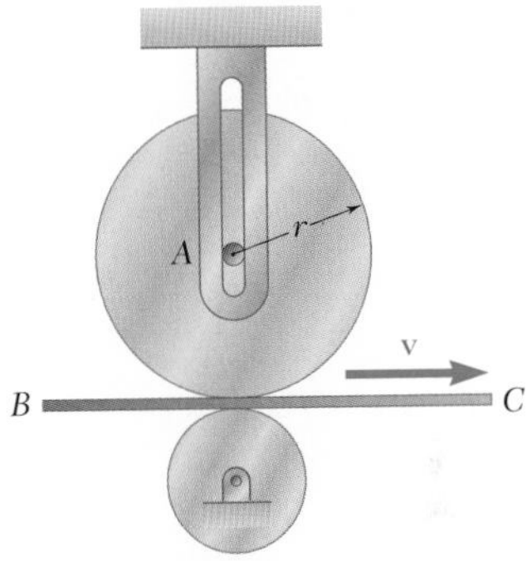

Fig. P17.7 and P17.8

17.8 Disk A is of constant thickness and is at rest when it is placed in contact with belt BC, which moves with a constant velocity **v**. Denoting by μ_k the coefficient of kinetic friction between the disk and the belt, derive an expression for the number of revolutions executed by the disk before it attains a constant angular velocity.

17.9 The 160-mm-radius brake drum is attached to a larger flywheel that is not shown. The total mass moment of inertia of the flywheel and drum is 20 kg · m^2 and the coefficient of kinetic friction between the drum and the brake shoe is 0.35. Knowing that the initial angular velocity of the flywheel is 360 rpm counterclockwise, determine the vertical force **P** that must be applied to the pedal C if the system is to stop in 100 revolutions.

17.10 Solve Prob. 17.9, assuming that the initial angular velocity of the flywheel is 360 rpm clockwise.

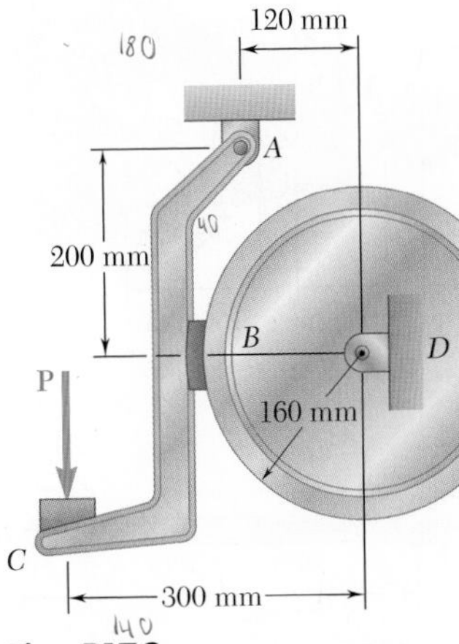

Fig. P17.9

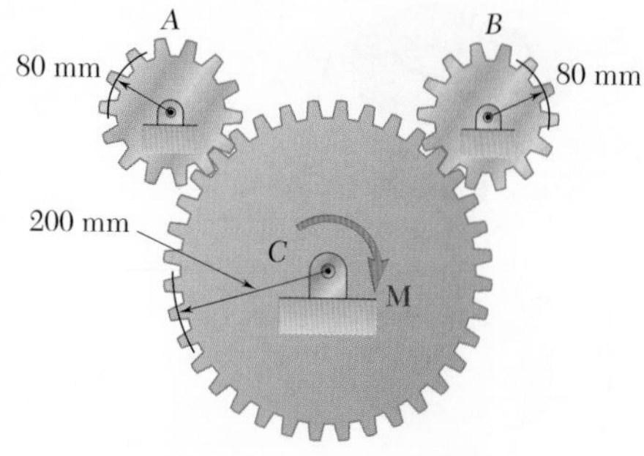

Fig. P17.11

17.11 Each of the gears A and B has a mass of 2.4 kg and a radius of gyration of 60 mm, while gear C has a mass of 12 kg and a radius of gyration of 150 mm. A couple $\mathbf{M}$ of constant magnitude 10 N · m is applied to gear C. Determine (*a*) the number of revolutions of gear C required for its angular velocity to increase from 100 to 450 rpm, (*b*) the corresponding tangential force acting on gear A.

17.12 Solve Prob. 17.11, assuming that the 10-N · m couple is applied to gear B.

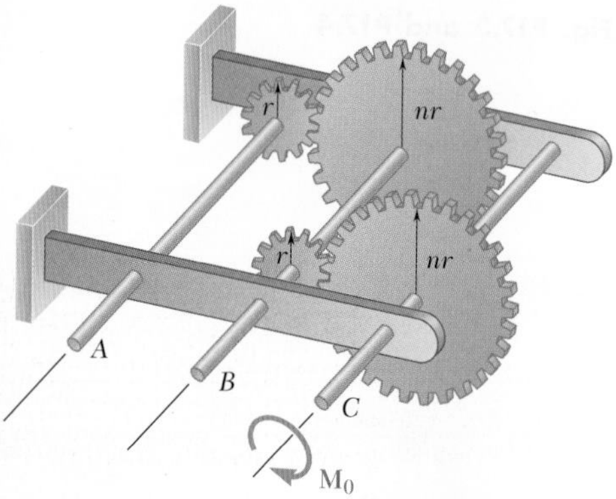

Fig. P17.13

17.13 The gear train shown consists of four gears of the same thickness and of the same material; two gears are of radius r, and the other two are of radius nr. The system is at rest when the couple $\mathbf{M}_0$ is applied to shaft C. Denoting by I_0 the moment of inertia of a gear of radius r, determine the angular velocity of shaft A if the couple $\mathbf{M}_0$ is applied for one revolution of shaft C.

17.14 The double pulley shown has a mass of 15 kg and a centroidal radius of gyration of 160 mm. Cylinder A and block B are attached to cords that are wrapped on the pulleys as shown. The coefficient of kinetic friction between block B and the surface is 0.2. Knowing that the system is at rest in the position shown when a constant force $\mathbf{P} = 200$ N is applied to cylinder A, determine (*a*) the velocity of cylinder A as it strikes the ground, (*b*) the total distance that block B moves before coming to rest.

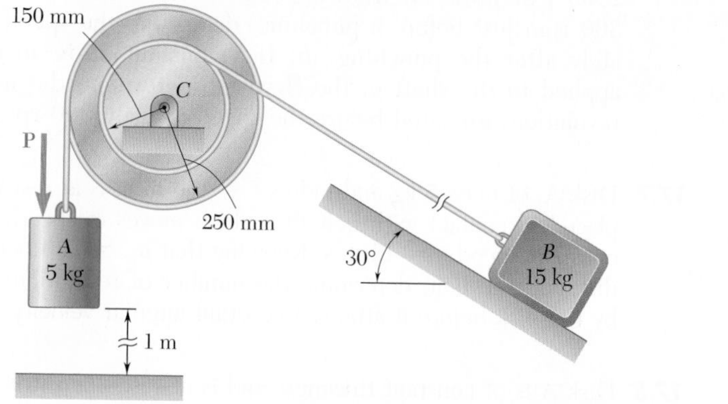

Fig. *P17.14*

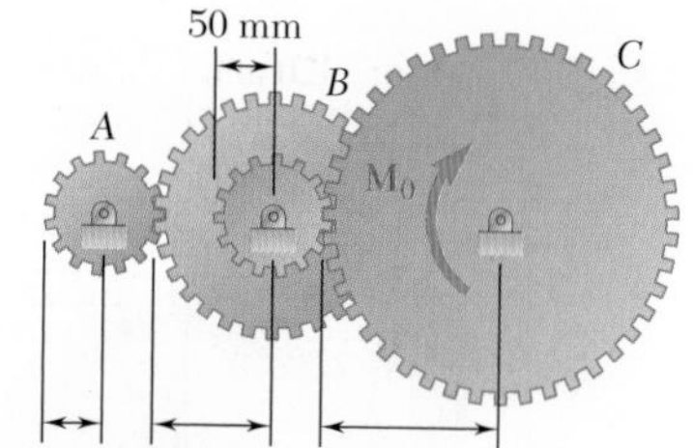

Fig. *P17.15*

17.15 Gear A has a mass of 1 kg and a radius of gyration of 30 mm; gear B has a mass of 4 kg and a radius of gyration of 75 mm; gear C has a mass of 9 kg and a radius of gyration of 100 mm. The system is at rest when a couple $\mathbf{M}_0$ of constant magnitude 4 N·m is applied to gear C. Assuming that no slipping occurs between the gears, determine the number of revolutions required for disk A to reach an angular velocity of 300 rpm.

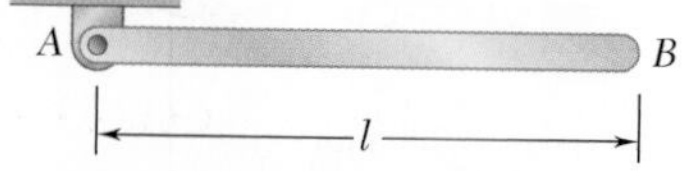

Fig. P17.16

17.16 A slender rod of length l and weight W is pivoted at one end as shown. It is released from rest in a horizontal position and swings freely. (*a*) Determine the angular velocity of the rod as it passes through a vertical position and determine the corresponding reaction at the pivot. (*b*) Solve part *a* for $W = 10$ N and $l = 1$ m.

17.17 A 15-kg slender rod AB is 2.5 m long and is pivoted about a point O which is 0.5 m from end B. The other end is pressed against a spring of constant $k = 300$ kN/m until the spring is compressed 40 mm. The rod is then in a horizontal position. If the rod is released from this position, determine its angular velocity and the reaction at the pivot O as the rod passes through a vertical position.

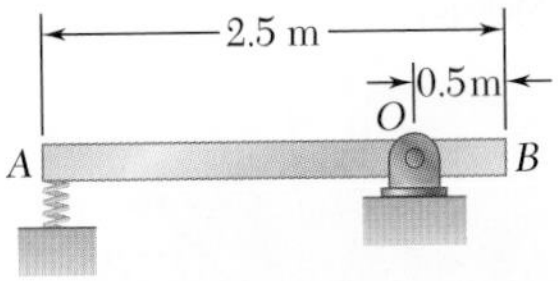

Fig. P17.17

17.18 and 17.19 A slender 4-kg rod can rotate in a vertical plane about a pivot at B. A spring of constant $k = 400$ N/m and of unstretched length 150 mm is attached to the rod as shown. Knowing that the rod is released from rest in the position shown, determine its angular velocity after it has rotated through 90°.

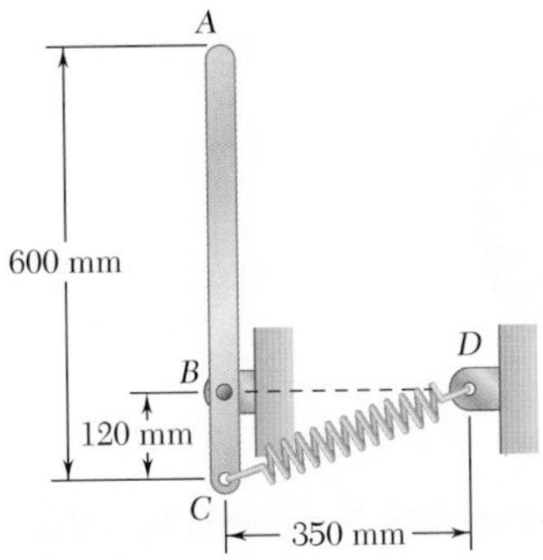

Fig. P17.18

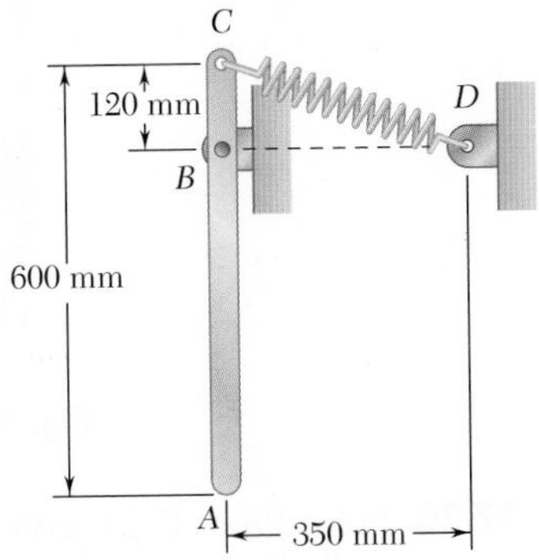

Fig. P17.19

17.20 An 80-kg gymnast is executing a series of full-circle swings on the horizontal bar. In the position shown he has a small and negligible clockwise angular velocity and will maintain his body straight and rigid as he swings downward. Assuming that during the swing the centroidal radius of gyration of his body is 0.4 m, determine his angular velocity and the force exerted on his hands after he has rotated through (*a*) 90°, (*b*) 180°.

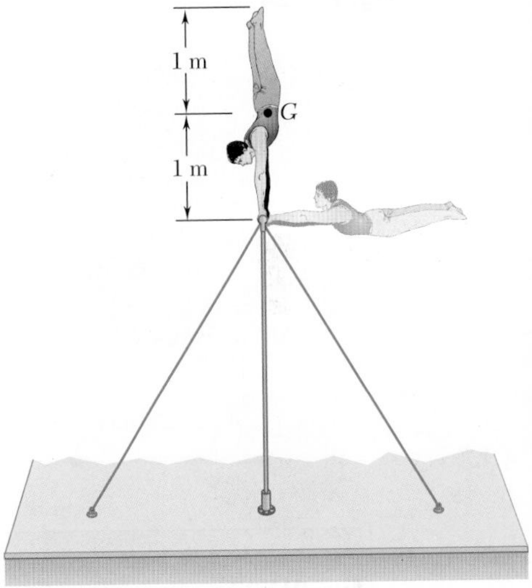

Fig. P17.20

17.21 A collar with a mass of 1 kg is rigidly attached at a distance $d = 300$ mm from the end of a uniform slender rod AB. The rod has a mass of 3 kg and is of length $L = 600$ mm. Knowing that the rod is released from rest in the position shown, determine the angular velocity of the rod after it has rotated through 90°.

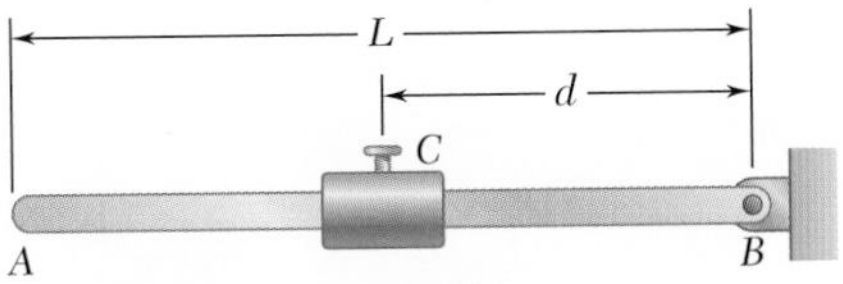

Fig. *P17.21* and *P17.22*

17.22 A collar with a mass of 1 kg is rigidly attached to a slender rod AB of mass 3 kg and length $L = 600$ mm. The rod is released from rest in the position shown. Determine the distance d for which the angular velocity of the rod is maximum after it has rotated through 90°.

17.23 Two identical slender rods AB and BC are welded together to form an L-shaped assembly. The assembly is pressed against a spring at D and released from the position shown. Knowing that the maximum angle of rotation of the assembly in its subsequent motion is 90° counterclockwise, determine the magnitude of the angular velocity of the assembly as it passes through the position where rod AB forms an angle of 30° with the horizontal.

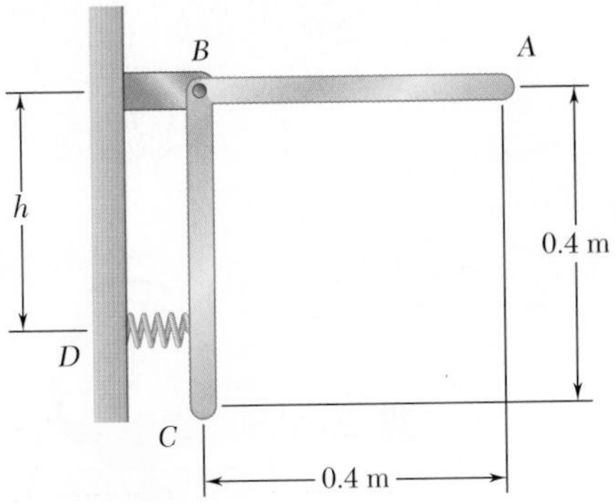

Fig. P17.23

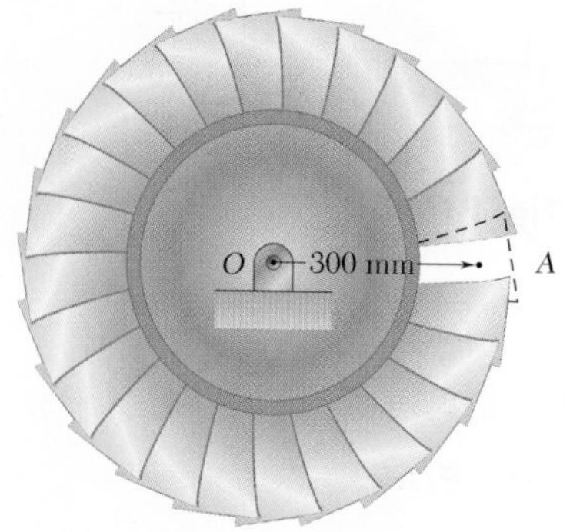

Fig. P17.24

17.24 The 30-kg turbine disk has a centroidal radius of gyration of 175 mm and is rotating clockwise at a constant rate of 60 rpm when a small blade of weight 0.5 N at point A becomes loose and is thrown off. Neglecting friction, determine the change in the angular velocity of the turbine disk after it has rotated through (*a*) 90°, (*b*) 270°.

17.25 A rope is wrapped around a cylinder of radius r and mass m as shown. Knowing that the cylinder is released from rest, determine the velocity of the center of the cylinder after it has moved downward a distance s.

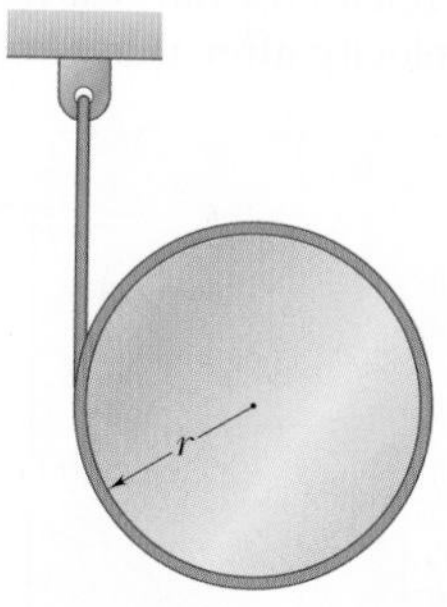

Fig. P17.25

17.26 Solve Prob. 17.25, assuming that the cylinder is replaced by a thin-walled pipe of radius r and mass m.

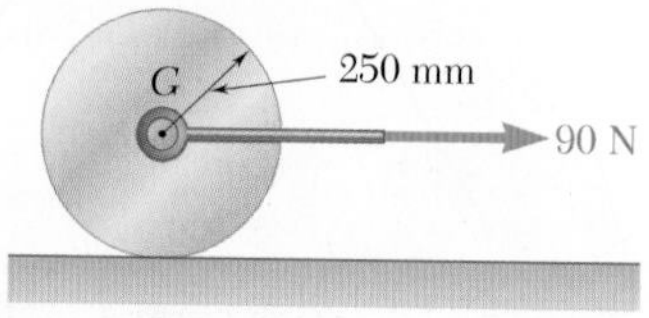

Fig. P17.27

17.27 A 20-kg uniform cylindrical roller, initially at rest, is acted upon by a 90-N force as shown. Knowing that the body rolls without slipping, determine (*a*) the velocity of its center G after it has moved 1.5 m, (*b*) the friction force required to prevent slipping.

17.28 A small sphere of mass m and radius r is released from rest at A and rolls without sliding on the curved surface to point B where it leaves the surface with a horizontal velocity. Knowing that $a = 1.5$ m and $b = 1.2$ m, determine (*a*) the speed of the sphere as it strikes the ground at C, (*b*) the corresponding distance c.

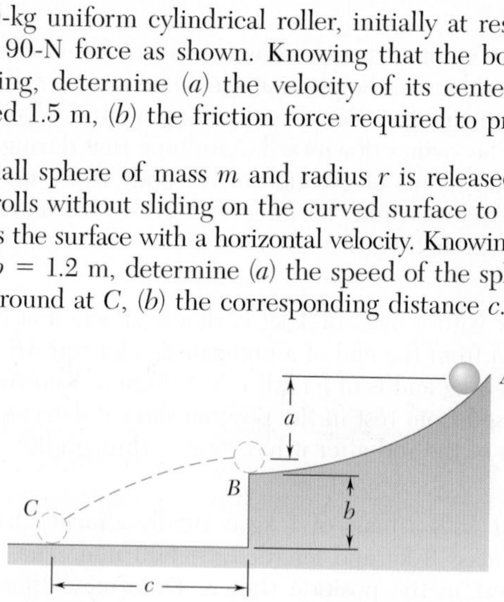

Fig. *P17.28*

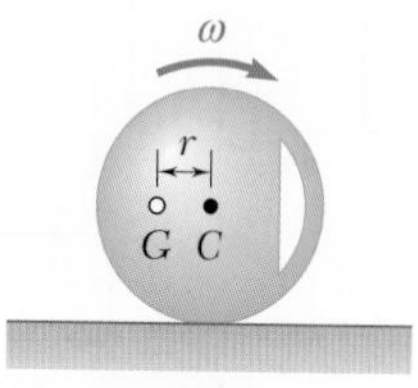

Fig. P17.29

17.29 The mass center G of a 3-kg wheel of radius $R = 180$ mm is located at a distance $r = 60$ mm from its geometric center C. The centroidal radius of gyration of the wheel is $\bar{k} = 90$ mm. As the wheel rolls without sliding, its angular velocity is observed to vary. Knowing that $\omega = 8$ rad/s in the position shown, determine (*a*) the angular velocity of the wheel when the mass center G is directly above the geometric center C, (*b*) the reaction at the horizontal surface at the same instant.

17.30 A half section of pipe of mass m and radius r is released from rest in the position shown. Knowing that the pipe rolls without sliding, determine (*a*) its angular velocity after it has rolled through 90°, (*b*) the reaction at the horizontal surface at the same instant. [*Hint:* Note that $GO = 2r/\pi$ and that, by the parallel-axis theorem, $\bar{I} = mr^2 - m(GO)^2$.]

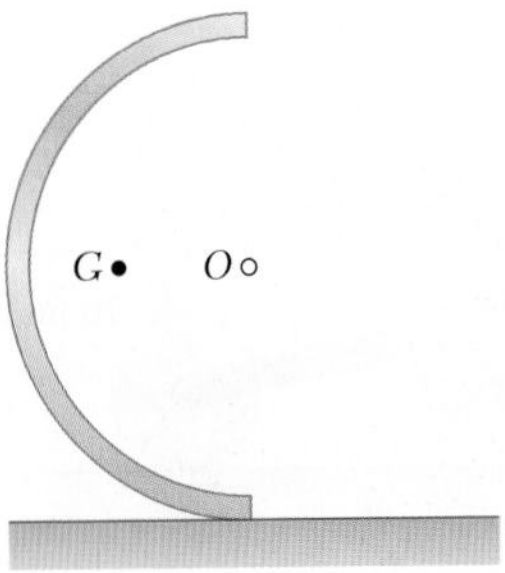

Fig. P17.30

17.31 A sphere of mass m and radius r rolls without slipping inside a curved surface of radius R. Knowing that the sphere is released from rest in the position shown, derive an expression for (*a*) the linear velocity of the sphere as it passes through B, (*b*) the magnitude of the vertical reaction at that instant.

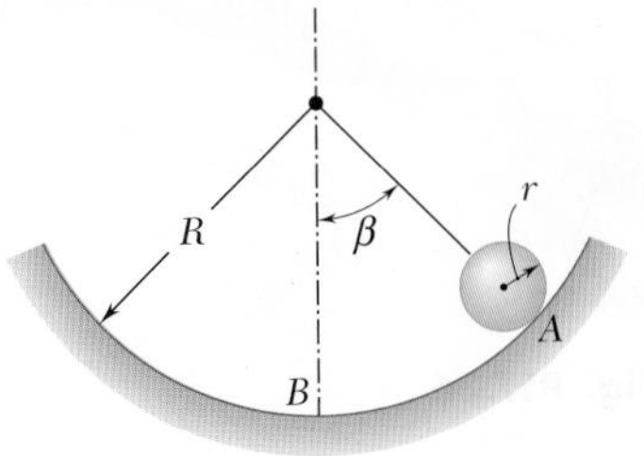

Fig. P17.31

17.32 Two uniform cylinders, each of mass m = 7 kg and radius r = 100 mm, are connected by a belt as shown. Knowing that the initial angular velocity of cylinder B is 30 rad/s counterclockwise, determine (*a*) the distance through which cylinder A will rise before the angular velocity of cylinder B is reduced to 5 rad/s, (*b*) the tension in the portion of belt connecting the two cylinders.

17.33 Two uniform cylinders, each of mass m = 7 kg and radius r = 100 mm, are connected by a belt as shown. If the system is released from rest, determine (*a*) the velocity of the center of cylinder A after it has moved through 1 m, (*b*) the tension in the portion of belt connecting the two cylinders.

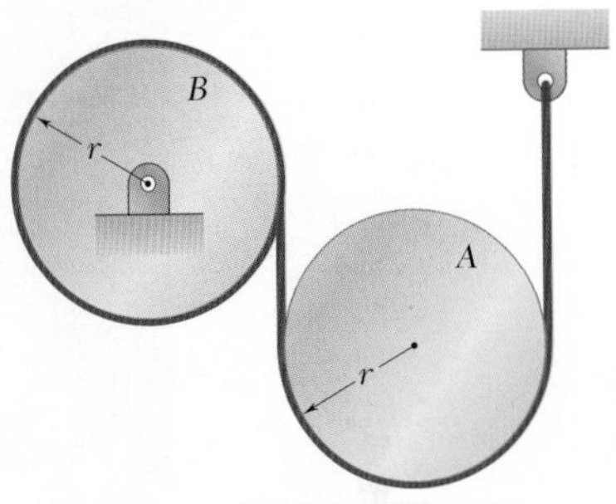

Fig. P17.32 and *P17.33*

17.34 A bar of mass m = 5 kg is held as shown between four disks each of mass m' = 2 kg and radius r = 75 mm. Knowing that the forces exerted on the disks are sufficient to prevent slipping and that the bar is released from rest, for each of the cases shown determine the velocity of the bar after it has moved through the distance h.

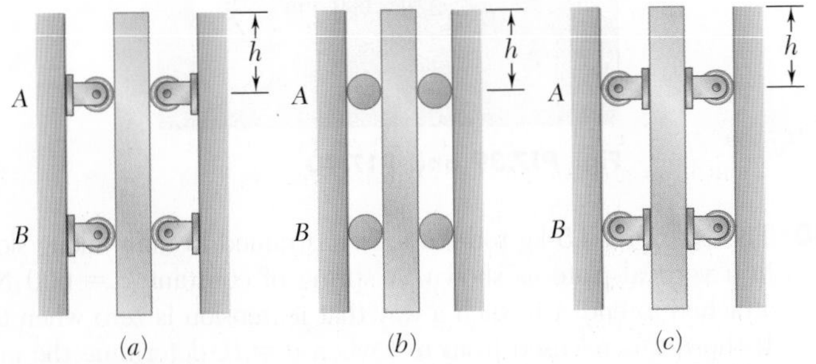

Fig. *P17.34*

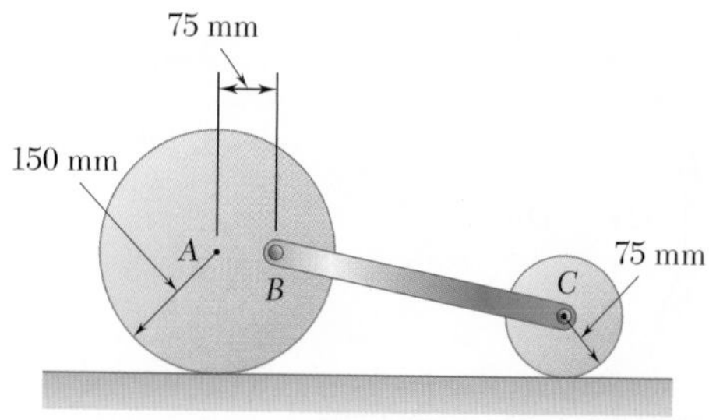

Fig. P17.35

17.35 The 5-kg rod BC is attached by pins to two uniform disks as shown. The mass of the 150-mm-radius disk is 6 kg and that of the 75-mm-radius disk is 1.5 kg. Knowing that the system is released from rest in the position shown, determine the velocity of the rod after disk A has rotated through 90°.

17.36 The motion of the uniform rod AB is guided by small wheels of negligible mass that roll on the surface shown. If the rod is released from rest when $\theta = 0$, determine the velocities of A and B when $\theta = 30°$.

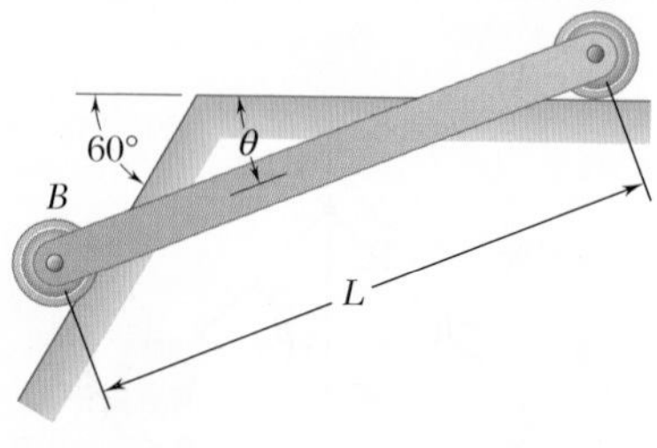

Fig. P17.36

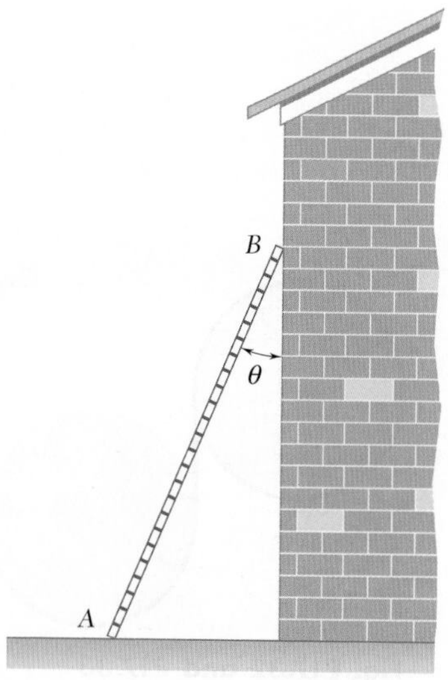

Fig. P17.37 and P17.38

17.37 A 5-m-long ladder has a mass of 15 kg and is placed against a house at an angle $\theta = 20°$. Knowing that the ladder is released from rest, determine the angular velocity of the ladder and the velocity of end A when $\theta = 45°$. Assume the ladder can slide freely on the horizontal ground and on the vertical wall.

17.38 A long ladder of length l, mass m, and centroidal mass moment of inertia $\bar{I}$ is placed against a house at an angle $\theta = \theta_0$. Knowing that the ladder is released from rest, determine the angular velocity of the ladder when $\theta = \theta_2$. Assume the ladder can slide freely on the horizontal ground and on the vertical wall.

17.39 The ends of a 4.5-kg rod AB are constrained to move along slots cut in a vertical plate as shown. A spring of constant $k = 600$ N/m is attached to end A in such a way that its tension is zero when $\theta = 0$. If the rod is released from rest when $\theta = 50°$, determine the angular velocity of the rod and the velocity of end B when $\theta = 0$.

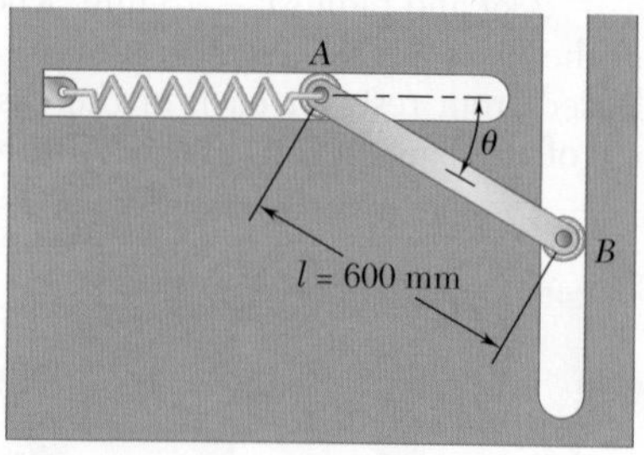

Fig. *P17.39* and P17.40

17.40 The ends of a 4.5-kg rod AB are constrained to move along slots cut in a vertical plate as shown. A spring of constant $k = 600$ N/m is attached to end A in such a way that its tension is zero when $\theta = 0$. If the rod is released from rest when $\theta = 0$, determine the angular velocity of the rod and the velocity of end B when $\theta = 30°$.

17.41 The motion of a slender rod of length R is guided by pins at A and B which slide freely in slots cut in a vertical plate as shown. If end B is moved slightly to the left and then released, determine the angular velocity of the rod and the velocity of its mass center (*a*) at the instant when the velocity of end B is zero, (*b*) as end B passes through point D.

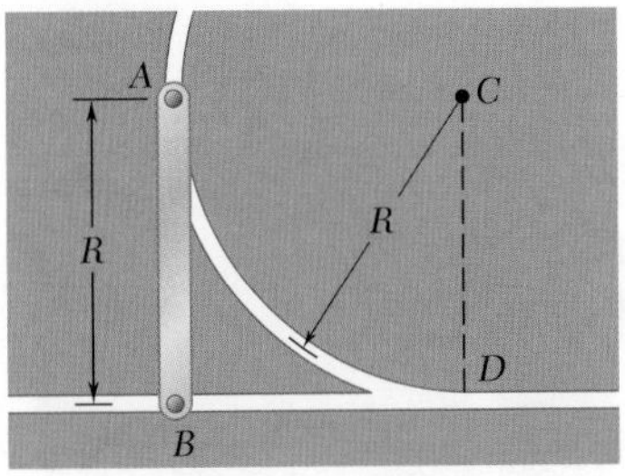

Fig. P17.41

17.42 Each of the two rods shown is of length $L = 1$ m and has a mass of 5 kg. Point D is connected to a spring of constant $k = 20$ N/m and is constrained to move along a vertical slot. Knowing that the system is released from rest when rod BD is horizontal and the spring connected to point D is initially unstretched, determine the velocity of point D when it is directly to the right of point A.`

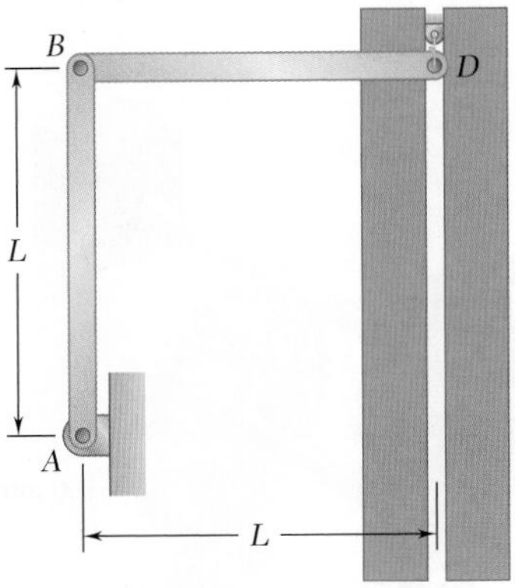

Fig. P17.42

17.43 The 4-kg rod AB is attached to a collar of negligible mass at A and to a flywheel at B. The flywheel has a mass of 16 kg and a radius of gyration of 180 mm. Knowing that in the position shown the angular velocity of the flywheel is 60 rpm clockwise, determine the velocity of the flywheel when point B is directly below C.

17.44 If in Prob. 17.43 the angular velocity of the flywheel is to be the same in the position shown and when point B is directly above C, determine the required value of its angular velocity in the position shown.

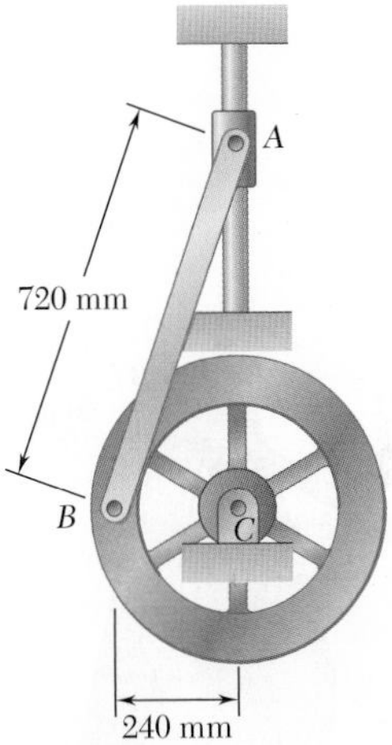

Fig. P17.43 and P17.44

17.45 The uniform rods AB and BC weigh 2.4 kg and 4 kg, respectively, and the small wheel at C is of negligible weight. If the wheel is moved slightly to the right and then released, determine the velocity of pin B after rod AB has rotated through 90°.

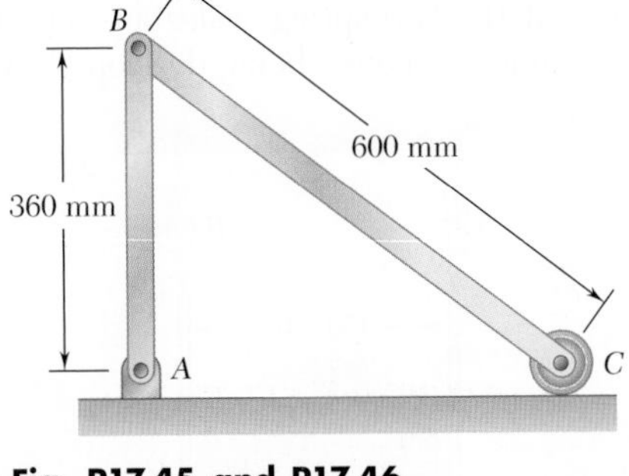

Fig. P17.45 and P17.46

17.46 The uniform rods AB and BC weigh 2.4 kg and 4 kg, respectively, and the small wheel at C is of negligible weight. Knowing that in the position shown the velocity of wheel C is 2 m/s to the right, determine the velocity of pin B after rod AB has rotated through 90°.

17.47 The 80-mm-radius gear shown has a mass of 5 kg and a centroidal radius of gyration of 60 mm. The 4-kg rod AB is attached to the center of the gear and to a pin at B that slides freely in a vertical slot. Knowing that the system is released from rest when $\theta = 60°$, determine the velocity of the center of the gear when $\theta = 20°$.

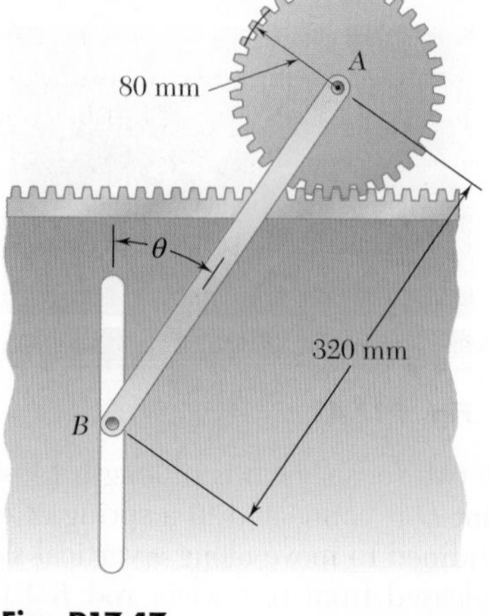

Fig. P17.47

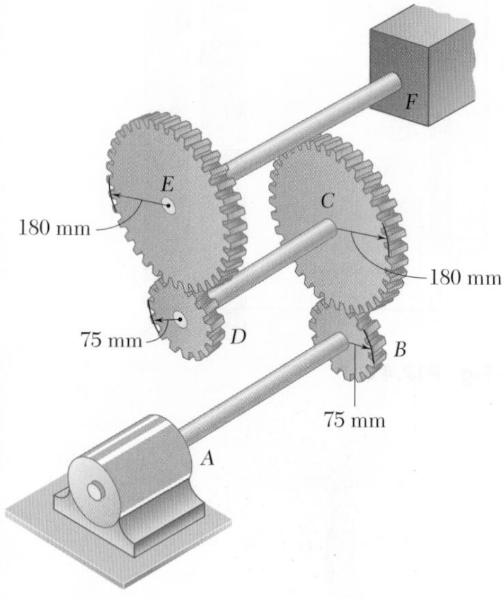

Fig. P17.49

17.48 Knowing that the maximum allowable couple that can be applied to a shaft is 2000 N · m, determine the maximum power (in kW) that can be transmitted by the shaft at (*a*) 180 rpm, (*b*) 480 rpm.

17.49 Three shafts and four gears are used to form a gear train which will transmit 7.5 kW from the motor at A to a machine tool at F. (Bearings for the shafts are omitted from the sketch.) Knowing that the frequency of the motor is 30 Hz, determine the magnitude of the couple which is applied to shaft (*a*) AB, (*b*) CD, (*c*) EF.

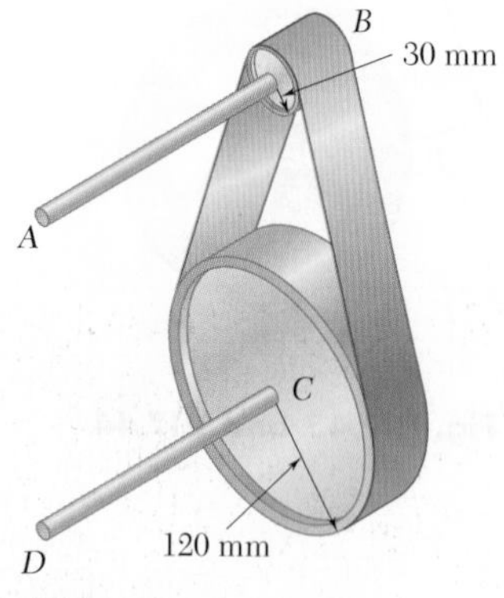

Fig. P17.50

17.50 The shaft-disk-belt arrangement shown is used to transmit 2.4 kW from point A to point D. Knowing that the maximum allowable couples that can be applied to shafts AB and CD are 25 N · m and 80 N · m, respectively, determine the required minimum speed of shaft AB.

17.51 The experimental setup shown is used to measure the power output of a small turbine. When the turbine is operating at 200 rpm, the readings of the two spring scales are 50 and 110 N, respectively. Determine the power being developed by the turbine.

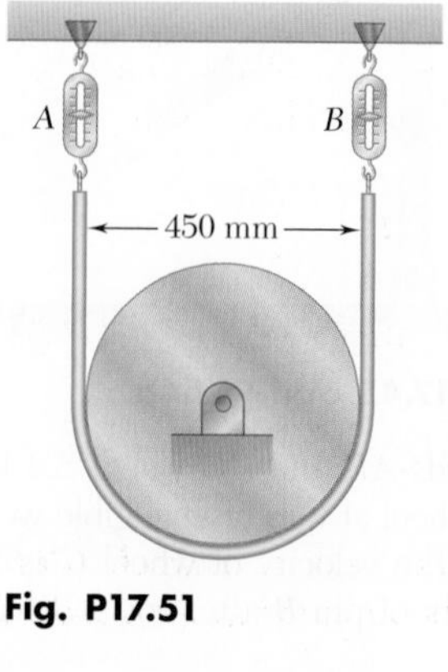

Fig. P17.51

17.8 PRINCIPLE OF IMPULSE AND MOMENTUM FOR THE PLANE MOTION OF A RIGID BODY

The principle of impulse and momentum will now be applied to the analysis of the plane motion of rigid bodies and of systems of rigid bodies. As was pointed out in Chap. 13, the method of impulse and momentum is particularly well adapted to the solution of problems involving time and velocities. Moreover, the principle of impulse and momentum provides the only practicable method for the solution of problems involving impulsive motion or impact (Secs. 17.11 and 17.12).

Considering again a rigid body as made of a large number of particles P_i, we recall from Sec. 14.9 that the system formed by the momenta of the particles at time t_1 and the system of the impulses of the external forces applied from t_1 to t_2 are together equipollent to the system formed by the momenta of the particles at time t_2. Since the vectors associated with a rigid body can be considered as sliding vectors, it follows (Sec. 3.19) that the systems of vectors shown in Fig. 17.6 are not only equipollent but truly *equivalent* in

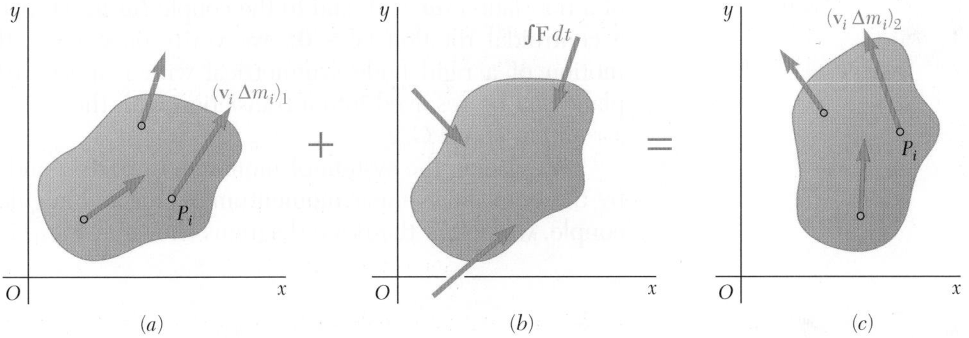

Fig. 17.6

the sense that the vectors on the left-hand side of the equals sign can be transformed into the vectors on the right-hand side through the use of the fundamental operations listed in Sec. 3.13. We therefore write

$$\textbf{Syst Momenta}_1 + \textbf{Syst Ext Imp}_{1\rightarrow 2} = \textbf{Syst Momenta}_2 \qquad (17.14)$$

But the momenta $\mathbf{v}_i\,\Delta m_i$ of the particles can be reduced to a vector attached at G, equal to their sum

$$\mathbf{L} = \sum_{i=1}^{n} \mathbf{v}_i\,\Delta m_i$$

and a couple of moment equal to the sum of their moments about G

$$\mathbf{H}_G = \sum_{i=1}^{n} \mathbf{r}'_i \times \mathbf{v}_i\,\Delta m_i$$

We recall from Sec. 14.3 that $\mathbf{L}$ and $\mathbf{H}_G$ define, respectively, the linear momentum and the angular momentum about G of the system

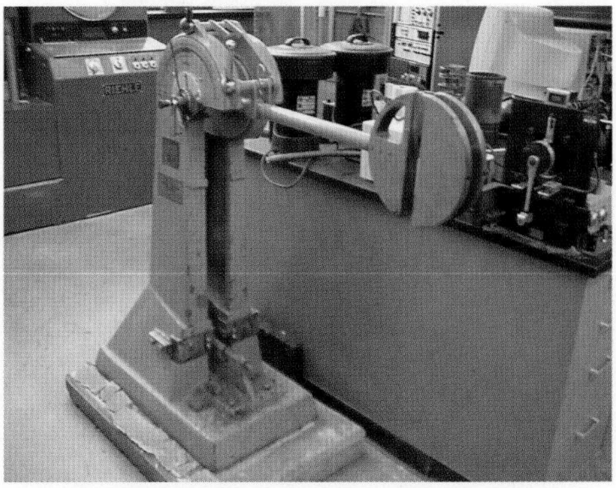

Photo 17.2 A Charpy impact test is used to determine the amount of energy absorbed by a material during impact by subtracting the final gravitation potential energy of the arm from its initial gravitational potential energy.

of particles forming the rigid body. We also note from Eq. (14.14) that $\mathbf{L} = m\overline{\mathbf{v}}$. On the other hand, restricting the present analysis to the plane motion of a rigid slab or of a rigid body symmetrical with respect to the reference plane, we recall from Eq. (16.4) that $\mathbf{H}_G = \overline{I}\boldsymbol{\omega}$. We thus conclude that the system of the momenta $\mathbf{v}_i\, \Delta m_i$ is equivalent to the *linear momentum vector* $m\overline{\mathbf{v}}$ attached at G and to the *angular momentum couple* $\overline{I}\boldsymbol{\omega}$ (Fig. 17.7). Observing that the

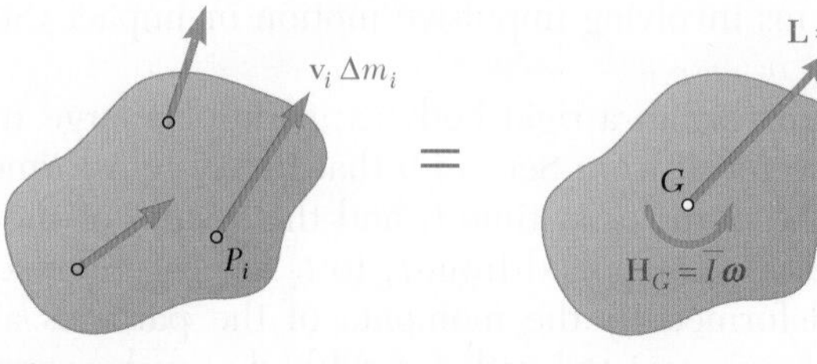

Fig. 17.7

system of momenta reduces to the vector $m\overline{\mathbf{v}}$ in the particular case of a translation ($\boldsymbol{\omega} = 0$) and to the couple $\overline{I}\boldsymbol{\omega}$ in the particular case of a centroidal rotation ($\overline{\mathbf{v}} = \mathbf{0}$), we verify once more that the plane motion of a rigid body symmetrical with respect to the reference plane can be resolved into a translation with the mass center G and a rotation about G.

Replacing the system of momenta in parts *a* and *c* of Fig. 17.6 by the equivalent linear momentum vector and angular momentum couple, we obtain the three diagrams shown in Fig. 17.8. This figure

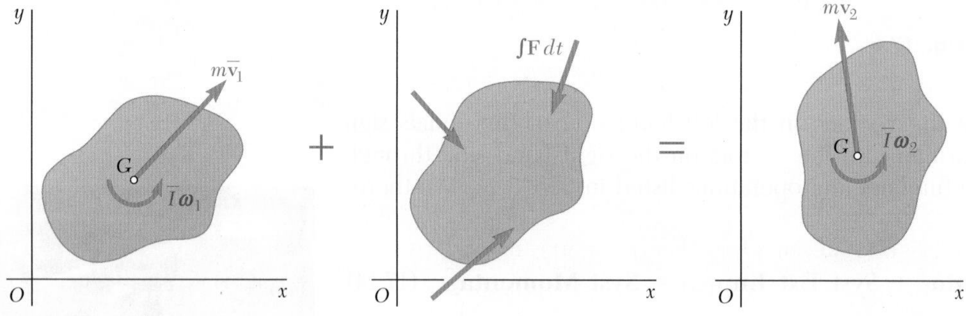

Fig. 17.8

expresses as a free-body-diagram equation the fundamental relation (17.14) in the case of the plane motion of a rigid slab or of a rigid body symmetrical with respect to the reference plane.

Three equations of motion can be derived from Fig. 17.8. Two equations are obtained by summing and equating the x and y *components* of the momenta and impulses, and the third equation is obtained by summing and equating the *moments* of these vectors *about any given point*. The coordinate axes can be chosen fixed in space, or allowed to move with the mass center of the body while maintaining a fixed direction. In either case, the point about which moments are taken should keep the same position relative to the coordinate axes during the interval of time considered.

In deriving the three equations of motion for a rigid body, care should be taken not to add linear and angular momenta indiscriminately. Confusion can be avoided by remembering that $m\bar{v}_x$ and $m\bar{v}_y$ represent the *components of a vector,* namely, the linear momentum vector $m\bar{\mathbf{v}}$, while $\bar{I}\omega$ represents the *magnitude of a couple,* namely, the angular momentum couple $\bar{I}\boldsymbol{\omega}$. Thus the quantity $\bar{I}\omega$ should be added only to the *moment* of the linear momentum $m\bar{\mathbf{v}}$, never to this vector itself nor to its components. All quantities involved will then be expressed in the same units, namely N · m · s or lb · ft · s.

Noncentroidal Rotation. In this particular case of plane motion, the magnitude of the velocity of the mass center of the body is $\bar{v} = \bar{r}\omega$, where $\bar{r}$ represents the distance from the mass center to the fixed axis of rotation and $\boldsymbol{\omega}$ represents the angular velocity of the body at the instant considered; the magnitude of the momentum vector attached at G is thus $m\bar{v} = m\bar{r}\omega$. Summing the moments about O of the momentum vector and momentum couple (Fig. 17.9)

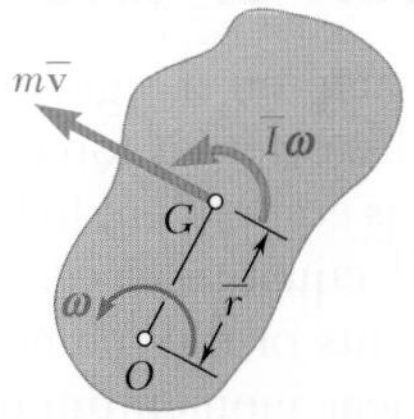

Fig. 17.9

and using the parallel-axis theorem for moments of inertia, we find that the angular momentum $\mathbf{H}_O$ of the body about O has the magnitude†

$$\bar{I}\omega + (m\bar{r}\omega)\bar{r} = (\bar{I} + m\bar{r}^2)\omega = I_O\omega \tag{17.15}$$

Equating the moments about O of the momenta and impulses in (17.14), we write

$$I_O\omega_1 + \sum \int_{t_1}^{t_2} M_O\,dt = I_O\omega_2 \tag{17.16}$$

In the general case of plane motion of a rigid body symmetrical with respect to the reference plane, Eq. (17.16) can be used with respect to the instantaneous axis of rotation under certain conditions. It is recommended, however, that all problems of plane motion be solved by the general method described earlier in this section.

†Note that the sum $\mathbf{H}_A$ of the moments about an arbitrary point A of the momenta of the particles of a rigid slab is, in general, *not* equal to $I_A\boldsymbol{\omega}$. (See Prob. 17.67.)

17.9 SYSTEMS OF RIGID BODIES

The motion of several rigid bodies can be analyzed by applying the principle of impulse and momentum to each body separately (Sample Prob. 17.6). However, in solving problems involving no more than three unknowns (including the impulses of unknown reactions), it is often convenient to apply the principle of impulse and momentum to the system as a whole. The momentum and impulse diagrams are drawn for the entire system of bodies. For each moving part of the system, the diagrams of momenta should include a momentum vector, a momentum couple, or both. Impulses of forces internal to the system can be omitted from the impulse diagram, since they occur in pairs of equal and opposite vectors. Summing and equating successively the x components, y components, and moments of all vectors involved, one obtains three relations which express that the momenta at time t_1 and the impulses of the external forces form a system equipollent to the system of the momenta at time t_2.† Again, care should be taken not to add linear and angular momenta indiscriminately; each equation should be checked to make sure that consistent units have been used. This approach has been used in Sample Prob. 17.8 and, further on, in Sample Probs. 17.9 and 17.10.

17.10 CONSERVATION OF ANGULAR MOMENTUM

When no external force acts on a rigid body or a system of rigid bodies, the impulses of the external forces are zero and the system of the momenta at time t_1 is equipollent to the system of the momenta at time t_2. Summing and equating successively the x components, y components, and moments of the momenta at times t_1 and t_2, we conclude that the total linear momentum of the system is conserved in any direction and that its total angular momentum is conserved about any point.

There are many engineering applications, however, in which *the linear momentum is not conserved* yet *the angular momentum* $\mathbf{H}_O$ of the system about a given point O *is conserved* that is, in which

$$(\mathbf{H}_O)_1 = (\mathbf{H}_O)_2 \tag{17.17}$$

Such cases occur when the lines of action of all external forces pass through O or, more generally, when the sum of the angular impulses of the external forces about O is zero.

Problems involving *conservation of angular momentum* about a point O can be solved by the general method of impulse and momentum, i.e., by drawing momentum and impulse diagrams as described in Secs. 17.8 and 17.9. Equation (17.17) is then obtained by summing and equating moments about O (Sample Prob. 17.8). As you will see later in Sample Prob. 17.9, two additional equations can be written by summing and equating x and y components and these equations can be used to determine two unknown linear impulses, such as the impulses of the reaction components at a fixed point.

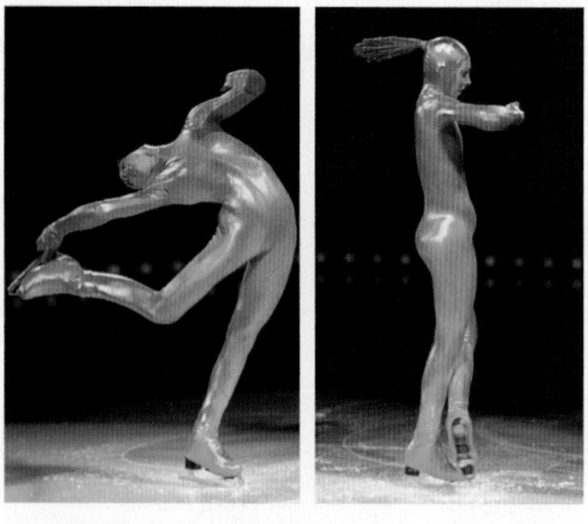

Photo 17.3 A figure skater at the beginning and at the end of a spin. By using the principle of conservation of angular momentum you will find that her angular velocity is much higher at the end of the spin.

†Note that as in Sec. 16.7, we cannot speak of *equivalent* systems since we are not dealing with a single rigid body.

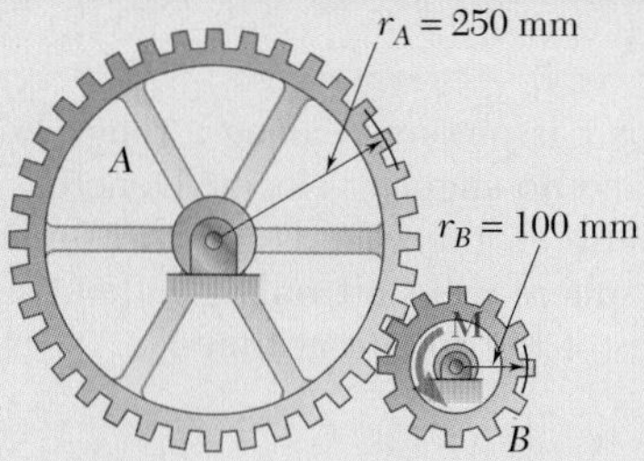

SAMPLE PROBLEM 17.6

Gear A has a mass of 10 kg and a radius of gyration of 200 mm, and gear B has a mass of 3 kg and a radius of gyration of 80 mm. The system is at rest when a couple $\mathbf{M}$ of magnitude 6 N · m is applied to gear B. (These gears were considered in Sample Prob. 17.2.) Neglecting friction, determine (a) the time required for the angular velocity of gear B to reach 600 rpm, (b) the tangential force which gear B exerts on gear A.

SOLUTION

We apply the principle of impulse and momentum to each gear separately. Since all forces and the couple are constant, their impulses are obtained by multiplying them by the unknown time t. We recall from Sample Prob. 17.2 that the centroidal moments of inertia and the final angular velocities are

$$\bar{I}_A = 0.400 \text{ kg} \cdot \text{m}^2 \qquad \bar{I}_B = 0.0192 \text{ kg} \cdot \text{m}^2$$
$$(\omega_A)_2 = 25.1 \text{ rad/s} \qquad (\omega_B)_2 = 62.8 \text{ rad/s}$$

Principle of Impulse and Momentum for Gear A. The systems of initial momenta, impulses, and final momenta are shown in three separate sketches.

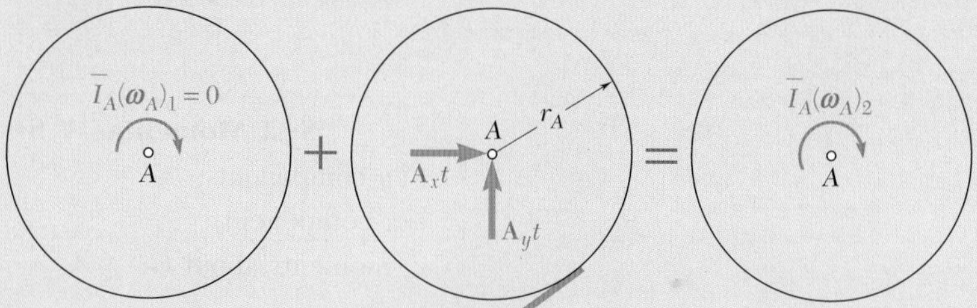

Syst Momenta$_1$ + Syst Ext Imp$_{1\to2}$ = Syst Momenta$_2$

$+\circlearrowleft$moments about A: $\quad 0 - Ftr_A = -\bar{I}_A(\omega_A)_2$

$$Ft(0.250 \text{ m}) = (0.400 \text{ kg} \cdot \text{m}^2)(25.1 \text{ rad/s})$$
$$Ft = 40.2 \text{ N} \cdot \text{s}$$

Principle of Impulse and Momentum for Gear B.

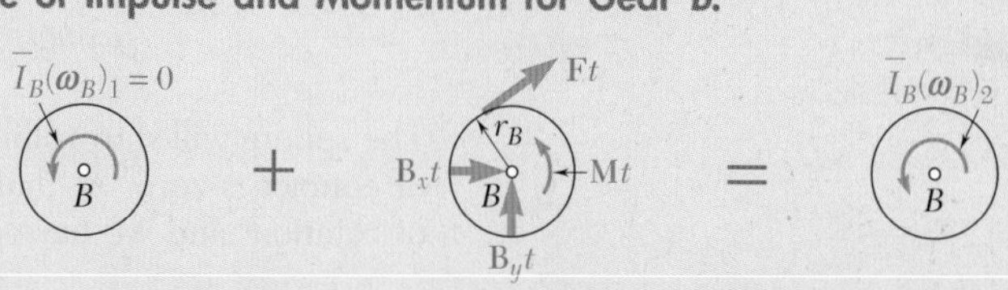

Syst Momenta$_1$ + Syst Ext Imp$_{1\to2}$ = Syst Momenta$_2$

$+\circlearrowleft$moments about B: $\quad 0 + Mt - Ftr_B = \bar{I}_B(\omega_B)_2$

$$+(6 \text{ N} \cdot \text{m})t - (40.2 \text{ N} \cdot \text{s})(0.100 \text{ m}) = (0.0192 \text{ kg} \cdot \text{m}^2)(62.8 \text{ rad/s})$$

$t = 0.871$ s ◀

Recalling that $Ft = 40.2$ N · s, we write

$$F(0.871 \text{ s}) = 40.2 \text{ N} \cdot \text{s} \qquad F = +46.2 \text{ N}$$

Thus, the force exerted by gear B on gear A is $\quad \mathbf{F} = 46.2 \text{ N} \swarrow$ ◀

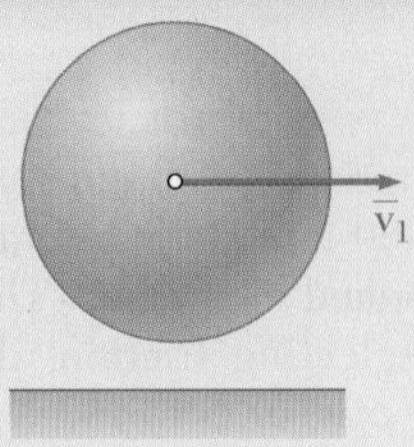

SAMPLE PROBLEM 17.7

A uniform sphere of mass m and radius r is projected along a rough horizontal surface with a linear velocity $\bar{\mathbf{v}}_1$ and no angular velocity. Denoting by μ_k the coefficient of kinetic friction between the sphere and the surface, determine (*a*) the time t_2 at which the sphere will start rolling without sliding, (*b*) the linear and angular velocities of the sphere at time t_2.

SOLUTION

While the sphere is sliding relative to the surface, it is acted upon by the normal force **N**, the friction force **F**, and its weight **W** of magnitude $W = mg$.

Principle of Impulse and Momentum. We apply the principle of impulse and momentum to the sphere from the time $t_1 = 0$ when it is placed on the surface until the time $t_2 = t$ when it starts rolling without sliding.

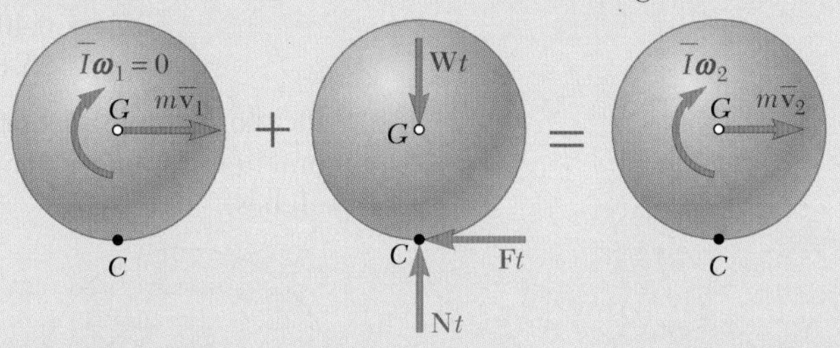

Syst Momenta$_1$ + Syst Ext Imp$_{1\to 2}$ = Syst Momenta$_2$

$+\uparrow y$ components: $\quad Nt - Wt = 0 \quad (1)$

$\xrightarrow{+} x$ components: $\quad m\bar{v}_1 - Ft = m\bar{v}_2 \quad (2)$

$+\downarrow$ moments about G: $\quad Ftr = \bar{I}\omega_2 \quad (3)$

From (1) we obtain $N = W = mg$. During the entire time interval considered, sliding occurs at point C and we have $F = \mu_k N = \mu_k mg$. Substituting this for F into (2), we write

$$m\bar{v}_1 - \mu_k mgt = m\bar{v}_2 \qquad \bar{v}_2 = \bar{v}_1 - \mu_k gt \tag{4}$$

Substituting $F = \mu_k mg$ and $\bar{I} = \frac{2}{5}mr^2$ into (3),

$$\mu_k mgtr = \tfrac{2}{5}mr^2\omega_2 \qquad \omega_2 = \frac{5}{2}\frac{\mu_k g}{r}t \tag{5}$$

The sphere will start rolling without sliding when the velocity $\mathbf{v}_C$ of the point of contact is zero. At that time, point C becomes the instantaneous center of rotation, and we have $\bar{v}_2 = r\omega_2$. Substituting from (4) and (5), we write

$$\bar{v}_2 = r\omega_2 \qquad \bar{v}_1 - \mu_k gt = r\left(\frac{5}{2}\frac{\mu_k g}{r}t\right) \qquad t = \frac{2}{7}\frac{\bar{v}_1}{\mu_k g} \quad ◀$$

Substituting this expression for t into (5),

$$\omega_2 = \frac{5}{2}\frac{\mu_k g}{r}\left(\frac{2}{7}\frac{\bar{v}_1}{\mu_k g}\right) \qquad \omega_2 = \frac{5}{7}\frac{\bar{v}_1}{r} \qquad \boldsymbol{\omega}_2 = \frac{5}{7}\frac{\bar{v}_1}{r} \downarrow \quad ◀$$

$$\bar{v}_2 = r\omega_2 \qquad \bar{v}_2 = r\left(\frac{5}{7}\frac{v_1}{r}\right) \qquad \bar{\mathbf{v}}_2 = \tfrac{5}{7}\,\bar{v}_1 \rightarrow \quad ◀$$

SAMPLE PROBLEM 17.8

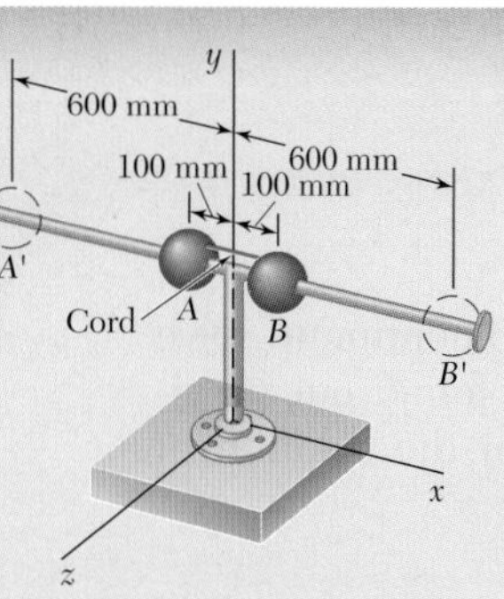

Two solid spheres of radius 100 mm, of mass 1 kg each, are mounted at A and B on the horizontal rod $A'B'$, which rotates freely about the vertical with a counterclockwise angular velocity of 6 rad/s. The spheres are held in position by a cord which is suddenly cut. Knowing that the centroidal moment of inertia of the rod and pivot is $I_R = 0.4 \text{ kg}\cdot\text{m}^2$, determine (*a*) the angular velocity of the rod after the spheres have moved to positions A' and B', (*b*) the energy lost due to the plastic impact of the spheres and the stops at A' and B'.

SOLUTION

a. Principle of Impulse and Momentum. In order to determine the final angular velocity of the rod, we will express that the initial momenta of the various parts of the system and the impulses of the external forces are together equipollent to the final momenta of the system. The final configuration (2) is just before the spleres hit the stop.

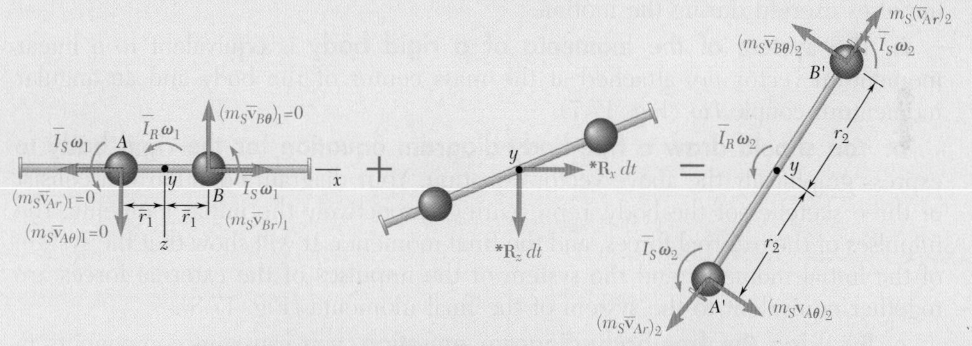

Syst Momenta$_1$ + Syst Ext Imp$_{1\to2}$ = Syst Momenta$_2$

We observe that the external forces consist of the weights and the reaction at the pivot have no moment about the y axis. We also observe that the r components of V_A and V_B do not contribute to the angular momentum about the y axis. Noting that $\bar{v}_{A\theta} = \bar{v}_{B\theta} = \bar{r}\omega$ we equate moments about the y axis: The final configuration (2) is first before the spheres hit the stop.

$$2(m_S\bar{r}_1\omega_1)\bar{r}_1 + 2\bar{I}_S\omega_1 + \bar{I}_R\omega_1 = 2(m_S\bar{r}_2\omega_2)\bar{r}_2 + 2\bar{I}_S\omega_2 + \bar{I}_R\omega_2$$

$$(2m_S\bar{r}_1^2 + 2\bar{I}_S + \bar{I}_R)\omega_1 = (2m_S\bar{r}_2^2 + 2\bar{I}_S + \bar{I}_R)\omega_2 \qquad (1)$$

which expresses that *the angular momentum of the system about the y axis is conserved.* We now compute

$$\bar{I}_S = \tfrac{2}{5}m_S a^2 = \tfrac{2}{5}(1 \text{ kg})(0.1 \text{ m})^2 = 0.004 \text{ kg}\cdot\text{m}^2$$

$$m_S\bar{r}_1^2 = (1 \text{ kg})(0.1 \text{ m})^2 = 0.01 \text{ kg}\cdot\text{m}^2 \quad m_S\bar{r}_2^2 = (1 \text{ kg})(0.6 \text{ m})^2 = 0.36 \text{ kg}\cdot\text{m}^2$$

Substituting these values, and $\bar{I}_R = 0.4$ and $\omega_1 = 6$ rad/s into (1):

$$0.428(6 \text{ rad/s}) = 1.128\omega_2 \qquad \omega_2 = 2.28 \text{ rad/s} \circlearrowleft \blacktriangleleft$$

b. Energy Lost. The kinetic energy of the system at any instant is

$$T = 2(\tfrac{1}{2}m_S\bar{v}^2 + \tfrac{1}{2}\bar{I}_S\omega^2) + \tfrac{1}{2}\bar{I}_R\omega^2 \qquad \bar{v}^2 = v_r^2 + v_\theta^2$$

Thus if $v_r = 0$, $T = \frac{1}{2}(2m_s\bar{r}^2 + 2\bar{I}_s + \bar{I}_R)\omega^2$

Let 1 denote the initial position, 2 denote the system just before the sphere hit the stop and 3 system just after the spheres have hit the stop.

We observe V_r of spheres at 1 = 0 as the chord is just cut. Since the impact is just plastic, V_r component of spheres at 3 = 0. The energy lost due to plastic impact = $T_2 - T_3$, as the potential energy of the system does not change.

From conservation of energy, $T_1 = T_2 \Rightarrow$Energy lost in impact $T_1 - T_3$ Recalling the numerical values found above, we have

$$T_1 = \tfrac{1}{2}(0.428)(6)^2 = 7.704 \text{ J} \qquad T_3 = \tfrac{1}{2}(1.128)(2.28)^2 = 2.932 \text{ J}$$

$$\text{Energy Lost} = T_1 - T_3 = 7.704 - 2.932 = 4.77 \text{ J} \blacktriangleleft$$

Note: By applying $T_1 = T_2$ we can find the radial velocity of the sphere at position 2.

SOLVING PROBLEMS ON YOUR OWN

In this lesson you learned to use the method of impulse and momentum to solve problems involving the plane motion of rigid bodies. As you found out previously in Chap. 13, this method is most effective when used in the solution of problems involving velocities and time.

1. The principle of impulse and momentum for the plane motion of a rigid body is expressed by the following vector equation:

$$\textbf{Syst Momenta}_1 + \textbf{Syst Ext Imp}_{1\rightarrow 2} = \textbf{Syst Momenta}_2 \qquad (17.14)$$

where **Syst Momenta** represents the system of the momenta of the particles forming the rigid body, and **Syst Ext Imp** represents the system of all the external impulses exerted during the motion.

a. The system of the momenta of a rigid body is equivalent to a linear momentum vector $m\overline{\mathbf{v}}$ attached at the mass center of the body and an angular momentum couple $\bar{I}\boldsymbol{\omega}$ (Fig. 17.7).

b. You should draw a free-body-diagram equation for the rigid body to express graphically the above vector equation. Your diagram equation will consist of three sketches of the body, representing respectively the initial momenta, the impulses of the external forces, and the final momenta. It will show that the system of the initial momenta and the system of the impulses of the external forces are together equivalent to the system of the final momenta (Fig. 17.8).

c. By using the free-body-diagram equation, you can sum components in any direction and sum moments about any point. When summing moments about a point, remember to include the *angular momentum* $\bar{I}\omega$ of the body, as well as the *moments* of the components of its *linear momentum.* In most cases you will be able to select and solve an equation that involves only one unknown. This was done in all the sample problems of this lesson.

2. In problems involving a system of rigid bodies, you can apply the principle of impulse and momentum to the system as a whole. Since internal forces occur in equal and opposite pairs, they will not be part of your solution [Sample Prob. 17.8].

3. Conservation of angular momentum about a given axis occurs when, for a system of rigid bodies, *the sum of the moments of the external impulses about that axis is zero.* You can indeed easily observe from the free-body-diagram equation that the initial and final angular momenta of the system about that axis are equal and, thus, that *the angular momentum of the system about the given axis is conserved.* You can then sum the angular momenta of the various bodies of the system and the moments of their linear momenta about that axis to obtain an equation which can be solved for one unknown [Sample Prob. 17.8].

PROBLEMS

CONCEPT QUESTIONS

17.CQ6 Slender bar A is rigidly connected to a massless rod BC in Case 1 and two massless cords in Case 2 as shown. The vertical thickness of bar A is negligible compared to L. If bullet D strikes A with a speed v_0 and becomes embedded in it, how will the speeds of the center of gravity of A immediately after the impact compare for the two cases?

a. Case 1 will be larger.
b. Case 2 will be larger.
c. The speeds will be the same.

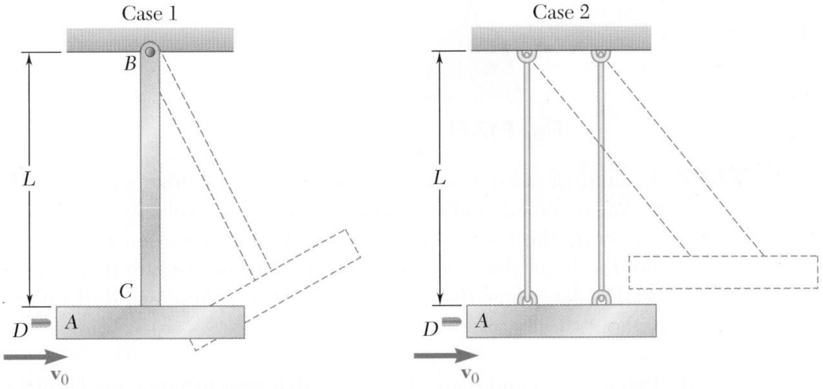

Fig. P17.CQ6

17.CQ7 A 1-m-long uniform slender bar AB has an angular velocity of 12 rad/s and its center of gravity has a velocity of 2 m/s as shown. About which point is the angular momentum of A smallest at this instant?

a. P_1
b. P_2
c. P_3
d. P_4
e. It is the same about all the points.

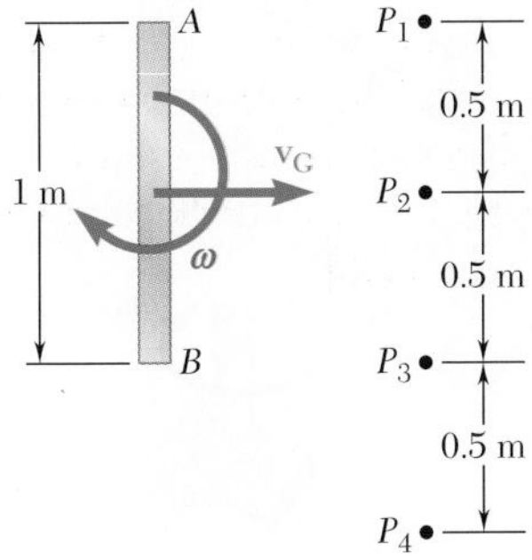

Fig. P17.CQ7

IMPULSE-MOMENTUM PRACTICE PROBLEMS

17.F1 The 350-kg flywheel of a small hoisting engine has a radius of gyration of 600 mm. If the power is cut off when the angular velocity of the flywheel is 100 rpm clockwise, draw an impulse-momentum diagram that can be used to determine the time required for the system to come to rest.

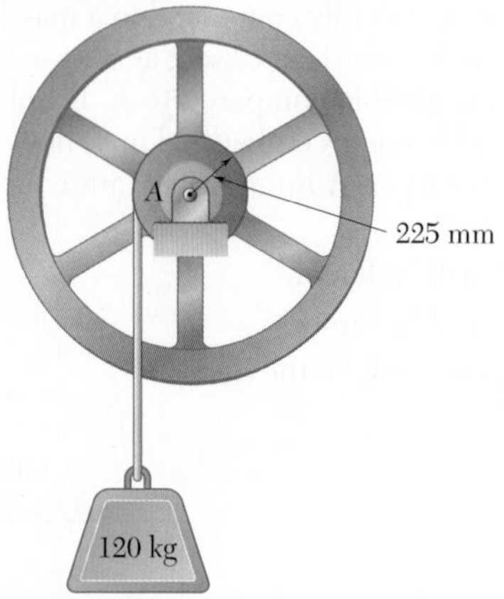

Fig. P17.F1

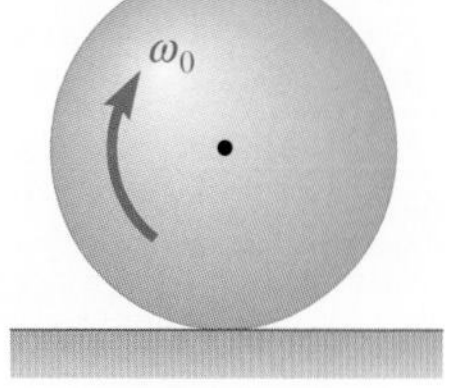

Fig. P17.F2

17.F2 A sphere of radius r and mass m is placed on a horizontal floor with no linear velocity but with a clockwise angular velocity ω_0. Denoting by μ_k the coefficient of kinetic friction between the sphere and the floor, draw the impulse-momentum diagram that can be used to determine the time t_1 at which the sphere will start rolling without sliding.

17.F3 Two panels A and B are attached with hinges to a rectangular plate and held by a wire as shown. The plate and the panels are made of the same material and have the same thickness. The entire assembly is rotating with an angular velocity ω_0 when the wire breaks. Draw the impulse-momentum diagram that is needed to determine the angular velocity of the assembly after the panels have come to rest against the plate.

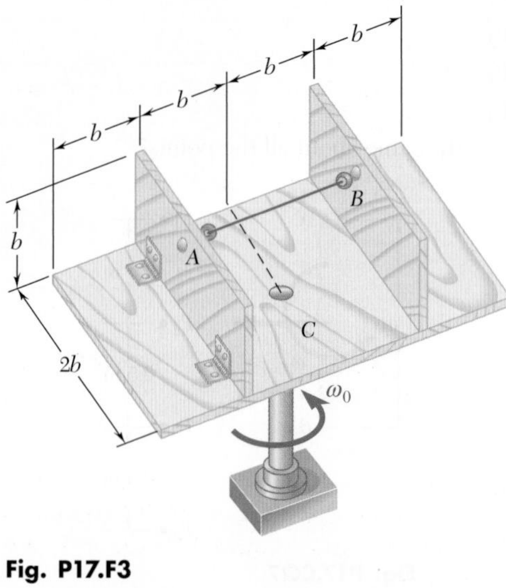

Fig. P17.F3

END-OF-SECTION PROBLEMS

17.52 The rotor of an electric motor has a mass of 25 kg, and it is observed that 4.2 min is required for the rotor to coast to rest from an angular velocity of 3600 rpm. Knowing that kinetic friction produces a couple of magnitude 1.2 N · m, determine the centroidal radius of gyration for the rotor.

17.53 A small grinding wheel is attached to the shaft of an electric motor which has a rated speed of 3600 rpm. When the power is turned off, the unit coasts to rest in 70 s. The grinding wheel and rotor have a combined mass of 3 kg and a combined radius of gyration of 50 mm. Determine the average magnitude of the couple due to kinetic friction in the bearings of the motor.

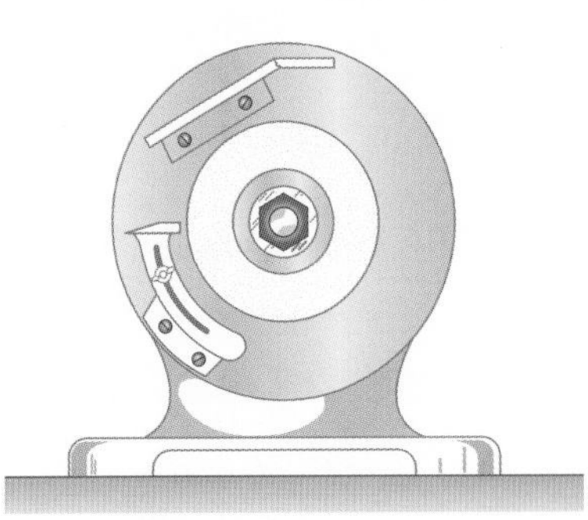

Fig. P17.53

17.54 A bolt located 50 mm from the center of an automobile wheel is tightened by applying the couple shown for 0.10 s. Assuming that the wheel is free to rotate and is initially at rest, determine the resulting angular velocity of the wheel. The wheel has a mass of 19 kg and has a radius of gyration of 250 mm.

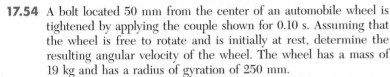

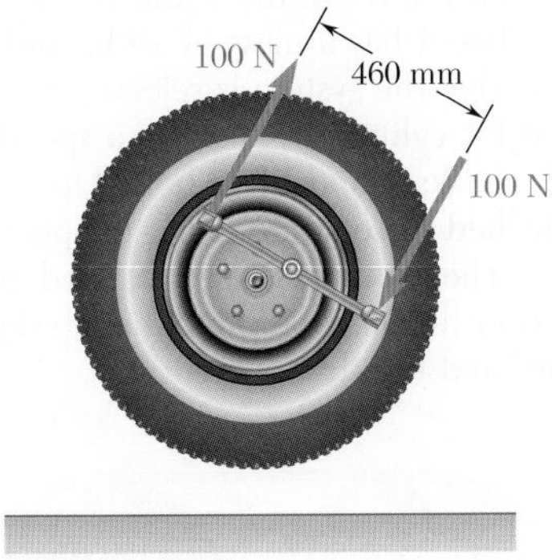

Fig. P17.54

17.55 Two disks of the same thickness and same material are attached to a shaft as shown. The 4-kg disk A has a radius $r_A = 100$ mm, and disk B has a radius $r_B = 150$ mm. Knowing that a couple **M** of magnitude 2.5 N·m is applied to disk A when the system is at rest, determine the time required for the angular velocity of the system to reach 960 rpm.

17.56 Two disks of the same thickness and same material are attached to a shaft as shown. The 3-kg disk A has a radius $r_A = 100$ mm, and disk B has a radius $r_B = 125$ mm. Knowing that the angular velocity of the system is to be increased from 200 rpm to 800 rpm during a 3-s interval, determine the magnitude of the couple **M** that must be applied to disk A.

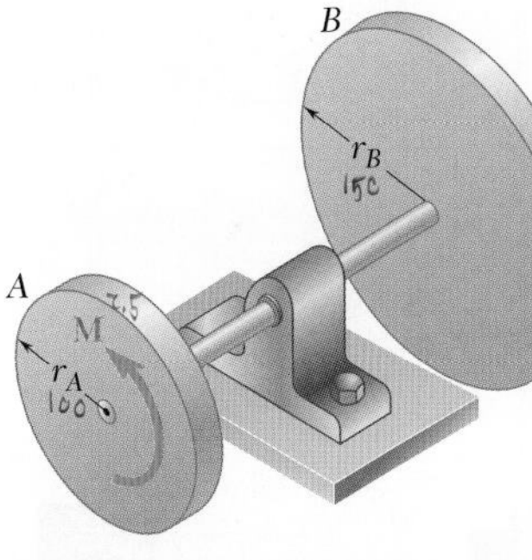

Fig. P17.55 and *P17.56*

17.57 A disk of constant thickness, initially at rest, is placed in contact with a belt that moves with a constant velocity **v**. Denoting by μ_k the coefficient of kinetic friction between the disk and the belt, derive an expression for the time required for the disk to reach a constant angular velocity.

17.58 Disk A, of mass 2.5 kg and radius $r = 100$ mm, is at rest when it is placed in contact with a belt which moves at a constant speed $v = 15$ m/s. Knowing that $\mu_k = 0.20$ between the disk and the belt, determine the time required for the disk to reach a constant angular velocity.

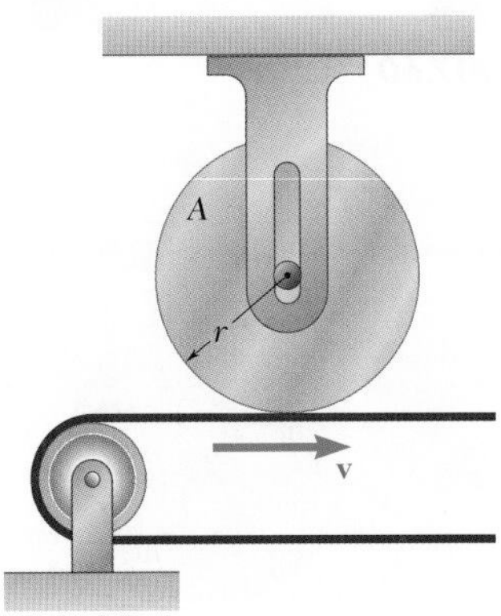

Fig. *P17.57* and P17.58

17.59 A cylinder of radius r and weight W with an initial counterclockwise angular velocity $\boldsymbol{\omega}_0$ is placed in the corner formed by the floor and a vertical wall. Denoting by μ_k the coefficient of kinetic friction between the cylinder and the wall and the floor, derive an expression for the time required for the cylinder to come to rest.

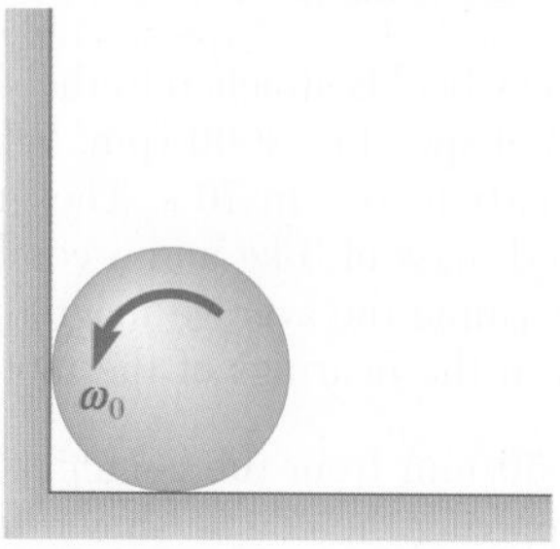

Fig. P17.59

17.60* and *17.61 Two uniform disks and two cylinders are assembled as indicated. Disk A has a mass of 10 kg and disk B has a mass of 6 kg. Knowing that the system is released from rest, determine the time required for cylinder C to have a speed of 0.5 m/s.

17.60 Disks A and B are bolted together and the cylinders are attached to separate cords wrapped on the disks.

17.61 The cylinders are attached to a single cord that passes over the disks. Assume that no slipping occurs between the cord and the disks.

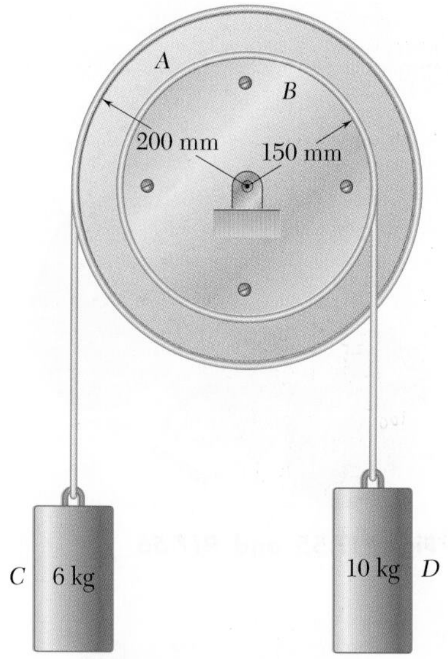

Fig. *P17.60*

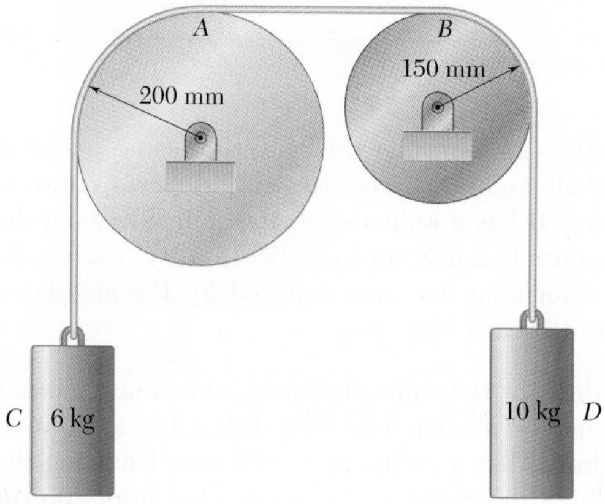

Fig. *P17.61*

17.62 Disk B has an initial angular velocity $\boldsymbol{\omega}_0$ when it is brought into contact with disk A which is at rest. Show that the final angular velocity of disk B depends only on ω_0 and the ratio of the masses m_A and m_B of the two disks.

17.63 The 4-kg disk A has a radius r_A = 150 mm and is initially at rest. The 5-kg disk B has a radius r_B = 200 mm and an angular velocity $\boldsymbol{\omega}_0$ of 900 rpm when it is brought into contact with disk A. Neglecting friction in the bearings, determine (*a*) the final angular velocity of each disk, (*b*) the total impulse of the friction force exerted on disk A.

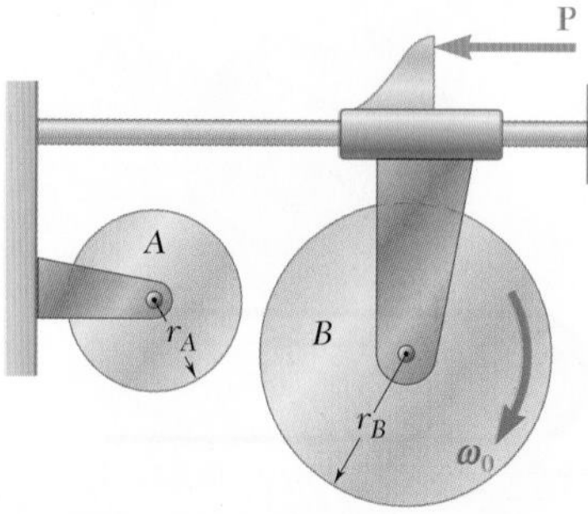

Fig. P17.62 and P17.63

17.64 A tape moves over the two drums shown. Drum A has a mass of 0.6 kg and has a radius of gyration of 20 mm, while drum B has a mass of 1.75 kg and has a radius of gyration of 30 mm. In the lower portion of the tape the tension is constant and equal to T_A = 4N. Knowing that the tape is initially at rest, determine (*a*) the required constant tension T_B if the velocity of the tape is to be v = 3 m/s after 0.24 s, (*b*) the corresponding tension in the portion of the tape between the drums.

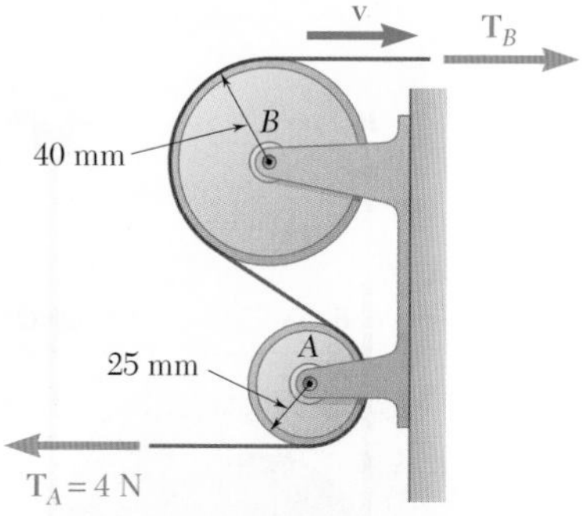

Fig. P17.64

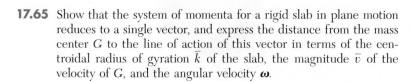

17.65 Show that the system of momenta for a rigid slab in plane motion reduces to a single vector, and express the distance from the mass center G to the line of action of this vector in terms of the centroidal radius of gyration $\bar{k}$ of the slab, the magnitude $\bar{v}$ of the velocity of G, and the angular velocity $\boldsymbol{\omega}$.

17.66 Show that, when a rigid slab rotates about a fixed axis through O perpendicular to the slab, the system of the momenta of its particles is equivalent to a single vector of magnitude $m\bar{r}\omega$, perpendicular to the line OG, and applied to a point P on this line, called the *center of percussion*, at a distance $GP = \bar{k}^2/\bar{r}$ from the mass center of the slab.

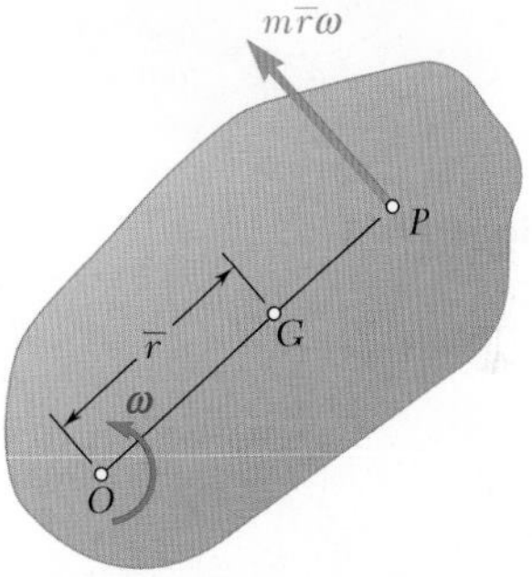

Fig. P17.66

17.67 Show that the sum $\mathbf{H}_A$ of the moments about a point A of the momenta of the particles of a rigid slab in plane motion is equal to $I_A\boldsymbol{\omega}$, where $\boldsymbol{\omega}$ is the angular velocity of the slab at the instant considered and I_A the moment of inertia of the slab about A, if and only if one of the following conditions is satisfied: (*a*) A is the mass center of the slab, (*b*) A is the instantaneous center of rotation, (*c*) the velocity of A is directed along a line joining point A and the mass center G.

17.68 Consider a rigid slab initially at rest and subjected to an impulsive force **F** contained in the plane of the slab. We define the *center of percussion* P as the point of intersection of the line of action of **F** with the perpendicular drawn from G. (*a*) Show that the instantaneous center of rotation C of the slab is located on line GP at a distance $GC = \bar{k}^2/GP$ on the opposite side of G. (*b*) Show that if the center of percussion were located at C the instantaneous center of rotation would be located at P.

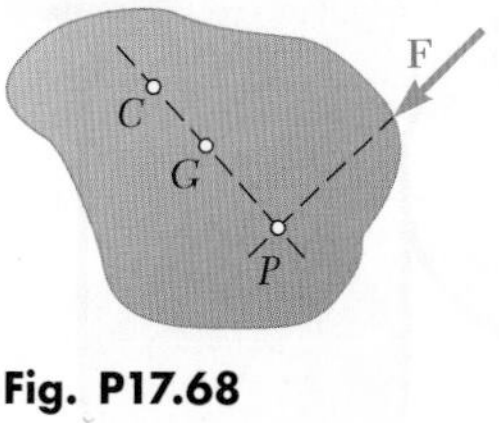

Fig. P17.68

17.69 A flywheel is rigidly attached to a 40-mm-radius shaft that rolls without sliding along parallel rails. Knowing that after being released from rest the system attains a speed of 150 mm/s in 30 s, determine the centroidal radius of gyration of the system.

Fig. P17.69

17.70 A wheel of radius r and centroidal radius of gyration $\bar{k}$ is released from rest on the incline shown at time $t = 0$. Assuming that the wheel rolls without sliding, determine (*a*) the velocity of its center at time t, (*b*) the coefficient of static friction required to prevent slipping.

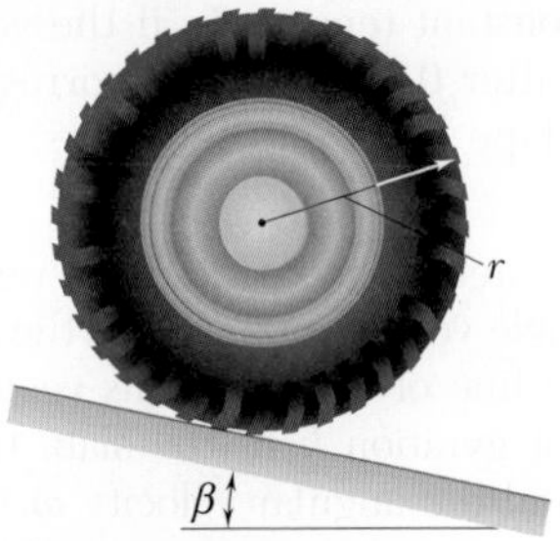

Fig. P17.70

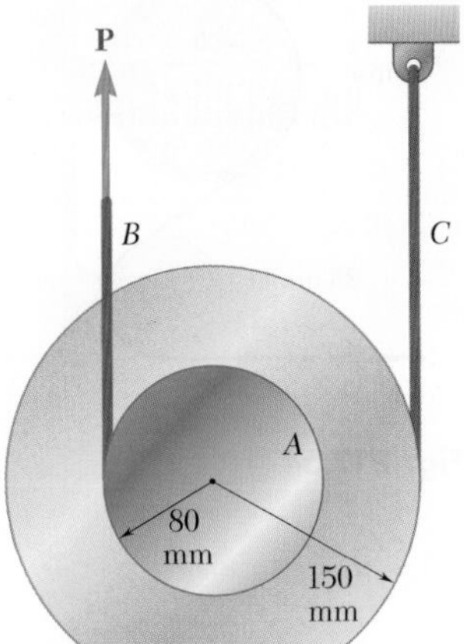

Fig. P17.71

17.71 The double pulley shown has a mass of 3 kg and a radius of gyration of 100 mm. Knowing that when the pulley is at rest, a force **P** of magnitude 24 N is applied to cord B, determine (*a*) the velocity of the center of the pulley after 1.5 s, (*b*) the tension in cord C.

17.72 and *17.73* A 240-mm-radius cylinder of mass 8 kg rests on a 3-kg carriage. The system is at rest when a force **P** of magnitude 10 N is applied as shown for 1.2 s. Knowing that the cylinder rolls without sliding on the carriage and neglecting the mass of the wheels of the carriage, determine the resulting velocity of (*a*) the carriage, (*b*) the center of the cylinder.

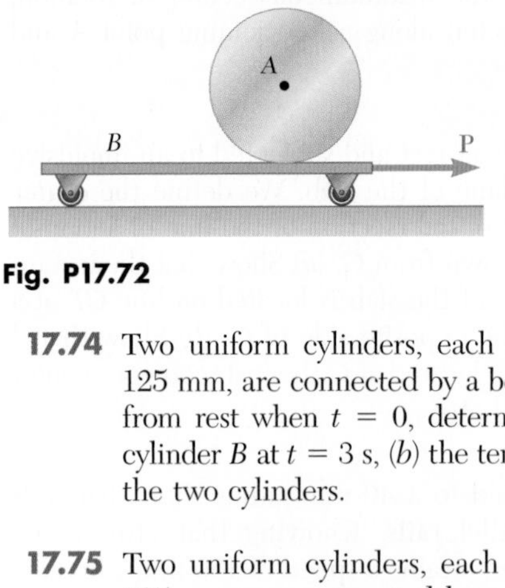

Fig. P17.72

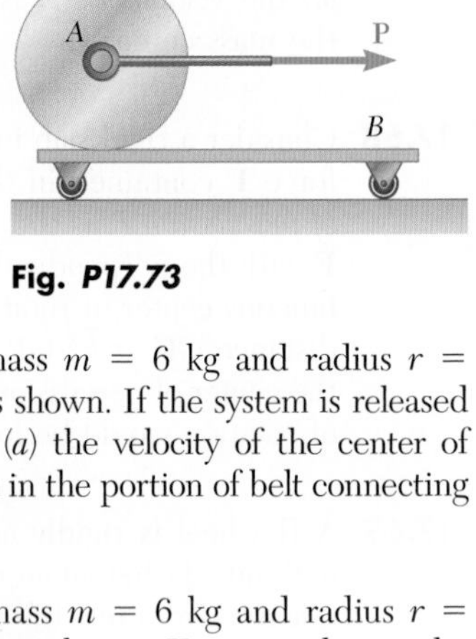

Fig. *P17.73*

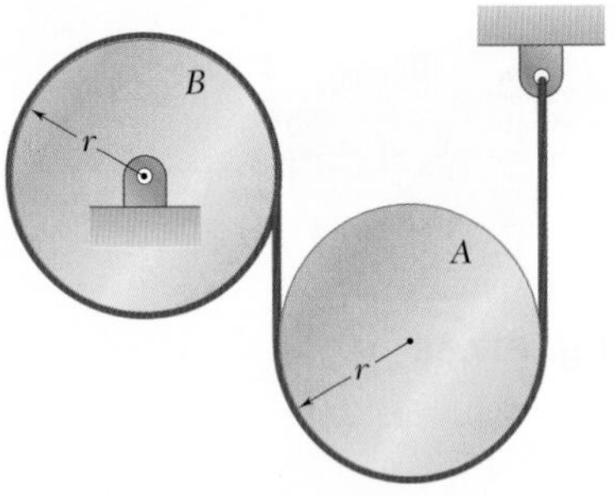

Fig. P17.74 and P17.75

17.74 Two uniform cylinders, each of mass $m = 6$ kg and radius $r = 125$ mm, are connected by a belt as shown. If the system is released from rest when $t = 0$, determine (*a*) the velocity of the center of cylinder B at $t = 3$ s, (*b*) the tension in the portion of belt connecting the two cylinders.

17.75 Two uniform cylinders, each of mass $m = 6$ kg and radius $r = 125$ mm, are connected by a belt as shown. Knowing that at the instant shown the angular velocity of cylinder A is 30 rad/s counterclockwise, determine (*a*) the time required for the angular velocity of cylinder A to be reduced to 5 rad/s, (*b*) the tension in the portion of belt connecting the two cylinders.

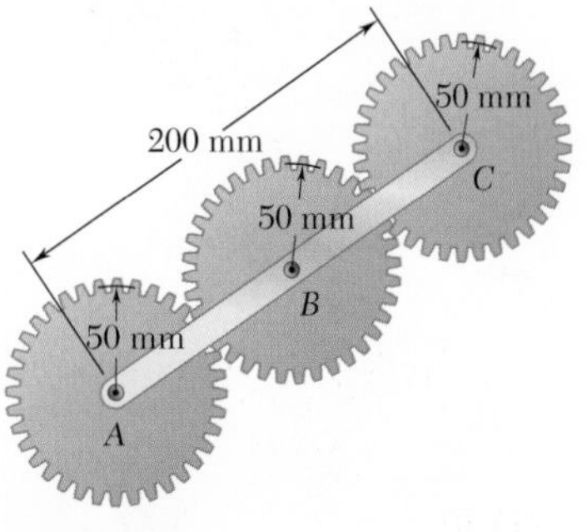

Fig. *P17.76*

17.76 In the gear arrangement shown, gears A and C are attached to rod ABC, which is free to rotate about B, while the inner gear B is fixed. Knowing that the system is at rest, determine the magnitude of the couple **M** which must be applied to rod ABC, if 2.5 s later the angular velocity of the rod is to be 240 rpm clockwise. Gears A and C have a mass of 1.25 kg each and may be considered as disks of radius 50 mm; rod ABC has a mass of 2 kg.

17.77 A sphere of radius r and mass m is projected along a rough horizontal surface with the initial velocities shown. If the final velocity of the sphere is to be zero, express (*a*) the required magnitude of $\boldsymbol{\omega}_0$ in terms of v_0 and r, (*b*) the time required for the sphere to come to rest in terms of v_0 and the coefficient of kinetic friction μ_k.

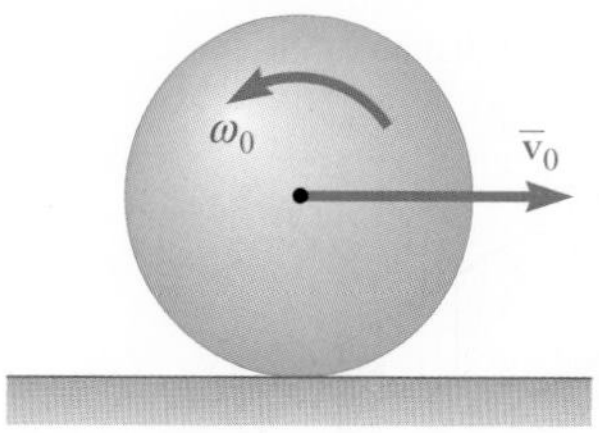

Fig. P17.77

17.78 A sphere of radius r and mass m is placed on a horizontal floor with no linear velocity but with a clockwise angular velocity $\boldsymbol{\omega}_0$. Denoting by μ_k the coefficient of kinetic friction between the sphere and the floor, determine (*a*) the time t_1 at which the sphere will start rolling without sliding, (*b*) the linear and angular velocities of the sphere at time t_1.

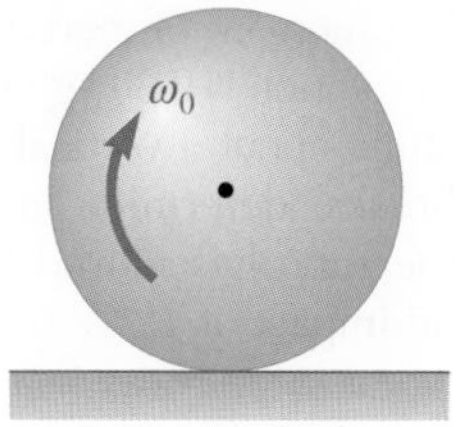

Fig. P17.78

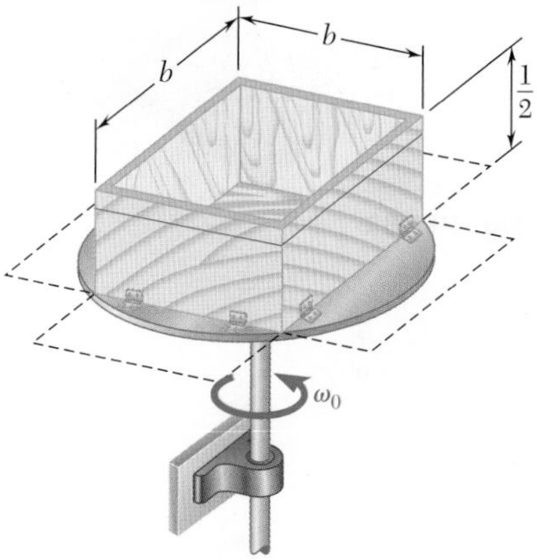

Fig. P17.79

17.79 Four rectangular panels, each of length b and height $\frac{1}{2}b$, are attached with hinges to a circular plate of diameter $\sqrt{2}\,b$ and held by a wire loop in the position shown. The plate and the panels are made of the same material and have the same thickness. The entire assembly is rotating with an angular velocity ω_0 when the wire breaks. Determine the angular velocity of the assembly after the panels have come to rest in a horizontal position.

17.80 A 1.25-kg disk of radius 100 mm is attached to the yoke BCD by means of short shafts fitted in bearings at B and D. The 0.75-kg yoke has a radius of gyration of 75 mm about the x axis. Initially the assembly is rotating at 120 rpm with the disk in the plane of the yoke ($\theta = 0$). If the disk is slightly disturbed and rotates with respect to the yoke until $\theta = 90°$, where it is stopped by a small bar at D, determine the final angular velocity of the assembly.

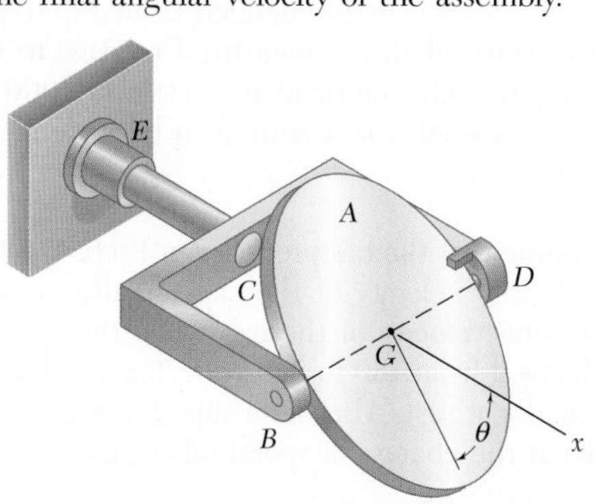

Fig. P17.80

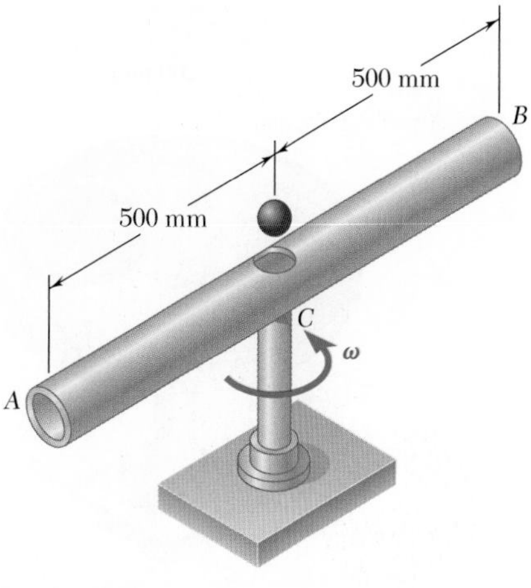

Fig. P17.81

17.81 Two 0.4-kg balls are to be put successively into the center C of the slender 2-kg tube AB. Knowing that when the first ball is put into the tube the initial angular velocity of the tube is 8 rad/s and neglecting the effect of friction, determine the angular velocity of the tube just after (*a*) the first ball has left the tube, (*b*) the second ball has left the tube.

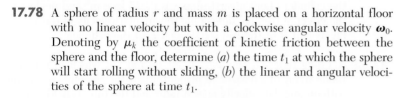

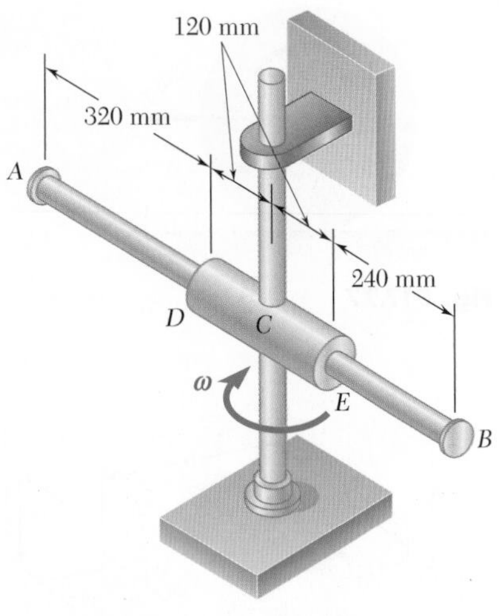

Fig. P17.82

17.82 A 3-kg rod of length 800 mm can slide freely in the 240-mm cylinder *DE*, which in turn can rotate freely in a horizontal plane. In the position shown the assembly is rotating with an angular velocity of magnitude $\omega = 40$ rad/s and end *B* of the rod is moving toward the cylinder at a speed of 75 mm/s relative to the cylinder. Knowing that the centroidal mass moment of inertia of the cylinder about a vertical axis is $0.025 \text{ kg} \cdot \text{m}^2$ and neglecting the effect of friction, determine the angular velocity of the assembly as end *B* of the rod strikes end *E* of the cylinder.

17.83 A 1.6-kg tube *AB* can slide freely on rod *DE* which in turn can rotate freely in a horizontal plane. Initially the assembly is rotating with an angular velocity $\omega = 5$ rad/s and the tube is held in position by a cord. The moment of inertia of the rod and bracket about the vertical axis of rotation is $0.30 \text{ kg} \cdot \text{m}^2$ and the centroidal moment of inertia of the tube about a vertical axis is $0.0025 \text{ kg} \cdot \text{m}^2$. If the cord suddenly breaks, determine (*a*) the angular velocity of the assembly after the tube has moved to end *E*, (*b*) the energy lost during the plastic impact at *E*.

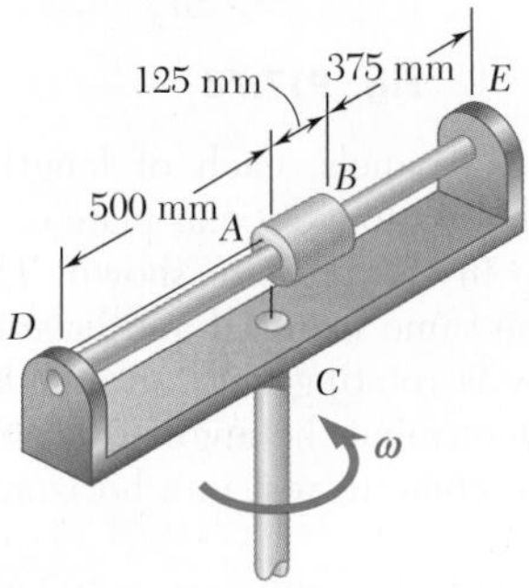

Fig. P17.83

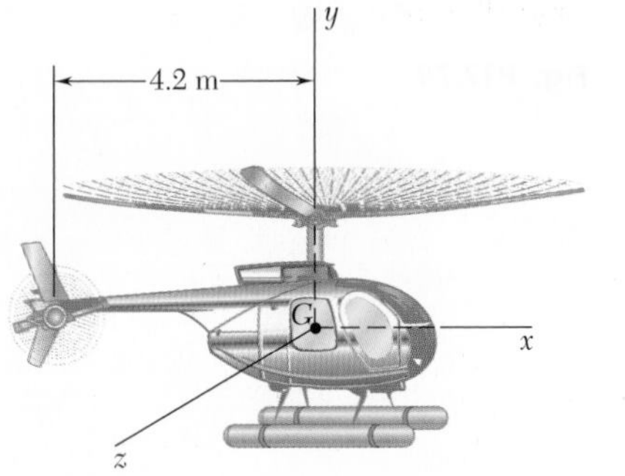

Fig. P17.84

17.84 In the helicopter shown, a vertical tail propeller is used to prevent rotation of the cab as the speed of the main blades is changed. Assuming that the tail propeller is not operating, determine the final angular velocity of the cab after the speed of the main blades has been changed from 180 to 240 rpm. (The speed of the main blades is measured relative to the cab, and the cab has a centroidal moment of inertia of $1000 \text{ kg} \cdot \text{m}^2$. Each of the four main blades is assumed to be a slender 4.2-m rod of mass 25 kg).

17.85 Assuming that the tail propeller in Prob. 17.84 is operating and that the angular velocity of the cab remains zero, determine the final horizontal velocity of the cab when the speed of the main blades is changed from 180 to 240 rpm. The cab has a mass of 625 kg and is initially at rest. Also determine the force exerted by the tail propeller if the change in speed takes place uniformly in 12 s.

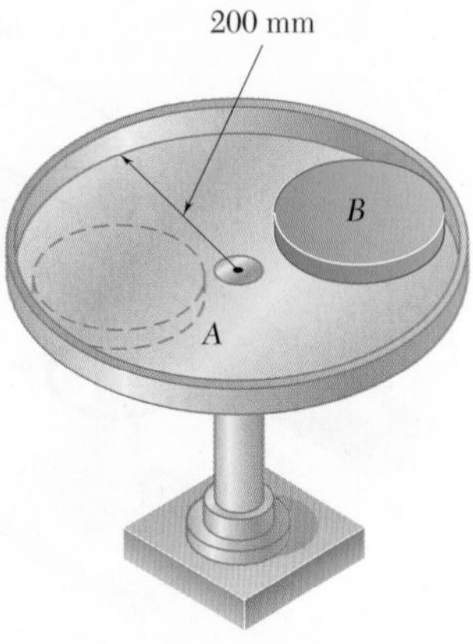

Fig. P17.86

17.86 The circular platform *A* is fitted with a rim of 200-mm inner radius and can rotate freely about the vertical shaft. It is known that the platform-rim unit has a mass of 5 kg and a radius of gyration of 175 mm with respect to the shaft. At a time when the platform is rotating with an angular velocity of 50 rpm, a 3-kg disk *B* of radius 80 mm is placed on the platform with no velocity. Knowing that disk *B* then slides until it comes to rest relative to the platform against the rim, determine the final angular velocity of the platform.

17.87 Two 4-kg disks and a small motor are mounted on a 6-kg rectangular platform which is free to rotate about a central vertical spindle. The normal operating speed of the motor is 240 rpm. If the motor is started when the system is at rest, determine the angular velocity of all elements of the system after the motor has attained its normal operating speed. Neglect the mass of the motor and of the belt.

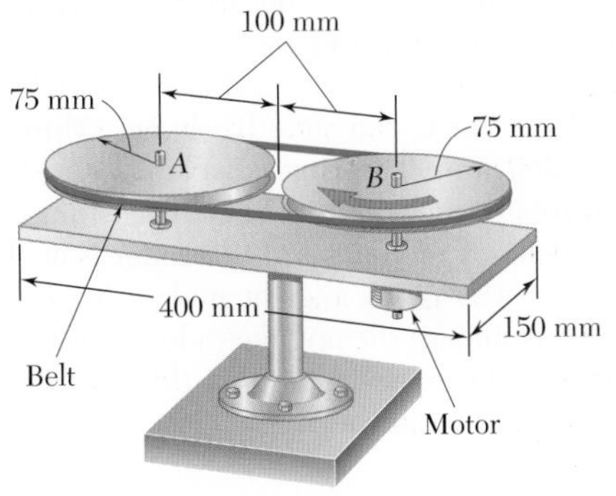

Fig. P17.87

17.88 The 4-kg rod AB can slide freely inside the 6-kg tube CD. The rod was entirely within the tube ($x = 0$) and released with no initial velocity relative to the tube when the angular velocity of the assembly was 5 rad/s. Neglecting the effect of friction, determine the speed of the rod relative to the tube when $x = 400$ mm.

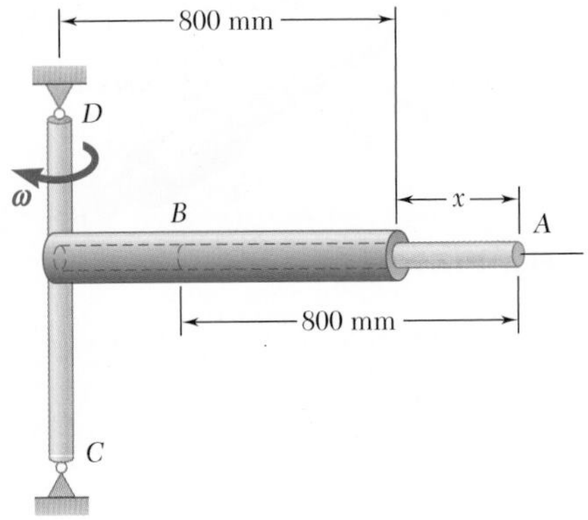

Fig. P17.88

17.89 A 1.8-kg collar A and a 0.7-kg collar B can slide without friction on a frame, consisting of the horizontal rod OE and the vertical rod CD, which is free to rotate about its vertical axis of symmetry. The two collars are connected by a cord running over a pulley that is attached to the frame at O. At the instant shown, the velocity $\mathbf{v}_A$ of collar A has a magnitude of 2.1 m/s and a stop prevents collar B from moving. The stop is suddenly removed and collar A moves toward E. As it reaches a distance of 0.12 m from O, the magnitude of its velocity is observed to be 2.5 m/s. Determine at that instant the magnitude of the angular velocity of the frame and the moment of inertia of the frame and pulley system about CD.

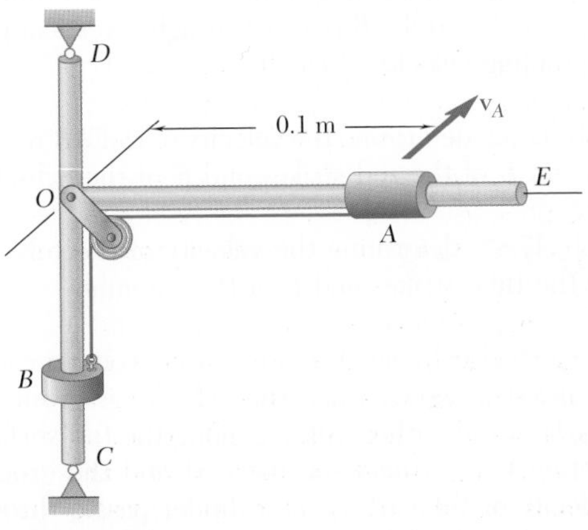

Fig. P17.89

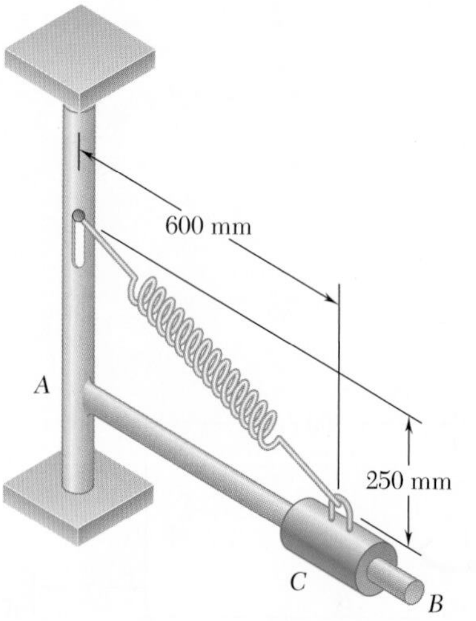

Fig. P17.90

17.90 A 3-kg collar C is attached to a spring and can slide on rod AB, which in turn can rotate in a horizontal plane. The mass moment of inertia of rod AB with respect to end A is 0.5 $kg \cdot m^2$. The spring has a constant $k = 3000$ N/m and an undeformed length of 250 mm. At the instant shown the velocity of the collar relative to the rod is zero and the assembly is rotating with an angular velocity of 12 rad/s. Neglecting the effect of friction, determine (*a*) the angular velocity of the assembly as the collar passes through a point located 180 mm from end A of the rod, (*b*) the corresponding velocity of the collar relative to the rod.

17.91 A small 2-kg collar C can slide freely on a thin ring of mass 3 kg and radius 250 mm. The ring is welded to a short vertical shaft, which can rotate freely in a fixed bearing. Initially the ring has an angular velocity of 35 rad/s and the collar is at the top of the ring ($\theta = 0$) when it is given a slight nudge. Neglecting the effect of friction, determine (*a*) the angular velocity of the ring as the collar passes through the position $\theta = 90°$, (*b*) the corresponding velocity of the collar relative to the ring.

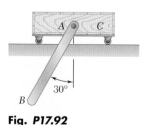

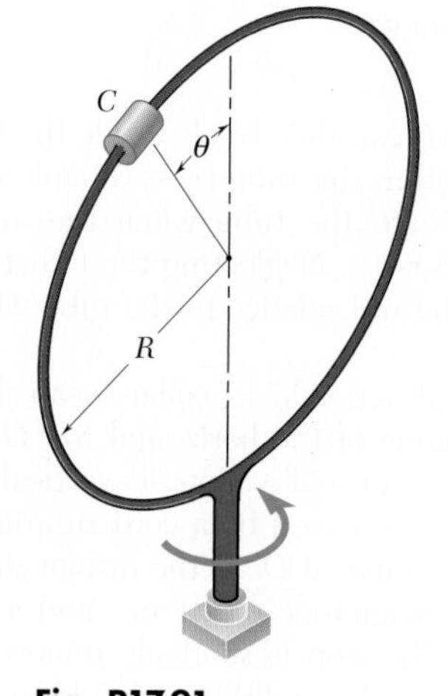

Fig. P17.91

Fig. P17.92

17.92 A uniform rod AB, of mass 7 kg and length 1.2 m, is attached to the 11-kg cart C. Knowing that the system is released from rest in the position shown and neglecting friction, determine (*a*) the velocity of point B as rod AB passes through a vertical position, (*b*) the corresponding velocity of cart C.

17.93 In Prob. 17.82, determine the velocity of rod AB relative to cylinder DE as end B of the rod strikes end E of the cylinder.

17.94 In Prob. 17.83, determine the velocity of the tube relative to the rod as the tube strikes end E of the assembly.

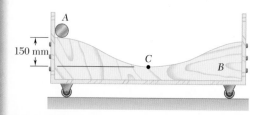

Fig. P17.95

17.95 The 3-kg steel cylinder A and the 5-kg wooden cart B are at rest in the position shown when the cylinder is given a slight nudge, causing it to roll without sliding along the top surface of the cart. Neglecting friction between the cart and the ground, determine the velocity of the cart as the cylinder passes through the lowest point of the surface at C.

17.11 IMPULSIVE MOTION

You saw in Chap. 13 that the method of impulse and momentum is the only practicable method for the solution of problems involving the impulsive motion of a particle. Now you will find that problems involving the impulsive motion of a rigid body are particularly well suited to a solution by the method of impulse and momentum. Since the time interval considered in the computation of linear impulses and angular impulses is very short, the bodies involved can be assumed to occupy the same position during that time interval, making the computation quite simple.

17.12 ECCENTRIC IMPACT

In Secs. 13.13 and 13.14, you learned to solve problems of *central impact,* i.e., problems in which the mass centers of the two colliding bodies are located on the line of impact. You will now analyze the *eccentric impact* of two rigid bodies. Consider two bodies which collide, and denote by $\mathbf{v}_A$ and $\mathbf{v}_B$ the velocities before impact *of the two points of contact* A and B (Fig. 17.10*a*). Under the impact, the two

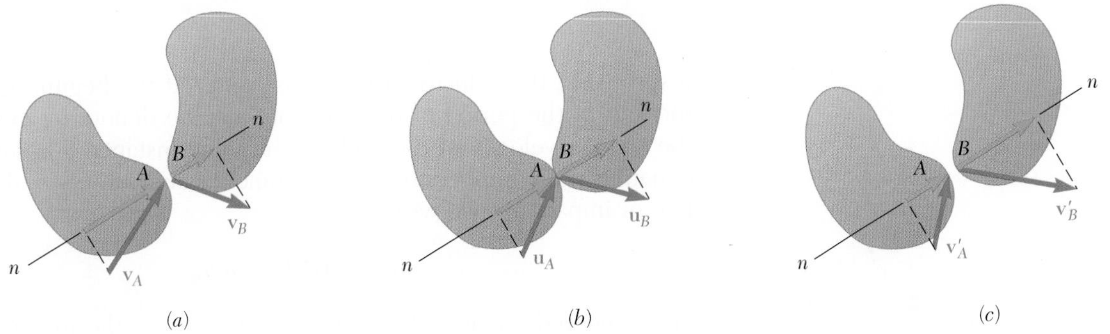

Fig. 17.10

bodies will *deform,* and at the end of the period of deformation, the velocities $\mathbf{u}_A$ and $\mathbf{u}_B$ of A and B will have equal components along the line of impact *nn* (Fig. 17.10*b*). A period of *restitution* will then take place, at the end of which A and B will have velocities $\mathbf{v}'_A$ and $\mathbf{v}'_B$ (Fig. 17.10*c*). Assuming that the bodies are frictionless, we find that the forces they exert on each other are directed along the line of impact. Denoting the magnitude of the impulse of one of these forces during the period of deformation by $\int P\,dt$ and the magnitude of its impulse during the period of restitution by $\int R\,dt$, we recall that the coefficient of restitution e is defined as the ratio

$$e = \frac{\int R\,dt}{\int P\,dt} \tag{17.18}$$

We propose to show that the relation established in Sec. 13.13 between the relative velocities of two particles before and after impact also holds between the components along the line of impact

Photo 17.4 When the rotating bat contacts the ball it applies an impulsive force to the ball requiring the method of impulse and momentum to be used to determine the final velocities of the ball and bat.

of the relative velocities of the two points of contact A and B. We propose to show, therefore, that

$$(v'_B)_n - (v'_A)_n = e[(v_A)_n - (v_B)_n] \tag{17.19}$$

It will first be assumed that the motion of each of the two colliding bodies of Fig. 17.10 is unconstrained. Thus the only impulsive forces exerted on the bodies during the impact are applied at A and B, respectively. Consider the body to which point A belongs and draw the three momentum and impulse diagrams corresponding to the period of deformation (Fig. 17.11). We denote by $\bar{\mathbf{v}}$ and $\bar{\mathbf{u}}$,

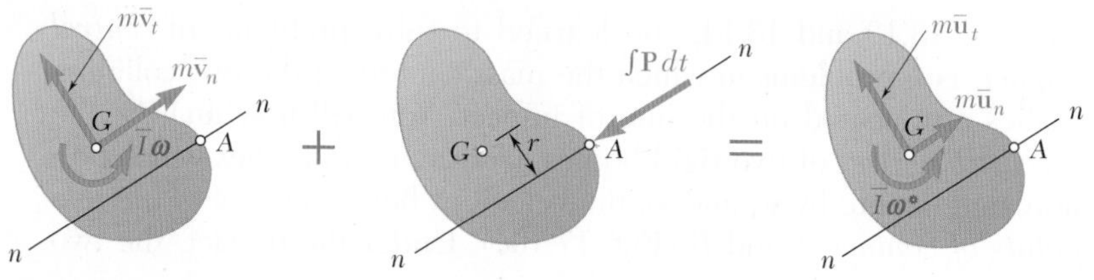

Fig. 17.11

respectively, the velocity of the mass center at the beginning and at the end of the period of deformation, and we denote by $\boldsymbol{\omega}$ and $\boldsymbol{\omega}^*$ the angular velocity of the body at the same instants. Summing and equating the components of the momenta and impulses along the line of impact nn, we write

$$m\bar{v}_n - \int P\,dt = m\bar{u}_n \tag{17.20}$$

Summing and equating the moments about G of the momenta and impulses, we also write

$$\bar{I}\omega - r\int P\,dt = \bar{I}\omega^* \tag{17.21}$$

where r represents the perpendicular distance from G to the line of impact. Considering now the period of restitution, we obtain in a similar way

$$m\bar{u}_n - \int R\,dt = m\bar{v}'_n \tag{17.22}$$

$$\bar{I}\omega^* - r\int R\,dt = \bar{I}\omega' \tag{17.23}$$

where $\bar{\mathbf{v}}'$ and $\boldsymbol{\omega}'$ represent, respectively, the velocity of the mass center and the angular velocity of the body after impact. Solving (17.20) and (17.22) for the two impulses and substituting into (17.18), and then solving (17.21) and (17.23) for the same two impulses and substituting again into (17.18), we obtain the following two alternative expressions for the coefficient of restitution:

$$e = \frac{\bar{u}_n - \bar{v}'_n}{\bar{v}_n - \bar{u}_n} \qquad e = \frac{\omega^* - \omega'}{\omega - \omega^*} \tag{17.24}$$

Multiplying by r the numerator and denominator of the second expression obtained for e, and adding respectively to the numerator and denominator of the first expression, we have

$$e = \frac{\bar{u}_n + r\omega^* - (\bar{v}'_n + r\omega')}{\bar{v}_n + r\omega - (\bar{u}_n + r\omega^*)} \tag{17.25}$$

Observing that $\bar{v}_n + r\omega$ represents the component $(v_A)_n$ along nn of the velocity of the point of contact A and that, similarly, $\bar{u}_n + r\omega^*$ and $\bar{v}'_n + r\omega'$ represent, respectively, the components $(u_A)_n$ and $(v'_A)_n$, we write

$$e = \frac{(u_A)_n - (v'_A)_n}{(v_A)_n - (u_A)_n} \tag{17.26}$$

The analysis of the motion of the second body leads to a similar expression for e in terms of the components along nn of the successive velocities of point B. Recalling that $(u_A)_n = (u_B)_n$, and eliminating these two velocity components by a manipulation similar to the one used in Sec. 13.13, we obtain relation (17.19).

If one or both of the colliding bodies is constrained to rotate about a fixed point O, as in the case of a compound pendulum (Fig. 17.12*a*), an impulsive reaction will be exerted at O (Fig. 17.12*b*).

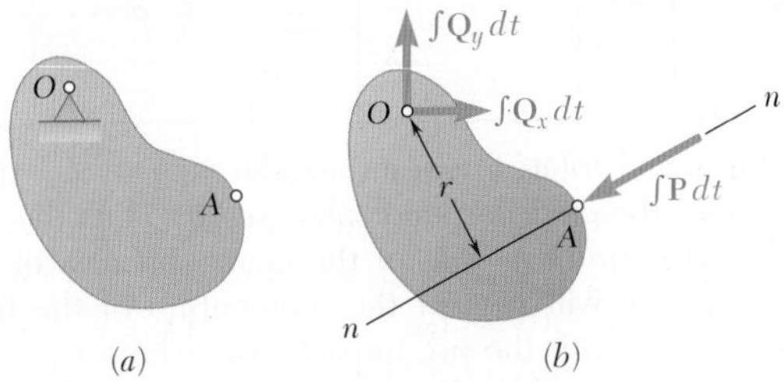

Fig. 17.12

Let us verify that while their derivation must be modified, Eqs. (17.26) and (17.19) remain valid. Applying formula (17.16) to the period of deformation and to the period of restitution, we write

$$I_O\omega - r\int P\,dt = I_O\omega^* \tag{17.27}$$
$$I_O\omega^* - r\int R\,dt = I_O\omega' \tag{17.28}$$

where r represents the perpendicular distance from the fixed point O to the line of impact. Solving (17.27) and (17.28) for the two impulses and substituting into (17.18), and then observing that $r\omega$, $r\omega^*$, and $r\omega'$ represent the components along nn of the successive velocities of point A, we write

$$e = \frac{\omega^* - \omega'}{\omega - \omega^*} = \frac{r\omega^* - r\omega'}{r\omega - r\omega^*} = \frac{(u_A)_n - (v'_A)_n}{(v_A)_n - (u_A)_n}$$

and check that Eq. (17.26) still holds. Thus Eq. (17.19) remains valid when one or both of the colliding bodies is constrained to rotate about a fixed point O.

In order to determine the velocities of the two colliding bodies after impact, relation (17.19) should be used in conjunction with one or several other equations obtained by applying the principle of impulse and momentum (Sample Prob. 17.10).

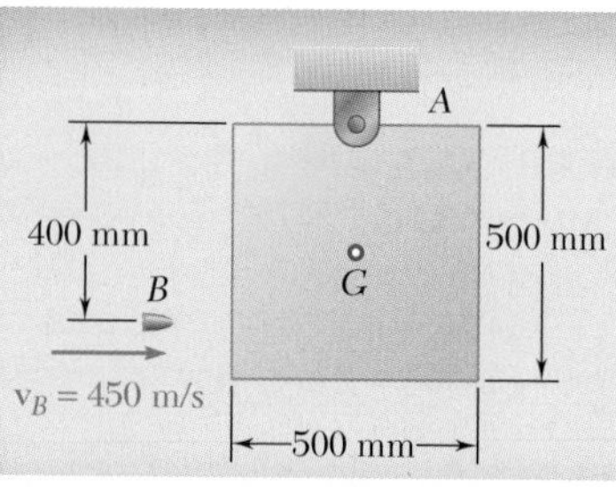

SAMPLE PROBLEM 17.9

A 25-g bullet B is fired with a horizontal velocity of 450 m/s into the side of a 10-kg square panel suspended from a hinge at A. Knowing that the panel is initially at rest, determine (*a*) the angular velocity of the panel immediately after the bullet becomes embedded, (*b*) the impulsive reaction at A, assuming that the bullet becomes embedded in 0.0006 s.

SOLUTION

Principle of Impulse and Momentum. We consider the bullet and the panel as a single system and express that the initial momenta of the bullet and panel and the impulses of the external forces are together equipollent to the final momenta of the system. Since the time interval $\Delta t = 0.0006$ s is very short, we neglect all nonimpulsive forces and consider only the external impulses $\mathbf{A}_x\, \Delta t$ and $\mathbf{A}_y\, \Delta t$. After the impact the bullet gets embedded in

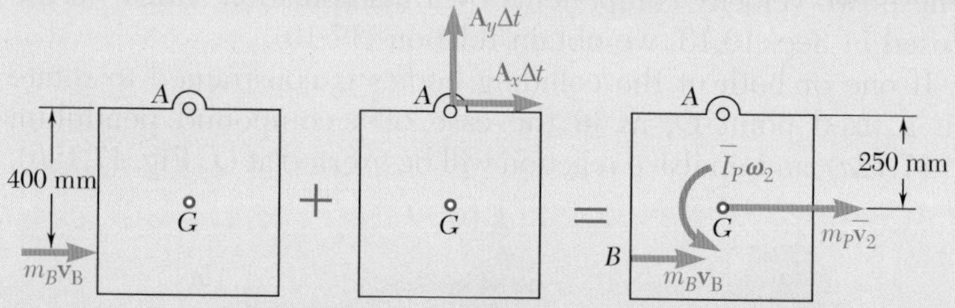

the panel. Since the panel rotates with an angular velocity ω_2, the velocity of the bullet is $(AB)\omega_2$ directed perpendicular to AB. Thus the velocity of the bullet is of the same order as that of the centre of mass of the panel. Since $m_{\text{bullet}} << m_{\text{panel}}$, we will neglect the momentum of the bullet after impact. This is in contrast with the pre-impact case, where $v_{\text{bullet}} >> v_G$. In case if m_{bullet} and m_{panel} comparable, the momentum of the bullet post impact can not be neglected.

Syst Momenta$_1$ + Syst Ext Imp$_{1\to 2}$ = Syst Momenta$_2$

+↺moments about A: $\quad m_B v_B(0.4 \text{ m}) + 0 = m_P \bar{v}_2(0.25 \text{ m}) + \bar{I}_P \omega_2 \quad (1)$

$\overset{+}{\rightarrow} x$ components: $\quad m_B v_B + A_x\, \Delta t = m_P \bar{v}_2 \quad (2)$

$+\uparrow y$ components: $\quad 0 + A_y\, \Delta t = 0 \quad (3)$

The centroidal mass moment of inertia of the square panel is

$$\bar{I}_P = \tfrac{1}{6} m_P b^2 = \frac{1}{6}(10 \text{ kg})(0.5 \text{ m})^2 = 0.417 \text{ kg} \cdot \text{m}^2$$

Substituting this value as well as the given data into (1) and noting that

$$\bar{v}_2 = (0.25 \text{ m})\omega_2$$

we write

$$(0.025)(450)(0.4) = (10)(0.25\omega_2)(0.25) + 0.417\omega$$

$$\omega_2 = 4.32 \text{ rad/s} \qquad \omega_2 = 4.32 \text{ rad/s}\circlearrowleft \blacktriangleleft$$

$$\bar{v}_2 = (0.25 \text{ m})\omega_2 = (0.25 \text{ m})(4.32 \text{ rad/s}) = 1.08 \text{ m/s}$$

Substituting $\bar{v}_2 = 1.08$ m/s, $\Delta t = 0.0006$ s, and the given data into Eq. (2), we have

$$(0.025)(450) + A_x(0.0006) = 10(1.08)$$

$$A_x = -750 \text{ N} \qquad \mathbf{A}_x = 750 \text{ N} \leftarrow \blacktriangleleft$$

From Eq. (3), we find $\quad A_y = 0 \qquad \mathbf{A}_y = 0 \blacktriangleleft$

SAMPLE PROBLEM 17.10

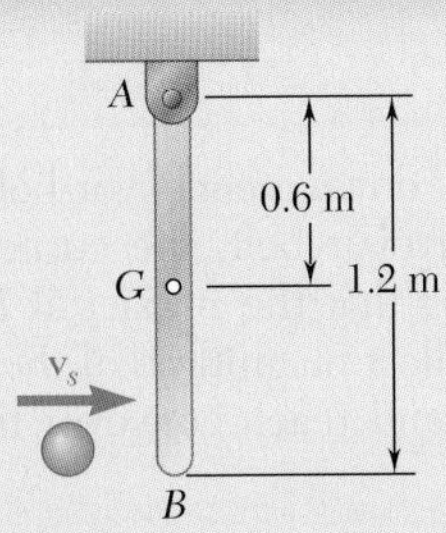

A 2-kg sphere moving horizontally to the right with an initial velocity of 5 m/s strikes the lower end of an 8-kg rigid rod *AB*. The rod is suspended from a hinge at *A* and is initially at rest. Knowing that the coefficient of restitution between the rod and the sphere is 0.80, determine the angular velocity of the rod and the velocity of the sphere immediately after the impact.

SOLUTION

Principle of Impulse and Momentum. We consider the rod and sphere as a single system and express that the initial momenta of the rod and sphere and the impulses of the external forces are together equipollent to the final momenta of the system. We note that the only impulsive force external to the system is the impulsive reaction at *A*.

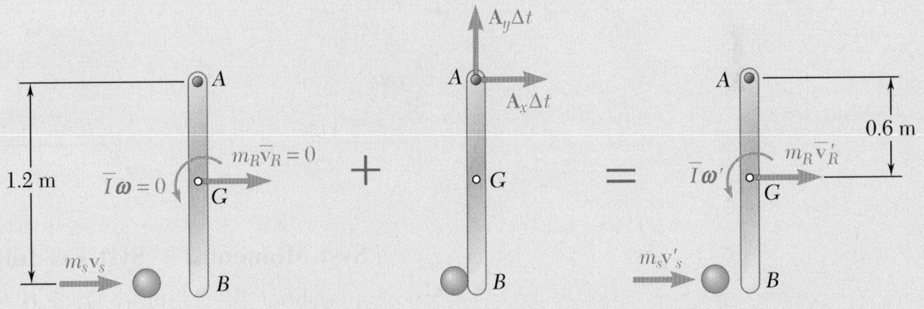

$$\textbf{Syst Momenta}_1 + \textbf{Syst Ext Imp}_{1\to2} = \textbf{Syst Momenta}_2$$

+↺moments about *A*:

$$m_s v_s(1.2\text{ m}) = m_s v'_s(1.2\text{ m}) + m_R\bar{v}'_R(0.6\text{ m}) + \bar{I}\omega' \tag{1}$$

Since the rod rotates about A, we have $\bar{v}'_R = \bar{r}\omega' = (0.6\text{ m})\omega'$. Also,

$$\bar{I} = \tfrac{1}{12}mL^2 = \tfrac{1}{12}(8\text{ kg})(1.2\text{ m})^2 = 0.96\text{ kg}\cdot\text{m}^2$$

Substituting these values and the given data into Eq. (1), we have

$$(2\text{ kg})(5\text{ m/s})(1.2\text{ m}) = (2\text{ kg})v'_s(1.2\text{ m}) + (8\text{ kg})(0.6\text{ m})\omega'(0.6\text{ m}) + (0.96\text{ kg}\cdot\text{m}^2)\omega'$$

$$12 = 2.4v'_s + 3.84\omega' \tag{2}$$

Relative Velocities. Choosing positive to the right, we write

$$v'_B - v'_s = e(v_s - v_B)$$

Substituting $v_s = 5$ m/s, $v_B = 0$, and $e = 0.80$, we obtain

$$v'_B - v'_s = 0.80(5\text{ m/s}) \tag{3}$$

Again noting that the rod rotates about *A*, we write

$$v'_B = (1.2\text{ m})\omega' \tag{4}$$

Solving Eqs. (2) to (4) simultaneously, we obtain

$\omega' = 3.21$ rad/s $\boldsymbol{\omega}' = 3.21$ rad/s ↺ ◀

$v'_s = -0.143$ m/s $\mathbf{v}'_s = -0.143$ m/s ← ◀

SAMPLE PROBLEM 17.11

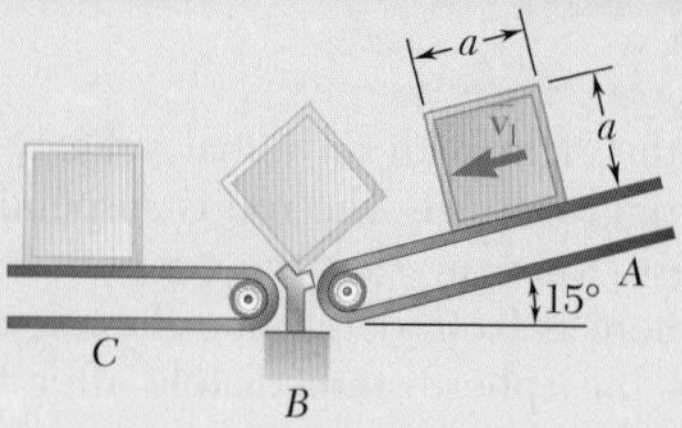

A square package of side a and mass m moves down a conveyor belt A with a constant velocity $\bar{\mathbf{v}}_1$. At the end of the conveyor belt, the corner of the package strikes a rigid support at B. Assuming that the impact at B is perfectly plastic, derive an expression for the smallest magnitude of the velocity $\bar{\mathbf{v}}_1$ for which the package will rotate about B and reach conveyor belt C.

SOLUTION

Principle of Impulse and Momentum. Since the impact between the package and the support is perfectly plastic, the package rotates about B during the impact. We apply the principle of impulse and momentum to the package and note that the only impulsive force external to the package is the impulsive reaction at B.

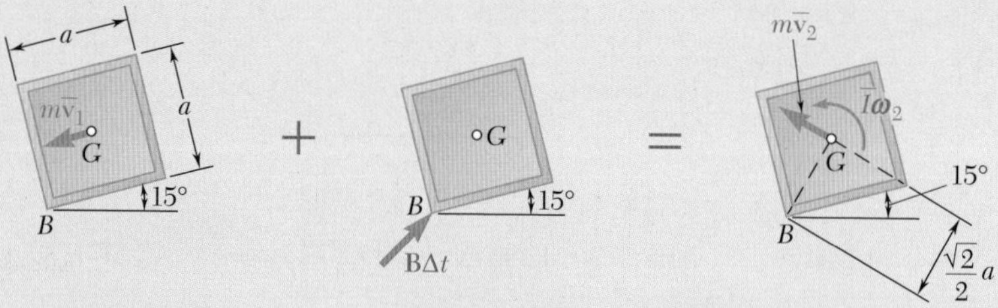

$$\textbf{Syst Momenta}_1 + \textbf{Syst Ext Imp}_{1\to 2} = \textbf{Syst Momenta}_2$$

$+\circlearrowleft$ moments about B: $\quad (m\bar{v}_1)(\tfrac{1}{2}a) + 0 = (m\bar{v}_2)(\tfrac{1}{2}\sqrt{2}a) + \bar{I}\omega_2 \qquad (1)$

Position 2

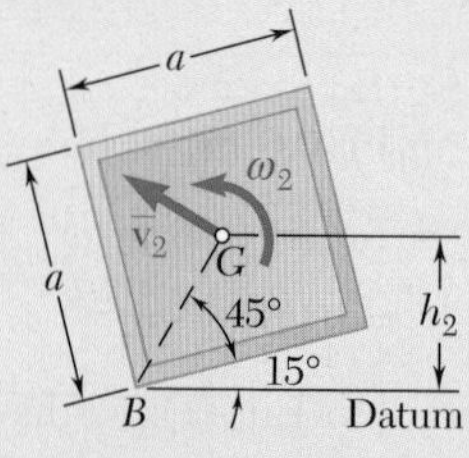

$GB = \frac{1}{2}\sqrt{2}a = 0.707a$

$h_2 = GB \sin(45° + 15°)$

$= 0.612a$

Since the package rotates about B, we have $\bar{v}_2 = (GB)\omega_2 = \frac{1}{2}\sqrt{2}a\omega_2$. We substitute this expression, together with $\bar{I} = \frac{1}{6}ma^2$, into Eq. (1):

$$(m\bar{v}_1)(\tfrac{1}{2}a) = m(\tfrac{1}{2}\sqrt{2}a\omega_2)(\tfrac{1}{2}\sqrt{2}a) + \tfrac{1}{6}ma^2\omega_2 \qquad \bar{v}_1 = \tfrac{4}{3}a\omega_2 \qquad (2)$$

Principle of Conservation of Energy. We apply the principle of conservation of energy between position 2 and position 3.

Position 2. $V_2 = Wh_2$. Recalling that $\bar{v}_2 = \frac{1}{2}\sqrt{2}a\omega_2$, we write

$$T_2 = \tfrac{1}{2}m\bar{v}_2^2 + \tfrac{1}{2}\bar{I}\omega_2^2 = \tfrac{1}{2}m(\tfrac{1}{2}\sqrt{2}a\omega_2)^2 + \tfrac{1}{2}(\tfrac{1}{6}ma^2)\omega_2^2 = \tfrac{1}{3}ma^2\omega_2^2$$

Position 3. Since the package must reach conveyor belt C, it must pass through position 3 where G is directly above B. Also, since we wish to determine the smallest velocity for which the package will reach this position, we choose $\bar{v}_3 = \omega_3 = 0$. Therefore $T_3 = 0$ and $V_3 = Wh_3$.

Position 3

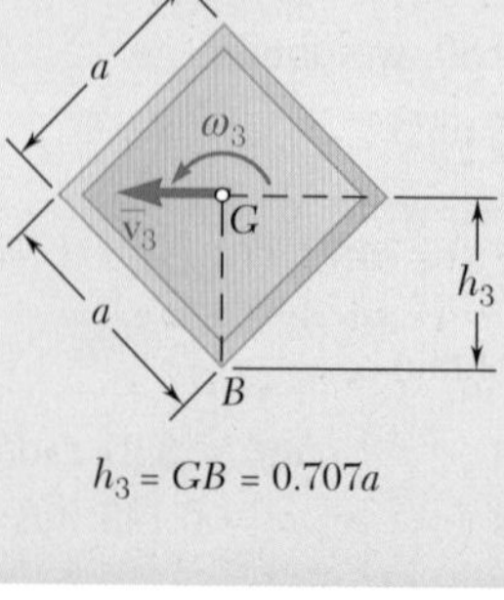

$h_3 = GB = 0.707a$

Conservation of Energy

$$T_2 + V_2 = T_3 + V_3$$
$$\tfrac{1}{3}ma^2\omega_2^2 + Wh_2 = 0 + Wh_3$$
$$\omega_2^2 = \frac{3W}{ma^2}(h_3 - h_2) = \frac{3g}{a^2}(h_3 - h_2) \qquad (3)$$

Substituting the computed values of h_2 and h_3 into Eq. (3), we obtain

$$\omega_2^2 = \frac{3g}{a^2}(0.707a - 0.612a) = \frac{3g}{a^2}(0.095a) \qquad \omega_2 = \sqrt{0.285g/a}$$

$$\bar{v}_1 = \tfrac{4}{3}a\omega_2 = \tfrac{4}{3}a\sqrt{0.285g/a} \qquad \bar{v}_1 = 0.712\sqrt{ga} \blacktriangleleft$$

SOLVING PROBLEMS ON YOUR OWN

This lesson was devoted to the *impulsive motion* and to the *eccentric impact of rigid bodies.*

1. Impulsive motion occurs when a rigid body is subjected to a very large force **F** for a very short interval of time Δt; the resulting impulse $\mathbf{F}\,\Delta t$ is both finite and different from zero. Such forces are referred to as *impulsive forces* and are encountered whenever there is an impact between two rigid bodies. Forces for which the impulse is zero are referred to as *nonimpulsive forces.* As you saw in Chap. 13, the following forces can be assumed to be nonimpulsive: the *weight* of a body, the force exerted by a *spring,* and any other force which is *known* to be small by comparison with the impulsive forces. Unknown reactions, however, *cannot be assumed* to be nonimpulsive.

2. Eccentric impact of rigid bodies. You saw that when two bodies collide, the velocity components along the line of impact of the *points of contact A and B* before and after impact satisfy the following equation:

$$(v'_B)_n - (v'_A)_n = e[(v_A)_n - (v_B)_n] \qquad (17.19)$$

where the left-hand member is the *relative velocity after the impact,* and the right-hand member is the product of the coefficient of restitution and the *relative velocity before the impact.*

This equation expresses the same relation between the velocity components of the points of contact before and after an impact that you used for particles in Chap. 13.

3. To solve a problem involving an impact you should use the *method of impulse and momentum* and take the following steps.

a. Draw a free-body-diagram equation of the body that will express that the system consisting of the momenta immediately before impact and of the impulses of the external forces is equivalent to the system of the momenta immediately after impact.

b. The free-body-diagram equation will relate the velocities before and after impact and the impulsive forces and reactions. In some cases, you will be able to determine the unknown velocities and impulsive reactions by solving equations obtained by summing components and moments [Sample Prob. 17.9].

c. In the case of an impact in which $e > 0$, the number of unknowns will be greater than the number of equations that you can write by summing components and moments, and you should supplement the equations obtained from the free-body-diagram equation with Eq. (17.19), which relates the relative velocities of the points of contact before and after impact [Sample Prob. 17.10].

d. During an impact you must use the method of impulse and momentum. However, *before and after the impact* you can, if necessary, use some of the other methods of solution that you have learned, such as the method of work and energy [Sample Prob. 17.11].

PROBLEMS

IMPULSE-MOMENTUM PRACTICE PROBLEMS

17.F4 A uniform slender rod AB of mass m is at rest on a frictionless horizontal surface when hook C engages a small pin at A. Knowing that the hook is pulled upward with a constant velocity $\mathbf{v}_0$, draw the impulse-momentum diagram that is needed to determine the impulse exerted on the rod at A and B. Assume that the velocity of the hook is unchanged and that the impact is perfectly plastic.

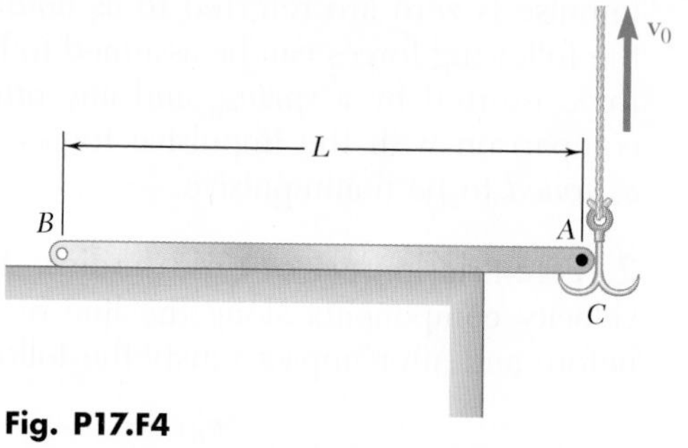

Fig. P17.F4

17.F5 A uniform slender rod AB of length L is falling freely with a velocity $\mathbf{v}_0$ when cord AC suddenly becomes taut. Assuming that the impact is perfectly plastic, draw the impulse-momentum diagram that is needed to determine the angular velocity of the rod and the velocity of its mass center immediately after the cord becomes taut.

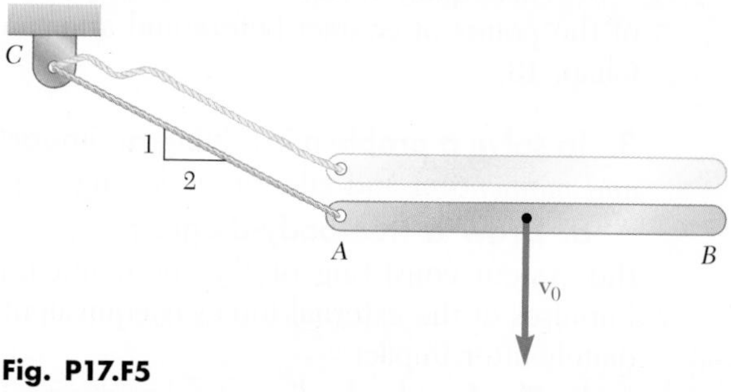

Fig. P17.F5

17.F6 A slender rod CDE of length L and mass m is attached to a pin support at its midpoint D. A second and identical rod AB is rotating about a pin support at A with an angular velocity $\boldsymbol{\omega}_1$ when its end B strikes end C of rod CDE. The coefficient of restitution between the rods is e. Draw the impulse-momentum diagrams that are needed to determine the angular velocity of each rod immediately after the impact.

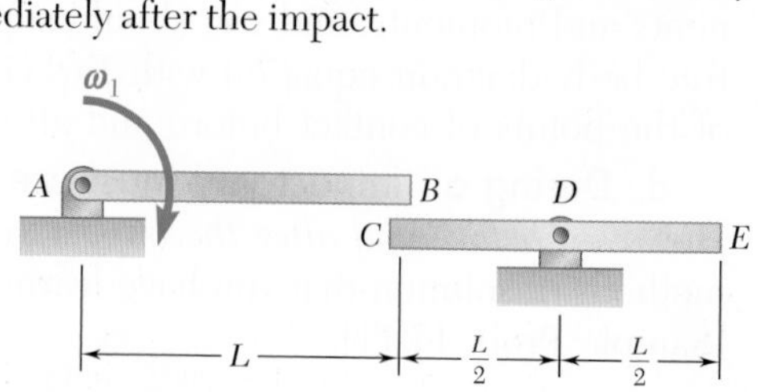

Fig. P17.F6

END-OF-SECTION PROBLEMS

17.96 At what height h above its center G should a billiard ball of radius r be struck horizontally by a cue if the ball is to start rolling without sliding?

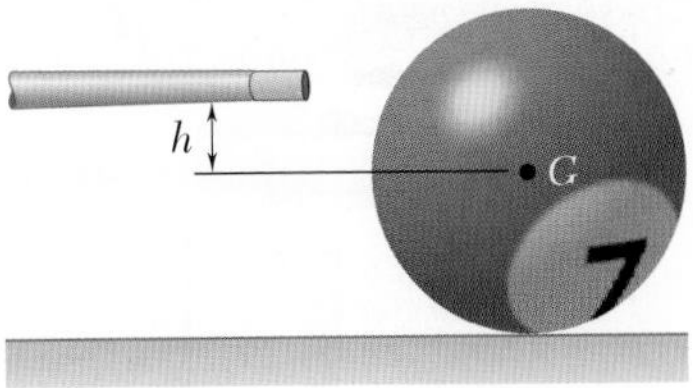

Fig. P17.96

17.97 A bullet of mass 40 g is fired with a horizontal velocity of 550 m/s into the lower end of a slender 7.5-kg bar of length $L = 800$ mm. Knowing that $h = 300$ mm and that the bar is initially at rest, determine (*a*) the angular velocity of the bar immediately after the bullet becomes embedded, (*b*) the impulsive reaction at C, assuming that the bullet becomes embedded in 0.001 s.

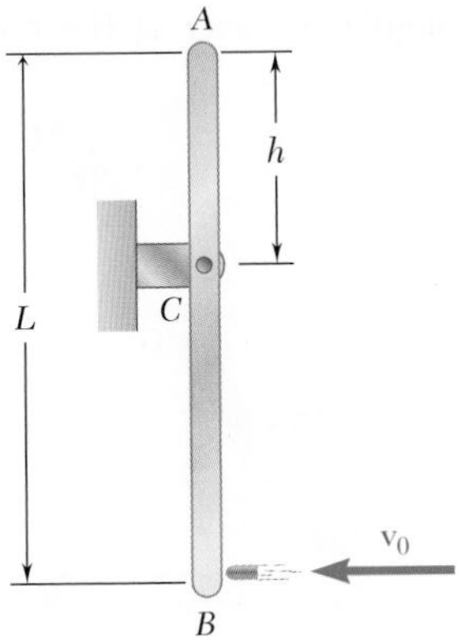

Fig. P17.97

17.98 In Prob. 17.97, determine (*a*) the required distance h if the impulsive reaction at C is to be zero, (*b*) the corresponding angular velocity of the bar immediately after the bullet becomes embedded.

17.99 An 8-kg wooden panel is suspended from a pin support at A and is initially at rest. A 2-kg metal sphere is released from rest at B and falls into a hemispherical cup C attached to the panel at a point located on its top edge. Assuming that the impact is perfectly plastic, determine the velocity of the mass center G of the panel immediately after the impact.

17.100 An 8-kg wooden panel is suspended from a pin support at A and is initially at rest. A 2-kg metal sphere is released from rest at B' and falls into a hemispherical cup C' attached to the panel at the same level as the mass center G. Assuming that the impact is perfectly plastic, determine the velocity of the mass center G of the panel immediately after the impact.

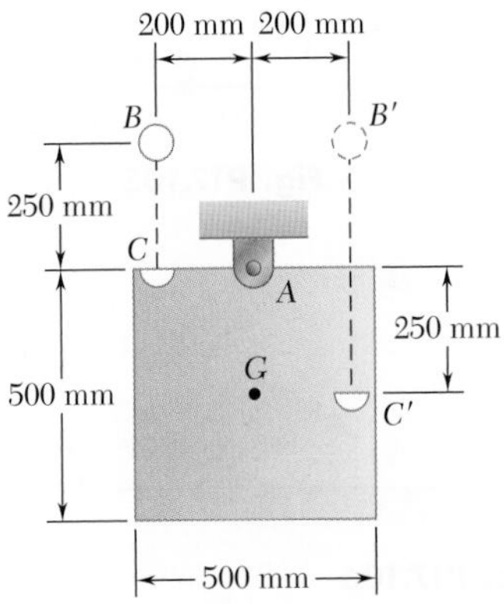

Fig. P17.99 and P17.100

17.101 A 45-g bullet is fired with a velocity of 400 m/s at $\theta = 30°$ into a 9-kg square panel of side $b = 200$ mm. Knowing that $h = 150$ mm and that the panel is initially at rest, determine (*a*) the velocity of the center of the panel immediately after the bullet becomes embedded, (*b*) the impulsive reaction at A, assuming that the bullet becomes embedded in 2 ms.

17.102 A 45-g bullet is fired with a velocity of 400 m/s at $\theta = 5°$ into a 9-kg square panel of side $b = 200$ mm. Knowing that the panel is initially at rest, determine (*a*) the required distance h if the horizontal component of the impulsive reaction at A is to be zero, (*b*) the corresponding velocity of the center of the panel immediately after the bullet becomes embedded.

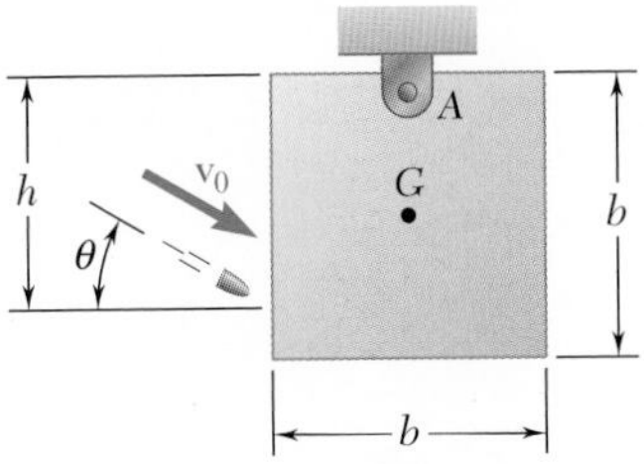

Fig. P17.101 and P17.102

17.103 Two uniform rods, each of mass m, form the L-shaped rigid body ABC which is initially at rest on the frictionless horizontal surface when hook D of the carriage E engages a small pin at C. Knowing that the carriage is pulled to the right with a constant velocity $\mathbf{v}_0$, determine immediately after the impact (*a*) the angular velocity of the body, (*b*) the velocity of corner B. Assume that the velocity of the carriage is unchanged and that the impact is perfectly plastic.

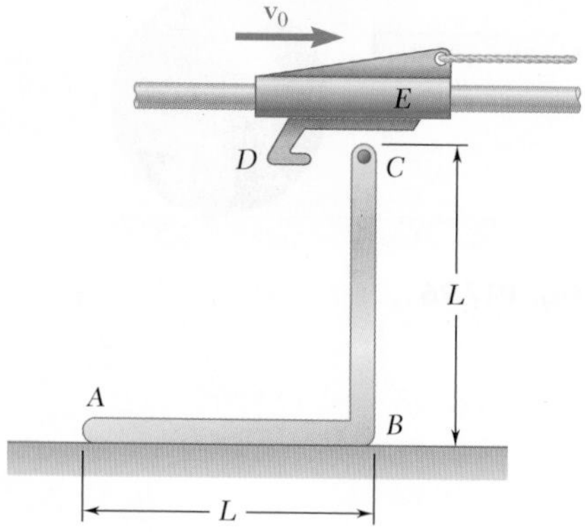

Fig. P17.103

17.104 The uniform slender rod AB of mass 2.5 kg and length 750 mm forms an angle $\beta = 30°$ with the vertical as it strikes the smooth corner shown with a vertical velocity $\mathbf{v}_1$ of magnitude 2.4 m/s and no angular velocity. Assuming that the impact is perfectly plastic, determine the angular velocity of the rod immediately after the impact.

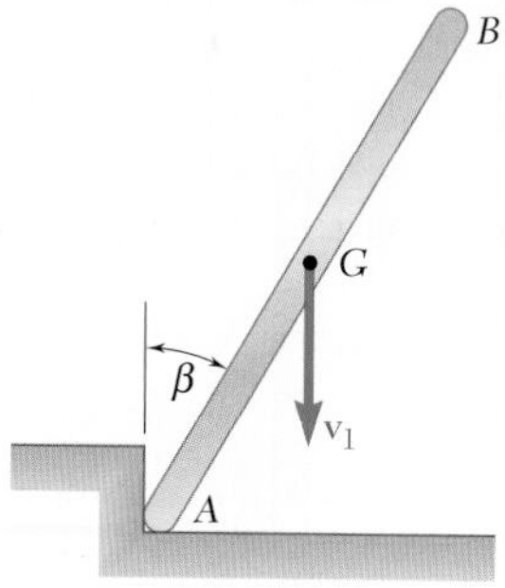

Fig. P17.104

17.105 A bullet of mass 40 g is fired with a horizontal velocity of 550 m/s into the 7.5-kg wooden rod AB of length L = 750 mm. The rod, which is initially at rest, is suspended by a cord of length L = 750 mm. Determine the distance h for which, immediately after the bullet becomes embedded, the instantaneous center of rotation of the rod is point C.

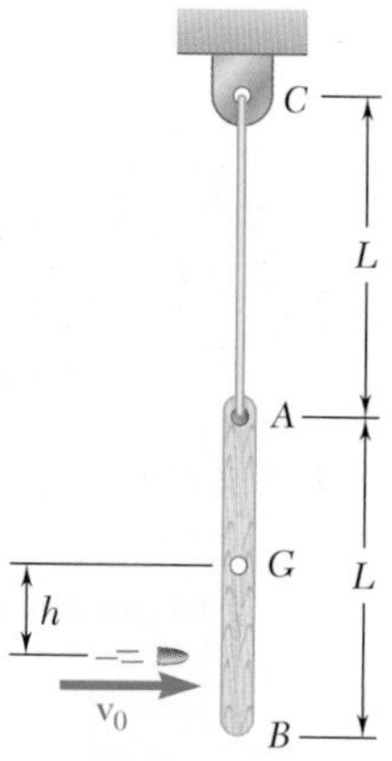

Fig. P17.105

17.106 A uniform sphere of radius r rolls down the incline shown without slipping. It hits a horizontal surface and, after slipping for a while, it starts rolling again. Assuming that the sphere does not bounce as it hits the horizontal surface, determine its angular velocity and the velocity of its mass center after it has resumed rolling.

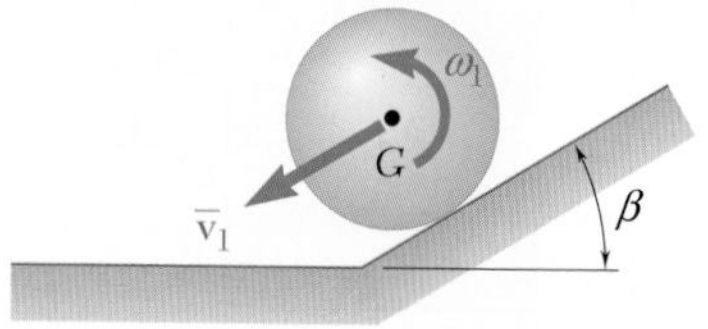

Fig. P17.106

17.107 A uniformly loaded rectangular crate is released from rest in the position shown. Assuming that the floor is sufficiently rough to prevent slipping and that the impact at B is perfectly plastic, determine the smallest value of the ratio a/b for which corner A will remain in contact with the floor.

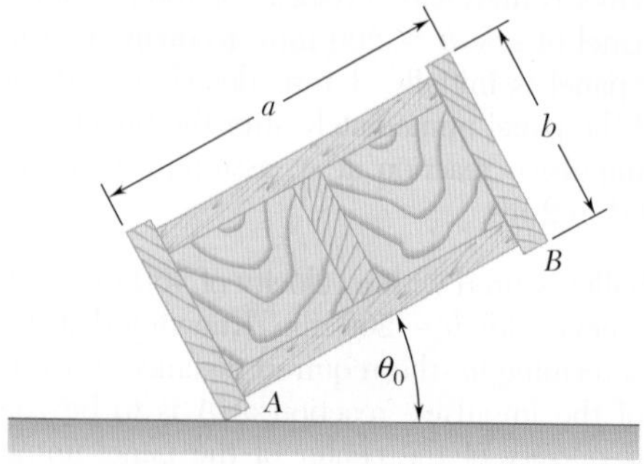

Fig. P17.107

17.108 A bullet of mass m is fired with a horizontal velocity $\mathbf{v}_0$ and at a height $h = \frac{1}{2}R$ into a wooden disk of much larger mass M and radius R. The disk rests on a horizontal plane and the coefficient of friction between the disk and the plane is finite. (*a*) Determine the linear velocity $\bar{\mathbf{v}}_1$ and the angular velocity ω_1 of the disk immediately after the bullet has penetrated the disk. (*b*) Describe the ensuing motion of the disk and determine its linear velocity after the motion has become uniform.

17.109 Determine the height h at which the bullet of Prob. 17.108 should be fired (*a*) if the disk is to roll without sliding immediately after impact, (*b*) if the disk is to slide without rolling immediately after impact.

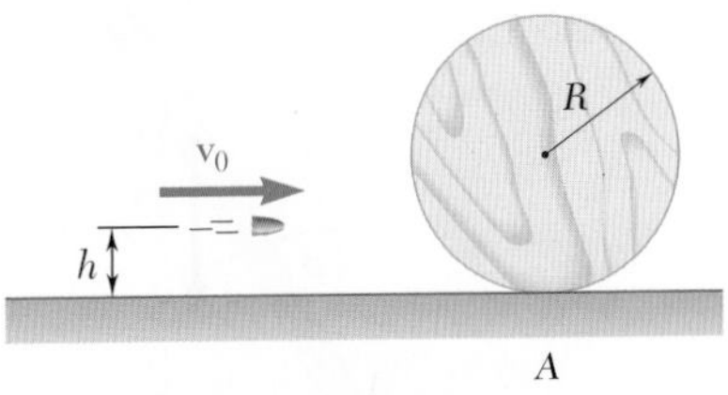

Fig. P17.108 and P17.109

17.110 A uniform slender bar of length $L = 200$ mm and mass $m = 0.5$ kg is supported by a frictionless horizontal table. Initially the bar is spinning about its mass center G with a constant angular speed $\omega_1 = 6$ rad/s. Suddenly latch D is moved to the right and is struck by end A of the bar. Knowing that the coefficient of restitution between A and D is $e = 0.6$, determine the angular velocity of the bar and the velocity of its mass center immediately after the impact.

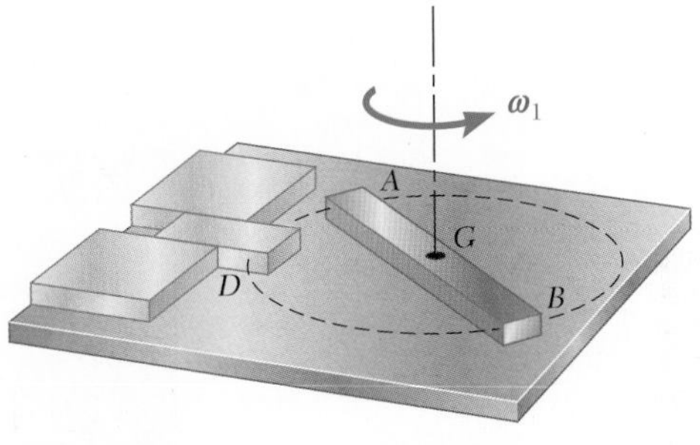

Fig. P17.110

17.111 A uniform slender rod of length L is dropped onto rigid supports at A and B. Since support B is slightly lower than support A, the rod strikes A with a velocity $\bar{\mathbf{v}}_1$ before it strikes B. Assuming perfectly elastic impact at both A and B, determine the angular velocity of the rod and the velocity of its mass center immediately after the rod (*a*) strikes support A, (*b*) strikes support B, (*c*) again strikes support A.

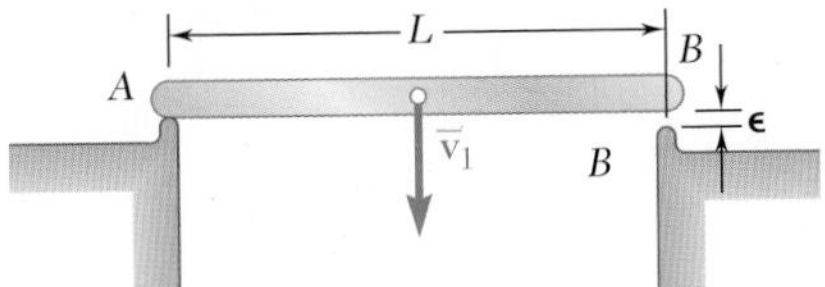

Fig. P17.111

17.112 The slender rod AB of length L forms an angle β with the vertical as it strikes the frictionless surface shown with a vertical velocity $\bar{\mathbf{v}}_1$ and no angular velocity. Assuming that the impact is perfectly plastic, derive an expression for the angular velocity of the rod immediately after the impact.

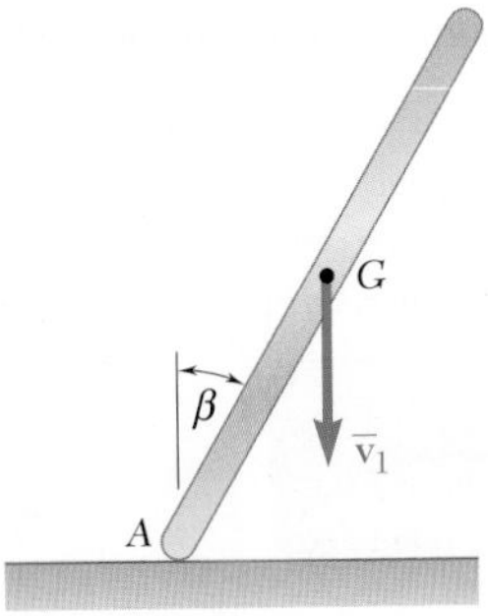

Fig. P17.112

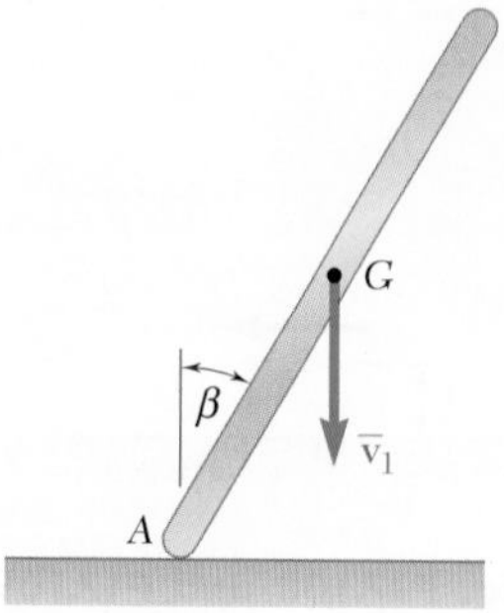

Fig. P17.113

17.113 The slender rod AB of length $L = 1$ m forms an angle $\beta = 30°$ with the vertical as it strikes the frictionless surface shown with a vertical velocity $\bar{\mathbf{v}}_1 = 2$ m/s and no angular velocity. Knowing that the coefficient of restitution between the rod and the ground is $e = 0.8$, determine the angular velocity of the rod immediately after the impact.

17.114 The trapeze/lanyard air drop (t/LAD) launch is a proposed innovative method for airborne launch of a payload-carrying rocket. The release sequence involves several steps as shown in (1) where the payload rocket is shown at various instances during the launch. To investigate the first step of this process, where the rocket body drops freely from the carrier aircraft until the 2-m lanyard stops the vertical motion of B, a trial rocket is tested as shown in (2). The rocket can be considered a uniform 1 × 7-m rectangle with a mass of 4000 kg. Knowing that the rocket is released from rest and falls vertically 2 m before the lanyard becomes taut, determine the angular velocity of the rocket immediately after the lanyard is taut.

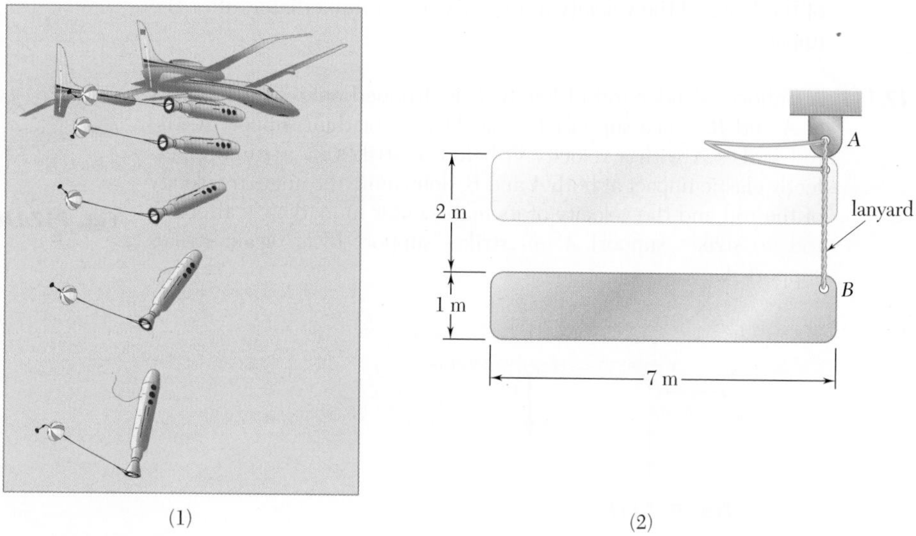

Fig. P17.114

17.115 The uniform rectangular block shown is moving along a frictionless surface with a velocity $\bar{\mathbf{v}}_1$ when it strikes a small obstruction at B. Assuming that the impact between corner A and obstruction B is perfectly plastic, determine the magnitude of the velocity $\bar{\mathbf{v}}_1$ for which the maximum angle θ through which the block will rotate will be 30°.

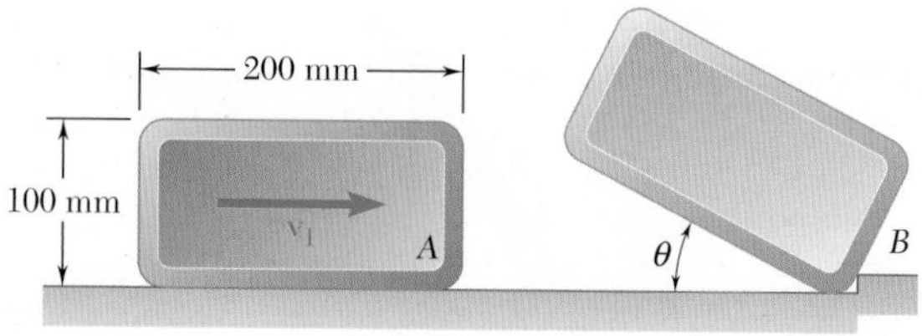

Fig. P17.115

17.116 A slender rod of length L and mass m is released from rest in the position shown. It is observed that after the rod strikes the vertical surface it rebounds to form an angle of 30° with the vertical. (*a*) Determine the coefficient of restitution between knob K and the surface. (*b*) Show that the same rebound can be expected for any position of knob K.

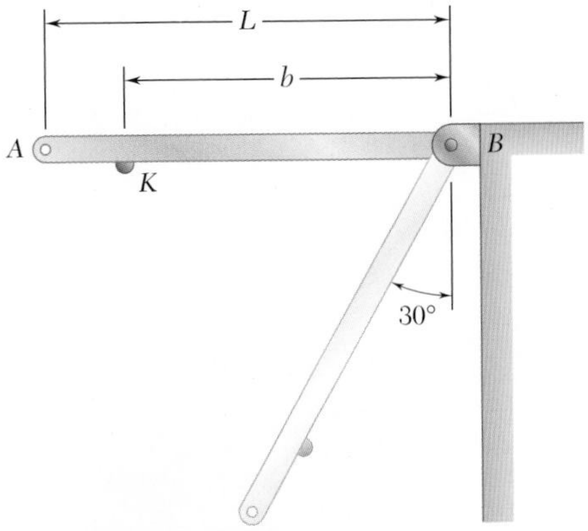

Fig. P17.116

17.117 A slender rod of mass m and length L is released from rest in the position shown and hits edge D. Assuming perfectly plastic impact at D, determine for $b = 0.6L$, (*a*) the angular velocity of the rod immediately after the impact, (*b*) the maximum angle through which the rod will rotate after the impact.

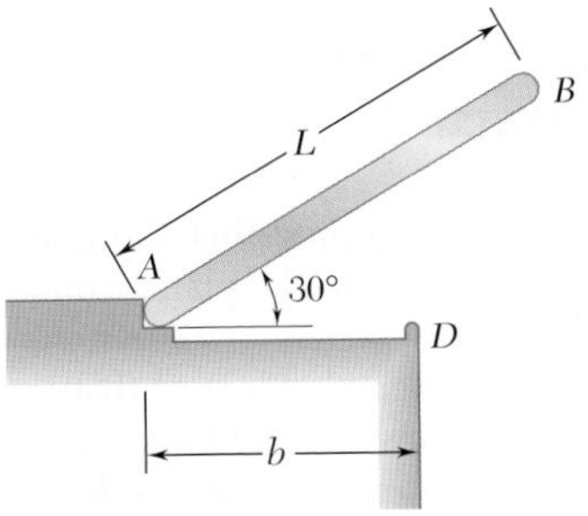

Fig. P17.117

17.118 A uniformly loaded square crate is released from rest with its corner D directly above A; it rotates about A until its corner B strikes the floor, and then rotates about B. The floor is sufficiently rough to prevent slipping and the impact at B is perfectly plastic. Denoting by $\boldsymbol{\omega}_0$ the angular velocity of the crate immediately before B strikes the floor, determine (*a*) the angular velocity of the crate immediately after B strikes the floor, (*b*) the fraction of the kinetic energy of the crate lost during the impact, (*c*) the angle θ through which the crate will rotate after B strikes the floor.

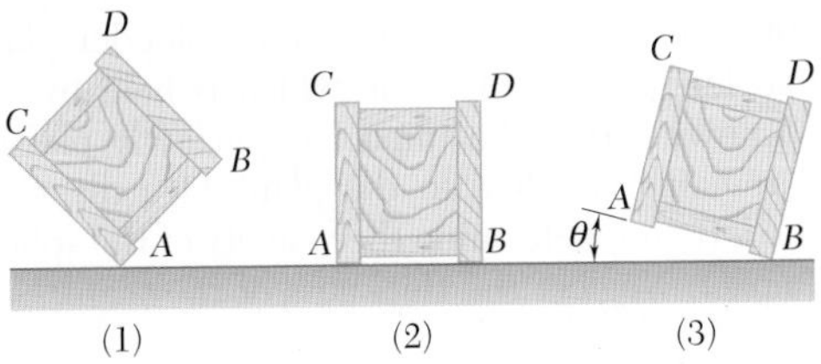

Fig. P17.118

17.119 A 30-g bullet is fired with a horizontal velocity of 350 m/s into the 8-kg wooden beam AB. The beam is suspended from a collar of negligible mass that can slide along a horizontal rod. Neglecting friction between the collar and the rod, determine the maximum angle of rotation of the beam during its subsequent motion.

17.120 For the beam of Prob. 17.119, determine the velocity of the 30-g bullet for which the maximum angle of rotation of the beam will be 90°.

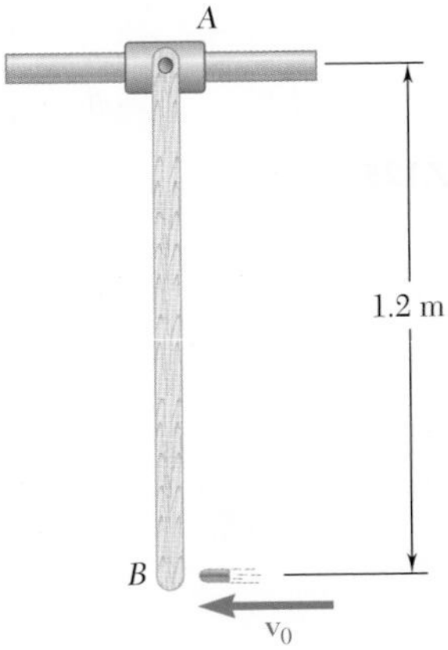

Fig. P17.119

17.121 The plank CDE has a mass of 15 kg and rests on a small pivot at D. The 55-kg gymnast A is standing on the plank at C when the 70-kg gymnast B jumps from a height of 2.5 m and strikes the plank at E. Assuming perfectly plastic impact and that gymnast A is standing absolutely straight, determine the height to which gymnast A will rise.

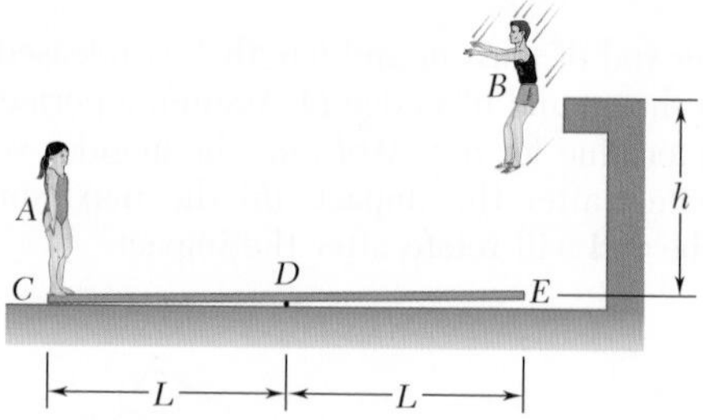

Fig. P17.121

17.122 Solve Prob. 17.121, assuming that the gymnasts change places so that gymnast A jumps onto the plank while gymnast B stands at C.

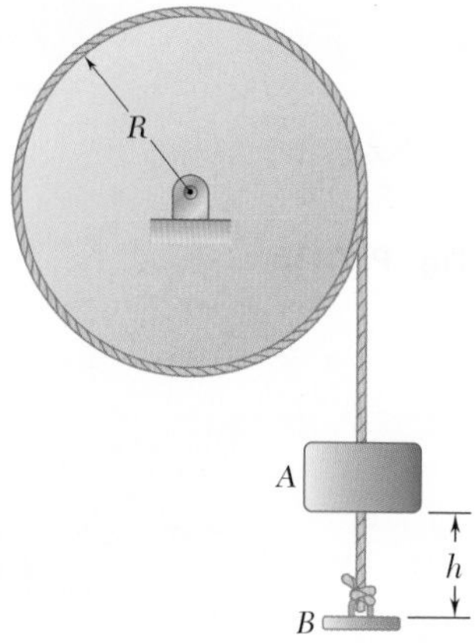

Fig. P17.123

17.123 A small plate B is attached to a cord that is wrapped around a uniform 4-kg disk of radius $R = 200$ mm. A 1.5-kg collar A is released from rest and falls through a distance $h = 300$ mm before hitting plate B. Assuming that the impact is perfectly plastic and neglecting the weight of the plate, determine immediately after the impact (*a*) the velocity of the collar, (*b*) the angular velocity of the disk.

17.124 Solve Prob. 17.123, assuming that the coefficient of restitution between A and B is 0.8.

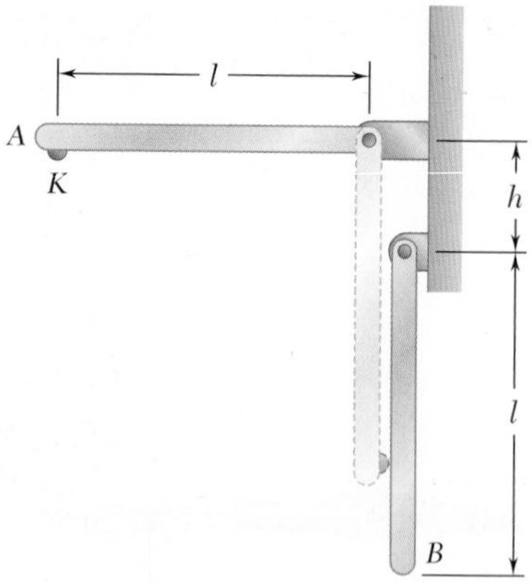

Fig. P17.125

17.125 Two identical slender rods may swing freely from the pivots shown. Rod A is released from rest in a horizontal position and swings to a vertical position, at which time the small knob K strikes rod B which was at rest. If $h = \frac{1}{2}l$ and $e = \frac{1}{2}$, determine (*a*) the angle through which rod B will swing, (*b*) the angle through which rod A will rebound.

17.126 A 2-kg solid sphere of radius $r = 40$ mm is dropped from a height $h = 200$ mm and lands on a uniform slender plank AB of mass 4 kg and length $L = 500$ mm which is held by two inextensible cords. Knowing that the impact is perfectly plastic and that the sphere remains attached to the plank at a distance $a = 40$ mm from the left end, determine the velocity of the sphere immediately after impact. Neglect the thickness of the plank.

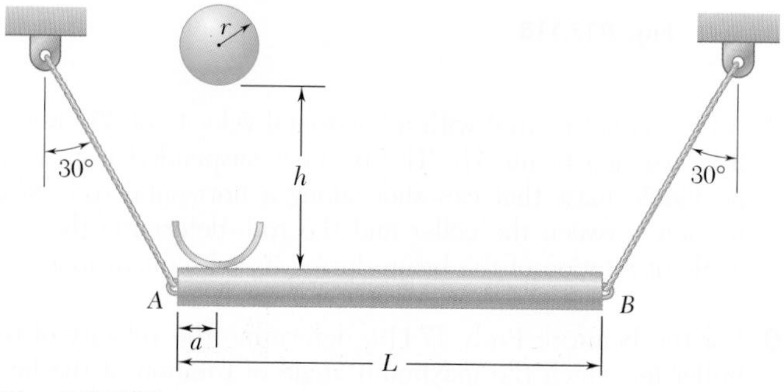

Fig. *P17.126*

17.127 and 17.128 Member ABC has a mass of 2.4 kg and is attached to a pin support at B. An 800-g sphere D strikes the end of member ABC with a vertical velocity $\mathbf{v}_1$ of 3 m/s. Knowing that $L = 750$ mm and that the coefficient of restitution between the sphere and member ABC is 0.5, determine immediately after the impact (*a*) the angular velocity of member ABC, (*b*) the velocity of the sphere.

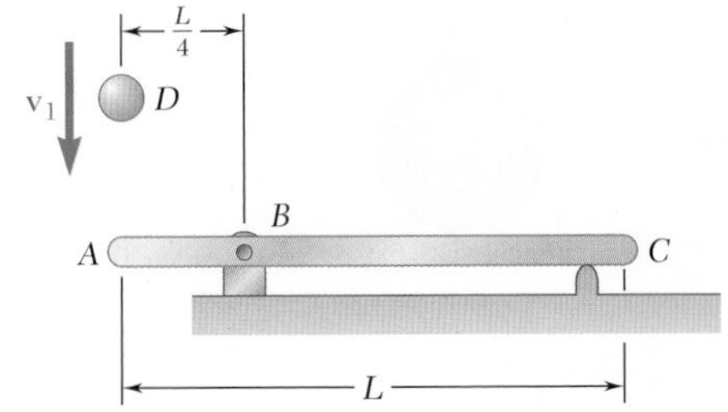

Fig. P17.127

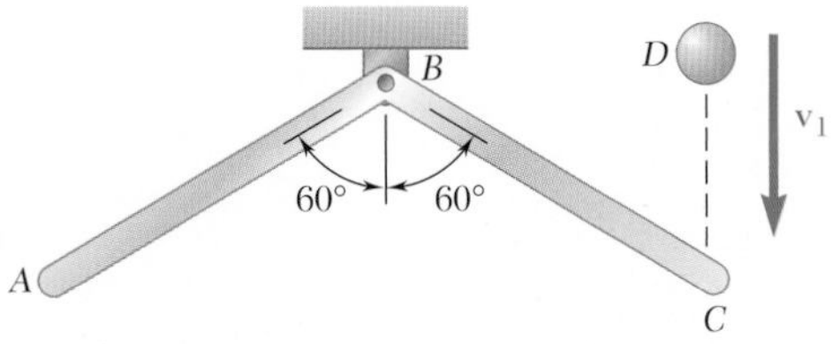

Fig. P17.128

17.129 Sphere A of mass $m_A = 2$ kg and radius $r = 40$ mm rolls without slipping with a velocity $\bar{\mathbf{v}}_1 = 2$ m/s on a horizontal surface when it hits squarely a uniform slender bar B of mass is $m_B = 0.5$ kg and length $L = 100$ mm that is standing on end and is at rest. Denoting by μ_k the coefficient of kinetic friction between the sphere and the horizontal surface, neglecting friction between the sphere and the bar, and knowing the coefficient of restitution between A and B is 0.1, determine the angular velocities of the sphere and the bar immediately after the impact.

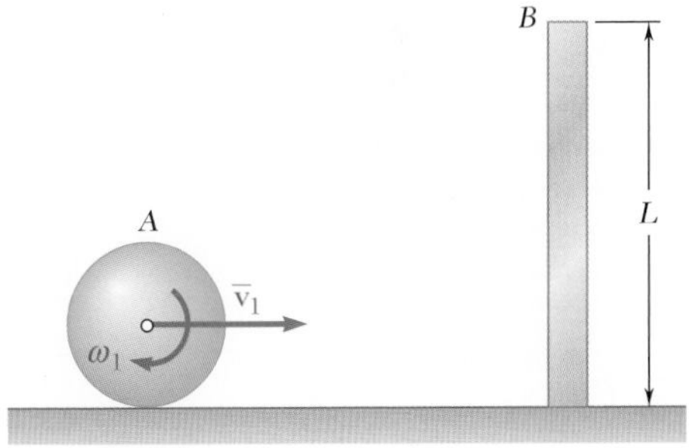

Fig. *P17.129*

17.130 A large 1.5-kg sphere with a radius $r = 100$ mm is thrown into a light basket at the end of a thin, uniform of mass 1 kg and length $L = 250$ mm as shown. Immediately before the impact the angular velocity of the rod is 3 rad/s counterclockwise and the velocity of the sphere is 0.5 m/s down. Assume the sphere sticks in the basket. Determine after the impact (*a*) the angular velocity of the bar and sphere, (*b*) the components of the reactions at A.

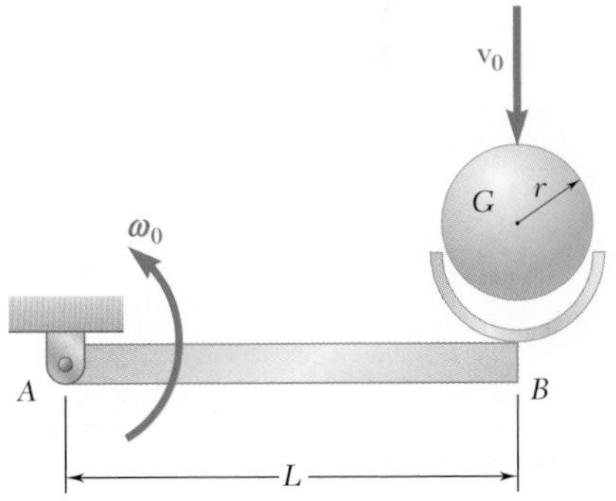

Fig. *P17.130*

17.131 A small rubber ball of radius r is thrown against a rough floor with a velocity $\bar{\mathbf{v}}_A$ of magnitude v_0 and a backspin $\boldsymbol{\omega}_A$ of magnitude ω_0. It is observed that the ball bounces from A to B, then from B to A, then from A to B, etc. Assuming perfectly elastic impact, determine the required magnitude ω_0 of the backspin in terms of $\bar{v}_0$ and r.

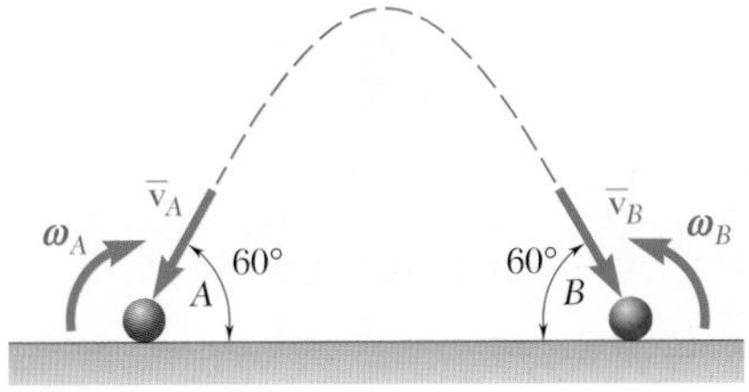

Fig. P17.131

17.132 Sphere A of mass m and radius r rolls without slipping with a velocity $\bar{\mathbf{v}}_1$ on a horizontal surface when it hits squarely an identical sphere B that is at rest. Denoting by μ_k the coefficient of kinetic friction between the spheres and the surface, neglecting friction between the spheres, and assuming perfectly elastic impact, determine (*a*) the linear and angular velocities of each sphere immediately after the impact, (*b*) the velocity of each sphere after it has started rolling uniformly.

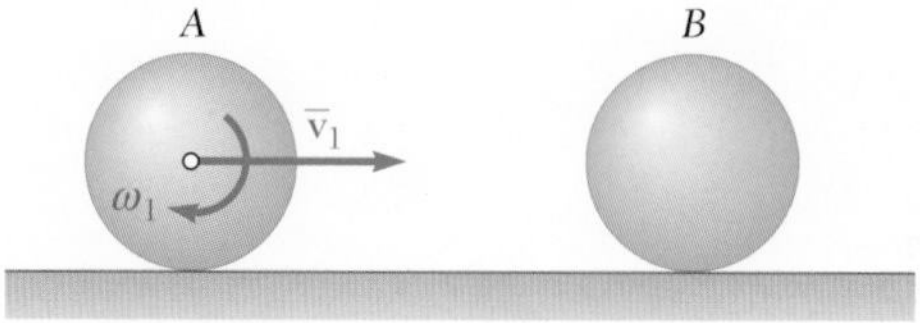

Fig. P17.132

17.133 In a game of pool, ball A is rolling without slipping with a velocity $\bar{\mathbf{v}}_0$ as it hits obliquely ball B, which is at rest. Denoting by r the radius of each ball and by μ_k the coefficient of kinetic friction between a ball and the table, and assuming perfectly elastic impact, determine (*a*) the linear and angular velocity of each ball immediately after the impact, (*b*) the velocity of ball B after it has started rolling uniformly.

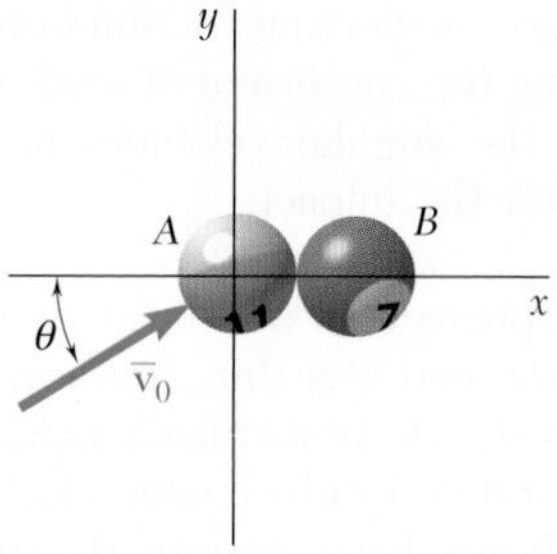

Fig. P17.133

17.134 Each of the bars AB and BC is of length $L = 400$ mm and mass $m = 1.2$ kg. Determine the angular velocity of each bar immediately after the impulse $\mathbf{Q}\Delta t = (1.5\text{ N}\cdot\text{s})\mathbf{i}$ is applied at C.

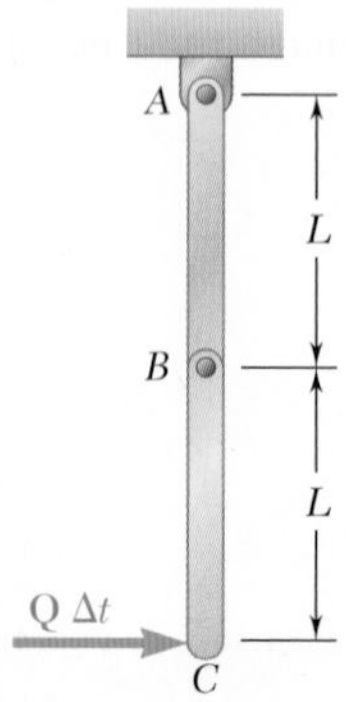

Fig. P17.134

REVIEW AND SUMMARY

In this chapter we again considered the method of work and energy and the method of impulse and momentum. In the first part of the chapter we studied the method of work and energy and its application to the analysis of the motion of rigid bodies and systems of rigid bodies.

Principle of work and energy for a rigid body

In Sec. 17.2, we first expressed the principle of work and energy for a rigid body in the form

$$T_1 + U_{1\rightarrow 2} = T_2 \qquad (17.1)$$

where T_1 and T_2 represent the initial and final values of the kinetic energy of the rigid body and $U_{1\rightarrow 2}$ represents the work of the *external forces* acting on the rigid body.

Work of a force or a couple

In Sec. 17.3, we recalled the expression found in Chap. 13 for the work of a force **F** applied at a point A, namely

$$U_{1\rightarrow 2} = \int_{s_1}^{s_2} (F \cos \alpha)\, ds \qquad (17.3')$$

where F was the magnitude of the force, α the angle it formed with the direction of motion of A, and s the variable of integration measuring the distance traveled by A along its path. We also derived the expression for the *work of a couple of moment* **M** applied to a rigid body during a rotation in θ of the rigid body:

$$U_{1\rightarrow 2} = \int_{\theta_1}^{\theta_2} M\, d\theta \qquad (17.5)$$

Kinetic energy in plane motion

We then derived an expression for the kinetic energy of a rigid body in plane motion [Sec. 17.4]. We wrote

$$T = \tfrac{1}{2}m\bar{v}^2 + \tfrac{1}{2}\bar{I}\omega^2 \qquad (17.9)$$

where $\bar{v}$ is the velocity of the mass center G of the body, ω is the angular velocity of the body, and $\bar{I}$ is its moment of inertia about an axis through G perpendicular to the plane of reference (Fig. 17.13) [Sample Prob. 17.3]. We noted that the kinetic energy of a rigid body in plane motion can be separated into two parts: (1) the kinetic energy $\frac{1}{2}m\bar{v}^2$ associated with the motion of the mass center G of the body, and (2) the kinetic energy $\frac{1}{2}\bar{I}\omega^2$ associated with the rotation of the body about G.

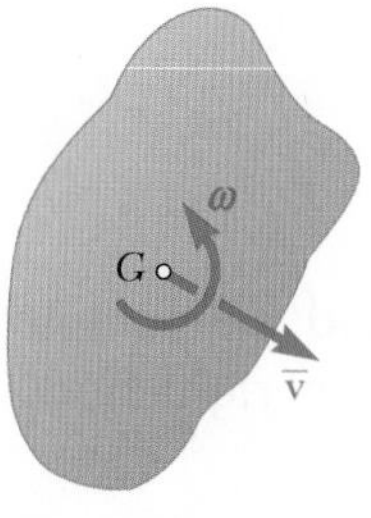

Fig. 17.13

For a rigid body rotating about a fixed axis through O with an angular velocity $\boldsymbol{\omega}$, we had

Kinetic energy in rotation

$$T = \tfrac{1}{2}I_O\omega^2 \tag{17.10}$$

where I_O was the moment of inertia of the body about the fixed axis. We noted that the result obtained is not limited to the rotation of plane slabs or of bodies symmetrical with respect to the reference plane, but is valid regardless of the shape of the body or of the location of the axis of rotation.

Systems of rigid bodies

Equation (17.1) can be applied to the motion of systems of rigid bodies [Sec. 17.5] as long as all the forces acting on the various bodies involved—internal as well as external to the system—are included in the computation of $U_{1\rightarrow 2}$. However, in the case of systems consisting of pin-connected members, or blocks and pulleys connected by inextensible cords, or meshed gears, the points of application of the internal forces move through equal distances and the work of these forces cancels out [Sample Probs. 17.1 and 17.2].

Conservation of energy

When a rigid body, or a system of rigid bodies, moves under the action of conservative forces, the principle of work and energy can be expressed in the form

$$T_1 + V_1 = T_2 + V_2 \tag{17.12}$$

which is referred to as the *principle of conservation of energy* [Sec. 17.6]. This principle can be used to solve problems involving conservative forces such as the force of gravity or the force exerted by a spring [Sample Probs. 17.4 and 17.5]. However, when a reaction is to be determined, the principle of conservation of energy must be supplemented by the application of d'Alembert's principle [Sample Prob. 17.4].

Power

In Sec. 17.7, we extended the concept of power to a rotating body subjected to a couple, writing

$$\text{Power} = \frac{dU}{dt} = \frac{M\,d\theta}{dt} = M\omega \tag{17.13}$$

where M is the magnitude of the couple and ω the angular velocity of the body.

The middle part of the chapter was devoted to the method of impulse and momentum and its application to the solution of various types of problems involving the plane motion of rigid slabs and rigid bodies symmetrical with respect to the reference plane.

Principle of impulse and momentum for a rigid body

We first recalled the *principle of impulse and momentum* as it was derived in Sec. 14.9 for a system of particles and applied it to the *motion of a rigid body* [Sec. 17.8]. We wrote

$$\textbf{Syst Momenta}_1 + \textbf{Syst Ext Imp}_{1\rightarrow 2} = \textbf{Syst Momenta}_2 \tag{17.14}$$

Next we showed that for a rigid slab or a rigid body symmetrical with respect to the reference plane, the system of the momenta of the particles forming the body is equivalent to a vector $m\bar{\mathbf{v}}$ attached at the mass center G of the body and a couple $\bar{I}\boldsymbol{\omega}$ (Fig. 17.14). The vector

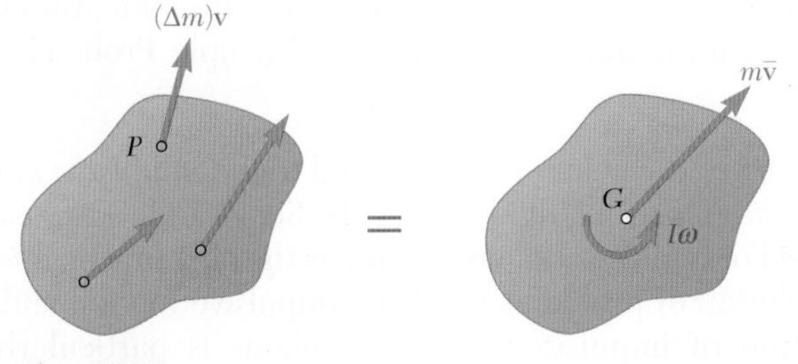

Fig. 17.14

$m\bar{\mathbf{v}}$ is associated with the translation of the body with G and represents the *linear momentum* of the body, while the couple $\bar{I}\boldsymbol{\omega}$ corresponds to the rotation of the body about G and represents the *angular momentum* of the body about an axis through G.

Equation (17.14) can be expressed graphically as shown in Fig. 17.15 by drawing three diagrams representing respectively the system of the initial momenta of the body, the impulses of the external forces acting on the body, and the system of the final momenta of the body.

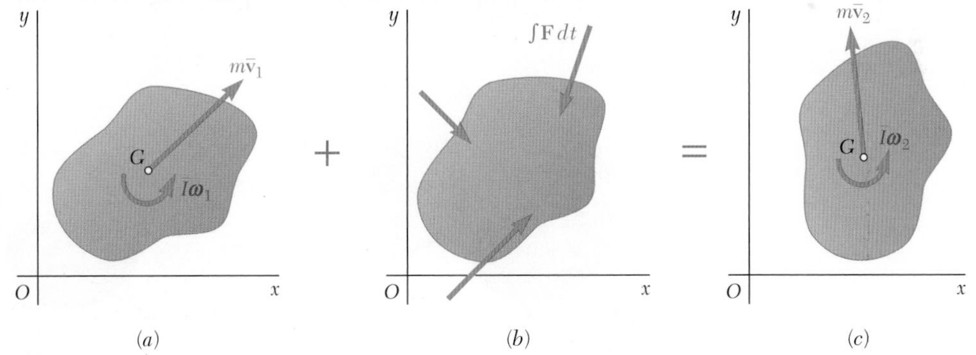

Fig. 17.15

Summing and equating respectively the *x components,* the *y components,* and the *moments about any given point* of the vectors shown in that figure, we obtain three equations of motion which can be solved for the desired unknowns [Sample Probs. 17.6 and 17.7].

In problems dealing with several connected rigid bodies [Sec. 17.9], each body can be considered separately [Sample Prob. 17.6], or, if no more than three unknowns are involved, the principle of impulse and momentum can be applied to the entire system, considering the impulses of the external forces only [Sample Prob. 17.8].

Conservation of angular momentum

When the lines of action of all the external forces acting on a system of rigid bodies pass through a given point O, the angular momentum of the system about O is conserved [Sec. 17.10]. It was suggested that problems involving conservation of angular momentum be solved by the general method described above [Sample Prob. 17.8].

Impulsive motion

The last part of the chapter was devoted to the *impulsive motion* and the *eccentric impact* of rigid bodies. In Sec. 17.11, we recalled that the method of impulse and momentum is the only practicable method for the solution of problems involving impulsive motion and that the computation of impulses in such problems is particularly simple [Sample Prob. 17.9].

Eccentric impact

In Sec. 17.12, we recalled that the eccentric impact of two rigid bodies is defined as an impact in which the mass centers of the colliding bodies are *not* located on the line of impact. It was shown that in such a situation a relation similar to that derived in Chap. 13 for the central impact of two particles and involving the coefficient of restitution e still holds, but that *the velocities of points A and B where contact occurs during the impact should be used.* We have

$$(v'_B)_n - (v'_A)_n = e[(v_A)_n - (v_B)_n] \tag{17.19}$$

where $(v_A)_n$ and $(v_B)_n$ are the components along the line of impact of the velocities of A and B before the impact, and $(v'_A)_n$ and $(v'_B)_n$ are their components after the impact (Fig. 17.16). Equation (17.19)

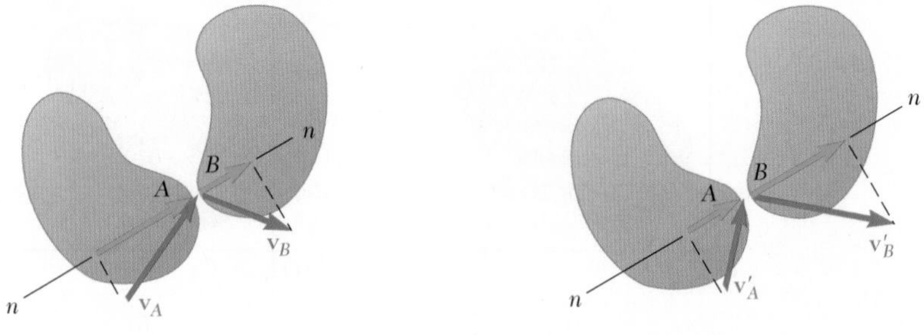

(*a*) Before impact (*b*) After impact

Fig. 17.16

is applicable not only when the colliding bodies move freely after the impact but also when the bodies are partially constrained in their motion. It should be used in conjunction with one or several other equations obtained by applying the principle of impulse and momentum [Sample Prob. 17.10]. We also considered problems where the method of impulse and momentum and the method of work and energy can be combined [Sample Prob. 17.11].

REVIEW PROBLEMS

17.135 A uniform disk of constant thickness and initially at rest is placed in contact with the belt shown, which moves at a constant speed $v = 25$ m/s. Knowing that the coefficient of kinetic friction between the disk and the belt is 0.15, determine (*a*) the number of revolutions executed by the disk before it reaches a constant angular velocity, (*b*) the time required for the disk to reach that constant angular velocity.

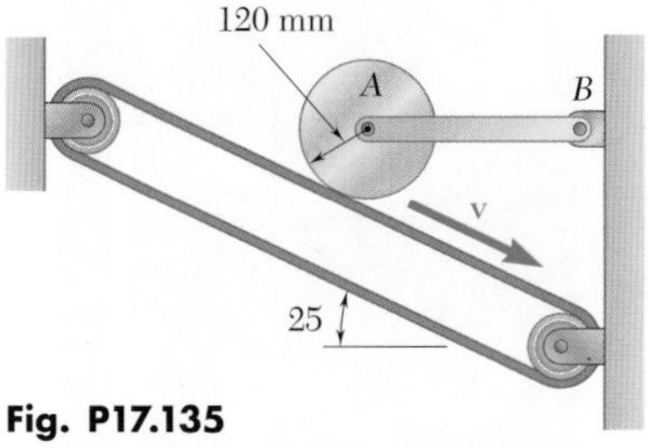

Fig. P17.135

17.136 The three friction disks shown are made of the same material and have the same thickness. It is known that disk *A* has a mass of 6 kg and that the radii of the disks are $r_A = 200$ mm, $r_B = 150$ mm, and $r_C = 100$ mm. The system is at rest when a couple $\mathbf{M}_0$ of constant magnitude 7.5 N · m is applied to disk *A*. Assuming that no slipping occurs between disks, determine the number of revolutions required for disk *A* to reach an angular velocity of 150 rpm.

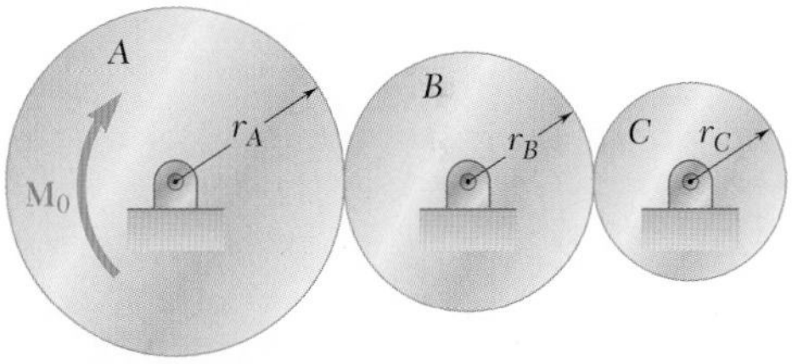

Fig. *P17.136*

17.137 A 300 × 400-mm rectangular plate is suspended by pins at *A* and *B*. The pin at *B* is removed and the plate swings freely about pin *A*. Determine (*a*) the angular velocity of the plate after it has rotated through 90°, (*b*) the maximum angular velocity attained by the plate as it swings freely.

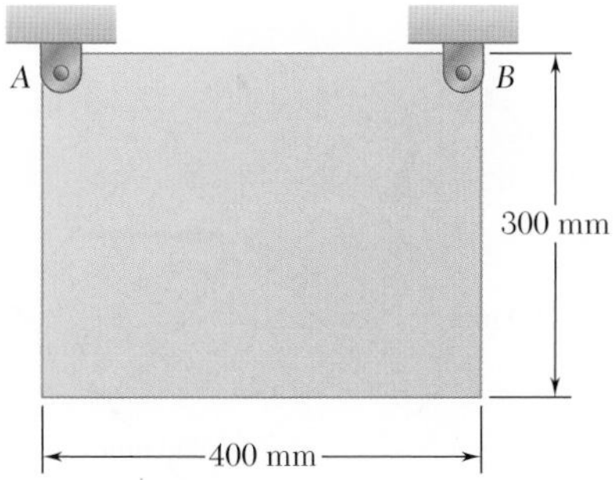

Fig. P17.137

17.138 The gear shown has a radius $R = 150$ mm and a radius of gyration $\bar{k} = 125$ mm. The gear is rolling without sliding with a velocity $\bar{\mathbf{v}}_1$ of magnitude 3 m/s when it strikes a step of height $h = 75$ mm. Because the edge of the step engages the gear teeth, no slipping occurs between the gear and the step. Assuming perfectly plastic impact, determine (*a*) the angular velocity of the gear immediately after the impact, (*b*) the angular velocity of the gear after it has rotated to the top of the step.

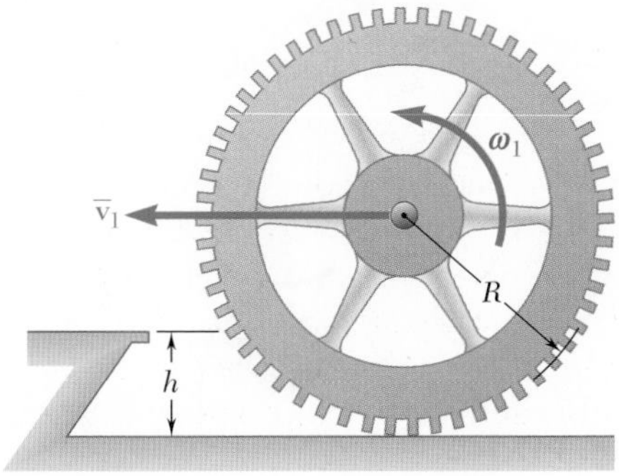

Fig. *P17.138*

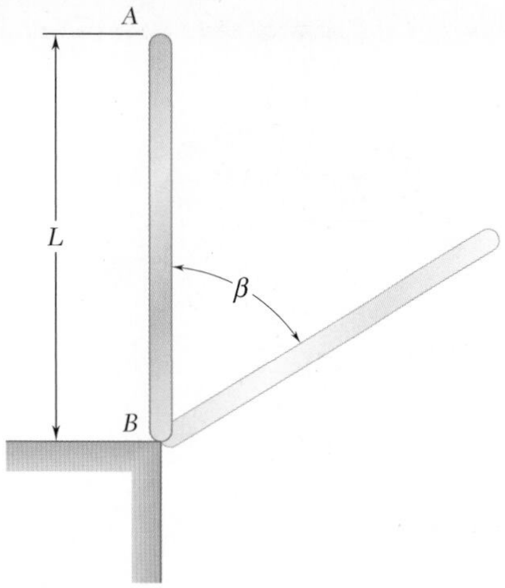

Fig. P17.139

17.139 A uniform slender rod is placed at corner B and is given a slight clockwise motion. Assuming that the corner is sharp and becomes slightly embedded in the end of the rod, so that the coefficient of static friction at B is very large, determine (*a*) the angle β through which the rod will have rotated when it loses contact with the corner, (*b*) the corresponding velocity of end A.

17.140 The motion of the slender 250-mm rod AB is guided by pins at A and B that slide freely in slots cut in a vertical plate as shown. Knowing that the rod has a mass of 2 kg and is released from rest when $\theta = 0$, determine the reactions at A and B when $\theta = 90°$.

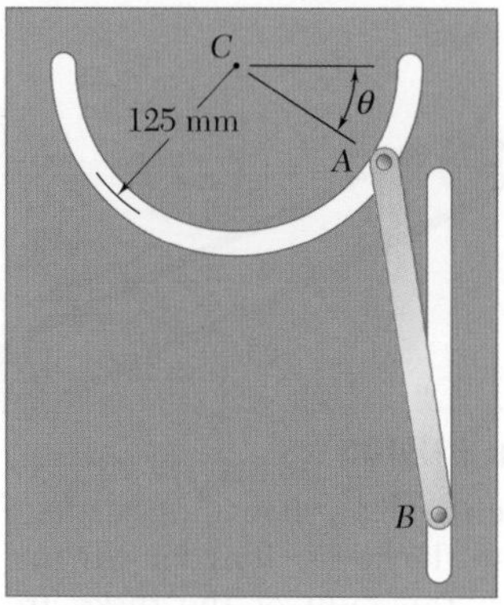

Fig. P17.140

17.141 A 35-g bullet B is fired horizontally with a velocity of 400 m/s into the side of a 3-kg square panel suspended from a pin at A. Knowing that the panel is initially at rest, determine the components of the reaction at A after the panel has rotated 45°.

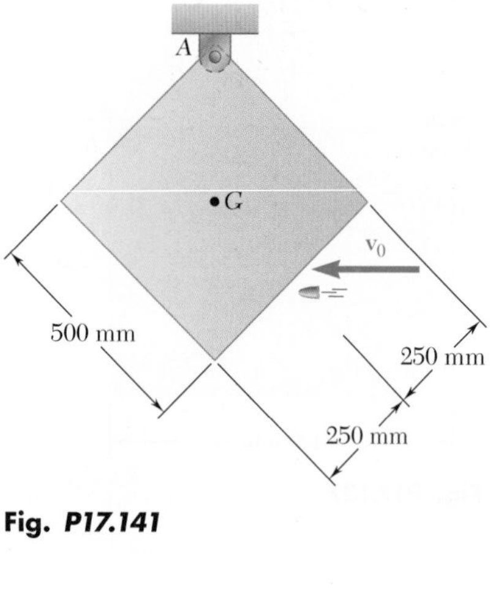

Fig. *P17.141*

17.142 Two panels A and B are attached with hinges to a rectangular plate and held by a wire as shown. The plate and the panels are made of the same material and have the same thickness. The entire assembly is rotating with an angular velocity $\boldsymbol{\omega}_0$ when the wire breaks. Determine the angular velocity of the assembly after the panels have come to rest against the plate.

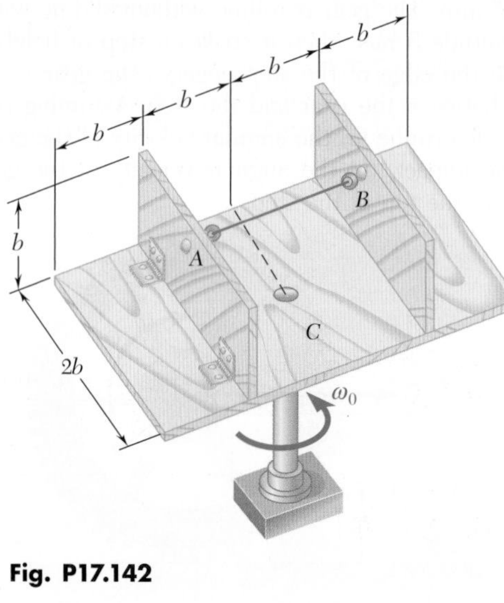

Fig. P17.142

17.143 Disks A and B are made of the same material and are of the same thickness; they can rotate freely about the vertical shaft. Disk B is at rest when it is dropped onto disk A, which is rotating with an angular velocity of 500 rpm. Knowing that disk A has a mass of 8 kg, determine (*a*) the final angular velocity of the disks, (*b*) the change in kinetic energy of the system.

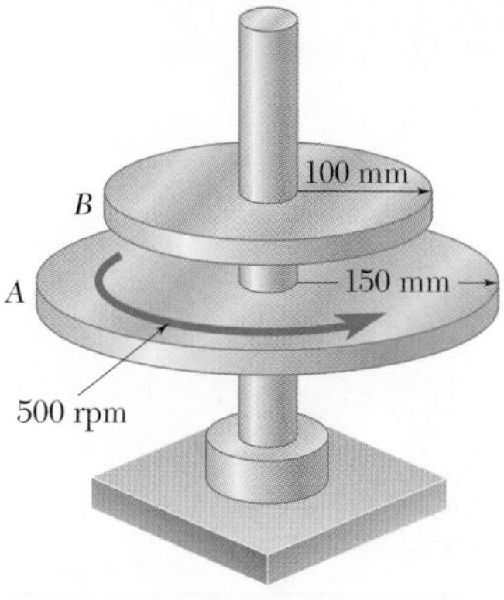

Fig. P17.143

17.144 A square block of mass m is falling with a velocity $\bar{\mathbf{v}}_1$ when it strikes a small obstruction at B. Knowing that the coefficient of restitution for the impact between corner A and the obstruction B is $e = 0.5$, determine immediately after the impact (*a*) the angular velocity of the block, (*b*) the velocity of its mass center G.

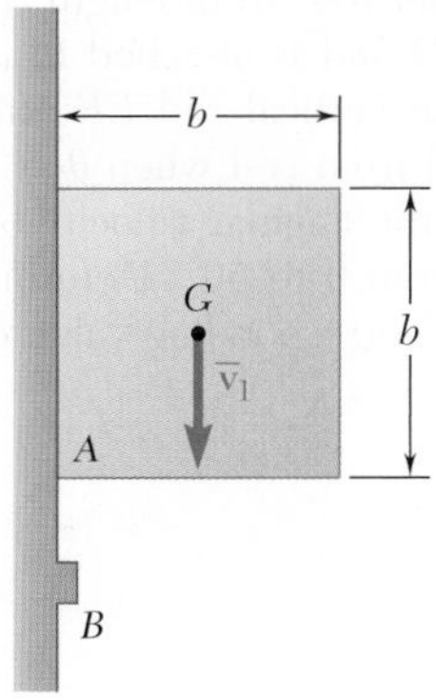

Fig. *P17.144*

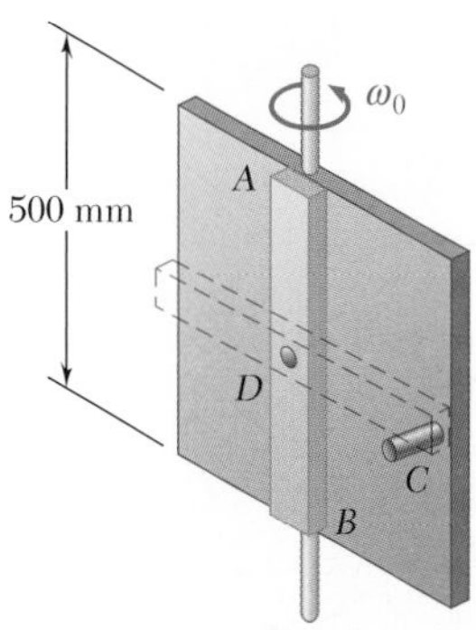

Fig. P17.145

17.145 A 3-kg bar AB is attached by a pin at D to a 4-kg square plate, which can rotate freely about a vertical axis. Knowing that the angular velocity of the plate is 120 rpm when the bar is vertical, determine (*a*) the angular velocity of the plate after the bar has swung into a horizontal position and has come to rest against pin C, (*b*) the energy lost during the plastic impact at C.

17.146 A slender rod CDE of length L and mass m is attached to a pin support at its midpoint D. A second and identical rod AB is rotating about a pin support at A with an angular velocity $\boldsymbol{\omega}_1$ when its end B strikes end C of rod CDE. Denoting by e the coefficient of restitution between the rods, determine the angular velocity of each rod immediately after the impact.

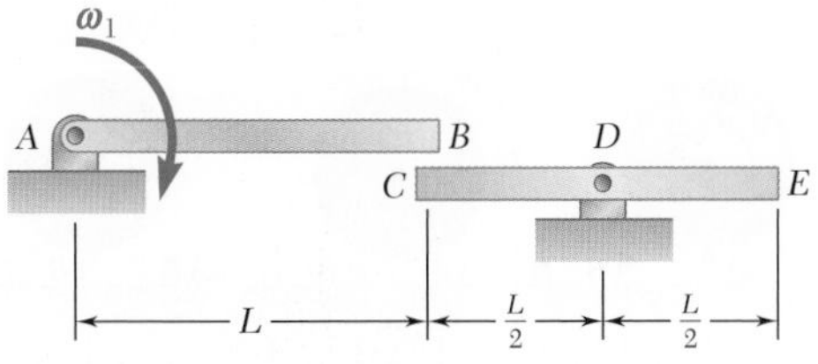

Fig. *P17.146*

COMPUTER PROBLEMS

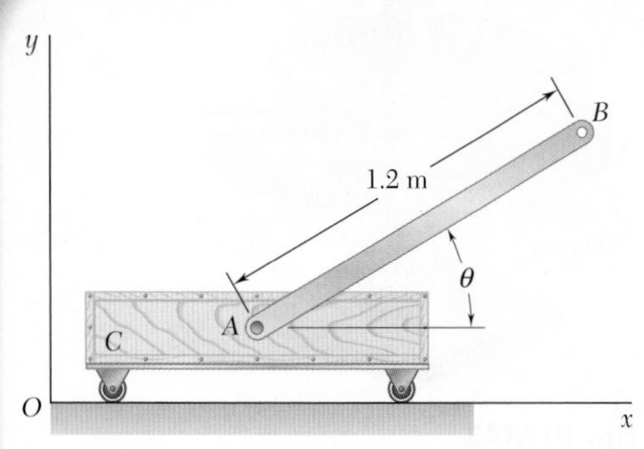

Fig. P17.C1

17.C1 Rod AB has a mass of 3 kg and is attached at A to a 5-kg cart C. Knowing that the system is released from rest when $\theta = 30°$ and neglecting friction, use computational software to determine the velocity of the cart and the velocity of end B of the rod for values of θ from $+30°$ to $-90°$. Determine the value of θ for which the velocity of the cart to the left is maximum and the corresponding value of the velocity.

17.C2 The uniform slender rod AB of length $L = 800$ mm and mass 5 kg rests on a small wheel at D and is attached to a collar of negligible mass that can slide freely on the vertical rod EF. Knowing that $a = 200$ mm and that the rod is released from rest when $\theta = 0$, use computational software to calculate and plot the angular velocity of the rod and the velocity of end A for values of θ from 0 to 50°. Determine the maximum angular velocity of the rod and the corresponding value of θ.

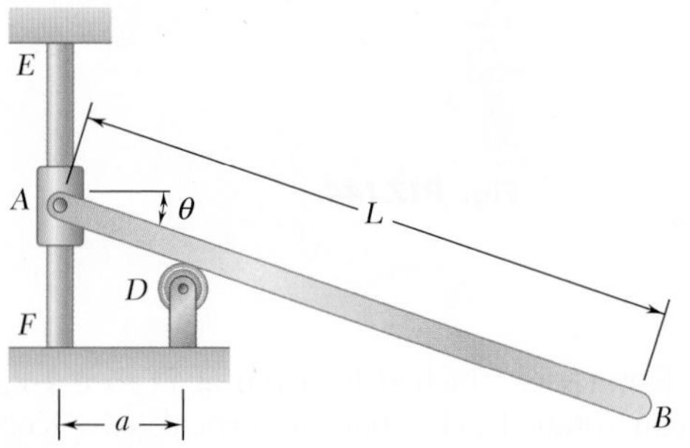

Fig. P17.C2

17.C3 A uniform 250-mm-radius sphere rolls over a series of parallel horizontal bars equally spaced at a distance d. As it rotates without slipping about a given bar, the sphere strikes the next bar and starts rotating about that bar without slipping, until it strikes the next bar, and so on. Assuming perfectly plastic impact and knowing that the sphere has an angular velocity $\boldsymbol{\omega}_0$ of 1.5 rad/s as its mass center G is directly above bar A, use computational software to calculate values of the spacing d from 25 to 150 mm (*a*) the angular velocity $\boldsymbol{\omega}_1$ of the sphere as G passes directly above bar B, (*b*) the number of bars over which the sphere will roll after leaving bar A.

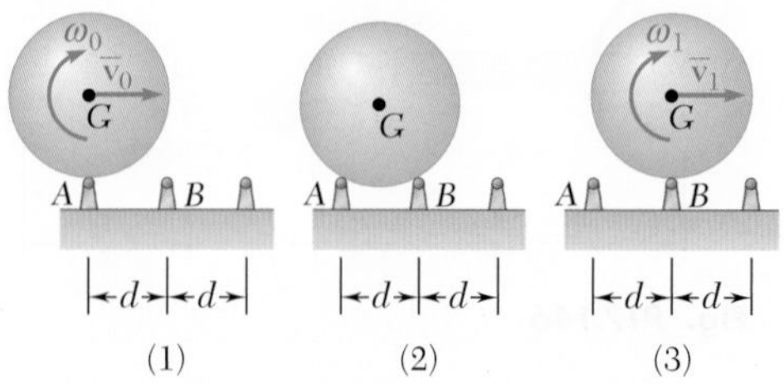

Fig. P17.C3

17.C4 Collar C has a mass of 2.5 kg and can slide without friction on rod AB. A spring of constant 750 N/m and an unstretched length r_0 = 500 mm is attached as shown to the collar and to the hub B. The total mass moment of inertia of the rod, hub, and spring is known to be 0.3 kg · m^2 about B. Initially the collar is held at a distance of 500 mm from the axis of rotation by a small pin protruding from the rod. The pin is suddenly removed as the assembly is rotating in a horizontal plane with an angular velocity $\boldsymbol{\omega}_0$ of 10 rad/s. Denoting by r the distance of the collar from the axis of rotation, use computational software to calculate and plot the angular velocity of the assembly and the velocity of the collar relative to the rod for values of r from 500 to 700 mm. Determine the maximum value of r in the ensuing motion.

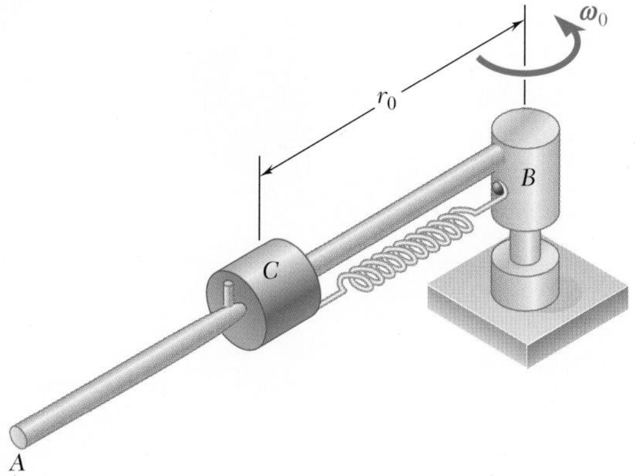

Fig. P17.C4

17.C5 Each of the two identical slender bars shown has a length L = 750 mm. Knowing that the system is released from rest when the bars are horizontal, use computational software to calculate and plot the angular velocity of rod AB and the velocity of point D for values of θ from 0 to 90°.

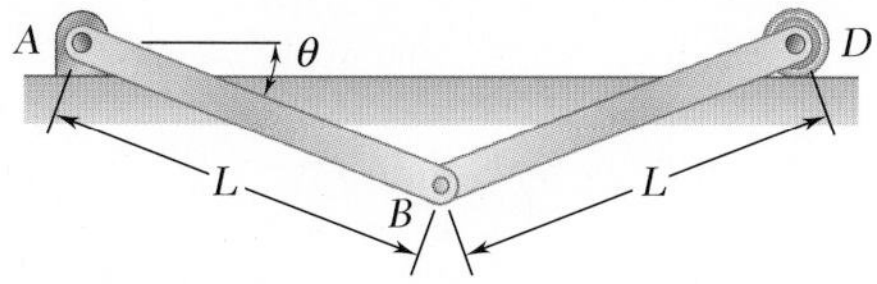

Fig. P17.C5

While the general principles that you learned in earlier chapters can be used again to solve problems involving the three-dimensional motion of rigid bodies, the solution of these problems requires a new approach and is considerably more involved than the solution of two-dimensional problems. One example is the determination of the forces acting on the space shuttle's robotic arm.

CHAPTER 18

Kinetics of Rigid Bodies in Three Dimensions

Chapter 18 Kinetics of Rigid Bodies in Three Dimensions

*18.1 INTRODUCTION

In Chaps. 16 and 17 we were concerned with the plane motion of rigid bodies and of systems of rigid bodies. In Chap. 16 and in the second half of Chap. 17 (momentum method), our study was further restricted to that of plane slabs and of bodies symmetrical with respect to the reference plane. However, many of the fundamental results obtained in these two chapters remain valid in the case of the motion of a rigid body in three dimensions.

For example, the two fundamental equations

$$\Sigma \mathbf{F} = m\bar{\mathbf{a}} \tag{18.1}$$

$$\Sigma \mathbf{M}_G = \dot{\mathbf{H}}_G \tag{18.2}$$

on which the analysis of the plane motion of a rigid body was based, remain valid in the most general case of motion of a rigid body. As indicated in Sec. 16.2, these equations express that the system of the external forces is equipollent to the system consisting of the vector $m\bar{\mathbf{a}}$ attached at G and the couple of moment $\dot{\mathbf{H}}_G$ (Fig. 18.1). However,

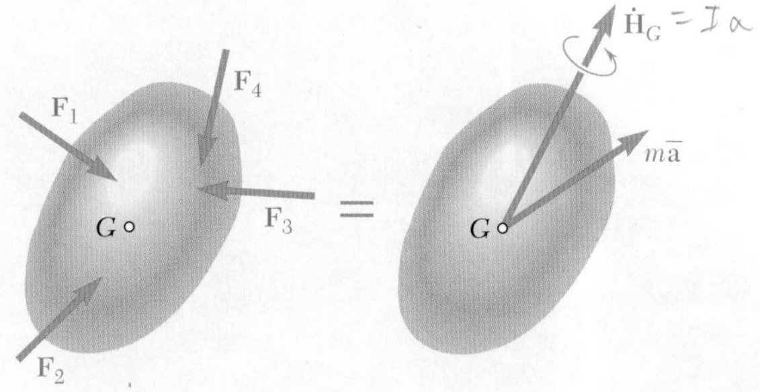

Fig. 18.1

the relation $\mathbf{H}_G = \bar{I}\boldsymbol{\omega}$, which enabled us to determine the angular momentum of a rigid slab and which played an important part in the solution of problems involving the plane motion of slabs and bodies symmetrical with respect to the reference plane, ceases to be valid in the case of nonsymmetrical bodies or three-dimensional motion. Thus in the first part of the chapter, in Sec. 18.2, a more general method for computing the angular momentum $\mathbf{H}_G$ of a rigid body in three dimensions will be developed.

Similarly, although the main feature of the impulse-momentum method discussed in Sec. 17.7, namely, the reduction of the momenta of the particles of a rigid body to a linear momentum vector $m\bar{\mathbf{v}}$ attached at the mass center G of the body and an angular momentum couple $\mathbf{H}_G$, remains valid, the relation $\mathbf{H}_G = \bar{I}\boldsymbol{\omega}$ must be discarded and replaced by the more general relation developed in Sec. 18.2 before this method can be applied to the three-dimensional motion of a rigid body (Sec. 18.3).

We also note that the work-energy principle (Sec. 17.2) and the principle of conservation of energy (Sec. 17.6) still apply in the case of the motion of a rigid body in three dimensions. However, the expression obtained in Sec. 17.4 for the kinetic energy of a rigid body in plane motion will be replaced by a new expression developed in Sec. 18.4 for a rigid body in three-dimensional motion.

In the second part of the chapter, you will first learn to determine the rate of change $\dot{\mathbf{H}}_G$ of the angular momentum $\mathbf{H}_G$ of a three-dimensional rigid body, using a rotating frame of reference with respect to which the moments and products of inertia of the body remain constant (Sec. 18.5). Equations (18.1) and (18.2) will then be expressed in the form of free-body-diagram equations, which can be used to solve various problems involving the three-dimensional motion of rigid bodies (Secs. 18.6 through 18.8).

The last part of the chapter (Secs. 18.9 through 18.11) is devoted to the study of the motion of the gyroscope or, more generally, of an axisymmetrical body with a fixed point located on its axis of symmetry. In Sec. 18.10, the particular case of the steady precession of a gyroscope will be considered, and, in Sec. 18.11, the motion of an axisymmetrical body subjected to no force, except its own weight, will be analyzed.

*18.2 ANGULAR MOMENTUM OF A RIGID BODY IN THREE DIMENSIONS

In this section you will see how the angular momentum $\mathbf{H}_G$ of a body about its mass center G can be determined from the angular velocity $\boldsymbol{\omega}$ of the body in the case of three-dimensional motion.

According to Eq. (14.24), the angular momentum of the body about G can be expressed as

$$\mathbf{H}_G = \sum_{i=1}^{n} (\mathbf{r}'_i \times \mathbf{v}'_i \, \Delta m_i) \tag{18.3}$$

where $\mathbf{r}'_i$ and $\mathbf{v}'_i$ denote, respectively, the position vector and the velocity of the particle P_i, of mass Δm_i, relative to the centroidal frame $Gxyz$ (Fig. 18.2). But $\mathbf{v}'_i = \boldsymbol{\omega} \times \mathbf{r}'_i$, where $\boldsymbol{\omega}$ is the angular

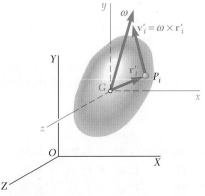

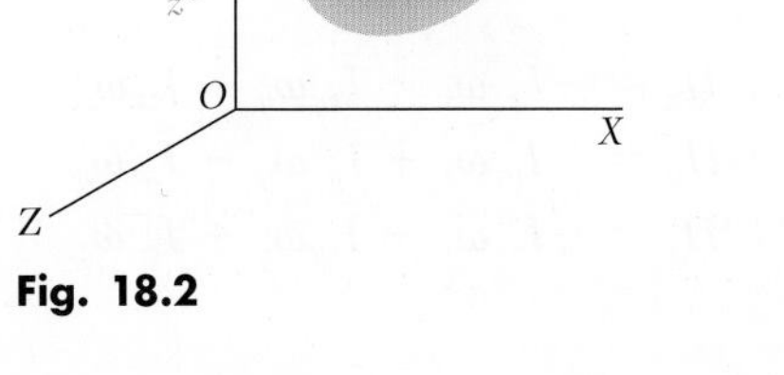

Fig. 18.2

velocity of the body at the instant considered. Substituting into (18.3), we have

$$\mathbf{H}_G = \sum_{i=1}^{n} [\mathbf{r}'_i \times (\boldsymbol{\omega} \times \mathbf{r}'_i)\, \Delta m_i]$$

Recalling the rule for determining the rectangular components of a vector product (Sec. 3.5), we obtain the following expression for the x component of the angular momentum:

$$\begin{aligned} H_x &= \sum_{i=1}^{n} [y_i(\boldsymbol{\omega} \times \mathbf{r}'_i)_z - z_i(\boldsymbol{\omega} \times \mathbf{r}'_i)_y]\, \Delta m_i \\ &= \sum_{i=1}^{n} [y_i(\omega_x y_i - \omega_y x_i) - z_i(\omega_z x_i - \omega_x z_i)]\, \Delta m_i \\ &= \omega_x \sum_i (y_i^2 + z_i^2)\, \Delta m_i - \omega_y \sum_i x_i y_i\, \Delta m_i - \omega_z \sum_i z_i x_i\, \Delta m_i \end{aligned}$$

Replacing the sums by integrals in this expression and in the two similar expressions which are obtained for H_y and H_z, we have

$$\begin{aligned} H_x &= \omega_x \int (y^2 + z^2)\, dm - \omega_y \int xy\, dm - \omega_z \int zx\, dm \\ H_y &= -\omega_x \int xy\, dm + \omega_y \int (z^2 + x^2)\, dm - \omega_z \int yz\, dm \\ H_z &= -\omega_x \int zx\, dm - \omega_y \int yz\, dm + \omega_z \int (x^2 + y^2)\, dm \end{aligned} \qquad (18.4)$$

We note that the integrals containing squares represent the *centroidal mass moments of inertia* of the body about the x, y, and z axes, respectively (Sec. 9.11); we have

$$\bar{I}_x = \int (y^2 + z^2)\, dm \qquad \bar{I}_y = \int (z^2 + x^2)\, dm \qquad \bar{I}_z = \int (x^2 + y^2)\, dm \qquad (18.5)$$

Similarly, the integrals containing products of coordinates represent the *centroidal mass products of inertia* of the body (Sec. 9.16); we have

$$\bar{I}_{xy} = \int xy\, dm \qquad \bar{I}_{yz} = \int yz\, dm \qquad \bar{I}_{zx} = \int zx\, dm \qquad (18.6)$$

Substituting from (18.5) and (18.6) into (18.4), we obtain the components of the angular momentum $\mathbf{H}_G$ of the body about its mass center:

$$\begin{aligned} H_x &= +\bar{I}_x\, \omega_x - \bar{I}_{xy}\omega_y - \bar{I}_{xz}\omega_z \\ H_y &= -\bar{I}_{yx}\omega_x + \bar{I}_y\, \omega_y - \bar{I}_{yz}\omega_z \\ H_z &= -\bar{I}_{zx}\omega_x - \bar{I}_{zy}\omega_y + \bar{I}_z\, \omega_z \end{aligned} \qquad (18.7)$$

The relations (18.7) show that the operation which transforms the vector $\boldsymbol{\omega}$ into the vector $\mathbf{H}_G$ (Fig. 18.3) is characterized by the array of moments and products of inertia

$$\begin{pmatrix} \bar{I}_x & -\bar{I}_{xy} & -\bar{I}_{xz} \\ -\bar{I}_{yx} & \bar{I}_y & -\bar{I}_{yz} \\ -\bar{I}_{zx} & -\bar{I}_{zy} & \bar{I}_z \end{pmatrix} \tag{18.8}$$

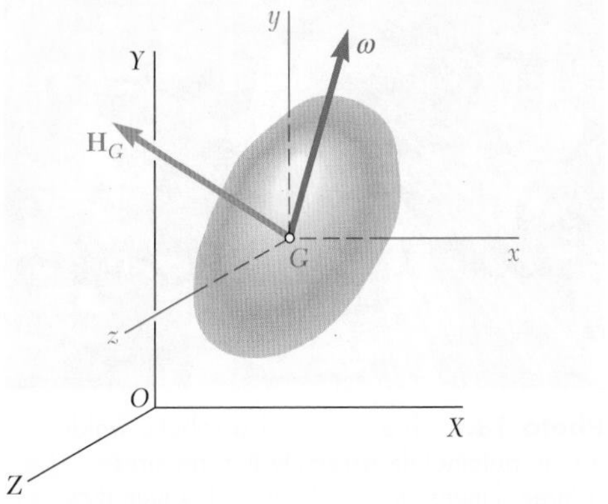

Fig. 18.3

The array (18.8) defines the *inertia tensor* of the body at its mass center G.† A new array of moments and products of inertia would be obtained if a different system of axes were used. The transformation characterized by this new array, however, would still be the same. Clearly, the angular momentum $\mathbf{H}_G$ corresponding to a given angular velocity $\boldsymbol{\omega}$ is independent of the choice of the coordinate axes. As was shown in Secs. 9.17 and 9.18, it is always possible to select a system of axes $Gx'y'z'$, called *principal axes of inertia*, with respect to which all the products of inertia of a given body are zero. The array (18.8) takes then the diagonalized form

$$\begin{pmatrix} \bar{I}_{x'} & 0 & 0 \\ 0 & \bar{I}_{y'} & 0 \\ 0 & 0 & \bar{I}_{z'} \end{pmatrix} \tag{18.9}$$

where $\bar{I}_{x'}$, $\bar{I}_{y'}$, $\bar{I}_{z'}$ represent the *principal centroidal moments of inertia* of the body, and the relations (18.7) reduce to

$$H_{x'} = \bar{I}_{x'}\omega_{x'} \qquad H_{y'} = \bar{I}_{y'}\omega_{y'} \qquad H_{z'} = \bar{I}_{z'}\omega_{z'} \tag{18.10}$$

We note that if the three principal centroidal moments of inertia $\bar{I}_{x'}$, $\bar{I}_{y'}$, $\bar{I}_{z'}$ are equal, the components $H_{x'}$, $H_{y'}$, $H_{z'}$ of the angular momentum about G are proportional to the components $\omega_{x'}$, $\omega_{y'}$, $\omega_{z'}$ of the angular velocity, and the vectors $\mathbf{H}_G$ and $\boldsymbol{\omega}$ are collinear. In general, however, the principal moments of inertia will be different, and the vectors $\mathbf{H}_G$ and $\boldsymbol{\omega}$ *will have different directions,* except when two of the three components of $\boldsymbol{\omega}$ happen to be zero, i.e., when $\boldsymbol{\omega}$ is directed along one of the coordinate axes. Thus, *the angular momentum* $\mathbf{H}_G$ *of a rigid body and its angular velocity* $\boldsymbol{\omega}$ *have the same direction if, and only if,* $\boldsymbol{\omega}$ *is directed along a principal axis of inertia.*‡

†Setting $\bar{I}_x = I_{11}$, $\bar{I}_y = I_{22}$, $\bar{I}_z = I_{33}$, and $-\bar{I}_{xy} = I_{12}$, $-\bar{I}_{xz} = I_{13}$, etc., we may write the inertia tensor (18.8) in the standard form

$$\begin{pmatrix} I_{11} & I_{12} & I_{13} \\ I_{21} & I_{22} & I_{23} \\ I_{31} & I_{32} & I_{33} \end{pmatrix}$$

Denoting by H_1, H_2, H_3 the components of the angular momentum $\mathbf{H}_G$ and by ω_1, ω_2, ω_3 the components of the angular velocity $\boldsymbol{\omega}$, we can write the relations (18.7) in the form

$$H_i = \sum_j I_{ij}\omega_j$$

where i and j take the values 1, 2, 3. The quantities I_{ij} are said to be the *components* of the inertia tensor. Since $I_{ij} = I_{ji}$, the inertia tensor is a *symmetric tensor of the second order.*

‡In the particular case when $\bar{I}_{x'} = \bar{I}_{y'} = \bar{I}_{z'}$, any line through G can be considered as a principal axis of inertia, and the vectors $\mathbf{H}_G$ and $\boldsymbol{\omega}$ are always collinear.

Photo 18.1 The design of a robotic welder for an automobile assembly line requires a three-dimensional study of both kinematics and kinetics.

Since this condition is satisfied in the case of the plane motion of a rigid body symmetrical with respect to the reference plane, we were able in Secs. 16.3 and 17.8 to represent the angular momentum $\mathbf{H}_G$ of such a body by the vector $\bar{I}\boldsymbol{\omega}$. We must realize, however, that this result cannot be extended to the case of the plane motion of a nonsymmetrical body, or to the case of the three-dimensional motion of a rigid body. Except when $\boldsymbol{\omega}$ happens to be directed along a principal axis of inertia, the angular momentum and angular velocity of a rigid body have different directions, and the relation (18.7) or (18.10) must be used to determine $\mathbf{H}_G$ from $\boldsymbol{\omega}$.

Reduction of the Momenta of the Particles of a Rigid Body to a Momentum Vector and a Couple at *G*. We saw in Sec. 17.8 that the system formed by the momenta of the various particles of a rigid body can be reduced to a vector $\mathbf{L}$ attached at the mass center G of the body, representing the linear momentum of the body, and to a couple $\mathbf{H}_G$, representing the angular momentum of the body about G (Fig. 18.4). We are now in a position to determine the vector $\mathbf{L}$

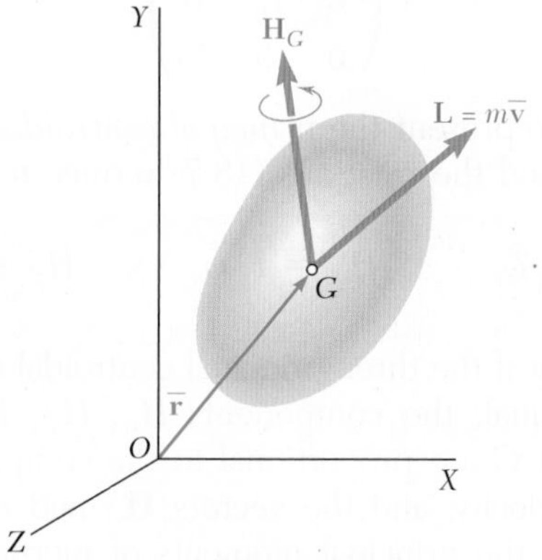

Fig. 18.4

and the couple $\mathbf{H}_G$ in the most general case of three-dimensional motion of a rigid body. As in the case of the two-dimensional motion considered in Sec. 17.8, the linear momentum $\mathbf{L}$ of the body is equal to the product $m\bar{\mathbf{v}}$ of its mass m and the velocity $\bar{\mathbf{v}}$ of its mass center G. The angular momentum $\mathbf{H}_G$, however, can no longer be obtained by simply multiplying the angular velocity $\boldsymbol{\omega}$ of the body by the scalar $\bar{I}$; it must now be obtained from the components of $\boldsymbol{\omega}$ and from the centroidal moments and products of inertia of the body through the use of Eq. (18.7) or (18.10).

We should also note that once the linear momentum $m\bar{\mathbf{v}}$ and the angular momentum $\mathbf{H}_G$ of a rigid body have been determined, its angular momentum $\mathbf{H}_O$ about any given point O can be obtained by adding the moments about O of the vector $m\bar{\mathbf{v}}$ and of the couple $\mathbf{H}_G$. We write

$$\mathbf{H}_O = \bar{\mathbf{r}} \times m\bar{\mathbf{v}} + \mathbf{H}_G \tag{18.11}$$

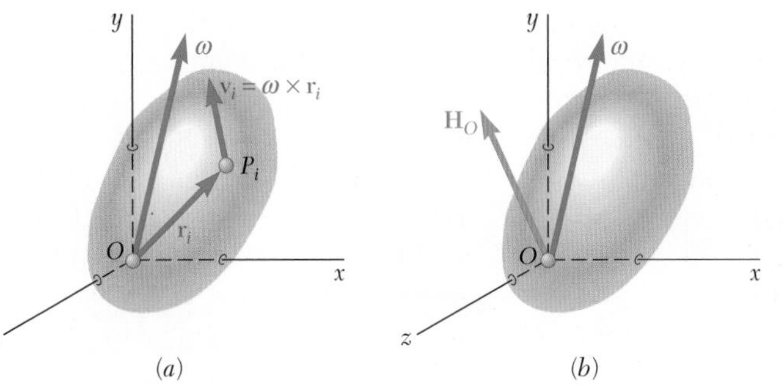

Fig. 18.5

Angular Momentum of a Rigid Body Constrained to Rotate about a Fixed Point. In the particular case of a rigid body constrained to rotate in three-dimensional space about a fixed point O (Fig. 18.5*a*), it is sometimes convenient to determine the angular momentum $\mathbf{H}_O$ of the body about the fixed point O. While $\mathbf{H}_O$ could be obtained by first computing $\mathbf{H}_G$ as indicated above and then using Eq. (18.11), it is often advantageous to determine $\mathbf{H}_O$ directly from the angular velocity $\boldsymbol{\omega}$ of the body and its moments and products of inertia with respect to a frame $Oxyz$ centered at the fixed point O. Recalling Eq. (14.7), we write

$$\mathbf{H}_O = \sum_{i=1}^{n} (\mathbf{r}_i \times \mathbf{v}_i \, \Delta m_i) \tag{18.12}$$

where $\mathbf{r}_i$ and $\mathbf{v}_i$ denote, respectively, the position vector and the velocity of the particle P_i with respect to the fixed frame $Oxyz$. Substituting $\mathbf{v}_i = \boldsymbol{\omega} \times \mathbf{r}_i$, and after manipulations similar to those used in the earlier part of this section, we find that the components of the angular momentum $\mathbf{H}_O$ (Fig. 18.5*b*) are given by the relations

$$\begin{aligned} H_x &= +I_x\,\omega_x - I_{xy}\omega_y - I_{xz}\omega_z \\ H_y &= -I_{yx}\omega_x + I_y\,\omega_y - I_{yz}\omega_z \\ H_z &= -I_{zx}\omega_x - I_{zy}\omega_y + I_z\,\omega_z \end{aligned} \tag{18.13}$$

where the moments of inertia I_x, I_y, I_z and the products of inertia I_{xy}, I_{yz}, I_{zx} are computed with respect to the frame $Oxyz$ centered at the fixed point O.

*18.3 APPLICATION OF THE PRINCIPLE OF IMPULSE AND MOMENTUM TO THE THREE-DIMENSIONAL MOTION OF A RIGID BODY

Before we can apply the fundamental equation (18.2) to the solution of problems involving the three-dimensional motion of a rigid body, we must learn to compute the derivative of the vector $\mathbf{H}_G$. This will be done in Sec. 18.5. The results obtained in the preceding section can, however, be used right away to solve problems by the impulse-momentum method.

Recalling that the system formed by the momenta of the particles of a rigid body reduces to a linear momentum vector $m\overline{\mathbf{v}}$

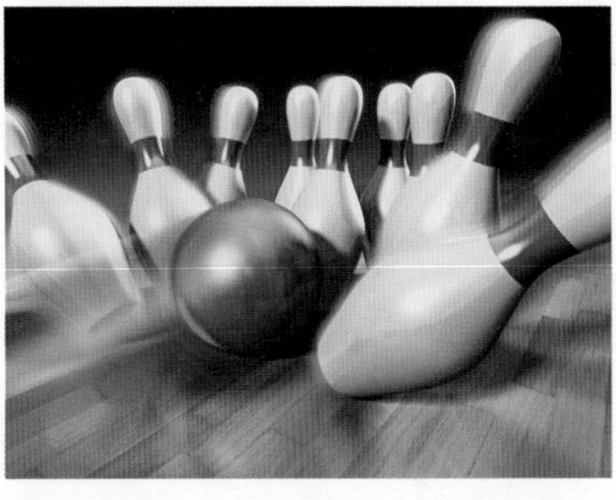

Photo 18.2 As a result of the impulsive force applied by the bowling ball, a pin acquires both linear momentum and angular momentum.

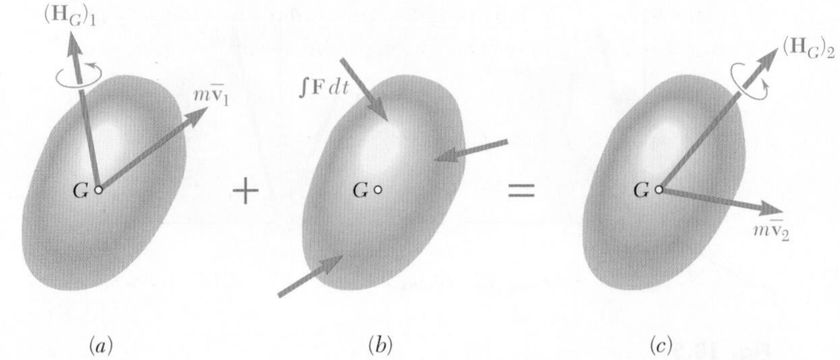

Fig. 18.6

attached at the mass center G of the body and an angular momentum couple $\mathbf{H}_G$, we represent graphically the fundamental relation

$$\textbf{Syst Momenta}_1 + \textbf{Syst Ext Imp}_{1\rightarrow 2} = \textbf{Syst Momenta}_2 \quad (17.14)$$

by means of the three sketches shown in Fig. 18.6. To solve a given problem, we can use these sketches to write appropriate component and moment equations, keeping in mind that the components of the angular momentum $\mathbf{H}_G$ are related to the components of the angular velocity $\boldsymbol{\omega}$ by Eqs. (18.7) of the preceding section.

In solving problems dealing with the motion of a body rotating about a fixed point O, it will be convenient to eliminate the impulse of the reaction at O by writing an equation involving the moments of the momenta and impulses about O. We recall that the angular momentum $\mathbf{H}_O$ of the body about the fixed point O can be obtained either directly from Eqs. (18.13) or by first computing its linear momentum $m\bar{\mathbf{v}}$ and its angular momentum $\mathbf{H}_G$ and then using Eq. (18.11).

*18.4 KINETIC ENERGY OF A RIGID BODY IN THREE DIMENSIONS

Consider a rigid body of mass m in three-dimensional motion. We recall from Sec. 14.6 that if the absolute velocity $\mathbf{v}_i$ of each particle P_i of the body is expressed as the sum of the velocity $\bar{\mathbf{v}}$ of the mass center G of the body and of the velocity $\mathbf{v}_i'$ of the particle relative to a frame $Gxyz$ attached to G and of fixed orientation (Fig. 18.7), the kinetic energy of the system of particles forming the rigid body can be written in the form

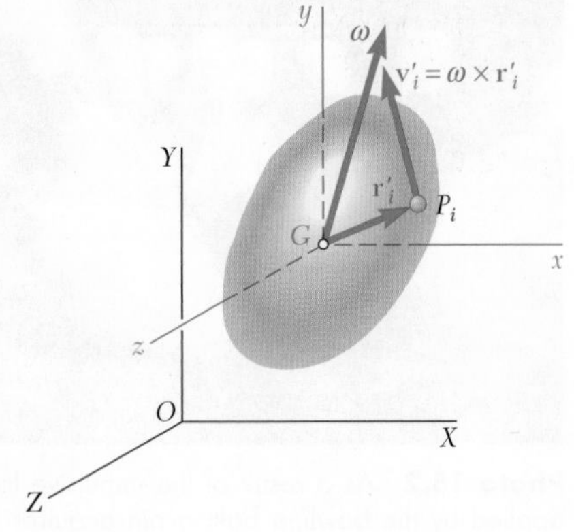

Fig. 18.7

$$T = \tfrac{1}{2}m\bar{v}^2 + \frac{1}{2}\sum_{i=1}^{n}\Delta m_i v_i'^2 \qquad (18.14)$$

where the last term represents the kinetic energy T' of the body relative to the centroidal frame $Gxyz$. Since $v_i' = |\mathbf{v}_i'| = |\boldsymbol{\omega} \times \mathbf{r}_i'|$, we write

$$T' = \frac{1}{2}\sum_{i=1}^{n}\Delta m_i v_i'^2 = \frac{1}{2}\sum_{i=1}^{n}|\boldsymbol{\omega} \times \mathbf{r}_i'|^2\,\Delta m_i$$

Expressing the square in terms of the rectangular components of the vector product, and replacing the sums by integrals, we have

$$\begin{aligned} T' &= \tfrac{1}{2}\int[(\omega_x y - \omega_y x)^2 + (\omega_y z - \omega_z y)^2 + (\omega_z x - \omega_x z)^2]\,dm \\ &= \tfrac{1}{2}[\omega_x^2\int(y^2 + z^2)\,dm + \omega_y^2\int(z^2 + x^2)\,dm + \omega_z^2\int(x^2 + y^2)\,dm \\ &\qquad - 2\omega_x\omega_y\int xy\,dm - 2\omega_y\omega_z\int yz\,dm - 2\omega_z\omega_x\int zx\,dm] \end{aligned}$$

or, recalling the relations (18.5) and (18.6),

$$T' = \tfrac{1}{2}(\bar{I}_x\omega_x^2 + \bar{I}_y\omega_y^2 + \bar{I}_z\omega_z^2 - 2\bar{I}_{xy}\omega_x\omega_y - 2\bar{I}_{yz}\omega_y\omega_z - 2\bar{I}_{zx}\omega_z\omega_x) \tag{18.15}$$

Substituting into (18.14) the expression (18.15) we have just obtained for the kinetic energy of the body relative to centroidal axes, we write

$$\begin{aligned} T = \tfrac{1}{2}m\bar{v}^2 + \tfrac{1}{2}(\bar{I}_x\omega_x^2 + \bar{I}_y\omega_y^2 + \bar{I}_z\omega_z^2 - 2\bar{I}_{xy}\omega_x\omega_y \\ - 2\bar{I}_{yz}\omega_y\omega_z - 2\bar{I}_{zx}\omega_z\omega_x) \end{aligned} \tag{18.16}$$

If the axes of coordinates are chosen so that they coincide at the instant considered with the principal axes x', y', z' of the body, the relation obtained reduces to

$$T = \tfrac{1}{2}m\bar{v}^2 + \tfrac{1}{2}(\bar{I}_{x'}\omega_{x'}^2 + \bar{I}_{y'}\omega_{y'}^2 + \bar{I}_{z'}\omega_{z'}^2) \tag{18.17}$$

where $\bar{\mathbf{v}}$ = velocity of mass center
$\boldsymbol{\omega}$ = angular velocity
m = mass of rigid body
$\bar{I}_{x'}, \bar{I}_{y'}, \bar{I}_{z'}$ = principal centroidal moments of inertia

The results we have obtained enable us to apply to the three-dimensional motion of a rigid body the principles of work and energy (Sec. 17.2) and conservation of energy (Sec. 17.6).

Kinetic Energy of a Rigid Body with a Fixed Point. In the particular case of a rigid body rotating in three-dimensional space about a fixed point O, the kinetic energy of the body can be expressed in terms of its moments and products of inertia with respect to axes attached at O (Fig. 18.8). Recalling the definition of the kinetic energy of a system of particles, and substituting $v_i = |\mathbf{v}_i| = |\boldsymbol{\omega} \times \mathbf{r}_i|$, we write

$$T = \frac{1}{2}\sum_{i=1}^{n} \Delta m_i v_i^2 = \frac{1}{2}\sum_{i=1}^{n} |\boldsymbol{\omega} \times \mathbf{r}_i|^2\,\Delta m_i \tag{18.18}$$

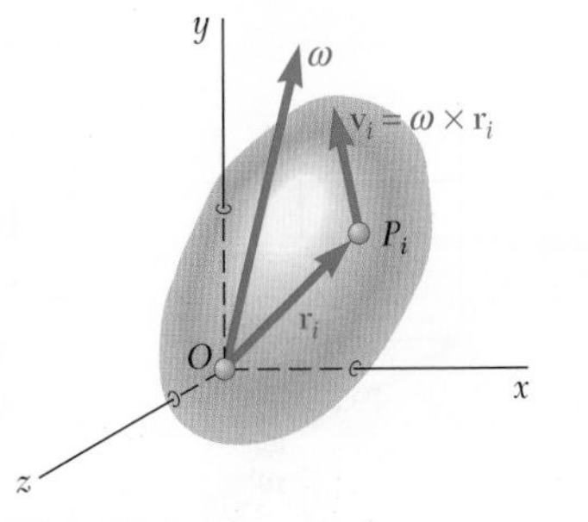

Fig. 18.8

Manipulations similar to those used to derive Eq. (18.15) yield

$$T = \tfrac{1}{2}(I_x\omega_x^2 + I_y\omega_y^2 + I_z\omega_z^2 - 2I_{xy}\omega_x\omega_y - 2I_{yz}\omega_y\omega_z - 2I_{zx}\omega_z\omega_x) \tag{18.19}$$

or, if the principal axes x', y', z' of the body at the origin O are chosen as coordinate axes,

$$T = \tfrac{1}{2}(I_{x'}\omega_{x'}^2 + I_{y'}\omega_{y'}^2 + I_{z'}\omega_{z'}^2) \tag{18.20}$$

SAMPLE PROBLEM 18.1

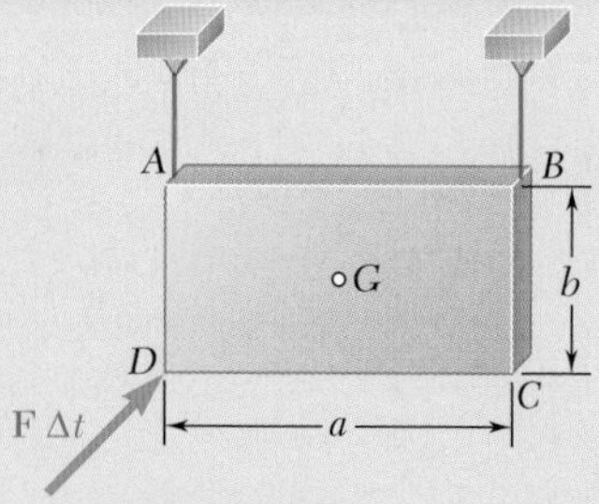

A rectangular plate of mass m suspended from two wires at A and B is hit at D in a direction perpendicular to the plate. Denoting by $\mathbf{F}\,\Delta t$ the impulse applied at D, determine immediately after the impact (*a*) the velocity of the mass center G, (*b*) the angular velocity of the plate.

SOLUTION

Assuming that the wires remain taut and thus that the components $\bar{v}_y$ of $\bar{\mathbf{v}}$ and ω_z of $\boldsymbol{\omega}$ are zero after the impact, we have

$$\bar{\mathbf{v}} = \bar{v}_x\mathbf{i} + \bar{v}_z\mathbf{k} \qquad \boldsymbol{\omega} = \omega_x\mathbf{i} + \omega_y\mathbf{j}$$

and since the x, y, z axes are principal axes of inertia,

$$\mathbf{H}_G = \bar{I}_x\omega_x\mathbf{i} + \bar{I}_y\omega_y\mathbf{j} \qquad \mathbf{H}_G = \tfrac{1}{12}mb^2\omega_x\mathbf{i} + \tfrac{1}{12}ma^2\omega_y\mathbf{j} \tag{1}$$

Principle of Impulse and Momentum. Since the initial momenta are zero, the system of the impulses must be equivalent to the system of the final momenta:

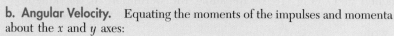

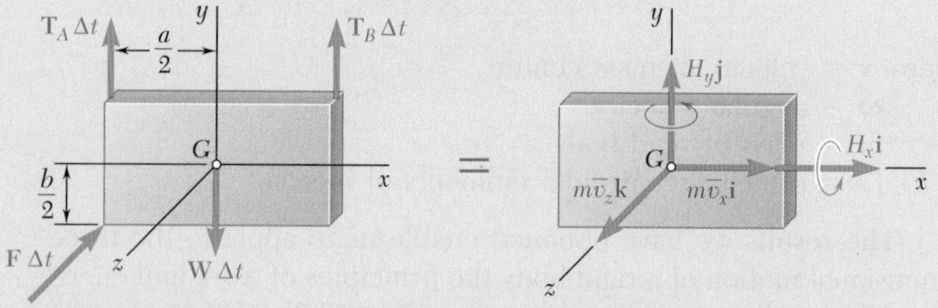

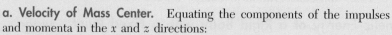

a. Velocity of Mass Center. Equating the components of the impulses and momenta in the x and z directions:

x components: $\quad 0 = m\bar{v}_x \qquad \bar{v}_x = 0$

z components: $\quad -F\,\Delta t = m\bar{v}_z \qquad \bar{v}_z = -F\,\Delta t/m$

$$\bar{\mathbf{v}} = \bar{v}_x\mathbf{i} + \bar{v}_z\mathbf{k} \qquad \bar{\mathbf{v}} = -(F\,\Delta t/m)\mathbf{k} \quad ◀$$

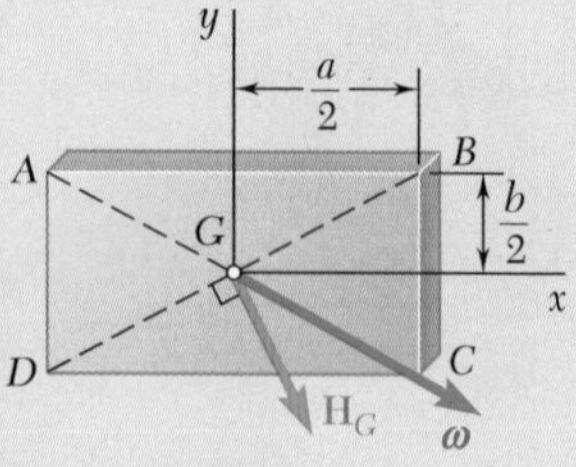

b. Angular Velocity. Equating the moments of the impulses and momenta about the x and y axes:

About x axis: $\quad \tfrac{1}{2}bF\,\Delta t = H_x$

About y axis: $\quad -\tfrac{1}{2}aF\,\Delta t = H_y$

$$\mathbf{H}_G = H_x\mathbf{i} + H_y\mathbf{j} \qquad \mathbf{H}_G = \tfrac{1}{2}bF\,\Delta t\mathbf{i} - \tfrac{1}{2}aF\,\Delta t\mathbf{j} \tag{2}$$

Comparing Eqs. (1) and (2), we conclude that

$$\omega_x = 6F\,\Delta t/mb \qquad \omega_y = -6F\,\Delta t/ma$$

$$\boldsymbol{\omega} = \omega_x\mathbf{i} + \omega_y\mathbf{j} \qquad \boldsymbol{\omega} = (6F\,\Delta t/mab)(a\mathbf{i} - b\mathbf{j}) \quad ◀$$

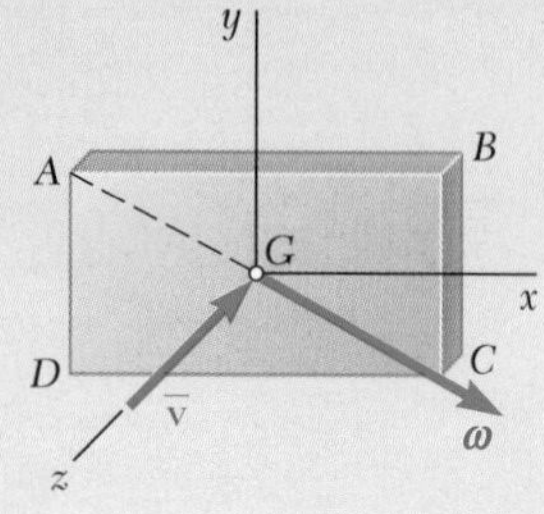

We note that $\boldsymbol{\omega}$ is directed along the diagonal AC.

Remark: Equating the y components of the impulses and momenta, and their moments about the z axis, we obtain two additional equations which yield $T_A = T_B = \tfrac{1}{2}W$. We thus verify that the wires remain taut and that our assumption was correct.

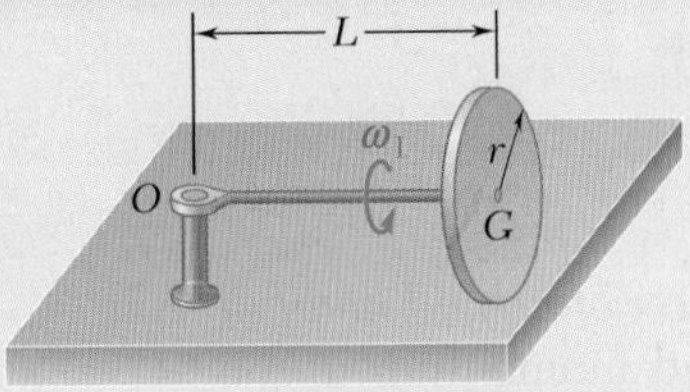

SAMPLE PROBLEM 18.2

A homogeneous disk of radius r and mass m is mounted on an axle OG of length L and negligible mass. The axle is pivoted at the fixed point O, and the disk is constrained to roll on a horizontal floor. Knowing that the disk rotates counterclockwise at the rate ω_1 about the axle OG, determine (*a*) the angular velocity of the disk, (*b*) its angular momentum about O, (*c*) its kinetic energy, (*d*) the vector and couple at G equivalent to the momenta of the particles of the disk.

SOLUTION

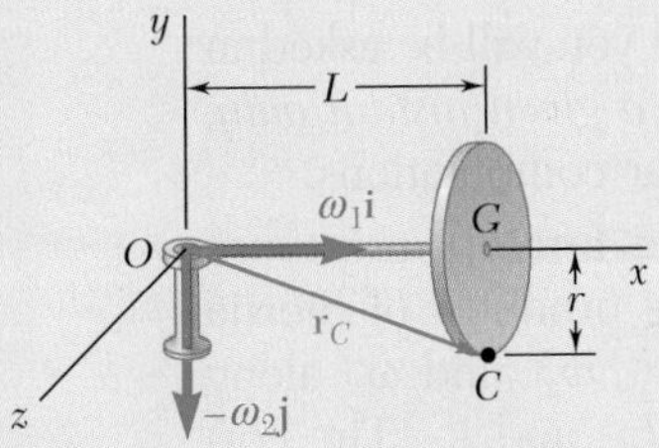

a. Angular Velocity. As the disk rotates about the axle OG it also rotates with the axle about the y axis at a rate ω_2 clockwise. The total angular velocity of the disk is therefore

$$\boldsymbol{\omega} = \omega_1\mathbf{i} - \omega_2\mathbf{j} \tag{1}$$

To determine ω_2 we write that the velocity of C is zero:

$$\mathbf{v}_C = \boldsymbol{\omega} \times \mathbf{r}_C = 0$$
$$(\omega_1\mathbf{i} - \omega_2\mathbf{j}) \times (L\mathbf{i} - r\mathbf{j}) = 0$$
$$(L\omega_2 - r\omega_1)\mathbf{k} = 0 \qquad \omega_2 = r\omega_1/L$$

Substituting into (1) for ω_2: $\quad \boldsymbol{\omega} = \omega_1\mathbf{i} - (r\omega_1/L)\mathbf{j}$ ◀

b. Angular Momentum about O. Assuming the axle to be part of the disk, we can consider the disk to have a fixed point at O. Since the x, y, and z axes are principal axes of inertia for the disk,

$$H_x = I_x\omega_x = (\tfrac{1}{2}mr^2)\omega_1$$
$$H_y = I_y\omega_y = (mL^2 + \tfrac{1}{4}mr^2)(-r\omega_1/L)$$
$$H_z = I_z\omega_z = (mL^2 + \tfrac{1}{4}mr^2)0 = 0$$
$$\mathbf{H}_O = \tfrac{1}{2}mr^2\omega_1\mathbf{i} - m(L^2 + \tfrac{1}{4}r^2)(r\omega_1/L)\mathbf{j} \quad ◀$$

c. Kinetic Energy. Using the values obtained for the moments of inertia and the components of $\boldsymbol{\omega}$, we have

$$T = \tfrac{1}{2}(I_x\omega_x^2 + I_y\omega_y^2 + I_z\omega_z^2) = \tfrac{1}{2}[\tfrac{1}{2}mr^2\omega_1^2 + m(L^2 + \tfrac{1}{4}r^2)(-r\omega_1/L)^2]$$
$$T = \tfrac{1}{8}mr^2\left(6 + \frac{r^2}{L^2}\right)\omega_1^2 \quad ◀$$

d. Momentum Vector and Couple at G. The linear momentum vector $m\bar{\mathbf{v}}$ and the angular momentum couple $\mathbf{H}_G$ are

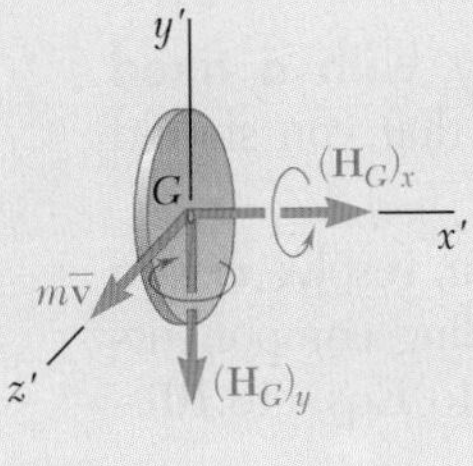

$$m\bar{\mathbf{v}} = mr\omega_1\mathbf{k} \quad ◀$$

and

$$\mathbf{H}_G = \bar{I}_{x'}\omega_x\mathbf{i} + \bar{I}_{y'}\omega_y\mathbf{j} + \bar{I}_{z'}\omega_z\mathbf{k} = \tfrac{1}{2}mr^2\omega_1\mathbf{i} + \tfrac{1}{4}mr^2(-r\omega_1/L)\mathbf{j}$$
$$\mathbf{H}_G = \tfrac{1}{2}mr^2\omega_1\left(\mathbf{i} - \frac{r}{2L}\mathbf{j}\right) \quad ◀$$

SOLVING PROBLEMS ON YOUR OWN

In this lesson you learned to compute the *angular momentum of a rigid body in three dimensions* and to apply the principle of impulse and momentum to the three-dimensional motion of a rigid body. You also learned to compute the *kinetic energy of a rigid body in three dimensions.* It is important for you to keep in mind that, except for very special situations, the angular momentum of a rigid body in three dimensions *cannot* be expressed as the product $\bar{I}\boldsymbol{\omega}$ and, therefore, *will not have the same direction as the angular velocity* $\boldsymbol{\omega}$ (Fig. 18.3).

1. To compute the angular momentum $\mathbf{H}_G$ of a rigid body about its mass center G, you must first determine the angular velocity $\boldsymbol{\omega}$ of the body with respect to a system of axes *centered at G and of fixed orientation.* Since you will be asked in this lesson to determine the angular momentum of the body *at a given instant only,* select the system of axes which will be most convenient for your computations.

a. If the principal axes of inertia of the body at G are known, use these axes as coordinate axes x', y', and z', since the corresponding products of inertia of the body will be equal to zero. Resolve $\boldsymbol{\omega}$ into components $\omega_{x'}$, $\omega_{y'}$, and $\omega_{z'}$ along these axes and compute the principal moments of inertia $\bar{I}_{x'}$, $\bar{I}_{y'}$, and $\bar{I}_{z'}$. The corresponding components of the angular momentum $\mathbf{H}_G$ are

$$H_{x'} = \bar{I}_{x'}\omega_{x'} \qquad H_{y'} = \bar{I}_{y'}\omega_{y'} \qquad H_{z'} = \bar{I}_{z'}\omega_{z'} \tag{18.10}$$

b. If the principal axes of inertia of the body at G are not known, you must use Eqs. (18.7) to determine the components of the angular momentum $\mathbf{H}_G$. These equations require prior computation of the *products of inertia* of the body as well as prior computation of its moments of inertia with respect to the selected axes.

c. The magnitude and direction cosines of $\mathbf{H}_G$ are obtained from formulas similar to those used in Statics [Sec. 2.12]. We have

$$H_G = \sqrt{H_x^2 + H_y^2 + H_z^2}$$

$$\cos\theta_x = \frac{H_x}{H_G} \qquad \cos\theta_y = \frac{H_y}{H_G} \qquad \cos\theta_z = \frac{H_z}{H_G}$$

d. Once $\mathbf{H}_G$ has been determined, you can obtain the angular momentum of the body *about any given point O* by observing from Fig. (18.4) that

$$\mathbf{H}_O = \bar{\mathbf{r}} \times m\bar{\mathbf{v}} + \mathbf{H}_G \tag{18.11}$$

where $\bar{\mathbf{r}}$ is the position vector of G relative to O, and $m\bar{\mathbf{v}}$ is the linear momentum of the body.

2. To compute the angular momentum $\mathbf{H}_O$ of a rigid body with a fixed point O, follow the procedure described in paragraph 1, except that you should now use axes centered at the fixed point O.

a. If the principal axes of inertia of the body at O are known, resolve $\boldsymbol{\omega}$ into components along these axes [Sample Prob. 18.2]. The corresponding components of the angular momentum $\mathbf{H}_G$ are obtained from equations similar to Eqs. (18.10).

b. If the principal axes of inertia of the body at O are not known, you must compute the products as well as the moments of inertia of the body with respect to the axes that you have selected and use Eqs. (18.13) to determine the components of the angular momentum $\mathbf{H}_O$.

3. To apply the principle of impulse and momentum to the solution of a problem involving the three-dimensional motion of a rigid body, you will use the same vector equation that you used for plane motion in Chap. 17,

$$\textbf{Syst Momenta}_1 + \textbf{Syst Ext Imp}_{1\rightarrow 2} = \textbf{Syst Momenta}_2 \qquad (17.14)$$

where the initial and final systems of momenta are each represented by a *linear-momentum vector* $m\bar{\mathbf{v}}$ and an *angular-momentum couple* $\mathbf{H}_G$. Now, however, these vector-and-couple systems should be represented in three dimensions as shown in Fig. 18.6, and $\mathbf{H}_G$ should be determined as explained in paragraph 1.

a. In problems involving the application of a known impulse to a rigid body, draw the free-body-diagram equation corresponding to Eq. (17.14). Equating the components of the vectors involved, you will determine the final linear momentum $m\bar{\mathbf{v}}$ of the body and, thus, the corresponding velocity $\bar{\mathbf{v}}$ of its mass center. Equating moments about G, you will determine the final angular momentum $\mathbf{H}_G$ of the body. You will then substitute the values obtained for the components of $\mathbf{H}_G$ into Eqs. (18.10) or (18.7) and solve these equations for the corresponding values of the components of the angular velocity $\boldsymbol{\omega}$ of the body [Sample Prob. 18.1].

b. In problems involving unknown impulses, draw the free-body-diagram equation corresponding to Eq. (17.4) and write equations which do not involve the unknown impulses. Such equations can be obtained by equating moments about the point or line of impact.

4. To compute the kinetic energy of a rigid body with a fixed point O, resolve the angular velocity $\boldsymbol{\omega}$ into components along axes of your choice and compute the moments and products of inertia of the body with respect to these axes. As was the case for the computation of the angular momentum, use the principal axes of inertia x', y', and z' if you can easily determine them. The products of inertia will then be zero [Sample Prob. 18.2], and the expression for the kinetic energy will reduce to

$$T = \tfrac{1}{2}(I_{x'}\omega_{x'}^2 + I_{y'}\omega_{y'}^2 + I_{z'}\omega_{x'}^2) \qquad (18.20)$$

If you must use axes other than the principal axes of inertia, the kinetic energy of the body should be expressed as shown in Eq. (18.19).

5. To compute the kinetic energy of a rigid body in general motion, consider the motion as the sum of a *translation with the mass center G and a rotation about G*. The kinetic energy associated with the translation is $\frac{1}{2}m\bar{v}^2$. If principal axes of inertia can be used, the kinetic energy associated with the rotation about G can be expressed in the form used in Eq. (18.20). The total kinetic energy of the rigid body is then

$$T = \tfrac{1}{2}m\bar{v}^2 + \tfrac{1}{2}(\bar{I}_{x'}\omega_{x'}^2 + \bar{I}_{y'}\omega_{y'}^2 + \bar{I}_{z'}\omega_{z'}^2) \qquad (18.17)$$

If you must use axes other than the principal axes of inertia to determine the kinetic energy associated with the rotation about G, the total kinetic energy of the body should be expressed as shown in Eq. (18.16).

PROBLEMS

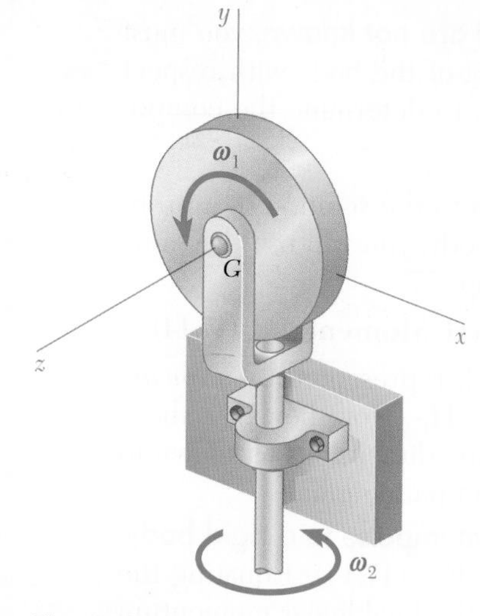

Fig. P18.1

18.1 A thin, homogeneous disk of mass m and radius r spins at the constant rate ω_1 about an axle held by a fork-ended vertical rod which rotates at the constant rate ω_2. Determine the angular momentum $\mathbf{H}_G$ of the disk about its mass center G.

18.2 A thin homogeneous square of mass m and side a is welded to a vertical shaft AB with which it forms an angle of 45°. Knowing that the shaft rotates with an angular velocity $\boldsymbol{\omega}$, determine the angular momentum of the plate about A.

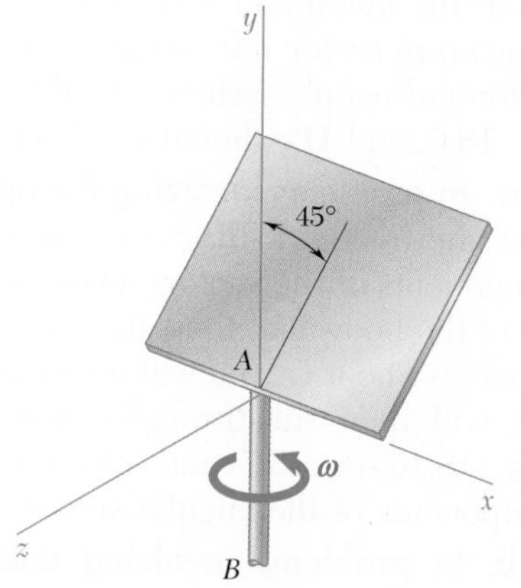

Fig. P18.2

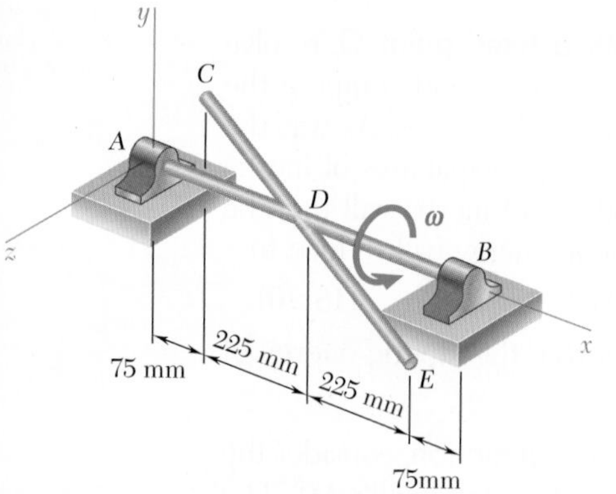

Fig. P18.3

18.3 Two uniform rods AB and CE, each of mass 1.5 kg and length 600 mm, are welded to each other at their midpoints. Knowing that this assembly has an angular velocity of constant magnitude $\omega = 12$ rad/s, determine the magnitude and direction of the angular momentum $\mathbf{H}_D$ of the assembly about D.

18.4 A homogeneous disk of mass $m = 3$ kg rotates at the constant rate $\omega_1 = 16$ rad/s with respect to arm ABC, which is welded to a shaft DCE rotating at the constant rate $\omega_2 = 8$ rad/s. Determine the angular momentum $\mathbf{H}_A$ of the disk about its center A.

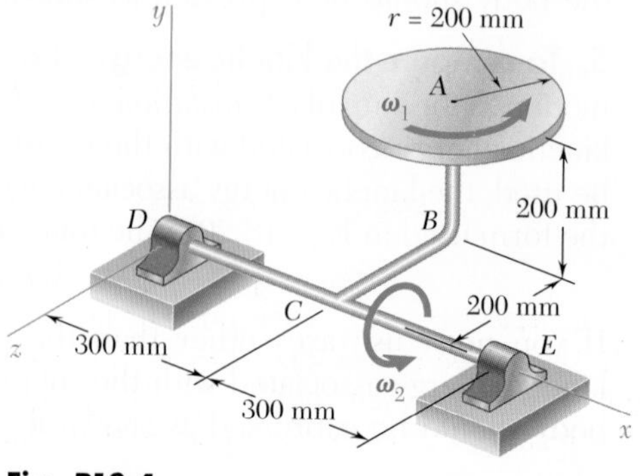

Fig. *P18.4*

18.5 A thin disk of mass $m = 4$ kg rotates at the constant rate $\omega_2 = 15$ rad/s with respect to arm ABC, which itself rotates at the constant rate $\omega_1 = 5$ rad/s about the y axis. Determine the angular momentum of the disk about its center C.

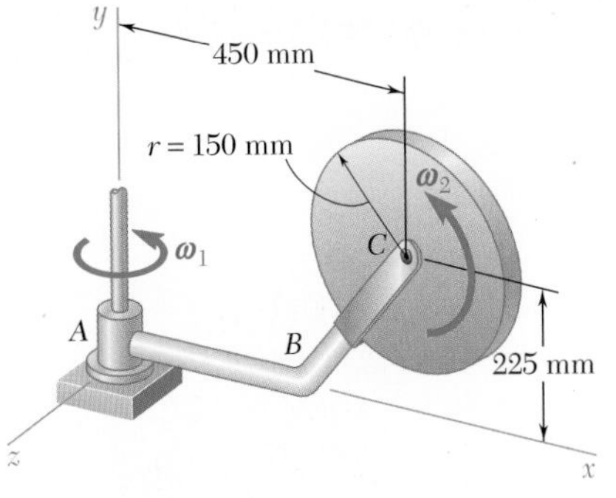

Fig. P18.5

18.6 A solid rectangular parallelepiped of mass m has a square base of side a and a length $2a$. Knowing that it rotates at the constant rate ω about its diagonal AC' and that its rotation is observed from A as counterclockwise, determine (*a*) the magnitude of the angular momentum $\mathbf{H}_G$ of the parallelepiped about its mass center G, (*b*) the angle that $\mathbf{H}_G$ forms with the diagonal AC'.

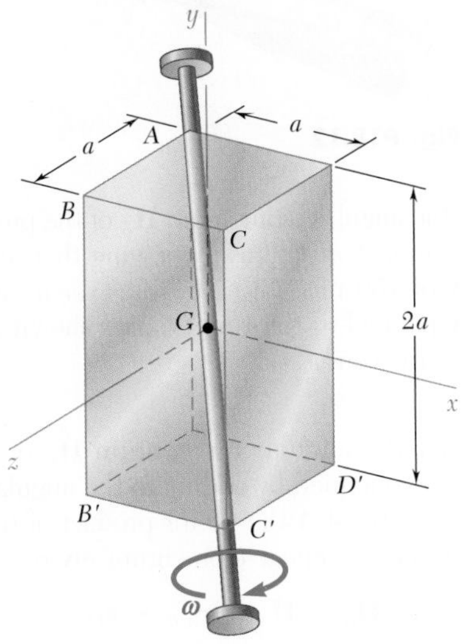

Fig. P18.6

18.7 Solve Prob. 18.6, assuming that the solid rectangular parallelepiped has been replaced by a hollow one consisting of six thin metal plates welded together.

18.8 A homogeneous disk of mass m and radius r is mounted on the vertical shaft AB. The normal to the disk at G forms an angle $\beta = 25°$ with the shaft. Knowing that the shaft has a constant angular velocity $\boldsymbol{\omega}$, determine the angle θ formed by the shaft AB and the angular momentum $\mathbf{H}_G$ of the disk about its mass center G.

18.9 Determine the angular momentum $\mathbf{H}_D$ of the disk of Prob. 18.6 about point D.

18.10 Determine the angular momentum of the disk of Prob. 18.5 about point A.

18.11 Determine the angular momentum $\mathbf{H}_O$ of the disk of Sample Prob. 18.2 from the expressions obtained for its linear momentum $m\bar{\mathbf{v}}$ and its angular momentum $\mathbf{H}_G$, using Eqs. (18.11). Verify that the result obtained is the same as that obtained by direct computation.

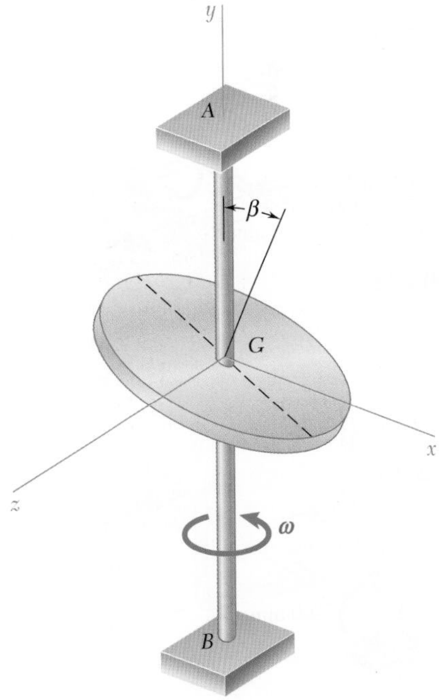

Fig. P18.8

18.12 The 100-kg projectile shown has a radius of gyration of 100 mm about its axis of symmetry Gx and a radius of gyration of 250 mm about the transverse axis Gy. Its angular velocity $\boldsymbol{\omega}$ can be resolved into two components; one component, directed along Gx, measures the *rate of spin* of the projectile, while the other component, directed along GD, measures its *rate of precession*. Knowing that $\theta = 6°$ and that the angular momentum of the projectile about its mass center G is $\mathbf{H}_G = (500 \text{ g} \cdot \text{m}^2/\text{s})\mathbf{i} - (10 \text{ g} \cdot \text{m}^2/\text{s})\mathbf{j}$, determine (*a*) the rate of spin, (*b*) the rate of precession.

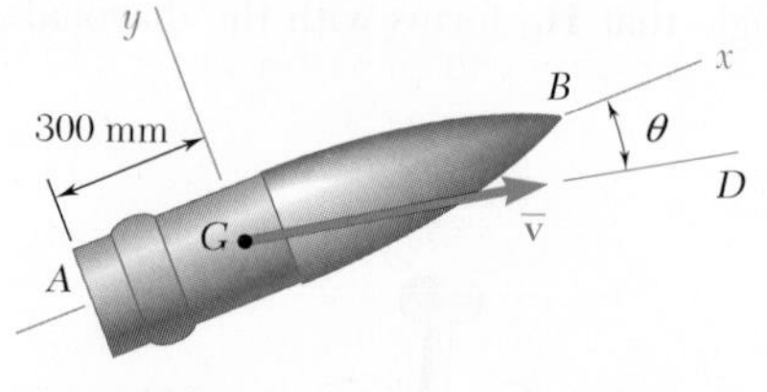

Fig. P18.12

18.13 Determine the angular momentum $\mathbf{H}_A$ of the projectile of Prob. 18.12 about the center A of its base, knowing that its mass center G has a velocity $\overline{\mathbf{v}}$ of 750 m/s. Give your answer in terms of components respectively parallel to the x and y axes shown and to a third axis z pointing toward you.

18.14 (*a*) Show that the angular momentum $\mathbf{H}_B$ of a rigid body about point B can be obtained by adding to the angular momentum $\mathbf{H}_A$ of that body about point A the vector product of the vector $\mathbf{r}_{A/B}$ drawn from B to A and the linear momentum $m\overline{\mathbf{v}}$ of the body:

$$\mathbf{H}_B = \mathbf{H}_A + \mathbf{r}_{A/B} \times m\overline{\mathbf{v}}$$

(*b*) Further show that when a rigid body rotates about a fixed axis, its angular momentum is the same about any two points A and B located on the fixed axis ($\mathbf{H}_A = \mathbf{H}_B$) if, and only if, the mass center G of the body is located on the fixed axis.

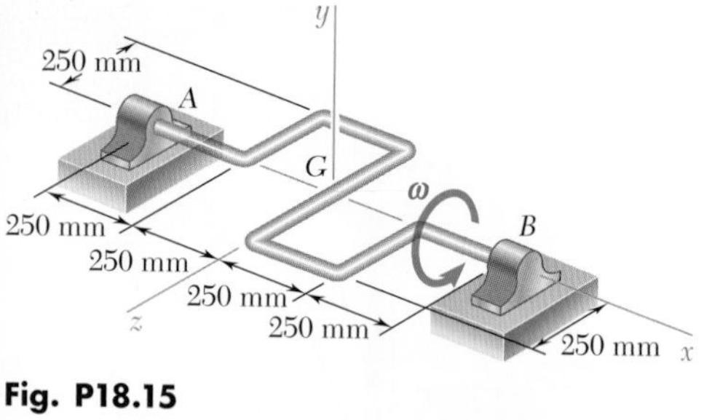

Fig. P18.15

18.15 A 5-kg rod of uniform cross section is used to form the shaft shown. Knowing that the shaft rotates with a constant angular velocity $\boldsymbol{\omega}$ of magnitude 12 rad/s, determine (*a*) the angular momentum $\mathbf{H}_G$ of the shaft about its mass center G, (*b*) the angle formed by $\mathbf{H}_G$ and the axis AB.

18.16 Determine the angular momentum of the shaft of Prob. 18.15 about (*a*) point A, (*b*) point B.

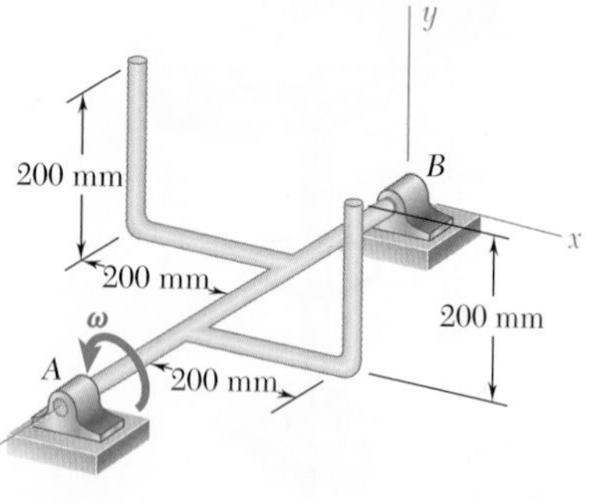

Fig. P18.17

18.17 Two L-shaped arms, each of mass 2 kg are welded at the third points of the 600 mm shaft AB. Knowing that shaft AB rotates at the constant rate $\omega = 240$ rpm, determine (*a*) the angular momentum of the body about A, (*b*) the angle formed by the angular momentum and shaft AB.

18.18 For the body of Prob. 18.17, determine (*a*) the angular momentum about B, (*b*) the angle formed by the angular momentum about shaft BA.

18.19 The triangular plate shown has a mass of 7.5 kg and is welded to a vertical shaft AB. Knowing that the plate rotates at the constant rate $\omega = 12$ rad/s, determine its angular momentum about (*a*) point C, (*b*) point A. (*Hint:* To solve part *b*, find $\bar{\mathbf{v}}$ and use the property indicated in part *a* of Prob. 18.14.)

18.20 The triangular plate shown has a mass of 7.5 kg and is welded to a vertical shaft AB. Knowing that the plate rotates at the constant rate $\omega = 12$ rad/s, determine its angular momentum about (*a*) point C, (*b*) point B. (See hint of Prob. 18.19.)

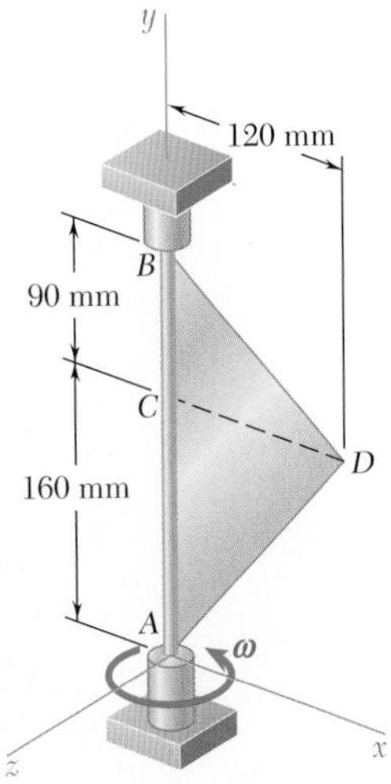

Fig. P18.19 and P18.20

18.21 One of the sculptures displayed on a university campus consists of a hollow cube made of six aluminum sheets, each 1.5×1.5 m, welded together and reinforced with internal braces of negligible weight. The cube is mounted on a fixed base at A and can rotate freely about its vertical diagonal AB. As she passes by this display on the way to a class in mechanics, an engineering student grabs corner C of the cube and pushes it for 1.2 s in a direction perpendicular to the plane ABC with an average force of 50 N. Having observed that it takes 5 s for the cube to complete one full revolution, she flips out her calculator and proceeds to determine the mass of the cube. What is the result of her calculation? (*Hint:* The perpendicular distance from the diagonal joining two vertices of a cube to any of its other six vertices can be obtained by multiplying the side of the cube by $\sqrt{2/3}$.)

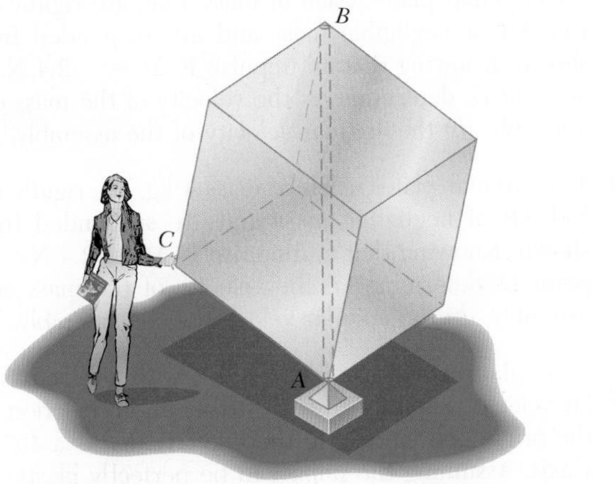

Fig. P18.21

18.22 If the aluminum cube of Prob. 18.21 were replaced by a cube of the same size, made of six plywood sheets with mass 8 kg each, how long would it take for that cube to complete one full revolution if the student pushed its corner C in the same way that she pushed the corner of the aluminum cube?

18.23 A uniform rod of total mass m is bent into the shape shown and is suspended by a wire attached at B. The bent rod is hit at D in a direction perpendicular to the plane containing the rod (in the negative z direction). Denoting the corresponding impulse by $\mathbf{F}\Delta t$, determine (*a*) the velocity of the mass center of the rod, (*b*) the angular velocity of the rod.

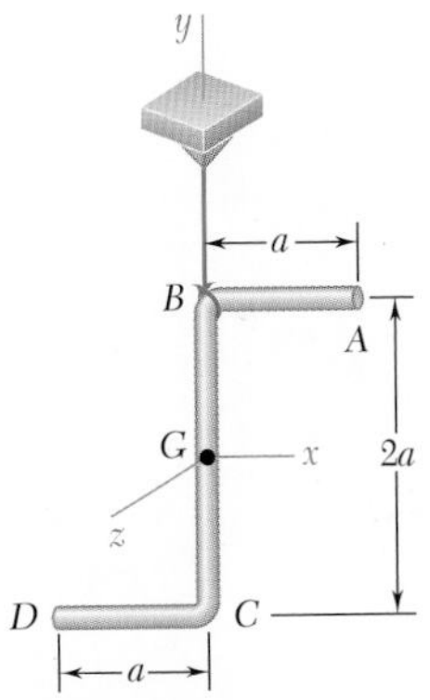

Fig. P18.23

18.24 Solve Prob. 18.23, assuming that the bent rod is hit at C.

18.25 Three slender rods, each of mass m and length $2a$, are welded together to form the assembly shown. The assembly is hit at A in a vertical downward direction. Denoting the corresponding impulse by $\mathbf{F}\,\Delta t$, determine immediately after the impact (*a*) the velocity of the mass center G, (*b*) the angular velocity of the rod.

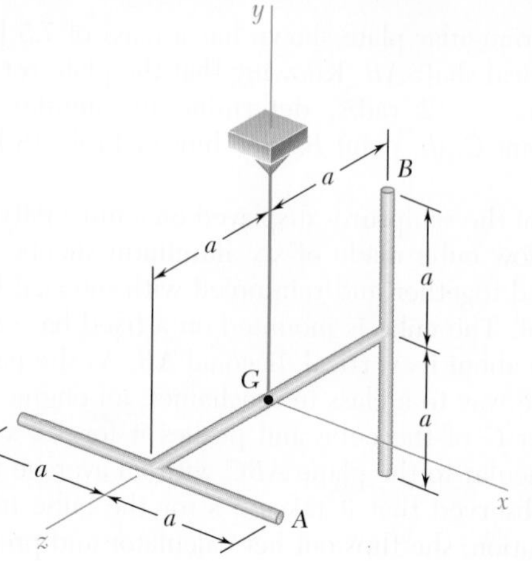

Fig. P18.25

18.26 Solve Prob. 18.25, assuming that the assembly is hit at B in a direction opposite to that of the x axis.

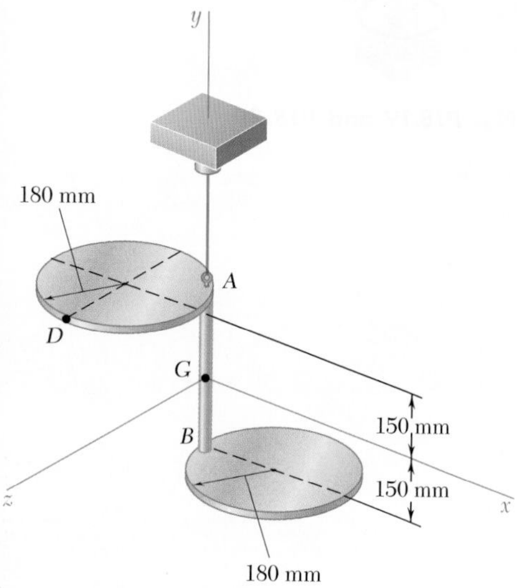

Fig. P18.27 and P18.28

18.27 Two circular plates, each of mass 4 kg, are rigidly connected by a rod AB of negligible mass and are suspended from point A as shown. Knowing that an impulse $\mathbf{F}\,\Delta t = -(2.4\ \text{N}\cdot\text{s})\mathbf{k}$ is applied at point D, determine (*a*) the velocity of the mass center G of the assembly, (*b*) the angular velocity of the assembly.

18.28 Two circular plates, each of mass 4 kg, are rigidly connected by a rod AB of negligible mass and are suspended from point A as shown. Knowing that an impulse $\mathbf{F}\,\Delta t = (2.4\ \text{N}\cdot\text{s})\mathbf{j}$ is applied at point D, determine (*a*) the velocity of the mass center G of the assembly, (*b*) the angular velocity of the assembly.

18.29 A circular plate of mass m is falling with a velocity $\bar{\mathbf{v}}_0$ and no angular velocity when its edge C strikes an obstruction. A line passing the origin and parallel to the line CG makes a 45° angle with the x-axis. Assuming the impact to be perfectly plastic ($e = 0$), determine the angular velocity of the plate immediately after the impact.

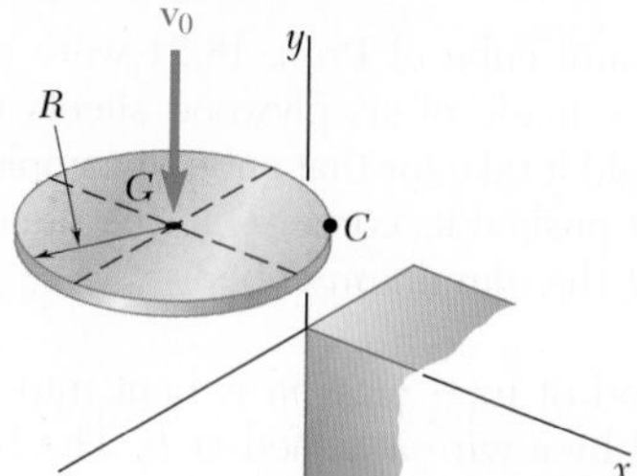

Fig. *P18.29*

18.30 For the plate of Prob. 18.29, determine (*a*) the velocity of its mass center G immediately after the impact, (*b*) the impulse exerted on the plate by the obstruction during the impact.

18.31 A square plate of side a and mass m supported by a ball-and-socket joint at A is rotating about the y axis with a constant angular velocity $\boldsymbol{\omega} = \omega_0\mathbf{j}$ when an obstruction is suddenly introduced at B in the xy plane. Assuming the impact at B to be perfectly plastic ($e = 0$), determine immediately after the impact (*a*) the angular velocity of the plate, (*b*) the velocity of its mass center G.

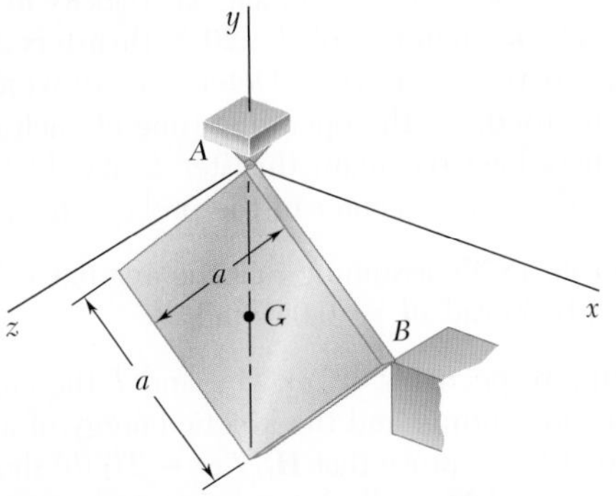

Fig. P18.31

18.32 Determine the impulse exerted on the plate of Prob. 18.31 during the impact by (*a*) the obstruction at B, (*b*) the support at A.

18.33 The coordinate axes shown represent the principal centroidal axes of inertia of a 1500-kg space probe whose radii of gyration are $k_x = 0.4$ m, $k_y = 0.45$ m, and $k_z = 0.375$ m. The probe has no angular velocity when a 150-g meteorite strikes one of its solar panels at point A with a velocity $\mathbf{v}_0 = (720\text{ m/s})\mathbf{i} - (900\text{ m/s})\mathbf{j} + (960\text{ m/s})\mathbf{k}$ relative to the probe. Knowing that the meteorite emerges on the other side of the panel with no change in the direction of its velocity, but with a speed reduced by 20 percent, determine the final angular velocity of the probe.

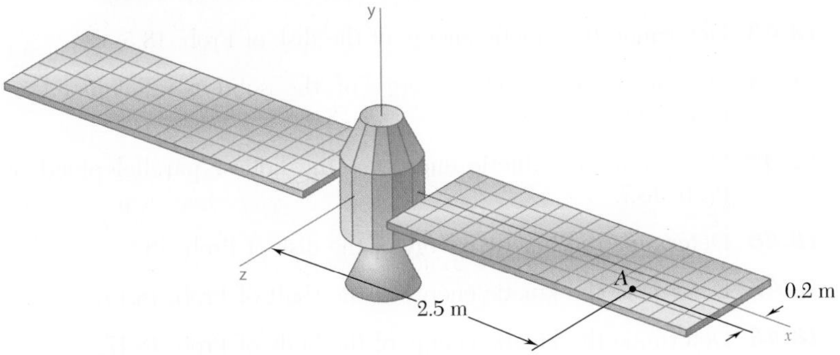

Fig. *P18.33* and *P18.34*

18.34 The coordinate axes shown represent the principal centroidal axes of inertia of a 1500-kg space probe whose radii of gyration are $k_x = 0.4$ m, $k_y = 0.45$ m, and $k_z = 0.375$ m. The probe has no angular velocity when a 150-g meteorite strikes one of its solar panels at point A and emerges on the other side of the panel with no change in the direction of its velocity, but with a speed reduced by 25 percent. Knowing that the final angular velocity of the probe is $\boldsymbol{\omega} = (0.05\text{ rad/s})\mathbf{i} - (0.12\text{ rad/s})\mathbf{j} + \omega_z\mathbf{k}$ and that the x component of the resulting change in the velocity of the mass center of the probe is -16 mm/s, determine (*a*) the component ω_z of the final angular velocity of the probe, (*b*) the relative velocity $\mathbf{v}_0$ with which the meteorite strikes the panel.

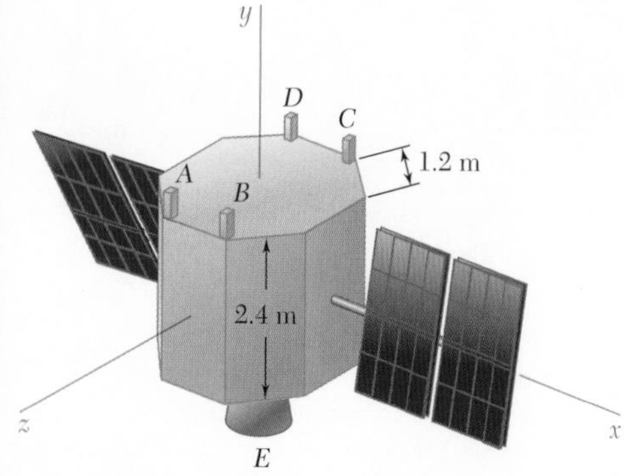

Fig. P18.35

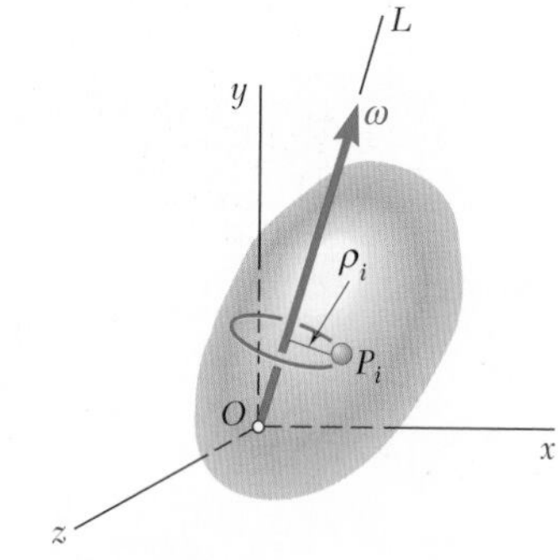

Fig. P18.38

18.35 A 2500-kg probe in orbit about the moon is 2.4 m high and has octagonal bases of sides 1.2 m. The coordinate axes shown are the principal centroidal axes of inertia of the probe, and its radii of gyration are $k_x = 0.98$ m, $k_y = 1.06$ m, and $k_z = 1.02$ m. The probe is equipped with a main 500-N thruster E and with four 20-N thrusters A, B, C, and D which can expel fuel in the positive y direction. The probe has an angular velocity $\boldsymbol{\omega} = (0.040 \text{ rad/s})\mathbf{i} + (0.060 \text{ rad/s})\mathbf{k}$ when two of the 20-N thrusters are used to reduce the angular velocity to zero. Determine (*a*) which of the thrusters should be used, (*b*) the operating time of each of these thrusters, (*c*) for how long the main thruster E should be activated if the velocity of the mass center of the probe is to remain unchanged.

18.36 Solve Prob. 18.35, assuming that the angular velocity of the probe is $\boldsymbol{\omega} = (0.060 \text{ rad/s})\mathbf{i} - (0.040 \text{ rad/s})\mathbf{k}$.

18.37 Denoting, respectively, by $\boldsymbol{\omega}$, $\mathbf{H}_O$, and T the angular velocity, the angular momentum, and the kinetic energy of a rigid body with a fixed point O, (*a*) prove that $\mathbf{H}_O \cdot \boldsymbol{\omega} = 2T$; (*b*) show that the angle θ between ω and $\mathbf{H}_O$ will always be acute.

18.38 Show that the kinetic energy of a rigid body with a fixed point O can be expressed as $T = \frac{1}{2} I_{OL}\omega^2$, where $\boldsymbol{\omega}$ is the instantaneous angular velocity of the body and I_{OL} is its moment of inertia about the line of action OL of $\boldsymbol{\omega}$. Derive this expression (*a*) from Eqs. (9.46) and (18.19), (*b*) by considering T as the sum of the kinetic energies of particles P_i describing circles of radius ρ_i about line OL.

18.39 Determine the kinetic energy of the disk of Prob. 18.1.

18.40 Determine the kinetic energy of the plate of Prob. 18.2.

18.41 Determine the kinetic energy of the assembly of Prob. 18.3.

18.42 Determine the kinetic energy of the disk of Prob. 18.4.

18.43 Determine the kinetic energy of the disk of Prob. 18.5.

18.44 Determine the kinetic energy of the solid parallelepiped of Prob. 18.6.

18.45 Determine the kinetic energy of the hollow parallelepiped of Prob. 18.7.

18.46 Determine the kinetic energy of the disk of Prob. 18.8.

18.47 Determine the kinetic energy of the shaft of Prob. 18.15.

18.48 Determine the kinetic energy of the body of Prob. 18.17.

18.49 Determine the kinetic energy of the triangular plate of Prob. 18.19.

18.50 Determine the kinetic energy imparted to the cube of Prob. 18.21.

18.51 Determine the kinetic energy lost when edge C of the plate of Prob. 18.29 hits the obstruction.

18.52 Determine the kinetic energy lost when the plate of Prob. 18.31 hits the obstruction at B.

18.53 Determine the kinetic energy of the space probe of Prob. 18.33 in its motion about its mass center after its collision with the meteorite.

18.54 Determine the kinetic energy of the space probe of Prob. 18.34 in its motion about its mass center after its collision with the meteorite.

*18.5 MOTION OF A RIGID BODY IN THREE DIMENSIONS

As indicated in Sec. 18.2, the fundamental equations

$$\Sigma \mathbf{F} = m\bar{\mathbf{a}} \quad (18.1)$$

$$\Sigma \mathbf{M}_G = \dot{\mathbf{H}}_G \quad (18.2)$$

remain valid in the most general case of the motion of a rigid body. Before Eq. (18.2) could be applied to the three-dimensional motion of a rigid body, however, it was necessary to derive Eqs. (18.7), which relate the components of the angular momentum $\mathbf{H}_G$ and those of the angular velocity $\boldsymbol{\omega}$. It still remains for us to find an effective and convenient way for computing the components of the derivative $\dot{\mathbf{H}}_G$ of the angular momentum.

Since $\mathbf{H}_G$ represents the angular momentum of the body in its motion relative to centroidal axes $GX'Y'Z'$ of fixed orientation (Fig. 18.9), and since $\dot{\mathbf{H}}_G$ represents the rate of change of $\mathbf{H}_G$ with respect to the same axes, it would seem natural to use components of $\boldsymbol{\omega}$ and $\mathbf{H}_G$ along the axes X', Y', Z' in writing the relations (18.7). But since the body rotates, its moments and products of inertia would change continually, and it would be necessary to determine their values as functions of the time. It is therefore more convenient to use axes x, y, z attached to the body, ensuring that its moments and products of inertia will maintain the same values during the motion. This is permissible since, as indicated earlier, the transformation of $\boldsymbol{\omega}$ into $\mathbf{H}_G$ is independent of the system of coordinate axes selected. The angular velocity $\boldsymbol{\omega}$, however, should still be *defined* with respect to the frame $GX'Y'Z'$ of fixed orientation. The vector $\boldsymbol{\omega}$ may then be *resolved* into components along the rotating x, y, and z axes. Applying the relations (18.7), we obtain the *components* of the vector $\mathbf{H}_G$ along the rotating axes. The vector $\mathbf{H}_G$, however, represents the angular momentum about G of the body *in its motion relative to the frame* $GX'Y'Z'$.

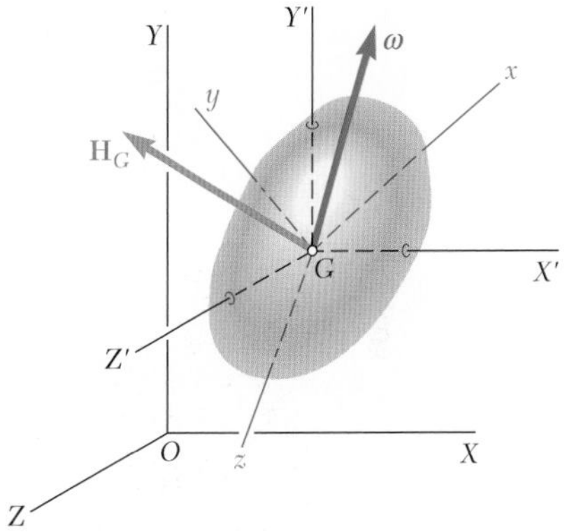

Fig. 18.9

Differentiating with respect to t the components of the angular momentum in (18.7), we define the rate of change of the vector $\mathbf{H}_G$ with respect to the rotating frame $Gxyz$:

$$(\dot{\mathbf{H}}_G)_{Gxyz} = \dot{H}_x\mathbf{i} + \dot{H}_y\mathbf{j} + \dot{H}_z\mathbf{k} \quad (18.21)$$

where $\mathbf{i}$, $\mathbf{j}$, $\mathbf{k}$ are the unit vectors along the rotating axes. Recalling from Sec. 15.10 that the rate of change $\dot{\mathbf{H}}_G$ of the vector $\mathbf{H}_G$ with respect to the frame $GX'Y'Z'$ is found by adding to $(\dot{\mathbf{H}}_G)_{Gxyz}$ the vector product $\boldsymbol{\Omega} \times \mathbf{H}_G$, where $\boldsymbol{\Omega}$ denotes the angular velocity of the rotating frame, we write

$$\dot{\mathbf{H}}_G = (\dot{\mathbf{H}}_G)_{Gxyz} + \boldsymbol{\Omega} \times \mathbf{H}_G \quad (18.22)$$

where $\mathbf{H}_G$ = angular momentum of body with respect to frame $GX'Y'Z'$ of fixed orientation

$(\dot{\mathbf{H}}_G)_{Gxyz}$ = rate of change of $\mathbf{H}_G$ with respect to rotating frame $Gxyz$, to be computed from the relations (18.7) and (18.21)

$\boldsymbol{\Omega}$ = angular velocity of rotating frame $Gxyz$

Substituting for $\dot{\mathbf{H}}_G$ from (18.22) into (18.2), we have

$$\Sigma \mathbf{M}_G = (\dot{\mathbf{H}}_G)_{Gxyz} + \mathbf{\Omega} \times \mathbf{H}_G \tag{18.23}$$

If the rotating frame is attached to the body, as has been assumed in this discussion, its angular velocity $\mathbf{\Omega}$ is identically equal to the angular velocity $\boldsymbol{\omega}$ of the body. There are many applications, however, where it is advantageous to use a frame of reference which is not actually attached to the body but rotates in an independent manner. For example, if the body considered is axisymmetrical, as in Sample Prob. 18.5 or Sec. 18.9, it is possible to select a frame of reference with respect to which the moments and products of inertia of the body remain constant, but which rotates less than the body itself.† As a result, it is possible to obtain simpler expressions for the angular velocity $\boldsymbol{\omega}$ and the angular momentum $\mathbf{H}_G$ of the body than could have been obtained if the frame of reference had actually been attached to the body. It is clear that in such cases the angular velocity $\mathbf{\Omega}$ of the rotating frame and the angular velocity $\boldsymbol{\omega}$ of the body are different.

*18.6 EULER'S EQUATIONS OF MOTION. EXTENSION OF D'ALEMBERT'S PRINCIPLE TO THE MOTION OF A RIGID BODY IN THREE DIMENSIONS

If the x, y, and z axes are chosen to coincide with the principal axes of inertia of the body, the simplified relations (18.10) can be used to determine the components of the angular momentum $\mathbf{H}_G$. Omitting the primes from the subscripts, we write

$$\mathbf{H}_G = \bar{I}_x\omega_x\mathbf{i} + \bar{I}_y\omega_y\mathbf{j} + \bar{I}_z\omega_z\mathbf{k} \tag{18.24}$$

where $\bar{I}_x$, $\bar{I}_y$, and $\bar{I}_z$ denote the principal centroidal moments of inertia of the body. Substituting for $\mathbf{H}_G$ from (18.24) into (18.23) and setting $\mathbf{\Omega} = \boldsymbol{\omega}$, we obtain the three scalar equations

$$\begin{aligned} \Sigma M_x &= \bar{I}_x\dot{\omega}_x - (\bar{I}_y - \bar{I}_z)\omega_y\omega_z \\ \Sigma M_y &= \bar{I}_y\dot{\omega}_y - (\bar{I}_z - \bar{I}_x)\omega_z\omega_x \\ \Sigma M_z &= \bar{I}_z\dot{\omega}_z - (\bar{I}_x - \bar{I}_y)\omega_x\omega_y \end{aligned} \tag{18.25}$$

These equations, called *Euler's equations of motion* after the Swiss mathematician Leonhard Euler (1707–1783), can be used to analyze the motion of a rigid body about its mass center. In the following sections, however, Eq. (18.23) will be used in preference to Eqs. (18.25), since the former is more general and the compact vectorial form in which it is expressed is easier to remember.

Writing Eq. (18.1) in scalar form, we obtain the three additional equations

$$\Sigma F_x = m\bar{a}_x \qquad \Sigma F_y = m\bar{a}_y \qquad \Sigma F_z = m\bar{a}_z \tag{18.26}$$

which, together with Euler's equations, form a system of six differential equations. Given appropriate initial conditions, these differential

†More specifically, the frame of reference will have no spin (see Sec. 18.9).

equations have a unique solution. Thus, the motion of a rigid body in three dimensions is completely defined by the resultant and the moment resultant of the external forces acting on it. This result will be recognized as a generalization of a similar result obtained in Sec. 16.4 in the case of the plane motion of a rigid slab. It follows that in three as well as two dimensions, two systems of forces which are equipollent are also equivalent; that is, they have the same effect on a given rigid body.

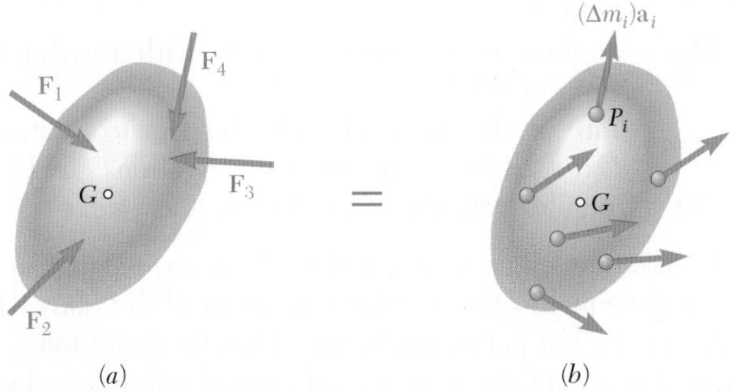

Fig. 18.10

Considering in particular the system of the external forces acting on a rigid body (Fig. 18.10*a*) and the system of the effective forces associated with the particles forming the rigid body (Fig. 18.10*b*), we can state that the two systems—which were shown in Sec. 14.2 to be equipollent—are also equivalent. This is the extension of d'Alembert's principle to the three-dimensional motion of a rigid body. Replacing the effective forces in Fig. 18.10*b* by an equivalent force-couple system, we verify that the system of the external forces acting on a rigid body in three-dimensional motion is equivalent to the system consisting of the vector $m\bar{\mathbf{a}}$ attached at the mass center G of the body and the couple of moment $\dot{\mathbf{H}}_G$ (Fig. 18.11), where $\dot{\mathbf{H}}_G$ is obtained from the relations (18.7) and (18.22). Note that the equivalence of the systems of vectors shown in Fig. 18.10 and in Fig. 18.11 has been indicated by *red* equals signs. Problems involving the three-dimensional motion of a rigid body can be solved by considering the free-body-diagram equation represented in Fig. 18.11 and writing appropriate scalar equations relating the components or moments of the external and effective forces (see Sample Prob. 18.3).

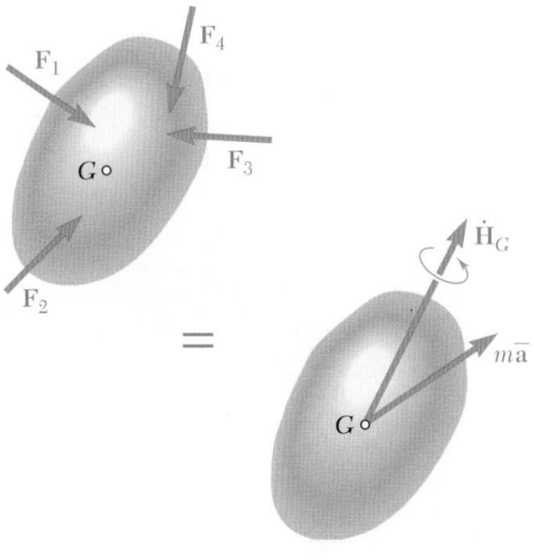

Fig. 18.11

*18.7 MOTION OF A RIGID BODY ABOUT A FIXED POINT

When a rigid body is constrained to rotate about a fixed point O, it is desirable to write an equation involving the moments about O of the external and effective forces, since this equation will not contain the unknown reaction at O. While such an equation can be obtained from Fig. 18.11, it may be more convenient to write it by considering the rate of change of the angular momentum $\mathbf{H}_O$ of the body about the fixed point O (Fig. 18.12). Recalling Eq. (14.11), we write

$$\Sigma \mathbf{M}_O = \dot{\mathbf{H}}_O \tag{18.27}$$

where $\dot{\mathbf{H}}_O$ denotes the rate of change of the vector $\mathbf{H}_O$ with respect to the fixed frame $OXYZ$. A derivation similar to that used in Sec. 18.5

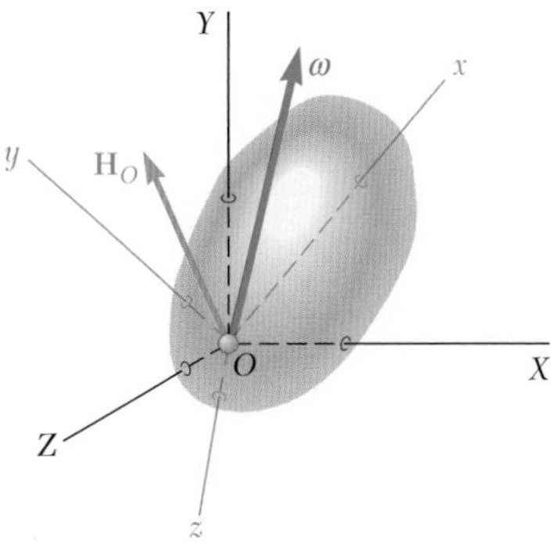

Fig. 18.12

Photo 18.3 The revolving radio telescope is an example of a structure constrained to rotate about a fixed point.

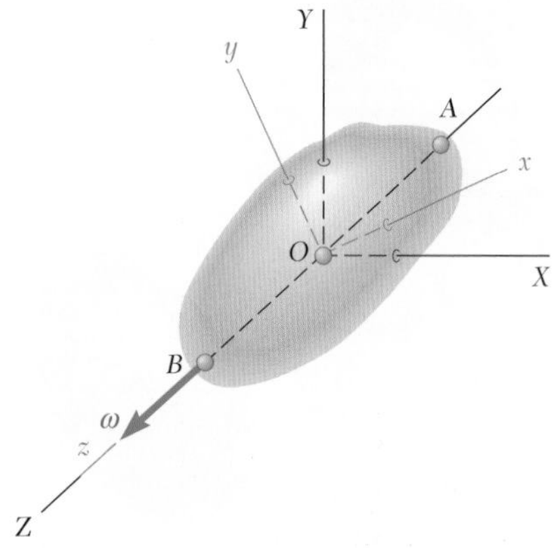

Fig. 18.13

enables us to relate $\dot{\mathbf{H}}_O$ to the rate of change $(\dot{\mathbf{H}}_O)_{Oxyz}$ of $\mathbf{H}_O$ with respect to the rotating frame $Oxyz$. Substitution into (18.27) leads to the equation

$$\Sigma\mathbf{M}_O = (\dot{\mathbf{H}}_O)_{Oxyz} + \mathbf{\Omega} \times \mathbf{H}_O \tag{18.28}$$

where $\Sigma\mathbf{M}_O$ = sum of moments about O of forces applied to rigid body

$\mathbf{H}_O$ = angular momentum of body with respect to fixed frame $OXYZ$

$(\dot{\mathbf{H}}_O)_{Oxyz}$ = rate of change of $\mathbf{H}_O$ with respect to rotating frame $Oxyz$, to be computed from relations (18.13)

$\mathbf{\Omega}$ = angular velocity of rotating frame $Oxyz$

If the rotating frame is attached to the body, its angular velocity $\mathbf{\Omega}$ is identically equal to the angular velocity $\boldsymbol{\omega}$ of the body. However, as indicated in the last paragraph of Sec. 18.5, there are many applications where it is advantageous to use a frame of reference which is not actually attached to the body but rotates in an independent manner.

*18.8 ROTATION OF A RIGID BODY ABOUT A FIXED AXIS

Equation (18.28), which was derived in the preceding section, will be used to analyze the motion of a rigid body constrained to rotate about a fixed axis AB (Fig. 18.13). First, we note that the angular velocity of the body with respect to the fixed frame $OXYZ$ is represented by the vector $\boldsymbol{\omega}$ directed along the axis of rotation. Attaching the moving frame of reference $Oxyz$ to the body, with the z axis along AB, we have $\boldsymbol{\omega} = \omega\mathbf{k}$. Substituting $\omega_x = 0$, $\omega_y = 0$, $\omega_z = \omega$ into the relations (18.13), we obtain the components along the rotating axes of the angular momentum $\mathbf{H}_O$ of the body about O:

$$H_x = -I_{xz}\omega \qquad H_y = -I_{yz}\omega \qquad H_z = I_z\omega$$

Since the frame $Oxyz$ is attached to the body, we have $\mathbf{\Omega} = \boldsymbol{\omega}$ and Eq. (18.28) yields

$$\begin{aligned}\Sigma\mathbf{M}_O &= (\dot{\mathbf{H}}_O)_{Oxyz} + \boldsymbol{\omega} \times \mathbf{H}_O\\ &= (-I_{xz}\mathbf{i} - I_{yz}\mathbf{j} + I_z\mathbf{k})\dot{\omega} + \omega\mathbf{k} \times (-I_{xz}\mathbf{i} - I_{yz}\mathbf{j} + I_z\mathbf{k})\omega\\ &= (-I_{xz}\mathbf{i} - I_{yz}\mathbf{j} + I_z\mathbf{k})\alpha + (-I_{xz}\mathbf{j} + I_{yz}\mathbf{i})\omega^2\end{aligned}$$

The result obtained can be expressed by the three scalar equations

$$\begin{aligned}\Sigma M_x &= -I_{xz}\alpha + I_{yz}\omega^2\\ \Sigma M_y &= -I_{yz}\alpha - I_{xz}\omega^2\\ \Sigma M_z &= I_z\alpha\end{aligned} \tag{18.29}$$

When the forces applied to the body are known, the angular acceleration α can be obtained from the last of Eqs. (18.29). The angular velocity ω is then determined by integration and the values obtained for α and ω substituted into the first two equations (18.29). These equations plus the three equations (18.26) which define the motion of the mass center of the body can then be used to determine the reactions at the bearings A and B.

It is possible to select axes other than the ones shown in Fig. 18.13 to analyze the rotation of a rigid body about a fixed axis. In many cases, the principal axes of inertia of the body will be found more advantageous. It is therefore wise to revert to Eq. (18.28) and to select the system of axes which best fits the problem under consideration.

If the rotating body is symmetrical with respect to the xy plane, the products of inertia I_{xz} and I_{yz} are equal to zero and Eqs. (18.29) reduce to

$$\Sigma M_x = 0 \qquad \Sigma M_y = 0 \qquad \Sigma M_z = I_z\alpha \tag{18.30}$$

which is in accord with the results obtained in Chap. 16. If, on the other hand, the products of inertia I_{xz} and I_{yz} are different from zero, the sum of the moments of the external forces about the x and y axes will also be different from zero, even when the body rotates at a constant rate ω. Indeed, in the latter case, Eqs. (18.29) yield

$$\Sigma M_x = I_{yz}\omega^2 \qquad \Sigma M_y = -I_{xz}\omega^2 \qquad \Sigma M_z = 0 \tag{18.31}$$

This last observation leads us to discuss the *balancing of rotating shafts*. Consider, for instance, the crankshaft shown in Fig. 18.14*a*, which is symmetrical about its mass center G. We first observe that when the crankshaft is at rest, it exerts no lateral thrust on its supports, since its center of gravity G is located directly above A. The shaft is said to be *statically balanced*. The reaction at A, often referred to as a *static reaction*, is vertical and its magnitude is equal to the weight W of the shaft. Let us now assume that the shaft rotates with a constant angular velocity $\boldsymbol{\omega}$. Attaching our frame of reference to the shaft, with its origin at G, the z axis along AB, and the y axis in the plane of symmetry of the shaft (Fig. 18.14*b*), we note that I_{xz} is zero and that I_{yz} is positive. According to Eqs. (18.31), the external forces include a couple of moment $I_{yz}\omega^2\mathbf{i}$. Since this couple is formed by the reaction at B and the horizontal component of the reaction at A, we have

$$\mathbf{A}_y = \frac{I_{yz}\omega^2}{l}\mathbf{j} \qquad \mathbf{B} = -\frac{I_{yz}\omega^2}{l}\mathbf{j} \tag{18.32}$$

Since the bearing reactions are proportional to ω^2, the shaft will have a tendency to tear away from its bearings when rotating at high speeds. Moreover, since the bearing reactions $\mathbf{A}_y$ and $\mathbf{B}$, called *dynamic reactions*, are contained in the yz plane, they rotate with the shaft and cause the structure supporting it to vibrate. These undesirable effects will be avoided if, by rearranging the distribution of mass around the shaft or by adding corrective masses, we let I_{yz} become equal to zero. The dynamic reactions $\mathbf{A}_y$ and $\mathbf{B}$ will vanish and the reactions at the bearings will reduce to the static reaction $\mathbf{A}_z$, the direction of which is fixed. The shaft will then be *dynamically as well as statically balanced*.

Photo 18.4 The forces exerted by a rotating automobile crankshaft on its bearings are the static and dynamic reactions. The crankshaft can be designed to be dynamically as well as statically balanced.

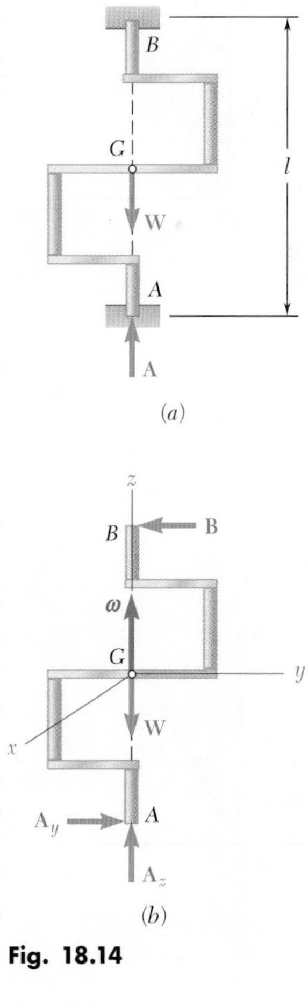

Fig. 18.14

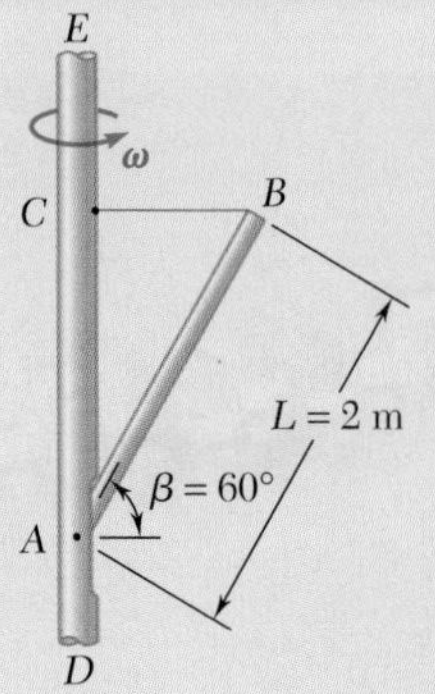

SAMPLE PROBLEM 18.3

A slender rod *AB* of length $L = 2$ m and mass $m = 20$ kg is pinned at *A* to a vertical axle *DE* which rotates with a constant angular velocity $\boldsymbol{\omega}$ of 15 rad/s. The rod is maintained in position by means of a horizontal wire *BC* attached to the axle and to the end *B* of the rod. Determine the tension in the wire and the reaction at *A*.

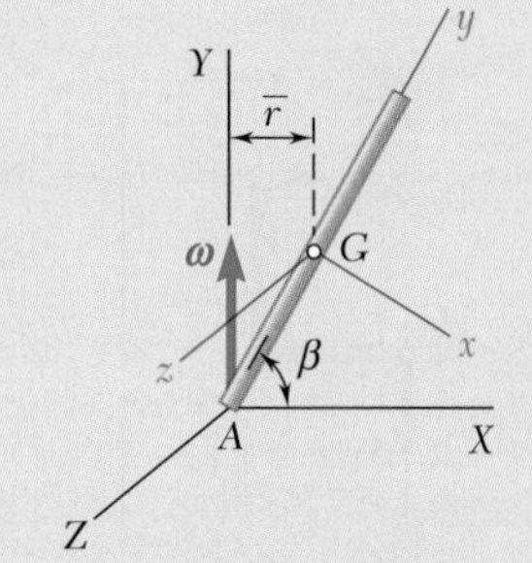

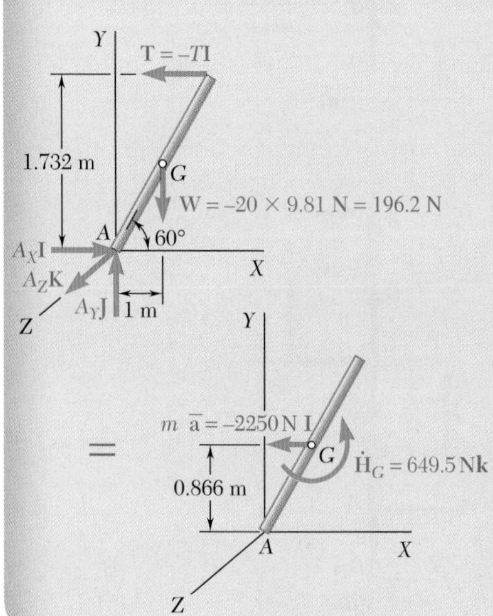

SOLUTION

The effective forces reduce to the vector $m\bar{\mathbf{a}}$ attached at *G* and the couple $\dot{\mathbf{H}}_G$. Since *G* describes a horizontal circle of radius $\bar{r} = \frac{1}{2}L\cos\beta$ at the constant rate ω, we have

$$\bar{\mathbf{a}} = \mathbf{a}_n = -\bar{r}\omega^2\mathbf{I} = -(\tfrac{1}{2}L\cos\beta)\omega^2\mathbf{I} = -(112.5\text{ m/s}^2)\mathbf{I}$$

$$m\bar{\mathbf{a}} = 20(-112.5\,\mathbf{I}) = -(2250\text{ N})\mathbf{I}$$

Determination of $\dot{\mathbf{H}}_G$. We first compute the angular momentum $\mathbf{H}_G$. Using the principal centroidal axes of inertia *x*, *y*, *z*, we write

$$\bar{I}_x = \tfrac{1}{12}mL^2 \qquad \bar{I}_y = 0 \qquad \bar{I}_z = \tfrac{1}{12}mL^2$$

$$\omega_x = -\omega\cos\beta \qquad \omega_y = \omega\sin\beta \qquad \omega_z = 0$$

$$\mathbf{H}_G = \bar{I}_x\omega_x\mathbf{i} + \bar{I}_y\omega_y\mathbf{j} + \bar{I}_z\omega_z\mathbf{k}$$

$$\mathbf{H}_G = -\tfrac{1}{12}mL^2\omega\cos\beta\,\mathbf{i}$$

The rate of change $\dot{\mathbf{H}}_G$ of $\mathbf{H}_G$ with respect to axes of fixed orientation is obtained from Eq. (18.22). Observing that the rate of change $(\dot{\mathbf{H}}_G)_{Gxyz}$ of $\mathbf{H}_G$ with respect to the rotating frame *Gxyz* is zero, and that the angular velocity $\boldsymbol{\Omega}$ of that frame is equal to the angular velocity $\boldsymbol{\omega}$ of the rod, we have

$$\dot{\mathbf{H}}_G = (\dot{\mathbf{H}}_G)_{Gxyz} + \boldsymbol{\omega}\times\mathbf{H}_G$$

$$\dot{\mathbf{H}}_G = 0 + (-\omega\cos\beta\,\mathbf{i} + \omega\sin\beta\,\mathbf{j})\times(-\tfrac{1}{12}mL^2\omega\cos\beta\,\mathbf{i})$$

$$\dot{\mathbf{H}}_G = \tfrac{1}{12}mL^2\omega^2\sin\beta\cos\beta\,\mathbf{k} = (649.5\text{ N}\cdot\text{m})\mathbf{k}$$

Equations of Motion. Expressing that the system of the external forces is equivalent to the system of the effective forces, we write

$\Sigma\mathbf{M}_A = \Sigma(\mathbf{M}_A)_{\text{eff}}$:

$$1.732\mathbf{J}\times(-T\mathbf{I}) + 1\mathbf{I}\times(-196.2\mathbf{J}) = 0.866\mathbf{J}\times(-2250\mathbf{I}) + 649.5\mathbf{K}$$

$$(1.732T - 196.2)\mathbf{K} = (1948.5 + 649)\mathbf{K} \qquad T = 1613\text{ N} \blacktriangleleft$$

$\Sigma\mathbf{F} = \Sigma\mathbf{F}_{\text{eff}}$: $\quad A_X\mathbf{I} + A_Y\mathbf{J} + A_Z\mathbf{K} - 1613\mathbf{I} - 196.2\mathbf{J} = -2250\mathbf{I}$

$$\mathbf{A} = -(697\text{ N})\mathbf{I} + (196.2\text{ N})\mathbf{J} \blacktriangleleft$$

Remark. The value of *T* could have been obtained from $\mathbf{H}_A$ and Eq. (18.28). However, the method used here also yields the reaction at *A*. Moreover, it draws attention to the effect of the asymmetry of the rod on the solution of the problem by clearly showing that both the vector $m\bar{\mathbf{a}}$ and the couple $\dot{\mathbf{H}}_G$ must be used to represent the effective forces.

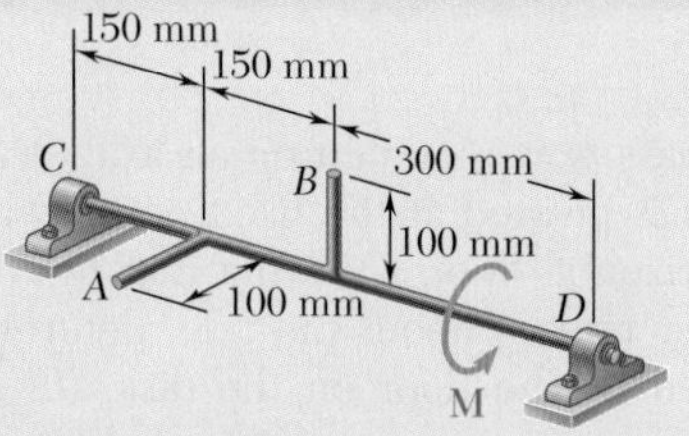

SAMPLE PROBLEM 18.4

Two 100-mm rods A and B, each of mass 300 g, are welded to shaft CD which is supported by bearings at C and D. If a couple $\mathbf{M}$ of magnitude equal to 6 N · m is applied to the shaft, determine the components of the dynamic reactions at C and D at the instant when the shaft has reached an angular velocity at 1200 rpm. Neglect the moment of inertia of the shaft itself.

SOLUTION

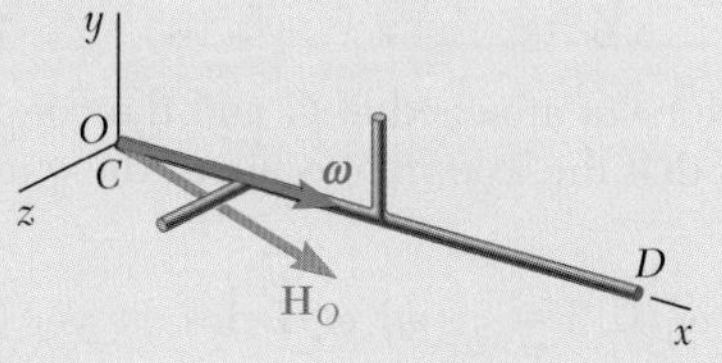

Angular Momentum About *O*. We attach to the body the frame of reference $Oxyz$ and note that the axes chosen are not principal axes of inertia for the body. Since the body rotates about the x axis, we have $\omega_x = \omega$ and $\omega_y = \omega_z = 0$. Substituting into Eqs. (18.13),

$$H_x = I_x\omega \qquad H_y = -I_{xy}\omega \qquad H_z = -I_{xz}\omega$$
$$\mathbf{H}_O = (I_x\mathbf{i} - I_{xy}\mathbf{j} - I_{xz}\mathbf{k})\omega$$

Moments of the External Forces About *O*. Since the frame of reference rotates with the angular velocity $\boldsymbol{\omega}$, Eq. (18.28) yields

$$\begin{aligned}\Sigma\mathbf{M}_O &= (\dot{\mathbf{H}}_O)_{Oxyz} + \boldsymbol{\omega} \times \mathbf{H}_O \\ &= (I_x\mathbf{i} - I_{xy}\mathbf{j} - I_{xz}\mathbf{k})\alpha + \omega\mathbf{i} \times (I_x\mathbf{i} - I_{xy}\mathbf{j} - I_{xz}\mathbf{k})\omega \\ &= I_x\alpha\mathbf{i} - (I_{xy}\alpha - I_{xz}\omega^2)\mathbf{j} - (I_{xz}\alpha + I_{xy}\omega^2)\mathbf{k} \qquad (1)\end{aligned}$$

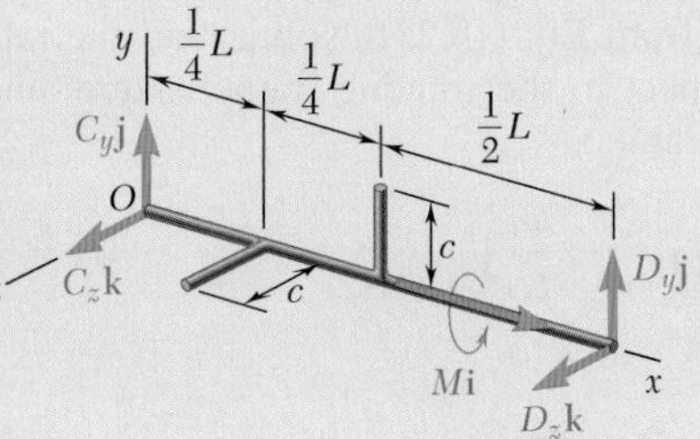

Dynamic Reaction at *D*. The external forces consist of the weights of the shaft and rods, the couple $\mathbf{M}$, the static reactions at C and D, and the dynamic reactions at C and D. Since the weights and static reactions are balanced, the external forces reduce to the couple $\mathbf{M}$ and the dynamic reactions $\mathbf{C}$ and $\mathbf{D}$ as shown in the figure. Taking moments about O, we have

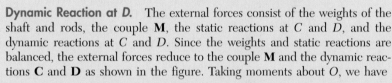

$$\Sigma\mathbf{M}_O = L\mathbf{i} \times (D_y\mathbf{j} + D_z\mathbf{k}) + M\mathbf{i} = M\mathbf{i} - D_zL\mathbf{j} + D_yL\mathbf{k} \qquad (2)$$

Equating the coefficients of the unit vector $\mathbf{i}$ in (1) and (2),

$$M = I_x\alpha \qquad M = 2(\tfrac{1}{3}mc^2)\alpha \qquad \alpha = 3M/2mc^2$$

Equating the coefficients of $\mathbf{k}$ and $\mathbf{j}$ in (1) and (2):

$$D_y = -(I_{xz}\alpha + I_{xy}\omega^2)/L \qquad D_z = (I_{xy}\alpha - I_{xz}\omega^2)/L \qquad (3)$$

Using the parallel-axis theorem, and noting that the product of inertia of each rod is zero with respect to centroidal axes, we have

$$I_{xy} = \Sigma m\bar{x}\bar{y} = m(\tfrac{1}{2}L)(\tfrac{1}{2}c) = \tfrac{1}{4}mLc$$
$$I_{xz} = \Sigma m\bar{x}\bar{z} = m(\tfrac{1}{4}L)(\tfrac{1}{2}c) = \tfrac{1}{8}mLc$$

Substituting into (3) the values found for I_{xy}, I_{xz}, and α:

$$D_y = -\tfrac{3}{16}(M/c) - \tfrac{1}{4}mc\omega^2 \qquad D_z = \tfrac{3}{8}(M/c) - \tfrac{1}{8}mc\omega^2$$

Substituting $\omega = 1200$ rpm $= 125.7$ rad/s, $c = 0.100$ m, $M = 6$ N · m, and $m = 0.300$ kg, we have

$$D_y = -129.8 \text{ N} \qquad D_z = -36.8 \text{ N} \quad ◀$$

Dynamic Reaction at *C*. Using a frame of reference attached at D, we obtain equations similar to Eqs. (3), which yield

$$C_y = -152.2 \text{ N} \qquad C_z = -155.2 \text{ N} \quad ◀$$

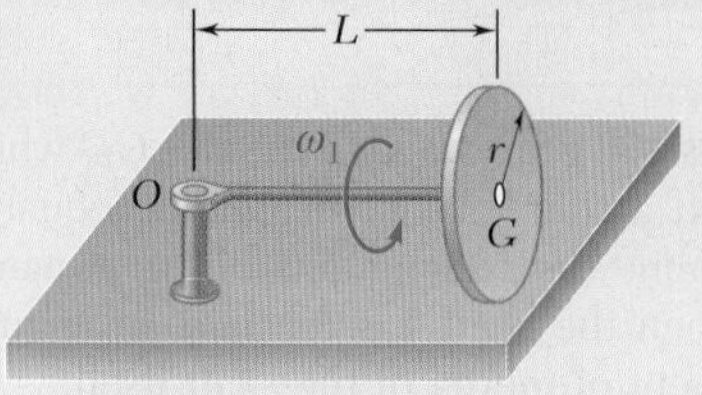

SAMPLE PROBLEM 18.5

A homogeneous disk of radius r and mass m is mounted on an axle OG of length L and negligible mass. The axle is pivoted at the fixed point O and the disk is constrained to roll on a horizontal floor. Knowing that the disk rotates counterclockwise at the constant rate ω_1 about the axle, determine (*a*) the force (assumed vertical) exerted by the floor on the disk, (*b*) the reaction at the pivot O.

SOLUTION

The effective forces reduce to the vector $m\bar{\mathbf{a}}$ attached at G and the couple $\dot{\mathbf{H}}_G$. Recalling from Sample Prob. 18.2 that the axle rotates about the y axis at the rate $\omega_2 = r\omega_1/L$, we write

$$m\bar{\mathbf{a}} = -mL\omega_2^2\mathbf{i} = -mL(r\omega_1/L)^2\mathbf{i} = -(mr^2\omega_1^2/L)\mathbf{i} \tag{1}$$

Determination of $\dot{\mathbf{H}}_G$. We recall from Sample Prob. 18.2 that the angular momentum of the disk about G is

$$\mathbf{H}_G = \tfrac{1}{2}mr^2\omega_1\left(\mathbf{i} - \frac{r}{2L}\mathbf{j}\right)$$

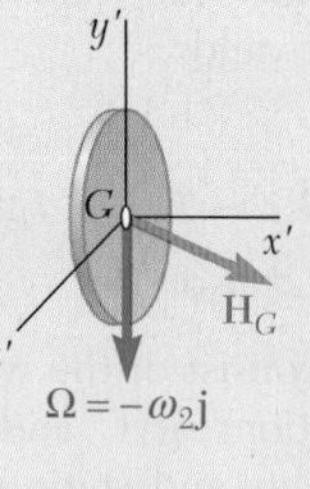

where $\mathbf{H}_G$ is resolved into components along the rotating axes x', y', z', with x' along OG and y' vertical. The rate of change $\dot{\mathbf{H}}_G$ of $\mathbf{H}_G$ with respect to axes of fixed orientation is obtained from Eq. (18.22). Noting that the rate of change $(\dot{\mathbf{H}}_G)_{Gx'y'z'}$ of $\mathbf{H}_G$ with respect to the rotating frame is zero, and that the angular velocity $\boldsymbol{\Omega}$ of that frame is

$$\boldsymbol{\Omega} = -\omega_2\mathbf{j} = -\frac{r\omega_1}{L}\mathbf{j}$$

we have

$$\begin{aligned}\dot{\mathbf{H}}_G &= (\dot{\mathbf{H}}_G)_{Gx'y'z'} + \boldsymbol{\Omega} \times \mathbf{H}_G \\ &= 0 - \frac{r\omega_1}{L}\mathbf{j} \times \tfrac{1}{2}mr^2\omega_1\left(\mathbf{i} - \frac{r}{2L}\mathbf{j}\right) \\ &= \tfrac{1}{2}mr^2(r/L)\omega_1^2\mathbf{k}\end{aligned} \tag{2}$$

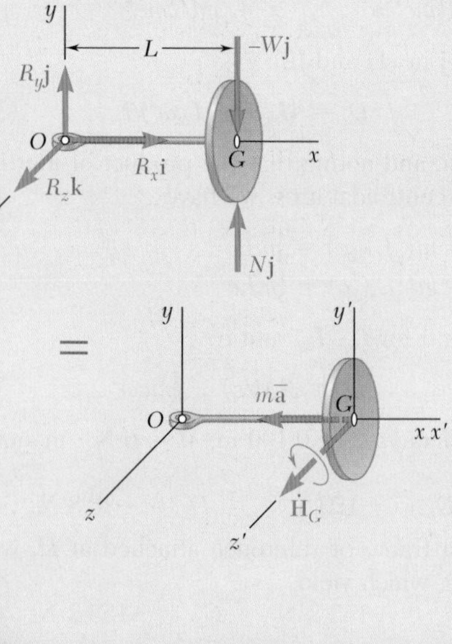

Equations of Motion. Expressing that the system of the external forces is equivalent to the system of the effective forces, we write

$\Sigma\mathbf{M}_O = \Sigma(\mathbf{M}_O)_{\text{eff}}$: $\quad L\mathbf{i} \times (N\mathbf{j} - W\mathbf{j}) = \dot{\mathbf{H}}_G$

$$(N - W)L\mathbf{k} = \tfrac{1}{2}mr^2(r/L)\omega_1^2\mathbf{k}$$

$$N = W + \tfrac{1}{2}mr(r/L)^2\omega_1^2 \qquad \mathbf{N} = [W + \tfrac{1}{2}mr(r/L)^2\omega_1^2]\mathbf{j} \tag{3}$$ ◀

$\Sigma\mathbf{F} = \Sigma\mathbf{F}_{\text{eff}}$: $\quad \mathbf{R} + N\mathbf{j} - W\mathbf{j} = m\bar{\mathbf{a}}$

Substituting for N from (3), for $m\bar{\mathbf{a}}$ from (1), and solving for $\mathbf{R}$, we have

$$\mathbf{R} = -(mr^2\omega_1^2/L)\mathbf{i} - \tfrac{1}{2}mr(r/L)^2\omega_1^2\mathbf{j}$$

$$\mathbf{R} = -\frac{mr^2\omega_1^2}{L}\left(\mathbf{i} + \frac{r}{2L}\mathbf{j}\right)$$ ◀

SOLVING PROBLEMS ON YOUR OWN

In this lesson you will be asked to solve problems involving the *three-dimensional motion of rigid bodies*. The method you will use is basically the same that you used in Chap. 16 in your study of the plane motion of rigid bodies. You will draw a free-body-diagram equation showing that the system of the external forces is equivalent to the system of the effective forces, and you will equate sums of components and sums of moments on both sides of this equation. Now, however, the system of the effective forces will be represented by the vector $m\bar{\mathbf{a}}$ and a couple vector $\dot{\mathbf{H}}_G$, the determination of which will be explained in paragraphs 1 and 2 below.

To solve a problem involving the three-dimensional motion of a rigid body, you should take the following steps:

1. Determine the angular momentum $\mathbf{H}_G$ of the body about its mass center G from its angular velocity $\boldsymbol{\omega}$ with respect to a frame of reference $GX'Y'Z'$ of fixed orientation. This is an operation you learned to perform in the preceding lesson. However, since the configuration of the body will be changing with time, it will now be necessary for you to use an auxiliary system of axes $Gx'y'z'$ (Fig. 18.9) to compute the components of $\boldsymbol{\omega}$ and the moments and products of inertia of the body. These axes may be rigidly attached to the body, in which case their angular velocity is equal to $\boldsymbol{\omega}$ [Sample Probs. 18.3 and 18.4], or they may have an angular velocity $\boldsymbol{\Omega}$ of their own [Sample Prob. 18.5].

Recall the following from the preceding lesson:

a. If the principal axes of inertia of the body at G are known, use these axes as coordinate axes x', y', and z', since the corresponding products of inertia of the body will be equal to zero. (Note that if the body is axisymmetric, these axes do not need to be rigidly attached to the body.) Resolve $\boldsymbol{\omega}$ into components $\omega_{x'}$, $\omega_{y'}$, and $\omega_{z'}$ along these axes and compute the principal moments of inertia $\bar{I}_{x'}$, $\bar{I}_{y'}$, and $\bar{I}_{z'}$. The corresponding components of the angular momentum $\mathbf{H}_G$ are

$$H_{x'} = \bar{I}_{x'}\omega_{x'} \qquad H_{y'} = \bar{I}_{y'}\omega_{y'} \qquad H_{z'} = \bar{I}_{z'}\omega_{z'} \tag{18.10}$$

b. If the principal axes of inertia of the body at G are not known, you must use Eqs. (18.7) to determine the components of the angular momentum $\mathbf{H}_G$. These equations require your prior computation of the *products of inertia* of the body, as well as of its moments of inertia, with respect to the selected axes.

(continued)

2. Compute the rate of change $\dot{\mathbf{H}}_G$ of the angular momentum $\mathbf{H}_G$ with respect to the frame $GX'Y'Z'$. Note that this frame has a *fixed orientation*, while the frame $Gx'y'z'$ you used when you calculated the components of the vector $\boldsymbol{\omega}$ was a *rotating frame*. We refer you to our discussion in Sec. 15.10 of the rate of change of a vector with respect to a rotating frame. Recalling Eq. (15.31), you will express the rate of change $\dot{\mathbf{H}}_G$ as follows:

$$\dot{\mathbf{H}}_G = (\dot{\mathbf{H}}_G)_{Gx'y'z'} + \boldsymbol{\Omega} \times \mathbf{H}_G \tag{18.22}$$

The first term in the right-hand member of Eq. (18.22) represents the rate of change of $\mathbf{H}_G$ with respect to the rotating frame $Gx'y'z'$. This term will drop out if $\boldsymbol{\omega}$—and, thus, $\mathbf{H}_G$—remain constant in both magnitude and direction when viewed from that frame. On the other hand, if any of the time derivatives $\dot{\omega}_{x'}$, $\dot{\omega}_{y'}$, and $\dot{\omega}_{z'}$ is different from zero, $(\dot{\mathbf{H}}_G)_{Gx'y'z'}$ will also be different from zero, and its components should be determined by differentiating Eqs. (18.10) with respect to t. Finally, we remind you that if the rotating frame is rigidly attached to the body, its angular velocity will be the same as that of the body, and $\boldsymbol{\Omega}$ can be replaced by $\boldsymbol{\omega}$.

3. Draw the free-body-diagram equation for the rigid body, showing that the system of the external forces exerted on the body is equivalent to the vector $m\bar{\mathbf{a}}$ applied at G and the couple vector $\dot{\mathbf{H}}_G$ (Fig. 18.11). By equating components in any direction and moments about any point, you can write as many as six independent scalar equations of motion [Sample Probs. 18.3 and 18.5].

4. When solving problems involving the motion of a rigid body about a fixed point O, you may find it convenient to use the following equation, derived in Sec. 18.7, which eliminates the components of the reaction at the support O,

$$\Sigma\mathbf{M}_O = (\dot{\mathbf{H}}_O)_{Oxyz} + \boldsymbol{\Omega} \times \mathbf{H}_O \tag{18.28}$$

where the first term in the right-hand member represents the rate of change of $\mathbf{H}_O$ with respect to the rotating frame $Oxyz$, and where $\boldsymbol{\Omega}$ is the angular velocity of that frame.

5. When determining the reactions at the bearings of a rotating shaft, use Eq. (18.28) and take the following steps:

a. Place the fixed point O at one of the two bearings supporting the shaft and attach the rotating frame $Oxyz$ to the shaft, with one of the axes directed along it. Assuming, for instance, that the x axis has been aligned with the shaft, you will have $\boldsymbol{\Omega} = \boldsymbol{\omega} = \omega\mathbf{i}$ [Sample Prob. 18.4].

b. Since the selected axes, usually, will not be the principal axes of inertia at *O*, you must compute the *products of inertia* of the shaft, as well as its moments of inertia, with respect to these axes, and use Eqs. (18.13) to determine $\mathbf{H}_O$. Assuming again that the x axis has been aligned with the shaft, Eqs. (18.13) reduce to

$$H_x = I_x\omega \qquad H_y = -I_{yx}\omega \qquad H_z = -I_{zx}\omega \tag{18.13'}$$

which shows that $\mathbf{H}_O$ *will not be directed along the shaft.*

c. To obtain $\dot{\mathbf{H}}_O$, substitute the expressions obtained into Eq. (18.28), and let $\boldsymbol{\Omega} = \boldsymbol{\omega} = \omega\mathbf{i}$. If the angular velocity of the shaft is constant, the first term in the right-hand member of the equation will drop out. However, if the shaft has an angular acceleration $\boldsymbol{\alpha} = \alpha\mathbf{i}$, the first term will not be zero and must be determined by differentiating with respect to t the expressions in (18.13′). The result will be equations similar to Eqs. (18.13′), with ω replaced by α.

d. Since point *O* coincides with one of the bearings, the three scalar equations corresponding to Eq. (18.28) can be solved for the components of the dynamic reaction at the other bearing. If the mass center G of the shaft is located on the line joining the two bearings, the effective force $m\bar{\mathbf{a}}$ will be zero. Drawing the free-body-diagram equation of the shaft, you will then observe that the components of the dynamic reaction at the first bearing must be equal and opposite to those you have just determined. If G is not located on the line joining the two bearings, you can determine the reaction at the first bearing by placing the fixed point O at the second bearing and repeating the earlier procedure [Sample Prob. 18.4]; or you can obtain additional equations of motion from the free-body-diagram equation of the shaft, making sure to first determine and include the effective force $m\bar{\mathbf{a}}$ applied at G.

e. Most problems call for the determination of the "dynamic reactions" at the bearings, that is, for the *additional forces* exerted by the bearings on the shaft when the shaft is rotating. When determining dynamic reactions, ignore the effect of static loads, such as the weight of the shaft.

PROBLEMS

18.55 Determine the rate of change $\dot{\mathbf{H}}_G$ of the angular momentum $\mathbf{H}_G$ of the disk of Prob. 18.1.

18.56 Determine the rate of change $\dot{\mathbf{H}}_A$ of the angular momentum $\mathbf{H}_A$ of the plate of Prob. 18.2, knowing that its angular velocity $\boldsymbol{\omega}$ remains constant.

18.57 Determine the rate of change $\dot{\mathbf{H}}_D$ of the angular momentum $\mathbf{H}_D$ of the assembly of Prob. 18.3.

18.58 Determine the rate of change $\dot{\mathbf{H}}_A$ of the angular momentum $\mathbf{H}_A$ of the disk of Prob. 18.4.

18.59 Determine the rate of change $\dot{\mathbf{H}}_C$ of the angular momentum $\mathbf{H}_C$ of the disk of Prob. 18.5.

18.60 Determine the rate of change $\dot{\mathbf{H}}_G$ of the angular momentum $\mathbf{H}_G$ of the disk of Prob. 18.8.

18.61 Determine the rate of change $\dot{\mathbf{H}}_D$ of the angular momentum $\mathbf{H}_D$ of the assembly of Prob. 18.3, assuming that at the instant considered the assembly has an angular velocity $\boldsymbol{\omega} = (12 \text{ rad/s})\mathbf{i}$ and an angular acceleration $\boldsymbol{\alpha} = -(96 \text{ rad/s}^2)\mathbf{i}$.

18.62 Determine the rate of change $\dot{\mathbf{H}}_D$ of the angular momentum $\mathbf{H}_D$ of the assembly of Prob. 18.3, assuming that at the instant considered the assembly has an angular velocity $\boldsymbol{\omega} = (12 \text{ rad/s})\mathbf{i}$ and an angular acceleration $\boldsymbol{\alpha} = (96 \text{ rad/s}^2)\mathbf{i}$.

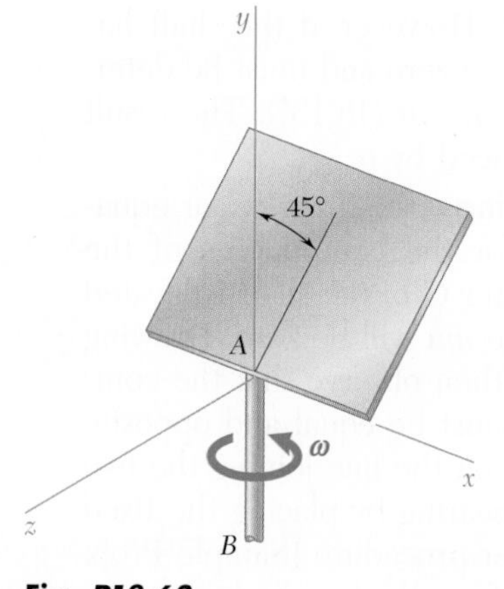

Fig. *P18.63*

18.63 A thin, homogeneous square of mass m and side a is welded to a vertical shaft AB with which it forms an angle of 45°. Knowing that the shaft rotates with an angular velocity $\boldsymbol{\omega} = \omega\mathbf{j}$ and an angular acceleration $\boldsymbol{\alpha} = \alpha\mathbf{j}$, determine the rate of change $\dot{\mathbf{H}}_A$ of the angular momentum $\mathbf{H}_A$ of the plate assembly.

18.64 Determine the rate of change $\dot{\mathbf{H}}_G$ of the angular momentum $\mathbf{H}_G$ of the disk of Prob. 18.8, assuming that at the instant considered the assembly has an angular velocity $\boldsymbol{\omega} = \omega\mathbf{j}$ and an angular acceleration $\boldsymbol{\alpha} = \alpha\mathbf{j}$.

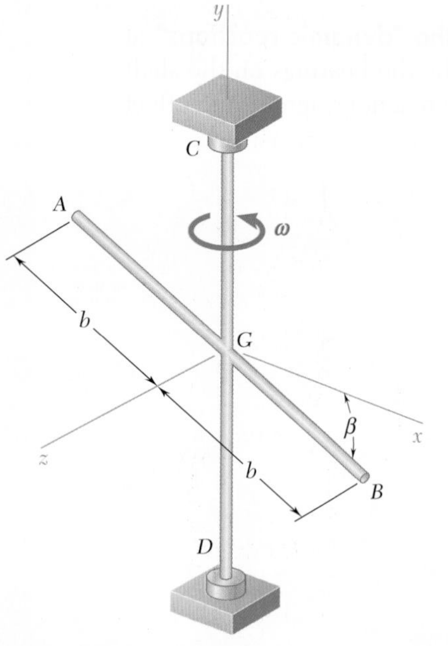

Fig. P18.65

18.65 A slender, uniform rod AB of mass m and a vertical shaft CD, each of length $2b$, are welded together at their midpoints G. Knowing that the shaft rotates at the constant rate ω, determine the dynamic reactions at C and D.

18.66 A thin homogeneous triangular plate of mass 2.5 kg is welded to a light vertical axle supported by bearings at A and B. Knowing that the plate rotates at the constant rate $\omega = 8$ rad/s, determine the dynamic reactions at A and B.

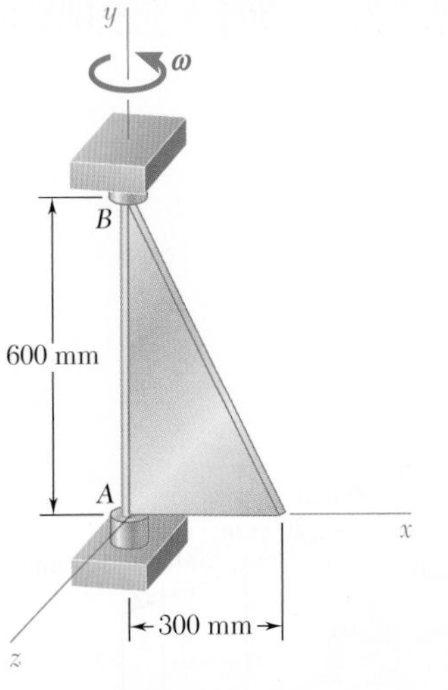

Fig. P18.66

18.67 The assembly shown consists of pieces of sheet aluminum of uniform thickness and of total mass 1.25 kg welded to a light axle supported by bearings at A and B. Knowing that the assembly rotates at the constant rate $\omega = 240$ rpm, determine the dynamic reactions at A and B.

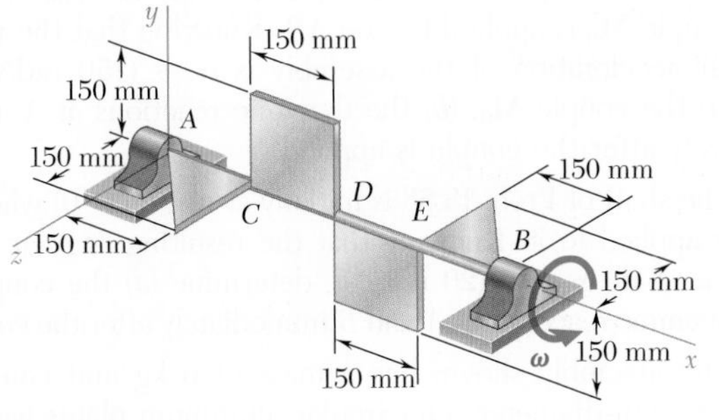

Fig. P18.67

18.68 The 8-kg shaft shown has a uniform cross section. Knowing that the shaft rotates at the constant rate $\omega = 12$ rad/s, determine the dynamic reactions at A and B.

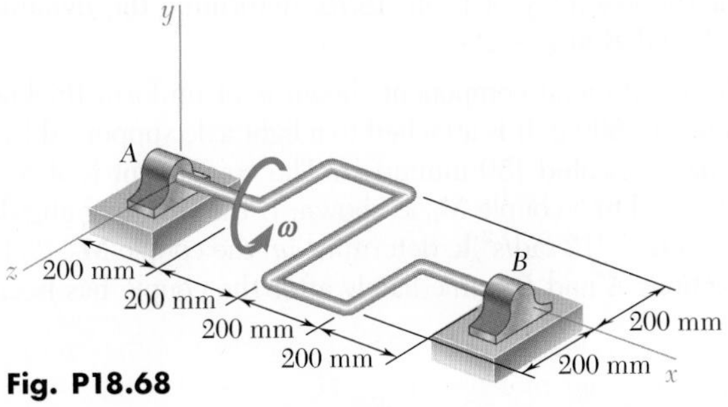

Fig. P18.68

18.69 After attaching the 18-kg wheel shown to a balancing machine and making it spin at the rate of 15 rev/s, a mechanic has found that to balance the wheel both statically and dynamically, he should use two corrective masses, a 170-g mass placed at B and a 56-g mass placed at D. Using a right-handed frame of reference rotating with the wheel (with the z axis perpendicular to the plane of the figure), determine before the corrective masses have been attached (*a*) the distance from the axis of rotation to the mass center of the wheel and the products of inertia I_{xy} and I_{zx}, (*b*) the force-couple system at C equivalent to the forces exerted by the wheel on the machine.

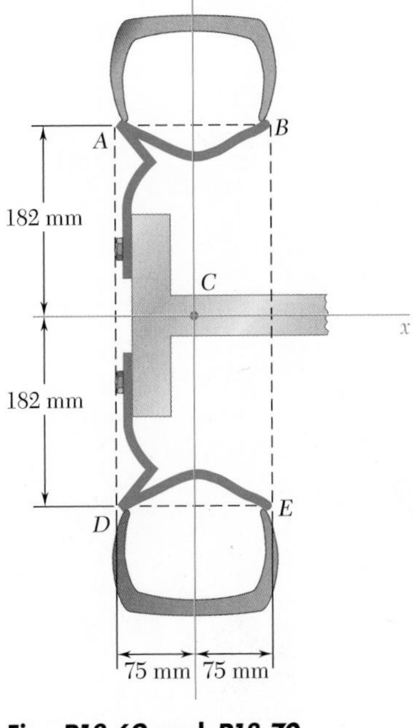

Fig. *P18.69* and *P18.70*

18.70 When the 18-kg wheel shown is attached to a balancing machine and made to spin at a rate of 12.5 rev/s, it is found that the forces exerted by the wheel on the machine are equivalent to a force-couple system consisting of a force $\mathbf{F} = (160 \text{ N})\mathbf{j}$ applied at C and a couple $\mathbf{M}_C = (14.7 \text{ N} \cdot \text{m})\mathbf{k}$, where the unit vectors form a triad which rotates with the wheel. (*a*) Determine the distance from the axis of rotation to the mass center of the wheel and the products of inertia I_{xy} and I_{zx}. (*b*) If only two corrective masses are to be used to balance the wheel statically and dynamically, what should these masses be and at which of the points A, B, D, or E should they be placed?

18.71 Knowing that the assembly of Prob. 18.65 is initially at rest ($\omega = 0$) when a couple of moment $\mathbf{M}_0 = M_0\mathbf{j}$ is applied to shaft CD, determine (*a*) the resulting angular acceleration of the assembly, (*b*) the dynamic reactions at C and D immediately after the couple is applied.

18.72 Knowing that the plate of Prob. 18.66 is initially at rest ($\omega = 0$) when a couple of moment $\mathbf{M}_0 = (0.75 \text{ N} \cdot \text{m})\mathbf{j}$ is applied to it, determine (*a*) the resulting angular acceleration of the plate, (*b*) the dynamics reactions A and B immediately after the couple has been applied.

18.73 The assembly of Prob. 18.67 is initially at rest ($\omega = 0$) when a couple $\mathbf{M}_0$ is applied to axle AB. Knowing that the resulting angular acceleration of the assembly is $\boldsymbol{\alpha} = (150 \text{ rad/s}^2)\mathbf{i}$, determine (*a*) the couple $\mathbf{M}_0$, (*b*) the dynamic reactions at A and B immediately after the couple is applied.

18.74 The shaft of Prob. 18.68 is initially at rest ($\omega = 0$) when a couple $\mathbf{M}_0$ is applied to it. Knowing that the resulting angular acceleration of the shaft is $\boldsymbol{\alpha} = (20 \text{ rad/s}^2)\mathbf{i}$, determine (*a*) the couple $\mathbf{M}_0$, (*b*) the dynamic reactions at A and B immediately after the couple is applied.

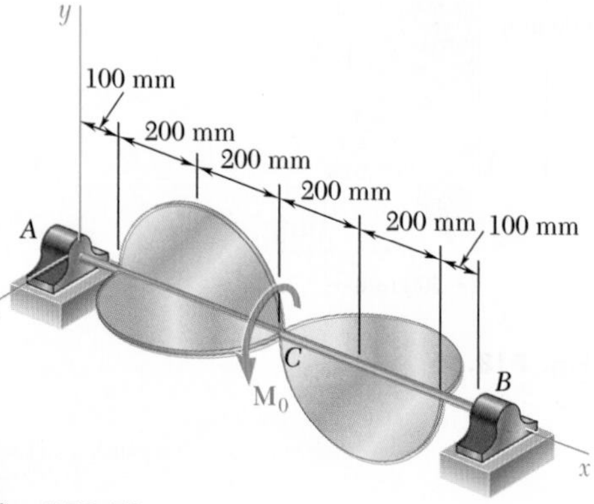

Fig. P18.75

18.75 The assembly shown has a mass of 6 kg and consists of 4 thin 400-mm-diameter semicircular aluminum plates welded to a light 1-m-long shaft AB. The assembly is at rest ($\omega = 0$) at time $t = 0$ when a couple $\mathbf{M}_0$ is applied to it as shown, causing the assembly to complete one full revolution in 2 s. Determine (*a*) the couple $\mathbf{M}_0$, (*b*) the dynamic reactions at A and B at $t = 0$.

18.76 For the assembly of Prob. 18.75, determine the dynamic reactions at A and B at $t = 2$ s.

18.77 The sheet-metal component shown is of uniform thickness and has a mass of 600 g. It is attached to a light axle supported by bearings at A and B located 150 mm apart. The component is at rest when it is subjected to a couple $\mathbf{M}_0$ as shown. If the resulting angular acceleration is $\boldsymbol{\alpha} = (12 \text{ rad/s}^2)\mathbf{k}$, determine (*a*) the couple $\mathbf{M}_0$, (*b*) the dynamic reactions A and B immediately after the couple has been applied.

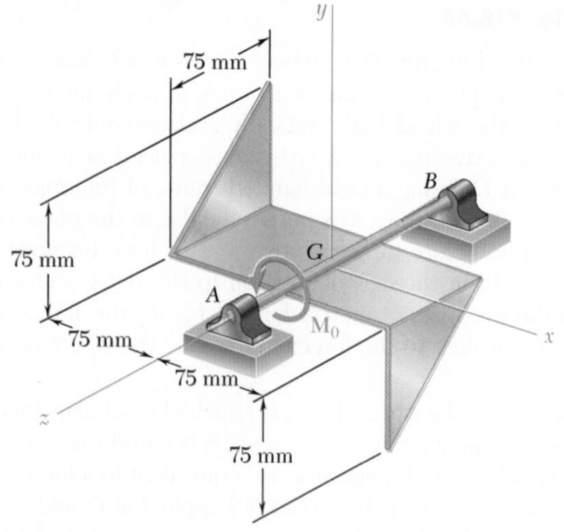

Fig. *P18.77*

18.78 For the sheet-metal component of Prob. 18.77, determine (*a*) the angular velocity of the component 0.6 s after the couple $\mathbf{M}_0$ has been applied to it, (*b*) the magnitude of the dynamic reactions at A and B at that time.

18.79 The blade of an oscillating fan and the rotor of its motor have a total mass of 300 g and a combined radius of gyration of 75 mm. They are supported by bearings at A and B, 125 mm apart, and rotate at the rate $\omega_1 = 1800$ rpm. Determine the dynamic reactions at A and B when the motor casing has an angular velocity $\boldsymbol{\omega}_2 = (0.6 \text{ rad/s})\mathbf{j}$.

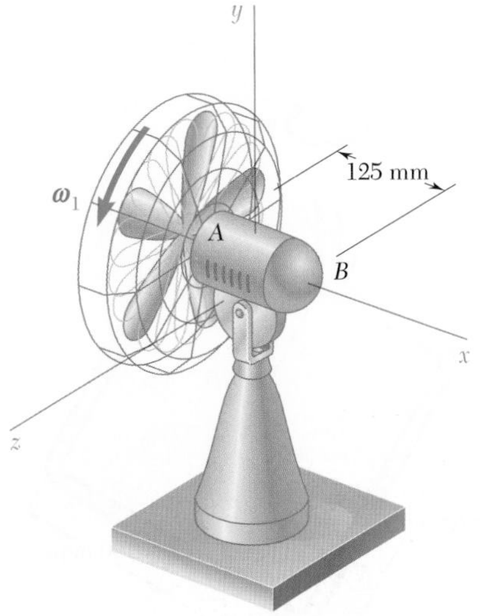

Fig. P18.79

18.80 The blade of a portable saw and the rotor of its motor have a total mass of 1.25 kg and a combined radius of gyration of 40 mm. Knowing that the blade rotates as shown at the rate $\omega_1 = 1500$ rpm, determine the magnitude and direction of the couple **M** that a worker must exert on the handle of the saw to rotate it with a constant angular velocity $\omega_2 = -(2.4 \text{ rad/s})\mathbf{j}$.

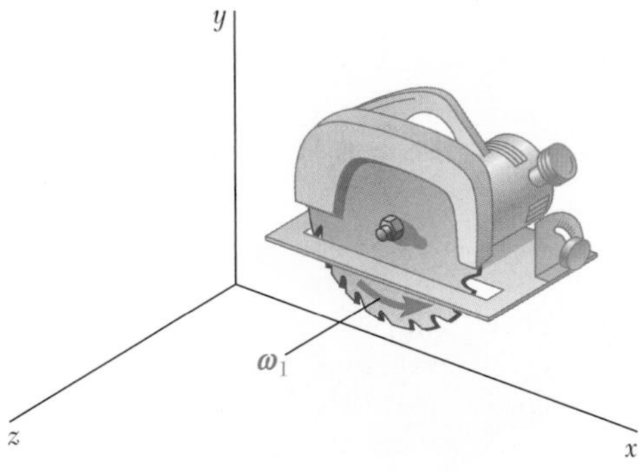

Fig. *P18.80*

18.81 The flywheel of an automobile engine, which is rigidly attached to the crankshaft, is equivalent to a 400-mm-diameter, 15-mm-thick steel plate. Determine the magnitude of the couple exerted by the flywheel on the horizontal crankshaft as the automobile travels around an unbanked curve of 200-m radius at a speed of 90 km/h, with the flywheel rotating at 2700 rpm. Assume the automobile to have (*a*) a rear-wheel drive with the engine mounted longitudinally, (*b*) a front-wheel drive with the engine mounted transversely. (Density of steel = 7860 kg/m^3.)

18.82 Each wheel of an automobile has a mass of 22 kg, a diameter of 575 mm, and a radius of gyration of 225 mm. The automobile travels around an unbanked curve of radius 150 m at a speed of 95 km/h. Knowing that the transverse distance between the wheels is 1.5 m, determine the additional normal force exerted by the ground on each outside wheel due to the motion of the car.

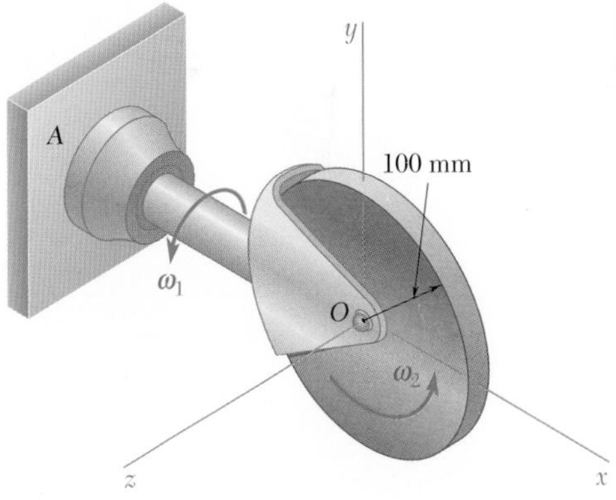

Fig. *P18.83*

18.83 The uniform, thin 2.5-kg disk spins at a constant rate $\omega_2 = 6$ rad/s about an axis held by a housing attached to a horizontal rod that rotates at the constant rate $\omega_1 = 3$ rad/s. Determine the couple which represents the dynamic reaction at the support A.

18.84 The essential structure of a certain type of aircraft turn indicator is shown. Each spring has a constant of 500 N/m, and the 200-g uniform disk of 40-mm radius spins at the rate of 10 000 rpm. The springs are stretched and exert equal vertical forces on yoke AB when the airplane is traveling in a straight path. Determine the angle through which the yoke will rotate when the pilot executes a horizontal turn of 750-m radius to the right at a speed of 800 km/h. Indicate whether point A will move up or down.

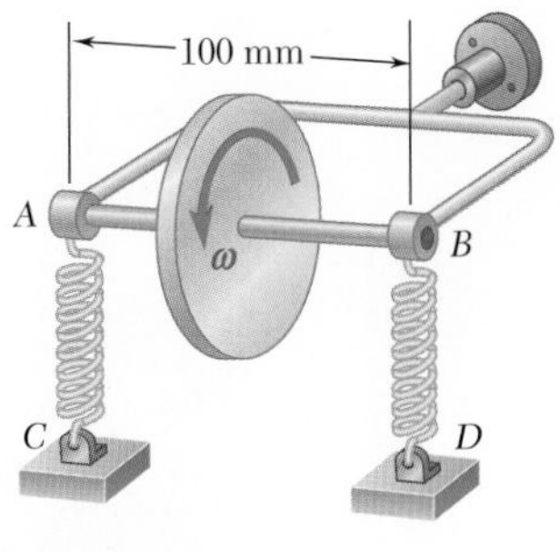

Fig. P18.84

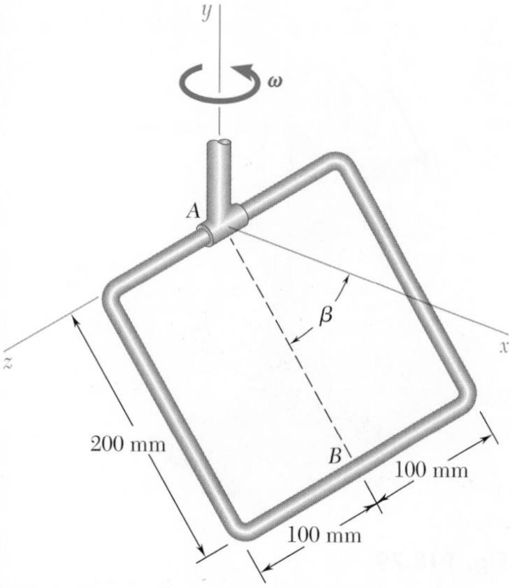

Fig. P18.85

18.85 A slender rod is bent to form a square frame of side 200 mm. The frame is attached by a collar at A to a vertical shaft which rotates with a constant angular velocity $\boldsymbol{\omega}$. Determine the value of ω for which line AB forms an angle $\beta = 48°$ with the horizontal x axis.

18.86 A uniform semicircular plate of radius 120 mm is hinged at A and B to a clevis which rotates with a constant angular velocity $\boldsymbol{\omega}$ about a vertical axis. Determine (*a*) the angle β that the plate forms with the horizontal x axis when $\omega = 15$ rad/s, (*b*) the largest value of ω for which the plate remains vertical ($\beta = 90°$).

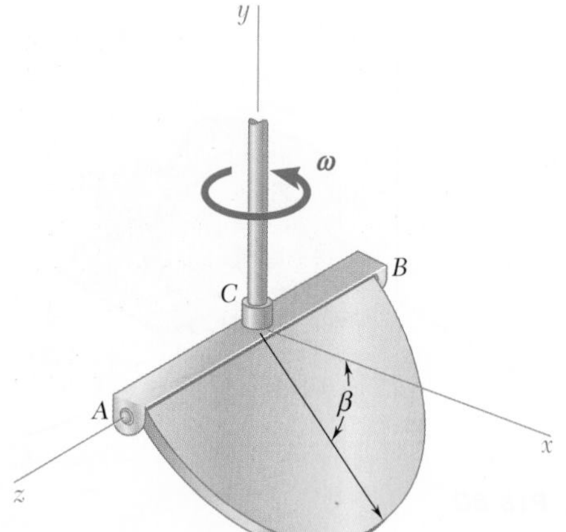

Fig. P18.86 and P18.87

18.87 A uniform semicircular plate of radius 120 mm is hinged at A and B to a clevis which rotates with a constant angular velocity $\boldsymbol{\omega}$ about a vertical axis. Determine the value of ω for which the plate forms an angle $\beta = 50°$ with the horizontal x axis.

18.88 and *18.89* The slender rod AB is attached by a clevis to arm BCD which rotates with a constant angular velocity $\boldsymbol{\omega}$ about the centerline of its vertical portion CD. Determine the magnitude of the angular velocity $\boldsymbol{\omega}$.

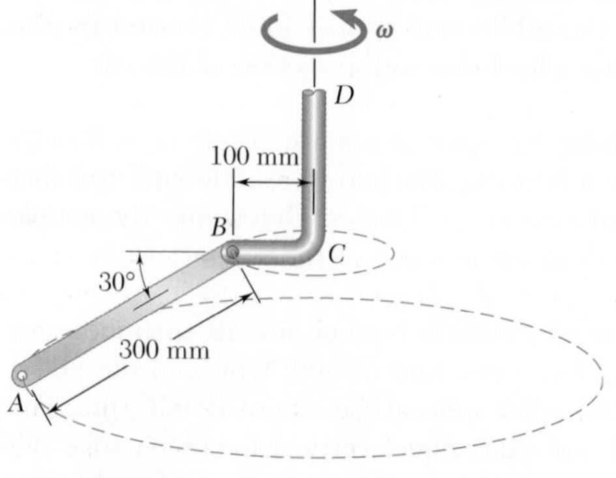

Fig. *P18.88*

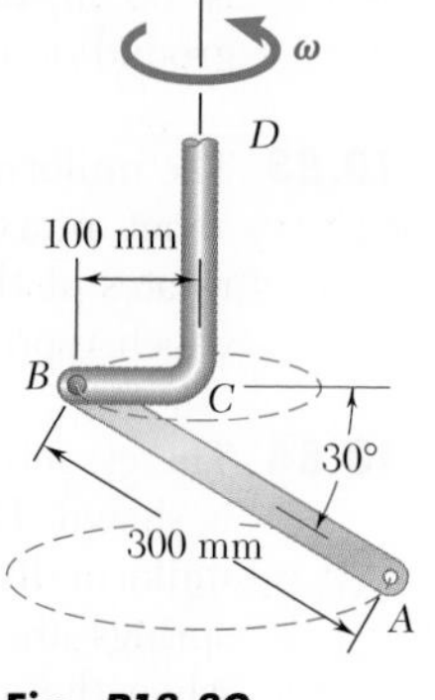

Fig. *P18.89*

18.90 The 950-g gear A is constrained to roll on the fixed gear B, but is free to rotate about axle AD. Axle AD, of length 400 mm and negligible mass, is connected by a clevis to the vertical shaft DE which rotates as shown with a constant angular velocity $\boldsymbol{\omega}_1$. Assuming that gear A can be approximated by a thin disk of radius 80 mm, determine the largest allowable value of ω_1 if gear A is not to lose contact with gear B.

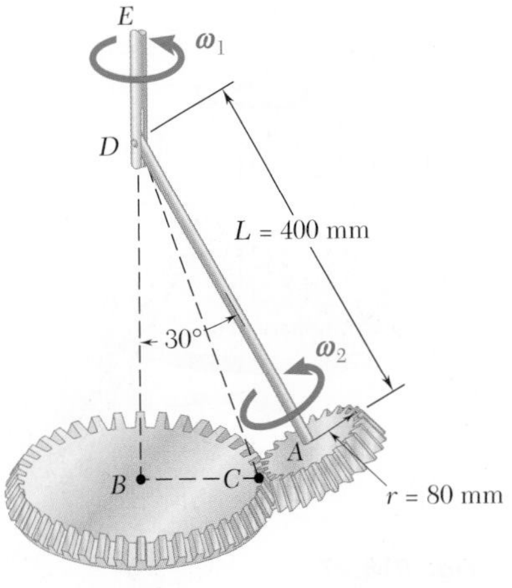

Fig. P18.90

18.91 Determine the force $\mathbf{F}$ exerted by gear B on gear A of Prob. 18.90 when shaft DE rotates with the constant angular velocity $\boldsymbol{\omega}_1 =$ 4 rad/s. (*Hint:* The force $\mathbf{F}$ must be perpendicular to the line drawn from D to C.)

18.92 The essential structure of a certain type of aircraft turn indicator is shown. Springs AC and BD are initially stretched and exert equal vertical forces at A and B when the airplane is traveling in a straight path. Each spring has a constant of 600 N/m and the uniform disk has a mass of 250 g and spins at the rate of 12 000 rpm. Determine the angle through which the yoke will rotate when the pilot executes a horizontal turn of 800-m radius to the right at a speed of 720 km/h. Indicate whether point A will move up or down.

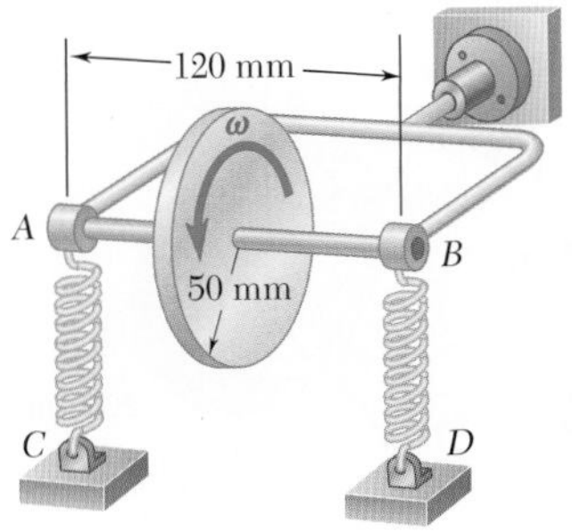

Fig. *P18.92*

18.93 The 250-g disk shown spins at the rate $\omega_1 = 750$ rpm, while axle AB rotates as shown with an angular velocity $\boldsymbol{\omega}_2$ of 6 rad/s. Determine the dynamic reactions at A and B.

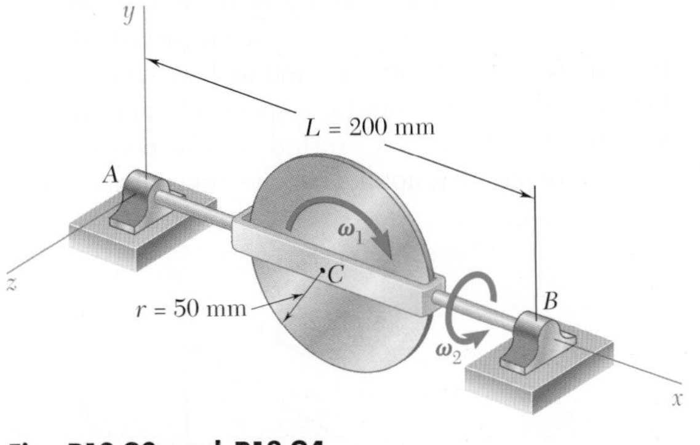

Fig. P18.93 and P18.94

18.94 The 250-g disk shown spins at the rate $\omega_1 = 750$ rpm, while axle AB rotates as shown with an angular velocity $\boldsymbol{\omega}_2$. Determine the maximum allowable magnitude of $\boldsymbol{\omega}_2$ if the dynamic reactions at A and B are not to exceed 1.25-N each.

18.95 Two disks, each of mass 5 kg and radius 100 mm, spin as shown at the rate $\omega_1 = 1500$ rpm about a rod AB of negligible mass which rotates about a vertical axis at the rate $\omega_2 = 45$ rpm. (*a*) Determine the dynamic reactions at C and D. (*b*) Solve part *a* assuming that the direction of spin of disk B is reversed.

18.96 Two disks, each of mass 5 kg and radius 100 mm, spin as shown at the rate $\omega_1 = 1500$ rpm about a rod AB of negligible mass which rotates about a vertical axis at a rate ω_2. Determine the maximum allowable value of ω_2 if the dynamic reactions at C and D are not to exceed 250 N each.

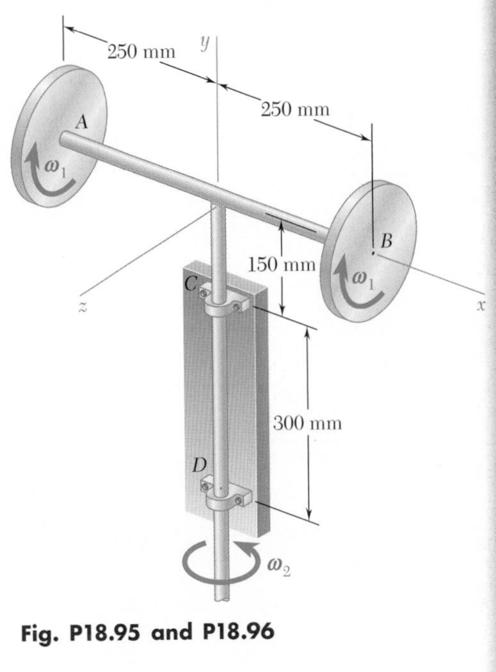

Fig. P18.95 and P18.96

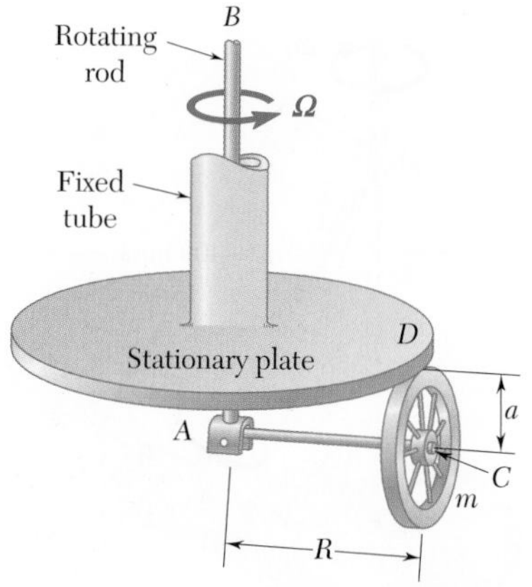

Fig. *P18.97*

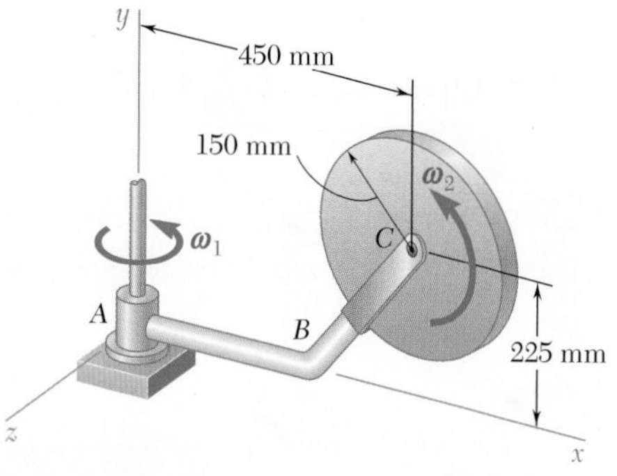

Fig. P18.99

18.97 A stationary horizontal plate is attached to the ceiling by means of a fixed vertical tube. A wheel of radius a and mass m is mounted on a light axle AC which is attached by means of a clevis at A to a rod AB fitted inside the vertical tube. The rod AB is made to rotate with a constant angular velocity Ω causing the wheel to roll on the lower face of the stationary plate. Determine the minimum angular velocity Ω for which contact is maintained between the wheel and the plate. Consider the particular cases (a) when the mass of the wheel is concentrated in the rim, (b) when the wheel is equivalent to a thin disk of radius a.

18.98 Assuming that the wheel of Prob. 18.97 has a mass of 4 kg, has a radius a = 100 mm, and a radius of gyration of 75 mm, and that R = 500 mm, determine the force exerted by the plate on the wheel when Ω = 25 rad/s.

18.99 A thin disk of mass m = 4 kg rotates with an angular velocity $\boldsymbol{\omega}_2$ with respect to arm ABC, which itself rotates with an angular velocity $\boldsymbol{\omega}_1$ about the y axis. Knowing that ω_1 = 5 rad/s and ω_2 = 15 rad/s and that both are constant, determine the force-couple system representing the dynamic reaction at the support at A.

18.100 An experimental Fresnel-lens solar-energy concentrator can rotate about the horizontal axis AB which passes through its mass center G. It is supported at A and B by a steel framework which can rotate about the vertical y axis. The concentrator has a mass of 30 Mg, a radius of gyration of 12 m about its axis of symmetry CD, and a radius of gyration of 10 m about any transverse axis through G. Knowing that the angular velocities $\boldsymbol{\omega}_1$ and $\boldsymbol{\omega}_2$ have constant magnitudes equal to 0.20 rad/s and 0.25 rad/s, respectively, determine for the position θ = 60° (a) the forces exerted on the concentrator at A and B, (b) the couple $M_2\mathbf{k}$ applied to the concentrator at that instant.

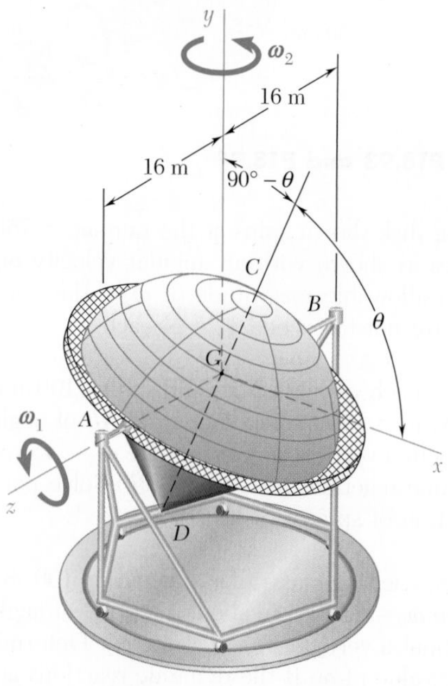

Fig. P18.100

18.101 A 3-kg homogeneous disk of radius 60 mm spins as shown at the constant rate $\omega_1 = 60$ rad/s. The disk is supported by the fork-ended rod AB, which is welded to the vertical shaft CBD. The system is at rest when a couple $\mathbf{M}_0 = (0.40\ \text{N} \cdot \text{m})\mathbf{j}$ is applied to the shaft for 2 s and then removed. Determine the dynamic reactions at C and D after the couple has been removed.

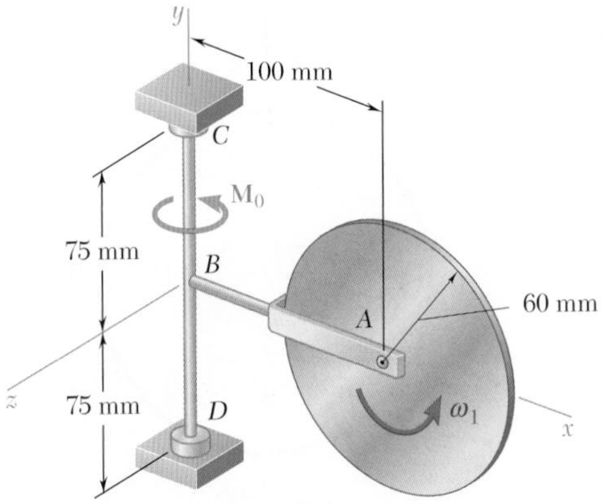

Fig. P18.101 and P18.102

18.102 A 3-kg homogeneous disk of radius 60 mm spins as shown at the constant rate $\omega_1 = 60$ rad/s. The disk is supported by the fork-ended rod AB, which is welded to the vertical shaft CBD. The system is at rest when a couple $\mathbf{M}_0$ is applied as shown to the shaft for 3 s and then removed. Knowing that the maximum angular velocity reached by the shaft is 18 rad/s, determine (*a*) the couple $\mathbf{M}_0$, (*b*) the dynamic reactions at C and D after the couple has been removed.

18.103 A 2.5-kg homogeneous disk of radius 80 mm rotates with an angular velocity $\boldsymbol{\omega}_1$ with respect to arm ABC, which is welded to a shaft DCE rotating as shown at the constant rate $\omega_2 = 12$ rad/s. Friction in the bearing at A causes ω_1 to decrease at the rate of 15 rad/s^2. Determine the dynamic reactions at D and E at a time when ω_1 has decreased to 50 rad/s.

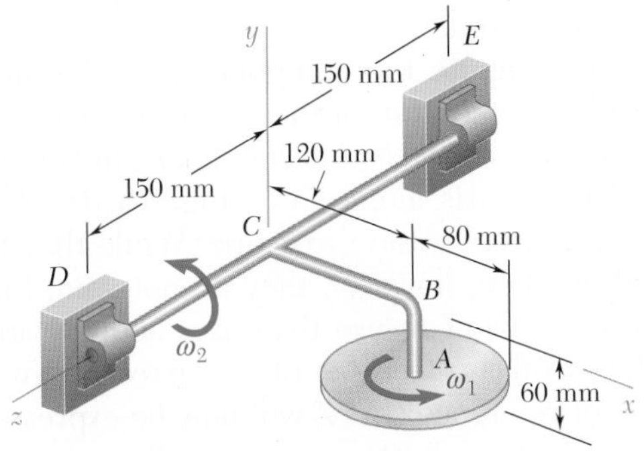

Fig. P18.103 and P18.104

18.104 A 2.5-kg homogeneous disk of radius 80 mm rotates at the constant rate $\omega_1 = 50$ rad/s with respect to arm ABC, which is welded to a shaft DCE. Knowing that at the instant shown, shaft DCE has an angular velocity $\boldsymbol{\omega}_2 = (12\ \text{rad/s})\mathbf{k}$ and an angular acceleration $\boldsymbol{\alpha}_2 = (8\ \text{rad/s}^2)\mathbf{k}$, determine (*a*) the couple which must be applied to shaft DCE to produce that acceleration, (*b*) the corresponding dynamic reactions at D and E.

18.105 For the disk of Prob. 18.99, determine (*a*) the couple $M_1\mathbf{j}$ which should be applied to arm ABC to give it an angular acceleration $\boldsymbol{\alpha}_1 = -(7.5\ \text{rad/s}^2)\mathbf{j}$ when $\omega_1 = 5$ rad/s, knowing that the disk rotates at the constant rate $\omega_2 = 15$ rad/s, (*b*) the force-couple system representing the dynamic reaction at A at that instant. Assume that ABC has a negligible mass.

****18.106*** A slender homogeneous rod AB of mass m and length L is made to rotate at a constant rate ω_2 about the horizontal z axis, while frame CD is made to rotate at the constant rate ω_1 about the y axis. Express as a function of the angle θ (*a*) the couple $\mathbf{M}_1$ required to maintain the rotation of the frame, (*b*) the couple $\mathbf{M}_2$ required to maintain the rotation of the rod, (*c*) the dynamic reactions at the supports C and D.

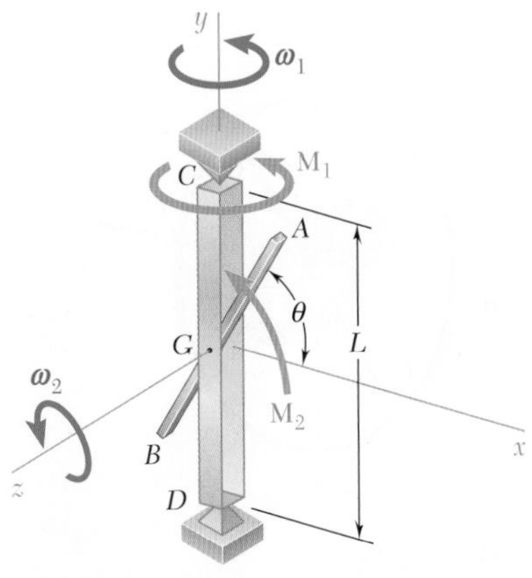

Fig. P18.106

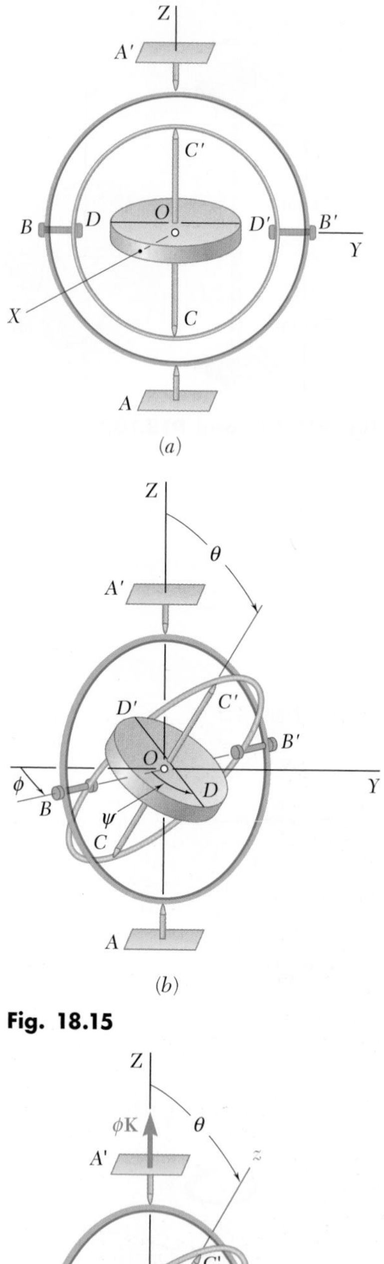

Fig. 18.15

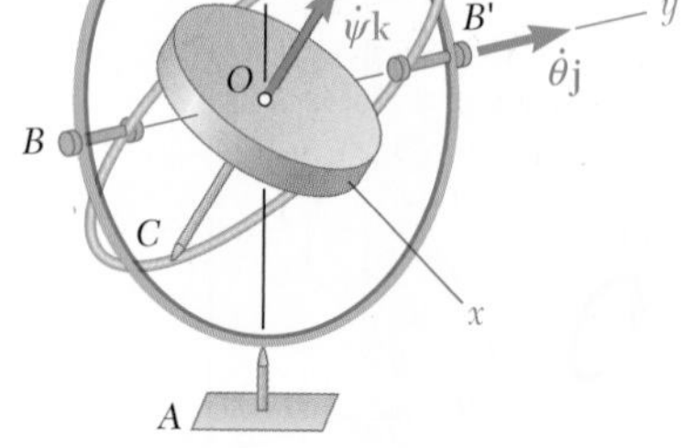

Fig. 18.16

*18.9 MOTION OF A GYROSCOPE. EULERIAN ANGLES

A *gyroscope* consists essentially of a rotor which can spin freely about its geometric axis. When mounted in a Cardan's suspension (Fig. 18.15), a gyroscope can assume any orientation, but its mass center must remain fixed in space. In order to define the position of a gyroscope at a given instant, let us select a fixed frame of reference $OXYZ$, with the origin O located at the mass center of the gyroscope and the Z axis directed along the line defined by the bearings A and A' of the outer gimbal. We will consider a reference position of the gyroscope in which the two gimbals and a given diameter DD' of the rotor are located in the fixed YZ plane (Fig. 18.15*a*). The gyroscope can be brought from this reference position into any arbitrary position (Fig. 18.15*b*) by means of the following steps: (1) a rotation of the outer gimbal through an angle ϕ about the axis AA', (2) a rotation of the inner gimbal through θ about BB', and (3) a rotation of the rotor through ψ about CC'. The angles ϕ, θ, and ψ are called the *Eulerian angles;* they completely characterize the position of the gyroscope at any given instant. Their derivatives $\dot{\phi}$, $\dot{\theta}$, and $\dot{\psi}$ define, respectively, the rate of *precession,* the rate of *nutation,* and the rate of *spin* of the gyroscope at the instant considered.

In order to compute the components of the angular velocity and of the angular momentum of the gyroscope, we will use a rotating system of axes $Oxyz$ *attached to the inner gimbal,* with the y axis along BB' and the z axis along CC' (Fig. 18.16). These axes are principal axes of inertia for the gyroscope. While they follow it in its precession and nutation, however, they do not spin; for that reason, they are more convenient to use than axes actually attached to the gyroscope. The angular velocity $\boldsymbol{\omega}$ of the gyroscope with respect to the fixed frame of reference $OXYZ$ will now be expressed as the sum of three partial angular velocities corresponding, respectively, to the precession, the nutation, and the spin of the gyroscope. Denoting by $\mathbf{i}$, $\mathbf{j}$, and $\mathbf{k}$ the unit vectors along the rotating axes, and by $\mathbf{K}$ the unit vector along the fixed Z axis, we have

$$\boldsymbol{\omega} = \dot{\phi}\mathbf{K} + \dot{\theta}\mathbf{j} + \dot{\psi}\mathbf{k} \tag{18.33}$$

Since the vector components obtained for $\boldsymbol{\omega}$ in (18.33) are not orthogonal (Fig. 18.16), the unit vector $\mathbf{K}$ will be resolved into components along the x and z axes; we write

$$\mathbf{K} = -\sin\theta\,\mathbf{i} + \cos\theta\,\mathbf{k} \tag{18.34}$$

and, substituting for $\mathbf{K}$ into (18.33),

$$\omega = -\dot{\phi}\sin\theta\,\mathbf{i} + \dot{\theta}\mathbf{j} + (\dot{\psi} + \dot{\phi}\cos\theta)\mathbf{k} \tag{18.35}$$

Since the coordinate axes are principal axes of inertia, the components of the angular momentum $\mathbf{H}_O$ can be obtained by multiplying the components of $\boldsymbol{\omega}$ by the moments of inertia of the rotor about the x, y, and z axes, respectively. Denoting by I the moment of inertia of the rotor about its spin axis, by I' its moment of inertia about a transverse axis through O, and neglecting the mass of the gimbals, we write

$$\mathbf{H}_O = -I'\dot{\phi} \sin\theta\,\mathbf{i} + I'\dot{\theta}\mathbf{j} + I(\dot{\psi} + \dot{\phi}\cos\theta)\mathbf{k} \qquad (18.36)$$

Recalling that the rotating axes are attached to the inner gimbal and thus do not spin, we express their angular velocity as the sum

$$\boldsymbol{\Omega} = \dot{\phi}\mathbf{K} + \dot{\theta}\mathbf{j} \qquad (18.37)$$

or, substituting for $\mathbf{K}$ from (18.34),

$$\boldsymbol{\Omega} = -\dot{\phi}\sin\theta\,\mathbf{i} + \dot{\theta}\mathbf{j} + \dot{\phi}\cos\theta\,\mathbf{k} \qquad (18.38)$$

Substituting for $\mathbf{H}_O$ and $\boldsymbol{\Omega}$ from (18.36) and (18.38) into the equation

$$\Sigma\mathbf{M}_O = (\dot{\mathbf{H}}_O)_{Oxyz} + \boldsymbol{\Omega} \times \mathbf{H}_O \qquad (18.28)$$

we obtain the three differential equations

$$\begin{aligned} \Sigma M_x &= -I'(\ddot{\phi}\sin\theta + 2\dot{\theta}\dot{\phi}\cos\theta) + I\dot{\theta}\,(\dot{\psi} + \dot{\phi}\cos\theta) \\ \Sigma M_y &= I'(\ddot{\theta} - \dot{\phi}^2\sin\theta\cos\theta) + I\dot{\phi}\sin\theta\,(\dot{\psi} + \dot{\phi}\cos\theta) \qquad (18.39) \\ \Sigma M_z &= I\frac{d}{dt}(\dot{\psi} + \dot{\phi}\cos\theta) \end{aligned}$$

The equations (18.39) define the motion of a gyroscope subjected to a given system of forces when the mass of its gimbals is neglected. They can also be used to define the motion of an *axisymmetrical body* (or body of revolution) attached at a point on its axis of symmetry, and to define the motion of an axisymmetrical body about its mass center. While the gimbals of the gyroscope helped us visualize the Eulerian angles, it is clear that these angles can be used to define the position of any rigid body with respect to axes centered at a point of the body, regardless of the way in which the body is actually supported.

Since the equations (18.39) are nonlinear, it will not be possible, in general, to express the Eulerian angles ϕ, θ, and ψ as analytical functions of the time t, and numerical methods of solution may have to be used. However, as you will see in the following sections, there are several particular cases of interest which can be analyzed easily.

Photo 18.5 A gyroscope can be used for measuring orientation and is capable of maintaining the same absolute direction in space.

*18.10 STEADY PRECESSION OF A GYROSCOPE

Let us now investigate the particular case of gyroscopic motion in which the angle θ, the rate of precession $\dot{\phi}$, and the rate of spin $\dot{\psi}$ remain constant. We propose to determine the forces which must be applied to the gyroscope to maintain this motion, known as the *steady precession* of a gyroscope.

Instead of applying the general equations (18.39), we will determine the sum of the moments of the required forces by computing the rate of change of the angular momentum of the gyroscope in the particular case considered. We first note that the angular velocity $\boldsymbol{\omega}$ of the gyroscope, its angular momentum $\mathbf{H}_O$, and the angular velocity $\boldsymbol{\Omega}$ of the rotating frame of reference (Fig. 18.17) reduce, respectively, to

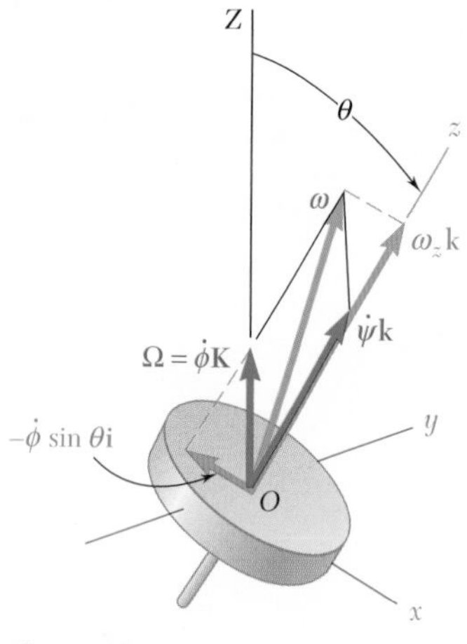

Fig. 18.17

$$\boldsymbol{\omega} = -\dot{\phi} \sin \theta \, \mathbf{i} + \omega_z \mathbf{k} \tag{18.40}$$
$$\mathbf{H}_O = -I'\dot{\phi} \sin \theta \, \mathbf{i} + I\omega_z \mathbf{k} \tag{18.41}$$
$$\boldsymbol{\Omega} = -\dot{\phi} \sin \theta \, \mathbf{i} + \dot{\phi} \cos \theta \, \mathbf{k} \tag{18.42}$$

where $\omega_z = \dot{\psi} + \dot{\phi} \cos \theta$ = rectangular component along spin axis of total angular velocity of gyroscope

Since θ, $\dot{\phi}$, and $\dot{\psi}$ are constant, the vector $\mathbf{H}_O$ is constant in magnitude and direction with respect to the rotating frame of reference and its rate of change $(\dot{\mathbf{H}}_O)_{Oxyz}$ with respect to that frame is zero. Thus Eq. (18.28) reduces to

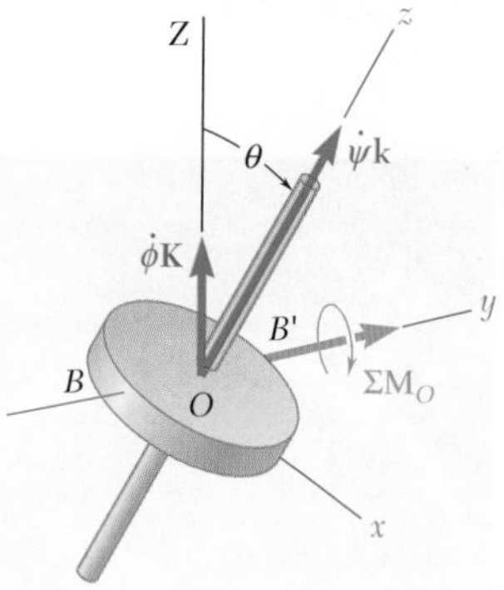

Fig. 18.18

$$\Sigma \mathbf{M}_O = \boldsymbol{\Omega} \times \mathbf{H}_O \tag{18.43}$$

which yields, after substitutions from (18.41) and (18.42),

$$\Sigma \mathbf{M}_O = (I\omega_z - I'\dot{\phi} \cos \theta)\dot{\phi} \sin \theta \, \mathbf{j} \tag{18.44}$$

Since the mass center of the gyroscope is fixed in space, we have, by (18.1), $\Sigma \mathbf{F} = 0$; thus, the forces which must be applied to the gyroscope to maintain its steady precession reduce to a couple of moment equal to the right-hand member of Eq. (18.44). We note that *this couple should be applied about an axis perpendicular to the precession axis and to the spin axis of the gyroscope* (Fig. 18.18).

In the particular case when the precession axis and the spin axis are at a right angle to each other, we have $\theta = 90°$ and Eq. (18.44) reduces to

$$\Sigma \mathbf{M}_O = I\dot{\psi}\dot{\phi} \, \mathbf{j} \tag{18.45}$$

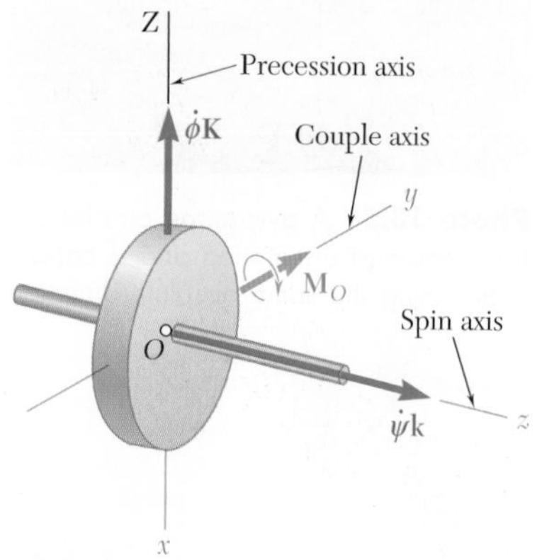

Fig. 18.19

Thus, if we apply to the gyroscope a couple $\mathbf{M}_O$ about an axis perpendicular to its axis of spin, the gyroscope will precess about an axis perpendicular to both the spin axis and the couple axis, in a sense such that the vectors representing the spin, the couple, and the precession, respectively, form a right-handed triad (Fig. 18.19).

Because of the relatively large couples required to change the orientation of their axles, gyroscopes are used as stabilizers in torpedoes and ships. Spinning bullets and shells remain tangent to their trajectory because of gyroscopic action. And a bicycle is easier to keep balanced at high speeds because of the stabilizing effect of its spinning wheels. However, gyroscopic action is not always welcome and must be taken into account in the design of bearings supporting rotating shafts subjected to forced precession. The reactions exerted by its propellers on an airplane which changes its direction of flight must also be taken into consideration and compensated for whenever possible.

*18.11 MOTION OF AN AXISYMMETRICAL BODY UNDER NO FORCE

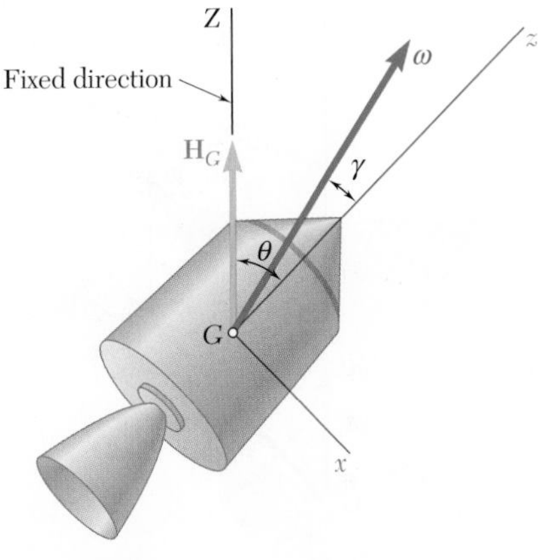

Fig. 18.20

In this section you will analyze the motion about its mass center of an axisymmetrical body under no force except its own weight. Examples of such a motion are furnished by projectiles, if air resistance is neglected, and by artificial satellites and space vehicles after burnout of their launching rockets.

Since the sum of the moments of the external forces about the mass center G of the body is zero, Eq. (18.2) yields $\dot{\mathbf{H}}_G = 0$. It follows that the angular momentum $\mathbf{H}_G$ of the body about G is constant. Thus, the direction of $\mathbf{H}_G$ is fixed in space and can be used to define the Z axis, or axis of precession (Fig. 18.20). Selecting a rotating system of axes $Gxyz$ with the z axis along the axis of symmetry of the body, the x axis in the plane defined by the Z and z axes, and the y axis pointing away from you, we have

$$H_x = -H_G \sin\theta \qquad H_y = 0 \qquad H_z = H_G \cos\theta \tag{18.46}$$

where θ represents the angle formed by the Z and z axes, and H_G denotes the constant magnitude of the angular momentum of the body about G. Since the x, y, and z axes are principal axes of inertia for the body considered, we can write

$$H_x = I'\omega_x \qquad H_y = I'\omega_y \qquad H_z = I\omega_z \tag{18.47}$$

where I denotes the moment of inertia of the body about its axis of symmetry and I' denotes its moment of inertia about a transverse axis through G. It follows from Eqs. (18.46) and (18.47) that

$$\omega_x = -\frac{H_G \sin\theta}{I'} \qquad \omega_y = 0 \qquad \omega_z = \frac{H_G \cos\theta}{I} \tag{18.48}$$

The second of the relations obtained shows that the angular velocity $\boldsymbol{\omega}$ has no component along the y axis, i.e., along an axis perpendicular to the Zz plane. Thus, the angle θ formed by the Z and z axes remains constant and *the body is in steady precession about the Z axis.*

Dividing the first and third of the relations (18.48) member by member, and observing from Fig. 18.21 that $-\omega_x/\omega_z = \tan\gamma$, we obtain the following relation between the angles γ and θ that the

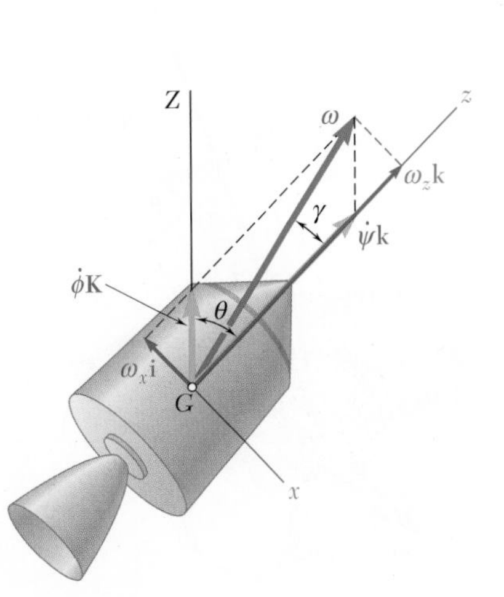

Fig. 18.21

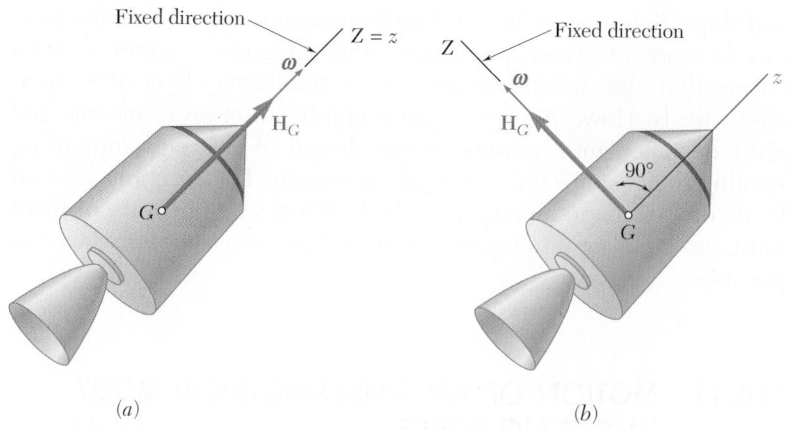

Fig. 18.22

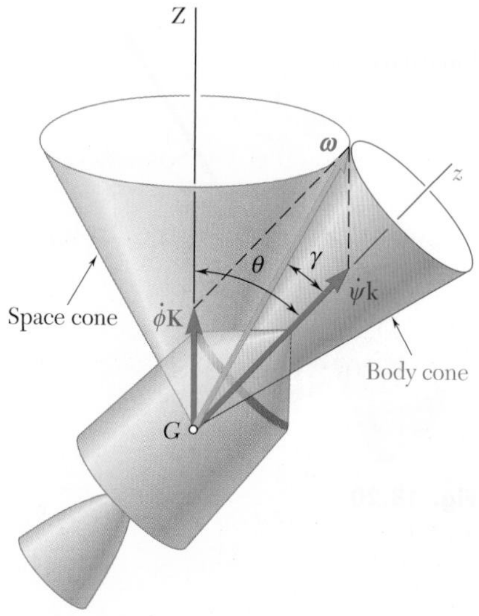

Fig. 18.23

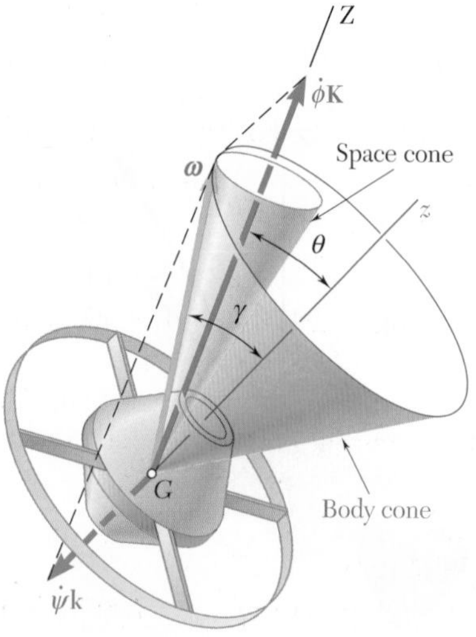

Fig. 18.24

vectors $\boldsymbol{\omega}$ and $\mathbf{H}_G$, respectively, form with the axis of symmetry of the body:

$$\tan \gamma = \frac{I}{I'} \tan \theta \qquad (18.49)$$

There are two particular cases of motion of an axisymmetrical body under no force which involve no precession: (1) If the body is set to spin about its axis of symmetry, we have $\omega_x = 0$ and, by (18.47), $H_x = 0$; the vectors $\boldsymbol{\omega}$ and $\mathbf{H}_G$ have the same orientation and the body keeps spinning about its axis of symmetry (Fig. 18.22*a*). (2) If the body is set to spin about a transverse axis, we have $\omega_z = 0$ and, by (18.47), $H_z = 0$; again $\boldsymbol{\omega}$ and $\mathbf{H}_G$ have the same orientation and the body keeps spinning about the given transverse axis (Fig. 18.22*b*).

Considering now the general case represented in Fig. 18.21, we recall from Sec. 15.12 that the motion of a body about a fixed point—or about its mass center—can be represented by the motion of a body cone rolling on a space cone. In the case of steady precession, the two cones are circular, since the angles γ and $\theta - \gamma$ that the angular velocity $\boldsymbol{\omega}$ forms, respectively, with the axis of symmetry of the body and with the precession axis are constant. Two cases should be distinguished:

1. $I < I'$. This is the case of an elongated body, such as the space vehicle of Fig. 18.23. By (18.49) we have $\gamma < \theta$; the vector $\boldsymbol{\omega}$ lies inside the angle ZGz; the space cone and the body cone are tangent externally; the spin and the precession are both observed as counterclockwise from the positive z axis. The precession is said to be *direct*.
2. $I > I'$. This is the case of a flattened body, such as the satellite of Fig. 18.24. By (18.49) we have $\gamma > \theta$; since the vector $\boldsymbol{\omega}$ must lie outside the angle ZGz, the vector $\dot{\psi}\mathbf{k}$ has a sense opposite to that of the z axis; the space cone is inside the body cone; the precession and the spin have opposite senses. The precession is said to be *retrograde*.

SAMPLE PROBLEM 18.6

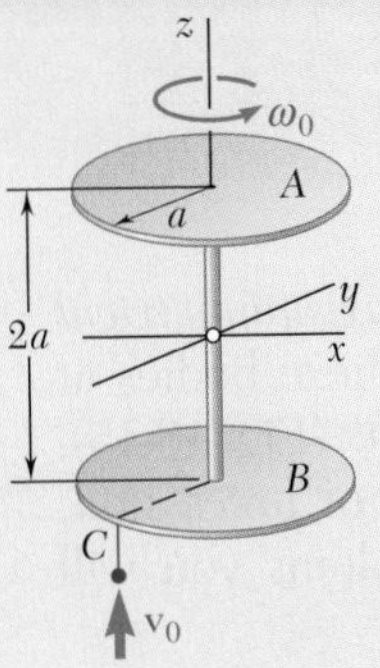

A space satellite of mass m is known to be dynamically equivalent to two thin disks of equal mass. The disks are of radius a = 800 mm and are rigidly connected by a light rod of length $2a$. Initially the satellite is spinning freely about its axis of symmetry at the rate ω_0 = 60 rpm. A meteorite, of mass $m_0 = m/1000$ and traveling with a velocity $\mathbf{v}_0$ of 2000 m/s relative to the satellite, strikes the satellite and becomes embedded at C. Determine (*a*) the angular velocity of the satellite immediately after impact, (*b*) the precession axis of the ensuing motion, (*c*) the rates of precession and spin of the ensuing motion.

SOLUTION

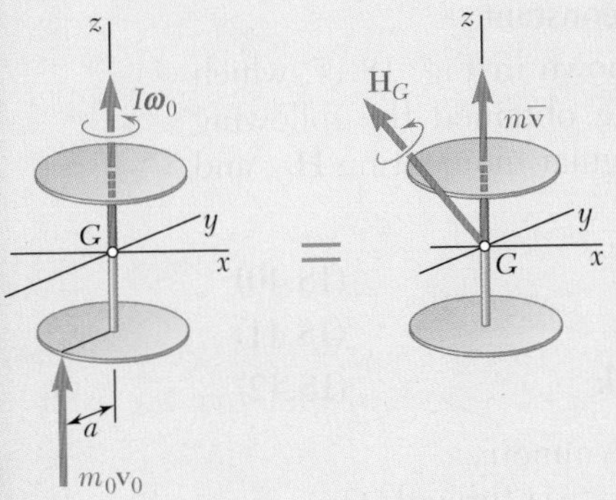

Moments of Inertia. We note that the axes shown are principal axes of inertia for the satellite and write

$$I = I_z = \tfrac{1}{2}ma^2 \qquad I' = I_x = I_y = 2[\tfrac{1}{4}(\tfrac{1}{2}m)a^2 + (\tfrac{1}{2}m)a^2] = \tfrac{5}{4}ma^2$$

Principle of Impulse and Momentum. We consider the satellite and the meteorite as a single system. Since no external force acts on this system, the momenta before and after impact are equipollent. Taking moments about G, we write

$$-a\mathbf{j} \times m_0v_0\mathbf{k} + I\omega_0\mathbf{k} = \mathbf{H}_G$$

$$\mathbf{H}_G = -m_0v_0a\mathbf{i} + I\omega_0\mathbf{k} \tag{1}$$

Angular Velocity After Impact. Substituting the values obtained for the components of $\mathbf{H}_G$ and for the moments of inertia into

$$H_x = I_x\omega_x \qquad H_y = I_y\omega_y \qquad H_z = I_z\omega_z$$

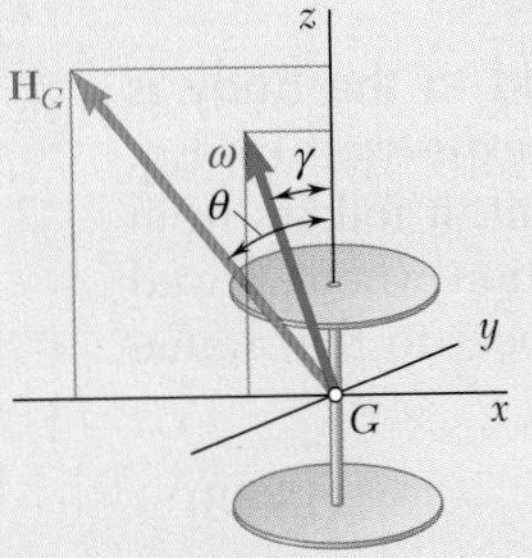

we write

$$-m_0v_0a = I'\omega_x = \tfrac{5}{4}ma^2\omega_x \qquad 0 = I'\omega_y \qquad I\omega_0 = I\omega_z$$

$$\omega_x = -\frac{4}{5}\frac{m_0v_0}{ma} \qquad \omega_y = 0 \qquad \omega_z = \omega_0 \tag{2}$$

For the satellite considered we have ω_0 = 60 rpm = 6.283 rad/s, $m_0/m = \frac{1}{1000}$, a = 0.800 m, and v_0 = 2000 m/s; we find

$$\omega_x = -2 \text{ rad/s} \qquad \omega_y = 0 \qquad \omega_z = 6.283 \text{ rad/s}$$

$$\omega = \sqrt{\omega_x^2 + \omega_z^2} = 6.594 \text{ rad/s} \qquad \tan\gamma = \frac{-\omega_x}{\omega_z} = +0.3183$$

$$\omega = 63.0 \text{ rpm} \qquad \gamma = 17.7° \blacktriangleleft$$

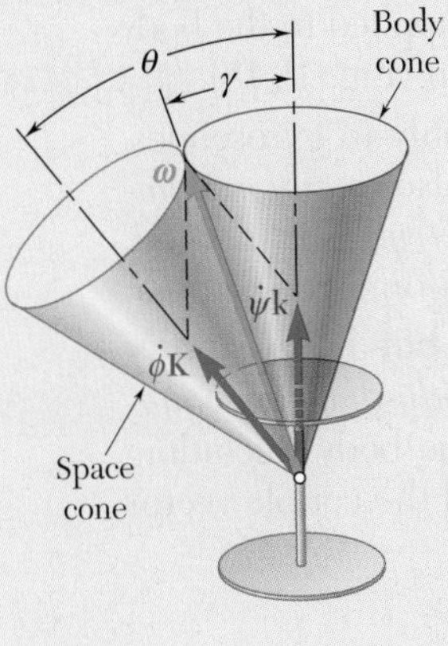

Precession Axis. Since in free motion the direction of the angular momentum $\mathbf{H}_G$ is fixed in space, the satellite will precess about this direction. The angle θ formed by the precession axis and the z axis is

$$\tan\theta = \frac{-H_x}{H_z} = \frac{m_0v_0a}{I\omega_0} = \frac{2m_0v_0}{ma\omega_0} = 0.796 \qquad \theta = 38.5° \blacktriangleleft$$

Rates of Precession and Spin. We sketch the space and body cones for the free motion of the satellite. Using the law of sines, we compute the rates of precession and spin.

$$\frac{\omega}{\sin\theta} = \frac{\dot{\phi}}{\sin\gamma} = \frac{\dot{\psi}}{\sin(\theta - \gamma)}$$

$$\dot{\phi} = 30.8 \text{ rpm} \qquad \dot{\psi} = 35.9 \text{ rpm} \blacktriangleleft$$

SOLVING PROBLEMS ON YOUR OWN

In this lesson we analyzed the motion of *gyroscopes* and of other *axisymmetrical bodies* with a fixed point O. In order to define the position of these bodies at any given instant, we introduced the three *Eulerian angles* ϕ, θ, and ψ (Fig. 18.15), and noted that their time derivatives define, respectively, the rate of *precession*, the rate of *nutation*, and the rate of *spin* (Fig. 18.16). The problems you will encounter fall into one of the following categories.

1. Steady precession. This is the motion of a gyroscope or other axisymmetrical body with a fixed point located on its axis of symmetry, in which the angle θ, the rate of precession $\dot{\phi}$, and the rate of spin $\dot{\psi}$ all remain constant.

a. Using the rotating frame of reference *Oxyz* shown in Fig. 18.17, which *precesses* with the body, *but does not spin* with it, we obtained the following expressions for the angular velocity $\boldsymbol{\omega}$ of the body, its angular momentum $\mathbf{H}_O$, and the angular velocity $\boldsymbol{\Omega}$ of the frame $Oxyz$:

$$\boldsymbol{\omega} = -\dot{\phi} \sin \theta \, \mathbf{i} + \omega_z \mathbf{k} \tag{18.40}$$

$$\mathbf{H}_O = -I'\dot{\phi} \sin \theta \, \mathbf{i} + I \, \omega_z \mathbf{k} \tag{18.41}$$

$$\boldsymbol{\Omega} = -\dot{\phi} \sin \theta \, \mathbf{i} + \dot{\phi} \cos \theta \, \mathbf{k} \tag{18.42}$$

where I = moment of inertia of body about its axis of symmetry
I' = moment of inertia of body about a transverse axis through O
ω_z = *rectangular* component of $\boldsymbol{\omega}$ along z axis = $\dot{\psi} + \dot{\phi} \cos \theta$

b. The sum of the moments about *O* of the forces applied to the body is equal to the rate of change of its angular momentum, as expressed by Eq. (18.28). But, since θ and the rates of change $\dot{\phi}$ and $\dot{\psi}$ are constant, it follows from Eq. (18.41) that $\mathbf{H}_O$ remains constant in magnitude and direction when viewed from the frame $Oxyz$. Thus, its rate of change is zero with respect to that frame and you can write

$$\Sigma \mathbf{M}_O = \boldsymbol{\Omega} \times \mathbf{H}_O \tag{18.43}$$

where $\boldsymbol{\Omega}$ and $\mathbf{H}_O$ are defined, respectively, by Eq. (18.42) and Eq. (18.41). The equation obtained shows that the moment resultant at O of the forces applied to the body is perpendicular to both the axis of precession and the axis of spin (Fig. 18.18).

c. Keep in mind that the method described applies, not only to gyroscopes, where the fixed point O coincides with the mass center G, but also *to any axisymmetrical body with a fixed point O located on its axis of symmetry*. This method, therefore, can be used to analyze the *steady precession of a top* on a rough floor.

d. When an axisymmetrical body has no fixed point, but is in steady precession about its mass center *G*, you should draw a *free-body-diagram equation* showing that the system of the external forces exerted on the body (including the body's weight) is equivalent to the vector $m\bar{\mathbf{a}}$ applied at G and the couple vector

$\dot{\mathbf{H}}_G$. You can use Eqs. (18.40) through (18.42), replacing $\mathbf{H}_O$ with $\mathbf{H}_G$, and express the moment of the couple as

$$\dot{\mathbf{H}}_G = \boldsymbol{\Omega} \times \mathbf{H}_G$$

You can then use the free-body-diagram equation to write as many as six independent scalar equations.

2. Motion of an axisymmetrical body under no force, except its own weight. We have $\Sigma\mathbf{M}_G = 0$ and, thus, $\dot{\mathbf{H}}_G = 0$; it follows that *the angular momentum* $\mathbf{H}_G$ is constant in magnitude and direction (Sec. 18.11). The body is in *steady precession* with the precession axis GZ directed along $\mathbf{H}_G$ (Fig. 18.20). Using the rotating frame $Gxyz$ and denoting by γ the angle that $\boldsymbol{\omega}$ forms with the spin axis Gz (Fig. 18.21), we obtained the following relation between γ and the angle θ formed by the precession and spin axes:

$$\tan \gamma = \frac{I}{I'} \tan \theta \tag{18.49}$$

The precession is said to be *direct* if $I < I'$ (Fig. 18.23) and *retrograde* if $I > I'$ (Fig. 18.24).

a. In many of the problems dealing with the motion of an axisymmetrical body under no force, you will be asked to determine the *precession axis* and the *rates of precession and spin* of the body, knowing the magnitude of its *angular velocity* $\boldsymbol{\omega}$ and the angle γ that it forms with the axis of symmetry Gz (Fig. 18.21). From Eq. (18.49) you will determine the angle θ that the precession axis GZ forms with Gz and resolve $\boldsymbol{\omega}$ into its two *oblique components* $\dot{\phi}\mathbf{K}$ and $\dot{\psi}\mathbf{k}$. Using the law of sines, you will then determine the rate of precession $\dot{\phi}$ and the rate of spin $\dot{\psi}$.

b. In other problems, the body will be subjected to *a given impulse* and you will first determine the resulting *angular momentum* $\mathbf{H}_G$. Using Eqs. (18.10), you will calculate the rectangular components of the angular velocity $\boldsymbol{\omega}$, its magnitude ω, and the angle γ that it forms with the axis of symmetry. You will then determine the *precession axis* and the *rates of precession and spin* as described above [Sample Prob. 18.6].

3. General motion of an axisymmetric body with a fixed point O located on its axis of symmetry, and subjected only to its own weight. This is a motion in which the angle θ is allowed to vary. At any given instant you should take into account the rate of precession $\dot{\phi}$, the rate of spin $\dot{\psi}$, *and the rate of nutation* $\dot{\theta}$, none of which will remain constant. An example of such a motion is the motion of a top, which is discussed in Probs. 18.137 and 18.138. The rotating frame of reference $Oxyz$ that you will use is still the one shown in Fig. 18.18, but this frame

(*continued*)

will now rotate about the y axis at the rate $\dot{\theta}$. Equations (18.40), (18.41), and (18.42), therefore, should be replaced by the following equations:

$$\boldsymbol{\omega} = -\dot{\phi} \sin \theta \, \mathbf{i} + \dot{\theta} \mathbf{j} + (\dot{\psi} + \dot{\phi} \cos \theta) \, \mathbf{k} \qquad (18.40')$$

$$\mathbf{H}_O = -I'\dot{\phi} \sin \theta \, \mathbf{i} + I'\dot{\theta} \mathbf{j} + I(\dot{\psi} + \dot{\phi} \cos \theta) \, \mathbf{k} \qquad (18.41')$$

$$\boldsymbol{\Omega} = -\dot{\phi} \sin \theta \, \mathbf{i} + \dot{\theta} \mathbf{j} + \dot{\phi} \cos \theta \, \mathbf{k} \qquad (18.42')$$

Since substituting these expressions into Eq. (18.44) would lead to nonlinear differential equations, it is preferable, whenever feasible, to apply the following conservation principles.

a. Conservation of energy. Denoting by c the distance between the fixed point O and the mass center G of the body, and by E the total energy, you will write

$$T + V = E: \qquad \tfrac{1}{2}(I'\omega_x^2 + I'\omega_y^2 + I\omega_z^2) + mgc \cos \theta = E$$

and substitute for the components of $\boldsymbol{\omega}$ the expressions obtained in Eq. (18.40′). Note that c will be positive or negative, depending upon the position of G relative to O. Also, $c = 0$ if G coincides with O; the *kinetic energy* is then conserved.

b. Conservation of the angular momentum about the axis of precession. Since the support at O is located on the Z axis, and since the weight of the body and the Z axis are both vertical and, thus, parallel to each other, it follows that $\Sigma M_Z = 0$ and, thus, that H_Z remains constant. This can be expressed by writing that the scalar product $\mathbf{K} \cdot \mathbf{H}_O$ is constant, where $\mathbf{K}$ is the unit vector along the Z axis.

c. Conservation of the angular momentum about the axis of spin. Since the support at O and the center of gravity G are both located on the z axis, it follows that $\Sigma M_z = 0$ and, thus, that H_z remains constant. This is expressed by writing that the coefficient of the unit vector $\mathbf{k}$ in Eq. (18.41′) is constant. Note that this last conservation principle cannot be applied when the body is restrained from spinning about its axis of symmetry, but in that case the only variables are θ and ϕ.

PROBLEMS

18.107 A solid cone of height 300 mm with a circular base of radius 100 mm, is supported by a ball-and-socket joint at A. Knowing that the cone is observed to precess about the vertical axis AC at the constant rate of 40 rpm in the sense indicated and that its axis of symmetry AB forms an angle $\beta = 40°$ with AC, determine the rate at which the cone spins about the axis AB.

18.108 A solid cone of height 300 mm with a circular base of radius 100 mm is supported by a ball-and-socket joint at A. Knowing that the cone is spinning about its axis of symmetry AB at the rate of 3000 rpm and that AB forms an angle $\beta = 60°$ with the vertical axis AC, determine the two possible rates of steady precession of the cone about the axis AC.

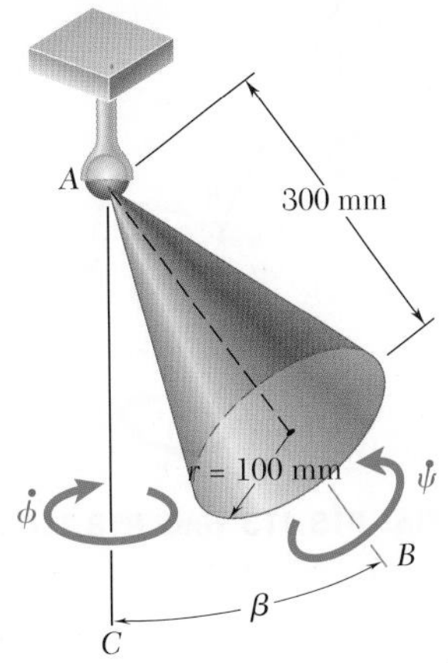

Fig. P18.107 and *P18.108*

18.109 The 85-g top shown is supported at the fixed point O. The radii of gyration of the top with respect to its axis of symmetry and with respect to a transverse axis through O are 21 mm and 45 mm, respectively. Knowing that $c = 37.5$ mm and that the rate of spin of the top about its axis of symmetry is 1800 rpm, determine the two possible rates of steady precession corresponding to $\theta = 30°$.

18.110 The top shown is supported at the fixed point O and its moments of inertia about its axis of symmetry and about a transverse axis through O are denoted, respectively, by I and I'. (*a*) Show that the condition for steady precession of the top is

$$(I\omega_z - I'\dot{\phi}\cos\theta)\dot{\phi} = Wc$$

where $\dot{\phi}$ is the rate of precession and ω_z is the rectangular component of the angular velocity along the axis of symmetry of the top. (*b*) Show that if the rate of spin $\dot{\psi}$ of the top is very large compared with its rate of precession $\dot{\phi}$, the condition for steady precession is $I\dot{\psi}\dot{\phi} \approx Wc$. (*c*) Determine the percentage error introduced when this last relation is used to approximate the slower of the two rates of precession obtained for the top of Prob. 18.109.

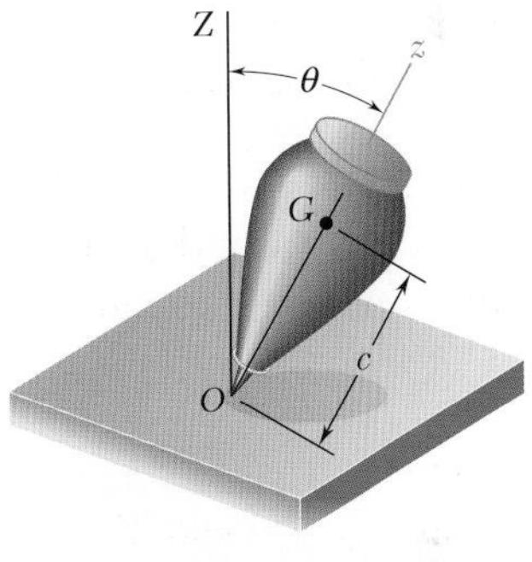

Fig. P18.109 and *P18.110*

18.111 A solid aluminum sphere of radius 100 mm is welded to the end of a 200 mm-long rod AB of negligible mass which is supported by a ball-and-socket joint at A. Knowing that the sphere is observed to precess about a vertical axis at the constant rate of 60 rpm in the sense indicated and that rod AB forms an angle $\beta = 30°$ with the vertical, determine the rate of spin of the sphere about line AB.

18.112 A solid aluminum sphere of radius 100 mm is welded to the end of a 200 mm-long rod AB of negligible mass which is supported by a ball-and-socket joint at A. Knowing that the sphere spins as shown about line AB at the rate of 700 rpm, determine the angle β for which the sphere will precess about a vertical axis at the constant rate of 60 rpm in the sense indicated.

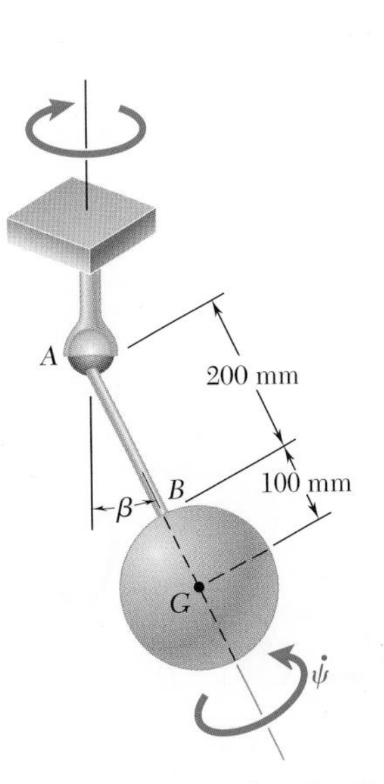

Fig. P18.111 and P18.112

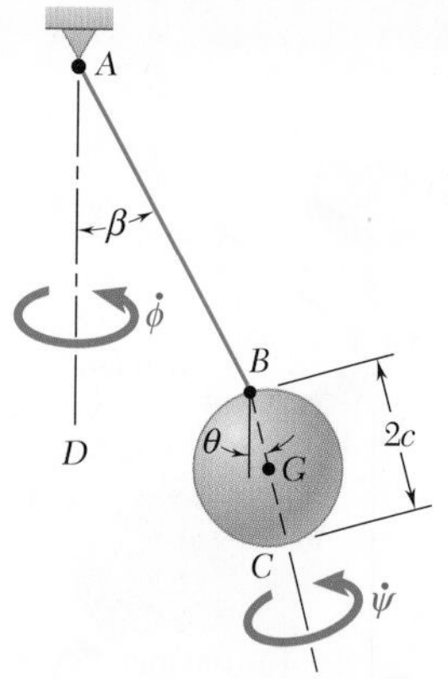

Fig. P18.113 and P18.114

18.113 A solid sphere of radius $c = 100$ mm is attached as shown to cord AB. The sphere is observed to precess at the constant rate $\dot{\phi} = 6$ rad/s about the vertical axis AD. Knowing that $\beta = 40°$, determine the angle θ that the diameter BC forms with the vertical when the sphere (*a*) has no spin, (*b*) spins about its diameter BC at the rate $\dot{\psi} = 50$ rad/s, (*c*) spins about BC at the rate $\dot{\psi} = -50$ rad/s.

18.114 A solid sphere of radius $c = 100$ mm is attached as shown to a cord AB of length 500 mm The sphere spins about its diameter BC and precesses about the vertical axis AD. Knowing that $\theta = 20°$ and $\beta = 35°$, determine (*a*) the rate of spin of the sphere, (*b*) its rate of precession.

18.115 A solid cube of side $c = 80$ mm is attached as shown to cord AB. It is observed to spin at the rate $\dot{\psi} = 40$ rad/s about its diagonal BC and to precess at the constant rate $\dot{\phi} = 5$ rad/s about the vertical axis AD. Knowing that $\beta = 30°$, determine the angle θ that the diagonal BC forms with the vertical. (*Hint:* The moment of inertia of a cube about an axis through its center is independent of the orientation of that axis.)

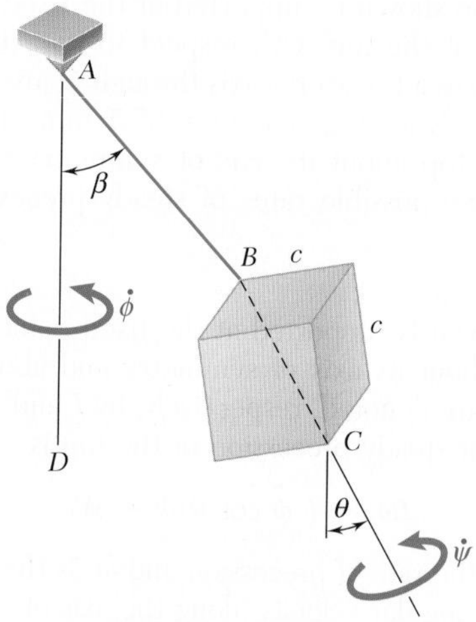

Fig. P18.115 and P18.116

18.116 A solid cube of side $c = 120$ mm is attached as shown to a cord AB of length 240 mm. The cube spins about its diagonal BC and precesses about the vertical axis AD. Knowing that $\theta = 25°$ and $\beta = 40°$, determine (*a*) the rate of spin of the cube, (*b*) its rate of precession. (See hint of Prob. 18.115.)

18.117 A high-speed photographic record shows that a certain projectile was fired with a horizontal velocity $\overline{\mathbf{v}}$ of 600 m/s and with its axis of symmetry forming an angle $\beta = 3°$ with the horizontal. The rate of spin $\dot{\psi}$ of the projectile was 6000 rpm, and the atmospheric drag was equivalent to a force $\mathbf{D}$ of 120 N acting at the center of pressure C_P located at a distance $c = 150$ mm from G. (*a*) Knowing that the projectile has a mass of 20 kg and a radius of gyration of 50 mm with respect to its axis of symmetry, determine its approximate rate of steady precession. (*b*) If it is further known that the radius of gyration of the projectile with respect to a transverse axis through G is 200 mm, determine the exact values of the two possible rates of precession.

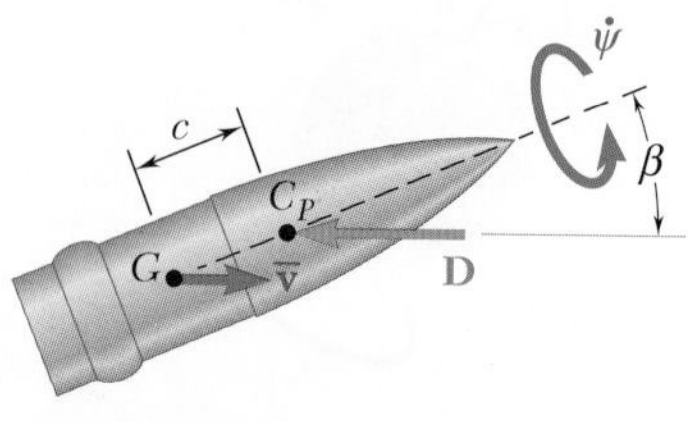

Fig. P18.117

18.118 If the earth were a sphere, the gravitational attraction of the sun, moon, and planets would at all times be equivalent to a single force **R** acting at the mass center of the earth. However, the earth is actually an oblate spheroid and the gravitational system acting on the earth is equivalent to a force **R** and a couple **M**. Knowing that the effect of the couple **M** is to cause the axis of the earth to precess about the axis *GA* at the rate of one revolution in 25 800 years, determine the average magnitude of the couple **M** applied to the earth. Assume that the average density of the earth is 5.51 g/cm^3, that the average radius of the earth is 6370 km, and that $\bar{I} = \frac{2}{5}mR^2$. (*Note:* This forced precession is known as the precession of the equinoxes and is not to be confused with the free precession discussed in Prob. 18.123.)

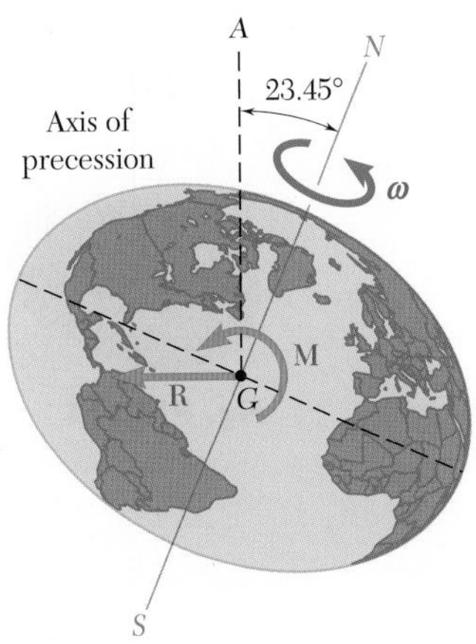

Fig. P18.118

18.119 Show that for an axisymmetrical body under no force, the rates of precession and spin can be expressed, respectively, as

$$\dot{\phi} = \frac{H_G}{I'}$$

and

$$\dot{\psi} = \frac{H_G \cos\theta(I' - I)}{II'}$$

where H_G is the constant value of the angular momentum of the body.

18.120 (*a*) Show that for an axisymmetrical body under no force, the rate of precession can be expressed as

$$\dot{\phi} = \frac{I\omega_z}{I' \cos\theta}$$

where ω_2 is the rectangular component of $\boldsymbol{\omega}$ along the axis of symmetry of the body. (*b*) Use this result to check that the condition (18.44) for steady precession is satisfied by an axisymmetrical body under no force.

18.121 Show that the angular velocity vector $\boldsymbol{\omega}$ of an axisymmetrical body under no force is observed from the body itself to rotate about the axis of symmetry at the constant rate

$$n = \frac{I' - I}{I'}\omega_z$$

where ω_2 is the rectangular component of $\boldsymbol{\omega}$ along the axis of symmetry of the body.

18.122 For an axisymmetrical body under no force, prove (*a*) that the rate of retrograde precession can never be less than twice the rate of spin of the body about its axis of symmetry, (*b*) that in Fig. 18.24 the axis of symmetry of the body can never lie within the space cone.

18.123 Using the relation given in Prob. 18.121, determine the period of precession of the north pole of the earth about the axis of symmetry of the earth. The earth may be approximated by an oblate spheroid of axial moment of inertia I and of transverse moment of inertia $I' = 0.9967I$. (*Note:* Actual observations show a period of precession of the north pole of about 432.5 mean solar days; the difference between the observed and computed periods is due to the fact that the earth is not a perfectly rigid body. The free precession considered here should not be confused with the much slower precession of the equinoxes, which is a forced precession. See Prob. 18.118.)

18.124 A coin is tossed into the air. It is observed to spin at the rate of 600 rpm about an axis GC perpendicular to the coin and to precess about the vertical direction GD. Knowing that GC forms an angle of 15° with GD, determine (*a*) the angle that the angular velocity $\boldsymbol{\omega}$ of the coin forms with GD, (*b*) the rate of precession of the coin about GD.

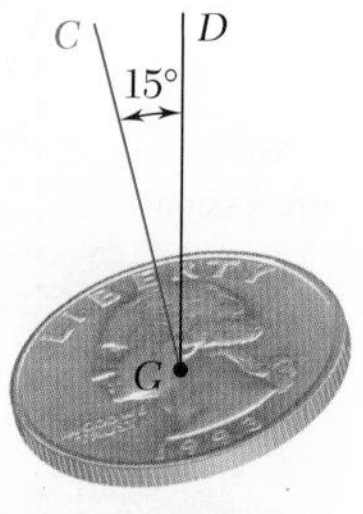

Fig. P18.124

18.125 The angular velocity vector of a football which has just been kicked is horizontal, and its axis of symmetry OC is oriented as shown. Knowing that the magnitude of the angular velocity is 200 rpm and that the ratio of the axis and transverse moments of inertia is $I/I' = \frac{1}{3}$, determine (*a*) the orientation of the axis of precession OA, (*b*) the rates of precession and spin.

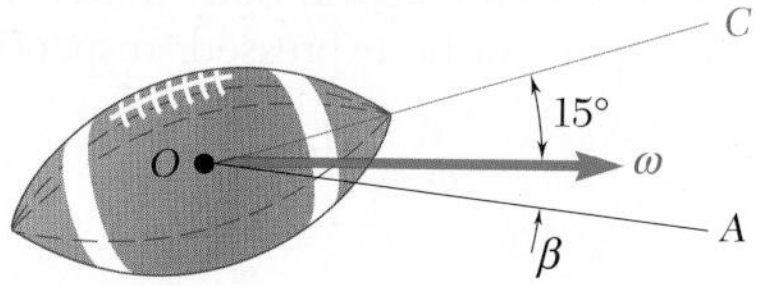

Fig. P18.125

18.126 The space capsule has no angular velocity when the jet at A is activated for 1 s in a direction parallel to the x axis. Knowing that the capsule has a mass of 1000 kg, that its radii of gyration are $\bar{k}_z = \bar{k}_y = 1.00$ m and $\bar{k}_z = 1.25$ m, and that the jet at A produces a thrust of 50 N, determine the axis of precession and the rates of precession and spin after the jet has stopped.

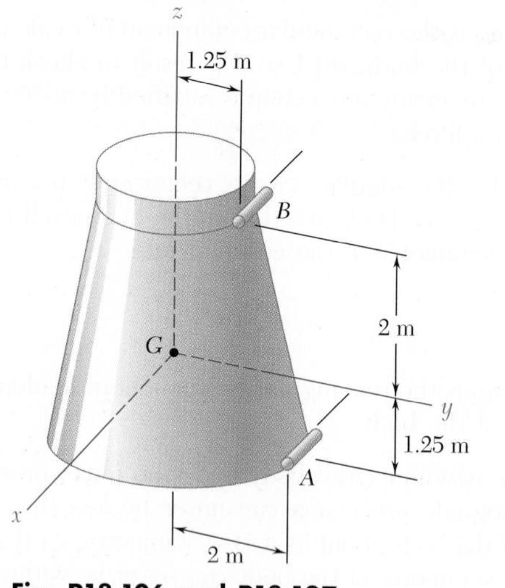

Fig. P18.126 and P18.127

18.127 The space capsule has an angular velocity $\boldsymbol{\omega} = (0.02 \text{ rad/s})\mathbf{j} + (0.10 \text{ rad/s})\mathbf{k}$ when the jet at B is activated for 1 s in a direction parallel to the x axis. Knowing that the capsule has a mass of 1000 kg, that its radii of gyration are $\bar{k}_x = \bar{k}_y = 1.00$ m and $\bar{k}_z = 1.25$ m, and that the jet at B produces a thrust of 50 N, determine the axis of precession and the rates of precession and spin after the jet has stopped.

18.128 Solve Sample Prob. 18.6, assuming that the meteorite strikes the satellite at C with a velocity $\mathbf{v}_0 = (2000 \text{ m/s})\mathbf{i}$.

18.129 A 400-kg geostationary satellite is spinning with an angular velocity $\boldsymbol{\omega}_0 = (1.5 \text{ rad/s})\mathbf{j}$ when it is hit at B by a 200 g meteorite traveling with a velocity $\mathbf{v}_0 = -(500 \text{ m/s})\mathbf{i} + (400 \text{ m/s})\mathbf{j} + (1250 \text{ m/s})\mathbf{k}$ relative to the satellite. Knowing that $b = 500$ mm and that the radii of gyration of the satellite are $\bar{k}_x = \bar{k}_z = 700$ mm and $\bar{k}_y = 800$ mm determine the precession axis and the rates of precession and spin of the satellite after the impact.

18.130 Solve Prob. 18.129, assuming that the meteorite hits the satellite at A instead of B.

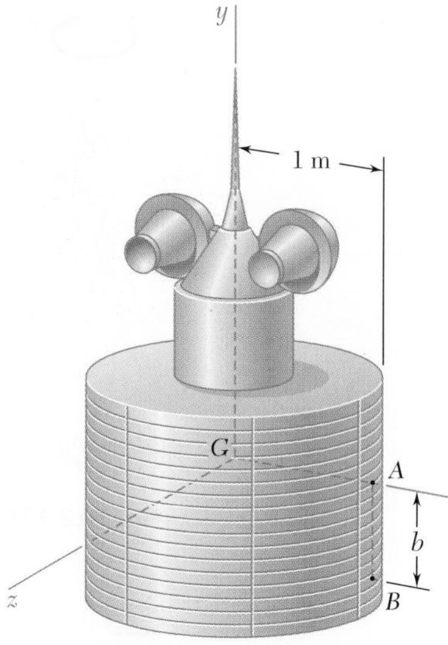

Fig. P18.129

18.131 A homogeneous rectangular plate of mass m and sides c and $2c$ is held at A and B by a fork-ended shaft of negligible mass which is supported by a bearing at C. The plate is free to rotate about AB, and the shaft is free to rotate about a horizontal axis through C. Knowing that, initially, $\theta_0 = 40°$, $\dot{\theta}_0 = 0$, and $\dot{\phi}_0 = 10$ rad/s, determine for the ensuing motion (*a*) the range of values of θ, (*b*) the minimum value of $\dot{\phi}$, (*c*) the maximum value of $\dot{\theta}$.

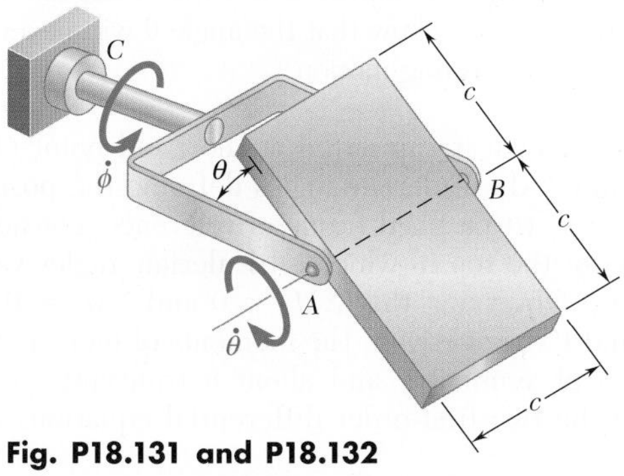

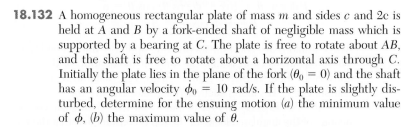

Fig. P18.131 and P18.132

18.132 A homogeneous rectangular plate of mass m and sides c and 2c is held at A and B by a fork-ended shaft of negligible mass which is supported by a bearing at C. The plate is free to rotate about AB, and the shaft is free to rotate about a horizontal axis through C. Initially the plate lies in the plane of the fork ($\theta_0 = 0$) and the shaft has an angular velocity $\dot{\phi}_0 = 10$ rad/s. If the plate is slightly disturbed, determine for the ensuing motion (*a*) the minimum value of $\dot{\phi}$, (*b*) the maximum value of $\dot{\theta}$.

18.133 A homogeneous disk of radius 180 mm is welded to a rod AG of length 360 mm and of negligible mass which is connected by a clevis to a vertical shaft AB. The rod and disk can rotate freely about a horizontal axis AC, and shaft AB can rotate freely about a vertical axis. Initially rod AG is horizontal ($\theta_0 = 90°$) and has no angular velocity about AC. Knowing that the maximum value $\dot{\phi}_m$ of the angular velocity of shaft AB in the ensuing motion is twice its initial value $\dot{\phi}_0$, determine (*a*) the minimum value of θ, (*b*) the initial angular velocity $\dot{\phi}_0$ of shaft AB.

18.134 A homogeneous disk of radius 180 mm is welded to a rod AG of length 360 mm and of negligible mass which is connected by a clevis to a vertical shaft AB. The rod and disk can rotate freely about a horizontal axis AC, and shaft AB can rotate freely about a vertical axis. Initially rod AG is horizontal ($\theta_0 = 90°$) and has no angular velocity about AC. Knowing that the smallest value of θ in the ensuing motion is 30°, determine (*a*) the initial angular velocity of shaft AB, (*b*) its maximum angular velocity.

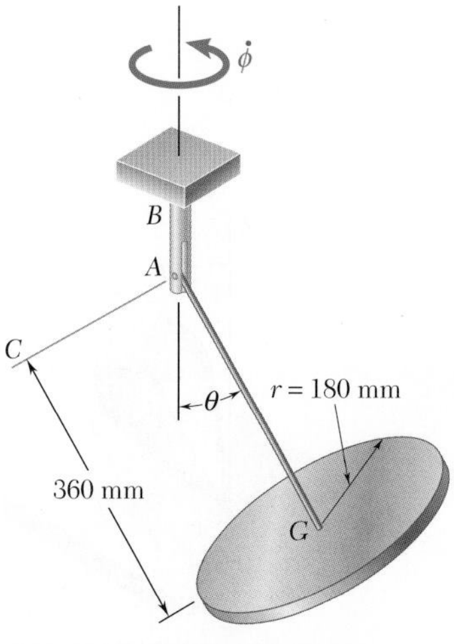

Fig. P18.133 and P18.134

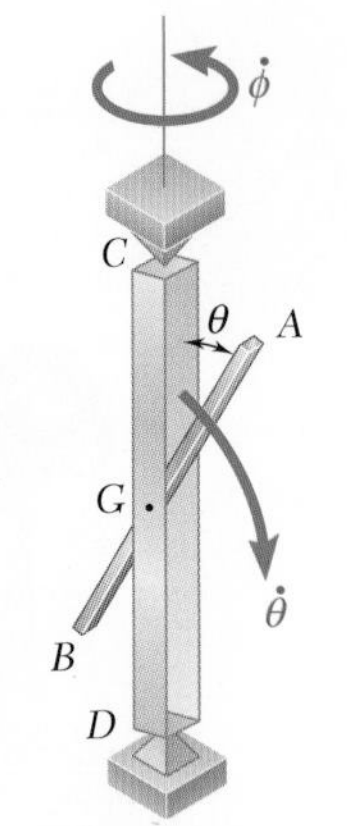

Fig. P18.135

18.135 The slender homogeneous rod AB of mass m and length L is free to rotate about a horizontal axle through its mass center G. The axle is supported by a frame of negligible mass which is free to rotate about the vertical CD. Knowing that, initially, $\theta = \theta_0$, $\dot{\theta} = 0$, and $\dot{\phi} = \dot{\phi}_0$, show that the rod will oscillate about the horizontal axle and determine (*a*) the range of values of angle θ during this motion, (*b*) the maximum value of $\dot{\theta}$, (*c*) the minimum value of $\dot{\phi}$.

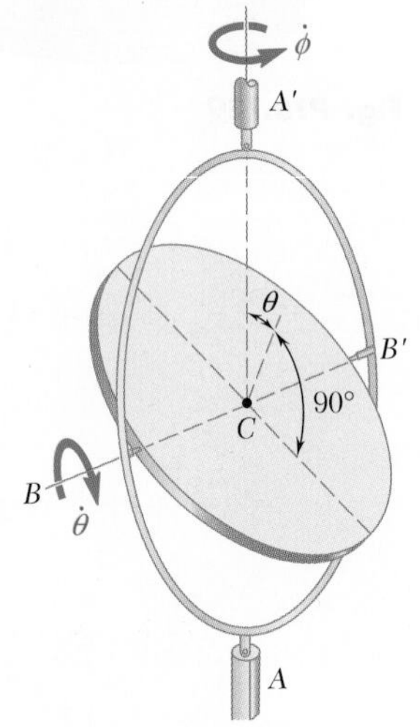

Fig. P18.136

18.136 The gimbal $ABA'B'$ is of negligible mass and may rotate freely about the vertical AA'. The uniform disk of radius a and mass m may rotate freely about its diameter BB', which is also the horizontal diameter of the gimbal. (*a*) Applying the principle of conservation of energy, and observing that, since $\Sigma M_{AA'} = 0$, the component of the angular momentum of the disk along the fixed axis AA' must be constant, write two first-order differential equations defining the motion of the disk. (*b*) Given the initial conditions $\theta_0 \neq 0$, $\dot{\phi}_0 \neq 0$, and $\dot{\theta}_0 = 0$, express the rate of nutation $\dot{\theta}$ as a function of θ. (*c*) Show that the angle θ will never be larger than θ_0 during the ensuing motion.

***18.137** The top shown is supported at the fixed point O. Denoting by ϕ, θ, and ψ the Eulerian angles defining the position of the top with respect to a fixed frame of reference, consider the general motion of the top in which all Eulerian angles vary.

(*a*) Observing that $\Sigma M_Z = 0$ and $\Sigma M_z = 0$, and denoting by I and I', respectively, the moments of inertia of the top about its axis of symmetry and about a transverse axis through O, derive the two first-order differential equations of motion

$$I'\dot{\phi}\sin^2\theta + I(\dot{\psi} + \dot{\phi}\cos\theta)\cos\theta = \alpha \qquad (1)$$
$$I(\dot{\psi} + \dot{\phi}\cos\theta) = \beta \qquad (2)$$

where α and β are constants depending upon the initial conditions. These equations express that the angular momentum of the top is conserved about both the Z and z axes, i.e., that the rectangular component of $\mathbf{H}_O$ along each of these axes is constant.

(*b*) Use Eqs. (1) and (2) to show that the rectangular component ω_z of the angular velocity of the top is constant and that the rate of precession $\dot{\phi}$ depends upon the value of the angle of nutation θ.

***18.138** (*a*) Applying the principle of conservation of energy, derive a third differential equation for the general motion of the top of Prob. 18.137.

(*b*) Eliminating the derivatives $\dot{\phi}$ and $\dot{\psi}$ from the equation obtained and from the two equations of Prob. 18.137, show that the rate of nutation $\dot{\theta}$ is defined by the differential equation $\dot{\theta}^2 = f(\theta)$, where

$$f(\theta) = \frac{1}{I'}\left(2E - \frac{\beta^2}{I} - 2mgc\cos\theta\right) - \left(\frac{\alpha - \beta\cos\theta}{I'\sin\theta}\right)^2 \qquad (1)$$

(*c*) Further show, by introducing the auxiliary variable $x = \cos\theta$, that the maximum and minimum values of θ can be obtained by solving for x the cubic equation

$$\left(2E - \frac{\beta^2}{I} - 2mgcx\right)(1 - x^2) - \frac{1}{I'}(\alpha - \beta x)^2 = 0 \qquad (2)$$

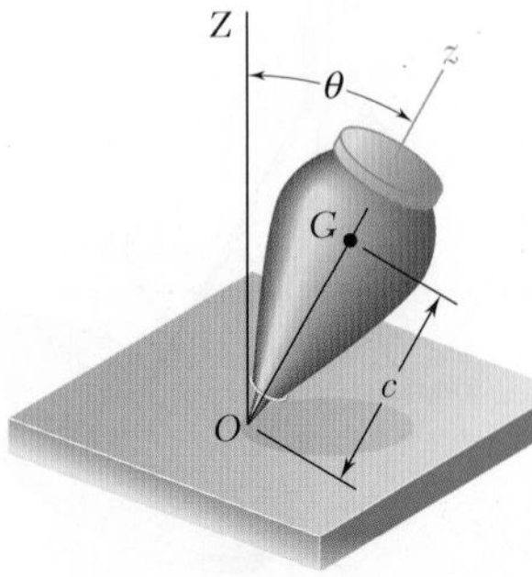

Fig. P18.137 and P18.138

***18.139** A solid cone of height 180 mm with a circular base of radius 60 mm is supported by a ball and socket at A. The cone is released from the position $\theta_0 = 30°$ with a rate of spin $\dot{\psi}_0 = 300$ rad/s, a rate of precession $\dot{\phi}_0 = 20$ rad/s, and a zero rate of nutation. Determine (*a*) the maximum value of θ in the ensuing motion, (*b*) the corresponding values of the rates of spin and precession. [*Hint:* Use Eq. (2) of Prob. 18.138; you can either solve this equation numerically or reduce it to a quadratic equation, since one of its roots is known.]

***18.140** A solid cone of height 180 mm with a circular base of radius 60 mm is supported by a ball and socket at A. The cone is released from the position $\theta_0 = 30°$ with a rate of spin $\dot{\psi}_0 = 300$ rad/s, a rate of precession $\dot{\phi}_0 = -4$ rad/s, and a zero rate of nutation. Determine (*a*) the maximum value of θ in the ensuing motion, (*b*) the corresponding values of the rates of spin and precession, (*c*) the value of θ for which the sense of the precession is reversed. (See hint of Prob. 18.139.)

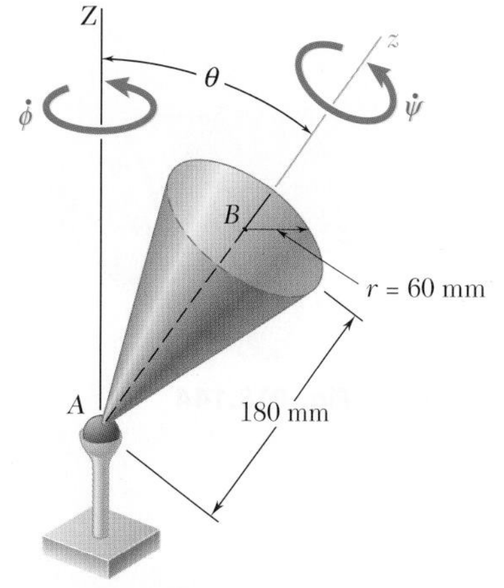

Fig. *P18.139* and *P18.140*

***18.141** A homogeneous sphere of mass m and radius a is welded to a rod AB of negligible mass, which is held by a ball-and-socket support at A. The sphere is released in the position $\beta = 0$ with a rate of precession $\dot{\phi}_0 = \sqrt{17g/11a}$ with no spin or nutation. Determine the largest value of β in the ensuing motion.

***18.142** A homogeneous sphere of mass m and radius a is welded to a rod AB of negligible mass, which is held by a ball-and-socket support at A. The sphere is released in the position $\beta = 0$ with a rate of precession $\dot{\phi} = \dot{\phi}_0$ with no spin or nutation. Knowing that the largest value of β in the ensuing motion is 30°, determine (*a*) the rate of precession $\dot{\phi}_0$ of the sphere in its initial position, (*b*) the rates of precession and spin when $\beta = 30°$.

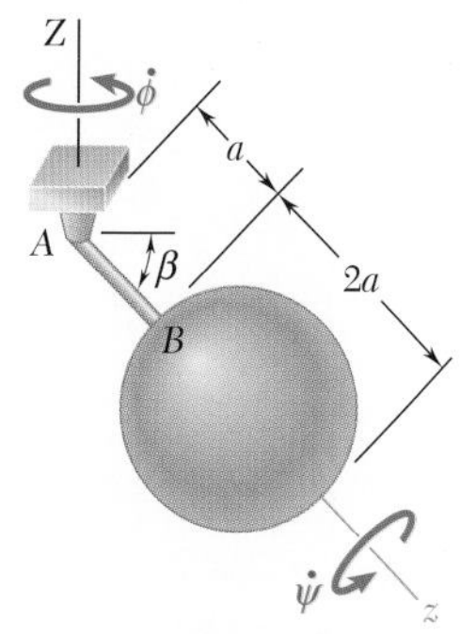

Fig. *P18.141* and *P18.142*

***18.143** Consider a rigid body of arbitrary shape which is attached at its mass center O and subjected to no force other than its weight and the reaction of the support at O.

(*a*) Prove that the angular momentum $\mathbf{H}_O$ of the body about the fixed point O is constant in magnitude and direction, that the kinetic energy T of the body is constant, and that the projection along $\mathbf{H}_O$ of the angular velocity $\boldsymbol{\omega}$ of the body is constant.

(*b*) Show that the tip of the vector $\boldsymbol{\omega}$ describes a curve on a fixed plane in space (called the *invariable plane*), which is perpendicular to $\mathbf{H}_O$ and at a distance $2T/H_O$ from O.

(*c*) Show that with respect to a frame of reference attached to the body and coinciding with its principal axes of inertia, the tip of the vector $\boldsymbol{\omega}$ appears to describe a curve on an ellipsoid of equation

$$I_x\omega_x^2 + I_y\omega_y^2 + I_z\omega_z^2 = 2T = \text{constant}$$

The ellipsoid (called the *Poinsot ellipsoid*) is rigidly attached to the body and is of the same shape as the ellipsoid of inertia, but of a different size.

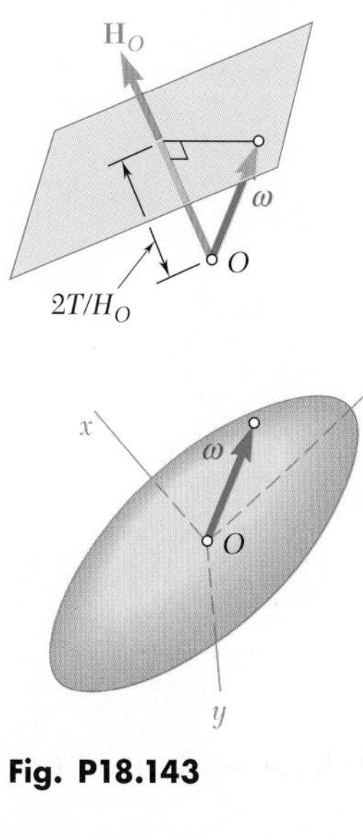

Fig. P18.143

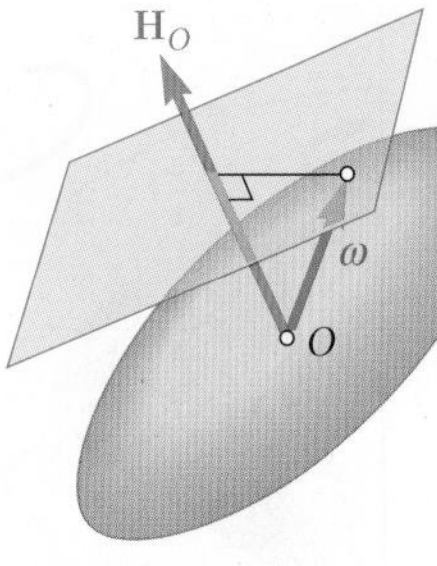

Fig. P18.144

***18.144** Referring to Prob. 18.143, (*a*) prove that the Poinsot ellipsoid is tangent to the invariable plane, (*b*) show that the motion of the rigid body must be such that the Poinsot ellipsoid appears to roll on the invariable plane. [*Hint:* In part *a*, show that the normal to the Poinsot ellipsoid at the tip of $\boldsymbol{\omega}$ is parallel to $\mathbf{H}_O$. It is recalled that the direction of the normal to a surface of equation $F(x, y, z) =$ constant at a point P is the same as that of **grad** F at point P.]

***18.145** Using the results obtained in Probs. 18.143 and 18.144, show that for an axisymmetrical body attached at its mass center O and under no force other than its weight and the reaction at O, the Poinsot ellipsoid is an ellipsoid of revolution and the space and body cones are both circular and are tangent to each other. Further show that (*a*) the two cones are tangent externally, and the precession is direct, when $I < I'$, where I and I' denote, respectively, the axial and transverse moment of inertia of the body, (*b*) the space cone is inside the body cone, and the precession is retrograde, when $I > I'$.

***18.146** Refer to Probs. 18.143 and 18.144.

(*a*) Show that the curve (called *polhode*) described by the tip of the vector $\boldsymbol{\omega}$ with respect to a frame of reference coinciding with the principal axes of inertia of the rigid body is defined by the equations

$$I_x\omega_x^2 + I_y\omega_y^2 + I_z\omega_z^2 = 2T = \text{constant} \qquad (1)$$
$$I_x^2\omega_x^2 + I_y^2\omega_y^2 + I_z^2\omega_z^2 = H_O^2 = \text{constant} \qquad (2)$$

and that this curve can, therefore, be obtained by intersecting the Poinsot ellipsoid with the ellipsoid defined by Eq. (2).

(*b*) Further show, assuming $I_x > I_y > I_z$, that the polhodes obtained for various values of H_O have the shapes indicated in the figure.

(*c*) Using the result obtained in part *b*, show that a rigid body under no force can rotate about a fixed centroidal axis if, and only if, that axis coincides with one of the principal axes of inertia of the body, and that the motion will be stable if the axis of rotation coincides with the major or minor axis of the Poinsot ellipsoid (z or x axis in the figure) and unstable if it coincides with the intermediate axis (y axis).

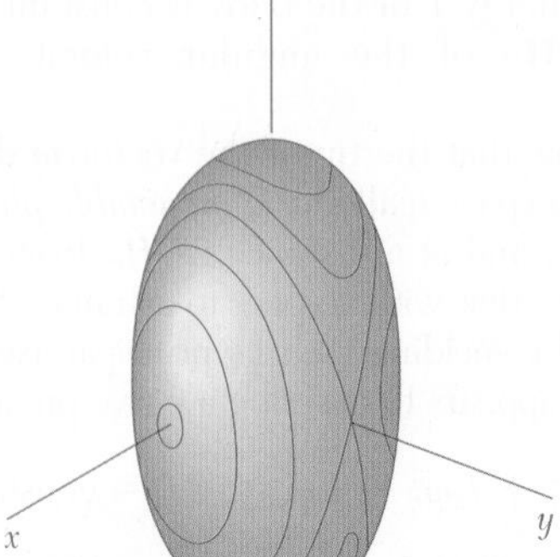

Fig. P18.146

REVIEW AND SUMMARY

This chapter was devoted to the kinetic analysis of the motion of rigid bodies in three dimensions.

Fundamental equations of motion for a rigid body

We first noted [Sec. 18.1] that the two fundamental equations derived in Chap. 14 for the motion of a system of particles

$$\Sigma \mathbf{F} = m\bar{\mathbf{a}} \tag{18.1}$$
$$\Sigma \mathbf{M}_G = \dot{\mathbf{H}}_G \tag{18.2}$$

provide the foundation of our analysis, just as they did in Chap. 16 in the case of the plane motion of rigid bodies. The computation of the angular momentum $\mathbf{H}_G$ of the body and of its derivative $\dot{\mathbf{H}}_G$, however, are now considerably more involved.

Angular momentum of a rigid body in three dimensions

In Sec. 18.2, we saw that the rectangular components of the angular momentum $\mathbf{H}_G$ of a rigid body can be expressed as follows in terms of the components of its angular velocity $\boldsymbol{\omega}$ and of its centroidal moments and products of inertia:

$$\begin{aligned} H_x &= +\bar{I}_x\,\omega_x - \bar{I}_{xy}\omega_y - \bar{I}_{xz}\omega_z \\ H_y &= -\bar{I}_{yx}\omega_x + \bar{I}_y\,\omega_y - \bar{I}_{yz}\omega_z \\ H_z &= -\bar{I}_{zx}\omega_x - \bar{I}_{zy}\omega_y + \bar{I}_z\,\omega_z \end{aligned} \tag{18.7}$$

If *principal axes of inertia* $Gx'y'z'$ are used, these relations reduce to

$$H_{x'} = \bar{I}_{x'}\omega_{x'} \qquad H_{y'} = \bar{I}_{y'}\omega_{y'} \qquad H_{z'} = \bar{I}_{z'}\omega_{z'} \tag{18.10}$$

We observed that, in general, *the angular momentum* $\mathbf{H}_G$ *and the angular velocity* $\boldsymbol{\omega}$ *do not have the same direction* (Fig. 18.25). They will, however, have the same direction if $\boldsymbol{\omega}$ is directed along one of the principal axes of inertia of the body.

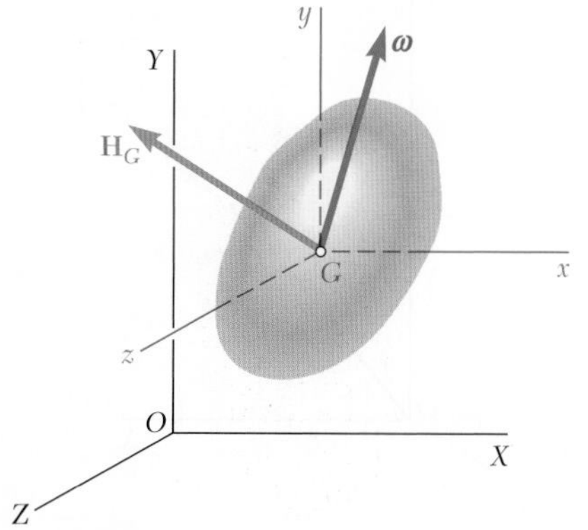

Fig. 18.25

Recalling that the system of the momenta of the particles forming a rigid body can be reduced to the vector $m\bar{\mathbf{v}}$ attached at G and the couple $\mathbf{H}_G$ (Fig. 18.26), we noted that, once the linear momentum $m\bar{\mathbf{v}}$ and the angular momentum $\mathbf{H}_G$ of a rigid body have been determined, the angular momentum $\mathbf{H}_O$ of the body about any given point O can be obtained by writing

Angular momentum about a given point

$$\mathbf{H}_O = \bar{\mathbf{r}} \times m\bar{\mathbf{v}} + \mathbf{H}_G \tag{18.11}$$

Rigid body with a fixed point

In the particular case of a rigid body *constrained to rotate about a fixed point O*, the components of the angular momentum $\mathbf{H}_O$ of the body about O can be obtained directly from the components of its angular velocity and from its moments and products of inertia with respect to axes through O. We wrote

$$\begin{aligned} H_x &= +I_x\,\omega_x - I_{xy}\omega_y - I_{xz}\omega_z \\ H_y &= -I_{yx}\omega_x + I_y\,\omega_y - I_{yz}\omega_z \\ H_z &= -I_{zx}\omega_x - I_{zy}\omega_y + I_z\,\omega_z \end{aligned} \tag{18.13}$$

Principle of impulse and momentum

The *principle of impulse and momentum* for a rigid body in three-dimensional motion [Sec. 18.3] is expressed by the same fundamental formula that was used in Chap. 17 for a rigid body in plane motion,

$$\textbf{Syst Momenta}_1 + \textbf{Syst Ext Imp}_{1\rightarrow 2} = \textbf{Syst Momenta}_2 \tag{17.4}$$

but the systems of the initial and final momenta should now be represented as shown in Fig. 18.26, and $\mathbf{H}_G$ should be computed from the relations (18.7) or (18.10) [Sample Probs. 18.1 and 18.2].

Kinetic energy of a rigid body in three dimensions

The *kinetic energy* of a rigid body in three-dimensional motion can be divided into two parts [Sec. 18.4], one associated with the motion of its mass center G and the other with its motion about G. Using principal centroidal axes x', y', z', we wrote

$$T = \tfrac{1}{2}m\bar{v}^2 + \tfrac{1}{2}(\bar{I}_{x'}\omega_{x'}^2 + \bar{I}_{y'}\omega_{y'}^2 + \bar{I}_{z'}\omega_{z'}^2) \tag{18.17}$$

where $\bar{\mathbf{v}}$ = velocity of mass center
$\boldsymbol{\omega}$ = angular velocity
m = mass of rigid body
$\bar{I}_{x'}, \bar{I}_{y'}, \bar{I}_{z'}$ = principal centroidal moments of inertia

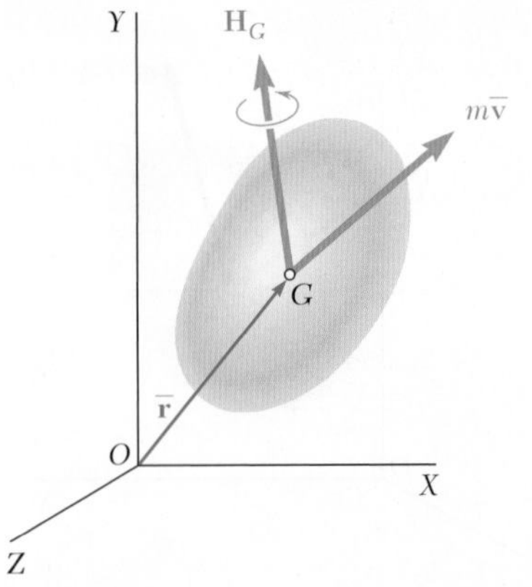

Fig. 18.26

We also noted that, in the case of a rigid body *constrained to rotate about a fixed point O*, the kinetic energy of the body can be expressed as

$$T = \frac{1}{2}(I_{x'}\omega_{x'}^2 + I_{y'}\omega_{y'}^2 + I_{z'}\omega_{z'}^2) \tag{18.20}$$

where the x', y', and z' axes are the principal axes of inertia of the body at O. The results obtained in Sec. 18.4 make it possible to extend to the three-dimensional motion of a rigid body the application of the *principle of work and energy* and of the *principle of conservation of energy*.

Using a rotating frame to write the equations of motion of a rigid body in space

The second part of the chapter was devoted to the application of the fundamental equations

$$\Sigma \mathbf{F} = m\bar{\mathbf{a}} \tag{18.1}$$

$$\Sigma \mathbf{M}_G = \dot{\mathbf{H}}_G \tag{18.2}$$

to the motion of a rigid body in three dimensions. We first recalled [Sec. 18.5] that $\mathbf{H}_G$ represents the angular momentum of the body relative to a centroidal frame $GX'Y'Z'$ of fixed orientation (Fig. 18.27)

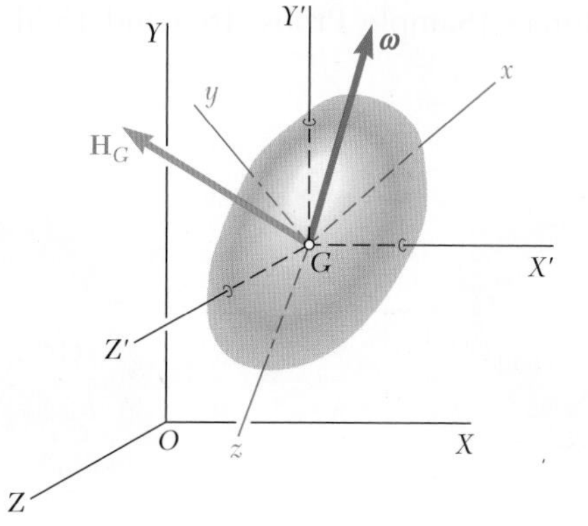

Fig. 18.27

and that $\dot{\mathbf{H}}_G$ in Eq. (18.2) represents the rate of change of $\mathbf{H}_G$ with respect to that frame. We noted that, as the body rotates, its moments and products of inertia with respect to the frame $GX'Y'Z'$ change continually. Therefore, it is more convenient to use a rotating frame $Gxyz$ when resolving $\boldsymbol{\omega}$ into components and computing the moments and products of inertia that will be used to determine $\mathbf{H}_G$ from Eqs. (18.7) or (18.10). However, since $\dot{\mathbf{H}}_G$ in Eq. (18.2) represents the rate of change of $\mathbf{H}_G$ with respect to the frame $GX'Y'Z'$ of fixed orientation, we must use the method of Sec. 15.10 to determine its value. Recalling Eq. (15.31), we wrote

$$\dot{\mathbf{H}}_G = (\dot{\mathbf{H}}_G)_{Gxyz} + \boldsymbol{\Omega} \times \mathbf{H}_G \tag{18.22}$$

where $\mathbf{H}_G$ = angular momentum of body with respect to frame $GX'Y'Z'$ of fixed orientation

$(\dot{\mathbf{H}}_G)_{Gxyz}$ = rate of change of $\mathbf{H}_G$ with respect to rotating frame $Gxyz$, to be computed from relations (18.7)

$\boldsymbol{\Omega}$ = angular velocity of the rotating frame $Gxyz$

Substituting for $\dot{\mathbf{H}}_G$ from (18.22) into (18.2), we obtained

$$\Sigma \mathbf{M}_G = (\dot{\mathbf{H}}_G)_{Gxyz} + \boldsymbol{\Omega} \times \mathbf{H}_G \tag{18.23}$$

If the rotating frame is actually attached to the body, its angular velocity $\boldsymbol{\Omega}$ is identically equal to the angular velocity $\boldsymbol{\omega}$ of the body. There are many applications, however, where it is advantageous to use a frame of reference which is not attached to the body but rotates in an independent manner [Sample Prob. 18.5].

Euler's equations of motion. D'Alembert's principle

Setting $\boldsymbol{\Omega} = \boldsymbol{\omega}$ in Eq. (18.23), using principal axes, and writing this equation in scalar form, we obtained *Euler's equations of motion* [Sec. 18.6]. A discussion of the solution of these equations and of the scalar equations corresponding to Eq. (18.1) led us to extend d'Alembert's principle to the three-dimensional motion of a rigid body and to conclude that the system of the external forces acting on the rigid body is not only equipollent, but actually *equivalent* to the effective forces of the body represented by the vector $m\bar{\mathbf{a}}$ and the couple $\dot{\mathbf{H}}_G$ (Fig. 18.28). Problems involving the three-dimensional motion of a rigid body can be solved by considering the free-body-diagram equation represented in Fig. 18.28 and writing appropriate scalar equations relating the components or moments of the external and effective forces [Sample Probs. 18.3 and 18.5].

Free-body-diagram equation

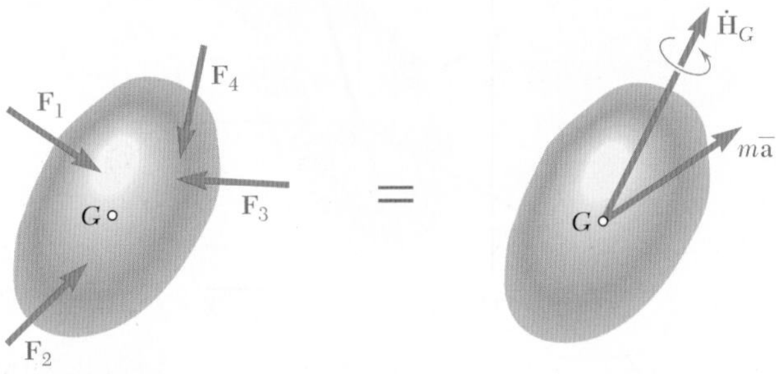

Fig. 18.28

Rigid body with a fixed point

In the case of a rigid body *constrained to rotate about a fixed point O*, an alternative method of solution, involving the moments of the forces and the rate of change of the angular momentum about point O, can be used. We wrote [Sec. 18.7]:

$$\Sigma \mathbf{M}_O = (\dot{\mathbf{H}}_O)_{Oxyz} + \boldsymbol{\Omega} \times \mathbf{H}_O \tag{18.28}$$

where $\Sigma \mathbf{M}_O$ = sum of moments about O of forces applied to rigid body
$\mathbf{H}_O$ = angular momentum of body with respect to fixed frame $OXYZ$
$(\dot{\mathbf{H}}_O)_{Oxyz}$ = rate of change of $\mathbf{H}_O$ with respect to rotating frame $Oxyz$, to be computed from relations (18.13)
$\boldsymbol{\Omega}$ = angular velocity of rotating frame $Oxyz$

This approach can be used to solve certain problems involving the rotation of a rigid body about a fixed axis [Sec. 18.8], for example, an unbalanced rotating shaft [Sample Prob. 18.4].

Motion of a gyroscope

In the last part of the chapter, we considered the motion of *gyroscopes* and other *axisymmetrical bodies*. Introducing the *Eulerian angles* ϕ, θ, and ψ to define the position of a gyroscope (Fig. 18.29), we observed that their derivatives $\dot{\phi}$, $\dot{\theta}$, and $\dot{\psi}$ represent, respectively, the rates of *precession, nutation,* and *spin* of the gyroscope [Sec. 18.9]. Expressing the angular velocity $\boldsymbol{\omega}$ in terms of these derivatives, we wrote

$$\boldsymbol{\omega} = -\dot{\phi} \sin\theta\, \mathbf{i} + \dot{\theta}\mathbf{j} + (\dot{\psi} + \dot{\phi}\cos\theta)\mathbf{k} \qquad (18.35)$$

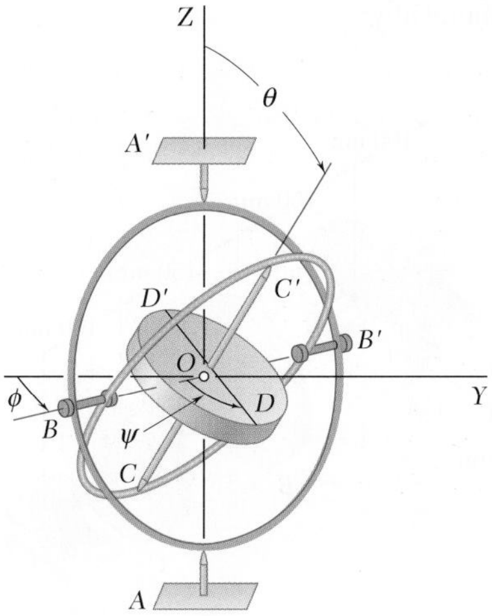

Fig. 18.29

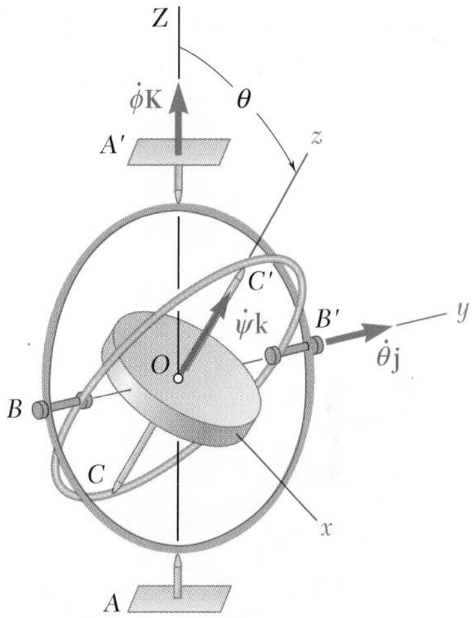

Fig. 18.30

where the unit vectors are associated with a frame *Oxyz* attached to the inner gimbal of the gyroscope (Fig. 18.30) and rotate, therefore, with the angular velocity

$$\boldsymbol{\Omega} = -\dot{\phi} \sin\theta\, \mathbf{i} + \dot{\theta}\mathbf{j} + \dot{\phi}\cos\theta\, \mathbf{k} \qquad (18.38)$$

Denoting by I the moment of inertia of the gyroscope with respect to its spin axis z and by I' its moment of inertia with respect to a transverse axis through O, we wrote

$$\mathbf{H}_O = -I'\dot{\phi} \sin\theta\, \mathbf{i} + I'\dot{\theta}\mathbf{j} + I(\dot{\psi} + \dot{\phi}\cos\theta)\mathbf{k} \qquad (18.36)$$

Substituting for $\mathbf{H}_O$ and $\boldsymbol{\Omega}$ into Eq. (18.28) led us to the differential equations defining the motion of the gyroscope.

Steady precession

In the particular case of the *steady precession* of a gyroscope [Sec. 18.10], the angle θ, the rate of precession $\dot{\phi}$, and the rate of spin $\dot{\psi}$ remain constant. We saw that such a motion is possible only if the moments of the external forces about O satisfy the relation

$$\Sigma\mathbf{M}_O = (I\omega_z - I'\dot{\phi}\cos\theta)\dot{\phi}\sin\theta\mathbf{j} \qquad (18.44)$$

i.e., if the external forces reduce to a couple of moment equal to the right-hand member of Eq. (18.44) and applied *about an axis perpendicular to the precession axis and to the spin axis* (Fig. 18.31). The chapter ended with a discussion of the motion of an axisymmetrical body spinning and precessing *under no force* [Sec. 18.11; Sample Prob. 18.6].

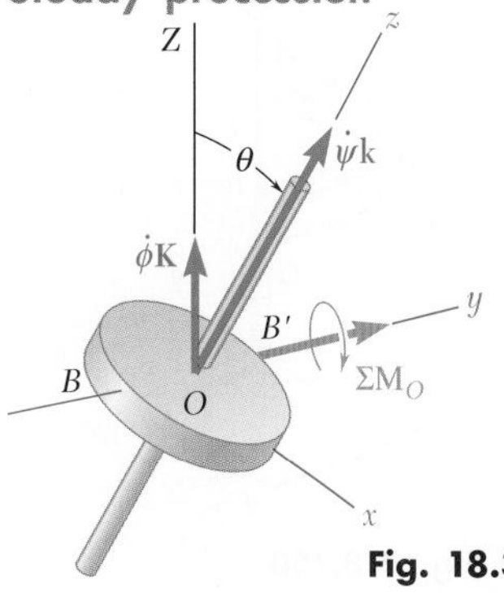

Fig. 18.31

REVIEW PROBLEMS

18.147 Three 12.5-kg rotor disks are attached to a shaft which rotates at 720 rpm. Disk A is attached eccentrically so that its mass center is 6 mm from the axis of rotation, while disks B and C are attached so that their mass centers coincide with the axis of rotation. Where should 1-kg masses be bolted to disks B and C to balance the system dynamically?

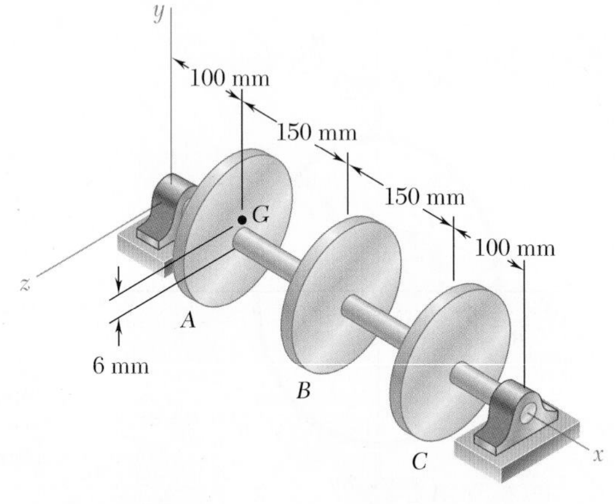

Fig. *P18.147*

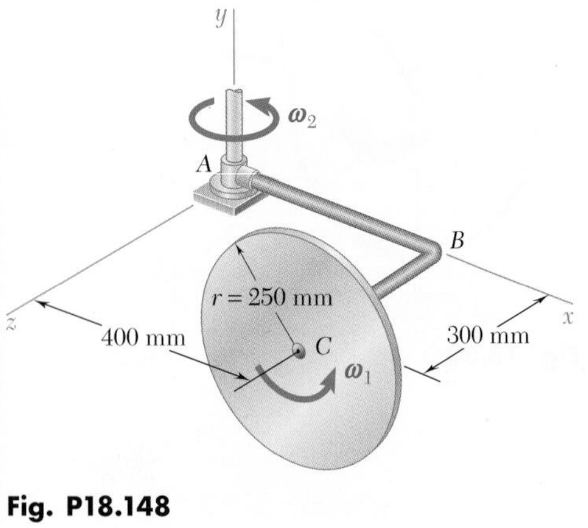

Fig. P18.148

18.148 A homogeneous disk of mass $m = 5$ kg rotates at the constant rate $\omega_1 = 8$ rad/s with respect to the bent axle ABC, which itself rotates at the constant rate $\omega_2 = 3$ rad/s about the y axis. Determine the angular momentum $\mathbf{H}_C$ of the disk about its center C.

18.149 A rod of uniform cross section is used to form the shaft shown. Denoting by m the total mass of the shaft and knowing that the shaft rotates with a constant angular velocity ω, determine (*a*) the angular momentum $\mathbf{H}_G$ of the shaft about its mass center G, (*b*) the angle formed by $\mathbf{H}_G$ and the axis AB, (*c*) the angular momentum of the shaft about point A.

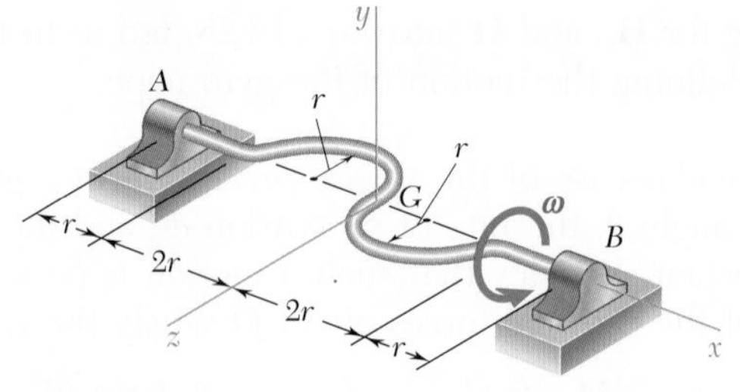

Fig. *P18.149*

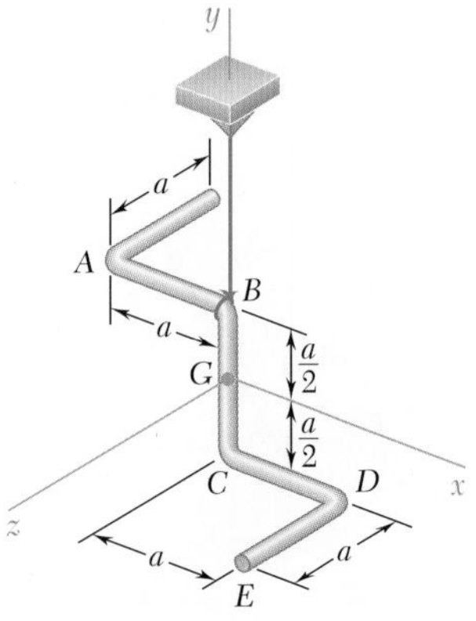

Fig. P18.150

18.150 A uniform rod of mass m and length $5a$ is bent into the shape shown and is suspended from a wire attached at point B. Knowing that the rod is hit at point A in the negative y direction and denoting the corresponding impulse by $-(F\,\Delta t)\mathbf{j}$, determine immediately after the impact (*a*) the velocity of the mass center G, (*b*) the angular velocity of the rod.

18.151 A four-bladed airplane propeller has a mass of 160 kg and a radius of gyration of 800 mm. Knowing that the propeller rotates at 1600 rpm as the airplane is traveling in a circular path of 600-m radius at 540 km/h, determine the magnitude of the couple exerted by the propeller on its shaft due to the rotation of the airplane.

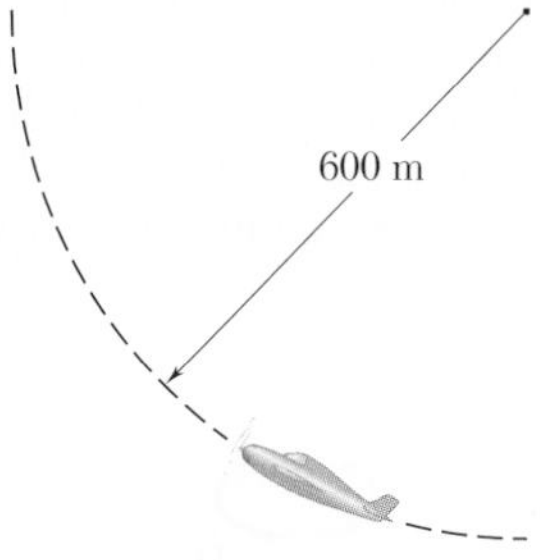

Fig. P18.151

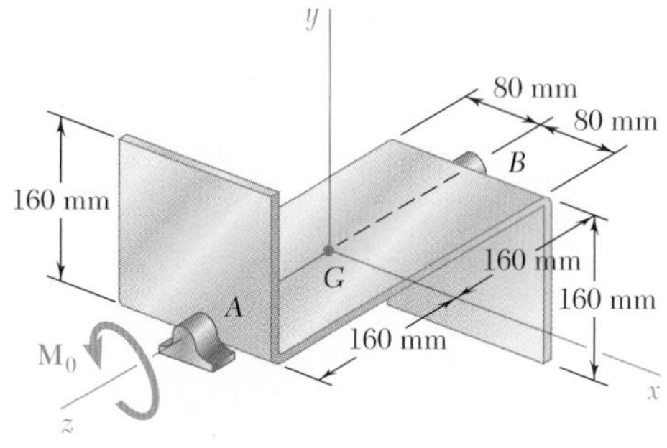

Fig. *P18.152*

18.152 A 2.4-kg piece of sheet steel with dimensions 160 × 640 mm was bent to form the component shown. The component is at rest ($\omega = 0$) when a couple $\mathbf{M}_0 = (0.8\ \text{N}\cdot\text{m})\mathbf{k}$ is applied to it. Determine (*a*) the angular acceleration of the component, (*b*) the dynamic reactions at *A* and *B* immediately after the couple is applied.

18.153 A homogeneous disk of mass $m = 3$ kg rotates at the constant rate $\omega_1 = 16$ rad/s with respect to arm *ABC*, which is welded to a shaft *DCE* rotating at the constant rate $\omega_2 = 8$ rad/s. Determine the dynamic reactions at *D* and *E*.

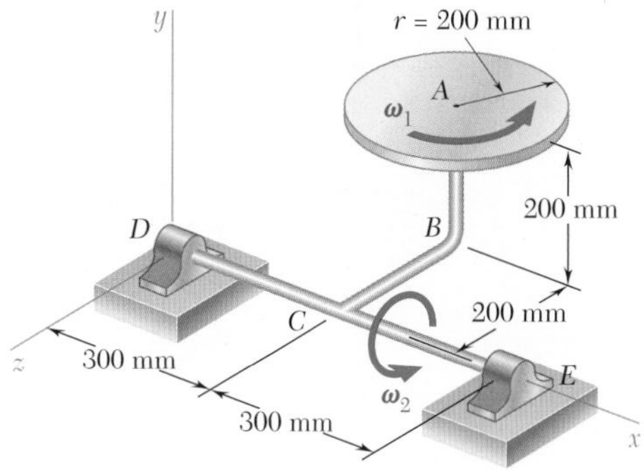

Fig. *P18.153*

18.154 A 48-kg advertising panel of length $2a = 2.4$ m and width $2b = 1.6$ m is kept rotating at a constant rate ω_1 about its horizontal axis by a small electric motor attached at *A* to frame *ACB*. This frame itself is kept rotating at a constant rate ω_2 about a vertical axis by a second motor attached at *C* to the column *CD*. Knowing that the panel and the frame complete a full revolution in 6 s and 12 s, respectively, express, as a function of the angle θ, the dynamic reaction exerted on column *CD* by its support at *D*.

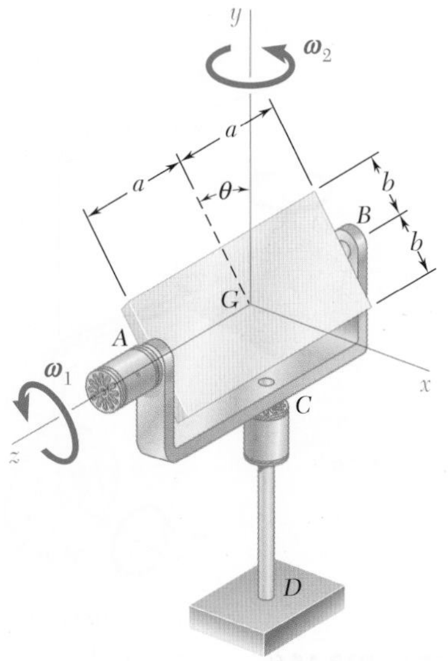

Fig. P18.154

18.155 A 2500-kg satellite is 2.4 m high and has octagonal bases of sides 1.2 m. The coordinate axes shown are the principal centroidal axes of inertia of the satellite, and its radii of gyration are $k_x = k_z = 0.90$ m and $k_y = 0.98$ m. The satellite is equipped with a main 500-N thruster E and four 20-N thrusters A, B, C, and D which can expel fuel in the positive y direction. The satellite is spinning at the rate of 36 rev/h about its axis of symmetry Gy, which maintains a fixed direction in space, when thrusters A and B are activated for 2 s. Determine (*a*) the precession axis of the satellite, (*b*) its rate of precession, (*c*) its rate of spin.

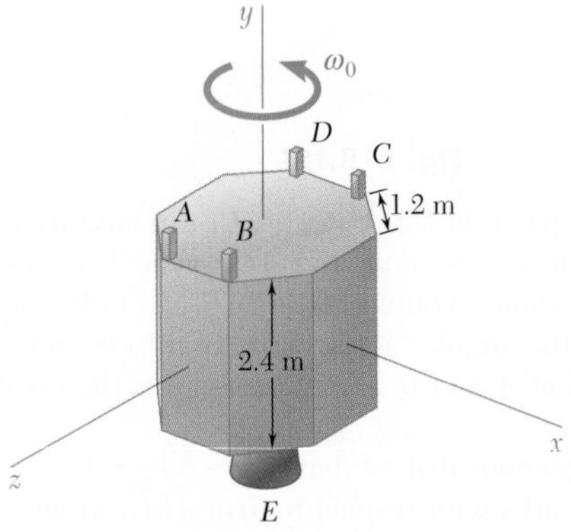

Fig. P18.155

18.156 A thin disk of mass $m = 4$ kg rotates with an angular velocity ω_2 with respect to arm OA, which itself rotates with an angular velocity ω_1 about the y axis. Determine (*a*) the couple $M_1\mathbf{j}$ which should be applied to arm OA to give it an angular acceleration $\boldsymbol{\alpha}_1 = (6 \text{ rad/s}^2)\mathbf{j}$ with $\omega_1 = 4$ rad/s, knowing that the disk rotates at the constant rate $\omega_2 = 12$ rad/s, (*b*) the force-couple system representing the dynamic reaction at O at that instant. Assume that arm OA has negligible mass.

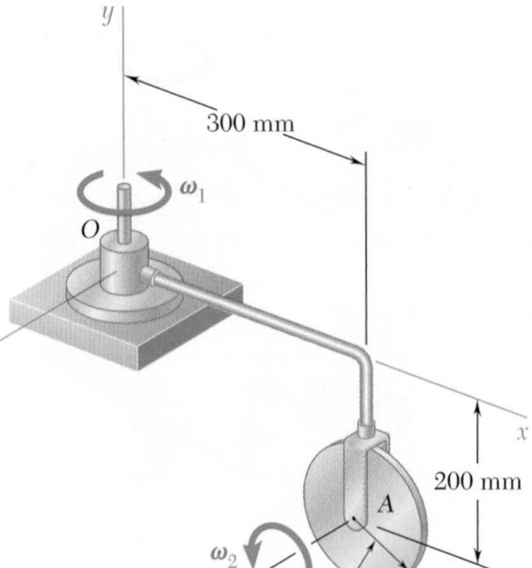

Fig. P18.156

18.157 A homogeneous disk of mass m is connected at A and B to a fork-ended shaft of negligible mass which is supported by a bearing at C. The disk is free to rotate about its horizontal diameter AB and the shaft is free to rotate about a vertical axis through C. Initially the disk lies in a vertical plane ($\theta_0 = 90°$) and the shaft has an angular velocity $\dot{\phi}_0 = 8$ rad/s. If the disk is slightly disturbed, determine for the ensuing motion (*a*) the minimum value of $\dot{\phi}$, (*b*) the maximum value of $\dot{\theta}$.

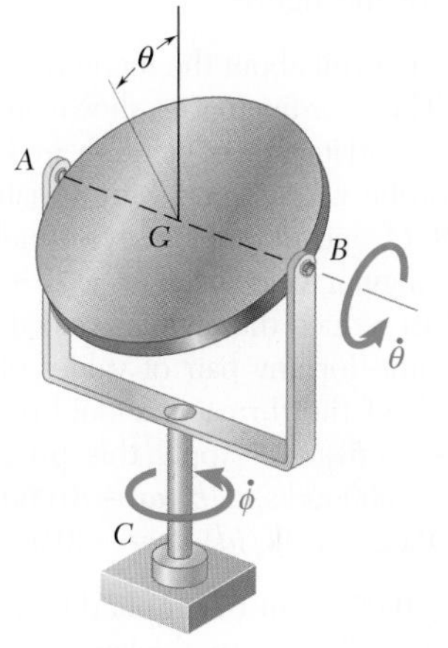

Fig. P18.157

18.158 The essential features of the gyrocompass are shown. The rotor spins at the rate $\dot{\psi}$ about an axis mounted in a single gimbal, which may rotate freely about the vertical axis AB. The angle formed by the axis of the rotor and the plane of the meridian is denoted by θ, and the latitude of the position on the earth is denoted by λ. We note that line OC is parallel to the axis of the earth, and we denote by $\boldsymbol{\omega}_e$ the angular velocity of the earth about its axis.

(*a*) Show that the equations of motion of the gyrocompass are

$$I'\ddot{\theta} + I\omega_z\omega_e \cos\lambda \sin\theta - I'\omega_e^2 \cos^2\lambda \sin\theta \cos\theta = 0$$

$$I\dot{\omega}_z = 0$$

where ω_z is the rectangular component of the total angular velocity $\boldsymbol{\omega}$ along the axis of the rotor, and I and I' are the moments of inertia of the rotor with respect to its axis of symmetry and a transverse axis through O, respectively.

(*b*) Neglecting the term containing ω_e^2, show that for small values of θ, we have

$$\ddot{\theta} + \frac{I\omega_z\omega_e \cos\lambda}{I'}\theta = 0$$

and that the axis of the gyrocompass oscillates about the north-south direction.

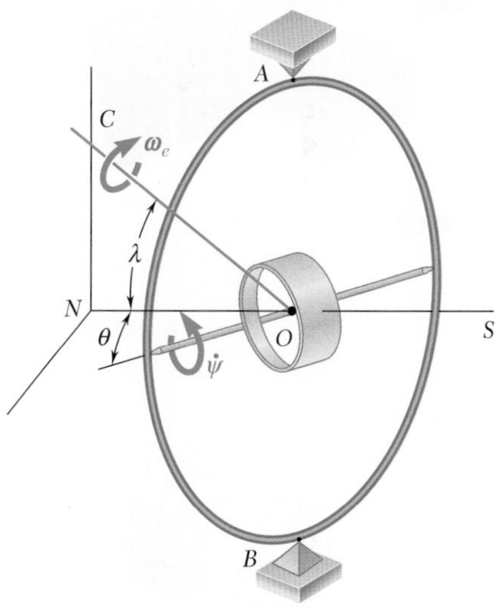

Fig. P18.158

COMPUTER PROBLEMS

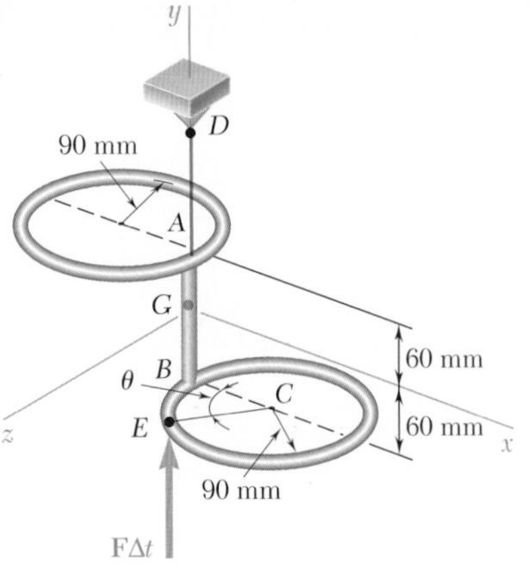

Fig. P18.C1

18.C1 A wire of uniform cross section weighing 60 g/m is used to form the wire figure shown, which is suspended from cord AD. An impulse $\mathbf{F}\,\Delta t = (2.5\text{ N}\cdot\text{s})\mathbf{j}$ is applied to the wire figure at point E. Use computational software to calculate and plot immediately after the impact, for values of θ from 0 to 180°, (*a*) the velocity of the mass center of the wire figure, (*b*) the angular velocity of the figure.

18.C2 A 2500-kg probe in orbit about the moon is 2.4 m high and has octagonal bases of sides 1.2 m. The coordinate axes shown are the principal centroidal axes of inertia of the probe, and its radii of gyration are $k_x = 0.98$ m, $k_y = 1.06$ m, and $k_z = 1.02$ m. The probe is equipped with a main 500-N thruster E and four 20-N thrusters A, B, C, and D that can expel fuel in the positive y direction. The probe has an angular velocity $\boldsymbol{\omega} = \omega_x\,\mathbf{i} + \omega_z\,\mathbf{k}$ when two of the 20-N thrusters are used to reduce the angular velocity to zero. Use computational software to determine for any pair of values of ω_x and ω_z less than or equal to 0.06 rad/s, which of the thrusters should be used and for how long each of them should be activated. Apply this program assuming ω to be (*a*) $\boldsymbol{\omega} = (0.040\text{ rad/s})\mathbf{i} + (0.060\text{ rad/s})\mathbf{k}$, (*b*) $\boldsymbol{\omega} = (0.060\text{ rad/s})\mathbf{i} - (0.040\text{ rad/s})\mathbf{k}$, (*c*) $\boldsymbol{\omega} = (0.06\text{ rad/s})\mathbf{i} + (0.02\text{ rad/s})\mathbf{k}$, (*d*) $\boldsymbol{\omega} = -(0.06\text{ rad/s})\mathbf{i} - (0.02\text{ rad/s})\mathbf{k}$.

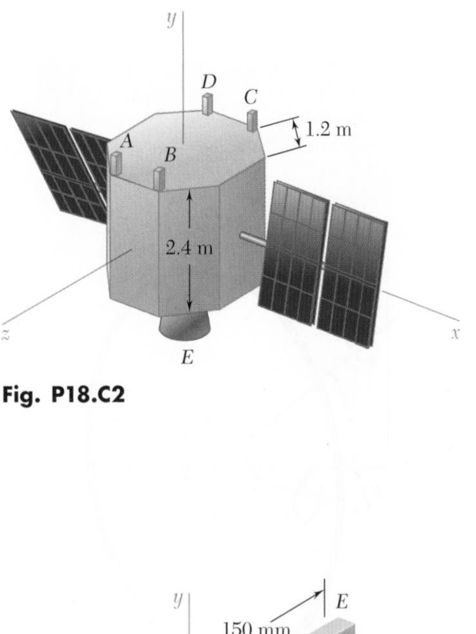

Fig. P18.C2

18.C3 A couple $\mathbf{M}_0 = (0.05\text{ N}\cdot\text{m})\mathbf{i}$ is applied to an assembly consisting of pieces of sheet aluminum of uniform thickness and of total mass 1.5 kg, which are welded to a light axle supported by bearings at A and B. Use computational software to determine the dynamic reactions exerted by the bearings on the axle at any time t after the couple has been applied. Resolve these reactions into components directed along y and z axes rotating with the assembly. (*a*) calculate and plot the components of the reactions from $t = 0$ to $t = 2$ s at 0.1-s intervals. (*b*) Determine the time at which the z components of the reactions at A and B are equal to zero.

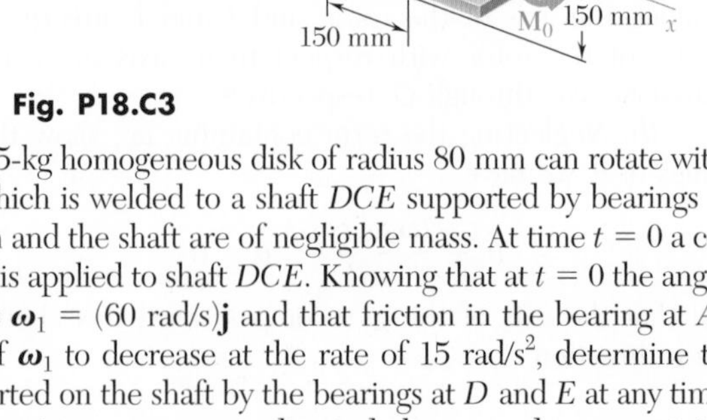

Fig. P18.C3

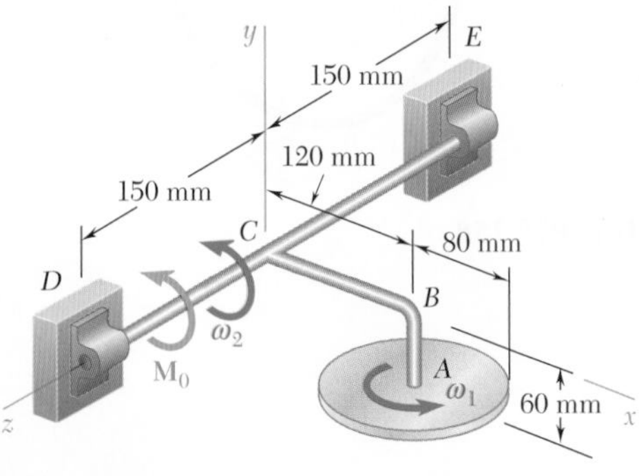

Fig. P18.C4

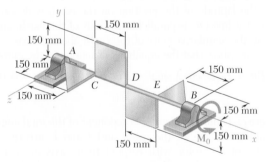

18.C4 A 2.5-kg homogeneous disk of radius 80 mm can rotate with respect to arm ABC, which is welded to a shaft DCE supported by bearings at D and E. Both the arm and the shaft are of negligible mass. At time $t = 0$ a couple $\mathbf{M}_0 = (0.5\text{ N}\cdot\text{m})\mathbf{k}$ is applied to shaft DCE. Knowing that at $t = 0$ the angular velocity of the disk is $\boldsymbol{\omega}_1 = (60\text{ rad/s})\mathbf{j}$ and that friction in the bearing at A causes the magnitude of $\boldsymbol{\omega}_1$ to decrease at the rate of 15 rad/s^2, determine the dynamic reactions exerted on the shaft by the bearings at D and E at any time t. Resolve these reactions into components directed along x and y axes rotating with the shaft. Use computational software (*a*) to calculate the components of the reactions from $t = 0$ to $t = 4$ s, (*b*) to determine the times t_1 and t_2 at which the x and y components of the reaction at E are respectively equal to zero.

18.C5 A homogeneous disk of radius 180 mm is welded to a rod AG of length 360 mm and of negligible mass which is connected by a clevis to a vertical shaft AB. The rod and disk can rotate freely about a horizontal axis AC, and shaft AB can rotate freely about a vertical axis. Initially rod AG forms a given angle θ_0 with the downward vertical and its angular velocity $\dot{\theta}_0$ about AC is zero. Shaft AB is then given an angular velocity $\dot{\phi}_0$ about the vertical. Use computational software (*a*) to calculate the minimum value θ_m of the angle θ in the ensuing motion and the period of oscillation in θ, that is, the time required for θ to regain its initial value θ_0, (*b*) to compute the angular velocity $\dot{\phi}$ of shaft AB for values of θ from θ_0 to θ_m. Apply this program with the initial conditions (*i*) $\theta_0 = 90°$, $\dot{\phi}_0 = 5$ rad/s, (*ii*) $\theta_0 = 90°$, $\dot{\phi}_0 = 10$ rad/s, (*iii*) $\theta_0 = 60°$, $\dot{\phi}_0 = 5$ rad/s. [*Hint:* Use the principle of conservation of energy and the fact that the angular momentum of the body about the vertical through A is conserved to obtain an equation of the form $\dot{\theta}^2 = f(\theta)$. This equation can be integrated by a numerical method.]

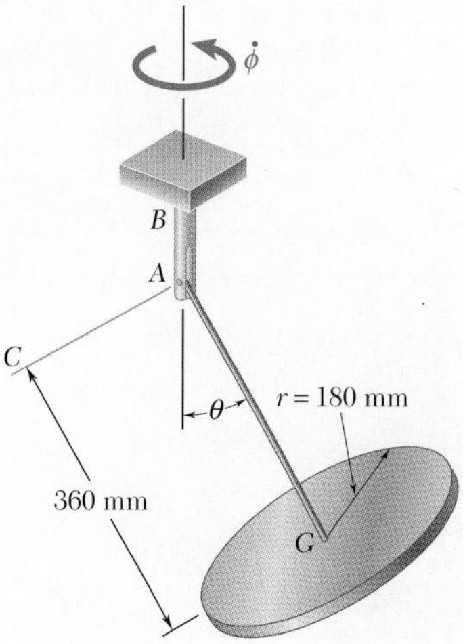

Fig. P18.C5

18.C6 A homogeneous disk of radius 180 mm is welded to a rod AG of length 360 mm and of negligible mass which is supported by a ball-and-socket joint at A. The disk is released in the position $\theta = \theta_0$ with a rate of spin $\dot{\psi}_0$, a rate of precession $\dot{\phi}_0$, and a zero rate of nutation. Use computational software (*a*) to calculate the minimum value θ_m of the angle θ in the ensuing motion and the period of oscillation in θ, that is, the time required for θ to regain its initial value θ_0, (*b*) to compute the rate of spin $\dot{\psi}$ and the rate of precession $\dot{\phi}$ for values of θ from θ_0 to θ_m, using 2° decrements. Apply this program with the initial conditions (*i*) $\theta_0 = 90°$, $\dot{\psi}_0 = 50$ rad/s, $\dot{\phi}_0 = 0$, (*ii*) $\theta_0 = 90°$, $\dot{\psi}_0 = 0$, $\dot{\phi}_0 = 5$ rad/s, (*iii*) $\theta_0 = 90°$, $\dot{\psi}_0 = 50$ rad/s, $\dot{\phi}_0 = 5$ rad/s, (*iv*) $\theta_0 = 90°$, $\dot{\psi}_0 = 10$ rad/s, $\dot{\phi}_0 = 5$ rad/s, (*v*) $\theta_0 = 60°$, $\dot{\psi}_0 = 50$ rad/s, $\dot{\phi}_0 = 5$ rad/s. [*Hint:* Use the principle of conservation of energy and the fact that the angular momentum of the body is conserved about both the Z and z axes to obtain an equation of the form $\dot{\theta}^2 = f(\theta)$. This equation can be integrated by a numerical method.]

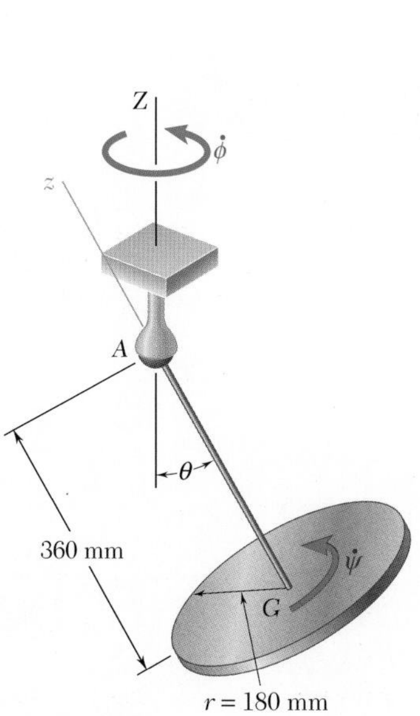

Fig. P18.C6

The Wind Damper inside of Taipei 101 helps protect against typhoons and earthquakes by reducing the effects of wind and vibrations on the building. Mechanical systems may undergo *free vibrations* or they may be subject to *forced vibrations.* The vibrations are *damped* when there is energy dissipation and *undamped* otherwise. This chapter is an introduction to many fundamental concepts in vibration analysis.

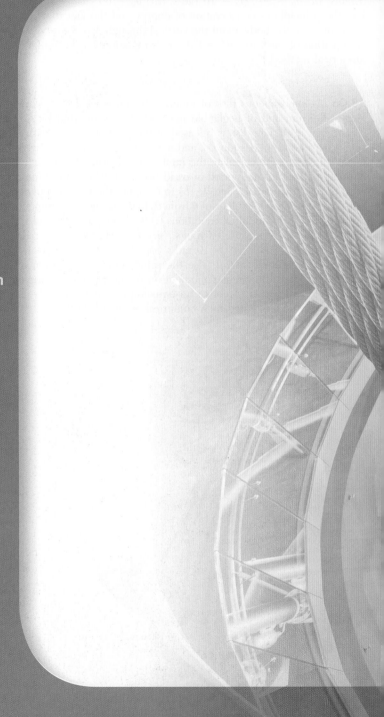

CHAPTER 19

Mechanical Vibrations

19.1 INTRODUCTION

A *mechanical vibration* is the motion of a particle or a body which oscillates about a position of equilibrium. Most vibrations in machines and structures are undesirable because of the increased stresses and energy losses which accompany them. They should therefore be eliminated or reduced as much as possible by appropriate design. The analysis of vibrations has become increasingly important in recent years owing to the current trend toward higher-speed machines and lighter structures. There is every reason to expect that this trend will continue and that an even greater need for vibration analysis will develop in the future.

The analysis of vibrations is a very extensive subject to which entire texts have been devoted. Our present study will therefore be limited to the simpler types of vibrations, namely, the vibrations of a body or a system of bodies with one degree of freedom.

A mechanical vibration generally results when a system is displaced from a position of stable equilibrium. The system tends to return to this position under the action of restoring forces (either elastic forces, as in the case of a mass attached to a spring, or gravitational forces, as in the case of a pendulum). But the system generally reaches its original position with a certain acquired velocity which carries it beyond that position. Since the process can be repeated indefinitely, the system keeps moving back and forth across its position of equilibrium. The time interval required for the system to complete a full cycle of motion is called the *period* of the vibration. The number of cycles per unit time defines the *frequency*, and the maximum displacement of the system from its position of equilibrium is called the *amplitude* of the vibration.

When the motion is maintained by the restoring forces only, the vibration is said to be a *free vibration* (Secs. 19.2 to 19.6). When a periodic force is applied to the system, the resulting motion is described as a *forced vibration* (Sec. 19.7). When the effects of friction can be neglected, the vibrations are said to be *undamped*. However, all vibrations are actually *damped* to some degree. If a free vibration is only slightly damped, its amplitude slowly decreases until, after a certain time, the motion comes to a stop. But if damping is large enough to prevent any true vibration, the system then slowly regains its original position (Sec. 19.8). A damped forced vibration is maintained as long as the periodic force which produces the vibration is applied. The amplitude of the vibration, however, is affected by the magnitude of the damping forces (Sec. 19.9).

VIBRATIONS WITHOUT DAMPING

19.2 FREE VIBRATIONS OF PARTICLES. SIMPLE HARMONIC MOTION

Consider a body of mass m attached to a spring of constant k (Fig. 19.1a). Since at the present time we are concerned only with the motion of its mass center, we will refer to this body as a particle. When the particle is in static equilibrium, the forces acting on it are its weight $\mathbf{W}$ and the force $\mathbf{T}$ exerted by the spring, of magnitude

$T = k\delta_{st}$, where δ_{st} denotes the elongation of the spring. We have, therefore,

$$W = k\delta_{st}$$

Suppose now that the particle is displaced through a distance x_m from its equilibrium position and released with no initial velocity. If x_m has been chosen smaller than δ_{st}, the particle will move back and forth through its equilibrium position; a vibration of amplitude x_m has been generated. Note that the vibration can also be produced by imparting a certain initial velocity to the particle when it is in its equilibrium position $x = 0$ or, more generally, by starting the particle from any given position $x = x_0$ with a given initial velocity $\mathbf{v}_0$.

To analyze the vibration, let us consider the particle in a position P at some arbitrary time t (Fig. 19.1*b*). Denoting by x the displacement OP measured from the equilibrium position O (positive downward), we note that the forces acting on the particle are its weight $\mathbf{W}$ and the force $\mathbf{T}$ exerted by the spring which, in this position, has a magnitude $T = k(\delta_{st} + x)$. Recalling that $W = k\delta_{st}$, we find that the magnitude of the resultant $\mathbf{F}$ of the two forces (positive downward) is

$$F = W - k(\delta_{st} + x) = -kx \tag{19.1}$$

Thus the *resultant* of the forces exerted on the particle is proportional to the displacement OP *measured from the equilibrium position*. Recalling the sign convention, we note that $\mathbf{F}$ is always directed *toward* the equilibrium position O. Substituting for F into the fundamental equation $F = ma$ and recalling that a is the second derivative $\ddot{x}$ of x with respect to t, we write

$$m\ddot{x} + kx = 0 \tag{19.2}$$

Note that the same sign convention should be used for the acceleration $\ddot{x}$ and for the displacement x, namely, positive downward.

The motion defined by Eq. (19.2) is called a *simple harmonic motion*. It is characterized by the fact that *the acceleration is proportional to the displacement and of opposite direction*. We can verify that each of the functions $x_1 = \sin(\sqrt{k/m}\,t)$ and $x_2 = \cos(\sqrt{k/m}\,t)$ satisfies Eq. (19.2). These functions, therefore, constitute two *particular solutions* of the differential equation (19.2). The *general solution* of Eq. (19.2) is obtained by multiplying each of the particular solutions by an arbitrary constant and adding. Thus, the general solution is expressed as

$$x = C_1x_1 + C_2x_2 = C_1 \sin\left(\sqrt{\frac{k}{m}}\,t\right) + C_2 \cos\left(\sqrt{\frac{k}{m}}\,t\right) \tag{19.3}$$

We note that x is a *periodic function* of the time t and does, therefore, represent a vibration of the particle P. The coefficient of t in the expression we have obtained is referred to as the *natural circular frequency* of the vibration and is denoted by ω_n. We have

$$\text{Natural circular frequency} = \omega_n = \sqrt{\frac{k}{m}} \tag{19.4}$$

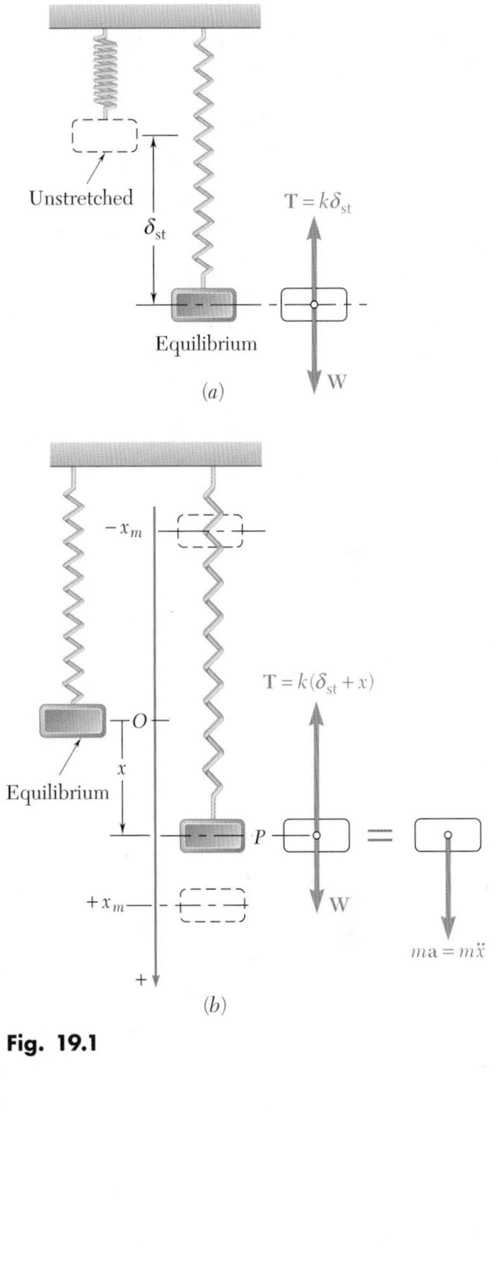

Fig. 19.1

Substituting for $\sqrt{k/m}$ into Eq. (19.3), we write

$$x = C_1 \sin \omega_n t + C_2 \cos \omega_n t \tag{19.5}$$

This is the general solution of the differential equation

$$\ddot{x} + \omega_n^2 x = 0 \tag{19.6}$$

which can be obtained from Eq. (19.2) by dividing both terms by m and observing that $k/m = \omega_n^2$. Differentiating twice both members of Eq. (19.5) with respect to t, we obtain the following expressions for the velocity and the acceleration at time t:

$$v = \dot{x} = C_1\omega_n \cos \omega_n t - C_2\omega_n \sin \omega_n t \tag{19.7}$$
$$a = \ddot{x} = -C_1\omega_n^2 \sin \omega_n t - C_2\omega_n^2 \cos \omega_n t \tag{19.8}$$

The values of the constants C_1 and C_2 depend upon the *initial conditions* of the motion. For example, we have $C_1 = 0$ if the particle is displaced from its equilibrium position and released at $t = 0$ with no initial velocity, and we have $C_2 = 0$ if the particle is started from O at $t = 0$ with a certain initial velocity. In general, substituting $t = 0$ and the initial values x_0 and v_0 of the displacement and the velocity into Eqs. (19.5) and (19.7), we find that $C_1 = v_0/\omega_n$ and $C_2 = x_0$.

The expressions obtained for the displacement, velocity, and acceleration of a particle can be written in a more compact form if we observe that Eq. (19.5) expresses that the displacement $x = OP$ is the sum of the x components of two vectors $\mathbf{C}_1$ and $\mathbf{C}_2$, respectively, of magnitude C_1 and C_2, directed as shown in Fig. 19.2*a*. As t varies, both vectors rotate clockwise; we also note that the magnitude of their resultant $\overrightarrow{OQ}$ is equal to the maximum displacement x_m. The simple harmonic motion of P along the x axis can thus be obtained by projecting on this axis the motion of a point Q describing an *auxiliary circle* of radius x_m *with a constant angular velocity* ω_n (which explains the name of natural *circular* frequency given to ω_n). Denoting by ϕ the angle formed by the vectors $\overrightarrow{OQ}$ and $\mathbf{C}_1$, we write

$$OP = OQ \sin (\omega_n t + \phi) \tag{19.9}$$

which leads to new expressions for the displacement, velocity, and acceleration of P:

$$x = x_m \sin (\omega_n t + \phi) \tag{19.10}$$

$$v = \dot{x} = x_m\omega_n \cos (\omega_n t + \phi) \tag{19.11}$$
$$a = \ddot{x} = -x_m\omega_n^2 \sin (\omega_n t + \phi) \tag{19.12}$$

The displacement-time curve is represented by a sine curve (Fig. 19.2*b*); the maximum value x_m of the displacement is called the *amplitude* of the vibration, and the angle ϕ which defines the initial position of Q on the circle is called the *phase angle*. We note from Fig. 19.2 that a full *cycle* is described as the angle $\omega_n t$ increases by 2π rad. The corresponding value of t, denoted by τ_n, is called the *period* of the free vibration and is measured in seconds. We have

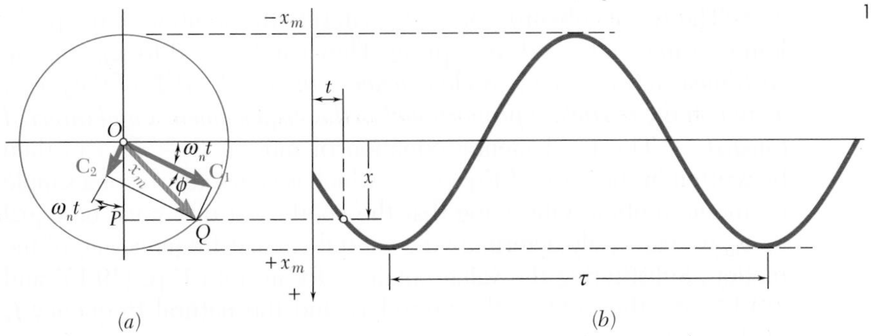

Fig. 19.2

$$\text{period} = \tau_n = \frac{2\pi}{\omega_n} \tag{19.13}$$

The number of cycles described per unit of time is denoted by f_n and is known as the *natural frequency* of the vibration. We write

$$\text{Natural frequency} = f_n = \frac{1}{\tau_n} = \frac{\omega_n}{2\pi} \tag{19.14}$$

The unit of frequency is a frequency of 1 cycle per second, corresponding to a period of 1 s. In terms of base units the unit of frequency is thus 1/s or s^{-1}. It is called a *hertz* (Hz) in the SI system of units. It also follows from Eq. (19.14) that a frequency of 1 s^{-1} or 1 Hz corresponds to a circular frequency of 2π rad/s. In problems involving angular velocities expressed in revolutions per minute (rpm), we have 1 rpm $= \frac{1}{60}\ s^{-1} = \frac{1}{60}$ Hz, or 1 rpm $= (2\pi/60)$ rad/s.

Recalling that ω_n was defined in (19.4) in terms of the constant k of the spring and the mass m of the particle, we observe that the period and the frequency are independent of the initial conditions and of the amplitude of the vibration. Note that τ_n and f_n depend on the *mass* rather than on the *weight* of the particle and thus are independent of the value of g.

The velocity-time and acceleration-time curves can be represented by sine curves of the same period as the displacement-time curve, but with different phase angles. From Eqs. (19.11) and (19.12), we note that the maximum values of the magnitudes of the velocity and acceleration are

$$v_m = x_m\omega_n \qquad a_m = x_m\omega_n^2 \tag{19.15}$$

Since the point Q describes the auxiliary circle, of radius x_m, at the constant angular velocity ω_n, its velocity and acceleration are equal, respectively, to the expressions (19.15). Recalling Eqs. (19.11) and (19.12), we find, therefore, that the velocity and acceleration of P can be obtained at any instant by projecting on the x axis vectors of magnitudes $v_m = x_m\omega_n$ and $a_m = x_m\omega_n^2$ representing, respectively, the velocity and acceleration of Q at the same instant (Fig. 19.3).

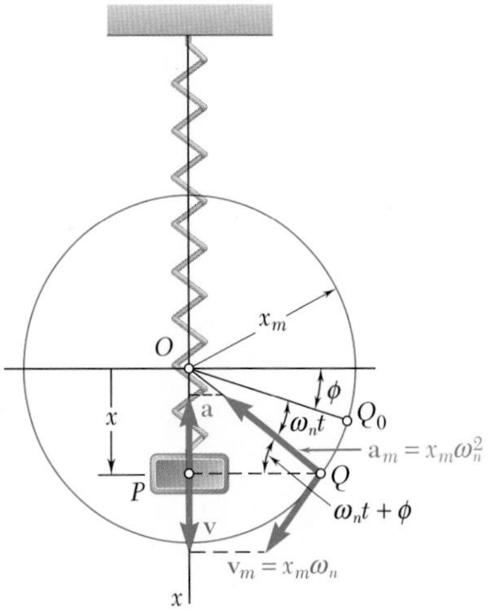

Fig. 19.3

The results obtained are not limited to the solution of the problem of a mass attached to a spring. They can be used to analyze the rectilinear motion of a particle *whenever the resultant* **F** *of the forces acting on the particle is proportional to the displacement x and directed toward O*. The fundamental equation of motion $F = ma$ can then be written in the form of Eq. (19.6), which is characteristic of a simple harmonic motion. Observing that the coefficient of x must be equal to ω_n^2, we can easily determine the natural circular frequency ω_n of the motion. Substituting the value obtained for ω_n into Eqs. (19.13) and (19.14), we then obtain the period τ_n and the natural frequency f_n of the motion.

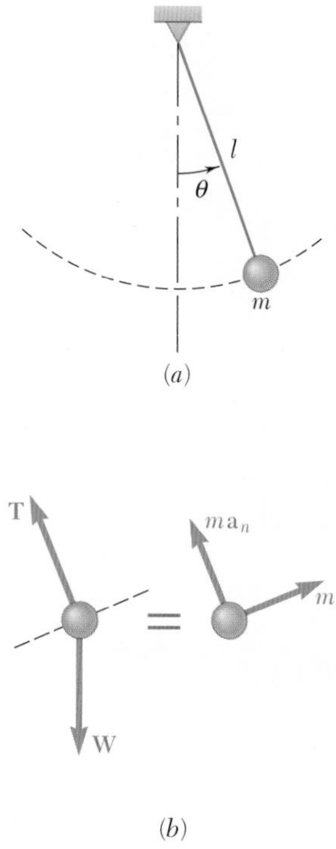

Fig. 19.4

19.3 SIMPLE PENDULUM (APPROXIMATE SOLUTION)

Most of the vibrations encountered in engineering applications can be represented by a simple harmonic motion. Many others, although of a different type, can be *approximated* by a simple harmonic motion, provided that their amplitude remains small. Consider, for example, a *simple pendulum*, consisting of a bob of mass m attached to a cord of length l, which can oscillate in a vertical plane (Fig. 19.4*a*). At a given time t, the cord forms an angle θ with the vertical. The forces acting on the bob are its weight **W** and the force **T** exerted by the cord (Fig. 19.4*b*). Resolving the vector $m\mathbf{a}$ into tangential and normal components, with $m\mathbf{a}_t$ directed to the right, i.e., in the direction corresponding to increasing values of θ, and observing that $a_t = l\alpha = l\ddot{\theta}$, we write

$$\Sigma F_t = ma_t: \qquad -W \sin \theta = ml\ddot{\theta}$$

Noting that $W = mg$ and dividing through by ml, we obtain

$$\ddot{\theta} + \frac{g}{l} \sin \theta = 0 \tag{19.16}$$

For oscillations of small amplitude, we can replace $\sin \theta$ by θ, expressed in radians, and write

$$\ddot{\theta} + \frac{g}{l} \theta = 0 \tag{19.17}$$

Comparison with Eq. (19.6) shows that the differential equation (19.17) is that of a simple harmonic motion with a natural circular frequency ω_n equal to $(g/l)^{1/2}$. The general solution of Eq. (19.17) can, therefore, be expressed as

$$\theta = \theta_m \sin (\omega_n t + \phi)$$

where θ_m is the amplitude of the oscillations and ϕ is a phase angle. Substituting into Eq. (19.13) the value obtained for ω_n, we get the following expression for the period of the small oscillations of a pendulum of length l:

$$\tau_n = \frac{2\pi}{\omega_n} = 2\pi\sqrt{\frac{l}{g}} \tag{19.18}$$

*19.4 SIMPLE PENDULUM (EXACT SOLUTION)

Formula (19.18) is only approximate. To obtain an exact expression for the period of the oscillations of a simple pendulum, we must return to Eq. (19.16). Multiplying both terms by $2\dot{\theta}$ and integrating from an initial position corresponding to the maximum deflection, that is, $\theta = \theta_m$ and $\dot{\theta} = 0$, we write

$$\left(\frac{d\theta}{dt}\right)^2 = \frac{2g}{l}(\cos\theta - \cos\theta_m)$$

Replacing $\cos\theta$ by $1 - 2\sin^2(\theta/2)$ and $\cos\theta_m$ by a similar expression, solving for dt, and integrating over a quarter period from $t = 0$, $\theta = 0$ to $t = \tau_n/4$, $\theta = \theta_m$, we have

$$\tau_n = 2\sqrt{\frac{l}{g}}\int_0^{\theta_m}\frac{d\theta}{\sqrt{\sin^2(\theta_m/2) - \sin^2(\theta/2)}}$$

The integral in the right-hand member is known as an *elliptic integral;* it cannot be expressed in terms of the usual algebraic or trigonometric functions. However, setting

$$\sin(\theta/2) = \sin(\theta_m/2)\sin\phi$$

we can write

$$\tau_n = 4\sqrt{\frac{l}{g}}\int_0^{\pi/2}\frac{d\phi}{\sqrt{1 - \sin^2(\theta_m/2)\sin^2\phi}} \tag{19.19}$$

where the integral obtained, commonly denoted by K, can be calculated by using a numerical method of integration. It can also be found in *tables of elliptic integrals* for various values of $\theta_m/2$.† In order to compare the result just obtained with that of the preceding section, we write Eq. (19.19) in the form

$$\tau_n = \frac{2K}{\pi}\left(2\pi\sqrt{\frac{l}{g}}\right) \tag{19.20}$$

Formula (19.20) shows that the actual value of the period of a simple pendulum can be obtained by multiplying the approximate value given in Eq. (19.18) by the correction factor $2K/\pi$. Values of the correction factor are given in Table 19.1 for various values of the amplitude θ_m. We note that for ordinary engineering computations the correction factor can be omitted as long as the amplitude does not exceed 10°.

TABLE 19.1 Correction Factor for the Period of a Simple Pendulum

θ_m	0°	10°	20°	30°	60°	90°	120°	150°	180°
K	1.571	1.574	1.583	1.598	1.686	1.854	2.157	2.768	∞
$2K/\pi$	1.000	1.002	1.008	1.017	1.073	1.180	1.373	1.762	∞

†See, for example, *Standard Mathematical Tables*, Chemical Rubber Publishing Company, Cleveland, Ohio.

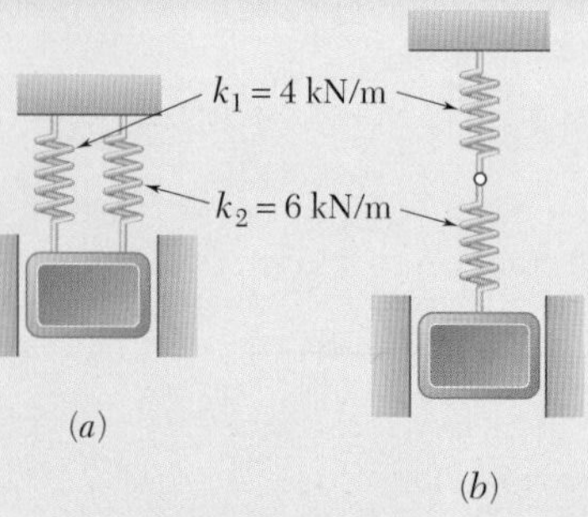

SAMPLE PROBLEM 19.1

A 50-kg block moves between vertical guides as shown. The block is pulled 40 mm down from its equilibrium position and released. For each spring arrangement, determine the period of the vibration, the maximum velocity of the block, and the maximum acceleration of the block.

SOLUTION

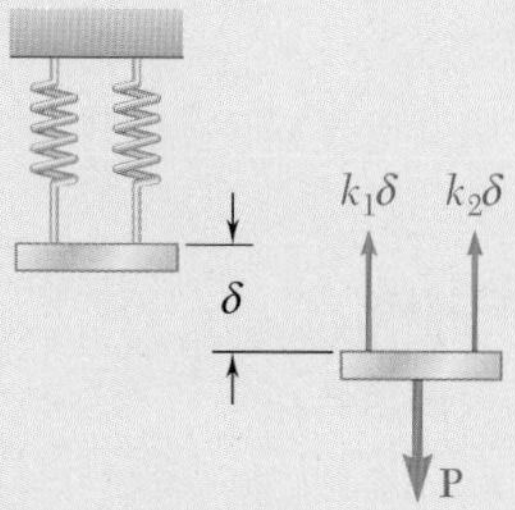

a. Springs Attached in Parallel. We first determine the constant k of a single spring equivalent to the two springs *by finding the magnitude of the force* **P** required to cause a given deflection δ. Since for a deflection δ the magnitudes of the forces exerted by the springs are, respectively, $k_1\delta$ and $k_2\delta$, we have

$$P = k_1\delta + k_2\delta = (k_1 + k_2)\delta$$

The constant k of the single equivalent spring is

$$k = \frac{P}{\delta} = k_1 + k_2 = 4 \text{ kN/m} + 6 \text{ kN/m} = 10 \text{ kN/m} = 10^4 \text{ N/m}$$

Period of Vibration: Since $m = 50$ kg, Eq. (19.4) yields

$$\omega_n^2 = \frac{k}{m} = \frac{10^4 \text{ N/m}}{50 \text{ kg}} \qquad \omega_n = 14.14 \text{ rad/s}$$

$$\tau_n = 2\pi/\omega_n \qquad \tau_n = 0.444 \text{ s} \blacktriangleleft$$

Maximum Velocity: $v_m = x_m\omega_n = (0.040 \text{ m})(14.14 \text{ rad/s})$

$$v_m = 0.566 \text{ m/s} \qquad \mathbf{v}_m = 0.566 \text{ m/s} \updownarrow \blacktriangleleft$$

Maximum Acceleration: $a_m = x_m\omega_n^2 = (0.040 \text{ m})(14.14 \text{ rad/s})^2$

$$a_m = 8.00 \text{ m/s}^2 \qquad \mathbf{a}_m = 8.00 \text{ m/s}^2 \updownarrow \blacktriangleleft$$

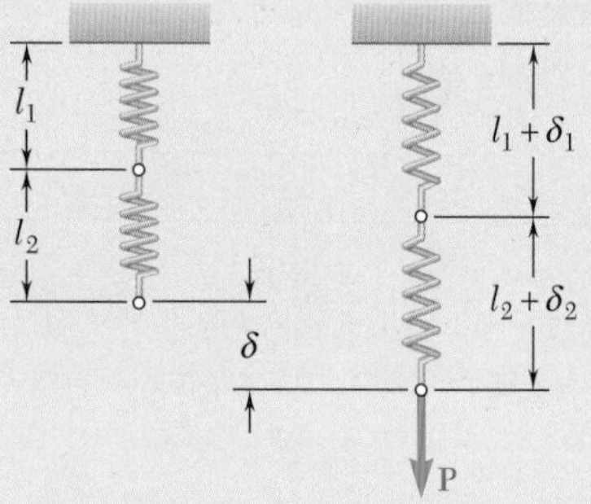

b. Springs Attached in Series. We first determine the constant k of a single spring equivalent to the two springs *by finding the total elongation* δ of the springs under a given static load **P**. To facilitate the computation, a static load of magnitude $P = 12$ kN is used.

$$\delta = \delta_1 + \delta_2 = \frac{P}{k_1} + \frac{P}{k_2} = \frac{12 \text{ kN}}{4 \text{ kN/m}} + \frac{12 \text{ kN}}{6 \text{ kN/m}} = 5 \text{ m}$$

$$k = \frac{P}{\delta} = \frac{12 \text{ kN}}{5 \text{ m}} = 2.4 \text{ kN/m} = 2400 \text{ N/m}$$

Period of Vibration: $\omega_n^2 = \dfrac{k}{m} = \dfrac{2400 \text{ N/m}}{50 \text{ kg}} \qquad \omega_n = 6.93$ rad/s

$$\tau_n = \frac{2\pi}{\omega_n} \qquad \tau_n = 0.907 \text{ s} \blacktriangleleft$$

Maximum Velocity: $v_m = x_m\omega_n = (0.040 \text{ m})(6.93 \text{ rad/s})$

$$v_m = 0.277 \text{ m/s} \qquad \mathbf{v}_m = 0.277 \text{ m/s} \updownarrow \blacktriangleleft$$

Maximum Acceleration: $a_m = x_m\omega_n^2 = (0.040 \text{ m})(6.93 \text{ rad/s})^2$

$$a_m = 1.920 \text{ m/s}^2 \qquad \mathbf{a}_m = 1.920 \text{ m/s}^2 \updownarrow \blacktriangleleft$$

SOLVING PROBLEMS ON YOUR OWN

This chapter deals with *mechanical vibrations,* i.e., with the motion of a particle or body oscillating about a position of equilibrium.

In this first lesson, we saw that a *free vibration* of a particle occurs when the particle is subjected to a force proportional to its displacement and of opposite direction, such as the force exerted by a spring (Fig. 19.1). The resulting motion, called a *simple harmonic motion,* is characterized by the differential equation

$$m\ddot{x} + kx = 0 \tag{19.2}$$

where x is the displacement of the particle, $\ddot{x}$ is its acceleration, m is its mass, and k is the constant of the spring. The solution of this differential equation was found to be

$$x = x_m \sin(\omega_n t + \phi) \tag{19.10}$$

where x_m = amplitude of the vibration
$\omega_n = \sqrt{k/m}$ = natural circular frequency (rad/s)
ϕ = phase angle (rad)

We also defined the *period* of the vibration as the time $\tau_n = 2\pi/\omega_n$ needed for the particle to complete one cycle, and the *natural frequency* as the number of cycles per second, $f_n = 1/\tau_n = \omega_n/2\pi$, expressed in Hz or s^{-1}. Differentiating Eq. (19.10) twice yields the velocity and the acceleration of the particle at any time. The maximum values of the velocity and acceleration were found to be

$$v_m = x_m\omega_n \qquad a_m = x_m\omega_n^2 \tag{19.15}$$

To determine the parameters in Eq. (19.10) you can follow these steps.

1. Draw a free-body diagram showing the forces exerted on the particle when the particle is at a distance x from its position of equilibrium. The resultant of these forces will be proportional to x and its direction will be opposite to the positive direction of x [Eq. (19.1)].

2. Write the differential equation of motion by equating to $m\ddot{x}$ the resultant of the forces found in step 1. Note that once a positive direction for x has been chosen, the same sign convention should be used for the acceleration $\ddot{x}$. After transposition, you will obtain an equation of the form of Eq. (19.2).

(continued)

3. Determine the natural circular frequency ω_n by dividing the coefficient of x by the coefficient of $\ddot{x}$ in this equation and taking the square root of the result obtained. Make sure that ω_n is expressed in rad/s.

4. Determine the amplitude x_m and the phase angle ϕ by substituting the value obtained for ω_n and the initial values of x and $\dot{x}$ into Eq. (19.10) and the equation obtained by differentiating Eq. (19.10) with respect to t.

Equation (19.10) and the two equations obtained by differentiating Eq. (19.10) twice with respect to t can now be used to find the displacement, velocity, and acceleration of the particle at any time. Equations (19.15) yield the maximum velocity v_m and the maximum acceleration a_m.

5. You also saw that for the small oscillations of a simple pendulum, the angle θ that the cord of the pendulum forms with the vertical satisfies the differential equation

$$\ddot{\theta} + \frac{g}{l}\theta = 0 \qquad (19.17)$$

where l is the length of the cord and where θ is expressed in radians [Sec. 19.3]. This equation defines again a *simple harmonic motion*, and its solution is of the same form as Eq. (19.10),

$$\theta = \theta_m \sin (\omega_n t + \phi)$$

where the natural circular frequency $\omega_n = \sqrt{g/l}$ is expressed in rad/s. The determination of the various constants in this expression is carried out in a manner similar to that described above. Remember that the velocity of the bob is tangent to the path and that its magnitude is $v = l\dot{\theta}$, while the acceleration of the bob has a tangential component $\mathbf{a}_t$, of magnitude $a_t = l\ddot{\theta}$, and a normal component $\mathbf{a}_n$ directed toward the center of the path and of magnitude $a_n = l\dot{\theta}^2$.

PROBLEMS

19.1 Determine the maximum velocity and maximum acceleration of a particle which moves in simple harmonic motion with an amplitude of 3 mm and a frequency of 20 Hz.

19.2 A particle moves in simple harmonic motion. Knowing that the amplitude is 300 mm and the maximum acceleration is 5 m/s^2, determine the maximum velocity of the particle and the frequency of its motion.

19.3 Determine the amplitude and maximum velocity of a particle which moves in simple harmonic motion with a maximum acceleration of 60 m/s^2 and a frequency of 40 Hz.

19.4 A 32-kg block is attached to a spring and can move without friction in a slot as shown. The block is in its equilibrium position when it is struck by a hammer which imparts to the block an initial velocity of 250 mm/s. Determine (*a*) the period and frequency of the resulting motion, (*b*) the amplitude of the motion and the maximum acceleration of the block.

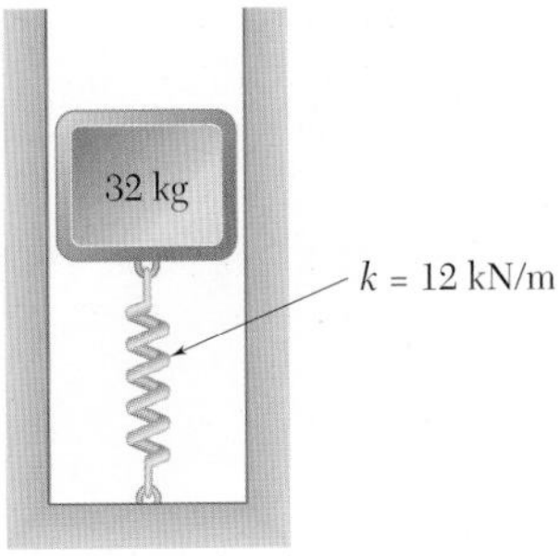

Fig. P19.4

19.5 A 15-kg block is supported by the spring shown. If the block is moved vertically downward from its equilibrium position and released, determine (*a*) the period and frequency of the resulting motion, (*b*) the maximum velocity and acceleration of the block if the amplitude of its motion is 50 mm.

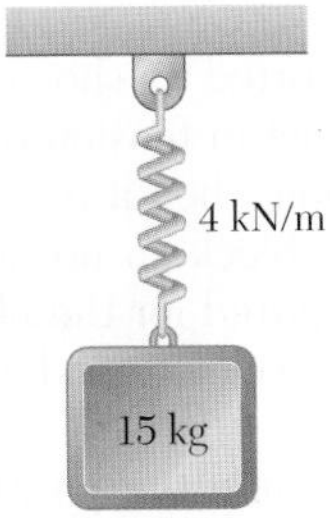

Fig. P19.5

19.6 An instrument package A is bolted to a shaker table as shown. The table moves vertically in simple harmonic motion at the same frequency as the variable-speed motor which drives it. The package is to be tested at a peak acceleration of 50 m/s^2. Knowing that the amplitude of the shaker table is 60 mm, determine (*a*) the required speed of the motor in rpm, (*b*) the maximum velocity of the table.

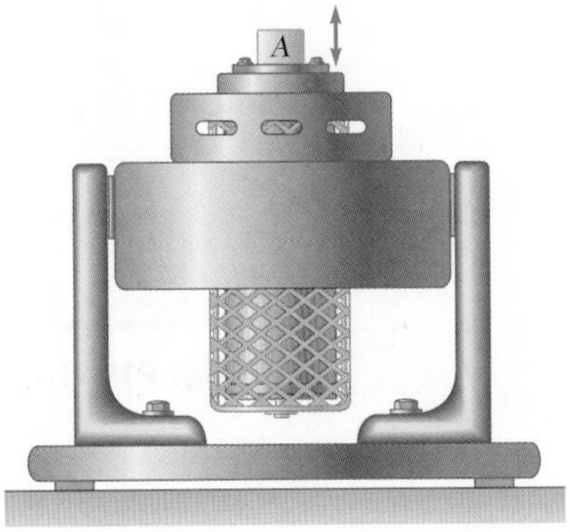

Fig. P19.6

19.7 A simple pendulum consisting of a bob attached to a cord oscillates in a vertical plane with a period of 1.3 s. Assuming simple harmonic motion and knowing that the maximum velocity of the bob is 0.4 m/s, determine (*a*) the amplitude of the motion in degrees, (*b*) the maximum tangential acceleration of the bob.

19.8 A simple pendulum consisting of a bob attached to a cord of length l = 800 mm oscillates in a vertical plane. Assuming simple harmonic motion and knowing that the bob is released from rest when θ = 6°, determine (*a*) the frequency of oscillation, (*b*) the maximum velocity of the bob.

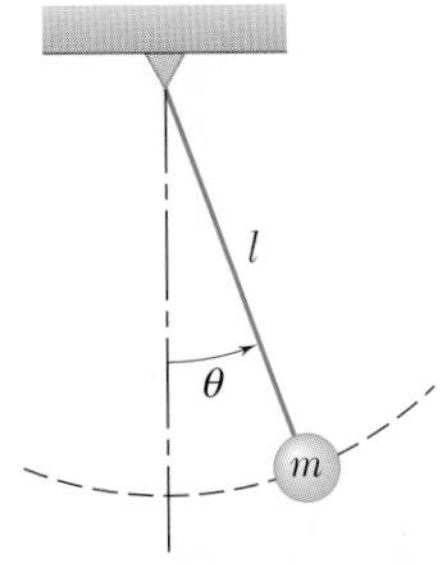

Fig. P19.7 and P19.8

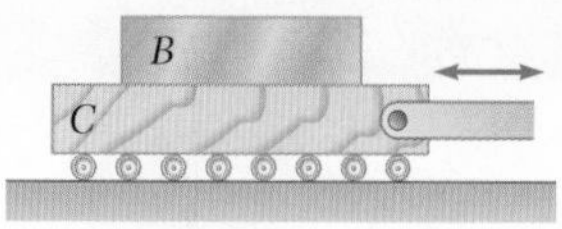

Fig. *P19.9*

19.9 An instrument package B is placed on the shaking table C as shown. The table is made to move horizontally in simple harmonic motion with a frequency of 3 Hz. Knowing that the coefficient of static friction is $\mu_s = 0.40$ between the package and the table, determine the largest allowable amplitude of the motion if the package is not to slip on the table.

19.10 A 5-kg fragile glass vase is surrounded by packing material in a cardboard box of negligible weight. The packing material has negligible damping and a force-deflection relationship as shown. Knowing that the box is dropped from a height of 1 m and the impact with the ground is perfectly plastic, determine (*a*) the amplitude of vibration for the vase, (*b*) the maximum acceleration the vase experiences in g's.

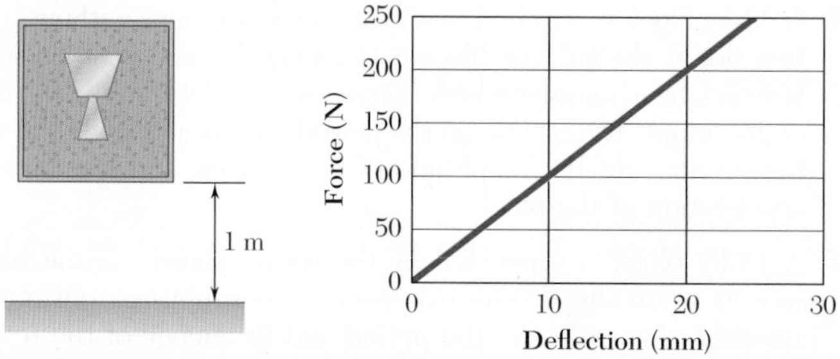

Fig. *P19.10*

Fig. P19.11

19.11 A 2-kg block is supported as shown by a spring of constant k = 400 N/m which can act in tension or compression. The block is in its equilibrium position when it is struck from below by a hammer which imparts to the block an upward velocity of 2.5 m/s. Determine (*a*) the time required for the block to move 100 mm upward, (*b*) the corresponding velocity and acceleration of the block.

19.12 In Prob. 19.11, determine the position, velocity, and acceleration of the block 0.90 s after it has been struck by the hammer.

19.13 The bob of a simple pendulum of length l = 800 mm is released from rest when $\theta = +5°$. Assuming simple harmonic motion, determine 1.6 s after release (*a*) the angle θ, (*b*) the magnitudes of the velocity and acceleration of the bob.

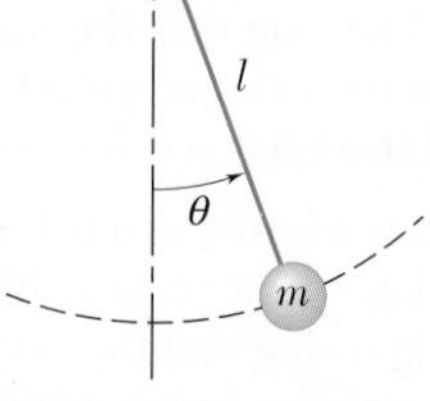

Fig. P19.13

19.14 A 150-kg electromagnet is at rest and is holding 100 kg of scrap steel when the current is turned off and the steel is dropped. Knowing that the cable and the supporting crane have a total stiffness equivalent to a spring of constant 200 kN/m, determine (*a*) the frequency, the amplitude, and the maximum velocity of the resulting motion, (*b*) the minimum tension which will occur in the cable during the motion, (*c*) the velocity of the magnet 0.03 s after the current is turned off.

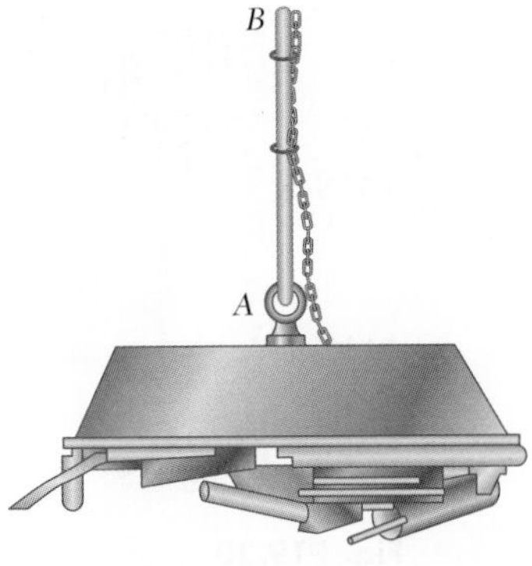

Fig. P19.14

19.15 A variable-speed motor is rigidly attached to beam BC. The rotor is slightly unbalanced and causes the beam to vibrate with a frequency equal to the motor speed. When the speed of the motor is less than 600 rpm or more than 1200 rpm, a small object placed at A is observed to remain in contact with the beam. For speeds between 600 rpm and 1200 rpm the object is observed to "dance" and actually to lose contact with the beam. Determine the amplitude of the motion of A when the speed of the motor is (*a*) 600 rpm, (*b*) 1200 rpm.

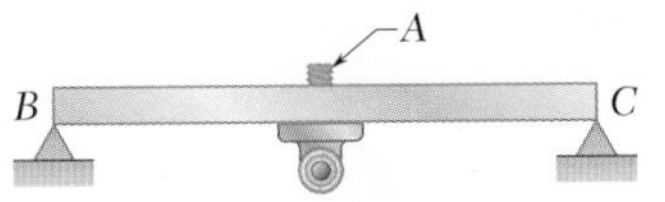

Fig. *P19.15*

19.16 A small bob is attached to a cord of length 1.2 m and is released from rest when $\theta_A = 5°$. Knowing that $d = 0.6$ m, determine (*a*) the time required for the bob to return to point A, (*b*) the amplitude θ_C.

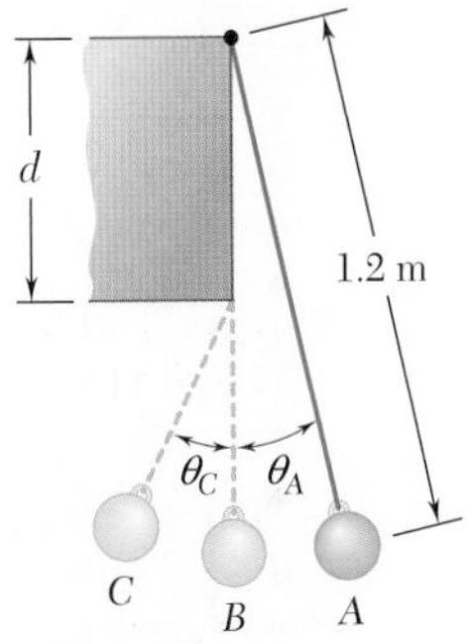

Fig. *P19.16*

19.17 A 5-kg block, attached to the lower end of a spring whose upper end is fixed, vibrates with a period of 6.8 s. Knowing that the constant k of a spring is inversely proportional to its length, determine the period of a 3-kg block which is attached to the center of the same spring if the upper and lower ends of the spring are fixed.

19.18 and 19.19 A 35-kg block is supported by the spring arrangement shown. The block is moved vertically downward from its equilibrium position and released. Knowing that the amplitude of the resulting motion is 45 mm, determine (*a*) the period and frequency of the motion, (*b*) the maximum velocity and maximum acceleration of the block.

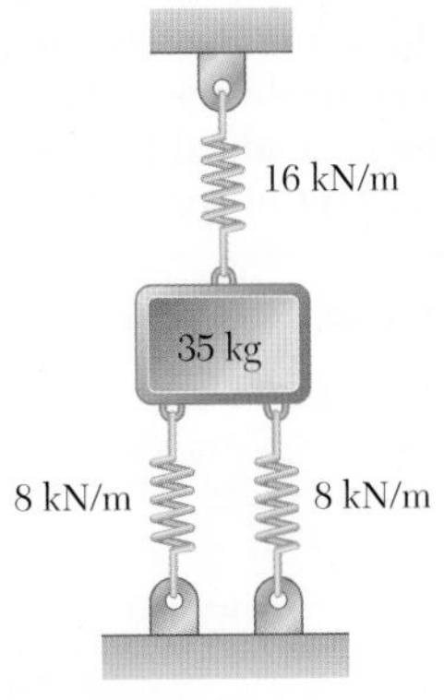

Fig. P19.18

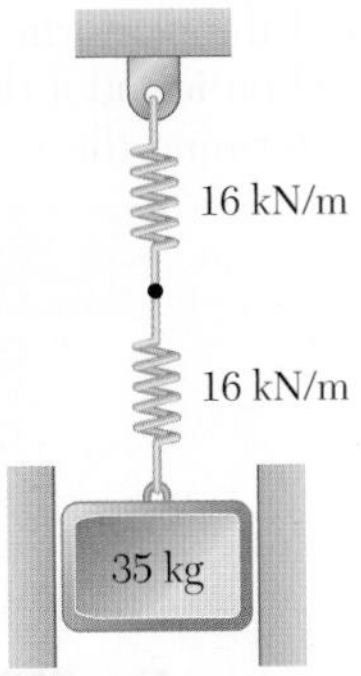

Fig. P19.19

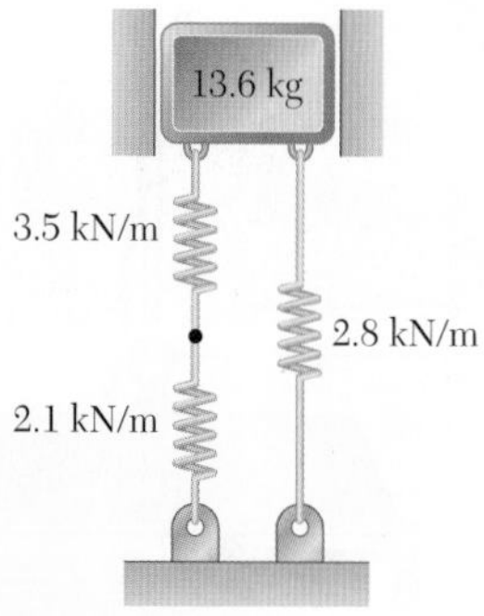

Fig. P19.20

19.20 A 13.6-kg block is supported by the spring arrangement shown. If the block is moved from its equilibrium position 44 mm vertically downward and released, determine (*a*) the period and frequency of the resulting motion, (*b*) the maximum velocity and acceleration of the block.

19.21 A 5-kg block, attached to the lower end of a spring whose upper end is fixed, vibrates with a period of 7.2 s. Knowing that the constant k of a spring is inversely proportional to its length (e.g., if you cut a 500 N/m spring in half, the remaining two springs each have a spring constant of 1000 N/m), determine the period of a 3-kg block which is attached to the center of the same spring if the upper and lower ends of the spring are fixed.

19.22 Block A of mass m is supported by the spring arrangement as shown. Knowing that the mass of the pulley is negligible and that the block is moved vertically downward from its equilibrium position and released, determine the frequency of the motion.

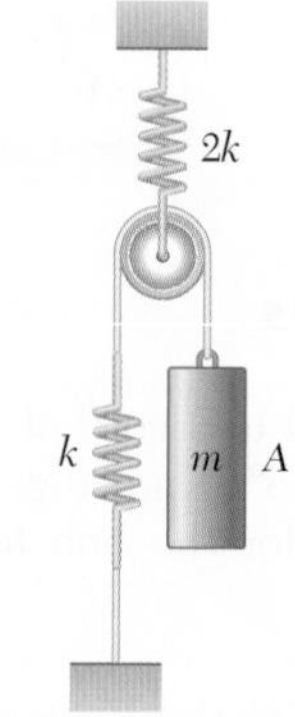

Fig. *P19.22*

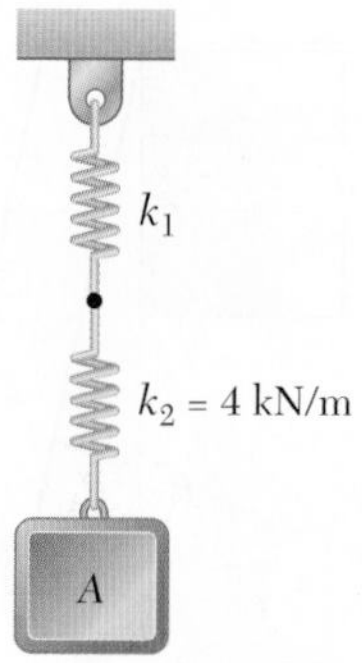

Fig. P19.23

19.23 The period of vibration of the system shown is observed to be 0.2 s. After the spring of constant $k_2 = 4$ kN/m is removed and block A is connected to the spring of constant k_1, the period is observed to be 0.12 s. Determine (*a*) the constant k_1 of the remaining spring, (*b*) the mass of block A.

Fig. P19.24

19.24 The period of vibration of the system shown is observed to be 0.8 s. If block A is removed, the period is observed to be 0.7 s. Determine (*a*) the mass of block C, (*b*) the period of vibration when both blocks A and B have been removed.

19.25 The 50-kg platform A is attached to springs B and D, each of which has a constant $k = 2$ kN/m. Knowing that the frequency of vibration of the platform is to remain unchanged when a 40-kg block is placed on it and a third spring C is added between springs B and D, determine the required constant of spring C.

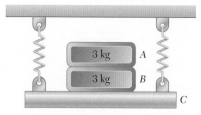

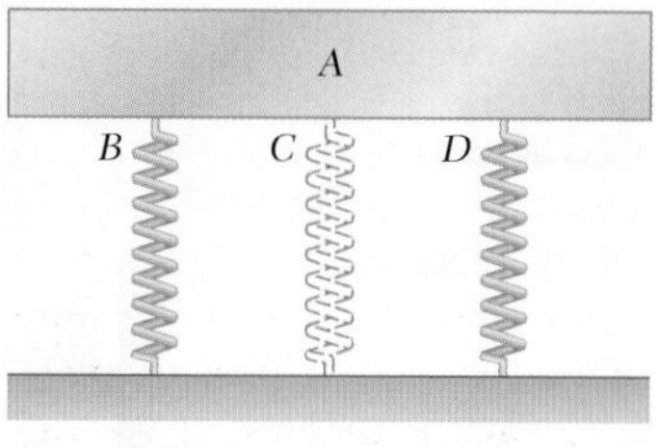

Fig. P19.25

19.26 The period of vibration for a barrel floating in salt water is found to be 0.58 s when the barrel is empty and 1.8 s when it is filled with 250 liters of crude oil. Knowing that the density of the oil is 900 kg/m^3, determine (*a*) the mass of the empty barrel, (*b*) the density of the salt water, ρ_{sw}. [*Hint*: The force of the water on the bottom of the barrel can be modeled as a spring with constant $k = \rho_{sw}gA$.]

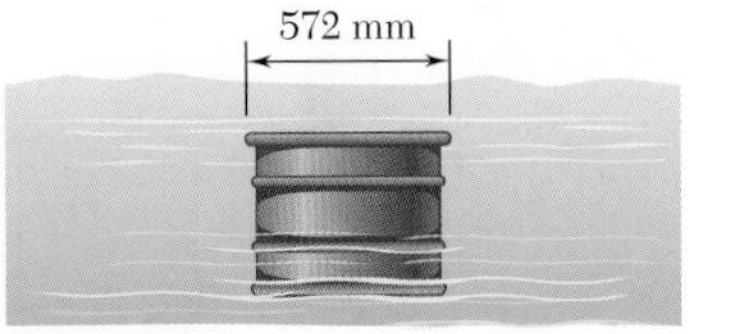

Fig. P19.26

19.27 From mechanics of materials it is known that for a cantilever beam of constant cross section a static load **P** applied at end B will cause a deflection $\delta_B = PL^3/3EI$, where L is the length of the beam, E is the modulus of elasticity, and I is the moment of inertia of the cross-sectional area of the beam. Knowing that $L = 3$ m, $E = 230$ GPa, and $I = 5 \times 10^6$ mm^4, determine (*a*) the equivalent spring constant of the beam, (*b*) the frequency of vibration of a 250-kg block attached to end B of the same beam.

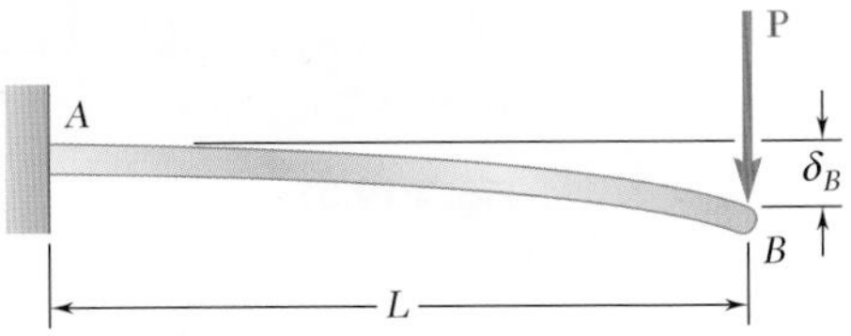

Fig. *P19.27*

19.28 From mechanics of materials it is known that when a static load **P** is applied at the end B of a uniform metal rod fixed at end A, the length of the rod will increase by an amount $\delta = PL/AE$, where L is the length of the undeformed rod, A is its cross-sectional area, and E is the modulus of elasticity of the metal. Knowing that $L = 450$ mm and $E = 200$ GPa and that the diameter of the rod is 8 mm, and neglecting the mass of the rod, determine (*a*) the equivalent spring constant of the rod, (*b*) the frequency of the vertical vibrations of a block of mass $m = 8$ kg attached to end B of the same rod.

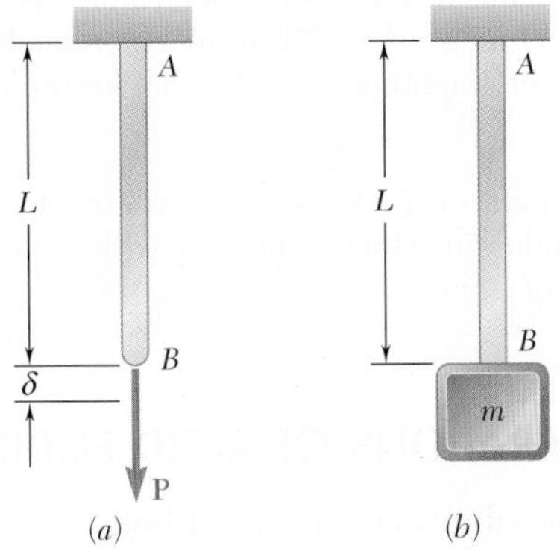

Fig. P19.28

19.29 Denoting by δ_{st} the static deflection of a beam under a given load, show that the frequency of vibration of the load is

$$f = \frac{1}{2\pi}\sqrt{\frac{g}{\delta_{st}}}$$

Neglect the mass of the beam, and assume that the load remains in contact with the beam.

19.30 A 40-mm deflection of the second floor of a building is measured directly under a newly installed 3500-kg piece of rotating machinery which has a slightly unbalanced rotor. Assuming that the deflection of the floor is proportional to the load it supports, determine (*a*) the equivalent spring constant of the floor system, (*b*) the speed in rpm of the rotating machinery that should be avoided if it is not to coincide with the natural frequency of the floor-machinery system.

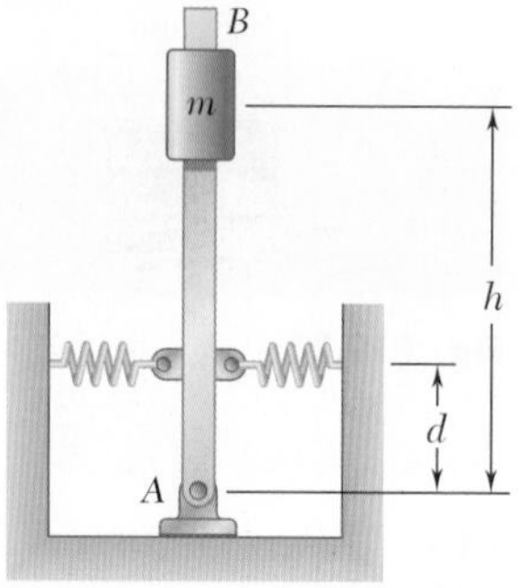

Fig. P19.31

19.31 If $h = 700$ mm and $d = 500$ mm and each spring has a constant $k = 600$ N/m, determine the mass m for which the period of small oscillations is (*a*) 0.50 s, (*b*) infinite. Neglect the mass of the rod and assume that each spring can act in either tension or compression.

19.32 The force-deflection equation for a nonlinear spring fixed at one end is $F = 5x^{1/2}$ where F is the force, expressed in newtons, applied at the other end and x is the deflection expressed in meters. (*a*) Determine the deflection x_0 if a 120-g block is suspended from the spring and is at rest. (*b*) Assuming that the slope of the force-deflection curve at the point corresponding to this loading can be used as an equivalent spring constant, determine the frequency of vibration of the block if it is given a very small downward displacement from its equilibrium position and released.

***19.33** Expanding the integrand in Eq. (19.19) of Sec. 19.4 into a series of even powers of sin φ and integrating, show that the period of a simple pendulum of length l may be approximated by the formula

$$\tau = 2\pi\sqrt{\frac{l}{g}}\left(1 + \tfrac{1}{4}\sin^2\frac{\theta_m}{2}\right)$$

where θ_m is the amplitude of the oscillations.

***19.34** Using the formula given in Prob. 19.33, determine the amplitude θ_m for which the period of a simple pendulum is $\frac{1}{2}$ percent longer than the period of the same pendulum for small oscillations.

***19.35** Using the data of Table 19.1, determine the period of a simple pendulum of length $l = 750$ mm (*a*) for small oscillations, (*b*) for oscillations of amplitude $\theta_m = 60°$, (*c*) for oscillations of amplitude $\theta_m = 90°$.

***19.36** Using the data of Table 19.1, determine the length in mm of a simple pendulum which oscillates with a period of 2 s and an amplitude of 90°.

19.5 FREE VIBRATIONS OF RIGID BODIES

The analysis of the vibrations of a rigid body or of a system of rigid bodies possessing a single degree of freedom is similar to the analysis of the vibrations of a particle. An appropriate variable, such as a distance x or an angle θ, is chosen to define the position of the body or system of bodies, and an equation relating this variable and its second derivative with respect to t is written. If the equation obtained is of the same form as (19.6), i.e., if we have

$$\ddot{x} + \omega_n^2 x = 0 \qquad \text{or} \qquad \ddot{\theta} + \omega_n^2 \theta = 0 \tag{19.21}$$

the vibration considered is a simple harmonic motion. The period and natural frequency of the vibration can then be obtained by identifying ω_n and substituting its value into Eqs. (19.13) and (19.14).

In general, a simple way to obtain one of Eqs. (19.21) is to express that the system of the external forces is equivalent to the system of the effective forces by drawing a free-body-diagram equation for an arbitrary value of the variable and writing the appropriate equation of motion. We recall that our goal should be *the determination of the coefficient* of the variable x or θ, *not* the determination of the variable itself or of the derivative $\ddot{x}$ or $\ddot{\theta}$. Setting this coefficient equal to ω_n^2, we obtain the natural circular frequency ω_n, from which τ_n and f_n can be determined.

The method we have outlined can be used to analyze vibrations which are truly represented by a simple harmonic motion, or vibrations of small amplitude which can be *approximated* by a simple harmonic motion. As an example, let us determine the period of the small oscillations of a square plate of side $2b$ which is suspended from the midpoint O of one of its sides (Fig. 19.5a). We consider the plate in an arbitrary position defined by the angle θ that the line OG forms with the vertical and draw a free-body-diagram equation to express that the weight $\mathbf{W}$ of the plate and the components $\mathbf{R}_x$ and $\mathbf{R}_y$ of the reaction at O are equivalent to the vectors $m\mathbf{a}_t$ and $m\mathbf{a}_n$ and to the couple $\bar{I}\boldsymbol{\alpha}$ (Fig. 19.5b). Since the angular velocity and angular acceleration of the plate are equal, respectively, to $\dot{\theta}$ and $\ddot{\theta}$, the magnitudes of the two vectors are, respectively, $mb\ddot{\theta}$ and $mb\dot{\theta}^2$, while the moment of the couple is $\bar{I}\ddot{\theta}$. In previous applications of this method (Chap. 16), we tried whenever possible to assume the correct sense for the acceleration. Here, however, we must assume the same positive sense for θ and $\ddot{\theta}$ in order to obtain an equation of the form (19.21). Consequently, the angular acceleration $\ddot{\theta}$ will be assumed positive counterclockwise, even though this assumption is obviously unrealistic. Equating moments about O, we write

$$+\circlearrowleft \qquad -W(b \sin \theta) = (mb\ddot{\theta})b + \bar{I}\ddot{\theta}$$

Noting that $\bar{I} = \frac{1}{12}m[(2b)^2 + (2b)^2] = \frac{2}{3}mb^2$ and $W = mg$, we obtain

$$\ddot{\theta} + \frac{3g}{5b} \sin \theta = 0 \qquad (19.22)$$

For oscillations of small amplitude, we can replace $\sin \theta$ by θ, expressed in radians, and write

$$\ddot{\theta} + \frac{3g}{5b} \theta = 0 \qquad (19.23)$$

Comparison with (19.21) shows that the equation obtained is that of a simple harmonic motion and that the natural circular frequency ω_n of the oscillations is equal to $(3g/5b)^{1/2}$. Substituting into (19.13), we find that the period of the oscillations is

$$\tau_n = \frac{2\pi}{\omega_n} = 2\pi\sqrt{\frac{5b}{3g}} \qquad (19.24)$$

The result obtained is valid only for oscillations of small amplitude. A more accurate description of the motion of the plate is obtained by comparing Eqs. (19.16) and (19.22). We note that the two equations are identical if we choose l equal to $5b/3$. This means that the plate will oscillate as a simple pendulum of length $l = 5b/3$ and the results of Sec. 19.4 can be used to correct the value of the period given in (19.24). The point A of the plate located on line OG at a distance $l = 5b/3$ from O is defined as the *center of oscillation* corresponding to O (Fig. 19.5a).

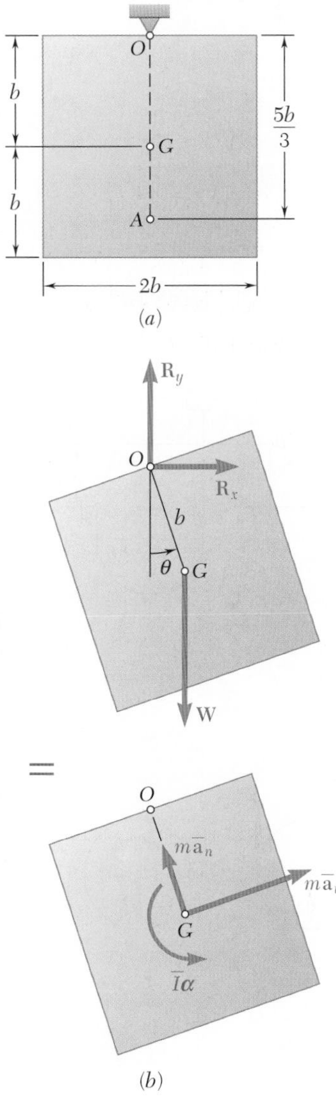

Fig. 19.5

SAMPLE PROBLEM 19.2

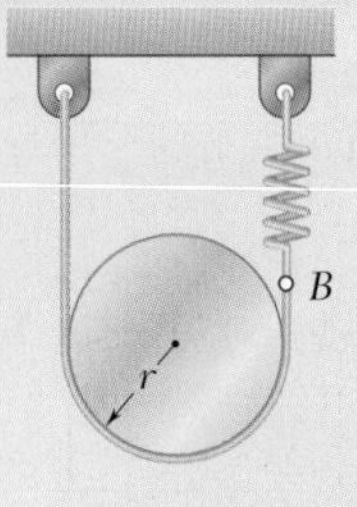

A cylinder of weight W and radius r is suspended from a looped cord as shown. One end of the cord is attached directly to a rigid support, while the other end is attached to a spring of constant k. Determine the period and natural frequency of the vibrations of the cylinder.

SOLUTION

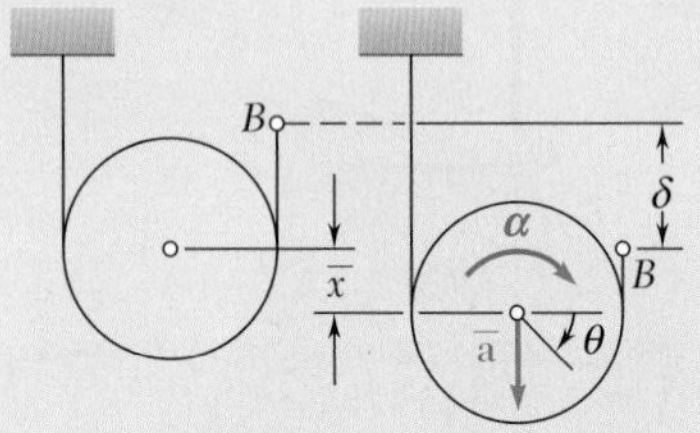

Kinematics of Motion. We express the linear displacement and the acceleration of the cylinder in terms of the angular displacement θ. Choosing the positive sense clockwise and measuring the displacements from the equilibrium position, we write

$$\bar{x} = r\theta \qquad \delta = 2\bar{x} = 2r\theta$$

$$\boldsymbol{\alpha} = \ddot{\theta} \downarrow \qquad \bar{a} = r\alpha = r\ddot{\theta} \qquad \bar{\mathbf{a}} = r\ddot{\theta} \downarrow \tag{1}$$

Equations of Motion. The system of external forces acting on the cylinder consists of the weight $\mathbf{W}$ and of the forces $\mathbf{T}_1$ and $\mathbf{T}_2$ exerted by the cord. We express that this system is equivalent to the system of effective forces represented by the vector $m\bar{\mathbf{a}}$ attached at G and the couple $\bar{I}\boldsymbol{\alpha}$.

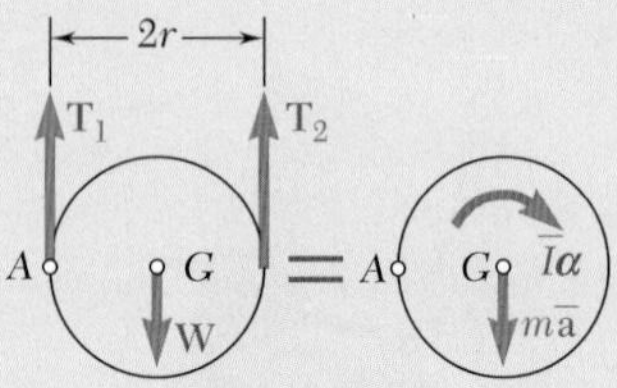

$$+\downarrow\Sigma M_A = \Sigma(M_A)_{\text{eff}}: \qquad Wr - T_2(2r) = m\bar{a}r + \bar{I}\alpha \tag{2}$$

When the cylinder is in its position of equilibrium, the tension in the cord is $T_0 = \frac{1}{2}W$. We note that for an angular displacement θ, the magnitude of $\mathbf{T}_2$ is

$$T_2 = T_0 + k\delta = \tfrac{1}{2}W + k\delta = \tfrac{1}{2}W + k(2r\theta) \tag{3}$$

Substituting from (1) and (3) into (2), and recalling that $\bar{I} = \frac{1}{2}mr^2$, we write

$$Wr - (\tfrac{1}{2}W + 2kr\theta)(2r) = m(r\ddot{\theta})r + \tfrac{1}{2}mr^2\ddot{\theta}$$

$$\ddot{\theta} + \frac{8}{3}\frac{k}{m}\theta = 0$$

The motion is seen to be simple harmonic, and we have

$$\omega_n^2 = \frac{8}{3}\frac{k}{m} \qquad \omega_n = \sqrt{\frac{8}{3}\frac{k}{m}}$$

$$\tau_n = \frac{2\pi}{\omega_n} \qquad \tau_n = 2\pi\sqrt{\frac{3}{8}\frac{m}{k}} \quad ◀$$

$$f_n = \frac{\omega_n}{2\pi} \qquad f_n = \frac{1}{2\pi}\sqrt{\frac{8}{3}\frac{k}{m}} \quad ◀$$

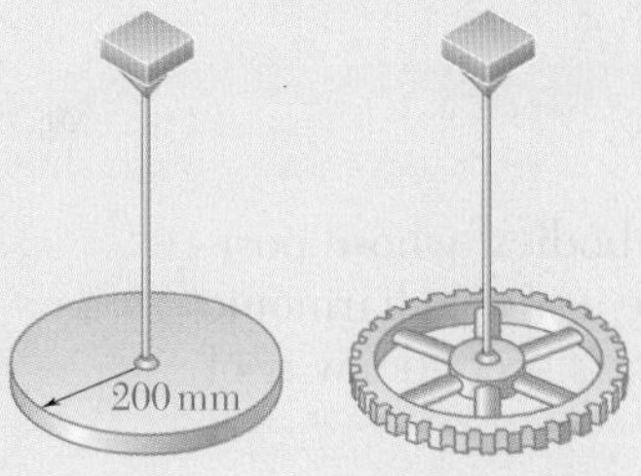

SAMPLE PROBLEM 19.3

A circular disk, of mass 10 kg and of radius 200 mm, is suspended from a wire as shown. The disk is rotated (thus twisting the wire) and then released; the period of the torsional vibration is observed to be 1.13 s. A gear is then suspended from the same wire, and the period of torsional vibration for the gear is observed to be 1.93 s. Assuming that the moment of the couple exerted by the wire is proportional to the angle of twist, determine (*a*) the torsional spring constant of the wire, (*b*) the centroidal moment of inertia of the gear, (*c*) the maximum angular velocity reached by the gear if it is rotated through 90° and released.

SOLUTION

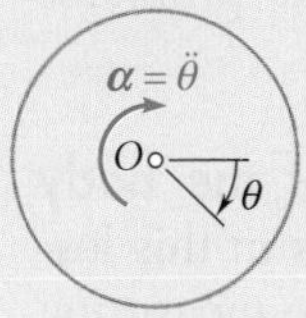

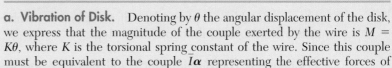

a. Vibration of Disk. Denoting by θ the angular displacement of the disk, we express that the magnitude of the couple exerted by the wire is $M = K\theta$, where K is the torsional spring constant of the wire. Since this couple must be equivalent to the couple $\bar{I}\boldsymbol{\alpha}$ representing the effective forces of the disk, we write

$$+\circlearrowleft \Sigma M_O = \Sigma (M_O)_{\text{eff}}: \qquad +K\theta = -\bar{I}\ddot{\theta}$$

$$\ddot{\theta} + \frac{K}{\bar{I}}\theta = 0$$

The motion is seen to be simple harmonic, and we have

$$\omega_n^2 = \frac{K}{\bar{I}} \qquad \tau_n = \frac{2\pi}{\omega_n} \qquad \tau_n = 2\pi\sqrt{\frac{\bar{I}}{K}} \tag{1}$$

For the disk, we have

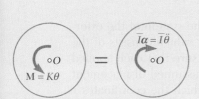

$$\tau_n = 1.13 \text{ s} \qquad \bar{I} = \tfrac{1}{2}mr^2 = \frac{1}{2}(10 \text{ kg})(0.2 \text{ m})^2 = 0.2 \text{ kg}\cdot\text{m}^2$$

Substituting into (1), we obtain

$$1.13 = 2\pi\sqrt{\frac{0.2}{K}} \qquad K = 6.18 \text{ N}\cdot\text{m/rad} \quad ◀$$

b. Vibration of Gear. Since the period of vibration of the gear is 1.93 s and $K = 6.183$ N·m/rad, Eq. (1) yields

$$1.93 = 2\pi\sqrt{\frac{\bar{I}}{6.183}} \qquad \bar{I}_{\text{gear}} = 0.583 \text{ kg}\cdot\text{m}^2 \quad ◀$$

c. Maximum Angular Velocity of Gear. Since the motion is simple harmonic, we have

$$\theta = \theta_m \sin \omega_n t \qquad \omega = \theta_m \omega_n \cos \omega_n t \qquad \omega_m = \theta_m \omega_n$$

Recalling that $\theta_m = 90° = 1.571$ rad and $\tau = 1.93$ s, we write

$$\omega_m = \theta_m \omega_n = \theta_m\left(\frac{2\pi}{\tau}\right) = (1.571 \text{ rad})\left(\frac{2\pi}{1.93 \text{ s}}\right)$$

$$\omega_m = 5.11 \text{ rad/s} \quad ◀$$

SOLVING PROBLEMS ON YOUR OWN

In this lesson you saw that a rigid body, or a system of rigid bodies, whose position can be defined by a single coordinate x or θ, will execute a simple harmonic motion if the differential equation obtained by applying Newton's second law is of the form

$$\ddot{x} + \omega_n^2 x = 0 \qquad \text{or} \qquad \ddot{\theta} + \omega_n^2 \theta = 0 \tag{19.21}$$

Your goal should be to determine ω_n, from which you can obtain the period τ_n and the natural frequency f_n. Taking into account the initial conditions, you can then write an equation of the form

$$x = x_m \sin(\omega_n t + \phi) \tag{19.10}$$

where x should be replaced by θ if a rotation is involved. To solve the problems in this lesson, you will follow these steps:

1. Choose a coordinate which will measure the displacement of the body from its equilibrium position. You will find that many of the problems in this lesson involve the rotation of a body about a fixed axis and that the angle measuring the rotation of the body from its equilibrium position is the most convenient coordinate to use. In problems involving the general plane motion of a body, where a coordinate x (and possibly a coordinate y) is used to define the position of the mass center G of the body, and a coordinate θ is used to measure its rotation about G, find kinematic relations which will allow you to express x (and y) in terms of θ [Sample Prob. 19.2].

2. Draw a free-body-diagram equation to express that the system of the external forces is equivalent to the system of the effective forces, which consists of the vector $m\bar{\mathbf{a}}$ and the couple $\bar{I}\boldsymbol{\alpha}$, where $\bar{a} = \ddot{x}$ and $\alpha = \ddot{\theta}$. Be sure that each applied force or couple is drawn in a direction consistent with the assumed displacement and that the senses of $\bar{\mathbf{a}}$ and $\boldsymbol{\alpha}$ are, respectively, those in which the coordinates x and θ are increasing.

3. Write the differential equations of motion by equating the sums of the components of the external and effective forces in the x and y directions and the sums of their moments about a given point. If necessary, use the kinematic relations developed in step 1 to obtain equations involving only the coordinate θ. If θ is a small angle, replace $\sin\theta$ by θ and $\cos\theta$ by 1, if these functions appear in your equations. Eliminating any unknown reactions, you will obtain an equation of the type of Eqs. (19.21). Note that in problems involving a body rotating about a fixed axis, you can immediately obtain such an equation by equating the moments of the external and effective forces about the fixed axis.

4. Comparing the equation you have obtained with one of Eqs. (19.21), you can identify ω_n^2 and, thus, determine the natural circular frequency ω_n. Remember that the object of your analysis is *not to solve* the differential equation you have obtained, *but to identify* ω_n^2.

5. Determine the amplitude and the phase angle ϕ by substituting the value obtained for ω_n and the initial values of the coordinate and its first derivative into Eq. (19.10) and the equation obtained by differentiating (19.10) with respect to t. From Eq. (19.10) and the two equations obtained by differentiating (19.10) twice with respect to t, and using the kinematic relations developed in step 1, you will be able to determine the position, velocity, and acceleration of any point of the body at any given time.

6. In problems involving torsional vibrations, the torsional spring constant K is expressed in $\text{N} \cdot \text{m/rad}$ or $\text{lb} \cdot \text{ft/rad}$. The product of K and the angle of twist θ, expressed in radians, yields the moment of the restoring couple, which should be equated to the sum of the moments of the effective forces or couples about the axis of rotation [Sample Prob. 19.3].

PROBLEMS

19.37 The uniform rod shown has mass 6 kg and is attached to a spring of constant $k = 700$ N/m. If end B of the rod is depressed 10 mm and released, determine (*a*) the period of vibration, (*b*) the maximum velocity of end B.

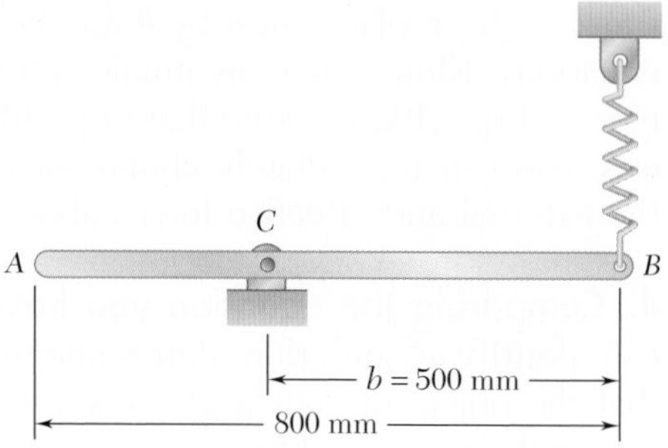

Fig. P19.37

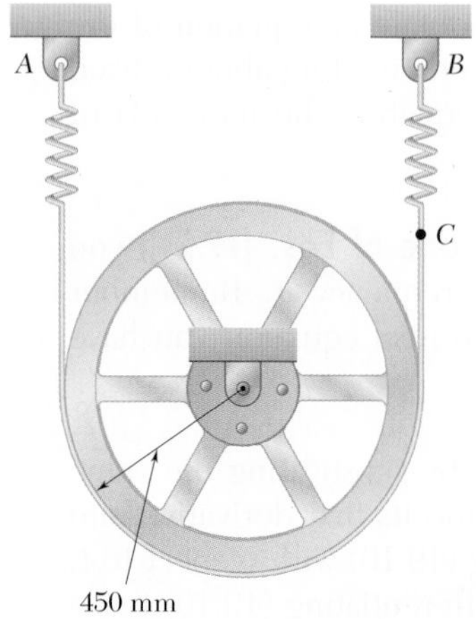

Fig. P19.38

19.38 A belt is placed around the rim of a 240-kg flywheel and attached as shown to two springs, each of constant $k = 15$ kN/m. If end C of the belt is pulled 40 mm down and released, the period of vibration of the flywheel is observed to be 0.5 s. Knowing that the initial tension in the belt is sufficient to prevent slipping, determine (*a*) the maximum angular velocity of the flywheel, (*b*) the centroidal radius of gyration of the flywheel.

19.39 An 8-kg uniform rod AB is hinged to a fixed support at A and is attached by means of pins B and C to a 12-kg disk of radius 400 mm. A spring attached at D holds the rod at rest in the position shown. If point B is moved down 25 mm and released, determine (*a*) the period of vibration, (*b*) the maximum velocity of point B.

19.40 Solve Prob. 19.39, assuming that pin C is removed and that the disk can rotate freely about pin B.

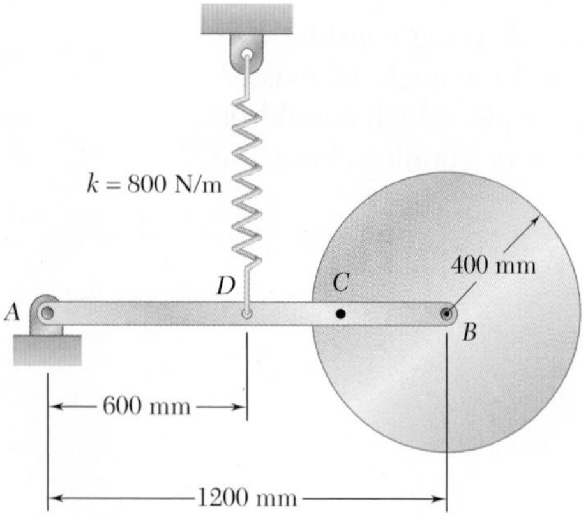

Fig. P19.39

19.41 A 7.5-kg slender rod AB is riveted to a 6-kg uniform disk as shown. A belt is attached to the rim of the disk and to a spring which holds the rod at rest in the position shown. If end A of the rod is moved 20 mm down and released, determine (*a*) the period of vibration, (*b*) the maximum velocity of end A.

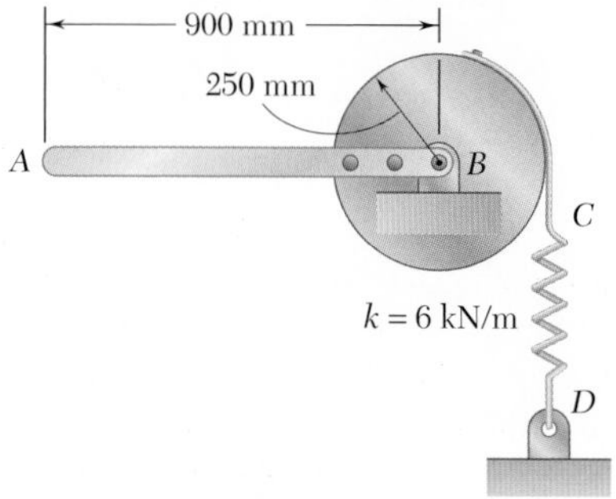

Fig. P19.41

19.42 A 15-kg uniform cylinder can roll without sliding on a 15°-incline. A belt is attached to the rim of the cylinder, and a spring holds the cylinder at rest in the position shown. If the center of the cylinder is moved 50 mm down the incline and released, determine (*a*) the period of vibration, (*b*) the maximum acceleration of the center of the cylinder.

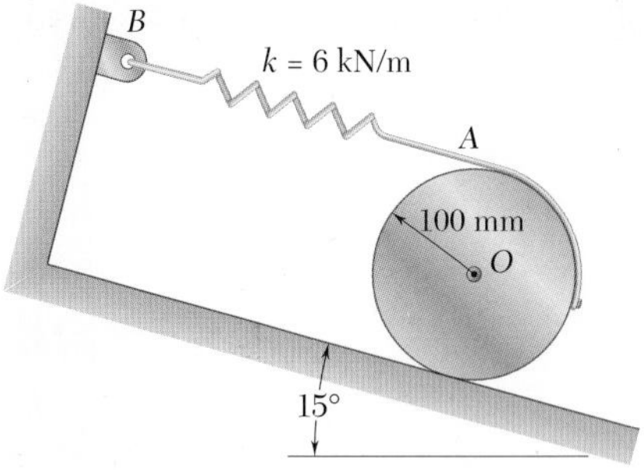

Fig. P19.42

19.43 A square plate of mass m is held by eight springs, each of constant k. Knowing that each spring can act in either tension or compression, determine the frequency of the resulting vibration if (*a*) the plate is given a small vertical displacement and released, (*b*) the plate is rotated through a small angle about G and released.

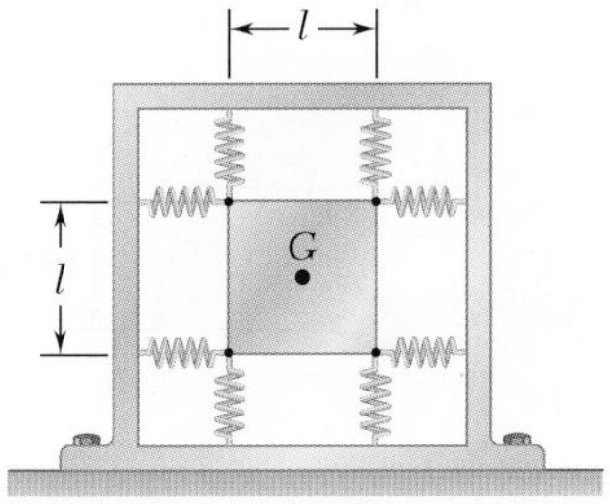

Fig. *P19.43*

19.44 Two small weights w are attached at A and B to the rim of a uniform disk of radius r and weight W. Denoting by τ_0 the period of small oscillations when $\beta = 0$, determine the angle β for which the period of small oscillations is $2\tau_0$.

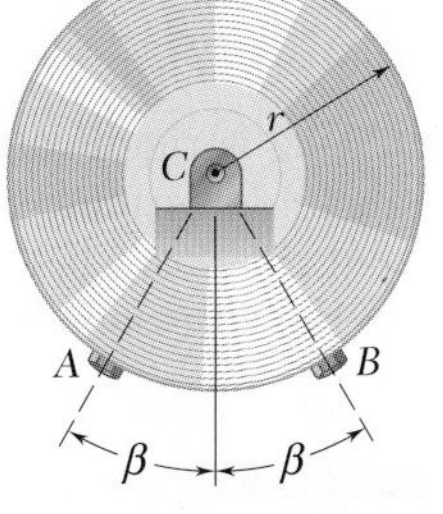

Fig. P19.44 and P19.45

19.45 Two 40-g weights are attached at A and B to the rim of a 1.5-kg uniform disk of radius $r = 100$ mm. Determine the frequency of small oscillations when $\beta = 60°$.

19.46 A three-blade wind turbine used for research is supported on a shaft so that it is free to rotate about O. One technique to determine the centroidal mass moment of inertia of an object is to place a known weight at a known distance from the axis of rotation and to measure the frequency of oscillations after releasing it from rest with a small initial angle. In this case, a mass of $m_{add} = 25$ kg is attached to one of the blades at a distance $R = 6$ m from the axis of rotation. Knowing that when the blade with the added weight is displaced slightly from the vertical axis, and the system is found to have a period of 7.6 s, determine the centroidal mass moment of inertia of the three-blade rotor.

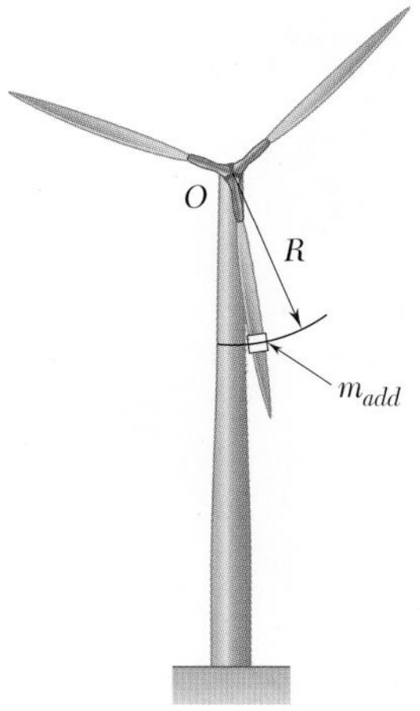

Fig. *P19.46*

19.47 A connecting rod is supported by a knife-edge at point A; the period of its small oscillations is observed to be 0.87 s. The rod is then inverted and supported by a knife edge at point B and the period of its small oscillations is observed to be 0.78 s. Knowing that $r_a + r_b = 250$ mm, determine (*a*) the location of the mass center G, (*b*) the centroidal radius of gyration $\overline{k}$.

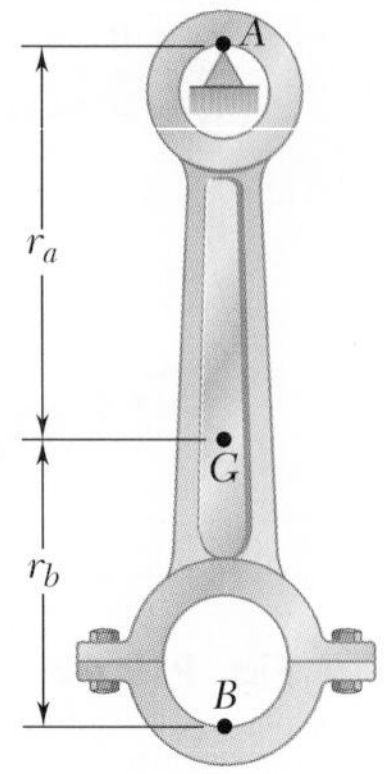

Fig. P19.47

19.48 A 75-mm-radius hole is cut in a 200-mm-radius uniform disk which is attached to a frictionless pin at its geometric center O. Determine (*a*) the period of small oscillations of the disk, (*b*) the length of a simple pendulum which has the same period.

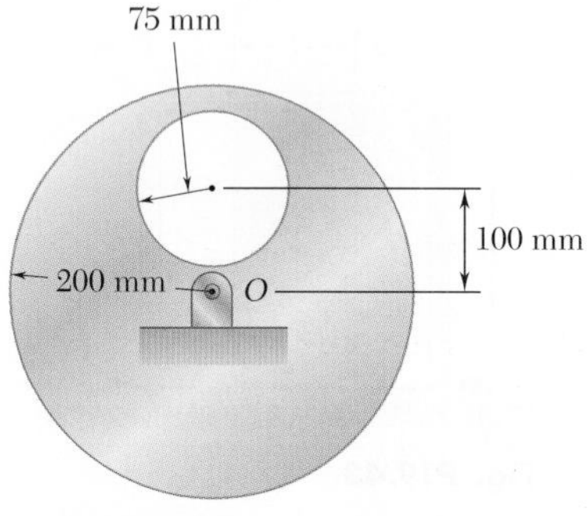

Fig. P19.48

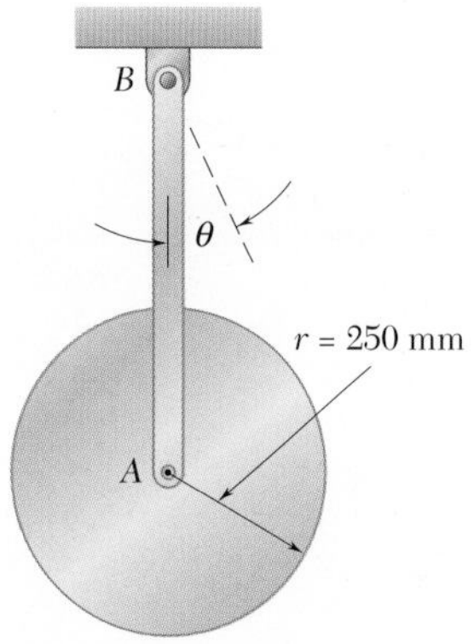

Fig. P19.49

19.49 A uniform disk of radius $r = 250$ mm is attached at A to a 650-mm rod AB of negligible mass which can rotate freely in a vertical plane about B. Determine the period of small oscillations (*a*) if the disk is free to rotate in a bearing at A, (*b*) if the rod is riveted to the disk at A.

19.50 A small collar of mass 1 kg is rigidly attached to a 3-kg uniform rod of length $L = 750$ mm. Determine (*a*) the distance d to maximize the frequency of oscillation when the rod is given a small initial displacement, (*b*) the corresponding period of oscillation.

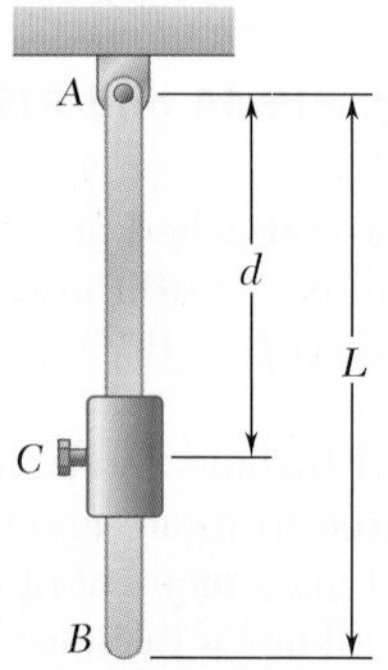

Fig. P19.50

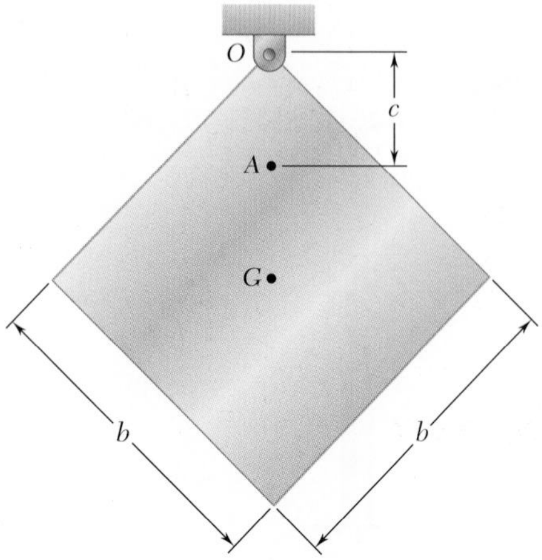

Fig. P19.51

19.51 For the uniform square plate of side $b = 300$ mm, determine (*a*) the period of small oscillations if the plate is suspended as shown, (*b*) the distance c from O to a point A from which the plate should be suspended for the period to be a minimum.

19.52 A *compound pendulum* is defined as a rigid slab which oscillates about a fixed point O, called the center of suspension. Show that the period of oscillation of a compound pendulum is equal to the period of a simple pendulum of length OA, where the distance from A to the mass center G is $GA = \bar{k}^2/\bar{r}$. Point A is defined as the center of oscillation and coincides with the center of percussion defined in Prob. 17.66.

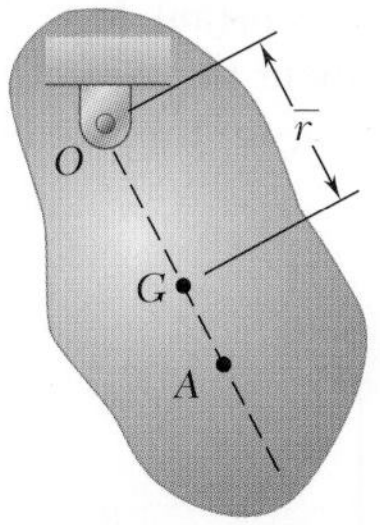

Fig. P19.52 and P19.53

19.53 A rigid slab oscillates about a fixed point O. Show that the smallest period of oscillation occurs when the distance $\bar{r}$ from point O to the mass center G is equal to $\bar{k}$.

19.54 Show that if the compound pendulum of Prob. 19.52 is suspended from A instead of O, the period of oscillation is the same as before and the new center of oscillation is located at O.

19.55 The 8-kg uniform bar AB is hinged at C and is attached at A to a spring of constant $k = 500$ N/m. If end A is given a small displacement and released, determine (*a*) the frequency of small oscillations, (*b*) the smallest value of the spring constant k for which oscillations will occur.

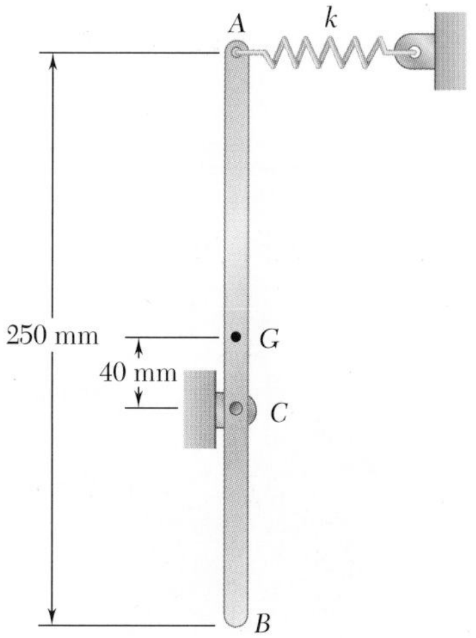

Fig. P19.55

19.56 Two uniform rods, each of mass $m = 12$ kg and length $L = 800$ mm, are welded together to form the assembly shown. Knowing that the constant of each spring is $k = 500$ N/m and that end A is given a small displacement and released, determine the frequency of the resulting motion.

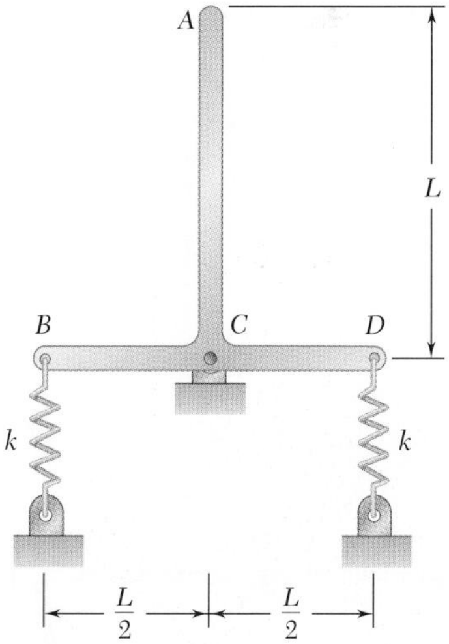

Fig. P19.56

19.57 A 20-kg uniform square plate is suspended from a pin located at the midpoint A of one of its 0.4 m edges and is attached to springs, each of constant $k = 1.6$ kN/m. If corner B is given a small displacement and released, determine the frequency of the resulting vibration. Assume that each spring can act in either tension or compression.

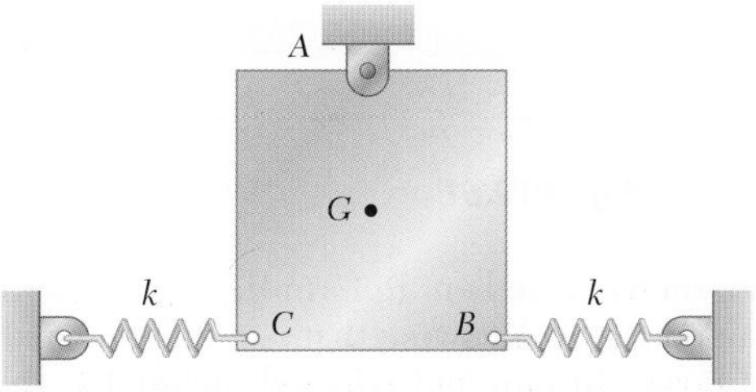

Fig. P19.57

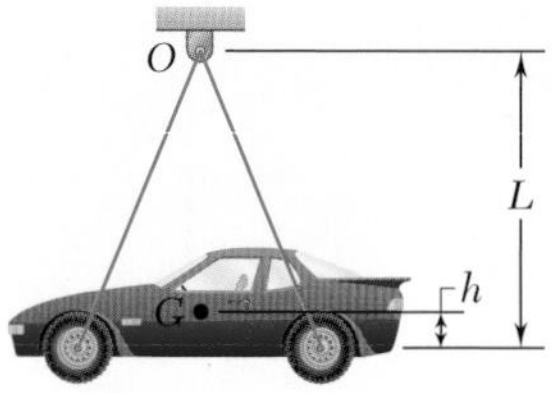

Fig. P19.58

19.58 A 1300-kg sports car has a center of gravity G located a distance h above a line connecting the front and rear axles. The car is suspended from cables that are attached to the front and rear axles as shown. Knowing that the periods of oscillation are 4.04 s when $L = 4$ m and 3.54 s when $L = 3$ m, determine h and the centroidal radius of gyration.

19.59 A 3-kg slender rod is suspended from a steel wire which is known to have a torsional spring constant $K = 2.25$ N · m/rad. If the rod is rotated through 180° about the vertical and released, determine (*a*) the period of oscillation, (*b*) the maximum velocity of end A of the rod.

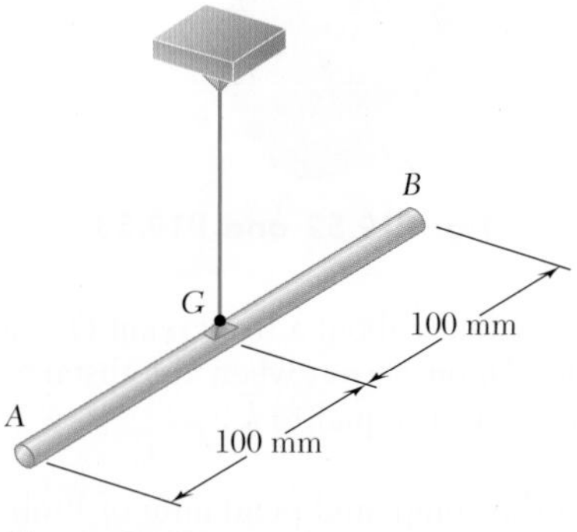

Fig. P19.59

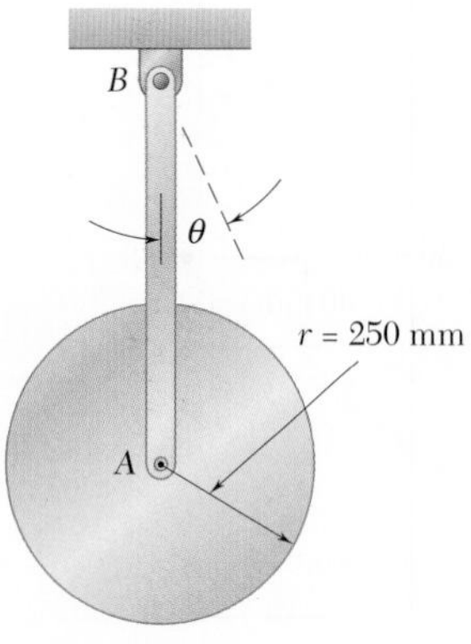

Fig. P19.60

19.60 A uniform disk of radius $r = 250$ mm is attached at A to a 650-mm rod AB of negligible mass which can rotate freely in a vertical plane about B. If the rod is displaced 2° from the position shown and released, determine the magnitude of the maximum velocity of point A, assuming that the disk is (*a*) free to rotate in a bearing at A, (*b*) riveted to the rod at A.

19.61 Two uniform rods, each of mass m and length l, are welded together to form the T-shaped assembly shown. Determine the frequency of small oscillations of the assembly.

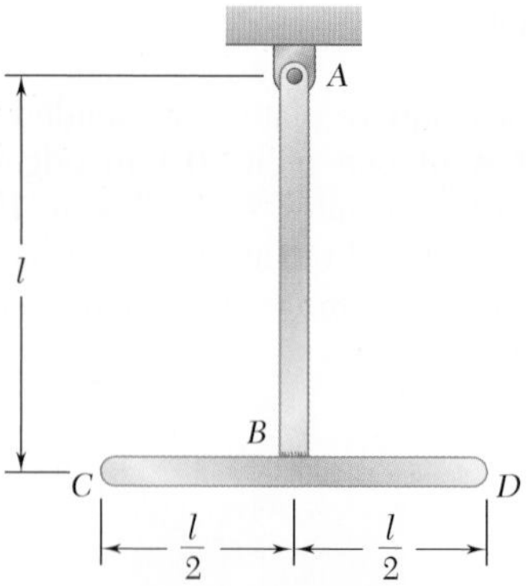

Fig. P19.61

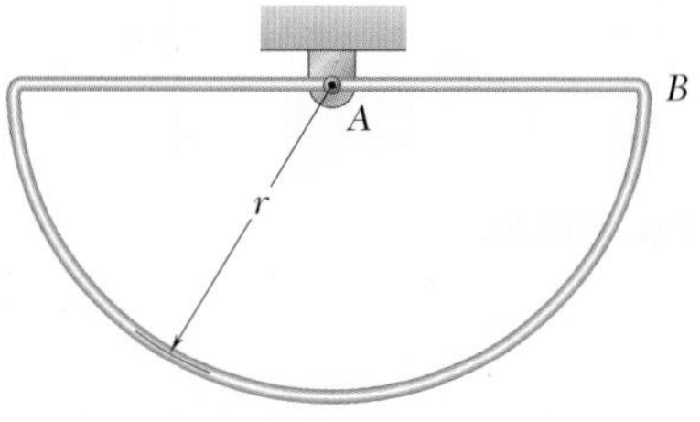

Fig. P19.62

19.62 A homogeneous wire bent to form the figure shown is attached to a pin support at A. Knowing that $r = 220$ mm and that point B is pushed down 20 mm and released, determine the magnitude of the velocity of B, 8 s later.

19.63 A horizontal platform P is held by several rigid bars which are connected to a vertical wire. The period of oscillation of the platform is found to be 2.2 s when the platform is empty and 3.8 s when an object A of unknown moment of inertia is placed on the platform with its mass center directly above the center of the plate. Knowing that the wire has a torsional constant $K = 27\ \text{N}\cdot\text{m/rad}$, determine the centroidal moment of inertia of object A.

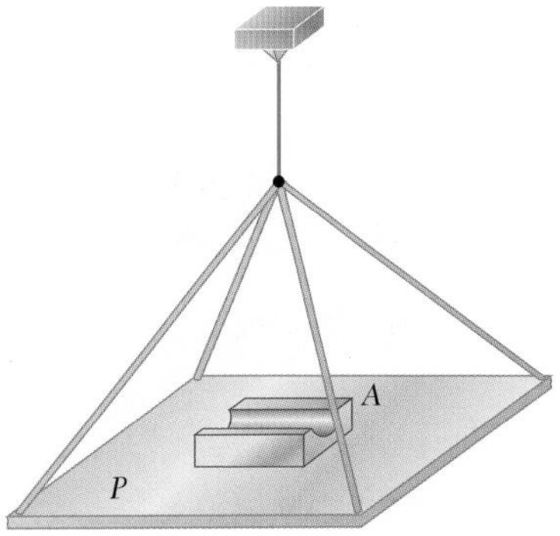

Fig. P19.63

19.64 A uniform disk of radius $r = 120$ mm is welded at its center to two elastic rods of equal length with fixed ends at A and B. Knowing that the disk rotates through an 8° angle when a 500-mN · m couple is applied to the disk and that it oscillates with a period of 1.3 s when the couple is removed, determine (*a*) the mass of the disk, (*b*) the period of vibration if one of the rods is removed.

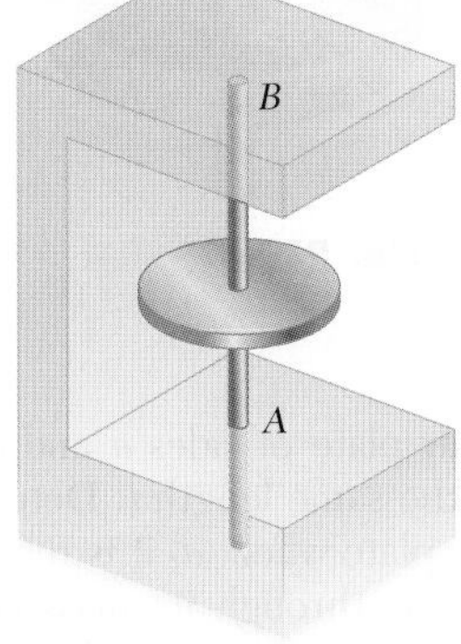

Fig. P19.64

19.65 A 5-kg uniform rod CD of length $l = 0.7$ m is welded at C to two elastic rods, which have fixed ends at A and B and are known to have a combined torsional spring constant $K = 24\ \text{N}\cdot\text{m/rad}$. Determine the period of small oscillations, if the equilibrium position of CD is (*a*) vertical as shown, (*b*) horizontal.

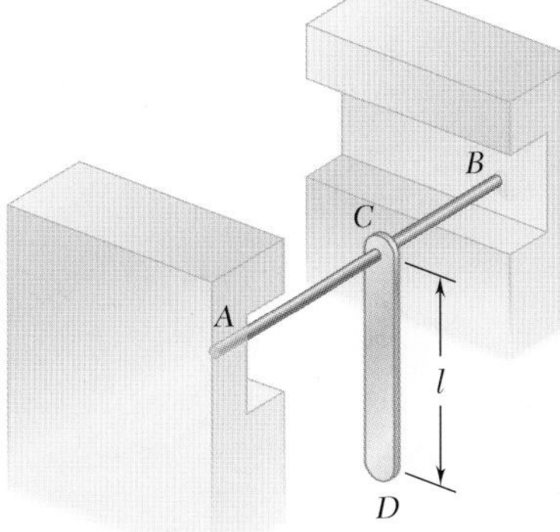

Fig. *P19.65*

19.66 A 1.8-kg uniform plate in the shape of an equilateral triangle is suspended at its center of gravity from a steel wire which is known to have a torsional constant $K = 35\ \text{mN}\cdot\text{m/rad}$. If the plate is rotated 360° about the vertical and then released, determine (*a*) the period of oscillation, (*b*) the maximum velocity of one of the vertices of the triangle.

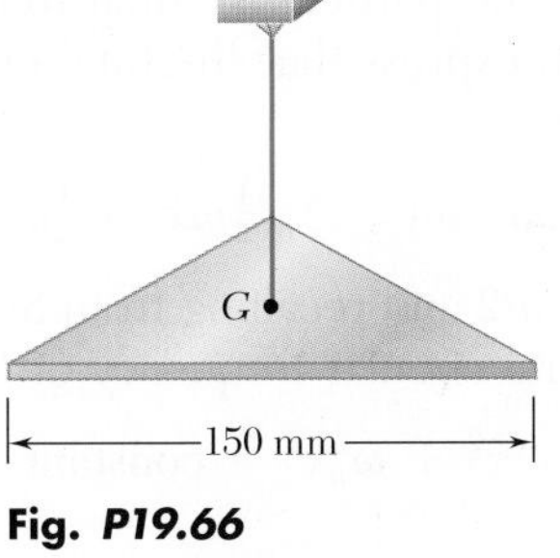

Fig. *P19.66*

19.67 A period of 6.00 s is observed for the angular oscillations of a 120-g gyroscope rotor suspended from a wire as shown. Knowing that a period of 3.80 s is obtained when a 30 mm-diameter steel sphere is suspended in the same fashion, determine the centroidal radius of gyration of the rotor. (density of steel = 7800 kg/m^3).

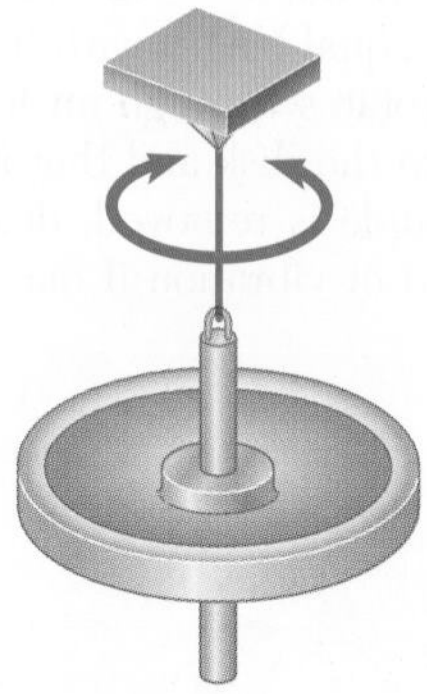

Fig. P19.67

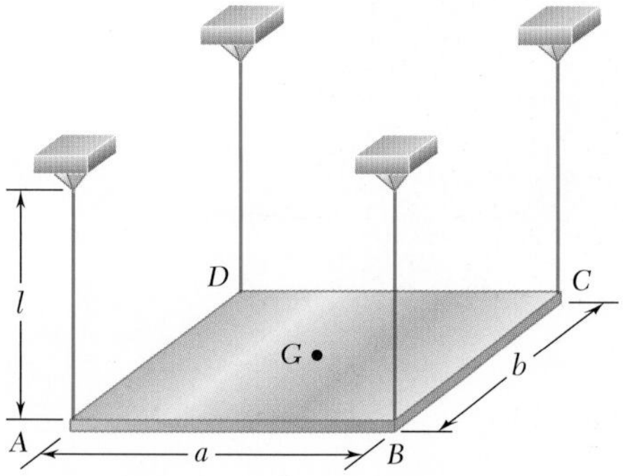

Fig. *P19.68*

19.68 A thin rectangular plate of sides a and b is suspended from four vertical wires of the same length l. Determine the period of small oscillations of the plate when (*a*) it is rotated through a small angle about a vertical axis through its mass center G, (*b*) it is given a small horizontal displacement in a direction perpendicular to AB, (*c*) it is given a small horizontal displacement in a direction perpendicular to BC.

19.6 APPLICATION OF THE PRINCIPLE OF CONSERVATION OF ENERGY

We saw in Sec. 19.2 that when a particle of mass m is in simple harmonic motion, the resultant **F** of the forces exerted on the particle has a magnitude proportional to the displacement x measured from the position of equilibrium O and is directed toward O; we write $F = -kx$. Referring to Sec. 13.6, we note that **F** is a *conservative force* and that the corresponding potential energy is $V = \frac{1}{2}kx^2$, where V is assumed equal to zero in the equilibrium position $x = 0$. Since the velocity of the particle is equal to $\dot{x}$, its kinetic energy is $T = \frac{1}{2}m\dot{x}^2$ and we can express that the total energy of the particle is conserved by writing

$$T + V = \text{constant} \qquad \tfrac{1}{2}m\dot{x}^2 + \tfrac{1}{2}kx^2 = \text{constant}$$

Dividing through by $m/2$ and recalling from Sec. 19.2 that $k/m = \omega_n^2$, where ω_n is the natural circular frequency of the vibration, we have

$$\dot{x}^2 + \omega_n^2 x^2 = \text{constant} \tag{19.25}$$

Equation (19.25) is characteristic of a simple harmonic motion, since it can be obtained from Eq. (19.6) by multiplying both terms by $2\dot{x}$ and integrating.

The principle of conservation of energy provides a convenient way for determining the period of vibration of a rigid body or of a system of rigid bodies possessing a single degree of freedom, once it has been established that the motion of the system is a simple harmonic motion or that it can be approximated by a simple harmonic motion. Choosing an appropriate variable, such as a distance x or an angle θ, we consider two particular positions of the system:

1. *The displacement of the system is maximum;* we have $T_1 = 0$, and V_1 can be expressed in terms of the amplitude x_m or θ_m (choosing $V = 0$ in the equilibrium position).
2. *The system passes through its equilibrium position;* we have $V_2 = 0$, and T_2 can be expressed in terms of the maximum velocity $\dot{x}_m$ or the maximum angular velocity $\dot{\theta}_m$.

We then express that the total energy of the system is conserved and write $T_1 + V_1 = T_2 + V_2$. Recalling from (19.15) that for simple harmonic motion the maximum velocity is equal to the product of the amplitude and of the natural circular frequency ω_n, we find that the equation obtained can be solved for ω_n.

As an example, let us consider again the square plate of Sec. 19.5. In the position of maximum displacement (Fig. 19.6*a*), we have

$$T_1 = 0 \qquad V_1 = W(b - b\cos\theta_m) = Wb(1 - \cos\theta_m)$$

or, since $1 - \cos\theta_m = 2\sin^2(\theta_m/2) \approx 2(\theta_m/2)^2 = \theta_m^2/2$ for oscillations of small amplitude,

$$T_1 = 0 \qquad V_1 = \tfrac{1}{2}Wb\theta_m^2 \tag{19.26}$$

As the plate passes through its position of equilibrium (Fig. 19.6*b*), its velocity is maximum and we have

$$T_2 = \tfrac{1}{2}m\bar{v}_m^2 + \tfrac{1}{2}\bar{I}\omega_m^2 = \tfrac{1}{2}mb^2\dot{\theta}_m^2 + \tfrac{1}{2}\bar{I}\dot{\theta}_m^2 \qquad V_2 = 0$$

or, recalling from Sec. 19.5 that $\bar{I} = \tfrac{2}{3}mb^2$,

$$T_2 = \tfrac{1}{2}(\tfrac{5}{3}mb^2)\dot{\theta}_m^2 \qquad V_2 = 0 \tag{19.27}$$

Substituting from (19.26) and (19.27) into $T_1 + V_1 = T_2 + V_2$, and noting that the maximum velocity $\dot{\theta}_m$ is equal to the product $\theta_m\omega_n$, we write

$$\tfrac{1}{2}Wb\theta_m^2 = \tfrac{1}{2}(\tfrac{5}{3}mb^2)\theta_m^2\omega_n^2 \tag{19.28}$$

which yields $\omega_n^2 = 3g/5b$ and

$$\tau_n = \frac{2\pi}{\omega_n} = 2\pi\sqrt{\frac{5b}{3g}} \tag{19.29}$$

as previously obtained.

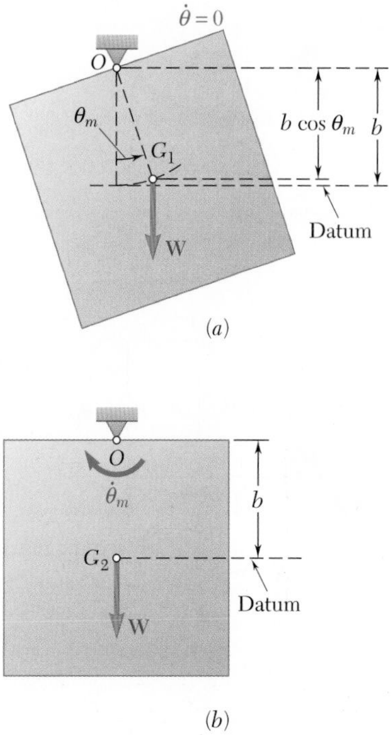

Fig. 19.6

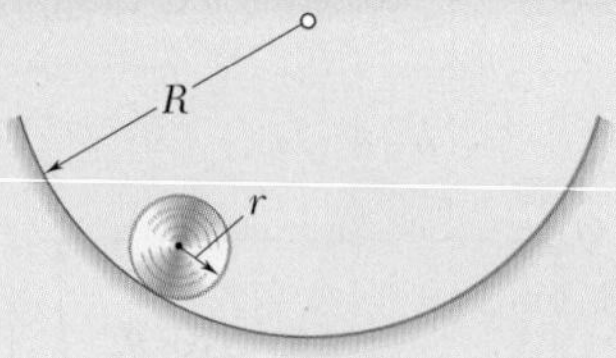

SAMPLE PROBLEM 19.4

Determine the period of small oscillations of a cylinder of radius r which rolls without slipping inside a curved surface of radius R.

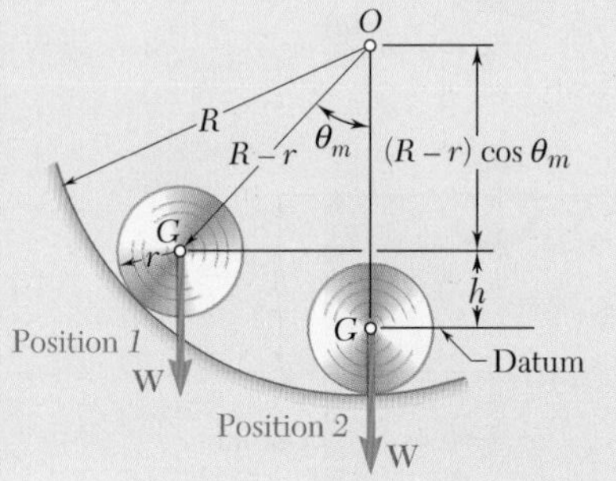

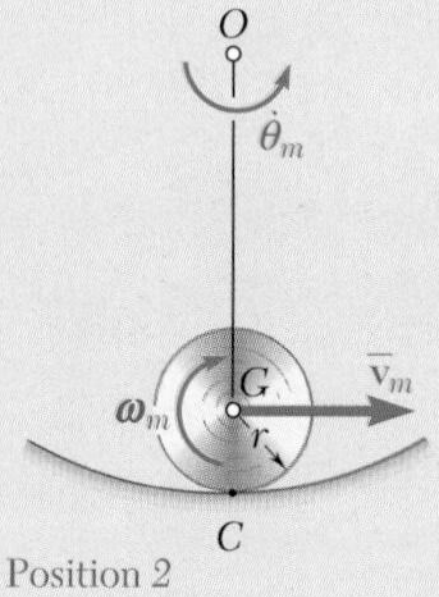

SOLUTION

We denote by θ the angle which line OG forms with the vertical. Since the cylinder rolls without slipping, we may apply the principle of conservation of energy between position *1*, where $\theta = \theta_m$, and position 2, where $\theta = 0$.

Position 1

Kinetic Energy. Since the velocity of the cylinder is zero, $T_1 = 0$.

Potential Energy. Choosing a datum as shown and denoting by W the weight of the cylinder, we have

$$V_1 = Wh = W(R - r)(1 - \cos\theta)$$

Noting that for small oscillations $(1 - \cos\theta) = 2\sin^2(\theta/2) \approx \theta^2/2$, we have

$$V_1 = W(R - r)\frac{\theta_m^2}{2}$$

Position 2. Denoting by $\dot{\theta}_m$ the angular velocity of line OG as the cylinder passes through position 2, and observing that point C is the instantaneous center of rotation of the cylinder, we write

$$\bar{v}_m = (R - r)\dot{\theta}_m \qquad \omega_m = \frac{\bar{v}_m}{r} = \frac{R - r}{r}\dot{\theta}_m$$

Kinetic Energy

$$\begin{aligned} T_2 &= \tfrac{1}{2}m\bar{v}_m^2 + \tfrac{1}{2}\bar{I}\omega_m^2 \\ &= \tfrac{1}{2}m(R - r)^2\dot{\theta}_m^2 + \tfrac{1}{2}(\tfrac{1}{2}mr^2)\left(\frac{R - r}{r}\right)^2\dot{\theta}_m^2 \\ &= \tfrac{3}{4}m(R - r)^2\dot{\theta}_m^2 \end{aligned}$$

Potential Energy

$$V_2 = 0$$

Conservation of Energy

$$T_1 + V_1 = T_2 + V_2$$

$$0 + W(R - r)\frac{\theta_m^2}{2} = \tfrac{3}{4}m(R - r)^2\dot{\theta}_m^2 + 0$$

Since $\dot{\theta}_m = \omega_n\theta_m$ and $W = mg$, we write

$$mg(R - r)\frac{\theta_m^2}{2} = \tfrac{3}{4}m(R - r)^2(\omega_n\theta_m)^2 \qquad \omega_n^2 = \frac{2}{3}\frac{g}{R - r}$$

$$\tau_n = \frac{2\pi}{\omega_n} \qquad \tau_n = 2\pi\sqrt{\frac{3}{2}\frac{R - r}{g}} \quad ◀$$

SOLVING PROBLEMS ON YOUR OWN

In the problems which follow you will be asked to use the *principle of conservation of energy* to determine the period or natural frequency of the simple harmonic motion of a particle or rigid body. Assuming that you choose an angle θ to define the position of the system (with $\theta = 0$ in the equilibrium position), as you will in most of the problems in this lesson, you will express that the total energy of the system is conserved, $T_1 + V_1 = T_2 + V_2$, between the position 1 of maximum displacement ($\theta_1 = \theta_m$, $\dot{\theta}_1 = 0$) and the position 2 of maximum velocity ($\dot{\theta}_2 = \dot{\theta}_m$, $\theta_2 = 0$). It follows that T_1 and V_2 will both be zero, and the energy equation will reduce to $V_1 = T_2$, where V_1 and T_2 are homogeneous quadratic expressions in θ_m and $\dot{\theta}_m$, respectively. Recalling that, for a simple harmonic motion, $\dot{\theta}_m = \theta_m \omega_n$ and substituting this product into the energy equation, you will obtain, after reduction, an equation that you can solve for ω_n^2. Once you have determined the natural circular frequency ω_n, you can obtain the period τ_n and the natural frequency f_n of the vibration.

The steps that you should take are as follows:

1. Calculate the potential energy V_1 of the system in its position of maximum displacement. Draw a sketch of the system in its position of maximum displacement and express the potential energy of all the forces involved (internal as well as external) in terms of the maximum displacement x_m or θ_m.

a. The potential energy associated with the weight W of a body is $V_g = W_y$, where y is the elevation of the center of gravity G of the body above its equilibrium position. If the problem you are solving involves the oscillation of a rigid body about a horizontal axis through a point O located at a distance b from G (Fig. 19.6), express y in terms of the angle θ that the line OG forms with the vertical: $y = b(1 - \cos\theta)$. But, for small values of θ, you can replace this expression with $y = \frac{1}{2}b\theta^2$ [Sample Prob. 19.4]. Therefore, when θ reaches its maximum value θ_m, and for oscillations of small amplitude, you can express V_g as

$$V_g = \tfrac{1}{2}Wb\theta_m^2$$

Note that *if G is located above O* in its equilibrium position (instead of below O, as we have assumed), the vertical displacement y will be negative and should be approximated as $y = -\frac{1}{2}b\theta^2$, which will result in a negative value for V_g. In the absence of other forces, the equilibrium position will be unstable, and the system will not oscillate. (See, for instance, Prob. 19.89.)

b. The potential energy associated with the elastic force exerted by a spring is $V_e = \frac{1}{2}kx^2$, where k is the constant of the spring and x its deflection. In problems involving the rotation of a body about an axis, you will generally have $x = a\theta$, where a is the distance from the axis of rotation to the point of the body

(*continued*)

where the spring is attached, and where θ is the angle of rotation. Therefore, when x reaches its maximum value x_m and θ reaches its maximum value θ_m, you can express V_e as

$$V_e = \tfrac{1}{2}kx_m^2 = \tfrac{1}{2}ka^2\theta_m^2$$

c. The potential energy V_1 of the system in its position of maximum displacement is obtained by adding the various potential energies that you have computed. It will be equal to the product of a constant and θ_m^2.

2. Calculate the kinetic energy T_2 of the system in its position of maximum velocity. Note that this position is also the equilibrium position of the system.

a. If the system consists of a single rigid body, the kinetic energy T_2 of the system will be the sum of the kinetic energy associated with the motion of the mass center G of the body and the kinetic energy associated with the rotation of the body about G. You will write, therefore,

$$T_2 = \tfrac{1}{2}m\bar{v}_m^2 + \tfrac{1}{2}\bar{I}\omega_m^2$$

Assuming that the position of the body has been defined by an angle θ, express $\bar{v}_m$ and ω_m in terms of the rate of change $\dot{\theta}_m$ of θ as the body passes through its equilibrium position. The kinetic energy of the body will thus be expressed as the product of a constant and $\dot{\theta}_m^2$. Note that if θ measures the rotation of the body about its mass center, as was the case for the plate of Fig. 19.6, then $\omega_m = \dot{\theta}_m$. In other cases, however, the kinematics of the motion should be used to derive a relation between ω_m and $\dot{\theta}_m$ [Sample Prob. 19.4].

b. If the system consists of several rigid bodies, repeat the above computation for each of the bodies, using the same coordinate θ, and add the results obtained.

3. Equate the potential energy V_1 of the system to its kinetic energy T_2,

$$V_1 = T_2$$

and, recalling the first of Eqs. (19.15), replace $\dot{\theta}_m$ in the right-hand term by the product of the amplitude θ_m and the circular frequency ω_n. Since both terms now contain the factor θ_m^2, this factor can be canceled and the resulting equation can be solved for the circular frequency ω_n.

PROBLEMS

19.69 A 1.8-kg collar A is attached to a spring of constant 800 N/m and can slide without friction on a horizontal rod. If the collar is moved 70 mm to the left from its equilibrium position and released, determine the maximum velocity and maximum acceleration of the collar during the resulting motion.

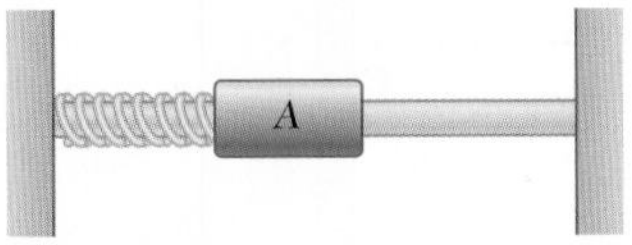

Fig. P19.69

19.70 Two blocks, each of mass 1.5 kg, are attached to links which are pin-connected to bar BC as shown. The weights of the links and bar are negligible, and the blocks can slide without friction. Block D is attached to a spring of constant $k = 800$ N/m. Knowing that block A is moved 12 mm from its equilibrium position and released, determine the magnitude of the maximum velocity of block D during the resulting motion.

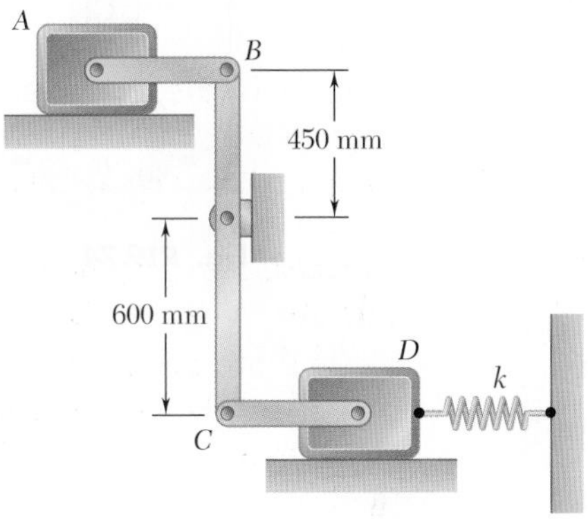

Fig. P19.70

19.71 A 400-g sphere A and a 300-g sphere C are attached to the ends of a rod AC of negligible weight which can rotate in a vertical plane about an axis at B. Determine the period of small oscillations of the rod.

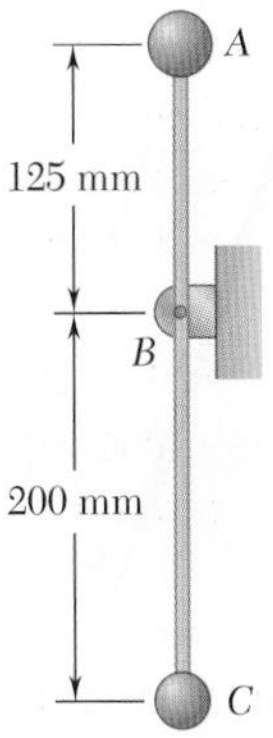

Fig. P19.71

19.72 Determine the period of small oscillations of a small particle which moves without friction inside a cylindrical surface of radius R.

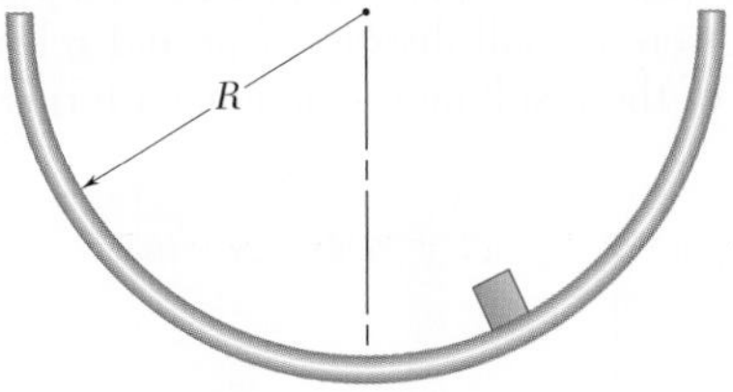

Fig. P19.72

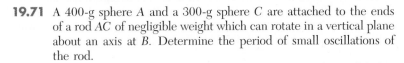

19.73 The inner rim of a 40-kg flywheel is placed on a knife edge, and the period of its small oscillations is found to be 1.26 s. Determine the centroidal moment of inertia of the flywheel.

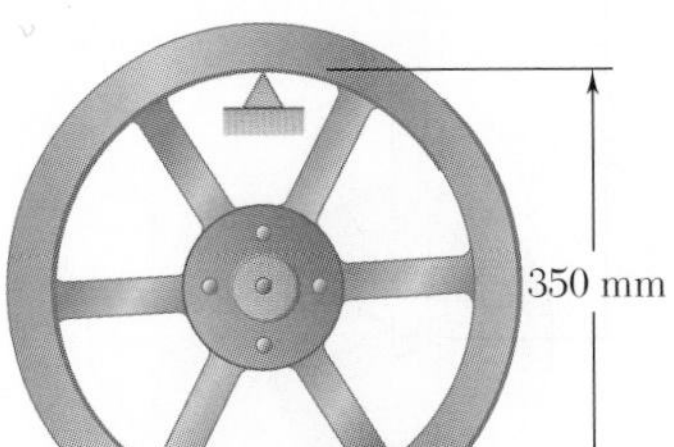

Fig. *P19.73*

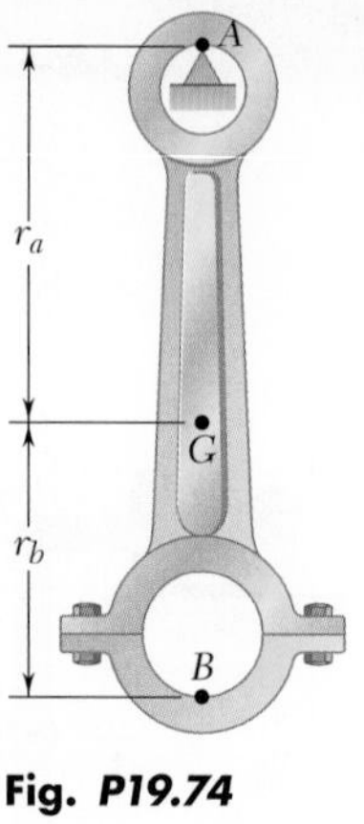

Fig. *P19.74*

19.74 A connecting rod is supported by a knife edge at point A; the period of its small oscillations is observed to be 1.03 s. Knowing that the distance r_a is 150 mm, determine the centroidal radius of gyration of the connecting rod.

19.75 A uniform rod AB can rotate in a vertical plane about a horizontal axis at C located at a distance c above the mass center G of the rod. For small oscillations determine the value of c for which the frequency of the motion will be maximum.

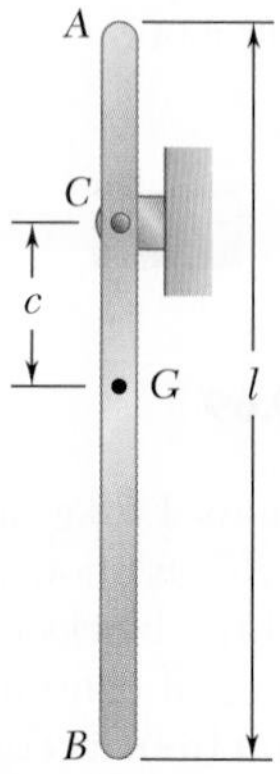

Fig. P19.75

19.76 A homogeneous wire of length $2l$ is bent as shown and allowed to oscillate about a frictionless pin at B. Denoting by τ_0 the period of small oscillations when $\beta = 0$, determine the angle β for which the period of small oscillations is $2\,\tau_0$.

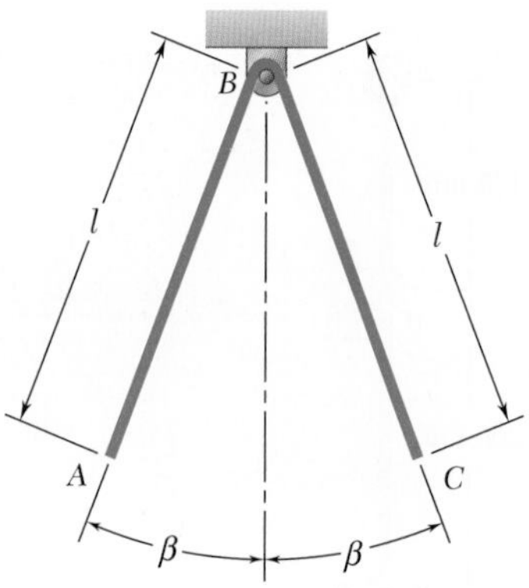

Fig. P19.76

19.77 A uniform disk of radius r and mass m can roll without slipping on a cylindrical surface and is attached to bar ABC of length L and negligible mass. The bar is attached to a spring of constant k and can rotate freely in the vertical plane about point B. Knowing that end A is given a small displacement and released, determine the frequency of the resulting oscillations in terms of m, L, k, and g.

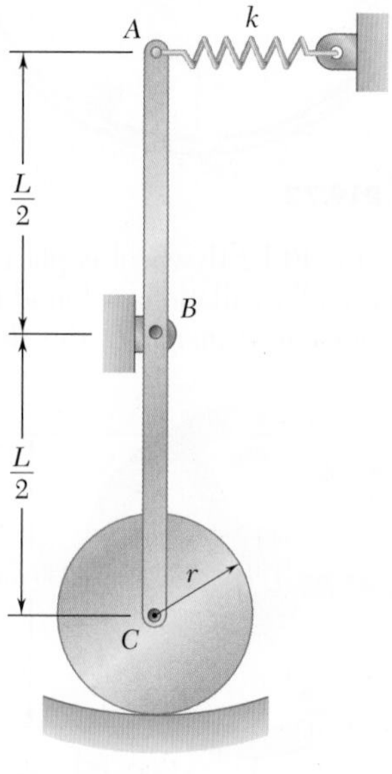

Fig. P19.77

19.78 Two uniform rods, each of mass $m = 600$ g and length $l = 200$ mm, are welded together to form the assembly shown. Knowing that the constant of each spring is $k = 120$ N/m and that end A is given a small displacement and released, determine the frequency of the resulting motion.

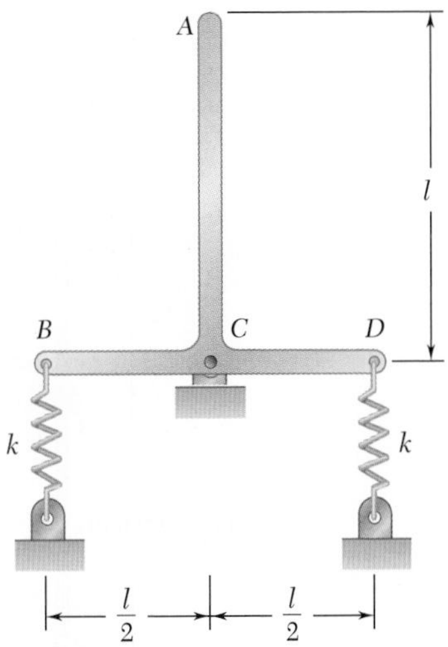

Fig. P19.78

19.79 A 7.5-kg uniform cylinder can roll without sliding on an incline and is attached to a spring AB as shown. If the center of the cylinder is moved 10 mm down the incline and released, determine (*a*) the period of vibration, (*b*) the maximum velocity of the center of the cylinder.

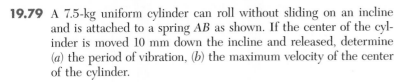

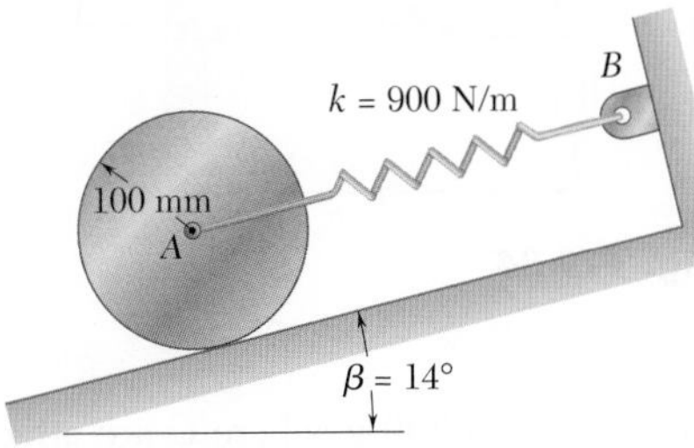

Fig. P19.79

19.80 A 3-kg slender rod AB is bolted to a 5-kg uniform disk. A spring of constant 280 N/m is attached to the disk and is unstretched in the position shown. If end B of the rod is given a small displacement and released, determine the period of vibration of the system.

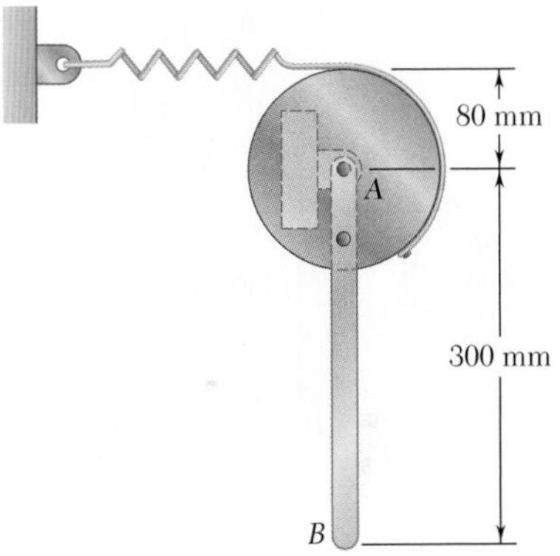

Fig. P19.80

19.81 A slender rod AB of mass m and length l is connected to two collars of negligible mass in a horizontal plane as shown. Collar A is attached to a spring of constant k. Knowing that the collars can slide freely on their respective rods and the system is in equilibrium in the position shown, determine the period of vibration if collar A is given a small displacement and released.

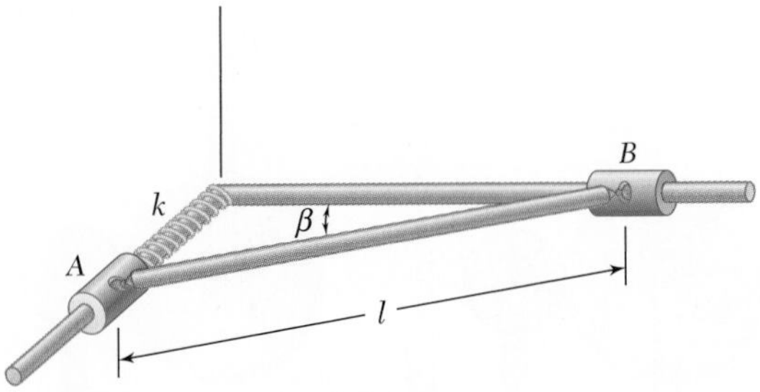

Fig. *P19.81* and *P19.82*

19.82 A slender rod AB of mass m and length l is connected to two collars of mass m_c in a horizontal plane as shown. Collar A is attached to a spring of constant k. Knowing that the collars can slide freely on their respective rods and the system is in equilibrium in the position shown, determine the period of vibration if collar A is given a small displacement and released.

19.83 An 800-g rod AB is bolted to a 1.2-kg disk. A spring of constant $k = 12$ N/m is attached to the center of the disk at A and to the wall at C. Knowing that the disk rolls without sliding, determine the period of small oscillations of the system.

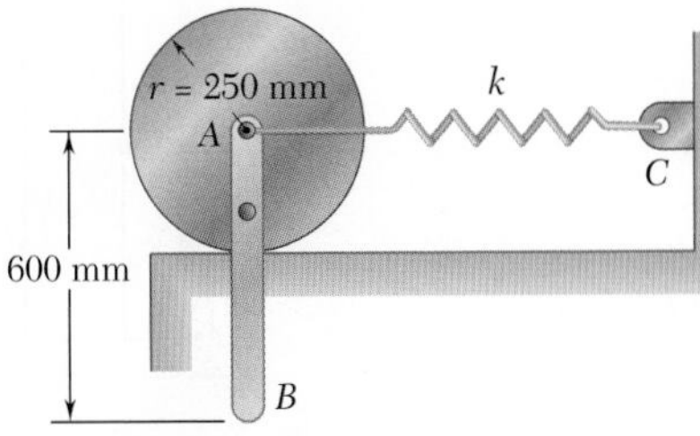

Fig. P19.83

19.84 Three identical rods are connected as shown. If $b = \frac{3}{4}l$, determine the frequency of small oscillations of the system.

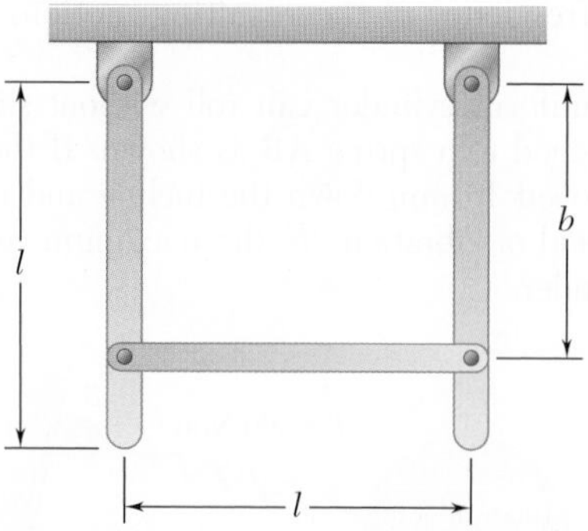

Fig. P19.84

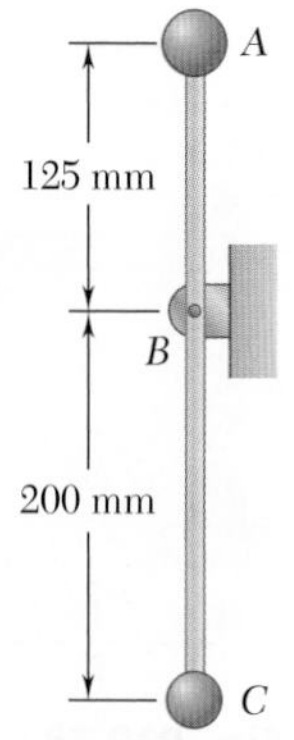

Fig. P19.85

19.85 A 400-g sphere A and a 300-g sphere C are attached to the ends of a 600-g rod AC which can rotate in a vertical plane about an axis at B. Determine the period of small oscillations of the rod.

19.86 A 5-kg uniform rod CD is welded at C to a shaft of negligible mass which is welded to the centers of two 10-kg uniform disks A and B. Knowing that the disks roll without sliding, determine the period of small oscillations of the system.

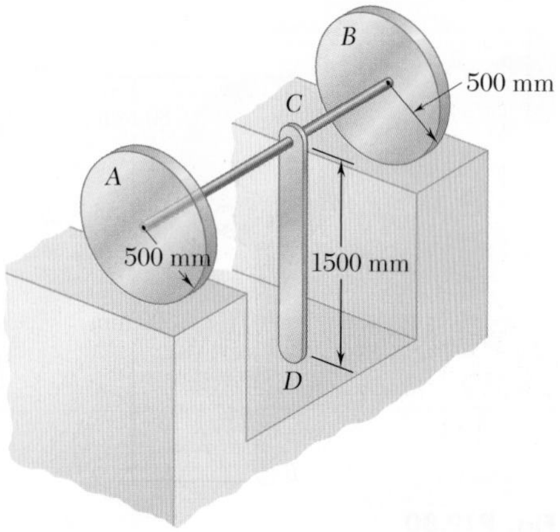

Fig. P19.86

19.87 and 19.88 Two uniform rods AB and CD, each of length l and mass m, are attached to gears as shown. Knowing that the mass of gear C is m and that the mass of gear A is $4m$, determine the period of small oscillations of the system.

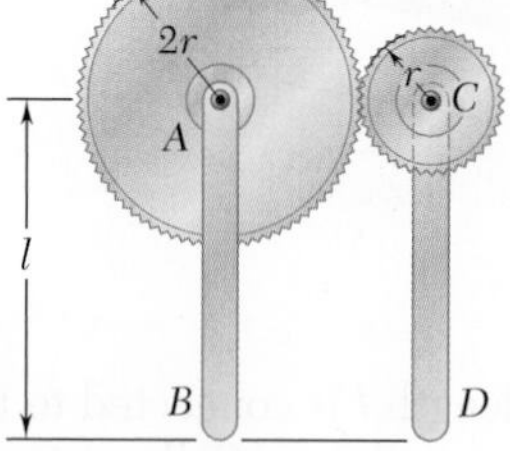

Fig. P19.87

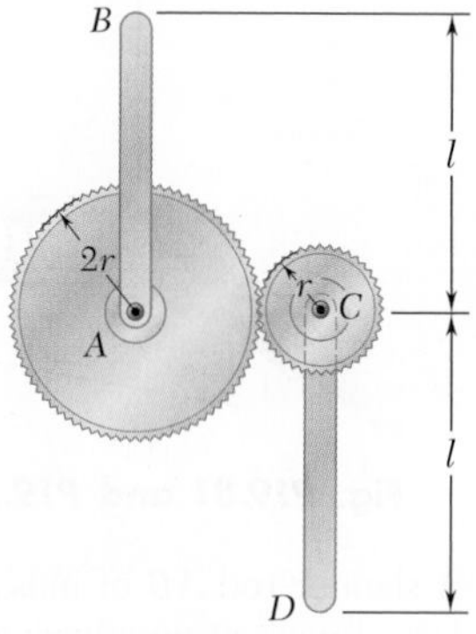

Fig. P19.88

19.89 An inverted pendulum consisting of a rigid bar ABC of length l and mass m is supported by a pin and bracket at C. A spring of constant k is attached to the bar at B and is undeformed when the bar is in the vertical position shown. Determine (*a*) the frequency of small oscillations, (*b*) the smallest value of a for which these oscillations will occur.

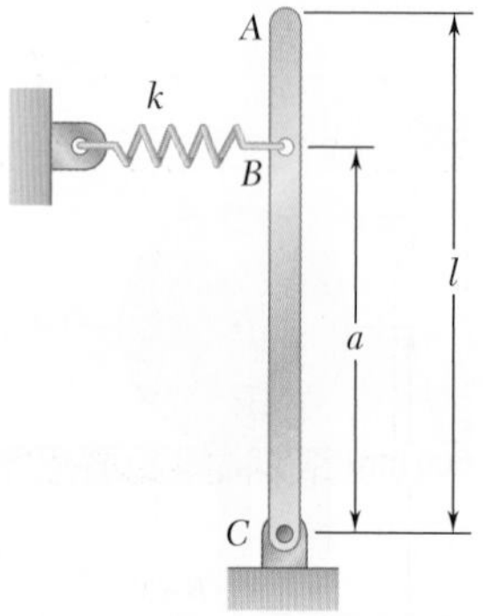

Fig. P19.89

19.90 Two 6-kg uniform disks are attached to the 10-kg rod AB as shown. Knowing that the constant of the spring is 6 kN/m and that the disks roll without sliding, determine the frequency of vibration of the system.

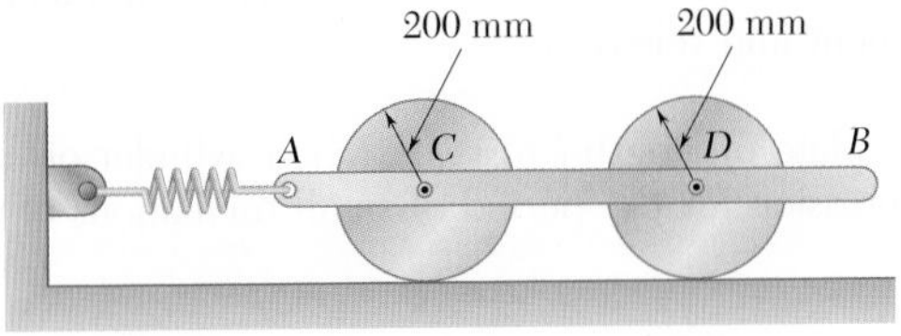

Fig. P19.90

19.91 The 10-kg rod AB is attached to two 4-kg disks as shown. Knowing that the disks roll without sliding, determine the frequency of small oscillations of the system.

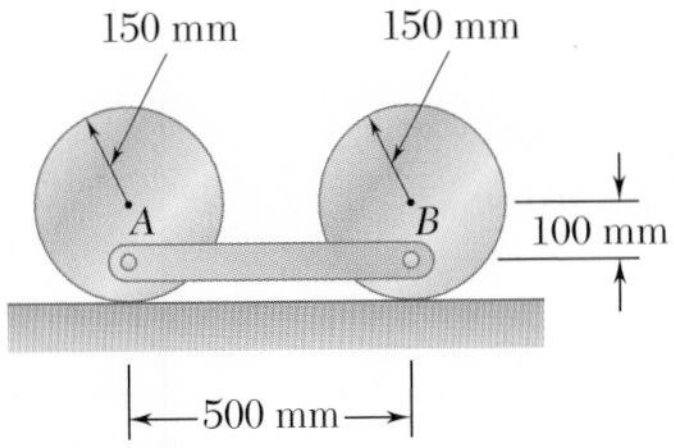

Fig. P19.91

19.92 A half section of a uniform cylinder of radius r and mass m rests on two casters A and B, each of which is a uniform cylinder of radius $r/4$ and mass $m/8$. Knowing that the half cylinder is rotated through a small angle and released and that no slipping occurs, determine the frequency of small oscillations.

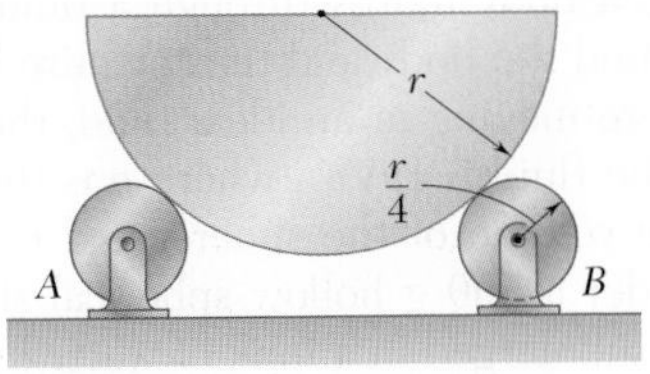

Fig. P19.92

19.93 The motion of the uniform rod AB is guided by the cord BC and by the small roller at A. Determine the frequency of oscillation when the end B of the rod is given a small horizontal displacement and released.

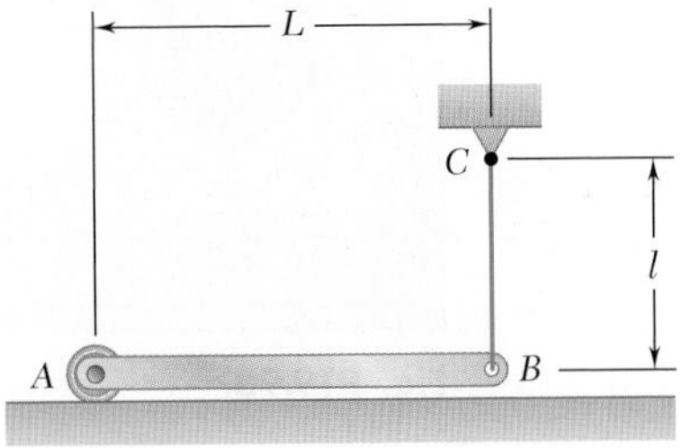

Fig. P19.93

19.94 A uniform rod of length L is supported by a ball-and-socket joint at A and by a vertical wire CD. Derive an expression for the period of oscillation of the rod if end B is given a small horizontal displacement and then released.

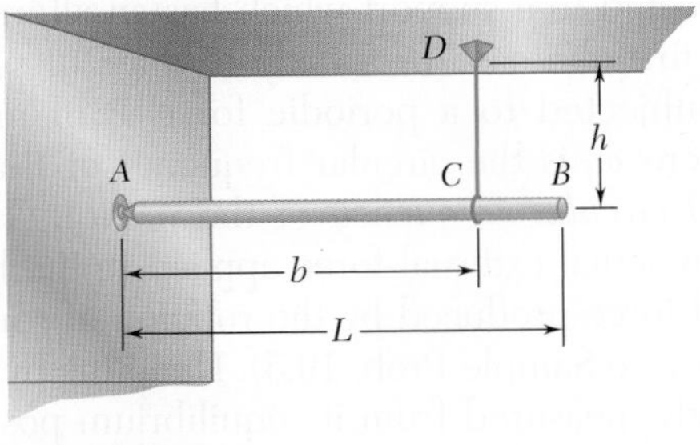

Fig. *P19.94*

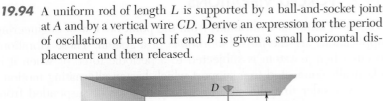

19.95 A section of uniform pipe is suspended from two vertical cables attached at A and B. Determine the frequency of oscillation when the pipe is given a small rotation about the centroidal axis OO' and released.

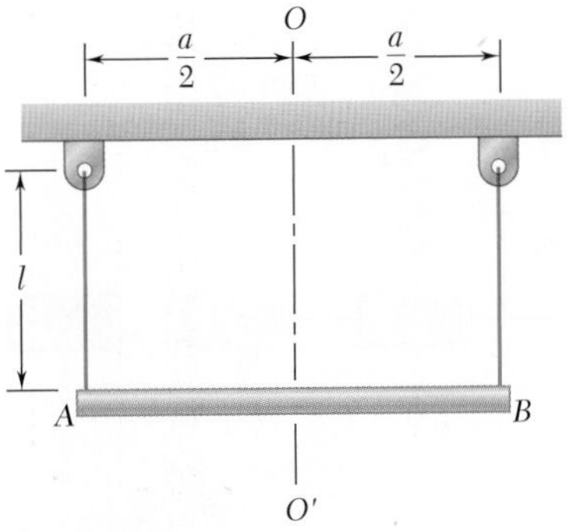

Fig. P19.95

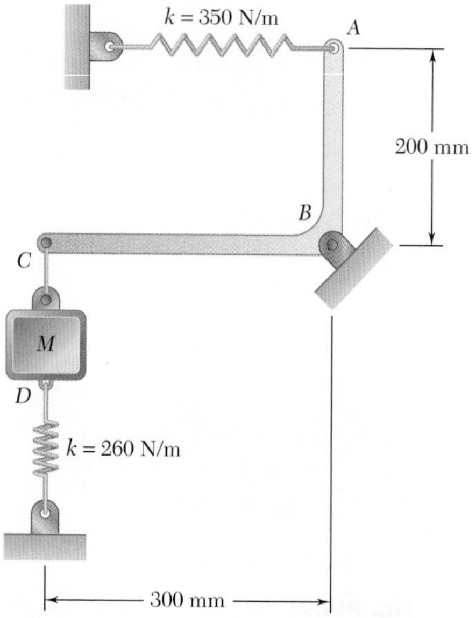

Fig. P19.96

19.96 A 0.6-kg uniform arm ABC is supported by a pin at B and is attached to a spring at A. It is connected at C to a 1.4-kg mass M which is attached to a spring. Knowing that each spring can act in tension or compression, determine the frequency of small oscillations of the system when the weight is given a small vertical displacement and released.

***19.97** A thin plate of length l rests on a half cylinder of radius r. Derive an expression for the period of small oscillations of the plate.

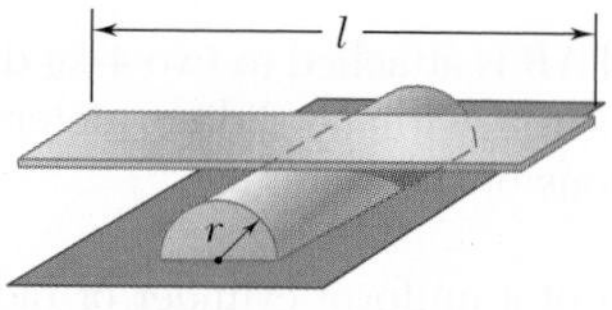

Fig. P19.97

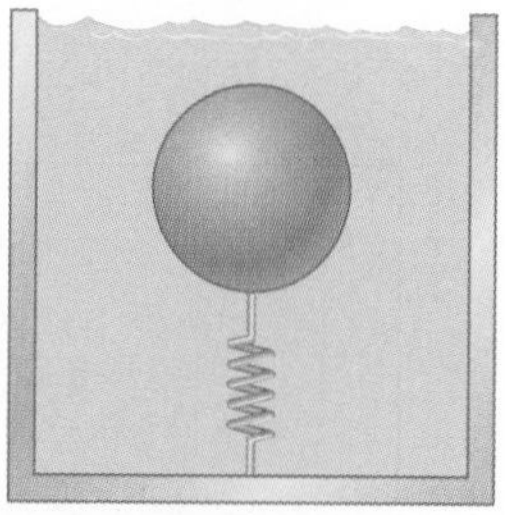

Fig. P19.98

***19.98** As a submerged body moves through a fluid, the particles of the fluid flow around the body and thus acquire kinetic energy. In the case of a sphere moving in an ideal fluid, the total kinetic energy acquired by the fluid is $\frac{1}{4}\rho V v^2$, where ρ is the mass density of the fluid, V is the volume of the sphere, and v is the velocity of the sphere. Consider a 500-g hollow spherical shell of radius 80 mm which is held submerged in a tank of water by a spring of constant 500 N/m. (*a*) Neglecting fluid friction, determine the period of vibration of the shell when it is displaced vertically and then released. (*b*) Solve part *a*, assuming that the tank is accelerated upward at the constant rate of 8 m/s^2.

19.7 FORCED VIBRATIONS

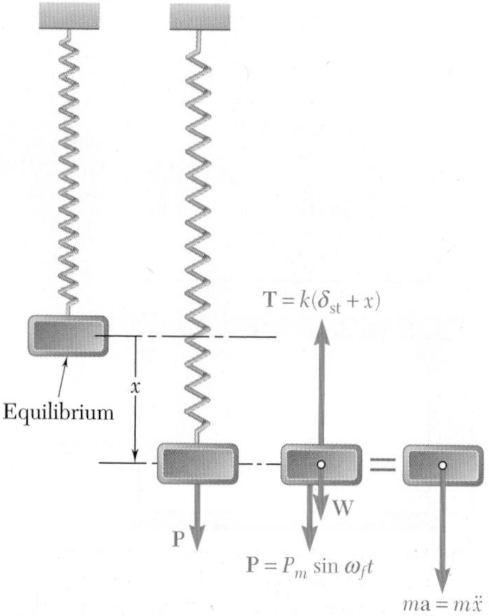

Fig. 19.7

The most important vibrations from the point of view of engineering applications are the *forced vibrations* of a system. These vibrations occur when a system is subjected to a periodic force or when it is elastically connected to a support which has an alternating motion.

Consider first the case of a body of mass m suspended from a spring and subjected to a periodic force $\mathbf{P}$ of magnitude $P = P_m \sin \omega_f t$, where ω_f is the circular frequency of $\mathbf{P}$ and is referred to as the *forced circular frequency* of the motion (Fig. 19.7). This force may be an actual external force applied to the body, or it may be a centrifugal force produced by the rotation of some unbalanced part of the body (see Sample Prob. 19.5). Denoting by x the displacement of the body measured from its equilibrium position, we write the equation of motion,

$$+\downarrow \Sigma F = ma: \qquad P_m \sin \omega_f t + W - k(\delta_{st} + x) = m\ddot{x}$$

Recalling that $W = k\delta_{st}$, we have

$$m\ddot{x} + kx = P_m \sin \omega_f t \tag{19.30}$$

Next we consider the case of a body of mass m suspended from a spring attached to a moving support whose displacement δ is equal to $\delta_m \sin \omega_f t$ (Fig. 19.8). Measuring the displacement x of the body from the position of static equilibrium corresponding to $\omega_f t = 0$, we find that the total elongation of the spring at time t is $\delta_{st} + x - \delta_m \sin \omega_f t$. The equation of motion is thus

$$+\downarrow \Sigma F = ma: \qquad W - k(\delta_{st} + x - \delta_m \sin \omega_f t) = m\ddot{x}$$

Recalling that $W = k\delta_{st}$, we have

$$m\ddot{x} + kx = k\delta_m \sin \omega_f t \qquad (19.31)$$

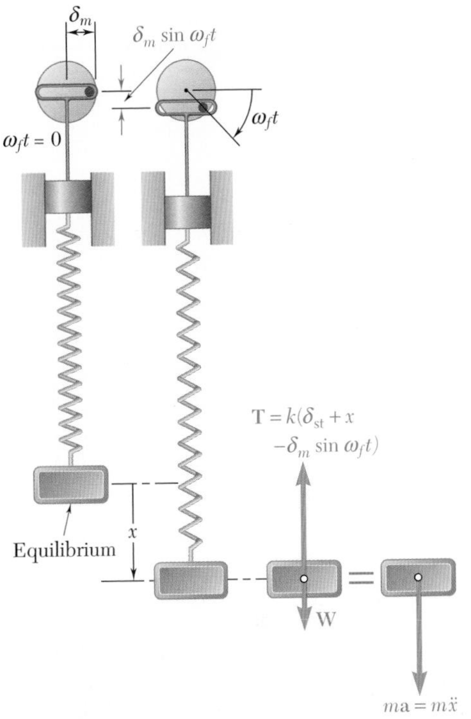

Fig. 19.8

We note that Eqs. (19.30) and (19.31) are of the same form and that a solution of the first equation will satisfy the second if we set $P_m = k\delta_m$.

A differential equation such as (19.30) or (19.31), possessing a right-hand member different from zero, is said to be *nonhomogeneous*. Its general solution is obtained by adding a particular solution of the given equation to the general solution of the corresponding *homogeneous* equation (with right-hand member equal to zero). A *particular solution* of (19.30) or (19.31) can be obtained by trying a solution of the form

$$x_{part} = x_m \sin \omega_f t \qquad (19.32)$$

Substituting x_{part} for x into (19.30), we find

$$-m\omega_f^2 x_m \sin \omega_f t + kx_m \sin \omega_f t = P_m \sin \omega_f t$$

which can be solved for the amplitude,

$$x_m = \frac{P_m}{k - m\omega_f^2}$$

Recalling from (19.4) that $k/m = \omega_n^2$, where ω_n is the natural circular frequency of the system, we write

$$x_m = \frac{P_m/k}{1 - (\omega_f/\omega_n)^2} \qquad (19.33)$$

Substituting from (19.32) into (19.31), we obtain in a similar way

$$x_m = \frac{\delta_m}{1 - (\omega_f/\omega_n)^2} \qquad (19.33')$$

The homogeneous equation corresponding to (19.30) or (19.31) is Eq. (19.2), which defines the free vibration of the body. Its general solution, called the *complementary function*, was found in Sec. 19.2:

$$x_{comp} = C_1 \sin \omega_n t + C_2 \cos \omega_n t \qquad (19.34)$$

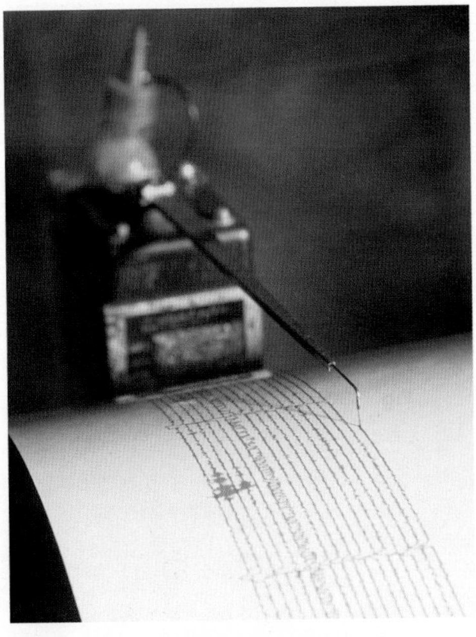

Photo 19.1 A seismometer operates by measuring the amount of electrical energy needed to keep a mass centered in the housing in the presence of strong ground shaking.

Adding the particular solution (19.32) to the complementary function (19.34), we obtain the *general solution* of Eqs. (19.30) and (19.31):

$$x = C_1 \sin \omega_n t + C_2 \cos \omega_n t + x_m \sin \omega_f t \qquad (19.35)$$

We note that the vibration obtained consists of two superposed vibrations. The first two terms in Eq. (19.35) represent a free vibration of the system. The frequency of this vibration is the *natural frequency* of the system, which depends only upon the constant k of the spring and the mass m of the body, and the constants C_1 and C_2 can be determined from the initial conditions. This free vibration is also called a *transient* vibration, since in actual practice it will soon be damped out by friction forces (Sec. 19.9).

The last term in (19.35) represents the *steady-state* vibration produced and maintained by the impressed force or impressed support movement. Its frequency is the *forced frequency* imposed by this force or movement, and its amplitude x_m, defined by (19.33) or (19.33′), depends upon the *frequency ratio* ω_f/ω_n. The ratio of the amplitude x_m of the steady-state vibration to the static deflection P_m/k caused by a force P_m, or to the amplitude δ_m of the support movement, is called the *magnification factor.* From (19.33) and (19.33′), we obtain

$$\text{Magnification factor} = \frac{x_m}{P_m/k} = \frac{x_m}{\delta_m} = \frac{1}{1 - (\omega_f/\omega_n)^2} \qquad (19.36)$$

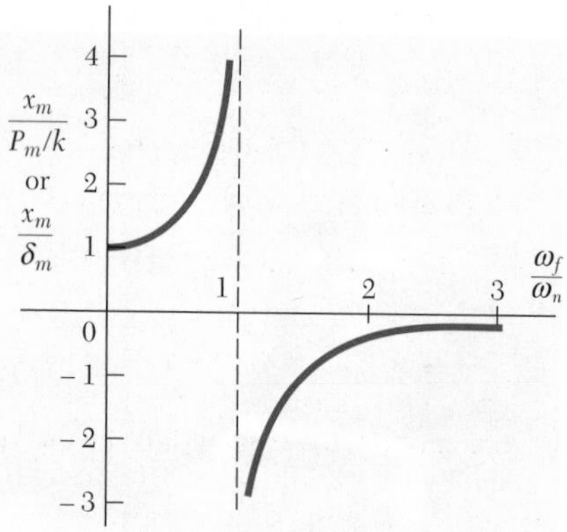

Fig. 19.9

The magnification factor has been plotted in Fig. 19.9 against the frequency ratio ω_f/ω_n. We note that when $\omega_f = \omega_n$, the amplitude of the forced vibration becomes infinite. The impressed force or impressed support movement is said to be in *resonance* with the given system. Actually, the amplitude of the vibration remains finite because of damping forces (Sec. 19.9); nevertheless, such a situation should be avoided, and the forced frequency should not be chosen too close to the natural frequency of the system. We also note that for $\omega_f < \omega_n$ the coefficient of $\sin \omega_f t$ in (19.35) is positive, while for $\omega_f > \omega_n$ this coefficient is negative. In the first case the forced vibration is *in phase* with the impressed force or impressed support movement, while in the second case it is 180° *out of phase.*

Finally, let us observe that the velocity and the acceleration in the steady-state vibration can be obtained by differentiating twice with respect to t the last term of Eq. (19.35). Their maximum values are given by expressions similar to those of Eqs. (19.15) of Sec. 19.2, except that these expressions now involve the amplitude and the circular frequency of the forced vibration:

$$v_m = x_m \omega_f \qquad a_m = x_m \omega_f^2 \qquad (19.37)$$

SAMPLE PROBLEM 19.5

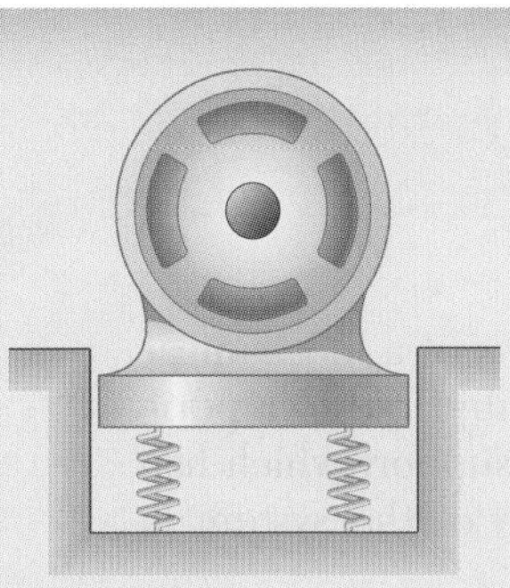

A motor of mass 200 kg is supported by four springs, each having a constant of 150 kN/m. The unbalance of the rotor is equivalent to a mass of 30 g located 150 mm from the axis of rotation. Knowing that the motor is constrained to move vertically, determine (*a*) the speed in rpm at which resonance will occur, (*b*) the amplitude of the vibration of the motor at a speed of 1200 rpm.

SOLUTION

a. Resonance Speed. The resonance speed is equal to the natural circular frequency ω_n (in rpm) of the free vibration of the motor. The mass of the motor and the equivalent constant of the supporting springs are

$$m = 200 \text{ kg}$$

$$k = 4(150 \text{ kN/m}) = 600{,}000 \text{ N/m}$$

$$\omega_n = \sqrt{\frac{k}{m}} = \sqrt{\frac{600{,}000}{200}} = 54.8 \text{ rad/s} = 523 \text{ rpm}$$

Resonance speed = 523 rpm ◀

b. Amplitude of Vibration at 1200 rpm. The angular velocity of the motor is

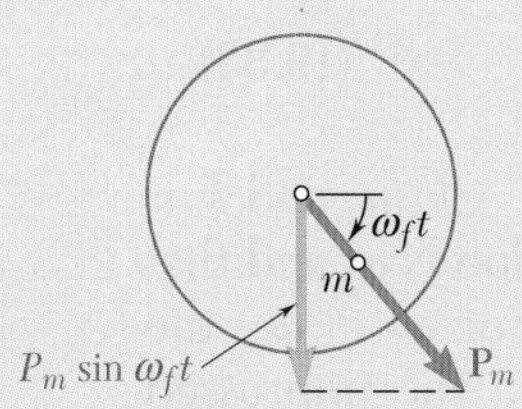

$$\omega = 1200 \text{ rpm} = 125.7 \text{ rad/s}$$

$$m = 0.03 \text{ kg}$$

The magnitude of the centrifugal force due to the unbalance of the rotor is

$$P_m = ma_n = mr\omega^2 = (0.03 \text{ kg})(0.15 \text{ m})(125.7 \text{ rad/s})^2 = 71.1 \text{ N}$$

The static deflection that would be caused by a constant load P_m is

$$\frac{P_m}{k} = \frac{71.1 \text{ N}}{600{,}000 \text{ N/m}} \times 1000 \text{ mm} = 0.1185 \text{ mm}$$

The forced circular frequency ω_f of the motion is the angular velocity of the motor,

$$\omega_f = \omega = 125.7 \text{ rad/s}$$

Substituting the values of P_m/k, ω_f, and ω_n into Eq. (19.33), we obtain

$$x_m = \frac{P_m/k}{1 - (\omega_f/\omega_n)^2} = \frac{0.1185 \text{ mm}}{1 - (125.7/54.8)^2} = -0.0278 \text{ mm}$$

$x_m = 0.0278$ mm (out of phase) ◀

Note. Since $\omega_f > \omega_n$, the vibration is 180° out of phase with the centrifugal force due to the unbalance of the rotor. For example, when the unbalanced mass is directly below the axis of rotation, the position of the motor is $x_m = 0.0278$ mm above the position of equilibrium.

SOLVING PROBLEMS ON YOUR OWN

This lesson was devoted to the analysis of the *forced vibrations* of a mechanical system. These vibrations occur either when the system is subjected to a periodic force **P** (Fig. 19.7), or when it is elastically connected to a support which has an alternating motion (Fig. 19.8). In the first case, the motion of the system is defined by the differential equation

$$m\ddot{x} + kx = P_m \sin \omega_f t \tag{19.30}$$

where the right-hand member represents the magnitude of the force **P** at a given instant. In the second case, the motion is defined by the differential equation

$$m\ddot{x} + kx = k\delta_m \sin \omega_f t \tag{19.31}$$

where the right-hand member is the product of the spring constant k and the displacement of the support at a given instant. You will be concerned only with the *steady-state* motion of the system, which is defined by a *particular solution* of these equations, of the form

$$x_{\text{part}} = x_m \sin \omega_f t \tag{19.32}$$

1. If the forced vibration is caused by a periodic force P, of amplitude P_m and circular frequency ω_f, the amplitude of the vibration is

$$x_m = \frac{P_m/k}{1 - (\omega_f/\omega_n)^2} \tag{19.33}$$

where ω_n is the *natural circular frequency* of the system, $\omega_n = \sqrt{k/m}$, and k is the spring constant. Note that the circular frequency of the vibration is ω_f and that the amplitude x_m does not depend upon the initial conditions. For $\omega_f = \omega_n$, the denominator in Eq. (19.33) is zero and x_m is infinite (Fig. 19.9); the impressed force **P** is said to be in *resonance* with the system. Also, for $\omega_f < \omega_n$, x_m is positive and the vibration is *in phase* with **P**, while, for $\omega_f > \omega_n$, x_m is negative and the vibration is *out of phase.*

a. In the problems which follow, you may be asked to determine one of the parameters in Eq. (19.33) when the others are known. We suggest that you keep Fig. 19.9 in front of you when solving these problems. For example, if you are asked to find the frequency at which the amplitude of a forced vibration has a given value, but you do not know whether the vibration is in or out of phase with respect to the impressed force, you should note from Fig. 19.9 that there can be two frequencies satisfying this requirement, one corresponding to a positive value of x_m and to a vibration in phase with the impressed force, and the other corresponding to a negative value of x_m and to a vibration out of phase with the impressed force.

b. Once you have obtained the amplitude x_m of the motion of a component of the system from Eq. (19.33), you can use Eqs. (19.37) to determine the maximum values of the velocity and acceleration of that component:

$$v_m = x_m\omega_f \qquad a_m = x_m\omega_f^2 \tag{19.37}$$

c. When the impressed force P is due to the unbalance of the rotor of a motor, its maximum value is $P_m = mr\omega_f^2$, where m is the mass of the rotor, r is the distance between its mass center and the axis of rotation, and ω_f is equal to the angular velocity ω of the rotor expressed in rad/s [Sample Prob. 19.5].

2. If the forced vibration is caused by the simple harmonic motion of a support, of amplitude δ_m and circular frequency ω_f, the amplitude of the vibration is

$$x_m = \frac{\delta_m}{1 - (\omega_f/\omega_n)^2} \tag{19.33'}$$

where ω_n is the *natural circular frequency* of the system, $\omega_n = \sqrt{k/m}$. Again, note that the circular frequency of the vibration is ω_f and that the amplitude x_m does not depend upon the initial conditions.

a. Be sure to read our comments in paragraphs 1, 1a, and 1b, since they apply equally well to a vibration caused by the motion of a support.

b. If the maximum acceleration a_m of the support is specified, rather than its maximum displacement δ_m, remember that, since the motion of the support is a simple harmonic motion, you can use the relation $a_m = \delta_m\omega_f^2$ to determine δ_m; the value obtained is then substituted into Eq. (19.33′).

PROBLEMS

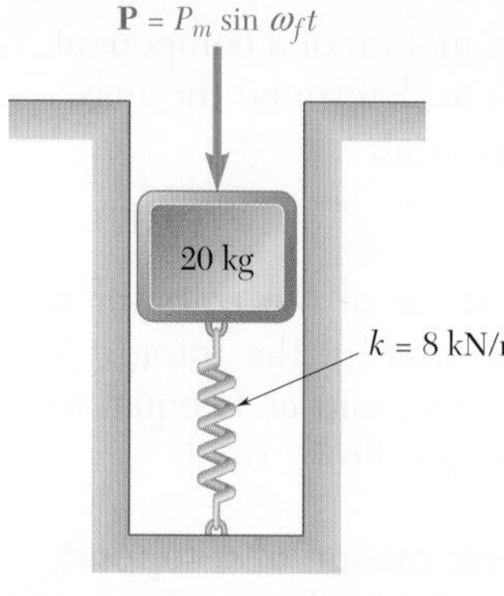

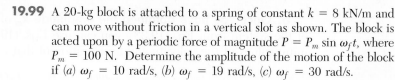

Fig. P19.99 and P19.100

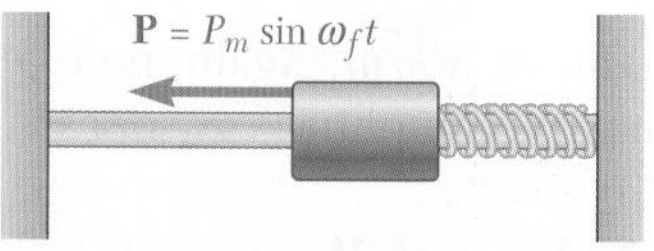

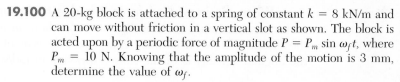

Fig. P19.101 and P19.102

19.99 A 20-kg block is attached to a spring of constant $k = 8$ kN/m and can move without friction in a vertical slot as shown. The block is acted upon by a periodic force of magnitude $P = P_m \sin \omega_f t$, where $P_m = 100$ N. Determine the amplitude of the motion of the block if (*a*) $\omega_f = 10$ rad/s, (*b*) $\omega_f = 19$ rad/s, (*c*) $\omega_f = 30$ rad/s.

19.100 A 20-kg block is attached to a spring of constant $k = 8$ kN/m and can move without friction in a vertical slot as shown. The block is acted upon by a periodic force of magnitude $P = P_m \sin \omega_f t$, where $P_m = 10$ N. Knowing that the amplitude of the motion is 3 mm, determine the value of ω_f.

19.101 A 5-kg can slide on a frictionless horizontal rod and is attached to a spring of constant k. It is acted upon by a periodic force of magnitude $P = P_m \sin \omega_f t$, where $P_m = 10$ N and $\omega_f = 5$ rad/s. Determine the value of the spring constant k knowing that the motion of the collar has an amplitude of 150 mm and is (*a*) in phase with the applied force, (*b*) out of phase with the applied force.

19.102 A collar of mass m which slides on a frictionless horizontal rod is attached to a spring of constant k and is acted upon by a periodic force of magnitude $P = P_m \sin \omega_f t$. Determine the range of values of ω_f for which the amplitude of the vibration exceeds two times the static deflection caused by a constant force of magnitude P_m.

19.103 A small 20-kg block A is attached to the rod BC of negligible mass which is supported at B by a pin and bracket and at C by a spring of constant $k = 2$ kN/m. The system can move in a vertical plane and is in equilibrium when the rod is horizontal. The rod is acted upon at C by a periodic force **P** of magnitude $P = P_m \sin \omega_f t$, where $P_m = 6$ N. Knowing that $b = 200$ mm, determine the range of values of ω_f for which the amplitude of vibration of block A exceeds 3.5 mm.

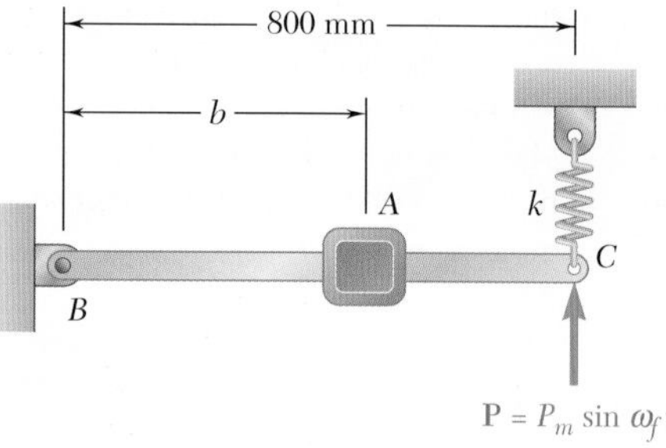

Fig. *P19.103*

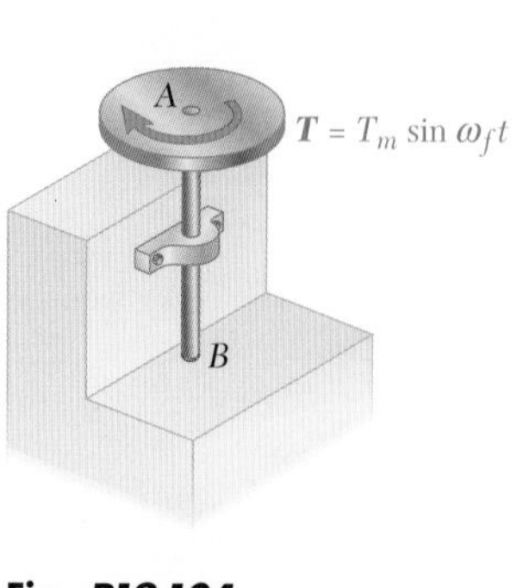

Fig. *P19.104*

19.104 An 8-kg uniform disk of radius 200 mm is welded to a vertical shaft with a fixed end at B. The disk rotates through an angle of 3° when a static couple of magnitude 50 N · m is applied to it. If the disk is acted upon by a periodic torsional couple of magnitude $T = T_m \sin \omega_f t$, where $T_m = 60$ N · m, determine the range of values of ω_f for which the amplitude of the vibration is less than the angle of rotation caused by a static couple of magnitude T_m.

19.105 An 8-kg block A slides in a vertical frictionless slot and is connected to a moving support B by means of a spring AB of constant $k = 1.6$ kN/m. Knowing that the displacement of the support is $\delta = \delta_m \sin \omega_f t$, where $\delta_m = 150$ mm, determine the range of values of ω_f for which the amplitude of the fluctuating force exerted by the spring on the block is less than 120 N.

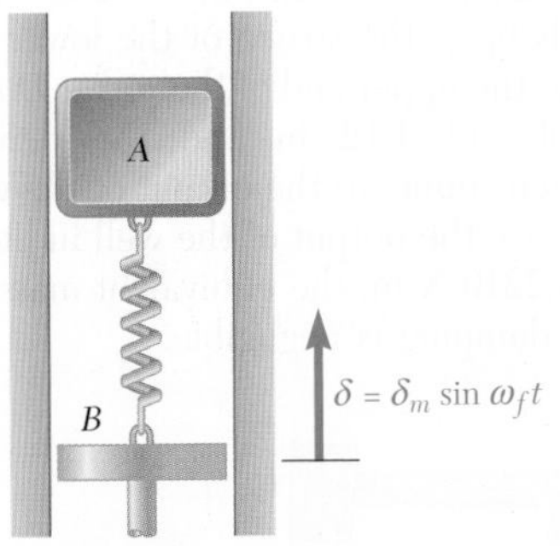

Fig. P19.105

19.106 A cantilever beam AB supports a block which causes a static deflection of 8 mm at B. Assuming that the support at A undergoes a vertical periodic displacement $\delta = \delta_m \sin \omega_f t$, where $\delta_m = 2$ mm, determine the range of values of ω_f for which the amplitude of the motion of the block will be less than 4 mm. Neglect the weight of the beam and assume that the block does not leave the beam.

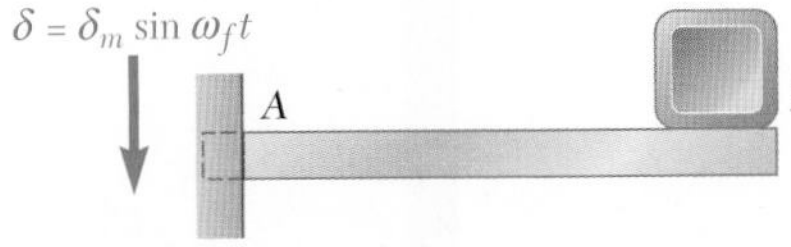

Fig. P19.106

19.107 Rod AB is rigidly attached to the frame of a motor running at a constant speed. When a collar of mass m is placed on the spring, it is observed to vibrate with an amplitude of 15 mm. When two collars, each of mass m, are placed on the spring, the amplitude is observed to be 18 mm. What amplitude of vibration should be expected when three collars, each of mass m, are placed on the spring? (Obtain two answers.)

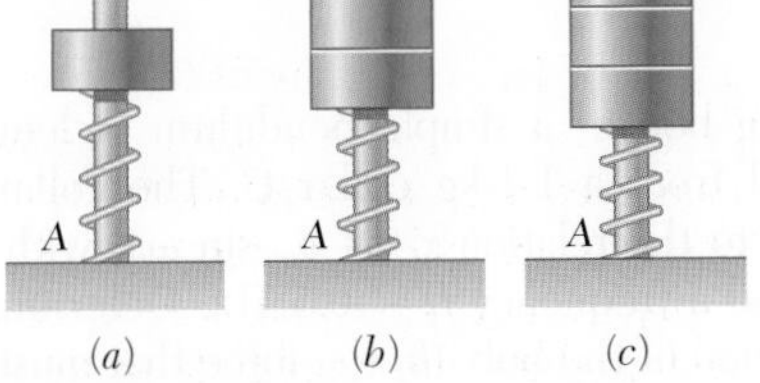

Fig. P19.107

19.108 The crude-oil-pumping rig shown is driven at 20 rpm. The inside diameter of the well pipe is 50 mm, and the diameter of the pump rod is 20 mm. The length of the pump rod and the length of the column of oil lifted during the stroke are essentially the same, and equal to 1.8 km. During the downward stroke, a valve at the lower end of the pump rod opens to let a quantity of oil into the well pipe, and the column of oil is then lifted to obtain a discharge into the connecting pipeline. Thus, the amount of oil pumped in a given time depends upon the stroke of the lower end of the pump rod. Knowing that the upper end of the rod at D is essentially sinusoidal with a stroke of 1.12 m and the density of crude oil is 800 kg/m^3, determine (*a*) the output of the well in liters/min if the shaft is rigid, (*b*) the output of the well in liters/min if the stiffness of the rod is 2210 N/m, the equivalent mass of the oil and shaft is 290 kg, and damping is negligible.

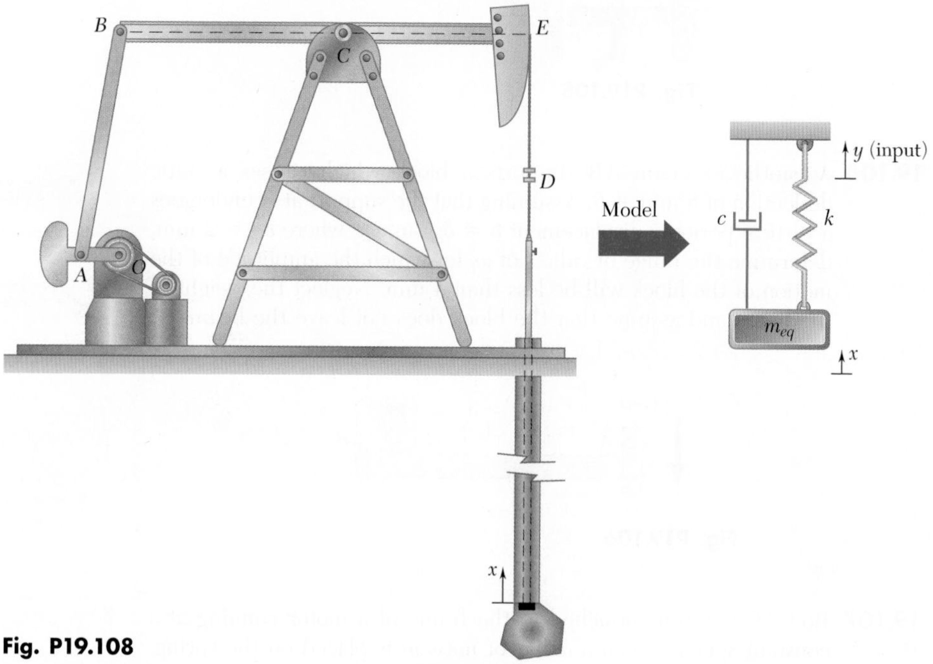

Fig. P19.108

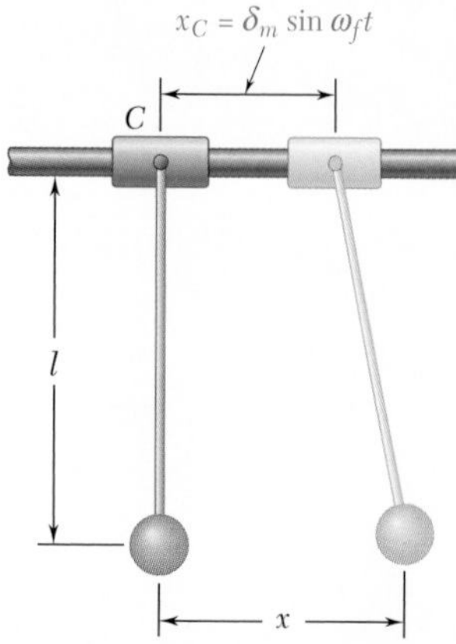

Fig. P19.109 and P19.110

19.109 A simple pendulum of length l is suspended from collar C which is forced to move horizontally according to the relation $x_C = \delta_m \sin \omega_f t$. Determine the range of values of ω_f for which the amplitude of the motion of the bob is less than δ_m. (Assume that δ_m is small compared with the length l of the pendulum.)

19.110 The 1.2-kg bob of a simple pendulum of length l = 600 mm is suspended from a 1.4-kg collar C. The collar is forced to move according to the relation $x_C = \delta_m \sin \omega_f t$, with an amplitude δ_m = 10 mm and a frequency f_f = 0.5 Hz. Determine (*a*) the amplitude of the motion of the bob, (*b*) the force that must be applied to collar C to maintain the motion.

19.111 An 8-kg block A slides in a vertical frictionless slot and is connected to a moving support B by means of a spring AB of constant $k = 120$ N/m. Knowing that the acceleration of the support is $a = a_m \sin \omega_f t$, where $a_m = 1.5$ m/s^2 and $\omega_f = 5$ rad/s, determine (*a*) the maximum displacement of block A, (*b*) the amplitude of the fluctuating force exerted by the spring on the block.

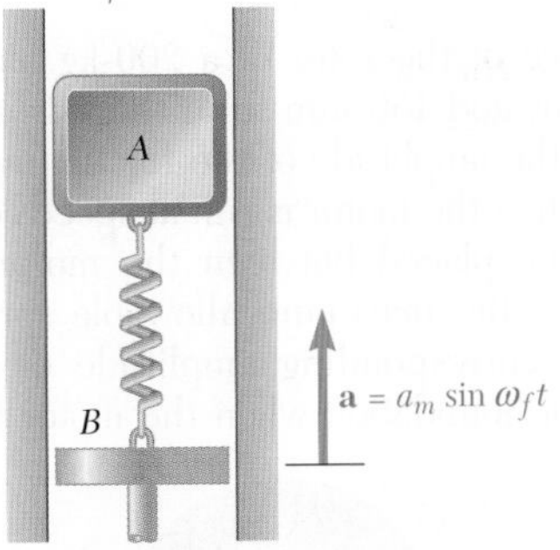

Fig. *P19.111*

19.112 A variable-speed motor is rigidly attached to a beam BC. When the speed of the motor is less than 600 rpm or more than 1200 rpm, a small object placed at A is observed to remain in contact with the beam. For speeds between 600 and 1200 rpm the object is observed to "dance" and actually to lose contact with the beam. Determine the speed at which resonance will occur.

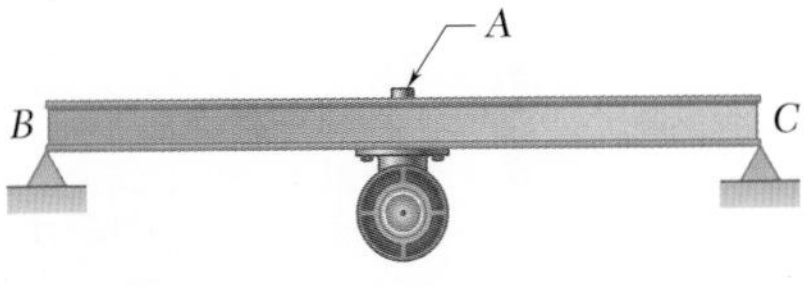

Fig. P19.112

19.113 A motor of mass M is supported by springs with an equivalent spring constant k. The unbalance of its rotor is equivalent to a mass m located at a distance r from the axis of rotation. Show that when the angular velocity of the motor is ω_f, the amplitude x_m of the motion of the motor is

$$x_m = \frac{r(m/M)(\omega_f/\omega_n)^2}{1 - (\omega_f/\omega_n)^2}$$

where $\omega_n = \sqrt{k/M}$.

19.114 As the rotational speed of a spring-supported 100-kg motor is increased, the amplitude of the vibration due to the unbalance of its 15-kg rotor first increases and then decreases. It is observed that as very high speeds are reached, the amplitude of the vibration approaches 3.3 mm. Determine the distance between the mass center of the rotor and its axis of rotation. (*Hint:* Use the formula derived in Prob. 19.113.)

19.115 A motor of mass 200 kg is supported by springs having a total constant of 240 kN/m The unbalance of the rotor is equivalent to a 30-g mass located 200 mm from the axis of rotation. Determine the range of allowable values of the motor speed if the amplitude of the vibration is not to exceed 1.5 mm.

19.116 As the rotational speed of a spring-supported motor is slowly increased from 300 to 500 rpm, the amplitude of the vibration due to the unbalance of its rotor is observed to increase continuously from 1.5 to 6 mm. Determine the speed at which resonance will occur.

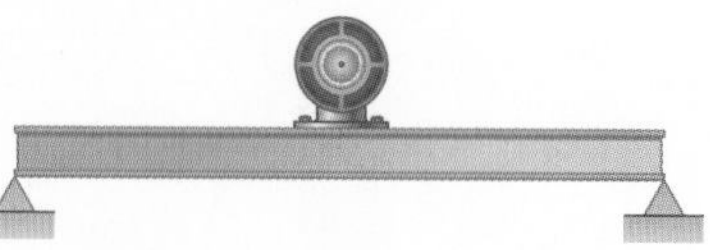

Fig. P19.117

19.117 A 180-kg motor is bolted to a light horizontal beam. The unbalance of its rotor is equivalent to a 28-g mass located 150 mm from the axis of rotation, and the static deflection of the beam due to the weight of the motor is 12 mm. The amplitude of the vibration due to the unbalance can be decreased by adding a plate to the base of the motor. If the amplitude of vibration is to be less than 60 μm for motor speeds above 300 rpm, determine the required mass of the plate.

19.118 The unbalance of the rotor of a 200-kg motor is equivalent to a 100-g mass located 150 mm from the axis of rotation. In order to limit to 1 N the amplitude of the fluctuating force exerted on the foundation when the motor is run at speeds of 100 rpm and above, a pad is to be placed between the motor and the foundation. Determine (*a*) the maximum allowable spring constant k of the pad, (*b*) the corresponding amplitude of the fluctuating force exerted on the foundation when the motor is run at 200 rpm.

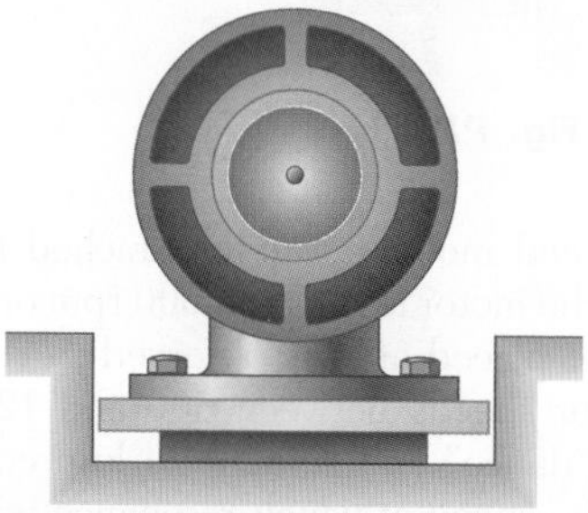

Fig. *P19.118*

19.119 A counter-rotating eccentric mass exciter consisting of two rotating 100-g masses describing circles of radius r at the same speed but in opposite senses is placed on a machine element to induce a steady-state vibration of the element. The total mass of the system is 300 kg, the constant of each spring is $k = 600$ kN/m, and the rotational speed of the exciter is 1200 rpm. Knowing that the amplitude of the total fluctuating force exerted on the foundation is 160 N, determine the radius r.

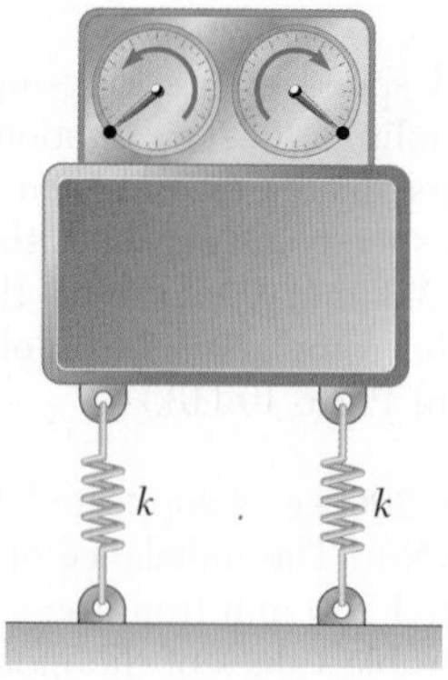

Fig. P19.119

19.120 A 180-kg motor is supported by springs of total constant 150 kN/m. The unbalance of the rotor is equivalent to a 28-g mass located 150 mm from the axis of rotation. Determine the range of speeds of the motor for which the amplitude of the fluctuating force exerted on the foundation is less than 20-N.

19.121 Figures (1) and (2) show how springs can be used to support a block in two different situations. In Fig. (1) they help decrease the amplitude of the fluctuating force transmitted by the block to the foundation. In Fig. (2) they help decrease the amplitude of the fluctuating displacement transmitted by the foundation to the block. The ratio of the transmitted force to the impressed force or the ratio of the transmitted displacement to the impressed displacement is called the *transmissibility*. Derive an equation for the transmissibility for each situation. Give your answer in terms of the ratio ω_f/ω_n of the frequency ω_f of the impressed force or impressed displacement to the natural frequency ω_n of the spring-mass system. Show that in order to cause any reduction in transmissibility, the ratio ω_f/ω_n must be greater than $\sqrt{2}$.

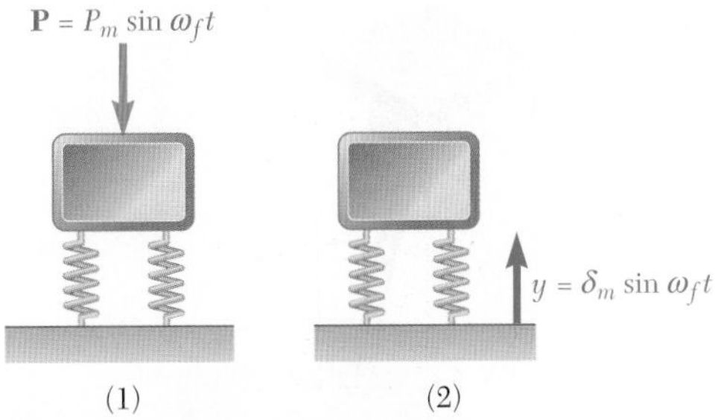

Fig. P19.121

19.122 A vibrometer used to measure the amplitude of vibrations consists essentially of a box containing a mass-spring system with a known natural frequency of 120 Hz. The box is rigidly attached to a surface which is moving according to the equation $y = \delta_m \sin \omega_f t$. If the amplitude z_m of the motion of the mass relative to the box is used as a measure of the amplitude δ_m of the vibration of the surface, determine (*a*) the percent error when the frequency of the vibration is 600 Hz, (*b*) the frequency at which the error is zero.

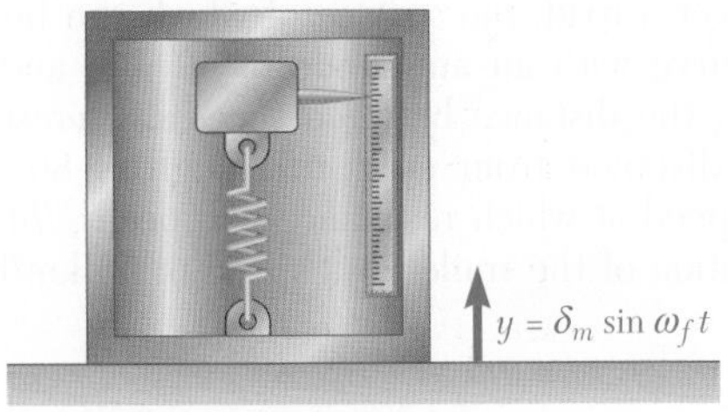

Fig. P19.122 and P19.123

19.123 A certain accelerometer consists essentially of a box containing a mass-spring system with a known natural frequency of 2200 Hz. The box is rigidly attached to a surface which is moving according to the equation $y = \delta_m \sin \omega_f t$. If the amplitude z_m of the motion of the mass relative to the box times a scale factor ω_n^2 is used as a measure of the maximum acceleration $\alpha_m = \delta_m \omega_f^2$ of the vibrating surface, determine the percent error when the frequency of the vibration is 600 Hz.

19.124 Block A can move without friction in the slot as shown and is acted upon by a vertical periodic force of magnitude $P = P_m \sin \omega_f t$, where $\omega_f = 2$ rad/s and $P_m = 20$ N. A spring of constant k is attached to the bottom of block A and to a 22-kg block B. Determine (*a*) the value of the constant k which will prevent a steady-state vibration of block A, (*b*) the corresponding amplitude of the vibration of block B.

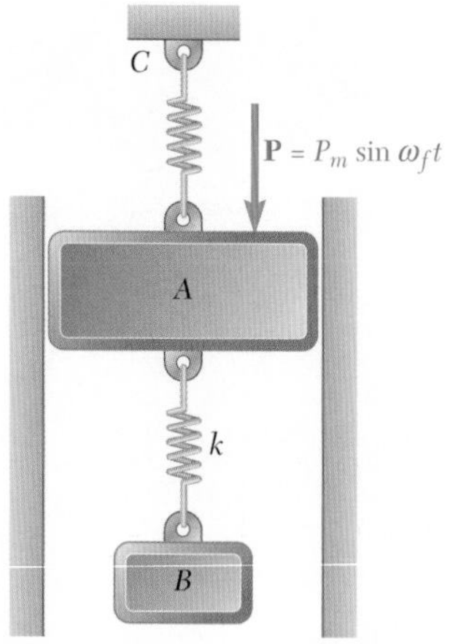

Fig. *P19.124*

19.125 A 30-kg disk is attached with an eccentricity $e = 0.15$ mm to the midpoint of a vertical shaft AB which revolves at a constant angular velocity ω_f. Knowing that the spring constant k for horizontal movement of the disk is 650 kN/m, determine (*a*) the angular velocity ω_f at which resonance will occur, (*b*) the deflection r of the shaft when $\omega_f = 1200$ rpm.

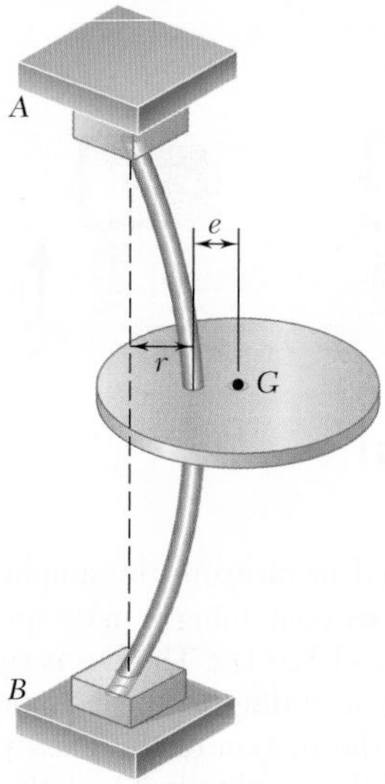

Fig. P19.125

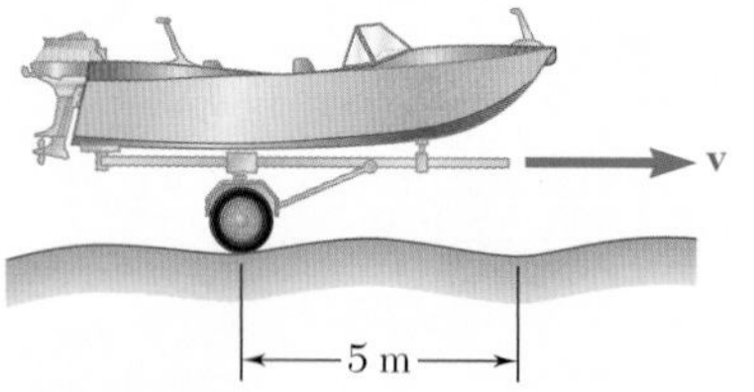

Fig. P19.126

19.126 A small trailer and its load have a total mass of 250 kg. The trailer is supported by two springs, each of constant 10 kN/m, and is pulled over a road, the surface of which can be approximated by a sine curve with an amplitude of 40 mm and a wavelength of 5 m (i.e., the distance between successive crests is 5 m and the vertical distance from crest to trough is 80 mm). Determine (*a*) the speed at which resonance will occur, (*b*) the amplitude of the vibration of the trailer at a speed of 50 km/h.

DAMPED VIBRATIONS

*19.8 DAMPED FREE VIBRATIONS

The vibrating systems considered in the first part of this chapter were assumed free of damping. Actually all vibrations are damped to some degree by friction forces. These forces can be caused by *dry friction,* or *Coulomb friction,* between rigid bodies, by *fluid friction* when a rigid body moves in a fluid, or by *internal friction* between the molecules of a seemingly elastic body.

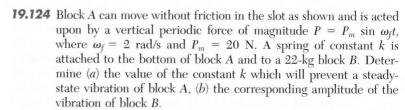

A type of damping of special interest is the *viscous damping* caused by fluid friction at low and moderate speeds. Viscous damping is characterized by the fact that the friction force is *directly proportional and opposite to the velocity* of the moving body. As an example, let us again consider a body of mass m suspended from a spring of constant k, assuming that the body is attached to the plunger of a dashpot (Fig. 19.10). The magnitude of the friction force exerted on the plunger by the surrounding fluid is equal to $c\dot{x}$, where the constant c, expressed in N · s/m or lb · s/ft and known as the *coefficient of viscous damping*, depends upon the physical properties of the fluid and the construction of the dashpot. The equation of motion is

$$+\downarrow \Sigma F = ma: \qquad W - k(\delta_{st} + x) - c\dot{x} = m\ddot{x}$$

Recalling that $W = k\delta_{st}$, we write

$$m\ddot{x} + c\dot{x} + kx = 0 \tag{19.38}$$

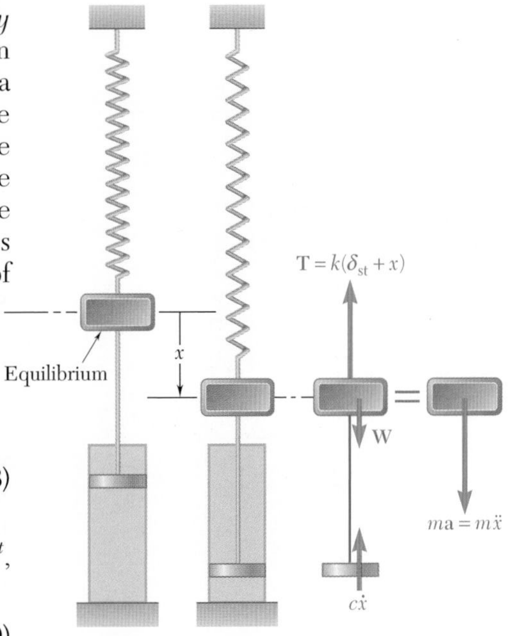

Fig. 19.10

Substituting $x = e^{\lambda t}$ into (19.38) and dividing through by $e^{\lambda t}$, we write the *characteristic equation*

$$m\lambda^2 + c\lambda + k = 0 \tag{19.39}$$

and obtain the roots

$$\lambda = -\frac{c}{2m} \pm \sqrt{\left(\frac{c}{2m}\right)^2 - \frac{k}{m}} \tag{19.40}$$

Defining the *critical damping coefficient* c_c as the value of c which makes the radical in Eq. (19.40) equal to zero, we write

$$\left(\frac{c_c}{2m}\right)^2 - \frac{k}{m} = 0 \qquad c_c = 2m\sqrt{\frac{k}{m}} = 2m\omega_n \tag{19.41}$$

where ω_n is the natural circular frequency of the system in the absence of damping. We can distinguish three different cases of damping, depending upon the value of the coefficient c.

1. *Heavy damping:* $c > c_c$. The roots λ_1 and λ_2 of the characteristic equation (19.39) are real and distinct, and the general solution of the differential equation (19.38) is

$$x = C_1 e^{\lambda_1 t} + C_2 e^{\lambda_2 t} \tag{19.42}$$

This solution corresponds to a nonvibratory motion. Since λ_1 and λ_2 are both negative, x approaches zero as t increases indefinitely. However, the system actually regains its equilibrium position after a finite time.

2. *Critical damping:* $c = c_c$. The characteristic equation has a double root $\lambda = -c_c/2m = -\omega_n$, and the general solution of (19.38) is

$$x = (C_1 + C_2 t)e^{-\omega_n t} \tag{19.43}$$

The motion obtained is again nonvibratory. Critically damped systems are of special interest in engineering applications since they regain their equilibrium position in the shortest possible time without oscillation.

3. *Light damping:* $c < c_c$. The roots of Eq. (19.39) are complex and conjugate, and the general solution of (19.38) is of the form

$$x = e^{-(c/2m)t}(C_1 \sin \omega_d t + C_2 \cos \omega_d t) \tag{19.44}$$

where ω_d is defined by the relation

$$\omega_d^2 = \frac{k}{m} - \left(\frac{c}{2m}\right)^2$$

Substituting $k/m = \omega_n^2$ and recalling (19.41), we write

$$\omega_d = \omega_n \sqrt{1 - \left(\frac{c}{c_c}\right)^2} \tag{19.45}$$

where the constant c/c_c is known as the *damping factor.* Even though the motion does not actually repeat itself, the constant ω_d is commonly referred to as the *circular frequency* of the damped vibration. A substitution similar to the one used in Sec. 19.2 enables us to write the general solution of Eq. (19.38) in the form

$$x = x_0 e^{-(c/2m)t} \sin (\omega_d t + \phi) \tag{19.46}$$

The motion defined by Eq. (19.46) is vibratory with diminishing amplitude (Fig. 19.11), and the time interval $\tau_d = 2\pi/\omega_d$ separating two successive points where the curve defined by Eq. (19.46) touches one of the limiting curves shown in Fig. 19.11 is commonly referred to as the *period of the damped vibration.* Recalling Eq. (19.45), we observe that $\omega_d < \omega_n$ and, thus, that τ_d is larger than the period of vibration τ_n of the corresponding undamped system.

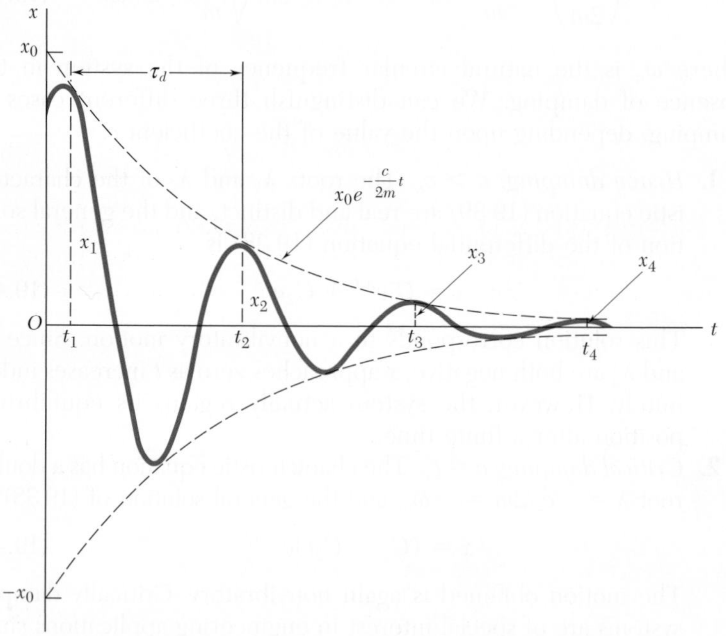

Fig. 19.11

*19.9 DAMPED FORCED VIBRATIONS

If the system considered in the preceding section is subjected to a periodic force **P** of magnitude $P = P_m \sin \omega_f t$, the equation of motion becomes

$$m\ddot{x} + c\dot{x} + kx = P_m \sin \omega_f t \tag{19.47}$$

The general solution of (19.47) is obtained by adding a particular solution of (19.47) to the complementary function or general solution of the homogeneous equation (19.38). The complementary function is given by (19.42), (19.43), or (19.44), depending upon the type of damping considered. It represents a *transient* motion which is eventually damped out.

Our interest in this section is centered on the steady-state vibration represented by a particular solution of (19.47) of the form

$$x_{\text{part}} = x_m \sin (\omega_f t - \varphi) \tag{19.48}$$

Substituting x_{part} for x into (19.47), we obtain

$$-m\omega_f^2 x_m \sin (\omega_f t - \varphi) + c\omega_f x_m \cos (\omega_f t - \varphi) + kx_m \sin (\omega_f t - \varphi) = P_m \sin \omega_f t$$

Making $\omega_f t - \varphi$ successively equal to 0 and to $\pi/2$, we write

$$c\omega_f x_m = P_m \sin \varphi \tag{19.49}$$

$$(k - m\omega_f^2)\, x_m = P_m \cos \varphi \tag{19.50}$$

Squaring both members of (19.49) and (19.50) and adding, we have

$$[(k - m\omega_f^2)^2 + (c\omega_f)^2]\, x_m^2 = P_m^2 \tag{19.51}$$

Solving (19.51) for x_m and dividing (19.49) and (19.50) member by member, we obtain, respectively,

$$x_m = \frac{P_m}{\sqrt{(k - m\omega_f^2)^2 + (c\omega_f)^2}} \qquad \tan \varphi = \frac{c\omega_f}{k - m\omega_f^2} \tag{19.52}$$

Recalling from (19.4) that $k/m = \omega_n^2$, where ω_n is the circular frequency of the undamped free vibration, and from (19.41) that $2m\omega_n = c_c$, where c_c is the critical damping coefficient of the system, we write

$$\frac{x_m}{P_m/k} = \frac{x_m}{\delta_m} = \frac{1}{\sqrt{[1 - (\omega_f/\omega_n)^2]^2 + [2(c/c_c)(\omega_f/\omega_n)]^2}} \tag{19.53}$$

$$\tan \varphi = \frac{2(c/c_c)(\omega_f/\omega_n)}{1 - (\omega_f/\omega_n)^2} \tag{19.54}$$

Photo 19.2 The automobile suspension shown consists essentially of a spring and a shock absorber, which will cause the body of the car to undergo *damped forced vibrations* when the car is driven over an uneven road.

Photo 19.3 This truck is experiencing damped forced vibration in the vehicle dynamics test shown.

Formula (19.53) expresses the magnification factor in terms of the frequency ratio ω_f/ω_n and damping factor c/c_c. It can be used to determine the amplitude of the steady-state vibration produced by an impressed force of magnitude $P = P_m \sin \omega_f t$ or by an impressed support movement $\delta = \delta_m \sin \omega_f t$. Formula (19.54) defines in terms of the same parameters the *phase difference* φ between the impressed force or impressed support movement and the resulting steady-state vibration of the damped system. The magnification factor has been plotted against the frequency ratio in Fig. 19.12 for various values of the damping factor. We observe that the amplitude of a forced vibration can be kept small by choosing a large coefficient of viscous damping c or by keeping the natural and forced frequencies far apart.

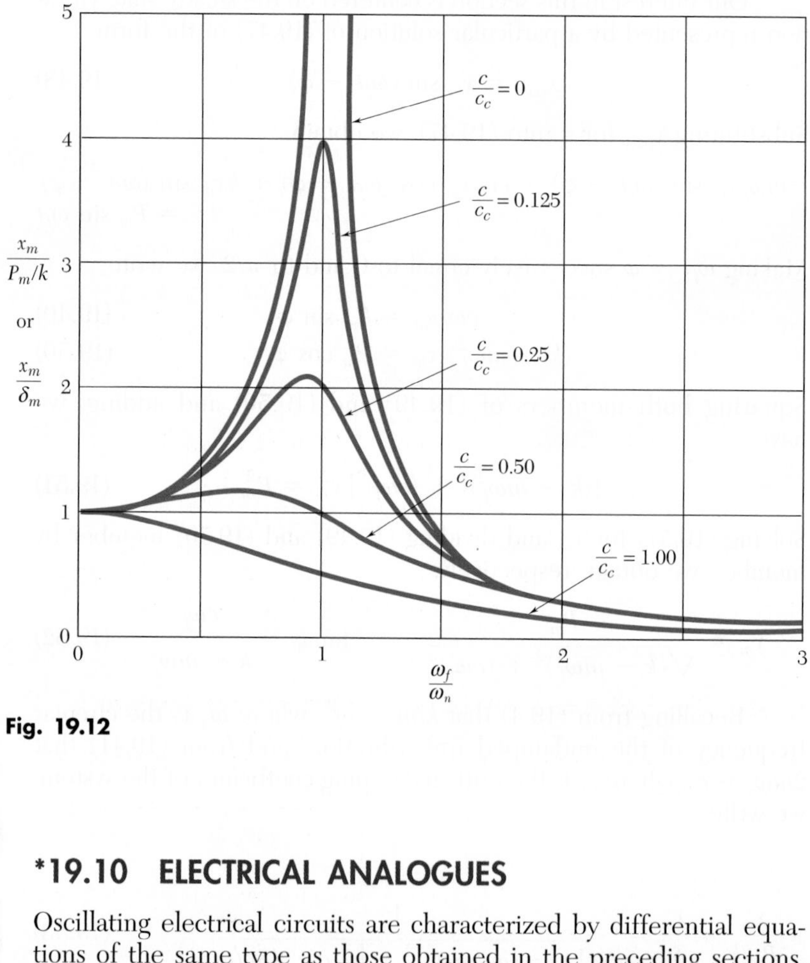

Fig. 19.12

*19.10 ELECTRICAL ANALOGUES

Oscillating electrical circuits are characterized by differential equations of the same type as those obtained in the preceding sections. Their analysis is therefore similar to that of a mechanical system, and the results obtained for a given vibrating system can be readily extended to the equivalent circuit. Conversely, any result obtained for an electrical circuit will also apply to the corresponding mechanical system.

Consider an electrical circuit consisting of an inductor of inductance L, a resistor of resistance R, and a capacitor of capacitance C, connected in series with a source of alternating voltage $E = E_m \sin \omega_f t$ (Fig. 19.13). It is recalled from elementary circuit theory† that if i denotes the current in the circuit and q denotes the electric charge on the capacitor, the drop in potential is $L(di/dt)$ across the inductor, Ri across the resistor, and q/C across the capacitor. Expressing that the algebraic sum of the applied voltage and of the drops in potential around the circuit loop is zero, we write

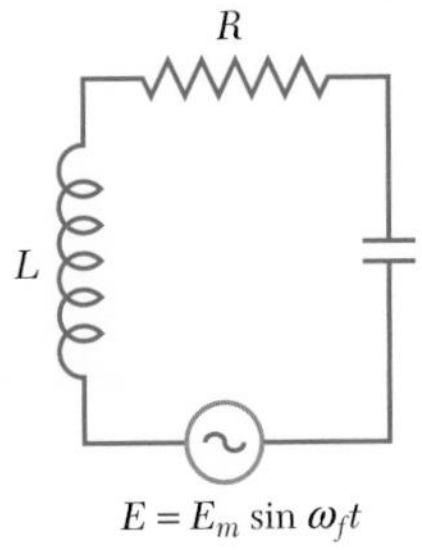

Fig. 19.13

$$E_m \sin \omega_f t - L\frac{di}{dt} - Ri - \frac{q}{C} = 0 \qquad (19.55)$$

Rearranging the terms and recalling that at any instant the current i is equal to the rate of change $\dot{q}$ of the charge q, we have

$$L\ddot{q} + R\dot{q} + \frac{1}{C}q = E_m \sin \omega_f t \qquad (19.56)$$

We verify that Eq. (19.56), which defines the oscillations of the electrical circuit of Fig. 19.13, is of the same type as Eq. (19.47), which characterizes the damped forced vibrations of the mechanical system of Fig. 19.10. By comparing the two equations, we can construct a table of the analogous mechanical and electrical expressions.

Table 19.2 can be used to extend the results obtained in the preceding sections for various mechanical systems to their electrical analogues. For instance, the amplitude i_m of the current in the circuit of Fig. 19.13 can be obtained by noting that it corresponds to the

TABLE 19.2 Characteristics of a Mechanical System and of Its Electrical Analogue

Mechanical System		Electrical Circuit	
m	Mass	L	Inductance
c	Coefficient of viscous damping	R	Resistance
k	Spring constant	$1/C$	Reciprocal of capacitance
x	Displacement	q	Charge
v	Velocity	i	Current
P	Applied force	E	Applied voltage

†See C. R. Paul, S. A. Nasar, and L. E. Unnewehr, *Introduction to Electrical Engineering*, 2nd ed., McGraw-Hill, New York, 1992.

maximum value v_m of the velocity in the analogous mechanical system. Recalling from the first of Eqs. (19.37) that $v_m = x_m \omega_f$, substituting for x_m from Eq. (19.52), and replacing the constants of the mechanical system by the corresponding electrical expressions, we have

$$i_m = \frac{\omega_f E_m}{\sqrt{\left(\frac{1}{C} - L\omega_f^2\right)^2 + (R\omega_f)^2}}$$

$$i_m = \frac{E_m}{\sqrt{R^2 + \left(L\omega_f - \frac{1}{C\omega_f}\right)^2}} \tag{19.57}$$

The radical in the expression obtained is known as the *impedance* of the electrical circuit.

The analogy between mechanical systems and electrical circuits holds for transient as well as steady-state oscillations. The oscillations of the circuit shown in Fig. 19.14, for instance, are analogous to the damped free vibrations of the system of Fig. 19.10. As far as the initial conditions are concerned, we should note that closing the switch S when the charge on the capacitor is $q = q_0$ is equivalent to releasing the mass of the mechanical system with no initial velocity from the position $x = x_0$. We should also observe that if a battery of constant voltage E is introduced in the electrical circuit of Fig. 19.14, closing the switch S will be equivalent to suddenly applying a force of constant magnitude P to the mass of the mechanical system of Fig. 19.10.

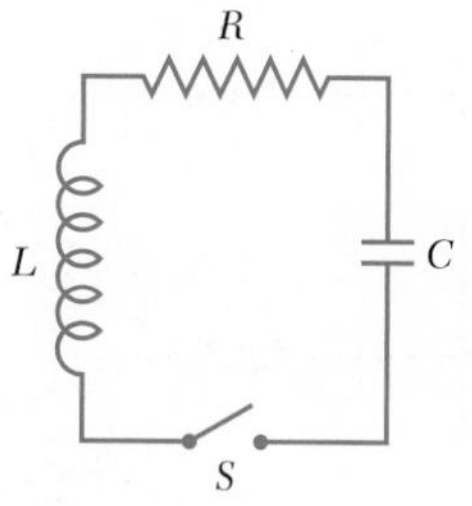

Fig. 19.14

The discussion above would be of questionable value if its only result were to make it possible for mechanics students to analyze electrical circuits without learning the elements of circuit theory. It is hoped that this discussion will instead encourage students to apply to the solution of problems in mechanical vibrations the mathematical techniques they may learn in later courses in circuit theory. The chief value of the concept of electrical analogue, however, resides in its application to *experimental methods* for the determination of the characteristics of a given mechanical system. Indeed, an electrical circuit is much more easily constructed than is a mechanical model, and the fact that its characteristics can be modified by varying the inductance, resistance, or capacitance of its various components makes the use of the electrical analogue particularly convenient.

To determine the electrical analogue of a given mechanical system, we focus our attention on each moving mass in the system and observe which springs, dashpots, or external forces are applied directly to it. An equivalent electrical loop can then be constructed to match each of the mechanical units thus defined; the various loops obtained in that way will together form the desired circuit. Consider, for instance, the mechanical system of Fig. 19.15. We observe that the mass m_1 is acted upon by two springs of constants k_1 and k_2 and by two dashpots characterized by the coefficients of viscous damping c_1 and c_2. The electrical circuit should therefore include a loop consisting of an inductor of inductance L_1 proportional to m_1, of two capacitors of capacitance C_1 and C_2 inversely proportional to k_1 and k_2, respectively, and of two resistors of resistance R_1 and R_2, proportional to c_1 and c_2, respectively. Since the mass m_2 is acted upon by the spring k_2 and the dashpot c_2, as well as the force $P = P_m \sin \omega_f t$, the circuit should also include a loop containing the capacitor C_2, the resistor R_2, the new inductor L_2, and the voltage source $E = E_m \sin \omega_f t$ (Fig. 19.16).

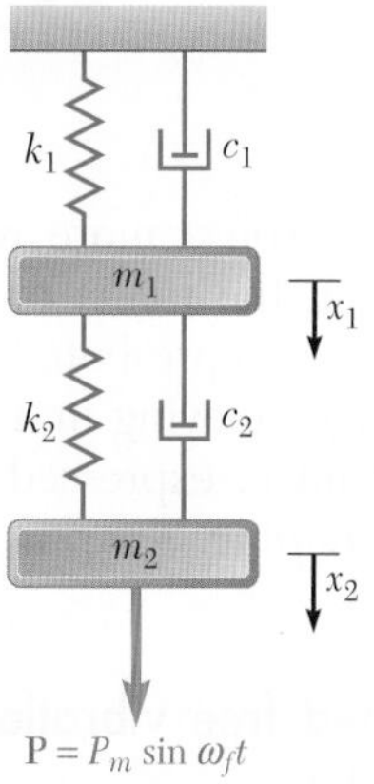

Fig. 19.15

To check that the mechanical system of Fig. 19.15 and the electrical circuit of Fig. 19.16 actually satisfy the same differential equations, the equations of motion for m_1 and m_2 will first be derived. Denoting, respectively, by x_1 and x_2 the displacements of m_1 and m_2 from their equilibrium positions, we observe that the elongation of the spring k_1 (measured from the equilibrium position) is equal to x_1, while the elongation of the spring k_2 is equal to the relative displacement $x_2 - x_1$ of m_2 with respect to m_1. The equations of motion for m_1 and m_2 are therefore

$$m_1\ddot{x}_1 + c_1\dot{x}_1 + c_2(\dot{x}_1 - \dot{x}_2) + k_1x_1 + k_2(x_1 - x_2) = 0 \quad (19.58)$$

$$m_2\ddot{x}_2 + c_2(\dot{x}_2 - \dot{x}_1) + k_2(x_2 - x_1) = P_m \sin \omega_f t \quad (19.59)$$

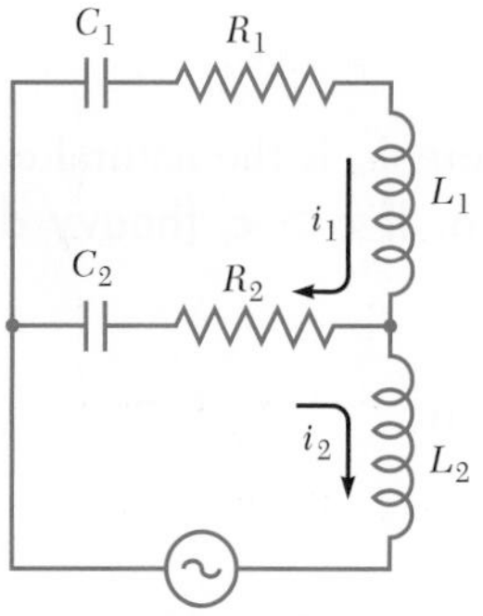

Fig. 19.16

Consider now the electrical circuit of Fig. 19.16; we denote, respectively, by i_1 and i_2 the current in the first and second loops, and by q_1 and q_2 the integrals $\int i_1\,dt$ and $\int i_2\,dt$. Noting that the charge on the capacitor C_1 is q_1, while the charge on C_2 is $q_1 - q_2$, we express that the sum of the potential differences in each loop is zero and obtain the following equations

$$L_1\ddot{q}_1 + R_1\dot{q}_1 + R_2(\dot{q}_1 - \dot{q}_2) + \frac{q_1}{C_1} + \frac{q_1 - q_2}{C_2} = 0 \quad (19.60)$$

$$L_2\ddot{q}_2 + R_2(\dot{q}_2 - \dot{q}_1) + \frac{q_2 - q_1}{C_2} = E_m \sin \omega_f t \quad (19.61)$$

We easily check that Eqs. (19.60) and (19.61) reduce to (19.58) and (19.59), respectively, when the substitutions indicated in Table 19.2 are performed.

SOLVING PROBLEMS ON YOUR OWN

In this lesson a more realistic model of a vibrating system was developed by including the effect of the *viscous damping* caused by fluid friction. Viscous damping was represented in Fig. 19.10 by the force exerted on the moving body by a plunger moving in a dashpot. This force is equal in magnitude to $c\dot{x}$, where the constant c, expressed in N · s/m or lb · s/ft, is known as the *coefficient of viscous damping*. Keep in mind that the same sign convention should be used for x, $\dot{x}$, and $\ddot{x}$.

1. Damped free vibrations. The differential equation defining this motion was found to be

$$m\ddot{x} + c\dot{x} + kx = 0 \qquad (19.38)$$

To obtain the solution of this equation, calculate the *critical damping coefficient* c_c, using the formula

$$c_c = 2m\sqrt{k/m} = 2m\omega_n \qquad (19.41)$$

where ω_n is the natural circular frequency of the *undamped* system.

a. If $c > c_c$ (heavy damping), the solution of Eq. (19.38) is

$$x = C_1 e^{\lambda_1 t} + C_2 e^{\lambda_2 t} \qquad (19.42)$$

where

$$\lambda_{1,2} = -\frac{c}{2m} \pm \sqrt{\left(\frac{c}{2m}\right)^2 - \frac{k}{m}} \qquad (19.40)$$

and where the constants C_1 and C_2 can be determined from the initial conditions $x(0)$ and $\dot{x}(0)$. This solution corresponds to a nonvibratory motion.

b. If $c = c_c$ (critical damping), the solution of Eq. (19.38) is

$$x = (C_1 + C_2 t)e^{-\omega_n t} \qquad (19.43)$$

which also corresponds to a nonvibratory motion.

c. If $c < c_c$ (light damping), the solution of Eq. (19.38) is

$$x = x_0 e^{-(c/2m)t} \sin(\omega_d t + \phi) \qquad (19.46)$$

where

$$\omega_d = \omega_n \sqrt{1 - \left(\frac{c}{c_c}\right)^2} \tag{19.45}$$

and where x_0 and ϕ can be determined from the initial conditions $x(0)$ *and* $\dot{x}(0)$. This solution corresponds to oscillations of decreasing amplitude and of period $\tau_d = 2\pi/\omega_d$ (Fig. 19.11).

2. Damped forced vibrations. These vibrations occur when a system with viscous damping is subjected to a periodic force **P** of magnitude $P = P_m \sin \omega_f t$ or when it is elastically connected to a support with an alternating motion $\delta = \delta_m \sin \omega_f t$. In the first case the motion is defined by the differential equation

$$m\ddot{x} + c\dot{x} + kx = P_m \sin \omega_f t \tag{19.47}$$

and in the second case by a similar equation obtained by replacing P_m with $k\delta_m$. You will be concerned only with the *steady-state* motion of the system, which is defined by a *particular solution* of these equations, of the form

$$x_{\text{part}} = x_m \sin (\omega_f t - \varphi) \tag{19.48}$$

where

$$\frac{x_m}{P_m/k} = \frac{x_m}{\delta_m} = \frac{1}{\sqrt{[1 - (\omega_f/\omega_n)^2]^2 + [2(c/c_c)(\omega_f/\omega_n)]^2}} \tag{19.53}$$

and

$$\tan \varphi = \frac{2(c/c_c)(\omega_f/\omega_n)}{1 - (\omega_f/\omega_n)^2} \tag{19.54}$$

The expression given in Eq. (19.53) is referred to as the *magnification factor* and has been plotted against the frequency ratio ω_f/ω_n in Fig. 19.12 for various values of the damping factor c/c_c. In the problems which follow, you may be asked to determine one of the parameters in Eqs. (19.53) and (19.54) when the others are known.

PROBLEMS

19.127 Show that in the case of heavy damping ($c > c_c$), a body never passes through its position of equilibrium O if it is (*a*) released with no initial velocity from an arbitrary position, (*b*) started from O with an arbitrary initial velocity.

19.128 Show that in the case of heavy damping ($c > c_c$), a body released from an arbitrary position with an arbitrary initial velocity cannot pass more than once through its equilibrium position.

19.129 In the case of light damping, the displacements x_1, x_2, x_3, shown in Fig. 19.11 may be assumed equal to the maximum displacements. Show that the ratio of any two successive maximum displacements x_n and x_{n+1} is a constant and that the natural logarithm of this ratio, called the *logarithmic decrement,* is

$$\ln\frac{x_n}{x_{n+1}} = \frac{2\pi(c/c_c)}{\sqrt{1 - (c/c_c)^2}}$$

19.130 In practice, it is often difficult to determine the logarithmic decrement of a system with light damping defined in Prob. 19.129 by measuring two successive maximum displacements. Show that the logarithmic decrement can also be expressed as $(1/k)\ \ln(x_n/x_{n+k})$, where k is the number of cycles between readings of the maximum displacement.

19.131 In a system with light damping ($c < c_c$), the period of vibration is commonly defined as the time interval $\tau_d = 2\pi/\omega_d$ corresponding to two successive points where the displacement-time curve touches one of the limiting curves shown in Fig. 19.11. Show that the interval of time (*a*) between a maximum positive displacement and the following maximum negative displacement is $\frac{1}{2}\tau_d$, (*b*) between two successive zero displacements is $\frac{1}{2}\tau_d$, (*c*) between a maximum positive displacement and the following zero displacement is greater than $\frac{1}{4}\tau_d$.

19.132 A loaded railroad car of mass 15,000 kg is rolling at a constant velocity $\mathbf{v}_0$ when it couples with a spring and dashpot bumper system (Fig. 1). The recorded displacement-time curve of the loaded railroad car after coupling is as shown (Fig. 2). Determine (a) the damping constant, (b) the spring constant. (*Hint:* Use the definition of logarithmic decrement given in 19.129.)

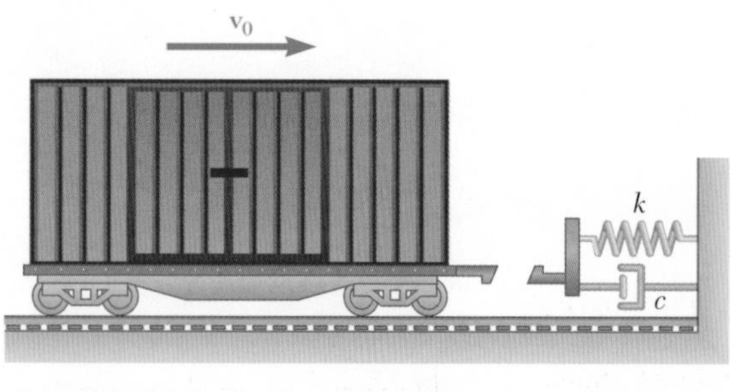

(1)

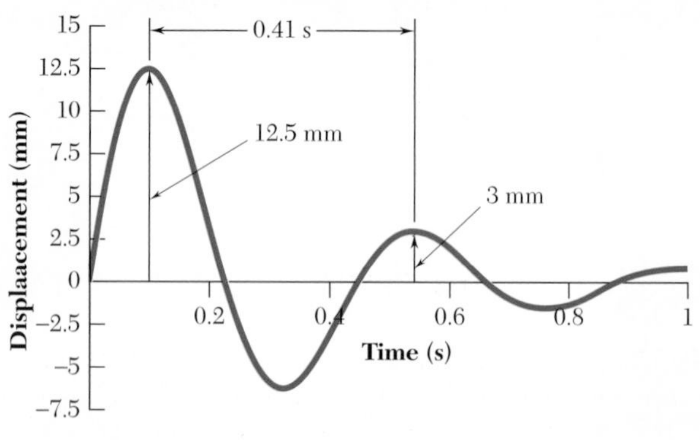

(2)

Fig. P19.132

19.133 A torsional pendulum has a centroidal mass moment of inertia of $0.3\ \text{kg}\cdot\text{m}^2$ and when given an initial twist and released is found to have a frequency of oscillation of 200 rpm. Knowing that when this pendulum is immersed in oil and when given the same initial condition it is found to have a frequency of oscillation of 180 rpm, determine the damping constant for the oil.

19.134 The barrel of a field gun has a mass of 750 kg and is returned into firing position after recoil by a recuperator of constant $c = 18\ \text{kN}\cdot\text{s/m}$. Determine (*a*) the constant k which should be used for the recuperator to return the barrel into firing position in the shortest possible time without any oscillation, (*b*) the time needed for the barrel to move back two-thirds of the way from its maximum-recoil position to its firing position.

19.135 A platform of mass 100 kg, supported by two springs each of constant $k = 50$ kN/m, is subjected to a periodic force of maximum magnitude equal to 625 N. Knowing that the coefficient of damping is 1600 N · s/m, determine (*a*) the natural frequency in rpm of the platform if there were no damping, (*b*) the frequency in rpm of the periodic force corresponding to the maximum value of the magnification factor, assuming damping, (*c*) the amplitude of the actual motion of the platform for each of the frequencies found in parts *a* and *b*.

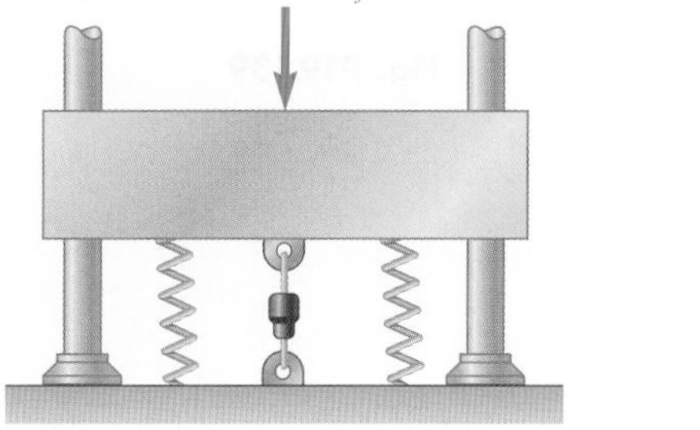

Fig. P19.135

19.136 A 4-kg block A is dropped from a height of 800 mm onto a 9-kg block B which is at rest. Block B is supported by a spring of constant $k = 1500$ N/m and is attached to a dashpot of damping coefficient $c = 230$ N · s/m. Knowing that there is no rebound, determine the maximum distance the blocks will move after the impact.

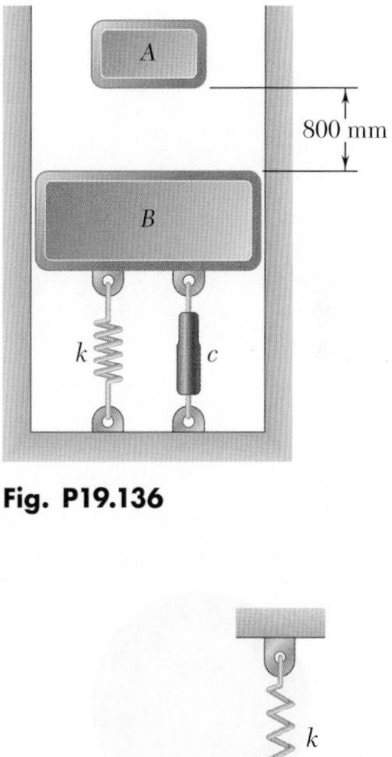

Fig. P19.136

19.137 A 3-kg slender rod AB is bolted to a 5-kg uniform disk. A dashpot of damping coefficient $c = 9$ N · s/m is attached to the disk as shown. Determine (*a*) the differential equation of motion for small oscillations, (*b*) the damping factor c/c_c.

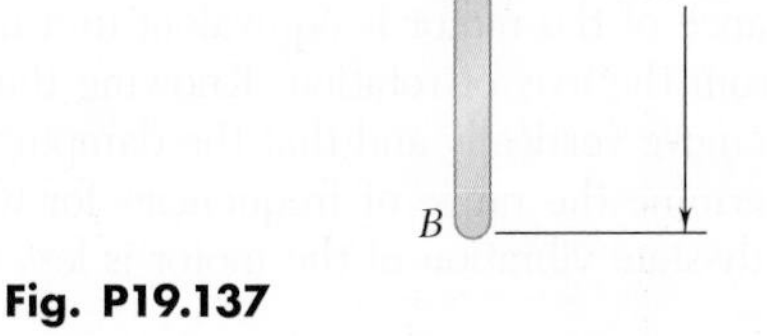

Fig. P19.137

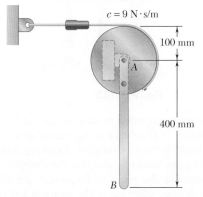

19.138 A uniform rod of mass m is supported by a pin at A and a spring of constant k at B and is connected at D to a dashpot of damping coefficient c. Determine in terms of m, k, and c, for small oscillations, (*a*) the differential equation of motion, (*b*) the critical damping coefficient c_c.

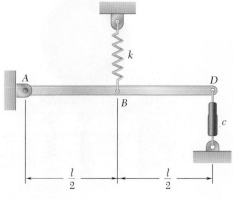

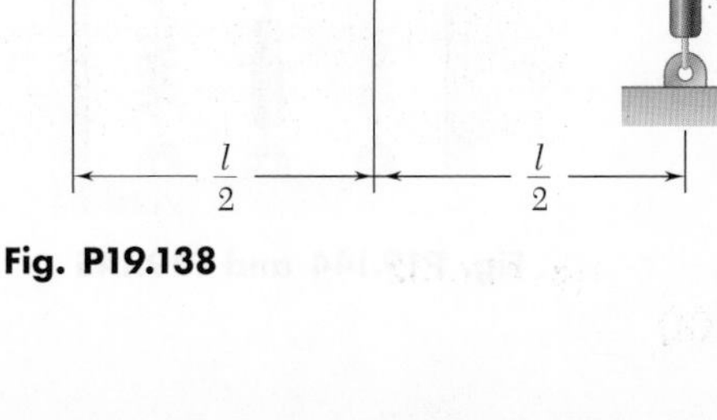

Fig. P19.138

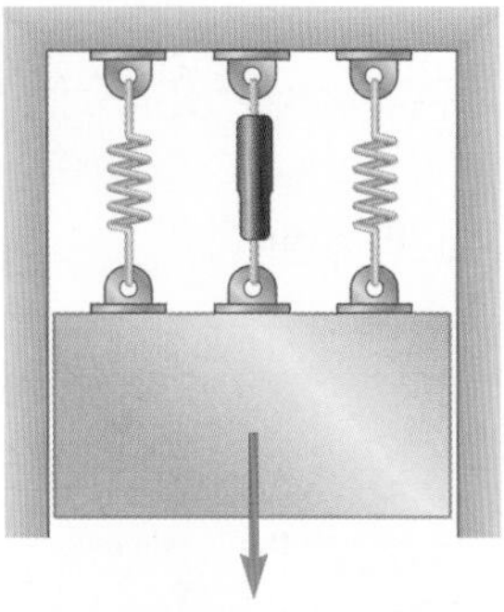

Fig. P19.139

19.139 A 500 kg machine element is supported by two springs, each of constant 50 kN/m. A periodic force of 150-N amplitude is applied to the element with a frequency of 2.8 Hz. Knowing that the coefficient of damping is 1.8 kN · s/m, determine the amplitude of the steady-state vibration of the element.

19.140 In Prob. 19.139, determine the required value of the constant of each spring if the amplitude of the steady-state vibration is to be 1.25 mm.

19.141 In the case of the forced vibration of a system, determine the range of values of the damping factor c/c_c for which the magnification factor will always decrease as the frequency ratio ω_f/ω_n increases.

19.142 Show that for a small value of the damping factor c/c_c, the maximum amplitude of a forced vibration occurs when $\omega_f \approx \omega_n$ and that the corresponding value of the magnification factor is $\frac{1}{2}(c/c_c)$.

19.143 A counter-rotating eccentric mass exciter consisting of two rotating 400-g masses describing circles of 150-mm radius at the same speed but in opposite senses is placed on a machine element to induce a steady-state vibration of the element and to determine some of the dynamic characteristics of the element. At a speed of 1200 rpm a stroboscope shows the eccentric masses to be exactly under their respective axes of rotation and the element to be passing through its position of static equilibrium. Knowing that the amplitude of the motion of the element at that speed is 15 mm and that the total mass of the system is 140 kg, determine (*a*) the combined spring constant k, (*b*) the damping factor c/c_c.

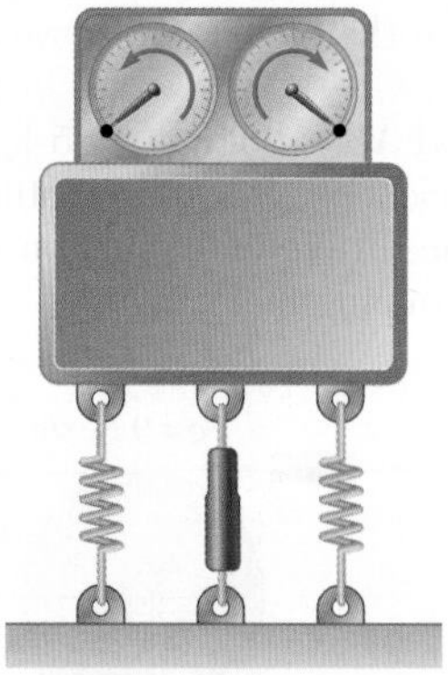

Fig. P19.143

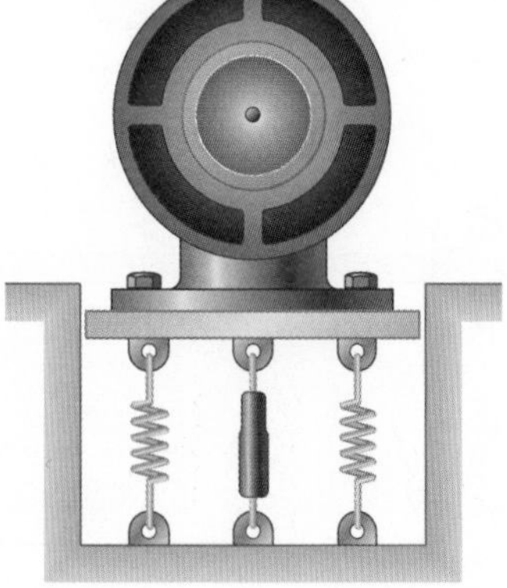

Fig. P19.144 and P19.145

19.144 A 15-kg motor is supported by four springs, each of constant 40 kN/m. The unbalance of the motor is equivalent to a mass of 20 g located 125 mm from the axis of rotation. Knowing that the motor is constrained to move vertically and that the damping factor c/c_c is equal to 0.4, determine the range of frequencies for which the amplitude of the steady-state vibration of the motor is less than 0.2 mm.

19.145 A 100-kg motor is supported by four springs, each of constant 90 kN/m, and is connected to the ground by a dashpot having a coefficient of damping c = 6500 N · s/m. The motor is constrained to move vertically, and the amplitude of its motion is observed to be 2.1 mm at a speed of 1200 rpm. Knowing that the mass of the rotor is 15 kg, determine the distance between the mass center of the rotor and the axis of the shaft.

19.146 A 50-kg motor is directly supported by a light horizontal beam which has a static deflection of 6 mm due to the weight of the motor. The unbalance of the rotor is equivalent to a mass of 100 g located 75 mm from the axis of rotation. Knowing that the amplitude of the vibration of the motor is 0.8 mm at a speed of 400 rpm, determine (*a*) the damping factor c/c_c, (*b*) the coefficient of damping c.

Fig. P19.146

19.147 A machine element is supported by springs and is connected to a dashpot as shown. Show that if a periodic force of magnitude $P = P_m \sin \omega_f t$ is applied to the element, the amplitude of the fluctuating force transmitted to the foundation is

$$F_m = P_m \sqrt{\frac{1 + [2(c/c_c)(\omega_f/\omega_n)]^2}{[1 - (\omega_f/\omega_n)^2]^2 + [2(c/c_c)(\omega_f/\omega_n)]^2}}$$

19.148 A 91-kg machine element supported by four springs, each of constant $k = 175$ N/m, is subjected to a periodic force of frequency 0.8 Hz and amplitude 89 N. Determine the amplitude of the fluctuating force transmitted to the foundation if (*a*) a dashpot with a coefficient of damping $c = 365$ N · s/m is connected to the machine element and to the ground, (*b*) the dashpot is removed.

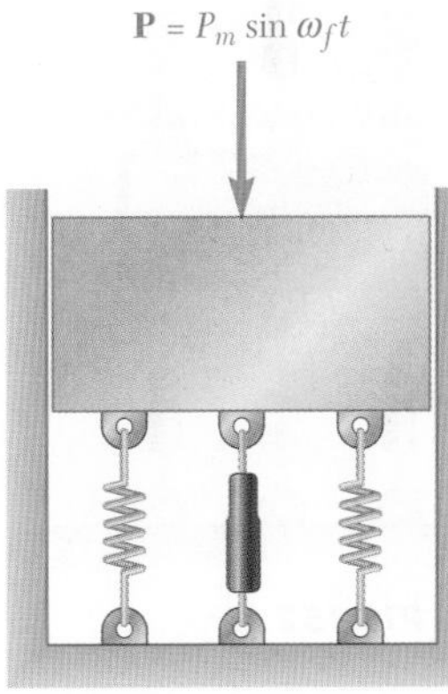

Fig. P19.147 and P19.148

19.149 A simplified model of a washing machine is shown. A bundle of wet clothes forms a mass m_b of 10 kg in the machine and causes a rotating unbalance. The rotating mass is 20 kg (including m_b) and the radius of the washer basket e is 0.25 m. Knowing the washer has an equivalent spring constant $k = 1000$ N/m and damping ratio $\zeta = c/c_c = 0.05$ and during the spin cycle the drum rotates at 250 rpm, determine the amplitude of the motion and the magnitude of the force transmitted to the sides of the washing machine.

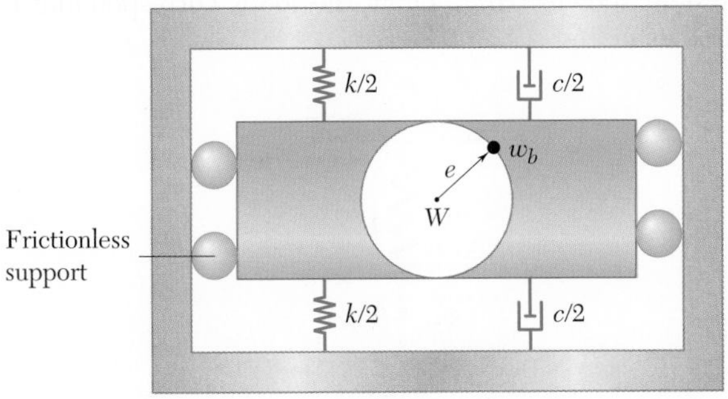

Fig. P19.149

***19.150** For a steady-state vibration with damping under a harmonic force, show that the mechanical energy dissipated per cycle by the dashpot is $E = \pi c x_m^2 \omega_f$, where c is the coefficient of damping, x_m is the amplitude of the motion, and ω_f is the circular frequency of the harmonic force.

***19.151** The suspension of an automobile can be approximated by the simplified spring-and-dashpot system shown. (*a*) Write the differential equation defining the vertical displacement of the mass m when the system moves at a speed v over a road with a sinusoidal cross section of amplitude δ_m and wave length L. (*b*) Derive an expression for the amplitude of the vertical displacement of the mass m.

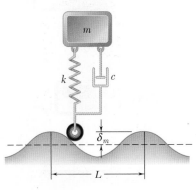

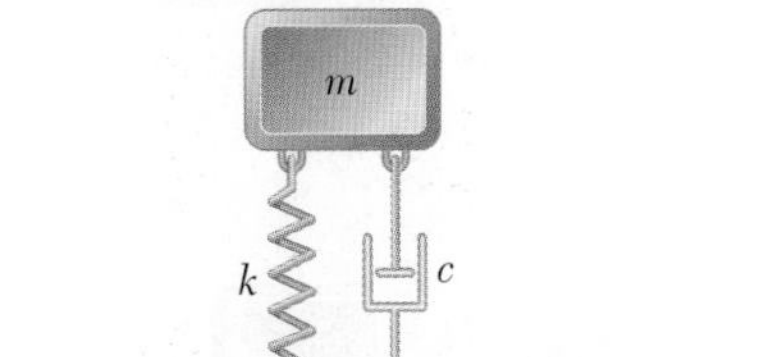

Fig. P19.151

***19.152** Two blocks A and B, each of mass m, are supported as shown by three springs of the same constant k. Blocks A and B are connected by a dashpot and block B is connected to the ground by two dashpots, each dashpot having the same coefficient of damping c. Block A is subjected to a force of magnitude $P = P_m \sin \omega_f t$. Write the differential equations defining the displacements x_A and x_B of the two blocks from their equilibrium positions.

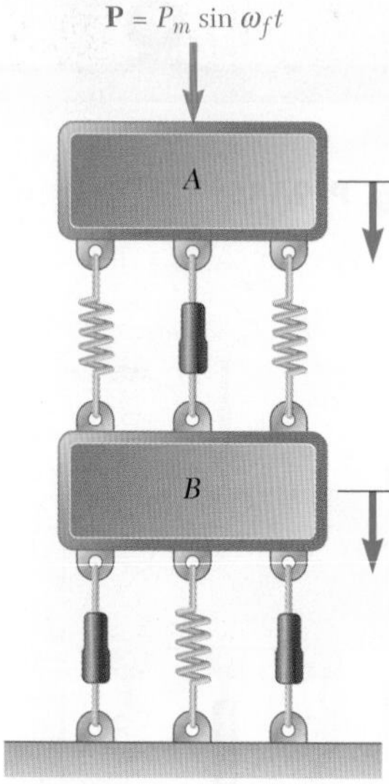

Fig. *P19.152*

19.153 Express in terms of L, C, and E the range of values of the resistance R for which oscillations will take place in the circuit shown when switch S is closed.

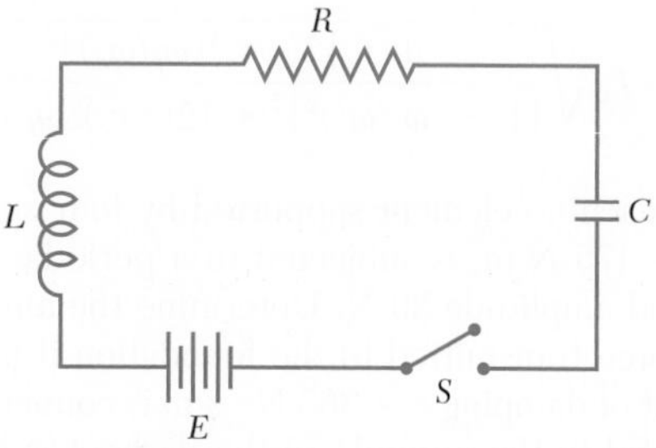

Fig. P19.153

19.154 Consider the circuit of Prob. 19.153 when the capacitor C is removed. If switch S is closed at time $t = 0$, determine (*a*) the final value of the current in the circuit, (*b*) the time t at which the current will have reached $(1 - 1/e)$ times its final value. (The desired value of t is known as the *time constant* of the circuit.)

19.155 and 19.156 Draw the electrical analogue of the mechanical system shown. (*Hint:* Draw the loops corresponding to the free bodies m and A.)

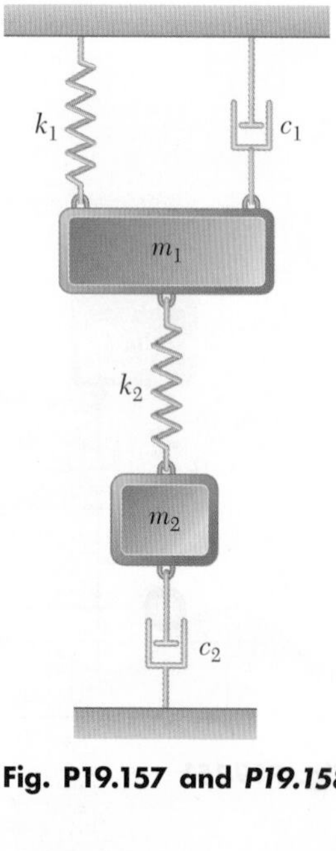

Fig. P19.157 and *P19.158*

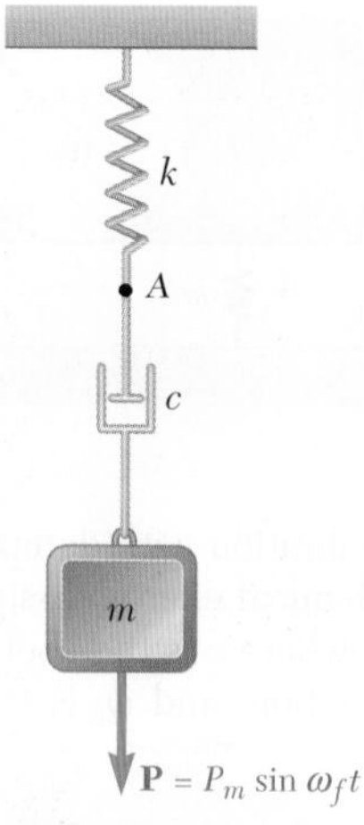

Fig. P19.155 and P19.156

19.157 and *19.158* Write the differential equations defining (*a*) the displacements of the mass m and of the point A, (*b*) the charges on the capacitors of the electrical analogue.

REVIEW AND SUMMARY

This chapter was devoted to the study of *mechanical vibrations,* i.e., to the analysis of the motion of particles and rigid bodies oscillating about a position of equilibrium. In the first part of the chapter [Secs. 19.2 through 19.7], we considered *vibrations without damping,* while the second part was devoted to *damped vibrations* [Secs. 19.8 through 19.10].

Free vibrations of a particle

In Sec. 19.2, we considered the *free vibrations of a particle,* i.e., the motion of a particle P subjected to a restoring force proportional to the displacement of the particle—such as the force exerted by a spring. If the displacement x of the particle P is measured from its equilibrium position O (Fig. 19.17), the resultant $\mathbf{F}$ of the forces acting on P (including its weight) has a magnitude kx and is directed toward O. Applying Newton's second law $F = ma$ and recalling that $a = \ddot{x}$, we wrote the differential equation

$$m\ddot{x} + kx = 0 \qquad (19.2)$$

or, setting $\omega_n^2 = k/m$,

$$\ddot{x} + \omega_n^2 x = 0 \qquad (19.6)$$

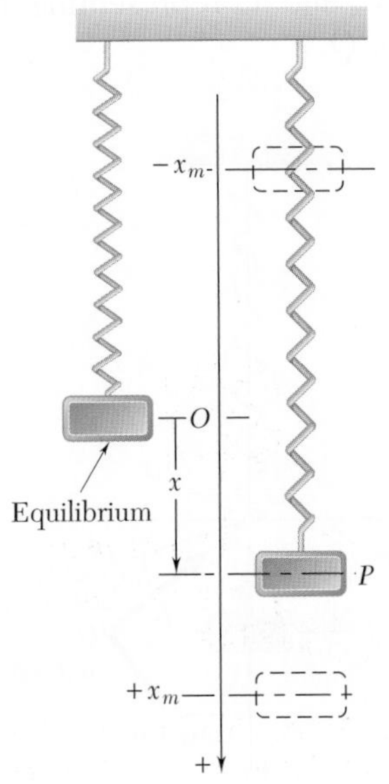

Fig. 19.17

The motion defined by this equation is called a *simple harmonic motion.*

The solution of Eq. (19.6), which represents the displacement of the particle P, was expressed as

$$x = x_m \sin (\omega_n t + \phi) \tag{19.10}$$

where x_m = amplitude of the vibration
$\omega_n = \sqrt{k/m}$ = natural circular frequency
ϕ = phase angle

The *period of the vibration* (i.e., the time required for a full cycle) and its *natural frequency* (i.e., the number of cycles per second) were expressed as

$$\text{Period} = \tau_n = \frac{2\pi}{\omega_n} \tag{19.13}$$

$$\text{Natural frequency} = f_n = \frac{1}{\tau_n} = \frac{\omega_n}{2\pi} \tag{19.14}$$

The velocity and acceleration of the particle were obtained by differentiating Eq. (19.10), and their maximum values were found to be

$$v_m = x_m\omega_n \qquad a_m = x_m\omega_n^2 \tag{19.15}$$

Since all the above parameters depend directly upon the natural circular frequency ω_n and thus upon the ratio k/m, it is essential in any given problem to calculate the value of the constant k; this can be done by determining the relation between the restoring force and the corresponding displacement of the particle [Sample Prob. 19.1].

It was also shown that the oscillatory motion of the particle P can be represented by the projection on the x axis of the motion of a point Q describing an auxiliary circle of radius x_m with the constant angular velocity ω_n (Fig. 19.18). The instantaneous values of the velocity and acceleration of P can then be obtained by projecting on the x axis the vectors $\mathbf{v}_m$ and $\mathbf{a}_m$ representing, respectively, the velocity and acceleration of Q.

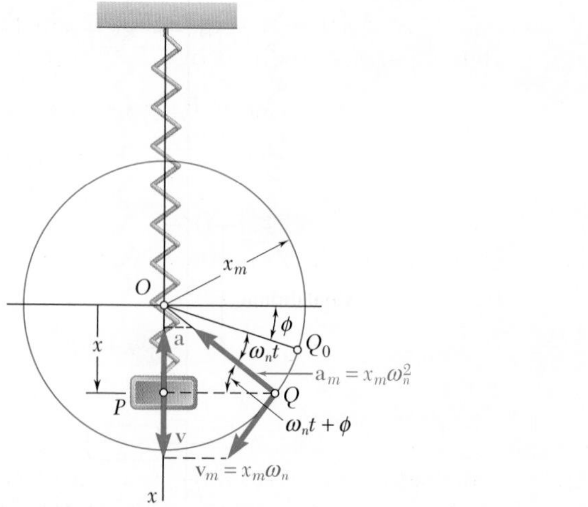

Fig. 19.18

Simple pendulum

While the motion of a *simple pendulum* is not truly a simple harmonic motion, the formulas given above can be used with $\omega_n^2 = g/l$ to calculate the period and natural frequency of the *small oscillations* of a simple pendulum [Sec. 19.3]. Large-amplitude oscillations of a simple pendulum were discussed in Sec. 19.4.

Free vibrations of a rigid body

The *free vibrations of a rigid body* can be analyzed by choosing an appropriate variable, such as a distance x or an angle θ, to define the position of the body, drawing a free-body-diagram equation to express the equivalence of the external and effective forces, and writing an equation relating the selected variable and its second derivative [Sec. 19.5]. If the equation obtained is of the form

$$\ddot{x} + \omega_n^2 x = 0 \qquad \text{or} \qquad \ddot{\theta} + \omega_n^2 \theta = 0 \qquad (19.21)$$

the vibration considered is a simple harmonic motion and its period and natural frequency can be obtained *by identifying* ω_n and substituting its value into Eqs. (19.13) and (19.14) [Sample Probs. 19.2 and 19.3].

Using the principle of conservation of energy

The *principle of conservation of energy* can be used as an alternative method for the determination of the period and natural frequency of the simple harmonic motion of a particle or rigid body [Sec. 19.6]. Choosing again an appropriate variable, such as θ, to define the position of the system, we express that the total energy of the system is conserved, $T_1 + V_1 = T_2 + V_2$, between the position of maximum displacement ($\theta_1 = \theta_m$) and the position of maximum velocity ($\dot{\theta}_2 = \dot{\theta}_m$). If the motion considered is simple harmonic, the two members of the equation obtained consist of homogeneous quadratic expressions in θ_m and $\dot{\theta}_m$, respectively.† Substituting $\dot{\theta}_m = \theta_m \omega_n$ in this equation, we can factor out θ_m^2 and solve for the circular frequency ω_n [Sample Prob. 19.4].

Forced vibrations

In Sec. 19.7, we considered the *forced vibrations* of a mechanical system. These vibrations occur when the system is subjected to a periodic force (Fig. 19.19) or when it is elastically connected to a support which has an alternating motion (Fig. 19.20). Denoting by ω_f the forced circular frequency, we found that in the first case, the motion of the system was defined by the differential equation

$$m\ddot{x} + kx = P_m \sin \omega_f t \qquad (19.30)$$

and that in the second case it was defined by the differential equation

$$m\ddot{x} + kx = k\delta_m \sin \omega_f t \qquad (19.31)$$

The general solution of these equations is obtained by adding a particular solution of the form

$$x_{\text{part}} = x_m \sin \omega_f t \qquad (19.32)$$

†If the motion considered can only be *approximated* by a simple harmonic motion, such as for the small oscillations of a body under gravity, the potential energy must be approximated by a quadratic expression in θ_m.

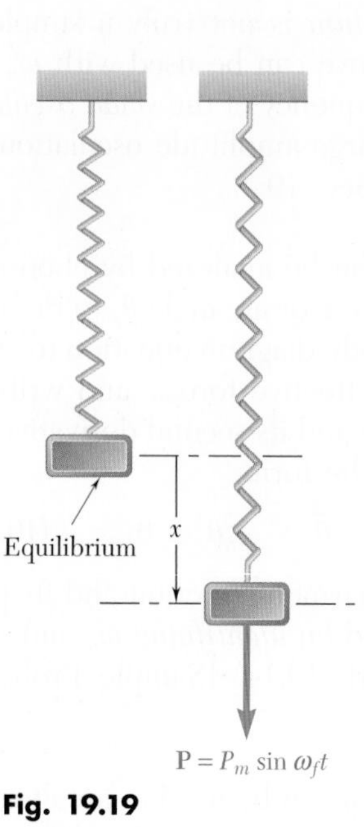

Fig. 19.19

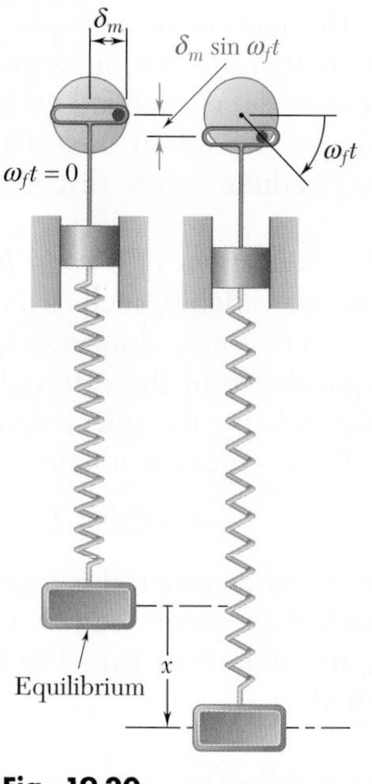

Fig. 19.20

to the general solution of the corresponding homogeneous equation. The particular solution (19.32) represents a *steady-state vibration* of the system, while the solution of the homogeneous equation represents a *transient free vibration* which can generally be neglected.

Dividing the amplitude x_m of the steady-state vibration by P_m/k in the case of a periodic force, or by δ_m in the case of an oscillating support, we defined the *magnification factor* of the vibration and found that

$$\text{Magnification factor} = \frac{x_m}{P_m/k} = \frac{x_m}{\delta_m} = \frac{1}{1 - (\omega_f/\omega_n)^2} \qquad (19.36)$$

According to Eq. (19.36), the amplitude x_m of the forced vibration *becomes infinite when* $\omega_f = \omega_n$, i.e., *when the forced frequency is equal to the natural frequency of the system.* The impressed force or impressed support movement is then said to be in *resonance* with the system [Sample Prob. 19.5]. Actually the amplitude of the vibration remains finite, due to damping forces.

Damped free vibrations

In the last part of the chapter, we considered the *damped vibrations* of a mechanical system. First, we analyzed the *damped free vibrations* of a system with *viscous damping* [Sec. 19.8]. We found that the motion of such a system was defined by the differential equation

$$m\ddot{x} + c\dot{x} + kx = 0 \qquad (19.38)$$

where c is a constant called the *coefficient of viscous damping*. Defining the *critical damping coefficient* c_c as

$$c_c = 2m\sqrt{\frac{k}{m}} = 2m\omega_n \qquad (19.41)$$

where ω_n is the natural circular frequency of the system in the absence of damping, we distinguished three different cases of damping, namely, (1) *heavy damping*, when $c > c_c$; (2) *critical damping*, when $c = c_c$; and (3) *light damping*, when $c < c_c$. In the first two cases, the system when disturbed tends to regain its equilibrium position without any oscillation. In the third case, the motion is vibratory with diminishing amplitude.

Damped forced vibrations

In Sec. 19.9, we considered the *damped forced vibrations* of a mechanical system. These vibrations occur when a system with viscous damping is subjected to a periodic force **P** of magnitude $P = P_m \sin \omega_f t$ or when it is elastically connected to a support with an alternating motion $\delta = \delta_m \sin \omega_f t$. In the first case, the motion of the system was defined by the differential equation

$$m\ddot{x} + c\dot{x} + kx = P_m \sin \omega_f t \qquad (19.47)$$

and in the second case by a similar equation obtained by replacing P_m by $k\delta_m$ in (19.47).

The *steady-state vibration* of the system is represented by a particular solution of Eq. (19.47) of the form

$$x_{\text{part}} = x_m \sin(\omega_f t - \varphi) \qquad (19.48)$$

Dividing the amplitude x_m of the steady-state vibration by P_m/k in the case of a periodic force, or by δ_m in the case of an oscillating support, we obtained the following expression for the magnification factor:

$$\frac{x_m}{P_m/k} = \frac{x_m}{\delta_m} = \frac{1}{\sqrt{[1 - (\omega_f/\omega_n)^2]^2 + [2(c/c_c)(\omega_f/\omega_n)]^2}} \qquad (19.53)$$

where $\omega_n = \sqrt{k/m}$ = natural circular frequency of undamped system
$c_c = 2m\omega_n$ = critical damping coefficient
c/c_c = damping factor

We also found that the *phase difference* φ between the impressed force or support movement and the resulting steady-state vibration of the damped system was defined by the relation

$$\tan \varphi = \frac{2(c/c_c)(\omega_f/\omega_n)}{1 - (\omega_f/\omega_n)^2} \qquad (19.54)$$

Electrical analogues

The chapter ended with a discussion of *electrical analogues* [Sec. 19.10], in which it was shown that the vibrations of mechnical systems and the oscillations of electrical circuits are defined by the same differential equations. Electrical analogues of mechanical systems can therefore be used to study or predict the behavior of these systems.

REVIEW PROBLEMS

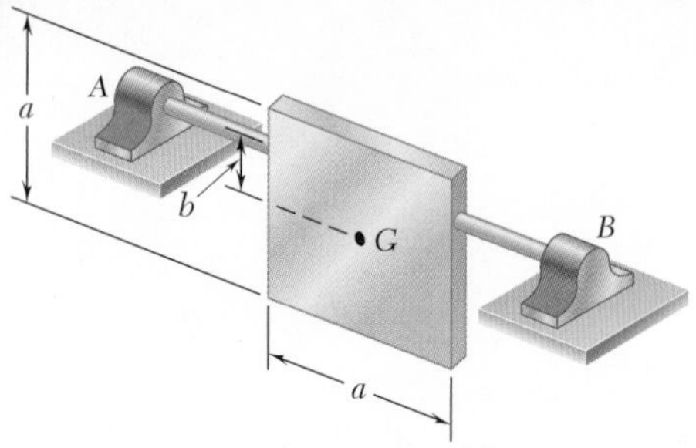

Fig. P19.159

19.159 A thin square plate of side a can oscillate about an axis AB located at a distance b from its mass center G. (*a*) Determine the period of small oscillations if $b = \frac{1}{2}a$ (*b*) Determine a second value of b for which the period of small oscillations is the same as that found in part *a*.

19.160 The period of vibration of the system shown is observed to be 0.6 s. After cylinder B has been removed, the period is observed to be 0.5 s. Determine (*a*) the mass of cylinder A, (*b*) the constant of the spring.

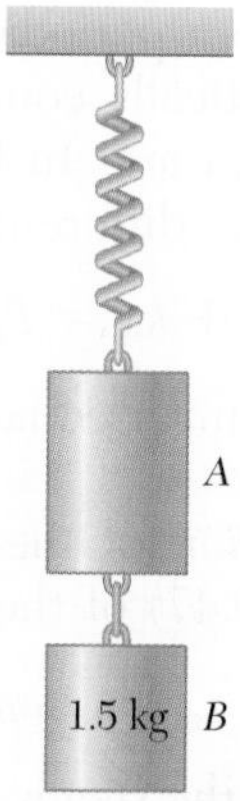

Fig. P19.160

19.161 Disks A and B have a mass of 15 kg and 6 kg, respectively, and a small 2.5-kg block C is attached to the rim of disk B. Assuming that no slipping occurs between the disks, determine the period of small oscillations of the system.

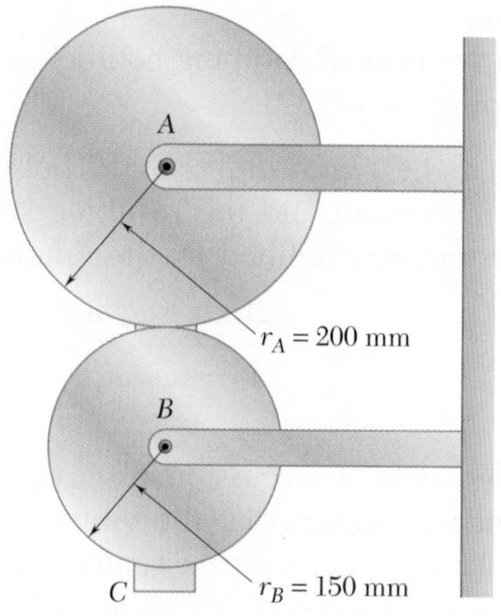

Fig. P19.161

19.162 For the uniform equilateral triangular plate of side l = 300 mm, determine the period of small oscillations if the plate is suspended from (*a*) one of its vertices, (*b*) the midpoint of one of its sides.

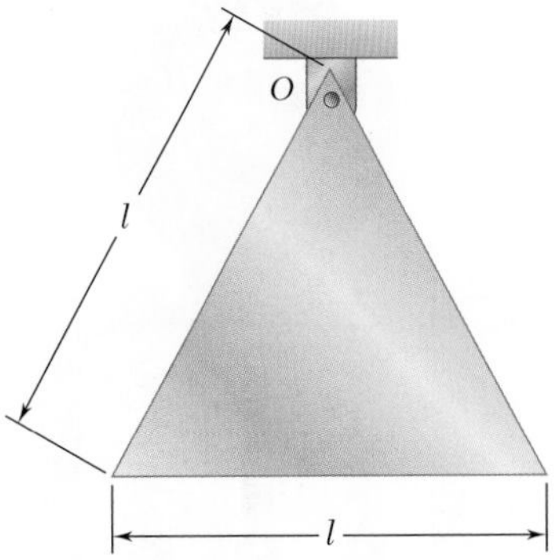

Fig. P19.162

19.163 An 20-g ball is connected to a paddle by means of an elastic cord AB of constant k = 7.5 N/m. Knowing that the paddle is moved vertically according to the relation $\delta = \delta_m \sin \omega_f t$, where δ_m = 200 mm, determine the maximum allowable circular frequency ω_f if the cord is not to become slack.

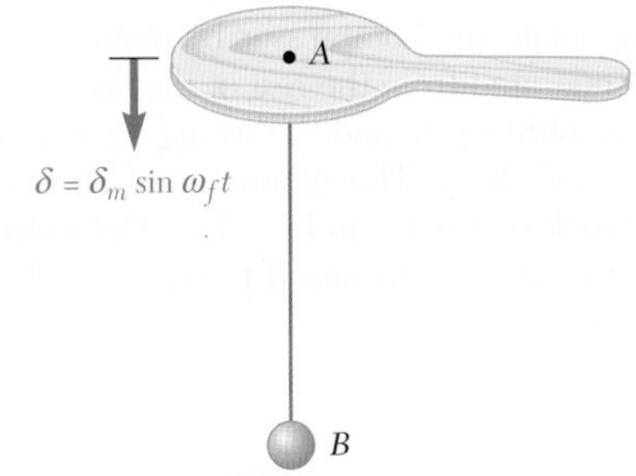

Fig. *P19.163*

19.164 The block shown is depressed 30 mm. from its equilibrium position and released. Knowing that after 10 cycles the maximum displacement of the block is 12.5 mm, determine (*a*) the damping factor c/c_c, (*b*) the value of the coefficient of viscous damping. (*Hint:* See Probs. 19.129 and 19.130.)

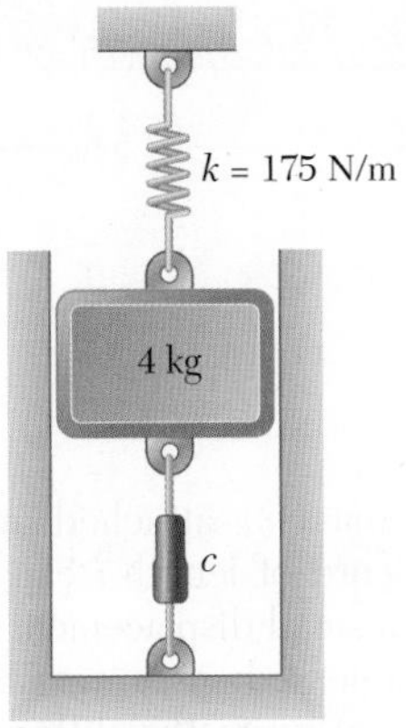

Fig. P19.164

19.165 A 2-kg uniform rod is supported by a pin at O and a spring at A and is connected to a dashpot at B. Determine (*a*) the differential equation of motion for small oscillations, (*b*) the angle that the rod will form with the horizontal 5 s after end B has been pushed 25 mm down and released.

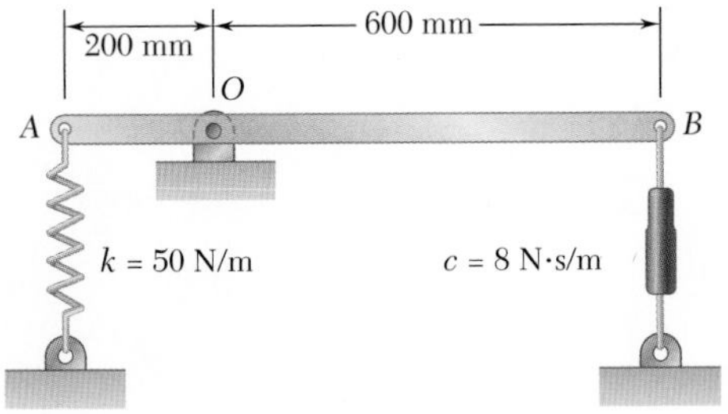

Fig. P19.165

19.166 A 400-kg motor supported by four springs, each of constant 150 kN/m, and a dashpot of constant $c = 6500$ N · s/m is constrained to move vertically. Knowing that the unbalance of the rotor is equivalent to a 23-g mass located at a distance of 100 mm from the axis of rotation, determine for a speed of 800 rpm (*a*) the amplitude of the fluctuating force transmitted to the foundation, (*b*) the amplitude of the vertical motion of the motor.

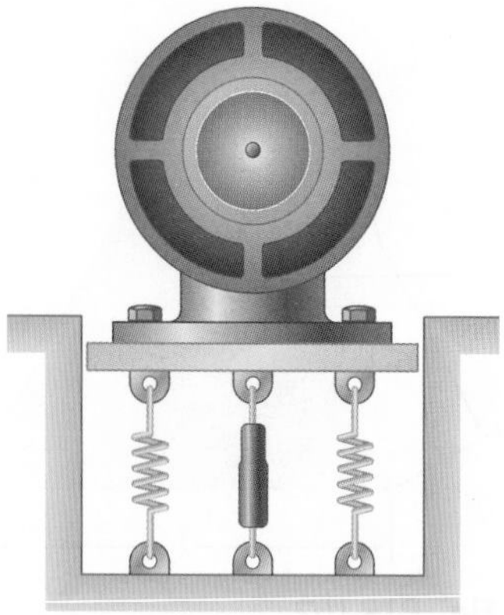

Fig. P19.166

19.167 The compressor shown has a mass of 250 kg and operates at 2000 rpm. At this operating condition, the force transmitted to the ground is excessively high and is found to be $mr\omega_f^2$, where mr is the unbalance and ω_f is the forcing frequency. To fix this problem, it is proposed to isolate the compressor by mounting it on a square concrete block separated from the rest of the floor as shown. The density of concrete is 2400 kg/m^3 and the spring constant for the soil is found to be 80×10^6 N/m. The geometry of the compressor leads to choosing a block that is 1.5 m by 1.5 m. Determine the depth h that will reduce the force transmitted to the ground by 75 percent.

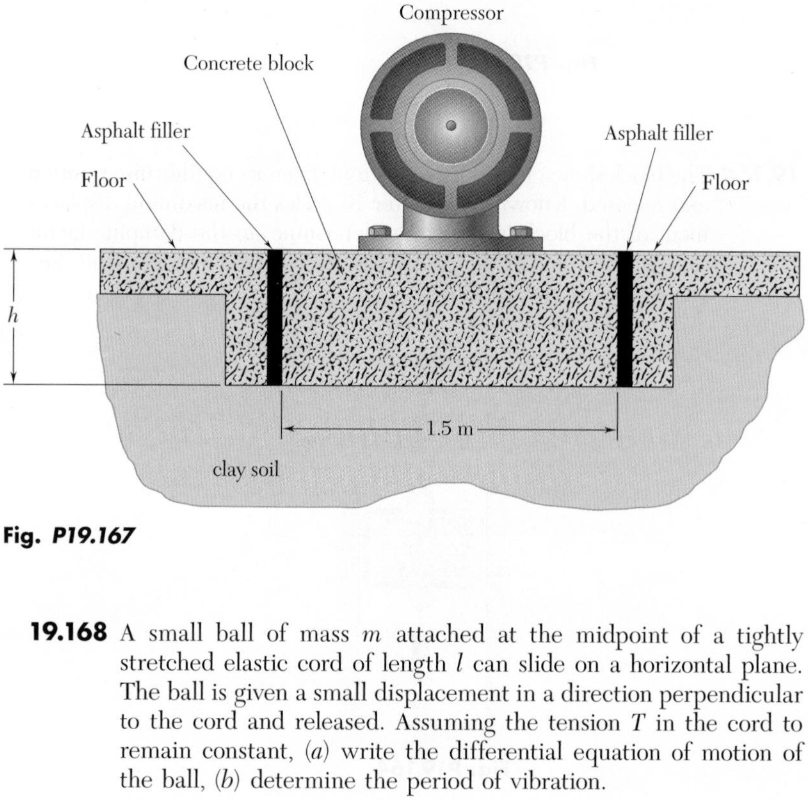

Fig. P19.167

19.168 A small ball of mass m attached at the midpoint of a tightly stretched elastic cord of length l can slide on a horizontal plane. The ball is given a small displacement in a direction perpendicular to the cord and released. Assuming the tension T in the cord to remain constant, (*a*) write the differential equation of motion of the ball, (*b*) determine the period of vibration.

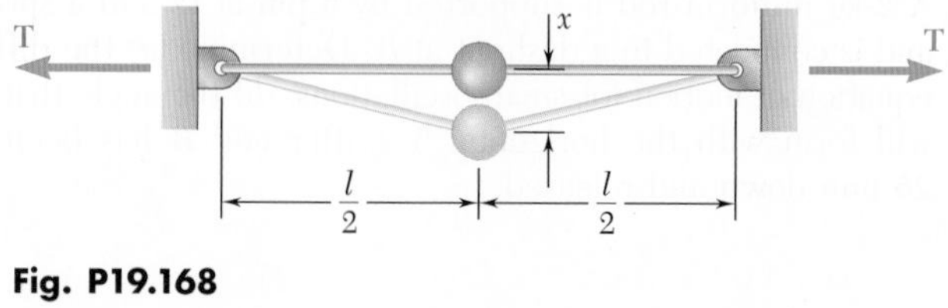

Fig. P19.168

19.169 A certain vibrometer used to measure vibration amplitudes consists essentially of a box containing a slender rod to which a mass m is attached; the natural frequency of the mass-rod system is known to be 5 Hz. When the box is rigidly attached to the casing of a motor rotating at 600 rpm, the mass is observed to vibrate with an amplitude of 1.5 mm relative to the box. Determine the amplitude of the vertical motion of the motor.

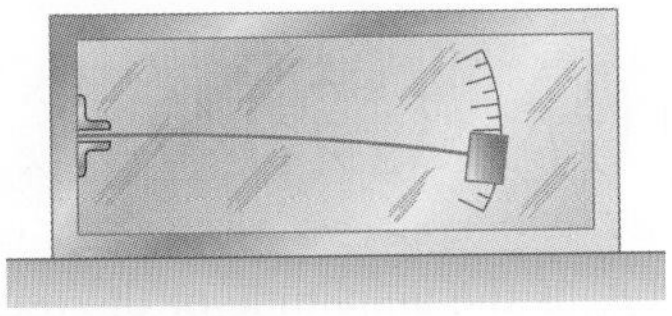

Fig. P19.169

19.170 If either a simple or a compound pendulum is used to determine experimentally the acceleration of gravity g, difficulties are encountered. In the case of the simple pendulum, the string is not truly weightless, while in the case of the compound pendulum, the exact location of the mass center is difficult to establish. In the case of a compound pendulum, the difficulty can be eliminated by using a reversible, or Kater, pendulum. Two knife edges A and B are placed so that they are obviously not at the same distance from the mass center G, and the distance l is measured with great precision. The position of a counterweight D is then adjusted so that the period of oscillation τ is the same when either knife edge is used. Show that the period τ obtained is equal to that of a true simple pendulum of length l and that $g = 4\pi^2 l/\tau^2$.

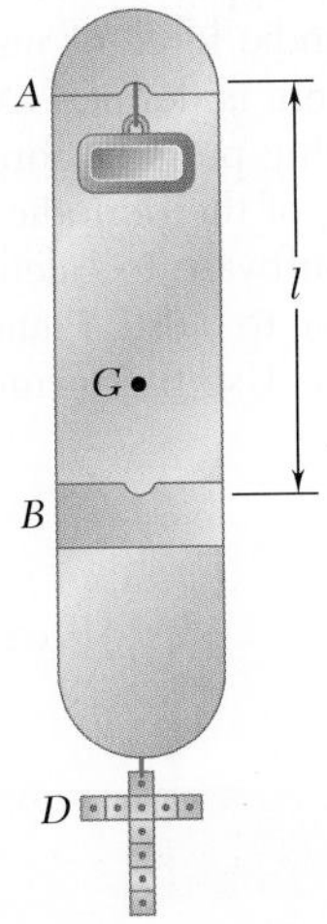

Fig. P19.170

COMPUTER PROBLEMS

19.C1 By expanding the integrand in Eq. (19.19) into a series of even powers of sin ϕ and integrating, it can be shown that the period of a simple pendulum of length l can be approximated by the expression

$$\tau_n = 2\pi\sqrt{\frac{l}{g}}\left[1 + \left(\frac{1}{2}\right)^2 c^2 + \left(\frac{1 \times 3}{2 \times 4}\right)^2 c^4 + \left(\frac{1 \times 3 \times 5}{2 \times 4 \times 6}\right)^2 c^6 + \cdots\right]$$

where $c = \sin \frac{1}{2}\theta_m$ and θ_m is the amplitude of the oscillations. Use computational software to calculate the sum of the series in brackets, using successively 1, 2, 4, 8, and 16 terms, for values of θ_m from 30 to 120° using 30° increments.

19.C2 The force-deflection equation for a class of nonlinear springs fixed at one end is $F = 5x^{1/n}$, where F is the magnitude, expressed in newtons, of the force applied at the other end of the spring and x is the deflection expressed in meters. Knowing that a block of mass m is suspended from the spring and is given a small downward displacement from its equilibrium position, use computational software to calculate and plot the frequency of vibration of the block for values of m equal to 0.2, 0.6, and 1.0 kg and values of n from 1 to 2. Assume that the slope of the force-deflection curve at the point corresponding to $F = mg$ can be used as an equivalent spring constant.

19.C3 A machine element supported by springs and connected to a dashpot is subjected to a periodic force of magnitude $P = P_m \sin \omega_f t$. The *transmissibility* T_m of the system is defined as the ratio F_m/P_m of the maximum value F_m of the fluctuating periodic force transmitted to the foundation to the maximum value P_m of the periodic force applied to the machine element. Use computational software to calculate and plot the value of T_m for frequency ratios ω_f/ω_n equal to 0.8, 1.4, and 2.0 and for damping factors c/c_c equal to 0, 1, and 2. (*Hint:* Use the formula given in Prob. 19.147.)

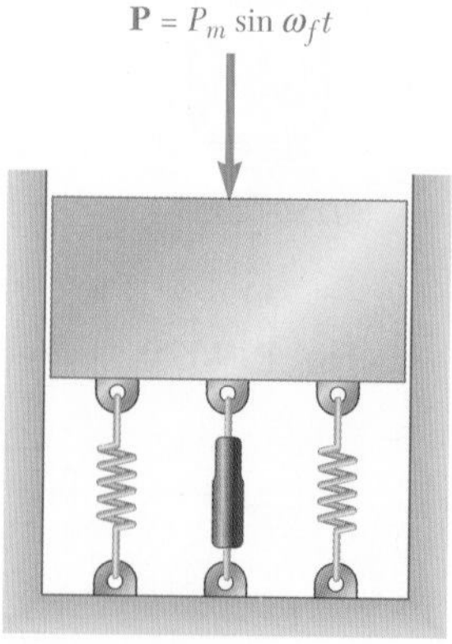

Fig. P19.C3

19.C4 A 15-kg motor is supported by four springs, each of constant 60 kN/m. The unbalance of the motor is equivalent to a mass of 20 g located 125 mm from the axis of rotation. Knowing that the motor is constrained to move vertically, use computational software to calculate and plot the amplitude of the vibration and the maximum acceleration of the motor for motor speeds of 1000 to 2500 rpm.

19.C5 Solve Prob. 19.C4, assuming that a dashpot having a coefficient of damping $c = 2.5\ \text{kN} \cdot \text{s/m}$ has been connected to the motor base and to the ground.

19.C6 A small trailer and its load have a total mass of 250 kg. The trailer is supported by two springs, each of constant 10 kN/m, and is pulled over a road, the surface of which can be approximated by a sine curve with an amplitude of 40 mm and a wave length of 5 m (i.e., the distance between successive crests is 5 m and the vertical distance from crest to trough is 80 mm). (*a*) Neglecting the mass of the wheels and assuming that the wheels stay in contact with the ground, use computational software to calculate and plot the amplitude of the vibration and the maximum vertical acceleration of the trailer for speeds of 10 to 80 km/h. (*b*) Determine the range of values of the speed of the trailer for which the wheels will lose contact with the ground.

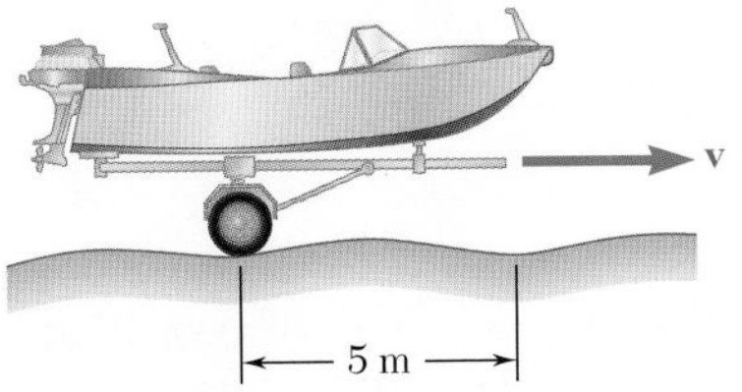

Fig. P19.C6

Appendix A

Some Useful Definitions and Properties of Vector Algebra

The following definitions and properties of vector algebra were discussed fully in Chaps. 2 and 3 of *Vector Mechanics for Engineers: Statics.* They are summarized here for the convenience of the reader, with references to the appropriate sections of the *Statics* volume. Equation and illustration numbers are those used in the original presentation.

A.1 ADDITION OF VECTORS (SECS. 2.3 AND 2.4)

Vectors are defined as *mathematical expressions possessing magnitude and direction, which add according to the parallelogram law.* Thus the sum of two vectors **P** and **Q** is obtained by attaching the two vectors to the same point *A* and constructing a parallelogram, using **P** and **Q** as two sides of the parallelogram (Fig. A.2). The diagonal that passes through *A* represents the sum of the vectors **P** and **Q**, and this sum is denoted by **P** + **Q**. Vector addition is *associative* and *commutative.*

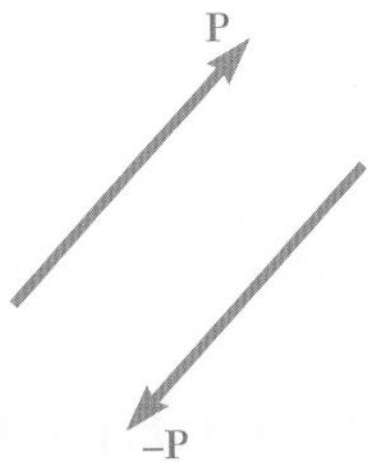

Fig. A.1

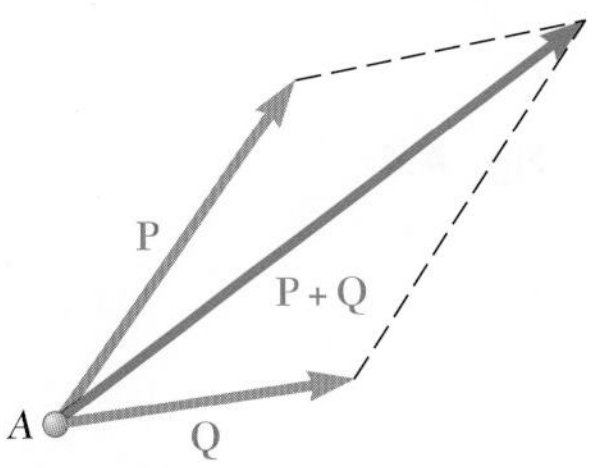

Fig. A.2

The *negative vector* of a given vector **P** is defined as a vector having the same magnitude *P* and a direction opposite to that of **P** (Fig. A.1); the negative of the vector **P** is denoted by −**P**. Clearly, we have

$$\mathbf{P} + (-\mathbf{P}) = 0$$

A.2 PRODUCT OF A SCALAR AND A VECTOR (SEC. 2.4)

The product $k\mathbf{P}$ of a scalar k and a vector $\mathbf{P}$ is defined as a vector having the same direction as $\mathbf{P}$ (if k is positive), or a direction opposite to that of $\mathbf{P}$ (if k is negative), and a magnitude equal to the product of the magnitude P and the absolute value of k (Fig. A.3).

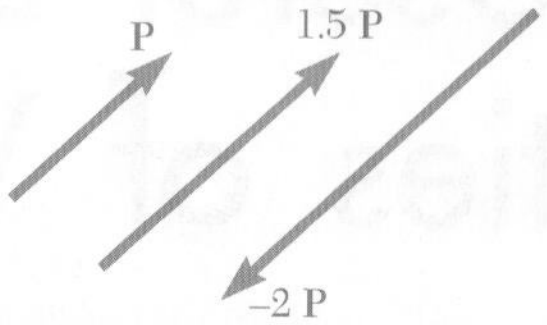

Fig. A.3

A.3 UNIT VECTORS. RESOLUTION OF A VECTOR INTO RECTANGULAR COMPONENTS (SECS. 2.7 AND 2.12)

The vectors **i**, **j**, and **k**, called *unit vectors*, are defined as vectors of magnitude 1, directed, respectively, along the positive x, y, and z axes (Fig. A.4).

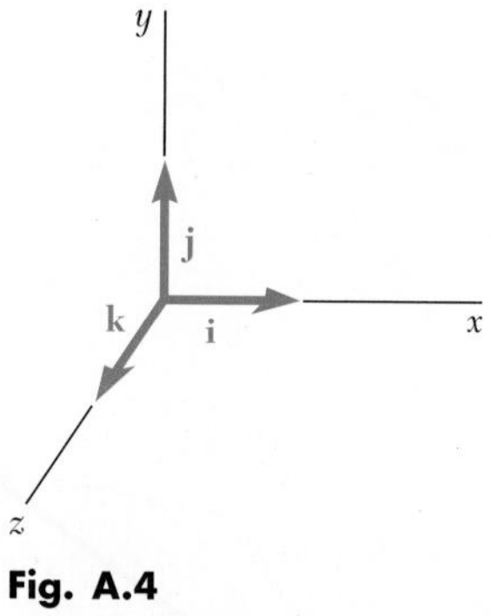

Fig. A.4

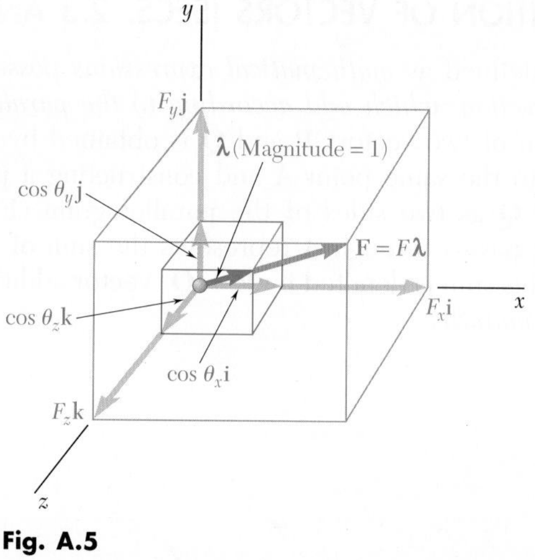

Fig. A.5

Denoting by F_x, F_y, and F_z, the scalar components of a vector $\mathbf{F}$, we have (Fig. A.5)

$$\mathbf{F} = F_x\mathbf{i} + F_y\mathbf{j} + F_z\mathbf{k} \tag{2.20}$$

In the particular case of a unit vector $\boldsymbol{\lambda}$ directed along a line forming angles θ_x, θ_y, and θ_z with the coordinate axes, we have

$$\boldsymbol{\lambda} = \cos\theta_x\mathbf{i} + \cos\theta_y\mathbf{j} + \cos\theta_z\mathbf{k} \tag{2.22}$$

A.4 VECTOR PRODUCT OF TWO VECTORS (SECS. 3.4 AND 3.5)

The vector product, or *cross product*, of two vectors $\mathbf{P}$ and $\mathbf{Q}$ is defined as the vector

$$\mathbf{V} = \mathbf{P} \times \mathbf{Q}$$

which satisfies the following conditions:

1. The line of action of **V** is perpendicular to the plane containing **P** and **Q** (Fig. A.6).
2. The magnitude of **V** is the product of the magnitudes of **P** and **Q** and of the sine of the angle θ formed by **P** and **Q** (the measure of which will always be 180° or less); we thus have
$$V = PQ \sin\theta \qquad (3.1)$$
3. The direction of **V** is obtained from the *right-hand rule.* Close your right hand and hold it so that your fingers are curled in the same sense as the rotation through θ which brings the vector **P** in line with the vector **Q**; your thumb will then indicate the direction of the vector **V** (Fig. A.6*b*). Note that if **P** and **Q** do not have a common point of application, they should first be redrawn from the same point. The three vectors **P**, **Q**, and **V**—taken in that order—are said to form a *right-handed triad.*

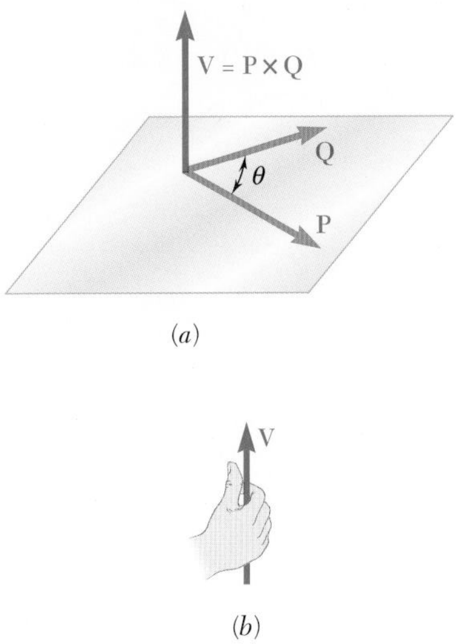

Fig. A.6

Vector products are *distributive* but *not commutative.* We have
$$\mathbf{Q} \times \mathbf{P} = -(\mathbf{P} \times \mathbf{Q}) \qquad (3.4)$$

Vector Products of Unit Vectors. It follows from the definition of the vector product of two vectors that
$$\begin{array}{lll} \mathbf{i} \times \mathbf{i} = 0 & \mathbf{j} \times \mathbf{i} = -\mathbf{k} & \mathbf{k} \times \mathbf{i} = \mathbf{j} \\ \mathbf{i} \times \mathbf{j} = \mathbf{k} & \mathbf{j} \times \mathbf{j} = 0 & \mathbf{k} \times \mathbf{j} = -\mathbf{i} \\ \mathbf{i} \times \mathbf{k} = -\mathbf{j} & \mathbf{j} \times \mathbf{k} = \mathbf{i} & \mathbf{k} \times \mathbf{k} = 0 \end{array} \qquad (3.7)$$

Rectangular Components of Vector Product. Resolving the vectors **P** and **Q** into rectangular components, we obtain the following expressions for the components of their vector product **V**:
$$\begin{aligned} V_x &= P_yQ_z - P_zQ_y \\ V_y &= P_zQ_x - P_xQ_z \\ V_z &= P_xQ_y - P_yQ_x \end{aligned} \qquad (3.9)$$

In determinant form, we have
$$\mathbf{V} = \mathbf{P} \times \mathbf{Q} = \begin{vmatrix} \mathbf{i} & \mathbf{j} & \mathbf{k} \\ P_x & P_y & P_z \\ Q_x & Q_y & Q_z \end{vmatrix} \qquad (3.10)$$

A.5 MOMENT OF A FORCE ABOUT A POINT (SECS. 3.6 AND 3.8)

The moment of a force **F** (or, more generally, of a vector **F**) about a point O is defined as the vector product
$$\mathbf{M}_O = \mathbf{r} \times \mathbf{F} \qquad (3.11)$$
where **r** denotes the *position vector* of the point of application A of **F** (Fig. A.7*a*).

According to the definition of the vector product of two vectors given in Sec. A.4, the moment $\mathbf{M}_O$ must be perpendicular to the plane containing O and the force **F**. Its magnitude is equal to
$$M_O = rF \sin\theta = Fd \qquad (3.12)$$

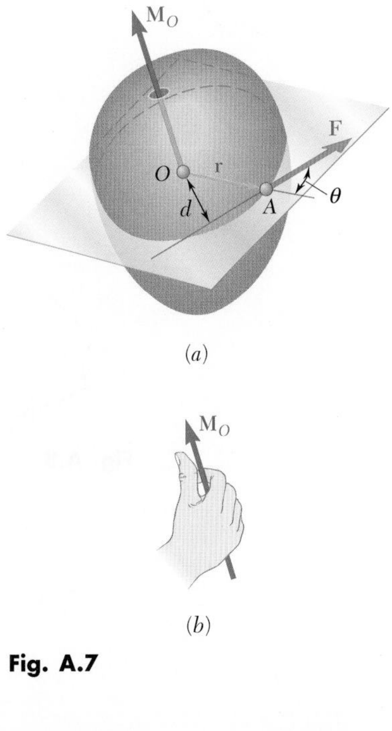

Fig. A.7

where d is the perpendicular distance from O to the line of action of **F**, and its sense is defined by the sense of the rotation which would bring the vector **r** in line with the vector **F**; this rotation should be viewed as *counterclockwise* by an observer located at the tip of $\mathbf{M}_O$. Another way of defining the sense of $\mathbf{M}_O$ is furnished by a variation of the *right-hand rule:* Close your right hand and hold it so that your fingers are curled in the sense of the rotation that **F** would impart to the rigid body about a fixed axis directed along the line of action of $\mathbf{M}_O$; your thumb will indicate the sense of the moment $\mathbf{M}_O$ (Fig. A.7*b*).

Rectangular Components of the Moment of a Force. Denoting by x, y, and z the coordinates of the point of application A of **F**, we obtain the following expressions for the components of the moment $\mathbf{M}_O$ of **F**:

$$\begin{aligned} M_x &= yF_z - zF_y \\ M_y &= zF_x - xF_z \\ M_z &= xF_y - yF_x \end{aligned} \tag{3.18}$$

In determinant form, we have

$$\mathbf{M}_O = \mathbf{r} \times \mathbf{F} = \begin{vmatrix} \mathbf{i} & \mathbf{j} & \mathbf{k} \\ x & y & z \\ F_x & F_y & F_z \end{vmatrix} \tag{3.19}$$

To compute the moment $\mathbf{M}_B$ about an arbitrary point B of a force **F** applied at A, we must use the vector $\mathbf{r}_{A/B} = \mathbf{r}_A - \mathbf{r}_B$ drawn from B to A instead of the vector **r**. We write

$$\mathbf{M}_B = \mathbf{r}_{A/B} \times \mathbf{F} = (\mathbf{r}_A - \mathbf{r}_B) \times \mathbf{F} \tag{3.20}$$

or, using the determinant form,

$$\mathbf{M}_B = \begin{vmatrix} \mathbf{i} & \mathbf{j} & \mathbf{k} \\ x_{A/B} & y_{A/B} & z_{A/B} \\ F_x & F_y & F_z \end{vmatrix} \tag{3.21}$$

where $x_{A/B}$, $y_{A/B}$, $z_{A/B}$ are the components of the vector $\mathbf{r}_{A/B}$:

$$x_{A/B} = x_A - x_B \qquad y_{A/B} = y_A - y_B \qquad z_{A/B} = z_A - z_B$$

A.6 SCALAR PRODUCT OF TWO VECTORS (SEC. 3.9)

The scalar product, or *dot product,* of two vectors **P** and **Q** is defined as the product of the magnitudes of **P** and **Q** and of the cosine of the angle θ formed by **P** and **Q** (Fig. A.8). The scalar product of **P** and **Q** is denoted by $\mathbf{P} \cdot \mathbf{Q}$. We write

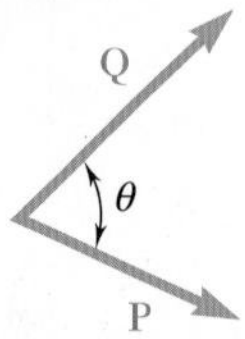

Fig. A.8

$$\mathbf{P} \cdot \mathbf{Q} = PQ \cos \theta \tag{3.24}$$

Scalar products are *commutative* and *distributive.*

Scalar Products of Unit Vectors. It follows from the definition of the scalar product of two vectors that

$$\begin{aligned} \mathbf{i} \cdot \mathbf{i} &= 1 \qquad & \mathbf{j} \cdot \mathbf{j} &= 1 \qquad & \mathbf{k} \cdot \mathbf{k} &= 1 \\ \mathbf{i} \cdot \mathbf{j} &= 0 \qquad & \mathbf{j} \cdot \mathbf{k} &= 0 \qquad & \mathbf{k} \cdot \mathbf{i} &= 0 \end{aligned} \tag{3.29}$$

Scalar Product Expressed in Terms of Rectangular Components. Resolving the vectors **P** and **Q** into rectangular components, we obtain

$$\mathbf{P} \cdot \mathbf{Q} = P_xQ_x + P_yQ_y + P_zQ_z \tag{3.30}$$

Angle Formed by Two Vectors. It follows from (3.24) and (3.29) that

$$\cos\theta = \frac{\mathbf{P} \cdot \mathbf{Q}}{PQ} = \frac{P_xQ_x + P_yQ_y + P_zQ_z}{PQ} \tag{3.32}$$

Projection of a Vector on a Given Axis. The projection of a vector **P** on the axis OL defined by the unit vector $\boldsymbol{\lambda}$ (Fig. A.9) is

$$P_{OL} = OA = \mathbf{P} \cdot \boldsymbol{\lambda} \tag{3.36}$$

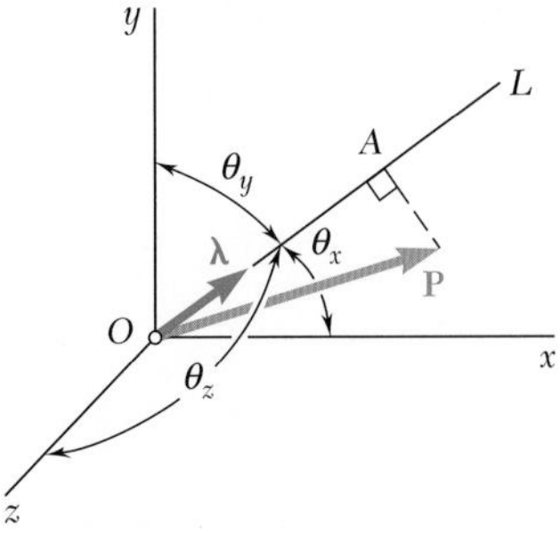

Fig. A.9

A.7 MIXED TRIPLE PRODUCT OF THREE VECTORS (SEC. 3.10)

The mixed triple product of the three vectors **S**, **P**, and **Q** is defined as the scalar expression

$$\mathbf{S} \cdot (\mathbf{P} \times \mathbf{Q}) \tag{3.38}$$

obtained by forming the scalar product of **S** with the vector product of **P** and **Q**. Mixed triple products are invariant under *circular permutations* but change sign under any other permutation:

$$\begin{aligned}\mathbf{S} \cdot (\mathbf{P} \times \mathbf{Q}) &= \mathbf{P} \cdot (\mathbf{Q} \times \mathbf{S}) = \mathbf{Q} \cdot (\mathbf{S} \times \mathbf{P}) \\ &= -\mathbf{S} \cdot (\mathbf{Q} \times \mathbf{P}) = -\mathbf{P} \cdot (\mathbf{S} \times \mathbf{Q}) = -\mathbf{Q} \cdot (\mathbf{P} \times \mathbf{S})\end{aligned} \tag{3.39}$$

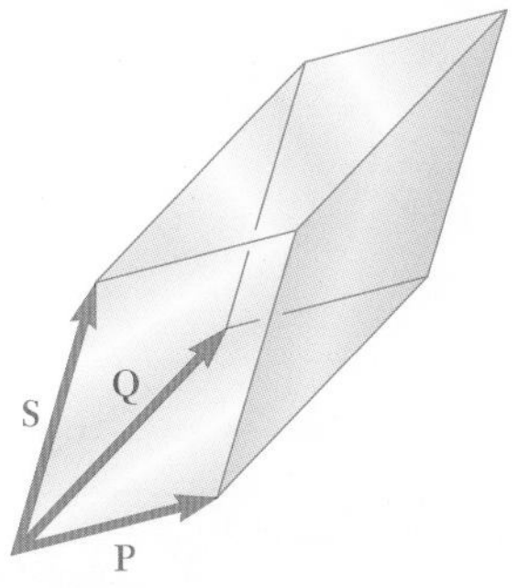

Fig. A.10

Mixed Triple Product Expressed in Terms of Rectangular Components. The mixed triple product of **S**, **P**, and **Q** may be expressed in the form of a determinant

$$\mathbf{S} \cdot (\mathbf{P} \times \mathbf{Q}) = \begin{vmatrix} S_x & S_y & S_z \\ P_x & P_y & P_z \\ Q_x & Q_y & Q_z \end{vmatrix} \tag{3.41}$$

The mixed triple product $\mathbf{S} \cdot (\mathbf{P} \times \mathbf{Q})$ measures the volume of the parallelepiped having the vectors **S**, **P**, and **Q** for sides (Fig. A.10).

A.8 MOMENT OF A FORCE ABOUT A GIVEN AXIS (SEC. 3.11)

The moment M_{OL} of a force **F** (or, more generally, of a vector **F**) about an axis OL is defined as the projection OC on the axis OL of the moment $\mathbf{M}_O$ of **F** about O (Fig. A.11). Denoting by $\boldsymbol{\lambda}$ the unit vector along OL, we have

$$M_{OL} = \boldsymbol{\lambda} \cdot \mathbf{M}_O = \boldsymbol{\lambda} \cdot (\mathbf{r} \times \mathbf{F}) \tag{3.42}$$

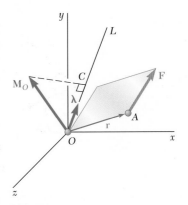

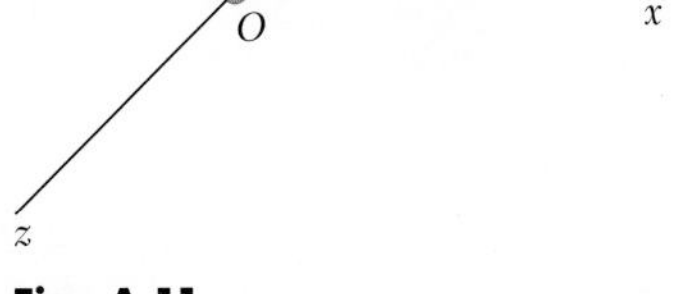

Fig. A.11

or, in determinant form,

$$M_{OL} = \begin{vmatrix} \lambda_x & \lambda_y & \lambda_z \\ x & y & z \\ F_x & F_y & F_z \end{vmatrix} \tag{3.43}$$

where $\lambda_x, \lambda_y, \lambda_z,$ = direction cosines of axis OL
x, y, z = coordinates of point of application of $\mathbf{F}$
F_x, F_y, F_z = components of force $\mathbf{F}$

The moments of the force $\mathbf{F}$ about the three coordinate axes are given by the expressions (3.18) obtained earlier for the rectangular components of the moment $\mathbf{M}_O$ of $\mathbf{F}$ about O:

$$\begin{aligned} M_x &= yF_z - zF_y \\ M_y &= zF_x - xF_z \\ M_z &= xF_y - yF_x \end{aligned} \tag{3.18}$$

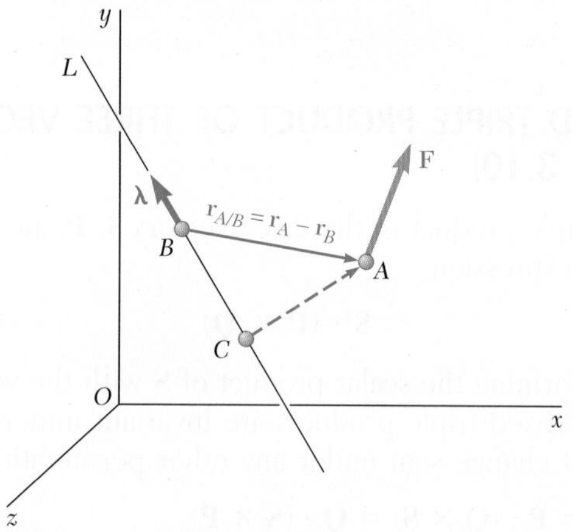

Fig. A.12

More generally, the moment of a force $\mathbf{F}$ applied at A about an axis which does not pass through the origin is obtained by choosing an arbitrary point B on the axis (Fig. A.12) and determining the projection on the axis BL of the moment $\mathbf{M}_B$ of $\mathbf{F}$ about B. We write

$$M_{BL} = \boldsymbol{\lambda} \cdot \mathbf{M}_B = \boldsymbol{\lambda} \cdot (\mathbf{r}_{A/B} \times \mathbf{F}) \tag{3.45}$$

where $\mathbf{r}_{A/B} = \mathbf{r}_A - \mathbf{r}_B$ represents the vector drawn from B to A. Expressing M_{BL} in the form of a determinant, we have

$$M_{BL} = \begin{vmatrix} \lambda_x & \lambda_y & \lambda_z \\ x_{A/B} & y_{A/B} & z_{A/B} \\ F_x & F_y & F_z \end{vmatrix} \tag{3.46}$$

where $\lambda_x, \lambda_y, \lambda_z$ = direction cosines of axis BL

$x_{A/B} = x_A - x_B, \; y_{A/B} = y_A - y_B, \; z_{A/B} = z_A - z_B$
F_x, F_y, F_z = components of force $\mathbf{F}$

It should be noted that the result obtained is independent of the choice of the point B on the given axis; the same result would have been obtained if point C had been chosen instead of B.

Appendix B

Moments of Inertia of Masses

MOMENTS OF INERTIA OF MASSES

B.1 MOMENT OF INERTIA OF A MASS

Consider a small mass Δm mounted on a rod of negligible mass which can rotate freely about an axis AA' (Fig. B.1*a*). If a couple is applied to the system, the rod and mass, assumed to be initially at rest, will start rotating about AA'. The details of this motion will be studied later in dynamics. At present, we wish only to indicate that the time required for the system to reach a given speed of rotation is proportional to the mass Δm and to the square of the distance r. The product $r^2\,\Delta m$ provides, therefore, a measure of the *inertia* of the system, that is, a measure of the resistance the system offers when we try to set it in motion. For this reason, the product $r^2\,\Delta m$ is called the *moment of inertia* of the mass Δm with respect to the axis AA'.

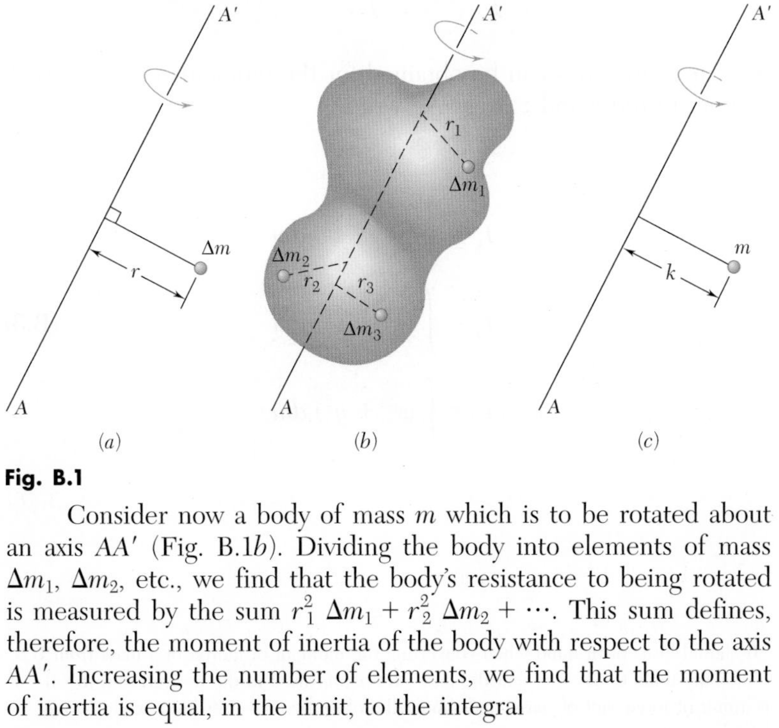

Fig. B.1

Consider now a body of mass m which is to be rotated about an axis AA' (Fig. B.1*b*). Dividing the body into elements of mass Δm_1, Δm_2, etc., we find that the body's resistance to being rotated is measured by the sum $r_1^2\,\Delta m_1 + r_2^2\,\Delta m_2 + \cdots$. This sum defines, therefore, the moment of inertia of the body with respect to the axis AA'. Increasing the number of elements, we find that the moment of inertia is equal, in the limit, to the integral

$$I = \int r^2\,dm \qquad \text{(B.1)}$$

The *radius of gyration* k of the body with respect to the axis AA' is defined by the relation

$$I = k^2 m \qquad \text{or} \qquad k = \sqrt{\frac{I}{m}} \tag{B.2}$$

The radius of gyration k represents, therefore, the distance at which the entire mass of the body should be concentrated if its moment of inertia with respect to AA' is to remain unchanged (Fig. B.1*c*). Whether it is kept in its original shape (Fig. B.1*b*) or whether it is concentrated as shown in Fig. B.1*c*, the mass m will react in the same way to a rotation, or *gyration*, about AA'.

If SI units are used, the radius of gyration k is expressed in meters and the mass m in kilograms, and thus the unit used for the moment of inertia of a mass is $kg \cdot m^2$. If U.S. customary units are used, the radius of gyration is expressed in feet and the mass in slugs (that is, in $lb \cdot s^2/ft$), and thus the derived unit used for the moment of inertia of a mass is $lb \cdot ft \cdot s^2$.†

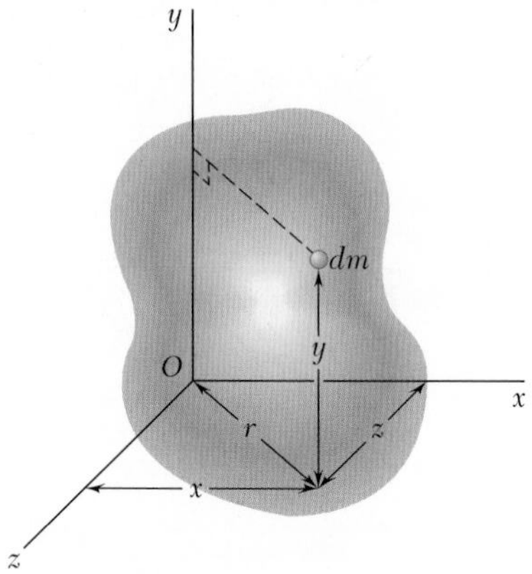

Fig. B.2

The moment of inertia of a body with respect to a coordinate axis can easily be expressed in terms of the coordinates x, y, z of the element of mass dm (Fig. B.2). Noting, for example, that the square of the distance r from the element dm to the y axis is $z^2 + x^2$, we express the moment of inertia of the body with respect to the y axis as

$$I_y = \int r^2\, dm = \int (z^2 + x^2)\, dm$$

Similar expressions can be obtained for the moments of inertia with respect to the x and z axes. We write

$$\begin{aligned} I_x &= \int (y^2 + z^2)\, dm \\ I_y &= \int (z^2 + x^2)\, dm \\ I_z &= \int (x^2 + y^2)\, dm \end{aligned} \tag{B.3}$$

Photo B.1 As you will discuss in your dynamics course, the rotational behavior of the camshaft shown is dependent upon the mass moment of inertia of the camshaft with respect to its axis of rotation.

†It should be kept in mind when converting the moment of inertia of a mass from U.S. customary units to SI units that the base unit *pound* used in the derived unit $lb \cdot ft \cdot s^2$ is a unit of force (*not* of mass) and should therefore be converted into newtons. We have

$$1\ lb \cdot ft \cdot s^2 = (4.45\ N)(0.3048\ m)(1\ s)^2 = 1.356\ N \cdot m \cdot s^2$$

or, since $1\ N = 1\ kg \cdot m/s^2$,

$$1\ lb \cdot ft \cdot s^2 = 1.356\ kg \cdot m^2$$

B.2 PARALLEL-AXIS THEOREM

Consider a body of mass m. Let $Oxyz$ be a system of rectangular coordinates whose origin is at the arbitrary point O, and $Gx'y'z'$ a system of parallel *centroidal axes,* that is, a system whose origin is at the center of gravity G of the body† and whose axes x', y', z' are parallel to the x, y, and z axes, respectively (Fig. B.3). Denoting by $\bar{x}, \bar{y}, \bar{z}$ the coordinates of G with respect to $Oxyz$, we write the following relations between the coordinates x, y, z of the element dm with respect to $Oxyz$ and its coordinates x', y', z' with respect to the centroidal axes $Gx'y'z'$:

$$x = x' + \bar{x} \qquad y = y' + \bar{y} \qquad z = z' + \bar{z} \tag{B.4}$$

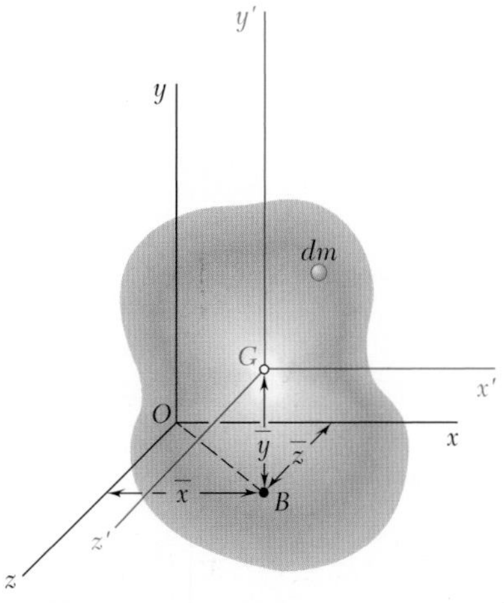

Fig. B.3

Referring to Eqs. (B.3), we can express the moment of inertia of the body with respect to the x axis as follows:

$$\begin{aligned} I_x &= \int (y^2 + z^2)\, dm = \int [(y' + \bar{y})^2 + (z' + \bar{z})^2]\, dm \\ &= \int (y'^2 + z'^2)\, dm + 2\bar{y} \int y'\, dm + 2\bar{z} \int z'\, dm + (\bar{y}^2 + \bar{z}^2) \int dm \end{aligned}$$

The first integral in this expression represents the moment of inertia $\bar{I}_{x'}$ of the body with respect to the centroidal axis x'; the second and third integrals represent the first moment of the body with respect to the $z'x'$ and $x'y'$ planes, respectively, and, since both planes contain G, the two integrals are *zero;* the last integral is equal to the total mass m of the body. We write, therefore,

$$I_x = \bar{I}_{x'} + m(\bar{y}^2 + \bar{z}^2) \tag{B.5}$$

and, similarly,

$$I_y = \bar{I}_{y'} + m(\bar{z}^2 + \bar{x}^2) \qquad I_z = \bar{I}_{z'} + m(\bar{x}^2 + \bar{y}^2) \tag{B.5'}$$

We easily verify from Fig. B.3 that the sum $\bar{z}^2 + \bar{x}^2$ represents the square of the distance OB, between the y and y' axes. Similarly, $\bar{y}^2 + \bar{z}^2$ and $\bar{x}^2 + \bar{y}^2$ represent the squares of the distance between the x and x' axes and the z and z' axes, respectively. Denoting by d the distance between an arbitrary axis AA' and the parallel centroidal axis BB' (Fig. B.4), we can, therefore, write the following general relation between the moment of inertia I of the body with respect to AA' and its moment of inertia $\bar{I}$ with respect to BB':

$$I = \bar{I} + md^2 \tag{B.6}$$

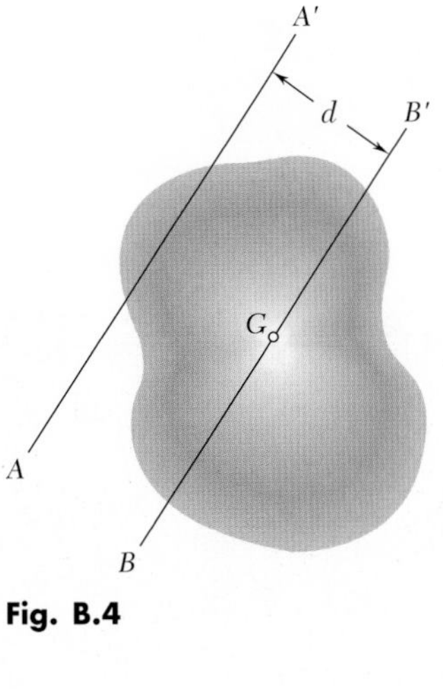

Fig. B.4

Expressing the moments of inertia in terms of the corresponding radii of gyration, we can also write

$$k^2 = \bar{k}^2 + d^2 \tag{B.7}$$

where k and $\bar{k}$ represent the radii of gyration of the body about AA' and BB', respectively.

†Note that the term *centroidal* is used here to define an axis passing through the center of gravity G of the body, whether or not G coincides with the centroid of the volume of the body.

B.3 MOMENTS OF INERTIA OF THIN PLATES

Consider a thin plate of uniform thickness t, which is made of a homogeneous material of density ρ (density = mass per unit volume). The mass moment of inertia of the plate with respect to an axis AA' *contained in the plane* of the plate (Fig. B.5a) is

$$I_{AA',\text{ mass}} = \int r^2\, dm$$

Since $dm = \rho t\, dA$, we write

$$I_{AA',\text{ mass}} = \rho t \int r^2\, dA$$

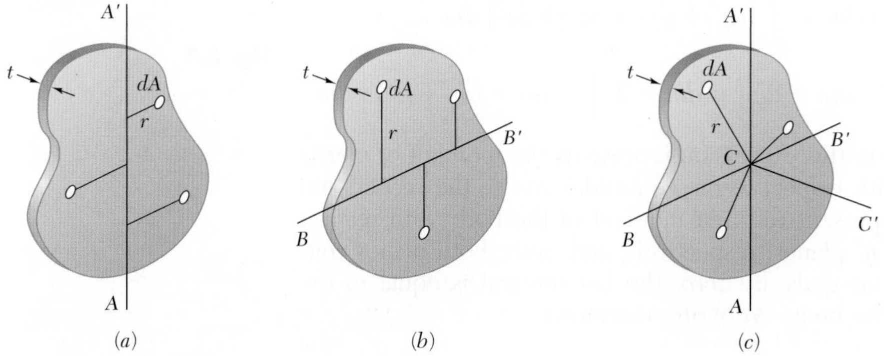

Fig. B.5

But r represents the distance of the element of area dA to the axis AA'; the integral is therefore equal to the moment of inertia of the area of the plate with respect to AA'. We have

$$I_{AA',\text{ mass}} = \rho t I_{AA',\text{ area}} \tag{B.8}$$

Similarly, for an axis BB' which is contained in the plane of the plate and is perpendicular to AA' (Fig. B.5b), we have

$$I_{BB',\text{ mass}} = \rho t I_{BB',\text{ area}} \tag{B.9}$$

Considering now the axis CC' which is *perpendicular* to the plate and passes through the point of intersection C of AA' and BB' (Fig. B.5c), we write

$$I_{CC',\text{ mass}} = \rho t J_{C,\text{ area}} \tag{B.10}$$

where J_C is the *polar* moment of inertia of the area of the plate with respect to point C.

Recalling the relation $J_C = I_{AA'} + I_{BB'}$ which exists between polar and rectangular moments of inertia of an area, we write the following relation between the mass moments of inertia of a thin plate:

$$I_{CC'} = I_{AA'} + I_{BB'} \tag{B.11}$$

Rectangular Plate. In the case of a rectangular plate of sides a and b (Fig. B.6), we obtain the following mass moments of inertia with respect to axes through the center of gravity of the plate:

$$I_{AA',\,\text{mass}} = \rho t I_{AA',\,\text{area}} = \rho t(\tfrac{1}{12}a^3b)$$
$$I_{BB',\,\text{mass}} = \rho t I_{BB',\,\text{area}} = \rho t(\tfrac{1}{12}ab^3)$$

Observing that the product ρabt is equal to the mass m of the plate, we write the mass moments of inertia of a thin rectangular plate as follows:

$$I_{AA'} = \tfrac{1}{12}ma^2 \qquad I_{BB'} = \tfrac{1}{12}mb^2 \tag{B.12}$$
$$I_{CC'} = I_{AA'} + I_{BB'} = \tfrac{1}{12}m(a^2 + b^2) \tag{B.13}$$

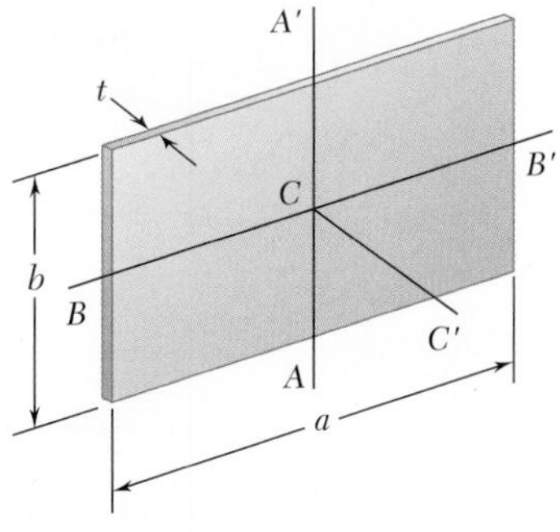

Fig. B.6

Circular Plate. In the case of a circular plate, or disk, of radius r (Fig. B.7), we write

$$I_{AA',\,\text{mass}} = \rho t I_{AA',\,\text{area}} = \rho t(\tfrac{1}{4}\pi r^4)$$

Observing that the product $\rho\pi r^2 t$ is equal to the mass m of the plate and that $I_{AA'} = I_{BB'}$, we write the mass moments of inertia of a circular plate as follows:

$$I_{AA'} = I_{BB'} = \tfrac{1}{4}mr^2 \tag{B.14}$$
$$I_{CC'} = I_{AA'} + I_{BB'} = \tfrac{1}{2}mr^2 \tag{B.15}$$

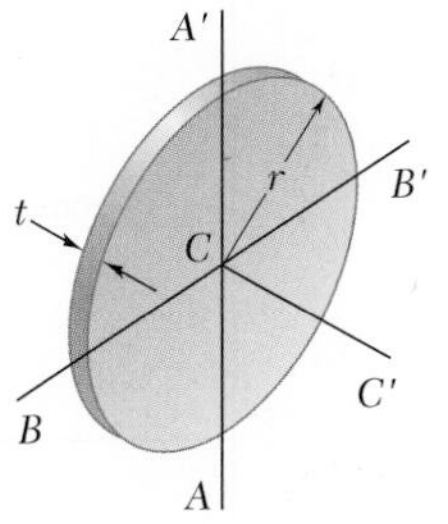

Fig. B.7

B.4 DETERMINATION OF THE MOMENT OF INERTIA OF A THREE-DIMENSIONAL BODY BY INTEGRATION

The moment of inertia of a three-dimensional body is obtained by evaluating the integral $I = \int r^2\,dm$. If the body is made of a homogeneous material of density ρ, the element of mass dm is equal to ρdV and we can write $I = \rho \int r^2\,dV$. This integral depends only upon the shape of the body. Thus, in order to compute the moment of inertia of a three-dimensional body, it will generally be necessary to perform a triple, or at least a double, integration.

However, if the body possesses two planes of symmetry, it is usually possible to determine the body's moment of inertia with a single integration by choosing as the element of mass dm a thin slab which is perpendicular to the planes of symmetry. In the case of bodies of revolution, for example, the element of mass would be a thin disk (Fig. B.8). Using formula (B.15), the moment of inertia of the disk with respect to the axis of revolution can be expressed as indicated in Fig. B.8. Its moment of inertia with respect to each of the other two coordinate axes is obtained by using formula (B.14) and the parallel-axis theorem. Integration of the expression obtained yields the desired moment of inertia of the body.

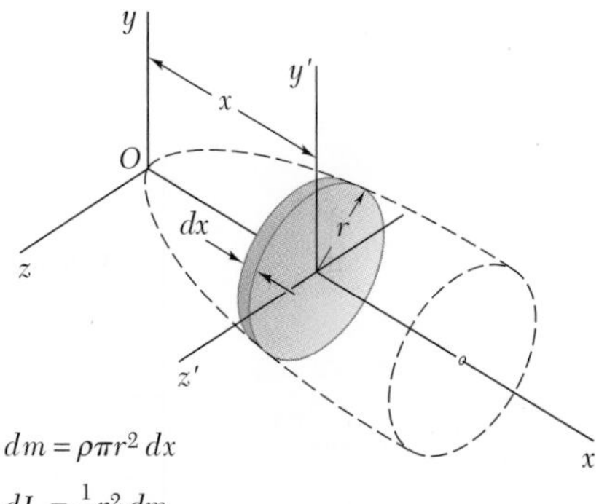

$dm = \rho\pi r^2\,dx$

$dI_x = \frac{1}{2}r^2\,dm$

$dI_y = dI_{y'} + x^2\,dm = \left(\frac{1}{4}r^2 + x^2\right)dm$

$dI_z = dI_{z'} + x^2\,dm = \left(\frac{1}{4}r^2 + x^2\right)dm$

Fig. B.8 Determination of the moment of inertia of a body of revolution.

B.5 MOMENTS OF INERTIA OF COMPOSITE BODIES

The moments of inertia of a few common shapes are shown in Fig. B.9. For a body consisting of several of these simple shapes, the moment of inertia of the body with respect to a given axis can be obtained by first computing the moments of inertia of its component parts about the desired axis and then adding them together. As was the case for areas, the radius of gyration of a composite body *cannot* be obtained by adding the radii of gyration of its component parts.

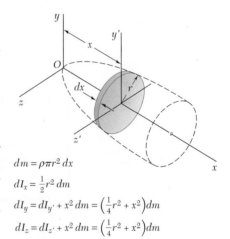

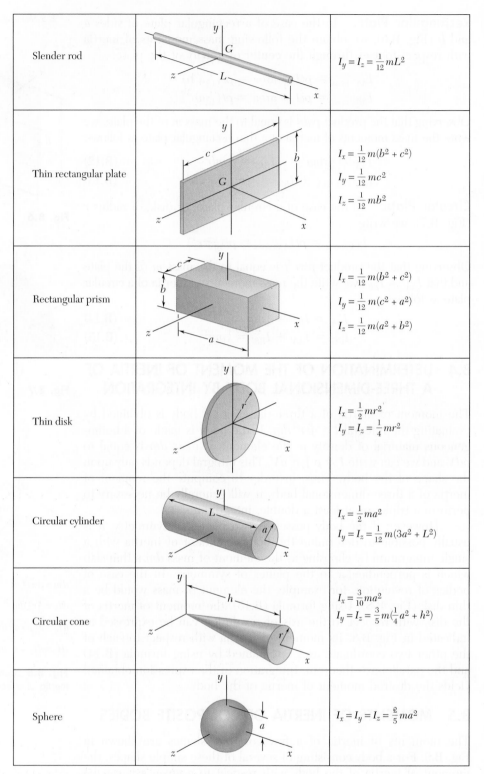

Slender rod	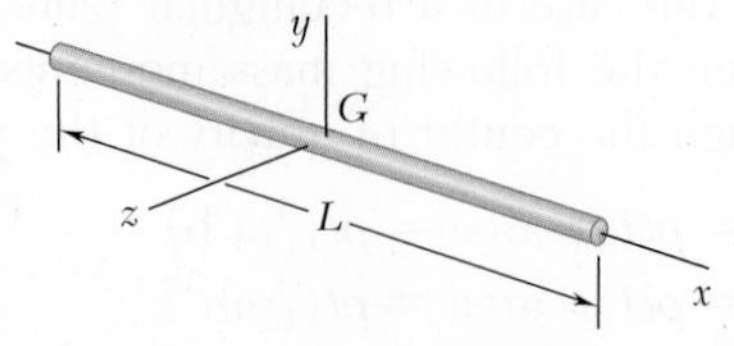	$I_y = I_z = \frac{1}{12}mL^2$
Thin rectangular plate	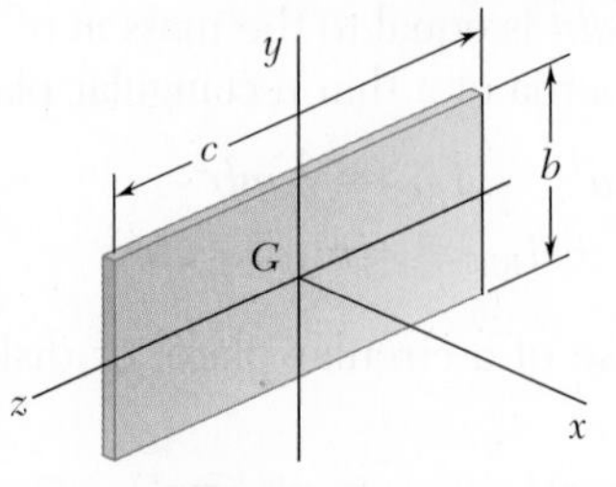	$I_x = \frac{1}{12}m(b^2 + c^2)$ $I_y = \frac{1}{12}mc^2$ $I_z = \frac{1}{12}mb^2$
Rectangular prism	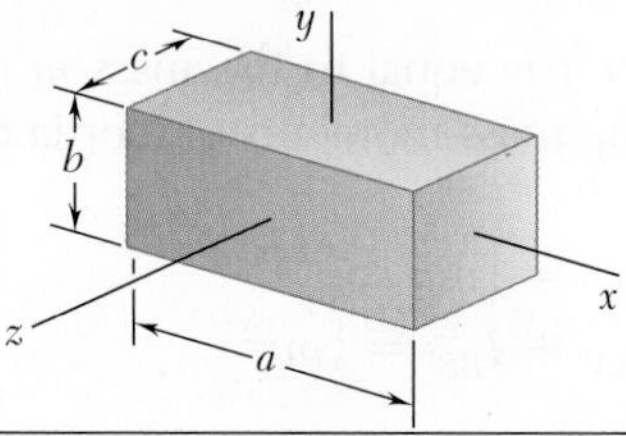	$I_x = \frac{1}{12}m(b^2 + c^2)$ $I_y = \frac{1}{12}m(c^2 + a^2)$ $I_z = \frac{1}{12}m(a^2 + b^2)$
Thin disk	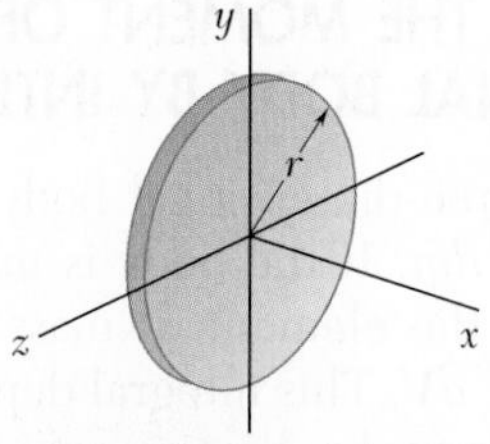	$I_x = \frac{1}{2}mr^2$ $I_y = I_z = \frac{1}{4}mr^2$
Circular cylinder	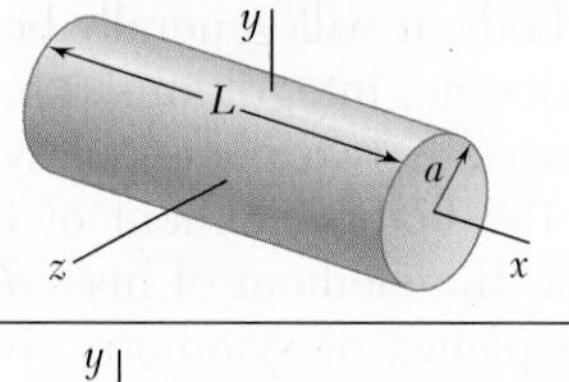	$I_x = \frac{1}{2}ma^2$ $I_y = I_z = \frac{1}{12}m(3a^2 + L^2)$
Circular cone	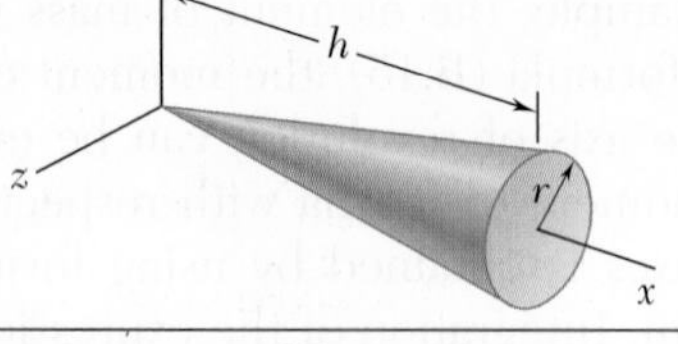	$I_x = \frac{3}{10}ma^2$ $I_y = I_z = \frac{3}{5}m(\frac{1}{4}a^2 + h^2)$
Sphere	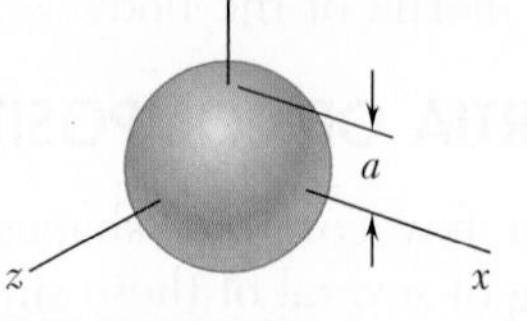	$I_x = I_y = I_z = \frac{2}{5}ma^2$

Fig. B.9 Mass moments of inertia of common geometric shapes.

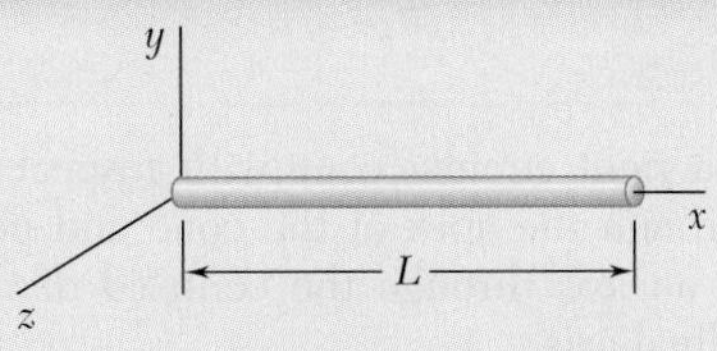

SAMPLE PROBLEM B.1

Determine the moment of inertia of a slender rod of length L and mass m with respect to an axis which is perpendicular to the rod and passes through one end of the rod.

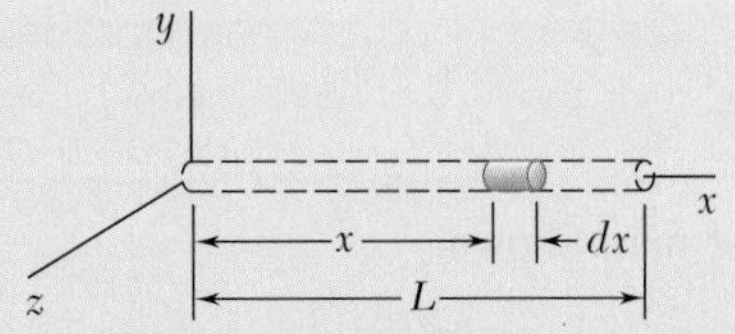

SOLUTION

Choosing the differential element of mass shown, we write

$$dm = \frac{m}{L}\,dx$$

$$I_y = \int x^2\,dm = \int_0^L x^2 \frac{m}{L}\,dx = \left[\frac{m}{L}\frac{x^3}{3}\right]_0^L \qquad I_y = \tfrac{1}{3}mL^2 \blacktriangleleft$$

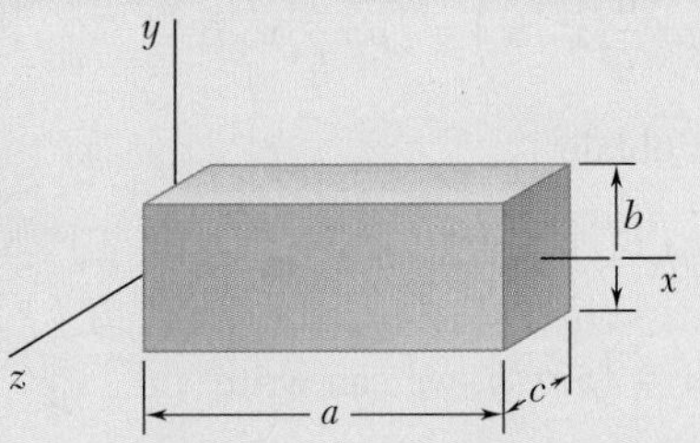

SAMPLE PROBLEM B.2

For the homogeneous rectangular prism shown, determine the moment of inertia with respect to the z axis.

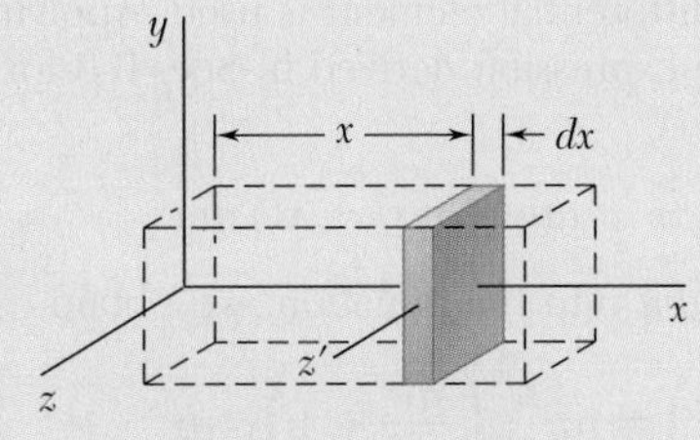

SOLUTION

We choose as the differential element of mass the thin slab shown; thus

$$dm = \rho bc\,dx$$

Referring to Sec. B.3, we find that the moment of inertia of the element with respect to the z' axis is

$$dI_{z'} = \tfrac{1}{12}b^2\,dm$$

Applying the parallel-axis theorem, we obtain the mass moment of inertia of the slab with respect to the z axis.

$$dI_z = dI_{z'} + x^2\,dm = \tfrac{1}{12}b^2\,dm + x^2\,dm = (\tfrac{1}{12}b^2 + x^2)\,\rho bc\,dx$$

Integrating from $x = 0$ to $x = a$, we obtain

$$I_z = \int dI_z = \int_0^a (\tfrac{1}{12}b^2 + x^2)\,\rho bc\,dx = \rho abc(\tfrac{1}{12}b^2 + \tfrac{1}{3}a^2)$$

Since the total mass of the prism is $m = \rho abc$, we can write

$$I_z = m(\tfrac{1}{12}b^2 + \tfrac{1}{3}a^2) \qquad I_z = \tfrac{1}{12}m(4a^2 + b^2) \blacktriangleleft$$

We note that if the prism is thin, b is small compared to a, and the expression for I_z reduces to $\frac{1}{3}ma^2$, which is the result obtained in Sample Prob. B.1 when $L = a$.

SAMPLE PROBLEM B.3

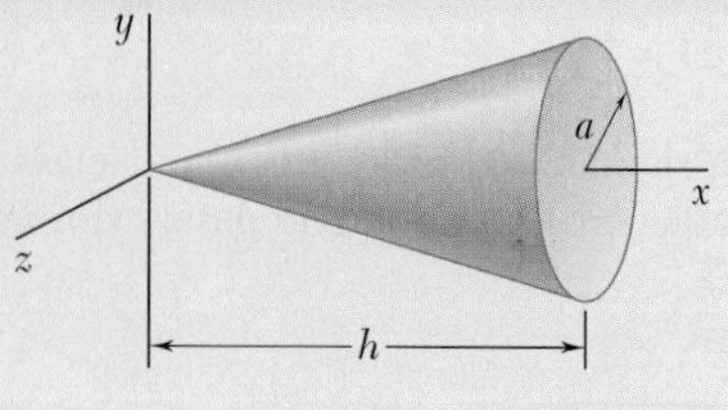

Determine the moment of inertia of a right circular cone with respect to (*a*) its longitudinal axis, (*b*) an axis through the apex of the cone and perpendicular to its longitudinal axis, (*c*) an axis through the centroid of the cone and perpendicular to its longitudinal axis.

SOLUTION

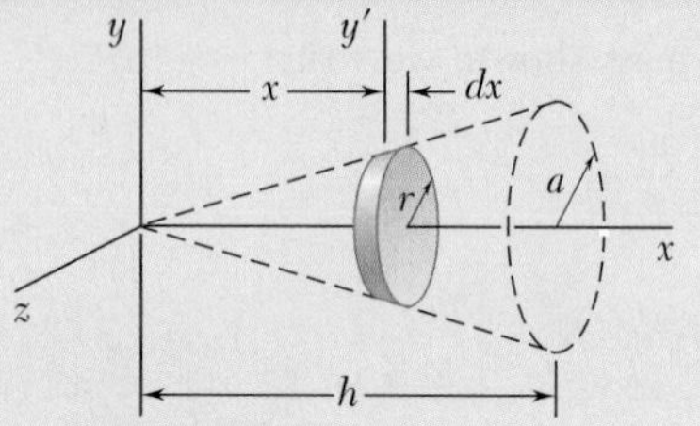

We choose the differential element of mass shown.

$$r = a\frac{x}{h} \qquad dm = \rho\pi r^2\,dx = \rho\pi\frac{a^2}{h^2}x^2\,dx$$

a. Moment of Inertia I_x. Using the expression derived in Sec. B.3 for a thin disk, we compute the mass moment of inertia of the differential element with respect to the x axis.

$$dI_x = \tfrac{1}{2}r^2\,dm = \tfrac{1}{2}\left(a\frac{x}{h}\right)^2\left(\rho\pi\frac{a^2}{h^2}x^2\,dx\right) = \tfrac{1}{2}\rho\pi\frac{a^4}{h^4}x^4\,dx$$

Integrating from $x = 0$ to $x = h$, we obtain

$$I_x = \int dI_x = \int_0^h \tfrac{1}{2}\rho\pi\frac{a^4}{h^4}x^4\,dx = \tfrac{1}{2}\rho\pi\frac{a^4}{h^4}\frac{h^5}{5} = \tfrac{1}{10}\rho\pi a^4 h$$

Since the total mass of the cone is $m = \tfrac{1}{3}\rho\pi a^2 h$, we can write

$$I_x = \tfrac{1}{10}\rho\pi a^4 h = \tfrac{3}{10}a^2(\tfrac{1}{3}\rho\pi a^2 h) = \tfrac{3}{10}ma^2 \qquad I_x = \tfrac{3}{10}ma^2 \quad ◀$$

b. Moment of Inertia I_y. The same differential element is used. Applying the parallel-axis theorem and using the expression derived in Sec. B.3 for a thin disk, we write

$$dI_y = dI_{y'} + x^2\,dm = \tfrac{1}{4}r^2\,dm + x^2\,dm = (\tfrac{1}{4}r^2 + x^2)\,dm$$

Substituting the expressions for r and dm into the equation, we obtain

$$dI_y = \left(\frac{1}{4}\frac{a^2}{h^2}x^2 + x^2\right)\left(\rho\pi\frac{a^2}{h^2}x^2\,dx\right) = \rho\pi\frac{a^2}{h^2}\left(\frac{a^2}{4h^2} + 1\right)x^4\,dx$$

$$I_y = \int dI_y = \int_0^h \rho\pi\frac{a^2}{h^2}\left(\frac{a^2}{4h^2} + 1\right)x^4\,dx = \rho\pi\frac{a^2}{h^2}\left(\frac{a^2}{4h^2} + 1\right)\frac{h^5}{5}$$

Introducing the total mass of the cone m, we rewrite I_y as follows:

$$I_y = \tfrac{3}{5}(\tfrac{1}{4}a^2 + h^2)\tfrac{1}{3}\rho\pi a^2 h \qquad I_y = \tfrac{3}{5}m(\tfrac{1}{4}a^2 + h^2) \quad ◀$$

c. Moment of Inertia $\bar{I}_{y''}$. We apply the parallel-axis theorem and write

$$I_y = \bar{I}_{y''} + m\bar{x}^2$$

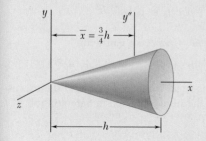

Solving for $\bar{I}_{y''}$ and recalling that $\bar{x} = \tfrac{3}{4}h$, we have

$$\bar{I}_{y''} = I_y - m\bar{x}^2 = \tfrac{3}{5}m(\tfrac{1}{4}a^2 + h^2) - m(\tfrac{3}{4}h)^2$$

$$\bar{I}_{y''} = \tfrac{3}{20}m(a^2 + \tfrac{1}{4}h^2) \quad ◀$$

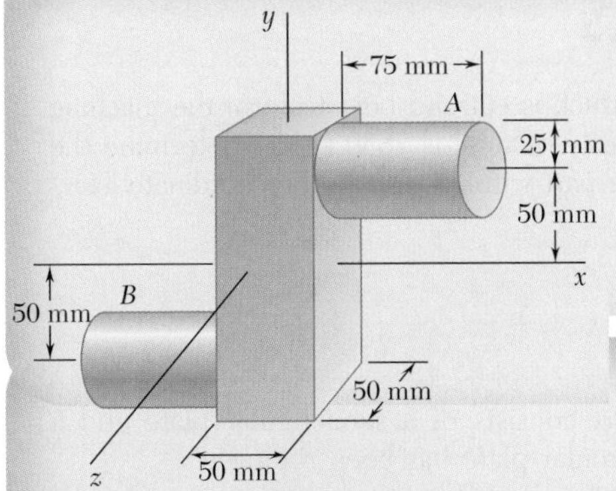

SAMPLE PROBLEM B.4

A steel forging consists of a 150 × 50 × 50 mm rectangular prism and two cylinders of diameter 50 mm and length 75 mm as shown. Determine the moments of inertia of the forging with respect to the coordinate axes, knowing that the density of steel is 7850 kg/m³.

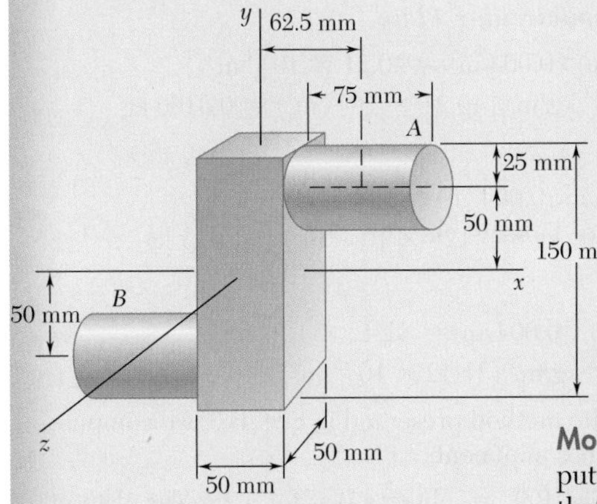

SOLUTION

Computation of Masses

Prism

$$V = (0.05\text{m})(0.05\text{m})(0.150\text{m}) = 3.75 \times 10^{-4}\text{ m}^3$$

$$m = (3.75 \times 10^{-4}\text{ m}^3)(7850\text{ kg/m}^3)$$

$$= 2.94\text{ kg}$$

Each Cylinder

$$V = \pi(0.025\text{m})^2(0.075\text{m}) = 1.473 \times 10^{-4}\text{ m}^3$$

$$m = (1.473 \times 10^{-4}\text{ m}^3)(7850\text{ kg/m}^3)$$

$$= 1.156\text{ kg}$$

Moments of Inertia. The moments of inertia of each component are computed from Fig. 9.28, using the parallel-axis theorem when necessary. Note that all lengths should be expressed in metres.

Prism

$$I_x = I_z = \frac{1}{12}(2.94\text{ kg})\left[\left(\frac{150}{1000}\text{m}\right)^2 + \left(\frac{50}{1000}\text{m}\right)^2\right] = 6.125 \times 10^{-3}\text{kg}\cdot\text{m}^2$$

$$I_y = \frac{1}{12}(2.94\text{ kg})\left[\left(\frac{50}{1000}\text{m}\right)^2 + \left(\frac{50}{1000}\text{m}\right)^2\right] = 1.225 \times 10^{-3}\text{kg}\cdot\text{m}^2$$

Each Cylinder

$$I_x = \tfrac{1}{2}ma^2 + m\bar{y}^2 = \tfrac{1}{2}(1.156\text{ kg})\left(\frac{25}{1000}\text{ m}\right)^2 + (1.156\text{ kg})\left(\frac{50}{1000}\right)^2 = 3.251 \times 10^{-3}\text{kg}\cdot\text{m}^2$$

$$I_y = \tfrac{1}{12}m\left(3a^2 + L^2\right) + m\bar{x}^2 = \tfrac{1}{12}(1.156\text{ kg})\left[3\times\left(\frac{25}{1000}\text{m}\right)^2 + \left(\frac{75}{1000}\text{m}\right)^2\right]$$

$$+(1.156\text{ kg})\left(\frac{62.5}{1000}\text{m}\right)^2 = 5.238 \times 10^{-3}\text{kg}\cdot\text{m}^2$$

$$I_z = \frac{1}{12}m\left(3a^2 + L^2\right) + m\left(\bar{x}^2 + \bar{y}^2\right) = \tfrac{1}{12}(1.156\text{ kg})\left[3\times\left(\frac{25}{1000}\text{m}\right)^2 + \left(\frac{75}{1000}\text{m}\right)^2\right]$$

$$+(1.156\text{ kg})\left[\left(\frac{62.5}{1000}\text{m}\right)^2 + \left(\frac{50}{1000}\text{m}\right)^2\right] = 8.128 \times 10^{-3}\text{kg}\cdot\text{m}^2.$$

Entire Body. Adding the values obtained,

$I_x = 6.125 \times 10^{-3} + 2(3.251 \times 10^{-3})$	$I_x = 12.63 \times 10^{-3}\text{ kg}\cdot\text{m}^2$ ◀
$I_y = 1.225 \times 10^{-3} + 2(5.238 \times 10^{-3})$	$I_y = 11.70 \times 10^{-3}\text{ kg}\cdot\text{m}^2$ ◀
$I_z = 6.125 \times 10^{-3} + 2(8.128 \times 10^{-3})$	$I_z = 22.4 \times 10^{-3}\text{ kg}\cdot\text{m}^2$ ◀

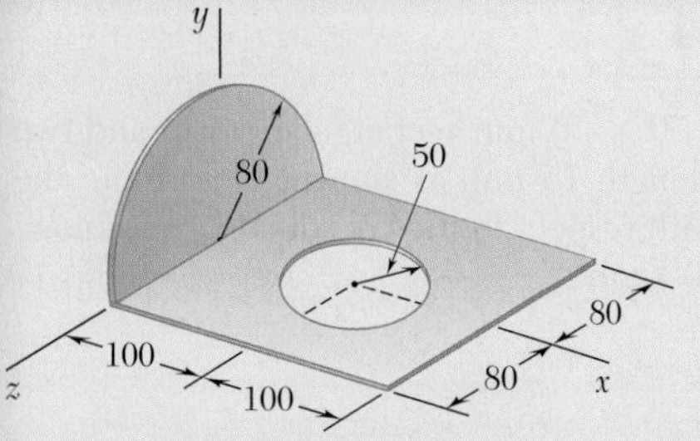

Dimensions in mm

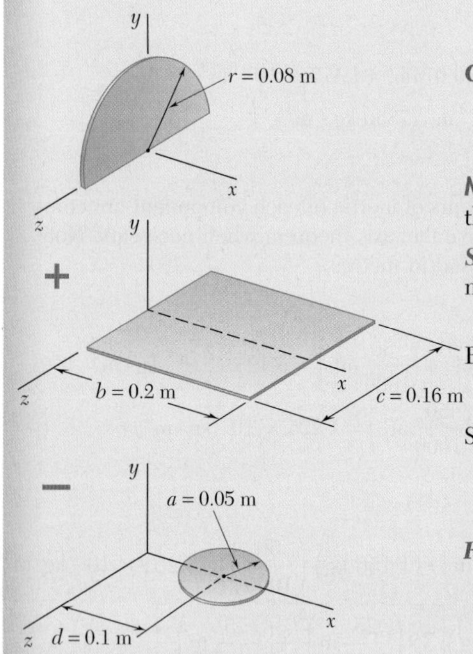

SAMPLE PROBLEM B.5

A thin steel plate which is 4 mm thick is cut and bent to form the machine part shown. Knowing that the density of steel is 7850 kg/m^3, determine the moments of inertia of the machine part with respect to the coordinate axes.

SOLUTION

We observe that the machine part consists of a semicircular plate and a rectangular plate from which a circular plate has been removed.

Computation of Masses. ***Semicircular Plate***

$$V_1 = \tfrac{1}{2}\pi r^2 t = \tfrac{1}{2}\pi(0.08\text{ m})^2(0.004\text{ m}) = 40.21 \times 10^{-6}\text{ m}^3$$
$$m_1 = \rho V_1 = (7.85 \times 10^3\text{ kg/m}^3)(40.21 \times 10^{-6}\text{ m}^3) = 0.3156\text{ kg}$$

Rectangular Plate

$$V_2 = (0.200\text{ m})(0.160\text{ m})(0.004\text{ m}) = 128 \times 10^{-6}\text{ m}^3$$
$$m_2 = \rho V_2 = (7.85 \times 10^3\text{ kg/m}^3)(128 \times 10^{-6}\text{ m}^3) = 1.005\text{ kg}$$

Circular Plate

$$V_3 = \pi a^2 t = \pi(0.050\text{ m})^2(0.004\text{ m}) = 31.42 \times 10^{-6}\text{ m}^3$$
$$m_3 = \rho V_3 = (7.85 \times 10^3\text{ kg/m}^3)(31.42 \times 10^{-6}\text{ m}^3) = 0.2466\text{ kg}$$

Moments of Inertia. Using the method presented in Sec. B.3, we compute the moments of inertia of each component.

Semicircular Plate. From Fig. B.9, we observe that for a circular plate of mass m and radius r

$$I_x = \tfrac{1}{2}mr^2 \qquad I_y = I_z = \tfrac{1}{4}mr^2$$

Because of symmetry, we note that for a semicircular plate

$$I_x = \tfrac{1}{2}(\tfrac{1}{2}mr^2) \qquad I_y = I_z = \tfrac{1}{2}(\tfrac{1}{4}mr^2)$$

Since the mass of the semicircular plate is $m_1 = \tfrac{1}{2}m$, we have

$$I_x = \tfrac{1}{2}m_1 r^2 = \tfrac{1}{2}(0.3156\text{ kg})(0.08\text{ m})^2 = 1.010 \times 10^{-3}\text{ kg}\cdot\text{m}^2$$
$$I_y = I_z = \tfrac{1}{4}(\tfrac{1}{2}mr^2) = \tfrac{1}{4}m_1 r^2 = \tfrac{1}{4}(0.3156\text{ kg})(0.08\text{ m})^2 = 0.505 \times 10^{-3}\text{ kg}\cdot\text{m}^2$$

Rectangular Plate

$$I_x = \tfrac{1}{12}m_2 c^2 = \tfrac{1}{12}(1.005\text{ kg})(0.16\text{ m})^2 = 2.144 \times 10^{-3}\text{ kg}\cdot\text{m}^2$$
$$I_z = \tfrac{1}{3}m_2 b^2 = \tfrac{1}{3}(1.005\text{ kg})(0.2\text{ m})^2 = 13.400 \times 10^{-3}\text{ kg}\cdot\text{m}^2$$
$$I_y = I_x + I_z = (2.144 + 13.400)(10^{-3}) = 15.544 \times 10^{-3}\text{ kg}\cdot\text{m}^2$$

Circular Plate

$$I_x = \tfrac{1}{4}m_3 a^2 = \tfrac{1}{4}(0.2466\text{ kg})(0.05\text{ m})^2 = 0.154 \times 10^{-3}\text{ kg}\cdot\text{m}^2$$
$$I_y = \tfrac{1}{2}m_3 a^2 + m_3 d^2$$
$$= \tfrac{1}{2}(0.2466\text{ kg})(0.05\text{ m})^2 + (0.2466\text{ kg})(0.1\text{ m})^2 = 2.774 \times 10^{-3}\text{ kg}\cdot\text{m}^2$$
$$I_z = \tfrac{1}{4}m_3 a^2 + m_3 d^2 = \tfrac{1}{4}(0.2466\text{ kg})(0.05\text{ m})^2 + (0.2466\text{ kg})(0.1\text{ m})^2$$
$$= 2.620 \times 10^{-3}\text{ kg}\cdot\text{m}^2$$

Entire Machine Part

$$I_x = (1.010 + 2.144 - 0.154)(10^{-3})\text{ kg}\cdot\text{m}^2 \qquad I_x = 3.00 \times 10^{-3}\text{ kg}\cdot\text{m}^2 \blacktriangleleft$$
$$I_y = (0.505 + 15.544 - 2.774)(10^{-3})\text{ kg}\cdot\text{m}^2 \qquad I_y = 13.28 \times 10^{-3}\text{ kg}\cdot\text{m}^2 \blacktriangleleft$$
$$I_z = (0.505 + 13.400 - 2.620)(10^{-3})\text{ kg}\cdot\text{m}^2 \qquad I_z = 11.29 \times 10^{-3}\text{ kg}\cdot\text{m}^2 \blacktriangleleft$$

SOLVING PROBLEMS ON YOUR OWN

In this lesson we introduced the *mass moment of inertia* and the *radius of gyration* of a three-dimensional body with respect to a given axis [Eqs. (B.1) and (B.2)]. We also derived a *parallel-axis theorem* for use with mass moments of inertia and discussed the computation of the mass moments of inertia of thin plates and three-dimensional bodies.

1. Computing mass moments of inertia. The mass moment of inertia I of a body with respect to a given axis can be calculated directly from the definition given in Eq. (B.1) for simple shapes [Sample Prob. B.1]. In most cases, however, it is necessary to divide the body into thin slabs, compute the moment of inertia of a typical slab with respect to the given axis—using the parallel-axis theorem if necessary—and integrate the expression obtained.

2. Applying the parallel-axis theorem. In Sec. B.2 we derived the parallel-axis theorem for mass moments of inertia

$$I = \bar{I} + md^2 \tag{B.6}$$

which states that the moment of inertia I of a body of mass m with respect to a given axis is equal to the sum of the moment of inertia $\bar{I}$ of that body with respect to the *parallel centroidal axis* and the product md^2, where d is the distance between the two axes. When the moment of inertia of a three-dimensional body is calculated with respect to one of the coordinate axes, d^2 can be replaced by the sum of the squares of distances measured along the other two coordinate axes [Eqs. (B.5) and (B.5′)].

3. Avoiding unit-related errors. To avoid errors, it is essential that you be consistent in your use of units. Thus, all lengths should be expressed in meters or feet, as appropriate, and for problems using U.S. customary units, masses should be given in $\text{lb} \cdot \text{s}^2/\text{ft}$. In addition, we strongly recommend that you include units as you perform your calculations [Sample Probs. B.4 and B.5].

4. Calculating the mass moment of inertia of thin plates. We showed in Sec. B.3 that the mass moment of inertia of a thin plate with respect to a given axis can be obtained by multiplying the corresponding moment of inertia of the area of the plate by the density ρ and the thickness t of the plate [Eqs. (B.8) through (B.10)]. Note that since the axis CC' in Fig. B.5*c* is *perpendicular to the plate*, $I_{CC', \text{mass}}$ is associated with the *polar* moment of inertia $J_{C, \text{area}}$.

Instead of calculating directly the moment of inertia of a thin plate with respect to a specified axis, you may sometimes find it convenient to first compute its moment of inertia with respect to an axis parallel to the specified axis and then apply the parallel-axis theorem. Further, to determine the moment of inertia of a thin plate with respect to an axis perpendicular to the plate, you may wish to first determine its moments of inertia with respect to two perpendicular in-plane axes

(continued)

and then use Eq. (B.11). Finally, remember that the mass of a plate of area A, thickness t, and density ρ is $m = \rho t A$.

5. Determining the moment of inertia of a body by direct single integration. We discussed in Sec. B.4 and illustrated in Sample Probs. B.2 and B.3 how single integration can be used to compute the moment of inertia of a body that can be divided into a series of thin, parallel slabs. For such cases, you will often need to express the mass of the body in terms of the body's density and dimensions. Assuming that the body has been divided, as in the sample problems, into thin slabs perpendicular to the x axis, you will need to express the dimensions of each slab as functions of the variable x.

a. In the special case of a body of revolution, the elemental slab is a thin disk, and the equations given in Fig. B.8 should be used to determine the moments of inertia of the body [Sample Prob. B.3].

b. In the general case, when the body is not of revolution, the differential element is not a disk, but a thin slab of a different shape, and the equations of Fig. B.8 cannot be used. See, for example, Sample Prob. B.2, where the element was a thin, rectangular slab. For more complex configurations, you may want to use one or more of the following equations, which are based on Eqs. (B.5) and (B.5′) of Sec. B.2.

$$dI_x = dI_{x'} + (\bar{y}_{el}^2 + \bar{z}_{el}^2)\, dm$$
$$dI_y = dI_{y'} + (\bar{z}_{el}^2 + \bar{x}_{el}^2)\, dm$$
$$dI_z = dI_{z'} + (\bar{x}_{el}^2 + \bar{y}_{el}^2)\, dm$$

where the primes denote the centroidal axes of each elemental slab, and where $\bar{x}_{el}$, $\bar{y}_{el}$, and $\bar{z}_{el}$ represent the coordinates of its centroid. The centroidal moments of inertia of the slab are determined in the manner described earlier for a thin plate: Referring to Fig. 9.12 on page 483, calculate the corresponding moments of inertia of the area of the slab and multiply the result by the density ρ and the thickness t of the slab. Also, assuming that the body has been divided into thin slabs perpendicular to the x axis, remember that you can obtain $dI_{x'}$ by adding $dI_{y'}$ and $dI_{z'}$ instead of computing it directly. Finally, using the geometry of the body, express the result obtained in terms of the single variable x and integrate in x.

6. Computing the moment of inertia of a composite body. As stated in Sec. B.5, the moment of inertia of a composite body with respect to a specified axis is equal to the sum of the moments of its components with respect to that axis. Sample Probs. B.4 and B.5 illustrate the appropriate method of solution. You must also remember that the moment of inertia of a component will be negative only if the component is *removed* (as in the case of a hole).

Although the composite-body problems in this lesson are relatively straightforward, you will have to work carefully to avoid computational errors. In addition, if some of the moments of inertia that you need are not given in Fig. B.9, you will have to derive your own formulas using the techniques of this lesson.

PROBLEMS

B.1 A thin plate of mass m is cut in the shape of an equilateral triangle of side a. Determine the mass moment of inertia of the plate with respect to (*a*) the centroidal axes AA' and BB', (*b*) the centroidal axis CC' that is perpendicular to the plate.

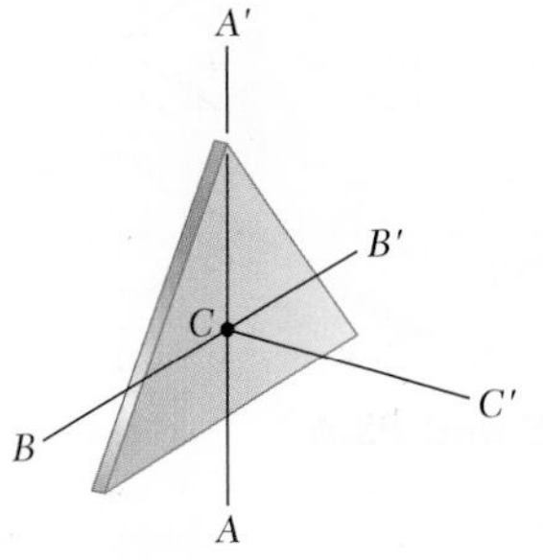

Fig. PB.1

B.2 The elliptical ring shown was cut from a thin, uniform plate. Denoting the mass of the ring by m, determine its mass moment of inertia with respect to (*a*) the centroidal axis BB', (*b*) the centroidal axis CC' that is perpendicular to the plane of the ring.

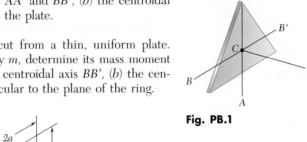

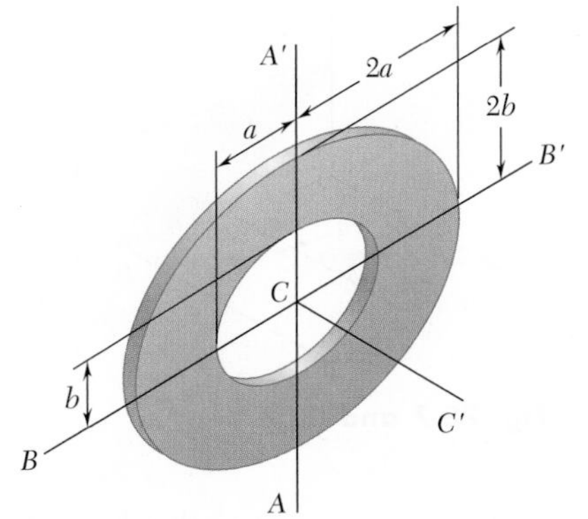

Fig. PB.2

B.3 A thin semicircular plate has a radius a and a mass m. Determine the mass moment of inertia of the plate with respect to (*a*) the centroidal axis BB', (*b*) the centroidal axis CC' that is perpendicular to the plate.

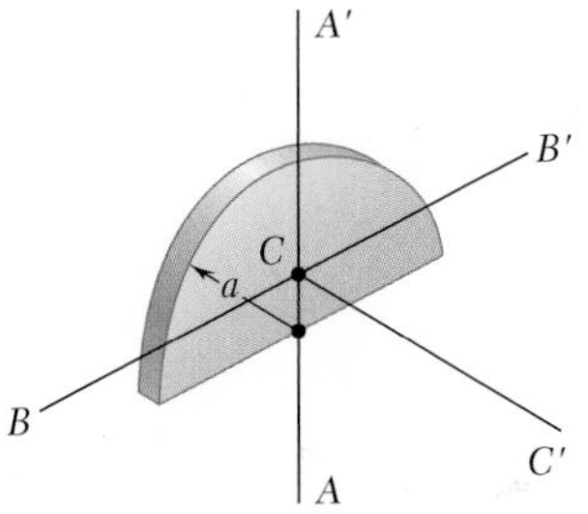

Fig. PB.3

B.4 The quarter ring shown has a mass m and was cut from a thin, uniform plate. Knowing that $r_1 = \frac{3}{4}r_2$, determine the mass moment of inertia of the quarter ring with respect to (*a*) the axis AA', (*b*) the centroidal axis CC' that is perpendicular to the plane of the quarter ring.

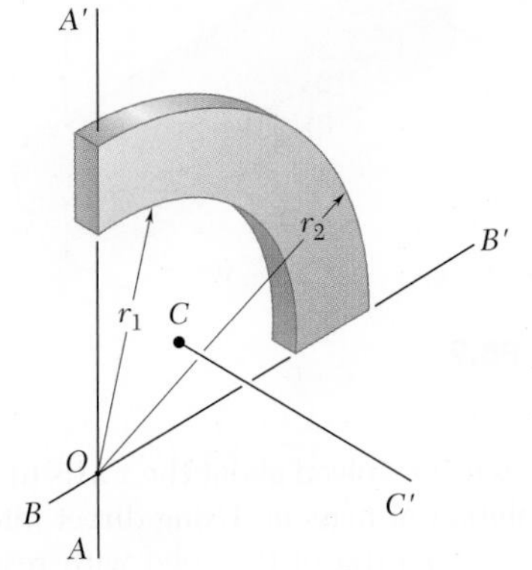

Fig. PB.4

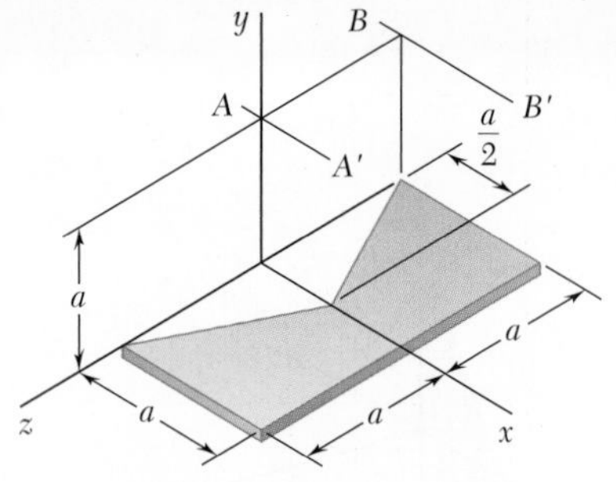

Fig. *PB.5* and *PB.6*

B.5 A piece of thin, uniform sheet metal is cut to form the machine component shown. Denoting the mass of the component by m, determine its mass moment of inertia with respect to (*a*) the x axis, (*b*) the y axis.

B.6 A piece of thin, uniform sheet metal is cut to form the machine component shown. Denoting the mass of the component by m, determine its mass moment of inertia with respect to (*a*) the axis AA', (*b*) the axis BB', where the AA' and BB' axes are parallel to the x axis and lie in a plane parallel to and at a distance a above the xz plane.

B.7 A thin plate of mass m was cut in the shape of a parallelogram as shown. Determine the mass moment of inertia of the plate with respect to (*a*) the x axis, (*b*) the axis BB', which is perpendicular to the plate.

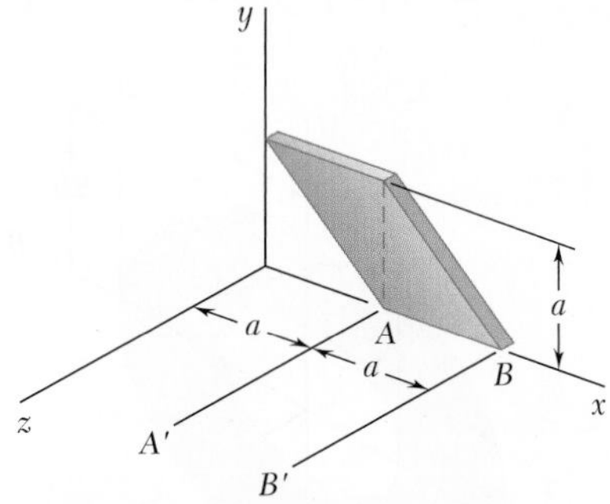

Fig. PB.7 and PB.8

B.8 A thin plate of mass m was cut in the shape of a parallelogram as shown. Determine the mass moment of inertia of the plate with respect to (*a*) the y axis, (*b*) the axis AA', which is perpendicular to the plate.

B.9 Determine by direct integration the mass moment of inertia with respect to the z axis of the right circular cylinder shown, assuming that it has a uniform density and a mass m.

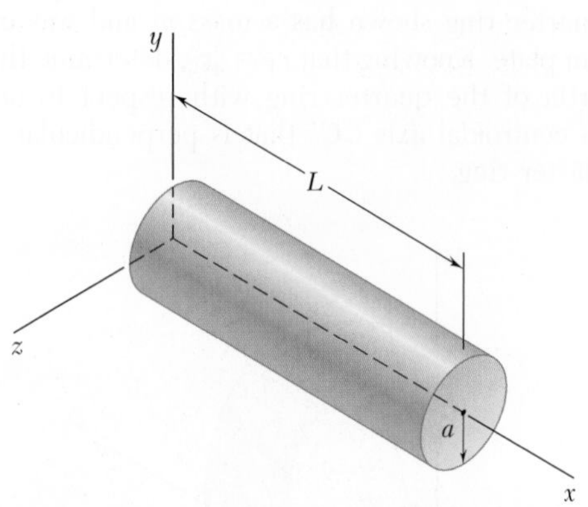

Fig. PB.9

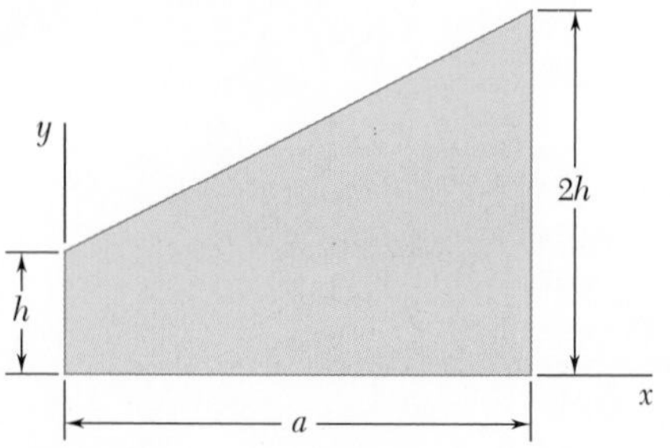

Fig. PB.10

B.10 The area shown is revolved about the x axis to form a homogeneous solid of revolution of mass m. Using direct integration, express the mass moment of inertia of the solid with respect to the x axis in terms of m and h.

B.11 The area shown is revolved about the x axis to form a homogeneous solid of revolution of mass m. Determine by direct integration the mass moment of inertia of the solid with respect to (*a*) the x axis, (*b*) the y axis. Express your answers in terms of m and the dimensions of the solid.

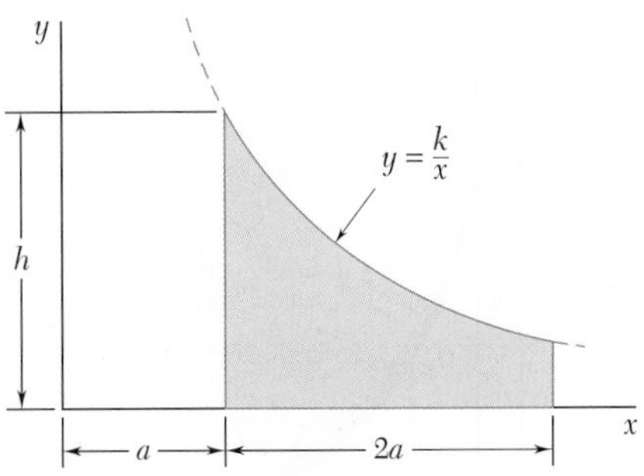

Fig. PB.11

B.12 Determine by direct integration the mass moment of inertia with respect to the x axis of the pyramid shown, assuming that it has a uniform density and a mass m.

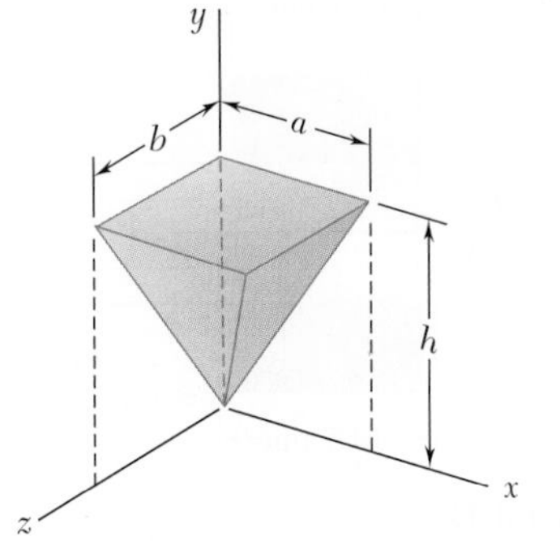

Fig. PB.12 and *PB.13*

B.13 Determine by direct integration the mass moment of inertia with respect to the y axis of the pyramid shown, assuming that it has a uniform density and a mass m.

B.14 Determine by direct integration the mass moment of inertia with respect to the y axis of the paraboloid shown, assuming that it has a uniform density and a mass m.

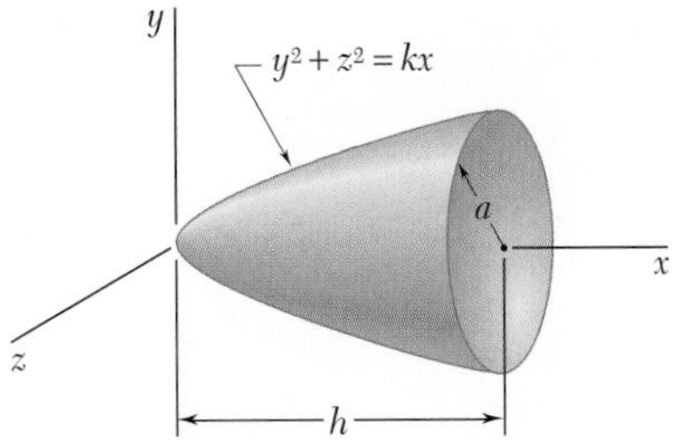

Fig. PB.14

B.15 A thin rectangular plate of mass m is welded to a vertical shaft AB as shown. Knowing that the plate forms an angle θ with the y axis, determine by direct integration the mass moment of inertia of the plate with respect to (*a*) the y axis, (*b*) the z axis.

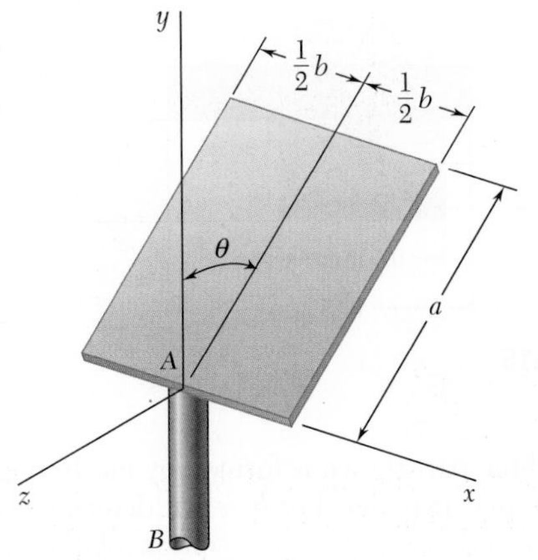

Fig. *PB.15*

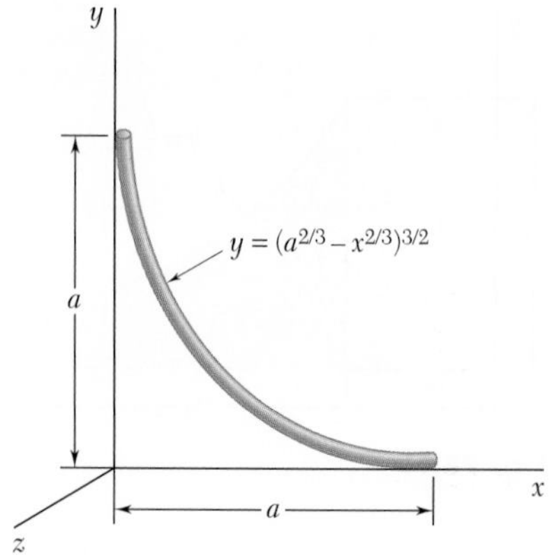

Fig. PB.16

***B.16** A thin steel wire is bent into the shape shown. Denoting the mass per unit length of the wire by m', determine by direct integration the mass moment of inertia of the wire with respect to each of the coordinate axes.

B.17 Shown is the cross section of an idler roller. Determine its mass moment of inertia and its radius of gyration with respect to the axis AA'. (The density of bronze is 8580 kg/m^3; of aluminum, 2770 kg/m^3; and of neoprene, 1250 kg/m^3.)

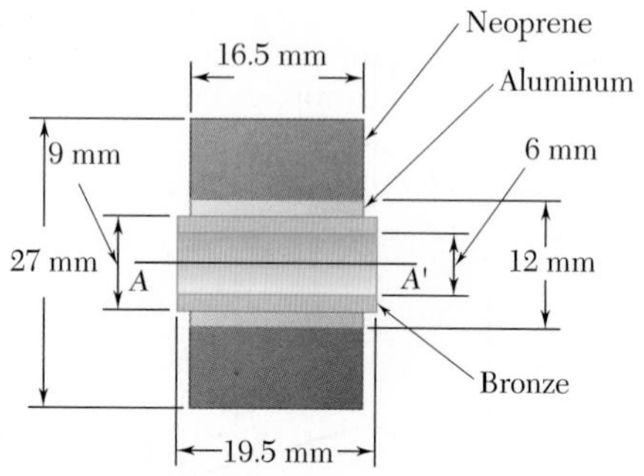

Fig. PB.17

B.18 Shown is the cross section of a molded flat-belt pulley. Determine its mass moment of inertia and its radius of gyration with respect to the axis AA'. (The density of brass is 8650 kg/m^3 and the density of the fiber-reinforced polycarbonate used is 1250 kg/m^3.)

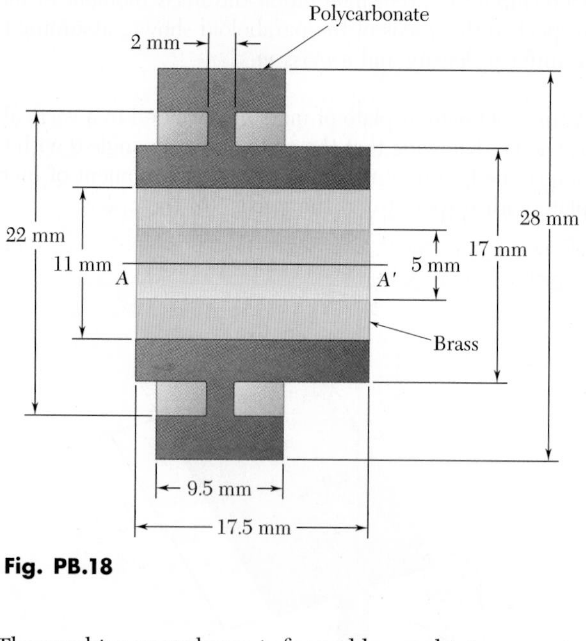

Fig. PB.18

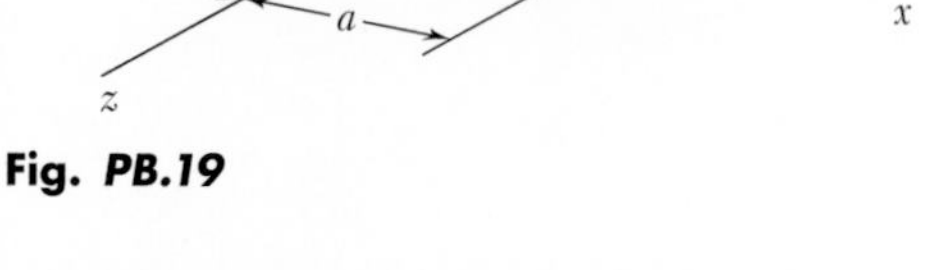

Fig. *PB.19*

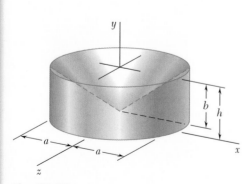

B.19 The machine part shown is formed by machining a conical surface into a circular cylinder. For $b = \frac{1}{2}h$, determine the mass moment of inertia and the radius of gyration of the machine part with respect to the y axis.

B.20 Given the dimensions and the mass m of the thin conical shell shown, determine the mass moment of inertia and the radius of gyration of the shell with respect to the x axis. (*Hint:* Assume that the shell was formed by removing a cone with a circular base of radius a from a cone with a circular base of radius $a + t$, where t is the thickness of the wall. In the resulting expressions, neglect terms containing t^2, t^3, etc. Do not forget to account for the difference in the heights of the two cones.)

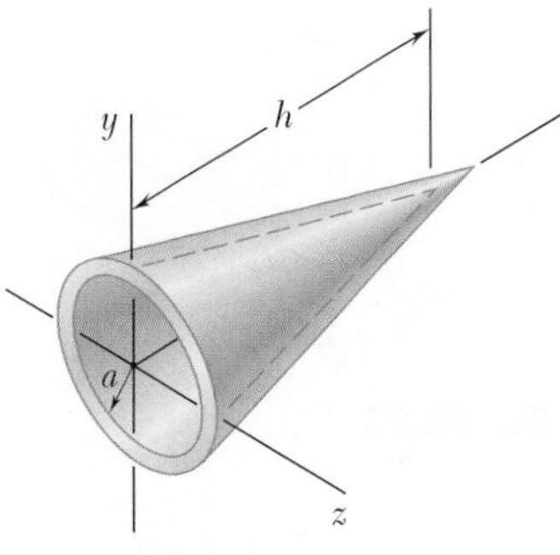

Fig. PB.20

B.21 A square hole is centered in and extends through the aluminum machine component shown. Determine (*a*) the value of a for which the mass moment of inertia of the component with respect to the axis AA', which bisects the top surface of the hole, is maximum, (*b*) the corresponding values of the mass moment of inertia and the radius of gyration with respect to the axis AA'. (The density of aluminum is 2770 kg/m^3.)

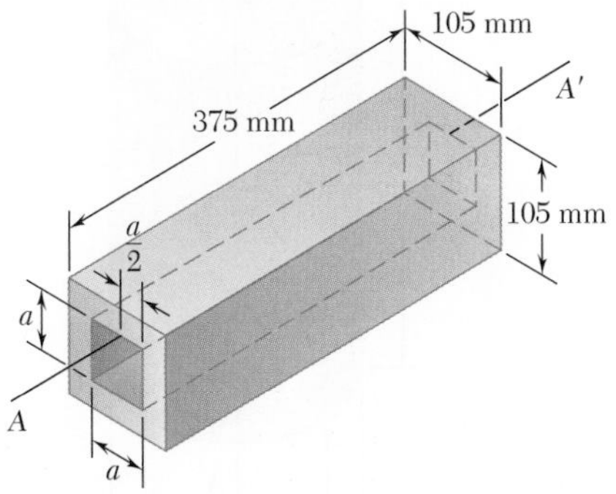

Fig. PB.21

B.22 The cups and the arms of an anemometer are fabricated from a material of density ρ. Knowing that the mass moment of inertia of a thin, hemispherical shell of mass m and thickness t with respect to its centroidal axis GG' is $5ma^2/12$, determine (*a*) the mass moment of inertia of the anemometer with respect to the axis AA', (*b*) the ratio of a to l for which the centroidal moment of inertia of the cups is equal to 1 percent of the moment of inertia of the cups with respect to the axis AA'.

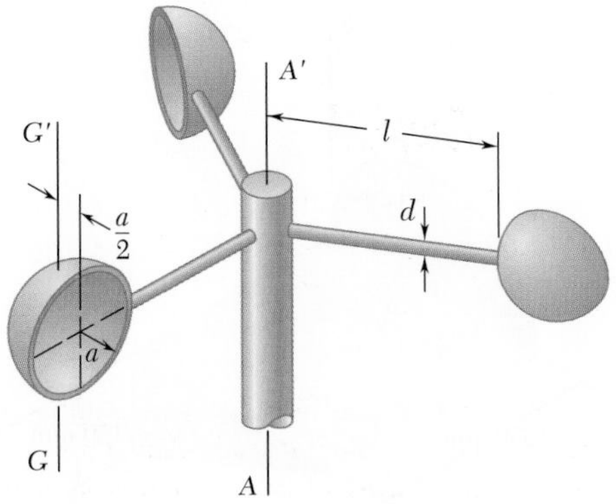

Fig. PB.22

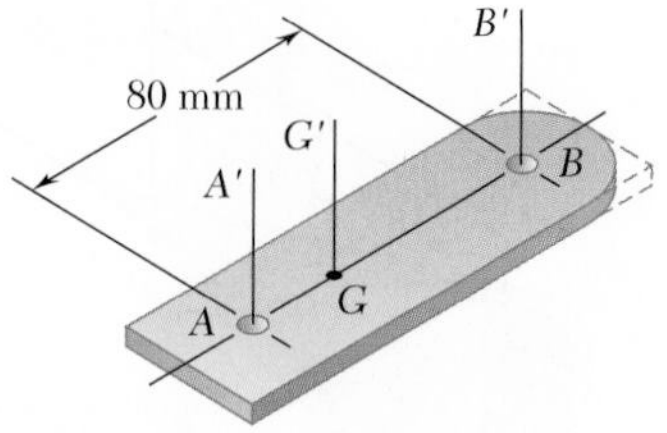

Fig. PB.23

B.23 After a period of use, one of the blades of a shredder has been worn to the shape shown and is of mass 0.18 kg. Knowing that the mass moments of inertia of the blade with respect to the AA' and BB' axes are $0.320 \text{ g} \cdot \text{m}^2$ and $0.680 \text{ g} \cdot \text{m}^2$, respectively, determine (*a*) the location of the centroidal axis GG', (*b*) the radius of gyration with respect to axis GG'.

B.24 Determine the mass moment of inertia of the 0.4-kg machine component shown with respect to the axis AA'.

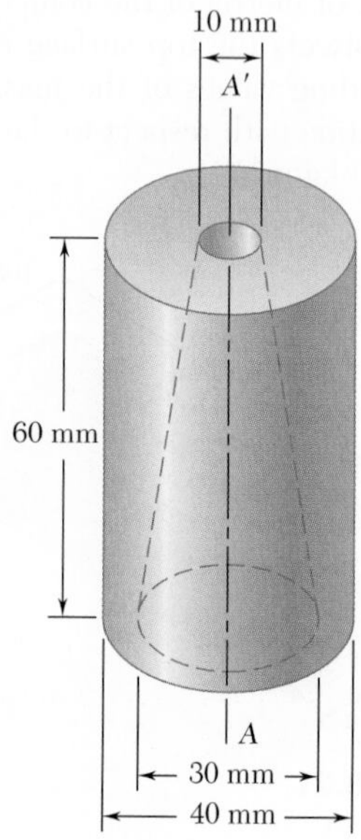

Fig. PB.24

B.25 and B.26 A 2-mm-thick piece of sheet steel is cut and bent into the machine component shown. Knowing that the density of steel is 7850 kg/m^3, determine the mass moment of inertia of the component with respect to each of the coordinate axes.

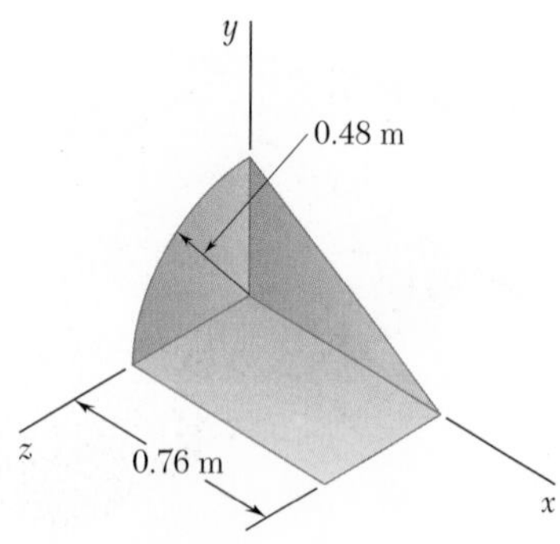

Fig. PB.25

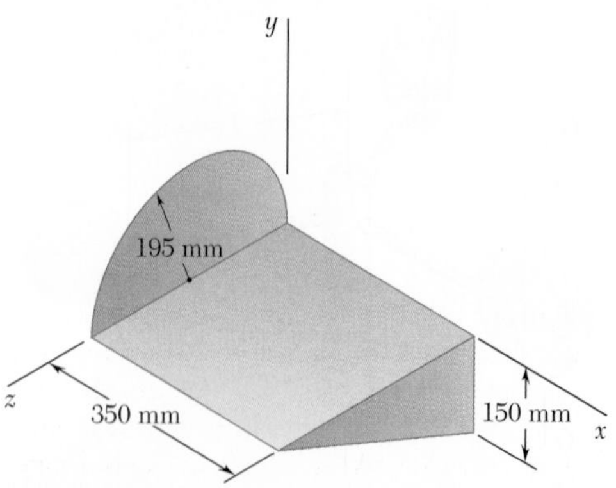

Fig. PB.26

B.27 A subassembly for a model airplane is fabricated from three pieces of 1.5-mm plywood. Neglecting the mass of the adhesive used to assemble the three pieces, determine the mass moment of inertia of the subassembly with respect to each of the coordinate axes. (The density of the plywood is 780 kg/m^3.)

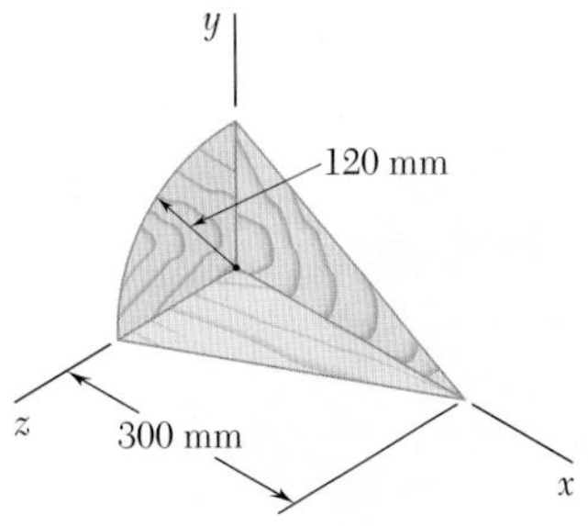

Fig. *PB.27*

B.28 The cover for an electronic device is formed from sheet aluminum that is 1.25 mm thick. Determine the mass moment of inertia of the cover with respect to each of the coordinate axes. (The density of aluminum is 2770 kg/m^3.)

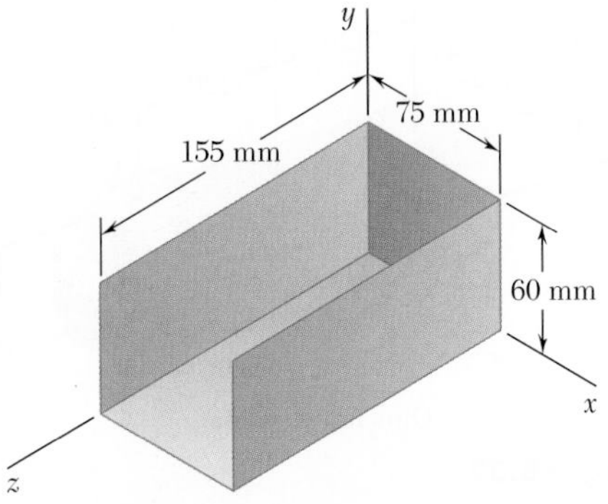

Fig. PB.28

B.29 A framing anchor is formed of 2 mm-thick galvanized steel. Determine the mass moment of inertia of the anchor with respect to each of the coordinate axes. (The density of galvanized steel is 7530 kg/m^3.)

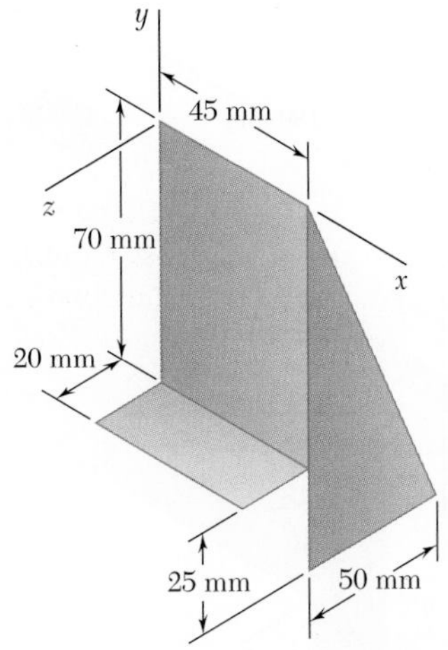

Fig. PB.29

***B.30** A farmer constructs a trough by welding a rectangular piece of 2-mm-thick sheet steel to half of a steel drum. Knowing that the density of steel is 7850 kg/m^3 and that the thickness of the walls of the drum is 1.8 mm, determine the mass moment of inertia of the trough with respect to each of the coordinate axes. Neglect the mass of the welds.

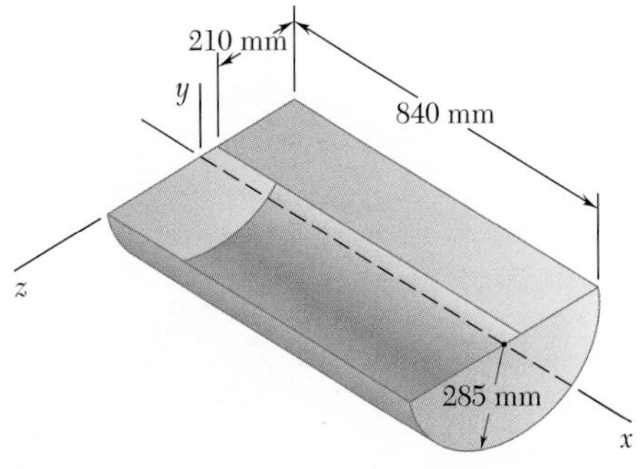

Fig. *PB.30*

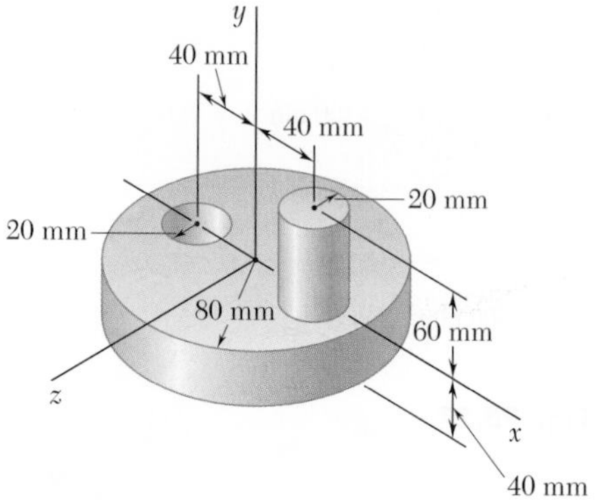

Fig. PB.31

B.31 The machine element shown is fabricated from steel. Determine the mass moment of inertia of the assembly with respect to (*a*) the x axis, (*b*) the y axis, (*c*) the z axis. (The density of steel is 7850 kg/m^3.)

B.32 Determine the mass moments of inertia and the radii of gyration of the steel machine element shown with respect to the x and y axes. (The density of steel is 7850 kg/m^3.)

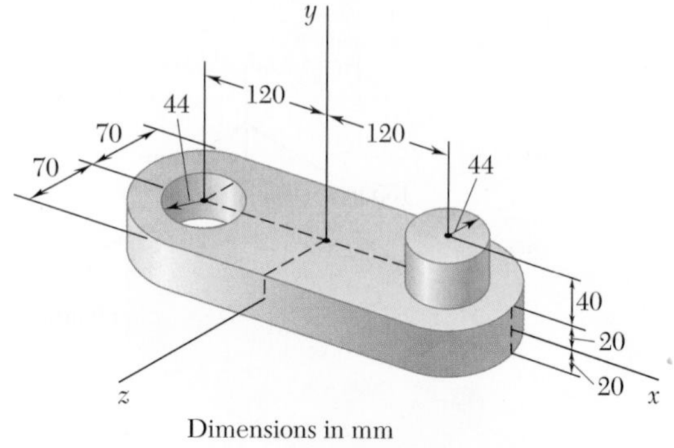

Fig. PB.32

B.33 Determine the mass moment of inertia of the steel machine element shown with respect to the y axis. (The density of steel is 7850 kg/m^3.)

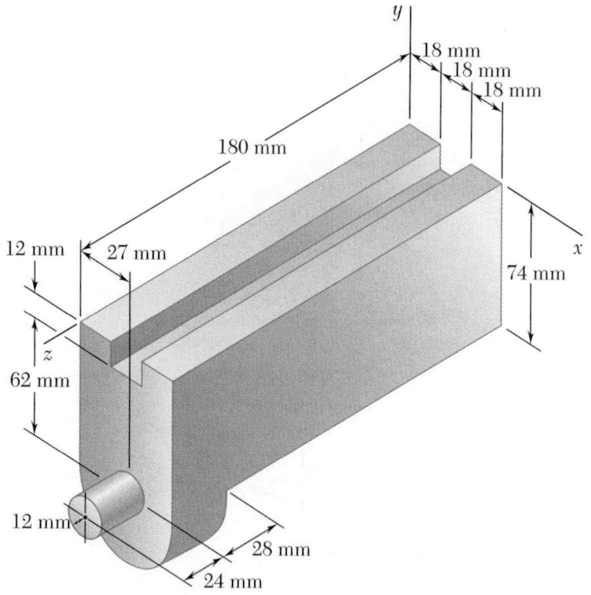

Fig. PB.33 and *PB.34*

B.34 Determine the mass moment of inertia of the steel machine element shown with respect to the z axis. (The density of steel is 7850 kg/m^3.)

B.35 Determine the mass moment of inertia of the steel fixture shown with respect to (*a*) the x axis, (*b*) the y axis, (*c*) the z axis. (The density of steel is 7850 kg/m^3.)

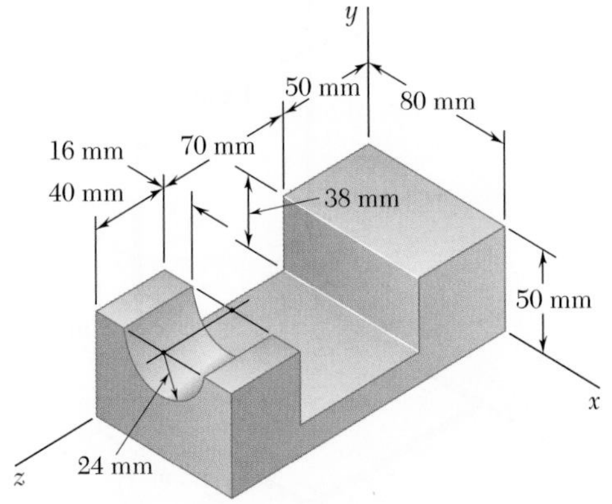

Fig. PB.35

B.36 Aluminum wire with mass per unit length of 0.05 kg/m is used to form the circle and the straight members of the figure shown. Determine the mass moment of inertia of the assembly with respect to each of the coordinate axes.

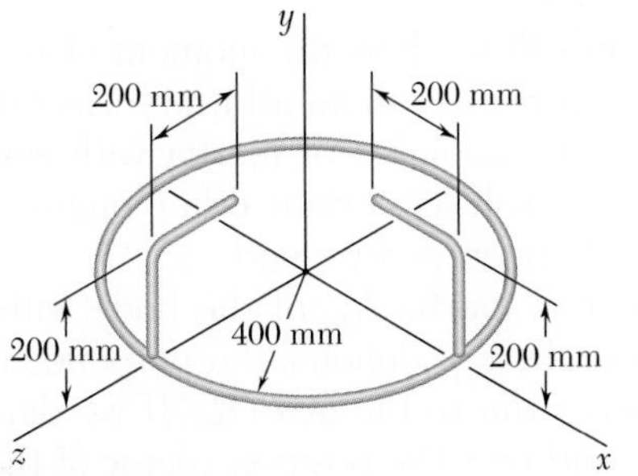

Fig. PB.36

B.37 The figure shown is formed of 3 mm-diameter steel wire. Knowing that the density of the steel is 7850 kg/m^3, determine the mass moment of inertia of the wire with respect to each of the coordinate axes.

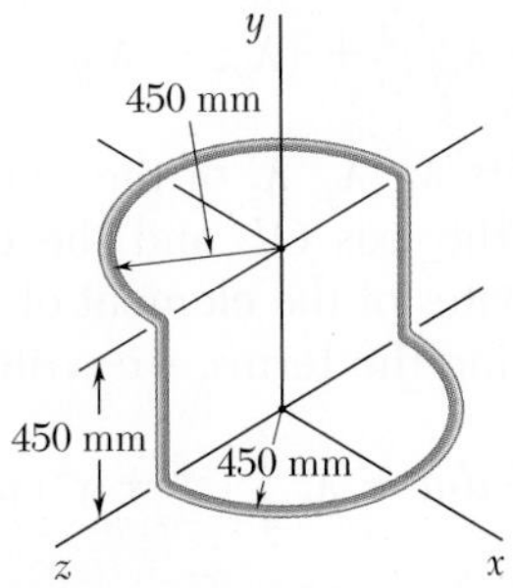

Fig. PB.37

B.38 A homogeneous wire with a mass per unit length of 0.056 kg/m is used to form the figure shown. Determine the mass moment of inertia of the wire with respect to each of the coordinate axes.

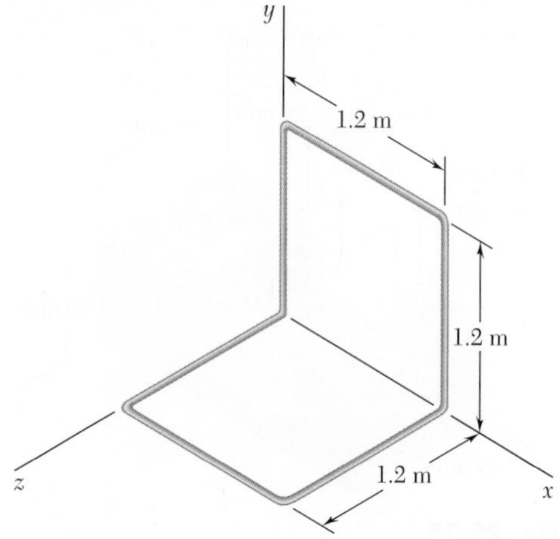

Fig. PB.38

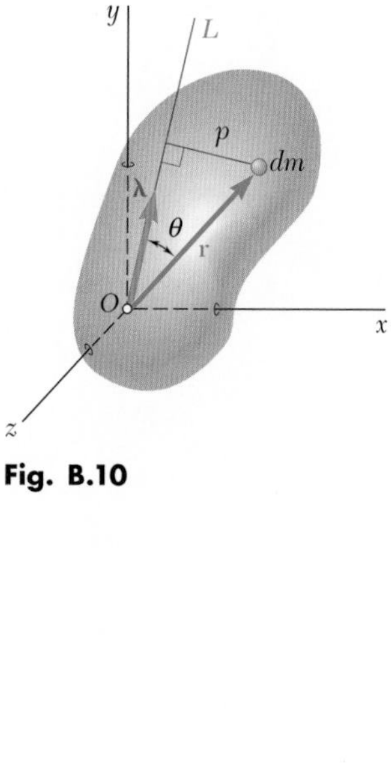

Fig. B.10

*B.6 MOMENT OF INERTIA OF A BODY WITH RESPECT TO AN ARBITRARY AXIS THROUGH O. MASS PRODUCTS OF INERTIA

In this section you will see how the moment of inertia of a body can be determined with respect to an arbitrary axis OL through the origin (Fig. B.10) if its moments of inertia with respect to the three coordinate axes, as well as certain other quantities to be defined below, have already been determined.

The moment of inertia I_{OL} of the body with respect to OL is equal to $\int p^2\,dm$, where p denotes the perpendicular distance from the element of mass dm to the axis OL. If we denote by $\boldsymbol{\lambda}$ the unit vector along OL and by $\mathbf{r}$ the position vector of the element dm, we observe that the perpendicular distance p is equal to $r \sin\theta$, which is the magnitude of the vector product $\boldsymbol{\lambda} \times \mathbf{r}$. We therefore write

$$I_{OL} = \int p^2\,dm = \int |\boldsymbol{\lambda} \times \mathbf{r}|^2\,dm \qquad \text{(B.16)}$$

Expressing $|\boldsymbol{\lambda} \times \mathbf{r}|^2$ in terms of the rectangular components of the vector product, we have

$$I_{OL} = \int [(\lambda_x y - \lambda_y x)^2 + (\lambda_y z - \lambda_z y)^2 + (\lambda_z x - \lambda_x z)^2]\,dm$$

where the components λ_x, λ_y, λ_z of the unit vector $\boldsymbol{\lambda}$ represent the direction cosines of the axis OL and the components x, y, z of $\mathbf{r}$ represent the coordinates of the element of mass dm. Expanding the squares and rearranging the terms, we write

$$I_{OL} = \lambda_x^2 \int (y^2 + z^2)\,dm + \lambda_y^2 \int (z^2 + x^2)\,dm + \lambda_z^2 \int (x^2 + y^2)\,dm$$
$$- 2\lambda_x\lambda_y \int xy\,dm - 2\lambda_y\lambda_z \int yz\,dm - 2\lambda_z\lambda_x \int zx\,dm \qquad \text{(B.17)}$$

Referring to Eqs. (B.3), we note that the first three integrals in (B.17) represent, respectively, the moments of inertia I_x, I_y, and I_z of the body with respect to the coordinate axes. The last three integrals in (B.17), which involve products of coordinates, are called the *products of inertia* of the body with respect to the x and y axes, the y and z axes, and the z and x axes, respectively. We write

$$I_{xy} = \int xy\, dm \qquad I_{yz} = \int yz\, dm \qquad I_{zx} = \int zx\, dm \tag{B.18}$$

Rewriting Eq. (B.17) in terms of the integrals defined in Eqs. (B.3) and (B.18), we have

$$I_{OL} = I_x\lambda_x^2 + I_y\lambda_y^2 + I_z\lambda_z^2 - 2I_{xy}\lambda_x\lambda_y - 2I_{yz}\lambda_y\lambda_z - 2I_{zx}\lambda_z\lambda_x \tag{B.19}$$

We note that the definition of the products of inertia of a mass given in Eqs. (B.18) is an extension of the definition of the product of inertia of an area (Sec. 9.8). Mass products of inertia reduce to zero under the same conditions of symmetry as do products of inertia of areas, and the parallel-axis theorem of mass products of inertia is expressed by relations similar to the formula derived for the product of inertia of an area. Substituting the expressions for x, y, and z given in Eqs. (B.4) into Eqs. (B.18), we find that

$$\begin{aligned} I_{xy} &= \bar{I}_{x'y'} + m\bar{x}\bar{y} \\ I_{yz} &= \bar{I}_{y'z'} + m\bar{y}\bar{z} \\ I_{zx} &= \bar{I}_{z'x'} + m\bar{z}\bar{x} \end{aligned} \tag{B.20}$$

where $\bar{x}, \bar{y}, \bar{z}$ are the coordinates of the center of gravity G of the body and $\bar{I}_{x'y'}$, $\bar{I}_{y'z'}$, $\bar{I}_{z'x'}$ denote the products of inertia of the body with respect to the centroidal axes x', y', z' (Fig. B.3).

*B.7 ELLIPSOID OF INERTIA. PRINCIPAL AXES OF INERTIA

Let us assume that the moment of inertia of the body considered in the preceding section has been determined with respect to a large number of axes OL through the fixed point O and that a point Q has been plotted on each axis OL at a distance $OQ = 1/\sqrt{I_{OL}}$ from O. The locus of the points Q thus obtained forms a surface (Fig. B.11). The equation of that surface can be obtained by substituting $1/(OQ)^2$ for I_{OL} in (B.19) and then multiplying both sides of the equation by $(OQ)^2$. Observing that

$$(OQ_z)\lambda_x = x \qquad (OQ)\lambda_y = y \qquad (OQ)\lambda_z = z$$

where x, y, z denote the rectangular coordinates of Q, we write

$$I_x x^2 + I_y y^2 + I_z z^2 - 2I_{xy}xy - 2I_{yz}yz - 2I_{zx}zx = 1 \tag{B.21}$$

The equation obtained is the equation of a *quadric surface*. Since the moment of inertia I_{OL} is different from zero for every axis OL, no point Q can be at an infinite distance from O. Thus, the quadric

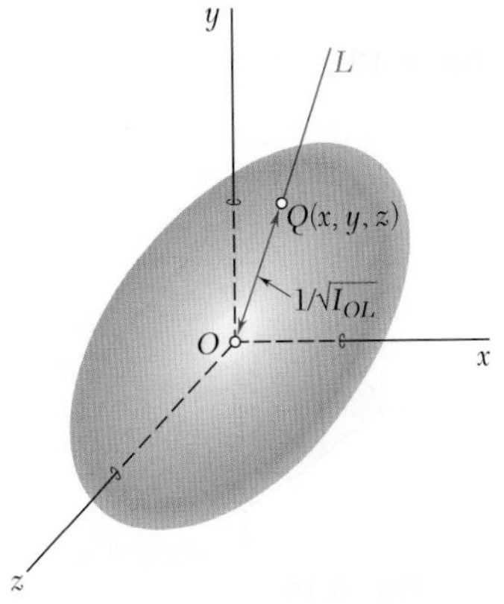

Fig. B.11

surface obtained is an *ellipsoid.* This ellipsoid, which defines the moment of inertia of the body with respect to any axis through O, is known as the *ellipsoid of inertia* of the body at O.

We observe that if the axes in Fig. B.11 are rotated, the coefficients of the equation defining the ellipsoid change, since they are equal to the moments and products of inertia of the body with respect to the rotated coordinate axes. However, the *ellipsoid itself remains unaffected,* since its shape depends only upon the distribution of mass in the given body. Suppose that we choose as coordinate axes the principal axes x', y', z' of the ellipsoid of inertia (Fig. B.12). The equation of the ellipsoid with respect to these coordinate axes is known to be of the form

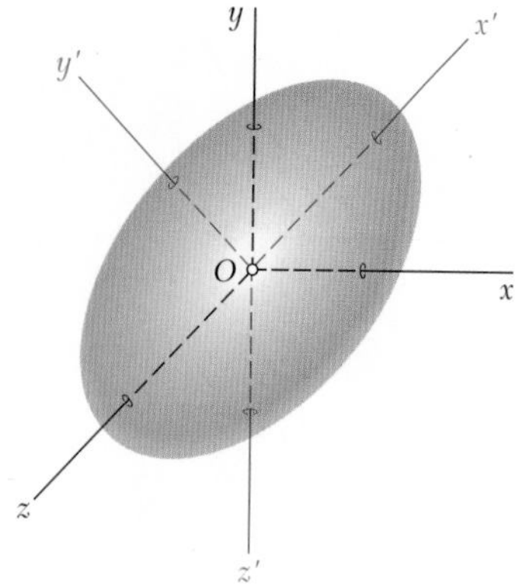

Fig. B.12

$$I_{x'}x'^2 + I_{y'}y'^2 + I_{z'}z'^2 = 1 \tag{B.22}$$

which does not contain any products of the coordinates. Comparing Eqs. (B.21) and (B.22), we observe that the products of inertia of the body with respect to the x', y', z' axes must be zero. The x', y', z' axes are known as the *principal axes of inertia* of the body at O, and the coefficients $I_{x'}$, $I_{y'}$, $I_{z'}$ are referred to as the *principal moments of inertia* of the body at O. Note that, given a body of arbitrary shape and a point O, it is always possible to find axes which are the principal axes of inertia of the body at O, that is, axes with respect to which the products of inertia of the body are zero. Indeed, whatever the shape of the body, the moments and products of inertia of the body with respect to x, y, and z axes through O will define an ellipsoid, and this ellipsoid will have principal axes which, by definition, are the principal axes of inertia of the body at O.

If the principal axes of inertia x', y', z' are used as coordinate axes, the expression obtained in Eq. (B.19) for the moment of inertia of a body with respect to an arbitrary axis through O reduces to

$$I_{OL} = I_{x'}\lambda_{x'}^2 + I_{y'}\lambda_{y'}^2 + I_{z'}\lambda_{z'}^2 \tag{B.23}$$

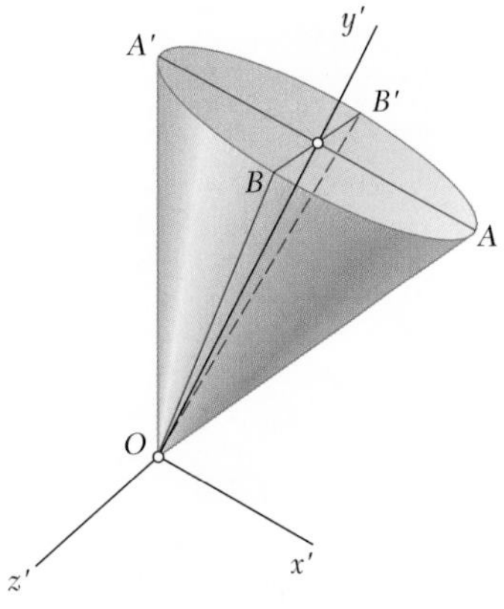

Fig. B.13

The determination of the principal axes of inertia of a body of arbitrary shape is somewhat involved and will be discussed in the next section. There are many cases, however, where these axes can be spotted immediately. Consider, for instance, the homogeneous cone of elliptical base shown in Fig. B.13; this cone possesses two mutually perpendicular planes of symmetry OAA' and OBB'. From the definition (B.18), we observe that if the $x'y'$ and $y'z'$ planes are chosen to coincide with the two planes of symmetry, all of the products of inertia are zero. The x', y', and z' axes thus selected are therefore the principal axes of inertia of the cone at O. In the case of the homogeneous regular tetrahedron $OABC$ shown in Fig. B.14, the line joining the corner O to the center D of the opposite face is a principal axis of inertia at O, and any line through O perpendicular to OD is also a principal axis of inertia at O. This property is apparent if we observe that rotating the tetrahedron through 120° about OD leaves its shape and mass distribution unchanged. It follows that the ellipsoid of inertia at O also remains unchanged under this rotation. The ellipsoid, therefore, is a body of revolution whose axis of revolution is OD, and the line OD, as well as any perpendicular line through O, must be a principal axis of the ellipsoid.

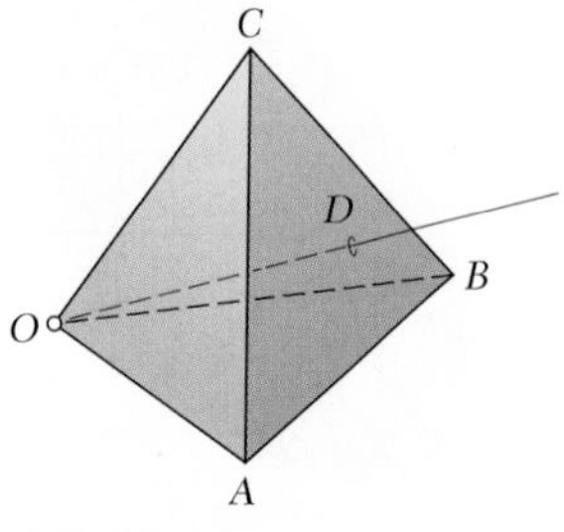

Fig. B.14

*B.8 DETERMINATION OF THE PRINCIPAL AXES AND PRINCIPAL MOMENTS OF INERTIA OF A BODY OF ARBITRARY SHAPE

The method of analysis described in this section should be used when the body under consideration has no obvious property of symmetry.

Consider the ellipsoid of inertia of the body at a given point O (Fig. B.15); let $\mathbf{r}$ be the radius vector of a point P on the surface of the ellipsoid and let $\mathbf{n}$ be the unit vector along the normal to that surface at P. We observe that the only points where $\mathbf{r}$ and $\mathbf{n}$ are collinear are the points P_1, P_2, and P_3, where the principal axes intersect the visible portion of the surface of the ellipsoid, and the corresponding points on the other side of the ellipsoid.

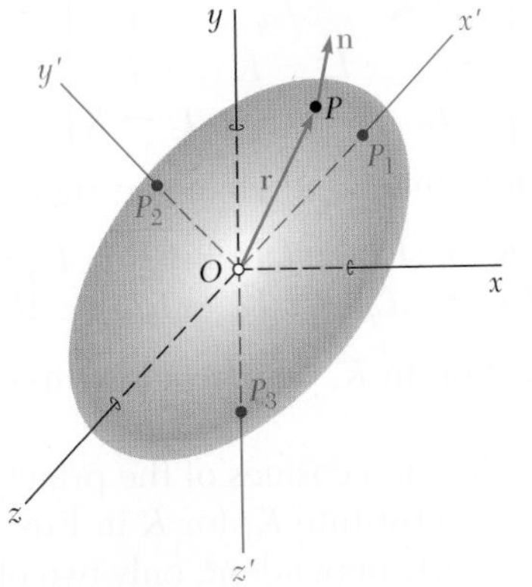

Fig. B.15

We now recall from calculus that the direction of the normal to a surface of equation $f(x, y, z) = 0$ at a point $P(x, y, z)$ is defined by the gradient ∇f of the function f at that point. To obtain the points where the principal axes intersect the surface of the ellipsoid of inertia, we must therefore write that $\mathbf{r}$ and ∇f are collinear,

$$\nabla f = (2K)\mathbf{r} \qquad \text{(B.24)}$$

where K is a constant, $\mathbf{r} = x\mathbf{i} + y\mathbf{j} + z\mathbf{k}$, and

$$\nabla f = \frac{\partial f}{\partial x}\mathbf{i} + \frac{\partial f}{\partial y}\mathbf{j} + \frac{\partial f}{\partial x}\mathbf{k}$$

Recalling Eq. (B.21), we note that the function $f(x, y, z)$ corresponding to the ellipsoid of inertia is

$$f(x, y, z) = I_x x^2 + I_y y^2 + I_z z^2 - 2I_{xy}xy - 2I_{yz}yz - 2I_{zx}zx - 1$$

Substituting for $\mathbf{r}$ and ∇f into Eq. (B.24) and equating the coefficients of the unit vectors, we write

$$\begin{aligned} I_x x - I_{xy} y - I_{zx} z &= Kx \\ -I_{xy} x + I_y y - I_{yz} z &= Ky \\ -I_{zx} x - I_{yz} y + I_z z &= Kz \end{aligned} \qquad \text{(B.25)}$$

Dividing each term by the distance r from O to P, we obtain similar equations involving the direction cosines λ_x, λ_y, and λ_z:

$$\begin{aligned} I_x\lambda_x - I_{xy}\lambda_y - I_{zx}\lambda_z &= K\lambda_x \\ -I_{xy}\lambda_x + I_y\lambda_y - I_{yz}\lambda_z &= K\lambda_y \\ -I_{zx}\lambda_x - I_{yz}\lambda_y + I_z\lambda_z &= K\lambda_z \end{aligned} \quad \text{(B.26)}$$

Transposing the right-hand members leads to the following homogeneous linear equations:

$$\begin{aligned} (I_x - K)\lambda_x - I_{xy}\lambda_y - I_{zx}\lambda_z &= 0 \\ -I_{xy}\lambda_x + (I_y - K)\lambda_y - I_{yz}\lambda_z &= 0 \\ -I_{zx}\lambda_x - I_{yz}\lambda_y + (I_z - K)\lambda_z &= 0 \end{aligned} \quad \text{(B.27)}$$

For this system of equations to have a solution different from $\lambda_x = \lambda_y = \lambda_z = 0$, its discriminant must be zero:

$$\begin{vmatrix} I_x - K & -I_{xy} & -I_{zx} \\ -I_{xy} & I_y - K & -I_{yz} \\ -I_{zx} & -I_{yz} & I_z - K \end{vmatrix} = 0 \quad \text{(B.28)}$$

Expanding this determinant and changing signs, we write

$$\begin{aligned} K^3 - (I_x + I_y + I_z)K^2 + (I_xI_y + I_yI_z + I_zI_x - I_{xy}^2 - I_{yz}^2 - I_{zx}^2)K \\ - (I_xI_yI_z - I_xI_{yz}^2 - I_yI_{zx}^2 - I_zI_{xy}^2 - 2I_{xy}I_{yz}I_{zx}) = 0 \end{aligned} \quad \text{(B.29)}$$

This is a cubic equation in K, which yields three real, positive roots K_1, K_2, and K_3.

To obtain the direction cosines of the principal axis corresponding to the root K_1, we substitute K_1 for K in Eqs. (B.27). Since these equations are now linearly dependent, only two of them may be used to determine λ_x, λ_y, and λ_z. An additional equation may be obtained, however, by recalling from Sec. 2.12 that the direction cosines must satisfy the relation

$$\lambda_x^2 + \lambda_y^2 + \lambda_z^2 = 1 \quad \text{(B.30)}$$

Repeating this procedure with K_2 and K_3, we obtain the direction cosines of the other two principal axes.

We will now show that *the roots K_1, K_2, and K_3 of Eq. (B.29) are the principal moments of inertia of the given body.* Let us substitute for K in Eqs. (B.26) the root K_1, and for λ_x, λ_y, and λ_z the corresponding values $(\lambda_x)_1$, $(\lambda_y)_1$, and $(\lambda_z)_1$ of the direction cosines; the three equations will be satisfied. We now multiply by $(\lambda_x)_1$, $(\lambda_y)_1$, and $(\lambda_z)_1$, respectively, each term in the first, second, and third equation and add the equations obtained in this way. We write

$$\begin{aligned} I_x^2(\lambda_x)_1^2 + I_y^2(\lambda_y)_1^2 + I_z^2(\lambda_z)_1^2 - 2I_{xy}(\lambda_x)_1(\lambda_y)_1 \\ - 2I_{yz}(\lambda_y)_1(\lambda_z)_1 - 2I_{zx}(\lambda_z)_1(\lambda_x)_1 = K_1[(\lambda_x)_1^2 + (\lambda_y)_1^2 + (\lambda_z)_1^2] \end{aligned}$$

Recalling Eq. (B.19), we observe that the left-hand member of this equation represents the moment of inertia of the body with respect to the principal axis corresponding to K_1; it is thus the principal moment of inertia corresponding to that root. On the other hand, recalling Eq. (B.30), we note that the right-hand member reduces to K_1. Thus K_1 itself is the principal moment of inertia. We can show in the same fashion that K_2 and K_3 are the other two principal moments of inertia of the body.

SAMPLE PROBLEM B.6

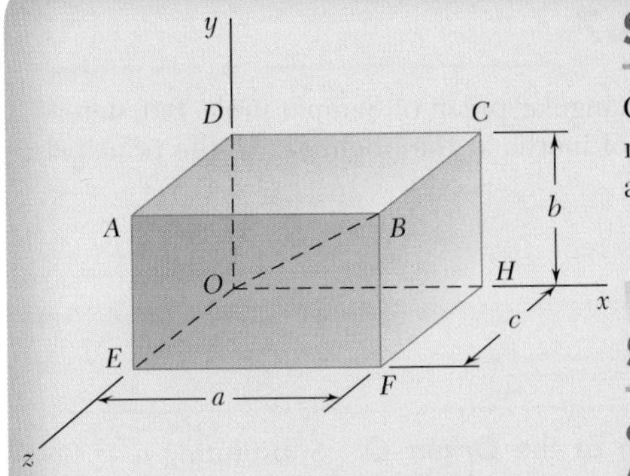

Consider a rectangular prism of mass m and sides a, b, c. Determine (*a*) the moments and products of inertia of the prism with respect to the coordinate axes shown, (*b*) its moment of inertia with respect to the diagonal OB.

SOLUTION

a. Moments and Products of Inertia with Respect to the Coordinate Axes. ***Moments of Inertia.*** Introducing the centroidal axes x', y', z', with respect to which the moments of inertia are given in Fig. B.9, we apply the parallel-axis theorem:

$$I_x = \bar{I}_{x'} + m(\bar{y}^2 + \bar{z}^2) = \tfrac{1}{12}m(b^2 + c^2) + m(\tfrac{1}{4}b^2 + \tfrac{1}{4}c^2)$$

$$I_x = \tfrac{1}{3}m(b^2 + c^2) \quad ◀$$

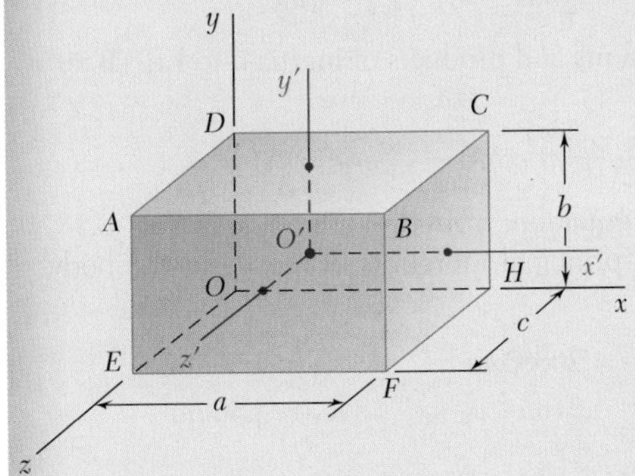

Similarly, $\quad I_y = \tfrac{1}{3}m(c^2 + a^2) \qquad I_z = \tfrac{1}{3}m(a^2 + b^2) \quad ◀$

Products of Inertia. Because of symmetry, the products of inertia with respect to the centroidal axes x', y', z' are zero, and these axes are principal axes of inertia. Using the parallel-axis theorem, we have

$$I_{xy} = \bar{I}_{x'y'} + m\bar{x}\bar{y} = 0 + m(\tfrac{1}{2}a)(\tfrac{1}{2}b) \qquad I_{xy} = \tfrac{1}{4}mab \quad ◀$$

Similarly, $\quad I_{yz} = \tfrac{1}{4}mbc \qquad I_{zx} = \tfrac{1}{4}mca \quad ◀$

b. Moment of Inertia with Respect to *OB*. We recall Eq. (B.19):

$$I_{OB} = I_x\lambda_x^2 + I_y\lambda_y^2 + I_z\lambda_z^2 - 2I_{xy}\lambda_x\lambda_y - 2I_{yz}\lambda_y\lambda_z - 2I_{zx}\lambda_z\lambda_x$$

where the direction cosines of OB are

$$\lambda_x = \cos\theta_x = \frac{OH}{OB} = \frac{a}{(a^2 + b^2 + c^2)^{1/2}}$$

$$\lambda_y = \frac{b}{(a^2 + b^2 + c^2)^{1/2}} \qquad \lambda_z = \frac{c}{(a^2 + b^2 + c^2)^{1/2}}$$

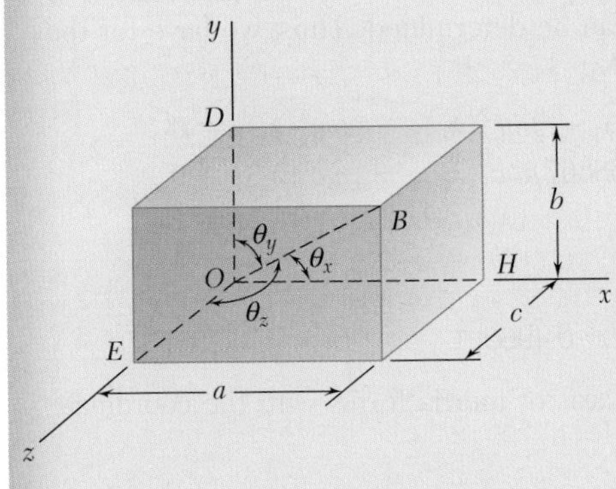

Substituting the values obtained for the moments and products of inertia and for the direction cosines into the equation for I_{OB}, we have

$$I_{OB} = \frac{1}{a^2 + b^2 + c^2}[\tfrac{1}{3}m(b^2 + c^2)a^2 + \tfrac{1}{3}m(c^2 + a^2)b^2 + \tfrac{1}{3}m(a^2 + b^2)c^2 - \tfrac{1}{2}ma^2b^2 - \tfrac{1}{2}mb^2c^2 - \tfrac{1}{2}mc^2a^2]$$

$$I_{OB} = \frac{m}{6}\frac{a^2b^2 + b^2c^2 + c^2a^2}{a^2 + b^2 + c^2} \quad ◀$$

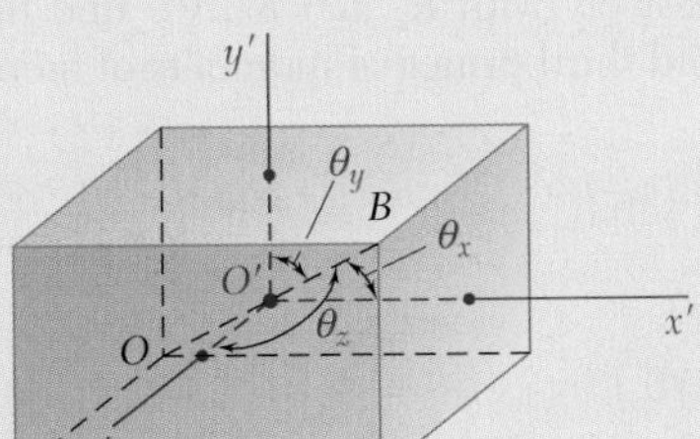

Alternative Solution. The moment of inertia I_{OB} can be obtained directly from the principal moments of inertia $\bar{I}_{x'}$, $\bar{I}_{y'}$, $\bar{I}_{z'}$, since the line OB passes through the centroid O'. Since the x', y', z' axes are principal axes of inertia, we use Eq. (B.23) to write

$$I_{OB} = \bar{I}_{x'}\lambda_x^2 + \bar{I}_{y'}\lambda_y^2 + \bar{I}_{z'}\lambda_z^2$$

$$= \frac{1}{a^2 + b^2 + c^2}\left[\frac{m}{12}(b^2 + c^2)a^2 + \frac{m}{12}(c^2 + a^2)b^2 + \frac{m}{12}(a^2 + b^2)c^2\right]$$

$$I_{OB} = \frac{m}{6}\frac{a^2b^2 + b^2c^2 + c^2a^2}{a^2 + b^2 + c^2} \quad ◀$$

SAMPLE PROBLEM B.7

If $a = 3c$ and $b = 2c$ for the rectangular prism of Sample Prob. B.6, determine (*a*) the principal moments of inertia at the origin O, (*b*) the principal axes of inertia at O.

SOLUTION

a. Principal Moments of Inertia at the Origin O. Substituting $a = 3c$ and $b = 2c$ into the solution to Sample Prob. B.6, we have

$$I_x = \tfrac{5}{3}mc^2 \qquad I_y = \tfrac{10}{3}mc^2 \qquad I_z = \tfrac{13}{3}mc^2$$
$$I_{xy} = \tfrac{3}{2}mc^2 \qquad I_{yz} = \tfrac{1}{2}mc^2 \qquad I_{zx} = \tfrac{3}{4}mc^2$$

Substituting the values of the moments and products of inertia into Eq. (B.29) and collecting terms yields

$$K^3 - (\tfrac{28}{3}mc^2)K^2 + (\tfrac{3479}{144}m^2c^4)K - \tfrac{589}{54}m^3c^6 = 0$$

We then solve for the roots of this equation; from the discussion in Sec. B.18, it follows that these roots are the principal moments of inertia of the body at the origin.

$$K_1 = 0.568867mc^2 \qquad K_2 = 4.20885mc^2 \qquad K_3 = 4.55562mc^2$$
$$K_1 = 0.569mc^2 \qquad K_2 = 4.21mc^2 \qquad K_3 = 4.56mc^2 \quad ◀$$

b. Principal Axes of Inertia at O. To determine the direction of a principal axis of inertia, we first substitute the corresponding value of K into two of the equations (B.27); the resulting equations together with Eq. (B.30) constitute a system of three equations from which the direction cosines of the corresponding principal axis can be determined. Thus, we have for the first principal moment of inertia K_1:

$$(\tfrac{5}{3} - 0.568867)mc^2(\lambda_x)_1 - \tfrac{3}{2}mc^2(\lambda_y)_1 - \tfrac{3}{4}mc^2(\lambda_z)_1 = 0$$
$$-\tfrac{3}{2}mc^2(\lambda_x)_1 + (\tfrac{10}{3} - 0.568867)mc^2(\lambda_y)_1 - \tfrac{1}{2}mc^2(\lambda_z)_1 = 0$$
$$(\lambda_x)_1^2 + (\lambda_y)_1^2 + (\lambda_z)_1^2 = 1$$

Solving yields

$$(\lambda_x)_1 = 0.836600 \qquad (\lambda_y)_1 = 0.496001 \qquad (\lambda_z)_1 = 0.232557$$

The angles that the first principal axis of inertia forms with the coordinate axes are then

$$(\theta_x)_1 = 33.2° \qquad (\theta_y)_1 = 60.3° \qquad (\theta_z)_1 = 76.6° \quad ◀$$

Using the same set of equations successively with K_2 and K_3, we find that the angles associated with the second and third principal moments of inertia at the origin are, respectively.

$$(\theta_x)_2 = 57.8° \qquad (\theta_y)_2 = 146.6° \qquad (\theta_z)_2 = 98.0° \quad ◀$$

and

$$(\theta_x)_3 = 82.8° \qquad (\theta_y)_3 = 76.1° \qquad (\theta_z)_3 = 164.3° \quad ◀$$

SOLVING PROBLEMS ON YOUR OWN

In this lesson we defined the *mass products of inertia* I_{xy}, I_{yz}, and I_{zx} of a body and showed you how to determine the moments of inertia of that body with respect to an arbitrary axis passing through the origin O. You also learned how to determine at the origin O the *principal axes of inertia* of a body and the corresponding *principal moments of inertia.*

1. Determining the mass products of inertia of a composite body. The mass products of inertia of a composite body with respect to the coordinate axes can be expressed as the sums of the products of inertia of its component parts with respect to those axes. For each component part, we can use the parallel-axis theorem and write Eqs. (B.20)

$$I_{xy} = \bar{I}_{x'y'} + m\bar{x}\,\bar{y} \qquad I_{yz} = \bar{I}_{y'z'} + m\bar{y}\,\bar{z} \qquad I_{zx} = \bar{I}_{z'x'} + m\bar{z}\,\bar{x}$$

where the primes denote the centroidal axes of each component part and where $\bar{x}$, $\bar{y}$, and $\bar{z}$ represent the coordinates of its center of gravity. Keep in mind that the mass products of inertia can be positive, negative, or zero, and be sure to take into account the signs of $\bar{x}$, $\bar{y}$, and $\bar{z}$.

a. From the properties of symmetry of a component part, you can deduce that two or all three of its centroidal mass products of inertia are zero. For instance, you can verify that for a thin plate parallel to the xy plane; a wire lying in a plane parallel to the xy plane; a body with a plane of symmetry parallel to the xy plane; and a body with an axis of symmetry parallel to the z axis, *the products of inertia* $\bar{I}_{y'z'}$ *and* $\bar{I}_{z'x'}$ *are zero.*

For rectangular, circular, or semicircular plates with axes of symmetry parallel to the coordinate axes; straight wires parallel to a coordinate axis; circular and semicircular wires with axes of symmetry parallel to the coordinate axes; and rectangular prisms with axes of symmetry parallel to the coordinate axes, *the products of inertia* $\bar{I}_{x'y'}$, $\bar{I}_{y'z'}$, *and* $\bar{I}_{z'x'}$ *are all zero.*

b. Mass products of inertia which are different from zero can be computed from Eqs. (B.18). Although, in general, a triple integration is required to determine a mass product of inertia, a single integration can be used if the given body can be divided into a series of thin, parallel slabs. The computations are then similar to those discussed in the previous lesson for moments of inertia.

2. Computing the moment of inertia of a body with respect to an arbitrary axis *OL*. An expression for the moment of inertia I_{OL} was derived in Sec. B.6 and is given in Eq. (B.19). Before computing I_{OL}, you must first determine the mass moments and products of inertia of the body with respect to the given coordinate axes as well as the direction cosines of the unit vector $\boldsymbol{\lambda}$ along OL.

(continued)

3. Calculating the principal moments of inertia of a body and determining its principal axes of inertia. You saw in Sec. B.7 that it is always possible to find an orientation of the coordinate axes for which the mass products of inertia are zero. These axes are referred to as the *principal axes of inertia* and the corresponding moments of inertia are known as the *principal moments of inertia* of the body. In many cases, the principal axes of inertia of a body can be determined from its properties of symmetry. The procedure required to determine the principal moments and principal axes of inertia of a body with no obvious property of symmetry was discussed in Sec. B.8 and was illustrated in Sample Prob. B.7. It consists of the following steps.

a. Expand the determinant in Eq. (B.28) and solve the resulting cubic equation. The solution can be obtained by trial and error or, preferably, with an advanced scientific calculator or with the appropriate computer software. The roots K_1, K_2, and K_3 of this equation are the principal moments of inertia of the body.

b. To determine the direction of the principal axis corresponding to K_1, substitute this value for K in two of the equations (B.27) and solve these equations together with Eq. (B.30) for the direction cosines of the principal axis corresponding to K_1.

c. Repeat this procedure with K_2 and K_3 to determine the directions of the other two principal axes. As a check of your computations, you may wish to verify that the scalar product of any two of the unit vectors along the three axes you have obtained is zero and, thus, that these axes are perpendicular to each other.

d. When a principal moment of inertia is approximately equal to a moment of inertia with respect to a coordinate axis, the calculated values of the corresponding direction cosines will be very sensitive to the number of significant figures used in your computations. Thus, for this case we suggest you express your intermediate answers in terms of six or seven significant figures to avoid possible errors.

PROBLEMS

B.39 Determine the mass products of inertia I_{xy}, I_{yz}, and I_{zx} of the steel fixture shown. (The density of steel is 7850 kg/m^3.)

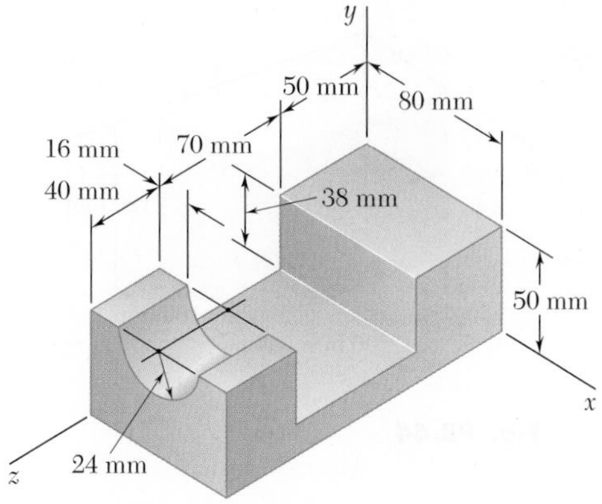

Fig. PB.39

B.40 Determine the mass products of inertia I_{xy}, I_{yz}, and I_{zx} of the steel machine element shown. (The density of steel is 7850 kg/m^3.)

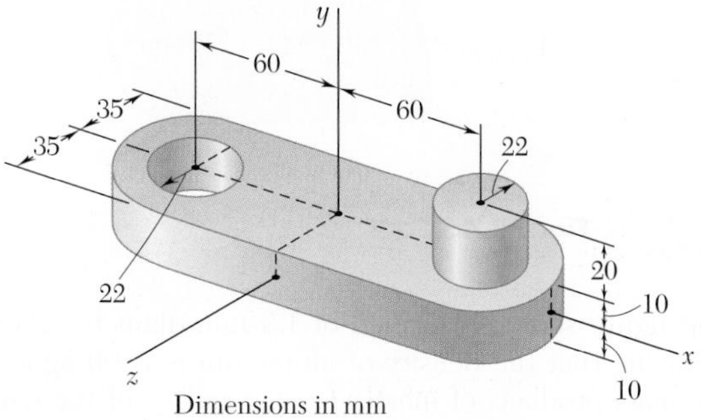

Fig. PB.40

B.41 and B.42 Determine the mass products of inertia I_{xy}, I_{yz}, and I_{zx} of the cast aluminum machine component shown. (The density of aluminum is 2770 kg/m^3)

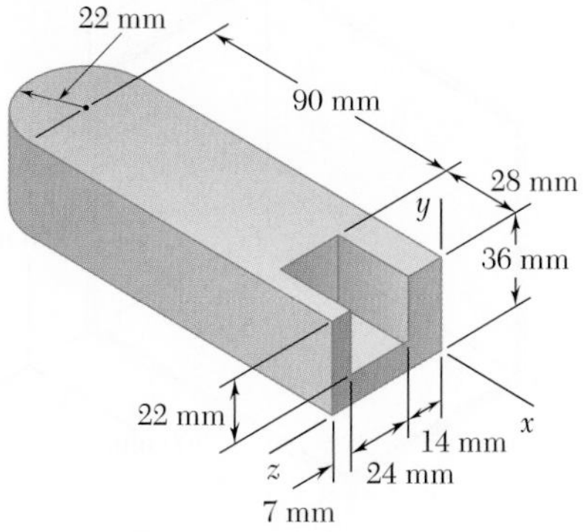

Fig. PB.41

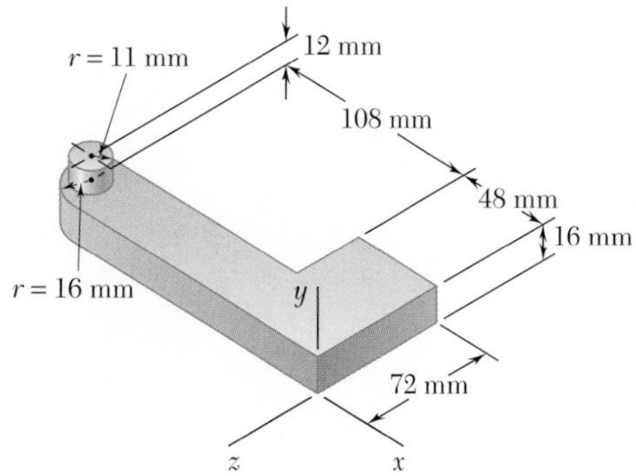

Fig. PB.42

B.43 through B.46 A section of sheet steel 2 mm thick is cut and bent into the machine component shown. Knowing that the density of steel is 7850 kg/m^3, determine the mass products of inertia I_{xy}, I_{yz}, and I_{zx} of the component.

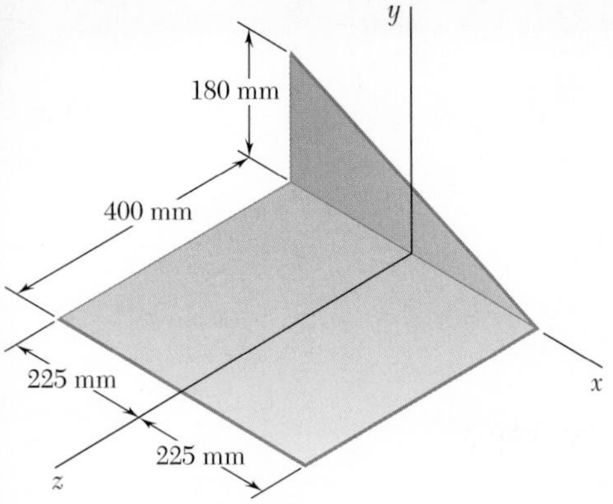

Fig. PB.43

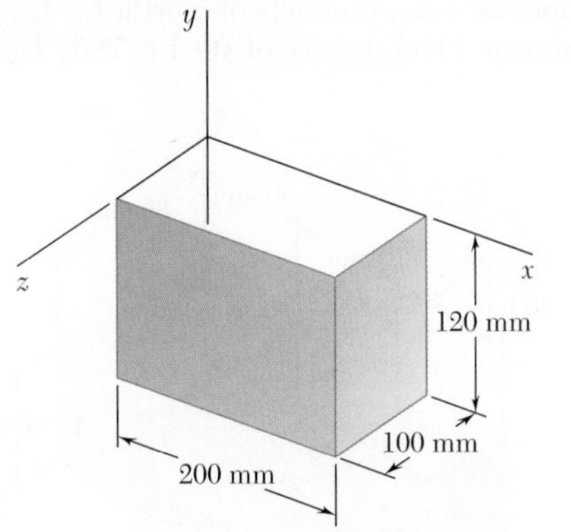

Fig. PB.44

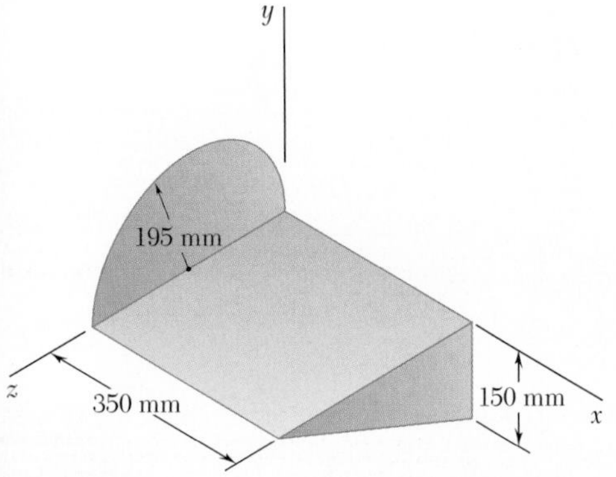

Fig. PB.45

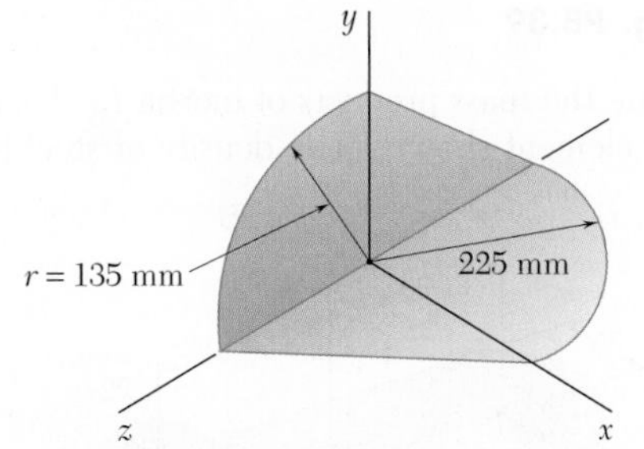

Fig. PB.46

B.47 The figure shown is formed of 1.5-mm-diameter aluminum wire. Knowing that the density of aluminum is 2800 kg/m^3, determine the mass products of inertia I_{xy}, I_{yz}, and I_{zx} of the wire figure.

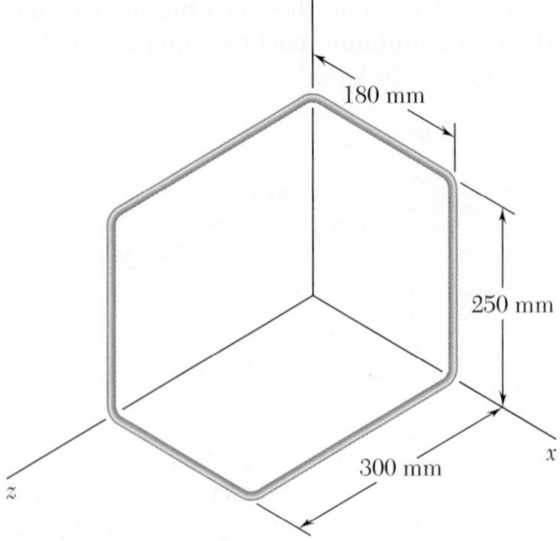

Fig. PB.47

B.48 Thin aluminum wire of uniform diameter is used to form the figure shown. Denoting by m' the mass per unit length of the wire, determine the mass products of inertia I_{xy}, I_{yz}, and I_{zx} of the wire figure.

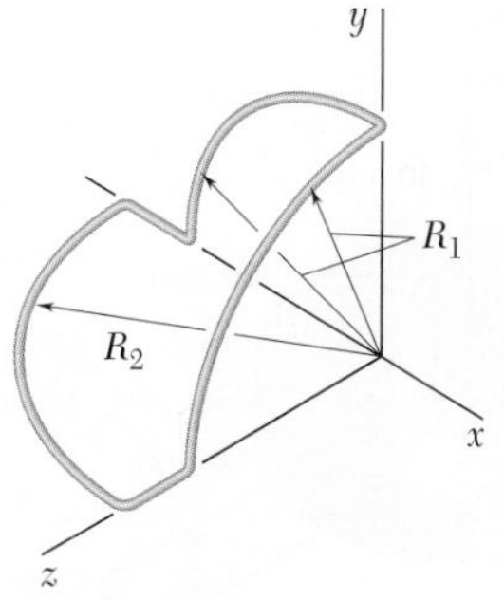

Fig. PB.48

B.49 and B.50 Brass wire with a weight per unit length w is used to form the figure shown. Determine the mass products of inertia I_{xy}, I_{yz}, and I_{zx} of the wire figure.

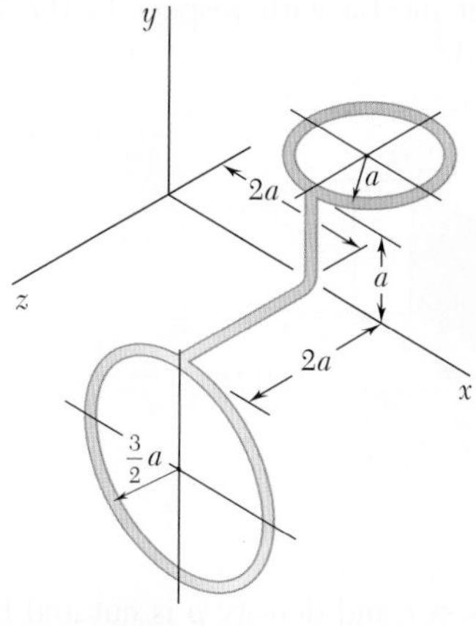

Fig. PB.49

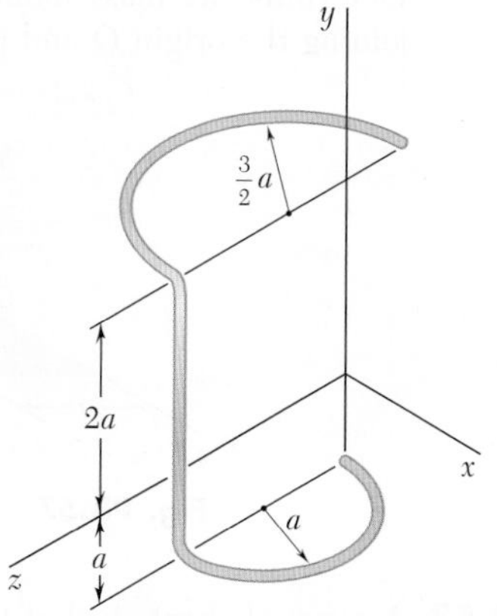

Fig. PB.50

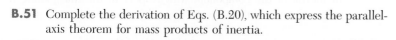

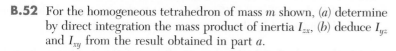

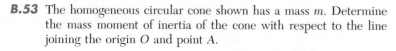

B.51 Complete the derivation of Eqs. (B.20), which express the parallel-axis theorem for mass products of inertia.

B.52 For the homogeneous tetrahedron of mass m shown, (*a*) determine by direct integration the mass product of inertia I_{zx}, (*b*) deduce I_{yz} and I_{xy} from the result obtained in part *a*.

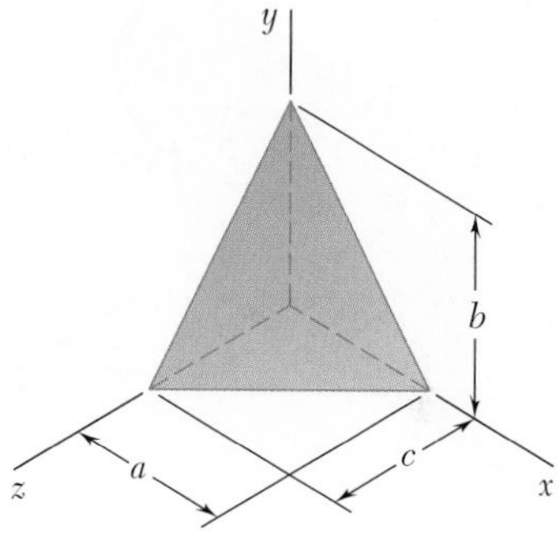

Fig. PB.52

B.53 The homogeneous circular cone shown has a mass m. Determine the mass moment of inertia of the cone with respect to the line joining the origin O and point A.

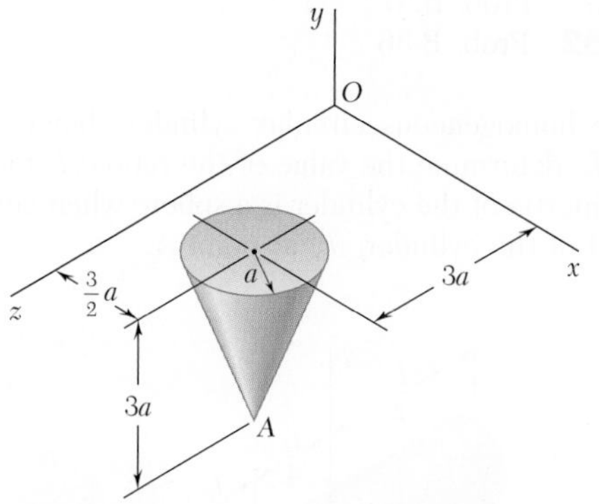

Fig. PB.53

B.54 The homogeneous circular cylinder shown has a mass m. Determine the mass moment of inertia of the cylinder with respect to the line joining the origin O and point A that is located on the perimeter of the top surface of the cylinder.

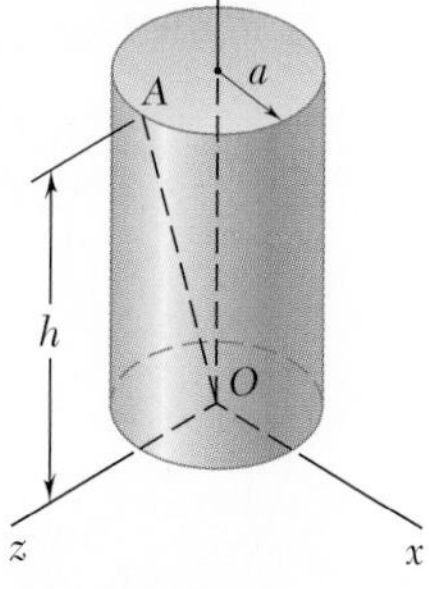

Fig. PB.54

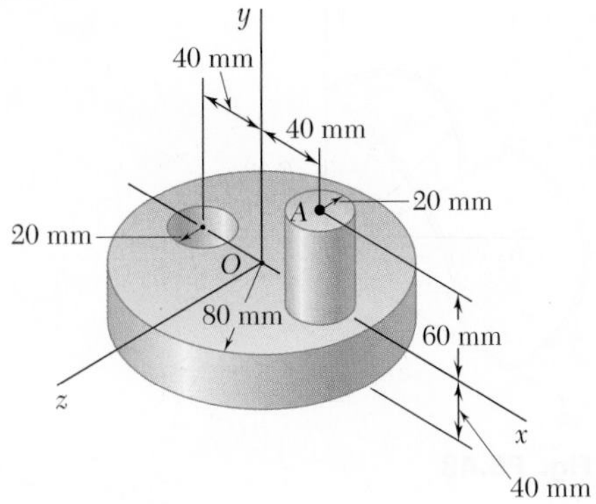

Fig. PB.55

B.55 Shown is the machine element of Prob. B.55. Determine its mass moment of inertia with respect to the line joining the origin O and point A.

B.56 Determine the mass moment of inertia of the steel fixture of Probs. B.35 and B.39 with respect to the axis through the origin that forms equal angles with the x, y, and z axes.

B.57 The thin bent plate shown is of uniform density and weight W. Determine its mass moment of inertia with respect to the line joining the origin O and point A.

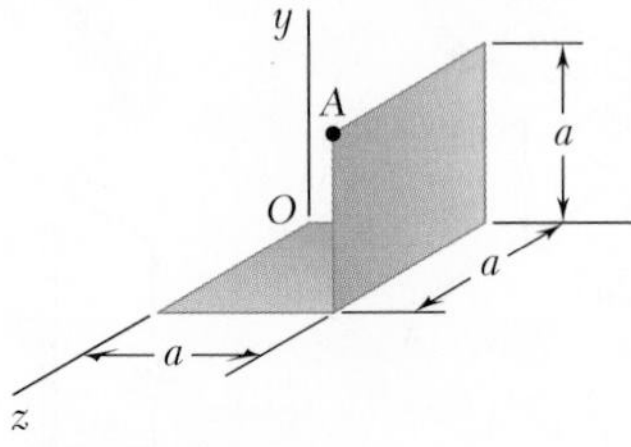

Fig. PB.57

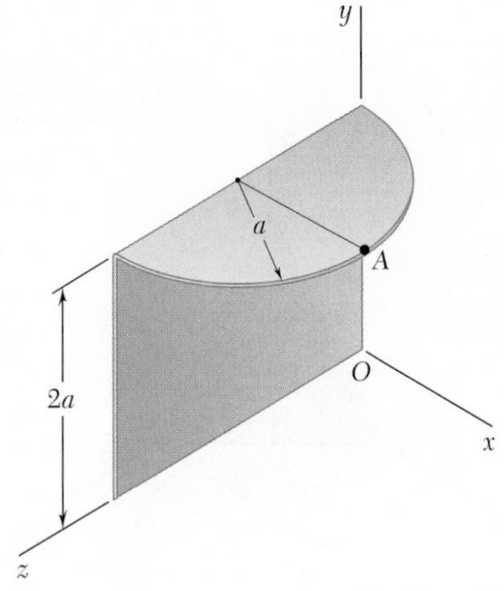

Fig. PB.58

B.58 A piece of sheet steel of thickness t and density ρ is cut and bent into the machine component shown. Determine the mass moment of inertia of the component with respect to the line joining the origin O and point A.

B.59 Determine the mass moment of inertia of the machine component of Probs. B.26 and B.45 with respect to the axis through the origin characterized by the unit vector $\lambda = (-4\mathbf{i} + 8\mathbf{j} + \mathbf{k})/9$.

B.60 through *B.62* For the wire figure of the problem indicated, determine the mass moment of inertia of the figure with respect to the axis through the origin characterized by the unit vector $\lambda = (-3\mathbf{i} - 6\mathbf{j} + 2\mathbf{k})/7$.

B.60 Prob. B.38
B.61 Prob. B.37
B.62 Prob. B.36

B.63 For the homogeneous circular cylinder shown, of radius a and length L, determine the value of the ratio a/L for which the ellipsoid of inertia of the cylinder is a sphere when computed (a) at the centroid of the cylinder, (b) at point A.

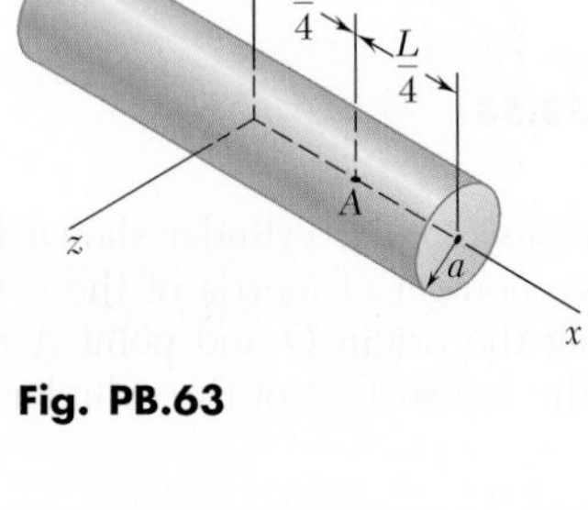

Fig. PB.63

B.64 For the rectangular prism shown, determine the values of the ratios b/a and c/a so that the ellipsoid of inertia of the prism is a sphere when computed (*a*) at point *A*, (*b*) at point *B*.

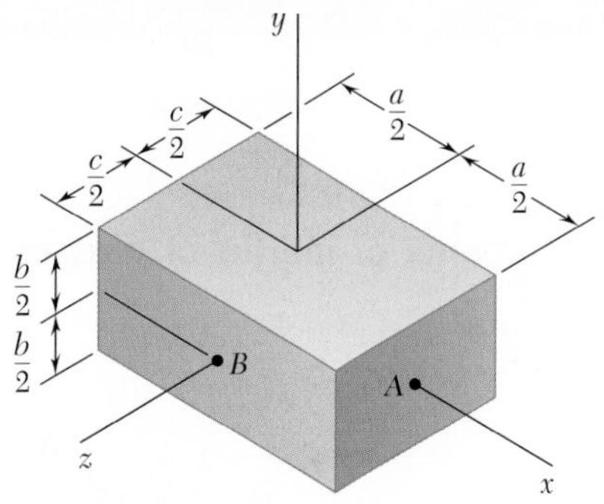

Fig. PB.64

B.65 For the right circular cone of Sample Prob. 9.11, determine the value of the ratio a/h for which the ellipsoid of inertia of the cone is a sphere when computed (*a*) at the apex of the cone, (*b*) at the center of the base of the cone.

B.66 Given an arbitrary body and three rectangular axes *x*, *y*, and *z*, prove that the mass moment of inertia of the body with respect to any one of the three axes cannot be larger than the sum of the mass moments of inertia of the body with respect to the other two axes. That is, prove that the inequality $I_x \leq I_y + I_z$ and the two similar inequalities are satisfied. Further, prove that $I_y \geq \frac{1}{2}I_x$ if the body is a homogeneous solid of revolution, where *x* is the axis of revolution and *y* is a transverse axis.

B.67 Consider a cube of mass *m* and side *a*. (*a*) Show that the ellipsoid of inertia at the center of the cube is a sphere, and use this property to determine the moment of inertia of the cube with respect to one of its diagonals. (*b*) Show that the ellipsoid of inertia at one of the corners of the cube is an ellipsoid of revolution, and determine the principal moments of inertia of the cube at that point.

B.68 Given a homogeneous body of mass *m* and of arbitrary shape and three rectangular axes *x*, *y*, and *z* with origin at *O*, prove that the sum $I_x + I_y + I_z$ of the mass moments of inertia of the body cannot be smaller than the similar sum computed for a sphere of the same mass and the same material centered at *O*. Further, using the result of Prob. B.66, prove that if the body is a solid of revolution, where *x* is the axis of revolution, its mass moment of inertia I_y about a transverse axis *y* cannot be smaller than $3ma^2/10$, where *a* is the radius of the sphere of the same mass and the same material.

***B.69** The homogeneous circular cylinder shown has a mass *m*, and the diameter *OB* of its top surface forms 45° angles with the *x* and *z* axes. (*a*) Determine the principal mass moments of inertia of the cylinder at the origin *O*. (*b*) Compute the angles that the principal axes of inertia at *O* form with the coordinate axes. (*c*) Sketch the cylinder, and show the orientation of the principal axes of inertia relative to the *x*, *y*, and *z* axes.

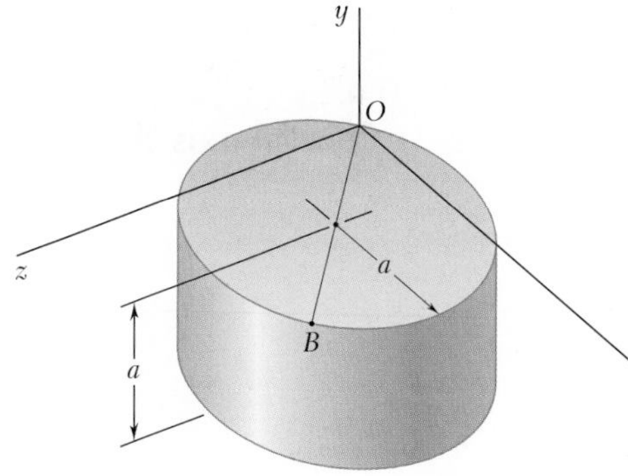

Fig. PB.69

B.70 through *B.74* For the component described in the problem indicated, determine (*a*) the principal mass moments of inertia at the origin, (*b*) the principal axes of inertia at the origin. Sketch the body and show the orientation of the principal axes of inertia relative to the *x*, *y*, and *z* axes.

***B.70** Prob. B.55
****B.71*** Probs. B.35 and B.39
***B.72** Prob. B.57
***B.73** Prob. B.58
****B.74*** Probs. B.38 and B.60–62

REVIEW AND SUMMARY

Moments of inertia of masses

This appendix was devoted to the determination of *moments of inertia of masses,* which are encountered in dynamics in problems involving the rotation of a rigid body about an axis. The mass moment of inertia of a body with respect to an axis AA' (Fig. B.16) was defined as

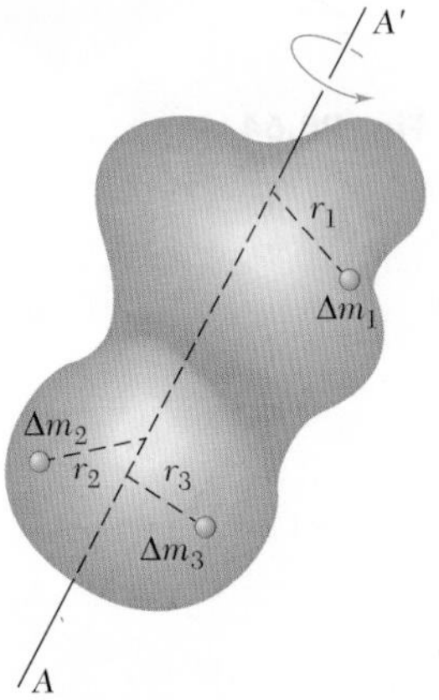

Fig. B.16

$$I = \int r^2\, dm \tag{B.1}$$

where r is the distance from AA' to the element of mass [Sec. B.1]. The *radius of gyration* of the body was defined as

$$k = \sqrt{\frac{I}{m}} \tag{B.2}$$

The moments of inertia of a body with respect to the coordinates axes were expressed as

$$\begin{aligned} I_x &= \int (y^2 + z^2)\, dm \\ I_y &= \int (z^2 + x^2)\, dm \\ I_z &= \int (x^2 + y^2)\, dm \end{aligned} \tag{B.3}$$

Parallel-axis theorem

We saw that the *parallel-axis theorem* also applies to mass moments of inertia [Sec. B.2]. Thus, the moment of inertia I of a body with respect to an arbitrary axis AA' (Fig. B.17) can be expressed as

$$I = \bar{I} + md^2 \tag{B.6}$$

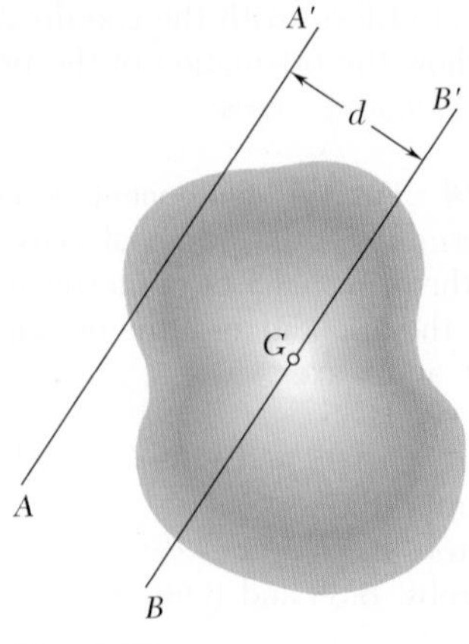

Fig. B.17

where $\bar{I}$ is the moment of inertia of the body with respect to the centroidal axis BB' which is parallel to the axis AA', m is the mass of the body, and d is the distance between the two axes.

Moments of inertia of thin plates

The moments of inertia of *thin plates* can be readily obtained from the moments of inertia of their areas [Sec. B.3]. We found that for a *rectangular plate* the moments of inertia with respect to the axes shown (Fig. B.18) are

$$I_{AA'} = \tfrac{1}{12}ma^2 \qquad I_{BB'} = \tfrac{1}{12}mb^2 \tag{B.12}$$

$$I_{CC'} = I_{AA'} + I_{BB'} = \tfrac{1}{12}m(a^2 + b^2) \tag{B.13}$$

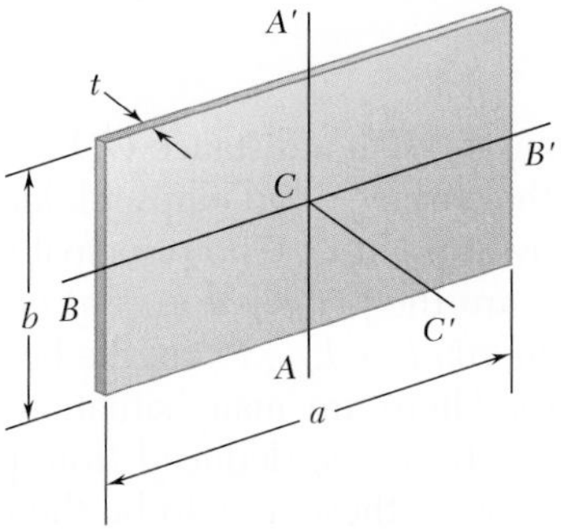

Fig. B.18

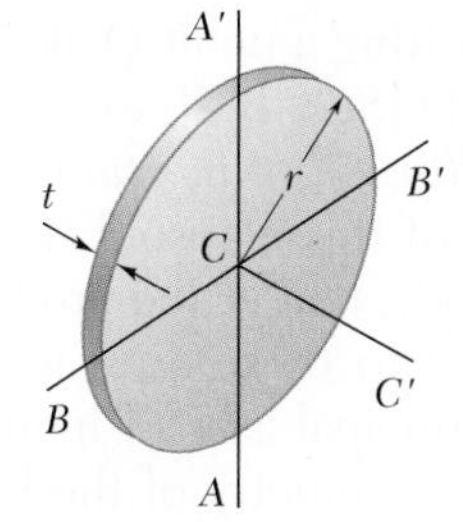

Fig. B.19

while for a *circular plate* (Fig. B.19) they are

$$I_{AA'} = I_{BB'} = \tfrac{1}{4}mr^2 \tag{B.14}$$

$$I_{CC'} = I_{AA'} + I_{BB'} = \tfrac{1}{2}mr^2 \tag{B.15}$$

Composite bodies

When a body possesses *two planes of symmetry,* it is usually possible to use a single integration to determine its moment of inertia with respect to a given axis by selecting the element of mass dm to be a thin plate [Sample Probs. B.2 and B.3]. On the other hand, when a body consists of *several common geometric shapes,* its moment of inertia with respect to a given axis can be obtained by using the formulas given in Fig. B.9 together with the parallel-axis theorem.

Moment of inertia with respect to an arbitrary axis

In the last portion of the chapter, we learned to determine the moment of inertia of a body *with respect to an arbitrary axis OL* which is drawn through the origin O [Sec. B.6]. Denoting by λ_x, λ_y, λ_z the components of the unit vector $\boldsymbol{\lambda}$ along OL (Fig. B.20) and introducing the *products of inertia*

$$I_{xy} = \int xy\,dm \qquad I_{yz} = \int yz\,dm \qquad I_{zx} = \int zx\,dm \tag{B.18}$$

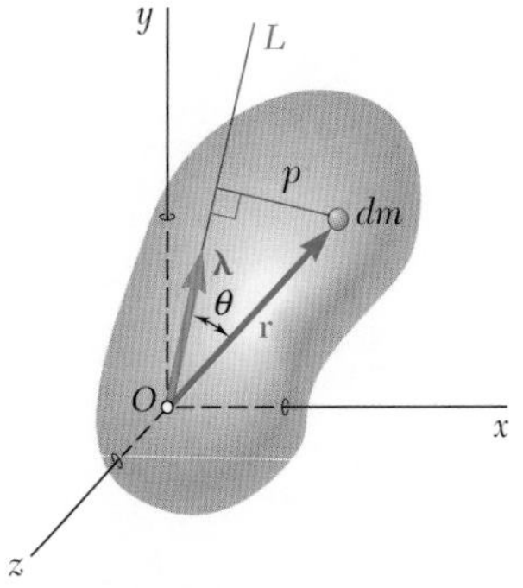

Fig. B.20

we found that the moment of inertia of the body with respect to OL could be expressed as

$$I_{OL} = I_x\lambda_x^2 + I_y\lambda_y^2 + I_z\lambda_z^2 - 2I_{xy}\lambda_x\lambda_y - 2I_{yz}\lambda_y\lambda_z - 2I_{zx}\lambda_z\lambda_x \tag{B.19}$$

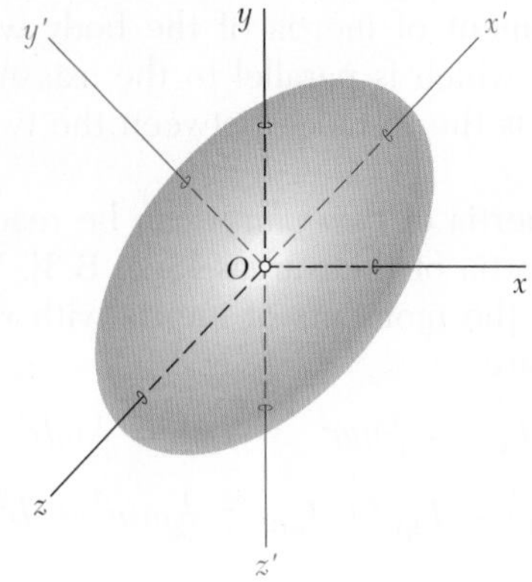

Fig. B.21

Ellipsoid of inertia

By plotting a point Q along each axis OL at a distance $OQ = 1\sqrt{I_{OL}}$ from O [Sec. B.7], we obtained the surface of an ellipsoid, known as the *ellipsoid of inertia* of the body at point O. The principal axes x', y', z' of this ellipsoid (Fig. B.21) are the *principal axes of inertia* of the body; that is, the products of inertia $I_{x'y'}$, $I_{y'z'}$, $I_{z'x'}$ of the body with respect to these axes are all zero. There are many situations when the principal axes of inertia of a body can be deduced from properties of symmetry of the body. Choosing these axes to be the coordinate axes, we can then express I_{OL} as

Principal axes of inertia
Principal moments of inertia

$$I_{OL} = I_{x'}\lambda_{x'}^2 + I_{y'}\lambda_{y'}^2 + I_{z'}\lambda_{z'}^2$$

where $I_{x'}$, $I_{y'}$, $I_{z'}$ are the *principal moments of inertia* of the body at O.

When the principal axes of inertia cannot be obtained by observation [Sec. B.7], it is necessary to solve the cubic equation

$$K^3 - (I_x + I_y + I_z)K^2 + (I_xI_y + I_yI_z + I_zI_x - I_{xy}^2 - I_{yz}^2 - I_{zx}^2)K \\ - (I_xI_yI_z - I_xI_{yz}^2 - I_yI_{zx}^2 - I_zI_{xy}^2 - 2I_{xy}I_{yz}I_{zx}) = 0$$

We found [Sec. B.8] that the roots K_1, K_2, and K_3 of this equation are the principal moments of inertia of the given body. The direction cosines $(\lambda_x)_1$, $(\lambda_y)_1$, and $(\lambda_z)_1$ of the principal axis corresponding to the principal moment of inertia K_1 are then determined by substituting K_1 into Eqs. (B.27) and solving two of these equations and Eq. (B.30) simultaneously. The same procedure is then repeated using K_2 and K_3 to determine the direction cosines of the other two principal axes.

Appendix C

Fundamentals of Engineering Examination

Engineers are required to be licensed when their work directly affects the public health, safety, and welfare. The intent is to ensure that engineers have met minimum qualifications involving competence, ability, experience, and character. The licensing process involves an initial exam, called the *Fundamentals of Engineering Examination*, professional experience, and a second exam, called the *Principles and Practice of Engineering*. Those who successfully complete these requirements are licensed as a *Professional Engineer*. The exams are developed under the auspices of the *National Council of Examiners for Engineering and Surveying*.

The first exam, the *Fundamentals of Engineering Examination*, can be taken just before or after graduation from a four-year accredited engineering program. The exam stresses subject material in a typical undergraduate engineering program, including dynamics. The topics included in the exam cover much of the material in this book. The following is a list of the main topic areas, with references to the appropriate sections in this book. Also included are problems that can be solved to review this material.

Kinematics (11.1–11.6; 11.9–11.14; 15.2–15.8)
Problems: 11.3, 11.5, 11.35, 11.61, 11.67, 11.97, 15.8, 15.29, 15.39, 15.61, 15.65, 15.87, 15.112, 15.142

Force, Mass, and Acceleration (12.1–12.6; 16.2–16.8)
Problems: 12.6, 12.8, 12.30, 12.32, 12.36, 12.45, 12.52, 12.55, 16.1, 16.3, 16.7, 16.26, 16.27, 16.50, 16.61, 16.63, 16.78, 16.85, 16.137

Work and Energy (13.1–13.6; 13.8; 17.1–17.7)
Problems: 13.5, 13.7, 13.17, 13.22, 13.40, 13.41, 13.47, 13.64, 13.66, 13.68, 17.1, 17.2, 17.16, 17.22

Impulse and Momentum (13.10–13.15; 17.8–17.12)
Problems: 13.119, 13.121, 13.129, 13.134, 13.146, 13.155, 13.160, 13.171, 17.53, 17.58, 17.70, 17.72, 17.96, 17.97, 17.111

Vibration (19.1–19.3; 19.5–19.7)
Problems: 19.1, 19.2, 19.11, 19.18, 19.23, 19.28, 19.49, 19.55, 19.63, 19.74, 19.83, 19.85, 19.101, 19.105, 19.116

Friction (Problems involving friction occur in each of the above subjects)

Photo Credits

CHAPTER 11

Opener: © NASA/Getty Images RF; **Photo 11.1:** U.S. Department of Energy; **Photo 11.2:** © Digital Vision/Getty RF; **Photo 11.3:** © Brand X Pictures/Jupiter Images RF; **Photo 11.4:** © Digital Vision/Getty RF; **Photo 11.5, Photo 11.6:** © Royalty-Free/Corbis.

CHAPTER 12

Opener: © Lester Lefkowitz/Corbis; **Photo 12.1:** © Royalty-Free/Corbis; **Photo 12.2:** © Brand X Pictures/PunchStock RF; **Photo 12.3:** © Royalty-Free/Corbis; **Photo 12.4:** © Russell Illig/Getty RF; **Photo 12.5:** © Royalty-Free/Corbis.

CHAPTER 13

Opener: © Tom Miles; **Photo 13.1, Photo 13.2:** © Scandia National Laboratories/Getty RF; **Photo 13.3:** © Andrew Davidhazy/RIT; **Photo 13.4:** © Tom McCarthy/Photolibrary.

CHAPTER 14

Opener: © XCOR; **Photo 14.1:** NASA; **Photo 14.2:** © Royalty-Free/Corbis; **Photo 14.3:** © Brand X Pictures/PunchStock RF.

CHAPTER 15

Opener: Courtesy A.P. Moller-Maersk; **(inset):** Courtesy of Wärtsilä Corporation; **Photo 15.1:** © Chris Hellier/Corbis; **Photo 15.2:** © Royalty-Free/Corbis; **Photo 15.3:** © Joseph Nettis/Stock Boston Inc.; **Photo 15.4:** © AGE Fotostock/Photolibrary; **Photo 15.5:** © George Tiedemann/NewSport/Corbis; **Photo 15.6:** © Royalty-Free/Corbis; **Photo 15.7:** Purdue University/Physics/PRIME Lab; **Photo 15.8:** © Northrop Grumman/Index Stock Imagery/Jupiter Images/Photolibrary; **Photo 15.9:** © Royalty-Free/Corbis.

CHAPTER 16

Opener, Photo 16.1: © Getty RF; **Photo 16.2:** Courtesy of Samsung Semiconductor, Inc.; **Photo 16.3:** © Tony Arruza/Corbis; **Photo 16.4:** © Robert E. Daemmrich.

CHAPTER 17

Opener: © AP Photo/Matt Dunham; **Photo 17.1:** © Richard McDowell/Alamy RF; **Photo 17.2:** Phillip Cornwell; **Photo 17.3:** © Photography by Leah; **Photo 17.4:** © Chuck Savage/Corbis.

CHAPTER 18

Opener: © Royalty-Free/Corbis; **Photo 18.1:** © age fotostock/SuperStock; **Photo 18.2:** © Matthias Kulka/Corbis; **Photo 18.3:** © Roger Ressmeyer/Corbis; **Photo 18.4:** Reprinted Courtesy of Caterpillar Inc.; **Photo 18.5:** © Lawrence Manning/Corbis RF.

CHAPTER 19

Opener: © Peter Tsai Photography; **Photo 19.1:** © Tony Freeman/Index Stock/Photolibrary; **Photo 19.2:** © The McGraw-Hill Companies, Inc., photo by Sabina Dowell; **Photo 19.3:** Courtesy of MTS Systems Corporation.

APPENDIX B

Photo B.1: Reprinted Courtesy of Caterpillar Inc.

Index

A

B

C

D

E

F

G

H

I

J

K

L

M

N

O

P

Q

R

S

T

U

V

W

Answers to Problems

Answers to problems with a number set in straight type are given on this and the following pages. Answers to problems set in italic are not listed.

CHAPTER 11

11.1 $x = 11.00$ m, $v = -8.00$ m/s, $a = -8.00$ m/s^2.

11.2 $t = 1.000$ s, $x_1 = 15.00$ m, $a_1 = -6.00$ m/s^2; $t = 2.00$ s, $x_2 = 14.00$ m, $a_2 = 6.00$ m/s^2.

11.3 (*a*) $x_1 = 102.9$ mm, $v_1 = -35.6$ mm/s, $a_1 = -11.40$ mm/s^2.
(*b*) $v_{max} = -36.1$ mm/s, $a_{max} = 72.1$ mm/s^2.

11.4 (*a*) $x_0 = 0$ mm, $v_0 = 960$ mm/s $\rightarrow$, $a_0 = 9220$ mm/s^2 $\leftarrow$.
(*b*) $x_{0.3} = 14.16$ mm $\leftarrow$, $v_{0.3} = 87.9$ mm/s $\rightarrow$, $a_{0.3} = 3110$ mm/s^2 or 3.11 m/s^2 $\rightarrow$.

11.5 $t = 0.667$ s, $x_{2/3} = 0.259$m, $v_{2/3} = -8.56$ m/s.

11.6 (*a*) $t = 2.00$ s, $t = 4.00$ s, (*b*) $x_3 = 10.00$ m, $m = 22.0$ m.

11.7 (*a*) $t = 1.000$ s and $t = 4.00$ s. (*b*) $x_{2.5} = +1.500$ m, Total distance = 24.5 m.

11.8 (*a*) $t = 6.00$ s, (*b*) $v_{10} = 144.0$ m/s, $a_0 = 48.0$ m/s^2, $m = 472$ m.

11.9 (*a*) $v_0 = 24.5$ m/s^2. (*b*) $t_f = 8.17$ s.

11.10 $v_7 = -825$ mm/s, $x_7 = 50.0$ mm, 2190 mm.

11.11 $x(t) = \dfrac{1}{108}t^4 + 10t + 24$.

$v(t) = \dfrac{1}{27}t^3 + 10$.

11.12 (*a*) $k = 6.00$ m/s^4. (*b*) $a = 6t^2$, $v = 2t^3 - 8$, $x = \dfrac{1}{2}t^4 - 8t + 8$.

11.13 $v = 42.4$ mm/s, $x = 199.5$ mm.

11.14 $v = -453$ mm/s, $x = 131.0$ mm.

11.15 $a = 800$ m/s^2 $\uparrow$.

11.16 (*a*) $a_0 = -729 \times 10^3$ m/s^2. (*b*) $t = 1.366 \times 10^{-3}$ s.

11.17 (*a*) $v = 1.962$ m/s. (*b*) $x = 0.591$ m.

11.18 $v_m = 167.1$ mm/s $\uparrow$, $a_m = 15.19$ m/s^2 $\uparrow$.

11.19 (*a*) $v = \pm 0.323$ m/s, (*b*) $x = -0.0801$ m, (*c*) $v_{max} = 1.004$ m/s.

11.20 $v_f = 10(\sqrt{l^2 + x_0^2} - l)$

11.21 (*a*) $x = 1.25$ m, (*b*) $t = 0.866$ s.

11.22 (*a*) $x = 22.5$ m, (*b*) $v = 38.4$ m/s.

11.23 $v = 5.88$ m/s.

11.24 (*a*) $v = 29.3$ m/s. (*b*) $t = 0.947$ s.

11.25 (*a*) $x = 3.33$ m. (*b*) $t = 2.22$ s. (*c*) $t = 1.667$ s.

11.26 (*a*) $k = 0.1457$ s/m. (*b*) $x = 145.2$ m. (*c*) $v_{max} = 6.86$ m/s.

11.27 (*a*) $a = -0.0525$ m/s^2. (*b*) $t_3 - t_1 = 6.17$ s.

11.28 (*a*) $x = 10.55$ km. (*b*) $a_0 = -2 \times 10^{-4}$ m/s^2. (*c*) $t = 49.9$ min.

11.29 (*a*) $y_{max} = 14.90 \times 10^3$ m. (*b*) $y_{max} = 41.6 \times 10^3$ m. (*c*) $y_{max} = -6.12 \times 10^{10}$ m. or $y_{max} \rightarrow \infty$

11.30 $v_e = 11{,}180$ m/s.

11.31 (*a*) $x_{3T} = 2.36\, v_0 T$, $a_{3T} = \dfrac{\pi v_0}{T}$. (*b*) $v_{ave} = 0.363\, v_0$.

11.33 (*a*) $\mathbf{v}_0 = 9.62$ m/s $\uparrow$. (*b*) $\mathbf{v} = 29.6.$ m/s $\downarrow$.

11.34 (*a*) $a = -0.417$ m/s^2. (*b*) $v = 18.00$ km/h.

11.35 (*a*) $a = 1.500$ m/s^2. (*b*) $t_1 - t_0 = 10.00$ s.

11.36 (*a*) $v_1 = 76.8$ m/s. (*b*) $y_{max} = 328$ m.

11.37 (*a*) $d = 2.40$ m. (*b*) $t_D = 2.06$ s.

11.38 (*a*) $a = 2.40 \text{ m/s}^2$. (*b*) $v_{max} = 12.96$ m/s. (*c*) $t_2 = 10.41$ s.
11.39 (*a*) $a_A = -2.10 \text{ m/s}^2$, $a_B = 2.06 \text{ m/s}^2$.
(*b*) Runner *B* should start to run 2.59 s before *A* reaches the exchange zone.
11.40 (*a*) $a_A = 1.563 \text{ m/s}^2$. (*b*) $a_B = 3.13 \text{ m/s}^2$.
11.41 (*a*) $(x_P)_{42} = 0.5$ km. (*b*) $v_M = 42.9$ km/h.
11.42 (*a*) $t_1 = 14.548$ s, $x_A = .203$ m.
(*b*) $v_A = 64.3$ km/h →. $v_B = 35.1$ km/h →.
11.43 (*a*) $a_A = -0.250 \text{ m/s}^2$, $a_B = 0.300 \text{ m/s}^2$.
(*b*) $t > 0 \Rightarrow t_{AB} = 20.8$ s. (*c*) $v_B = 85.5$ km/h.
11.44 (*a*) $t = 1.330$ s. (*b*) 4.68 m below the man.
11.45 (*a*) $t_1 = 13.08$ s. (*b*) $v_{B/A} = 128.3$ m/s ↑.
11.46 (*a*) $a_A = 3.11 \text{ m/s}^2$. (*b*) $t_{AB} = 7.61$ s. (*c*) $d = 279$ m
11.47 (*a*) $\mathbf{v}_C = 8.00$ m/s ↑. (*b*) $\mathbf{v}_W = 4.00$ m/s ↑. (*c*) $\mathbf{v}_{C/E} = 12.00$ m/s ↑. (*d*) $\mathbf{v}_{W/E} = 8.00$ m/s ↑.
11.48 (*a*) $\mathbf{a}_E = 0.800 \text{ m/s}^2$ ↑, $\mathbf{a}_C = 1.600 \text{ m/s}^2$ ↓.
(*b*) $(\mathbf{v}_E)_5 = 4.00$ m/s ↑.
11.49 (*a*) $\mathbf{v}_B = 2.00$ m/s ↑. (*b*) $\mathbf{v}_D = 2.00$ m/s ↓. (*c*) $\mathbf{v}_{C/D} = 8.00$ m/s ↑.
11.50 (*a*) $\mathbf{a}_A = 20 \text{ m/s}^2$ →, $\mathbf{a}_B = 6.67 \text{ m/s}^2$ ↓.
(*b*) $\mathbf{v}_B = 13.33$ m/s ↓, $\mathbf{y}_B - (\mathbf{y}_B)_0 = 13.33$ m ↓.
11.51 (*a*) $\mathbf{v}_A = 200$ mm/s →. (*b*) $\mathbf{v}_C = 600$ mm/s →.
(*c*) $\mathbf{v}_D = 200$ mm/s ←. (*d*) $\mathbf{v}_{C/A} = 400$ mm/s →.
11.52 (*a*) $\mathbf{a}_A = 13.33 \text{ mm/s}^2$ ←, $\mathbf{a}_B = 20.0 \text{ mm/s}^2$ ←.
(*b*) $\mathbf{a}_D = 13.33 \text{ mm/s}^2$→. (*c*) $\mathbf{v}_B = 70.0$ mm/s→. $\mathbf{x}_B - (\mathbf{x}_B)_0 = 440$ mm→.
11.53 (*a*) $\mathbf{a}_A = 0.0500 \text{ m/s}^2$ ↑ $\mathbf{a}_B = 0.250 \text{ m/s}^2$↓.
(*b*) $\mathbf{v}_B = 0.1500$ m/s↓ $\mathbf{y}_B - (\mathbf{y}_B)_0 = 0.450$ m↓.
11.54 (*a*) $\mathbf{v}_L = 16.67$ mm/s↑. (*b*) $\mathbf{v}_{B/L} = 16.67$ mm/s↑.
11.55 (*a*) $t = 1.500$ s. (*b*) $t = 3.00$ s. (*c*) Change in position = 250 mm ↑.
11.56 (*a*) $t = 1.000$ s. (*b*) $\Delta_{yC} = 75.0$ mm. ↓.
11.57 (*a*) $\mathbf{a}_B = 240 \text{ mm/s}^2$ ↓, $\mathbf{a}_A = 345 \text{ mm/s}^2$ ↑.
(*b*) $(\mathbf{v}_C)_0 = 130.0$ mm/s→, $(\mathbf{v}_A)_0 = 43.3$ mm/s ↑.
⊂*c*) $\mathbf{x}_C - (\mathbf{x}_C)_0 = 728$ mm →.
11.58 (*a*) $(\mathbf{v}_C)_0 = 10.00$ mm/s →. (*b*) $\mathbf{a}_C = 6.00 \text{ mm/s}^2$→,
$\mathbf{a}_A = 2.00 \text{ mm/s}^2$↑. (*c*) $\mathbf{y}_A - (\mathbf{y}_A)_0 = 175.0$ mm↑.
11.59 (*a*) $\mathbf{v}_C = 120.$ mm/s↓. (*b*) $\mathbf{y}_D - (\mathbf{y}_D)_0 = 125.0$ mm↑.
11.60 (*a*) $\mathbf{a}_A = 20.0 \text{ mm/s}^2$↓. $\mathbf{a}_B = 40.0 \text{ mm/s}^2$↓.
(*b*) $\mathbf{y}_D - (\mathbf{y}_D)_0 = 375$ mm↓.
11.61 $d = 1812$ m
11.62 (*a*) $v_{max} = +10.00$ m/s (*b*) $x_{max} = +33.0$ m
11.63 (*b*) Total distance traveled = 1383 m. (*c*) $(t_{x=0})_1 = 9.00$ s, $(t_{x=0})_2 = 49.5$ s.
11.64 (*b*) $x_{max} = 420$ m. (*c*) $t_1 = 10.69$ s, $t_2 = 40.0$ s.
11.65 $t_{total} = 10.50$ s.
11.66 (*a*) $t_{total} = 44.8$ s. (*b*) $\mathbf{a}_{initial} = 103.3 \text{ m/s}^2$ ↑.
11.67 (a) $t_1 = 5$ min (*b*) $v_2 = 6$ km/h (*c*) $a_{final} = -0.0444 \text{ m/s}^2$
11.68 $t_{cycle} = 10.5$ s
11.69 (*a*) $t_1 = 0.6$ s. (*b*) $v_{1.4} = 0.20$ m/s, $x_{1.4} = 2.84$ m.
11.70 (*a*) $v_{20} = 60.0$ m/s, $x_{20} = 1194$ m. (*b*) $v_{average} = 59.25$ m/s.
11.71 (*a*) $t_A = 52.2$ s, $t_B = 52.0$ s, $\Delta x = 1.879$ m.
11.72 $t_F = 9.20$ s.
11.73 $t_F = 8.54$ s, $v_m = 86.9$ km/h.
11.74 $x_{B/A} = 39.4$ m.
11.75 $t = 5.67$ s.
11.76 (*a*) $(x_{A/B})_{min} = 8.73$ m. (*b*) $t = 14.25$ s.
11.77 $v_{0.2} = 2.13$ m/s, $v_{0.3} = 2.73$ m/s, $v_{0.4} = 2.93$ m/s; $x_{0.3} = 0.1137$ m, $x_{0.2} = -0.1327$ m.
11.78 (*a*) $t_1 = 18.00$ s. (*b*) $x_{18} = 178.8$ m. (*c*) $v_{ave} = 34.7$ km/h.
11.79 (*a*) $t_{min} = 5$ min 11.95 s. (*b*) $v_{ave} = 28.9$ km/h.
11.80 (*a*) $t_{min} = 2.04$ s. (*b*) $v_{max} = 0.392$ m/s, $v_{ave} = 0.196$ m/s.

11.81 (*a*) $v_0 = 1.914$ m/s. (*b*) $x = 0.840$ m.
11.82 (*a*) $v = 117.3$ km/h. (*b*) $x = 660$ m.
11.83 (*a*) $t = 3.00$ s. (*b*) $x = 68.0$ m.
11.84 (*a*) $a = 4070$ mm/s^2. (*b*) $a = 2860$ mm/s^2.

11.85 $x_0 + v_0 t + \frac{1}{2}at^2$ Q.E.D.

11.86 $x_{14} = +32$ m
11.87 (*a*) $t_1 = 2.40$ s (*b*) $x = 9.60$ m
11.88 (*a*) $x = -12.50$ m (*b*) $x_{max} = 420$ m
11.89 (*a*) $v = 6.28$ m/s ∠ 37.2°. (*b*) $x = 7.49$ m.
11.90 (*a*) $\mathbf{v} = 67.1$ mm/s ∠ 63.4°, $\mathbf{a} = 256$ mm/s^2 ∠ 69.4°.
(*b*) $\mathbf{v} = 8.29$ mm/s ∠ 36.2°, $\mathbf{a} = 336$ mm/s^2 ∠ 86.6°.

11.91 (*a*) $\mathbf{v} = -(4\pi \text{ m/s})\mathbf{i}$, $\mathbf{a} = -(4\pi^2 \text{ m/s}^2)\mathbf{j}$. (*b*) $y = \frac{1}{8}x^2 - 1$ (Parabola).

11.92 (*a*) $v_{min} = 50$ mm/s, $v_{max} = 150$ mm/s.
(*b*) $t = 2n\pi$ s, $x = 200\, n\pi$ mm, $y = 50.0$ mm; $\theta_{v_{min}} = 0 \rightarrow$, $t = (2n + 1)\pi$ s,
$x = 100(2n + 1)\pi$ mm, $y = 150$ mm, $\theta_{v_{max}} = 0 \rightarrow$

11.93 (*a*) $\mathbf{r} = 20$ mm↑, $\mathbf{v} = 43.4$ mm/s ∠ 46.3°, $\mathbf{a} = 743$ mm/s^2 ∠ 85.4°
(*b*) $\mathbf{r} = 18.10$ mm ∠ 6.01°, $\mathbf{v} = 5.65$ mm/s ∠ 31.08°, $\mathbf{a} = 70.3$ mm/s^2 ∠ 86.9°

11.94 (*a*) $1 - t^2 = 0$ or $t = 1$ s (*b*) $2t = 0$ or $t = 0$

11.95 $v = \sqrt{R^2\left(1+\omega_n^2 t^2\right) + c^2}$; $a = R\omega_n\sqrt{4+\omega_n^2 t^2}$

11.96 $\left(\frac{y}{A}\right)^2 - \left(\frac{x}{A}\right)^2 - \left(\frac{z}{B}\right)^2 = 1$ Q.E.D.,

(*a*) $v = 3$ m/s, $a = 3.61$ m/s^2. (*b*) $t = 3.82$ s.
11.97 $d = 353$ m.
11.98 (*a*) $d = 330$ m. (*b*) $h = 149.9$ m.
11.99 (*a*) 115.3 km/h $\leq v_0 \leq$ 148.0 km/h.
(*b*) $\alpha = 6.66°$; $\alpha = 4.05°$.
11.100 4.65 m/s $\leq v_0 \leq$ 10.58 m/s.
11.101 $0 \leq d \leq 0.5173$ m.
11.102 0.244 m $< h <$ 0.386 m.
11.103 (*a*) $y_C > 2.43$ m (height of net) ⇒ ball clears net.
(*b*) $b = (16.01 - 9.00)$ m $= 7.01$ m from the net.
11.104 $d = 233$ m
11.105 $v_0 = 6.93$ m/s.
11.106 16.20 m/s $< v_0 <$ 21.0 m/s.
11.107 (*a*) $v_0 = 8.96$ m/s. (*b*) $v_0 = 8.87$ m/s.
11.108 37.7 m/s $< v_0 <$ 44.3 m/s.
11.109 0.678 m/s $< v_0 <$ 1.211 m/s.
11.110 10.95 m/s $< v_0 <$ 11.93 m/s.
11.111 (*a*) $\alpha = 10.38°$. (*b*) $\theta = 9.74°$.
11.112 (*a*) $\alpha = 4.17°$. (*b*) $y_{max} = 285$ m. (*c*) $t = 15.89$ s.
11.113 (*a*) $\alpha_{max} = 14.66°$. (*b*) $t_{enter} = 0.1161$ s.
11.114 (*a*) $d_{max} = 4.98$ m. (*b*) $\alpha = 23.8°$.
11.115 (*a*) $\alpha = 45°$, $d_{max} = 6.52$ m. (*b*) $\alpha = 58.2°$, $d_{max} = 5.84$ m
11.116 $\alpha = 26.1°$, $(v_0)_{min} = 2.94$ m/s
11.117 $\mathbf{v}_{A/B} = 5.05$ m/s ∠ 55.8°.
11.118 $\mathbf{v}_A = 125$ mm/s↑, $\mathbf{v}_B = 75$ mm/s↓, $\mathbf{v}_C = 175$ mm/s↓.
11.119 (*a*) $\mathbf{v}_{B/A} = 91.0$ ft/s ∠ 47.0°. (*b*) $\mathbf{r}_{B/A} = 364$ ft ∠ 47.0°.
(*c*) Distance between autos = 293 ft.
11.120 $\mathbf{v}_R = 3.20$ km/h ∠ 17.8°.
11.121 (*a*) $\mathbf{v}_{B/A} = 405$ km/h ∠ 11.53°. (*b*) $\mathbf{v}_A = 379$ km/h ∠ 76.17°.
(*c*) $\Delta\mathbf{r}_{C/B} = 100$ km ∠ 40°
11.122 (*a*) $\mathbf{v}_P = 80.7$ mm/s ∠ 21.7° (*b*) $\mathbf{v}_P = 218$ mm/s ∠ 53.4°
(*a*) $v_P = 80.7$ mm/s ∠ 21.7°. (*b*) $v_P = 218$ mm/s ∠ 53.4°

11.123 (*a*) $\mathbf{v}_B$ = 175.3 mm/s ∠ 60.6°. (*b*) $\mathbf{a}_B$ = 292 mm/s^2 ∠ 60.6°.
11.124 (*a*) $\mathbf{v}_B$ = 213.194 mm/s ∠ 54.1°. (*b*) $\mathbf{a}_B$ = 159.9 mm/s^2 ∠ 54.1°.
11.125 (*a*) d = 0.979 m. (*b*) $\mathbf{v}_{B/D}$ 12.55 m/s ∠ 86.5°.
11.126 (*a*) $\mathbf{a}_C$ = 0.835 mm/s^2 ∠ 75°. (*b*) $\mathbf{v}_C$ = 8.35 mm/s ∠ 75°.
11.127 (*a*) $\mathbf{v}_B$ = 2.59 m/s ∠ 15°. (*b*) $\mathbf{v}_B$ = 1.011 m/s ∠ 15°.
11.128 $\mathbf{v}_{B/A}$ = 3.18 m/s ∠ 81.4°.
11.129 v_R = 5.96 m/s ∠ 82.8°, v_R = 5.96 m/s ∠ 82.8°
11.130 $\mathbf{v}_w$ = 17.49 km/h ∠ 59.0°
11.131 $\mathbf{v}_W$ = 15.79 km/h ∠ 26.0°.
11.132 $\mathbf{v}_D$ = 0.341 m/s ∠ 2.07°, $\mathbf{v}_D$ = 0.341 m/s ∠ 2.07°.
11.133 ρ = 500 m.
11.134 $(v_{\max})_{AB}$ = 97.6 km/h.
11.135 a_n = 444 m/s^2.
11.136 (*a*) d = 250 m. (*b*) v = 82.9 km/h.
11.137 a = 0.690 m/s^2.
11.138 (*a*) a = 10.20 mm/s^2. (*b*) t = 25.2 s.
11.139 (*a*) x_1 = 178.9 m. (*b*) a_t = 1.118 m/s^2.
11.140 (*a*) v_B = 83.8 km/h. (*b*) $|a|$ = 3.71 m/s^2
11.141 (*a*) $\mathbf{v}_{B/A}$ = 189.5 km/h ∠ 54.0°. (*b*) $\mathbf{a}_{B/A}$ = 21.8 m/s^2 ∠ 5.3°.
11.142 (*a*) $\mathbf{v}_{B/A}$ = 501 km/h ∠ 68.9°. (*b*) $\mathbf{a}_{B/A}$ = 82.2 m/s^2 ∠ 25.1°
11.143 (*a*) $\mathbf{v}_A$ = 7.99 m/s ∠ 40°. (*b*) ρ_B = 3.82 m.
11.144 (*a*) ρ_A = 1.575 m. (*b*) ρ_B = 0.0271 m.
11.145 (*a*) ρ_A = 281 m. (*b*) ρ_B = 209 m.
11.146 (*a*) ρ_B =27.6 m. (*b*) ρ_C = 34.0 m.
11.147 (*a*) ρ_A = 0.634 m. (*b*) ρ_B = 9.07 m.
11.148 (*a*) $\mathbf{v}_A$ = 14.01 m/s ∠ 36.9°. (*b*) ρ_B = 12.80 m
11.149 $\mathbf{v}_B$ = 18.17 m/s ∠ 4.04° and $\mathbf{v}_B$ = 18.17 m/s ∠ 4.04°.
11.150 (*a*) $\rho_{\min} = \rho_B = \dfrac{v_0^2 \cos^2\alpha}{g}$ Q.E.D. (*b*) $\rho_C = \dfrac{\rho_{\min}}{\cos^3\theta}$ Q.E.D.
11.151 $\rho = \dfrac{R^2 + c^2}{2\omega_n R}$.
11.152 ρ = 2.50 m.
11.153 r = 149.8 Gm.
11.154 r = 1425 Gm.
11.155 $v_{\text{circ.}} = 25.8 \times 10^3$ km/h.
11.156 $v_{\text{circ.}} = 12.56 \times 10^3$ km/h.
11.157 $v_{\text{circ.}} = 153.3 \times 10^3$ km/h.
11.158 h_2 = 542 km.
11.159 t_{orbit} = 1.606 h.
11.160 T_C = 54.6 h.
11.161 (*a*) $\mathbf{v}_A = -(25\text{ mm/s})\mathbf{e}_r - (250\text{ mm/s})\mathbf{e}_\theta$
(*b*) $\mathbf{a}_B = -(490\text{ mm/s}^2)\mathbf{e}_r + (100\text{ mm/s}^2)\mathbf{e}_\theta$
(*c*) $\mathbf{a}_{B/OA} = (10\text{ mm/s}^2)\mathbf{e}_r$.
11.162 (*a*) $\mathbf{v}_B = -(150\text{ mm/s})\mathbf{e}_r$
(*b*) $\mathbf{a}_B = (2000\pi\text{ mm/s}^2)\mathbf{e}_\theta$
(*c*) $\mathbf{a}_{B/OA} = 0$.
11.163 (*a*) $\mathbf{v} = (2\pi\text{ m/s})\mathbf{e}_\theta$, $\mathbf{a} = -(4\pi^2\text{ m/s}^2)\mathbf{e}_r$.
(*b*) $\mathbf{v} = -\left(\dfrac{\pi}{2}\text{ m/s}\right)\mathbf{e}_r + (\pi\text{ m/s})\mathbf{e}_\theta$. $\mathbf{a} = -\left(\dfrac{\pi^2}{2}\text{ m/s}^2\right)\mathbf{e}_r - (\pi^2\text{ m/s}^2)\mathbf{e}_\theta$
11.164 (*a*) $v = 2abt$, $a = 2ab\sqrt{1 + 4b^2t^4}$
(*b*) $\rho = a$.
11.165 (*a*) $\mathbf{v} = bk\mathbf{e}_\theta$, $\mathbf{a} = -\dfrac{1}{2}bk^2\mathbf{e}_r$.
(*b*) $\mathbf{v} = 2bk\mathbf{e}_r + 2bk\mathbf{e}_\theta$, $\mathbf{a} = 2bk^2\mathbf{e}_r + 4bk^2\mathbf{e}_\theta$.
11.166 (*a*) a = constant (*b*) Thus, $\mathbf{a}$ is directed toward A.

11.167 (*a*) $v = -\dfrac{b\dot{\theta}}{\cos^2\theta}$, $v = -\dfrac{b\dot{\theta}}{\cos^2\theta}$

(*b*) $a = -\dfrac{b}{\cos^2\theta}(\ddot{\theta} + 2\dot{\theta}^2\tan\theta)$, $a = -\dfrac{b}{\cos^2\theta}(\ddot{\theta} + 2\dot{\theta}^2\tan\theta)$

11.168 $v = \dfrac{d\dot{\theta}\tan\beta\sec\beta}{(\tan\beta\cos\theta - \sin\theta)^2}$, $v = \dfrac{d\dot{\theta}\tan\beta\sec\beta}{(\tan\beta\cos\theta - \sin\theta)^2}$

11.169 $\dot{r} = 120$ m/s , $\dot{\theta} = -0.0900$ rad/s , $\ddot{r} = 34.8$ m/s^2 , $\ddot{\theta} = -0.0156$ rad/s^2

11.170 (*a*) $\dot{r} = -\dfrac{1}{2}d\omega$, $\dot{\theta} = \dfrac{1}{2}\omega$ (*b*) $\ddot{r} = -\dfrac{\sqrt{3}}{4}d\omega^2$, $\ddot{\theta} = 0$

11.171 v_{ave} = 185.7 km/h.
11.172 v_{ave} = 103.5 km/h, β = 51.2°.

11.173 $v = \dfrac{b}{\theta^2}\sqrt{1+\theta^2}\,\dot{\theta}$

11.174 $v\ v = e^{b\theta}\sqrt{1+b^2}\,\dot{\theta}$

11.175 $a = \dfrac{b\omega^2}{\theta^3}\sqrt{4+\theta^4}$

11.176 $a = (1 + b^2)w^2e^{b\theta}$.

11.177 $v = 2\pi\sqrt{A^2 + n^2B^2\cos^2 2\pi nt}$, $a = 4\pi^2\sqrt{A^2 + n^4B^2\sin^2 2\pi nt}$

11.178 $\dot{r} = h\dot{\phi}\sin\theta$ Q.E.D.

11.179 (*a*) $v = \sqrt{A^2 + B^2}$. (*b*) $v = 2\pi A$,

(a) $a = \sqrt{(1+16\pi^2)A^2 + B^2}$. (b) $a = 4\pi^2 A$.

11.180 $\beta = \tan^{-1}\left[\dfrac{R(2+\omega_n^2t^2)}{c\sqrt{4+\omega_n^2t^2}}\right]$

11.181 (*a*) θ_x = 90°, θ_y = 123.7°, θ_z = 33.7°.
(*b*) θ_x = 103.4°, θ_y = 134.3°, θ_z = 47.4°.
11.182 (*a*) t = 1.00 s and t = 4.00 s. (*b*) x = 1.50 m, d = 24.5 m
11.183 A = −36.8 m^2, k = 1.832 s^{-2}.
11.184 v_{18} = 36 m/s, x_{18} = 52 m, d = 164 m
11.185 (*a*) $\mathbf{v}_{B/A}$ = 111.4 km/h ∠ 10.50°. (*b*) $r_{B/A}$ = 2.96 km.
11.186 (a) $\mathbf{a}_{C/A}$ = 400 mm/s^2↑ (b) $\mathbf{v}_C$ = 894 m/s ∠ 63.4°
11.187 (*a*) a_B = −50 mm/s^2 or $\mathbf{a}_B$ = 50.0 mm/s^2 ↑, a_C = 75 mm/s^2 or $\mathbf{a}_C$ = 75.0 mm/s^2 ↓. (*b*) t = 0.667 s.
(*c*) $\mathbf{y}_C - (\mathbf{y}_C)_0$ = 16.67 mm↑.
11.188 (*a*) v_0 = 38.1 m/s, d = 20.4 m. (*b*) v_0 = 41.1 m/s, d = 29.6 m.
11.189 (*a*) α = 22.4°, $\mathbf{a}_B$ = 0.964 m/s^2 ∢ 22.4°. (*b*) $\mathbf{v}_B$ = 1.929 m/s^2 ∢ 22.4°.
11.190 a_{max} = 3.94 m/s^2.
11.191 (*a*) v_0 = 7.10 m/s. (*b*) ρ = 31.0 m.
11.192 (*a*) $\mathbf{v}_B$ = (−2.00 m/s)$\mathbf{e}_r$ + (−3.00 m/s)$\mathbf{e}_\theta$, $\mathbf{v}_B$ = 3.61 m/s ∠ 59.0°
(*b*) $\mathbf{a}_B$ = (−4.00 m/s^2)$\mathbf{e}_r$ + (2.50 m/s^2)$\mathbf{e}_\theta$, $\mathbf{a}_B$ = 4.72 m/s^2 ∠ 29.3°
(*c*) ρ = 2.76 m
11.193 (*a*) $\mathbf{v}$ = (−31.5 m/s)$\mathbf{e}_r$ + (10.5 m/s)$\mathbf{e}_\theta$, v = 33.2 m/s ∠ 6.57°
(*b*) $\mathbf{a}$ = (−3.74 m/s^2)$\mathbf{e}_r$ + (4.59 m/s^2)$\mathbf{e}_\theta$, a = 5.92 m/s^2 ∠ 25.9°
(*c*) $\mathbf{d}$ = 125.5 m

CHAPTER 12

12.1 (*a*) W = 3.24 N. (*b*) m = 2.00 kg.
12.2 (*a*) ϕ = 0°: W = (5 kg)(9.78 m/s^2) = 48.9 N
ϕ = 45°: W = (5 kg)(9.8059 m/s^2) = 49.0 N
ϕ = 60°: W = (5 kg)(9.8189 m/s^2) = 49.1 N
(*b*) m = 5.000 kg.

12.3 $L = 2.84 \times 10^6$ kg · m/s.
12.4 (*a*) $W = 66.8$ N. (*b*) $F_s = 73.6$ N, $m_w = 6.81$ kg.
12.5 $x_{\text{upgrade}} = 0.470$ km.
12.6 (*a*) $v_0 = 6.67$ m/s. (*b*) $\mu_k = 0.0755$.
12.7 $\mathbf{a} = 0.414$ m/s² ⦪ 15°.
12.8 (*a*) $v = 110.5$ km/h. (*b*) $v = 85.6$ km/h. (*c*) $v = 69.9$ km/h.
12.9 (*a*) $x_f - x_0 = 40.1$ m. (*b*) $x_f - x_0 = 47.0$ m.
12.10 (*a*) $t = 2.22$ s. (*b*) $x = 3.32$ m.
12.11 $x = 51.0$ m.
12.12 (*a*) $x = 234$ m. (*b*) $P = 3.33$ kN (tension).
12.13 (*a*) $\mathbf{a}_A = 2.49$ m/s²→, $\mathbf{a}_B = 0.831$ m/s²↓.
(*b*) $T = 74.8$ N
12.14 (*a*) $\mathbf{a}_A = 0.698$ m/s²→, $\mathbf{a}_B = 0.233$ m/s²↓.
(*b*) $T = 79.8$ N
12.15 (*a*) (1): $(\mathbf{a}_A)_1 = 3.27$ m/s² ↓, (2): $(\mathbf{a}_A)_2 = 4.91$ m/s² ↓, (3): $(\mathbf{a}_A)_3 = 0.228$ m/s² ↓.
(*b*) (1): $(v_A)_1 = 4.43$ m/s ↓, (2): $(v_A)_2 = 5.42$ m/s ↓, (3): $(v_A)_3 = 1.170$ m/s ↓.
(*c*) (1): $t_1 = 1.835$ s, (2): $t_2 = 1.223$ s, (3): $t_3 = 26.3$ s.
12.16 $\mathbf{a}_A = 0.997$ ft/s² ⦣ 15°, $\mathbf{a}_B = 1.619$ ft/s² ⦣ 15°.
12.17 (*a*) $F = 6816$ N. (*b*) $F_{AB} = 5060$ N.
12.18 (*a*) $\mathbf{a}_B = 0.986$ m/s² ⦩ 25°. (*b*) $T = 51.7$ N.
12.19 (*a*) $\mathbf{a}_B = 1.794$ m/s² ⦩ 25°. (*b*) $T = 58.2$ N.
12.20 $\mathbf{a}^2 = 6.63$ m/s²←. (*b*) $x_{p/\text{belt}} = 0.321$ m→
12.21 $\mu_s > 0.300$
12.22 (a) $\mathbf{a}_T = 0.309$ m/s²→ (*b*) $\mathbf{a}_T = 4.17$ m/s²→
12.23 $\mathbf{a}_1 = 19.53$ m/s² ⦣ 65°, $\mathbf{a}_2 = 4.24$ m/s² ⦤ 65°.
12.24 $x_1 = 1.598$ km.
12.25 $x = 0.347\dfrac{m_0 v_0^2}{F_0}$.
12.26 $x = \dfrac{Pt}{k} - \dfrac{kv}{m}$, which is linear.
12.27 answer: $v = \sqrt{\dfrac{k}{m}}\left(\sqrt{\ell^2 + x_0^2} - \ell\right)$
12.28 (*a*) $P = 10.00$ N. (*b*) $T_{AD} = 103.1$ N.
12.29 (*a*) $\mathbf{a} = 2.73$ m/s², $T = 88.6$ N.
(*b*) $\mathbf{a} = 3.77$ m/s²←, $T = 75.5$ N. (*c*) $\mathbf{a} = 3.77$ m/s²←; $T = 75.5$ N.
12.30 (*a*) $T = 33.6$ N. (*b*) $\mathbf{a}_A = 4.76$ m/s²→,
$\mathbf{a}_B = 3.08$ m/s²↓, $\mathbf{a}_C = 1.401$ m/s²←.
12.31 (*a*) $m_C = 1.426$ kg. (*b*) $\mathbf{a}_A = 1.503$ m/s²→,
$\mathbf{a}_B = 0.348$ m/s²→, $\mathbf{a}_C = 1.759$ m/s²↓.
12.32 (*a*) $\mathbf{a}_A = \mathbf{a}_B = \mathbf{a}_D = 1.073$ m/s²↓ and $a_C = -4(1.0728 \text{ m/s}^2)$ or $\mathbf{a}_C = 4.29$ m/s²↑
(*b*) $T_1 = 6(9.81 + 4(1.0728))$ or $T_1 = 84.6$ N
12.33 (*a*) $\mathbf{v}_{D/A} = 8.33$ m/s↓. (*b*) $\mathbf{v}_{C/D} = 7.41$ m/s↑
12.34 (*a*) $\mathbf{a}_A = 2.80$ m/s²←. (*b*) $\mathbf{a}_{B/A} = 8.32$ m/s² ⦩ 25°.
12.35 (*a*) $\mathbf{a}_B = 5.94$ m/s² ⦪ 75.6°. (*b*) $\mathbf{v}_{B/A} = 3.74$ m/s ⦪ 20°.
12.36 (*a*) $\theta = 49.9°$. (*b*) $T_{AB} = 6.85$ N.
12.37 (*a*) $T_{BC} = 80.4$ N. (*b*) $v_A = 2.30$ m/s.
12.38 $v = 3.47$ m/s.
12.39 $v \geq 3.01$ m/s, $3.01 \text{ m/s} \leq v \leq 3.96$ m/s.
12.40 $3.01 \text{ m/s} \leq v \leq 3.85$ m/s.
12.41 $1.121 \text{ m/s} < v_D < 1.663$ m/s
12.42 $2.72 \text{ m/s} < v_C \leq 3.84$ m/s.
12.43 $0.536 \text{ m/s} \leq v \leq 4.10$ m/s.
12.44 (*a*) $T_{BA} = 533$ N. (*b*) $T_{BA} = 659$ N.
12.45 (*a*) $\rho = 201$ m. (*b*) N = 612 N↑.
12.46 (*a*) $T_{BA} = 131.7$ N. (*b*) $T_{BA} = 88.4$ N.
12.47 (*a*) $a_t = 4.63$ m/s². (*b*) $a_t = 1.962$ m/s². (*c*) $a_t = 0.1842$ m.s².
12.48 $24.1° < \theta < 155.9°$.
12.49 (*a*) $N = 2.905$ N. (*b*) $\theta = 13.09°$.

12.50 $(\mathbf{F}_{pilot})_B = 1126$ N $\measuredangle$ 25.6°.
12.51 (*a*) $v_0 = 12.19$ m/s. (*b*) $N = 2290$ N.
12.52 (*a*) $\theta = 43.5°$. (*b*) $\mu = 0.408$. (*c*) $v = 121.8$ km/h
12.53 (*a*) $F_s = 0.1631W$. (*b*) $\phi = 9.06°$.
12.54 $\phi = 3.52°$.
12.55 $v = 7.67$ m/s.
12.56 (*a*) $v = 12.00$ m/s. (*b*) $F_{tuft} = 2.05 \times 10^{-3}$ N.
12.57 $\mu_s = 0.236$.
12.58 $r = 468$ mm.
12.59 $2.36 \text{ m/s} < \mathbf{v} < 4.99$ m/s.
12.60 (*a*) $\mathbf{F} = 9.3617$ N $\measuredangle$ 80°. (*b*) $\mathbf{F} = 5.413$ N $\measuredangle$ 40°.
12.61 $\mu_s = 0.400$.
12.62 (*a*) $(\mu_s)_{min} = 0.204$. (*b*) $\theta = 11.53°$, $\theta = 168.5°$
12.63 (*a*) $(v_B)_{max} = 0.900$ m/s. (*b*) $\theta = 19.29°$, $\theta = 160.7°$
12.64 $\delta = \dfrac{eV\ell L}{mdv_0^2}$.
12.65 $\dfrac{d}{\ell} > 1.054\sqrt{\dfrac{eV}{mv_0^2}}$.
12.66 (*a*) $F_r = -10.73$ N, $F_\theta = 0.754$ N.
(*b*) $F_r = -4.44$ N, $F_\theta = 1.118$ N.
12.67 (*a*) $F_r = -0.49$ N, $F_\theta = 0.1$ N.
(*b*) $F_r = -0.249$ N, $F_\theta = -0.0250$ N.
12.68 $t = 2.00$ s.
12.69 (*a*) $\mathbf{a}_{B/rod} = 72$ m/s^2 radically outward. (*b*) $F_\theta = 1.25$ N.
12.70 (*a*) $F_r = -50.8$ N, $F_\theta = 8.10$ N
(*b*) $\mathbf{P} = 26.6$ N $\measuredangle$ 70°, $\mathbf{Q} = 54.1$ N $\measuredangle$ 40°
12.71 (*a*) $T = 126.6$ N. (*b*) $\mathbf{a}_A = 5.48$ m/s^2→. (*c*) $\mathbf{a}_B = 4.75$ m/s^2↑.
12.72 (*a*) $T = 142.7$ N. (*b*) $\mathbf{a}_A = 6.18$ m/s^2→. (*c*) $\mathbf{a}_B = 4.10$ m/s^2↓.
12.73 (*a*) $v_r = -4.00$ m/s, $v_\theta = 3.00$ m/s.
(*b*) $a_r = -60.0$ m/s^2, $a_\theta = -80.0$ m/s^2.
(*c*) $F_\theta = -20.0$ N.
12.74 $v_r = v_0 \dfrac{\sin 2\theta}{\sqrt{\cos 2\theta}}, \quad v_\theta = v_0\sqrt{\cos 2\theta}$.
12.75 (*a*) $v = \dfrac{v_0 r}{r_0}, \quad F_r = \dfrac{mv_0^2 r}{r_0^2}$. (*b*) $\rho = \dfrac{r^3}{r_0^2}$.
12.76 $v = \dfrac{v_0}{\cos^2\theta}$.
12.77 (*a*) $F_t = 0$. (*b*) $F_t = \dfrac{8mv_0^2}{r_0}$.
12.78 $M = 6.01 \times 10^{24}$ kg.
12.79 $r = \left(\dfrac{g\tau^2 R^2}{4\pi^2}\right)^{1/3}$ Q.E.D., $r = 383 \times 10^3$ km.
12.80 (*a*) $h = 35{,}800$ km. (*b*) $v = 3.07$ km/s.
12.81 $r = \left(\dfrac{g\tau^2 R^2}{4\pi^2}\right)^{1/3}$ Q.E.D., $g_{Jupiter} = 24.8$ m/s^2.
12.82 (*a*) $M = 1.998 \times 10^{30}$ kg. (*b*) $g = 276$ m/s^2.
12.83 (*a*) $R = 60{,}000$ km. (*b*) $M = 5.62 \times 10^{24}$ kg.
12.84 (*a*) $M = 86.9 \times 10^{24}$ kg. (*b*) $r_T = 436{,}000$ km.
12.85 (*a*) $F = 1684$ N. (*b*) $r_M = 2510$ km. (*c*) $g_{moon} = 1.62$ m/s^2.
12.86 (*a*) $(v_B)_{TR} = 1551$ m/s. (*b*) $\Delta v_B = -15.8$ m/s.
12.87 $v = 5000$ m/s.
12.88 (*a*) 1632 m/s. (*b*) $\Delta v_A = 2600$ m/s.
12.89 $\Delta v_D = 18.33$ m/s.
12.90 (*a*) $(a_A)_r = 0$, $(a_A)_\theta = 0$. (*b*) $(a_{collar/rod})_A = 38.4$ m/s^2. (*c*) $(v_B)_\theta = 0.800$ m/s.

12.91 (*a*) $(a_B)_r = 0$, $(a_B)_\theta = 0$. (*b*) $\ddot{r}_B = 20.0\ \text{m/s}^2$. (*c*) $(v_A)_f = 0.931$ m/s.
12.92 (*a*) $(v_\theta)_2 = 1.200$ m/s. (*b*) $T = 8.38$ N, $a_{A/\text{rod}} = 3.11\ \text{m/s}^2$ radially inward.
12.93 (*a*) $\ell_1^3 \sin^3\theta_1 \tan\theta_1 = \ell_2^3 \sin^3\theta_2 \tan\theta_2$. (*b*) $\theta_2 = 43.6°$.
12.96 $F = -\dfrac{mh^2 r}{r_0^4}$.
12.97 F is proportional to $\dfrac{1}{r^5}$ $F \times \dfrac{1}{r^5}$ Q.E.D.
12.98 $\varepsilon = 2.33$.
12.99 $v_{max} = 10.42$ km/s.
12.100 (*a*) $(v_A)_{par} = 10.13$ km/s. (*b*) $|\Delta v_A| = 2.97$ km/s.
12.101 $\varepsilon = 1.147$.
12.103 $\beta_{min} = \sqrt{\dfrac{2}{2+\alpha}}$.
12.104 (*a*) $v_0 = 8.00 \times 10^3$ m/s. (*b*) $\Delta v_B = 127$ m/s.
12.105 (*a*) $\Delta v_A = 159.0$ m/s. (*b*) $\Delta v_B = 70.1$ m/s. (*c*) $\Delta v_C = 2370$ m/s.
12.106 (*a*) $r_B = 71.8 \times 10^3$ km. (*b*) $\Delta v_B = 247$ m/s, $\Delta v_C = 2010$ m/s.
12.107 (*a*) $v_A = 16.3 \times 10^3$ m/s. (*b*) $|\Delta v_A| = 582$ m/s, $|\Delta v_B| = 1210$ m/s.
12.108 $r_{max} = 5.31 \times 10^9$ km.
12.109 $\tau = 91.8 \times 10^3$ yr.
12.110 $\tau_{AB} = 122.6$ h.
12.111 $\tau = 95.5$ min.
12.112 $\tau = 4.95$ h.
12.113 $t_{BC} = 50$ min 55 s.
12.114 $\phi = \cos^{-1}[(1 - n\beta^2)/(1 - \beta^2)]$.
12.115 (*a*) $v_C = 4$ km/s. (*b*) $\varepsilon = 0.684$.
12.116 $\measuredangle AOB = 130.3°$.
12.117 (*a*) $\measuredangle AOB = 14.37°$. (*b*) $v_B = 59.8$ km/s.
12.118 $\dfrac{1}{\rho} = \dfrac{1}{2}\left(\dfrac{1}{r_0} + \dfrac{1}{r_1}\right)$.
12.119 (*a*) $\varepsilon = \dfrac{r_1 - r_0}{r_1 + r_0}$. (*b*) $r_1 = 609 \times 10^{12}$ m.
12.120 $\left(\dfrac{\tau_1}{\tau_2}\right)^2 = \left(\dfrac{a_1}{a_2}\right)^3$.
12.121 $h = \sqrt{GMa(1-\varepsilon^2)}$.
12.122 (*a*) $\mathbf{v}_A = 1.091$ m/s→. (*b*) $\mathbf{v}_B = 0.545$ m/s←.
12.123 $a = 1.141$ m
12.124 (*a*) $\mathbf{a}_A = 6.24\ \text{m/s}^2$ ⦣ 30°. (*b*) $\mathbf{a}_{B/A} = 5.40\ \text{m/s}^2$→.
12.125 (*a*) $\mathbf{a}_{B/A} = 0.363\ \text{m/s}^2$ ←. (*b*) $T_{CD} = 1145$ N.
12.126 (*a*) $|a_t| = 5.79\ \text{m/s}^2$. (*b*) $|a_t| = 2.45\ \text{m/s}^2$. (*c*) $|a_t| = 0.230\ \text{m/s}^2$.
12.127 (*a*) $F_r = (5.76\ \text{N}) \tan^2\theta \sec\theta$, $F_\theta = (5.76\ \text{N}) \tan\theta \sec\theta$.
(*b*) $\mathbf{P} = (5.76\ \text{N}) \tan\theta \sec^2\theta$ ⦪ θ, $\mathbf{Q} = (5.76\ \text{N}) \tan^2\theta \sec^2\theta$ →.
12.128 (*a*) $(\mu_s)_{min} = 0.454$, The direction of the impending motion is downward.
(*b*) $(\mu_s)_{min} = 0.1796$, The direction of the impending motion is downward.
(c) $(\mu_s)_{min} = 0.218$, The direction of the impending motion is upward.
12.129 (*a*) $N = 539$ N. (*b*) $\rho = 47.1$ m.
12.130 $\tau = (24\pi/G\rho)^{1/2}$.
12.131 (*a*) $\tau_{AB} = 43$ min 56.6 sec. (*b*) $\tau_{circ} = 1$h 29 min 48 sec.
12.132 $\varepsilon = 104.8$.
12.133 (*a*) $r = 0.5$ m, $F_H = 0$. (*b*) $r = 0.270$ m, $F_H = -84.1$ N.

CHAPTER 13

13.1 $T = 10.11$ GJ.
13.2 (*a*) $T_2 = 225$ N·m, $h = 45.9$ m. (*b*) $T_2 = 225$ N·m, $h = 276$ m.
13.3 (*a*) $T_1 = 128$ N·m, (*b*) $T_2 = 75.1$ N·m, 33.7 m. (*c*) $h = 34.3$ m.
13.4 $T = 2.36$ GJ

13.5 $v = 4.05$ m/s.
13.6 $d = 2.99$ m.
13.7 (*a*) $v_2 = 112.2$ km/h. (*b*) $v_2 = 91.6$ km/h.
13.8 (*a*) $v_2 = 15.34$ m/s. (*b*) $v_3 = 59.9$ m/s.
13.9 (*a*) $\mu_k = 0.20$. (*b*) $v_3 = 1.103$ m/s $\leftarrow$.
13.10 (*a*) $v_2 = 15.54$ m/s. (*b*) $h = 27.3$ m. (*c*) $v_4 = 23.1$ m/s.
13.11 $d = 6.71$ m.
13.12 (*a*) $v_C = 2.90$ m/s. (*b*) $x_{belt} = 0.893$ m.
13.13 $\mathbf{v}_0 = 4.61$ m/s $\nearrow$ 15°.
13.14 $\mathbf{v}_0 = 3.87$ m/s $\nearrow$ 15°.
13.15 (*a*) $d = 57.8$ m. (*b*) $\mathbf{H} = 154$ N $\rightarrow$.
13.16 (*a*) $F_t = 7.41$ kN. (*b*) $F_c = 5.56$ kN (tension)
13.17 (*a*) $x = 40.576$ m. (*b*) $F_{AB} = 95.1$ kN (tension);
$F_{BC} = 42.3$ kN (tension).
13.18 (*a*) $x = 91.3$ m. (*b*) $F_{AB} = 95.1$ kN (compression);
$F_{BC} = 42.3$ kN (compression).
13.19 (*a*) $E_p = 45.7$ J.
(*b*) $T_A = 83.2$ N, $T_B = 60.3$ N.
13.20 (*a*) $\mathbf{v}_B = 2.34$ m/s $\leftarrow$. (*b*) $d = 235$ mm.
13.21 (*a*) $d = 11.88$ m. (*b*) $T = 5.05$ kN.
13.22 (*a*) $\mathbf{v}_B = 1.510$ m/s $\leftarrow$. (*b*) $F = 96.0$ N.
13.23 (*a*) $\mathbf{v}_B = 1.218$ m/s $\leftarrow$. (*b*) $F = 91.0$ N.
13.24 $v_A = 1.190$ m/s.
13.25 (*a*) $\mathbf{v}_2 = 4.43$ m/s $\nearrow$ 23.6°.
(*b*) $\mathbf{v}_3 = 6.26$ m/s $\nearrow$ 23.6°.
13.26 (*a*) $v^2 = 3.29$ m/s. (*b*) $h = 1.533$ m.
13.27 (*a*) $v_2 = 3.29$ m/s. (*b*) $h = 1.472$ m.
13.28 (*a*) $d = 559$ mm. (*b*) $d = 218$ mm.
13.29 (*a*) $\mathbf{x}_m = 18.75$ mm $\downarrow$. (*b*) $\mathbf{v}_m = 0.217$ m/s $\updownarrow$.
13.30 (*a*) $v = 0.597$ m/s. (*b*) $v_{max} = 0.617$ m/s.
13.31 $k_2 = 3110$ N/m.
13.32 $v = 0.759\sqrt{\dfrac{paA}{m}}$.
13.33 (*a*) $d = 3.72$ m. (*b*) $a_D = 120$ m/s^2.
13.34 (*a*) $v_m = 162.8$ mm/s. (*b*) $\mathbf{a}_m = 14.72$ m/s^2 $\uparrow$.
13.35 (*a*) $x_m = 32.7 \times 10^{-3}$ m $= 32.7$ mm, $\mathbf{F}_m = 98.1$ N $\uparrow$.
(*b*) $x_m = 30.4$ mm, $\mathbf{F}_m = 104.9$ N $\uparrow$.
13.36 $\dfrac{h_n}{h_u} = \dfrac{1}{1 - \dfrac{(v_0^2 - v^2)}{(2g_m R_m)}}$.
13.37 (*a*) $P = 0.0314\%$. (*b*) $P = 25.3\%$.
13.38 $h_m = 364$ m.
13.39 $\theta = 14.00°$.
13.40 (*a*) $v_0 = \sqrt{3gl}$. (*b*) $v_0 = \sqrt{2gl}$.
13.41 (*a*) $T = 1.5\,mg$. (*b*) $T = 2.5\,mg$.
13.42 (*a*) $\mathbf{N}_B = 5490$ N $\uparrow$, $\mathbf{N}_D = 785$ N $\downarrow$. (*b*) $\rho = 24$ m.
13.43 15 m $\leq h \leq$ 17.25 m.
13.44 $v_B = 2.30$ m/s.
13.45 (*a*) $\theta = -28.5°$. (*b*) $x = 1.261$ m.
13.46 (*a*) Average power = 57.2 kW. (*b*) Motor capacity = $\dfrac{229\text{ kW}}{0.85}$ = 269 kW.
13.47 (*a*) $(P_P)_A = 2.75$ kW. (*b*) $(P_E)_A = 3.35$ kW.
13.48 $F = 14.8$ kN.
13.49 (*a*) $P_W = 25.4$ W. (*b*) $P_B = 148.2$ W.
13.50 (*a*) $P(\text{kw}) = 0.278 \times 10^{-6}\,\dfrac{mgb}{\eta}$.

13.51 (*a*) Power = 109.0 kJ/s = 109.0 kW, Power = $\dfrac{(109.0\text{ kW})}{(0.7457\text{ kW/hp})}$ = 146.2 hp.

(*b*) kW Power = 530, 200 J = 530 kW, hp Power = $\dfrac{(530\text{ kW})}{(0.7457\text{ kW/hp})}$ = 711 hp.

13.52 (*a*) P_0 = 375 kW. (*b*) v = 5.79 km/h.

13.53 Power = (132.5 kW/s)t, Power =375 kW.

13.54 (*a*) P = 58.9 kW. (*b*) P = 52.9 kW.

13.55 (*a*) $k_e = k_1k_2/k_1 + k_2$. (*b*) $k_e = k_1 + k_2$.

13.56 (*a*) $\delta = v_0\sqrt{m(k_1+k_2)/k_1k_2}$. (*b*) $\delta = v_0\sqrt{\dfrac{m}{k_1+k_2}}$.

13.57 (*a*) v_B = 4.22 m/s. (*b*) v_D = 4.42 m/s.

13.58 v_2 = 1.481 m/s ↓.

13.59 $\mathbf{v}_2$ = 3.19 m/s ↔.

13.60 (*a*) v_A = 87.2 m/s. (*b*) v_C = 105.8 m/s.

13.61 v_2 = 21.9 m/s.

13.62 (*a*) k = 7060 N/m. (*b*) d = 12.72 m.

13.63 (*a*) $y_m = y_{st}\left(1+\sqrt{\dfrac{2h}{y_{st}}}\right)$.

13.64 (*a*) $\mathbf{v}_B$ = 2.48 m/s ←. (*b*) $\mathbf{v}_C$ = 1.732 m/s ↑.

13.65 (*a*) v_B = 3.31 m/s. (*b*) v_E = 3.90 m/s.

13.66 (*a*) θ = 43.2°. (*b*) $\mathbf{v}_A$= 2.43 m/s ↓.

13.67 v_2 = 2.11 m/s.

13.68 x = 0.269 m.

13.69 x = 0.1744 m.

13.70 N = 731 N.

13.71 731 N, 5520 N.

13.72 $\mathbf{v}_B$ = 4.18 m/s ←, $\mathbf{N}$ = 63.2 N ↑.

13.73 (*a*) v_C = 2.86 m/s. (*b*) $\mathbf{F}$ = –(38.1N)$\mathbf{i}$ + (38.6 N) $\mathbf{j}$.

13.74 1: (*a*) v_0 = 7.99 m/s. (*b*) $\mathbf{N}_C$ = 5.89 N ←.
2: (*a*) v_0 = 7.67 m/s. (*b*) $\mathbf{N}_C$ = 3.92 N ←.

13.75 (*a*) v_C = 3.836 m/s > 3.5 m/s. (*b*) v_0 = 7.83 m/s.

13.76 Loop 1: (*a*) $v_0 = \sqrt{5gr}$ ←. (*b*) $N_B = 3W$ →.
Loop 2: (*a*) $v_0 = \sqrt{4gr}$ ←. (*b*) $\mathbf{N} = 2W$ →.

13.77 y_B = 0.448 m.

13.78 (*a*) $\phi = \tan^{-1}\left(\dfrac{2.0642}{6-y}\right)$.

(*b*) N = 292 N, θ_x = 85.4°, θ_y = 71.6°, θ_z = 161.0°.

13.79 $\dfrac{\partial F_x}{\partial y}=\dfrac{\partial F_y}{\partial x},\ \dfrac{\partial F_x}{\partial z}=\dfrac{\partial F_z}{\partial y}\ \dfrac{\partial F_z}{\partial x}=\dfrac{\partial Fx}{\partial z}.$

13.80 $V = -\ln xyz$.

13.81 (*a*) $U_{ABCA} = \dfrac{\pi k a^2}{4}$. (*b*) U_{ABCA} = 0.

13.82 (*a*)
$$P_x=-\frac{\partial V}{\partial x}=-\frac{\partial[-(x^2+y^2+z^2)^{1/2}]}{\partial x}=x(x^2+y^2+z^2)^{-1/2}$$
$$P_y=-\frac{\partial V}{\partial y}=-\frac{\partial[-(x^2+y^2+z^2)^{1/2}]}{\partial y}=y(x^2+y^2+z^2)^{-1/2}$$
$$P_z=-\frac{\partial V}{\partial z}=-\frac{\partial[-(x^2+y^2+z^2)^{1/2}]}{\partial z}=z(x^2+y^2+z^2)^{-1/2}.$$

(*b*) $U_{OABD} = a\sqrt{3}$, $\Delta V_{OD} = -a\sqrt{3}$.

13.83 (*a*) $U_{OD} = \sqrt{3a}$. (*b*) $U_{OABDO} = \sqrt{3a} - \sqrt{3a} = 0$.

13.84 (*b*) $V = \dfrac{1}{(x^2+y^2+z^2)^{1/2}}$.

13.85 (*a*) $\dfrac{T_1}{m} = 62.5$ MJ/kg. (*b*) $V_{esc} = 11.18$ km/s.

13.86 (*a*) $v_A = 9.56$ km/s. (*b*) $v_B = 2.39$ km/s.

13.87 (*a*) $\Delta E_{300} = 90.4$ GJ. (*b*) $\Delta E_E = 208$ GJ.

13.88 (*a*) $\dfrac{\Delta E}{m} = 2.81 \times 10^6 \dfrac{\text{J}}{\text{kg}}$. (*b*) $\dfrac{\Delta E}{m} = 1.35 \times 10^6 \dfrac{\text{J}}{\text{kg}}$.

13.89 $v_B = 25.1$ Mm/h.

13.90 $v_B = 6.48$ km/s.

13.91 $\dfrac{M_B}{M_{sun}} = 1.334 \text{ x } 10^9$.

13.92 $V_g = WR\left(\dfrac{y}{R}\right) = Wy$. (*b*) $V_g = Wy - \dfrac{Wy^2}{R}$.

13.93 $v_r = \pm 3.87$ m/s, $v_\theta = 1.000$ m/s.

13.94 (*a*) $r_m = 0.720$ m. (*b*) $v_2 = 0.834$ m.

13.95 (*a*) $v_A' = 1.713$ m/s $\measuredangle$ 37.8°. (*b*) $v_A' = 0.316$ m/s.

13.96 (*a*) $v_A = 0.919$ m/s. (*b*) $v_B = 8.27$ m/s.

13.97 (*a*) $v_A = 7.35$ m/s. (*b*) $v_A = 11.02$ m/s.

13.98 $r_{max} = 66{,}700$ km.

13.99 (*a*) $r_{max} = 1.661$ m, $r_{min} = 0.339$ m. (*b*) $v_{max} = 26.6$ m/s, $v_{min} = 5.21$ m/s.

13.100 $v_A = 27.6 \times 10^3$ km/h.

13.101 (*a*) $\Delta v_A = 2430$ m/s. (*b*)) $\Delta v_B = 1470$ m/s.

13.102 (*a*) $\Delta v_A = 8{,}880$ m/s. (*b*) $v_B = 6860$ m/s.

13.103 $\Delta v_A = 14.20$ km/s.

13.104 (*a*) $v_A - v_E = 32.695 - 29.758 = 2.94$ km/s.
(*b*) $v_A - v_M = 24.115 - 21.471 = 2.64$ km/s.

13.105 (*a*) $\Delta v_A = 726$ m/s, $\Delta v_B = 135.7$ m/s. (*b*) $\dfrac{E}{m} = 12.90 \times 10^6 \text{ m}^2/\text{s}^2$

13.106 (*a*) $v_B = 7.35$ km/s. (*b*) $\phi_B = 45.0°$.

13.107 $\gamma = 68.9°$.

13.108 $r_{min} = (1 - \sin\alpha) r_0$.

13.109 $v_C = 1555$ m/s, $\phi = 79.3°$.

13.110 (*a*) $v_A = 2450$ m/s. (*b*) $v_B = 2960$ m/s.

13.111 $v_B = 9560$ m/s, $\phi_B = 57.4°$.

13.112 $v_p^2 = \dfrac{2GM}{r_A + r_p}\left(\dfrac{r_A}{r_P}\right)(3)$, $v_A^2 = \dfrac{2GM}{r_A + r_p}\left(\dfrac{r_P}{r_A}\right)$ Q.E.D.

13.113 $E = -\dfrac{GMm}{r_A + r_P}$.

13.114 (*a*) $\alpha = \sin\theta \sqrt{\dfrac{2(n-1)}{n^2 - \sin^2\theta}}$ Q.E.D. (*b*) $\alpha = \sin 90° \sqrt{\dfrac{2(n-1)}{n^2 - \sin^2 90°}} = \sqrt{\dfrac{2}{n+1}}$.

13.115 (*b*) $v_{esc}\sqrt{\dfrac{\alpha}{1+\alpha}} < v_0 < v_{esc}\sqrt{\dfrac{1+\alpha}{2+\alpha}}$,

13.118 (*a*) $h = r_{min} v_{max}$, $\dfrac{E}{m} = \dfrac{1}{m}\left(\dfrac{1}{2} m v_{max}^2 - \dfrac{GMn}{r_{min}}\right) = \dfrac{1}{2} v_{max}^2 - \dfrac{GM}{r_{min}}$.

(*b*) $\dfrac{1}{r_{min}} = \dfrac{GM}{h^2}\left[1 + \sqrt{1 + \dfrac{2E}{m}\left(\dfrac{h}{GM}\right)^2}\right]$.

(*d*) 1. Hyperbola if $\varepsilon > 1$, that is, if $E > 0$,
2. Parabola if $\varepsilon = 1$, that is, if $E = 0$,
3. Ellipse if $\varepsilon < 1$, that is, if $E < 0$.

13.119 $t = 4$ min 19 s.

13.120 (*a*) $t = 3.40$ s. (*b*) $t = 25.5$ s.

13.121 $F_n = 83.3$ N.

13.122 $t = 0.603$ s.

13.123 $t = 5.51$ s.
13.124 (*a*) $F = 7940$ N. (*b*) $t = 3.00$ s.
13.125 $\mu_s = 0.260$.
13.126 (*a*) $t = 11.42$ s. (*b*) $\mathbf{v} = -(125.5 \text{ m/s})\mathbf{j} - (194.5 \text{ m/s})\mathbf{k}$.
13.127 $t = 2.53$ s.
13.128 (*a*) $t_{0-20} = 2.61$ s. (*b*) $t_{0-400} = 11.95$ s.
13.129 (*a*) $t_{1-2} = 14.42$ s. (*b*) $F_C = 3470$N (tension).
13.130 (*a*) $t_{1-2} = 28.8$ s. (*b*) $F_C = 12{,}480$ N (tension).
13.131 (*a*) $t_{1-2} = 19.60$ s (*b*) $Q = 10.20$ kN (compression).
13.132 (*a*) $t = 0.549$ s. (*b*) $T = 56.8$ N.
13.133 (*a*) $v = 3.92$ m/s. (*b*) $F_C = 39.2$ N.
13.134 (*a*) $F_{ave} = 18.18$ kN. (*b*) $F_m = 36.4$ kN.
13.135 (*a*) $v_2 = 70.5$ m/s. (b) $t_3 = 7.99$ s.
13.136 $p_0 = 223$ Mpa.
13.137 (*a*) $t_1 = 4.71$ s. (*b*) $v_m = 18.42$ m/s. (*c*) $t_m = 18.76$ s.
13.138 (*a*) $t_1 = 7.0632$ s. (*b*) $\mathbf{v}_m = 3.14 \text{ m/s} \rightarrow$. (*c*) $t_s = 13.36$ s.
13.139 $F_{AV} = 350$ N.
13.140 $F_m = 7.23$ kN.
13.141 $P_v = 6.21$ W.
13.142 $P_H = 2.68$ kN.
13.143 (*a*) $v_2 = 20.0$ m/s. (*b*) $R_{ave} = 40.0$ kN.
13.144 $\mathbf{F}' = 445 \text{ kN} \measuredangle 40.1°$.
13.145 (*a*) $v' = 1.333 \text{ km/h} \leftarrow$. (*b*) $t = 0.1888$ s.
13.146 (*a*) A was going faster. (*b*) $v_A = 115.\ 2$ km/h.
13.147 $F_{AV} = 65.0$ kN.
13.148 (*a*) $T_1 - T_2 = 13.50 \text{ N} \cdot \text{m}$, $F\Delta t = 4.5 \text{ N} \cdot \text{S}$.
(*b*) $T_1 - T_2 = 11.57 \text{ N} \cdot \text{m}$, $F(\Delta t) = 3.86 \text{ N} \cdot \text{m}$.
13.149 $v_0 = 208$ m/s.
13.150 (*a*) $\mathbf{v}_2 = 0.875 \text{ m/s} \leftarrow$. (*b*) $\mathbf{v}_2' = -0.0714 \text{ m/s} \leftarrow$.
13.151 (*a*) $v'_{plate} = 1.694 \text{ m/s} \downarrow$. (*b*) Energy lost = $(1.3272 - 1.1653)\text{J} = 0.1619$ J.
13.152 $v_0 = 1.650$ m/s.
13.153 (*a*) $\mathbf{v}' = 3.64 \text{ m/s} \downarrow$. (*b*) $R_x\Delta t = 5.31 \text{ N} \cdot \text{S}$, $R_y\Delta t = 9.11 \text{ N} \cdot \text{S}$.
13.154 (*a*) $F\Delta t = 162.7 \text{ N} \cdot \text{s}$, $E = 441$ J. (*b*) $F\Delta t = 130.2 \text{ N} \cdot \text{s}$, $E = 353$ J.
13.155 (*a*) $\mathbf{v}_A = 0.594 \text{ m/s} \leftarrow$, $\mathbf{v}_B = 1.156 \text{ m/s} \rightarrow$. (*b*) $T_1 - T_2 = 2.99$ J.
13.156 $T_1 - T_2 = (1 - e^2)mv^2$.
13.157 $0.728 \leq e \leq 0.762$.
13.158 (*a*) $m_B = 1.500$ kg. (*b*) $1.000 \text{ kg} \leq m_B \leq 3.00$ kg.
13.159 $m_B = 4.47$ kg.
13.160 $\mathbf{v}_C' = 0.150 \text{ m/s} \leftarrow$, $\mathbf{v}_A' = 1.013 \text{ m/s} \leftarrow$, $\mathbf{v}_B'' = 0.338 \text{ m/s} \leftarrow$.
13.161 (*a*) $v_A' = \dfrac{v_0(1-e)}{2}$, $v_B' = \dfrac{v_0(1+e)}{2}$.

(*b*) $v_C' = \dfrac{v_0(1+e)^2}{4}$, $v_B'' = \dfrac{v_0(1-e^2)}{4}$.

(*c*) $v_n' = \dfrac{v_0(1+e)^{(n-1)}}{2^{(n-1)}}$. (*d*) $v_B' = 0.815\, v_0$.

13.162 (*a*) $\mathbf{v}_A' = 1.288 \text{ m/s} \leftarrow$, $\mathbf{v}_B' = 0.312 \text{ m/s} \rightarrow$, $\mathbf{v}_C' = 1.512 \text{ m/s} \rightarrow$.
(*b*) $\mathbf{v}_A''' = 0.956 \text{ m/s} \leftarrow$, $\mathbf{v}_B''' = 0.0296 \text{ m/s} \leftarrow$, $\mathbf{v}_C'' = 1.552 \text{ m/s} \rightarrow$.
13.163 $\mathbf{v}_C = 0.294 \text{ m/s} \leftarrow$.
13.164 (*b*) $\bar{v}_A = 0.711\, v_0 \measuredangle 39.3°$, $\bar{v}_B = 0.636 v_0 \measuredangle 45°$.
13.165 (*a*) $\theta = 70.2°$. (*b*) $\mathbf{v}_B' = 0.322 \text{ m/s} \rightarrow$.
13.166 $\mathbf{v}_A' = 6.37 \text{ m/s} \measuredangle 77.2°$, $\mathbf{v}_B' = 1.802 \text{ m/s} \measuredangle 40°$.
13.167 $\mathbf{v}_A' = 1.322 \text{ m/s} \measuredangle 70.9°$, $\mathbf{v}_B' = 3.85 \text{ m/s} \measuredangle 27.0°$.
13.168 (*a*) $\mathbf{v}_A = 0.878\, v_0 \measuredangle 24.2°$, $v_B = 0.412\, v_0 \measuredangle 61.0°$.
13.169 $e = 0.837$.
13.170 (*a*) $y = 15.69$ m. (*b*) $x = 73.8$ m.
13.171 $d = 15.94$ m.

13.172 $h = 0.156$ m.
13.173 (a) $\alpha = 22.5°$. (b) $\alpha = 21.3°$.
13.174 (a) $v_B = 37.9$ km/h. (b) $e = 0.1902$.
13.175 (a) $h = 0.294$ m. (b) $x = 54.4$ mm.
13.176 (a) $e = 0.258$. (b) $v_0 = 4.34$ m/s.
13.177 (a) $v'_B = 2.90$ m/s. (b) $\Delta T_L = 100.5$ J.
13.178 (a) $e_{AB} = 0.724$, $e_{BC} = 0.288$. (b) $x = 293$ mm.
13.179 (a) $h = 0.0240$ m. (b) $k = 817$ N/m.
13.180 (a) $e = 1.000$. (b) $h = 0.0667$ m. (c) $\Delta = 0.0762$ m.
13.181 (a) $t = 0.08495$ s. (b) $x_C = 7.08$ mm, $x_A = 28.3$ mm.
13.182 (a) $\mathbf{v}'_A = 0$, $\mathbf{v}'_B = 0$. (b) $\mathbf{v}'_A = 1.201$ m/s $\leftarrow$, $\mathbf{v}'_B = 0.400$ m/s $\rightarrow$.
13.183 (a) $h = 401$ mm. (b) $F\Delta t = 4.10$ N · s.
13.184 (a) $\mathbf{v}'_A = 0.300$ m/s $\uparrow$, $\mathbf{v}'_B = 0.209$ m/s $\measuredangle$ 10.38°. (b) $h = 6.72$ mm.
13.185 $h_B = 86.6$ mm.
13.186 (a) $e = 0.923$. (b) $h_1 = 1.278$ m.
13.187 (a) $\theta = 62.7°$. (b) $T_{\text{lost}} = 0.1400 v_0^2$.
13.188 (a) $\mathbf{v}'_A = 0.690$ m/s $\measuredangle$ 88.9°, $\mathbf{v}'_B = 1.053$ m/s $\rightarrow$.
13.189 (a) $\mathbf{v}'_A = 0.330$ m/s $\leftarrow$, $\mathbf{v}'_B = 0.991$ m/s $\rightarrow$. (b) $\Delta x = 36.2$ mm.
13.190 $N = 877$ N.
13.191 $E_L = 2.38$ J.
13.193 $v_2 = 3.13$ m/s.
13.194 $v_0 = 3970$ m/s.
13.195 (a) $\mathbf{N}_A = 13.31$ N $\rightarrow$. (b) $\mathbf{N}_B = 4.49$ N$\downarrow$. (c) $\mathbf{N}_C = 13.31$ N $\leftarrow$.
13.196 $h = 0.707\,a$.
13.197 $\nearrow_{30}$ $d_A = 0.21675$ m $= 217$ mm, $d_B = 69.1$ mm.
13.198 (a) $v'_A = v'_B = v'_C = 1.368$ m/s. (b) $d = 0.668$ m. (c) $x = 1.049$ m.
13.199 $\mathbf{a}_n = 28.8$ m/s^2, $N = 157.2$ N.
13.200 (a) $v_2 = 2.94$ m/s. (b) $T = 16.14$ N.
13.201 $\mathbf{v}_0 = 4.64$ m/s $\rightarrow$.

CHAPTER 14

14.1 (a) $\mathbf{v}' = 4.46$ m/s $\leftarrow$. (b) $\mathbf{v}'' = 0.409$ m/s $\leftarrow$.
14.2 (a) $\mathbf{v}_A = 306$ m/s $\leftarrow$. (b) $\mathbf{v}'' = 0.409$ m/s $\leftarrow$.
14.3 (a) $\mathbf{v}_1 = 1.125$ m/s$\rightarrow$, $\mathbf{v}_2 = 1.417$ m/s$\rightarrow$. (b) $\mathbf{v}'_1 = 0.8889$ m/s$\rightarrow$, $\mathbf{v}'_2 = 1.417$ m/s
14.4 (a) $m = 25.2$ g. (b) $\mathbf{v}_1 = 271$ m/s $\rightarrow$.
14.5 (a) $\mathbf{v}_2 = 0.875$ m/s $\leftarrow$. (b) $\mathbf{v}'_2 = -0.0714$ m/s $\leftarrow$.
14.6 (a) $\mathbf{v}_2 = 2.88$ m/s $\leftarrow$. (b) $\mathbf{v}'_2 = 2.93$ m/s $\leftarrow$.
14.7 (a) $\mathbf{v}_1 = 3.79$ km/h $\rightarrow$, $\mathbf{v}_2 = 2.77$ km/h $\rightarrow$.
(b) $\mathbf{v}_1 = 5.54$ km/h $\rightarrow$, $\mathbf{v}_2 = 2.77$ km/h $\rightarrow$.
(c) $\mathbf{v}_1 = 5.54$ km/h $\rightarrow$, $\mathbf{v}_2 = 3.60$ km/h $\rightarrow$.
14.8 (a) $\mathbf{v}'_C = 0.901$ m/s $\rightarrow$. (b) $\mathbf{v}''_A = 0.807$ m/s $\rightarrow$.
14.9 $\mathbf{H}_O = -(4.80\text{ kg}\cdot\text{m}^2/\text{s})\mathbf{j} + (9.60\text{ kg}\cdot\text{m}^2/\text{s})\mathbf{k}$.
14.10 (a) $\bar{\mathbf{r}} = (1.867\text{ m})\mathbf{i} + (1.533\text{ m})\mathbf{j} + (0.667\text{ m})\mathbf{k}$.
(b) $m\bar{\mathbf{v}} = (12.00\text{ kg}\cdot\text{m/s})\mathbf{i} + (28.0\text{ kg}\cdot\text{m/s})\mathbf{j} + (14.00\text{ kg}\cdot\text{m/s})\mathbf{k}$.
(c) $\mathbf{H}_G = -(2.80\text{ kg}\cdot\text{m}^2/\text{s})\mathbf{i} + (13.33\text{ kg}\cdot\text{m}^2/\text{s})\mathbf{j} - (24.3\text{ kg}\cdot\text{m}^2/\text{s})\mathbf{k}$.
14.11 (a) $v_x = \frac{1}{4}(4v_y - 5.5)$ m/s, $v_z = \frac{1}{4}(3v_y - 2.125)$ m/s. (b) $\mathbf{H}_O = (-13.63\text{ N}\cdot\text{m}\cdot\text{s})\mathbf{i}$.
14.12 (a) $v_z = \frac{1}{4}(3v_y + 11.5)$ m/s, $v_x = \frac{1}{3}(3v_y + 11.5)$ m/s.
(b) $\mathbf{H}_O = -\left(v_y + \frac{28}{3}\right)\text{N}\times\text{m}\times\text{s}$.
14.13 $\mathbf{H}_O = -(31.2\text{ kg}\cdot\text{m}^2/\text{s})\mathbf{i} - (64.8\text{ kg}\cdot\text{m}^2/\text{s})\mathbf{j} + (48.0\text{ kg}\cdot\text{m}^2/\text{s})\mathbf{k}$.
14.14 (a) $\bar{\mathbf{r}} = (0.600\text{ m})\mathbf{i} + (1.400\text{ m})\mathbf{j} + (1.525\text{ m})\mathbf{k}$.
(b) $m\bar{\mathbf{v}} = -(26.0\text{ kg}\cdot\text{m/s})\mathbf{i} + (14.00\text{ kg}\cdot\text{m/s})\mathbf{j} + (14.00\text{ kg}\cdot\text{m/s})\mathbf{k}$.
(c) $\mathbf{H}_G = -(29.5\text{ kg}\cdot\text{m}^2/\text{s})\mathbf{i} - (16.75\text{ kg}\cdot\text{m}^2/\text{s})\mathbf{j} + (3.20\text{ kg}\cdot\text{m}^2/\text{s})\mathbf{k}$.

14.15 $\mathbf{r}_B = (114.4 \text{ m})\mathbf{i} - (76.1 \text{ m})\mathbf{j} + (8.75 \text{ m})\mathbf{k}$.
14.16 $\mathbf{r}_C = (1180 \text{ m})\mathbf{i} + (140 \text{ m})\mathbf{j} + (155 \text{ m})\mathbf{k}$.
14.17 43.1 m (east), 113.0 m (up).
14.18 $x_B = 253$ m, $y_B = 6.61$ m, $z_B = -4.67$ m.
14.19 $x_P = 17.40$ m, $y_P = 16.50$ m.
14.20 (*a*) $t_P = 2.00$ s. (*b*) $v_A = 144.0$ km/h.
14.21 $\mathbf{r}_p = (26.0 \text{ m})\mathbf{i} + (125.4 \text{ m})\mathbf{k}$.
14.22 (*a*) $\mathbf{v}'_A = 2.46 \text{ m/s} \rightarrow$. (*b*) $\theta = 36.6°$, $v_C = 3.30$ m/s, $v_D = 2.77$ m/s.
14.23 $v_A = 2.98$ m/s, $v_B = 1.437$ m/s.
14.24 $v_A = 431$ m/s, $v_B = 395$ m/s, $v_C = 528$ m/s.
14.25 $v_A = 646$ m/s, $v_B = 789$ m/s, $v_C = 176$ m/s.
14.26 $v_A = 919$ m/s, $v_B = 717$ m/s, $v_C = 619$ m/s.
14.31 $F_f d = 2.97$ J, Loss = 24.3 J, Loss = 3007 J.
14.32 (*a*) $m = 25.2$ g. (*b*) $\mathbf{v}_1 = 271$ m/s. (*a*) energy lost = 1622 J. (*b*) energy lost = 917 J.
14.33 $T_1 - T_0 = 600$ J, $T'_2 + T'_1 = 703$ J
14.34 Loss = 4.86 J, Loss = 3.28 J, Loss = 0.688 J.
14.35 (*a*) $\dfrac{E_A}{E_B} = \dfrac{m_B}{m_A}$. (*b*) $E_A = 180.0$ kJ, $E_B = 320$ kJ.
14.36 $\dfrac{S_A}{S_B} = \dfrac{m_B^2}{m_A^2}$.
14.37 (*a*) $v_B = \dfrac{m_A v_0}{m_A + m_B} \rightarrow$. (*b*) $h = \dfrac{m_A}{m_A + m_B} \dfrac{v_0^2}{2g}$.
14.38 $\mathbf{v}_A = 4.11$ m/s ∠ 46.9°, $\mathbf{v}_B = 17.39$ m/s ∠ 16.7°.
14.39 (*a*) $\mathbf{v}_{B/A} = 2.86$ m/s ∠ 30°. (*b*) $\mathbf{v}_A = 0.929 \text{ m/s} \leftarrow$.
14.40 $\mathbf{v}_B = 1.485 \text{ m/s} \rightarrow$, $\mathbf{v}_A = 0.990 \text{ m/s} \leftarrow$.
14.41 $v_A = 3.54$ m/s, $v_B = 1.768$ m/s, $v_C = 3.06$ m/s.
14.42 $v_A = 2.50$ m/s, $v_B = 3.06$ m/s, $v_C = 3.06$ m/s.
14.43 (*a*) $\mathbf{v_C} = 0$, $\mathbf{v}_A = 0.250v_0$ ∠ 60°, $\mathbf{v}_B = 0.901v_0$ ∠ 13.9°.
(*b*) Fraction of kinetic energy lost = $\dfrac{1}{8}$.
14.44 (*a*) $\mathbf{v}_A = 0.200v_0 \leftarrow$, $\mathbf{v}_B = 0.693v_0$ ∠ 30°, $\mathbf{v}_C = 0.693v_0$ ∠ 30°.
(*b*) $\mathbf{v}_A = 0.250v_0$ ∠ 60°, $\mathbf{v}_B = 0.866v_0$ ∠ 30°, $\mathbf{v}_C = 0.433v_0$ ∠ 30°.
14.45 (*a*) $\mathbf{L} = (8.75 \text{ kg} \cdot \text{m/s})\mathbf{i}$, $\mathbf{H}_G = -(0.500 \text{ kg} \cdot \text{m}^2/\text{s})\mathbf{k}$.
(*b*) $\mathbf{v}'_A = (1.50 \text{ m/s})\mathbf{i}$, $\mathbf{v}'_B = (5.00 \text{ m/s})\mathbf{i}$.
14.46 $\mathbf{v}_A = (60.0 \text{ m/s})\mathbf{i} + (60.0 \text{ m/s})\mathbf{j} + (390 \text{ m/s})\mathbf{k}$.
14.47 (*a*) $v_C = 3.50$ m/s, $v_D = 1.750$ m/s. (*b*) $\dfrac{(T_1 - T_2)}{T_1} = 0.786$.
14.48 $x_{B_0} = 181.7$ mm, $y_{B_0} = 0$, $z_{B_0} = 139.4$ mm.
14.49 $(v_B)_y = -0.6$ m/s, $\mathbf{v}_C = (0.6 \text{ m/s})\mathbf{i} + (0.6 \text{ m/s})\mathbf{j}$.
14.50 (*a*) $v_D = 0.500v_0$. (*b*) $u = 0.750v_0$. (*c*) $\dfrac{T_1 - T_2}{T_1} = 0.1875$.
14.51 (*a*) $\mathbf{v}_B = 2.40$ m/s ∠ 53.1°, $\mathbf{v}_C = 2.56 \text{ m/s} \rightarrow$. (*b*) $c = 1.059$ m.
14.52 (*a*) $\mathbf{v}_A = 2.40 \text{ m/s}\downarrow$, $\mathbf{v}_B = 3.00$ m/s ∠ 53.1°. (*b*) $a = 1.864$ m.
14.53 (*a*) $\mathbf{v}_A = 2.56 \text{ m/s}\uparrow$, $\mathbf{v}_B = 4.24$ m/s ∠ 31.9°. (*b*) $a = 2.34$ m.
14.54 (*a*) $\overline{\mathbf{v}}_0 = (2.4 \text{ m/s})\mathbf{i} + (1.8 \text{ m/s})\mathbf{j}$. (*b*) Length of cord = 600 mm.
(*c*) Original rate of spin = 20.0 rad/s.
14.55 (*a*) $\mathbf{v}_A = 0.693 \text{ m/s}\uparrow$, $\mathbf{v}_B = 0.693 \text{ m/s}\downarrow$, $\mathbf{v}_C = 1.200 \text{ m/s} \rightarrow$. (*b*) $d = 200$ mm.
14.56 (*a*) $v_0 = 0.600 \text{ m/s} \rightarrow$. (*b*) $l = 240$ mm. (*c*) $\dot{\theta} = 5.00$ rad/s ↻.
14.57 $P = \rho(A_1 - A_2)v_1^2$.
14.58 $P = \rho A_2 v_2^2 - \rho A_1 v_1^2 \cos\theta$.
14.59 $P = 312$ N.
14.60 $V = 4.18$ m/s.
14.61 (*a*) $F_z = 14.8$ kN. (*b*) $F_x = 27.7$ kN.
14.62 $\mathbf{F}_x = 90.6 \text{ N} \leftarrow$.
14.63 $R = 4040 \text{ N} \uparrow$.

14.64 $D_x = 329$ N, $D_y = 0$, $C_x = -203$ N, $C_y = 271$ N.
14.65 $\mathbf{C} = 161.7$ N ↑, $\mathbf{D}_x = 154.8$ N →, $\mathbf{D}_y = 170.2$ N ↑.
14.66 (*a*) $v_A = 61.1$ m/s. (*b*) $\mathbf{R} = 59.2$ N ⦪ 49.0°.
14.67 $C_x = 90.0$ N, $C_y = 2360$ N, $D_x = 0$, $D_y = 2900$ N.
14.68 (*a*) $\mathbf{A} = qv_0 \leftarrow$. (*b*) $\mathbf{B} = \sqrt{2gh}$ ⦪ 30°.
14.69 $\dfrac{dm}{dt} = 100$ kg/s.
14.70 $D = 36.9$ kN.
14.71 $\mathbf{F} = 33.6$ kN ←.
14.72 $W = 41.4$ kN.
14.73 $h = 1.096$ m.
14.74 (*a*) $F = 46100$ N, $d = 1.205$ m. (*b*) $F = 32200$ N, $d = 3.45$ m.
14.75 (*a*) $v_1 = 760$ km/h. (*b*) $v_2 = 562$ km/h.
14.76 (*a*) $\mathbf{a} = 3.03$ m/s^2 ⦣ 18°. (*b*) $v_A = 922$ km/h.
14.77 (*a*) $v = 30.6$ m/s. (*b*) $Q = 96.1$ m^3/s. (*c*) $\dfrac{dT}{dt} = 55{,}100$ N · m/s.
14.78 (*a*) $\dfrac{dT}{dt} = 3.234$ MW. (*b*) $\eta = 0.464$.
14.79 $d_2 = 213$ m.
14.80 (*a*) Propulsion power = 12 MW. (*b*) Total power = 21.6 MW.
(*c*) Mechanical efficiency = 0.556.
14.81 (*a*) $v_A = \dfrac{1}{2}V$. (*b*) $(P_{\text{out}})_{\max} = \dfrac{1}{4}\rho A v_A^3(1-\cos\theta)$. (*c*) $\eta = 2\left(1 - \dfrac{V}{v_A}\right)\dfrac{V}{v_A}(1-\cos\theta)$.
14.82 $d = \dfrac{D}{\sqrt{2}}$.
14.83 (*a*) $P = qv$. (*b*) $\dfrac{\Delta T}{\Delta t} = \dfrac{1}{2}\dfrac{\Delta U}{\Delta t}$ Q.E.D. ,
The other half of the work of **P** is dissipated into heat by friction as the gravel slips on the belt before reaching the speed v.
14.84 $Q = b\sqrt{\dfrac{1}{2}gd_1d_2(d_1+d_2)}$.
14.85 $Q = 17.58$ m^3/s.
14.86 (*a*) $P = \dfrac{m}{l}(v^2 + gy)$. (*b*) $\mathbf{R} = mg\left(1 - \dfrac{y}{l}\right)\uparrow$.
14.87 (*a*) $P = \dfrac{mgy}{l}$. (*b*) $\mathbf{R} = \dfrac{m}{l}[g(L-y) + v^2]\uparrow$.
14.88 $v = \sqrt{gh}\tanh\left(\dfrac{\sqrt{gh}}{L}t\right)$.
14.89 $v_{\max} = 3.37$ m/s.
14.90 $U = 1.485$ m/s.
14.91 $\dfrac{dm}{dt} = 533$ kg/s.
14.92 Total thrust = 3.03 MN.
14.93 (*a*) $\mathbf{a} = 31.9$ m/s^2 ↑. (*b*) $\mathbf{a} = 240$ m/s^2 ↑.
14.94 (*a*) $(a_t)_0 = 90.0$ m/s^2. (*b*) $v_1 = 35.9 \times 10^3$ km/h.
14.95 $v_m = 7930$ m/s.
14.96 (*a*) $v_1 = 1800$ m/s. (*b*) $v_2 = 9240$ m/s.
14.97 $w_{\text{fuel}} = 23200$ N.
14.98 $\Delta v = 1151$ m/s.
14.99 $x = 30.8$ km.
14.100 (*a*) $h = 173.9$ km. (*b*) $v = 6.38$ km/s.
14.101 $h = 186.8$ km.
14.102 (*a*) $h_1 = 31.2$ km. (*b*) $h_2 = 197.5$ km.
14.103 $\eta = \dfrac{2v}{(u+v)}$.

14.104 $\eta = \dfrac{2uv}{(u^2 + v^2)}$.

14.105 (a) $\mathbf{v}_A = \mathbf{v}_B = \mathbf{v}_C = 0.667$ km/h →. (b) $\mathbf{v}_A = \mathbf{v}_B = 2.8$ km/h ← $\mathbf{v}_C = 7.6$ km/h →.

14.106 (a) $v_f = 1.595$ m/s. (b) $x = 0.370$ m.

14.107 (a) $\mathbf{v}' = 5.20$ km/h →. (b) $\mathbf{v}_f = 4.00$ km/h →.

14.108 Components of $\mathbf{v}_B$: 146.3 km/h 60.5 km/h 22.9 km/h.

14.109 (a) $\mathbf{v}_A = 3.11$ m/s. (b) $h = 1.356$ m.

14.110 $\mathbf{v}_A = 4.81$ m/s →, $\mathbf{v}_B = 1.602$ m/s ←.

14.111 $\mathbf{v}_B = 42.4$ km/h, $\mathbf{v}_C = 84.7$ km/h.

14.112 $\mathbf{M}_A = 46.0$ N · m ↓, $\mathbf{A} = 274$ N ⦩ 20°.

14.113 (a) $m'l = \dfrac{1-\beta}{\beta} m$. (b) $|F_{\max}| = \dfrac{(1-\beta) m v_0^2}{\beta l}$.

14.114 (a) $m_0 + qt_L = m_0 e^{qL/m_0 v_0}$. (b) $v_L = v_0 e^{-qL/m_0 v_0}$.

14.115 $\omega = 414$ rpm.

14.116 Case 1: (a) $\mathbf{a} = 0.333\, g$↓. (b) $\mathbf{v} = 0.817\sqrt{gl}$ ↓.

Case 2: (a) $\mathbf{a} = \dfrac{gy}{l}$ ↓. (b) $\mathbf{v} = \sqrt{gl}$ ↓.

CHAPTER 15

15.1 (a) $\omega = 29.6$ rad/s. (b) $\theta = 32.2$ rev.

15.2 (a) $\theta = 0$, $\omega = 0$, $\alpha = 0$. (b) $\theta = 6.00$ rad, $\omega = 4.71$ rad/s, $\alpha = -3.70$ rad/s^2.

15.3 (a) $\theta = 0$, $\omega = 0.1000$ rad/s, $\alpha = -0.250$ rad/s^2.
(b) $\theta = 0.211$ rad, $\omega = 0.0472$ rad/s, $\alpha = -0.01181$ rad/s^2.
(c) $\theta = 0.400$ rad, $\omega = 0$, $\alpha = 0$.

15.4 (a) $\alpha = -3.01$ rad/s^2. (b) $\theta = 13{,}80$ rev.

15.5 (a) $\theta = 150$ rev. (b) $\theta = 2100$ rev.

15.6 (a) $\omega_{\max} = 0.855$ rad/s. (b) $\theta = 3.71°$.

15.7 $\omega = 26.8$ rad/s.

15.8 (a) $k = 4.00$ s^{-2}. (b) $\omega = 5.29$ rad/s.

15.9 (a) $\theta = 12.73$ rev. (b) $t = \infty$. (c) $t = 18.42$ s.

15.10 $\mathbf{v}_C = -(0.45$ m/s$)\mathbf{i} - (1.2$ m/s$)\mathbf{j} + (1.5$ m/s$)\mathbf{k}$, $\mathbf{a}_C = (12.60$ m/s$^2)\mathbf{i} + (7.65$ m/s$^2)\mathbf{j}$ $+ (9.90$ m/s$^2)\mathbf{k}$.

15.11 $\mathbf{v}_B = (0.75$ m/s$)\mathbf{i} + (1.5$ m/s$)\mathbf{k}$, $\mathbf{a}_B = (12.75$ m/s$^2)\mathbf{i} + (11.25$ m/s$^2)\mathbf{j} + (3$ m/s$^2)\mathbf{k}$.

15.12 $\mathbf{v}_F = -(0.4$ m/s$)\mathbf{i} - (1.4$ m/s$)\mathbf{j} - (0.7$ m/s$)\mathbf{k}$,
$\mathbf{a}_F = (8.4$ m/s$^2)\mathbf{i} + (3.3$ m/s$^2)\mathbf{j} - (11.4$ m/s$^2)\mathbf{k}$.

15.13 $\mathbf{v}_H = -(0.4$ m/s$)\mathbf{i} + (0.7$ m/s$)\mathbf{k}$.
$\mathbf{a}_H = -(2$ m/s$^2)\mathbf{i} - (6.5$ m/s$^2)\mathbf{j} - (3$ m/s$^2)\mathbf{k}$.

15.14 $\mathbf{v}_E = (3.12$ m/s$)\mathbf{i} + (2.88$ m/s$)\mathbf{j} + (1.200$ m/s$)\mathbf{k}$,
$\mathbf{a}_E = -(81.1$ m/s$^2)\mathbf{i} + (74.9$ m/s$^2)\mathbf{j} + (31.2$ m/s$^2)\mathbf{k}$.

15.15 $\mathbf{v}_E = (3.12$ m/s$)\mathbf{i} + (2.88$ m/s$)\mathbf{j} + (1.200$ m/s$)\mathbf{k}$,
$\mathbf{a}_B = -(73320$ mm/s$^2)\mathbf{i} + (82080$ mm/s$^2)\mathbf{j} + (34200$ mm/s$^2)\mathbf{k}$
$= -(73.3$ m/s$^2)\mathbf{i} + (82.1$ m/s$^2)\mathbf{j} + (34.2$ m/s$^2)\mathbf{k}$.

15.16 $v = 10{,}750$ km/h, $a = 5.95 \times 10^{-4}$ m/s^2.

15.17 (a) $v = 465$ m/s, $a = 0.0339$ m/s^2. (b) $v = 356$ m/s, $a = 0.0260$ m/s^2.
(c) $v = a = 0$.

15.18 (a) $\boldsymbol{\omega} = 2.50$ rad/s ↺, $\boldsymbol{\alpha} = 1.500$ rad/s^2 ↻. (b) $\mathbf{a}_B = 771$ mm/s^2 ⦩ 76.5°.

15.19 12.00 rad/s^2 ↺ or ↻.

15.20 (a) $a_B = 90.05$ m/s^2. (b) $a_B = 1440$ m/s^2.

15.21 (a) $v_C = 6.28$ m/s, $a_C = 15.79$ m/s^2. (b) $v_9 = 0.628$ m/s, $a_C = 15.80$ m/s^2.

15.22 left: $t = 3.49$ s; middle: $t = 6.98$ s; right: $t = 13.96$ s.

15.23 (a) $\mathbf{v}_C = 0.15$ m/s →, $\mathbf{a}_C$ 0.45 m/s^2 ←.
(b) $\mathbf{a}_B = 1.273$ m/s^2 ⦩ 45°.

15.24 (a) $\boldsymbol{\omega}_B = 300$ rpm ↺, $\boldsymbol{\omega}_C = 100$ rpm ↻. (b) $\mathbf{a}_B = 49.3$ m/s^2 ←,
$\mathbf{a}_C = 16.45$ m/s^2 →.

15.25 (*a*) A: $\boldsymbol{\omega}_A = 15.00$ rad/s ↺; *B*: $\boldsymbol{\omega}_B = 7.50$ rad/s ↺.
(*b*) A: $\mathbf{a}_A = 22.5$ m/s^2 ↑; *B*: $\mathbf{a}_B = 11.25$ m/s^2 ↓.
15.26 (*a*) *C*: $\omega_C = 120$ rpm; *B*: $\omega_B = 275$ rpm.
(*b*) A: $\mathbf{a}_A = 23.7$ m/s^2 ↑; *B*: $\mathbf{a}_B = 19.90$ m/s^2 ↓.
15.27 (*a*) $\boldsymbol{\omega}_B = 10$ rad/s ↺. (*b*) A: $\mathbf{a}_A = 7.50$ m/s^2 ↓; B: $\mathbf{a}_B = 3.00$ m/s^2 ↓.
(*c*) $\mathbf{a}_D = 4.00$ m/s^2 ↓.
15.28 (*a*) $\boldsymbol{\alpha}_B = 0.400$ rad/s^2 ↻. (*b*) $\theta_B = 1.528$ rev.
15.29 (*a*) $\theta = 2.75$ rev. (*b*) $\mathbf{v}_B = 1.710$ m/s ↓, $\Delta y_B = 3.11$ m ↓.
(*c*) $\mathbf{a}_D = 849$ mm/s^2 $\measuredangle$ 32.0°.
15.30 (*a*) $\mathbf{v}_A = 1.152$ m/s ↑, $\mathbf{y}_A = 2.30$ m ↑. (*b*) $\mathbf{v}_B = 1.728$ m/s ↓, $\mathbf{y}_B = 3.46$ m ↓.
15.31 (*a*) $\theta_A = 15.28$ rev. (*b*) $t_f = 10.14$ s.
15.32 $\boldsymbol{\alpha}_A = 5.41$ rad/s^2 ↺, $\boldsymbol{\alpha}_B = 1.466$ rad/s^2 ↺.
15.33 (*a*) $t = 10.39$ s. (*b*) $\boldsymbol{\omega}_A = 413$ rpm ↻, $\boldsymbol{\omega}_B = 248$ rpm ↺.
15.34 (*a*) $t = 19.06$ s. (*b*) $\boldsymbol{\omega}_A = 341$ rpm ↻, $\boldsymbol{\omega}_B = 455$ rpm ↺.
15.35 (*a*) $\boldsymbol{\alpha}_A = 2.36$ rad/s^2 ↻, $\boldsymbol{\alpha}_B = 4.19$ rad/s^2 ↻. (*b*) $t = 6.00$ s.
15.36 $\mathbf{a} = \dfrac{b\omega_0^2}{2\pi} \rightarrow$.
15.37 $\alpha = \dfrac{bv^2}{2\pi r^3}$ ↻.
15.38 $\mathbf{v}_C = 0$, $\mathbf{v}_B = 44.4$ m/s →, $\mathbf{v}_D = 42.9$ m/s $\measuredangle$ 15.0°, $\mathbf{v}_E = 31.4$ m/s $\measuredangle$ 45.0°.
15.39 (*b*) $\mathbf{v}_A = 160.4$ mm/s ↑. (*a*) $\boldsymbol{\omega}_{AB} = 0.378$ rad/s ↻.
15.40 (*a*) $\boldsymbol{\omega}_{AB} = 1.173$ rad/s ↺. (*b*) $\mathbf{v}_A = 0.998$ m/s $\measuredangle$ 25°.
15.41 (*b*) $\mathbf{v}_A = 1.963$ m/s ↓. (*a*) $\boldsymbol{\omega}_{AB} = 3.62$ rad/s ↺.
15.42 (*b*) $\mathbf{v}_B = 1.328$ m/s $\measuredangle$ 30°. (*a*) $\boldsymbol{\omega}_{AB} = 2.54$ rad/s ↻.
15.43 (*a*) $\boldsymbol{\omega}_{ABC} = 1.175$ rad/s ↺. (*b*) $\mathbf{v}_B = 0.449$ m/s $\measuredangle$ 59.1°.
15.44 (*a*) $\boldsymbol{\omega} = 2$ rad/s ↻. (*b*) $\mathbf{v}_A = (120 \text{ mm/s})\mathbf{i} + (660 \text{ mm/s})\mathbf{j}$.
15.45 (*a*) $\mathbf{v}_B = -(240 \text{ mm/s})\mathbf{i} + (300 \text{ mm/s})\mathbf{j}$.
(*b*) $y = -60$ mm, $x = 150$ mm.
15.46 (*a*) $\boldsymbol{\omega} = -(5.00 \text{ rad/s})\mathbf{k} = 5.00$ rad/s ↻. (*b*) $\mathbf{v}_A = (150 \text{ mm/s})\mathbf{i} + (300 \text{ mm/s})\mathbf{j}$.
15.47 (*a*) $\boldsymbol{\omega} = 4.00$ rad/s ↻. (*b*) $\mathbf{v}_B = -(100 \text{ mm/s})\mathbf{i}$.
15.48 (*a*) $\boldsymbol{\omega}_B = \boldsymbol{\omega}_C = \boldsymbol{\omega}_D = \dfrac{1}{2}\,\omega_A$ ↺. (*b*) $\boldsymbol{\omega}_S = \dfrac{1}{4}\,\omega_A$ ↻.
15.49 (*a*) $\boldsymbol{\omega}_B = \boldsymbol{\omega}_C = \boldsymbol{\omega}_D = 5\pi$ rad/s ↻. (*b*) $\boldsymbol{\omega}_S = 6.5\pi$ rad/s ↻ $= 195$ rmp ↻.
15.50 (*a*) $\boldsymbol{\omega}_B = 48$ rad/s ↻. (*b*) $\mathbf{v}_D = 3.39$ m/s $\measuredangle$ 45°.
15.51 (*a*) $v_C = 5.65$ m/s. (*b*) $\omega_D = 9000$ rpm, (*c*) $n = 1500$.
15.52 (*a*) $r = 1.5$ mm. (*b*) $a_n = 0.457 \times 10^{-3}$ mm/s^2. (*c*) $a_n = 2.74 \times 10^{-3}$ mm/s^2.
15.53 (*a*) $\boldsymbol{\omega}_A = 200$ rad/s ↺. (*b*) $\boldsymbol{\omega}_B = 24.0$ rad/s ↻.
15.54 (*a*) $\boldsymbol{\omega}_A = 104.0$ rad/s ↺. (*b*) $\boldsymbol{\omega}_B = 120.0$ rad/s ↻.
15.55 (*a*) $\boldsymbol{\omega}_{AD} = (6.00 \text{ rad/s})\mathbf{k} = 6.00$ rad/s ↺.
(*b*) $\mathbf{v}_B = (360 \text{ mm/s})\mathbf{i} - (672 \text{ mm/s})\mathbf{j} = 762$ mm/s $\measuredangle$ 61.8°.
15.56 (*a*) $\mathbf{v}_A = 540$ mm/s →. (*b*) $\mathbf{v}_B = 457$ mm/s $\measuredangle$ 61.8°.
15.57 (*a*) $\mathbf{v}_B = \dfrac{r\omega_D}{\cos\theta} \rightarrow$. (*b*) $\boldsymbol{\omega}_{AB} = \omega_D = \tan^2\theta$ ↺.
15.58 (*a*) $\boldsymbol{\omega}_D = 3.02$ rad/s ↻. (*b*) $\boldsymbol{\omega}_{AB} = 0.657$ rad/s ↺.
15.59 (*a*) $\omega_{BD} = 1.500$ rad/s ↺. (*b*) $\mathbf{v}_D = 450$ mm/s ↑.
(*c*) $\mathbf{v}_M = -(168.8 \text{ mm/s})\mathbf{i} + (225 \text{ mm/s})\mathbf{j} = 281$ mm/s $\measuredangle$ 53.1°.
15.60 (*a*) $\mathbf{v}_B = 497$ mm/s ←.
15.61 (*a*) $\mathbf{v}_P = 0$, $\boldsymbol{\omega}_{BD} = 39.3$ rad/s ↺.
(*b*) $\omega_{BD} = 0$, $\mathbf{v}_P = 6.28$ m/s ↓.
15.62 $\mathbf{v}_P = 6.52$ m/s ↓, $\omega_{BD} = 20.8$ rad/s ↺.
15.63 (*a*) $\boldsymbol{\omega}_{BD} = 12.00$ rad/s ↺. (*b*) $\mathbf{v}_M = 3.90$ m/s $\measuredangle$ 67.4°.

15.64 $\boldsymbol{\omega}_{DE} = 2.55$ rad/s ↻, $\boldsymbol{\omega}_{BD} = 0.955$ rad/s ↺.
15.65 $\boldsymbol{\omega}_{DE} = 6.4$ rad/s ↻, $\boldsymbol{\omega}_{BD} = 5.2$ rad/s ↻.
15.66 $\boldsymbol{\omega}_{DE} = 5.00$ rad/s ↻, $\boldsymbol{\omega}_{BD} = 5.00$ rad/s ↺, (*b*) $\mathbf{v}_F = 2.90$ m/s →.
15.67 (*b*) $\omega_{AB} = 2.00$ rad/s ↻, $\boldsymbol{\omega}_{DE} = 1.464$ rad/s ↻, $\mathbf{v}_F = 1.039$ m/s →.
15.68 (*a*) $\boldsymbol{\omega}_{BD} = 3.33$ rad/s ↺. (*b*) $\mathbf{v}_F = 2.00$ m/s ⦩ 56.3°.
15.69 (*a*) $h = 1.500$ m. (*b*) $\mathbf{v}_M = 5.00$ m/s ↓.
15.70 $\mathbf{v}_E = 369$ mm/s $\mathbf{i}$ = 369 mm/s →.
15.71 (*a*) $\mathbf{v}_B = 338$ mm/s ←, $\omega_{AB} = 0$.
(*b*) $\mathbf{v}_B = 710$ mm/s ←, $\omega_{AB} = 2.37$ rad/s ↻.
15.72 $\omega_C = \left(1 - \dfrac{r_A}{r_C}\right)\omega_{ABC}$.
15.73 $v_A = 10.2$ m/s, $v_B = 24.0$ ft/s, Point *C* lies 40 mm to the left of *G*.
15.74 (*a*) *C* lies 0.25 m to the right of *A*. (*b*) $\mathbf{v}_A = 100.0$ mm/s ↑.
15.75 $z = 2.75$ m.
15.76 (*a*) $\boldsymbol{\omega} = 3.00$ rad/s ↺. (*b*) $\mathbf{v}_A = 300$ mm/s ←.
(*c*) Cord wound per second = 180.0 mm.
15.77 (*a*) $\boldsymbol{\omega} = 3.00$ rad/s ↻. (*b*) $\mathbf{v}_A = 180$ mm/s →.
(*c*) Cord unwound per second = 300 mm.
15.78 (*a*) The instantaneous center lies 50 mm to the right of the axle.
(*b*) $\mathbf{v}_B = 750$ mm/s ↓, $\mathbf{v}_D = 1.950$ m/s ↑.
15.79 (*a*) $OC = 25$ mm. (*b*) $\mathbf{v}_D = 420$ mm/s ↑.
15.80 (*a*) Gear A: 300 mm left of *A*.
Gear *C*: 600 mm left of *C*.
(*b*) $\boldsymbol{\omega}_A = 4.00$ rad/s ↻, $\boldsymbol{\omega}_C = 2.00$ rad/s ↺.
15.81 (*a*) $\boldsymbol{\omega} = 2.40$ rad/s ↺.
(*b*) $\mathbf{v}_A = 240$ mm/s = 0.24 m/s ↑, $\mathbf{v}_D = 0.433$ m/s ⦨ 33.7°.
15.82 (*a*) $\boldsymbol{\omega} = 0.389$ rad/s ↺. (*b*) $\mathbf{v}_D = 1.164$ m/s ⦨ 59.2°.
15.83 (*a*) $\boldsymbol{\omega} = 2.89$ rad/s ↻. (*b*) $\mathbf{v}_D = 2080$ m/s ⦩ 73.9°.
15.84 (*a*) $\boldsymbol{\omega}_{BCD} = 2.84$ rad/s ↺. (*b*) $\mathbf{v}_E = 1.817$ m/s ⦪ 82.4°.
15.85 (*a*) $\boldsymbol{\omega}_{BDE} = 3.26$ rad/s ↻, $\boldsymbol{\omega}_{AB} = 6.78$ rad/s ↺.
15.86 (*a*) $\boldsymbol{\omega} = 5.13$ rad/s ↺. (*b*) $\mathbf{v}_D = 0.924$ m/s ←. (*c*) $\mathbf{v}_A = 1.870$ m/s ⦫ 34.7°.
15.87 (*a*) $\boldsymbol{\omega}_{AD} = 4.27$ rad/s ↻. (*b*) $\mathbf{v}_D = 1.330$ m/s ↓. (*c*) $\mathbf{v}_A = 1.557$ m/s ⦨ 34.7°.
15.88 (*a*) $\omega = \dfrac{v_A}{l}\cdot\dfrac{\sin\beta}{\cos(\beta-\theta)}$. (*b*) $v_B = v_A\dfrac{\cos\theta}{\cos(\beta-\theta)}$.
15.89 (*a*) $\boldsymbol{\omega} = 4.42$ rad/s ↺. (*b*) $\mathbf{v}_B = 3.26$ m/s ⦨ 50°.
15.90 (*a*) $\boldsymbol{\omega} = 0.9$ rad/s ↻, $\mathbf{v}_F = 439$ mm/s ⦨ 79.6°.
(*b*) $\mathbf{v}_G = 411$ mm/s ⦩ 20.5°.
15.91 $\boldsymbol{\omega}_{AB} = 1.000$ rad/s ↻. (*b*) $\mathbf{v}_B = 1.039$ m/s →.
15.92 (*a*) $\boldsymbol{\omega}_{ABD} = 1.579$ rad/s ↻. (*b*) $\mathbf{v}_A = 699$ mm/s ⦨ 78.3°.
15.93 (*a*) $\boldsymbol{\omega}_{ABD} = 2.25$ rad/s ↺. (*b*) $\mathbf{v}_A = 996$ mm/s ⦫ 78.3°.
15.94 (*a*) $\boldsymbol{\omega}_{AB} = 1.920$ rad/s ↻, $\boldsymbol{\omega}_{BD} = 1.200$ rad/s ↻.
(*b*) $\mathbf{v}_B = 750$ mm/s ⦫ 73.7°.
15.95 $\boldsymbol{\omega}_{AB} = 0.9$ rad/s ↻. (*a*) $\boldsymbol{\omega}_{DE} = 0.338$ rad/s ↻. (*b*) $\mathbf{v}_E = 78.8$ mm/s ←.
15.96 (*a*) $\boldsymbol{\omega}_{DE} = 5.00$ rad/s ↺. (*b*) $\mathbf{v}_E = 3.00$ m/s ↓.
15.97 (*a*) $\mathbf{v}_D = 1260$ mm/s ↑. (*b*) $\boldsymbol{\omega}_{AB} = 1.250$ rad/s ↺.
15.98 (*a*) $\boldsymbol{\omega}_{DE} = 2.5$ rad/s ↻, $\boldsymbol{\omega}_{AB} = 1.177$ rad/s ↻.
(*b*) $\mathbf{v}_A = 735$ mm/s ←.
15.99 The *space centrode* is a quarter circle of 15 in. radius centered at *O*, The *body centrode* is a semicircle of 7.5 in. radius centered midway between *A* and *B*.
15.100 *space centrode*: lower rack.
body centrode: circumference of gear.

15.101 $\mathbf{v}_B = 497$ mm/s ←.
15.102 $\boldsymbol{\omega}_{BD} = 0.955$ rad/s ↺, $\boldsymbol{\omega}_{DE} = 2.55$ rad/s ↺.
15.103 $\boldsymbol{\omega}_{BD} = 5.2$ rad/s ↺, $\boldsymbol{\omega}_{DE} = 6.4$ rad/s ↺.
15.104 $\mathbf{v}_C = 0$, $\mathbf{v}_B = 44.4$ m/s →, $\mathbf{v}_D = 42.9$ m/s ∠ 15.0°, $\mathbf{v}_E = 31.4$ m/s ∠ 45.0°.
15.105 (*a*) $\boldsymbol{\alpha} = 0.833$ rad/s^2 ↺, (*b*) $\mathbf{a}_C = 1.083$ m/s^2 ↑.
15.106 $\mathbf{a}_A = 2.50$ m/s^2 ↑, $\mathbf{a}_B = 0.100$ m/s^2 ↑.
15.107 (*a*) $\mathbf{a}_G = 0.9$ m/s^2 →. (*b*) $\mathbf{a}_B = 1.8$ m/s^2 ←.
15.108 (*a*) $\mathbf{a} = 0$ at 0.6 m from A. (*b*) $\mathbf{a} = 2.4$ m/s^2 → at 0.2 m from A.
15.109 (*a*) $\mathbf{a}_A = 2.56$ m/s^2 ↓. (*b*) $\mathbf{a}_D = 4.62$ m/s^2 ∠ 16.1°.
15.110 (*a*) $\mathbf{a}_D = 2.88$ m/s^2 ←. (*b*) $\mathbf{a}_E = 3.60$ m/s^2 ←.
15.111 (*a*) $\mathbf{a}_B = 1430$ m/s^2 ↓. (*b*) $\mathbf{a}_C = 1430$ m/s^2 ↓, (*c*) $\mathbf{a}_D = 1430$ m/s^2 ∠ 60°.
15.112 (*a*) $\mathbf{a}_A = 315$ mm/s^2 ∠ 64.7°. (*b*) $\mathbf{a}_B = 301$ mm/s^2 ∠ 67.4°.
15.113 $\mathbf{a}_B = 2000$ mm/s^2 ↑, $\mathbf{a}_A = 1415$ mm/s^2 ∠ 58.0°,
$\mathbf{a}_C = 4300$ mm/s^2 ∠ 25.8°
15.114 $\mathbf{a}_A = 1200$ mm/s^2 ↑, $\mathbf{a}_B = 2140$ mm/s^2 ∠ 69.4°,
$\mathbf{a}_C = 2070$ mm/s^2 ∠ 65.0°.
15.115 $\boldsymbol{\alpha}_A = 96.0$ rad/s^2 ↻, $\mathbf{a}_A = 2.40$ m/s^2 ←.
$\boldsymbol{\alpha}_B = 48.0$ rad/s^2 ↻, $\mathbf{a}_B = 1.200$ m/s^2 ←.
15.116 (*a*) $\mathbf{a}_A = 1.442$ m/s^2 ∠ 33.7°. (*b*) $\mathbf{a}_B = 3.44$ m/s^2 ∠ 35.5°.
(*c*) $\mathbf{a}_C = 3.20$ m/s^2 ↑.
15.117 $\mathbf{a}_D = 742$ mm/s^2 ∠ 46.0°.
15.118 (*a*) $a_T = 3080$ mm/s^2. (*b*) $a_E = 9250$ mm/s^2.
15.119 $\omega_{AB} = -1.6525$ rad/s, $\mathbf{a}_A = (12.98\text{ m/s}^2)\mathbf{i} = 12.98$ m/s^2 →.
15.120 $\mathbf{a}_P = 148.3$ m/s^2 ↓.
15.121 $\mathbf{a}_P = 296$ m/s^2 ↑.
15.122 $\mathbf{a}_D = 1558$ m/s^2 ∠ 45°. $\mathbf{a}_E = 337$ m/s^2 ∠ 45°.
15.123 $\mathbf{a}_D = 59.8$ m/s^2 ↑. $\mathbf{a}_D = 190.6$ m/s^2 ↑
15.125 $\mathbf{a}_D = 17350$ mm/s^2 ←.
15.126 (*a*) $\boldsymbol{\alpha}_{AB} = 0.718$ rad/s^2 ↺. (*b*) $\mathbf{a}_B = 125.0$ mm/s^2 →
15.127 $\mathbf{a}_D = 1.745$ m/s^2 ∠ 68.2°.
15.128 $\mathbf{a}_E = 1.296$ m/s^2 ←.
15.129 (*a*) $\boldsymbol{\alpha}_{DE} = \omega_0^2$ ↻. (*b*) $\mathbf{a}_D = \sqrt{2}l\omega_0^2$ ∠ 45°.
15.130 $\mathbf{a}_C = 9.60$ m/s^2 →.
15.131 (*a*) $\alpha_{BD} = 10.75$ rad/s^2 ↻. (*b*) $\alpha_{DE} = 2.30$ rad/s^2 ↻.
15.132 (*a*) $\boldsymbol{\alpha}_{BD} = 4.18$ rad/s^2 ↺. (*b*) $\boldsymbol{\alpha}_{DE} = 2.43$ rad/s^2 ↺.
15.133 (*a*) $\alpha_{BD} = 8.15$ rad/s^2 ↻. (*b*) $\alpha_{DE} = 0.896$ rad/s^2 ↺.
15.134 (*a*) $\boldsymbol{\alpha}_{BD} = 3.70$ rad/s^2 ↺. (*b*) $\boldsymbol{\alpha}_{DE} = 3.70$ rad/s^2 ↺.
15.135 (*a*) $\boldsymbol{\alpha}_{DE} = 16.53$ rad/s^2 ↻.
(*b*) $\mathbf{a}_F = -(4800\text{ mm/s}^2)\mathbf{i} - (620\text{ mm/s}^2)\mathbf{j} = 4840$ mm/s^2 ∠ 7.36°.
15.136 $\mathbf{v}_D = 1.382$ m/s ↓. $\mathbf{a}_D = 0.695$ m/s^2 ↓.
15.137 (*a*) $\mathbf{r}_C = \mathbf{r}_A + \dfrac{\boldsymbol{\omega}\times\mathbf{v}_A}{\omega^2}$, (*b*) $\mathbf{a}_A = \dfrac{\alpha}{\omega}\mathbf{v}_A + \boldsymbol{\omega}\times\mathbf{v}_A$.
15.138 $v_B = b\omega\cos\theta$, $a_B = b\alpha\cos\theta - b\omega^2\sin\theta$.
15.139 $\omega = \dfrac{v_B\sin\beta}{l\cos\theta}$.
15.140 $\alpha = \left[\dfrac{v_B\sin\beta}{l}\right]^2 \dfrac{\sin\theta}{\cos^3\theta}$.
15.141 $v_x = v\left(1-\cos\dfrac{vt}{r}\right)$. $v_y = v\sin\dfrac{vt}{r}$.
15.142 $\omega = \dfrac{bv_A}{b^2+x_A^2}$ ↻, $\alpha = \dfrac{2bx_Av_A^2}{\left(b^2+x_A^2\right)^2}$ ↻.
15.143 $(v_B)_x = v_A - \dfrac{lb^2v_A}{\left(b^2+x_A^2\right)^{3/2}}$ →, $(v_B)_y = \dfrac{lbx_Av_A}{\left(b^2+x_A^2\right)^{3/2}}$ ↑.

15.144 $\omega_{BD} = \dfrac{b(b + l\cos\theta)}{l^2 + b^2 + 2bl\cos\theta}\omega$ ↻,

$\mathbf{v}_E = \dfrac{bl\sin\theta}{l^2 + b^2 + 2bl\cos\theta}\omega$ ↖ $\tan^{-1}\left(\dfrac{b\sin\theta}{l + b\cos\theta}\right)$.

15.145 $\alpha_{BD} = \dfrac{bl(l^2 - b^2)\sin\theta}{l^2 + b^2 + 2bl\cos\theta}\omega^2$ ↺.

15.146 (*a*) $\boldsymbol{\omega} = \dfrac{v_0}{b}\sin^2\theta$ ↻. (*b*) $\mathbf{v}_E = \dfrac{v_0 l}{b}\sin^2\theta\cos\theta \rightarrow + \dfrac{v_0 l}{b}\sin^3\theta \uparrow$.

(*a*) $\boldsymbol{\alpha} = \dfrac{2v_0^2}{b^2}\sin^3\theta\cos\theta$ ↺.

15.147 $\boldsymbol{\omega} = \dfrac{v_0}{r}\dfrac{\sin^2\theta}{\cos\theta}$ ↺, $\boldsymbol{\alpha} = \left(\dfrac{v_0}{r}\right)^2(1 + \cos^2\theta)\tan^3\theta$ ↺.

15.148 $(v_P)_x = r\omega\left(\cos\dfrac{r\omega t}{R - r} - \cos\omega t\right)$, $(v_P)_y = r\omega\left(\sin\dfrac{r\omega t}{R - r} + \sin\omega t\right)$.

15.149 The path is the y axis. $\mathbf{v} = (R\omega\sin\omega t)\mathbf{j}$, $\mathbf{a} = (R\omega^2\cos\omega t)\mathbf{j}$.

15.150 $\mathbf{v}_P = 2.40$ m/s ↖ 73.9°.

15.151 $\mathbf{v}_P = 2.87$ m/s ↖ 44.8°.

15.152 (*a*) $\boldsymbol{\omega}_{BE} = 1.815$ rad/s ↻. (*b*) $\mathbf{v}_{P/BE} = 410$ mm/s ↖ 20°.

15.153 (*a*) $\boldsymbol{\omega}_{BD} = 5.16$ rad/s ↻. (*b*) $\mathbf{v}_{P/AD} = 1.399$ m/s ↘ 60°.

15.154 (*a*) $\boldsymbol{\omega}_{BD} = 3.81$ rad/s ↻, $\mathbf{v}_{P/F} = 6.53$ m/s ↗ 16.26°.
(*b*) $\boldsymbol{\omega}_{BD} = 3.00$ rad/s ↻, $\mathbf{v}_{P/F} = 4.00$ m/s →.

15.155 (*a*) $\boldsymbol{\omega}_{BD} = 14.00$ rad/s ↻. (*b*) $\mathbf{v}_{E/F} = -8.40$ m/s ↓.

15.156 (*a*) $\boldsymbol{\omega}_{EF} = 12.00$ rad/s ↺. (*b*) $\mathbf{v}_{D/EF} = 4.80$ m/s ↑.

15.157 $\mathbf{v}_P = 2.26$ m/s ↙ 74.6°.

15.158 $\mathbf{a}_1 = r\omega^2\mathbf{i} - 2\omega u\mathbf{j}$, $\mathbf{a}_2 = 2\omega u\mathbf{i} - r\omega^2\mathbf{j}$, $\mathbf{a}_3 = -\left(r\omega^2 + \dfrac{u^2}{r} + 2\omega u\right)\mathbf{i}$,
$\mathbf{a}_4 = (r\omega^2 + 2\omega u)\mathbf{j}$.

15.159 $\mathbf{a}_1 = r\omega^2\mathbf{i} + 2\omega u\mathbf{j}$, $\mathbf{a}_2 = -2\omega u\mathbf{i} - r\omega^2\mathbf{j}$, $\mathbf{a}_3 = \left(2\omega u - r\omega^2 - \dfrac{u^2}{r}\right)\mathbf{i}$,
$\mathbf{a}_4 = (r\omega^2 - 2\omega u)\mathbf{j}$.

15.160 $\mathbf{a}_P = 15.47$ m/s² ↙ 77.3°.

15.161 (*a*) $\mathbf{a}_C = 1.750 \times 10^{-3}$ m/s² west. (*b*) $\mathbf{a}_C = 1.341 \times 10^{-3}$ m/s² west.
(*c*) $\mathbf{a}_C = 1.341 \times 10^{-3}$ m/s² west.

15.162 $\mathbf{a}_C = 0.0234$ m/s² west.

15.163 (*a*) $\mathbf{v}_D = (0.659$ m/s$)\mathbf{i} - (0.470$ m/s$)\mathbf{j} = 0.809$ m/s ↖ 35.5°.
(*b*) $\mathbf{a}_D = -(0.657$ m/s²$)\mathbf{i} + (1.593$ m/s²$)\mathbf{j} = 1.723$ m/s² ↘ 67.6°.

15.164 (*a*) $\mathbf{v}_B = 0.520$ m/s ↖ 82.6°. (*b*) $\mathbf{a}_B = 50.0$ mm/s² ↘ 9.8°.

15.165 (*a*) $\mathbf{v}_B = 0.520$ m/s ↖ 37.4°. (*b*) $\mathbf{a}_B = 50.0$ mm/s² ↖ 69.8°.

15.166 (*a*) $\mathbf{a}_D = -(1.32$ m/s$)\mathbf{j} - (2.70$ m/s$)\mathbf{k}$. (*b*) $\mathbf{a}_P = -(1.32$ m/s²$)\mathbf{j}$.

15.167 (*a*) $\mathbf{a}_D = (2.4$ m/s$)\mathbf{i} - (2.70$ m/s$)\mathbf{k}$. (*b*) $\mathbf{a}_P = (2.4$ m/s²$)\mathbf{i}$.

15.168 (1) $\mathbf{a}_1 = 303$ mm/s² →; (2) $\mathbf{a}_1 = 168.5$ mm/s² ↙ 57.7°.

15.169 (3) $\mathbf{a}_3 = 483$ mm/s² ←; (4) $\mathbf{a}_4 = 168.5$ mm/s² ↘ 57.7°.

15.170 $\mathbf{v}_W = 0.750$ m/s ↗ 71.3°, $\mathbf{a}_W = 2.13$ m/s² ↙ 61.9°.

15.171 $\boldsymbol{\omega} = -(2.79$ rad/s$)\mathbf{k} = 2.79$ rad/s ↻, $\boldsymbol{\alpha} = -(2.13$ rad/s²$)\mathbf{k} = 2.13$ rad/s² ↻.

15.172 $a_P = 2.60$ m/s².

15.173 (*a*) $\mathbf{a}_A = 0.621$ m/s² ↑. (*b*) $\mathbf{a}_B = 1.767$ m/s² ↖ 30.6°. (*c*) $\mathbf{a}_C = 2.42$ m/s² ↓.

15.174 (*a*) $\mathbf{a}_A = 0.621$ m/s² ↑. (*b*) $\mathbf{a}_B = 1.733$ m/s² ↖ 53.9°.
(*c*) $\mathbf{a}_C = 2.62$ m/s² ↙ 67.6°.

15.175 (*a*) $\boldsymbol{\alpha}_{BD} = 8.09$ rad/s² ↻. (*b*) $\mathbf{a}_{P/F} = 8.43$ m/s² ↙ 16.26°.

15.176 $\boldsymbol{\omega}_{BP} = 7.86$ rad/s ↺, $\boldsymbol{\alpha}_{BP} = 81.1$ rad/s² ↺.

15.177 $\boldsymbol{\omega}_S = 3.81$ rad/s ↻, $\alpha_S = 81.4$ rad/s² ↻.

15.178 $\boldsymbol{\omega}_S = 1.526$ rad/s ↻, $\boldsymbol{\alpha}_S = 57.6$ rad/s² ↻.

15.179 $\alpha_P = 43.0$ rad/s² ↻.

15.180 $\alpha_P = 47.0$ rad/s² ↻.

15.181 (*a*) $\boldsymbol{\omega} = 3.85$ rad/s ↺. (*b*) $\mathbf{u} = 2.31$ m/s ∠ 30°.
(*c*) $\mathbf{a}_P = 16.03$ m/s^2 ⦣ 46.1°.

15.182 (*a*) $\boldsymbol{\omega} = 3.85$ rad/s ↻. (*b*) $\mathbf{u} = 2.31$ m/s ∠ 30°, $\mathbf{a}_P = 16.03$ m/s^2 ∠ 46.1°.

15.183 $\mathbf{a}_P = -(36.8 \text{ m/s}^2)\mathbf{i} + (36.0 \text{ m/s}^2)\mathbf{j} = 51.5$ m/s^2 ⦣ 44.4°.

15.184 (*a*) $\boldsymbol{\omega} = (0.480 \text{ rad/s})\mathbf{i} - (1.600 \text{ rad/s})\mathbf{j} + (0.600 \text{ m/s})\mathbf{k}$.
(*b*) $\mathbf{v}_A = (400 \text{ mm/s})\mathbf{i} + (300 \text{ mm/s})\mathbf{j} + (480 \text{ mm/s})\mathbf{k}$.

15.185 (*a*) $\boldsymbol{\omega} = -(0.400 \text{ rad/s})\mathbf{i} - (0.360 \text{ rad/s})\mathbf{k}$.
(*b*) $\mathbf{v}_A = (100 \text{ mm/s})\mathbf{i} - (90 \text{ mm/s})\mathbf{j} + (120 \text{ mm/s})\mathbf{k}$.

15.186 (*a*) $\boldsymbol{\omega} = (1.5 \text{ rad/s})\mathbf{i} - (3.5 \text{ rad/s})\mathbf{j} - (3.0 \text{ rad/s})\mathbf{k}$.
(*b*) $\mathbf{v}_D = (640 \text{ mm/s})\mathbf{i} - (360 \text{ mm/s})\mathbf{j} + (740 \text{ mm/s})\mathbf{k}$.

15.187 $\boldsymbol{\alpha} = -(9.87 \text{ rad/s}^2)\mathbf{k}$.

15.188 $\boldsymbol{\alpha} = (118.4 \text{ rad/s}^2)\mathbf{i}$.

15.189 $\boldsymbol{\alpha} = (230 \text{ rad/s}^2)\mathbf{i} - (2.5 \text{ rad/s}^2)\mathbf{k}$.

15.190 (*a*) $\boldsymbol{\alpha} = -(2260 \text{ rad/s}^2)\mathbf{k}$. (*b*) $\boldsymbol{\alpha} = (2260 \text{ rad/s}^2)\mathbf{j}$.

15.191 (*a*) $\boldsymbol{\omega}_A = \omega_1\mathbf{j} + \left(\frac{R}{r}\right)\omega_1\mathbf{k}$. (*b*) $\boldsymbol{\alpha}_A = \frac{R}{r}\omega_1^2\mathbf{i}$.

15.192 (*a*) $\boldsymbol{\omega}_A = \omega_1\mathbf{j} + \frac{R}{r}(\omega_1 - \omega_2)\mathbf{k}$. (*b*) $\boldsymbol{\alpha}_A = \frac{R}{r}\omega_1(\omega_1 - \omega_2)\mathbf{i}$.

15.193 (*a*) $\mathbf{v}_A = -(0.600 \text{ m/s})\mathbf{i} + (0.750 \text{ m/s})\mathbf{j} - (0.600 \text{ m/s})\mathbf{k}$.
(*b*) $\mathbf{a}_A = -(6.15 \text{ m/s}^2)\mathbf{i} - (3.00 \text{ m/s}^2)\mathbf{j}$.

15.194 (*a*) $\boldsymbol{\omega} = -(0.1745 \text{ rad/s})\mathbf{i} - (0.524 \text{ rad/s})\mathbf{j}$. (*b*) $\boldsymbol{\alpha} = -(0.0914 \text{ rad/s}^2)\mathbf{k}$.
(*c*) $\mathbf{v}_P = -(1.818 \text{ m/s})\mathbf{i} + (0.605 \text{ m/s})\mathbf{j} - (3.49 \text{ m/s})\mathbf{k}$,
$\mathbf{a}_P = -(1.366 \text{ m/s}^2)\mathbf{i} - (0.0609 \text{ m/s}^2)\mathbf{j} - (1.055 \text{ m/s}^2)\mathbf{k}$.

15.195 (*a*) $\alpha = -(20.0 \text{ rad/s}^2)\mathbf{j}$. (*b*) $\mathbf{a}_P = -(1.6 \text{ m/s}^2)\mathbf{i} + (4 \text{ m/s}^2)\mathbf{k}$.
(*c*) $\mathbf{a}_P = -(4.1 \text{ m/s}^2)\mathbf{j}$.

15.196 $\mathbf{a}_P = -(1.386 \text{ m/s}^2)\mathbf{i} - (2.05 \text{ m/s}^2)\mathbf{j} + (3.46 \text{ m/s}^2)\mathbf{k}$.

15.197 (*a*) $\boldsymbol{\omega} = (8.00 \text{ rad/s}^2)\mathbf{i}$. (*b*) $\boldsymbol{\alpha} = -(19.20 \text{ rad/s}^2)\mathbf{k}$.
(*c*) $\mathbf{a}_C = -(1.103 \text{ m/s}^2)\mathbf{i} - (2.005 \text{ m/s}^2)\mathbf{j}$.

15.198 (*a*) $\boldsymbol{\alpha} = (0.0375 \text{ rad/s}^2)\mathbf{i}$.
(*b*) $\mathbf{v}_C = -(0.1434 \text{ m/s})\mathbf{i} + (0.204 \text{ m/s})\mathbf{j} - (0.1229 \text{ m/s})\mathbf{k}$.
(*c*) $\mathbf{a}_C = -(0.0696 \text{ m/s}^2)\mathbf{i} - (0.0359 \text{ m/s}^2)\mathbf{j} + (0.0430 \text{ m/s}^2)\mathbf{k}$.

15.199 (*a*) $\boldsymbol{\omega} = (28.4 \text{ rad/s})\mathbf{i} + (5.24 \text{ rad/s})\mathbf{j}$. (*b*) $\boldsymbol{\omega}_{FH} = (25.8 \text{ rad/s})\mathbf{i}$.

15.200 (*a*) $\boldsymbol{\alpha} = 135.1 \text{ rad/s}^2\mathbf{k}$. (*b*) $\mathbf{a}_1 = (5.8 \text{ m/s}^2)\mathbf{i} - (232 \text{ m/s}^2)\mathbf{j}$.

15.201 (*a*) $\boldsymbol{\omega} = (0.75 \text{ rad/s})\mathbf{i} + (1.5 \text{ rad/s})\mathbf{j}$. (*b*) $\mathbf{v}_A = (300 \text{ mm/s})\mathbf{i} - (150 \text{ mm/s})\mathbf{j}$.
(*c*) $\mathbf{v}_C = (60 \text{ mm/s})\mathbf{i} - (30 \text{ mm/s})\mathbf{j} - (90 \text{ mm/s})\mathbf{k}$.

15.202 $\alpha = (1.125 \text{ rad/s})\mathbf{k}$. (*b*) $\mathbf{a}_C = -(225 \text{ mm/s}^2)\mathbf{i} + (180 \text{ mm/s}^2)\mathbf{j} - (112.5 \text{ mm/s}^2)\mathbf{k}$.

15.203 $\mathbf{v}_A = -(667 \text{ mm/s})\mathbf{j}$.

15.204 $\mathbf{v}_A = (210 \text{ mm/s})\mathbf{k}$.

15.205 $\mathbf{v}_C = -(34.5 \text{ mm/s})\mathbf{i}$.

15.206 $\mathbf{v}_A = -(750 \text{ mm/s})\mathbf{j}$.

15.207 $\mathbf{v}_A = (0.914 \text{ m/s})\mathbf{j}$.

15.208 $\mathbf{v}_A = (12.78 \text{ mm/s})\mathbf{j}$.

15.209 $\mathbf{v}_A = (4.66 \text{ mm/s})\mathbf{j}$.

15.210 $\boldsymbol{\omega}_{EG} = \frac{\omega_2}{\cos 25°}(-\sin 25°\mathbf{j} + \cos 25°\mathbf{k})$.

15.211 $\boldsymbol{\omega}_{EG} = \omega_1 \cos 25° \, (-\sin 25°\mathbf{i} + \cos 25°\mathbf{k})$.

15.212 (*a*) $\boldsymbol{\omega}_{AB} = (0.348 \text{ rad/s})\mathbf{i} - (0.279 \text{ rad/s})\mathbf{j} + (1.089 \text{ rad/s})\mathbf{k}$.
(*b*) $\mathbf{v}_A = -(750 \text{ mm/s})\mathbf{j}$.

15.213 (*a*) $\boldsymbol{\omega}_{BC} = (1.463 \text{ rad/s})\mathbf{i} + (0.1052 \text{ rad/s})\mathbf{j} + (0.0841 \text{ rad/s})\mathbf{k}$.
(*b*) $\mathbf{v}_C = -(34.5 \text{ mm/s})\mathbf{i}$.

15.214 $\mathbf{a}_B = -(510 \text{ mm/s}^2)\mathbf{k}$.

15.215 $\mathbf{a}_C = -(191.6 \text{ mm/s}^2)\mathbf{i}$.

15.216 $\mathbf{a}_A = -(1125 \text{ mm/s}^2)\mathbf{j}$.

15.217 $\mathbf{v}_A = (0.914 \text{ m/s})\mathbf{j}$, $\mathbf{a}_A = (4100 \text{ mm/s}^2)\mathbf{j}$.

15.218 $\mathbf{a}_A = -(9.51 \text{ mm/s}^2)\mathbf{j}$.

15.219 $\mathbf{a}_A = -(8.76 \text{ mm/s}^2)\mathbf{j}$.

15.220 (*a*) $\mathbf{v}_C = (684 \text{ mm/s})\mathbf{i} + (1879 \text{ mm/s})\mathbf{j} - (1410 \text{ mm/s})\mathbf{k}$.
(*b*) $\mathbf{a}_C = -(11.75 \text{ mm/s}^2)\mathbf{i} + (2.74 \text{ m/s}^2)\mathbf{j} - (4.10 \text{ m/s}^2)\mathbf{k}$.

15.221 (*a*) $\mathbf{v}_D = (1.434 \text{ mm/s})\mathbf{i} + (1.879 \text{ m/s})\mathbf{j} - (1.410 \text{ m/s})\mathbf{k}$.
(*b*) $\mathbf{a}_D = -(11.75 \text{ m/s}^2)\mathbf{i} + (2.74 \text{ m/s}^2)\mathbf{j} - (6.35 \text{ m/s}^2)\mathbf{k}$.
15.222 $\mathbf{v}_B = -(1.215 \text{ m/s})\mathbf{i} + (1.620 \text{ m/s})\mathbf{k}$. $\mathbf{a}_B = -(30.4 \text{ m/s}^2)\mathbf{j}$.
15.223 $\mathbf{v}_C = -(1.215 \text{ m/s})\mathbf{i} - (1.080 \text{ m/s})\mathbf{j} + (1.620 \text{ m/s})\mathbf{k}$.
$\mathbf{a}_C = (19.44 \text{ m/s}^2)\mathbf{i} - (30.4 \text{ m/s}^2)\mathbf{j} - (12.96 \text{ m/s}^2)\mathbf{k}$.
15.224 (*a*) $\mathbf{v}_D = (1.2 \text{ m/s})\mathbf{i} + (0.5 \text{ m/s})\mathbf{j} - (1.2 \text{ m/s})\mathbf{k}$.
(*b*) $\mathbf{a}_D = -(7.2 \text{ m/s}^2)\mathbf{i} - (14.4 \text{ m/s}^2)\mathbf{k}$.
15.225 (*a*) $\mathbf{v}_C = (0.800 \text{ m/s})\mathbf{i} - (0.600 \text{ m/s})\mathbf{j} - (0.675 \text{ m/s})\mathbf{k}$,
$\mathbf{a}_C = -(2.025 \text{ m/s}^2)\mathbf{j} + (3.60 \text{ m/s}^2)\mathbf{k}$.
(*a*) $\mathbf{v}_C = (0.8 \text{ m/s})\mathbf{i} - (0.6 \text{ m/s})\mathbf{j}$, $\mathbf{a}_C = (3.60 \text{ m/s}^2)\mathbf{k}$.
15.226 (*a*) $\mathbf{v}_D = (0.480 \text{ m/s})\mathbf{i} + (1.50 \text{ m/s})\mathbf{j} + (2.64 \text{ m/s})\mathbf{k}$.
(*b*) $\mathbf{a}_D = -(22.8 \text{ m/s}^2)\mathbf{j} + (15 \text{ m/s}^2)\mathbf{k}$.
15.227 (*a*) $\mathbf{v}_D = (0.750 \text{ m/s})\mathbf{i} + (1.299 \text{ m/s})\mathbf{j} - (1.732 \text{ m/s})\mathbf{k}$.
(*b*) $\mathbf{a}_D = (27.1 \text{ m/s}^2)\mathbf{i} + (5.63 \text{ m/s}^2)\mathbf{j} - (15.00 \text{ m/s}^2)\mathbf{k}$.
15.228 (*a*) $\mathbf{v}_D = (129.9 \text{ mm/s})\mathbf{i} + (75.0 \text{ mm/s})\mathbf{j} + (86.6 \text{ mm/s})\mathbf{k}$.
(*b*) $\mathbf{a}_D = (45.0 \text{ mm/s}^2)\mathbf{i} - (112.6 \text{ mm/s}^2)\mathbf{j} + (60.0 \text{ mm/s}^2)\mathbf{k}$.
15.229 (*a*) $\mathbf{v}_D = (0.750 \text{ m/s})\mathbf{i} + (1.299 \text{ m/s})\mathbf{j} - (1.732 \text{ m/s})\mathbf{k}$.
(*b*) $\mathbf{a}_D = -(28.6 \text{ m/s}^2)\mathbf{i} + (3.03 \text{ m/s}^2)\mathbf{j} - (10.67 \text{ m/s}^2)\mathbf{k}$.
15.230 (*a*) $\mathbf{v}_C = (0.800 \text{ m/s})\mathbf{i} - (0.600 \text{ m/s})\mathbf{j} - (0.675 \text{ m/s})\mathbf{k}$.
(*b*) $\mathbf{v}_C = (0.800 \text{ m/s})\mathbf{i} - (0.600 \text{ m/s})\mathbf{j}$,
$\mathbf{a}_C = -(1.600 \text{ m/s}^2)\mathbf{i} + (1.200 \text{ m/s}^2)\mathbf{j} + (3.60 \text{ m/s}^2)\mathbf{k}$.
15.231 (*a*) $\boldsymbol{\omega} = \omega_1\mathbf{j} + \frac{R}{r}(\omega_1 - \omega_2)\mathbf{k}$. (*b*) $\boldsymbol{\alpha} = \omega_1(\omega_1 - \omega_2)\frac{R}{r}\mathbf{j}$.
15.232 $\mathbf{a}_P = -(1.386 \text{ m/s}^2)\mathbf{i} - (2.05 \text{ m/s}^2)\mathbf{j} + (3.46 \text{ m/s}^2)\mathbf{k}$.
15.233 (*a*) $\boldsymbol{\alpha} = (0.0375 \text{ rad/s}^2)\mathbf{i}$.
(*b*) $\mathbf{v}_C = -(0.143 \text{ m/s})\mathbf{i} + (0.205 \text{ m/s})\mathbf{j} - (0.123 \text{ m/s})\mathbf{k}$.
(*c*) $\mathbf{a}_C = -(0.0696 \text{ m/s}^2)\mathbf{i} - (0.0358 \text{ m/s}^2)\mathbf{j} + (0.0430 \text{ m/s}^2)\mathbf{k}$.
15.234 $\mathbf{v}_C = (0.600 \text{ m/s})\mathbf{j} - (0.585 \text{ m/s})\mathbf{k}$.
$\mathbf{a}_C = -(4.76 \text{ m/s}^2)\mathbf{i}$.
15.235 $\mathbf{v}_D = (0.600 \text{ m/s})\mathbf{i} - (0.225 \text{ m/s})\mathbf{k}$.
$\mathbf{a}_D = -(0.675 \text{ m/s}^2)\mathbf{i} + (3.00 \text{ m/s}^2)\mathbf{j} - (3.60 \text{ m/s}^2)\mathbf{k}$.
15.236. $\mathbf{v}_B = -(0.428 \text{ m/s})\mathbf{i} + (1.175 \text{ m/s})\mathbf{j} + (0.585 \text{ m/s})\mathbf{k}$.
$\mathbf{a}_B = (0.381 \text{ m/s}^2)\mathbf{i} + (0.1069 \text{ m/s}^2)\mathbf{j} - (0.1283 \text{ m/s}^2)\mathbf{k}$.
15.237 (*a*) $\boldsymbol{\alpha}_{BCD} = -(0.0012 \text{ rad/s}^2)\mathbf{k}$.
(*b*) $\mathbf{v}_D = -(0.169 \text{ m/s})\mathbf{i} + (0.225 \text{ m/s})\mathbf{j} + (0.230 \text{ m/s})\mathbf{k}$.
(*c*) $\mathbf{a}_D = (0.0138 \text{ m/s}^2)\mathbf{i} - (0.0092 \text{ m/s}^2)\mathbf{j} + (0.0141 \text{ m/s}^2)\mathbf{k}$.
15.238 (*a*) $\boldsymbol{\alpha} = -(0.27 \text{ rad/s}^2)\mathbf{i}$.
(*b*) $\mathbf{v}_D = (156 \text{ mm/s})\mathbf{i} - (90 \text{ mm/s})\mathbf{j} - (420 \text{ mm/s})\mathbf{k}$.
(*c*) $\mathbf{a}_D = -(293 \text{ mm/s}^2)\mathbf{i} - (70.1 \text{ mm/s}^2)\mathbf{j} - (187 \text{ mm/s}^2)\mathbf{k}$.
15.239 $\mathbf{v}_B = (1.299 \text{ m/s})\mathbf{i} - (1.828 \text{ m/s})\mathbf{j} + (1.633 \text{ m/s})\mathbf{k}$,
$\mathbf{a}_B = (0.817 \text{ m/s}^2)\mathbf{i} - (0.826 \text{ m/s}^2)\mathbf{j} - (0.956 \text{ m/s}^2)\mathbf{k}$.
$\mathbf{v}_B = (1.299 \text{ m/s})\mathbf{i} - (1.828 \text{ m/s})\mathbf{j} + (1.633 \text{ m/s})\mathbf{k}$,
$\mathbf{a}_B = (0.817 \text{ m/s}^2)\mathbf{i} - (0.826 \text{ m/s}^2)\mathbf{j} - (0.956 \text{ m/s}^2)\mathbf{k}$.
15.240 (*a*) $\mathbf{a}_A = (0.9 \text{ m/s}^2)\mathbf{i} - (0.1333 \text{ m/s}^2)\mathbf{j}$.
(*b*) $\mathbf{a}_B = (0.500 \text{ m/s}^2)\mathbf{i} - (0.300 \text{ m/s}^2)\mathbf{k}$.
15.241 (*a*) $\mathbf{a}_C = -(0.1 \text{ m/s}^2)\mathbf{i} + (0.1333 \text{ m/s}^2)\mathbf{j}$.
(*b*) $\mathbf{a}_D = (0.500 \text{ m/s}^2)\mathbf{i} + (0.300 \text{ m/s}^2)\mathbf{k}$.
15.242 $\mathbf{v}_A = -(5.04 \text{ m/s})\mathbf{i} - (1.200 \text{ m/s})\mathbf{k}$.
$\mathbf{a}_A = -(9.60 \text{ m/s}^2)\mathbf{i} - (25.9 \text{ m/s}^2)\mathbf{j} + (57.6 \text{ m/s}^2)\mathbf{k}$.
15.243 $\mathbf{v}_B = -(0.720 \text{ m/s})\mathbf{i} - (1.200 \text{ m/s})\mathbf{k}$,
$\mathbf{a}_B = -(9.60 \text{ m/s}^2)\mathbf{i} + (25.9 \text{ m/s}^2)\mathbf{j} - (11.52 \text{ m/s}^2)\mathbf{k}$.
15.244 (*a*) $\mathbf{a}_B = r\omega_2^2 \sin 30°\mathbf{j} - (r\omega_2^2 \cos 30° + 2r\omega_1\omega_2)\mathbf{k}$.
(*b*) $\mathbf{a}_B = -r(\omega_1^2 + \omega_2^2 + 2\omega_1\omega_2 \cos 30°)\mathbf{i} + r\omega_1^2 \cos 30°\mathbf{k}$.
(*c*) $\mathbf{a}_B = -r\omega_2^2 \sin 30°\mathbf{j} + r(2\omega_1^2 \cos 30° + \omega_2^2 \cos 30° + 2\omega_1\omega_2)\mathbf{k}$.
15.245 (*a*) $\mathbf{v}_E = (0.610 \text{ m/s})\mathbf{k}$, $\mathbf{a}_E = -(0.880 \text{ m/s}^2)\mathbf{i} + (1.170 \text{ m/s}^2)\mathbf{j}$.
(*b*) $\mathbf{v}_F = (5.20 \text{ m/s})\mathbf{i} - (0.390 \text{ m/s})\mathbf{j} - (1.000 \text{ m/s})\mathbf{k}$,
$\mathbf{a}_F -(4.00 \text{ m/s}^2)\mathbf{i} - (3.25 \text{ m/s}^2)\mathbf{k}$.
15.246 (*a*) $\mathbf{v}_G = (1.390 \text{ m/s})\mathbf{k}$, $\mathbf{a}_G = (7.12 \text{ m/s}^2)\mathbf{i} - (1.170 \text{ m/s}^2)\mathbf{j}$.
(*b*) $\mathbf{v}_H = (0.520 \text{ m/s})\mathbf{i} - (0.390 \text{ m/s})\mathbf{j} + (1.000 \text{ m/s})\mathbf{k}$,
$\mathbf{a}_H = (4.00 \text{ m/s}^2)\mathbf{i} - (3.25 \text{ m/s}^2)\mathbf{k}$.

15.247 Method 1: (*a*) $\mathbf{v}_A = (0.78\text{ m/s})\mathbf{i} - (0.72\text{ m/s})\mathbf{j} + (0.76\text{ m/s})\mathbf{k}$.
(*b*) $\mathbf{a}_A = (0.64\text{ m/s}^2)\mathbf{i} - (1.392\text{ m/s}^2)\mathbf{j} - (1.824\text{ m/s}^2)\mathbf{k}$.
Method 2: (*a*) $\mathbf{v}_A = (0.78\text{ m/s})\mathbf{i} - (0.72\text{ m/s})\mathbf{j} + (0.76\text{ m/s})\mathbf{k}$.
(*b*) $\mathbf{a}_A = (0.64\text{ m/s}^2)\mathbf{i} - (1.392\text{ m/s}^2)\mathbf{j} - (1.824\text{ m/s}^2)\mathbf{k}$.
15.248 (*a*) $\mathbf{a}_B = 6.00\text{ m/s}^2$. (*b*) $\mathbf{a}_B = 9.98\text{ m/s}^2$. (*c*) $\mathbf{a}_B = 60.0\text{ m/s}^2$.
15.249 (*a*) $\boldsymbol{\alpha} = 3.00\text{ rad/s}^2$ ↻. (*b*) $t_1 = 4.00$ s.
15.250 $\boldsymbol{\omega}_A = 1701$ rmp ↻, $\boldsymbol{\omega}_B = 1573$ rmp ↺.
15.251 (*a*) $\boldsymbol{\omega}_{AB} = 11.11$ rad/s ↻. (*b*) $\boldsymbol{\omega}_B = 33.3$ rad/s ↻. (*c*) $\mathbf{a}_1 = 29.6\text{ m/s}^2 \rightarrow$.
15.252 $\boldsymbol{\alpha}_{BD} = 306\text{ rad/s}^2$ ↺, $\boldsymbol{\alpha}_{DE} = 737\text{ rad/s}^2$ ↺.
15.253 (*a*) $\boldsymbol{\alpha}_{DE} = 1080\text{ rad/s}^2$ ↻. (*b*) $\mathbf{a}_D = 137.9\text{ m/s}^2$ ⦩ 64.9°.
15.254 (*a*) $\alpha_{AB} = 7.90\text{ rad/s}^2$ ↻. (*b*) $\boldsymbol{\alpha}_{BP} = 134.6\text{ rad/s}^2$ ↺.
15.255 $\mathbf{a}_P = 49.4\text{ m/s}^2$ ⦪ 26.0°.
15.256 (*a*) $\mathbf{v}_D = (0.45\text{ m/s})\mathbf{k}$, $\mathbf{a}_D = (4.05\text{ m/s}^2)\mathbf{i}$.
(*b*) $\mathbf{v}_F = -(1.35\text{ m/s})\mathbf{k}$, $\mathbf{a}_F = -(6.75\text{ m/s}^2)\mathbf{i}$.
15.257 (*a*) AE: $u_1 \rightarrow = 2.00\text{ m/s} \rightarrow$, BD: u_2 ∠ 60° = 0.
(*b*) AE: $\dot{u}_1 = 23.5\text{ m/s}^2 \leftarrow$, BD: $\dot{u}_2$ ∠ 60° $= 46.2\text{ m/s}^2$ ⦫ 60°.
15.258 $\mathbf{v}_C = (1.000\text{ m/s})\mathbf{k}$.
15.259 Method 1: $\mathbf{v}_A = (0.600\text{ m/s})\mathbf{i} - (0.400\text{ m/s})\mathbf{j} + (0.300\text{ m/s})\mathbf{k}$.
$\mathbf{a}_A = (0.400\text{ m/s}^2)\mathbf{i} - (1.500\text{ m/s}^2)\mathbf{j} - (0.300\text{ m/s}^2)\mathbf{k}$.
Method 2: $\mathbf{v}_A = (0.600\text{ m/s})\mathbf{i} - (0.400\text{ m/s})\mathbf{j} + (0.300\text{ m/s})\mathbf{k}$.
$\mathbf{a}_A = (0.400\text{ m/s}^2)\mathbf{i} - (1.500\text{ m/s}^2)\mathbf{j} - (0.300\text{ m/s}^2)\mathbf{k}$.

CHAPTER 16

16.1 $\mathbf{a}_{max} = 3.57\text{ m/s}^2 \leftarrow$.
16.2 (*a*) $\mathbf{C} = 3.43$ N ∠ 20°. (*b*) $\mathbf{B} = 24.4$ N ⦩ 73.4°.
16.3 (*a*) $\bar{\mathbf{a}} = 7.85\text{ m/s}^2$. (*b*) $\bar{\mathbf{a}} = 3.74\text{ m/s}^2$. (*c*) $\bar{\mathbf{a}} = 4.06\text{ m/s}^2 \rightarrow$.
16.4 (*a*) $\bar{\mathbf{a}} = 3.20\text{ m/s}^2 \rightarrow$. (*b*) $\mathbf{A} = 3.82$ N↑, $\mathbf{B} = 20.71$ N↑.
16.5 (*a*) $\mathbf{a}_A = \bar{\mathbf{a}} = 4.09\text{ m/s}^2 \rightarrow$. (*b*) $T = 42.5$ N.
16.6 (*a*) 5270 N↑. (*b*) 4120 N (compression).
16.7 (*a*) $\mathbf{a} = 0.337g$ ∠ 30°. (*b*) $\dfrac{h}{d} = 4.00$.
16.8 (*a*) $\mathbf{a} = 0.252g$ ⦫ 30°. (*b*) $\dfrac{h}{d} = 4.00$.
16.9 (*a*) $\bar{\mathbf{a}} = 5.00\text{ m/s}^2 \rightarrow$. (*b*) $0.311\text{ m} \leq h \leq 1.489$ m.
16.10 (*a*) $\bar{\mathbf{a}} = 2.55\text{ m/s}^2 \rightarrow$. (*b*) $h \leq 1.047$ m.
16.11 $m_C = 195.9$ kg.
16.12 $F = 229$ N.
16.13 (*a*) $\bar{\mathbf{a}} = 8.50\text{ m/s}^2$ ⦪ 60°. (*b*) $F = 87.8$ N.
16.14 (*a*) $\bar{\mathbf{a}} = 4.91\text{ m/s}^2$ ⦫ 30°. (*b*) $T_{BE} = 11.43$ N, $T_{AD} = 31.0$ N.
16.15 $\theta = 51.3°$.
16.16 $F_{CF} = +14.90$ N compression, $F_{BE} = +52.7$ N compression.
16.17 (*a*) $\bar{\mathbf{a}} = 8.50\text{ m/s}^2$ ⦪ 60°. (*b*) $\mathbf{A} = 11.17$ N↑.
16.18 $\mathbf{C}_y = -194.0$ N↓, $\mathbf{B}_y = 194.0$ N↓.
16.19 (*a*) $\bar{\mathbf{a}} = 9.29\text{ m/s}^2$ ⦪ 83.8°. (*b*) $\mathbf{B} = 6.63$ N ∠ 30°,
$\mathbf{A} = 2.60$ N ∠ 30°.
16.20 $\mathbf{a}_b = 5.18\text{ m/s}^2$ ⦪ 67.1°, $\mathbf{a}_P = 9.55\text{ m/s}^2$ ⦪ 30°.
16.21 $V_{max} = 14.49$ N, $M_{max} = 3.26$ N · m.
16.22 $\mathbf{C}_y = -194.0$ N↓, $\mathbf{a}_y = 61.544\text{ m/s}^2$; $|M|_{max} = 36.4$ N · m, $V_B = -194.0$ N.
16.25 $\theta = 5230$ rev.
16.26 $M = 127.2$ N · m.
16.27 $\theta = 75.1$ rev.
16.28 $\theta = 86.4$ rev.

16.29 t = 74.5 s.
16.30 $\boldsymbol{\alpha}$ = 20.4 rad/s^2 ↻.
16.31 $\boldsymbol{\alpha}$ = 32.7 rad/s^2 ↺.
16.32 $\bar{I} = 112.1\ \text{kg} \cdot \text{m}^2$.
16.33 (*a*) $\mathbf{a}_A$ = 1.784 m/s^2↓. (*b*) v_A = 2.31 m/s.
16.34 (1): (*a*) $\boldsymbol{\alpha}$ = 10 rad/s^2 ↺. (*b*) $\boldsymbol{\omega}$ = 15.49 rad/s ↺.
(2): (*a*) $\boldsymbol{\alpha}$ = 7.97 rad/s^2 ↺. (*b*) $\boldsymbol{\omega}$ = 13.83 rad/s ↺.
(3): (*a*) $\boldsymbol{\alpha}$ = 4.52 rad/s^2 ↺. (*b*) $\boldsymbol{\omega}$ = 10.42 rad/s ↺.
(4): (*a*) $\boldsymbol{\alpha}$ = 6.62 rad/s^2 ↺. (*b*) $\boldsymbol{\omega}$ = 12.61 rad/s ↺.
16.35 (*a*) $\boldsymbol{\alpha}_A$ = 15.14 rad/s^2 ↺. (*b*) $\mathbf{F}_A$ = 21.8 N↗.
16.36 (*a*) $\boldsymbol{\alpha}_A$ = 6.06 rad/s^2 ↻. (*b*) $\mathbf{F}_{AC}$ = 11.28 N ↗.
16.37 t = 59.4 s.
16.38 (*a*) $\mathbf{a}_D$ = 0.218 m/s^2↓. (*b*) $\mathbf{a}_E$ = 0.164 m/s^2↑.
16.39 (*a*) Slipping occurs between disk *B* and the belt.
$\boldsymbol{\alpha}_B$ = 9.81 rad/s^2 ↻.
There is no slipping between *A* and the belt.
$\boldsymbol{\alpha}_A$ = 65.5 rad/s^2 ↺.
16.40 (*a*) $\boldsymbol{\alpha}_A$ = 16.00 rad/s^2 ↺. (*b*) $\boldsymbol{\alpha}_B$ = 8.00 rad/s^2 ↻.
16.41 (*a*) $\boldsymbol{\alpha}_A$ = 12.50 rad/s^2 ↺, $\boldsymbol{\alpha}_B$ = 33.3 rad/s^2 ↺.
(*b*) $\boldsymbol{\omega}_A$ = 240 rpm ↻, $\boldsymbol{\omega}_B$ = 320 rpm ↺.
16.42 (*a*) $\boldsymbol{\alpha}_A$ = 12.50 rad/s^2 ↺, $\boldsymbol{\alpha}_B$ = 33.3 rpm ↺.
(*b*) $\boldsymbol{\omega}_A$ = 90.0 rpm ↺, $\boldsymbol{\omega}_B$ = 120.0 rpm ↻.
16.43 (*a*) $\boldsymbol{\alpha}_B$ = 38.2 rad/s^2 ↺, $\mathbf{R}_B$ = 5.4936 N←; $\boldsymbol{\alpha}_A$ = 9.16 rad/s^2 ↺.
(*b*) $\mathbf{C}$ = 54.9 N↑, $\mathbf{M}_C$ = 2.64 N · m ↺.
16.44 (*b*) $\omega_A = \omega_0/(1 + m_B/m_A)$ ↻.
16.45 (*a*) $\boldsymbol{\alpha}_A$ = 104.2 rad/s^2 ↺, $\boldsymbol{\alpha}_B$ = 20.85 rad/s^2 ↻, $\boldsymbol{\alpha}_C$ = 10.43 rad/s^2 ↺.
(*b*) $\boldsymbol{\omega}_A$ = 120.0 rpm ↻, $\boldsymbol{\omega}_C$ = 60.0 rpm ↺, $\boldsymbol{\omega}_B$ = 120.0 rpm ↻.
16.46 $d = \dfrac{\bar{k}^2\alpha}{\bar{a}}$.
16.48 (*a*) $\mathbf{a}_A$ = 4.36 m/s^2 →. (*b*) $\mathbf{a}_B$ = 2.18 m/s^2 ←.
16.49 (*a*) Thus, *P* is located $\dfrac{1}{3}$m from end *A*. (*b*) $\mathbf{a}_A$ = 2.18 m/s^2 →.
16.50 (*a*) $\mathbf{a}_A$ = 2.50 m/s^2 →. (*b*) $\mathbf{a}_B$ = 0.
16.51 $x_1 = \pi_r$ Q.E.D.
16.52 (*a*) $\bar{\mathbf{a}}$ = 0, $\boldsymbol{\alpha}$ = −(1.200 rad/s^2)**j**. (*b*) $\bar{\mathbf{a}}$ = −(0.1350 m/s^2)**i**, $\boldsymbol{\alpha}$ = −(0.900 rad/s^2)**j**.
16.53 (*a*) $\mathbf{a}_E$ = (3.50 m/s^2)**k**. (*b*) $\mathbf{a}_B$ = (1.728 m/s^2)**i** + (3.5 m/s^2)**k**.
16.54 (*a*) $\boldsymbol{\alpha}$ = −(8 rad/s^2)**j**. (*b*) $\mathbf{a}_B$ = (0.6 m/s^2)**i** + (2.6 m/s^2)**k**.
16.55 $\boldsymbol{\alpha} = \dfrac{rg}{\bar{k}^2}$ ↻.
16.56 (*a*) T = 0.865 N. (*b*) α = 72.1 rad/s^2 ↻.
16.57 $\mathbf{a}_A$ = 0.885 m/s^2↓, $\mathbf{a}_B$ = 2.60 m/s^2 ↑.
16.58 (*a*) $\boldsymbol{\alpha}$ = 0.741 rad/s^2 ↺. (*b*) $\bar{\mathbf{a}}$ = 0.857 m/s^2 ↑.
16.59 (*a*) T_B = 2800 N. (*b*) $\boldsymbol{\alpha}$ = 15.11 rad/s^2 ↻.
16.60 T_A = 1802 N, T_B = 1590 N.
16.61 T_A = 1378 N, T_B = 1855 N.
16.62 (*a*) $\boldsymbol{\alpha}_A$ = 22.4 rad/s^2 ↺, $\boldsymbol{\alpha}_B$ = 44.8 rad/s^2 ↻.
(*b*) T_{AB} = 19.62 N. (*c*) $\mathbf{v}_A$ = 2.37 m/s↓.
16.63 (*a*) $\boldsymbol{\alpha} = \dfrac{3g}{L}$ ↻. (*b*) $\mathbf{a}_A$ = g↑. (*c*) $\mathbf{a}_B$ = 2g↓.
16.64 (*a*) $\mathbf{a}_G = \dfrac{2g}{3}$↓, $\boldsymbol{\alpha} = \dfrac{2g}{L}$. (*b*) $\mathbf{a}_A = \dfrac{g}{3}$↑. (*c*) $\mathbf{a}_B = \dfrac{5g}{3}$↓.
16.65 (*a*) $\boldsymbol{\alpha} = \dfrac{g}{L}$ ↻. (*b*) $\mathbf{a}_A$ = 0.866g ←. (*c*) $\mathbf{a}_B$ = 1.323g ↗ 49.1°.
16.66 (*a*) $\mathbf{a}_A = \dfrac{1}{4}g$↑. (*b*) $\mathbf{a}_B = \dfrac{5}{4}g$↓.

16.67 (*a*) $\mathbf{a}_A = \dfrac{1}{2}g\uparrow$. (*b*) $\mathbf{a}_B = \dfrac{3}{2}g\downarrow$.

16.68 (*a*) $\mathbf{a}_A = \dfrac{1}{2}g(\mathbf{i}-\mathbf{j}) + \dfrac{3g(a+b)a}{a^2+b^2}\mathbf{j}$.

(*b*) $\mathbf{a}_B = \dfrac{1}{2}g(\mathbf{i}-\mathbf{j}) - \dfrac{3g(a+b)a}{a^2+b^2}\mathbf{j}$.

16.69 $t_1 = \dfrac{2}{7}\dfrac{\overline{v}_0}{\mu_k g}$, $\omega_1 = \dfrac{5}{7}\dfrac{\overline{v}_0}{r}$ ↓, $\mathbf{v}_1 = \dfrac{5}{7}\overline{v}_0 \rightarrow$.

16.70 (*a*) $\omega_0 = \dfrac{v_0}{r}$ ↰. (*b*) $t_1 = \dfrac{v_0}{\mu_k g}$. (*c*) $s_1 = \dfrac{v_0^2}{2\mu_k g}$.

16.71 (*a*) t_1 = 1.718 s. (*b*) $\overline{v}_1$ = 3.31 m/s. (*c*) s_1 = 7.14 m.

16.72 (*a*) t_1 = 1.980 s. (*b*) $\overline{v}_1$ = 3.06 m/s. (*c*) s_1 = 7.98 m.

16.73 (*a*) $t_1 = \dfrac{2}{7}\dfrac{v_1}{\mu_k g}$. (*b*) $\overline{\mathbf{v}} = \dfrac{2}{7}v_1 \rightarrow$, $\boldsymbol{\omega} = \dfrac{5}{7}\dfrac{v_1}{r}$ ↰.

16.74 $t_1 = \dfrac{2v_1}{5\mu_k g}$, $v_0 = \dfrac{2}{5}v_1$, $\mathbf{s} = \dfrac{2}{25}\dfrac{v_1^2}{\mu_k g} \leftarrow$.

16.75 $GP = \dfrac{\overline{k}^2}{\overline{r}}$ (Q.E.D.).

16.76 (*a*) $\boldsymbol{\alpha}$ = 107.1 rad/s² ↓. (*b*) $\mathbf{C}_y$ = 39.2 N↑, $\mathbf{C}_x$ = 21.4 N←.

16.77 (*a*) $\overline{r}$ = 150 mm. (*b*) $\boldsymbol{\alpha}$ = 125 rad/s² ↓.

16.78 (*a*) $\boldsymbol{\alpha}$ = 12.00 rad/s² ↓. (*b*) $\mathbf{A}_y$ = 19.62 N ↑, $\mathbf{A}_x$ = 4.00 N←.

16.79 (*a*) $\boldsymbol{\alpha}$ = 8.00 rad/s² ↓. (*b*) h = 0.667 m.

16.80 (*a*) $T = w\left(lz - \dfrac{z^2}{2}\right)\omega^2$. (*b*) T = 5.09 N.

16.81 R = 4.55 N.

16.82 (*a*) $\mathbf{a}_D$ = 2.50g↓. (*b*) $\mathbf{C} = \dfrac{3}{8}mg\uparrow$.

16.83 R = 13.64 kN →.

16.84 (*a*) $\mathbf{a}_B = \dfrac{3}{2}g\downarrow$. (*b*) $\mathbf{A} = \dfrac{1}{4}mg\uparrow$.

16.85 (*a*) $\mathbf{a}_B = \dfrac{9}{7}g\downarrow$. (*b*) $\mathbf{C} = \dfrac{4}{7}mg\uparrow$.

16.86 (*a*) $\mathbf{P}$ = 46.8 N↑. (*b*) $\mathbf{A}_x$ = 360 N ←, $\mathbf{A}_y$ = 39.0 N↑.

16.87 (*a*) $\boldsymbol{\alpha}$ = 43.6 rad/s² ↓. (*b*) $\mathbf{C}_x$ = 21.0 N ←, $\mathbf{C}_y$ = 54.6 N↑.

16.88 (*a*) $\boldsymbol{\alpha}$ = 16.88 rad/s² ↰. (*b*) $\mathbf{M}$ = 8.49 N · m ↰.

16.89 $\mathbf{C}$ = 150.1 N ∠ 83.2°.

16.90 (*a*) $\boldsymbol{\alpha}$ = 9.29 rad/s² ↰. (*b*) $\mathbf{M}$ = 10.13 N · m ↰.

16.91 (*a*) M = 99.4 N-m ↰. (*b*) $\mathbf{C}$ = 30.0 N ∠ 30°.

16.92 $\Sigma M_C = I_C\alpha$ (Q.E.D.).

16.93 Referring to the first diagram, we note that this will occur only when Points *G*, *O*, and *C* lie in a straight line. (Q.E.D.).

16.94 $\overline{a} = \dfrac{r^2}{r^2 + \overline{k}^2} g \sin\beta$.

16.95 (*a*) $x_{C/P} = \dfrac{1}{2}\left(\dfrac{1}{6}9.81 \text{ m/s}^2\right)\sin 10° (4 \text{ s})^2 = 2.27$ m.

(*b*) $x_{S/C} = \dfrac{1}{2}\left(\dfrac{1}{21}9.81 \text{ m/s}^2\right)\sin 10° (4 \text{ s})^2 = 0.649$ m.

16.96 (*a*) $\boldsymbol{\alpha}$ = 11.91 rad/s² ↓. (*b*) $\overline{\mathbf{v}}$ = 2.98 m/s ∠ 15°.

16.97 P = 303 N.

16.98 (*a*) $\boldsymbol{\alpha}$ = 17.78 rad/s² ↓, $\overline{\mathbf{a}}$ = 2.13 m/s² →. (*b*) $(\mu_s)_{\min}$ = 0.122.

16.99 (*a*) $\boldsymbol{\alpha}$ = 26.7 rad/s² ↓, $\overline{\mathbf{a}}$ = 3.20 m/s² →. (*b*) $(\mu_s)_{\min}$ = 0.0136.

16.100 (*a*) $\boldsymbol{\alpha}$ = 8.89 rad/s² ↓, $\overline{\mathbf{a}}$ = 1.067 m/s² →. (*b*) $(\mu_s)_{\min}$ = 0.231.

16.101 (*a*) $\boldsymbol{\alpha} = 8.89 \text{ rad/s}^2$ ↻, $\bar{\mathbf{a}} = 1.067 \text{ m/s}^2 \leftarrow$. (*b*) $(\mu_s)_{\min} = 0.165$.
16.102 (*a*) $\boldsymbol{\alpha} = 17.78 \text{ rad/s}^2$ ↻, $\bar{\mathbf{a}} = 2.13 \text{ m/s}^2$. (*b*) $(\mu_s)_{\min} = 0.122$.
16.103 (*a*) $\boldsymbol{\alpha} = 26.7 \text{ rad/s}^2$ ↻, $\bar{\mathbf{a}} = 3.20 \text{ m/s}^2 \rightarrow$. (*b*) $(\mu_s)_{\min} = 0.0136$.
16.104 (*a*) $\boldsymbol{\alpha} = 8.89 \text{ rad/s}^2$ ↻, $\bar{\mathbf{a}} = 1.067 \text{ m/s}^2 \rightarrow$. (*b*) $(\mu_s)_{\min} = 0.231$.
16.105 (*a*) $\boldsymbol{\alpha} = 8.89 \text{ rad/s}^2$ ↻, $\bar{\mathbf{a}} = 1.067 \text{ m/s}^2 \leftarrow$. (*b*) $(\mu_s)_{\min} = 0.165$.
16.106 (*a*) $\mathbf{a}_B = 3.53 \text{ m/s}^2 \rightarrow$. (*b*) $\mathbf{a}_A = 1.176 \text{ m/s}^2 \rightarrow$. (*c*) $\mathbf{x}_{B/A} = 0.294 \text{ m} \rightarrow$.
16.107 (*a*) $\mathbf{a}_B = 2.06 \text{ m/s}^2 \rightarrow$. (*b*) $\mathbf{a}_A = 1.176 \text{ m/s}^2 \rightarrow$. (*c*) $\mathbf{x}_{A/B} = 0.1103 \text{ m} \rightarrow$.
16.108 (*a*) $\boldsymbol{\alpha}_C = 72.4 \text{ rad/s}^2$ ↺. (*b*) $\mathbf{a}_B = 7.24 \text{ m/s}^2 \downarrow$.
16.109 (*a*) $\mathbf{a}_A = \mathbf{a}_B = 1.923 \text{ m/s}^2 \leftarrow$. (*b*) $\mathbf{D}_x = 4.33 \text{ N} \leftarrow$.
16.110 (*a*) $\boldsymbol{\alpha} = 17.70 \text{ rad/s}^2$ ↺ (*b*) $\mathbf{F} = 4.42 \text{ N} \angle 5°$, $\mathbf{N} = 48.9 \text{ N} \angle 85°$.
16.111 (*a*) $\mu_{\min} = 0.298$. (*b*) $\mathbf{a}_B = 0.536g \rightarrow$.
16.112 (*a*) $\mu_{\min} = 0.322$. (*b*) $\mathbf{a}_B = 0.566g$.
16.113 $\mathbf{P} = 8.26 \text{ N} \leftarrow$.
16.114 (*a*) $\boldsymbol{\alpha} = \frac{1}{8}\frac{g}{r}$ ↻. (*b*) $(\mathbf{a}_B)_x = \frac{1}{8}g \rightarrow, (\mathbf{a}_B)_y = \frac{1}{8}g \downarrow$.
16.115 $\alpha = \frac{g}{2r}\frac{m_B \sin\theta}{m_h + m_B(1+\cos\theta)}$.
16.116 $\mathbf{P} = 16.48 \text{ N} \angle 70.5°$, $\mathbf{M}_P = 0.228 \text{ N}\cdot\text{m}$ ↻.
16.117 (*a*) $\boldsymbol{\alpha} = 11.11 \text{ rad/s}^2$ ↻. (*b*) $\mathbf{A} = 37.7 \text{ N} \uparrow$. (*c*) $\mathbf{B} = 28.2 \text{ N} \rightarrow$.
16.118 (*a*) $\mathbf{P} = 97.8 \text{ N}\uparrow$. (*b*) $A = 60.3 \text{ N}\uparrow$.
16.119 (*a*) $\boldsymbol{\alpha} = 10.62 \text{ rad/s}^2$ ↺. (*b*) $\mathbf{B} = 4.25 \text{ N} \leftarrow$.
16.120 $T = \frac{mg \sin\theta}{1 + 3\sin\theta}$.
16.121 (*a*) $\boldsymbol{\alpha} = 10.405 \text{ rad/s}^2$ ↻. (*b*) $T = 36.8$ N. (*c*) $N = 61.3 \text{ N}\uparrow$.
16.122 (*a*) $\boldsymbol{\alpha} = 16.99 \text{ rad/s}^2$ ↺. (*b*) $\mathbf{A} = 25.5 \text{ N}\rightarrow$.
16.123 (*a*) general plane motion. (*b*) $T_{BF} = 65.2$ N, $T_{CG} = 0$.
16.124 $\mathbf{A} = 6.40 \text{ N} \leftarrow$.
16.125 (a) $\mathbf{P} = 2.0 \text{ N} \leftarrow$. (*b*) $\mathbf{B} = 15.19 \text{ N} \angle 20°$.
16.126 (*a*) $\boldsymbol{\alpha} = 9.56 \text{ rad/s}^2$ ↻. (*b*) $\mathbf{B} = 8.80 \text{ N} \angle 20°$.
16.127 $\mathbf{D} = 171.7 \text{ N} \rightarrow$.
16.128 $\mathbf{D} = 60.0 \text{ N} \rightarrow$.
16.129 $\mathbf{D} = 25.9 \text{ N} \angle 60°$.
16.130 $\mathbf{A} = 31.6 \text{ N} \uparrow$, $\mathbf{B} = 15.89 \text{ N} \leftarrow$.
16.131 (*a*) $\mathbf{a}_G = 9.36 \text{ m/s}^2 \angle 27.1°$. (*b*) $\mathbf{N} = 278 \text{ N}\uparrow$.
16.132 $\omega_f = 2.43$ rad/s.
16.133 $\mathbf{a}_A = 1.360 \text{ m/s}^2 \rightarrow$.
16.134 $\mathbf{D} = 1.618 \text{ N} \leftarrow$.
16.135 (*a*) $\mathbf{M} = 36.3 \text{ N}\cdot\text{m}$ ↺. (*b*) $\mathbf{C} = 231 \text{ N} \leftarrow + 524 \text{ N}\downarrow$.
16.136 (*a*) $\mathbf{M} = 82.3 \text{ N}\cdot\text{m}$ ↺. (*b*) $\mathbf{C} = 147.2 \text{ N} \leftarrow + 479 \text{ N}\downarrow$.
16.137 $\mathbf{B} = 805 \text{ N} \leftarrow$, $\mathbf{D} = 426 \text{ N} \rightarrow$.
16.138 $\mathbf{B} = 525 \text{ N} \angle 38.1°$, $\mathbf{D}' = 322 \text{ N} \angle 15.7°$.
16.139 $\mathbf{B} = 8.61 \text{ N}\uparrow$, $\mathbf{A}_y = 8.13 \text{ N}\uparrow$.
16.140 $\mathbf{A}_x = 3.24 \text{ N} \rightarrow$.
16.141 $\mathbf{B}_x = 120 \text{ N} \rightarrow$, $\mathbf{B}_y = 88.2 \text{ N}\uparrow$, $\text{M} = 15 \text{ N}\cdot\text{m}$ ↺.
16.142 $\mathbf{B}_y = 190 \text{ N} \rightarrow$, $\mathbf{B}_y = 104.9 \text{ N}\uparrow$, $\text{M} = 25.0 \text{ N}\cdot\text{m}$ ↺.
16.143 (*a*) $M_{\max} = \frac{mgL}{27}$ at $\frac{L}{3}$ from A.
16.144 $\mathbf{a}_A = \frac{2}{7}\frac{P}{m} \rightarrow, \mathbf{a}_B = \frac{22}{7}\frac{P}{m} \leftarrow$.
16.145 (*a*) $\mathbf{a}_C = 5.63 \text{ m/s}^2 \angle 25°$. (*b*) $\boldsymbol{\alpha} = 7.66 \text{ rad/s}^2$ ↻.
16.146 (*a*) $\mathbf{a}_A = 13.55 \text{ m/s}^2\downarrow$. (*b*) $\mathbf{a}_B = 2.34 \text{ m/s}^2\downarrow$.
16.147 (*a*) $\mathbf{a}_A = 1.950 \text{ m/s}^2 \rightarrow$. (*b*) $\boldsymbol{\alpha} = 34.6 \text{ rad/s}^2$ ↺.
***16.148** (*a*) $\mathbf{a}_A = 5.38 \text{ m/s}^2 \angle 20°$. (*b*) $\boldsymbol{\alpha} = 35.6 \text{ rad/s}^2$ ↺.
16.149 $\boldsymbol{\alpha}_{AB} = 11.43 \text{ rad/s}^2$ ↻, $\boldsymbol{\alpha}_{BC} = 57.1 \text{ rad/s}^2$ ↺.
16.150 (*a*) $b = 227$ mm. (*b*) $\boldsymbol{\alpha} = 7.27 \text{ rad/s}^2$ ↺.

16.151 M_{max} = 10.39 lb · in. located 20.8 in. below A.

16.152 $\boldsymbol{\alpha}_A = \frac{2}{5}\frac{g}{r}$ ⤾, $\boldsymbol{\alpha}_B = \frac{2}{5}\frac{g}{r}$ ⤵, $T = \frac{1}{5}mg$, $\mathbf{a}_B = \frac{4}{5}g\downarrow$.

16.153 s = 5.45 m.

16.154 x = 5.12 m, μ_{req} = 0.09 < 0.30. The create does not slide.

16.155 (*a*) $\boldsymbol{\alpha}_{disk}$ = 80.0 rad/s² ⤾. (*b*) $\mathbf{a}_A$ = 1.250 m/s² ↑.

16.156 (*a*) $\bar{a} = \frac{2\mu}{1+3\mu}g$. (*b*) $\bar{a} = g$.

16.157 (*a*) $\boldsymbol{\alpha}$ = 0.513 $\frac{g}{L}$ ⤵. (*b*) $\mathbf{N}_A$ = 0.912 mg↑. (*c*) $\mathbf{F}_A$ = 0.241 mg →.

16.158 (*a*) $\boldsymbol{\alpha}$ = 1.519 $\frac{g}{L}$ ⤵. (*b*) $\bar{\mathbf{a}}$ = 0.260g↓. (*c*) $\mathbf{A}$ = 0.740 mg↑.

16.159 (*a*) $\bar{\mathbf{a}} = \frac{5}{9}g\downarrow$. (*b*) $\bar{\mathbf{a}} = \frac{9}{8}g\downarrow$. (*c*) $\bar{\mathbf{a}} = \frac{13}{17}g\downarrow$.

16.160 (1): (*a*) $\boldsymbol{\alpha}$ = 1.2 $\frac{g}{c}$ ⤵. (*b*) $\bar{\mathbf{a}}$ = 0.671g ⦣ 63.4°.

(2): (*a*) $\alpha = \frac{24}{17}\frac{g}{c}$ ⤵. (*b*) $\bar{\mathbf{a}} = \frac{12}{17}g\downarrow$.

(3): (*a*) $\boldsymbol{\alpha} = 2.4\frac{g}{c}$ ⤵, (*b*) $\bar{\mathbf{a}} = 0.5g\downarrow$.

16.161 (*a*) $\boldsymbol{\alpha}$ = 0. (*b*) $\mathbf{C}_x$ = 0, $\mathbf{C}$ = 62.0 N↑.

16.162 (*a*) $\boldsymbol{\alpha}$ = 51.2 rad/s² ⤵. (*b*) $\mathbf{A}$ = 21.0 N↑.

16.163 (*a*) $\boldsymbol{\alpha}$ = 59.8 rad/s² ⤵. (*b*) $\mathbf{A}$ = 20.4 N↑.

16.164 (*a*) $\mathbf{a}_C$ = 1.8g ↓. (*b*) $\mathbf{B}$ = 0.2 mg ↑.

CHAPTER 17

17.1 $\bar{k}$ = 188.1 mm.

17.2 M = 157.1 N · m.

17.3 r_B = 98.8 mm.

17.4 n = 0.760.

17.5 (*a*) I = 32.5 kg · m². (*b*) θ = 13.26 rev

17.6 (*a*) ω_2 = 293 rpm. (*b*) θ = 15.92 rev.

17.7 θ = 19.47 rev.

17.8 $\theta = \frac{v^2}{8\pi r\mu_k g}$ rev.

17.9 $\mathbf{P}$ = 417 N ↓.

17.10 $\mathbf{P}$ = 480 N ↓.

17.11 (*a*) θ_C = 6.35 rev. (*b*) F_t = 7.14 N.

17.12 (*a*) θ_C = 2.54 rev. (*b*) F_t = 17.86 N.

17.13 $\omega_A = \frac{2n}{n^2+1}\sqrt{\frac{\pi M_0}{\bar{I}_0}}$.

17.14 (*a*) $\mathbf{v}_A$ = 4.79 m/s ↑. (*b*) $s_B + s_B'$ = 1.936 m

17.15 θ_A = 1.063 rev

17.16 (*a*) $\omega_2 = \sqrt{\frac{3g}{l}}$ ⤵, $\mathbf{A} = \frac{5}{2}W$ ↑. (*b*) ω_2 = 5.42 rad/s ⤵, $\mathbf{A}$ = 25.0 N ↑.

17.17 (*a*) ω_2 = 3.995 rad/s⤵. (*b*) $\mathbf{R}$ = 32.4 N ↓.

17.18 ω_2 = 11.13 rad/s ⤾.

17.19 ω_2 = 3.27 rad/s ⤵.

17.20 (*a*) ω_2 = 4.11 rad/s ⤵, $\mathbf{R}$ = 1357 N ⦬ 4.57°.

(*b*) ω_3 = 5.82 rad/s ⤵, $\mathbf{R}$ = 3490 N ↑.

17.21 $\bar{\omega}$ = 7.23 rad/s ⤾.

17.22 d = 181.7 mm.

17.23 ω_2 = 7.09 rad/s.

17.24 (a) $\Delta\omega = -0.250$ rpm. (b) $\Delta\omega = 0.249$ rpm.

17.25 $\overline{\mathbf{v}} = \sqrt{\dfrac{4gs}{3}} \downarrow$

17.26 $\overline{\mathbf{v}} = \sqrt{gs} \downarrow$

17.27 (a) $\mathbf{v}_G = 3.00$ m/s $\rightarrow$. (b) $\mathbf{F}_f = 30.0$ N $\leftarrow$.

17.28 (a) $v_C = 6.68$ m/s. (b) $c = 2.27$ m.

17.29 (a) $\omega_2 = 5.00$ rad/s. (b) N = 24.9 N$\uparrow$.

17.30 (a) $\omega_2 = 1.324\sqrt{\dfrac{g}{r}}$ $\circlearrowleft$, (b) $\mathbf{A} = 2.12\ mg\uparrow$.

17.31 (a) $\overline{v}_2 = \sqrt{\dfrac{10}{7}g(R-r)(1-\cos\beta)}$.

(b) $N = \dfrac{1}{7}mg[17 - 10\cos\beta]\uparrow$.

17.32 (a) $h = 0.390$ m. (b) $P = 19.62$ N.

17.33 (a) $\overline{\mathbf{v}}_A = 2.37$ m/s $\downarrow$. (b) $P = 19.62$ N.

17.34 (a) $v = 1.054\sqrt{gh}$. (b) $v = 1.500\sqrt{gh}$ (c) $v = 1.237\sqrt{gh}$.

17.35 $\mathbf{v}_{AB} = 292$ mm/s $\rightarrow$.

17.36 $\mathbf{v}_A = 0.775\sqrt{gL} \leftarrow$, $\mathbf{v}_B = 0.775\sqrt{gL} \measuredangle\ 60°$.

17.37 $\omega = 1.170$ rad/s $\circlearrowright$, $\mathbf{v}_A = 5.07$ m/s $\leftarrow$.

17.38 $\omega = \sqrt{3g(\cos\theta_0 - \cos\theta_2)/L}\ \circlearrowright$.

17.39 $\omega_2 = 3.67$ rad/s $\circlearrowleft$, $\mathbf{v}_B = 2.20$ m/s $\uparrow$.

17.40 $\boldsymbol{\omega} = 6.47$ rad/s, $\mathbf{v}_B = 3.36$ m/s.

17.41 (a) $\boldsymbol{\omega}_2 = 1.225\sqrt{\dfrac{g}{R}}$ $\circlearrowright$, $\mathbf{v}_R = 0.612\sqrt{gR} \measuredangle\ 60°$. (b) $\boldsymbol{\omega}_3 = 0$, $\mathbf{v}_3 = \sqrt{gR} \rightarrow$.

17.42 $\mathbf{v}_D = 2.69$ m/s $\downarrow$.

17.43 $\boldsymbol{\omega}_2 = 84.7$ rpm $\circlearrowright$.

17.44 $\boldsymbol{\omega}_1 = 110.8$ rpm $\circlearrowright$.

17.45 $\mathbf{v}_B = 3.25$ m/s $\downarrow$.

17.46 $\mathbf{v}_B = 4.43$ m/s $\downarrow$.

17.47 $\mathbf{v}_A = 0.770$ m/s $\leftarrow$.

17.48 (a) 37.7 kW (b) 100.5 kW.

17.49 (a) $M_{AB} = 39.8$ N · m. (b) $M_{CD} = 95.5$ N · m. (c) $M_{EF} = 229$ N · m.

17.50 min $\omega_{AB} = 1146$ rpm.

17.51 Power = 0.283 kW.

17.52 $\overline{k} = 179.1$ mm.

17.53 $M = 0.0404$ N · m.

17.54 $\omega = 3.87$ rad/s.

17.55 $t = 4.88$ s.

17.56 $M = 1.081$ N · m.

17.57 $t = \dfrac{v}{2\mu_k g}$

17.58 $t = 3.82$ s.

17.59 $t = \dfrac{\overline{I}\omega_0}{(F_A + F_B)r} = \dfrac{(1+\mu_k^2)\overline{I}\omega_0}{\mu_k(1+\mu_k)Wr}$ $\quad t = \dfrac{1+\mu_k^2}{2\mu_k(1+\mu_k)}\dfrac{r\omega_0}{g}$.

17.60 $t = 2.07$ s.

17.61 $t = 0.663$ s.

17.62 $\omega_B = \dfrac{\omega_0}{1 + \dfrac{m_A}{m_B}}$.

17.63 (a) $\boldsymbol{\omega}_A = 667$ rpm $\circlearrowleft$, $\boldsymbol{\omega}_B = 500$ rpm $\circlearrowright$. (b) $\mathbf{F}t = 20.9$ N · s $\uparrow$.

17.64 (a) $T_B = 21.1$ N. (b) $T_{AB} = 8.80$ N.

17.65 $\mathbf{X} = m\overline{\mathbf{v}}$, $d = \dfrac{\overline{k}^2\omega}{\overline{v}}$.

17.66 $X = m\overline{r}\omega$, $(GP) = \dfrac{\overline{k}^2}{\overline{r}}$.

17.68 (*a*) $GC = \dfrac{\bar{k}^2}{GP}$, (*b*) $d_C = d_p$ or $GC' = GP$.
17.69 $\bar{k} = 900$ mm.
17.70 (*a*) $\bar{\mathbf{v}} = \dfrac{r^2 gt \sin\beta}{r^2 + \bar{k}^2} \measuredangle \beta$. (*b*) $\mu_s \geq \dfrac{\bar{k}^2 \tan\beta}{r^2 + \bar{k}^2}$
17.71 (*a*) $\bar{\mathbf{v}} = 2.55$ m/s↑. (*b*) $Q = 10.53$ N
17.72 (*a*) $\mathbf{v}_B = 2.12$ m/s →. (*b*) $\mathbf{v}_A = 0.706$ m/s →.
17.73 (*a*) $\mathbf{v}_B = 0.706$ m/s →. (*b*) $\mathbf{v}_A = 0.235$ m/s →.
17.74 (*a*) $\bar{\mathbf{v}}_A = 8.41$ m/s ↓. (*b*) $P = 16.82$ N.
17.75 (*a*) $t = 0.557$ s. (*b*) $P = 16.82$ N.
17.76 $\mathbf{M} = 0.444$ N · m ↺.
17.77 (*a*) $\omega_0 = \dfrac{5}{2}\dfrac{\bar{v}_0}{r}$ (*b*) $t = \dfrac{\bar{v}_0}{\mu_k g}$
17.78 (*a*) $t_1 = \dfrac{2}{7}\dfrac{r\omega_0}{\mu_k g}$. (*b*) $\mathbf{v}_2 = \dfrac{2}{7} r\omega_0$ →, $\boldsymbol{\omega}_2 = \dfrac{2}{7}\omega_0$ ↺.
17.79 $\boldsymbol{\omega}_1 = 0.614\ \omega_0$ ↻.
17.80 $\omega_2 = 84.2$ rpm.
17.81 (*a*) $\omega_2 = 5.00$ rad/s. (*b*) $\omega_2 = 3.13$ rad/s.
17.82 $\omega_2 = 18.07$ rad/s.
17.83 (*a*) 2.54 rad/s. (*b*) $T_1 - T_2 = 1.902$ J.
17.84 $\Omega_2 = -22.2$ rpm.
17.85 (*a*) $v_2 = 1.019$ m/s. (*b*) $F = 61.6$ N.
17.86 $\omega_2 = 37.2$ rpm.
17.87 $\omega_A = 212$ rpm ↺, $\omega_B = 212$ rpm ↺, $\omega_P = 27.9$ rpm ↻.
17.88 $v_r = 2.51$ m/s.
17.89 $\omega = 18.83$ rad/s, $I = 0.0508$ kg · m^2.
17.90 (*a*) $\omega_2 = 31.7$ rad/s. (*b*) $(v_r)_2 = 5.64$ m/s.
17.91 (*a*) $\omega_2 = 15.00$ rad/s. (*b*) $v_y = 6.14$ m/s.
17.92 (*a*) $\mathbf{v}_C = 0.503$ m/s ←. (*b*) $\mathbf{v}_B = 2.08$ m/s →.
17.93 $(v_r)_2 = 7.45$ m/s.
17.94 $(\bar{v}_r)_2 = 1.542$ m/s.
17.95 $\mathbf{v}_B = 0.607$ m/s ←.
17.96 $h = \dfrac{2}{5} r$.
17.97 (*a*) $\boldsymbol{\omega} = 22.7$ rad/s ↺. (*b*) $\mathbf{C} = 4540$ N →.
17.98 (*a*) $h = 267$ mm. (*b*) $\omega = 21.5$ rad/s.
17.99 $\bar{\mathbf{v}}_2 = 242$ mm/s →.
17.100 $\bar{\mathbf{v}}_2 = 302$ mm/s ←.
17.101 (*a*) $\mathbf{v}_G = 2.16$ m/s →. (*b*) $\mathbf{A} = 4.87$ kN ∠ 66.9°.
17.102 (*a*) $h = 158$ mm. (*b*) $\mathbf{v}_G = 1.992$ m/s →.
17.103 (*a*) $\omega = 0.900\ v_0/L$ ↺. (*b*) $\mathbf{v}_B = 0.100\ v_0$ →.
17.104 $\omega = 2.40$ rad/s ↺.
17.105 $h = 41.7$ mm.
17.106 $\omega_2 = \dfrac{1}{7}(2 + 5\cos\beta)\bar{v}_1/r$ ↻, $\bar{\mathbf{v}}_2 = \dfrac{1}{7}(2 + 5\cos\beta)\bar{v}_1$ ←.
17.107 $\dfrac{a^2}{b^2} = 2\ \dfrac{a}{b} = \sqrt{2}\ \dfrac{a}{b} = 1.414$.
17.108 (*a*) $\bar{\mathbf{v}}_1 = \dfrac{mv_0}{M}$ →, $\omega_1 = \dfrac{mv_0}{MR}$ ↻. (*b*) $\bar{\mathbf{v}}_2 = \dfrac{mv_0}{3M}$ →.
17.109 (*a*) $h = \dfrac{3}{2}R$. (*b*) $h = R$.
17.110 $\omega_2 = 1.200$ rad/s ↻, $\mathbf{v}_2 = 0.240$ m/s ↑.
17.111 (*a*) $\omega_2 = \dfrac{3v_1}{L}$ ↺, $\bar{\mathbf{v}}_2 = \dfrac{1}{2}v_1$ ↓. (*b*) $\boldsymbol{\omega}_3 = \dfrac{3v_1}{L}$ ↻, $\bar{\mathbf{v}}_3 = \dfrac{1}{2}v_1$ ↑.
(*c*) $\omega_4 = 0$, $\bar{\mathbf{v}}_4 = v_1$ ↑.

17.112 $\omega = \dfrac{6\sin\beta}{1+3\sin^2\beta}\dfrac{v_1}{L}$.

17.113 $\omega = 6.17$ rad/s ↻.

17.114 $\omega' = 1.336$ rad/s ↻.

17.115 $\mathbf{v}_1 = 2.38$ m/s.

17.116 $e = 0.366$.

17.117 (*a*) $\omega_3 = 0.437\sqrt{\dfrac{g}{L}}$ ↻. (*b*) $\theta = 5.12°$.

17.118 (*a*) $\omega = \dfrac{1}{4}\omega_0$ ↻. (*b*) $\dfrac{15}{16}$. (*c*) $\theta = 1.50°$.

17.119 $\theta_m = 55.9°$.

17.120 $v_0 = 528$ m/s.

17.121 $h_2 = 725$ mm.

17.122 $h_2 = 447$ mm.

17.123 (*a*) $\mathbf{v}_2 = 1.040$ m/s ↓. (*b*) $\omega_2 = 5.20$ rad/s ↻.

17.124 (*a*) $\mathbf{v}'_A = 0.0318$ m/s ↑. (*b*) $\omega' = 9.55$ rad/s ↻.

17.125 $\theta_B = 50.2°$. (*b*) $\theta_A = 16.26°$.

17.126 $\boldsymbol{\omega} = 0.60716$ rad/s ↺, $\mathbf{v}_s^1 = 0.314$ m/s ⦣ 23.9°.

17.127 (*a*) $\omega' = 3.00$ rad/s ↺. (*b*) $v'_D = 0.938$ m/s ↑.

17.128 (*a*) $\boldsymbol{\omega}_2 = 2.60$ rad/s ↻. (*b*) $\mathbf{v}_D = 1.635$ m/s ⦣ 53.4°.

17.129 $\omega_1 = 50$ rad/s ↻, $\boldsymbol{\omega}_A = 50.0$ rad/s ↻, $\boldsymbol{\omega}_B = 19.27$ rad/s ↺.

17.130 (*a*) $\boldsymbol{\omega}' = 0.922$ rad/s ↻.
(*b*) $\boldsymbol{\alpha} = 36.1766$ rad/s^2 ↻, $\mathbf{A}_x = 5.21$ N →, $\mathbf{A}_y = 6.31$ N ↑.

17.131 $\omega_0 = \dfrac{5}{4}\dfrac{v_0}{r}$.

17.132 (*a*) $\mathbf{v}_A = 0$, $\boldsymbol{\omega}_A = \dfrac{v_1}{r}$ ↻, $\mathbf{v}_B = v_1$ →, $\omega_B = 0$.

(*b*) $\mathbf{v}'_A = \dfrac{2}{7}v_1$ →, $\mathbf{v}'_B = \dfrac{5}{7}v_1$ →.

17.133 (*a*) $v_A = (v_0\sin\theta)\mathbf{j}$, $v_B = (v_0\cos\theta)\mathbf{i}$, $\boldsymbol{\omega}_A = \left(\dfrac{v_0}{r}\right)(-\sin\theta\mathbf{i} + \cos\theta\mathbf{j})$, $\boldsymbol{\omega}_B = 0$.

(*b*) $\mathbf{v}'_B = \dfrac{5}{7}(v_0\cos\theta)\mathbf{i}$.

17.134 $\boldsymbol{\omega}_{AB} = 2.68$ rad/s ↻, $\boldsymbol{\omega}_{BC} = 13.39$ rad/s ↺.

17.135 (*a*) $\theta = 118.7$ rev. (*b*) $t = 7.16$ s.

17.136 $\theta = 0.569$ rev.

17.137 (*a*) $\boldsymbol{\omega}_2 = 3.43$ rad/s ↻. (*b*) $\boldsymbol{\omega}_3 = 4.85$ rad/s ↻.

17.138 $\boldsymbol{\omega}_2 = 14.10$ rad/s ↺, $\boldsymbol{\omega}_3 = 12.66$ rad/s ↺.

17.139 (*a*) $\beta = 53.1°$. (*b*) $\mathbf{v}_A = 1.095\sqrt{gL}$ ⦣ 53.1°.

17.140 $\mathbf{B} = 43.9$ N →, $\mathbf{A} = 100.1$ N ↑.

17.141 $\mathbf{A}_x = 189.7$ N →, $\mathbf{A}_y = 7.36$ N ↑.

17.142 $\omega_1 = \dfrac{7}{9}\omega_2$.

17.143 (*a*) $\omega_2 = 418$ rpm. (*b*) $\Delta T = -20.4$ J.

17.144 (*a*) $\boldsymbol{\omega}_2 = 1.125\dfrac{v_1}{b}$ ↻. (*b*) $\overline{\mathbf{v}}_2 = 0.566$ m/s ⦣ 6.34°

17.145 (*a*) $\omega_2 = 68.6$ rpm. (*b*) $T_1 - T_2 = 2.82$ J.

17.146 $(\omega_{CE})_2 = \omega_1(1+e)$ ↺, $(\omega_{AB})_2 = \dfrac{1}{2}\omega_1(1-e)$ ↻.

CHAPTER 18

18.1 $\mathbf{H}_G = \dfrac{1}{4}mr^2\omega_2\mathbf{j} + \dfrac{1}{2}mr^2\omega_1\mathbf{k}$.

18.2 $\mathbf{H}_A = \dfrac{ma^2\omega}{12}(3\mathbf{j} + 2\mathbf{k})$.

18.3 $\mathbf{H}_D = 0.357\ \text{kg}\cdot\text{m}^2/\text{s}$, $\theta_x = 48.6°$, $\theta_y = 41.4°$, $\theta_z = 90°$.
18.4 $\mathbf{H}_A = (0.24\ \text{kg}\cdot\text{m}^2/\text{s})\mathbf{i} + (0.96\ \text{kg}\cdot\text{m}^2/\text{s})\mathbf{j}$.
18.5 $\mathbf{H}_C = (0.1125\ \text{kg}\cdot\text{m}^2/\text{s})\mathbf{j} + (0.675\ \text{kg}\cdot\text{m}^2/\text{s})\mathbf{k}$.
18.6 *(a)* $\mathbf{H}_G = 0.276\ ma^2\omega$. *(b)* $\theta = 25.2°$.
18.7 *(a)* $\mathbf{H}_G = 0.432\ ma^2\omega$. *(b)* $\theta = 20.2°$.
18.8 $\theta = 11.88°$.
18.9 $\mathbf{H}_D = (2.16\ \text{kg}\cdot\text{m}^2/\text{s})\mathbf{i} - (0.48\ \text{kg}\cdot\text{m}^2/\text{s})\mathbf{j} + (1.440\ \text{kg}\cdot\text{m}^2/\text{s})\mathbf{k}$.
18.10 $\mathbf{H}_A = -(2.03\ \text{kg}\cdot\text{m}^2/\text{s})\mathbf{i} + (4.16\ \text{kg}\cdot\text{m}^2/\text{s})\mathbf{j} + (0.675\ \text{kg}\cdot\text{m}^2/\text{s})\mathbf{k}$.
18.12 *(a)* $\omega_s = 0.485$ rad/s. *(b)* $\omega_P = 0.01531$ rad/s.
18.13 $\mathbf{H}_A = (0.500\ \text{kg}\cdot\text{m}^2/\text{s})\mathbf{i} - (0.100\ \text{kg}\cdot\text{m}^2/\text{s})\mathbf{j} - (2350\ \text{kg}\cdot\text{m}^2/\text{s})\mathbf{k}$.
18.15 *(a)* $\mathbf{H}_G = (1.563\ \text{kg}\cdot\text{m}^2/\text{s})\mathbf{i} - (0.938\ \text{kg}\cdot\text{m}^2/\text{s})\mathbf{k}$. *(b)* 31.0.
18.16 *(a)* $\mathbf{H}_A = (1.563\ \text{kg}\cdot\text{m}^2/\text{s})\mathbf{i} - (0.938\ \text{kg}\cdot\text{m}^2/\text{s})\mathbf{k}$.
(b) $\mathbf{H}_B = (1.563\ \text{kg}\cdot\text{m}^2/\text{s})\mathbf{i} - (0.938\ \text{kg}\cdot\text{m}^2/\text{s})\mathbf{k}$.
18.17 *(a)* $\mathbf{H}_A = -(3.02\ \text{kg}\cdot\text{m}^2/\text{s})\mathbf{i} + (3.02\ \text{kg}\cdot\text{m}^2/\text{s})\mathbf{j} + (6.70\ \text{kg}\cdot\text{m}^2/\text{s})\mathbf{k}$.
(b) $\theta = 147.5°$.
18.18 *(a)* $\mathbf{H}_B = -(3.02\ \text{kg}\cdot\text{m}^2/\text{s})\mathbf{i} + (3.02\ \text{kg}\cdot\text{m}^2/\text{s})\mathbf{j} + (6.70\ \text{kg}\cdot\text{m}^2/\text{s})\mathbf{k}$.
(b) $\theta = 32.5°$.
18.19 (a) $\mathbf{H}_C = (0.063\ \text{kg}\cdot\text{m}^2/\text{s})\mathbf{i} + (0.216\ \text{kg}\cdot\text{m}^2/\text{s})\mathbf{j}$.
(b) $\mathbf{H}_A = -(0.513\ \text{kg}\cdot\text{m}^2/\text{s})\mathbf{i} + (0.216\ \text{kg}\cdot\text{m}^2/\text{s})\mathbf{j}$.
18.20 (a) $\mathbf{H}_C = (0.063\ \text{kg}\cdot\text{m}^2/\text{s})\mathbf{i} + (0.216\ \text{kg}\cdot\text{m}^2/\text{s})\mathbf{j}$.
(b) $\mathbf{H}_B = (0.387\ \text{kg}\cdot\text{m}^2/\text{s})\mathbf{i} + (0.216\ \text{kg}\cdot\text{m}^2/\text{s})\mathbf{j}$.
18.21 $m = 93.6$ kg.
18.22 $t = 2.57$ s.
18.23 (a) $\overline{\mathbf{v}} = -\dfrac{F\Delta t}{m}\mathbf{k}$. (b) $\boldsymbol{\omega} = \dfrac{12}{7}\dfrac{F\Delta t}{ma}(-\mathbf{i} - 5\mathbf{j})$.
18.24 (a) $\overline{\mathbf{v}} = \dfrac{F\Delta t}{m}\mathbf{k}$. (b) $\boldsymbol{\omega} = \dfrac{12}{7}\dfrac{F\Delta t}{ma}(2\mathbf{i} + 3\mathbf{j})$.
18.25 *(a)* $\overline{\mathbf{v}} = 0$. *(b)* $\boldsymbol{\omega} = (3F\Delta t/8ma)(\mathbf{i} - 4\mathbf{k})$.
18.26 *(a)* $\overline{\mathbf{v}} = -(F\Delta t/m)\mathbf{i}$. *(b)* $\boldsymbol{\omega} = (3F\Delta t/8ma)(\mathbf{j} + 4\mathbf{k})$.
18.27 *(a)* $\overline{\mathbf{v}} = -(0.300\ \text{m/s})\mathbf{k}$. *(b)* $\boldsymbol{\omega} = -(0.962\ \text{rad/s})\mathbf{i} - (0.577\ \text{rad/s})\mathbf{j}$.
18.28 *(a)* $\overline{\mathbf{v}} = (0.300\ \text{m/s})\mathbf{j}$.
(b) $\boldsymbol{\omega} = -(3.46\ \text{rad/s})\mathbf{i} + (1.923\ \text{rad/s})\mathbf{j} - (0.857\ \text{rad/s})\mathbf{k}$.
18.29 $\boldsymbol{\omega} = \dfrac{2\sqrt{2}}{5}\dfrac{v_0}{R}(\mathbf{i} + \mathbf{k})$.
18.30 (a) $\overline{\mathbf{v}} = -\dfrac{4}{5}\overline{v}_0\mathbf{j}$. (b) $\mathbf{C}\Delta t = \dfrac{1}{5}m\overline{v}_0\mathbf{j}$
18.31 *(a)* $\boldsymbol{\omega} = \dfrac{1}{8}\omega_0(-\mathbf{i} + \mathbf{j})$. *(b)* $\overline{\mathbf{v}} = 0.0884a\omega_0\mathbf{k}$.
18.32 *(a)* $(F\Delta t)\mathbf{k} = 0.1031\ ma\omega_0\mathbf{k}$. *(b)* $\mathbf{A}\Delta t = -0.01473\ ma\omega_0\mathbf{k}$.
18.33 $\boldsymbol{\omega} = (0.0225\ \text{rad/s})\mathbf{i} - (0.223\ \text{rad/s})\mathbf{j} - (0.320\ \text{rad/s})\mathbf{k}$.
18.34 *(a)* $\omega_z = -0.711$ rad/s.
(b) $\mathbf{v}_0 = -(640\ \text{m/s})\mathbf{i} - (1600\ \text{m/s})\mathbf{j} + (338\ \text{m/s})\mathbf{k}$.
18.35 *(a)* Use thrusters *C* and *B*. *(b)* $\Delta t_C = 8.16$ s, $\Delta t_B = 4.84$ s. *(c)* $\Delta t_E = 0.520$ s.
18.36 *(a)* Use thrusters *D* and *A*. *(b)* $\Delta t_D = 6.82$ s, $\Delta t_A = 1.848$ s. *(c)* $\Delta t_E = 0.347$ s.
18.38 *(a)* $T = \dfrac{1}{2}I_{OL}\omega^2$. (b) $T = \dfrac{1}{2}I_{OL}\omega^2$.
18.39 $T = \dfrac{1}{8}mr^2(\omega_2^2 + 2\omega_1^2)$.
18.40 $T = \dfrac{1}{8}ma^2\omega^2$.
18.41 $T = 1.417$ J
18.42 $T = 16.32$ N · m.
18.43 $T = 15.47$ J.
18.44 $T = 0.1250\ ma^2\omega^2$.
18.45 $T = 0.203\ ma^2\omega^2$.
18.46 $T = 0.228\ mr^2\omega^2$.
18.47 $T = 9.38$ J.
18.48 $T = 84.2$ N · m.

18.49 $T = 1.296$ J.
18.50 $T = 46.2$ J.
18.51 $T_0 - T = \frac{1}{10} m\bar{v}_0^2$
18.52 $T_0 - T_1 = \frac{7}{192} ma^2\omega_0^2$.
18.53 $T = 18.40$ N · m.
18.54 $\omega_z = -0.711$ rad/s, $T' = 55.8$ N · m.
18.55 $\dot{\mathbf{H}}_G = \frac{1}{2} mr^2\omega_1\omega_2\mathbf{i}$.
18.56 $\dot{\mathbf{H}}_A = \frac{1}{6} ma^2\omega^2\mathbf{i}$.
18.57 $\dot{\mathbf{H}}_D = (3.21\ \text{N}\cdot\text{m})\mathbf{k}$.
18.58 $\dot{\mathbf{H}}_A = (7.68\ \text{N}\cdot\text{m})\mathbf{k}$.
18.59 $\dot{\mathbf{H}}_C = (3.38\ \text{N}\cdot\text{m})\mathbf{i}$.
18.60 $\dot{\mathbf{H}}_G = -0.0958mr^2\omega^2\mathbf{k}$.
18.61 $\dot{\mathbf{H}}_D = -(1.890\ \text{N}\cdot\text{m})\mathbf{i} - (2.14\ \text{N}\cdot\text{m})\mathbf{j} + (3.21\ \text{N}\cdot\text{m})\mathbf{k}$.
18.62 $\dot{\mathbf{H}}_D = (1.890\ \text{N}\cdot\text{m})\mathbf{i} - (2.14\ \text{N}\cdot\text{m})\mathbf{j} + (3.21\ \text{N}\cdot\text{m})\mathbf{k}$.
18.63 $\dot{\mathbf{H}}_A = \frac{ma^2}{12}(2\omega^2\mathbf{i} + 3\alpha\mathbf{j} + 2\alpha\mathbf{k})$.
18.64 $\dot{\mathbf{H}}_G = mr^2(0.0958\alpha\mathbf{i} + 0.455\alpha\mathbf{j} - 0.0958\omega^2\mathbf{k})$.
18.65 $\mathbf{C} = \frac{1}{6} mb\omega^2 \sin\beta \cos\beta\mathbf{i}, \mathbf{D} = -\frac{1}{6} mb\omega^2 \sin\beta \cos\beta\mathbf{i}$.
18.66 $\mathbf{A} = -(12.00\ \text{N})\mathbf{i}$, $\mathbf{B} = -(4.00\ \text{N})\mathbf{i}$.
18.67 $\mathbf{A} = -(4.93\ \text{N})\mathbf{j} - (4.11\ \text{N})\mathbf{k}$, $\mathbf{B} = (4.93\ \text{N})\mathbf{j} + (4.11\ \text{N})\mathbf{k}$.
18.68 $\mathbf{A} = (14.4\ \text{N})\mathbf{k}$, $\mathbf{B} = -(14.4\ \text{N})\mathbf{k}$.
18.69 (*a*) $\bar{y} = -1.153$ mm, $\bar{z} = 0$; $I_{xy} = -3.08 \times 10^{-3}\ \text{kg}\cdot\text{m}^2$, $I_{zx} = 0$.
(b) $\mathbf{F} = -(184.3\ \text{N})\mathbf{j}$, $\mathbf{M}_C = -(27.4\ \text{N}\cdot\text{m})\mathbf{k}$.
18.70 (a) $\bar{r} = 1.441$ mm, $I_{xy} = 2.38 \times 10^{-3}\ \text{kg}\cdot\text{m}^2$, $I_{xz} = 0$.
(b) $m_E = 158.5$ g, $m_A = 16.05$ g
18.71 (*a*) $\alpha = \frac{3M_0}{mb^2\cos^2\beta}$. (*b*) $\mathbf{C} = \left(\frac{M_0}{2b}\right)\tan\beta\mathbf{k}, \mathbf{D} = -\left(\frac{M_0}{2b}\right)\tan\beta\mathbf{k}$.
18.72 $\boldsymbol{\alpha} = (20.0\ \text{rad/s}^2)\mathbf{j}$, $\mathbf{B} = -(1.250\ \text{N})\mathbf{k}$, $\mathbf{A} = -(3.75\ \text{N})\mathbf{k}$.
18.73 (*a*) $M_0 = (1.172\ \text{N}\cdot\text{m})\mathbf{i}$.
(*b*) $\mathbf{A} = -(0.977\ \text{N})\mathbf{j} + (1.172\ \text{N})\mathbf{k}$, $\mathbf{B} = (0.977\ \text{N})\mathbf{j} - (1.172\ \text{N})\mathbf{k}$.
18.74 (*a*) $M_0 = (2.67\ \text{N}\cdot\text{m})\mathbf{i}$. (*b*) $\mathbf{B} = -(2.00\ \text{N})\mathbf{j}$, $\mathbf{A} = (2.00\ \text{N})\mathbf{j}$.
18.75 (*a*) $M_0 = (0.1885\ \text{N}\cdot\text{m})\mathbf{i}$.
(*b*) $\mathbf{A} = -(0.1600\ \text{N})\mathbf{j} + (0.1600\ \text{N})\mathbf{k}$, $\mathbf{B} = (0.1600\ \text{N})\mathbf{j} - (0.1600\ \text{N})\mathbf{k}$.
18.76 $\mathbf{A} = -(2.17\ \text{N})\mathbf{j} - (1.851\ \text{N})\mathbf{k}$, $\mathbf{B} = -(2.17\ \text{N})\mathbf{j} - (1.851\ \text{N})\mathbf{k}$.
18.77 (*a*) $M_0 = 24.8 \times 10^{-3}$ N · m.
(*b*) $\mathbf{A} = (7.50 \times 10^{-3}\text{N})\mathbf{i} + (15.00 \times 10^{-3}\text{N})\mathbf{j}$,
$\mathbf{B} = -(7.50 \times 10^{-3}\text{N})\mathbf{i} - (15.00 \times 10^{-3}\text{N})\mathbf{j}$
18.78 (*a*) $\boldsymbol{\omega} = (7.20\ \text{rad/s})\mathbf{k}$
(*b*) $\mathbf{A} = -(57.3 \times 10^{-3}\text{N})\mathbf{i} + (47.3 \times 10^{-3}\text{N})\mathbf{j}$,
$\mathbf{B} = (57.3 \times 10^{-3}\text{N})\mathbf{i} - (47.3 \times 10^{-3}\text{N})\mathbf{j}$
18.79 $\mathbf{A} = (1.527\ \text{N})\mathbf{j}$, $\mathbf{B} = -(1.527\ \text{N})\mathbf{j}$.
18.80 $\mathbf{M} = -(0.754\ \text{N}\cdot\text{m})\mathbf{i}$.
18.81 (*a*) $\mathbf{M}' = 10.47$ N · m. (*b*) $\mathbf{M}' = 10.47$ N · m.
18.82 $\mathbf{F}_y = 24.0$ N↑.
18.83 $\mathbf{M} = -(0.225\ \text{N}\cdot\text{m})\mathbf{j}$.
18.84 $\theta = 1.138°$↓; Point *A* moves *up*.
18.85 $\omega = 8.90$ rad/s.
18.86 (*a*) $\beta = 38.1°$. (*b*) $\omega = 11.78$ rad/s.
18.87 $\omega = 13.46$ rad/s.
18.88 $\omega = 7.89$ rad/s.

18.89 $\omega = 12.49$ rad/s.
18.90 $\omega_2 = 5.45$ rad/s.
18.91 2.11 N ∠ 18.7°.
18.92 *A* will move up;

$$\text{Angle of rotation} = \frac{x}{GA} = \frac{1.3635 \times 10^{-3}\text{ m}}{0.06\text{ m}} = 0.022725\text{ rad} = 1.30°.$$

18.93 $\mathbf{A} = (0.736\text{ N})\,\mathbf{K}$, $\mathbf{B} = -(0.736\text{ N})\mathbf{K}$.
18.94 $\omega_2 = 10.19$ rad/s.
18.95 (*a*) $\mathbf{C} = -(123.4\text{ N})\mathbf{i}$; $\mathbf{D} = (123.4\text{ N})\mathbf{i}$. (*b*) $\mathbf{C} = \mathbf{D} = 0$.
18.96 $\omega_2 = 91.2$ rpm.
18.97 (a) $\Omega = \sqrt{g/a}$. (b) $\Omega = \sqrt{2g/a}$.
18.98 $\mathbf{D} = 101.4$ N↓.
18.99 $\mathbf{A} = -(45\text{ N})\mathbf{i}$, $\mathbf{M}_A = (3.38\text{ N}\cdot\text{m})\mathbf{i} + (10.13\text{ N}\cdot\text{m})\mathbf{k}$.
18.100 (*a*) $\mathbf{A} = (1.786\text{ kN})\mathbf{i} + (143.5\text{ kN})\mathbf{j}$, $\mathbf{B} = -(1.786\text{ kN})\mathbf{i} + (150.8\text{ kN})\mathbf{j}$.
(*b*) $M_2\mathbf{k} = -(35.7\text{ kN}\cdot\text{m})\mathbf{k}$.
18.101 $\mathbf{C} = -(89.8\text{ N})\mathbf{i} + (52.8\text{ N})\mathbf{k}$,
$\mathbf{D} = -(89.8\text{ N})\mathbf{i} - (52.8\text{ lb})\mathbf{k}$.
18.102 (*a*) $M_0 = (0.1962\text{ N}\cdot\text{m})\mathbf{j}$.
(*b*) $\mathbf{C} = -(48.6\text{ N})\mathbf{i} + (38.9\text{ N})\mathbf{k}$, $\mathbf{D} = -(48.6\text{ N})\mathbf{i} - (38.9\text{ N})\mathbf{k}$.
18.103 $\mathbf{D} = -(22.0\text{ N})\mathbf{i} + (26.8\text{ N})\mathbf{j}$, $\mathbf{E} = -(21.2\text{ N})\mathbf{i} - (5.20\text{ N})\mathbf{j}$.
18.104 (*a*) $M_0 = (0.392\text{ N}\cdot\text{m})\mathbf{k}$.
(*b*) $\mathbf{D} = -(21.0\text{ N})\mathbf{i} + (28.0\text{ N})\mathbf{j}$, $\mathbf{E} = -(21.0\text{ N})\mathbf{i} - (4.00\text{ N})\mathbf{j}$.
18.105 (*a*) $\mathbf{M}_1\mathbf{j} = -(624\text{ N}\cdot\text{m})\mathbf{j}$.
(*b*) $\mathbf{A} = -(45.0\text{ N})\mathbf{i} + (13.50\text{ N})\mathbf{k}$, $\mathbf{M}_A = (6.41\text{ N}\cdot\text{m})\mathbf{i} + (10.13\text{ N}\cdot\text{m})\mathbf{k}$.
18.106 (a) $\mathbf{M}_1 = -\frac{1}{12}mL^2\omega_1\omega_2\sin 2\theta\,\mathbf{j}$.

(b) $\mathbf{M}_2 = \frac{1}{24}mL^2\omega_1^2\sin 2\theta\,\mathbf{k}$, $\mathbf{C} = \frac{1}{6}mL\omega_1\omega_2\sin^2\theta\,\mathbf{k}$; $\mathbf{D} = -\frac{1}{6}mL\omega_1\omega_2\sin^2\theta\,\mathbf{k}$.
18.107 $\dot{\psi} = 1141$ rpm.
18.108 Solving the quadratic equations, $\dot{\phi} = 2.2064$ rad/s, -38.1104 rad/s
$\dot{\phi} = 21.1$ rpm, -364 rpm
18.109 $\dot{\varphi} = 45.9$ rpm, 533 rpm.
18.110 (*a*) $Wc = (I\omega_z - I'\dot{\varphi}\cos\theta)\dot{\varphi}$. (*b*) $Wc \approx I\dot{\psi}\dot{\varphi}$. (c) %error = –7.95%.
18.111 $\dot{\psi} = 50.9$ rpm.
18.112 $\beta = 72.0°$.
18.113 (*a*) $\theta = 40.0°$. (*b*) $\theta = 3.63°$. (*c*) $\theta = -75.1°$.
18.114 (*a*) $\dot{\psi} = 49.0$ rad/s. (*b*) $\dot{\varphi} = 4.63$ rad/s.
18.115 $\theta = 23.7°$.
18.116 (*a*) $\dot{\psi} = 52.7$ rad/s. (*b*) $\dot{\varphi} = 6.44$ rad/s.
18.117 (*a*) $\dot{\varphi} \approx 5.47$ rpm. (*b*) $\dot{\varphi} = 5.55$ rpm, 395 rpm.
18.118 $M = 21.7 \times 10^{21}$ N · m.
18.121 $n = \frac{I' - I}{I'}\omega_z$.
18.122 The axis lies outside the space cone.
18.123 period = 302 days
18.124 (*a*) $\beta = 13.19°$. (*b*) $|\dot{\varphi}| = 1242$ rpm (retrograde).
18.125 (*a*) $\beta = 23.8°$. (*b*) *precession:* $\dot{\varphi} = 82.6$ rpm; *spin:* $\dot{\psi} = 128.8$ rpm.
18.126 (*a*) $\theta = 32.005°$. (*b*) $\dot{\varphi} = 1.126$ rpm; $\dot{\psi} = 0.344$ rpm.
18.127 $\theta = 52.001°$; $\dot{\varphi} = 0.1523$ rad/s, $\dot{\psi} = 0.0338$ rad/s.
18.128 (*a*) $\omega = 109.4$ rpm; $\gamma_x = 90°$, $\gamma_y = 100.05°$, $\gamma_z = 10.05°$.
(*b*) $\theta_x = 90°$, $\theta_y = 113.9°$, $\theta_z = 23.9°$.
(*c*) precession: $\dot{\varphi} = 47.1$ rpm; spin: $\dot{\psi} = 64.6$ rpm.
18.129 $\theta_x = 132.3°$, $\theta_y = 43.8°$, $\theta_z = 80.7°$;
$\dot{\varphi} = 0.947$ rad/s, $\dot{\psi} = 0.1602$ rad/s, Since $\gamma > \theta$, precession is retograde.
18.130 $\theta_x = 90.0°$, $\theta_y = 30.8°$, $\theta_z = 59.2°$;
$\dot{\varphi} = 0.796$ rad/s, $\dot{\psi} = 0.1602$ rad/s, Since $\gamma > \theta$, precession is retograde.

18.131 (*a*) $40° < \theta < 140°$. (*b*) $\dot{\phi}_{min} = 5.31$ rad/s. (*c*) $\dot{\phi}_{max} = 5.58$ rad/s.

18.132 (*a*) $\dot{\phi}_{min} = 2.00$ rad/s. (*b*) $\dot{\phi}_{max} = 4.47$ rad/s.

18.133 (*a*) $\theta_m = 41.2°$. (*b*) $\dot{\varphi}_0 = 6.21$ rad/s.

18.134 (*a*) $\dot{\varphi}_0 = 4.76$ rad/s. (*b*) $\dot{\varphi}_m = 14.09$ rad/s.

18.135 (*a*) $\theta_0 \le \theta \le 180° - \theta_0$. (*b*) $\dot{\theta}_{max} = \dot{\phi}_0 \sin\theta_0 \cos\theta_0$.
(*c*) $\dot{\phi}_{min} = \dot{\phi}_0 \sin^2\theta_0$.

18.136 (*a*) $(1+\cos^2\theta)\dot{\phi}^2 + \dot{\theta}^2 = \text{constant}, \quad \dot{\phi}(1+\cos^2\theta) = \text{constant}$.

(*b*) $\dot{\theta} = \dot{\phi}_0 \sqrt{\dfrac{(1+\cos^2\theta_0)(\cos^2\theta - \cos^2\theta_0)}{1+\cos^2\theta}}$.

(*c*) $\theta \le \theta_0$.

18.137 (*a*) $I'\dot{\phi}\sin^2\theta + I(\dot{\psi} + \dot{\phi}\cos\theta)\cos\theta = \alpha, \quad I(\dot{\psi} + \dot{\phi}\cos\theta) = \beta$.

(*b*) $\omega_z = \dfrac{\beta}{I} = \text{constant}, \ \dot{\phi} = \dfrac{\alpha - \beta\cos\theta}{I'\sin^2\theta}$.

18.138 (*a*) $\dfrac{1}{2}I'(\dot{\varphi}\sin\theta)^2 + \dfrac{1}{2}I'\dot{\theta}^2 + \dfrac{1}{2}I\omega_z^2 + mgc\cos\theta = E$. .

(*b*) $f(\theta) = \dfrac{1}{I'}\left(2E - \dfrac{\beta^2}{I} - 2mgc\cos\theta\right) - \left(\dfrac{\alpha - \beta\cos\theta}{I'\sin\theta}\right)^2$.

(c) $\left(2E - \dfrac{\beta^2}{I} - 2mgcx\right)(1-x^2) - \dfrac{1}{I'}(\alpha - \beta x)^2 = 0$.

18.139 (*a*) $\theta_{max} = 47.0°$. (*b*) precession: $\dot{\phi} = 15.25$ rad/s; spin: $\dot{\psi} = 307$ rad/s.

18.140 (*a*) $\theta_{max} = 76.3°$. (*b*) precession: $\dot{\phi} = 9.62$ rad/s; spin: $\dot{\psi} = 294$ rad/s.
(*c*) $\theta = 36.5°$.

18.141 $\beta_{max} = 27.5°$

18.142 (*a*) $\dot{\phi}_0 = \sqrt{\dfrac{15}{11}\dfrac{g}{a}}$. (*b*) $\dot{\phi} = 2\sqrt{\dfrac{20}{33}\dfrac{g}{a}}, \quad \dot{\psi} = \sqrt{\dfrac{20}{33}\dfrac{g}{a}}$.

18.143 (a) (1) $\mathbf{H}_O$ = constant, (2) T = constant, (3) $\omega \cos\beta$ = constant.

18.147 On *B*: 150 mm below shaft. On *C*: 75 mm above shaft.

18.148 $\mathbf{H}_C = (0.234 \text{ kg}\cdot\text{m}^2/\text{s})\mathbf{j} + (1.250 \text{ kg}\cdot\text{m}^2/\text{s})\mathbf{k}$.

18.149 (a) $\mathbf{H}_G = mr^2\omega(0.379\mathbf{i} - 0.483\mathbf{k})$. (b) $\theta = 51.9°$. (c) $\mathbf{H}_A = mr^2\omega(0.379\mathbf{i} - 0.483\mathbf{k})$.

18.150 (*a*) $\bar{\mathbf{v}} = 0$. (*b*) $\boldsymbol{\omega} = \left(\dfrac{F\Delta t}{ma}\right)(2.50\mathrm{i} - 1.454\mathrm{j} + 2.19\mathrm{k})$.

18.151 4.29 kN · m.

18.152 (a) $\alpha = 52.1$ rad/s^2. (b) $\mathbf{A} = -(2.50 \text{ N})\mathbf{i}$, $\mathbf{B} = (2.50 \text{ N})\mathbf{i}$

18.153 $\mathbf{D} = -(32.0 \text{ N})\mathbf{j} + (19.20 \text{ N})\mathbf{k}$, $\mathbf{E} = -(6.40 \text{ N})\mathbf{j} + (19.20 \text{ N})\mathbf{k}$.

18.154 $\mathbf{R} = 0$; $\mathbf{M}_D = (11.23 \text{ N}\cdot\text{m})\cos^2\theta\mathbf{i} +$
$(11.23 \text{ N}\cdot\text{m})\sin\theta\cos\theta\mathbf{j} - (2.81 \text{ N}\cdot\text{m})\sin\theta\cos\theta\mathbf{k}$.

18.155 (*a*) $\theta_x = 52.5°$, $\theta_y = 37.5°$, $\theta_z = 90°$.
(*b*) $\dot{\varphi} = 53.8$ rev/h. (*c*) $\dot{\psi} = 6.68$ rev/h.

18.156 (*a*) $(\mathbf{M}_O)_y = (3.12 \text{ N}\cdot\text{m})\mathbf{j}$. (*b*) $\mathbf{R} = -(19.20 \text{ N})\mathbf{i} - (7.20 \text{ N})\mathbf{k}$;
$\mathbf{M}_O = (2.40 \text{ N}\cdot\text{m})\mathbf{i} - (3.84 \text{ N}\cdot\text{m})\mathbf{k}$

18.157 (*a*) $\dot{\varphi}_{min} = 4.00$ rad/s. (*b*) $\dot{\theta}_{max} = 5.66$ rad/s.

18.158 (*a*) $I'\ddot{\theta} + (-I'\omega_e\cos\lambda\sin\theta)(\omega_e\cos\lambda\cos\theta) - (-\omega_e\cos\lambda\sin\theta)I\omega_z = 0$,

$$I'\ddot{\theta} + I\omega_z\omega_e\cos\lambda\sin\theta - I'\omega_e^2\cos^2\lambda\sin\theta\cos\theta = 0$$

$I\dot{\omega}_z = 0$ Q.E.D.

(b) $\ddot{\theta} + \dfrac{I\omega_z\omega_e\cos\lambda}{I'}\theta = 0$ Q.E.D.

CHAPTER 19

19.1 $v_m = 0.377$ m/s, $a_m = 47.3$ m/s^2.
19.2 f_n 0.650 Hz, $v_m = 1.225$ m/s.
19.3 $x_m = 0.950$ mm, $v_m = 239$ mm/s.
19.4 (*a*) $\tau_n = 0.324$ s, $f_n = \dfrac{1}{\tau_n} = \dfrac{1}{0.324} = 3.08$ Hz. (*b*) $x_m = 12.91$ mm, $a_m = 4.84$ m/s^2.
19.5 (*a*) $\tau_n = 0.222$ s, $f_n = \dfrac{1}{\tau_n} = \dfrac{1}{0.22214} = 4.50$ Hz.
(*b*) $v_m = 1.414$ m/s, $a_m = 40.0$ m/s^2.
19.6 (*a*) speed = 276 rpm. (*b*) $v_{max} = 1.732$ m/s.
19.7 (*a*) $\theta_m = 11.29°$. (*b*) $(a_t)_m = 1.933$ m/s^2.
19.8 (*a*) $f_n = 0.557$ Hz. (*b*) $v_m = 293$ mm/s.
19.9 $x_m = 11.04$ mm.
19.10 (*a*) $x_m = 99.0$ mm. (*b*) $a_m = 20.2$ *g*.
19.11 (*a*) $t = 0.046$ s. (*b*) $\mathbf{v} = 2.06$ m/s ↑, $\mathbf{a} = 20.0$ m/s^2↓.
19.12 $\mathbf{x} = 0.0284$ m↑, $\mathbf{v} = 2.47$ m/s↑, $\mathbf{a} = 5.69$ m/s^2↓.
19.13 (*a*) $\theta = 0.06786$ rad $= 3.89°$.
(*b*) $v = l\dot{\theta} = (0.800\text{ m})(0.19223\text{ rad/s}) = 0.1538$ m/s, $a = 0.666$ m/s^2.
19.14 (*a*) amplitude $x_m = 4.91$ mm, frequency $f_n = 5.81$ Hz,
maximum velocity $v_m = 0.1791$ m/s.
(*b*) $T_{min} = 491$ N, (c) $\dot{\mathbf{x}} = 0.1592$ m/s↑.
19.15 (*a*) $x_m = 2.48$ mm. (*b*) $x_m = 0.621$ mm.
19.16 (*a*) $\tau = 1.876$ s. (*b*) $\theta_C = 7.07°$.
19.17 $\tau_{n2} = 2.63$ s.
19.18 (*a*) $\tau_n = \dfrac{2\pi}{\omega_n} = \dfrac{2\pi}{30.23} = 0.208$ s, $f_n = \dfrac{1}{\tau_n} = 4.81$ Hz.
(*b*) $v_{max} = 1.361$ m/s, $a_{max} = 41.1$ m/s^2.
19.19 (*a*) $\tau_n = \dfrac{2\pi}{\sqrt{\dfrac{k}{m}}} = \dfrac{2\pi}{\sqrt{\dfrac{8\times 10^3}{35}}} = 0.416$ s, $f_n = \dfrac{1}{\tau_n} = \dfrac{1}{0.416} = 2.41$ Hz.
(*b*) $v_{max} = 0.680$ m/s, $a_{max} = 10.29$ m/s^2.
19.20 (*a*) $t_n = 0.361$ s, $f_n = 2.77$ Hz. (*b*) $v_{max} = 0.765$ m/s, $a_{max} = 13.30$ m/s^2.
19.21 $\tau_{n2} = 2.79$ s.
19.22 $\omega_n = \sqrt{\dfrac{k}{2m}}$.
19.23 (*a*) $k_1 = 7.11$ kN/m. (*b*) $W_A = 25.4$ N.
19.24 (*a*) $m_C = 6.80$ kg. (*b*) $\tau_3 = 0.583$ s.
19.25 $k_C = 5.20$ kN/m.
19.26 (*a*) $m_b = 26.1$ kg. (*b*) $\rho_{sw} = 1213$ kg/m^3.
19.27 (*a*) $k_e = 127.8$ kN/m. (*b*) $f_n = 3.60$ Hz.
19.28 (*a*) $k_e = 22.3$ MN/m. (*b*) $f_n = 266$ Hz.
19.29 $f_n = \dfrac{\omega_n}{2\pi} = \dfrac{1}{2\pi}\sqrt{\dfrac{g}{\delta_{st}}}$.
19.30 (*a*) $k_e = 858$ N/mm. (*b*) Speed = 149.5 rpm.
19.31 (*a*) $m = 3.56$ kg. (*b*) $m = 43.7$ kg.
19.32 (*a*) $x_0 = 616$ mm. (*b*) $f_n = 0.449$ Hz.
19.33 $\tau_n = 2\pi\sqrt{\dfrac{l}{g}}\left(1 + \dfrac{1}{4}\sin^2\dfrac{\theta_m}{2}\right)$.
19.34 $\theta_m = 16.26°$.
19.35 (*a*) $\tau_n = 1.737$ s. (*b*) $\tau_n = 1.864$ s. (*c*) $\tau_n = \dfrac{2(1.854)(1.737\text{ s})}{\pi} = 2.05$ s.
19.36 $l = 713$ mm.
19.37 (*a*) $\tau = 0.293$ s. (*b*) $v_m = 0.215$ m/s.
19.38 (*a*) $\dot{\theta}_m = 1.117$ rad/s. (*b*) $\bar{k} = 400$ mm.
19.39 (*a*) $\tau_n = 1.740$ s. (*b*) $(v_B)_{max} = 90.3$ mm/s.

19.40 (*a*) τ_n = 1.701 s. (*b*) $(v_B)_{max}$ = 92.3 mm/s.
19.41 (*a*) τ = 0.483 s. (*b*) v_m = 260 m/s.
19.42 (*a*) τ_n = 0.1924 s. (*b*) $(a_0)_{max}$ = 53.3 m/s^2.
19.43 (*a*) $f = 0.318\sqrt{\frac{k}{m}}$. (*b*) $f = 0.551\sqrt{\frac{k}{m}}$.
19.44 β = 75.5°.
19.45 f_n = 0.346 Hz.
19.46 $\overline{I}$ = 1255 kg · m^2.
19.47 (*a*) r_a = 153.7 mm, r_b = 96.31 mm.
19.48 (*a*) $\tau_n = \frac{2\pi}{2.753} = 2.28$ s. (*b*) l = 1.294 m.
19.49 (*a*) τ_n = 1.617 s. (*b*) τ_n = 1.676 s.
19.50 (*a*) d = 227 mm. (*b*) τ_n = 1.352 s.
19.51 (*a*) τ_{n0} = 1.067 s. (*b*) c = 89.7 mm.
19.52 $GA = l - \overline{r} = \frac{\overline{k}^2}{\overline{r}}$ Q.E.D.
19.53 $\overline{r} = \overline{k}$ Q.E.D.
19.55 (*a*) f_n = 2.21 Hz. (*b*) k = 115.3 N/m.
19.56 f_n = 0.945 Hz.
19.57 f_n = 3.24 Hz.
19.58 h = 0.1123 m, $\overline{k}$ = 0.808 m.
19.59 (*a*) τ_n = 0.419 s. (*b*) $(v_A)_m$ = 4.71 m/s.
19.60 (*a*) $(v_A)_m$ = 0.0881 m/s. (*b*) $(v_A)_m$ = 0.0851 m/s.
19.61 $f = 0.1638\sqrt{\frac{g}{l}}$.
19.62 v_B = 82.2 mm/s.
19.63 $\overline{I}_A$ = 6.57 kg · m^2.
19.64 (*a*) m = 21.3 kg. (*b*) τ = 1.838 s.
19.65 (*a*) τ_n = 0.885 s. (*b*) τ_n = 1.159 s.
19.66 (*a*) τ = 1.951 s. (*b*) v_m = 1.752 m/s.
19.67 $\overline{k}$ = 14.36 mm.
19.68 $\tau_n = 2\pi\sqrt{\frac{\frac{1}{12}m(a^2+b^2)l}{mg\frac{1}{4}(a^2+b^2)}} = 2\pi\sqrt{\frac{l}{3g}}$. (*b*) $\tau_n = 2\pi\sqrt{\frac{l}{g}}$. (*c*) $\tau_n = 2\pi\sqrt{\frac{l}{g}}$.
19.69 $\dot{x}_m = 1.476$ m/s, $\dot{x}_m = 31.1$ m/s^2.
19.70 $|v_D|_m$ = 0.296 m/s.
19.71 τ_n = 3.18 s.
19.72 $\tau_n = \frac{2\pi}{\omega_n} = 2\pi\sqrt{\frac{R}{g}}$.
19.73 $\overline{I}$ = 1.537 kg · m^2.
19.74 $\overline{k}$ = 130.6 mm.
19.75 $c = \frac{l}{\sqrt{12}}$.
19.76 β = 75.5°.
19.77 $f_n = \frac{1}{2\pi}\sqrt{\frac{2k}{3m} + \frac{4g}{3L}}$.
19.78 f_n = 2.75 Hz.
19.79 (*a*) $\tau_n = \frac{2\pi}{\omega_n} = \frac{2\pi}{\sqrt{\frac{2}{3}\frac{(900\text{ N/m})}{(7.5\text{ kg})}}} = \frac{2\pi}{\sqrt{80}} = 0.702$ s.

(*b*) v_m = (0.01 m/s) $\sqrt{80}$ = 0.0894 m/s.
19.80 τ_n = 0.821 s.

19.81 $\tau = 2\pi\sqrt{m/3k\cos^2\beta}$.

19.82 $\tau = 2\pi\sqrt{\left(\frac{m}{3}+m_C\right)/k\cos^2\beta}$.

19.83 $\tau_n = 1.327$ s.

19.84 $f_n = 0.1899\sqrt{\frac{g}{l}}$.

19.85 $\tau_n = 1.737$ s.

19.86 $\tau_n = 3.06$ s.

19.87 $\tau_n = \frac{2\pi}{\sqrt{\omega_n}} = 2\pi\sqrt{\frac{12r^2+2l^2}{3gl}}$.

19.88 $\tau_n = \frac{2\pi}{\omega_n} = 2\pi\sqrt{\frac{60r^2+10l^2}{9gl}}$.

19.89 (*a*) $f = 2\pi\sqrt{(6ka^2-3mgl)/2ml^2}$. (*b*) $a_{\min} = \sqrt{\frac{mgl}{2k}}$.

19.90 $f = 2.33$ Hz.

19.91 $f_n = 0.918$ Hz.

19.92 $f_n = 0.1312\sqrt{\frac{g}{r}}$.

19.93 $f = 0.1125\sqrt{\frac{g}{l}}$.

19.94 $\tau_n = 2\pi\sqrt{\frac{2hL}{3bg}}$.

19.95 $f_n = \frac{1}{2\pi}\sqrt{\frac{3g}{l}}$.

19.96 $f = 2.59$ Hz.

19.97 $\tau_n = \frac{\pi l}{\sqrt{3gr}}$.

19.98 (*a*) $\tau_n = 0.352$ s. (*b*) $\tau_n = 0.352$ s.

19.99 (*a*) $x_m = 166.7$ mm (in-phase). (*b*) $x_m = 128.2$ mm (in-phase). (*c*) $x_m = 10.00$ mm (out-of-phase).

19.100 $\omega_f = 15.28$ rad/s, $\omega_f = 23.8$ rad/s.

19.101 (*a*) $k = 191.7$ N/m. (*b*) $k = 58.3$ N/m.

19.102 Range: $\sqrt{\frac{k}{2m}} < \omega_f < \sqrt{\frac{3k}{2m}}$.

19.103 35.5 rad/s $< \omega_f <$ 44.1 rad/s.

19.104 $\omega_f > 109.3$ rad/s.

19.105 $\omega_f < 8.16$ rad/s.

19.106 $\omega_f < 24.8$ rad/s, $\omega_f > 42.9$ rad/s.

19.107 $(x_m)_3 = 22.5$ mm, $(x_m)_3 = -5.63$ mm.

19.108 (*a*) 36.9 liters/min. (*b*) 87.1 liters/min.

19.109 $\omega_f > \sqrt{\frac{2g}{l}}$.

19.110 (*a*) $x_m = 25.2$ mm. (*b*) $F = -0.437 \sin \pi t$ (N).

19.111 (*a*) $x_m = 90.0$ mm. (*b*) $F_m = 18.00$ N.

19.112 $\omega_n = 651$ rpm.

19.113 $x_m = \frac{r\left(\frac{m}{M}\right)\left(\frac{\omega_f}{\omega_n}\right)^2}{1-\left(\frac{\omega_f}{\omega_n}\right)^2}$ Q.E.D.

19.114 $r = 22$ mm.

19.115 $\omega_f < 328$ rpm, $\omega_f > 334$ rpm.

19.116 $\omega_n = 783$ rpm.
19.117 $\Delta M = 39.1$ kg.
19.118 (*a*) $k = 8.292$ kN/m. (*b*) $|F_m| = 0.687$ N.
19.119 $r = 149.3$ mm.
19.120 (*a*) $\omega_f < 254$ rpm, $\omega_f > 304$ rpm.
19.121 Transmissibility $= \dfrac{(P_T)_m}{P_m} = \dfrac{1}{1-\left(\frac{\omega_f}{\omega_n}\right)^2}$, Transmissibility $= \dfrac{x_m}{\delta_m} = \dfrac{1}{1-\left(\frac{\omega_f}{\omega_n}\right)^2}$,

$\dfrac{\omega_f}{\omega_n} > \sqrt{2}$ Q.E.D.

19.122 (*a*) Error = 4.17%. (*b*) $f_n = 84.9$ Hz.
19.123 Error = 8.04%.
19.124 (*a*) $k = 88.0$ N/m. (*b*) $x_m = -0.227$ m.
19.125 (*a*) $\omega_n = \omega_f = 1406$ rpm. (*b*) $r = 0.403$ mm.
19.126 (*a*) $v = 25.6$ km/h. (*b*) $x_m = -14.25$ mm.
19.129 $\ln\dfrac{x_n}{x_{n+1}} = \dfrac{2\pi\left(\frac{c}{c_c}\right)}{\sqrt{1-\left(\frac{c}{c_c}\right)^2}}$ Q.E.D.
19.130 $\text{log decrement} = \dfrac{1}{k}\ln\dfrac{x_n}{x_{n+k}}$ Q.E.D.
19.132 (*a*) $c = 104.4$ kN · s/m. (*b*) $k = 3.70 \times 10^6$ N/m.
19.133 $C = 5.48$ N · m · s.
19.134 (*a*) $k = \dfrac{\left(\frac{c_c}{2}\right)^2}{m} = \dfrac{\left(\frac{18000}{2}\right)^2}{750} = 108$ kN/m. (*b*) $t = 0.1908$ s.
19.135 (*a*) $f = 302$ rpm. (*b*) $f = 282$ rpm. (*c*) $x_m = 12.03$ mm, $x_m = 12.77$ mm.
19.136 Total distance = 0.02619 + 0.0307 = 0.0569 m = 56.9 mm.
19.137 $\dfrac{c}{c_c} = 0.0431.$
19.138 (*a*) $\ddot{\theta} + \left(\dfrac{3c}{m}\right)\dot{\theta} + \left(\dfrac{3k}{4m}\right)\theta = 0.$

(*b*) $c_c = \sqrt{\dfrac{km}{3}}.$

19.139 $x_m = 2.37$ mm.
19.140 $\dfrac{k}{2} = 135.3$ kN/m.
19.141 $\dfrac{c}{c_c} \geq 0.707.$
19.142 $\dfrac{x_m}{\left(\frac{P_m}{k}\right)} = \dfrac{1}{2}\dfrac{c_c}{c}.$
19.143 (*a*) $k = 2.21$ Mn/m. (*b*) $\dfrac{c}{c_c} = 0.0286.$
19.144 $f_f > 30.8$ Hz and $f_f < 15.85$ Hz.
19.145 $e = 13.01$ mm.
19.146 (*a*) $k = 2.21$ Mn/m. (*b*) $\dfrac{c}{c_c} = 0.0286.$
19.147 $F_m = \dfrac{P_m\sqrt{1+\left[2\left(\frac{c}{c_c}\right)\left(\frac{\omega_f}{\omega_n}\right)\right]^2}}{\sqrt{\left[1-\left(\frac{\omega_f}{\omega_n}\right)^2\right]^2+\left[2\left(\frac{c}{c_c}\right)\left(\frac{\omega_f}{\omega_n}\right)\right]^2}}$ Q.E.D.
19.148 (*a*) $F_m = 71.8$ N. (*b*) $F_m = 39.0$ N.
19.149 (*a*) $x_m = 134.8$ mm. (*b*) $F_m = 143.7$ N.
19.150 $E = \pi c x_m^2 \omega_f$ Q.E.D.

19.151 (*a*) $m\frac{d^2x}{dt}+c\frac{dx}{dt}+kx=(k\sin\omega_f t+c\omega_f\cos\omega_f t)\delta_m$.

(*b*) $x = x_m \sin(\omega_f t - \varphi + \psi)$ (where analogous to Equations (19.52))

$$x_m=\frac{\delta_m\sqrt{k^2+(c\omega_f)^2}}{\sqrt{(k-m\omega_f^2)^2+(c\omega_f)^2}},\ \tan\varphi=\frac{c\omega_f}{k-m\omega_f^2},\ \tan\psi=\frac{c\omega_f}{k}.$$

19.152 $m\ddot{x}_A+c(\dot{x}_A-\dot{x}_B)+2k(x_A-x_B)=P_m\sin\omega_f t,\ m\ddot{x}_B+3c\dot{x}_B-c\dot{x}_A+3kx_B-2kx_A=0.$

19.153 $R<2\sqrt{\frac{L}{C}}$.

19.154 (*a*) $i_{\text{final}}=\frac{E}{R}$. (*b*) $t=\frac{L}{R}$.

19.157 (*a*) $c\frac{d}{dt}(x_A-x_m)+kx_A=0,\ m\frac{d^2x_m}{dt^2}+c\frac{d}{dt}(x_m-x_A)=P_m\sin\omega_f t.$

(*b*) $R\frac{d}{dt}(q_A-q_m)+\left(\frac{1}{C}\right)q_n=0,\ L\frac{d^2q_m}{dt^2}+R\frac{d}{dt}(q_m-q_A)=E_m\sin\omega_f t.$

19.158 (*a*) $m_1\frac{d^2x_1}{dt^2}+c_1\frac{dx_1}{dt}+k_1x_1+k_2(x_1-x_2)=0,\ m_2\frac{d^2x_2}{dt^2}+c_2\frac{dx_2}{dt}+k_2(x_2-x_1)=0.$

(*b*) $L_1\frac{d^2q_1}{dt^2}+R_1\frac{dq_1}{dt}+\frac{q_1}{C_1}+\frac{(q_1-q_2)}{C_2}=0,\ L_2\frac{d^2q_2}{dt^2}+R_2\frac{dq_2}{dt}+\frac{q_2-q_1}{C_2}=0.$

19.159 $\tau_n=2\pi\sqrt{\frac{2a}{3g}}$. (*b*) $b = 0.1667a$.

19.160 (*a*) $W_A = 33.4$ N. (*b*) $k = 538$ N/m.
19.161 $\tau_n = 1.772$ s.
19.162 (*a*) $\tau_n = 2.28$ s. (*b*) $l = 1.294$ m.
19.163 $\omega_f < 6.59$ rad/s.
19.164 (*a*) $\frac{c}{c_c}=0.01393$. (*b*) $c = 0.737$ N · s/m.

19.165 (*a*) $0.18667\ddot{\theta}+2.88\dot{\theta}+2\theta=0$.

19.166 (*a*) $F_m = 5.75$ N. (*b*) $x_m = 0.00710$ mm.
19.167 $h = 1.642$ m.

19.168 (*a*) $m\ddot{x}+(2T)\left(\frac{2x}{l}\right)=0$. (*b*) $\tau_n=\pi\sqrt{\frac{ml}{T}}$.

19.169 $x_m = 1.125$ mm.

SI Prefixes

Multiplication Factor	Prefix†	Symbol
$1\,000\,000\,000\,000 = 10^{12}$	tera	T
$1\,000\,000\,000 = 10^{9}$	giga	G
$1\,000\,000 = 10^{6}$	mega	M
$1\,000 = 10^{3}$	kilo	k
$100 = 10^{2}$	hecto‡	h
$10 = 10^{1}$	deka‡	da
$0.1 = 10^{-1}$	deci‡	d
$0.01 = 10^{-2}$	centi‡	c
$0.001 = 10^{-3}$	milli	m
$0.000\,001 = 10^{-6}$	micro	μ
$0.\,000\,000\,001 = 10^{-9}$	nano	n
$0.000\,000\,000\,001 = 10^{-12}$	pico	p
$0.000\,000\,000\,000\,001 = 10^{-15}$	femto	f
$0.000\,000\,000\,000\,000\,001 = 10^{-18}$	atto	a

† The first syllable of every prefix is accented so that the prefix will retain its identity. Thus, the preferred pronunciation of kilometer places the accent on the first syllable, not the second.

‡ The use of these prefixes should be avoided, except for the measurement of areas and volumes and for the nontechnical use of centimeter, as for body and clothing measurements.

Principal SI Units Used in Mechanics

Quantity	Unit	Symbol	Formula
Acceleration	Meter per second squared	. . .	m/s^2
Angle	Radian	rad	†
Angular acceleration	Radian per second squared	. . .	rad/s^2
Angular velocity	Radian per second	. . .	rad/s
Area	Square meter	. . .	m^2
Density	Kilogram per cubic meter	. . .	kg/m^3
Energy	Joule	J	$N \cdot m$
Force	Newton	N	$kg \cdot m/s^2$
Frequency	Hertz	Hz	s^{-1}
Impulse	Newton-second	. . .	$kg \cdot m/s$
Length	Meter	m	‡
Mass	Kilogram	kg	‡
Moment of a force	Newton-meter	. . .	$N \cdot m$
Power	Watt	W	J/s
Pressure	Pascal	Pa	N/m^2
Time	Second	s	‡
Velocity	Meter per second	. . .	m/s
Volume, solids	Cubic meter	. . .	m^3
Liquids	Liter	L	$10^{-3}\ m^3$
Work	Joule	J	$N \cdot m$

† Supplementary unit (1 revolution = 2π rad = 360°).
‡ Base unit.